Third Edition

Handbook of
Photosynthesis

BOOKS IN SOILS, PLANTS, AND THE ENVIRONMENT

Soil Biochemistry, Volume 1, edited by A. D. McLaren and G. H. Peterson

Soil Biochemistry, Volume 2, edited by A. D. McLaren and J. Skujins

Soil Biochemistry, Volume 3, edited by E. A. Paul and A. D. McLaren

Soil Biochemistry, Volume 4, edited by E. A. Paul and A. D. McLaren

Soil Biochemistry, Volume 5, edited by E. A. Paul and J. N. Ladd

Soil Biochemistry, Volume 6, edited by Jean-Marc Bollag and G. Stotzky

Soil Biochemistry, Volume 7, edited by G. Stotzky and Jean-Marc Bollag

Soil Biochemistry, Volume 8, edited by Jean-Marc Bollag and G. Stotzky

Soil Biochemistry, Volume 9, edited by G. Stotzky and Jean-Marc Bollag

Organic Chemicals in the Soil Environment, Volumes 1 and 2, edited by C. A. I. Goring and J. W. Hamaker

Humic Substances in the Environment, M. Schnitzer and S. U. Khan

Microbial Life in the Soil: An Introduction, T. Hattori

Principles of Soil Chemistry, Kim H. Tan

Soil Analysis: Instrumental Techniques and Related Procedures, edited by Keith A. Smith

Soil Reclamation Processes: Microbiological Analyses and Applications, edited by Robert L. Tate III and Donald A. Klein

Symbiotic Nitrogen Fixation Technology, edited by Gerald H. Elkan

Soil–Water Interactions: Mechanisms and Applications, Shingo Iwata
and Toshio Tabuchi with Benno P. Warkentin

Soil Analysis: Modern Instrumental Techniques, Second Edition, edited by Keith A. Smith

Soil Analysis: Physical Methods, edited by Keith A. Smith and Chris E. Mullins

Growth and Mineral Nutrition of Field Crops, N. K. Fageria, V. C. Baligar, and Charles Allan Jones

Semiarid Lands and Deserts: Soil Resource and Reclamation, edited by J. Skujins

Plant Roots: The Hidden Half, edited by Yoav Waisel, Amram Eshel, and Uzi Kafkafi

Plant Biochemical Regulators, edited by Harold W. Gausman

Maximizing Crop Yields, N. K. Fageria

Transgenic Plants: Fundamentals and Applications, edited by Andrew Hiatt

Soil Microbial Ecology: Applications in Agricultural and Environmental Management, edited by F. Blaine Metting, Jr.

Principles of Soil Chemistry, Second Edition, Kim H. Tan

Water Flow in Soils, edited by Tsuyoshi Miyazaki

Handbook of Plant and Crop Stress, edited by Mohammad Pessarakli

Genetic Improvement of Field Crops, edited by Gustavo A. Slafer

Agricultural Field Experiments: Design and Analysis, Roger G. Petersen

Mechanisms of Plant Growth and Improved Productivity: Modern Approaches, edited by Amarjit S. Basra

Selenium in the Environment, edited by W. T. Frankenberger, Jr. and Sally Benson

Plant–Environment Interactions, edited by Robert E. Wilkinson

Handbook of Plant and Crop Physiology, edited by Mohammad Pessarakli

Handbook of Phytoalexin Metabolism and Action, edited by M. Daniel and R. P. Purkayastha

Soil–Water Interactions: Mechanisms and Applications, Second Edition, Revised and Expanded, Shingo Iwata, Toshio Tabuchi, and Benno P. Warkentin

Stored-Grain Ecosystems, edited by Digvir S. Jayas, Noel D. G. White, and William E. Muir

Agrochemicals from Natural Products, edited by C. R. A. Godfrey

Seed Development and Germination, edited by Jaime Kigel and Gad Galili

Nitrogen Fertilization in the Environment, edited by Peter Edward Bacon

Phytohormones in Soils: Microbial Production and Function, William T. Frankenberger, Jr. and Muhammad Arshad

Handbook of Weed Management Systems, edited by Albert E. Smith

Soil Sampling, Preparation, and Analysis, Kim H. Tan

Soil Erosion, Conservation, and Rehabilitation, edited by Menachem Agassi

Plant Roots: The Hidden Half, Second Edition, Revised and Expanded, edited by Yoav Waisel, Amram Eshel, and Uzi Kafkafi

Photoassimilate Distribution in Plants and Crops: Source–Sink Relationships, edited by Eli Zamski and Arthur A. Schaffer

Mass Spectrometry of Soils, edited by Thomas W. Boutton and Shinichi Yamasaki

Handbook of Photosynthesis, edited by Mohammad Pessarakli

Chemical and Isotopic Groundwater Hydrology: The Applied Approach, Second Edition, Revised and Expanded, Emanuel Mazor

Fauna in Soil Ecosystems: Recycling Processes, Nutrient Fluxes, and Agricultural Production, edited by Gero Benckiser

Soil and Plant Analysis in Sustainable Agriculture and Environment, edited by Teresa Hood and J. Benton Jones, Jr.

Seeds Handbook: Biology, Production, Processing, and Storage, B. B. Desai, P. M. Kotecha, and D. K. Salunkhe

Modern Soil Microbiology, edited by J. D. van Elsas, J. T. Trevors, and E. M. H. Wellington

Growth and Mineral Nutrition of Field Crops, Second Edition, N. K. Fageria, V. C. Baligar, and Charles Allan Jones

Fungal Pathogenesis in Plants and Crops: Molecular Biology and Host Defense Mechanisms, P. Vidhyasekaran

Plant Pathogen Detection and Disease Diagnosis, P. Narayanasamy

Agricultural Systems Modeling and Simulation, edited by Robert M. Peart and R. Bruce Curry

Agricultural Biotechnology, edited by Arie Altman

Plant–Microbe Interactions and Biological Control, edited by Greg J. Boland and L. David Kuykendall

Handbook of Soil Conditioners: Substances That Enhance the Physical Properties of Soil, edited by Arthur Wallace and Richard E. Terry

Environmental Chemistry of Selenium, edited by William T. Frankenberger, Jr., and Richard A. Engberg

Principles of Soil Chemistry, Third Edition, Revised and Expanded, Kim H. Tan

Sulfur in the Environment, edited by Douglas G. Maynard

Soil–Machine Interactions: A Finite Element Perspective, edited by Jie Shen and Radhey Lal Kushwaha

Mycotoxins in Agriculture and Food Safety, edited by Kaushal K. Sinha and Deepak Bhatnagar

Plant Amino Acids: Biochemistry and Biotechnology, edited by Bijay K. Singh

Handbook of Functional Plant Ecology, edited by Francisco I. Pugnaire and Fernando Valladares

Handbook of Plant and Crop Stress, Second Edition, Revised and Expanded, edited by Mohammad Pessarakli

Plant Responses to Environmental Stresses: From Phytohormones to Genome Reorganization, edited by H. R. Lerner

Handbook of Pest Management, edited by John R. Ruberson

Microbial Endophytes, edited by Charles W. Bacon and James F. White, Jr.

Plant–Environment Interactions, Second Edition, edited by Robert E. Wilkinson

Microbial Pest Control, Sushil K. Khetan

Soil and Environmental Analysis: Physical Methods, Second Edition, Revised and Expanded, edited by Keith A. Smith and Chris E. Mullins

The Rhizosphere: Biochemistry and Organic Substances at the Soil–Plant Interface, Roberto Pinton, Zeno Varanini, and Paolo Nannipieri

Woody Plants and Woody Plant Management: Ecology, Safety, and Environmental Impact, Rodney W. Bovey

Metals in the Environment, M. N. V. Prasad

Plant Pathogen Detection and Disease Diagnosis, Second Edition, Revised and Expanded, P. Narayanasamy

Handbook of Plant and Crop Physiology, Second Edition, Revised and Expanded, edited by Mohammad Pessarakli

Environmental Chemistry of Arsenic, edited by William T. Frankenberger, Jr.

Enzymes in the Environment: Activity, Ecology, and Applications, edited by Richard G. Burns and Richard P. Dick

Plant Roots: The Hidden Half, Third Edition, Revised and Expanded, edited by Yoav Waisel, Amram Eshel, and Uzi Kafkafi

Handbook of Plant Growth: pH as the Master Variable, edited by Zdenko Rengel

Biological Control of Major Crop Plant Diseases edited by Samuel S. Gnanamanickam

Pesticides in Agriculture and the Environment, edited by Willis B. Wheeler

Mathematical Models of Crop Growth and Yield, Allen R. Overman and Richard Scholtz

Plant Biotechnology and Transgenic Plants, edited by Kirsi-Marja Oksman Caldentey and Wolfgang Barz

Handbook of Postharvest Technology: Cereals, Fruits, Vegetables, Tea, and Spices, edited by Amalendu Chakraverty, Arun S. Mujumdar, G. S. Vijaya Raghavan, and Hosahalli S. Ramaswamy

Handbook of Soil Acidity, edited by Zdenko Rengel

Humic Matter in Soil and the Environment: Principles and Controversies, edited by Kim H. Tan

Molecular Host Plant Resistance to Pests, edited by S. Sadasivam and B. Thayumanayan

Soil and Environmental Analysis: Modern Instrumental Techniques, Third Edition, edited by Keith A. Smith and Malcolm S. Cresser

Chemical and Isotopic Groundwater Hydrology, Third Edition, edited by Emanuel Mazor

Agricultural Systems Management: Optimizing Efficiency and Performance, edited by Robert M. Peart and W. David Shoup

Physiology and Biotechnology Integration for Plant Breeding, edited by Henry T. Nguyen and Abraham Blum

Global Water Dynamics: Shallow and Deep Groundwater, Petroleum Hydrology, Hydrothermal Fluids, and Landscaping, , edited by Emanuel Mazor

Principles of Soil Physics, edited by Rattan Lal

Seeds Handbook: Biology, Production, Processing, and Storage, Second Edition, Babasaheb B. Desai

Field Sampling: Principles and Practices in Environmental Analysis, edited by Alfred R. Conklin

Sustainable Agriculture and the International Rice–Wheat System, edited by Rattan Lal, Peter R. Hobbs, Norman Uphoff, and David O. Hansen

Plant Toxicology, Fourth Edition, edited by Bertold Hock and Erich F. Elstner

Drought and Water Crises: Science, Technology, and Management Issues, edited by Donald A. Wilhite

Soil Sampling, Preparation, and Analysis, Second Edition, Kim H. Tan

Climate Change and Global Food Security, edited by Rattan Lal, Norman Uphoff, B. A. Stewart, and David O. Hansen

Handbook of Photosynthesis, Second Edition, edited by Mohammad Pessarakli

Environmental Soil-Landscape Modeling: Geographic Information Technologies and Pedometrics, edited by Sabine Grunwald

Water Flow in Soils, Second Edition, Tsuyoshi Miyazaki

Biological Approaches to Sustainable Soil Systems, edited by Norman Uphoff,
 Andrew S. Ball, Erick Fernandes, Hans Herren, Olivier Husson, Mark Laing,
 Cheryl Palm, Jules Pretty, Pedro Sanchez, Nteranya Sanginga, and Janice Thies

Plant–Environment Interactions, Third Edition, edited by Bingru Huang

Biodiversity in Agricultural Production Systems, edited by Gero Benckiser
 and Sylvia Schnell

Organic Production and Use of Alternative Crops, Franc Bavec and Martina Bavec

Handbook of Plant Nutrition, edited by Allen V. Barker and David J. Pilbeam

Modern Soil Microbiology, Second Edition, edited by Jan Dirk van Elsas,
 Janet K. Jansson, and Jack T. Trevors

Functional Plant Ecology, Second Edition, edited by Francisco I. Pugnaire
 and Fernando Valladares

*Fungal Pathogenesis in Plants and Crops: Molecular Biology and Host Defense
 Mechanisms,* Second Edition, P. Vidhyasekaran

Handbook of Turfgrass Management and Physiology, edited by Mohammad Pessarakli

Soils in the Humid Tropics and Monsoon Region of Indonesia, Kim H. Tan

Handbook of Agricultural Geophysics, edited by Barry J. Allred, Jeffrey J. Daniels,
 and M. Reza Ehsani

Environmental Soil Science, Third Edition, Kim H. Tan

Principles of Soil Chemistry, Fourth Edition, Kim H. Tan

Handbook of Plant and Crop Stress, Second Edition, edited by Mohammad Pessarakli

Handbook of Plant and Crop Physiology, Third Edition, edited by Mohammad Pessarakli

Humic Matter in Soil and the Environment: Principles and Controversies, Second Edition,
 Kim H. Tan

Handbook of Photosynthesis, Third Edition, Mohammad Pessarakli

Third Edition

Handbook of
Photosynthesis

Edited by
Mohammad Pessarakli

The University of Arizona
School of Plant Sciences
Tucson, Arizona, USA

CRC Press
Taylor & Francis Group
Boca Raton London New York

CRC Press is an imprint of the
Taylor & Francis Group, an **informa** business

CRC Press
Taylor & Francis Group
6000 Broken Sound Parkway NW, Suite 300
Boca Raton, FL 33487-2742

First issued in paperback 2021

© 2016 by Taylor & Francis Group, LLC
CRC Press is an imprint of Taylor & Francis Group, an Informa business

No claim to original U.S. Government works

Version Date: 20160126

ISBN 13: 978-1-03-209800-5 (pbk)
ISBN 13: 978-1-4822-3073-4 (hbk)

Visit the Taylor & Francis Web site at
http://www.taylorandfrancis.com

and the CRC Press Web site at
http://www.crcpress.com

To the memory of my beloved parents, Fatemeh and Vahab, who regretfully did not live to see this work and my other works, which in no small part resulted from their gift of many years of unconditional love to me.

Contents

Preface...xvii
Acknowledgments...xxi
Editor xxiii
Contributors xxv

SECTION I   Principles of Photosynthesis

Chapter 1   Biogenesis of Thylakoid Membranes: Correlation of Structure and Function.......................3

Łucja Kowalewska and Agnieszka Mostowska

Chapter 2   Nature of Light from the Perspective of a Biologist: What Is a Photon? 17

Randy Wayne

Chapter 3   Adaptation and Regulation of Photosynthetic Apparatus in Response to Light....................53

Selene Casella, Fang Huang, and Lu-Ning Liu

Chapter 4   Estimation of Mesophyll Conductance and Its Applications 65

Naomi Kodama and Wataru Yamori

SECTION II   Biochemistry of Photosynthesis

Chapter 5   Development of Chloroplast: Biogenesis, Senescence, and Regulations 77

Basanti Biswal and Jitendra K. Pandey

Chapter 6   Physicochemical Properties of Chlorophylls and Bacteriochlorophylls 95

Masami Kobayashi, Yuhta Sorimachi, Daisuke Fukayama, Hirohisa Komatsu, Terumitsu Kanjoh,
Katsuhiro Wada, Masanobu Kawachi, Hideaki Miyashita, Mayumi Ohnishi-Kameyama, and Hiroshi Ono

Chapter 7   Thylakoid Ndh Complex 149

Bartolomé Sabater and Mercedes Martín

Chapter 8   Redox Metabolism in Photosynthetic Organisms 159

Sergio A. Guerrero, Diego G. Arias, and Alberto A. Iglesias

Chapter 9   Regulation of Chlorophyll Metabolism in Plants...173

Koichi Kobayashi and Tatsuru Masuda

Chapter 10 Carbonic Anhydrases of Higher Plant Thylakoids and Their Participation in Photosynthesis 193

L.K. Ignatova and B.N. Ivanov

Chapter 11 Hybrid Interfaces for Electron and Energy Transfer Based on Photosynthetic Proteins 201

   R. Roberto Tangorra, Alessandra Antonucci, Francesco Milano, Simona la Gatta,
   Gianluca M. Farinola, Angela Agostiano, Roberta Ragni, and Massimo Trotta

SECTION III Molecular Aspects of Photosynthesis: Photosystems, Photosynthetic Enzymes and Genes

Chapter 12 Quality Control of Photosystem II: Role of Structural Changes of Thylakoid Membranes and FtsH
   Proteases in High Light Tolerance and Recovery from Photoinhibition 223

   Miho Yoshioka-Nishimura

Chapter 13 Cytochrome c_6-Like Proteins in Cyanobacteria, Algae, and Higher Plants 229

   Alejandro Torrado and Fernando P. Molina-Heredia

SECTION IV Atmospheric and Environmental Factors Affecting Photosynthesis

Chapter 14 Role of the LHCII Organization for the Sensitivity of the Photosynthetic Apparatus to Temperature
   and High Light Intensity 243

   E.L. Apostolova and A.G. Dobrikova

Chapter 15 Impact of Solar Ultraviolet (280–400 nm) Exclusion on Photosynthesis in C_3 and C_4 Plants 257

   K.N. Guruprasad and Sunita Kataria

Chapter 16 What Role Does UVB Play in Determining Photosynthesis? 275

   Brian R. Jordan, Å. Strid, and J.J. Wargent

Chapter 17 Small Fruit Crop Responses to Leaf Intercellular CO_2, Primary Macronutrient Limitation and Crop Rotation ... 287

   Hong Li

SECTION V Photosynthetic Pathways in Various Crop Plants

Chapter 18 Photosynthesis in Nontypical C_4 Species: C_4 Cycles without Kranz Anatomy and C_4 Crassulacean Acid
   Metabolism Transitions 307

   María Valeria Lara and Carlos Santiago Andreo

SECTION VI Photosynthesis in Lower and Monocellular Plants

Chapter 19 Structural and Functional Studies of the Cytochrome bc_1 Complex from Photosynthetic Purple Bacteria 327

   Fei Zhou, Lothar Esser, Chang-An Yu, and Di Xia

Chapter 20 Photosynthetic Apparatus in Cyanobacteria and Microalgae 349

Isabella Moro, Nicoletta La Rocca, and Nicoletta Rascio

Chapter 21 Excess Light and Limited Carbon: Two Problems with Which Cyanobacteria and Microalgae Cope 369

Nicoletta La Rocca, Isabella Moro, and Nicoletta Rascio

Chapter 22 Rethinking the Limitations of Photosynthesis in Cyanobacteria.. 397

Yehouda Marcus

Chapter 23 Induction and Relaxation of Bacteriochlorophyll Fluorescence in Photosynthetic Bacteria............................ 405

Péter Maróti

Chapter 24 Photosynthesis in Eukaryotic Algae with Secondary Plastids.. 425

Christian Wilhelm and Reimund Goss

SECTION VII   *Photosynthesis in Higher Plants*

Chapter 25 Current Methods in Photosynthesis Research.. 447

Jutta Papenbrock, Yelena Churakova, Bernhard Huchzermeyer, and Hans-Werner Koyro

Chapter 26 Forage Crops and Their Photosynthesis.. 465

M. Anowarul Islam and Albert T. Adjesiwor

Chapter 27 Photosynthetic Competition between Forest Trees 475

Piotr Robakowski

Chapter 28 Functional Traits and Plasticity of Plants.. 487

Elena Masarovičová, Mária Májeková, and Ivana Vykouková

SECTION VIII   *Photosynthesis and Plant/Crop Productivity and Photosynthetic Products*

Chapter 29 Carbon Photoassimilation and Photosynthate Partitioning in Plants.. 509

Carlos M. Figueroa, Claudia V. Piattoni, Karina E.J. Trípodi, Florencio E. Podestá, and Alberto A. Iglesias

SECTION IX   *Photosynthesis and Plant Genetics*

Chapter 30 Genetic Factors Affecting Photosynthesis 539

A.K. Joshi and Shree P. Pandey

SECTION X   Photosynthetic Activity Measurements and Analysis of Photosynthetic Pigments

Chapter 31 Quantifying Growth Nondestructively Using Whole-Plant CO_2 Exchange Is a Powerful Tool for Phenotyping 571

Evangelos D. Leonardos and Bernard Grodzinski

Chapter 32 Nonubiquitous Carotenoids in Higher Plants: Presence, Role in Photosynthesis, and Guidelines for Identification 589

Raquel Esteban and José Ignacio García-Plazaola

SECTION XI   Photosynthesis and Its Relationship with Plant Nutrient Elements

Chapter 33 Role of Phosphorus in Photosynthetic Carbon Assimilation and Partitioning 603

Anna M. Rychter, Idupulapati M. Rao, and Juan Andrés Cardoso

SECTION XII   Photosynthesis under Environmental Stress Conditions

Chapter 34 Photosynthesis in Plants under Stressful Conditions 629

Rama Shanker Dubey

Chapter 35 Effects of Salinity and Drought Stress on Photosynthesis, Growth, and Development of Ornamental Plants 651

Hyun-Sug Choi, Xiaoya Cai, and Mengmeng Gu

Chapter 36 Photosynthetic Strategies of Desiccation-Tolerant Organisms 663

Beatriz Fernández-Marín, Andreas Holzinger, and José Ignacio García-Plazaola

Chapter 37 Drought Tolerance of Photosynthesis 683

Katya Georgieva and Gergana Mihailova

Chapter 38 Photosynthesis under Heat Stress 697

Muhammad Farooq, Abdul Rehman, Abdul Wahid, and Kadambot H.M. Siddique

SECTION XIII   Photosynthesis in the Past, Present, and Future Prospective

Chapter 39 Evolutionary Ecology of s-Triazine-Resistant Plants: Pleiotropic Photosynthetic Reorganization in the Chloroplast Chronomutant 705

Jack Dekker

Chapter 40 Next Evolution of Agriculture: A Review of Innovations in Plant Factories 723

Merrill F. Brandon, Na Lu, Toshitaka Yamaguchi, Michiko Takagaki, Toru Maruo, Toyoki Kozai, and Wataru Yamori

Chapter 41 Strategies for Optimizing Photosynthesis with Biotechnology to Improve Crop Yield 741

Wataru Yamori, Louis J. Irving, Shunsuke Adachi, and Florian A. Busch

Chapter 42 Efficiency of Light-Emitting Diodes for Future Photosynthesis...761

Soleyman Dayani, Parisa Heydarizadeh, and Mohammad Reza Sabzalian

Index...785

Preface

Because photosynthesis has probably been given more attention than any other physiological processes in plant physiology, there have been hundreds of articles published on this topic since the first and the second editions of this book were published in 1997 and 2005, respectively. Numerous new ideas on photosynthesis have emerged during the last decade that not only have captured the attention of the experts and researchers on this topic but also have attracted interested individuals from various disciplines to photosynthesis. Therefore, I felt it not only necessary but also urgent that this book be revised and some of these interesting recent and relevant findings be included in the new volume, the third edition of the *Handbook of Photosynthesis*. In revising the book, I have eliminated most of the old chapters and included new ones. The few previous chapters that are included in the new volume have been extensively revised.

Over 90% of the material in the new edition of the *Handbook of Photosynthesis* is entirely new. The other less than 10% of the old material have been updated and substantially modified. Therefore, overall, about 95% of this book is new, and a totally new volume has emerged. From the total 42 chapters, only 5 are from the second edition, but substantially revised and updated. The other 37 chapters are entirely new. Therefore, the new volume looks like a new book. Like the first and the second editions, the new edition of the *Handbook of Photosynthesis* is the unique, most comprehensive, and complete collection of the topics on photosynthesis.

Photosynthesis is by far the most spectacular physiological process in plant growth and productivity. Due to this fact, the study of photosynthesis has captivated plant physiologists, botanists, plant biologists, horticulturalists, agronomists, agriculturalists, crop growers, and most recently, plant molecular and cellular biologists around the world.

From an aesthetic perspective, I thought that it would be wonderful to include many of the remarkable findings on photosynthesis in a single inclusive volume. In such an album, selected sources could be surveyed on this most magnificent subject. With the abundance of research on photosynthesis available at present, an elegantly prepared exhibition of the knowledge on photosynthesis is indeed in order. Accordingly, one mission of this collection is to provide an array of information on photosynthesis in a single and unique volume. Ultimately, this unique and comprehensive source of intelligence will both attract beginning students and stimulate further exploration by their educators. Furthermore, since more books, papers, and articles are currently available on photosynthesis than on any other plant physiological processes, preparation of a single volume by inclusion of the most recent and relevant issues and information on this subject can be appreciably useful and substantially helpful to those seeking specific information.

I see from a scientific perspective that the novelty of photosynthesis and its attraction in various disciplines have resulted in a voluminous, but somewhat scattered, database. However, none of the available sources comprehensively discusses the topic. The sources are either too specific or too general in scope. Therefore, a balanced presentation of the information on this subject is necessary. Accordingly, another main objective of this collection is to provide a balanced source of information on photosynthesis.

Now, more than ever, the excessive levels and exceedingly high accumulation rates of CO_2 due to the industrialization of nations have drawn the attention of scientists around the globe. If the current accumulation rates of carbon dioxide with the consequence of the imbalance between the atmospheric O_2 and CO_2 continue, all living organisms, including human beings and animals, will be endangered. The only natural mechanism known to utilize the atmospheric CO_2 is photosynthesis by green plants. Therefore, another purpose of preparing this volume is to gather the most useful and relevant issues on photosynthesis of selected plant species. In this regard, we must consider plant species with the most efficient photosynthetic pathways to reduce the excess atmospheric CO_2 concentrations. The use of such plants will result in balanced O_2 and CO_2 concentrations and will reduce toxic levels of atmospheric carbon dioxide. This higher consumption of atmospheric CO_2 by plants through the photosynthetic process not only reduces the toxic levels of CO_2 but will also result in more biomass production and higher crop yields.

To adequately cover many of the issues related to photosynthesis and for the advantage of easy accessibility to desired information, the volume has been divided into several sections. Each section includes one or more chapters that are closely related to each other.

Like any other physiological process, photosynthesis differs greatly among various plant species, particularly between C_3 and C_4 plants, whether growing under normal or stressful conditions. Therefore, examples of plants with various photosynthetic rates and different responses are presented in different chapters and included in this collection.

Now, it is well established that any plant species during its life cycle, at least once, is subjected to environmental stress. Since any stress alters the normal course of plant growth and development, metabolism, and other physiological processes, photosynthesis is also subject to this alteration and severely affected under stressful conditions. Therefore, a portion of this volume discusses plant photosynthesis under stressful conditions.

As with other fields, accessibility of knowledge is among the most critical of factors involved with photosynthesis. For this reason, as many of the photosynthetic factors as possible are included in this handbook. To further facilitate accessibility of the desired information on photosynthetic processes covered in this collection, the volume has been divided into 13 sections, including Principles of Photosynthesis; Biochemistry of Photosynthesis; Molecular Aspects of

Photosynthesis: Photosystems, Photosynthetic Enzymes, and Genes; Atmospheric and Environmental Factors Affecting Photosynthesis; Photosynthetic Pathways in Various Crop Plants; Photosynthesis in Lower and Monocellular Plants; Photosynthesis in Higher Plants; Photosynthesis and Plant/Crop Productivity and Photosynthetic Products; Photosynthesis and Plant Genetics; Photosynthetic Activity Measurements and Analysis of Photosynthetic Pigments; Photosynthesis and Its Relationship with Plant Nutrient Elements; Photosynthesis under Environmental Stress Conditions; and Photosynthesis in the Past, Present, and Future Prospective. Although the sections are interrelated, each serves independently to facilitate understanding of the material presented therein. Each section also enables the reader to acquire confidence in his or her learning and use the information offered. Each of these sections consists of one or more chapters to discuss, independently, as many aspects of photosynthesis as possible.

The section Principles of Photosynthesis consists of four chapters, as follows: Biogenesis of Thylakoid Membranes—Correlation of Structure and Function; Nature of Light from the Perspective of a Biologist: What Is a Photon?; Adaptation and Regulation of Photosynthetic Apparatus in Response to Light; and Estimation of Mesophyll Conductance and Its Applications. These chapters address various aspects of photosynthesis.

The section Biochemistry of Photosynthesis contains seven chapters, including Development of Chloroplast: Biogenesis, Senescence, and Regulations; Physicochemical Properties of Chlorophylls and Bacteriochlorophylls; The Thylakoid Ndh Complex; Redox Metabolism in Photosynthetic Organisms; Regulation of Chlorophyll Metabolism in Plants; Carbolic Anhydrases of Higher Plant Thylakoids and Their Participation in Photosynthesis; and Hybrid Interfaces for Electron Transfer and Energy Transfer Based on Photosynthetic Proteins.

The section Molecular Aspects of Photosynthesis: Photosystems, Photosynthetic Enzymes, and Genes presents detailed information on these aspects of photosynthesis in the following two chapters: Quality Control of Photosystem II: The Role of Structural Changes of Thylakoid Membranes and FtsH Proteases in High Light Tolerance and Recovery from Photoinhibition; and Cytochrome c_6-Like Proteins in Cyanobacteria, Algae, and Higher Plants.

Due to the recent climatic changes and increase in CO_2 levels, which have substantial impacts on photosynthesis, plant photosynthetic responses to these changes must be considered for crop production under these conditions. Therefore, a section on Atmospheric and Environmental Factors Affecting Photosynthesis, consisting of four chapters, presents recent information on this subject. These chapters are as follows: Role of LHCII Organization for the Sensitivity of the Photosynthetic Apparatus to Temperature and High Light Treatment; Impact of Solar Ultraviolet (280–400 nm) Exclusion on Photosynthesis in C_3 and C_4 Plants; What Role Does UVB Play in Photosynthesis?; and Small Fruit Crop Responses to Leaf Intercellular CO_2 Primary Macronutrient Limitation and Crop Rotation.

In the section Photosynthetic Pathways in Various Crop Plants, detailed information is given in a chapter entitled Photosynthesis in Nontypical C_4 Species: C_4 Cycles without Kranz Anatomy and C_4 Crassulacean Acid Metabolism Transitions.

The section Photosynthesis in Lower and Monocellular Plants contains six chapters, as follows: Structural and Functional Studies of the Cytochrome bc_1 Complex from Photosynthetic Purple Bacteria; Photosynthetic Apparatus in Cyanobacteria and Microalgae; Excess Light and Limited Carbon: Two Problems that Cyanobacteria and Microalgae Have to Cope with; Rethinking the Limitations of Photosynthesis in Cyanobacteria; Induction and Relaxation of Bacteriochlorophyll Fluorescence in Photosynthetic Bacteria; and Photosynthesis in Eukaryotic Algae with Secondary Plastids.

The section Photosynthesis in Higher Plants includes four chapters, as follows: Current Methods in Photosynthesis Research; Forage Crops and Their Photosynthesis; Photosynthetic Competition between Forest Trees; and Functional Traits and Plasticity of Plants.

The section Photosynthesis and Plant/Crop Productivity and Photosynthetic Products contains one chapter, entitled Carbon Photoassimilation and Photosynthate Partitioning in Plants, which presents detailed information on this subject.

The section Photosynthesis and Plant Genetics comprehensively discusses genetics aspects of photosynthesis in a chapter entitled Genetic Factors Affecting Photosynthesis.

The section Photosynthetic Activity Measurements and Analysis of Photosynthetic Pigments contains two chapters, as follows: Quantifying Growth Nondestructively Using Whole-Plant CO_2 Exchange Is a Powerful Tool for Phenotyping; and Nonubiquitous Carotenoids in Higher Plants: Presence, Role in Photosynthesis, and Guidelines for Identification.

The section Photosynthesis and Its Relationship with Plant Nutrient Elements includes a chapter entitled Role of Phosphorus in Photosynthetic Carbon Assimilation and Partitioning.

Since plants and crops, like other living things, at one time or another during their life cycle, encounter biotic or abiotic stressful conditions, a section entitled Photosynthesis under Environmental Stress Conditions is included in this volume that presents detailed information on the photosynthetic responses of plants under stressful conditions. This section contains five chapters, as follows: Photosynthesis in Plants under Stressful Conditions; Effects of Salinity and Drought Stress on Photosynthesis, Growth, and Development of Ornamental Plants; Photosynthetic Strategies of Desiccation-Tolerant Organisms; Drought Tolerance of Photosynthesis; and Photosynthesis under Heat Stress. These chapters are devoted to photosynthesis under various stresses.

Finally, a section entitled Photosynthesis in the Past, Present, and Future Perspective is included in this volume that consists of four chapters, as follows: Evolutionary Ecology of s-Triazine-Resistant Plants: Pleiotropic Photosynthetic Reorganization in the Chloroplast Chronomutant; The Next Evolution of Agriculture: A Review of Innovations in Plant Factories; Strategies for Optimizing Photosynthesis with

Biotechnology to Improve Crop Yield; and Efficiency of Light-Emitting Diodes for Future Photosynthesis. These chapters present information on the evolution of photosynthesis and its future perspectives.

Numerous tables, figures, and illustrations are included in the volume to facilitate understanding and comprehension of the presented information throughout the text. Thousands of references have been used to prepare this unique collection.

Thousands of index words are also included in this volume to further increase accessibility to the desired information.

Mohammad Pessarakli, PhD
Professor
University of Arizona
Tucson, Arizona, USA

Acknowledgments

I would like to express my appreciation for the secretarial and administrative assistance that I received from the secretarial and administrative staff of the School of Plant Sciences, College of Agriculture and Life Sciences, University of Arizona. The encouraging words of several of my colleagues, which are always greatly appreciated, have certainly been a driving force for the successful completion of this project.

In addition, I sincerely acknowledge Randy Brehm (senior editor, Taylor and Francis Group, CRC Press), whose professionalism, patience, hard work, and proactive methods helped in the completion of this project as well as my previous book projects. This job would not have been completed as smoothly and rapidly without her valuable support and efforts.

I am indebted to Jill Jurgensen (senior project coordinator, Taylor and Francis Group, CRC Press) for her professional and careful handling of this volume as well as my previous books. I would also like to acknowledge the eye for detail, sincere efforts, and hard work put in by the copy editor and the project editor.

The collective efforts and invaluable contributions of several experts in the field of photosynthesis made it possible to produce this unique source, which presents comprehensive information on this subject. Each and every one of these contributors and their contributions are greatly appreciated.

Last, but not least, I thank my wife, Vinca, a high school science teacher, and my son, Mahdi Pessarakli, MD, who supported me during the course of this work.

Editor

Dr. Mohammad Pessarakli is a professor in the School of Plant Sciences, College of Agriculture and Life Sciences, at the University of Arizona, Tucson, Arizona, United States. His work at the University of Arizona includes research and extension services as well as teaching courses in turfgrass science, management, and stress physiology. He is the editor of the *Handbook of Plant and Crop Stress*, the *Handbook of Plant and Crop Physiology* (both titles published formerly by Marcel Dekker, Inc., and currently Taylor and Francis Group, CRC Press), the *Handbook of Photosynthesis*, and *Handbook of Turfgrass Management and Physiology*. Dr. Pessarakli has written 20 book chapters; is an editorial board member of the *Journal of Plant Nutrition, Communications in Soil Science and Plant Analysis, Advances in Plants & Agriculture Research Journal*, and the *Journal of Agricultural Technology*; a member of the book review committee of the Crop Science Society of America; and a reviewer for the *Crop Science, Agronomy, Soil Science Society of America, and HortScience Journals*. He is the author or coauthor of 185 journal articles and 55 trade magazine articles. Dr. Pessarakli is an active member of the Agronomy Society of America, Crop Science Society of America, and Soil Science Society of America, among others. He is an executive board member of the American Association of University Professors (AAUP), Arizona chapter. Dr. Pessarakli is a well-known internationally recognized scientist and scholar and an esteemed member (invited) of *Sterling Who's Who, Marques Who's Who, Strathmore Who's Who, Madison Who's Who,* and *Continental Who's Who* as well as numerous honor societies (i.e., Phi Kappa Phi, Gamma Sigma Delta, Pi Lambda Theta, Alpha Alpha Chapter). He is a certified professional agronomist and certified professional soil scientist (CPAg/SS), designated by the American Registry of the Certified Professionals in Agronomy, Crop Science, and Soil Science. Dr. Pessarakli is a United Nations consultant in agriculture for underdeveloped countries. He earned a BS degree (1977) in environmental resources in agriculture and an MS degree (1978) in soil management and crop production from Arizona State University, Tempe, and a PhD degree (1981) in soil and water science from the University of Arizona, Tucson. Dr. Pessarakli's research work on environmental stress and his expertise on plants and crops are internationally recognized.

For more information about Dr. Pessarakli, please visit

http://ag.arizona.edu/pls/faculty/pessarakli.htm
http://cals.arizona.edu/spls/people/faculty

Contributors

Shunsuke Adachi
Global Innovation Research Organization
Tokyo University of Agriculture and Technology
Fuchu, Tokyo, Japan

and

Japan Science and Technology Agency (JST), PRESTO
Kawaguchi, Saitama, Japan

Albert T. Adjesiwor
Department of Plant Sciences
University of Wyoming
Laramie, Wyoming

Angela Agostiano
Department of Chemistry
Università degli Studi di Bari "Aldo Moro"
and
Istituto per i Processi Chimico Fisici
Consiglio Nazionale delle Ricerche
Bari, Italy

Carlos Santiago Andreo
Centro de Estudios Fotosintéticos y Bioquímicos (CEFOBI),
    CONICET
Facultad de Ciencias Bioquímicas y Farmacéuticas
Universidad Nacional de Rosario
Suipacha, Rosario, Argentina

Alessandra Antonucci
Department of Chemistry
Università degli Studi di Bari "Aldo Moro"
Bari, Italy

E.L. Apostolova
Institute of Biophysics and Biomedical Engineering
Bulgarian Academy of Sciences
Sofia, Bulgaria

Juan B. Arellano
Instituto de Recursos Naturales y Agrobiología de
    Salamanca (IRNASA-CSIC)
Salamanca, Spain

Diego G. Arias
Instituto de Agrobiotecnología del Litoral (IAL, UNL
    CONICET)
and
Facultad de Bioquímica y Ciencias Biológicas, UNL
Centro Científico Tecnológico Santa Fe
Santa Fe, Argentina

Basanti Biswal
Laboratory of Biochemistry and Molecular Biology
School of Life Sciences
Sambalpur University
Odisha, India

Merrill F. Brandon
Center for Environment, Health and Field Sciences
Chiba University
Kashiwa, Chiba, Japan

Florian A. Busch
Research School of Biology
Australian National University
Canberra, Australia

Xiaoya Cai
Department of Horticultural Sciences
Texas A&M AgriLife Extension
College Station, Texas

Juan Andrés Cardoso
Centro Internacional de Agricultura Tropical (CIAT)
Cali, Colombia

Selene Casella
Institute of Integrative Biology
University of Liverpool
Liverpool, United Kingdom

Hyun-Sug Choi
Department of Horticulture
Catholic University of Daegu
Gyeongsan, Republic of South Korea

Yelena Churakova
Department of Pathology
Weill Cornell Medicine
New York, New York

Soleyman Dayani
Department of Agricultural Biotechnology
Payame Noor University (PNU)
Tehran, Iran

Jack Dekker
New Weed Biology Laboratory
Ames, Iowa

A.G. Dobrikova
Institute of Biophysics and Biomedical Engineering
Bulgarian Academy of Sciences
Sofia, Bulgaria

Rama Shanker Dubey
Department of Biochemistry
Faculty of Science
Banaras Hindu University
Varanasi, India

Lothar Esser
Laboratory of Cell Biology
Center for Cancer Research
National Cancer Institute
National Institutes of Health
Bethesda, Maryland

Raquel Esteban
Institute of Agrobiotechnology
IdAB-CSIC-UPNA-Government of Navarre
Pamplona, Spain

Gianluca M. Farinola
Department of Chemistry
Università degli Studi di Bari "Aldo Moro"
Bari, Italy

Muhammad Farooq
Department of Agronomy
University of Agriculture
Faisalabad, Pakistan

and

The UWA Institute of Agriculture
The University of Western Australia
Crawley, Western Australia, Australia

and

College of Food and Agricultural Sciences
King Saud University
Riyadh, Saudi Arabia

Beatriz Fernández-Marín
Functional Plant Biology
Institute of Botany
University of Innsbruck
Innsbruck, Austria

Carlos M. Figueroa
Instituto de Agrobiotecnología del Litoral (IAL, UNL
  CONICET)
and
Facultad de Bioquímica y Ciencias Biológicas, UNL
Centro Científico Tecnológico Santa Fe
Santa Fe, Argentina

Daisuke Fukayama
Division of Materials Science
Faculty of Pure and Applied Science
University of Tsukuba
Tsukuba, Ibaraki, Japan

José Ignacio García-Plazaola
Department of Plant Biology and Ecology
Universidad del País Vasco (UPV/EHU)
Bilbao, Spain

Katya Georgieva
Institute of Plant Physiology and Genetics
Bulgarian Academy of Sciences
Sofia, Bulgaria

Reimund Goss
Department of Plant Physiology
Institute of Biology
Faculty of Biosciences, Pharmacy and Psychology
University of Leipzig
Leipzig, Germany

Bernard Grodzinski
Department of Plant Agriculture
University of Guelph
Guelph, Ontario, Canada

Mengmeng Gu
Department of Horticultural Sciences
Texas A&M AgriLife Extension
College Station, Texas

Sergio A. Guerrero
Instituto de Agrobiotecnología del Litoral
  (IAL, UNL CONICET)
and
Facultad de Bioquímica y Ciencias Biológicas, UNL
Centro Científico Tecnológico Santa Fe
Santa Fe, Argentina

K.N. Guruprasad
School of Life Sciences
Devi Ahilya University
Indore, Madhya Pradesh, India

Parisa Heydarizadeh
Department of Agronomy and Plant Breeding
College of Agriculture
Isfahan University of Technology
Isfahan, Iran

Andreas Holzinger
Functional Plant Biology
Institute of Botany
University of Innsbruck
Innsbruck, Austria

Fang Huang
Institute of Integrative Biology
University of Liverpool
Liverpool, United Kingdom

Bernhard Huchzermeyer
Department of Biology
Northeastern University
Boston, Massachusetts

Alberto A. Iglesias
Instituto de Agrobiotecnología del Litoral
    (IAL, UNL CONICET)
and
Facultad de Bioquímica y Ciencias Biológicas, UNL
Centro Científico Tecnológico Santa Fe
Santa Fe, Argentina

L.K. Ignatova
Institute of Basic Biological Problems of Russian Academy
    of Sciences
Pushchino, Russia

Louis J. Irving
Graduate School of Life and Environmental Sciences
University of Tsukuba
Tsukuba, Ibaraki, Japan

M. Anowarul Islam
Department of Plant Sciences
University of Wyoming
Laramie, Wyoming

B.N. Ivanov
Institute of Basic Biological Problems of Russian Academy
    of Sciences
Pushchino, Russia

Brian R. Jordan
Faculty of Agriculture and Life Sciences
Lincoln University
Lincoln, New Zealand

A.K. Joshi
Department of Genetics and Plant Breeding
Institute of Agricultural Sciences
Banaras Hindu University
Varanasi, India

Terumitsu Kanjoh
Division of Materials Science
Faculty of Pure and Applied Science
University of Tsukuba
Tsukuba, Ibaraki, Japan

Sunita Kataria
School of Life Sciences
Devi Ahilya University
Indore, Madhya Pradesh, India

Masanobu Kawachi
National Institute for Environmental Studies
Tsukuba, Ibaraki, Japan

Koichi Kobayashi
Department of Life Sciences
Graduate School of Arts and Sciences
University of Tokyo
Komaba, Tokyo, Japan

Masami Kobayashi
Division of Materials Science
Faculty of Pure and Applied Science
University of Tsukuba
Tsukuba, Ibaraki, Japan

Naomi Kodama
Swiss Federal Institute for Forest, Snow and Landscape
    Research (WSL)
Birmensdorf, Switzerland

and

School of Human Science and Environment
Hyogo University
Shinzaike-honcho, Himeji, Japan

Hirohisa Komatsu
Division of Materials Science
Faculty of Pure and Applied Science
University of Tsukuba
Tsukuba, Ibaraki, Japan

Łucja Kowalewska
Department of Anatomy and Cytology of Plants
Faculty of Biology
University of Warsaw
Warsaw, Poland

Hans-Werner Koyro
Institut für Pflanzenökologie
Justus-Liebig-Universität
Gießen, Germany

Toyoki Kozai
Japan Plant Factory Association
Chiba University
Kashiwa, Chiba, Japan

Simona la Gatta
Department of Chemistry
Università degli Studi di Bari "Aldo Moro"
Bari, Italy

Nicoletta La Rocca
Department of Biology
University of Padova
Padova, Italy

María Valeria Lara
Centro de Estudios Fotosintéticos y Bioquímicos (CEFOBI),
    CONICET
Facultad de Ciencias Bioquímicas y Farmacéuticas
Universidad Nacional de Rosario
Suipacha, Rosario, Argentina

Evangelos D. Leonardos
Department of Plant Agriculture
University of Guelph
Guelph, Ontario, Canada

Hong Li
Environment and Plant Protection Institute
Chinese Academy of Tropical Agricultural Science
Haikou, Hainan, China

and

Nova Scotia Institute of Agrologists
Truro, Nova Scotia, Canada

Lu-Ning Liu
Institute of Integrative Biology
University of Liverpool
Liverpool, United Kingdom

Na Lu
Center for Environment, Health and Field Sciences
Chiba University
Kashiwa, Chiba, Japan

Mária Májeková
Department of Soil Science
Faculty of Natural Sciences
Comenius University
Bratislava, Slovak Republic

Yehouda Marcus
Department of Plant Molecular Biology & Ecology
George S. Wise Faculty of Life Sciences
Tel Aviv University
Tel Aviv, Israel

Péter Maróti
Department of Medical Physics
University of Szeged
Hungary

Mercedes Martín
Department of Life Sciences
University of Alcalá
Alcalá de Henares, Madrid, Spain

Toru Maruo
Center for Environment, Health and Field Sciences
Chiba University
Kashiwa, Chiba, Japan

Elena Masarovičová
Department of Soil Science
Faculty of Natural Sciences
Comenius University
Bratislava, Slovak Republic

Tatsuru Masuda
Department of General Systems Studies
Graduate School of Arts and Sciences
University of Tokyo
Komaba, Tokyo, Japan

Barry J. Micallef
Department of Plant Agriculture
University of Guelph
Guelph, Ontario, Canada

Gergana Mihailova
Institute of Plant Physiology and Genetics
Bulgarian Academy of Sciences
Sofia, Bulgaria

Francesco Milano
Istituto per i Processi Chimico Fisici
Consiglio Nazionale delle Ricerche
Bari, Italy

Hideaki Miyashita
Graduate School of Human and Environment Studies
Kyoto University
Kyoto, Japan

Fernando P. Molina-Heredia
Instituto de Bioquímica Vegetal y Fotosíntesis
Universidad de Sevilla & CSIC
Sevilla, Spain

Isabella Moro
Department of Biology
University of Padova
Padova, Italy

Agnieszka Mostowska
Department of Anatomy and Cytology of Plants
Faculty of Biology
University of Warsaw
Warsaw, Poland

K. Razi Naqvi
Department of Physics
Norwegian University of Science and Technology
Trondheim, Norway

Mayumi Ohnishi-Kameyama
National Food Research Institute, NARO
Tsukuba, Ibaraki, Japan

Hiroshi Ono
National Food Research Institute, NARO
Tsukuba, Ibaraki, Japan

Jitendra K. Pandey
Laboratory of Biochemistry and Molecular Biology
School of Life Sciences
Sambalpur University
Odisha, India

Shree P. Pandey
Department of Biological Sciences
Indian Institute of Science Education and Research Kolkata
Mohanpur Campus
Nadia, West Bengal, India

Jutta Papenbrock
Institut für Botanik
Gottfried Wilhelm Leibniz Universität
Hannover, Germany

Claudia V. Piattoni
Instituto de Agrobiotecnología del Litoral
    (IAL, UNL CONICET)
and
Facultad de Bioquímica y Ciencias Biológicas, UNL
Centro Científico Tecnológico Santa Fe
Santa Fe, Argentina

Florencio E. Podestá
Centro de Estudios Fotosintéticos y Bioquímicos
    (CEFOBI, CONICET)
Facultad de Ciencias Bioquímicas y Farmacéuticas (UNR)
Suipacha, Rosario, Argentina

Roberta Ragni
Department of Chemistry
Università degli Studi di Bari "Aldo Moro"
Bari, Italy

Idupulapati M. Rao
Centro Internacional de Agricultura Tropical (CIAT)
Cali, Colombia

Nicoletta Rascio
Department of Biology
University of Padova
Padova, Italy

Abdul Rehman
Department of Agronomy
University of Agriculture
Faisalabad, Pakistan

Piotr Robakowski
Poznan University of Life Sciences
Faculty of Forestry
Department of Forest Sites and Ecology
Poznan, Poland

Anna M. Rychter
Institute of Experimental Plant Biology and Biotechnology
Faculty of Biology
University of Warsaw
Warsaw, Poland

Bartolomé Sabater
Department of Life Sciences
University of Alcalá
Alcalá de Henares, Madrid, Spain

Mohammad Reza Sabzalian
Department of Agronomy and Plant Breeding
College of Agriculture
Isfahan University of Technology
Isfahan, Iran

Kadambot H.M. Siddique
The Institute of Agriculture
The University of Western Australia
Crawley, Western Australia, Australia

Yuhta Sorimachi
Division of Materials Science
Faculty of Pure and Applied Science
University of Tsukuba
Tsukuba, Ibaraki, Japan

Å. Strid
School of Science and Technology
Örebro University
Örebro, Sweden

Michiko Takagaki
Center for Environment, Health and Field Sciences
Chiba University
Kashiwa, Chiba, Japan

R. Roberto Tangorra
Department of Chemistry
Università degli Studi di Bari "Aldo Moro"
Bari, Italy

Alejandro Torrado
Instituto de Bioquímica Vegetal y Fotosíntesis
Universidad de Sevilla & CSIC
Seville, Spain

Karina E.J. Trípodi
Centro de Estudios Fotosintéticos y Bioquímicos
   (CEFOBI, CONICET)
Facultad de Ciencias Bioquímicas y Farmacéuticas (UNR)
Suipacha, Rosario, Argentina

Massimo Trotta
Istituto per i Processi Chimico Fisici
Consiglio Nazionale delle Ricerche
Bari, Italy

Ivana Vykouková
Department of Soil Science
Faculty of Natural Sciences
Comenius University
Bratislava, Slovak Republic

Katsuhiro Wada
Division of Materials Science
Faculty of Pure and Applied Science
University of Tsukuba
Tsukuba, Ibaraki, Japan

Abdul Wahid
Department of Botany
University of Agriculture
Faisalabad, Pakistan

J.J. Wargent
Institute of Agriculture and Environment
Massey University
Palmerston North, New Zealand

Randy Wayne
Laboratory of Natural Philosophy
Section of Plant Biology
School of Integrative Plant Science
Cornell University
Ithaca, New York

Christian Wilhelm
Department of Plant Physiology
Institute of Biology
Faculty of Biosciences, Pharmacy and Psychology
University of Leipzig
Leipzig, Germany

Di Xia
Laboratory of Cell Biology
Center for Cancer Research
National Cancer Institute
National Institutes of Health
Bethesda, Maryland

Toshitaka Yamaguchi
Center for Environment, Health and Field Sciences
Chiba University
Kashiwa, Chiba, Japan

Wataru Yamori
Center for Environment, Health and Field Sciences
Chiba University
Kashiwa, Chiba, Japan

and

Japan Science and Technology Agency (JST), PRESTO
Kawaguchi, Saitama, Japan

Miho Yoshioka-Nishimura
Graduate School of Natural Science and Technology
Okayama University
Okayama, Japan

Chang-An Yu
Department of Biochemistry
Oklahoma State University
Stillwater, Oklahoma

Fei Zhou
Laboratory of Cell Biology
Center for Cancer Research
National Cancer Institute
National Institutes of Health
Bethesda, Maryland

Section I

Principles of Photosynthesis

1 Biogenesis of Thylakoid Membranes
Correlation of Structure and Function

Łucja Kowalewska and Agnieszka Mostowska

CONTENTS

1.1 Introduction 3
1.2 Biogenesis of Chloroplasts as a Process Highly Coordinated with Plant Development 3
1.3 Chloroplast Biogenesis at the Structural Level...4
1.4 Changes in Proteins, Lipids, and Photosynthetic Pigments during Chloroplast Biogenesis.................6
1.5 Changes in Gene Expression during Chloroplast Biogenesis..7
1.6 Correlation of Structure and Function during Chloroplast Biogenesis 8
    1.6.1 Model of Mutual Relations between Membrane Composition and Structure during Chloroplast Differentiation ... 8
        1.6.1.1 Microscopic Analysis of Subsequent Chloroplast Biogenesis Stages...............................8
        1.6.1.2 Appearance of CP Complexes during Chloroplast Biogenesis...10
        1.6.1.3 Functional Changes of Photosynthetic Apparatus during Biogenesis 10
        1.6.1.4 Immunodetection of Proteins during Biogenesis...12
1.7 Conclusions...12
Acknowledgments...13
References...13

1.1 INTRODUCTION

The development of photosynthesis was one of the most fundamental evolutionary processes that shaped recent living organisms. It is generally believed that the ability of oxygen photosynthesis was developed several billion years ago by ancestors of the present cyanobacteria and, later, by the endosymbiosis implemented in algae and higher plants [1,2]. Chloroplasts, which are endosymbionts of the latter, are highly specialized organelles that perform photosynthesis [3].

Light-dependent photosynthetic machinery is localized within thylakoids, a well-developed system of chloroplast internal membranes [4,5]. Because of the unique architecture of pigment–protein complexes (chlorophyll–protein [CP] complexes), chloroplasts trap solar energy with higher efficiency than man-made photovoltaic systems.

Rearrangements of the thylakoid membranes are crucial for photosynthetic efficiency. Therefore, attempts to develop artificial photosynthetic systems based on the structure and function of photosynthesis [6] depend on the interdisciplinary research on the structure, biochemistry, and functionality of the thylakoid membranes and also on their molecular regulation. The understanding of the mechanism of chloroplast membranes' biogenesis at different levels of the chloroplast organization is important for development of this area of plant biology. It is important to remember that chloroplast biogenesis can be modulated by different environmental factors such as temperature, light intensity, etc. Biogenesis can vary depending on plant tissue, organ, and, of course, the plant species.

1.2 BIOGENESIS OF CHLOROPLASTS AS A PROCESS HIGHLY COORDINATED WITH PLANT DEVELOPMENT

Chloroplast biogenesis is a process highly coordinated with plant development in which the chloroplast precursors, which are unable to perform photosynthesis, differentiate into fully mature and functional chloroplasts. Chloroplasts develop from proplastids, small organelles in leaf primordia, which possess small vesicles containing little or no photosynthetic complexes. In light, during natural photomorphogenesis, proplastids transform to mature chloroplasts with a developed thylakoid network and photosynthetic capacity. In the dark, proplastids differentiate into etioplasts. The development of chloroplasts through the stage of etioplasts often proceeds during the initial plant ontogenesis, when the first stages of seed germination and seedling growth take place without light (Figure 1.1). This process, called scotomorphogenesis, is a frequent and natural process, which does not disturb plant growth nor the development of differentiating chloroplasts; it can be considered as a form of adaptation of the angiosperm plants to the initial growth process under the ground [7].

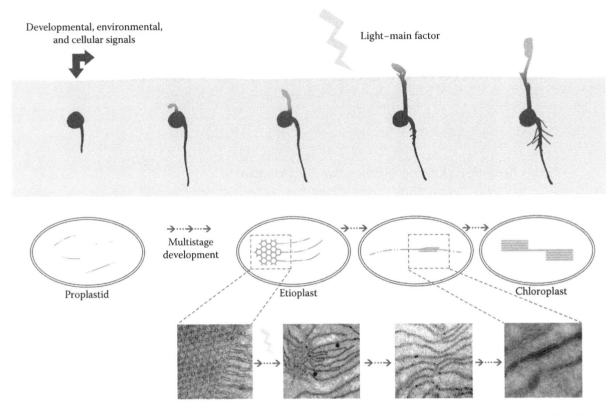

FIGURE 1.1 **(See color insert.)** Chloroplast biogenesis in leaves of a hypogeal germinating plant coordinated with seedling development.

Following illumination, young seedlings "switch" to photomorphogenesis [8], and etioplasts differentiate into chloroplasts (Figure 1.1). During this process, the tubular system of the prolamellar body (PLB), a characteristic feature of etioplasts, transforms into a linear system of lamellae arranged parallel to each other, and the first grana appear [9 and literature therein]. Such photomorphogenic transition, a so-called de-etiolation or greening process, is completed within one to several days depending on the species and light conditions [9].

Differentiation of mature photosynthetically active chloroplasts is a highly coordinated process, which requires cooperation of different cell compartments [10] (Figure 1.2). The main processes that take place during differentiation of chloroplasts are exogenous and endogenous signals, especially light perception and subsequent expression of nuclear and plastid genes; biosynthesis of proteins, lipids, and pigments; import of appropriate pigment-binding proteins into developing chloroplasts; insertion of these proteins into the thylakoid membranes; and finally, protein assembly into functional complexes (Figure 1.2). At the very end, the thylakoid membrane structure in mature chloroplasts is arranged into a 3-D network of cylindrical grana linked to each other by the stroma lamellae [9 and literature therein].

In spite of the fact that the main steps of chloroplast differentiation were already described in several plant species, all data were scattered: in some papers, the ultrastructural differentiation of chloroplasts during greening was described [8,11–13], while in other papers proteomic and functional analyses were done [14–16].

Below we present collected data on different levels of chloroplast biogenesis.

1.3 CHLOROPLAST BIOGENESIS AT THE STRUCTURAL LEVEL

When the first stages of seed germination and seedling growth take place without light, the proplastids develop into etioplasts. The main and a very characteristic feature of etioplasts are PLBs with a regular, paracrystalline structure. Together with prothylakoids (PTs), flattened porous membranes, PLBs are precursors of the chloroplast thylakoid membranes [11,12]. The reasons for the formation and maintaining of such regular structure in the darkness and its degradation in light are still not fully understood. Protochlorophyllide–protochlorophyllide oxidoreductase–NADPH (Pchlide:LPOR:NADPH) complex is probably responsible for keeping the paracrystalline structure intact. According to Solymosi and coworkers [17,18] and Solymosi and Schoefs [7], the PLB, because of its maximum packing of membranes, should play a crucial role in chloroplast differentiation. However, mutual interaction of the complex components, interaction with the membrane lipids, and localization in PLBs are still far from being understood [19]. Apart from Pchlide bound to the LPOR and NADPH, the PLB usually contains two main carotenoids (lutein and violaxanthin), proteins, and lipids [20 and literature therein].

Light triggers transformation of etioplasts into chloroplasts starting photomorphogenetic processes at structural, molecular, and physiological levels. An important function of the PLB

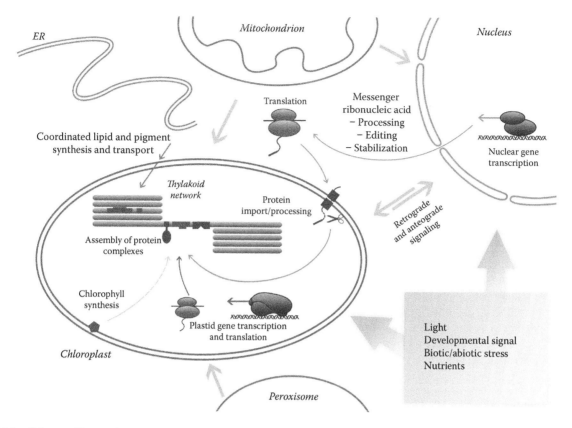

FIGURE 1.2 Scheme of internal and external factors that influence chloroplast biogenesis with main processes occurring during thylakoid network formation.

is also protection of the membrane components from light in developing chloroplasts. Within PLB, Pchlide and chlorophyllide (Chlide) are bound into complexes that efficiently dissipate energy, preventing reactive oxygen species (ROS) formation [21]. In spite of these considerations, the direct role of PLBs in thylakoid formation in light remains unclear [14]. The role of the PLBs in etioplast–chloroplast transformation is discussed broadly in the review of Adam and coworkers [9]. Recently, Grzyb and coworkers [22] have shown, on isolated wheat PLBs, that Pchlide photoreduction was followed by the disruption of the PLB lattice, and the formation of vesicles around the PLBs was observed. Data on the isolated PLBs obtained by transmission electron microscope (TEM) were correlated with atomic force microscopy (AFM) results [22]. It is important that PLBs are present not only in etioplasts formed in darkness. Paracrystalline PLBs can coexist with grana in etiochloroplasts, when chloroplasts that are not fully developed are transferred to darkness for a short time. The question arises as to what the role of PLBs is in functionally active chloroplasts.

Chloroplast development from proplastids has not often been a subject of studies [23]. This process was well described during the development of the monocotyledon leaf. Such leaves exhibit a basipetal differentiation; this means that the youngest cells, usually meristematic cells, are located in the basal part of the leaf with chloroplasts of the youngest development stage—proplastids—while the oldest ones are at the apical part [24,25]. Because of the special pattern of monocotyledon leaf development, all chloroplast developmental stages

can thus be found within a single leaf. Proplastids are the simplest plastids. They are small (1–2 µm) spherical organelles with many invaginations from the internal membrane of the proplastid envelope and a small number of vesicles originating from it [25]. After illumination, a system of thylakoids differentiates, leading to the final stage where fully developed chloroplasts have a sophisticated system of appressed (grana thylakoids) and nonappressed thylakoids (stroma lamellae) joined together by margin regions of a different enzymatic activity. Grana size and the ratio of the grana thylakoids to the stroma thylakoids are characteristic of given species of higher plants and depend on environmental factors [5,26].

The spatial structure of fully differentiated chloroplasts has been a subject of investigations for many years, starting from models based on the analysis of the TEM images of Gunning and Steer [23] and, more recently, by Mustardy and coworkers [27]. The arrangement of the thylakoid membrane of the mature chloroplasts based on electron tomography has been presented recently by several authors [28–30]. According to their data the thylakoid membranes are arranged into a 3-D network, with cylindrical grana linked to each other by the stroma lamellae [9 and literature therein]. It seems that the quasi-helical model, based on the Paolillo model [31], presents the most probable structure of the mature thylakoid grana [27,30].

The reconstruction of the spatial structure of the *in vivo* chloroplast membranes of isolated, intact, fully differentiated chloroplasts, based on chlorophyll (Chl) autofluorescence and by stacking of the optical slices, was performed by our group in the confocal laser scanning microscopy (CLSM). These results will be discussed in Section 1.6.

1.4 CHANGES IN PROTEINS, LIPIDS, AND PHOTOSYNTHETIC PIGMENTS DURING CHLOROPLAST BIOGENESIS

Chloroplast differentiation during plant ontogenesis is related not only to the formation of the sophisticated system of the internal membranes but also to the intensive protein, pigment, and lipid synthesis. Data on the proteomic analysis of plastid membranes during their development are scattered and connected mainly with either the dark stage of etioplasts or the fully mature stage of differentiated chloroplasts. Detailed proteomic analysis was performed on isolated PLBs of rice [15] and wheat [14]. Among 64 proteins extracted from wheat leaf PLBs are those related with pigment biosynthesis, the Calvin cycle, chaperon synthesis, and other proteins connected with the light photosynthetic reactions. This is a strong argument that the PLB membranes are direct precursors of thylakoids during light differentiation [14].

Photosynthetic pigments located in PLBs and their changes during chloroplast differentiation from etioplasts in light are relatively well known (Figure 1.3). PLBs contain two spectral forms of the LPOR–Pchlide complexes with different absorption maxima: LPOR–Pchlide at 640 nm and POR–Pchlide at 650 nm [32]. LPOR is one of the major proteins of the PLB membranes and is responsible for the NADPH-dependent reduction of Pchlide to Chlide [32]. Analysis of the low-temperature fluorescence (77 K) spectra revealed three major types of Pchlide: a nonphotoconvertible form, Pchlide 628–633 nm, and two photoconvertible forms, Pchlide 640–645 nm and Pchlide 650–657 nm [33–35]. These long-wavelength forms are bound to PLBs, while the short-wavelength form, unbound to LPOR, is mainly found in PTs [36]. It was demonstrated that the *in vivo* association of LPOR with PLBs and PT membranes requires NADPH and ATP but not the ternary Pchlide:LPOR:NADPH complex [37]. Light leads to a conversion of the photoconvertible Pchlide to Chlide, which is accompanied by the emission's maximum shift from 650 nm to 678 nm, then to 684 nm, and after longer light exposure, to 672 nm (Shibata shift) [38]. The Shibata shift is attributed to the release of Chlide from LPOR. This release leads

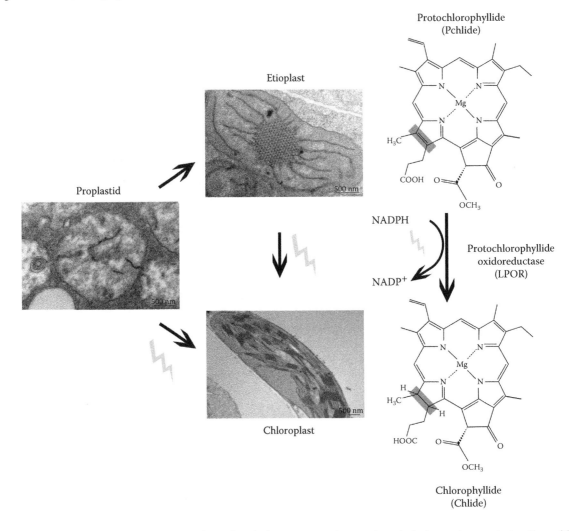

FIGURE 1.3 Scheme of basic plastid form transformation during scotomorphogenesis and photomorphogenesis together with only one light-dependent reaction in the chlorophyll biosynthesis pathway.

to a conversion of the PLB membranes to the thylakoid membrane system [32]. Conversion of Pchlide to Chlide has been well described; both photoactive and nonphotoactive Pchlide forms were identified in different plant species during growth in the natural photoperiod after each dark phase [38]. Apart from serving other functions, the PLB is also a reservoir of Pchlide and its reduced and esterified forms: Chlide and Chl. As long as the PLB structure is regular, chlorophyll synthesis is protected against proteolysis [14,36].

It is known that carotenoid biosynthesis undergoes dynamic changes during plant ontogenesis, especially during germination, photomorphogenesis, fruit development, and photosynthesis. Carotenoids also play an essential role in the photosynthetic apparatus function; they are also important for the formation of PLBs. Generally four carotenoids accumulate in chloroplasts of higher plants: lutein, β-carotene, violaxanthin, and neoxanthin [39–41]. Lutein is the most abundant both in chloroplasts and etioplasts. Therefore, its role in the formation of PLBs needs to be elucidated. It was already demonstrated, in lutein-deficient *Arabidopsis* mutants, that perturbations in carotenoid biosynthesis result in a disturbance in PLB formation [41]. The *ccr2 Arabidopsis* mutant, which is a carotenoid-isomerase mutant, exhibits a substantial reduction in lutein accumulation, and in darkness the *ccr2* seedlings lack PLBs [41]. However, it is not clear whether it is connected with accumulation of the *cis* carotenes or due to a loss of activity of the carotenoid isomerase [41]. It was shown, with the help of the *ccr1* mutant (carotenoid chloroplast regulatory mutant), that the lutein level in *Arabidopsis* is regulated by a chromatin-modifying enzyme (histone methyltransferase SDG8), but still, understanding of the regulatory mechanisms of carotenoid biosynthesis and composition is not sufficient [39]. It should be pointed out that even fundamental changes in the carotenoid composition affect neither total carotenoid content nor chlorophyll content, which proves the easy compensation for one carotenoid by another and the adjustability of the photosynthetic apparatus of higher plants [42].

Glycoglycerolipids are important structural lipids of the plastid membranes both in dark and in light; therefore, their role in PLB formation, structure, and transformation as well as in the early steps of grana formation during early biogenesis of chloroplasts should be clarified. The thylakoid membrane lipids mainly comprise two classes of glycoglycerolipids: monogalactosyldiacylglycerol (MGDG), around 50% of total membrane lipids, and digalactosyldiacylglycerol (DGDG), around 30% of total membrane lipids; both classes of glycoglycerolipids are necessary for photosynthesis [43–45]. MGDG is the main polar component of the PLB and, together with the Pchlide:LPOR:NADPH complex, plays an important role in the formation and stabilization of PLBs [46]. An important role of glycoglycerolipids was shown in *Arabidopsis* mutants with a reduced level of MGDG (*mgd-1*) and DGDG (*dgd-1*). The *mgd-1* mutants had 50% chlorophyll and a distorted chloroplast ultrastructure compared with wild plants, while the *dgd-1* mutants exhibited a distorted protein composition of both photosystems together with the level of the xanthophyll cycle pigments [46,47]. Recent results of

Mysliwa-Kurdziel and coworkers [19] shed light on the role of lipids in Pchlide aggregation in the etioplast membranes in the *in vitro* system simulating natural conditions. These results strongly pointed to the role of glycoglycerolipids of the PLB *in vivo* in regulating the proportion between Pchlide aggregates and monomers, which is important for the proper assembly of the ternary photoactive Pchlide:LPOR:NADPH complexes and for an efficient Pchlide-to-Chlide photoreduction.

The chloroplast structure of higher plants is determined by the protein composition of the CP complexes, their distribution in the thylakoid membranes, and the membrane lipids [48]. Regarding biogenesis of CP complexes, photosystem I (PSI) and photosystem II (PSII) were analyzed in different species [49,50], in some cases also during etioplast–chloroplast transformation [51]. The first step in CP-complex organization is the insertion of an anchor protein in the thylakoid membrane, which is crucial for the location of other components of certain complexes in such a membrane. Such anchor proteins are the main subunits of complexes: D2 for PSII, PetB for Cytb_6f, and PsaB for PSI [52]. In the absence of these proteins, the translation of further components is inhibited [53]. Such a process is regulated by the mechanism referred to as control by epistasy of synthesis (CES).

CP-complex components are of both nuclear and chloroplast origin. Therefore, the analysis of the CP complexes deals with different steps and levels of their formation, such as light signal perception, regulation of the transcription and translation, transport to the plastid and within a plastid, balance between protein synthesis and degradation, and also interaction between proteins and lipids [20,54].

The chloroplast structure is also determined by the arrangement of the photosynthetic complexes within the thylakoid membranes [55]. The structure and arrangement of appressed and nonappressed thylakoids within chloroplasts depend on different features, e.g., the lateral separation of light-harvesting complex II (LHC II) and PSII from PSI and ATP synthase. They are also characteristic for the plant species, although they can change under different environmental conditions [5,26,55–57].

1.5 CHANGES IN GENE EXPRESSION DURING CHLOROPLAST BIOGENESIS

One of the processes that proceed in parallel to structural changes during chloroplast biogenesis in light is the accumulation of the gene transcripts regulated by light. It was demonstrated by microarray analysis that hundreds of *Arabidopsis* genes are regulated by different light intensities and colors at the transcriptional level. Moreover, a large number of identified *Arabidopsis* mutants with distorted photomorphogenesis showed how complicated the process of gene regulation is [58]. Investigations performed on the well-recognized *Arabidopsis hy5* mutant with distorted chloroplast development proved that HY5 is one of the main modulators coordinating the perception of light signals and regulating the expression of genes induced by light [58].

Many mutations influence chloroplast development and the formation of the thylakoid network, but it is difficult to point out their direct impact on each of these processes [25]. Mutation *vipp1* is an example, which manifests by the inhibition of the thylakoid and vesicle formation in *Arabidopsis* chloroplasts [59]. In the case of *Synechocystis* mutants without the Vipp1 protein (vesicle inducing protein in plastid), the formation of the thylakoid membranes was not disturbed, but the inhibition of PSI biogenesis and not that of PSII was noticed [60]. Such discrepancy does not allow us to arrive at a clear conclusion on the regulation role of the Vipp1 protein [61]. Various factors play different roles in the organization of the CP complexes. One of the main factors necessary for the integration of the LHC proteins with the thylakoid membrane is the ALB3 protein (ALBINO 3) [62], while proteins Ycf3, Ycf4, and Ycf37 interact with the PSI core proteins, influencing the proper assembly of the PSI complex [52].

Chloroplast biogenesis is regulated by a coordinated mechanism of joined action of nuclear and chloroplast gene expression. A "retrograde" mechanism transmitting signals from a plastid to the nucleus coordinates the expression of the nuclear genes encoding plastid proteins with the metabolism and the plastid stage of development [63,64]. Such a mechanism provides an assembly of complex subunits, encoded by both the nucleus and the plastid in a proper stoichiometry; this mechanism also coordinates the direct reorganization of the complex as a response to a changing environment [63]. Mg-protoporphyrin IX (Mg-ProtoIX), a intermediate of chlorophyll biosynthesis, is one of the signaling factors that accumulates in plastid due to stress conditions, later is exported from the plastid to the cytoplasm, and then inhibits expression of the photosynthetic genes. Proper coordination of nuclear and plastid gene expression is crucial for chloroplast biogenesis because most of the multicomponent complexes are of both nuclear and plastid origin [65]. Such coordination is regulated at the level of transcript maturation, posttranscriptional modifications, splicing, regulation of translation, and organization of proteins in the functional complexes [54 and literature therein]. Some *Arabidopsis* mutants have not only disturbed development but also the expression of some specific genes after transfer from darkness to light or vice versa. For example, pea mutant *lip1*, after re-etiolation, exhibits a higher transcript level of some photosynthetic genes as compared to wild type plants [66], while *Arabidopsis* mutant *ab13* maintains the proper paracrystalline PLB structure after transfer from darkness to light [67]. In spite a lot of data on mutants with disturbed chloroplast biogenesis [64], not much directly related to plant development can be found.

Concluding the chloroplast biogenesis is a multistage process that proceeds on many organizational levels. It requires cooperation of both cellular and external factors acting at a certain time of the plant ontogenesis. From external factors, a crucial role is played by light, which can modify both the transcriptional and translational level and also the stability of newly synthesized proteins. Both chloroplast and cellular internal factors can also modulate many processes, from the transduction of signals to the organization of the thylakoid membrane architecture together with the arrangement of the CP complexes.

Many important questions on chloroplast biogenesis remain unanswered: What are the relations between chloroplast development and plant photomorphogenesis? How do the interactions between organelles influence chloroplast biogenesis? Does the retrograde signaling influence chloroplast differentiation directly? Are there any control points of the chloroplast biogenesis?

1.6 CORRELATION OF STRUCTURE AND FUNCTION DURING CHLOROPLAST BIOGENESIS

1.6.1 MODEL OF MUTUAL RELATIONS BETWEEN MEMBRANE COMPOSITION AND STRUCTURE DURING CHLOROPLAST DIFFERENTIATION

The biogenesis of chloroplasts has been a subject of studies for many years [7,20,64,68 and literature therein], but the direct correlation of simultaneous changes at the biochemical, structural, and functional levels was studied only recently [16,69].

It is important to find changes in the photosynthetic protein levels, the functionality of photosynthetic apparatus and CP-complex presence, and the structure of etioplasts and proplastids transforming into mature chloroplasts taking place at the same time intervals.

Complex analysis of all these parameters was performed by our group [69] in exactly the same stages of plant ontogenesis to correlate at the same time composition, structure, and function of developing chloroplasts. Although particular biochemical, physiological, and structural aspects of chloroplast biogenesis were demonstrated in the literature, their mutual relations in the developmental stages have never been shown.

Such analysis has been done on the so-called etioplast model imitating the natural initial seedling growth under the soil surface. In this model, the chloroplast with typical stroma and grana thylakoids differentiate from etioplast with a regular paracrystalline PLB and porous PTs.

Another model, the so-called proplastid model, in which a plant after germination grows in light, did not give reliable results, because it is hard to get synchronized stages of chloroplast development. Therefore, this model was used for selected analysis of some points in time. In both models, natural conditions of day and night were imitated (16 h photoperiod). The study was performed on two closely related plant species: pea (*Pisum sativum* L.) and bean (*Phaseolus coccineus* L.), which differ in thylakoid membrane arrangement. There are large appressed thylakoid regions in peas and less distinguished appressed regions in bean chloroplasts [26,55,56,69].

1.6.1.1 Microscopic Analysis of Subsequent Chloroplast Biogenesis Stages

Transmission electron microscopy was the basis for determination of the main structural stages of biogenesis for all correlations. TEM detail analysis during time development allowed choosing of the crucial stages of inner membrane transformations for bean and pea plastids. In this way, the time within the

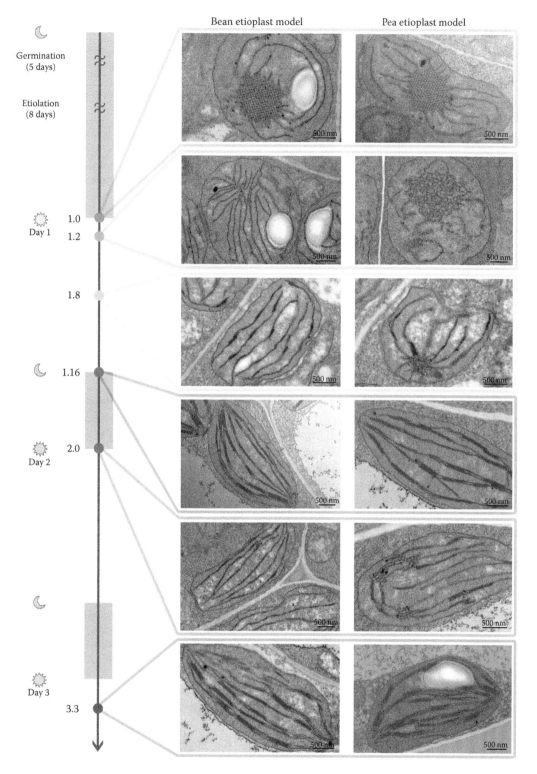

FIGURE 1.4 TEM images of subsequent stages of chloroplast biogenesis during day/night growth in the etioplast model. Each stage is presented with the corresponding sample collection point on the timeline of seedling germination and growth. Numbers refer to day, number, and hours of illumination, e.g., 1.2 denotes two hours of illumination during the first light growth. These numbers are given in parentheses in the text.

day/night cycle was assigned to particular moments of structural changes in both species.

In the etioplast model, the same several stages of chloroplast development in characteristic times for both species are shown in Figure 1.4; more detailed analysis together with measured distances between adjacent thylakoids in the

granum can be found in the work of Rudowska and coworkers [69].

After 8 days of etiolation (1.0), the plastids of both species had typical features of etioplasts: well-developed PLBs joined with PTs. In both species, PLBs were of the tetrahedral closed type only. Plastoglobules of low electrodensity were localized both inside PLBs and at their peripheries. Exposure to light caused a gradual degradation of PLBs (1.2). The process proceeded faster in bean plastids as compared to peas. The degree of PLB degradation in peas after 2 h of illumination was comparable with the state of PLBs in beans after 1 h of illumination only. The formation of the first grana was observed after 8 h of illumination (1.8) in both species. However, in pea chloroplasts, strongly degraded PLBs were observed, while in beans, no remnants of PLBs were visible. Subsequent 8 h of illumination (1.16) caused complete PLB degradation in peas.

After 8 h of darkness, meaning after the first night of the day/night cycle, in plastids of both species, small but well-formed grana were observed (2.0). After this dark period, a partial reconstruction of PLBs could be observed in peas only. Moreover, the distance between adjacent grana membranes increased in both species, which means that the membranes were less appressed than earlier (1.16).

After the second night of the experiment, no recrystallization of PLBs was seen in both species. The last observed stage was after 3 h of light during the third day of the experiment (3.3). At this stage, well-developed grana and starch grains were seen. The distance between adjacent grana was similar in both species and was similar to the one in mature chloroplasts. Generally, the ultrastructure of this last analyzed stage was comparable with that in mature chloroplasts except for smaller grana and smaller chloroplast dimensions [26,55,56].

A similar ultrastructural analysis was performed also for the proplastid model, that is, without the initial etiolation. The detailed process of proplastid differentiation into chloroplast is presented in Figure 1.5. No homogeneous plastid population was observed in pea nor in bean seedlings; this made it hard to follow changes in the plastid structure during the subsequent illumination. The majority of plastids possessed well-formed paracrystalline PLB, so they were etioplasts, in fact. Degradation of PLBs was faster than in the etioplast model during the next hours of development. After 2 h of illumination, there were not even remnants of PLBs. Moreover, the first appressed membranes appeared faster than in the etioplast model, as early as after 2 h of light treatment. The first appressed membranes did not mean that fully developed grana appeared faster as well. The grana stacks observed in the final stage of the experiment, that is, after 3 h of illumination during the third day of the experiment (3.3), contained fewer thylakoids than those in the etioplast model. For further investigations, only the etioplast model, characterized by a highly synchronized biogenesis process in the total plastid population, was used. It this way, a clearer picture of the time development studies was given.

Structural differences between the bean and pea chloroplasts can be observed also when they are fully developed. The in vivo structure of the mature chloroplasts was examined in these two plants species. For the first time, the 3-D view of the overall structure of the mature chloroplast was obtained by our group from chlorophyll autofluorescence with the help of a confocal laser scanning microscope [55,69]. The images show round red bodies with red intensive fluorescence in the range of 660–700 nm separated from each other by dark nonfluorescent areas. The red fluorescence comes from the LHC supercomplex LHCII–PSII, dimers/monomers of PSII, and free trimers of LHCII localized in grana [70]. At room temperature, the Chl autofluorescence in the nonappressed regions of supercomplex LHCI–PSI with maximum fluorescence around 739 nm is very low, and visualization is not possible [71–73]. We recovered the 3-D structures using computer assistance from hundreds of fluorescence images from different focal depths. These reconstructed structures show the spatial layout of the Chl fluorescence.

In other words, the 3-D computer models demonstrated different arrangements of the grana and stroma thylakoids of bean and pea mature chloroplasts [55,70]. In pea chloroplasts, appressed domains that were large and separated from each other were shown, while in bean chloroplasts, less distinguished appressed regions existed [26,55,56]. The structural differences were a consequence of the size differences between the PSII and PSI complexes and also the quantitative and/or qualitative composition of microdomains [55,74,75]. Thus, the structure and arrangement of appressed and nonappressed thylakoids within chloroplasts are characteristic for the plant species.

1.6.1.2 Appearance of CP Complexes during Chloroplast Biogenesis

The analysis of the particular CP complexes' appearance during the greening of pea and bean seedlings was performed with the help of chlorophyll fluorescence at low (77 K) temperature [69]. This method enables registration of the fluorescence from PSII and PSI.

In the case of the etioplast model, no CP-complex fluorescence coming from the CP complexes was observed in etiolated seedlings of both species; only Pchlide:POR:NADPH complex with a fluorescence maximum of 655 nm was found. Formation of the CP complexes proceeded in a similar way in both species. At first, the bands characteristic of the LHCII antenna complexes and the PSII core complexes appeared, followed by bands characteristic of the PSI–LHCI complexes [69]. The only difference between the species was slower changes in peas during the first 4 h of de-etiolation. In fully developed thylakoids of beans and peas, differences in the band intensities and their absorption maxima were observed, indicating different CP-complex organization in these two species This is consistent with our previous results [70].

1.6.1.3 Functional Changes of Photosynthetic Apparatus during Biogenesis

Analysis of the photosynthetic apparatus functional changes was performed, by measurement of chlorophyll a fluorescence in vivo [69]. The measurements were performed on bean and pea leaves adapted to darkness of etiolated and de-etiolated seedlings with the help of a Dual-PAM fluorometer (Waltz). Membranes of etiolated seedlings of both examined species exhibited no photochemical activity. The normal induction curves with proper induction points and subsequent regeneration were observed in beans during the first day of experiment after 16 h of light, while

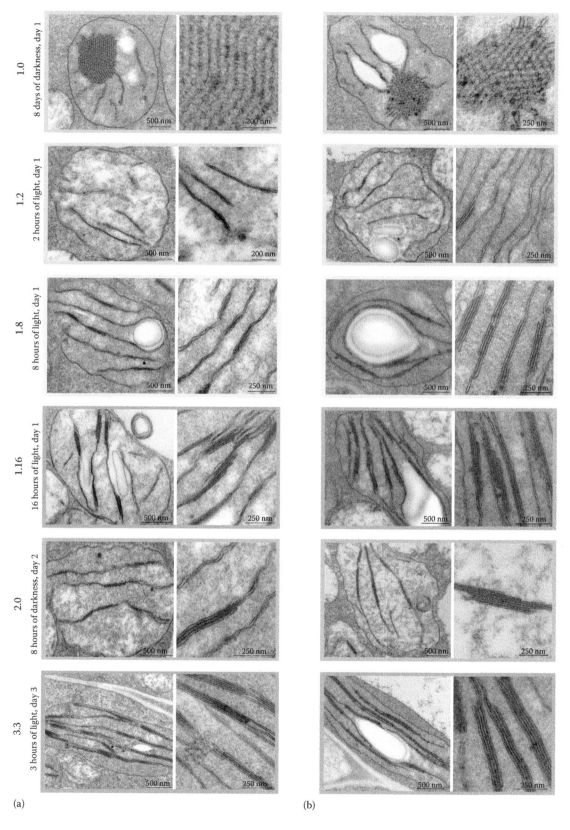

FIGURE 1.5 TEM images of subsequent stages of chloroplast biogenesis during day/night growth in the proplastid model (view on the whole plastid and magnification of characteristic thylakoid arrangements). (a) Bean and (b) pea proplastid models.

in peas, these changes were observed later, on the next day after 8 h of light. The first symptoms of PSI activity were recorded in peas after 8 h of light, and in beans, after 10 h.

The process of CP-complex formation in the thylakoid membranes and their photochemical functionality proceeded faster in beans than in peas.

1.6.1.4 Immunodetection of Proteins during Biogenesis

The studies of the protein composition changes in the cell protein extract were performed using the Western blot technique with the help of antibodies for selected proteins: the PSI and PSII core proteins and the LHCI and LHCII antenna proteins. We present the appearance of these proteins during subsequent hours of the experiment (Table 1.1). Detailed results of the Western blot analysis are given in Ref. [69]. In beans, the Lhcb1 protein of the PSII complex antenna was the first protein that appeared, as early as after 6 h of illumination in leaf extracts. Later, after 12 h of illumination, the Lhcb1 protein registered, and 2 h later, the reaction center protein D2 was found. On the second day of the experiment, the PSII core protein CP43 was found; later, Lhcb4 and, finally, Lhcb3 were registered in the bean leaf extracts. In the case of the PSI complex, Lhca1 was the first antenna protein, and 2 h later, a core PSI protein PsaA was found in the bean leaf extracts. Other antenna proteins, Lhca2 and Lhca3, were registered on the second day of the experiment.

In contrast to beans, the first protein detected in the pea leaf extract was a core PSI protein, PsaA, already found in etiolated plants. At the same time, antenna proteins Lhca2, 2 h later, Lhca3, and 6 h later, Lhca1 were localized. The first proteins from the PSII complex appeared after 8 h of illumination: antenna protein Lhcb2 and PSII reaction center protein D2. Later, after 12 h of light, a PSII core protein, CP43, and also antenna Lhcb1 and Lhcb4 appeared, and finally, Lhcb3 registered on the next day of the experiment.

In short, in extracts from the bean leaves, the main PSII antenna proteins (Lhcb1 and Lhcb2) were localized earlier than the core proteins, and the PSI core proteins, at roughly

TABLE 1.1
Appearance of Photosynthetic Proteins during Chloroplast Biogenesis

Time Point of Protein Appearance	Photosynthetic Protein	
	Bean	Pea
1.0	–	PsaA
1.1	–	Lhca2
1.2	–	Lhca3
1.4	Lhcb1	Lhca1, Lhcb2, D2
1.6	Lhca1, PsaA, Lhcb2	–
1.8	–	Lhcb1, Lhcb4, CP43
1.12	Lhca3	–
1.14	Lhca2, D2	Lhcb3
2.0	Lhcb4, CP43	–
3.3	Lhcb3	–

the same time as the PSI antenna proteins. In the case of peas, however, the core and antenna PSII proteins were localized at the same time, but the PSI core proteins were localized earlier than the corresponding antenna proteins. An unexpected result was the localization of one of the PSI core proteins—PsaA—in the protein extracts of etiolated seedlings, without any exposition to light. These results require further investigation and suggest the existence of a light-independent biosynthesis pathway of this protein.

1.7 CONCLUSIONS

In this review, the correlation of the structural changes of the internal plastid membranes with their functionality together with the photosynthetic protein appearance was presented in two plants species—peas and beans. Both species are good examples to compare because they have the same hypogeal type of germination; they have different low-temperature resistance (the pea is a chilling-tolerant plant, and the bean is chilling sensitive); and they differ in thylakoid membrane organization [69,70].

Studies on the relation between the structural changes and the membrane functionality of developing chloroplasts demonstrated that the formation of active chlorophyll–protein complexes and the structural differentiation of the pea chloroplasts were delayed in comparison to beans during the first day of photomorphogenesis; however, during further processing, these differences were not visible. The structural changes of the bean chloroplast biogenesis during the day/night cycle were sequential—the first stacked membranes were observed after full PLB degradation. In the case of peas, grana formation took place in the presence of a degrading PLB [69]. After-darkness reformation of PLBs observed in peas means that PLB recrystallization can take place also in the case of other plant species. Thus, the chloroplast development in a continuous illumination differs as compared to that proceeding in the day/night cycle. Most data concerning the structural level of chloroplast development come from the continuous light illumination conditions [8,11–13]. Therefore, they are not comparable with our data obtained in periodical light conditions.

As opposed to differences in the structural changes, especially in PLB reformation, we showed that protein biosynthesis as well as photosynthetic complex formation proceeded gradually and irreversibly in peas and beans in spite of periods of darkness. As mentioned before, many proteins connected with the light photosynthetic reactions were localized in the PLB membranes in darkness, which strongly indicates that they are "prepared" to start photosynthetic activity as soon as the PLB tubules transform directly to thylakoids during light differentiation [14]. We demonstrated that both core complexes of PSI and PSII were localized in the chloroplast membranes before the first appressed thylakoids were formed; that means that their presence in the thylakoid membranes was not sufficient for the first grana formation [69]. We also showed that the presence of the main LHCII proteins is necessary for a regular grana arrangement. Finally, it was demonstrated that 3 days in the day/night cycle is sufficient for etioplast transformation to a fully developed and functionally mature chloroplast. This study points out that in order to find correlations and create a global picture of chloroplast biogenesis, it is necessary to analyze different levels of plastid organization at the same time.

These studies also showed that is important to visualize chloroplast development within the total plant ontogenesis and that it is important to preserve the natural day/night cycle photoperiod in order to analyze chloroplast biogenesis; otherwise, the pattern of chloroplast development is modified.

ACKNOWLEDGMENTS

TEM imaging was performed in the Laboratory of Electron Microscopy, Nencki Institute of Experimental Biology, on a JEM 1400 (JEOL Co., Japan) electron microscope. This equipment was installed for the project, sponsored by the EU Structural Funds: Centre of Advanced Technology BIM as an equipment purchase for the Laboratory of Biological and Medical Imaging. This work was supported by National Research Centre Poland, Grant 2014/13/B/NZ3/00413.

REFERENCES

1. Margulis, L. 1970. Recombination of non-chromosomal genes in *Chlamydomonas*: Assortment of mitochondria and chloroplasts? *J. Theor. Biol.* 26(2): 337–342.
2. Keeling, P.J. 2004. Diversity and evolutionary history of plastids and their hosts. *Am. J. Bot.* 10: 1481–1493.
3. Lopez-Juez, E. 2009. Steering the solar panel: Plastids influence development. *New Phytol.* 182(2): 287–290.
4. Albertsson, P.A. 2001. A quantitative model of the domain structure of the photosynthetic membrane. *Trends Plant Sci.* 6: 349–354.
5. Dekker, J.P., Boekema, E.J. 2005. Supramolecular organization of thylakoid membrane proteins in green plants. *Biochim. Biophys. Acta* 1706: 12–39.
6. Janssen, P.J.D., Lambreva, M.D., Plumere, N., Bartolucci, C., Antonaccia, A., Buonasera, K., Frese, R.N., Scognamiglio, V., Rea, G. 2014. Photosynthesis at the forefront of a sustainable life. *Front. Chem.* 2: 36.
7. Solymosi, K., Schoefs, B. 2010. Etioplast and etio-chloroplast formation under natural conditions: The dark side of chlorophyll biosynthesis in angiosperms. *Photosynth. Res.* 105(2): 143–166.
8. Lohmanova, G., Zdrahal, Z., Konecna, H., Koukalova, S., Malbeck, J., Soucek, P., Valkova, M., Kiran, N.S., Brzobohaty, B. 2008. Cytokinin-induced photomorphogenesis in dark-grown *Arabidopsis*: A proteomic analysis. *J. Exp. Bot.* 59: 3705–3719.
9. Adam, Z., Charuvi, D., Tsabari, O., Knopf, R.R., Reich, Z. 2011. Biogenesis of thylakoid networks in angiosperms: Knowns and unknowns. *Plant Mol. Biol.* 76: 221–234.
10. Nevo, R., Charuvi, D., Tsabari, O., Reich, Z. 2012. Composition, architecture and dynamics of the photosynthetic apparatus in higher plants. *Plant J.* 70: 157–176.
11. Mostowska, A. 1986. Changes induced on the prolamellar body of pea seedlings by white, red and blue low intensity light. *Protoplasma* 131: 166–173.
12. Mostowska, A. 1986. Thylakoid and grana formation during the development of pea chloroplast, illuminated by white, red and blue low intensity light. *Protoplasma* 134: 88–94.
13. Gunning, B.E. 2001. Membrane geometry of "open" prolamellar bodies. *Protoplasma* 215(1–4): 4–15.
14. Blomqvist, L.A., Ryberg, M., Sundqvist, C. 2008. Proteomic analysis of highly purified prolamellar bodies reveals their significance in chloroplast development. *Photosynth. Res.* 96(1): 37–50.
15. von Zychlinski, A., Kleffman, T., Krishnamurthy, N., Sjolander, K., Baginsky, S., Gruissem, W. 2005. Proteome analysis of the rice etioplast: Metabolic and regulatory network and novel protein functions. *Mol. Cell Proteomics* 4: 1072–1084.
16. Kleffmann, T., von Zychlinski, A., Russenberger, D., Hirsch-Hoffmann, M., Gehrig, P., Gruissem, W., Baginsky, S. 2007. Proteome dynamics during plastid differentiation in rice. *Plant Physiol.* 143(2): 912–923.
17. Solymosi, K., Mysliwa-Kurdziel, B., Boka, K., Strzalka, K., Boddi, B. 2006. Disintegration of the prolamellar body structure at high concentrations of Hg^{2+}. *Plant Biol. (Stuttg)* 8: 627–635.
18. Solymosi, K., Vitanyi, B., Hideg, E., Boddi, B. 2007. Etiolation symptoms in sunflower (*Helianthus annuus*) cotyledons partially covered by the pericarp of the achene. *Ann. Bot.-London* 99: 857–867.
19. Mysliwa-Kurdziel, B., Kruk, J., Strzalka, K. 2013. Protochlorophyllide in model systems—An approach to *in vivo* conditions. *Biophys. Chem.* 175–176: 28–38.
20. Pogson, B.J., Albrecht, V. 2011. Genetic dissection of chloroplast biogenesis and development: An overview. *Plant Physiol.* 155(4): 1545–1551.
21. Schoefs, B., Franck, F. 2003. Protochlorophyllide reduction: Mechanisms and evolutions. *Photochem. Photobiol.* 78(6): 543–557.
22. Grzyb, J.M., Solymosi, K., Strzalka, K., Mysliwa-Kurdziel, B. 2013. Visualization and characterization of prolamellar bodies with atomic force microscopy. *J. Plant Physiol.* 170: 1217–1227.
23. Gunning, E.S., Steer, M.W. 1975. *Ultrastructure and the Biology of Plant Cells.* Edward Arnold, London.
24. Rascio, N., Orsenigo, M. 1976. Chloroplast fine structure in the japonica-2 maize mutant exposed to continuous illumination. 2. The white tissues. *Cytobios* 16(63–64): 183–191.
25. Vothknecht, U., Westhoff, P. 2001. Biogenesis and origin of thylakoid membranes. *Biochim. Biophys. Acta* 154: 91–101.
26. Garstka, M., Drozak, A., Rosiak, M., Venema, J.H., Kierdaszuk, B., Simeonova, E., van Hasselt, P.R., Dobrucki, J., Mostowska, A. 2005. Light-dependent reversal of dark chilling induced changes in chloroplast structure and arrangement of chlorophyll–protein complexes in bean thylakoid membranes. *Biochim. Biophys. Acta* 1710: 13–23.
27. Mustardy, L., Buttle, K., Steinbach, G., Garab, G. 2008. The three-dimensional network of the thylakoid membranes in plants: Quasihelical model of the granum–stroma assembly. *Plant Cell* 20: 2552–2557.
28. Shimoni, E., Rav-Hon, O., Ohad, I., Brumfeld, V., Reich, Z. 2005. Three dimensional organization of higher-plant chloroplast thylakoid membranes revealed by electron tomography. *Plant Cell* 17: 2580–2586.
29. Daum, B., Nicastro, D., Austin, J.R., McIntosh, J.R., Kuhlbrandt, W. 2010. Arrangement of photosystem II and ATP synthase in chloroplast membranes of spinach and pea. *Plant Cell* 22: 1299–1312.

30. Austin, J.R., Staehelin, L.A. 2011. Three-dimensional architecture of grana and stroma thylakoids of higher plants as determined by electron tomography. *Plant Physiol.* 155: 1601–1611.

31. Paolillo, D.J. Jr. 1970. The three-dimensional arrangement of intergranal lamellae in chloroplasts. *J. Cell Sci.* 6: 243–255.

32. Selstam, E., Schelin, J., Brain, T., Williams, W.P. 2002. The effects of low pH on the properties of protochlorophyllide oxidoreductase and the organization of prolamellar bodies of maize (*Zea mays*). *Eur. J. Biochem.* 269: 2336–2346.

33. Boddi, B., Kis-Petik, K., Kaposi, A.D., Fidy, J., Sundqvist, C. 1998. The two spectroscopically different short wavelength protochlorophyllide forms in pea epicotyls are both monomeric. *Biochim. Biophys. Acta.* 1365(3): 531–540.

34. Mysliwa-Kurdziel, B., Amirjani, M.R., Strzalka, K., Sundqvist, C. 2003. Fluorescence lifetimes of protochlorophyllide in plants with different proportions of short-wavelength and long-wavelength protochlorophyllide spectral forms. *Photochem. Photobiol.* 78: 205–212.

35. Mysliwa-Kurdziel, B., Stecka, A., Strzalka, K. 2012. Initial stages of angiosperm greening monitored by low-temperature fluorescence spectra and fluorescence lifetimes. *Methods Mol. Biol.* 875: 231–239.

36. Abdelkader, A.F., Aronsson, H., Solymosi, K., Boddi, B., Sundqvist, C. 2007. High salt stress induced swollen prothylakoids in dark-grown wheat and alters both prolamellar body transformation and reformation after irradiation. *J. Exp. Bot.* 58: 2553–2564.

37. Engdahl, S., Aronsson, H., Sundqvist, C., Timko, M.P., Dahlin, C. 2001. Association of the NADPH:protochlorophyllide oxidoreductase (POR) with isolated etioplast inner membranes from wheat. *Plant J.* 27(4): 297–304.

38. Schoefs, B., Franck, F. 2008. The photoenzymatic cycle of NADPH: Protochlorophyllide oxidoreductase in primary bean leaves (*Phasoleus vulgaris*) during the first days of photoperiodic growth. *Photosynth. Res.* 96: 15–26.

39. Cazzonelli, C.I., Cuttriss, A.J., Cossetto, S.B., Pye, W., Crisp, P., Whelan, J., Finnegan, E.J., Turnbull, C., Pogson, B.J. 2009. Regulation of carotenoid composition and shoot branching in *Arabidopsis* by a chromatin modifying histone methyltransferase, SDG8. *Plant Cell* 21: 39–53.

40. Cazzonelli, C.I., Pogson, B.J. 2010. Source to sink: Regulation of carotenoid biosynthesis in plants. *Trends Plant Sci.* 15: 266–274.

41. Cuttriss, A., Chubb, A., Alawady, A., Grimm, B., Pogson, J.B. 2007. Regulation of lutein biosynthesis and prolamellar body formation in *Arabidopsis*. *Funct. Plant Biol.* 34: 663–372.

42. Pogson, B., McDonald, K.A., Truong, M., Britton, G., DellaPenna, D. 1996. *Arabidopsis* carotenoid mutants demonstrate that lutein is not essential for photosynthesis in higher plants. *Plant Cell* 8(9): 1627–1639.

43. Kobayashi, K., Narise, T., Sonoike, K., Hashimoto, H., Sato, N., Kondo, M., Nishimura, M., Sato, M., Toyooka, K., Sugimoto, K., Wada, H., Masuda, T., Ohta, H. 2013. Role of galactolipid biosynthesis in coordinated development of photosynthetic complexes and thylakoid membranes during chloroplast biogenesis in *Arabidopsis*. *Plant J.* 73: 250–261.

44. Kobayashi, K., Fujii, S., Sasaki, D., Baba, S., Ohta, H., Masuda, T., Wada, H. 2014. Transcriptional regulation of thylakoid galactolipid biosynthesis coordinated with chlorophyll biosynthesis during the development of chloroplasts in *Arabidopsis*. *Front. Plant Sci.* 5: 272.

45. Xu, C., Fan, J., Cornish, A.J., Benning, C. 2008. Lipid trafficking between the endoplasmic reticulum and the plastid in *Arabidopsis* requires the extraplastidic TGD4 protein. *Plant Cell* 20: 2190–2204.

46. Aronsson, H., Schottler, M.A., Kelly, A.A., Sundqvist, C., Dormann, P., Karim, S., Jarvis, P. 2008. Monogalactosyldiacylglycerol deficiency in *Arabidopsis* affects pigment composition in the prolamellar body and impairs thylakoid membrane energization and photoprotection in leaves. *Plant Physiol.* 148(1): 580–592.

47. Aronsson, H. 2008. The galactolipid monogalactosyldiacylglycerol (MGDG) contributes to photosynthesis-related processes in *Arabidopsis thaliana*. *Plant Signal. Behav.* 12: 1093–1095.

48. Rumak, I., Gieczewska, K., Koziol-Lipinska, J., Kierdaszuk, B., Mostowska, A., Garstka, M. 2008. Arrangement of chlorophyll–protein complexes determines chloroplast structure. *Photosynthesis*. *Energy from the Sun* (Allen J.F., Gantt E., Golbeck J.H., Osmond B., eds.) pp. 795–797, Springer, Netherlands.

49. Minai, L., Wostrikoff, K., Wollman, F.-A., Choquet, Y. 2006. Chloroplast biogenesis of photosystem II cores involves a series of assembly-controlled steps that regulate translation. *Plant Cell* 18: 159–117.

50. Zak, E., Norling, B., Maitra, R., Huang, F., Andersson, B., Pakrasi, H.B. 2001. The initial steps of biogenesis of cyanobacterial photosystems occur in plasma membranes. *Proc. Natl. Acad. Sci. U.S.A.* 98: 13443–13448.

51. Baena-Gonzalez, E., Aro, E.-M. 2002. Biogenesis, assembly and turnover of photosystem II units. *Phil. Trans. R. Soc. Lond B* 357: 1451–1460.

52. Rochaix, J.D. 2011. Assembly of the photosynthetic apparatus. *Plant Physiol.* 155(4): 1493–1500.

53. Choquet, Y., Vallon, O. 2000. Synthesis, assembly and degradation of thylakoid membrane proteins. *Biochimie* 82(6–7): 615–634.

54. Lyska, D., Meierhoff, K., Westhoff, P. 2013. How to build functional thylakoid membranes: From plastid transcription to protein complex assembly. *Planta* 237(2): 413–428.

55. Rumak, I., Gieczewska, K., Kierdaszuk, B., Gruszecki, W.I., Mostowska, A., Mazur, R., Garstka, M. 2010. 3-D modelling of chloroplast structure under (Mg^{2+}) magnesium ion treatment. Relationship between thylakoid membrane arrangement and stacking. *Biochim. Biophys. Acta* 1797: 1736–1748.

56. Garstka, M., Venema, J.H., Rumak, I., Gieczewska, K., Rosiak, M., Koziol-Lipinska, J., Kierdaszuk, B., Vredenberg, W.J., Mostowska, A. 2007. Contrasting effect of dark chilling on chloroplast structure and arrangement of chlorophyll–protein complexes in pea and tomato plants with a different susceptibility to non-freezing temperature. *Planta* 226: 1165–1181.

57. Mostowska, A. 1997. Environmental factors affecting chloroplasts. *Handbook of Photosynthesis* (Pessarakli M. ed.) pp. 407–426. Marcel Dekker, New York.

58. Chang, J., Li, Y.-H., Chen, L.-T., Chen, W.-C., Hsieh, W.-P., Shin, J., Jene, W.-N., Chou, S.-J., Choi, G., Somerville, S., Wu, S.-H. 2008. LZF1, a HY5-regulated transcriptional factor, functions in *Arabidopsis* de-etiolation. *Plant J.* 54: 205–219.

59. Aseeva, E., Ossenbuhl, F., Sippel, C., Cho, W.K., Stein, B., Eichacker, L.A., Meurer, J., Wanner, G., Westhoff, P., Soll, J., Vothknecht, U.C. 2007. Vipp1 is required for basic thylakoid membrane formation but not for the assembly of thylakoid protein complexes. *Plant Physiol. Biochem.* 45(2): 119–128.

60. Zhang, S., Shen, G., Li, Z., Golbeck, J.H., Bryant, D.A. 2014. Vipp1 is essential for the biogenesis of photosystem I but not thylakoid membranes in *Synechococcus* sp. PCC 7002. *J. Biol. Chem.* 289(23): 15904–15914.

61. Rutgers, M., Schroda, M. 2013. A role of VIPP1 as a dynamic structure within thylakoid centers as sites of photosystem biogenesis? *Plant Signal. Behav.* 8(11): e27037.

62. Sundberg, E., Slagter, J.G., Fridborg, I., Cleary, S.P., Robinson, C., Coupland, G. 1997. ALBINO3, an *Arabidopsis* nuclear gene essential for chloroplast differentiation, encodes a chloroplast protein that shows homology to proteins present in bacterial membranes and yeast mitochondria. *Plant Cell* 9(5): 717–730.

63. Ankele, E., Kindgren, P., Pesquet, E., Strand. 2007. A. *In vivo* visualization of Mg-protoporphyrin IX, a coordinator of photosynthetic gene expression in the nucleus and the chloroplast. *Plant Cell* 19: 1964–1979.

64. Waters, M.T., Langdale, J.A. 2009. The making of a chloroplast. *EMBO J.* 28: 2861–2873.

65. Ruckle, M.E., DeMarco, S.M., Larkin, R.M. 2007. Plastid signals remodel light signaling networks are essentials for efficient chloroplast biogenesis in *Arabidopsis. Plant Cell* 10: 3944–3960.

66. Frances, S., Thompson, W.F. 1997. The dark-adaptation response of the de-etiolated pea mutant *lip1* is modulated by external signals and endogenous programs. *Plant Physiol.* 115: 23–28.

67. Rohde, A., De Rycke, R., Beeckman, T., Engler, G., Montagu, M.V., Boerjan, W. 2000. ABI3 affects plastid differentiation in dark-grown *Arabidopsis* seedlings. *Plant Cell* 12: 35–52.

68. Jarvis, P., Lopez-Juez, E. 2013. Biogenesis and homeostasis of chloroplasts and other plastids. *Nat. Rev. Mol. Cell Biol.* 12: 787–802.

69. Rudowska, L., Gieczewska, K., Mazur, R., Garstka, M., Mostowska, A. 2012. Chloroplast biogenesis—Correlation between structure and function. *Biochim. Biophys. Acta.* 1817(8): 1380–1387.

70. Rumak, I., Mazur, R., Gieczewska, K., Koziol-Lipinska, J., Kierdaszuk, B., Michalski, W.P., Shiell, B.J., Venema, J.H., Vredenberg, W.J., Mostowska, A., Garstka, M. 2012. Correlation between spatial (3D) structure of pea and bean thylakoid membranes and arrangement of chlorophyll–protein complexes. *BMC Plant Biol.* 12: 72.

71. Mehta, M., Sarafis, V., Critchley, C. 1999. Thylakoid membrane architecture. *Aust. J. Plant Physiol.* 26: 709–771.

72. Vacha, F., Adamec, F., Valenta, J., Vacha, M. 2007. Spatial location of photosystem pigment–protein complexes in thylakoid membranes of chloroplasts of *Pisum sativum* studied by chlorophyll fluorescence. *J. Lumin.* 122–123: 301–303.

73. Hasegawa, M., Shiina, T., Terazima, M., Kumazaki, S. 2010. Selective photosystems in chloroplasts inside plant leaves observed by near-infrared laser-based fluorescence spectral microscopy. *Plant Cell Physiol.* 51: 225–238.

74. Danielsson, R., Albertsson, P.-A., Mamedov, F., Styring, S. 2004. Quantification of photosystem I and II in different parts of the thylakoid membrane from spinach. *Biochim. Biophys. Acta* 1608: 53–61.

75. Danielsson, R., Suorsa, M., Paakkarinen, V., Albertsson, P.-A., Styring, S., Aro, E.-M., Mamedov, F. 2006. Dimeric and Monomeric Organization of Photosystem II. Distribution of five distinct complexes in the different domains of the thylakoid membrane. *J. Biol. Chem.* 281: 14241–14249.

2 Nature of Light from the Perspective of a Biologist
What Is a Photon?

Randy Wayne

CONTENTS

2.1 Introduction 17
2.2 Quantum Mechanical Photon and the Wave–Particle Duality.. 17
2.3 Binary Photon.. 23
2.4 Uncertainty Principle... 34
2.5 A Test of the Binary Photon: Questioning the Relativity of Space and Time... 36
2.6 The Real World: Mathematical or More?.. 42
2.7 Summary of the Properties of a Binary Photon 42
Acknowledgments.. 43
References.. 43

The light which makes the plants grow and which gives us warmth has the double characteristics of waves and particles, and is found to exist ultimately of photons. Having carried the analysis of the universe as far as we are able, there thus remains the proton, the electron, and the photon—these three. And, one is tempted to add, the greatest of these is the photon, for it is the life of the atom.

Arthur Compton (1929)

2.1 INTRODUCTION

Isaac Newton (1730) asked, "Are not gross Bodies and Light convertible into one another, and may not Bodies receive much of their Activity from the Particles of Light which enter their Composition?" Photosynthesis is the process by which plants and other autotrophic organisms transform the rapidly flowing radiant energy of sunlight into stable and stored chemical energy (Herschel 1833; Mayer 1845; Boltzmann 1886; Franck and Wood 1936; Franck and Herzfeld 1941; Oppenheimer 1941; Arnold and Oppenheimer 1950; Calvin 1959; Arnon 1961; Clayton 1971; Kamen 1985; Laible et al. 1994; Campbell and Norman 1998; Jagendorf 1998; Fuller 1999; Govindjee 2000; Feher 2002; Monteith and Unsworth 2008; Nobel 2009; Wayne 2009b). Photosynthesis, the basic process that feeds the world, begins when the pigments in the antenna complex capture the sunlight and transfer the energy to the pair of chlorophyll molecules that make up the reaction center of a photosystem. The chlorophyll molecules in the reaction center undergo a photochemical charge separation that initiates a sequence of oxidation–reduction reactions that generate an electrochemical potential gradient across the photosynthetic membrane. These electrochemical events facilitate the fixation of carbon dioxide and the evolution of oxygen. These life-sustaining energy conversion processes are initiated by the absorption of a particle of light now known as a photon; but what is a photon?

2.2 QUANTUM MECHANICAL PHOTON AND THE WAVE–PARTICLE DUALITY

Albert Einstein (1905a) described the quantum of light (*Lichtquanten*) like so:

it seems to me that the observations regarding 'black-body radiation,' photoluminescence, production of cathode rays by ultraviolet light, and other groups of phenomena associated with the production or conversion of light can be understood better if one assumes that the energy of light is discontinuously distributed in space. According to this assumption to be contemplated here, when a light ray is spreading from a point, the energy is not distributed continuously over ever-increasing spaces, but consists of a finite number of energy quanta that are localized in points in space, move without dividing, and can be absorbed or generated only as a whole.

Radiant energy quanta are currently known as photons (from φώτο, the Greek word for light), a name coined independently and with a myriad of meanings by such polymaths as Leonard T. Troland (1916, 1917), John Joly (1921), René Wurmser (1925a,b), Frithiof Wolfers (1926), Gilbert Lewis (1926a,b), and others (Kidd et al. 1989; Kragh 2014).

On the centenary of the publication of Einstein's paper entitled On a Heuristic Point of View Concerning the Production and Transformation of Light, John Rigden (2005) wrote,

What makes a physics paper revolutionary? Perhaps the most important requirement is that it contains a 'big idea'. Next, the big idea must contradict the accepted wisdom of its time. Third, physicists capable of judging the intrinsic merit of the big idea typically reject it until they are forced to accept it. Finally, the big idea must survive and eventually become part of the woodwork of physics... Einstein's...paper...meets these criteria.

Einstein's mathematical point-like photon differed from Newton's light particle in that the former lacked extension, while the latter had both bigness and sidedness. Many of Einstein's contemporaries, including Max Planck (1920), Niels Bohr (1922), Hendrik Lorentz (1924), and Robert Millikan (1924) did not accept Einstein's model of a mathematical point-like photon since it could not explain the interference of nearby light beams (Einstein 1909c; Stuewer 1989, 2006; Miller 1994; Campos 2004; Rigden 2005). In fact, Einstein "outplancked Planck in not only accepting quantization, but in conceiving of light quanta as actual small packets or particles of energy transferable to single electrons in toto" (Davisson 1937). Einstein's light quantum lacked the spatial extension given to the wavelength of light that is necessary to explain interference and diffraction (Young 1807) that can be observed in soap bubbles, peacock feathers, and the beautiful iridescent blue colors found in a variety of plants, including the leaves of the spike moss, *Selaginella willdenowii*, the leaves of the fern, *Danaea nodosa*, the fruits of *Elaeocarpus angustifolius*, and the petals of the "Queen of the Night" tulip (Lee 2007; Vignolini et al. 2013).

An intuitive description and explanation of interference depends on the wave-like characteristics of light. Classically, the flux of energy or intensity of light depends on the instantaneous amplitude (A) of a monochromatic plane light wave with wavelength λ and frequency ν. The sinusoidally varying amplitude of a light wave is given by

$$A = A_o \sin\left[2\pi\left(\frac{z}{\lambda} \pm \nu t\right)\right] \tag{2.1}$$

where A_o is the maximum amplitude of the wave and may represent the electric field or the magnetic field. The speed $c = \dfrac{z}{t}$ of the wave is equal to the product of λ and ν. The negative sign inside the argument represents a sinusoidal plane wave moving along the z-axis to the right, and a positive sign represents a sinusoidal plane wave moving along the z-axis to the left. The flux of energy or intensity (I, in W/m^2) of the light wave is proportional to the square of the time average of the amplitude of the electric field (in V/m) and not related to the wavelength or frequency. The intensity is given by

$$I = c\varepsilon_o A^2 = \frac{c\varepsilon_o}{2}A_o^2 \tag{2.2}$$

where $\sin^2\left[2\pi\left(\frac{z}{\lambda} \pm \nu t\right)\right] = \dfrac{1}{2}$. Interference effects result when light waves from two sources meet in a given space at the same time. The intensity of the interfering waves depends on the square of the sum of the amplitudes of two (or more) waves and *not* on the sum of the squares:

$$I = c\varepsilon_o(A_1 + A_2 + ... + A_n)^2$$
$$\neq c\varepsilon_o(A_1)^2 + c\varepsilon_o(A_2)^2 + ... + c\varepsilon_o\langle A_n\rangle^2 \tag{2.3}$$

Consequently, waves can both destructively and constructively interfere.

Interestingly, a one-dimensional point-like particle of polychromatic white light can be mathematically modelled by summing an infinite number of plane waves with an infinite number of wavelengths. A larger polychromatic particle of light known as a wave packet can be modelled by summing a group of plane waves with slightly different wavelengths (de Broglie 1924; Bohr 1928; Darwin 1931). Such a particle-like wave packet can be created experimentally with a pinhole and a rapid shutter (Bohm 1979).

Newton's particulate theory of light could not explain the colors of soap bubbles and peacock feathers observed by himself and by Robert Hooke (1665) and the diffraction of light described and named by Francesco Maria Grimaldi (1665). However, these phenomena could be explained at the turn of the nineteenth century by Thomas Young (1804, 1807; Anon 1804) in terms of the interference of light waves. By the end of the nineteenth century, as a result of the successes of James Clerk Maxwell's (1865) electromagnetic wave theory and Heinrich Hertz's (1893) demonstration that electromagnetic waves obey the laws of optics, the wave theory of light (Huygens 1690), which itself had been marginalized by the particulate theory of light, was resurrected and improved, and its proponents relegated Newton's particulate theory of light to the sidelines (Stokes 1884). At the *fin de siècle*, Albert Michelson (1903) triumphantly wrote, "The more important fundamental laws and facts of physical science have all been discovered, and these are now so firmly established that the possibility of their ever being supplanted in consequence of new discoveries is exceedingly remote." However, Lord Kelvin (1904) recognized that there were *nineteenth century clouds* over the wave theory of light created in part by the results of the Michelson–Morley experiment (Michelson and Morley 1887). Some of these clouds would blow over following the introduction of the quantum mechanical, mathematical point-like photon, which could explain the photoelectric effect (Einstein 1905a), while others would remain since it was not possible to describe and explain interference in terms of the mathematical point-like quantum mechanical photon.

Experiments performed in the later part of the nineteenth century by Hertz and Philipp Lenard led to the idea that there was more to the description of the energy of light than just the intensity given by the wave theory. While doing research to experimentally verify Maxwell's electromagnetic wave theory that predicted the propagation of electromagnetic waves through space (Yang 2014), Hertz

(1887) discovered serendipitously that the ultraviolet light produced by the spark gap powered by an oscillating high-voltage coil that he used to transmit electromagnetic waves enhanced the ability of the receiver, which was a copper wire loop with a gap, to produce a spark (Klassen 2011). The presence of a spark in the gap of the receiver that was unconnected to the transmitter was proof that the electromagnetic waves had been transmitted from the transmitter to the receiver through space. Although Hertz hoped that he would be able to see the spark produced in the receiver better when he put it in the dark, he found that when he covered the receiver, the spark it produced was much weaker. The ultraviolet light from the transmitter sparks enhanced spark production in the receiver. This ultraviolet light-induced production of an electric spark became known as the photoelectric effect. The photoelectric effect is a physical analogue of the charge separation that takes place in the photosynthetic reaction center.

Lenard (1900, 1902), who had been an assistant of Hertz, placed the spark gap in a vacuum tube that allowed him to produce a photocurrent instead of a spark in response to ultraviolet irradiation. By moving the actinic spark light closer to the metal, Lenard showed that the magnitude of the photocurrent produced across the spark gap in the vacuum tube, which was a measure of the number of ejected photoelectrons, was a function of the ultraviolet light intensity that fell upon the metal cathode.

Lenard found that he could accelerate or retard the photoelectrons ejected from the metal by applying electrical energy in forward bias and reverse bias mode, respectively, between the negatively charged cathode from which the electrons were emitted and the positively charged anode to which the electrons traveled. He placed an electric field in the reverse bias mode so that it diminished the kinetic energy of the emitted photoelectrons. When the electric field was small, the photoelectrons had high kinetic energy, and when the electric field was large, the photoelectrons had reduced kinetic energy. At one electric field strength, the photoelectrons had zero kinetic energy. Lenard realized that the photoelectrons could only be ejected from the metal atoms if their kinetic energy exceeded the binding energy. Lenard equated the binding energy with the electrical energy that would produce a photoelectron with zero kinetic energy. Lenard found that he could not cause the photoelectrons to be ejected at the threshold electrical potential nor could he cause the ejected photoelectrons to gain additional kinetic energy by increasing the intensity of the actinic light provided by the arc lamp by either moving it closer or increasing the current that flowed through it. However, he did find that the kinetic energy of the ejected photoelectrons did depend on the type of light he used. Lenard (1902) suggested that the spectral composition (i.e., frequency) of the light determined the characteristics of the photoelectrons that were ejected from the atom by ultraviolet light (Thomson 1908; Wheaton 1978, 1983)—a prediction that was confirmed by Millikan (1950) in 1912.

Looking at Lenard's (1902) experimental results, Einstein (1905a) realized that the photoelectric effect could be understood better if the energy of light was discontinuously distributed in space. Einstein wrote,

> According to the conception that the exciting light consists of energy quanta of energy $\dfrac{R}{N}\beta v$, the production of cathode rays by light can be conceived in the following way. The body's surface layer is penetrated by energy quanta whose energy is converted at least partially to kinetic energy of electrons. The simplest possibility is that a light quantum transfers its entire energy to a single electron; we will assume that this can occur. However, we will not exclude the possibility that the electrons absorb only a part of the energy of the light quanta. An electron provided with kinetic energy in the interior of the body will have lost a part of its kinetic energy by the time it reaches the surface. In addition, it will have to be assumed that in leaving the body, each electron has to do some work P (characteristic of the body).

Einstein presented a heuristic equation to describe the photoelectric effect like so:

$$KE = \frac{R}{N}\beta v - P \qquad (2.4)$$

where KE is the kinetic energy of the ejected photoelectron, P is the amount of work that must be done by the quantum of light just to overcome the attractive force between the electron and the nucleus; $\dfrac{R}{N}$ is the ratio of the universal gas constant to Avogadro's number and is equal to Boltzmann's constant; and β is the ratio of Planck's constant to Boltzmann's constant. Consequently, $\dfrac{R}{N}\beta v = h v$. By changing P to W to stand for the work function, the modern form of Einstein's equation for the kinetic energy $KE = \dfrac{1}{2}mv^2$ of the photoelectron becomes

$$KE = h v - W \qquad (2.5)$$

Einstein (1905a) wrote that

> As far as I can see, our conception does not conflict with the properties of the photoelectric effect observed by Mr. Lenard. If each energy quantum of the exciting light transmits its energy to electrons independent of all others, then the velocity distribution of the electrons, i.e., the quality of the cathode rays produced, will be independent of the intensity of the exciting light; on the other hand, under otherwise identical circumstances, the number of electrons leaving the body will be proportional to the intensity of the exciting light.

"After ten years of testing and changing and learning and sometimes blundering," Millikan (1916, 1924) provided the experimental proof using the photoelectric effect that quantitatively confirmed the validity of Einstein's equation describing "the bold, not to say the reckless, hypothesis of an electro-magnetic light corpuscle of energy $h v$, which energy was transferred upon absorption to an electron." The slope of the line that related the kinetic energy of the photoelectrons ejected from sodium and lithium metal to the frequency of the incident ultraviolet and visible light was equal to Planck's constant, and the product of the *x-intercept* and Planck's constant was equal to the work function (Millikan 1914, 1916, 1924, 1935). William Duane and Franklin Hunt (1915) designed an experiment that was the reverse of the photoelectric effect and showed that, consistent with Equation 2.5, the energy of the x-rays emitted from a metal in a vacuum tube was proportional to the kinetic energy of the electrons that were used to bombard the metal. Their results supported Einstein's hypothesis concerning the proportionality between the energy of a photon and frequency of light.

Charles D. Ellis (1921, 1926; Ellis and Skinner 1924a,b) extended Millikan's experiments on the photoelectric effect to the x-ray range and showed that the slope of the graph that related the kinetic energy of photoelectrons to the frequency of incident x-rays was the same for different metals. This supported the idea that Planck's constant was a property of all photons.

These experimental confirmations of Einstein's heuristic proposal that the energy of a photon was related to its wavelength or frequency, but not its amplitude, were quite a blow to the wave theory of light (Einstein 1931), although Millikan (1924) was not convinced as he expressed in his Nobel Lecture,

...the general validity of Einstein's equation is, I think, now universally conceded, and to that extent the reality of Einstein's light-quanta may be considered as experimentally established. But the conception of localized [point-like] light-quanta out of which Einstein got his equation must still be regarded as far from being established...It may be said then without hesitation that it is not merely the Einstein equation which is having extraordinary success at the moment, but the Einstein conception as well. But until it can account for the facts on interference and the other effects which have seemed thus far to be irreconcilable with it, we must withhold our full assent. Possible the recent steps taken by Duane, Compton, Epstein and Ehrenfest may ultimately bear fruit in bringing even interference under the control of localized light-quanta. But as yet the path is dark.

Additional support for Einstein's point-like quantum of light came from experiments done by Arthur Compton using x-rays. Compton (1923) scattered x-rays from the electrons of graphite (carbon) and measured the wavelength of the scattered x-rays with an x-ray diffraction grating spectrometer. He discovered that the wavelength of the scattered x-rays was longer than the wavelength of the incident x-rays. Compton realized that if x-rays were considered to be particles with energy and linear momentum,* and if both energy and linear

momentum were conserved in a collision between a photon and an electron, as they are in collisions between massive particles, then the wavelength of the x-rays scattered from a recoiling electron would be greater than the wavelength of the incident x-rays. Compton found that the red shift in the wavelength of the scattered radiation was also consistent with the Doppler effect since the recoiling electron was actually moving away from the incident and scattered x-ray photons. The interpretation of the Compton effect was a double bonus for Einstein since Compton also found that the recoil of the electron caused by the high energy photons could only be explained by taking into consideration Einstein's (1905b) special theory of relativity.

Chandrasekhara V. Raman (1930) provided further support for the particulate nature of light by performing experiments that were an optical analogue of the Compton effect. Raman showed that long wavelength light described by ultraviolet, visible, and infrared wavelengths was scattered by the vibrating electrons of molecules as if the light had a particulate nature. Depending on the direction of movement of the electrons, the incident light could lose or gain energy and linear momentum resulting in a lengthening or shortening of the wavelength (Wayne 2014a). Likewise, x-rays can gain energy and linear momentum from interacting with electrons moving toward them, which results in a shortening of their wavelength in a process known as the inverse Compton or the Sunyaev–Zel'dovich effect (Rybicki and Lightman 1979; Shu 1982).

For nearly a century, the widely accepted quantum mechanical model has described the photon as a point-like elementary particle or wave packet characterized by the following four quantities: speed, energy, linear momentum, and angular momentum (Jeans 1914, 1924; Jordan 1928; Darwin 1931; Heitler 1944; Weinberg 1975; Feynman 1979; Loudon 1983; Zeilinger et al. 2005; Bialynicki-Birula 2006). The speed (c) of a photon in free space is defined as a constant equal to 2.99792458×10^8 m/s (Jaffe 1960; Livingston 1973). The speed of light is related to two other constants of nature: the electrical permittivity of the vacuum ($\varepsilon_o = 8.854187817 \times 10^{-12}$ F m^{-1}) and the magnetic permeability of the vacuum ($\mu_o = 4\pi \times 10^{-7}$ H m^{-1}) by the following equation:

$$c = \frac{1}{\sqrt{\varepsilon_o \mu_o}} \qquad (2.6)$$

The energy (E) of a photon is given by

$$E = \frac{hc}{\lambda} = \hbar ck \qquad (2.7)$$

where h is Planck's constant ($6.62606957 \times 10^{-34}$ J s), $\hbar$ or h-bar is the reduced Planck's constant $\hbar = \frac{h}{2\pi} = 1.055 \times 10^{-34}$ J s, λ is the wavelength of the photon, and k is the angular wave

---

* When the mass is constant and invariant, the linear momentum (Leibnitz's dead force or *vis mortua*) is equal to the derivative of the kinetic energy (Leibnitz's living force or *vis viva*) with respect to velocity:

$$\frac{dKE}{dv} = \frac{d\frac{1}{2}mv^2}{dv} = mv.$$

number of the photon $= \frac{2\pi}{\lambda}$. The wavelength of a photon is inversely proportional to its energy:

$$\lambda = \frac{hc}{E} \qquad (2.8)$$

The proportionality constant between energy and wavelength is hc (= 1.99×10^{-25} J m). The wavelength of the quantum mechanical photon represents only a number and not spatial wave-like properties. Since the frequency (ν) of the quantum mechanical photon is equal to the ratio of its speed to its wavelength as given by the dispersion relation $\nu = \frac{c}{\lambda}$, the energy of a photon in free space that is traveling at a speed c is also given by

$$E = h\nu = \hbar\omega \qquad (2.9)$$

where h is the proportionality constant between the energy of a photon and its frequency. The angular frequency ω equals $2\pi\nu$, and the dispersion relation is $c = \frac{\omega}{k} = \frac{2\pi\nu}{2\pi/\lambda} = \nu\lambda$. Energy is a scalar quantity that only has magnitude and is easy to work with algebraically. Linear momentum, on the other hand, is more difficult to work with since it is a vector quantity that has both direction and magnitude. This was especially true in the early years of the fledgling field of quantum theory, and linear momentum had not been included in Einstein's (1905a) original concept of the quantum of light.

The linear momentum (p) of a massive body is equal to the product of the mass (m) of the body and its velocity (v). Johannes Stark (1909) took the unidirectional nature of light propagation into serious consideration and stated that the linear momentum (p) of a photon is parallel to the direction of propagation and is related to its energy (E) in the following manner:

$$p = \frac{E}{c} \qquad (2.10)$$

where the speed of light is a constant that relates the linear momentum of a photon to its energy. Consequently,

$$p = \frac{h\nu}{c} = \frac{h}{\lambda} = \hbar k \qquad (2.11)$$

The fact that the linear momentum of light is capable of exerting a radiation pressure was already predicted by electromagnetic wave theory (Maxwell 1873; Poynting 1904) and experimentally measured (Lebedew 1901; Nichols and Hull 1903a,b). Moreover, the fact that energy and linear momentum are conserved in collisions between photons and electrons supports the particulate nature of the photon and also suggests that the photon has some kind of mass associated

with it. Since the linear momentum of a photon is inversely proportional to its wavelength, photons in the x-ray range (λ = 0.01–10 nm) have very large linear momenta. Since photons propagate at the speed of light ($v = c$), the linear momentum can be considered to be given by

$$p = mv = mc \qquad (2.12)$$

And since $p = mc$ and $E = pc$, then

$$E = mcc \qquad (2.13)$$

which is more commonly written as the world's most famous equation:

$$E = mc^2 \qquad (2.14)$$

This equation states that mass and energy are transformable. It is helpful in understanding many high energy processes. One such process is the transformation of the mass of protons into the lesser mass of helium nuclei with the attendant release of radiant energy that occurs in the core of the sun (Bethe 1967), and that makes photosynthesis on earth possible.

In addition to linear momentum, each photon has angular momentum (L), a three-dimensional vector quantity that is even more difficult to work with than linear momentum and was a latecomer to quantum theory. The angular momentum of each and every photon is given by the following equation:

$$L = \frac{h}{2\pi} = \hbar \qquad (2.15)$$

where $\hbar$ is the product of energy and time (Schuster and Nicholson 1924). The angular momentum of a photon was determined by Beth (1936) by measuring the torque exerted on a birefringent crystal by polarized light. Interestingly, the angular momentum, which like linear momentum is also a vector quantity, is unique in terms of conserved quantities in that it is the only conserved property shared by all photons, independent of their frequency and wavelength. The angular momentum* of a photon is related to its total energy (E) by the following relationships:

$$L = \hbar = \frac{h}{2\pi} = \frac{h\nu}{2\pi\nu} = \frac{E}{2\pi\nu} = \frac{E}{\omega} \qquad (2.16)$$

---

* Historically, there has been contention concerning the relation between rotational motion and spin (Tomonaga 2007). According to Landau and Lifshitz (1958), "in quantum mechanics, some 'intrinsic' angular momentum must be ascribed to an elementary particle, regardless of its motion in space. This property of elementary particles is peculiar to quantum theory..., and hence is essentially incapable of a classical interpretation. In particular, it would be wholly meaningless to imagine the 'intrinsic' angular momentum of an elementary particle as being the result of its rotation about 'its own axis', if only because we cannot ascribe any finite dimensions to an elementary particle."

The quantum mechanical photon is characterized by its contradictory and seemingly irreconcilable particle-like properties such as mass and linear momentum and wave-like properties such as wavelength and frequency. Max Born (1963) described particle–wave duality like so:

The ultimate origin of this difficulty lies in the fact (or philosophical principle) that we are compelled to use the words of common language when we wish to describe a phenomenon, not by logical or mathematical analysis, but by a picture appealing to the imagination. Common language has grown by everyday experience and can never surpass these limits. Classical physics has restricted itself to the use of concepts of this kind; by analysing visible motions it has developed two ways of representing them by elementary processes: moving particles and waves. There is no other way of giving a pictorial description of motions—we have to apply it even in the region of atomic processes, where classical physics breaks down. Every process can be interpreted either in terms of corpuscles or in terms of waves, but on the other hand it is beyond our power to produce proof that it is actually corpuscles or waves with which we are dealing, for we cannot simultaneously determine all the other properties which are distinctive of a corpuscle or of a wave, as the case may be. We can therefore say that the wave and corpuscular descriptions are only to be regarded as complementary ways of viewing one and the same objective process, a process which only in definite limiting cases admits of complete pictorial interpretation. It is just the limited feasibility of measurements that defines the boundaries between our concepts of a particle and a wave. The corpuscular description means at the bottom that we carry out the measurements with the object of getting exact information about momentum and energy relations (e.g. the Compton effect), while experiments which amount to determinations of place and time we can always picture to ourselves in terms of the wave representation....

It seems to me that the longer the wavelength of a photon, the better the wave model describes its interactions with matter, and the shorter the wavelength of the photon, the better a mathematical point describes its interactions with matter. In his Nobel Lecture, Arthur Compton (1927) offered these thoughts:

An examination of the spectrum of the secondary X-rays shows that the primary beam has been split into two parts... one of the same wavelength and the other of increased wavelength. When different primary wavelengths are used, we find always the same difference in wavelength between these two components; but the relative intensity of the two components changes. For the longer wavelengths the unmodified ray has the greater energy, while for the shorter wavelengths the modified ray is predominant. In fact when hard γ-rays are employed, it is not possible to find any radiation of the original wavelength. Thus in the wavelength of secondary radiation we have a gradually increasing departure from the classical electron theory of scattering as we go from the optical region to the region of X-rays and γ-rays.... According to the classical theory, an electromagnetic wave is scattered when it sets the electrons which it traverses into forced oscillations, and these oscillating electrons reradiate the energy which they receive. In order to account for the change in

wavelength of the scattered rays, however, we have had to adopt a wholly different picture of the scattering process.... Here we do not think of the X-rays as waves but as light corpuscles, quanta, or, as we may call them, photons. Moreover, there is nothing here of the forced oscillation pictured on the classical view, but a sort of elastic collision, in which the energy and momentum are conserved.... Thus we see that as a study of the scattering of radiation is extended into the very high frequencies of X-rays, the manner of scattering changes. For the lower frequencies the phenomena could be accounted for in terms of waves. For these higher frequencies we can find no interpretation of the scattering except in terms of the deflection of corpuscles or photons of radiation. Yet it is certain that the two types of radiation, light and X-rays, are essentially the same kind of thing. We are thus confronted with the dilemma of having before us a convincing evidence that radiation consists of waves, and at the same time that it consists of corpuscles.... Thus by a study of X-rays as a branch of optics we have found in X-rays all of the well-known wave characteristics of light, but we have found also that we must consider these rays as moving in directed quanta. It is these changes in the laws of optics when extended to the realm of X-rays that have been in large measure responsible for the recent revision of our ideas regarding the nature of the atom and of radiation.

Neither the quantum mechanical model of a mathematical point-like photon nor the classical model of light as an infinite plane wave is sufficient on their own to explain all the observable interactions of light with matter. William Henry Bragg (1922) described the situation in 1921 like so:

On Mondays, Wednesdays and Fridays, we use the wave theory; on Tuesdays, Thursdays and Saturdays we think in streams of flying quanta or corpuscles. That is after all a very proper attitude to take. We cannot state the whole truth since we have only partial statements, each covering a portion of the field. When we want to work in any one portion of the field or other, we must take out the right map. Some day we shall piece all the maps together.

In 1938, Einstein and Leopold Infeld asked,

But what is light really? Is it a wave or a shower of photons? Once before we put a similar question when we asked: is light a wave or a shower of light corpuscles? At that time there was every reason for discarding the corpuscular theory of light and accepting the wave theory, which covered all phenomena. Now, however, the problem is much more complicated. There seems no likelihood for forming a consistent description of the phenomena of light by a choice of only one of the two languages. It seems as though we must use sometimes the one theory and sometimes the other, while at times we may use either. We are faced with a new kind of difficulty. We have two contradictory pictures of reality; separately neither of them fully explains the phenomena of light, but together they do.

In his *own obituary*, Einstein (1949) wrote,

The double nature of radiation (and of material corpuscles) is a major property of reality, which has been interpreted by quantum-mechanics in an ingenious and amazingly successful fashion. This interpretation, which is looked upon as essentially final by almost all contemporary physicists, appears to me as only a temporary way out....

While Einstein saw the Copenhagen interpretation of the wave–particle duality of light as a temporary fix, Niels Bohr (1934, 1958, 1963; see Jammer 1966) saw it as a fundamental aspect of reality when he wrote, "we are compelled to acknowledge... a new trait which is not describable in terms of spatiotemporal pictures... [and we must envision processes] which are incompatible with the properties of mechanical models...and which defy the use of ordinary space–time models." While the irreconcilability of the wave–particle duality and the principle of complementarity has become an *idola tribus* (R. Bacon 1267; F. Bacon 1620) among almost all contemporary physicists, perhaps it is possible to take the best parts of both theories to get a synthetic and realistic model of a photon that can describe both gamma rays and radio waves. Such a theory should be approximated by the quantum mechanical mathematical point-like photon in the gamma ray region and by the wave theory that describes infinite plane waves in the radio wave region of the spectrum.

2.3 BINARY PHOTON

In the quantum mechanical, mathematical point-like model of the photon, there is no indication of how the photon can transfer the electromagnetic force from an emitter to an absorber (Lehnert 2006, 2008). Here I will present a model of a photon that has bigness and sidedness as Newton (1730) would say. The extension beyond that of a mathematical point allows the carrier of the electromagnetic force to possess an electric dipole moment and a magnetic moment. I will derive the finite transversal dimension of the photon from its angular momentum, linear momentum, and energy. I will also describe why I think that the photon is not an elementary particle but is divisible—being composed of two component parts that oscillate and rotate in such a way to generate wave-like behavior. Perhaps such wave-like behavior is what allows a single photon to interfere with itself when subject to an obstruction (Taylor 1909; Tsuchiya et al. 1985). Notable physicists such as William Bragg (1907a,b,c, 1911, 1933; Bragg and Madsen 1908), Louis de Broglie (1924, 1932a,b,c, 1933, 1934a,b,c,d, 1939; de Broglie and Winter 1934), Pascual Jordan (1935, 1936a,b,c, 1937a,b; Jordan and Kronig 1936), and others (Kronig 1935a,b,c, 1936; Scherzer 1935; Born and Nagendra Nath 1936a,b; Fock 1936, 1937; Nagendra Nath 1936; Sokolow 1937; Pryce 1938; Rao 1938; Greenberg and Wightman 1955; Case 1957; Rosen and Singer 1959; Barbour et al. 1963; Ferretti 1964; Perkins 1965, 1972; Ruderfer 1965, 1971; Broido 1967; Bandyopadhyay and Ray Choudhuri 1971; Inoue et al. 1972; Sarkar et al. 1975; Clapp 1980; Dvoeglazov 1998, 1999; Varlamov 2002; Beswick and Rizzo 2008) have

proffered, modified, or refuted models of a binary photon* composed of two semiphotons.

Some particles, such as neutral mesons that were once thought to be elementary, have turned out to be composite particles (Dirac 1933; Fermi and Yang 1949). I start with the assumption that the photon may not be an elementary particle, but a binary structure consisting of two semiphotons[†]—one a particle of matter and the other an antiparticle of antimatter (Wayne 2009a). I have defined matter as having a positive mass and antimatter as having a negative mass (Ginzburg and Wayne 2012; Wayne, 2012c, 2013b). Negative mass is a legitimate (Belletête and Paranjape 2013; Mbarek and Paranjape 2014) although an unwelcomed (Dirac 1930, 1931; Djerassi and Hoffmann 2001) concept in physics. The cosmologist Hermann Bondi (1957) characterized many of its properties. The particle and antiparticle that make up a binary photon are conjugate in that they have equal and opposite mass (M), charge (C), and sense of rotation or parity (P). The sums of two masses or two charges that are equal in magnitude but opposite in sign are zero. Thus, a binary photon in free space is massless and charge-neutral, as is required (Okun 2006; Altschul 2008; Olive et al. 2014). Although the binary photon is neutral as a result of being composed of two conjugate[‡] semiphotons, it can form an electric dipole moment and a magnetic moment, which one could argue is a *sine qua non* for a photon to carry the electromagnetic force. Moreover, since the senses of rotation and the signs of the masses are opposite, the angular momenta of the two particles do not cancel each other but add to each other such that the binary photon has angular momentum ($L = \hbar$).

By contrast, the standard model of physics defines the conjugate particles of matter and antimatter as differing in charge (C), sense of rotation or parity (P), and direction in time (T), which gives CPT symmetry (Feynman 1987). According to Richard Feynman (1985), "Every particle in nature has an amplitude to move backwards in time, and therefore has an anti-particle..." Assuming that time is most accurately described as being unidirectional (Wayne 2012a, 2013b), I define the conjugate particles of matter and antimatter as differing in charge (C), sense of rotation or parity (P), and mass (M), which gives CPM symmetry (Wayne 2012c).

In order to travel at the speed of light, according to de Broglie (1930), the photon in free space must be massless, even

---

* Sing to the tune of Mack the Knife (Pais 1986):

Und Herr Jordan	Mister Jordan
Nimmt Neutrinos	Takes neutrinos
Und daraus baut	And from those he
Er das Licht	Builds the light
Und sie fahren	And in pairs they
Stets in Paaren	Always travel
Ein Neutrino	One neutrino's
Sieht man nicht	Out of sight

[†] Edwin Salpeter came to Cornell to work on the model of a binary photon with Hans Bethe (personal communication).

[‡] From the Latin word *conjugare* meaning yoked together, united, or married and from the mathematical meaning of changing the sign from positive to negative or negative to positive.

though it has energy $E = \dfrac{hc}{\lambda}$, linear momentum $p = \dfrac{h}{\lambda}$, and angular momentum ($L = \hbar$) that can be observably transferred to any object with which it interacts. However, given the measured energy ($E = mc^2$) and linear momentum ($p = mc$) of a photon, the observable photon must by necessity also have a measurable mass (Haas 1928; Ruark and Urey 1930; O'Leary 1964; Young 1976) when it interacts with either matter or antimatter. The mass transferred to the object is given by the following equation:

$$m = \pm \frac{h\nu}{c^2} \qquad (2.17)$$

where the + sign describes the mass of a photon interacting with matter, and the − sign describes the mass of a photon interacting with antimatter. I assume that measurements made with an equal number of matter and antimatter detectors that would separately give a positive or a negative mass, respectively, when added together would give a vanishing photon mass.

Newton's Second Law was written only for bodies with positive mass, which was reasonable because no other substance besides matter was known. I have generalized Newton's Second Law to include masses that are positive and negative (Wayne 2009a). According to the generalized Second Law of Newton, the ratio* of the inertial force (F) to acceleration (a) of a body is given by

$$m = \frac{F}{a} \qquad (2.18)$$

where mass (m) is a scalar quantity with sign and magnitude, and force and acceleration are vector quantities with magnitude and direction in space. The vector of acceleration is parallel to the force vector for a positive mass, and the two vectors are antiparallel for a negative mass. Specifically, a positive mass will accelerate toward an attractive force, and a negative mass will accelerate away from an attractive force (Figure 2.1). A positive mass will accelerate away from a repulsive force, and a negative mass will accelerate toward a repulsive force.

How do particles of negative and positive mass interact with themselves and with each other? At the onset, if we consider the particles to have mass but not charge, then we can use Newton's Law of Gravitation in a generalized version to describe the causal force and Newton's (1687) Second Law in a generalized version to determine how any two particles, with masses of arbitrary sign, respond to the causal force and accelerate relative to each other (Wayne 2009a).

* The vector division is done with vectors that have direction in one-dimensional vector space where their magnitudes are described by real numbers and their directions are either parallel or antiparallel.

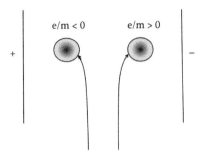

FIGURE 2.1 In an electric field, a particle, such as an electron with a charge-to-mass (e/m) ratio less than zero, accelerates toward an attractive force and bends toward the positive plate. A negative mass electron (= positron), with a charge-to-mass ratio greater than zero, accelerates away from the positive plate.

By equating the causal gravitational force (F_g) to the responsive inertial force (F_i) we get

$$\frac{G}{r^2} m_1 m_2 \, \hat{r} = F_g = F_i = m_2 g \qquad (2.19)$$

where r is the distance between the two masses, $\hat{r}$ is the unit vector from m_2 to m_1, G is the gravitational constant (6.673003×10^{11} m^3 kg^{-1} s^{-2}), m_1 is the mass of a large body like the earth or the sun, m_2 is the test mass, and g is the acceleration due to gravity of the test mass relative to the large body (Figure 2.2). The test mass accelerates toward the large body when $g > 0$, and the test body accelerates away from the large body when $g < 0$. When $F_g > 0$, there are like masses and the gravitational force is attractive. When $F_g < 0$, there are unlike masses and the gravitational force is repulsive. The relationship between the gravitational force and the acceleration for any combination of masses can be obtained by plugging masses of various signs into the above equation.

For example, when the mass of a large body such as the earth is positive, there will be an attractive force ($F_g > 0$) between it and a positive test mass. Consequently, the positive test mass will accelerate toward the large positive mass ($g > 0$). When the mass of a large body is positive, there will be a

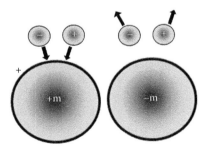

FIGURE 2.2 The direction of acceleration of positive and negative test masses relative to a large body composed of positive or negative mass. Positive and negative test masses accelerate toward a large body composed of positive mass, while positive and negative test masses accelerate away from a large body composed of negative mass.

repulsive force ($F_g < 0$) between it and a negative test mass. Consequently, the negative test mass will accelerate toward the large positive mass ($g > 0$). Recent tests on the effect of gravity on antimatter supports this conjecture (ALPHA Collaboration and A. E. Charman 2013).

When the mass of a large body is negative, there will be a repulsive force ($F_g < 0$) between it and a positive test mass. Consequently, the positive test mass will accelerate away from the large negative mass ($g < 0$). When the mass of a large body is negative, there will be an attractive force ($F_g > 0$) between it and a negative test mass. Consequently, the negative test mass will accelerate away from the large positive mass ($g > 0$).

Now for the interesting part that is relevant for the binary photon. If the magnitudes of the masses of a negative mass particle and a positive mass particle are the same, the positive mass particle will accelerate away from the negative mass particle ($g < 0$), and the negative mass particle will accelerate toward the positive mass particle ($g > 0$). Consequently, the negative mass particle will chase the positive mass particle (Figure 2.3). Since $\frac{G}{r^2} m_1 m_2$ is the same for the two semiphotons but the signs of $\hat{r}$ are opposite, the force exerted by each semiphoton on the other is equal and opposite, and the propagating binary photon obeys Newton's Third Law.

I suggest that the gravitational force between the two conjugate semiphotons that make up the binary photon provides the motive force that causes a photon to move. While this is the only dynamic answer I know of to the question "what causes light to move?" it contradicts the widely held assumption that the gravitational force, which is the weakest of the four fundamental forces (e.g., strong, weak, electromagnetic, gravitational), is unimportant when it comes to subatomic distances (Yang 1957; Dirac 1964). The proposed involvement of the gravitational force in binding the two conjugate semiphotons of the binary photon together and in propelling the binary photon through Euclidean space and Newtonian time may provide insight to explore the connection sought by Faraday (1846), Maxwell (1865), and Einstein (Pais 1982) between the gravitational and electromagnetic fields.

If the conjugate semiphotons that constitute the binary photon only had the properties of mass, the binary photon would accelerate to infinite velocity. Consequently, the conjugate particle and antiparticle that make up the binary photon must also have charge that could interact with the electric permittivity (ε_o) and magnetic permeability (μ_o) of the vacuum in

Direction of propagation

FIGURE 2.3 The propagation of conjugate particles composed of positive and negative mass. The negative mass particle chases the positive mass particle, and the positive mass particle accelerates away from the negative mass particle.

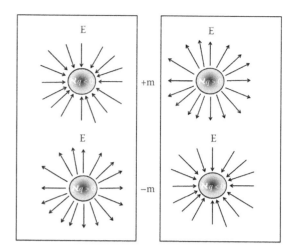

FIGURE 2.4 The electric field lines that radiate from a semiphoton. The two semiphotons on the top have a positive mass and thus are the leading semiphotons. The one on the right has a positive ($\aleph q > 0$) charge and the electric field lines point outwardly, and the one on the left has a negative ($\aleph q < 0$) charge and the electric field lines point inwardly. The two semiphotons on the bottom have a negative mass and thus are the trailing semiphotons. The one on the right has a negative ($\aleph q < 0$) charge and the electric field lines point inwardly, and the one on the left has a positive ($\aleph q > 0$) charge and the electric field lines point outwardly. The two semiphotons on the left are conjugate particles that make one type of binary photon, and the two semiphotons on the right are conjugate particles that make another type.

order to constrain the velocity of the photon to the speed of light (or the reciprocal of the square root of the product of ε_o and μ_o). The existence of charge within a photon seems reasonable since the photon is the carrier of the electromagnetic force. However, the electric field radiating from the charges of the particle and antiparticle must be equal in magnitude and opposite in sign to ensure that the charge of the binary photon is neutral overall (de Broglie 1934d). The direction of the electric field that radiates from a charge depends on both the sign of the charge and the sign of the mass* (Figure 2.4). The gravitational force-induced movement of the charged particles causes a magnetic field according to Ampere's law and an oppositely directed electromotive force according to Faraday's and Lenz's laws that is responsible for reducing the velocity of the binary photon to the speed of light

$$c = \frac{1}{\sqrt{\varepsilon_o \mu_o}}$$. The prophetic Michael Faraday (1846) wrote,

"Neither accepting nor rejecting the hypothesis of an ether, or the corpuscular, or any other view that may be entertained of the nature of light; and, as far as I can see, nothing being

---

* See Wayne (2012c) for the complete equations of symmetry that include a coefficient $\aleph$ that keeps track of the sign of the mass where $\aleph$ is +1 for positive mass and −1 for negative mass. The electric fields generated by the charges cancel when ($\aleph q$)m of the leading photon equals ($\aleph q$)m of the trailing photon. In terms of the electric field, a negatively charged electron with negative mass is equivalent to a positively charged electron (positron) with positive mass, and in both cases, the electric field lines point away from the charge.

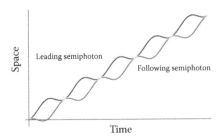

FIGURE 2.5 The positive and negative mass semiphotons oscillate toward and away from the center of gravity as the center of gravity of the binary photon propagates at the speed of light. It is easy to visualize the particle–wave duality when there is not *one* particle but *two* particles that oscillate and can form an oscillating wave. The laws of electromagnetism predict that as the leading particle accelerates away from the negative mass particle as a result of the gravitational force, the leading particle will generate a greater magnetic field, which will produce an electromotive force on itself. This self-induction will put an electromagnetic brake on the leading particle so that the trailing particle can catch up to it. Before the trailing particle catches up to it, the leading particle again accelerates when the gravitational force becomes greater than the electromagnetic braking force that weakens as the leading particle slows down. The combined effects of the gravitational motive force and the electromagnetic braking force result in a longitudinal wave.

FIGURE 2.6 The longitudinal wave formed by the binary photon moving through space and time. This is a graph of Equation 3.20.

($\phi_{following}$) semiphotons travelling along the z-axis as a function of time is shown in Figure 2.6 and given by the following formulae:

really known of a ray of light more than of a line of magnetic or electric force, or even a line of gravitating force."

I assume that the center of gravity of the binary photon, which can be considered to be its rest frame, propagates at the speed of light c along the z-axis as a function of time (Figure 2.5). As a result of the gravitational force on a moving charge inducing an oppositely directed electromotive force, the binary photon may have internal longitudinal motions* that were predicted by Wilhelm Röntgen (1896) and George FitzGerald (1896) and consistent with Einstein's (1909a) "oscillation energy of frequency ν [that] can occur only in quanta of magnitude $h\nu$." Indeed de Broglie (1924) wrote, "Naturally, the light quantum must have an internal binary symmetry corresponding to the symmetry of an electromagnetic wave…." I have described the predicted sinusoidal oscillations with an antisymmetric normal mode using wave equations. The positions of the leading ($\phi_{leading}$) and following

$$\begin{matrix} \phi_{leading}(t) \\ \phi_{following}(t) \end{matrix} = \begin{matrix} ct + \dfrac{\lambda}{4}\left(1 - \cos(2\pi\nu t)\right) \\ ct - \dfrac{\lambda}{4}\left(1 - \cos(2\pi\nu t)\right) \end{matrix} \hat{z} \qquad (2.20)$$

In order for the semiphotons with mass $\dfrac{\hbar\omega}{2c^2} = \dfrac{hc}{2\lambda c^2}$ to oscillate in a sinusoidal manner with angular frequency ($\omega = 2\pi\nu$), there must be a restoring force characterized by a spring constant[†] (K in N/m). The angular frequency of the oscillator is related to the spring constant according to the following formula:

$$\omega = 2\pi\nu = \sqrt{\dfrac{K}{m}} \qquad (2.21)$$

Solving for K, we find that the spring constant that provides the restoring force to the semiphoton is equal to the ratio of a constant ($2\pi^2 hc$) to the cube of the wavelength:

$$K = \dfrac{2\pi^2 hc}{\lambda^3} \qquad (2.22)$$

The longer the wavelength is, the lesser the spring constant becomes, and the more the binary photon approaches a floppy wave. On the other hand, the shorter the wavelength is, the greater the spring constant becomes, and the more the binary photon approaches a *hard* mathematical point. The spring constant[‡] is 2.9×10^9 N/m for a 0.01 nm x-ray binary photon, 3921.1 N/m for a 1 nm x-ray binary photon, 6.1×10^{-5} N/m for a

---

* While the center of gravity of a wave packet moves with a group velocity equal to the speed of light, the particles formed by a wave packet do not all move at the same velocity (de Broglie 1924; French and Taylor 1978). The particles at the front of the wave packet that represent the short wavelengths move with a phase velocity greater than the speed of light, and the particles at the back of the wave packet that represent the long wavelengths move with a phase velocity less than the speed of light. Consequently, the wave packet spreads over time. Also, according to quantum electrodynamics (QED), light has an amplitude to go faster and slower than the vacuum speed of light (Feynman 1985). In a binary photon, the velocities of the semiphotons are greater and less than the speed of light but are coupled in a harmonic oscillator so that the binary photon does not smear out while the center of gravity moves with a velocity equal to the speed of light. The longitudinal oscillation could explain the oscillation in radiation pressure (Einstein 1909b). Longitudinal polarization has been observed experimentally (Wang et al. 2008; Ye et al. 2013).

---

† The spring constant is a one-dimensional property related to flexural stiffness (in N m²), which is a two-dimensional property that is important for accessing the mechanical properties of the photosynthetic leaf blade and its supporting petiole (Niklas 1992).

‡ As a reference, the spring constant of a binary photon of visible light is similar to the spring constants of the neutrophil microvilli and the elastic cytoplasm, which are 4×10^{-5} N/m (Shao et al. 1998; Hochmuth 2000) and 10^{-5} N/m (Guo et al. 2014), respectively.

400 nm visible binary photon, 2.1×10^{-5} N/m for a 500 nm visible binary photon, 1.8×10^{-5} N/m for a 600 nm visible binary photon, 14.5×10^{-20} N/m for a 3 cm microwave binary photon, and 2.9×10^{-24} N/m for a 1 m radio wavelength binary photon.

The velocities of the leading ($v_{leading}$) and following ($v_{following}$) semiphotons along the direction of propagation as a function of time are obtained by differentiating Equation 2.20 and are given by the following formulae:

$$\begin{matrix} v_{leading}(t) \\ v_{following}(t) \end{matrix} = \begin{matrix} c + \dfrac{\pi c}{2}\sin(2\pi\nu t) \\ c - \dfrac{\pi c}{2}\sin(2\pi\nu t) \end{matrix} \quad [\hat{z}] \quad (2.23)$$

Heretofore, the wave–particle duality of the quantum mechanical photon has been unintuitive. Friedrich Hund (1974) wrote, "one way of explaining quantum theory in physical terms these days consists in regarding it as a completely non-intuitive unification or two intuitive pictures, i.e., classical particles and classical waves of fields." By considering the photon to be a binary photon composed of two conjugate particles, instead of an elementary particle, it becomes possible to visualize simultaneously the wave and particle nature of the photon or what Arthur Eddington (1928) and Charles Galton Darwin, Charles Darwin's grandson, called *wavicles*. The simultaneous visualization of the wave-like and particle-like properties was an unrealized goal of Erwin Schrödinger's (1933) wave mechanics.

The longitudinal wave propagating along the z-axis with a maximal spatial extension of λ and an average spatial extension of $\dfrac{\lambda}{2}$ is possible if the photon is composed of two particles as opposed to one. Consequently, the binary photons that make up radio waves (1 m–100 km) and microwaves (1 mm–1 m) are predicted to be very long, and binary photons that make up gamma rays (<0.01 nm) and x-rays (0.01–10 nm) are predicted to be very short—approximating a mathematical point. The binary photons that make up the visible light effective in photosynthesis (Engelmann 1882) are predicted to be intermediate in length.

The possibility that a real photon has transverse extension in addition to longitudinal extension comes from an intuitive and mechanical understanding of angular momentum as a mechanical property (Oberg et al. 2000) that means something more than just a number. John Nicholson (1912, 1913) interpreted Planck's constant as a "natural unit of angular momentum" when he realized that the characteristic absorption and emission spectra of atoms would be intelligible if "the angular momentum of an atom can only rise or fall by discrete amounts when electrons leave or return."

Niels Bohr (1913) applied Nicholson's idea of quantized angular momentum to Ernest Rutherford's (1911) planetary model of the atom and wrote:

In any molecular system consisting of positive nuclei and electrons in which the nuclei are at rest relative to each other

and the electrons move in circular orbits, the angular momentum of every electron round the centre of its orbit will in the permanent state of the system be equal to $h/2\pi$, where h is Planck's constant.

Realizing that the planets orbited the sun in elliptical orbits as Newton showed was required by a central force, Arnold Sommerfeld (1923) suggested that electrons also orbit the nucleus in elliptical orbits. In addition, Sommerfeld suggested that angular momentum, which was then known as the moment of momentum or impulse moment (Ruark and Urey 1927), must not only characterize the atomic system but also be conserved when the atom emits a photon. Sommerfeld wrote,

…in the process of emission…, we demanded…the conservation of energy. The energy that is made available by the atom should be entirely accounted for in the energy of radiation ν, which is, according to the quantum theory of the oscillator, equal to $h\nu$. With the same right, we now demand the conservation of momentum and of moment of momentum: if in a change of configuration of the atom, its momentum or moment of momentum alters, then these quantities are to be reproduced entirely and unweakened in the momentum and moment of momentum of the radiation.

The significance of Planck's constant as a natural unit of angular momentum was also emphasized by Linus Pauling, Sommerfeld's student. Pauling and E. Bright Wilson Jr. (1935) wrote

…h, is a new constant of nature; it is called Planck's constant…Its dimensions (energy × time) are those of the old dynamical quantity called action; they are such that the product of h and frequency ν (with dimensions sec^{-1}) has the dimensions of energy. The dimensions of h are also those of angular momentum, and…just as hν is a quantum of radiant energy of frequency ν, so is $h/2\pi$ a natural unit or quantum of angular momentum.

The selection rules that successfully describe and explain the absorption and emission spectra of atoms and molecules, including chlorophyll, are based on the conservation of angular momentum (Hund 1974; French and Taylor 1978). In the absorption process, a unit of angular momentum is gained by the absorber, and in the emission process, a unit of angular momentum is lost by the emitter. While the unit of angular momentum carried to or carried away from the substance has a magnitude of $\hbar$, the direction reverses, and thus the sign of the angular momentum changes, between absorption and emission.

What would the radius of the binary photon be in order for it to have its observed angular momentum? While this question cannot be answered using current quantum mechanics (Landau and Lifshitz 1958), to answer this question, I went back to Niels Bohr's correspondence principle, which sets a classical quantity equal to a quantum quantity. Classically, the angular momentum of a particle is equal to $mvr\Gamma$, where m is the mass of body, v is its angular velocity, r is its radius, and

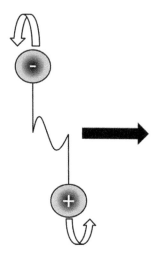

FIGURE 2.7 The rotational motion of the semiphotons is superimposed on the oscillating translational motion.

Γ is a dimensionless geometric factor between 0 and 1 that equals 1 for a point mass at the end of a massless string of radius r. For simplicity (and no better reason), I will let $\Gamma = 1$, which describes the movement of a mass at the end of a massless string. The rotational motion will be superimposed on the oscillating translational motion (Figure 2.7).

The mass of each semiphoton that composes the binary photon is one-half of the total mass of the binary photon and is given by

$$m = \frac{h\nu}{2c^2} \qquad (2.24)$$

Using the correspondence principle where v is the angular velocity and r is the radius of each semiphoton that composes the binary photon, we get

$$L = \frac{\hbar}{2} = \frac{h}{4\pi} = mvr \qquad (2.25)$$

for a semiphoton with angular momentum equal to $\frac{\hbar}{2}$.

We can calculate the radius of the semiphoton from Equation 2.25 by letting $v = 2\pi vr$ and inserting the mass $m = \frac{h\nu}{2c^2}$ of that semiphoton to get

$$\frac{h}{4\pi} = \frac{h\nu}{2c^2} 2\pi vr^2 \qquad (2.26)$$

After cancelling and rearranging, we get

$$r^2 = \frac{c^2}{(2\pi)^2 v^2} \qquad (2.27)$$

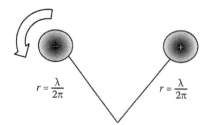

FIGURE 2.8 The radius of the binary photon can be determined from the angular momentum of the binary photon ($\hbar$), the angular momentum of the semiphoton $\frac{\hbar}{2}$, the mass of the binary photon $\frac{h\nu}{c^2}$, and the mass of a semiphoton $\frac{h\nu}{2c^2}$ using the correspondence principle.

Since according to the dispersion relation, $\frac{c^2}{v^2} = \lambda^2$, we get

$$r^2 = \frac{\lambda^2}{(2\pi)^2} \qquad (2.28)$$

And after taking the square root of both sides, we get

$$r = \frac{\lambda}{2\pi} \qquad (2.29)$$

That is, the radius of the binary photon is equal to the wavelength of light divided by 2π (Figure 2.8), and the circumference ($2\pi r$) is equal to the wavelength. The radius of the binary photon is identical to the radius of the semiphoton, since for the binary photon, the angular momentum is equal to $\frac{h}{2\pi}$ and the mass is equal to $\frac{h\nu}{c^2}$. The diameter (d) of a cylinder- or needle-like binary photon is approximately equal to one-third of its wavelength*:

$$d = 2r = \frac{\lambda}{\pi} = 0.32\lambda \qquad (2.30)$$

This equation, which is based on the strong assumptions that the binary photon has energy, linear momentum, and angular momentum, all of which have mechanical consequences, and the arbitrary assumption concerning the geometry of the binary photon, describes the transverse extension or *bigness* of a binary photon with a given wavelength. Likewise, J. J. Thomson (1925) proposed that the photon was a vibrating ring-shaped Faraday *tube of force* where the circumference was equal to the wavelength of light and the diameter of the ring was equal to $\frac{\lambda}{\pi}$. Although I hope to elucidate the form of the binary photon eventually, the arbitrariness

---

* This explains why the lateral resolution of optical systems, including those used for superresolution microscopy, is approximately three times greater than the axial resolution (Wayne 2014a).

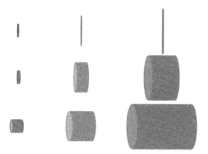

FIGURE 2.9 The predicted three-dimensional forms and relative sizes of oscillating binary photons with a wavelength of 400 nm (left), 500 nm (center), and 600 nm (right).

of the geometrical assumption probably does not introduce a great error since by considering the photon to be a single Newtonian corpuscle, and using similar reasoning, Zu (2008) calculated the diameter of a photon to be 0.5 λ. Previously, Ludwik Silberstein (1922; Mees 1922; Silberstein and Trivelli 1922) obtained a similar diameter by modeling the interaction of photons with photographic silver grains, and Bo Lehnert (2006, 2008, 2013) also derived a similar diameter by revising the assumptions of quantum electrodynamic theory.

When the wavelength of a binary photon approaches zero, so does its diameter, and the *bigness* of the binary photon, or perhaps its smallness, approaches the size of a mathematical point. When the wavelength of a binary photon approaches infinity, so does its diameter, and the *bigness* of the binary photon approaches infinity and can be described as an infinite plane wave. A binary photon of monochromatic 500 nm light has an *average* length of 250 nm and a diameter of 159.2 nm. This is why two *close* binary photons can interfere or a single binary photon can interfere with itself. The *bigness* of a binary photon with a wavelength of 400 nm is smaller; and the *bigness* of a binary photon with a wavelength of 600 nm is larger than the bigness of a binary photon with a wavelength of 500 nm (Figure 2.9).

The size of a photon can be used to derive Planck's black-body radiation law (Shanks 1956) where real space replaces phase space. Support for the predicted three-dimensional size of the binary photon, in which the radius and average length are given by $\dfrac{\lambda}{2\pi}$ and $\dfrac{\lambda}{2}$, comes from the ability to predict the relationship between the number densities of photons of given wavelengths and the temperature of a blackbody cavity with a constant volume (Wayne 2014b*).

In order for the binary photon to have a nonvanishing angular momentum that is equal to $\dfrac{h}{2\pi}$, the two semiphotons, with masses of opposite signs, have to rotate perpendicular to the axis of propagation with opposite senses. Using the

calculated radius, I have incorporated the rotation of the two semiphotons that make up the binary photon into the wave equation that describes the time-varying positions (ϕ) of the two semiphotons:

$$
\begin{array}{c} \phi_{\text{leading}}(t) \\ \phi_{\text{following}}(t) \end{array} =
\begin{array}{ccc}
\dfrac{\lambda}{2\pi}\cos(2\pi vt) & -\dfrac{\lambda}{2\pi}\sin(2\pi vt) & ct + \dfrac{\lambda}{4}(1-\cos(2\pi vt)) \\
-\dfrac{\lambda}{2\pi}\cos(2\pi vt) & -\dfrac{\lambda}{2\pi}\sin(2\pi vt) & ct - \dfrac{\lambda}{4}(1-\cos(2\pi vt))
\end{array}
\begin{array}{c} \hat{x} \\ \hat{y} \\ \hat{z} \end{array}
$$

(2.31)

The positions of the two semiphotons with respect to time as they spiral along the propagation axis are shown in Figure 2.10. The velocities (v) of the semiphotons with respect to time are given by the following formulae:

$$
\begin{array}{c} v_{\text{leading}}(t) \\ v_{\text{following}}(t) \end{array} =
\begin{array}{ccc}
-c\sin(2\pi vt) & -c\cos(2\pi vt) & c + \dfrac{\pi c}{2}\sin(2\pi vt) \\
c\sin(2\pi vt) & -c\cos(2\pi vt) & c - \dfrac{\pi c}{2}\sin(2\pi vt)
\end{array}
\begin{array}{c} \hat{x} \\ \hat{y} \\ \hat{z} \end{array}
$$

(2.32)

The motions of the semiphotons are evocative of Maxwell's (1861) mechanical interpretation of the luminous ether composed of particles and vortices. The binary photon is also reminiscent of two vibrating rings or strings moving through an observable Euclidean space and Newtonian time. The binary photon has been described as a three-dimensional version of string theory. Actual string theory, according to Michio Kaku (1994), expounds that "the laws of nature become simpler and more elegant when expressed in higher dimensions." Kaku writes

String theory is such a promising candidate for physics because it gives a simple origin of the symmetries found in particle physics as well as general relativity...The heterotic string consists of a closed string that has two types of vibrations, clockwise and counterclockwise, which are treated differently. The clockwise vibrations live in a ten-dimensional space. The counterclockwise live in a 26-dimensional space, of which 16 dimensions have been compactified... The heterotic string owes its name to the fact that the clockwise and the counterclockwise vibrations live in two different dimensions but are combined to produce a single superstring theory. That is why it is named after the Greek word for heterosis, which means 'hybrid vigor.'

The biophysical approach to nature assumes that a form–function relationship exists. Although this is not necessarily true (Niklas and Spatz 2012), the biophysicist in me thinks that the structure of the binary photon has hybrid vigor in performing the function of electromagnetic energy transfer from the sun to the chloroplast through Euclidean space and Newtonian time.

Since photons are the carrier of the electromagnetic force (Fermi 1932), it is only natural that they generate electric

---

* Although the final equation given in Wayne (2014b) is correct, there is a factor of 2 error in calculating the cross-sectional area of a photon that was cancelled out by unnecessarily taking the polarization of the photon into consideration.

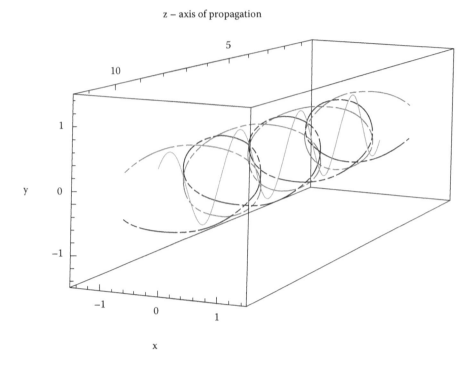

z – axis of propagation

FIGURE 2.10   The anatomy of the binary photon showing the positions of the leading (dashed dark line) and following (dashed light line) semi-photons that compose a binary photon with respect to time as predicted by Equation 3.31. The solid line is the mean position of the two semiphotons.

and magnetic fields described by Faraday's, Ampere's, and Lenz' laws (Jackson 1999). Since the two semiphotons carry charge ($\aleph q$), Coulomb's law predicts that they each generate a time-varying, three-dimensional electric field ($\mathfrak{E}$) inside the binary photon. Positively charged particles ($\aleph q > 0$) generate electric fields that point away from the source, while negatively charged particles ($\aleph q < 0$) generate electric fields* that point toward the source.

The electric field can be calculated from the addition of the principal inward normal unit vectors from each semiphoton. As the two conjugate particles of the binary photon rotate, their electric fields are superimposed inside and diverge outside the binary photon. This geometry ensures quantization of a real time-varying electric field in real space. At 0° (N) and 180° (S), the electric field vectors inside the binary photon destructively interfere and at 90° (E) and 270° (W), the electric field vectors inside the binary photon constructively interfere to give a linearly polarized wave equivalent to Faraday's (1846) *line of electric force*. I claim that, in the binary photon, only linearly polarized light can result from the two charges with opposite polarity rotating with opposite senses. By contrast, Dirac (1958) claimed that the electric field of a single photon

was circularly polarized. This is a testable difference between the model of the binary photon and the quantum mechanical point-like photon. The azimuth of polarization of the electric field of the binary photon depends on the azimuth of the line between the two particles of the binary photon when they are maximally separated and the dipole moment is greatest. The transverse electric field in the *x–y* plane is not confined to one point along the *y*-axis but oscillates up and down along the *y*-axis (Figure 2.11) osculating perpendicular to the solid curve shown in Figure 2.10. This is the three-dimensional extension of Maxwell's (1873) planar electric wave. Consistent with the wave theory of light, the electric fields of two binary photons constructively or destructively interfere in a manner that depends on the phase of the three spatial components of each binary photon. Each linearly polarized binary photon has at least two isomers—one with a parallel magnetic moment and one with an antiparallel magnetic moment (see below). Perhaps entanglement (Ismail et al. 2014) is related to racemic mixtures of binary photons.

While the positions of the semiphotons determine the polarization of the electric field, the velocities (v) of the semiphotons determine the polarization of the magnetic field ($\mathfrak{B}$). The three-dimensional form of the magnetic field of the binary photon depends on the time-varying three-dimensional velocities, which can be determined relative to the principal unit tangent vectors of the two moving charges. Because the products of the charge and the velocity of each conjugate semiphoton have the same sign, the magnetic fields they generate add together. The superposition of the magnetic fields is maximal in the *y–z* plane. The magnetic field oscillates perpendicular to the electric field, being greatest when

---

* Coulomb's law only applies to a mathematical point that cannot blow apart. Assuming that the semiphotons are not mathematical points, and the circumference has width, we ask in the spirit of Henri Poincaré, what stops the charge of a semiphoton from repelling itself and splitting into fragments? I assume that the charge is indivisible and that the mass of the charged particle provides the Poincaré force necessary to hold the charge within a small volume. As a result, the electrical potential decreases exponentially with distance in a manner analogous to the Yukawa potential (de Broglie 1962).

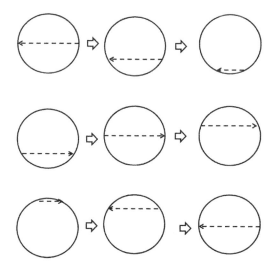

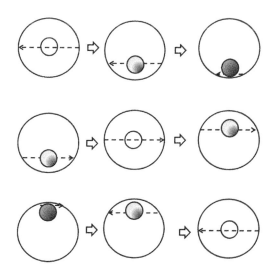

FIGURE 2.11 Transverse sections of a binary photon over time. The azimuth of polarization of the electric field (dashed line) of the binary photon is determined from the superposition of the electric field vectors of the two conjugate particles that make up the binary photon. The electric dipole moment is greatest when the two semi-photons are maximally separated in the x–y plane and minimally separated in the y–z plane. The transverse electric field in the x–y plane is not confined to one point on the y-axis but oscillates up and down the y-axis. The plane of the oscillating electric field is perpendicular to the solid sinusoidal curve shown in Figure 2.10.

FIGURE 2.12 Transverse sections of a binary photon over time. The magnetic field formed by the superposition of the magnetic fields generated by the two semiphotons is orthogonal to the electric field and oriented either parallel or antiparallel to the axis of propagation. The strength of the magnetic field is indicated by the relative darkness of the circles. The oscillating magnetic field in the y–z plane is parallel to the plane of the yellow sinusoid curve shown in Figure 2.10. The electromagnetic field is quantized inside the binary photon, and the strength of the magnetic field alternates in time with the electric field. The electric dipole moment is weakest when the magnetic dipole moment is strongest. This occurs when the two semiphotons are minimally separated in the x–y plane and maximally separated in the y–z plane. The Poynting vector, or the flux of energy density, is obtained by rotating the right hand from the plane of the electric field to the plane of the magnetic field.

the electric dipole moment is weakest and weakest when the electric dipole moment is greatest (Figure 2.12). The magnetic field of the binary photon is a three-dimensional extension of Maxwell's (1873) planar magnetic wave that was predicted by Evans and Vigier (1994). The Poynting vector, which directly gives the energy density flux and from which the radiation pressure can be derived, can be obtained for the binary photon from the principal planes of the electric and magnetic fields using the right-hand rule. The three-dimensional magnetic field may be useful in explaining the effects of magnets on light (Weinberger 2008).

After finding an effect of magnetism on the azimuth of polarization of light propagating through glass, Faraday (1846) wrote

> Thus is established, I think for the first time, a true, direct relation and dependence between light and the magnetic and electric forces; and thus a great addition made to the facts and considerations which tend to prove that all natural forces are tied together, and have one common origin…the powers of nature is…manifested by particular phenomena in particular forms, is here further identified and recognized, by the direct relation of its form of light to its forms of electricity and magnetism.

Michael Faraday (see Thompson 1901), John Kerr (1877, 1878), Pieter Zeeman (1903), and others (Rikken and van Tiggelen 1996; 't Hooft and van der Mark 1996; van Tiggelen and Rikken 2002) found that a magnetic field could influence the polarization of light, but it is not clear whether the magnet acts on the light itself and/or on the electrons of the

material that influence the propagation of light. On the other hand, Röntgen (1896) found that x-rays were not deflected by a magnet and used the fact that cathode rays but not x-rays could be bent by a magnetic field to distinguish the newly discovered x-rays from cathode rays. This distinction was also used by George P. Thomson (1928, 1938) to confirm that the diffraction pattern he saw was due to electrons and not x-rays. Does the fact that the deflection of x-ray was not detected mean that x-rays do not have a magnetic moment? Should the results be extrapolated to mean that photons do not have a magnetic moment? It could be argued *a priori* that as the carrier of the electromagnetic force, the photon should have a magnetic moment, and based on their analysis of a gamma photon produced by the annihilation of an electron and a positron, Sahin and Saglam (2009) derived a formula to calculate the magnetic moment of a photon. Although the photon is usually supposed to lack a magnetic moment (Jackson 1999; Karpa and Weitz 2006; Altschul 2008), this may be an unjustified and unintended consequence of the assumption that the photon is a massless $m = \dfrac{\hbar\omega}{c^2} = 0$ mathematical point. Using the model of the binary photon, I predict that all photons have a magnetic moment, and the formula I derive below is identical to that derived by Sahin and Saglam (2009).

According to the model of the binary photon, the magnetic moment (μ) of a semiphoton with charge ($\aleph q$) and mass $\dfrac{\hbar\omega}{2c^2}$ is related to its angular momentum $\dfrac{\hbar}{2}$ by the following equation:

$$= \frac{\aleph q}{2m} L = \frac{\aleph q}{2m} \frac{\hbar}{2} = \frac{\aleph q \hbar 2c^2}{4\hbar\omega} = \frac{\aleph q c^2}{2\omega} \quad (2.33)$$

Since the conjugate particles that make up a binary photon have opposite charge ($\aleph q$) and opposite spinning frequency (ω), then $\dfrac{\aleph q c^2}{2\omega} = \dfrac{\aleph q c^2}{2\omega}$ for the leading and trailing semiphotons, respectively. Thus, the magnetic moment for the binary photon does not vanish; but it is twice as great as the magnetic moment of each individual particle. The magnetic moment (μ) of the binary photon is equal to

$$= \pm \left| \frac{\aleph q c\lambda}{2\pi} \right| = \pm \left| \frac{\aleph q c^2}{\omega} \right| \quad (2.34)$$

The orientation of the magnetic moment depends on the composition of the binary photon (Table 2.1). When the leading semiphoton with positive mass has a positive charge ($\aleph q > 0$) and a clockwise spin (the trailing semiphoton would have a negative mass, a negative charge [$\aleph q < 0$], and an anticlockwise spin), the magnetic moment is antiparallel to the vector of propagation (Class I), and when the leading semiphoton with positive mass has a negative charge ($\aleph q < 0$) and a clockwise spin (the trailing semiphoton would have a negative mass, a positive charge [$\aleph q > 0$], and an anticlockwise spin), the magnetic moment is parallel to the vector of propagation (Class II).

Both of these binary photons (Classes I and II) have an angular momentum that is antiparallel to the axis of propagation. Symmetry predicts that there may also be two other binary photons with an angular momentum that is parallel to the axis of propagation (Classes III and IV). However, it is also possible that for binary photons that travel at the speed of light, nature favors one isomer over the other as it does in the case of neutrinos and antineutrinos. All neutrinos have left-handed helicity with spin antiparallel to the propagation axis, and all antineutrinos have right-handed helicity with spin parallel to the propagation axis (Lee 1957; Lee and Yang 1957; Goldhaber et al. 1958; Griffiths 1987; Solomey 1997; Bilenky 2013).

Equation 2.34 predicts that the magnetic moment of a binary photon is proportional to the wavelength and inversely proportional to the angular frequency. The magnetic moment of a binary photon with a wavelength of 0.01 nm is 7.61×10^{-23} A m²; the magnetic moment of a binary photon with a wavelength of 400 nm is 1.45×10^{-21} A m²; the magnetic moment of a binary photon with a wavelength of 500 nm is 1.81×10^{-21} A m²; the magnetic moment of a binary photon with a wavelength of 600 nm is 2.17×10^{-21} A m²; and the magnetic moment of a binary photon with a wavelength of 1 m is 8.62×10^{-14} A m².

The predicted proportional relationship between the magnitude of the magnetic moment and the wavelength indicates that long wavelength binary photons are more likely to be bent by a magnetic field than x-rays. However, symmetry predicts that a beam of natural light with both parallel and antiparallel magnetic moments will be broadened by a magnetic field, while a beam with only one orientation of the magnetic moment will be bent (Figure 2.13). Perhaps the x-rays observed by Röntgen were broadened but not bent. Experimental tests of the magnetic moment of light could reify or falsify the model of the binary photon.

TABLE 2.1

Four Possible Classes of Binary Photons

Class	Symmetry	Leading Semiphoton	Following Semiphoton	Angular Momentum (L)	Magnetic Moment (μ)
I	M	+m	−m	Antiparallel	Antiparallel
	C	$\aleph q > 0$	$\aleph q < 0$		
	P	CW	ACW		
II	M	+m	−m	Antiparallel	Parallel
	C	$\aleph q < 0$	$\aleph q > 0$		
	P	CW	ACW		
III	M	+m	−m	Parallel	Parallel
	C	$\aleph q > 0$	$\aleph q < 0$		
	P	ACW	CW		
IV	M	+m	−m	Parallel	Antiparallel
	C	$\aleph q < 0$	$\aleph q > 0$		
	P	ACW	CW		

Note: Mass (M), charge (C), parity (P), clockwise (CW), and anticlockwise (ACW). Parallel and antiparallel is relative to the propagation vector.

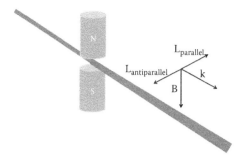

FIGURE 2.13 Predicted effect of a magnetic field on binary photons. If the binary photon has a magnetic moment, a magnetic field will induce a torque on it. The torque exerted on the binary photon will depend on the orientation of the magnetic moment. As a result of the magnetic field, there will be a broadening of the beam in the direction parallel or antiparallel to the magnetic field lines.

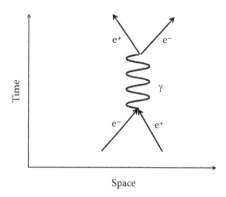

FIGURE 2.14 Feynman diagram of the annihilation of an electron (e⁻) and positron (e⁺) to form a gamma ray photon (γ) and the production or creation of an electron and a positron from a gamma ray photon.

Eugene Wigner (1967) assumed that elementary particles were symmetrical in terms of right and left. Lee and Yang (1956) questioned the assumption that parity was a conserved quantity, and Wu et al. (1957) and Garwin et al. (1957) showed that parity was not a conserved quantity.

In the standard model of physics, symmetry includes real particles of matter, real particles of antimatter, and the virtual particles that pop in and out of the vacuum (Lee 1988). In order to balance the positive energy of matter and antimatter, the vacuum was endowed by Dirac (1930) with an infinite number of particles with negative energy that could give rise to virtual particles (Feynman 1949a,b, 1987). "A virtual particle," according to David Kaiser (2005), "is one that has borrowed energy from the vacuum, briefly shimmering into existence literally from nothing. Virtual particles must pay back the borrowed energy quickly, popping out of existence again, on a time scale set by Werner Heisenberg's uncertainty principle." The uncertainty principle, according to quantum field theory, allows photons to develop internal structures that give rise to fermion–antifermion pairs for a short period of time that carry the same quantum numbers as the photon itself (Przybycień 2003; Lehnert 2008). Perhaps each binary photon propagating through an empty and vacuous vacuum is actually composed of a fermion–antifermion pair that conserves the energy, linear momentum, and angular momentum of the binary photon and can be produced when the binary photon experiences an electric field that is great enough to split it. If so, the binary photon serves as "a unification between the charges (and thus of the forces) by…a single entity, of which the various charges are components in the sense that they can be transformed one into the other" (Salam 1979).

Indeed, pair production is known to occur when a photon with a very short wavelength enters the strong electric field of an atom (Figure 2.14). During pair production, a photon (γ) with energy of 1.02 MeV undergoes an internal conversion to form an electron (e⁻), which is a particle, and a positron (e⁺), which is an antiparticle (Curie and Joliot 1933; Rose and Uhlenbeck 1935; Leone and Robotti 2010). Pair production, in general, results when a photon with sufficient energy (≥ $2mc^2$)

is transformed into a particle (mc^2) and its antiparticle (mc^2) both measured by detectors made of matter to have mass m. Conversely, when an antiparticle such as a positron collides with a particle such as an electron, they annihilate each other and are transformed into high-energy photons in a process known as pair annihilation. Positrons are not *other-worldly* and have been used to visualize photosynthesis in the leaves of *Cannabis sativa* (Kawachi et al. 2006).

It turned out that CPT, the product of the signs of charge (C), parity (P), and time (T), was a conserved quantity (Lee 1988). Wayne (2012c) suggested that the sign of mass (M) might be a more realistic indicator than the sign of time, and it is CPM not CPT that is conserved. CPM theory allows all symmetries to be satisfied with real particles of matter and real particles of antimatter and a vacuum that has been swept clean of everything except its electric permittivity and magnet permeability.

The photon, according to the standard model of physics, is a gauge boson that carries the electromagnetic force (Glashow 1979; Salam 1979; Weinberg 1979). The binary photon could be considered to be a boson with spin ±1 composed of two conjugate fermions with spin ±1/2 (de Broglie 1934d).

I claim that the photon cannot be a mathematical point since the presence of two rotating particles ensures that the binary photon is longer and wider than a mathematical point. The extension allows the formation of an electric dipole moment (±|ℵ$q\lambda$|) and a magnetic moment $\pm\left|\dfrac{\aleph q\lambda c}{2\pi}\right|$, two characteristics that I presume are necessary for the carrier of the electromagnetic force. The model of the binary photon follows Franks Lloyd Wright's (1953) dictum *Form and Function Are One*, which was inspired by his love of design and experience with the natural world.

Robert Hooke (1665) learned long ago that a mathematical point is an idealization that is not found in nature. He wrote in his *Micrographia*,

As in Geometry, the most natural way of beginning is from a Mathematical point; so is the same method in Observations and Natural history the most genuine, simple, and instructive….

And in Physical Enquiries, we must endeavour to follow Nature in the more plain and easie ways she treads in the most simple and uncompounded bodies, to trace her steps, and to be acquainted with her manner of walking there, before we venture our selves into the multitude of meanders she has in bodies of a more complicated nature; lest, being unable to distinguish and judge our way, we quickly lose both Nature our guide, and our selves too, and are left to wander in the labyrinth of groundless opinions; wanting both judgment, that light, and experience, that clew, which should direct our proceedings. We will begin these our Inquiries therefore with the Observations of Bodies of the most simple nature first, and so gradually proceed to those of a more compounded one. In prosecution of which method, we shall begin with a Physical point; of which the Point of a Needle is commonly reckon'd for one; and is indeed, for the most part, made so sharp, that the naked eye cannot distinguish and parts of it…. But if view'd with a very good Microscope, we may find that the top of a Needle…appears a broad, blunt, and very irregular end….

Indeed, Einstein (in Campos 2004) wrote to Hendrik Lorentz in 1909 stating that, "I am not at all of the opinion that one should think of light as being composed of mutually independent quanta localized in relatively small spaces."

Just as a plant systematist has to weigh the advantages and disadvantages of lumping two taxa into one taxon or splitting one taxon into two, so must the biophysical plant biologist weigh the value and limitations of the binary photon and the quantum mechanical, mathematical point-like photon as the carrier of the electromagnetic force that separates charge in the reaction center that results in the evolution of oxygen and the fixation of carbon dioxide, two key events that make the contemplation of the photon possible.

2.4 UNCERTAINTY PRINCIPLE

The uncertainty principle originated when Werner Heisenberg (1927) realized the difficulty one would have trying to use just one photon to determine the position and momentum of a subatomic particle such as an electron at an instant in time without disturbing it. Pierre-Simon Laplace (1814) had written,

> We ought then to regard the present state of the universe as the effect of its anterior state and as the cause of the one which is to follow. Given for one instant an intelligence which could comprehend all the forces by which nature is animated and the respective situation of the beings who compose it an intelligence sufficiently vast to submit these data to analysis it would embrace in the same formula the movements of the greatest bodies of the universe and those of the lightest atom; for it, nothing would be uncertain and the future, as the past, would be present to its eyes.

Knowledge of the position and momentum of an electron would allow a Laplacian super being to predict all future movements of the electron with deterministic physical laws. The fundamental nature of chance and statistics given by the uncertainty principle of quantum mechanics would not allow such determinism in Euclidean space and Newtonian

time (Jordan 1927; Heisenberg 1933, 1974; Frayn 2000). Ralph Lillie (1927), Eddington (1928), Bohr (1934), Compton (1935), Dingle (1937), Schrödinger (1945), Heitler (1963), Hawking and Mlodinow (2010), and Heisenberg's son Martin Heisenberg (2009) have discussed the relationship between physical indeterminism and the beliefs in free will and the freedom of the human mind. To me, free will is a fact (Wayne 2010c), and the determinacy found in the binary photon suggests that the source of free will must be sought outside of quantum mechanics.

In principle, the electron can be localized best with a microscope by using the shortest wavelength of the illuminating gamma rays. However, as $\lambda \to 0$, the hard gamma ray photons, with great linear momentum, send the electron flying, disturbing its location. On the other hand, if one observed the subatomic particle with soft gamma rays with longer wavelengths, the electron would not be as disturbed, but as a result of diffraction, the localization would be coarser. Likewise the linear momentum of the moving electron can be determined by measuring the Doppler shift of the scattered gamma rays as described by the Compton effect. But to get the most accurate measure of the electron's linear momentum, short wavelength gamma rays, which give the greatest Doppler shift, should be used. However, the short wavelength gamma rays cause great recoil and disturb the original position of the electron. One could determine the linear momentum with softer gamma rays with longer wavelength, which do not disturb the electron as much, but the Doppler shift is reduced and the linear momentum measurement like that for the position will be less sharp. It is impossible to accurately measure the position and linear momentum of an electron at the same time, and thus, the description of a quantum system at a given instant of time is accurately described by a smaller number of quantities than a classical system, where differential equations can be used to determine the instantaneous velocity of a particle at a given position.

The uncertainty is a consequence of the particulate nature of light because even one photon, the minimum unit possible, disturbs the localization and linear momentum of an electron (Bohr 1928; Heisenberg 1930; Darwin 1931; Hawking 1999). Heisenberg (1927) realized that the mutually incompatible requirement for longer and shorter wavelengths is a general principle that results in incomplete knowledge of the electron in principle. Since position and linear momentum were two canonically linked variables in quantum mechanics, he suggested that there was a fundamental limit to knowledge. Heisenberg (1927) wrote, "At the instant of the determination of its position—i.e., the instant at which the light quantum is diffracted by the electron—the electron discontinuously changes its impulse. That change will be more pronounced, the smaller the wavelength of the light used, i.e., the more precise the position determination is to be." That is, the measurement of position disturbs the simultaneous measurement of linear momentum.

To describe the reciprocal relationship between the canonical variables of quantum mechanics that result in incomplete knowledge, Heisenberg introduced the principle of *Umbestimmtheit*, which could stand for the principle of indeterminacy, indefiniteness, or uncertainty in the following forms (Ruark and Urey 1930; Pauling and Wilson 1935; Mott and Sneddon 1948):

$$\Delta p \Delta z \sim h \qquad (2.35)$$

and

$$\Delta p \Delta z = \hbar \qquad (2.36)$$

Subsequently, the uncertainty relation has been presented in alternative but not equivalent forms such as

$$\Delta p \Delta z \geq \frac{\hbar}{2} \qquad (2.37)$$

where the relationship is derived from the mathematical structure of the quantum theory, and Δ represents the uncertainty due to the standard deviation (Kennard 1927: Richtmyer and Kennard 1942, 1947; Richtmyer et al. 1955, 1969; Brehm and Mullin 1989; Griffiths 2005; Serway et al. 2005), and

$$\Delta p \Delta z = h \qquad (2.38)$$

where Δ represents the uncertainty due to the wave nature of light (Slater and Frank 1933; Slater 1951; Brehm and Mullin 1989; Serway et al. 2005).

The uncertainty principle, which replaced the principle of causality, undergirds the principle of complementarity touted by Bohr's (1934) Copenhagen School that treats quantum mechanics as a complete theory and emphasizes the particle-like *or* wave-like properties of light and the necessity of chance. By contrast, the binary photon is a melting pot for particle(s)-like *and* (as opposed to *or*) wave light properties that welcomes causality. The unity in diversity displayed in the binary photon provides a way of describing the heretofore hidden variables (Bohm 1952a,b; Bohm and Vigier 1954) within the photon that would seem to cause a mathematical point-like photon to scatter in a probabilistic manner. De Broglie (1957) wrote

It is possible that looking into the future to a deeper level of physical reality we will be able to interpret the laws of probability and quantum physics as being the statistical results of the development of completely determined values of variables which are at present hidden from us.

The way the transverse electric and longitudinal magnetic fields oscillate above and below the axis of propagation is one such hidden variable. In this way, the binary photon provides a challenge to the fundamental nature of the principle of uncertainty, the principle that has led to the counterintuitive elevation of chance and the promotion of paradoxical interpretations of reality supported by the maxim *shut up and calculate* (Mermin 1989, 2004; Tegmark 2007). Indeed, Eddington

(1928) wrote that "if we could understand it [$qp - pq = \frac{ih}{2\pi}$, the root of the Uncertainty Principle] we should not think it so fundamental."

The time-varying position and extension of the binary photon may provide the hidden variables that allow the complete description of a process. A precisely defined state of the linear momentum* and the position of the binary photon can be calculated in principle from Equations 2.31 and 2.32 and the initial conditions. The product of the velocity variation $\frac{\pi c}{2} + \frac{\pi c}{2} = \pi c$ along the axis of propagation of the binary photon and its mass $\frac{hc}{\lambda c^2}$ gives its variation in linear momentum $\Delta p = \Delta \frac{hc}{\lambda c^2} \pi c$. The product of the variation in linear momentum and the variation in the length ($\Delta z = \Delta \lambda$) of the binary photon along the axis of propagation results in an equation comparable to the uncertainty relation:

$$(\Delta \lambda) \; \Delta \frac{hc}{\lambda c^2} \pi c \; = \pi h \qquad (2.39)$$

Since the two rotating semiphotons are in a plane including the propagation axis only twice during a cycle $\frac{2}{2\pi}$, then the product of the length variation and the momentum variation in the plane that includes the axis of propagation is

$$(\Delta \lambda) \; \Delta \frac{hc}{\lambda c^2} \pi c \; \frac{2}{2\pi} \; = h \qquad (2.40)$$

If $\Delta \lambda$ is related to the electric field and $\Delta \frac{hc}{\lambda c^2} \pi c$ is related to the magnetic field, it may be possible to visualize the hidden variables experimentally by mapping the electric and magnetic fields in a standing wave formed in Lecher (1890) wires. In principle, a linear wire antenna could be used to map the electric field, and a circular wire loop antenna could be used to map the magnetic field of a binary photon.

Could a knowledge of the phase of the binary photon, which is in principle knowable, tell us how much of the momentum of an incident photon will be transferred to an electron whose position is being located (McQuarrie et al. 2010)? Could the phase of the binary photon, which is in principle knowable,

---

* The direction of the linear momentum vector depends on the sign of the mass of the semiphoton and its velocity. Thus, oscillating semiphotons in a binary photon have linear momentum vectors that point in the same direction at any given time during the oscillation. The linear momentum increases as the semiphoton with positive mass moves in the direction of propagation and decreases as the semiphoton with positive mass moves antiparallel to the direction of propagation. In order for linear momentum to be conserved in a harmonic oscillator, the *kinetic* linear momentum must be transformed into *potential* linear momentum (~ the spring constant) just as the kinetic energy is transformed into potential energy.

determine whether a photon is reflected from or transmitted through an interface (Feynman 1985)? Could the phase of the two semiphotons be the hidden variables proposed by Max Born (1926), long-searched for, and often poo-pooed (von Neumann 1932; Bohm 1957; Belinfante 1973; Pinch 1977; Peat 1997)? Such an interpretation would provide support for the idea that Heisenberg's (1927) uncertainty principle is not a foundational principle that "once and for all establishes the invalidity of the law of causality." Schrödinger (in Heisenberg 1927) described "quantum mechanics as a formal theory of frightening, even revulsive un-intuitiveness and abstraction." The calculable and predictive but paradoxical nature of quantum mechanics may, in part, result from considering the photon as a mathematical point-like elementary particle subject to statistical laws that hide important *real-world* parameters instead of a pair of particles, with theoretically knowable time-varying momenta and distance, and electric and magnetic fields that interact causally with matter. Thus, even though the act of observation would have an effect on atomic and subatomic particles (Park 1992), the cause and effect relation could be knowable in principle. Perhaps this is what Einstein meant when he wrote to Born (2005) on December 4, 1926: "Quantum mechanics is certainly imposing. But an inner voice tells me that it is not yet the real thing. The theory says a lot, but does not really bring us any closer to the secret of the 'old one.' I, at any rate, am convinced that He is not playing at dice."

2.5 A TEST OF THE BINARY PHOTON: QUESTIONING THE RELATIVITY OF SPACE AND TIME

I have recently put the model of the binary photon to a test by describing and explaining the observed magnitude of the gravitational deflection of starlight—the *experimentum crucis* in favor of the general theory of relativity, in terms of the binary photon (Wayne 2012b,d).

By assuming that the gravity was not a Newtonian force that influenced massive objects directly, but that mass influenced the movement of mathematical point-like objects by warping an interdependent space–time through which they moved, Einstein (1916, 1920) predicted that starlight would be bent by the sun twice as much as was predicted by Johann von Soldner using the Newtonian model that gravity is a force that interacts with massive particles and that light itself was a particle with translational motion only (Jaki 1978).

Following the horrors of World War I, there was a favorable eclipse that allowed the deflection of starlight to be measured in the heavens (Figure 2.15). Dyson et al. (1920) found that "the results of the expeditions to Sobral and Principe can leave little doubt that a deflection of light takes place in the neighbourhood of the sun and that is of the amount demanded by EINSTEIN'S generalized theory of relativity." Following the observation of the signs in the heavens, Einstein became an instant celebrity. According to Subramanya Chandrasekhar (1983), Rutherford told him on May 29, 1919,

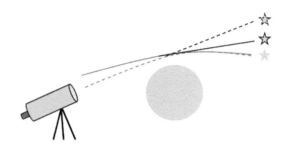

FIGURE 2.15 The deflection of starlight. As a result of the gravitational attraction of the sun, starlight composed of photons is deflected (dashed line) as it passes close to the sun. The star is assumed to exist in a direction parallel to the telescope axis. As a result of gravity, the source of the starlight (star without outline) appears to be displaced away from the sun (star with dashed outline). The observed *double* deflection is predicted equally well by (1) the general theory of relativity and (2) the binary photon theory. If the binary photon did not have rotational motion, the translational energy would be twice as large, and the starlight would be bent half as much (solid line). Since the starlight would be deflected half as much, the star would appear to be closer to the sun (star with solid outline).

The war had just ended, and the complacency of the Victorian and Edwardian times had been shattered. The people felt that all their values and all their ideals had lost their bearings. Now, suddenly, they learnt that an astronomical prediction by a German scientist had been confirmed...by British astronomers. Astronomy had always appealed to public imagination; and an astronomical discovery, transcending worldly strife, struck a responsive chord. The meeting of the Royal Society, at which the results of the British expeditions were reported, was headlined in all the British papers: and the typhoon of publicity crossed the Atlantic. From that point on, the American press played Einstein to the maximum.

The *New York Times* (1919) reported that "if those English scientists are right in feeling that the theory is strongly supported we may be forced to conclude after all that our world is in just a topsy-turvy condition, and that we must learn the theory of relativity to understand it." Unfortunately, they also reported that "As all common folk are suavely informed by the President of the Royal Society that Dr. Einstein's deductions from the behavior of light observed during an eclipse cannot be put in language comprehensible to them, they are under no obligation to worry their heads, already tired by contemplation of so many other hard problems...."

How did Einstein the iconoclast that overturned Newton become an icon himself and *Time* magazine's *Person of the Century* (Golden 1999)? According to Pais (1994), in the wake of the horrors of World War I, Einstein "carried a message of a new world order in the universe"; and Einstein knew how to use language. Everyone knows what *space* and *warp* mean, yet hardly anyone understands what *warped space* is. Einstein himself said to a Dutch newspaper in 1921, "It is the mystery of the non-understanding that appeals to them...."

The general theory of relativity that posited that a relative and interdependent space–time directed the movement of a mathematical point like light quantum became accepted by the scientific community. In appreciation, Einstein (1923) won the Nobel Prize in Physics for 1921 "for his services to Theoretical Physics, and especially for his discovery of the law of the photoelectric effect" and gave his lecture on the fundamental ideas and the problems of the theory of relativity.

Is it possible that space is Euclidean and time is Newtonian and that the hidden properties revealed in the binary photon could explain the double deflection? Since the binary photon has angular momentum and radial extension, it must have rotational motion, which means that it must have rotational energy. If the binary photon had infinite translational energy, it would not be deflected by the sun, and if it had vanishing translational energy, it would fall into the sun. But, if the total energy of a binary photon ($E = h\nu$) is equipartitioned between the translational energy and the rotational energy, then the binary photon would have one-half of the expected translational energy. Consequently, the deflection of starlight would be twice a great as that which von Soldner predicted for a particle that has translational energy only (Wayne 2012b). The deflection of starlight composed of binary photons would also be equal to that predicted by Einstein (1916, 1920).

That is, the binary photon model, which assumes that the binary photon rotates as it translates through absolute space and time, gives the same prediction as Einstein's general theory of relativity. This means that the interpretation of the *experimentum crucis* that gave support for the relative and interdependent nature of space–time proffered by the general theory of relativity depends on the model of the photon. If the photon is a mathematical point whose energy cannot be partitioned into translational and rotational energy, then space and time must be relative and interdependent. However, if the photon is a binary compound with extension and its total energy is equipartitioned between its translational energy and its rotational energy, then space must be Euclidean and time must be Newtonian.

In the same paper in which I offered this interpretation of the deflection of starlight, I also offered a quantitatively accurate interpretation of the gravitational red shift. According to the general theory of relativity, the warping of space–time results in a reddening of the photons emitted by a star. According to the binary photon theory, the reddening results because the binary photon loses energy as it does work against the gravitational binding energy of a star. If the star is so massive, the reddening will be so extreme that the massive star would appear black in Euclidean space and Newtonian time (Wayne 2012b). According to the theory of general relativity, the atomic clocks of the global positioning system that emit photons of a given frequency must be adjusted to take into consideration the warping of space–time by the earth. According to the binary photon theory, the decrease in the frequency or *clock ticks* of the binary photons moving away from the earth and the increase in the frequency or *clock ticks* of the binary photons moving toward the earth result from the loss or gain in the energy of the binary photon due to the work done as it propagates against or along the gradient in gravitational binding energy (Wayne 2012b). The other successes of

Einstein's theories of relativity are also understandable and explainable in terms of Euclidean space and Newtonian time.

According to Einstein (1923), the speed of light is a fundamental universal constant that relates relative space to an interdependent relative time and "to harmonize the relativity principle with the light principle, the assumption that an absolute time (agreeing for all inertial frames) exists, had to be abandoned." That is, the speed of light is a universal absolute that relegates space and time to relative geometrical quantities. The following equation expresses the relationship between absolute and relative quantities:

$$ds^2 = dx^2 + dy^2 + dz^2 - c^2dt^2 \tag{2.41}$$

where ds is a line element or world line in a Minkowski four-dimensional space–time. According to the theory of relativity, the square of the line element (ds^2) and the square of the speed of light (c^2) are constant for all observers, while the square of the distance in space ($dx^2 + dy^2 + dz^2$) and the square of the duration of time (dt^2) are relative quantities that depend on the velocity of the observer or the mass of an object—both of which warp space–time.

I asked myself if there could there be a heretofore hidden property of light itself that is relative when it moves through absolute space and time. Could Einstein have discounted such a property of light when he concentrated on its speed? The answer is, yes! The spatial extension of the binary photon allows one to see the fundamental nature of the wave-like properties of the binary photon that are subject to the Doppler effect expanded to the second order. The Doppler effect was discovered by Gregor Mendel's physics teacher, Christian Doppler (Baksalary and Styan 2009; Wayne 2013a) and I expanded it to second order.

Doppler (1842) guessed that the color of binary stars might be caused by their movement toward or away from an observer (Andrade 1959; Hujer 1963; Gill 1965; Toman 1984; Schuster 2005). Following the introduction of the rapidly moving steam locomotive, Christophorus Buijs Ballot (1845) tested Doppler's wave theory acoustically by placing musicians on a railroad train that traveled 40 mph past musically trained observers. The stationary observers perceived the notes played by the horn players to be a half-note sharper when the train approached and a half-note flatter when the train receded. Three years later, John Scott Russell (1848) noticed that when he was on a moving train, the pitch of the whistle of a stationary train was higher when the train moved toward it and lower when the train moved away. Further support for the Doppler effect came when Hermann Vogel (1876) quantified the increase and decrease in the pitch of a train whistle as the train approached or receded by matching the tone on a violin.

Following the rise of chemical spectroscopy (Roscoe 1869; Kirchhoff and Bunsen 1860), Ernst Mach (1860, 1873) and Hippolyte Fizeau (1870) proposed that the radial velocity of objects could be ascertained by observing the Doppler shift in the spectral lines that identified each chemical. The value of the Doppler effect on determining the velocity of objects was confirmed in the heavens (Huggins 1868; Slipher 1913) and in the laboratory (Bélopolsky 1901; Stark 1906; Galitzin and Wilip 1907). The cited acoustic and optical phenomena demonstrated the first-order Doppler effect. I have derived the Doppler effect expanded to the second order by starting with Maxwell's second-order wave equations (Wayne 2010b).

Einstein tried to reformulate Maxwell's wave equation so that it would take into consideration two inertial frames moving relative to each other but was unsuccessful (Wertheimer 1959). Consequently, he concluded that Maxwell's wave equation, as it was written with its single explicit velocity (c), was a fundamental law of physics valid in all inertial frames and that the speed of light was invariant. I have reformulated Maxwell's wave equation so that it takes into consideration the changes in the spatial and temporal characteristics of electromagnetic waves observed when there is relative motion between the inertial frame that includes the source and the inertial frame that includes the observer. My reformulation of Maxwell's wave equation is based on the primacy of the Doppler effect expanded to the second order, which is experienced by all waves (Wayne 2010b). Since, for any solution to the second-order wave equation in the form of $\Psi = \Psi_o e^{2\pi i \frac{1}{\lambda} z - vt}$, $\dfrac{1}{\lambda}$ and (z) as well as (v) and (t) are complementary pairs $\dfrac{r}{\lambda}$ and vt, it is only a matter of taste which members of the pairs $\dfrac{r}{\lambda}$, v or (z, t) one assumes to depend on the relative velocity of the source and the observer, and which members of the pairs one assumes to be invariant. Einstein chose z and t to be velocity-dependent and $\dfrac{1}{\lambda}$ and v to be invariant in all inertial frames, and I chose $\dfrac{1}{\lambda}$ and v to be velocity-dependent and z and t to be invariant in all inertial frames. The Doppler-based relativistic wave equation is given below in two equivalent forms—the first emphasizing symmetry and the second, which is Equation 2.42 multiplied by $\dfrac{c\sqrt{c - v\cos\theta}}{c\sqrt{c - v\cos\theta}} = 1$, emphasizing the similarity with the Lorentz transformation:

$$\frac{\partial^2 \Psi}{\partial t^2} = cc \frac{\sqrt{c - v\cos\theta}}{\sqrt{c + v\cos\theta}} \,^2\Psi \tag{2.42}$$

$$\frac{\partial^2 \Psi}{\partial t^2} = cc \frac{1 - \dfrac{v}{c}\cos\theta}{\sqrt{1 - \dfrac{v^2\cos^2\theta}{c^2}}} \,^2\Psi \tag{2.43}$$

The magnitude of the relative velocity of the source and the observer is given by v; θ is the angle subtending the velocity

vector originating at the source and the wave vector originating at the source and pointing toward the observer; c is the speed of light through the vacuum and is equal to $\dfrac{1}{\sqrt{\varepsilon_o \,_o}}$; and c' is the product of the frequency (v_{source}) of the source in its inertial frame and the wavelength ($\lambda_{observer}$) observed in any inertial frame. When the source and the observer are receding from each other, $\theta = \pi$ radians, and when the source and the observer are approaching each other, $\theta = 0$ radians. The following equation is a general plane wave solution to the second-order relativistic wave equation given above:

$$\Psi = \Psi_o e^{2\pi i \left(\frac{z}{\lambda_{observer}} - v_{source} \frac{1 - \frac{v}{c}\cos\theta}{\sqrt{1 - \frac{v^2\cos^2\theta}{c^2}}} t \right)} \tag{2.44}$$

Solving the relativistic wave equation for the speed of the wave c results in the following relativistic dispersion relation:

$$c = \lambda_{observer} v_{source} \frac{1 + \dfrac{v}{c}\cos\theta}{\sqrt{1 - \dfrac{v^2\cos^2\theta}{c^2}}} = 2.99 \times 10^8 \text{ m/s} \tag{2.45}$$

indicating that the speed of light (c) is equal to 2.99×10^8 m/s and is independent of the velocity of the observer. When v vanishes, the source and the observer are in the same inertial frame, and the relativistic dispersion relation reduces to the standard dispersion relation $c = \lambda_{source} v_{source}$. After replacing v_{source} with $\dfrac{c}{\lambda_{source}}$, Equation 2.45 transforms into a simple, perspicuous, and lucid relativistic equation that describes the new relativistic Doppler effect:

$$\lambda_{observer} = \lambda_{source} \frac{1 - \dfrac{v}{c}\cos\theta}{\sqrt{1 - \dfrac{v^2\cos^2\theta}{c^2}}} \tag{2.46}$$

and the effect of relative velocity on the wavelength of the observed light.

The Doppler effect expanded to second order differs from the first-order Doppler effect in that the denominator in the first-order Doppler effect is unity. Consequently, as a result of the first-order Doppler effect, at any relative velocity, the *average* wavelength of light observed by or colliding with an observer or object from the front and the back is unchanged and predicted to be (Page 1918)

$$\lambda_{observer} = \lambda_{source} \frac{1}{2} \left[\left(1 - \frac{v}{c}\right) + \left(1 + \frac{v}{c}\right) \right] = \lambda_{source}. \tag{2.47}$$

By contrast, when the Doppler effect is expanded to the second order, at any relative velocity, the *average* wavelength of light observed by or colliding with an observer or object from the front and the back will change and will be given by:

$$\lambda_{observer} = \lambda_{source} \frac{1}{2} \left[\frac{1 - \frac{v}{c}}{\sqrt{1 - \frac{v^2 \cos^2 \theta}{c^2}}} + \frac{1 + \frac{v}{c}}{\sqrt{1 - \frac{v^2 \cos^2 \theta}{c^2}}} \right] = \frac{\lambda_{source}}{\sqrt{1 - \frac{v^2 \cos^2 \theta}{c^2}}}$$

(2.48)

The equation, which describes the new relativistic Doppler effect, differs from Einstein's relativistic Doppler effect equation by having a cosine term in both the numerator and the denominator. The cosine term describes the dependence of the first-order and second-order velocity-dependent spatial properties of electromagnetic waves on the component of the velocity relative to the propagation vector. Unlike Einstein's relativistic Doppler effect, where the term in the denominator describes the relativity of time independent of the propagation vector, the new relativistic Doppler effect shown here does not predict a transverse Doppler effect when $\theta = \frac{\pi}{2}$ since at this angle, $\cos \theta = 0$. The Doppler effect expanded to the second order will cause a velocity-dependent change in the observed length ($L_{observer}$) of the binary photon according to the following equation:

$$L_{observer} = \lambda_{source} \frac{1 - \frac{v}{c} \cos \theta}{\sqrt{1 - \frac{v^2 \cos^2 \theta}{c^2}}}$$

(2.49)

and a velocity-dependent change in its observed cross-sectional area ($A_{observer}$) according to the following equation:

$$A_{observer} = \pi r^2 = \pi \frac{\lambda_{source}^2}{(2\pi)^2} \frac{\left(1 - \frac{v}{c} \cos \theta\right)^2}{\sqrt{1 - \frac{v^2 \cos^2 \theta}{c^2}}} = \frac{\lambda_{source}^2}{4\pi} \frac{\left(1 - \frac{v}{c} \cos \theta\right)^2}{\sqrt{1 - \frac{v^2 \cos^2 \theta}{c^2}}}$$

(2.50)

I assume that the velocity-induced change in the cross-sectional area of the binary photon is a result of the equipartition of energy between the longitudinal motion and the rotational motion. Indeed, the fact that the entropy (S) of a photon is $2.60k$, where k is Boltzmann's constant, indicates that a photon has approximately 36 microstates (Ω) among which to share the entropy (Wayne 2015):

$$S = k \ln \Omega$$

(2.51)

Red-shifted and blue-shifted binary photons are shown in Figure 2.16. Curiously, even though the Doppler effect is

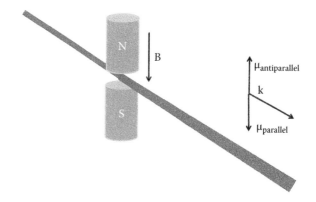

FIGURE 2.16 Red-shifted and blue-shifted binary photons. The Doppler effect will cause a velocity-dependent change in the length and cross-sectional area of the oscillating binary photon.

readily perceived when there is relative motion, whether one is looking at the water waves produced by a swimming swan, the water waves striking a cattail, the sound waves produced by the siren on a fire truck, or the light coming from a distant galaxy, standard theories rarely, if ever, include the Doppler effect as a primary consideration in the study and description of relative motion. The analyses done by my colleagues and me (Wayne 2010a,b, 2012a, 2013c, 2015; Maers and Wayne 2011; Maers et al. 2013) are unique in that we incorporate the relativistic Doppler effect *ab initio*. When expanded to the second order, the inclusion of the Doppler effect makes it possible to unify many aspects of mechanics, electrodynamics, and optics that are usually treated separately. Indeed, the Doppler effect expanded to the second order combined with absolute time also provides alternative derivations of results familiar from the special theory of relativity describing the relativity of simultaneity and why charged particles cannot exceed the vacuum speed of light. It also describes the optics of moving bodies, the mass equivalent of energy, and allows the combination of Newton's second law with the second law of thermodynamics to produce a fundamental, relativistic, and irreversible law of motion.

Einstein lived at a time when fast moving coal-powered trains and telegraphic communication made time seem as if it were relative (Galison 2003). Imagine someone living at that time who was 1000 miles away telling you that their train or a telegram was going to arrive at 12 o'clock noon. Which 12 o'clock noon, the noon of the person telling you or the noon of the person waiting for the train or the telegram? The confusion led to the creation of standard time (Blaise 2000). Before the creation of standard time in 1884, there was local time or solar time where each community reckoned 12 o'clock noon to be the time that the sun was highest in the sky at that location.

In his book *Relativity: The Special and the General Theory*, Einstein (1920) used a train analogy developed by David Comstock (1910) to describe the foundations of the special theory of relativity to a general audience in a non-mathematical manner.

Einstein (1920) demonstrated that time is relative by comparing the observations of a person on "a very long train

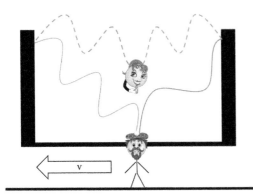

FIGURE 2.17 The observer in the railroad car midway between the lights sees two identical lights come on simultaneously, while the observer on the platform, midway between the two lights and moving backward relative to the railroad car, sees the light from the back come on before the light from the front.

travelling along the rails with the constant velocity *v*" with the observations made by a person on a "railway embankment." He asked the reader to picture an observer in a railroad car midway between light sources at the back of the railroad car and at the front of the railroad car (Figure 2.17). This observer would see the lights come on simultaneously. By contrast, an observer standing on the railway embankment, who is moving backward at velocity *v* relative to the train, would see the light at the back of the railroad car come on before the light at the front of the railroad car comes on. Since there was only one simultaneous event observed by the person on the train, but two nonsimultaneous events observed by the person on the embankment, Einstein concluded that time was relative and the time reckoned depended on the relative velocity of the observer.

Working at a time when transformations between local times and standard time were being made by engineers and telegraph operators, Einstein was immersed in the relativity of time. Combined with the fact that he considered light to be a mathematical point where wavelength and frequency were just numbers that represented momentum and energy, Einstein considered the relativity of time to be a more reasonable explanation than the relativity of wavelength and frequency (i.e., color) due to the Doppler effect. By contrast, I am immersed in a time of Doppler radar, Doppler weather, Doppler ultrasound, and Doppler MRI (Doviak and Zrnić 1993; Maulik 1997; Baksalary and Styan 2009), and as a child of the 1960s, how could I not appreciate the train metaphor in terms of the Doppler effect and the relativity of color?

While there is a lack of clarity as to whether color is described by wavelength or by frequency (Johnsen 2012), the color of light can be described equally well in terms of wavelength and frequency (Wayne 2014c). According to the Doppler theory (Wayne 2010b), if the person in the railroad car midway between the lights at the back and front of the railway car sees the lights come on simultaneously, he or she would see them to be the same color. By contrast the person

on the embankment would see the light at the back of the train to be bluer and the light at the front of the train to be redder as a result of the Doppler effect expanded to the second order and the relative motion between the train and the person on the railway embankment. While the velocities of the blue-shifted and red-shifted light are the same and equal to *c*, the speed of light in free space, the amplitude, energy, or probability of finding a photon (Born 1954; Bloch 1976) described by the blue-shifted wave arrives at the observer before the amplitude, energy, or probability of seeing a photon described by the red-shifted wave arrives at the observer. Consequently, the person on the platform would not observe the two lights coming on simultaneously, but because of the difference in the wavelengths that results from the Doppler effect, the person on the railway embankment would observe the blue-shifted light from the back before observing the red-shifted light from the front.

The Doppler effect experienced by the binary photon can also be used to describe and explain the electrodynamics of moving bodies and why particles with a charge and/or a magnetic moment cannot go faster than the speed of light (Wayne 2010a; Figure 2.18). When an electron is accelerated through an electric field in a cavity, it moves through a photon gas. According to Planck's blackbody radiation law, the greater the temperature of a cavity, the greater the number of photons in the cavity and the shorter their wavelength. This means that at any temperature greater than absolute zero, which, according

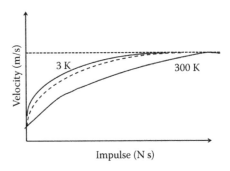

FIGURE 2.18 Particles cannot exceed the speed of light. The reason why a particle with a charge and/or a magnetic moment cannot exceed the speed of light is explained by the special theory of relativity by saying that the duration of time that the electron experiences the accelerating field gets shorter and shorter as the particle gets faster and faster, and consequently, it accelerates less and less. As the particle approaches the speed of light, the duration of time is so short that the particle can no longer accelerate. According to the special theory of relativity, the particle is only moving through a relative and interdependent space–time. The Doppler theory also explains why particles with a charge and/or a magnetic moment, the only kind of particles that can interact with binary photons, the carrier of the electromagnetic force, cannot exceed the speed of light. Special relativity makes no predictions about the effect of temperature on the velocity–impulse relation, while the Doppler theory predicts that as the temperature increases, the impulse needed to accelerate a particle to a given velocity will be greater as a result of the increased velocity-dependent optomechanical counterforce caused by the binary photons on the charged particle.

to the third law of thermodynamics developed by Walther Nernst, is unattainable, there will be photons. This means that there will be binary photons in any space through which a particle with charge and/or magnetic moment moves.

If a particle is moving through a photon gas, then the binary photons that scatter from the front of the moving particle will be blue-shifted as a result of the Doppler effect expanded to the second order, and the binary photons that scatter from the back of the moving particle will be red-shifted (Figure 2.19). The binary photons that collide with the back of the moving particle can also be considered to be red-shifted as a result of the Compton effect, and the binary photons that collide with the front of the moving particle can also be considered to be blue-shifted as a result of the inverse Compton or Sunyaev-Zel'dovich effect (Rybicki and Lightman 1979; Shu 1982).

Since the energy and linear momentum of binary photons are inversely proportional to their wavelength, the blue-shifted binary photons that collide with or scatter from the front of a moving particle will push the particle backward more than the red-shifted binary photons that collide with or scatter from the back of the moving particle will push the particle forwards. The faster the particle moves, the greater the difference is between the wavelengths of the binary photons hitting the front and back of the moving particle, and the greater the optomechanical counterforce is provided by the binary photons through which the particle moves. As the electron approaches the speed of light, the counterforce approaches the accelerating force. Since the acceleration of the electron is proportional to the difference between the accelerating force and the counterforce, when the counterforce equals the accelerating force, acceleration is no longer possible. This means light itself prevents a particle with charge and/or magnetic moment from moving faster than the speed of light.

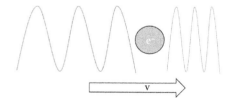

FIGURE 2.19 According to the Doppler theory, at any temperature greater than absolute zero, the particle is moving through a photon gas described by Planck's blackbody radiation law. As the particle moves through the photon gas, it experiences the photons through which it moves as being Doppler shifted. The binary photons that strike the leading side of the particle are blue-shifted by the Doppler effect, and the binary photons that strike the trailing side of the particle are red-shifted. The greater the particle velocity, the greater the difference between the blue-shifted and red-shifted binary photons. Since the linear momentum of the binary photons is inversely proportional to their wavelength, the binary photons through which the particles move exert an optomechanical counterforce on the moving particle. In this way, light itself prevents particles with a charge and/or a magnetic moment from exceeding the speed of light. Only particles with a charge and/or a magnetic moment are able to interact with the binary photons, the carriers of the electromagnetic force.

Friction in physics is considered to be fundamentally negligible and unimportant (Einstein and Infeld 1938). However, a biophysical plant biologist knows that a frictional counterforce is experienced by anything that moves, including a substrate diffusing toward an enzyme (Wayne 2009b), the thylakoids moving through the stroma during chloroplast biogenesis (Paolillo Jr. and Reighard 1967), nuclear-encoded proteins passing through the chloroplast envelope (Jarvis and López-Juez 2013), proteins trafficking through plastid stromules (Hanson and Sattarzadeh 2013), chloroplasts moving through the cell (Kadota et al. 2009; Wada 2013), and leaves tracking the movement of the sun (Koller 2011). By extrapolation, I have found that at any temperature above absolute zero, friction is inevitable and that the binary photons have the properties necessary to provide the optomechanical counterforce that prevents particles, with a charge and/or magnetic moment that makes them capable of interacting with photons, from exceeding the speed of light.

According to the optomechanical model of how binary photons limit the speed of a moving particle to that of light, the greater the temperature of the space through which the particle moves, the greater the number of binary photons and the greater the optomechanical counterforce or the resistance to acceleration. Consequently, the optomechanical counterforce hypothesis is testable since the counterforce exerted on the moving particle increases with temperature. If the speed in which a particle is accelerated by an impulse *is not* temperature dependent, then the special theory of relativity gives a better explanation of the limiting speed of particles. If the speed in which a particle is accelerated by an impulse *is* temperature dependent, the theory of the optomechanical counterforce provided by Doppler-shifted binary photons gives a better explanation of the limiting speed of particles. I look forward to someone measuring the impulse–velocity relationship at 3 and 300 K in a linear accelerator. According to the optomechanical counterforce theory, the impulse needed to accelerate a particle to a given velocity should be 10,000 times greater at 300 K than at 3 K (Wayne 2010a).

When we look at the development of the photosynthetic system (Majeran et al. 2010) and the *adaptive walk* taken in the evolutionary history of photosynthetic plants (Niklas 1997), there seems to be an undeniable arrow of time. Yet, according to the standard model of physics, time is an illusion because the fundamental equations of physics do not have an arrow of time. According to Brian Greene (2004), "Even though experience reveals over and over again that there is an arrow of how events unfold in time, this arrow seems not to be found in the fundamental laws of physics." However, the reversibility of time *is* the foundational assumption and *only* equations that are quadratic in time (t^2) are allowed to be called fundamental. This is why the second law of thermodynamics, which according to me foundationally describes and explains the observed unidirectional arrow of time, is not considered to be a fundamental law of physics.

By taking into consideration the optomechanical counterforce produced by Doppler-shifted binary photons, I have been able to combine Newton's second law of motion with the second law of thermodynamics to produce a fundamental, relativistic, and irreversible law of motion (Wayne 2012a). It states that processes are irreversible because Doppler-shifted binary photons that collide with any moving object radiate away at the speed of light. These binary photons cannot be rounded up to reverse the natural process.

2.6 THE REAL WORLD: MATHEMATICAL OR MORE?

Is it possible to come up with laws of physics that coincide with the visual world? In his Nobel lecture, Heisenberg (1933) stated that "The impossibility of harmonizing the Maxwellian theory with the pronouncedly visual concepts expressed in the hypothesis of light quanta subsequently compelled research workers to the conclusion that radiation phenomena can only be understood by largely renouncing their immediate visualization." One could no longer ask where is a given photon in space and time? Because, according to Walter Heitler (1944) "there are no indication that, for instance, the idea of the 'position of a light quantum' (or the 'probability for the position') has any simple physical meaning." More recently, David Griffiths (2005) wrote,

> The particle wasn't really anywhere. It was the act of measurement that forced the particle to 'take a stand' (though how and why it decided on the point C we dare not ask). Jordan said it most starkly, 'Observations not only disturb what is to be measured, they produce it.... We compel (the particle) to assume a definite position.' This view (the so-called Copenhagen interpretation), is associated with Bohr and his followers. Among physicists it has always been the most widely accepted position.

According to Armstrong (1983), the photon is not a physical unit with any form of localization but more like coefficients in a Fourier series until it is commanded into existence.

Mathematics seems to have trumped other forms of knowledge about the natural world. James Jeans (1945) wrote that "the history of theoretical physics is a record of the clothing of mathematical formulae which were right, or very nearly right, with physical interpretations which were often very badly wrong." In *The Mysterious Universe*, Jeans (1934) wrote,

> Lapsing back again into the crudely anthropomorphic language we have already used, we may say that we have already considered with disfavour the possibility of the universe having been planned by a biologist or an engineer; from the intrinsic evidence of his creation, the Great Architect of the Universe now begins to appear as a pure mathematician.

I think that the current mathematical models in physics that consider all particles fundamentally as mathematical points, matter as being fundamentally massless, friction to be a fiction, and space and time to be fundamentally an illusion are too simplistic in their assumptions, and, because of this, they may be misleading when it comes to describing the real world. Consequently, I am endeavoring to create a realistic theory

of the photon, which inevitably creates friction as a result of the Doppler effect expanded to the second order, where space and time are real-world quantities defined by common sense and only approximated by mathematical equations (Synge 1951, 1970). My point of view contrasts with the mathematical physicists who think that the mathematical equations are fundamentally real and anything less abstract is accidental and misleading (Tegmark 2007).

After reading Hermann Weyl's book *Space, Time and Matter*, Felix Bloch (1976) told Heisenberg "that space was simply the field of linear operations." Heisenberg replied, "Nonsense, space is blue and birds fly through it." Heisenberg was warning Bloch that "it was dangerous for a physicist to describe Nature in terms of idealized abstractions too far removed from the evidence of actual observation." Einstein also thought that idealized abstractions provided inadequate pictures of the world. When Max Born's wife Hedwig asked Einstein, "Do you believe that everything can be pictured in a scientific manner?" Einstein answered, "Yes, it is conceivable but it would be of no use. It would be an inadequate means of expression—like representing a Beethoven symphony in terms of curves of air pressure" (Born 1965).

Wholistic, intuitive, aural, and visual interpretations of reality contrast with the current orthodox interpretation of reality where reality is completely described mathematically by the foundational principles of uncertainty and relativity, and consequently, events do not really take place in a cause and effect manner over time in three-dimensional space. According to Bohr (1934), the commonsense yet illusional view of reality prevails because most people do not have experience with velocities that are comparable to the speed of light and with objects as small as atoms (Miller 1994). The binary photon allows for a physical and mathematical description of the real world, capable of visual imagery, and consistent with common sense and intuition, where time differs from space, friction is not a fiction, and all effects require a cause.

2.7 SUMMARY OF THE PROPERTIES OF A BINARY PHOTON

The importance of plants in transforming the energy of light to the requirements for life was recognized by Julius Robert Mayer (1845), the founder of the first law of thermodynamics (Tyndall 1915). Mayer wrote,

> Nature undertakes the task of storing up the light which streams earthward—of condensing the most volatile of all powers into a rigid form, and thus preserving it for our use. She has overspread the earth with organisms which while living take into them the solar light, and by the appropriation of its energy generate incessantly chemical forces. These organisms are plants. The vegetable world constitutes the reservoir in which the fugitive solar rays are deposited, and rendered ready for useful application. With this economical provision the existence of the human race is also inseparably connected. The reducing action exerted by solar light on both inorganic and organic substances is well known. This reduction takes place most copiously in full sunlight, less copiously in the

shade, being entirely absent in darkness, and even in candle-light. The reduction is a conversion of one form of energy into another—of mechanical effect into chemical tension.

Given the importance of light to plant life in terms of photosynthesis, photomorphogenesis (Wayne and Hepler 1984, 1985), and photomovement (Wayne et al. 1991), I became interested in the nature of the photon. The nature of the light quantum has been questioned ever since Einstein proposed it in 1905. On December 12, 1951, Einstein wrote to his friend Michele Besso: "All the fifty years of conscious brooding have brought me no closer to the answer to the question, 'What are light quanta?' Of course, today, every rascal thinks he knows the answer, but he is deluding himself" (see Klein 1970). The ideas on the nature of the photon that I present here are incomplete and still in progress. While they are the best I have to offer, some may be wrong. Arthur Schuster (1898), the person who first came up with the idea of antimatter, reminds us that as scientists, occasionally we should think about the unknown and perhaps even the unknowable. I hope my ideas developed from by background as a biophysical plant biologist have stimulated you to think about the photon. Here is a summary of my conclusions:

- The photon is not an elementary particle but a composite particle composed of two semiphotons and a boson composed of two fermions. The mass of the boson is not unique but depends on the frequency of the photon.
- The semiphotons are conjugate particles. One semiphoton has positive mass and the other has negative mass. The positive mass semiphoton is equivalent to a particle (matter), and the negative mass semiphoton is equivalent to an antiparticle (antimatter).
- A binary photon cannot occupy a single mathematical point and thus by necessity it must have extension.
- The gravitational force between the two conjugate particles provides the motive force that causes the negative mass semiphoton to chase the positive mass semiphoton unidirectionally in space and time. This is why light moves.
- As the carrier of the electromagnetic force, the binary photon must carry charge yet remain neutral. To remain electrically neutral, the semiphotons have opposite charges. The charges of the semiphotons confine the speed of the center of gravity of the binary photon to the speed of light.
- The two semiphotons rotate with opposite senses around the axis of propagation in a manner that gives the binary photon one unit of angular momentum and a magnetic moment. In so doing, the semiphotons generate a transverse linearly polarized electric field and electric dipole moment that is equivalent to Faraday's electric line of force. The rotation of the semiphotons also results in a three-dimensional and predominantly longitudinal magnetic field, equivalent to Faraday's magnetic line of force. The binary photon, with its electric dipole moment and orthogonal magnetic moment, is fit to be the gauge boson that carries the electromagnetic force.

- The internal structure of the binary photon, represented by the time-varying electric and magnetic fields, may provide the hidden variables or the variables that were hidden to the founders of quantum mechanics that call into question the fundamental nature of the uncertainty principle.
- The model of the binary photon has been tested in that it is able to predict the double deflection of starlight in Euclidean space and Newtonian time as well as the general theory of relativity does for a mathematical point-like photon in warped space–time.
- By postulating that the Doppler effect expanded to the second order is fundamental and that the wavelength and frequency of light are relative and space and time are absolute, the relativity of simultaneity as well as the reason why a particle with a charge and/or magnetic moment cannot exceed the speed of light can be described and explained in terms of the binary photon moving through Euclidean space and Newtonian time. Thus, the postulate of an interdependent and relative space–time may be superfluous, and the foundational value of the special and general theories of relativity may be called into question by the binary photon.
- The binary photon, with its time-varying electrical dipole and magnetic moments, is fit to initiate photochemical charge separation that leads to the photosynthetic fixation of carbon dioxide, the evolution of oxygen, and life as we know it.

ACKNOWLEDGMENTS

I thank my teachers, family, friends, and colleagues in plant biology, physics, biological and environmental engineering, mathematics, history, and science and technology studies for their contributions to the ideas presented here.

REFERENCES

ALPHA Collaboration and A. E. Charman. 2012. Description and first application of a new technique to measure the gravitational mass of antihydrogen. *Nature Communications* 4:1745.

Altschul, B. 2008. Astrophysical bounds on the photon charge and magnetic moment. *Astroparticle Physics* 29:290–298.

Andrade, E. N. da C. 1959. Doppler and the Doppler effect. *Endeavour* 18:14–19.

Anon. 1804. Dr Young's Bakerian lecture. *Edinburgh Review* 5:97–102.

Armstrong, H. 1982. No place for a photon? *American Journal of Physics* 51:103–104.

Arnold, W. and J. R. Oppenheimer. 1950. Internal conversion in the photosynthetic mechanism of blue-green algae. *Journal of General Physiology* 33:423–435.

Arnon, D. I. 1961. Cell-free photosynthesis and the energy conversion process. In *A Symposium on Light and Life*, eds. W. D. McElroy and B. Glass. Baltimore: Johns Hopkins University Press, pp. 489–569.

Bacon, F. [1620]. 1960. *The New Organon: Or True Directions Concerning the Interpretation of Nature*. New York: Liberal Arts Press.

Bacon, R. [1267]. 1985. *Four Stumbling Blocks to Truth*. Dallas: Somesuch Press.

Baksalary, O. M. and G. P. H. Styan. 2009. A philatelic introduction to the Doppler effect and some of its applications. *Erwin Schrödinger Institute of Mathematical Physics* 4(2):4–11.

Bandyopadhyay, P. and P. Ray Choudhuri. 1971. The photon as a composite state of a neutrino–antineutrino pair. *Physical Review D* 3:1378–1381.

Barbour, I. M., A. Bietti and B. F. Touschek. 1962. A remark on the neutrino theory of light. *Nuovo Cimento* 28:452–454.

Belinfante, F. J. 1972. *A Survey of Hidden-Variables Theories*. New York: Pergamon Press.

Belletête, J. and M. B. Paranjape. 2012. On negative mass. Available at http://arxiv.org/pdf/1304.1566v1.pdf.

Bélopolsky, A. 1901. On an apparatus for the laboratory demonstration of the Doppler–Fizeau principle. *Astrophysical Journal* 13:15–24.

Beswick, J. A. and C. Rizzo. 2008. Structure of the photon and magnetic field induced birefringence and dichroism. *Annales de la Fondation Louis de Broglie* 33:31–42.

Beth, R. A. 1936. Mechanical detection and measurement of the angular momentum of light. *Physical Review* 10:115–125.

Bethe, H. 1967. Energy production in stars. Nobel Lecture, December 11, 1967. Available at http://www.nobelprize.org/nobel_prizes/physics/laureates/1967/bethe-lecture.html.

Bialynicki-Birula, I. 2006. Photon as a quantum particle. *Acta Physica Polonica B* 37:935–946.

Bilenky, S. M. 2012. Neutrino. History of a unique particle. *The European Physical Journal H* 38:345–404.

Blaise, C. 2000. *Time Lord: Sir Sandford Fleming and the Creation of Standard Time*. New York: Random House.

Bloch, F. 1976. Heisenberg and the early days of quantum mechanics. *Physics Today* 29:23–27.

Bohm, D. 1952a. A suggested interpretation of the quantum theory in terms of "hidden" variables. I. *Physical Review* 85:166–179.

Bohm, D. 1952b. A suggested interpretation of the quantum theory in terms of "hidden" variables. II. *Physical Review* 85:180–192.

Bohm, D. 1957. *Causality and Chance in Modern Physics*. New York: Harper & Brothers.

Bohm, D. 1979. *Quantum Theory*. Mineola: Dover.

Bohm, D. and J. P. Vigier. 1954. Model of the causal interpretation of quantum theory in terms of a fluid with irregular fluctuations. *Physical Review* 96:208–216.

Bohr, N. 1912. On the constitution of atoms and molecules. *Philosophical Magazine* 26:1–25, 476–502, 856–875.

Bohr, N. 1922. The structure of the atom. Nobel Lecture, December 11, 1922. Available at http://www.nobelprize.org/nobel_prizes/physics/laureates/1922/bohr-lecture.html.

Bohr, N. 1928. The quantum postulate and the recent development of atomic theory. *Nature* 121 (Supplement):580–590.

Bohr, N. 1934. *Atomic Theory and the Description of Nature*. Cambridge: Cambridge University Press.

Bohr, N. 1958. *Atomic Physics and Human Knowledge*. New York: John Wiley & Sons.

Bohr, N. 1962. *Essays 1958–1962 on Atomic Physics and Human Knowledge*. Bungay: Richard Clay and Co.

Boltzmann, L. [1886]. 1905. Der zweite Hauptsatz der mechanischen Wärmetheorie. In *Populäre Schriften*. Leipzig: Johann Ambrosius Barth, pp. 25–50.

Bondi, H. 1957. Negative mass in general relativity. *Reviews of Modern Physics* 29:423–428.

Born, M. 1926. Quantenmechanik der Stoßvorgange. *Zeitschrift für Physik* 38:803–827.

Born, M. 1954. The statistical interpretation of quantum mechanics. Nobel Lecture, December 11, 1954. Available at http://www.nobelprize.org/nobel_prizes/physics/laureates/1954/born-lecture.pdf.

Born, M. 1962. *Atomic Physics*, 7th Edition. New York: Hafner Publishing.

Born, M. 1965. Erinnerungen an Einstein. *Physikalische Blätter* 21(7):297–306.

Born, M. 2005. *The Born–Einstein Letters. Friendship, Politics and Physics in Uncertain Times*. New York: Macmillan.

Born, M. and N. S. Nagendra Nath. 1936a. The neutrino theory of light. *Proceedings of the Indian Academy of Sciences* 3:318–377.

Born, M. and N. S. Nagendra Nath. 1936b. The neutrino theory of light. II. *Proceedings of the Indian Academy of Sciences* 4:611–620.

Bragg, W. H. 1907a. A comparison of some forms of electric radiation. *Transactions, Proceedings and Report, Royal Society of South Australia* 31:79–92.

Bragg, W. H. 1907b. The nature of Röntgen rays. *Transactions, Proceedings and Report, Royal Society of South Australia* 31:94–98.

Bragg, W. H. 1907c. On the properties and nature of various electric radiations. *Philosophical Magazine* 6th Series 14:429–449.

Bragg, W. H. 1911. Corpuscular radiation. *Report of the British Association for the Advancement of Science*, 80th Meeting at Portsmouth, August 31–September 7, 1911, 340–341.

Bragg, W. 1922. Electrons and ether waves. The Robert Boyle Lecture at Oxford University for the year 1921. *Scientific Monthly* 14:153–160.

Bragg, W. 1932. *The Universe of Light*. New York: Macmillan.

Bragg, W. H. and J. P. V. Madsen. 1908. An experimental investigation of the nature of γ rays. *Proceedings of the Physical Society of London* 21:261–275.

Brehm, J. J. and W. J. Mullin. 1989. *Introduction to the Structure of Matter*. New York: John Wiley & Sons.

Broido, M. M. 1967. Photon as a composite particle in quantum electrodynamics. *Physical Review* 157:1444–1447.

Buijs Ballot, C. H. D. 1845. Alustische Versuche auf der Niederl Eisenbahn, nebst gelegentliche Bemerkungen zur Theorie des Herrn. Prof. Doppler. *Annalen der Physik* 2nd Series 66:321–351.

Calvin, M. 1959. Energy reception and transfer in photosynthesis. *Reviews of Modern Physics* 31:147–156.

Campbell, G. S. and J. M. Norman. 1998. *An Introduction to Environmental Biophysics*, 2nd Edition. New York: Springer.

Campos, R. A. 2004. Still shrouded in mystery: The photon in 1925. Available at http://arxiv.org/ftp/physics/papers/0401/0401044.pdf.

Case, K. M. 1957. Composite particle of zero mass. *Physical Review* 106:1316–1320.

Chandrasekhar, S. 1982. *Eddington, the Most Distinguished Astrophysicist of His Time*. New York: Cambridge University Press, p. 28.

Clapp, R. E. 1980. Nonlocal structures: Bilocal photon. *International Journal of Theoretical Physics* 19:31–88.

Clayton, R. K. 1971. *Light and Living Matter, Volume 2: The Biological Part*. New York: McGraw-Hill.

Compton, A. H. 1922. A quantum theory of the scattering of X-rays by light elements. *Physical Review* 21:483–502.

Compton, A. H. 1927. X-rays as a branch of optics. Nobel Lecture, December 12, 1927.

Compton, A. H. 1929. What things are made of. *Scientific American* 140:110–113, 234–236.

Compton, A. H. 1935. *The Freedom of Man*. New Haven: Yale University Press:

Comstock, D. F. 1910. The principle of relativity. *Science* 31:767–772.

Curie, I. and F. Joliot. 1932. Électrons de matérialisation et de transmutation. *Journal de Physique et le Radium* 4:494–500.

Darwin, C. G. 1931. *The New Conceptions of Matter*. New York: Macmillan.

Davisson, C. J. 1937. The discovery of electron waves. Nobel Lecture, December 13, 1937.

de Broglie, L. 1924. A tentative theory of light quanta. *Philosophical Magazine* 6th Series 47:446–458.

de Broglie, L. 1930. *An Introduction to the Study of Wave Mechanics*. London: Methuen.

de Broglie, L. 1932a. Sur une analogie entre l'électron de Dirac et l'onde électromagnétique. *Comptes Rendus Hebdomadaires des Séances de l'Académie des Sciences* 195:536–537.

de Broglie, L. 1932b. Remarques sur le moment-magnétique et le moment de rotation de l'électron. *Comptes Rendus Hebdomadaires des Séances de l'Académie des Sciences* 195:577–588.

de Broglie, L. 1932c. Sur le champ électromagnétique de l'onde lumineuse. *Comptes Rendus Hebdomadaires des Séances de l'Académie des Sciences* 195:862–864.

de Broglie, L. 1932. Sur la densité de l'énergie dans la théorie de la lumiére. *Comptes Rendus Hebdomadaires des Séances de l'Académie des Sciences* 197:1377–1380.

de Broglie, L. 1934a. Sur la nature du photon. *Comptes Rendus Hebdomadaires des Séances de l'Académie des Sciences* 198:135–138.

de Broglie, L. 1934b. L'équation d'ondes du photon. *Comptes Rendus Hebdomadaires des Séances de l'Académie des Sciences* 199:445–448.

de Broglie, L. 1934c. Sur l'expression de la densité dans la nouvelle théorie du photon. *Comptes Rendus Hebdomadaires des Séances de l'Académie des Sciences* 199:1165–1168.

de Broglie, L. 1934d. *A New Conception of Light*. Translated by D. H. Delphenich. Paris: Hermann.

de Broglie, L. 1939. *Matter and Light. The New Physics*. New York: W. W. Norton.

de Broglie, L. 1957. Forward. In *Causality and Chance in Modern Physics*, ed. D. Bohm. New York: Harper & Brothers, pp. ix–xi.

de Broglie, L. 1962. *New Perspectives in Physics*. Edinburgh: Oliver & Boyd, pp. 8–9.

de Broglie, L. and J. Winter. 1934. Sur le spin du photon. *Comptes Rendus Hebdomadaires des Séances de l'Académie des Sciences* 199:813–816.

Dingle, H. 1937. *Through Science to Philosophy*. Oxford: Clarendon Press.

Dirac, P. A. M. 1930. A theory of electrons and protons. *Proceedings of the Royal Society* 126A:360–365.

Dirac, P. A. M. 1931. Quantised singularities in the electromagnetic field. *Proceedings of the Royal Society* 133A:60–72.

Dirac, P. A. M. 1932. Theory of electrons and positrons. Nobel Lecture, December 12, 1932. Available at http://www.nobelprize.org/nobel_prizes/physics/laureates/1933/dirac-lecture.pdf.

Dirac, P. A. M. 1958. *The Principles of Quantum Mechanics*, 4th Edition (Revised). Oxford: Oxford University Press.

Dirac, P. A. M. [1964]. 2001. *Lectures on Quantum Mechanics*. Mineola: Dover.

Djerassi, C. and R. Hoffmann. 2001. *Oxygen*. Weinheim: Wiley-VCH.

Doppler, C. [1842]. 1992. On the coloured light of the double stars and certain other stars of the heavens. In *The Search for Christian Doppler*, ed. A. Eden. New York: Springer-Verlag, pp. 103–134.

Doviak, R. J. and D. S. Zrnić. 1992. *Doppler Radar and Weather Observations*, 2nd Edition. New York: Dover Publications.

Duane, W. and F. L. Hunt. 1915. On X-ray wave lengths. *Physical Review* 6:166–171.

Dvoeglazov, V. V. 1998. A note on the neutrino theory of light. *Speculations in Science and Technology* 21:111–115.

Dvoeglazov, V. V. 1999. Speculations on the neutrino theory of light. *Annales de la Fondation Louis de Broglie* 24:111–127.

Dyson, F. W., A. S. Eddington and C. Davidson. 1920. A determination of the deflection of light by the sun's gravitational field, from observations made at the total eclipse of May 29, 1919. *Philosophical Transactions of the Royal Society of London* 220A:291–332.

Eddington, A. S. 1928. *The Nature of the Physical World*. New York: McMillan, p. 201.

Einstein, A. [1905a]. 1989. On a heuristic point of view concerning the production and transformation of light. Doc. 14. In *The Collected Papers of Albert Einstein, Vol. 2. The Swiss Years: Writings, 1900–1909*. English Translation by A. Beck. Princeton: Princeton University Press, pp. 86–102.

Einstein, A. [1905b]. 1989. On the electrodynamics of moving bodies. Doc. 22. In *The Collected Papers of Albert Einstein, Vol. 2. The Swiss Years: Writings, 1900–1909*. English Translation by A. Beck. Princeton: Princeton University Press, pp. 140–171.

Einstein, A. 1909a. On the present status of the radiation problem. Doc 56. In *The Collected Papers of Albert Einstein, Vol. 2. The Swiss Years: Writings, 1900–1909*. English Translation by A. Beck. Princeton: Princeton University Press, pp. 357–375.

Einstein, A. 1909b. On the development of our views concerning the nature and constitution of radiation. Doc 60. In *The Collected Papers of Albert Einstein, Vol. 2. The Swiss Years: Writings, 1900–1909*. English Translation by A. Beck. Princeton: Princeton University Press, pp. 379–394.

Einstein, A. 1909c. "Discussion" following lecture version of "On the development of our views concerning the nature and constitution of radiation." Doc 61. In *The Collected Papers of Albert Einstein, Vol. 2. The Swiss Years: Writings, 1900–1909*. English Translation by A. Beck. Princeton: Princeton University Press, pp. 395–398.

Einstein, A. [1916]. 1997. The foundation of the General Theory of Relativity. Doc. 30, In *The Collected Papers of Albert Einstein, Vol. 6. The Swiss Years: Writings, 1914–1917*. English Translation by A. Engel. Princeton: Princeton University Press, pp. 146–200.

Einstein, A. 1920. *Relativity. The Special and the General Theory*. A Popular Exposition by Albert Einstein. Authorized Translation by R. W. Lawson. New York: Henry Holt.

Einstein, A. 1922. Fundamental ideas and the problems of the theory of relativity. Nobel Lecture, July 11, 1922. Available at http://www.nobelprize.org/nobel_prizes/physics/laureates/1921/einstein-lecture.pdf.

Einstein, A. 1931. Professor Einstein at the California Institute of Technology. *Science* 73:375–379.

Einstein, A. 1949. Autobiographical Notes. In *Albert Einstein: Philosopher-Scientist*, ed. P. A. Schilpp. New York: Tudor Publishing, pp. 3–95.

Einstein, A. and L. Infeld. 1938. *The Evolution of Physics from Early Concepts to Relativity and Quanta*. New York: Simon and Schuster, p. 278.

Ellis, C. D. 1921. The magnetic spectrum of the β-rays excited by γ-rays. *Proceedings of the Royal Society of London* 99A:261–271.

Ellis, C. D. 1926. The light-quantum hypothesis. *Nature* 117:895–897.

Ellis, C. D. and H. W. B. Skinner. 1924a. A re-investigation of the β-ray spectrum of Radium B and Radium C. *Proceedings of the Royal Society of London* 105A:165–184.

Ellis, C. D. and H. W. B. Skinner. 1924b. An interpretation of β-ray spectra. *Proceedings of the Royal Society of London* 105A:185–198.

Engelmann, T. W. [1882]. 1955. On the production of oxygen by plant cells in a microspectrum. In *Great Experiments in Biology*, eds. M. L. Gabriel and S. Fogel. Englewood Cliffs: Prentice-Hall, pp. 166–170.

Evans, M. and J.-P. Vigier. 1994. *The Enigmatic Photon. Volume 1: The Field B$^{(3)}$*. Dordrecht: Kluwer Academic Publishers.

Faraday, M. 1846. Experimental researches in electricity. Nineteenth and Twentieth Series. *Philosophical Transactions of the Royal Society* 136:1–40.

Feher, G. 2002. My road to biophysics: Picking flowers on the way to photosynthesis. *Annual Review of Biophysics and Biomolecular Structure* 31:1–44.

Fermi, E. 1932. Quantum theory of radiation. *Review of Modern Physics* 4:87–132.

Fermi, E. and C. N. Yang. 1949. Are mesons elementary particles? *Physical Review* 76:1739–1742.

Ferretti, B. 1964. A comment on the neutrino-theory of light. *Nuovo Cimento* 33:264–266.

Feynman, R. P. 1949a. The theory of positrons. *Physical Review* 76:749–759.

Feynman, R. P. 1949b. Space-time approach to quantum electrodynamics. *Physical Review* 76:769–789.

Feynman, R. 1979. QED: Photons—corpuscles of light. The Sir Douglas Robb Lectures, University of Auckland. Available at http://www.youtube.com/watch?v=eLQ2atfqk2c.

Feynman, R. 1985. *QED: The Strange Theory of Light and Matter*. Princeton: Princeton University Press, p. 98. Available at http://www.vega.org.uk/video/subseries/8.

Feynman, R. P. 1987. The reason for antiparticles. In *Elementary Particles and the Laws of Physics*, eds. R. MacKenzie and P. Doust. Cambridge: Cambridge University Press, pp. 1–59.

FitzGerald, G. F. 1896. On the longitudinal component in light. *Philosophical Magazine* 5th Series 42:260–271.

Fizeau, H. 1870. Des effets du mouvement sur le ton des vibrations sonores et sur la longueur d'onde des rayons de lumière. *Annales de Chimie et de Physique Séries* 4. 19:211–221.

Fock, V. 1936. Inconsistency of the neutrino theory of light. *Nature* 140:1011–1012.

Fock, V. 1937. The neutrino theory of light. *Nature* 140:112.

Franck, J. and R. W. Wood. 1936. Fluorescence of chlorophyll in its relation to photochemical processes in plants and organic solutions. *The Journal of Physical Chemistry* 4:551–560.

Franck, J. and K. F. Herzfeld. 1941. Contribution to a theory of photosynthesis. *The Journal of Physical Chemistry* 45:978–1025.

Frayn, M. 2000. *Copenhagen*. New York: Random House. Available at https://www.youtube.com/watch?v=4hGAq2kc6u0.

French, A. P. and E. F. Taylor. 1978. *An Introduction to Quantum Physics*. New York: W.W. Norton.

Fuller, R. C. 1999. Forty years of microbial photosynthesis research: Where is came from and what it led to. *Photosynthesis Research* 62:1–29.

Galison, P. 2002. *Einstein's Clocks, Poincaré's Maps*. New York: W. W. Norton.

Galitzin, P. B. and J. Wilip. 1907. Experimental test of Doppler's principle for light-rays. *Astrophysical Journal* 26:49–58.

Garwin, R. L., L. M. Lederman and M. Weinrich. 1957. Observations of the failure of conservation of parity and charge conjugation in meson decays: The magnetic moment of the free muon. *Physical Review* 105:1415–1417.

Gill, T. P. 1965. *The Doppler Effect: An Introduction to the Theory of the Effect*. New York: Academic Press.

Ginzburg, B.-Z. and R. Wayne. 2012. Symmetry and the order of events in time: The asymmetrical order of events in time in a reversible energy converter. *Turkish Journal of Physics* 36:155–162.

Glashow, S. L. 1979. Towards a unified theory—threads in a tapestry. Nobel Lecture, December 8, 1979. Available at http://www.nobelprize.org/nobel_prizes/physics/laureates/1979/glashow-lecture.pdf.

Golden, F. 1999. Albert Einstein. *Time Magazine* December 31, 1999. Available at http://content.time.com/time/magazine/article/0,9171,993017,00.html (accessed November 13, 2014).

Goldhaber, M., L. Grodzins and A. W. Sunyar. 1958. Helicity of neutrinos. *Physical Review* 109:1015–1017.

Govindjee. 2000. Milestones in photosynthetic research. In *Probing Photosynthesis: Mechanisms, Regulation and Adaptation*, eds. M. Yunus, U. Pathre and P. Mohanty. London: Taylor & Francis, pp. 9–39.

Greenberg, O. W. and A. S. Wightman. 1955. Re-examination of the neutrino theory of light. *Physical Review* 99:675.

Greene, B. 2004. *The Fabric of the Cosmos: Space, Time, and the Texture of Reality*. New York: Vintage Books.

Griffiths, D. 1987. *Introduction to Elementary Particles*. New York: John Wiley & Sons.

Griffiths, D. J. 2005. *Introduction to Quantum Mechanics*, 2nd Edition. Upper Saddle River: Pearson.

Grimaldi, F. M. 1665. *Physico-mathesis de Lumine*. Bononiae: V. Benatij.

Guo, M., A. J. Ehrlicher, M. H. Jensen, M. Renz, J. R. Moore, R. D. Goldman, J. Lippincott-Schwartz, F. C. Mackintosh and D. A. Weitz. 2014. Probing the stochastic, motor-driven properties of the cytoplasm using force spectrum microscopy. *Cell* 158:822–932.

Haas, A. 1928. *Wave Mechanics and the New Quantum Theory*. London: Constable.

Hanson, M. R. and A. Sattarzadeh. 2012. Trafficking of proteins through plastid stromules. *The Plant Cell* 25:2774–2782.

Hawking, S. 1999. Does God play dice? Public Lecture. Available at http://www.hawking.org.uk/does-god-play-dice.html, http://www.hawking.org.uk/lectures.html (accessed November 2, 2014).

Hawking, S. and L. Mlodinow. 2010. *The Grand Design*. New York: Bantam Books.

Heisenberg, W. 1927. Über den anschaulichen Inhalt der quantentheoretischen Kinematik und Mechanik. *Zeitschrift für Physik* 43:172–198.

Heisenberg, W. 1930. *The Physical Principles of the Quantum Theory*. New York: Dover.

Heisenberg, W. 1932. The development of quantum mechanics. Nobel Lecture, December 11, 1932. Available at http://www.nobelprize.org/nobel_prizes/physics/laureates/1932/heisenberg-lecture.pdf.

Heisenberg, W. [1974]. 1982. Encounters and conversations with Albert Einstein. In *Tradition in Science*. New York: Seabury Press, 107–122.

Heisenberg, M. 2009. Is free will an illusion? *Nature* 459:164–165.

Heitler, W. 1944. *The Quantum Theory of Radiation*, 2nd Edition. Oxford: Oxford University Press.

Heitler, W. 1962. *Man and Science*. Edinburgh: Oliver and Boyd.

Herschel, J. F. W. 1832. *A Treatise on Astronomy*. London: Longman, Rees, Orme, Brown, Green & Longman.

Hertz, H. 1887. Ueber einen Einfluss des ultravioletten Lichtes auf die electrische Entladung. *Annalen der Physik* 267:983–1000.

Hertz, H. [1893]. 1962. *Electric Waves*. New York: Dover.

Hochmuth, R. M. 2000. Micropipette aspiration of living cells. *Journal of Biomechanics* 33:15–22.

Hooke, R. 1665. *Micrographia*. London: Jo. Martyn and Ja. Allestry.

Huggins, W. 1868. Further observations on the spectra of some of the stars and nebulae, with an attempt to determine therefrom whether these bodies are moving towards or from the earth, also observations on the spectra of the sun and of comet II. *Philosophical Transactions of the Royal Society of London* 158:529–564.

Hujer, K. 1962. Christian Doppler in Prague. *Royal Astronomical Society of Canada Journal* 57(4):177–180.

Hund, F. 1974. *The History of Quantum Theory*. New York: Barnes & Noble, p. 141.

Huygens, C. [1690]. 1945. *Treatise on Light*. Chicago: University of Chicago Press.

Inoue, H., T. Tajima and S. Tanaka. 1972. On the neutrino theory of light. *Progress in Theoretical Physics* 48:1338–1362.

Ismail, Y., A. R. Mirza, A. Forbes and F. Petruccione. 2014. Characterization of a polarization based entanglement photon source. *African Review of Physics* 9:217–226.

Jackson, D. J. 1999. *Classical Electrodynamics*, 3rd Edition. New York: John Wiley & Sons.

Jaffe, B. 1960. *Michelson and the Speed of Light*. Garden City: Doubleday.

Jagendorf, A. T. 1998. Chance, luck and photosynthesis research: An inside story. *Photosynthesis Research* 57:215–229.

Jaki, S. L. 1978. Johann Georg von Soldner and the gravitational bending of light, with an English translation of his essay on it published in 1801. *Foundations of Physics* 8:927–950.

Jammer, M. 1966. *The Conceptual Development of Quantum Mechanics*. New York: McGraw-Hill.

Jarvis, P. and E. López-Juez. 2012. Biogenesis and homeostasis of chloroplasts and other plastids. *Nature Reviews Molecular Cell Biology* 14:787–802.

Jeans, J. H. 1914. *Report on Radiation and the Quantum Theory*. London: "The Electrician" Printing and Publishing Co.

Jeans, J. H. 1924. *Report on Radiation and the Quantum Theory*, 2nd Edition. London: Fleetwood Press.

Jeans, J. H. 1934. *The Mysterious Universe*, 2nd Edition. Cambridge: Cambridge University Press, p. 122.

Jeans, J. H. 1945. *Physics & Philosophy*. Cambridge: Cambridge University Press, pp. 16, 175–176, 190.

Johnsen, S. 2012. *The Optics of Light. A Biologist's Guide to Light in Nature*. Princeton: Princeton University Press.

Joly, J. 1921. A quantum theory of colour vision. *Proceedings of the Royal Society* 92B:219–232.

Jordan, P. 1927. Philosophical foundations of quantum theory. *Nature* 119:566–569.

Jordan, P. 1928. Die Lichtquantenhypotheses. Entwicklung und gegenwärtiger Stand. *Ergebnisse der exakten Naturwissenschaften* 7:158–208.

Jordan, P. 1935. Zur neutrinotheorie des Lichtes. *Zeitschrift für Physik* 93:464–472.

Jordan, P. 1936a. Lichtquant und Neutrino. *Zeitschrift für Physik* 98:759–767.

Jordan, P. 1936b. Zur Herleitung der Vertauschungsregeln in der Neutrinotheorie des Lichtes. *Zeitschrift für Physik* 99:109–112.

Jordan, P. 1936c. Beiträge zur Neutrinotheorie des Lichts. I. *Zeitschrift für Physik* 102:243–252.

Jordan, P. 1937a. Beiträge zur Neutrinotheorie des Lichts. II. *Zeitschrift für Physik* 105:114–121.

Jordan, P. 1937b. Beiträge zur Neutrinotheorie des Lichts. III. *Zeitschrift für Physik* 105:229–231.

Jordan, P. and R. L. de Kronig. 1936. Lichtquant und Neutrino. II. Dreidimensionales Strahlungsfeld. *Zeitschrift für Physik* 100:569–582.

Kadota, A., N. Yamada, N. Suetsugu, M. Hirose, C. Saito, K. Shoda, S. Ichikawa, T. Kagawa, A. Nakano and M. Wada. 2009. Short actin-based mechanism for light-directed chloroplast movement in *Arabidopsis*. *Proceedings of the National Academy of Sciences of the United States of America* 106:13106–13111.

Kaiser, D. 2005. Physics and Feynman's diagrams. *American Scientist* 93:156–165.

Kaku, M. 1994. *Hyperspace. A Scientific Odyssey through Parallel Universes, Time Warps, and the Tenth Dimension*. New York: Anchor Books.

Kamen, M. 1985. *Radiant Science, Dark Politics: A Memoir of the Nuclear Age*. Berkeley: University of California Press.

Karpa, L. and M. Weitz. 2006. A Stern–Gerlach experiment for slow light. *Nature Physics* 2:332–335.

Kawachi, N., K. Sakamoto, S. Ishii, S. Fujimaki, N. Suzui, N. S. Ishioka and S. Matsuhashi. 2006. Kinetic analysis of carbon-11-labeled carbon dioxide for studying photosynthesis in a leaf using positron emitting tracer imaging system. *IEEE Transactions on Nuclear Science* 53(5):2991–2997.

Kelvin, Lord. 1904. *Baltimore Lecture on Molecular Dynamics and the Wave Theory of Light*. London: C. J. Clay and Sons.

Kennard, E. H. 1927. Zur Quantenmechanik einfacher Bewegungstypen. *Zeitschrift für Physik* 44:326–352.

Kerr, J. 1877. On the rotation of the plane of polarization by reflection from the pole of a magnet. *Philosophical Magazine* 5th Series 3:321–342.

Kerr, J. 1878. On the reflection of polarized light from the equatorial surface of a magnet. *Philosophical Magazine* 5th Series 5:161–177.

Kidd, R., J. Ardini and A. Anton. 1989. Evolution of the modern photon. *American Journal of Physics* 57:27–35.

Kirchhoff, G and R. Bunsen. [1860]. 1901. Chemical analysis by spectral observations. In *The Laws of Radiation and Absorption; Memoirs by Prévost, Stewart, and Kirchhoff and Bunsen*, ed. D. B. Brace. New York: Amer. Book Co., pp. 99–126.

Klassen, S. 2011. The photoelectric effect: Reconstructing the story for the physics classroom. *Science & Education* 20:719–731.

Klein, M. J. 1970. The first phase of the Bohr–Einstein dialogue. *Historical Studies in the Physical Sciences* 2:1–39.

Koller, D. 2011. *The Restless Plant*. Cambridge: Harvard University Press.

Kragh, H. 2014. Photon: New light on an old name. Available at http://arxiv.org/abs/1401.0292.

Kronig, R. de L. 1935a. Zur Neutrinotheorie des Lichtes. *Physica* 2:491–498.

Kronig, R. de L. 1935b. Zur Neutrinotheorie des Lichtes II. *Physica* 2:854–860.

Kronig, R. de L. 1935c. Zur Neutrinotheorie des Lichtes III. *Physica* 2:968–980.

Kronig, R. de L. 1936. On a relativistically invariant formulation of the neutrino theory of light. *Physica* 3:1120–1132.

Laible, P. D., W. Zipfel and T. G. Owens. 1994. Excited state dynamics in chlorophyll-based antennae: The role of transfer equilibrium. *Biophysical Journal* 66:844–860.

Landau, L. D. and E. M. Lifshitz. 1958. *Quantum Mechanics Non-Relativistic Theory*. London: Pergamon Press, p. 186.

Laplace, P. S. Marquis de [1814]. 1902. *A Philosophical Essay on Probabilities*. Translated form the Sixth French Edition by F. W. Truscott and F. L. Emory. New York: John Wiley & Sons, p. 4.

Lebedew, P. 1901. Untersuchungen über die Druckkräfte des Lichtes. *Annalen der Physik* 4th Series 6:433–458.

Lecher, E. 1890. Eine Studie über elektrische Resonanzerscheinungen. *Annalen der Physik* 3rd Series 41:850–870.

Lee, T. D. 1957. Weak interactions and nonconservation of parity. Nobel Lecture, December 11, 1957. Available at http://www.nobelprize.org/nobel_prizes/physics/laureates/1957/lee-lecture.pdf.

Lee, T. D. 1988. *Symmetries, Asymmetries, and the World of Particles.* Seattle: University of Washington Press.

Lee, D. 2007. *Nature's Palette. The Science of Plant Color.* Chicago: University of Chicago Press.

Lee, T. D. and C. N. Yang. 1956. Question of parity conservation in weak interactions. *Physical Review* 104:254–258.

Lee, T. D. and C. N. Yang. 1957. Parity nonconservation and a two-component theory of the neutrino. *Physical Review* 105:1671–1675.

Lehnert, B. 2006. Photon physics of revised electromagnetics. *Progress in Physics* 2:78–85.

Lehnert, B. 2008. Wave-particle properties and pair formation of the photon. In *International Workshop on the Frontiers of Modern Plasma Physics*, The Abdus Salam International Centre for Theoretical Physics, July 14–25, 2008.

Lehnert, B. 2013. On angular momentum and rest mass of the photon. *Journal of Plasma Physics* 79:1133–1135.

Lenard, P. 1900. Erzeugung von Kathodenstrahlen durch ultraviolettes Licht. *Annalen der Physik* 307:359–375.

Lenard, P. 1902. Ueber die lichtelektrische Wirkung. *Annalen der Physik* 8:149–198.

Leone, M. and N. Robotti. 2010. Frédéric Joliot, Irène Curie and the early history of the positron (1932–33). *European Journal of Physics* 31:975–987.

Lewis, G. N. 1926a. The nature of light. *Proceedings of the National Academy of Sciences of the United States of America* 12:22–29.

Lewis, G. N. 1926b. The conservation of photons. *Nature* 118:874–875.

Lillie, R. S. 1927. Physical indeterminism and vital action. *Science* 66:139–144.

Livingston, D. M. 1973. *The Master of Light. A Biography of Albert A. Michelson.* Chicago: University of Chicago Press.

Lorentz, H. A. 1924. The radiation of light. *Nature* 113:608–611.

Loudon, R. 1983. *The Quantum Theory of Light*, 2nd Edition. Oxford: Clarendon Press.

Mach, E. 1860. Über die Änderung des Tones und der Färbe durch Bewegung. *Sitzungsberichte der kaiserlichen Akademie der Wissenschaften. Math.-Naturwissen Klasse, Vienna* 41: 543–560.

Mach, E. 1873. *Beiträge zur Doppler'schen Theorie der Ton- und Fäbenänderung durch Bewegung.* Prague: J. G. Calve.

Maers, A. F. and R. Wayne. 2011. Rethinking the foundations of the theory of special relativity: Stellar aberration and the Fizeau experiment. *African Physical Review* 5:7–40. Available at http://arxiv.org/abs/1105.2305.

Maers, A., R. Furnas, M. Rutzke and R. Wayne. 2013. The Fizeau experiment: Experimental investigations of the relativistic Doppler effect. *African Review of Physics* 8:297–312. Available at http://www.aphysrev.org/index.php/aphysrev/article/view/762/321.

Majeran, W., G. Friso, L. Ponnala, B. Connolly, M. Huang, E. Reidel, C. Zhang, Y. Asakura, N. H. Bhuiyan, Q. Sun, R. Turgeon and K. J. van Wijk. 2010. Structural and metabolic transitions of C4 leaf development and differentiation defined by microscopy and quantitative proteomics in maize. *The Plant Cell* 22:3509–3542.

Maulik, D. ed. 1997. *Doppler Ultrasound in Obstetrics & Gynecology.* New York: Springer-Verlag.

Maxwell, J. C. 1861. On physical lines of force. *Philosophical Magazine* 4th Series 21:338–348.

Maxwell, J. C. 1865. A dynamical theory of the electromagnetic field. *Philosophical Transactions of the Royal Society of London* 155:459–512.

Maxwell, J. C. 1873. *A Treatise on Electricity and Magnetism*, Vol. II. Oxford: Clarendon Press.

Mayer, J. R. [1845]. 1869. On organic motion and nutrition. In *Heat Considered as a Mode of Motion*, J. Tyndall. New York: D. Appleton, pp. 522–529.

Mbarek, S. and M. B. Pranjape. 2014. Negative mass bubbles in de Sitter space-time. Available at http://arxiv.org/pdf/1407.1457v1.pdf.

McQuarrie, D. A., P. A. Rock and E. B. Gallogly. 2010. *General Chemistry*, 4th Edition. Mill Valley: University Science Books.

Mees, C. E. K. 1922. Darts of light. What twentieth century physics has done to the wave theory of classical optics. *Scientific American* 126:336.

Mermin, N. D. 1989. What's wrong with this pillow? *Physics Today* 42(4):9–11.

Mermin, N. D. 2004. Could Feynman have said this? *Physics Today* 57(5):10–11.

Michelson, A. A. 1903. *Light Waves and their Uses.* Chicago: University of Chicago Press, pp. 23–24.

Michelson, A. A. and E. W. Morley. 1887. On the relative motion of the earth and the luminiferous ether. *American Journal of Science* 34:333–345.

Miller, A. I. 1994. *Early Quantum Electrodynamics: A Source Book.* Cambridge: Cambridge University Press.

Millikan, R. A. 1914. A direct determination of "h." *Physical Review* 4:73–75.

Millikan, R. A. 1916. A direct photoelectric determination of Planck's "h." *Physical Review* 7:355–388.

Millikan, R. A. 1924. The electron and the light-quant from the experimental point of view. Nobel Lecture, May 23, 1924. Available at http://www.nobelprize.org/nobel_prizes/physics/laureates/1923/millikan-lecture.html.

Millikan, R. A. 1935. *Electrons (+ and −), Protons, Photons, Neutrons, and Cosmic Rays.* Chicago: University of Chicago Press.

Millikan, R. A. 1950. The experimental proof of the existence of the photon—Einstein's photoelectric equation. In: *The Autobiography of Robert A. Millikan.* New York: Prentice-Hall, pp. 100–107.

Monteith, J. and M. Unsworth. 2008. *Principles of Environmental Physics*, 3rd Edition. Amsterdam: Elsevier/Academic Press.

Mott, N. F. and I. N. Sneddon. 1948. *Wave Mechanics and Its Applications.* Oxford: Clarendon Press.

Nagendra Nath, N. S. 1936. Neutrinos and light quanta. *Proceedings of the Indian Academy of Sciences* 3:448.

Newton, I. 1687. *Philosophiæ Naturalis Principia Mathematica.* London: Jussu Societatis Regiae.

Newton, I. [1730]. 1979. *Opticks.* New York: Dover.

New York Times. 1919. Eclipse showed gravity variation, November 9; Lights all askew in the heavens, November 10; Amateurs will be resentful, November 11; They have already a geometry, November 11; Sir Isaac finds a defender, November 11; Don't worry over new light theory, November 16; Jazz in scientific world, November 16; Light and logic, November 16; Nobody need be offended, November 18; A new physics based on Einstein, November 25; Bad times for the learned, November 26.

Nichols, E. F. and G. F. Hull. 1903a. The pressure due to radiation *Physical Review* 17:26–50.

Nichols, E. F. and G. F. Hull. 1903b. The pressure due to radiation. *Physical Review* 17:91–104.

Nicholson, J. W. 1912. On the constitution of the solar corona. II. *Monthly Notices of the Royal Astronomical Society* 72:677–692.

Nicholson, J. W. 1913. The theory of radiation. *Nature* 92:199.

Niklas, K. J. 1992. *Plant Biomechanics. An Engineering Approach to Plant Form and Function.* Chicago: University of Chicago Press.

Niklas, K. J. 1997. *The Evolutionary Biology of Plants.* Chicago: University of Chicago Press.

Niklas, K. J. and H.-C. Spatz. 2012. *Plant Physics.* Chicago: University of Chicago Press.

Nobel, P. S. 2009. *Physicochemical and Environmental Plant Physiology,* 4th Edition. Amsterdam: Elsevier/Academic Press.

Oberg, E., F. D. Jones, H. L. Horton and H. H. Ryffell. 2000. *Machinery's Handbook,* 26th Edition. New York: Industrial Press.

Okun, L. B. 2006. Photon: History, mass, charge. *Acta Physica Polonica B* 37:565–573.

O'Leary, A. J. 1964. Redshift and deflection of photons by gravitation: A comparison of relativistic and Newtonian treatments. *American Journal of Physics* 32:52–55.

Olive, K. A. et al. (Particle Data Group). 2014. *Chinese Physics C38*: 090001. Available at http://pdg.lbl.gov, http://pdg.lbl.gov/2014/listings/rpp2014-list-photon.pdf.

Oppenheimer, J. R. 1941. Internal conversion in photosynthesis. Proceedings of the American Physical Society Meeting in Pasadena, California June 18–20, 1941. *Phys. Rev.* 60:158.

Page, L. 1918. Is a moving star retarded by the reaction of its own radiation? *Proceedings of the National Academy of Sciences of the United States of America* 4:47–49.

Pais, A. 1982. *'Subtle is the Lord...' The Science and the Life of Albert Einstein.* Oxford: Clarendon Press.

Pais, A. 1986. *Inward Bound.* Oxford: Clarendon Press, p. 419.

Pais, A. 1994. *Einstein Lived Here.* Oxford: Clarendon Press, pp. 148–150.

Paolillo Jr., D. J. and J. A. Reighard. 1967. On the relationship between mature structure and ontogeny in the grana of chloroplasts. *Canadian Journal of Botany* 45:773–782.

Park, D. 1992. *Introduction to the Quantum Theory,* 3rd Edition. Mineola: Dover.

Pauling, L. and E. B. Wilson Jr. 1935. *Introduction to Quantum Mechanics.* New York: McGraw-Hill.

Peat, F. D. 1997. *Infinite Potential. The Life and Times of David Bohm.* Reading, MA: Addison-Wesley.

Perkins, W. A. 1965. Neutrino theory of photons. *Physical Review* 137B:1291–1301.

Perkins, W. A. 1972. Statistics of a composite photon formed of two fermions. *Physical Review* 5D:1375–1384.

Pinch, T. J. 1977. What does a proof do if it does not prove? In *The Social Production of Scientific Knowledge,* eds. E. Mendelsohn, P. Weingart and R. Whitley. Dordrecht: D. Reidel, pp. 171–215.

Planck, M. 1920. The genesis and present state of development of the quantum theory. Nobel Lecture, June 2, 1920. Available at http://www.nobelprize.org/nobel_prizes/physics/laureates/1918/planck-lecture.html.

Poynting, J. H. 1904. Radiation in the solar system: Its effect on temperature and its pressure on small bodies. *Philosophical Transactions of the Royal Society of London* 202A:525–552.

Pryce, M. H. L. 1938. On the neutrino theory of light. *Proceedings of the Royal Society* 165A:247–271.

Przybycień, M. 2003. Study of the photon structure at LEP. Available at http://home.agh.edu.pl/~mariuszp/hab.pdf.

Raman, C. V. 1930. The molecular scattering of light. Nobel Lecture, December 11, 1930. Available at http://www.nobelprize.org/nobel_prizes/physics/laureates/1930/raman-lecture.pdf.

Rao, B. S. M. 1938. Question of invariance in the neutrino theory of light. *Proceedings of the Indian Academy of Sciences Section A* 7:293–295.

Richtmyer, F. K. and E. H. Kennard. 1942. *Introduction to Modern Physics,* 3rd Edition. New York: McGraw-Hill.

Richtmyer, F. K. and E. H. Kennard. 1947. *Introduction to Modern Physics,* 4th Edition. New York: McGraw-Hill.

Richtmyer, F. K., E. H. Kennard and T. Lauritsen. 1955. *Introduction to Modern Physics,* 5th Edition. New York: McGraw-Hill.

Richtmyer, F. K., E. H. Kennard and J. N. Cooper. 1969. *Introduction to Modern Physics,* 6th Edition. New York: McGraw-Hill.

Rigden, J. S. 2005. Einstein's revolutionary paper. *Physics World* 18(4):18–19.

Rikken, G. L. J. A. and B. A. ven Tiggelen. 1996. Observation of magnetically induced transverse diffusion of light. *Nature* 381:54–55.

Röntgen, W. C. 1896. On a new kind of rays. *Nature* 53:274–276.

Roscoe, H. 1869. *Spectrum Analysis.* London: Macmillan.

Rose, M. E. and G. E. Uhlenbeck. 1935. The formation of electron-positron pairs by internal conversion of γ-radiation. *Physical Review* 48:211–223.

Rosen, N. and P. Singer. 1959. The photon as a composite particle. *Bulletin of the Research Council of Israel* 8F:51–62.

Ruark, A. E. and H. C. Urey. 1927. The impulse moment of the light quantum. *Proceedings of the National Academy of Sciences of the United States of America* 13:763–771.

Ruark, A. E. and H. C. Urey. 1930. *Atoms, Molecules and Quanta.* New York: McGraw-Hill.

Ruderfer, M. 1965. Phasor-neutrino theory of electromagnetism. *Electronics & Power* 11(5):164–165.

Ruderfer, M. 1971. On the neutrino theory of light. *American Journal of Physics* 39:116.

Russell, J. S. 1848. On certain effects produced on sound by the rapid motion of the observer. Notices and Abstracts of Communications to the British Association for the Advancement of Science at the Swansea Meeting, August 1848. *British Association Reports* 18:37–38.

Rutherford, E. 1911. The scattering of α and β particles by matter and the structure of the atom. *Philosophical Magazine* 6th Series 21:669–688.

Rybicki, G. B. and A. P. Lightman. 1979. *Radiative Processes in Astrophysics.* New York: Wiley.

Sahin, G. and M. Saglam. 2009. Calculation of the magnetic moment of the photon. XXVI International Conference on Photon, Electronic and Atomic Collisions. *Journal of Physics: Conference Series* 194:022006.

Salam, A. 1979. Gauge unification of fundamental forces. Nobel Lecture, December 8, 1979. Available at http://www.nobelprize.org/nobel_prizes/physics/laureates/1979/salam-lecture.pdf.

Sarkar, H., B. Bhattacharya and P. Bandyopadhyay. 1975. Neutrino theory of light from the viewpoint of the Bethe–Salpeter equation. *Physical Review D* 11:935–938.

Scherzer, O. 1935. Zur Neutrinotheorie des Lichts. *Zeitschrift für Physics* 97:725–739.

Schrödinger, E. 1926. An undulatory theory of the mechanics of atoms and molecules. *Physical Review* 28:1049–1070.

Schrödinger, E. 1933. The fundamental idea of wave mechanics. Nobel Lecture, December 12, 1933. Available at http://www.nobelprize.org/nobel_prizes/physics/laureates/1933/schrodinger-lecture.pdf.

Schrödinger, E. 1945. *What is Life? The Physical Aspect of the Living Cell*. Cambridge: Cambridge University Press.

Schuster, A. 1898. Potential matter—a holiday dream. *Nature* 58:367.

Schuster, P. M. 2005. *Moving the Stars: Christian Doppler, His Life, His Works and Principle. Christian Doppler and World After*. Pöllauberg: Living Edition.

Schuster, A. and J. W. Nicholson. 1924. *An Introduction to the Theory of Optics*, 3rd Edition. London: Edward Arnold.

Serway, R. A., C. J. Moses and C. A. Moyer. 2005. *Modern Physics*, 3rd Edition. Belmont: Brooks/Cole.

Shanks, D. 1956. Monochromatic approximation of blackbody radiation. *American Journal of Physics* 24:244–246.

Shao, J.-Y., H. P. Ting-Beall and R. M. Hochmuth. 1998. Static and dynamic lengths of neutrophil microvilli. *Proceedings of the National Academy of Sciences of the United States of America* 95:6797–6802.

Shu, F. H. 1982. *The Physical Universe: An Introduction to Astronomy*. Sausalito: University Science Books.

Silberstein, L. 1922. Quantum theory of photographic exposure. *Philosophical Magazine* 6th Series 44:257–273.

Silberstein, L. and A. P. H. Trivelli. 1922. Quantum theory of photographic exposure (Second paper). *Philosophical Magazine* 6th Series 44:956–968.

Slater, J. C. 1951. *Quantum Theory of Matter*. New York: McGraw-Hill.

Slater, J. C. and N. H. Frank. 1933. *Introduction to Theoretical Physics*. New York: McGraw-Hill.

Slipher, V. M. 1913. The radial velocity of the Andromeda Nebula. *Lowell Observatory BulletinLowell Observatory Bulletin* 2:56–57.

Sokolow, A. 1937. Neutrino theory of light. *Nature* 139:1071.

Solomey, N. 1997. *The Elusive Neutrino*. New York: Scientific American Library.

Sommerfeld, A. 1923. *Atomic Structures and Spectral Lines*. Translated from the 3rd German Edition. London: Methuen.

Stark, J. 1906. Über die Lichtemission der Kanalstrahlen in Wasserstoff. *Annalen der Physik* 21:401–456.

Stark, J. 1909. Zur experimentellen Entscheidung zwischen Ätherwellen- und Lichtquantenhypothese. I. Röntgenstrahlung. *Physikalische Zeitschrift* 10:902–1913.

Stokes, G. G. 1884. *First Course on the Nature of Light*. London: Macmillan, p. 32.

Stuewer, R. H. 1989. Non-Einsteinian interpretations of the photoelectric effect. In *Historical and Philosophical Perspective of Science*, ed. R. H. Stuewer. New York: Gordon and Breach Science Publishers, 246–263.

Stuewer, R. H. 2006. Einstein's revolutionary light-quantum hypothesis. *Acta Physica Polonica B* 37:543–558.

Synge, J. L. 1951. *Science, Sense and Nonsense*. New York: W. W. Norton.

Synge, J. L. 1970. *Talking about Relativity*. New York: American Elsevier.

Taylor, G. I. 1909. Interference fringes with feeble light. *Proceedings of the Cambridge Philosophical Society* 15:114–115.

Tegmark, M. 2007. Shut up and calculate. Available at http://arxiv.org/pdf/0709.4024v1.pdf.

Thomson, G. P. 1928. Experiments on the diffraction of cathode rays. *Proceedings of the Royal Society of London* 117A:600–609.

Thomson, G. P. 1938. Electronic waves. Nobel Lecture, June 7, 1938. Available at http://www.nobelprize.org/nobel_prizes/physics/laureates/1937/thomson-lecture.html.

Thomson, J. J. 1908. On the ionization of gases by ultra-violet light and on the evidence as to the structure of light afforded by its electrical effects. *Proceedings of the Cambridge Philosophical Society* 14:417–424.

Thomson, J. J. 1925. *The Structure of Light*. Cambridge: Cambridge University Press.

Thompson, S. P. 1901. *Michael Faraday. His Life and Work*. London: Cassell.

t Hooft, G. W. and M. B. van der Mark. 1996. Light bent by magnets. *Nature* 381:27–28.

Toman, K. 1984. Christian Doppler and the Doppler effect. *Eos* 65:7–10.

Tomonaga, S.-I. 1997. *The Story of Spin*. Chicago: University of Chicago Press.

Troland, L. T. 1916. Apparent brightness; its conditions and properties. *Transactions of the Illuminating Engineering Society* 11:947–975.

Troland, L. T. 1917. On the measurement of visual stimulation intensities. *Journal of Experimental Psychology* 2:1–33.

Tsuchiya, T. E. Enuzuka, T. Kurono and M. Hosoda. 1985. Photon-counting imaging and its application. *Advances in Electronics and Electron Physics* 64A:21–31.

Tyndall, J. 1915. *Heat. A Mode of Motion*, 6th Edition. New York: D. Appleton.

van Tiggelen, B. A. and G. L. J. A. Rikken. 2002. Manipulating light with a magnetic field. In: *Optical Properties of Nanostructured Random Media*, Topics Appl. Phys. Vol. 82, ed. V. M. Shalaev, pp. 275–302.

Varlamov, V. V. 2002. About algebraic foundations of Majorana-Oppenheimer quantum electrodynamics and de Broglie-Jordan neutrino theory of light. *Annales de la Fondation Louis de Broglie* 27:273.

Vignolini, S., E. Moyround, B. J. Glover and U. Steiner. 2013. Analysing photonic structures on plants. *Journal of the Royal Society Interface* 10:20130394.

Vogel, H. C. 1876. Ueber die Veränderung de Tonhöhe bei Bewegung eines tönenden Korpers. *Annalen der Physik* 2nd Series 158:287–306.

von Neumann, J. [1932]. 1955. *Mathematical Foundations of Quantum Mechanics*. Princeton: Princeton University Press.

Wada, M. 2013. Chloroplast movement. *Plant Science* 210:177–182.

Wang, H., L. Shi, B. Luk'yanchuk, C. Sheppard and C. T. Chong. 2008. Creation of a needle of longitudinally polarized light in vacuum using binary optics. *Nature Photonics* 2:501–505.

Wayne, R. 2009a. *Light and Video Microscopy*. Amsterdam: Elsevier/Academic Press.

Wayne, R. 2009b. *Plant Cell Biology: From Astronomy to Zoology*. Amsterdam: Elsevier/Academic Press.

Wayne, R. 2010a. Charged particles are prevented from going faster than the speed of light by light itself: A biophysical cell biologist's contribution to physics. *Acta Physica Polonica B* 41:1001–1027. Available at http://arxiv.org/abs/1103.3697.

Wayne, R. 2010b. The relativity of simultaneity: An analysis based on the properties of electromagnetic waves. *African Physical Review* 4:43–55.

Wayne, R. 2010c. Letter to the Editor: In defense of free will. *Cornell Daily Sun* May 13. Available at http://cornellsun.com/blog/2010/05/13/letter-to-the-editor-in-defense-of-free-will/

Wayne, R. 2012a. A fundamental, relativistic, and irreversible law of motion: A unification of Newton's Second Law of motion and the Second Law of Thermodynamics. *African Review of Physics* 7:115–134.

Wayne, R. 2012b. Rethinking the concept of space-time in the general theory of relativity: The deflection of starlight and the gravitational red shift. *African Review of Physics* 7:183–201.

Wayne, R. 2012c. Symmetry and the order of events in time: A proposed identity of negative mass with antimatter. *Turkish Journal of Physics* 36:165–177.

Wayne, R. 2012d. Taking the mechanics out of space-time and putting it back into quantum mechanics. *FQXi Forum*. Available at http://fqxi.org/community/forum/topic/1402 (accessed November 14, 2014).

Wayne, R. 2013a. Identifying the individual chromosomes of maize. In *Perspectives On Nobel Laureate Barbara McClintock's Publications (1926–1984): A Companion Volume*, Vol. 1, ed. L. B. Kass. Ithaca: Internet-First University Press, pp. 1.77–71.86.

Wayne, R. 2013b. Symmetry and the order of events in time: Description of a reversible thermal energy converter composed of negative mass. *Turkish Journal of Physics* 37:1–21.

Wayne, R. 2013c. The relationship between the optomechanical Doppler force and the magnetic vector potential. *African Review of Physics* 8:283–296.

Wayne, R. 2014a. *Light and Video Microscopy*, 2nd Edition. Amsterdam: Elsevier/Academic Press.

Wayne, R. 2014b. Evidence that photons have extension in space. *Turkish Journal of Physics* 38:17–25.

Wayne, R. 2014c. Deriving the Snel-Descartes law for a single photon. *Turkish Journal of Physics* 38:26–38.

Wayne, R. 2015. The equivalence of mass and energy: Blackbody radiation in uniform translational motion. *African Review of Physics* 10:1–9.

Wayne, R. and P. K. Hepler. 1984. The role of calcium ions in phytochrome-mediated germination of spores of *Onoclea sensibilis* L. *Planta* 160:12–20.

Wayne, R. and P. K. Hepler. 1985. Red light stimulates an increase in intracellular calcium in the spores of *Onoclea sensibilis*. *Plant Physiology* 77:8–11.

Wayne, R., A. Kadota, M. Watanabe and M. Furuya. 1991. Photomovement in *Dunaliella salina*: Fluence rate-response curves and action spectra. *Planta* 184:515–524.

Weinberg, S. 1975. Light as a fundamental particle. *Physics Today* 32(6):32–37.

Weinberg, S. 1979. Conceptual foundations of the unified theory of weak and electromagnetic interactions. Nobel Lecture, December 8, 1979. Available at http://www.nobelprize.org/nobel_prizes/physics/laureates/1979/weinberg-lecture.pdf.

Weinberger, P. 2008. John Kerr and his effects found in 1877 and 1878. *Philosophical Magazine Letters* 88:897–907.

Wertheimer, W. 1959. *Productive Thinking*. New York: Harper & Brothers, p. 216.

Wheaton, B. R. 1978. Philipp Lenard and the photoelectric effect, 1889–1911. *Historical Studies in the Physical Sciences* 9:299–322.

Wheaton, B. R. 1983. *The Tiger and the Shark. Empirical Roots of Wave-Particle Dualism*. Cambridge: Cambridge University Press.

Wigner, E. 1967. *Symmetries and Reflections, Scientific Essays of Eugene P. Wigner*. Bloomington: Indiana University Press.

Wolfers, F. 1926. Une action probable de la matière sur les quanta de radiation. *Comptes Rendus Hebdomadaires des Séances de l'Académie des Sciences* 183:276–277.

Wright, F. L. 1953. *The Future Architecture*. New York: Horizon Press.

Wu, C. S., E. Ambler, R. W. Hayward, D. D. Hoppes and R. P. Hudson. 1957. Experimental test of parity conservation in beta decay. *Physical Review* 105:1413–1415.

Wurmser, R. 1925a. La rendement énergétique de la photosynthèse chlorophylliene. *Annales de Physiologie et de Physicochimie Biologique* 1:47–63.

Wurmser, R. 1925b. Sur l'activité des diverses radiations dans la photosynthèse. *Comptes Rendus Hebdomadaires des Séances de l'Académie des Sciences* 181:374–375.

Yang, C. N. 1957. The law of parity conservation and other symmetry laws of physics. Nobel Lecture, December 11, 1957. Available at http://www.nobelprize.org/nobel_prizes/physics/laureates/1957/yang-lecture.pdf.

Yang, C. N. 2014. The conceptual origins of Maxwell's equations and gauge theory. *Physics Today* 67(11):45–51.

Ye, H., C.-W. Qiu, K. Huang, J. Teng, B. Luk'yanchuk and S. P. Yeo. 2013. Creation of a longitudinally polarized subwavelength hotspot with an ultra-thin planar lens: vectorial Rayleigh–Sommerfeld method. *Laser Physics Letters* 10:065004.

Young, T. 1804. Reply to the animadversions of the Edinburgh reviewers on some papers published in the Philosophical Transactions. London: Joseph Johnson.

Young, T. 1807. *A Course of Lectures on Natural Philosophy and the Mechanical Arts*. London: Joseph Johnson.

Young, R. A. 1976. Thinking of the photon as a quantum-mechanical particle. *American Journal of Physics* 44:1043–1046.

Zeeman, P. 1903. Light radiation in a magnetic field. Nobel Lecture, May 2, 1903.

Zeilinger, A., G. Weihs, T. Jennewein and M. Aspelmeyer. 2005. Happy centenary, photon. *Nature* 433:230–238.

Zu, D. 2008. The classical structure model of a single photon and classical point of view with regard to wave-particle duality of photon. *Electromagnetics Research Letters* 1:109–118.

3 Adaptation and Regulation of Photosynthetic Apparatus in Response to Light

Selene Casella, Fang Huang, and Lu-Ning Liu

CONTENTS

3.1 Photosynthetic Apparatus: Composition and Organization 53
   3.1.1 Light-Harvesting Complexes 54
   3.1.2 Reaction Centers 54
   3.1.3 Cytochrome and ATP Synthase... 55
   3.1.4 Diversity of Photosynthetic Apparatus Organization.. 55
3.2 Light Effects on the Photosynthetic Stoichiometry 55
   3.2.1 Light Intensity... 55
   3.2.2 Light Quality... 56
   3.2.3 Light–Dark Cycle 57
3.3 Light Effects on the Organization of Photosynthetic Machinery.. 57
   3.3.1 Photoadaptation of the Photosynthetic Apparatus... 57
   3.3.2 Light-State Transitions... 58
3.4 Regulation of Electron Transport Pathways 60
3.5 Photoprotection... 61
   3.5.1 Nonphotochemical Quenching and Orange Carotenoid Protein 61
   3.5.2 Photoprotection of PBsome 61
3.6 Summary 62
Acknowledgments.. 62
References... 62

Solar light can be efficiently captured and converted into chemical energy by all phototrophs, including photosynthetic bacteria, cyanobacteria, algae, and vascular plants, through their specific photosynthetic machinery. The major environmental challenge to photosynthetic organisms is the variability in the light. To cope with the rapidly fluctuating light environment (specifically in intensity and wavelength distribution), the phototrophs have evolved complex regulatory systems to adapt to changing light conditions. Advanced understanding of the adaptive mechanisms developed by photosynthetic organisms will inform the bioengineering of phototrophs to enhance the utilization of solar energy, for the production of biofuel and commodity products from light, CO_2, and water. In this chapter, we will summarize recent studies on the molecular basis underlying the physiological adaptation and regulation of photosynthetic machinery, in particular, in cyanobacteria, toward irradiance variation.

3.1 PHOTOSYNTHETIC APPARATUS: COMPOSITION AND ORGANIZATION

The primary reactions of photosynthesis are mediated by a series of photosynthetic complexes associated with or embedded in the photosynthetic membranes. These pigment–protein complexes can be classified into several groups according to their functions: light-harvesting antenna complexes, photosynthetic reaction centers (RCs), the cytochrome (Cyt) complex, and ATP synthase (ATPase). They are structurally and functionally linked in order through the photosynthetic electron transport chain (Figure 3.1). Light energy captured by the light-harvesting antenna is rapidly and efficiently transferred to the RCs to drive the transmembrane charge separation. The electrons are then transferred to the (plasto)quinone pool and subsequently to the Cyt enzymes. The electron transfer reactions are coupled to the formation of an electrochemical gradient across the photosynthetic membranes, which is essential for driving the ATP synthesis.

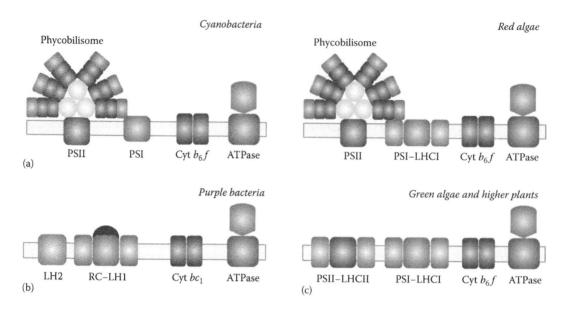

Cyanobacteria

Phycobilisome

Red algae

Phycobilisome

(a)   PSII   PSI   Cyt $b_6 f$ ATPase   PSII   PSI–LHCI   Cyt $b_6 f$ ATPase

Purple bacteria

Green algae and higher plants

(b)   LH2   RC–LH1   Cyt bc_1 ATPase   (c)   PSII–LHCII   PSI–LHCI   Cyt $b_6 f$ ATPase

FIGURE 3.1 **(See color insert.)** Variety of the photosynthetic apparatus in different photosynthetic organisms. (a) Cyanobacterial and red algal photosynthetic apparatus are composed of light-harvesting phycobilisomes, chlorophyll *a*-containing PSII and PSI, Cyt $b_6 f$, and ATPase. (b) Purple bacteria contain type II reaction centers with two major types of membrane light-harvesting complexes, LH1 and LH2. (c) The photosynthetic apparatus of green algae and higher plants consists of PSI and PSII with their membrane-embedded antennae, LHCI and LHCII, respectively, as well as Cyt $b_6 f$ and ATPase.

3.1.1 Light-Harvesting Complexes

The first step in the process of photosynthesis is the absorption of light photons by an array of antenna pigment–protein complexes, termed light-harvesting complexes (LHCs). The spectral properties and macromolecular conformations of photosynthetic antenna complexes vary dramatically depending on the different origins. The photosynthetic apparatus of the anoxygenic purple bacteria presents the simplest configurations (Cogdell et al. 2006; Liu et al. 2011; Liu and Scheuring 2013). Most purple photosynthetic bacteria synthesize two types of LHCs classified according to their *in vivo* absorbance, B875 (LH1) and B800-850 (LH2) complexes. Such antenna complexes are generally composed of two polypeptides (α and β subunits), two or three *bacteriochlorophyll* (*BChl*) molecules, and some carotenoids. Green algae and higher plants contain integral LHCI and LHCII as the peripheral antenna proteins associated with photosystem I (PSI) and photosystem II (PSII) supercomplexes, respectively (Croce and van Amerongen 2014). The migration of LHCs between PSII and PSI in the thylakoid membrane is essential to balancing the excitation energy between the two photosystems during state transitions (Minagawa 2013).

Phycobilisomes (PBsomes) are the major light-harvesting antenna complexes in cyanobacteria and red algae (Adir 2005; Liu et al. 2005b; Watanabe and Ikeuchi 2013). They are aggregations of water-soluble phycobiliproteins (PBPs) and linker polypeptides (Liu et al. 2005b), and serve as external antenna macrocomplexes associated with the stromal surfaces of thylakoid membranes (Arteni et al. 2008; Liu et al. 2008a). Red algae also have an intrinsic antenna LHCI-like

complex, functionally associated with PSI (Wolfe et al. 1994). PBsomes consist of two structural domains: the inner domain contains three cylinders that are arranged in a triangular PBsome core, and the peripheral domain contains six rodlike structures that radiate from the core. Both the core and the rods of the PBsome are composed of stacked PBP hexamers. This domain is composed predominantly of allophycocyanins (APCs), whereas the peripheral rods are mainly composed of phycocyanins (PCs) and phycoerythrins (PEs). Energy transfer in PBsomes is expected to progress from PE ($\lambda_{max} = 545 – 565$ nm) (Liu et al. 2005a) stepwise to PC ($\lambda_{max} = 620$ nm), APC ($\lambda_{max} = 650$ nm), and eventually, chlorophylls (Chls). It is evident that the presence of PBsomes extends the absorbance range covered by PSII and PSI. The stepwise energy transfer within the PBsomes may probably also play a photoprotective role (Liu et al. 2008b). A key physiological importance of PBsomes is reflected in light-state transitions, which will be discussed in detail below.

3.1.2 Reaction Centers

Photosynthetic RCs are pigment–protein complexes that convert the excitation energy from antenna complexes into chemical potential energy (Olson and Blankenship 2004). The key reactions of photosynthesis occur in two homologous types of RCs: (1) RCI type in some anoxygenic photosynthetic bacteria, such as green sulfur bacteria and heliobacteria, and (2) RCII type in other anoxygenic photosynthetic bacteria, for instance, purple bacteria and green filamentous bacteria; RCI and RCII coexist in all oxygenic photosynthetic organisms, i.e., cyanobacteria, algae, and plants (Hohmann-Marriott and Blankenship 2011). These two types of RCs probably share a common evolutionary ancestor because of their similar structures and cofactor arrangements of the electron transfer domains (Schubert et al. 1998). The cyanobacterial PSI complex presents a trimeric structure, and PSII complexes exist as dimers (Mazor et al. 2014; Suga et al. 2015).

3.1.3 Cytochrome and ATP Synthase

Various cytochromes are mainly responsible for the electron transfer released from the primary processes of charge separation in RCs. The Cyt $b_6 f$ complex is a dimeric membrane-intrinsic complex located in the thylakoid membranes. It is essential for both photosynthetic and respiratory electron transfer chains (Berry et al. 2009). In the linear electron transfer scheme of oxygenic photosynthesis, Cyt $b_6 f$ receives electrons from PSII by plastoquinol and passes them to PSI by reducing plastocyanin or Cyt c_6. This results in proton release in the lumen, generating a proton electrochemical gradient across the membrane. Cyt $b_6 f$ can switch from linear electron transfer between both PSs to a cyclic mode of electron transfer around PSI (Shikanai 2014). The cytochrome c oxidoreductases (Cyt bc_1) are multisubunit enzymes existing in a broad variety of organisms, including the purple nonsulfur photosynthetic bacteria (Berry et al. 2009). They are components of both cyclic photosynthetic and mitochondria-like linear respiratory electron transport chains.

ATPase is a large protein complex, catalyzing the synthesis of ATP from ADP and inorganic phosphate driven by a flux of protons across the membrane down the proton gradient generated by electron transfer. ATPases are located in the plasma membrane and photosynthetic membrane of bacteria,

the chloroplast thylakoid membrane in algae and plants, and the mitochondrial inner membrane in plants and animals.

3.1.4 Diversity of Photosynthetic Apparatus Organization

The physiological arrangement and functional coordination of these photosynthetic constituents are fundamental to efficient light capture and energy transfer mechanisms. In order to adapt to diverse habitats, photosynthetic organisms have developed distinct photosynthetic machinery to regulate energy conversion, as shown in Figure 3.1. Purple photosynthetic bacteria contain one type of RC and two types of LHCs, LH1 and LH2. Cyanobacteria, red algae, green algae, and higher plants consist of PSI, PSII, and various membrane or extramembrane light-harvesting antenna complexes. Distinct from green algae and higher plants, cyanobacteria and red algae utilize PBsomes to capture light for PSI and PSII. In addition, cyanobacterial thylakoid membranes house both photosynthetic and respiratory electron transport chains (Liu et al. 2012), and some complexes are shared by the two electron transport pathways.

3.2 LIGHT EFFECTS ON THE PHOTOSYNTHETIC STOICHIOMETRY

3.2.1 Light Intensity

The ratio of antenna quantity to photosynthetic RCs has been demonstrated to depend on light intensity during cell growth. To retain efficient light absorbance to RCs, additional 30% PBsome antenna complexes are synthesized under low-light conditions compared to high-light conditions (Figure 3.2)

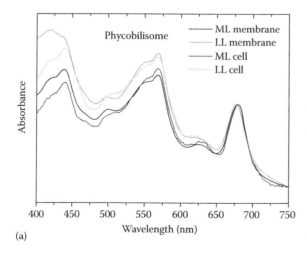
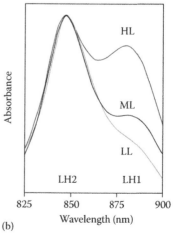

FIGURE 3.2 (See color insert.) (a) Room-temperature absorption spectra of whole cells and isolated PBsome thylakoid membranes from the unicellular red alga *Porphyridium cruentum* grown under low light (LL, 6 W/m²) and moderate light (ML, 15 W/m²). (From Liu, L.N. et al., *J. Biol. Chem.*, 283, 34946–34953, 2008.) Spectra were normalized to the chlorophyll absorption band at 682 nm. Results showed that the amount of PBsomes (500–650 nm) is dependent upon the light intensity. (b) The absorption spectrum of purple photosynthetic membranes adapted to high light (HL), ML, and LL. The absorption ratios of LH2 (845 nm) to the core complex (LH1, 880 nm) are 1.16, 1.86, and 2.6, respectively. (From Liu, L.N. et al., *J. Mol. Biol.*, 393, 27–35, 2009; Scheuring, S., and J.N. Sturgis, *Science*, 309, 484–487, 2005.) Results indicated that the ratio of antenna to RC varies according to the light intensity.

(Liu et al. 2008a). Similarly, in the photosynthetic membranes from the purple photosynthetic bacterium *Rhodospirillum photometricum* adapted to high light, ~3.5 LH2s are present per core complex, whereas under low-light conditions, ~7 LH2s per core were observed, and the moderate-light-adapted membranes have an intermediate LH2/core ratio of 4.8 (Liu et al. 2009b; Scheuring and Sturgis 2005).

In cyanobacteria, the ratio of PSII to PSI is variable according to the light intensity and spectral quality. The switch from low light to high light suppresses the PSI biosynthesis, resulting in an increase of the PSII/PSI ratio. This responsive effect has been shown to be triggered by the redox state of the electron transport chain (Murakami and Fujita 1991). Furthermore, two genes have been implicated in the regulation of the PSII/PSI ratio. Inactivation of a gene encoding for a putative sensory histidine kinase, *rppA*, leads to phenotypic changes consistent with a role in transducing redox signals to changes in PSII and PSI gene expression (Li and Sherman 2000). Inactivation of *pmgA* specifically abolishes the PSII/PSI ratio change in response to high light (Hihara et al. 1998). Both RppA and PmgA are excellent candidates for redox signal transduction proteins.

3.2.2 Light Quality

The acclimation of photosynthetic organisms to changes in light wavelength is ubiquitous and may be best characterized by the complex process of complementary chromatic adaptation (CCA) (Grossman et al. 1993; Kehoe and Gutu 2006). In many freshwater, marine, and soil cyanobacterial species whose PBsomes contain both PE and PC, the ratio of PC to PE in PBsomes varies in response to light spectra (Figure 3.3). Green light (optimally 540 nm) promotes PE biosynthesis, whereas red light (optimally 650 nm) elevates PC biosynthesis. CCA leads to the optimized absorbance of PBsomes to capture the most abundant wavelength of light in the green-to-red spectral region.

CCA regulation of PC and PE synthesis is predominantly at the transcriptional level for PE and PC genes (*cpeBA* and *cpc2*). The β and α subunits of PE are encoded by the *cpeBA* operon, which is highly upregulated by green light (Federspiel and Grossman 1990). The β and α subunits of "inducible PC" as well as three corresponding linker proteins are encoded by a large transcription unit, *cpc2* (*cpcB2A2H2I2D2*) (Conley et al. 1988). Its expression is highly upregulated by red light. Maintaining the expression of these operons is not essential once steady-state CCA is obtained (Oelmuller et al. 1989). The changes in RNA levels of PBsome components after an inductive light treatment are relatively rapid (Oelmuller et al. 1988). *cpc2* RNA reaches a maximum level 2 h after a shift from green to red light and drops to undetectable levels 2 h after a shift from red to green light, while *cpeBA* RNA reaches a maximum level 8 h after a shift from red to green light and drops to undetectable levels more than 14 h after a shift from green to red light. In contrast to the rapid response at the transcriptional level, PE and PC protein levels altered more slowly, requiring a few days to fully shift between the red- and green-light steady states. The regulatory mechanisms governing these responses are different for *cpc2* versus *cpeBA*. Two light-response pathways controlling the PBsome biosynthesis during CCA, an Rca system and a Cgi system, have been identified (Kehoe and Gutu 2006).

The initial light signaling for CCA is not well characterized. A phytochrome would be a potential candidate for a red-light sensor. However, the phytochrome responding to green light has not been found. If CCA is controlled by a single photoreceptor, it should contain a novel pigment or another green-absorbing chromophore. There is a precedent in the purple photosynthetic bacterium *R. centenum* for a

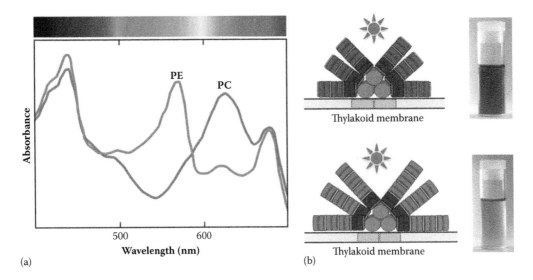

(a)   (b)

FIGURE 3.3 **(See color insert.)** (a) Whole-cell absorption spectra of the cyanobacterium *Fremyella diplosiphon* cells grown in green and red light. The phycoerythrin (PE) and phycocyanin (PC) absorption peaks are indicated. (From Kehoe, D.M., and A. Gutu, *Annu. Rev. Plant Biol.*, 57, 127–150, 2006.) (b) Changes of *F. diplosiphon* PBsomes in composition and structure induced by green and red light.

phytochrome-like protein that binds a second pigment (Jiang et al. 1999). Alternatively, a second photoreceptor, possibly a rhodopsin, could be involved (Hoff et al. 1995).

3.2.3 LIGHT–DARK CYCLE

The daily light–dark cycle controls rhythmic changes in the behavior and physiology of most species, such as cyanobacteria, plants, animals, and fungi. The so-called circadian rhythm displays an endogenous and entrainable oscillation of about 24 h. Cyanobacteria exhibit a self-sustained circadian rhythm that results in temporal changes in gene expression patterns, even in the absence of environmental cues (Dong and Golden 2008; Dong et al. 2010; Golden 1995). Correct circadian regulation maximizes photosynthesis, which is carried out specifically in plant chloroplasts, and productivity, by increasing Chl content, modifying the stoichiometry of photosynthetic complexes, and enhancing photosynthetic carbon fixation (Dodd et al. 2005). A recent study has described how the nuclear-encoded clock regulates the expression of key chloroplast-associated genes (Noordally et al. 2013). Based on observations of rhythms in delayed fluorescence (Gould et al. 2009), with a readout of the chemical state of PSII, the clock is likely to have profound effects on the structure and function of thylakoid membranes. However, how the circadian clock controls the photosynthetic performance at the biochemical and structural levels remains uncharacterized.

3.3 LIGHT EFFECTS ON THE ORGANIZATION OF PHOTOSYNTHETIC MACHINERY

Light variability not only determines the optimization of photosynthetic stoichiometry but also results in, inevitably, the reconfiguration of photosynthetic apparatus organization, for regulating the energy transduction under diverse light conditions.

3.3.1 PHOTOADAPTATION OF THE PHOTOSYNTHETIC APPARATUS

To understand the dynamic organization of photosynthetic complexes and the mechanisms of photosynthetic processes, it is essential to assess the *in situ* assembly and distribution of membrane proteins in native photosynthetic membranes under their physiological conditions. Recently, atomic force microscopy (AFM) has matured as a unique and powerful tool for directly assessing the supramolecular organization of photosynthetic complexes in their native environment at submolecular resolution (Liu and Scheuring 2013). The tight arrangement of LHCs has been observed in AFM topographs of the photosynthetic membranes in a unicellular red alga, *Porphyridium cruentum* (Figure 3.4) (Liu et al. 2008a). Under moderate light, PBsomes are randomly distributed and tightly clustered on photosynthetic membranes. In contrast, under low light, increasing amounts of PBsomes form densely packed rows on the membrane surface. The presence of dense

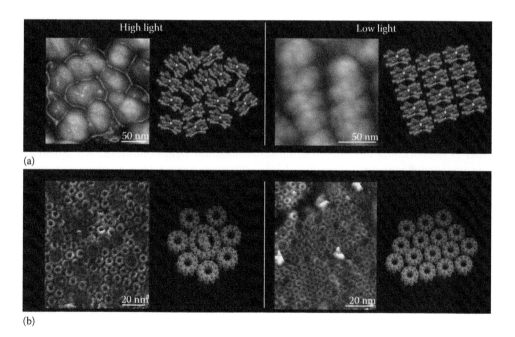

(a)

(b)

FIGURE 3.4 (See color insert.) (a) AFM images presenting the adaptation of the organization of PBsomes on the red algal thylakoid membrane. (From Liu, L.N. et al., *J. Biol. Chem.*, 283, 34946–34953, 2008.) The PBsomes show a random distribution and clustering in high light, whereas low-light illumination induces the synthesis of more PBsomes and facilitates the formation of parallel rows. The structural models of the photosynthetic apparatus are based on the AFM topographs. (b) AFM images showing the organization of photosynthetic membranes from the purple photosynthetic bacterium *Rhodospirillum photometricum* adapted to high light and low light. LH2-rich antenna membrane domains were observed in the low-light-adapted photosynthetic membrane of *R. photometricum*. (From Liu, L.N., and S. Scheuring, *Trends Plant Sci.*, 18, 277–286, 2013; Scheuring, S., and J.N. Sturgis, *Science*, 309, 484–487, 2005.)

antenna domains might be a general regulatory mechanism for light trapping when photons are relatively rare. Such structural constraints would enhance photosynthetic electron transfer under diverse light conditions (Liu et al. 2009b).

A similar response has also been recorded in the thylakoid membrane of higher plants (Dekker and Boekema 2005; Kirchhoff 2014). Electron microscopy data on the dark- and light-adapted *Arabidopsis thaliana* thylakoids indicated that the granal thylakoid lumen significantly expands under light stress (Kirchhoff et al. 2011). This light-induced expansion may moderate the restrictions imposed on protein diffusion in the lumen in the dark.

AFM imaging has also characterized comprehensively how the organization of the *Rsp. photometricum* photosynthetic membrane is modulated toward light variation (Liu et al. 2009b; Scheuring and Sturgis 2005). In high-light-adapted membranes, ~3.5 LH2s are present per core complex, whereas after low-light growth, ~7 LH2s per core were recorded. Two different types of protein assemblies in the bacterial photosynthetic membranes were identified: core–LH2 domains and paracrystalline LH2-rich domains (Figure 3.4). Additional LH2 incorporated into the membrane segregated in paracrystalline antenna domains, whereas the domains with core complexes seemed architecturally unaffected. The two domains have distinct roles to optimize the photosynthetic activity during light intensity change: core–LH2 domains maintain efficient harvesting, trapping, and transmission of solar energy; LH2-rich domains enhance light capture when only few photons are available but do not perturb the photosynthetically active core assemblies. The membrane adapted to medium-light conditions exhibited the intermediate composition and organization of the photosynthetic apparatus (Liu et al. 2009b). Similar protein assembly patterns have been found in other species, for instance, *Phaeospirillum molischianum* (Gonçalves et al. 2005), *Rhodopseudomonas palustris* (Scheuring et al. 2006), and *Rhodobacter sphaeroides* (Adams and Hunter 2012). In *Rps. palustris* photosynthetic membranes of high-light-adapted cells, the core complexes also segregate into hexagonally packed paracrystalline domains, reminiscent of the assembly found in *Blastochloris viridis* (Scheuring et al. 2004). The dense packing of the large paracrystalline LH2 domains may limit quinone diffusion and, therefore, favor quinone distribution in the proximity of the cores.

3.3.2 LIGHT-STATE TRANSITIONS

State transitions are rapid adaptive responses to changes in light quality. Illumination conditions that lead to excess excitation energy of PSII compared to PSI induce a transition to state 2, in which more absorbed excitation energy is diverted to PSI. When PSI is overexcited relative to PSII, it induces a transition to state 1, in which more energy is transferred to PSII. Thus, state transitions act as a mechanism to balance excitation of the two photosystems under changing light regimes. State transitions have been extensively characterized in green alga *Chlamydomonas reinhardtii* and higher plant *A. thaliana*. The processes of the state transitions involve the LHCII migration, the molecular reorganization

of photosystem supercomplexes, the identification of LHCII kinase, the mapping of phosphorylated residues in LHCII, and the involvement of Cyt $b_6 f$ in the control of LHCII phosphorylation (Minagawa 2013). It has been shown that the state transitions are regulated by the redox state of plastoquinone (PQ), an electron carrier located between two photosystems (Mullineaux and Allen 1990), and involve posttranslational modifications by phosphorylation of LHCII. Under different light conditions, for example, when PSII is excited, the redox state of the PQ pool is more reduced. The more reduced PQ pool induces the activation of an LHCII kinase and LHCII phosphorylation, resulting in the LHCII movement from PSII to PSI (state 2). Conversely, when the PQ pool is oxidized, the LHCII kinase is inactive, and LHCII is dephosphorylated by phosphatase and moves back from PSI to PSII (state 1). Thus, the PQ redox-regulated reversible phosphorylation of LHCII promotes state transitions and acts to redistribute absorbed excitation energy in response to different light conditions.

In cyanobacteria, there is no specific light-harvesting antenna for PSI, and the PBsomes serve as the major antenna for both photosystems. State transitions regulate the excitation energy transfer from the PBsomes to PSII or PSI. The structural basis of state transitions in cyanobacteria is still controversial (Figure 3.5). One hypothesis is that of *mobile PBsomes*, which suggests that state transitions may involve the physical association and disassociation of PBsomes between PSII and PSI, and thus, the energy redistribution between PSII and PSI (Allen and Holmes 1986). About 80% of all PBsomes were found to connect with PSI, and energy is transferred via PBsomes independently to PSII and PSI (Rakhimberdieva et al. 2001). Another model is that of *energy spillover*, which proposes that PBsomes can only associate with PSII and that excess Chl *a*-absorbed excitation energy may be redistributed from PSII to PSI (Biggins and Bruce 1989; Bruce et al. 1989). The redistribution of excitation energy absorbed by Chl is independent of the redistribution of excitation energy absorbed by the PBsomes. Both changes are triggered by the same environmental light conditions. An updated model states that PBsomes are capable of physically interacting with both PSII and PSI. Instead of the long-range movement, redistribution of PBsomes between PSI and PSII in the local membrane region might be essential to the state transitions (McConnell et al. 2002).

Studies using fluorescence recovery after photobleaching (FRAP) based on live-cell imaging using a confocal fluorescence microscope have indicated that the PBsomes are mobile along the surface of a thylakoid membrane (Mullineaux et al. 1997) and that the diffusion of PBsomes from RC to RC is required for state transitions (Joshua and Mullineaux 2004) and nonphotochemical quenching (NPQ) (Joshua et al. 2005). A recent study further demonstrated that state transitions have an important regulatory function in mesophilic red algae, but this process is replaced by NPQ in thermophilic red algae (Kana et al. 2014). However, FRAP experiments have revealed a partial fluorescence recovery in wholly bleached cells of the red alga *P. cruentum* and, more interestingly,

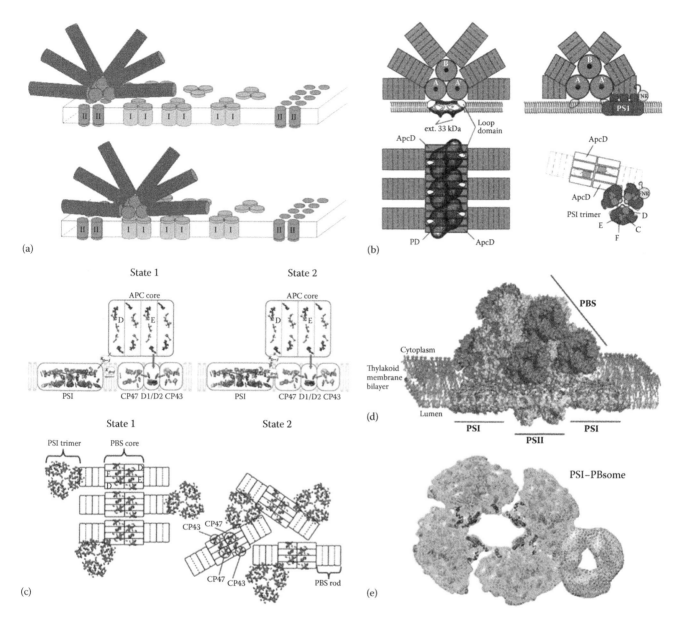

FIGURE 3.5 Models of the association between PBsome and PSI and PSII. (a) Mobile PBsomes, suggesting that PBsomes have a loose association with PSI and PSII and that the movement of PBsomes between the two photosystems is essential to state transitions. (From Mullineaux, C.W. et al., *Nature*, 390, 421–424, 1997; Bald, D. et al., *Photosynth. Res.*, 49, 103–118, 1996.) (b) Possible coupling of PBsome with PSII and PSI. The PBsome core can come into contact with a PSII dimer. Tilted packing of PBsomes and PSII is essential to driving the connection. PBsomes can also interact with PSI trimers through the rods, to form a PSI–PBsome supercomplex under the state 2 condition. (The model is adapted from Bald, D. et al., *Photosynth. Res.*, 49, 103–118, 1996. From McConnell, M.D. et al., *Plant Physiol.*, 130, 1201–1212, 2002.) (c) A schematic model showing the organization of a cyanobacterial thylakoid membrane, showing possible associations between the PBsome core and PSII and PSI complexes in states 1 and 2. (From McConnell, M.D. et al., *Plant Physiol.*, 130, 1201–1212, 2002; Liu, H. et al., *Science*, 342, 1104–1107, 2013.) Energy transfer from the core to PSII occurs from the ApcE subunit to D1/D2. Energy transfer from the core to PSI is via the ApcD subunit. Energy "spillover" from PSII to PSI is assumed to occur from CP47 to PSI. APC core cylinders are associated with PSII dimers, and PC rods interact with PSI trimers. In state 2, trimeric PSI is in close contact with both the PsbB (CP47) protein of PSII and the ApcD subunits of the core. In state 1, the PBsome–PSII supercomplexes are organized into rows. One PBsome–PSII supercomplex is shown to remain, coupling with one PSI trimer via ApcD, to form the PBsome–PSII–PSI supercomplex. (d) A structural model of the PBsome–PSII–PSI photosynthetic megacomplex, depicting that the PBsome core fully covers and close-couples with the PSII dimer, whereas the PSI is associated with ApcD through a side-on orientation. (From Liu, H. et al., *Science*, 342, 1104–1107, 2013; Watanabe, M. et al., *PNAS*, 11, 2512–2517, 2014.) (e) A structural model of the PBsome–CpcL–PSI supercomplex isolated from the cyanobacterium *Anabaena* (Watanabe et al. 2014), describing the CpcL–PBsome rods, which specifically bind at the periphery of the PSI tetramers. FNR, ferredoxin; PBS, PBsome.

immobilized PBsome complexes *in vitro* (Liu et al. 2009a). The observations might suggest that the fluorescence recovery recorded during FRAP experiments could be ascribed to the intrinsic photophysics of the bleached PBsomes *in situ*, rather than the diffusion of PBsome complexes on the thylakoid membranes. Furthermore, AFM images on the native thylakoid membrane of *P. cruentum* showed significant crowding of PBsome complexes (Figure 3.4) (Liu et al. 2008a). Under such a crowd circumstance, the rapid and long-range movement of PBsomes may be significantly restricted by steric hindrance, taking into account the large size of individual PBsomes, their dense lateral packing membrane surface, as well as the limited free vertical spacing between opposite thylakoid layers. In addition, given the fact that PSII and PSI are mixed in the thylakoid membrane (Mustardy et al. 1992), the dense coverage of PBsomes on the thylakoid membranes may denote the structural association between the PBsomes and both photosystems underneath.

Recently, the existence of PBsome–photosystem supercomplexes has been proved (Figure 3.5). Using a chemical cross-linking strategy, a protein megacomplex composed of a PBsome, PSII, and PSI from a cyanobacterium, *Synechocystis* sp. PCC6803, has been isolated (Liu et al. 2013). This provided evidence about the presence of PBsome–PSII–PSI supercomplexes *in vivo*. Time-resolved fluorescence spectroscopy further demonstrated that the PBsome could deliver excitations to the RCs of either PSI or PSII, although the energy transfer from the PBsome to PSII is efficient, whereas that from the PBsome to PSI is slow. Another work characterized a supercomplex PBsome–CpcL–PSI isolated from a cyanobacterium, *Anabaena* (Watanabe et al. 2014). Within the supercomplex, PSI is organized into tetramers (a dimer of dimers). The PBsome subcomplex, CpcL–PBsome rods, specifically binds at the periphery of the PSI pseudotetramers.

3.4   REGULATION OF ELECTRON TRANSPORT PATHWAYS

The organization of photosynthetic complexes regulates the electron transport pathway and efficiency. All photosynthetic membranes that have been analyzed exhibit a dense packing of multicomponent photosynthetic complexes (Liu and Scheuring 2013). On the one hand, protein crowding is favorable for excitation energy transfer between complexes; on the other hand, it significantly reduces the lipid content and space between protein complexes and, as a consequence, probably membrane fluidity, which is required for the diffusion of hydrophobic electron/proton transport carriers (i.e., quinone molecules). Therefore, it represents an obstacle for efficient cyclic electron transduction between RCs and Cyt bc_1 complexes in membranes. Analysis of the molecular environment and long-range protein organization proposed a continuous "lipid area network" for long-range quinone diffusion throughout the photosynthetic membrane of the purple photosynthetic bacterium *R. photometricum* (Liu et al. 2009b). Recent studies on the distribution and dynamics of respiratory components in the plasma membrane of *Escherichia coli* revealed that

respiratory complexes are concentrated in mobile domains in the membrane (Llorente-Garcia et al. 2014). Different complexes are concentrated in separate domains, with no significant colocalization and, therefore, no supercomplexes. This is another indication of a rapid and long-range quinone diffusion that serves to shuttle electrons between islands of distinct electron transport complexes in the membrane.

Photosynthetic electron transfer induced by light excitation modulates the redox state of electron transport components. A number of cyanobacterial responses are known to be triggered by changes in the redox state of PQ or the Cyt b_6f complex, and thioredoxin, which accepts electrons from PSI. Light-harvesting regulation can act to control the balance of linear and cyclic electron transport, and therefore, the balance of proton-motive force and reducing power as photosynthetic outputs. Switches that remove electrons from the photosynthetic electron transport chain are also known as *electron valves*: they serve to prevent dangerous overreduction of the electron transport chain (Liu et al. 2012). There is scope for short-scale posttranslational mechanisms to switch between cyclic and linear electron transport. One example is the regulation of the cyclic electron transport pathway involving complex I under different light intensities (Liu et al. 2012). Fluorescence microscopy images of the fluorescently tagged complex I in *Synechococcus elongatus* PCC7942 showed that the larger-scale distribution of complex I in the thylakoid membrane is controlled in response to a redox switch triggered by light intensity changes. Oxidation of the PQ pool induces the clustering of complex I in segregated thylakoid membrane zones, whereas reduction of the PQ pool induces a posttranslational switch in the distribution of respiratory complexes to a state in which it is more evenly dispersed in the membrane. Complex II (succinate dehydrogenase) showed a similar change in distribution under the same conditions. This switch in the distribution of respiratory complexes correlates with a major change in the probability that electrons from the respiratory complexes are transferred to a PSI rather than to a terminal oxidase (Liu et al. 2012). The switch provides a mechanism to promote cyclic electron transport when the reduction of the PQ pool indicates an adequate supply of electrons in the cell. Although many questions about the mechanism remain to be addressed, the observation indicates that the distribution of electron transport complexes in the membrane at the submicron scale is under physiological control and plays a crucial role in controlling pathways of electron flux.

Another example of cyanobacterial electron valves is the flavodiiron (Flv) proteins Flv1–4: they are cytoplasmic proteins that take electrons from the photosynthetic electron transport chain and divert them to alternative acceptors. Flv1 and Flv3 form a heterodimer that takes electrons from the acceptor side of PSI and uses them to reduce oxygen (Allahverdiyeva et al. 2013). An Flv2/Flv4 heterodimer takes electrons from the acceptor side of PSII, passing them to an unknown acceptor (Zhang et al. 2012). The regulation of the activities of Flv proteins is not known, and they presumably act only as an electron transport switch on slow timescales.

3.5 PHOTOPROTECTION

Light not only is the basic driving force for photosynthesis but can also be destructive, particularly when the light-harvesting antennae capture excess photons after photosynthetic electron transport saturation. The photosynthetic apparatus has developed appropriate physiological mechanisms to modulate the absorbance of excitation energy while avoiding the potentially phototoxic effects of excess photons (Bailey and Grossman 2008).

3.5.1 NONPHOTOCHEMICAL QUENCHING AND ORANGE CAROTENOID PROTEIN

High levels of solar radiation can increase the production of reactive oxygen species and cause damage of photosynthetic membranes and pigment–protein complexes. Cyanobacteria have evolved a protective mechanism, NPQ, to dissipate excess PBsome-absorbed energy as heat (El Bissati et al. 2000). In contrast to plants and eukaryotes, cyanobacteria lack both pH-dependent quenching and the xanthophyll cycle (Gorbunov et al. 2011). NPQ in cyanobacteria is triggered by strong blue light, which excites both PBsomes and Chls. Recent studies have shown that the NPQ in cyanobacteria is mediated by a 35 kDa water-soluble orange carotenoid protein (OCP) (Kerfeld et al. 2003; Kirilovsky and Kerfeld 2013; Wilson et al. 2006). As a high-light sensor, OCP is directly involved in the fluorescence quenching of PBsomes and possibly in the regulation of energy transfer between the PBsomes and photosystems (Kirilovsky and Kerfeld 2013). OCP contains a single bound carotenoid (3′-hydroxyechinenone), which can change the conformation between its orange (OCP^O) and red forms (OCP^R) (Kirilovsky and Kerfeld 2013). The photoactivated OCP^R binds to the PBsome core, where it takes excitation energy from the phycobilins and converts it to

heat in order to prevent photodamage of the RCs at high light. The reversal of OCP-induced energy quenching (conversion of OCP^R back to OCP^O) depends on a second cytoplasmic protein, the fluorescence recovery protein (FRP), which binds to the OCP and weakens its association with the PBsome (Gwizdala et al. 2013).

3.5.2 PHOTOPROTECTION OF PBSOME

The photoprotection of PBsomes from excess excitation energy remains poorly characterized. A study using single-molecule spectroscopy imaging on purified PBsomes from the red alga *P. cruentum* elucidated an energetic decoupling in PBsomes with respect to intense light (Figure 3.6) (Liu et al. 2008b). Strong green light was able to induce the fluorescence decrease of PBsomes and the fluorescence increase of the peripheral PE in the PBsome at the first stage of photobleaching. This indicates that excess photon energy can be dissipated from the peripheral PE in the PBsome to minimize the photodamage of RCs. This process may serve as a photoprotective mechanism ascribed to the PBsomes under strong light illumination. It is corroborated with high-light-induced reorganization (Six et al. 2007; Stoitchkova et al. 2007) and photodegradation of PBsomes (Rinalducci et al. 2008). The photoprotective role of PE has also been characterized in marine cyanobacteria (Wyman et al. 1985). The chromophore variety and increasing abundance extend the absorbance spectrum and enhance the absorption capacity, enabling photosynthetic organisms to survive in various environments. The energetic decoupling of PBsomes occurring under high light indicates a novel physiological role of the chromophore variety: creating a multistep photoprotection mechanism to effectively prevent photodamage of photosynthetic RCs in response to excess excitation energy (Liu et al. 2008b).

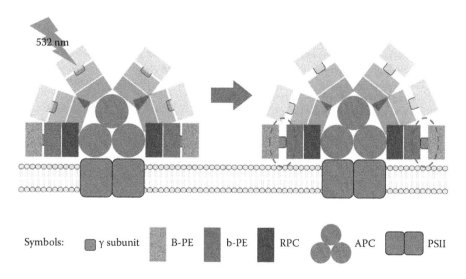

FIGURE 3.6 (**See color insert.**) Schematic model of light-induced decoupling of PBsomes, indicating a possible photoprotective mechanism of the PBsome. Green open circles present the potential decoupling sites within the PBsome. B-PE, B-phycoerythrin; b-PE, b-phycoerythrin; RPC, R-phycocyanin.

3.6 SUMMARY

By harvesting solar energy and converting it into chemical energy, the phototrophs play essential roles in maintaining life on Earth. Variability in the light environment presents major challenges to photosynthetic organisms. To survive in such a fluctuating environment, the phototrophs have evolved regulatory and photoprotective mechanisms to optimize the organization and efficiency of the photosynthetic apparatus. Cyanobacteria are the most important contributors to the global energy production and carbon cycle. For a sustainable future of our society, cyanobacteria are also promising industrial organisms for the production of fuels and metabolic chemicals. Advanced understanding of the photoadaptation/photoacclimation process of cyanobacteria will be of fundamental necessity to the development of biotechnology and bioengineering of photosynthetic microorganisms.

ACKNOWLEDGMENTS

L.-N. Liu acknowledges the support of a university research fellowship (UF120411) and a research grant (RG130442) from the Royal Society.

REFERENCES

Adams, P. G., and C. N. Hunter. 2012. Adaptation of intracytoplasmic membranes to altered light intensity in *Rhodobacter sphaeroides*. *Biochim Biophys Acta* 1817:1616–27.

Adir, N. 2005. Elucidation of the molecular structures of components of the phycobilisome: Reconstructing a giant. *Photosynth Res* 85:15–32.

Allahverdiyeva, Y. et al. 2013. Flavodiiron proteins Flv1 and Flv3 enable cyanobacterial growth and photosynthesis under fluctuating light. *Proc Natl Acad Sci U S A* 110:4111–16.

Allen, J. F., and N. G. Holmes. 1986. A general model for regulation of photosynthetic unit function by protein phosphorylation. *FEBS Lett* 202:175–81.

Arteni, A. A. et al. 2008. Structure and organization of phycobilisomes on membranes of the red alga *Porphyridium cruentum*. *Photosynth Res* 95:169–74.

Bailey, S., and A. Grossman. 2008. Photoprotection in cyanobacteria: Regulation of light harvesting. *Photochem Photobiol* 84:1410–20.

Bald, D. et al. 1996. Supramolecular architecture of cyanobacterial thylakoid membranes: How is the phycobilisome connected with the photosystems? *Photosynth Res* 49:103–18.

Berry, E. et al. 2009. Structural and mutational studies of the cytochrome bc 1 complex. In *The Purple Phototrophic Bacteria*, edited by C. Neil Hunter et al., 425–50. Springer, Dordrecht.

Biggins, J., and D. Bruce. 1989. Regulation of excitation energy transfer in organisms containing phycobilins. *Photosynth Res* 20:1–34.

Bruce, D. et al. 1989. State transitions in a phycobilisome-less mutant of the cyanobacterium *Synechococcus* sp. PCC 7002. *Biochim Biophys Acta* 974:66–73.

Cogdell, R. J. et al. 2006. The architecture and function of the light-harvesting apparatus of purple bacteria: From single molecules to in vivo membranes. *Q Rev Biophys* 39:227–324.

Conley, P. B. et al. 1988. Molecular characterization and evolution of sequences encoding light-harvesting components in the chromatically adapting cyanobacterium Fremyella diplosiphon. *J Mol Biol* 199:447–65.

Croce, R., and H. van Amerongen. 2014. Natural strategies for photosynthetic light harvesting. *Nat Chem Biol* 10:492–501.

Dekker, J. P., and E. J. Boekema. 2005. Supramolecular organization of thylakoid membrane proteins in green plants. *Biochim Biophys Acta* 1706:12–39.

Dodd, A. N. et al. 2005. Plant circadian clocks increase photosynthesis, growth, survival, and competitive advantage. *Science* 309:630–3.

Dong, G., and S. S. Golden. 2008. How a cyanobacterium tells time. *Curr Opin Microbiol* 11:541–6.

Dong, G. et al. 2010. Simplicity and complexity in the cyanobacterial circadian clock mechanism. *Curr Opin Genet Dev* 20: 619–25.

El Bissati, K. et al. 2000. Photosystem II fluorescence quenching in the cyanobacterium Synechocystis PCC 6803: Involvement of two different mechanisms. *Biochim Biophys Acta* 1457:229–42.

Federspiel, N. A., and A. R. Grossman. 1990. Characterization of the light-regulated operon encoding the phycoerythrin-associated linker proteins from the cyanobacterium Fremyella diplosiphon. *J Bacteriol* 172:4072–81.

Golden, S. S. 1995. Light-responsive gene expression in cyanobacteria. *J Bacteriol* 177:1651–4.

Gonçalves, R. P. et al. 2005. Architecture of the native photosynthetic apparatus of *Phaeospirillum molischianum*. *J Struct Biol* 152:221–8.

Gorbunov, M. Y. et al. 2011. A kinetic model of non-photochemical quenching in cyanobacteria. *Biochim Biophys Acta* 1807: 1591–9.

Gould, P. D. et al. 2009. Delayed fluorescence as a universal tool for the measurement of circadian rhythms in higher plants. *Plant J* 58:893–901.

Grossman, A. R. et al. 1993. The phycobilisome, a light-harvesting complex responsive to environmental conditions. *Microbiol Rev* 57:725–49.

Gwizdala, M. et al. 2013. Characterization of the Synechocystis PCC 6803 fluorescence recovery protein involved in photoprotection. *Biochim Biophys Acta* 1827:348–54.

Hihara, Y. et al. 1998. A novel gene, pmgA, specifically regulates photosystem stoichiometry in the cyanobacterium Synechocystis species PCC 6803 in response to high light. *Plant Physiol* 117:1205–16.

Hoff, W. D. et al. 1995. Rhodopsin(s) in eubacteria. *Biophys Chem* 56:193–9.

Hohmann-Marriott, M. F., and R. E. Blankenship. 2011. Evolution of photosynthesis. *Annu Rev Plant Biol* 62:515–48.

Jiang, Z. et al. 1999. Bacterial photoreceptor with similarity to photoactive yellow protein and plant phytochromes. *Science* 285: 406–9.

Joshua, S., and C. W. Mullineaux. 2004. Phycobilisome diffusion is required for light-state transitions in cyanobacteria. *Plant Physiol* 135:2112–19.

Joshua, S. et al. 2005. Involvement of phycobilisome diffusion in energy quenching in cyanobacteria. *Plant Physiol* 138: 1577–85.

Kana, R. et al. 2014. Phycobilisome mobility and its role in the regulation of light harvesting in red algae. *Plant Physiol* 165:1618–31.

Kehoe, D. M., and A. Gutu. 2006. Responding to color: The regulation of complementary chromatic adaptation. *Annu Rev Plant Biol* 57:127–50.

Kerfeld, C. A. et al. 2003. The crystal structure of a cyanobacterial water-soluble carotenoid binding protein. *Structure* 11: 55–65.

Kirchhoff, H. 2014. Structural changes of the thylakoid membrane network induced by high light stress in plant chloroplasts. *Philos Trans R Soc Lond B Biol Sci* 369:20130225.

Kirchhoff, H. et al. 2011. Dynamic control of protein diffusion within the granal thylakoid lumen. *Proc Natl Acad Sci U S A* 108:20248–53.

Kirilovsky, D., and C. A. Kerfeld. 2013. The orange carotenoid protein: A blue-green light photoactive protein. *Photochem Photobiol Sci* 12:1135–43.

Li, H., and L. A. Sherman. 2000. A redox-responsive regulator of photosynthesis gene expression in the cyanobacterium Synechocystis sp. strain PCC 6803. *J Bacteriol* 182:4268–77.

Liu, L. N., and S. Scheuring. 2013. Investigation of photosynthetic membrane structure using atomic force microscopy. *Trends Plant Sci* 18:277–86.

Liu, L. N. et al. 2005a. One-step chromatography method for efficient separation and purification of R-phycoerythrin from *Polysiphonia urceolata*. *J Biotechnol* 116:91–100.

Liu, L. N. et al. 2005b. Characterization, structure and function of linker polypeptides in phycobilisomes of cyanobacteria and red algae: An overview. *Biochim Biophys Acta—Bioenergetics* 1708:133–42.

Liu, L. N. et al. 2008a. Watching the native supramolecular architecture of photosynthetic membrane in red algae: Topography of phycobilisomes and their crowding, diverse distribution patterns. *J Biol Chem* 283:34946–53.

Liu, L. N. et al. 2008b. Light-induced energetic decoupling as a mechanism for phycobilisome-related energy dissipation in red algae: a single molecule study. *PLoS One* 3:e3134.

Liu, L. N. et al. 2009a. FRAP analysis on red alga reveals the fluorescence recovery is ascribed to intrinsic photoprocesses of phycobilisomes than large-scale diffusion. *PLoS One* 4:e5295.

Liu, L. N. et al. 2009b. Quinone pathways in entire photosynthetic chromatophores of *Rhodospirillum photometricum*. *J Mol Biol* 393:27–35.

Liu, L. N. et al. 2011. Forces guiding assembly of light-harvesting complex 2 in native membranes. *Proc Natl Acad Sci U S A* 108:9455–9.

Liu, L. N. et al. 2012. Control of electron transport routes through redox-regulated redistribution of respiratory complexes. *Proc Natl Acad Sci U S A* 109:11431–6.

Liu, H. et al. 2013. Phycobilisomes supply excitations to both photosystems in a megacomplex in cyanobacteria. *Science* 342: 1104–7.

Llorente-Garcia, I. et al. 2014. Single-molecule in vivo imaging of bacterial respiratory complexes indicates delocalized oxidative phosphorylation. *Biochim Biophys Acta—Bioenergetics* 1837:811–24.

Mazor, Y. et al. 2014. Crystal structures of virus-like photosystem I complexes from the mesophilic cyanobacterium Synechocystis PCC 6803. *Elife* 3:e01496.

McConnell, M. D. et al. 2002. Regulation of the distribution of chlorophyll and phycobilin-absorbed excitation energy in cyanobacteria. A structure-based model for the light state transition. *Plant Physiol* 130:1201–12.

Minagawa, J. 2013. Dynamic reorganization of photosynthetic supercomplexes during environmental acclimation of photosynthesis. *Front Plant Sci* 4:513.

Mullineaux, C. W., and J. F. Allen. 1990. State 1–State 2 transitions in the cyanobacterium *Synechococcus* 6301 are controlled by the redox state of electron carriers between Photosystems I and II. *Photosynth Res* 23:297–311.

Mullineaux, C. W. et al. 1997. Mobility of photosynthetic complexes in thylakoid membranes. *Nature* 390:421–4.

Murakami, A., and Y. Fujita. 1991. Regulation of photosystem stoichiometry in the photosynthetic system of the cyanophyte Synechocystis PCC 6714 in response to light-intensity. *Plant Cell Physiol* 32:223–30.

Mustardy, L. et al. 1992. Photosynthetic membrane topography: Quantitative in situ localization of photosystems I and II. *Proc Natl Acad Sci U S A* 89:10021–5.

Noordally, Z. B. et al. 2013. Circadian control of chloroplast transcription by a nuclear-encoded timing signal. *Science* 339:1316–19.

Oelmuller, R. et al. 1988. Changes in accumulation and synthesis of transcripts encoding phycobilisome components during acclimation of Fremyella diplosiphon to different light qualities. *Plant Physiol* 88:1077–83.

Oelmuller, R. et al. 1989. Role of protein synthesis in regulation of phycobiliprotein mRNA abundance by light quality in Fremyella diplosiphon. *Plant Physiol* 90:1486–91.

Olson, J. M., and R. E. Blankenship. 2004. Thinking about the evolution of photosynthesis. *Photosynth Res* 80:373–86.

Rakhimberdieva, M. G. et al. 2001. Interaction of phycobilisomes with photosystem II dimers and photosystem I monomers and trimers in the cyanobacterium spirulina platensis. *Biochemistry* 40:15780–8.

Rinalducci, S. et al. 2008. Generation of reactive oxygen species upon strong visible light irradiation of isolated phycobilisomes from Synechocystis PCC 6803. *Biochim Biophys Acta* 1777:417–24.

Scheuring, S., and J. N. Sturgis. 2005. Chromatic adaptation of photosynthetic membranes. *Science* 309:484–7.

Scheuring, S. et al. 2004. Watching the photosynthetic apparatus in native membranes. *Proc Natl Acad Sci U S A* 101:11293–7.

Scheuring, S. et al. 2006. The photosynthetic apparatus of *Rhodopseudomonas palustris*: Structures and organization. *J Mol Biol* 358:83–96.

Schubert, W. D. et al. 1998. A common ancestor for oxygenic and anoxygenic photosynthetic systems: A comparison based on the structural model of photosystem I. *J Mol Biol* 280:297–314.

Shikanai, T. 2014. Central role of cyclic electron transport around photosystem I in the regulation of photosynthesis. *Curr Opin Biotechnol* 26:25–30.

Six, C. et al. 2007. UV-induced phycobilisome dismantling in the marine picocyanobacterium *Synechococcus* sp. WH8102. *Photosynth Res* 92:75–86.

Stoitchkova, K. et al. 2007. Heat- and light-induced reorganizations in the phycobilisome antenna of Synechocystis sp. PCC 6803. Thermo-optic effect. *Biochim Biophys Acta* 1767:750–6.

Suga, M. et al. 2015. Native structure of photosystem II at 1.95 Å resolution viewed by femtosecond X-ray pulses. *Nature* 517:99–103.

Watanabe, M., and M. Ikeuchi. 2013. Phycobilisome: Architecture of a light-harvesting supercomplex. *Photosynth Res* 116:265–76.

Watanabe, M. et al. 2014. Attachment of phycobilisomes in an antenna–photosystem I supercomplex of cyanobacteria. *Proc Natl Acad Sci U S A* 111:2512–17.

Wilson, A. et al. 2006. A soluble carotenoid protein involved in phycobilisome-related energy dissipation in cyanobacteria. *Plant Cell* 18:992–1007.

Wolfe, G. R. et al. 1994. Evidence for a common origin of chloroplasts with light-harvesting complexes of different pigmentation. *Nature* 367:566–8.

Wyman, M. et al. 1985. Novel role for phycoerythrin in a marine cyanobacterium, *Synechococcus* strain DC2. *Science* 230:818–20.

Zhang, P. et al. 2012. Operon flv4-flv2 provides cyanobacterial photosystem II with flexibility of electron transfer. *Plant Cell* 24:1952–71.

4 Estimation of Mesophyll Conductance and Its Applications

Naomi Kodama and Wataru Yamori

CONTENTS

4.1   Introduction 65
    4.1.1 CO_2 Diffusional Pathways 65
4.2   Theory and Method of Analyses of Conductance 66
    4.2.1 A/C_c Curve-Fitting Methods 66
    4.2.2   Chlorophyll Fluorescence Method 67
        4.2.2.1   Constant J Method 67
        4.2.2.2   Variable J Method 67
    4.2.3   Isotope Method 68
        4.2.3.1   Theory of Carbon Isotope Discrimination during Photosynthesis 68
4.3   Applications to Investigate Mesophyll Conductance against Environmental Response 69
    4.3.1   Dynamic Analyses of Mesophyll Conductance to Investigate the Short-Term Variations under Variable
            Environmental Conditions 69
    4.3.2   Analyses of Limiting Step in Photosynthesis at Various Leaf Temperatures 69
        4.3.2.1   Response of CO_2 Assimilation Rate to CO_2 69
        4.3.2.2   Response of Chloroplast Electron Transport Rate to CO_2 71
4.4   Concluding Remarks 71
Acknowledgments.. 71
References.. 71

4.1   INTRODUCTION

4.1.1 CO_2 DIFFUSIONAL PATHWAYS

CO_2 travels from the atmosphere through stomata to the carboxylation sites where CO_2 is fixed (Figure 4.1). During this diffusional process, CO_2 travels through the leaf boundary layer, stomata, mesophyll cell wall, mesophyll plasmalemma, and cytosol and chloroplast membranes. Carbonic anhydrase catalyzes the reversible hydration of CO_2 to HCO_3^- in the aqueous phase (i.e., chloroplast, cytosol, plasma membrane) and is thought to maintain the supply of CO_2 to ribulose 1,5-bisphosphate carboxynase/oxynase (RuBisCO) by speeding up the dehydration of HCO_3^-. In the chloroplast, RuBisCO catalyzes the carboxylation of ribulose-1,5-bisphosphate (RuBP) by CO_2 and produces 3-phosphoglyceric acid (PGA). ATP and NADPH produced by chloroplast electron transport are used to produce sugars and starch, as well as for the regeneration of RuBP from PGA in the Calvin–Benson cycle. During CO_2 diffusion from the atmosphere into mesophyll, each step is influenced by multiple factors, and depending on the environmental variables, the diffusional conductance at each step could change independently or simultaneously [1–4]. It is therefore important to investigate the diffusional conductance steps separately.

The main resistances between the pathways are (1) the leaf surface, (2) through the stomata, and (3) within the leaf mesophyll. Gradients in CO_2 concentration promote diffusion to the sites of carboxylation. Stomatal conductance to CO_2 diffusion (g_s) is regularly measured using leaf gas exchange techniques, and many models describing variability in g_s have been suggested so far [5,6]. On the other hand, variability in the conductance to CO_2 diffusion from leaf intercellular air spaces to the sites of carboxylation within the leaf (g_m) is less well understood, both in the short and long term and at different developmental stages. Recent studies, however, have shown that mesophyll conductance has a more important role than previously thought and thus needs to be included in photosynthetic models [7,8]. Moreover, g_m has been found to vary in response to current and previous irradiance [9,10], leaf anatomy and age [11], developmental stage of the leaf [12], photosynthetic rate [13], growth temperature [14–16], nitrogen availability [17], water availability [18], carbonic anhydrase activity [19], aquaporin activity [20,21], and leaf internal CO_2 concentration [22,23]. Furthermore, with the aid of recent technological advances, some studies have revealed that g_m can respond to current environmental variables in the short term [24].

So far, three different techniques have been suggested to estimate g_m and are described in the following sections. Two of the three techniques require variables to be measured together with conventional gas exchange measurements,

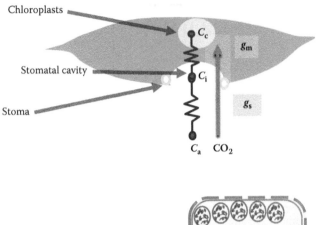

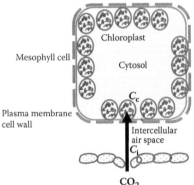

FIGURE 4.1 **(See color insert.)** A model of components of a photosynthetic CO_2 diffusion pathway.

and mesophyll diffusional conductance is estimated based on the previously developed theory of photosynthesis. Each theory has been well studied and developed, but still requires assumptions. Conventional gas exchange and combined gas exchange with fluorescence methods enable a limited number of measurements per day, and it was technically impossible to measure short-term variations in g_m using these methods. Recently, a diode laser coupled with leaf gas exchange measurements proved to be an innovative technique for understanding the mechanisms of rapid changes in CO_2 diffusion from the atmosphere to leaves during photosynthesis [3,23]. This combination of techniques has enabled researchers to measure dynamic changes in CO_2 isotope discrimination on a short time scale and thus estimate the limiting factors of CO_2 assimilation rates during transient environmental change. However, the underlying mechanisms of such changes have not been clarified. Additionally, the response time of mesophyll and stomatal conductance to different environmental variables may vary, and thus, more frequent measurements are required than previous studies have allowed. Recent technological developments and modifications to the sampling sequence now allow more frequent measurement of CO_2 isotopologues, using diode lasers such as lead-salt or quantum cascade lasers [25–27]. By deploying this technique, we can discern the independent behavior of stomatal and mesophyll diffusional conductance in response to different environmental variables.

4.2 THEORY AND METHOD OF ANALYSES OF CONDUCTANCE

Figure 4.2 summarizes the response of photosynthetic rate to CO_2 and the candidates for the limiting steps of CO_2 assimilation in C_3 plants. In C_3 species, CO_2 assimilation is classically considered to be limited by the capacities of RuBP carboxylation, RuBP regeneration, or Pi regeneration (Figure 4.2) [28,29]. At low CO_2 concentrations, the CO_2 assimilation rate is limited by carboxylation of RuBP. In the process of RuBP carboxylation, CO_2 diffusion (via stomatal conductance and mesophyll conductance [16]) and RuBisCO activity (i.e., RuBisCO amount, RuBisCO kinetics, and RuBisCO activation [30,31]) can affect the CO_2 assimilation rate. On the other hand, at high CO_2 concentrations, the CO_2 assimilation rate is limited by the regeneration of RuBP. The RuBP regeneration rate is determined by either the chloroplast electron transport capacity to generate NADPH and ATP [32] or the activity of Calvin cycle enzymes involved in the regeneration of RuBP (e.g., SBPase [33]). P_i regeneration limits CO_2 assimilation under some conditions, such as high CO_2 concentrations and/ or low temperatures [29,34].

4.2.1 A/C_c CURVE-FITTING METHODS

g_m may be estimated from analysis of a photosynthetic CO_2 response curve (A/C_i curve) using traditional gas exchange approaches over a range of CO_2 concentration between 0 and 2000 µmol/mol under saturating light. This method takes between 30 and 120 min to complete a single g_m estimate; it is not suitable to measure short-term responses. The

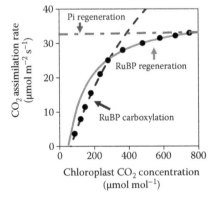

FIGURE 4.2 CO_2 response of CO_2 assimilation rate and the candidates for the limiting steps of CO_2 assimilation in C_3 plants. At low CO_2 concentrations, the CO_2 assimilation rate is limited by RuBP carboxylation, where CO_2 diffusion (stomatal conductance and mesophyll conductance) and Rubisco activity (i.e., Rubisco amount, Rubisco kinetics, and Rubisco activation) would affect the CO_2 assimilation rate. At high CO_2 concentrations, the CO_2 assimilation rate is limited by RuBP regeneration. RuBP regeneration rate in turn is determined by the chloroplast electron transport capacity to generate NADPH and ATP, or the activity of Calvin cycle enzymes involved in regeneration of RuBP or Pi regeneration capacity.

method is usually used for comparison between species. Photosynthetic rate (A) within the RuBisCO-limited phase can be described by

$$A = \frac{(C_c - \Gamma^*)V_{c\max}}{C_c + K_c \left(1 + \dfrac{O_i}{K_o}\right)} - R_d \qquad (4.1)$$

and within the ribulose biphosphate (RuBP) regeneration limited phase by

$$A = J_{\max} \frac{C_c - \Gamma^*}{4C_c + 8\Gamma} - R_d. \qquad (4.2)$$

At steady state,

$$A = g_m(C_i - C_c), \qquad (4.3)$$

where $V_{c\max}$ (μmol m^{-2} s^{-1}) is the maximum rate of RuBP carboxylation on the basis of leaf area; $J_{\max}$ (μmol m^{-2} s^{-1}) is the rate of electron transport; K_c and K_o are Michaelis–Menten constants for CO_2 and O_2, respectively; and Γ^* is the CO_2 compensation point. K_c, K_o, and Γ^* are determined by using a Bernacchi-type temperature-dependent method [14]. O_i is the oxygen concentration in the atmosphere (usually given as 210,000 ppm). R_d is mitochondrial respiration. Solving Equation 4.1 (for $V_{c\max}$ estimation) and Equation 4.2 (for $J_{\max}$ estimation) for C_c and substituting in Equation 4.3 gives a nonrectangular hyperbola equation according to von Caemmerer and Evans [35]. By fitting the nonrectangular hyperbola, $V_{c\max}$, $J_{\max}$, and R_d are given. Then, g_m is estimated by using the data, either RuBisCO-limited data only [36] or RuBP-limited data, or using a nonrectangular hyperbola curve-fitting, which is described in Refs. [36,37].

g_m can be also estimated from only the RuBisCO-limited curve using Equation 4.1 at the CO_2 compensation point according to Ref. [38]. In this case, the chloroplastic CO_2 concentration at CO_2 compensation point (C^*) can be given as follows [39]:

$$C_c = \Gamma^* - R_d/g_m \qquad (4.4)$$

then

$$g_m = R_d/(\Gamma^* - C^*). \qquad (4.5)$$

4.2.2 Chlorophyll Fluorescence Method

4.2.2.1 Constant J Method

In the constant J method, it is assumed that absorption and potential differences in light partitioning between photosystem II and photosystem I are constant across the leaf [40].

The response of chlorophyll a fluorescence to CO_2 is determined simultaneously with gas exchange by an integrated fluorescence chamber head (LI-6400, LI-6400-40 Leaf Chamber Fluorometer, LI-COR). After measurements of the quantum yield of photosystem II (Φ_{PSII}) [41], the rate of linear electron transport (J_{flu}) is determined as

$$J_{flu} = \phi_{PSII} \cdot PPFD \cdot \alpha \cdot f \qquad (4.6)$$

where f is the fraction of absorbed light reaching PSII (assumed to be 0.5 for C_3 plants [42]), $PPFD$ is the incident photon flux density, and α is the leaf absorptance (often assumed to be 0.85). The photochemical efficiency of photosystem II (Φ_{PSII}) is determined by measurement of steady-state fluorescence (F_s) and maximum fluorescence with a saturating pulse of light (F_m') according to Genty, Briantais, and Baker [41]:

$$\phi_{PSII} = (F_m' - F_s)/F_m'. \qquad (4.7)$$

The electron transport rate calculated from the fluorescence measurement (J_{flu}) is calculated from above-mentioned method.

C_c is then estimated by using the following [43]:

$$C_c = \Gamma^* (J_{flu} + (A + R_d))/J_{flu} - 4(A + R_d). \qquad (4.8)$$

Finally, g_m is estimated:

$$C_c = C_i - (A/g_m). \qquad (4.9)$$

4.2.2.2 Variable J Method

Another method to estimate g_m using chlorophyll fluorescence measurement and gas exchange is the variable J method [44,45]. In this method, the measurement is done under a single condition (e.g., under atmospheric condition). The measurements should be conducted independently under the same condition to estimate the electron transport rates. The electron transport rates are estimated through conventional gas exchange measurement and the fluorescence method. From the gas exchange measurement, the electron transport rate (J) is estimated from the following equation, which is supported by assimilation rate (A). R_d, C_c, and Γ^* are day respiration rate, leaf internal CO_2 concentration, and CO_2 photocompensation point, respectively.

$$J = (A + R_d)(4C_c + 8\Gamma^*)/(C_c - \Gamma^*). \qquad (4.10)$$

The electron transport (J_{flu}) is also measured from the fluorescence measurement, and using Equation 4.6 above, J_{flu} is estimated. $V_{c\max}$ and $J_{\max}$ are estimated using the A/C_c curves based on the temperature dependence of kinetic RuBisCO parameters according to Ref. [14]. By combining Equations 4.8 and 4.10, g_m is estimated directly. g_m and C_c are then accounted for from the instantaneous measurement described in Refs. [43,46].

4.2.3 ISOTOPE METHOD

4.2.3.1 Theory of Carbon Isotope Discrimination during Photosynthesis

CO_2 containing the heavier 13-carbon isotope (about 1.1% in the atmosphere) was discriminated by each physical and chemical step during photosynthesis. $^{13}CO_2$ diffuses more slowly than $^{12}CO_2$ and interacts more slowly with RuBisCO during carboxylation in C_3 plants. Therefore, the carbon isotope composition of C_3 plant biomass is more depleted in ^{13}C compared to CO_2 in the atmosphere.

A theory has been developed to predict carbon isotope discrimination processes [47], and the model has been modified [48,49]:

$$\Delta^{13}C = a_b \frac{C_a - C_s}{C_a} + a \frac{C_s - C_i}{C_a} + (b_s + a_l)\frac{C_i - C_c}{C_a} + b\frac{C_c}{C_a} - \frac{f\Gamma^*}{C_a} - e\frac{R_d}{A + R_d} \cdot \frac{C_c - \Gamma^*}{C_a}. \tag{4.11}$$

Thereafter, Farquhar and Cernusak [50] have further modified the model as follows:

$$\Delta^{13}C = \frac{1}{1-t}\, a_b \frac{C_a - C_s}{C_a} + a_s \frac{C_s - C_i}{C_a}$$

$$+ \frac{1+t}{1-t}\, a_m \frac{C_i - C_c}{C_a} + b\frac{C_c}{C_a} - \frac{\alpha_b}{\alpha_e}\cdot e\frac{R_d}{A + R_d}$$

$$\cdot \frac{C_c - \Gamma^*}{C_a} - \frac{\alpha_b}{\alpha_f}\cdot \frac{f\Gamma^*}{C_a}, \tag{4.12}$$

where b (29‰), a_b (2.9‰), a_s (4.4‰), and a_m (1.17‰) are fractionations associated with carboxylation, diffusion through the leaf boundary layer and stomata, and combined fractionations during dissolution into solution (1.1‰ at 25°C) and diffusion into leaf water (0.7‰), respectively. α_b, α_e, and α_f are $^{13}CO_2$ fractionation factors for carboxylation, photorespiration, and photorespiration with respect to assimilation rate, which are defined as $\alpha_b = 1 + b$, $\alpha_e = 1 + e'$, and $\alpha_f = 1 + f$, respectively. C_a, C_s, C_i, and C_c are the CO_2 concentrations of the ambient air, at the leaf surface, in the substomatal cavities, and at the sites of carboxylation, respectively; A is the photosynthetic rate; e' (0.5‰, taking into account both isotopic effects [48] and the ^{13}C disequilibrium of 27‰ between growth and measurement of CO_2, and thus between ambient air and CO_2, from the CO_2 cylinder used in this study [51]) and f (16.2‰ [52]) are fractionations associated with mitochondrial respiration in light and photorespiration, respectively; and Γ^* is the CO_2 compensation point in the absence of mitochondrial respiration (R_d).

The following equation is the carbon isotope discrimination if g_m is infinite [50]:

$$\Delta^{13}C_i = \frac{1}{1-t}\, a_b \frac{C_a - C_s}{C_a} + a_s \frac{C_s - C_i}{C_a}$$

$$+ \frac{1+t}{1-t}\, b\frac{C_i}{C_a} - \frac{\alpha_b}{\alpha_e}\cdot e\frac{R_d}{A + R_d}\cdot \frac{C_i - \Gamma^*}{C_a} - \frac{\alpha_b}{\alpha_f}\cdot \frac{f\Gamma^*}{C_a}. \tag{4.13}$$

Finally, t is a *ternary effect*, which describes the influence of water vapor added by transpiration on discrimination [50]. The ternary effect is expressed as

$$t = \frac{\alpha_{ac}E}{2g_{ac}}, \tag{4.14}$$

where

$$\frac{1}{\alpha_{ac}} = \frac{1}{(1+\bar{a})}. \tag{4.15}$$

$\bar{a}$ is the weighted fractionation across the boundary layer and stomata (2.9‰ and 4.4‰, respectively, for $^{13}CO_2$). g_{ac} is the conductance to diffusion of CO_2 in air from the atmosphere to the intercellular air space. C_c in the above equation can be replaced [50] with $C_i - Ar_m/P$, where r_m is the resistance to CO_2 in the mesophyll (the inverse of g_m) and P is the atmospheric pressure in the cuvette, $P = p_a/C_a$ (p_a is the ambient partial pressure of CO_2).

The measured isotope value ($\Delta^{13}C_{obs}$) is used to calculate the actual discrimination as follows [53]:

$$\Delta^{13}C_{obs} = \frac{\xi(\delta^{13}C_o - \delta^{13}C_e)}{1 + \delta^{13}C_o - \xi(\delta^{13}C_o - \delta^{13}C_e)}, \tag{4.16}$$

where $\xi = C_e/(C_e - C_o)$. Note that C_e and C_o are the CO_2 concentrations of the air streams after removal of water vapor by a Nafion drying tube. CO_2 concentration of the leaf chamber inlet air stream (measured using the infrared gas analyzer on the photosynthesis system) was expressed on a dry air basis using:

$$\Delta^{13}C_i - \Delta^{13}C_{obs} = \frac{1+t}{1-t}\, b - a_m - \frac{\alpha_b}{\alpha_e}\cdot e\frac{R_d}{A + R_d}\, r_m \cdot \frac{A}{Pc_a} \tag{4.17}$$

and g_m can be estimated using [50]

$$g_m = \frac{\frac{1+t}{1-t}\, b - a_m - \frac{\alpha_b}{\alpha_e}\cdot e\frac{R_d}{A + R_d}\cdot \frac{A}{Pc_a}}{\Delta^{13}C_i - \Delta^{13}C_{obs}}. \tag{4.18}$$

Carbon isotope discrimination can be measured using isotope ratio mass spectrometric analysis (IRMS) of CO_2 from a leaf gas exchange chamber (e.g., [53]). CO_2 of a known isotope composition is mixed to the required concentration and plumbed to the inlet of the leaf chamber. Inlet CO_2 concentration and isotope composition are assumed to be constant over the experiment, meaning that only the outlet concentration and isotope composition need to be measured. Mesophyll conductance is estimated from Equations 4.13, 4.17, and 4.18. Because IRMS usually requires high CO_2 concentrations for precise measurements, the air stream exiting the leaf chamber is collected over a period of minutes to hours, and the CO_2 is purified using cryogenic traps [2]. Care must be taken to avoid leaks and contamination in the leaf chamber, during gas collection, and during CO_2 purification. Further, the low temporal resolution limits experiments to comparisons between species or genotypes, with no information on dynamic responses.

More recently, carbon isotope discrimination has been measured online and in real time using a stable isotope tunable diode laser absorption spectrometer (e.g., the Campbell Scientific TGA100A or TGA100B or Aerodyne QC TILDAS ISO) directly plumbed to a leaf gas exchange system (e.g., [54–56]). The concentrations of individual isotopologues ($^{12}CO_2$ and $^{13}CO_2$) are measured at a frequency of 10 Hz (10 times per second), but averages over 15–20 s are usually calculated to attain the required precision for g_m estimation (0.1/mL or better). The laser absorption spectrometer instruments require frequent calibration, using cylinders or gas mixtures of known isotope composition and absolute concentration, usually every 7–9 min. The required frequency of calibration limits the temporal resolution of g_m estimation to once every 2–5 min, but this resolution allows dynamic responses of g_m to changes in environmental conditions inside the leaf chamber to be quantified, or for two or more chambers to be plumbed and sampled sequentially. The direct plumbing of the leaf gas exchange system to isotope measurements eliminates sample-handling errors during purification, which provides significant advantages over off-line IRMS measurement.

4.3 APPLICATIONS TO INVESTIGATE MESOPHYLL CONDUCTANCE AGAINST ENVIRONMENTAL RESPONSE

4.3.1 DYNAMIC ANALYSES OF MESOPHYLL CONDUCTANCE TO INVESTIGATE THE SHORT-TERM VARIATIONS UNDER VARIABLE ENVIRONMENTAL CONDITIONS

From the continuous measurement of changes in isotopologues of CO_2 in response to environmental variables, it has become possible to track the changes in carbon and oxygen isotope fractionation during photosynthesis. $\delta^{13}C$ of CO_2 discrimination during photosynthesis ($\Delta^{13}C$) has been reported to be related to mesophyll diffusional conductance and photosynthetic assimilation rate [56]. This setup of the instrument enables us to track the changes in the discrimination ($\Delta^{13}C$) and photosynthetic parameters. The method helps us to understand the mechanisms of the variations of g_s and g_m

during the transient period in response to alterations in some environmental factors, such as irradiance or temperature.

4.3.2 ANALYSES OF LIMITING STEP IN PHOTOSYNTHESIS AT VARIOUS LEAF TEMPERATURES

In C_3 plants, analyses of the response to CO_2 of (1) CO_2 assimilation rate (Figure 4.3a) and (2) chloroplast electron transport rate estimated from chlorophyll fluorescence (Figure 4.3b) can determine the limiting step of CO_2 assimilation by RuBP carboxylation or RuBP regeneration under various measurement conditions [10,57,58]. It should be noted that there is an error in the description of how we could analyze the limiting step of CO_2 assimilation in the figure legends in previous papers [10,57,58].

Using independent methods, such as the A/C_c response curve and chlorophyll fluorescence measurement (Figure 4.3), it is possible to analyze which process limits CO_2 assimilation rate, RuBP carboxylation, or RuBP regeneration by two independent methods: (1) response of CO_2 assimilation rate to CO_2 and (2) response of chloroplast electron transport rate to CO_2 (Figure 4.4).

4.3.2.1 Response of CO_2 Assimilation Rate to CO_2
The chloroplast CO_2 concentration at which the transition from RuBP carboxylation to RuBP regeneration limitation occurs ($C_{transition}$) is determined as [10,57,58]

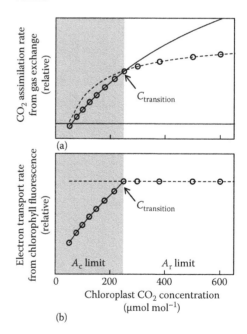

(a)

(b)

FIGURE 4.3 CO_2 response of net CO_2 assimilation rate from gas-exchange (a) and electron transport rate from chlorophyll fluorescence (b). Net CO_2 assimilation rate and electron transport rate is shown as open circles. Solid line shows RuBP carboxylation-limited A (A_c), and dotted line shows RuBP regeneration-limited A (A_r). Arrows shows the chloroplast CO_2 concentration ($C_{transition}$) at which the transition from A_c to A_r limitation occurs. Shaded area shows that A_{400} is limited by A_c, whereas others show that A_{400} is limited by A_r.

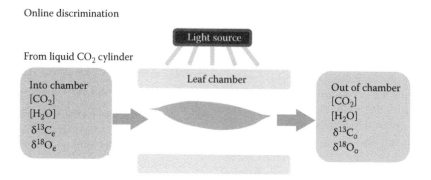

FIGURE 4.4 **(See color insert.)** A model of measurement set up with leaf chamber.

$$C_{\text{transition}} = \frac{K_c(1 + /K_o)J_{\max}/4V_{c\max} - 2\Gamma *}{1 - J_{\max}/4V_{c\max}}. \quad (4.19)$$

The C_c measured under growth CO_2 conditions (e.g., 400 µmol/mol) is compared to $C_{\text{transition}}$ (Figure 4.5). If C_c for

CO_2 assimilation rate at growth CO_2 conditions (e.g., A_{400}) is less than $C_{\text{transition}}$, it indicates a limitation by RuBP carboxylation, whereas if C_c for CO_2 assimilation rate at growth CO_2 conditions (e.g., A_{400}) is greater than $C_{\text{transition}}$, it indicates a limitation by RuBP regeneration. At $C_{\text{transition}}$, a colimitation by RuBP carboxylation and RuBP regeneration exists.

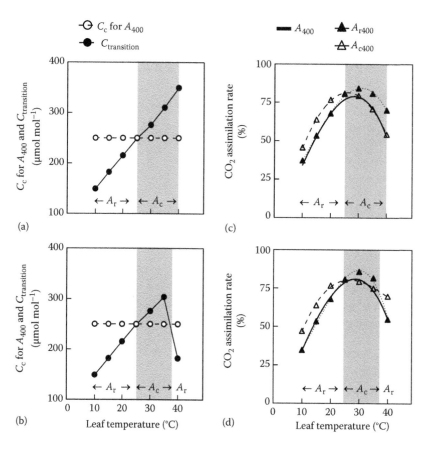

FIGURE 4.5 Differences in the limiting step of CO_2 assimilation by RuBP carboxylation (A_c) or RuBP regeneration (A_r). (a, b) Temperature dependence of chloroplast CO_2 concentration ($C_{\text{transition}}$) at which the transition from A_c to A_r limitation occurs and C_c for CO_2 assimilation rate measured at 400 µmol mol^{-1} CO_2 concentration (A_{400}). C_c for A_{400} above the $C_{\text{transition}}$ indicates that CO_2 assimilation is limited by A_r, whereas C_c for A_{400} less than $C_{\text{transition}}$ indicates that CO_2 assimilation is limited by A_c. (c, d) Temperature dependence of A_{400}, A_c and A_r. A_{400} can be expressed using the minimum rate of two partial reactions, A_c or A_r. Shaded area shows that A_{400} is limited by A_c, whereas others show that A_{400} is limited by A_r. In the case of (c), A_{400} at lower temperature is limited by A_r whereas at higher temperature by A_c. On the other hand, in the case of (b) and (d), A_{400} at lower and higher temperature is limited by A_r whereas at around the optimum temperature of CO_2 assimilation, A_{400} is limited by A_c.

4.3.2.2 Response of Chloroplast Electron Transport Rate to CO_2

The response of chlorophyll a fluorescence to CO_2 is determined simultaneously with gas exchange by an integrated fluorescence chamber head (LI-6400, LI-6400-40 Leaf Chamber Fluorometer, LI-COR). After measurements of the quantum yield of photosystem II (Φ_{PSII}) [41], the rate of linear electron transport (J_{flu}) is determined using Equation 4.6. J_{flu} is shown to be constant at high CO_2 concentration (Figure 4.3b). CO_2 assimilation rate at growth CO_2 conditions (e.g., A_{400}) is limited by RuBP regeneration rate if J_{flu} at growth CO_2 concentrations (e.g., 400 μmol/mol) is similar to the constant J_f at high CO_2 concentration. On the other hand, if J_{flu} at growth CO_2 concentration (e.g., 400 μmol/mol) is lower than the constant J_{flu} at high CO_2 concentration, this indicates that CO_2 assimilation rate at growth CO_2 conditions (e.g., A_{400}) is limited by RuBP carboxylation. The J_{flu}–C_c curve is analyzed to obtain the chloroplast CO_2 concentration at which the electron transport rate (J_{flu}) is constant ($C_{transition}$ [J_{flu}–C_c curve]), since the $C_{transition}$ (J_{flu}–C_c curve) indicates the transition point from RuBP carboxylation to RuBP regeneration limitation, as with analyses of response of CO_2 assimilation rate to CO_2.

Figure 4.5 shows an example of under what temperatures RuBP carboxylation or RuBP regeneration limits CO_2 assimilation rate at an ambient CO_2 concentration of 400 μmol/mol (A_{400}). The temperature response of CO_2 assimilation can be described with a parabolic curve having an optimum temperature, and thus, the CO_2 assimilation rate is inhibited at both low and high temperatures [59,60]. At current levels of atmospheric CO_2, the control of the temperature response of photosynthesis is typically considered to be a mixture of limitations by RuBP carboxylation and RuBP regeneration (and sometimes P_i regeneration) [59]. Figure 4.5c and d shows a similar temperature response of A_{400}; however, the limiting step of A_{400} is different, especially at high temperature. In Figure 4.5a and c, A_{400} is limited by RuBP regeneration at low temperature, whereas at high temperature, A_{400} is limited by RuBP carboxylation. On the other hand, in Figure 4.5b and d, A_{400} is limited by RuBP regeneration at low and high temperature, whereas at around the optimum temperature of CO_2 assimilation, A_{400} is limited by RuBP carboxylation. It has been shown that the limiting step of CO_2 assimilation differs depending on plant species (e.g., cold-tolerant plants versus cold-sensitive plants [10,61]) and growth condition, including growth temperature [16,30,62] and growth nitrogen concentration [58]. Understanding the limiting step of photosynthesis at various environmental conditions is of immense importance for identifying a biomolecular target for enhancing leaf photosynthesis [63].

4.4 CONCLUDING REMARKS

Work to establish the techniques and calculations to estimate mesophyll conductance is still underway. There are still potential artifacts associated with different methods; for example, there are potential errors when using the fluorescence method due to possible gradients in quantum yields through the leaves [64], and there are also artifacts in both the fluorescence methods and the isotope methods associated with respiratory and photorespiratory fluxes [65,66], as well as possible issues associated with bias in gas exchange measurements [67]. However, continuous measurement using a diode laser would be advantageous in investigating the rapid response under fluctuating environmental conditions and revealing the mechanisms behind them. Besides, technical developments will advance to optimize the collection of more data more effectively than conventional techniques. The technique will help us to understand water use efficiency between species or genotypes to identify the most efficient species or genotypes that are more beneficial under harsh environmental conditions, such as drought.

ACKNOWLEDGMENTS

First of all, we would like to thank to Prof. Margaret M. Barbour for giving updated information regarding the laser spectroscopy methods and the method to estimate the mesophyll conductance. This work was supported by the New Technology Development Foundation (N.K.); a Grant-in-Aid for Scientific Research for young researchers from the Japan Society for the Promotion of Science (scientific research no. 26850009, N.K.); the Japan Science and Technology Agency, Precursory Research for Embryonic and Technology (PRESTO) (W.Y.); and a Grant-in-Aid for Scientific Research from the Japan Society for the Promotion of Science (scientific research no. 25891005, W.Y.).

REFERENCES

1. Flexas, J. et al., Mesophyll conductance to CO_2: Current knowledge and future prospects. *Plant Cell and Environment*, 2008. **31**(5): pp. 602–621.
2. Hanba, Y.T., S.I. Miyazawa, and I. Terashima, The influence of leaf thickness on the CO_2 transfer conductance and leaf stable carbon isotope ratio for some evergreen tree species in Japanese warm-temperate forests. *Functional Ecology*, 1999. **13**(5): pp. 632–639.
3. Douthe, C. et al., Mesophyll conductance to CO_2, assessed from online TDL-AS records of $^{13}CO_2$ discrimination, displays small but significant short-term responses to CO_2 and irradiance in Eucalyptus seedlings. *Journal of Experimental Botany*, 2011. **62**(15): pp. 5335–5346.
4. Kodama, N. et al., Short-term dynamics of the carbon isotope composition of CO_2 emitted from a wheat agroecosystem— Physiological and environmental controls. *Plant Biology*, 2011. **13**(1): pp. 115–125.
5. Collatz, G.J. et al., Physiological and environmental-regulation of stomatal conductance, photosynthesis and transpiration—A model that includes a laminar boundary-layer. *Agricultural and Forest Meteorology*, 1991. **54**(2–4): pp. 107–136.
6. Leuning, R. et al., Leaf nitrogen, photosynthesis, conductance and transpiration—Scaling from leaves to canopies. *Plant Cell and Environment*, 1995. **18**(10): pp. 1183–1200.
7. Epron, D. et al., Limitation of net CO_2 assimilation rate by internal resistance to CO_2 transfer in the leaves of 2 tree species (Fagus-sylvatica L and Castanea-sativa Mill). *Plant Cell and Environment*, 1995. **18**(1): pp. 43–51.

8. Flexas, J. et al., Mesophyll diffusion conductance to CO_2: An unappreciated central player in photosynthesis. *Plant Science*, 2012. **193**: pp. 70–84.

9. Niinemets, U., The controversy over traits conferring shade-tolerance in trees: Ontogenetic changes revisited. *Journal of Ecology*, 2006. **94**(2): pp. 464–470.

10. Yamori, W., J.R. Evans, and S. Von Caemmerer, Effects of growth and measurement light intensities on temperature dependence of CO_2 assimilation rate in tobacco leaves. *Plant Cell and Environment*, 2010. **33**(3): pp. 332–343.

11. Evans, J.R. et al., The relationship between CO_2 transfer conductance and leaf anatomy in transgenic tobacco with a reduced content of RUBISCO. *Australian Journal of Plant Physiology*, 1994. **21**(4): pp. 475–495.

12. Hanba, Y.T. et al., Effects of leaf age on internal CO_2 transfer conductance and photosynthesis in tree species having different types of shoot phenology. *Australian Journal of Plant Physiology*, 2001. **28**(11): pp. 1075–1084.

13. Evans, J.R., Leaf anatomy enables more equal access to light and CO_2 between chloroplasts. *New Phytologist*, 1999. **143**(1): pp. 93–104.

14. Bernacchi, C.J. et al., Temperature response of mesophyll conductance. Implications for the determination of Rubisco enzyme kinetics and for limitations to photosynthesis in vivo. *Plant Physiology*, 2002. **130**(4): pp. 1992–1998.

15. Warren, C.R. and M.A. Adams, Internal conductance does not scale with photosynthetic capacity: Implications for carbon isotope discrimination and the economics of water and nitrogen use in photosynthesis. *Plant Cell and Environment*, 2006. **29**(2): pp. 192–201.

16. Yamori, W. et al., Effects of internal conductance on the temperature dependence of the photosynthetic rate in spinach leaves from contrasting growth temperatures. *Plant and Cell Physiology*, 2006. **47**(8): pp. 1069–1080.

17. Warren, C.R., The photosynthetic limitation posed by internal conductance to CO_2 movement is increased by nutrient supply. *Journal of Experimental Botany*, 2004. **55**(406): pp. 2313–2321.

18. Warren, C.R., Soil water deficits decrease the internal conductance to CO_2 transfer but atmospheric water deficits do not. *Journal of Experimental Botany*, 2008. **59**(2): pp. 327–334.

19. Makino, A. et al., Distinctive response of ribulose-1,5-biphosphate carboxylase and carbonic anhydrase in wheat leaves to nitrogen nutrition and their possible relationship to CO_2 transfer resistance. *Plant Physiology*, 1992. **100**(4): pp. 1737–1743.

20. Hanba, Y.T. et al., Overexpression of the barley aquaporin HvPIP2;1 increases internal CO_2 conductance and CO_2 assimilation in the leaves of transgenic rice plants. *Plant and Cell Physiology*, 2004. **45**(5): pp. 521–529.

21. Flexas, J. et al., Tobacco aquaporin NtAQP1 is involved in mesophyll conductance to CO_2 in vivo. *Plant Journal*, 2006. **48**(3): pp. 427–439.

22. Flexas, J. et al., Keeping a positive carbon balance under adverse conditions: Responses of photosynthesis and respiration to water stress. *Physiologia Plantarum*, 2006. **127**(3): pp. 343–352.

23. Tazoe, Y. et al., Light and CO_2 do not affect the mesophyll conductance to CO_2 diffusion in wheat leaves. *Journal of Experimental Botany*, 2009. **60**(8): pp. 2291–2301.

24. Tazoe, Y. et al., Using tunable diode laser spectroscopy to measure carbon isotope discrimination and mesophyll conductance to CO_2 diffusion dynamically at different CO_2 concentrations. *Plant Cell and Environment*, 2011. **34**(4): pp. 580–591.

25. Nelson, D.D. et al., New method for isotopic ratio measurements of atmospheric carbon dioxide using a 4.3 mu m pulsed quantum cascade laser. *Applied Physics B-Lasers and Optics*, 2008. **90**(2): pp. 301–309.

26. Tuzson, B. et al., High precision and continuous field measurements of delta C-13 and delta O-18 in carbon dioxide with a cryogen-free QCLAS. *Applied Physics B-Lasers and Optics*, 2008. **92**(3): pp. 451–458.

27. Wada, R. et al., Observation of carbon and oxygen isotopic compositions of CO_2 at an urban site in Nagoya using Mid-IR laser absorption spectroscopy. *Atmospheric Environment*, 2011. **45**(5): pp. 1168–1174.

28. Farquhar, G.D., S.V. Caemmerer, and J.A. Berry, A biochemial-model of photosynthetic CO_2 assimilation in leaves of C3 species. *Planta*, 1980. **149**(1): pp. 78–90.

29. Sharkey, T.D., Photosynthesis in intact leaves of C3 plants: Physics, physiology and rate limitations. *The Botanical Review*, 1985. **51**(1): pp. 53–105.

30. Yamori, W., K. Noguchi, and I. Terashima, Temperature acclimation of photosynthesis in spinach leaves: Analyses of photosynthetic components and temperature dependencies of photosynthetic partial reactions. *Plant Cell and Environment*, 2005. **28**(4): pp. 536–547.

31. Yamori, W. et al., Effects of Rubisco kinetics and Rubisco activation state on the temperature dependence of the photosynthetic rate in spinach leaves from contrasting growth temperatures. *Plant Cell and Environment*, 2006. **29**(8): pp. 1659–1670.

32. Yamori, W. et al., Cyclic electron flow around photosystem I via chloroplast NAD(P)H dehydrogenase (NDH) complex performs a significant physiological role during photosynthesis and plant growth at low temperature in rice. *Plant Journal*, 2011. **68**(6): pp. 966–976.

33. Raines, C.A., Transgenic approaches to manipulate the environmental responses of the C3 carbon fixation cycle. *Plant Cell and Environment*, 2006. **29**(3): pp. 331–339.

34. Sage, R.F. and T.D. Sharkey, The effect of temperature on the occurrence of O_2 and CO_2 insensitive photosynthesis in field-grown plants. *Plant Physiology*, 1987. **84**(3): pp. 658–664.

35. von Caemmerer, S. and J.R. Evans, Determination of the average partial-pressure of CO_2 in chloroplasts from leaves of several C3 plants. *Australian Journal of Plant Physiology*, 1991. **18**(3): pp. 287–305.

36. Ethier, G.J. and N.J. Livingston, On the need to incorporate sensitivity to CO_2 transfer conductance into the Farquhar–von Caemmerer–Berry leaf photosynthesis model. *Plant Cell and Environment*, 2004. **27**(2): pp. 137–153.

37. Sharkey, T.D. et al., Fitting photosynthetic carbon dioxide response curves for C-3 leaves. *Plant Cell and Environment*, 2007. **30**(9): pp. 1035–1040.

38. von Caemmerer, S., W.P. Quick, and R. Kennedy, Rubisco: Physiology in vivo, in *Photosynthesis: Physiology and metabolism*, R.C. Leegood, T.D. Sharkey, and S. von Caemmerer, Editors. 2000, Kluwer Academic Publishers, Dordrecht, Netherlands, pp. 85–113.

39. Peisker, M. and H. Apel, Inhibition by light of CO_2 evolution from dark respiration: Comparison of two gas exchange methods. *Photosynthesis Research*, 2001. **70**(3): pp. 291–298.

40. Laisk, A. and F. Loreto, Determining photosynthetic parameters from leaf CO_2 exchange and chlorophyll fluorescence—Ribulose-1,5-bisphosphate carboxylase oxygenase specificity factor, dark respiration in the light, excitation distribution between photosystems, alternative electron transport rate, and mesophyll diffusion resistance. *Plant Physiology*, 1996. **110**(3): pp. 903–912.

41. Genty, B., J.M. Briantais, and N.R. Baker, The relationship between the quantum yield of photosynthetic electron-transport and quenching of chlorophyll fluorescence. *Biochimica Et Biophysica Acta*, 1989. **990**(1): pp. 87–92.

42. Ogren, E. and J.R. Evans, Photosynthetic light-response curves. 1. The influence of CO_2 partial-pressure and leaf inversion. *Planta*, 1993. **189**(2): pp. 182–190.

43. Harley, P.C. et al., Modeling photosynthesis of cotton grown in elevated CO_2. *Plant Cell and Environment*, 1992. **15**(3): pp. 271–282.

44. Di Marco, G. et al., Fluorescence parameters measured concurrently with net photosynthesis to investigate chloroplastic carbon dioxide concentration in leaves of QUERCUS-ILEX L. *Journal of Plant Physiology*, 1990. **136**(5): pp. 538–543.

45. Harley, P.C. et al., Theoretical considerations when estimating the mesophyll conductance to CO_2 flux by analysis of the response of photosynthesis to CO_2. *Plant Physiology*, 1992. **98**(4): pp. 1429–1436.

46. Grassi, G. and F. Magnani, Stomatal, mesophyll conductance and biochemical limitations to photosynthesis as affected by drought and leaf ontogeny in ash and oak trees. *Plant Cell and Environment*, 2005. **28**(7): pp. 834–849.

47. Farquhar, G.D., M.H. Oleary, and J.A. Berry, On the relationship between carbon isotope discrimination and the intercellular carbon-dioxide concentration in leaves. *Australian Journal of Plant Physiology*, 1982. **9**(2): pp. 121–137.

48. Brugnoli, E. et al., Correlation between the carbon isotope discrimination in leaf starch and sugars of C-3 plants and the ratio of intercellular and atmospheric partial pressures of carbon-dioxide. *Plant Physiology*, 1988. **88**(4): pp. 1418–1424.

49. Lanigan, G.J. et al., Carbon isotope fractionation during photorespiration and carboxylation in senecio. *Plant Physiology*, 2008. **148**(4): pp. 2013–2020.

50. Farquhar, G.D. and L.A. Cernusak, Ternary effects on the gas exchange of isotopologues of carbon dioxide. *Plant Cell and Environment*, 2012. **35**(7): pp. 1221–1231.

51. Wingate, L. et al., Variations in ^{13}C discrimination during CO_2 exchange by Picea sitchensis branches in the field. *Plant Cell and Environment*, 2007. **30**(5): pp. 600–616.

52. Evans, J.R. and S. Von Caemmerer, Temperature response of carbon isotope discrimination and mesophyll conductance in tobacco. *Plant Cell and Environment*, 2013. **36**(4): pp. 745–756.

53. Evans, J.R. et al., Carbon isotope discrimination measured concurrently with gas-exchange to investigate CO_2 diffusion in leaves of higher-plants. *Australian Journal of Plant Physiology*, 1986. **13**(2): pp. 281–292.

54. Barbour, M.M. et al., A new measurement technique reveals rapid post-illumination changes in the carbon isotope composition of leaf-respired CO_2. *Plant Cell and Environment*, 2007. **30**(4): pp. 469–482.

55. Barbour, M.M. et al., Variability in mesophyll conductance between barley genotypes, and effects on transpiration efficiency and carbon isotope discrimination. *Plant Cell and Environment*, 2010. **33**(7): pp. 1176–1185.

56. Kodama, N. et al., Spatial variation in photosynthetic CO_2 carbon and oxygen isotope discrimination along leaves of the monocot triticale (Triticum x Secale) relates to mesophyll conductance and the Peclet effect. *Plant Cell and Environment*, 2011. **34**(9): pp. 1548–1562.

57. Yamori, W. and S. von Caemmerer, Effect of Rubisco activase deficiency on the temperature response of CO_2 assimilation rate and Rubisco activation state: Insights from transgenic tobacco with reduced amounts of Rubisco activase. *Plant Physiology*, 2009. **151**(4): pp. 2073–2082.

58. Yamori, W., T. Nagai, and A. Makino, The rate-limiting step for CO_2 assimilation at different temperatures is influenced by the leaf nitrogen content in several C-3 crop species. *Plant Cell and Environment*, 2011. **34**(5): pp. 764–777.

59. Yamori, W., K. Hikosaka, and D.A. Way, Temperature response of photosynthesis in C-3, C-4, and CAM plants: Temperature acclimation and temperature adaptation. *Photosynthesis Research*, 2014. **119**(1–2): pp. 101–117.

60. Way, D.A. and W. Yamori, Thermal acclimation of photosynthesis: On the importance of adjusting our definitions and accounting for thermal acclimation of respiration. *Photosynthesis Research*, 2014. **119**(1–2): pp. 89–100.

61. Yamori, W. et al., Cold-tolerant crop species have greater temperature homeostasis of leaf respiration and photosynthesis than cold-sensitive species. *Plant and Cell Physiology*, 2009. **50**(2): pp. 203–215.

62. Yamori, W., K. Noguchi, and I. Terashima, Mechanisms of temperature acclimation of photosynthesis. *Plant and Cell Physiology*, 2006. **47**: p. S4.

63. Yamori, W., Improving photosynthesis to increase food and fuel production by biotechnological strategies in crops. *Journal of Plant Biochemistry & Physiology*, 2013. **1**: p. 113.

64. Evans, J.R., Potential errors in electron transport rates calculated from chlorophyll fluorescence as revealed by a multilayer leaf model. *Plant and Cell Physiology*, 2009. **50**(4): pp. 698–706.

65. Tholen, D. et al., Variable mesophyll conductance revisited: Theoretical background and experimental implications. *Plant Cell and Environment*, 2012. **35**(12): pp. 2087–2103.

66. Tholen, D., G. Ethier, and B. Genty, Mesophyll conductance with a twist. *Plant Cell and Environment*, 2014. **37**(11): pp. 2456–2458.

67. Gu, L. and Y. Sun, Artefactual responses of mesophyll conductance to CO_2 and irradiance estimated with the variable J and online isotope discrimination methods. *Plant Cell and Environment*, 2014. **37**(5): pp. 1231–1249.

Section II

Biochemistry of Photosynthesis

5 Development of Chloroplast
Biogenesis, Senescence, and Regulations

Basanti Biswal and Jitendra K. Pandey

CONTENTS

5.1   Introduction 77
5.2   Chloroplast Genome: Structure, Organization, and Dynamics for Gene Expression during Chloroplast Development 79
5.3   Biogenesis of Thylakoids and Stromal Enzyme Complexes during Leaf Development: Assembly of Structural Components and Regulation 79
    5.3.1   Organization and Assembly of Photosystem II, a Well-Examined Thylakoid Complex 80
    5.3.2   Assembly of Photosystem I, Cytochrome b_6f Complex, and ATP Synthase 81
    5.3.3   Possible Mechanism of Assembly of Thylakoid Complexes 81
    5.3.4   Rubisco, an Important Stromal Enzyme: Synthesis, Assembly, and Regulation by Chaperones 81
5.4   Changes in Chloroplasts during Leaf Senescence: Formation of Gerontoplasts 82
    5.4.1   Ultrastructural Changes of Chloroplasts 82
    5.4.2   Degradation of Photosynthetic Pigments 82
        5.4.2.1   Tracing the Path of Chlorophyll Degradation: Enzymes Involved and the Role of Pheophorbide a Oxygenase 82
        5.4.2.2   Carotenoid Degradation 84
        5.4.2.3   Coordinated Breakdown of Pigment–Protein Complexes during Leaf Senescence: Possible Regulation 84
    5.4.3   Disassembly of Thylakoid Complexes and Loss in Primary Photochemical Reactions 84
    5.4.4   Decline in Rubisco Activity and Loss of the Enzyme Protein: Participation of Proteases and the Vesicular Transport System 84
5.5   Signals Regulating Chloroplast Biogenesis and Senescence during Leaf Development 85
    5.5.1   Sugar Signaling 85
        5.5.1.1   Depletion of Sugars and Induction of Senescence 85
        5.5.1.2   Induction of Senescence by Excess Sugars 86
        5.5.1.3   Sugar Sensors and Mechanism of Signal Transduction 87
        5.5.1.4   Cross-Talk of Sugar with Nitrogen, Phytohormones, and ROS in Signaling Network 87
    5.5.2   Role of ROS in Signaling Network of Chloroplast Development 87
        5.5.2.1   Reactive Oxygen Species Production 87
        5.5.2.2   ROS and Retrograde Signaling 88
        5.5.2.3   Interaction of ROS with Sugar and Phytohormone Signalings 88
5.6   Conclusions: Knowns, Unknowns, and the Future 89
Acknowledgments 90
References 90

5.1   INTRODUCTION

The biology of plastids in green plants has a long history, and the literature in the area is vast. A book by Wise and Hoober (2006) provides relevant information on different forms of plastids, their origin, specific structural features, functions, and locations in plants. Chloroplast is recognized as the most important form of plastid in green plants and algae because of its function in photosynthesis, which sustains life on the planet Earth. In addition to synthesis of pigments and proteins, chloroplasts in higher plants also participate in the biosynthesis of several essential cellular metabolites, including amino acids, fatty acids, and phytohormones. The organelle also plays a role in the network of sulfur metabolism (Biswal et al. 2008). Existing literature suggests chloroplast as a sensor of abiotic stress and modulator of plant stress response (Biswal et al. 2011). The organelle is also known to play a dominant role in regulating leaf and whole plant development (López-Zuez and Pyke 2005; Inaba and Ito-Inaba 2010). Importantly, the development of the plastid is tightly coupled to the developmental program of the plants, specifically in monocarpic plants (Krupinska et al. 2013). In this context, the mechanism of development of chloroplast and the regulatory system associated with it are emerging as a fascinating area of study in plant science. As such, the development of different plastid forms in plants is complex. The diversity of different plastid forms is primarily because of the diversity in their developmental programs as determined by their position; location (e.g., cells, tissues, and organs); environmental condition; and interaction with neighboring cells (Biswal et al. 2003; Jarvis and López-Juez 2013). Currently, excellent reviews are available on the nature and mechanism of chloroplast development and existing knowledge gaps in our understanding of the regulatory systems associated with the developmental programs (Croce and Amerongen 2011; Fischer 2012; Flores-Pérez and Jarvis 2013; Jarvis and López-Juez 2013; Juvany et al. 2013; Nickelsen and Rengstl 2013; Rochaix 2014; Pogson et al. 2015; Suga et al. 2015). This review will focus only on the pathway of development of chloroplast from proplastid with subsequent transformation of chloroplast to gerontoplast (senescing chloroplast). Gerontoplast is rather a new entry to the plastid family in green plants (Sitte 1977). This plastid with characteristic structural features plays a role in nutrient recycling (Thomas 2013). Figure 5.1 provides a quick introduction to the theme of the review. The development of chloroplast covers three major phases, namely, the buildup phase in emerging leaves with gradual transformation of proplastids to young chloroplasts, which are subsequently transformed to mature chloroplasts. The chloroplasts in fully green mature leaves are capable of capture, assimilation, and storage of carbon with high efficiency. Transformation of mature chloroplast to gerontoplast is the terminal phase of the developmental program, leading to leaf yellowing followed by necrosis and death (Thomas 2013). The genetic regulation and the regulatory network with several internal and environmental factors that determine these developmental transitions are complex and are not fully resolved. Among the environmental factors, light is recognized as the most important one that modulates the plastid development in the photomorphogenic network, mediated by well-identified photoreceptors (Biswal et al. 2003). Reports available on several internal factors modulating plastid development are plenty (Wise and Hoober 2006; Biswal et al. 2013). Phytohormones are well reported to play a critical role during plant development (Biswal et al. 2003). Development-dependent signaling and organellar interactions also influence plastid biogenesis and senescence during leaf development (Biswal et al. 2013). Undoubtedly, the nucleus is one of the important organelles, which codes for most of the plastid proteins, both structural and regulatory. The other organelles, including mitochondria, peroxisomes, and the

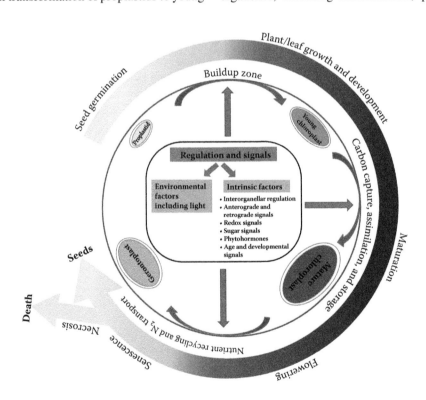

FIGURE 5.1 **(See color insert.)** The major events associated with formation and demolition of chloroplast and associated regulatory mechanisms during leaf development and senescence. The interaction between the internal factors and environmental cues, including light in a complex signaling network, regulates three major developmental phases, namely, the "buildup zone," when the proplastid is transformed to young chloroplast; the formation of mature chloroplast, the phase referred to as the "carbon capturing and assimilation zone"; and the final phase of development, in which mature chloroplasts are converted to gerontoplasts during leaf senescence. This phase is physiologically important for nutrient relocation and nitrogen transport. Necrosis is the last event leading to death. (Modified from Springer Science+Business Media: *Plastid Development in Leaves During Growth and Senescence*, ed. B. Biswal, K. Krupinska, and U.C. Biswal, 3–16, Advances in Photosynthesis and Respiration, vol. 36, The dynamic role of chloroplasts in integrating plant growth and development, 2013, Krupinska, K. et al.)

endoplasmic reticulum, are, in recent years, known to play a no less important role in regulating plastid development, through an interconnected metabolic network (Pogson and Albrecht 2011; Biswal et al. 2013).

Figure 5.1 depicts the developmental changes of the plastids as a marker of whole plant development. The best example of coupling between chloroplast development and the developmental program of the plant is the gradual visible color changes of leaves from light green to green and finally to yellow, reflecting the plastid-specific developmental changes in the pigments during the transformation of proplastid to chloroplast and finally to gerontoplast. In spite of development of several sophisticated and more accurate markers, the level of chlorophyll is still being used as a reliable indicator of plant development, supporting the view of tight coupling between the developmental programs of the organelle with leaf/plant development, and this review, therefore, very often uses chloroplast development and leaf/plant development interchangeably.

5.2 CHLOROPLAST GENOME: STRUCTURE, ORGANIZATION, AND DYNAMICS FOR GENE EXPRESSION DURING CHLOROPLAST DEVELOPMENT

Chloroplasts normally contain several copies of their DNA. The DNA has a circular structure and ranges in size from 100 to 200 kbps. The plastid genome is sequenced and mapped in many plant systems (Biswal et al. 2003; Kanamaru and Sugita 2013). The details of plastid DNA, and its association with RNA and proteins of organized structures referred to as plastid nucleoids, were recently described by Powikrowska et al. (2014). The plastid nucleoids are dynamic and significantly change in their structure and composition during development of the organelle (Liere and Börner 2013; Powikrowska et al. 2014). The nucleoids and their interaction with other factors including environmental cues regulate plastid genome replication and gene expression (Powikrowska et al. 2014).

The plastid gene expression is mostly regulated by nuclear-encoded proteins, but plastid-specific regulations also play a role in the process. Expression of plastid genes by nuclear-encoded proteins is regulated both at transcriptional and post-transcriptional levels. Chloroplast of higher plants uses two distinct types of RNA polymerases, namely, nuclear-encoded RNA polymerase (NEP) and plastid-encoded RNA polymerase (PEP). The coordinated interaction between PEP and NEP appears to regulate a highly complex pattern of plastid transcription during development of the organelle (Kanamaru and Sugita 2013). Nuclear control of chloroplast development is well examined through mutational studies (Biswal et al. 2003). Many nuclear mutants were identified that block synthesis of proteins encoded by the organelle genome (Barkan et al. 1995; Goldschmidt-Clermont 1998). For example, a nuclear mutant of *Chlamydomonas*, a green alga, has been shown to lack the ability to synthesize the large subunit (RbcL) of

Rubisco encoded by the plastid gene in spite of the synthesis of the small subunit (RbcS) encoded by the nuclear gene and other plastid proteins (Hong and Spreitzer 1994). The specific effect of the nuclear gene product on the synthesis of the RbcL may suggest that the signal from the nuclear genome has a target site on the plastid for the expression of specific gene(s). Analysis of the nuclear mutants also reveals the control of nuclear gene products on the accumulation of other proteins of the photosystem II (PSII) reaction center (Goldschmidt-Clermont 1998).

In addition to nuclear control, the plastid gene expression is also known to be regulated by its own developmental program (Kapoor et al. 1994). The accumulation of transcripts for the synthesis of proteins associated with the core complex of the reaction centers of photosystem I (PSI) and PSII is significantly influenced by the aging and functional status of developing chloroplast (Kapoor et al. 1994). The tissue and organ specificity is another factor assumed to control plastid gene expression (Taylor 1989; Kapoor et al. 1993; Biswal et al. 2003). Since plastid biogenesis in developing leaves involves several proteins of dual genetic origin, the proteins, which are encoded by nuclear genes, are targeted to chloroplasts through a complex transport system located in the plastid envelope (Biswal et al. 2003; Flores-Pérez and Jarvis 2013; Ling et al. 2013).

5.3 BIOGENESIS OF THYLAKOIDS AND STROMAL ENZYME COMPLEXES DURING LEAF DEVELOPMENT: ASSEMBLY OF STRUCTURAL COMPONENTS AND REGULATION

Thylakoid complexes are composed of several protein and nonprotein components that are assembled into the functional structures through a tightly regulated network, as shown in Figure 5.2. The lipids that constitute the bilayers of thylakoids most likely are derived from the inner envelope through vesicular budding (Block et al. 2013). Proteins encoded by nuclear and plastid genomes and the pigments coordinately assemble in the target site. The assembly process is regulated by several intrinsic factors, including phytohormones and modulation by other organelles. Light is the most important environmental cue, which has been extensively reported to play a dominant role in the assembly of thylakoid complexes and Rubisco through the action of several photo receptors (Biswal et al. 2003). There are four major thylakoid complexes, namely, PSI and PSII with their light-harvesting systems, cytochrome b_6f (Cytb_6f) complex, and ATP synthase. These complexes, in addition to protein subunits, consist of several redox components and metal ions, and some have different types of pigments (Ke 2001). PSII is a well-studied complex, and its structural details are well known (Umena et al. 2011). It is located in appressed grana regions of thylakoids. On the other hand, PSI and ATP synthase are primarily distributed in stroma lamellae, and

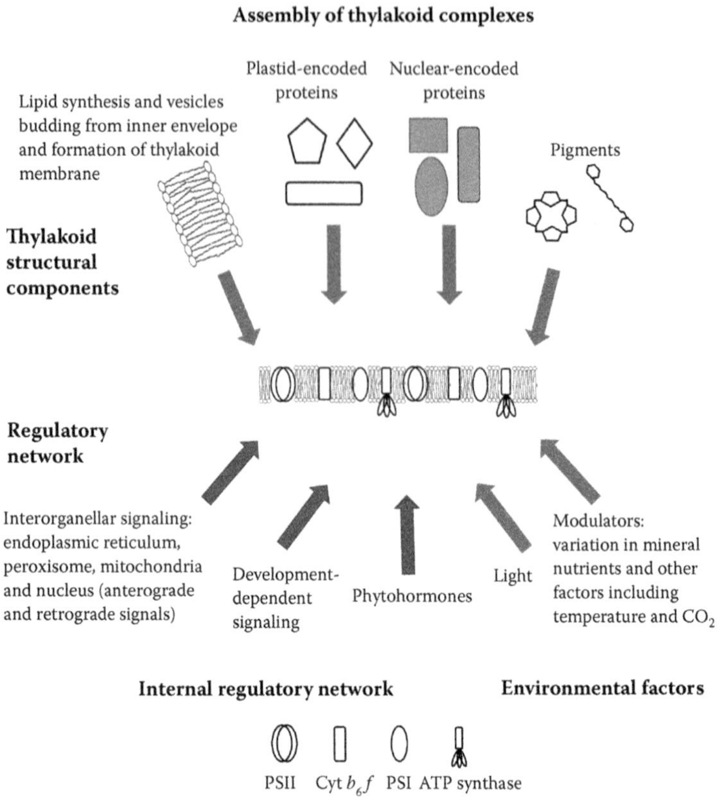

Assembly of thylakoid complexes

FIGURE 5.2 Scheme showing the assembly of different structural components of thylakoid. The possible regulatory events, both internal and environmental cues, are also shown. (Modified from Springer Science+Business Media: *Plastid Development in Leaves During Growth and Senescence*, ed. B. Biswal, K. Krupinska, and U.C. Biswal, 17–35, Advances in Photosynthesis and Respiration, vol. 36, Chloroplast development: Time, dissipative structures and fluctuations, 2013, Raval, M.K. et al.)

$Cytb_6f$, the other major thylakoid complex, is more or less uniformly distributed in the thylakoids.

The structures of most of the individual complexes are known now. For example, the structures at 1.9 Å atomic resolution (Umena et al. 2011; Suga et al. 2015) of PSII and at 3.3 Å resolution of PSI (Amunts et al. 2010) reveal the complex association of proteins with the cofactors that make these complexes functionally active. These structural details not only add to our understanding of the structure and function of the complexes but also provide information about the possible mechanism for the formation of the complexes during chloroplast biogenesis in developing leaves. During assembly of the thylakoid complexes and Rubisco, a major stromal enzyme complex, the proteins of individual complex, which are encoded by nuclear genome, are synthesized in the cytoplasm and targeted to chloroplasts through a complex transport machine consisting of a translocon at the outer envelope membrane of chloroplasts (TOC) and a translocon at the inner envelope membrane of chloroplasts (TIC). Both the translocons are hetero-oligomeric protein complexes that transport nuclear-encoded proteins in a tightly regulated manner. The structural features of translocons involving receptors, channels, motors, and regulators are largely known. Because of the differential location, the mechanisms of targeting are different for different complexes (Biswal et al. 2003; Flores-Pérez and Jarvis 2013; Ling et al. 2013).

5.3.1 Organization and Assembly of Photosystem II, a Well-Examined Thylakoid Complex

Among the individual complexes, the assembly of PSII has been widely studied in recent years (Nyitrai 1999; Biswal et al. 2003; Nield and Barber 2006; Nickelsen and Rengstl 2013). PSII usually exists in dimeric form. A monomer of PSII contains at least 27–28 subunits (Dekker and Boekema 2005). The major intrinsic protein subunits of the PSII complex are D1, D2, cytochrome b559, CP43, and CP47. These proteins of the core complex are inserted on thylakoids with other proteins and nonprotein components in a definite sequence (Biswal et al. 2003). The extrinsic proteins of molecular weights 33, 23, and 16 kDa are encoded by nuclear genes, synthesized in cytoplasm as high-molecular-weight precursors, processed and transported through the chloroplast envelope and subsequently to the thylakoid membrane (Biswal et al. 2003). Finally, the proteins are targeted to the lumen and are attached to the intrinsic core complex to form the functional complex. It appears that formation of several subassembly complexes determines the sequence and stability of the complex during the assembly process (Cai et al. 2010; Rochaix 2011). Assembly of PSII requires several cofactors, including a number of low-molecular-weight proteins, redox components, metal ions, and photosynthetic pigments

(Rochaix 2011). The stability and assembly of the photosystem largely depend on the coordinated synthesis of proteins and pigments, including chlorophylls and different forms of carotenoids (Biswal et al. 2003, 2013; see review by Nickelsen and Rengstl 2013). It is proposed that some of the protein subunits may remain stable in the absence of other subunits of the complex but cannot have a proper orientation on lamellar bilayer membranes (Sutton et al. 1987). The synthesis, regulation, and assembly of both intrinsic and extrinsic proteins and their final insertion to the PSII core complex were examined in detail in both *in vitro* and *in vivo* conditions (for a review, see Biswal et al. 2003).

5.3.2 Assembly of Photosystem I, Cytochrome b_6f Complex, and ATP Synthase

PSI complex, in addition to several protein subunits, contains a large number of cofactors, including redox centers. (Ke 2001; Biswal et al. 2003; Hihara and Sonoike 2013). The photosystem is composed of 19 protein subunits in higher plants (Croce 2012), and the assembly process is known to be regulated primarily by the nuclear gene products (Biswal et al. 2003).

Through several mutational studies and biochemical analyses, Cytb_6f complex, which interlinks both the photosystems, and ATP synthase, responsible for ATP synthesis through a highly sophisticated motor device mechanism, appear to follow more or less a similar pattern of the assembly process, as demonstrated during assembly of the photosystems (Biswal et al. 2003). Steps like heme attachment, synthesis, and binding of the iron–sulfur centers and other cofactors modulate the assembly of the Cytb_6f complex (Biswal et al. 2003).

5.3.3 Possible Mechanism of Assembly of Thylakoid Complexes

The molecular mechanisms of assembly of thylakoid and stromal complexes largely remain unclear. Extensive genetic and biochemical studies, however, provide some basic information to elucidate the involvement of several steps involved in the assembly process. The coordinated regulation of synthesis and insertion of protein subunits of individual complexes of the thylakoid membrane is critically discussed by Rochaix (2011). The insertion of an anchor protein subunit on the membrane may be the initial step in the assembly process. The nature of recognition of the anchoring protein by thylakoid membrane on the appropriate site, however, remains obscured. The attachment of the anchoring protein subsequently facilitates synthesis and association of other protein subunits of the complex in a sequence (Rochaix 2011). It is likely that the anchoring proteins regulate the next protein(s) to be attached to the complex in the sequence. This mechanism explains the controlled insertion of the right protein at the right time of the assembly process during complex formation. During assembly of PSII complex, D2, the intrinsic

subunit of the photosystem, is recognized as an anchoring protein. The insertion of the protein is followed by the insertion of other subunits in the core complex, including D1 and antenna proteins (Cai et al. 2010; Rochaix 2011). Similarly, PsaB of PSI complex is initially inserted, followed by insertion of other subunits, including PsaA of the complex in a sequence (Rochaix 2011; Hihara and Sonoike 2013). This pattern of sequential assembly of protein subunits is most likely followed during assembly of other thylakoid complexes.

5.3.4 Rubisco, an Important Stromal Enzyme: Synthesis, Assembly, and Regulation by Chaperones

Rubisco, an important stromal enzyme of the Calvin cycle, has been extensively studied from various angles (Andersson and Backlund 2008; Bracher et al. 2011; Durão et al. 2015). It is the major source of nitrogen in green plants. Very often, the enzyme is also used as a model to examine the coordinated interaction and regulation of nuclear and plastid genes. Its structure–function relationship and regulation are reviewed (Spreitzer and Salvucci 2002; Andersson and Backlund 2008). The enzyme is composed of equal numbers of RbcL subunits encoded by plastid genes and RbcS subunits encoded by nuclear genes. The RbcS subunits are synthesized as precursors in the cytoplasm, processed, and targeted to the organelle, where they bind with RbcL subunits and take up a hexadecameric form of the holoenzyme (Biswal et al. 2003). The picture of the regulatory network associated with the coordinated synthesis of both the subunits so far remains hazy, although several internal and external factors are known to participate in the regulation process. Optimization and controlled synthesis of both RbcL and RbcS encoded by different organellar genomes are mostly managed by the interorganellar communication systems. Among the environmental factors, light control of the enzyme synthesis primarily through photoreceptors like phytochrome and blue light receptors is well known (Biswal et al. 2003).

Although the precise nature and sequence of various steps of the assembly process still remain unclear, the modulation of the steps by different chaperones has been extensively examined in the past (Roy and Gilson 1997; Roy 2013). The discovery of the crystallographic structure of a chaperone-bound assembly intermediate by Bracher et al. (2011) significantly adds toward the understanding of the assembly process of the enzyme. The chaperones facilitate the aggregation of RbcL and stabilize it by making RbcL–RbcX intermediate complex. The chaperone is assumed to modulate the geometry of the aggregation and stability of RbcL subunits that favor the formation of the enzyme complex with RbcS subunits encoded by nuclear genes. The initial steps of the assembly process involve the chaperone-induced dimerization of RbcL subunits, which leads to the formation of $RbcL_8RbcX_8$ complex (Figure 5.3). In the final steps of the assembly, the RbcS subunits join in the assembly process, and their binding

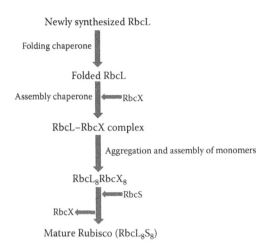

Newly synthesized RbcL

Folding chaperone

Folded RbcL

Assembly chaperone ← RbcX

RbcL–RbcX complex

Aggregation and assembly of monomers

$RbcL_8RbcX_8$

RbcX ← RbcS

RbcX ←

Mature Rubisco ($RbcL_8S_8$)

FIGURE 5.3 Possible steps during the assembly of ribulose bisphosphate carboxylase/oxygenase (Rubisco). The chaperone-mediated initial steps result in the aggregation of a large subunit (RbcL) with a specific chaperone, namely, RbcX, forming an $RbcL_8RbcX_8$ intermediate. The final event involves in the insertion of RbcS subunits to the intermediate. On binding RbcS, RbcX is released, which results in the formation of mature Rubisco. (Modified from Springer Science+Business Media: *Plastid Development in Leaves During Growth and Senescence*, ed. B. Biswal, K. Krupinska, and U.C. Biswal, 117–129, Advances in Photosynthesis and Respiration, vol. 36, Rubisco assembly: A research memoir, 2013, Roy, H.)

knocks out the chaperone from the intermediate complex, leading to completion of the assembly process (Bracher et al. 2011; Roy 2013).

5.4 CHANGES IN CHLOROPLASTS DURING LEAF SENESCENCE: FORMATION OF GERONTOPLASTS

Senescence in green leaves is the terminal phase of the organ's development. The dismantling and degradation of cellular organelles of the leaf are part of the senescence program (Biswal and Biswal 1988). Chloroplast is the first organelle to show extensive degradative symptoms, including alterations in structural fabrics of the thylakoid membrane with degradation of pigments and proteins of the organelle. Some of the cellular organelles, including the nucleus and mitochondria, remain relatively stable till the final stage of leaf senescence. Studies in the area of the molecular biology of plant senescence have unambiguously defined plant senescence as a genetic program of plant development, and the process is tightly regulated by the senescence-associated genes (*SAGs*) and environmental cues (Biswal et al. 2003; Buchanan-Wollaston et al. 2005; Guiboileau et al. 2010; Fischer 2012; Biswal et al. 2013; Kusaba et al. 2013; Ay et al. 2014; Penfol and Buchanan-Wollaston 2014). The execution and completion of the senescence program need energy. Because of this, the stability of mitochondria to supply energy is absolutely essential for completion of the process (Biswal and Biswal 1988; Biswal et al. 2012a).

The early events associated with the degradative changes in chloroplasts during leaf senescence are physiologically significant, specifically for nutrient recycling (Biswal et al. 2003; Guiboileau et al. 2010; Juvany et al. 2013).

5.4.1 ULTRASTRUCTURAL CHANGES OF CHLOROPLASTS

Senescence-induced ultrastructural modifications and changes in molecular composition of thylakoids have been extensively examined by electron microscopy, x-ray diffraction, immunological techniques, and absorption and fluorescence techniques in different plant systems (Biswal and Biswal 1988; Biswal et al. 2003; Lichtenthaler 2013; Mulisch and Krupinska 2013). The gradual and regulated disorganization of thylakoid membranes as probed by electron microscopy appears to be sequential, starting with the unstacking of grana thylakoids as the first event, followed by the formation of loose and elongated lamellae. These loose lamellae subsequently undergo massive degradation with the concomitant formation of plastoglobuli, the degradation products of thylakoids (Biswal and Biswal 1988; Lichtenthaler 2013). The plastoglobuli are observed to increase in size and number with the progress of senescence. These plastoglobuli are composed primarily of lipids and are possibly transported to cytoplasm through the chloroplast envelope for further degradation (Krupinska et al. 2012; Mulisch and Krupinska 2013). The details of the sequential changes in the ultrastructure of thylakoids are shown in Figure 5.4.

5.4.2 DEGRADATION OF PHOTOSYNTHETIC PIGMENTS

Reports published so far on the enzymatic degradation of individual pigments are extensively reviewed (Biswal 1995; Matile et al. 1999; Biswal et al. 2003; Hörtensteiner and Kräutler 2011).

5.4.2.1 Tracing the Path of Chlorophyll Degradation: Enzymes Involved and the Role of Pheophorbide a Oxygenase

The degradation of chlorophyll has been accepted as one of the major events of thylakoid disorganization during leaf senescence (Biswal et al. 2003; Hörtensteiner and Kräutler 2011; Hörtensteiner 2013; Thomas and Ougham 2014). In fact, the loss of chlorophyll, which gradually changes the leaf color from green to yellow, is considered a visible syndrome of senescence. The degradation of the pigments is mediated through different steps with sequential participation of enzymatic and nonenzymatic reactions that lead to the formation of several intermediates and chlorophyll catabolites. The degradation pathway is likely to involve several cellular compartments, namely, chloroplasts, cytoplasm, and the central vacuole. The central vacuole may be considered as the terminal compartment for the degradation of the pigments (Hörtensteiner 2013). The precise nature of communications between the compartments and the transport of degradative intermediates remains unclear. The individual steps of the degradation pathway are

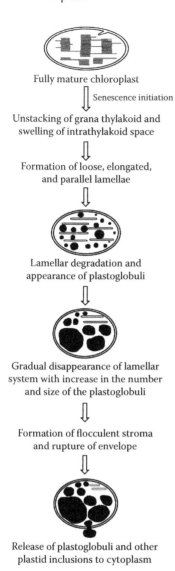

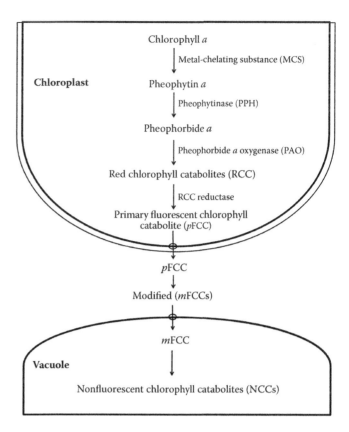

FIGURE 5.5 Pheophorbide *a* oxygenase (PAO) pathway of chlorophyll degradation during leaf senescence, showing several chlorophyll catabolites, the enzymes involved, and their location. The final events associated with the degradation pathway possibly occur in the vacuole. (Modified from Biswal, U.C. et al., *Chloroplast Biogenesis: From Proplastid to Gerontoplast*, Kluwer Academic Publishers, Dordrecht, Netherlands, 2003; Modified from Springer Science+Business Media: *Plastid Development in Leaves During Growth and Senescence*, ed. B. Biswal, K. Krupinska, and U.C. Biswal, 363–392, Advances in Photosynthesis and Respiration, vol. 36, The pathway of chlorophyll degradation: Catabolites, enzymes and pathway regulation, 2013, Hörtensteiner, S.)

FIGURE 5.4 Structure and organization of thylakoids in the mature chloroplasts and their degradation with the progress of senescence. The unstacking of grana thylakoids and the subsequent formation of loose, elongated, and parallel lamellae are the few initial steps. Subsequently, the lamellar systems are rapidly degraded with simultaneous appearance of plastoglobuli, which increase in size and number and thereafter are released to cytoplasm. (Modified from Springer Science+Business Media: *Photosynthesis: Plastid Biology, Energy Conversion and Carbon Assimilation*, ed. J.J. Eaton-Rye, B.C. Tripathy, and T.D. Sharkey, 217–230. Advances in Photosynthesis and Respiration, vol. 34, Leaf senescence and transformation of chloroplasts to gerontoplasts, 2012, Biswal, B. et al.; Modified from Springer Science+Business Media: *Plastid Development in Leaves During Growth and Senescence*, ed. B. Biswal, K. Krupinska, and U.C. Biswal, 337–361, Advances in Photosynthesis and Respiration, vol. 36, Plastoglobuli, thylakoids, chloroplast structure and development of plastids, 2013, Lichtenthaler, H.K.; Modified from Springer Science+Business Media: *Plastid Development in Leaves During Growth and Senescence*, ed. B. Biswal, K. Krupinska, and U.C. Biswal, 307–335, Advances in Photosynthesis and Respiration, vol. 36, Ultrastructural analyses of senescence associated dismantling of chloroplasts revisited, 2013, Mulisch, M., and K. Krupinska.)

shown in Figure 5.5. The pathway is commonly known as the pheophorbide *a* oxygenase (PAO) pathway (Hörtensteiner 2013), because the PAO is the key enzyme that is primarily responsible for opening the macrocycle ring of the pigment for further degradation and production of terminal chlorophyll catabolites. Chlorophyll *a*, recognized by the PAO pathway, is the substrate for initiation of pigment degradation. Chlorophyll *b* during senescence is converted to chlorophyll *a* through a complex interconversion system (Hörtensteiner 2013). The enzymatic steps subsequently participating in the degradation pathway are shown in Figure 5.5.

The reports of initial steps of chlorophyll breakdown especially during Mg dechelation and dephytylation during senescence are rather confusing (Hörtensteiner 2013). Chlorophyll *a* is suggested to be converted to pheophytin *a* through metal-chelating substance (MCS), and pheophytin *a* subsequently produces pheophorbide *a* by pheophytinase (PPH). The next step of the chlorophyll degradation pathway involves the participation of PAO, which, in combination with another enzyme, red chlorophyll catabolite reductase (RCCR), is responsible for the opening of the ring structure of the pigment, resulting in red chlorophyll catabolite (RCC) production. The cleavage of the ring results in the loss of green color of the pigment. The enzyme is specific to the senescence process. The product RCC, in a series of reactions, is converted to fluorescent chlorophyll catabolites (FCCs), which are subsequently modified and converted to nonfluorescent chlorophyll catabolites (NCCs).

5.4.2.2 Carotenoid Degradation

Carotenoids are relatively stable during senescence (Biswal and Biswal 1984; Biswal 1995). Although slow, the loss of pigments is distinctly marked with the progress of senescence. Not much is known about the enzymes that participate in the degradation of carotenoids in spite of the reports available on qualitative changes of the pigments, like the formation of carotenoid esters and epoxides (Biswal et al. 1994). Kusaba et al. (2009) suggested the possible conversion of lutein to lutein 3-acetate during senescence of rice leaves. The possibility of enzymatic participation, identification of the enzymes, and their regulation for quantitative loss of these pigments is described by Biswal et al. (2003).

5.4.2.3 Coordinated Breakdown of Pigment–Protein Complexes during Leaf Senescence: Possible Regulation

The general kinetic pattern of loss in pigments and pigment-binding membrane proteins remains more or less the same, suggesting a common point in their degradation mechanism (Biswal 1995). Since these pigments exist in the form of complexes with proteins, dislocation or breakdown of any individual component may lead to the destabilization of the complex, which is the prerequisite for enzymatic degradation of individual components. It appears that the structural status of different pigment–protein complexes may play a key role in coordinating the loss of photosynthetic pigments and proteins during senescence. The possibility of senescence-induced modification in the structure of the light-harvesting protein complex and a change in the topology of the pigments on the protein with consequent loss of pigments has been proposed in the chloroplasts of wheat leaves (Joshi et al. 1994). Although the precise nature of destabilization of pigment–protein complex still remains unclear, the possible regulatory factors for the degradation have been recently suggested (Hörtensteiner 2013).

5.4.3 Disassembly of Thylakoid Complexes and Loss in Primary Photochemical Reactions

The complexes of the thylakoid membrane that participate in the electron transport systems and phosphorylations are demonstrated to be destabilized during leaf senescence, most likely in an ordered sequence (Biswal and Biswal 1988; Tang et al. 2005). In most of the plant systems, leaf senescence is demonstrated to cause earlier and rapid loss of photochemical activities associated with PSII compared to PSI activities (Biswal et al. 2003). There could be several

factors contributing to the rapid degradation of the PSII of chloroplasts. The loss in oxygen evolution associated with PSII but retention of PSII activity measured by 2,6-dichlorophenol indophenol photoreduction in chloroplasts with an exogenous electron donor like diphenyl carbazide during leaf senescence may suggest relative instability of the oxygen-evolving system compared to the reaction center core complex of the photosystem (Biswal and Biswal 1988; Mohapatra et al. 2013). The exact nature of senescence-induced loss in the oxygen-evolving capacity of chloroplasts is not known. Senescence-induced distortion of the Mn cluster and release of Mn as observed by Margulies (1971) may be a factor directly affecting oxygen evolution. The loss of Mn may be the consequence of the senescence-induced loss of a 33 kDa extrinsic protein that is known to stabilize Mn binding to the PSII reaction center core complex of the photosystem. The loss of this extrinsic protein as immunologically analyzed has been demonstrated during leaf senescence of *Festuca pratensis* (Nock et al. 1992). Experiments conducted during leaf senescence of barley also suggest a parallel loss of extrinsic proteins and a decline in oxygen evolution (Choudhury and Imaseki 1990). It is assumed that the loss of the proteins may lead to destabilization of the topology of Mn on PSII proteins, leading to the inactivation of the oxygen-evolving system. The reaction center core complex may start showing signs of deterioration, contributing to the total loss of PSII photochemistry during the final phase of senescence. The core complex may be damaged by either quantitative loss of reaction center proteins (Biswal et al. 1994; Prakash et al. 2001a) or their structural modifications (Joshi et al. 1993). Senescence-induced loss and disorganization of the light-harvesting system may be another factor contributing to the loss in the primary photochemistry of the photosystem (Prakash et al 2001b).

The photochemical reactions associated with PSI decline during senescence. However, the rate of decline is relatively slow compared to PSII. The decline is attributed to the inactivation and loss of plastocyanin and NADP reductase (Biswal and Biswal 1988). The quantitative loss and/or inactivation of some of the components of the intersystem electron transport chain, namely, plastoquinones associated with PSII and plastocyanins with PSI, two important shuttling molecules between the photosystems, is likely to cause the loss in the primary photochemistry of thylakoids during senescence (Biswal and Biswal 1988; Grover 1993; Mae et al. 1993). The precise nature of dismantling of ATP synthase is not known, although senescence-induced loss in photophosphorylation and loss of some of the protein subunits of the complex are reported (Biswal et al. 2003).

5.4.4 Decline in Rubisco Activity and Loss of the Enzyme Protein: Participation of Proteases and the Vesicular Transport System

Senescence-induced changes in the activities of many stromal enzymes in different plant systems are known (Lauriere 1983;

Biswal et al. 2003) Rubisco, a major enzyme of the stroma, is known to rapidly degrade during the initial phase of senescence (Lauriere 1983; Biswal and Biswal 1988; Grover 1993; Kato and Sakamoto 2013). The loss in the enzyme activity may be attributed primarily to the quantitative loss of the enzyme protein. The possibility of qualitative change of the enzyme resulting in its inactivation cannot be ignored (Biswal et al. 2003). It seems logical to suggest a senescence-induced alteration in the turnover rate of the enzyme. The turnover may preferentially shift more toward degradation than synthesis, thereby causing a loss in the level of the enzyme protein. The complete degradation of the protein may be mediated by senescence-induced activity of specific proteases located both inside the chloroplast and outside the organelle (Kato and Sakamoto 2013).

The literature on the mechanism of Rubisco degradation in stroma during senescence is still limited. Currently, new ideas are emerging on the degradation of this protein during different developmental phases of green plants, including senescence (Kato and Sakamoto 2013). A review by Kato and Sakamoto (2013) suggests several important proteases responsible for chloroplast protein degradation through multiple pathways that involve the activities of the proteases of chloroplasts and other cellular organelles. It is quite likely that initial steps for degradation of Rubisco occur in chloroplast itself, followed by transport of the transient products to cytoplasm and finally to the vacuole through the autophagic pathway. The initial steps of degradation of stromal proteins may involve the damage to the proteins induced by reactive oxygen species (ROS). Senescence-induced ROS production by thylakoids in light is possible. The ROS can oxidize the proteins, including Rubisco, within chloroplasts to fragments (Avila-Ospina et al. 2014), and these damaged enzyme proteins/fragments may be transported to cytoplasm through vesicular bodies (Costa et al. 2013; Wada and Ishida 2013), which deliver these products to the central vacuole for their final degradation.

5.5 SIGNALS REGULATING CHLOROPLAST BIOGENESIS AND SENESCENCE DURING LEAF DEVELOPMENT

The signaling network that regulates chloroplast development is not yet fully resolved. Although the literature on the environmental and intrinsic signals in regulating chloroplast development is rich, a clear picture of the interaction between these two types of signals and their impact on development is yet to emerge (Biswal et al. 2003; Pogson and Albrecht 2011; Fischer 2012; Jarvis and López-Zuez 2013). In addition to environmental factors, intrinsic factors like hormones, sugars, ROS, and interorganellar and intercellular interactions play pivotal roles in regulating plastid biogenesis and senescence (Pogson and Albrecht 2011; Juvany et al. 2013; Thomas 2013). This review will focus primarily on the current views on the induction and progression of chloroplast degradation during leaf senescence regulated by sugar and ROS signaling systems.

5.5.1 SUGAR SIGNALING

Historically, sugars are known as a cellular energy carrier and have an extensive cellular metabolic network. The metabolite provides organic carbon skeletons for the synthesis of essential structural and regulatory molecules of cells and also sustains cellular osmotic balance. In addition, sugars are currently recognized as an important group of signaling molecules that participate in a very intricate signaling network regulating plant development (Sheen 2014). The work during last two decades strongly supports involvement of sugar signaling in the regulation of plant development, senescence, and stress responses (see reviews by Wingler et al. 2006; Eveland and Jackson 2013; Granot et al. 2013; Thomas 2013; Lastdrager et al. 2014; Ruan 2014; Sheen 2014). On the other hand, the role of sugars in regulating senescence appears to be intricate, and its molecular mechanism is not precisely understood so far, in spite of serious attempts in this area of study currently made by several laboratories (see reviews by van Doorn 2008; Biswal et al. 2012a; Sheen 2014). It is the level of the cellular sugars that is believed to be sensed by sensors, followed by a sugar signaling cascade for developmental responses. The question of what really signals the induction of senescence in green leaves, low or high levels of sugar, remains unanswered so far. There is a strong body of evidence that indicates the induction of senescence by depletion of sugars, which is opposite to the view of induction of the process by their accumulation (van Doorn 2008). These two contradictory views have not been rationalized so far. On the other hand, Zwack and Rashotte (2013) have tried to explain sugar-induced (lower or higher) modulation of senescence through a source–sink relationship.

5.5.1.1 Depletion of Sugars and Induction of Senescence

In green plants, chloroplasts, through photosynthesis, play a pivotal role in regulating plant development and aging (Biswal et al. 2003; Krupinska et al. 2013). In fact, an age-dependent decline in photosynthesis has been suggested as a possible factor for induction of senescence in green leaves (Mohapatra et al. 2010; Biswal et al. 2012a). The loss of photosynthesis as a signal for the induction of senescence in green leaves is evident during sequential senescence in the canopy of individual plants (see reviews by Smart 1994; Biswal 1999; Biswal et al. 2003). The lower leaves in a canopy receive light that is different in quality and quantity when compared to the light received by the upper leaves in the canopy of the plant body. The light transmitted through and reflected from the upper leaves is enriched with the far-red component with a loss in photosynthetically active radiation. This may result in the downregulation of photosynthesis, which causes induction of senescence. Reports on repression of photosynthesis-associated genes (PAGs) and, consequently loss in photosynthesis and induction of SAGs, support this proposition (Biswal et al. 2012a,b). The precise nature of signaling molecules that interplay between the decline in photosynthesis and induction of senescence, however, still remains unclear, although the sizable literature available indicates the involvement of sugars as the signaling molecules (see review by Biswal et al. 2012a). It is proposed that a decline in photosynthesis and, consequently, loss of sugars may be a signal that initiates induction of senescence. This hypothesis is supported by the observation on induction of senescence by dark-induced depletion of sugars and its suppression by addition of exogenous sugars (van Doorn 2008). Dark treatment also causes expression of several *SAGs* (Fujiki et al. 2001, 2005; Lee et al. 2004; Buchanan-Wollaston et al. 2005), which are repressed by addition of sugars (Fujiki et al. 2001; Lee et al. 2004). These findings may, therefore, assign a link between age-dependent decline in photosynthesis, the consequent loss of sugars, and induction of senescence. We have examined the possibility of whether loss of photosynthesis leading to sugar depletion could be a factor for induction and execution of the senescence program in *Arabidopsis* leaves. The question that is addressed is as follows: how do senescing leaves reprogram sugar metabolism with the loss of photosynthetic production of sugars? The loss of sugars may act as a signal for induction of catabolic events associated with the breakdown of cellular macromolecules including starch,

lipids, proteins, and cell wall polysaccharides to produce sugars. In a sugar depletion environment, as discussed earlier, several *SAGs* participating in the catabolic network are expressed, and some of them, including dark-inducible gene 2 (*din2*), are expressed late during senescence (Fujiki et al. 2001, 2005); *din2* codes for β-glucosidase is responsible for the breakdown of cell wall polysaccharides (Biswal et al. 2012a). Since the cell wall remains intact till the late phase of senescence and *din2* coding for β-glucosidase is expressed late, it is quite possible that the breakdown of cell wall polysaccharides to sugars mediated by cell wall hydrolases is the last event for the remobilization and completion of the energy-dependent senescence program (Figure 5.6) (Fujiki et al. 2005; Mohapatra et al. 2010; Patro et al. 2014). In addition, we have also demonstrated a significant enhancement of other cell wall hydrolases that break down cell wall polysaccharides at the late phase of senescence (Pandey and Biswal unpublished).

5.5.1.2 Induction of Senescence by Excess Sugars

Contrary to the view of senescence induction by depletion of sugars, accumulation of excess sugars has been shown to

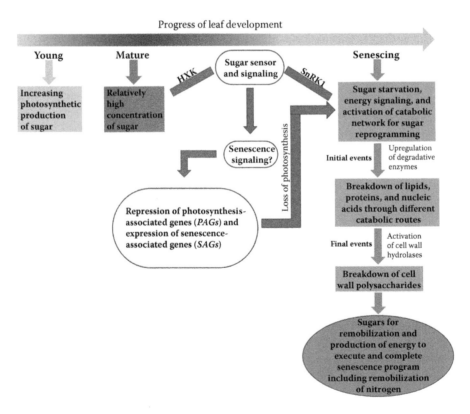

FIGURE 5.6 (**See color insert.**) Changes in sugar reprogramming during leaf development and senescence. The figure shows variation in the concentration of sugar during development and involvement of regulatory mechanisms mediated by sugar sensors for induction of senescence and cellular sugar homeostasis. Hexokinase (HXK) appears to initiate a senescence signaling cascade through a repression of photosynthesis-associated genes (*PAGs*) and expression of senescence-associated genes (*SAGs*). The loss of sugars creates a sugar starvation environment, which is recognized by another sugar sensor, namely, sucrose nonfermenting-1-related protein kinase 1 (SnRK1), which activates a catabolic network that participates in the degradation of starch, lipids, proteins, and nucleic acids to provide sugars for completion of the energy-dependent senescence program, including nutrient recycling. The final event of sugar reprogramming possibly includes the degradation of cell wall polysaccharides to provide sugars to energy-deficient senescing leaves. (Modified from Biswal, B. et al., Photosynthetic regulation of senescence in green leaves: Involvement of sugar signalling, in *Photosynthesis—Overviews on Recent Progress and Future Perspectives*, eds. S. Itoh, P. Mohanty, and K.N. Guruprasad, 245–260, IK International Publishing House Pvt, Ltd., New Delhi, India, 2012.)

induce the process in several plant systems (see reviews by van Doorn 2008; Biswal et al. 2012a).

5.5.1.3 Sugar Sensors and Mechanism of Signal Transduction

Through several mutational studies and biochemical analyses, some sugar sensors are identified, and their action on downstream components are partially known (Sheen 2014). It is likely that the sensors, of both high and low sugars, can sense the cellular sugar levels and act as a switch system for induction and repression of genes through a complex cellular signaling network during plant senescence for sugar homeostasis.

Two excellent current reviews are available describing different types of sugar sensors that appropriately sense the cellular sugar status and induce characteristic gene expression and metabolism for sugar reprogramming necessary during different phases of plant development and stress response (Thomas 2013; Sheen 2014). On the basis of current literature, Figure 5.6 depicts the possible involvement of sugar sensors and signaling during leaf development and initiation and progress of senescence, finally leading to necrosis. The rate of photosynthesis in green leaves varies with the development of the organ and the variation in the rate modulates of the level of cellular sugars. The gradual increase in the concentration of sugars after leaf emergence is likely to reach a peak when the leaves are fully mature, obviously with a high efficiency of photosynthesis. The relatively high concentration of sugars in mature leaves is likely to act as a signal for inhibition of photosynthesis by downregulating *PAGs* mediated by a feedback mechanism. Most likely, hexokinase (HXK), a sugar sensor, plays a role here for repression of *PAGs* through a glucose-repressive feedback loop. The repression leads to downregulation of photosynthesis and induction of *SAGs*. The loss in photosynthesis and, consequently, the decline in the level of cellular sugar, create an energy-deficient environment, which is sensed by another major sugar sensor, the sucrose nonfermenting-1-related protein kinase 1 (SnRK1), which is well studied by Sheen's group in Harvard Medical School in Boston (Baena-Gonzalez et al. 2007). The sensor is reported to activate several catabolic events by upregulating the relevant transacting factors to induce sugar reprogramming to meet the challenges of energy deficiency in cells. The catabolic network induced by the sensor may include several degradative enzymes responsible for the breakdown of proteins, lipids, and other macromolecules, converting them to sugars. In the absence of photosynthesis, production of sugars from other sources is absolutely necessary for completion of the energy-dependent senescence program, including nutrient recycling.

5.5.1.4 Cross-Talk of Sugar with Nitrogen, Phytohormones, and ROS in Signaling Network

The cross-talk of sugar signaling with nitrogen metabolism, the hormone regulatory network, and ROS signaling is not fully dissected yet, although nitrogen metabolism is extensively reported to play a dominant role in modulating sugar signaling in plants. Experiments with addition of sugars and low nitrogen supply suggest induction of senescence in leaves (Biswal et al. 2012a). One of the reasons for the induction of senescence could be attributed to accumulation of sugars in absence of its utility in a weak nitrogen metabolism sink. The reversal of low nitrogen-induced senescence by addition of exogenous nitrogen supports the proposition that the C/N ratio in leaves determines the nature of cross-talk between sugar and nitrogen signaling that regulates plant growth and development (Biswal et al. 2012a). In recent years, a scenario is emerging on cross-talk between sugars and phytohormones that participates in a signaling network regulating plant development. A close interaction between sugars and hormones like ethylene, abscisic acid (promoter of leaf senescence), and cytokinins (senescence retardant) is suggested to participate in the regulatory network (Lara et al. 2004; van Doorn 2004, 2008; Cho et al. 2010; Zwack and Rashotte 2013; Sheen 2014). Although the precise interaction between sugars and phytohormones is yet to be resolved, sugar-induced variation in the sensitivity of the hormones is a possibility (Biswal et al. 2012a). On the other hand, a review by Sheen (2014) has critically discussed, with available current literature, a close link between HXK-mediated sugar signaling and several phytohormones (Rolland et al. 2006; Huang et al. 2008; Ramon et al. 2008; Cho et al. 2010; Karve et al. 2012; Hsu et al. 2014).

5.5.2 ROLE OF ROS IN SIGNALING NETWORK OF CHLOROPLAST DEVELOPMENT

ROS are currently recognized as the major players of the signaling network that participate in the leaf developmental program starting right from leaf emergence till senescence (Juvany et al. 2013). Chloroplasts in green leaves are the major organelles that not only produce molecular oxygen in light but also generate several ROS with oxygen as the primary substrate. The presence of photosynthetic pigments capable of absorbing light and a well-organized long-chain electron transport system with several redox components potentially capable of producing ROS makes the organelle sensitive to the production of ROS.

5.5.2.1 Reactive Oxygen Species Production

Depending on the developmental status and environmental conditions, thylakoid complexes, primarily pigment-binding complexes like PSI and PSII of the photosynthetic organelle, can produce ROS. PSII, on development of excitation and redox pressure under certain developmental and environmental conditions, is capable of producing ^3Chl, which subsequently reacts with O_2 to produce $^1O_2$. On the other hand, under certain conditions, ferredoxin associated with PSI receives electrons from the electron transport system and converts O_2 to O_2^- radicals, which finally leads to formation of other oxy free radicals (Biswal et al. 2011). ROS with high oxidizing potential are well reported to oxidize and damage membrane lipids, proteins, and nucleic acids (Foyer and Noctor 2005; Biswal et al. 2011). Normally, the mature leaves in favorable environmental conditions maintain a cellular homeostasis with a perfect balance between the inputs (light energy absorbed) as the source and the outputs (energy utilization in the sink) (Biswal

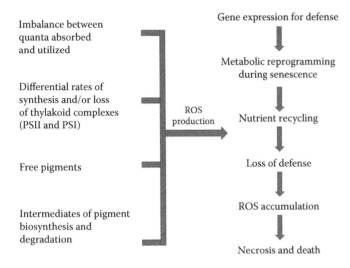

FIGURE 5.7 Chloroplast-related factors responsible for the production of highly toxic reactive oxygen species (ROS) and their damaging and signaling effects for defense during leaf development, senescence, and necrosis.

et al. 2011). There are several channels for utilization of energy absorbed, including the Calvin cycle for producing organic carbons as a major energy sink. However, the balance between the energy input and output is likely to be modulated during different phases of leaf development and environmental stresses (Biswal et al. 2011; Jarvis and López-Juez 2013). The energy imbalance may result in production of ROS. There are several other developmental factors that could contribute to produce ROS (Figure 5.7). For example, sequential appearance of PSI, PSII, and other electron transport carriers during plastid development and differential loss of PSI and PSII during senescence are very likely to change the energy balance and redox homeostasis (Biswal et al. 2011) (Figure 5.7). The selective damage to any component in the electron transport system during senescence similarly may result in the loss of redox homeostasis, leading to ROS production (Biswal et al. 2003). In addition, during either development or senescence, free unbound chlorophylls or the production of their intermediates during both synthesis and degradation are potentially photodynamic in nature and are likely to produce ROS in the presence of light. The question that has been addressed in last several years is as follows: how do the green plants counter this oxidative environment during different transitions of plant development? In addition to other factors, ROS-induced expression of several genes that very effectively counter this oxidative environment through production of several antioxidant systems, including the enzymes encoded by nuclear genes through the retrograde signaling network, is well reported (Foyer and Noctor 2005; Biswal et al. 2011).

5.5.2.2 ROS and Retrograde Signaling

ROS, through an intricate signaling network, regulate the induction of a very effective defense system, including antioxidant enzymes encoded by nuclear genes through well-known retrograde signaling to counter the toxic action of free

radicals during different phases of leaf development (Juvany et al. 2013). In addition to ROS-scavenging enzymes, including superoxide dismutase, peroxidase, and catalase, other effective antioxidants include glutathione, ascorbate, tocopherols, and several pigments (Foyer and Noctor 2005).

The molecular biology of ROS-induced antioxidant systems and their protective mechanisms against an oxidative environment are largely known (Apel and Hirt 2004; Foyer and Noctor 2005; Biswal et al. 2011; Jarvis and López-Juez 2013). But the interplay between ROS chemistry and the signaling response for the development of antioxidants is not fully resolved during leaf development in spite of the availability of models proposed by several laboratories (see reviews by Foyer and Noctor 2005; Biswal et al. 2011). A review by Juvany et al. (2013) provides a critical discussion about the developmental-dependent ROS production and induction of different kinds of antioxidants to meet the challenges of the oxidative environment during both leaf emergence and senescence. The retrograde signaling, however, is not straightforward. Chloroplast-linked retrograde signaling is modulated by redox signalings generated by other cellular organelles including mitochondria (Pogson et al. 2008). Mitochondria, with distinct features of an energy transduction system and several redox components potentially active for ROS production, can interfere in the retrograde signaling generated in the chloroplasts (Pogson et al. 2008). Because of the participation of several other cellular organelles in addition to chloroplasts, the retrograde signaling responsible for induction of nuclear genes for the production of antioxidants becomes complex. A second factor that adds to the complexity of the ROS signaling cascade is the specificity of different types of ROS and their differential signal transduction pathways. Some of the ROS may act through secondary messenger systems for the expression of nuclear genes for defense (Pogson et al. 2008).

5.5.2.3 Interaction of ROS with Sugar and Phytohormone Signalings

The complexity of ROS signalings becomes distinct when discussed in the background of its cross-talk with the signaling network of sugars, phytohormones, and other signaling elements. A variation in the level of cellular sugars is reported for ROS production and signaling during plant development and stress response (Bolouri-Moghaddam et al. 2010). A review by Juvany et al. (2013) provides another interesting discussion on the possible modulation of ROS production by the action of different photoreceptors, specifically during leaf emergence. These types of interactions make ROS signaling complex and do not permit us to get a clear picture of ROS signaling during plant development.

No doubt, reactive oxygen free radicals damage different cellular organelles and develop the antioxidant-mediated defense systems against oxidative damage. Nevertheless, ROS play a positive role in regulating leaf development, including initiation of senescence (Fischer 2012; Juvany et al. 2013). The degradation of cellular macromolecules mediated by the free radicals is absolutely required for nutrient recycling for completion of the senescence program. For example, literature is now available on the initial steps of the degradation pathway of stromal proteins including Rubisco induced by ROS. The oxygen free radicals are likely to cleave Rubisco into several fragments, the initial step in the degradation pathway of the major stromal enzyme (Avila-Ospina et al. 2014). The fragments in the degradation pathway finally move to a vacuole possibly through an autophagic and vacuolar delivery system (Avila-Ospina et al. 2014). On one hand, ROS induce antioxidants necessary against an oxidative environment, and on the other hand, free radicals are also necessary for degradation pathways essential for completion of developmental programs, including senescence. The balance between the two appears to be development dependent and is greatly influenced by environmental fluctuations. The precise regulation of an appropriate balance is absolutely required during plant development, and the mechanisms of the balance remain, so far, unknown. What we know is that the failure of induction of the antioxidant system and accumulation of ROS during the terminal phase of senescence lead to collapse of the balance and, finally, death (Juvany et al. 2013).

5.6 CONCLUSIONS: KNOWNS, UNKNOWNS, AND THE FUTURE

Rapid progress made in plant molecular biology, genomics, proteomics, and development of highly sophisticated techniques, especially in structural biology, has significantly contributed to our understanding of the events associated with chloroplast biogenesis and senescence. In spite of the large accumulation of data, a clear picture of plastid development and regulations has not emerged yet. There are several unknowns that need to be addressed in future.

1. The crystallographic structures of some of the thylakoid and stromal complexes have recently become known. Thylakoid complexes contain several nonprotein components, including pigments, redox components, and metal clusters in addition to proteins. We have a fairly good understanding of the topology of the nonprotein components on the proteins. The structures at atomic resolution also reveal the possible geometry of the individual complex. However, the mechanism of sequential and temporal assembly and integration of the nonprotein components on proteins of the individual complex to form the final functional complex during the formation of the chloroplasts is poorly understood. Importantly, the device that regulates the exact location and topology of nonprotein components on the protein surface has not been resolved. Several other factors, not necessarily associated with the structure and organization of the complexes, are currently characterized, and these factors possibly facilitate the correct targeting of the various components of the complex, but their mechanistic action in the assembly process is not fully dissected.

2. Lack of sufficient knowledge on the targeting of nuclear-encoded proteins to chloroplast is a major hurdle to having a clear understanding of protein transport during chloroplast biogenesis. The transport of the proteins occurs through a highly complex and sophisticated machine located on the plastid envelope. In spite of vast accumulation of data, we fail to understand the mechanistic action of TOC and TIC involved in targeting nuclear-encoded proteins to chloroplast. The directional movement of proteins to the correct location in chloroplast and the protein processing during transport largely remain unclear. The receptors, channels, motors, and regulators of both transporters located on the plastid envelope are characterized, but the integrated mechanism of their action for transport of individual proteins to different locations during assembly process still remains unclear.

3. Protease-mediated degradation of plastid proteins, both thylakoidal and stromal, is a common feature during both biogenesis and senescence of the organelle. But the exact mechanism of proteases and their location for complete degradation of individual proteins of the organelle is not elucidated specifically during senescence. There are a large number of current excellent papers on the involvement of the central vacuole as the terminal organelle of the degradation pathway. The initial events associated with the degradation pathway are suggested to occur in chloroplast, followed by transport of degradative or structurally damaged intermediates through different kinds of vesicular systems, like Rubisco-containing bodies (RCB) and senescence-associated vacuole (SAV), to vacuoles for completion of the degradation process. The confusion and contradictions about the nature of inclusions (degradative intermediates) of these vesicular bodies that are docked to the central vacuole need clarification in the future.

4. The involvement of sugar signaling in regulating chloroplast-to-gerontoplast transition during leaf senescence is discussed in this review with citation of recent literature. The level of cellular sugars is known to induce a signaling cascade. The question that has been raised in the review is the following: what really initiates the induction of the senescence process, high or low level of sugars? Both are reported to initiate senescence. The reports obviously are contradictory and need a clarification in the future. Second, the mechanism of the signaling at high or low sugar levels is yet to be understood. Through mutational and biochemical analyses, two major sugar sensors, namely, HXK and SnRK1, participating in the developmental program are identified and characterized. HXK is likely to regulate the senescence process through downregulation of photosynthetic genes and, consequently, loss in photosynthesis. On the other hand, SnRK1-mediated

transcription reprogramming is reported to induce several catabolic events for the production of sugars in a sugar-deficient environment for sugar reprogramming. Our knowledge in understanding of upstream and downstream regulations induced by both sugar sensors is still limited, and the coordination between these two sensors in the plant developmental signaling network remains unclear.

5. The ROS, in spite of their damaging effect on chloroplast, regulate the expression of several genes responsible for the development of defense mechanisms and therefore participate in the organellar developmental program. Most of the data published have emphasized that ROS originating from plastid induces the retrograde signaling during development of the organelle. But the interference of ROS production by other cellular organelles like mitochondria in the signaling network has not been seriously examined. Second, it is likely that different ROS follow different signaling pathways. The possible implication of the cross-talk between these pathways in regulating plastid development has not so far been addressed.

6. The participation of phytohormones, sugars, and ROS-induced signalings in the regulation of plant development and senescence are mostly examined separately. This review has, however, described their possible cross-talk. But the exact nature of their integration in the signaling network in initiating and executing different developmental programs largely remains unknown and needs attention in the future.

ACKNOWLEDGMENTS

The authors are very thankful to Professor U.C. Biswal for his suggestions and critical review during the preparation of the manuscript. They also thank Drs. M.K. Raval, P.N. Joshi, and Lalitendu Nayak for their comments during the preparation of the figures. The authors thank the Council of Scientific and Industrial Research (CSIR), New Delhi, for financial support (Emeritus Scientist project no. 21 (0886)/12-EMR II).

REFERENCES

Amunts, A., H. Toporik, A. Borovikova, and N. Nelson. 2010. Structure determination and improved model of plant photosystem I. *J. Biol. Chem.* 285:3478–86.

Andersson, I. and A. Backlund. 2008. Structure and function of Rubisco. *Plant Physiol. Biochem.* 46:275–91.

Apel, K. and H. Hirt. 2004. Reactive oxygen species: Metabolism, oxidative stress, and signal transduction. *Annu. Rev. Plant Biol.* 55:373–99.

Avila-Ospina, L., M. Moison, K. Yoshimoto, and C. Masclaux-Daubresse. 2014. Autophagy, plant senescence, and nutrient recycling. *J. Exp. Bot.* 65:3799–811.

Ay, N., B. Janack, and K. Humbeck. 2014. Epigenetic control of plant senescence and linked processes. *J. Exp. Bot.* 65:3875–87.

Baena-Gonzalez, E., F. Rolland, J.M. Thevelein, and J. Sheen. 2007. A central integrator of transcription networks in plant stress and energy signalling. *Nature* 448:938–42.

Barkan, A., R. Voelker, J. Mendel-Hartvig, D. Johnson, and M. Walker. 1995. Genetic analysis of chloroplast biogenesis in higher plants. *Physiol. Plant.* 93:163–70.

Biswal, B. 1995. Carotenoid catabolism during leaf senescence and its control by light. *J. Photochem. Photobiol. B* 30:3–13.

Biswal, B. 1999. Senescence-associated genes of leaves. *J. Plant Biol.* 26:43–50.

Biswal, B. and U.C. Biswal. 1984. Photocontrol of leaf senescence. *Photochem. Photobiol.* 39:875–9.

Biswal, B. and U.C. Biswal. 1988. Ultrastructural modifications and biochemical changes during senescence of chloroplasts. *Int. Rev. Cytol.* 113:271–321.

Biswal, B., L.J. Rogers, A.J. Smith, and H. Thomas. 1994. Carotenoid composition and its relationship to chlorophyll and D1 protein during leaf development in a normally senescing cultivar and a stay green mutant of *Festuca pratensis. Phytochemistry* 37:1257–62.

Biswal, B., U.C. Biswal, and M.K. Raval. 2003. *Chloroplast biogenesis: From proplastid to gerontoplast.* Dordrecht: Kluwer Academic Publishers.

Biswal, B., M.K. Raval, U.C. Biswal, and P.N. Joshi. 2008. Response of photosynthetic organelles to abiotic stress: Modulation by sulphur metabolism. In *Sulfur assimilation and abiotic stress in plants*, eds. N.A. Khan, S. Singh, and S. Umar, 167–91. Berlin, Heidelberg: Springer.

Biswal, B., P.N. Joshi, M.K. Raval, and U.C. Biswal. 2011. Photosynthesis, a global sensor of environmental stress in green plants: Stress signaling and adaptation. *Curr. Sci.* 101:47–56.

Biswal, B., P.K. Mohapatra, M.K. Raval, and U.C. Biswal. 2012a. Photosynthetic regulation of senescence in green leaves: Involvement of sugar signalling. In *Photosynthesis—overviews on recent progress and future perspectives*, eds. S. Itoh, P. Mohanty, and K.N. Guruprasad, 245–60. New Delhi, India: IK International Publishing House Pvt, Ltd.

Biswal, B., P.K. Mohapatra, U.C. Biswal, and M.K. Raval. 2012b. Leaf senescence and transformation of chloroplasts to gerontoplasts. In *Photosynthesis: Plastid biology, energy conversion and carbon assimilation*, eds. J.J. Eaton-Rye, B.C. Tripathy, and T.D. Sharkey, 217–30. Advances in Photosynthesis and Respiration, Vol. 34, Dordrecht: Springer.

Biswal, B., K. Krupinska, and U.C. Biswal (ed.). 2013. *Plastid development in leaves during growth and senescence.* Advances in Photosynthesis and Respiration, Vol. 36, Dordrecht: Springer.

Block, M.A., E. Dubots, and E. Maréchal. 2013. Glycerolipid biosynthesis and chloroplast biogenesis. In *Plastid development in leaves during growth and senescence*, eds. B. Biswal, K. Krupinska, and U.C. Biswal, 131–54. Advances in Photosynthesis and Respiration, Vol. 36, Dordrecht: Springer.

Bolouri-Moghaddam, M.R., K.L. Roy, L. Xiang, F. Rolland, and W.V.D. Ende. 2010. Sugar signalling and antioxidant network connections in plant cells. *FEBS J.* 277:2022–37.

Bracher, A., A. Starling-Windhof, F.U. Hartl, and M. Hayer-Hartl. 2011. Crystal structure of a chaperone-bound assembly intermediate of form I Rubisco. *Nat. Struct. Mol. Biol.* 18:875–80.

Buchanan-Wollaston, V., T. Page, E. Harrison et al. 2005. Comparative transcriptome analysis reveals significant differences in gene expression and signalling pathways between developmental and dark/starvation-induced senescence in *Arabidopsis. Plant J.* 42:567–85.

Cai, W., J. Ma, W. Chi et al. 2010. Cooperation of LPA3 and LPA2 is essential for photosystem II assembly in *Arabidopsis. Plant Physiol.* 154:109–20.

Cho, Y.H., J. Sheen, and S.D. Yoo. 2010. Low glucose uncouples hexokinase-dependent sugar signaling from stress and defense hormone abscisic acid and C_2H_4 responses in *Arabidopsis*. *Plant Physiol*. 152:1180–2.

Choudhury, N.K. and H. Imaseki. 1990. Loss of photochemical functions of thylakoid membranes and PS 2 complex during senescence of barley leaves. *Photosynthetica* 24:436–45.

Costa, M.L., D.E. Martínez, F.M. Gomez, C.A. Carrión, and J.J. Guiamet. 2013. Chloroplast protein degradation: Involvement of senescence-associated vacuoles. In *Plastid development in leaves during growth and senescence*, eds. B. Biswal, K. Krupinska, and U.C. Biswal, 417–33. Advances in Photosynthesis and Respiration, Vol. 36, Dordrecht: Springer.

Croce, R. 2012. Chlorophyll-binding proteins of higher plants and cyanobacteria. In *Photosynthesis: Plastid biology, energy conversion and carbon assimilation*, eds. J.J. Eaton-Rye, B.C. Tripathy, and T.D. Sharkey, 127–49. Advances in Photosynthesis and Respiration, Vol. 34, Dordrecht: Springer.

Croce, R. and H.V. Amerongen. 2011. Light-harvesting and structural organization of Photosystem II: From individual complexes to thylakoid membrane. *J. Photochem. Photobiol. B* 104:142–53.

Dekker, J.P. and E.J. Boekema. 2005. Supramolecular organization of thylakoid membrane proteins in green plants. *Biochim. Biophys. Acta* 1706:12–39.

Durão, P., H. Aigner, P. Nagy et al. 2015. Opposing effects of folding and assembly chaperones on evolvability of Rubisco. *Nat. Chem. Biol.* 11:148–55.

Eveland, A.L. and D.P. Jackson. 2012. Sugars, signalling, and plant development. *J. Exp. Bot.* 63:3367–77.

Fischer, A.M. 2012. The complex regulation of senescence. *Crit. Rev. Plant Sci.* 31:124–47.

Flores-Pérez, Ú. and P. Jarvis. 2013. Molecular chaperone involvement in chloroplast protein import. *Biochim. Biophys. Acta* 1833:332–40.

Foyer, C.H. and G. Noctor. 2005. Redox homeostasis and antioxidant signaling: A metabolic interface between stress perception and physiological responses. *Plant Cell* 17:1866–75.

Fujiki, Y., Y. Yoshikawa, T. Sato et al. 2001. Dark-inducible genes from *Arabidopsis thaliana* are associated with leaf senescence and repressed by sugars. *Physiol. Plant.* 111:345–52.

Fujiki, Y., Y. Nakagawa, T. Furumoto et al. 2005. The response to darkness of the late-responsive dark-inducible genes is positively regulated by leaf age and negatively by calmodulin antagonist–sensitive signalling in *Arabidopsis thaliana*. *Plant Cell Physiol.* 46:1741–6.

Goldschmidt-Clermont, M. 1998. Coordination of nuclear and chloroplast gene expression in plant cells. *Int. Rev. Cytol.* 177:115–80.

Granot, D., R. David-Schwartz, and G. Kelly. 2013. Hexose kinases and their role in sugar-sensing and plant development. *Front. Plant Sci.* 4:44.

Grover, A. 1993. How do senescing leaves lose photosynthetic activity? *Curr. Sci.* 64:226–33.

Guiboileau, A., R. Sormani, C. Meyer, and C. Masclaux-Daubresse. 2010. Senescence and death of plant organs: Nutrient recycling and developmental regulation. *C. R. Biol.* 333:382–91.

Hihara, Y. and K. Sonoike. 2013. Organization and assembly of photosystem I. In *Plastid development in leaves during growth and senescence*, eds. B. Biswal, K. Krupinska, and U.C. Biswal, 101–16. Advances in Photosynthesis and Respiration, Vol. 36, Dordrecht: Springer.

Hong, S. and R.J. Spreitzer. 1994. Nuclear mutation inhibits expression of the chloroplast gene that encodes the large subunit of ribulose-1,5-bisphosphate carboxylase/oxygenase. *Plant Physiol.* 106:673–8.

Hörtensteiner. S. 2013. The pathway of chlorophyll degradation: Catabolites, enzymes and pathway regulation. In *Plastid development in leaves during growth and senescence*, eds. B. Biswal, K. Krupinska, and U.C. Biswal, 363–92. Advances in Photosynthesis and Respiration, Vol. 36, Dordrecht: Springer.

Hörtensteiner, S. and B. Kräutler. 2011. Chlorophyll breakdown in higher plants. *Biochim. Biophys. Acta* 1807:977–88.

Hsu, Y.F., Y.C. Chen, Y.C. Hsiao et al. 2014. AtRH57, a DEAD-box RNA helicase, is involved in feedback inhibition of glucose-mediated abscisic acid accumulation during seedling development and additively affects pre-ribosomal RNA processing with high glucose. *Plant J.* 77:119–35.

Huang, Y., C.Y. Li, K.D. Biddle, and S.I. Gibson. 2008. Identification, cloning and characterization of sis7 and sis10 sugar-insensitive mutants of *Arabidopsis*. *BMC Plant Biol.* 8:104.

Inaba, T. and Y. Ito-Inaba. 2010. Versatile roles of plastids in plant growth and development. *Plant Cell Physiol.* 51:1847–53.

Jarvis, P. and E. López-Juez. 2013. Biogenesis and homeostasis of chloroplasts and other plastids. *Nat. Rev. Mol. Cell Biol.* 14:787–802.

Joshi, P.N., N.K. Ramaswamy, M.K. Raval et al. 1993. Alteration in photosystem II photochemistry of thylakoids isolated from senescing leaves of wheat seedlings. *J. Photochem. Photobiol. B* 20:197–202.

Joshi, P.N., B. Biswal, G. Kulandaivelu, and U.C. Biswal. 1994. Response of senescing wheat leaves to ultraviolet A light: Changes in energy transfer efficiency and PS II photochemistry. *Radiat. Environ. Biophys.* 33:167–76.

Juvany, M., M. Müller, and S. Munné-Bosch. 2013. Photo-oxidative stress in emerging and senescing leaves: A mirror image? *J. Exp. Bot.* 64:3087–98.

Kanamaru, K. and M. Sugita. 2013. Dynamic features of plastid genome and its transcriptional control in plastid development. In *Plastid development in leaves during growth and senescence*, eds. B. Biswal, K. Krupinska, and U.C. Biswal, 189–213. Advances in Photosynthesis and Respiration, Vol. 36, Dordrecht: Springer.

Kapoor, S., S.C. Maheshwari, and A.K. Tyagi. 1993. Organ specific expression of plastid-encoded genes in rice involves both quantitative and qualitative changes in m-RNAs. *Plant Cell Physiol.* 34:943–7.

Kapoor, S., S.C. Maheshwari, and A.K. Tyagi. 1994. Developmental and light dependent cues interact to establish steady-state levels of transcripts for photosynthesis-related genes (psbA, psbD, psaA and rbcL) in rice (*Oryza sativa* L.). *Curr. Genet.* 25:362–6.

Karve, A., X. Xia, and B.D. Moore. 2012. *Arabidopsis* hexokinase-like1 and hexokinase1 form a critical node in mediating plant glucose and ethylene responses. *Plant Physiol.* 158:1965–75.

Kato, Y. and W. Sakamoto. 2013. Plastid protein degradation during leaf development and senescence: Role of proteases and chaperones. In *Plastid development in leaves during growth and senescence*, eds. B. Biswal, K. Krupinska, and U.C. Biswal, 453–77. Advances in Photosynthesis and Respiration, Vol. 36, Dordrecht: Springer.

Ke, B. (ed.). 2001. *Photosynthesis photobiochemistry and photobiophysics*. Advances in Photosynthesis and Respiration, Vol. 10, Dordrecht: Kluwer Academic Publishers.

Krupinska, K., M. Mulisch, J. Hollmann et al. 2012. An alternative strategy of dismantling of the chloroplasts during leaf senescence observed in a high-yield variety of barley. *Physiol. Plant.* 144:189–200.

Krupinska, K., U.C. Biswal, and B. Biswal. 2013. The dynamic role of chloroplasts in integrating plant growth and development. In *Plastid development in leaves during growth and senescence*, eds. B. Biswal, K. Krupinska, and U.C. Biswal, 3–16. Advances in Photosynthesis and Respiration, Vol. 36, Dordrecht: Springer.

Kusaba, M., T. Maoka, R. Morita, and S. Takaichi. 2009. A novel carotenoid derivative, lutein 3-acetate, accumulates in senescent leaves of rice. *Plant Cell Physiol.* 50:1573–7.

Kusaba, M., A. Tanaka, and R. Tanaka. 2013. Stay-green plants: What do they tell us about the molecular mechanism of leaf senescence. *Photosynth. Res.* 117:221–34.

Lara, M.E.B., M.-C.G. Garcia, T. Fatima et al. 2004. Extracellular invertase is an essential component of cytokinin-mediated delay of senescence. *Plant Cell* 16:1276–87.

Lastdrager, J., J. Hanson, and S. Smeekens. 2014. Sugar signals and the control of plant growth and development. *J. Exp. Bot.* 65:799–807.

Lauriere, C. 1983. Enzymes and leaf senescence. *Physiol. Veg.* 21:1159–77.

Lee, E.-J., N. Koizumi, and H. Sano. 2004. Identification of genes that are up-regulated in concert during sugar depletion in *Arabidopsis*. *Plant Cell Environ.* 27:337–45.

Lichtenthaler, H.K. 2013. Plastoglobuli, thylakoids, chloroplast structure and development of plastids. In *Plastid development in leaves during growth and senescence*, eds. B. Biswal, K. Krupinska, and U.C. Biswal, 337–61. Advances in Photosynthesis and Respiration, Vol. 36, Dordrecht: Springer.

Liere, K. and T. Börner. Development-dependent changes in the amount and structural organization of plastid DNA. In *Plastid development in leaves during growth and senescence*, eds. B. Biswal, K. Krupinska, and U.C. Biswal, 215–37. Advances in Photosynthesis and Respiration, Vol. 36, Dordrecht: Springer.

Ling, Q., R. Trösch, and P. Jarvis. 2013. The ins and outs of chloroplast protein transport. In *Plastid development in leaves during growth and senescence*, eds. B. Biswal, K. Krupinska, and U.C. Biswal, 239–80. Advances in Photosynthesis and Respiration, Vol. 36, Dordrecht: Springer.

López-Juez, E. and K.A. Pyke. 2005. Plastids unleashed: Their development and their integration in plant development. *Int. J. Dev. Biol.* 49:557–77.

Mae, T., H. Thomas, A.P. Gay, A. Makino, and J. Hidema. 1993. Leaf development in *Lolium temulentum*: Photosynthesis and photosynthetic proteins in leaves senescing under different irradiances. *Plant Cell Physiol.* 34:391–9.

Margulies, M.M. 1971. Electron transport properties of chloroplasts from aged bean leaves and their relationships to the manganese content of the chloroplasts. In *Proceedings of the second international congress on photosynthesis research*, eds. G. Forti, M. Avron, and A. Melandri, 539–45. The Hague: Dr. W. Junk Publishers.

Matile, P., S. Hörtensteiner, and H. Thomas. 1999. Chlorophyll degradation. *Annu. Rev. Plant Physiol. Plant Mol. Biol.* 50:67–95.

Mohapatra, P.K., L. Patro, M.K. Raval, N.K. Ramaswamy, U.C. Biswal, and B. Biswal. 2010. Senescence induced loss in photosynthesis enhances cell wall β-glucosidase activity. *Physiol. Plant.* 138:346–55.

Mohapatra, P.K., P.N. Joshi, N.K. Ramaswamy et al. 2013. Damage of photosynthetic apparatus in the senescing basal leaf of *Arabidopsis thaliana*: A plausible mechanism of inactivation of reaction centre II. *Plant Physiol. Biochem.* 62:116–21.

Mulisch, M. and K. Krupinska. 2013. Ultrastructural analyses of senescence associated dismantling of chloroplasts revisited. In *Plastid development in leaves during growth and senescence*, eds. B. Biswal, K. Krupinska, and U.C. Biswal, 307–35. Advances in Photosynthesis and Respiration, Vol. 36, Dordrecht: Springer.

Nickelsen, J. and B. Rengstl. 2013. Photosystem II assembly: From cyanobacteria to plants. *Annu. Rev. Plant Biol.* 64:609–35.

Nield, J. and J. Barber. 2006. Refinement of the structural model for the photosystem II supercomplex of higher plants. *Biochim. Biophys. Acta* 1757:353–61.

Nock, L.P., L.J. Rogers, and H. Thomas. 1992. Metabolism of protein and chlorophyll in leaf tissue of *Festuca pratensis* during chloroplast assembly and senescence. *Phytochemistry* 31:1465–70.

Nyitrai, P. 1997. Development of functional thylakoid membranes: Regulation by light and hormones. In *Handbook of photosynthesis*, ed. M. Pessarakli, 391–406. New York: Marcel Dekker.

Patro, L., P.K. Mohapatra, U.C. Biswal, and B. Biswal. 2014. Dehydration induced loss of photosynthesis in *Arabidopsis* leaves during senescence is accompanied by the reversible enhancement in the activity of cell wall β-glucosidase. *J. Photochem. Photobiol. B* 137:49–54.

Penfol, C.A. and V. Buchanan-Wollaston. 2014. Modelling transcriptional networks in leaf senescence. *J. Exp. Bot.* 65:3859–73.

Pogson, B.J. and V. Albrecht. 2011. Genetic dissection of chloroplast biogenesis and development: An overview. *Plant Physiol.* 155:1545–51.

Pogson, B.J., N.S. Woo, B. Forster, and I.D. Small. 2008. Plastid signalling to the nucleus and beyond. *Trends Plant Sci.* 13:602–9.

Pogson, B.J., D. Ganguly, and V. Albrecht-Borth. 2015. Insights into chloroplast biogenesis and development. *Biochim. Biophys. Acta.* 1847:1017–24.

Powikrowska, M., S. Oetke, P.E. Jensen, and K. Krupinska. 2014. Dynamic composition, shaping and organization of plastid nucleoids. *Front. Plant Sci.* 5:424.

Prakash, J.S.S., M.A. Baig, and P. Mohanty. 2001a. Differential changes in the steady state levels of thylakoid membrane proteins during senescence in *Cucumis sativus* cotyledons. *Z. Naturforsch. C* 56:585–92.

Prakash, J.S.S., M.A. Baig, and P. Mohanty. 2001b. Senescence induced structural reorganization of thylakoid membranes in *Cucumis sativus* cotyledons; LHC II involvement in reorganization of thylakoid membranes. *Photosynth. Res.* 68:153–61.

Ramon, M., F. Rolland, and J. Sheen. 2008. Sugar sensing and signaling. *The Arabidopsis Book* 6:e0117.

Raval, M.K., B.K. Mishra, B. Biswal, and U.C. Biswal. 2013. Chloroplast development: Time, dissipative structures and fluctuations. In *Plastid development in leaves during growth and senescence*, eds. B. Biswal, K. Krupinska, and U.C. Biswal, 17–35. Advances in Photosynthesis and Respiration, Vol. 36, Dordrecht: Springer.

Rochaix, J.D. 2011. Assembly of the photosynthetic apparatus. *Plant Physiol.* 155:1493–500.

Rochaix, J.D. 2014. Regulation and dynamics of the light-harvesting system. *Annu. Rev. Plant Biol.* 65:287–309.

Rolland, F., E. Baena-Gonzalez, and J. Sheen. 2006. Sugar sensing and signaling in plants: Conserved and novel mechanisms. *Annu. Rev. Plant Biol.* 57:675–709.

Roy, H. 2013. Rubisco assembly: A research memoir. In *Plastid development in leaves during growth and senescence*, eds. B. Biswal, K. Krupinska, and U.C. Biswal, 117–29. Advances in Photosynthesis and Respiration, Vol. 36, Dordrecht: Springer.

Roy, H. and M. Gilson. 1997. Rubisco and the chaperonins. In *Handbook of Photosynthesis*, ed. M. Pessarakli, 295–304. New York: Marcel Dekker.

Ruan, Y.L. 2014. Sucrose metabolism: Gateway to diverse carbon use and sugar signaling. *Annu. Rev. Plant Biol.* 65:33–67.

Sheen, J. 2014. Master regulators in plant glucose signaling networks. *J. Plant Biol.* 57:67–79.

Sitte, P. 1977. Chromoplasten—Bunte Objekte der modernen Zellbiologie. *BIUZ* 7:65–74.

Smart, C.M. 1994. Gene expression during leaf senescence. *New Phytol.* 126:419–48.

Spreitzer, R.J. and M.E. Salvucci. 2002. Rubisco: Structure, regulatory interactions, and possibilities for a better enzyme. *Annu. Rev. Plant Biol.* 53:449–75.

Suga, M., F. Akita, K. Hirata et al. 2015. Native structure of photosystem II at 1.95 Å resolution viewed by femtosecond X-ray pulses. *Nature* 517:99–103.

Sutton, A., L.E. Sieburth, and J. Bennett. 1987. Light dependent accumulation and localization of photosystem II proteins in maize. *Eur. J. Biochem.* 164:571–8.

Tang, Y., X. Wen, and C. Lu. 2005. Differential changes in degradation of chlorophyll–protein complexes of photosystem I and photosystem II during flag leaf senescence of rice. *Plant Physiol. Biochem.* 43:193–201.

Taylor, W.C. 1989. Regulatory interactions between nuclear and plastid genomes. *Annu. Rev. Plant Physiol. Plant Mol. Biol.* 40:211–33.

Thomas, H. 2013. Senescence, ageing and death of the whole plant. *New Phytol.* 197:696–711.

Thomas, H. and H. Ougham. 2014. The stay-green trait. *J. Exp. Bot.* 65:3889–900.

Umena, Y., K. Kawakami, J.R. Shen, and N. Kamiya. 2011. Crystal structure of oxygen-evolving photosystem II at a resolution of 1.9 Å. *Nature* 473:55–60.

van Doorn, W.G. 2004. Is petal senescence due to sugar starvation? *Plant Physiol.* 134:35–42.

van Doorn, W.G. 2008. Is the onset of senescence in leaf cells of intact plants due to low or high sugar levels? *J. Exp. Bot.* 59:1963–72.

Wada, S. and H. Ishida. 2013. Autophagy of chloroplasts during leaf senescence. In *Plastid development in leaves during growth and senescence*, eds. B. Biswal, K. Krupinska, and U.C. Biswal, 435–51. Advances in Photosynthesis and Respiration, Vol. 36, Dordrecht: Springer.

Wingler, A., S. Purdy, J.A. MacLean, and N. Pourtau. 2006. The role of sugars in integrating environmental signals during the regulation of leaf senescence. *J. Exp. Bot.* 57:391–9.

Wise, R.R. and J.K. Hoober (ed.). 2006. *The structure and function of plastids*. Advances in Photosynthesis and Respiration, Vol. 23, Dordrecht: Springer.

Zwack, P.J. and A.M. Rashotte. 2013. Cytokinin inhibition of leaf senescence. *Plant Signal. Behav.* 8:e24737.

6 Physicochemical Properties of Chlorophylls and Bacteriochlorophylls

Masami Kobayashi, Yuhta Sorimachi, Daisuke Fukayama,
Hirohisa Komatsu, Terumitsu Kanjoh, Katsuhiro Wada, Masanobu Kawachi,
Hideaki Miyashita, Mayumi Ohnishi-Kameyama, and Hiroshi Ono

CONTENTS

6.1   Introduction 96
   6.1.1   Chlorophylls *a*, *b*, *d*, *f* and DV-Chl *a* 96
   6.1.2   Chls c_1 and c_2 96
   6.1.3   BChls *a*, *b*, and *g* 96
   6.1.4   Chls *a′*, *d′* and DV-Chl *a′* 96
   6.1.5   BChls *a′* and *g′* 97
   6.1.6   Phe *a*, BPhes *a* and *b* 98
   6.1.7 8^1-OH-Chl *a* and BChl 663 100
   6.1.8   [Zn]-BChl *a* 100
6.2   High-Performance Liquid Chromatography.. 100
   6.2.1   Chls *a*, *b*, *d*, *f* and Their Derivatives 100
   6.2.2   BChls *a*, *b*, *g* and Their Derivatives 101
   6.2.3   DV-Chls *a*, *a′*, *b* and Divinyl Phe *a* 101
6.3   Absorption Spectra 102
   6.3.1   Chls *a*, *b*, *d*, *f* and DV-Chl *a* 103
   6.3.2   Phes *a*, *b*, *d*, *f* and Divinyl Pheophytin *a* 108
   6.3.3   BChls *a*, *b*, *g* and Their Epimers 109
   6.3.4   BPhe *a* and BPhe *b* 109
   6.3.5   [Zn]-BChl *a* 110
6.4   Fluorescence Spectra 110
   6.4.1   Fluorescence Wavelength Maxima 110
   6.4.2   Fluorescence Quantum Yield: ϕ_F 111
   6.4.3   Fluorescence Lifetime: τ 111
6.5   CD Spectra 114
   6.5.1   Chls *a′*, *b′*, *d′*, *f′*, DV-Chl *a′* and Their Pheophytins 114
   6.5.2   BChls *g* and *g′* 114
   6.5.3   [Mg]- and [Zn]-BChl *a* 116
6.6   Mass Spectra 116
   6.6.1   APCI-Mass Spectra: Chls *a*, *b*, *d*, *f* and Their Phes 116
   6.6.2   FAB-Mass Spectra 117
      6.6.2.1   Chl *a* and Chl *d* 117
      6.6.2.2   BChl 663: Chl *a* Esterified with Δ2,6-Phytadienol, Chl *a*Δ2,6$_{PD}$ 117
      6.6.2.3   [Mg]-, [2H]-, and [Zn]-BChl *a* 119
   6.6.3   MALDI-Mass Spectra 119
6.7   NMR Spectra 120
   6.7.1   Chls *a*, *b*, *d*, *f* and Divinyl Chl *a* 120
      6.7.1.1 ^{1}H-NMR Spectra 120
      6.7.1.2 ^{13}C-NMR Spectra 122
      6.7.1.3   Two-Dimensional NMR 124
   6.7.2   Chls *a* and *a′* 126
   6.7.3   BChl 663 (Chl *a*Δ2,6-Phytadienol) 126
   6.7.4   [Mg]-BChl *a*, [Zn]-BChl *a*, and [2H]-BChl *a* (BPhe *a*) 134

6.8   Redox Potentials 138
    6.8.1   Chls *a*, *b*, *d*, *f* and DV-Chl *a* 138
    6.8.2   [M]-Chl *a*, [Zn]-BChl *a*, and [2H]-BChl *a* 139
Acknowledgments 142
Abbreviations 142
References 142

6.1   INTRODUCTION

Analysis by a combination of high-performance liquid chromatography, absorption, fluorescence, circular dichroism, mass and nuclear magnetic resonance spectrometry has been used to investigate the structure of chlorophylls. Here, we show several spectroscopic and redox properties of unique chlorophylls, e.g., chlorophylls *d* and *f*, divinyl chlorophyll *a*, and [Zn]-bacteriochlorophyll *a*, and compare them with those of the well-known chlorophyll *a* and bacteriochlorophyll *a*.

The various *naturally occurring* chlorophylls (Chls) to be treated here are Chls *a*, *a'*, *b*, c_1, c_2, *d*, *d'*, *f*, divinyl chlorophylls (DV-Chls) *a*, *a'*, and pheophytin (Phe) *a* found in oxygenic photosynthesis; bacteriochlorophylls (BChls) *a*, *b*, and bacteriopheophytins (BPhes) *a*, *b* in purple bacteria; BChls *g*, *g'*, and 8^1-OH-Chl *a* in heliobacteria; BChl *a'* and BChl 663 in green sulfur bacteria; and novel Zn-containing BChl *a* in a unique photosynthetic bacterium *Acidiphilium rubrum*. In this chapter, we briefly present their short history and some examples of high-performance liquid chromatography (HPLC) traces, absorption, fluorescence, circular dichroism (CD), mass (MS), nuclear magnetic resonance (NMR) spectra, and redox potentials.

6.1.1   CHLOROPHYLLS *a*, *b*, *d*, *f* AND DV-CHL *a*

In 1818, the term *chlorophyll* (Chl), the green (Greek *chloros*) of leaf (Greek *phyllon*), was introduced for the pigments extracted from leaves with organic solvents (Pelletier and Caventou 1818). In 1903, a Russian botanist Tsvet (Tswett) (in Russian meaning "color") separated leaf pigments by chromatography (from Greek *chroma* and *graphein* meaning "color" and "to write", respectively) into the blue Chl *a*, the green Chl *b*, and several yellow to orange carotenoids.

In 1943, Chl *d* (Figure 6.1) was first reported as a minor pigment in several red macroalgae (Manning and Strain 1943). In 1996, a unique cyanobacterium, *Acaryochloris marina*, was isolated from colonial ascidians containing Chl *d* as the dominant chlorophyll (Miyashita et al. 1996; Ohashi et al. 2008; a review by Kobayashi et al. 2013).

In 2010, Chl *f* (Figure 6.1) was first reported in a methanolic extract of Shark Bay stromatolites incubated under near-infrared light for the initial purpose of the isolation of new Chl *d*-containing phototrophs (Chen et al. 2010). At around the same time, a Chl *f*-producing cyanobacterium, strain KC1, was discovered from Lake Biwa, the largest freshwater lake in Japan; Chl *f* was induced only when incubated far red light (Kobayashi et al. 2013; Miyashita et al. 2014).

In 1983, DV-Chl *a* (Figure 6.1) was found in algae fractions with the particles of less-than-1 μm in surface waters of the

tropical Atlantic Ocean (Gieskes and Kraay 1983, 1986). In 1988, DV-Chl *a* was found in a small free-living marine prochlorophyte, *Prochlorococcus marinus* (Chisholm et al. 1988).

6.1.2   CHLS c_1 AND c_2

Chls c_1 and c_2 are widely distributed and abundant in the chromophyte algae. They have the fully unsaturated porphyrin macrocycle and do not carry an esterifying long chain (Figure 6.2). Chls c_1 and c_2 differ by an 8-ethyl and 8-vinyl substituents, respectively.

6.1.3   BCHLS *a*, *b*, AND *g*

BChl *a* (Figure 6.3) is the most popular bacteriochlorophyll in photosynthetic bacteria and the only bacteriochlorophyll in most *Rhodospirillales*. BChl *b* was first isolated from *Rhodopseudomonas viridis* (Eimhjellen et al. 1963); BChl *b* differs from BChl *a* by the presence of an 8-ethylidene group on ring II (Figure 6.3) (Brockmann and Kleber 1970; Scheer et al. 1974; Risch 1981). In 1981, BChl *g* (Figure 6.3) was found in *Heliobacterium chlorum*, isolated from a soil sample collected in front of the Biology Department, Indiana University, using an incorrectly prepared culture medium for other anoxygenic bacteria (Gest 1994), and the molecular structure was determined by Brockmann and Lipinski (1983).

Isomerization (intramolecular proton transfer) of BChl *g* on ring II easily takes place to yield Chl *a* esterified with farnesol (Brockmann and Lipinsky 1983; Michalski et al. 1987; Kobayashi et al. 1998a), indicating that BChl *g* is the most likely candidate for the ancestor of Chl *a* in oxygenic photosynthesis (Figure 6.4): similar reactions were seen in BChl *b* yielding [3-acetyl]-Chl *a* but not being found as a naturally occurring pigment (Steiner et al. 1983; Kobayashi et al. 1998b). Heliobacteria are taxonomically closer to the cyanobacteria than the other groups of anoxygenic photosynthetic bacteria (Xiong et al. 2000), supporting the notion of Chl *a* arising from BChl *g* (Figure 6.4).

6.1.4   CHLS *a'* *d'* AND DV-CHL *a'*

The 13^2-epimer of Chl *a*, Chl *a'* ("a-prime") (Figure 6.1), was first reported in 1942 (Strain and Manning 1942). In 1988, it was proposed that Chl *a'* constitutes P700 as a heterodimer of Chl *a/a'* (Figure 6.5) (Kobayashi et al. 1988), and the idea has been confirmed in 2001 (Jordan et al. 2001). For more details, see Chapter 4 in Kobayashi et al. (2006a).

Chl *d'* (Figure 6.1) was also detected in *A. marina* as a minor component, while Chl *a'* was absent (Akiyama et al.

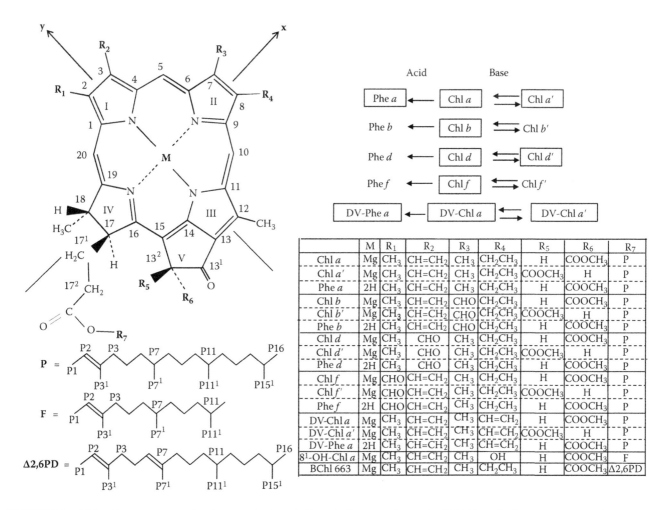

	M	R_1	R_2	R_3	R_4	R_5	R_6	R_7
Chl a	Mg	CH_3	$CH=CH_2$	CH_3	CH_2CH_3	H	$COOCH_3$	P
Chl a'	Mg	CH_3	$CH=CH_2$	CH_3	CH_2CH_3	$COOCH_3$	H	P
Phe a	2H	CH_3	$CH=CH_2$	CH_3	CH_2CH_3	H	$COOCH_3$	P
Chl b	Mg	CH_3	$CH=CH_2$	CHO	CH_2CH_3	H	$COOCH_3$	P
Chl b'	Mg	CH_3	$CH=CH_2$	CHO	CH_2CH_3	$COOCH_3$	H	P
Phe b	2H	CH_3	$CH=CH_2$	CHO	CH_2CH_3	H	$COOCH_3$	P
Chl d	Mg	CH_3	CHO	CH_3	CH_2CH_3	H	$COOCH_3$	P
Chl d'	Mg	CH_3	CHO	CH_3	CH_2CH_3	$COOCH_3$	H	P
Phe d	2H	CH_3	CHO	CH_3	CH_2CH_3	H	$COOCH_3$	P
Chl f	Mg	CHO	$CH=CH_2$	CH_3	CH_2CH_3	H	$COOCH_3$	P
Chl f'	Mg	CHO	$CH=CH_2$	CH_3	CH_2CH_3	$COOCH_3$	H	P
Phe f	2H	CHO	$CH=CH_2$	CH_3	CH_2CH_3	H	$COOCH_3$	P
DV-Chl a	Mg	CH_3	$CH=CH_2$	CH_3	$CH=CH_2$	H	$COOCH_3$	P
DV-Chl a'	Mg	CH_3	$CH=CH_2$	CH_3	$CH=CH_2$	$COOCH_3$	H	P
DV-Phe a	2H	CH_3	$CH=CH_2$	CH_3	$CH=CH_2$	H	$COOCH_3$	P
8^1-OH-Chl a	Mg	CH_3	$CH=CH_2$	CH_3	OH	H	$COOCH_3$	F
BChl 663	Mg	CH_3	$CH=CH_2$	CH_3	CH_2CH_3	H	$COOCH_3$	$\Delta2,6PD$

FIGURE 6.1 Molecular structures and carbon numbering of chlorophylls, according to the IUPAC numbering system. Naturally occurring chlorophylls are designated by squares.

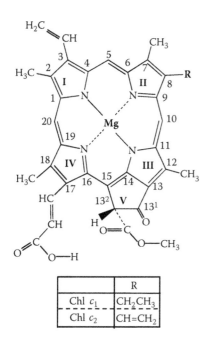

	R
Chl c_1	CH_2CH_3
Chl c_2	$CH=CH_2$

FIGURE 6.2 Molecular structures and carbon numbering of Chl c_1 and Chl c_2, according to the IUPAC numbering system.

2001) (see Figure 6.6). P740, the primary electron donor of PS I in *A. marina*, was proposed to be a Chl d/d' heterodimer (Akiyama et al. 2002, 2004; Kobayashi et al. 2005, 2007, 2013; Ohashi et al. 2008), just like the Chl a/a' for P700 (Figure 6.5). For more details, see a review by Kobayashi et al. (2013). DV-Chl a' is found in a free-living marine prochlorophyte, *P. marinus* (Fujinuma et al. 2012; Komatsu et al. 2014). Note that neither Chl b' nor Chl f' has been detected as naturally occurring species.

6.1.5 BChls a' and g'

As seen in Figure 6.5, it has been also shown that P798 consists of a homodimer of BChl g', (BChl $g')_2$, in the reaction center (RC) of heliobacteria (Kobayashi et al. 1991), and that P840 consists of a homodimer of BChl a' in green sulfur bacteria (Kobayashi et al. 1992, 2000).

Here we introduce our hypothesis about the evolution of the PS I-type RCs based on the chlorophyll structures (Figure 6.4). The prime-type chlorophylls, BChl a' in green sulfur bacteria, BChl g' in heliobacteria, Chl a' in Chl a-type PS I, and Chl d' in Chl d-type PS I, function as the special pairs, either as homodimers, (BChl $a')_2$ and (BChl $g')_2$ in

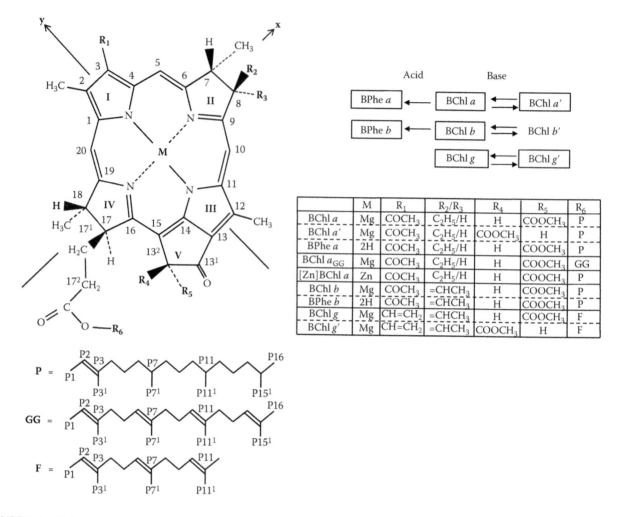

	M	R$_1$	R$_2$/R$_3$	R$_4$	R$_5$	R$_6$
BChl a	Mg	COCH$_3$	C$_2$H$_5$/H	H	COOCH$_3$	P
BChl a'	Mg	COCH$_3$	C$_2$H$_5$/H	COOCH$_3$	H	P
BPhe a	2H	COCH$_3$	C$_2$H$_5$/H	H	COOCH$_3$	P
BChl a_{GG}	Mg	COCH$_3$	C$_2$H$_5$/H	H	COOCH$_3$	GG
[Zn]BChl a	Zn	COCH$_3$	C$_2$H$_5$/H	H	COOCH$_3$	P
BChl b	Mg	COCH$_3$	=CHCH$_3$	H	COOCH$_3$	P
BPhe b	2H	COCH$_3$	=CHCH$_3$	H	COOCH$_3$	P
BChl g	Mg	CH=CH$_2$	=CHCH$_3$	H	COOCH$_3$	F
BChl g'	Mg	CH=CH$_2$	=CHCH$_3$	COOCH$_3$	H	F

FIGURE 6.3 Molecular structures and carbon numbering of bacteriochlorophylls. Naturally occurring chlorophylls are designated by squares.

anoxygenic organisms, or heterodimers, Chl a/a' and Chl d/d' in oxygenic photosynthesis (Figure 6.5). BChl g/g' may be a convincing ancestor of Chl a/a', because the BChl $g/g' \to$ Chl a/a' conversion takes place spontaneously under mild conditions *in vitro* (Kobayashi et al. 1998a). Further, a Chl $a/a' \to$ Chl d/d' conversion also occurs with ease *in vitro* (Kobayashi et al. 2005; Koizumi et al. 2005), supporting the succession from the Chl a-type cyanobacteria to *A. marina* (Figure 6.4). Chl f is produced in very small amounts in a Chl a-type special cyanobacterium, only when cultivated under far red light, suggesting that Chl f appeared after the birth of Chl a. As mentioned later, the primary electron acceptors, A_0, are all Chl a-derivatives even in anoxygenic PS I-type RCs without exception (Figure 6.5).

Note that Chls b', f' and BChl b' are not found in natural photosynthesis. Chls b and f function only as antenna pigments, and BChl b function in not PS I- but PS II-type RC.

6.1.6 Phe a, BPhes a and b

In 1974, Phe a (Figure 6.1), a demetallated Chl a, was first postulated to be the primary electron acceptor in PS II (Figure 6.5) (van Gorkom 1974), and the idea was experimentally

confirmed by Klimov et al. (1977a,b). In 1975, BPhe a was found to function as a primary electron acceptor in the RC of purple bacteria (Figure 6.3) (Fajer et al. 1975; Kaufmann et al. 1975; Parson et al. 1975; Rockley et al. 1975), and shortly thereafter, BPhe b was also found to perform the same function (Figure 6.5) (Klimov et al. 1977c). In 1986, BPhe a was also found to function in green filamentous bacteria (Figure 6.3) (Kirmaier et al. 1986; Shuvalov et al. 1986a,b). In 2001, the primary electron acceptor of PS II in *A. marina* was first found to be Phe a (Figure 6.1) (Akiyama et al. 2001) and later supported by Tomo et al. (2007). In 2012, DV-Phe a was found in *P. marinus* (Fujinuma et al. 2012; Komatsu et al. 2014). For more details, see Chapter 4 in Kobayashi et al. (2006a) and reviews by Ohashi et al. (2008) and Kobayashi et al. (2013).

It is of interest to note that Phe a as well as Chl a', d' and Chl d are artifacts easily produced *in vitro* (see Figure 6.4): Phe a is readily produced from Chl a under acidic conditions, primed chlorophylls from nonprimed ones by epimerization under basic conditions, and Chl d from Chl a under oxidative conditions. These artifacts, however, function as key components in natural photosynthesis, while other artifacts, Phes b, d, f and Chls b', f', are not naturally occurring pigments.

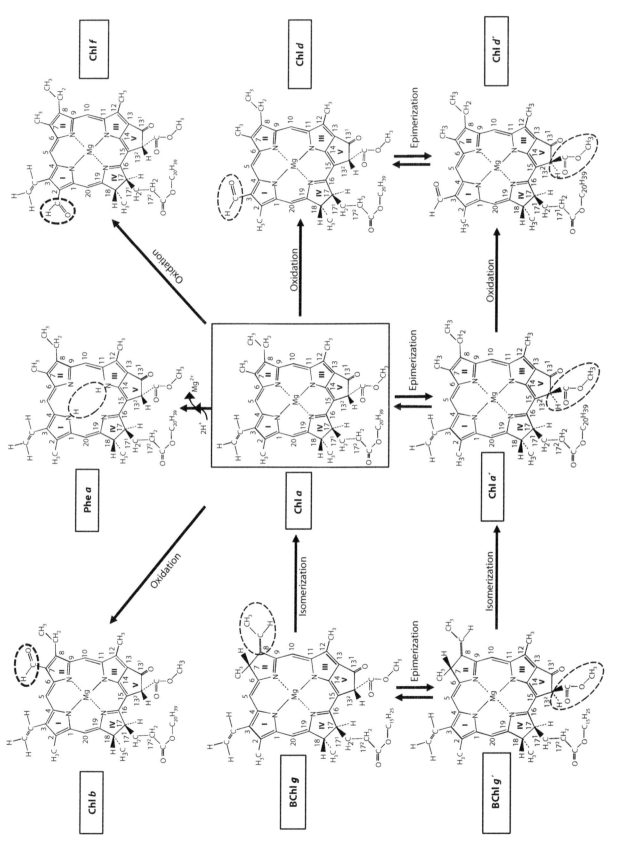

FIGURE 6.4 Bacteriochlorophyll *g/g'* and their derivatives functioning in natural photosynthesis.

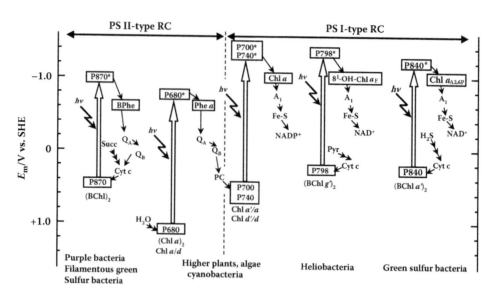

FIGURE 6.5 Schematic comparison of photosynthetic electron transport in PS I-type RC and PS II-type RC. Components are placed according to their estimated or approximate midpoint potentials. The arrows indicate the direction of electron flow. In order to simplify the figure, some primary electron donors, P970, P850, and P865, are omitted here: P970 and P850 are the primary electron donors of BChl *b* and Zn-BChl *a* containing purple bacteria, respectively; P865 is the primary electron donor of green filamentous bacteria.

6.1.7 8¹-OH-CHL *a* AND BCHL 663

The A_0 in the RC of heliobacteria showing a bleaching around 670 nm (Fuller et al. 1985; Nuijs et al. 1985a) was first thought to be a BChl *c*- or Chl *a*-like pigment, but the molecular structure was clarified to be 8¹-OH-Chl *a* esterified with farnesol (Figure 6.1) (Van de Meent et al. 1991).

The A_0 in the RC of green sulfur bacteria showed a bleaching around 670 nm and was tentatively suggested to involve BPhe *c* (Nuijs et al. 1985b), and was designated BChl 663 after its absorption maximum in the HPLC eluent (Braumann et al. 1986); it was considered to be a lipophilic form of BChl *c*. Later, BChl 663 was thought to be an isomer of Chl *a* (van de Meent et al. 1992), and finally it was identified as Chl *a* esterified at C17³ with Δ2,6-phytadienol possessing an additional double bond between P6- and P7- carbons of the phytol carbon chain (Figure 6.1) (Kobayashi et al. 2000).

It is interesting to note that in general it appears that the A_0 molecules in type-1 RCs are Chl *a* derivatives without exception (Figure 6.5), suggesting that the type-1 RCs are derived from a common ancestor. Chl *a*-type pigments are thought to be necessary to reduce $NADP^+$ or NAD^+ through an Fe–S center: the reduction potentials of pheophytins are not sufficiently negative to reduce $NAD(P)^+$, as mentioned in Section 6.8.

6.1.8 [Zn]-BCHL *a*

Natural Chls and BChls were previously assumed, without question, to be magnesium (Mg) complexes of substituted chlorin or bacteriochlorin macrocycles with one obvious and important exception, pheophytins (see Figures 6.1 and 6.3). In 1996, however, a novel Zn-containing BChl was discovered in an aerobic bacterium *A. rubrum* growing in a low pH habitat (Wakao et al. 1996). Its structure was identical to that of BChl *a*

esterified with phytol, but the central metal was Zn not Mg (Figure 6.3) (Akiyama et al. 1998a,b; Kobayashi et al. 1998c; Chapter 4 in Kobayashi et al. 2006a).

6.2 HIGH-PERFORMANCE LIQUID CHROMATOGRAPHY

In the late 1970s, the HPLC technique was applied to the separation of plant pigments. In many cases, the reversed-phase HPLC was preferred (Eskins et al. 1977; Schoch 1978; Shoaf 1978) and is still the main option to date. In that system, however, an eluent gradient is usually required for simultaneous separation of (B)Chls and (B)Phes, and the gradient system is unfavorable for quantitative analysis, since the molar absorptivities of pigments strongly depend on solvents. In this context, an isocratic eluent system seemed to be favorable. In 1978, a simultaneous separation of Chls and Phes by normal-phase HPLC was attained by an isocratic procedure (Iriyama et al. 1978), and then the isocratic normal-phase HPLC was established as a powerful tool for chlorophyll analysis (Watanabe et al. 1984). In this chapter, we present the *isocratic* normal-phase and reversed-phase HPLC separation of (B)Chls and (B)Phes.

6.2.1 CHLS *a*, *b*, *d*, *f* AND THEIR DERIVATIVES

As illustrated in Figure 6.6f, authentic eight Chls and four Phes are clearly separated by isocratic normal-phase HPLC. One can easily see that *Synechocystis* sp. PCC6803 possesses minor Phe *a* and Chl *a'* as well as major Chl *a* (Figure 6.6a), and that *Chlorella vulgaris* has also Chl *b* (Figure 6.6b). In Figure 6.6c, *A. marina* has three minor chlorophylls, Phe *a*, Chl *a*, and Chl *d'*, as well as the major Chl *d* (Akiyama et al. 2001); Phe *a* (not Phe *d*) functions as the primary electron acceptor in PS II, Chl *a* (not Chl *d*) as the primary electron

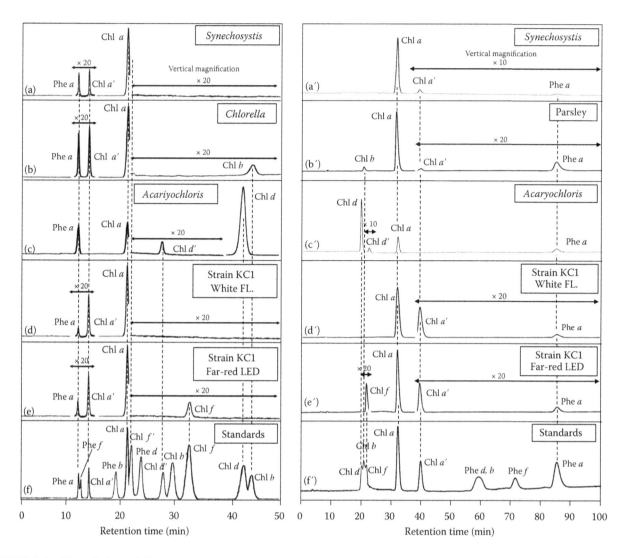

FIGURE 6.6 Normal-phase (left) and reversed-phase (right) HPLC profiles for acetone/methanol extracts of (a, a′) *Synechocystis* sp. PCC6803, (b) *Chlorella vulgaris*, (b′) parsley, (c, c′) *A. marina*, (d, d′) the cyanobacterium strain KC1 grown under white fluorescent light, (e, e′) the cyanobacterium strain KC1 grown under far red LED light, and (f) a standard mixture of Chls and Phes. Detection wavelength is 670 nm. Normal-phase HPLC (left): column, YMC-pak SIL (250 × 4.6 mm i.d.); temperature, 277 K; eluent, degassed hexane/2-propanol/methanol (100/0.7/0.2, v/v/v) at a flow rate of 0.9 mL min⁻¹. Reversed-phase HPLC (right): column, Kaseisorb LC ODS 2000-3 (250 × 4.6 mm i.d.); temperature, 277 K; eluent: degassed ethanol/methanol/2-propanol/water (86/13/1/3, v/v/v/v) at a flow rate of 0.4 mL min⁻¹.

acceptor in PS I, and Chl *d′* as the primary electron donor P740 as a heterodimer of Chl *d/d′*, like the Chl *a/a′* in P700 (Figure 6.5). Chlorophyll *f* is detected only when the strain KC1 is incubated under far red light (Figure 6.6d, e).

Chls *a*, *a′* and Phes *f*, *a* are well separated by isocratic reversed-phase HPLC, but the separation of both Chls *b*, *d*, *f* and Phes *b*, *d* is poor (Figure 6.6f′). One had better pay attention when analyzing Chl *b*-containing algae incubated under far red light, because induction of 3-desmethyl-3-formyl-Chl *b* is not negligible.

6.2.2 BChls *a*, *b*, *g* and Their Derivatives

As seen in Figure 6.7, BChls and BPhes are well separated by isocratic normal-phase (left) and reversed-phase (right) HPLC. A typical purple bacteria, *Rhodobacter sphaeroides*, possesses BChl *a* and BPhe *a* as a minor pigment (Figure 6.7a, a′). Figure 6.7b, b′ shows that *A. rubrum* has [Zn]-BChl *a* as

well as two minor pigments, BPhe *a* and BChl *a* ([Mg]-BChl *a*), where [Mg]-BChl *a* functions only as antenna in this bacterium. In *Rhodospirillum rubrum*, esterifying alcohol for BChl *a* is geranylgeraniol, but that of BPhe *a* is phytol (Figure 6.7c, c′). In *R. viridis*, *b*-type pigments, instead of BChl *a* and BPhe *a*, are present and function (Figure 6.7d, d′). One of the PS I-type photosynthetic bacteria, *Heliobacillus mobilis*, has BChl *g* and its epimer, BChl *g′*. Note that PS II-type photosynthetic bacteria (Figure 6.7a–d, a′–d′) have metal-free BPhes but no primed pigments and that PS I-type bacteria (Figure 6.7e and e′) have primed pigment, BChl *g′*, but no BPhes.

6.2.3 DV-Chls *a*, *a′*, *b* and Divinyl Phe *a*

As described above, the isocratic HPLC systems showed remarkable separations of (B)Chls and (B)Phes, but exhibit low separation of DV-Chls and MV-Chls as well as DV- and MV-Phes.

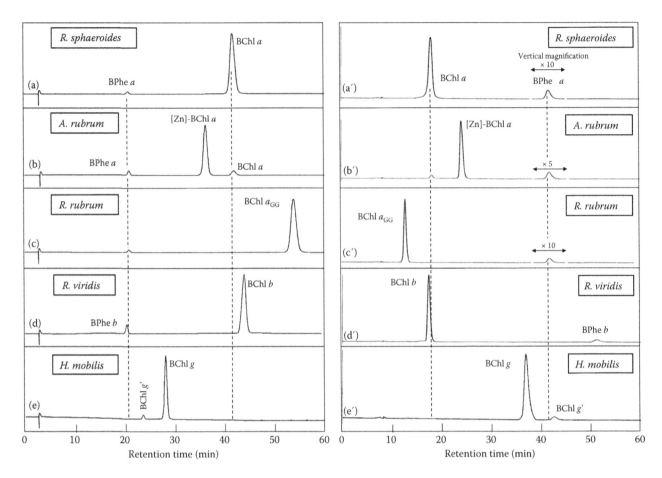

FIGURE 6.7 Normal-phase (left) and reversed-phase (right) HPLC profiles for acetone/methanol extracts of (a, a') *R. sphaeroides*, (b, b') *A. rubrum*, (c, c') *R. rubrum*, (d, d') *R. viridis*, (e, e') *H. mobilis*. Detection wavelength is 765 nm. Normal-phase HPLC (left): column, YMC-pak SIL (250 × 4.6 mm i.d.); temperature, 277 K; eluent, degassed hexane/2-propanol/methanol (100/0.8/0.3, v/v/v) at a flow rate of 1.3 mL min^{-1}. Reversed-phase HPLC (right): column, Kaseisorb LC ODS 2000-3 (250 × 4.6mm i.d.); temperature: 277 K; eluent: degassed ethanol/methanol/2-propanol/water (86/13/1/3, v/v/v/v) at a flow rate of 0.4 mL min^{-1}.

For example, typical isocratic normal-phase HPLC traces for acetone/methanol extract from cells of *Prochlorococcus* sp. NIES-3376 and pigment standard (MV-Phe *a*, Chl *a'*, Chl *a*, Chl *b'*, Chl *b*; DV-Phe *a*, Chl *a'*, Chl *a*) are shown in Figure 6.8a and b, respectively. DV-Chl *b* and MV-Chl *b* are clearly separated, while DV-Chl *a* and MV-Chl *a* are partially separated (Komatsu et al. 2014). Neither a primed pair (DV- and MV-Chls *a'*) nor metal-free Chls (DV- and MV-Phes *a*) are separated. Isocratic C18 reversed-phase HPLC also shows poor separation (data not shown).

In contrast, both a primed pair (DV- and MV-Chls *a'*) and a metal-free pair (DV- and MV-Phes *a*) were well separated by a C22-based reversed-phase HPLC column with simple isocratic eluent mode in a single run (Figure 6.8c and d) (Fujinuma et al. 2012; Komatsu et al. 2014). DV-Chl *a* and MV-Chl *a* were also nicely separated, but only the resolution of DV-Chl *b* and MV-Chl *b* was poor. One should note that both MV-Chl *a'* and MV-Phe *a* were not detected at all in *Prochlorococcus* sp. NIES-3376, but that both DV-Chl *a'* and DV-Phe *a* exit as the minor components in this picoplankton (Figure 6.8c). In *Prochlorococcus* sp. NIES-3376, DV-Chl *a'* and DV-Phe *a* must function as the primary donor of PS I and

the primary electron acceptor in PS II, respectively, instead of MV-Chl *a'* and MV-Phe *a* functioning in normal cyanobacterial RCs (Fujinuma et al. 2012).

The simple isocratic eluent HPLC systems with the combination of normal- and reversed-phase HPLC columns are found to be favorable and powerful tools for assessing the distribution of the phytoplankton in marine water and freshwater, since the former and the latter have the high selectivity to resolve DV-Chls *b* and *a*, the marker pigments for the prokaryote *Prochlorococcus*, from MV-Chl *b* and *a*, respectively. For more details, see Komatsu et al. (2014).

6.3 ABSORPTION SPECTRA

The absorption spectrum is the simplest, most useful, and extensively used analytical property to characterize (B)Chls and (B)Phes. The absorption maxima and the corresponding molar absorption coefficients (ε) in several organic solvents are listed in Table 6.1. Absorption spectra of (B)Chls show the electronic transitions along the *x*-axis of the (B) Chl running through the two nitrogen (N) atoms of rings II and IV, and along the *y*-axis through the N atoms of rings I

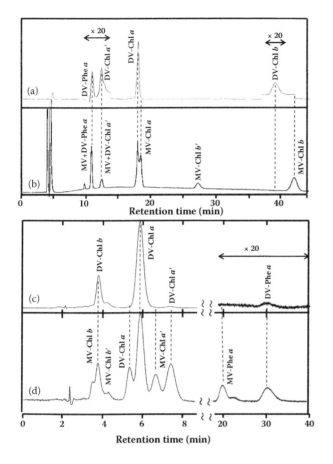

FIGURE 6.8 Normal-phase (top) and reversed-phase (bottom) HPLC traces for (a, c) acetone/methanol extracts of *Prochlorococcus* sp. NIES-3376 and (b, d) a mixture of authentic MV-Chl *b*, MV-Chl *b'*, MV-Chl *a*, DV-Chl *a*, MV-Chl *a'*, DV-Chl *a'*, MV-Phe *a*, and DV-Phe *a*. Detection wavelength is 665 nm. Normal-phase HPLC: column, YMC-pak SIL (250 × 4.6 mm i.d.); temperature, 277 K; eluent, degassed hexane/2-propanol/methanol (100/0.7/0.2, v/v/v) at a flow rate of 0.9 mL min^{-1}. Reversed-phase HPLC (right): column, Senshupak DOCOSIL SP100 (250 × 4.6 mm i.d.); temperature, 277 K; eluent, degassed ethanol/methanol/2-propanol (86/13/1, v/v/v) at a flow rate of 1.4 mL min^{-1}.

and III (see Figures 6.1 through 6.3). The two main absorption bands in the blue and red regions are called Soret and Q bands, respectively, and arise from $\pi \rightarrow \pi^*$ transitions of four frontier orbitals (Weiss 1978; Petke et al. 1979; Hanson 1991). The fundamental macrocycles of (B)Chls are readily distinguishable as porphyrins (Figure 6.2), chlorins (Figure 6.1), and bacteriochlorins (Figure 6.3) by their absorption spectra in organic solvents; for example, absorption spectra of Chl c_1 (porphyrin), Chl a (chlorin), and BChl a (bacteriochlorin) are illustrated in Figure 6.9. For more details, see Chapter 6 in Kobayashi et al. (2006b).

6.3.1 Chls *a, b, d, f* and DV-Chl *a*

Absorption spectra of Chls *a, b, d,* and *f* in four kinds of solvent measured at room temperature are shown in Figure 6.10.

TABLE 6.1

Absorption Properties of Chlorophylls in Diethylether at Room Temperature

Compound	λ_{max}, Blue (nm) (ε[10^3 M^{-1}cm^{-1}])	λ_{max}, Red (nm) (ε[10^3 M^{-1}cm^{-1}])	Reference
Chl *a*	430	662	Kim (1967)
	(147,1)	(112.9)	Kim (1967)
	428.4	660.3	Watanabe et al. (1984)
	(115)	(89.8)	Watanabe et al. (1984)
	429.1[b]	661.6[b]	Watanabe et al. (1984)
	(100)[b]	(81.3)[b]	Watanabe et al. (1984)
	432.5[d]	665.4[d]	Watanabe et al. (1984)
	(101)[d]	(79.7)[d]	Watanabe et al. (1984)
	–	663[b]	Kobayashi (1989)
	–	(81.3)[b]	Kobayashi (1989)
	432.5[d]	665.5[d]	Kobayashi et al. (2000)
	429.0	661.0	Kobayashi et al. (2000)
	431.0[b]	662.5[b]	Kobayashi et al. (2000)
	431.5[e]	664.0[e]	Kobayashi et al. (2000)
	431.5[g]	664.5[g]	Kobayashi et al. (2000)
	433.0[c]	666.0[c]	Kobayashi et al. (2000)
	431.5[c]	665.7[c]	Vladkova (2000)
	431[g]	665[g]	Vladkova (2000)
	429.6[b]	661.7[b]	Vladkova (2000)
	433[c]	668[c]	Takeuchi and Amao (2005)
	429.0	660.9	Kobayashi et al. (2006b)
	(1.00)[a]	(0.775)[a]	Kobayashi et al. (2006b)
	429.8[b]	662.2[b]	Tomo et al. (2009)
	–	665[c]	Chen et al. (2010)
	431.3[b]	662.2[b]	Kobayashi et al. (2013)
	(1.00)[a,b]	(0.828)[a,b]	Kobayashi et al. (2013)
	432.5[c]	665.8[c]	Kobayashi et al. (2013)
	(0.944)[a,c]	(1.00)[a,c]	Kobayashi et al. (2013)
	432.5[d]	665.3[d]	Kobayashi et al. (2013)
	(1.00)[a,d]	(0.785)[a,d]	Kobayashi et al. (2013)
	429.0	660.9	Komatsu et al. (2014)
	431.3[b]	662.2[b]	Komatsu et al. (2014)
	432.5[c]	665.8[c]	Komatsu et al. (2014)
	432.5[d]	665.3[d]	Komatsu et al. (2014)
	430.6[b]	662.4[b]	This work
	(1.00)[a,b]	(0.828)[a,b]	This work
Phe *a*	409	667	Kim (1967)
	(151.5)	(73.1)	Kim (1967)
	408.4	667.9	Watanabe et al. (1984)
	(107)	(52.6)	Watanabe et al. (1984)
	409.2[b]	665.9[b]	Watanabe et al. (1984)
	(104)[b]	(46.0)[b]	Watanabe et al. (1984)
	414.8[d]	671.6[d]	Watanabe et al. (1984)
	(108)[d]	(53.1)[d]	Watanabe et al. (1984)
	–	668[b]	Kobayashi (1989)
	–	(46)[b]	Kobayashi (1989)
	408.4	667.3	Kobayashi et al. (2006b)
	(1.00)[a]	(0.497)[a]	Kobayashi et al. (2006b)

(Continued)

TABLE 6.1 (CONTINUED)
Absorption Properties of Chlorophylls in Diethylether at Room Temperature

Compound	λ_{max}, Blue (nm) ($\varepsilon[10^3$ $M^{-1}cm^{-1}]$)	λ_{max}, Red (nm) ($\varepsilon[10^3$ $M^{-1}cm^{-1}]$)	Reference
	408.9[b]	665.4[b]	Kobayashi et al. (2013)
	(1.00)[a,b]	(0.440)[a,b]	Kobayashi et al. (2013)
	409.2[c]	665.7[c]	Kobayashi et al. (2013)
	(1.00)[a,c]	(0.464)[a,c]	Kobayashi et al. (2013)
	414.5[d]	670.8[d]	Kobayashi et al. (2013)
	(1.00)[a,d]	(0.798)[a,d]	Kobayashi et al. (2013)
	408.4	667.3	Komatsu et al. (2014)
	408.9[b]	665.4[b]	Komatsu et al. (2014)
	409.2[c]	665.7[c]	Komatsu et al. (2014)
	414.5[d]	670.8[d]	Komatsu et al. (2014)
DV-Chl a	436	661	Bazzaz (1981)
	435	660	Goericke and Repeta (1992)
	435.6[b]	661.8[b]	Tomo et al. (2009)
	436.1	659.9	Komatsu et al. (2014)
	(1.00)[a]	(0.721)[a]	Komatsu et al. (2014)
	438.3[b]	661.9[b]	Komatsu et al. (2014)
	(1.00)[a,b]	(0.731)[a,b]	Komatsu et al. (2014)
	443.0[c]	666.4[c]	Komatsu et al. (2014)
	(1.00)[a,c]	(0.873)[a,c]	Komatsu et al. (2014)
	440.4[d]	665.2[d]	Komatsu et al. (2014)
	(1.00)[a,d]	(0.749)[a,d]	Komatsu et al. (2014)
DV-Phe a	416.9	666.8	Komatsu et al. (2014)
	(1.00)[a]	(0.431)[a]	Komatsu et al. (2014)
	417.3[b]	665.3[b]	Komatsu et al. (2014)
	(1.00)[a,b]	(0.484)[a,b]	Komatsu et al. (2014)
	416.5[c]	666.1[c]	Komatsu et al. (2014)
	(1.00)[a,c]	(0.555)[a,c]	Komatsu et al. (2014)
	423.3[d]	670.8[d]	Komatsu et al. (2014)
	(1.00)[a,d]	(0.497)[a,d]	Komatsu et al. (2014)
Chl b	453	645	Kim (1967)
	(192.6)	(68.3)	Kim (1967)
	451.9	641.9	Watanabe et al. (1984)
	(159)	(56.7)	Watanabe et al. (1984)
	455.8[b]	644.6[b]	Watanabe et al. (1984)
	(136)[b]	(47.6)[b]	Watanabe et al. (1984)
	457.9[d]	646.2[d]	Watanabe et al. (1984)
	(152)[d]	(56.2)[d]	Watanabe et al. (1984)
	452.4	642.5	Kobayashi et al. (2006b)
	(1.00)[a]	(0.355)[a]	Kobayashi et al. (2006b)
	–	652[c]	Chen et al. (2010)
	454.0	644.0	Kobayashi et al. (2013)
	(1.00)[a]	(0.361)[a]	Kobayashi et al. (2013)
	458.7[b]	646.0[b]	Kobayashi et al. (2013)
	(1.00)[a,b]	(0.355)[a,b]	Kobayashi et al. (2013)
	469.4[c]	652.2[c]	Kobayashi et al. (2013)
	(1.00)[a,c]	(0.355)[a,c]	Kobayashi et al. (2013)
	458.3[d]	646.5[d]	Kobayashi et al. (2013)
	(1.00)[a,d]	(0.364)[a,d]	Kobayashi et al. (2013)

(Continued)

TABLE 6.1 (CONTINUED)
Absorption Properties of Chlorophylls in Diethylether at Room Temperature

Compound	λ_{max}, Blue (nm) ($\varepsilon[10^3$ $M^{-1}cm^{-1}]$)	λ_{max}, Red (nm) ($\varepsilon[10^3$ $M^{-1}cm^{-1}]$)	Reference
Phe b	434	653	Kim (1967)
	(244.0)	(47.5)	Kim (1967)
	432.7	654.6	Watanabe et al. (1984)
	(172)	(34.8)	Watanabe et al. (1984)
	433.8[b]	653.3[b]	Watanabe et al. (1984)
	(153)[b]	(29.3)[b]	Watanabe et al. (1984)
	439.5[d]	656.7[d]	Watanabe et al. (1984)
	(152)[d]	(32.3)[d]	Watanabe et al. (1984)
	433.2	654.5	Kobayashi et al. (2006b)
	(1.00)[a]	(0.202)[a]	Kobayashi et al. (2006b)
	433.4	654.4	Kobayashi et al. (2013)
	(1.00)[a]	(0.193)[a]	Kobayashi et al. (2013)
	434.4[b]	653.3[b]	Kobayashi et al. (2013)
	(1.00)[a,b]	(0.195)[a,b]	Kobayashi et al. (2013)
	436.0[c]	653.9[c]	Kobayashi et al. (2013)
	(1.00)[a,c]	(0.245)[a,c]	Kobayashi et al. (2013)
	440.0[d]	656.9[d]	Kobayashi et al. (2013)
	(1.00)[a,d]	(0.216)[a,d]	Kobayashi et al. (2013)
Chl c_1	446.1[b]	629.1[b]	Jeffrey (1972)
	(8.89)[a,b]	(1.00)[a,b]	Jeffrey (1972)
Chl c_2	444.6[b]	629.6[b]	Jeffrey (1972)
	(8.62)[a,b]	(1.00)[a,b]	Jeffrey (1972)
Chl d	447	688	Smith and Benitez (1955)
	(87.6)	(98.9)	Smith and Benitez (1955)
	447	688	French (1960)
	(87.6)	(98.5)	French (1960)
	390, 445	686	Miyashita et al. (1997)
	392[b], 447[b]	688[b]	Miyashita et al. (1997)
	400[c], 455[c]	697[c]	Miyashita et al. (1997)
	445.6	686.2	Kobayashi et al. (2006b)
	(0.853)[a]	(1.00)[a]	Kobayashi et al. (2006b)
	–	696[c]	Chen et al. (2010)
	394.3[b], 451.7[b]	691.4[b]	Kobayashi et al. (2013)
	(0.559)[a,b] (0.826)[a,b]	(1.00)[a,b]	Kobayashi et al. (2013)
	400.8[c], 455.5[c]	698.1[c]	Kobayashi et al. (2013)
	(0.735)[a,c] (0.706)[a,c]	(1.00)[a,c]	Kobayashi et al. (2013)
	394.2[d], 450.2[d]	692.7[d]	Kobayashi et al. (2013)
	(0.532)[a,d], (0.885)[a,d]	(1.00)[a,d]	Kobayashi et al. (2013)
	390.3, 445.3	685.8	This work
	(0.525)[a], (0.853)[a]	(1.00)[a]	This work

(Continued)

TABLE 6.1 (CONTINUED)
Absorption Properties of Chlorophylls in Diethylether at Room Temperature

Compound	λ_{max}, Blue (nm) (ε[10^3 M^{-1}cm^{-1}])	λ_{max}, Red (nm) (ε[10^3 M^{-1}cm^{-1}])	Reference
Phe d	421	692	Smith and Benitez (1955)
	(84.9)	(72.2)	Smith and Benitez (1955)
	421	692	French (1960)
	(84.9)	(72.2)	French (1960)
	382.7, 421.3	692.0	Kobayashi et al. (2006b)
	(0.881)[a], (1.00)[a]	(0.911)[a]	Kobayashi et al. (2006b)
	383.3, 421.0	691.9	Kobayashi et al. (2013)
	(0.869)[a], (1.00)[a]	(0.918)[a]	Kobayashi et al. (2013)
	383.7[b], 421.5[b]	691.0[b]	Kobayashi et al. (2013)
	(0.888)[a,b], (1.00)[a,b]	(0.761)[a,b]	Kobayashi et al. (2013)
	384.0[c], 410.7[c]	693.1[c]	Kobayashi et al. (2013)
	(1.00)[a,c], (0.964)[a,c]	(0.637)[a,c]	Kobayashi et al. (2013)
	387.8[d], 428.8[d]	697.3[d]	Kobayashi et al. (2013)
	(0.802)[a,d], (1.00)[a,d]	(0.915)[a,d]	Kobayashi et al. (2013)
Chl f	406[c]	706[c]	Chen et al. (2010)
	(1.00)[a,d]	(0.527)[a,d]	Chen et al. (2010)
	406.7[c]	708.3[c]	Akutsu et al. (2011)
	(0.904)[a,c]	(1.00)[a,c]	Akutsu et al. (2011)
	395.6, 440.5	695.2	Kobayashi et al. (2013)
	(0.657)[a] (0.648)[a]	(1.00)[a]	Kobayashi et al. (2013)
	398.2[b], 442.0[b]	701.0[b]	Kobayashi et al. (2013)
	(0.780)[a,c], (0.576)[a,b]	(1.00)[a,b]	Kobayashi et al. (2013)
	400.9[d], 444.0[d]	700.9[d]	Kobayashi et al. (2013)
	(0.668)[a,d], (0.658)[a,d]	(1.00)[a,d]	Kobayashi et al. (2013)
Phe f	409.3	696.9	Kobayashi et al. (2013)
	(1.00)[a]	(0.727)[a]	Kobayashi et al. (2013)
	409.3[b]	697.9[b]	Kobayashi et al. (2013)
	(1.00)[a,b]	(0.610)[a,b]	Kobayashi et al. (2013)
	410.0[c]	699.8[c]	Kobayashi et al. (2013)
	(1.00)[a,c]	(0.561)[a,c]	Kobayashi et al. (2013)
	415.0[d]	701.9[d]	Kobayashi et al. (2013)
	(1.00)[a,d]	(0.776)[a,d]	Kobayashi et al. (2013)
[Ba]-Chl a	–	678[b]	Kobayashi (1989)
	–	(35.5)[b]	Kobayashi (1989)
[Mn(II)]-Chl a	–	660[i]	Kobayashi (1989)
	–	(47.7)[i]	Kobayashi (1989)
[Mn(III)]-Chl a	–	675[b]	Kobayashi (1989)
	–	(28.2)[b]	Kobayashi (1989)
[Zn]-Chl a	–	655[b]	Kobayashi (1989)
	–	(83)[b]	Kobayashi (1989)
	421[c]	662[c]	Takeuchi and Amao (2005)
[Cd]-Chl a	–	660[b]	Kobayashi (1989)
	–	(64.8)[b]	Kobayashi (1989)
[Ni]-Chl a	–	648[b]	Kobayashi (1989)
	–	(54.2)[b]	Kobayashi (1989)
[Co(II)]-Chl a	–	652[j]	Kobayashi (1989)
	–	(58.1)[i]	Kobayashi (1989)
[Co(III)]-Chl a	–	651[b]	Kobayashi (1989)
	–	(38.4)[b]	Kobayashi (1989)
[Sn(IV)]-Chl a	–	653[b]	Kobayashi (1989)
	–	(46.4)[b]	Kobayashi (1989)
[Fe(II)]-Chl a	–	665[i]	Kobayashi (1989)
	–	(81.8)[i]	Kobayashi (1989)
[Fe(III)]-Chl a	–	621[b]	Kobayashi (1989)
	–	(27.4)[b]	Kobayashi (1989)
[Pb]-Chl a	–	676[b]	Kobayashi (1989)
	–	(36)[b]	Kobayashi (1989)
[Cu]-Chl a	414	648	Kim (1967)
	(108.3)	(58.5)	Kim (1967)
	412[b]	649[b]	Kim (1967)
	(108.8)[b]	(58.4)[b]	Kim (1967)
	412[c]	650[c]	Kim (1967)
	(96.6)[c]	(55.5)[c]	Kim (1967)
	416[d]	654[d]	Kim (1967)
	(104.9)[d]	(58.6)[d]	Kim (1967)
	–	651[b]	Kobayashi (1989)
	–	(64.2)[b]	Kobayashi (1989)
[Ag]-Chl a	–	647[b]	Kobayashi (1989)
	–	(47.1)[b]	Kobayashi (1989)
[Hg]-Chl a	–	678[b]	Kobayashi (1989)
	–	(34.7)[b]	Kobayashi (1989)
[Pd]-Chl a	–	631[b]	Kobayashi (1989)
	–	(102.6)[b]	Kobayashi (1989)
[VO]-Chl a	–	676[b]	Kobayashi (1989)
	–	(28.7)[b]	Kobayashi (1989)
BChl a	358	770	Kim (1967)
	(75.2)	(98.7)	Kim (1967)
	–	770.5	Connoly et al. (1982b)
	–	771.5[c]	Connoly et al. (1982b)
	–	782.5[d]	Connoly et al. (1982b)
	–	775.5[e]	Connoly et al. (1982b)
	356	771	Oelze (1985)

(Continued)

TABLE 6.1 (CONTINUED)
Absorption Properties of Chlorophylls in Diethylether at Room Temperature

Compound	λ_{max}, Blue (nm) $(\varepsilon[10^3\ \text{M}^{-1}\text{cm}^{-1}])$	λ_{max}, Red (nm) $(\varepsilon[10^3\ \text{M}^{-1}\text{cm}^{-1}])$	Reference
	(76)	(96)	Oelze (1985)
	–	770.0[b]	Becker et al. (1991)
	–	770.5[c]	Becker et al. (1991)
	357	771	Scheer and Hartwich (1995)
	357	771	Hartwich et al. (1998)
	(73.3)	(91.0)	Hartwich et al. (1998)
	357.0	771.0	Kobayashi et al. (1998c)
	(0.767)[a]	(1.00)[a]	Kobayashi et al. (1998c)
	358.5[b]	769.0[b]	Kobayashi et al. (1998c)
	(0.947)[a,b]	(1.00)[a,b]	Kobayashi et al. (1998c)
	361.0[d]	781.0[d]	Kobayashi et al. (1998c)
	(0.790)[a,d]	(1.00)[a,d]	Kobayashi et al. (1998c)
	362.5[e]	775.5[e]	Kobayashi et al. (1998c)
	(0.848)[a,e]	(1.00)[a,e]	Kobayashi et al. (1998c)
	361[h]	781[h]	Musewald et al. (1998)
	357.2	771.1	Permentier et al. (2000)
	(72.9)	(97.0)	Permentier et al. (2000)
	–	781.8[d]	Permentier et al. (2000)
	–	(90.8)[d]	Permentier et al. (2000)
	–	770.6[b]	Permentier et al. (2000)
	–	(69.3)[b]	Permentier et al. (2000)
	–	776.5[e]	Permentier et al. (2000)
	–	(65.1)[e]	Permentier et al. (2000)
	–	773.9[g]	Permentier et al. (2000)
	–	(59.6)[g]	Permentier et al. (2000)
	–	771.0[c]	Permentier et al. (2000)
	–	(54.8)[c]	Permentier et al. (2000)
	–	770.5	Akiyama (2001)
BChl 663	432.5[d]	665.5[d]	Kobayashi et al. (2000)
	429.0	661.0	Kobayashi et al. (2000)
	431.0[b]	662.5[b]	Kobayashi et al. (2000)
	432.0[e]	664.5[e]	Kobayashi et al. (2000)
	432.0[g]	665.0[g]	Kobayashi et al. (2000)
	432.5[c]	665.5[c]	Kobayashi et al. (2000)
BPhe a	357	750	Kim (1967)
	(120.8)	(73.1)	Kim (1967)
	354	748	Oelze (1985)
	(106)	(67.6)	Oelze (1985)
	–	746.5[b]	Becker et al. (1991)
	–	746.5[c]	Becker et al. (1991)
	357	749	Scheer and Hartwich (1995)
	356	750	Hartwich et al. (1998)
	(113)	(67.5)	Hartwich et al. (1998)
	356.5	748.5	Kobayashi et al. (1998c)
	(1.00)[a]	(0.63)[a]	Kobayashi et al. (1998c)
	357.0[b]	746.0[b]	Kobayashi et al. (1998c)
	(1.00)[a,b]	(0.482)[a,b]	Kobayashi et al. (1998c)
	362.0[d]	758.0[d]	Kobayashi et al. (1998c)
	(1.00)[a,d]	(0.717)[a,d]	Kobayashi et al. (1998c)
	358.5[e]	750.0[e]	Kobayashi et al. (1998c)
	(1.00)[a,e]	(0.521)[a,e]	Kobayashi et al. (1998c)
	–	749.0	Akiyama (2001)
BChl b	368[b]	795[b]	Jensen et al. (1964)
	(1.00)[a]	(0.96)[a]	Jensen et al. (1964)
	372	791	Oelze (1985)
	(77.3)	(106.0)	Oelze (1985)
	371.1	795.5	Kobayashi et al. (2006b)
	(0.702)[a]	(1.00)[a]	Kobayashi et al. (2006b)
BPhe b	368[b]	775[b]	Jensen et al. (1964)
	(1.00)[a]	(0.48)[a]	Jensen et al. (1964)
	366	776	Oelze (1985)
	(1.00)[a]	(0.678)[a]	Oelze (1985)
	367.9	778.5	Kobayashi et al. (2006b)
	(1.00)[a]	(0.696)[a]	Kobayashi et al. (2006b)
BChl c[f]	432	660	Stanier and Smith (1960)
	(139)[d]	(89)[d]	Stanier and Smith (1960)
BChl d[f]	425	650	Stanier and Smith (1960)
	(114)[e]	(88)[e]	Stanier and Smith (1960)
BChl e[f]	456-459[b]	646-648[b]	Gloe et al. (1975)
	(1.00)[a]	(0.24-0.32)[a]	Gloe et al. (1975)
BChl g	365, 405	766	Michalski et al. (1987)
	364.0, 404.0	767.2	Kobayashi et al. (1991)
	(90)	(96)	Kobayashi et al. (1991)
	364.8[b], 404.8[b]	761.6[b]	Kobayashi et al. (1991)
	(100)[b], (89)[b]	(75)[b]	Kobayashi et al. (1991)
	366.4[d], 409.6[d]	776.0[d]	Kobayashi et al. (1991)
	(98)[d], (95)[d]	(100)[d]	Kobayashi et al. (1991)
	364.4	767.4	Kobayashi et al. (2006b)
	(0.938)[a]	(1.00)[a]	Kobayashi et al. (2006b)
[Cd]-BChl a	359	760	Scheer and Hartwich (1995)
[Ni]-BChl a	336	780	Scheer and Hartwich (1995)
[Co]-BChl a	336	764	Scheer and Hartwich (1995)
[Mn(III)]-BChl a	363	769	Scheer and Hartwich (1995)
[Cu]-BChl a	425	666	Kim (1967)
	(40.5)	(35.5)	Kim (1967)
	428[b]	665[b]	Kim (1967)
	(42.0)[b]	(36.1)[b]	Kim (1967)
	432[d]	672[d]	Kim (1967)
	(38.0)[d]	(36.4)[d]	Kim (1967)

(Continued)

TABLE 6.1 (CONTINUED)
Absorption Properties of Chlorophylls in Diethylether at Room Temperature

Compound	λ_{max}, Blue (nm) ($\varepsilon[10^3$ M^{-1}cm^{-1}])	λ_{max}, Red (nm) ($\varepsilon[10^3$ M^{-1}cm^{-1}])	Reference
	342	772	Scheer and Hartwich (1995)
[Zn]-BChl a	353	762	Scheer and Hartwich (1995)
	353 (58.9)	762 (67.7)	Hartwich et al. (1998) Hartwich et al. (1998)
	353.0 (0.801)[a]	762.0 (1.00)[a]	Kobayashi et al. (1998c) Kobayashi et al. (1998c)
	355[h]	773[h]	Musewald et al. (1998)
	–	762.5	Akiyama (2001)
	353.5 (0.788)[a]	763.2 (1.00)[a]	Kobayashi et al. (2006b) Kobayashi et al. (2006b)
[Pd]-BChl a	329	755	Scheer and Hartwich (1995)
	334[h]	762[h]	Musewald et al. (1998)

[a] Relative values.
[b] In acetone.
[c] In methanol.
[d] In benzene.
[e] In 2-propanol.
[f] Mixture of several homologues.
[g] In ethanol.
[h] In toluene.
[i] In butyronitrile.

As compared to Chl a, Chl b shows red-shifted Soret bands and blue-shifted weak Q_Y bands, while the Q_Y bands of Chls d and f are intensified and shifted to longer wavelengths. The Q_X exhibits practically no intensity. The ratios of Soret/Q_Y band intensities show remarkable differences, e.g., in diethyl ether,

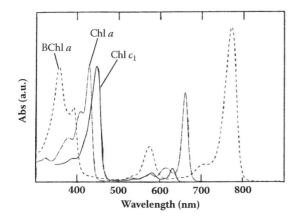

FIGURE 6.9 Absorption spectra of Chl a, Chl c_1, and BChl a in diethyl ether. Soret-band maxima are arbitrarily scaled to a common height. (The spectrum of Chl c_1 is reproduced from Jeffrey et al., 1997.)

about 1.3 in Chl a, about 2.8 in Chl b, about 0.85 in Chl d, and about 0.65 in Chl f (Table 6.1). Note that the Soret/Q_Y band ratios in Chl b are more than 2, and those in Chls d and f are below 1 in all solvents in Figure 6.10, whereas the ratios of Chl a in three solvents excluding methanol are around 1.3, but slightly below 1 in methanol. So one can easily distinguish Chl b from Chls a, d, and f, also Chl a from Chls d and f, by their absorption spectra in any solvents used here.

It is somewhat difficult to distinguish Chl f from Chl d, when one roughly compares the absorption spectrum of Chl f in diethyl ether with that of Chl d in methanol, because their spectral shapes are very similar. In contrast, in diethyl ether, one can easily distinguish Chl f from Chl d without spectrophotometer, because Chl f looks blue-green as Chl a, while Chl d looks light-green as Chl b, indicating that the naked eye is often powerful for color judgment.

The Soret bands include several intense bands. In diethyl ether and benzene, the Soret band of Chl f is clearly split into two bands, most probably the so-called B-bands (longer wavelength) and η-bands (shorter wavelength), while Chl d shows such a split not in those solvents but in methanol, and hence one can easily distinguish them, comparing their optical spectra in the same solvents, e.g., diethyl ether.

Absorption spectral shapes of DV-Chl a and MV-Chl a are very similar to each other, except that MV-Chl a in methanol shows a broadening of the Soret band and its smaller intensity than that of the Q_Y band. As compared to MV-Chl a, DV-Chl a shows the slightly but significantly red-shifted Soret band and the very small intensity reduction of the Q_Y band, while little changes are seen in the Q_Y band wavelengths in all solvents examined here. The Q_X exhibits practically no intensity.

Inductive effects on the absorption wavelengths and intensities of the Q_Y bands of chlorophylls strongly depend on the nature and position of substituent(s) on the macrocycle, due to the presence of two different electronic transitions polarized in the x and y directions (the axes of transition moments are depicted in Figures 6.1 through 6.3; Gouterman 1961; Gouterman et al. 1963; Weiss 1978; Petke et al. 1979; Hanson 1991, Kobayashi et al. 2006b). Replacement of the electron-donating group, –CH$_3$, on ring II of Chl a by the electron-withdrawing group, –CHO, yielding Chl b, causes the blue shift and significant intense reduction of the Q_Y band (Figure 6.10). In contrast, replacement of –CH$_3$ on ring I of Chl a by –CHO, yielding to Chl f, causes the red shift and intensity increase of the Q_Y band (Figure 6.10). A similar phenomenon is clearly seen in Chl d, where –CH=CH$_2$ on ring I of Chl a is replaced with –CHO. These results indicate that it is a general feature that substitution by the electron-withdrawing group on ring II causes the blue shift and intensity reduction of the Q_Y band and that the same substitution on ring I leads to the opposite, namely, the red shift and intensity increase of the Q_Y band. Moreover, it looks that substitution on ring I by the electron-withdrawing group generates the well-split Soret band, while showing heavy dependence on solvent as described above (Kobayashi et al. 2013).

A similar but weak tendency is seen in DV-Chl a, where a ←CH$_2$–CH$_3$ group on ring II of MV-Chl a is replaced with

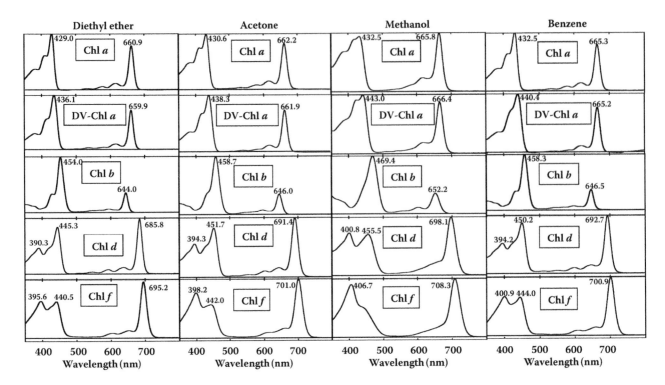

FIGURE 6.10 Comparison of the absorption spectra of Chls *a*, *b*, *d*, *f*, and DV-Chl *a* in diethyl ether, acetone, methanol, and benzene at room temperature. Spectra are scaled to the Soret- or Q_Y-band maximum.

a –CH=CH₂ moiety; the red shift of the Soret band was seen but smaller (about 7 nm), and the negligibly small intensity reduction of the Q_Y band was observed, supporting the general feature that substitution by the electron-withdrawing group on ring II causes the intensity reduction of the Q_Y band as well as the red shift of the Soret band (Komatsu et al. 2014). The Q_X(0.0) band (longer wavelength) of DV-Phe *a* looks unclear. DV-Phe *a* also showed the red shift of the Soret band by about 10 nm and the very slight reduction of the Q_Y band in intensity compared to MV-Phe *a*.

In methanol, when compared with DV-Chl *a*, MV-Chl *a* shows the broad Soret band and the intensity of the Soret band is smaller than that of the Q_Y band, while such features are not seen in MV-Phe *a*. The broad Soret band may be caused by some interactions of the central metal, Mg, of MV-Chl *a* with a methanol molecule(s) or a neighboring MV-Chl *a* molecule. It is not clear whether similar tendency will be observed in MV-Chls *b*, *d*, and *f* in methanol compared with DV species, and further investigations are needed.

6.3.2 Phes *a*, *b*, *d*, *f* and Divinyl Pheophytin *a*

The free base related to (B)Chl is called (B)Phe, and we emphasize that in natural oxygenic photosynthesis, only Phe *a* functions, and Phes *b*, *d*, and *f* are not functional (Figures 6.1 and 6.5). In general, the more structured shape and red-shifted Soret band of Chls distinguishes them from the corresponding Phes. In contrast to Chls, the η bands in the Soret band was poorly resolved in any Phes except Phe *d* (Figure 6.11). Removal of the central Mg increases deviation from planarity

and reduces the molecular symmetry, thus increasing Soret and Q_X transition. The Soret/Q_Y band ratios noticeably increased by pheophytinization; in diethyl ether, Phe *b* shows the highest value of around 5, Phe *a* the second highest value of about 2, and Phe *d* the lowest near 1 (compare Figure 6.11 with Figure 6.10; see also Table 6.1). Therefore, contamination of pheophytins in a Chl sample is often noticed from the optical spectra.

As seen in Figure 6.11, Phe *b* can be easily distinguished from Phe *a* by its blue-shifted Q_Y band, red-shifted Soret band, and its marginally higher Soret/Q_Y band ratio. Phes *d* and *f* can be distinguished from Phes *a* and *b* by their red-shifted Q_Y band and intense Q_Y bands, i.e., the Soret/Q_Y band ratios in Phes *d* and *f* are not high and almost the same as those seen in Chl *a*. We should also pay attention to Phe *f*, because in methanol its optical shape is somewhat similar to Phe *a*, although they can be distinguished by the Q_Y wavelength difference. We must emphasize again that Phes possess relatively strong and characteristic Q_X bands in the region of 490–570 nm; the Q_X bands in Phes *a* and *d* are better resolved to the Q_X(0,0) and Q_X(1,0) transitions. Pheophytin *d* also shows significantly well splits at the Soret and Q_X bands in all solvents illustrated in Figure 6.11. In contrast, the Q_X band corresponding to the Q_X(1,0) transition (shorter wavelength) of Phe *f* looks unclear. However, the Q_X (0,0) band (longer wavelength) of DV-Phe *a* looks unclear. DV-Phe *a* also showed the red shift of the Soret band by about 10 nm and the very slight reduction of the Q_Y band in intensity compared to Phe *a* (MV-Phe *a*). It is of interest to note that in diethyl ether, Phe *d* assumes a pale pink color, while both Phes *b* and *f* show a dull color. These

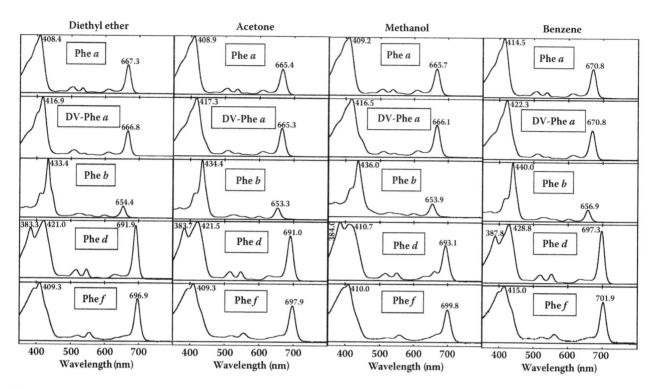

FIGURE 6.11 Comparison of the absorption spectra of Phes *a*, *b*, *d*, *f* and DV-Phe *a* in diethyl ether, acetone, methanol, and benzene at room temperature. Spectra are scaled to the Soret- or Q_Y-band maximum.

characteristics will help us to discern Phes from Chls, and among Phes.

6.3.3 BChls *a*, *b*, *g* and Their Epimers

Absorption spectra of BChl *a*, BChl *b*, and BChl *g*, which possess a bacteriochlorin macrocyle, are shown in Figure 6.12: bacteriochlorin Q_Y bands are much red-shifted relative to the chlorins (Figure 6.10). BChls *a* and *g* show very similar absorption spectra and have shorter wavelength for Q_Y maxima than BChl *b* and is easily distinguished from BChl *a* and BChl *g*. BChl *g* is distinguished from BChl *a* by its better-split Soret band and by its higher Soret/Q_y band ratio (about 0.90 in BChl *g* and about 0.75 in BChl *a*).

In Figure 6.12, the absorption spectra of BChl *a*/*a'*, *b*/*b'*, and *g*/*g'* in a *n*-hexane/2-propanol/methanol solution (100/2/0.3, v/v/v) show marked differences in the Q_X band region and slight differences in the Soret bands (Kobayashi 1996; Takahashi et al. 2005; Kobayashi et al. 2006b). Similar differences are seen in benzene but not in diethyl ether and acetone. The corresponding Mg-free BChls, BPhe *a*/*a'*, *b*/*b'*, and *g*/*g'*, are not distinguishable by absorption spectra, strongly suggesting that central metal, Mg, of BChls should play a key role in the optical differences observed here.

The Q_X band of BChl *a* (Evans and Katz 1975) and metal-substituted BChl *a* (Hartwich et al. 1998; Noy et al. 2000) is very sensitive to the coordination state of the central metal, shifting to the red with an increasing number of ligands. Therefore, the differences seen in Figure 6.12 between the C13²-stereoisomers of BChls are very likely related to a higher proportion of 6-coordinated Mg (two axial ligands) in the *normal* BChls, while the *prime* BChls are almost exclusively 5-coordinated (only one axial ligand). Such considerations can explain why the difference appeared only in certain solvents and disappeared in the corresponding BPhes, which cannot have axial ligands because of the lack of Mg. The axial ligand for *normal* BChls is methanol in Figure 6.12 and probably water in benzene.

In contrast, Chl *a*/*a'*, Chl *b*/*b'*, *d*/*d'*, *f*/*f'*, and their pheophytin pairs cannot be distinguished by their absorption spectra (Watanabe et al. 1984; Furukawa et al. 2000; Akiyama et al. 2001; Kobayashi et al. 2013), indicating that $\pi \rightarrow \pi^*$ transition along the *x*-axis in bacteriochlorin with Mg as the central metal is much more strongly affected by the C13²-stereochemistry on the macrocycle. For more details, see Takahashi et al. (2005).

6.3.4 BPhe *a* and BPhe *b*

As seen in Figure 6.13c, BPhe *a* can be easily distinguished from BChl *a* ([Mg]-BChl *a* in this figure) by their shape of their Soret bands, Soret/Q_y band ratios (about 1.6 in BPhe *a* and about 0.75 in BChl *a*), and wavelengths of Q_X maxima (about 520 nm in BPhe *a* and 570 nm in BChl *a*). Similar tendency is seen between BPhe *b* and BChl *b*; the Soret/Q_y band ratios (about 1.4 in BPhe *b* and 0.7 in BChl *b*) and Q_X wavelengths (about 530 nm in BPhe *b* and 580 nm in BChl *b*) show remarkable differences, although the shape of their Soret bands are very similar (Figure 6.13a). Two Q_X bands of Phe *a* in Figure 6.13b are conspicuous for shape, as described in Section 6.3.2.

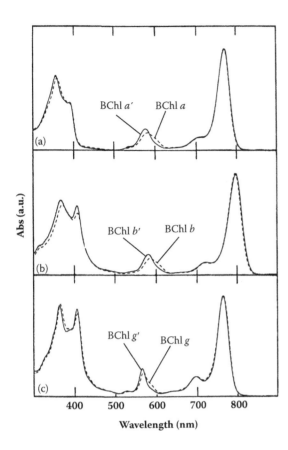

FIGURE 6.12 Absorption spectra of (a) BChls *a/a'*, (b) BChls *b/b'*, and (c) BChls *g/g'* in *n*-hexane/2-propanol/methanol (100/2/0.3, v/v/v). Q_Y-band maxima are arbitrarily scaled to a common height.

6.3.5 [Zn]-BChl *a*

The shape of the absorption spectrum of [Zn]-BChl *a* (Figure 6.3) is remarkably similar to that of BChl *a* (Figure 6.13c); thus, [Zn]-BChl *a* in *A. rubrum* was first mistaken for *normal* [Mg]-BChl *a* (Wakao et al. 1993; Kishimoto et al. 1995). The close similarity between absorption spectra as seen in Figure 6.13e explains the delayed discovery until 1996 of [Zn]-BChl *a* (Wakao et al. 1996); the absorption maxima of [Zn]-BChl *a* are slightly blue-shifted compared to those of BChl *a* in the four solvents listed in Table 6.1. The pink-purple color of [Zn]-BChl *a* contrasts with the blue-purple of BChl *a* in organic solvents resulting from the blue shift of the Q_X band in [Zn]-BChl *a* relative to BChl *a*. Note that the pink-purple color of [Zn]-BChl *a* changes to blue-purple when dried.

The absorption spectrum of artificially prepared [Zn]-Chl *a* is deceptively similar to that of [Mg]-Chl *a* (Figure 6.13b): they cannot be distinguished visually. Even if [Zn]-Chl *a* participated in natural photosynthesis, it would be difficult to recognize by its spectrum or coloring, alone.

Figure 6.14 shows the linear correlation for the absorption maxima, $h\nu_{max}$, of synthesized, unnatural [M]-BChls *a* and [M]-Chls *a*, as well as natural BChl *a* ([Mg]-BChl *a*), [Zn]-BChl *a*, BPhe *a* ([2H]-BChl *a*), Chl *a* ([Mg]-Chl *a*), and Phe *a* ([2H]-Chl *a*), against Pauling electronegativity (E_N) values. These values are interpreted as the electron density change

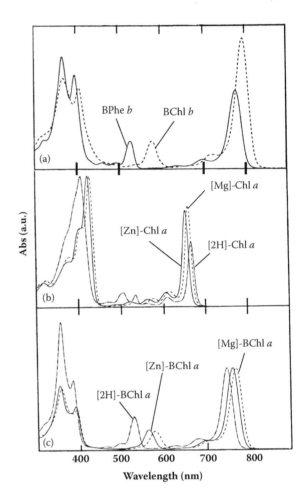

FIGURE 6.13 Absorption spectra of (a) BChl *b* and BPhe *b*; (b) Chl *a* ([Mg]-Chl *a*), Phe *a* ([2H]-Chl *a*), and [Zn]-Chl *a*; and (c) BChl *a* ([Mg]-BChl *a*), BPhe *a* ([2H]-BChl *a*), and [Zn]-BChl *a* in diethyl ether. Q_X-band maxima in (a), Soret-band maxima in (b), and Q_Y-band maxima in (c) are arbitrarily scaled to a common height, respectively.

in the bacteriochlorin and chlorin π-system, due to the inductive effect of the central metal (Watanabe and Kobayashi 1991; Geskes et al. 1995; Hartwich et al. 1998; Kobayashi et al. 1998b; Noy et al. 1998; Kobayashi et al. 1999a,b). Figure 6.14 shows that the change of $h\nu_{max}$ between [M]-BChls *a* and [M]-Chls *a* by metal exchange is small, indicating that the difference in light energy harvested would not be a primary reason to select Mg as the central metal of most photosynthetic pigments (Kobayashi et al. 1999a,b).

6.4 FLUORESCENCE SPECTRA

6.4.1 FLUORESCENCE WAVELENGTH MAXIMA

Analytical techniques based on fluorescence detection have the advantages of high sensitivity and selectivity. The fluorescence emission spectra of (B)Chls and (B)Phes bear a mirror-image relationship to their absorption, and peak fluorescence occurs at a longer wavelength than that of the corresponding absorption spectrum (Karukstis 1991; Hall and Rao 1994). Typical emission spectra of Chls *a, b, d, f,*

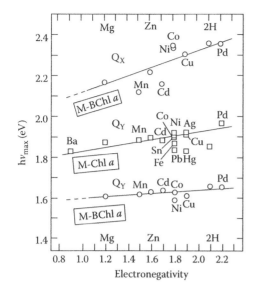

FIGURE 6.14 Correlation of the Q_X and Q_Y transition energy maxima, $h\nu_{max}$, of a series of [M]-BChls a in diethyl ether and the Q_Y transition maxima of [M]-chlorophylls a in acetone with the Pauling electronegativity (E_N) of the central metal. M is divalent except for Mn(III), Sn(IV), and Fe(III). Mg, Zn, and 2H naturally found pigments are indicated on the x-axis. (Adapted from Kobayashi M et al., *Current Topics in Plant Biology,* 1: 17–35, 1999.)

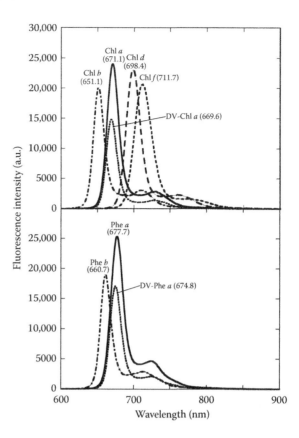

FIGURE 6.15 Emission spectra of Chls a, b, d, f, DV-Chl a, Phes a, b, and DV-Phe a in benzene.

DV-Chl a, Phes a, b, and DV-Phe a in benzene are illustrated in Figure 6.15. The wavelengths of the fluorescence maxima of several (B)Chls and (B)Phes are listed in Table 6.2. The fluorescence maximum of Chl a in benzene is observed at 671 nm, which is at 665 nm: such red shift is called Stoke's shift (Table 6.2). One can easily see that the emission band of Chl b nicely overlaps with the Q_Y absorption band of Chl a (see Figures 6.10 and 6.15). Note that Stoke's shift of Chl f in benzene is about 11 nm, which is significantly larger than those observed in Chls a, b, d and DV-Chl a (about 6, 5, 5, and 4 nm, respectively).

Emission spectra of seven kinds of [M]-Chls a in benzene are also illustrated in Figure 6.16, where Stoke's shifts are about 7, 6, 8, 11, 10, 5, and 7 nm (M = Mg, 2H, Zn, Sn(IV), Cd, Ba and Pd, respectively).

6.4.2 Fluorescence Quantum Yield: ϕ_F

The fluorescence quantum yields of several (B)Chls and (B)Phes are also listed in Table 6.2. Chl b exhibits remarkably small fluorescence quantum yield, $\phi_F = 0.16$, in benzene, while Chls a, d, and f and DV-Chl a show almost the same and higher values (0.39, 0.36, 0.39, and 0.37), suggesting that the presence of formyl moiety at ring I has little effects on ϕ_F, but the same moiety at ring II decreases ϕ_F drastically. As illustrated in Figure 6.17, a weak linear correlation for ϕ_F versus ε or Soret/Q_Y intensity ratio among Chls a, b, d, and f DV-Chl a or among Phes a, b, and DV-Phe a is seen.

As seen in Table 6.2, (B)Chls participating in natural photosynthesis are all fluorescent. However, only five [M]-BChls a

among the following 7, where M = Mg, Zn, Cd, Pd, Ni, Cu, or 2H (Connolly et al. 1982b; Teuchner et al. 1997), and seven [M]-Chls a among the following 16, where M = Mg, Zn, Cd, Pd, Ni Cu, 2H, VO, Mn(III), Fe(III), Co, Ag, Sn(IV), Hg, Pb, or Ba, fluoresce in organic solvents (Watanabe et al. 1985; Watanabe and Kobayashi 1988; Leupold et al. 1990; Kobayashi et al. 1998a,b). Figure 6.18 shows a correlation for ϕ_F between [M]-Chls a in benzene and [M]-BChls a in diethyl ether. [Mg]-, [Zn]-, and [2H]-(B)Chls a are strongly fluorescent, [Sn(IV)]-Chl a and [Cd]-(B)Chl a show weak fluorescence, and [Ba]-Chl a and [Pd]-(B)Chl a exhibit very weak fluorescence, probably due to the heavy atom effect. The values of ϕ_F of [Sn(IV)]-BChl a and [Ba]-BChl a have not been measured yet, but may be predicted from the line in Figure 6.18.

6.4.3 Fluorescence Lifetime: τ

The fluorescence lifetime of (B)Chls and (B)Phes is also listed in Table 6.2. Chl b shows remarkably short fluorescence lifetime, $\tau = 2.96$, in benzene, whereas Chls a, d, and f have much longer values (5.39, 5.54, and 5.15), which indicates that a formyl moiety at ring I gives no influence on τ, and the moiety at ring II extremely shortens the lifetime. We should note that Phe b presents almost the same long lifetime, 5.31, as those of Chls a, d, and f as well as Phe a (6.11), strongly suggesting that there seems little effect of a formyl group at ring II of pheophytins on τ. The lifetimes of BChl a and BPhe a

TABLE 6.2

Fluorescence Properties of Natural Chlorophylls in Diethyl Ether at Room Temperature

Compound	λ_{max} (nm)	ϕ_F	τ_F (ns)	Stokes Shift (nm)	Reference
Chl a	668	–	–	7	French et al. (1956)
	–	0.33	–	–	Latimer et al. (1956)
	–	0.32[d]	–	–	Latimer et al. (1956)
	–	0.32	–	–	Weber and Teale (1957)
	–	0.29–0.325[c]	–	–	Weber and Teale (1957)
	–	0.30[a]	–	–	Weber and Teale (1957)
	–	0.225[b]	–	–	Weber and Teale (1957)
	–	0.23[d]	–	–	Weber and Teale (1957)
	–	0.34	5.2	–	Tomita and Rabinowitch (1962)
	–	0.45[d]	6.8[d]	–	Tomita and Rabinowitch (1962)
	–	0.51[c]	7.8[c]	–	Tomita and Rabinowitch (1962)
	–	–	6.3[b]	–	Butler and Norris (1963)
	–	0.35–0.39[e]	–	–	Gradyushko et al. (1970)
	666	–	–	5	Boardman and Thorne (1971)
	668	–	–	7	White et al. (1972)
	668[a], 731[a]	0.24[a]	–	–	Jeffrey (1972)
	–	–	6.12–6.90[d]	–	Kaplanova and Cermak (1981)
	–	–	6.19–7.30[d]	–	Kaplanova and Cermak (1981)
	–	–	6.09	–	Connolly et al. (1982a)
	–	–	6.05[f]	–	Connolly et al. (1982a)
	–	–	5.44[d]	–	Connolly et al. (1982a)
	–	0.33	6.09	–	Moog et al. (1984)
	664[b]	0.30–0.33[b]	5.5 ± 0.3[b]	–	Jabben et al. (1986)
	670[c]	0.28–0.30[c]	–	5	Kobayashi (1989)
	–	0.32	6.0	–	Leupold et al. (1990)
	674[d]	–	6.0	8	Vladkova (2000)
	674[b]	–	6.15	–	Vladkova (2000)
	669[a]	–	6.15	7	Vladkova (2000)
	–	–	5.9[d]	–	Takeuchi and Amao (2005)
	671.1[c]	0.39[c]	5.39[c]	5.8	This work
DV–Chl a	665	0.36 ± 0.04	6.0 ± 0.2	5	Steglich et al. (2003)
	669.6[c]	0.37[c]	5.50[c]	4.4	This work
Phe a	672.5			5.2	Smith and Benitez (1955)
	673	–	–	6	French et al. (1956)
	–	0.175[a]	–	–	Weber and Teale (1957)
	672	–	–	5	Boardman and Thorne (1971)
	673	–	–	6	White et al. (1972)
	674	0.22–0.24[b]	–	7	Kobayashi (1989)
	673[b]	–	–	–	Chen et al. (2009)
	677.7[c]	0.28[c]	6.11[c]	6.9	This work
DV–Phe a	674.8[c]	0.28[c]	6.00[c]	4.0	This work
Chl b	648			6	French et al. (1956)
	–	0.16	–	–	Latimer et al. (1956)
	–	0.084	–	–	Latimer et al. (1956)
	–	0.117	–	–	Weber and Teale (1957)
	–	0.11–0.122[c]	–	–	Weber and Teale (1957)
	–	0.09[a]	–	–	Weber and Teale (1957)
	–	0.095[b]	–	–	Weber and Teale (1957)
	–	0.10[d]	–	–	Weber and Teale (1957)
	–	0.17	4.0	–	Tomita and Rabinowitch (1962)
	–	0.26[d]	6.0[d]	–	Tomita and Rabinowitch (1962)
	–	0.28[c]	6.4[c]	–	Tomita and Rabinowitch (1962)
	646	–	–	4	Boardman and Thorne (1971)

(Continued)

TABLE 6.2 (CONTINUED)
Fluorescence Properties of Natural Chlorophylls in Diethyl Ether at Room Temperature

Compound	λ_{max} (nm)	ϕ_F	τ_F (ns)	Stokes Shift (nm)	Reference
	649	–	–	7	White et al. (1972)
	652[a], 710[a]	0.09[a]	–	6	Jeffrey (1972)
	649	0.12	3.5 ± 0.3	7	Jabben et al. (1986)
	651.1[c]	0.16[c]	2.96[c]	4.6	This work
DV–Chl b	650.5	0.10 ± 0.01	2.9 ± 0.2	–	Steglich et al. (2003)
Phe b	661			7	Smith and Benitez (1955)
	661	–	–	7	French et al. (1956)
	658	–	–	4	Boardman and Thorne (1971)
	661	–	–	7	White et al. (1972)
	660.7[c]	0.20[c]	5.31[c]	3.8	This work
Chl c_1	632				Wilhelm (1987)
	633[a], 694[a]	0.16[a]	–	–	Jeffrey (1972)
Chl c_2	632	–	–	–	Wilhelm (1987)
	635[a], 696[a]	0.15[a]	–	–	Jeffrey (1972)
	638[b], 700[b]	0.08[b]	–	–	Jeffrey (1972)
Chl d	693	–	–	7	Manning and St rain (1943)
	696	–	–	10	Smith and Benitez (1955)
	699	–	–	13	French et al. (1956)
	695	–	–	9	Goedheer (1966)
	698.4[c]	0.36[c]	5.54[c]	4.7	This work
Phe d	701	–	–	9	French et al. (1956)
Chl f	711.7[c]	0.39[c]	5.15[c]	10.8	This work
[Zn]–Chl a	663[c]	0.21–0.23[c]	–	–	Kobayashi (1989)
	–	–	5.2[d]	–	Takeuchi and Amao (2005)
[Cd]–Chl a	670[c]	0.040–0.043[c]	–	–	Kobayashi (1989)
[Pd]–Chl a	638[c]	0.0025[c]	–	–	Kobayashi (1989)
[Sn(IV)]–Chl a	664[c]	0.096–0.110[c]	–	–	Kobayashi (1989)
[Ba]–Chl a	683[c]	0.0071–0.0077[c]	–	–	Kobayashi (1989)
BChl a	782.0	0.196	3.21	11	Connolly et al. (1982b)
	785.5[a]	0.189[a]	3.14[a]	16.5	Connolly et al. (1982b)
	793.5[c]	0.190[c]	2.73[c]	12.5	Connolly et al. (1982b)
	795.0[b]	0.144[b]	2.50[b]	–	Connolly et al. (1982b)
	793.0[d]	0.129[d]	2.32[d]	17.5	Connolly et al. (1982b)
	793.5[e]	0.152[e]	2.54[e]	–	Connolly et al. (1982b)
	793[d]	0.11 ± 0.01[d]	2.32[d]	18	Becker et al. (1991)
	795[e]	0.14 ± 0.01[e]	2.54[e]	20	Becker et al. (1991)
	786[a]	0.16 ± 0.01[a]	3.14[a]	17	Becker et al. (1991)
	782	0.19	3	11	Teuchner et al. (1997)
	800[f]	0.189[f]	2.6[f]	–	Musewald et al. (1998)
BPhe a	760	–	–	11	Smith and Benitez (1955)
	764.5	0.126	2.54	15.5	Connolly et al. (1982b)
	766[d]	0.08 ± 0.02[d]	2.25[d]	20	Becker et al. (1991)
	763[a]	0.10 ± 0.01[a]	2.55[a]	17	Becker et al. (1991)
[Zn]–BChl a	777	0.14	2.6	14	Teuchner et al. (1997)
	790[f]	0.110[f]	2.1[f]	–	Musewald et al. (1998)
[Cd]–BChl a	775	0.03	0.8	–	Teuchner et al. (1997)
[Pd]–BChl a	762	0.004	<0.2	–	Teuchner et al. (1997)
	–	0.005[f]	0.065[f]	–	Musewald et al. (1998)
[Ni]–BChl a	–	–	–	–	Teuchner et al. (1997)
[Cu]–BChl a	–	–	–	–	Teuchner et al. (1997)

Note: In diethyl ether at room temperature.

[a] In acetone.

[b] In ethanol.

[c] In benzene.

[d] In methanol.

[e] In propanol.

[f] In toluene.

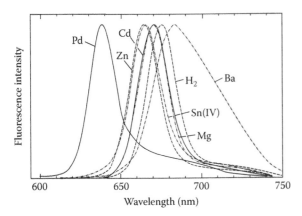

FIGURE 6.16 Emission spectra of [M]-Chls *a* in benzene: M = Mg, 2H, Zn, Sn(IV), Cd, Ba, and Pd. Emission maxima are arbitrarily scaled to a common height.

are in the range of 2.0–3.6, significantly smaller than those of Chls *a*, *d*, and *f*, but almost the same as that of Chl *b*. As illustrated in Figure 6.17, a weak linear correlation for τ versus ε or Soret/Q_Y intensity ratio is also seen. These findings provide insight into the fluorescence behavior of chlorophylls, and further theoretical explanation is expected.

6.5 CD SPECTRA

CD spectra are very useful for distinction between the primed chlorophyll, e.g., Chl *a'*, and the corresponding nonprimed one, Chl *a*, although the absorption characteristics of the primed derivatives (Chls *a'*, *b'*, *d'*, *f'*, DV-Chl *a'*, Phes *a'*, *b'*, *d'*, *f'*, and DV-Phe *a'*) are identical with those of the nonprimed ones (Wolf and Scheer 1973; Weiss 1978; Watanabe et al. 1984; Kobayashi et al. 2006b, 2013; Komatsu et al. 2014).

6.5.1 CHLS *a'*, *b'*, *d'*, *f'*, DV-CHL *a'*
AND THEIR PHEOPHYTINS

The CD spectra of Chl *a/a'*, *b/b'*, *d/d'*, *f/f'* and DV-Chl *a/a'* in benzene are illustrated in Figure 6.19. For a given pair of epimers, the CD spectra are considerably different, although the absorption spectra are practically identical with each other (Kobayashi et al. 2013; Komatsu et al. 2014). For each of Chls *a'*, *b'*, *d'*, *f'* and DV-Chl *a'*, an intense negative CD is associated with $Q_Y(0,0)$ and a well-defined weakly negative satellite with $Q_Y(1,0)$, although weakly positive satellite with $Q_Y(1,0)$ for DV-Chl *a'*. On the other hand, the nonprimed species, Chls *a*, *b*, *d*, *f* and DV-Chl *a*, show complicated, very weak negative, and/or positive activities at these transitions. In Figure 6.20, all pheophytins show negative activities, and primed ones reveal stronger and blue-shifted signals compared to the nonprimed ones, although the absorption spectra of the primed derivatives are also identical with those of the nonprimed ones. The findings suggest that the Q_Y maximum transition consists of at least two bands, and shorter wavelength band shows stronger activity in primed Phes and longer wavelength band stronger in nonprimed Phes.

A series of Q_X transitions occur in the *valley* of the absorption spectrum. The positive CD activities are derived from the $Q_X(0,0)$ absorption satellites (bands III) of Chl *a'*, Phe *a'*, Chl *d'*, Phes *d*, *d'*, Chl *f'*, Phes *f*, *f'*, and DV-Phe *a'*, respectively, while Chl *f* and Phe *a* exhibit weakly negative activities at 609 and 535 nm, respectively. DV-Chl *a* and DV-Phe *a* showed practically no intensity in this region. Note that although the absorption peak associated with $Q_X(0,0)$ transition of DV-Phe *a'* is very small as compared with that of MV-Phe *a'*, the corresponding CD signal of DV-Phe *a'* is strong enough to be clearly seen. The CD activity associated with the $Q_X(0,1)$ absorption satellites (band IV) at the shorter wavelength is very weak and vague in all the pigments shown in Figures 6.19 and 6.20.

Band B in the Soret absorption consists of two nearly degenerate electronic transitions, $B_X(0,0)$ and $B_Y(0,0)$. All the primed derivatives gave single and strongly positive CD spectra at this absorption peak, suggesting that the two transitions contribute to CD spectra in a similar manner (Watanabe et al. 1984). In contrast, the CD spectra of nonprimed species, except Chl *f* and Phes, apparently reflect the existence of the two transitions: they show a maximum and a minimum with the center wavelength roughly coinciding with the Soret absorption maximum. Different feature of CD spectrum for Chl *f* among Chls may come from its characteristically splitted Soret absorption arising from B-bands and so-called η-bands in benzene. Primed Chls exhibit relatively intense negative CD spectra in the near ultraviolet at η-bands of the Soret region (Weiss 1978; Petke et al. 1979), whereas nonprimed ones exhibit positive activities. Phes *a* and *f* also show positive CD spectra at η-bands in the near ultraviolet region, but such a tendency is not clear in Phe *b*, and Phe *d* exhibited negative activity, although all the primed Phes show negative activity and Phe *d'* showed the most intense activity.

The positive CD spectra at η-bands of DV-Chl *a* and DV-Phe *a* are significantly weaker than those of MV-Chl *a* and MV-Phe *a*. Relatively a little weaker CD intensity seen in DV-Chl *a* and DV-Phe *a* may be caused by the configuration difference at C8 between DV- and MV-derivatives, because the C8-ethyl moiety in MV-Chl *a* was reported to be perpendicular to the ring plane, but the direction of the C8-vinyl side chain was almost coplanar to the ring plane in DV-Chl *a*, based on the density functional theory (DFT) calculations (Tomo et al. 2009).

6.5.2 BCHLS *g* AND *g'*

One can distinguish, with some difficulty, BChl *g'* from BChl *g* by their absorption spectra in benzene (Figure 6.21a), in particular, the shape of the Q_X band, while they are hardly distinguishable in other solvents, as mentioned in Section 6.3.3. Like Chl *a/a'*, however, the CD spectra of BChl *g* and *g'* are distinctly different from each other (Figure 6.21b), and the two epimers can be readily identified. We had better note that the BChl *g/g'* pair shows excellent mirror-image CD spectra typical for compounds with a single asymmetric carbon atom, as in D- and L-alanine ($CH_3C^*H(NH_2)COOH$), whereas such

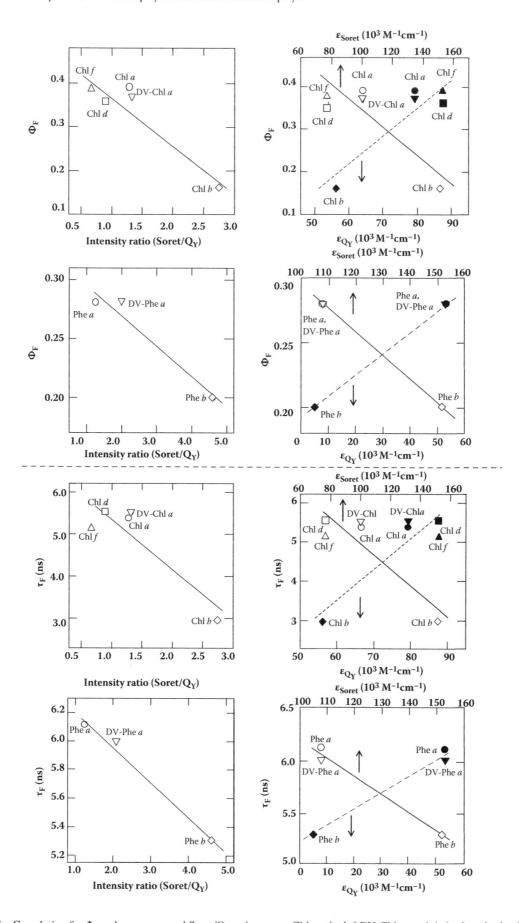

FIGURE 6.17 Correlation for Φ_F and τ versus ε and Soret/Q_Y ratio among Chls a, b, d, f, DV-Chl a, and their pheophytins in benzene.

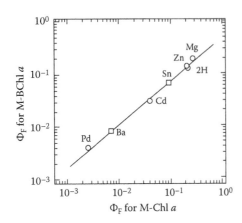

FIGURE 6.18 Correlation for fluorescence quantum yield, Φ_F, between [M]-chlorophylls a in benzene and [M]-BChls a in diethyl ether. The values of Φ_F of [Sn(IV)]-BChl a and [Ba]-BChl a are predicted values only. (Adapted from Kobayashi M et al., *Current Topics in Plant Biology*, 1: 17–35, 1999.)

a nice mirror image is not seen in other epimer pairs (see Figures 6.19 and 6.20).

6.5.3 [Mg]- AND [Zn]-BCHL a

The CD spectrum of Zn-BChl a is similar in shape to that of BChl a, although the positions of the peaks differed (Figure 6.22), indicating that the stereochemistry of C7, C8, c13^2, C17, and C18 in Zn-BChl a is the same as in BChl a (Kobayashi et al. 1998a).

6.6 MASS SPECTRA

Chlorophylls in natural photosynthesis are sometimes present in very small amounts, and hence the use of mass spectrometry (MS) can be advantageous since only minute samples are required. Mass spectra can provide accurate and useful information not only on molecular weights and elemental compositions but also on the nature of functional groups attached to the macrocycle (e.g., phytol) and of the central metal (see reviews by Smith 1975; Hunt and Michalski 1991; Porra and Scheer 2000; Kobayashi et al. 2006b, 2013).

In mass spectrometric analysis, a molecular ion (M$^{+\cdot}$) and/ or a protonated molecular ion ([M+H]$^+$) are produced, which may subsequently undergo fragmentation. A variety of ionization techniques are used (see reviews by Hunt and Michalski 1991; Hoffmann and Stroobant 2002): atmospheric pressure chemical ionization (APCI), electron ionization (EI), electrospray ionization (ESI), fast atom bombardment (FAB), field desorption (FD), laser desorption (LD), matrix-assisted laser desorption/ionization (MALDI), and plasma desorption (PD) have all been used. Some typical examples of Chls and BChls, which have been measured by APCI-, FAB- and MALDI- mass spectrometry, are presented here.

The characteristic feature of the mass spectra of chlorophylls is the way in which the fragments are split into two separate groups: the highest intense mass fragment is due to

the molecular ion, M$^+$, and one of the other strong fragment is due to the subsequent loss at C17^3 of the long alcohol chain, e.g., phytol, [M—phytyl(C$_{20}$H$_{39}$) + H]$^+$.

6.6.1 APCI-MASS SPECTRA: CHLS a, b, d, f AND THEIR PHES

Typical APCI mass spectra are illustrated in Figure 6.23. Chl a (C$_{55}$H$_{72}$MgN$_4$O$_5$, monoisotopic mass; 892.535; hereafter, the value in the bracket shows the monoisotopic mass of the molecule or the ion) gives the protonated molecule ([M+H]$^+$) at m/z 893.2 producing the dominant fragment ion at m/z 615.1. The mass difference 278 between [M+H]$^+$ and the product ion corresponds to C$_{20}$H$_{38}$. This suggests the presence of a phytyl chain in Chl a. The other product ions at m/z 583.0 and m/z 555.2 corresponding to [M+H-278-32]$^+$ and [M+H-278-60]$^+$, respectively, are supposed to be the results of the loss of carboxymethyl group followed by the cleavage of phytol. The losses of 278, 310, and 338 from the precursor ion in MS/MS spectra are seen in all the pigments examined here, which reveals the presence of a phytyl chain.

Chl d (C$_{54}$H$_{70}$MgN$_4$O$_6$, 894.515) gives the protonated molecule ([M+H]$^+$) at m/z 895.2 and the prominent fragment ion at m/z 617.1. Though Chls b and f (both C$_{52}$H$_{70}$MgN$_4$O$_6$, 906.515) are eluted at different LC retention times, they show the same mass and MS/MS spectral patterns, [M+H]$^+$ at m/z 907.2 and the dominant product ion at m/z 629.1, which correspond to [C$_{52}$H$_{71}$MgN$_4$O$_6$]$^+$ (907.522) and [M-C$_{20}$H$_{38}$+H]$^+$ (629.225). These results suggest that Chl f also possesses a phytyl long chain such as Chls a, b, and d, and that most probably one –CH$_3$ moiety of Chl a is substituted for the –CHO group in Chl f like Chl b, yielding [2-formyl]-Chl a, [12-formyl]-Chl a, or [18-formyl]-Chl a (Kobayashi et al. 2013; Miyashita et al. 2014).

DV-Chl a (C$_{55}$H$_{70}$MgN$_4$O$_5$, monoisotopic mass; 890.520) gives the protonated molecule ([M+H]$^+$) at m/z 891.3 producing the dominant fragment ion at m/z 613.1. The mass difference 278 between [M+H]$^+$ and the product ion corresponds to C$_{20}$H$_{38}$, supporting the presence of a phytyl chain in DV-Chl a. The other product ions at m/z 581.2 and m/z 553.3 corresponding to [M+H-278-32]$^+$ and [M+H-278-60]$^+$, respectively, are the results of the loss of carboxymethyl group followed by the cleavage of phytol. The losses of 278, 310, and 338 from the precursor ion in MS/MS spectra reveal the presence of a phytyl chain. All the observed values are 2 Da smaller than those observed in MV-Chl a, due to the replacement of –CH$_2$CH$_3$ with –CH=CH$_2$ in DV-Chl a (Komatsu et al. 2014).

As seen in Figure 6.23 (right), the corresponding pheophytins prepared by acid treatment clearly showed the absence of magnesium (Figure 6.1). For example, [M+H]$^+$ of Phe a is observed at m/z 871.3, which is 22 Da smaller than that of Chl a, showing the substitution of Mg with two H atoms by pheophytinization (see Figure 6.1). Phes b, f (C$_{52}$H$_{74}$N$_4$O$_6$, 884.545), Phe d (C$_{54}$H$_{72}$N$_4$O$_6$, 872.545), and DV-Phe a show the similar pattern, supporting that all of them do not possess Mg as central metal.

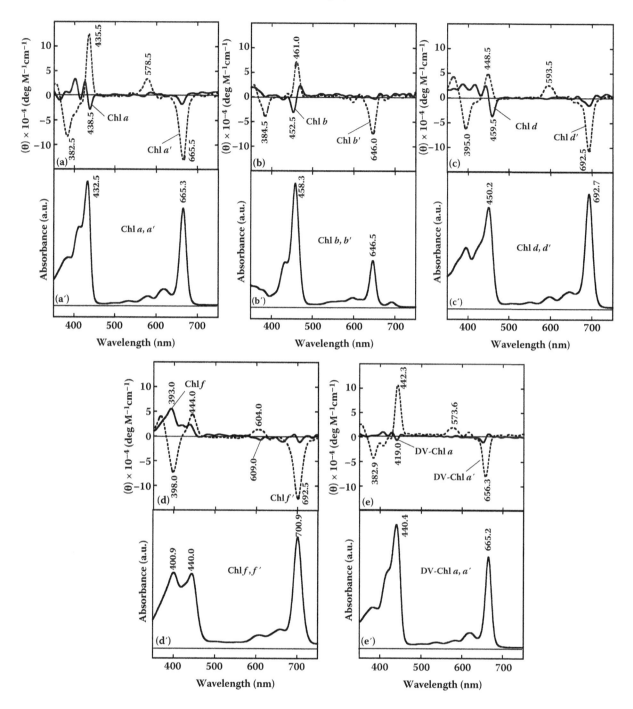

FIGURE 6.19 CD spectra for (a) Chls a/a', (b) Chls b/b', (c) Chls d/d', (d) Chls f/f', (e) DV-Chls a/a', and (a'–e') the corresponding absorption spectra in benzene at room temperature. [θ] denotes the molar ellipticity.

6.6.2 FAB-Mass Spectra

6.6.2.1 Chl a and Chl d

As shown in Figure 6.24, a typical FAB-mass spectrum of Chl d ($C_{54}H_{70}N_4O_6Mg$, calcd. 894.5146) with m-nitrobenzyl alcohol as a matrix has a molecular ion peak [M+H]$^+$ at m/z 895, which is 2 mass units higher than the corresponding peak [M+H]$^+$ at m/z 893 for Chl a ($C_{55}H_{72}N_4O_5Mg$, calcd. 892.5353). The intense peaks at m/z 614 and 616 were derived from loss

of the esterifying phytol from Chl a and Chl d, respectively (Kobayashi et al. 2005).

6.6.2.2 BChl 663: Chl a Esterified with Δ2,6-Phytadienol, Chl aΔ2,6$_{PD}$

The reaction center (RC) component, A$_0$, of green sulfur bacteria was designated BChl 663 after its absorption maximum in the HPLC eluent, as mentioned in Section 6.1.7. A typical FAB-mass spectrum of BChl 663 (Figure 6.25, left) with

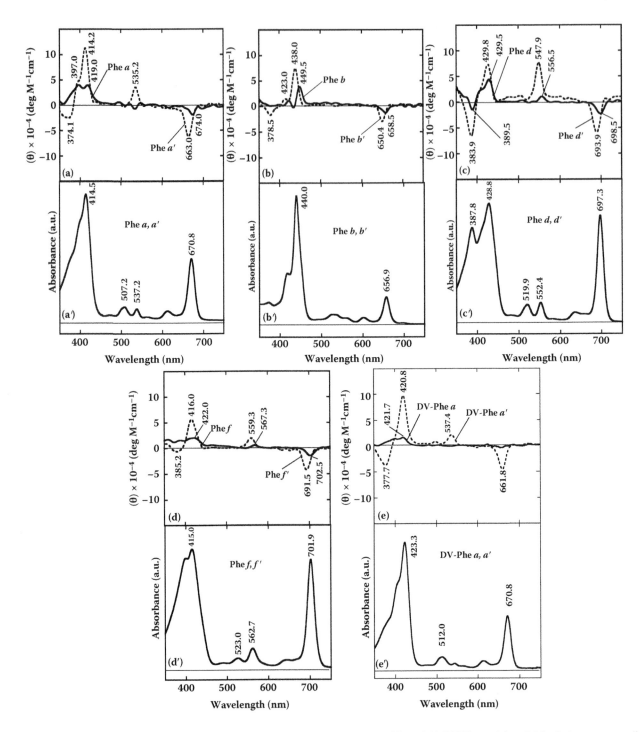

FIGURE 6.20 CD spectra for (a) Phes *a*/*a*′, (b) Phes *b*/*b*′, (c) Phes *d*/*d*′, (d) Phes *f*/*f*′, and (e) DV-Phes *a*/*a*′, and (a′–e′) the corresponding absorption spectra in benzene at room temperature. [θ] denotes the molar ellipticity.

m-nitrobenzyl alcohol as a matrix has molecular ion peaks [M]⁺ and [M+H]⁺ at *m/z* 890.6 and 891.6, respectively, which are 2.0 mass units smaller than the [M]⁺ and [M+H]⁺ peaks at *m/z* 892.6 and 893.6 of Chl *a*$_P$ (C$_{55}$H$_{72}$N$_4$O$_5$Mg, calcd. 892.5353). FAB-MS spectra of BChl 663 and Chl *a*$_P$ both show an identical intense mass peak at *m/z* 614.3 indicating that the 2.0 mass units difference between Chl *a*$_P$ and BChl 663 is due to a difference in the esterifying alcohol. The

high-resolution mass measurement of BChl 663 gives a value of *m/z* 890.5191, consistent with the calculated value of Chl *a* esterified with phytadienol (Chl *a*$_{PD}$: C$_{55}$H$_{70}$N$_4$O$_5$Mg, calcd. 890.5197) (Kobayashi et al. 2000). Thus, MS analysis gave unequivocal evidence for the presence of an additional C=C double bond in the phytyl chain of BChl 663 (Figure 6.1). The position of this double bond, however, could not be clarified by MS, and awaited NMR analysis.

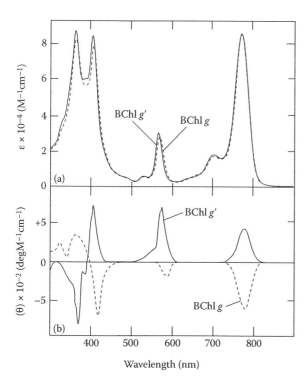

FIGURE 6.21 (a) Absorption spectra and (b) CD spectra of BChls g and g' in benzene.

6.6.2.3 [Mg]-, [2H]-, and [Zn]-BChl a

The FAB-mass spectrum of [Zn]-BChl a ($C_{55}H_{74}N_4O_6{}^{64}Zn$; mol wt = 950) shows a molecular ion [M]$^+$ at m/z 950.5, while the [M]$^+$ ion of BChl a ($C_{55}H_{74}N_4O_6Mg$) is 39.9 mass units lower at m/z 910.6, which is consistent with the mass difference between ^{64}Zn (63.9) and Mg (24.0) atoms (Figure 6.25, right) (Kobayashi et al. 1998a). An intense fragment at m/z 672.2 corresponds to the loss of the phytyl group ($C_{20}H_{39}$ = 279 m/z) from [Zn]-BChl a [950.5 − 279 + H]$^+$. The same dephytylated fragment occurs 39.9 mass units lower at m/z 632.3 with [Mg]-BChl a due to the mass difference between ^{64}Zn and Mg. The FAB-mass spectrum of BPhe a ([2H]-BChl a) shows an M$^{+\cdot}$ ion at m/z 888, which coincides with M$^{+\cdot}$ of BPhe a ($C_{55}H_{74}N_4O_62H$), while the fragment at m/z 610 arises by loss of the phytyl group [888 − 279 + H]$^+$.

The two satellite peaks at m/z 952.5 and 954.5 near M$^{+\cdot}$ of [Zn]-BChl a ($C_{55}H_{74}N_4O_6{}^{64}Zn$) at m/z 950.5 arise from the isotopic molecular ions, $C_{55}H_{74}N_4O_6{}^{66}Zn$ and $C_{55}H_{74}N_4O_6{}^{68}Zn$, respectively: the natural abundance is ^{64}Zn(48.6%), ^{66}Zn(27.9%), and ^{68}Zn(18.8%). The high-resolution mass spectrum of [Zn]-BChl a gives a major peak at m/z 950.4935, which can only be explained by one rational formula, $C_{55}H_{74}N_4O_6{}^{64}Zn$ (calcd. 950.4899) (Kobayashi et al. 1998a).

6.6.3 MALDI-MASS SPECTRA

Recently, MALDI-mass is often used for chlorophyll analyses. A typical example of a MALDI-mass spectrum of Chl f is shown in Figure 6.26a, where the MALDI-mass spectrum has

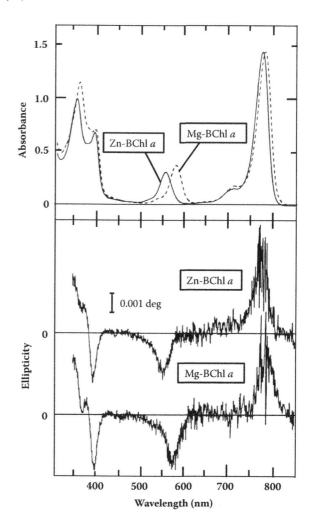

FIGURE 6.22 (Top) Absorption spectra of Zn-BChl a purified from *A. rubrum* and Mg-BChl a from *Rba. sphaeroides*, and (bottom) their CD spectra in benzene.

the molecular ion at m/z 906, and the major fragment ion at m/z 628 is due to the loss of a phytyl (Willows et al. 2013); the other major ion at m/z 657 is due to the terthiophene matrix. The other major fragment ions at m/z 597.1 and 569.1 seen in the APCI-mass spectrum (Figure 6.26b) are not clear in the MALDI-mass spectrum (Figure 6.26a).

We had better pay attention to the selection of matrix when we use MALDI-mass for chlorophyll analysis. For example, we show MALDI-mass spectra of Chl a and BChl a with three different conventional matrices, 3-hydroxy-2-picolinic acid (HPA), 2,5-dihydroxybenzoic acid (DHB), and 4-hydoxy-α-cyanocinnamic acid (CHCA) (Figure 6.27). As seen in Figure 6.27a, a′, nice MALDI-mass spectra of both Chl a and BChl a are obtained when HPA is used as a matrix, while the minor fragment ion at m/z 870.8 for Chl a due to the loss of Mg is also detected. The loss of Mg is remarkable, when DHB (Figure 6.27b, b′) or CHCA (Figure 6.27c, c′) is selected, in particular, a molecular ion [M]$^+$ is not observed when CHCA is used (Figure 6.27c, c′). It is of interest to note that all phytol-free fragments at m/z 614 for Chl a and m/z 632.6 for BChl a

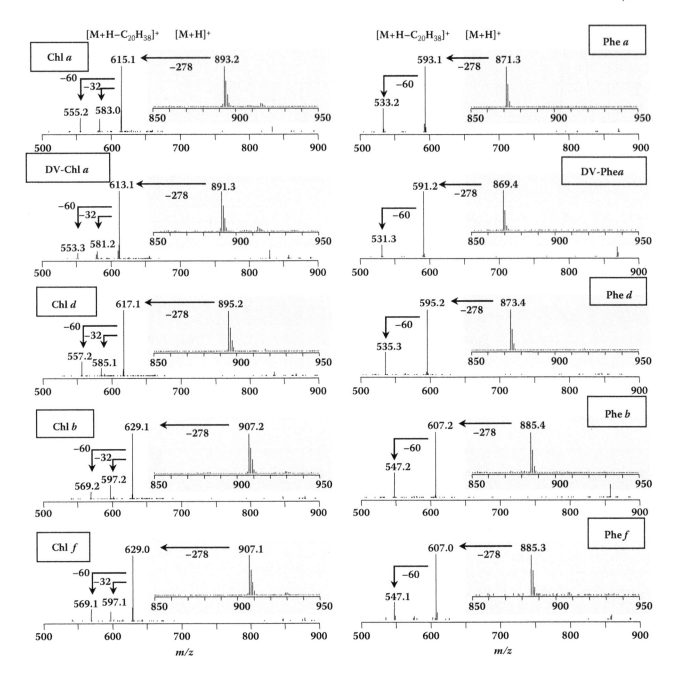

FIGURE 6.23 APCI-mass spectra of Chls *a*, *b*, *d*, *f*, DV-Chl *a* (left column), Phes *a*, *b*, *d*, *f*, and DV-Phe *a* (right column). Each mass spectrum of the chlorophyll fraction is shown in the shaded square. MS/MS spectra of the protonated molecules ([M+H]$^+$) of Chls *a*, *b*, *d*, *f*, and DV-Chl *a* give product ions of [M+H-278]$^+$, [M+H-278-32]$^+$, and [M+H-278-60]$^+$. Pheophytins *a*, *b*, *d*, *f*, and DV-Phe *a* give product ions of [M+H-278]$^+$ and [M+H-278-60]$^+$.

show the presence of Mg. In Figure 6.27, DHB is the most unfavorable matrix for Chl *a*.

6.7 NMR SPECTRA

NMR spectroscopy can provide precise information about Chl structure. Coupled use of NMR with HPLC, absorption, CD, and mass spectrometries has not only definitively identified the structures of several major naturally occurring Chls but also assisted recent studies on minor Chl pigments, present in minute quantities, such as electron donors and acceptors in the RC.

6.7.1 CHLS *a*, *b*, *d*, *f* AND DIVINYL CHL *a*

6.7.1.1 ¹H-NMR Spectra

As observed in one-dimensional ¹H-NMR spectra of Chls *a*, *b*, *d*, and *f* (Figure 6.28, Table 6.3), marked differences are seen in the signals arising from the formyl group (Kobayashi et al. 2013; Willows et al. 2013). Each low-field singlet signal characteristic of the formyl moiety observed around 11 ppm in the spectra of Chls *b* (7^1), *d* (3^1), and *f* (2^1) is absent from the spectrum of Chl *a*. Similarly, double doublet signal of 3^1-H vinylic proton at 8–8.5 ppm in the spectra of Chls *a*, *b*, and *f* is

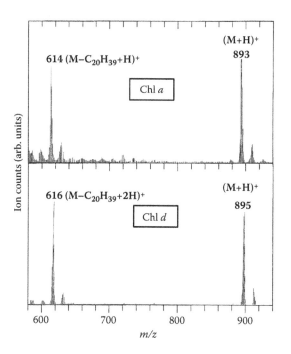

not seen in the spectrum of Chl d. These results reconfirm that Chl d is 3-desvinyl-3-formyl Chl a ([3-formyl]-Chl a).

Here we note that the 3^1-H vinylic proton shows a large downfield shift in Chl f (8.534 ppm) and that is slightly upfield-shifted in Chl b (8.043 ppm), as compared to Chl a (8.162 ppm), suggesting that Chl f should be formylated along the y-axis and that the interaction between –CH=CH$_2$ and –CHO in Chl f is rather strong than that in Chl b. The –CHO substitution position in Chl f is hence most probably at C2, next to the 3-vinyl group.

The pair signals of 3^2- and $3^{2'}$-H vinylic protons are well resolved in the spectra of Chls a (6.242 and 6.028 ppm) and b (6.302 and 6.055 ppm), while the corresponding pair signals in Chl f show low resolution (6.365 and 6.324 ppm), suggesting that the environment of 3-vinyl moiety in Chls a and d is very similar to each other, but is profoundly different from that in Chl f, most probably due to the presence of formyl moiety at the neighboring C2 in Chl f. The ^{1}H singlet signals for 7^1-CH$_3$ in Chls a, d, and f are at 3.3 ppm and absent from the spectrum of Chl b, while the ^{1}H signal for 7^1-CHO in Chl b appears on a much lower field at 11.3 ppm. Another marked difference seen in Chl f spectrum is the disappearance of the singlet signal of 2^1-CH$_3$ proton; the corresponding signals are observed at 3.3–3.72 ppm in the spectra of Chls a, b, and d,

FIGURE 6.24 FAB-mass spectra of Chl a and Chl d. Matrix: m-nitrobenzyl alcohol; acceleration voltage: 10 kV. (Adapted from Kobayashi M et al., *Photosynth. Res.*, 84: 201–207, 2005.)

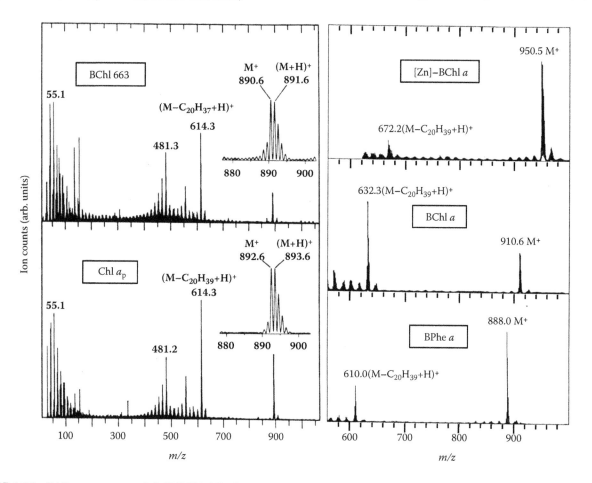

FIGURE 6.25 FAB-mass spectra of (left) BChl 663, Chl a_p and (right) [Zn]-BChl a, BChl a ([Mg]-BChl a), and BPhe a ([2H]-BChl a). Matrix: m-nitrobenzyl alcohol; acceleration voltage: 10 kV. (Adapted from Kobayashi M et al., *Anal. Chim. Acta*, 365: 199–203, 1998; Kobayashi M et al., *Photosynth. Res.*, 63: 269–280, 2000.)

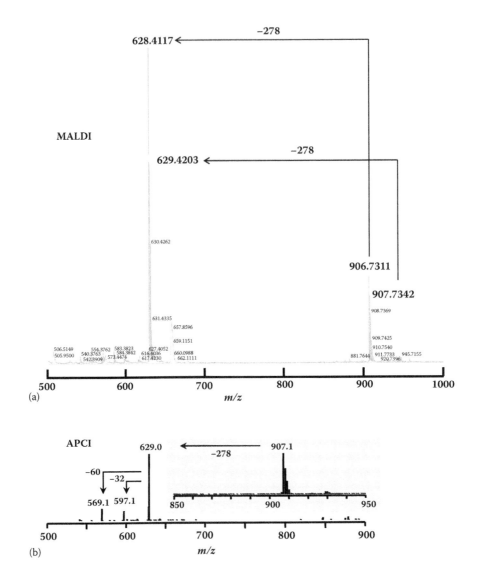

FIGURE 6.26 (a) MALDI- and (b) APCI-mass spectra of Chl *f*. Matrix for (a) is terthiophene. (a: Adapted from Willows RD et al., *Org. Lett.*, 15: 1588–1590, 2013; b: Adapted from Kobayashi M et al., Physicochemical properties of chlorophylls in oxygenic photosynthesis—Succession of co-factors from anoxygenic to oxygenic photosynthesis. In *Photosynthesis*, ed. by Dubinsky Z, Intech, Croatia, Chapter 3, pp. 47–90, 2013.)

implying that 2^1-CH$_3$ of Chl *a* is substituted in Chl *f* for some other moiety, most probably –CHO. No other change is clear in the one-dimensional ^{1}H-NMR spectra.

As observed in one-dimensional ^{1}H-NMR spectra of MV-Chl *a* and DV-Chl *a* (Figure 6.28, Table 6.3), marked differences are seen in the signals arising from the ethyl and vinyl groups (Komatsu et al. 2014). Two high-field signals characteristic of the ethyl group observed at 3.817 ppm (q, 8^1-H) and 1.696 ppm (t, 8^2-H) in the spectra of MV-Chls *a* were absent from the spectrum of DV-Chl *a*. Instead, the signals derived from 8-vinylic protons were observed at 8.180 ppm (dd, 8^1-H), 6.237 ppm (d, 8^2-H), and 6.019 ppm (d, 8^2-H), in the spectrum of DV-Chl *a*, as well as the signals of 3-vinylic protons at 8.291 ppm (dd, 3^1-H), 6.438 ppm (d, 3^2-H), and 6.266 ppm (d, 3^2-H); the latter signals in the spectrum of MV-Chls *a* appeared at 8.162 ppm (dd, 3^1-H), 6.242 ppm (dd, 3^2-H), and 6.028 ppm (dd, 3^2-H), respectively. The results confirm that

8-CH$_2$CH$_3$ of Chl *a* is substituted by –CH=CH$_2$ in DV-Chl *a*. No other clear changes are observed in the one-dimensional ^{1}H-NMR spectra.

6.7.1.2 ^{13}C-NMR Spectra

In the one-dimensional ^{13}C-NMR spectra of Chls *a*, *b*, *d*, and *f* (Figure 6.29, Table 6.4), marked differences are noted in the range of 0–20, 120–140, and 180–200 ppm, relating to the –CH$_3$, –CH=CH$_2$, and –CHO moieties (Kobayashi et al. 2013). Compared to Chl *a*, in the spectrum of Chl *f*, the ^{13}C signal of 2^1-CH$_3$ is absent; the 7^1-CH$_3$ and $3^{1,2}$-CH=CH$_2$ carbon signals appear at 11, 130, and 126 ppm, respectively, similar to Chl *a* (11, 131, and 120 ppm). Further, a new carbon signal appears at 189 ppm, close to the signals of –CHO in Chls *b* (188 ppm) and *d* (190 ppm), supporting the presence of –CHO at C2 in Chl *f*.

The signals of 7^1-CH$_3$ of Chls *a*, *d*, and *f* show almost the same chemical shifts, 11.15, 11.24, and 11.17 ppm, respectively,

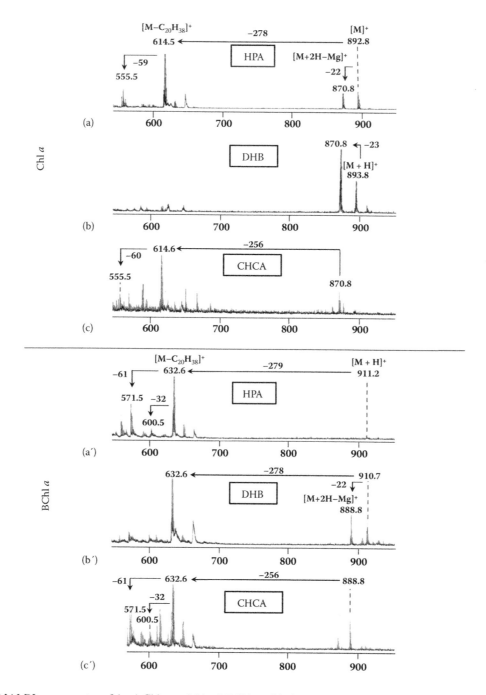

FIGURE 6.27 MALDI-mass spectra of (a–c) Chl *a* and (a′–c′) BChl *a* with three different matrixes, 3-hydroxy-2-picolinic acid (HPA), 2,5-dihydroxybenzoic acid (DHB), and 4-hydoxy-α-cyanocinnamic acid (CHCA).

indicating that the interaction between the –CHO and 7^1-CH_3 moieties in Chls *d* and *f* are negligibly small; hence, the –CHO substituent is not so close to the 7^1-CH_3 moiety in Chls *d* and *f*. The chemical shifts of 3^2-vinyl carbons in Chls *a* and *b* are almost identical (120.3–120.8 ppm), but Chl *f* shows a slight but significant downfield shift (126.3 ppm), suggesting that the formyl substituent in Chl *f* is positioned very close to the 3-vinyl group, most probably at the next C2. The signals of –CHO moiety of Chls *d* and *f* exhibit almost the same chemical shift (189.59 and 189.27 ppm), but a slight upfield C-formyl signal is observed at 188.62 ppm in Chl *b*, indicating that the

environment of –CHO in Chls *d* and *f* is very similar, supporting that the –CHO moieties of Chls *d* and *f* are positioned at the same ring I, while that of Chl *b* is at ring II.

In the ^{13}C-NMR spectra of MV-Chl *a* and DV-Chl *a* (Figure 6.29, Table 6.4), marked differences are noted in the range of 0–20 and 120–140 ppm, relating to the –CH_2CH_3 and –$CH=CH_2$ moieties. Compared to MV-Chl *a*, the ^{13}C signals of 8-CH_2CH_3 were absent in the spectrum of DV-Chl *a*, while new 8-$CH=CH_2$ carbon signals appeared at 131 ppm (8^1-C) and 121 ppm (8^2-C), respectively, as well as the 3-$CH=CH_2$ carbon signals (130 ppm [3^1-C] and 124 ppm [3^2-C]), similar

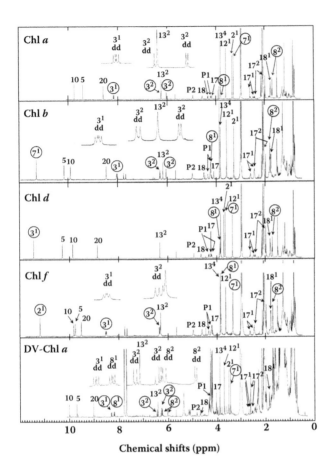

Chemical shifts (ppm)

FIGURE 6.28 ^{1}H-NMR spectra of Chls a, b, d, f, and DV-Chl a measured in acetone-d_6 at 273 K. Signals corresponding to ^{1}H atoms of the macrocycle are labeled with the numbers of the corresponding carbons. The peak at 2.06–2.11 ppm is acetone and 3.07–3.10 ppm is H$_2$O, respectively. (Adapted from Kobayashi M et al., Physicochemical properties of chlorophylls in oxygenic photosynthesis—Succession of co-factors from anoxygenic to oxygenic photosynthesis. In *Photosynthesis*, ed. by Dubinsky Z, Intech, Croatia, Chapter 3, pp. 47–90, 2013; Komatsu H et al., *Photomed. Photobiol.*, 36: 59–69, 2014.)

to Chl a (131 ppm [3^1-C] and 120 ppm [3^2-C]). Further remarkable differences are not seen in the one-dimensional ^{13}C-NMR spectra.

6.7.1.3 Two-Dimensional NMR

6.7.1.3.1 NOESY

Two-dimensional NMR spectra provide further information about a molecule than one-dimensional NMR spectra. NOESY is one of several types of two-dimensional NMR, where the nuclear Overhauser effect (NOE) between nuclear spins is used to establish the correlations. The cross peaks in the two-dimensional spectrum connect resonances from spins that are spatially close.

Here, we will trace the coherent correlations from meso-20-H, because Chl f possesses –CHO most probably at C2 near to C20 (Kobayashi et al. 2013; Willows et al. 2013). Coherent correlations can be easily traced from 20-H on

the NOESY spectrum of Chl a (Figure 6.30), where the signal of 20-H at 8.582 ppm shows three cross peaks with the signals of 18^1-H at 1.771 ppm, 2^1-H at 3.343 ppm, and 18-H at 4.572 ppm. Good coherent correlations can also be traced from meso-5-H and meso-10-H: (1) the signal of 5-H at 9.410 ppm shows three cross peaks with the signal of 7^1-H at 3.300 ppm, 3^2-H at 6.242 ppm, and 3^1-H at 8.162 ppm; and (2) the signal of 10-H at 9.749 ppm shows two cross peaks with the signal of 8^2-H at 1.696 ppm and 12^1-H at 3.619 ppm. In Chl b and Chl d, similar nice correlations are seen (Figure 6.30).

As shown in Figure 6.30, coherent correlations can be easily traced from 3^1-H of –CH=CH$_2$ on the NOESY spectrum of DV-Chl a, where the signal of 3^1-H at 8.291 ppm on the NOESY spectrum of DV-Chl a shows two cross peaks with the signals of 3^2-H at 6.438 and 6.266 ppm, respectively, while the signal of 3^1-H at 8.162 ppm on the MV-Chl a spectrum also possesses two cross peaks with the corresponding signal in the similar region, at 6.242 and 6.028 ppm. As expected, good coherent correlations can also be traced from 8^1-H, but the position of cross peaks is completely different from each other; in DV-Chl a, the signal of 8^1-H of –CH=CH$_2$ at 8.180 ppm shows two cross peaks with the signal of 8^2-H at 6.237 and 6.019 ppm, respectively, whereas in MV-Chl a, the signal of 8^1-H of –CH$_2$CH$_3$ at 3.817 ppm exhibits a single cross peak with the signal of 8^2-H at 1.696 ppm, and hence its cross peak position appears to be a faraway area of the two cross peaks of DV-Chl a.

6.7.1.3.2 HSQC

HSQC is a two-dimensional inverse correlation technique that allows for the determination of connectivity between two different nuclear species, and HSQC is selective for direct coupling. As illustrated in the ^{1}H-^{13}C HSQC spectra of Chls a, b, d, f and DV-Chl a (Figure 6.31), all substituents on the macrocycle show the corresponding cross peaks. For example, three cross peaks associated with 3^1-3^1 and 3^2-3^2 of –CH=CH$_2$ in both DV-Chl a and MV-Chl a appeared in the same area (in the lower left corner of Figure 6.31), but the area and number of 8^1-8^1 and 8^2-8^2 cross peaks were completely different, where three cross peaks in DV-Chl a are in the lower left corner, whereas two cross peaks in MV-Chl a are in the upper right corner of Figure 6.31. Other cross peaks are seen in a similar area. The results support that 8-CH$_2$CH$_3$ of Chl a is replaced with –CH=CH$_2$ in DV-Chl a. Similarly, one methyl group of Chl a is replaced with a formyl moiety in Chls b, d, and f.

6.7.1.3.3 HMBC

HMBC is also a two-dimensional inverse correlation method that allows for the determination of connectivity between two different nuclear species like HSQC, but HMBC gives longer range coupling (2-4 bond coupling) than HSQC.

Three meso-Hs in Chl a exhibit one to three cross peaks as seen in Figure 6.32. The formyl-H in Chls b, d, and f shows two to five cross peaks. For example, in the ^{1}H-^{13}C HMBC spectrum of Chl f, the ^{1}H-signal for 2^1-CHO has three cross

TABLE 6.3
^{1}H-Chemical Shifts of Chl *a*, *b*, *d*, *f*, and DV-Chl *a* in Acetone-d_6 at 273K[a,c]

IUPAC No. of Carbon Atom	Chl *a*[a]	BChl 663[b,i]	DV-Chl *a*[c]	Chl *b*[a]	Chl *d*[a]	Chl *f*[a]
2^1	3.343 (3.36)[b,i](s)	3.36(s)	3.511	3.316 (3.40)[d,j] (s) (3.22)[e,m]	3.724 (3.68)[f](s)	11.215 (11.35)[g,k](s), (11.165)[h,l](s)
3	–	–	–	–	–	–
3^1	8.162 (8.18)[b,i](dd)	8.18(q)	8.291(dd)	8.043 (7.95)[d,j] (dd) (7.85)[e,m]	11.460 (11.40)[f](s)	8.534(dd), (8.312)[h,l](dd)
3^2	6.242, 6.028(dd) (6.24, 6.03)[b,i]	6.24, 6.03	6.438(d), 6.266(d)	6.302, 6.055(dd) (6.25, 6.04)[d,j] (6.15, 5.98)[e,m]	–	6.324(dd), 6.365(dd), (6.266, 6.227)[h,l](d)
4	–	–	–	–	–	–
5	9.410 (9.40)[b,i](s)	9.39(s)	9.679	10.192 (10.04)[d,j] (s) (9.87)[e,m]	10.294 (10.20)[f](s)	9.770 (9.79)[g,k](s), (9.615)[h,l](s)
7^1	3.300 (3.30)[b,i](s)	3.30(s)	3.439	11.305 (11.22)[d,j] (s) (10.92)[e,m]	3.365 (3.33)[f](s)	3.351(s), (3.29)[h,l](s)
8	–	–	–	–	–	–
8^1	3.817 (3.82)[b,i](q)	3.81(q)	8.180(dd)	4.243	3.876 (3.86)[f](q)	3.754(q), (3.757)[h,l](q)
8^2	1.696 (1.69)[b,i](t)	1.70(t)	6.237(d), 6.019(d)	1.815	1.723 (1.73)[f](t)	1.705(t), (1.698)[h,l](t)
10	9.749 (9.75)[b,i](s)	9.74(s)	9.995	9.934 (9.64)[d,j] (s) (9.55)[e,m]	9.873 (9.8)[f](s)	9.838 (9.86)[g,k](s), (9.615)[h,l](s)
11	–	–	–	–	–	–
12	–	–	–	–	–	–
12^1	3.619 (3.61)[b,i](s)	3.61(s)	3.709(s)	3.606 (3.65)[d,j] (s) (3.52)[e,m]	3.668 (3.65)[f](s)	3.637(s), (3.671)[h,l](s)
13	–	–	–	–	–	–
13^2	6.234 (6.24)[b,i](s)	6.24(s)	6.406	6.189 (6.19)[d,j] (s) (6.10)[e,m]	6.335 (6.28)[f](s)	6.318(s), (6.219)[h,l](s)
13^3	–	–	–	–	–	–
13^4	3.829 (3.83)[b,i](s)	3.84(s)	3.882	3.842 (4.02)[d,j] (s) (3.95)[e,m]	3.851 (3.83)[f](s)	3.887(s), (3.847)[h,l](s)
17	4.175 (4.16)[b,i]	4.17	4.261	4.128 (4.15)[e,m]	4.242 (4.25)[f]	4.230, (4.115)[h,l](d)
17^1	2.589, 2.461 (2.60, 2.45)[b,i]	2.61, 2.47	2.422, 2.699	2.43, 2.593 (~2.35)[e,m]	2.484, 2.622 (2.63, 2.46)[f]	2.467, 2.632, (2.25, 2.487)[h,l]
17^2	2.431, 2.159 (2.35, 2.05)[b,i]	2.47, 2.10	2.280, 2.538	2.08, 2.44 (~2.35)[e,m]	1.98, 2.418 (2.48, 2.21)[f]	2.08, 2.47, (2.36, 1.913)[h,l]
18	4.572 (4.57)[b,i](q)	4.57(q)	4.691	4.524(q) (4.45)[e,m]	4.660 (4.63)[f](q)	4.634(q), (4.46)[h,l](q)
18^1	1.772, 1.762 (d) (1.77, 1.76)[b,i]	1.78, 1.77(dd)	1.855, 1.845(d)	1.768, 1.759 (1.78)[d,j](d)	1.812, 1.802 (1.82)[f](d)	1.800, 1.791(d), (1.644)[h,l] (d)
20	8.582 (8.58)[b,i](s)	8.58(s)	9.009	8.480 (8.20)[d,j] (s) (8.18)[e,m]	8.867 (8.81)[f](s)	9.533 (9.77)[g,k](s), (9.42)[h,l] (s)
P1	4.342, 4.224 (4.33, 4.21)[b,i]	4.33, 4.19	4.277, 4.228	4.364, 4.247	4.343, 4.227 (4.36, 4.26)[f]	4.361, 4.263, (4.44)[h,l](dd)
P2	4.955 (4.95)[b,i]	4.95(t)	4.829	4.980	4.944 (5.04)[f]	4.987, (5.125)[h,l](t)
P3	–	–	–	–	–	–
P3^1	1.509 (1.51)[b,i]	1.53(s)	1.52	1.519	1.505 (1.54)[f]	1.525
P4	1.822 (1.82)[b,i]	1.86	1.85	1.845	1.832 (1.85)[f]	1.845, (1.88)[h,l]
P5	1.31	1.95	1.25	1.31	1.30	1.195
P6	0.97, 1.17 (0.94–1.02)[b,i]	5.01(t)	1.01–0.93	0.98, 1.18	1.97, 1.16 (1.02, 1.20)[f]	0.97, 1.17
P7	1.31	–	1.33	1.33	1.31 (1.34)[f]	1.324, (1.3)[h,l]
P7^1	0.811, 0.803 (0.81, 0.80)[b,i]	1.49(s)	0.81, 0.81	0.785, 0.777	0.778, 0.770 (0.79)[f]	0.785, 0.777

(Continued)

TABLE 6.3 (CONTINUED)

^{1}H-Chemical Shifts of Chl *a*, *b*, *d*, *f*, and DV-Chl a in Acetone-d_6 at 273K[a,c]

IUPAC No. of Carbon Atom	Chl a[a]	BChl 663[b,i]	DV-Chl a[c]	Chl b[a]	Chl d[a]	Chl f[a]
P8	1.01, 1.23 (0.94–1,02)[b,i]	1.88	1.01–0.93	1.02, 1.22	1.01, 1.22 (1.23, 1.04)[f]	1.01, 1.22
P9	1.15, 1.28	1.32	1.15	1.15, 1.28	1.14, 1.28 (1.23)[f]	1.15, 1.27
P10	1.01, 1.23 (0.94–1.02)[b,i]	0.99 or 1.02	1.01–0.93	1.02, 1.22	1.01, 1.22 (1.23, 1.04)[f]	1.01, 1.22
P11	1.31	1.33	1.31	1.32	1.32 (1.34)[f]	1.32, (1.3)[h,l]
P11^1	0.783, 0.774 (0.79, 0.78)[b,i]	0.80, 0.79(d)	0.79, 0.78	0.809, 0.801	0.806, 0.797 (0.81)[f]	0.807, 0.798
P12	1.01, 1.23 (0.94–1.02)[b,i]	0.99 or 1.02	1.01–0.93	1.02, 1.22	1.01, 1.22 (1.23, 1.04)[f]	1.01, 1.22
P13	1.23, 1.28	1.27	1.18	1.23, 1.28	1.23, 1.28 (1.30)[f]	1.23, 1.28
P14	1.12	1.12	1.12	1.12	1.12 (1.13)[f]	1.12, (1.1)[h,l]
P15	1.500, (1.50)[b,i]	1.51	1.58	1.489	1.497 (1.51)[f]	1.495, (1.465)[h,l]
P15^1	0.854, 0.845 (0.86, 0.84)[b,i]	0.85, 0.84(d)	0.85, 0.85	0.851, 0.842	0.850, 0.842 (0.85)[f]	0.849, 0.841
P16	0.854, 0.845 (0.86, 0.84)[b,i]	0.85, 0.84(d)	0.85, 0.85	0.851, 0.842	0.850, 0.842 (0.85)[f]	0.849, 0.841, (0.82)[h,l](d)

[a] From Kobayashi M et al., Physicochemical properties of chlorophylls in oxygenic photosynthesis—Succession of co-factors from anoxygenic to oxygenic photosynthesis. In *Photosynthesis*, ed. by Dubinsky Z, Intech, Croatia, Chapter 3, pp. 47–90, 2013.

[b] From Kobayashi M et al., *Photosynth. Res.*, 63: 269–280, 2000.

[c] From Komatsu H et al., *Photomed. Photobiol.*, 36: 59–69, 2014.

[d] From Wu S-M and Rebeiz CA, *J. Biol. Chem.*, 260: 3632–3634, 1985.

[e] From Katz JJ and Brown CE, *Bull. Magn. Reson.*, 5: 3–49, 1983.

[f] From Miyashita H et al., *Plant Cell Physiol.*, 38: 274–281, 1997.

[g] From Chen M et al., *Science*, 329: 1318–1319, 2010.

[h] From Willows RD et al., *Org. Lett.*, 15: 1588–1590, 2013.

[i] 263 K.

[j] In CDCl$_3$.

[k] In CD$_2$Cl$_2$/d_5-pyridine(97/3, v/v).

[l] In CDCl$_3$/d_5-pyridine(99.5/0.5, v/v).

[m] In CD$_3$Cl/CD$_3$OD.

peaks: one is with the signal for 3-vinylic carbons, one is with the signal for 2C carbon, and the last is with the signal for 2^1-CHO carbon. Conclusively, the Chl *f*-like pigment isolated from the strain KC1 has been identified as Chl *f*, namely, 2-desmethyl-2-formyl-Chl *a* ([2-CHO]-Chl *a*).

In the ^{1}H-^{13}C HMBC spectrum of MV- and DV-Chls *a*, the 3^1-H signal for –CH=CH$_2$ had one cross peak associated with the signal of 2-C, and the corresponding 3^2-H signal had two cross peaks with the signals for 3-C and 3^1-C carbons. In the spectrum of DV-Chl *a*, the signal of an 8^1-vinilic hydrogen shows a cross peak with the signal of 8-C, while one of the signals of two 8^2-vinilic hydrogens has a cross peak with the signal of 8^1-C. In contrast, the 8^1-H signal for –CH$_2$–CH$_3$ in MV-Chl *a* exhibits four cross peaks: one is associated with the signal for 7-C, one is with the signal for 8-C, one is with the signal for 9-C, and the last is with the signal for 8^2-C carbon. On the other hand, the 8^2-H signal has two cross peaks: one is with 8-C and the other with 8^1-C, respectively.

6.7.2 Chls *a* and *a*′

The fingerprint region of ^{1}H-NMR spectra for Chl *a* and Chl *a*′ in acetone-d_6 is shown in Figure 6.33. Signals of *meso*-H (5, 10, and 20) on a macrocycle and P2-H on a phytyl chain are clearly downfield-shifted in Chl *a*′. On the other hand, 13^2-H shows upfield shift in Chl *a*′. Such a difference in tertiary structure is not recognized by other analytical methods. The ^{13}C chemical shift effects are in agreement with those for the ^{1}H-NMR of the epimers (Lötjönen and Hynninen 1983; Abraham and Rowan 1991).

6.7.3 BChl 663 (Chl *a*Δ2,6-Phytadienol)

BChl 663 has been identified as Chl *a* esterified with Δ-phytadienol, Chl *a*$_{PD}$, by absorption, fluorescence, CD, and mass spectroscopy, but these methods cannot locate the two C=C double bonds in the long-chain alcohol at C17^3.

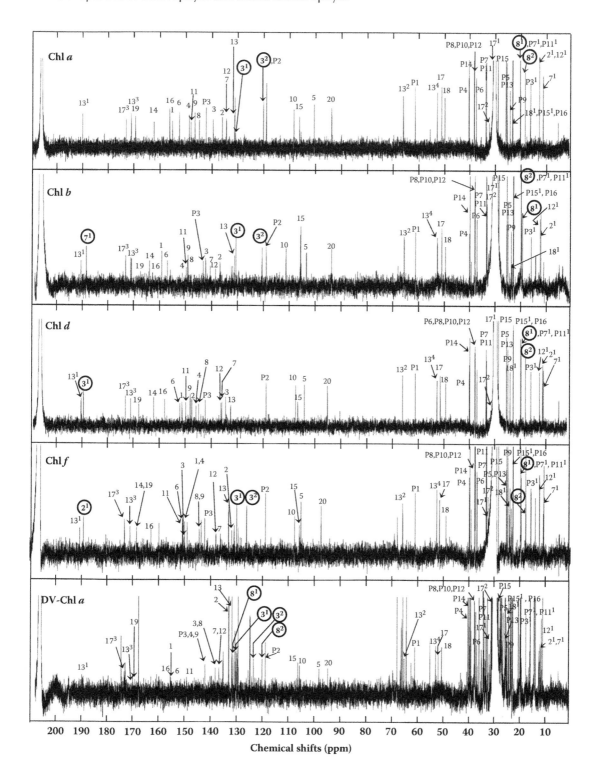

FIGURE 6.29 ^{13}C-NMR spectra of Chls *a*, *b*, *d*, *f*, and DV-Chl *a* measured in acetone-d_6 at 273 K. Signals corresponding to ^{13}C atoms of the molecules are labeled. The peak at 30–32 ppm is acetone. (Adapted from Kobayashi M et al., Physicochemical properties of chlorophylls in oxygenic photosynthesis—Succession of co-factors from anoxygenic to oxygenic photosynthesis. In *Photosynthesis*, ed. by Dubinsky Z, Intech, Croatia, Chapter 3, pp. 47–90, 2013; Komatsu H et al., *Photomed. Photobiol.*, 36: 59–69, 2014.)

TABLE 6.4

^{13}C-Chemical Shifts of Chl *a*, *b*, *d*, *f*, and DV-Chl *a* in Acetone-*d*$_6$ at 273 K[a,c]

IUPAC No. of Carbon Atoms	Chl *a*[a]	BChl 663[b]	DV-Chl *a*[c]	Chl *b*[a]	Chl *d*[a]	Chl *f*[a]
1	155.47 (155.46)[b,g]	155.45	155.66	159.43 (157.54)[d]	151.47 (151.81)[e]	150.20 (149.5)[f,i]
2	136.28 (136.24)[b,g]	136.24	133.35	136.98 (136.98)[d]	147.66 (147.33)[e]	131.19 (130.3)[f,i]
2^1	12.65 (12.70)[b,g]	12.70	12.38	12.33 (12.64)[d]	12.58 (11.75)[e]	189.27 (189.2)[f,i]
3	139.76 (139.68)[b,g]	139.68	138.46	142.53 (141.06)[d]	134.77 (135.12)[e]	150.5 (150.1)[f,i]
3^1	131.30 (131.29)[b,g]	131.28	129.92	130.83 (130.94)[d]	189.59 (189.54)[e]	130.22 (129.0)[f,i]
3^2	120.33 (120.33)[b,g]	120.33	123.67	120.80 (120.80)[d]	–	126.36 (126.1)[f,i]
4	148.96 (148.99)[b,g]	148.97	142.36	150.15 (150.05)[d]	146.00 (146.36)[e]	150.20
5	100.58 (100.51)[b,g]	100.52	98.42	103.63 (103.59)[d]	106.24 (104.41)[e]	105.39 (104.8 or 107.0)[f,i]
6	152.78 (152.80)[b,g]	152.78	152.11	157.25 (149.45)[d]	152.27 (152.55)[e]	150.85 (150.1)[f,i]
7	134.68 (134.65)[b,g]	134.67	135.39	140.76 (131.22)[d]	136.19 (136.31)[e]	136.49 (135.7)[f,i]
7^1	11.15 (11.16)[b,g]	11.16	12.75	188.62 (187.69)[d]	11.25 (11.39)[e]	11.17 (11.2)[f,i]
8	145.08 (145.02)[b,g]	145.03	138.46	148.32 (155.79)[d]	145.02 (145.13)[e]	144.59 (148.4)[f,i]
8^1	19.91 (19.88)[b,g]	19.88	130.898	19.51 (19.55)[d]	19.94 (20.12)[e]	19.87 (19.6)[f,i]
8^2	18.12 (18.18)[b,g]	18.18	120.399	19.91 (19.55)[d]	18.12 (18.12)[e]	18.07 (17.4)[f,i]
9	146.74 (146.74)[b,g]	146.73	142.359	149.29 (143.41)[d]	148.28 (148.64)[e]	144.59 (143.7)[f,i]
10	108.50 (108.53)[b,g]	108.52	106.50	111.53 (111.37)[d]	107.68 (107.8)[e]	107.82 (104.8 or 107.0)[f,i]
11	148.34 (148.34)[b,g]	148.32	148.78	149.94 (149.45)[d]	149.97 (150.35)[e]	151.02 (150.4)[f,i]
12	134.57 (134.46)[b,g]	134.48	135.52	138.66 (138.87)[d]	136.50 (136.63)[e]	138.17 (138.5)[f,i]
12^1	12.65 (12.66)[b,g]	12.64	11.74	12.47 (12.38)[d]	12.79 (12.89)[e]	12.82 (13.0)[f,i]
13	131.53 (131.41)[b,g]	131.42	131.63	132.13 (132.78)[d]	132.76 (133.04)[e]	133.35 (132.5)[f,i]
13^1	190.30 (190.37)[b,g]	190.36	190.1	190.66 (190.62)[d]	190.49 (190.50)[e]	190.68 (190)[f,i]
13^2	66.00 (65.95)[b,g]	65.95	65.43	65.77 (65.91)[d]	66.19 (66.47)[e]	66.13 (65.4)[f,i]
13^3	171.32 (171.36)[b,g]	171.35	170.3	171.04 (171.05)[d]	171.20 (171.33)[e]	171.15 (170.2)[f,i]
13^4	52.67 (52.71)[b,g]	52.72	53.08	52.79 (52.84)[d]	52.79 (52.84)[e]	52.83 (52.8)[f,i]
14	162.55 (162.58)[b,g]	162.55	167.98	164.11 (164.38)[d]	162.30 (162.68)[e]	168.67 (162.6)[f,i]
15	106.35 (106.27)[b,g]	106.29	106.9	105.94 (105.96)[d]	106.82 (107.04)[e]	106.03 (104.6)[f,i]
16	156.50 (156.54)[b,g]	156.49	157.40	160.38 (159.70)[d]	158.12 (158.60)[e]	163.04 (162.6)[f,i]
17	50.98 (50.92)[b,g]	50.90	52.08	51.14 (51.45)[d]	51.54 (52.05)[e]	51.74 (51.0)[f,i]
17^1	30.28 (30.03)[b,g]	30.02	29.51	30.28 (31.28)[d]	30.66 (30.64)[e]	30.90 (29.7)[f,i]
17^2	30.66 (30.09)[b,g]	30.09	30.66	30.66 (30.18)[d]	31.27 (31.37)[e]	30.35 (30.5)[f,i]
17^3	173.29 (173.39)[b,g]	173.33	173.18	173.35 (173.42)[d]	173.26 (173.45)[e]	173.30 (172.9)[f,i]
18	49.73 (49.69)[b,g]	49.71	50.70	49.65 (50.03)[d]	49.44 (49.80)[e]	49.26 (48.5)[f,i]
18^1	23.01 (23.88)[b,g]	23.89	24.37	23.68 (23.76)[d]	24.13 (24.25)[e]	24.20 (23.5)[f,i]
19	169.69 (167.74)[b,g]	169.71	169.02	168.03 (171.46)[d]	168.09 (168.34)[e]	168.67 (167.8)[f,i]
20	93.78 (93.79)[b,g]	93.78	94.91	93.86 (93.96)[d]	95.26 (95.36)[e]	97.60 (96.8)[f,i]
P1	61.31 (61.32)[b,g]	61.28	61.31	61.42	61.33 (61.67)[e]	61.35 (61.4)[f,i]
P2	120.17 (119.12)[b,g]	119.24	119.11	119.19	119.13 (119.53)[e]	119.18 (117.7)[f,i]
P3	142.46 (142.48)[b,g]	142.08	142.36	143.42	142.49 (142.70)[e]	142.49 (142.8)[f,i]
P3^1	16.11 (16.10)[b,g]	16.23	16.12	16.16	16.11 (16.40)[e]	16.13
P4	40.19 (40.19)[b,g]	39.96	40.21	40.23	40.19 (40.46)[e]	40.20 (39.7)[f,i]
P5	25.54 (25.52)[b,g]	26.63	25.60	25.57	25.53 (25.86)[e]	25.55
P6	37.11 (38.01, 37.94, 37.87 or 37.11)[b,g]	124.35	37.13	37.14	38.11 (37.40)[e]	37.11 (37.2)[f,i]
P7	33.43 (33.44)[b,g]	135.94	33.44	33.44	33.22 (33.47)[e]	33.50 (32.6)[f,i]
P7^1	20.03 (20.02)[b,g]	15.84	20.03	20.03	19.98 (20.19)[e]	20.02
P8	37.94 (38.01, 37.94, 37.87 or 37.11)[b,g]	40.50	38.02, 37.96, 37.89	37.95	37.94 (38.21)[e]	37.94 (37.2)[f,i]
P9	25.04 (25.06)[b,g]	25.82	25.05	25.06	25.04 (25.25)[e]	25.04
P10	38.01 (38.01, 37.94, 37.87 or 37.11)[b,g]	37.90 or 37.11	38.02, 37.96, 37.89	38.02	38.01 (38.26)[e]	38.01 (37.2)[f,i]
P11	33.22 (33.23)[b,g]	33.28	33.24	33.25	33.42 (33.65)[e]	33.42 (32.6)[f,i]
P11^1	19.98 (19.98)[b,g]	19.98	19.86	19.99	20.02 (20.23)[e]	19.98

(Continued)

TABLE 6.4 (CONTINUED)

^{13}C-Chemical Shifts of Chl *a*, *b*, *d*, *f*, and DV-Chl *a* in Acetone-d_6 at 273 K[a,c]

IUPAC No. of Carbon Atoms	Chl *a*[a]	BChl 663[b]	DV-Chl *a*[c]	Chl *b*[a]	Chl *d*[a]	Chl *f*[a]
P12	37.87 (38.01, 37.94, 37.87 or 37.11)[b,g]	37.90 or 37.11	38.02, 37.96, 37.89	37.89	37.88 (38.14)[e]	37.88 (37.2)[f,i]
P13	25.49 (25.52)[b,g]	25.52	25.49	25.49	25.49 (25.64)[e]	25.48
P14	40.00 (39.98)[b,g]	39.96	40.01	40.00	40.00 (40.27)[e]	40.00 (39.2)[f,i]
P15	28.65 (28.66)[b,g]	28.67	28.66	28.65	28.65 (28.85)[e]	28.65 (28.0)[f,i]
P15^1	23.01 (23.02)[b,g]	23.02, 22.91	23.01[h]	23.01[h]	23.01[h] (23.15)[e]	23.00[h]
P16	22.90[h] (22.91)[b,g]	23.02, 22.91	22.91[h]	22.91[h]	22.90[h] (23.07)[e]	22.90[h] (22.7)[f,i]

[a] From Kobayashi M et al., Physicochemical properties of chlorophylls in oxygenic photosynthesis—Succession of co-factors from anoxygenic to oxygenic photosynthesis. In *Photosynthesis*, ed. by Dubinsky Z, Intech, Croatia, Chapter 3, pp. 47–90, 2013.

[b] From Kobayashi M et al., *Photosynth. Res.*, 63: 269–280, 2000.

[c] From Komatsu H et al., *Photomed. Photobiol.*, 36: 59–69, 2014.

[d] From Risch N and Brockmann H, *Tetrahed. Lett.*, 24: 173–176, 1983.

[e] From Miyashita H et al., *Plant Cell Physiol.*, 38: 274–281, 1997.

[f] From Willows RD et al., *Org. Lett.*, 15: 1588–1590, 2013.

[g] 263 K.

[h] Assignment interchangeable.

[i] In CDCl$_3$/d_5-pyridine (99.5/0.5, v/v).

In the one-dimensional ^{1}H-NMR spectra, identical signals corresponding to all ^{1}H atoms of the Chl a_P macrocycle are observed for BChl 663 at identical chemical shifts (Figure 6.34a,b) (Kobayashi et al. 2000), demonstrating that the macrocycles of BChl 663 and Chl a_P are identical. There are marked differences in the signals from the long alcohol chains of each pigment. However, a triplet signal (signal A) present in the spectrum of BChl 663 but absent in the spectrum of Chl a_P appeared at 5.01 ppm on the low-field side of signal P2 in the spectrum of BChl 663: this suggests that the environment of the proton is very similar to that of P2. Similarly, a complex signal B appeared at 1.95 ppm on the low-field side of signal P4 at 1.86 ppm. The intensity of signal P4 (denoted as signal C) in the spectrum of BChl 663 is twice that in Chl a_P, suggesting two more protons in Chl a_P in an environment similar to that of P4. Another marked difference is the disappearance of one of the doublet peaks from the Chl a_P spectrum, either at P7^1 or P11^1, at the high-field end of the spectrum, and the concomitant appearance of a singlet signal D at 1.49 ppm (near peak P3^1) in the spectrum of BChl 663. These results are consistent with FAB-MS (Figure 6.25) results, indicating an additional double bond in the phytyl chain of BChl 663. In the ^{13}C-NMR spectra, marked differences are also noted in the range of 15–40 and 120–145 ppm, relating to the esterifying alcohol (data not shown).

In the ^{1}H-^{1}H DQF-COSY spectrum for BChl 663 (Figure 6.34c, d), the signals at 1.86 ppm (signal C) assigned to signal P4 overlapping with another unknown proton signal showed a cross peak with the signal at 1.95 ppm (signal B), situated at the low-field side of signal P4. The signal at 1.95 ppm had a cross peak again with the signal at 5.01 ppm (signal A), the low-field side signal to signal P2, which had a small cross peak with the peak at 1.49 ppm (signal D), the new singlet peak in BChl 663. This small cross peak indicates that the carbon atom, to which the proton atom at 5.01 ppm (signal A) is attached, is positioned at the opposite side of a double bond linking to a methyl group as indicated by the disappearance of one of the doublet peaks at 0.79 ppm, as mentioned above. The methyl group should be P7^1 or P11^1; thus, the position of the double bond must be at P7 or P11.

If a double bond is present between P11 and P12, the peak at 1.49 ppm D is P11^1, and hence the peak at 5.01 ppm A is P12. The peak at 1.95 ppm B is P13, and the peak C at 1.86 ppm, overlapping signal P4, is P14. Conversely, if the double bond is situated between P10 and P11, the peak C is assigned to be P8 linking to P7; if between P7 and P8, the peak C is P10 linking to P11; and if between P6 and P7, the peak C is P4 linking to P3. The coherent correlation is broken on the high-field part of the peak C (Figure 6.34d); the next carbon atom should have no proton. FAB-mass spectrum (Figure 6.25) shows that only one double bond is increased, which consists of the peak A being present at the opposite side of the increased double bond to the attaching methyl group (P7^1). Therefore, the high-field signal in the peak C is P4 connecting to P3, which is forming a double bond with P2. The peak B is P5, and the peak A is P6, which is linked to P7 to form the new double bond. The peak D is then P7^1, while the remaining doublet peak at 0.79 ppm is P11^1.

The low-field part of the peak C overlapping with signal P4 shows a cross peak at 1.32 ppm. A coherent correlation can be traced from the terminal protons of P11^1 on the ^{1}H-^{1}H DQF-COSY spectrum (Figure 6.34d). The signal P11^1 has a cross peak with the signal P11 at 1.33 ppm, which has cross peaks with signals P10 plus P12 around 1.0 ppm. They have

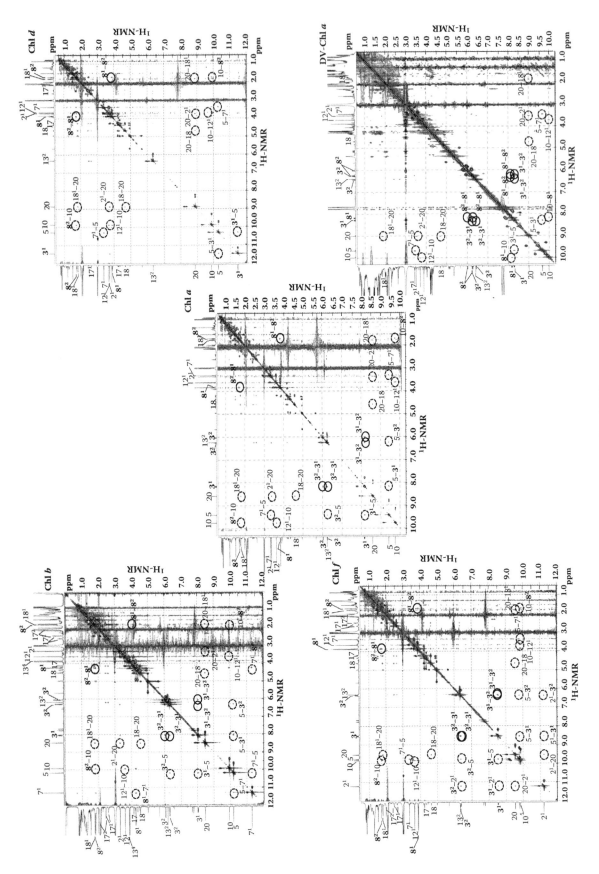

FIGURE 6.30 ¹H-¹H NOESY spectra of Chls *a*, *b*, *d*, *f*, and DV-Chl *a* measured in acetone-*d₆* at 273 K.

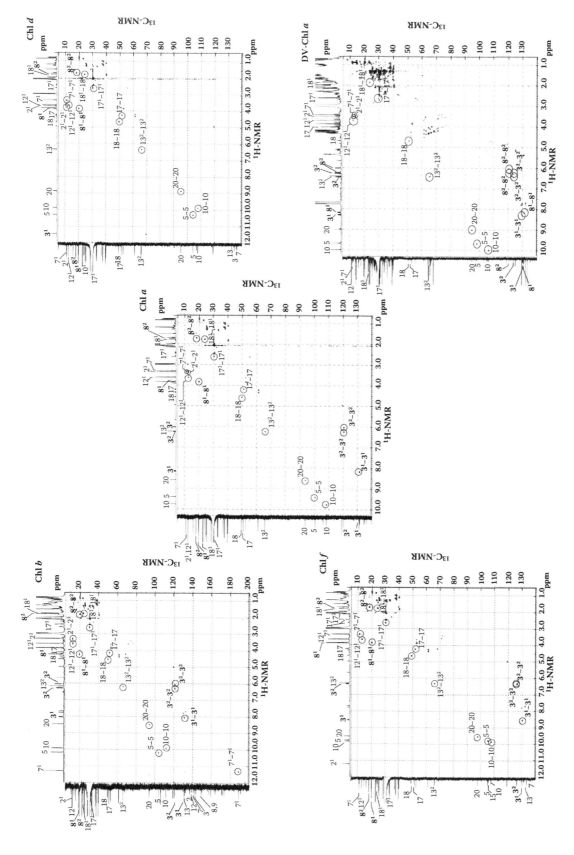

FIGURE 6.31 ^1H-^{13}C HSQC spectra of Chls a, b, d, f, and DV-Chl a measured in acetone-d_6 at 273 K.

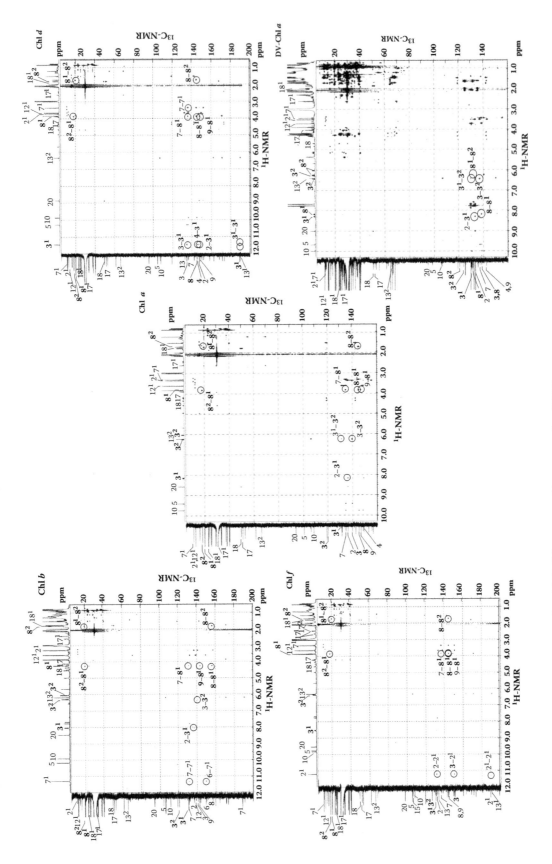

FIGURE 6.32 ^{1}H-^{13}C HMBC spectra of Chls a, b, d, f, and DV-Chl a measured in acetone-d_6 at 273 K.

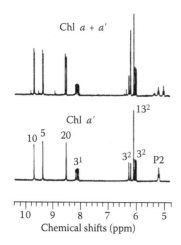

FIGURE 6.33 The fingerprint region of ¹H-NMR spectra of (top) a mixture of Chls *a* and *a'*, and (bottom) pure Chl *a'* in acetone-d_6 at 263 K.

cross peaks with the signal P9 at 1.32 ppm; the correlation returns to the large cross peak with which the low-field part of the peak C correlated. The correlation stops here, indicating that this signal is due to protons attaching to the carbon atom next to that without protons. Therefore, the peak at 1.88 ppm overlapping to the signal P4 is due to P8, which is in a very similar environment to P4.

These relationships are seen on the HMBC spectra as compared with the HMQC spectra. Based on the HMQC spectra (Figure 6.35a), ¹³C-signals corresponding to the peak D, the peak A, the peak B, the high-field part of the peak C, and the low-field part of the peak C situate at 15.8, 124.4, 26.6, 40.0, and 40.5 ppm, respectively.

In the HMBC spectrum (Figure 6.35b), since the ¹H-signal of the high-field part of the peak C has cross peaks with the ¹³C-signals P2 at 119 ppm and P3 at 142 ppm, this signal is the signal P4. It also shows a cross peak with the ¹³C-signal at 26.6 ppm (indicated by α) originating from the peak B;

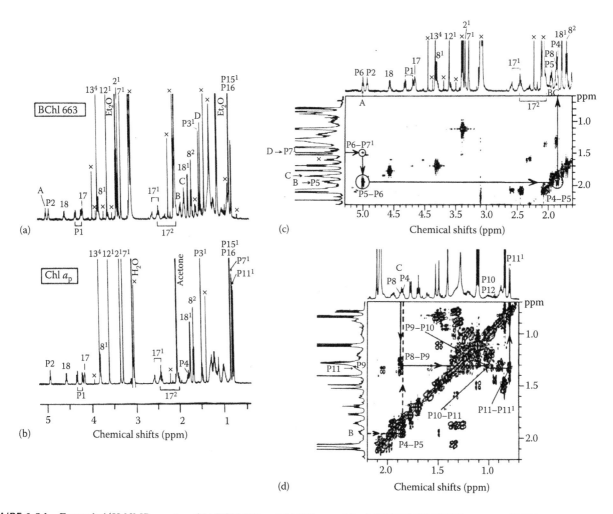

FIGURE 6.34 Expanded ¹H-NMR spectra of (a) BChl 663 and (b) Chl a_p, and (c, d) ¹H-¹H DQF-COSY spectra of BChl 663 in acetone-d_6 at 263 K. (c) Lines between cross peaks show the correlation of carbon chain, and circles on the cross peaks indicate the relationship between the signals. (d) The broken lines show the correlation from P5 to P4, which stops, while the solid lines show the correlation from P8 to P11. Characteristic signals observed in the BChl 663 spectrum are indicated by A, B, C, and D. (Adapted from Kobayashi M et al., *Photosynth. Res.*, 63: 269–280, 2000.)

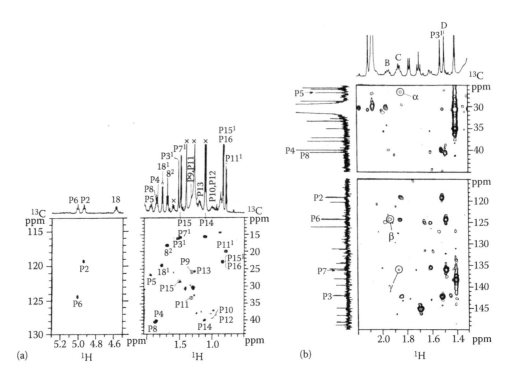

FIGURE 6.35 (a) ¹H-¹³C HMQC and (b) ¹H-¹³C HMBC spectra of BChl 663 in acetone-d_6 at 263 K. ¹H-¹³C HMBC spectra of BChl 663 indicating phytyl carbon atom region (top) and double-bond carbon atom region (bottom). Symbol α indicates the cross peak between the ¹H-signal P4 and the ¹³C-signal P5, β the 1H-signal P5 and the ¹³C-signal P6, and γ the ¹H-signal P8 and the ¹³C-signal P7. (Adapted from Kobayashi M et al., *Photosynth. Res.,* 63: 269–280, 2000.)

therefore, the peak B is signal P5. The ¹H-signal of the peak B, the signal P5, has a cross peak with the ¹³C-signal at 124 ppm (indicated by β) in the double-bond region corresponding to the peak A, and the peak A is P6. The ¹³C-signal P6 shows a cross peak with the singlet ¹H-signal at the high-field side to the signal P3¹. Thus, the peak D is P7¹, which has a cross peak with the ¹³C-signal at 136 ppm in the double-bond region due to P7. These results indicate that the double bond is present between P6 and P7.

The ¹³C-signal corresponding to the ¹H-signal P4 is positioned at 40.0 ppm on the HMQC spectrum (Figure 6.35a, right), which has a cross peak with the ¹H-signal P3¹ in the HMBC spectrum (Figure 6.35b, top). On the other hand, the ¹³C-signal corresponding to the low-field part of the peak C is located at 40.5 ppm on the HMQC spectrum (Figure 6.35a, right), which shows a cross peak with the ¹H-signal P7¹, the peak D (Figure 6.35b, top). Therefore, the low-field part of the peak C is the signal P8, which has, as expected, a cross peak with the ¹³C-signal P7 at 136 ppm (Figure 6.35b, bottom; indicated by γ).

Comparing the ¹³C-signals of geranylgeraniol esterified to BChl *c* (Fages et al. 1990), the ¹³C-signals from P1 to P8 for BChl 663 are positioned to within 1 ppm of those for geranylgeranyl except for P3 (1.12 ppm difference), while the signals from P9 to P16 are quite different between the two. This supports the finding in the current investigation that the structure of the esterifying alcohol for BChl 663 is *trans*-Δ2,6-phytadienol (Figure 6.1).

6.7.4 [Mg]-BCHL *a*, [Zn]-BCHL *a*, AND [2H]-BCHL *a* (BPHE *a*)

A series of NMR spectra of [Zn]-BChl *a* are summarized in Figure 6.36, Tables 6.5 and 6.6. The ¹H-¹H DQF-COSY spectrum shows coherent correlations between carbon 7¹ and 8², and between carbon 17³ and 18¹; the solid lines show the correlation from 7¹ to 8² and the broken lines from 17³ to 18¹ (Figure 6.36b). The ¹H-¹³C HMQC spectrum shows the relationship between proton and carbon signals of the macrocycle (Figure 6.36c).

Changed electron distribution on the [M]-Chl macrocycle due to the inductive effect of the central metal can be seen in ¹H-NMR spectra. A fair correlation is seen between the electronegativity (E_N) of the central metal and the chemical shifts of ¹H-NMR for [Mg]-BChl *a* (i.e., BChl *a*), [Zn]-BChl *a*, and [2H]-BChl *a* (i.e., BPhe *a*) in acetone-d_6 (Figure 6.37) as observed with absorption maxima (see Figure 6.13c). Clear downfield shifts of *meso*-H (at 5-, 10- and 20-), 2-Me, 12-Me, and 13²-H on removal of Zn and Mg are observed, although other protons remain almost unchanged or show even slightly upfield shifts. The protons exhibiting large downfield shifts on the removal of metals are located on rings I, III, and V (Kobayashi et al. 1998a,b, 1999). Interestingly, the shifts reported for ¹³C-NMR for Chls (Boxer et al. 1974) are in the opposite direction to the ¹H-NMR signal changes for BChls.

The ¹H-NMR chemical shift can be interpreted on the basis of ring current and in both localized diamagnetic and

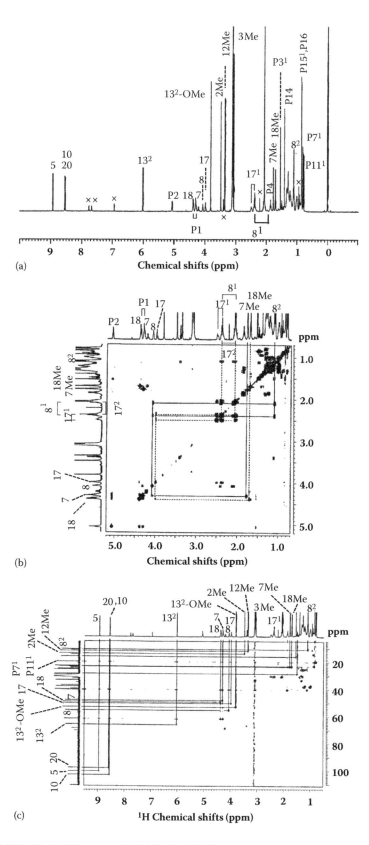

FIGURE 6.36 (a) ^{1}H-, (b) ^{1}H-^{1}H DQF-COSY, and (c) ^{1}H-^{13}C HMQC NMR spectra of Zn-BChl a purified from $A.$ $rubrum.$ Measurements were carried out in acetone-d_6 at 263 K. (a) The signals at 2.06–2.11 ppm is due to acetone and 3.10–3.15 ppm is that of water. The symbol x indicates the signals of impurity.

TABLE 6.5
^{1}H-Chemical Shifts of BChl a, BChl b, BPhe a, BPhe g, and Zn-BChl a in Acetone-d_6 at 263 K[a]

IUPAC No. of Carbon Atoms	BChl a[a]	BPhe a[a]	Zn-BChl a[a]	BChl b[c,g]	BPhe g[f,h]
2^1	3.41. (3.40[b]. 3.33[c,g]. 3.41[d])	3.59. (3.5)[b]	3.49	3.34	3.19(s)
3^1	–	–	–	–	7.71(dd)
3^2	3.02. (3.01[b]. 3.00[c,g]. 3.02[d])	3.06. (2.07)[b]	3.09	2.99	6.11, 6.06(dd)
5	8.81. (8.81[b]. 9.23[c,g]. 8.81[d])	9.12. (9.05)[b]	8.95	9.41	8.22(s)
7	4.32. (4.24[b]. 4.10[c,g]. 4.10[d])	4.21. (4.35)[b]	4.30	4.93(dd). (4.93(qq))[c,g]	~5.0
7^1	1.76. (1.73[b].1.58(d)[c,g]. 1.76[d])	1.81. (1.8)[b]	1.77	1.66. (1.66)[c,g](d)	~1.65
8	4.10. (4.03[b]. 3.86[c,g]. 4.32[d])	4.10. (4.07)[b]	4.07	–	–
8^1	2.5–2.2. (2.08[b]. ~2.5[c,g]. 2.5–2.2[d])	2.6–2.3. (2.07)[b]	2.40, 2.09	6.84(dd). (6.84(qq))[c,g]	6.94
8^2	1.10. (1.09[b]. 1.10[d])	1.12. (1.09)[b]	1.10	2.01. (2.01)[c,g](d)	2.27(d)
10	8.43. (8.40[b]. 9.50[c,g]. 8.43[d])	8.81. (8.66)[b]	8.55	8.93	8.60(s)
12^1	3.30. (3.29[b]. 3.44[c,g]. 3.30[d])	3.57. (3.20)[b]	3.36	3.45	3.38(s)
13^2	5.90. (5.92[b]. 6.44[c,g]. 5.90[d])	6.17. (6.10)[b]	6.02	6.43	5.96(s)
13^4	3.76. (3.76[b]. 3.66(d)[c,g]. 3.76[d])	3.84. (3.76)[b]	3.81	3.66	3.83(s)
17	4.00. (3.92[b]. 4.10[c,g]. 4.00[d])	4.01. (3.99)[b]	3.99	4.1	4.15
17^1	2.5–2.2. (2.37[b]. ~2.5[c,g]. 2.5–2.2[d])	2.6–2.3. (2.4)[b]	2.50, 2.37	~2.5	2.0-2.7
17^2	2.20. (2.45[b]. ~2.5[c,g]. 2.20[d])	1.81. (2.30)[b]	2.02	~2.5	2.0-2.7
18	4.40. (4.32[b]. 4.21[c,g]. 4.40[d])	4.41. (4.41)[b]	4.39	4.21	~4.45
18^1	1.67. (1.65[b]. 1.41(d)[c,g]. 1.67[d])	1.74. (1.70)[b]	1.70	1.41(d)	1.81(d)
20	8.38. (8.36[b]. 8.28[c,g]. 8.38[d])	8.73. (8.73)[b]	8.57	8.39	8.06(s)
P1	4.42. 4.36. (4.39[b]. 4.42[d]. 4.36[d])	4.40–4.33. (4.26)[b]	4.38, 4.29	–	–
P2	5.14. (5.13[b]. 5.14[d])	4.97. (5.20)[b]	5.06	–	–
$P3^1$	1.58. (1.58)[d]	1.48	1.54	–	–
P4	(1.91)[b]	(1.93)[b]	–	–	–
P5	(1.35)[b]	(1.45)[b]	–	–	–
P6	(1.09)[b]	(1.17. 1.04)[b]	–	–	–
P7	(0.89)[b]	(1.35. 1.26)[b]	–	–	–
$P7^1$	–	0.74, 0.72	0.81, 0.79	–	–
P8	(1.09)[b]	(1.09)[b]	–	–	–
P9	(1.28)[b]	(1.21)[b]	–	–	–
P10	(1.13)[b]	(1.09)[b]	–	–	–
P11	(0.89)[b]	(1.28)[b]	–	–	–
$P11^1$	–	0.74, 0.72	0.81, 0.79	–	–
P12	(1.13)[b]	(1.09)[b]	–	–	–
P13	(1.21)[b]	(1.21)[b]	–	–	–
P14	(1.11)[b]	(1.04)[b]	–	–	–
P15	(1.17. 1.04)[b]	(1.25)[b]	–	–	–
$P15^1$	–	0.81	0.85	–	–
P16	0.81. (0.84)[b]	0.81. (0.84)[b]	0.85	–	–
NH	–	–0.82	–	–	–

[a] From Kobayashi M et al., *Photomed. Photobiol.*, 20: 75–80, 1998.

[b] From Egorova-Zachernyuk T et al., *Magn. Reson. Chem.*, 46: 1074–1083, 2008.

[c] From Katz JJ and Brown CE, *Bull. Magn. Reson.*, 5: 3–49, 1983.

[d] From Brereton RG and Sanders JKM, *J. Chem. Soc. Perkin Trans.*, 1: 423–430, 1983.

[e] From Scheer H et al., *J. Am. Chem. Soc.*, 96: 3714–3716, 1974.

[f] From Brockmann H Jr and Lipinski A, *Arch. Microbial.*, 136: 17–19, 1983.

[g] In pyridine-d_5.

[h] In CDCl$_3$.

TABLE 6.6
^{13}C-Chemical Shifts of BChl *a*, BPhe *a*, and Zn-BChl *a* in Acetone-d_6

IUPAC No. of Carbon Atoms	BChl *a*[a]	BPhe *a*[a]	Zn-BChl *a*[d]
1	151.2, (149.57[b,e], 166.87[c,e], 166.64[c,f])	139.7	142.110
2	142.0, (142.86[b,e], 124.25[c,e], 125.13[c,f])	138.5	150.712
2^1	13.5, (13.76[b,e], 13.76[c,e], 13.63[c,e])	13.9	13.9
3	137.7, (129.97[b,e], 129.97[c,e], 129.15[c,f])	135.0	137.267
3^1	199.3, (200.01[b,e], 200.01[c,e], 199.92[c,f])	199.2	199.641
3^2	32.9, (33.18[b,e], 33.18[c,e], 32.47[c,e])	34.0	33.431
4	150.0, (149.91[b,e], 149.91[c,e], 149.41[c,f])	138.1	149.779
5	99.9, (99.48[b,e], 99.48[c,e], 99.30[c,f])	97.9	98.941
6	168.9, (166.87[b,e], 149.57[c,e], 149.07[c,f])	172.4	168.563
7	48.3, (48.76[b,e], 48.76[c,e], 47.56[c,f])	49.6	47.748
7^1	23.4, (23.53[b,e], 23.53[c,e], 22.98[c,f])	23.7	23.023
8	55.8, (56.05[b,e], 56.05[c,e], 54.86[c,f])	55.4	55.047
8^1	30.8, (31.07[b,e], 31.07[c,e], 30.32[c,f])	30.7	–
8^2	10.5, (10.86[b,e], 10.86[c,e], 10.66[c,e])	11.5	10.5
9	158.5, (158.69[b,e], 158.69[c,e], 158.62[c,f])	164.3	158.654
10	102.4, (101.89[b,e], 101.89[c,e], 102.13[c,f])	100.2	101.569
11	149.5, (150.74[b,e], 150.74[c,e], 150.44[c,f])	139.3	149.222
12	124.0, (124.25[b,e], 142.86[c,e], 141.49[c,f])	121.3	124.361
12^1	11.9, (12.13[b,e], 12.13[c,e], 11.88[c,f])	11.5	11.9
13	130.6, (137.492, 137.493, 136.783)	129.2	130.115
13^1	189.0, (189.75[b,e], 189.75[c,e], 188.44[c,f])	189.3	189.202
13^2	65.7, (66.062, 66.063, 64.513)	65.5	65.211
13^3	171.6, (172.38[b,e], 172.38[c,e], 170.60[c,f])	170.2	171.165
13^4	52.3, (52.67[b,e], 52.673[c,e], 52.63[c,f])	52.8	–
14	160.8, (160.64[b,e], 160.64[c,e], 160.37[c,f])	148.7	160.011
15	109.7, (109.42[b,e], 109.42[c,e], 108.07[c,f])	110.3	109.466
16	152.0, (152.19[b,e], 152.19[c,e], 151.23[c,f])	158.7	153.076
17	50.5, (51.07[b,e], 51.07[c,e], 49.46[c,f])	51.4	50.058
17^1	30.5, (30.74[b,e], 30.74[c,e], 29.56[c,f])	31.3	–
17^2	29.4, (31.29[b,e], 31.29[c,e], 30.63[c,f])	31.5	–
17^3	173.4, (173.96[b,e], 173.96[c,e], 174.01[c,f])	173.0	173.349
18	49.5, (50.03[b,e], 50.03[c,e], 49.07[c,f])	50.9	48.981
18^1	22.8, (23.46[b,e], 23.46[c,e], 22.88[c,f])	22.9	23.023
19	167.3, (168.17[b,e], 168.17[c,e], 168.75[c,f])	171.7	167.76
20	96.3, (96.22[b,e], 96.22[c,e], 95.30[c,f])	97.2	96.449
P1	61.7, (61.71[b,e], 61.71[c,e], 61.88[c,f])	61.7	61.34
P2	119.3, (119.45[b,e], 119.45[c,e], 116.77[c,f])	119.4	119.167
P3	142.4, (143.06[b,e], 143.06[c,e], 143.50[c,f])	143.0	142.555
P3^1	(16.27[b,e], 16.27[c,e], 16.17[c,f])	–	16.1
P4	40.2, (40.41[b,e], 40.41[c,e], 39.83[c,f])	40.2	39.976
P5	25.3, (25.75[b,e], 25.75[c,e], 25.06[c,f])	25.3	25.511
P6	37.6, (37.28[b,e], 37.28[c,e], 36.75[c,f])	37.6	37.863
P7	33.3, (33.38[b,e], 33.38[c,e], 32.67[c,e])	33.2	33.431
P7^1	(20.10[b,e], 20.10[c,e], 19.73[c,f])	–	20.022
P8	38.5, (38.09[b,e], 38.09[c,e], 37.37[c,f])	37.9	37.863
P9	24.9, (25.15[b,e], 25.15[c,e], 24.47[c,f])	24.9	–
P10	38.0, (38.15[b,e], 38.15[c,e], 37.43[c,f])	38.5	37.863
P11	31.0, (33.55[b,e], 33.55[c,e], 32.78[c,f])	33.2	33.431
P11^1	(20.04[b,e], 20.04[c,e], 19.66[c,f])	–	20.022

(Continued)

TABLE 6.6 (CONTINUED)
^{13}C-Chemical Shifts of BChl *a*, BPhe *a*, and Zn-BChl *a* in Acetone-d_6

IUPAC No. of Carbon Atoms	BChl *a*[a]	BPhe *a*[a]	Zn-BChl *a*[d]
P12	37.8, (38.04[b,e], 38.04[c,e], 37.29[c,f])	38.5	37.863
P13	25.5, (25.54[b,e], 25.54[c,e], 24.78[c,f])	25.0	37.863
P14	39.8, (40.15[b,e], 40.15[c,e], 39.36[c,f])	39.8	39.976
P15	28.6, (28.75[b,e], 28.75[c,e], 27.96[c,f])	28.6	–
P15^1	(23.00[b,e], 23.00[c,e], 22.62[c,f])	–	23.023
P16	22.9, (22.92[b,e], 22.92[c,e], 22.62[c,f])	22.8	23.023

[a] From Egorova-Zachernyuk T et al., *Magn. Reson. Chem.*, 46: 1074–1083, 2008.

[b] From Facelli JC, *J. Phys. Chem. B*, 102: 2111–2116, 1998.

[c] From Okazaki T and Kajiwara M, *Chem. Pharm. Bull.*, 43: 1311–1317, 1995.

[d] From this work.

[e] In acetone-d_6: $CD_3OD = 4:1$.

[f] In $CDCl_3$.

remote paramagnetic terms. The effect of the ring current in the macrocycle is not related to the observed downfield shift of ^{1}H-NMR signals: the ring current has to shift ^{1}H-NMR signals upfield with regard to the change of the redox potential on introduction of a more electronegative metal (see Figure 6.37). Taking into account the fact that the large shifts are observed primarily on the same rings where the downfield shift of ^{13}C-NMR occurs, a possible explanation is that the enhanced anti-bonding level with the less electronegative metal coordination increases electron density not only on the π-system but also on the protons of rings I, III, and V. Consequently, the magnetic field shielding increases, and an anisotropic inductive effect from metal to rings I and III takes place (Kobayashi et al. 1998a,b, 1999a).

A nonuniform electron distribution on the macrocycle caused by the anisotropic inductive effect may produce forces of different magnitude along an axis through rings I, III, and IV to those along an axis through rings II and IV, and make a bend in the macrocycle planarity. The force is larger in [Mg]-BChl *a* than in [Zn]-BChl *a*, resulting in large instability in the former rather than the latter. In [M]-Chls *a*, increased electrons by the coordination of less electron negative metals may cause even ring II, as well as rings I and III, to participate in the formation of the π-system. The presence of nonbonding electrons on ring II induces an excessive distortion on the macrocycle, which results in less stability even in [Zn]-Chl *a* compared with [Zn]-BChl *a*, which accounts for the stronger resistance of [Zn]-BChl *a* than [Mg]-Chl *a* to demetallation (see Chapter 4 in Kobayashi et al. 2006a).

A unique ^{13}C-NMR and ^{15}N-NMR study of metal-free Chl derivatives showed the changes of free electrons on the

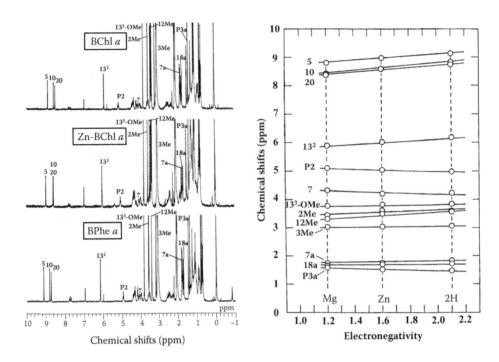

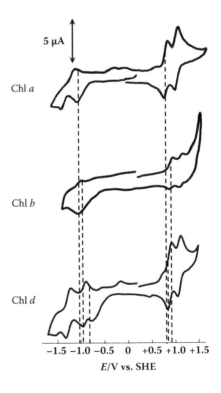

FIGURE 6.37 ¹H-NMR spectra (a) and correlation of the ¹H-NMR chemical shifts (b) among BChl *a* ([Mg]-BChl *a*), [Zn]-BChl *a*, and BPhe *a* ([2H]-BChl *a*) in acetone-d_6 at 263 K with E_N of the central metal. (Adapted from Kobayashi M et al., *Curr. Topics Plant Biol.*, 1: 17–35, 1999.)

macrocycle associated with the inductive effect. The changes in free electron density in rings I, III, and V, or rings II and IV by the substitution of central metal and solvents closely correlate with the electronegativity of the central metal and refractive index of solvents.

6.8 REDOX POTENTIALS

To understand the charge separation in the RC, electrochemical characterization of chlorophylls and bacteriochlorophylls is of crucial importance. In this section, the redox potentials of (B)Chls, (B)Phes, and [M]-(B)Chls *a in vitro* are presented.

6.8.1 CHLS *a, b, d, f* AND DV-CHL *a*

Typical cyclic voltammogram (CV) and square wave voltammogram (SWV) for Chl *a* in acetonitrile are illustrated in Figures 6.38 and 6.39. The anodic sweep of CV for Chl *d* in acetonitrile showed that $E_{red}^2 = -1.20$, $E_{red}^1 = -0.88$, $E_{ox}^1 = +0.93$, and $E_{ox}^2 = +1.12$ V versus SHE, and the cathodic sweep showed that $E_{red}^2 = -1.32$, $E_{red}^1 = -0.94$, $E_{ox}^1 = +0.84$, and $E_{ox}^2 = +1.06$ V versus SHE (Figure 6.38), resulting in $E_{red}^2 = -1.27$, $E_{red}^1 = -0.91$, $E_{ox}^1 = +0.88$, and $E_{ox}^2 = +1.09$ V versus SHE. The values agreed well with the redox potentials obtained from the SWV: $E_{red}^2 = -1.27$, $E_{red}^1 = -0.91$, $E_{ox}^1 = +0.88$, and $E_{ox}^2 = +1.09$ V versus SHE (Figure 6.39). The four potentials in anodic and cathodic sweep of SWV (data not shown) are identical to each other, indicating that the four redox reactions are reversible.

Table 6.7 summarizes the redox potentials for several (B) Chls. Chl *d* shows higher oxidation potentials than Chl *a*, lower than Chl *b*, and much lower than Phe *a*, Phe *b*, and Phe *d*. Chl *f*

exhibits higher oxidation potentials than Chls *a, d*, and lower than Chl *b*. The very similar absorption spectra of DV-Chl *a* and MV-Chl *a* suggest that the redox potential difference between them is too much small for CV. As expected, redox potential difference between DV-Chl *a* and MV-Chl *a* cannot

FIGURE 6.38 Cyclic voltammograms of Chls *a, b*, and *d* in acetonitrile.

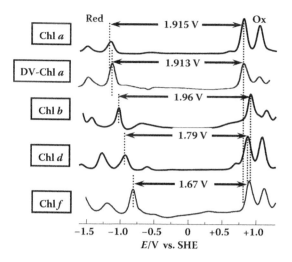

FIGURE 6.39 Square wave voltammograms of Chls a, b, d, f, and DV-Chl a in acetonitrile.

be clarified by CV, but SWV shows that DV-Chl a has slightly more positive oxidation potentials than MV-Chl a only by 8 mV (Figure 6.39). The results can be explained by the inductive effect of substituent groups on the macrocycle, because the redox potentials of chlorophylls are sensibly affected by the nature of substituent groups on the π-electron system (Fuhrhop 1975; Watanabe and Kobayashi 1991; Kobayashi et al. 2007).

The $-$CHO substituent on Chls b, d, and f is an electron-withdrawing group ($\rightarrow$CHO) and hence reduces the electronic density in the π-system of chlorophyll. The replacements of $-$CH$_3$ at C7 or C2 of Chl a by $\rightarrow$CHO to yield Chl b or Chl f cause the macrocycle to be electron poor, thus rendering the molecule less oxidizable (E_{ox}^1: Chl b, f > Chl a). Similarly, replacement of $-$CH=CH$_2$ at C3 of Chl a by $\rightarrow$CHO to yield Chl d makes the first oxidation potential, E_{ox}^1, more positive than that of Chl a (E_{ox}^1: Chl d > Chl a). Therefore, the E_{ox}^1 order becomes Chls b, d, f > Chl a (see Figure 6.40). When one pays attention to the group of $-$CH$_3$ at C7 of Chl d and the group of $-$CH=CH$_2$ at C3 of Chl b or C7 of Chl f, the $-$CH$_3$ group is more electron-donating ($\leftarrow$CH$_3$), thus making the macrocycle of Chl d more electron rich, and hence its oxidation potential less positive (Chls b, f > d); the E_{ox}^1 order results in Chls b, f > Chl d > Chl a. As expected from the inductive effect of substituent groups, Chls b and f show almost the same E_{ox}^1 values. As expected from the inductive effect of substituent groups, replacement of $\leftarrow$CH$_2$–CH$_3$ at C8 by $-$CH=CH$_2$ causes the macrocycle to be electron poor, thus rendering the molecule less oxidizable (first oxidation potential, E_{ox}^1: DV-Chl a > MV-Chl a). Slightly less negative first reduction potential, E_{red}^1, observed in DV-Chl a by 10 mV can be also explained in a similar way, namely, replacement of $\leftarrow$CH$_2$–CH$_3$ by $-$CH=CH$_2$ causes the macrocycle to be electron poor, thus rendering the molecule more easily reduced. Consequently, as seen in Figure 6.40, the E_{ox}^1 order results in Chl b > Chl f > Chl d $\gg$ DV-Chl a $\geq$ Chl a; a

little higher oxidation potential of Chl b than that of Chl f, 20 mV, cannot be explained from the primitive way used here.

The redox behavior of a compound is related to the energy levels of its molecular orbitals: E_{ox}^1 is intimately related to the highest occupied molecular orbital (HOMO) and E_{red}^1 to the lowest unoccupied molecular orbital (LUMO) (Hanson 1991; Watanabe and Kobayashi 1991). The order of HOMO energy levels well parallels the E_{ox}^1 values, while the correlation between the LUMO levels and the E_{red}^1 values is less conspicuous (see Figure 7 in Watanabe and Kobayashi 1991).

As clearly seen in Figure 6.40, the order of absolute values of the first reduction potentials, E_{red}^1, of Chls a, b and DV-Chl a is Chl a $\geq$ DV-Chl a > Chl b, which can be well explained by substituent inductive effect, like E_{ox}^1 as mentioned above. This simple rule, however, does not hold for Chls d and f; their values of E_{red}^1 are remarkably more positive than those of Chls a and b; compared to Chl a, Chl b is harder to oxidize by 135 mV and easier to reduce by 90 mV, namely, similar in degree, while Chls d and f are peculiarly easier to reduce by 200 and 360 mV. The findings indicate that the inductive effect of substituent groups at ring I conspicuously appears for E_{red}^1. Anyway, this irregularity may come from the fact that the LUMO energy levels are less correlated with the E_{red}^1 values.

The primary redox potential difference, $\Delta E = E_{ox}^1 - E_{red}^1$, seen in Figures 6.39 and 6.40 can be taken as an index for the Q_Y excitation energy (Hanson 1991; Watanabe and Kobayashi 1991). In Figure 6.40, for example, ΔE for Chl a is 1.915 eV, which corresponds to the Q_Y excitation wavelengths to a certain extent, 661–666 nm for Chl a in Figure 6.10. Almost the same value of ΔE for DV-Chl a and MV-Chl a correlates to the practically identical Q_Y excitation wavelengths as seen in Figure 6.10. Similarly, $\Delta E = 1.96$, 1.79, and 1.67 eV for Chls b, d, and f (Figures 6.39 and 6.40) and are also correlated to the Q_Y wavelengths, 644–652, 686–698, and 695–708 nm (Figure 6.10). Pheophytins also behave in a similar fashion.

In 1959, the domination of inductive effects of the central metal over a conjugative macrocycle has first been formulated (Gouterman 1959). The redox potential of chlorophyll shows a systematic shift with the electronegativity of the central metal, and such a trend is rationalized in terms of an electron density decrease in the chlorin π-system by the presence of an electron negative metal in the center of chlorophylls (Hanson 1991; Watanabe and Kobayashi 1991; Noy et al. 1998). Inspection of Table 6.7 demonstrates that such a trend is essential for the pair of Chls and Phes; the electron negativity of 2.2 for H is significantly higher than that of 1.2 for Mg, which renders Phes more difficult to oxidize than the corresponding Chls.

6.8.2 [M]-Chl a, [Zn]-BChl a, and [2H]-BChl a

A linear correlation for the primary redox potentials of M-BChls a and M-Chls a against the Pauling electronegativity (E_N) of the central metal is observed in Figure 6.41. The results are interpreted as an electron density change in the bacteriochlorin and chlorin π-system due to the inductive effect caused by the central metal (Watanabe and Kobayashi 1991).

TABLE 6.7

Redox Potentials of Chlorophylls and Bacteriochlorophylls

Compound	Solvent	Electrolyte	E_{red}^2	E_{red}^1	E_{ox}^1	E_{ox}^2	$E_{ox}^1 - E_{red}^1$	Reference
Chl a	AN	NaClO$_4$	–	–1.04	+0.76	+0.99	1.80	Watanabe and Kobayashi (1991)
	BuN	TBAP	–1.40	–0.91	+0.87	+1.10	1.78	Watanabe and Kobayashi (1991)
	DMSO	TPAP	–1.30	–0.88	–	–	–	Felton et al. (1964)
	PrN	LiClO$_4$	–	–	+0.76	+1.01	–	Stanienda (1965)
	DMF	TBAP	–1.37	–0.90	+0.83	+1.04	1.73	Saji and Bard (1977)
	CH$_2$I$_2$	TBAP	–	–	+0.86	+1.10	–	Wasielewski et al. (1981)
	DMF	NaClO$_4$	–1.33	–0.92	+0.84	+1.02	1.76	Watanabe and Kobayashi (1991)
	PrN	TBAP, TPAP	–	–	+0.82	–	–	Fajer et al. (1982)
	PrN	TBAC	–	–	+0.75	–	–	Fajer et al. (1982)
	CH$_2$Cl$_2$	TBAP, TPAP	–	–	+0.80	–	–	Fajer et al. (1982)
	CH$_2$Cl$_2$	TBAC	–	–	+0.74	–	–	Fajer et al. (1982)
	THF	TBAP, TPAP	–	–	+0.93	–	–	Fajer et al. (1982)
	CH$_2$Cl$_2$	TBAP	–	–	+0.80	–	–	Maggiora et al. (1985)
	AN	?	–	–0.58[a]	+0.61[a]	–	1.19	Takeuchi and Amao (2005)
	AN	TBAP	–1.454	–1.110	+0.805	+1.035	1.915	Komatsu et al. (2014)
DV–Chl a	AN	TBAP	–1.447	–1.100	+0.813	+1.044	1.913	Komatsu et al. (2014)
Chl b	AN	NaClO$_4$	–	–	+0.89	+1.14	–	Watanabe and Kobayashi (1991)
	PrN	LiClO$_4$	–	–	+0.89	+1.11	–	Stanienda (1965)
	DMF	NaClO$_4$	–	–0.87	+0.96	+1.10	1.83	Watanabe and Kobayashi (1991)
	AN	TBAP	–1.41	–1.02	+0.94	+1.15	1.96	Kobayashi et al. (2013)
Chl d	AN	TBAP	–1.27	–0.91	+0.88	+1.09	1.79	Kobayashi et al. (2013)
Chl f	AN	TBAP	–1.12	–0.75	+0.92	+1.13	1.67	Kobayashi et al. (2013)
Zn–Chl a	BuN	TBAP	–	–0.87	+1.00	–	1.87	Watanabe and Kobayashi (1991)
	AN	?	–	–0.67[a]	+0.60[a]	–	1.27	Takeuchi and Amao (2005)
Ni–Chl a	BuN	TBAP	–	–0.66	+1.22	–	1.88	Watanabe and Kobayashi (1991)
Co–Chl a	BuN	TBAP	–	–1.33	+1.31	–	2.64	Watanabe and Kobayashi (1991)
Fe–Chl a	BuN	TBAP	–	–0.69	+1.53	–	2.22	Watanabe and Kobayashi (1991)
Cu–Chl a	BuN	TBAP	–	–0.70	+1.10	–	1.80	Watanabe and Kobayashi (1991)
Ag–Chl a	BuN	TBAP	–	–	+1.31	–	–	Watanabe and Kobayashi (1991)
Hg–Chl a	BuN	TBAP	–	–0.96	+1.14	–	2.10	Watanabe and Kobayashi (1991)
Pd–Chl a	BuN	TBAP	–	–0.64	+1.24	–	1.88	Watanabe and Kobayashi (1991)
Pd–Chl a	BuN	TBAP	–	–0.64	+1.24	–	1.88	Watanabe and Kobayashi (1991)
Mn(III)–Chl a	BuN	TBAP	–	–0.58	+1.46	–	2.04	Watanabe and Kobayashi (1991)
Phe a	BuN	TBAP	–0.99	–0.70	+1.28	+1.60	1.98	Watanabe and Kobayashi (1991)
	PrN	LiClO$_4$	–	–	+1.41	+1.64	–	Stanienda (1965)
	DMF	NaClO$_4$	–0.86	–0.56	–	–	–	Watanabe and Kobayashi (1991)
	CH$_2$Cl$_2$	TBAP	–	–0.78	–	–	–	Bucks et al. (1982)
	CH$_2$Cl$_2$	TEAC	–	–0.64	–	–	–	Bucks et al. (1982)
	AN	TBAP	–1.00	–0.75	+1.14	+1.49	1.89	Kobayashi et al. (2013)
Phe b	PrN	LiClO$_4$	–	–	+1.23	+1.52	–	Stanienda (1965)
	AN	TBAP	–1.05	–0.64	+1.25	+1.58	1.89	Kobayashi et al. (2013)
Phe d	AN	TBAP	–0.87	–0.63	+1.21	+1.50	1.84	Kobayashi et al. (2013)
BChl a	CH$_2$Cl$_2$	TPAP	–	–0.86	+0.64	–	1.50	Fajer et al. (1975)
	Dry AN	TBAP, TBABF$_4$	–1.04	–0.75	+0.66	+1.02	1.41	Cotton and Van Duyne (1979)
	Wet AN	TBAP, TBABF$_4$	–1.08	–0.74	+0.66	+1.02	1.50	Cotton and Van Duyne (1979)
	Dry AN + THF	TBAP, TBABF$_4$	–1.11	–0.77	+0.66	+1.08	1.43	Cotton and Van Duyne (1979)
	THF	TBAP, TBABF$_4$	–1.17	–0.83	+0.76	+1.07	1.59	Cotton and Van Duyne (1979)
	CH$_2$Cl$_2$	TBAP, TBABF$_4$	–1.33	–0.84	+0.66	+1.04	1.51	Cotton and Van Duyne (1979)
	CH$_2$Cl$_2$ + THF	TBAP, TBABF$_4$	–1.19	–0.79	+0.66	+1.03	1.45	Cotton and Van Duyne (1979)
	MeOH	TBAP	–	–	0.56[a]	–	–	Cotton and Heald (1987)
	THF	TBAPF$_6$	–	–1.02	+0.57	–	1.59	Geskes et al. (1995)
	AN + DHF	TBAF	–	–0.88	+0.48	–	1.36	Noy et al. (1998)
BChl b	DMF	?	–	–0.70	–	–	–	Fajer et al. (1976, 1978)

(Continued)

TABLE 6.7 (CONTINUED)
Redox Potentials of Chlorophylls and Bacteriochlorophylls

Compound	Solvent	Electrolyte	E_{red}^2	E_{red}^1	E_{ox}^1	E_{ox}^2	$E_{ox}^1 - E_{red}^1$	Reference
BChl c^b	CH_2Cl_2	TPAP	–	–1.03	–	–	–	Fajer et al. (1983)
BChl d^b	DMF	TPAP	–	–1.00	–	–	–	Fajer et al. (1983)
BChl e^b	DMF	TPAP	–	–0.95	–	–	–	Fajer et al. (1983)
Mn(III)–BChl a	THF	TBAPF$_6$	–	–1.00	+0.91	–	1.91	Geskes et al. (1995)
Zn–BChl a	THF	TBAPF$_6$	–	–0.93	+0.60	–	1.53	Geskes et al. (1995)
	AN + DHF	TBAF	–	–0.83	+0.59	–	1.42	Noy et al. (1998)
	AN + DHF	TBAF	–	–0.83	+0.59	–	1.42	Noy et al. (1998)
Cd–BChl a	THF	TBAPF$_6$	–	–0.87	+0.65	–	1.52	Geskes et al. (1995)
	AN + DHF	TBAF	–	–0.80	+0.58	–	1.38	Noy et al. (1998)
Ni–BChl a	THF	TBAPF$_6$	–	–0.73	+0.58	–	1.31	Geskes et al. (1995)
	AN + DHF	TBAF	–	–0.64	+0.67	–	1.31	Noy et al. (1998)
Co–BChl a	THF	TBAPF$_6$	–	–1.32	+0.56	–	1.88	Geskes et al. (1995)
Cu–BChl a	THF	TBAPF$_6$	–	–0.87	+0.64	–	1.51	Geskes et al. (1995)
	AN + DHF	TBAF	–	–0.68	+0.67	–	1.35	Noy et al. (1998)
Pd–BChl a	THF	TBAPF$_6$	–	–0.72	+0.88	–	1.60	Geskes et al. (1995)
	AN + DHF	TBAF	–	–0.64	+0.82	–	1.46	Noy et al. (1998)
BPhe a	BuN	TPAP	–	–0.54	–	–	–	Fajer et al. (1975)
	DMF	TPAP	–	–0.51	–	–	–	Fajer et al. (1975)
	CH_2Cl_2	TPAP	–	–0.58	–	–	–	Fajer et al. (1975)
	THF	TBAP, TBABF$_4$	–0.93	–0.60	+1.12	–	1.72	Cotton and Van Duyne (1979)
	CH_2Cl_2	TBAP, TBABF$_4$	–0.86	–0.61	+1.02	–	1.63	Cotton and Van Duyne (1979)
	THF	TBAPF$_6$	–	–0.79	+0.88	–	1.67	Geskes et al. (1995)
	AN + DHF	TBAF	–	–0.67	+0.88	–	1.55	Noy et al. (1998)
BPhe b	DMF	?	–	–0.50	–	–	–	Fajer et al. (1976)
	CH_2Cl_2	?	–	–0.56	–	–	–	Fajer et al. (1976)
BPhe $c^{b,c}$	DMF	TPAP	–	–0.79	–	–	–	Fajer et al. (1983)
BPhe $d^{b,c}$	DMF	TPAP	–	–0.81	–	–	–	Fajer et al. (1983)
BPhe $e^{b,c}$	DMF	TPAP	–	–0.76	–	–	–	Fajer et al. (1983)

[a] Ag/AgCl.
[b] Homologous mixtures.
[c] Methyl ester derivatives.

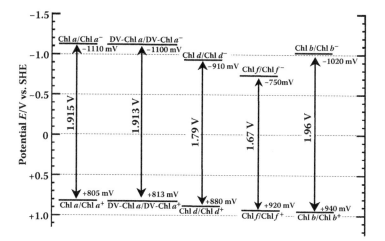

FIGURE 6.40 Schematic comparison of redox potentials of Chls a, b, d, and f in acetonitrile.

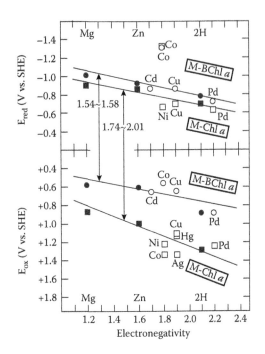

FIGURE 6.41 Correlation of the first redox potentials of a series of M(I1)-BChls *a* (○ and ●) in tetrahydrofuran (Geskes et al. 1995; Scheer and Hartwich 1995) and M(II)-Chls *a* (□ and ■) in butyronitrile with electronegativity, E_N, of the central metal. Metal-centered redox potentials are omitted. The vertical arrows denote the values of $E_{ox}^1 - E_{red}^1$.

The effective charge values calculated for the central metal in several M-Chls *a* are +0.50 (Mg), +0.32 (Zn), +0.29 (Cu), and +0.026 (Pd) in units of elementary charge (Kobayashi et al. 1998d). The increase in the π-electron density in the macrocycle in the order of Mg > Zn > Cu > Pd antiparallels the order of E_N (Mg = 1.2, Zn ≈ 1.6, Cu = 1.9, Pd = 2.2). In the plot of E_{red} against E_N (Figure 6.41), Co-BChl *a* and Co- Chl *a* show a large deviation from the straight line. One reason may be the relative ease of metal redox in Co as compared with other metals and its effect on the ring redox potential.

Interestingly, the difference in the E_{red} values between M-BChls *a* and M-Chls *a* is small, and this indicates little change in LUMO levels (Petke et al. 1980; Watanabe and Kobayashi 1991). In contrast, the E_{ox}^1 values of M-BChls *a* are more negative than those of M-Chls *a*, mainly resulting from the large difference in the HOMO levels, due to the reduction of the C7=C8 double bond (Weiss 1972, 1978; Spangler et al. 1977; Watanabe and Kobayashi 1991; Geskes et al. 1995; Scheer and Hartwich 1995). These are qualitatively in line with the following. M-BChls *a* and M-Chls *a* in the singlet excited state can reduce the primary electron acceptors, BPhe *a* and Phe *a*, that have similar reduction potentials both *in vitro* and *in vivo*. On the other hand, the marked difference in the E_{ox}^1 values between M-BChls *a* and M-Chls *a* reflects the difference in the substance acting as the electron donor *in vivo*, namely, H_2S or organic substances for the former and H_2O for the latter.

Note that the difference in E_{ox}^1 between Zn-BChl *a* and Mg-BChl *a* is small. This may enable *A. rubrum* to keep

the photosynthetic polypeptide changes to a minimum (Nagashima et al. 1997). This hypothesis is supported partly by the *in vitro* exchange experiments of Mg-BChl *a* to Zn-BChl *a* in the purple bacterial RC by Scheer and Hartwich (1995). A chance of finding Zn-Chl *a* in nature may be small, since its oxidation potential is considerably more positive compared to Mg-Chl *a*, and hence, large polypeptide changes for the protection against oxidation damage are required, although a high oxidation potential is advantageous for water oxidation.

ACKNOWLEDGMENTS

We thank Prof. Yoshihiro Shiraiwa, Dr. Koji Iwamoto, Dr. Yutaka Hanawa (Univ. Tsukuba), and Hamamatsu Photonics Co. Ltd. for their helpful technical support.

ABBREVIATIONS

A. marina: *Acaryochloris marina*
APCI: atmospheric pressure chemical ionization
A. rubrum: *Acidiphilium rubrum*
BChl: bacteriochlorophyll
BPhe: bacteriopheophytin
CD: circular dichroism
Chl: chlorophyll
COSY: correlation spectroscopy
DQF-COSY: double quantum filtered correlation spectroscopy
DV-Chl: divinyl chlorophyll
DV-Phe: divinyl pheophytin
F: farnesol
FAB: fast atom bombardment
H. mobilis: *Heliobacillus mobilis*
HMBC: heteronuclear multiple-bond correlation
HSQC: heteronuclear single-quantum coherence
HPLC: high-performance liquid chromatography
MALDI: matrix-assisted laser desorption ionization
NMR: nuclear magnetic resonance
NOESY: nuclear Overhauser and exchange spectroscopy
P: phytol
P. marinus: *Prochlorococcus marinus*
Phe: pheophytin
Δ2,6PD: Δ2,6-phytadienol
R. rubrum: *Rhodospirillum rubrum*
R. sphaeroides: *Rhodobacter sphaeroides*
R. viridis: *Rhodopseudomonas viridis*
RC: reaction center

REFERENCES

Abraham RJ and Rowan AE (1991) Nuclear magnetic resonance spectroscopy of chlorophyll. In *Chlorophylls*, ed. by Scheer H, CRC Press, Boca Raton, FL, pp. 797–834.

Akiyama M, Kobayashi M, Kise H, Hara M, Wakao N and Shimada K (1998a) Pigment composition of the reaction center complex isolated from an acidophilic bacterium *Acidiphilium rubrum* grown at pH 3.5. *Photomed. Photobiol.*, **20**: 85–87.

Akiyama M, Kobayashi M, Kise H, Takaichi S, Watanabe T, Shimada K, Iwaki M, Itoh S, Ishida N, Koizumi M, Kano H, Wakao N and Hiraishi A (1998b) *Acidiphilium rubrum* and zinc-bacteriochlorophyll, part 1: Molecular structure of the zinc-containing bacteriochlorophyll. In *Photosynthesis: Mechanisms and Effects*, ed. by Garab, G, Kluwer Academic Publishers, Dordrecht, vol. **2**, pp. 731–734.

Akiyama M, Miyashita H, Watanabe T, Kise H, Miyachi S and Kobayashi M (2001) Detection of chlorophyll *d'* and pheophytin *a* in a chlorophyll *d*-dominating oxygenic photosynthetic prokaryote *Acaryochloris marina*. *Anal. Sci.*, **17**: 205–208.

Akiyama M, Miyashita H, Kise H, Watanabe T, Mimuro M, Miyachi S and Kobayashi M (2002) Quest for minor but key chlorophyll molecules in photosynthetic reaction centers—Unusual pigment composition in the reaction centers of a chlorophyll *d*-dominated cyanobacterium *Acaryochloris marina*. *Photosynth. Res.*, **74**: 97–107.

Akiyama M, Gotoh T, Kise H, Miyashita H, Mimuro M and Kobayashi M (2004) Stoichiometries of chlorophyll *d'*/PSI and chlorophyll *a*/PSII in a chlorophyll *d*-dominated cyanobacterium *Acaryochloris marina*. *Jpn. J. Phycol.*, **52**: 67–72.

Akutsu S, Fujinuma D, Furukawa H, Watanabe T, Ohnishi-Kameyama M, Ono H, Ohkubo S, Miyashita H and Kobayashi M (2011) Pigment analysis of a chlorophyll *f*-containing cyanobacterium strain KC1 isolated from Lake Biwa. *Photomed. Photobiol.*, **33**: 35–40.

Bazzaz MB (1981) New chlorophyll chromophores isolated from a chlorophyll deficient mutant of maize. *Photobiochem. Photobiophys.*, **2**: 199–207.

Becker M, Nagarajan V and Parson WW (1991) Properties of the excited-singlet states of bacteriochlorophyll *a* and bacteriopheophytin *a* in polar solvents. *J. Am. Chem. Soc.*, **113**: 6840–6848.

Boardman NK and Thorne SW (1971) Sensitive fluorescence method for the determination of chlorophyll *a*/chlorophyll *b* rations. *Biochim. Biophys. Acta*, **253**: 222–231.

Boxer SG, Closs GL and Katz JJ (1974) The effect of magnesium coordination on the ^{13}C and ^{15}N magnetic resonance spectra of chlorophyll *a*. The relative energies of nitrogen nπ* states as deduced from a complete assignment of chemical shifts. *J. Am. Chem. Soc.*, **96**: 7058–7066.

Braumann T, Vasmel H, Grimme LH and Amesz J (1986) Pigment composition of the photosynthetic membrane and reaction center of the green bacterium *Prosthecochloris aestuarii*. *Biochim. Biophys. Acta*, **848**: 83–91.

Brereton RG and Sanders JKM (1983) Co-ordination and aggregation of bacteriochlorophyll *a*: An n. m. r. and electronic absorption study. *J. Chem. Soc. Perkin Trans.*, **1**: 423–430.

Brockmann H Jr and Kleber I (1970) Bacteriochlorophyll b. *Tetrahedron Lett.*, **25**: 2195–2198.

Brockmann H Jr and Lipinski A (1983) Bacteriochlorophyll g. A new bacteriochlorophyll from *Heliobacterium chlorum*. *Arch. Microbial.*, **136**: 17–19.

Bucks RR, Netzel TL, Fujita I and Boxer SG (1982) Picosecond spectroscopic study of chlorophyll-based models for the primary photochemistry of photosynthesis. *J. Phys. Chem.*, **86**: 1947–1955.

Butler WL and Norris KH (1963) Lifetime of the long-wavelength chlorophyll fluorescence. *Biochim. Biophys. Acta*, **66**: 72–77.

Chen K, Preuβ A, Hackbarth S, Wacker M, Langer K and Röder B (2009) Novel photosensitizer-protein nanoparticles for photodynamic therapy: Photophysical characterization and *in vitro* investigations. *J. Photochem. Photobiol. B, Biology*, **96**: 66–74.

Chen M, Schliep M, Willows RD, Cai Z-L, Neilan BA and Scheer H (2010) A red-shifted chlorophyll. *Science*, **329**: 1318–1319.

Chisholm SW, Olson RJ, Zettler ER, Goericke R, Waterbury JB and Welschmeyer NA (1988) A novel free-living prochlorophyte abundant in the oceanic euphotic zone. *Nature*, **334**: 340–343.

Connolly JS, Janzen AF and Samuel EB (1982a) Fluorescence lifetimes of chlorophyll *a*: Solvent, concentration and oxygen dependence. *Photochem. Photobiol.*, **36**: 559–563.

Connolly JS, Samuel EB and Janzen AF (1982b) Effects of solvent on the fluorescence properties of bacteriochlorophyll *a*. *Photochem. Photobiol.*, **36**: 565–574.

Cotton TM and Heald RL (1987) Solvent effects on the heterogeneous electron-transfer kinetics of bacteriochlorophyll *a* and bacteriopheophytin *a* at platinum and gold electrodes. *J. Phys. Chem.*, **91**: 3891–3898.

Cotton TM and Van Duyne RP (1979) An electrochemical investigation of the redox properties of bacteriochlorophyll and bacteriopheophytin in aprotic solvents. *J. Am. Chem. Soc.*, **101**: 7605–7612.

Egorova-Zachernyuk T, van Rossum B, Erkelens C and de Groot H (2008) Characterisation of uniformly 13C, 15N labelled bacteriochlorophyll *a* and bacteriopheophytin *a* in solution and in solid state: Complete assignment of the ^{13}C, ^{1}H and ^{15}N chemical shifts. *Magn. Reson. Chem.*, **46**: 1074–1083.

Eimhjellen KE, Aasmundrud O and Jensen A (1963) A new bacterial chlorophyll. *Biochem. Biophys. Res. Commun.*, **10**: 232–236.

Eskins K, Scholfield CR, Dutton HJ (1977) High-performance liquid chromatography of plant pigments. *J. Chromatogr.*, **135**: 217–220.

Evans TA and Katz JJ (1975) Evidence for 5- and 6-coordinated magnesium in bacteriochlorophyll *a* from visible absorption spectroscopy. *Biochim. Biophys. Acta*, **396**: 414–426.

Facelli JC (1998) Density functional theory calculations of the structure and the ^{15}N and ^{13}C chemical shifts of methyl bacteriopheophorbide *a* and bacteriochlorophyll *a*. *J. Phys. Chem. B*, **102**: 2111–2116.

Fages F, Griebenow N, Griebenow K, Holzwarth AR and Schaffner K (1990) Characterization of light-harvesting pigments of *Chloroflexus aurantiacus*. Two new chlorophylls: Oleyl (octadec-9-enyl) and cetyl (hexadecanyl) bacteriochlorophyllides-*c*. *J. Chem. Soc. Perkin Trans.*, **1**: 2791–2797.

Fajer J, Brune DC, Davis MS, Forman A and Spaulding LD (1975) Primary charge separation in bacterial photosynthesis: Oxidized chlorophylls and reduced pheophytin. *Proc. Natl. Acad. Sci. U.S.A.*, **72**: 4956–4960.

Fajer J, Davis MS, Brune DC, Spaulding LD, Borg DC and Forman A (1976) Chlorophyll radicals and primary events. *Brookhaben Symp. Biol.*, **28**: 74–104.

Fajer J, Davis MS, Brune DC, Forman A and Thornber JP (1978) Optical and paramagnetic identification of a primary electron acceptor in bacterial photosynthesis. *J. Am. Chem. Soc.*, **100**: 1918–1920.

Fajer J, Fujita I, Davis MS, Forman A, Hanson LK and Smith KM (1982) Photosynthetic energy transduction: Spectral and redox characteristics of chlorophyll radicals in vitro and in vivo. In *Electrochemical and Spectrochemical Studies of Biological Redox Components*, ed. by Kadish KM, American Chemical Society, Washington, DC, chapter 21, pp. 489–513.

Fajer J, Fujita I, Forman A, Hanson LK, Craig GW, Goff DA, Kehres LA and Smith KM (1983) Anion radicals of bacteriochlorophylls *c*, *d* and *e*. Likely electron acceptors in the primary photochemistry of green and brown photosynthetic bacteria. *J. Am. Chem. Soc.*, **105**: 3837–3843.

Felton R, Sherman GM and Linschitz H (1964) Formation of phase test intermediate of chlorophyll by electrolytic reduction. *Nature*, **203**: 637–639.

French CS (1960) The chlorophylls *in vivo* and *in vitro*. In *Handbuch der Pflanzenphysiologie/Encyclopedia of Plant Physiology*, ed. by Ruhland W, Springer-Verlag, Berlin, Heidelberg, vol. **5**, pp. 252–297.

French CS, Smith JHC, Virgin HI and Airth RL (1956) Fluorescence spectrum curves of chlorophylls, pheophytins, phycoerythrins, phycocyanins and hypericin. *Plant Physiol.*, **31**: 369–374.

Fuhrhop JH (1975) Reversible reactions of porphyrins and metalloporphyrins and electrochemistry. In *Porphyrins and Metalloporphyrins*, ed. by Smith KM, Elsevier, Amsterdam, chapter 14.

Fujinuma D, Akutsu S, Komatsu H, Watanabe T, Miyashita H, Iwamoto K, Shiraiwa Y, Islam MR, Koike H, Kawachi M, Kobayashi M (2012) Detection of divinyl chlorophyll a' and divinyl pheophytin a as minor key components in a marine picoplankton *Prochlorococcus* sp. RCC315. *Photomed. Photobiol.*, **34**: 47–52.

Fuller RC, Sprague SG, Gest H and Blankenship RE (1985) A unique photosynthetic reaction center from *Heliobacterium chlorum*. *FEBS Lett.*, **182**: 345–349.

Furukawa H, Oba T, Tamiaki H and Watanabe T (2000) Effect of $C13^2$-stereochemistry on the molecular properties of chlorophylls. *Bull. Chem. Soc. Jpn.*, **73**: 1341–1351.

Geskes C, Hartwich G, Scheer H, Mäntele W and Heinze J (1995) An electrochemical and spectroelectrochemical investigation of metal-substituted bacteriochlorophyll a. *J. Am. Chem. Soc.*, **117**: 7776–7783.

Gest H (1994) Discovery of the heliobacteria. *Photosynth. Res.*, **41**: 17–21.

Gieskes WW and Kraay GW (1983) Unknown chlorophyll a derivatives in the North Sea and the tropical Atlantic Ocean revealed by HPLC analysis. *Limnol. Oceanogr.*, **28**: 757–766.

Gieskes WW and Kraay GW (1986) Floristic and physiological differences between the shallow and the deep nanoplankton community in the euphotic zone of the open tropical Atlantic revealed by HPLC analysis of pigments. *Mar. Biol.*, **91**: 567–576.

Gloe A, Pfenning N, Brockmann H Jr and Trowitzsch W (1975) A new bacteriochlorophyll from brown-colored chlorobiaceae. *Arch. Microbiol.*, **102**: 103–109.

Goedheer JC (1966) Visible absorption and fluorescence of chlorophyll and its aggregates in solution. In *The Chlorophylls*, ed. by Vernon LP and Seely GR, Academic Press, New York, pp. 147–184.

Goericke R and Repeta DJ (1992) The pigments of *Prochlorococcus marinus*: The presence of divinyl chlorophyll a and b in a marine rokaryote. *Limnol. Oceanogr.*, **37**: 425–433.

Gouterman M (1959) Study of the effects of substitution on the absorption spectra of porphin. *J. Chem. Phys.*, **30**: 1139–1161.

Gouterman M (1961) Spectra of porphyrins. *J. Mol. Spectrosc.*, **6**: 138–163.

Gouterman M, Wagniere GH and Snyder LC (1963) Spectra of porphyrins: Part II. Four orbital model. *J. Mol. Spectrosc.*, **11**: 108–127.

Gradyushko AT, Sevchenko AN, Solovyov KN and Tsvirko MP (1970) Energetics of photophysical processes in chlorophyll-like molecules. *Photochem. Photobiol.*, **11**: 387–400.

Hall DO and Rao KK (1994) *Photosynthesis*, 5th edition, Cambridge University Press, Cambridge, England.

Hanson LK (1991) Molecular orbital theory on monomer pigments. In *Chlorophylls*, ed. by Scheer H, CRC Press, Boca Raton, FL, pp. 993–1014.

Hartwich G, Fiedor L, Simonin I, Cmiel E, Schäfer W, Noy D, Scherz A and Scheer H (1998) Metal-substituted bacteriochlorophylls. 1. Preparation and influence on metal and coordination on spectra. *J. Am. Chem. Soc.*, **120**: 3675–3683.

Hoffmann E and Stroobant V (2002) *Mass Spectrometry: Principles and Applications*, 2nd edition, John Wiley & Sons, Chichester.

Hunt JE and Michalski TJ (1991) Desorption-ionization mass spectrometry of chlorophylls. In *Chlorophylls*, ed. by Scheer H, CRC Press, Boca Raton, FL, pp. 835–853.

Iriyama K, Yoshiura M and Shiraki M (1978) Micro-method for the qualitative and quantitative analysis of photosynthetic pigments using high-performance liquid chromatography. *J. Chromatogr.*, **154**: 302–305.

Jabben M, Garcia NA, Braslavsky SE and Schaffner K (1986) Photophysical parameters of chlorophylls a and b Fluorescence and laser-induced optoacoustic measurements. *Photochem. Photobiol.*, **43**: 127–131.

Jeffrey SW (1972) Preparation and some properties of crystalline chlorophyll c_1 and c_2 from marine algae. *Biochim. Biophys. Acta*, **279**: 15–33.

Jeffrey SW, Mantoura RFC and Wright SW (1997) *Phytoplankton Pigments in Oceanography*. UNESCO Publishing, Paris, France.

Jensen A, Aasmundrud O and Eimhjellen KE (1964) Chlorophylls of photosynthetic bacteria. *Biochim. Biophys. Acta*, **88**: 466–479.

Jordan P, Fromme P, Witt HT, Klukas O, Saenger W and Krauβ N (2001) Three-dimensional structure of cyanobacterial photosystem I at 2.5 Å resolution. *Nature*, **411**: 909–917.

Kaplanova M and Cermak K (1981) Effect of reabsorption on the concentration dependence of fluorescence lifetimes of chlorophyll a. *J. Photochem.*, **15**: 313–319.

Karukstis KK (1991) Chlorophyll fluorescence as a physiological probe of the photosynthetic apparatus. In *Chlorophylls*, ed. by Scheer H, CRC Press, Boca Raton, FL, pp. 769–795.

Katz JJ and Brown CE (1983) Nuclear magnetic resonance spectroscopy of chlorophylls and corrins. *Bull. Magn. Reson.*, **5**: 3–49.

Kaufmann KJ, Dutton PL, Netzel TL, Leigh JS and Rentzepis PM (1975) Picosecond kinetics of events leading to reaction center bacteriochlorophyll oxidation. *Science*, **188**: 1301–1304.

Kim SW (1967) Copper replacement of magnesium in the chlorophylls and bacteriochlorophyll. *Z. Naturforschg.*, **22b**: 1054–1064.

Kirmaier C, Blankenship RE and Holten D (1986) Formation and decay of radical-pair P^+I^- in *Chloroflexus aurantiacus* reaction centers. *Biochim. Biophys. Acta*, **850**: 275–285.

Kishimoto N, Fukaya F, Inagaki K, Sugio T, Tanaka H and Tano T (1995) Distribution of bacteriochlorophyll a among aerobic and acidophilic bacteria and light-enhanced CO_2-incorporation in *Acidiphilium rubrum*. *FEMS Microbiol. Ecol.*, **16**: 291–296.

Klimov VV, Klevanik AV, Shuvalov VA and Krasnovsky AA (1977a) Reduction of pheophytin in the primary light reaction of photosystem II. *FEBS Lett.*, **82**: 183–186.

Klimov VV, Allkhverdiev SI, Demeter S and Krasnovsky AA (1977b) Photoreduction of pheophytin in photosystem 2 of chloroplasts with respect to the redox potential of the medium. *Dokl. Akad. Nauk SSSR*, **249**: 227–230.

Klimov VV, Shuvalov VA, Krakhmaleva IN, Klevanik AV and Krasnovskii AA (1977c) Photoreduction of bacteriopheophytin b in the primary light reaction of *Rhodopseudomonas viridis* chromatophores. *Biokhimiya* **42**: 519–530.

Kobayashi M (1989) Study on the molecular mechanism of photosynthetic reaction centers. Thesis, University of Tokyo, Tokyo, Japan.

Kobayashi M (1996) Study of precise pigment composition of photosystem I-type reaction centers by means of normal-phase HPLC. *J. Plant Res.*, **109**: 223–230.

Kobayashi M, Watanabe T, Nakazato M, Ikegami I, Hiyama T, Matsunaga T and Murata N (1988) Chlorophyll a'/P700 and pheophytin a/P680 stoichiometries in higher plants and cyanobacteria determined by HPLC analysis. *Biochim. Biophys. Acta*, **936**: 81–89.

Kobayashi M, van de Meent EJ, Erkelens C, Amesz J, Ikegami I and Watanabe T (1991) Bacteriochlorophyll g epimer as a possible reaction center component of heliobacteria. *Biochim. Biophys. Acta*, **1057**: 89–96.

Kobayashi M, van de Meent EJ, Oh-Oka H, Inoue K, Itoh S, Amesz J and Watanabe T (1992) Pigment composition of heliobacteria and green sulfur bacteria. In *Research in Photosynthesis*, ed. by Murata N, Kluwer Academic Publishers, Dordrecht, vol. **1**, pp. 393–396.

Kobayashi M, Hamano T, Akiyama M, Watanabe T, Inoue K, Oh-Oka H, Amesz J, Yamamura M and Kise H (1998a) Light-independent isomerization of bacteriochlorophyll g to chlorophyll a catalyzed by weak acid *in vitro. Anal. Chim. Acta*, **365**: 199–203.

Kobayashi M, Yamamura M, Akutsu S, Miyake J, Hara M, Akiyama M and Kise H (1998b) Successfully controlled isomerization and pheophytinization of bacteriochlorophyll b by weak acid in the dark *in vitro. Anal. Chim. Acta*, **361**: 285–290.

Kobayashi M, Akiyama M, Yamamura M, Kise H, Takaichi S, Watanabe T, Shimada K, Iwaki M, Itoh S, Ishida N, Koizumi M, Kano H, Wakao N and Hiraishi A (1998c) Structural determination of the novel Zn-containing bacteriochlorophyll in *Acidiphilium rubrum. Photomed. Photobiol.*, **20**: 75–80.

Kobayashi M, Akiyama M, Yamamura M, Kise H, Ishida N, Koizumi M, Kano H and Watanabe T (1998d) *Acidiphilium rubrum* and zinc-bacteriochlorophyll, part2: Physicochemical comparison of zinc-type chlorophylls and other metallochlorophylls. In *Photosynthesis: Mechanisms and Effects*, ed. by Garab, G, Kluwer Academic Publishers, Dordrecht, vol. **2**, pp. 735–738.

Kobayashi M, Akiyama M, Watanabe T and Kano H (1999a) Exotic chlorophylls as key components of photosynthesis. *Curr. Top. Plant Biol.*, **1**: 17–35.

Kobayashi M, Akiyama M, Yamamura M, Kise H, Wakao N, Ishida N, Koizumi M, Kano H and Watanabe T (1999b) Comparison of physicochemical properties of metallobacteriochlorophylls and metallochlorophylls. *Z. Physik. Chem.*, **213**: 207–214.

Kobayashi M, Oh-Oka H, Akutsu S, Akiyama M, Tominaga K, Kise H, Nishida F, Watanabe T, Amesz J, Koizumi M, Ishida N and Kano H (2000) The primary electron acceptor of green sulfur bacteria, bacteriochlorophyll 663, is chlorophyll a esterified with Δ2,6-phytadienol. *Photosynth. Res.*, **63**: 269–280.

Kobayashi M, Watanabe S, Gotoh T, Koizumi H, Itoh Y, Akiyama M, Shiraiwa Y, Tsuchiya T, Miyashita H, Mimuro M, Yamashita T and Watanabe T (2005) Minor but key chlorophylls in Photosystem II. *Photosynth. Res.*, **84**: 201–207.

Kobayashi M, Akiyama M, Kise H and Watanabe T (2006a) Unusual tetrapyrrole pigments of photosynthetic antennae and reaction centers: Specially-tailored chlorophylls. In *Chlorophylls and Bacteriochlorophylls: Biochemistry, Biophysics, Functions and Applications*, ed. by Grimm B, Porra RJ, Rüdiger W and Scheer H, Springer, Dordrecht, pp. 55–66.

Kobayashi M, Akiyama M, Kano H and Kise H (2006b) Spectroscopy and structure determination. In *Chlorophylls and Bacteriochlorophylls: Biochemistry, Biophysics, Functions and Applications*, ed. by Grimm B, Porra RJ, Rüdiger W and Scheer H, Springer, Dordrecht, pp. 79–94.

Kobayashi M, Ohashi S, Iwamoto K, Shiraiwa Y, Kato Y and Watanabe T (2007) Redox potential of chlorophyll d *in vitro. Biochim. Biophys. Acta*, **1767**: 596–602.

Kobayashi M, Akutsu S, Fujinuma D, Furukawa D, Komatsu H, Hotota Y, Kato Y, Kuroiwa Y, Watanabe T, Ohnishi-Kameyama M, Ono H, Ohkubo S and Miyashita H (2013) Physicochemical properties of chlorophylls in oxygenic photosynthesis— Succession of co-factors from anoxygenic to oxygenic photosynthesis. In *Photosynthesis*, ed. by Dubinsky Z, Intech, Croatia, chapter 3, pp. 47–90.

Koizumi H, Itoh Y, Hosoda S, Akiyama M, Hoshino T, Shiraiwa Y and Kobayashi M (2005) Serendipitous discovery of Chl d formation from Chl a with papain. *Sci. Technol. Adv. Mat.*, **6**: 551–557.

Komatsu H, Fujinuma D, Akutsu S, Fukayama D, Sorimachi Y, Kato Y, Kuroiwa Y, Watanabe T, Miyashita H, Iwamoto K, Shiraiwa Y, Ohnishi-Kameyama M, Ono H, Koike H, Sato M, Kawachi M and Kobayashi M (2014) Physicochemical properties of divinyl chlorophylls a, a' and divinyl pheophytin a compared with those of monovinyl derivatives. *Photomed. Photobiol.*, **36**: 59–69.

Latimer P, Bannister TT and Rabinowitch E (1956) Quantum yields of fluorescence of plant pigments. *Science*, **124**: 585–586.

Leupold D, Struck A, Stiel H, Teuchner K, Oberländer S and Scheer H (1990) Excited-state properties of 20-chloro-chlorophyll a. *Chem. Phys. Lett.*, **170**: 478–484.

Lötjönen S and Hynninen PH (1983) Carbon-13 NMR spectra of chlorophyll a, chlorophyll a', pyrochlorophyll a and the corresponding pheophytins. *Org. Magn. Reson.*, **21**: 757–765.

Maggiora LL, Petke JD, Gopal D, Iwamoto RT and Maggiora GM (1985) Experimental and theoretical studies of chiff base chlorophylls. *Photochem. Photobiol.*, **42**: 69–75.

Manning WM and Strain HH (1943) Chlorophyll d, a green pigment of red algae. *J. Biol. Chem.*, **151**: 1–19.

Michalski TJ, Hunt JE, Bowman MK, Smith U, Bardeen K, Gest H, Norris JR and Katz JJ (1987) Bacteriopheophytin g: Properties and some speculations on a possible primary role for bacteriochlorophylls b and g in the biosynthesis of chlorophylls. *Proc. Natl. Acad. Sci. U.S.A.*, **84**: 2570–2574.

Miyashita H, Ikemoto H, Kurano N, Adachi K, Chihara M and Miyachi S (1996) Chlorophyll d as a major pigment. *Nature*, **383**: 402.

Miyashita H, Adachi K, Kurano N, Ikemoto H, Chihara M and Miyachi S (1997) Pigment composition of a novel oxygenic photosynthetic prokaryote containing chlorophyll d as the major chlorophyll. *Plant Cell Physiol.*, **38**: 274–281.

Miyashita H, Ohkubo S, Komatsu H, Sorimachi Y, Fukayama D, Fujinuma D, Akutsu S and Kobayashi M (2014) Discovery of chlorophyll d in *Acaryochloris marina* and chlorophyll f in a unicellular cyanobacterium, Strain KC1, Isolated from Lake Biwa. *J. Phys. Chem. Biophys.*, **4**: 10004–10012.

Moog RS, Kuki A, Fayer MD and Boxer SG (1984) Excitation transport and trapping in a synthetic chlorophyllide substituted hemoglobin: Orientation of the chlorophyll S_1 transition dipole. *Biochemistry*, **23**: 1564–1571.

Musewald C, Hartwich G, Pöllinger-Dammer F, Lossau H, Scheer H and Michel-Beyerle ME (1998) Time-resolved spectral investigation of bacteriochlorophyll a and its transmetalated derivatives [Zn]-bacteriochlorophyll a and [Pd]-bacteriochlorophyll a. *J. Phys. Chem. B*, **102**: 8336–8342.

Nagashima KVP, Matsuura K, Wakao N, Hiraishi A and Shimada K (1997) Nucleotide sequences of genes cording for photosynthetic reaction centers and light-harvesting proteins of *Acidiphilium rubrum* and related aerobic acidophilic bacteria. *Plant Cell Physiol.*, **38**: 1249–1258.

Noy D, Fiedor L, Hartwich G, Scheer H and Scherz A (1998) Metal-substituted bacteriochlorophylls. 2. Changes in redox potentials and electronic transition energies are dominated by intramolecular electrostatic interactions. *J. Am. Chem. Soc.*, **120**: 3684–3693.

Noy D, Yerushalmi R, Brumfeld V, Ashur I, Scheer H, Baldridge KK and Scherz A (2000) Optical absorption and computational studies of [Ni]-bacteriochlorophyll-*a*. New insight into charge distribution between metal and ligands. *J. Am. Chem. Soc.,* **122**: 3937–3944.

Nuijs AM, van Dorssen RJ, Duysens LNM and Amesz J (1985a) Excited states and primary photochemical reaction in the photosynthetic bacterium *Heliobacterium chlorum. Proc. Natl. Acad. Sci. U.S.A.,* **82**: 6865–6868.

Nuijs AM, Vasmel H, Joppe HLP, Duysens LNM and Amesz J (1985b) Excited states and primary charge separation in the pigment system of the green photosynthetic bacterium *Prosthecochloris aestuarii* as studied by picosecond absorbance difference spectroscopy. *Biochim. Biophys. Acta,* **807**: 24–34.

Oelze J (1985) Analysis of bacteriochlorophylls. *Methods Microbiol.,* **18**: 257–284.

Ohashi S, Miyashita H, Okada N, Iemura T, Watanabe T and Kobayashi M (2008) Unique photosystems in *Acaryochloris marina. Photosynth. Res.,* **98**: 141–149.

Okazaki T and Kajiwara M (1995) Studies on the biosynthesis of corrinoids and porphyrinoids. X. biosynthetic studies of bacteriochlorophyll-*a* in *Rhodopseudomonas spheroids*: Incorporation of ^{2}H-, ^{13}C- and ^{15}N-labeled substrates and origin of the hydrogen atoms. *Chem. Pharm. Bull.,* **43**: 1311–1317.

Parson WW, Clayton RK and Cogdell RJ (1975) Excited states of photosynthetic reaction centers at low redox potentials. *Biochim. Biophys. Acta,* **387**: 265–278.

Pelletier PJ and Caventou JB (1818) Sur la matière verte des feuilles. *Ann. Chim. Phys.,* **9**: 194–196.

Permentier HP, Schmidt KA, Kobayashi M, Akiyama M, Hager-Braun C, Neerken S, Miller M and Amesz J (2000) Composition and optical properties of reaction centre core complexes from the green sulfur bacteria *Prosthecochloris aestuarii* and *Chlorobium tepidum. Photosynth. Res.,* **64**: 27–39.

Petke JD, Maggiora G, Shipman L and Christoffersen RE (1979) Stereoelectronic properties of photosynthetic and related systems—v. *ab initio* configuration interaction calculations on the ground and lower excited singlet and triplet states of ethyl chlorophyllide a and ethyl pheophorbide *a. Photochem. Photobiol.,* **30**: 203–223.

Petke JD, Maggiora GM, Shipman LL and Christoffersen RE (1980) Stereoelectronic properties of photosynthetic and related systems—vii. Ab initio quantum mechanical characterization of the electronic structure and spectra of chlorophyllide *a* and bacteriochlorophyllide *a* cation radicals. *Photochem. Photobiol.,* **31**: 243–257.

Porra RJ and Scheer H (2000) ^{18}O and mass spectrometry in chlorophyll research: Derivation and loss of oxygen atoms at the periphery of the chlorophyll macrocycle during biosynthesis, degradation and adaptation. *Photosynth. Res.,* **66**: 159–175.

Risch N (1981) Bacteriochlorophyll *b*. Determination of its configuration by nuclear overhauser effect difference spectroscopy. *J. Chem. Res. (S),* **4**: 116–117.

Risch N and Brockmann H (1983) Chlorophyll *b*. Totalzuordnung des ^{13}C-NMR-Spektrums. *Tetrahedron Lett.,* **24**: 173–176.

Rockley MG, Windsor MW, Cogdell RJ and Parson WW (1975) Picosecond detection of an intermediate in the photochemical reaction of bacterial photosynthesis. *Proc. Natl. Acad. Sci. U.S.A.,* **72**: 2251–2255.

Saji T and Bard AJ (1977) Electrogenerated chemiluminescence. 29. The electrochemistry and chemiluminescence of chlorophyll a in *N,N*-dimethylformamide solutions. *J. Am., Chem. Soc.,* **99**: 2235–2240.

Scheer H and Hartwich G (1995) Bacterial reaction centers with modified tetrapyrrol chromophores. In *Anoxygenic Photosynthetic Bacteria,* ed. by Blankenship RE, Madigan MT and Bauer CE, Kluwer Academic Publishers, Dordrecht, pp. 649–663.

Scheer H, Svec WA, Cope BT, Studier MH, Scott RG and Katz JJ (1974) Structure of bacteriochlorophyll *b. J. Am. Chem. Soc.,* **96**: 3714–3716.

Schoch S (1978) The esterification of chlorophyllide *a* in greening bean leaves. *Z. Naturforsch.,* **33c**: 712–714.

Shoaf WT (1978) Rapid method for the separation of chlorophylls *a* and *b* by high-pressure liquid chromatography. *J. Chromatogr.,* **152**: 247–249.

Shuvalov VA, Vasmel H, Amesz J and Duysens LNM (1986a) Picosecond spectroscopy of the charge separation in reaction centers of *Chloroflexus aurantiacus* with selective excitation of the primary electron donor. *Biochim. Biophys. Acta,* **851**: 361–368.

Shuvalov VA, Amesz J and Duysens LNM (1986b) Picosecond spectroscopy of isolated membranes of the photosynthetic green sulfur bacterium *Prosthecochloris aestuarii* upon selective excitation of the primary electron donor. *Biochim. Biophys. Acta,* **851**: 1–5.

Smith KM (1975) Mass spectrometry of porphyrins and metalloporphyrins. In *Porphyrins and Metalloporphyrins,* ed. by Smith KM, Elsevier Science Amsterdam, pp. 381–398.

Smith JHC and Benitez A (1955) Chlorophylls: Analysis in plant materials. In *Moderne Methoden der Pflanzenanalyse,* ed. by Paech K and Tracey MV, Springer-Verlag, Berlin, Heidelberg, vol. **4**, pp. 142–196.

Spangler D, Maggiora GM, Shipman LL and Christoffersen RE (1977) Stereoelectronic properties of photosynthetic and related systems. 2. ab initio quantum mechanical ground state characterization of magnesium porphine, magnesium chlorin, and ethyl chlorophyllide a. *J. Am. Chem. Soc.,* **99**: 7478–7489.

Stanienda A (1965) Elektrochemische untersuchungen der chlorophylle (a,b) und phäophytine (a,b). *Z. Phys. Chem.,* **229**: 257–272.

Steglich C, Mullineaux CW, Teuchner K, Hess WR and Lokstein H (2003) Photophysical properties of *Prochlorococcus marinus* SS120 divinyl chlorophylls and phycoerythrin in vitro and in vivo. *FEBS Lett.,* **553**: 79–84.

Steiner R, Cmiel E and Scheer H (1983) Chemistry of bacteriochlorophyll *b*: Identification of some (photo)oxidation products. *Z. Naturforsch,* **38c**: 748–752.

Strain HH and Manning WM (1942) Isomerization of chlorophylls *a* and *b. J. Biol. Chem.,* **146**: 275–276.

Takahashi K, Itoh Y, Akiyama M, Watanabe T, Inoue K, Oba T, Umetsu M and Kobayashi M (2005) Delicate distinction between absorption spectra of 'normal' and 'prime' bacteriochlorophylls. In *Photosynthesis: Fundamental Aspects to Global Perspectives,* ed. by van der Est A and Bruce D, Alliance Communications Group, Lawrence, KS, pp. 46–48.

Takeuchi Y and Amao Y (2005) Light-harvesting properties of zinc complex of chlorophyll-*a* from *spirulina* in surfactant micellar media. *Biometals,* **18**: 15–21.

Teuchner K, Stiel H, Leupold D, Scherz A, Noy D, Simonin I, Hartwich G and Scheer H (1997) Fluorescence and excited state absorption in modified pigments of bacterial photosynthesis a comparative study of metal-substituted bacteriochlorophylls a. *J. Lumin.,* **72–74**: 612–614.

Tomita G and Rabinowitch E (1962) Excitation energy transfer between pigments in photosynthetic cells. *Biophys. J.,* **2**: 483–499.

Tomo T, Okubo T, Akimoto S, Yokono M, Miyashita H, Tsuchiya T, Noguchi T and Mimuro M (2007) Identification of the special pair of photosystem II in a chlorophyll *d*-dominated cyanobacterium. *Proc. Natl. Acad. Sci. U.S.A.*, **104**: 7283–7288.

Tomo T, Akimoto S, Ito H, Tsuchiya T, Fukuya M, Tanaka A and Mimuro M (2009) Replacement of chlorophyll with di-vinyl chlorophyll in the antenna and reaction center complexes of the cyanobacterium *Synechocystis* sp. PCC 6803: Characterization of spectral and photochemical properties. *Biochim. Biophys. Acta*, **1787**: 191–200.

Van de Meent EJ, Kobayashi M, Erkelens C, Van Veelen PA, Amesz J and Watanabe T (1991) Identification of 8^1-hydroxychlorophyll *a* as a functional reaction center pigment in heliobacteria. *Biochim. Biophys. Acta*, **1058**: 356–362.

Van de Meent EJ, Kobayashi M, Erkelens C, Van Veelen PA, Otte SCM, Inoue K, Watanabe T and Amesz J (1992) The nature of the primary electron acceptor in green sulfur bacteria. *Biochim. Biophys. Acta*, **1102**: 371–378.

Van Gorkom HJ (1974) Identification of the reduced primary electron acceptor of photosystem II as a bound semiquinone anion. *Biochim. Biophys. Acta*, **347**: 439–442.

Vladkova R (2000) Chlorophyll a self-assembly in polar solvent-water mixtures. *Photochem. Photobiol.*, **71**: 71–83.

Wakao N, Shiba T, Hiraishi A, Ito M and Sakurai Y (1993) Distribution of bacteriochlorophyll *a* in species of the genus *Acidiphilium*. *Curr. Microbiol.*, **27**: 277–279.

Wakao N, Yokoi N, Isoyama N, Hiraishi A, Shimada K, Kobayashi M, Kise H, Iwaki M, Itoh S, Takaichi S and Sakurai Y (1996) Discovery of natural photosynthesis using Zn-containing bacteriochlorophyll in an aerobic bacterium *Acidiphilium rubrum*. *Plant Cell Physiol.*, **37**: 889–893.

Wasielewski MR, Norris JR, Shipman LL, Lin C-P and Svec WA (1981) Monomeric chlorophyll *a* enol: Evidence for its possible role as the primary electron donor in photosystem I of plant photosynthesis. *Proc. Natl. Acad. Sci. U.S.A.*, **78**: 2957–2961.

Watanabe T and Kobayashi M (1988) Chlorophylls as functional molecules in photosynthesis—Molecular composition *in vivo* and physical chemistry *in vitro*. *Chem. Soc. Japan*, **4**: 383–395.

Watanabe T and Kobayashi M (1991) Electrochemistry of chlorophylls. In *Chlorophylls*, ed. by Scheer H, CRC Press, Boca Raton, FL, pp. 287–315.

Watanabe T, Hongu A, Honda K, Nakazato M, Konno M and Saitoh S (1984) Preparation of chlorophylls and pheophytins by isocratic liquid chromatography. *Anal. Chem.*, **56**: 251–256.

Watanabe T, Machida K, Suzuki H, Kobayashi M and Honda K (1985) Photoelectrochemistry of metallochlorophylls. *Coord. Chem. Rev.*, **64**: 207–224.

Weber G and Teale FWJ (1957) Determination of the absolute quantum yield of fluorescent solutions. *Trans. Faraday Soc.*, **53**: 646–655.

Weiss C Jr (1972) The pi electron structure and absorption spectra of chlorophylls in solution. *J. Mol. Spectrosc.*, **44**: 37–80.

Weiss C (1978) Electronic absorption spectra of chlorophylls. In *The Porphyrins, vol. III, Physical Chemistry, Part A*, ed. by Dolphin D, Academic Press, New York, pp. 211–223.

White RC, Jones ID, Gibbs E and Butler LS (1972) Fluorometric estimation of chlorophylls, chlorophyllides, pheophytins and pheophorbides in mixtures. *J. Agric. Food Chem.*, **20**: 773–778.

Wilhelm C (1987) Purification and identification of chlorophyll c_1 from the green alga *Mantoniella squamata*. *Biochim. Biophys. Acta*, **892**: 23–29.

Willows RD, Li Y, Scheer H and Chen M (2013) Structure of chlorophyll *f*. *Org. Lett.*, **15**: 1588–1590.

Wolf H and Scheer H (1973) Stereochemistry and chiroptic properties of pheophorbides and related compounds. *Ann. N. Y. Acad. Sci.*, **206**: 549–567.

Wu S-M and Rebeiz CA (1985) Chloroplast biogenesis. Molecular structure of chlorophyll *b* (E489 F666). *J. Biol. Chem.*, **260**: 3632–3634.

Xiong J, Fischer WM, Inoue K, Nakahara M and Bauer CE (2000) Molecular evidence for the early evolution of photosynthesis. *Science*, **289**: 1724–1730.

7 Thylakoid Ndh Complex

Bartolomé Sabater and Mercedes Martín

CONTENTS

7.1 Introduction 149
7.2 Reaction Catalyzed by the Ndh Complex 149
7.3 Nuclear *Ndh* Genes and the Structure and Specificity of the Ndh Complex 151
7.4 Role of the Ndh Complex in Photosynthesis 153
7.5 Regulation of the Thylakoid Ndh Complex 154
    7.5.1 Transcriptional and Posttranscriptional Control of *Ndh* Gene Expression 154
    7.5.2 Regulation of the Activity of the Ndh Complex by Phosphorylation 154
7.6 Concluding Remarks and Further Prospects 155
Acknowledgments 155
References 155

7.1 INTRODUCTION

The plastid DNA of most of higher plants contains 11 *Ndh* genes homologous to those encoding polypeptides of mitochondrial complex I (NADH dehydrogenase, EC 1.6.5.3) (Ohyama et al. 1986; Shinozaki et al. 1986; Maier et al. 1995). Soon after their discovery, it was found that the *Ndh* genes are transcribed (Kanno and Hirai 1993; Martínez et al. 1997) and their encoded polypeptides (NDH) were immunoidentified in chloroplasts (Berger et al. 1993; Martín et al. 1996). The thylakoid Ndh complex containing the polypeptide products of plastid *Ndh* genes was first purified from peas (Burrows et al. 1998) and later from barley (Casano et al. 2000). Subsequently, a number of NDH polypeptides have been identified by proteomics in chloroplastic Ndh complex preparations (Darie et al. 2005; Rumeau et al. 2005; Shimizu et al. 2008; Sirpiö et al. 2009; Suorsa et al. 2009).

Except for Charophyceae and several Prasinophyceae, in the evolutionary lineage leading to higher plants, the plastid DNAs of eukaryotic algae lack *ndh* genes that are present in most photosynthetic land plants (Martín and Sabater 2010). Some *ndh* genes are present as pseudogenes or are completely lost/absent in parasitic nonphotosynthetic species of the *Cuscuta*, *Epiphagus*, and *Orobanche* genera and of the Orchidaceae family (Braukmann et al. 2013; Barret et al. 2014), which suggests that the thylakoid Ndh complex, encoded by the 11 plastid *ndh* genes and a still-unknown number of nuclear genes, has a role in the photosynthesis of land plants. However, the plastid DNAs of the photosynthetic gymnosperms Pinaceae and Gnetales and of a few species scattered among angiosperm genera, families, and orders (e.g., *Erodium*, Ericaceae, Alismatales) lack *ndh* genes (Braukmann et al. 2009, 2013; Blazier et al. 2011; Braukmann and Stefanovíc 2012; Peredo et al. 2013), which suggests that the role of the *ndh* genes in photosynthesis could be dispensable under certain environments.

Despite intense research since 2000, uncertainties remain as to the structure and function of the thylakoid Ndh complex. Hence, the absence of *ndh* genes in the plastid DNAs of several plants raises doubts regarding the functional role of the Ndh complex in photosynthesis. On the other hand, the molecular characterization of the Ndh complex is difficult due to its instability and low abundance in chloroplasts. In addition, the investigation of nuclear genes encoding putative additional polypeptides of the Ndh complex frequently produced ambiguous and even conflicting results. Nevertheless, the combination of molecular, genetic, and functional investigations and the comparison with evolutionarily related complexes provide valuable insights on the structure and function of the thylakoid Ndh complex, which will be reviewed here.

7.2 REACTION CATALYZED BY THE Ndh COMPLEX

In a purified or partially purified preparation of the thylakoid Ndh complex, the enzymatic activity is assayed spectrophotometrically as NADH:ferricyanide oxidoreductase (K_m for NADH, 26 µM) and as NADH:nitroblue tetrazolium oxidoreductase in zymograms (Cuello et al. 1995a; Corneille et al. 1998; Sazanov et al. 1998; Martín et al. 2004, 2009; Rumeau et al. 2005). Despite former affirmations (Endo et al. 1998), the activity is negligible with NADPH as an electron donor. Figure 7.1a shows that the solubilized preparations of tobacco thylakoids produce a number of NADH:nitroblue tetrazolium oxidoreductase bands after native electrophoresis. Bands correspond to different NADH dehydrogenases and diaphorases that have NADH:nitroblue tetrazolium oxidoreductase activity. Among them, the low-intensity band corresponding to the NADH dehydrogenase of the Ndh complex is unambiguously identified because, after subsequent transfer, it was the only band recognized by both NDH-F and NDH-J antibodies. Similarly, the thylakoid Ndh complex activity can

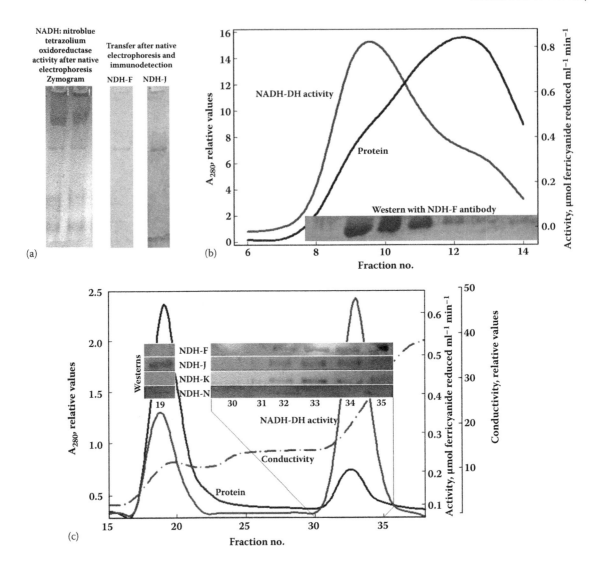

FIGURE 7.1 (a) Zymograms show NADH:nitroblue tetrazolium oxidoreductase bands after native electrophoresis of solubilized preparations of tobacco thylakoid as described by Martín et al. (2009). After transfer, only the band migrating some 30% was recognized by both NDH-F and NDH-J antibodies and corresponded to the thylakoid Ndh complex. (b) Elution profile in Sephadex G-200 of proteins and NADH:ferricyanide oxidoreductase activities of solubilized preparations of barley thylakoids. Aliquots of the fractions were separated by SDS-PAGE, and after transfer, assay with NDH-F antibody revealed the presence of the thylakoid Ndh complex in fractions 9, 10, and 11. (c) When fractions were analyzed by SDS-PAGE and Western blot with different NDH antibodies, the elution, by increasing ionic strength of the thylakoid proteins retained in a Q Sepharose column, revealed low activity and weakly retained subcomplex eluting at fraction 19 (including NDH-J and NDH-N but not NDH-F and NDH-K subunits), and high-activity complete Ndh complex eluting at higher ionic strength at around fraction 33.

be distinguished from the different NADH:ferricyanide oxidoreductase activities eluted from a Sephadex G-200 column (Figure 7.1b) by sodium dodecyl sulfate polyacrylamide gel electrophoresis (SDS-PAGE) and Western blot with an NDH-F antibody, as shown in fractions 9, 10, and 11 of the chromatography. A more complex pattern of activity is observed after ion-exchange chromatography. When the thylakoid proteins retained in a Q Sepharose column were eluted with increasing ionic strength (Figure 7.1c), detection with NDH antibodies revealed a main band of activity around fraction 33, which is recognized by all NDH antibodies tested (for NDH-F, NDH-J, NDH-K, and NDH-N subunits). A minor band of activity in fraction 19 contains NDH-J and NDH-N but not NDH-F and

NDH-K subunits and must correspond to the hydrophilic subcomplex detached from the complete Ndh complex. Despite its low abundance (1 Ndh complex per 100 to 200 photosystems) (Sazanov et al. 1998; Casano et al. 2000), several subunits of the Ndh complex have been identified by proteomics in purified preparations of the Ndh complex (Rumeau et al. 2005). Figure 7.2 shows four tryptic peptides identified in the barley Ndh complex immunopurified as described by Casano et al. (2000).

Mitochondrial complex I catalyzes the oxidation of NADH with ubiquinone, which suggests that the thylakoid Ndh complex could use oxidized plastoquinone (PQ) as an electron acceptor. Accordingly, purified preparations of the Ndh complex use oxidized PQ as an electron acceptor (in addition to

Tryptic peptides of the NDH-H subunit identified in purified preparation of the thylakoid Ndh complex of barley	101 IR**VIMLELSR** IASHLLWLGP FMADLGAQTP FFYIFRER**EL IYDLFEAATG**
	151 **MR**MMHNYFRI GGVAADLPYG WIDKCLDFCD YFLRGVVEYQ QLITQNPIFL
	201 ERVEGVGFIS GEEAVNWGLS GPMLRASGIQ WDLRKVDPYE SYNQFDWKVQ
	251 WQKEGDSLAR YLVRVGEMSE SIKIIQQAIE **KIPGGPYENL EVRR**FKKEK
	301 SEWNDFEYKF LGKKPSPNFE LSRQELYVRV EAPKGELGIY LVGDDSLFPW
	351 RWKIRPPGFI NLQILPQLVK KMK**LADIMTI LGSIDIIMGE VDR**

FIGURE 7.2 After bidimensional electrophoresis (del Riego et al. 2006) of the Ndh complex of *Hordeum vulgare*, proteomic analysis detected four tryptic peptides (bold) in one protein spot corresponding to the NDH-H subunit.

ferricyanide and nitroblue tetrazoilum), as demonstrated in *in vitro* spectrophotometric assays in a partially hydrophobic media (Casano et al. 2000). *In vivo*, the comparison of the fluorescence of chlorophyll of *wt* and transgenic tobaccos defective in *ndh* genes also points to oxidized PQ as an electron acceptor because there is a positive correlation between the chlorophyll fluorescence and the levels of reduced PQ (Burrows et al. 1998; Kofer et al. 1998; Shikanai et al. 1998; Martín et al. 2004, 2009; Wang and Portis 2007). In fact, chlorophyll fluorescence transitorily increases after relative high-to-minimum light transition (postillumination fluorescence) presumably due to the slightly higher levels of reduced PQ in plants containing the Ndh complex than in *ndh*-defective plants, which do not show the fluorescence increase (Figure 7.3a). Hence, the postillumination fluorescence increase is commonly assumed (Serrot et al. 2012) as a test of the reduction of PQ by the Ndh complex *in vivo*. A sudden increase of light intensity produces a burst of chlorophyll fluorescence that decays within a few seconds. The increase of fluorescence is lower and the persistence of high fluorescence is shorter in *ndh*-defective plants than in *wt* (Figure 7.3b), which again suggests that the Ndh complex catalyzes the transfer of electrons from NADH to oxidized PQ.

7.3 NUCLEAR *Ndh* GENES AND THE STRUCTURE AND SPECIFICITY OF THE Ndh COMPLEX

By comparison with the minimum 14 core polypeptides of the homologous respiratory complex I (Moparthi et al. 2014),

the 11 plastid-encoded polypeptides of the Ndh complex constitute the hydrophobic thylakoid-inserted arm (P module: NDH-A, NDH-B, NDH-C, NDH-D, NDH-E, NDH-F, and NDH-G subunits) and part of the bridge between the two arms (Q module: NDH-H, NDH-I, NDH-J, and NDH–K subunits) of the *L*-shaped Ndh complex (Figure 7.4), which is located at the stromal thylakoid (Berger et al. 1993; Quiles et al. 2000; Albertsson 2001; Lennon et al. 2003; Casano et al. 2004). Four nuclear-encoded subunits (NDH-L, NDH-M, NDH-N, and NDH-O, not shown in Figure 7.4) (Rumeau et al. 2005; Shimizu et al. 2008; Sirpiö et al. 2009) are probably part of the bridge between the two arms. Other nuclear-encoded potential subunits of the Ndh complex (CRR3, CRR7, NDF1, NDF2, NDF6, etc.) seem associated with the *Arabidopsis* Ndh complex, but the lack of a three-dimensional, high-resolution, crystal structure of the Ndh complex hampers the identification of the electron donor site and subunit topology (Suorsa et al. 2009). A molecular weight of between 400 and 500 kDa is estimated for the Ndh complex. Frequently, *Arabidopsis* nuclear mutants defective in the thylakoid Ndh complex are affected in subunit assembly, plastid *ndh* transcript processing, and, in general, processes that can have pleiotropic effects in several chloroplast functions (Meurer et al. 1996).

It was frequently assumed that the Ndh complex includes additional not-yet-identified nuclear-encoded subunits (N module: NuoE, NuoF and NuoG) harboring the electron donor binding site in the hydrophilic arm oriented toward the stroma (Figure 7.4). However, no such genes have been found, which raises doubts on the nature of the electron donor (Suorsa et al. 2009). As an alternative, reduced ferredoxin has

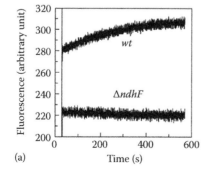

(a)

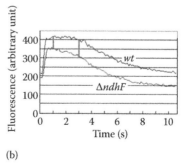

(b)

FIGURE 7.3 (a) Chlorophyll fluorescence increases after relative high-to-minimum light transition in *wt* but not in Δ*ndhF* tobacco deficient in the Ndh complex. Traces correspond to fluorescence readings every 0.1 s during the 9 min following the final 0.15–0.1 μmol photon m^{-2} s^{-1} photosynthetic active radiation (PAR) transition as described by Serrot et al. (2012). (b) When the intensity of the incident light suddenly increases from 61 to 870 μmol photon m^{-2} s^{-1} PAR, the continuous record shows a burst of chlorophyll fluorescence that is higher and more persistent in *wt* than in Δ*ndhF* tobacco (compare vertical gray bars at 1 and 3 min).

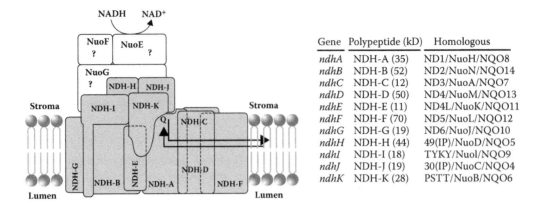

Gene	Polypeptide (kD)	Homologous
ndhA	NDH-A (35)	ND1/NuoH/NQO8
ndhB	NDH-B (52)	ND2/NuoN/NQO14
ndhC	NDH-C (12)	ND3/NuoA/NQO7
ndhD	NDH-D (50)	ND4/NuoM/NQO13
ndhE	NDH-E (11)	ND4L/NuoK/NQO11
ndhF	NDH-F (70)	ND5/NuoL/NQO12
ndhG	NDH-G (19)	ND6/NuoJ/NQO10
ndhH	NDH-H (44)	49(IP)/NuoD/NQO5
ndhI	NDH-I (18)	TYKY/NuoI/NQO9
ndhJ	NDH-J (19)	30(IP)/NuoC/NQO4
ndhK	NDH-K (28)	PSTT/NuoB/NQO6

FIGURE 7.4 Proposed organization of the Ndh complex in the thylakoid. (Modified from Martín, M. et al., *Biochim. Biophys. Acta*, 1787, 920–928, 2009.) The 11 subunits encoded in chloroplast DNA are shaded. The table on the right shows the correspondence of nomenclature of genes and polypeptides encoded in chloroplast DNA with homologous subunits in respiratory complex I and other evolutionarily related complexes.

been proposed as an electron donor, which implies that the Ndh complex endows the thylakoid with a new route for cyclic photosynthetic electron transport in addition to the commonly accepted model in which ferredoxin directly donates electrons to the PQ/*cyt.b₆f* intermediary electron pool (Kurisu et al. 2003). In this line, a protein, CRR31, which could bind ferredoxin and be required for the *in vivo* activity of the Ndh complex, has been described in *Arabidopsis* (Yamamoto et al. 2011). The CRR31-dependent activity of the Ndh complex has been assayed in membranous fractions of *Arabidopsis* and uses NADPH in the presence of ferredoxin for the reduction of PQ. The ferredoxin-dependent activity is estimated by the increase of chlorophyll fluorescence. However, CRR31 accumulates in thylakoids independently of the Ndh complex, and the increase of fluorescence (around 0.15 arbitrary unit) was rather low when compared to the 200-fold higher values usually measured *in vivo* (Figure 7.3). A similar argument was used by Nandha et al. (2007) to discard the involvement of the PGR5 protein in cyclic electron transport.

Although the possibility of a singular ferredoxin-dependent Ndh complex in *Arabidopsis* cannot be ruled out, spectrophotometric and zymogram assays of activity, combined with immunodetection with antibodies raised against proteins encoded by chloroplast *ndh* genes, indicate that NADH is the electron donor in most, if not all, plants (Cuello et al. 1995a; Corneille et al. 1998; Sazanov et al. 1998; Casano et al. 2000; Díaz et al. 2007; Martín et al. 2009; Serrot et al. 2012). A comparison with evolutionarily related complexes could be useful to settle the questions of the apparent absence of NADH-binding subunits in the Ndh complex and the nature of the electron donor. The 11 *ndh* genes encoded in chloroplasts are homologous to those encoding the 11 subunits of the basic complex I found both in archaeal and eubacterial kingdoms (Moparthi et al. 2014). The 11-subunit complex I (such as that of *Bacillus cereus*) would be the common ancestor of all present-day complex I-like activities. It has the functional properties of the full-sized complex I and probably binds different electron donors (Moparthi et al. 2014). Assuming that

the thylakoid Ndh complex is similar to the 11-subunit complex I of *B. cereus*, polypeptides NDH-A, NDH-B, NDH-C, NDH-D, NDH-E, NDH-F, and NDH-G integrating the P module and polypeptides NDH-H, NDH-I, NDH-J, and NDH-K of the Q module (compare with Figure 7.4) would be sufficient for full activity and could use diverse electron donors in different organisms/organelles. Lacking the N module present in the mitochondrial complex I, the Q module of the thylakoid Ndh complex could have evolved to bind NADH as an electron donor in most plants.

Similarly to respiratory complex I (Friedrich and Böttcher 2004; Euro et al. 2008) and membrane transhydrogenases (Mathiesen and Hägerhäall 2002), and in contrast to the type II NAD(P)H dehydrogenase of *Chlamydomonas* (Desplats et al. 2009), the thylakoid Ndh complex has H⁺ pumping activity, probably involved in energy conversion, linked to the transmembrane helices of the NDH-A, NDH-B, NDH-D, and NDH-F subunits. Accordingly, the NDH-F subunit contains histidine, lysine, and phenylalanine (at positions 349, 353, and 357, respectively) (Casano et al. 2004), which are conserved in the homologous plastid and respiratory (NuoL) polypeptides analyzed (Figure 7.4) within the same transmembrane helix. The histidine and probably another conserved histidine at position 345 may be involved in the H⁺ pumping channel.

According to the cyanobacterial origin of chloroplasts, nucleotide sequences of *ndh* genes and their organization in plastid DNA show a high similitude with those of the homologous genes in the cyanobacteria *Synechocystis* (Martín and Sabater 2010). There are several copies of some *ndh*-like genes in *Synechocystis* organized in different operons, which agrees with the presence of several Ndh-like complexes in cyanobacteria that have different functions, such as those in cyclic electron transport around photosystem I (PSI) and respiration and CO_2 uptake (Prommeenate et al. 2004; Battchikova et al. 2005; Ogawa and Mi 2007; Ma and Mi 2008). It must be remembered that cyanobacteria, in addition to the photosynthetic electron transport chain, have a bacterial-like respiratory chain. In this regard, some results suggest (Schultze et al. 2009) the presence of an NADH-specific respiratory complex (NDH-2) in the plasma membrane as well as a similar NDH-2 complex and an NADPH-specific complex (NDH-1) in the thylakoid membrane of *Synechocystis*. More investigations are necessary on the structural and functional properties of the complexes containing *ndh*-encoded subunits in cyanobacteria to permit a comparison with the Ndh complex of chloroplasts.

7.4 ROLE OF THE Ndh COMPLEX IN PHOTOSYNTHESIS

The oxidation of P700 in far-red light is faster in *ndh*-defective mutants than in *wt* tobacco (Bart and Krause 2002), which suggests that the Ndh complex competes with P700* to reduce PQ and is hardly compatible with the role of the Ndh complex in providing additional ferredoxin-mediated cyclic electron transport. The behavior of *ndh*-deficient transgenic tobaccos and their responses to environmental and developmental factors suggest another role of the Ndh complex in cyclic electron transport, which depends on its NADH:PQ oxidoreductase activity.

Under field conditions, leaves receive light that fluctuates intensely and rapidly (Külheim et al. 2002), disrupting the balance of the rates of NADPH production and consumption, which would strongly affect the redox level of electron transporters. As pointed by Heber and Walker (1992), the redox levels of the components of photosynthetic electron transport must be middle-poised. Otherwise, when they are overreduced or overoxidized, the rate of cyclic electron transport would be too low to transport extra protons to the thylakoid lumen to maintain the appropriate proton gradient (ΔpH). This gradient is required for photophosphorylation and for the dissipation of excess light energy through zeaxanthin (Eskling et al. 2001), reducing the formation of singlet oxygen. Under fluctuating light, the Ndh complex acts as a valve feeding electrons (Figure 7.5) to poise the redox level of the intermediaries that optimizes the rate of cyclic electron transport (Casano et al. 2000; Joët et al. 2002) when the electron supply from photosystem II (PSII) is very low.

Hence, the Ndh complex would catalyze the first stage of the chlororespiratory reaction chain in which, at high light intensity, the Mehler reaction and superoxide dismutase (SOD) and peroxidase (PX) activities drain excess electrons (Martín et al. 2004; Rumeau et al. 2007; Sabater and Martín 2013) to ensure the fine-tuning of the redox level of the cyclic electron transporters. Thus, in addition to the photoinhibition of PSII, the generation of $O_2^{\cdot-}$ and H_2O_2 would be a less harmful alternative than the formation of $^1O_2$ under transitory high light. Once light intensity returns to a low level, the recovery of photoinhibited PSII could last minutes. The feeding of electrons from NADH by the thylakoid Ndh complex maintains the redox poising of cyclic electron transporters and the transmembrane ΔpH (Casano et al. 2000; Sabater and Martín 2013). In *Chlamydomonas*, which lacks *ndh* genes, a nuclear-encoded NAD(P)H dehydrogenase (Nda2) seems to play a similar role as the Ndh complex of higher plants (Desplats et al. 2009), contributing to balancing the redox level of cyclic electron transport. Therefore, the reactions draining and feeding electrons take place alternatively and successively in chlororespiration. Accordingly, net photosynthetic rates are impaired in *ndh*-defective transgenic tobaccos under rapidly fluctuating light intensities (Martín et al. 2009), and the Ndh complex is coordinately induced with SOD and PX under rapid light changes and stress conditions in general (Casano et al. 1994, 1999, 2001; Martín et al. 1996). When a leaf ages, the induction of SOD progressively decreases, and the induction of the Ndh complex is not compensated for by that of reactive oxygen species (ROS) scavenging activities (Casano et al. 1999; Abarca et al. 2001), resulting in an increase of ROS and programmed leaf cell death (Zapata et al. 2005; Sabater and Martín 2013).

Significantly, the impairment of photosynthesis in *ndh*-defective transgenic tobacco is especially pronounced at high CO_2 concentrations and with sudden increases of light intensity (Figure 7.6). Therefore, the functional role of the Ndh complex seems related to the adaptation of photosynthesis to land environments where frequent and rapid changes of light intensity as well as temperature and water availability produce different stresses related to the redox level of photosynthetic

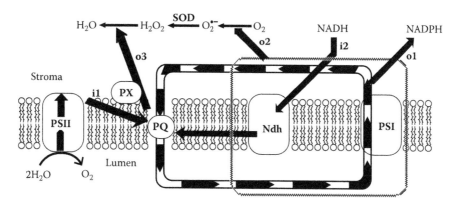

FIGURE 7.5 Redox poising of cyclic electron transporters by chlororespiratory reactions. i1 and i2 are influxes of electrons from, respectively, photosystem II (PSII) and NADH (through the thylakoid Ndh complex). o1, o2, and o3 are different outfluxes of electrons to, respectively, NADP+, the Mehler reaction, and plastoquinol peroxidase.

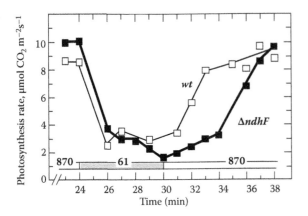

FIGURE 7.6 Transgenic Ndh-defective ($\Delta ndhF$) tobacco shows a delay in reaching full photosynthetic rate with respect to *wt* when light intensity is suddenly increased from 61 to 870 µmol photon m^{-2} s^{-1} PAR. Net photosynthesis was measured at different times in the leaf section as described by Martín et al. (2009), except that CO_2 concentration was 600 ppm. After 24 min acclimation at 870 µmol photon m^{-2} s^{-1} PAR, a low-intensity light (61 µmol photon m^{-2} s^{-1} PAR) was applied for 6 min, after which light was again raised to 870 µmol photon m^{-2} s^{-1} PAR (see numbers on inserted bar).

electron transporters (Martín et al. 2004, 2009, 2015; Martín and Sabater 2010).

7.5 REGULATION OF THE THYLAKOID Ndh COMPLEX

NADH, the substrate of the Ndh complex, can be generated in chloroplasts from NAD^+ by the activities of several enzymes such as malate dehydrogenase, pyruvate dehydrogenase, and NAD^+-dependent glyceraldehide-3-phosphate dehydrogenase. Due to the redox potential of pairs of $NADH/NAD^+$ (−0.32 V) and reduced PQ (PQH_2)/oxidized PQ (around 0 V), the reaction catalyzed by the Ndh complex ($NADH + H^+ + PQ \rightarrow NAD^+ + PQH_2$) will always proceed forward, and only the low activity of the complex as compared with the rapid turnover of PQH_2 prevents high levels of PQH_2 in chloroplasts. In this context, the inhibition of activity by the products NAD^+ and PQH_2 could be a possible mechanism of control that must be investigated. The rate of postillumination chlorophyll fluorescence increase provides an indirect, semiquantitative indication of the activity of the Ndh complex *in vivo* (Martín et al. 2009; Serrot et al. 2012) in response to developmental and environmental factors, but it does not provide information on the mechanisms involved. Zymograms and/or spectrophotometric assays show that the NADH dehydrogenase activity of the Ndh complex increases under different stress conditions; during senescence (Casano et al. 1999, 2000, 2001; Lascano et al. 2003; Martín et al. 2004); and when leaves are treated with methyl jasmonate (Cuello et al. 1995b).

The level of the Ndh complex has usually been determined by Western blotting with antibodies against polypeptides encoded by plastid *ndh* genes and, in a similar manner to its activity, increases in response to different stresses (Martín

et al. 1996, 2004; Casano et al. 1999; Lascano et al. 2003) and during leaf senescence (Martín et al. 1996; Zapata et al. 2005). The stress-related increases of levels of the Ndh complex and its activity can be reproduced by treating the leaves with hydrogen peroxide (H_2O_2) (Casano et al. 2001; Lascano et al. 2003), which suggests that H_2O_2 is a signal mediating the responses of the Ndh complex to stress. Important information has been provided by investigations on the mechanisms and factors that affect the posttranscriptional processing of the *ndh* genes and the posttranslational modification of the Ndh complex.

7.5.1 TRANSCRIPTIONAL AND POSTTRANSCRIPTIONAL CONTROL OF *NDH* GENE EXPRESSION

Of the 11 *ndh* genes, 6 (*ndhH*, *ndhA*, *ndhI*, *ndhG*, *ndhE*, and *ndhD*) map in the small single-copy region of the plastid DNA constituting the *ndhH-D* operon, which also includes the gene *psaC* (encoding a polypeptide of the photosystem I complex, PSI) between genes *ndhE* and *ndhD*. The *ndhC*, *ndhK*, and *ndhJ* genes are grouped in another transcriptional unit (ndhC-J operon) in the large single-copy region of plastid DNA. The *ndhF* gene is located in the small single-copy region, and *ndhB* is duplicated in the inverted repeated regions; both are transcribed as single independent units (Shinozaki et al. 1986; Maier et al. 1995; Del Campo et al. 2000, 2002; Serrot et al. 2008). There is evidence (Favory et al. 2005) that the plastid RNA polymerase requires the nuclear-encoded sigma4 factor for the transcription of the *ndhF* gene, which in turn would stimulate the transcription of the other plastid *ndh* genes. The primary transcripts of the *ndh* operons are processed to obtain monocystronic translatable transcripts in reactions carried out by the products of nuclear genes, as described for other operons in plastid DNA (Monde et al. 2000; Schmitz-Linneweber et al. 2001; Till et al. 2001). Accordingly, the levels of the mRNAs of *ndhB* and *ndhF* genes increase under stress conditions and senescence (Martínez et al. 1997; Casano et al. 2001). Transcripts of the *ndhH-D* and *ndhC-J* operons show a complex profile with multiple bands after Northern blot (Martínez et al. 1997; Del Campo et al. 2000, 2002, 2006; Serrot et al. 2008), which is affected by different factors. Under stress conditions and in senescence, the intensity of large polycistronic bands decreases, whereas the intensity of small monocistronic translatable bands increases. Therefore, rather than at the transcriptional level, the synthesis of NDH polypeptides is controlled in chloroplasts at the posttranscriptional level, involving stages of C-to-U editing and intron and intergenic splicing (Del Campo et al. 2000, 2002, 2006; Serrot et al. 2008) that depend on the products of nuclear genes, the details of which are beyond the scope of this revision.

7.5.2 REGULATION OF THE ACTIVITY OF THE NDH COMPLEX BY PHOSPHORYLATION

The NDH-F subunit of the thylakoid Ndh complex may be phosphorylated at the threonine at position 181 (Lascano et al. 2003), in a hydrophilic sequence that is oriented toward the stromal side of the thylakoid between the two hydrophobic transmembrane arms (Casano et al. 2004; Martín et al. 2009). The homologous ND5/NuoL subunit of respiratory complex I lacks Thr-181, which is conserved, together with a large sequence of contiguous amino acids, in the plastid NDH-F of Charophyceae and Prasinophyceae algae, bryophytes, gymnosperms, and angiosperms. The phosphorylation of NDH-F seems dependent on Ca^{2+}. It is stimulated under stress conditions and by treatment with H_2O_2 and results in an increase of the NADH dehydrogenase activity of the Ndh complex (Lascano et al. 2003). Site-directed transgenic tobaccos in which Thr-181 is substituted by other amino acids demonstrate the role of phosphorylation in activation, which seems related to the negative charge of the phosphate because activation is mimicked in the transgenic T181D in which Thr-181 is substituted by aspartic acid (Martín et al. 2009).

7.6 CONCLUDING REMARKS AND FURTHER PROSPECTS

A large variety of complexes exist in different organisms and organelles that are homologous to respiratory complex I. Although the function of all complexes is usually related to redox reactions and ion transport across membranes and there are well-established evolutionary relationships among the genes encoding their polypeptides, the complexes differ significantly in subunit composition and functions. In addition, the instability of the complexes and the hydrophobicity of a majority of the subunits create formidable difficulties in their structural and functional characterization.

When compared with homologous complexes, the thylakoid Ndh complex has the added difficulties of its low levels, which make the purification necessary for an accurate characterization laborious, and its absence in several photosynthetic organisms, which can give rise to doubts on its function in photosynthesis and dispensability. However, in recent years, investigations with purified thylakoid Ndh complex and with nuclear and chloroplastic mutants are clarifying the function of the complex in photosynthesis, its dispensability under certain environments, and its comparison with evolutionarily related complexes. Overwhelming evidence indicates that the Ndh complex is required for optimal cyclic electron transport and that it is related to the protective response against different stresses. It is also generally agreed that it catalyzes the transfer of electrons to oxidized PQ. The main remaining discussion refers to the nature of the electron donor (NADH or reduced ferredoxin), a question that is related to uncertainties about the structure and subunit composition of the complex and to the role of the complex in cyclic electron transport. Most evidence indicates that NADH is the electron donor. As the subunits of the N module that bind NADH in mitochondrial complex I have not yet been identified in plastids, the possibility exists that the thylakoid Ndh complex has a similar structure to the elemental (11 subunits) core complex I of *B. cereus* and catalyzes essentially the same reaction.

With NADH as the electron donor, the Ndh complex participates in the chlororespiratory electron transport chain, which balances the redox level of transporters under rapidly fluctuating light intensities (Figure 7.5). The appropriate redox level is required both to optimize the rate of cyclic electron transport and to maintain the proton gradient across the thylakoid. Therefore, the function of the thylakoid Ndh complex in photosynthesis is particularly relevant when, under fluctuating light, linear electron transport must be supplemented with cyclic electron transport to maintain the proton gradient necessary for photophosphorylation and to reduce the formation of singlet oxygen. The supplementation of linear with cyclic electron transport should be specially required under stress conditions and at high concentrations of CO_2, which explains the low photosynthetic performance of *ndh*-deficient plants at high CO_2 concentrations and fluctuating light and suggests that *ndh* genes could be dispensable under a prolonged mild environment and/or low CO_2 concentrations.

Further investigations should definitively clarify the structure and electron donor of the Ndh complex. The regulation of the expression of the *ndh* genes, in particular, the mechanism and control of the posttranscriptional processing, also remains a promising field of research, which is also relevant to ascertaining the functional role of the Ndh complex. The regulation of the Ndh complex, including posttranslational modification, should also be an active field of research to understand its functional relevance and the factors that affect its activity *in vivo*.

ACKNOWLEDGMENTS

We thank Guillermo del Riego for help performing chromatography, electrophoresis, and proteomic analyses of the Ndh complex and Patricia H. Serrot for helpful discussion, reading, and corrections of the manuscript.

REFERENCES

Abarca, D., M. Martín, and B. Sabater. 2001. Differential leaf stress responses in young and senescent plants. *Physiol. Plant.* 113:409–415.

Albertsson, P. A. 2001. A quantitative model of the domain structure of the photosynthetic membrane. *Trends Plant Sci.* 6:349–354.

Barret, C. F., J. V. Freudenstein, J. Li, D. R. Mayfield-Jones, L. Perez, J. C. Pires, and C. Santos. 2014. Investigating the path of plastid genome degradation in an early-transitional clade of heterotrophic orchids, and implications for heterotrophic angiosperms. *Mol. Biol. Evol.* 31:3095–3112.

Bart, C. and G. H. Krause. 2002. Study of tobacco transformants to asses the role of chloroplastic NAD(P)H dehydrogenase in photoprotection of photosystems I and II. *Planta* 216:273–279.

Battchikova, N., P. Zhang, S. Rudd, T. Ogawa, and E. M. Aro. 2005. Identification of NdhL and Ssl1690 (NdhO) in NDH-1L and NDH-1M complexes of *Synechocystis* sp. PCC 6803. *J. Biol. Chem.* 280:2587–2595.

Berger, S., U. Ellersiek, P. Westhoff, and K. Steinmuller. 1993. Studies on the expression of NDH-H, a subunit of the NAD(P)H-plastoquinone-oxidoreductase of higher plant chloroplasts. *Planta* 190:25–31.

Blazier, J. C., M. M. Guisinger, and R. K. Jansen. 2011. Recent loss of plastid-encoded *ndh* genes within *Erodium* (Geraniaceae). *Plant Mol. Biol.* 76:263–272. doi:10.1007/s11103-011-9753-5.

Braukmann, T. W. A. and S. Stefanovíc. 2012. Plastid genome evolution in mycoheterotrophic Ericaceae. *Plant Mol. Biol.* 79:5–20. doi:10.1007/s11103-012-9884-3.

Braukmann, T. W. A., M. Kuzmina, and S. Stefanovíc. 2009. Loss of all *ndh* genes in Gnetales and conifers: Extent and evolutionary significance for seed plant phylogeny. *Curr. Genet.* 55:323–337. doi:10.1007/s00294-009-0249-7.

Braukmann, T. W. A., M. Kuzmina, and S. Stefanovíc. 2013. Plastid genome evolution across the genus Cuscuta (Convolvulaceae): Two clades within subgenus Grammica exhibit extensive gene loss. *J. Exp. Bot.* 64:977–989. doi:10.1093/jxb/ers391.

Burrows, P. A., L. A. Sazanov, Z. Svab, P. Maliga, and P. J. Nixon. 1998. Identification of a functional respiratory complex in chloroplasts through analysis of tobacco mutants containing disrupted plastid ndh genes. *EMBO J.* 17:868–876.

Casano, L. M., M. Martín, and B. Sabater. 1994. Sensitivity of superoxide dismutase transcript levels and activities to oxidative stress is lower in mature-senescent than in young barley leaves. *Plant Physiol.* 106:1033–1039.

Casano, L. M., M. Martín, J. M. Zapata, and B. Sabater. 1999. Leaf-age and paraquat concentration-dependent effects on the levels of enzymes protecting against photooxidative stress. *Plant Sci.* 149:13–22.

Casano, L. M., M. Martín, J. M. Zapata, and B. Sabater. 2000. Chlororespiration and poising of cyclic electron transport: Plastoquinone as electron transporter between thylakoid NADH dehydrogenase and peroxidase. *J. Biol. Chem.* 275:942–948.

Casano, L. M., M. Martín, and B. Sabater. 2001. Hydrogen peroxide mediates the induction of chloroplast Ndh complex under photooxidative stress in barley. *Plant Physiol.* 125:1450–1458.

Casano, L. M., H. R. Lascano, M. Martín, and B. Sabater. 2004. Topology of the plastid Ndh complex and its NDH-F subunit in thylakoid membranes. *Biochem. J.* 382:145–155.

Corneille, S., L. Courmac, G. Guedeney, M. Havaux, and G. Peltier. 1998. Reduction of the plastoquinone pool by exogenous NADH and NADPH in higher plant chloroplasts. Characterization of a NAD(P)H–plastoquinone oxidoreductase activity. *Biochim. Biophys. Acta* 1363:59–69.

Cuello, J., M. J. Quiles, M. E. Albacete, and B. Sabater. 1995a. Properties of a large complex of NADH dehydrogenase from barley leaves. *Plant Cell Physiol.* 36:265–271.

Cuello, J., M. J. Quiles, J. Rosauro, and B. Sabater, B. 1995b. Effects of growth regulators and light on chloroplast NAD(P)H dehydrogenase activities of senescent barley leaves. *Plant Growth Regul.* 17:225–232.

Darie, C. C., M. L. Biniossek, V. Winter, B. Mutschler, and W. Haehnel. 2005. Isolation and structural characterization of the Ndh complex from mesophyll and bundle sheath chloroplasts of *Zea mays*. *FEBS J.* 272:2705–2716.

Del Campo, E. M., B. Sabater, and M. Martín. 2000. Transcripts of the ndhH-D operon of barley plastids: Possible role of unedited site III in splicing of the ndhA intron. *Nucleic Acids Res.* 28:1092–1098.

Del Campo, E. M., B. Sabater, and M. Martín. 2002. Post-transcriptional control of chloroplast gene expression: Accumulation of stable psaC mRNA is due to downstream RNA cleavages in the ndhD gene. *J. Biol. Chem.* 277:36457–36464.

Del Campo, E. M., B. Sabater, and M. Martín. 2006. Characterization of the 50- and 30-ends of mRNAs of ndhH, ndhA and ndhI genes of the plastid ndhH-D operon. *Biochimie* 88:347–357.

del Riego, G., L. M. Casano, M. Martín, and B. Sabater. 2006. Multiple phosphorylation sites in the β subunit of thylakoid ATP synthase. *Photosynth. Res.* 89:11–18.

Desplats, C., F. Mus, S. Cuiné, E. Billon, L. Courmac, and G. Peltier. 2009. Characterization of the Nda2, a plastoquinone-reducing type II NAD(P)H dehydrogenase in *Chlamydomonas* chloroplasts. *J. Biol. Chem.* 284:4148–4157.

Díaz, M., V. de Haro, R. Muñoz, and M. J. Quiles. 2007. Chlororespiration is involved in the adaptation of Brassica plants to heat and high light intensity. *Plant Cell Environ.* 30:1578–1585.

Endo, T., T. Shikanai, F. Sato, and K. Asada. 1998. NAD(P)H dehydrogenase-dependent, antymicin A–sensitive electron donation to plastoquinone tobacco chloroplasts. *Plant Cell Physiol.* 39:1226–1231.

Eskling, M., A. Emanuelsson, and H. E. Akerlund. 2001. Enzyme and mechanisms for violaxanthin–zeaxanthin conversions. In *Regulation of Photosynthesis*, eds. E. M. Aro, and B. Andersson, 433–452. Dordrecht: Kluwer.

Euro, L., G. Belevich, M. I. Verkhovsky, M. Wikström, and M. Verkhovskaya. 2008. Conserved lysine residues of the membrane subunit NuoM are involved in energy conversion by the proton-pumping NADH:ubiquinone oxidoreductase (Complex I). *Biochim. Biophys. Acta* 1777:1166–1172. doi:10.1016/j.bbabio.2008.06.001.

Favory, J. J., M. Kobayshi, K. Tanaka, G. Peltier, J. Kreis, J. G. Valay, and S. Lerbs-Mache. 2005. Specific function of a plastid sigma factor for ndhF gene transcription. *Nucleic Acids Res.* 33:5991–5999.

Friedrich, T. and B. Böttcher. 2004. The gross structure of the respiratory complex I: A Lego system. *Biochim. Biohys. Acta* 1608:1–9. doi:10.1016/j.bbabio.2003.10.002.

Heber, U. and D. Walker. 1992. Concerning a dual function of coupled cyclic electron transport in leaves. *Plant Physiol.* 100:1621–1626. doi:10.1104/pp.100.4.1621.

Joët, T., L. Cournac, G. Peltier, and M. Havaux. 2002. Cyclic electron flow around photosystem I in C3 plants. In vivo control by the redox state of chloroplast and involvement of the NADH-dehydrogenase complex. *Plant Physiol.* 128:760–769.

Kanno, A. and A. Hirai. 1993. A transcription map of the chloroplast genome from rice (*Oryza sativa*). *Curr Genet.* 23:166–174.

Kofer, W., H. U. Koop, G. Wanner, and K. Steinmüller. 1998. Mutagenesis of the genes encoding subunits A, C, H, I, J, and K of the plastid (NAD(H)-plastoquinone-oxidoreductase in tobacco by polyethylene glycol-mediated plastome transformation. *Mol. Gen. Gen.* 258:166–173.

Külheim, C., J. Ågren, and S. Jansson. 2002. Rapid regulation of light harvesting and plant fitness in the field. *Science* 297:91–93. doi:10.1126/science.1072359.

Kurisu, G., H. Zhang, J. L. Smith, and W. A. Cramer. 2003. Structure of the cytochrome b6f complex of oxygenic photosynthesis: Tuning the cavity. *Science* 302:1009–1014.

Lascano, H. R., L. M. Casano, M. Martín, and B. Sabater. 2003. The activity of the chloroplastic Ndh complex is regulated by phosphorylation of the NDH-F subunit. *Plant Physiol.* 132:256–262. doi:10.1104/pp.103.020321.

Lennon, A. M., P. Prommeenate, and P. J. Nixon. 2003. Location, expression and orientation of the putative chlororespiratory enzymes, Ndh and IMMUTANS, in higher plant plastids. *Planta* 218:254–260.

Ma, W. and H. Mi. 2008. Effect of exogenous glucose on the expression and activity of NADPH dehydrogenase complexes in the cyanobacterium *Synechocystis* sp. strain PCC 6803. *Plant Physiol. Biochem.* 46:775–779.

Maier, R. M., K. Neckermann, G. L. Igloi, and H. Kossel. 1995. Complete sequence of the maize chloroplast genome: Gene content, hotspots of divergence and fine tuning of genetic information by transcript editing. *J. Mol. Biol.* 251:614–628.

Martín, M. and B. Sabater. 2010. Plastid *ndh* genes in plant evolution. *Plant Physiol. Biochem.* 48:636–645. doi:10.1016/j.plaphy.2010.04.009.

Martín, M., L. M. Casano, and B. Sabater. 1996. Identification of the product of ndhA gene as a thylakoid protein synthesized in response to photooxidative treatment. *Plant Cell Physiol.* 37:293–298.

Martín, M., L. M. Casano, J. M. Zapata, A. Guéra, E. M. Del Campo, C. Schmitz-Linneweber, R. M. Maier, and B. Sabater. 2004. Role of thylakoid Ndh complex and peroxidase in the protection against photo-oxidative stress: Fluorescence and enzyme activities in wild-type and *ndhF*-deficient tobacco. *Physiol. Plant.* 122:443–452.

Martín, M., H. T. Funk, P. H. Serrot, P. Poltnigg, and B. Sabater. 2009. Functional characterization of the thylakoid Ndh complex phosphorylation by site-directed mutations in the *ndhF* gene. *Biochim. Biophys. Acta* 1787:920–928. doi:10.1016/jbbabio.2009.03.001.

Martín, M., D. Marín, P. H. Serrot, and B. Sabater. 2015. The rise of the photosynthetic rate when light intensity increases is delayed in ndh gene-defective tobacco at high but not at low CO_2 concentrations. *Front Plant Sci.* 6:34. doi:10.3389/fpls.2015.00034.

Martínez, P., C. Lopez, M. Roldan, B. Sabater, and M. Martin. 1997. Plastid DNA of five ecotypes of *Arabidopsis thaliana*: Sequence of *ndhG* gene and maternal inheritance. *Plant Sci.* 123:113–122.

Mathiesen, C. and C. Hägerhäall. 2002. Transmembrane topology of the NuoL, M and N subunits of NADH:quinone oxidoreductase and their homologues among membrane bound hydrogenases and bona fide antiporters. *Biochim. Biophys. Acta* 1556:121–132.

Meurer, J., A. Berger, and P. Westhoff. 1996. A nuclear mutant of *Arabidopsis* with impaired stability on distinct transcripts of the plastid *psbB*, *psbD/C*, *ndhH*, and *ndhC* operons. *Plant Cell* 8:1193–1207.

Monde, R. A., J. C. Greene, and D. B. Stern. 2000. The sequence and secondary structure of the 3′-UTR affect 3′-end maturation, RNA accumulation, and translation in tobacco chloroplasts. *Plant Mol. Biol.* 44:529–542.

Moparthi, V. K., B. Kumar, Y. Al-Eryani, E. Sperling, K. Górecki, T. Drakenberg, and C. Hägerhäll. 2014. Functional role of the MrpA- and MrpD-homologous protein subunits in enzyme complexes evolutionary related to respiratory chain complex. *Biochim. Biophys. Acta* 1837:178–185. doi:/10.1016/j.bbabio.2013.09.012.

Nandha, B., G. Finazzi, P. Joliot, S. Hald, and G. N. Johnson. 2007. The role of PGR5 in the redox poising of photosynthetic electron transport. *Biochim. Biophys. Acta* 1767:1252–1259. doi:10.1016/j.bbabio.2007.07.007.

Ogawa, T. and H. Mi. 2007. Cyanobacterial NADPH dehydrogenase complexes. *Photosynth. Res.* 93:69–77.

Ohyama, K., H. Fukuzawa, T. Kohchi, H. Shirai, T. Sano, S. Sano, K. Umesono, Y. Shiki, M. Takeuchi, Z. Chang, S. Aota, N. H. Inokuchi, and H. Ozeki. 1986. Chloroplast gene organization deduced from complete sequence of liverwort *Marchantia polymorpha* chloroplast DNA. *Nature* 322:572–574.

Peredo, E. L., U. M. King, and D. H. Les. 2013. The plastid genome of *Najas flexilis*: Adaptation to submersed environments is accompanied by the complete loss of the NDH Complex in an Aquatic Angiosperm. *PLos One* 8(7):e68591. http://www.plosone.org/article/info%3Adoi%2F10.1371%2Fjournal.pone.0068591. doi:10.1371/journal.pone.0068591.

Prommeenate, P., A. M. Lennon, C. Market, M. Hippler, and P. Nixon. 2004. Subunit composition of NDH-1 complexes of *Synechocystis* sp. PCC 6803. *J. Biol. Chem.* 279:28165–28173.

Quiles, M. J., A. García, and J. Cuello. 2000. Separation by blue-native PAGE and identification of the whole NAD(P)H dehydrogenase complex from barley stroma thylakoids. *Plant Physiol. Biochem.* 38:225–232.

Rumeau, D., N. Becuwe-Linka, A. Beyly, M. Louwagie, J. Garin, and G. Peltier. 2005. New subunits NDH-M, -N, and –O, encoded by nuclear genes, are essential for plastid Ndh complex functioning in higher plants. *Plant Cell* 17:219–232.

Rumeau, D., G. Peltier, and L. Courmac. 2007. Chlororespiration and cyclic electron flow around PSI during photosynthesis and plant stress response. *Plant Cell Environ.* 30:1041–1051.

Sabater, B. and M. Martín. 2013. Hypothesis: Increase of the ratio singlet oxygen plus superoxide radical to hydrogen peroxide changes stress defence response to programmed leaf death. *Front. Plant Sci.* 4:479. http://journal.frontiersin.org/Journal/10.3389/fpls.2013.00479/full. doi:10.3389/fpls.2013.00479.

Sazanov, L. A., P. A. Burrows, and P. J. Nixon. 1998. The plastid *ndh* genes code for an NADH-specific dehydrogenase: Purification and characterization of a mitochondrial-like complex I from pea thylakoid membranes. *Proc. Natl. Acad. Sci. U.S.A.* 95:1319–1324.

Schmitz-Linneweber, C., M. Tillich, R. G. Herrmann, and R. M. Maier. 2001. Heterologous, splicing-dependent RNA editing in chloroplasts: Allotetraploidy provides trans-factors. *EMBO J.* 20:4874–4883.

Schultze, M., B. Forberich, S. Rexroth, G. N. Dyczmons, M. Roegener, and J. Appel. 2009. Localization of cytochrome b6f complexes implies an incomplete respiratory chain in cytoplasmic membranes of the cyanobacterium *Synechocystis* sp. PCC 6803. *Biochim. Biophys. Acta* 1787:1479–1485.

Serrot, P. H., B. Sabater, and M. Martín. 2008. Expression of the *ndhCKJ* operon of barley and editing at the 13th base of the mRNA of the *ndhC* gene. *Biol. Plant.* 52:347–350.

Serrot, P. H., B. Sabater, and M. Martín. 2012. Activity, polypeptide and gene identification of thylakoid Ndh complex in trees: Potential physiological relevance of fluorescence assays. *Physiol. Plant.* 146:110–120. doi:10.1111/j.1399-3054.2012.01598.x.

Shikanai, T., T. Endo, T. Hashimoto, Y. Yamada, K. Asada, and A. Yokota. 1998. Directed disruption of the tobacco *ndhB* gene impairs cyclic electron flow around photosystem. *Proc. Natl. Acad. Sci. U.S.A.* 95:9705–9709.

Shimizu, H., L. Peng, F. Myouga, R. Motohashi, K. Shinozaki, and T. Shikanai. 2008. CRR23/NdhL is a subunit of the chloroplast NAD(P)H dehydrogenase complex in *Arabidopsis. Plant Cell Physiol.* 49:835–842.

Shinozaki, K., M. Ohme, M. Tanaka, T. Wakasugi, N. Hayashida, T. Matsubayashi, N. Zaita, J. Chunwongse, J. Obokata, K. Yamaguchi-Shinozaki, C. Ohto, K. Torazawa, B. Y. Meng, M. Sugita, H. Deno, T. Kamogashira, K. Yamada, J. Kusuda, F. Takaiwa, A. Kato, N. Tohdoh, H. Shimada, and M. Sugiura. 1986. The complete nucleotide sequence of the tobacco chloroplast genome: Its gene organization and expression. *EMBO J.* 5:2043–2049.

Sirpiö, S., Y. Allahverdiyeva, M. Holmstrom, A. Khrouchtchova, A. Haldrup, N. Battchikova, and E. M. Aro. 2009. Novel nuclear-encoded subunits of the chloroplast NAD(P)H dehydrogenase complex. *J. Biol. Chem.* 284:905–912.

Suorsa, M., S. Sirpiö, and E. M. Aro. 2009. Towards characterization of the chloroplast NAD(P)H dehydrogenase complex. *Mol. Plant* 2:1127–1140.

Till, B., C. Schmitz-Linneweber, R. Williams-Carrier, and A. Barkan. 2001. CRS1 is a novel group II intron splicing factor that was derived from a domain of ancient origin. *RNA* 7:1227–1238.

Wang, D. and A. R. Portis. 2007. A novel nucleus-encoded chloroplast protein, PIFI, is involved in NAD(P)H dehydrogenase complex-mediated chlororespiratory electron transport in *Arabidopsis*. *Plant Physiol.* 144:1742–1752.

Yamamoto, H., L. Peng, Y. Fukao, and T. Shikanai. 2011. A Src homology 3 domain-like fold protein forms a ferredoxin binding site for the chloroplast NADH dehydrogenase-like complex of *Arabidopsis*. *Plant Cell* 23:1480–1493. doi:10.1105/tpc.110.080291.

Zapata, J. M., A. Guéra, A. Esteban-Carrasco, M. Martín, and B. Sabater. 2005. Chloroplasts regulate leaf senescence: Delayed senescence in transgenic *ndhF*-defective tobacco. *Cell Death Differ.* 12:1277–1284.

8 Redox Metabolism in Photosynthetic Organisms

Sergio A. Guerrero, Diego G. Arias, and Alberto A. Iglesias

CONTENTS

8.1 Reducing Power Generation and Partitioning in Plants 159
    8.1.1 NADPH Production in Plastids 159
    8.1.2 NADPH Metabolism in the Cytosol 161
8.2 Redox Molecules and Enzymatic Systems 163
    8.2.1 Glutathione 163
    8.2.2 Ascorbate 163
    8.2.3 Thioredoxin System 164
    8.2.4 Glutathione-Dependent System 164
    8.2.5 Peroxiredoxin 165
    8.2.6 Superoxide Dismutase 165
    8.2.7 Glutathione S-Transferase 166
    8.2.8 Ascorbate Peroxidase 166
8.3 Systems for Repairing Oxidative Damage in Plants 166
8.4 Concluding Remarks 167
Acknowledgments 168
References 168

8.1 REDUCING POWER GENERATION AND PARTITIONING IN PLANTS

The reducing equivalents required for anabolism as well as for detoxifying and regenerative systems in plant cells derive (directly or indirectly) from both ferredoxin (Fd) and NADPH. The chloroplast protein Fd is the final acceptor of the photosynthetic electron transfer through the thylakoid membrane and, in its reduced form, is linked to conversion of NADP+ to NADPH, thus generating the primary products that conserve part of the energy from sunlight as reducing power (Rochaix 2011). NADPH and ATP are the main products of the light phase of photosynthesis that then are used for carbon assimilation during the synthetic phase (Cortassa et al. 2012; Iglesias and Podestá 2005). Plants exhibit the following key characteristics: (1) cells are highly compartmentalized, with the occurrence of plastids, and (2) there are two different kinds of tissues, with characteristic metabolic scenarios: photosynthetic or heterotrophic. Consequently, specific routes take place in the different cells to partition photosynthetically generated reducing equivalents intracellularly and intercellularly; the whole picture is schematized in Figure 8.1.

8.1.1 NADPH PRODUCTION IN PLASTIDS

Plant cells are characterized by a high degree of compartmentalization with the existence of different kinds of plastids: chloroplasts (present in photosynthetic tissues) as well

as amyloplasts, leucoplasts, or chromoplasts (found in cells of heterotrophic tissues) (Hudák 1997). These diverse plastids have distinctive specialized functions and metabolic capacities that also change with time. It is worth considering the particular characteristics of chloroplasts, where metabolic fluxes exhibit substantial differences during the light or the dark period. In higher plants, the primary production of NADPH in plastids is associated with NADP+ reduction in light (photoreduction) by the photosynthetic electron flow, or the generation of the reduced nucleotide by the oxidative pentose phosphate pathway (OPPP) in the dark (in chloroplasts) or in nonphotosynthetic tissues.

During the light period, chloroplasts have an active photosynthetic activity upholding NADPH as one key metabolite. Using solar energy, the photosynthetic electron transport drives reducing equivalents from water to NADP+ ($E^{\circ\prime}_{O2/H2O} = 815$ mV and $E^{\circ}_{NADP^+/NADPH} = -340$ mV, $E^{\circ\prime}$ being the midpoint oxidation–reduction potential relative to the standard hydrogen electrode at pH 7.0). In fact, NADPH and ATP are the main products of the light phase of photosynthesis, and they constitute the molecular way by which the energy from sunlight is converted into cellular reducing and energetic power, respectively (Cortassa et al. 2012; Iglesias and Podestá 2005). The entire process of the light phase takes place, physiologically, by coupling two successive photochemical reactions involving photosystem II and photosystem I (PSII and PSI; Figure 8.1). These photosystems are linked by means of an electron transport chain that includes the integral membrane assemblage

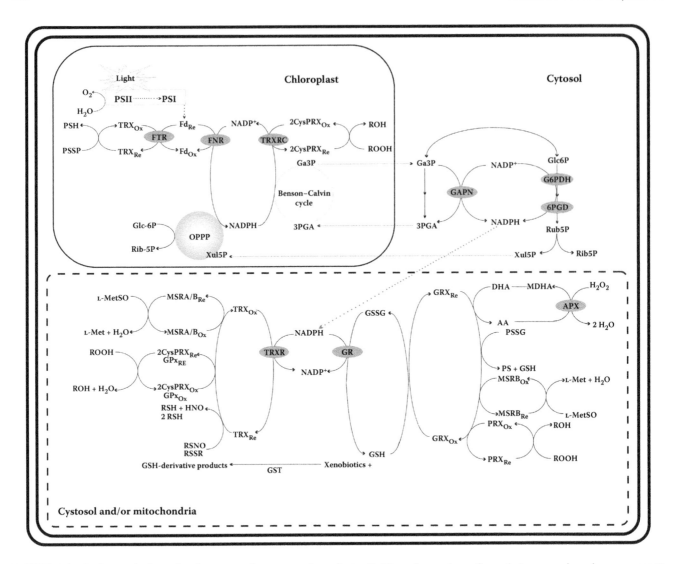

FIGURE 8.1 Redox equivalents flux between redox systems in a plant cell. The scheme shows fluxes between main redox components in plants. 2CysPRX, typical two-cysteine peroxiredoxin; 3PGA, 3-phosphoglycerate; 6GPD, 6-phosphogluconate dehydrogenase; AA, ascorbic acid; APX, ascorbate peroxidase; DHA, dehydroascorbate; Fd, ferredoxin; FNR, ferredoxin–NADP oxidoreductase; FTR, ferredoxin–thioredoxin oxide-reductase, G6PDH, Glc6P dehydrogenase; Ga3P, glyceraldehyde 3-phosphate; GAPN, nonphosphorylating glyceraldehyde 3-phosphate dehydrogenase; Glc6P, glucose 6-phosphate; GPX, glutathione peroxidase; GR, glutathione reductase; GRX, glutaredoxin; GSH, reduced glutathione; GSSG, oxidized glutathione; GST, glutathione S-transferase; H_2O, water; H_2O_2, hydrogen peroxide; HNO, nitroxyl; L-Met, L-methionine; L-MetSO, L-methionine sulfoxide; MDHA, monodehydroascorbate; MSRA, A-type methionine sulfoxide reductase; MSRB, B-type methionine sulfoxide reductase; O_2, molecular oxygen; OPPP, oxidative pentose phosphate pathway; PRX, peroxiredoxin; PSH, thiol protein; PSI, photosynthetic system I; PSII, photosynthetic system II; PSSG, glutathionyl protein; PSSP, disulfide protein; Rib5P, ribose 5-phosphate; ROH, alcohol; ROOH, peroxide; RSH, low-molecular-weight thiol; RSNO, S-nitrosothiols; RSSR, low-molecular-weight disulfide; Rub5P, ribulose 5-phosphate; TRX, thioredoxin; TRXR, thioredoxin reductase; TRXRC, thioredoxin reductase C; Xul5P, xylulose 5-phosphate.

know as the cytochrome $b_6 f$ complex. At the terminal stage, Fd reduces $NADP^+$ to NADPH via an enzyme known as Fd-$NADP^+$ oxidoreductase. Electron transfer from water to $NADP^+$ is coupled to proton translocation across the thylakoid membrane, thus generating the electrochemical potential gradient energetically required for ATP synthesis mediated by the F_0–F_1 complex of the ATP synthase (Rochaix 2011).

During the dark period, as well as in nonphotosynthetic tissues, NAPDH production in plastids is directly linked to the OPPP (Figure 8.1) through oxidation, interconversion, and rearrangement of different sugar-phosphate metabolites

(Brownleader 1997). In fact, the OPPP can be analyzed as composed by two operative phases: one oxidative phase (phase 1), and one nonoxidative phase (phase 2). Phase 1 is oxidative because glucose 6-phosphate (Glc6P) is oxidized/decarboxylated to ribulose 5-phosphate (Rub5P) in three enzymatic steps with the net production of two molecules of NADPH. The first reaction of this metabolic route is the oxidation of Glc6P by one specific dehydrogenase (Glc6PDHase, EC 1.1.1.49) to form 6-phosphoglucono-δ-lactone and NADPH. The lactone is hydrolyzed by a specific lactonase to yield the acid 6-phosphogluconate, which then undergoes oxidation and decarboxylation by a specific dehydrogenase (6PGDHase, EC 1.1.1.44) to form Rub5P and a second molecule of NADPH (Nelson and Cox 2004). The second phase consists of a set of reactions routing the interconversion between phosphorylated three-, four-, five-, six-, and seven-carbon sugars, involving (between others) the enzymes ribose 5-phosphate isomerase (Rib5P isomerase), Rub5P 3-epimerase, transaldolase, and transketolase (Debnam and Emes 1999). Phase 2 regenerates Glc6P, thus cycling the pathway and connecting it with glycolysis. Unlike what happens in animal and yeast cells, plants have a duplication of the OPPP: with a complete pathway inside the chloroplast and another one (incomplete) in the cytosol (Anderson and Advani 1970; Bailey-Serres et al. 1992; Brownleader 1997; Eicks et al. 2002; Kruger and von Schaewen 2003; Schnarrenberger et al. 1973, 1995). Notwithstanding the physical separation, both routes are interconnected by specific metabolite transporters present in the chloroplast envelope (mainly glyceraldehyde 3-phosphate [Ga3P] and xylulose 5-phosphate [Xyl5P]; Figure 8.1) (Eicks et al. 2002; Neuhaus and Emes 2000; Weber et al. 2005).

Many of the enzymes catalyzing reversible reactions of the OPPP also play a part in the Benson–Calvin cycle (Debnam and Emes 1999). Distinct is the case for Glc6PDHase, involved in the irreversible and specific step of oxidation of Glc6P and constituting a relevant control point for the regulation of the hexose-P substrate partitioning between the OPPP and glycolysis. Detailed studies carried out on chloroplastidic Glc6PDHase evidenced that its activity is under different levels of regulatory control. The NADPH/NADP$^+$ ratio, pH, Mg^{2+}, and concentration of Glc6P can affect the enzyme activity (Brownleader 1997). Moreover, the enzyme is activated in the dark (low NADPH/NADP$^+$ ratio) by a redox mechanism involving thioredoxin (TRX) (Buchanan 1980; Scheibe 1991; Wenderoth et al. 1997). The activity of several chloroplastidic enzymes is known to be regulated by reversible thiol–disulfide interchange (Buchanan 1980; Scheibe 1991). During the photosynthetic light period, the electron transport leads to the reductive activation of several stromal target enzymes (fructose 1,6-bisphosphatase, phosphoribulokinase, NADP-dependent Ga3P dehydrogenase, and others) through covalent redox modification mediated by the Fd–TRX system. Chloroplastidic Glc6PDHase is regulated in an opposite way by the same system, being inactive in its reduced state (light period) (Wenderoth et al. 1997; Wendt et al. 1999). This regulation prevents futile cycles, like simultaneous carbohydrate synthesis in the Benson–Calvin cycle and catabolism by the OPPP. Thus, in line with its physiological role in chloroplasts, Glc6PDHase is active to generate NADPH only during the dark phase, when supply of the reduced dinucleotide by the photosynthetic electron flow ceases (Wenderoth et al. 1997).

8.1.2 NADPH METABOLISM IN THE CYTOSOL

The production of NADPH in the cytosol of plant cells exhibits a higher degree of complexity than that observed in animals and yeasts. The passage of this reduced intermediate from the chloroplast to the cytosol is indirect, via transport

of key metabolites and metabolic conversions, one of which includes the OPPP. The extent to which the OPPP operates in the cytosol of higher plants is a matter of discussion. Both Glc6PDHase (Schnarrenberger et al. 1973, 1975) and 6PGDHase (Bailey-Serres and Nguyen 1992; Bailey-Serres et al. 1992) are found in the cytosol of many plant cells, but the organization and functioning of the reversible nonoxidative phase of the metabolic path are far less clear. Different studies support the absence of Rub5P 3-epimerase, Rib5P isomerase, transketolase, and transaldolase in the cytosol of leaf cells from spinach, peas, and maize (Debnam and Emes 1999; Schnarrenberger et al. 1995). Similar results were reported by Kruger and von Schaewen (2003), who analyzed the complete *Arabidopsis thaliana* genome looking for the predicted subcellular localization of multiple copies of genes encoding for enzymes of the OPPP. Such an analysis identified putative cytosolic and plastidic forms of Glc6PDHase, 6PGDHase, Rib5P isomerase, and Rub5P 3-epimerase, whereas transketolase and transaldolase are predicted only as plastidic enzymes.

Thus, *Arabidopsis* has the genetic capacity to convert cytosolic Rub5P to Rib5P and Xyl5P, but the production of other intermediates such as fructose 6-phosphate and triose phosphate only occurs within plastids (Kruger and von Schaewen 2003). The group of M.J. Emes arrived at the same metabolic scenario (Hutchings et al. 2005) after analyzing the subcellular distribution of OPPP in *Brassica napus* embryos by measuring activity of the different enzymes. Given the observations regarding the cytosolic nonoxidative phase 2 of the OPPP, the extent to which the cycle is operative is an open question (Averill et al. 1998). To further metabolize cytosolic pentose-P, the nonoxidative part of the OPPP would require transport of the metabolite across the plastid envelope, and in such a way, a phosphate translocator protein with specificity to counter the exchange of Xul5P, triose-P, and Pi has been identified in the inner membrane of the chloroplast envelope (Eicks et al. 2002). Thus, as illustrated in Figure 8.1, the transport of Xyl5P would represent a key connection between chloroplast and cytosol playing a critical role of the functioning of the OPPP in both compartments.

As it happens in plastids, the key regulatory step of the cytosolic OPPP seems to be that catalyzed by Glc6PDHase. However, the characterization of the regulatory properties for this cytosolic enzyme is scarce. An early report indicated that both chloroplastidic and cytosolic Glc6PDHases are inactivated in response to light and dithiothreitol (DTT) (Anderson et al. 1974). Later, it was pointed out that the two cytosolic isoforms of G6PDHase exhibit differences in modulation, with one being slightly activated by reductants and the other one behaving insensitive to redox signaling (Hauschild and von Schaewen 2003; Hutchings et al. 2005; Wakao and Benning 2005; Wakao et al. 2008). Current data are not clear enough to define the exact role played by cytosolic forms of this enzyme. Thus, in *Arabidopsis*, single and double null mutants of cytosolic Glc6PDHases indicate the occurrence of alternative, compensatory mechanisms to supply NADPH in the cytosol (Wakao et al. 2008). Since the double mutant showed increased seed oil content and mass, it was concluded that a metabolic change possibly takes place with increase in the flux of carbon through glycolysis.

Besides the incomplete OPPP, plant cells have other enzymes able to generate NADPH in the cytosol. Worthy of mention and analysis in some detail are the nonphotosynthetic form of NADP-malic enzyme (NADP-ME, EC 1.1.1.40) and the nonphosphorylating Ga3P dehydrogenase (GAPN, EC 1.2.1.9). NADP-ME catalyzes the oxidative decarboxylation of L-malate to produce pyruvate, CO_2, and NADPH in the presence of a divalent cation (mainly Mg^{2+}) (Drincovich et al. 2001). The cytosolic (nonphotosynthetic) isoform of NADP-ME has been characterized with respect to kinetics, regulation, and function (Detarsio et al. 2008; Drincovich et al. 2001; Maurino et al. 1996, 2001; Smirnoff and Wheeler 2000). The enzyme is involved in different physiological processes in the following ways: as pH control by balancing malate levels and also by providing NADPH for plant defense response (to oxidative conditions), and for synthesis of lignin (Smirnoff and Wheeler 2000). Genetically transformed *Arabidopsis* plants overexpressing rice cytosolic NADP-ME exhibit an increased NADPH/$NADP^+$ ratio in the cytosol, which confers salt tolerance to transgenic seedlings (Cheng and Long 2007; Liu et al. 2007).

As specified, the other main enzyme involved in the generation of reducing power in the form of NADPH in the cytosol of plant cells is GAPN. This enzyme catalyzes the irreversible oxidation of Ga3P to 3-phosphoglycerate (3PGA), being highly specific for the cofactor and thus generating NADPH according to the following reaction (Habenicht 1997; Iglesias 1989):

$$Ga3P + NADP^+ + H_2O \rightarrow 3PGA + NADPH + 2H^+$$

In this way, the enzyme uses as substrate a primary product of the Benson–Calvin cycle that, after being exported from the chloroplast, serves to increase the level of reducing equivalents in the form of the reduced dinucleotide in the cytosol. Structural analysis determined that GAPN is a member of the aldehyde dehydrogenase superfamily (Habenicht 1997), and it is only found in some specialized eubacteria (Boyd et al. 1995; Brown and Wittenberger 1971; Fourrat et al. 2007; Habenicht 1997; Iddar et al. 2005), archaebacteria (Brunner et al. 1998), and the cytosol of green algae and higher plants (Arnon et al. 1954; Bustos and Iglesias 2002; Gomez Casati et al. 2000; Iddar et al. 2002; Iglesias 1989; Iglesias and Losada 1988; Iglesias et al. 1987; Jacob and D'Auzac 1972; Marchal et al. 2001; Mateos and Serrano 1992; Pupillo and Faggiani 1979). The enzyme is encoded by a single gene (*gapN*), and so far, its active form has been characterized as a homotetramer with a molecular mass of ~200 kDa (Bustos and Iglesias 2002; Gomez Casati et al. 2000; Habenicht 1997; Iddar et al. 2002).

Early studies (Kelly and Gibbs 1973) established that, in photosynthetic cells, GAPN is involved in a shuttle system for the export of NADPH (photogenerated) from the chloroplast to the cytosol. This transport system involves the interexchange of triose phosphate and Pi between both cell compartments via the triose phosphate/Pi translocator

of the chloroplast envelope. In plants accumulating acyclic polyols (such as celery), the enzyme plays an additional role by supplying the NADPH necessary for the synthesis of the reduced sugars (Gao and Loescher 2000; Rumpho et al. 1983). On the other hand, in heterotrophic cells (from plants and bacteria), the enzyme would couple the production of NADPH required by anabolism with glycolysis (Habenicht 1997). The presence of GAPN in the cytosol determines a distinctive feature for glycolysis in plant cells by establishing an alternative to oxidation of triose phosphate (Figure 8.1) (Iglesias 1989; Plaxton 1996). In this way, Ga3P can be metabolized to 3PGA either by the two enzymatic steps of phosphorylating NAD-dependent Ga3PDHase (EC 1.2.1.12) and 3PGA kinase (EC 2.7.2.3) or via the single reaction catalyzed by GAPN. By the first route, the products are NADH and ATP, whereas the alternative path renders NADPH but not ATP (Iglesias 1989). From a bioenergetic point of view, this branch point in glycolysis is relevant, and it should be regulated in order to effectively modulate the production of energetic and reductive power within the cell (Plaxton 1996). In agreement with the latter analysis, it has been demonstrated that in heterotrophic plant cells, GAPN undergoes posttranslational regulation by phosphorylation (Bustos and Iglesias 2002). The phosphorylated enzyme is able to interact with 14-3-3 regulatory proteins to render a protein complex with altered kinetic properties (the enzyme is less active and more sensitive to regulation by nucleotides and pyrophosphate) (Bustos and Iglesias 2003, 2005).

Studies carried out in *Arabidopsis* and wheat have shown that the functionality of GAPN is of pivotal importance for carbon and energy metabolism in plants under both physiological and oxidative stress conditions (Bustos et al. 2008; Rius et al. 2006). An *Arabidopsis* null mutant of GAPN was found to have altered morphology of the siliques, reduced carbon flux through glycolysis, lower capacity for CO_2 fixation, and increased oxidative stress (Rius et al. 2006). The absence of GAPN elicited a parallel increase of Glc6PDHase activity. The concurrence of these results with others from studies with cytosolic Glc6PDHase *Arabidopsis* mutants (Wakao et al. 2008) supports the statement that the disruption of one pathway generating NADPH is counterbalanced by boosting up levels of enzymes involved in alternative routes operating within the cytosol.

With this background, it could be concluded that NADPH production in the cytosol of plant cells is not strictly dependent on a particular system, but diverse metabolic alternatives are possible (Figure 8.1). Rather, the provision of reducing equivalents (photosynthetically generated) in the cytosol can be visualized as an integrated complex metabolic network. In the latter scenario, different primary metabolic pathways interact in order to assure supply by reducing power demands to the cell and/or to ameliorate oxidative stress situations. It is worth taking into account that the universal provider of reducing equivalents in plant cells is NADPH, which primarily is generated photosynthetically in the chloroplast and then is exported indirectly (the dinucleotide cannot pass the chloroplast envelope), via transport of different metabolites, to the different cell compartments and to other plant cells.

8.2 REDOX MOLECULES AND ENZYMATIC SYSTEMS

Besides NADPH, which, as discussed, is the key redox metabolite providing reducing equivalents in different cell compartments and tissues, many other molecules participate in the redox metabolism in plants. These molecules are linked to different enzymatic systems that orchestrate the maintenance of redox balance under physiological conditions and also participate in mechanisms allowing plant defense to cope with oxidant conditions. The following sections discuss in detail the major redox molecules in plant cells and their respective functions and characteristics.

8.2.1 GLUTATHIONE

The tripeptide γ-glutamyl-cysteinyl-glycine (or reduced glutathione [GSH]) is a moderate reducing agent ($E^{\circ\prime}_{GSSG/GSH} = -240$ mV) that exists interchangeably with the oxidized form (GSSG) and plays a critical role in many cells (Banerjee 2008). Some plant species contain homologue forms of GSH in which the carboxy terminal glycine of the tripeptide is replaced by other amino acids (Noctor and Foyer 1998). Glutathione reductase (GR, EC 1.8.1.7), a pyridine nucleotide disulfide reductase, is the enzyme responsible for the transfer of electrons from NADPH to GSSG to generate GSH (Smirnoff 2005). GSH biosynthesis takes place in two ATP-dependent steps catalyzed by glutamate cysteine ligase and γ-glutamyl-transpeptidase. GSH is also found in other eukaryotes as well as in Gram-negative bacteria and in a reduced number of Gram-positive prokaryotes (Banerjee 2008). In eukaryotes, most (~90%) of the intracellular pool of GSH is found in the cytoplasm, and the rest is distributed in mitochondria, the endoplasmic reticulum, and the nucleus (Noctor and Foyer 1998). In plant cells, biosynthesis of GSH takes place in the cytosol and the chloroplast (Smirnoff 2005), and it constitutes the major nonprotein thiol.

The physiological implicance of GSH in plants may be divided into two groups: sulfur metabolism and antioxidant defense (Scheibe 1991; Smirnoff 2005). The tripeptide is a product of the primary sulfur metabolism, contributing to the transport and storage of reduced sulfur and being an important antioxidant and redox buffer (Banerjee 2008; Buchanan and Balmer 2005). GSH is also critical for detoxification of heavy metals (such as cadmium and nickel) since it is necessary for phytochelatin synthesis (Abhilash et al. 2009; Noctor and Foyer 1998). Furthermore, GSH is a substrate providing reducing equivalents to glutathione S-transferases (GST) (Kulinskii and Kolesnichenko 2009; Smirnoff 2005), glutathione peroxidase (GPX) (Dietz et al. 2006), and phospholipid hydroperoxide glutathione peroxidase (PHGPX) (Smirnoff 2005), thus being respectively involved in the detoxification of xenobiotics and in the reduction of H_2O_2 and lipidic

peroxide. Its capacity to be an electron donor for glutaredoxins (GRXs) (Lillig et al. 2008) makes GSH useful in mediating the mixed-disulfide reduction of proteins (Figure 8.1).

8.2.2 ASCORBATE

Ascorbate (L-*threo*-hex-2-enono-1,4-lactone, also known as vitamin C) is well known for its ability to detoxify free radicals, being most abundant between water-soluble small molecules playing an antioxidant role in plants (Smirnoff 2005). Oxidation of ascorbate renders the monodehydro derivative, this reaction having a moderately positive standard redox potential ($E^{\circ\prime} = 280$ mV), after which this compound constitutes a very suitable one-electron donor for many oxygenases and hydroxylases (Noctor and Foyer 1998). Given the values of pK for ascorbic acid (pK_1 = 4.2 and pK_2 = 11.8), the monoanion form of ascorbate is predominant at cellular conditions (Banerjee 2008), and the enediol group loses one electron to produce the monodehydroascorbate (MDHA) radical and can also be further oxidized to dehydroascorbate (DHA) (Banerjee 2008). The MDHA radical is stabilized by delocalization of the unpaired electron within the conjugated structure of the five-atom lactone ring containing the enediol group (Noctor and Foyer 1998). The latter stability of the product of one-electron oxidation is the basis of the biological function of ascorbate as an antioxidant and scavenger of free radicals. However (and paradoxically), the anion has also properties of a prooxidant agent in the reduction of metal ions such as Fe^{3+}, and the generation of Fe^{2+} can then catalyze the production of ·OH from H_2O_2 by the Fenton reaction (Banerjee 2008; Halliwell 2006).

All plants and animals (except primates and guinea pigs) are able to synthesize ascorbic acid (Noctor and Foyer 1998). Levels of ascorbate in photosynthetic and heterotrophic plant tissues can reach millimolar concentrations. Synthesis occurs through the Smirnoff–Wheeler–Running pathway (Linster and Clarke 2008; Smirnoff and Wheeler 2000), where the relatively rare sugar L-galactose undergoes two-step oxidation in reactions catalyzed by NAD-dependent L-galactose dehydrogenase and L-galactono-1,4-lactone dehydrogenase. Guanosine diphosphate–mannose (GDP-mannose) is the source of L-galactose via the intermediate metabolite GDP-L-galactose, which itself originates through epimerization from GDP-mannose (Noctor and Foyer 1998; Smirnoff and Wheeler 2000). Catabolism of ascorbate produces oxalate, tartrate, and threonine, although the enzymes involved this degradative pathway have not been identified at present (Noctor and Foyer 1998).

The salvage of ascorbate is performed by direct reduction of DHA by two molecules of GSH (Figure 8.1), a reaction that has been demonstrated in cell-free systems (Banerjee 2008; Meyer 2008; Noctor and Foyer 1998). Plants also contain proteins with GSH-dependent DHA reductase activity (Meyer 2008; Smirnoff and Pallanca 1996). Besides, electrons coming from the water photolysis by PSII can be transferred to MDHA via Fd (Smirnoff 2000). Ascorbate peroxidase (APX) (Smirnoff 2005) and ascorbate oxidase (AO) are two enzymes identified in plants for ascorbate oxidation (Dawson et al. 1975). APX is found in all subcellular compartments (Dawson et al. 1975), and it catalyzes the ascorbate-dependent reduction of H_2O_2 to H_2O (Shigeoka et al. 2002), which is critical for the antioxidant function of ascorbate in plants. AO is a glycoprotein pertaining to the blue copper oxidase family that localizes at the apoplastic level (Pignocchi and Foyer 2003), being particularly active in expanding cucurbit fruit tissues (Arrigoni 1994).

8.2.3 Thioredoxin System

TRXs are small redox proteins (usually ~12 kDa) having the conserved motif WCG/PPC as the active site. The two cysteine residues in the motif undergo redox changes, and thus, TRXs exist either in the reduced (dithiol) or in the oxidized (disulfide) form (Banerjee 2008). Reduced TRX (linked to the reducing power of NADPH and Fd systems) rapidly reacts with disulfide substrates mainly constituted by target proteins (Smirnoff 2005). The redox potential (which is important for the catalytic capacity) of plant TRXs ranges between −285 and −350 mV (Buchanan and Balmer 2005). The number of genes encoding for TRX in plants is outstandingly high, and many of them are implicated in photosynthetic regulation. In the *A. thaliana* genome, genes were identified that encode for at least 22 TRX isoforms that are classified in six different groups: *f*, *m*, *x*, and *y* in chloroplasts; *o* in mitochondria; and *h* in other cell compartments (cytosol, mitochondria, and endoplasmic reticulum) as well as outside the cell (Gelhaye et al. 2005; Montrichard et al. 2009). The functionality of many enzymes, receptors, and transcription factors may be controlled via the disulfide reductase activity of TRX. For example, photosynthetic enzymes are regulated by specific TRXs that are reduced during the light period, as TRX_m and TRX_f, and are specific for target enzymes like malate dehydrogenase and fructose bisphosphatase, respectively (Smirnoff 2005).

TRXs are the main thiol–disulfide proteins in plants, and they have a common tertiary polypeptide structure known as the TRX fold (Gelhaye et al. 2005). This domain is also found in many other thiol proteins, including GRX, GST, GPX, peroxiredoxin (PRX), and protein disulfide isomerase (Martin 1995). Through rapid thiol–disulfide interchange reactions, the reduced TRX can act to directly reduce protein disulfides (Bindoli et al. 2008). The action mechanism can be divided into three stages. First, reduced TRX noncovalently docks to the target protein through a hydrophobic interaction surface and hydrogen bonds. Then, the thiolate of the N-term cysteine (in the CXXC active domain) nucleophilically attacks the target protein performing a thiol–disulfide exchange. Finally, the C-term cysteine cleaves the transient protein–protein complex disulfide intermediate to render oxidized TRX and the reduced target protein (Banerjee 2008; Buchanan and Balmer 2005).

TRXs are involved in two redox systems found in different plant cell compartments, as schematized in Figure 8.1 (Smirnoff 2005). The Fd system, located in chloroplasts, is formed by Fd, FTR, and TRX *f*, *m*, *x*, and *y* (Gelhaye et al.

2005; Schurmann and Jacquot 2000). The system is involved in the electron flow by thiol–disulfide exchange intermediates from Fd (reduced in the light via PSI) to target proteins following this sequence: light → Fd → FTR → TRX → target protein (Buchanan and Balmer 2005; Scheibe 1991). In the dark, the system is oxidized by either O_2, GSSG, or reactive oxygen species (ROS) (Montrichard et al. 2009). The cytosolic TRX system (also found in mitochondria) is NADPH dependent and includes a low-molecular-weight TRX reductase (TRXR), plus a FAD-containing and NADPH-dependent TRXR (Gelhaye et al. 2005). In *A. thaliana* two distinct genes code for similar forms of TRXR (A and B) (Meyer et al. 2005, 2008) that have been localized in the cytoplasm or in mitochondria (Meyer et al. 2005). These TRXRs transfer electrons by a thiol–disulfide exchange mechanism from NADPH to TRX_h (in the cytoplasm) or TRX_h and TRX_o (in mitochondria) following this sequence: NADPH → TRXR → TRX → target protein (Buchanan and Balmer 2005; Gelhaye et al. 2005; Smirnoff 2005). In chloroplasts, a modified type of TRXR, named TRXRC, has been identified that has a TRX domain in the C-term (Kirchsteiger et al. 2009), exhibits a complete dependence on NADPH, and is able to transfer reducing equivalents from NADPH to the plastidic 2CysPRX called BAS1 (Alkhalfioui et al. 2007; Moon et al. 2006). Figure 8.1 details other potential functions of the TRX system (besides its documented involvement in dark/light regulation), such as exertion of responses against environmental stresses (Gelhaye et al. 2005) and ROS detoxification mediating redox interchange with several PRX isoforms, GPX, and MetSO via methionine sulfoxide reductase A/B (MSRA/B) proteins (Montrichard et al. 2009).

8.2.4 Glutathione-Dependent System

The homeostasis redox intracellular in most organisms is dependent on GSH, a key low-molecular-weight component involved in the protection of plant cells from oxidative stress (Smirnoff 2005). GSH undergoes redox interexchange being converted to its oxidized form, GSSG, which is reduced again enzymatically by glutathione reductase (GR) at the expense of NADPH (Noctor and Foyer 1998). GR is a flavoenzyme belonging to the disulfide oxidoreductase family (Rouhier et al. 2008) that usually conforms to a homodimeric structure arrangement of the active site (Rybus-Kalinowska et al. 2009). Three domains can be visualized in each monomer of GR: the FAD, the $NADP^+$, and the interface domain, with the binding site for GSSG being formed by a bridge between the interface domain of one subunit and the FAD domain of the adjacent subunit (Banerjee 2008; Rybus-Kalinowska et al. 2009).

GSSG can also produce glutathionylation or protein thiolation after reacting with cysteine residues of many proteins to form protein–glutathione mixed disulfides (PSSG) (Dalle-Donne et al. 2007; Rouhier et al. 2008). The reversion of the latter process is mediated by glutaredoxins (GRXs or thioltransferases), which are GSH-dependent proteins (Dalle-Donne et al. 2007) that belong to the thiol–disulfide oxidoreductase enzyme family (Lillig et al. 2008). NADPH, GR, GSH, and GRX constitute the GSH system, which often operates in parallel with the TRX system to regulate the redox homeostasis in plant cells. Many enzymes are inactivated or activated after thiolation/dethiolation, and thus, the GSH system is involved in modulation of some metabolic pathways and cell signalling (Buchanan and Balmer 2005).

GRXs are small proteins of 10 to 24 kDa able to catalyze the reduction of proteins that are thiolated by GSH (PSSG). Approximately 30 GRX isoforms were found in *A. thaliana* (Meyer et al. 2008). The reduction of PSSG is carried out by GSH, which is oxidized to GSSG and recycled to GSH via the recycling system of NADPH and GR, thus establishing the following electron transfer path: NADPH → GR → GSH → GRX (Smirnoff 2005). GRX can reduce PSSG or low-molecular-weight dithiols (GSSR) via either a monothiol or a dithiol mechanism (Lillig et al. 2008). Three main classes of GRXs are distinguished according to their respective redox-active center: CPYC, CGFS, and CCXC/S types (Rouhier et al. 2004), with the latter type only found in higher plants (Meyer et al. 2008). The redox potential of GRXs is between −190 and −230 mV, with these proteins being involved in different cellular processes. They behave as electron carriers in the GSH-dependent synthesis of deoxyribonucleotides by ribonucleotide reductase (Lillig et al. 2008) and participate in antioxidant defense, performing reduction of DHA (Holmgren and Aslund 1995) and serving as an electron donor for several PRXs (Meyer et al. 2008) and for MSRB (Tarrago et al. 2009b). GRXs were also found to bind iron–sulfur clusters and to deliver them to enzymes on demand (Rouhier et al. 2008). Moreover, in *Arabidopsis*, GRXs are involved in flower development and salicylic acid signaling (Meyer et al. 2008; Rouhier et al. 2008).

8.2.5 PEROXIREDOXIN

PRXs are a family of thiol-based antioxidant proteins ubiquitously found in nature that catalyze the reduction of peroxides at the expenses of thiol substrates, according to the following general reaction (Banerjee 2008):

$$2\ RSH + R'OOH \rightarrow RSSR + R'OH + H_2O$$

Between the peroxides that can be reduced, a few are of mention: H_2O_2; a wide variety of organic hydroperoxides (from *tert*-butyl hydroperoxide, cumene hydroperoxide, and fatty acid peroxides to complex phosphatidyl choline peroxides); and peroxynitrite radical. PRXs pertain to the group of proteins having the TRX fold and TPXC motif where the active site cysteine residue localizes (Dietz 2003).

From a mechanistic analysis, three classes of PRXs are differentiated: typical 2-CysPRX, atypical 2-CysPRX, and 1-CysPRX (Poole 2007; Rouhier and Jacquot 2002). All these proteins exhibit the same basic catalytic mechanism where the oxidant substrate oxidizes a redox-active cysteine (the peroxidatic cysteine or R-S_PH) forming the sulfenic acid (R-S_POH) derivative. The regeneration process of the oxidized cysteine

is what differentiates the three mechanistic classes. The typical and atypical 2-CysPRXs have a second reactive cysteine, named resolving cysteine (R-S_RH), forming a disulfide bond with the R-S_POH (to generate R-S_P-S_R-R) before reduction. In typical 2-CysPRX, the R-S_RH is on a partner subunit with respect to the R-S_PH (Dietz et al. 2006) and gives rise to an intersubunit disulfide, whereas in the atypical 2-CysPRX, both cysteine residues (the R-S_RH and the R-S_PH) pertain to the same monomer, forming an intrasubunit disulfide (Smirnoff 2005). In the 1-CysPRX, the disulfide bond formation prior to reduction is bypassed (Banerjee 2008). The mechanism for recycling 2-CysPRX involves TRX, GRX, or another CXXC-containing redox module protein (Dietz 2003) such as CDSP32 (Rey et al. 2005), TRXRC (Kirchsteiger et al. 2009), or cyclophilin (Dietz 2007). The ways of 1-Cys PRX reduction are still quite unclear, although several studies demonstrated that both ascorbate and GSH act as reducing substrates (Dietz 2003).

The *A. thaliana* genome contains 10 *prx* genes belonging to four different groups: one 1-CysPRX, one typical 2-CysPRX, and two atypical 2CysPRXs (peroxiredoxin Q [PRXQ] and type II PRX) (Smirnoff 2005). Some of the PRX transcripts have been detected in all analyzed plant tissues, though the expression of each individual member of the *prx* gene family may differ significantly. The accumulation of transcripts encoding chloroplastidic PRX shows a correlation with chlorophyll both in tissue distribution and during leaf development (Dietz 2003). It has been established that PRXs play roles alternative to the antioxidant functions previously ascribed to these and other redox proteins (Rouhier and Jacquot 2002). Numerous works evidenced that H_2O_2 is an abundant and ubiquitous molecule serving as a signal to regulate important cellular functions (Neill et al. 2002; Noctor 2006; Vranova et al. 2002). The activity of PRXs is central to the regulation of such functions since it modulates levels of the peroxide. It has been suggested that modulation of the peroxidase and chaperone activities of PRXs is relevant for the H_2O_2-mediated signal transduction in plants (Hall et al. 2009). Moreover, it has been established that the activity and other properties of some PRXs (particularly the typical 2-CysPRX) are regulated by different modifications such as overoxidation (Banerjee 2008), nitrosylation (Romero-Puertas et al. 2007), and/or phosphorylation (Aran et al. 2009). Therefore, PRXs are key proteins for modulation of reactive oxygen species regulatory networks and critically determines the intensity of the response.

8.2.6 SUPEROXIDE DISMUTASE

Another enzymatic machinery that is relevant as an antioxidant defense in photosynthetic cells is represented by superoxide dismutase (SOD), a metalloprotein that catalyzes dismutation of superoxide anion $\left(O_2^-\right)$ into O_2 and H_2O_2 (Smirnoff 2005). Even when in biological systems, O_2^- can undergo a fast dismutation (velocity ~10^5 M^{-1} s^{-1} at pH 7.0); the very high turnover number exhibited by SOD increases the process velocity by four orders of magnitude. In fact, ultimately, the reaction becomes limited only by the collision frequency between the enzyme and the substrate (Banerjee 2008). These characteristics determine that, *in vivo*, SOD stands for the first line of defense against ROS, protecting cells from oxidative destruction (Hancock 1997). The biological importance of SOD is established mainly because the enzyme activity minimizes the fast reaction of O_2^- with other biological radicals, which generates more toxic species (for example, the reaction between O_2^- and nitric oxide produces nitrite peroxide) (Valderrama et al. 2007).

Different forms of SOD, distinguished by the metal cofactor specificity (CuZnSOD, MnSOD, and FeSOD), are found in plants. The CuZnSOD variant is a homodimer of a molecular mass of ~32 kDa containing one Cu^{2+} and one Zn^{2+} per subunit (Tainer et al. 1983) and localized in the cytosol as well as organelles (Hart et al. 1999). In this enzyme form, Cu^{2+} belongs to the catalytic center, while Zn^{2+} plays a role of structural support (Hart et al. 1999). MnSOD is a 23 kDa homotetrameric protein that is found in the mitochondrial matrix in animals and plants (Jackson et al. 1978). FeSODs are constitutive enzymes that can be found in plastids (Bridges and Salin 1981). There are three FeSODs in *A. thaliana*: FeSOD1, FeSOD2, and FeSOD3. FeSOD2 and FeSOD3 are arranged in a heteromeric protein complex in the chloroplast nucleoids (Myouga et al. 2008). This heterocomplex defends chloroplast nucleoids against oxidative stress and is essential for chloroplast development. In addition, the overexpression of FeSOD modifies the regulation of photosynthesis at low CO_2 partial pressures or following exposure to the prooxidant herbicide methyl viologen (Arisi et al. 1998).

8.2.7 GLUTATHIONE S-TRANSFERASE

GSTs catalyze the nucleophilic attack of GSH on different electrophilic substrates (Smirnoff 2005). This enzyme plays different physiological roles such as detoxification of xenobiotics, catalysis for specific reactions in a number of biosynthetic and catabolic pathways, and defense against oxidative stress by reducing ROOH and DHA (Banerjee 2008). It has been determined that some GSTs are active as GPXs and also that these two proteins interact, forming a complex, although the possible involvement of the latter complex in reactions related to stress signaling is uncertain at the present time (Noctor and Foyer 1998). All GSTs so far described form homodimers of 50 kDa, and in *Arabidopsis*, six major groups of the enzyme were identified: φ-, τ-, ζ-, ω-, λ-, and θ-GST (Dixon et al. 2002). Although the six types of GSTs are able to function as dehydroascorbic acid reductase (DHAR), this activity was particularly recognized for the ω and λ forms. The ω-GST is a DHAR protein with an active site determined by the CPFC/S motif (Dixons et al. 2002; Smirnoff 2005), and thus, it regenerates ascorbate from DHA using GSH as an electron donor (Figure 8.1). Between λ-GSTs, two members are distinguishable because they lack DHAR activity but are active as GSH-dependent thiol transferases, suggesting that they could function in dethiolation of *S*-glutathionylated proteins accumulated under oxidative stress situations (Dixon et al. 2002; Rouhier et al. 2008).

8.2.8 ASCORBATE PEROXIDASE

APX is a hemoprotein that catalyzes the ascorbate-associated reduction of H_2O_2:

$$Ascorbate + H_2O_2 \rightarrow dehydroascorbate + 2\ H_2O$$

This is a unique enzyme found mainly in plants, algae, and some protozoa (Noctor and Foyer 1998; Wilkinson et al. 2002). In plants, the enzyme is present in almost every cell and compartment and is actively involved in the detoxification of H_2O_2 as part of the ascorbate–glutathione or Asada–Halliwell–Foyer pathway (Hiner et al. 2002). Different genes coding for APXs are identified in *Arabidopsis*, with four of them corresponding to chloroplastic isoenzymes (Kubo et al. 1992). Two stromal APXs are dually targeted to the stroma and the mitochondrial intermembrane space (Smirnoff 2000), two are localized in cytoplasm (Santos et al. 1996), whereas the other two are proposed to be targeted to peroxisomes and glyoxysomes (Hoshi and Heinemann 2001; Nito et al. 2001). Finally, there are two microsomal APXs that might bind to the external surface of glyoxysomes or be transported into peroxisomes (Lisenbee et al. 2003). Three forms of chloroplastic APX were identified: thylakoid, soluble (localized in the lumen), and stromal APX (Smirnoff 2000). APX function is dependent on the bioavailability of ascorbate and, in some cases, GSH (Smirnoff 2005). Cellular pools of these antioxidants are maintained in reduced form by enzymatic systems such as GR, DHAR, GRX, and GST (Buchanan and Balmer 2005). In addition, the enzyme presents higher affinity for H_2O_2 than for organic peroxides, making it a suitable candidate to participate in the removal of the former for signaling purposes (Apel and Hirt 2004; Reddy et al. 2009). Studies with transgenic plants have demonstrated that APX is an important defense enzyme implicated in the elimination of H_2O_2 (Smirnoff 2005). APX functions as component of the H_2O_2 regulation network of plants, controlling levels of the peroxide used for cellular signaling (Pauly et al. 2006). Thus, in concert with other H_2O_2 regulation enzymes of the cell, APX balances different cellular systems that generate H_2O_2 in plants (for example, NADPH oxidase in pathogen response) and also controls levels of H_2O_2 used for signaling during biotic or abiotic stress (Apel and Hirt 2004; Kotchoni and Gachomo 2006).

8.3 SYSTEMS FOR REPAIRING OXIDATIVE DAMAGE IN PLANTS

Reactive species of oxygen (ROS) or of nitrogen (RNS) can modify different macromolecules to generate cellular damage (Moskovitz 2005). For this reason, cells have many protective systems helping to eliminate or minimize injurious molecular species or to repair the oxidative damage caused on cellular components (Banerjee 2008). Particularly, oxidation of proteins can generate conformational changes and, in some cases, loss of function (Friguet 2006). The amino acids most susceptible to oxidation are histidine, tryptophan, tyrosine, cysteine, and methionine, the latter two being the most sensitive (Moller et al. 2007). Oxidation of methionine to methionine sulfoxide (MetSO) produces a mixture of epimers on the sulfur atom (Met(S)SO and Met(R)SO) (Boschi-Muller et al. 2008), affecting the biological function of the oxidized protein (Friguet 2006). Consequently, the occurrence of a system for reducing MetSO to Met does matter in the cell. Such a reaction is catalyzed by the enzyme MSR, which is present in almost all organisms (Boschi-Muller et al. 2008). The enzyme protects the organism from oxidative damage, being essentially involved in resistance against abiotic and biotic stress in plants and animals (Bechtold et al. 2004; Moskovitz et al. 1998).

At present, two nonrelated classes of MSRs were described in several organisms: MSRA, which is stereospecific to the *S* isomer of the sulfur atom in the sulfoxide group (Boschi-Muller et al. 2008), and MSRB, which is specific to the *R* isomer (Boschi-Muller et al. 2008). MSRA and MSRB share a similar reaction mechanism (Boschi-Muller et al. 2008; Kauffmann et al. 2005), where the catalytic cysteine thiolate (situated in the N-terminal part of MSRA and the C-terminal portion of MSRB) first attacks the sulfoxide and releases methionine with generation of a sulfenic acid intermediate on the cysteine moiety. It follows the attack by the resolving cysteine on the sulfenic acid intermediate to generate an intramolecular disulfide bond. At last, TRX (or other electron donors such as GRX or GSH) reduces the intramolecular disulfide by restoring MSR (Tarrago et al. 2009b). Most MSRBs also contain a single zinc atom that supports a structural role in these proteins (Kauffmann et al. 2005). In addition, some MSRBs lack the resolving cysteine, suggesting that the sulfenic acid intermediate may be directly reduced by TRX or GSH (linking to GRX) (Vieira Dos Santos et al. 2007) or CDSP32 (Rey et al. 2005).

In plants, multiple forms of MSRA and MSRB have been identified with localization in different cellular compartments (Tarrago et al. 2009a), which makes it so that reduction of oxidized methionine residues in proteins is required independently in different subcellular places (Rouhier et al. 2006, 2007). The functional importance of this process was evidenced after results showing that the overexpression of MSRB2 in *A. thaliana* transgenic plants generated enhanced tolerance to cellular oxidative damage during long nights (Bechtold et al. 2004; Romero et al. 2004). In addition, MSRB7 reverses the oxidation of glutathione S-transferase F2/3 (GSTF2/3) and confers tolerance of *A. thaliana* to oxidative stress (Lee et al. 2014). The MSRB2 protein is a defense regulator against oxidative stress and pathogen attack (via the regulation of cell redox status) in pepper and tomato plants (Oh et al. 2010).

There has been evidence of the existence of a mechanism by which plants can perceive when H_2O_2 is associated with methionine (Emes 2009). The data indicate that chemical oxidation of methionine residues by H_2O_2 takes place at key hydrophobic positions within canonical phosphorylation motifs in numerous enzymes, which inhibits protein kinases binding. This modification of the kinases (including

the calcium-dependent protein kinase and adenosine monophosphate [AMP]-activated protein kinase families) can be reversed by MSR *in vivo* (Hardin et al. 2009). This latter mechanism is directly linked to oxidative signals through changes in protein phosphorylation, thus enhancing the action of redox signalling in plants.

8.4 CONCLUDING REMARKS

Plants, like all aerobic organisms, benefit from the redox potential of oxygen, using it in metabolic pathways. However, the use of oxygen also produces problems associated with deleterious effects of oxidant conditions. Antioxidants and ROS/reactive nitrogen species (RNS)/reactive sulphur species (RSS) are important interacting systems with different functions in higher plants, which ensure high redox flexibility to the organism. Antioxidants are not passive intermediates in this crosstalk, but rather, they function as critical signal components that constitute a dynamic metabolic interface between cell stress perception and physiological responses (Scandalios 2005). Normal metabolic conditions produce ROS and RNS as by-products, and to neutralize their undesirable effects, plants developed effective defense mechanisms, including antioxidant enzymes and free radical scavengers. Under certain situations, like biotic or abiotic stress, as well as over the course of development, an induction of ROS and RNS synthesis takes place, these act as signaling molecules, with incoming complex downstream effects on both primary and secondary metabolism. Circumstances like this could result in different situations; the cell responds to the ROS/RNS signal, making appropriate adjustments and returning to physiological conditions. Otherwise, ROS/RNS production exceeds the capacity of antioxidant systems, cells are not able to return to homeostasis, oxidative damage prevails, and genetically programmed cell suicide events are triggered.

It is now well established that most cells can adapt to oxidative stress by altering global gene expression patterns, including transcription and translation of genes encoding antioxidants as well as other metabolic enzymes, and/or by posttranslational changes operating on proteins that could modify key regulators of redox responses (Grant 2008). In plants, H_2O_2 has been recognized as a second messenger for signals triggered by ROS, and advances have been made in the understanding of metabolic changes operating in consequence. Also, the role of nitric oxide and the problematic nature of RNS have been clearly evidenced. However, areas related to regulation of gene expression, modulation of enzyme activity, and coordination/redirectioning of metabolic fluxes requiring intensive research to reach a complete and comprehensive view of how higher plants maintain levels of oxidative species under control. The use of postgenomics tools, including proteomics and metabolomics, is coming to be utilized in this field and is critical for advancements in integrative outlooks. The challenge is complex, but finding responses to the different open questions foresees relevant consequences. This is a critical prerequisite in the design and establishment of manageable procedures to improve plant productivity under different environmental scenarios. In an overall view, it is worth taking into account that independently of the different mechanisms to control redox homeostasis and to cope with oxidation, the generation of reducing equivalents in plants is primarily a photosynthetic product. In such a way, photoproduction and partitioning of NADPH is a key master strategy for feeding the different enzymes and proteins involved in redox metabolism.

ACKNOWLEDGMENTS

Work in our laboratory was granted by Agencia Nacional de Promoción de la Ciencia y la Tecnología (ANPCyT), Consejo Nacional de Investigaciones Científicas y Tecnológicas (CONICET), and Universaidad Nacional del Litoral (UNL). Sergio A. Guerrero, Diego G. Arias y Alberto A. Iglesias are career investigators from CONICET.

REFERENCES

Abhilash, P. C., Jamil, S. et al. 2009. Transgenic plants for enhanced biodegradation and phytoremediation of organic xenobiotics. *Biotechnol Adv* 27(4): 474–88.

Alkhalfioui, F., Renard, M. et al. 2007. Unique properties of NADP-thioredoxin reductase C in legumes. *J Exp Bot* 58(5): 969–78.

Anderson, L. E. and Advani, V. R. 1970. Chloroplast and cytoplasmic enzymes: Three distinct isoenzymes associated with the reductive pentose phosphate cycle. *Plant Physiol* 45(5): 583–5.

Anderson, L. E., Ng, T. C. et al. 1974. Inactivation of pea leaf chloroplastic and cytoplasmic glucose 6-phosphate dehydrogenases by light and dithiothreitol. *Plant Physiol* 53(6): 835–9.

Apel, K. and Hirt, H. 2004. Reactive oxygen species: Metabolism, oxidative stress, and signal transduction. *Annu Rev Plant Biol* 55: 373–99.

Aran, M., Ferrero, D. S. et al. 2009. Typical 2-Cys peroxiredoxins—Modulation by covalent transformations and noncovalent interactions. *FEBS J* 276(9): 2478–93.

Arisi, A. C., Cornic, G. et al. 1998. Overexpression of iron superoxide dismutase in transformed poplar modifies the regulation of photosynthesis at low CO_2 partial pressures or following exposure to the prooxidant herbicide methyl viologen. *Plant Physiol* 117(2): 565–74.

Arnon, D. I., Rosenberg, L. L. et al. 1954. A new glyceraldehyde phosphate dehydrogenase from photosynthetic tissues. *Nature* 173: 1132–4.

Arrigoni, O. 1994. Ascorbate system in plant development. *J Bioenerg Biomembr* 26(4): 407–19.

Averill, R. H., Bailey-Serres, J. et al. 1998. Co-operation between cytosolic and plastidic oxidative pentose phosphate pathways revealed by 6-phosphogluconate dehydrogenase-deficient genotypes of maize. *Plant J* 14(4): 449–57.

Bailey-Serres, J. and Nguyen, M. T. 1992. Purification and characterization of cytosolic 6-phosphogluconate dehydrogenase isozymes from maize. *Plant Physiol* 100(3): 1580–3.

Bailey-Serres, J., Tom, J. et al. 1992. Expression and distribution of cytosolic 6-phosphogluconate dehydrogenase isozymes in maize. *Biochem Genet* 30(5–6): 233–46.

Banerjee, R. 2008. *Redox Biochemistry*. Hoboken, NJ: Wiley-Interscience.

Bechtold, U., Murphy, D. J. et al. 2004. *Arabidopsis* peptide methionine sulfoxide reductase2 prevents cellular oxidative damage in long nights. *Plant Cell* 16(4): 908–19.

Bindoli, A., Fukuto, J. M. et al. 2008. Thiol chemistry in peroxidase catalysis and redox signaling. *Antioxid Redox Signal* 10(9): 1549–64.

Boschi-Muller, S., Gand, A. et al. 2008. The methionine sulfoxide reductases: Catalysis and substrate specificities. *Arch Biochem Biophys* 474(2): 266–73.

Boyd, D. A., Cvitkovitch, D. G. et al. 1995. Sequence, expression, and function of the gene for the nonphosphorylating, NADP-dependent glyceraldehyde-3-phosphate dehydrogenase of Streptococcus mutans. *J Bacteriol* 177(10): 2622–7.

Bridges, S. M. and Salin, M. L. 1981. Distribution of iron-containing superoxide dismutase in vascular plants. *Plant Physiol* 68(2): 275–8.

Brown, A. T. and Wittenberger, C. L. 1971. The occurrence of multiple glyceraldehyde-3-phosphate dehydrogenases in cariogenic streptococci. *Biochem Biophys Res Commun* 43(1): 217–24.

Brownleader, M. D., Harborne, J. B., Dey, P. M. 1997. Carbohydrate metabolism: Primary metabolism of monosaccharides. In *Plant Biochemistry*, eds. Dey, P. M. and Harborne, J. B., 111–41. London: Elsevier.

Brunner, N. A., Brinkmann, H. et al. 1998. NAD$^+$-dependent glyceraldehyde-3-phosphate dehydrogenase from *Termoproteus tenax*. The first identified archeal member of the aldehyde dehydrogenase superfamily is a glycolytic enzyme with unusual regulatory properties. *J Biol Chem* 273: 6149–56.

Buchanan, B. B. 1980. Role of light in the regulation of chloroplast enzymes. *Annu Rev Plant Physiol* 31(1): 341–74.

Buchanan, B. B. and Balmer, Y. 2005. Redox regulation: A broadening horizon. *Annu Rev Plant Biol* 56: 187–220.

Bustos, D. M. and Iglesias, A. A. 2002. Non-phosphorylating glyceraldehyde-3-phosphate dehydrogenase is post-translationally phosphorylated in heterotrophic cells of wheat (*Triticum aestivum*). *FEBS Lett* 530(1–3): 169–73.

Bustos, D. M. and Iglesias, A. A. 2003. Phosphorylated non-phosphorylating glyceraldehyde-3-phosphate dehydrogenase from heterotrophic cells of wheat interacts with 14-3-3 proteins. *Plant Physiol* 133(4): 2081–8.

Bustos, D. M. and Iglesias, A. A. 2005. A model for the interaction between plant GAPN and 14-3-3zeta using protein–protein docking calculations, electrostatic potentials and kinetics. *J Mol Graph Model* 23(6): 490–502.

Bustos, D. M., Bustamante, C. A. et al. 2008. Involvement of non-phosphorylating glyceraldehyde-3-phosphate dehydrogenase in response to oxidative stress. *J Plant Physiol* 165(4): 456–61.

Cheng, Y. and Long, M. 2007. A cytosolic NADP-malic enzyme gene from rice (*Oryza sativa* L.) confers salt tolerance in transgenic *Arabidopsis*. *Biotechnol Lett* 29(7): 1129–34.

Cortassa, S., Aon, M. A. et al. 2012. *An Introduction to Metabolic and Cellular Engineering*, 2nd Edition. New York: World Scientific Publishing Co. Pte. Ltd.

Dalle-Donne, I., Rossi, R. et al. 2007. S-glutathionylation in protein redox regulation. *Free Radic Biol Med* 43(6): 883–98.

Dawson, C. R., Strothkamp, K. G. et al. 1975. Ascorbate oxidase and related copper proteins. *Ann N Y Acad Sci* 258: 209–20.

Debnam, P. M. and Emes, M. J. 1999. Subcellular distribution of enzymes of the oxidative pentose phosphate pathway in root and leaf tissues. *J Exp Bot* 50(340): 1653–61.

Detarsio, E., Maurino, V. G. et al. 2008. Maize cytosolic NADP-malic enzyme (ZmCytNADP-ME): A phylogenetically distant isoform specifically expressed in embryo and emerging roots. *Plant Mol Biol* 68(4–5): 355–67.

Dietz, K. J. 2003. Plant peroxiredoxins. *Annu Rev Plant Biol* 54: 93–107.

Dietz, K. J. 2007. The dual function of plant peroxiredoxins in antioxidant defence and redox signaling. *Subcell Biochem* 44: 267–94.

Dietz, K. J., Jacob, S. et al. 2006. The function of peroxiredoxins in plant organelle redox metabolism. *J Exp Bot* 57(8):
1697–709.

Dixon, D. P., Davis, B. G. et al. 2002. Functional divergence in the
glutathione transferase superfamily in plants. Identification of
two classes with putative functions in redox homeostasis in
Arabidopsis thaliana. J Biol Chem 277(34): 30859–69.

Drincovich, M. F., Casati, P. et al. 2001. NADP-malic enzyme from
plants: A ubiquitous enzyme involved in different metabolic
pathways. *FEBS Lett* 490(1–2): 1–6.

Eicks, M., Maurino, V. et al. 2002. The plastidic pentose phosphate
translocator represents a link between the cytosolic and the
plastidic pentose phosphate pathways in plants. *Plant Physiol*
128(2): 512–22.

Emes, M. J. 2009. Oxidation of methionine residues: The missing link
between stress and signalling responses in plants. *Biochem J*
422(2): e1–2.

Fourrat, L., Iddar, A. et al. 2007. Cloning, gene expression and
characterization of a novel bacterial NAD-dependent non-
phosphorylating glyceraldehyde-3-phosphate dehydrogenase
from Neisseria meningitidis strain Z2491. *Mol Cell Biochem*
305(1–2): 209–19.

Friguet, B. 2006. Oxidized protein degradation and repair in ageing
and oxidative stress. *FEBS Lett* 580(12): 2910–16.

Gao, Z. and Loescher, W. H. 2000. NADPH supply and mannitol
biosynthesis. Characterization, cloning, and regulation of the
non-reversible glyceraldehyde-3-phosphate dehydrogenase in
celery leaves. *Plant Physiol* 124(1): 321–30.

Gelhaye, E., Rouhier, N. et al. 2005. The plant thioredoxin system.
Cell Mol Life Sci 62(1): 24–35.

Gomez Casati, D. F., Sesma, J. I. et al. 2000. Structural and kinetic
characterization of NADP-dependent, non-phosphorylating
glyceraldehyde-3-phosphate dehydrogenase from celery
leaves. *Plant Sci* 154(2): 107–15.

Grant, C. M. 2008. Metabolic reconfiguration is a regulated response
to oxidative stress. *J Biol* 7(1): 1.

Habenicht, A. 1997. The non-phosphorylating glyceraldehyde-
3-phosphate dehydrogenase: Biochemistry, structure, occurrence and evolution. *Biol Chem* 378(12): 1413–19.

Hall, A., Karplus, P. A. et al. 2009. Typical 2-Cys peroxiredoxins—
Structures, mechanisms and functions. *FEBS J* 276(9): 2469–77.

Halliwell, B. 2006. Reactive species and antioxidants. Redox biology is a fundamental theme of aerobic life. *Plant Physiol*
141(2): 312–22.

Hancock, J. T. 1997. Superoxide, hydrogen peroxide and nitric oxide
as signalling molecules: Their production and role in disease.
Br J Biomed Sci 54(1): 38–46.

Hardin, S. C., Larue, C. T. et al. 2009. Coupling oxidative signals to protein phosphorylation via methionine oxidation in
Arabidopsis. Biochem J 422(2): 305–12.

Hart, P. J., Balbirnie, M. M. et al. 1999. A structure-based mechanism for copper–zinc superoxide dismutase. *Biochemistry*
38(7): 2167–78.

Hauschild, R. and von Schaewen, A. 2003. Differential regulation
of glucose-6-phosphate dehydrogenase isoenzyme activities in
potato. *Plant Physiol* 133(1): 47–62.

Hiner, A. N., Raven, E. L. et al. 2002. Mechanisms of compound I
formation in heme peroxidases. *J Inorg Biochem* 91(1): 27–34.

Holmgren, A. and Aslund, F. 1995. Glutaredoxin. *Methods Enzymol*
252: 283–92.

Hoshi, T. and Heinemann, S. 2001. Regulation of cell function by
methionine oxidation and reduction. *J Physiol* 531(Pt 1): 1–11.

Hudák, J. 1997. Chapter 2: Photosynthetic apparatus. In *Handbook of
Photosynthesis*, ed. Pressarakli, M. New York: Marcel Dekker.

Hutchings, D., Rawsthorne, S. et al. 2005. Fatty acid synthesis and the
oxidative pentose phosphate pathway in developing embryos of
oilseed rape (*Brassica napus* L.). *J Exp Bot* 56(412): 577–85.

Iddar, A., Valverde, F. et al. 2002. Expression, purification, and
characterization of recombinant nonphosphorylating NADP-
dependent glyceraldehyde-3-phosphate dehydrogenase from
Clostridium acetobutylicum. Protein Expr Purif 25(3): 519–26.

Iddar, A., Valverde, F. et al. 2005. Widespread occurrence of non-
phosphorylating glyceraldehyde-3-phosphate dehydrogenase
among gram-positive bacteria. *Int Microbiol* 8(4): 251–8.

Iglesias, A. A. 1989. On the metabolism of triose-phosphates in photosynthetic cells. Their involvement on the traffic of ATP and
NADPH. *Biochem Educ* 18: 2–5.

Iglesias, A. A. and Losada, M. 1988. Purification and kinetic and
structural properties of spinach leaf NADP-dependent non-
phosphorylating glyceraldehyde-3-phosphate dehydrogenase.
Arch Biochem Biophys 260(2): 830–40.

Iglesias, A. A. and Podestá, F. E. 2005. Photosynthate formation and
partitioning in crop plants. In *Handbook of Photosynthesis*,
2nd Edition, Pessarakli, M., 525–45. Boca Raton, FL: CRC
Press.

Iglesias, A. A., Serrano, A. et al. 1987. Purification and properties
of NADP-dependent glyceraldehyde-3-phosphate dehydrogenase from green alga *Chlamydomonas reinhardtii. Biochem
Biophys Acta* 925: 1–10.

Jackson, C., Dench, J. et al. 1978. Subcellular localisation and identification of superoxide dismutase in the leaves of higher plants.
Eur J Biochem 91(2): 339–44.

Jacob, J. L. and D'Auzac, J. 1972. [Glyceraldehyde-3-phosphate
dehydrogenase from the latex of Hevea brasiliensis. Comparative
study with its phosphorylating homologue]. *Eur J Biochem*
31(2): 255–65.

Kauffmann, B., Aubry, A. et al. 2005. The three-dimensional structures of peptide methionine sulfoxide reductases: Current
knowledge and open questions. *Biochim Biophys Acta* 1703(2):
249–60.

Kelly, G. J. and Gibbs, M. 1973. A mechanism for the indirect transfer of photosynthetically reduced nicotinamide adenine dinucleotide phosphate from chloroplasts to the cytoplasm. *Plant
Physiol* 52(6): 674–6.

Kirchsteiger, K., Pulido, P. et al. 2009. NADPH thioredoxin reductase C controls the redox status of chloroplast 2-Cys peroxiredoxins in *Arabidopsis thaliana. Mol Plant* 2(2): 298–307.

Kotchoni, S. O. and Gachomo, E. W. 2006. The reactive oxygen species network pathways: An essential prerequisite for perception of pathogen attack and the acquired disease resistance in
plants. *J Biosci* 31(3): 389–404.

Kruger, N. J. and von Schaewen, A. 2003. The oxidative pentose
phosphate pathway: Structure and organisation. *Curr Opin
Plant Biol* 6(3): 236–46.

Kubo, A., Saji, H. et al. 1992. Cloning and sequencing of a cDNA
encoding ascorbate peroxidase from *Arabidopsis thaliana.
Plant Mol Biol* 18(4): 691–701.

Kulinskii, V. I. and Kolesnichenko, L. S. 2009. [Glutathione system.
I. Synthesis, transport, glutathione transferases, glutathione
peroxidases]. *Biomed Khim* 55(3): 255–77.

Lee, S. H., Li, C. W. et al. 2014. MSRB7 reverses oxidation of
GSTF2/3 to confer tolerance of *Arabidopsis thaliana* to oxidative stress. *J Exp Bot* 65(17): 5049–62.

Lillig, C. H., Berndt, C. et al. 2008. Glutaredoxin systems. *Biochim
Biophys Acta* 1780(11): 1304–17.

Linster, C. L. and Clarke, S. G. 2008. L-Ascorbate biosynthesis in
higher plants: The role of VTC2. *Trends Plant Sci* 13(11):
567–73.

Lisenbee, C. S., Heinze, M. et al. 2003. Peroxisomal ascorbate peroxidase resides within a subdomain of rough endoplasmic reticulum in wild-type *Arabidopsis* cells. *Plant Physiol* 132(2): 870–82.

Liu, S., Cheng, Y. et al. 2007. Expression of an NADP-malic enzyme gene in rice (*Oryza sativa* L.) is induced by environmental stresses; over-expression of the gene in *Arabidopsis* confers salt and osmotic stress tolerance. *Plant Mol Biol* 64(1–2): 49–58.

Marchal, S., Cobessi, D. et al. 2001. Chemical mechanism and substrate binding sites of NADP-dependent aldehyde dehydrogenase from Streptococcus mutans. *Chem Biol Interact* 130–132(1–3): 15–28.

Martin, J. L. 1995. Thioredoxin—A fold for all reasons. *Structure* 3(3): 245–50.

Mateos, M. I. and Serrano, A. 1992. Occurrence of phosphorylating and non-phosphorylating NADP+-dependent glyceraldehyde-3-phosphate dehydrogenases in photosynthetic organisms. *Plant Sci* 84(2): 163–70.

Maurino, V. G., Drincovich, M. F. et al. 1996. NADP-malic enzyme isoforms in maize leaves. *Biochem Mol Biol Int* 38(2): 239–50.

Maurino, V. G., Saigo, M. et al. 2001. Non-photosynthetic 'malic enzyme' from maize: A constituvely expressed enzyme that responds to plant defence inducers. *Plant Mol Biol* 45(4): 409–20.

Meyer, A. J. 2008. The integration of glutathione homeostasis and redox signaling. *J Plant Physiol* 165(13): 1390–403.

Meyer, Y., Reichheld, J. P. et al. 2005. Thioredoxins in *Arabidopsis* and other plants. *Photosynth Res* 86(3): 419–33.

Meyer, Y., Siala, W. et al. 2008. Glutaredoxins and thioredoxins in plants. *Biochim Biophys Acta* 1783(4): 589–600.

Moller, I. M., Jensen, P. E. et al. 2007. Oxidative modifications to cellular components in plants. *Annu Rev Plant Biol* 58: 459–81.

Montrichard, F., Alkhalfioui, F. et al. 2009. Thioredoxin targets in plants: The first 30 years. *J Proteomics* 72(3): 452–74.

Moon, J. C., Jang, H. H. et al. 2006. The C-type *Arabidopsis* thioredoxin reductase ANTR-C acts as an electron donor to 2-Cys peroxiredoxins in chloroplasts. *Biochem Biophys Res Commun* 348(2): 478–84.

Moskovitz, J. 2005. Methionine sulfoxide reductases: Ubiquitous enzymes involved in antioxidant defense, protein regulation, and prevention of aging-associated diseases. *Biochim Biophys Acta* 1703(2): 213–19.

Moskovitz, J., Flescher, E. et al. 1998. Overexpression of peptide-methionine sulfoxide reductase in Saccharomyces cerevisiae and human T cells provides them with high resistance to oxidative stress. *Proc Natl Acad Sci U S A* 95(24): 14071–5.

Myouga, F., Hosoda, C. et al. 2008. A heterocomplex of iron superoxide dismutases defends chloroplast nucleoids against oxidative stress and is essential for chloroplast development in *Arabidopsis*. *Plant Cell* 20(11): 3148–62.

Neill, S. J., Desikan, R. et al. 2002. Hydrogen peroxide and nitric oxide as signalling molecules in plants. *J Exp Bot* 53(372): 1237–47.

Nelson, D. L. and Cox, M. M. 2004. *Lehninger Principles of Biochemistry*. New York: W. H. Freeman.

Neuhaus, H. E. and Emes, M. J. 2000. Nonphotosynthetic metabolism in plastids. *Annu Rev Plant Physiol Plant Mol Biol* 51: 111–40.

Nito, K., Yamaguchi, K. et al. 2001. Pumpkin peroxisomal ascorbate peroxidase is localized on peroxisomal membranes and unknown membranous structures. *Plant Cell Physiol* 42(1): 20–7.

Noctor, G. 2006. Metabolic signalling in defence and stress: The central roles of soluble redox couples. *Plant Cell Environ* 29(3): 409–25.

Noctor, G. and Foyer, C. H. 1998. Ascorbate and glutathione: Keeping active oxygen under control. *Annu Rev Plant Physiol Plant Mol Biol* 49: 249–79.

Oh, S. K., Baek, K. H. et al. 2010. CaMsrB2, pepper methionine sulfoxide reductase B2, is a novel defense regulator against oxidative stress and pathogen attack. *Plant Physiol* 154(1): 245–61.

Pauly, N., Pucciariello, C. et al. 2006. Reactive oxygen and nitrogen species and glutathione: Key players in the legume–Rhizobium symbiosis. *J Exp Bot* 57(8): 1769–76.

Pignocchi, C. and Foyer, C. H. 2003. Apoplastic ascorbate metabolism and its role in the regulation of cell signalling. *Curr Opin Plant Biol* 6(4): 379–89.

Plaxton, W. C. 1996. The organization and regulation of plant glycolysis. *Annu Rev Plant Physiol Plant Mol Biol* 47: 185–214.

Poole, L. B. 2007. The catalytic mechanism of peroxiredoxins. *Subcell Biochem* 44: 61–81.

Pupillo, P. and Faggiani, R. 1979. Subunit structure of three glyceraldehyde 3-phosphate dehydrogenases of some flowering plants. *Arch Biochem Biophys* 194(2): 581–92.

Reddy, R. A., Kumar, B. et al. 2009. Molecular cloning and characterization of genes encoding Pennisetum glaucum ascorbate peroxidase and heat-shock factor: Interlinking oxidative and heat-stress responses. *J Plant Physiol* 166(15): 1646–59.

Rey, P., Cuine, S. et al. 2005. Analysis of the proteins targeted by CDSP32, a plastidic thioredoxin participating in oxidative stress responses. *Plant J* 41(1): 31–42.

Rius, S. P., Casati, P. et al. 2006. Characterization of an *Arabidopsis thaliana* mutant lacking a cytosolic non-phosphorylating glyceraldehyde-3-phosphate dehydrogenase. *Plant Mol Biol* 61(6): 945–57.

Rochaix, J. D. 2011. Reprint of: Regulation of photosynthetic electron transport. *Biochim Biophys Acta* 1807(8): 878–86.

Romero, H. M., Berlett, B. S. et al. 2004. Investigations into the role of the plastidial peptide methionine sulfoxide reductase in response to oxidative stress in *Arabidopsis*. *Plant Physiol* 136(3): 3784–94.

Romero-Puertas, M. C., Laxa, M. et al. 2007. S-nitrosylation of peroxiredoxin II E promotes peroxynitrite-mediated tyrosine nitration. *Plant Cell* 19(12): 4120–30.

Rouhier, N. and Jacquot, J. P. 2002. Plant peroxiredoxins: Alternative hydroperoxide scavenging enzymes. *Photosynth Res* 74(3): 259–68.

Rouhier, N., Gelhaye, E. et al. 2004. Plant glutaredoxins: Still mysterious reducing systems. *Cell Mol Life Sci* 61(11): 1266–77.

Rouhier, N., Vieira Dos Santos, C. et al. 2006. Plant methionine sulfoxide reductase A and B multigenic families. *Photosynth Res* 89(2–3): 247–62.

Rouhier, N., Kauffmann, B. et al. 2007. Functional and structural aspects of poplar cytosolic and plastidial type a methionine sulfoxide reductases. *J Biol Chem* 282(5): 3367–78.

Rouhier, N., Lemaire, S. D. et al. 2008. The role of glutathione in photosynthetic organisms: Emerging functions for glutaredoxins and glutathionylation. *Annu Rev Plant Biol* 59: 143–66.

Rumpho, M. E., Edwards, G. E. et al. 1983. A pathway for photosynthetic carbon flow to mannitol in celery leaves: Activity and localization of key enzymes. *Plant Physiol* 73(4): 869–73.

Rybus-Kalinowska, B., Zwirska-Korczala, K. et al. 2009. [Activity of antioxidative enzymes and concentration of malondialdehyde as oxidative status markers in women with non-auto-immunological subclinical hyperthyroidism]. *Endokrynol Pol* 60(3): 199–202.

Santos, M., Gousseau, H. et al. 1996. Cytosolic ascorbate peroxidase from *Arabidopsis thaliana* L. is encoded by a small multigene family. *Planta* 198(1): 64–9.

Scandalios, J. G. 2005. Oxidative stress: Molecular perception and transduction of signals triggering antioxidant gene defenses. *Braz J Med Biol Res* 38(7): 995–1014.

Scheibe, R. 1991. Redox-modulation of chloroplast enzymes: A common principle for individual control. *Plant Physiol* 96(1): 1–3.

Schnarrenberger, C., Oeser, A. et al. 1973. Two isoenzymes each of glucose-6-phosphate dehydrogenase and 6-phosphogluconate dehydrogenase in spinach leaves. *Arch Biochem Biophys* 154(1): 438–48.

Schnarrenberger, C., Tetour, M. et al. 1975. Development and intracellular distribution of enzymes of the oxidative pentose phosphate cycle in radish cotyledons. *Plant Physiol* 56(6): 836–40.

Schnarrenberger, C., Flechner, A. et al. 1995. Enzymatic evidence for a complete oxidative pentose phosphate pathway in chloroplasts and an incomplete pathway in the cytosol of spinach leaves. *Plant Physiol* 108(2): 609–14.

Schurmann, P. and Jacquot, J. P. 2000. Plant thioredoxin systems revisited. *Annu Rev Plant Physiol Plant Mol Biol* 51: 371–400.

Shigeoka, S., Ishikawa, T. et al. 2002. Regulation and function of ascorbate peroxidase isoenzymes. *J Exp Bot* 53(372): 1305–19.

Smirnoff, N. 2000. Ascorbate biosynthesis and function in photoprotection. *Philos Trans R Soc Lond B Biol Sci* 355(1402): 1455–64.

Smirnoff, N. 2005. *Antioxidants and Reactive Oxygen Species in Plants*. Oxford: Blackwell Publishing.

Smirnoff, N. and Pallanca, J. E. 1996. Ascorbate metabolism in relation to oxidative stress. *Biochem Soc Trans* 24(2): 472–8.

Smirnoff, N. and Wheeler, G. L. 2000. Ascorbic acid in plants: Biosynthesis and function. *Crit Rev Biochem Mol Biol* 35(4): 291–314.

Tainer, J. A., Getzoff, E. D. et al. 1983. Structure and mechanism of copper, zinc superoxide dismutase. *Nature* 306(5940): 284–7.

Tarrago, L., Laugier, E. et al. 2009a. Protein-repairing methionine sulfoxide reductases in photosynthetic organisms: Gene organization, reduction mechanisms, and physiological roles. *Mol Plant* 2(2): 202–17.

Tarrago, L., Laugier, E. et al. 2009b. Regeneration mechanisms of *Arabidopsis thaliana* methionine sulfoxide reductases B by glutaredoxins and thioredoxins. *J Biol Chem* 284(28): 18963–71.

Valderrama, R., Corpas, F. J. et al. 2007. Nitrosative stress in plants. *FEBS Lett* 581(3): 453–61.

Vieira Dos Santos, C., Laugier, E. et al. 2007. Specificity of thioredoxins and glutaredoxins as electron donors to two distinct classes of *Arabidopsis* plastidial methionine sulfoxide reductases B. *FEBS Lett* 581(23): 4371–6.

Vranova, E., Inze, D. et al. 2002. Signal transduction during oxidative stress. *J Exp Bot* 53(372): 1227–36.

Wakao, S. and Benning, C. 2005. Genome-wide analysis of glucose-6-phosphate dehydrogenases in *Arabidopsis*. *Plant J* 41(2): 243–56.

Wakao, S., Andre, C. et al. 2008. Functional analyses of cytosolic glucose-6-phosphate dehydrogenases and their contribution to seed oil accumulation in *Arabidopsis*. *Plant Physiol* 146(1): 277–88.

Weber, A. P., Schwacke, R. et al. 2005. Solute transporters of the plastid envelope membrane. *Annu Rev Plant Biol* 56: 133–64.

Wenderoth, I., Scheibe, R. et al. 1997. Identification of the cysteine residues involved in redox modification of plant plastidic glucose-6-phosphate dehydrogenase. *J Biol Chem* 272(43): 26985–90.

Wendt, U. K., Hauschild, R. et al. 1999. Evidence for functional convergence of redox regulation in G6PDH isoforms of cyanobacteria and higher plants. *Plant Mol Biol* 40(3): 487–94.

Wilkinson, S. R., Obado, S. O. et al. 2002. Trypanosoma cruzi expresses a plant-like ascorbate-dependent hemoperoxidase localized to the endoplasmic reticulum. *Proc Natl Acad Sci U S A* 99(21): 13453–8.

9 Regulation of Chlorophyll Metabolism in Plants

Koichi Kobayashi and Tatsuru Masuda

CONTENTS

9.1   Introduction 173
9.2   Overview of Chl Biosynthesis in Plants 174
    9.2.1   ALA Biosynthesis...174
    9.2.2   Common Steps...174
    9.2.3   Mg Branch 176
    9.2.4   Chl Cycle 177
    9.2.5   Fe Branch..177
9.3   Localization of Enzymes Involved in Chl Biosynthesis...177
    9.3.1   Subcellular Localization of Biosynthetic Enzymes...177
    9.3.2   Complex Formation of Biosynthetic Enzymes 178
9.4   Regulation of Chl Biosynthesis 179
    9.4.1   Transcriptional Regulation...179
        9.4.1.1   Transcription Factors Involved in Chl biosynthesis..179
        9.4.1.2   Developmental and Environmental Cues 182
    9.4.2   Posttranslational Regulation 184
        9.4.2.1   Regulation of Enzyme Activities 184
        9.4.2.2   Redox Regulation..184
        9.4.2.3   Regulation by Protein Stability...185
9.5   Conclusions...186
Acknowledgments...186
References..186

9.1   INTRODUCTION

Photosynthetic organisms contain some form of the light-absorbing pigment chlorophyll (Chl). Photosynthetic organisms that perform oxygenic photosynthesis, such as cyanobacteria, algae, and plants, synthesize Chl a. A group of cyanobacteria (prochlorophytes), green algae, and plants also contain Chl b. Photosynthetic bacteria, which perform anaerobic photosynthesis, produce a variety of derivatives called bacteriochlorophylls (Bchls). In the order of their discovery, Chls and Bchls are named a–f and a–g, respectively. Chls contain a tetrapyrrole ring structure like that found in heme (Figure 9.1). However, Chls bind an Mg atom in the center of its ring structure, while hemes bind an Fe atom. In addition, a hydrophobic phytol chain is attached to the ring structure of Chl that makes the molecule extremely nonpolar. All Chls are noncovalently bound to various photosystem subunits in the thylakoid membrane.

Chls absorb light at wavelengths below 480 nm and between 550 and 700 nm (Figure 9.2). In photosynthesis, Chls in light-harvesting complexes capture light energy and transfer it to the reaction centers where charge separation takes place. Since only Chl a is a constituent of the reaction centers, this pigment is regarded as the central photosynthetic

pigment. Chl b, in which the methyl group at the C_7 position of Chl a is replaced by a formyl group (Figure 9.1), has absorption properties substantially different from those of Chl a. Chl b binds to light-harvesting antenna complexes (Figure 9.2) and enhances the utilization of light energy in photosynthesis by transferring the absorbed light energy to Chl a very efficiently. In plants, the Chl a/Chl b ratio is about 3:1.

As light-excitable pigments, Chls and their intermediates can cause severe photodamage and cell death under some conditions, because they generate free radicals and reactive oxygen species (ROS), primarily singlet oxygen, with unregulated excitation by light (op den Camp et al. 2003). In addition, molecular genetic studies suggest signaling functions of Chl intermediates that control nuclear gene expression (Nott et al. 2006; Kleine, Voigt, and Leister 2009) or trigger light-independent cell death (Hirashima, Tanaka, and Tanaka 2009). Therefore, Chl metabolism needs to be tightly regulated by various environmental, developmental, and physiological factors.

In this chapter, we discuss the regulation of Chl metabolism in plants, mainly in angiosperms. Interested readers should conduct several comprehensive reviews on the tetrapyrrole pathway (Eckhardt, Grimm, and Hortensteiner 2004; Tanaka

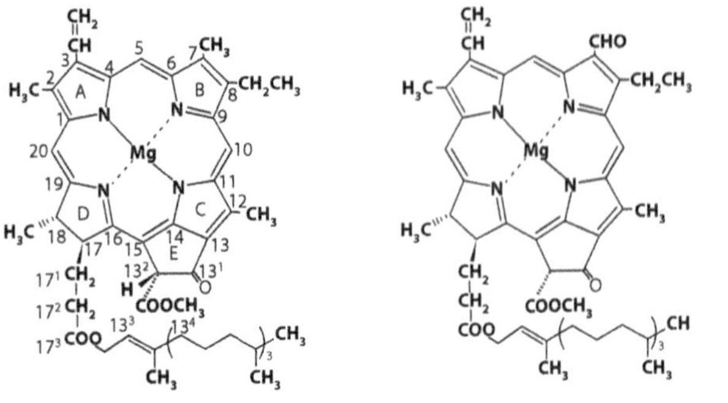

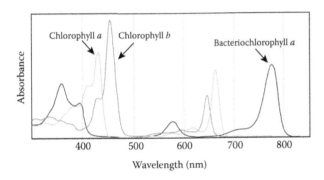

FIGURE 9.1 Structures of chlorophyll (Chl) *a* and *b*. Chl is a tetrapyrrole macrocycle containing Mg^{2+}, a phytol chain, and an isocyclic fifth ring. The five rings in Chls are lettered A through E, and the substituent positions on the macrocycle are numbered clockwise, beginning with ring A. In Chl *b*, the methyl group at the C_7 position of Chl *a* is replaced by a formyl group.

Chlorophyll *a*   Chlorophyll *b*
(Chl *a*)   (Chl *b*)

Chlorophyll *a*   Chlorophyll *b*   Bacteriochlorophyll *a*

Absorbance

400   500   600   700   800

Wavelength (nm)

FIGURE 9.2 **(See color insert.)** Absorption spectra of Chls. The absorption spectra of pigments dissolved in nonpolar solvents are shown for Chl *a* and *b* and Bchl *a*. Note that the spectra of these pigments show substantial shifts in absorbance *in vivo*, where they are associated with specific proteins.

and Tanaka 2007; Masuda and Fujita 2008; Mochizuki et al. 2010; Tanaka, Kobayashi, and Masuda 2011) and the putative roles in signaling (Nott et al. 2006; Kleine, Voigt, and Leister 2009).

9.2 OVERVIEW OF CHL BIOSYNTHESIS IN PLANTS

Figure 9.3 shows an overview of the Chl biosynthetic pathway, and Table 9.1 shows a list of enzymes involved in this pathway. Enzymes involved in this pathway are exclusively nuclear encoded, with the exception of *chlL*, *chlN*, and *chlB* genes encoding subunits of dark-operative (light-independent) protochlorophyllide (Pchlide) reductase (DPOR), which reside in the chloroplast genome in nonangiosperms and green and red algae. In addition, glutamyl-tRNA (tRNA^{Glu}), a substrate of glutamyl-tRNA synthetase (GluRS), is plastid encoded. Most enzymes are monomeric

or homomultimeric enzymes composed of a single subunit, with the exception of some multisubunit enzyme complexes, described later. However, it is becoming apparent that enzymes for Chl biosynthesis are forming multiple protein complexes for substrate channeling and regulatory purposes (see Section 9.3.2). Some enzymes seem to be encoded by single-copy genes, and others, in small families of differentially expressed genes.

Chl biosynthesis is composed of four distinct sections, as follows: 5-aminolevulinic acid (ALA) biosynthesis (Section 9.2.1), the common steps (Section 9.2.2), the Mg branch (Section 9.2.3), and the Chl cycle (Section 9.2.4); it branches to the Fe branch (Section 9.2.5) after the common steps.

9.2.1 ALA BIOSYNTHESIS

Chl biosynthesis starts with ALA synthesis, the first committed precursor of the entire tetrapyrrole biosynthetic pathway. Photosynthetic eukaryotes and many bacteria synthesize ALA from glutamate (Glu) via the so-called C_5 pathway consisting of three enzymatic steps. On the other hand, in nonphotosynthetic eukaryotes and α-proteobacteria, ALA is synthesized by ALA synthase in a single-step condensation of succinyl-CoA and glycine, which is the so-called Shemin pathway. In photosynthetic eukaryotes, Glu is first activated by coupling with plastid-encoded tRNA^{Glu} by GluRS. Thus, this step is shared with plastidic protein biosynthesis. Charged Glu-tRNA^{Glu} is then reduced to Glu 1-semialdehyde (GSA) in a NADPH-dependent manner by Glu-tRNA reductase (GluTR), which is one of the most important regulatory steps in Chl biosynthesis (see Section 9.4.2). GSA is further metabolized to ALA by GSA aminotransferase (GSAT), catalyzing intramolecular transfer of an amino group of GSA. ALA synthesis is the rate-limiting step of the entire tetrapyrrole biosynthetic pathway and is strictly controlled at the transcriptional and posttranslational levels by multiple factors (see Section 9.4).

9.2.2 COMMON STEPS

Chl and heme share the same biosynthetic pathway from ALA to the last common intermediate, protoporphyrin IX (Proto). This common pathway is highly conserved in virtually all living organisms except some bacteria and many Archaea, which bypass the pathway from coproporphyrinogen III to Proto through the precorrin 2-dependent pathway (Storbeck et al. 2010; Kobayashi, Masuda et al. 2014). In the common pathway, two ALA molecules are condensed by ALA dehydratase (ALAD) to form porphobilinogen (PBG), the first monopyrrole. PBG deaminase (PBGD) (also known as hydroxymethylbilane [HMB] synthase) condenses four PBG molecules to form the linear tetrapyrrole, HMB. Uroporphyrinogen III (Urogen III) synthase (UROS) catalyzes the cyclization of HMB to form the first macrocyclic tetrapyrrole, Urogen III. Siroheme, a cofactor of nitrite and sulfite reductases, is formed in a three-step reaction from Urogen III, while in the main branch of tetrapyrrole

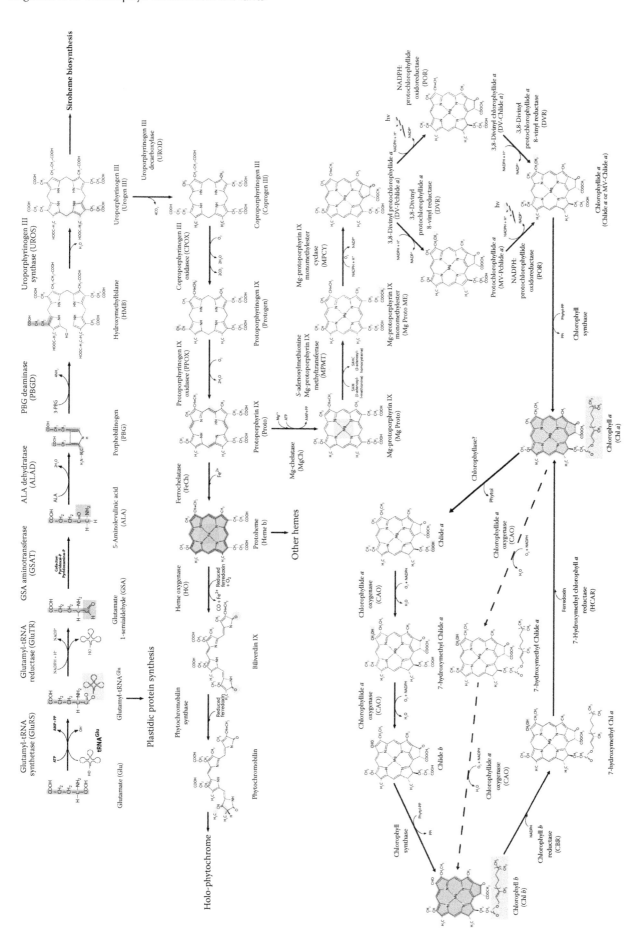

FIGURE 9.3 Biosynthesis of Chls and hemes.

TABLE 9.1

Enzymes Involved in Chl and Heme Biosynthetic Pathway in Angiosperms

Step[a]	Enzyme Name	Abbreviation	Gene Name	Localization
A	Glutamyl-tRNA synthase	GluRS		S
	Glutamyl-tRNA reductase	GluTR	HEMA	S (T)[e]
	Glutamate 1-semialdehyde aminotransferase	GSAT	GSA (HEML)	S
C	5-Aminolevulinic acid dehydratase (porphobilinogen synthase)	ALAD	ALAD (HEMB)	S
	Porphobilinogen deaminase (hydroxymethylbilane synthase)	PBGD	PBGD (HEMC)	S
	Uroporphyrinogen III synthase	UROS	UROS (HEMD)	S
	Uroporphyrinogen III decarboxylase	UROD	UROD (HEME)	S
	Coproporphyrinogen III oxidase	CPOX	HEMF (HEMN)[b]	S
	Protoporphyrinogen IX oxidase	PPOX	PPO (HemY)[c]	E, T (MT?)[f]
M	Mg-chelatase	MgCh		
	I subunit		CHLI	S
	D subunit		CHLD	S
	H subunit		CHLH	S, E, T
	Mg-protoporphyrin IX methyltransferase	MPMT	CHLM	E, T
	Mg-protoporphyrin IX monomethylester cyclase	MPCY		
	Subunit 1		CHL27	E, T
	Subunit 2		LCAA/YCF54	?
	NADPH protochlorophyllide oxidoreductase	POR	POR[d]	E, T
	Divinyl reductase	DVR	DVR	E, T
	Chlorophyll synthase		CHLG	T
Y	Chlorophyllide a oxygenase	CAO	CAO	T?[g]
	Chlorophyll b reductase	CBR		
	Subunit 1		NOL	E
	Subunit 2		NYC1	T?[g]
	7-Hydroxymethyl chlorophyll a reductase	HCAR		T?[h]
F	Ferrochelatase	FeCh	FC	E, T (MT?)[f]
	Heme oxygenase	HO	HO (HY1)	S
	Phytochromobilin synthase		HY2	S

[a] Step indicates each section of the biosynthetic pathway: ALA biosynthesis (A), the common steps (C), the Mg branch (M), the Chl cycle (C), and the Fe branch (F).

[b] HEMN is an oxygen-independent type of CPOX, the function of which is unknown in angiosperms.

[c] Three types of PPO have been reported: HemG, HemJ, and HemY. Plants use the HemY-type PPO enzyme, which requires oxygen as terminal electron acceptor and a flavin cofactor.

[d] Many gymnosperms, ferns, mosses, cyanobacteria, and photosynthetic bacteria have a light-independent type of Pchlide oxidoreductase (DPOR).

[e] It is proposed that a minor portion of GluTR1 is anchored to thylakoid membranes by FLU and/or GluTR1-binding protein.

[f] Mitochondrial localizations of PPOX and FeCh are proposed in some plants.

[g] Proteomic analyses fail to detect CAO and NYC1. Considering their functions, these enzymes are presumed to be localized in thylakoid membranes.

[h] Although the enzyme activity is detected, HCAR has not been identified yet. Considering the function, this enzyme is presumed to be localized in thylakoid membranes.

biosynthesis, Urogen III decarboxylase (UROD) and coproporphyrinogen III oxidase (CPOX) catalyze the two sequential reactions of the formation of protoporphyrinogen IX (Protogen). Protogen oxidase (PPOX) catalyzes the removal of six electrons of Protogen to form Proto. At this step, the tetrapyrrole macrocycle is fully conjugated and consequently exhibits a red color.

9.2.3 Mg Branch

At the end of the common steps, the tetrapyrrole biosynthetic pathway branches into two distinct pathways, the Mg and Fe branch. Whereas the Fe branch leads to the formation of heme

and its derivatives, the catalytic reactions of the Mg branch result in Chl formation.

In the Mg branch, an Mg^{2+} is inserted into the Proto macrocycle by Mg-chelatase (MgCh) to form Mg Proto in an ATP-dependent manner. MgCh consists of the three subunits I, D, and H, the average molecular weights of which are 40, 70, and 140 kDa, respectively. These subunits are designated CHLI, CHLD, and CHLH in Chl-producing organisms, and BchI, BchD, and BchH in Bchl-producing organisms, respectively. Each I and D subunit forms a hexameric ring that interacts with the catalytically active H subunit that catalyzes the Mg^{2+} insertion into Proto. Catalysis by MgCh proceeds with a two-step reaction: an activation step, followed by an Mg^{2+}

insertion step. At the first activation step, I and D form an I–D–Mg–ATP complex without ATP hydrolysis, and at the second insertion step, this complex binds to the Mg–H–Proto complex. Here, ATP is hydrolyzed by the ATPase activity of the I subunit to form Mg Proto.

Subsequently, Mg Proto is methylated by Mg Proto methyltransferase (MPMT), in which a methyl group from S-adenosyl-L-methionine (SAM) is transferred to Mg Proto to form Mg Proto monomethylester (ME). Mg Proto ME cyclase (MPCY) then catalyzes the formation of a fifth isocyclic ring by incorporating an atomic oxygen to Mg Proto ME, forming 3,8-divinyl-protochlorophyllide (DV-Pchlide). Although two types of MPCY, anaerobic and aerobic, have been identified in photosynthetic organisms, higher plants contain only the aerobic MPCY. This is one of the least understood steps in Chl biosynthesis. MPCY is assumed to consist of three subunits (Rzeznicka et al. 2005), in which CHL27 (Tottey et al. 2003) and low Chl accumulation A (LCAA)/YCF54 (Albus et al. 2012; Hollingshead et al. 2012) are identified to be essential subunits for the cyclase activity. Inactivation of the cyanobacterial $ycf54$ gene causes a significant reduction of Chl levels with concomitant accumulation of Mg Proto ME, and only trace amounts of Pchlide, indicating that YCF54 is essential for the activity and/or stability of MPCY (Hollingshead et al. 2012). LCAA is a tobacco homolog of YCF54, which might have an additional role in the feedback control of ALA biosynthesis (Albus et al. 2012). Since YCF54 is similar to the photosystem II assembly factor (Psb28), this protein may be involved in coordinating Chl biosynthesis and photosystem biogenesis.

The D ring of DV-Pchlide is then reduced by Pchlide oxidoreductase (POR) to form 3,8-divinyl-chlorophyllide (DV-Chlide). Two different types of enzymes have been identified for this reaction. Light-dependent POR absolutely requires light for catalysis. Since angiosperms contain only this type of enzyme, they cannot produce Chl in the dark. However, many gymnosperms, ferns, mosses, cyanobacteria, and photosynthetic bacteria have the ability to synthesize Chl in the dark, since they have another type of POR, a dark-operative (light-independent) DPOR. DPOR is an oxygen-sensitive nitrogenase-like enzyme consisting of L, N, and B subunits. The L protein is the reductant component, and the NB proteins are the catalytic component; these are functionally equivalent to the Fe protein and MoFe protein of nitrogenase, respectively (Fujita and Bauer 2003). The 8-vinyl group of DV-Chlide is then reduced by the DV-reductase (DVR) to form 3-vinyl Chlide a (MV-Chlide a). This enzyme can also reduce DV-Pchlide; however, the reaction efficiency with this substrate is substantially lower than that with DV-Chlide in $vivo$ and in $vitro$ (Nagata, Tanaka, and Tanaka 2007).

In the last step of the Mg branch, 17-propionate on the D ring of MV-Chlide is esterified with a long-chain polyisoprenol (geranylgeraniol or phytol) by Chl synthase to synthesize Chl a. This enzyme is also found to recycle Chlide a during Chl turnover (Lin et al. 2014). Phytol is provided from geranylgeranyl pyrophosphate (GGPP), which is produced via isopentenyl pyrophosphate (IPP) in the nonmevalonate 2-C-methyl-D-erythriotol-4-phosphate (MEP) pathway in plastids. In plants, the geranylgeranyl side chain is stepwise

reduced by geranylgeranyl reductase (GGR) before or after attachment to a Chl moiety. Reduced GGPP availability by chemical inhibition of the MEP pathway results in a light-dependent lethal phenotype, suggesting that a stoichiometric imbalance of GGPP and Chl precursors can cause photooxidative stresses on plants (Kim et al. 2013).

9.2.4 CHL CYCLE

Among oxygenic phototrophs, only land plants, green algae, and a few groups of cyanobacteria (prochlorophytes) possess Chl b. In these organisms, Chlide a is converted to Chlide b by Chlide a oxygenase (CAO), a Rieske-type monooxygenase. By an in $vitro$ assay, it is shown that CAO performs a two-step oxygenation of Chlide a as a substrate, in which 7-hydroxymethyl Chlide a is produced as the intermediate (Oster et al. 2000). However, it is also assumed that CAO is able to directly convert Chl a to Chl b in $vivo$ (Tanaka and Tanaka 2011), suggesting two alternative routes for Chl a to Chl b conversion. Chl a is directly converted into Chl b, or Chl a is first dephytylated to form Chlide a, and subsequently, Chlide a is converted into Chlide b by CAO and finally conjugated with phytol to form Chl b (Figure 9.2). For dephytylation of Chl a, an enzyme, chlorophyllase, is proposed to be responsible (Tsuchiya et al. 2003), but involvement of this enzyme is disputed by several studies (Schelbert et al. 2009; Lin et al. 2014). Chl b can be reversibly converted to Chl a through 7-hydroxymethyl Chl a. These two-step reactions are catalyzed by Chl b reductase (CBR) and 7-hydroxymethyl Chl a reductase (HCAR).

9.2.5 FE BRANCH

In the Fe branch, ferrochelatase (FeCh) catalyzes the insertion of Fe^{2+} into Proto to form protoheme (heme b). Other hemes, such as heme a and heme c, are synthesized from protoheme. Protoheme is also converted into biliverdin IXα by heme oxygenase (HO). In land plants, biliverdin IXα is then reduced to $3E$-phytochromobilin, the chromophore of phytochromes (PHYs), by phytochromobilin synthase, whereas it is converted into phycobilins in cyanobacteria and various algae (Terry, Linley, and Kohchi 2002).

9.3 LOCALIZATION OF ENZYMES INVOLVED IN CHL BIOSYNTHESIS

9.3.1 SUBCELLULAR LOCALIZATION OF BIOSYNTHETIC ENZYMES

Enzymes involved in Chl biosynthesis are exclusively localized in plastids. The predicted localization of each enzyme is summarized in Table 9.1. Since early Chl intermediates are hydrophilic, it is generally considered that the enzymes involved in the ALA biosynthesis and early enzymes in the common steps are localized in the soluble stroma, while the enzymes involved in the later steps, which react with more hydrophobic intermediates, are plastid membrane bound (Masuda and Fujita 2008;

Mochizuki et al. 2010; Tanaka, Kobayashi, and Masuda 2011). The ALA synthesizing activity is biochemically detected in the stromal fraction of barley and other plants (Gough and Kannangara 1976). One of two isoforms of GluTR (GluTR1 encoded by *HEMA1*) is likely anchored to thylakoid membranes in the dark through an interaction with the membrane-bound fluorescent (FLU) protein for negative regulation (see Section 9.4.2). In addition, another GluTR-binding protein is identified to be involved in anchoring a minor portion of GluTR1 to thylakoid membranes (see Section 9.3.2).

The subcellular localization of plastid proteins, including enzymes involved in Chl biosynthesis, is examined by proteomic analysis (Joyard et al. 2009). The first two enzymes involved in ALA biosynthesis (GluRS and GluTR) were undetectable in the proteomic assay; however, other enzymes until the CPOX steps of the common pathway are detected in stroma (Joyard et al. 2009). Except for the subunits of MgCh, most enzymes used later than PPOX were detected in both inner envelope and thylakoid membranes (Joyard et al. 2009). In MgCh, CHLI and CHLD subunits are considered to be stroma localized (Nakayama et al. 1995; Masuda and Fujita 2008), while the catalytic H subunits conditionally bind to envelope membranes in an Mg^{2+}- and porphyrin-dependent manner (Gibson et al. 1996; Nakayama et al. 1998; Adhikari et al. 2009). The dual localization of enzymes in envelope and thylakoid membranes suggests that Chl biosynthesis may take place in both membranes. However, considering the large volume of the thylakoid membrane and the final destination of Chl, it is likely that the thylakoid membrane is the major site for Chl biosynthesis.

By analogy of mitochondrial heme biosynthesis in animals and yeast, it has been disputed that heme biosynthesis also takes place in mitochondria of higher plants (Papenbrock et al. 2001; Cornah et al. 2002; Mochizuki et al. 2010). For PPOX, three nonhomologous isofunctional enzymes, HemG, HemJ, and HemY, have been identified among organisms (Kato et al. 2010), and eukaryotes possess only HemY-type PPOX (Kobayashi, Masuda et al. 2014). Phylogenetic analysis suggests that the main PPOX isoform PPO1 in plants originated from cyanobacteria and diversified ubiquitously within Plantae, whereas PPO2, which is related to the HemY homologs in *Chloroflexus* species, originated in land plants after divergence from a green algal lineage (Kobayashi, Masuda et al. 2014). Although PPO1 is exclusively localized to plastids, PPO2 is detected in mitochondria in tobacco (Lermontova and Grimm 2006) and spinach (Watanabe et al. 2001). Meanwhile, a green alga, *Chlamydomonas reinhardtii*, possesses only the PPO1-type PPOX, which is exclusively found in the chloroplast (van Lis et al. 2005), so the PPOX activity in the chloroplast is sufficient for this alga. It is notable that in addition to having a role as a metabolic enzyme, PPOX in the plastid is identified as a regulator of RNA editing (Zhang et al. 2014), indicating that mitochondrial PPOX may have a distinct function. Similarly, the presence of mitochondrial FeCh activity is reported in peas (Cornah et al. 2002) and tobacco (Papenbrock et al. 2001). In cucumber (Masuda et al. 2003) and *Arabidopsis* (Woodson, Perez-Ruiz, and Chory 2011), two isoforms of FeCh are localized within plastids, but not in mitochondria. Phylogenetic analysis of FeCh suggests that plants and green algal enzymes are of cyanobacterial origin, while red algal enzymes are of α-proteobacterial origin (Obornik and Green 2005). Consistent with this assumption, FeCh is detected in the chloroplast in *C. reinhardtii* (van Lis et al. 2005) and in mitochondria in a red alga, *Cyanidioschyzon merolae*, (Watanabe et al. 2013) and an apicomplexan parasite, *Plasmodium* (Sato and Wilson 2003). Therefore, considering that plants are descendent from the green algal lineage, it is reasonable to assume that plastids are the major site of heme biosynthesis. On the other hand, mitochondrial heme biosynthesis in plants is questionable considering the minor presence of last two enzymes in mitochondria and that trafficking of tetrapyrrole intermediates from plastid to mitochondria should be necessary (Cornah et al. 2002; Tanaka, Kobayashi, and Masuda 2011).

9.3.2 COMPLEX FORMATION OF BIOSYNTHETIC ENZYMES

Since most Chl intermediates are highly reactive and, especially, fluorescent intermediates photodynamically produce ROS, the metabolic pathway must be arranged to flow intermediates smoothly by enzyme complexes. In fact, a mega complex of enzymes called the *Chl biosynthetic center* is proposed based on physiological and biochemical data (Shlyk 1971). Some enzymes involved in Chl biosynthesis actually form complexes to enable the efficient channeling of substrates. From the determined protein structures, a homodimer of V-shaped GluTR is proposed to form a complex with a homodimer of GSAT, which may prevent the release of the highly reactive aldehyde moiety of GSA by direct channeling of this intermediate from GluTR to GSAT (Moser et al. 2001). Although physical interactions between GluTR and GSAT have been shown using recombinant proteins (Luer et al. 2005; Nogaj and Beale 2005), their direct interaction *in vivo* has not yet been proven. In addition, complex formation of the enzymes in common steps has been proposed from determined structures (Koch et al. 2004). Similarly, since the MPMT activity is stimulated by the H subunit of MgCh in photosynthetic bacteria, it is proposed that Mg Proto may be channeled directly to MPMT from MgCh through the H subunit (Hinchigeri, Hundle, and Richards 1997; Shepherd and Hunter 2004; Alawady et al. 2005), although such protein–protein interaction of the H subunit and MPMT has not been demonstrated in higher plants.

From the regulatory point of view, *HEMA1*-encoded GluTR1, which is expressed in photosynthetically active tissues and required for Chl synthesis, interacts with the membrane-localized FLU protein in the dark. FLU downregulates the GluTR1 activity in response to accumulating Pchlide *a*, the substrate of POR (Meskauskiene et al. 2001; Meskauskiene and Apel 2002). Interestingly, FLU does not bind to *HEMA2*-encoded GluTR2, which is expressed in all of the *Arabidopsis* tissues. FLU was identified by the analysis of the *Arabidopsis* conditional fluorescence (*flu*) mutant in which ALA synthesis is not repressed in darkness, resulting in dramatic accumulation of Pchlide *a* (Meskauskiene et al. 2001). FLU is a ~27 kDa protein localized in plastid membranes and contains two tetratricopeptide repeat motifs that are presumed to be involved in protein–protein interactions. Further analysis shows that GluTR1 forms a complex with CHL27, POR, and FLU in darkness, but not in the light, proposing that GluTR1 is inactivated in darkness through binding to FLU, CHL27, and POR, whereas in the light, GluTR1 is released from this complex and becomes active (Kauss et al. 2012). In addition, it is reported that a small portion of GluTR1 is also attached to the thylakoid membrane by a GluTR-binding protein (GluBP) (Czarnecki et al. 2012). GluBP spatially separates GluTR1 from a soluble GluTR1 fraction and protects it from the FLU-dependent inactivation in darkness. It is proposed that GluBP contributes to subcompartmentation of ALA biosynthesis by maintaining a portion of GluTR1 at the thylakoid membrane, which flows ALA into heme biosynthesis (Czarnecki et al. 2012). However, as subsequent enzymes in the common steps are shared for heme biosynthesis, the mechanism of the spatial separation is still not clear.

MgCh itself is a multisubunit enzyme and may form a complex with the regulatory GUN4 protein. In fact, it is shown that GUN4 interacts with the CHLH subunit of MgCh (Sobotka et al. 2008; Zhou et al. 2012). GUN4 was first identified as the defective gene in the *genome uncoupled 4* (*gun4*) mutant, which exhibits aberrant regulation of retrograde signal transduction (Larkin et al. 2003). GUN4 binds to both Proto and Mg Proto and stimulates the MgCh activity (Davison et al. 2005; Verdecia et al. 2005). Recombinant *Synechocystis* CHLH proteins carrying *gun*-related mutations can bind Proto and Mg Proto and form the MgCh complex, but inhibits the MgCh activity. Addition of GUN4 to the recombinant complex led to recovery of the MgCh activity (Davison and Hunter 2011).

The LIL3 protein, a member of the light-harvesting protein family, has been reported to interact with GGR encoded by *CHLP* to stabilize and/or activate this enzyme (Tanaka et al. 2010; Takahashi et al. 2014). In addition, during de-etiolation of barley leaves, LIL3 forms two pigment–protein complexes containing either Chl *a* or protochlorophyll *a*, which may be involved in the transfer of Chl *a* to pigment–protein complexes (Reisinger, Ploscher, and Eichacker 2008). Similarly, in cyanobacteria, Chl synthase forms a protein complex including the high-light-inducible protein HliD, the putative assembly factor for photosystem II Ycf39, and the YidC/Alb3 insertase (Chidgey et al. 2014). It is proposed that this complex coordinates the binding of Chls to apoproteins for producing photosynthetic pigment–protein complexes with a minimal risk of accumulating phototoxic unbound Chls. Identification of complexes of enzymes involved in the final steps of Chl biosynthesis, pigment–protein apoproteins, and related regulatory proteins is becoming more important to understanding the mechanism of assembly of photosynthetic complexes.

9.4 REGULATION OF CHL BIOSYNTHESIS

9.4.1 TRANSCRIPTIONAL REGULATION

Regulated gene expression plays a central role in the coordination of each step of Chl biosynthesis with the synthesis of cognate proteins and formation and maintenance of the photosynthetic machinery. Genes involved in Chl biosynthesis (Chl genes) are strictly regulated by various environmental and endogenous cues. Light is a major driver of photosynthesis-associated gene expression, and the light-signaling pathway plays an essential role in the expression of Chl genes. The expression of Chl genes is also regulated by endogenous factors such as hormones, the circadian clock, and developmental programs. Various transcription factors function in the regulation of Chl genes to coordinate Chl metabolism with formation of the photosynthetic machinery in the chloroplast.

9.4.1.1 Transcription Factors Involved in Chl biosynthesis

9.4.1.1.1 HY5

Reflecting the necessity of light in photosynthesis, light signal transduction plays a central role in transcriptional regulation of Chl biosynthesis. LONG HYPOCOTYL5 (HY5) is one of the pivotal transcription factors in the nucleus functioning downstream of photoreceptors. The HY5 protein is degraded in the dark through CONSTITUTIVE PHOTOMORPHOGENIC1 (COP1), which is a ubiquitin E3 ligase controlling the abundance of several light-signaling components, in conjunction with DE-ETIOLATED1 (DET1) (Pepper et al. 1994; Osterlund et al. 2000; Lau and Deng 2012). In the light, photoreceptors mediate accumulation of the HY5 protein in part by excluding COP1 from the nucleus. HY5 regulates many light-associated events, including photomorphogenesis and chloroplast development (Bae and Choi 2008). A genome-wide chromatin immunoprecipitation-chip (ChIP-chip) analysis revealed that many Chl genes (*GluRS, URO2, PPO1, CHLH, GUN4, CHL27, DVR, PORC, CAO, CHLP,* and *HO1*) are putative direct targets of HY5 together with various photosynthesis-associated nuclear genes (Lee et al. 2007). Although *HEMA1* is not found as the direct HY5 target in the ChIP-chip analysis, loss of the *HY5* function results in a substantial decrease in responses of *HEMA1* expression to blue, red, and far-red light (McCormac and Terry 2002), indicating that *HEMA1* is also under the HY5 regulation. Consistent with the role of HY5 in transcriptional upregulation of key Chl genes, *hy5* mutations partially inhibit Chl accumulation during photomorphogenesis (Holm et al. 2002; Lee et al. 2007). The role of HY5 in Chl biosynthesis is more evident in roots. The expression of key Chl genes is intensely downregulated in *hy5* roots, leading to the absence of Chls in the mutant roots (Oyama, Shimura, and Okada 1997; Usami et al. 2004; Kobayashi et al. 2012). By contrast, *cop1* and *det1* mutants fail to degrade their target proteins, including HY5, resulting in strong Chl accumulation in roots in the light as well as photomorphogenic growth in the dark (Chory and Peto 1990; Deng and Quail 1992). The data indicate that HY5 upregulates nuclear genes associated with Chl biosynthesis and photosynthesis, but its activity is repressed by the COP1/DET1 components in roots or in the dark.

HY5 directly binds to G-box (CACGTG), a well-characterized light-responsive *cis* element, in the promoters of target genes and upregulates them in a light-dependent manner (Chattopadhyay et al. 1998). In fact, the HY5-targeted genes identified by the ChIP-chip analysis frequently possess the G-box on their promoter regions (Lee et al. 2007). Furthermore, in silico searches for cis elements in key Chl genes (CHLH, CHL27, and CHLP) and their coexpressed genes associated with photosynthesis revealed significant enrichment of the G-box sequence in their promoter regions (Kobayashi, Obayashi, and Masuda 2012). A deletion of the G-box sequence from the promoter of the CHLH gene resulted in decreased CHLH transcription, particularly in the root, indicating the importance of the G-box for its expression, presumably through HY5 binding (Kobayashi, Obayashi, and Masuda 2012). Moreover, a ChIP analysis targeting CHLH revealed that HY5 binds to a G-box-containing region of the CHLH promoter (Toledo-Ortiz et al. 2014). In addition to CHLH, other light-responsive genes, PORC, PHYTOENE, SYNTHASE, and LHCA4, also possess the G-box-containing region where HY5 binds (Toledo-Ortiz et al. 2014). Loss of HY5 impaired light induction of these genes, reflecting a role of HY5 in coordinate upregulation of these genes in response to light. This may be applicable to other coexpressed genes related to photosynthesis that have G-box elements in their promoter regions.

9.4.1.1.2 PIF

In contrast to HY5, a small subset of basic helix-loop-helix transcription factors designated PHYTOCHROME-INTERACTING FACTORs (PIFs) negatively regulates Chl biosynthesis under light signaling (Leivar and Quail 2011). PIFs are involved in broad cellular processes as a signaling hub integrating light, hormone, and other multiple developmental signals. PIFs (PIF1, PIF3, PIF4, and PIF5) accumulate in the nucleus in the dark and repress photomorphogenic responses, including Chl biosynthesis. However, in the light, these factors are phosphorylated and subsequently degraded via the ubiquitin–proteasome system in a PHY-dependent manner (Leivar and Quail 2011).

Out of six well-characterized PIF proteins, at least four PIFs (PIF1, PIF3, PIF4, and PIF5) function redundantly to downregulate Chl genes in the dark. In the pif1 and pif3 single mutants, expression of key Chl genes such as HEMA1, CHLH, and GUN4 was derepressed, resulting in excessive production of Pchlide in the dark (Stephenson, Fankhauser, and Terry 2009). Consequently, light exposure to the dark-grown pif1 and pif3 mutants caused Pchlide-mediated photooxidation and photobleaching. A similar phenomenon was observed in a pif5 mutant (Shin et al. 2009). Moreover, quadruple mutations (pifQ) in PIF1, PIF3, PIF4, and PIF5 caused global upregulation of Chl genes along with many photosynthesis-associated nuclear genes (Leivar et al. 2009; Shin et al. 2009), demonstrating that PIFs repress Chl biosynthesis in a redundant manner.

As is the case in HY5, PIFs specifically bind to the G-box element in target gene promoters (Leivar and Quail 2011). A ChIP analysis revealed that PIF3 binds to the G-box region of CHLH in addition to those of other photosynthesis-associated

nuclear genes (LHCB2.2, PSBQ, and PSAE1) (Liu et al. 2013). Another ChIP analysis showed that PIF1 binds to the G-box region of CHLH, PORC, and LHCA4 (Toledo-Ortiz et al. 2014). Although PIF3 and HY5 both positively regulate anthocyanin biosynthesis genes by directly binding to different regions of the gene promoters (Shin, Park, and Choi 2007), PIFs (PIF1 and PIF4) and HY5 act antagonistically to their target genes involved in photosynthesis and Chl and carotenoid biosynthesis, presumably in the same G-box-containing region (Toledo-Ortiz et al. 2014). Furthermore, regulation of photosynthesis-associated nuclear genes by PIF3 involves histone deacetylation of target gene promoters by HISTONE DEACETYLASE15 (HDA15) (Liu et al. 2013). In the dark, the HDA15 protein interacting with PIF3 binds to the G-box regions in the target gene promoters and decreases their transcriptional activities via histone deacetylation. Upon illumination, light-activated PHYs induce the PIF3 degradation and dissociation of HDA15 from its target genes. Subsequently, histone acetyltransferases (HATs) such as TAF1 (HAF2) and GCN5, which are involved in light-regulated gene expression and photomorphogenic development, may induce expression of photosynthesis-associated genes via histone acetylation (Benhamed et al. 2006). Because these HATs are genetically associated with HY5, HY5 may recruit these factors to trigger the expression of photosynthesis-associated genes in response to light.

PIFs are also involved in light-regulated gene repositioning within the Arabidopsis nucleus (Feng et al. 2014). The Chl a/b-binding proteins (CAB) locus, including LHCB1.1, LHCB1.2, and LHCB1.3, was repositioned rapidly from the nuclear interior to the nuclear periphery in response to light before full transcriptional activation of the locus. Meanwhile, in the pifQ mutant as well as cop1 and det1 mutants, the CAB locus was positioned at the nuclear periphery even in the dark, suggesting that COP1, DET1, and the PIFs are required to retain the CAB locus in the nuclear interior in the dark. Because the loci for CHLH, PLASTOCYANIN, and RBCS1A, encoding the small subunit of ribulose-1,5-bisphosphate carboxylase, which are distributed throughout the Arabidopsis genome, were also repositioned in response to light as observed in the CAB locus, the gene locus reposition in the nucleus may be a common induction mechanism of photosynthesis-associated nuclear genes, including key Chl genes, by light.

9.4.1.1.3 GLK

Reflecting the strong need for coordination of tetrapyrrole biosynthesis with the development of the photosynthetic machinery, key Chl genes form a tight coexpression network with other nuclear photosynthesis-associated genes (Masuda and Fujita 2008; Kobayashi, Obayashi, and Masuda 2012). Golden2-like (GLK) transcription factors are involved in the positive regulation of these coexpressed genes during chloroplast biogenesis. Deficiencies of two GLK homologs GLK1 and GLK2 in Arabidopsis resulted in strong downregulation of nuclear photosynthesis-associated genes in leaves, especially those associated with Chl biosynthesis and light harvesting (Fitter et al. 2002). By contrast, a transient induction of *GLKs* in the *Arabidopsis glk1 glk2* double mutant caused immediate upregulation of Chl genes and other nuclear photosynthesis-associated genes, suggesting that the primary inducible genes after the *GLK* induction are direct targets of GLK factors (Waters et al. 2009). Indeed, a ChIP analysis revealed that GLK1 can bind to the promoter regions of its primary targets, including key Chl genes (*HEMA1, CHLH, GUN4, CHLM, CHL27, PORA, PORB, PORC*, and *CAO*) (Waters et al. 2009). With the exception of *PORA* and *PORB*, these key Chl genes form a tight coexpression network with other GLK-targeted photosynthetic genes (Masuda and Fujita 2008; Kobayashi, Obayashi, and Masuda 2012), implying that GLKs are pivotal regulators responsible for the transcriptional regulation of these coexpressed genes. Moreover, stable overexpression of GLKs strongly upregulates not only their target genes but also other genes involved in chloroplast biogenesis in nonphotosynthetic tissues, which results in ectopic chloroplast biogenesis there (Nakamura et al. 2009; Powell et al. 2012; Kobayashi, Sasaki et al. 2013). These data suggest that the GLK family is one of the key regulators determining chloroplast differentiation during organ development.

An *in silico* analysis of the promoter sequences in GLK-targeted genes identified a putative GLK-recognition *cis* element (CCAATC) (Waters et al. 2009). Moreover, *cis*-element searches in key Chl genes and their coexpressed genes revealed enrichment of the GLK-recognition element in their promoter regions (Kobayashi, Obayashi, and Masuda 2012). The GLK-recognition element coexists with G-box in the promoter regions of the coexpressed genes with significant frequency, suggesting a possibility of interactions of GLKs with G-box binding factors such as HY5 and PIFs.

9.4.1.1.4 Other Transcription Factors

Two paralogous GATA transcription factors, GNC (GATA, NITRATE-INDUCIBLE, CARBON METABOLISM INVOLVED) and CGA1 (CYTOKININ-RESPONSIVE GATA TRANSCRIPTION FACTOR1), play a role in regulation of chloroplast biogenesis. Loss of function of GNC and CGA1 decreased Chl content in leaves, whereas overexpression of either gene enhanced Chl accumulation and chloroplast production in leaves, roots, and hypocotyls (Mara and Irish 2008; Richter et al. 2010; Hudson et al. 2011; Chiang et al. 2012; Hudson et al. 2013). Whereas the expression of *HEMA1, GUN4, PORB*, and *PORC* were decreased in the *gnc cga1* double mutants, these genes were upregulated in the overexpression lines of these factors (Hudson et al. 2011). However, in contrast to the direct regulation of Chl genes by HY5, PIFs, and GLKs, the GATA factors upregulated these genes an indirect manner, and their regulatory pathways remain unknown (Hudson et al. 2011). GNC and CGA1 are actively expressed in the light but are downregulated in the dark (Naito et al. 2007; Richter et al. 2010). Because PIFs downregulate GNC and CGA1 via direct binding to their promoter regions (Richter et al. 2010), PIF may regulate Chl biosynthesis and chloroplast biogenesis in part through these GATA factors.

FAR-RED ELONGATED HYPOCOTYL3 (FHY3) and its homolog FAR-RED IMPAIRED RESPONSE1 (FAR1) are transposase-derived transcription factors functioning in PHYA signaling. Mutant analyses revealed that FHY3 is required for the upregulation of *HEMA1* in response to far-red and red light irradiation (McCormac and Terry 2002). FHY3 also contributes to the expression of *GUN4* and *CHLH* under far-red light (Stephenson and Terry 2008). A ChIP-sequence analysis in *Arabidopsis* showed that *HEMA1*, but not *GUN4* and *CHLH*, is a putative direct target of FHY3 (Ouyang et al. 2011). In addition, FHY3 directly binds to the promoter region of *HEMB1* encoding an ALAD through the FHY3/FAR1 binding site (CACGCGC), thereby upregulating *HEMB1* expression (Tang et al. 2012). Meanwhile, PIF1 partially represses FHY3 activity by a physical interaction with the DNA-binding domain of FHY3 (Tang et al. 2012). Although HY5 also physically interacts with FHY3 and FAR1 through their respective DNA-binding domains (Li et al. 2010), the role of these protein interactions in the regulation of Chl genes remains elusive.

Scarecrow-like (SCL) transcription factors SCL6, SCL22, and SCL27 are involved in regulation of Chl synthesis in addition to the proliferation of meristematic cells and polar organization in *Arabidopsis* (Wang et al. 2010; Ma et al. 2014). A triple mutant for these factors showed upregulation of *PORA, PORB, PORC*, and *CAO* along with increased Chl accumulation in leaves. SCL27 can directly bind to the GT *cis* elements of the *PORC* promoter and suppress *PORC* expression. MicroRNAs (miR171s) target *SCL* transcripts for degradation and thereby inhibit their function. Whereas constitutive overexpression of miR171c decreased *SCL* transcripts and increased leaf Chl levels, overexpression of miR171-resistant SCL27 reduced the expression of Chl genes and Chl accumulation. Thus, the balance between SCLs and miR171s in part determines the level of Chl biosynthesis in leaves. Because SCL27 can upregulate the miR171a expression by binding to its promoter region, these two factors form a feedback regulatory loop, which may contribute to fine-tuning of Chl biosynthesis during leaf development. In addition, DELLAs, a family of nuclear proteins functioning downstream of gibberellin signaling as negative regulators, can interact with SCL27 and inhibit its DNA binding to the *PORC* promoter, suggesting a link between gibberellin signaling and regulation of Chl biosynthesis.

Key Chl genes are under circadian regulation (Matsumoto et al. 2004). TIMING OF CAB EXPRESSION1 (TOC1) is one of the core components of the circadian clock machinery. TOC1 is activated in the evening and prevents the induction of morning-expressed genes at night (Hsu and Harmer 2014). TOC1 acts as a general transcriptional repressor that directly binds to the promoter regions of its target genes (Gendron et al. 2012). One of the targets of TOC1 is *CHLH*. *CHLH* expression is high in the day and low at night (Papenbrock et al. 1999; Matsumoto et al. 2004). However, loss of function of TOC1 increased *CHLH* expression at night, whereas gain of function of TOC1 repressed *CHLH* expression constantly (Legnaioli, Cuevas, and Mas 2009). Indeed, TOC1 binding to the *CHLH* promoter was antiphasic to *CHLH* expression, reflecting the direct regulation of *CHLH* expression by TOC1 in response to the circadian clock.

9.4.1.2 Developmental and Environmental Cues

9.4.1.2.1 Transcriptional Regulation during Photomorphogenesis

Angiosperm seedlings germinated in the dark differentiae etioplasts, which accumulate Pchlide *a* with POR and NADPH in prolamellar bodies, instead of chloroplasts in cotyledon cells. Expression of most Chl genes is repressed in the dark, in which PIFs play a pivotal role, as described earlier. Gibberellin signaling is involved in the PIF-dependent repression of Chl genes. DELLAs can inhibit the DNA binding activity of PIFs through a physical interaction (de Lucas et al. 2008; Feng et al. 2008). However, gibberellins accumulated in the dark induce degradation of DELLA proteins and thereby promote PIF activities to repress Chl genes (Cheminant et al. 2011). Retaining the key Chl gene loci in the nuclear interior through COP1, DET, and the PIFs may also contribute to repression of these genes in the dark (Feng et al. 2014). In addition, the positive regulator HY5 is inactivated in the dark through COP1/DET1-mediated degradation (Osterlund et al. 2000), and *GLKs* are downregulated in transcriptional levels (Waters et al. 2009), leading to low transcriptional activities of light-inducible Chl genes in the dark.

Meanwhile, *PORA* and *PORB* transcripts are actively synthesized in etiolated *Arabidopsis* seedlings to form prolamellar bodies in etioplasts (Armstrong et al. 1995; Matsumoto et al. 2004). COP1 plays an essential role in the expression of the *PORs* in the dark. The expression of *PORA* and *PORB* was substantially decreased in dark-grown *cop1* seedlings, resulting in the loss of prolamellar bodies in cotyledon plastids (Sperling et al. 1998). Ethylene signaling is shown to be involved in the COP1-mediated regulation of *POR* expression in the dark. An ethylene-inducible transcription factor, EIN3/EIL1, which is stabilized in the dark by ethylene and COP1 signaling, binds to the promoter regions of both *PORA* and *PORB* and upregulates them (Zhong et al. 2009, 2010). DELLA proteins also upregulate *PORA* and *PORB* expression in the dark and during initial light exposure in a PIF-independent manner (Cheminant et al. 2011).

At the onset of chloroplast biogenesis upon illumination, substantial amounts of Chl are newly synthesized concomitantly with other photosynthetic components. Light-activated PHYs induce degradation of PIFs via phosphorylation and subsequent ubiquitination, leading to derepression of Chl genes (Leivar and Quail 2011). Changes in gibberellin signaling upon illumination may also inhibit PIF activities through DELLAs. Inactivation of PIFs along with COP1 and DET1 may induce repositioning of key Chl gene loci to the nuclear periphery, although the relationship between the reposition of light-inducible gene loci and their upregulation remains unknown (Feng et al. 2014). At the same time, photoreceptors induce HY5 accumulation through inactivation of COP1 and upregulate the expression of *GLKs*, which results in the

strong induction of key Chl genes. Signaling of phytohormones cytokinin and strigolactone can also mediate HY5 accumulation by inhibiting COP1 activity (Vandenbussche et al. 2007; Tsuchiya et al. 2010). In fact, a cytokinin treatment to etiolated *Arabidopsis* seedlings activates the expression of *HEMA1*, *CHLH*, and *CHL27*, whereas double mutations in cytokinin receptors (*ahk2* and *ahk3*) reduce their expression levels in the light (Hedtke et al. 2011; Kobayashi, Fujii, Sasaki et al. 2014).

Meanwhile, *PORA* and *PORB* transcripts that accumulate in etiolated seedlings decrease rapidly after illumination. In this process, the degradation of EIN3 through inactivation of ethylene signaling and COP1 triggers downregulation of *PORA* and *PORB* (Zhong et al. 2009, 2010). By contrast, the *PORC* transcript, which is undetectable in the dark, accumulates after illumination in a light intensity–dependent manner in *Arabidopsis* (Oosawa et al. 2000; Su et al. 2001; Matsumoto et al. 2004). In gymnosperms and liverworts, which use DPOR in the dark, the expression of *PORs* is induced by light together with other photosynthesis-related genes (Takio et al. 1998; Skinner and Timko 1999), although the signaling pathway is largely unknown.

9.4.1.2.2 Transcriptional Regulation by the Endogenous Circadian Rhythm

The circadian day/night clock strictly regulates expression of key Chl genes (Papenbrock et al. 1999; Matsumoto et al. 2004; Stephenson, Fankhauser, and Terry 2009). The core circadian clock machinery in *Arabidopsis* is mainly composed of the evening-phased factor TOC1 and two morning-phased transcription factors CCA1 (CIRCADIAN CLOCK-ASSOCIATED1) and LHY (LATE ELONGATED HYPOCOTYL), which form a transcriptional feedback loop to generate circadian rhythms (Hsu and Harmer 2014). As described earlier, *CHLH* expression is under direct regulation by the core clock component TOC1 (Legnaioli, Cuevas, and Mas 2009). A ChIP-sequence analysis confirms the direct binding of TOC1 to the *CHLH* promoter, although other circadian-regulated genes involved in Chl biosynthesis are not found in the TOC1 target list (Huang et al. 2012).

PIFs play a crucial role in the circadian control of key Chl genes, and the periodic expression of *HEMA1*, *GUN4*, and *CHLH* is largely perturbed in *pif1* and *pif3* (Stephenson, Fankhauser, and Terry 2009). PIFs function in the output from the circadian clock to downregulate Chl biosynthesis along with formation of photosynthetic complexes in response to the day/night cycle.

In contrast to PIFs, HY5 functions to upregulate key Chl genes during circadian oscillation. The *hy5* mutation depresses expression peaks of *CHLH* and *PORC* at daytime (Toledo-Ortiz et al. 2014). HY5 can physically interact with CCA1 and act synergistically on *Lhcb1.1* expression (Andronis et al. 2008), suggesting that interaction of HY5 and CCA1 plays a role in the circadian expression of photosynthesis-associated nuclear genes. In addition to the role in the output of light signaling to control Chl biosynthesis, HY5 functions in gating PHY signaling to the circadian clock with FHY3 and FAR1

(Li et al. 2011). Thus, HY5 may be involved in circadian regulation in various phases.

In addition to key Chl genes, the expression of *GLK2* is also under circadian control (Fitter et al. 2002). Considering that GLK2 functions as a direct upregulator of key Chl genes (Waters et al. 2009), the oscillated *GLK2* expression could affect the circadian rhythms of target Chl genes. *GNC* and *CGA1* are also strongly upregulated by light following periods of darkness and are subject to circadian regulation with the expression peak at predawn (Manfield et al. 2007). However, the involvement of these factors in circadian regulation of Chl genes is unclear.

9.4.1.2.3 Transcriptional Coordination of Chl Biosynthesis with Chloroplast Functionality

Coordination of Chl biosynthesis with formation of the photosynthetic machinery is essential to prevent photooxidative reactions by free Chl and its intermediates and derivatives. The coexpression network of Chl genes with other photosynthesis-associated genes is one of the central mechanisms to synchronize the synthesis of Chl and cognate proteins (Masuda and Fujita 2008; Kobayashi, Obayashi, and Masuda 2012). Indeed, perturbed transcriptional regulation of Chl biosynthesis in PIF mutants (Huq et al. 2004; Leivar et al. 2009; Shin et al. 2009; Stephenson, Fankhauser, and Terry 2009) or overexpressors of *CHLH* or *HEMA1* (Shin et al. 2009; Schmied, Hedtke, and Grimm 2011) resulted in photooxidative damage to leaves, showing the importance of transcriptional regulation of key Chl genes.

Photosynthetic Chl–protein complexes are embedded in a lipid matrix of thylakoid membranes. Transcriptional coordination of Chl biosynthesis with thylakoid membrane biogenesis has been characterized in mutants defective in thylakoid lipid biosynthesis. Defects in the synthesis of galactolipids, the major lipid components in the thylakoid membrane, lead to strong downregulation of Chl genes along with other photosynthesis-associated genes (Kobayashi, Narise et al. 2013; Fujii et al. 2014). Deficiency of another major thylakoid lipid, phosphatidylglycerol, also resulted in the downregulation of Chl genes (Kobayashi, Fujii, Sato et al. 2015). However, the repressed Chl gene expression in thylakoid lipid-deficient mutants was restored in response to thylakoid biogenesis via alternative lipid biosynthetic pathways (Kobayashi, Narise et al. 2013; Kobayashi, Fujii, Sato et al. 2015). These data indicate a coupled regulation of Chl genes with formation of the thylakoid lipid bilayer to coordinate Chl biosynthesis with thylakoid membrane biogenesis. Indeed, Chl biosynthesis and thylakoid lipid biosynthesis are transcriptionally coordinated with each other in response to developmental and environmental cues (Kobayashi, Fujii, Sasaki et al. 2014). This regulation may involve plastid-to-nucleus retrograde signaling, as described next.

When the chloroplast function is severely disrupted, photosynthesis-associated nuclear genes, including most Chl genes, are downregulated to prevent further photooxidative damages. Various *Arabidopsis* mutants that fail to downregulate photosynthesis-associated nuclear genes in response to chloroplast dysfunction have been identified and characterized to reveal the plastid-to-nucleus retrograde signaling pathway (Nott et al. 2006). Key Chl genes, which are downregulated by chloroplast defects in wild type, showed constant expression in gun mutants defective in the plastid signaling with chloroplast dysfunction (Strand et al. 2003; Moulin et al. 2008), indicating that these genes are under strict plastid-signaling control.

One of the factors playing a central role in plastid signaling is a plastidic pentatricopeptide repeat protein, GUN1, which integrates multiple plastid signals to downregulate photosynthesis-associated nuclear genes in response to chloroplast defects. In a currently proposed model (Koussevitzky et al. 2007; Sun et al. 2011), GUN1 activated by chloroplast dysfunction somehow induces proteolytic cleavage of PTM, a chloroplast envelope-bound plant homeodomain transcription factor with transmembrane domains. The processed PTM accumulates in the nucleus and upregulates the ABA INSENSITIVE4 (ABI4) transcription factor through histone modifications. Subsequently, ABI4 binds to the CCAC motif of the CCACGT box and inhibits *LHCB* expression presumably by preventing the binding of G-box transcription factors. However, the involvement of ABI4 in regulation of Chl genes needs to be elucidated.

The expression of *GLK1* and *GLK2*, which are direct upregulators of key Chl genes, is also regulated in response to chloroplast functionality through plastid signaling. GUN1 downregulates *GLK1* expression when chloroplasts are dysfunctional and thereby represses the expression of GLK-regulated genes, including key Chl genes, independently of ABI4 signaling (Kakizaki et al. 2009; Waters et al. 2009). GLKs control Chl synthesis downstream of the retrograde signaling pathway at the transcriptional level to optimize synthesis of Chl under varying environmental and developmental conditions.

From an analysis of the *LHCB* expression, HY5 is proposed to repress its target genes through cryptochrome (CRY) when chloroplasts are dysfunctional (Ruckle, DeMarco, and Larkin 2007). In this model, a plastid signal converts HY5 from a positive to a negative regulator of photosynthesis-associated nuclear genes independently of GUN1 signaling, resulting in downregulation of HY5-targeted genes. Indeed, inhibition of the light-signaling pathway by mutations in CRY1 or HY5 caused strong photodamage under high-light conditions due to dysregulation of photosynthesis-associated genes. Introduction of the *gun1* mutation in these light-signaling mutants further enhanced the light-dependent growth defect (Ruckle, DeMarco, and Larkin 2007). Thus, the dual regulation of photosynthesis-associated genes by the repressive light signaling through CRY1 and HY5 and the GUN1 signaling is likely to be important to control the chloroplast biogenesis in response to light conditions and chloroplast functionality. The flexible conversion of light-signaling pathways from positive to negative regulators may facilitate fine-tuning of the biosynthesis of Chl and cognate proteins in diverse light environments, although the mechanism converting light-signaling pathways remains elusive.

9.4.2 Posttranslational Regulation

It is known that plants exert multiple levels of the posttranslational regulation over tetrapyrrole metabolism. These regulatory mechanisms may promptly adjust the rate of tetrapyrrole biosynthesis depending on environmental factors and plant growth and development. Posttranslational regulation of tetrapyrrole biosynthesis is also summarized by Czarneki and Grimm (2012).

9.4.2.1 Regulation of Enzyme Activities

ALA biosynthesis is the first regulatory point and subject to regulation by the end products. GluTR is regarded as the central regulator of Chl biosynthesis, and its activity is regulated by multiple posttranslational mechanisms in addition to transcriptional regulation. Heme is a potent feedback inhibitor of GluTR and can suppress ALA formation, which means that heme biosynthesis affects overall tetrapyrrole biosynthesis, including Chl biosynthesis (Beale 1999). It is reported that heme directly binds to GluTR and inhibits its activity at a site distinct from the catalytic center (Pontoppidan and Kannangara 1994; Vothknecht, Kannangara, and von Wettstein 1996; Vothknecht, Kannangara, and von Wettstein 1998) and the FLU binding site of the enzyme (see Section 9.3.2) (Goslings et al. 2004). In fact, an impairment of heme catabolism in hy1 and hy2 mutants in the Fe branch leads to repression of GluTR activity (Terry and Kendrick 1999), reflecting the feedback inhibitory function of heme in vivo. Moreover, a mutation (ulf3) that suppresses the flu phenotype is identified in the HO1 locus encoding a major HO in Arabidopsis, suggesting that the accumulation of heme in the ulf3 mutant suppresses GluTR activity independently of the action of FLU, which may result in a block of the excessive Pchlide accumulation in flu (Goslings et al. 2004).

In addition, several experiments show a strong inverse correlation between Pchlide accumulation and ALA synthesis, implying that high Pchlide accumulation induces downregulation of Chl biosynthesis at the level of ALA biosynthesis (Fluhr et al. 1975; Huang and Castelfranco 1989). This idea is consistent with the fact that feeding of ALA to dark-grown seedlings, which can bypass the negative feedback regulation of ALA synthesis, resulted in a substantial accumulation of Pchlide and other tetrapyrrole intermediates (Terry and Kendrick 1999). The activity of GluTR1 is found to be downregulated by the membrane-localized FLU protein in response to dark-accumulating Pchlide (Meskauskiene et al. 2001). As described, FLU is not likely associated with HEMA2-encoded GluTR2 (Meskauskiene and Apel 2002). GluTR2 expressed under control of the HEMA1 promoter in the hema1 mutant continuously synthesized ALA and accumulated tetrapyrrole intermediates in the dark, resulting in leaf necrosis after illumination (Apitz et al. 2014). The data confirm that, unlike GluTR1, GluTR2 is independent of the posttranslational regulation by FLU. Meanwhile, low GluTR2 accumulation despite the high HEMA2 expression under the HEMA1 promoter in the dark implies the existence of different posttranscriptional regulation for both GluTR

isoforms (Apitz et al. 2014). FLU-like proteins (FLPs) have been also identified in C. reinhardtii (Falciatore et al. 2005) and a gymnosperm, Picea abies (Demko et al. 2009), both of which barely accumulate Pchlide in the dark owing to the DPOR activity. Chlamydomonas FLPs can interact with both the Chlamydomonas GluTR and the Arabidopsis GluTR1 and partially complement the flu mutation in Arabidopsis. Chlamydomonas mutants with reduced FLP levels accumulate high levels of Pchlide, Proto, and Mg Proto in the dark, and cannot survive after transfer to high-light conditions, reflecting a crucial role of FLPs in regulation of Chl biosynthesis (Falciatore et al. 2005).

During daily growth conditions, MgCh activity increased and decreased in the day and at night, respectively, whereas FeCh activity showed the opposite profile (Papenbrock et al. 1999). This inverse relationship would be in part caused by the changes in the stromal concentrations of ATP and Mg^{2+}. Although ATP is essential for MgCh activity (Gibson et al. 1995; Reid and Hunter 2004), it inhibits FeCh activity (Cornah et al. 2002). MgCh activity is also stimulated by increased Mg^{2+} concentration in the light (Reid and Hunter 2004), suggesting that light/dark cycles influence MgCh activity through varying ATP/ADP ratios and Mg^{2+} concentrations in plastids. In fact, in chloroplasts, ATP/ADP ratios increase from 1 in the dark to 4 in the light (Usuda 1988), while Mg^{2+} concentrations increase from 0.5 mM in the dark to 2 mM in the light (Ishijima et al. 2003). The required Mg^{2+} concentration for the full activation of MgCh decreased in vitro in the presence of GUN4 (Davison et al. 2005), so GUN4 would play an important role in the Mg^{2+}-induced MgCh activation. In addition, Mg^{2+} influences the localization of the H subunit (CHLH) within plastids; in 1 mM Mg^{2+}, CHLH is mainly present in the stroma fraction, whereas in 5 mM Mg^{2+}, a significant portion of the CHLH protein is associated with membranes, particularly those from the envelope fraction (Gibson et al. 1996; Nakayama et al. 1998). Translocation of CHLH from the stroma to the envelope may enable the substrate binding of CHLH. Similarly, it has been found that CHLH and GUN4 are translocated to the membrane in a porphyrin-dependent manner, suggesting that GUN4 contributes to MgCh activity not only by interacting with this enzyme but also by facilitating the interaction of the H subunit to chloroplast envelope membranes (Adhikari et al. 2011).

9.4.2.2 Redox Regulation

In chloroplasts, regulation of the activity of a number of enzymes involved in photosynthetic reactions is coupled to photosynthetic electron transport. The most extensively studied example of this kind of regulation is the thioredoxin (Trx) system, in which a light-induced change in enzyme activity is linked to the redox state of a disulfide bond located on the enzyme molecule. Trx is a class of ubiquitous small proteins containing a redox-reactive cysteine pair within the active site and has critical regulatory roles in various biological processes, including Chl biosynthesis and photosynthesis. In Arabidopsis plastids, two stromal types of Trx, f-type (Trx-f) and m-type (Trx-m), designated according to their first identified target proteins, have been initially studied. Thereafter, several other types of Trxs have been reported, which are classified according to their structures, target proteins, and localizations (Chibani et al. 2010). Single Trx-deficient mutants generally show no obvious growth defects, so Trx proteins may have functional redundancy in terms of their capacity to reduce thiol groups of the same target proteins. The thiol-reducing system that controls the redox state of chloroplast enzymes is driven by light and the ferredoxin-Trx-dependent and/or the NADPH-Trx-dependent pathway.

In the former pathway, the electrons flow through thiol–disulfide exchanges from ferredoxin to Trx. This reaction is catalyzed by ferredoxin-Trx reductase (FTR). Ferredoxin is reduced by photosynthetic electron transport and reacts with oxidized Trx through FTR activity. Reduced Trx can transfer electrons to the respective target proteins and reduce regulatory intramolecular disulfide bonds, thereby altering their conformation and enzymatic activities. In the dark, the regulatory sulfhydryl group of Trx-targeted enzymes become oxidized and deactivated. In plastids, a number of enzymes involved in the Calvin cycle, oxidative pentose phosphate cycle, and fatty acid biosynthesis are known to be activated by reduced Trx in the light and deactivated in the dark. Using Trx-affinity chromatography, several proteins involved in Chl biosynthesis (GSAT, UROD, ALAD, CHLI, and PPOX) have been identified as potential targets of Trx (Balmer et al. 2003). Among them, MgCh is known to be regulated by the redox state (Stenbaek and Jensen 2010). Continuous addition of dithiothreitol (DTT) is required for MgCh activity *in vitro*, indicating that thiols are essential for its catalytic activity (Jensen, Reid, and Hunter 2000). Inhibition of MgCh activity by a thiol modifier, *N*-ethylmaleimide (NEM), shows that cysteine residues in the I and H subunits may be involved in this redox regulation (Jensen, Reid, and Hunter 2000). Indeed, Trx reduces the CHLI subunits of *Arabidopsis* MgCh, stimulates their ATPase activity (Ikegami et al. 2007; Kobayashi et al. 2008), and increases the total MgCh activity *in vitro* and *in vivo* (Luo et al. 2012).

In the latter pathway, the electrons are transferred from NADPH to target proteins via NADPH-dependent Trx reductase (NTR). This enzyme contains a flavin-adeninediphosphate-binding (FAD-binding) domain and a double cysteine-forming peptide motif in the catalytic center. In *Arabidopsis*, three NTR isoforms have been identified; NTRC is exclusively localized to the plastid as one of the main redox regulators in photosynthetic and nonphotosynthetic plastids (Serrato et al. 2004; Kirchsteiger et al. 2012), while other NTR isoforms (NTRA and NTRB) are dual-targeted to the cytosol and mitochondria (Reichheld et al. 2007). NTRC is a unique enzyme containing a C-terminal TRX domain and acts as a bifunctional enzyme with NTR/TRX activity for the reduction of specific target proteins. NTRC is able to reduce 2-Cys peroxiredoxins (2-Cys PRXs), which are small thiol-containing proteins with antioxidative properties (Moon et al. 2006). Based on the NADPH dependency, its ability to reduce hydrogen peroxide-scavenging 2-Cys PRXs, and the severe phenotype of the *ntrc* mutant when exposed to long dark periods,

NTRC was initially proposed to be part of a hydrogen peroxide-scavenging system particularly important during nighttime, when reduced ferredoxin is limited but NADPH can be regenerated via the oxidative pentose phosphate pathway (Perez-Ruiz et al. 2006).

The activity of MgCY is stimulated *in vitro* by addition of catalase and ascorbic acid, probably by removing hydrogen peroxide produced during the reaction (Bollivar and Beale 1996). NTRC in combination with 2-Cys PRXs can substitute catalase in stimulation of MgCY activity, suggesting that NTRC is important for protection of the MgCY reaction (Stenbaek et al. 2008). In fact, the *Arabidopsis ntrc* mutant accumulates Mg Proto ME, a substrate of MgCY (Stenbaek et al. 2008). In addition, high levels of Mg Proto, the substrate of MPMT, accumulated in *ntrc* under ALA feeding, indicating that MPMT may also be a target of hydrogen peroxide inhibition (Stenbaek et al. 2008). Latter analysis of the *ntrc* mutant shows that protein levels of GluTR, MPMT, and POR are decreased in the mutant in a posttranslational manner, suggesting that NTRC affects the stability and activity of these enzymes (Richter and Grimm 2013; Richter et al. 2013). Moreover, NTRC interacts with the CHLI subunit in *Arabidopsis* and reduces it *in vitro* and *in vivo*. ATPase activity of the CHLI subunit is stimulated by NTRC more efficiently than by chloroplast Trxs (Perez-Ruiz et al. 2014). It is noted that NTRC failed to stimulate total MgCh activity (Stenbaek and Jensen 2010).

9.4.2.3 Regulation by Protein Stability

Chl *b* synthesizing activity is feedback-regulated through the stability of CAO. Except for the N-terminal transit peptide, the CAO protein sequence can be divided into three parts: A, B, and C domains (Nagata et al. 2004). The A domain is responsible for the regulation of CAO stability, while the C domain has a catalytic function and is conserved among nearly all CAO sequences. The B domain is considered to be a linker of the A and C domains (Sakuraba et al. 2007). Although overexpression of the gene encoding the full-length CAO protein in *Arabidopsis* failed to accumulate its translated product due to protein destabilization, removal of the A domain led to a significant accumulation of the truncated CAO protein. The full-length CAO protein was destabilized in a Chl *b* dependent manner, so the A domain may sense the presence of Chl *b* and regulate the accumulation of the CAO protein in the plastids (Yamasato et al. 2005). A chloroplast Clp protease was found to be involved in the destabilization of the CAO protein in response to Chl *b* accumulation (Nakagawara et al. 2007). A specific amino acid (degron) sequence in the A domain, which may be recognized by the Clp protease, is identified to be essential for the destabilization of CAO. Although the entire A domain was stabilized by the elimination of Chl *b* in the CAO-deficient *chlorina1* mutant, the degron sequence alone was unstable in the mutant (Sakuraba et al. 2009). This indicates that the A-domain residues other than the degron sequence are required to respond to the change in Chl *b* levels. The A domain may shield the degron sequence from the Clp protease in the absence of Chl *b*, whereas, in the presence of Chl *b*, a conformational change in the A domain may occur to expose the degron sequence, resulting in the proteolytic digestion of CAO by Clp protease (Tanaka, Ito, and Tanaka 2010; Tanaka, Kobayashi, and Masuda 2011; Tanaka and Tanaka 2011). This feedback mechanism may provide fine and prompt regulation of CAO activity to coordinate the pigment supply and its assembly into the photosynthetic apparatus.

9.5 CONCLUSIONS

While Chls allow cyanobacteria, algae, and plants to harvest sunlight and perform oxygenic photosynthesis, these highly reactive tetrapyrrole molecules at the same time cause a threat of photooxidative stresses to these organisms. To achieve efficient photosynthesis while preventing photooxidative damage by Chls and their intermediates, these oxygenic phototrophs have developed various regulatory mechanisms of Chl metabolism in addition to photoprotective systems in photosynthesis. The regulation of plant Chl metabolism covers a wide range of mechanisms, including transcriptional regulation in response to environmental and endogenous cues; feedback control of enzymes by tetrapyrrole products and intermediates; posttranslational regulation by cofactors, redox state, and level of protein stability; and formation of multiple protein complexes for substrate channeling and regulation. The strict regulation of Chl metabolism at multiple levels enables plants to coordinate this potentially harmful process with the formation of photosynthetic machinery and photoprotective systems and to fine-tune the metabolic flow to the varying need for Chl in response to exogenous and endogenous factors.

In plants, almost all of the genes and enzymes directly involved in Chl biosynthesis have been identified during the last few decades. In addition, genetic, biochemical, and molecular biological studies have identified key regulatory factors and processes involved in Chl biosynthesis. However, a number of questions on Chl metabolism, especially those on the regulatory mechanism and the interaction of Chl metabolism and other cellular processes, still need to be answered. Moreover, the mechanism of how Chl molecules newly synthesized or recycled are safely assembled with photosynthetic complexes is another challenging area of research.

ACKNOWLEDGMENTS

The authors would like to acknowledge support from Grants-in-Aid for Scientific Research (nos. 24570042, 22370016, and 24770055).

REFERENCES

Adhikari, N. D., R. Orler, J. Chory, J. E. Froehlich, and R. M. Larkin. 2009. Porphyrins promote the association of GENOMES UNCOUPLED 4 and a Mg-chelatase subunit with chloroplast membranes. *J. Biol. Chem.* 284:24783–96.

Adhikari, N. D., J. E. Froehlich, D. D. Strand, S. M. Buck, D. M. Kramer, and R. M. Larkin. 2011. GUN4–porphyrin complexes bind the ChlH/GUN5 subunit of Mg-Chelatase and promote chlorophyll biosynthesis in Arabidopsis. *Plant Cell* 23:1449–67.

Alawady, A., R. Reski, E. Yaronskaya, and B. Grimm. 2005. Cloning and expression of the tobacco CHLM sequence encoding Mg protoporphyrin IX methyltransferase and its interaction with Mg chelatase. *Plant Mol. Biol.* 57:679–91.

Albus, C. A., A. Salinas, O. Czarnecki, S. Kahlau, M. Rothbart, W. Thiele, W. Lein, R. Bock, B. Grimm, and M. A. Schottler. 2012. LCAA, a novel factor required for magnesium protoporphyrin monomethylester cyclase accumulation and feedback control of aminolevulinic acid biosynthesis in tobacco. *Plant Physiol.* 160:1923–39.

Andronis, C., S. Barak, S. M. Knowles, S. Sugano, and E. M. Tobin. 2008. The clock protein CCA1 and the bZIP transcription factor HY5 physically interact to regulate gene expression in Arabidopsis. *Mol. Plant* 1:58–67.

Apitz, J., J. Schmied, M. J. Lehmann, B. Hedtke, and B. Grimm. 2014. GluTR2 complements a hema1 mutant lacking glutamyl-tRNA reductase 1, but is differently regulated at the post-translational level. *Plant Cell Physiol.* 55:645–57.

Armstrong, G. A., S. Runge, G. Frick, U. Sperling, and K. Apel. 1995. Identification of NADPH:protochlorophyllide oxidoreductases A and B: A branch pathway for light-dependent chlorophyll biosynthesis in Arabidopsis thaliana. *Plant Physiol.* 108:1505–17.

Bae, G. and G. Choi. 2008. Decoding of light signals by plant phytochromes and their interacting proteins. *Annu. Rev. Plant Biol.* 59:281–311.

Balmer, Y., A. Koller, G. del Val, W. Manieri, P. Schurmann, and B. B. Buchanan. 2003. Proteomics gives insight into the regulatory function of chloroplast thioredoxins. *Proc. Natl. Acad. Sci. U.S.A.* 100:370–5.

Beale, S. I. 1999. Enzymes of chlorophyll biosynthesis. *Photosynth. Res.* 60:43–73.

Benhamed, M., C. Bertrand, C. Servet, and D. X. Zhou. 2006. Arabidopsis GCN5, HD1, and TAF1/HAF2 interact to regulate histone acetylation required for light-responsive gene expression. *Plant Cell* 18:2893–903.

Bollivar, D. W. and S. I. Beale. 1996. The chlorophyll biosynthetic enzyme Mg-protoporphyrin IX monomethyl ester (oxidative) cyclase (Characterization and partial purification from *Chlamydomonas reinhardtii* and *Synechocystis* sp. PCC 6803). *Plant Physiol.* 112:105–14.

Chattopadhyay, S., P. Puente, X.-W. Deng, and N. Wei. 1998. Combinatorial interaction of light-responsive elements plays a critical role in determining the response characteristics of light-regulated promoters in *Arabidopsis*. *Plant J.* 15:69–77.

Cheminant, S., M. Wild, F. Bouvier, S. Pelletier, J.-P. Renou, M. Erhardt, S. Hayes, M. J. Terry, P. Genschik, and P. Achard. 2011. DELLAs regulate chlorophyll and carotenoid biosynthesis to prevent photooxidative damage during seedling deetiolation in Arabidopsis. *Plant Cell* 23:1849–60.

Chiang, Y.-H., Y. Zubo, W. Tapken, H. J. Kim, A. Lavanway, L. Howard, M. Pilon, J. Kieber, and G. E. Schaller. 2012. Functional characterization of the GATA transcription factors GNC and CGA1 reveals their key role in chloroplast development, growth, and division in Arabidopsis. *Plant Physiol.* 160:332–48.

Chibani, K., J. Couturier, B. Selles, J. P. Jacquot, and N. Rouhier. 2010. The chloroplastic thiol reducing systems: Dual functions in the regulation of carbohydrate metabolism and regeneration of antioxidant enzymes, emphasis on the poplar redoxin equipment. *Photosynth. Res.* 104:75–99.

Chidgey, J. W., M. Linhartová, J. Komenda, P. J. Jackson, M. J. Dickman, D. P. Canniffe, P. Koník, J. Pilny, C. N. Hunter, and R. Sobotka. 2014. A cyanobacterial chlorophyll synthase–HliD complex associates with the Ycf39 protein and the YidC/Alb3 insertase. *Plant Cell* 26:1267–79.

Chory, J. and C. A. Peto. 1990. Mutations in the DET1 gene affect cell-type-specific expression of light-regulated genes and chloroplast development in Arabidopsis. *Proc. Natl. Acad. Sci. U.S.A.* 87:8776–80.

Cornah, J. E., J. M. Roper, D. Pal Singh, and A. G. Smith. 2002. Measurement of ferrochelatase activity using a novel assay suggests that plastids are the major site of haem biosynthesis in both photosynthetic and non-photosynthetic cells of pea (*Pisum sativum* L.). *Biochem. J.* 362:423–32.

Czarnecki, O. and B. Grimm. 2012. Post-translational control of tetrapyrrole biosynthesis in plants, algae, and cyanobacteria. *J. Exp. Bot.* 63:1675–87.

Czarnecki, O., B. Hedtke, M. Melzer, M. Rothbart, A. Richter, Y. Schroter, T. Pfannschmidt, and B. Grimm. 2012. An Arabidopsis GluTR binding protein mediates spatial separation of 5-aminolevulinic acid synthesis in chloroplasts. *Plant Cell* 23:4476–91.

Davison, P. A. and C. N. Hunter. 2011. Abolition of magnesium chelatase activity by the gun5 mutation and reversal by Gun4. *FEBS Lett.* 585:183–6.

Davison, P. A., H. L. Schubert, J. D. Reid, C. D. Iorg, A. Heroux, C. P. Hill, and C. N. Hunter. 2005. Structural and biochemical characterization of Gun4 suggests a mechanism for its role in chlorophyll biosynthesis. *Biochemistry* 44:7603–12.

de Lucas, M., J.-M. Davière, M. Rodríguez-Falcón, M. Pontin, J. M. Iglesias-Pedraz, S. Lorrain, C. Fankhauser, M. A. Blázquez, E. Titarenko, and S. Prat. 2008. A molecular framework for light and gibberellin control of cell elongation. *Nature* 451:480–4.

Demko, V., A. Pavlovic, D. Valková, L. Slováková, B. Grimm, and J. Hudák. 2009. A novel insight into the regulation of light-independent chlorophyll biosynthesis in Larix decidua and Picea abies seedlings. *Planta* 230:165–76.

Deng, X. W. and P. H. Quail. 1992. Genetic and phenotypic characterization of cop1 mutants of *Arabidopsis thaliana*. *Plant J.* 2:83–95.

Eckhardt, U., B. Grimm, and S. Hortensteiner. 2004. Recent advances in chlorophyll biosynthesis and breakdown in higher plants. *Plant Mol. Biol.* 56:1–14.

Falciatore, A., L. Merendino, F. Barneche, M. Ceol, R. Meskauskiene, K. Apel, and J. D. Rochaix. 2005. The FLP proteins act as regulators of chlorophyll synthesis in response to light and plastid signals in Chlamydomonas. *Genes Dev.* 19:176–87.

Feng, S., C. Martinez, G. Gusmaroli, Y. Wang, J. Zhou, F. Wang, L. Chen, L. Yu, J. M. Iglesias-Pedraz, S. Kircher, E. Schäfer, X. Fu, L.-M. Fan, and X.-W. Deng. 2008. Coordinated regulation of Arabidopsis thaliana development by light and gibberellins. *Nature* 451:475–9.

Feng, C. M., Y. Qiu, E. K. Van Buskirk, E. J. Yang, and M. Chen. 2014. Light-regulated gene repositioning in Arabidopsis. *Nat. Commun.* 5:3027.

Fitter, D. W., D. J. Martin, M. J. Copley, R. W. Scotland, and J. A. Langdale. 2002. GLK gene pairs regulate chloroplast development in diverse plant species. *Plant J.* 31:713–27.

Fluhr, R., E. Harel, S. Klein, and E. Meller. 1975. Control of delta-aminolevulinic acid and chlorophyll accumulation in greening maize leaves upon light–dark transitions. *Plant Physiol.* 56:497–501.

Fujii, S., K. Kobayashi, Y. Nakamura, and H. Wada. 2014. Inducible knockdown of MONOGALACTOSYLDIACYLGLYCEROL SYNTHASE1 reveals roles of galactolipids in organelle differentiation in Arabidopsis cotyledons. *Plant Physiol.* 166:1436–49.

Fujita, Y. and C. E. Bauer. 2003. The light-independent protochlorophyllide reductase: A nitrogenase-like enzyme catalyzing a key reaction for greening in the dark. In *The Porphyrin Handbook*, edited by K. M. Kadish, K. M. Smith and R. Guilard. Amsterdam: Elsevier Science.

Gendron, J. M., J. L. Pruneda-Paz, C. J. Doherty, A. M. Gross, S. E. Kang, and S. A. Kay. 2012. Arabidopsis circadian clock protein, TOC1, is a DNA-binding transcription factor. *Proc. Natl. Acad. Sci. U.S.A.* 109:3167–72.

Gibson, L. C., R. D. Willows, C. G. Kannangara, D. von Wettstein, and C. N. Hunter. 1995. Magnesium-protoporphyrin chelatase of *Rhodobacter sphaeroides*: Reconstitution of activity by combining the products of the bchH, -I, and -D genes expressed in *Escherichia coli*. *Proc. Natl. Acad. Sci. U.S.A.* 92:1941–4.

Gibson, L. C., J. L. Marrison, R. M. Leech, P. E. Jensen, D. C. Bassham, M. Gibson, and C. N. Hunter. 1996. A putative Mg chelatase subunit from Arabidopsis thaliana cv C24. Sequence and transcript analysis of the gene, import of the protein into chloroplasts, and in situ localization of the transcript and protein. *Plant Physiol.* 111:61–71.

Goslings, D., R. Meskauskiene, C. Kim, K. P. Lee, M. Nater, and K. Apel. 2004. Concurrent interactions of heme and FLU with Glu tRNA reductase (HEMA1), the target of metabolic feedback inhibition of tetrapyrrole biosynthesis, in dark- and light-grown Arabidopsis plants. *Plant J.* 40:957–67.

Gough, S. P. and C. G. Kannangara. 1976. Synthesis of δ-amino levulinic acid by isolate plastids. *Carlsberg Res. Commun.* 41:183–90.

Hedtke, B., A. Alawady, A. Albacete, K. Kobayashi, M. Melzer, T. Roitsch, T. Masuda, and B. Grimm. 2011. Deficiency in riboflavin biosynthesis affects tetrapyrrole biosynthesis in etiolated Arabidopsis tissue. *Plant Mol. Biol.* 78:77–93.

Hinchigeri, S. B., B. Hundle, and W. R. Richards. 1997. Demonstration that the BchH protein of Rhodobacter capsulatus activates S-adenosyl-L-methionine:magnesium protoporphyrin IX methyltransferase. *FEBS Lett.* 407:337–42.

Hirashima, M., R. Tanaka, and A. Tanaka. 2009. Light-independent cell death induced by accumulation of pheophorbide a in Arabidopsis thaliana. *Plant Cell Physiol.* 50:719–29.

Hollingshead, S., J. Kopecna, P. J. Jackson, D. P. Canniffe, P. A. Davison, M. J. Dickman, R. Sobotka, and C. N. Hunter. 2012. Conserved chloroplast open-reading frame ycf54 is required for activity of the magnesium protoporphyrin monomethylester oxidative cyclase in Synechocystis PCC 6803. *J. Biol. Chem.* 287:27823–33.

Holm, M., L. G. Ma, L. J. Qu, and X. W. Deng. 2002. Two interacting bZIP proteins are direct targets of COP1-mediated control of light-dependent gene expression in Arabidopsis. *Genes Dev.* 16:1247–59.

Hsu, P. Y. and S. L. Harmer. 2014. Wheels within wheels: The plant circadian system. *Trends Plant Sci.* 19:240–9.

Huang, L. and P. A. Castelfranco. 1989. Regulation of 5-aminolevulinic acid synthesis in developing chloroplasts 1. Effect of light/dark treatments *in vivo* and *in organello*. *Plant Physiol.* 90: 996–1002.

Huang, W., P. Perez-Garcia, A. Pokhilko, A. J. Millar, I. Antoshechkin, J. L. Riechmann, and P. Mas. 2012. Mapping the core of the Arabidopsis circadian clock defines the network structure of the oscillator. *Science* 336:75–9.

Hudson, D., D. Guevara, M. W. Yaish, C. Hannam, N. Long, J. D. Clarke, Y.-M. Bi, and S. J. Rothstein. 2011. GNC and CGA1 modulate chlorophyll biosynthesis and glutamate synthase (GLU1/Fd-GOGAT) expression in Arabidopsis. *PLoS One* 6:e26765.

Hudson, D., D. R. Guevara, A. J. Hand, Z. Xu, L. Hao, X. Chen, T. Zhu, Y. M. Bi, and S. J. Rothstein. 2013. Rice cytokinin GATA transcription Factor1 regulates chloroplast development and plant architecture. *Plant Physiol.* 162:132–44.

Huq, E., B. Al-Sady, M. Hudson, C. Kim, K. Apel, and P. H. Quail. 2004. Phytochrome-interacting factor 1 is a critical bHLH regulator of chlorophyll biosynthesis. *Science* 305:1937–41.

Ikegami, A., N. Yoshimura, K. Motohashi, S. Takahashi, P. G. N. Romano, T. Hisabori, K.-I. Takamiya, and T. Masuda. 2007. The CHLI1 subunit of Arabidopsis thaliana magnesium chelatase is a target protein of the chloroplast thioredoxin. *J. Biol. Chem.* 282:19282–91.

Ishijima, S., A. Uchibori, H. Takagi, R. Maki, and M. Ohnishi. 2003. Light-induced increase in free Mg2+ concentration in spinach chloroplasts: Measurement of free Mg2+ by using a fluorescent probe and necessity of stromal alkalinization. *Arch. Biochem. Biophys.* 412:126–32.

Jensen, P. E., J. D. Reid, and C. N. Hunter. 2000. Modification of cysteine residues in the ChlI and ChlH subunits of magnesium chelatase results in enzyme inactivation. *Biochem. J.* 352:435–41.

Joyard, J., M. Ferro, C. Masselon, D. Seigneurin-Berny, D. Salvi, J. Garin, and N. Rolland. 2009. Chloroplast proteomics and the compartmentation of plastidial isoprenoid biosynthetic pathways. *Mol. Plant* 2:1154–80.

Kakizaki, T., H. Matsumura, K. Nakayama, F. S. Che, R. Terauchi, and T. Inaba. 2009. Coordination of plastid protein import and nuclear gene expression by plastid-to-nucleus retrograde signaling. *Plant Physiol.* 151:1339–53.

Kato, K., R. Tanaka, S. Sano, A. Tanaka, and H. Hosaka. 2010. Identification of a gene essential for protoporphyrinogen IX oxidase activity in the cyanobacterium Synechocystis sp. PCC6803. *Proc. Natl. Acad. Sci. U.S.A.* 107:16649–54.

Kauss, D., S. Bischof, S. Steiner, K. Apel, and R. Meskauskiene. 2012. FLU, a negative feedback regulator of tetrapyrrole biosynthesis, is physically linked to the final steps of the Mg(++)-branch of this pathway. *FEBS Lett.* 586:211–16.

Kim, S., H. Schlicke, K. Van Ree, K. Karvonen, A. Subramaniam, A. Richter, B. Grimm, and J. Braam. 2013. Arabidopsis chlorophyll biosynthesis: An essential balance between the methylerythritol phosphate and tetrapyrrole pathways. *Plant Cell* 25:4984–93.

Kirchsteiger, K., J. Ferrandez, M. B. Pascual, M. Gonzalez, and F. J. Cejudo. 2012. NADPH thioredoxin reductase C is localized in plastids of photosynthetic and nonphotosynthetic tissues and is involved in lateral root formation in Arabidopsis. *Plant Cell* 24:1534–48.

Kleine, T., C. Voigt, and D. Leister. 2009. Plastid signalling to the nucleus: messengers still lost in the mists? *Trends Genet.* 25:185–92.

Kobayashi, K., N. Mochizuki, N. Yoshimura, K. Motohashi, T. Hisabori, and T. Masuda. 2008. Functional analysis of Arabidopsis thaliana isoforms of the Mg-chelatase CHLI subunit. *Photochem. Photobiol. Sci.* 7:1188–95.

Kobayashi, K., S. Baba, T. Obayashi, M. Sato, K. Toyooka, M. Keränen, E.-M. Aro, H. Fukaki, H. Ohta, K. Sugimoto, and T. Masuda. 2012. Regulation of root greening by light and auxin/cytokinin signaling in Arabidopsis. *Plant Cell* 24:1081–95.

Kobayashi, K., T. Obayashi, and T. Masuda. 2012. Role of the G-box element in regulation of chlorophyll biosynthesis in Arabidopsis roots. *Plant Signal. Behav.* 7:922–6.

Kobayashi, K., T. Narise, K. Sonoike, H. Hashimoto, N. Sato, M. Kondo, M. Nishimura, M. Sato, K. Toyooka, K. Sugimoto, H. Wada, T. Masuda, and H. Ohta. 2013. Role of galactolipid biosynthesis in coordinated development of photosynthetic complexes and thylakoid membranes during chloroplast biogenesis in Arabidopsis. *Plant J.* 73:250–61.

Kobayashi, K., D. Sasaki, K. Noguchi, D. Fujinuma, H. Komatsu, M. Kobayashi, M. Sato, K. Toyooka, K. Sugimoto, K. K. Niyogi, H. Wada, and T. Masuda. 2013. Photosynthesis of root chloroplasts developed in Arabidopsis lines overexpressing GOLDEN2-LIKE transcription factors. *Plant Cell Physiol.* 54:1365–77.

Kobayashi, K., S. Fujii, D. Sasaki, S. Baba, H. Ohta, T. Masuda, and H. Wada. 2014. Transcriptional regulation of thylakoid galactolipid biosynthesis coordinated with chlorophyll biosynthesis during the development of chloroplasts in Arabidopsis. *Front. Plant Sci.* 5:272.

Kobayashi, K., T. Masuda, N. Tajima, H. Wada, and N. Sato. 2014. Molecular phylogeny and intricate evolutionary history of the three isofunctional enzymes involved in the oxidation of protoporphyrinogen IX. *Genome Biol. Evol.* 6:2141–55.

Kobayashi, K., S. Fujii, M. Sato, K. Toyooka, and H. Wada. 2015. Specific role of phosphatidylglycerol and functional overlaps with other thylakoid lipids in Arabidopsis chloroplast biogenesis. *Plant Cell Rep.* 34:631–42.

Koch, M., C. Breithaupt, R. Kiefersauer, J. Freigang, R. Huber, and A. Messerschmidt. 2004. Crystal structure of protoporphyrinogen IX oxidase: A key enzyme in haem and chlorophyll biosynthesis. *EMBO J.* 23:1720–8.

Koussevitzky, S., A. Nott, T. C. Mockler, F. Hong, G. Sachetto-Martins, M. Surpin, J. Lim, R. Mittler, and J. Chory. 2007. Signals from chloroplasts converge to regulate nuclear gene expression. *Science* 316:715–19.

Larkin, R. M., J. M. Alonso, J. R. Ecker, and J. Chory. 2003. GUN4, a regulator of chlorophyll synthesis and intracellular signaling. *Science* 299:902–6.

Lau, O. S. and X. W. Deng. 2012. The photomorphogenic repressors COP1 and DET1: 20 years later. *Trends Plant Sci.* 17:584–93.

Lee, J., K. He, V. Stolc, H. Lee, P. Figueroa, Y. Gao, W. Tongprasit, H. Zhao, I. Lee, and X. W. Deng. 2007. Analysis of transcription factor HY5 genomic binding sites revealed its hierarchical role in light regulation of development. *Plant Cell* 19:731–49.

Legnaioli, T., J. Cuevas, and P. Mas. 2009. TOC1 functions as a molecular switch connecting the circadian clock with plant responses to drought. *EMBO J.* 28:3745–57.

Leivar, P. and P. H. Quail. 2011. PIFs: Pivotal components in a cellular signaling hub. *Trends Plant Sci.* 16:19–28.

Leivar, P., J. M. Tepperman, E. Monte, R. H. Calderon, T. L. Liu, and P. H. Quail. 2009. Definition of early transcriptional circuitry involved in light-induced reversal of PIF-imposed repression of photomorphogenesis in young Arabidopsis seedlings. *Plant Cell* 21:3535–53.

Lermontova, I. and B. Grimm. 2006. Reduced activity of plastid protoporphyrinogen oxidase causes attenuated photodynamic damage during high-light compared to low-light exposure. *Plant J.* 48:499–510.

Li, J., G. Li, S. Gao, C. Martinez, G. He, Z. Zhou, X. Huang, J. H. Lee, H. Zhang, Y. Shen, H. Wang, and X. W. Deng. 2010. Arabidopsis transcription factor ELONGATED HYPOCOTYL5 plays a role in the feedback regulation of phytochrome A signaling. *Plant Cell* 22:3634–49.

Li, G., H. Siddiqui, Y. B. Teng, R. C. Lin, X. Y. Wan, J. G. Li, O. S. Lau, X. H. Ouyang, M. Q. Dai, J. M. Wan, P. F. Devlin, X. W. Deng, and H. Y. Wang. 2011. Coordinated transcriptional regulation underlying the circadian clock in Arabidopsis. *Nat. Cell Biol.* 13:616–22.

Lin, Y. P., T. Y. Lee, A. Tanaka, and Y. Y. Charng. 2014. Analysis of an Arabidopsis heat-sensitive mutant reveals that chlorophyll synthase is involved in reutilization of chlorophyllide during chlorophyll turnover. *Plant J.* 80:14–26.

Liu, X., C. Y. Chen, K. C. Wang, M. Luo, R. Tai, L. Yuan, M. Zhao, S. Yang, G. Tian, Y. Cui, H. L. Hsieh, and K. Wu. 2013. PHYTOCHROME INTERACTING FACTOR3 associates with the histone deacetylase HDA15 in repression of chlorophyll biosynthesis and photosynthesis in etiolated Arabidopsis seedlings. *Plant Cell* 25:1258–73.

Luer, C., S. Schauer, K. Mobius, J. Schulze, W. D. Schubert, D. W. Heinz, D. Jahn, and J. Moser. 2005. Complex formation between glutamyl-tRNA reductase and glutamate-1-semialdehyde 2,1-aminomutase in Escherichia coli during the initial reactions of porphyrin biosynthesis. *J. Biol. Chem.* 280:18568–72.

Luo, T., T. Fan, Y. Liu, M. Rothbart, J. Yu, S. Zhou, B. Grimm, and M. Luo. 2012. Thioredoxin redox regulates ATPase activity of magnesium chelatase CHLI subunit and modulates redox-mediated signaling in tetrapyrrole biosynthesis and homeostasis of reactive oxygen species in pea plants. *Plant Physiol.* 159:118–30.

Ma, Z., X. Hu, W. Cai, W. Huang, X. Zhou, Q. Luo, H. Yang, J. Wang, and J. Huang. 2014. Arabidopsis miR171-targeted scarecrow-like proteins bind to GT cis-elements and mediate gibberellin-regulated chlorophyll biosynthesis under light conditions. *PLoS Genet.* 10:e1004519.

Manfield, I. W., P. F. Devlin, C. H. Jen, D. R. Westhead, and P. M. Gilmartin. 2007. Conservation, convergence, and divergence of light-responsive, circadian-regulated, and tissue-specific expression patterns during evolution of the Arabidopsis GATA gene family. *Plant Physiol.* 143:941–58.

Mara, C. D. and V. F. Irish. 2008. Two GATA transcription factors are downstream effectors of floral homeotic gene action in Arabidopsis. *Plant Physiol.* 147:707–18.

Masuda, T. and Y. Fujita. 2008. Regulation and evolution of chlorophyll metabolism. *Photochem. Photobiol. Sci.* 7:1131–49.

Masuda, T., T. Suzuki, H. Shimada, H. Ohta, and K. Takamiya. 2003. Subcellular localization of two types of ferrochelatase in cucumber. *Planta* 217:602–9.

Matsumoto, F., T. Obayashi, Y. Sasaki-Sekimoto, H. Ohta, K. Takamiya, and T. Masuda. 2004. Gene expression profiling of the tetrapyrrole metabolic pathway in Arabidopsis with a mini-array system. *Plant Physiol.* 135:2379–91.

McCormac, A. C. and M. J. Terry. 2002. Loss of nuclear gene expression during the phytochrome A-mediated far-red block of greening response. *Plant Physiol.* 130:402–14.

Meskauskiene, R. and K. Apel. 2002. Interaction of FLU, a negative regulator of tetrapyrrole biosynthesis, with the glutamyl-tRNA reductase requires the tetratricopeptide repeat domain of FLU. *FEBS Lett.* 532:27–30.

Meskauskiene, R., M. Nater, D. Goslings, F. Kessler, R. op den Camp, and K. Apel. 2001. FLU: A negative regulator of chlorophyll biosynthesis in *Arabidopsis thaliana*. *Proc. Natl. Acad. Sci. U.S.A.* 98:12826–31.

Mochizuki, N., R. Tanaka, B. Grimm, T. Masuda, M. Moulin, A. G Smith, A. Tanaka, and M. J. Terry. 2010. The cell biology of tetrapyrroles: A life and death struggle. *Trends Plant Sci.* 15:488–98.

Moon, J. C., H. H. Jang, H. B. Chae, J. R. Lee, S. Y. Lee, Y. J. Jung, M. R. Shin, H. S. Lim, W. S. Chung, D. J. Yun, K. O. Lee, and S. Y. Lee. 2006. The C-type Arabidopsis thioredoxin reductase ANTR-C acts as an electron donor to 2-Cys peroxiredoxins in chloroplasts. *Biochem. Biophys. Res. Commun.* 348:478–84.

Moser, J., W. D. Schubert, V. Beier, I. Bringemeier, D. Jahn, and D. W. Heinz. 2001. V-shaped structure of glutamyl-tRNA reductase, the first enzyme of tRNA-dependent tetrapyrrole biosynthesis. *EMBO J.* 20:6583–90.

Moulin, M., A. C. McCormac, M. J. Terry, and A. G. Smith. 2008. Tetrapyrrole profiling in Arabidopsis seedlings reveals that retrograde plastid nuclear signaling is not due to Mg-protoporphyrin IX accumulation. *Proc. Natl. Acad. Sci. U.S.A.* 105:15178–83.

Nagata, N., S. Satoh, R. Tanaka, and A. Tanaka. 2004. Domain structures of chlorophyllide a oxygenase of green plants and Prochlorothrix hollandica in relation to catalytic functions. *Planta* 218:1019–25.

Nagata, N., R. Tanaka, and A. Tanaka. 2007. The major route for chlorophyll synthesis includes [3,8-divinyl]-chlorophyllide a reduction in Arabidopsis thaliana. *Plant Cell Physiol.* 48:1803–8.

Naito, T., T. Kiba, N. Koizumi, T. Yamashino, and T. Mizuno. 2007. Characterization of a unique GATA family gene that responds to both light and cytokinin in Arabidopsis thaliana. *Biosci. Biotechnol. Biochem.* 71:1557–60.

Nakagawara, E., Y. Sakuraba, A. Yamasato, R. Tanaka, and A. Tanaka. 2007. Clp protease controls chlorophyll b synthesis by regulating the level of chlorophyllide a oxygenase. *Plant J.* 49:800–9.

Nakamura, H., M. Muramatsu, M. Hakata, O. Ueno, Y. Nagamura, H. Hirochika, M. Takano, and H. Ichikawa. 2009. Ectopic overexpression of the transcription factor OsGLK1 induces chloroplast development in non-green rice cells. *Plant Cell Physiol.* 50:1933–49.

Nakayama, M., T. Masuda, N. Sato, H. Yamagata, C. Bowler, H. Ohta, Y. Shioi, and K. Takamiya. 1995. Cloning, subcellular localization and expression of CHL1, a subunit of magnesium-chelatase in soybean. *Biochem. Biophys. Res. Commun.* 215:422–8.

Nakayama, M., T. Masuda, T. Bando, H. Yamagata, H. Ohta, and K. Takamiya. 1998. Cloning and expression of the soybean chlH gene encoding a subunit of Mg-chelatase and localization of the Mg2+ concentration-dependent ChlH protein within the chloroplast. *Plant Cell Physiol.* 39:275–84.

Nogaj, L. A. and S. I. Beale. 2005. Physical and kinetic interactions between glutamyl-tRNA reductase and glutamate-1-semialdehyde aminotransferase of Chlamydomonas reinhardtii. *J. Biol. Chem.* 280:24301–7.

Nott, A., H. S. Jung, S. Koussevitzky, and J. Chory. 2006. Plastid-to-nucleus retrograde signaling. *Annu. Rev. Plant Biol.* 57:739–59.

Obornik, M. and B. R. Green. 2005. Mosaic origin of the heme biosynthesis pathway in photosynthetic eukaryotes. *Mol. Biol. Evol.* 22:2343–53.

Oosawa, N., T. Masuda, K. Awai, N. Fusada, H. Shimada, H. Ohta, and K. Takamiya. 2000. Identification and light-induced expression of a novel gene of NADPH–protochlorophyllide oxidoreductase isoform in *Arabidopsis thaliana*. *FEBS Lett.* 474:133–6.

op den Camp, R. G. L., D. Przybyla, C. Ochsenbein, C. Laloi, C. Kim, A. Danon, D. Wagner, E. Hideg, C. Göbel, I. Feussner, M. Nater, and K. Apel. 2003. Rapid induction of distinct stress responses after the release of singlet oxygen in Arabidopsis. *Plant Cell* 15:2320–32.

Oster, U., R. Tanaka, A. Tanaka, and W. Rudiger. 2000. Cloning and functional expression of the gene encoding the key enzyme for chlorophyll b biosynthesis (CAO) from Arabidopsis thaliana. *Plant J.* 21:305–10.

Osterlund, M. T., C. S. Hardtke, N. Wei, and X. W. Deng. 2000. Targeted destabilization of HY5 during light-regulated development of Arabidopsis. *Nature* 405:462–6.

Ouyang, X., J. Li, G. Li, B. Li, B. Chen, H. Shen, X. Huang, X. Mo, X. Wan, R. Lin, S. Li, H. Wang, and X. W. Deng. 2011. Genome-wide binding site analysis of FAR-RED ELONGATED HYPOCOTYL3 reveals its novel function in Arabidopsis development. *Plant Cell* 23:2514–35.

Oyama, T., Y. Shimura, and K. Okada. 1997. The Arabidopsis HY5 gene encodes a bZIP protein that regulates stimulus-induced development of root and hypocotyl. *Genes Dev.* 11:2983–95.

Papenbrock, J., H.-P. Mock, E. Kruse, and B. Grimm. 1999. Expression studies in tetrapyrrole biosynthesis: Inverse maxima of megnesium chelatase and ferrochelatase activity during cyclic photoperiods. *Planta* 208:264–73.

Papenbrock, J., S. Mishra, H. P. Mock, E. Kruse, E. K. Schmidt, A. Petersmann, H. P. Braun, and B. Grimm. 2001. Impaired expression of the plastidic ferrochelatase by antisense RNA synthesis leads to a necrotic phenotype of transformed tobacco plants. *Plant J.* 28:41–50.

Pepper, A., T. Delaney, T. Washburn, D. Poole, and J. Chory. 1994. *DET1*, a negative regulator of light-mediated development and gene expression in Arabidopsis, encodes a novel nuclear-localized protein. *Cell* 78:109–16.

Perez-Ruiz, J. M., M. C. Spinola, K. Kirchsteiger, J. Moreno, M. Sahrawy, and F. J. Cejudo. 2006. Rice NTRC is a high-efficiency redox system for chloroplast protection against oxidative damage. *Plant Cell* 18:2356–68.

Perez-Ruiz, J. M., M. Guinea, L. Puerto-Galan, and F. J. Cejudo. 2014. NADPH thioredoxin reductase C is involved in redox regulation of the Mg-chelatase I subunit in Arabidopsis thaliana chloroplasts. *Mol. Plant* 7:1252–5.

Pontoppidan, B. and C. G. Kannangara. 1994. Purification and partial characterization of barley glutamyl-tRNA^Glu reductase, the enzyme that directs glutamate to chlorophyll synthesis. *Eur. J. Biochem.* 225:529–37.

Powell, A. L. T., C. V. Nguyen, T. Hill, K. L. Cheng, R. Figueroa-Balderas, H. Aktas, H. Ashrafi, C. Pons, R. Fernández-Muñoz, A. Vicente, J. Lopez-Baltazar, C. S. Barry, Y. Liu, R. Chetelat, A. Granell, A. Van Deynze, J. J. Giovannoni, and A. B. Bennett. 2012. Uniform ripening encodes a Golden 2-like transcription factor regulating tomato fruit chloroplast development. *Science* 336:1711–15.

Reichheld, J. P., M. Khafif, C. Riondet, M. Droux, G. Bonnard, and Y. Meyer. 2007. Inactivation of thioredoxin reductases reveals a complex interplay between thioredoxin and glutathione pathways in Arabidopsis development. *Plant Cell* 19:1851–65.

Reid, J. D. and C. N. Hunter. 2004. Magnesium-dependent ATPase activity and cooperativity of magnesium chelatase from Synechocystis sp. PCC6803. *J. Biol. Chem.* 279:26893–9.

Reisinger, V., M. Ploscher, and L. A. Eichacker. 2008. Lil3 assembles as chlorophyll-binding protein complex during deetiolation. *FEBS Lett.* 582:1547–51.

Richter, A. S. and B. Grimm. 2013. Thiol-based redox control of enzymes involved in the tetrapyrrole biosynthesis pathway in plants. *Front. Plant Sci.* 4:371.

Richter, R., C. Behringer, I. K. Muller, and C. Schwechheimer. 2010. The GATA-type transcription factors GNC and GNL/CGA1 repress gibberellin signaling downstream from DELLA proteins and PHYTOCHROME-INTERACTING FACTORS. *Genes Dev.* 24:2093–104.

Richter, A. S., E. Peter, M. Rothbart, H. Schlicke, J. Toivola, E. Rintamaki, and B. Grimm. 2013. Posttranslational influence of NADPH-dependent thioredoxin reductase C on enzymes in tetrapyrrole synthesis. *Plant Physiol.* 162:63–73.

Ruckle, M. E., S. M. DeMarco, and R. M. Larkin. 2007. Plastid signals remodel light signaling networks and are essential for efficient chloroplast biogenesis in Arabidopsis. *Plant Cell* 19:3944–60.

Rzeznicka, K., C. J. Walker, T. Westergren, C. G. Kannangara, D. von Wettstein, S. Merchant, S. P. Gough, and M. Hansson. 2005. Xantha-1 encodes a membrane subunit of the aerobic Mg-protoporphyrin IX monomethyl ester cyclase involved in chlorophyll biosynthesis. *Proc. Natl. Acad. Sci. U.S.A.* 102:5886–91.

Sakuraba, Y., A. Yamasato, R. Tanaka, and A. Tanaka. 2007. Functional analysis of N-terminal domains of Arabidopsis chlorophyllide a oxygenase. *Plant Physiol. Biochem.* 45:740–9.

Sakuraba, Y., R. Tanaka, A. Yamasato, and A. Tanaka. 2009. Determination of a chloroplast degron in the regulatory domain of chlorophyllide a oxygenase. *J. Biol. Chem.* 284:36689–99.

Sato, S. and R. J. M. Wilson. 2003. Proteobacteria-like ferrochelatase in the malaria parasite. *Curr. Genet.* 42:292–300.

Schelbert, S., S. Aubry, B. Burla, B. Agne, F. Kessler, K. Krupinska, and S. Hortensteiner. 2009. Pheophytin pheophorbide hydrolase (pheophytinase) is involved in chlorophyll breakdown during leaf senescence in Arabidopsis. *Plant Cell* 21:767–85.

Schmied, J., B. Hedtke, and B. Grimm. 2011. Overexpression of HEMA1 encoding glutamyl-tRNA reductase. *J. Plant Physiol.* 168:1372–9.

Serrato, A. J., J. M. Pérez-Ruiz, M. C. Spínola, and F. J. Cejudo. 2004. A novel NADPH thioredoxin reductase, localized in the chloroplast, which deficiency causes hypersensitivity to abiotic stress in Arabidopsis thaliana. *J. Biol. Chem.* 279:43821–7.

Shepherd, M. and C. N. Hunter. 2004. Transient kinetics of the reaction catalysed by magnesium protoporphyrin IX methyltransferase. *Biochem. J.* 382:1009–13.

Shin, J., E. Park, and G. Choi. 2007. PIF3 regulates anthocyanin biosynthesis in an HY5-dependent manner with both factors directly binding anthocyanin biosynthetic gene promoters in Arabidopsis. *Plant J.* 49:981–94.

Shin, J., K. Kim, H. Kang, I. S. Zulfugarov, G. Bae, C.-H. Lee, D. Lee, and G. Choi. 2009. Phytochromes promote seedling light responses by inhibiting four negatively-acting phytochrome-interacting factors. *Proc. Natl. Acad. Sci. U.S.A.* 106:7660–5.

Shlyk, A. A. 1971. Biosynthesis of chlorophyll *b. Annu. Rev. Plant Physiol.* 22:169–84.

Skinner, J. S. and M. P. Timko. 1999. Differential expression of genes encoding the light-dependent and light-independent enzymes for protochlorophyllide reduction during development in loblolly pine. *Plant Mol. Biol.* 39:577–92.

Sobotka, R., U. Duhring, J. Komenda, E. Peter, Z. Gardian, M. Tichy, B. Grimm, and A. Wilde. 2008. Importance of the cyanobacterial Gun4 protein for chlorophyll metabolism and assembly of photosynthetic complexes. *J. Biol. Chem.* 283:25794–802.

Sperling, U., F. Franck, B. van Cleve, G. Frick, K. Apel, and G. A. Armstrong. 1998. Etioplast differentiation in *Arabidopsis*: Both PORA and PORB restore the prolamellar body and photoactive protochlorophyllide-F655 to the *cop1* photomorphogenic mutant. *Plant Cell* 10:283–96.

Stenbaek, A. and P. E. Jensen. 2010. Redox regulation of chlorophyll biosynthesis. *Phytochemistry* 71:853–9.

Stenbaek, A., A. Hansson, R. Wulff, M. Hansson, K. Dietz, and P. Jensen. 2008. NADPH-dependent thioredoxin reductase and 2-Cys peroxiredoxins are needed for the protection of Mg-protoporphyrin monomethyl ester cyclase. *FEBS Lett.* 582:2773–8.

Stephenson, P. and M. Terry. 2008. Light signalling pathways regulating the Mg-chelatase branchpoint of chlorophyll synthesis during de-etiolation in Arabidopsis thaliana. *Photochem. Photobiol. Sci.* 7:1243–52.

Stephenson, P. G., C. Fankhauser, and M. J. Terry. 2009. PIF3 is a repressor of chloroplast development. *Proc. Natl. Acad. Sci. U.S.A.* 106:7654–9.

Storbeck, S., S. Rolfes, E. Raux-Deery, M. J. Warren, D. Jahn, and G. Layer. 2010. A novel pathway for the biosynthesis of heme in archaea: Genome-based bioinformatic predictions and experimental evidence. *Archaea* 2010:175050.

Strand, A., T. Asami, J. Alonso, J. R. Ecker, and J. Chory. 2003. Chloroplast to nucleus communication triggered by accumulation of Mg-protoporphyrinIX. *Nature* 421:79–83.

Su, Q., G. Frick, G. Armstrong, and K. Apel. 2001. POR C of *Arabidopsis thaliana*: A third light- and NADPH-dependent protochlorophyllide oxidoreductase that is differentially regulated by light. *Plant Mol. Biol.* 47:805–13.

Sun, X., P. Feng, X. Xu, H. Guo, J. Ma, W. Chi, R. Lin, C. Lu, and L. Zhang. 2011. A chloroplast envelope-bound PHD transcription factor mediates chloroplast signals to the nucleus. *Nat. Commun.* 2:477.

Takahashi, K., A. Takabayashi, A. Tanaka, and R. Tanaka. 2014. Functional analysis of light-harvesting-like protein 3 (LIL3) and its light-harvesting chlorophyll-binding motif in Arabidopsis. *J. Biol. Chem.* 289:987–99.

Takio, S., N. Nakao, T. Suzuki, K. Tanaka, I. Yamamoto, and T. Satoh. 1998. Light-dependent expression of protochlorophyllide oxidoreductase gene in the liverwort, *Marchantia paleacea* var. *diptera*. *Plant Cell Physiol.* 39:665–9.

Tanaka, R. and A. Tanaka. 2007. Tetrapyrrole biosynthesis in higher plants. *Annu. Rev. Plant Biol.* 58:321–46.

Tanaka, R. and A. Tanaka. 2011. Chlorophyll cycle regulates the construction and destruction of the light-harvesting complexes. *Biochim. Biophys. Acta* 1807:968–76.

Tanaka, R., H. Ito, and A. Tanaka. 2010. Regulation and functions of the chlorophyll cycle. In *The Chloroplast: Basics and Applications*, edited by C. A. Rebeiz, C. Benning, H. J. Bohnert, J. K. Daniell, J. K. Hoober, H. K. Lishtenthaler, A. R. Portis and B. C. Tripathy. New York: Springer.

Tanaka, R., M. Rothbart, S. Oka, A. Takabayashi, K. Takahashi, M. Shibata, F. Myouga, R. Motohashi, K. Shinozaki, B. Grimm, and A. Tanaka. 2010. LIL3, a light-harvesting-like protein, plays an essential role in chlorophyll and tocopherol biosynthesis. *Proc. Natl. Acad. Sci. U.S.A.* 107:16721–5.

Tanaka, R., K. Kobayashi, and T. Masuda. 2011. Tetrapyrrole metabolism in *Arabidopsis thaliana*. *Arabidopsis Book* 9:e0145.

Tang, W., W. Wang, D. Chen, Q. Ji, Y. Jing, H. Wang, and R. Lin. 2012. Transposase-derived proteins FHY3/FAR1 interact with PHYTOCHROME-INTERACTING FACTOR1 to regulate chlorophyll biosynthesis by modulating HEMB1 during deetiolation in Arabidopsis. *Plant Cell* 24:1984–2000.

Terry, M. J. and R. E. Kendrick. 1999. Feedback inhibition of chlorophyll synthesis in the phytochrome chromophore-deficient aurea and yellow-green-2 mutants of tomato. *Plant Physiol.* 119:143–52.

Terry, M. J., P. J. Linley, and T. Kohchi. 2002. Making light of it: The role of plant haem oxygenases in phytochrome chromophore synthesis. *Biochem. Soc. Trans.* 30:604–9.

Toledo-Ortiz, G., H. Johansson, K. P. Lee, J. Bou-Torrent, K. Stewart, G. Steel, M. Rodríguez-Concepción, and K. J. Halliday. 2014. The HY5-PIF regulatory module coordinates light and temperature control of photosynthetic gene transcription. *PLoS Genet.* 10:e1004416.

Tottey, S., M. A. Block, M. Allen, T. Westergren, C. Albrieux, H. V. Scheller, S. Merchant, and P. E. Jensen. 2003. Arabidopsis CHL27, located in both envelope and thylakoid membranes, is required for the synthesis of protochlorophyllide. *Proc. Natl. Acad. Sci. U.S.A.* 100:16119–24.

Tsuchiya, T., T. Suzuki, T. Yamada, H. Shimada, T. Masuda, H. Ohta, and K. Takamiya. 2003. Chlorophyllase as a serine hydrolase: Identification of a putative catalytic triad. *Plant Cell Physiol.* 44:96–101.

Tsuchiya, Y., D. Vidaurre, S. Toh, A. Hanada, E. Nambara, Y. Kamiya, S. Yamaguchi, and P. McCourt. 2010. A small-molecule screen identifies new functions for the plant hormone strigolactone. *Nat. Chem. Biol.* 6:741–9.

Usami, T., N. Mochizuki, M. Kondo, M. Nishimura, and A. Nagatani. 2004. Cryptochromes and phytochromes synergistically regulate Arabidopsis root greening under blue light. *Plant Cell Physiol.* 45:1798–808.

Usuda, H. 1988. Adenine nucleotide levels, the redox state of the NADP system, and assimilatory force in nonaqueously purified mesophyll chloroplasts from maize leaves under different light intensities. *Plant Physiol.* 88:1461–8.

van Lis, R., A. Atteia, L. A. Nogaj, and S. I. Beale. 2005. Subcellular localization and light-regulated expression of protoporphyrinogen IX oxidase and ferrochelatase in *Chlamydomonas reinhardtii*. *Plant Physiol.* 139:1946–58.

Vandenbussche, F., Y. Habricot, A. S. Condiff, R. Maldiney, D. Van der Straeten, and M. Ahmad. 2007. HY5 is a point of convergence between cryptochrome and cytokinin signalling pathways in *Arabidopsis thaliana*. *Plant J.* 49:428–41.

Verdecia, M. A., R. M. Larkin, J. L. Ferrer, R. Riek, J. Chory, and J. P. Noel. 2005. Structure of the Mg-chelatase cofactor GUN4 reveals a novel hand-shaped fold for porphyrin binding. *PLoS Biol.* 3:e151.

Vothknecht, U. C., C. G. Kannangara, and D. von Wettstein. 1996. Expression of catalytically active barley glutamyl tRNA^{Glu} reductase in *Escherichia coli* as a fusion protein with glutathione S-transferase. *Proc. Natl. Acad. Sci. U.S.A.* 93:9287–91.

Vothknecht, U., C. Kannangara, and D. von Wettstein. 1998. Barley glutamyl tRNAGlu reductase: mutations affecting haem inhibition and enzyme activity. *Phytochemistry* 47:513–19.

Wang, L., Y. X. Mai, Y. C. Zhang, Q. Luo, and H. Q. Yang. 2010. MicroRNA171c-targeted SCL6-II, SCL6-III, and SCL6-IV genes regulate shoot branching in Arabidopsis. *Mol. Plant* 3:794–806.

Watanabe, N., F. Che, M. Iwano, S. Takayama, S. Yoshida, and A. Isogai. 2001. Dual targeting of spinach protoporphyrinogen oxidase II to mitochondria and chloroplasts by alternative use of two in-frame initiation codons. *J. Biol. Chem.* 276:20474–81.

Watanabe, S., M. Hanaoka, Y. Ohba, T. Ono, M. Ohnuma, H. Yoshikawa, S. Taketani, and K. Tanaka. 2013. Mitochondrial localization of ferrochelatase in a red alga Cyanidioschyzon merolae. *Plant Cell Physiol.* 54:1289–95.

Waters, M. T., P. Wang, M. Korkaric, R. G. Capper, N. J. Saunders, and J. A. Langdale. 2009. GLK transcription factors coordinate expression of the photosynthetic apparatus in Arabidopsis. *Plant Cell* 21:1109–28.

Woodson, J. D., J. M. Perez-Ruiz, and J. Chory. 2011. Heme synthesis by plastid ferrochelatase I regulates nuclear gene expression in plants. *Curr. Biol.* 21:897–903.

Yamasato, A., N. Nagata, R. Tanaka, and A. Tanaka. 2005. The N-terminal domain of chlorophyllide a oxygenase confers protein instability in response to chlorophyll B accumulation in Arabidopsis. *Plant Cell* 17:1585–97.

Zhang, F., W. Tang, B. Hedtke, L. Zhong, L. Liu, L. Peng, C. Lu, B. Grimm, and R. Lin. 2014. Tetrapyrrole biosynthetic enzyme protoporphyrinogen IX oxidase 1 is required for plastid RNA editing. *Proc. Natl. Acad. Sci. U.S.A.* 111:2023–8.

Zhong, S., M. Zhao, T. Shi, H. Shi, F. An, Q. Zhao, and H. Guo. 2009. EIN3/EIL1 cooperate with PIF1 to prevent photo-oxidation and to promote greening of Arabidopsis seedlings. *Proc. Natl. Acad. Sci. U.S.A.* 106:21431–6.

Zhong, S., H. Shi, Y. Xi, and H. Guo. 2010. Ethylene is crucial for cotyledon greening and seedling survival during de-etiolation. *Plant Signal. Behav.* 5:739–42.

Zhou, S., A. Sawicki, R. D. Willows, and M. Luo. 2012. C-terminal residues of oryza sativa GUN4 are required for the activation of the ChlH subunit of magnesium chelatase in chlorophyll synthesis. *FEBS Lett.* 586:205–10.

10 Carbonic Anhydrases of Higher Plant Thylakoids and Their Participation in Photosynthesis

L.K. Ignatova and B.N. Ivanov

CONTENTS

10.1 Introduction 193
10.2 Thylakoid Carbonic Anhydrases 193
    10.2.1 Membrane-Bound Carbonic Anhydrases 193
        10.2.1.1 Carbonic Anhydrases near PSII 194
        10.2.1.2 Carbonic Anhydrase near PSI 196
    10.2.2 Soluble Lumenal Carbonic Anhydrase 196
10.3 Possible Functions of Thylakoid Carbonic Anhydrases 197
10.4 Conclusions 198
Acknowledgments 198
References 198

10.1 INTRODUCTION

Several isoforms of carbonic anhydrases (CAs), soluble and membrane-bound ones, were discovered in the chloroplasts of higher plants. By now, two soluble CAs, α-CA1 and β-CA1, are known to be situated in the chloroplast stroma, and one soluble CA belonging possibly to the β-CA family was found in the thylakoid lumen. α-CA4 was identified among the proteins of the thylakoid membrane. At the same time, at least three sources of CA activity in the thylakoid membrane were revealed. The thylakoid membrane fragments enriched with photosystem II (PSII membranes) possess two such sources. One of them was detected as a low-molecular-mass protein in the course of the native electrophoresis of the PSII membranes treated with n-dodecyl-β-maltoside. The other was disclosed among high-molecular-mass proteins of the PSII core complex. Besides the different molecular mass, these CAs demonstrate different sensitivity to the different specific CA inhibitors, sulfonamides. One source of CA activity was revealed in the fragments of thylakoid membrane enriched with photosystem I (PSI). This CA activity was equally susceptible to both lipophilic and hydrophilic sulfonamides. The possible functions of thylakoid CAs are discussed.

Carbonic anhydrase (CA) catalyzes reversible hydration of carbon dioxide according to the equation $CO_2 + H_2O \leftrightarrow HCO_3^- + H^+$. The reaction may occur spontaneously, but it is very slow. CAs are enzymes with an extremely high turnover rate (up to 10^{-6} s). The enzyme is found in all living organisms, from prokaryotes to animals. All CAs are divided into several families based on conservative nucleotide sequences (Hewett-Emmett and Tashian 1996). *Arabidopsis thaliana* has 19 genes belonging to three distinct CA families: α-, β-, and γ-CAs. Some cellular compartments contain several isoforms of CA. The cytoplasm of Arabidopsis cells contains two CAs of the β family: β-CA2 and β-CA3 (specific nomenclature was proposed by Fabre et al. in 2007). Chloroplast stroma contains a significant amount of β-CA1 (known since the 1940s) and α-CA1 (recently discovered) (Villarejo et al. 2005). Exact localization of β-CA5 in chloroplasts is unknown (Fabre et al. 2007), and α-CA4 has been found in thylakoid membranes (Friso et al. 2004). A number of research teams around the world discovered CA activity (1) in two sites of the pigment–protein complex of photosystem II (PSII) in both C_4 (Lu and Stemler 2002) and C_3 plants (Lu et al. 2005; Ignatova et al. 2006, 2011); (2) adjacent to photosystem I (PSI) complex (Pronina et al. 2002; Ignatova et al. 2006, 2011); (3) and in the lumenal compartment of thylakoids (Rudenko et al. 2007; Fedorchuk et al. 2014). The nature of these CAs is unknown. Three membrane-bound CAs (two in PSII-enriched fragments of thylakoid membranes [PSII membranes] and one in PSI-enriched fragments [PSI membranes]) and three soluble CAs (two in the stroma and one in the lumen) have been found in chloroplasts of higher plants as of this writing. This review will focus on thylakoid CAs.

10.2 THYLAKOID CARBONIC ANHYDRASES

10.2.1 MEMBRANE-BOUND CARBONIC ANHYDRASES

Membrane-bound CA activity in higher plants has been found in thylakoid membranes of beans (Komarova et al. 1982) and peas (Vaklinova et al. 1982). For many years, this CA activity has been disputed as a contamination of membrane preparations by highly active and abundant stromal β-CA1.

Later, during the 1980s and 1990s, a lot of experimental data have been obtained proving that properties of membrane-bound thylakoid CA differ from soluble stromal CA. The CA activity of thylakoid membranes from peas was inhibited by diuron and hydroxylamine, whereas soluble CA has been insensitive to them (Vaklinova et al. 1982). The redox state of the medium has been found to have an effect on the CA activity of thylakoid membranes (Moubarak-Milad and Stemler 1994). Some specific inhibitors of CAs, namely, acetazolamide and azide in submicromolar concentration, have been found to have unusual stimulating effects on thylakoid CA activity (Moskvin et al. 1995). Thylakoid CA had higher affinity to CO_2 than stromal CA; K_m of thylakoid CA was determined as 9 mM, whereas soluble stromal CA had a K_m of 20 mM (Ignatova et al. 1998). Dehydration activity of thylakoid CA depended on pH, with maximum activity at 6.8–7.0, but soluble CA had no pH dependence (Ignatova et al. 1998). Comprehensive analysis of facts in support of the existence of membrane-bound CA has been presented in the review by A. Stemler in 1997. Finally, it has been shown that antibodies against soluble CA from spinach showed strong cross-reaction with soluble CA of pea chloroplasts but not with thylakoids exhibiting similar CA activity (Moskvin et al. 2004).

In the early 2000s, the first evidence of more than one membrane-bound CA in thylakoids started to appear. Isolated pigment–protein complexes of PSII from maize mesophyll have been found to have two sources of CA activity with different properties (Lu and Stemler 2002). Both PSII and PSI membranes from peas had CA activity, but only PSII membranes showed cross-reaction with antibodies against Cah3—lumenal α-CA—from *Chlamydomonas reinhardtii* (Pronina et al. 2002). The CA activity of thylakoids increased after disruption of membranes by either detergents, freeze–thaw, or sonication (Moskvin et al. 2004). Loosening of thylakoid membranes by incubation of thylakoids with progressively increasing amounts of Triton X-100 showed two distinct maximums of CA activity at 0.3 and 1.0 triton/chlorophyll ratios (Rudenko et al. 2006, 2007), suggesting the existence of at least two membrane-bound CAs.

10.2.1.1 Carbonic Anhydrases near PSII

A link between PSII and CA activity of thylakoids has been found first by researchers in Bulgaria (Vaklinova et al. 1982). As mentioned, CA activity strongly associated with thylakoid membranes from peas was inhibited by diuron and hydroxylamine—inhibitors of PSII. It was noticed that many properties of PSII and CA were similar—sulfonamides (specific inhibitors of CA) and some anions capable of inhibiting CA also inhibited photochemical activity of PSII (Stemler 1985, 1986, 1997). PSII membranes from peas, spinach, and beans exhibited CA activity that was greatly increased by detachment of outer (PsbO, PsbP, and PsbQ) proteins of

oxygen-evolving complex (OEC) (Moskvin et al. 1999). In another study, however, the CA activity of PSII membranes dropped after removal of proteins of OEC (Dai et al. 2001). McConnell et al. (2007) studied the CA activity of PSII membranes from spinach, which was high enough but varied greatly between preparations. However, after removal of proteins of OEC, CA activity was present in neither the soluble protein fraction nor the core complex (McConnell et al. 2007). Such mixed experimental results may be explained both by the different methods used to measure CA activity and by the presence of more than one source of CA activity with different properties in PSII membranes.

Lu and Stemler (2002) were first to discover two sources of CA activity with different properties in isolated pigment–protein complexes of PSII from maize mesophyll. The first one (called extrinsic CA) had easily measurable activity by various methods and was able to move into solution after treatment with 1 M $CaCl_2$. The other one (intrinsic CA activity) was tightly bound to the membrane and exhibited only hydration activity (Lu and Stemler 2002). However, studies performed with other objects showed dehydration activity of internal CA too (Moskvin et al. 1999). Studies of CA activity of PSII membranes from three species of higher plants showed that CA activity increases after loosening of the membranes with Triton X-100 (Table 10.1). The same effect was observed with PSII membranes from wheat (Khristin et al. 2004).

The presence of two sources of CA activity (high molecular mass and low molecular mass) can be observed visually in gel. Figure 10.1a shows 8% polyacrylamide gel after native electrophoresis of PSII membranes from Arabidopsis treated with n-dodecyl-β-maltoside (DM). The gel was stained with bromothymol blue, a pH indicator, which became yellow at the place where CA activity was located after dipping the gel into CO_2-enriched distilled water. CA activity showed itself in two bands: the bottom part of the gel, i.e., the low-molecular-mass region of DM-solubilized proteins, and in the top part of the gel, the green band where core proteins of PSII (CP43 and CP49) are found (Ignatova, unpublished data).

TABLE 10.1

CA Activity (Measured as the Rate of CO_2 Hydration) of PSII Membranes from Arabidopsis, Pea, and Spinach Leaves after 20 min Incubation at 0°C with Triton X-100 Taken at a Triton/Chlorophyll Ratio of 1 (w/w)

Source of PSII Membranes*	CA Activity, μmol H⁺(mg Chlorophyll min)⁻¹	
	Without Triton X-100	**After Incubation with Triton X-100**
Spinach	82.1 ± 16.3	109.9 ± 16.6
Pea	78.0 ± 7.6	97.6 ± 4.1
Arabidopsis	77.9 ± 18.1	119.2 ± 13.8

* PSII membranes were isolated as described by Ignatova et al. (2006, 2011).

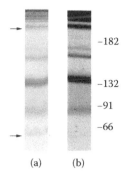

FIGURE 10.1 Polyacrylamide gel (8%) after native electrophoresis of the Arabidopsis PSII membranes treated by n-dodecyl-β-maltoside (DM) at a DM/chlorophyll ratio of 10. (Electrophoresis performed according to Peter and Thornber 1991.) (a) Gel stained with bromothymol blue. (b) Gel stained with Coomassie. The arrows show positions of the bands possessing CA activity. The positions of molecular mass markers (kDa) are shown on the right.

Sometimes, yellow staining of high-molecular-mass CA activity could be obscured by the natural green color of the band. CA activity of that band was confirmed by measuring CA activity of eluates of the band in the hydration direction (Ignatova et al. 2011). The two bands in Figure 10.1 contain two CA activities with different properties: CA activity of the low-molecular-mass band was increased by 10^{-7} M acetazolamide, and CA activity of the high-molecular-mass band was inhibited by the same 10^{-7} M acetazolamide (Ignatova et al. 2011). Such distribution of CA activity with the same effects of acetazolamide was observed after electrophoresis of PSII membranes from peas and Arabidopsis (Ignatova et al. 2011). Low-molecular-mass CA activity in both peas and Arabidopsis was inhibited by a nanomolar concentration of a specific lipophilic inhibitor, ethoxzolamide (Ignatova et al. 2006, 2011). High-molecular-mass CA activity was barely sensitive to ethoxzolamide (Ignatova, unpublished data).

Low-molecular-mass CA in our studies can be compared with the extrinsic CA of PSII-enriched membranes from maize, and high-molecular-mass CA, with intrinsic CA (Lu and Stemler 2002). These two carriers of CA activity differ by molecular mass, their place in the pigment–protein complex of PSII, and their reaction to sulfonamide inhibitors. They also showed diverse responses to some anions and cations. Activity of extrinsic CA increased three times in presence of 5 mM Cl⁻ and dropped sharply in 30 mM Cl⁻. Intrinsic CA activity increased gradually up to 0.4 M Cl⁻ after detachment of outer proteins of OEC (Lu and Stemler 2007). This increase of activity was unusual for CAs, whose activity was inhibited by Cl⁻ (Graham et al. 1984). Magnesium and zinc inhibited activity of high-molecule-mass CA, whereas calcium and especially manganese increased its activity (Moskvin et al. 2004; Lu and Stemler 2007).

Data about the nature of low-molecular-mass CA activity of PSII are scarce and contradictory. Lu and Stemler (2002) believe, based on cross-reaction of antibodies against PsbO and against Cah3 from *C. reinhardtii*, that manganese-stabilizing

protein PsbO is the protein that exhibits CA activity. However, later, it turned out that antibodies against Cah3 recognized at least one additional protein that had a similar molecular mass as Cah3, and therefore, their results should be used with caution (Mitra et al. 2005). Lu et al. (2005) obtained additional experimental data supporting the hypothesis that PsbO is responsible for the CA activity of external CA: a product of cDNA of the PsbO gene expressed in *Escherichia coli* showed CA activity, had similar thermostability as PsbO, and exhibited the highest CA activity in the presence of manganese. Other researchers found that CA activity is inherent to all three hydrophilic proteins of water-splitting complex, PsbO, PsbP, and PsbQ (Shitov et al. 2009). The CA activity of these proteins except PsbQ was also increased by manganese, so that authors refer to them as a special group of Mn^{2+}-inducible CAs (Shitov et al. 2009).

These experimental facts were critically analyzed by Bricker and Frankel (2011), who offered their explanations of the observed facts. If the PsbO protein indeed functionally exhibits CA activity, then it is a truly unique CA with extraordinarily novel properties. Based on the molecular structure of the protein, it is unlikely that it possesses CA activity. The most unusual feature is the stimulation of CA activity by manganese. So far, all known CAs have zinc in the active center (excepting a recently identified cadmium-containing CA of marine diatoms) (Lane et al. 2005; Xu et al. 2008). Another explanation of CA activity associated with the PsbO component was derived from a contaminating "authentic" CA. All CAs exhibit extremely high k_{cat}, and very small amounts of a contaminating CA could yield the observed CA activity reported for the PsbO protein. Finally, it is possible that the PsbO protein could artifactually bind manganese and that this protein-bound manganese could exhibit CA activity (Bricker and Frankel 2011). It was shown that manganese-substituted human CA I exhibits 7–10% activity of the normal zinc-containing enzyme (Lanir et al. 1975).

We believe that low-molecular-mass CA of PSII is a CA of the α family. Firstly, low-molecular-mass CA of PSII was very sensitive to ethoxzolamide (see Section 10.2.1.1). It is known that the CAs of the α family are the most sensitive to sulfonamides (Karlsson et al. 1995). Secondly, CA was stimulated by acetazolamide. We discovered this unusual effect in 1995 with pea thylakoids (Moskvin et al. 1995). Sodium azide had a similar effect. Subsequent studies showed that this property is inherent to low-molecular-mass CA of PSII membranes, and is retained after extraction of protein from the corresponding band (Ignatova et al. 2006). The underlying mechanism of this stimulation was understood only recently (Ilies et al. 2004). CAs from mammals (belonging to the α family) were found to exhibit stimulation of their activity by various compounds, including azoles, one of which is the specific inhibitor of CA acetazolamide. The limiting stage of CA reaction is the ejection of a proton from the reaction center (Silverman 1991). Possibly, the low-molecular-mass CA in PSII has such a structure in the vicinity of the active center that acetazolamide at low concentrations is able to fulfill the role of proton shuttle.

CA belonging to the α family (α-CA4 according Fabre et al. 2007) was found among proteins of thylakoid membranes (Friso et al. 2004).

Even less data are available on the nature of internal, high-molecular-mass, CA activity of PSII membranes. The presence of an inhibitor of CAs, p-aminomethylbenzensulfon-amide (mafenide), at the start line in the course of native electrophoresis of PSII membranes solubilized by DM decreased the amount of PSII core complex in the gel (Khristin et al. 2004). The elution of PSII core complex from the column with immobilized mafenide occurred only by either mafenide or ethoxzolamide. The above results led to a conclusion that membrane-bound CA activity associated with PSII was situated in the core complex (Khristin et al. 2004). According to Lu and Stemler (2007), intrinsic CA activity is provided by $CaMn_4$ complex and its immediate neighbors. We agree that CA activity may be provided not by one of the proteins but by an interface between two or more proteins, like in γ-CAs, which have a reaction center formed by amino acids from different subunits. We assume that high-molecule-mass CA activity may be carried out by PSII proteins PsbA, PsbD, and PsbC. It is possible that the reaction center of this complex contains some metal other than zinc, which is found in almost all known CAs. Replacement of zinc with cobalt did not cause loss of function in animal α-CA2 (Wells et al. 1974), and replacement to Fe(II) in γ-CA from *Methanosarcina thermophila* even increased CA activity in anaerobic conditions (Tripp et al. 2004). Figure 10.2 shows the supposed location of PSII CAs.

10.2.1.2 Carbonic Anhydrase near PSI

Another source of CA activity of thylakoid membranes was found in PSI membranes (Figure 10.2). Both PSI and PSII membranes from peas exhibited CA activity, but only PSII membranes showed cross-reaction with antibodies against Cah3 from *C. reinhardtii* (Karlsson et al. 1995; Pronina et al. 2002). CA activity when loosening PSI membranes with detergent Triton X-100 was represented by a bell-shaped curve with a maximum at a Triton/chlorophyll ratio of 0.3

(Rudenko et al. 2007); the same ratio was found for PSI membranes from Arabidopsis (Ignatova et al. 2011). The CA activity of PSI membranes was equally sensitive to sulfonamides acetazolamide and ethoxzolamide, with I_{50} equal 10^{-6} M (Ignatova et al. 2006).

10.2.2 Soluble Lumenal Carbonic Anhydrase

It was established earlier that there is soluble CA inside thylakoid lumen (Rudenko et al. 2007). Later, this fact was confirmed with thylakoids from Arabidopsis (Figure 10.2) (Fedorchuk et al. 2014). This CA was isolated from a fraction of lumenal proteins, obtained from Arabidopsis thylakoids washed from stromal and cytoplasmic soluble CAs. It was obtained from both wild-type plants and a knockout mutant lacking the *At3g01500* gene encoding stromal β-CA1. The CA activity of lumenal proteins was even higher in the mutant. This CA was isolated in pure form; after native electrophoresis, it had a molecular mass of 132 kDa. It was sensitive to ethoxzolamide, similar to stromal β-CA1, with I_{50} of 5×10^{-6}M (Fedorchuk et al. 2014), which is usual for a CA of the β family (Karlsson et al. 1995). The CA activity of lumenal proteins increased after addition of dithiothreitol, which is also usual for β-CAs that have relatively high levels of sulfur-containing amino acids, cysteine and methionine, compared to animal CAs belonging to α-CAs (Graham et al. 1984). These facts together allow us to assume that soluble lumenal CA may belong to the β family. Besides soluble β-KA1 in stroma, β-KA5 was found in chloroplasts by the method of green fluorescent protein (GFP) fusion (Fabre et al. 2007), which allows us to suggest that the soluble CA lumen may be β-KA5.

A high content of a protein in some organelles does not necessarily correlate with its importance for metabolism. For example, 100-fold reduction of the most abundant β-CA1 did not lead to a decrease in photosynthesis (Price et al. 1994). The level of expression of *At4g33580* encoding β-CA5 is two to three orders of magnitude lower than *At3g01500* encoding stromal β-CA1. Despite that, the knockout mutant lacking *At4g33580* shows severe growth retardation (D.V. Moroney,

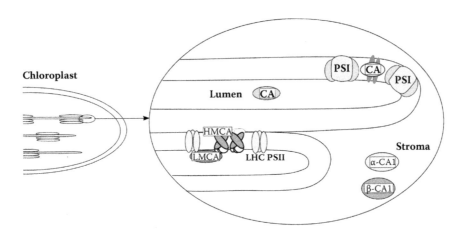

FIGURE 10.2 The tentative scheme of CA arrangement in the thylakoids of higher plants. HMCA, high-molecule-mass CA activity of PSII; LHC, light harvesting complex of PSII; LMCA, low-molecule-mass CA of PSII.

personal communication). β-CA5 is the only CA that, with its shortage in the cell, causes changes in phenotype, and its exact localization is still unknown to this day. This indicates that β-CA5 is one of the key enzymes and is necessary for normal plant development.

10.3 POSSIBLE FUNCTIONS OF THYLAKOID CARBONIC ANHYDRASES

The possible functions of thylakoid CAs were discussed for the first time by A. Stemler (1997). It is obvious that the existence of several CAs in chloroplast, an organelle whose main function is photosynthesis, indicates their possible involvement in that process; however, the nature of this link is unclear yet. The unique feature of CA as an enzyme is that all participants of the catalyzed reaction are key cell metabolites, which may be involved in regulation of central metabolic processes (photosynthesis, respiration, photorespiration) including multiple cell compartments (Igamberdiev and Roussel 2012). The catalyzed reaction involves protons, which play not only an energetic but also a regulatory role in photosynthesis. CA increases the rate of changes of proton concentration in a compartment; moreover, CO_2 may pass into the membrane and stay there because the solubility of CO_2 in lipids is three to eight times higher than in water (Mitz 1979).

Most of the experimental data and proposed theories about the role of thylakoid CAs are limited to processes taking place around PSII where water is oxidized by water-oxidizing complex, and electrons taken from water are transferred to terminal acceptors of PSII complex: Q_A and Q_B. It is known that bicarbonate ions are needed to maintain the physiological rate of electron transfer on the acceptor side of PSII (Wydrzynski and Govindjee 1975). There are two bicarbonate pools linked to the membrane (Stemler 1977). The first, and the smallest, contains one bicarbonate ion per 380–400 chlorophyll molecules, which is approximately equal to the number of reaction centers of PSII. Removal of this bicarbonate leads to more than 90% suppression of oxygen evolution (Stemler 1977). In recent years, Klimov et al. (Klimov et al. 1997; Klimov and Baranov 2001) obtained more data that support the hypothesis that bicarbonate is also needed for normal functioning of the donor side of PSII. The authors assume the presence of CA, possibly involved in bicarbonate exchange. It is unclear which product of CA reaction plays a defining role on the donor and acceptor sides of PSII. On the acceptor side, it is most likely to be bicarbonate that facilitates electron transfer between Q_A and Q_B, which is under the control of nonheme iron. In that case, protonation of Q_B by CA in that site, as was suggested by many researchers (Stemler 1979; van Rensen et al. 1988; Govindjee and van Rensen 1997), would be an associated reaction. High-molecular-mass CA, described in the previous sections, could be that CA.

On the donor side of PSII, capture of protons produced by the OEC may be needed for prevention of photoinhibition. In this case, proton binding is the main role of CA, whereas bicarbonate is just a useful acceptor of protons. A similar role is proposed for Cah3 bound to the lumenal side of the thylakoid membrane in green alga *C. reinhardtii*. The absence of this CA leads to lower sustainability under high light (Villarejo et al. 2002).

The second bicarbonate pool mentioned earlier is quantitatively close to chlorophyll concentration and may be removed without disabling PSII activity. It is possible that this pool plays a role in proton uptake by thylakoids coupled with photosynthetic electron transfer, because the rate of proton uptake increased if bicarbonate was added to thylakoids (Podorvanov et al. 2005). CA inhibitors such as acetazolamide and ethoxzolamide decreased the amount of protons consumed by thylakoids under illumination, the amount of bound bicarbonate, and the buffer capacity of thylakoids (Podorvanov et al. 2005). Podorvanov et al. (2005) showed that acetazolamide and ethoxzolamide had no effect on the rate of photosynthetic electron transfer but substantially decreased the rate of photophosphorylation that is coupled to that electron transfer (Zolotareva 2008; Onoiko et al. 2010). The authors assume the existence of CA in the thylakoid lumen, which enables participation of the bicarbonate pool in proton uptake and transport, contributing to photophosphorylation (Zolotareva 2008). The existence of lumenal CA was proved in another study (Fedorchuk et al. 2014) (see Section 10.2.2), where the authors proposed that it facilitates stabilization of lumenal pH by increasing the rate of proton flow from the sites of production into the lumen and proton transport within the lumen. Both assumptions may be combined into one where lumenal CA increases the rate of proton transfer to proton channels of ATP synthase in the thylakoid membrane.

Previously, α-CA4 has been found among membrane-bound proteins in thylakoids of Arabidopsis (Friso et al. 2004). We found that the effective quantum yield of PSII under saturating light and CO_2 was higher in knockout mutants lacking the gene *At4g20990* encoding α-CA4, than in wild type, whereas nonphotochemical quenching of fluorescence (NPQ) was 30–50% lower (Zhurikova et al. 2015). The value of NPQ indicates the extent of changes in photosynthetic apparatus that lead to dissipation of absorbed light energy into heat. In our experiments, NPQ changes were mostly the changes of qE component of NPQ, which depend on proton concentration in the lumen. Protons play a role in the NPQ process in two ways: causing a conformational change in light-harvesting antenna, predominantly by protonation of PsbS protein (Crouchman et al. 2006; Johnson and Ruban 2011), and initiating the process of de-epoxidation of violaxanthin by violaxanthin de-epoxidase (VDE) (Gilmore and Yamamoto 1992). In addition, knockout mutants showed higher starch content in chloroplasts, two to three times higher than wild type (Zhurikova et al. 2015). It appears that the lack of α-CA4 decreased the supply of protons for PbsS and VDE, thus increasing ATP synthesis and, consequently, starch synthesis rates. The lack of α-CA4 also had an effect on stability of PSII complex during isolation, namely, in order to isolate functional PSII-containing membranes, a very high concentration (up to 10 mM) of Mg^{2+} in isolation medium was needed (Ignatova, unpublished data); this indicates that normally, α-CA4 may be present in stacked granal regions of thylakoid membranes. Experimental data listed here suggest direct involvement of CA in regulating the functional activity of peripheral light-harvesting complexes of PSII antenna.

Functions of CA that was found near PSI have not been studied as thoroughly as those near PSII. In 1984, one prolific hypothesis was suggested, that CA provides CO_2 for Rubisco by taking advantage of low pH in the lumen of thylakoids (Pronina and Semenenko 1984). According to this hypothesis, membrane-bound CA (existence of several CAs was not known at the time) increases the rate of conversion of bicarbonate to CO_2, which diffuses freely through membrane and is used by Rubisco in stroma. It was shown theoretically that spontaneous dehydration of bicarbonate is not enough to explain the rates of photosynthesis found in nature (Raven 1997). However, this scheme requires additional assumptions, namely, how to avoid rapid conversion of CO_2 into HCO_3^- in stroma, where there is not only high pH but also a high amount of β-CA1. We suggested the scheme that involves protons from thylakoid lumen in conversion of HCO_3^- into CO_2 in stroma (Ivanov et al. 2007). In essence, this scheme involves the presence of CA in thylakoid membrane so that it is located near a proton channel and uses protons transferred through this channel in a way similar to ATP synthase. In order to limit the diffusion of CO_2 molecules produced by this CA, its reaction centers should be closely associated with Rubisco. It was shown that a significant portion of Rubisco molecules is associated with thylakoid membranes (Anderson et al. 1996) and that CA is part of the supermolecular complex containing Rubisco (Jebanathirajah and Coleman 1998). Some experimental data were obtained (Moskvin et al. 2000) that are consistent with the above hypothesis and led to its establishment; specifically, it was shown that stimulation of the dehydration CA activity of thylakoids, in conditions that favor acidification of the lumen in the light, could be inhibited by ethoxzolamide.

Thylakoid CAs can be involved in adaptation to stress factors. It is well known that stress factors have a negative effect on photosynthesis, especially on PSII, which is involved in water oxidation and provides electrons for the photosynthetic electron transport chain, whereas PSI is substantially more tolerant. It was established that salt stress and drought lead to increased concentration of one of the thylakoid CAs, associated with PSII (Lazova et al. 2012). In addition, it was shown that α-CA4 plays a role in adaptation to water stress (Li et al. 2008).

10.4 CONCLUSIONS

Despite the fact that CA has been studied for almost 100 years now, this enzyme is still a mystery. Decoding of the genome helps us to find the number of isoforms of the enzyme in the organism, but ascertainment of the localization and specific function of each isoform is a hard task. It is even harder if the same cellular compartment contains several isoforms. Location of six CAs of the α-family *A. thaliana* is still unknown, and it is possible that other CAs may be found in chloroplasts. Variability of the enzyme in its numerous isoforms suggests variability of its functions in the cell. Finding out which functions are attributed to each of the discovered isoforms is the goal for future studies, and the most efficient way, apparently, is the use of knockout mutants.

ACKNOWLEDGMENTS

The authors express their gratitude to Dr. J.V. Moroney of Louisiana State University (USA) for providing the seeds of the knockout mutant and Dr. I.A. Naidov for technical help. This work was supported in part by a grant from the Russian Foundation for Basic Research (project 15-04-03883).

REFERENCES

Anderson, L.E., Gibbons, J.T., and Wang, X. 1996. Distribution of ten enzymes of carbon metabolism in pea (*Pisum sativum*) chloroplasts. *Int J Plant Sci* 157:525–538.

Bricker, T.M. and Frankel, L.K. 2011. Auxiliary functions of the PsbO, PsbP and PsbQ proteins of higher plant photosystem II: A critical analysis. *J Photochem Photobiol B* 104:165–178.

Crouchman, S., Ruban, A., and Horton, P. 2006. PsbS enhances non-photochemical fluorescence quenching in the absence of zeaxanthin. *FEBS Lett* 580:2053–2058.

Dai, X., Yu, Y., Zhang, R., Yu, X., He, P., and Xu, C. 2001. Relationship among photosystem II carbonic anhydrase, extrinsic polypeptides and manganese cluster. *Chin Sci Bull* 46:406–408.

Fabre, N., Reiter, I.M., Becuwe-Linka, N., Genty, B., and Rumeau, D. 2007. Characterization and expression analysis of genes encoding alpha- and beta-carbonic anhydrases in Arabidopsis. *Plant Cell Environ* 30:617–629.

Fedorchuk, T.P., Rudenko, N.N., Ignatova, L.K., and Ivanov, B.N. 2014. The presence of soluble carbonic anhydrase in the thylakoid lumen of chloroplasts from Arabidopsis leaves. *J. Plant Physyol* 171:903–906.

Friso, G., Giacomelli, L., Ytterberg, A.J. et al. 2004. In-depth analysis of the thylakoid membrane proteome of *Arabidopsis thaliana* chloroplasts: New proteins, new functions, and a plastid proteome database. *Plant Cell* 16:478–499.

Gilmore, A.M. and Yamamoto, H.Y. 1992. Linear models relating xanthophylls and lumen acidity to non-photochemical fluorescence quenching: Evidence that antheraxanthin explains zeaxanthin-independent quenching. *Photosynth Res* 35:67–78.

Govindjee, X.C. and van Rensen, J.J.S. 1997. On the requirement of bound bicarbonate for photosystem II activity. *Z. Naturforsch., C-A J. Biosciences* 52:24–32.

Graham, D., Reed, M.R., Patterson, B.D., Hockley, D.G., and Dwyer, M.R. 1984. Chemical properties, distribution and physiology of plant and algal carbonic anhydrase. *Ann N Y Acad Sci* 429:222–237.

Hewett-Emmett, D. and Tashian, R.E. 1996. Functional diversity, conservation, and convergence in the evolution of the alpha-, beta-, and gamma-carbonic anhydrase gene families. *Mol Phylogen Evol* 5:50–77.

Igamberdiev, A.U. and Roussel, M.R. 2012. Feedforward non-Michaelis–Menten mechanism for CO_2 uptake by Rubisco: Contribution of carbonic anhydrases and photorespiration to optimization of photosynthetic carbon assimilation. *BioSystems* 107:158–166.

Ignatova, L.K., Moskvin, O.V., Romanova, A.K., and Ivanov, B.N. 1998. Carbonic anhydrases in the C3-plant leaf cell. *Aust J Plant Physiol* 25:673–677.

Ignatova, L.K., Rudenko, N.N., Khristin, M.S., and Ivanov, B.N. 2006. Heterogeneous origin of carbonic anhydrase activity of thylakoid membranes. *Biochemistry (Moscow)* 71:525–532.

Ignatova, L.K., Rudenko, N.N., Mudrik, V.A., Fedorchuk, T.P., and Ivanov, B.N. 2011. Carbonic anhydrase activity in *Arabidopsis thaliana* thylakoid membrane and fragments enriched with PSI or PSII. *Photosynth Res* 110:89–98.

Ilies, M., Scozzafava, A., and Supuran, C.T. 2004. Carbonic anhydrase activators. In *Carbonic anhydrase: Its inhibitors and activators*, eds. C.T. Supuran, A. Scozzafava, and G. Conway. Boca Raton, FL: CRC Press.

Ivanov, B.N., Ignatova, L.K., and Romanova, A.K. 2007. Diversity in forms and functions of carbonic anhydrase in terrestrial higher plants. *Russ J Plant Physiol* 54:143–162.

Jebanathirajah, J.A. and Coleman, J.R. 1998. Association of carbonic anhydrase with a Calvin cycle enzyme complex in *Nicotiana tabacum*. *Planta* 203:177–182.

Johnson, M.P. and Ruban, A.V. 2011. Restoration of rapidly reversible photoprotective energy dissipation in the absence of PsbS protein by enhanced pH. *J Biol Chem* 286:19973–19981.

Karlsson, J., Hiltonen, T., Husic, H.D., Ramazanov, Z., and Samuelsson, G. 1995. Intracellular carbonic anhydrase of *Chlamydomonas reinhardtii*. *Plant Physiol* 109:533–539.

Khristin, M.S., Ignatova, L.K., Rudenko, N.N., Ivanov, B.N., and Klimov, V.V. 2004. Photosystem II associated carbonic anhydrase activity in higher plant is situated in core complex. *FEBS Lett* 577:305–308.

Klimov, V.V. and Baranov, S.V. 2001. Bicarbonate requirement for the water-oxidizing complex of photosystem II. *Biochim Biophys Acta* 1503:187–196.

Klimov, V.V., Hulsebosch, R.J., Allakhverdiev, S.I., Wincencjusz, Y., and van Gorkom, B. 1997. Bicarbonate may be required for ligation of manganese in the oxygen-evolving complex of photosystem II. *Biochemistry* 51:16277–16281.

Komarova, Y.M., Doman, N.G., and Shaposhnikov, G.L. 1982. Two forms of carbonic anhydrase from bean chloroplasts. *Biochemistry (Moscow)* 47:856–862.

Lane, T.W., Saito, M.A., George, G.N., Pikering, I.J., Prince, R.C., and Morel, F.M.M. 2005. A cadmium enzyme from a marine diatom. *Nature* 435:42.

Lanir, A., Gradstajn, S., and Navon, G. 1975. Temperature and frequency dependence of solvent proton relaxation rates in solutions of manganese(II) carbonic anhydrase. *Biochemistry* 14:242–248.

Lazova, G., Ignatova, L., and Baydanova, V. 2012. Drought and salinity stress: Changes in hydratase and dehydratase activities of thylakoid-associated carbonic anhydrase in pea seedlings. *J Plant Sci* 173:7–15.

Li, Y., Zhu, Y., Liu, Y. et al. 2008. Genome-wide identification of osmotic stress response gene in *Arabidopsis thaliana*. *Genomics* 92:488–493.

Lu, Y.-K. and Stemler, A.J. 2002. Extrinsic photosystem II carbonic anhydrase in maize mesophyll chloroplasts. *Plant Physiol* 128:643–649.

Lu, Y.-K. and Stemler, A.J. 2007. Differing responses of the two form of photosystem II carbonic anhydrase to chloride, cations, and pH. *Biochim Biophys Acta* 1767:633–638.

Lu, Y.-K., Theg, S.M., and Stemler, A.J. 2005. Carbonic anhydrase activity of the photosystem II OEC33 protein from pea. *Plant Cell Physiol* 46:1944–1953.

McConnell, I.L., Badger, M.R., Wydrzynski, T., and Hillier, W. 2007. A quantitative assessment of the carbonic anhydrase activity in photosystem II. *Biochim Biophys Acta* 1767:639–647.

Mitra, M., Mason, C.B., Xiao, Y., Ynalvez, R.A., Lato, S.M., and Moroney, J.V. 2005. The carbonic anhydrase gene families of *Chlamydomonas reinhardtii*. *Can J Bot* 83:1–15.

Mitz, M.A. 1979. CO_2 Biodynamics: A new concept of cellular control. *J Theor Biol* 80:537–551.

Moskvin, O.V., Ignatova, L.K., Ovchinnikova, V.I., and Ivanov, B.N. 1995. Membrane associated carbonic anhydrase of pea thylakoids. *Biochemistry (Moscow)* 60:859–864.

Moskvin, O.V., Razguljayeva, A.Y., Shutova, T.V., Khristin, M.S., Ivanov, B.N., and Klimov, V.V. 1999. Carbonic anhydrase activity of different photosystem II preparation. In *Photosynthesis: Mechanism and effects*, vol. II, ed. G. Garab, 1201–1204. Dordrecht: Kluwer.

Moskvin, O.V., Ivanov, B.N., Ignatova, L.K., and Kollmeier, M.A. 2000. Light-induced stimulation of carbonic anhydrase activity in pea thylakoids. *FEBS Lett* 470:375–377.

Moskvin, O.V., Shutova, T.V., Khristin, M.S. et al. 2004. Carbonic anhydrase activity in pea thylakoids: A photosystem II core complex–associated carbonic anhydrase. *Photosynth Res* 79:93–100.

Moubarak-Milad, M. and Stemler, A. 1994. Oxidation–reduction potential dependence of photosystem II carbonic anhydrase in maize thylakoids. *Biochemistry* 33:4432–4438.

Onoiko, E.V., Polishchuck, A.V., and Zolotareva, E.K. 2010. The stimulation of photophosphorylation in isolated spinach chloroplasts by exogenous bicarbonate: The role of carbonic anhydrase. *Rep Nat Acad Sci Ukr* 10:160–165.

Peter, G.F. and Thornber, J.P. 1991 Biochemical composition and organization of higher plant photosystem II light-harvesting pigment proteins. *J Biol Chem* 266:16746–16754.

Podorvanov, V.V., Zolotareva, E.K., and Chernoshtan, A.A. 2005. Role of bicarbonate in light-dependent proton uptake in isolated chloroplasts. *Physiol Biochem Cult Plants* 37:326–332.

Price, G.D., von Caemmerer, S., Evans, J.R. et al. 1994. Specific reduction of chloroplast carbonic anhydrase activity by antisense RNA in transgenic tobacco plants has a minor effect on photosynthetic CO_2 assimilation. *Planta* 193:331–340.

Pronina, N.A. and Semenenko, V.E. 1984. Localization of membrane-bound and soluble forms of carboanhydrase in Chlorella cells. *Sov Plant Physiol* 31:241–251.

Pronina, N.A., Allakhverdiev, S.I., Kupriyanova, E.V., Klyachko-Gurvich, G.L., and Klimov, V.V. 2002. Carbonic anhydrase in subchloroplast particles of pea plants. *Russ J Plant Physiol* 49:303–310.

Raven, J.A. 1997. CO_2 concentrating mechanisms: A role for thylakoid lumen acidification? *Plant Cell Environ* 20:147–154.

Rudenko, N.N., Ignatova, L.K., Kamornitskaya, V.B., and Ivanov, B.N. 2006. Pea leaf thylakoids contain several carbonic anhydrases. *Dokl Biochem Biophys* 408:155–157.

Rudenko, N.N., Ignatova, L.K., and Ivanov, B.N. 2007. Multiple sources of carbonic anhydrase activity in pea thylakoids: Soluble and membrane-bound forms. *Photosynth Res* 91:81–89.

Shitov, A.V., Pobeguts, O.V., Smolova, T.N., Allakhverdiev, S.I., and Klimov, V.V. 2009. Manganese dependent carboanhydrase activity of photosystem II proteins. *Biochemistry (Moscow)* 74:509–517.

Silverman, D.N. 1991. The catalytic mechanism of carbonic anhydrase. *Can J Bot* 69:1070–1078.

Stemler, A. 1977. The binding of bicarbonate ions to washed chloroplast grana. *Biochim Biophys Acta* 460:511–522.

Stemler, A. 1979. A dynamic interaction between the bicarbonate ligand and photosystem II reaction center complexes in chloroplasts. *Biochim Biophys Acta* 545:36–45.

Stemler, A. 1985. Carbonic anhydrase: Molecular insights applied to photosystem II in thylakoid membranes. In *Inorganic carbon uptake by aquatic photosynthetic organisms*, eds. W.J. Lucas, and J.A. Berry, 377–387. Rockville, MD: American Society of Plant Physiologists.

Stemler, A. 1986. Carbonic anhydrase associated with thylakoids and photosystem II particles from maize. *Biochim Biophys Acta* 850:97–107.

Stemler, A. 1997. The case for chloroplast thylakoid carbonic anhydrase. *Physiol Plant* 99:348–353.

Tripp, B.C., Bell, C.B. III, Cruz, F., Krebs, C., and Ferry, J.G. 2004. A role for iron in an ancient carbonic anhydrase. *J Biol Chem* 279:6683–6687.

Vaklinova, S.G., Goushtina, L.M., and Lazova, G.N. 1982. Carboanhydrase activity in chloroplasts and chloroplast fragments. *C R Acad Bulg Sci* 35:1721–1724.

van Rensen, J.J., Tonk, W.J.M., and Bruijn, S.M. 1988. Involvement of bicarbonate in the protonation of the secondary quinone electron acceptor of Photosystem II via the non-haem iron of the quinone–iron acceptor complex. *FEBS Lett* 226:347–351.

Villarejo, A., Shutova, T.V., Moskvin, O.V., Forssen, M., Klimov, V.V., and Samuelsson, G. 2002. A photosystem II associated carbonic anhydrase regulates the efficiency of photosynthetic oxygen evolution. *EMBO J* 21:1930–1938.

Villarejo, A., Buren, S., Larsson, S. et al. 2005. Evidence for a protein transported through the secretory pathway en route to the higher plant chloroplast. *Nat Cell Biol* 7:1224–1231.

Wells, J.W., Kandel, S.I., Kandel, M., and Gornall, A.G. 1974. The esterase activity of bovine carbonic anhydrase B above pH 9. Reversible and covalent inhibition by acetazolamide. *J Biol Chem* 250:3522–3530.

Wydrzynski, T. and Govindjee. 1975. A new site of bicarbonate effect in photosystem II of photosynthesis; evidence from chlorophyll fluorescence transients in spinach chloroplasts. *Biochim Biophys Acta* 387:403–408.

Xu, Y., Feng, L., Jeffrey, P.D., Shi, Y., and Morel, F.M. 2008. Structure and metal exchange in the cadmium carbonic anhydrase of marine diatoms. *Nature* 452:56–61.

Zhurikova, E.M., Ignatova, L.K., Semenova, G.A., Rudenko, N.N., Mudrik, V.A., and Ivanov, B.N. 2015. Effect of gene knockout alpha-carbonic anhydrase 4 on photosynthetic characteristics of *Arabidopsis thaliana* and starch accumulation in leaves. *Russ J Plant Physiol* 62:564–569.

Zolotareva, E.K. 2008. Proton regulation of energy transformation processes in thylakoid membranes. PhD diss., Kharkov National Univ., Ukraine.

11 Hybrid Interfaces for Electron and Energy Transfer Based on Photosynthetic Proteins

*R. Roberto Tangorra, Alessandra Antonucci, Francesco Milano, Simona la Gatta,
Gianluca M. Farinola, Angela Agostiano, Roberta Ragni, and Massimo Trotta*

CONTENTS

11.1 Introduction 201
11.2 Photosynthesis 202
    11.2.1 Photochemical Energy Conversion..202
    11.2.2 Q-Type RC 202
        11.2.2.1 Spatial Organization 202
        11.2.2.2 Some Spectroscopic Properties..204
        11.2.2.3 Electron Transfer Cascade 204
        11.2.2.4 The Photocycle...204
    11.2.3 Light-Harvesting Complexes 205
        11.2.3.1 The Spatial Organization 205
        11.2.3.2 Some Spectroscopic Properties..206
        11.2.3.3 The Energy Transfer 206
11.3 Artificial Photosynthesis...207
    11.3.1 Basic Architecture 207
    11.3.2 Photoconversion Efficiency versus Structural Complexity 208
11.4 Nano-Biohybrid (or Biohybrid) Photoconverter 208
    11.4.1 Hybrid Interfaces for Electron Transfer..210
        11.4.1.1 Electrostatic Adsorption 210
        11.4.1.2 Physisorption Techniques 210
        11.4.1.3 Covalent Binding 210
        11.4.1.4 Selective Anchoring...211
        11.4.1.5 Interface of RC with Carbon Substrates 211
        11.4.1.6 RC–NP Interface 212
    11.4.2 Hybrid Interfaces for Energy Transfer..212
        11.4.2.1 RC–NP Interface 212
        11.4.2.2 RC-Organic 213
11.5 Conclusions...216
Acknowledgments..216
References..216

11.1 INTRODUCTION

Sunlight is the most abundant energy source for the planet earth [1]: in 1 h, Earth receives the amount of energy needed by mankind in 1 year. Only photosynthetic organisms, however, are able to use solar energy, trapping solar energy and converting it in highly valuable energetic compounds as biomass. Photosynthesis can hence be considered as one of the cornerstone biological processes in the biosphere, since life on earth, directly or indirectly, depends on it. Modern civilization has a further reason to cherish photosynthesis as the energy stored in fossil fuels originates from sunlight captured by plants during the Carboniferous period. The careless use of fossil energy reserves is posing severe problems to the well-being of mankind's future generations [2].

Photosynthesis has so many implications in the survival of the planet that botanists, biologists, geneticists, chemists, and physicists set the ambitious goal of replicating it in the laboratory, as the Italian photochemist Giacomo Ciamician [3] wishfully stated in the early twentieth century:

On the arid lands there will spring up industrial colonies without smoke and without smokestacks; forests of glass tubes will extend over the plains and glass buildings will rise everywhere; inside of these will take place the photochemical processes that hitherto have been the guarded secret of the plants, but that will have been mastered by human industry which will know how to make them bear even more abundant fruit than nature, for nature is not in a hurry and mankind is. And if in a distant future the supply of coal becomes completely exhausted, civilization will not be checked by that, for life and civilization will continue as long as the sun shines! If our black and nervous civilization, based on coal, shall be followed by a quieter civilization based on the utilization of solar energy, that will not be harmful to progress and to human happiness.

This excerpt maintains its strength today, as long as we substitute oil or uranium for coal within the text [4]! The need for understanding the *guarded secret of the plants* has spurred a large effort in the scientific community, and the deep understanding gathered at the molecular level of natural photosynthesis is inspiring what is known as artificial photosynthesis by assembling artificial systems based on molecular and supramolecular architectures [5–7].

Ideal biomimetic systems must efficiently harvest the sunlight, with the help of suitable antennas, and convert the energy in a stable charge-separated state with a lifetime long enough to allow ancillary chemistry to take place [8]. The combination of these requirements, finalized in the photosynthetic process, has not yet been fully attained: whereas efficient light harvesting and energy transfer have been obtained in artificial systems, some important limits in the lifetime of charge separated states are still present.

Parallel to artificial photosynthesis, a novel strategy involving functional components of the photosynthetic apparatus is developing [9]. The underlying idea is to take advantage of outstanding performances of photosynthetic proteins in harvest and convert sunlight, assembling hybrid systems interfacing them to relatively simple biomimetic components. In this chapter, several recent studies in this field are reviewed, offering an outlook to strategies employed to assemble photosynthetic proteins in hybrid biological devices.

11.2 PHOTOSYNTHESIS

Earth is colonized by photosynthetic organisms, all using the same basic strategy to drive their metabolism: light is absorbed by specialized photosynthetic units (PSUs), where solar energy is converted into chemical energy. The energy stored in the light-dependent reactions is then used in the so-called dark phase for the synthesis of adenosine triphosphate (ATP) and nicotinamide adenine dinucleotide phosphate (NADPH) eventually employed for fixing carbon dioxide and synthesize carbohydrates in the Calvin cycle (see Figure 11.1) [10].

11.2.1 PHOTOCHEMICAL ENERGY CONVERSION

PSUs of photosynthetic organisms share a basic architecture, composed of an efficient light collecting system, the

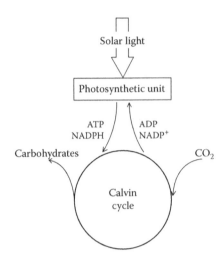

FIGURE 11.1 Schematic representation of the photosynthetic process. Sunlight is absorbed by the PSU formed by the association of light-harvesting complexes (LHCs) and the reaction center (RC). Here the energy associated to the electromagnetic radiation is used for synthesizing ATP and NADPH, high content energy molecules which, in turn, fuel the capture of atmospheric CO_2 (or other carbon sources) and the subsequent carbohydrate synthesis in the Calvin cycle.

light-harvesting complexes (LHCs), which acts as antenna harvesting light and funneling it to the second and more important component, the reaction center (RC), where photons are converted into chemical reducing power to be used in the Calvin cycle. The RCs play a pivotal role in photosynthesis as they generate, within a few hundred picoseconds from photon absorption, charge-separated states that span the membrane. The yield of this conversion is close to unity, an efficiency resulting from the careful design of protein composition and structural organization refined in billion years of evolution. RCs have little variability within photosynthetic organisms and can be divided into two main groups, classified on the basis of the electron acceptor, the chemical species carrying the electron of the charge-separated state, as schematically shown in Figure 11.2 [11].

11.2.2 Q-TYPE RC

11.2.2.1 Spatial Organization

The structure and properties of Q-type (PSII-like) RCs from purple bacteria have been intensively investigated since the early 1960s. A landmark of bacterial photosynthesis is the successful crystallization and subsequent solution of the three-dimensional structure of the RC from the purple bacteria *R. viridis* in 1984. The first crystallization of a membrane protein was awarded with the Nobel Prize in Chemistry in 1988 to a team of German researchers based at the Max-Planck Institute [12]. It was such a breakthrough that research in photosynthesis is occasionally divided in the periods BC/AC, i.e., before crystallization/after crystallization [13]. The wealth of information presently available on the RC makes this protein a model system to exploit in investigating non-physiological application of photosynthesis. It lacks indeed

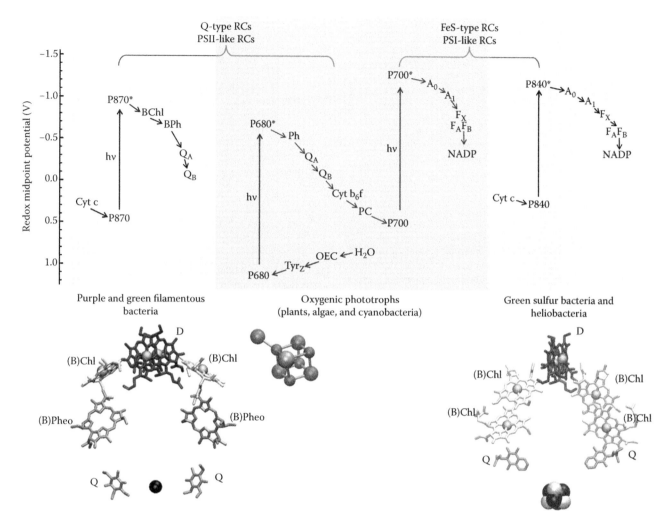

FIGURE 11.2 The photosynthetic reaction centers classified according to the final electron acceptor, either a quinone (Q-type RC) or an iron–sulfur cluster (FeS-type RC). (Adapted from Allen, J.P. and J.C. Williams, *FEBS Lett*, 438(1–2), 5–9, 1998.) Q-type RCs, on the left, present in purple and some green bacteria, among which are the *Rhodobacter sphaeroides* and *Rhodopseudomonas* (now reclassified as *Blastochloris*) *viridis*, show chemical composition and structural organization very similar to composition and structure of one of the reaction centers present in cyanobacteria, algae, and plants, the RC of photosystem II. For this reason, this class of RCs is also called PSII-like. A dramatically important difference between PSII-like and PSII RCs is the presence of a protein complex specialized in the oxidation of water to produce, as by-product, molecular oxygen. This complex, called oxygen evolving system (OEC), is shown in correspondence of the left end of the Z-scheme shown at the center of the figure. Similarly, FeS-RCs are very similar to the RCs of photosystem I and hence are called PSI-like RCs. Q-type and FeS-type RCs are considered evolutionary precursors of PSII and PSI RCs, respectively [12]. PSII and PSI RCs have later teamed up in the above-mentioned Z-scheme. This evolution has eventually produced enough oxygen on the planet Earth to shape the oxygen-based life still enduring on the planet [9]. Interestingly, the two types of RC show striking similarities since (1) the primary electron donor is composed by two chlorophylls or two bacteriochlorophylls associated in a functional dimer (often called special pair P or dimer [D]); (2) the electrons are passed to intermediate monomeric (bacterio)chlorin-based electron acceptors, either (bacterio) chlorophylls ((B)Chl) or (bacterio)pheophytins ((B)Pheo); (3) the cofactors are arranged in two (quasi) symmetric branches; and (4) both types present quinone molecules (Q), which represent the final electron acceptors in Q-type RCs, while are intermediate electron acceptors in FeS-type RCs. The alkyl chains have been truncated in the structures for the sake of clarity. The upper scheme depicts the electron transfer steps between cofactors initiated by photon ($h\nu$) absorption along with the midpoint potentials of each cofactor. Vertical lines represent the absorption of light, while other lines show the electron transfer steps. Cyt c: cytochrome c; P870 and P870*: special pair absorbing at 870 nm and its excited state; BChl: monomeric bacteriochlorophylls (monomeric chlorophylls are omitted); BPh and Ph (bacterio) pheophytin; Q_A and Q_B are either ubiquinones (in PSII-like) or plastoquinones (in PSII); Tyr_Z is a tyrosine residue donating electrons to the OEC; P680 and P680*: special pair absorbing at 680 nm and its excited state; Cyt b_6f: cytochrome b_6f; PC: plastocyanine; P700 and P700*: special pair absorbing at 700 nm and its excited state; P840 and P840*: special pair absorbing at 840 nm and its excited state; A_0: monomeric chlorophyll; A_1: phylloquinone; F_X and F_AF_B: consecutive iron–sulfur centers; NADP: nicotinamide adenine dinucleotide phosphate. Structures were created with the VMD [13] software and using crystallographic coordinates retrieved from the Protein Data Bank files 2J8C (*R. sphaeroides* [14]) and 1JB0 (*Synechococcus elongatus* [15]).

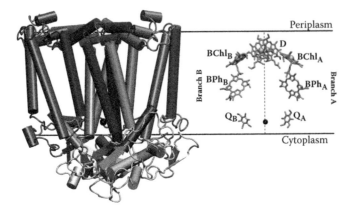

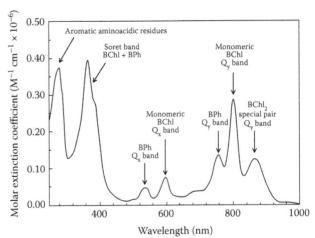

FIGURE 11.3 Spatial organization of subunits and cofactors of the photosynthetic RC from *R. sphaeroides* strain R26 (PDB ID: 1AIJ [16]). Main chains are shown as cartoons (L: dark gray; M: medium gray; H: light gray), and the iron is shown as a black sphere. Cofactors are shown on the right. Four bacteriochlorophylls a (BChl), two of which strongly coupled to form a BChl dimer (P870 or D) that behaves as a single species and functions as a primary electron donor, are located on the periplasmic side; two bacteriopheophytins (BPh), consisting of a Mg-free bacteriochlorin ring, sit in the middle of the protein; two ubiquinones (UQ$_{10}$) with a 10 isoprenyl units side chain and one non-heme Fe^{2+} are located in the close proximity of the cytoplasmic site. Subscripts A and B indicate cofactors located in the corresponding branches. Mg and Fe ions are represented as gray and black spheres, respectively. (From Tangorra et al., *Photochem. Photobiol. Sci.*, 14, 1844–1852, 2015. DOI: Reproduced by permission of the Royal Society of Chemistry.)

FIGURE 11.4 Optical spectrum of purified RC of *R. sphaeroides* strain R26. The spectrum contains several peaks associated with electronic transitions of the protein cofactors. The peak centered at 280 nm is mostly associated to the aromatic amino acids (plus a small contribution due to quinones). The intensity ratio between the absorption at 280 and at 802 nm is used to assess the purity of the protein. The ratio between the absorption peaks at 865 and 770 nm is instead used to assess protein integrity as, in degraded proteins, the bacteriochlorophyll dimer tends to lose the central magnesium ion (lowering the 865 nm peak) and converts in bacteriopheophytin (increasing the 770 nm peak). The optical spectrum of the RC from *R. sphaeroides* wild type (strain 2.4.1) is very similar to the above reported one, but features three peaks ranging between 400 and 500 nm and associated to a molecule of the carotenoid spheroidene, lacking instead in R26.

the chemical and structural complexity of PSII RCs, and its functioning is fully understood. The spatial organization of the RC from *R. sphaeroides* is described in some detail to offer an example of the main features owned by such proteins. It is an integral membrane protein formed by three subunits M (34 kDa), L (31 kDa), and H (28 kDa) surrounding a series of noncovalently bound cofactors (see Figure 11.3) and embedded in the cytoplasmic membrane. The LM complex forms an elliptic central *core* with the main axes of 40 and 70 Å. The two subunits are structurally very similar and are composed of five transmembrane α-helices arranged around a twofold pseudosymmetry axle oriented perpendicular to the membrane plane. The hydrophilic H subunit is located on the cytoplasmic side of the membrane and contains a single transmembrane helix. The cofactors located within the protein scaffolding, responsible for the electron transfer reactions, are organized in two branches, called A and B, arranged around the above-mentioned pseudosymmetry axle. Structural details are shown in Figure 11.3.

11.2.2.2 Some Spectroscopic Properties

The presence of bacteriochlorophylls, bacteriopheophytins, and quinones, along with the aromatic amino acids within the protein, results in a rich optical spectrum of the isolated RC, showing intense peaks from near ultraviolet (260 nm) to near infrared (1000 nm) [18]. The electronic transitions responsible for the optical absorption peaks of the RC are shown in Figure 11.4.

11.2.2.3 Electron Transfer Cascade

The absorption of a photon promotes the dimer D (P870 in *R. sphaeroides*) in its excited state within a handful of femtoseconds. This elementary photochemical event is followed by a cascade of electron transfer reactions and, as the electron moves from first to last cofactor, a charge-separated state is eventually produced in the protein, having the positive charge sitting on the oxidized dimer and the electron sitting on the newly formed semiquinone. The formation of this *nanocondenser* has a yield close to unit, i.e., almost each absorbed photon converts into a charge-separated state [19]. Such yet unmatched yield is a direct consequence of the finely tuned spatial organization of cofactors and amino acid residues that causes forward electron transfer to be at least three orders of magnitude faster than the wasteful, energy dissipative charge recombination reaction [20]. The counterpart of such high efficiency is the loss of more than 60% of the excited D* state energy eventually stored in the D$^+$Q$_B^-$ state (see Figure 11.5). The overall result of the photoconversion is anyway an outstanding nanocondenser having charges separated by 25–30 Å, having a free energy, relative to the ground state, close to 0.5 eV and decaying with a constant of roughly 1 s [21–23].

11.2.2.4 The Photocycle

Under physiological conditions, i.e., when the RC is embedded in the cytoplasmic membrane, the oxidized dimer D$^+$

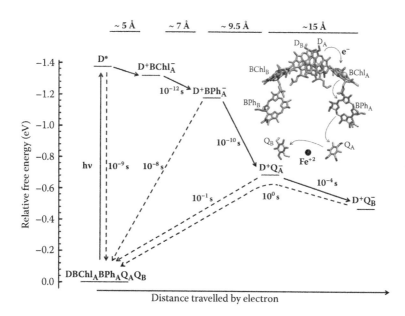

FIGURE 11.5 Energetic scheme of the cofactors involved in the cascade of electron transfer reactions initiated by the absorption of a photon and producing the final charge separated state $D^+Q_B^-$. Arrows indicate the forward (solid line) and backward (dashed line) electron transfer reactions. All involved cofactors are shown in the ground state, while after electron absorption, only the cofactors involved in the new state are shown. The inset shows the three-dimensional arrangement of the cofactors (symbols are explained in Figure 11.2). The upper row indicates the approximate distance travelled by the electron in moving from one cofactor to the next one along the A branch, virtually the sole functioning branch, of the RC. (From Kleinfeld, D. et al., *Biochimica et Biophysica Acta (BBA)—Bioenergetics*, 766(1): 126–140, 1984.)

formed after photon absorption is re-reduced by an external electron donor, the reduced form of cytochrome c_2 in the specific case of *R. sphaeroides*. The dimer is now prone to absorb a second photon and transfer a second electron that reaches the semiquinone formed after absorption of the first photon. The quinone is now doubly reduced and eventually picks up two protons from the cytoplasm milieu, forming a quinol molecule that leaves the binding site and reaches the membrane.

The quinol is replaced by a quinone molecule arriving from a pool sitting in the membrane. Meanwhile the oxidized dimer is re-reduced by a second cytochrome molecule. This cyclic system, known as RC photocycle and shown in Figure 11.6, is continuously fed by the reduced cytochrome and the oxidized quinone originated by the neighboring enzymatic complex cytochrome bc_1.

In isolated systems, the photocycle can be driven as long as suitable pools of exogenous electron donors (e.g., cytochrome c, ferrocene, ferrocyanide [24]) and quinone are present, while it would come to a stop once one of the exogenous pools comes to exhaustion.

11.2.3 LIGHT-HARVESTING COMPLEXES

The RC must fulfill the metabolic needs of the photosynthetic organisms even under dim light. Light-harvesting or antennae complexes operate to this purpose. Structure and composition of these complexes are very dependent on the photosynthetic organism. A detailed description of antennae complexes is beyond the scope of this chapter and can be found in the literature [25–28]. Here, a summary of the LHCs of the purple nonsulfur bacteria, to which *R. sphaeroides* belongs, will be presented [29].

11.2.3.1 The Spatial Organization

Among purple bacteria, the best investigated antenna system belongs to the bacterium *Rhodopseudomonas acidophila*, whose three-dimensional structure has been obtained by x-ray crystallography at 2.0 Å resolution [30,31] and is shown

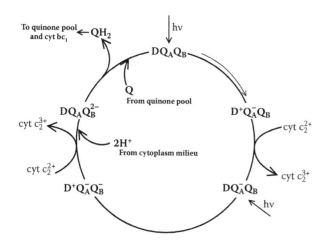

FIGURE 11.6 Simplified version of the photocycle of a Q-type RC. (Modified from Feher, G. et al., *Nature*, 339: 111–116, 1989.) In this scheme, only the primary electron donor (D) and the final electron acceptors (Q_A and Q_B) are shown. cyt c_2^{2+} and cyt c_2^{3+} represent the reduced and oxidized forms of the physiological electron donor to the oxidized dimer.

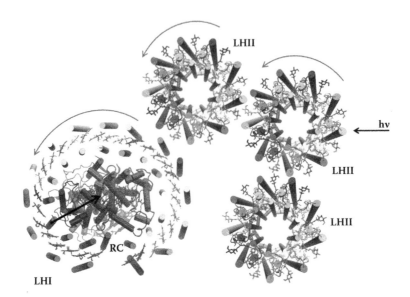

FIGURE 11.7 Spatial organization the PSU of the purple photosynthetic bacteria as obtained from x-ray crystallography. Each heterodimer formed by two transmembrane α-helices, three bacteriochlorophylls, and, for LHII, one carotenoid has the same color. The RC–LHI supercomplex is obtained from *Rhodopseudomonas palustris* [32] and deposited in the protein databank with the code 1PYH. The LHII complex is obtained from *R. acidophila* [32] and deposited in the protein databank with the code 2FKW. Gray arrows indicate the energy transfer, which eventually takes excitation (thick black line) to the dimer of the RC sketched in the center of LHI.

in Figure 11.7, where the RC is surrounded by protein complex in turn in close contact with a few smaller complexes forming the bacterial PSU.

The two complexes are composed of two types of subunits, α and β, associated into a heterodimer to form a α/β minimal unit and are structurally arranged in a toroidal proteic scaffold in which pigments are arranged. The complex LHII is formed of nine heterodimers and allocates two rings of bacteriochlorophyll molecules: one ring contains 18 molecules in an almost circular face-to-face arrangement close to the periplasmic side, and another set of nine Bchl monomeric molecules all lying in a plane perpendicular to the previous ring and sitting in the middle of the bilayer. LHII allocates, next to BChl, also one carotenoid molecule per heterodimer. Carotenoids play a double important role contributing to the harvesting of light in the 400–500 nm region and protecting the entire assembly against photooxidation by quenching the singlet oxygen molecules produced when the bacterium is contemporarily exposed to light and oxygen. This carotenoid is absent in *R. sphaeroides* R26, making this carotenoidless mutant very sensitive to oxidative damage if contemporarily exposed to light and oxygen [29,33,34]. LHI is the largest toroid, which consists of 15 α/β heterodimers containing a single elliptical ring of bacteriochlorophylls. LHI surrounds the RC to which directly funnels the harvested energy [32].

11.2.3.2 Some Spectroscopic Properties

The antenna complexes are the main responsible for the color of photosynthetic bacteria [29]. Since the beginning of bacterial photosynthesis research, *R. sphaeroides* has been investigated for its pigment composition, and several mutations, discernible on the base of the color of their colony, have been discovered. An early example of the spectral variability found

in five classes of mutants is shown in Figure 11.8 [35]. The blue-green mutant is optically very similar to the so-called strain R26 later on characterized and heavily investigated [33] to clarify the molecular basis of bacterial photosynthesis [36].

The absorption spectrum of photosynthetic bacteria strongly depends on several factors, the most important being the growth conditions; indeed, pigmented cells appear at low oxygen partial pressure and under illumination to reach maximal pigment concentration in cells cultivated anaerobically in the light [37]. Under such conditions, the pigment concentration, i.e., the color intensity of the cells, decreases as light intensity increases. This effect is a direct consequence of the role played by the LHCs, which substantially increases the optical cross section of the RC to ensure maximum photon absorption. In presence of intense illumination, this requirement disappears and the LH complex biosynthesis is frustrated. Changes in the relative amount of LH complexes synthesized under illumination are clearly reflected in the optical spectrum of the wild type [38]. At low light intensities, the absorption at 850 nm (LHII) is far more intense than the 870 nm peak (LHI) (Figure 11.8). The LHI peak intensity decreases as light increases, and the 870 nm peak eventually emerges as a separate peak under very high light intensities. Oxygen inhibits the pigment synthesis but does not prevent the bacterium to start its photosynthetic growth once the oxygen partial pressure has dropped and light is again shone.

11.2.3.3 The Energy Transfer

LHCs surround the Q-type RCs of purple bacteria as sketched in Figure 11.7. Light collected by the LHI and LHII antennas is funneled to the RC, where excitation reaches the unexcited bacteriochlorophyll molecules before any detrimental

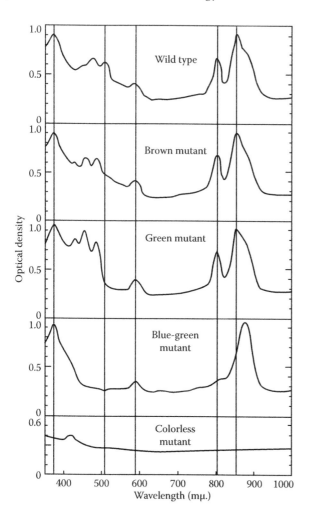

FIGURE 11.8 Optical spectra of whole cells of several mutants of *R. sphaeroides* grown photosynthetically. The wild type, also known as strain 2.4.1, presents the typical absorption bands of the LHC components. The BChls associated to LHI have an absorption peak appearing as shoulder (870 nm) of the bluemost peak (850 nm) of the BChls associated to LHII. The LHII complex has a second absorption peak (800 nm). Two further BChls peaks (merging LHI and LHII contributions) are present at 590 and 390 nm. The region from 450 to 550 nm presents the carotenoid peaks associated to the LHI complex. The cytochrome Soret band is hidden under the carotenoids. Brown and green mutants show a substantially unaltered spectrum in the BChls regions, while consistent differences are found in the carotenoid region, indicating differences in the LHI composition compared to 2.4.1. Substantial spectral differences are instead shown by the blue-green mutant both in the NIR and in the carotenoid regions. These changes are due to the absence of LHI complex in the bacterium. The colorless mutant shown at the bottom lacks all photosynthetic pigments, and hence the cytochrome peak is visible. This mutant cannot grow photosynthetically. (Reprinted from Griffiths, M. and R.Y. Stanier, *Journal of General Microbiology*, 14(3): 698–715, 1956. With permission.)

decay process takes place. The probability that excitation consequent to the absorption of a photon by the LH complexes produces a charge separation in the RC, i.e., the quantum yield or efficiency, is rather high—close to unity. Such a high value, descending from the spatial organization of the photosystem composed by RC–LHI–LHII, is the product of a kinetic competition between the dissipation of energy, i.e., excited bacteriochlorophylls returning to ground state by radiative decay (fluorescence), taking place in the nanoseconds timescale, and the transfer of excitation to RC, where energy is trapped by the charge separation, taking place in the picoseconds timescale [39,40]. This excitation transfer takes place nonradiatively, i.e., without photon emission, and is typically called Förster (or fluorescence) resonance energy transfer (FRET) [41]. It depends on three main parameters: (1) the overlapping of the emission band of the donor and the absorption band of the acceptor, (2) the orientation factor that accounts how properly oriented the electric dipoles of the donor and the acceptor molecules are and (3) the distance between the acceptor and the donor molecules. The distance at which energy transfer efficiency is 50% is called Förster's radius. The photosynthetic complexes of algae, plants, and bacteria are all structurally organized to favor the FRET over other energy transfer mechanisms, ensuring maximum energy transfer efficiency.

11.3 ARTIFICIAL PHOTOSYNTHESIS

Photosynthetic process inspires researchers around the world to fabricate artificial photosynthetic systems to be implemented in devices for solar energy conversion. Such devices should (1) efficiently absorb photons to generate electron excited states, (2) trap the excited state and convert it into stable charge separated state, (3) efficiently inject charges into an external circuitry and restore the initial condition to reinitiate the entire process, and (4) possess an interface to couple the electron flow with the production of energy-rich chemical species. A schematic view is shown in Figure 11.9.

11.3.1 Basic Architecture

The complex organization of a photosynthetic apparatus is out of reach for the synthetic chemistry of today, and probably of tomorrow in synthetic chemistry, but natural and synthetic photoconverters share the same overall architecture shown in Figure 11.9. Even though the fight against wasteful charge recombination reaction has not been won yet (see Figure 11.5), artificial photosynthesis has improved during the last 15 years, reaching appreciable performances [42] by introducing a plethora of artificial light-harvesting structures and RCs mimicking the photosynthetic electron transfer [5,6,8,42–45].

The minimum model for generating a photoinduced charge separation consists of a bifunctional molecule composed of electron–donor and electron–acceptor moieties. To efficiently absorb light, at least one of the moieties must be a chromophore [46,47]. The understanding of the parameters involved in the donor–acceptor interaction has improved in time [48], and more elaborate molecules have been introduced to bypass the competition between charge separation and charge recombination. Indeed, if charge separation occurs in a single step (as in dyads), the competition with charge recombination favors the latter, wasting the excitation energy. Spatial separation of the charges in the hole–electron couple (the trick

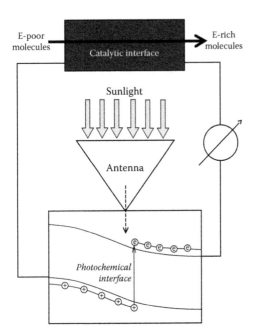

FIGURE 11.9 Schematic representation of a device able to convert sunlight into chemical energy. The light energy from the sun is absorbed and collected by the large array of antennae (triangle) that surrounds the photochemical interface (rectangular box) acting as *p-n* junction semiconductor. After the excitation energy is funneled to the photochemical interface (dotted line), charge separation (production of an electron (e)/hole (+) pair) takes place followed by consecutive electron transfer processes (solid lines). The emerging electrons reach the e-donating catalytic site where drive reductive processes are performed (producing, e.g., molecular hydrogen or methanol from carbon dioxide). Electrons produced in this device can also possibly be used in power systems.

used by the RC; see Figure 11.5) increases the lifetime of the charge-separated state [49–51]. Triad arrangement composed, for example, of porphyrin (P), quinone (Q), and carotenoid (C) shows a charge-separation yield of 4%, an energy conversion efficiency of 2%, and a charge-separated state lifetime of 300 ns [47]. In a pentad arrangement based on the previous triad, added with two further electron transfer steps, the yield reaches 83%, the energy conversion efficiency reaches 50%, and the lifetime of the charge separated state increases up to 55 μs [52] (see Figure 11.10).

To improve performance in artificial photosynthesis, the challenge of increasing complexity was tackled, and several examples of fairly elaborated molecular assemblies, including dendrimersome and fullerene-based systems, have been synthesized [43,53–56].

11.3.2 PHOTOCONVERSION EFFICIENCY VERSUS STRUCTURAL COMPLEXITY

Photoconversion efficiency improves with the increase in structural complexity and, to date, artificial photosynthetic devices do not reach the dazzling efficiency of natural apparatus. This is graphically illustrated in Figure 11.11 where the photoconversion efficiency is represented as a function of

structural complexity, with the dyads representing the simplest and less efficient system, while photosystem I and, even more, photosystem II, where the OEC is located in oxygenic photosynthesis, represent the most complex and efficient systems. Throughout the chapter, the notations PSI and PSII refer to the whole photosystems I and II. PSI and PSII contain the Q-type and FeS-type RCs, respectively, together with a large number of other functionally needed subunits [17,56].

Chemical robustness is also included to show that biological materials are far more perishable than the organic counterpart. Even though the scales used in Figure 11.11 are somehow arbitrary, an unquestionable gap between the best artificial system and biological photosynthesis does exist. A clear tendency is to fill this gap by climbing the hill (or mountain) of complexity, figuring out better and better organic molecules. This tendency is well documented in the literature, and the performance of new molecules is steadily improving [57].

An alternative approach that has been explored is the so-called molecular maquettes, a minimal assembly in which nonnative peptides are transformed into proteins with native-like properties [58,59]. The idea of using molecular maquettes sits on the search for minimal polypeptide assembly, which performs the property sought, without carrying the parts of the natural protein irrelevant to the pursued activity. This idea was applied to some protein cases, e.g., for artificial heme-containing proteins, for artificial Q-type RCs [60,61], and for the higher plants OEC [62].

Assembling biological and nonbiological materials represents a further opportunity to fill the gray gap shown in Figure 11.11 by improving the energy conversion efficiency. The basic idea is to couple the RCs with one (or more) synthetic moiety for performing either effective light-harvesting or ancillary redox chemistry. Strictly speaking, this does not represent a fully artificial photosynthetic system, being based on a photoenzyme, and will be discussed in Section 11.4 as a nano-biohybrid (or biohybrid) photoconverter.

11.4 NANO-BIOHYBRID (OR BIOHYBRID) PHOTOCONVERTER

Photosynthetic hybrid systems are a hot research topic since the beginning of this century, as photoelectrons generated with very high efficiency can be used either in redox chemistry or in electric circuits. Preliminary demonstrations were given for PSI, PSII, and Q-type RCs used as photosensors [63,64], as gating elements for phototransistors [65], as optoelectronic components [66,67], and as biosensors for the detection of herbicides [68,69] or other environmental pollutants [70]. Particularly attractive is the application of photoenzymes in photovoltaic devices [71–74] and light-powered fuel cells for hydrogen generation [75–77] or water-splitting reactions [78].

These and the future challenging applications require the selection of suitable bio-nanocomposite materials that ensure full functionality of the photoenzyme. Several aspects of the

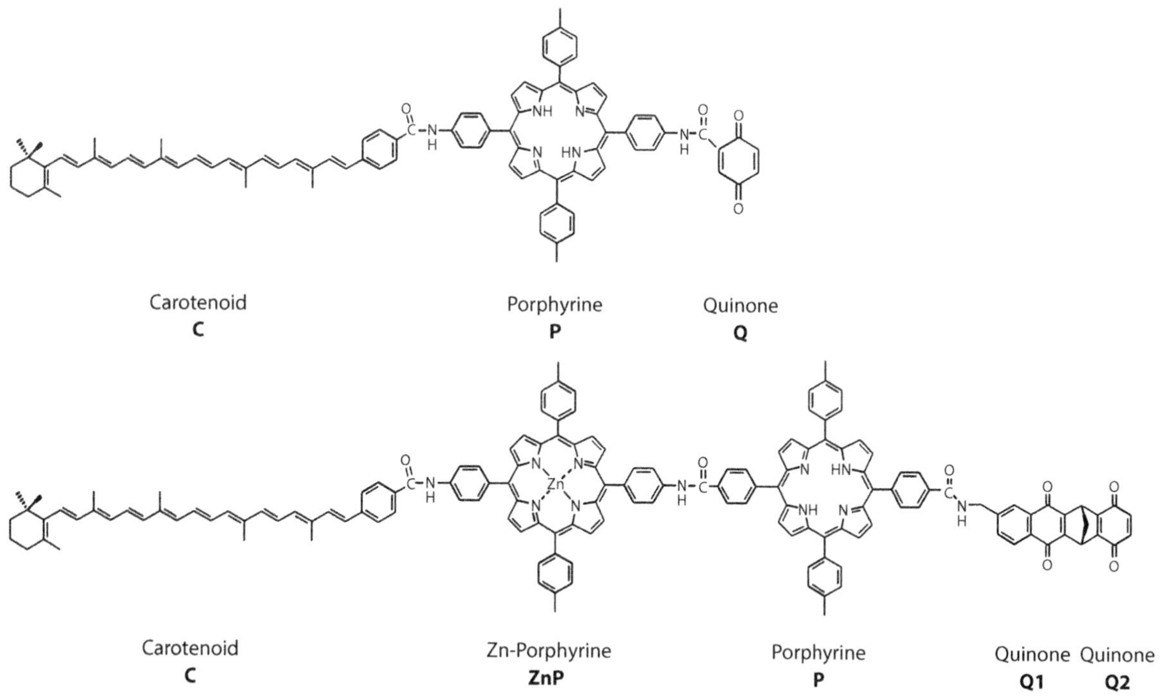

FIGURE 11.10 Porphyrin, quinone, and carotenoid based triad and pentad. In the triad, the central porphyrin absorbs photons in the visible region and to become excited. Excitation is transferred to the quinone forming the semiquinone radical anion and the carotenoid radical cation forming the charge-separated state $C^{\cdot+}-Q^{\cdot-}$. The pentad contains two more electron transfer steps and is performing more than triad: the central ZnP absorbs photons to form the state $C-ZnP^*-P-Q_1-Q_2$. The excitation energy is transferred to the neighbor porphyrin P forming the singlet state. From P, a photoelectron is shuttled toward the first electron acceptor, the quinone Q1, forming the electron hole system $C-ZnP-P^+-Q1^{\cdot-}-Q2$. The following step is performed by the secondary electron acceptor (Q2) and electron donor (C) to establish the final charge separated state. $C^{\cdot+}-ZnP-P-Q1-Q2^{\cdot-}$. (Redrawn from Balzani, V. et al., *ChemSusChem*, 1(1–2): 26–58, 2008.)

integration between proteins and nanocomposite materials must be addressed to properly design and efficiently integrate *in vitro* hybrid system, including membrane protein manipulability, biocompatibility of connecting interfaces, and effectiveness of the electron or energy transfer among components. Indeed, efficient handling of the nanoscale communication still faces many open challenges in designing hybrid schemes, constructing stable conjugates, and understanding protein–material interaction mechanism [79].

Photosynthetic proteins are interfaced to a wide range of materials, using several strategies, to improve the light-harvesting capability or to produce integrated optoelectronic devices, energy conversion systems, and biosensors. In this chapter, we wish to offer an overview of these applications showing technical difficulties and limitations.

Two possible (see Figure 11.12) photodriven reaction categories will be discussed: (1) electron transfer among a metal, a semiconductor, or carbon-based substrates and photosynthetic

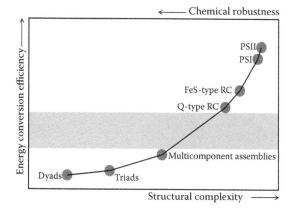

FIGURE 11.11 Photoconversion efficiency versus structural complexity and chemical robustness.

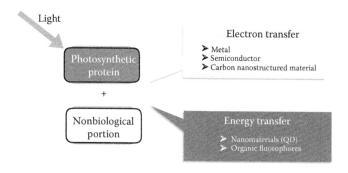

FIGURE 11.12 Photophysical processes involved in photosynthetic protein-based hybrid systems. (Redrawn from Kim, Y. et al., *Nanotechnology*, 25(34): 342001, 2014.)

proteins, and (2) energy transfer between biomimetic light-harvesting moieties and a photosynthetic protein [80].

11.4.1 Hybrid Interfaces for Electron Transfer

Efficient electron transfer between a metal or semiconductor substrate and a photosynthetic protein is a key goal in photoelectric power [81] and requires a suitable strategy [72,82,83]. The solid structural knowledge of RCs, PSI, and PSII is being used to design several immobilization techniques: (1) electrostatic adsorption, (2) physisorption, (3) affinity binding, (4) covalent binding, and (5) *plugging* (Figure 11.13) [84]. These techniques will be briefly outlined in the following paragraphs. Furthermore, Sections 11.4.1.5 and 11.4.1.6 describe two special hybrids that involve the use of photosynthetic reaction center with carbon-based and nanoparticle (NP) substrates.

11.4.1.1 Electrostatic Adsorption

Electrostatic absorption is a simple and generally applicable procedure realized by a preliminary step in which protein and substrate are charged followed by either a direct or layer-by-layer protein deposition. Charging the substrate can be achieved by simple electrostatic interaction, while the charge of the protein descends from its isoelectric point and can be set choosing the appropriate pH value of the solubilizing solution [24,85]. The electrostatic assembly forms by simple deposition [86].

The layer-by-layer assembly offers the possibility to fabricate multilayer films of (bio)organic and inorganic materials on a variety of substrates [71,73,87]. Thylakoid enriched in PSII was immobilized on quartz substrate by alternating protein to the polyelectrolyte polyethylenimine layers [88]. Multilayered (up to 24 layers) films containing ordered RC were also obtained by electrostatic adsorption using positively charged polydimethyldiallylammonium chloride (PDDA) and negatively charged protein [87].

11.4.1.2 Physisorption Techniques

Physisorption or physical sorption involves direct protein deposition on the bare substrate surface to form a thin film and can be effectively achieved by drop casting or spin coating [63,89–91]. Dense PSI monolayers deposited on several different substrates incorporated into stand-alone stable biohybrid photoelectrochemical cell were successfully used to convert light into electrical energy [71,72].

Direct immobilization by laser printing has been alternatively used to successfully immobilize thylakoid membranes from fresh spinach leaves on screen-printed gold electrodes [92,93] for applications in herbicide biosensors.

Physisorption lacks selective protein-oriented adhesion on bare surfaces and, furthermore, direct contact with the substrate might introduce conformational changes in the protein, modifying the active centers, the specific binding surfaces, and/or the redox properties of cofactors. Notwithstanding these potential drawbacks, several proofs that different photosynthetic proteins remain functional on bare electrode surfaces have been produced [63,89,91,94,95]. Also, introducing cysteine tags on the protein surface favors oriented adhesion on bare surface, especially onto gold electrodes. Surprisingly, high photocurrent was registered when RCs of *R. sphaeroides* and RC–LH1 complexes from *R. acidophila* were adhered to bare gold electrodes [96].

11.4.1.3 Covalent Binding

A stable and robust interaction between protein and substrate can be obtained by covalent binding. Protein can be readily functionalized by bioconjugating primary amines, sulfhydryls, carboxyl, and carbonyl groups [97]. The primary amines of lysine residues are a common target for bioconjugation as they are abundant and mostly located on the outer part of proteins. The sulfhydryl group of cysteines is less abundant than lysines, hence offering higher chances to chemically modify specific sites. The single cysteine residue present on the RC surface is used to immobilize the protein,

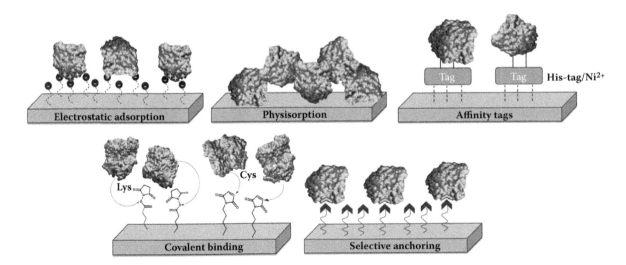

FIGURE 11.13 Schematic overview of protein immobilization techniques onto biomimetic material substrates. (Modified from Kim, Y. et al., *Nanotechnology*, 25(34): 342001, 2014.)

in the proximity of the electron acceptor quinone, to a carbon electrode surface [98] producing photocurrents. Similar strategy is also employed to prepare ultrathin ordered films based on spontaneous molecular assembly of bifunctional linkers, interacting by hydrophobic interactions, anchored by alkylthiol–gold surface affinity with the surface substrate at one termination and covalently bound to the protein with the other termination [83]. Alkylthiols–gold surface can be used to construct ordered RC array on nanocomposite electrodes for photocurrent generation [99–102].

Self-assembled monolayers (SAMs) of bifunctional linkers offer a simple way to assemble and orient the protein on electrode surface preventing its alteration or degradation [78]. Indeed, *R. sphaeroides* RCs adhered to bare-Pt electrode in the presence of cytochrome c produce a photocurrent whose intensity can be increased by at least a factor of 10 when the electrode is previously coated with an alkanethiol SAM [103]. Care should, however, be taken in using SAMs, as they tend to strongly impair the rate of electron transfer between the electrode surface and RC redox cofactors [98,104].

11.4.1.3.1 Affinity Tags

Alternatively to bioconjugation, genetically engineered RC bearing a polyhistidine tag can be bound via a Ni^{2+}–nitrilotriacetic acid (Ni–NTA) complex, a well-known method to chelate His-tagged proteins. This method, widely used for protein purification procedure, has been successfully adapted to anchor proteins to electrode surfaces [102,104–106]. By opportunely inserting the His-Tag chain on the protein, a preferential RC orientation can be achieved and a consequent cathodic or anodic photocurrent can be generated [105]. By expressing the His-Tag at the C-terminal end of the M subunit, the bacteriochlorophyll dimer will face the electrode, and electrons will flow toward the protein [104]; on the other hand, by expressing it on the H subunit, the quinones will face the electrode and electrons will leave the protein and flow toward the substrate [101].

11.4.1.4 Selective Anchoring

Anchoring proteins is a handy alternative to other interfacing procedures: bacterial Q-type RCs, in particular, can be hooked to a substrate using either the quinone or the cytochrome binding sites. Cytochromes have a high affinity to the RC whose periplasmic site presents a cytochrome docking site very close to the dimer D. When bound to the RC, reduced cytochrome donates an electron to the oxidized dimer (see Section 11.2.2.4). The high-affinity cytochrome c and RC can be exploited to anchor the protein to the substrate surface [99,103]. Cytochrome c has been also used for assembling the PSI complex on a thiol-modified gold surface [107], where it acts as an assembly template of an oriented and dense protein layer, and as a wiring agent to shuttle electrons from the electrode to the PSI. A noteworthy example of selective anchoring is the plugging strategy in RC [108], based on the affinity of the buried quinone binding pocket toward the quinoid moiety. A first attempt of RC immobilization on Pt electrode using the plugging strategy attempted to insert a naphthoquinone-based linker in the ubiquinone binding pocket of the bacterial RC [109]. As a consequence of the lower affinity of the quinone-binding pocket to naphthoquinones as compared to physiological ubiquinone, the attempt was not very successful. However, it paved the way to successful PSI immobilization on gold NPs, achieved by molecular wires of vitamin K1 (phylloquinone A_1) plugged in the quinone-binding site of the protein [110,111].

11.4.1.5 Interface of RC with Carbon Substrates

Interesting substrates to which photosynthetic proteins can be interfaced for photoinduced charge transfer are the carbon-based ones, including the nanostructures fullerenes, carbon nanotubes, and graphene. The initial attempt used freshly cleaved pyrolytic graphite electrode with RCs immobilized by covalent binding either to a lysine or cysteine residue [98]. The graphite electrode was further activated with a layer of UQ_{10} as the redox mediator. Photocurrent could be obtained if RC binding occurs via cysteines, while additional UQ_{10} is required in the solution if RC binding occurs via lysines. A similar approach was used more recently using highly ordered pyrolytic graphite (HOPG) as a substrate [112]. *R. sphaeroides* Q-type RC generates photocurrent densities up to $7 \, \mu A \cdot cm^{-2}$, which is among the highest values reported to date [113]. Carbon electrodes were also used in connection with genetically engineered poly-His tagged RCs using a pyrene-NTA linker [104]. Two possible RC orientations (D-side or Q-side facing the electrode) were obtained, and in both cases, electrons flow in the same direction: in Q-side RCs, the photocurrent is anodic, while in P-side RCs, the photocurrent is cathodic.

Carbon structures are also employed to facilitate electron transfer to noncarbon electrodes. His-tagged RCs were immobilized on Au-covered ITO electrode via a Ni^{2+}–NTA linker. The dimer faces the electrode, generating a cathodic photocurrent. A 60 nm layer of fullerene (C_{60}) is used to cover the electrode improving the photocurrent [114]. Multiwalled or single-walled carbon nanotubes (MWCNTs or SWCNTs, respectively) were also used in connection with RCs [89,94,115] by either physisorption (driven by hydrophobic interaction) or chemical cross-linking by using the conducting polymer poly(3-thiophene acetic acid) [116] (Figure 11.14). Encapsulation of a genetically engineered RC with specifically synthesized organic molecular linkers inside carbon nanotubes and bound to the inner tube walls with unidirectional orientation was also reported [117].

Graphene-based electrodes, composed of 2D layers of sp^2-hybridized carbon atoms, have very high carrier mobility, elasticity, and thermal conductivity, and are transparent over the visible spectrum. These extraordinary electronic, mechanical, and optical properties of graphene are appealing for applications ranging from solar cells to transistors. So far, there are a few examples of coupling graphene with photosynthetic materials, dealing with plant PSI attached via physisorption [118,119] and bacterial RCs covalently attached via click chemistry [120].

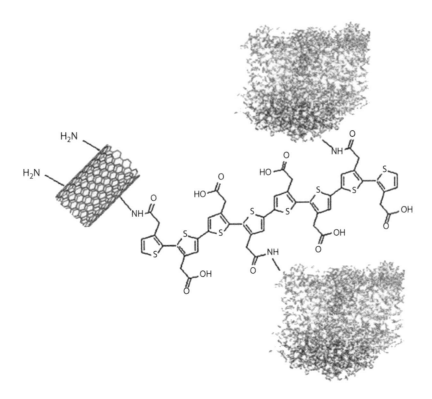

FIGURE 11.14 SWCNTs used as a substrate of the bacterial Q-type RC from *R. sphaeroides*. The conducting poly(3-thiophene acetic acid) ensures covalent binding and electric connection between SWNT and RC. (Reprinted from Szabó, T. et al., *Physica Status Solidi B*, 249(12): 2386–2389, 2012. With permission.)

11.4.1.6 RC–NP Interface

NPs exhibit unique electronic, photonic, and catalytic properties that can also be tuned by opportunely selecting their size. Today, NP and nanomaterial synthesis is so advanced that it is possible to accurately tailor them for specific purposes including patterning and assembling in hybrid nanobiomaterials [121–123]. Metal or semiconductor NPs having size comparable to the photosynthetic proteins can easily be obtained, making these novel materials proper candidates for bio-hybrid systems. These outstanding properties were exploited in several NP-photosynthetic protein hybrid systems, in which electron mechanism and enhanced catalytic activity are well documented [80].

Nanocrystals (NCs) have high surface-active area, which is a consequence of the nanodimension, thus making them suitable for charge transfer from redox proteins, including photosynthetic RCs [106,124]. Semiconducting and conducting NPs, e.g., CdS and Au NCs, were coelectropolymerized on an electrode surface, and the resulting composites [125] were used in conjunction with PSII for photocurrent generation [126]. Alignment of PSI on electrode surfaces was also achieved on Pt nanoclusters [127], which play the double role of enhancing the electron transfer between the F_B iron–sulfur cluster of PSI (see Figure 11.2) and surface electrode, and facilitating the alignment of the protein on the electrode.

A very intriguing result was reported [75,128] where a PSI and Pt NPs are covalently connected by a bifunctional linker, namely, 1,6-hexanedithiol or 1,4-benzenedithiol. The linker is

short enough (1–2 nm) to allow electron injection from PSI to the NP before any intraprotein charge recombination occurs. As a result, H_2 production takes place at the Pt NP (Figure 11.15a).

Molecular wires formed by NCs were employed to extract electron from PSI and deliver them to the gate of a field-effect transistor (FET, see Figure 11.15b), with the aim of generating a PSI-bio-photosensor for use in imaging devices [111,129,130].

Finally, NPs can be employed in hybrid systems as ancillary catalytic site for exploiting the charge-separated state. Early example used hybrids prepared by electrostatic adsorption of Pt or Au nanoclusters directly on the plant organelles (the thylakoids). The experiment showed that the charge-separated state of PSI injects photoexcited electrons into the NP, catalyzing the subsequent electrocatalytic reduction of protons from the acid aqueous solution to form molecular hydrogen [131].

11.4.2 HYBRID INTERFACES FOR ENERGY TRANSFER

The chemical complexity of photosynthetic proteins ensures their high-performing standards but is accompanied by intrinsically low chemical robustness. In the following, new interfaces recently introduced to increase chemical robustness in hybrid systems devoted to light harvesting will be presented.

11.4.2.1 RC–NP Interface

NPs, used in electron-transfer interfaces (see Section 11.4.1.6), are also exploited in hybrid system as optical antenna and

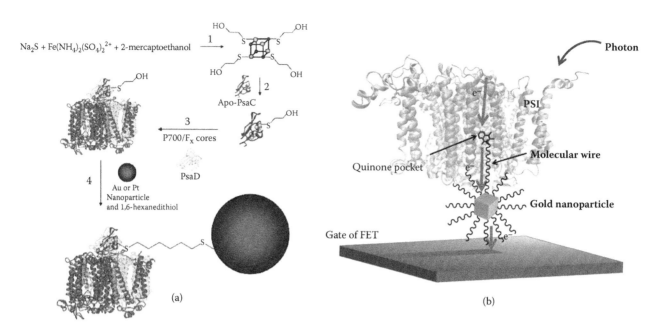

FIGURE 11.15 (a) Scheme of the assembly of PS I to Au or Pt NP. Construction procedure of the PSI/1,6-hexanedithiol/NP bioconjugate. In (1), a [4Fe–4S] cluster is assembled in the solution starting from sodium sulfide, ferrous ammonium sulfate, and 2-mercaptoethanol. In (2), an *ad hoc* engineered subunit from PSI (Apo-PsaC) displaced the [4Fe-4S] cluster, and then (3) bound to the FeS-type RC. In (4), the Apo–PsaC–FeS-type RC is covalently linked to the Pt or Au (gray sphere) NP to eventually produce molecular hydrogen [75]. (Reprinted with permission from Grimme, R.A. et al., *Journal of the American Chemical Society*, 130(20): 6308. Copyright 2008 American Chemical Society.) (b) Schematic representation of the biophotosensor concept based on PSI gold NP hybrid coupled with FET. (Redrawn from Terasaki, N. et al., *Biochimica et Biophysica Acta (BBA)—Bioenergetics*, 1767(6): 653–659, 2007.)

light concentrators [86]. NPs can achieve high light concentration ability at the nanometer scale, and hence are very suitable for designing efficient light-harvesting devices. This is particularly true for quantum dots (QDs), semiconductor NCs small enough to exhibit quantum mechanical properties [132]. Electronic and optical absorption properties of QDs can be finely tuned from ultraviolet (UV) to visible and even to near infrared (NIR) by conveniently selecting chemical composition, size, or shape [133]. QDs are perfect as energy donors because their size is comparable with Förster radius (see Section 11.4.1.6), and hence very short donor–acceptor separation distances can be achieved, with consequent high FRET efficiencies. Furthermore, their wide absorption bands allow QDs to be excited at wavelengths where no direct excitation of the acceptor occurs [134]. QDs are not soluble in water, and their use in conjunction with biological materials [135] requires the use of proper linker molecules surrounding the QDs and ensuring water solubility.

Silver QDs were proposed and modelled as light harvesters for photosynthetic proteins [124,136], and the goodness of the model was indeed confirmed in the PSI–silver QD complex. A dramatic increase of roughly 20 times in the excitation rate of PSI demonstrates that silver QDs can efficiently be used for PSI excitation [137].

Experimental proofs of efficient energy transfer from a QD to a Q-type bacterial RC were obtained by using CdSe and CdTe NPs artificial antenna [138]. Hybrid complexes CdSe and CdTe–RC were obtained by electrostatically assembling the QDs, with diameter ranging between 2.0 and 4.6 nm, to

the bacterial RC from *R. sphaeroides* (Figure 11.16). The CdTe–RC hybrid complex was found to be more efficient than the CdSe complex because of its shorter radius and ligand lengths, resulting in a shorter QD–RC distance. In both QD–RC hybrid complexes, the efficient quenching of QDs' exciton emission and a corresponding enhancement of the photoluminescence emission of the BChl dimer D in the RC suggested an efficient QD to RC FRET.

Energy transfer study from CdSe QDs with 3 nm diameters to His$_6$-tagged PSI from *Chlamydomonas reinhardtii* [139] was performed. For PSI adsorption, CdSe QDs were previously coated with ZnS and trioctylphosphine oxide to ensure solubilization. The QD emitting at 575 nm was found to work as an artificial antenna and efficiently extends the absorption spectral range of the PSI. Commercially available water-soluble CdSe/ZnS QDs (5.2 nm in diameter) coated with hydrophilic capping were used to form QD–PSII hybrids with the NP functioning as an efficient electron donor to the PSII [140]. In both cases, the energy transfer between the QDs and the photosynthetic portion of the hybrid was found to take place via resonance energy transfer (FRET).

11.4.2.2 RC-Organic

Alternatively to bulky RC-organic interface (see Section 11.4.2.1), an increase in the absorption of RCs can be achieved by using *ad hoc* synthesized or commercial organic chromophores as light-harvesting systems. The chromophores can be covalently bound to specific amino acid residues of the protein [141] to form an organic–biological hybrid. To effectively

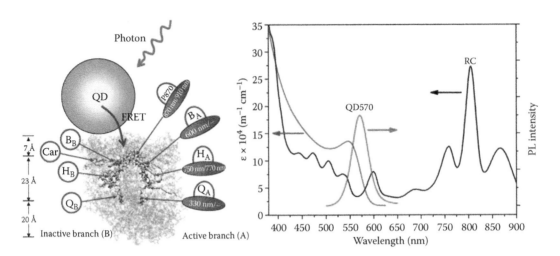

FIGURE 11.16 Schematic representation of the RC–CdTe QD hybrid system in which protein cofactors absorption/photoluminescence maxima are shown. BChl special pair (P870), BChl monomer (B), bacteriopheophytin (H), and quinone (Q) are indicated for the sole active branch (A). Photons are absorbed by both the RC and the QD. An exciton from the QD is transferred to the RC by FRET. Car = carotenoid. Energy is transferred from the CdTe QD to the RC thanks to the good overlapping of donor emission and acceptor absorption spectra as shown on the right. (Reprinted from Nabiev, I. et al., *Angewandte Chemie—International Edition*, 49(40): 7217–7221, 2010. With permission.)

improve the RC light-harvesting and subsequent the energy transfer efficiency, the chromophore must fulfill a few requirements: (1) allow a fine-tuning of spectroscopic properties for a large spectral overlap between its emission bands and RC absorption bands; (2) possess an appreciable water solubility and (3) an adequate fluorescence quantum yield, even in aqueous mean; (4) must tightly bound to the selected amino acid targets; and (5) result smaller than the protein size to avoid any steric hindrance [142]. Recently, the organic fluorophore belonging to the class of aryleneethynylenes (AE; see Figure 11.17a) with all the above features was synthesized and covalently bound to the Q-type RC of *R. sphaeroides* strain R26 to function as an antenna [143].

The conjugated backbone of AE absorbs light at 450 nm in correspondence to a minimal absorption of the RC, and presents a large Stokes shift, with an emission maximum at 602 nm, corresponding to an RC absorption peak (Figure 11.17b). The presence of the n-hexyl chains and triple bonds allows, respectively, favoring the intercalation of the dye skeleton with surfactant molecules and modulating excitation and emission wavelength without posing severe steric hindrance. The succinimidyl ester group was introduced to react selectively to lysine RC residues.

The effective AE bioconjugation to RC was confirmed by matrix-assisted laser desorption/ionization-time of flight (MALDI-ToF) measurements and sodium dodecyl sulfate

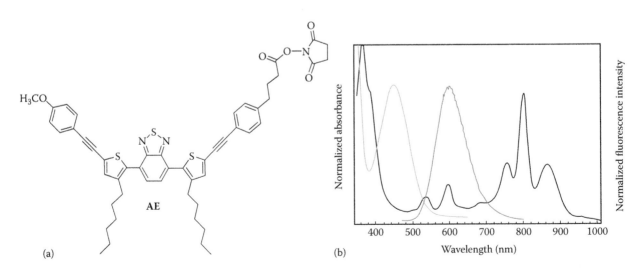

FIGURE 11.17 (a) Chemical structure of the organic fluorophore AE activated with the succinimidyl ester group. (b) Absorption spectra of RC (black) and AE (light gray) superimposed with the emission spectrum of AE (dark gray). (Adapted from Milano, F. et al., *Angewandte Chemie*, 124(44): 11181–11185, 2012.)

polyacrylamide gel electrophoresis (SDS-PAGE). A calculated average of four AE molecules per RC was found. The actual AE ability to transfer energy to the RC without modifying its energetics and activity was checked by recording, at the wavelength of AE maximum absorption, the enzymatic RC photocycle in presence of electron donor cytochrome c²⁺ and by detecting the formation of photoinduced charge-separated state (Figure 11.18b) in native and hybrid RCs. On the other hand, steady-state emission and fluorescence decay measurements (Figure 11.18a) showed a drastic emission quenching and a reduction of fluorescence lifetime of AE attached to RC, suggesting that an efficient dye-to-RC energy transfer by Förster mechanism can occur. It was found that, at 450 nm, at the AE absorption maximum, the hybrid system outperforms the protein by a factor of three according to the photocycle assay, and by three to five times according to charge separation and steady-state fluorescence measurements, confirming that the organic dye bioconjugation to RC is not detrimental to the protein activity, but, more importantly, enhances the RC ability to drive its photochemical reactions in specific wavelength ranges.

To gain insight in FRET mechanism and to explore the needed geometric design and spectral requirements for efficient energy transfer between an organic artificial antenna and RC, Dutta et al. [144] reported a bright study of covalent binding of several commercial Alexa Fluor dyes to a genetically modified RC. To make more specific the covalent reaction, a triple *R. sphaeroides* 2.4.1 mutant presenting three unique cysteine targets near the dimer D was engineered. These residues were covalently attached to three Alexa Fluor dyes (AF647, AF660, and AF750) functionalized with a maleimide group. The bioconjugation of each dye resulted in an increase in the photochemical RC activity and in the amount of charge separated state at the specific dye absorption wavelengths.

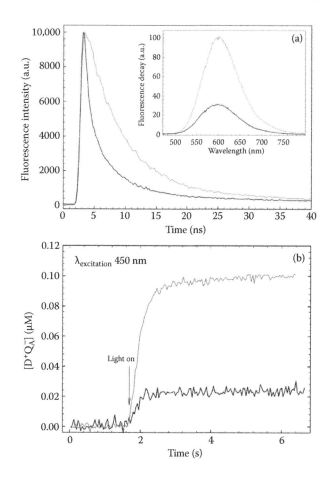

FIGURE 11.18 (a) Fluorescence decay and steady-state emission of AE–RC (dark gray) and AE (black). (b) Charge-separated state concentration of AE–RC (dark gray) and RC (black) at excitation wavelength of 450 nm. (Adapted from Milano, F. et al., *Angewandte Chemie*, 124(44): 11181–11185, 2012.)

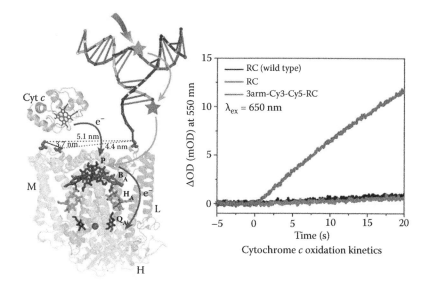

FIGURE 11.19 **(See color insert.)** RC from the purple bacterium, *R. sphaeroides* 2.4.1, bioconjugated to a 3arm-DNA construct sequences. The two stars on strands 2 and 3 represent the positions of the two dye molecules. Because of the presence of three Cys residues, up to three 3arm-DNA junctions (and hence three pairs of dyes) can be conjugated to the RC. (Reprinted with permission from Dutta, P.K. et al., *Journal of the American Chemical Society*, 136(47): 16618–16625. Copyright 2014 American Chemical Society.)

Also, in this case, it was possible to demonstrate energy transfer from the Alexa Fluors to the RC by Förster energy-transfer mechanism, and an enhancement in the RC photochemical activity.

Similar organic fluorophores (Cy3 and Cy5 or AF660 and AF750) have been employed to form hybrid with the same triple RC mutant, where the artificial light-harvesting systems are anchored to a 3arm-DNA nanostructure, resulting in a geometrically well-defined antenna complex [145] as shown in Figure 11.19. The complex was obtained by attaching two different pairs of dyes (Cy3 and Cy5, or AF660 and AF750) to strands 2 and 3 of a 3arm-DNA nanostructure. Strand 1 is instead bioconjugated to a cysteine residue of the mutant RC by a *N*-succinimidyl-3-(2-pyridyldithio)propionate crosslinker (Figure 11.19). Also in this case a significant increase in the amount of cytochrome c oxidized during the photocycle was found, showing that the biohybrid device outperforms the bare RC.

Finally, Förster resonance energy transfer from organic compounds to photosynthetic LHCs was also reported [146]. Energy transfer from a conjugated polymer blend (poly(9,9-dioctylfluorenyl-2,7-diyl):poly(2-methoxy-5-(2-ethylhexyloxy)-1,4-phenylenevinylene) to a LHII obtained from the purple bacteria *R. palustris* was demonstrated, observing a 30% reduction of the fluorescence lifetime of the polymer emission as compared to the pure polymer layer.

11.5 CONCLUSIONS

A brief overview of the state-of-the-art in assembling hybrid systems based on photosynthetic proteins is presented. Two main functional purposes of the hybrid interfaces have been described, namely, for electron and energy transfer, and literature has been reviewed and discussed according to the functional purpose.

ACKNOWLEDGMENTS

This work was financed by Università degli Studi di Bari "Aldo Moro" (IDEA Giovani Ricercatori 2011 project: "BIOEXTEND: Extending enzymatic properties by bioconjugation of enzymes with fluorescent organic oligomers"), PON 02_00563_3316357 Molecular Nanotechnology for Health and Environment MAAT and COST Action TD1102 "PHOTOTECH."

REFERENCES

1. Rosner, R., *MacMillan Encyclopedia of Physics*, Vol. 4. 1996, New York: Simon & Schuster.
2. Armaroli, N. and V. Balzani, The future of energy supply: Challenges and opportunities. *Angewandte Chemie-International Edition in English*, 2007. **46**(1–2): pp. 52–66.
3. Ciamician, G., The photochemistry of the future. *Science*, 1912. **36**(926): pp. 385–394.
4. Armaroli, N. and V. Balzani, *Energy for a Sustainable World: From the Oil Age to a Sun-Powered Future*. 2010, Darmstadt, Germany: Wiley-VCH.
5. Alstrum-Acevedo, J.H., M.K. Brennaman, and T.J. Meyer, Chemical approaches to artificial photosynthesis. 2. *Inorganic Chemistry*, 2005. **44**(20): pp. 6802–6827.
6. Barber, J. and P.D. Tran, From natural to artificial photosynthesis. *Journal of the Royal Society, Interface*, 2013. **10**(81): p. 20120984.
7. Balzani, V., A. Credi, and M. Venturi, Photochemical conversion of solar energy. *ChemSusChem*, 2008. **1**(1–2): pp. 26–58.
8. Maróti, P. and M. Trotta, Artificial photosynthetic systems, in *CRC Handbook of Organic Photochemistry and Photobiology, Third Edition—Two Volume Set*, A. Griesbeck, M. Oelgemöller, and F. Ghetti, Editors. 2012, Boca Raton, FL: CRC Press, pp. 1289–1324.
9. Hohmann-Marriott, M.F. and R.E. Blankenship, Evolution of photosynthesis. *Annual Review of Plant Biology*, 2011. **62**(1): pp. 515–548.
10. Blankenship, R.E., *Molecular Mechanisms of Photosynthesis*. 2002, Oxford: Blackwell Science.
11. Allen, J.P. and J.C. Williams, Photosynthetic reaction centers. *FEBS Letters*, 1998. **438**(1–2): pp. 5–9.
12. Michel, H., O. Epp, and J. Deisenhofer, Pigment-protein interactions in the photosynthetic reaction centre from Rhodopseudomonas viridis. *EMBO Journal*, 1986. **5**(10): pp. 2445–2451.
13. Feher, G., Three decades of research in bacterial photosynthesis and the road leading to it: A personal account. *Photosynthesis Research*, 1998. **55**(1): pp. 1–40.
14. Blankenship, R.E., Origin and early evolution of photosynthesis. *Photosynthesis Research*, 1992. **33**: pp. 91–111.
15. Humphrey, W., A. Dalke, and K. Schulten, VMD: Visual molecular dynamics. *Journal of Molecular Graphics & Modelling*, 1996. **14**(1): pp. 33–38.
16. Stowell, M.H.B. et al., Light-induced structural changes in photosynthetic reaction center: Implications for mechanism of electron-proton transfer. *Science*, 1997. **276**(5313): pp. 812–816.
17. Jordan, P. et al., Three-dimensional structure of cyanobacterial photosystem I at 2.5 A resolution. *Nature*, 2001. **411**(6840): pp. 909–917.
18. Feher, G., Some chemical and physical properties of a bacterial reaction center particle and its primary photochemical reactants. *Photochemistry and Photobiology*, 1971. **14**(3): pp. 373–387.
19. Wraight, C.A. and R.K. Clayton, The absolute quantum efficiency of bacteriochlorophyll photooxidation in reaction centres of Rhodopseudomonas spheroides. *Biochimica et Biophysica Acta (BBA)*, 1974. **333**(2): pp. 246–260.
20. Feher, G. et al., Structure and function of bacterial photosynthetic reaction centres. *Nature*, 1989. **339**: pp. 111–116.
21. Mavelli, F. et al., The binding of quinone to the photosynthetic reaction centers: Kinetics and thermodynamics of reactions occurring at the QB-site in zwitterionic and anionic liposomes. *European Biophysics Journal*, 2014. **43**(6–7): pp. 301–315.
22. Wraight, C.A., Proton and electron transfer in the acceptor quinone complex of photosynthetic reaction centers from Rhodobacter sphaeroides. *Frontiers in Bioscience*, 2004. **9**: pp. 309–337.
23. Kleinfeld, D., M.Y. Okamura, and G. Feher, Electron transfer in reaction centers of Rhodopseudomonas sphaeroides. I. Determination of the charge recombination pathway of D+QAQ−B and free energy and kinetic relations between Q−AQB and QAQ−B. *Biochimica et Biophysica Acta (BBA)—Bioenergetics*, 1984. **766**(1): pp. 126–140.

24. Milano, F. et al., Mechanism of quinol oxidation by ferricenium produced by light excitation in reaction centers of photosynthetic bacteria. *Journal of Physical Chemistry B*, 2007. **111**(16): pp. 4261–4270.

25. Croce, R. and H. van Amerongen, Natural strategies for photosynthetic light harvesting. *Nature Chemical Biology*, 2014. **10**(7): pp. 492–501.

26. Neilson, J.A. and D.G. Durnford, Structural and functional diversification of the light-harvesting complexes in photosynthetic eukaryotes. *Photosynthesis Research*, 2010. **106**(1–2): pp. 57–71.

27. Rochaix, J.D., Assembly of the photosynthetic apparatus. *Plant Physiology*, 2011. **155**(4): pp. 1493–1500.

28. Rochaix, J.D., Regulation and dynamics of the light-harvesting system. *Annual Review of Plant Biology*, 2014. **65**: pp. 287–309.

29. Cogdell, R.J. et al., The structural basis of light-harvesting in purple bacteria. *FEBS Letters*, 2003. **555**(1): pp. 35–39.

30. McDermott, G. et al., Crystal structure of an integral membrane light harvesting complex from photosynthetic bacteria. *Nature*, 1995. **374**: pp. 517–521.

31. Papiz, M.Z. et al., The structure and thermal motion of the B800-850 LH2 complex from Rps. acidophila at 2.0 (A) over-circle resolution and 100 K: New structural features and functionally relevant motions. *Journal of Molecular Biology*, 2003. **326**(5): pp. 1523–1538.

32. Roszak, A.W. et al., Crystal structure of the RC-LH1 core complex from Rhodopseudomonas palustris. *Science*, 2003. **302**(5652): pp. 1969–1972.

33. Clayton, R.K. and C. Smith, Rhodopseudomonas spheroides: High catalase and blue-green double mutants. *Biochemical and Biophysical Research Communications*, 1960. **3**: pp. 143–145.

34. Cogdell, R.J. and A.R. Crofts, Analysis of the pigment content of an antenna pigment-protein complex from three strains of Rhodopseudomonas sphaeroides. *Biochimica et Biophysica Acta (BBA)*, 1978. **502**(3): pp. 409–416.

35. Griffiths, M. and R.Y. Stanier, Some mutational changes in the photosynthetic pigment system of Rhodopseudomonas spheroides. *Journal of General Microbiology*, 1956. **14**(3): pp. 698–715.

36. Arnold, W. and R.K. Clayton, The first step in photosynthesis: Evidence for its electronic nature. *Proceedings of the National Academy of Sciences of the United States of America*, 1960. **46**(6): pp. 769–776.

37. Timpmann, K. et al., Efficiency of light harvesting in a photosynthetic bacterium adapted to different levels of light. *Biochimica et Biophysica Acta (BBA)*, 2014. **1837**(10): pp. 1835–1846.

38. Woronowicz, K. et al., Differential assembly of polypeptides of the light-harvesting 2 complex encoded by distinct operons during acclimation of Rhodobacter sphaeroides to low light intensity. *Photosynthesis Research*, 2012. **111**(1–2): pp. 125–8.

39. Law, C.J. et al. How purple bacteria harvest light energy, in *Energy Harvesting Materials*, L.A. David, Editor. 2005, Singapore: World Scientific, pp. 65–95.

40. Şener, M.K. and K. Schulten, Physical principles of efficient excitation transfer in light harvesting, in *Energy Harvesting Materials*, D.L. Andrews, Editor. 2005, Singapore: World Scientific, pp. 1–26.

41. Lakowicz, J.R., *Principles of Fluorescence Spectroscopy*, 3rd edition. 2006, New York: Springer.

42. Gust, D., T.A. Moore, and A.L. Moore, Realizing artificial photosynthesis. *Faraday Discussions*, 2012. **155**: pp. 9–26.

43. Kremer, A. et al., Versatile bisethynyl[60]fulleropyrrolidine scaffolds for mimicking artificial light-harvesting photoreaction centers. *Chemistry-A European Journal*, 2015. **21**(3): pp. 1108–1117.

44. Moore, T.A. et al., Photodriven charge separation in a carotenoporphyrinquinone triad. *Nature*, 1984. **307**: pp. 630–632.

45. Hambourger, M. et al., Biology and technology for photochemical fuel production. *Chemical Society Reviews*, 2009. **38**(1): pp. 25–35.

46. Kurreck, H. and M. Huber, Model reactions for photosynthesis-photoinduced charge and energy-transfer between covalently-linked porphyrin and quinone units. *Angewandte Chemie-International Edition in English*, 1995. **34**(8): pp. 849–866.

47. Gust, D. et al., Mimicking the photosynthetic triplet energy-transfer relay. *Journal of the American Chemical Society*, 1993. **115**(13): pp. 5684–5691.

48. Holten, D., D.F. Bocian, and J.S. Lindsey, Probing electronic communication in covalently linked multiporphyrin arrays. A guide to the rational design of molecular photonic devices. *Accounts of Chemical Research*, 2002. **35**(1): pp. 57–69.

49. Guldi, D.M. and H. Imahori, Supramolecular assemblies for electron transfer. *Journal of Porphyrins and Phthalocyanines*, 2004. **8**(7): pp. 976–983.

50. Gust, D., T.A. Moore, and A.L. Moore, Mimicking photosynthetic solar energy transduction. *Accounts of Chemical Research*, 2000. **34**(1): pp. 40–48.

51. Wasielewski, M.R. et al., Supramolecular structures for modeling photosynthetic reaction center function. *Photosynthesis Research*, 1992. **34**(1): p. 91.

52. Gust, D. and T.A. Moore, Intramolecular photoinduced electron transfer reactions of porphyrins, in *The Porphyrin Handbook: Electron Transfer*, K.M. Kaddis, K.M. Smith, and R. Guillard, Editors. 2000, San Diego, CA: Academic Press, pp. 153–190.

53. Yoosaf, K. et al., A supramolecular photosynthetic model made of a multiporphyrinic array constructed around a C60 core and a C60-imidazole derivative. *Chemistry*, 2014. **20**(1): pp. 223–231.

54. Accorsi, G. et al., The electronic properties of a homoleptic bisphosphine Cu(I) complex A joint theoretical and experimental insight. *Journal of Molecular Structure-Theochem*, 2010. **962**(1–3): pp. 7–14.

55. McConnell, I., G. Li, and G.W. Brudvig, Energy conversion in natural and artificial photosynthesis. *Chemistry & Biology*, 2010. **17**(5): pp. 434–447.

56. Broser, M. et al. Crystal structure of monomeric photosystem II from Thermosynechococcus elongatus at 3.6-a resolution. *The Journal of Biological Chemistry*, 2010. **285**(34): pp. 26255–26262.

57. Sherman, B.D. et al., Evolution of reaction center mimics to systems capable of generating solar fuel. *Photosynthesis Research*, 2014. **120**(1–2): pp. 59–70.

58. Gibney, B.R. et al., Iterative protein redesign. *Journal of the American Chemical Society*, 1999. **121**(21): pp. 4952–4960.

59. Koder, R.L. and P.L. Dutton, Intelligent design: The de novo engineering of proteins with specified functions. *Dalton Transactions*, 2006(25): pp. 3045–3051.

60. Fry, B.A. et al., Designing protein maquettes for interprotein and transmembrane electron transfer. *Biophysical Journal*, 2013. **104**(2): p. 489a.

61. Noy, D., C.C. Moser, and P.L. Dutton, Design and engineering of photosynthetic light-harvesting and electron transfer using length, time, and energy scales. *Biochimica et Biophysica Acta (BBA)—Bioenergetics*, 2006. **1757**(2): pp. 90–105.

62. Conlan, B. et al., Photo-catalytic oxidation of a di-nuclear manganese centre in an engineered bacterioferritin 'reaction centre'. *Biochimica et Biophysica Acta (BBA)—Bioenergetics*, 2009. **1787**(9): pp. 1112–1121.

63. Frolov, L. et al., Fabrication of a photoelectronic device by direct chemical binding of the photosynthetic reaction center protein to metal surfaces. *Advanced Materials*, 2005. **17**(20): pp. 2434–2437.

64. Frolov, L. et al., Photoelectric junctions between GaAs and photosynthetic reaction center protein. *Journal of Physical Chemistry C*, 2008. **112**(35): pp. 13426–13430.

65. Nishihara, H. et al., Construction of redox- and photo-functional molecular systems on electrode surface for application to molecular devices. *Coordination Chemistry Reviews*, 2007. **251**(21–24): pp. 2674–2687.

66. Wang, F., X.Q. Liu, and I. Willner, Integration of photo-switchable proteins, photosynthetic reaction centers and semi-conductor/biomolecule hybrids with electrode supports for optobioelectronic applications. *Advanced Materials*, 2013. **25**(3): pp. 349–377.

67. Yehezkeli, O. et al., Photosynthetic reaction center-functionalized electrodes for photo-bioelectrochemical cells. *Photosynthesis Research*, 2014. **120**(1): pp. 71–85.

68. Giardi, M.T. and E. Pace, Photosynthetic proteins for technological applications. *Trends in Biotechnology*, 2005. **23**(5): pp. 257–263.

69. Swainsbury, D.J. et al., Evaluation of a biohybrid photoelectrochemical cell employing the purple bacterial reaction centre as a biosensor for herbicides. *Biosensors & Bioelectronics*, 2014. **58**: pp. 172–178.

70. Sanders, C.A., M. Rodriguez, and E. Greenbaum, Stand-off tissue-based biosensors for the detection of chemical warfare agents using photosynthetic fluorescence induction. *Biosensors & Bioelectronics*, 2001. **16**(7–8): pp. 439–446.

71. Ciesielski, P.N. et al., Enhanced photocurrent production by photosystem I multilayer assemblies. *Advanced Functional Materials*, 2010. **20**(23): pp. 4048–4054.

72. Ciesielski, P.N. et al., Photosystem I-based biohybrid photoelectrochemical cells. *Bioresource Technology*, 2010. **101**(9): pp. 3047–3053.

73. Yehezkeli, O. et al., Photosystem I (PSI)/Photosystem II (PSII)-based photo-bioelectrochemical cells revealing directional generation of photocurrents. *Small*, 2013. **9**(17): pp. 2970–2978.

74. Takshi, A. et al., A photovoltaic device using an electrolyte containing photosynthetic reaction centers. *Energies*, 2010. **3**(11): pp. 1721–1727.

75. Grimme, R.A. et al., Photosystem I/molecular wire/metal nanoparticle bioconjugates for the photocatalytic production of H-2. *Journal of the American Chemical Society*, 2008. **130**(20): pp. 6308–6309.

76. Lubner, C.E. et al., Wiring photosystem I for electron transfer to a tethered redox dye. *Energy & Environmental Science*, 2011. **4**(7): pp. 2428–2434.

77. Krassen, H. et al., Photosynthetic hydrogen production by a hybrid complex of photosystem I and [NiFe]-hydrogenase. *ACS Nano*, 2009. **3**(12): pp. 4055–4061.

78. Kato, M. et al., Covalent immobilization of oriented photosystem II on a nanostructured electrode for solar water oxidation. *Journal of the American Chemical Society*, 2013. **135**(29): pp. 10610–10613.

79. Whyburn, G.P., Y.J. Li, and Y. Huang, Protein and protein assembly based material structures. *Journal of Materials Chemistry*, 2008. **18**(32): pp. 3755–3762.

80. Kim, Y. et al., Hybrid system of semiconductor and photosynthetic protein. *Nanotechnology*, 2014. **25**(34): p. 342001.

81. Pace, R.J., An integrated artificial photosynthesis model, in *Artificial Photosynthesis*, A. Collings and C. Critchley, Editors. 2005, Weinheim: Wiley-VCH. pp. 13–34.

82. Badura, A. et al., Wiring photosynthetic enzymes to electrodes. *Energy & Environmental Science*, 2011. **4**(9): pp. 3263–3274.

83. Lu, Y.D. et al., Photosynthetic reaction center functionalized nano-composite films: Effective strategies for probing and exploiting the photo-induced electron transfer of photosensitive membrane protein. *Biosensors & Bioelectronics*, 2007. **22**(7): pp. 1173–1185.

84. LeBlanc, G. et al., Photosystem I protein films at electrode surfaces for solar energy conversion. *Langmuir*, 2014. **30**(37): pp. 10990–11001.

85. Liu, J.G. et al., Characterization of photosystem I from spinach: Effect of solution pH. *Photosynthesis Research*, 2012. **112**(1): pp. 63–70.

86. Rakovich, A. et al., Linear and nonlinear optical effects induced by energy transfer from semiconductor nanoparticles to photosynthetic biological systems. *Journal of Photochemistry and Photobiology C-Photochemistry Reviews*, 2014. **20**: pp. 17–32.

87. Zhao, J. et al., Photoelectric conversion of photosynthetic reaction center in multilayered films fabricated by layer-by-layer assembly. *Electrochimica Acta*, 2002. **47**(12): pp. 2013–2017.

88. Ventrella, A. et al., Photosystem II based multilayers obtained by electrostatic layer-by-layer assembly on quartz substrates. *Journal of Bioenergetics and Biomembranes*, 2014. **46**(3): pp. 221–228.

89. Dorogi, M. et al., Stabilization effect of single-walled carbon nanotubes on the functioning of photosynthetic reaction centers. *The Journal of Physical Chemistry B*, 2006. **110**(43): pp. 21473–21479.

90. Szabó, T. et al., Photosynthetic reaction centers/ITO hybrid nanostructure. *Materials Science and Engineering: C*, 2013. **33**(2): pp. 769–773.

91. Yaghoubi, H. et al., The role of gold-adsorbed photosynthetic reaction centers and redox mediators in the charge transfer and photocurrent generation in a bio-photoelectrochemical cell. *The Journal of Physical Chemistry C*, 2012. **116**(47): pp. 24868–24877.

92. Touloupakis, E. et al., A photosynthetic biosensor with enhanced electron transfer generation realized by laser printing technology. *Analytical and Bioanalytical Chemistry*, 2012. **402**(10): pp. 3237–3244.

93. Boutopoulos, C. et al., Direct laser immobilization of photosynthetic material on screen printed electrodes for amperometric biosensor. *Applied Physics Letters*, 2011. **98**(9): p. 093703.

94. Magyar, M. et al., Long term stabilization of reaction center protein photochemistry by carbon nanotubes. *Physica Status Solidi (b)*, 2011. **248**(11): pp. 2454–2457.

95. Magis, G.J. et al., Light harvesting, energy transfer and electron cycling of a native photosynthetic membrane adsorbed onto a gold surface. *Biochimica et Biophysica Acta (BBA)—Biomembranes*, 2010. **1798**(3): pp. 637–645.

96. den Hollander, M.J. et al., Enhanced photocurrent generation by photosynthetic bacterial reaction centers through molecular relays, light-harvesting complexes, and direct protein-gold interactions. *Langmuir*, 2011. **27**(16): pp. 10282–10294.

97. Hermanson, G.T., *Bioconjugate Techniques*, 2nd edition. 2008, London: Academic Press.

98. Katz, E., Application of bifunctional reagents for immobilization of proteins on a carbon electrode surface: Oriented immobilization of photosynthetic reaction centers. *Journal of Electroanalytical Chemistry*, 1994. **365**(1–2): pp. 157–164.

99. Yaghoubi, H. et al., Hybrid wiring of the rhodobacter sphaeroides Reaction center for applications in bio-photoelectrochemical solar cells. *The Journal of Physical Chemistry C*, 2014. **118**(41): pp. 23509–23518.

100. Kong, J.L. et al., Direct electrochemistry of cofactor redox sites in a bacterial photosynthetic reaction center protein. *Journal of the American Chemical Society*, 1998. **120**(29): pp. 7371–7372.

101. Nakamura, C. et al., Self-assembling photosynthetic reaction centers on electrodes for current generation. *Applied Biochemistry and Biotechnology*, 2000. **84–86**: pp. 401–408.

102. Trammell, S.A. et al., Orientated binding of photosynthetic reaction centers on gold using Ni-NTA self-assembled monolayers. *Biosensors & Bioelectronics*, 2004. **19**(12): pp. 1649–1655.

103. Lebedev, N. et al., Conductive wiring of immobilized photosynthetic reaction center to electrode by cytochrome c. *Journal of the American Chemical Society*, 2006. **128**(37): pp. 12044–12045.

104. Trammell, S.A. et al., Effect of protein orientation on electron transfer between photosynthetic reaction centers and carbon electrodes. *Biosensors & Bioelectronics*, 2006. **21**(7): pp. 1023–1028.

105. Kondo, M. et al., Photocurrent and electronic activities of oriented-his-tagged photosynthetic light-harvesting/reaction center core complexes assembled onto a gold electrode. *Biomacromolecules*, 2012. **13**(2): pp. 432–438.

106. Terasaki, N. et al., Photocurrent generation properties of Histag-photosystem II immobilized on nanostructured gold electrode. *Thin Solid Films*, 2008. **516**(9): pp. 2553–2557.

107. Stieger, K.R. et al., Advanced unidirectional photocurrent generation via cytochrome c as reaction partner for directed assembly of photosystem I. *Physical Chemistry Chemical Physics*, 2014. **16**(29): pp. 15667–15674.

108. Xiao, Y. et al., "Plugging into enzymes": Nanowiring of redox enzymes by a gold nanoparticle. *Science*, 2003. **299**(5614): pp. 1877–1881.

109. Katz, E.Y. et al., Coupling of photoinduced charge separation in reaction centers of photosynthetic bacteria with electron transfer to a chemically modified electrode. *Biochimica et Biophysica Acta (BBA)—Bioenergetics*, 1989. **976**(2–3): pp. 121–128.

110. Terasaki, N. et al., Plugging a molecular wire into photosystem I: Reconstitution of the photoelectric conversion system on a gold electrode. *Angewandte Chemie-International Edition in English*, 2009. **48**(9): pp. 1585–1587.

111. Terasaki, N. et al., Bio-photosensor: Cyanobacterial photosystem I coupled with transistor via molecular wire. *Biochimica et Biophysica Acta (BBA)—Bioenergetics*, 2007. **1767**(6): pp. 653–659.

112. Takshi, A., J.D. Madden, and J.T. Beatty, Diffusion model for charge transfer from a photosynthetic reaction center to an electrode in a photovoltaic device. *Electrochimica Acta*, 2009. **54**(14): pp. 3806–3811.

113. Mahmoudzadeh, A. et al., Photocurrent generation by direct electron transfer using photosynthetic reaction centres. *Smart Materials and Structures*, 2011. **20**(9): p. 094019.

114. Das, R. et al., Integration of photosynthetic protein molecular complexes in solid-state electronic devices. *Nano Letters*, 2004. **4**(6): pp. 1079–1083.

115. Ham, M.H. et al., Photoelectrochemical complexes for solar energy conversion that chemically and autonomously regenerate. *Nature Chemistry*, 2010. **2**(11): pp. 929–936.

116. Szabó, T. et al., Charge stabilization by reaction center protein immobilized to carbon nanotubes functionalized by amine groups and poly(3-thiophene acetic acid) conducting polymer. *Physica Status Solidi (b)*, 2012. **249**(12): pp. 2386–2389.

117. Lebedev, N. et al., Increasing efficiency of photoelectronic conversion by encapsulation of photosynthetic reaction center proteins in arrayed carbon nanotube electrode. *Langmuir*, 2008. **24**(16): pp. 8871–8876.

118. Gunther, D. et al., Photosystem I on graphene as a highly transparent, photoactive electrode. *Langmuir*, 2013. **29**(13): pp. 4177–4180.

119. Darby, E. et al., Photoactive films of photosystem I on transparent reduced graphene oxide electrodes. *Langmuir*, 2014. **30**(29): pp. 8990–8994.

120. Tangorra, R.R. et al., Photoactive film by covalent immobilization of a bacterial photosynthetic protein on reduced graphene oxide surface. *MRS Proceedings*, 2015. **1717**: mrsf14-1717-a03-01.

121. Curri, M.L. et al., Inorganic nanocrystals: Patterning and assembling, in *Encyclopedia of Inorganic Chemistry*, R. Bruce King, Editor. 2006, Chichester: John Wiley & Sons.

122. Gaponenko, S.V., *Optical Properties of Semiconductor Nanocrystals*. 1998, Cambridge: Cambridge University Press.

123. Katz, E. and I. Willner, Integrated nanoparticle-biomolecule hybrid systems: Synthesis, properties, and applications. *Angewandte Chemie-International Edition in English*, 2004. **43**(45): pp. 6042–6108.

124. Govorov, A.O. and I. Carmeli, Hybrid structures composed of photosynthetic system and metal nanoparticles: Plasmon enhancement effect. *Nano Letters*, 2007. **7**(3): pp. 620–625.

125. Yildiz, H.B., R. Tel-Vered, and I. Willner, Solar cells with enhanced photocurrent efficiencies using oligoaniline-crosslinked Au/CdS nanoparticles arrays on electrodes. *Advanced Functional Materials*, 2008. **18**(21): pp. 3497–3505.

126. Yehezkeli, O. et al., Integrated photosystem II-based photo-bioelectrochemical cells. *Nature Communications*, 2012. **3**: p. 742.

127. Yehezkeli, O. et al., Generation of photocurrents by Bis-aniline-cross-linked Pt nanoparticle/photosystem I composites on electrodes. *The Journal of Physical Chemistry B*, 2010. **114**(45): pp. 14383–14388.

128. Grimme, R.A., C.E. Lubner, and J.H. Golbeck, Maximizing H2 production in photosystem I/dithiol molecular wire/platinum nanoparticle bioconjugates. *Dalton Transactions*, 2009(45): pp. 10106–10113.

129. Terasaki, N. et al., Photosensor based on an FET utilizing a biocomponent of photosystem I for use in imaging devices. *Langmuir*, 2009. **25**(19): pp. 11969–11974.

130. Terasaki, N. et al., Fabrication of novel photosystem I–gold nanoparticle hybrids and their photocurrent enhancement. *Thin Solid Films*, 2006. **499**(1–2): pp. 153–156.

131. Greenbaum, E., Platinized chloroplasts: A novel photocatalytic material. *Science (Washington DC)*, 1985. **230**(4732): pp. 1373–1375.

132. Reed, M.A. et al., Observation of discrete electronic states in a zero-dimensional semiconductor nanostructure. *Physical Review Letters*, 1988. **60**(6): pp. 535–537.

133. Bera, D. et al., Quantum dots and their multimodal applications: A review. *Materials*, 2010. **3**(4): pp. 2260–2345.

134. Prasad, P.N., *Nanophotonics*. 2004, Hoboken, NJ: John Wiley & Sons.

135. Corricelli, M. et al., Biotin-decorated silica coated PbS nano-crystals emitting in the second biological near infrared window for bioimaging. *Nanoscale*, 2014. **6**(14): pp. 7924–7933.

136. Govorov, A.O., Enhanced optical properties of a photosynthetic system conjugated with semiconductor nanoparticles: The role of Förster transfer. *Advanced Materials*, 2008. **20**(22): pp. 4330–4335.

137. Kim, I. et al., Metal nanoparticle plasmon-enhanced light-harvesting in a photosystem I thin film. *Nano Letters*, 2011. **11**(8): pp. 3091–3098.

138. Nabiev, I. et al., Fluorescent quantum dots as artificial antennas for enhanced light harvesting and energy transfer to photosynthetic reaction centers. *Angewandte Chemie-International Edition in English*, 2010. **49**(40): pp. 7217–7221.

139. Jung, H. et al., Optical and electrical measurement of energy transfer between nanocrystalline quantum dots and photosystem I. *Journal of Physical Chemistry B*, 2010. **114**(45): pp. 14544–14549.

140. Maksimov, E.G. et al., Hybrid system based on quantum dots and photosystem 2 core complex. *Biochemistry-Moscow*, 2012. **77**(6): pp. 624–630.

141. Osvath, S., J.W. Larson, and C.A. Wraight, Site specific labeling of Rhodobacter sphaeroides reaction centers with dye probes for surface pH measurements. *Biochimica et Biophysica Acta (BBA)*, 2001. **1505**(2–3): pp. 238–247.

142. Ragni, R. et al., Bursting photosynthesis: Designing ad-hoc fluorophores to complement the light harvesting capability of the photosynthetic reaction center. *MRS Proceedings*, 2014. **1689**, mrss14-1689-z06-01. doi:10.1557/opl.2014.617.

143. Milano, F. et al., Enhancing the light harvesting capability of a photosynthetic reaction center by a tailored molecular fluorophore. *Angewandte Chemie-International Edition in English*, 2012. **124**(44): pp. 11181–11185.

144. Dutta, P.K. et al., Reengineering the optical absorption cross-section of photosynthetic reaction centers. *Journal of the American Chemical Society*, 2014. **136**(12): pp. 4599–4604.

145. Dutta, P.K. et al., A DNA-directed light-harvesting/reaction center system. *Journal of the American Chemical Society*, 2014. **136**(47): pp. 16618–16625.

146. Buczynska, D. et al., Energy transfer from conjugated polymer to bacterial light-harvesting complex. *Applied Physics Letters*, 2012. **101**(17): p. 173703.

Section III

Molecular Aspects of Photosynthesis:
Photosystems, Photosynthetic Enzymes and Genes

12 Quality Control of Photosystem II

Role of Structural Changes of Thylakoid Membranes and FtsH Proteases in High Light Tolerance and Recovery from Photoinhibition

Miho Yoshioka-Nishimura

CONTENTS

12.1 Introduction 223
12.2 Thylakoid Membranes 223
    12.2.1 Structure and Composition 223
    12.2.2 Thylakoid Membrane Lipids 224
12.3 Structural Changes of the Thylakoid Membranes 224
    12.3.1 Unstacking of the Thylakoids 224
    12.3.2 Shrinkage and Swelling of the Thylakoids 224
    12.3.3 Regulation of Grana Formation 225
12.4 The Role of Thylakoid Membranes in Quality Control of Photosystem II 225
12.5 Concluding Remarks 225
Acknowledgment 227
References 227

12.1 INTRODUCTION

Plants are influenced by various environmental conditions, such as the quantity of light, ambient temperature, and water and nutrient levels in the soil. Although sunlight is essential for photosynthesis, intense light easily causes oxidative damage to photosynthetic functions. Most proteins that participate in the photochemical reactions of photosynthesis are embedded in the thylakoid membranes, and therefore, it is important to consider the mechanisms of photosynthesis in terms of the structure and function of thylakoid membranes. Recently, more information about the structural changes of thylakoid membranes has become available. This chapter describes how structural changes of the thylakoid membranes contribute to regulation of photosynthesis under light stress.

12.2 THYLAKOID MEMBRANES

12.2.1 STRUCTURE AND COMPOSITION

Plants have chloroplasts, which are green, oval-shaped organelles, 5–10 μm in diameter and 2–3 μm in thickness. Chloroplasts are made up of three types of membranes—the outer (envelope) membrane, the inner (envelope) membrane, and the thylakoid membrane. The thylakoid membranes exist inside of the chloroplasts and are well known as the site of photochemical reactions. In the thylakoids, many proteins related to photosynthesis and pigments such as chlorophylls and carotenoids are present. Thylakoid membranes have flat, sack-like structures and are piled up to form grana. It has been reported that a grana disk is 300–600 nm in diameter and the width of the partition gap between two adjacent membranes in the grana is about 3.5 nm [1,2]. The grana occupy about 80% of the thylakoids and are interconnected by stroma thylakoids, forming stable thylakoid networks. The grana stacks create large surface areas, and this contributes to efficient light harvesting in the chloroplasts.

The protein content is different between the grana and stroma thylakoids. Photosystem II (PSII) and the light-harvesting chlorophyll protein complexes of PSII (LHCII) are abundant in the stacked grana, whereas photosystem I (PSI), the light-harvesting chlorophyll protein complexes of PSI (LHCI), and ATPases exist in the unstacked regions, such as the stroma thylakoids, grana margins, and grana end membranes. Cytochrome b_6/f is localized in both stacked and unstacked thylakoid regions [3,4]. The reason for the heterogeneous distribution of proteins in the thylakoid membranes can be explained by results obtained from x-ray crystal structure analysis of the protein complexes in the thylakoids and cryo-electron microscopy of the thylakoid membranes. The surfaces of both the PSII and LHCII complexes exposed to the stromal side are comparatively flat [5], and therefore, these complexes can exist in the tightly stacked grana regions. Conversely, PSI, LHCI, and ATPase protrude into the stromal side [6,7], which prevents them from existing in the narrow gaps between membranes in the stacked grana.

12.2.2 Thylakoid Membrane Lipids

The lipids of higher-plant thylakoid membranes consist of phosphatidylglycerol (PG), monogalactosyldiacylglycerol (MGDG), digalactosyldiacylglycerol (DGDG), and sulfoquinovosyl diacylglycerol (SQDG). PG is a phospholipid, MGDG and DGDG are galactolipids, and SQDG is an acidic glycolipid. It is notable that 80% of the thylakoid membranes are made up of galactolipids, although other cellular membranes such as the endoplasmic reticulum and mitochondria consist mainly of phospholipids. This characteristic lipid composition of thylakoid membranes is conserved from cyanobacteria to higher plants and has become a basis for the endosymbiotic theory, which explains the origin of the chloroplasts.

Because most photosynthetic proteins are embedded in the thylakoid membranes, the thylakoids are crowded with protein molecules and large protein complexes [8,9]. Among the membrane lipids, there are bound lipids (or boundary lipids) tightly associated with membrane proteins and free lipids (or bulk lipids) that diffuse freely within the membranes. Membrane fluidity depends on the degree of fatty acid saturation of the free lipids; a higher degree of fatty acid unsaturation increases membrane fluidity. Various photosynthetic processes are supported by membrane lipids providing fluidity in the membranes, which is important, in particular, for protein–protein interactions in the thylakoids.

12.3 STRUCTURAL CHANGES OF THE THYLAKOID MEMBRANES

12.3.1 Unstacking of the Thylakoids

The thylakoid membranes of higher-plant chloroplasts are tightly stacked and show the typical grana structure in the dark. When illuminated, the grana stacks loosen, and the distance between two adjacent membranes in the grana increases [2,10,11]. After treatment of spinach thylakoids with a 0.5% solution of the nonionic surface-active agent digitonin, a grana membrane fraction can be obtained as a precipitate by centrifugation at 10,000 × g for 30 min [12]. By comparing the chlorophyll contents in the precipitates, the level of thylakoid membrane stacking before and after illumination was determined [10]. The degree of grana stacking was reduced approximately by half after 1 h of strong illumination (light intensity, 1000 μmol photons m^{-2} s^{-1}) of the thylakoid membranes. When the thylakoids were put into dark conditions, the thylakoid stacking recovered. These results show that the strong light-induced unstacking of the thylakoid membranes is a reversible process.

It is possible that the interaction of the van der Waals force between thylakoid membranes and the surface charge density of the membrane are involved in the stacking of thylakoid membranes [13]. Because thylakoid stacking is dependent on the concentration of monovalent or divalent cations and the pH in the suspension, we can control thylakoid stacking *in vitro* by changing these factors. When NaCl and $MgCl_2$ were added to a thylakoid suspension, thylakoid stacking increased from 10% in the no-salt control to 50% in salt-added samples [10].

The physiological meaning of the strong light-induced unstacking of the thylakoid membrane is apparent when we look at the generation of reactive oxygen species (ROS) in thylakoids under light stress. When plants are subjected to excessive light, the acceptor side of PSII is overreduced, and the plants produce ROS near PSII. It was revealed by the electron paramagnetic resonance (EPR) spin trapping method that more hydroxyl radicals are produced in stacked thylakoids than in unstacked thylakoids under strong illumination [10]. Hydroxyl radicals are the most oxidative molecular species among ROS, which oxidize proteins, carbohydrates, and lipids. The stimulated production of ROS in stacked thylakoid membranes is probably due to the uneven distribution of PSI and PSII in the thylakoid membranes. A delay of electron transport between PSII and PSI is expected because PSI and PSII are separated into two membrane domains, namely, the stroma thylakoids and the grana, respectively, and this delay of electron transport from PSII may promote ROS generation under high light. Thylakoid unstacking enables PSI and PSII to migrate freely, and this would reduce the delay of electron transport and suppress ROS generation. Thus, higher plants should avoid further damage of the thylakoid membranes by unstacking of the thylakoids under high light.

12.3.2 Shrinkage and Swelling of the Thylakoids

Other structural changes of the thylakoid membranes, including *shrinkage* and *swelling* of the thylakoids, have been reported [14]. Shrinkage of the thylakoids results in reduction of the grana diameter, which leads to an increase in the unstacked region. The diameter of the grana is actually reduced by about 20%, from 370 to 300 nm, by 1 h of strong-light illumination [11]. By contrast, it was reported that shrinkage of the thylakoid membranes under strong light is not seen in an *Arabidopsis thaliana* knockout mutant of the kinase genes *stn7* and *stn8*, which participate in phosphorylation of LHCII and the PSII core proteins [11]. The phosphorylation level of the proteins that exist in PSII and the grana is possibly related to the regulation of thylakoid shrinkage.

The thylakoid lumen swells during illumination. In *A. thaliana*, voltage-gated chloride channels were identified in the thylakoid membranes [15–18]. Chloride influx by light-induced proton motive force probably causes thylakoid swelling. It was reported that the height of the lumen expanded from 4.7 to 9.2 nm under illumination at 500 µmol photons $m^{-2} s^{-1}$ [2].

12.3.3 REGULATION OF GRANA FORMATION

An increased number of grana stacks are found in plants grown in the shade [19]. Grana stacks may develop under limited light conditions to collect more light energy. It is suggested that thylakoid protein phosphorylation controls grana formation. Longer grana regions were observed in the *stn8* mutant and *stn7 stn8* double mutant of *A. thaliana* lacking thylakoid membrane protein phosphorylation [20]. PSII core phosphatase (PBCP), which works against STN8 kinase, was also shown to be involved in grana formation. The number of grana layers was reduced in the *pbcp* mutant [21]. In *A. thaliana*, a protein called CURVATURE THYLAKOID1 (CURT1) exists in the grana margins that regulates the formation of grana membranes [22]. Grana margins were almost lost in the mutant lacking the CURT1 protein, whereas its overexpression resulted in increased grana stacks. Grana formation is controlled by multiple genes, suggesting that plants deal with various environmental stresses by expression of these genes and changing the structure of the grana.

12.4 THE ROLE OF THYLAKOID MEMBRANES IN QUALITY CONTROL OF PHOTOSYSTEM II

PSII splits water into protons and oxygen molecules using light energy. ROS arise in the process of photochemical reactions and damage photosynthetic proteins. In particular, the D1 protein of PSII is easily damaged by strong light, which leads to a decrease in photosynthesis activity. It is suggested that strong-light and high-temperature stresses cause lipid peroxidation in the thylakoid membranes and the generation of singlet oxygen [23]. Singlet oxygen damages the D1 protein and other photosynthetic proteins in the thylakoid membranes. Lipid peroxidation affects the whole thylakoid membrane, and the mobility of thylakoid membranes is impaired by the aggregation of damaged proteins [24,25]. Decreased thylakoid mobility will also reduce the function of other thylakoid proteins that have not yet been damaged. Because the accumulation of damaged D1 proteins and the aggregation of damaged and neighboring proteins may ultimately lead to cell death, these protein aggregates are promptly removed, and newly synthesized D1 proteins are inserted into the PSII complexes. This PSII repair cycle maintains the homeostasis of photosynthetic systems under light stress.

FtsH proteases that bind to the thylakoids with two transmembrane helices degrade the photodamaged D1 protein. Since FtsH protease has a large, hydrophilic portion about 6.5 nm in height exposed to the stromal side in its C-terminal end, it is too large for the grana stack, where the partition gap between adjacent membranes is 3–4 nm. Although it has been supposed that FtsH proteases exist exclusively in the stroma thylakoids, it was recently found that many FtsH proteases exist in the grana margins and close to the PSII complexes [26]. This suggests that FtsH proteases move only a short distance to reach damaged D1 proteins. However, the grana membrane is densely packed with semicrystalline arrays of PSII complexes and LHCII [4], and thus, membrane mobility is low in the grana. Thylakoid membrane proteins may be able to move more freely once the thylakoids are unstacked because the restriction on the movement of the protein complexes imposed by grana stacking is removed. Indeed, in transmission electron microscopy (TEM) observations, the grana of spinach leaves exposed to high light (light intensity, 2,000 µmol photons $m^{-2} s^{-1}$) showed partial unstacking (Figure 12.1), especially at both ends of the thylakoid membrane, which increased the area of the grana margins [27]. Additionally, it was found by tomographic analysis of TEM images of grana that thylakoid membrane stacking is loosened under high light so that the membrane proteins move more freely [27]. To investigate the distribution of FtsH protease before and after illumination, immunoelectron microscopy with FtsH antibody was performed (Figure 12.2). After strong illumination, FtsH proteases that had been found only in the stroma thylakoids and grana margins were detected in the grana stacks as well [27]. The unstacking of thylakoid membranes enables FtsH proteases to move to the grana stacks, promotes prompt assembly of the FtsH proteases, and stimulates removal of damaged D1 proteins from the PSII complexes. Thus, structural changes of thylakoid membranes are an important factor in the quality control of PSII under light stress. Plants may change the structure of their thylakoid membranes flexibly to cope with various environmental stresses.

12.5 CONCLUDING REMARKS

Structural changes of thylakoid membranes reduce photo-oxidative damage to the photosynthetic proteins, by facilitating the migration of membrane proteins in the thylakoids. The increased areas of grana margins, which are necessary for swift repair of PSII, result from a combination of unstacking, shrinkage, and swelling of the thylakoid membranes. Free lateral diffusion of proteins provided by structural changes of the thylakoid membranes helps avoid aggregation of crowded membrane proteins in the thylakoids [24,25] and makes it possible to reduce damage under high light. To further understand the quality control of PSII, investigations of the structures and dynamics of thylakoid membranes are as important as those about the molecular process of the photosynthesis reaction. More concrete knowledge about the structure and function of thylakoid membranes might be acquired by real-time tracing of structural changes of the thylakoid membranes *in vivo*. Progress in visualization techniques for thylakoid membranes by microscopy is also important for the progress of future photosynthesis research.

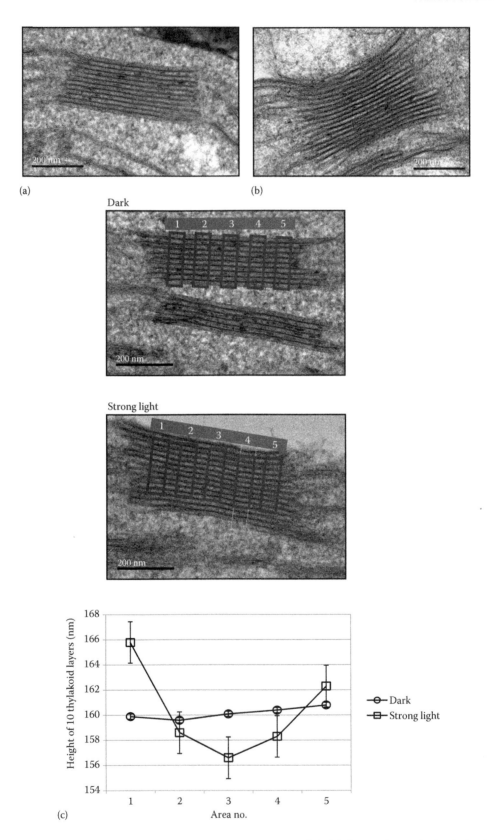

FIGURE 12.1 Structural changes of the thylakoid membranes. TEM images of thylakoid membranes in (a) dark-adapted or (b) light-illuminated spinach leaves. Scale bar = 200 nm. The spinach leaves were kept at 4°C in the dark prior to the light treatment, which was performed at a light intensity of 2000 μmol photons m^{-2} s^{-1} for 60 min. The height of 10 thylakoid membrane layers was measured in each area shown with rectangles (nos. 1–5) in the TEM images. Areas nos. 1 and 5 correspond to the grana margins. Area no. 3 represents the grana core. (c) The results are shown at the bottom of the figure. Circles and squares indicate the dark-adapted and strong light-treated samples, respectively. The data are means ± SD ($n = 10$). (Reprinted with permission from Yoshioka-Nishimura, M. et al., *Plant Cell Physiol.*, 55, 7, 1255–1265, 2014.)

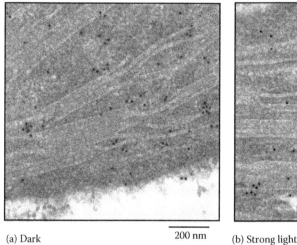
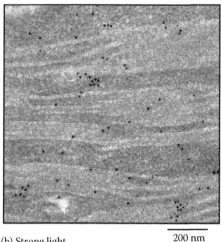

(a) Dark   200 nm   (b) Strong light   200 nm

FIGURE 12.2 Immunogold labeling of FtsH proteases in the thylakoid membranes. (a) Dark control. (b) TEM images of thylakoids subjected to strong illumination (1000 µmol photons m^{-2} s^{-1}) for 30 min. A primary antibody against FtsH2/8 and a secondary antibody attached to 10 nm gold particles were used to visualize the location of FtsH proteases. Scale bar = 200 nm. (Reprinted with permission from Yoshioka-Nishimura, M. et al., *Plant Cell Physiol.*, 55, 7, 1255–1265, 2014.)

ACKNOWLEDGMENT

I thank Prof. Yasusi Yamamoto for his constant support and encouragement.

REFERENCES

1. Daum, B. et al., Arrangement of photosystem II and ATP synthase in chloroplast membranes of spinach and pea. *Plant Cell*, 2010. **22**(4): pp. 1299–1312.
2. Kirchhoff, H. et al., Dynamic control of protein diffusion within the granal thylakoid lumen. *Proceedings of the National Academy of Sciences of the United States of America*, 2011. **108**(50): pp. 20248–20253.
3. Albertsson, P.A., A quantitative model of the domain structure of the photosynthetic membrane. *Trends in Plant Science*, 2001. **6**(8): pp. 349–354.
4. Dekker, J.P. and E.J. Boekema, Supramolecular organization of thylakoid membrane proteins in green plants. *Biochimica et Biophysica Acta-Bioenergetics*, 2005. **1706**(1–2): pp. 12–39.
5. Nield, J. and J. Barber, Refinement of the structural model for the photosystem II supercomplex of higher plants. *Biochimica et Biophysica Acta-Bioenergetics*, 2006. **1757**(5–6): pp. 353–361.
6. Amunts, A. and N. Nelson, Functional organization of a plant photosystem I: Evolution of a highly efficient photochemical machine. *Plant Physiology and Biochemistry*, 2008. **46**(3): pp. 228–237.
7. Junge, W., H. Sielaff, and S. Engelbrecht, Torque generation and elastic power transmission in the rotary F0F1-ATPase. *Nature*, 2009. **459**(7245): pp. 364–370.
8. Kirchhoff, H., Molecular crowding and order in photosynthetic membranes. *Trends in Plant Science*, 2008. **13**(5): pp. 201–207.
9. Mullineaux, C.W., Factors controlling the mobility of photosynthetic proteins. *Photochemistry and Photobiology*, 2008. **84**(6): pp. 1310–1316.
10. Khatoon, M. et al., Quality control of photosystem II: Thylakoid unstacking is necessary to avoid further damage to the D1 protein and to facilitate D1 degradation under light stress in spinach thylakoids. *The Journal of Biological Chemistry*, 2009. **284**(37): pp. 25343–25352.
11. Herbstova, M. et al., Architectural switch in plant photosynthetic membranes induced by light stress. *Proceedings of the National Academy of Sciences of the United States of America*, 2012. **109**(49): pp. 20,130–20,135.
12. Chow, W.S. et al., The stacking of chloroplast thylakoids. Effects of cation screening and binding, studied by the digitonin method. *Archives of Biochemistry and Biophysics*, 1980. **201**(1): pp. 347–355.
13. Chow, W.S. et al., Granal stacking of thylakoid membranes in higher plant chloroplasts: The physicochemical forces at work and the functional consequences that ensue. *Photochemical & Photobiological Sciences*, 2005. **4**(12): pp. 1081–1090.
14. Kirchhoff, H., Architectural switches in plant thylakoid membranes. *Photosynthesis Research*, 2013. **116**(2–3): pp. 481–487.
15. Schonknecht, G. et al., A voltage-dependent chloride channel in the photosynthetic membrane of a higher-plant. *Nature*, 1988. **336**(6199): pp. 589–592.
16. Hechenberger, M. et al., A family of putative chloride channels from Arabidopsis and functional complementation of a yeast strain with a CLC gene disruption. *Journal of Biological Chemistry*, 1996. **271**(52): pp. 33632–33638.
17. De Angeli, A. et al., CLC-mediated anion transport in plant cells. *Philosophical Transactions of the Royal Society B—Biological Sciences*, 2009. **364**(1514): pp. 195–201.
18. Spetea, C. and B. Schoefs, Solute transporters in plant thylakoid membranes: Key players during photosynthesis and light stress. *Communicative & Integrative Biology*, 2010. **3**(2): pp. 122–129.
19. Anderson, J.M., Photoregulation of the composition, function, and structure of thylakoid membranes. *Annual Review of Plant Physiology and Plant Molecular Biology*, 1986. **37**: pp. 93–136.

20. Fristedt, R. et al., Phosphorylation of photosystem II controls functional macroscopic folding of photosynthetic membranes in Arabidopsis. *Plant Cell*, 2009. **21**(12): pp. 3950–3964.

21. Samol, I. et al., Identification of a photosystem II phosphatase involved in light acclimation in Arabidopsis. *Plant Cell*, 2012. **24**(6): pp. 2596–2609.

22. Armbruster, U. et al., Arabidopsis CURVATURE THYLAKOID1 proteins modify thylakoid architecture by inducing membrane curvature. *Plant Cell*, 2013. **25**(7): pp. 2661–2678.

23. Chan, T. et al., Quality control of photosystem II: Lipid peroxidation accelerates photoinhibition under excessive illumination. *PLoS One*, 2012. **7**(12): p. e52100.

24. Yamamoto, Y. et al., Quality control of photosystem II: Reversible and irreversible protein aggregation decides the fate of photosystem II under excessive illumination. *Frontiers in Plant Science*, 2013. **4:** p. 433.

25. Yamamoto, Y. et al., Quality control of PSII: Behavior of PSII in the highly crowded grana thylakoids under excessive light. *Plant and Cell Physiology*, 2014. **55**(7): pp. 1206–1215.

26. Yoshioka, M. et al., Quality control of photosystem II: FtsH hexamers are localized near photosystem II at grana for the swift repair of damage. *The Journal of Biological Chemistry*, 2010. **285**(53): pp. 41972–41981.

27. Yoshioka-Nishimura, M. et al., Quality control of photosystem II: Direct imaging of the changes in the thylakoid structure and distribution of FtsH proteases in spinach chloroplasts under light stress. *Plant and Cell Physiology*, 2014. **55**(7): pp. 1255–1265.

13 Cytochrome c_6-Like Proteins in Cyanobacteria, Algae, and Higher Plants

Alejandro Torrado and Fernando P. Molina-Heredia

CONTENTS

13.1 Introduction 229
13.2 Photosynthesis and Respiration in Cyanobacteria 230
    13.2.1 Cytochrome c_6 and Plastocyanin 230
    13.2.2 Cytochrome c_M 232
13.3 Green Algae and Higher-Plant Cyt c_6-Like Proteins 232
13.4 Cyanobacterial Cyt c_6-Like Proteins 232
13.5 Concluding Remarks 236
References 237

13.1 INTRODUCTION

This chapter presents new functional information about cytochrome (Cyt) c_6 and its isoforms found in cyanobacteria, as well as recent data about the related higher-plant Cyt c_{6A} and cyanobacterial Cyt c_M. Cyt c_6 is the protein responsible for the electron transfer from Cyt b_6f complex to photosystem I (PSI) in the thylakoid lumen of cyanobacteria and some green algae, replacing plastocyanin (Pc) under copper deficiency conditions (De la Rosa et al. 2002; Hervas et al. 2003). This protein has been thought to be involved in photosynthesis (Kerfeld et al. 1999; Kerfeld and Krogmann 1998), respiration (Peschek 1999), and anoxygenic photosynthesis, in which Cyt c_6 could transport electrons between quinones and some sulfoferric clusters during anaerobic oxidation of sulfur (Garlick et al. 1977; Padan 1979). However, it remains unclear whether these processes involve one or several isoforms (Ki 2005). New systems for genome sequencing have uncovered three genes that could encode up to three possible isoforms of Cyt c_6 in some cyanobacteria, such as in the case of *Anabaena variabilis* or *Nostoc* sp. PCC 7119. The presence of isogenes opens the door to the possibility that each one could perform a different metabolic function within a cyanobacterium. The first to be found, the *petJ* gene, encodes for the native Cyt c_6, widely studied in several organisms and present in all cyanobacterial genomes sequenced. The second one encodes a Cyt c_6-like protein (herein after Cyt c_{6-2}), which cannot oxidize Cyt b_6f complex but can reduce, with low efficiency, PSI (Reyes-Sosa et al. 2011), and the third one encodes for Cyt c_{6-3}, found only in heterocyst-forming filamentous cyanobacteria (Torrado et al. 2015).

In recent years, a lot of work has been done to specifically investigate the mechanism of PSI reduction by Pc and Cyt c_6. However, in cyanobacteria, there are important aspects of the respiratory and photosynthetic processes that remain unknown. For example, it has been described that

Synechocystis sp. PCC 6803, which only presents one isoform of Cyt c_6, can grow photoautotrophically in the absence of both Cyt c_6 and Pc (Ardelean et al. 2002; Duran et al. 2006; Zhang et al. 1994). To explain this observation, numerous assumptions were made. Some authors suggested the existence of a third (although inefficient) electron donor to PSI (Metzger et al. 1995). Other authors proposed the formation of a supercomplex between Cyt b_6f and PSI, to allow direct electron transfer without the presence of soluble redox mediators (Schmetterer 1994). Also, under diazotrophic conditions (absence of combined nitrogen in the medium), the photosynthetic electron flow is disengaged at the Cyt b_6f complex level. To support the high PSI activity during the fixation of atmospheric nitrogen, the existence of an alternative electron carrier has also been suggested (Misra et al. 2003). On the other hand, in photosynthetic eukaryotes, the respiratory and photosynthetic chains are localized in different cell organelles, whereas in cyanobacteria, both are located in the same membrane system and share a number of components (plastoquinone, Pc or Cyt c_6, and Cyt b_6f). In these organisms, Cyt c_6 and Pc could alternatively transport electrons from the Cyt b_6f complex to PSI or Cyt c oxidase (Scheme 13.1), although it is not well established (Moser et al. 1991; Nicholls et al. 1991).

Another soluble cytochrome has been reported in cyanobacteria, Cyt c_M, whose function remains unknown. It is able to donate electrons to the terminal oxidase with a much higher efficiency than Cyt c_6 and Pc (Bernroitner et al. 2009). In contrast, Cyt c_M is unable to obtain electrons from Cyt b_6f as well as incapable of efficiently reducing PSI (Molina-Heredia et al. 2002). All this makes plausible the possible existence of an alternative electron donor, which could act as respiratory Cyt c in cyanobacteria, thus avoiding interference between the two electron transport chains.

This chapter concludes with a general reflection about the possible function of the isoforms of Cyt c_6 found in cyanobacteria.

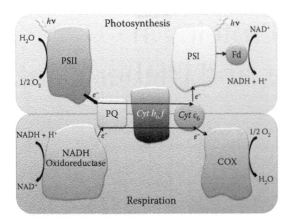

SCHEME 13.1 Photosynthesis and respiration in cyanobacteria. Cyt b_6f complex, and redox partners plastoquinone (PQ) and Cyt c_6, are shared in photosynthesis and respiration. COX, cytochrome c oxidase; Fd, ferredoxin; PSI, photosystem I; PSII, photosystem II.

13.2 PHOTOSYNTHESIS AND RESPIRATION IN CYANOBACTERIA

Cyanobacteria (formerly called blue-green algae) can perform oxygenic photosynthesis and respiration simultaneously in the same cellular compartment, inside a single prokaryotic cell, and also, many species are able to fix atmospheric nitrogen (Koike 1996; Scherer et al. 1988).

Over the years, photosynthesis has been widely studied, and information about it has been revised; it is the metabolic process responsible for the appearance of oxygen in the Earth's atmosphere (Gantt 1994; Lyons et al. 2014). This process can be divided into two phases: a first, light-dependent phase, in which reducing power and energy-rich compounds are generated, and a second phase, in which the ATP and reducing power produced are used to reduce CO_2, nitrates, and sulfates, thus assimilating the bioelements C, N, and S, in order to synthesize carbohydrates, amino acids, and other substances. The light-dependent phase is carried out in the so-called electron transport chains that perform the photosynthetic apparatus of these organisms. The photosynthetic apparatus is constituted by a series of protein complexes in membranes and a series of electron carrier molecules of lipidic nature (plastoquinones) or protein nature (Cyt c_6, Pc, ferredoxin [Fd], flavodoxin [Fld]). There are three membrane-embedded redox complexes in the photosynthetic electron transport chain (Scheme 13.1): (1) photosystem II (PSII), which takes electrons from the water; (2) Cyt b_6f complex, which accepts electrons from PSII and donates them to PSI; and (3) PSI, which reduces the protein Fd, whose redox center is a sulfoferric group, or alternatively, under iron deficiency conditions, Fld, whose prosthetic group is a flavin. Both photosystems contain molecules of chlorophyll (Chl), which are excited by solar light, capturing the energy needed to transport electrons from water to the Fd or Fld. In the two photosystems, the reaction center consists of a dimer of Chl molecules, P700 in PSI and P680 in PSII. The electron transport along the photosynthetic electron chain creates a proton (H^+) gradient that is used by ATP

synthase to produce ATP from ADP and phosphate. In addition, in photosynthesis, the function of Cyt c_6 is to transport electrons from Cyt b_6f complex to PSI. In some green algae and cyanobacteria, this function is carried out by Pc, which is replaced by Cyt c_6 in copper deficiency conditions. In higher plants, Pc is the only electron carrier to PSI (Molina-Heredia et al. 2003; Weigel et al. 2003).

Respiratory electron transport is the final stage of catabolism. The electrons derived from the oxidation of stored metabolites are sequentially transported to an external electron acceptor. The acceptor is usually oxygen, although many bacteria can make use of alternative electron acceptors in anoxic environments. In cyanobacteria, respiratory electron transport takes place mainly in thylakoid membranes, simultaneously with photosynthetic electron transport, with which it shares a number of membrane complexes and soluble electron carriers (plastoquinone, Cyt b_6f complex, and Pc or Cyt c_6). Recently, Conrad W. Mullineaux (2014) has revised information on the coexistence of both a respiratory and photosynthetic electron transfer chain in cyanobacteria. As shown in Scheme 13.1, in the thylakoid membrane from cyanobacteria, there are different pathways of electron transport. In any case, in cyanobacteria, the Cyt b_6f complex works as both Cyt b_6f from chloroplasts and Cyt bc_1 from mitochondria in photosynthetic eukaryotes, transporting electrons toward PSI or cytochrome c oxidase (COX), alternatively. Cyt b_6f, PSI, and COX are integral membrane complexes that are connected by the soluble electron carriers Cyt c_6 or Pc.

13.2.1 CYTOCHROME C_6 AND PLASTOCYANIN

Cyt c_6 is a well-known soluble electron carrier between the two membrane-bound complexes Cyt b_6f and PSI in oxygenic photosynthesis: it is present in all known cyanobacteria and in most green algae (Schmetterer and Pils 2004). Throughout evolution from cyanobacteria to higher plants, Cyt c_6 has been replaced by Pc (De la Rosa et al. 2002). However, in several cyanobacteria, both proteins are present, and Cyt c_6 substitutes for Pc under copper-deficient conditions (Davis et al. 1980; Sandmann 1985). When the same organism synthesizes both Cyt c_6 and Pc, both proteins are similar in terms of size, isoelectric point, and redox potential. The two metalloproteins are acidic in green algae, as well as the Pc of plants, but in cyanobacteria, they can be acidic, neutral, or basic (Hervas et al. 1995, 2003; Ho and Krogmann 1984). Organisms that synthesize both Cyt c_6 and Pc tend to regulate the synthesis of one or the other metalloprotein at the initiation of transcription (Bovy et al. 1992b; Nakamura et al. 1992; Zhang et al. 1994). There are some exceptions: for example, in the particular case of the green alga *Chlamydomonas reinhardtii*, the promoter of the gene that encodes Cyt c_6 is strongly inhibited in the presence of copper in the medium. However, both genes are transcribed in copper deficiency, i.e., both Cyt c_6 and preapo-Pc are synthesized, but preapo-Pc is quickly degraded in absence of copper (Hill and Merchant 1992; Merchant et al. 1991).

Cyt c_6 is a soluble hemeprotein of about 85 amino acids with a molecular mass close to 10 kDa and with a redox potential of approximately +350 mV (Campos et al. 1993; Hervas et al. 2003). It is a class I c-type cytochrome with a low-spin heme group. The heme appears covalently attached to the polypeptide chain by the N-terminal sequence consensus Cys-X-X-Cys-His (Campos et al. 1993; Kerfeld et al. 1995; Moore and Pettigrew 1990). Cyt c_6 is encoded by the *petJ* gene, which is nuclear in eukaryotes. Cyt c_6 is synthesized in the cytoplasm in the form of preapoprotein and is translocated to the chloroplasts and/or thylakoid lumen, where the transit peptide is removed and the heme group is covalently attached to the polypeptide chain (Bovy et al. 1992a; Howe and Merchant 1994; Merchant and Bogorad 1987). In the amino-acidic sequence of Cyt c_6, there are three highly conserved regions; two of them, from 9 to 21 and 56 to 65, contain the two axial ligands of heme: His18 and Met58. Another significant area comprises the residues 68 to 71 and is highly conserved in cytochromes from eukaryotes but not in those from cyanobacteria (Cohn et al. 1989; Kerfeld et al. 1995). The secondary structure of Cyt c_6 consists of four α-helices connected by three loops (Beissinger et al. 1998; Frazao et al. 1995; Kerfeld et al. 1995; Schnackenberg et al. 1999). The heme group is located in a hydrophobic pocket surrounded by 23 amino acids, of which 13 are unchanged or variable. Of the heme group, only 6% are exposed to solvent: atoms in the C and D pyrroles (Frazao et al. 1995). Heme is covalently attached to the polypeptide through thioether links by residues Cys15 and Cys18 (Frazao et al. 1995).

Pc is a copper protein that contains a metal atom attached to a single polypeptide chain of about 100 amino acids (approximately 10.5 kDa molecular mass). The nature and structure of the ligands of the copper atom make this protein blue. It is a type I blue copper protein, with a maximum absorption peak at 600 nm, which is due to the charge transfer that occurs on the (Cys84) S–Cu link, between the ligand and the metal, as a result of their orbital overlap. The spectral characteristics are similar to other copper proteins, such as azurin, pseudoazurin, and stelacyanin (Redinbo et al. 1994). The *petE* gene, which encodes for Pc, is nuclear in plants and green algae (Grossman et al. 1982; Smeekens et al. 1985). Pc is synthesized in the cytoplasm as a precursor with a transit peptide. Preapo-Pc is transported to thylakoid lumen of chloroplast (in eukaryotes) or directly through the thylakoid membrane (in cyanobacteria) (Merchant et al. 1990; Van der Plas et al. 1989). The Pcs from plants and green algae have a high homology between them, while those of cyanobacteria present a lesser conservation of sequences. The most conserved residues are the four ligands of the copper atom (His37, His87, Cys84, and Met92) and Tyr83. The net charge of the protein varies greatly from one organism to another. Pc may be acidic, with an isoelectric point close to 4, in plants and green algae, or basic or neutral in some cyanobacteria, such as *Synechocystis*, with an isoelectric point from 5.6 to 8.8 (Diaz et al. 1994; Medina et al. 1993; Molina-Heredia et al. 1998). A comparison of known Pc structures indicates that all have an identical basic folding, with eight chains connected by seven loops. Seven chains have β configuration, whereas one is usually a small α-helix. The overall form of the protein is a flattened cylinder or β-barrel with

approximate dimensions of 40 × 28 × 32 Å, with the copper atom located in the so-called north pole of the molecule (Guss et al. 1986). This copper atom, which is located in a pocket formed by highly conserved hydrophobic residues, is close to the surface of the protein, but it is not exposed to the solvent. Two nitrogen atoms from the imidazole rings of residues His37 and His87 and two atoms of sulfur from residues Cys84 and Met92 coordinate the copper atom; only His87 is exposed to the solvent (Guss and Freeman 1983; Redinbo et al. 1994). Pc presents a redox potential of + 370 mV at a pH of 7, which is probably due to the special shape of the active site, a distorted tetrahedron (Garrett et al. 1984). It is important to distinguish between these two types of Pc reduced structures because at low pH values, the molecule is kinetically inactive for its oxidation by inorganic complexes (Segal and Sykes 1978; Sykes 1985). The region flanking the area of His87 in Pc, which is exposed to the solvent, is called the *hydrophobic face* or *north pole*. In eukaryotes, in this area, the conserved residues are Leu12, Ala33, Gly34, Phe35, Pro36, Gly89, and Ala90 (Guss et al. 1992; Redinbo et al. 1994). However, in cyanobacteria, positions 33, 34, and 35 are variable. There are many data obtained through chemical and directed mutagenesis, indicating that the hydrophobic area in Pc, and in particular, His87, is responsible for the electron transfer to PSI (Anderson et al. 1987; Haehnel et al. 1994; Hope 2000; Nordling et al. 1991; Sigfridsson et al. 1996). Around the Tyr83 residue, which is located 19 Å from the copper atom, there is another possible site of interaction (He et al. 1991). Tyr83 is exposed to solvent and surrounded (in the Pc of higher plants) by negatively charged residues (Glu and Asp) in positions 42–45 and 59–61 (Redinbo et al. 1993, 1994). The two acidic areas at positions 42–45 and 59–61 are the so-called negative face or east side (Durell et al. 1990). In green algae, the negative face presents one less charge in areas 59–61, and two acidic residues appear in additional positions 53–85; this is a small change in the distribution but not in the amount of negative residues surrounding Tyr83 (Collyer et al. 1990; Redinbo et al. 1993, 1994). The east side of Pc is highly conserved in plants and green algae, where it is negative, but in cyanobacteria, it can be basic or neutral, as in *Synechocystis* (Briggs et al. 1990; Redinbo et al. 1994). In Cyt c_6, there also appears two regions equivalent to the electrostatic and hydrophobic areas present in Pc (Molina-Heredia et al. 1999). In addition, in cyanobacteria, both Pc and Cyt c_6 present a single arginyl residue, strictly conserved, located between electrostatic and hydrophobic areas, which is essential for the interaction of both proteins with PSI (Molina-Heredia et al. 2001).

As Cyt c_6 and Pc show similar physicochemical properties and carry out the same physiological function, we would expect both proteins to present a similar structure. In fact, even though Cyt c_6 is a hemeprotein with four α-helices and Pc is a copper protein whose secondary structure is based on β-sheets, their molecular weights and redox potentials are similar. In addition, when both proteins are isolated from the same organism, their isoelectric points and surface distribution of charges are similar (Ho and Krogmann 1984; Navarro et al. 1997; Ullmann et al. 1997). Both the heme group from Cyt c_6 and the His87 from Pc (ligand of the copper atom where the electron transfer is supposed to occur) are surrounded by hydrophobic residues. Cyt c_6 and the Pc from eukaryotic organisms present a negative face, but in cyanobacteria, both proteins can be neutral or present a positive side.

13.2.2 Cytochrome c_M

Cyt c_M is a soluble class I Cyt c, homologous to Cyt c_6, found in cyanobacteria (Malakhov et al. 1994). In fact, it has been found in all cyanobacteria whose genome has been sequenced; however, its function remains unknown. Little is known about this metalloprotein, and only the Cyt c_M from the cyanobacterium *Synechocystis* sp. PCC 6803 has been studied. In these studies, it has been shown that Cyt c_M messenger RNA (mRNA) levels are very low and that mutants lacking Cyt c_M grow normally under photoautotrophic conditions, displaying rates of photosynthesis and respiration that are comparable to the wild type (Malakhov et al. 1994; Manna and Vermaas 1997). On the other hand, under stress conditions (low temperature and high light intensity), when the synthesis of the two soluble metalloproteins Cyt c_6 and Pc is repressed, the Cyt c_M expression increases drastically (Malakhov et al. 1999). In these conditions Cyt c_M could play an important role as a protector against photoinduced stress and/or oxidative stress (Malakhov et al. 1999). Cyt c_M from *Synechocystis* contains a transit peptide to thylakoid lumen homologous to those from Cyt c_6. This indicates that it is probably located in the same cell compartment as Pc and Cyt c_6 (Molina-Heredia et al. 2002). Metzger et al. (1995) suggested that Cyt c_M could serve as a photosynthetic electron carrier, alternative to Cyt c_6 and Pc. However, kinetic analysis of PSI reduction by Cyt c_M, Cyt c_6, and Pc disproved this hypothesis, since the bimolecular constant determined by the reaction was up to 100 times lower with Cyt c_M than with its physiological electron donors Pc and Cyt c_6 (Molina-Heredia et al. 2002). Also the redox potential from Cyt c_M is +150 mV, 200 mV lower than that of Cyt f, so it is unlikely, from a thermodynamic point of view, that Cyt c_M could be capable of accepting electrons from Cyt $b_6 f$ complex (Molina-Heredia et al. 2002). In addition, Cyt c_M does not substitute Pc or Cyt c_6, because it is unable to keep electrons from Cyt $b_6 f$ complex and donate them to PSI (Molina-Heredia et al. 2002). Manna and Vermaas (1997) have suggested that Cyt c_M could operate in the respiratory electron transport, transferring electrons to a caa3-type Cyt c oxidase. This suggestion is based on the analysis of electron transport in *Synechocystis* sp. 6803 Cyt c_M deletion mutants, in which it was impossible to obtain a double mutant Cyt c_M/Cyt c_6. These authors also observed that it was not possible to eliminate the PSI activity in a Cyt c_M deletion mutant, indicating that Cyt c_M might be necessary for the respiratory electron transfer chain. Subsequently, Bernroitner et al. (2009) suggested that Cyt c_6 could be capable of donating electrons to terminal oxidases, with an efficiency higher than Cyt c_6 and Pc, although this is not completely confirmed. All these data make plausible the existence of an alternative electron donor that could play the role of respiratory Cyt c in

cyanobacteria, avoiding interferences between both respiratory and photosynthetic electron transport chains. However, nothing is known yet about the possible electron donor to Cyt c_M. Be that as it may, all these data lead us to conclude that the functional role of Cyt c_M would be related to the respiratory electron transfer chain.

13.3 GREEN ALGAE AND HIGHER-PLANT CYT c_6-LIKE PROTEINS

In 2002, Wastl et al. (2002) reported the identification of an *Arabidopsis thaliana* gene with clear homology to Cyt c_6. They gave evidence for the transcription of this gene in *A. thaliana* by expressed sequence tag (EST) sequencing. In addition, this gene was also found in other higher plants and even in green algae (Wastl et al. 2002; Worrall et al. 2008). Thus, both Cyt c_6 and Cyt c_{6A} are present in some green algae. Gupta et al. (2002) described knockout experiments that indicated that plants lacking either Pc or Cyt c_{6A} were viable, but those lacking both were not. They also reported that heterologously expressed plant Cyt c_6 could functionally replace Pc in reconstitution experiments *in vitro* using inside-out thylakoids and in measuring oxygen evolution. Therefore, they concluded that Cyt c_{6A} is the functional counterpart of Pc. However, Weigel et al. (2003) showed that inactivation of the two Pc genes of *Arabidopsis* resulted in plants unable to grow photoautotrophically, even when the Cyt c_{6A} protein was overexpressed at the same time. Furthermore, *in vitro*, the laser flash-induced kinetic analysis demonstrated the lack of reactivity of Cyt c_{6A} toward PSI (Molina-Heredia et al. 2003). In addition, Cyt c_{6A} $E_{m,7}$ was estimated at +150 mV, 200 mV lower than those of Cyt c_6, Pc, or Cyt f; with this, from a thermodynamic point of view, it is difficult to think that it will be able to capture electrons from the Cyt $b_6 f$ complex.

What is the possible function of plant Cyt c_{6A}? What we know for sure is that this protein is not involved in photosynthetic electron transfer (Molina-Heredia et al. 2003; Weigel et al. 2003). Some authors speculated about the possibility of regulatory functions (Wastl et al. 2002; Weigel et al. 2003; Worrall et al. 2008), but to date, its function remains unknown. The only truth is that it is present in many higher plants and green algae. In addition, although it is a protein related to Cyt c_6, other authors associate it evolutionarily with Cyt c_M (De la Rosa et al. 2006) or Cyt c_{6-2} (Zatwarnicki et al. 2014).

13.4 CYANOBACTERIAL CYT c_6-LIKE PROTEINS

As we have already mentioned in the introduction, a near-copy of the *petJ* gene, encoding for Cyt c_6, has been found in all cyanobacteria whose genomes are sequenced. However, in a large number of cyanobacteria, a second copy is present (Cyt c_{6-2}) (Bialek et al. 2008), and recently, we have found a third differentiated form present only in heterocyst-forming filamentous cyanobacteria (Cyt c_{6-3}) (Torrado et al. 2015). We must bear in mind that sequencing data are always preliminary data and they do not serve to distinguish between paralogous and orthologous proteins. For this reason, it is important to study the newfound isoforms of Cyt c_6 to determine if they are true isoforms of the same protein or if they constitute new paralogous proteins that have evolved to a specific new function.

In higher plants, it is common to find two different functional isoforms of Pc, which play the same function with a similar efficiency (Abdel-Ghany 2009). Nevertheless, to date, in green algae, only the presence of a single copy of a functional Cyt c_6 is described. For this reason, the presence in some cyanobacteria of two or three possible copies of Cyt c_6 can be explained from a different point of view. They can be orthologous proteins, as in the case of higher plants' Pc, or they can be paralogous proteins that have evolved from a common ancestor to carry out different functions. In this section, we will review all the information available on these isoforms of Cyt c_6 to try to explain whether they are orthologous or paralogous proteins and, in the second case, what their metabolic function might be. As far as these isoforms from Cyt c_6 are only found in cyanobacteria, in order to study their possible functions, we must begin by describing some fundamental details of this bacterial group.

With all the kinds of metabolism present in prokaryotic organisms, cyanobacteria are a uniquely qualified group for performing an oxygenic photosynthesis similar to higher plants (Blankenship 1992), due to the presence of a complete photosynthetic complex integrated in the thylakoid membrane. Thylakoid membranes are also the site of a principal respiratory electron transport chain. Thus, in cyanobacteria, both respiratory and photosynthetic electron transport chains are located in the same membrane system, and as we have discussed in the introduction, they share some elements. Nevertheless, the cytoplasmic membrane does not contain a functional photosynthetic electron transport chain but a second respiratory chain has been found (Schmetterer and Pils 2004). A particularity of these organisms is the use of atmospheric nitrogen and a wide range of combined nitrogen, such as nitrate, nitrite, ammonium, urea, and amino acids as a nitrogen source (Flores and Herrero 1994; Guerrero and Lara 1987; Stewart 1980). For the atmospheric

nitrogen fixation, some cyanobacteria have developed a truly differentiated cell: the heterocyst. This cell is developed in the absence of a combined nitrogen source, and it is specialized for the nitrogen fixation process, spatially separated from the photosynthesis process. This prevents the irreversible inactivation of nitrogenase by molecular oxygen (Wolk 1982).

In the cyanobacterium *Nostoc* sp. PCC 7119, these three isoforms are found, so we have centered our research on this strain. To elucidate the possible function of the three isoforms of Cyt c_6, the first step is to analyze and compare their amino acid sequences (Figure 13.1). The three cytochromes contain a typical transit peptide to the periplasmic space/thylakoid lumen of 25 amino acids. This transit peptide consists of three well-defined regions: a positively charged N-terminus region, a central hydrophobic region, and a C-terminus end comprising the consensus motif AxA, a specific cleavage site for peptidases. In this context, the close similarity between the transit peptides of the three proteins suggests that both Cyt c_{6-2} and Cyt c_{6-3} could, in fact, be located in the same cellular compartment as Cyt c_6. Both Cyt c_6 and Pc are located inside the thylakoid lumen (Obinger et al. 1990), and also, a periplasmic Cyt c_6 was identified as an electron donor to cytoplasmic membrane-bound COX (Serrano et al. 1990). The occurrence of Pc in the periplasmic space has not been proved yet. One significative sequence difference is that Cyt c_{6-3} contains two additional regions that do not appear in Cyt c_6 or Cyt c_{6-2}, an intermediate region (LLKY) and a final one (NLEKE) (Figure 13.2); also, the presence of a glutamine at position 54 in the sequence of a protein belonging to the Cyt c family makes its redox potential 100 mV higher than if that position were occupied by any other residue (Worrall et al. 2007). Cyt c_6, whose $E_{m,7}$ is +337 mV (Molina-Heredia et al. 1998), and Cyt c_{6-3}, whose $E_{m,7}$ is +300 mV (Torrado et al. 2015), possesses a glutamine at position 54. However, it is not present in Cyt c_{6-2}, whose $E_{m,7}$ is +199 mV (Reyes-Sosa et al. 2011). Bialek et al. (2008) performed a comparative study of the sequence of Cyt c_6 and Cyt c_{6-2}, concluding that Cyt c_{6-2} constitutes a new branch separated from Cyt c_6. In addition, they divided Cyt c_6 into two groups, which they called

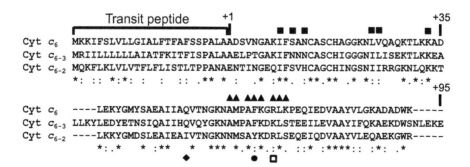

FIGURE 13.1 Alignment of the amino acid sequences of precytochromes c_6, c_{6-2}, and c_{6-3} from *Nostoc*. The box encompasses the transit peptide. The surface residues from Cyt c_6 involved in hydrophobic or electrostatic interaction with PSI are indicated by as squares or triangles, respectively. The rhomb indicates the glutamine residue at position 50 in Cyt c_6 and 54 in Cyt c_{6-3}, responsible for the increase by 100 mV of the redox potential in the Cyt c protein family. The circle indicates the tyrosine 61 from Cyt c_{6-2}, conserved in all Cyt c_{6-2}. The white square indicates the arginyl residue conserved in all Cyt c_6.

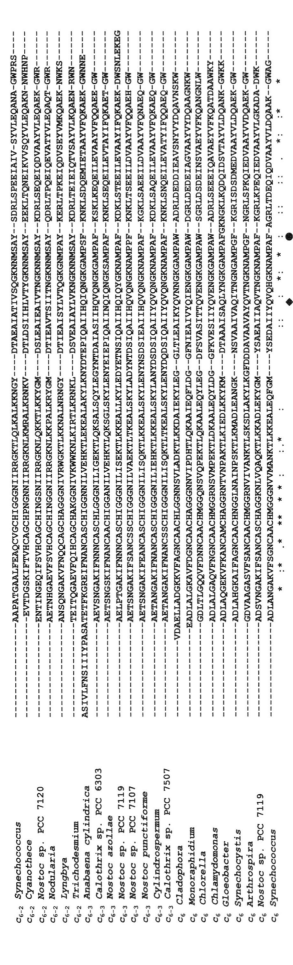

FIGURE 13.2 Sequence alignment of several cytochromes c_6, c_{6-2}, and c_{6-3} of cyanobacteria. The rhomb indicates the glutamine residue at position 50 in Cyt c_6, responsible for the increase by 100 mV of the redox potential in the Cyt c protein family. The circle indicates a low variable residue conserved in c_{6-2} as tyrosine and c_6/c_{6-3} as phenylalanine.

Cyt c_{6B} and Cyt c_{6C}. In similar studies, Torrado et al. (2015) included the third isoform, Cyt c_{6-3}. If we compare these cytochromes with other Cyt c_6-type cytochromes studied, some interesting correlations are found (Figure 13.2). As stated by Bialek et al. (2008), some residues are conserved in the cyanobacterial Cyt c_6 family. The residue in position 61 is conserved in Cyt c_6 and Cyt c_6-like proteins; in this position, phenylalanine appears in Cyt c_6 and in Cyt c_{6-3}, whereas in Cyt c_{6-2}, it is occupied by tyrosine (Bialek et al. 2008, 2014). The residue at position 54, as we discussed before, is crucial to determine the redox potential. Notwithstanding the extensive taxonomic work done by Bialek et al. (2008), Cyt c_{6-3} is not included in any group. The evolutionary origin of Cyt c_{6-3} has been studied by Torrado et al. (2015), where the amino acid sequences of Cyt c_{6-3} were compared with other soluble monoheme cytochromes with His–Met axial metal coordination of photosynthetic organisms: Cyt c_6, Cyt c_{6-2}, and Cyt c_M from cyanobacteria; Cyt c_2 from anoxygenic bacterial Cyt c_{6A} from higher plants and green algae; and respiratory Cyt c from algal and higher plants' mitochondria. As seen in Figure 13.3, Cyt c_{6-3} constitutes, by itself, a different clade independent of the other proteins.

In the photosynthetic electron flux from the heterocyst-forming cyanobacteria *Nostoc* sp. PCC 7119, Cyt c_6 and Pc, whose $E_{m,7}$ are +337 and +355 mV, respectively (Molina-Heredia et al. 1998), take electrons from Cyt f, whose $E_{m,7}$ is +334 mV (Albarran et al. 2005), and donate them to the photooxidized PSI, whose E_m (P700) is +500 mV (Brettell 1997). The $E_{m,7}$ values of Cyt c_{6-2} described are +199 mV in *Nostoc* and +155 mV in *Synechococcus* (Bialek et al. 2014; Reyes-Sosa et al. 2011). This redox potential is different but closer to those reported for other related cytochromes at the same pH, such as Cyt c_M, whose $E_{m,7}$ is +150 mV (Molina-Heredia et al. 2002), and higher plants' Cyt c_{6A}, with an $E_{m,7}$ of +140 mV (Molina-Heredia et al. 2003). With such E_m, it is clear that Cyt c_{6-2} cannot accept electrons from Cyt f. In addition, as Cyt c_{6-2} is not able to obtain electrons from Cyt b_6f complex, it is clear that it is a paralogous form of Cyt c_6. To date, it is not known what the electron donor to Cyt c_{6-2} may be. This must be further investigated. However, the E_m from *Nostoc* Cyt c_{6-3} is +300 mV at a pH of 7.0 and + 345 mV at the more physiological pH of 4.5 (Torrado et al. 2015). This E_m is similar to those from Cyt c_6, Pc, and Cyt f, making it plausible that Cyt c_{6-3} captures electrons from Cyt b_6f complex. Then, if Cyt c_{6-3} captures electrons from Cyt b_6f complex, which is its target?

Cyt c_6 interacts with PSI and Cyt b_6f using the same regions (Diaz-Moreno et al. 2005; Molina-Heredia et al. 1999). The first one comprehends a hydrophobic surface, located around the region through which the heme group is exposed to solvent, providing a contact surface for the electron transfer. The second one is a charged region, responsible for driving long-distance electrostatic interactions with PSI and Cyt b_6f complexes (De la Rosa et al. 2006). The hydrophobic site is located approximately between residues 9 and 29, and the electrostatic site, between residues 57 and 65 of the mature protein (De la Rosa et al. 2006). As shown

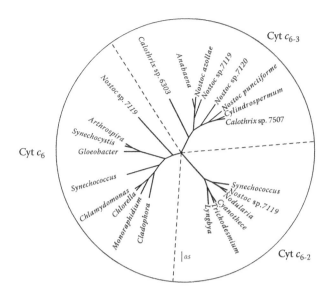

FIGURE 13.3 Phylogenetic tree of cytochromes c_6 and c_6-like from several cyanobacteria. Trees were constructed by protein maximum likelihood with PROTML (MOLPHY). (From Adachi, J., and Hasegawa, M., *Comput. Sci. Monogr.*, 27, 1–77, 1992.) The following sequences were aligned: Cyt c_{6-3} (c_{6-3}) of *Anabaena cylindrical* PCC 7122 (GenBank gi 440681468), *Calothrix* sp. PCC 6303 (GenBank gi 428297476), *Calothrix* sp. PCC 7507 (GenBank gi 427719567), *Cylindrospermum stagnale* PCC 7417 (GenBank gi 434404902), *Nostoc azollae* 0708 (GenBank gi 298490297), *Nostoc punctiforme* PCC 73102 (GenBank gi 186683068), *Nostoc* sp. PCC 7107 (GenBank gi 427710665), and *Nostoc* sp. PCC 7119 (TrEMBL HG316543); Cyt c_6 (c_6) of *Arthrospira platensis* NIES-39 (GenBank gi YP_005072220.1), *Chlamydomonas reinhardtii* (GenBank gi 117924), *Chlorella vulgaris* (GenBank gi 30578153), *Cladophora glomerata* (GenBank gi 24636293), *Gloeobacter violaceus* (GenBank gi 37521549), *Monoraphidium braunii* (GenBank gi 729268), *Nostoc* sp. PCC 7119 (EMBL AJ002361), *Synechococcus elongatus* (GenBank gi 25014058), and *Synechocystis* sp. PCC 6803 (EMBL P46445); and Cyt c_{6-2} (c_{6-2}) of *Cyanothece* sp. CCY0110 (GenBank gi 126660190), *Lyngbya* sp. PCC 8106 (GenBank gi 119484682), *Nodularia spumigena* CCY9414 (GenBank gi 119509989), *Nostoc* sp. PCC 7119 (EMBL AM902496), *Synechococcus elongatus* PCC 6301 (GenBank gi 56751577), and *Trichodesmium erythraeum* IMS101 (GenBank gi 113478003). Branch lengths reflect the estimated number of substitutions per site.

in Figures 13.1 and 13.2, these regions are also highly conserved in both Cyt c_{6-2} and Cyt c_{6-3}. In addition, in position 64 of Cyt c_6 (68 in Cyt c_{6-3}), an arginine residue appears, strictly conserved in all known cyanobacteria, which is essential for efficient interaction with PSI (Molina-Heredia et al. 2001). In Pc, there is a strictly conserved arginine residue that performs the same function (Molina-Heredia et al. 2001). This arginine residue is conserved in Cyt c_{6-2}, which is able to interact with PSI (Reyes-Sosa et al. 2011), but is not conserved in Cyt c_{6-3}. Anyhow, there is experimental evidence for the fact that Cyt c_{6-2} is able to interact with PSI, although with an efficiency significantly lower than Cyt c_6 or Pc (Figure 13.4) (Reyes-Sosa et al. 2011). However, even if *Nostoc* Cyt c_{6-3} could capture electrons from Cyt b_6f complex, the laser flash-induced kinetic analysis demonstrated the

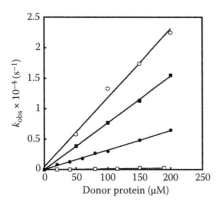

FIGURE 13.4 Dependence upon electron donor protein concentration of the observed rate constant (k_{obs}) for PSI reduction by Cyt c_6 (white circles), Pc (black squares), Cyt c_{6-2} (black circles), and Cyt c_{6-3} (white squares). The oscilloscope traces obtained with the Cyt c_6 molecule fit to biphasic kinetics, whereas those corresponding to Cyt c_{6-2}, Cyt c_{6-3}, and Pc fit to single exponential curves. Temperature was 25°C, and pH was 7.5.

lack of reactivity of Cyt c_{6-3} toward the photosystem (Figure 13.4) (Torrado et al. 2015).

So far, we have seen that Cyt c_{6-2} is not able to capture electrons from Cyt $b_6 f$ complex; however, it would donate them to PSI at low efficiency compared to Cyt c_6. On the other hand, Cyt c_{6-3} would be able to get electrons from Cyt $b_6 f$ complex, but it is not able to donate them to PSI. To check if Cyt c_{6-2} or Cyt c_{6-3} may have a role in the respiratory electron transport chain, Torrado et al. (2015) analyzed the Cyt c oxidase-dependent O_2 uptake activities in the presence of Cyt c_6, Cyt c_{6-2}, Cyt c_{6-3}, and Pc from *Nostoc*. *Nostoc* is able to grow photoautotrophically and chemoheterotrophically in darkness, in both cases with combined nitrogen or with dinitrogen as the nitrogen source. In *Nostoc*, there are also several terminal oxidases (Pils et al. 2004). For all this, we studied the O_2 uptake activities in membranes from cells grown in chemoheterotrophic conditions, photoautotrophic conditions with combined nitrogen, and photoautotrophic conditions with dinitrogen. In the last case, we purified membranes from whole cells (vegetative and heterocyst cells) and from aisled heterocysts. As shown in Table 13.1, *Nostoc* Cyt c_6 reacted efficiently with all membrane preparations. Surprisingly, the reaction of *Nostoc* Pc with the heterocyst membranes was incredibly low, on the order of nonspecific protein, for those experiment conditions. This makes us think that *Nostoc* Pc is not the main partner in respiration processes of heterocysts. The reactivity of Cyt c_{6-2} is similar to those from the nonspecific electron donor *Arabidopsis* Pc, so we can conclude that it is not an efficient electron donor to terminal oxidases. The case of Cyt c_{6-3} is different from the previous one and very significant: it only reacts efficiently with membranes from heterocyst cells. What could be the difference between the membranes from heterocyst cells and the others? Valladares et al. (2007) described three different heme-copper-type terminal respiratory oxidases (COX) in the related strain *Anabaena* sp. PCC 7120. COX1 is expressed in vegetative cells independently of the nitrogen source. In contrast, COX2 and COX3 are expressed under atmospheric nitrogen fixation conditions, specifically in the development of mature heterocysts

TABLE 13.1

Oxygen Uptake Rate of Several Redox Proteins with Membranes of *Nostoc* sp. PCC 7119

	O_2 Uptake Rate (nmol/mL·min)		
Protein	**BG11**	**BG11 – NO_3^-**	**Heterocyst Membranes**
c_6	11.2	4.5	11.4
c_{6-2}	0.0	1.5	3.3
c_{6-3}	0.0	1.6	11.6
Pc	6.5	3.2	2.2
Pc*	0.0	2.0	3.2

Note: Membranes were obtained from synchronic cultures of *Nostoc* sp. PCC 7119 in BG11 media, BG11 media without combined nitrogen and, from this last one, membranes of isolated heterocysts. Pc*, *Arabidopsis thaliana* Pc.

(Valladares et al. 2007). Thus, Cyt c_{6-3} could be a specific electron donor to the terminal oxidases specifically expressed in the heterocyst cells.

13.5 CONCLUDING REMARKS

Cyt c_6, whose principal function is to transfer electrons from Cyt $b_6 f$ complex to PSI (photosynthesis) or terminal oxidases (respiration), is present in all studied cyanobacteria. In some cyanobacteria, that function is carried out by Pc, and Cyt c_6 replaces it in copper deficiency conditions. Something similar happens in green algae. On the other hand, in higher plants, Cyt c_6 has been lost through evolution and has been completely replaced by Pc.

The recent sequencing of genomes of several photosynthetic organisms has allowed codifying Cyt c_6-like proteins' genes to be found in some cyanobacteria, green algae, and higher plants. In all sequenced cyanobacteria, Cyt c_M has been found, but its function remains unknown. This protein is not able to capture electrons from Cyt $b_6 f$ complex or donate them to PSI, and it is thought that it could be involved in respiration. In green algae and higher plants, even in green algae that have Cyt c_6, the related protein Cyt c_{6A} appears, which, as in cyanobacteria with Cyt c_M, does not interact efficiently with the Cyt $b_6 f$ complex nor with PSI, and it is not known what its functional role may be. The case of cyanobacteria is even more complex. They contain not only Cyt c_6 and Cyt c_M; in many of them, Cyt c_{6-2} is present, and in addition, in heterocyst-forming filamentous cyanobacteria, there is another Cyt c_6-like protein, Cyt c_{6-3}. Cyt c_{6-2} is capable of transferring electrons to PSI with significant kinetics, but is not able to capture them from Cyt $b_6 f$ complex. The very low concentration of Cyt c_{6-3} detected in cells, with respect to the Cyt c_6, makes us think that they may have a regulatory function. Cyt c_{6-3}, unlike the other Cyt c_6-like proteins, is able to accept electrons from Cyt $b_6 f$ complex. However, it is not able to donate them to PSI. On the other hand, Cyt c_{6-3} is able to transfer electrons to specific oxidases expressed in diazotrophic conditions. Therefore, we think that Cyt c_{6-3} may be involved in the formation of the heterocyst, or in any process

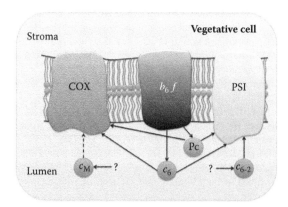

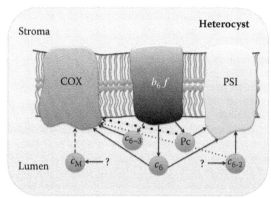

SCHEME 13.2 Proposed electron transfer flux in photosynthesis and respiration from *Nostoc*. Represented are two different states depending on the cellular type, vegetative cell, and heterocyst. In vegetative cells, the solid line represents the electron flux with a high level of interaction, like the COX and PSI with their common partners Pc and Cyt c_6, and the dashed line represents a probable interaction. In the case of Cyt c_{6-2} and Cyt c_M, the electron donor to these proteins remains unknown. In the heterocyst cells, the situation is different. The dotted lines represent a low level of interaction, which is the case for Cyt c_{6-2} and Pc with COX. Despite all that, the interaction of Cyt c_{6-3} with COX is high, at the same level as Cyt c_6.

related to the fixation of atmospheric nitrogen in heterocyst-forming cyanobacteria, but further studies are required to make a firm assertion.

Having taken everything into account, we present a scheme (Scheme 13.2) of the electron flux in photosynthesis and respiration in heterocyst-forming filamentous cyanobacteria, pending confirmation by further research. As we can see in the scheme, there are still several questions to be solved in order to compose a complete picture of the respiratory/photosynthesis flux in these organisms.

REFERENCES

Abdel-Ghany, S. E. (2009). Contribution of plastocyanin isoforms to photosynthesis and copper homeostasis in *Arabidopsis thaliana* grown at different copper regimes. *Planta, 229*(4), 767–779. doi:10.1007/s00425-008-0869-z.

Adachi, J. and Hasegawa, M. (1992). MOLPHY: Programs for molecular phylogenetics, I—PROTML: Maximum likelihood inference of protein phylogeny. *Comput Sci Monogr, 27*, 1–77.

Albarran, C., Navarro, J. A., Molina-Heredia, F. P., Murdoch Pdel, S., De la Rosa, M. A., and Hervas, M. (2005). Laser flash-induced kinetic analysis of cytochrome *f* oxidation by wild-type and mutant plastocyanin from the cyanobacterium *Nostoc* sp. PCC 7119. *Biochemistry, 44*(34), 11601–11607. doi:10.1021/bi050917g.

Anderson, G. P., Sanderson, D. G., Lee, C. H., Durell, S., Anderson, L. B., and Gross, E. L. (1987). The effect of ethylenediamine chemical modification of plastocyanin on the rate of cytochrome *f* oxidation and P-700+ reduction. *Biochim Biophys Acta, 894*(3), 386–398.

Ardelean, I., Matthijs, H. C., Havaux, M., Joset, F., and Jeanjean, R. (2002). Unexpected changes in photosystem I function in a cytochrome c_6-deficient mutant of the cyanobacterium *Synechocystis* PCC 6803. *FEMS Microbiol Lett, 213*(1), 113–119.

Beissinger, M., Sticht, H., Sutter, M., Ejchart, A., Haehnel, W., and Rosch, P. (1998). Solution structure of cytochrome c_6 from the thermophilic cyanobacterium *Synechococcus elongatus*. *EMBO J, 17*(1), 27–36. doi:10.1093/emboj/17.1.27.

Bernroitner, M., Tangl, D., Lucini, C., Furtmuller, P. G., Peschek, G. A., and Obinger, C. (2009). Cyanobacterial cytochrome *c*(M): Probing its role as electron donor for Cu(A) of cytochrome *c* oxidase. *Biochim Biophys Acta, 1787*(3), 135–143. doi:10.1016/j.bbabio.2008.12.003.

Bialek, W., Nelson, M., Tamiola, K., Kallas, T., and Szczepaniak, A. (2008). Deeply branching c_6-like cytochromes of cyanobacteria. *Biochemistry, 47*(20), 5515–5522. doi:10.1021/bi701973g.

Bialek, W., Krzywda, S., Zatwarnicki, P., Jaskolski, M., Kolesinski, P., and Szczepaniak, A. (2014). Insights into the relationship between the haem-binding pocket and the redox potential of c_6 cytochromes: Four atomic resolution structures of c_6 and c_6-like proteins from *Synechococcus* sp. PCC 7002. *Acta Crystallogr D Biol Crystallogr, 70*(Pt 11), 2823–2832. doi:10.1107/S1399004714013108.

Blankenship, R. E. (1992). Origin and early evolution of photosynthesis. *Photosynth Res, 33*(2), 91–111. doi:10.1007/BF00039173.

Bovy, A., de Vrieze, G., Borrias, M., and Weisbeek, P. (1992a). Isolation and sequence analysis of a gene encoding a basic cytochrome *c*-553 from the cyanobacterium *Anabaena* SP. PCC 7937. *Plant Mol Biol, 19*(3), 491–492.

Bovy, A., de Vrieze, G., Borrias, M., and Weisbeek, P. (1992b). Transcriptional regulation of the plastocyanin and cytochrome *c*553 genes from the cyanobacterium *Anabaena* species PCC 7937. *Mol Microbiol, 6*(11), 1507–1513.

Brettell, K. (1997). Electron transfer and arrangement of the redox cofactors in photosystem I. *Biochim Biophys Acta, 1318*, 322–373.

Briggs, L. M., Pecoraro, V. L., and McIntosh, L. (1990). Copper-induced expression, cloning, and regulatory studies of the plastocyanin gene from the cyanobacterium *Synechocystis* sp. PCC 6803. *Plant Mol Biol, 15*(4), 633–642.

Campos, A. P., Aguiar, A. P., Hervas, M., Regalla, M., Navarro, J. A., Ortega, J. M., and Teixeira, M. (1993). Cytochrome c_6 from *Monoraphidium braunii*. A cytochrome with an unusual heme axial coordination. *Eur J Biochem, 216*(1), 329–341.

Cohn, C. L., Hermodson, M. A., and Krogmann, D. W. (1989). The amino acid sequence of cytochrome *c*553 from *Microcystis aeruginosa*. *Arch Biochem Biophys, 270*(1), 219–226.

Collyer, C. A., Guss, J. M., Sugimura, Y., Yoshizaki, F., and Freeman, H. C. (1990). Crystal structure of plastocyanin from a green alga, *Enteromorpha prolifera*. *J Mol Biol, 211*(3), 617–632. doi:10.1016/0022-2836(90)90269-R.

Davis, D. J., Krogmann, D. W., and Pietro, A. S. (1980). Electron donation to photosystem I. *Plant Physiol, 65*(4), 697–702.

De la Rosa, M. A., Navarro, J. A., Diaz-Quintana, A., De la Cerda, B., Molina-Heredia, F. P., Balme, A., and Hervas, M. (2002). An evolutionary analysis of the reaction mechanisms of photosystem I reduction by cytochrome *c*(6) and plastocyanin. *Bioelectrochemistry, 55*(1–2), 41–45.

De la Rosa, M. A., Molina-Heredia, F. P., Hervas, M., and Navarro, J. A. (2006). Convergent evolution of cytochrome c_6 and plastocyanin. In J. H. Goldbeck (Ed.), *Advances in photosynthesis and respiration* (Vol. 24, pp. 683–694). Dordrecht: Springer.

Diaz, A., Hervas, M., Navarro, J. A., De La Rosa, M. A., and Tollin, G. (1994). A thermodynamic study by laser-flash photolysis of plastocyanin and cytochrome c_6 oxidation by photosystem I from the green alga *Monoraphidium braunii*. *Eur J Biochem, 222*(3), 1001–1007.

Diaz-Moreno, I., Diaz-Quintana, A., Molina-Heredia, F. P., Nieto, P. M., Hansson, O., De la Rosa, M. A., and Karlsson, B. G. (2005). NMR analysis of the transient complex between membrane photosystem I and soluble cytochrome c_6. *J Biol Chem, 280*(9), 7925–7931. doi:10.1074/jbc.M412422200.

Duran, R. V., Hervas, M., De la Cerda, B., De la Rosa, M. A., and Navarro, J. A. (2006). A laser flash-induced kinetic analysis of in vivo photosystem I reduction by site-directed mutants of plastocyanin and cytochrome c_6 in *Synechocystis* sp. PCC 6803. *Biochemistry, 45*(3), 1054–1060. doi:10.1021/bi052090w.

Durell, S. R., Labanowski, J. K., and Gross, E. L. (1990). Modeling of the electrostatic potential field of plastocyanin. *Arch Biochem Biophys, 277*(2), 241–254.

Flores, E. and Herrero, A. (1994). Assimilatory nitrogen metabolism and its regulation. In D. A. Bryant (Ed.), *The molecular biology of cyanobacteria* (pp. 487–517). Dordrecht: Kluwer Academic Publishers.

Frazao, C., Soares, C. M., Carrondo, M. A., Pohl, E., Dauter, Z., Wilson, K. S., and Sheldrick, G. M. (1995). Ab initio determination of the crystal structure of cytochrome c_6 and comparison with plastocyanin. *Structure, 3*(11), 1159–1169.

Gantt, E. (1994). Supramolecular membrane organization. In D. A. Bryant (Ed.), *The molecular biology of cyanobacteria* (pp. 119–138). Dordrecht: Kluwer Academic Publishers.

Garlick, S., Oren, A., and Padan, E. (1977). Occurrence of facultative anoxygenic photosynthesis among filamentous and unicellular cyanobacteria. *J Bacteriol, 129*(2), 623–629.

Garrett, T. P., Clingeleffer, D. J., Guss, J. M., Rogers, S. J., and Freeman, H. C. (1984). The crystal structure of poplar apoplastocyanin at 1.8-Å resolution. The geometry of the copper-binding site is created by the polypeptide. *J Biol Chem, 259*(5), 2822–2825.

Grossman, A. R., Bartlett, S. G., Schmidt, G. W., Mullet, J. E., and Chua, N. H. (1982). Optimal conditions for post-translational uptake of proteins by isolated chloroplasts. In vitro synthesis and transport of plastocyanin, ferredoxin–NADP+ oxidoreductase, and fructose-1,6-bisphosphatase. *J Biol Chem, 257*(3), 1558–1563.

Guerrero, M. G. and Lara, C. (1987). Assimilation of inorganic nitrogen. In C. Fay (Ed.), *The cyanobacteria* (pp. 163–186). New York: Elsevier Science.

Gupta, R., He, Z., and Luan, S. (2002). Functional relationship of cytochrome *c*(6) and plastocyanin in *Arabidopsis*. *Nature, 417*(6888), 567–571. doi:10.1038/417567a.

Guss, J. M. and Freeman, H. C. (1983). Structure of oxidized poplar plastocyanin at 1.6 Å resolution. *J Mol Biol, 169*(2), 521–563.

Guss, J. M., Harrowell, P. R., Murata, M., Norris, V. A., and Freeman, H. C. (1986). Crystal structure analyses of reduced (CuI) poplar plastocyanin at six pH values. *J Mol Biol, 192*(2), 361–387.

Guss, J. M., Bartunik, H. D., and Freeman, H. C. (1992). Accuracy and precision in protein structure analysis: Restrained least-squares refinement of the structure of poplar plastocyanin at 1.33 Å resolution. *Acta Crystallogr B, 48*(Pt 6), 790–811.

Haehnel, W., Jansen, T., Gause, K., Klosgen, R. B., Stahl, B., Michl, D., and Herrmann, R. G. (1994). Electron transfer from plastocyanin to photosystem I. *EMBO J, 13*(5), 1028–1038.

He, S., Modi, S., Bendall, D. S., and Gray, J. C. (1991). The surface-exposed tyrosine residue Tyr83 of pea plastocyanin is involved in both binding and electron transfer reactions with cytochrome *f*. *EMBO J, 10*(13), 4011–4016.

Hervas, M., Navarro, J. A., Diaz, A., Bottin, H., and De la Rosa, M. A. (1995). Laser-flash kinetic analysis of the fast electron transfer from plastocyanin and cytochrome c_6 to photosystem I. Experimental evidence on the evolution of the reaction mechanism. *Biochemistry, 34*(36), 11321–11326.

Hervas, M., Navarro, J. A., and De La Rosa, M. A. (2003). Electron transfer between membrane complexes and soluble proteins in photosynthesis. *Acc Chem Res, 36*(10), 798–805. doi:10.1021/ar020084b.

Hill, K. L. and Merchant, S. (1992). In vivo competition between plastocyanin and a copper-dependent regulator of the *Chlamydomonas reinhardtii* cytochrome *c*(6) gene. *Plant Physiol, 100*(1), 319–326.

Ho, K. K. and Krogmann, D. W. (1984). Electron donors to P700 in cyanobacteria and algae. An instance of unusual genetic variability. *Biochim Biophys Acta, 766*, 310–316.

Hope, A. B. (2000). Electron transfers amongst cytochrome *f*, plastocyanin and photosystem I: Kinetics and mechanisms. *Biochim Biophys Acta, 1456*(1), 5–26.

Howe, G. and Merchant, S. (1994). Role of heme in the biosynthesis of cytochrome c_6. *J Biol Chem, 269*(8), 5824–5832.

Kerfeld, C. A. and Krogmann, D. W. (1998). Photosynthetic cytochromes *c* in cyanobacteria, algae, and plants. *Annu Rev Plant Physiol Plant Mol Biol, 49*, 397–425. doi:10.1146/annurev.arplant.49.1.397.

Kerfeld, C. A., Anwar, H. P., Interrante, R., Merchant, S., and Yeates, T. O. (1995). The structure of chloroplast cytochrome c_6 at 1.9 Å resolution: Evidence for functional oligomerization. *J Mol Biol, 250*(5), 627–647. doi:10.1006/jmbi.1995.0404.

Kerfeld, C. A., Ho, K. K., and Krogmann, D. W. (1999). The cytochrome *c* of cyanobacteria. In G. Peschek, W. Löffelhardt, and G. Schmetterer (Eds.), *The prototrophic prokaryotes* (pp. 259–268). New York: Kluwer Academic/Plenum Publishers.

Ki, H. K. (2005). Cytochrome c_6 genes in cyanobacteria and higher plants. In M. Pessarakli (Ed.), *Handbook of photosynthesis* (Second Edition ed.). Boca Raton, FL: Taylor & Francis Ltd./CRC Press.

Koike, H. (1996). Respiration and photosynthetic electron transport system in cyanobacteria—Recent advances. *J Sci Ind Res, 55*, 564–582.

Lyons, T. W., Reinhard, C. T., and Planavsky, N. J. (2014). The rise of oxygen in Earth's early ocean and atmosphere. *Nature, 506*, 307–315.

Malakhov, M. P., Wada, H., Los, D. A., Semenenko, V. E., and Murata, N. (1994). A new type of cytochrome *c* from *Synechocystis* PCC 6803. *J Plant Physiol, 144*, 259–264.

Malakhov, M. P., Malakhova, O. A., and Murata, N. (1999). Balanced regulation of expression of the gene for cytochrome c_M and that of genes for plastocyanin and cytochrome c_6 in *Synechocystis*. *FEBS Lett, 444*(2–3), 281–284.

Manna, P. and Vermaas, W. (1997). Lumenal proteins involved in respiratory electron transport in the cyanobacterium *Synechocystis* sp. PCC 6803. *Plant Mol Biol, 35*(4), 407–416.

Medina, M., Diaz, A., Hervas, M., Navarro, J. A., Gomez-Moreno, C., De la Rosa, M. A., and Tollin, G. (1993). A comparative laser-flash absorption spectroscopy study of *Anabaena* PCC 7119 plastocyanin and cytochrome c_6 photooxidation by photosystem I particles. *Eur J Biochem, 213*(3), 1133–1138.

Merchant, S. and Bogorad, L. (1987). The Cu(II)-repressible plastidic cytochrome *c*. Cloning and sequence of a complementary DNA for the pre-apoprotein. *J Biol Chem, 262*(19), 9062–9067.

Merchant, S., Hill, K., Kim, J. H., Thompson, J., Zaitlin, D., and Bogorad, L. (1990). Isolation and characterization of a complementary DNA clone for an algal pre-apoplastocyanin. *J Biol Chem, 265*(21), 12372–12379.

Merchant, S., Hill, K., and Howe, G. (1991). Dynamic interplay between two copper-titrating components in the transcriptional regulation of cyt c_6. *EMBO J, 10*(6), 1383–1389.

Metzger, S. U., Pakrasi, H. B., and Whitmarsh, J. (1995). Characterization of a double deletion mutant that lacks cytochrome c_6 and cytochrome c_M in *Synechocystis* 6803. In P. Mathis (Ed.), *Photosynthesis: From light to biosphere* (pp. 823–826). Dordrecht: Kluwer Academic Publishers.

Misra, H. S., Khairnar, N. P., and Mahajan, S. K. (2003). An alternate photosynthetic electron donor system for PSI supports light dependent nitrogen fixation in a non-heterocystous cyanobacterium, *Plectonema boryanum. J Plant Physiol, 160*(1), 33–39. doi:10.1078/0176-1617-00846.

Molina-Heredia, F. P., Hervas, M., Navarro, J. A., and De la Rosa, M. A. (1998). Cloning and correct expression in *Escherichia coli* of the petE and petJ genes respectively encoding plastocyanin and cytochrome c_6 from the cyanobacterium *Anabaena* sp. PCC 7119. *Biochem Biophys Res Commun, 243*(1), 302–306. doi:10.1006/bbrc.1997.7953.

Molina-Heredia, F. P., Diaz-Quintana, A., Hervas, M., Navarro, J. A., and De La Rosa, M. A. (1999). Site-directed mutagenesis of cytochrome *c*(6) from *Anabaena* species PCC 7119. Identification of surface residues of the hemeprotein involved in photosystem I reduction. *J Biol Chem, 274*(47), 33565–33570.

Molina-Heredia, F. P., Hervas, M., Navarro, J. A., and De la Rosa, M. A. (2001). A single arginyl residue in plastocyanin and in cytochrome *c*(6) from the cyanobacterium *Anabaena* sp. PCC 7119 is required for efficient reduction of photosystem I. *J Biol Chem, 276*(1), 601–605. doi:10.1074/jbc.M007081200.

Molina-Heredia, F. P., Balme, A., Hervas, M., Navarro, J. A., and De la Rosa, M. A. (2002). A comparative structural and functional analysis of cytochrome c_M cytochrome c_6 and plastocyanin from the cyanobacterium *Synechocystis* sp. PCC 6803. *FEBS Lett, 517*(1–3), 50–54.

Molina-Heredia, F. P., Wastl, J., Navarro, J. A., Bendall, D. S., Hervas, M., Howe, C. J., and De La Rosa, M. A. (2003). Photosynthesis: A new function for an old cytochrome? *Nature, 424*(6944), 33–34. doi:10.1038/424033b.

Moore, G. R. and Pettigrew, G. W. (1990). *Cytochromes c: Evolutionary, structural and physicochemical aspects*. Berlin: Springer-Verlag.

Moser, D., Nicholls, P., Wastyn, M., and Peschek, G. (1991). Acidic cytochrome c_6 of unicellular cyanobacteria is an indispensable and kinetically competent electron donor to cytochrome oxidase in plasma and thylakoid membranes. *Biochem Int, 24*(4), 757–768.

Mullineaux, C. W. (2014). Co-existence of photosynthetic and respiratory activities in cyanobacterial thylakoid membranes. *Biochim Biophys Acta, 1837*(4), 503–511. doi:10.1016/j.bbabio.2013.11.017.

Nakamura, M., Yamagishi, M., Yoshizaki, F., and Sugimura, Y. (1992). The syntheses of plastocyanin and cytochrome *c*-553 are regulated by copper at the pre-translational level in a green alga, *Pediastrum boryanum. J Biochem, 111*(2), 219–224.

Navarro, J. A., Hervas, M., and De la Rosa, M. A. (1997). Co-evolution of cytochrome c_6 and plastocyanin, mobile proteins transferring electrons from cytochrome $b_6 f$ to photosystem I. *J Biol Inorg Chem, 2*, 11–22.

Nicholls, P., Obinger, C., Niederhauser, H., and Peschek, G. A. (1991). Cytochrome *c* and *c*-554 oxidation by membranous *Anacystis nidulans* cytochrome oxidase. *Biochem Soc Trans, 19*(3), 252S.

Nordling, M., Sigfridsson, K., Young, S., Lundberg, L. G., and Hansson, O. (1991). Flash-photolysis studies of the electron transfer from genetically modified spinach plastocyanin to photosystem I. *FEBS Lett, 291*(2), 327–330.

Obinger, C., Knepper, J. C., Zimmermann, U., and Peschek, G. A. (1990). Identification of a periplasmic *c*-type cytochrome as electron donor to the plasma membrane-bound cytochrome oxidase of the cyanobacterium *Nostoc* MAC. *Biochem Biophys Res Commun, 169*(2), 492–501.

Padan, E. (1979). Facultative anoxygenic photosynthesis in cyanobacteria. *Annu Rev Plant Physiol, 30*, 27–40.

Peschek, G. A. (1999). Photosynthesis and respiration in cyanobacteria. Bioenergetic significance and molecular interactions. In G. A. Peschek, W. Löffelhardt and G. Schmetterer (Eds.), *The phototrophic prokaryotes* (pp. 201–209). New York: Kluwer Academic/Plenum Publishers.

Pils, D., Wilken, C., Valladares, A., Flores, E., and Schmetterer, G. (2004). Respiratory terminal oxidases in the facultative chemoheterotrophic and dinitrogen fixing cyanobacterium *Anabaena variabilis* strain ATCC 29413: Characterization of the COX2 locus. *Biochim Biophys Acta, 1659*(1), 32–45. doi:10.1016/j.bbabio.2004.06.009.

Redinbo, M. R., Cascio, D., Choukair, M. K., Rice, D., Merchant, S., and Yeates, T. O. (1993). The 1.5-Å crystal structure of plastocyanin from the green alga *Chlamydomonas reinhardtii*. *Biochemistry, 32*(40), 10560–10567.

Redinbo, M. R., Yeates, T. O., and Merchant, S. (1994). Plastocyanin: Structural and functional analysis. *J Bioenerg Biomembr, 26*(1), 49–66.

Reyes-Sosa, F. M., Gil-Martinez, J., and Molina-Heredia, F. P. (2011). Cytochrome c_6-like protein as a putative donor of electrons to photosystem I in the cyanobacterium *Nostoc* sp. PCC 7119. *Photosynth Res, 110*(1), 61–72. doi:10.1007/s11120-011-9694-5.

Sandmann, G. (1985). Consequences of iron deficiency on photosynthetic and respiratory electron transport in blue-green algae. *Photosynth Res, 6*(3), 261–271. doi:10.1007/BF00049282.

Scherer, S., Almon, H., and Boger, P. (1988). Interaction of photosynthesis, respiration and nitrogen fixation in cyanobacteria. *Photosynth Res, 15*(2), 95–114. doi:10.1007/BF00035255.

Schmetterer, G. (1994). Cyanobacterial respiration. In D. A. Bryant (Ed.), *The molecular biology of cyanobacteria* (pp. 409–435). Dordrecht: Kluwer Academic Publishers.

Schmetterer, G. and Pils, D. (2004). Cyanobacterial respiration. In D. Zannoni (Ed.), *Respiration in archaea and bacteria* (pp. 261–278). Dordrecht: Springer.

Schnackenberg, J., Than, M. E., Mann, K., Wiegand, G., Huber, R., and Reuter, W. (1999). Amino acid sequence, crystallization and structure determination of reduced and oxidized cytochrome c_6 from the green alga *Scenedesmus obliquus*. *J Mol Biol, 290*(5), 1019–1030. doi:10.1006/jmbi.1999.2944.

Segal, M. G. and Sykes, A. G. (1978). Kinetic studies on 1:1 electron-transfer reactions involving glue copper proteins. *J Am Chem Soc, 100*, 251–262.

Serrano, A., Gimenez, P., Scherer, S., and Böger, P. (1990). Cellular localization of cytochrome $c553$ in the N_2-fixing cyanobacterium *Anabaena variabilis*. *Arch Microbiol, 154*, 614–618.

Sigfridsson, K., Young, S., and Hansson, O. (1996). Structural dynamics in the plastocyanin-photosystem 1 electron-transfer complex as revealed by mutant studies. *Biochemistry, 35*(4), 1249–1257. doi:10.1021/bi9520141.

Smeekens, S., van Binsbergen, J., and Weisbeek, P. (1985). The plant ferredoxin precursor: Nucleotide sequence of a full length cDNA clone. *Nucleic Acids Res, 13*(9), 3179–3194.

Stewart, W. D. (1980). Some aspects of structure and function in N_2-fixing cyanobacteria. *Annu Rev Microbiol, 34*, 497–536. doi:10.1146/annurev.mi.34.100180.002433.

Sykes, A. G. (1985). Structure and electron-transfer reactivity of the blue copper protein plastocyanin. *Chem Soc Rev, 14*, 283–315.

Torrado, A., Valladares, A., Puerto-Galán, L., Navarro, J. A., and Molina-Heredia, F. P. (2015). Heterocyst-forming filamentous cyanobacteria contains a new and specific cytochrome c_6 isoform. Manuscript submitted for publication.

Ullmann, G. M., Hauswald, M., Jensen, A., Kostic, N. M., and Knapp, E. W. (1997). Comparison of the physiologically equivalent proteins cytochrome c_6 and plastocyanin on the basis of their electrostatic potentials. Tryptophan 63 in cytochrome c_6 may be isofunctional with tyrosine 83 in plastocyanin. *Biochemistry, 36*(51), 16187–16196. doi:10.1021/bi971241v.

Valladares, A., Maldener, I., Muro-Pastor, A. M., Flores, E., and Herrero, A. (2007). Heterocyst development and diazotrophic metabolism in terminal respiratory oxidase mutants of the cyanobacterium *Anabaena* sp. strain PCC 7120. *J Bacteriol, 189*(12), 4425–4430. doi:10.1128/JB.00220-07.

Van der Plas, J., Bovy, A., Kruyt, F., de Vrieze, G., Dassen, E., Klein, B., and Weisbeek, P. (1989). The gene for the precursor of plastocyanin from the cyanobacterium *Anabaena* sp. PCC 7937: Isolation, sequence and regulation. *Mol Microbiol, 3*(3), 275–284.

Wastl, J., Bendall, D. S., and Howe, C. J. (2002). Higher plants contain a modified cytochrome $c(6)$. *Trends Plant Sci, 7*(6), 244–245.

Weigel, M., Varotto, C., Pesaresi, P., Finazzi, G., Rappaport, F., Salamini, F., and Leister, D. (2003). Plastocyanin is indispensable for photosynthetic electron flow in *Arabidopsis thaliana*. *J Biol Chem, 278*(33), 31286–31289. doi:10.1074/jbc.M302876200.

Wolk, C. P. (1982). Heterocysts. In B. A. Carr (Ed.), *The biology of cyanobacteria* (pp. 359–386). Oxford: Blackwell Scientific Publications.

Worrall, J. A., Schlarb-Ridley, B. G., Reda, T., Marcaida, M. J., Moorlen, R. J., Wastl, J., and Howe, C. J. (2007). Modulation of heme redox potential in the cytochrome c_6 family. *J Am Chem Soc, 129*(30), 9468–9475. doi:10.1021/ja072346g.

Worrall, J. A., Luisi, B. F., Schlarb-Ridley, B. G., Bendall, D. S., and Howe, C. J. (2008). Cytochrome c_{6A}: Discovery, structure and properties responsible for its low haem redox potential. *Biochem Soc Trans, 36*(Pt 6), 1175–1179. doi:10.1042/BST0361175.

Zatwarnicki, P., Barciszewski, J., Krzywda, S., Jaskolski, M., Kolesinski, P., and Szczepaniak, A. (2014). Cytochrome $c(6B)$ of *Synechococcus* sp. WH 8102—crystal structure and basic properties of novel $c(6)$-like family representative. *Biochem Biophys Res Commun, 443*(4), 1131–1135. doi:10.1016/j.bbrc.2013.10.167.

Zhang, L., Pakrasi, H. B., and Whitmarsh, J. (1994). Photoautotrophic growth of the cyanobacterium *Synechocystis* sp. PCC 6803 in the absence of cytochrome $c553$ and plastocyanin. *J Biol Chem, 269*(7), 5036–5042.

Section IV

Atmospheric and Environmental Factors Affecting Photosynthesis

14 Role of the LHCII Organization for the Sensitivity of the Photosynthetic Apparatus to Temperature and High Light Intensity

E.L. Apostolova and A.G. Dobrikova

CONTENTS

14.1 Introduction 243
14.2 Organization and Functions of the LHCII–PSII Supercomplex 243
    14.2.1 Organization of the PSII Core Complex and OEC...244
    14.2.2 Organization and Functions of the LHCII Complex. Role of the Lipids and Xanthophylls 244
14.3 Photosynthetic Response to Temperature Stress 245
    14.3.1 Structural and Functional Changes under High Temperature..245
        14.3.1.1 Structural Changes in Thylakoid Membranes 245
        14.3.1.2 Changes in the LHCII–PSII Supercomplex...246
        14.3.1.3 Influence of Lipid Composition on Temperature Sensitivity 247
        14.3.1.4 Role of the LHCII Organization for Temperature Sensitivity of the Photosynthetic Apparatus..........247
    14.3.2 Structural and Functional Changes in the Photosynthetic Apparatus under Low Temperature 248
        14.3.2.1 Structural Changes...248
        14.3.2.2 Functional Changes..248
14.4 Photosynthetic Response to High Light Intensity 249
    14.4.1 Structural and Functional Changes in the Photosynthetic Apparatus under High Light Intensity 249
    14.4.2 Photosynthetic Response to High Light Intensity Depending on the Temperature 250
14.5 Conclusions..251
References..251

14.1 INTRODUCTION

Temperature and high light intensity are among the main environmental factors that limit growth and development of the photosynthetic organisms, and their survival depends on the ability to adapt and respond to environmental changes. It is well known that the photosynthetic apparatus is very sensitive to abiotic stress factors. Temperatures above and below normal physiological range, as well as high light intensity, provoke oxidative stress causing primary damage in the photosynthetic apparatus and an inhibition of the photosynthesis (Wise and Naylor 1987; Hodgson and Raison 1991; Nie et al. 1992; Prochazkova and Wilhelmova 2010). The changes in photosynthetic membranes depend on growth conditions and plant species. A multilevel of adaptation strategies exists to help photosynthetic organisms to cope with altered environmental conditions, ensuring high levels of survival (Ruban 2009). A central role in the adaptation of plants and green algae plays the chlorophyll *a/b* binding light-harvesting complex of photosystem II (LHCII) by regulating the efficiency of light harvesting in photosynthesis (for a review, see Ruban et al. 2012). The primary target of action of stress factors in plants and algae is the photosystem II (PSII) complex since one of the most sensitive sites is the oxygen evolving complex (OEC) (Hakala et al. 2005; Murata et al. 2007; Tyystjärvi 2008; Mathur et al. 2014). This chapter is focused on the role of different amounts and organization of LHCII complex on the sensitivity of the photosynthetic apparatus to temperature and high light intensity.

14.2 ORGANIZATION AND FUNCTIONS OF THE LHCII–PSII SUPERCOMPLEX

PSII is a large multisubunit pigment–protein complex, which acts as a light-driven water–plastoquinone oxidoreductase in thylakoid membranes. It uses sunlight to split water into molecular oxygen, protons, and electrons (Andersson and Styring 1991; Dekker and Boekema 2005; Kern and Renger 2007). Using photons in the visible spectrum, PSII carries out: (1) a charge separation in the reaction center leading to the formation of the radical ion pair $P_{680}^{+\bullet} Q_A^{-\bullet}$; (2) reduction of plastoquinone to plastoquinol at the Q_B site via a two-step reaction sequence with $Q_A^{-\bullet}$ as a reductant at the acceptor side; and (3) O_2 evolution at the manganese-containing catalytic cluster of OEC at the donor side (Kern and Renger 2007). The linear electron transfer reaction in PSII catalyzes the light-induced water plastoquinone oxidation–reduction pass with high quantum deficiency. In addition to the main linear electron transport, cyclic electron transport is proposed, which may protect PSII against photoinhibition by preventing over-reduction of Q_A and Q_B on the acceptor side and accumulation of long-lived P_{680}^{+} on the donor side (Minagawa and Takahashi 2004). A variable number of the peripheral antenna proteins can associate with PSII core complex to form the so-called LHCII–PSII supercomplex (Minagawa and Takahashi 2004; Dekker and Boekema 2005; Nield and Barber 2006).

14.2.1 ORGANIZATION OF THE PSII CORE COMPLEX AND OEC

The PSII core complex is composed of a D1/D2 heterodimer and a few intrinsic low molecular weight polypeptides, and is surrounded by the core antenna chlorophyll *a* binding proteins, CP43 and CP47, which play an important role in excitation energy transfer from the peripheral antenna to the PSII reaction center in higher plants and green algae (Dekker and Boekema 2005). Furthermore, they also play an important role in maintaining the structural integrity and oxygen evolving capacity of PSII. The core complex is organized as a dimer in the stacked, appressed regions of the thylakoid membranes (Dekker and van Grondelle 2000; Broess et al. 2008). The OEC contains three extrinsic proteins of 33, 23, and 17 kDa associated with the luminal surface of PSII core complex (Ferreira et al. 2004; Nield and Barber 2006). The water oxidation on the PSII donor side is catalyzed by a cluster of four Mn ions of OEC (Andersson and Styring 1991; Nugent et al. 2001; Nield and Barber 2006), and the liberated electrons are transferred to the reaction center chlorophyll, P_{680}, via the immediate redox-active tyrosine residue, Y_Z (Tyr-161), located on the D1 protein.

14.2.2 ORGANIZATION AND FUNCTIONS OF THE LHCII COMPLEX. ROLE OF THE LIPIDS AND XANTHOPHYLLS

In higher plants and algae, the LHCII is composed of six polypeptides (encoded by the *Lhcb 1–6* genes) (Jansson 1994; Jansson et al. 1997). Two types of LHCII associated with PSII can be distinguished: the so-called major that bind 70% of PSII chlorophyll and minor LHCII (see Kouřil et al. 2012 and references therein). The major (peripheral) LHCII (*Lhcb 1–3*) is assembled as multiple trimers (oligomers) in thylakoid membranes, while the minor LHCII usually occurs in monomeric form and includes CP29, CP26, and CP24 (*Lhcb 4–6*) proteins (Jansson 1994; Caffarri et al. 2004; Nield and Barber 2006; van Amerongen and Croce 2008; Kouřil et al. 2012). There are a few proposed models for the LHCII protein

arrangement in PSII supercomplexes, which show differences in the stoichiometry of the proteins (Jansson 1994; Dekker and Boekema 2005; Nield and Barber 2006; Broess et al. 2008).

PSII supercomplexes in granum usually contain two to four copies of trimeric LHCII complexes (Broess et al. 2008; Kouřil et al. 2012). The role of LHCII has been recognized in stabilizing the membrane structure and in particular in stacking of adjacent membranes in granum for the segregation of PSII and LHCII into stacked grana regions from PSI and ATPase in stroma lamellae (Chow et al. 1991; Garab and Mustardy 1999; Stoitchkova et al. 2006). Moreover, peripheral LHCII also plays an important role in regulation of light absorption and energy transfer to the reaction centers at limiting sunlight intensity and dissipates the excess energy at high intensity for photoprotection (Horton et al. 1991, 2005; Krüger et al. 2013; Cui et al. 2014); therefore any damage of this complex influences the function of the photosynthetic apparatus. It has also been proposed the important role of PsbS protein in the remodeling of LHCII–PSII supercomplexes under elevated light conditions and the regulation of the energy balance in the thylakoid membrane (Dong et al. 2015). Studies with *Arabidopsis* mutants showed that lack of minor LHCII polypeptides (CP29, CP26, and CP24) *in vivo* causes impairment of the PSII organization and negatively affects electron transport rates and photoprotection capacity (Yakushevska et al. 2003; Dall'Osto et al. 2014).

Wentworth et al. (2004) studied the biological significance of LHCII trimerization and found that, when compared to monomers, LHCII trimers have increased thermal stability and a decreased structural flexibility, and the pigment configuration and protein conformation in trimers are adapted for efficient light harvesting and enhanced protein stability. Furthermore, trimers regulate the energy dissipation by modulating the development of the quenched state of the complex (Wentworth et al. 2004). The dynamic properties of the LHCII allow excess energy to be dissipated as heat and to regulate the energy redistribution between the two photosystems (Horton et al. 1996, 2008).

Xanthophylls are an important factor for controlling and stabilizing the conformation of LHCII. It has been shown that phototransformation of violaxanthin in xanthophyll cycle correlates with structural changes in LHCII (Horton et al. 1996). It has also been found that structural differences in the xanthophylls affect the tertiary and quaternary structures of LHCII (reviewed in Ruban et al. 2012). The more hydrophobic xanthophylls (such as zeaxanthin) were found to favor aggregated states of LHCII, while the more polar xanthophylls (such as violaxanthin) favored fluorescent LHCII conformations. On the other hand, it has been observed that operation of the xanthophyll cycle is associated with a low level of lipid peroxidation (Sarry et al. 1994). Tardy and Havaux (1997), based on the investigations on the relationship among xanthophyll cycle, lipid phase, and stability of the membrane under stress, proposed a hypothesis that this cycle is the efficient stabilizing protector under potentially harmful light and temperature conditions. It has also been suggested that the rapid adjustment of the PSII to heat may involve a modification of the interaction between violaxanthin and LHCII (Havaux and Tardy 1996). The carotenoids of the xanthophyll cycle may be also involved in the protection of PSII complex against heat stress (Havaux et al. 1996).

The lipid environment of the LHCII complexes plays a significant role in determining the structural flexibility of the isolated macroaggregates of LHCII, and those lamellar aggregates are capable of incorporating substantial amounts of different thylakoid lipids (Simidjiev et al. 1998). Monogalactosyldiacylglycerol (MGDG) may facilitate the optimal packing of large intrinsic proteins within the bilayer structure, digalactosyldiacylglycerol (DGDG) is essential for the formation of 3D and large 2D crystals, and phosphatidylglycerol (PG) is tightly bound to the polypeptide chain and can participate directly in the formation of LHCII trimers and in subunit–subunit interactions (Nußberger et al. 1993; Simidjiev et al. 1998). Simidjiev et al. (2000) also demonstrated that the addition of the nonbilayer lipid MGDG induces the transformation of isolated, disordered macroaggregates of LHCII into stacked lamellar aggregates with a long-range chiral order of the complexes. Recently, it has been concluded that the amount of the main thylakoid lipid MGDG associated to the LHCII complexes correlates with the amount of LHCII-associated violaxanthin (Schaller et al. 2010, 2011). On the other hand, the formation of oligomeric LHCII is found to be accompanied by an increase in *trans*-16:1 content of PG (Gray et al. 2005).

In a DGDG-deficient *Arabidopsis* mutant, a decreased ratio of the PSII to PSI has been established, which is a result of a decrease in the amount of the PSII and a concomitant increase in the amount of the PSI (Härtel et al. 1997). This decrease in the PSII amount is compensated with an increase in the amount of peripheral (major) LHCII relative to the inner antenna complex of PSII. Recently, it has been reported that DGDG plays important roles in the overall organization of thylakoid membranes especially at elevated temperatures (Krumova et al. 2010). The authors also showed that DGDG deficiency prevents the formation of the chirally organized macrodomains containing the major LHCII and leads to decreased thermal stability. Moreover, the thermal stability and structure of the chloroplasts are also influenced by unsaturation of the membrane lipids. It has been reported that the decrease in the degree of lipid unsaturation in *Arabidopsis* mutants is accompanied by a decreased amount of appressed membranes, chlorophyll, protein, and LHCII content, and directly affects the thermal stability of photosynthetic membranes (Hugly et al. 1989; Tsvetkova et al. 1994, 1995).

14.3 PHOTOSYNTHETIC RESPONSE TO TEMPERATURE STRESS

Exposure of the photosynthetic organisms to temperature above and below the normal environmental conditions results in reversible and irreversible modification of the thylakoid membrane structure and composition, and strongly influences photosynthetic efficiency (Carpentier 1999; Mathur et al. 2014). The plants from different geographic zones could

be adapted to the cool or hot environmental temperatures. During the acclimation to environmental factors, the LHCII content in the photosynthetic membranes often changes, which affects the ultrastructure of the grana and functions of the photosynthetic apparatus (Anderson 1999). The extent of the damage of the photosynthetic apparatus caused by exposure to temperature stress differs depending on the plant species, the state of growth, the temperature, the rate of temperature change, as well as the duration of the temperature treatment (Gounaris et al. 1983; Havaux and Tardy 1997; Kóta et al. 2002; Sung et al. 2003). A number of reports also indicated that plants exposed to high-temperature stress show reduced chlorophyll biosynthesis (Ristic et al. 2007; Efeoglu and Terzioglu 2009).

14.3.1 STRUCTURAL AND FUNCTIONAL CHANGES UNDER HIGH TEMPERATURE

High-temperature stress is one of the most prominent abiotic stresses affecting plants in view of global warming. Photosynthesis is highly sensitive to high-temperature stress and is often inhibited before other cell functions are impaired. The responses to high-temperature stress are as follows: destacking of thylakoid membranes, alteration in membrane lipid composition, production of secondary metabolites, increased amount of antioxidant enzymes, damage of D1 protein, dissociation of the extrinsic proteins of OEC, disorganization of Mn clusters, enhanced cyclic electron flow around PSI, reactive oxygen species production, inactivation of Rubisco activity, etc. (for a review, see Mathur et al. 2014). Threshold temperatures for structural and functional changes in thylakoid membranes strongly depend on growth temperature of photosynthetic organisms.

14.3.1.1 Structural Changes in Thylakoid Membranes

Exposure of higher plant chloroplasts to temperatures above 35°C led to major changes in their structural and functional properties, as O_2 evolution, CO_2 fixation, photophosphorylation, and PSII-mediated electron transport are all inhibited while PSI-mediated electron transport is stimulated. At the same time, thylakoid membranes become destacked (Gounaris et al. 1983, 1984; Thomas et al. 1984; Semenova 2004). The temperature treatment at 35–40°C causes complete destacking of thylakoids accompanied by disruption of chlorophyll–protein complexes of both photosystems, the formation of cylindrical inverted micelles, and vesiculation of thylakoid membranes (Gounaris et al. 1983; Semenova 2004). At temperatures around 50°C and above, packing of thylakoids in grana is changed and formed strands of a pseudograna with increased membrane stacking, since the membrane–membrane interactions in these pseudogranas are of another origin than in the normal grana (Semenova 2004).

It is well known that LHCII takes part in temperature-induced structural changes (Takeuchi and Thornber 1994; Horton et al. 1996; Cseh et al. 2000; Allen and Forsberg 2001; Mohanty et al. 2002). The changes in the structural organization of thylakoid membranes above 40°C are accompanied by dissociation of LHCII from PSII core complex and its diffusion from the granal to the stromal regions (Sundby et al. 1986; Yamane et al. 1997). Pastenes and Horton (1996) have shown that high temperatures, which lead to reversible inhibition of photosynthesis, influence the redistribution of the excitation energy between the two photosystems and the ratio between two subpopulations of PSII – PSIIα and PSIIβ. In addition, Mathur et al. (2011b) reported that subsequent increase in temperature from 25°C to 45°C in wheat leaves led to a decrease in the proportion of PSIIα centers (located in grana), while PSIIβ and PSIIγ centers in stroma thylakoids showed an increase. Other results (Mohanty et al. 2002) have shown that the temperature treatment *in vivo* (around 42°C) influences the pigment–protein and pigment–pigment interaction in LHCII and leads to the increase in the migration of LHCII from stacked to unstacked thylakoid membranes.

Temperature-induced structural rearrangements in isolated thylakoid membranes as well as the temperature stability of membrane complexes were revealed by differential scanning calorimetry and circular dichroism spectroscopy (Cseh et al. 2000; Dobrikova et al. 2003). Several thermo-induced structural transitions are occurring in thylakoid membranes: the first begins with the unstacking of the membranes followed closely by a lateral disassembly of the LHCII macrodomains in the granum, after which the trimers of LHCII are transformed into monomers during heating around 60°C (Dobrikova et al. 2003). The temperature of the trimer-to-monomer transition for isolated LHCII trimers was found to be 55°C (Garab et al. 2002). Moreover, investigations with circular dichroism spectroscopy on intact thylakoid membranes showed that the chiral macrodomains in granal thylakoid membranes (i.e., aggregates of the main LHCII) displayed high stability below 40°C, but they were gradually disassembled between 50°C and 60°C, and that the thermal stability of chiral macrodomains strongly depends on the ionic strength of the reaction media (Cseh et al. 2000).

14.3.1.2 Changes in the LHCII–PSII Supercomplex

The most sensitive component to high temperature in photosynthetic membranes is the LHCII–PSII supercomplex (Berry and Björkman 1980; Mathur et al. 2014). It has been reported that the PSII reaction center, the oxygen-evolving complex, and the light-harvesting complexes are initially disrupted by high temperature (Tiwari et al. 2007). Cytochrome *b*559 and plastoquinone (PQ) are also affected, while PSI exhibits higher stability than PSII (Yamane et al. 1995). Zhang and Sharkey (2009) proposed that the heat alters the redox balance away from PSII toward PSI and that regulation of the cyclic electron transport around PSI protects the photosynthetic apparatus from heat-induced damage.

Temperature deactivation of PSII includes functional separation of LHCII from PSII core complex (Schreiber and Berry 1977; Gounaris et al. 1984; Sundby et al. 1986; Vani et al. 2001; Mohanty et al. 2002) and damage of the donor (Nash et al. 1985; Enami et al. 1994) and the acceptor side of PSII (Bukov et al. 1990; Cao and Govindjee 1990; Yamane et al. 1998). Havaux and Strasser (1990) showed that the

photosynthetic oxygen evolution was irreversibly inhibited after short exposure to heat (42°C) of pea leaves. In addition, Shi et al. (1998) reported that the inactivation of the oxygen evolution is not associated with any major protein secondary structural changes. Temperatures up to 42°C have no inhibitory effect on the acceptor side of PSII as the electron transfer from primary (Q_A) to the secondary (Q_B) electron acceptor of PSII and the fraction of Q_B-non-reducing PSII centers remain unchanged (Havaux 2003). Inhibition of the electron flow from Q_A to Q_B was registered after heating at around 50°C (Yamane et al. 1998). It has been shown that high-temperature treatment of thylakoid membranes in the dark leads also to the degradation of D1 protein, as maximal degradation is occurring at 45°C concomitant with the release of 23 kDa fragment (Singh and Singhal 1999).

It has been reported that the antenna size, maximal fluorescence (Fm), and primary photochemistry of PSII (Fv/Fm) decreased, while the dissipation in the form of heat increased in response to high-temperature stress (45°C) (Mathur et al. 2011a). Recent investigations with wheat cultivars revealed the influence of the heat treatment (40°C) on the primary photochemistry of PSII and quantum yield of PSII at different growth stages (Haque et al. 2014). Our recent investigation on isolated thylakoid membranes from pea showed that the temperatures higher than 45°C lead to strong inhibition of the photochemistry of PSII in light- and dark-adapted state, which is a result of changes in the ratio of the photochemical and nonphotochemical processes (Dankov et al. 2015).

Vani et al. (2001) showed that *in vivo* exposure of rice seedling to elevated temperature (40°C) leads to inhibition of their photochemical functions as a result of disorganization of membrane structure and of some thylakoid membrane complexes without any changes in protein or pigment content of thylakoid membranes. The authors suggested that the loss in quantum yield accompanied by a decrease in the number of PSII active reaction centers could be due to possible uncoupling of light-harvesting antenna from the reaction center brought by heat treatment and/or decrease in the core antenna proteins, CP43 and CP47.

The high temperature alters the energy distribution in LHCII–PSII supercomplex and energy redistribution between two photosystems (Joshi et al. 1995; Apostolova and Dobrikova 2010). The increase in the energy transfer to PSI strongly depends on the LHCII organization. It has been shown that the increase is smaller in the membranes with a smaller degree of LHCII oligomerization (Apostolova and Dobrikova 2010). Reduction in the PQ pool is observed as a consequence of high-temperature stress (Pshybytko et al. 2008), which induces the increase in the photosynthetic capacity of PSI. Moreover, reduced PQ molecules are known to be efficient scavengers of the singlet oxygen that is generated in PSII.

The OEC is the most sensitive component of the LHCII–PSII supercomplex to heat stress and determines the heat sensitivity of the photosynthetic apparatus (Nishiyama et al. 1993). The changes in OEC are connected with the release of manganese atoms, which catalyze the O_2 evolution, from the

PSII core causing loss of the oxygen-evolving activity (Nash et al. 1985). Enami et al. (1994) suggested that the release of the Mn-stabilizing 33 kDa extrinsic protein of the OEC occurs first followed by liberation of manganese atoms and inhibition of oxygen evolution. It was supposed that 33 kDa protein releases from PSII core complex during heat treatment up to 40°C, becomes loosely bound, and rebinds to the complex when the chloroplast membranes were cooled down to 25°C (Yamane et al. 1998). At temperatures higher than 50°C, the 33 kDa protein is denatured as the changes are irreversible. Electron paramagnetic resonance study of manganese (Mn) release from thylakoids and PSII-enriched membranes under heat stress revealed that little loss of Mn ions up to 42°C leads to a significant decrease in the PSII activity (Tiwari et al. 2007).

It is well known that chloride ions are very important for normal function of PSII, and their loss causes inactivation of the OEC (Homann 1988). Chloride ions not only take part in the process of the oxygen evolution but also preserve PSII from temperature-induced damage (Coleman et al. 1988). It has been suggested that there are two sites of chloride binding, high-affinity (tightly bound) and low-affinity (loosely bound) binding sites (Lindberg and Andersson 1996; Olesen and Andersson 2003). The high-affinity site of chloride is affected earlier (around 37°C), while the low-affinity site is affected at higher temperature (42.5°C) for thylakoid membranes (Tiwari et al. 2007). Barra et al. (2005) showed that, at 47°C, releases of the 18 kDa protein is the main cause behind the loss of essential Ca ions from the Mn_4Ca complex of OEC. A slight change in the conformation of the bound 18 kDa protein may also facilitate the access of Y_D from the bulk and may favor a rapid reduction of it in the dark at comparatively lower temperatures (35–45°C) (Tiwari et al. 2008).

14.3.1.3 Influence of Lipid Composition on Temperature Sensitivity

Several studies revealed that the temperature denaturation of PSII is accompanied with big changes in the lipid bilayer of thylakoid membranes (Berry and Björkman 1980; Raison et al. 1982; Gounaris et al. 1983, 1984). Increase in the temperature leads to increase in the molecule movement (fluidity) of the membrane lipids (Raison et al. 1982) and subsequent formation of nonbilayer lipid structure (Gounaris et al. 1983). It has been shown that at temperatures above 45–55°C, phase separation of the nonbilayer-forming lipids from the bulk phase occurs, which results in the breakdown and vesiculation of the thylakoids (Gounaris et al. 1983, 1984). The main consequences of these lipid changes are destabilization of lipid–protein interactions and alteration of the structure and functions of PSII. It has also been proposed that nonbilayer-forming lipids are involved in packing of the LHCII complex, which correlates with the reported conversion of the major LHCII from trimeric to monomeric forms during heat stress (Takeuchi and Thornber 1994).

Many authors suggested that lipid composition plays an important role for temperature stability of the thylakoid membranes (Raison et al. 1982; Yordanov 1992; Wada et al. 1994;

Tsvetkova et al. 1995). Aminaka et al. (2006) suggested that one of the major factors that change the thermostability of PSII is the fluidity of the membranes. Mutants deficient in fatty acid desaturation have correspondingly reduced levels of polyunsaturated lipids, which resulted in enhanced tolerance to high temperature (Hugly et al. 1989; Gombos et al. 1994a; Wada et al. 1994; Tsvetkova et al. 1995). On the other hand, the changes in lipid composition affect the chloroplast ultrastructure, the membrane proteins, the chlorophyll content, as well as photosynthetic functions (Hugly et al. 1989; Gombos et al. 2002; Hagio et al. 2002; Sakurai et al. 2007; Domonkos et al. 2008; Dankov et al. 2009b, 2011). Analysis of the chlorophyll–protein complexes in a mutant of *Arabidopsis* revealed that reduced levels of polyunsaturated lipids result in a small decrease in the amount of light-harvesting complex (Hugly et al. 1989). The authors suggested that changes of the LHCII in this mutant also influence the thermostability of the LHCII–PSII supercomplex.

14.3.1.4 Role of the LHCII Organization for Temperature Sensitivity of the Photosynthetic Apparatus

Developing thylakoid membranes during greening, mutants with different amount and organization of LHCII, as well as manipulation of the LHCII size with exogenous polyamines are used by several authors as model systems for studying the role of antenna size of PSII in stress tolerance of the photosynthetic apparatus (Havaux and Tardy 1997; Kóta et al. 2002; Sfichi et al. 2004; Dankov et al. 2011). In our previous study, we have compared the effects of high temperature on the pea mutants with different organization of the LHCII (Apostolova and Dobrikova 2010). Data showed that the temperature-induced rearrangement of pigment–protein complexes in the thylakoid membranes with smaller PSII antenna size is stronger than that in the membranes with bigger antenna size. The changes in the membrane organization under abiotic stress in turn affect the oxygen evolution as the degree of inhibition depends on the organization of the LHCII–PSII supercomplex. Damage effect of heat stress on the oxygen evolution parameters decreases with an increase in the amount and oligomerization of LHCII (Apostolova and Dobrikova 2010).

The development of thylakoid membranes of greening plant seedlings is connected with the monotonic accumulation of chlorophylls and light-harvesting chlorophyll *a/b* binding polypeptides associated with PSII apoproteins (Thorne and Boardman 1971; Hoober and Eggink 1999; Kóta et al. 2002). LHCII is known to be in the monomeric form in the early greening phases (Dreyfuss and Thornber 1994) and associates into trimers or larger aggregates during development (Kóta et al. 2002). It has been supposed that formation of higher-order oligomeric LHCII complexes is related to the grana stacking formation in chloroplasts. Some authors reported an increase in the number of thylakoids per granum during the greening period (Duysen et al. 1980; Gal et al. 1997). The appearance of redox-controlled LHCII phosphorylation induced by light was detected only after 24–36 h of greening, which suggests correlating with the segregation of the thylakoid protein components and the development of grana stacks possessing at least four thylakoids per stack (Duysen et al. 1980; Gal et al. 1997). Investigations of developing thylakoid membranes using Fourier transform infrared and spin label electron paramagnetic resonance spectroscopic techniques suggested that the greening is accompanied with the reorganization of the membrane protein assemblies and alteration on the protein–lipid interactions (Kóta et al. 2002). The thermal stability of the oxygen evolution of developing thylakoid membranes increases during greening (Apostolova and Dobrikova 2010). It has also been shown that 24 h greening barley leaves have similar thermal sensitivity as barley mutant *Chlorina f2* (with the absence of the oligomeric forms of LHCII) but smaller than 36 and 48 h greening plants. Therefore, it can be concluded that a larger amount of the LHCII stabilizes the function of the OEC during heat stress.

Environmental temperature has a significant impact on the lipid and the carotenoid composition of the thylakoid membranes (Haldimann 1996; Kłodawska et al. 2012). Carotenoids are integral components of the photosynthetic apparatus and play a crucial role in the photoprotection function and in the stabilization of the LHCII trimers as well as in the assembly of LHCII monomers (Formaggio et al. 2001; Standfuss et al. 2005). Investigation on the temperature sensitivity of the thylakoid membranes from pea plants treated with fluridone (inhibitor of the carotenoid biosynthesis) revealed that the heat-induced damage of the PSIIα centers as well as on the donor side of all PSII centers (PSIIα and PSIIβ) is accentuated by fluridone (Dankov et al. 2015). Considering that the fluridone treatment influences LHCII organization (Dankov et al. 2009a), it can be supposed that one of the reasons for the increased heat sensitivity in fluridone-treated plants is a result from the modification of LHCII.

14.3.2 STRUCTURAL AND FUNCTIONAL CHANGES IN THE PHOTOSYNTHETIC APPARATUS UNDER LOW TEMPERATURE

It is well known that the structure and the dynamics of the thylakoid membranes directly affect the functioning of the photosynthetic apparatus, especially at low temperatures, where the lipid-related physical parameters (e.g., fluidity) change steeply. Photosynthetic rate is affected by both chilling and freezing regimes, although the impact is more intense in the latter (Ploschuk et al. 2014).

14.3.2.1 Structural Changes

Huner et al. (2011) demonstrated that chloroplast ultrastructure and pigment content are altered upon growth and development at cold-hardening temperatures. Changes in the chloroplast ultrastructure by growth at low temperature are indicated by a higher frequency of small grana stacks, but no significant differences were observed in the number of grana per chloroplast. Chlorophyll per plastid increased upon growth at cold temperatures, as well as β-carotene and the xanthophylls (Huner et al. 2011). Furthermore, Elfman et al. (2011) reported more oligomers of the LHCII in samples prepared from thylakoid membranes of warm-grown than those

from membranes of cold-grown rye plants. They concluded that the pigment–protein interactions in LHCII–PSII supercomplexes are altered upon growth and development at cold temperatures. Earlier observations of Huner et al. (1987) also demonstrated that low developmental temperatures modulate the LHCII organization. The authors showed that the oligomeric forms of LHCII predominate in rye thylakoid membranes developed in warm temperature (20°C), while in cold-hardening plants, monomeric forms predominate. In addition, DSC measurements revealed decreased thermostability of the membrane protein complexes of the thylakoid membranes isolated from cold-hardening plants (Huner et al. 1987). Ottander et al. (1995) showed that the content of D1 protein and LHCII declines under low-temperature stress. The authors also revealed that the recovery of the photosynthesis under favorable conditions is completed with an increase in the epoxidation state of xanthophylls and light-harvesting polypeptides, followed by recovery of D1 content.

PG, which is very important for oligomerization of the complexes of the photosynthetic apparatus (Domonkos et al. 2008), seems to play an exceptional role in the chilling resistance of plants. It is known that the degree of unsaturation of the PG fatty acyl chains exhibits a strong correlation with the cold resistance of plants (Murata and Yamaya 1984; Szalontai et al. 2003). This correlation is very interesting since PG makes up only 5–10% of the total plant lipid content. It has been suggested (Murata 1983; Murata and Yamaya 1984; Murata and Nishida 1990) that molecular species of the chloroplast PG containing a combination of saturated fatty acid (16:0 and 18:0) or 16:1-*trans* at both the sn-1 and sn-2 positions of the glycerol backbone (high-melting-point PG molecular species) confers chilling sensitivity on plants. Murata and Yamaya (1984) proposed that the presence of high-melting-point PG molecular species in the chloroplast membranes induces a phase transition at chilling temperatures and that the phase separation within the membranes is a direct cause of chilling injuries. Roughan (1985) investigated the fatty acid composition of PG in a number of plant species and found, in broad terms, a correlation between the chilling sensitivity and the percentage of saturated transmono-unsaturated PG molecular species. Therefore, the *cis*-unsaturated molecular species of PG in chloroplasts are crucial components in low-temperature tolerance (Ariizumi et al. 2002).

Membrane stability (i.e., conformation of membrane proteins and the packing/phase properties of membrane lipids) is primarily affected by low-temperature stress. Szalontai et al. (2012) noted that temperatures lower than the growth temperature lead to the increase in the membrane fluidity.

14.3.2.2 Functional Changes

The changes in the photosynthetic apparatus under chilling regime influence the fluorescence parameters measured (coefficient of the photochemical quenching, the quantum efficiency of PSII in the light adapted state, electron transport rate). Only lethal freezing regimes produced stable depletions in the water status, the chlorophyll content, and the primary photochemistry of PSII (Ploschuk et al. 2014). The sensitivity of the photosynthesis depends on both the suppression of photosynthesis and the slowed rate of replacement of damaged proteins of the photosynthetic apparatus.

Investigations of the influence of freezing on the isolated pea thylakoid membranes with different organization of the LHCII–PSII supercomplex revealed low-temperature-induced changes in the acceptor side of PSII, which depend on the oligomerization of LHCII. The increased oligomerization of LHCII preserves the capacity for oxidation of Q_B from artificial electron acceptors and plastoquinone during freezing (Dankov et al. 2009c). On the other hand, the determination of the kinetic parameters of the oxygen evolution under continuous illumination revealed that freezing-induced changes in thylakoid membranes are strongly related to the PSIIα centers, while the PSIIβ centers are not significantly affected.

In conclusion, based on existing literature data, it can be assumed that the amount of oligomeric forms of the LHCII and cis-unsaturated molecular species of PG in chloroplasts are crucial components in low-temperature resistance of the photosynthetic organisms.

14.4 PHOTOSYNTHETIC RESPONSE TO HIGH LIGHT INTENSITY

Strong light intensities cause a decrease in photosynthetic efficiency, growth, and productivity, a phenomenon referred to as photoinhibition. The primary site of photodamage is the multiprotein complex of PSII (Aro et al. 1993; Murata et al. 2007). The susceptibility of plants to photoinhibition depends on plant species and growth light condition. The term *photoinhibition of PSII complex* refers to a multistep process initiated and sustained by light excitation resulting in the inactivation, disassembly, and proteolysis of PSII components. The extent of the decline represents the imbalance between the quicker rate of damage and the slower rate of repair processes, i.e., synthesis of the proteins to form an active PSII (Murata et al. 2007; Mulo et al. 2008; Takahashi and Murata 2008). Strong visible light has two effects: a direct effect on photodamage and an inhibitory effect on the repair of PSII via the production of reactive oxygen species (Murata et al. 2007).

Recent studies showed that light absorption by the manganese cluster in the oxygen-evolving complex of PSII causes primary photodamage, whereas excess light absorbed by light-harvesting complexes acts to cause inhibition of the PSII through the generation of reactive oxygen species (Tyystjärvi 2008; Takahashi and Badger 2011). A number of studies (Keren and Krieger-Liszkay 2011; Ohad et al. 2011) proposed two hypotheses for the mechanisms that lead to photoinactivation of PSII: (1) The *excess-energy hypothesis* states that excess energy absorbed by chlorophylls, neither utilized in photosynthesis nor dissipated harmlessly in nonphotochemical quenching, leads to PSII photoinactivation (Ögren et al. 1984; Demmig and Björkman 1987; Osmond 1994); and (2) the *Mn hypothesis* (also termed the *two-step hypothesis*) states that light absorption by the Mn cluster in PSII is the primary effect that leads to dissociation of Mn, followed by damage to the reaction center by light absorption by chlorophylls (Hakala et

al. 2005; Ohnishi et al. 2005). Oguchi et al. (2011) proposed that both mechanisms operate in the leaf with the relative contribution depending on growth conditions or plant species.

14.4.1 STRUCTURAL AND FUNCTIONAL CHANGES IN THE PHOTOSYNTHETIC APPARATUS UNDER HIGH LIGHT INTENSITY

Cseh et al. (2000) showed that LHCII-containing macrodomains are susceptible to prolonged high light illumination, which induces similar irreversible disassembly to that observed at high temperatures. It is well known that light and heat stresses cause a significant change in the structure of the thylakoids and the PSII complexes. Especially prominent are those changes detected in the thylakoids from higher plants at the electron microscopic level under stress conditions, i.e., unstacking of the thylakoids, which may be either a reversible or an irreversible process, depending on the stresses imposed on the thylakoids (Yamamoto et al. 2008). The exposure of spinach thylakoids to light stress (1000 µE m^{-2} s^{-1} at 25°C) leads to significant unstacking of thylakoids, which is not recovered to the original stacking level during the subsequent incubation at 25°C in the dark (Yamamoto et al. 2008). The authors suggested that important for the irreversible membrane unstacking are the changes in the distribution of protein complexes such as LHCII in the thylakoid membranes, as well as the structural changes of the individual proteins in PSII and changes in the lipid environments.

On the other hand, it has been demonstrated that the illumination of isolated pea thylakoid membranes with strong light leads to an increase in the energy transfer from PSII to PSI, which could be a result of the disconnection of LHCII from PSII (Apostolova 2013). It has also been shown that these changes in the energy transfer do not depend on LHCII oligomerization and/or amount of anionic lipids. Chen et al. (2011) suggested that the higher tolerance to photoinhibition in wheat hybrids is associated with its higher capacity for antioxidative defense metabolism and the xanthophyll cycle, as the content of violaxanthin is decreased significantly, whereas the contents of zeaxanthin and antheraxanthin are increased considerably during high light treatments.

Depending on light growth conditions, up to two more LHCII trimers per PSII core can be present in the thylakoids (Dekker and Boekema 2005). Kouřil et al. (2013) demonstrated that the acclimation to different light intensities is accompanied by a regulation of the amount of LHC proteins and the PSII/PSI ratio. Under high light, the amount of LHCII trimers is reduced, whereas the PSII content increases compared to PSI. It has also been shown that acclimation to high light leads to a reduction of the antenna size (from 4.8 to 3.8 LHCII trimers per PSII core dimer), whereas the opposite effect is observed upon acclimation to low light (7.4 LHCII trimers per PSII core dimer). In line with these results, the Chl *a/b* ratio is also altered in high light plants (Chl *a/b* = 3.4) compared to normal plants (Chl *a/b* = 3.0), whereas under low light conditions, plants have a larger PSII antenna (Chl *a/b* = 2.7) (Kouřil et al. 2013).

Despite many studies, the role of light-harvesting antenna size of PSII in the susceptibility of the photosynthetic apparatus to high light treatment is not fully understood. Some investigators suggested that photoinhibition depends on the PSII antenna size (Stroch et al. 2004; Kim et al. 2009), while others (Tyystjärvi et al. 1994) claimed that it does not depend on the LHCII size. These differences could be a result of different organization of the photosynthetic apparatus and conditions of the treatments.

Recent investigations with pea mutant plants, which have different ratios of oligomeric to monomeric forms of LHCII (LHCIIo/LHCIIm), revealed that the oligomeric forms of LHCII in the pea mutants protect the PSII complexes against photoinhibition (Dankov et al. 2011; Apostolova 2013). Therefore, the sensitivity of the photosynthetic apparatus to high light treatment, i.e., the parameters of the PAM chlorophyll fluorescence and the photosynthetic oxygen evolution, depends on the amount and the structural organization of LHCII as well as the lipid composition. In addition, the LHCII oligomerization correlates with the amount of two anionic lipids, sulfoquinovosyl diacylglycerol (SQDG) and PG (Dankov et al. 2009b, 2011), which are essential for the function and stability of PSII (Hagio et al. 2002; Domonkos et al. 2008). It has also been shown that the degree of unsaturation of the membrane lipids (Gombos et al. 1994a; Kanervo et al. 1995; Moon et al. 1995; Tasaka et al. 1996) affects the rate of photoinhibition depending on light intensity, temperature, and time of treatment (Aro et al. 1990, 1993). Our previous investigations demonstrated that the sensitivity of the primary photochemistry of PSII (F_v/F_m) and the oxygen evolution to high light stress decreases with the increase in the oligomerization of LHCII proteins, accompanied by decreased amount of anionic lipids in pea mutant plants (Dankov et al. 2009b, 2011). On the other hand, data also showed that the degree of LHCII oligomerization and the lipid composition do not influence the degree of recovery of the PSII photochemistry after excess light exposure (Dankov et al. 2011).

The sensitivity to high light was also studied using chlorophyll b-less mutants of barley Chlorina f2 (Stroch et al. 2004) and Arabidopsis thaliana (Kim et al. 2009). It has been found that the strongly reduced amount of LHCII in Chlorina f2 is accompanied with an increased amount of PG and SQDG (Bolton et al. 1978). On the other hand, the plants grown at low light, having more stacked membranes and higher antenna size, are more susceptible to photoinhibition, but they have no differences in the lipid class composition (Chapman et al. 1986). Therefore, the controversial results for the role of antenna size in photoinhibitory damage obtained with mutant plants and plants grown at different light conditions could be due to the alterations in the lipid composition, which affect susceptibility of the photosynthetic apparatus to high light treatment.

It has also been shown that the MGDG lipid is required for the structural organization of thylakoids and the formation of grana (Jarvis et al. 2000), as well as for the photoprotection of thylakoid membranes toward light stress conditions (Aronsson et al. 2008). Schaller et al. (2010) recently suggested that an essential factor for the violaxanthin de-epoxidase activity under saturating light conditions is the presence of MGDG and that the concentration of LHCII-associated violaxanthin correlates with the amount of the main thylakoid lipid MGDG associated to the LHCII complexes.

14.4.2 PHOTOSYNTHETIC RESPONSE TO HIGH LIGHT INTENSITY DEPENDING ON THE TEMPERATURE

High light intensity and temperature stress cause significant changes in the structure of the thylakoids and strong influence of the LHCII–PSII complex. Furthermore, lower thermal stability is coupled with lower light stability, and temperature during light treatment influences the light-induced damage of the photosynthetic apparatus. Earlier investigation with wheat plants showed that high light intensity leads to enhancement of the thermal injury of the photosynthesis (Al-Khatib and Paulsen 1989). When the light stress is superimposed on heat stress, the photoinhibition of PSII is much stronger compared to the case when plants are not exposed to heat stress (see Yamamoto et al. 2008).

At the same time, it has been shown that low temperature protects intact isolated thylakoid membranes of higher plants against photoinhibition (Tyystjarvi et al. 1994). Previously, it has been shown that high light illumination at low temperature does not induce any lateral rearrangement in the location of PSII in thylakoid membranes, and the PSII electron transport is less sensitive to high light at low temperatures (Aro et al. 1990). On the other hand, during cold photoinhibition, the degradation of the D1 has been shown to be slow (Tyystjarvi et al. 1994), but the recovery is remarkably impaired (Gombos et al. 1994b). In addition, recent study on the isolated thylakoid membranes revealed that the decreased amount of anionic lipids, accompanied by an increase in the oligomeric forms of LHCII, stabilized the PSII complex against low-temperature photoinhibition (Apostolova 2013). The in vivo investigations (on leaf disks) of barley wild type and Chlorina mutants (chlorophyll-b deficient mutants with a strong decrease in the trimeric forms of LHCII) showed that mutants are more sensitive to heat and light stress than the wild type (Georgieva et al. 2003). In addition, authors also showed that high light intensity has a protective effect on the photosynthetic activity when barley plants were treated with high temperature. On the other hand, the study of Königer et al. (1998) showed that shade Alocasia macrorrhiza plants are quite tolerant to high-temperature stress in combination with low light intensity.

Other investigations showed that at temperatures between 5°C and 15°C, the LHCII macrodomains in the grana thylakoid membranes are not susceptible to high light, while the preillumination at 35–40°C almost completely destroyed these macrodomains (Cseh et al. 2000). Exposure of non-hardened Arabidopsis leaves to high light stress at 5°C resulted in a decrease in both PSII (45%) and PSI (35%) photochemical efficiencies compared to nontreated plants. In contrast, cold-acclimated leaves exhibited only 35% and 22% decrease in PSII and PSI photochemistry, respectively, under the same conditions (Ivanov et al. 2012). Involvement of alternative electron transport pathways in inducing greater resistance of both PSII and PSI to high light stress in cold-acclimated plants has been proposed. In addition, Krol et al. (2011) showed that 20°C seedlings shifted to 5°C were twofold more tolerant to photoinhibition than 20°C unshifted control seedlings, as the tolerance to photoinhibition correlates with anthocyanin accumulation in 20°C grown seedlings shifted to 5°C.

14.5 CONCLUSIONS

Plants have evolved different mechanisms for protection to environmental stress factors. The supramolecular organization of the pigment–protein complexes in the thylakoid membranes and their flexibility play an important role in various regulatory mechanisms of green plant photosynthesis (Vener 2007; Kargul and Barber 2008; Yamamoto et al. 2008; Kouřil et al. 2012; Zivcak et al. 2014). The alterations in the organization of LHCII–PSII supercomplex in turn modify the electric properties of the thylakoid membranes, the degree of grana staking, and the functions of the photosynthetic apparatus (Apostolova et al. 2006; Apostolova and Misra 2014). The comparison of the sensitivity of the photosynthetic membranes with different organization of the LHCII–PSII supercomplex in mutants (*Pisum sativum, Hordeun vulgaris, A. thaliana*) and greening plants suggests that the oligomerization of LHCII plays a key role on the sensitivity of the photosynthetic apparatus to temperature and high light intensity (Figure 14.1). In addition, the lipid composition and unsaturation, as well as carotenoids, also participate in the membrane dynamics-based protection against the abiotic stress by influencing LHCII oligomerization. Therefore, it can be concluded that the LHCII organization in the photosynthetic membranes has regulatory significance and lays the foundation for a number of studies on the dynamics of the photosynthetic apparatus and its role in the adaptation of plants.

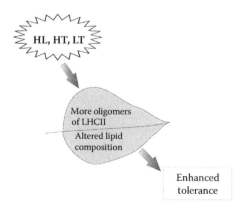

FIGURE 14.1 Sensitivity of the photosynthetic apparatus to high light (HL), high-temperature (HT), and low-temperature (LT) stresses depending on LHCII–PSII supercomplex organization in mutants and greening plants.

REFERENCES

Al-Khatib K. and G.M. Paulsen. 1989. Enhancement of thermal injury to photosynthesis in wheat plants and thylakoids by high light intensity. *Plant Physiol.* 90(3):1041–1048.

Allen J.F. and J. Forsberg. 2001. Molecular recognition in thylakoid structure and function. *Trends Plant Sci.* 6:317–326.

Anderson J. 1999. Insights into the consequences of grana stacking of thylakoid membranes in vascular plants: A personal perspective. *Aust. J. Plant Physiol.* 26(7):625–639.

Andersson B. and S. Styring. 1991. Photosystem II: Molecular organization, function and acclimation. *Curr. Top. Bioenerg.* 16:1–81.

Apostolova E.L. 2013. Effect of high-light on photosynthetic apparatus with different content of anionic lipids and organization of light-harvesting complex of photosystem II. *Acta Physiol. Plant.* 35(3):975–978.

Apostolova E.L. and A.G. Dobrikova. 2010. Effect of high temperature and UV-A radiation on the photosystem II. In: *Handbook of Plant and Crop Stress*, 3rd edition, ed. M. Pessarakli, 577–591. CRC Press, Boca Raton, FL.

Apostolova E.L. and A.N. Misra. 2014. Alterations in structural organization affect the functional ability of photosynthetic apparatus. In: *Handbook of Plant and Crop Physiology*, ed. M. Pessarakli, 103–118. Marcel Dekker, New York.

Apostolova E.L., A.G. Dobrikova, P.I. Ivanova, I.B. Petkanchin and S.G. Taneva. 2006. Relationship between the organization of the PSII supercomplex and the functions of the photosynthetic apparatus. *J. Photochem. Photobiol. B* 83:114–122.

Ariizumi T., S. Kishitani, R. Inatsugi, I. Nishida, N. Murata and K. Toriyama. 2002. An increase in unsaturation of fatty acids in phosphatidylglycerol from leaves improves the rates of photosynthesis and growth at low temperatures in transgenic rice seedlings. *Plant Cell Physiol.* 43(7):751–758.

Aro E.-M., T. Hundal, I. Carlberg and B. Andersson. 1990. In vitro studies on light-induced inhibition of Photosystem II and D_1-protein degradation at low temperatures. *Biochim. Biophys. Acta* 1019:269–275.

Aro E.-M., I. Virgin and B. Andersson. 1993. Photoinhibition of photosystem II. Inactivation, protein damage and turnover. *Biochim. Biophys. Acta* 1143:113–134.

Aronsson H., M.A. Schottler, A.A. Kelly et al. 2008. Monogalactosyldiacylglycerol deficiency in *Arabidopsis* affects pigment composition in the prolamellar body and impairs thylakoid membrane energization and photoprotection in leaves. *Plant Physiol.* 148(1):580–592.

Barra M., M. Haumann and H. Dau. 2005. Specific loss of the extrinsic 18 kDa protein from Photosystem II upon heating to 47°C causes inactivation of oxygen evolution likely due to Ca release from the Mn complex. *Photosynth. Res.* 84:231–237.

Berry J. and O. Björkman. 1980. Photosynthetic response and adaptation to temperature in higher plants. *Annu. Rev. Plant Physiol.* 31:491–543.

Bolton P., J. Wharte and J.L. Harwood. 1978. The lipid composition of a barley mutant lacking chlorophyll *b*. *Biochem. J.* 174:67–72.

Broess K., G. Trinkunas, A. van Hoek, R. Croce and H. van Amerongen. 2008. Determination of the excitation migration time in Photosystem II: Consequences for the membrane organization and charge separation parameters. *Biochim. Biophys. Acta* 1777(5):404–409.

Bukov N.G., S.C. Sabat and P. Mohanty. 1990. Analysis of chlorophyll a fluorescence changes in weak light and heat-treated *Amaranthus* chloroplasts. *Photosynth. Res.* 23:81–87.

Caffarri S., R. Croce, L. Cattivelli and R. Bassi. 2004. A look within LHCII: Differential analysis of the Lhcb1–3 complexes building the major trimeric antenna complex of higher-plant photosynthesis. *Biochemistry* 43:9467–9476.

Cao J. and Govindjee. 1990. Chlorophyll a fluorescence transitions as an indicator of active and inactive photosystem II in thylakoid membranes. *Biochim. Biophys. Acta* 1015:180–188.

Carpentier R. 1999. The effect of high temperature stress on the photosynthetic apparatus. In: *Handbook of Plant and Crop Stress*, 2nd edition, ed. M. Pessarakli, 337–348. Marcel Dekker, New York.

Chapman D., J. De Felice and J. Barber. 1986. Polar lipid composition of chloroplast thylakoids isolated from leaves grown under different lighting conditions. *Photosynth. Res.* 8:257–265.

Chen X., W. Li, Q. Lu et al. 2011. The xanthophyll cycle and antioxidative defense system are enhanced in the wheat hybrid subjected to high light stress. *J. Plant Physiol.* 168(15):1828–1836.

Chow W.S., C. Miller and J.M. Anderson. 1991. Surface charges the heterogeneous lateral distribution of the two photosystems, and thylakoid stacking. *Biochim. Biophys. Acta* 1057:69–77.

Coleman W., Govindjee and H.S. Gutowsky. 1988. The effect of chloride on the thermal inactivation of oxygen evolution. *Photosynth. Res.* 16:261–276.

Cseh Z., S. Rajagopal, T. Tsonev, M. Busheva, E. Papp and G. Garab. 2000. Thermooptic effect in chloroplast thylakoid membranes. Thermal and light stability of pigment arrays with different levels of structural complexity. *Biochemistry* 39:15250–15257.

Cui Z., Y. Wang, A. Zhang and L. Zhang. 2014. Regulation of reversible dissociation of LHCII from PSII by phosphorylation in plants. *Am. J. Plant Sci.* 5:241–249.

Dall'Osto L., C. Ünlü, S. Cazzaniga and H. van Amerongen. 2014. Disturbed excitation energy transfer in *Arabidopsis thaliana* mutants lacking minor antenna complexes of photosystem II. *Biochim. Biophys. Acta* 1837(12):1981–1988.

Dankov K., M. Busheva, D. Stefanov and E. Apostolova. 2009a. Relationship between the degree of carotenoid depletion and function of the photosynthetic apparatus. *J. Photochem. Photobiol. B* 96:49–56.

Dankov K., A. Dobrikova, B. Bogos, Z. Gombos and E. Apostolova. 2009b. The role of anionic lipids in LHCII organization and in photoinhibition of the photosynthetic apparatus. *Comp. Rend. Acad. Bulg. Sci.* 62(8):941–948.

Dankov K., S. Taneva and E.L. Apostolova. 2009c. Freeze-thaw damage of photosynthetic apparatus. Effect of the organization of LHCII_PSII supercomplex. *Comp. Rend. Acad. Bulg. Sci.* 62(9):1103–1110.

Dankov K.G., A.G. Dobrikova, B. Ugly, B. Bogos, Z. Gombos and E.L. Apostolova. 2011. LHCII organization and thylakoid lipids affect the sensitivity of the photosynthetic apparatus to high-light treatment. *Plant Physiol. Biochem.* 49:629–635.

Dankov K., G. Rashkov, A.N. Misra and E.L. Apostolova. 2015. Temperature sensitivity of photosystem II in isolated thylakoid membranes from fluridone treated pea leaves. *Turk. J. Bot.* 39:420–428.

Dekker J.P. and E.J. Boekema. 2005. Supramolecular organization of thylakoid membrane proteins in green plants. *Biochim. Biophys. Acta* 1706:12–39.

Dekker J.P. and R. van Grondelle. 2000. Primary charge separation in photosystem II. *Photosynth. Res.* 63:195–208.

Demmig B. and O. Björkman. 1987. Comparison of the effect of excessive light on chlorophyll fluorescence (77K) and photon yield of O2 evolution in leaves of higher-plants. *Planta* 171:171–184.

Dobrikova A.G., Z. Várkonyi, S.B. Krumova et al. 2003. Structural rearrangements in chloroplast thylakoid membranes revealed by differential scanning calorimetry and circular dichroism spectroscopy. Thermo-optic effect. *Biochemistry* 42:11272–11280.

Domonkos I., H. Laczkó-Dobos and Z. Gombos. 2008. Lipid-assisted protein-protein interactions that support photosynthetic and other cellular activities. *Progr. Lipid Res.* 6:422–435.

Dong L., W. Tu, K. Liu, R. Sun, C. Liu, K. Wang and C. Yang. 2015. The PsbS protein plays important roles in photosystem II supercomplex remodeling under elevated light conditions. *J. Plant Physiol.* 172:33–41.

Dreyfuss B.W. and J.P. Thornber. 1994. Assembly of the light-harvesting complexes (LHCs) of photosystem II (monomeric LHC IIb complexes are intermediates in the formation of oligomeric LHC IIb complexes). *Plant Physiol.* 106(3):829–839.

Duysen M.E., T.P. Freeman and R.D. Zabrocki. 1980. Light and the correlation of chloroplast development and coupling of phosphorylation to electron transport. *Plant Physiol.* 65:880–883.

Efeoglu B. and S. Terzioglu. 2009. Photosynthetic responses of two wheat varieties to high temperature. *Eur. Asia J. BioSci.* 3:97–106.

Elfman B., N.P.A. Huner, M. Griffith, M. Krol, W.G. Hopkins and D.B. Hayden. 2011. Growth and development at cold-hardening temperatures. Chlorophyll–protein complexes and thylakoid membrane polypeptides. *Can. J. Bot.* 62(1):61–67.

Enami I., M. Kitimura, Y. Isokova, H. Ohta and S. Katoh. 1994. Is primary cause of inactivation of oxygen evolution in spinach PSII membranes release extrinsic 33 kDa protein or Mn? *Biochim. Biophys. Acta* 1186:52–58.

Ferreira K.N., T.M. Iverson, K. Maghlaoui, J. Barber and S. Iwata. 2004. Architecture of the photosynthetic oxygen-evolving center. *Science* 303:1831–1838.

Formaggio E., G. Cinque and R. Bassi. 2001. Functional architecture of the major light harvesting complex from higher plants. *J. Mol. Biol.* 314:1157–1166.

Gal A., H. Zer and I. Ohad. 1997. Redox-controlled thylakoid protein phosphorylation. News and views. *Physiol. Plant.* 100, 4:869–885.

Garab G. and L. Mustardy. 1999. Role of LHCII-containing macrodomains in the structure, function and dynamics of grana. *Aust. J. Plant Physiol.* 26:649–658.

Garab G., Z. Cseh, L. Kovács et al. 2002. Light-induced trimer to monomer transition in the main light-harvesting antenna complex of plants: Thermo-optic mechanism. *Biochemistry* 41:15121–15129.

Georgieva K., I. Fedina, L. Maslenkova and V. Peeva. 2003. Response of the *chlorina* barley mutants to heat stress under low and high light. *Funct. Plant Biol.* 30:515–524.

Gombos Z., H. Wada, E. Hideg and N. Murata. 1994a. The unsaturation of membrane lipids stabilizes photosynthesis against heat stress. *Plant Physiol.* 104:563–567.

Gombos Z., H. Wada and N. Murata. 1994b. The recovery of photosynthesis from low-temperature photoinhibition is accelerated by the unsaturation of membrane lipids: A mechanism of chilling tolerance. *Proc. Natl. Acad. Sci. U.S.A.* 91:8787–8791.

Gombos Z., Z. Várkonyi, M. Hagio et al. 2002. Phosphatidylglycerol requirement for the function of electron acceptor plastoquinone Q(B) in the photosystem II reaction center. *Biochemistry* 41:3796–3802.

Gounaris K., A.R.R. Brain, P.J. Quinn and W.P. Williams. 1983. Structural and functional changes associated with heat-induced phase-separations of non-bilayer lipids in chloroplast thylakoid membranes. *FEBS Lett.* 153:47–52.

Gounaris K., A.R.R. Brain, P.J. Quinn and W.P. Williams. 1984. Structural reorganization of chloroplast thylakoid membranes in response to heat stress. *Biochim. Biophys. Acta* 766:198–208.

Gray G.R., A.G. Ivanov, M. Król et al. 2005. Temperature and light modulate the trans-delta3-hexadecenoic acid content of phosphatidylglycerol: Light-harvesting complex II organization and non-photochemical quenching. *Plant Cell Physiol.* 46(8):1272–1282.

Hagio M., I. Sakurai, S. Sato, T. Kato, S. Tabata and H. Wada. 2002. Phosphatidylglycerol is essential for the development of thylakoid membranes in *Arabidopsis thaliana. Plant Cell Physiol.* 43:1456–1464.

Hakala M., I. Tuominen, M. Keranen, T. Tyystjärvi and E. Tyystjärvi. 2005. Evidence for the role of the oxygen-evolving manganese complex in photoinhibition of Photosystem II. *Biochim. Biophys. Acta* 1706:68–80.

Haldimann P. 1996. Effects of changes in growth temperature on photosynthesis and carotenoid composition in *Zea mays* leaves. *Physiol. Plant.* 97:554–562.

Haque M.S., K.H. Kjaer, E. Rosenqvist, D.K. Sharma and C.-O. Ottosen. 2014. Heat stress and recovery of photosystem II efficiency in wheat (*Triticum aestivum* L.) cultivars acclimated to different growth temperatures. *Environ. Exp. Bot.* 99:1–8.

Härtel H., H. Lokstein, P. Dorman, B. Grima and C. Benning. 1997. Changes in the composition of the photosynthetic apparatus in the galactolipid-deficient *dgd1* mutant of *Arabidopsis thaliana. Plant Physiol.* 115:1175–1184.

Havaux M. 2003. Characterization of thermal damage to the photosynthetic electron transport systems in potato leaves. *Plant Sci.* 94:19–33.

Havaux M. and R.J. Strasser. 1990. Protection of photosystem II by light in heat-stressed pea leaves. *Z. Naturforsch.* 45c:1133–1141.

Havaux M. and F. Tardy. 1996. Temperature-dependent adjustment of the thermal stability of photosystem II in vivo: Possible involvement of xanthophylls cycle pigments. *Planta* 198:324–333.

Havaux M. and F. Tardy. 1997. Thermostability and photostability of photosystem II in leaves of the *Chlorina* f2 barley mutant deficient in light-harvesting chlorophyll *a/b* protein complex. *Plant Physiol.* 113:913–923.

Havaux M., F. Tardy, J. Revenel, D. Chanu and P. Parot. 1996. Thylakoid membrane stability to heat stress studied by flash spectroscopy measurements of the electrochromic shift in intact potato leaves: Influence of the xanthophylls content. *Plant Cell Environ.* 19:1359–1368.

Hodgson R.A. and J.K. Raison. 1991. Lipid peroxidation and superoxide dismutase activity in relation to photoinhibition induced by chilling in moderate light. *Planta* 183:222–228.

Homann P.H. 1988. Chloride relations of photosystem II membrane preparations depleted of, and resupplied with, 17 and 23 kDa extrinsic polypeptides. *Photosynth. Res.* 15:205–220.

Hoober J.K. and L.L. Eggink. 1999. Assembly of light-harvesting complex II and biogenesis of thylakoid membranes in chloroplasts. *Photosynth. Res.* 61:197–215.

Horton P., A.V. Ruban, D. Rees, A.A. Pascal, G. Noctor and A.J. Young. 1991. Control of the light-harvesting function of chloroplast membranes by aggregation of the LHCII chlorophyll–protein complex. *FEBS Lett.* 292(1):1–4.

Horton P., A.V. Ruban and R.G. Walters. 1996. Regulation of light harvesting in green plants. *Annu. Rev. Plant Physiol. Plant Mol. Biol.* 47:655–684.

Horton P., M. Wentworth and A. Ruban. 2005. Control of the light harvesting function of chloroplast membranes: The LHCII-aggregation model for non-photochemical quenching. *FEBS Lett.* 579:4201–4206.

Horton P., M.P. Johnson, M.L. Perez-Bueno, A.Z. Kiss and A.V. Ruban. 2008. Photosynthetic acclimation: Does the dynamic structure and macro-organisation of photosystem II in higher plant grana membranes regulate light harvesting states? *FEBS J.* 275(6):1069–1079.

Hugly S., L. Kunst, J. Browse and C.R. Somerville. 1989. Enhanced thermal tolerance of photosynthesis and altered chloroplast ultrastructure in a mutant of *Arabidopsis* deficient in lipid desaturation. *Plant Physiol.* 90(3):1134–1142.

Huner N.P.A., M. Krol, J. Williams et al. 1987. Low temperature development induces a specific decrease in trans-3-hexadecenoic acid content which influences LHCII organization. *Plant Physiol.* 84:12–18.

Huner N.P.A., B. Elfman, M. Krol and A. McIntosh. 2011. Growth and development at cold-hardening temperatures. Chloroplast ultrastructure, pigment content, and composition. *Can. J. Bot.* 62(1):53–60.

Ivanov A.G., D. Rosso, L.V. Savitch et al. 2012. Implications of alternative electron sinks in increased resistance of PSII and PSI photochemistry to high light stress in cold-acclimated *Arabidopsis thaliana. Photosynth. Res.* 113(1–3):191–206.

Jansson S. 1994. The light-harvesting chlorophyll a/b binding proteins. *Biochim. Biophys. Acta* 1184:1–19.

Jansson S., H. Stefansson, U. Nystrom, P. Gustafsson and P.-A. Albertsson. 1997. Antenna protein composition of PS I and PS II in thylakoid sub-domains. *Biochim. Biophys. Acta* 1320:297–309.

Jarvis P., P. Dormann, C.A. Peto, J. Lutes, C. Benning and J. Chory. 2000. Galactolipid deficiency and abnormal chloroplast development in the Arabidopsis MGD synthase 1 mutant. *Proc. Natl. Acad. Sci. U.S.A.* 97:8175–8179.

Joshi M.K., T.S. Desai and P. Mohanty. 1995. Temperature dependent alterations in the pattern of photochemical and non-photochemical quenching and associated changes in the photosystem II conditions of the leaves. *Plant Cell Physiol.* 36:1221–1227.

Kanervo E., E.-M. Aro and N. Murata. 1995. Low unsaturation level of thylakoid membrane lipids limits turnover of the D1 protein of photosystem II at high irradiance. *FEBS Lett.* 364:239–242.

Kargul J. and J. Barber. 2008. Photosynthetic acclimation: Structural reorganisation of light harvesting antenna—role of redox dependent phosphorylation of major and minor chlorophyll a/b binding proteins. *FEBS J.* 275:1056–1068.

Keren N. and A. Krieger-Liszkay. 2011. Photoinhibition: Molecular mechanisms and physiological significance. *Physiol. Plant.* 142(1):1–5.

Kern J. and G. Renger. 2007. Photosystem II: Structure and mechanism of the water: Plastoquinone oxidoreductase. *Photosynth. Res.* 94(2–3):183–202.

Kim E.H., X.P. Li, R. Razeghifard et al. 2009. The multiple roles of light-harvesting chlorophyll a/b-protein complexes define structure and optimize function of *Arabidopsis* chloroplasts: a study using two chlorophyll b-less mutants. *Biochim. Biophys. Acta* 1787:973–984.

Kłodawska K., P. Malec, M. Kis, Z. Gombos and K. Strzałka. 2012. EPR study of thylakoid membrane dynamics in mutants of the carotenoid biosynthesis pathway of *Synechocystis sp.* PCC6803. *Acta Biochim. Pol.* 59:87–90.

Königer M., G.C. Harris and R.W. Pearcy. 1998. Interaction between photon flux density and elevate temperatures on photoinhibition in *Alocasia macrorrhiza. Planta* 205:214–222.

Kóta Z., L.I. Horvath, M. Droppa, G. Horvath, T. Farkas and T. Pali. 2002. Protein assembly and heat stability in developing thylakoid membranes during greening. *Proc. Natl. Acad. Sci. U.S.A.* 99(19):12149–12154.

Kouřil R., J.P. Dekker and E.J. Boekema. 2012. Supramolecular organization of photosystem II in green plants. *Biochim. Biophys. Acta* 1817:2–12.

Kouřil R., E. Wientjes, J.B. Bultema, R. Croce and E.J. Boekema. 2013. High-light vs. low-light: Effect of light acclimation on photosystem II composition and organization in *Arabidopsis thaliana*. *Biochim. Biophys. Acta* 1827(3):411–419.

Krol M., G.R. Gray, N.P.A. Huner, V.M. Hurry, G. Öquist and L. Malek. 2011. Low-temperature stress and photoperiod affect an increased tolerance to photoinhibition in Pinus banksiana seedlings. *Can. J. Bot.* 73(8):1119–1127.

Krüger T.P., C. Ilioaia, M.P. Johnson et al. 2013. The specificity of controlled protein disorder in the photoprotection of plants. *Biophys. J.* 105(4):1018–1026.

Krumova S.B., S.P. Laptenok, L. Kovács et al. 2010. Digalactosyldiacylglycerol-deficiency lowers the thermal stability of thylakoid membranes. *Photosynth. Res.* 105:229–242.

Lindberg K. and L.E. Andersson. 1996. A one-site, two-state, model for binding of anions in photosystem II. *Biochemistry* 35:14259–14267.

Mathur S., A. Jajoo, P. Mehta and S. Bharti. 2011a. Analysis of elevated temperature-induced inhibition of photosystem II by using chlorophyll *a* fluorescence induction kinetics in wheat leaves (*Triticum aestivum*). *Plant Biol.* 13:1–6.

Mathur S., S.I. Allakhverdiev and A. Jajoo. 2011b. Analysis of high temperature stress on the dynamics of antenna size and reducing side heterogeneity of photosystem II in wheat (*Triticum aestivum*). *Biochim. Biophys. Acta* 1807:22–29.

Mathur S., D. Agrawal and A. Jajoo. 2014. Photosynthesis: Response to high temperature stress. *J. Photochem. Photobiol. B* 137:116–126.

Minagawa J. and Y. Takahashi. 2004. Structure, function and assembly of photosystem II and its light-harvesting proteins. *Photosynth. Res.* 82:241–263.

Mohanty P., B. Vani and J.S.S. Prakash. 2002. Elevate temperature treatment induced alteration in thylakoid membrane organization and energy distribution between two photosystems in *Pisum sativum*. *Z. Naturforsch.* 57c:836–842.

Moon B.Y., S.I. Higashi, Z. Gombos and N. Murata. 1995. Unsaturation of the membrane lipids of chloroplasts stabilizes the photosynthetic machinery against low temperature photoinhibition in transgenic tobacco plants. *Proc. Natl. Acad. Sci. U.S.A.* 92:6219–6223.

Mulo P., S. Sirpio, M. Suorsa and E.M. Aro. 2008. Auxiliary proteins involved in the assembly and sustenance of photosystem II. *Photosynth. Res.* 98:489–501.

Murata N. 1983. Molecular-species composition of phosphatidylglycerols from chilling-sensitive and chilling-resistant plants. *Plant Cell Physiol.* 24:81–86.

Murata N. and J. Yamaya. 1984. Temperature-dependent phase behavior of phosphatidylglycerols from chilling-sensitive and chilling-resistant plants. *Plant Physiol.* 74(4):1016–1024.

Murata N. and I. Nishida. 1990. Lipids in relation to chilling sensitivity of plants. In: *Chilling Injury of Horticultural Crops*, ed. C. Wang, 181–199. CRC Press, Boca Raton, FL.

Murata N., S. Takahashi, Y. Nishiyama and S. Allakhverdiev. 2007. Photoinhibition of photosystem II under environmental stress. *Biochim. Biophys. Acta* 1767:414–421.

Nash D., M. Miyao and N. Murata. 1985. Heat inactivation of oxygen evolution in photosystem II particles and its acceleration by chloride depletion and exogenous manganese. *Biochim. Biophys. Acta* 807:127–133.

Nie G.-Y., S.P. Long and N.R. Baker. 1992. The effects of development at sub-optimal growth temperatures on photosynthetic capacity and susceptibility to chilling-dependent photoinhibition in *Zea mays*. *Physiol. Plant.* 85:554–560.

Nield J. and J. Barber. 2006. Refinement of the structural model for the Photosystem II supercomplex of higher plants. *Biochim. Biophys. Acta* 1757:353–361.

Nishiyama Y., E. Kovács, C.B. Lee et al. 1993. Photosynthetic adaptation to high temperature associated with thylakoid membranes of *Synechococcus* PCC7002. *Plant Cell Physiol.* 34:337–343.

Nugent J.H.A., A.M. Rich and C.M.W. Evans. 2001. Photosynthetic water oxidation: Towards a mechanism. *Biochim. Biophys. Acta* 1503:138–146.

Nußberger S., K. Dörr, D.N. Wang and W. Kühlbrandt. 1993. Lipid–protein interactions in crystals of plant light-harvesting complex. *J. Mol. Biol.* 234:347–356.

Ögren E., G. Öquist and J.E. Höllgren. 1984. Photoinhibition of photosynthesis in *Lemna gibba* as induced by the interaction between light and temperature. 1. Photosynthesis in vivo. *Physiol. Plant.* 62:181–186.

Oguchi R., I. Terashima, J. Kou and W.S. Chow. 2011. Operation of dual mechanisms that both lead to photoinactivation of photosystem II in leaves by visible light. *Physiol. Plant.* 142:47–55.

Ohad I., A. Berg, S.M. Berkowicz, A. Kaplan and N. Keren. 2011. Photoinactivation of photosystem II: Is there more than one way to skin a cat? *Physiol. Plant.* 142:79–86.

Ohnishi N., S.I. Allakhverdiev, S. Takahashi et al. 2005. Two-step mechanism of photodamage to photosystem II: Step 1 occurs at the oxygen-evolving complex and step 2 occurs at the photochemical reaction center. *Biochemistry* 44:8494–8499.

Olesen K. and L.E. Andersson. 2003. The function of the chloride in photosynthetic oxygen evolution. *Biochemistry* 42:2025–2035.

Osmond C.B. 1994. What is photoinhibition? Some insights from comparisons of shade and sun plants. In: *Photoinhibition of Photosynthesis: From Molecular Mechanisms to the Field*, eds. Baker NR and JR Bowyer, 1–24. BIOS Scientific Publishing Ltd, Oxford.

Ottander C., D. Campbell and G. Öquist. 1995. Seasonal changes in photosystem II organization and pigment composition in *Pinus sylvestris*. *Planta* 197:176–183.

Pastenes C. and P. Horton. 1996. Effect of high temperature on photosynthesis in beans 1. Oxygen evolution and chlorophyll fluorescence. *Plant Physiol.* 112:1245–1251.

Ploschuk E.L., L.A. Bado, M. Salinas, D.F. Wasser, L.B. Windauer and P. Insausti. 2014. Photosynthesis and fluorescence responses of Jatropha curcas to chilling and freezing stress during early vegetative stages. *Environ. Exp. Bot.* 102:18–26.

Prochazkova D. and N. Wilhelmova. 2010. Antioxidant protection during abiotic stresses. In: *Handbook of Plant and Crop Stress*, 3rd edition, Chapter 6, ed. M. Pessarakli, 139–155. CRC Press, Boca Raton, FL.

Pshybytko N.L., J. Kruk, L.F. Kabashnikova and K. Strzalka. 2008. Function of plastoquinone in heat stress reactions of plants. *Biochim. Biophys. Acta* 1777:1393–1399.

Raison J.K., G.S. Pike and J.A. Berry. 1982. Growth temperature-induced alterations in thermophylic properties of *Nerium oleander* membrane lipids. *Plant Physiol.* 70:215–218.

Ristic Z., U. Bukovnik and P.V.V. Prasad. 2007. Correlation between heat stability of thylakoid membranes and loss of chlorophyll in winter wheat under heat stress. *Grop Sci.* 47:2067–2073.

Roughan P.G. 1985. Phosphatidylglycerol and chilling sensitivity in plants. *Plant Physiol.* 77(3):740–746.

Ruban A.V. 2009. Plants in light. *Commun. Integr. Biol.* 2:50–55.

Ruban A.V., M.P. Johnson and C.D. Duffy. 2012. The photoprotective molecular switch in the photosystem II antenna. *Biochim. Biophys. Acta* 1817(1):167–181.

Sakurai I., N. Mizusawa, S. Ohashi, M. Kobayashi and H. Wada. 2007. Effects of the lack of phosphatidylglycerol on the donor side of photosystem II. *Plant Physiol.* 144:1336–1346.

Sarry J.E., J.-L. Montillet, Y. Sauvaire and M. Havaux. 1994. The protective function of the xanthophyll cycle in photosynthesis. *FEBS Lett.* 353(2):147–150.

Schaller S., D. Latowski, M. Jemioła-Rezemińska and C. Wilhelm. 2010. The main thylakoid membrane lipid monogalactosyldiacylglycerol (MGDG) promotes the de-epoxidation of violaxanthin associated with the light-harvesting complex of photosystem II (LHCII). *Biochim. Biophys. Acta* 1797:414–424.

Schaller S., D. Latowski, M. Jemioła-Rezemińska et al. 2011. Regulation of LHCII aggregation by different thylakoid membrane lipids. *Biochim. Biophys. Acta* 1807(3):326–335.

Schreiber U. and J.A. Berry. 1977. Heat-induced changes in chlorophyll fluorescence in intact leaves correlated with damage of photosynthetic apparatus. *Planta* 136:233–238.

Semenova G.A. 2004. Structural reorganization of thylakoid systems in response to heat treatment. *Photosynthetica* 42:521–527.

Sfichi L., N. Ioannidis and K. Kotzabasis. 2004. Thylakoid-associated polyamines adjust the UVB-sensitivity of the photosynthetic apparatus by means of LHCII changes. *Photochem. Photobiol.* 80(3):499–506.

Shi H., L. Xiong, K. Yang, C. Tang, T. Kuang and N. Zhao. 1998. Protein secondary structure and conformation changes of photosystem II during heat denaturation studied by Fourier transform-infrared spectroscopy. *J. Mol. Struc.* 446:137–147.

Simidjiev I., V. Barzda, L. Mustárdy and G. Garab. 1998. Role of thylakoid lipids in the structural flexibility of lamellar aggregates of the isolated light-harvesting chlorophyll a/b complex of photosystem II. *Biochemistry* 37(12):4169–4173.

Simidjiev I., S. Stoylova, H. Amenitsch et al. 2000. Self-assembly of large, ordered lamellae from non-bilayer lipids and integral membrane proteins *in vitro. Proc. Natl. Acad. Sci. U.S.A.* 97(4):1473–1476.

Singh A.K. and G.S. Singhal. 1999. Specific degradation of D1 protein during exposure of thylakoid membranes to high temperature in the dark. *Photosynthetica* 36:433–440.

Standfuss J., A.C.T. van Scheltinga, M. Lamborghini and W. Kühlbrandt. 2005. Mechanisms of photoprotection and non-photochemical quenching in pea light-harvesting complex at 2.5 Å resolution. *EMBO J.* 24:919–928.

Stoitchkova K., M. Busheva, E. Apostolova and A. Andreeva. 2006. Changes in the energy distribution in mutant thylakoid membranes of pea with modified pigment content II. Changes due to magnesium ions concentration. *J. Photochem. Photobiol. B* 83:11–20.

Stroch M., M. Cajanek, J. Kalina and V. Spunda. 2004. Regulation of the excitation energy utilization in the photosynthetic apparatus of *chlorina f2* barley mutant grown under different irradiances. *J. Photochem. Photobiol. B* 75:41–50.

Sundby C., A. Melis, P. Maenpaa and B. Andersson. 1986. Temperature dependent changes in the antenna size of photosystem II. Reversible conversion of photosystem IIα to photosystem IIβ. *Biochim. Biophys. Acta* 851:475–483.

Sung D.-Y., F. Kaolan, K.-J. Lee and C.L. Guy. 2003. Acquired tolerance to temperature extremes. *Trends Plant Sci.* 8:179–187.

Szalontai B., Z. Kota, H. Nonaka and N. Murata. 2003. Structural consequences of genetically engineered saturation of the fatty acids of phosphatidylglycerol in tobacco thylakoid membranes. An FTIR study. *Biochemistry* 42:4292–4299.

Szalontai B., I. Domonkos and Z. Gombos. 2012. The role of membrane structure in acclimation to low-temperature stress. In: *Photosynthesis: Plastid Biology, Energy Conversion and Carbon Assimilation, Advances in Photosynthesis and Respiration*, vol. 34, eds. J.J. Eaton-Rye, B.C. Tripathy and T.D. Sharkey, 233–250. Springer Science+Business Media B.V., Dordrecht.

Takahashi S. and M.R. Badger. 2011. Photoprotection in plants: A new light on photosystem II damage. *Trends Plant Sci.* 16:53–60.

Takahashi S. and N. Murata. 2008. How do environmental stresses accelerate photoinhibition. *Trends Plant Sci.* 13:178–182.

Takeuchi T. and J.P. Thornber. 1994. Heat-induced alterations in thylakoid membrane protein composition in barley. *Aust. J. Plant Physiol.* 21:759–770.

Tardy F. and M. Havaux. 1997. Thylakoid membrane fluidity and thermostability during operation of xanthophyll cycle in higher-plant chloroplasts. *Biochim. Biophys. Acta* 1330:179–193.

Tasaka Y., Z. Gombos, Y. Nishiyama et al. 1996. Targeted mutagenesis of acyl-lipid desaturases in *Synechocystis*: evidence for the important roles of polyunsaturated membrane lipids in growth, respiration and photosynthesis. *EMBO J.* 15:6416–6425.

Thomas P.G., P.J. Quinn and W.P. Williams. 1984. Temperature-induced changes in the structure and function of pea chloroplasts and their relation to chloroplast membrane organisation. In: *Advances in Photosynth. Res.*, ed. C. Sybesma, 35–38. Nijhoff/Junk Publishers, The Hague.

Thorne S.W. and N.K. Boardman. 1971. Formation of chlorophyll *b* and the fluorescence properties and photochemical activities of isolated plastids from greening pea seedlings. *Plant Physiol.* 47:252–261.

Tiwari A., A. Jajoo, S. Bharti and P. Mohanty. 2007. Differential response of chloride binding sites to elevated temperature: A comparative study in spinach thylakoids and PSII-enriched membranes. *Photosynth. Res.* 93:123–132.

Tiwari A., A. Jajoo and S. Bharti. 2008. Heat induced changes in the EPR signal of tyrosine D$\left(Y_D^{ox}\right)$: A possible role of Cytochrome b559. *J. Bioenerg. Biomembr.* 40:237–243.

Tsvetkova N.M., A.P. Brain and P.J. Quinn. 1994. Structural characteristics of thylakoid membranes of *Arabidopsis* mutants deficient in lipid fatty acid desaturation. *Biochim. Biophys. Acta* 1192(2):263–271.

Tsvetkova N.M., E.L. Apostolova, A.P.R. Brain, W.P. Willams and P.J. Quinn. 1995. Factors influencing PS II particle array formation in *Arabidopsis thaliana* chloroplasts and the relationship of such arrays to the thermostability of PS II. *Biochim. Biophys. Acta* 1228:201–210.

Tyystjärvi E. 2008. Photoinhibition of photosystem II and photodamage of the oxygen evolving manganese cluster. *Coord. Chem. Rev.* 252:361–376.

Tyystjärvi E., R. Kettunen and E.-M. Aro. 1994. The rate constant of photoinhibition in vitro is independent of the antenna size of photosystem II but depends on temperature. *Biochim. Biophys. Acta* 1186:177–185.

van Amerongen H. and R. Croce. 2008. Structure and function of photosystem II Light harvesting proteins (Lhcb) of higher plants. In: *Primary Processes of Photosynthesis*, ed. G. Renger, pp. 329–368. Royal Society of Chemistry, Cambridge.

Vani B., P. Saradhi and P. Mohanty. 2001. Alteration of chloroplast structure and thylakoid membrane composition due to in vivo heat treatment of rice seedlings: Correlation with the functional changes. *J. Plant Physiol.* 158:583–592.

Vener A.V. 2007. Environmentally modulated phosphorylation and dynamics of proteins in photosynthetic membranes. *Biochim. Biophys. Acta* 1767:449–457.

Wada H., Z. Gombos and N. Murata. 1994. Contribution of membrane lipids to the ability of the photosynthetic machinery to tolerate temperature stress. *Proc. Natl. Acad. Sci. U.S.A.* 91:4273–4277.

Wentworth M., A.V. Ruban and P. Horton. 2004. The functional significance of the monomeric and trimeric states of the photosystem II light harvesting complexes. *Biochemistry* 43(2):501–509.

Wise R.R. and A.W. Naylor. 1987. Chilling-enhanced photooxidation. Evidence for the role of singlet oxygen and endogenous antioxidants. *Plant Physiol.* 83:278–282.

Yakushevska A.E., W. Keegstra, E.J. Boekema et al. 2003. The structure of photosystem II in Arabidopsis: Localization of the CP26 and CP29 antenna complex. *Biochemistry* 42(3):608–613.

Yamamoto Y., R. Aminaka, M. Yoshioka et al. 2008. Quality control of photosystem II: impact of light and heat stresses. *Photosynth. Res.* 98:589–608.

Yamane Y., Y. Kashino, H. Koike and K. Satoh. 1995. Effects of high temperatures on photosynthetic systems in higher plants. 1. Causes of the increase in the fluorescence Fo level. In: *Photosynthesis: From Light to Biosphere*, vol. IV, ed. P. Mathis, 849–852. Kluwer Academic Publishers, Dordrecht.

Yamane Y., Y. Kashino, H. Koike and K. Satoh. 1997. Increase of Fo level and reversible inhibition of photosystem II reaction center by high temperature treatments in higher plants. *Photosynth. Res.* 52:57–64.

Yamane Y., Y. Kashino, H. Koike and K. Satoh. 1998. Effect of high temperature on the photosynthetic system in spinach: Oxygen evolving activities, fluorescence characteristics and denaturation processes. *Photosynth. Res.* 57:51–59.

Yordanov I. 1992. Response of photosynthetic apparatus to temperature stress and molecular mechanisms of its adaptations. *Photosynthetica* 26:517–531.

Zhang R. and T.D. Sharkey. 2009. Photosynthetic electron transport and proton flux under moderate heat stress. *Photosynth. Res.* 100:29–43.

Zivcak M., M. Brestic, H.M. Kalaji and Govindjee. 2014. Photosynthetic responses of sun- and shade-grown barley leaves to high light: Is the lower PSII connectivity in shade leaves associated with protection against excess of light? *Photosynth. Res.* 119:339–354.

15 Impact of Solar Ultraviolet (280–400 nm) Exclusion on Photosynthesis in C_3 and C_4 Plants

K.N. Guruprasad and Sunita Kataria

CONTENTS

15.1 Introduction 257
    15.1.1 Stratospheric Ozone Depletion 257
    15.1.2 Solar UV Exclusion 258
    15.1.3 Effects of Solar UV Exclusion on C_3 and C_4 Plants 259
15.2 Effects of Solar UV Exclusion on Photosynthetic Performance 259
    15.2.1 Leaf Area and Biomass Accumulation 259
    15.2.2 Photosynthetic Pigments 260
    15.2.3 Chlorophyll Fluorescence 262
    15.2.4 Gas Exchange Parameters 264
    15.2.5 C_3 and C_4 Carbon Metabolism 265
        15.2.5.1 Carbonic Anhydrase, Rubisco, and PEP Carboxylase 265
15.3 Effect of Solar UV Exclusion on Crop Yield 268
15.4 Solar UV Sensitivity of C_3 and C_4 Plants 269
15.5 Conclusions and Future Perspective 270
Acknowledgments 271
References 271

15.1 INTRODUCTION

15.1.1 STRATOSPHERIC OZONE DEPLETION

Plants use sunlight for photosynthesis and, as a consequence, are exposed to solar ultraviolet (UV) radiation (280–400 nm). Solar UV radiation is a part of the solar electromagnetic spectrum, which is generally divided into three classes: UV-A (315–400 nm), UV-B (280–315 nm), and UV-C (≤280 nm). The UV-C is completely absorbed by the Earth's atmosphere, while UV-B and UV-A can reach the surface of the Earth. The level of UV-B radiation reaching the Earth's surface is mainly influenced by the stratospheric ozone, which is the primary UV-B absorbing component. The ozone layer directly absorbs about 90% of the UV-B radiation (McKenzie et al. 2007). The level of UV-A reaching the Earth's surface is independent of ozone concentration, since it is not attenuated by ozone, and it causes negligible damage to biological systems (Caldwell and Flint 1997; Solomon 2008).

UV-B is an important component of the environment, acting as an ecophysiological factor. In global agriculture, due to the threat to productivity, now the effects of UV-B radiation on plants is of major concern to plant biologists. A global depletion of the stratospheric ozone layer, largely due to the release of chlorofluorocarbons (CFCs) caused by human activities, has resulted in an increase of solar UV-B radiation at the Earth's surface. In spite of the current efforts to restrict the production of ozone-depleting substances, thinning of the stratospheric ozone layer and increasing penetration of UV-B radiation to the Earth's surface will still continue for decades (de la Rosa et al. 2001). As an effect of climate change, the ozone layer is not expected to recover until 2070, due to the decrease of temperature in the stratosphere (Caldwell et al. 2007).

In most of the tropical regions of the world, including India, the solar UV-B background level is often high and poses an environmental challenge. A substantial part of India lies in the low ozone belt and is consequently expected to receive a high flux of UV-B radiation (Sahoo et al. 2005). In the tropical latitudes, particularly high flux of ambient UV-B potentially impacts the photosynthetic performance of the tropical crop plants. An elevation in the flux of UV-B (280–315 nm) is an important atmospheric stress and is detrimental to plant growth and photosynthesis (Reddy et al. 2013). Various aspects of photosynthesis are affected by UV-B (Jansen et al. 2010; Kataria et al. 2014). The most common consequences of exposure to enhanced UV-B radiation on the photosynthetic functions are as follows:

- Decreased CO_2 fixation and oxygen evolution (Renger et al. 1989; Cicek et al. 2012)
- Impairment of photosystem II (PSII) and, to a lesser extent, photosystem I (PSI) (Teveni 2000; Tyystjärvi 2008)
- Reduction in dry weight, secondary sugars, starch, and total chlorophyll (Chl) (Basiouny et al. 1978; Ines et al. 2007)
- Decrease in ribulose-1,5-biphosphate carboxylase/oxygenase (Rubisco) activity (Strid et al. 1990; Allen et al. 1998; Yu et al. 2013)
- Inactivation of ATP synthase (Zhang et al. 1994)

Conversely, UV-A (315–400 nm) is considered less harmful than UV-B radiation. In general, by supplementation of UV-A with photosynthetically active radiation (PAR, 400–700 nm), the enhanced levels of pigments such as Chl, carotenoids, and UV-absorbing compounds, including antioxidants, has been reported to stimulate growth (Shiozaki et al. 1999; Helsper et al. 2003). Flint and Caldwell (1996) reported that UV-A irradiation alleviates the damaging effect of UV-B.

15.1.2 Solar UV Exclusion

Plants usually are subjected to combined stresses in the natural environment, and the stress factors cause several changes in plant metabolism and morphology. Many early studies examining UV-B effects on terrestrial plants were conducted indoors using growth chambers or greenhouses, in which plants were exposed to unnaturally high UV-B from lamps against a background of low UV-A and PAR (400–700 nm) (Caldwell et al. 1994; Caldwell and Flint 1997; Krizek and Mirecki 2004).

Thus, outdoors field studies that use visible background irradiance provided by sunlight are necessary to realistically evaluate the biological effects of solar UV radiation. The two most widely used approaches in outdoor field studies are the attenuation approach and the enhancement approach. Most attenuation approaches use specific filters that either absorb or transmit UV-B or UV-A+B from the natural solar spectrum. These filters were wrapped on specially designed UV exclusion chambers made up of iron mesh. The chambers held filters that excluded both UV-B and UV-A, excluded only UV-B, transmitted all UV (filter control [FC]), or lacked filters (open control [OC]) (Amudha et al. 2005; Kataria et al. 2013). The transmission spectra of these filters are given in Figure 15.1. Experiments using UV exclusion by specific filters to remove much of the radiation at a shorter wavelength from the solar spectrum are most suited to assessing the effects of current UV-B radiation at particular latitudes (Rousseaux et al. 2004).

An assessment of the impact of solar UV (280–400 nm) at a given latitude is possible by growing plants after the exclusion of this radiation by using specific filters. At a particular place, the amount of UV-B radiation reaching the Earth's surface mainly depends upon its latitude as well as on the ozone profile above it. The plants in tropical regions of the world are exposed to high ambient levels of UV-B radiation compared to the temperate zones (Sahoo et al. 2005). Different laboratories that have carried out studies using the UV exclusion approach have generally focused on UV-B impacts on plant growth and morphology, with varying results. At temperate locations, many species often fail to respond to UV-B exclusion (Cybulski and Peterjohn 1999). Solar UV-B exclusion does not cause any significant effect on the naturally occurring plants of the alpine plant community (Caldwell 1968). By excluding solar UV-B in field experiments carried out in Japan, a transient reduction in biomass accumulation in tomato plants was obtained (Tezuka et al. 1993), whereas at a high-elevation site in Colorado,

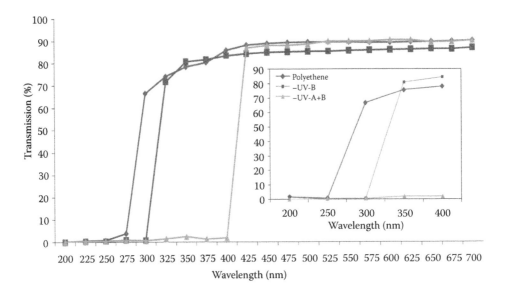

FIGURE 15.1 Transmission spectra of UV cutoff filters and polyethene filter used in growth chambers for raising plants under field conditions. The insert shows the transmission of all the filters at every 50 nm from 200 to 400 nm.

UV-B exclusion did not cause any effect of on final biomass in plants of potato, radish, and wheat (Becwar et al. 1982). Such exclusion studies in radish have shown an enhancement in leaf growth but a decrease in the yield (Zavala and Botto 2002). An increase in biomass and yield by exclusion of solar UV-B has been observed in tropical crops like *Cyamopsis* and *Vigna* (Amudha et al. 2005), soybean (Guruprasad et al. 2007; Baroniya et al. 2011), cotton (Dehariya et al. 2012), and *Amaranthus* (Kataria and Guruprasad 2014).

Exclusion of UV-B and UV-A+B from solar radiation causes an increase in root biomass in *Vaccinium ulginosum* (Rinnan et al. 2005). The solar UV-B radiation also showed its impact on the process of nitrogen metabolism, as has been observed through the changes in nitrate reductase enzyme activity. The nitrate reductase activity was increased in the leaves of common beans, barley, wheat, and pea plants grown in the absence of solar UV-B radiation (Pal et al. 2006; Moussa and Khodary 2008; Kataria and Guruprasad 2015). Thus, findings on the exclusion of solar UV indicate that the ambient solar UV-B currently is a significant stress factor for plants and, thus, even small changes in ozone depletion may have important biological consequences.

15.1.3 Effects of Solar UV Exclusion on C$_3$ and C$_4$ Plants

C$_3$ and C$_4$ plant species have evolved in different climates. C$_3$ plants are believed to have a temperate origin, and C$_4$ plants have evolved in tropical and arid environments (Ward et al. 1999). For their climatic requirements, they differ from each other both structurally and functionally (Ward et al. 1999; Nayyar 2003). A comparison between C$_3$ and C$_4$ plants is shown in Box 15.1. C$_3$ plant is one in which CO$_2$ is fixed into a compound containing three carbon atoms before entering the Calvin cycle of photosynthesis. A C$_4$ plant fixes CO$_2$ into a compound containing four carbon atoms.

Most of the previous studies concerning the responses of C$_3$ and C$_4$ plants to UV-B radiation have been carried out under enhanced UV-B provided by lamp supplementation. Far fewer studies have investigated the effects of exclusion of solar UV radiation on C$_3$ and C$_4$ plants (Pal et al. 1997, 2006; Amudha et al. 2005; Kataria et al. 2013). UV exclusion studies have

indicated the profound influence of the small constituent of UV-B in the solar spectrum on growth, biomass accumulation, and carbon and nitrogen fixation. Exclusion of solar UV enhanced plant height, internodal length, fresh and dry weight of plants, and number of leaves in C$_3$ plants like radish (Zavala and Botto 2002), soybean (Chouhan et al. 2008; Baroniya et al. 2011), mung bean (Pal et al. 1997), cucumber (Krizek et al. 1997; Krizek and Mirecki 2004), cotton (Dehariya et al. 2012), pea (Pal et al. 2006), and wheat (Kataria and Guruprasad 2012a), and C$_4$ plants like sorghum (Kataria and Guruprasad 2012b; Kataria et al. 2013), maize (Shine and Guruprasad 2012), and *Amaranthus* (Kataria et al. 2013; Kataria and Guruprasad 2014). Exclusion of UV also promotes nodulation and increases leghemoglobin content and nitrogenase activity in the root nodules of soybean and *Trigonella* (Chouhan et al. 2008; Singh 2011; Sharma and Guruprasad 2012; Baroniya et al. 2014).

The UV components present in the solar spectrum inhibit the photomorphogenetic pattern, and exclusion of UV-B and UV-A+B from the solar radiation causes significant enhancement in morphological and physiological parameters in C$_3$ and C$_4$ plants (Kataria et al. 2013). The inference from the reviewed literature concerning the UV exclusion studies are that the plant sensitivities to an ambient level of UV-B and UV-A radiation differ among the cultivars within a species and plants vary greatly in their response to an ambient level of UV radiation (Baroniya et al. 2011; Kataria and Guruprasad 2012a,b, 2014). The effects of solar UV exclusion on C$_3$ and C$_4$ monocot and dicot plants suggested that most of the UV-affected plants were mainly broad-leaved C$_3$ plants like cotton, cucumber, soybean, etc. This could be due to their morphological and anatomical differences from narrow-leaved C$_4$ plants like maize and sorghum. This chapter mainly focuses on the impacts of exclusion of solar UV on photosynthesis in C$_3$ and C$_4$ plants.

15.2 EFFECTS OF SOLAR UV EXCLUSION ON PHOTOSYNTHETIC PERFORMANCE

15.2.1 Leaf Area and Biomass Accumulation

Plant leaves are of prime importance as solar radiation falls directly on the foliar surface, and thus, in relation to the incidence of UV-B radiation, the leaf orientation may be

---

BOX 15.1 COMPARISON BETWEEN C$_3$ AND C$_4$ PLANTS

C$_3$ Plants	C$_4$ Plants
1. Photosynthesis occurs in mesophyll cells.	Photosynthesis occurs in both mesophyll and bundle sheath cells.
2. The carbon dioxide acceptor is Rubisco.	The carbon dioxide acceptor is phosphoenolpyruvate (PEP) carboxylase.
3. Kranze anatomy absent.	Kranze anatomy present.
4. First stable carbon compound is 3C compound called 3-phosphoglyceric acid (PGA).	First stable carbon compound is 4C compound called oxaloacetic acid (OAA).
5. Photorespiratory loss is high.	Photorespiration does not take place.
6. Examples of C$_3$ plants are wheat, rice, cotton, cucumber, soybean, etc.	Examples of C$_4$ plants are sorghum, maize, Amaranthus, sugarcane, etc.

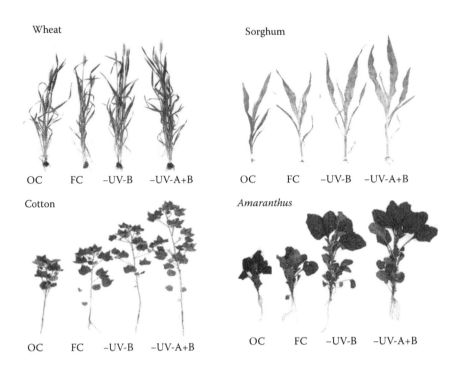

FIGURE 15.2 **(See color insert.)** Photographs showing effect of exclusion of UV-B and UV-A+B on growth and biomass accumulation of wheat, cotton, *Amaranthus*, and sorghum plants.

an important factor. The enhancements in the area of the leaf, specific leaf weight (SLW), and biomass accumulation are linked to the enhanced rate of photosynthesis. The effects of solar UV exclusion on C_3 (cotton and wheat) and C_4 (*Amaranthus* and sorghum) plants are shown in Figure 15.2. Exclusion of UV-B and UV-A+B enhanced both the leaf area and biomass accumulation in C_3 and C_4 plants (Figure 15.2). However, the C_3 and C_4 monocotyledonous (wheat and sorghum) plants showed greater enhancements in leaf area and biomass accumulation after the exclusion of solar UV-B, while C_3 and C_4 dicotyledonous (cotton and *Amaranthus*) plants showed enhancement after the exclusion of solar UV-A+B (Figure 15.3a–d). Exclusion of solar UV-B enhances the leaf area and biomass accumulation in several plant species: *Cymopsis* (Amudha et al. 2005), soybean (Guruprasad et al. 2007; Baroniya et al. 2011), cucumber (Krizek and Mirecki 2004), cotton (Dehariya et al. 2012), radish (Zavala and Botto 2002), pea (Pal et al. 2006), mung bean (Pal et al. 1997), barley (Mazza et al. 1999), and *Trigonella* (Sharma and Guruprasad 2012). This increase in biomass accumulation after exclusion of ambient UV-B may be due to more branches and leaves in dicots or more tillers in monocots like in barley, wheat, cotton, *Amaranthus*, and sorghum (Mazza et al. 1999; Coleman and Day 2004; Kataria and Guruprasad 2012a, 2014).

The SLW (the leaf dry weight per unit leaf area) and the leaf weight ratio (LWR, leaves' biomass allocation to the plant biomass) were also significantly increased by exclusion of solar UV-B and UV-A+B in the varieties of *Amaranthus tricolor* and *Sorghum bicolor* (Kataria and Guruprasad 2012b, 2014). The increase in SLW and LWR after UV exclusion indicates a larger biomass partition to leaves and increased leaf thickness. Similar results were observed in other UV-B exclusion studies, like in

Vaccinium uliginosum (Albert et al. 2010), whereas the plants grown in ambient UV conditions showed increased specific leaf area and decreased SLW, indicating that ambient UV-B radiation decreases leaf thickness, as reported in other UV-B enhancement and supplementation studies (Correia et al. 1999; Singh et al. 2009). This decrease may be important in plant dry weight reduction because decreased photosynthesis rates have been correlated with an increase in specific leaf area, contributing to lower relative growth rates (Poorter 1989). Increased specific leaf area under supplemental UV-B led to more damage of mesophyll tissues, thus significantly affecting the rate of photosynthesis (Correia et al. 1999). Thus, the UV components present in the solar spectrum seem to inhibit growth and biomass accumulation. However, UV-free-grown C_3 and C_4 monocot plants like wheat, maize, and sorghum showed only small differences in leaf area and biomass accumulation as compared with C_3 and C_4 dicot plants like cotton, pea, soybean, and *Amaranthus* (Pal et al. 2006; Guruprasad et al. 2007; Shine and Guruprasad 2012; Kataria et al. 2013). Solar UV exclusion studies suggested that leaf orientation may be one of the reasons for monocots of either C_3 or C_4 being less sensitive to ambient UV-B radiation than dicots of either C_3 or C_4; monocot leaves are nearly vertical, whereas dicot leaves are horizontal in orientation. The effects of supplementary UV-B radiation were more severe in artificially constrained horizontal leaves than in the near-vertical leaves of two rice cultivars (He et al. 1993).

15.2.2 Photosynthetic Pigments

Photosynthesis is the most important plant process, and it is necessary for the production of biomass. Photosynthesis is dependent on the light-harvesting properties of the chlorophyll

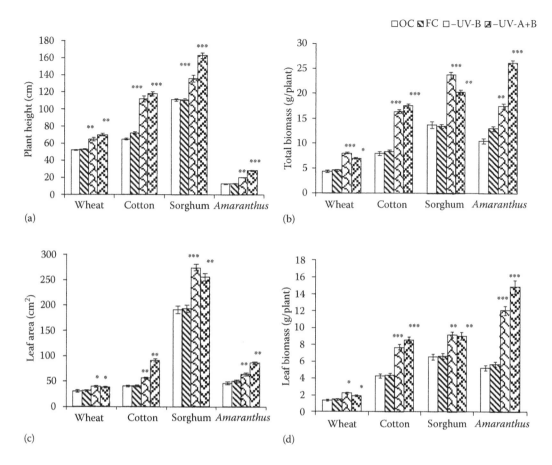

FIGURE 15.3 Effect of exclusion of UV-B and UV-A+B on (a) plant height, (b) total biomass, (c) leaf area, and (d) leaf biomass of cotton, wheat, *Amaranthus*, and sorghum plants. Changes are depicted as the absolute values of each treatment compared to the respective filter control. The vertical bars indicate mean ± standard error (SE) ($n = 15$). Values are significantly different ($^*P < .05$, $^{**}P < .01$, $^{***}P < .001$) from filter control (Newman–Keulis multiple comparison test).

(Chl). Hence, the measurements of photosynthetic pigment composition are known as useful indicators of UV-B tolerance or sensitivity (Strid et al. 1994). The changes in photosynthetic pigments after exclusion of solar UV in C_3 and C_4 plants are shown in Table 15.1. The exclusion of solar UV-B and UV-A+B caused enhancement in the photosynthetic pigments like Chl a, Chl b, total Chl, and carotenoids per unit fresh weight of leaves in C_3 and C_4 plants. The C_3 and C_4 dicot plant species like cotton and *Amaranthus* showed significant increase in these pigments after exclusion of solar UV as compared to the monocot plant species like wheat and sorghum (Dehariya et al. 2012; Kataria and Guruprasad 2012b, 2014; Kataria et al. 2013).

The UV exclusion studies suggested that the presence of UV components in the solar spectrum caused reduction in the photosynthetic pigments (Dehariya et al. 2012; Kataria and Guruprasad 2012b, 2014; Kataria et al. 2013). The pronounced decrease in the amount of photosynthetic pigments by ambient UV could affect photosynthesis and reduce both biomass accumulation and economic yield. The reduction in total Chl concentration induced by UV-B radiation was probably due to the destruction of the structure of chloroplasts, inhibited synthesis of new Chl, and increased degradation of

Chls (Sakaki et al. 1983; Sedej and Gabercik 2008). In contrast, on removal of the UV stress, under exclusion of UV, the inhibition is overcome, or degradation processes might have been stopped, resulting in an accumulation of these pigments (Amudha et al. 2005). The increase in total Chl level by the exclusion of solar UV was due to the enhanced synthesis of Chl a and Chl b both, though the extent of promotion was greater in Chl b than in Chl a, which can cause a decrease in the ratio of Chl a/b after UV exclusion in C_3 and C_4 plants (Kataria et al. 2013). Chl a/b may reflect the relative ratio of stacked regions to unstacked regions, which were inversely proportional to the stacking degree of thylakoids (Ibanez et al. 2008). Thus, an increase of Chl a/b ratio in plants grown under ambient UV may indicate that UV-B radiation causes the decomposition of stacked regions of thylakoids (Jiang et al. 1991).

In higher plants, carotenoids play an important role against UV-B damage. Carotenoids scavenge the singlet oxygen species formed during intense light, and they are involved in the light harvesting and protection of Chls from photooxidative destruction (Middleton and Teramura 1993). We have found a significant increase in carotenoid contents in plants grown under UV-B and UV-A+B filtered radiation (Table 15.1). The

TABLE 15.1

Photosynthesis Pigments (mg/g leaf fresh weight) in C_3 and C_4 Plants after Exclusion of Solar UV-B and UV-A+B Radiation

Plant Species	Treatment	Chl a	Chl b	Total Chl	Carotenoids
1. Cotton	OC	1.186	0.300	1.487	154
	FC	1.080	0.256	1.199	152
	–UV-B	1.342 (124)	0.425 (166)	1.768 (147)	168 (110.5)
	–UV-A+B	**1.597 (148)**	**0.500 (195)**	**1.941 (162)**	**180 (118.4)**
2. Wheat	OC	3.421	0.740	4.15	228
	FC	3.720	0.700	4.42	225
	–UV-B	**4.940 (132)**	**0.850 (121)**	**5.79 (131)**	**298 (132)**
	–UV-A+B	4.510 (121)	0.920 (132)	5.44 (123)	240 (107)
3. *Amaranthus*	OC	0.944	0.112	1.057	69.84
	FC	0.887	0.121	1.008	65.96
	–UV-B	1.034 (117)	0.175 (145)	1.209 (120)	76.50 (116)
	–UV-A+B	**1.240 (140)**	**0.247 (204)**	**1.488 (148)**	**88.23 (134)**
4. Sorghum	OC	2.310	0.410	2.720	163
	FC	2.440	0.560	3.010	177
	–UV-B	**2.950 (121)**	**0.744 (134)**	**3.695 (123)**	**234 (132)**
	–UV-A+B	2.924 (120)	0.683 (123)	3.590 (120)	204 (115)

Source: Data modified from Kataria et al., 2013.

Note: Data are the mean ± SEM of 15 plants ($n = 15$). The numbers in parentheses are percent changes with reference to respective filter controls. Bold data represents the maximum promotion in the parameters examined by the removal of UV-B or UV-A+B radiation.

carotenoid content was lower in the plants grown under ambient UV radiation, and exclusion of solar UV significantly enhanced the carotenoids (Dehariya et al. 2012; Kataria and Guruprasad 2012b). Similarly, the levels of carotenoids were decreased under enhanced UV radiation (Singh et al. 2011; Ibrahim et al. 2013). The reduction in carotenoid content under ambient UV radiation may result either from inhibition of synthesis or from breakdown of the pigments or their precursors.

15.2.3 Chlorophyll Fluorescence

Photosynthetic performance in intact plants is conveniently monitored by a powerful technique—Chl fluorescence analysis. The polyphasic rise of the Chl a fluorescence transient has been analyzed by the JIP test and has applications in investigating *in vivo* the "vitality" of plants and the response of the photosynthetic apparatus to different stresses (Srivastava and Strasser 1996; Christen et al. 2007), like high temperature (Mathur et al. 2011), UV-B (Albert et al. 2005, 2008), and salinity stress (Mehta et al. 2010). The JIP test is named after the basic steps in the fluorescence transient when plotted on a logarithmic time scale. When plotted on a logarithmic time scale, the fluorescence rise of the Kautsky transient is polyphasic and can be divided into O-J-I-P steps, where O is the minimal fluorescence (Fo), P is the peak at about 500 ms (Fp), and J and I are inflections (Strasser and Strasser 1995). This test of OJIP transients (the so-called JIP test) allows us to translate the fluorescence transient measurements into several

phenomenological and biophysical parameters that provide information on the PSII functioning of the photosynthetic organism (Stirbet and Govindjee 2011). The OJIP transient has been used for the characterization of the photochemical quantum yield of PSII photochemistry as well as the electron transport activity (Stirbet and Govindjee 2011; Chen et al. 2008). The OJIP transient originates from the O phase corresponding to minimal fluorescence yield, mostly from Chl a of PSII light-harvesting complex (Butler 1978). The J phase represents the maximum of Q_A^- (Hsu 1993), and the following fluorescence raises the JI transient and characterizes the closure of the remaining PSII open centers, resulting in the accumulation of $Q_A^- Q_B^-$ (Strasser and Strasser 1995). The maximum yield of Chl a fluorescence is attained when the plastoquinone (PQ) pool becomes reduced; therefore, the IP fluorescence step is a consequence of $Q_B^{2-} Q_A^-$ (Strasser and Strasser 1995).

Effects of exclusion of solar UV on Chl a fluorescence induction kinetics in the leaves of C_3 (wheat and cotton) and C_4 (sorghum and *Amaranthus*) plants are shown in Figure 15.4. The exclusion of solar UV-B and UV-A+B enhanced the intensity of fluorescence yield in the JIP phase; however, the differences in the polyphasic increase in fluorescence transients became more evident at the IP phase in C_3 (wheat and cotton) and C_4 (sorghum and *Amaranthus*) plants growing under solar UV exclusion (Kataria et al. 2013). Exclusion of UV-B radiation caused a decrease of F_o and an increase of F_m, resulting in an increase of the F_v/F_m value (Table 15.2). Kataria et al. (2013) reported that F_v/F_m does not show a

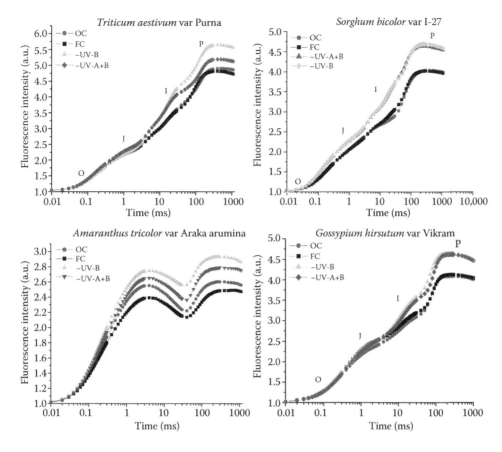

FIGURE 15.4 Fluorescence emission transient of leaves of wheat, cotton, *Amaranthus*, and sorghum plants normalized at Fo. FC, filter control, OC; open control; –UV-A+B, UV-A+B-excluded plants; –UV-B, UV-B-excluded plants. O-J-I-P, where O is the minimal fluorescence (Fo), P is the peak at about 500 ms (Fp), and J and I are inflections.

significant difference between UV-excluded plants and plants grown under ambient UV, which shows that exclusion of UV radiation did not influence the number of quanta absorbed per unit time (Mehta et al. 2010).

Among the constellation of JIP expressions, one of the most sensitive parameters is the performance index (PI$_{ABS}$). The solar UV exclusion studies established that the PI$_{ABS}$ is a much more sensitive indicator of UV stress than the well-known F_v/F_m ratio (Dehariya et al. 2012; Shine and Guruprasad 2012; Kataria et al. 2013; Kataria and Guruprasad 2014). Moreover, PI$_{ABS}$ allows broader analysis of photosynthetic performance, such as the relationship between photon absorption efficiency and capture of excited energy in PSII, as well as analysis of the density of active reaction centre (RC) and the probability that excited energy moves an electron further than Q_A^- (Goncalves and Santos Junior 2005). Therefore, the PI$_{ABS}$ helps to estimate the vitality of the plants with high resolution. Differences in response pattern in quantum efficiencies and phenomenological fluxes are integrated in the performance indexes. Their enhancement clearly signifies an improved overall processing of light energy per cross section of leaf sample in UV-excluded plants, indicating a negative impact of ambient UV on all of these parameters in C₃ (wheat and cotton) and C₄ (maize, sorghum, and *Amaranthus*) plants (Shine and Guruprasad 2012;

Kataria et al. 2013). A negative impact of ambient UV-B on these parameters was also observed in short-term and long-term UV exclusion experiments with high-Arctic plants (Albert et al. 2005, 2008, 2011).

A phenomenological leaf model (generated by Biolyzer HP3 software) in UV-excluded C₃ (wheat and cotton) and C₄ (sorghum and *Amaranthus*) plants depicts more active reaction centers per unit area (Figure 15.5). In this model, open circles represent the active reaction center, and UV-B- and UV-A+B-excluded plants had more active reaction centers combined with higher efficiency of electron transport, indicated by broader width of the arrow in the leaf models of C₃ (wheat and cotton) and C₄ (sorghum and *Amaranthus*) (Figure 15.5).

Exclusion of solar UV significantly enhanced leaf biomass and total biomass in C₃ and C₄ plant species (Amudha et al. 2005; Kataria and Guruprasad 2012a,b, 2014; Shine and Guruprasad 2012; Kataria et al. 2013). Quantum electron transport (ETo/ABS) and performance index (PI$_{ABS}$) are related to the productivity of photosynthetic metabolites, and hence, they offer a diagnostic tool for biomass production capability. Thus, ambient UV caused a significant decrease in quantum yield (φ_{Eo}) and estimated electron fluxes per cross section of the leaves of C₃/C₄ plant species, which may lead to the decrease in carbon uptake in these plants. The reduction in

TABLE 15.2

Values of Different Structural and Functional Parameters Provided by the JIP Test in C_3 and C_4 Plant Species

C_3/C_4 Plants	F_o	F_m	F_v/F_m	PI_{ABS}	ETo/ABS
			Cotton		
OC	422	1306	0.755	3.280 ± 0.13	0.400
FC	405	1266	0.757	2.671 ± 0.18	0.401
–UV-B	422	1554	0.784	$\mathbf{4.915 \pm 0.181^{***}}$	**0.471**
–UV-A/B	397	1442	0.784	$\mathbf{6.143 \pm 0.181^{***}}$	**0.468**
			Amaranthus		
OC	282	734	0.616	0.033 ± 0.009	0.0450
FC	268	668	0.599	0.053 ± 0.008	0.0590
–UV-B	287	841	0.659	$\mathbf{0.099 \pm 0.081^{***}}$	**0.0810**
–UV-A/B	293	817	0.641	$\mathbf{0.074 \pm 0.061^{***}}$	**0.0685**
			Wheat		
OC	214	1049	0.796	8.468 ± 0.15	0.509
FC	215	1036	0.792	7.313 ± 0.13	0.510
–UV-B	204	1150	0.823	$\mathbf{15.38 \pm 0.281^{***}}$	**0.579**
–UV-A/B	217	1126	0.807	$8.747 \pm 0.191^{***}$	0.527
			Sorghum		
OC	198	799	0.752	3.942 ± 0.18	0.442
FC	201	808	0.751	3.981 ± 0.25	0.443
–UV-B	223	1046	0.787	$\mathbf{5.161 \pm 0.191^{***}}$	**0.472**
–UV-A/B	213	988	0.784	$4.954 \pm 0.141^{***}$	0.469

Note: Values are significantly different (***$P < .001$) from filter control (Newman–Keulis multiple comparison test). Data in bold represents the maximum promotion in the parameters examined by the removal of UV-B and UV-A+B radiation. ETo/ABS, quantum yield of electron transport; F_m, maximum fluorescence at 1 s; F_o, initial fluorescence at 20 µs; F_v/F_m, maximum quantum yield of primary photochemistry; PI_{ABS}, performance index based on absorption.

net CO_2 uptake and PSI efficiency was found in tree seedlings when exposed to solar UV-B radiation (Krause et al. 2003).

The relationship between log PI_{ABS} (the driving force for photosynthesis) and ETo/ABS can be considered a typical property of a plant's ability to transform light energy into chemical energy (NADH), which is consequently directed into metabolic reactions in the biochemical process of photosynthesis (Hermans et al. 2003). The relationship between log PI_{ABS} and ETo/ABS is shown in Figure 15.6. It reveals that the efficiency in transforming light energy to chemical energy is higher in C_3 (wheat and cotton) and C_4 (sorghum and *Amaranthus*) plants grown under solar UV exclusion. This plot also emphasizes the need for the presence of the UV-A part of the solar spectrum to achieve the highest efficiency in case of wheat (Figure 15.6). Thus, UV exclusion enhances the ability of C_3 (wheat and cotton) and C_4 (sorghum and *Amaranthus*) plants to convert light energy to chemical energy that can be used to reduce CO_2 to carbohydrate, which can lead to larger accumulation of biomass. Thus, these results suggest that leaves of UV-excluded C_3 and C_4 plants species have higher reducing power with higher efficiency of electron transport and more active reaction centers.

15.2.4 GAS EXCHANGE PARAMETERS

Photosynthesis is the basis of plant growth, and improving photosynthesis can contribute towards greater food security in the coming decades as world population increases. Exclusion of solar UV components causes enhancement in the rate of photosynthesis in C_3 (wheat and cotton) and C_4 (Sorghum and *Amaranthus*) plants along with a simultaneous increase in stomatal conductance (Figure 15.7). This indicates that the presence of solar UV components reduced the stomatal conductance, which caused a reduction in the net rate of photosynthesis. A higher photosynthetic capacity was found in UV-B-free-grown common bean plants (up to 33%) against only a marginal increase (12%) in barley plants grown under similar conditions (Moussa and Khodary 2008). The rate of photosynthesis was found to increase by 58% and 55% in cotton and *Amaranthus*, respectively, with the exclusion of solar UV-B, while increases of 46% and 26% were seen in wheat and sorghum, respectively, after UV-B exclusion (Figure 15.7a) as compared with the plants grown under ambient solar UV radiation (+UV-A and UV-B). A similar observation of enhancement in the net rate of photosynthesis and concomitant increase in

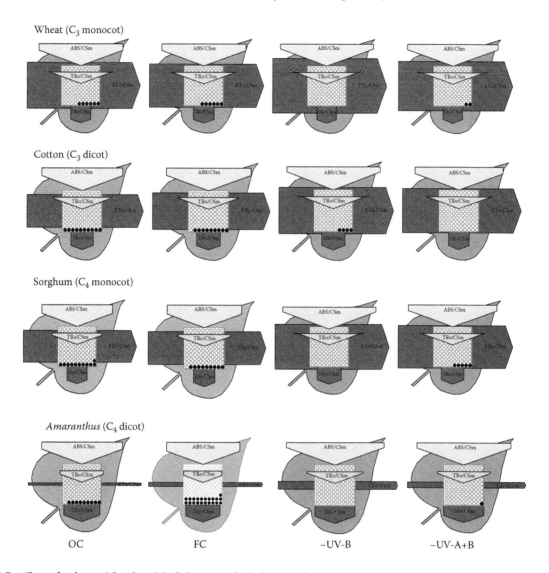

Wheat (C₃ monocot)

Cotton (C₃ dicot)

Sorghum (C₄ monocot)

Amaranthus (C₄ dicot)

OC   FC   –UV-B   –UV-A+B

FIGURE 15.5   (See color insert.) Leaf model of phenomenological energy fluxes per excited cross section (CS) of wheat, cotton, sorghum, and *Amaranthus* leaves showing the effect of exclusion of UV-B and UV-A+B. The arrows indicate fluxes for light absorbance (ABS), excitation energy trapping (TRo), energy dissipation (DIo), and electron transport (ETo) beyond QA⁻. ABS/CSm, absorption flux CS approximated by F_m; DIo/CSm, dissipated energy per CS; ETo/CSm, electron transport flux per CS; TRo/CSm, trapped energy per CS. Each relative value is represented by the size of the proper parameters (arrow), the width of each arrow denotes the relative size of the fluxes or the antenna, empty circles represent reducing Q_A reaction centers (active), and full black circles represent nonreducing Q_A reaction centers (inactive or silent).

stomatal conductance after exclusion of UV-B has been made in *Populus* (Schumaker et al. 1997), maize and mung bean (Pal et al. 1997), wheat and pea (Pal et al. 2006), sorghum (Kataria and Guruprasad 2012b), and *A. tricolor* (Kataria and Guruprasad 2014). Thus, the primary effect of ambient UV on photosynthetic performance is due to the reduced rate of photosynthesis and fixation of CO_2 (Kataria and Guruprasad 2012b, 2014; Kataria et al. 2013; Baroniya et al. 2014)

A decreased rate of photosynthesis has been shown in other UV-B exclusion experiments (Xiong and Day 2001; Albert et al. 2008) and in plants exposed to realistic supplemental UV-B irradiance in the field (Keiller and Holmes 2001; Keiller et al. 2003). The lower photosynthesis in ambient UV-B radiation could potentially be caused by effects on PSII performance (source side) or effects on performance of the Calvin cycle, CO_2 diffusion changes, etc. (sink side). Some

studies suggested that PSII is the primary target of UV-B damage, which reduces PSII activity (Strid et al. 1990; Melis et al. 1992) and the abundance of D1 (Aro et al. 1993; Jansen et al. 1996), while other studies have shown that photosynthetic inhibition can occur without any measurable effect on PSII (Middleton and Teramura 1993; Allen et al. 1999).

15.2.5   C₃ AND C₄ CARBON METABOLISM

15.2.5.1   Carbonic Anhydrase, Rubisco, and Phosphoenolpyruvate Carboxylase (PEP Case)

Since the quantum yield of electron transport was reduced in plants grown under ambient UV radiation, it is probable that the linear flow of electrons beyond PSII is also decreased,

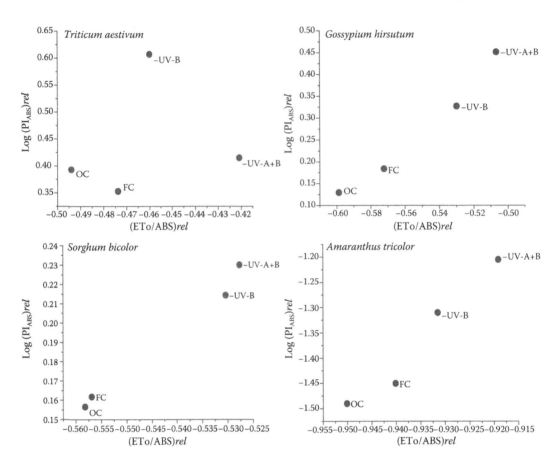

FIGURE 15.6 Correlation between the driving force $(DF_{ABS})rel = \log(PI_{ABS})rel$ as a function of the relative yield of electron transport $(ETo/ABS)rel$ in wheat, cotton, sorghum, and *Amaranthus* showing the effect of exclusion of solar UV-B and UV-A+B.

leading to a reduced proportion of electrons available for the Calvin cycle. Rubisco competitively binds with CO_2 or O_2. However, binding with O_2 triggers photorespiration vis-à-vis reduction in photosynthesis. Carbonic anhydrase (CA) plays an important role in the acceleration of carbon assimilation, by catalyzing the reversible interconversion of CO_2 and HCO_3^-. In leaves, CA represents 1–20% of total soluble protein, and its abundance is next only to Rubisco in chloroplast. In C_3 plants, CA activities are largely restricted in stroma of mesophyll chloroplasts (Tsuzuki et al. 1985), where it is believed to facilitate the diffusion of CO_2 across the chloroplast envelope (Majeau and Coleman 1991). Close association of CA with Rubisco increases the availability of CO_2 at the site of carboxylation (Werdan et al. 1972).

Rubisco is the primary catalyst for the assimilation of atmospheric CO_2 into the biosphere. Rubisco is a key enzyme in the C_3 cycle of photosynthetic fixation of CO_2 and a remarkably abundant protein; it constitutes up to 65% of total soluble leaf proteins in C_3 plant species, and in C_4 plant species, the percentage is lower (about 50%), since the enzyme is only in bundle sheath cells (Furbank and Taylor 1995; Hollosy 2002). The rate of photosynthesis and biomass accumulation largely depends on the quantity and activity of Rubisco (Lorimer 1981).

In C_4 photosynthetic carbon metabolism, the initial carboxylation reaction is catalyzed by phosphoenolpyruvate carboxylase (PEPcase). An important aspect of C_4 pathways is that the photosynthetic reactions are divided between the mesophyll cells and bundle sheath cells; Rubisco and the C_3 photosynthetic carbon reduction cycle are located exclusively in the chloroplast of bundle sheath cells, whereas initial assimilation of atmospheric CO_2 takes place in the cytosol of mesophyll cells (Hatch 1987) by PEPcase. PEPcase utilizes bicarbonate rather than CO_2 as the inorganic substrate (O'Leary 1982). In C_4 plants, at the site of carboxylation, a continuous supply of HCO_3^- is provided via association of CA with PEPcase (Rathnam and Das 1975). This is supported by the fact that CA of C_4 leaves is largely or exclusively confined to the cytosol of mesophyll cells, while bundle sheath cells contain little or no CA activity (Burnell and Hatch 1988). To sustain this process, atmospheric CO_2 entering the mesophyll cells must be rapidly converted to HCO_3^-, and this reaction is the critical first step of C_4 photosynthesis (Hatch and Burnell 1990). In C_4 plants, however, most of the CA activity is found in mesophyll cells, where PEP carboxylase is also located (Hatch and Burnell 1990). Coexistence and feeding of Rubisco in C_3 plants and PEPcase in C_4 plants with CA as a carbon source assumes the special significance of CA as an efficient biochemical marker for carbon sequestration and environmental amelioration in the current global warming scenario linked with elevated CO_2 concentration (Tiwari et al. 2005).

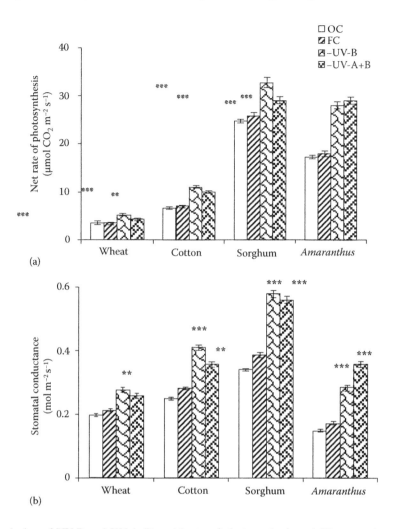

(a)

(b)

FIGURE 15.7 Effect of exclusion of UV-B and UV-A+B on (a) rate of photosynthesis and (b) stomatal conductance in wheat, cotton, *Amaranthus*, and sorghum. The vertical bars indicate mean ± SE ($n = 15$). Values are significantly different ($^{**}P < .01$, $^{***}P < .001$) from filter control (Newman–Keulis multiple comparison test). Nonsignificant terms are not shown.

We have measured the activity of CA, Rubisco, and PEPcase in C_3 and C_4 monocot/dicot crop species (Figure 15.8a–c). Solar UV exclusion leads to a remarkable increase in the activity of CA, Rubisco, and PEPcase in C_3 and C_4 monocot/dicot crop species (Figure 15.8a–c). This was accompanied by an enhancement in the total soluble proteins (Figure 15.8d). Kataria et al. (2013) first reported the enhancement in the activity of Rubisco by the fixation of $^{14}CO_2$ in C_3 and C_4 monocot/dicot crop species by the exclusion of solar UV components. In the presence of solar UV radiation, two possible effects might be responsible for the reduced activity of Rubisco. One of the reasons could be degradation of Rubisco subunits upon exposure to UV-B, as shown for higher plants and macroalgae (Bischof et al. 2002; Pedro et al. 2009). The second reason might be that wavelengths effective in suppressing gene expression of key proteins involved in photosynthesis (like rbcL encoding for large subunit [LSU] of Rubisco and *psbA* for the D1 proteins) are excluded (Jordan et al. 1992; Mackerness et al. 1999). Damage to enzyme proteins such as Rubisco could also result from the active oxygen species formed under ambient UV-B radiation (Shine and Guruprasad 2012).

Previous studies on UV effects on higher plants have stressed the central role of Rubisco in inhibition of photosynthesis under UV exposure (Vu et al. 1984; Jordan et al. 1992; Allen et al. 1997). Both the activity and content of the enzyme may decline during UV exposure, and thus, CO_2 assimilation drops. A large reduction in the expression and abundance of both large and small subunits of Rubisco would contribute to depressing photosynthesis and yield (Kulandaivelu et al. 1991; Pfündel 2003). On the other hand, UV exclusion enhanced the expression and abundance of small subunits of Rubisco (14 kDa) in cotton plants (Dehariya et al. 2012) and large subunits of Rubisco (55 kDa) in soybean (Singh 2011).

Carbon fixation in *A. tricolor* (C_4 dicot plant) and *S. bicolor* (C_4 monocot plant) is catalyzed by PEPcase. The enhanced activity of CA and PEPcase was found in C_3/C_4 plants like cotton, *Amaranthus*, wheat, and Sorghum after exclusion of ambient UV-B and UV-A+B, although in C_4 plants like *Amaranthus* and Sorghum, the extent of promotion in PEP carboxylase activity was higher as compared to C_3 plants-cotton and wheat (Figure 15.8c). The increase in CA activity by exclusion of solar UV radiation may enhance the fixation of

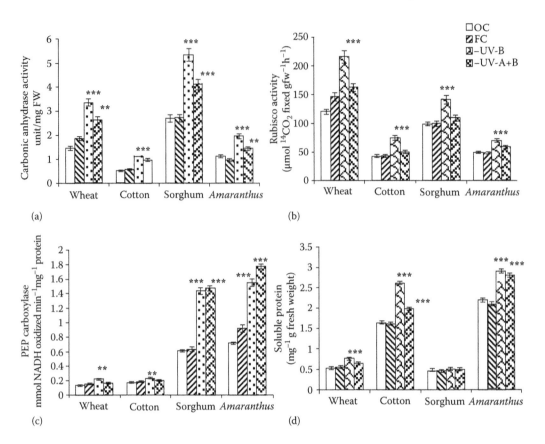

FIGURE 15.8 Effect of exclusion of UV-B and UV-A+B on (a) carbonic anhydrase, (b) Rubisco activity, (c) PEPcase, and (d) soluble protein in wheat, cotton, sorghum, and *Amaranthus*. The vertical bars indicate mean ± SE ($n = 3$). Values are significantly different ($^{**}P < .01$, $^{***}P < .001$) from filter control (Newman–Keulis multiple comparison test). Nonsignificant terms are not shown.

CO_2 via PEP carboxylase as HCO_3^- is the active CO_2 species for PEPcase (Hatch 1987). Enhanced or supplemental UV-B radiation inhibited PEPcase activities in many plant species (Vu et al. 1982; Jordan et al. 1992; He et al. 1994; Correia et al. 1999). Consistent with these results, the activity of the enzymes involved in CO_2 assimilation like CA and Rubisco were also suppressed by enhanced UV-B and solar UV-B in many plants, and it may be due to protein destruction or enzyme inactivation (Allen et al. 1997; Takeuchi et al. 2002; Xu et al. 2008; Kataria et al. 2013). Moreover, the decrease in PEP carboxylase (Vu et al. 1981) and Rubisco content and activity would contribute to depressing photosynthesis and yield (Takeuchi et al. 2002; Pfündel 2003; Yu et al. 2013). The high levels of UV radiation may possibly lead to degradation of the CA protein, as found for Rubisco and D1 protein (Bischof et al. 2002; Bouchard et al. 2005; Wu and Gao 2009).

In C_3 plants, PEPcase is thought to carry out various functions, including supplying carbon skeletons to the tricarboxylic acid (TCA) cycle, operating in malate homeostasis during stress, supplying carbon skeletons to allow ammonium assimilation, and regulating stomatal conductance. PEPcase also appears to play a role in the extension of cotton fibers (Li et al. 2010), and it is proposed that PEPcase activity allows malate production and, therefore, the increased turgor that is required for fiber elongation. In wheat, PEPcase is relatively abundant

in the meristematic and vascular cells, and the abundance of transcripts encoding PEPcase increased during salt and drought stress (Gonzalez et al. 2003). Thus, enhancement in the net rate of photosynthesis owing to solar UV exclusion by CA, Rubisco, and PEPcase in C_3/C_4 plants channelized the additional CO_2 fixation towards the yield of these plants.

15.3 EFFECT OF SOLAR UV EXCLUSION ON CROP YIELD

The increase in biomass, enhanced rate of photosynthesis, and increased activity of enzymes related to C_3 and C_4 carbon fixation by solar UV exclusion was also reflected in the yield parameters of C_3/C_4 plants. The yield parameters of C_3/C_4 plants—cotton, wheat, *Amaranthus*, and sorghum—are shown in Figure 15.9. Exclusion of UV-A along with UV-B enhanced the yield parameters in terms of weight of total bolls and fibers in cotton; number of leaves and foliage yield (fresh weight of leaves) in *Amaranthus*; and number of ears/panicles, grains, and grain yield per plant in wheat and sorghum (Figure 15.9). However, a considerable variation in the response of these parameters exists amongst the crops. In wheat and sorghum, enhancement was greater after UV-B exclusion, whereas in cotton and *Amaranthus*, the yield parameters were enhanced to a greater extent after exclusion of UV-A+B (Kataria et al. 2013).

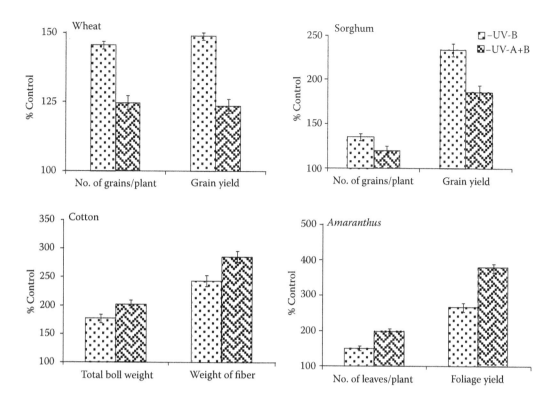

FIGURE 15.9 Effect of exclusion of UV-B and UV-A+B on yield parameters of wheat, cotton, sorghum, and *Amaranthus*. Changes are depicted in percent change as compared to the filter control of each treatment.

Removal of UV from solar radiation caused a significant increase in yield attributes, like number of pods and seeds and weight of seeds in *Cymopsis* (Amudha et al. 2005) and soybean (Baroniya et al. 2011). Germ et al. (2005) found that exclusion of UV-B from the solar radiation led to more than double the yield of pumpkin fruits. Thus, ambient UV caused reduction in the crop yield, due to alterations in plant vegetative and reproductive growth, e.g., plant stunting, flower suppression and/or delay of flowering, and lower pod set (Amudha et al. 2005). A high level of UV-B radiation delays flowering, thereby reducing the yield (Kakani et al. 2003), while exclusion of UV-B by covering plants with a Mylar sheet increased

the flowering in alpine plants (Caldwell 1968). A delay in the onset of flowering was observed in the ambient light-grown *Cyamopsis* plants, and this delay in flowering may be due to the impact of solar UV radiation on the biosynthesis of gibberellins, as found *in Hyoscyamus niger* (Saile-Mark 1993) and bush beans (Mark and Tevini 1997).

15.4 SOLAR UV SENSITIVITY OF C_3 AND C_4 PLANTS

The data obtained from the solar UV exclusion studies showed the sensitivity of crops to ambient UV-A and UV-B radiation. Sensitivity to UV, radiation, however varies considerably between the plant species. Sensitivity indices have been established as useful indicators of plant sensitivity to solar UV and enhanced UV-B radiation (Saile-Mark and Tevini 1997; Li et al. 2002; Kataria and Guruprasad 2012a). Excluding UV (280–400 nm) significantly increased plant height, leaf area, biomass accumulation, and yield in the varieties of C_3 and C_4 plants. We have determined the UV sensitivity index of C_3 (cotton/wheat) and C_4 (*Amaranthus*/sorghum) to an ambient level of UV radiation after the exclusion of solar UV radiation. Figure 15.10 illustrates a UV sensitivity index (UV SI) of C_3 and C_4 plants. A UV SI of C_3 and C_4 plant species was calculated in terms of plant height, total dry weight accumulation, and grain/seed/foliage yield in the presence and absence of solar UV-B and UV-A+B components, and it could reflect the overall sensitivity of the crop species to the current level of UV radiation (Figure 15.10). Kataria and Guruprasad

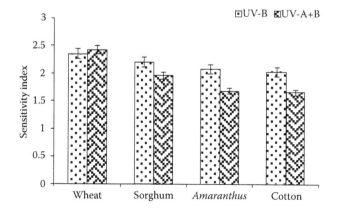

FIGURE 15.10 UV sensitivity index for C_3 and C_4 plants to current level of UV radiation by the exclusion of solar UV-B and UV-A+B.

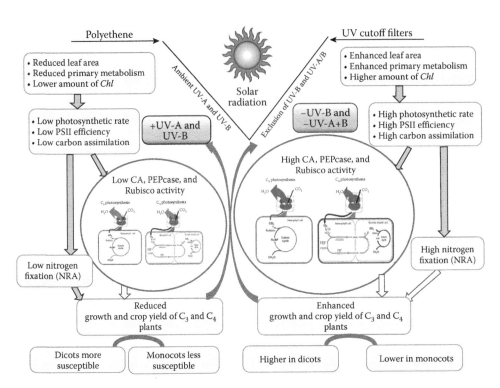

FIGURE 15.11 (See color insert.) A model to summarize the growth, photosynthetic, and yield responses to solar UV-B and UV-A+B exclusion for C_3 and C_4 plants. NRA, nitrate reductase activity.

(2012a) reported that when the UV SI of the crop species is significantly less than 3, that crop is more sensitive to solar UV, whereas when it is nearer to 3, the crop is less sensitive to solar UV. UV SI calculated after exclusion of solar UV-A and UV-B radiation indicated wheat (UV SI, 2.42) and sorghum (UV SI, 1.96) to be the least sensitive and *Amaranthus* (UV SI, 1.680) and cotton (UV SI, 1.653) to be more UV sensitive (Figure 15.10). According to UV SI, the sensitivity of C_3 and C_4 plants to an ambient level of UV (280–400 nm) radiation has the following sequence: wheat > sorghum > *Amaranthus* > cotton. Thus, cotton is the most sensitive, and wheat is the least sensitive crop to solar UV radiation. UV exclusion studies seem to be more appropriate to identify the most sensitive or insensitive crop to the ambient level of solar UV radiation.

15.5 CONCLUSIONS AND FUTURE PERSPECTIVE

In conclusion, UV exclusion findings emphasize that ambient UV radiation is a significant stress factor for crops in tropical climatic conditions. Exclusion of UV radiation is beneficial from the agricultural standpoint to improve the growth, biomass production, photosynthetic performance, and yield in C_3 and C_4 plants (Figure 15.11). Improvement in crop yield by solar UV exclusion is due to an increase in the overall processing of light energy per leaf sample, and as a consequence, higher amounts of ATP and NADPH are produced. Thus, the capacity for ribulose-1,5-bisphosphate and phosphophenol pyruvate regeneration increases. Exclusion of solar UV caused a remarkable increase in the activity of Rubisco and

PEPcase and channelized the additional fixation of carbon toward the improvement of yield (Figure 15.11). However, C_3 and C_4 dicots (cotton and *Amaranthus*) are more sensitive than C_3 and C_4 monocots (wheat and sorghum) to ambient UV (280–400 nm) radiation (Figure 15.10). This indicates that dicot plants are more sensitive to ambient solar UV radiation as compared to monocots, while there is no difference between C_3 and C_4 plants. Moreover, the leaf orientation of dicots and monocots is different. Thus, wheat, rice, sorghum, and maize, which are monocots and have nonhorizontal leaves, are less sensitive than dicots like cotton, soybean, pea, cucumber, and *Amaranthus*, which have broad, horizontal leaves.

The enhancement in photosynthetic efficiency of plants in the absence of solar UV components not only fixes atmospheric carbon dioxide rapidly but also reduces global warming. In addition to this, solar UV exclusion experiments for sequestration of atmospheric carbon are environmentally and biologically nonhazardous. Solar UV exclusion neither causes any change in the existing environment nor adds artificial chemical or foreign genes. Solar UV radiation will be more advantageous and challenging if studied in combination with other abiotic stresses, including increased atmospheric carbon dioxide, temperature, drought, and nitrogen deposition. Further investigations are needed to realize these possibilities.

Another important area of interest will be elucidating the significance of UV impact on the photosynthetic apparatus in relation to damage caused at the level of nucleic acids. A significant downregulation of genes associated with photosynthesis was found in response to supplementary/enhanced UV-B, but the changes in gene expression after exclusion of solar UV have to be investigated. Understanding the mechanism of how solar UV radiation modifies gene expression and results in the enhancement of overall photosynthetic performance in C₃ and C₄ plants will require further exploration.

ACKNOWLEDGMENTS

Financial support by the Women Scientists-A Scheme (SR/WOS-A/LS-101/2009 and LS-674/2012-G) given by Department of Science and Technology, New Delhi to Dr. S. Kataria is thankfully acknowledged.

REFERENCES

Albert, K.R., T.N. Mikkelsen and H. Ro-Poulsen. 2005. Effects of ambient versus reduced UV-B radiation on high arctic *Salix arctica* assessed by measurements and calculations of chlorophyll-*a* fluorescence parameters from fluorescence transients. *Physiologia Plantarum* 124: 208–26.

Albert, K.R., T.N. Mikkelsen and H. Ro-Poulsen. 2008. Ambient UV-B radiation decreases photosynthesis in high arctic *Vaccinium uliginosum*. *Physiologia Plantarum* 133: 199–210.

Albert, K.R., T.N. Mikkelsen, H. Ro-Poulsen, A. Michelsen, M.F. Arndal, L. Bredahl, K.B. Hakansson, K. Boesgaard and N.M. Schmidt. 2010. Improved UV-B screening capacity does not prevent negative effects of ambient UV irradiance on PSII performance in High Arctic plants. Results from a six year UV exclusion study. *Journal of Plant Physiology* 167: 1542–49.

Albert, K.R., T.N. Mikkelsen, H. Ro-Poulsen, M.F. Arndal and A. Michelsen. 2011. Ambient UV-B radiation reduces PSII performance and net photosynthesis in high Arctic *Salix arctica*. *Environmental Experimental Botany* 72: 439–47.

Allen, D.J., I.F. Mckee, P.K. Farage and N.R. Baker. 1997. Analysis of limitations to CO₂ assimilation on exposure of leaves of two *Brassica napus* cultivars to UV-B. *Plant Cell Environment* 20: 633–40.

Allen, D.J., S. Nogués and N.R. Baker. 1998. Ozone depletion and increased UV-B radiation: Is there a real threat to photosynthesis? *Journal of Experimental Botany* 49: 1775–78.

Allen, D.J., S. Nogués, J.I.L. Morison, P.D. Greenslade, A.R. McLeod and N.R. Baker. 1999. A thirty percent increase in UV-B has no impact on photosynthesis in well-watered and droughted pea plants in the field. *Global Change Biology* 5: 235–44.

Amudha, P., M. Jayakumar and G. Kulandaivelu. 2005. Impacts of ambient solar UV (280–400 nm) radiation on three tropical legumes. *Journal of Plant Biology* 48: 284–91.

Aro, E.M, I. Virgin and B. Andersson. 1993. Photoinhibition of photosystem II. Inactivation, protein damage and turnover. *Biochimica et Biophysica Acta* 1143: 113–34.

Baroniya, S.S., S. Kataria, G.P. Pandey and K.N. Guruprasad. 2011. Intraspecific variation in sensitivity to ambient ultraviolet-B radiation in growth and yield characteristics of eight soybean cultivars grown under field conditions. *Brazilian Journal of Plant Physiology* 23: 197–202.

Baroniya, S.S., S. Kataria, G.P. Pandey and K.N. Guruprasad. 2014. Growth, photosynthesis and nitrogen metabolism in soybean varieties after exclusion of UV-B and UV-A/B components of solar radiation. *The Crop Journal* 2: 388–97.

Basiouny, C.J., T.K. Van and R.H. Biggs. 1978. Some morphological and biochemical characteristics of C₃ and C₄ plants irradiated with UV-B. *Physiologia Plantarum* 42: 29–32.

Becwar, M.R., F.D. Moore and M.J. Burke. 1982. Effect of deletion and enhancement of ultraviolet-B (280–315) radiation on plant growth at 3000m elevation. *American Society for Horticultural Science* 107: 771–74.

Bischof, K., G. Krabs. C. Wiencke and D. Hanelt. 2002. Solar ultraviolet radiation affects the activity of ribulose-1,5-bisphosphate carboxylase-oxygenase and the composition of photosynthetic and xanthophyll cycle pigments in the intertidal green alga *Ulva lactuca* L. *Planta* 215: 502–09.

Bouchard, J.N., D.A. Campbell and S. Roy. 2005. Effects of UV-B radiation on the D1 protein repair cycle of natural phytoplankton communities from three latitudes (Canada, Brazil, and Argentina). *Journal of Phycology* 41: 273–86.

Burnell, J.N. and M.D. Hatch. 1988. Low bundle sheath carbonic anhydrase is apparently essential for effective C₄ pathway operation. *Plant Physiology* 86: 1252–56.

Butler, W.L. 1978. Energy distribution in the photochemical apparatus of photosynthesis. *Annual Review of Plant Physiology* 29: 345–78.

Caldwell, M.M. 1968. Solar ultraviolet radiation as an ecological factor for alpine plants. *Ecological Monograph* 38: 243–68.

Caldwell, M.M., J.F. Bornman, C.L. Ballare, S.D. Flint and G. Kulandaivelu. 2007. Terrestrial ecosystems, increased solar ultraviolet radiation, and interactions with other climate change factors. *Photochemistry Photobiology Science* 6: 252–66.

Caldwell, M.M., S.D. Flint and P.S. Searles. 1994. Spectral balance and UV-B sensitivity of soybean: A field experiment. *Plant Cell Environment* 17: 267–76.

Caldwell, M.M. and S.D. Flint. 1997. Uses of biological spectral weighting functions and the need of scaling for the ozone reduction problem. *Plant Ecology* 128: 66–76.

Chen, S. G., C. Y. Yin, X. B. Dai, S. Qiang and X. M. Xu. 2008. Action of tenuazonic acid, a natural phytotoxin, on photosystem II of spinach. *Environmental and Experimental Botany* 62: 279–89.

Chouhan, S., K. Chauhan, S. Kataria and Guruprasad, K.N. 2008. Enhancement in leghemoglobin content of root nodules by exclusion of UV-A and UV-B radiation in soybean. *Journal of Plant Biology* 51: 132–38.

Christen, D., S. Schönmanna, M. Jermini, R.J. Strasser and G. Defago. 2007. Characterization and early detection of grapevine (*Vitis vinifera*) stress responses to esca disease by in situ chlorophyll fluorescence and comparison with drought stress. *Environmental and Experimental Botany* 60: 504–14.

Cicek, N., I. Fedina, H. Cakirlar, M. Velitchkova and K. Georgieva. 2012. The role of short-term high temperature pre-treatment on the UV-B tolerance of barley cultivars. *Turkish Journal of Agriculture and Forestry* 36: 153–65.

Coleman, R.S. and T.A. Day. 2004. Response of cotton and sorghum to several levels of subambient solar UV-B radiation: A test of the saturation hypothesis. *Physiologia Plantarum* 122: 362–72.

Correia, C.M., E.L.V. Areal, M.S. Torres-Pereira, J.M.G. and Torres-Pereira.1999. Intraspecific variation in sensitivity to ultraviolet-B radiation in maize grown under field conditions II. Physiological and biochemical aspects. *Field Crops Research* 62: 97–105.

Cybulski, W.J. and W.T. Peterjohn. 1999. Effects of ambient UV-B radiation on the above-ground biomass of seven temperate-zone plant species. *Plant Ecology* 145: 175–81.

Dehariya, P., S. Kataria, G.P. Pandey and K.N.Guruprasad 2012. Photosynthesis and yield in cotton (*Gossypium hirsutum* L.) var. vikram after exclusion of ambient solar UV-B/A. *Acta Physiologia Plantarum* 34: 1133–44.

de la Rosa, T.M., R. Julkunen-Tiitto, T. Lehto and P.J. Aphalo. 2001. Secondary metabolites and nutrient concentrations in silver birch seedlings under five levels of daily UV-B exposure and two relative nutrient addition rates. *New Phytology* 150: 121–31.

Flint, S.D. and M.M. Caldwell. 1996. Scaling plant ultraviolet spectral responses from laboratory action spectra to field spectral weighing factors. *Journal of Plant Physiology* 148: 107–14.

Furbank, R.T. and W.C. Taylor. 1995. Regulation of photosynthesis in C_3 and C_4 plants: A molecular approach. *American Society of Plant Physiology* 7: 797–807.

Germ, M., I. Kreft and J. Osvald. 2005. Influence of UV-B exclusion and selenium treatment on photochemical efficiency of photosystem II, yield and respiratory potential in pumpkins (*Cucurbita pepo* L.). *Plant Physiology and Biochemistry* 43: 445–48.

Goncalves, J.F.C. and U.M. Santos Junior. 2005. Utilization of the chlorophyll *a* fluorescence technique as a tool for selecting tolerant species to environments of high irradiance. *Brazilian Journal of Plant Physiology* 17: 307–13.

Gonzalez, M. C., R. Sanchez and F.J. Cejudo. 2003. Abiotic stresses affecting water balance induce phosphoenolpyruvate carboxylase expression in roots of wheat seedlings. *Planta* 216: 985–92.

Guruprasad, K., S. Bhattacharjee, S. Kataria, S. Yadav, A. Tiwari, S. Baroniya, A. Rajiv and P. Mohanty. 2007. Growth enhancement of soybean (Glycine max) upon exclusion of UV-B and UV-A components of solar radiation: Characterization of photosynthetic parameters in leaves. *Photosynthetic Research* 94: 299–306.

Hatch, M.D. 1987. C_4 photosynthesis: A unique blend of modified biochemistry, anatomy and ultrastructure. *Biochimica et Biophysica Acta* 895: 81–96.

Hatch, M.D. and J.N. Burnell. 1990. Carbonic anhydrase activity in leaves and its role in the first step of C_4 photosynthesis. *Plant Physiology* 93: 825–28.

He, J., L.K. Huang, W.S. Chow, M.I. Whitecross and J.M. Anderson. 1993. Effects of supplementary ultraviolet-B radiation on rice and pea plants. *Australian Journal of Plant Physiology* 20: 129–42.

He, J., L.K. Huang and M.I., Whitecross. 1994. Chloroplast ultra structure changes in *Pisum sativum* associated with supplementary ultraviolet (UV-B) radiation. *Plant Cell and Environment* 17: 771–75.

Helsper, J.P.F., C.H. Ric de Vos, F.M. Maas, H.H. Jonker, H. C. Van den Broeck, W. Jordi, C. Sander, L.C. Paul Keizer. and H.C.M. Schapendonk. 2003. Response of selected antioxidants and pigments in tissues of *Rosa hybrida* and *Fuchsia hybrida* to supplemental UV-A exposure. *Physiologia Plantarum* 117: 171–78.

Hermans, C., M. Smeyers, R.M. Rodriguez, M. Eyletters, R.J. Strasser and J.P. Delhaye. 2003. Quality assessment of urban trees: A comparative study of physiological characterization, airborne imaging and on site fluorescence monitoring by the OJIP-test. *Journal of Plant Physiology* 160: 81–90.

Hollosy, F. 2002. Effects of ultraviolet radiation on plant cells. *Micron* 33: 179–97.

Hsu, B.D. 1993. Evidence for the contribution of S state transitions of oxygen evolution to the initial phase of fluorescence induction. *Photosynthesis Research* 36: 81–88.

Ibanez, S., M. Rosa, M. Hilal, J.A. Gonzalez and F.E. Prado. 2008. Leaves of *Citrus aurantifolia* exhibit a different sensibility to solar UV-B radiation according to development stage in relation to photosynthetic pigments and UV-B absorbing compounds production. *Journal of Photochemistry Photobiology B: Biology* 90: 163–69.

Ibrahim, M.M., A.A. Alsahli and A.A. Al-Ghamdi. 2013. Cumulative abiotic stresses and their effect on the antioxidant defense system in two species of wheat, *Triticum durum* Desf. and *Triticum aestivum* L. *Archives of Biological Sciences Belgrade* 65: 1423–33.

Ines, C., F.F. Terezinha and L.D. Anne. 2007. Growth and physiological responses of sunflower plants exposed to ultraviolet-B radiation. *Sci. Rural* 37: 85–90.

Jansen, M.A.K., B.L. Martreta and M. Koornneef. 2010. Variations in constitutive and inducible UV-B tolerance; dissecting photosystem II protection in *Arabidopsis thaliana* accessions. *Physiologia Plantarum* 138: 22–34.

Jansen, M.A.K., V. Gaba, B.M. Greenberg, A.K. Mattoo and M. Edelman. 1996. Low threshold levels of ultraviolet-B in a background of photosynthetically active radiation trigger rapid degradation of the D2 protein of photosystem II. *Plant Journal* 9: 693–99.

Jiang, M.Y., J.H. Jing and S.T. Wang. 1991. Effect of osmotic stress on photosynthetic pigment and level of membrane-lipidperoxidation in rice seedlings. *Acta Universitatis Agriculture Boreali-Occidentalis* 19: 79–84.

Jordan, B.R., J. He, W.S. Chow and J.M. Anderson. 1992. Changes in mRNA levels and polypeptide subunits of ribulose-1, 5-bisphosphate carboxylase in response to supplemental UV-B radiation. *Plant Cell Environment* 15: 91–98.

Kakani, V.G., K.R. Reddy, D. Zhao and K. Sailaja. 2003. Field crop responses to ultraviolet-B radiation: A review. *Agricultural and Forest Meteorology* 120: 191–218.

Kataria, S. and K.N. Guruprasad. 2012a. Solar UV-B and UV-A/B exclusion effects on intraspecific variations in crop growth and yield of wheat varieties. *Field Crops Research* 125: 8–13.

Kataria, S. and K.N. Guruprasad. 2012b. Intraspecific variations in growth, yield and photosynthesis of sorghum varieties to ambient UV (280–400 nm) radiation. *Plant Science* 196: 85–92.

Kataria, S. and K.N. Guruprasad. 2014. Exclusion of solar UV components improves growth and performance of *Amaranthus tricolor* varieties. *Scientia Horticulturae* 174: 36–45.

Kataria, S., K.N. Guruprasad, S. Ahuja, B. Singh. 2013. Enhancement of growth, photosynthetic performance and yield by exclusion of ambient UV components in C_3 and C_4 plants. *Journal of Photochemistry and Photobiology B: Biology* 127: 140–52.

Kataria, S., A. Jajoo and K. N. Guruprasad. 2014. Impact of increasing Ultraviolet-B (UV-B) radiation on photosynthetic processes. *Journal of Photochemistry and Photobiology B: Biology* 137: 55–66.

Kataria, S. and K.N. Guruprasad. 2015. Exclusion of solar UV Radiation improves photosynthetic performance and yield of Wheat varieties. Accepted in *Plant Physiology and Biochemistry*.

Keiller, D.R. and M.G. Holmes. 2001. Effects of long term exposure to elevated UV-B radiation on the photosynthetic performance of five broad leaved tree species. *Photosynthetic Research* 67: 229–40.

Keiller, D.R., S.A.H. Mackerness and M.G. Holmes. 2003. The action of a range of supplementary ultraviolet (UV) wavelengths on photosynthesis in *Brassica napus* L. in the natural environment: Effects on PS-II, CO_2 assimilation and level of chloroplast proteins. *Photosynthetic Research* 75: 139–50.

Krause, G.H., E. Grube, A. Virgo and K. Winter. 2003. Sudden exposure to solar UV-B radiation reduces net CO_2 uptake and photosystem-I efficiency in shade-acclimated tropical tree seedlings. *Plant Physiology* 131: 745–52.

Krizek, D.T. and R.M. Mirecki. 2004. Evidence for phytotoxic effects of cellulose acetate in UV exclusion studies. *Environmental and Experimental Botany* 51: 33–43.

Krizek, D.T., R.M. Mirecki and S.J. Britz. 1997. Inhibitory effects of ambient levels of solar UV-A and UV-B radiation on growth of cucumber. *Physiologia Plantarum* 100: 886–93.

Kulandaivelu, G., N. Neduchezhian and K. Annamalainathan. 1991. Ultraviolet-B (280–320 nm) radiation induced changes in photochemical activities and polypeptide components of C₃ and C₄ chloroplasts. *Photosynthetica* 25: 333–39.

Li, X.R., L. Wang and Y.L. Ruan. 2010. Developmental and molecular physiological evidence for the role of phosphoenolpyruvate carboxylase in rapid cotton fibre elongation. *Journal of Experimental Botany* 61: 287–95.

Li, Y., Y.Q. Zu, H.Y. Chen, J.J. Chen, J.L. Yang and Z.D. Hu. 2000. Intraspecific responses in crop growth and yield of 20 wheat cultivars to enhanced ultraviolet-B radiation under field conditions. *Field Crops Research* 67: 25–33.

Li, Y., Y.Q. Zu, J.J. Chen and H.Y. Chen. 2002. Intraspecific responses in crop growth and yield of 20 soybean cultivars to enhanced ultraviolet-B radiation under field conditions. *Field Crops Research* 78: 1–8.

Lorimer, G.H. 1981. The carboxylation and oxygenation of ribulose-1,5-bisphosphate: The primary event in photosynthesis and photorespiration. *Annual Review of Plant Physiology* 32: 349–82.

Mackerness, S.A.H., S.L. Surplus, P. Blake, C.F. John, V. Buchanan-Wollaston, B.R. Jordan and B. Thomas. 1999. UV-B induced stress and changes in gene expression in Arabidopsis thaliana: Role of signaling pathways controlled by jasmonic acid, ethylene and reactive oxygen species. *Plant Cell Environment* 22: 1413–23.

Majeau, N. and J.R. Coleman. 1991. Isolation and characterization of a cDNA coding for pea chloroplastic carbonic anhydrase. *Plant Physiology* 95: 264–68.

Mark, S.M. and M. Tevini. 1997. Effects of solar UV-B radiation on growth, flowering and yield of central and southern European bush bean cultivars (*Phaseolus vulgaris* L.). *Plant Ecology* 128: 114–25.

Mathur, S., A. Jajoo, P. Mehta and S. Bharti. 2011. Analysis of elevated temperature-induced inhibition of photosystem II by using chlorophyll *a* fluorescence induction kinetics in wheat leaves (*Triticum aestivum*). *Plant Biology* 13: 1–6.

Mazza, C.A., D. Battista, A.M. Zima, M. Scwarcberg-Bracchitta, C. Giordano, A. Acevedo, A.L. Scopel and C.L. Ballare. 1999. The effects of solar ultraviolet-B radiation on the growth and yield of barley are accompanied by increased DNA damage and antioxidant responses. *Plant Cell Environment* 22: 61–70.

McKenzie, R.L., P.J. Aucamp, A.F. Bais, L.O. Björn and M. Ilyas. 2007. Changes in biologically-active ultraviolet radiation reaching the Earth's surface. *Photochemistry and Photobiology Science* 6: 218–31.

Mehta, P., A. Jajoo, S. Mathur and S. Bharti. 2010. Chlorophyll *a* fluorescence study revealing effects of high salt stress on Photosystem II in wheat leaves. *Plant Physiology and Biochemistry* 48: 16–20.

Melis, A., J.A. Nemson, M.A. and Harrison. 1992. Damage to functional components and partial degradation of photosystem II reaction centre proteins upon chloroplast exposure to ultraviolet-B radiation. *Biochimica et Biophysica Acta* 1109: 312–20.

Middleton, E.M. and A.H. Teramura. 1993. The role of flavonol glycosides and carotenoids in protecting soybean from ultraviolet-B damage. *Plant Physiology* 103: 741–52.

Moussa, H.R. and S.D.K. Khodary. 2008. Changes in growth and ¹⁴CO₂ fixation of *Hordeum vulgare* and *Phaseolus vulgaris* induced by UV-B radiation. *Journal of Agriculture and Social Sciences* 4: 59–64.

Nayyar, H. 2003. Accumulation of osmolytes and osmotic adjustment in water-stressed wheat (*Triticum aestivum*) and maize (*Zea mays*) as affected by calcium and its antagonists. *Environmental Experimental Botany* 50: 253–64.

O'Leary, M. H. 1982. Phosphoenolpyruvate carboxylase: An enzymologist's view. *Annual Review of Plant Physiology* 33: 297–315.

Pal, M., A. Sharma, Y.P. Abrol and U.K. Sengupta. 1997. Exclusion of solar UV-B radiation from normal spectrum on growth of mung bean and maize. *Agriculture, Ecosystems and Environment* 61: 29–34.

Pal, M., P.H. Zaidi, S.R. Voleti and A. Raj. 2006. Solar UV-B exclusion effect on growth and photosynthetic characteristics of wheat and pea. *Journal of New Seeds* 8: 19–34.

Pedro, J.A., M.V. Elina, M.R. Tania and L. Tarja. 2009. Does Supplemental UV-B radiation affect gas exchange and Rubisco activity of *Betula pendula* Roth. seedlings grown in forest soil under greenhouse conditions? *Plant Ecology and Diversity* 2: 37–43.

Pfündel, E. 2003. Action of UV and visible radiation on chlorophyll fluorescence from dark adapted grape leaves (*Vitis vinifera L.*). *Photosynthesis Research* 75: 29–39.

Poorter, H. 1989. Interspecific variation in relative growth rate: On ecological causes and physiological consequences. In *Causes and Consequences of Variation in Growth Rate and Productivity of Higher Plants*, ed. H. Lambers, M.L. Cambridge, H. Konings, and T.L. Pons, 45–68. SPB Academic Publishing, The Hague.

Rathnam, C.K.M. and V.S.R. Das. 1975. Aspartate-type C-4 photosynthetic carbon metabolism in leaves of *Eleusine coracana* GAERTN. *Z. Pflanzenphysiology* 74: 377–93.

Reddy, R.K., S.K. Singh, S. Koti, V.G. Kakani, D., Zhao, W. Gao and V.R. Reddy. 2013. Quantifying the effects of corn growth and physiological responses to Ultraviolet-B radiation for modeling. *Agronomy Journal* 105: 1367–77.

Renger, G., M. Volkar, H.J. Eckort, R. Fromona, S. Hohm-Veit and P. Graber. 1989. On the mechanism of photosystem II deterioration by UV-B irradiation. *Photochemistry and Photobiology* 49: 97–105.

Rinnan, R., M.M. Keinanen, A. Kasurinen, J. Asikainen, T.K. Kekki, T. Holopainen, H. Ro-Poulasen, T.N. Mikkelsens and A. Michelse. 2005. Ambient ultraviolet radiation in the Arctic reduces root biomass and alters microbial community composition but has no effects on microbial biomass. *Global Change Biology* 11: 564–74.

Rousseaux, C.M., S.D. Flint, P.S. Searles and M.M. Caldwell. 2004. Plant responses to current solar ultraviolet-B radiation and supplemented solar ultraviolet-B radiation simulating ozone depletion: An experimental comparison. *Photochemistry and Photobiology* 80: 224–30.

Sahoo, A., S. Sarkar, R.P. Singh, M. Kafatos and M.E. Summers. 2005. Declining trend of total ozone column over the northern parts of India. *International Journal of Remote Sensing* 26: 3433–40.

Saile-Mark, M. 1993. Zur Beteiligung von Phytohormone an Wachstum und Blutenbildung verschiedener Bohnenkulturvarietaten (*Vigna vulgaris* L.) in Abhangigkeit von artifizeller und solarer UV-B Strahlung. *Karlsruher Beiträge zur Entwicklungs- und Okophysiologie der Pflanzen*, 13: 1–152.

Saile-Mark, M. and M. Tevini. 1997. Effects of solar UV-B radiation on growth, flowering and yield of central and southern European bush bean cultivars (*Phaseolus vulgaris* L.). *Plant Ecology* 128: 115–25.

Sakaki, T., N. Kondo and K. Sugahara. 1983. Breakdown of photosynthetic pigments and lipids in spinach leaves with ozone fumigation: Role of active oxygens. *Physiologia Plantarum* 59: 28–34.

Schumaker, M.A., J.H. Bassman, R. Robberecht and G.K. Radamaker. 1997. Growth, leaf anatomy and physiology of *Populus* clones in response to solar ultraviolet-B radiation. *Tree Physiology* 17: 617–26.

Sedej, T. and A. Gabercik. 2008. The effects of enhanced UV-B radiation on physiological activity and growth of Norway spruce planted outdoors over 5 Years. *Trees* 22: 423–35.

Sharma, S. and K.N. Guruprasad. 2012. Enhancement of root growth and nitrogen fixation in Trigonella by UV-exclusion from solar radiation. *Plant Physiology and Biochemistry* 61: 97–102.

Shine, M.B. and K.N. Guruprasad. 2012. Oxyradicals and PS II activity in maize leaves in the absence of UV components of solar spectrum. *Journal of Bioscience* 37: 703–12.

Shiozaki, N., I. Hattori, R. Gojo and T. Tezuka. 1999. Activation of growth and nodulation in a symbiotic system between pea plants and leguminous bacteria by near-UV radiation. *Journal of Photochemistry and Photobiology B: Biology* 50: 33–37.

Singh, P. 2011. Impact of exclusion of solar UV components on photosynthesis, nitrogen fixation and growth of soybean. *Ph.D.Thesis*, Devi Ahilya Vishwavidyalaya, Indore, India.

Singh, R., S. Singh, R. Tripathi and S.B. Agrawal. 2011. Supplemental UV-B radiation induced changes in growth, pigments and antioxidant pool of bean (*Dolichos lablab*) under field conditions. *Journal of Environmental Biology* 32: 139–45.

Singh, S., R. Kumari, M. Agrawal and S.B. Agrawal. 2009. Modification of growth and yield responses of *Amaranthus tricolor* L. to sUV-B under varying mineral nutrient supply. *Scientia Horticulturae* 120: 173–80.

Solomon, K.R. 2008. Effects of ozone depletion and UV-B radiation on humans and the environment. *Atmosphere-Ocean* 46: 185–202.

Srivastava, A. and R.J. Strasser. 1996. Stress and stress management of land plants during a regular day. *Journal of Plant Physiology* 148: 445–55.

Stirbet, A. and Govindjee. 2011. On the relation between the Kautsky effect (chlorophyll *a* fluorescence induction) and photosystem II: Basics and applications of the OJIP fluorescence transient. *Journal of Photochemistry and Photobiology B* 104: 236–57.

Strasser, B.J. and R.J. Strasser. 1995. Measuring fast fluorescence transients to address environmental questions: The JIP-Test. In *Photosynthesis: From Light to Biosphere,* ed. P. Mathis, 977–980. Kluwer Academic Publishers, Dordrecht.

Strid, A., W.S. Chow and J.M. Anderson. 1990. Effects of supplementary ultraviolet-B radiation on photosynthesis of in *Pisum sativum*. *Biochimca Biophysica Acta* 1020: 260–68.

Strid, A., W.S. Chow and J.M. Anderson. 1994. UV-B damage and protection at the molecular level in plants. *Photosynthesis Research* 39: 475–89.

Takeuchi, A., T. Yamaguchi, J. Hidema, A. Strid and T. Kumagai. 2002. Changes in synthesis and degradation of Rubisco and LHCII with leaf age in rice (*Oryza sativa* L.) growing under supplementary UV-B radiation. *Plant Cell Environment* 25: 695–706.

Tevini, M. 2000. UV-B effects on plants. In *Environmental Pollution and Plant Responses*, ed. S.B. Agrawal, and M. Agrawal, pp. 83–97. Lewis Publishers, Boca Raton, FL.

Tezuka, T., T. Hotta and I. Wanatabe. 1993. Growth promotion of tomato and radish plants by solar UV radiation reaching the earth's surface. *Photochemistry Photobiol*ogy 19: 61–66.

Tiwari, A., P. Kumar, S. Singh and S.A. Ansari. 2005. Carbonic anhydrase in relation to higher plants. *Photosynthetica* 43: 1–11.

Tsuzuki, M., S., Miyachi and G.E. Edwards. 1985. Localization of carbonic anhydrase in mesophyll cells of terrestrial C_3 plants in relation to CO_2 assimilation. *Plant Cell Physiology* 26: 881–91.

Tyystjärvi, E. 2008. Photoinhibition of Photosystem II and photodamage of the oxygen evolving manganese cluster. *Coordination Chemistry Reviews* 252: 361–76.

Vu, C.V., L.H.J. Allen and L.A. Garrard. 1981. Effects of supplemental UV-B radiation on growth and leaf photosynthetic reactions of soybean (*Glycine max*). *Physiologia Plantarum* 52: 353–62.

Vu, C.V., L.H. Allen and L.A. Garrard. 1982. Effects of supplemental UV-B radiation on primary photosynthetic carboxylating enzymes and soluble proteins in leaves of C_3 and C_4 crop plants. *Physiologia Plantarum* 55: 11–16.

Vu, C.V., Allen L.H. and L.A. Garrard. 1984. Effects of UV-B radiation (280–320 nm) on ribulose-1, 5-bisphosphate carboxylase in pea and soybean. *Environmental and Experimental Botany* 24: 131–43.

Ward, J.K., D.T. Tissue, R.B. Thomas and B.R. Strain. 1999. Comparative responses of model C_3 and C_4 plants to drought in low and elevated CO_2. *Global Change Biology* 5: 857–67.

Werdan, K. and H.W. Heldt. 1972. Accumulation of bicarbonate in intact chloroplasts following a pH gradient. *Biochimica et Biophysica Acta* 283: 430–41.

Wu, H. and K. Gao. 2009. Ultraviolet radiation stimulated activity of extracellular carbonic anhydrase in the marine diatom *Skeletonema costatum*. *Functional Plant Biology* 36: 137–43.

Xiong, F.S. and T.A. Day. 2001. Effect of solar ultraviolet-B radiation during springtime ozone depletion on photosynthesis and biomass production of antarctic vascular plants. *Plant Physiology* 125: 738–51.

Xu, C., S. Natarajan and J.H. Sullivan. 2008. Impact of solar ultraviolet-B radiation on the antioxidant defense system in soybean lines differing in flavonoids content. *Environmental and Experimental Botany* 63: 39–48.

Yu, G.H., W. Li, Z.Y. Yuan, H.Y. Cui, C.G. Lv, Z.P. Gao, B. Han, Y.Z. Gong and G.X. Chen. 2013. The effects of enhanced UV-B radiation on photosynthetic and biochemical activities in super high-yield hybrid rice *Liangyoupeijiu* at the reproductive stage. *Photosynthetica* 51: 33–44.

Zavalla. J.A. and J.F. Botto. 2002. Impact of solar UV-B radiation on seedling emergence, chlorophyll fluorescence, and growth and yield of radish (*Raphanus sativus*). *Functional Plant Biology* 29: 797–804.

Zhang, X., L. Hu. Henkow, B.R. Jordan and A. Strid. 1994. The effects of ultraviolet-B radiation on the CF0F1-ATPase. *Biochimica et Biophysica Acta* 1185: 295–302.

16 What Role Does UVB Play in Determining Photosynthesis?

Brian R. Jordan, Å. Strid, and J.J. Wargent

CONTENTS

16.1 Introduction 275
16.2 Experimental Factors Affecting the Outcome of UVB Responses 276
16.3 UVB and Plant Physiological Responses 276
16.4 Effects of UVB on Biochemistry and Bioenergetics of Photosynthesis 277
    16.4.1 General Aspects 277
    16.4.2 Proteins Involved in the Photosynthetic Light and Dark Reactions 278
    16.4.3 Lipids and Membranes 279
    16.4.4 Reactive Oxygen Species and Various Radicals...279
16.5 Molecular-Level Responses to UVB and Implications for Photosynthesis 280
16.6 Conclusions...281
References..282

16.1 INTRODUCTION

Photosynthesis is a key component of plant biology. It is an essential driver of primary carbohydrate biosynthesis and other aspects of metabolism such as the assimilation of nitrogen into organic compounds within the chloroplast (Blankenship 2002; Forde and Lea 2007). Photosynthesis is a sophisticated process involving complex bioenergetics and a wide range of different molecules, such as proteins, lipids, and light absorbing chromophores. Photosynthesis is therefore sensitive to regulation and perturbation by biotic and abiotic stimuli. One major abiotic environmental stimulus is UVB radiation (280–315 nm), which is a typical component of the light environment. UVB wavelengths are frequently absorbed by component molecules of photosynthesis (Strid et al. 1990, 1994; Jordan 1996), and such wavelengths have the potential to initiate loss of function and a cascade of damaging consequences, such as the production of reactive oxygen species (ROS) (Mackerness et al. 1999b). UVB is readily absorbed by DNA, causing damaging lesions (Jordan 1996; Taylor et al. 1996). In addition, UVB regulates nonspecific changes to gene expression and also acts through a specific photoreceptor molecule that can alter gene expression (see below). In terms of photosynthesis itself, there is a requirement for substantial gene expression activity to establish the complex biostructure of the chloroplast (with its contrasting soluble and membrane components) and to maintain component parts by constant turnover, for example, D1 and D2 polypeptide production. Overall, there is substantial potential for UVB-mediated changes to the photosynthetic process.

The understanding of UVB regulation of plant cell function has advanced substantially with the characterization of UVR8 (UV RESISTANCE LOCUS 8) as a UVB photoreceptor (Rizzini et al. 2011; Christie et al. 2012; Wu et al. 2012). A specific low UVB fluence response perceived by UVR8 is mediated through transcription factors such as HY5 (ELONGATED HYPOCOTYL 5) and the regulatory photomorphogenic protein COP1 (CONSTITUTIVELY PHOTOMORPHOGENIC 1) (Jenkins and Brown 2007; Stracke et al. 2010; Cloix et al. 2012). In addition to the UVR8 pathway, a nonspecific signaling network exists that overlaps with other known signal transduction pathways, commonly including ROS, jasmonic acid, and salicylic acid (Surplus et al. 1998; Mackerness et al. 1999a; Jordan 2002). This response pathway has usually been related to higher ultraviolet (UV) fluences and potentially damaging stress (Jenkins and Brown 2007; Brown and Jenkins 2008; Jenkins 2009; Hideg et al. 2013).

The wide range of phenotypic effects initiated by UVB has been recently reinterpreted in terms of the perceived potential for damaging consequences (Ballaré et al. 2012; Hideg et al. 2013; Wargent and Jordan 2013). It is now recognized that plants are rarely damaged under field conditions when exposed to solar UVB, due to a suite of adaptations at the cellular and physiological level that can provide efficient protection. Thus, photosynthesis itself may not be particularly downregulated (Allen et al. 1998). These defense mechanisms involve DNA repair, antioxidant production, physiological adjustments, and the synthesis of compounds that strongly absorb UVB radiation (Jordan 1996; Wargent and Jordan 2013). In fact, there are strong indications that UVB may actually be of ultimate benefit to the plant. In this chapter, we will explore the role that UVB plays in photosynthetic function, and the physiological, biochemical, and molecular mechanisms that are involved.

16.2 EXPERIMENTAL FACTORS AFFECTING THE OUTCOME OF UVB RESPONSES

Since UVB is absorbed by many of the proteinaceous macromolecules of the chloroplast, the metabolic state of a plant prior to UVB exposure (e.g., high-light acclimated, low-light acclimated, low-UV acclimated, subjected to biotic or abiotic stress, grown outdoors/in a glasshouse/in controlled environment [CE] chambers, and so on) is of great consequence for responses subsequently observed. For instance, the pre-experimental state will have a profound influence on the proteome of the plant tissue and therefore on the different proteins and protein complexes that can be affected by UVB. Also, the history of the experimental plants will determine the potential state of protection against UVB. For instance, prior acclimation of plants to low levels of UVB has been shown to protect against damage to photosystem II (PSII) (the D1 protein) when plants are subjected to sudden acute levels of UV (Wilson and Greenberg 1993). This protection is at least partially conferred by increased content of flavonoids and sinapates, both being protective UVB-absorbing pigments (Booij-James et al. 2000), in addition to the capacity for DNA photorepair and photoreactivation (Jansen et al. 1998).

The multitude of UVB absorbers in plant cells are of different molecular origin and have different molecular surroundings. This means that the actual UVB wavelengths absorbed will differ with each chromophore. This also means that by using different types of UVB radiation sources (Aphalo et al. 2012), differing from each other with respect to the quality and quantity of the radiation they emit, different cellular components will be targeted. Broadband UV sources will nonspecifically provoke damage to more molecules than narrow-band UV sources. From a mechanistic point of view, and to pinpoint as few targets as possible in each experiment, an extremely narrow-band UVB laser source could ideally be employed for such studies (Aphalo et al. 2012; Czégény et al. 2014) and to specifically record action spectra for the different types of responses observed.

Most significantly, the level of photosynthetically active radiation (PAR; 400–700 nm), concomitant with the UVB exposure, also profoundly influences the level of UVB-induced lesions (Krizek 2004). An optimal high PAR level will lead to a smaller degree of damage compared to low PAR (see below). Such understanding of how high PAR may protect gene expression, photosynthesis, and growth from inhibition by UVB is at best fragmentary (e.g., increased capacity for UV screening, photoreactivation of DNA, and repair of other cellular components) and deserves increased future attention, since there are likely other hitherto undiscovered mechanisms for protection against UVB by high visible light.

Overall, for the correct interpretation and comparison of experimental results from different laboratories, the status of environmental factors and, in particular, the PAR regimen, quantity and quality of UV radiation, and pre-experimental history need to be reported in detail.

16.3 UVB AND PLANT PHYSIOLOGICAL RESPONSES

UVB radiation is an energetic component of sunlight and is capable of initiating a broad range of physiological and biochemical changes in plants. In terms of the consequences for photosynthetic competency and performance, this breadth of phenotypic responses to UV could conceivably alter numerous aspects of photosynthetic function, and many studies have described varied responses on that basis. There are several archetypal responses to UV exposure: firstly, reductions in leaf expansion and/or hypocotyl length (Gonzalez et al. 1998; Ruhland and Day 2000; Searles et al. 2001), and induction of phenylpropanoid or "sunscreening" metabolism (Krizek et al. 1998; Bassman 2004). In addition, other physiological responses have been observed in several studies, including increases in leaf thickness (Staxen and Bornman 1994; Rozema et al. 1997), and increased leaf surface wax content (Cen and Bornman 1990). The underlying mechanisms that drive some of these common UV responses remain poorly understood, with the exception of phenylpropanoid metabolism, which is regulated in part by the UVR8-COP1-HY5 UVB signal transduction pathway (Brown et al. 2005). Nonetheless, such physiological changes have likely implications for photosynthesis, yet the complex, multifaceted nature of UV photomorphogenesis has created historical barriers to our understanding of the impact of realistic UV exposure for photosynthetic function. As discussed earlier, decreases in photosynthetic performance have been typically observed in CE studies, or in so called square-wave supplementation studies in outdoor conditions, where seasonal and climatic variation can affect observations. Ultimately, both types of experimental approach often include sudden UV exposure treatments, in addition to the potential for unbalanced UV/PAR ratios and wavelength compositions that do not follow closely the solar spectrum on the earth's surface. For example, in a meta-analysis of earlier UV-response studies, Searles et al. (2001) note that predominant historical observations have often showed some reduction in CO_2 fixation rates, often using experimental systems such as those described above. However, Searles and colleagues also observed that in some studies, CO_2 fixation rates increased in response to UV treatment.

Resolving a robust model of UV-photosynthetic response requires an understanding of the effects of physiological (e.g., inhibition of leaf expansion), biochemical (e.g., sun screening metabolism), and signal transduction (e.g., UVR8-COP1-HY5) responses upon the various photosynthetic processes. Studies to date that have indicated a positive effect of UV exposure upon photosynthesis have predictably focused on specific responses or have been unable to fully separate those different aspects of photosynthetic function, including direct photochemical effects, photoprotection of photosynthetic apparatus, and changes to photosynthesizing volume per unit of plant biomass/area. Those studies have been carried out in a range of plant species; for example, when *Lactuca sativa* (lettuce) seedlings were grown under photoselective films, which allowed increased transmission of solar UVB and UVA radiation under outdoor conditions for 13 days, net photosynthetic rate increased over that period of exposure (Wargent et al. 2011), with no reduction in the maximum photochemical efficiency of PSII (F_v/F_m) noted. Those authors also observed similar results when *L. sativa* seedlings of a similar age were consistently exposed to a supplementary UV dose of 10 kJ m^{-2} day^{-1} in indoor, CE conditions for 10 days. Moreover, maximum light-saturated photosynthetic rate was significantly higher in UV-treated plants in CE conditions, but there was no difference in the quantum yield of photosynthesis or light compensation point. Also, in controlled conditions, marked recovery of F_v/F_m values was observed in UV-pretreated plants that were then subsequently exposed to 24 h of high daytime temperature and PAR flux (35°C and 1000 μmol m^{-2} s^{-1}, respectively) by the time of solar dawn the day after stress exposure took place.

Increased net photosynthesis rates have also been observed in longer-term studies of deciduous tree species (Sprtova et al. 2003; Yang and Yao 2008), which some authors have attributed to increases in leaf thickness and potential accumulation of foliar phenolic compounds (Bolink et al. 2001), thus providing protection against photoinhibition in response to PAR exposure. Wargent et al. (2011) observed that a certain level of increased leaf thickness in *L. sativa* did not fully explain those increases in net photosynthesis discussed above but did not observe any UV-dependent relationship between increases in phenolic compounds and F_v/F_m values in field-treated plants.

Photoprotection could be mediated by phenolic compounds induced by UV exposure, but there is still somewhat limited mechanistic evidence for this, albeit there are several studies at a descriptive level (Xu and Gao 2010), including suggestion that photoprotection mediated by UV could be linked with antioxidant metabolism (Agati and Tattini 2010). Few studies to date have attempted to quantify changes in the metabolite pool simultaneously with alterations in photosynthetic performance or related gene expression. Wargent et al. (2015) observed a concomitant increase in P_{max} in UV-exposed *L. sativa* seedlings, with a decrease in F_v/F_m, which dissipated by the sixth day of UV exposure. Both of these responses were parallel with an increase in leaf UV shielding due to increased screening by a pool of secondary metabolites identified by broad metabolomics analysis, including quercetin and chlorogenic acid. Despite the description of parallel increases in specific secondary metabolites, we still have very limited information regarding any specific mechanism that leads to such direct increases in net photosynthesis for particular periods of time following UV exposure, usually very quickly after the onset of UV exposure. There is a paucity of studies that have integrated primary and secondary metabolism responses to UV, but it was suggested by Kusano and colleagues (2011) that UV exposure drove changes in tricarboxylic acid (TCA) intermediates, leading to increased carbon availability for phenolic biosynthesis and foliar shielding to UV. Clearly, such responses warrant further investigation in the future.

It is possible that manipulation of UVB photomorphogenesis may represent a future tool for improvement of photosynthetic performance of crops. Studies to date have revealed different insights into potential photosynthetic enhancement of crops including, but not limited to, wheat, rice, soybean, and various vegetables. Those enhancements in crop species include the aforementioned photoprotection from PAR wavelengths, increases in antioxidant activity, and increases in net photosynthetic rate (as discussed by Wargent and Jordan 2013). In terms of experimental conditions, there has been large variation between studies, which can limit meaningful subsequent meta-analysis or experimental follow-up. For example, when UVB dose is weighted with the Caldwell (1971) biological spectral weighting function (BSWF), daily UV doses in the above studies vary between 5.0 and 10.3 kJ m^{-2} day^{-1}, and some studies have been routinely carried out in field conditions, where UV irradiance and total dose may not have been measured at all. Equally, experimental duration, intraspecific cultivar variation, and environmental variables such as PAR flux will vary. Such responses could also impact on plant responses to other abiotic factors, such as limited water availability; prior exposure to UVB has been observed to reduce plant sensitivity to drought conditions (Manetas et al. 1996; Nogues et al. 1998) and may also increase water use efficiency (Poulson et al. 2002, 2006). Equally, authors have noted that stomatal patterning can be altered in response to UV (Kakani et al. 2003; Wargent et al. 2009), yet those studies suggesting almost instantaneous change in parameters such as P_{max} are not likely related to time-dependent developmental change. In summary, despite such indications of positive photosynthetic regulation by UVB in many crop species, a robust understanding of underpinning mechanisms remains largely elusive, and technologies to exploit any UV-mediated photosynthetic improvements are poorly developed at the present time.

In conclusion, plants in nature do not routinely display symptoms of chronic photosynthetic stress in response to UV wavelengths. Yet at the same time, UV photomorphogenesis drives a suite of physiological and biochemical changes in plants, to which many responses are likely to impact on photosynthetic performance. The concept of UV exposure underpinning agronomically valuable outcomes for photosynthetic performance represents an exciting new paradigm in UV plant photobiology, and our ability to harness UV for desirable traits in crop species will rest upon further investigation of the underlying mechanisms responsible for UV-photosynthetic interactions.

16.4 EFFECTS OF UVB ON BIOCHEMISTRY AND BIOENERGETICS OF PHOTOSYNTHESIS

16.4.1 GENERAL ASPECTS

Proteins and their function in metabolism provide the links between plant gene expression, and plant physiology and morphology. Compared with the numerous recent studies describing gene expression and physiology of plants in response to

UVB, less attention has been devoted to contemporary studies of how UVB exposure affects chloroplast proteins, membranes, lipids, other cellular proteins that affect chloroplast function, or the metabolism and bioenergetics within the chloroplast. However, numerous historical studies have focused on these aspects (Strid et al. 1994), including whole-plant studies; those focused on isolated plant tissue or chloroplast preparations as well as on different types of algae or cyanobacteria. As with most studies concerning the role of UVB in plants, both the quantity and the quality of the experimental UVB vary between reports, as does the ratio of UVB to PAR. Generally, these studies reveal that UV radiation indeed affects the structural integrity and function of plant cell components, at least at high fluence rates and high UVB/PAR ratios. As discussed above, such deleterious changes may occur after direct absorption of the energetic radiation by the biomolecules, or, alternatively or in addition, ROS may arise as a result of UV absorption by cellular content, which in turn will chemically affect cellular processes.

16.4.2 Proteins Involved in the Photosynthetic Light and Dark Reactions

Not surprisingly, most structural and functional studies of plant proteins show damaging and inhibitory effects as the dose rates of UVB increase (Jordan 1996). This is most likely due to the fact that aromatic amino acid residues of proteins or double bonds of cofactors bound to proteins act as UV chromophores absorbing radiation energy (Schmid 2001). This energy is then available for chemical reactions of a specific or nonspecific nature. Proteins may also be the targets for reactions with short-lived radicals or ROS yielding adducts with altered chemical and functional properties (Gill and Tuteja 2010; Hideg et al. 2013). Given the fact that plant cells contain high concentrations of many small and large molecules absorbing in the UVB region of the electromagnetic spectrum, some of which are actually protecting the cell (Strid et al. 1994), it is not surprising that the mechanistic studies that have so far been published have identified damage to a large number of cellular components at the molecular and/or functional level. In fact, a large number of the functionally important proteins that have been studied have also been shown to be damaged by UVB, provided the doses applied have been high enough to overwhelm defense or the sample preparation has been on an organizational level low enough to exclude the UV-protective components. For example, of the main protein complexes involved in the light reactions of photosynthesis, PSII (Bornman 1989; Vass 2012), the cytochrome b/f complex (Strid et al. 1990), photosystem I (PSI) (Krause et al. 2003), and ATP synthase (Strid et al. 1990; Zhang et al. 1994) have all been shown to be affected by UVB in one form or the other. Enzymes participating in the dark reaction, such as RUBISCO (Vu et al. 1984; Strid et al. 1990; Jordan et al. 1992; Wilson et al. 1995; Ferreira et al. 1996; Desimone et al. 1998; Gerhardt et al. 1999) and sedoheptulose-1,7-bisphosphatase (SBPase) (Allen et al. 1998) are also vulnerable. In fact, the content of the latter enzyme is specifically downregulated in

Brassica napus leaves as a result of U-B exposure, a specificity of degradation that is not shared with the other enzymes of the ribulose-1,5-bisphosphate-regenerative part of the Calvin cycle (Allen et al. 1998). The mechanism for the UVB-specific degradation of SBPase has not been investigated in any detail, although induction of senescence-related proteases has been suggested (Allen et al. 1998).

Large numbers of previous studies have implicated PSII and RUBISCO as the photosynthetic components most easily damaged by UVB. For instance, there have been dozens of studies devoted to the functional and structural investigations of UVB damage to PSII (Bornman 1989; Vass et al. 2005; Vass 2012). As early as the 1970s, studies of the integrity of PSII photochemistry as a function of supplementary UVB radiation were carried out and showed that photosystem-linked activities were prone to damage (Brandle et al. 1977; Iwanzik et al. 1983; Tevini and Pfister 1985; Greenberg et al. 1989; Strid et al. 1990; Babu et al. 1999; Krause et al. 1999). Different studies isolated different targets within PSII and associated components and cofactors as specific UVB absorbers. These included the water-splitting complex (particularly the manganese cluster S_2 and S_3 states) and the TyrZ on the donor side (Renger et al. 1989; Post et al. 1996; Vass et al. 1996; Larkum et al. 2001; Szilárd et al. 2007), the D1 and D2 proteins (Tevini and Pfister 1985; Greenberg et al. 1989; Trebst and Depka 1990; Jansen et al. 1996; Salter et al. 1997) of the core complex (Iwanzik et al. 1983; Renger et al. 1989; Ihle 1997), as well as the quinone acceptors (Trebst and Depka 1990; Melis et al. 1992; Jansen et al. 1993; Rodrigues et al. 2006). The action of an ensemble of PSII chromophores situated on both the donor and acceptor sides has also been suggested for the sites of damage (Hideg et al. 1993; Vass et al. 1996, 1999; Babu et al. 1999; van Rensen et al. 2007), and a sequential process was formulated by Vass et al. (2005). This process, starting with the damage to the water-splitting Mn cluster, continuing with impairment of the quinone electron acceptors and the tyrosine(s) on the donor side, eventually triggers the breakdown of the D1 and D2 proteins of the PSII core.

Large UVB-dependent decreases in total RUBISCO carboxylation activity were found at the same time as the degree of *in vivo* activation of the RUBISCO enzyme increased (Strid et al. 1990; Huang et al. 1993). UVB can modify RUBISCO to induce formation of a high-molecular-weight adduct (65–70 kDa in green plants, compared to the large subunit [LSU] size of 57 kDa and the small subunit [SSU] size of 14 kDa) (Wilson et al. 1995; Ferreira et al. 1996; Bischof et al. 2000), probably consisting of a covalently linked LSU and SSU heterodimer (Ferreira et al. 1996). As is the case with the UVR8 photoreceptor, tryptophan seems to be the main absorber of UV radiation in RUBISCO (Greenberg et al. 1996), leading to LSU/SSU dimer formation after tryptophan photolysis (Caldwell 1993; Gerhardt et al. 1999). Not surprisingly, actual degradation of RUBISCO protein is also enhanced by UVB treatment (Bischof et al. 2000; Takeuchi et al. 2002). For both tryptophan photolysis and RUBISCO degradation, ROS have been suggested as triggers (Caldwell 1993; Desimone et al. 1998; Kataria et al. 2014), although there is currently no direct evidence. Finally, there have been a large number of studies examining the influence of different supplementary UVB light regimens on the chlorophyll content of many different plant species and algae (Häder and Häder 1989; Strid and Porra 1992; Jordan et al. 1994). Commonly, a decrease is observed due to chlorophyll breakdown (Häder and Häder 1989; Strid and Porra 1992; Jordan et al. 1994), which, in addition to PSII and RUBISCO damage, could likely be deleterious to the net photosynthetic rate.

16.4.3 Lipids and Membranes

Chloroplast and thylakoid ultrastructure can be negatively affected by UVB in plants (Brandle et al. 1977; He et al. 1994), in *Chlorophyta* (Malanga et al. 1997), and in *Rhodophyta* (Schmidt et al. 2012), possibly as a result of widespread UVB-induced lipid peroxidation (Kramer et al. 1991; Takeuchi et al. 1995; Malanga et al. 1997; Hideg et al. 2003; Lidon and Ramalho 2011) at high UVB levels. However, more interesting is the increased ion permeability of the thylakoid membrane conferred by UVB at low doses (Chow et al. 1992; Strid et al. 1994; Mewes and Richter 2002), resulting in uncoupling of the membrane-bound bioenergetic processes of the chloroplast. This change in ion permeability is manifested as an increased rate of relaxation of the single-turnover flash-induced electrochromic shift of photosynthetic pigments (Chow et al. 1992; van Hasselt et al. 1996). The thylakoid membrane permeability is achieved at levels of UVB as low as those needed to accomplish expression of genes involved in synthesis of UV-protective pigment (i.e., at "signaling" doses), which means UV levels about 20-fold lower than those needed to affect PSII photochemistry, or about 10-fold lower than doses inhibiting RUBISCO activity (Strid et al. 1994, 1996a). Moreover, in contrast to many other effects of UVB, the change in the electrochromic shift relaxation was more or less independent of the PAR during UV exposure (Strid et al. 1996a). Also, photooxidation does not seem to be the cause of such uncoupling, judging by the independence of the effect on the oxygen concentration (between 2% and 60% oxygen) (van Hasselt et al. 1996). Furthermore, using the thiobarbituric acid method, no lipid peroxidation was detected under the low-dose UVB conditions that led to increased rates of electrochromic shift relaxation (Strid et al. 1996a). However, the temperature dependency of the relaxation showed that the effect is due to a chemical reaction, not a direct temperature-independent physical reaction between UVB quanta and membrane components (Strid et al. 1996b). Therefore, the molecular nature of this effect of UVB on thylakoid membrane integrity is as yet unidentified and is worth following up. Finally, the oxygen independence and temperature dependence of the UVB-induced relaxation of the electrochromic shift is similar to the thylakoid membrane permeabilization accomplished during photoinhibitory conditions (Tjus and Andersson 1993), and it would be worthwhile to investigate whether such phenomena share a common molecular mechanism.

16.4.4 Reactive Oxygen Species and Various Radicals

Under normal physiological conditions H_2O_2 generation in PSI (Mehler 1951) and singlet oxygen production in PSII (Telfer et al. 1994) are commonplace. Chloroplast stroma lack enzymes such as catalase and superoxide dismutase, and therefore, the ascorbate/glutathione-linked enzymes of the water–water cycle are used to detoxify hydrogen peroxide, whereas singlet oxygen is rapidly detoxified directly within PSII by carotenoids (Asada 2006). ROS and non-oxygen-centered radicals can also be produced after exposure of plants to environmental cues, e.g., as the result of absorption of UVB by the chloroplast content. For instance, hydrogen peroxide, superoxide anion, hydroxyl radicals, peroxy radicals, and carbon-centered radicals (Hideg and Vass 1996; Hideg et al. 2002; Majer et al. 2014) have all been found to be induced by UVB, and some of these are long-lived (Hideg and Vass 1996).

However, ROS are also physiologically used by plant cells as signaling molecules (Pitzschke et al. 2006). ROS produced after UV exposure will be detoxified by the enzymatic and small molecule antioxidants as long as the ROS amounts do not overwhelm plant defense capacity. The term "eustress" has been coined to describe the ROS that is produced and effectively removed by the antioxidants, whereas "distress" is the suggested term that describes ROS produced in amounts exceeding the saturation of the constitutive and inducible antioxidative defense measures (Hideg et al. 2013). In principle, the terms *eustress* and *distress* do not necessarily have to be linked to UVB-induced ROS but could also refer to any alteration in plant cells that is triggered by UVB and that potentially may be damaging to a plant process, such as photosynthesis.

As was discussed earlier, although ambient UVB can be a damaging agent to photosynthesis in certain plant species under certain conditions, as shown in UVB exclusion studies using a number of agricultural species in the tropics (Kataria et al. 2013), or for a dwarf *Salix* species in the arctic (Albert et al. 2005), in most cases, ambient UVB is not a limiting factor for photosynthesis or for growth (Allen et al. 1998). In a recent study, it was shown that UVB-induced antioxidant defense measures in plants were primarily directed toward neutralization of hydrogen peroxide (Majer et al. 2014). In agreement with this, Czégény et al. (2014) showed that UVB radiation can synergistically transform excess hydrogen peroxide through a photochemical mechanism yielding the very toxic hydroxyl radical. Since hydrogen peroxide normally can be produced in chloroplasts, photosynthetically linked components and reactions may be particularly vulnerable to this type of UVB-dependent sensitizing mechanism once the H_2O_2 scavenging capacity has been overwhelmed. Therefore, to avoid UVB-induced hydroxyl radical formation from hydrogen peroxide, it seems logical that screening of UVB radiation and scavenging of hydrogen peroxide are priorities for protecting photosynthesis at both ambient and enhanced UVB levels, which is also the case (Jansen et al. 1998; Majer et al. 2014). Thus, ambient UVB levels would lead to eustress with respect to ROS, but this eustress is kept under control of the antioxidative system, leading to no significant negative effects on plant performance in most cases.

16.5 MOLECULAR-LEVEL RESPONSES TO UVB AND IMPLICATIONS FOR PHOTOSYNTHESIS

In the early 1990s, Jordan et al. (1992, 1994) showed that genes for photosynthetic proteins were downregulated by UVB radiation. Nuclear-encoded genes for photosynthetic proteins were downregulated to a greater degree than those encoded by the chloroplast. This downregulation could have been due to nonspecific DNA damage (Jordan 1996; Taylor et al. 1996). However, in these experiments, simultaneous upregulation of genes for the flavonoid pathway, such as chalcone synthase, was detected (Strid 1993; Jordan et al. 1994). This strongly suggested that a more specific mechanism of regulation was involved. Furthermore, the changes to gene expression were dependent on the developmental stage of the tissue (Jordan et al. 1994; Mackerness et al. 1998b). Thus, etiolated tissue, which logically should be most susceptible to UVB (as it has less protective pigmentation), was least affected, and older mature green tissue, most susceptible (see discussion below). This aspect has been largely unexplored but could be as simple as the physical properties of the tissue. For instance, the anticlinal cell walls are known to allow passage of UVB, and these may be more transmissible in older tissue (Day et al. 1992, 1993). Of particular interest was the observation that high PAR applied at the same time as UVB exposure protected the levels of transcripts compared to UVB with low PAR (Jordan et al. 1992; Mackerness et al. 1996). This is consistent with previous findings showing high PAR protection at the physiological level (Warner and Caldwell 1983; Caldwell et al. 1994; Jordan 1996). One possible mechanism that could be involved is photorepair through enzymes such as photolyase. However, it was shown, using high-pressure sodium lamps to provide PAR but not support photoreactivation, that photolyase activity was not involved (Mackerness et al. 1996). Experiments were therefore carried out to determine what aspect of the PAR provided the protective response (Mackerness et al. 1996). From these experiments, chloroplast electron transport was shown to be involved in protection, and more specifically, photophosphorylation was found to be essential. It is still unknown how this may work. However, because ATP from photophosphorylation is needed for the transit of peptides from the cytoplasm, any reduction of photophosphorylation may prevent the passage of peptides into the chloroplasts. Indeed, photophosphorylation, particularly CF_0F_1-ATP synthase function, has been shown to be affected by UVB (Zhang et al. 1994). In turn, the accumulating transcripts in the cytoplasm will be degraded, and an overall decline will be seen. Thus, the high light ameliorates the apparent UVB-dependent downregulation by maintaining a high level of photophosphorylation. Another potential mechanism that could reduce the RNA transcripts is feedback repression from carbohydrates. In a series of

experiments, Mackerness et al. (1997a) tested this possibility. It was shown that under UVB and low light, glucose levels declined (sucrose and starch remained the same until longer exposure). Effects of carbohydrate were minimal under UVB and high irradiance. Because UVB effects on transcripts are most marked at low-light conditions, when glucose levels are reduced, carbohydrate feedback is not a likely explanation for UVB-induced downregulation of gene activity.

It has also been shown that free radical generators and antioxidant feeding mimicked UVB-induced changes in gene expression (Henkow et al. 1996; Mackerness et al. 1998b). This strongly suggested that the signal transduction pathway from UVB involved ROS. However, experiments with mutants and inhibitors identified at least three signal transduction pathways that involved ROS and altered gene expression (Surplus et al. 1998; Mackerness et al. 1999a; Mackerness 2000). These pathways involved intermediates such as jasmonic acid, ethylene, and salicylic acid. Significantly, the signal transduction pathway that was involved in the downregulation of photosynthetic gene activity did not involve these compounds. In contrast to other pathways, the ROS was not generated through NADPH or through NO but did include hydrogen peroxide downstream (Mackerness et al. 2001; Jordan 2002). It is possible that a signaling concentration of hydrogen peroxide was transformed to hydroxyl radicals by UVB, which, in turn, led to downregulation (Czégény et al. 2014).

The transition of an etiolated seedling into a mature green plant capable of photosynthesis requires light to regulate gene activity and involves changes in ultrastructure and biochemistry. Under normal irradiance, genes for photosynthetic proteins, such as CAB and RUBISCO, increase in activity until the tissue is green and the chloroplasts fully formed. This gene expression is controlled via a number of photoreceptors, notably phytochrome. However, the role of UVB in the development of the photosynthetic apparatus during de-etiolation is less well understood. In a series of experiments, Mackerness et al. (1999b) studied the response of genes for photosynthetic proteins exposed to supplemental UVB radiation during various development stages, including etiolated tissue. As noted above, in green tissue, the levels of gene expression declined substantially after UVB exposure. In contrast, when etiolated tissue was exposed to supplemental UVB, there was no difference from plants under PAR alone; that is, the gene expression increased throughout de-etiolation. One possible explanation for this was that members of multigene families played different roles during different stages of development. To test this idea, all the members of the CAB family in pea were investigated in plants exposed to supplemental UVB radiation during de-etiolation (Mackerness et al. 1998a). These experiments showed that differences between members of the CAB family could not account for the different response of green and etiolated tissue. The experiments did show, however, that the members of the gene family could be put into different groups based on their response to UVB. Findings by Liu and White (1998) also suggest the involvement of intact and functional chloroplasts in the downregulation of nuclear-encoded genes, such as CAB. A potential mechanism may involve the chloroplast signal that is generated through ROS (Taylor 1989). However, using *Arabidopsis gun* (genome uncoupled) mutants, no involvement of the chloroplast signal was found (Jordan et al. 1998). To analyze the transition from etiolated to fully green tissue in the context of gene regulation, the signal transduction pathways involving ROS were investigated. There was a clear distinction in the ROS activity between etiolated tissue and that of green tissue. Etiolated tissue consistently had higher levels of ROS than did green tissue. At 54 h of greening of etiolated tissue, the levels of ROS-induced downregulation were the same in the green and etiolated material (Mackerness et al. 1999b).

The recent discovery of UVR8 as the UVB photoreceptor has introduced a new level of complexity into the potential signal transduction pathways that may change gene expression (Jenkins and Brown 2007). UVR8 works independently of other known photoreceptors, such as phytochrome and cryptochrome (Jenkins and Brown 2007). UVR8 is a specific photoreceptor that works at low fluence of UVB (Jenkins 2009), and this separates UVR8 from the more nonspecific changes to gene expression that takes place at high fluence (e.g., ROS). UVR8 exists as a dimer and becomes a monomer after exposure to UVB. The monomer forms a heterodimer with COP1 and is thought to be involved in chromatin remodeling (Jenkins and Brown 2007) in the nucleus. These changes in chromatin are also associated with the promoter of HY5 (Jenkins and Brown 2007). Recently, HY5 has been shown to bind to the promoters of UVB-responsive genes in a UVR8-dependent manner and to be enhanced by UVB (Binkert et al. 2014). Furthermore, the cis-acting element T/G^{HY5}-box in the HY5 promoter itself binds HY5 and is UVB responsive. Light-driven changes in gene expression are well established to act at the level of histone modification within the chromatin (Wu 2014), and so, it is possible that UVR8 may control transcription for photosynthetic proteins through this mechanism. However, it is also well established that translation and posttranslational modifications may be involved (Mackerness et al. 1997b; Wu 2014). So does UVR8 control UVB-mediated gene expression for photosynthetic proteins? There have been few studies so far on this question. Davey et al. (2012) showed that preservation of photosynthetic function as measured by maximum quantum yield and efficiency of PSII did require UVR8. In addition, UVB increased the levels of psbD-BLRP transcripts encoding the PS11 D2 protein. This increase was mediated by the UVR8-mediated RNA polymerase sigma factor SIG5. The D1 transcripts were not, however, reduced, although the D1 protein did decline, suggesting possible translation control (Mackerness et al. 1997b; Wu 2014). Davey and colleagues (2012) did not fully resolve a role for the nuclear-bound light stress-induced EARLY LIGHT INDUCED PROTEIN 1 (ELIP1), which is expressed via UVR8 in response to UVB (Sävenstrand et al. 2004). It is likely that further hitherto undescribed mechanisms may provide links between UVB-specific signaling and photosynthetic function.

16.6 CONCLUSIONS

A major change in our perception of how UVB can affect plants has been brought about by the isolation and characterization

of UVR8 as a specific UVB photoreceptor (Rizzini et al. 2011; Christie et al. 2012; Wu et al. 2012). These findings have in turn enhanced the concept that UVB is more than just a damaging radiation but can provide UVB-mediated benefits to plants (Wargent and Jordan 2013). However, through this revision in our understanding, there has been little focus on UVB and photosynthesis. To some extent, the potential impact of UVB on photosynthesis "has gone out of fashion" and is still considered to have little relevance to events in the natural environment. Consequently, there remain many outstanding questions that need to be reevaluated regarding UVB and photosynthesis (Table 16.1). Undoubtedly, much of the evidence for effects of UVB on photosynthesis has come from studies carried out in CE facilities and under relatively extreme conditions (see comments above). These conditions have frequently involved relatively high UVB and low PAR and have often been dismissed as unrealistic (Allen et al. 1998). However, there is now sufficient evidence from more realistic field environments that is indicative of an influence of UVB on photosynthetic function (Kataria et al. 2014 and references therein) and that may even indicate beneficial effects. In addition, a recent investigation on *Vitis vinifera* (Liu et al. 2015) looking at the UVR8 pathway in relation to flavonoid biosynthesis in both vineyards and CEs strongly suggests a number of similar responses. Therefore, CE experiments may well still provide an insight into the possible UVB-induced mechanisms that impact upon photosynthesis in the natural environment.

A major shortfall in our knowledge is how UVR8 is involved in photosynthesis. At the physiological level, UVR8 seems to be involved in functions such as stomatal opening and changes in photosynthetic activity (e.g., chlorophyll fluorescence) and molecular biology. There are, however, many questions left unanswered. For instance, the signal transduction pathway involved in the downregulation of gene expression for photosynthetic proteins remains unclear. It does not seem to involve a high-fluence signalling pathway involving common intermediates, such as jasmonic acid, salicylic acid, or ethylene (Mackerness et al. 1999a; Agrawal et al. 2009). So

---

TABLE 16.1
UVB and Photosynthesis: Outstanding Issues

Understanding:

- The role of strong light in amelioration of UVB effects on photosynthesis
- The signal transduction chain that leads to downregulation of gene expression for photosynthetic proteins
- The relationship between PAR and UVB effects, e.g., leaf thickness
- How UVB alters changes in gene expression through development
- The role of UVB-induced reactive oxygen in photosynthesis
- The molecular mechanisms involved in the changes induced by low UVB fluence to membrane integrity and the relationship to photoinhibition

is UVR8 the method of perception? In addition, does UVR8 have a role in ameliorating the protective influence of high PAR? A reevaluation of what has been learnt from CE experiments may well be very appropriate as greater understanding of UVB signal transduction from UVR8 develops. Another aspect that requires further investigation is the impact of UVB on photosynthesis during development, particularly in perennial commercial crops. It is well known from CE trials that the response to UVB is dependent on the developmental stage of the plant. Such results could be reevaluated during development phases within field environments, particularly in relation to the UVR8-induced response. This is important because the gene expression of UVR8 is thought to be constitutive, but recent studies in *V. vinifera* clearly show that UVR8 expression is dependent on the developmental stage of the vines (Liu et al. 2015). Therefore, there are many questions to be answered in relation to UVB regulation of photosynthesis in the context of UVR8 and development. Other observations also require further investigation, such as the fact that high PAR may change leaf morphology and photosynthetic function, as opposed to low PAR. These changes include thicker leaves, less granal stacking, and increased photosynthesis activity. Interestingly, UVB can cause similar changes, and so, there may be common signal pathways between PAR and UVB. Also, recent results indicate a role of UVB in the breakdown of hydrogen peroxide to hydroxyl radicals (Czégény et al. 2014). Such a UV-induced reaction could have deleterious physiological consequences in the chloroplast of plants whose H_2O_2 antioxidant capacity is already saturated, such as those under multiple stress conditions. In conclusion, further developing our understanding of photosynthetic form and function in the context of the newly elucidated knowledge of UVB photomorphogenesis offers hugely exciting scientific and agricultural possibilities.

REFERENCES

Agati, G. and Tattini, M. (2010). Multiple functional roles of flavonoids in photoprotection. *New Phytologist*. 186: 786–793.

Agrawal, S.B., Singh, S. and Agrawal, M. (2009). Ultraviolet-B induced changes in gene expression and antioxidants in plants. *Advances in Botanical Research*. 52: 47–86.

Albert, K.R., Mikkelsen, T.N. and Ro-Poulsen, H. (2005). Effects of ambient versus reduced UV-B radiation on high arctic *Salix arctica* assessed by measurements and calculations of chlorophyll a fluorescence parameters from fluorescence transients. *Physiologia Plantarum*. 124: 208–226.

Allen, D.J., Nogués, S. and Baker, N.R. (1998). Ozone depletion and increased UV-B radiation: Is there a real threat to photosynthesis? *Journal of Experimental Botany*. 49: 1775–1788.

Aphalo, P.J., Albert, A., McLeod, A. et al. (2012). Manipulating UV radiation. In: *Beyond the visible: A handbook of best practice in plant UV photobiology. COST action F0906 'UV4growth'*, eds. Aphalo, P.J., Albert, A., Björn, L.O., McLeod, A.R., Robson, T.M. and Rosenqvist, E. Helsinki University, Division of Plant Biology, Helsinki, pp. 35–70.

Asada, K. (2006). Production and scavenging of reactive oxygen species in chloroplasts and their functions. *Plant Physiology*. 141: 391–396.

Babu, T.S., Jansen, M.A.K., Greenberg, B.M. et al. (1999). Amplified degradation of the photosystem II D1 and D2 proteins under a mixture of PAR and UV-B: Dependence on photosystem II electron transport. *Photochemistry and Photobiology*. 69: 553–559.

Ballaré, C.L., Mazza, C.A., Austin, T.A. and Pierik, R. (2012). Canopy light and plant health. *Plant Physiology*. 160: 145–155.

Bassman, J.H. (2004). Ecosystem consequences of enhanced solar ultraviolet radiation: Secondary plant metabolites as mediators of multiple trophic interactions in terrestrial plant communities. *Photochemistry and Photobiology*. 79: 382–398.

Binkert, M., Kozma-Bognar, L., Terecskei, K., Veylder, L.D., Nagy, F. and Ulm, R. (2014). UV-B-responsive association of the *Arabidopsis* bZip transcription factor ELONGATED HYPOCOTYL5 with target genes, including its own promoter. *Plant Cell*. 26: 4200–4213.

Bischof, K., Hanelt, D. and Wiencke, C. (2000). Effect of ultraviolet radiation on photosynthesis and related enzyme reactions of marine macroalgae. *Planta*. 211: 555–562.

Blankenship, R.E. (2002). *Molecular mechanisms of photosynthesis*. Blackwell Science, Oxford.

Bolink, E.M., van Schalkwijk, I., Posthumus, F. and van Hasselt, P.R. (2001). Growth under UV-B radiation increases tolerance to high-light stress in pea and bean plants. *Plant Ecology*. 154: 149–156.

Booij-James, I.S., Dube, S.K., Jansen, M.A.K., Edelman, M. and Mattoo, A. (2000). Ultraviolet-B radiation impacts light-mediated turnover of the photosystem II reaction center heterodimer in *Arabidopsis* mutants altered in phenolic metabolism. *Plant Physiology*. 124: 1275–1283.

Bornman, J.F. (1989). New trends in photobiology: Target sites of UV-B radiation in photosynthesis of higher plants. *Journal of Photochemistry and Photobiology B: Biology*. 4: 145–158.

Brandle, J.R., Campbell, W.F., Sisson, W.B. and Caldwell, M.M. (1977). Net photosynthesis, electron transport capacity, and ultrastructure of *Pisum sativum* L. exposed to ultraviolet-B radiation. *Plant Physiology*. 60: 165–169.

Brown, B.A. and Jenkins, G.I. (2008). UV-B signaling pathways with different fluence-rate response profiles are distinguished in mature *Arabidopsis* leaf tissue by requirement for UVR8, HY5, and HYH. *Plant Physiology*. 146(2): 576–588.

Brown, B.A., Cloix, C., Jiang, G.H. et al. (2005). A UV-B-specific signaling component orchestrates plant UV protection. *Proceedings of the National Academy of Sciences of the United States of America*. 102: 18225–18230.

Caldwell, C.R. (1993). Ultraviolet-induced photodegradation of cucumber (*Cucumis sativus* L.) microsomal and soluble protein tryptophanyl residues in vitro. *Plant Physiology*. 101: 947–953.

Caldwell, M.M. (1971). *Solar UV radiation and the growth and development of higher plants*. Academic Press, New York.

Caldwell, M.M., Flint, S. and Searles, P.S. (1994). Spectral balance and UV-B sensitivity of soybean: A field experiment. *Plant, Cell and Environment*. 17: 267–276.

Cen, Y.P. and Bornman, J.F. (1990). The response of bean-plants to UV-B radiation under different irradiances of background visible-light. *Journal of Experimental Botany*. 41: 1489–1495.

Chow, W.S., Strid, Å. and Anderson, J.M. (1992). Short-term treatment of pea plants with supplementary ultraviolet-B radiation: Recovery time-courses of some photosynthetic functions and components. In: *Research in photosynthesis*, vol. 4, ed. Murata, N. Kluwer Academic Publishers, Dordrecht, pp. 361–364.

Christie, J.M., Arvai, A.S., Baxter, K.J. et al. (2012). Plant UVR8 photoreceptor senses UV-B by tryptophan-mediated disruption of cross-dimer salt bridges. *Science.* 335: 1492–1496.

Cloix, C., Kaiserli, E., Heilmann, M. et al. (2012). C-terminal region of the UV-B photoreceptor UVR8 initiates signaling through interaction with the COP1 protein. *Proceedings of the National Academy of Sciences of the United States of America.* 109: 16366–16370.

Czégény, G., Wu, M., Dér, A., Eriksson, L.A., Strid, Å. and Hideg, É. (2014). Hydrogen peroxide contributes to the ultraviolet-B (280–315 nm) induced oxidative stress of plant leaves through multiple pathways. *FEBS Letters.* 588: 2255–2261.

Davey, M.P., Susanti, N.I., Wargent, J.J. et al. (2012). The UV-B photoreceptor UVR8 promotes photosynthetic efficiency in *Arabidopsis thaliana* exposed to elevated levels of UV-B. *Photosynthesis Research.* 114: 121–131.

Day, T.A., Vogelmann, T.C. and DeLucia, E.H. (1992). Are some plant life forms more effective than others in screening out ultraviolet-B radiation? *Oecologia.* 92: 513–519.

Day, T.A., Martin, G. and Vogelmann, T.C. (1993). Penetration of UV-B radiation in foliage: Evidence that the epidermis behaves as a non-uniform filter. *Plant, Cell and Environment.* 16: 735–741.

Desimone, M., Wagner, E. and Johanningmeier, U. (1998). Degradation of active-oxygen-modified ribulose-1,5-bisphosphate carboxylase/oxygenase by chloroplastic proteases requires ATP hydrolysis. *Planta.* 205: 459–466.

Ferreira, R.M.B., Franco, E. and Teixeira, A.R.N. (1996). Covalent dimerization of ribulose bisphosphate carboxylase subunits by UV radiation. *Biochemical Journal.* 318: 227–234.

Forde, B.G. and Lea, P.J. (2007). Glutamate in plants: Metabolism, regulation, and signalling. *Journal of Experimental Botany.* 58: 2339–2358.

Gerhardt, K.E., Wilson, M.I. and Greenberg, B.M. (1999). Tryptophan photolysis leads to a UVB-induced 66 kDa photoproduct of ribulose-1,5-bisphosphate carboxylase/oxygenase (Rubisco) in vitro and in vivo. *Photochemistry and Photobiology.* 70: 49–56.

Gill, S.S. and Tutega, N. (2010). Reactive oxygen species and antioxidant machinery in abiotic stress tolerance in crop plants. *Plant Physiology and Biochemistry.* 48: 909–930.

Gonzalez, R., Mepsted, R., Wellburn, A.R. and Paul, N.D. (1998). Non-photosynthetic mechanisms of growth reduction in pea (*Pisum sativum* L.) exposed to UV-B radiation. *Plant Cell and Environment.* 21: 23–32.

Greenberg, B.M., Gaba, V., Canaani, O., Malkin, S., Mattoo, A.K. and Edelman, M. (1989). Separate photosensitizers mediate degradation of the 32-kDa photosystem II reaction center protein in the visible and UV spectral regions. *Proceedings of the National Academy of Sciences of America.* 86: 6617–6620.

Greenberg, B.M., Wilson, M.I., Gerhardt, K.E. and Wilson, K.E. (1996). Morphological and physiological responses of *Brassica napus* to ultraviolet-B radiation: Photomodification of ribulose-1,5-bisphosphate carboxylase/oxygenase and potential acclimation processes. *Journal of Plant Physiology.* 148: 78–85.

Häder, D.-P. and Häder, M.A. (1989). Effects of solar and artificial radiation on motility and pigmentation in *Cyanophora paradoxa. Archives Microbiology.* 152: 453–457.

He, J., Huang, L.-K. and Whitecross, M.I. (1994). Chloroplast ultrastructure changes in *Psium sativum* associated with supplementary ultraviolet (UV-B) radiation. *Plant, Cell and Environment.* 17: 771–775.

Henkow, L., Strid, Å., Berglund, T., Rydström, J. and Ohlsson, A.B. (1996). Alteration of gene expression in tissue cultures and leaves of *Pisum sativum* caused by the oxidative stress-inducing agent 2,2′-azobis (2-amidinopropane) dihydrochloride. *Physiologia Plantarum.* 96: 6–12.

Hideg, É. and Vass, I. (1996). UV-B induced free radical production in plant leaves and isolated thylakoid membranes. *Plant Science.* 115: 251–260.

Hideg, É., Sass, L., Barbato, R. and Vass, I. (1993). Inactivation of photosynthetic oxygen evolution by UV-B irradiation: A thermoluminescence study. *Photosynthesis Research.* 38: 455–462.

Hideg, É., Barta, C., Kálai, T., Vass, I., Hideg, K. and Asada, K. (2002). Detection of singlet oxygen and superoxide with fluorescent sensors in leaves under stress by photoinhibition or UV radiation. *Plant Cell Physiology.* 43: 1154–1164.

Hideg, É., Nagy, T., Oberschall, A., Dudits, D. and Vass, I. (2003). Detoxification function of aldose/aldehyde reductase during drought and ultraviolet-B (280–320 nm) stresses. *Plant, Cell and Environment.* 26: 513–522.

Hideg, É., Jansen, M. and Strid, Å. (2013). UV-B radiation, ROS and stress; inseparable companions or loosely linked associates? *Trends in Plant Science.* 18: 107–115.

Huang, L.K., He, J., Chow, W.S., Whitecross, M.I. and Anderson, J.M. (1993). Responses of detached rice leaves (*Oryza sativa* L.) to moderate supplementary ultraviolet-B radiation allow early screening for relative sensitivity to ultraviolet-B irradiation. *Australian Journal of Plant Physiology.* 20: 285–297.

Ihle, C. (1997). Degradation and release from the thylakoid membrane of photosystem II subunits after UV-B irradiation of the liverwort *Conocephalum conicum. Photosynthesis Research.* 54: 73–78.

Iwanzik, W., Tevini, M., Dohnt, G., Voss, M., Weiss, W., Gräber, P. and Renger, G. (1983). Action of UV-B radiation on photosynthetic primary reactions in spinach chloroplasts. *Physiologia Plantarum.* 58: 401–407.

Jansen, M.A.K., Gaba, V., Greenberg, B.M., Mattoo, A.K. and Edelman, M. (1993). UV-B driven degradation of the D1 reaction-center of photosystem II proceeds via plastosemiquinone. In: *Photosynthetic responses to the environment,* eds. Yamamoto, H.Y. and Smith, C.M. Amer. Soc. Plant Physiol., Washington, DC, pp. 142–149.

Jansen, M.A.K., Gaba, V., Greenberg, B.M., Mattoo, A.K. and Edelman, M. (1996). Low threshold levels of ultraviolet-B in a background of photosynthetically active radiation trigger rapid degradation of the D2 protein of photosystem-II. *Plant Journal.* 9: 693–699.

Jansen, M.A.K., Gaba, V. and Greenberg, B.M. (1998). Higher plants and UV-B radiation: Balancing damage, repair and acclimation. *Trends in Plant Science.* 3: 131–135.

Jenkins, G.I. (2009). Signal transduction in response to UV-B radiation. *Annual Review of Plant Biology.* 60: 407–431.

Jenkins, G.I. and Brown, B.A. (2007). UV-B perception and signal transduction. In: *Annual plant reviews volume 30: Light and plant development,* eds. Whitelam, G.C. and Halliday, K.J. Blackwell Publishing, Oxford, pp. 155–182.

Jordan, B.R. (1996). The effects of ultraviolet-B radiation on plants: A molecular perspective. In: *Advances in botanical research.* Academic Press, London, pp. 97–162.

Jordan, B.R. (2002). Review: Molecular response of plant cells to UV-B stress. *Functional Plant Biology.* 29: 909–916.

Jordan, B.R., He, J., Chow, W.S. and Anderson, J.M. (1992). Changes in mRNA levels and polypeptide subunits of ribulose 1,5-bisphosphate carboxylase in response to supplementary ultraviolet-B radiation. *Plant, Cell and Environment.* 15: 91–98.

Jordan, B.R., James, P., Strid, Å. and Anthony, R. (1994). The effect of supplementary UV-B radiation on gene expression and pigment composition in etiolated and green pea leaf tissue: UV-B-induced changes are gene-specific and dependent upon the developmental stage. *Plant, Cell and Environment*. 17: 45–54.

Jordan, B.R., James, P. and Mackerness, S.A.-H. (1998). Factors affecting UV-B induced changes in *Arabidopsis thaliana* gene expression: Role of development, protective pigments and the chloroplast signal. *Plant and Cell Physiology*. 39: 769–778.

Kakani, V.G., Reddy, K.R., Zhao, D. and Mohammed, A.R. (2003). Effects of ultraviolet-B radiation on cotton (*Gossypium hirsutum* L.) morphology and anatomy. *Annals of Botany*. 91: 817–826.

Kataria, S., Guruprasad, K.N., Ahuja, S. and Singh, B. (2013). Enhancement of growth, photosynthetic performance and yield by exclusion of ambient UV components in C_3 and C_4 plants. *Journal of Photochemistry and Photobiology B: Biology*. 127: 140–152.

Kataria, S., Jajoo, A. and Guruprasad, K.N. (2014). Impact of increasing ultraviolet-B (UV-B) radiation on photosynthetic processes. *Journal of Photochemistry and Photobiology B: Biology*. 137: 55–66.

Kramer, G.F., Norman, H.A., Krizek, D.T. and Mirecki, R.M. (1991). Influence of UV-B radiation on polyamines, lipid peroxidation and membrane lipids in cucumber. *Phytochemistry*. 30: 2101–2108.

Krause, G.H., Schmude, C., Garden, H. Koroleva, O.Y. and Winter, K. (1999). Effects of solar ultraviolet radiation on the potential efficiency of photosystem II in leaves of tropical plants. *Plant Physiology*. 121: 1349–1358.

Krause, G.H., Grube, E., Virgo, A. and Winter, K. (2003). Sudden exposure to solar UV-B radiation reduces net CO_2 uptake and photosystem I efficiency in shade-acclimated tropical tree seedlings. *Plant Physiology*. 131: 745–752.

Krizek, D.T. (2004). Invited review: Influence of PAR and UV-A in determining plant sensitivity and photomorphogenic responses to UV-B radiation. *Photochemistry and Photobiology*. 79: 307–315.

Krizek, D.T., Britz, S.J. and Mirecki, R.M. (1998). Inhibitory effects of ambient levels of solar UV-A and UV-B radiation on growth of cv. New Red Fire lettuce. *Physiologia Plantarum*. 103: 1–7.

Kusano, M., Tohge, T., Fukushima, A. et al. (2011). Metabolomics reveals comprehensive reprogramming involving two independent metabolic responses of *Arabidopsis* to UV-B light. *Plant Journal*. 67: 354–369.

Larkum, A.W.D., Karge, M., Reifarth, F., Eckert, H.-J., Post, A. and Renger, G. (2001). Effect of monochromatic UV-B radiation on electron transfer reactions of photosystem II. *Photosynthesis Research*. 68: 49–60.

Lidon, F.C. and Ramalho, J.C. (2011). Impact of the UV-B irradiation on photosynthetic performance and chloroplast membrane components in *Oryza sativa* L. *Journal of Photochemistry and Photobiology B: Biology*. 104: 457–466.

Liu, L. and White, M.J. (1998). UV-B responses of nuclear genes encoding light harvesting complex 2 proteins in pea are altered by norflurazon- and photobleaching-induced chloroplast changes. *Physiologia Plantarum*. 102: 128–138.

Liu, L., Gregan, S., Winefield, C. and Jordan, B. (2015). From UVR8 to flavonol synthase: UV-B-induced gene expression in Sauvignon blanc grape berry. *Plant, Cell and Environment*. 38: 905–919.

Mackerness, S. (2000). Plant responses to ultraviolet-B (UV-B: 280–320 nm) stress: What are the key regulators? *Plant Growth Regulation*. 32: 27–39.

Mackerness, S.A.H., Butt, J.P. and Jordan, B.R. (1996). Amelioration of ultraviolet-B-induced down-regulation of mRNA transcripts for chloroplast proteins, by high irradiance, is mediated by photosynthesis. *Journal of Plant Physiology*. 148: 100–106.

Mackerness, S.A.-H., Surplus, S.L., Jordan, B.R. and Thomas, B. (1997a). Ultraviolet-B effects on transcript levels for photosynthetic genes are not mediated through carbohydrate metabolism. *Plant, Cell and Environment*. 20: 1431–1437.

Mackerness, S.A.H., Thomas, B. and Jordan, B.R. (1997b). The effect of supplementary UV-B radiation on mRNA transcripts, translation and stability of chloroplast proteins and pigment formation in *Pisum sativum* L. *Journal of Experimental Botany*. 48: 729–738.

Mackerness, S.A.-H., Liu, L., Thomas, B., Thompson, W.F., Jordan, B.R. and White, M.J. (1998a). Individual members of the chlorophyll a/b-binding protein gene family in pea show differential responses to UV-B radiation. *Physiologia Plantarum*. 103: 377–384.

Mackerness, S.A.-H., Surplus, S.L., Jordan, B.R. and Thomas, B. (1998b). Effects of supplementary UV-B radiation on photosynthetic transcripts at different stages of leaf development and light levels in pea: Role of active oxygen species and antioxidant enzymes. *Photochemistry and Photobiology*. 68: 88–96.

Mackerness, S., Surplus, S.L., Blake, P. et al. (1999a) Ultraviolet-B-induced stress and changes in gene expression in *Arabidopsis thaliana*: Role of signalling pathways controlled by jasmonic acid, ethylene and reactive oxygen species. *Plant, Cell and Environment*. 22: 1413–1423.

Mackerness, S.A.-H., Jordan, B.R. and Thomas, B. (1999b). Reactive oxygen species in the regulation of photosynthetic genes by ultraviolet-B radiation (UV-B: 280–320nm) in green and etiolated buds of pea (*Pisum sativum* L.). *Journal of Photochemistry and Photobiology B: Biology* 48: 180–188.

Mackerness, S., John, C.F., Jordan, B. and Thomas, B. (2001). Early signaling components in ultraviolet-B responses: Distinct roles for different reactive oxygen species and nitric oxide. *FEBS Letters*. 489: 237–242.

Majer, P., Czégény, G., Sándor, G., Dix, P.J. and Hideg, É. (2014). Antioxidant defence in UV-irradiated tobacco leaves is centred on hydrogen-peroxide neutralization. *Plant Physiology and Biochemistry*. 82: 239–243.

Malanga, G., Calmanovici, G. and Puntarulo, S. (1997). Oxidative damage to chloroplasts from *Chlorella vulgaris* exposed to ultraviolet-B radiation. *Physiologia Plantarum*. 101: 455–462.

Manetas, Y., Petropoulou, Y., Stamatakis, K., Nikolopoulos, D., Levizou, E., Psaras, G. and Karabourniotis, G. (1996). Beneficial effects of enhanced UV-B radiation under field conditions: Improvement of needle water relations and survival capacity of *Pinus pinea* L seedlings during the dry Mediterranean summer. *International Workshop Entitled UV-B and Biosphere*, Kluwer Academic Publishers, Wageningen, pp. 100–108.

Mehler, A.H. (1951). Studies on reactions of illuminated chloroplasts. I. Mechanism of the reduction of oxygen and other Hill reagents. *Archives of Biochemistry and Biophysics*. 33: 65–77.

Melis, A., Nemson, J.A. and Harrison, M.A. (1992). Damage to functional components and partial degradation of photosystem II reaction center proteins upon chloroplast exposure to ultraviolet-B radiation. *Biochimica et Biophysica Acta*. 1100: 312–320.

Mewes, H. and Richter, M. (2002). Supplementary ultraviolet-B radiation induces a rapid reversal of the diadinoxanthin cycle in the strong light-exposed diatom *Phaeodactylum tricornutum*. *Plant Physiology*. 130: 1527–1535.

Nogues, S., Allen, D.J., Morison, J.L. and Baker, N.R. (1998). Ultraviolet-B radiation effects on water relations, leaf development, and photosynthesis in droughted pea plants. *Plant Physiology.* 117: 173–181.

Pitzschke, A., Forzani, C. and Hert, H. (2006). Reactive oxygen species signaling in plants. *Antioxidants & Redox Signaling.* 8: 1757–1764.

Post, A., Lukins, P.B., Walker, P.J. and Larkum, A.W.D. (1996). The effects of ultraviolet irradiation on P680+ reduction in PS II core complexes measured for individual S-states and during repetitive cycling of the oxygen-evolving complex. *Photosynthesis Research.* 49: 21–27.

Poulson, M.E., Donahue, R.A., Konvalinka, J. and Boeger, M.R.T. (2002). Enhanced tolerance of photosynthesis to high-light and drought stress in *Pseudotsuga menziesii* seedlings grown in ultraviolet-B radiation. *Tree Physiology.* 22: 829–838.

Poulson, M.E., Boeger, M.R.T. and Donahue, R.A. (2006). Response of photosynthesis to high light and drought for *Arabidopsis thaliana* grown under a UV-B enhanced light regime. *Photosynthesis Research.* 90: 79–90.

Renger, G., Völker, M., Eckert, H.J., Fromme, R., Hohm-Veit, S. and Gräber, P. (1989). On the mechanism of photosystem II deterioration by UV-B irradiation. *Photochemistry and Photobiology.* 49: 97–105.

Rizzini, L., Favory, J.-J., Cloix, C. et al. (2011). Perception of UV-B by the Arabidopsis UVR8 protein. *Science.* 332: 103–106.

Rodrigues, G.C., Jansen, M.A.K., van den Noort, M.E. and van Rensen, J.J.S. (2006). Evidence for the semireduced primary quinone electron acceptor of photosystem II being a photosensitizer for UVB damage to the photosynthetic apparatus. *Plant Science.* 170: 283–290.

Rozema, J., Chardonnens, A., Tosserams, M., Hafkenscheid, R. and Bruijnzeel, S. (1997). Leaf thickness and UV-B absorbing pigments of plants in relation to an elevational gradient along the Blue Mountains, Jamaica. *Plant Ecology.* 128: 150–159.

Ruhland, C.T. and Day, T.A. (2000). Effects of ultraviolet-B radiation on leaf elongation, production and phenylpropanoid concentrations of *Deschampsia antarctica* and *Colobanthus quitensis* in Antarctica. *Physiologia Plantarum.* 109: 244–251.

Salter, A.H., Koivuniemi, A. and Strid, Å. (1997). UV-B and UV-C irradiation of PSII core and reaction centre particles in vitro. Identification of different fragments of the D1 protein and of multiple cleavage sites depending upon irradiation wavelength and particle type. *Plant Physiology and Biochemistry.* 35: 809–817.

Sävenstrand, H., Olofson, M., Samuelsson, M. and Strid, Å. (2004). Induction of early light-inducible protein gene expression in *Pisum sativum* after exposure to low levels of UV-B irradiation and other environmental stresses. *Plant Cell Reports.* 22: 532–536.

Schmid, F.-X. (2001). Biological macromolecules: UV-visible spectrometry. In: *eLS.* John Wiley & Sons, Chichester.

Schmidt, É.C., Pereira, B., Pontes, C.L.M. et al. (2012). Alterations in architecture and metabolism induced by ultraviolet radiation-B in the carragenophyte *Chondacanthus teedei* (Rhodophyta, Gigartinales). *Protoplasma.* 249: 353–367.

Searles, P.S., Flint, S.D. and Caldwell, M.M. (2001). A meta analysis of plant field studies simulating stratospheric ozone depletion. *Oecologia.* 127: 1–10.

Sprtova, M., Spunda, V., Kalina, J. and Marek, M.V. (2003). Photosynthetic UV-B response of beech (*Fagus sylvatica* L.) saplings. *Photosynthetica.* 41: 533–543.

Staxen, I. and Bornman, J.F. (1994). A morphological and cytological study of petunia-hybrida exposed to UV-B radiation. *Physiologia Plantarum.* 91: 735–740.

Stracke, R., Favory, J.-J., Gruber, H. et al. (2010). The *Arabidopsis* bZIP transcription factor HY5 regulates expression of the PFG1/MYB12 gene in response to light and ultraviolet-B radiation. *Plant, Cell and Environment.* 33: 88–103.

Strid, Å. (1993). Increased expression of defence genes in *Pisum sativum* after exposure to supplementary ultraviolet-B radiation. *Plant Cell Physiology.* 34: 949–953.

Strid, Å. and Porra, R.J. (1992). Alterations in pigment content in leaves of *Pisum sativum* after exposure to supplementary UV-B. *Plant Cell Physiology.* 33: 1015–1023.

Strid, Å., Chow, W.S. and Anderson, J.M. (1990). Effects of supplementary UV-B radiation on photosynthesis in *Pisum sativum.* *Biochimica et Biophysica Acta.* 1020: 260–268.

Strid, Å., Chow, W.S. and Anderson, J.M. (1994). UV-B damage and protection at the molecular level in plants. *Photosynthesis Research.* 39: 475–489.

Strid, Å., Chow, W.S. and Anderson, J.M. (1996a). Changes in the relaxation of electrochromic shifts of photosynthetic pigments and in the levels of mRNA transcripts in leaves of *Pisum sativum* as a result of exposure to supplementary UV-B radiation. The dependency on the intensity of the photosynthetically active radiation. *Plant Cell Physiology.* 37: 61–67.

Strid, Å., Chow, W.S. and Anderson, J.M. (1996b). Temperature-dependency of changes in the relaxation of electrochromic shifts, of chlorophyll fluorescence, and in the levels of mRNA transcripts in detached leaves from *Pisum sativum* exposed to supplementary UV-B radiation. *Plant Science.* 115: 199–206.

Surplus, S.L., Jordan, B.R., Murphy, A.M., Carr, J.P., Thomas, B. and Mackerness, S.A.H. (1998). Ultraviolet-B-induced responses in *Arabidopsis thaliana*: Role of salicylic acid and reactive oxygen species in the regulation of transcripts encoding photosynthetic and acidic pathogenesis-related proteins. *Plant, Cell and Environment.* 21: 685–694.

Szilárd, A., Sass, L., Deák, Z. and Vass, I. (2007). The sensitivity of photosystem II to damage by UV-B radiation depends on the oxidation state of the water-splitting complex. *Biochimica et Biophysica Acta.* 1767: 786–882.

Takeuchi, Y., Fukumoto, R., Kasahara, H. Sakaki, T. and Kitao, M. (1995). Peroxidation of lipids and growth inhibition induced by UV-B irradiation. *Plant Cell Report.* 14: 566–570.

Takeuchi, A. Yamaguchi, T., Hidema, J., Strid, Å. and Kumagai, T. (2002). Changes in synthesis and degradation of Rubisco and LHCII with leaf age in rice (*Oryza sativa* L.) growing under supplementary UV-B radiation. *Plant, Cell and Environment.* 25: 695–706.

Taylor, W.C. (1989). Regulatory interactions between nuclear and plastid genomes. *Annual Review of Plant Physiology and Plant Molecular Biology.* 40: 211–233.

Taylor, R.M., Nikaido, O., Jordan, B.R., Rosamond, J., Bray, C.M. and Tobin, A.K. (1996). Ultraviolet-B-induced DNA lesions and their removal in wheat (*Triticum aestivum* L.) leaves. *Plant, Cell and Environment.* 19: 171–181.

Telfer, A., Bishop, S.M., Phillips, D. and Barber, J. (1994). Isolated photosynthetic reaction center of photosystem II as a sensitizer for the formation of singlet oxygen. *Journal of Biological Chemistry.* 269: 13244–13253.

Tevini, M. and Pfister, K. (1985). Inhibition of photosystem II by UV-B radiation. *Zeitschrift für Naturforschung.* 40c: 129–133.

Tjus, S.E. and Andersson, B. (1993). Loss of the trans-thylakoid proton gradient is an early event during photoinhibitory illumination of chloroplast preparations. *Biochimica et Biophysica Acta.* 1183: 315–322.

Trebst, A. and Depka, B. (1990). Degradation of the D-1 protein subunit of photosystem II in isolated thylakoids by UV light. *Zeitschrift für Naturforschung.* 45c: 765–771.

van Hasselt, P.R., Chow, W.S. and Anderson, J.M. (1996). Short-term treatment of pea leaves with supplementary UV-B at different oxygen concentrations: Impacts on chloroplast and plasma membrane bound processes. *Plant Science.* 120: 1–9.

van Rensen, J.J.S., Vredenberg, W.J. and Rodrigues, G.C. (2007). Time sequence of the damage to the acceptor and donor sides of photosystem II by UV-B radiation as evaluated by chlorophyll a fluorescence. *Photosynthesis Research.* 94: 291–297.

Vass, I. (2012). Molecular mechanism of photodamage in the photosystem II complex. *Biochimica et Biophysica Acta.* 1817: 209–217.

Vass, I., Sass, L., Spetea, C., Bakou, A., Ghanotakis, D.F. and Petrouleas, V. (1996). UV-B-induced inhibition of photosystem II electron transport studied by EPR and chlorophyll fluorescence. Impairment of donor and acceptor side components. *Biochemistry.* 35: 8964–8973.

Vass, I., Kirilovsky, D. and Etienne, A.-L. (1999). UV-B radiation-induced donor- and acceptor-side modifications of photosystem II in the cyanobacterium *Synechocystis* sp. PCC 6803. *Biochemistry.* 38: 12786–12794.

Vass, I., Szilard, A. and Sicora, C. (2005). Adverse effects of UV-B light on the structure and function of the photosynthetic apparatus. In: *Handbook of photosynthesis*, 2nd edition, ed. Pessarakli, M. CRC Press, Boca Raton, FL, pp. 827–843.

Vu, C.V., Allen Jr., L.H. and Garrard, L.A. (1984). Effects of enhanced UV-B radiation (280–320 nm) on ribulose-1,5-bisphosphate carboxylase in pea and soybean. *Environmental and Experimental Botany.* 24: 131–143.

Wargent, J.J. and Jordan, B.R. (2013). From ozone depletion to agriculture: Understanding UV-B in sustainable crop production. *New Phytologist.* 197: 1058–1076.

Wargent, J.J., Gegas, V.C., Jenkins, G.I., Doonan, J.H. and Paul, N.D. (2009). UVR8 in *Arabidopsis thaliana* regulates multiple aspects of cellular differentiation during leaf development in response to ultraviolet B radiation. *New Phytologist.* 183: 315–326.

Wargent, J.J., Elfadly, E.M., Moore, J.P. and Paul, N.D. (2011). Increased exposure to UV-B radiation during early development leads to enhanced photoprotection and improved long-term performance in *Lactuca sativa. Plant Cell and Environment.* 34: 1401–1413.

Wargent, J.J., Nelson, B.C.W., McGhie, T.K. and Barnes, P.W. (2015). Acclimation to UV-B radiation and visible light in *Lactuca sativa* involves up-regulation of photosynthetic performance and orchestration of metabolome-wide responses. *Plant, Cell and Environment.* 38: 929–940.

Warner, C.W. and Caldwell, M.M. (1983). Influence of photon flux density in the 400–700 nm waveband on inhibition of photosynthesis by UV-B (280–320 nm) irradiation in soybean leaves: Separation of indirect and immediate effects. *Photochemistry and Photobiology.* 38: 341–346.

Wilson, M.I. and Greenberg, B.M. (1993). Protection of the D1 photosystem II reaction center protein from degradation in ultraviolet radiation following adaptation of *Brassica napus* L. to growth in ultraviolet-B. *Photochemistry and Photobiology.* 57: 556–563.

Wilson, M.I., Ghosh, S., Gerhardt, K.E. et al. (1995). In vivo photomodification of ribulose 1,5-bisphosphate carboxylase oxygenase holoenzyme by ultraviolet-B radiation. Formation of a 66-kilodalton variant of the large subunit. *Plant Physiology.* 109: 221–229.

Wu, S.-H. (2014). Gene expression regulation in photomorphogenesis from the perspective of the central dogma. *Annual Reviews of Plant Biology.* 65: 311–333.

Wu, D., Hu, Q., Yan, Z. et al. (2012). Structural basis of ultraviolet-B perception by UVR8. *Nature.* 484: 214–219.

Xu, J.T. and Gao, K.S. (2010). UV-A enhanced growth and UV-B induced positive effects in the recovery of photochemical yield in *Gracilaria lemaneiformis* (Rhodophyta). *Journal of Photochemistry and Photobiology B—Biology.* 100: 117–122.

Yang, Y.Q. and Yao, Y. (2008). Photosynthetic responses to solar UV-A and UV-B radiation in low-and high-altitude populations of *Hippophae rhamnoides. Photosynthetica.* 46: 307–311.

Zhang, J., Henkow, L., Hu, X., Jordan, B.R. and Strid, Å. (1994). The effects of ultraviolet-B radiation on the CF_oF_1-ATPase. *Biochimica et Biophysica Acta.* 1185: 295–302.

17 Small Fruit Crop Responses to Leaf Intercellular CO_2, Primary Macronutrient Limitation and Crop Rotation

Hong Li

CONTENTS

17.1 Introduction 287
    17.1.1 Carbon Dioxide, Primary Macronutrients, and Crop Rotation 288
    17.1.2 Limited Primary Macronutrient Supply as an Environmental Abiotic Stress Factor 289
17.2 Case Study: Limited K–N Macronutrient Supplies Inducing Leaf Intercellular C_i to Enhance Plant Carboxylation, Water Use Efficiency, Early Fruit Formation, and Leaf/Fruit K^+ Holding, but Higher Leaf C_i Inhibiting NO_3^- Ion Retention 290
    17.2.1 Limited K/N Macronutrient Supplies Resulting in Higher Leaf C_i to Enhance Plant Photosynthesis, Carboxylation, and Water Use Efficiency 291
    17.2.2 Limited K/N Macronutrients Advance Berry Formation and Marketable Yields 293
    17.2.3 Higher Leaf C_i Inhibits Leaf/Fruit NO_3^- Retention, but Not K^+ Ions and TDS 294
    17.2.4 Colimitation of Reduced Macronutrient K/N Inputs on Strawberry Fruit Productivity and Mineral Nutrition Quality 295
17.3 Case Study: Phosphorus Macronutrient Limitation and Small Fruit Crop Development 296
    17.3.1 Strawberry Plants Cope with Limited Macronutrient P Supplies in Nursery Production ...297
    17.3.2 Strawberry Plant P Optimization Requires a Substantial Acquisition of Plant N..............299
    17.3.3 Excessive P Inputs Could Prohibit Plant P/N Uptake to Hurt Vegetative Propagation.......301
17.4 Conclusions..302
Acknowledgments..302
Abbreviations 302
References..302

17.1 INTRODUCTION

Many environmental abiotic stress factors and farming practices can contribute to influence plant physiological development and crop productivity. Environmental variations in CO_2, water, radiation, and nutrients are among the most significant abiotic stress factors to influence crop performance. Farming practices such as nutrient supply and crop rotation are especially important to ensure healthy crop production. This paper is focused on understanding small fruit crop physiological responses to primary macronutrient limitation under different crop rotation conditions through summarizing the results of two separate field studies conducted in Nova Scotia, Canada. The field studies have investigated the physiological development of strawberry (*Fragaria* × *ananassa* Duch.) crops under reduced primary macronutrient supplies in different types of crop rotation regimes. It is concluded that strawberry plants treated with 25%-reduced K–N inputs could promote strawberry K/N assimilation to render the plants healthier with significantly higher leaf intercellular CO_2 concentrations, plant photosynthesis,

carboxylation, and water use efficiency. These physiological advantages could lead to advance berry formation and produce significantly higher marketable yields ($P < 0.05$). Higher leaf intercellular CO_2 concentrations inhibited leaf/fruit NO_3^- ion retention, but this inhibition did not occur in retention of leaf/fruit K^+ ions and total dissolved solids. Also, 50%-reduced macronutrient P inputs could result in a suitable nursery plant P/N retention ratio (0.14–0.17) for optimum nursery formation in wheat–ryegrass–ryegrass–strawberry (WRRS) and corn–ryegrass–ryegrass–strawberry (CRRS) rotation systems. High P inputs could hurt the strawberry nursery crops by prohibition of plant P/N uptake. Overall, maximum NPK deficiency (control plots) could cause plant stress, and unlimited macronutrient supplies could constrain more fruit formation and nutritional attributes than reduced NPK inputs. Limited primary macronutrient supplies would be a profitable strategy to enhance plant macronutrient assimilation and physiological development while reducing uses of nonrenewable fertilizers and environmental risk to promote sustainable, low-input small fruit crop production.

17.1.1 CARBON DIOXIDE, PRIMARY MACRONUTRIENTS, AND CROP ROTATION

Carbon dioxide (CO_2) is the source of carbon required for plant growth. Crops need CO_2 as the primary raw materials to produce organic compounds to construct tissues (Larios et al., 2004; Larbi et al., 2006). Plant photosynthesis involves the input of CO_2 and water with radiant energy and the presence of chlorophyll as the catalyst, as shown by the photosynthesis equation $6CO_2 + 6H_2O + \nu \rightarrow C_6H_{12}O_6 + 6O_2$. Pores in leaves let CO_2 enter. The outputs of the production are carbohydrates and oxygen. Enriching the air with CO_2 can enhance plant growth and development based on the photosynthesis process equation (Taiz and Zeiger, 2002; Sicher and Bunce, 2008; Lawlor and Tezara, 2009).

The direct effects of increased CO_2 concentrations on crop nutrient use and crop productivity are significantly positive, quantified on different types of crops (Viktor and Cramer, 2003; Leakey et al., 2009; Bloom et al., 2010; Foyer et al., 2011). Elevated CO_2 could alleviate any plant depression of photosynthesis, and hence elevated CO_2 transported in xylem sap can be fixed in photosynthetic cells in plant leaves and branches to enhance C assimilation, photosynthetic N use, and leaf area, shoot, and root productivities (Cramer et al., 2005; McGuire et al., 2009; He et al., 2010). When the increase in CO_2 in the air was doubled, the increased plant growth resulting from raising CO_2 level was estimated to vary between 22% and 41% (Ainsworth and Rogers, 2007). Elevated CO_2 can diminish crop leaf stomatal activity to reduce water use (Perez-Lopez et al., 2012).

Negative effects of CO_2 on plant physiological and nutritional aspects have also been reported. The effects of increased CO_2 in the air on plant growth can refer to the changes in elevated temperature and evaporation that can lead to the decline of growth conditions for most conventional crops (Taiz and Zeiger, 2002; Geissler et al., 2009). Crop photosynthesis and transpiration are primarily limited by light but not by CO_2 level (Tartachnyk and Blanke, 2007). High CO_2 concentrations could inhibit the assimilation of nitrate in grain crop (Bloom et al., 2010) and NO_3^- ions retention in rapid top-growth small fruit crop (Li et al., 2013). Also, elevated CO_2 reduces stomatal and metabolic limitations on photosynthesis under salinity conditions (Perez-Lopez et al., 2012).

As an essential nutrient for plant growth, carbon and oxygen are absorbed by plants from the air for plant growth (Taiz and Zeiger, 2002). The primary macronutrients, nitrogen (N), phosphorus (P), and potassium (K), are absorbed in large quantities, representing 0.2–4% of plant tissues on a dry matter weight basis. The other secondary macronutrients (Ca, Mg, and S) are less important, and micronutrients (B, Cl, Fe, Zn, Cu, Mn, Mo) represent much smaller quantities (<0.2%) in plant tissues (Taiz and Zeiger, 2002). Plant N assimilation is a fundamental biological process, and N determines the synthesis of amino acids and therefore proteins and vitamins in plants (Gastal and Lemaire, 2002; Davis, 2009; Li, 2012; Bloom et al., 2012). Nitrogen is the most required nutrient for plant growth, and it occurs in all parts of plants in substantial

amounts that have marked effects on crop productivity and quality components (Li et al., 2002, 2003, 2006a, 2011; Lea and Azevedo, 2006; Swarbreck et al., 2011). Nitrogen remobilization and translocation determine yield and quality of fruit crops (Nestby et al. 2005; Nava et al., 2008; Li et al., 2013) and nonfruit crops including cotton (Li et al., 2002; Li and Lascano 2011), potato (Li et al., 2003, 2006a), rice (Tabuchi et al., 2007), broccoli (Li et al., 2011), cauliflower (Li, 2012), wheat and maize (Bloom et al., 2010, 2012), and barley (Dordas, 2012). Amino acids and proteins can only be built from NH_4^+ ions and therefore NO_3^- ions must be reduced (Swarbreck et al., 2011).

In primary macronutrient management, phosphorus (P) is considered as the *energizer* for general health of plants and one of the most important nutrients to help store and transfer energy during the photosynthesis process (Schachtman et al., 1998; Hariprasad and Niranjana, 2009; Li et al., 2010b). Phosphorus is responsible for the conversion of light energy to chemical energy (ATP) during photosynthesis for plant growth (Smith et al., 2003), and it is needed as a structural component of the nucleic acid, deoxyribonucleic nucleic acid (DNA), and ribose nucleic acid (RNA) (Sinegani and Rashidi, 2011). Plants acquire P nutrients as phosphate (P_i) anions from soil solution, and P_i is the key player in all metabolic processes including energy transfer, biosynthesis, and respiration (Schachtman et al., 1998; Raghothama and Karthikeyan, 2005; Reich et al., 2009). Plant P fertilizer use efficiency is usually lower than N–K nutrients, and total plant P can make up about 0.2% of the whole plant dry matter (Schachtman et al., 1998; Li et al., 2010b). It is reported that excessive P inputs may not hurt the crops, but it can build up residual soil P levels and increase P_i movement from the soil to the environment (Rehm et al., 1998). But recent study has found that too much P inputs can hurt the nursery crop by non-optimal plant P/N uptake and its ratios (Preusch et al., 2004; Li et al., 2014).

Macronutrient potassium (K) is also essential to all plants, and K^+ ions are outstanding in the formation of plant proteins and carbohydrates, and the regulation of photosynthesis (Pettigrew, 2008; Lester et al., 2010; Li et al., 2013). K^+ is also the major cation creating cell turgor responsible for plant water and solute movement that influence yield formation of many crops (Khayyat et al., 2007; Pettigrew, 2008). Potassium occurs in all parts of plants in substantial amounts comparable to plant N holding levels (Nava et al., 2008; Li et al., 2013). Healthy food generally carries high amounts of K^+ ions that make up most of the cell fluid cations at about 4 g per liter (He and MacGregor, 2008; Lester et al., 2010). K^+ ions can interact with plant N elements in crop yield formation, and when plant N is deficient, the impact of K nutrition on plant N use is more critical (Nava et al., 2008; Li et al., 2013).

Plant nutrient uses are dependent on CO_2 concentrations to impact plant photosynthesis and plant growth (Pettersson and McDonald, 1994; Warren, 2004; Reich et al., 2006, 2009; Baslam et al., 2012). When tomato plant seedlings were supplied with both NO_3^- and NH_4^+, there was a significant increase in biomass accumulation in response to increased CO_2 (Viktor and Cramer, 2003). The effects of increased atmospheric CO_2 levels were mainly on the stimulation of C assimilation, photosynthetic N use efficiency, plant photosynthesis, and the improvement of soluble sugar concentrations (Leakey et al., 2009). Greater supplies of CO_2 concentrations could improve the incorporation of N into amino acids in the grass roots, which made greater use of inorganic C to build organic compounds, leading to amino acid and protein synthesis (Swarbreck et al., 2011). Also, elevated CO_2 can impair the beneficial effect of mycorrhizal fungi on minerals (P, Fe, and Cu) and phytochemical quality in short-season vegetable crop (Baslam et al., 2012).

Crop rotation, the practice of growing different types of crops in the same area in sequential seasons, is considered as one of the most efficient measures to increase plant nutrient use efficiency and crop productivity (Six et al., 2006; Anderson, 2010; Li et al., 2014). Crop rotation can balance soil nutrients and prevent disease and pest incidents by using the concepts that different crops have different nutrient requirements and crops in different botanical families tend to avoid the same pest and disease problems (Stanger and Lauer, 2008; Kong et al., 2011). Cover crops such as ryegrass used in crop rotation can build healthier soil, fewer soil-borne problems, and more beneficial insects (Stanger and Lauer, 2008; Murugan and Kumar, 2013). Cover crops' roots and top-growth contribute nutrients and organic matter to the soil after they are tilled under (Anderson, 2010; Kong et al., 2011). Yet, lengthy rotations are needed to build healthier soils because of the times necessary for roots and top residues of previous crops to balance soil macronutrients and stimulate microorganisms for helping suppress disease organisms (Stanger and Lauer, 2008; Anderson, 2010).

17.1.2 LIMITED PRIMARY MACRONUTRIENT SUPPLY AS AN ENVIRONMENTAL ABIOTIC STRESS FACTOR

Plant stress is defined as an external factor that exerts a disadvantageous influence on a plant, and stress tolerance is the plant's capability to cope with an unfavorable environment and an attack (Mittler, 2006; Li, 2009). Plant abiotic and biotic stress is related to unfavorable and environmental constraints (Gutschick, 1999; Li et al., 2006b; Laughlin and Abella, 2007; Li, 2009). Plant abiotic stress is about the determination of negative impacts on plant development that resulted from nonliving environmental stressors (Li et al., 2006b, 2008; Laughlin and Abella, 2007). Plants show some degree of stress when exposed to unfavorable environments, and these stresses can collectively contribute to affect plant development, plant nutrient assimilation, crop productivity, and food quality (Taiz and Zeiger, 2002; Li, 2008). It is estimated that up to 82% of potential crop yields are lost due to abiotic stress annually (Taiz and Zeiger, 2002; Mittler, 2006).

Soil nutrient limitation, among the most important plant abiotic stressors such as high temperatures, CO_2 variation, soil acidity, salinity, and water shortage can affect plant physiological development and crop productivity (Reich et al., 2006; Li et al., 2006b, 2008, 2011; Keutgen and Pawelzik, 2009; Kong et al., 2011; Li and Lascano, 2011). Limited primary

macronutrient supply can result in crop stress because crops become unhealthy in the soil environment that is deficient in required nutritional elements (Kong et al., 2011; Li et al., 2014). Crop macronutrient stress can affect negatively plant development leading to the decline in crop yield (Perez-Lopez et al., 2012; Li et al., 2013). Increase in crop growth is reflected in the harvestable yields with rising CO_2 mechanisms and minerals (Ainsworth and Rogers, 2007; Baslam et al., 2012). Negative consequences in crop physiological aspects are especially more pronounced when the effects of changes in CO_2 levels are combined with these from other plant abiotic stressors (Warren, 2004; Geissler et al., 2009; Baslam et al., 2012).

Plant stress from nutrient limitation is indicated by reduced leaf photosynthesis, diminished plant macronutrient assimilation, and low nitrate reduction (Pettersson and McDonald, 1994; Gastal and Lemaire, 2002; Foyer et al., 2011). Macronutrient N availability can have a stimulatory effect on plant nitrate reduction (or nitrate reductase activity) that can control plant physiological activity (Gallais and Hirel, 2004; Sicher and Bunce, 2008). Nitrogen absorbed by plant roots is generally in the form of nitrate $\left(NO_3^- \right)$, and assimilated NO_3^- in plants needs to be incorporated into organic forms, which can stimulate continuously both plant photosynthesis process and assimilation of external NO_3 for plant development (Taghavi and Babalar, 2007; Swarbreck et al., 2011; Dordas, 2012). As most of the assimilated NO_3^- is finally to be allocated to photosynthetic components such as chlorophyll, low rates of NO_3 reduction in crops can diminish plant acclimation capacity to changes in the environment, resulting in low crop productivity (Taiz and Zeiger, 2002; Sicher and Bunce, 2008).

The effects of CO_2 concentrations on plant growth depended on types of species and sufficiency of mineral nutrition (Viktor and Cramer, 2003; Warren, 2004; Long, 2006; Sicher and Bunce, 2008). Close relationships have been determined for different species between photosynthetic capacity and leaf P concentrations (Smith et al., 2003; Reich et al., 2009). Nitrogen input rates, leaf respiration, and plant N use were related to plant N use efficiency (Hirel et al., 2007; Foyer et al., 2011). It is estimated that more than 50% of leaf N components are associated with plant photosynthesis (Lea and Azevedo, 2006; Reich et al., 2006; Hirel et al., 2007). Also, as plant organic compounds are produced during the periods of NO_3 reduction and incorporation in the photosynthetic process, plant N status and its metabolic adjustments can affect glutamine synthetase and crop productivity (Hirel et al., 2007; Foyer et al., 2011; Swarbreck et al., 2011).

There is evidence that N limitation can constrain plant responses to the elevated ambient CO_2 concentrations (Reich et al., 2006), and CO_2 enrichment can inhibit crop assimilation of nitrate $\left(NO_3^- \right)$ into organic N compounds (Bloom et al., 2010). However, the overall impacts of CO_2 and macronutrient limitation on plant physiological development and macronutrient translocation are not well understood. The influence of limited macronutrient supplies in crop macronutrient uptake and crop productivity in crop rotation systems is still little known. It is import to understand how crops cope with environmental macronutrient stress, which maintain or even improve crop yield at lowered fertilizer inputs (Kong et al., 2011; Li and Lascano, 2011; Swarbreck et al., 2011; Li et al., 2013, 2014).

17.2 CASE STUDY: LIMITED K–N MACRONUTRIENT SUPPLIES INDUCING LEAF INTERCELLULAR C_i TO ENHANCE PLANT CARBOXYLATION, WATER USE EFFICIENCY, EARLY FRUIT FORMATION, AND LEAF/FRUIT K^+ HOLDING, BUT HIGHER LEAF C_i INHIBITING NO_3^- ION RETENTION

Small fruit crops, such as strawberries, blueberries, grapes, raspberries, and blackberries, usually produced on small perennial crops, are all known as high-value cash crops (Rieger, 2005). Among the popular small fruit crops, strawberry (*Fragaria* × *ananassa* Duch.) of the *Rosaceae* family is one of the most important small fruit crops grown worldwide (Pritts and Handley, 1998; Rieger, 2005). Strawberry is a rapid top-growth plant that required chilling and is usually the first fruit to be harvested in the year in North America. The crop is especially sensitive to environmental changes in nutrition, water, and temperature (Pritts and Handley, 1998; Preusch et al., 2004; Li et al., 2010a).

While little information is available about overlay effects of plant CO_2 and primary macronutrient limitations on strawberry crop development, macronutrient assimilation, and fruit yield formation, it is reported that photosynthetic contributions by young green fruits to their daily C budget could be significant for various fruit species including oranges (Moreshet and Green, 1980), apples (Jones, 1981), avocadoes (Blanke and Whiley, 1995), and tomatoes (Mingo et al., 2003). Individual effects of atmospheric CO_2 concentrations, temperatures, radiation, nutrients, or water/salt levels could determine strawberry yield formation and fruit quality (Itani et al., 1998; Tagliavini et al., 2005; Eshghi and Tafazoli, 2007; Grant et al., 2010; Li et al., 2010a).

In North America, strawberry plants are perennial, growing in the field during hot and cold seasons, and the crops adapt to grow in various soil environments (Pritts and Handley, 1998; Rieger, 2005). In strawberry fruit production in Canada, usually a 6-year crop rotation is practiced to ensure a healthy soil environment for fruit growth (Li et al., 2010a). As a rapid top-growth crop, strawberry plants require a sufficient acquisition of macronutrients for photosynthesis needs to meet their rapid top development and fruit formation. Growers tend to supply with high rates of macronutrients and irrigation to ensure rapid top growth and fruit formation of the strawberry (Pritts and Handley, 1998; Li et al., 2013). However, high inputs have become unsustainable because of high cost of chemical fertilizers and other materials leading to lower production incomes and environmental issues (Swarbreck et al., 2011). A promising solution to the problems mentioned above is the adoption

of low-input strategy for sustainable and profitable crop production (Kong et al., 2011).

Given the increased atmospheric CO_2 concentrations as the most certain predicted environmental change (Pettersson and McDonald, 1994; Swarbreck et al., 2011) and the central roles of primary macronutrients in crop production (Gastal and Lemaire, 2002; Hirel et al., 2007), understanding how crops can cope with primary macronutrient limitation is critically important. Li et al. (2013) have demonstrated that strawberry crops could cope with macronutrient limitation to understand the mechanisms underlying small fruit crop stress tolerance to adverse environmental conditions.

The 2-year (2009–2010) study of strawberry plant responses to limited K/N macronutrient supplies was conducted on a well-certified farm, Millen Farms, in Nova Scotia, eastern Atlantic Canada (45°23'42"N, 63°32'16"W). The Millen Farms were registered in the North American Small Fruit Production Program. The crop management practice was a 6-year, three-crop rotation regime, which consisted of 2 years of perennial ryegrass (*Lolium perenne* L.), 3 years of strawberry, and 1 year corn (*Zea mays* L.) to ensure clean soil for strawberry production. The soil was a sandy loam, classified as a Spodosol in the American Soil Classification Systems.

The hypothesis was that adequate macronutrient limitation could enhance plant assimilation of macronutrients, which could render the plants healthier that might induce significantly higher leaf intercellular CO_2 levels to enhance crop photosynthesis. The macronutrient experimental treatments consisted of five rates of K_2O fertilizers (0, 6, 12, 18, and 24 kg ha^{-1}) and five rates of N fertilizers (0, 5, 10, 15, and 20 kg ha^{-1}), which represented 0%, 25%, 50%, 75%, and 100%, respectively, of the K/N macronutrient recommendations for strawberry fruit production in loamy soils in North America (Pritts and Handley, 1998). The full (or nonreduced) inputs (100% of K/N macronutrient requirements for strawberry) were the highest rates, i.e., K_2O 24 kg ha^{-1} (K_{24}) and N 20 kg ha^{-1}(N_{20}). No P nutrient applications were required based on soil testing. A split-plot design was used to arrange the K and N treatments with three blocks in the field. The K treatments were applied in the main plots (15 units) using granular potassium chloride (KCl, 0-0-60). The N treatments were in the subplots (75 units) randomly distributed in each main plot using granular NH_4NO_3–Ca (27.5-0-0).

The strawberry cultivar was "Mira", a June-bearing-type crop. The planting stocks were commercially certified, well-rooted, cold-stored, and disease-free. The information of strawberry plant/soil management (including plant transplanting, nutrient applications, mulching, irrigation, runner thinning, winter protection, fruit bearing care, etc.) and plant physiological and nutritional data collation (Figure 17.1) have been reported by Li et al. (2013).

The case study revealed three novel results: (1) strawberry plants treated with adequately reduced K/N macronutrient inputs resulted in significantly higher leaf intercellular C_i concentrations to enhance plant photosynthesis, carboxylation, and water use efficiency to advance berry formation and improve marketable yields; (2) higher C_i concentrations

FIGURE 17.1 **(See color insert.)** Limited K/N macronutrient study field views (irrigated June-bearing strawberry cv. *Mira*) initial flowering (a) and initial fruit ripening (b); strawberry plants at full blooming on June 3 (c), full bearing on June 26 (d), early maturity on July 2 (e), full maturity on July 13 (f); marketable fruit yields for the 50%-limited K/N treatments plot on July 13 (g); and higher marketable fruit yields for the 25%-limited K/N treatments plot determined on the same day (July 13) (h).

were found to inhibit leaf NO_3^- ion retention and its translocation to berries, but higher C_i concentrations enhance leaf K^+ ion retention and translocation; and (3) there was a colimitation of reduced K/N inputs on C_i, leaf K^+/NO_3^- retention and translocation, marketable fruit yield, and fruit nutritional quality attributes. The newly reported results from this case study could be summarized in the following.

17.2.1 LIMITED K/N MACRONUTRIENT SUPPLIES RESULTING IN HIGHER LEAF C_i TO ENHANCE PLANT PHOTOSYNTHESIS, CARBOXYLATION, AND WATER USE EFFICIENCY

As referred to by Li et al. (2013), the in-season weekly strawberry physiological development, simultaneously determined during flowering, bearing, and fruit maturity stages (Figure 17.1), was characterized by the increases in plant photosynthesis, instantaneous carboxylation efficiency (CE), and leaf water use efficiency of photosynthesis (WUE). At full flowering stage, leaf intercellular CO_2 concentration (C_i), stomatal conductance (g_s), net photosynthesis rate (P_r), and transpiration rate per unit leaf area (T_r) showed that the plant responses to the reduced K/N macronutrient treatments were significant (Table 17.1). The C_i/C_a ratios varied between 0.51 and 0.92 with higher C_i/C_a ratios (0.80–0.92) determined in the

25%-reduced K/N input treatments. Leaf transpiration rates were significantly lower in the reduced input treatments (mean 8.13 µmol m^{-2}·s^{-1}) compared to the control (9.05 µmol m^{-2}·s^{-1}) (ANOVA, $F_{1,74} = 3.92$, $P < 0.032$). For instantaneous CE, determined using the equation CE = P_r/C_i, the highest CE value (0.126 mol m^{-2}·s^{-1}) was measured at the reduced rate K_{15}, which was significantly higher than the CE value for the control K_0 (0.096 mol m^{-2}·s^{-1}). The CE value at the K_{18} level was also significantly higher (0.121 mol m^{-2}·s^{-1}) compared to the other treatments (HSD = 0.016 mol m^{-2}·s^{-1}, α = 0.05).

The leaf water use efficiency WUE, determined using the equation WUE = P_r/T_r, had the lowest value in the control (2.50 µmol mmol^{-1}). The highest WUE value (3.84 µmol mmol^{-1}) was measured at the 25%-reduced N_{15} level, which was also significantly higher than the WUE grand mean value (3.11 µmol mmol^{-1}, $n = 75$, Table 17.1). Among all the physiological variables, the root mean square error (RMSE) and the standard deviation (SD) values were high for C_i. Also, only the distribution of the C_i data was skewed (kurtosis value > 3, Table 17.1), showing the great variation in leaf intercellular CO_2 concentrations.

Both leaf C_i and P_r variables were strongly induced by the reduced K/N input treatments (Li et al., 2013). The C_i and P_r values increased up to the 25%-reduced input levels (K_{18} and N_{15}, Figure 17.2). Higher leaf P_r rates were associated with higher leaf C_i concentrations, statistically separated by the K treatments

TABLE 17.1

Effects and Interactions of the K and N Macronutrient Treatments on Strawberry Physiological Development Determined at First Fruit Maturity in Early July

		Strawberry Leaf Physiological Variables[a]					
Sources of Variation	df	C_i	g_s	P_r	C_i/C_a	CE	WUE
Block (B)	2	10.7**	7.01**	12.9**	6.43**	7.56**	8.79**
K_2O rates (K)	4	3.14*	3.74*	3.32*	3.26*	3.79*	3.65*
B × K	8	0.51 ns	0.93 ns	0.73 ns	1.03 ns	0.98 ns	1.14 ns
N rates (N)	4	3.10*	3.65*	3.27*	3.34*	3.75*	3.23*
K × N	16	0.72 ns	0.78 ns	0.45 ns	0.43 ns	0.54 ns	0.62 ns
		Contrasts[b]					
K_0 vs. K_6, K_{12}, K_{18}, and K_{24}	1	3.53*	3.57*	4.06*	3.41*	3.82*	3.74*
N_0 vs. N_5, N_{10}, N_{15}, and N_{20}	1	3.85*	3.62*	4.16*	3.86*	4.51**	3.35*
N_{15} vs. N_5, N_{10}, and N_{20}	1	0.74 ns	0.58 ns	3.23*	1.72 ns	1.96 ns	1.28 ns
		Model					
F value	74	2.77**	2.23*	2.83**	2.98**	3.29**	4.12**
R^2		0.70**	0.62*	0.68**	0.69*	0.79**	0.81**
Mean[c]		251.8	3.02	27.3	0.7	0.109	3.11
RMSE[c]		16.7	1.89	2.63	0.09	0.008	0.42
SD[c]		22.5	1.96	2.97	0.08	0.043	1.01
Kurtosis		3.32	0.77	0.29	2.89	0.34	2.41

Source: Li, H. et al., *Photosynthesis Research* 115:101–14, 2013.

[a] *, **, ***, and ns: Significant at $P < 0.05$, $P < 0.01$, $P < 0.001$, and nonsignificant at $P < 0.05$, respectively.

[b] K_0, K_6, K_{12}, K_{18}, and K_{24} were the K_2O rates at 0, 6, 12, 18, and 24 kg ha^{-1}, and N_0, N_5, N_{10}, N_{15}, and N_{20} were the N rates at 0, 5, 10, 15, and 20 kg ha^{-1}, respectively. Each variable $n = 75$.

[c] Mean, SD (standard deviation), and RMSE (root mean square error). C_i, leaf intracellular CO_2 concentrations in µmol mol^{-1}; g_s, leaf stomatal conductance to CO_2 diffusion in mmol m^{-2} s^{-1}; P_r, leaf photosynthesis rate in µmol fixed CO_2 m^{-2} s^{-1}; C_i/C_0, leaf intracellular CO_2 concentrations to ambient air CO_2 concentrations ratio; CE, carboxylation efficiency (P_r/C_i) in mol m^{-2} s^{-1}; WUE, water use efficiency to photosynthesis (P_r/T_r) in µmol mmol^{-1}.

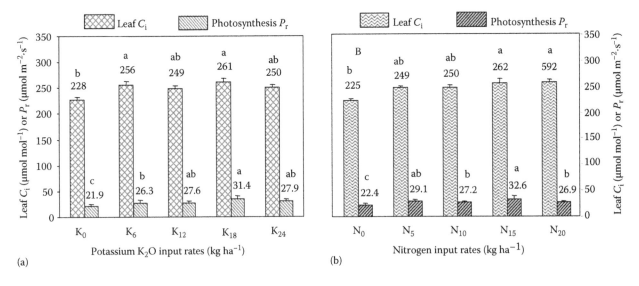

FIGURE 17.2 Comparison of means of strawberry leaf intercellular CO_2 concentrations (C_i) and net photosynthesis rates (P_r) related by K treatments (a) and by N treatments (b). Treatments K_0, K_6, K_{12}, K_{18}, and K_{24} represent the K_2O rates at 0, 6, 12, 18, and 24 kg ha^{-1} and N_0, N_5, N_{10}, N_{15}, and N_{20} represent the N rates at 0, 5, 10, 15, and 20 kg N ha^{-1}, respectively. Each bar represents the mean and standard error; $n = 15$. The honestly significant difference HSD = 26.3 mg L^{-1} for C_i, and HSD = 4.3 µmol m^{-2}·s^{-1} ($\alpha = 0.05$) for P_r. Means with the same letters were not significantly different at $P < 5\%$. Each variable $n = 75$. (From Li, J. et al., *Photosynthesis Research* 115:101–14, 2013.)

(Figure 17.2a) and by the N treatments (Figure 17.2b). The contrast analysis showed that the differences in C_i and P_r were significantly different between the control (K/N macronutrient at maximal deficiency) and other treatments (Table 17.1).

Correlations among C_0, C_i, C_i/C_0, CE, and WUE were significant ($0.62 < r < 0.84$, $P < 0.0001$). Leaf temperature (LT) and g_s were also significantly related to CE and WUE ($0.40 < r < 0.54$, $P < 0.0004$). Leaf C_i concentrations were significantly correlated with LT ($r = 0.47$, $P < 0.0001$), and the best-fit model for these two variables was a quadratic trendline (Figure 17.3a). The significant correlation between leaf C_i and leaf temperature meant that increases in leaf CO_2 could trap heat to increase leaf temperature for high carboxylation in temperate summer. The positive relation between leaf CO_2 and leaf temperature could be explained by the fact that increases in air CO_2 concentrations could trap more heat in the air. The lower leaf transpiration (8.03 µmol m^{-2}·s^{-1}) corresponding to higher C_i in the control (9.1 µmol m^{-2}·s^{-1}) suggested that rising leaf intercellular CO_2 could reduce plant transpiration (Li et al., 2013).

The regression model was a logarithmic model for leaf P_r vs. g_s (Figure 17.3b). Pores in leaves let CO_2 enter, and plants also lose water through the pores in leaves. Plants transpire water vapor to cool itself, and plant water movement is active through leaf stomates. Under higher CO_2 conditions, plants do not need to open widely their pores, which can reduce water loss (Ainsworth and Rogers, 2007); however, in humid environments, as in the study areas, plants have to lose water through the stomates, resulting in the increase in leaf P_r with increases in g_s activity. Relationships of strawberry leaf C_i, CE, and WUE relations in this cool, humid area could be described by the best-fit power models in the work of Li et al. (2013) as follows:

$$CE = 0.0003 \, C_i^{1.0951} \quad R^2 = 0.44^{**} \quad n = 75 \quad (17.1)$$

$$WUE = 0.0000006 \, C_i^{2.316} \quad R^2 = 0.75^{**} \quad n = 75 \quad (17.2)$$

17.2.2 Limited K/N Macronutrients Advance Berry Formation and Marketable Yields

Strawberry plants treated with adequately reduced K/N input rates could advance berry formation and improve marketable yields from high plant carboxylation and water use efficiency by higher leaf C_i concentrations (Li et al., 2013). The data showed that higher C_i/C_0 ratios (0.80–0.92) corresponding to higher P_r, CE, and WUE would lead to early fruit formation and high yield. Strawberry fruit maturity could advance significantly by 6 days for all other K–N treatments compared to the control plots (K_0 and N_0). Total marketable yield varied in the range 2207 ± 134 g m^{-2}, and the yields significantly increased at the 25%-reduced K treatment (K_{18}). The lower marketable yields were comparable with the 75%-reduced input (K_6) and the nonreduced input (K_{24}) treatments, showing the inefficiency of full inputs.

The impacts of soil N nutrients at maximal deficiency (N_0) on strawberry fruit productivity were shown by the lowest marketable fruit yields measured at each harvest date for the control plot N_0. When N nutrient was the most deficient (N_0), constant fruit yields at the N_0 level for all K treatments meant nonresponses of strawberry fruit growth to the K inputs (Figure 17.4a), indicating the limitation of K nutrition on fruit formation. When K nutrient was the most deficient (K_0), the linear increase in fruit yields meant the strong responses of strawberry fruit growth to the N inputs (Figure 17.4b), indicating the greater effects of N nutrients on fruit formation than K nutrients (Li et al., 2013).

The interactions of the K/N treatments on marketable fruit yields were both linear and quadratic (Figure 17.4). In relation to the K treatments, the plotted curves showed that except for the N_0, fruit yields increased and finally intercepted at the reduced and nonreduced K levels (Figure 17.4a). In relation to the N treatments, the fruit yields increased with all the K treatments including the K_0 level, and the plotted patterns were intercepted only at the reduced N levels (N_5, N_{10}, and N_{15}) (Figure 17.4b). The interaction curves showed that

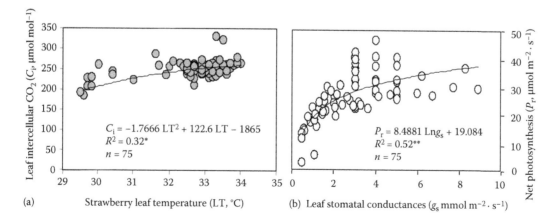

FIGURE 17.3 Scatted regression patterns and the best-fit models for strawberry leaf intercellular CO_2 concentrations C_i vs. leaf temperature LT (a), and leaf photosynthesis rate P_r vs. leaf stomatal conductance g_s (b). The data labels represent means. Each variable $n = 75$. (From Li, J. et al., *Photosynthesis Research* 115:101–14, 2013.)

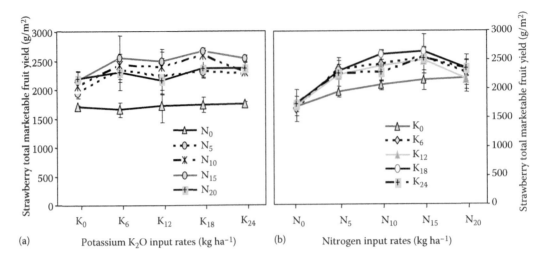

FIGURE 17.4 Interaction patterns of the K input treatments related to the N treatments (a), and the N input treatments related to the K treatments on strawberry total marketable fruit yields (b). Each bar represents the values of means and standard errors ($n = 15$). (From Li, J. et al., *Photosynthesis Research* 115:101–14, 2013.)

strawberry fruit formation responded more strongly to the reduced N treatments than the reduced K treatments, which showed a difference in strawberry responses to the types of macronutrients.

17.2.3 HIGHER LEAF C_i INHIBITS LEAF/FRUIT NO_3^- RETENTION, BUT NOT K^+ IONS AND TDS

Variation in CO_2 concentrations could lead to changes in the mineral and phytochemical compounds of plant tissues (Baslam et al., 2012). When the study has determined together the variables of leaf C_i, leaf/fruit K^+, NO_3^-, and total dissolved solids (TDS), the data showed that higher leaf C_i could inhibit leaf/fruit NO_3^- retention and translocation, but it did not affect K^+ and TDS holding and partitioning within leaves and fruits (Li et al., 2013). These results provided new insights to compare with the reports that leaf N concentrations per unit leaf mass could decrease by 13% in elevated CO_2 environment (Ainsworth and Rogers, 2007), and air CO_2 enrichment inhibited the assimilation of NO_3 into organic N compounds (Bloom et al., 2010).

In the study, leaf sap K^+, NO_3^-, and TDS levels, simultaneously analyzed in newly matured leaves and fresh fruits in each harvest day (every 3 days during the ripening period), varied with the K/N treatments (Figure 17.5). Leaf sap K^+ and NO_3^- concentrations were 2150 ± 183 and 576 ± 33 μg g^{-1} (mean ± standard error), respectively, and leaf TDS levels varied in the range $6.53 \pm 0.7\%$ among the treatments. Although leaf sap K^+ concentrations were correlated with sap NO_3^- concentrations ($r = 0.42$, $P < 0.01$), leaf NO_3^- concentrations tended to decline, but leaf K^+ holding remained similarly high with the increase in the K/N input rates. There was a significant increase in leaf K^+, NO_3^-, and TDS concentrations at the N_5 level compared to the control (maximum N nutrient deficiency). However, leaf sap K^+ concentrations increased with K treatment rates up to the 25%-reduced level (K_{18}, Figure 17.5a,b). Leaf TDS per unit

leaf area increased to 6.3% (2-year mean) in the reduced input plots compared to the control (5.4%), and the difference was significant (HSD = 0.8%, graph not shown).

The significantly lower leaf K^+/NO_3^- ion retention in the control plots (K_0 and N_0) resulted from the impacts of soil K/N deficiency. The increases in leaf K^+ ions (Figure 17.5a,b) corresponding to the declines in leaf NO_3^- concentrations (Figure 17.5c,d) with the reduced K/N inputs suggested that higher leaf C_i concentrations could inhibit leaf NO_3^- retention, but this inhibition did not occur in leaf K^+ ion retention. As the retention of mineral K^+ and NO_3^- ions in leaves was a temporary storage for eventual translocation to the fruit sinks, inhibition in leaf NO_3^- ion retention from higher leaf CO_2 diffusion would have caused the difference in leaf K^+/NO_3^- ratios (Li et al., 2013).

The temporal patterns of fresh fruit sap K^+, NO_3^-, and TDS concentrations, simultaneously analyzed on each harvest date, were related with leaf physiological variables and leaf sap K^+, NO_3^-, and TDS concentrations (Li et al., 2013). Fruit NO_3^- ions were negatively correlated with leaf PAR and leaf temperatures ($-0.38 < r < -0.37$). Leaf sap K^+ ions were 2.65 times higher than fruit K^+ concentrations (mean 2150 μg g^{-1} vs. 802 μg g^{-1}). Leaf sap NO_3^- ions were about two times higher than fruit NO_3^- concentrations (mean 589 μg g^{-1} vs. 292 μg g^{-1}). The leaf K^+/NO_3^- ratio averaged 3.7:1 (2150/589) compared to the fruit K^+/NO_3^- ratio 3.1:1 (802/292) in berries that were picked 9 days later. The driving force of K^+/NO_3^- translocation from leaves to fruits could be the needs of macronutrients for fruit formation from leaf sources, as shown by the N translocation patterns within cauliflower plants (Li, 2012). As the K^+/NO_3^- ratio difference diminished in the fruits, fruit NO_3^- ion retention could become stronger during the nutrient translocation process from leaves to the fruits (Li et al., 2013).

Changes in TDS with time in fresh berries showed also similar patterns as in temporal holdings of K^+ and NO_3^- ion concentrations in fruits. The TDS was significantly higher ($9.71 \pm 0.18\%$) in early matured berries than in berries that

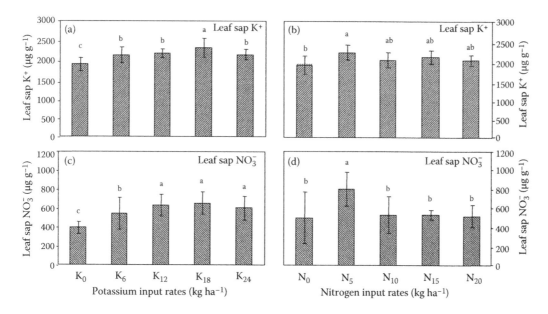

FIGURE 17.5 Strawberry cv. *Mira* leaf sap K$^+$ concentrations (a and b) and leaf sap NO$_3^-$ concentrations (c and d) related to the K and N input treatments, respectively. Each bar represents the values of means and standard errors with $n = 15$. The honestly significant difference HSD = 195 µg g^{-1} for K$^+$ ($\alpha = 0.05$) and HSD = 46 µg g^{-1} for NO$_3^-$ ($\alpha = 0.05$). Means with the same letters were not significantly different at $P < 5\%$. (From Li, J. et al., *Photosynthesis Research* 115:101–14, 2013.)

matured 9 days later (5.33% ± 0.16%) (ANOVA, $F = 5.34$, $df = 1$, 74, $P = 0.001$). The fruit TDS holding patterns corresponding to the leaf TDS levels meant the translocations of photosynthate energy for fruit formation. Leaf sap TDS was also positively correlated with leaf sap K$^+$ and NO$_3^-$ (leaf K$^+$ vs. TDS, $r = 0.54$, $P < 0.01$; leaf NO$_3^-$ vs. TDS, $r = 0.40$, $P < 0.01$, $n = 75$). Fruit TDS and fruit K$^+$ and NO$_3^-$ concentrations were positive (fruit K$^+$ vs. TDS, $r = 0.34$, $P < 0.05$; fruit NO$_3^-$ vs. TDS, $r = 0.32$, $P < 0.05$). Their significant correlation coefficients meant that both leaf/fruit K$^+$ and NO$_3^-$ ion fluxes were important mineral ions in leaf/fruit dissolved solids or photosynthates.

17.2.4 COLIMITATION OF REDUCED MACRONUTRIENT K/N INPUTS ON STRAWBERRY FRUIT PRODUCTIVITY AND MINERAL NUTRITION QUALITY

Colimitation of the reduced K/N inputs on strawberry fruit productivity and berry mineral nutrition quality could be shown by the quadratic relationships among C_i, T_r, CE, WUE, marketable fruit yield, and leaf/fruit nutritional attributes (K$^+$, NO$_3^-$, and TDS) influenced by the K/N macronutrient limitation (Li et al., 2013). The total fruit marketable yields were quadratically related to the leaf C_i, as it was demonstrated that higher fruit yields were associated with higher C_i concentrations varying between 240 and 330 µmol m^{-2}·s^{-1} (Figure 17.6a). Fruit TDS variable also showed a trend to increase quadratically with leaf C_i concentrations (Figure 17.6b).

Higher C_i/C_0 ratios (0.80–0.92), P_r (27.6–32.6 µmol fixed CO$_2$ m^{-2}·s^{-1}), CE (0.109–0.121 mol m^{-2}·s^{-1}), and WUE (3.37–3.84 µmol mmol^{-1}) corresponded to higher leaf TDS, which means higher photosynthate products for fruit formation from

high photosynthesis, carboxylation, and water use efficiency. Leaf T_r, CE, and WUE would explain 32–89% of the variations in total marketable fruit yield (Y_{st}). Their significant relations could be described by the following quadratic regression equations:

$$Y_{st} = -13.875T_r^2 + 301.27T_r - 669.12 \quad R^2 = 0.32^{**}, \quad n = 75 \tag{17.3}$$

$$Y_{st} = -8.02 \times 10^6 CE^2 + 1.94 \times 10^5 CE - 9283$$
$$R^2 = 0.53^{**}, \quad n = 75 \tag{17.4}$$

$$Y_{st} = -4.11 \times 10^2 WUE^2 + 3.09 \times 10^3 WUE - 3356$$
$$R^2 = 0.89^{**}, \quad n = 75 \tag{17.5}$$

Higher leaf C_i, photosynthesis, carboxylation, and water use efficiency for strawberry plants treated with the 25%-reduced K/N inputs (K$_{18}$ and N$_{15}$) would mean that the plants were healthier and therefore more productive (Li et al., 2013). Healthier plants should demand more CO$_2$ for its photosynthesis needs to enhance crop growth as the correlations among C_i, CE, WUE, and strawberry fruit yield were positively significant (Equations 17.3–17.5). The significant 6-day advance in berry fruit maturity and higher fruit yields at the 25%-reduced K/N levels resulted from higher leaf C_i, P_r, CE, and WUE efficiency. These results further suggested that adequate macronutrient limitation could enhance plants to adequately assimilate macronutrients, which could render the crops healthier for inducing high leaf intercellular CO$_2$ levels to enhance crop productivity.

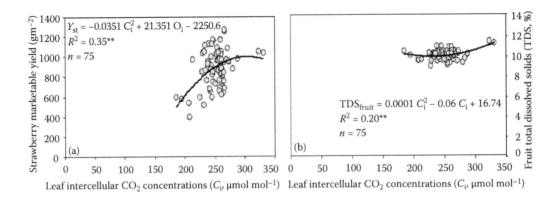

FIGURE 17.6 Correlation relationships of strawberry marketable fruit yield Y_{st} vs. leaf intercellular CO_2 concentrations C_i (a); fruit TDS vs. leaf intercellular CO_2 concentrations C_i (b). Data were measured at early fruit ripening stage (8 July). Each variable $n = 75$. (From Li, J. et al., *Photosynthesis Research* 115:101–14, 2013.)

17.3 CASE STUDY: PHOSPHORUS MACRONUTRIENT LIMITATION AND SMALL FRUIT CROP DEVELOPMENT

Profitable strawberry fruit production is primarily relying on highly healthy, cold-stored, well-rooted planting stocks (Stewart et al., 2005; Li et al., 2010b). Strawberry crops are very susceptible to infections of many fungus and bacterial diseases that can result in reduced yields and economic loss (Rieger, 2005). High-quality (cold-stored, well-rooted) planting stocks are therefore very important for profitable production of high-quality strawberries. Adequate macronutrient inputs and crop rotation practices have been among the most useful measures to ensure both successful nursery and fruit production (Pritts and Handley, 1998; Stewart et al., 2005; Li et al., 2014).

Strawberry plants require a high acquisition of macronutrients for the need of its rapid reproduction (Preusch et al., 2004; Li et al., 2010b). In strawberry production in Atlantic Canada regions, Nova Scotia is characterized by a cool summer season (seasonal average temperature 14±5°C) influenced by the Atlantic Ocean. Such temperature ranges are well suitable for reproduction of cold-stored, well-rooted, high-quality strawberry planting stocks. Nursery plants are grown from *in vitro* propagated, disease-free mother stocks that are acclimatized in sterilized greenhouses. Healthy daughter plants are usually harvested in early fall, and then nursery crops are exported across North America for transplanting for fruit production. The crop rotation regime in nursery for strawberry is commonly a 4-year cycle with wheat, corn, and ryegrass to ensure healthy soil for healthy, disease-free nursery reproduction (Li et al., 2014).

In primary macronutrient management for strawberry nursery, phosphorus is considered as one of the most important nutrients in helping store and transfer energy for rapid growth of strawberry nursery plants. Highly variable P availability in the soils has been one of the challenges in strawberry nursery plant production (Stewart et al., 2005; Li et al., 2010b). Yet there is a lack of knowledge about the effects of limited and unlimited P nutrient inputs on this small fruit nursery crop and plant P/N nutrition uptake.

Phosphorus is in the nitrogen family, and P ions can interact with N elements in plants (Li et al., 2011, 2014). A 2-year study of limited P macronutrient efficiency in strawberry nursery production was conducted on two registered, certified farms in center Cobequid Bay Coast in Nova Scotia, Canada. One of the registered, certified farms was the Kelly Farm in Annapolis Valley (Li et al., 2010b), and the other certified farm was the Millen Farm where the previous K/N macronutrient study was conducted (Li et al., 2013). The responses of strawberry nursery plants to P macronutrient limitation were examined with a total of eight cultivars including V1 = "Strawberry Festival," V2 = "Darselect," V3 = "Mesabi," V4 = "V151," V5 = "Seneca," V6 = "Serenity," V7 = "K93-20," and V8 = "Jewel" (Li et al., 2010b, 2014).

The cultivar "Strawberry Festival" was a newly certified cultivar in Florida. Its responses to the reduced P inputs were separately reported because the crop was tested on the Millen Farm in two types of crop rotation systems: one was the wheat–ryegrass–ryegrass–strawberry (WRRS) system, and the other was corn–ryegrass–ryegrass–strawberry (CRRS) system. The soils at both experimental systems were classified as an imperfectly drained Orthic Humo–Ferric Podzols, known as Spodosols in the American Soil Classification System (Li et al., 2014).

The experimental treatments consisted of three mother-stock P (MSP) rates (0, 15, and 30 kg P_2O_5 ha^{-1}) and five daughter-plant P (DPP) fertilizer rates (0, 30, 60, 90, and 120 kg P_2O_5 ha^{-1}). The P treatments were with small increments (15 kg ha^{-1} in the MSP and 30 kg ha^{-1} in the DPP). The split P treatments could be combined into three groups: MSP-0 (0 + 0, 0 + 30, 0 + 60, 0 + 90, and 0 + 120), MSP-15 (15 + 0, 15 + 30, 15 + 60, 15 + 90, and 15 + 120), and MSP-30 (30 + 0, 30 + 30, 30 + 60, 30 + 90, and 30 + 120) kg P_2O_5 ha^{-1}. Based on the regional recommendation rate (120 kg P_2O_5 ha^{-1}) for strawberry, there were 1 control treatment (maximum soil P deficiency), 10 treatments at the reduced P levels (i.e., 12.5–87.5% of the regional recommendation rate), and 4 treatments at the unlimited P levels (i.e., 100–125% of the recommendation, 0 + 120, 15 + 120, 30 + 90, and 30 + 120 kg ha⁻¹).

The MSP and DPP treatments were arranged with three blocks in a split-plot design at both rotation systems. The MSP rates were randomly applied in the main plots at the first runner emerging stage, and then the DPP rates were randomly arranged in the subplots at first daughter-plant initiation stage. Both MSP and DPP treatments were applied using di-ammonium phosphate (DAP, $(NH_4)_2HPO_4$, 18–46–0) by side dress. The crop requirements for N and K macronutrients were supplied in uniform rates with 105 kg N ha⁻¹ and 165 kg K_2O ha⁻¹ in all plots for both experiments.

Strawberry mother-stock transplanting and crop growth cares including irrigation, weeding, disease, and insect control were done based on the small fruit nursery certification programs. In-season nursery development and whole-plant P and N holdings were assessed using samples taken at different growth stages in each plot, detailed in the work of Li et al. (2014). Nursery plant productivity was assessed using determinations including whole-plant biomass, whole-plant P/N uptake dynamics, tri-leaves, runners, runner-tip lengths, and daughter-plant numbers (Figure 17.7).

The P limitation study revealed four novel findings through the assessments:

1. Strawberry plants could cope with macronutrient P limitation by adequate assimilation of P nutrients to enhance nursery formation.
2. Strawberry nursery growth required adequate plant P/N retention ratios that were related to adequately reduced MSP/DPP inputs.
3. Optimizing plant P nutrition needed a substantial N acquisition to stimulate strawberry nursery formation.

4. Loading the soils with excessive P inputs would hurt the nursery crops by prohibiting plant P/N uptake and vegetative propagation (Li et al., 2014).

17.3.1 STRAWBERRY PLANTS COPE WITH LIMITED MACRONUTRIENT P SUPPLIES IN NURSERY PRODUCTION

Strawberry nursery growth stages include leaf-runner expansions from mother-stock crowns, leafing at runner nodes, daughter-plant rooting stages, etc. Mother stocks usually grow into taller, larger canopies in center rows, and daughter plants, grown from runners, are of young, small sizes tensely surrounding the mother-stock crowns (Figure 17.7).

As referred to by Li et al. (2014), the effects of the limited P treatments were significant on strawberry nursery development, as shown by the significant differences in daughter-plant numbers at harvest (Table 17.2). In the WRRS rotation system, the highest nursery productivities (16–17 daughter plants per mother stock) and optimal daughter plant/runner (D/R) ratios (1.00–1.20) were measured in the limited P treatments. More dead runners were also related to higher DPP levels (Table 17.2). The MSP-15 treatments with lower DPP rates (30–60 kg ha⁻¹) tended to have a higher daughter-plant number. The linear and quadratic interactions between the MSP and DPP treatments were significant on daughter plants, runners, and D/R ratios. The experimental model was significant on daughter plants and runners (Table 17.2).

Total tri-leaf numbers (mean 26–30 leaves) per mother stock were significantly different among the MSP treatments in the CRRS rotation system (ANOVA $F = 3.58$, $df = 2$, $P < 0.0431$). Runner-tip lengths varied in the range 1.02 ± 0.24 m, and the MSP main effects were significant on runner expansion

FIGURE 17.7 **(See color insert.)** Strawberry nursery plant propagation and methods of determination of strawberry nursery productivity. Positioning of mother-stock crowns, runners, and daughter plants at first daughter-plant growth stage in the fields in mid-July (a) and overview positioning of strawberry nursery production systems in late September (b).

TABLE 17.2

Analysis of Variance (ANOVA) and Contrasts of the MSP and DPP Treatments on Strawberry Nursery Productivity, Determined in the WRRS Rotation System

Year	Phosphorus Sources	Daughter Plants	Daughter/Runners	Whole-Plant Biomass	Dead Runners
WRRS	0 + 0	8	0.56	251	1
	0 + 30	14	0.72	315	1
	0 + 60	13	0.93	254	1
	0 + 90	12	0.77	274	3
	0 + 120	10	0.57	249	4
	15 + 0	14	0.87	300	1
	15 + 30	16	0.96	263	1
	15 + 60	17	1.00	276	2
	15 + 90	15	0.90	288	2
	15 + 120	15	0.57	280	3
	30 + 0	16	1.20	302	1
	30 + 30	15	0.92	339	2
	30 + 60	16	1.04	265	2
	30 + 90	14	0.68	354	1
	30 + 120	8	0.96	200	1
ANOVA					
Block (B)	2	2.24 ns	1.97 ns	0.25 ns	0.52 ns
MSP	2	4.24*	2.97*	3.85*	0.88 ns
B × MSP	4	2.23 ns	2.42 ns	2.94*	1.88 ns
DPP	4	2.26 ns	2.43 ns	0.36 ns	3.73**
MSP × DPP	8	1.03 ns	0.86 ns	1.54 ns	1.44 ns
MSP Contrasts					
MSP Linear	1	7.02**	5.52*	5.89*	0.28 ns
MSP Quadratic	1	8.03**	5.82*	4.51*	0.16 ns
DPP Contrasts					
DPP Linear	1	2.72 ns	3.89*	0.30 ns	5.51*
DPP Quadratic	1	1.36 ns	2.28 ns	0.69 ns	5.23*
DPP Cubic	1	1.94 ns	1.24 ns	0.35 ns	2.88 ns
MSP × DPP Contrasts					
MSPlin. × DPPlin.	1	1.42 ns	4.20*	0.56 ns	1.06 ns
MSPqua. × DPPlin.	1	0.91 ns	3.56*	0.35 ns	0.46 ns
MSPlin. × DPPqua.	1	2.12 ns	3.20 ns	0.66 ns	0.32 ns
MSPqua. × DPPqua.	1	1.48 ns	2.61 ns	0.58 ns	0.15 ns
Model	44	1.98*	1.82 ns	1.63 ns	1.75 ns
R^2		63	60	58	59
Mean (g/plant)		13.5	0.85	284	1.7
Stand. dev. (g/plant)		2.6	0.26	42	1.3
RMSE (g/plant)[a]		3.6	0.25	38	1.1

Source: Li, H. et al., *Journal of Plant Nutrition and Soil Science* 177:260–70, 2014.

[a] RMSE: Root mean square error. $n = 45$.

(ANOVA $F = 3.79$, $df = 2$, $P < 0.0372$). The D/R ratios varied between 1.04 and 2.03 (about 8–16 runners and 15–20 daughter plants per mother stock). Daughter-plant formation responded positively to the MSP main effects (ANOVA $F = 10.25$, $df = 2$, $P < 0.0006$) and the SPP main effects (ANOVA $F = 4.76$, $df = 4$, $P < 0.0057$). The slowest growth of daughter plants was associated with the maximum soil P deficiency (the control plot 0 + 0) and the highest, nonreduced input rate (30 + 120 kg ha⁻¹) (Table 17.2).

High leaf/runner (L/R) ratios (1.6–2.2) meant that plant energy was mainly for leaf and runner expansion needs in the early season. High daughter plant/runner (D/R) ratios demonstrated that more plant energy was used for daughter-plant formation in the late season. There was a significant difference in nursery productivity between the WRSS and CRRS rotation systems. The difference was significant in leaf numbers (ANOVA $F = 5.09$, $df = 1$, $P < 0.0287$), runners (ANOVA $F = 44.2$, $P < 0.0001$), marketable daughter plants (ANOVA

$F = 31.84$, $P < 0.0001$), D/R ratios (ANOVA $F = 68.22$, $P < 0.0001$), runner-tip length (ANOVA $F = 137.9$, $P < 0.0001$), root mass (ANOVA $F = 42.1$, $P < 0.0001$), and whole-plant biomass (ANOVA $F = 41.36$, $P < 0.0001$).

In-season temporal development patterns of the nursery formation were affected by the MSP treatments (Figure 17.8a). The nursery productivity was similar (2.2–2.8 daughter plants) among the MSP treatments in mid-August, and it was until the second daughter-plant rooting period that the MSP-15 had the significantly higher daughter-plant numbers (15.3 ± 1.7) than in the control MSP-0 (11.7 ± 3.2). There was no difference in daughter plants among MSP-30 (13.7 ± 2.8), MSP-15, and MSP-0. In-season temporal development of whole-plant biomass showed an S-pathway, with the significantly higher whole-plant biomass determined in the MSP-15 and MSP-30 treatments compared to the MSP-0 treatment (Figure 17.8b).

The combined effects of the MSP–DPP treatments on nursery productivity were characterized by the significantly fast growth of daughter plants for the >50%-reduced MSP–DPP rate (15 + 30 kg ha⁻¹). From early rooting stage (July) through harvest (late September), 50%-reduced MSP-DPP treatments performed better than the other MSP-0 and MSP-30 combined levels (Figure 17.9). There were consistent increases in runners, daughter-plants, and whole-plant P/N retention at the 37.5–62.5%-reduced input rates (15 + 0, 15 + 30, 15 + 60 kg P_2O_5 ha⁻¹) (Figure 17.9). Strawberry plants could cope well with macronutrient P limitation with rapid top growth, tested in two different rotation environments. The reduced inputs 15 + 30 and 15 + 60 kg ha⁻¹ could be recommended to be applied at the first runner expansion stage and the first daughter-plant rooting stage.

Too many runners (18–22 per mother stock) could result in reduced whole-plant N/P accumulation ratios. The lowest D/R ratio associated with the unlimited input rate (15 + 120 kg P_2O_5 ha⁻¹) would be due to the fact that this treatment produced too many runners (26 runners) that could have taken more energy from the plants, which reduce the growth of daughter plants. Too many runners could diminish daughter-plant development (Table 17.1) because too many runners

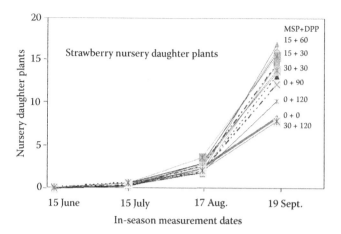

FIGURE 17.9 Comparisons of strawberry nursery daughter-plant formation patterns by P treatment rates and types, i.e., MSP) and DPP. The dates 15 June, 15 July, 17 August, and 19 September corresponded to the first runner expansion stage, the first daughter-plant rooting stage, the second daughter-plant rooting stage, and the harvest date, respectively. Each point represents the mean values ($n = 9$). (From Li, H. et al., *Journal of Plant Nutrition and Soil Science* 177:260–70, 2014.)

might take energy from the plants to leave little energy for daughter plants. Runner thinning should be considered in nursery production (Li et al., 2014).

17.3.2 STRAWBERRY PLANT P OPTIMIZATION REQUIRES A SUBSTANTIAL ACQUISITION OF PLANT N

Rapid top-growth strawberry nursery plants needed a substantial macronutrient P/N uptake to ensure a healthy, rapid development, tested in different rotation environments (Li et al., 2014). Optimizing plant P nutrition required a substantial acquisition of plant N to stimulate nursery plant formation. As shown, total P content in strawberry whole plants (including mother stocks, tri-leaves, runners, and daughter plants) varied in the range 0.21 ± 0.06% (mean ± SD, $n = 45$) among the MSP–DPP treatments in the WRRS system. With a dry

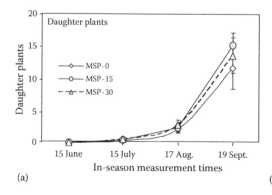

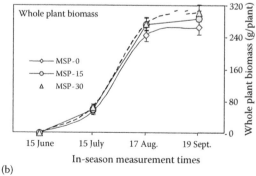

(a)   (b)

FIGURE 17.8 Temporal development patterns of strawberry nursery daughter-plant numbers (a) and whole-plant biomass (b) by MSP treatments. The times of measurements corresponded to the first runner expansion stage (15 June), the first daughter-plant rooting stage (15 July), the second daughter-plant rooting stage (17 August, September), and the harvest date (19 September). Each point represents the means and standard errors ($n = 15$). (From Li, H. et al., *Journal of Plant Nutrition and Soil Science* 177:260–70, 2014.)

matter weight varying between 80 and 102 g/plant, total P uptake was significantly higher at the MSP-15 level (HSD = 0.018 g/plant, $\alpha = 0.05$).

The whole-plant P uptake was 1.79 ± 0.24, 2.28 ± 0.29, and 2.08 ± 0.32 mg/g, respectively, on a dry matter basis for the MSP-0, MSP-15, and MSP-30 treatments, which was significantly different (HSD = 0.12 mg/g, $\alpha = 0.05$). The main effects of the MSP treatments were significant on plant P% content (ANOVA $F = 68.05$, $df = 2$, $P = 0.0001$) and total plant P uptake (ANOVA $F = 34.48$, $df = 2$, $P = 0.0001$). The main effects of the DPP treatments were also significant on plant P% (ANOVA $F = 8.54$, $df = 4$, $P = 0.0002$). The ANOVA model was significant for plant P% content (ANOVA $F = 10.14$, $df = 20, 24$, $P = 0.0002$) and total plant P uptake (ANOVA $F = 4.37$, $df = 20, 24$, $P = 0.0002$).

Strawberry nursery plants were especially in need of P nutrition for its rapid top growth, shown by the significant effects and interactions of the MSP and DPP treatments on the nursery plant (Table 17.2). The significant increases in strawberry runners and daughter plants related to plant total P accumulation further showed that plant P nutrition could significantly help capture and convert energy into useful plant compounds for the needs of plant leaf expansion and top growth.

Optimization of plant P nutrition in the nursery plants required a sufficient acquisition of plant N as quadratic increases in plant P were associated with whole-plant N accumulation (Li et al., 2014). A sufficient acquisition of P nutrient could need a higher concentration (seven times) of plant N to form an adequate plant P/N nutrition ratio for rapid top growth of nursery plants. The nursery whole-plant total N% varied in the range $1.79 \pm 0.24\%$ ($n = 45$) on a dry matter basis (total N holding in the whole plants was within 13.2–23.1 mg g⁻¹ dry matter or 1.08–1.94 g/plant) in the WRRS system. The peak of plant N accumulation appeared also in the MSP-15 level. The main effects of the MSP treatments were significant on plant N% (ANOVA $F = 7.73$, $df = 2$, $P < 0.0026$) and plant total N uptake (ANOVA $F = 31.78$, $df = 2$, $P < 0.0001$). The main effects of the DPP treatments were also significant on plant N% (ANOVA $F = 6.32$, $df = 2$, $P < 0.0013$) and plant

total N uptake (ANOVA $F = 3.92$, $df = 2$, $P < 0.0139$). The linear interactions (MSP × DPP) were significant on strawberry plant N% (ANOVA $F = 12.16$, $df = 1$, $P < 0.0019$) and total plant N accumulation (ANOVA $F = 60.91$, $df = 1$, $P < 0.0001$).

In the CRRS system, the in-season patterns of strawberry plant P uptake were comparable with the trends determined in the WRRS system. The total plant P content varied in the range $0.25 \pm 0.04\%$ among the P treatments. Plant total P retentions were 0.18 ± 0.03, 0.25 ± 0.05, and 0.21 ± 0.06 g/plant for MSP-0, MSP-15, and MSP-30 treatments, respectively, which was significantly different (HSD = 0.031 g/plant, $\alpha = 0.05$). On a dry matter basis, the P uptakes were 2.36, 2.71, and 2.49 mg/g for the MSP-0, MSP-15, and MSP-30 treatments, respectively. Both linear and quadratic interactions (MSP × DPP) were significant on plant P accumulation (ANOVA, $F = 30.62$, $df = 1$, $P < 0.0001$; $F = 8.19$, $df = 1$, $P < 0.0086$).

The significantly higher N accumulation (1.7–1.9 g/plant or 19–23 mg g⁻¹ dry matter) in the CRRS system was measured in the limited MSP and DPP treatments at the medium levels (i.e., 15 + 30 and 15 + 60 kg ha⁻¹). The lowest plant N accumulation was measured in the plots with the highest P input rates. The main effects of the MSP treatments were significant on plant N% content (ANOVA $F = 5.28$, $df = 2$, $P < 0.0126$) and plant N holding (ANOVA $F = 19.95$, $df = 2$, $P < 0.0001$).

Nursery plant total P uptake was estimated to be 14.3–29.3 kg P_2O_5 ha⁻¹, and total N uptake could be converted to 30.3–53.4 kg N ha⁻¹ in the WRRS system. Whole-plant total P and N holdings were higher in the CRRS system (19.4–32.4 kg P_2O_5 ha⁻¹ and 44.2–54.3 kg N ha⁻¹), and their differences in total N uptake were significant.

Whole-plant P/N nutrition ratios ranged within 0.14 ± 0.03, showing that strawberry nursery plant N retention was 7.14 times higher than plant P retention. There was a significant regression relationship between whole-plant P holding and whole-plant N retention (Figure 17.10). Plant P retention tended to increase linearly with the increase in the whole-plant N holding, and the highest plant P retention (0.31 g/plant) corresponded to the N holding of 2.28 g/plant, leading to the mean N/P ratio (0.14). Whole-plant P retention increased with the increase in plant N holding up to 2.2 g N/plant with a

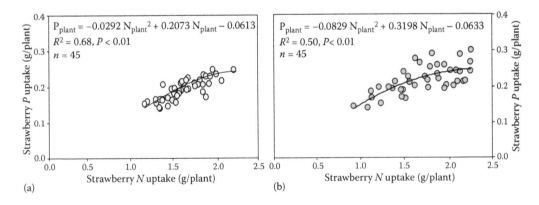

FIGURE 17.10 Regression relationships between plant P (P_{plant}) and plant N (N_{plant}) retention in strawberry whole nursery plants, determined in the WRRS rotation system (a) and in the CRRS rotation system (b). (From Li, H. et al., *Journal of Plant Nutrition and Soil Science* 177:260–70, 2014.)

significant R^2 value (0.50–0.68, Figure 17.10). The plant P/N ratios (0.12–0.13) were related to the 50–70% limited P input rates (15 + 30 and 15 + 60 kg/ha), where high daughter-plant numbers were measured (Figure 17.10).

As both P and N elements are major nutrients in plant functions, it is difficult to separate plant P and N roles in broad variation in plant development (Reich et al., 2009; Li et al., 2014). In strawberry nursery reproduction, low plant P concentrations could have limited crop leaf expansion and nursery plant rooting formation because significantly lower daughter-plant numbers were determined when plant P accumulation was inferior to 0.18 g/plant. This dependence could be explained by leaf P/N nutrition influence in leaf expansion because P/N-rich compounds play an important role in the biochemical fixation process (Reich et al., 2009).

17.3.3 Excessive P Inputs Could Prohibit Plant P/N Uptake to Hurt Vegetative Propagation

Loading the soils with excessive P inputs would also hurt the crops by prohibiting plant P/N uptake and nursery vegetative propagation (Li et al., 2014). This knowledge has provided new insights looking into the negative effects of excessive P inputs on both agronomical and environmental aspects. The impacts of excess P treatments on strawberry nursery data showed that less P/N retention (Figure 17.8) and the slowest growth of daughter plants (Table 17.2) were associated with the maximum P deficiency (the control plot) and the highest total P inputs (30 + 120 kg ha^{-1}). These results suggested that too much P in the soil could prohibit nursery plant vegetative propagation and P/N macronutrient uptake to leave more P into the environment. This result was different from the report that excess P supplies could generally not hurt the crops, but it can build up soil P levels (Rehm, 1998).

Hurting the nursery crops by affecting runner expansion from the excessive P inputs could occur in the highest MSP rate (MSP-30) applied at the early vegetative stage. This level of high inputs was associated with more dead runners (Table 17.2), meaning it might cause runner lesion to affect runner expansion.

The nursery leaf expansion was equally related to leaf P/N retention. The tri-leaves increased with the linear effects of plant P nutrition, but leaf numbers tended to decrease with plant P retention higher than 0.28 g/plant and plant N holding higher than 2.0 g/plant. The effects of whole-plant P and N nutrition on nursery plant tri-leaf formation were linear and quadratic, described by the following regression equations ($n = 45$):

$$\text{Tri-leaves} = -136.03 \, P_{plant}^2 + 115.35 P_{plant} + 10.35$$
$$R^2 = 0.33, \, P < 0.05 \qquad (17.6)$$

$$\text{Tri-leaves} = -38.36 \, N_{plant}^2 + 113.3 N_{plant} - 52.58$$
$$R^2 = 0.39, \, P < 0.01 \qquad (17.7)$$

Daughter-plant numbers, referred to as the nursery productivity, increased also with linear and quadratic effects of plant P/N ratios ($P < 0.05$). By counting overlay patterns of regression plot curves among tri-leaves, runners, and well-rooted daughter plants with plant P/N variables, significantly higher nursery plant development was dependent on nutrition holding in the ranges of 0.18–0.29 g P/plant (Figure 17.11a) and 1.5–2.1 g N/plant (Figure 17.11b). The whole plant P/N nutrition ranging from 0.18 to 0.29 g P and 1.3 to 2.1 g N per plant would lead to the production of adequate daughter/runner (D/R) ratios (D/R 1.4–1.7), i.e., 17–26 daughter plants to 12–16 runners per mother stock. Limited P inputs ranging from 37.5% to 62.5% of regular recommendation can be used for controlling the formation of too many runner numbers. It was concluded that loading the soils with excessive P inputs would also hurt the crops by prohibiting plant P/N assimilation and vegetative propagation. These results have implication in reducing the uses of nonrenewal, chemical fertilizers to promote low-input horticultural management for small fruit crops.

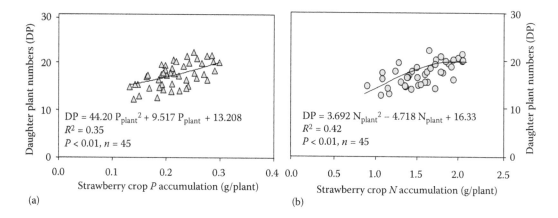

DP = 44.20 P$_{plant}^2$ + 9.517 P$_{plant}$ + 13.208
$R^2 = 0.35$
$P < 0.01$, $n = 45$

DP = 3.692 N$_{plant}^2$ − 4.718 N$_{plant}$ + 16.33
$R^2 = 0.42$
$P < 0.01$, $n = 45$

(a)   (b)

FIGURE 17.11 Regression relationships between strawberry nursery daughter-plants (DP) vs. plant P retention (P$_{plant}$, a), and DP vs. plant N retention (N$_{plant}$, b), determined in the CRRS rotation system.

17.4 CONCLUSIONS

Adequate primary K/N macronutrient limitation could regulate plant assimilation of macronutrients to render the plants healthier with reduced uses of nonrenewal, chemical fertilizers. With 25%-reduced K/N inputs, strawberry crops would be healthier, capable of inducting significantly higher leaf intercellular CO_2 concentrations to enhance crop net photosynthesis, carboxylation, water use efficiency, and fruit formation. Plant N concentrations at its most deficient levels were more crucial to small fruit crops than plant K concentrations at its most deficient levels. When N was most deficient, the stressed crops had no responses to any K inputs. When K was most deficient, strawberry plants responded significantly to the reduced N inputs. Yet, higher leaf intercellular CO_2 could inhibit plant NO_3 ion retention, but this inhibition did not occur in leaf/fruit K^+ retention. Also, healthy strawberry nursery formation was associated with substantial P/N uptake and plant P/N holding ratios from adequately reduced P inputs. With 50%-reduced P inputs, it could promote adequate plant P/N assimilation, faster top growth, leaf-runner expansion, and daughter-plant formation in 4-year crop rotation systems. Macronutrient P was the single controlled factor influencing strawberry nursery plant N uptake in crop rotation systems to produce healthy, well-rooted planting stocks for profitable strawberry fruit production. The results have implication in improving low-input horticultural management. Limited primary macronutrient supplies would be a profitable strategy to enhance plant macronutrient assimilation and physiological development while reducing chemical fertilizer usage, production costs, and environmental risk to promote sustainable, low-input small fruit crop production.

ACKNOWLEDGMENTS

The author thanks the support of the Fundamental Research Funds for the Chinese Academy of Tropical Agricultural Science (#1630032015005, #2013hzs1J006) and Hainan Soil Improvement Key Techniques Research and Demonstration Program (#HNGDg12015). The author thanks greatly the support of the Technology Development Program—Nova Scotia Department of Agriculture, North American Strawberry Growers Association, Millen Farms, and Kelly Farms in Nova Scotia, Canada.

ABBREVIATIONS

C_0: ambient environment CO_2 concentration
C_i: leaf intercellular CO_2 concentration
CE: instantaneous carboxylation efficiency
CRRS: corn–ryegrass–ryegrass–strawberry rotation
DPP: daughter-plant phosphorus
g_s: stomatal conductance
LT: leaf temperature
MSP: mother-stock phosphorus
P_r: net photosynthesis rate
PAR: photosynthetically active radiation
T_r: transpiration rate per unit leaf area

TDS: total dissolved solids
WRRS: wheat–ryegrass–ryegrass–strawberry rotation
WUE: water use efficiency of photosynthesis

REFERENCES

Ainsworth, E.A. and A. Rogers. 2007. The response of photosynthesis and stomatal conductance to rising CO_2: Mechanisms and environmental interactions. *Plant, Cell and Environment* 30:258–70.

Anderson, R.L. 2010. A rotation design to reduce weed density in organic farming. *Renewable Agriculture and Food Systems* 25:189–95.

Baslam, M., I. Garmendia, and N. Goicoechea. 2012. Elevated CO_2 may impair the beneficial effect of arbuscular mycorrhizal fungi on the mineral and phytochemical quality of lettuce. *Annals of Applied Biology* 161:180–91.

Blanke, M.M. and A.W. Whiley. 1995. Bioenergetics, respiration cost and water relations of developing avocado fruit. *Journal of Plant Physiology* 145:87–92.

Bloom, A.J., M. Burger, J.S.R. Asensio, and A.B. Cousins. 2010. Carbon dioxide enrichment inhibits nitrate assimilation in wheat and *Arabidopsis. Science* 328:899–903.

Bloom, A.J., L. Randall, A.R. Taylor, and W.K. Silk. 2012. Deposition of ammonium and nitrate in the roots of maize seedlings supplied with different nitrogen salts. *Journal of Experimental Botany* 63:1997–2006.

Cramer, M.D., M.W. Shane, and H. Lambers. 2005. Physiological changes in white lupin associated with variation in root-zone CO_2 concentration and cluster-root P mobilization. *Plant, Cell and Environment* 28:1203–17.

Davis, D.R. 2009. Declining fruit and vegetable nutrient composition: What is the evidence? *HortScience* 44:15–19.

Dordas, C. 2012. Variation in dry matter and nitrogen accumulation and remobilization in barley as affected by fertilization, cultivar, and source-sink relations. *European Journal of Agronomy* 37:31–42.

Eshghi, S. and E. Tafazoli. 2007. Changes in mineral nutrition levels during floral transition in strawberry (*Fragaria × ananassa* Duch.). *International Journal of Agricultural Research* 2: 180–84.

Foyer, C.H., G. Noctor, and M. Hodges. 2011. Respiration and nitrogen assimilation: Targeting mitochondria-associated metabolism as a means to enhance nitrogen use efficiency. *Journal of Experimental Botany* 62:1467–82.

Gallais, A. and B. Hirel. 2004. An approach to the genetics of NUE in maize. *Journal of Experimental Botany* 55:295–306.

Gastal, F. and Lemaire, G. 2002. N uptake and distribution in crops: An agronomical and ecophysiological perspective. *Journal of Experimental Botany* 53:789–99.

Geissler, N., S. Hussin, and H.-W. Koyro. 2009. Elevated atmospheric CO_2 concentration ameliorates effects of NaCl salinity on photosynthesis and leaf structure of *Aster tripolium* L. *Journal of Experimental Botany* 60:137–51.

Grant, O.M., A.W. Johnson, M.J. Davies, C.M. James, and D.M. Simpson. 2010. Physiological and morphological diversity of cultivated strawberry (*Fragaria × ananassa*) in response to water deficit. *Environmental and Experimental Botany* 68:264–72.

Gutschick, V.P. 1999. Biotic and abiotic consequences of differences in leaf structure. *New Phytology* 143:3–18.

Hariprasadm, P. and S.R. Niranjana. 2009. Isolation and characterization of phosphate solubilizing rhizobacteria to improve plant health of tomato. *Plant and Soil* 316:13–24.

He, F.J. and G.A. MacGregor. 2008. Beneficial effects of potassium on human health. *Physiologia Plantarum* 133:725–35.

He, J., P.T. Austin, and S.K. Lee. 2010. Effects of elevated root zone CO_2 and air temperature on photosynthetic gas exchange, nitrate uptake, and total reduced nitrogen content in aeroponically grown lettuce plants. *Journal of Experimental Botany* 61:3959–69.

Hirel, B., F. Chardon, and J. Durand. 2007. The contribution of molecular physiology to the improvement of nitrogen use efficiency in crops. *Journal of Crop Science and Biotechnology* 10:129–36.

Itani, Y., Y. Yoshida, and Y. Fujime. 1998. Effects of CO_2 enrichment on growth, yield and fruit quality of strawberry grown with rockwool. *Environmental Control and Biology* 36:125–29.

Jones, H.G. 1981. Carbon dioxide exchange of developing apple (*Malus pumila* Mill.) fruits. *Journal of Experimental Botany* 131:1203–10.

Keutgen, A.J. and E. Pawelzik. 2009. Impacts of NaCl stress on plant growth and mineral nutrient assimilation in two cultivars of strawberry. *Environmental and Experimental Botany* 65:170–76.

Khayyat, M., E. Tafazoli, S. Eshghi, M. Rahemi, and S. Rajaee. 2007. Salinity, supplementary calcium and potassium effects on fruit yield and quality of strawberry (*Fragaria* × *ananassa* Duch.). *Journal of Agronomy and Environmental Science* 2:539–44.

Kong, A.Y.Y., K.M. Scow, A.L. Cordova-Kreylos, W.E. Holmes, and J. Six. 2011. Microbial community composition and carbon cycling within soil microenvironments of conventional, low-input, and organic cropping systems. *Soil Biology and Biochemistry* 43:20–30.

Larbi, A., A. Abadia, J. Abadıa, and F. Morales. 2006. Down co-regulation of light absorption, photochemistry, and carboxylation in Fe-deficient plants growing in different environments. *Photosynthesis Research* 89:113–26.

Larios, B., E. Aguera, P. Cabello, J.M. Maldonado, and P. de la Haba. 2004. The rate of CO_2 assimilation controls the expression and activity of glutamine synthetase through sugar formation in sunflower (*Helianthus annuus* L.) leaves. *Journal of Experimental Botany* 55:69–75.

Laughlin, D.C. and S.R. Abella. 2007. Abiotic and biotic factors explain independent gradients of plant community composition in ponderosa pine forests. *Ecological Modeling* 205:231–40.

Lawlor, D.W. and W. Tezara. 2009. Causes of decreased photosynthetic rate and metabolic capacity in water-deficient leaf cells: A critical evaluation of mechanisms and integration of processes. *Annals of Botany* 103:561–79

Lea, P.J. and Azevedo, R.A. 2006. Nitrogen use efficiency. I: Uptake of nitrogen from the soil. *Annals of Applied Biology* 149:243–47.

Leakey, A.D.B., E.A. Ainsworth, C.J. Bernacchi, A. Rogers, S.P. Long, and D.R. Ort. 2009. Elevated CO_2 effects on plant carbon, nitrogen, and water relations: Six important lessons from FACE. *Journal of Experimental Botany* 60:2859–76.

Lester, G.E., J.L. Jifon, and D.J. Makus. 2010. Impact of potassium nutrition on postharvest fruit quality: Melon (*Cucumis melo* L.) case study. *Plant and Soil* 335:117–31.

Li, H. 2009. Citrus tree abiotic and biotic stress and implication of simulation and modeling tools in tree management. *Tree and Forestry Science and Biotechnology* 3:66–78.

Li, H. 2012. Nitrogen leaf-stem source and head sink relations in three cauliflower hybrid cultivars. *Acta Horticulturae* 938:219–26.

Li, H. and R.J. Lascano. 2011. Deficit irrigation for enhancing sustainable water use: Comparison of cotton nitrogen uptake and prediction of lint yield in a multivariate autoregressive state-space model. *Environmental and Experimental Botany* 71:224–31.

Li, H., R.J. Lascano, J. Booker, L.T. Wilson, K.F. Bronson, and E. Segarra. 2002. State-space description of field heterogeneity: Water and nitrogen use in cotton. *Soil Science Society of America Journal* 66:585–95.

Li, H., L.E. Parent, A. Karam, and C. Tremblay. 2003. Efficiency of soil and fertilizer nitrogen of a sod-potato system in the humid, acid and cool environment. *Plant and Soil* 251:23–36.

Li, H., L.E. Parent, and A. Karam. 2006a. Simulation modeling of soil and plant nitrogen use in a potato cropping system in the humid and cool environment. *Agriculture, Ecosystem and Environment* 115:248–60.

Li, H., J.P. Syvertsen, R.J. Stuart, C.W. McCoy, and A. Schumann. 2006b. Water stress and root injury from simulated flooding and *Diaprepes* root weevil feeding in citrus. *Soil Science* 171:138–51.

Li, H., W.A. Payne, G.J. Michels, and C.M. Rush. 2008. Reducing plant abiotic and biotic stress: Drought and attacks of greenbugs, corn leaf aphids and virus disease in dryland sorghum. *Environmental Experimental Botany* 63:305–16.

Li, H., T. Li, R.J. Gordon, S. Asiedu, and K. Hu. 2010a. Strawberry plant fruiting efficiency and its correlation with solar irradiance, temperature and reflectance water index variation. *Environmental and Experimental Botany* 68:165–74.

Li, H., R. Huang, T. Li, and K. Hu. 2010b. Ability of nitrogen and phosphorus assimilation of seven strawberry cultivars in a northern Atlantic coastal soil. *Proceedings of 19th World Congress of Soil Science*, August 1–6, Brisbane, Australia, Paper 0739, pp. 1–4.

Li, H., H.M. Jiang, and T. Li. 2011. Broccoli plant nitrogen, phosphorus and water relations at field scale and in various growth media. *International Journal of Vegetable Science* 17:1–21.

Li, H., T.X. Li, G. Fu, and P. Katulanda. 2013. Induced leaf intercellular CO_2, photosynthesis, potassium and nitrate retention and strawberry early fruit formation under macronutrient limitation. *Photosynthesis Research* 115:101–14.

Li, H., T.X. Li, G. Fu, and K. Hu. 2014. How strawberry plants cope with limited phosphorus supply: Nursery-crop formation and phosphorus and nitrogen uptake dynamics. *Journal of Plant Nutrition and Soil Science* 177:260–70.

Long, S.P. 2006. Increased C availability at elevated carbon dioxide concentration improves N assimilation in a legume. *Plant, Cell and Environment* 29:1651–58.

McGuire, M.A., J.D. Marshall, and R.O. Teskey. 2009. Assimilation of xylem-transported ^{13}C-labelled CO_2 in leaves and branches of sycamore (*Platanus occidentalis* L.). *Journal of Experimental Botany* 60:3809–17.

Mingo, M.D., M.A. Bacon, and W.J. Davies. 2003. Non-hydraulic regulation of fruit growth in tomato plants (*Lycopersicon esculentum* cv. Solairo) growing on drying soil. *Journal of Experimental Botany* 54:1205–12.

Mittler, R. 2006. Abiotic stress, the field environment and stress combination. *Trends in Plant Science* 11:15–19.

Moreshet, S. and G.C. Green. 1980. Photosynthesis and diffusion conductance of the Valencia orange fruit under field conditions. *Journal of Experimental Botany* 120:15–27.

Murugan, R. and S. Kumar. 2013. Influence of long-term fertilisation and crop rotation on changes in fungal and bacterial residues in a tropical rice-field soil. *Biology and Fertility of Soils* 49:847–56.

Nava, G., A.R. Dechen, and G.R. Nachtigall. 2008. Nitrogen and potassium fertilization affect apple fruit quality in southern Brazil. *Communication of Soil Science and Plant Analysis* 39:96–107.

Nestby, R., F. Lieten, D. Pivot, L.C. Raynal, and M. Tagliavini. 2005. Influence of mineral nutrients on strawberry fruit quality and their accumulation in plant organs: A review. *International Journal of Fruit Science* 5:139–56.

Perez-Lopez, U., A. Robredo, M. Lacuesta, A. Mena-Petite, and A. Munoz-Rueda. 2012. Elevated CO_2 reduces stomatal and metabolic limitations on photosynthesis caused by salinity in *Hordeum vulgare*. *Photosynthesis Research* 111:269–83.

Pettersson, R. and J.S. McDonald. 1994. Effects of nitrogen supply on the acclimation of photosynthesis to elevated CO_2. *Photosynthesis Research* 39:389–400.

Pettigrew, W.T. 2008. Potassium influences on yield and quality production for maize, wheat, soybean and cotton. *Physiologia Plantarum* 133:670–81.

Preusch, P.L., F. Takeda, and T.J. Tworkoski. 2004. N and P uptake by strawberry plants grown with composed poultry litter. *Scientifica Horticulturae* 102:91–103.

Pritts, M. and D. Handley. 1998. *Strawberry Production Guide for the Northeast, Midwest and Eastern Canada*. Natural Research Agricultural Engineer Service (NRAES), Ithaca, NY.

Raghothama, K.G. and A.S. Karthikeyan. 2005. Phosphate acquisition. *Plant and Soil* 274:37–49.

Rehm, G., M. Schmitt, J. Lamb, G. Randall, and L. Busman. 1998. Phosphorus in agricultural environment. Document FO-06288-GO. Univ. Minnesota. Minneapolis, MN.

Reich, P.B., S.E. Hobbie, T. Lee, D.S. Ellsworth, J.B. West, J.M.H. Knops, S. Naeem, and J. Trost. 2006. Nitrogen limitation constrains sustainability of ecosystem response to CO_2. *Nature* 440:922–25.

Reich, P.B., J. Oleksyn, and I.J. Wright. 2009. Leaf phosphorus influences the photosynthesis-nitrogen relation: A cross-biome analysis of 314 species. *Oecologia* 160:207–12.

Rieger, M. 2005. *Introduction to Fruit Crops*. Haworth Food & Agricultural Products Press, New York.

Schachtman, D.P., R.J. Reid, and S.M. Ayling. 1998. Phosphorus uptake by plants: From soil to cell. *Plant Physiology* 116:447–58.

Sicher, R.C. and J.A. Bunce. 2008. Growth, photosynthesis, nitrogen partitioning and responses to CO_2 enrichment in a barley mutant lacking NADH-dependent nitrate reductase activity. *Physiologia Plantarum* 134:31–40.

Sinegani, A.A. and S.T. Rashidi. 2011. Changes in phosphorus fractions in the rhizosphere of some crop species under glasshouse conditions. *Journal of Plant Nutrition and Soil Science* 174:899–907.

Six, J., S.D. Frey, R.K. Thiet, and K.M. Batten. 2006. Bacterial and fungal contributions to carbon sequestration in agroecosystems. *Soil Science Society of America Journal* 70:555–69.

Smith, F.W., S.R. Mudge, A.L. Rae, and D. Glassop. 2003. Phosphate transport in plants. *Plant and Soil* 248:71–83.

Stanger, T.F. and J. Lauer. 2008. Corn grain yield response to crop rotation and nitrogen over 35 years. *Agronomy Journal* 100:643–50.

Stewart, L., I. Hamel, C. Hogue, and R.P. Moutoglis. 2005. Response of strawberry to inoculation with arbuscular mycorrhizal fungi under very high soil phosphorus conditions. *Mycorrhiza* 15:612–19.

Swarbreck, S.M., M. Defoin-Platel, M. Hindle, M. Saqi, and D.Z. Habash. 2011. New perspectives on glutamine synthetase in grasses. *Journal of Experimental Botany* 62:1511–22.

Tabuchi, M., T. Abiko, and T. Yamaya. 2007. Assimilation of ammonium ions and reutilization of nitrogen in rice (*Oryza sativa* L.). *Journal of Experimental Botany* 58:2319–27.

Taghavi, T.S. and M. Babalar. 2007. The effect of nitrate and plant size on nitrate uptake and in vitro nitrate reductase activity in strawberry (*Fragaria* × *ananassa* cv. Selva). *Scientifica Horticulturae* 112:393–98.

Tagliavini, M., E. Baldi, P. Lucchi, M. Antonelli, G. Sorrenti, G. Baruzzi, and W. Faedi. 2005. Dynamics of nutrients uptake by strawberry plants (*Fragaria* × *ananassa* Dutch.) grown in soil and soilless culture. *European Journal of Agronomy* 23:15–25.

Taiz, L. and E. Zeiger. 2002. *Plant Physiology*, 3rd ed. Sinauer, Sunderland, MA.

Tartachnyk, I.I. and M.M. Blanke. 2007. Photosynthesis and transpiration of tomato and CO_2 fluxes in a greenhouse under changing environmental conditions in winter. *Annals of Applied Biology* 150:149–56.

Viktor, A. and M.D. Cramer. 2003. Variation in root-zone CO_2 concentration modifies isotopic fractionation of carbon and nitrogen in tomato seedlings. *New Phytologist* 157:45–54.

Warren, C. 2004. The photosynthetic limitation posed by internal conductance to CO_2 movement is increased by nutrient supply. *Journal of Experimental Botany* 55:2313–21.

Section V

Photosynthetic Pathways in Various Crop Plants

18 Photosynthesis in Nontypical C$_4$ Species
C$_4$ Cycles without Kranz Anatomy and C$_4$ Crassulacean Acid Metabolism Transitions

María Valeria Lara and Carlos Santiago Andreo

CONTENTS

18.1 Introduction 307
18.2 Induction of a C$_4$-like Mechanism in Submersed Aquatic Plants of the Hydrocharitaceae Family 308
    18.2.1 The Case of *Egeria densa*..308
    18.2.2 The Case of *Hydrilla verticillata*...310
18.3 C$_4$ Photosynthesis in the Chenopodiaceae Family 311
    18.3.1 The Case of *Bienertia cycloptera* and *Bienertia sinuspersici*.......................................311
    18.3.2 The Case of *Suaeda aralocaspica*..313
18.4 Study of the Transition from C$_4$ Photosynthesis to Crassulacean Acid-like Metabolism in the *Portulaca* Genus.........314
    18.4.1 The Case of *Portulaca oleracea*..315
    18.4.2 The Case of *Portulaca grandiflora* 318
18.5 Concluding Remarks and Perspectives for Biotechnological Approaches to Improve C$_3$ Photosynthesis 319
References..319

18.1 INTRODUCTION

All plants and kinds of eukaryotic photoautotrophs use the same basic pathway for photosynthetic CO_2 fixation: the C$_3$ cycle (alternatively called photosynthetic carbon reduction cycle or Calvin and Benson cycle). In this pathway, ribulose bisphosphate carboxylase–oxygenase (RuBisCO) catalyzes the entry of CO_2 into the cycle. At ambient CO_2 and O_2 conditions, the enzyme also acts as an oxygenase incorporating O_2 into the photorespiratory carbon oxidation cycle with the resultant loss of the fixed carbon [1]. To overcome the effect of O_2 on RuBisCO, some plants have developed ways to increase the level of CO_2 at the location of RuBisCO in the plant, decreasing in this way the oxygenation reaction and, thus, the carbon flux through the photorespiratory carbon oxidation cycle. Among the different photosynthetic modes are the C$_4$ cycle and the crassulacean acid metabolism (CAM), which are evolutionary derived from C$_3$ photosynthesis [2]. The C$_4$ photosynthesis requires the coordination of biochemical functions between two types of cells and the cell type-specific expression of the enzymes involved [1,3–5]. In these plants, atmospheric CO_2 is first incorporated into C$_4$ acids in the mesophyll cells by phosphoenolpyruvate carboxylase. These C$_4$ acids are then transported to bundle sheath cells, where they are decarboxylated, and the released CO_2 is incorporated into the C$_3$ cycle. The C$_4$ system is more efficient under some

environmental conditions as it increases the concentration of CO_2 in bundle sheath cells, suppressing the oxygenase activity of RuBisCO and, thus, photorespiration. On the other hand, CAM is a metabolic adaptation to arid environments: stomata are closed during much of the day and opened at night. Malic acid is accumulated in the vacuoles of mesophyll cells at night as a result of fixation of CO_2 by the phosphoenolpyruvate carboxylase [6]. During the day, malic acid is decarboxylated, and the released CO_2 is refixed in the C$_3$ cycle [4]. Compared with C$_4$ plants, leaves of CAM plants have a simple inner structure [7].

Another mechanism found among photosynthetic organisms that eliminates the O_2 inhibition of photosynthesis is the one present in unicellular and multicellular algae [8], and cyanobacteria. In this case, the concentration of inorganic carbon in the site of RuBisCO is the result of different transporters located at the plasma membrane or at the chloroplast envelope and carbonic anhydrase [9–14].

Although most C$_4$ plants present Kranz anatomy and C$_4$ biochemical features in a constitutive manner, many variations as well as transitions from/to other photosynthetic modes have been described. These nontraditional C$_4$ plants can be grouped as follows: (1) submersed aquatic monocot species like those belonging to the Hydrocharitaceae that show induction of a C$_4$-like metabolism without Kranz anatomy under conditions of high temperatures and light intensities [15–17];

(2) C_4 photosynthetic plants belonging to the Chenopodiaceae where C_4 photosynthesis functions within a single photosynthetic cell, though lacking the Kranz anatomy [18–19]; and (3) C_4 succulent species of *Portulaca* that exhibit, under water stress, transition to a CAM-like metabolism (as in *Portulaca oleracea*) [20] or induction of a CAM-cycling metabolism compartmentalized in a different cell type while the C_4 pathway is also operating (*Portulaca grandiflora*) [21].

18.2 INDUCTION OF A C_4-LIKE MECHANISM IN SUBMERSED AQUATIC PLANTS OF THE HYDROCHARITACEAE FAMILY

Submersed aquatic macrophytes are a large group of phototrophs, which include nonvascular plants, primitive vascular plants, and angiosperms. The supply of dissolved inorganic carbon species in water can be limiting because of the high diffusive resistance in water [22]. In this respect, submersed aquatic autotrophs exhibit plasticity in relation to photosynthesis in aspects such as biochemistry, physiology, and anatomy [23] and have developed mechanisms to cope with limiting CO_2 and high O_2 concentrations, such as CO_2 concentrating mechanisms and the ability to use HCO_3^- in photosynthesis, and/or the presence of carbonic anhydrase in the apoplast [23,24]. In general, submersed aquatic macrophytes display unique characteristics related to their environment, such as low photosynthetic rates [25], low light requirements, very high $K_m\left(CO_2/HCO_3^-\right)$ values [26], and the requirement of high CO_2 levels to saturate photosynthesis [22,24,25].

In a variety of submersed aquatic macrophytes, low CO_2 compensation points are induced by submergence and growth under stress conditions of low CO_2 levels, high temperatures, and long photoperiods [15,23,27–31]. At least four members of the monocot family Hydrocharitaceae, *Hydrilla verticillata*, *Elodea canadensis*, *Egeria densa*, and *Ottelia alismoides* [17], have an appreciable Kranz-less C_4-acid metabolism in the light. C_4-like metabolism has also been proposed for other species like *Myriophyllum spicatum* [25], *Cabomba caroliniana* [27], *Saggitaria subulata*, and *Orcuttia viscida* [32] belonging to other families. Due to the great interest in improving C_3 photosynthesis in terrestrial plants, the attention to the understanding of the C_4 facultative aquatic species has been renewed [33].

18.2.1 THE CASE OF *EGERIA DENSA*

Among the higher aquatic plants, *E. densa* has been preferred material for a number of different studies in plant physiology. Its leaves contain a single longitudinal vascular bundle, and the blade consists of two layers of cell only, allowing studies of the whole undamaged organ in a natural environment. In this species, heterogeneity is reduced to a minimum; all leaf cells are in direct contact with the external medium and at the same developmental stage, and thus, in similar physiological conditions. These properties, together with the leaf polarity displayed by *E. densa*, represent an advantage for different

kinds of research and make this species one of the model organisms of the plant kingdom for experiments, such as electrophysiology [34–37].

E. densa is a submersed rooted aquatic dioecious herb [38]. It is a common waterweed, which occurs in streams, ponds, and lakes. The slender stems of *Egeria* are usually a foot or two long but can be much longer. The small leaves are strap shaped, about 1 in. long and 1/4 in. wide. The leaf margins have very fine saw teeth that require a magnifying lens to see. Leaves occur in whorls of three to six around the stem. The flowers are on short stalks about 1 in. above the water. Flowers have three white petals and are about 3/4 in. across.

Electrophysiological studies in *E. densa* reveal that this species has leaf pH polarity and values of electric potential difference at the plasmalemma of up to ca. −300 mV in the light [37]. Under strong illumination, leaf cells acidify the medium on the abaxial side of the leaf and alkalize the adaxial side of the leaf. This acidification is mediated by a H^+-ATPase located in the plasmalemma [39,40] and controlled by the photosynthetic process, apparently by redox regulation [34]. The leaf polarity is proposed to be used for bicarbonate utilization, not only in species of the Hydrocharitaceae but also in *Potamogeton* species [41]. At high light intensities and low dissolved carbon concentration, *Egeria* generates a low pH at the abaxial leaf side for CO_2 uptake. To balance the loss of the H^+ from the symplasm, there is an OH^- efflux at the upper side of the leaf together with the K^+ (Ca^{+2}) flux from the abaxial to the adaxial solution. In this way, the photosynthetic reduction of HCO_3^- by these so-called polar plants produces one OH^- for each CO_2 assimilated. The acidification in the lower side of the leaf results in a shift in the equilibrium from HCO_3^- to CO_2, with CO_2 then entering the abaxial cells by passive diffusion [42]. By this means, *E. densa* cells have extra CO_2 available in conditions when the concentration of this gas is limiting for photosynthesis.

Two photosynthetic modes have been described for *E. densa* depending on the environmental conditions. The high photorespiration state with C_3 photosynthesis is characterized by a high CO_2 compensation point, which occurs in conditions of high CO_2 availability, and the second one is a low photorespiration state, which is induced after conditions of high light, temperature, O_2, probably UVB radiation, and low CO_2 [16,43,44]. The first studies in *E. densa* [45] showed radiolabeled inorganic carbon fixation into malate and aspartate, especially at low pH and in the short term. Further reports also indicated malate as the major product of short-term ^{14}C labeling in *Egeria* [46] and *E. canadensis* Michx [47]. Later, the Calvin cycle was found as the primary carboxylation mechanism, responsible for over 90% of the ^{14}C initially incorporated [46]. Although these primary studies in *E. densa* produced conflicting results, it was then established that low CO_2 levels influence the products formed, with malate increasing at the expense of the Calvin cycle intermediates [46,48–51].

Then, a CO_2 compensation point of 43 μL CO_2/L (typical of terrestrial C_3 plants) was described when plants were incubated at conditions of low temperature (12°C) and light [27].

But when incubated under high light and temperature (30°C), conditions that cause a decrease of gases' solubility in water, a value of 17 µL CO_2/L was observed. Maximal RuBisCO activity (76.0 and 70.6 µmol mg Chl^{-1} h^{-1}) in both conditions shows no correlation between the decrease in compensation point and the RuBisCO activity.

After 23 days under high light (300 µmol m^{-2} s^{-1}) and temperature (30°C), the activities of phosphoenolpyruvate carboxylase (PEPC) and NADP-malic enzyme (NADP-ME) increased 3.7-fold and 3-fold, respectively, with respect to the levels in plants under low light (30 µmol m^{-2} s^{-1}) and temperature (12°C) [16], due to increased protein levels. In contrast, RuBisCO content remained constant; thus, the PEPC/RuBisCO ratio was increased but to a lesser extent than that observed in *H. verticillata* [6,27].

Purified NADP-ME from induced *E. densa* leaves showed physical and kinetic properties similar to those of the enzyme from terrestrial C₃ plants. Thus, the increase in the amount of a C₃ NADP-ME type after the induction period may facilitate the maintenance of high rates of decarboxylation of malate and the delivery of CO_2 to RuBisCO [16].

In the case of PEPC, two immunoreactive bands of 108 and 115 kDa were expressed in plants kept under low light and temperature, and after 23 days of induction, the lower-molecular-mass form was clearly induced, while the level of the other isoform was not affected by the treatment. The purified inducible 108 kDa isoform exhibited low K_m for phosphoenolpyruvate and showed a hyperbolic response as a function of this substrate [16]. Moreover, the estimated K_m value for HCO_3^- (7.7 µM) is lower than all reported values for the

enzyme from different C₄ species [52]. Thus, this *E. densa* PEPC isoform not only has a high affinity for its substrates but also is induced under conditions of low CO_2 availability [16]. In addition, it shows differential phosphorylation during the day, with the phosphorylation/dephosphorylation process highly regulated during the induction of this Kranz-less C₄-acid metabolism in *E. densa* [53].

Cellular fractionation followed by Western blot indicated that PEPC is located in the cytosol of the photosynthetic cells of *E. densa* leaves, whereas NADP-ME and RuBisCO are located in the chloroplasts [16]. In this way, the specific compartmentation of these enzymes is very important for delivering inorganic carbon from the cytosol to the chloroplasts via C₄ acids. The chloroplast is the site of CO_2 generation from C₄ acids and, consequently, of the concentrating mechanism.

In summary, *E. densa* responds to the decrease in CO_2, which takes place under conditions of high light and temperature, by inducing an ancient-like isoform of NADP-ME similar to the one present in C₃ terrestrial species [54]. The regulation of PEPC in *E. densa* induced plants appears to be similar to the one occurring in PEPC from C₄ species, so the 108 kDa isoform is probably participating in a C₄-like mechanism, with a similar regulation to that of PEPC from C₄ plants. In this way, high activity and kinetic and regulatory properties of C₄ enzymes such as PEPC and NADP-ME, and labeling of malate, together with a decrease in the compensation point are evidence that a C₄-like photosynthetic system operates in *E. densa*, when transferred to conditions of low CO_2 availability (Figure 18.1). Finally, in *E. densa*,

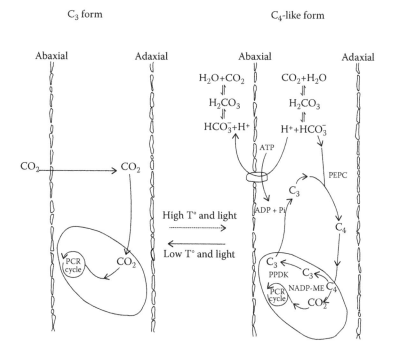

FIGURE 18.1 Proposed photosynthetic mechanisms operating in *Egeria densa* and *Hydrilla verticillata* under conditions of low (induced by high light and temperature) and high (induced by low light and temperature [T°]) CO_2 availability. The intracellular localization of the main C₄ photosynthetic enzymes is shown. NADP-ME, NADP-malic enzyme; PCR cycle, photosynthetic carbon reduction cycle; PEPC, phosphoenolpyruvate carboxylase; PPDK, pyruvate orthophosphate dikinase.

both the C_4-like mechanism and the pH polarity mechanism contribute to photosynthetic performance under CO_2-limiting circumstances.

18.2.2 THE CASE OF *HYDRILLA VERTICILLATA*

H. verticillata is a submersed, usually rooted, aquatic perennial herb with slender ascending stems up to 30 ft. long, heavily branched. The stems from slender rhizomes are often tipped with a small tuber. Leaves are whorled, three to eight per whorl, bearing coarse teeth along the margins. Fleshy axillary buds are often formed at the leaf axils, up to 2 in. long, with three sepals and three petals, each about 0.3 in. long, whitish or translucent, floating on the water surface. Male flowers detached and free-floating at maturity are white to reddish brown, about 2 mm long. Flowers release floating pollen from stamens when they pop open on the water surface.

In 1924, it was shown that *H. verticillata* was more acidic in summer than in winter and had a higher malate content, and that malate could replace CO_2 in driving photosynthetic O_2 evolution [55]. Later, it was indicated that its CO_2 compensation point was modified depending on growing conditions [56] and correlated with changes in PEPC activity. Plants grown in winter or under a short photoperiod accompanied by low temperature (25°C) exhibited higher CO_2 compensation points than those grown in summer or under a longer photoperiod and higher temperatures (27°C). The low-compensation-point plants incorporated 60% $^{14}CO_2$ into malate and aspartate, with only 16% in sugar phosphates [51]. Moreover, in *H. verticillata* plants, low CO_2 levels influenced the products formed, with malate increasing at the expense of the Calvin cycle intermediates. However, pulse-chase labeling experiments indicated that malate was turned over at a low rate [51].

In high-compensation-point plants, photorespiration, as a percentage of net photosynthesis, was equivalent to that in terrestrial C_3 plants, while plants with decreasing CO_2 compensation point were associated with reduced photorespiration and increased net photosynthesis rates [27]. In this last group, the PEPC/RuBisCO activity ratio was higher with respect to the high compensation group, and plants exhibited increases in activity of pyruvate orthophosphate dikinase, pyrophosphatase, adenylate kinase, NAD- and NADP-malate dehydrogenase, NAD- and NADP-MEs, and aspartate and alanine aminotransferases [27]. In contrast, the activities of the photorespiratory enzymes phosphoglycolate phosphatase and glycolate oxidase, and of phosphoglycerate phosphatase, showed no change. The decrease in the CO_2 compensation point reflects at least increased (or refixed) fixation via PEPC. In addition, fixation of CO_2 in the dark and diurnal fluctuations in the level of titratable acidity were observed in *H. verticillata* [29,51]. It was then established that malate formation occurred in both photorespiratory states, but reduced photorespiratory states resulted when malate was utilized in the light [15].

It is interesting to note that induction of the C_4-like cycle takes place in the same leaves without the development of

new leaves [31]. Sections of leaves show no evidence of Kranz anatomy and reveal the existence of two layers. The adaxial surface was composed of larger cells than the abaxial side and had prominent vacuoles. RuBisCO was found in the chloroplasts of both photosynthetic-type cells to the same extent. PEPC was mainly cytosolic, in both types of cells [56,57]. Subcellular fractionation of leaves under s low CO_2 compensation point showed that NADP-ME, pyruvate orthophosphate dikinase, and most of the NADP-malate dehydrogenase were chloroplastic, while NAD-ME was mitochondrial [57]. Since intercellular differentiation of fixation events, found in plants with Kranz anatomy, does not occur with low CO_2 compensation points in *H. verticillata* plants, intracellular separation of C_3 and C_4 fixation events may account for the low photorespiration state [56]. In this way, a C_4-like cycle concentrating CO_2 takes place in the chloroplast, with PEPC fixing inorganic carbon in the cytosol and C_4 acids moving to the chloroplasts and being decarboxylated by NADP-ME so as to supply CO_2 to RuBisCO [57] (Figure 18.1). An interesting feature of *Hydrilla* is that the chloroplasts show grana stacks, in contrast to NADP-ME Kranz plants; also, they are presumably not deficient in photosystem II or NADP, and leaves are not prone to photoinhibition. In this way, oxaloacetate or aspartate, rather than malate, is probably the major imported acid to the chloroplasts [55]. After a 12-day induction period, the increases in the activities of C_4 enzymes (PEPC, aspartate and alanine aminotransferases, NADP-ME, and pyruvate phosphate dikinase), accompanied by constant levels of RuBisCO [58], are consistent with a C_4-like cycle concentrating CO_2 in leaves of *H. verticillata* [57].

The contribution of PEPC to the low photorespiratory gas-exchange features was demonstrated by inhibition of the enzyme, which resulted in increases in the CO_2 compensation point, O_2 inhibition of photosynthesis, and CO_2 evolution in the dark by 51% [59]. This also indicates that the use of HCO_3^- is not responsible for concentrating CO_2. As the relation between PEPC activity and the CO_2 compensation point was nonlinear, the expression of low photorespiratory gas-exchange characteristics might require the subsequent induction of other C_4 photosynthetic enzymes [59]. The increase in PEPC activity was correlated with *de novo* synthesis of the protein. The 110 kDa enzyme from C_4-type leaf extracts was 14-fold more active than that in C_3-type leaves and had daytime values 53% higher than at night, resembling the enzyme from C_4 obligated plants [55]. During the day the enzyme was less sensitive to malate inhibition as accounted for by higher I_{50} values and glucose-6-phosphate acting as a positive effector of the enzyme. In contrast, while this form exhibited upregulation in the light, PEPC from C_3 leaves showed no light activation and lower rates, and it was virtually insensitive to malate inhibition [55]. Three full-length cDNAs encoding PEPCs were isolated from *H. verticillata*. Hvpepc4 was exclusively expressed in leaves during C_4 induction, with kinetic data from the expressed protein consistent with a C_4 form of the enzyme, and thus the one involved in photosynthesis. While it lacks the C_4 signature serine of terrestrial plant C_4 isoforms [60], it contains the putative C_4 determinant

Lys349. *Hvpepc*3 (probably involved in the dark fixation) and *Hvpepc*5 (expressed in roots) have Arg, a common feature of C$_3$ sequences [61]. In phylogenetic analyses, the three sequences grouped with C$_3$, nongraminaceous C$_4$, and CAM PEPCs but not with the graminaceous C$_4$, and formed a clade with a gymnosperm, which is consistent with *H. verticillata* PEPC predating that of other C$_4$ angiosperms [60].

Regarding NADP-ME, three genes encoding the enzyme were identified. *Hvnadp-me1* encodes a plastidic enzyme with kinetic properties intermediate between photosynthetic and nonphotosynthetic isoforms. This isoform, although constitutive, is induced during the light period in the C$_4$ leaves and would act as the photosynthetic decarboxylase when the C$_4$ cycle is induced. On the other hand, HVNADP-ME2 and HVNADP-ME3 are cytosolic enzymes, with HVNADP-ME2 highly expressed in C$_3$ leaves and HVNADP-ME3 equally expressed in C$_3$ and C$_4$ leaves [55,62,63].

Taken together, a C$_4$-like cycle occurs in *H. verticillata* when grown under summer conditions (high temperature and light), which results in limiting levels of inorganic carbon in water [29]. Gas-exchange and biochemical modifications demonstrate this shift from C$_3$ to C$_4$-like photosynthesis in the leaves [23], despite lacking Kranz anatomy [30,55]. Direct measurements of internal inorganic carbon pools in C$_3$ and C$_4$ *H. verticillata* leaves [30] demonstrate that the C$_4$ cycle concentrates the CO$_2$ in this species, even in a low-pH medium, where HCO$_3^-$ is negligible. A CO$_2$ flux mechanism facilitates the entrance of inorganic carbon to the cells in both C$_3$ and C$_4$ modes, by providing access to the HCO$_3^-$ pool in the medium. In this way, this plant is a HCO$_3^-$ user and becomes pH polarized in the light [63,64]. However, when inorganic carbon is limiting, the effectiveness of this carbon flux mechanism is diminished, and the CO$_2$ concentrating mechanism in the chloroplasts based on a C$_4$-like cycle minimizes CO$_2$ losses from respiration in *H. verticillata* [55] (Figure 18.1). In addition, net photosynthetic rates at limiting CO$_2$ concentrations are substantially greater when shoots are in the C$_4$ mode [27,65]. However the C$_4$ quantum yield is half of that in the C$_3$ shoots [65], demonstrating that the C$_4$ cycle in *Hydrilla* has a substantial energy cost to it, perhaps more than in a terrestrial C$_4$ plant [55].

18.3 C$_4$ PHOTOSYNTHESIS IN THE CHENOPODIACEAE FAMILY

The Chenopodiaceae family has C$_3$, C$_4$, and C$_3$–C$_4$ intermediate metabolisms species [66–70], with the most C$_4$ species among any of the dicot families [71]. There are interesting anatomical variations in the photosynthetic apparatus of leaves/stems and cotyledons of representative species of the tribe Salsoleae [72]. Apart from the Kranz-like organs with mesophyll cells and bundle sheath cells, in many species, a circular anatomy that includes water storage and subepidermal cells have been described [72]. In addition, the species *Bienertia cycloptera*, *Bienertia sinuspersici*, and *Suaeda aralocaspica* show appreciable C$_4$ photosynthesis in a single chlorenchyma cell [18,19,73–75]. In spite of C$_4$ photosynthesis

being accomplished by more than a dozen biochemical and anatomical combinations that arose independently [76,77], all C$_4$ plants share the common initial step of phosphoenolpyruvate carboxylation. Nevertheless, Kranz anatomy is not essential for C$_4$ photosynthesis in terrestrial plants [76], as is the case in *B. cycloptera*, *B. sinuspersici*, and *S. aralocaspica* [78].

18.3.1 THE CASE OF *BIENERTIA CYCLOPTERA* AND *BIENERTIA SINUSPERSICI*

Initially thought of as a monotypic genus of the tribe Suaedeae within the Chenopodiaceae family [78], *Bienertia* is a genus with three species: *B. cycloptera*, *B. sinuspersici* [79], and *Bienertia kavirense* [80]. They are Irano-Turanian floristic elements distributed in the central Iranian deserts and subdeserts with northern and southern radiation in the Persian Gulf countries, and Central Asia *Bienertia* grows as a halophytic annual on clay, silt, or sandy alluvial soils in depressions and along lagoons. Climatic conditions are marked by hot, sunny, and dry summers and by winter with no or little to moderate frost and mean precipitation of ca. 100–200 mm [81].

B. cycloptera plants are 10–60 cm in height and richly branched from the base. Leaves (2–15 × 2–5 × 1.5–2.5 mm) are usually glaucous, narrow to broad oblong, slightly narrow at the base, and obtuse at the apex. Young leaves are covered by small, short-stalked vesicular hairs [78]. *B. kavirense* are taller (30–100 cm height). Leaves from this species are glaucous and covered by a powdery bloom. They are oblong and flat (23–28 × 9–11 × 2–2.5 mm) [80]. In the case of *B. sinuspersici*, plants are up to 130 cm height and show fleshy oblong and glaucous leaves (27.0 × 8.15 × 1.7–2.2 mm) without petioles; their surface is covered with a powderlike bloom [79].

Five variants of Kranz anatomy, Atriplicoid, Kochioid, Salsoloid, Kranz-Suaedoid [67], and Conospermoid [81], were identified in the family. In the case of *Bienertia*, it represents a new leaf type named Bienertiod, and in the overall layout, it resembles most of the C$_3$ species of *Suaeda*, except for the cytological structure and arrangement of the chlorenchyma and its strict separation from the water storage tissue [81] (Figure 18.2). The epidermis is one layered with large individual cells in which chloroplasts are missing, except for the small and slightly sunken guard cells. The chlorenchyma consists of one to three (depending on the species) layers of radially elongated cells that are arranged in short rows between the epidermis and the aqueous tissue. Cells are unique in containing two cytoplasmic compartments, a peripheral cytoplasmic layer with scattered chloroplasts and a large globular cytoplasmic body located in the center of the cell, completely surrounded by the vacuole, densely packed with starch-producing chloroplasts, and joined by the nucleus [81]. Both compartments are connected by cytoplasmic channels that contain a few chloroplasts with a development of grana intermediate between that of both cytoplasmic compartments [19]. The sheetlike channels are proposed as the route for metabolites as well as the pathway of leakages of CO$_2$ from the central compartment [82]. The central compartment is filled with mitochondria, with an extensive system of tubules and lamellae, which encircle granal chloroplasts. In contrast, the peripheral compartment lacks mitochondria and has agranal chloroplasts that exhibit lower granal index, density of appressed thylakoids, and ratio of appressed to nonappressed thylakoids than the chloroplasts in the central compartment [19]. It has been shown in *B. sinuspersici* and *B. cycloptera* as well as in *S. aralocaspica* that a complex cytoskeletal system maintains the organization of the compartments [83]. Even after long light exposure periods, starch was absent in the chloroplasts of the peripheral cytoplasmic compartment [81]. The aqueous tissue together with the embedded vascular bundles accounts for about two-thirds to three-fourths of the leaf volume (Figure 18.2). It consists of one to two layers of large cells with thin walls and thin peripheral cytoplasmic layers without chloroplasts. The vascular system consists of one large central bundle and a varying number of smaller secondary bundles [81].

Studies involving leaf anatomy observation and the finding that after 10 s of $^{14}CO_2$ exposure, 45% of the total radioactivity was present in sugar phosphate and only 19% and 11% of malate and aspartate, respectively, indicated that the plant was exclusively C_3 in *B. cycloptera* [68]. However, later, $\delta^{13}C$ isotope values of −15.4‰ [70] and −14.3‰ [66] were discovered, which are similar to those of C_4 plants with Kranz anatomy. After that, Freitag and Stichler [81] measured $\delta^{13}C$ isotope discrimination values in *B. cycloptera* leaves under a wide range of conditions. They found values of −13.4‰ to −15.5‰ in leaves from natural habitats (typical of C_4 species), but values ranging from −15.5‰ to −21.1‰ were measured in newly formed leaves in the greenhouse. In addition, Voznesenskaya et al. [19], using mature leaves, found values (−13.5‰) similar to those reported for plants collected in the natural habitat and also reported more negative values for younger leaves. These results suggest that *Bienertia* is a C_4/C_3 facultative species and explain the differences in the results reported by Glagoleva et al. (1992, [68]). With respect to the photosynthetic carbon labeling pattern, considering that the diffusion pathway is

shorter within a single cell than between mesophyll and bundle sheath cells, Voznesenskaya et al. [19] used a shorter pulse (3 s) to catch the C_4 acids. Under this experimental condition, 50% of the initial products were in malate and aspartate and 13% in 3-phosphoglycerate.

Analysis of the main photosynthetic enzymes showed that RuBisCO was mainly located in the chloroplasts of the central compartment and only a weak signal was found in the chloroplasts of the peripheral compartment. Although PEPC was found throughout the cytosol, the most intensive labeling was found in the peripheral cytoplasm [19]. Regarding PEPC, biochemical studies reveal that PEPC from *B. sinuspersici* shows kinetic and regulatory features similar to those of PEPC from C_4 species [84]. These enzymes have levels similar to those in C_4 Chenopods as well as exhibited a PEPC/RuBisCO ratio greater than 1. In contrast to RuBisCO, pyruvate orthophosphate dikinase (PPDK) was located in the peripheral chloroplasts. Again, activity and immunoreactive protein levels were similar to those found in the C_4 plant *Salsola laricina*. Higher activity levels of NAD-ME relative to NADP-ME were measured in *B. cycloptera* and *B. sinuspersici*. While the former was immunodetected, the NADP-dependent enzyme was not. Immunoelectron microscopy indicated that NAD-ME and glycine decarboxylase were localized in the mitochondria of the central compartment [19]. Isolation of peripheral chloroplasts and organelles from the central compartment of *B. sinuspersici* revealed that peripheral chloroplasts possessed a transporter for pyruvate and thus are capable of importing this metabolite for the reaction catalyzed by PPDK. In addition, it was also shown that while both types of chloroplasts photoreduce 3-phosphoglycerate, a completely functional Calvin cycle occurs only in chloroplasts from the central compartment [85].

Its photosynthetic response to varying CO_2, at 21% versus 3% for O_2, is typical of C_4 plants with Kranz anatomy. In addition, O_2 did not inhibit photosynthesis in *Bienertia*. The lack of nighttime CO_2 fixation indicated this species is not a CAM-type [19]. Studies regarding CO_2 assimilation rates and leakiness as a function of irradiation also reveal that the response of *B. sinuspersici* is similar to that of C_4 species with Kranz anatomy [86,87].

Based on carbon isotope values, anatomical studies (although lacking Kranz anatomy) [18,88], products of $^{14}CO_2$ fixation, immunolocalization studies, and facultative C_3/C_4 expression [81], it can be concluded that *B. cycloptera and B. sinuspersici* exhibit a novel solution to C_4 photosynthesis without Kranz anatomy. This metabolism is carried out in a single chlorenchyma cell through spatial compartmentation of dimorphic chloroplasts, other organelles, and photosynthetic enzymes in distinct positions in the cell that mimic the spatial separation of Kranz anatomy [19] (Figure 18.2). In this way, the peripheral and central cytoplasmic compartments may be functionally equivalent in photosynthesis to palisade and Kranz cells in the Kranz-type C_4 plants [81]. The vacuole appears to be the resistant barrier minimizing CO_2 efflux, and the cytoplasmic strands are the channels for metabolite flux [77]. CO_2 would be initially fixed by PEPC localized in the

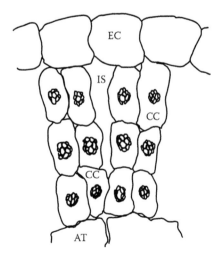

FIGURE 18.2 Diagram of various cell types in a cross section of *Bienertia cycloptera*. AT, aqueous tissue; CC, chlorenchyma cell; EC, epidermal cell; IS, intercellular space.

peripheral compartment. Then C$_4$ acids would be transported through the cytoplasmic channel to the mitochondria in the central cytoplasmic compartment, where NAD-ME acting as decarboxylase would provide RuBisCO in the neighboring chloroplasts with CO$_2$. Three carbon compounds would then return to the peripheral cytoplasmic compartment where PPDK located in the chloroplasts would regenerate phosphoenolpyruvate. As this last enzyme is not found in the central cytoplasmic compartment, and thus, the presence of phosphoenolpyruvate is unlikely to occur there, PEPC in the central compartment probably would not be involved in this metabolism [19].

B. cycloptera is also remarkable for its ability to adapt its photosynthetic machinery to environmental conditions as a facultative C$_4$/C$_3$ species. Leaves developed under conditions of high light and temperature carry out C$_4$ photosynthesis, but under conditions of low light and temperature, C$_3$ photosynthesis is performed [81].

18.3.2 The Case of *Suaeda aralocaspica*

S. aralocaspica Bunge is a Central Asian floristic element with a scattered distribution through the semideserts from the northeastern Caspian lowlands to western China and southern Mongolia. It belongs to the strongest halophytes, and it is found in monospecific stands in the foremost zone of higher vegetation around salt lakes and deep depressions. The habitat is usually flooded in spring and covered with a thick salt crust during summer and autumn [81].

S. aralocaspica is a succulent species with unusual chlorenchyma. Plants are 4–50 cm in height and richly branched from the base. Succulent glabrous leaves (10–17 × 1.5–2.2 × 0.5–1.1 mm) are isolateral, with a pronounced tendency toward a centric arrangement of tissues [81].

Similar to the C$_4$ Salsoloid type, *S. aralocaspica* has particular cytological features, with only a single layer of unusual palisade-shaped chlorenchyma cells located between the central water storage tissue and the hypodermal cells (Figure 18.3) [18]. The pattern of vascularization with numerous small peripheral bundles attached to the inner surface of the chlorenchyma makes *S. aralocaspica* similar to the Salsolid type of C$_4$ chenopods [81].

Hypodermal cells are large, with thin walls and no chloroplasts. The central aqueous tissue comprises one-third of the volume of the cell, and it is constituted by large polyhedral cells with thin walls, a thin protoplast, and few or no chloroplasts [81].

Chlorenchyma cells have a cylindrical shape in the upper two-thirds to three-fourths, but the basal parts of the cells are trapezoid. The layer is compact to the center of the leaf and has intercellular spaces in the outermost part of the chlorenchyma (Figure 18.3). The cell wall is thicker in the basal parts and has perfect straight orientation. The protoplast also has differential features in the upper and basal parts of the chlorenchyma cells [81]. The basal parts possess a much smaller central vacuole, a higher number of chloroplasts per unit volume, chloroplasts with abundant starch grains, a higher

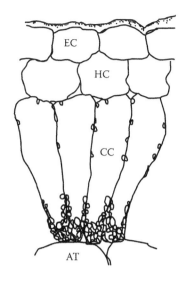

FIGURE 18.3 Diagram of various cell types in a cross section of *Borszczowia aralocaspica*. AT, aqueous tissue; CC, chlorenchyma cell; EC, epidermal cell; HC, hypodermal cell.

density of cytoplasm, and firmer attachment of plasmalemma to the cell walls. The nucleus is invariably located at or just above the border of the two macrocompartments of the chlorenchyma cells [81]. Large mitochondria are in the proximal position (from the vascular bundle) [18], and chloroplasts in the distal part of the cell lack grana and are without starch, while those in the proximal part have grana and contain starch. In this way, *S. aralocaspica* has leaf anatomy similar to C$_4$ *Salsola*, but it has dimorphic chloroplasts in a single photosynthetic cell instead of in two cell types with Kranz anatomy [18].

S. aralocaspica carbon isotope ratios of −13.78‰ [81] and −13.37‰ δ^{13}C [18] match not only those described for C$_4$ plants (−10‰ to −14‰, 124) but also those reported for 198 C$_4$ chenopods (−9.27‰ to −15.06‰) [66]. In contrast, the C$_3$ species *Suaeda heterophylla* exhibited values of −25.34‰ and −27.28‰ [18] as in this plant, RuBisCO fixes atmospheric CO$_2$. In this way, although lacking Kranz anatomy, *S. aralocaspica* probably fixes atmospheric CO$_2$ by PEPC, which does not discriminate against ^{13}CO$_2$.

Voznesenskaya et al. [18] found biochemical compartmentation in the chlorenchyma cells for the carbon assimilation enzymes. RuBisCO was found in the chloroplasts in the proximal part of the cell, while PPDK was located in the chloroplasts in the distal cytosol. RuBisCO was also detected in the few and small chloroplasts of water storage and hypodermal cells. Phosphoenolpyruvate carboxylase was visualized in all the cytosol. NAD-ME was located in the mitochondria in the proximal part of the cells.

The activities of RuBisCO, phosphoenolpyruvate carboxylase, NAD-ME, and PPDK are high enough to support the measured rates of CO$_2$ assimilation and, except for NAD-ME, are similar to the values measured in the C$_4$ plant *S. laricina*.

As the response of photosynthesis to varying CO$_2$ and O$_2$ is a diagnostic tool for discriminating C$_3$ versus C$_4$ photosynthesis, the response of *S. aralocaspica* against different O$_2$

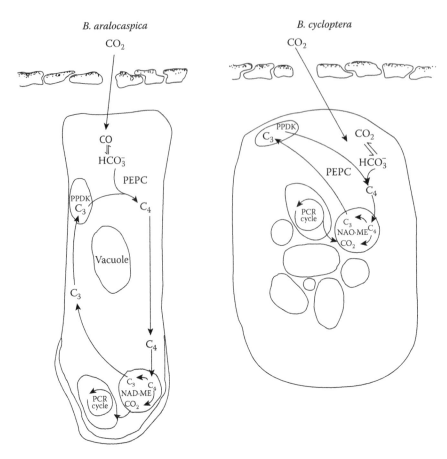

FIGURE 18.4 Proposed scheme of C_4 photosynthesis within a single cell in *Borszczowia aralocaspica* and *Bienertia cycloptera*. The intracellular localization of the main C_4 photosynthetic enzymes is shown within the chlorenchyma cell. NAD-ME, NAD-malic enzyme; PCR cycle, photosynthetic carbon reduction cycle; PEPC, phosphoenolpyruvate carboxylase; PPDK, pyruvate orthophosphate dikinase.

concentrations was evaluated, with the finding that photorespiration is restricted in this species [18]. While ambient levels of O_2 (21%) inhibited photosynthesis in C_3 *S. heterophylla*, it did not have effect on photosynthesis in the C_4 species *S. laricina* or *S. aralocaspica*.

Regarding the biochemical pathway, *S. aralocaspica* presents a unique pattern of enzyme distribution, with some of the C_4 enzymes present to the same extent as in C_4 plants. Gas-exchange analysis and carbon isotope composition indicate that *S. aralocaspica* behaves as a C_4 plant. Overall, *Suaeda* is a C_4 plant, although it lacks Kranz anatomy, in which the distal and proximal cytosolic compartments would probably act as the mesophyll and bundle sheath in C_4 plants (Figure 18.4). In this novel photosynthetic metabolism, atmospheric CO_2 enters the cell at the distal part thorough the thin cell wall and is fixed by PEPC. Then C_4 acids would be transported to the proximal part of the cell, where the decarboxylation in the mitochondria by the action of NAD-ME provides RuBisCO with CO_2. In this proximal part of the cell, CO_2 leakage is restricted by the thicker cell wall. Thus, the radial arrangement of the elongated chlorenchyma cell is of great importance to prevent the CO_2 efflux [77]. As PPDK is located in the chloroplasts in the distal cytosol, phosphoenolpyruvate would be regenerated in this part of the cell, providing the substrate for PEPC. Distal generation of phosphoenolpyruvate and the

thick wall at the proximal part of the cell make it improbable that PEPC in the proximal cytosol fixes atmospheric CO_2.

18.4 STUDY OF THE TRANSITION FROM C_4 PHOTOSYNTHESIS TO CRASSULACEAN ACID-LIKE METABOLISM IN THE *PORTULACA* GENUS

In general, leaves of a plant fix CO_2 through one type of photosynthetic pathway. However, some plants can shift their photosynthetic mode depending on the age or environment, like the shift from C_3 photosynthesis to CAM [89,90]. Current mechanistic understanding of the induction of CAM is largely based on studies with facultative species that present a transition from C_3 photosynthesis. In contrast, the shift from C_4 photosynthesis to CAM has been described only in some succulent C_4 plants belonging to the genus *Portulaca* [20,21,91–95]. *P. oleracea*, *P. grandiflora*, and *Portulaca mundula* were reported to express CAM characteristics when subject to drought-stress conditions or short photoperiods [20,21,91–96].

The photosynthetic diversity of the Portulacaceae is considerable despite the small size of the family, with species C_3, C_4 and facultative CAM plants. Moreover, the occurrence of the decarboxylating enzymes reinforces this diversity.

P. grandiflora is a NADP-ME species [94], *P. oleracea* is a NAD-ME subtype [97], and *Portulacalaria afra* is a phosphoenolpyruvate carboxykinase plant [98].

CAM induction is a complex reaction needing coordination of many changes in metabolism [99]. The study of the transition from one type of photosynthesis to another is of great interest since it can provide insight into the different kinds of expression regulation of the enzymes involved in each type of photosynthesis. For example, the transition from C$_4$ to CAM photosynthesis may involve the expression of different isoenzymes or the occurrence of an opposite regulation of the same enzyme, like phosphoenolpyruvate carboxylase, which is opposite-regulated in both metabolisms [100,101]. In this way, investigations were performed in *Portulaca* species (*P. oleracea, P. grandiflora*, and *P. mundula*) under different stress conditions, mainly drought stress [20,21,91–95,102; D'Andrea et al., unpublished results].

Different metabolic modifications can be induced in *Portulaca* due to water stress: transition to a CAM-like metabolism (as in *P. oleracea*) [20] or induction of a CAM-cycling metabolism compartmentalized in a different cell type while the C$_4$ pathway is also operating (*P. grandiflora*) [21]. Inducible CAM may be a physiological feature in leaves of water-stressed *Portulaca* spp. that could account for their relatively long duration in habitats characterized by extended dry periods. In addition, comparative studies indicated that *P. mundula* exhibited a higher degree of acid fluctuations and higher water use efficiency in comparison with *P. oleracea* and *P. grandiflora*; also, it took more time to effect an appreciable decrease in the shoot water potential, and net CO$_2$ exchange rates were much less affected [93]. This shows, as this species tends to grow in the most arid environments, a correlation between photosynthetic mode and habitat in which the species is found, thus indicating that, indeed, the transition from C$_4$ metabolism to a type of CAM is an adaptation to drought stress. Moreover, due to the biochemical similarities between CAM and C$_4$ plants, both exploiting a common set of enzymes and facilitating CO$_2$ capture and concentration around the active site of RuBisCO, the transition between both pathways should not be surprising.

In this way, *Portulaca* spp. constitute attractive biological systems for the study of the molecular and biochemical modifications underlying the shift from C$_4$ photosynthesis to CAM and offering an alternative field of investigation on the different CO$_2$ fixation mechanisms [20]. The capacity for acid metabolism may be dependent on endogenous (species and developmental stage) as well as on environmental conditions such as photoperiod, temperature, and light intensity [94].

18.4.1 THE CASE OF *PORTULACA OLERACEA*

P. oleracea, commonly known as purslane, pigweed, fatweed, pulse, little hogweed, and verdolaga, is found in very diverse places such as gardens, cultivated fields, or agriculturally poor habitats like roadsides and out-rocks, but it is usually found in hot, dry-tendency, and high-light-intensities environments [103]. This herb also has been used in traditional medicine [104] and as an edible plant for human and livestock feed. In addition, during the last decade, it has received great attention

since it is a rich source of antioxidants like glutathione, α-tocopherol, β-carotene, and flavonoids, and it is one of the richest source of Ω-3 fatty acids [105–107].

On the other hand it is considered a weed, in part because of its process and pattern of growth [103]. *P. oleracea* uses a wide variety of photoperiods, and capsule numbers are correlated with amounts of light received. This weed is widely tolerant of high salt concentrations and different light intensities, temperature regimes, and soil types, and the plant produces adequate levels of capsules over a wide range of these factors [108]. In this way, the plasticity shown by *P. oleracea* with respect to its habitats would be just a feature of its flexibility with respect to its photosynthetic metabolism.

Early studies have classified *P. oleracea* as a C$_4$ plant, although the stage of leaf development was one of the most important factors determining the operation of C$_4$ photosynthesis. Young and mature leaves fixed ^{14}CO$_2$ primarily into organic and amino acids, and less than 2% of the ^{14}CO$_2$ fixed appeared in phosphorylated compounds. Young leaves produced more malate than aspartate, whereas mature leaves produced more aspartate than malate [109]. In addition, young and mature leaves exhibited typical C$_4$ plant light/dark ^{14}CO$_2$ evolution ratios [110]. In contrast, senescent leaves had a relatively large amount of C$_3$ photosynthesis, as accounted for by a quantitative shift of primary products toward phosphorylated compounds (18% in phosphoglyceric acid) with a concomitant reduction of the label residing in malate and aspartate [109]. Senescent leaves had photorespiration ratios similar to C$_3$ plants. In this state, *P. oleracea* leaves had a smaller absolute amount of ribulose-1,5-bisphosphate, PEPC (10% of mature leaves), and RuBisCO (27% of mature leaves) than mature leaves [110].

The C$_4$ dicot *P. oleracea* possesses succulent leaves with branched venation and composed of various types of cells (Figure 18.5). Bundle sheath cells are around the vascular bundles, and their chloroplasts are located in a centripetal

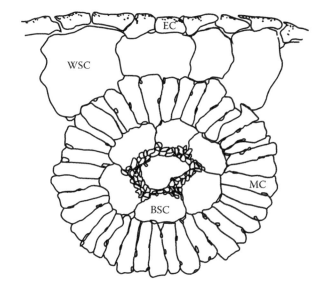

FIGURE 18.5 Diagram of various cell types in a cross section of *Portulaca oleracea*. BSC, bundle sheath cell; EC, epidermal cell; MC, mesophyll cell; WSC, water storage cell.

position (toward the vascular tissue). Mesophyll cells completely surround the bundle sheath cells. A third type of cell constitutes two or three layers between the epidermis and the parenchyma. These cells, called water storage cells, are particularly large, are vacuolated, and contain few small chloroplasts [20]. Total chlorophyll content is low, but as leaves of *P. oleracea* are relatively succulent, it is still within the reported range for several other C_4 plants [109].

Studies on a wide range of environments in which *P. oleracea* grows, the succulence of its leaves, its particularly high water use efficiency, and its location within a family that possesses species exhibiting C_3, C_4, or CAM metabolism led to investigation of the possible occurrence of CAM or facultative CAM in this species by analyzing diurnal acid fluctuation, CO_2 gas exchange, and leaf resistance under various photoperiods and water regimes [91]. Data strongly suggested that CAM occurred in *P. oleracea* leaves and stems. Under short photoperiods or water stress, *P. oleracea* presented a CAM-like pattern of acid fluctuation accompanied by low nocturnal leaf resistance, in the first case, and this feature plus net dark CO_2 uptake and daytime CO_2 release, in the second [91]. Later, Koch and Kennedy [92] studied the occurrence of CAM under natural environmental conditions but protected from the rain. This study confirmed the previous results obtained in drought-stressed plants as well as established that malate was the predominant compound labeled during the night, with some citrate and aspartate [92]. Under natural environment conditions, CAM contributed to the carbon balance and water retention in the C_4 dicot *P. oleracea*. Stomatal closure in the light reduced water loss from plants, with insoluble compounds being synthesized from the CO_2 assimilated during the night. Not only drought but also photoperiod, developmental state, and diurnal temperature changes appear to be important in CAM expression in *P. oleracea* [92]. Similar results have been reported by Kraybill and Martin [93]. In either case, the diurnal fluctuation of titratable acidity, accounted for by malic acid, suggests that a CAM-like metabolism is effectively induced in *P. oleracea* upon water-stress treatment.

Later, the induction of a CAM-like metabolism in *P. oleracea* was studied in plants under drought-stress conditions and a 12 h photoperiod [20]. After 23 days of withholding water, a diurnal change in titratable acidity, evidenced by a sevenfold to eightfold increase in the Δ titratable (day/night) acidity, was found. The stressed leaves presented an increase in malate levels of almost two orders of magnitude between the end of the day and night periods relative to well-watered plants, suggesting the induction of a CAM-like metabolism. This was also confirmed by limited or no CO_2 assimilation in the light and slow CO_2 rates of uptake in the dark of *P. oleracea* stressed plants.

Drought had an important effect on the leaf structure of *P. oleracea*, with the distortion of the mesophyll cells and the displacement of its chloroplasts toward the bundle sheath cells, probably due to water loss, accompanied by a general reduction in the chloroplast amount through the tissue. Partial collapse of the water storage cells and loss of organization of this tissue were also observed [20].

In C_4 and CAM plants, PEPC is regulated by a mechanism of phosphorylation/dephosphorylation [111–113]. In the phosphorylated state, feedback inhibition by L-malate is severely diminished [114]. The regulatory properties of the enzyme were also studied in *P. oleracea* under both conditions. The enzyme presented a subunit mass per subunit of 110 kDa and exhibited changes in the isoelectric point and electrophoretic mobility of the native enzyme [20]. *In vivo* phosphorylation and native isoelectrofocusing studies indicated that PEPC activity and regulation are modified upon drought-stress treatment in a way that allows *P. oleracea* to perform a CAM-like metabolism. Stressed leaves contained less (lower specific activity and lower enzyme content on a protein basis) of a non-cooperative form of PEPC, with different kinetic properties such as higher affinity for phosphoenolpyruvate and more sensitivity to malate inhibition. In summary, the regulation of the enzyme from *P. oleracea* control plants appears to be similar to that occurring in C_4 species, while that of the enzyme from drought-stressed plants resembles that in the CAM type [20].

As the function of a C_4 and CAM metabolism implies a different spatial and temporal participation of the enzymes involved in the CO_2 fixation, the study of the decarboxylating systems, RuBisCO and PPDK, in both tissue conditions was of great interest. This species is a NAD-ME-subtype C_4 plant, and this enzyme is upregulated by light in this photosynthetic mode [115]. However, as in the case of maize, phosphoenolpyruvate carboxykinase may contribute, along with NAD-ME, to the concentrating mechanism [116]. After 23 days of drought stress, a general decrease in the photosynthetic metabolism was found, as accounted for by the decrease in the net CO_2 fixation [20] and in the activity of enzymes related to that metabolism such as RuBisCO, PEPC, PPDK, phosphoenolpyruvate carboxykinase, and NAD-ME. In contrast, NADP-ME shows no activity or immunoreactive level variation. Western blot analysis indicated that changes in the activities were correlated with the levels of immunoreactive proteins, except for phosphoenolpyruvate carboxykinase activity, in which putative phosphorylation of the enzyme would be responsible for the changes in activity. PPDK showed two immunoreactive bands whose levels varied not only between day and night but also with the water status; a higher level of immunoreactive protein was found in control samples than in stressed plants, with the lower molecular mass band being more abundant and possibly being involved in phosphoenolpyruvate regeneration in control plants performing C_4 photosynthesis. In contrast, as the levels of the higher-molecular-mass isoform were higher in stressed leaves, this isoform could be related to the operation of CAM [20].

In situ immunolocalization studies were also conducted. RuBisCO is located in the bundle sheath chloroplasts of *P. oleracea* leaves, while NAD-ME is found in the mitochondria of the same cells. In contrast, PEPC was found in the cytosol of mesophyll and water storage cells. Although the localization of the enzymes was the same for control and stressed plants, the amount of immunogold particles varied in the same way as the amount of protein previously determined. The main change was observed for PEPC, which exhibited higher levels of immunocomplexes in the cytosol of water storage cells in stressed samples in comparison with control leaves [20].

Regarding carbohydrate metabolism-related enzymes, different patterns of assimilate partitioning are described among CAM species. As in *Mesembryanthemun crystallinum* [4], in *P. oleracea* there have been reports of changes in the day/night activities and in the level of immunoreactive protein of some of the enzymes involved in glycolysis and gluconeogenesis and some enzymes involved in the generation of reduction power when CAM is induced. While some enzymes (ATP-dependent phosphofructokinase, PP$_i$-dependent phosphofructokinase, NAD- and NADP-malate dehydrogenase, and nonphosphorylating glyceraldehyde-3-phosphate dehydrogenase) increased under stress conditions, others (such as aldolase and glucose-6-phosphate dehydrogenase) decreased. In this way, these changes could be of great importance in the establishment of CAM in *P. oleracea* [20].

In control *P. oleracea* plants under well-watered conditions, a C$_4$ photosynthetic metabolism operates with NAD-ME as major decarboxylating enzyme, accompanied by phosphoenolpyruvate carboxykinase (Figure 18.6). Additionally, as

PEPC was also found in the water storage cells, these cells could also contribute to the primary CO$_2$ fixation by producing oxaloacetate. This then could be transformed into malate by malate dehydrogenase and then transported to the bundle sheath cells through the mesophyll cell, thus increasing the malate pool given by the latter cells.

An inducible CAM may be a physiological feature in leaves of water-stressed *P. oleracea* that could account for the relatively long duration of this species in habitats characterized by extended dry periods [91]. Under these periods, a general decrease of CO$_2$ would occur. A spatial separation of the primary and secondary carbon fixation would be accompanied by a temporal separation of these processes. According to this, during the night, CO$_2$ would be fixed through PEPC in the cytosol of mesophyll cells yielding oxaloacetate (OAA), which then would be kept in the vacuoles in the form of malate. In the following light period, the released malate would be transported to the bundle sheath cells, where NAD-ME, together with other decarboxylating enzymes, would provide RuBisCO with CO$_2$. Probably, a cytosolic CAM-specific PPDK could participate in the phosphoenolpyruvate regeneration. Although the role of water

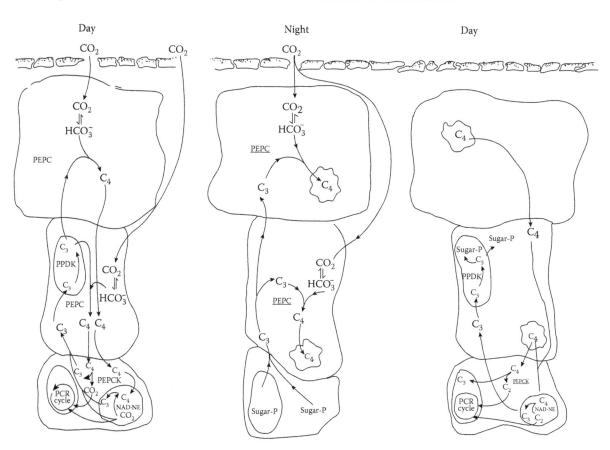

FIGURE 18.6 Proposed photosynthetic metabolisms operating in *Portulaca oleracea* in well-irrigated (C$_4$ metabolism) and in water-stressed plants (crassulacean acid-like metabolism). From top to bottom, water storage and mesophyll and bundle sheath cells are shown. The intracellular localization of the main C$_4$ photosynthetic enzymes is shown. NAD-ME, NAD-malic enzyme; PCR cycle, photosynthetic carbon reduction cycle; PEPC, phosphoenolpyruvate carboxylase; PEPCK, phosphoenolpyruvate carboxykinase; PPDK, pyruvate orthophosphate dikinase.

storage cells is still uncertain, enhanced activity of these cells could take place under this condition through the generation and storage of malate. Alternatively, these cells act as water reservoirs for mesophyll cells and bundle sheath cells for conditions of limited water supply, and the presence of PEPC in water storage cells reveals a possible remaining ancestral CAM metabolism.

Other physiological and biochemical approaches also shown that in CAM-like plants, chlorophyll fluorescence parameters were transitorily affected and nonradiative energy dissipation mechanisms were induced to protect the photosynthetic apparatus. Moreover, an increase in pigments (flavonoids and betalains) as well in the reactive species-scavenging enzymes ascorbate peroxidase and peroxidase would also prevent damages to photosynthetic machinery [102]. Metabolic analysis of polar compounds highlighted a clear metabolic shift, when a CAM-like metabolism is induced and also when there is a reversion to C_4 metabolism after rewatering. Rises in N-containing metabolites, such as free amino acids and urea, and pinitol would also help to cope with water deficit. Arginase and urease activity assays also demonstrated that both enzymes work together to modulate urea synthesis [102]. Recovery of C_4 metabolism was accounted for by CO_2 assimilation pattern and malate levels. In these plants, greater levels of glycerol and polyamines were also detected. Together, the studies conducted show that *P. oleracea* deploys multiple strategies, from induction of several metabolites to the transitory development of a CAM-like metabolism, to cope with drought [102].

18.4.2   The Case of *Portulaca grandiflora*

P. grandiflora Hook is an ornamental dicot that grows best in sandy soil originally from northern South America [117]. It has been reported as C_4 plant based on measurement of $\delta^{13}C$ [118] and biochemical studies [97]. The occurrence of *P. grandiflora* in an open strand community with xeric and high-light conditions, its succulent cylindrical leaves having a large volume of water tissue and a low ratio of surface area to volume (which aids water retention), and its phylogenetic position (it belongs to Portulacaceae) motivated the investigation of the possible occurrence of CAM in this species, especially under drought-stress conditions.

Ku et al. measured [94] greater diurnal acid fluctuations and malic acid concentration in stems and leaves of well-watered plants of *P. grandiflora* than in drought-stressed individuals. These parameters were much reduced under severe drought conditions. In contrast, Kraybill and Martin [93] indicated that no significant diurnal malic acid fluctuations occurred under well-watered conditions, but under drought stress, significant diurnal malic acid fluctuations were measured. They also found that no net CO_2 uptake took place under drought-stressed conditions and suggested that this species underwent CAM cycling. Subsequent research from Guralnick and Jackson [119] confirmed that *P. grandiflora* maintained high organic acid levels and large diurnal acid fluctuation when watered stressed.

Later, diurnal fluctuation in acidity in leaves, which was not accompanied by a net gain or loss of CO_2 at night and a decrease in net CO_2 fixation during the day, showed the upregulation of the CAM pathway after water stress [21]. After 8 days of water stress, the leaves presented CAM-cycling activity, which suggested that this feature may occur completely in water storage tissue. Based on diurnal fluctuations in acidity in stems and the absence of net carbon gain during the day or night, as the stems lack stomata, it was also proposed that CAM-idling photosynthesis could take place in stems, which may have an important role in recycling carbon and conserving water to support photosynthetic activity in leaves during water stress [21].

P. grandiflora has a slightly different Kranz leaf anatomy (Figure 18.7). The succulent, cylindrical leaves of this dicot species possess three distinct green cell types: bundle sheath cells in radial arrangement around the vascular bundles, mesophyll cells in an outer layer adjacent to the bundle sheath cells, and water storage cells in the leaf center. This last type of cell constitutes the water tissue and contains scattered chloroplasts and a large vacuole. Unlike typical Kranz leaf anatomy, the mesophyll cells do not surround the bundle sheath tissue but occur only in the area between the bundle sheath and the epidermis [93,94]. While bundle sheath cells around the peripheral vascular bundles possess agranal chloroplasts that are centripetally oriented, as is typical of NADP-ME plants [120,121], mesophyll cells have granal chloroplasts [121]. Nevertheless, granal development is suppressed in the chloroplasts of the innermost water storage cells adjacent to the central vascular bundle. Chloroplasts in the water storage cells just adjacent to the central vascular bundle show a tendency to be located near the vascular bundles [121].

When leaf anatomy is examined in plants under drought stress, collapse of the water storage cells is observed, with C_4 tissue still hydrated and functioning. Thus, under limited water

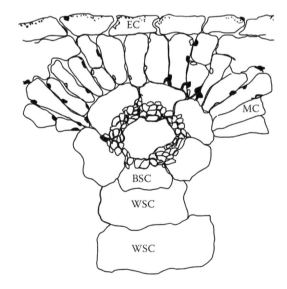

FIGURE 18.7   Diagram of various cell types in a cross section of *Portulaca grandiflora*. BSC, bundle sheath cell; EC, epidermal cell; MC, mesophyll cell; WSC, water storage cell.

supply, the water storage tissue transfers water to the mesophyll and bundle sheath cells [21]. Similar to *P. oleracea* mesophyll cells, the chloroplasts of cortical cells of *P. grandiflora* became clustered in comparison to well-watered leaves [21].

Immunogold labeling revealed that RuBisCO was accumulated in bundle sheath chloroplasts and in the chloroplasts of water storage cells adjacent to the central vascular bundle [121]. Labeling was markedly reduced in mesophyll chloroplasts and in the chloroplasts of the remaining water storage cells. A gradient in labeling of RuBisCO and granal development is not observed in the remaining water storage cells. Thus, accumulation of RuBisCO and the suppression of granal development are restricted to the chloroplasts in the cells adjacent to the vascular bundles [121]. *In situ* immunolocalization in well-watered leaves showed that PEPC was located in mesophyll and water storage cells, with lower density in the latter. In stems, RuBisCO and PEPC were located in cortical cells. Tissue printing was performed and revealed an increase in PEPC in the water storage tissue of leaves and the cortex of stems [21].

Western blot analysis showed that stressed leaves had a slight decrease in the C$_4$-CAM pathway proteins (NADP-ME and PPDK), while a new isoform of NADP-ME appeared, and an increase of PEPC was observed after 10 days of water-stress treatment. RuBisCO content remained constant during the water-stress period in leaves as well as in stems. Stems exhibited increases in the levels of PEPC, NADP-ME, and PPDK when subject to drought stress [21].

Regarding the biochemical pathways operating *P. grandiflora* leaves, they have shown CAM-cycling activity, while the stems exhibit CAM-idling metabolism, both upregulating the CAM pathways under water stress. Guralnick et al. [21] proposed that both C$_4$ and CAM metabolisms operate simultaneously in stressed plants as the diurnal course of acid fluctuation commences in leaves as the light period begins, as does CO$_2$ uptake. Nevertheless, the photosynthetic role of the innermost water storage cells is questionable because of their distance from mesophyll cells and sparse occurrence of chloroplasts in these cells [121]. However, some kind of activity may take place in the outermost water storage cells, as accounted for by the presence of RuBisCO in these cells and the increase of PEPC protein in the same cells after drought-stress treatment [21].

18.5 CONCLUDING REMARKS AND PERSPECTIVES FOR BIOTECHNOLOGICAL APPROACHES TO IMPROVE C$_3$ PHOTOSYNTHESIS

Photosynthesis constitutes the basis on which all ecosystems on earth function. Most plants are C$_3$ plants, in which the first product of photosynthesis is a three-carbon compound. C$_3$ plants include many important crops such as wheat, rice, and soybean. C$_4$ plants, in which the first product is a four-carbon compound, have evolved independently, with at least 66 independent origins, and are found in at least 19 families [122]. These distinctive properties enable the capture of CO$_2$ and its

concentration in the vicinity of RuBisCO. In addition, photosynthetic organs of C$_4$ plants show alteration in their anatomy and ultrastructure [3], and thus, C$_4$ photosynthesis in terrestrial plants was thought to require Kranz anatomy because the cell wall between mesophyll and bundle sheath cells restricts leakage of CO$_2$. The discovery in chenopods that shows that C$_4$ photosynthesis functions efficiently in individual cells containing both the C$_4$ and the C$_3$ cycles [18], together with the occurrence of a single-celled C$_4$-type CO$_2$ concentrating mechanism in the aquatic plants, such as *E. densa* and *H. verticillata* [15,16], reveals that even though the vast majority of C$_4$ plants exhibit Kranz anatomy, it would not be essential for the C$_4$ photosynthesis. These discoveries provide new inspiration for efforts to convert C$_3$ crops into C$_4$ plants because the anatomical changes required for C$_4$ photosynthesis might be less stringent than previously thought [123]. In this respect, many attempts to enhance photosynthesis in C$_3$ species have been conducted [33,124–126].

In summary, since Kortschak and coworkers [127] performed ^{14}CO$_2$ pulse-chase experiments with sugarcane and found a different pattern than expected, giving the basis for the discovery of the C$_4$ metabolism, much progress has been made in the study of CO$_2$ concentration mechanisms. C$_4$ photosynthesis is more plastic than previously thought; even C$_4$ plants with Kranz anatomy, such as *P. oleracea* and *P. grandiflora*, can shift to a kind of CAM when subject to drought-stress conditions. Nevertheless, as science keeps in motion, new variations in the photosynthetic modes are found, and not everything is as previously stated or thought. For the lucky ones who are studying the photosynthetic mechanisms, there is still a long way to go to the complete characterization of photosynthesis, and a lot of work awaits.

REFERENCES

1. Edwards GE, Walker DA. *C$_3$, C$_4$: mechanisms, and cellular and environmental regulation of photosynthesis.* Oxford: Blackwell Scientific, 1983.
2. Ehleringer JR, Monson RK. Evolutionary and ecological aspects of photosynthetic pathway variation. *Annu Rev Ecol Syst* 1993; 24: 411–439.
3. Hatch MD. C$_4$ photosynthesis: A unique blend of modified biochemistry, anatomy and ultrastructure. *Biochim Biophys Acta* 1987; 895: 81–106.
4. Holtum JAM, Winter K. Activity of enzymes of carbon metabolism in *Mesembryanthemun crystallinum* by salt stress. *Plant Cell Physiol* 1982; 37: 257–262.
5. Raghavendra AS, Sage RF. C$_4$ photosynthesis and related CO$_2$ concentrating mechanisms introduction. In: Raghavendra AS, Sage RF, eds. *C$_4$ photosynthesis and related CO$_2$ concentrating mechanisms.* Advances in photosynthesis and respiration, vol. 32. Dordrecht: Springer, 2011: 17–25.
6. Cushman JC, Bohnert HJ. Crassulacean acid metabolism: Molecular genetics. *Annu Rev Plant Physiol Plant Mol Biol* 1999; 50: 305–332.
7. Kluge M, Ting IP. *Crassulacean acid metabolism. Analysis of an ecological adaptation.* New York: Springer, 1978.
8. Raven JA. CO$_2$-concentrating mechanisms: A direct role for thylakoid lumen acidification? *Plant Cell Environ* 1997; 2: 147–154.

9. Moroney JV, Somanchi A. How do algae concentrate CO$_2$ to increase the efficiency of photosynthetic carbon fixation? *Plant Physiol* 1999; 119: 9–16.

10. Badger MR, Spalding MH. CO$_2$ acquisition, concentration and fixation in cyanobacteria and algae. In: Leegood RC, Sharkey TD, von Caemmerer S, eds. *Photosynthesis: Physiology and metabolism*. Dordrecht: Kluwer Academic Publishers, 2000: 369–397.

11. Giordano M, Beardall J, Raven JA. CO$_2$ concentrating mechanisms in algae: Mechanisms, environmental modulation, and evolution. *Annu Rev Plant Biol* 2005; 56: 99–131.

12. Price GD, Badger MR, Woodger FJ, Long BM. Advances in understanding the cyanobacterial CO$_2$-concentrating-mechanism (CCM): Functional components, Ci transporters, diversity, genetic regulation and prospects for engineering into plants. *J Exp Bot* 2008; 59: 1441–1461.

13. Badger MR, Price GD. CO$_2$ concentrating mechanisms in cyanobacteria: Molecular components, their diversity and evolution. *J Exp Bot* 2003; 383: 609–622.

14. Raven JA, Giordano M, Beardall J, Maberly SC. Algal evolution in relation to atmospheric CO$_2$: Carboxylases, carbon-concentrating mechanisms and carbon oxidation cycles. *Phil Trans R Soc B* 2012; 367: 493–507.

15. Salvucci ME, Bowes G. Two photosynthetic mechanisms mediating the low photorespiratory state in submersed aquatic angiosperms. *Plant Physiol* 1983; 73: 488–496.

16. Casati P, Lara MV, Andreo CS. Induction of a C$_4$-like mechanism of CO$_2$ fixation in *Egeria densa*, a submersed aquatic species. *Plant Physiol* 2000; 123: 1611–1622.

17. Zhang Y, Yin L, Jiang H-S, Li W, Gontero B, Maberly SC. Biochemical and biophysical CO$_2$ concentrating mechanisms in two species of freshwater macrophyte within the genus Ottelia (Hydrocharitaceae). *Photosynth Res* 2014; 121: 285–297.

18. Voznesenskaya EV, Franceschi VR, Kiirats O, Freitag H, Edwards GE. Kranz anatomy is not essential for terrestrial C$_4$ plant photosynthesis. *Nature* 2001; 414: 543–546.

19. Voznesenskaya EV, Franceschi VR, Kiirats O, Artyusheva EG, Freitag H, Edwards GE. Proof of C$_4$ photosynthesis without Kranz anatomy in *Bienertia cycloptera* (Chenopodiaceae). *Plant J* 2002; 31: 649–662.

20. Lara MV, Disante KB, Podestá FE, Andreo CS, Drincovich MF. Induction of a crassulacean acid like metabolism in the C$_4$ succulent plant, *Portulaca oleracea* L.: Physiological and morphological changes are accompanied by specific modifications in phosphoenolpyruvate carboxylase. *Photosynth Res* 2003; 77: 241–254.

21. Guralnick LJ, Edwards GE, Ku MSB, Hockema B, Franceschi VR. Photosynthetic and anatomical characteristics in the C$_4$-crassulacean acid metabolism-cycling plant, *Portulaca grandiflora*. *Funct Plant Biol* 2002; 29: 763–773.

22. Madsen JS, Sand-Jensen K. Photosynthetic carbon assimilation in aquatic macrophytes. *Aquat Bot* 1991; 41: 5–40.

23. Bowes G, Salvucci ME. Plasticity in the photosynthetic carbon metabolism of submersed aquatic macrophytes. *Aquat Bot* 1989; 34: 233–266.

24. Raven JA. Exogenous inorganic carbon sources in plant photosynthesis. *Biol Rev* 1970; 45: 167–221.

25. Van TK, Haller WT, Bowes G. Comparison of the photosynthetic characteristics of three submersed aquatic plant. *Plant Physiol* 1976; 58: 761–768.

26. Maberly SC. Photosynthesis by *Fontinalis antipyretica*: II. Assessment of environmental factors limiting photosynthesis and production. *New Phytol* 1985; 100: 141–155.

27. Salvucci ME, Bowes G. Induction of reduced photorespiratory activity in submerse and amphibious aquatic macrophytes. *Plant Physiol* 1981; 67: 335–340.

28. Salvucci ME, Bowes G. Ethoxyzolamide repression of the low photorespiration state in two submersed angiosperms. *Planta* 1983; 158: 27–34.

29. Holaday AS, Salvucci ME, Bowes G. Variable photosynthesis/photorespiration ratios in *Hydrilla* and other submersed aquatic macrophyte species. *Can J Bot* 1983; 61: 229–236.

30. Reiskind JB, Madsen TV, van Ginkel LC, Bowes G. Evidence that inducible C$_4$-type photosynthesis is a chloroplastic CO$_2$-concentrating mechanism in *Hydrilla*, a submersed monocot. *Plant Cell Environ* 1997; 20: 211–220.

31. Bowes G. Single cell C$_4$ photosynthesis in aquatic plants. In: Raghavendra AS, Sage RF, eds. *C$_4$ photosynthesis and related CO$_2$ concentration mechanisms*. Dordrecht: Springer, 2011: 80.

32. Bowes G, Rao SK, Reiskind JB. Photosynthetic acclimation of rice to global climate changes: Will a same-cell C$_4$ system help? In: Mew TW, Brar DS, Peng S, Dawe D, Hardly B, eds. *Rice science: Innovations and impact for livelihood*. Philippines: IRRI, 2003: 659–672.

33. von Caemmerer S, Edwards GE, Koteyeva N, Cousins AB. Single cell C$_4$ photosynthesis in aquatic and terrestrial plants: A gas exchange perspective. *Aquat Bot* 2014; 118: 71–80.

34. Elzenga JTM, Prins HBA. Light-induced polar pH changes in leaves of *Elodea canadensis*. *Plant Physiol* 1989; 91: 62–67.

35. Prins HBA, Elzenga JTM. Bicarbonate utilization: Function and mechanism. *Aquatic Bot* 1989; 34: 59–83.

36. Rascio N, Mariani P, Tommasini E, Bodner M, Larcher W. Photosynthetic strategies in leaves and stems of *Egeria densa*. *Planta* 1991; 185: 297–303.

37. Buschmann P, Sack H, Köhler AE, Dahse I. Modeling plasmalemma ion transport of the aquatic plant *Egeria densa*. *J Membrane Biol* 1996; 154: 109–118.

38. Watson L, Dallwitz MJ. The families of flowering plants: Descriptions, illustrations, identification, and information retrieval. Version: May 11, 2015. Available at http://delta-int key.com/angio/.

39. Miedema H, Prins HBA. pH-dependent proton permeability of the plasma membrane is a regulating mechanism of polar transport through the submerged leaves of *Potamogeton lucens*. *Can J Bot* 1991; 69: 1116–1122.

40. Miedema H, Staal M, Prins HBA. pH-induced proton permeability changes of plasma vesicles. *J Membr Biol* 1996; 152: 159–167.

41. van Ginkel LC, Prins HBA. Bicarbonate utilization and pH polarity. The response of photosynthetic electron transport to carbon limitation in *Potamogeton lucens* leaves. *Can J Bot* 1998; 76: 1018–1024.

42. Staal M, Elzenga JTM, Prins HBA. ^{14}C fixation by leaves and leaf cell protoplasts of the submerged aquatic angiosperm *Potamogeton lucens* L.: Carbon dioxide or bicarbonate? *Plant Physiol* 1989; 90: 1035–1040.

43. Casati P, Lara MV, Andreo CS. Regulation of enzymes involved in C$_4$ photosynthesis and the antioxidant metabolism by UV-B radiation in *Egeria densa*, a submersed aquatic species. *Photosynth Res* 2000; 71: 251–264.

44. Lara MV, Casati P, Andreo CS. CO$_2$ concentration mechanisms in *Egeria densa*, a submersed aquatic species. *Physiol Plant* 2002; 115: 487–495.

45. Brown JMA, Dromgoole FI, Towsey MW, Browse J. Photosynthesis and photorespiration in aquatic macrophytes. In: Bieleski RL, Ferguson AR, Cresswell MM, eds. *Mechanism of regulation of plant growth*. Wellington: The Royal Society of New Zealand, 1974: 243–249.

46. Browse JA, Dromgoole FI, Brown JMA. Photosynthesis in the aquatic macrophyte *Egeria densa*. I. $^{14}CO_2$ fixation at natural CO_2 concentrations. *Aust J Plant Physiol* 1977; 4: 169–176.

47. DeGroote D, Kennedy RA. Photosynthesis in *Elodea canadensis* Michx. Four carbon acid synthesis. *Plant Physiol* 1977; 59: 1133–1135.

48. Browse JA, Dromgoole FI, Brown JMA. Photosynthesis in the aquatic macrophyte *Egeria densa*. II. The effects of inorganic carbon conditions on ^{14}C-fixation. *Aust J Plant Physiol* 1979; 6: 1–9.

49. Browse JA, Dromgoole FI, Brown JMA. Photosynthesis in the aquatic macrophyte *Egeria densa*. III. Gas exchange studies. *Aust J Plant Physiol* 1979; 6: 1133–1135.

50. Browse JA, Brown JMA, Dromgoole FI. Malate synthesis and metabolism during photosynthesis in *Egeria densa* Planch. *Aquat Bot* 1980; 8: 295–305.

51. Holaday AS, Bowes G. C$_4$ metabolism and dark CO_2 fixation in a submersed aquatic macrophyte (*Hydrilla verticillata*). *Plant Physiol* 1980; 89: 1231–1237.

52. Bauwe H. An efficient method for the determination of K_m values for HCO_3^- of phospho*enol*pyruvate carboxylase. *Planta* 1986; 169: 356–360.

53. Lara MV, Casati P, Andreo CS. *In vivo* phosphorylation of phospho*enol*pyruvate carboxylase in *Egeria densa*, a submersed aquatic species. *Plant Cell Physiol* 2001; 42: 141–145.

54. Drincovich MF, Casati P, Andreo CS. NADP-malic enzyme from plants: A ubiquitous enzyme involved in different metabolic pathways. *FEBS Lett* 2001; 460: 1–6.

55. Bowes G, Rao SK, Estavillo GM, Reiskind JB. C$_4$ mechanisms in aquatic angiosperms: Comparisons with terrestrial C$_4$ systems. *Funct Plant Biol* 2002; 29: 379–392.

56. Reiskind JB, Berg RH, Salvucci ME, Bowes G. Immunogold localization of primary carboxylases in leaves of aquatic and a C$_3$–C$_4$ intermediate species. *Plant Sci* 1989; 61: 43–52.

57. Magnin NC, Cooley BA, Reiskind JB, Bowes G. Regulation and localization of key enzymes during the induction of Kranz-less, C$_4$ type photosynthesis in *Hydrilla verticillata*. *Plant Physiol* 1997; 115: 1681–1689.

58. Ascencio J, Bowes G. Phosphoenolpyruvate carboxylase in *Hydrilla* plants with varying CO_2 compensation points. *Photosynth Res* 1983; 4: 151–170.

59. Spencer WE, Wetzel RG, Teeri J. Photosynthetic phenotype plasticity and the role of phosphoenolpyruvate carboxylase in *Hydrilla verticillata*. *Plant Sci* 1996; 118: 1–9.

60. Rao SK, Magnin NC, Reiskind JB, Bowes J. Photosynthetic and other phosphoenolpyruvate carboxylase isoforms in the single-cell, facultative C$_4$ system of *Hydrilla verticillata*. *Plant Physiol* 2002; 130: 876–886.

61. Blässing OE, Weshoff P, Svensson P. Evolution of C$_4$ phosphoenolpyruvate carboxylase in *Flaveria*, a conserved serine residue in the carboxyl-terminal part of the enzyme is a major determinant of C$_4$-specific characteristics. *J Biol Chem* 2000; 275: 27917–27923.

62. Estavillo GE, Rao SK, Reiskind JB, Bowes G. Molecular studies of an inducible C$_4$-type photosynthetic system: NADP-ME isoforms. *Plant Mol Biol Rep* 2000; 18: S21–S23.

63. Estavillo GM, Rao SK, Reiskind JB, Bowes G. Characterization of the NADP malic enzyme gene family in the facultative, single-cell C$_4$ monocot *Hydrilla verticillata*. *Photosynth Res* 2007; 94: 43–57.

64. Prins HBA, Snel JFH, Zanastra PE, Helder RJ. The mechanism of bicarbonate assimilation by the polar leaves of *Potamogeton* and *Elodea*. CO_2 concentrations at the leaf surface. *Plant Cell Environ* 1982; 5: 207–214.

65. Spenser W, Teeri J, Wetzel RG. Acclimation of photosynthetic phenotype to environmental heterogeneity. *Ecology* 1994; 75: 301–314.

66. Akhani H, Trimborn P, Zeigler H. Photosynthetic pathways in Chenopodeaceae from Africa, Asia and Europe with their ecological, phytogeographical and taxonomical importance. *Plant Syst Evol* 1997; 206: 187–221.

67. Carolin RC, Jacobs SWL, Vesk M. Leaf structure in Chenopodiaceae. *Bot Jahrb Syst Pflanzegesh Planzengeogr* 1975; 95: 226–255.

68. Glagoleva TA, Chulanovskaya MV, Pakhomova MV, Voznesenskaya EV, Gamaley YV. Effect of salinity on the structure of the assimilatory organs and C labelling patterns in C$_3$ and C$_4$ plants of Ararat plain. *Photosynthetica* 1992; 26: 363–369.

69. Pyankov VI, Voznesenskaya EV, Kuz'min AN, Ku MSB, Ganko E, Franceschi VR, Black CC Jr, Edwards GE. Occurrence of C$_3$ and C$_4$ photosynthesis in cotyledons and leaves of *Salsola* species (Chenopodiaceae). *Photosynth Res* 2000; 63: 69–84.

70. Winter K. C$_4$ plants of high biomass in arid regions of Asia: Occurrence of C$_4$ photosynthesis in Chenopodiaceae and Polygonaceae from the Middle East and USSR. *Oecologia* 1981; 48: 100–106.

71. Sage RF, Li MR, Monson RK. The taxonomic distribution of C$_4$ photosynthesis. In: Sage RF, Monson RK, eds. *C$_4$ plant biology*. New York: Academic Press, 1999: 551–584.

72. Voznesenskaya EV, Franceschi VR, Pyankov VI, Edwards GE. Anatomy, chloroplast structure and compartmentation of enzymes relative to photosynthetic mechanisms in leaves and cotyledons of species in the tribe Salsolaeae (Chenopodiaceae). *J Exp Bot* 1999; 50: 1779–1795.

73. Edwards GE, Franceschi VR, Voznesenskaya EV. Single-cell C$_4$ photosynthesis versus the dual-cell (Kranz) paradigm. *Annu Rev Plant Biol* 2004; 55: 173–196.

74. Edwards GE, Voznesenskaya EV. C$_4$ photosynthesis: Kranz forms and single-cell C$_4$ in terrestrial plants. In: Raghavendra AS, Sage RF, eds. *C$_4$ photosynthesis and related CO_2 concentrating mechanisms*. Dordrecht: Springer, 2011: 29–61.

75. Sharpe R, Offermann S. One decade after the discovery of single-cell C$_4$ species in terrestrial plants: What did we learn about the minimal requirements of C$_4$ photosynthesis? *Photosynth Res* 2014; 119: 169–180.

76. Sage RF. Environmental and evolutionary preconditions for the origin and diversification of the C$_4$ photosynthesis syndrome. *Plant Biol* 2001; 3: 202–213.

77. Sage RF. C$_4$ photosynthesis in terrestrial plants does not require Kranz anatomy. *Trends Plant Sci* 2002; 7: 283–285.

78. Akhani H, Ghobadnejhad M, Hashemi SMH. Ecology, biogeography and pollen morphology of *Bienertia cycloptera* Bung ex Boiss. (Chenopodiaceae), an enigmatic C$_4$ plant without Kranz anatomy. *Plant Biol* 2003; 5: 167–178.

79. Akhani H, Barroca J, Koteeva N, Voznesenskaya E, Franceschi V, Edwards G, Ghaffari SM, Ziegler H. *Bienertia sinuspersici* (Chenopodiaceae): A new species from Southwest Asia and discovery of a third terrestrial C$_4$ plant without Kranz anatomy. *System Bot* 2005; 30: 290–301.

80. Akhani H, Chatrenoor T, Dehghani M, Khoshravesh R, Mahdavi P, Matinzadeh Z. A new species of *Bienertia* (Chenopodiaceae) from Iranian salt deserts: A third species of the genus and discovery of a fourth terrestrial C$_4$ plant without Kranz anatomy. *Plant Biosyst* 2012; 146: 550–559.

81. Freitag H, Stichler W. *Bienertia cycloptera* Bunge ex Boiss., Chenopodiaceae, another C$_4$ plant without Kranz tissues. *Plant Biol* 2002; 4: 121–132.

82. Voznesenskaya EV, Koteyeva NK, Chuong SDX, Akhani H, Edwards GE, Franceschi VR. Differentiation of cellular and biochemical features of the single-cell C_4 syndrome during leaf development in *Bienertia cycloptera* (Chenopodiaceae). *Am J Bot* 2005; 92: 1784–1795.

83. Chuong SDX, Franceschi VR, Edwards GE. The cytoskeleton maintains organelle partitioning required for single-cell C_4-photosynthesis in Chenopodiaceae species. *Plant Cell* 2006; 18: 2207–2223.

84. Lara MV, Choung SDX, Akhani H, Andreo CS, Edwards GE. Species having C_4 single cell type photosynthesis in family Chenopodiaceae evolved a photosynthetic phosphoenolpyruvate carboxylase like that of C_4 Kranz type species. *Plant Physiol* 2006; 142: 673–684.

85. Offermann S, Okita TW, Edwards GE. Resolving the compartmentation and function of C_4 photosynthesis in single-cell C_4 species *Bienertia sinuspersici*. *Plant Physiol* 2011; 155: 1612–1628.

86. Edwards GE, Voznesenskaya E, Smith M, Koteyeva N, Park YI, Park JH, Kiirats O, Okita TW, Chuong SDX. Breaking the Kranz paradigm in terrestrial C_4 plants: Does it hold promise for C_4 rice. In: Sheehy JE, Mitchell PL, Hardy B, eds. *Charting new pathways to C_4 rice*. Philippines: IRRI, 2007: 249–274.

87. King JL, Edwards GE, Cousins AB. The efficiency of the CO_2-concentrating mechanisms during single cell C_4 photosynthesis. *Plant Cell Environ* 2012; 35: 513–523.

88. Freitag H, Stichler W. A remarkable new leaf type with unusual photosynthetic tissue in a central asiatic genus of Chenopodiaceae. *Plant Biol* 2000; 2: 154–160.

89. Osmond CB. Crassulacean acid metabolism. A curiosity in context. *Annu Rev Plant Physiol* 1978; 29: 379–414.

90. Winter K, Smith JAC. An introduction to crassulacean acid metabolism. biochemical principles and ecological diversity. In: Winter K, Smith JAC, eds. *Crassulacean acid metabolism: Biochemistry, ecology and evolution*. Ecological studies, vol. 114. New York: Springer-Verlag, 1996: 1–10.

91. Koch K, Kennedy RA. Characteristics of crassulacean acid metabolism in the succulent C_4 Dicot, *Portulaca oleracea* L. *Plant Physiol* 1980; 65: 193–197.

92. Koch KE, Kennedy RA. Crassulacean acid metabolism in the succulent C_4 dicot, *Portulaca oleracea* L under natural environmental conditions. *Plant Physiol* 1982; 69: 757–761.

93. Kraybill AA, Martin CE. Crassulacean acid metabolism in three species of the C_4 genus *Portulaca*. *Int J Plant Sci* 1996; 157: 103–109.

94. Ku S, Shie Y, Reger B, Black CC. Photosynthetic characteristics of *Portulaca grandiflora*, a succulent C_4 dicot. *Plant Physiol* 1981; 68: 1073–1080.

95. Mazen AMA. Changes in levels of phosphoenolpyruvate carboxylase with induction of crassulacean acid metabolism (CAM)-like behavior in the C_4 plant *Portulaca oleracea*. *Physiol Plant* 1996; 98: 111–116.

96. Gurlanick LJ, Jackson MD. Crassulacean acid metabolism activity in the family Portulacaceae. *Plant Physiol* 1993; 102 (suppl): 139.

97. Gutierrez M, Gracen VE, Edwards GE. Biochemical and cytological relationships in C_4 plants. *Planta* 1974; 119: 279–300.

98. Guralnick LJ, Ting IP. Seasonal patterns of water relations and enzyme activity in the facultative CAM plant *Portulacalaria afra* (L.) Jacq. *Plant Cell Environ* 1988; 811–818.

99. Cushman J, Bohnert H. Molecular genetics of crassulacean acid metabolism. *Plant Physiol* 1997; 113: 667–676.

100. Chollet R, Vidal J, O'Leary MH. Phosphoenolpyruvate carboxylase: A ubiquitous, highly regulated enzyme in plants. *Annu Rev Plant Physiol Plant Mol Biol* 1996; 47: 273–298.

101. Nimmo HG. The regulation of phosphoenolpyruvate carboxylase in CAM plants. *Trends Plant Sci* 2000; 2: 75–80.

102. D'Andrea RM, Andreo CS, Lara MV. Deciphering the mechanisms involved in *Portulaca oleracea* (C_4) response to drought: Metabolic changes including crassulacean acid-like metabolism induction and reversal upon re-watering. *Physiol Plant* 2014; 152: 414–430.

103. Zimmerman CA. Growth characteristics of weediness in *Portulaca oleracea* L. *Ecology* 1976; 57: 964–974.

104. Simopoulos AP, Norman HA, Gillapsy JE, Duke JA. Common purslane: A source of omega-3 fatty acids and antioxidants. *J Am Coll Nutr* 1992; 11: 374–382.

105. Xiang L, Xing D, Wang W, Wang R, Ding Y, Du L. Alkaloids from *Portulaca oleracea* L. *Phytochemistry* 2005; 66: 2595–2601.

106. Liu L, Howe P, Zhou Y-F, Xu Z-Q, Hocart C, Zhang R. Fatty acids and β-carotene in Australian purslane (*Portulaca oleracea*) varieties. *J Chromatography A* 2000; 893: 207–213.

107. Simopoulos AP. Omega-3 fatty acids and antioxidants in edible wild plants. *Biol Res* 2004; 37: 263–277.

108. Yang Y, Chen J, Liu Q, Ben C, Todd CD, Shi J, Yang Y, Hu X. Comparative proteomic analysis of the thermotolerant plant *Portulaca oleracea* acclimation to combined high temperature and humidity stress. *J Proteome Res* 2012; 11: 3605–3623.

109. Kennedy RA, Laetsch WM. Relationship between leaf development and primary photosynthetic products in the C_4 plant *Portulaca oleracea*. *Planta* 1973; 115: 113–124.

110. Kennedy RA. Relationship between leaf development, carboxylase enzyme activities and photorespiration in the C_4 plant *Portulaca oleracea* L. *Planta* 1976; 128: 149–154.

111. Vidal J, Chollet R. Regulatory phosphorylation of phosphoenolpyruvate carboxylase. *Trends Plant Sci* 1997; 2: 230–237.

112. Izui K, Matsumura H, Furumoto F, Kai Y. Phosphoenolpyruvate carboxylase: A new era of structural biology. *Annu Rev Plant Biol* 2004; 55: 69–84.

113. Lepiniec L, Thomas M, Vidal J. From enzyme activity to plant biotechnology: 30 years of research on phosphoenolpyruvate carboxylase. *Plant Physiol Biochem* 2003; 41: 533–539.

114. Andreo CS, González DH, Iglesias A. Higher plant phosphoenolpyruvate carboxylase: Structure and regulation. *FEBS Lett* 1987; 213: 1–8.

115. Lara MV, Drincovich MF, Andreo CS. Induction of a crassulacean acid like metabolism in the C_4 succulent plant, *Portulaca oleracea* L.: Study of enzymes involved in carbon fixation and carbohydrate metabolism. *Plant Cell Physiol* 2004; 45: 618–626.

116. Walker RP, Acheson RM, Técsi LI, Leegood RC. Phosphoenolpyruvate carboxykinase in C_4 plants: Its role and regulation. *Aust J Plant Physiol* 1997; 24: 459–468.

117. Bailey LH, Bailey EZ. *Hortus third: A concise dictionary of plants cultivated in the United States and Canada*. New York: Macmillan, 1976.

118. Troughthon JH, Card KA, Hendy CH. Photosynthetic pathways and carbon isotope discrimination by plants. *Carnegie Inst Wash Year Book* 1974; 73: 768–780.

119. Guralnick LJ, Jackson MD. The occurrence and phylogenetics of crassulacean acid metabolism in the Potulacaceae. *Int J Plant Sci* 2001; 162: 257–262.

120. Nishioka D, Brisibe EA, Miyake H, Taniguchi T. Ultrastructural observations in the suppression of granal development in bundle sheath chloroplasts of NAD-ME type C_4 monocot and dicot species. *Jap J Crop Sci* 1993; 62: 621–627.

121. Nishioka D, Miyake H, Taniguchi T. Suppression of granal development and accumulation of Rubisco in different bundle sheath chloroplasts of the C$_4$ succulent plant *Portulaca grandiflora*. *Annals Bot* 1996; 77: 629–637.

122. Sage RF, Sage TL, Kocacinar F. Photorespiration and the evolutions of C$_4$ photosynthesis. *Annu Rev Plant Biol* 2012; 63: 19–47.

123. Häusler RE, Hirsch H-J, Kreuzaler F, Peterhánsenl C. Overexpression of C$_4$-cycle enzymes in transgenic C$_3$ plants: A biotechnological approach to improve C$_3$-photosynthesis. *J Exp Bot* 2002; 53: 591–607.

124. Leegood RC. C$_4$ photosynthesis: Principles of CO$_2$ concentration and prospects for its introduction into C$_3$ plants. *J Exp Bot* 2002; 53: 581–590.

125. Miyao M, Masumoto C, Miyazawa S-I, Fukayama H. Lessons from engineering a single-cell C$_4$ photosynthetic pathway into rice. *J Exp Bot* 2011; 62: 3021–3029.

126. Zhu XG, Long SP, Ort DR. Improving photosynthetic efficiency for greater yield. *Annu Rev Plant Biol* 2010; 61: 235–261.

127. Kortschak HP, Hartt CK, Burr GO. Carbon dioxide fixation in sugarcane leaves. *Plant Physiol* 1965; 40: 209–213.

Section VI

Photosynthesis in Lower and Monocellular Plants

19 Structural and Functional Studies of the Cytochrome bc_1 Complex from Photosynthetic Purple Bacteria

Fei Zhou, Lothar Esser, Chang-An Yu, and Di Xia

CONTENTS

19.1 Introduction 327
    19.1.1 Reactions Catalyzed by Cyt bc_1 Complexes 328
    19.1.2 Role of Cyt bc_1 in the Growth of Photosynthetic Bacteria 328
    19.1.3 Photosynthetic Purple Bacteria Offer a Convenient System for Genetic Manipulation of Cyt bc_1 328
19.2 Structural Analysis of Cyt bc_1 from Photosynthetic Purple Bacteria 329
    19.2.1 Purification, Crystallization, and Subunit Composition of bc_1 from *R. sphaeroides* 330
    19.2.2 Crystal Structures of Bacterial bc_1 from *R. sphaeroides* 330
        19.2.2.1 Structure of the Cyt b Subunit 330
        19.2.2.2 Structure of the Cyt c_1 Subunit 331
        19.2.2.3 Structure of the Iron–Sulfur Protein Subunit 333
    19.2.3 The Role of Subunit IV of $Rsbc_1$ 333
19.3 Mechanism of bc_1 Function in Light of Structural Information 334
    19.3.1 Q-Cycle Mechanism and Bifurcated ET at the Quinol Oxidation Site 334
    19.3.2 Controlled Movement of the Extrinsic Domain of ISP Provides a Possible Mechanism for the Bifurcated ET at the Q_P Site 334
    19.3.3 Experimental Verification of SAMICS 336
        19.3.3.1 ISP-ED Mobility Is Necessary for ET Function of bc_1 336
        19.3.3.2 Mutagenesis in bc_1 Supports the Presence of a Control Mechanism for the ISP-ED Conformation Switch 336
    19.3.4 Presence or Absence of a Ubisemiquinone Free Radical as a Reaction Intermediate at the Quinol Oxidation Site 336
19.4 Regulation of bc_1 Activity in Photosynthetic Purple Bacteria 338
    19.4.1 Inhibition of Bacterial bc_1 Activity by Zn^{2+} 338
    19.4.2 Effect of O_2 on the Electron Transfer Activity of bc_1 338
    19.4.3 Interaction of bc_1 Complexes with Cytosolic Proteins 339
19.5 Origin of Superoxide Generation 340
    19.5.1 Mutations That Suppress or Enhance Superoxide Generation in Bacterial bc_1 340
    19.5.2 Ubisemiquinol as a Source for Superoxide Formation 341
    19.5.3 Reduced Heme b_L as a Source for Superoxide Formation 342
    19.5.4 Role of Oxygen in Mediating Superoxide Production 342
19.6 Concluding Remarks 343
Acknowledgments 343
Abbreviations 343
References 343

19.1 INTRODUCTION

As part of the mitochondrial respiratory chain and the photosynthetic apparatus of nonoxygenic purple bacteria, the cytochrome bc_1 complex couples the oxidation of quinol to proton translocation across the membrane, contributing to the proton motive force essential for a variety of cellular activities such as ATP synthesis. With the accumulation of information from biochemical, genetic, and structural investigations, an understanding of the mechanism by which the complex performs its biological function and the way it is inhibited has been obtained at an unprecedented level of detail. With their unique genetic background, photosynthetic bacteria afford an indispensible and convenient system that has been used extensively in functional and structural studies of the cytochrome bc_1 complex. In this chapter, we review contributions from many laboratories including our own and point out certain challenges that lie ahead.

19.1.1 REACTIONS CATALYZED BY CYT bc_1 COMPLEXES

The cytochrome bc_1 complex (ubiquinol–cytochrome c oxidoreductase, complex III, cyt bc_1, or bc_1) is an essential energy-transducing electron transfer (ET) complex in mitochondria and in many aerobic and photosynthetic bacteria (Brandt and Trumpower 1994; Crofts 2004; Trumpower and Gennis 1994; Wikstrom et al. 1981). It catalyzes the ET reaction from ubiquinol (QH_2) to cytochrome c (cyt c) with concomitant generation of a cross-membrane proton gradient and membrane potential that is essential for cellular ATP supply. A second well-documented side reaction of bc_1 is its ability to generate superoxide anion $\left(O_2^{\cdot-}\right)$ under various conditions (Boveris and Chance 1973). The physiological relevance of $O_2^{\cdot-}$ generated by the respiratory chain is under intensive investigation.

19.1.2 ROLE OF CYT bc_1 IN THE GROWTH OF PHOTOSYNTHETIC BACTERIA

Photosynthesis converts sunlight into chemical energy and is considered one of the most important reactions on planet Earth. In anoxygenic, facultative photosynthetic purple bacteria such as *Rhodobacter sphaeroides* and *Rhodobacter capsulatus*, the presence of light and oxygen induces the production of modular machineries for both photosynthetic and respiratory growth (Figure 19.1). In fact, these bacteria can switch from one growth mode to another easily, acclimating to aerobic, anaerobic, or photosynthetic conditions. Under continuous light exposure, respiration is slowed or inhibited due to the inhibitory buildup of membrane potential or possibly, by competition between the photosynthetic reaction center (RC) and the cyt c oxidase for reduced cyt c_2 (Oh and Kaplan 2001). The bacterial photosynthetic apparatus consists of membrane-bound light-harvesting or antenna complexes (LHCs), photosynthetic RCs, cyt bc_1 complexes, and the soluble electron carrier cyt c_2. The presence of oxygen stimulates the development of the machinery for cellular respiration, which includes NADH and succinate dehydrogenases, the cyt bc_1 complex, and various terminal oxidases (Figure 19.1). Cyt bc_1 is at the intersection of the two processes: phototropic and chemotropic ET.

Under light conditions and in the absence of oxygen, all photosynthetic bacteria rely on a functional cyt bc_1 complex for survival, making it essential for photosynthetic growth. To ensure an efficient collection of light energy, especially under low light conditions, the photosynthetic apparatus of purple bacteria usually contains two types of LHCs: the RC-encircling LH-1 and peripheral LH-2. After light absorption by the LHCs, the excitation energy is transferred to the RC, where a charge separation occurs between donor (bacteriochlorophyll [Bchl]) and acceptor (bacteriophaeophytin [BPh]) molecules. The resulting high-energy electron is ultimately used to reduce bound quinone molecules (Q). The photooxidized primary electron donor, a dimer of Bchl molecules, is reduced by the high-potential electron carrier cyt c_2, which is in turn reduced by the cyt bc_1 complex (Figure 19.1). To complete the cyclic ET, the bc_1 complex oxidizes QH_2 and then reduces mobile cyt c_2 (Vermeglio and Joliot 1999). When switched from photosynthetic growth to aerobic growth, expressions of the photosynthetic apparatus, as shown, for example, by reduction of photopigments, are downregulated, while those for aerobic growth are upregulated (Pemberton et al. 1998). Under aerobic conditions, electrons flow downhill energetically through various respiratory complexes, ending in molecular oxygen. This ET path is coupled to proton pumping across the membrane, contributing to the proton motive force.

19.1.3 PHOTOSYNTHETIC PURPLE BACTERIA OFFER A CONVENIENT SYSTEM FOR GENETIC MANIPULATION OF CYT bc_1

The dependence of photosynthetic bacteria on cyt bc_1 under anaerobic or photosynthetic growth conditions makes them sensitive to both bc_1 inhibitors and mutations that render the complex nonfunctional. This is in sharp contrast to growth under aerobic conditions, where respiration is supported by alternate oxidases in the absence of bc_1. The key difference, compared to the alternate oxidase pathway, is that cyt bc_1 couples the oxidation of ubiquinol more efficiently to proton translocation. As it turns out, it is this growth-mode switch feature that renders photosynthetic bacteria useful tools in the study of respiratory components. Furthermore, the bc_1 complex in a photosynthetic membrane can be functionally coupled to the photosynthetic RC, using flash photolysis to study electron flow through the enzyme and to examine the effects of various amino acid substitutions. Numerous mutations have been generated in the cyt b subunit, in the Rieske iron–sulfur protein (ISP) subunit, and in the cyt c_1 subunit, illuminating many important aspects of the bc_1 function (Gennis et al. 1993).

The purple bacterium *R. capsulatus* contains a single 4.4 Mb chromosome, whereas *R. sphaeroides* contains two circular chromosomes, chromosome I (CI) with 3 Mb and chromosome II (CII) with 0.9 Mb (Suwanto and Kaplan 1989b). *R. sphaeroides* additionally possesses five endogenous plasmids, which account for a total of 0.45 Mb of the genome (Suwanto and Kaplan 1989a). In *R. sphaeroides* and *R. capsulatus*, genes encoding cyt bc_1 and other photosynthetic components are located on CI (Mouncey et al. 2000; Pemberton et al. 1998); they are located in operons called *pet*ABC and *fbc*FBC, respectively, for *R. capsulatus* and *R. sphaeroides* (Gennis et al. 1993). The size of the genes coding for the three core subunits, cyt b, cyt c_1, and ISP, of $Rsbc_1$ complex is only 3.3 Kb (Yun et al. 1990). The supernumerary subunit of $Rsbc_1$, the subunit IV, is encoded by the *fbc*Q gene, which is approximately 278 Kb away from the *fbc*FBC operon (Tso et al. 2000; Yu et al. 1999).

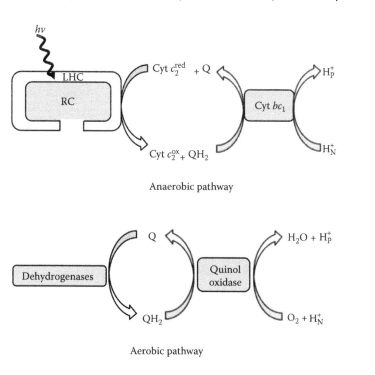

Anaerobic pathway

Aerobic pathway

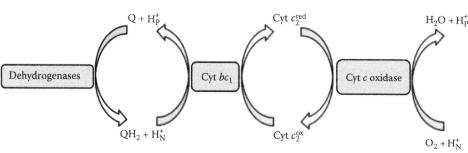

Semiaerobic pathway

FIGURE 19.1 Photosynthetic and respiratory machineries in photosynthetic purple bacteria. Under photosynthetic and anaerobic growth conditions (anaerobic pathway), the photosynthetic apparatus, consisting of the light-harvesting complex (LHC), photosynthetic reaction center (RC), cyt bc_1 complex, quinone (Q), quinol (QH$_2$), and oxidized $\left(\text{cyt } c_2^{\text{ox}}\right)$ and reduced cyt c_2 $\left(\text{cyt } c_2^{\text{red}}\right)$, performs a cyclic reaction among these components to sustain growth. The RC oxidizes cyt c_2^{red} and reduces Q to QH$_2$, whereas the cyt bc_1 oxidizes QH$_2$ to produce cyt c_2^{red}. Under aerobic growth conditions (aerobic pathway), the respiratory chain can bypass cyt bc_1 with QH$_2$, produced by various dehydrogenases, to be oxidized by various quinol oxidases, passing electrons to terminal oxidant O$_2$ and pumping protons across the membrane. When the oxygen level is low (semiaerobic pathway), cyt bc_1 is again used in the oxidation of QH$_2$ and reduction of cyt c_2^{ox}. The cyt c_2^{red} is used as a substrate by cyt c oxidase for proton pumping.

The value of photosynthetic purple bacteria in studies of cyt bc_1 has been demonstrated through decades of structural and functional investigations. Most advantageous to the bacterial systems is their ability to perform site-directed mutagenesis (Mouncey et al. 2000). Compared to eukaryotic systems, purple bacteria have a very simple subunit composition. These bacteria are easy to grow under a variety of different conditions, allowing control of the expression of specific proteins and making possible molecular engineering of the bc_1 complex. Unlike in eukaryotic systems, where the gene encoding the cyt b subunit is located in the mitochondrial genome, making molecular manipulation a challenge, application of site-directed mutagenesis in purple bacteria has been well developed and widely used. In the *R. sphaeroides*

strain *BC17*, the *fbc*FBC and *fbc*Q operons have been deleted from the chromosome (Yun et al. 1990). Therefore, without the bc_1 complex, this strain cannot grow photosynthetically. However, it can survive aerobically, due to the alternate bc_1-independent ET pathways. By inserting *fbc*FBCQ genes into the low-copy-number expression vector pRKD418, the bc_1 complex can be overexpressed in *BC17*, permitting photosynthetic growth.

19.2 STRUCTURAL ANALYSIS OF CYT bc_1 FROM PHOTOSYNTHETIC PURPLE BACTERIA

Several decades of intensive studies of bc_1 with biophysical, biochemical, and molecular genetic approaches have generated large amounts of information (Crofts and Wang 1989; di Rago and Colson 1988; Musatov and Robinson 1994; Schagger et al. 1995), leading to a basic understanding of the mechanism by which bc_1 complexes carry out the ET-coupled proton pumping function. Perhaps one of the most significant advances in elucidating the detailed mechanism of bc_1 function came from x-ray crystallographic investigations of mitochondrial and bacterial bc_1 complexes. Since the first report of the bovine mitochondrial bc_1 structure in 1997 (Xia et al. 1997), crystal structures of bc_1 from different sources and in different forms have been reported (Berry et al. 2004; Esser et al. 2004, 2006, 2008; Hunte et al. 2000; Iwata et al. 1998; Kim et al. 1998; Kleinschroth et al. 2011; Kurisu et al. 2003; Stroebel et al. 2003; Xia et al. 2007, 2008; Zhang et al. 1998b). In this chapter, focus will be given to structural studies of cyt bc_1 complexes from photosynthetic bacteria R. capsulatus and R. sphaeroides.

19.2.1 Purification, Crystallization, and Subunit Composition of bc_1 from R. sphaeroides

Subunit compositions of cyt bc_1 from different organisms vary. In high vertebrates such as *Homo sapiens* or *Bos taurus*, bc_1 complexes are composed of as many as 11 different protein subunits, whereas in bacteria such as R. capsulatus or R. sphaeroides, the number of subunits is much smaller. Consequently, bacterial bc_1 is considerably less stable and has lower activity (Ljungdahl et al. 1987). Essential to the function of all bc_1 complexes are the four redox prosthetic groups: the low- and high-potential hemes b_L (heme b_{566}) and b_H (heme b_{562}), heme c_1, and the high-potential [2Fe-2S] cluster (iron–sulfur cluster [ISC]), which are found in three highly conserved subunits—cyt b, cyt c_1, and ISP, respectively (Trumpower and Gennis 1994). Subunits that are not involved in the ET function are known as supernumerary subunits. These subunits contain no redox prosthetic groups, and their roles in bc_1 function are not entirely clear. For example, in bovine bc_1 there are eight supernumerary subunits (I, II, VI–XI) (Oudshoorn et al. 1987; Schagger et al. 1986). By contrast, both bc_1 complexes of R. sphaeroides (Rsbc$_1$) and R. capsulatus bc_1 (Rcbc$_1$) were believed to have one supernumerary subunit when they were first purified (Ljungdahl et al. 1987). The subunit IV of Rsbc$_1$ was shown subsequently to be an integral part of Rsbc$_1$ (Andrews et al. 1990; Purvis et al. 1990; Yu et al. 1999), whereas that of Rcbc$_1$ turned out to be an artifact (Robertson et al. 1993).

Procedures for isolating bacterial bc_1 from different species can be found in the literature (Berry et al. 2004; Ljungdahl et al. 1987; Yu and Yu 1982). For the purpose of producing large amounts of Rsbc$_1$ suitable for crystallographic studies, a hexahistidine tag was introduced to the C-terminus of the cyt c_1 subunit. R. sphaeroides strain BC17 cells bearing the mutated pRKD418-fbcFBC$_{6H}$Q plasmid (Mather et al. 1995) were grown photosynthetically at 30°C in an enriched Sistrom medium (Tian et al. 1997). Cells were harvested when the optical density of the culture, measured at a wavelength of 600 nm (OD$_{600}$), reached 1.8-2.0 units. Chromatophore membranes were prepared from

cell pellets as described previously (Yu and Yu 1991) and stored in the presence of 20% glycerol at −80°C. To purify the hexahistidine-tagged Rsbc$_1$ complex, frozen chromatophores thawed on ice were solubilized with n-dodecyl ß-D-maltoside (ß-DDM) to a final concentration of 0.56 mg detergent/nmol cyt b. After removing unsolubilized material by centrifugation, the supernatant was subjected to a nickel-nitrilotriacetic (Ni-NTA) affinity chromatography. Fractions containing purified bc_1 were combined, concentrated, and stored in the presence of 10% glycerol at −80°C (Xiao et al. 2014). Proteins prepared this way can be readily crystallized (Esser et al. 2008; Xia et al. 2008). The presence of the bc_1-specific inhibitor stigmatellin, the use of the amino acid histidine, and a mixture of ß-octyl glucopyranoside (ß-OG) and sucrose monocaprate are important factors for obtaining high-quality crystals.

19.2.2 Crystal Structures of Bacterial bc_1 from R. sphaeroides

Crystal structures of Rsbc$_1$ were determined to better than 2.4 Å resolution in the presence of various inhibitors and substrates (Esser et al. 2008). In these structures, only the three core subunits were present. The structure of Rsbc$_1$ revealed dimeric association of the three subunits, cyt b, cyt c_1, and ISP (Figure 19.2a). The assembled three-subunit Rsbc$_1$ resembles closely that of corresponding subunits in bovine mitochondrial bc_1 (B. taurus bc_1 [Btbc$_1$] (Figure 19.2b). As in mitochondrial bc_1, the extrinsic domain of the ISP subunit (ISP-ED) in Rsbc$_1$ crosses over to form an active site with the cyt b subunit of an adjacent monomer. Unlike mitochondrial cyt bc_1 (Mtbc$_1$) complexes, there is a lack of supernumerary subunits on the cytoplasmic side of the membrane in bacterial bc_1 (Figure 19.2b). A proposed function for these additional subunits of mitochondrial bc_1 is to shield the portion of the enzyme complex, which is embedded in the membrane bilayer against proton and/or water leakage (Esser et al. 2008; Gurung et al. 2005).

19.2.2.1 Structure of the Cyt b Subunit

The cyt b subunit of Rsbc$_1$ has eight transmembrane (TM) helices (A through H), forming two helical bundles (A–E and F–H) (Figure 19.3a). The two heme groups, b_L and b_H, reside within the first bundle. Extra-membranous loops connect pairs of TM helices, and those that are longer than 20 residues are the AB, CD, DE, and EF loops. The quinol oxidation site (Q$_P$ site) near the periplasmic side of the membrane (positive side or P-side) and the quinone reduction (Q$_N$) site on the opposite side (cytoplasmic side, negative side, or N-side) can be identified with bound stigmatellin and ubiquinone, respectively.

The sequence of the cyt b subunit of Rsbc$_1$ is considerably longer (approximately 15%) compared to the corresponding bovine subunit. Most of the insertions and extensions are concentrated on the negative side of the membrane, suggesting that their function(s) were eventually taken over by the two large core subunits in Mtbc$_1$ (Figures 19.2b and 19.3a). Bacterial cyt b has significant extensions at both its N-terminus (helix a0) and its C-terminus (helix i). In the crystal structures of all

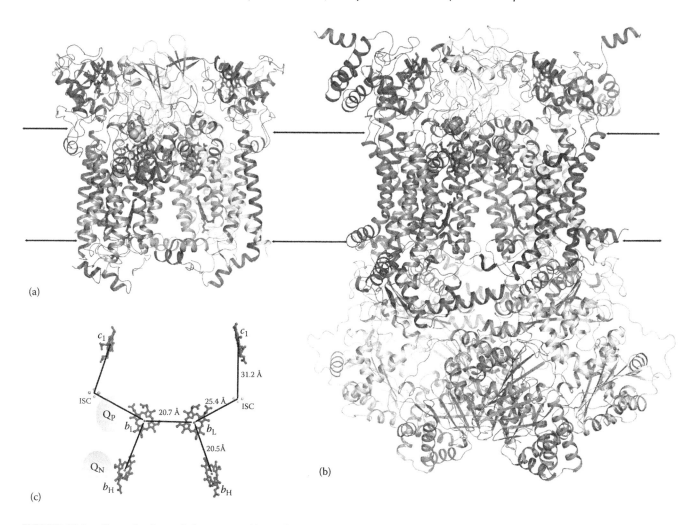

FIGURE 19.2 **(See color insert.)** Structures of bacterial and mitochondrial cyt bc_1 complexes. (a) Ribbon representation of the structure of dimeric cyt bc_1 from the photosynthetic purple bacterium *R. sphaeroides*. Only the three core subunits are present in the structure: cyt b (green), cyt c_1 (blue), and iron–sulfur protein (ISP, yellow). Heme moieties are shown as stick models in red. The 2Fe-2S clusters are shown as spheres. Structural elements that are colored pink are insertions, as compared to bovine bc_1. (b) Structure of dimeric cyt bc_1 from bovine heart mitochondria. All eleven different subunits are present, and the three core subunits are the same color as those of the bacterial complex. The membrane bilayer is delineated by the two parallel black lines. (c) Arrangement of prosthetic heme moieties and ISC groups in dimeric *Rsbc$_1$*. Distances between pairs of these groups are given, and the two active sites, Q_P and Q_N, are indicated and labeled.

bacterial forms, the first ordered residue (as judged by electron density) for cyt b starts at position 3 and ends at residue 430. This suggests that the 15 remaining C-terminal residues are highly flexible and can in fact be deleted without affecting the function of the enzyme in *R. sphaeroides* (Liu et al. 2004). The extra helices a0, de, and i enable cyt b to interact with a lipid molecule modeled as phosphatidylethanolamine (PE), as shown in Figure 19.2a, apparently creating additional shielding for enhanced membrane impermeability.

There is an insertion of 18 residues (310–327), denoted the ef1 helix, which is located near the positive side of the membrane. This is the only insertion on the positive side of the membrane, providing a unique structural element in bacterial bc_1 complexes (Figure 19.3a). As revealed in high-resolution crystal structures of *Rsbc$_1$*, the N-terminal side of the ef1 helix interacts with lipid-like molecules (deduced from the residual density that is not shown in Figure 19.3a). These densities so far have defied attempts to be modeled as lipids or detergents. However, it has been shown that deletion of the ef1 helix or as little as the point mutation S322A significantly impairs the function of bc_1 (Esser et al. 2006). Since mitochondrial cyt b does not have the ef1 helix, its direct involvement in catalysis or structural integrity can be ruled out. This protruding element may be needed to interact with membrane lipids, providing help to orient the complex in the membrane environment, as seen in Figure 19.3a.

19.2.2.2 Structure of the Cyt c_1 Subunit

The cyt c_1 subunit folds in a manner similar to that of its mitochondrial counterpart, having a C-terminal TM helix and an N-terminal periplasmic globular domain that features the C36–X–X–C39–H40 motif characteristic of c-type cytochromes, with the heme iron atom being coordinated by the side chains of H40 and M185 as fifth and sixth axial ligands, respectively. Compared

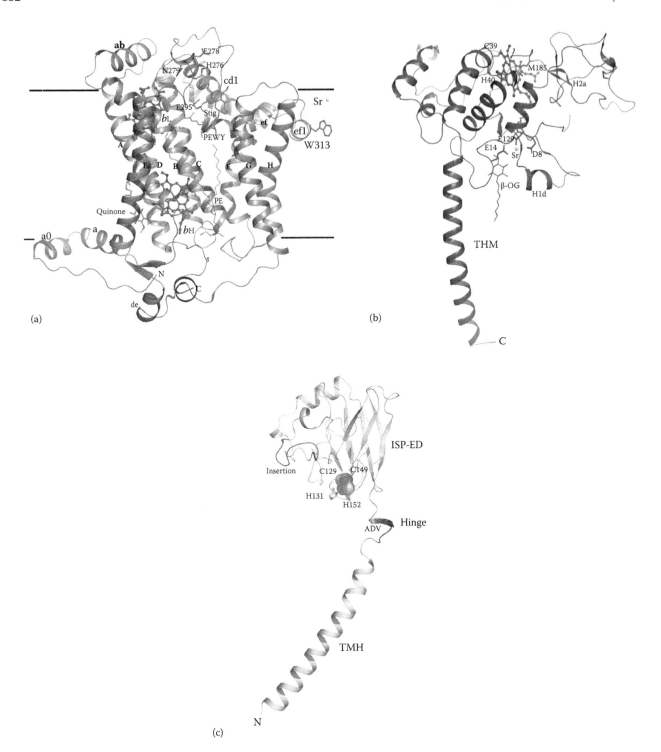

FIGURE 19.3 Structures of monomeric cyt *b*, cyt *c₁*, and ISP subunits of *Rsbc₁*. (a) Ribbon diagram of the structure of the cyt *b* subunit. The eight TM helices are labeled sequentially from A to H. The two heme moieties, b_L and b_H, are shown as ball-and-stick models. The ef1 insertion features a protruding W313 residue that interacts with the neighboring W313. Both tryptophan residues are involved in interaction with a bound Sr^{2+} ion as part of crystal contacts. The bound lipid molecule was identified in the crystal structure as PE, and is shown in a stick model. Labeled are the highly conserved cd1 helix and the PEWY motif. The Q_P site occupant is stigmatellin, whereas a quinone molecule was found in the Q_N site. Structures of monomeric cyt *b*, cyt *c₁*, and ISP subunits of *Rsbc₁*. (b) Ribbon diagram of the structure of cyt *c₁* subunit. Insertions compared to the sequence of *Btbc₁* are the a0, de and ef1 secondary structure elements. The heme moiety is shown as a ball-and-stick model, which is covalently linked by C36 and C39, shown as stick models. Residues H40 and M185 serve as axial ligands for the *c₁* heme and are also shown as stick models. The Sr^{2+} binding site is surrounded by residues E14, E129, and D8, which are given as stick models. The bound detergent ß-OG is also shown. (c) Ribbon representation of the structure of the ISP subunit. The only insertion with respect to the *Btbc₁* is shown in pink. The ISC is rendered as a spherical model, and its four ligands are given as stick models and labeled. The conserved ala-asp-val (ADV) sequence motif is part of the hinge between ISP-ED and the N-terminal transmembrane helix (TMH).

to mitochondrial cyt c_1, the sequences of bacterial cyt c_1 suffer a loss of two major structural elements and a gain of two smaller insertions (Figure 19.3b). First, the C36-X-X-C39-H40 signature motif for the c-type heme is perfectly conserved, including the remote M185. Residue R107 is also conserved; it forms a salt bridge with one of the propionate groups of the heme and is located on the surface of cyt c_1 that interacts transiently with the moving ISP-ED. A superposition of coordinates (Cα positions) of cyt c_1 between $Rsbc_1$ and $Btbc_1$ reveals a very high degree of preservation of structural elements, in particular, the position of the important heme redox group as well as the C-terminal TM helix. Two insertions have occurred in mitochondrial cyt c_1 after position 76 and 94 (bovine sequence), both of which contribute to a closing up of cyt c_1, including the establishment of a direct bridge between the two cyt c_1 units of the homodimeric bc_1 complex (Esser et al. 2008).

A prominent loop is inserted into the sequences of *R. sphaeroides* and *R. capsulatus* containing a stabilizing disulfide bridge C145-C169. Compared to mitochondrial bc_1, $Rsbc_1$ lacks the supernumerary subunit (subunit 8) located next to cyt c_1, which is a disulfide-linked helical hairpin guarding the perimeter of cyt c_1. Thus, the insertion at this position suggests a compensatory function. Another highlighted feature of bacterial cyt c_1 is the largely helical insertion between residues 109 and 127 (Figure 19.3b). This element, designated h1d helix, has been deleted from mitochondrial forms of cyt c_1 without any discernable compensation. The high degree of flexibility of this loop rendered it "invisible" in the structure of *R. capsulatus* cyt c_1, but it could be modeled with high-temperature factors in $Rsbc_1$. Here, the lateral displacement of the helix away from the globular cyt c_1 core revealed a cavity that appears to provide space for a lipid head group of a phosphatidylethanolamine (PE). However, the high degree of disorder in this region prevented it from being included in the final model. While the exact nature of the function of this loop awaits further investigation, it may be another demonstration that the three core subunits of bacterial bc_1 not only fulfill their redox catalytic functions but also deal with proper partitioning of the complex into lipids, a task that appears to be shared to a large degree with supernumerary subunits in mitochondria.

A strontium ion (Sr^{2+}) from the crystallization additive has been shown to bind to the side chains of D8, E14, and E129 and the backbone carbonyl oxygen atom of V9 in *R. sphaeroides* cyt c_1 (Figure 19.3b). This site is conserved in $Rcbc_1$ but is not in mitochondrial enzymes, where this site is structurally maintained, not the electrical charge, as the residues D8 and E14 have been replaced by neutral, polar, or even basic ones. Only vestiges of the formerly acidic site remain; residue E129 was replaced by D125 (bovine) or D189 (yeast), forming hydrogen bonds to the indole group of W12 (or W76), which replaced F11 in bacterial sequences. While the Sr^{2+} ion is most likely not a physiological substrate, it was shown that the presence of Ca^{2+} ions can prevent the binding of subunit IV of $Rsbc_1$, possibly through blocking the acidic cluster of residues meant to bind a basic portion of $Rsbc_1$'s only supernumerary unit. The purpose of this site in $Rcbc_1$, which is not known to require an additional subunit, remains unclear.

19.2.2.3 Structure of the Iron–Sulfur Protein Subunit

The ISP subunit has a C-terminal periplasmic globular domain (extrinsic domain, ISP-ED), which connects through a flexible hinge region to its N-terminal TM helix (Figure 19.3c). The ISP-ED is predominantly a ß-structure consisting of three ß-sheets arranged in three parallel layers with the 2Fe-2S cluster located at the apex of the ISP-ED between the second and third ß-sheets. There is an apparently flexible linker or hinge between the TM helix and ISP-ED, which provides flexibility for ISP-ED to relay the electron from the Q_P site to cyt c_1. The conserved A42-D43-V44 motif in the hinge adopts an α-helical conformation, whereas the corresponding element in the bovine bc_1 structure is a random coil. The ISP is well conserved among species, but structure-based sequence alignment highlights one ~12-residue long globular insertion in the bacterial ISP featuring three ß-turns and an inverse γ-turn (Figure 19.3c). Despite the remoteness of this element from the ISC (20–25 Å), an interruption in its hydrogen bonding network by more than one point mutation results in a failure to integrate this subunit into bc_1 (Xiao et al. 2004).

19.2.3 THE ROLE OF SUBUNIT IV OF $RSBC_1$

The subunit IV is a 124-amino acid residue peptide and is predicted to have a single membrane-spanning helix. It was shown to be an integral part of the $Rsbc_1$, providing structural stability to the complex of *R. sphaeroides* and involved in substrate ubiquinol (QH_2) binding (Yu and Yu 1991). Topological studies using subunit IV-specific antibodies found the N-terminus of the subunit to be located in the cytosol (Yu et al. 1999). A subunit IV knockout mutant gave rise to a photosynthetically competent phenotype only after a period of adaptation (Chen et al. 1994). The deletion of subunit IV also changed K_m values for QH_2 binding from 3 to 13 µM, suggesting involvement of subunit IV in QH_2 binding. After binding of the photoaffinity-labeled QH_2 derivative [³H]azido-Q to $Rsbc_1$, radioactivity was detected in subunits I (cyt b) and IV, supporting the notion that these two subunits are responsible for Q binding in the complex (Yu and Yu 1987).

Although purified $Rsbc_1$, both wild type and mutant, contained subunit IV, this subunit was absent from the complex in crystal structures, indicating that the crystallization medium (including polyethyleneglycol 400 [PEG400], detergents, etc.) must have caused its detachment (Xia et al. 2008). A sodium dodecyl sulfonate-polyacrylamide gel electrophoresis (SDS-PAGE) gel revealed the presence of subunit IV in solution, but no detectable amount was found in crystals (data not shown). It is not uncommon to lose a supernumerary subunit during crystallization of mitochondrial bc_1 complexes (Huang et al. 2005; Hunte et al. 2000). To test whether subunit IV is indirectly required for the crystallization of $Rsbc_1$, we purified the Δ-subIV mutant and subjected it to the same crystallization conditions. Crystals grew readily, demonstrating that subunit IV is not required for crystallization.

In the absence of an experimental structure for subunit IV, molecular modeling based on subunit 7 of the $Btbc_1$ structure was conducted (Yu et al. 1999). Subsequent mutagenesis to verify the model identified W79 as an important residue both for structural stability and substrate-binding functions of subunit IV.

19.3 MECHANISM OF bc_1 FUNCTION IN LIGHT OF STRUCTURAL INFORMATION

The chemical reaction catalyzed by the cyt bc_1 complex can be summarized in the following equation:

$$QH_2 + 2c^{3+} + 2H_N^+ \xrightarrow{bc_1} Q + 2c^{2+} + 4H_P^+$$

where QH_2 and Q are reduced and oxidized ubiquinol, respectively; c^{2+} and c^{3+} represent reduced and oxidized cyt c, respectively; and H_N^+ and H_P^+ are protons on the negative and positive side of the membrane, respectively. In one complete turnover, one ubiquinol molecule is oxidized, two molecules of cyt c are reduced, and four protons are translocated to the positive side of the membrane. For every electron transferred from quinol to cyt c, two protons are translocated across the membrane (2H$^+$/e stoichiometry).

19.3.1 Q-CYCLE MECHANISM AND BIFURCATED ET AT THE QUINOL OXIDATION SITE

The reaction stoichiometry shown in the previous equation can be explained by a modified Q-cycle mechanism (Figure 19.4a) (Crofts et al. 1983; Garland et al. 1975; Trumpower 1976). This mechanism defines two distinct binding sites for ubiquinol (Q_P or quinol oxidation) and ubiquinone (Q_N) near the electrochemically positive (P) and negative (N) sides of the membrane, respectively. Each site is associated with a b-type heme (b_L or b_H) necessary for the obligatory electronic coupling of the oxidation of quinol on one side to the reduction of quinone on the other. In one complete cycle, two molecules of quinol are oxidized in succession, and their four protons are delivered to the positive side. Simultaneously, one quinone molecule is reduced, requiring the uptake of two protons from the negative side. This means that in one half-cycle, a single electron enters the 2Fe-2S cluster of the ISP and is delivered to the final electron acceptor within bc_1, cyt c_1. At this point, a net imbalance of charge exists, as only one electron from quinol was removed to the P-side aqueous phase but both of its protons enter the P-side (protons exit bc_1 on the positive side of the membrane). The remaining electron takes a transverse path to the N-side of the membrane through the cyt b hemes. It transiently reduces the proximal heme b_L, followed by the remote b_H, and ends up on a quinone (Q) or semiquinone radical $\left(SQ_N^\bullet\right)$ at the Q_N site. Protons coming from the N-side aqueous phase of the membrane compensate for the sequential arrival of negative charges at the Q_N site (Figure 19.4a). Blockage of electron flow within this elegant circuit was demonstrated using inhibitors targeting specific sites of cyt bc_1. Antimycin, for example, blocks ET from the b_H heme to substrate Q, whereas myxothiazol and stigmatellin inhibit ET to the low- and high-potential chains, respectively (Figure 19.4a).

The most important feature of the Q-cycle mechanism is the mandatory separation of the two electrons of substrate ubiquinol at the Q_P site, forcing the first electron into the high-potential chain formed by the ISP, cyt c_1, and aqueous substrate cyt c_2 and the second electron into the low-potential chain consisting of the b_L and b_H hemes. The bifurcated ET is obligatory, as antimycin A, a Q_N site inhibitor, completely inhibits bc_1 activity; it ensures the 2H$^+$/e reaction stoichiometry.

19.3.2 CONTROLLED MOVEMENT OF THE EXTRINSIC DOMAIN OF ISP PROVIDES A POSSIBLE MECHANISM FOR THE BIFURCATED ET AT THE Q_P SITE

Binding of inhibitors has a profound effect on the spectra of b-type hemes and on the redox potential of the ISP (Link et al. 1993; von Jagow and Link 1986). Inhibitor binding was also recognized to be a function of the redox state of the enzyme (Brandt and von Jagow 1991). However, these phenomena were difficult to explain in the absence of structural information. In 1998, less than a year after the first bc_1 structure was published, Kim et al. (1998) reported crystallographic observations concerning the conformational switch of ISP-ED in response to binding of different types of bc_1 inhibitors to the Q_P site. One type of inhibitor, including stigmatellin, undecyl hydroxy dioxobenzothiazole (UHDBT), and famoxadone, led to fixation of the ISP-ED conformation at the Q_P site of cyt b (b-site), and another type, including myxothiazol, azoxystrobin, and methoxyacrylate (MOA) stilbene, mobilized ISP-ED. These two types of inhibitors were named P_f- and P_m-type inhibitors, respectively, according to their ability to fix or mobilize the ISP-ED conformation (Esser et al. 2004). High-resolution crystallographic data also showed that these two types of inhibitors bind to two different sites, the P_f or P_m sites, within the Q_P pocket, driving movement of the highly conserved cd1 helix in a bidirectional fashion (Esser et al. 2006). Since the cd1 helix forms part of the ISP-ED binding surface, this bidirectional movement was hypothesized to modulate the binding affinity of ISP-ED, leading to its conformation switch (Esser et al. 2006; Xia et al. 2013).

Elucidating the mechanism that enforces the bifurcated ET turned out to be challenging, as evidenced by many proposed mechanisms (Brandt 1996; Brandt and von Jagow 1991; Crofts et al. 2006; Ding et al. 1992; Junemann et al. 1998; Link 1997; Osyczka et al. 2005). The first crystal structures of bovine bc_1 (Xia et al. 1997) showed the ISP-ED to be disordered in both native and inhibitor-bound (antimycin A or myxothiazol) structures, demonstrating its mobility. However, the ISC of ISP-ED left an unmistakable signal in the anomalous difference Fourier map, placing it very far (~31 Å) away from the cyt c_1 heme iron (Figure 19.2c). This distance is incompatible with the observed turnover rate of bc_1 reaction, as it would prohibit efficient ET between ISP and cyt c_1. In subsequent crystal structures, ISP-ED was found in different locations in different crystal forms, leading to the proposal that ET between ISP and cyt c_1 can be accomplished by domain movement (Iwata et al. 1998; Zhang et al. 1998b). It was later realized that a stochastic

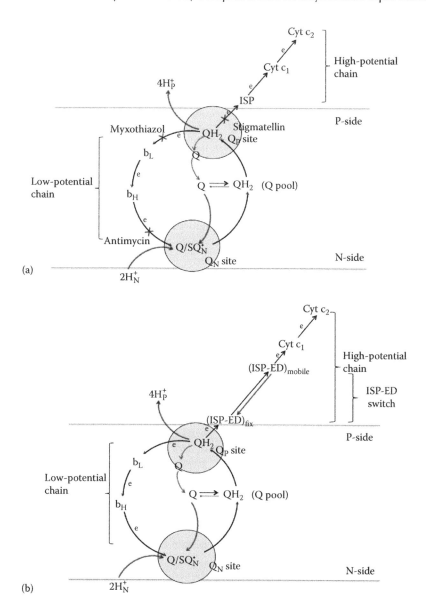

FIGURE 19.4 The modified Q-cycle mechanism of cyt bc_1 function and bifurcated ET at the quinol oxidation (Q_P) site. (a) The modified Q-cycle mechanism defines a Q_P site and a quinone reduction (Q_N) site. The Q_P site is connected to a high-potential chain and a low-potential chain. The essence of the Q-cycle mechanism is the bifurcated ET at the Q_P site, where the first electron from QH_2 is sent to the high-potential chain, which can be inhibited by stigmatellin, and the second electron of QH_2 follows the low-potential chain, which can be blocked by myxothizol. The second electron ultimately reaches the Q_N site, and this reaction can be inhibited by antimycin. As a result of the Q-cycle, four protons are deposited into the positive side of the membrane, and two protons are taken from the negative side of the membrane. (b) Surface-affinity modulated ISP conformation switch (SAMICS). The ISP-ED undergoes a conformation switch from a fixed conformation that places its ISC 31 Å from the cyt c_1 heme to a mobile conformation that allows the ISC to be close to cyt c_1. Due to the long intersubunit distance, the one-electron carrier ISP-ED in the fixed conformation does not allow ET to c_1 heme, forcing the resulting highly reactive second electron to the low-potential chain. Once the second electron arrives at the Q_N site, the ISP-ED switches to the mobile conformation, allowing ET to cyt c_1.

movement of ISP-ED is not a satisfactory model for the bifurcated ET of bc_1 function (Esser et al. 2006).

The essence of the operation of bc_1 is the separation of the electrons from substrate quinol, which is initiated after the binding of the substrate and the docking of the mobile ISP-ED at the Q_P site of cyt b, bringing the high-potential ISC into close proximity with the substrate. Experimental observations of the ISP-ED conformation switch in the presence of two different types of inhibitors suggested a mechanism for

the bifurcated ET at the Q_P site. As shown in Figure 19.4b, ISP-ED must be in a fixed conformation when substrate QH_2 enters the Q_P pocket, occupying first the P_f site. It remains fixed in place during the first ET to the one-electron carrier ISC. At the same time, the two protons are transferred to their respective receptors, the glutamate of the P294-E295-W296-Y297 (PEWY) motif and a histidine ligand (H131) to ISC. Because of the 31 Å separation between the ISC and heme c_1, ET between these two redox partners will not occur as long as the reduced ISP-ED remains fixed at this position. The resulting highly unstable ubisemiquinone at the Q_P site $\left(SQ_P^{\bullet}\right)$ is, however, electronically coupled with hemes b_L and b_H, allowing the second electron to reduce the Q_N site occupant. Once the second electron reaches the Q_N site, the now fully oxidized quinone substrate Q moves to the P_m site, releasing ISP-ED from the b-site to deliver the first electron to cyt c_1. This hypothesis is known as the surface-affinity modulated ISP-ED conformation switch mechanism [SAMICS] (Esser et al. 2006; Xia et al. 2007).

19.3.3 Experimental Verification of SAMICS

Photosynthetic purple bacteria have played critical roles in experimental verification of hypotheses derived from structural observations, as they represent the most convenient and efficient way to introduce mutations into bc_1 complexes. As mentioned earlier (Figure 19.1), cyt bc_1 is required for bacteria under photosynthetic growth conditions. During the process of photosynthetic energy conversion, electrons are shuttled back and forth between RCs and cyt bc_1 by cyt c_2 and quinol/quinone. Importantly, there is no exogenous electron donor or final acceptor in this cyclic ET pathway, and light energy is converted to the proton motif force (pmf) for ATP synthesis. However, the quinol oxidation activity catalyzed by bc_1 can be bypassed when photosynthetic bacteria are grown under aerobic conditions, which is an important feature that allows an impaired or defective bc_1 to be experimentally examined (Gennis et al. 1993; Thony-Meyer 1997).

19.3.3.1 ISP-ED Mobility Is Necessary for ET Function of bc_1

Mutations were introduced to cyt b and ISP of bc_1 complexes from *R. sphaeroides, R. capsulatus,* and baker's yeast to determine whether the mobility of ISP-ED is a critical and vital aspect of bc_1 function. The possibility of a mobile ISP-ED arose immediately following the first structure determination of $Btbc_1$ (Kim et al. 1998; Xia et al. 1997), which placed the ISC at a distance too far from its oxidation partner cyt c_1 for rapid ET. With a globular head domain connected to a single TM helix through a flexible "neck" or hinge region, the ISP has all the structural features required to swivel between cyt b and cyt c_1, carrying one electron at a time. Mutations that modify the neck region of ISP to make it more rigid or flexible, or mutations that alter the length (deletion or insertion mutants), clearly inhibit or even abolish the enzymatic activity of bc_1 (Darrouzet et al. 2000a, 2000b; Obungu et al. 1998; Tian et al. 1998, 1999). Mutations were also introduced to cross-link ISP-ED to the cyt b subunit with a double cystein mutation (Ma et al. 2008; Xiao et al. 2000). The mutation sites, one in cyt b and one in the ISP-ED, are spatially close so that a disulfide bridge could be formed under oxidizing conditions or broken under reduction or alkylation conditions. The cross-linking experiment demonstrated unequivocally the dependence of bc_1 activity on the mobility of ISP-ED. Moreover, manipulation of the disulfide bridge provided an effective and reversible on–off switch for the mobile ISP head domain.

19.3.3.2 Mutagenesis in bc_1 Supports the Presence of a Control Mechanism for the ISP-ED Conformation Switch

A requirement of the SAMICS mechanism is that the ISP-ED remains fixed at the Q_P site or the b-position until the transfer of the second electron and protons from ubiquinol is completed, which explains the high fidelity of the bifurcation of the electron pathway. This hypothesis thus predicts that mutations in residues of the cyt b subunit lining the surface of the ISP-ED binding site would significantly alter the dynamics of ISP-ED movement, affecting bc_1 activity. The ISP-ED binding surface at the b-site consists of residues from the cd1 helix and EF loop with a total of 13 residues, 5 on the cd1 helix and 8 on the EF loop, that have their side chains exposed to the binding surface. The mutations introduced into these residues produced effects that agreed well with the prediction of altered ET kinetics between the ISP and cyt c_1 with minimal impact on substrate binding. For instance, mutations W142R/T/K/S (W141 in $Btbc_1$) were introduced to $Scbc_1$ (Bruel et al. 1995) and T160Y/S (T144 in $Btbc_1$) to $Rsbc_1$ (Mather et al. 1995), both on the cd1 helix, resulting in a dramatic reduction in bc_1 activity and altered cyt c_1 reduction kinetics; yet substrate binding was not affected. Similar observations were made for residues on the EF loop as in K329A of *R. sphaeroides* (K287 $Btbc_1$) (Esser et al. 2006), L286 and I292 (I268 $Btbc_1$) of *R. sphaeroides* (Rajagukguk et al. 2007), K329 of *R. sphaeroides* (Crofts et al. 1995), Y279 of yeast (Wenz et al. 2007), and Y302 of *R. capsulatus* (Lee et al. 2011b).

A number of these mutants were purified, and ET rates between ISP and cyt c_1 were measured by rapid photooxidation of a laser-excitable ruthenium cluster (Millett and Durham 2009). For example, residue I292 on the EF loop was mutated to I292L/M/V/E/R, and the ET rate between ISP and c_1 dropped precipitously for all mutants, when compared to the wild-type bc_1. In particular, the I292A and I292M mutants behaved in a manner similar to bc_1 inhibited by the P_f inhibitor famoxadone (Rajagukguk et al. 2007). The SAMICS hypothesis also predicts that limiting the movement of the cd1 helix leads to a modification in the activity of the complex depending on the size of the substitution, as seen in S155 of *R. sphaeroides* (Tian et al. 1997) and in G158 of *R. capsulatus* (Ding et al. 1995).

19.3.4 Presence or Absence of a Ubisemiquinone Free Radical as a Reaction Intermediate at the Quinol Oxidation Site

As discussed, we have a basic structural understanding on the bifurcation of electrons from substrate QH_2 at the Q_P site, where one electron follows the high-potential chain (ISP to cyt c_1 to cyt c_2) and the other is guided through the low-potential chain (b_L, b_H, and quinone). Exactly how the Q_P site functions to separate the two electrons at higher temporal resolutions remains controversial. A critical point seems to be the presence or absence of a ubisemiquinone $\left(SQ_P^{\bullet}\right)$ as a reaction intermediate, as represented by the "sequential"

(Crofts et al. 2003; Hong et al. 1999; Link 1997) and "concerted" mechanisms (Hunte et al. 2003; Snyder et al. 2000; Trumpower 2002; Zhu et al. 2007), respectively (Figure 19.5). In the sequential oxidation mechanism, the first electron of QH_2 is transferred to the ISC and then to heme c_1, and the resulting $SQ_P^•$ is used to reduce heme b_L and then b_H. In the concerted oxidation mechanism, the two electrons from QH_2 are transferred to ISC and heme b_L near simultaneously, and thus, no $SQ_P^•$ intermediate forms or can be detected.

However, the search for the free radical at the Q_P site, which should be sensitive to Q_P site inhibitors such as myxothiazol or stigmatellin, has been mostly disappointing, leading

to proposals that feature a highly unstable or short-lived $SQ_P^•$ (Crofts and Wang 1989; Junemann et al. 1998). Naturally, the failure to detect $SQ_P^•$ under physiological conditions at the Q_P site (Junemann et al. 1998; Yang et al. 2008) raises concerns regarding the validity of the sequential oxidation mechanism, even though several plausible explanations for the absence of $SQ_P^•$ have been offered, such as antiferromagnetic coupling between $SQ_P^•$ and ISC (Link 1997) or $SQ_P^•$ being rapidly oxidized by heme b_L (Crofts et al. 2003).

The unstable ubisemiquinone intermediate obligates the reduction of the low-potential chain by the second electron to occur almost simultaneously with the reduction of the

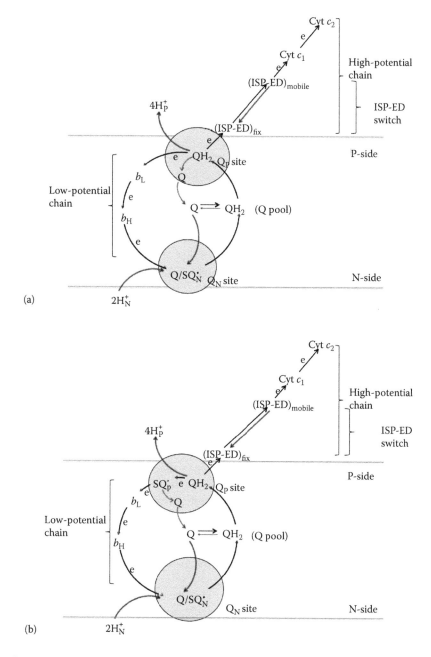

FIGURE 19.5 Concerted versus sequential ubiquinol oxidation. (a) The concerted quinol oxidation mechanism features a bifurcated ET that does not allow $SQ_P^•$ to exist for more than a picosecond under physiological conditions. (b) The sequential quinol oxidation mechanism mandates the existence of $SQ_P^•$ as part of the bifurcated ET mechanism.

high-potential chain by the first electron. In fact, Osyczka et al. (2005) suggested that the two ET events should be termed *concerted* ET if they occur within 1 ps of each other, and they argued that the concerted mechanism presents the best means to avoid short-circuit, wasteful ET reactions, because the bc_1 complexes, like other enzymes, are fully reversible (Osyczka et al. 2004).

Taking advantage of the unique electron paramagnetic resonance (EPR) signatures of the ISC (Orme-Johnson et al. 1974), hemes b_L, b_H, and c_1 (de Vries et al. 1979; Salerno 1984) and ubisemiquinone radicals (Ohnishi and Trumpower 1980; Yu et al. 1980), investigators carefully studied the pre-steady-state kinetics of the ISC and heme b_L reduction by QH_2 in bovine bc_1, using an ultra-fast microfluidic mixer and a freeze-quenching device coupled with EPR spectroscopy (Zhu et al. 2007). The results showed that heme b_L and the ISC have similar reduction kinetics, with a $t_{1/2}$ of 250 μs, which is consistent with the concerted bifurcated oxidation mechanism, at least at the time resolution of the instrument (60 μs). Thus, pre-steady-state kinetic analysis of the reduction of cyt b_L and ISP in the same sample supports the concerted scheme for the bifurcated oxidation of quinol at the Q_P site (Snyder et al. 2000; Zhu et al. 2007).

Although the concerted mechanism explains why the proposed $SQ_P^{\bullet}$ is not detected, it was argued that the similar reduction rates observed in heme b_L and ISC could be due to the inadequate (60 μs) time resolution of the instrumentation available, which failed to detect the tiny amount of highly unstable ubisemiquinones (Crofts et al. 2008). Various approaches to stabilize the proposed SQ_P have been employed. In one case, using a b_H heme knockout bc_1 under the conditions of multiple laser flashes and alkaline pH, Zhang et al. (2007) reported detection of radicals that are sensitive to stigmatellin but not to myxothiazol at an estimated 1% stoichiometric level. In another report, ubisemiquinone radicals were detected under anaerobic and oxidant-induced cyt b reduction conditions (Cape et al. 2007). However, using a mutant complex lacking heme b_L designed to maximize the amount of $SQ_P^{\bullet}$, Yang et al. (2008) were unable to detect ubisemiquinol signal by EPR, indicating that $SQ_P^{\bullet}$ is not stabilized under these conditions. Although it is difficult to determine the ISC reduction kinetics, so far, there is no evidence showing that the ISC receives an electron from QH_2 before the cyt b hemes do.

19.4 REGULATION OF bc_1 ACTIVITY IN PHOTOSYNTHETIC PURPLE BACTERIA

As an integral part of both photosynthetic and respiratory machineries, the cyt bc_1 complex of photosynthetic purple bacteria has its activity regulated by membrane potential (pmf) (Bechmann and Weiss 1991), by the concentration of substrates (QH_2 and cyt c_2) (Bechmann and Weiss 1991), and by lipid molecules (Schagger et al. 1990; Yu et al. 1984). Additionally, many other factors were reported to be capable of regulating bc_1 activity, although the physiological significance of these effects is not fully appreciated.

19.4.1 INHIBITION OF BACTERIAL bc_1 ACTIVITY BY Zn^{2+}

Zinc ions have been shown to inhibit cellular respiration (Skulachev et al. 1967), and cyt bc_1 was one of the inhibition sites (Kleiner and von Jagow 1972). Both mitochondrial and bacterial bc_1 complexes are inhibited by Zn^{2+} (Kleiner 1974; Lee et al. 2011a). Kinetic analysis of the redox reactions of the cytochromes indicated that Zn^{2+} affects the activity of the complex at the Q_P site. The crystalline chicken bc_1 complex specifically binds Zn^{2+} at two sites in each monomer (Berry et al. 2000). Both are close to the stigmatellin-binding site and are candidates for the inhibitory site. One binding site is actually in the hydrophobic channel between the Q_P site and the bulk lipid phase, and may interfere with quinone binding. The other is in a hydrophilic area between cytochromes b and c_1 and might interfere with the egress of protons from the Q_P site to the intermembrane aqueous medium. No zinc was bound near the putative proteolytic active site of subunits 1 and 2 (homologous to mitochondrial processing peptidase) under these conditions.

A comparative x-ray absorption fine-structure spectroscopy (XAFS) study of Zn^{2+} binding to both mitochondrial and bacterial cyt bc_1 complexes showed quite different spectral features exhibited by the two complexes (Giachini et al. 2007). While the XAFS spectra suggested tetrahedral ligand coordination for the $Mtbc_1$, the spectra for *R. capsulatus* are consistent with octahedral coordination. By aligning the crystal structures of the bacterial and avian enzymes, a group of residues was found to be located in the region homologous to that of the more hydrophobic site in avian bc_1. This cluster included the H276, D278, E295, and N279 residues of the cyt b subunit (Figure 19.3a). More detailed biochemical and mutagenesis data suggested that Zn^{2+} inhibits bacterial bc_1 function by blocking the proton exiting pathway involving E295, because mutation E295V significantly reduced K_i value by Zn^{2+} (Lee et al. 2011a).

19.4.2 EFFECT OF O_2 ON THE ELECTRON TRANSFER ACTIVITY OF bc_1

Facultative photosynthetic purple bacteria are able to acclimate to different growth conditions by switching from aerobic to anaerobic or to photosynthetic growth mode depending upon the level of oxygen. Because these bacteria usually grow more favorably under aerobic conditions (Hunter 1950), it is reasonable to speculate that the level of oxygen may have an effect on the function of cellular respiratory chain components. Since molecular oxygen is the terminal electron acceptor for the respiratory chain, much attention concerning the interaction between molecular oxygen and respiratory chain complexes has been focused on cyt c oxidase. However, an increasing number of experiments have indicated that oxygen has an effect on the cyt bc_1 function as well. It was shown that the ET activity of mitochondrial cyt bc_1 increases in response to elevated oxygen levels (Zhang et al. 1998a). In a study that was more systematic, with the oxygen levels carefully controlled, Zhou et al. (2012) compared ET activities of bc_1 from photosynthetic bacteria in the presence and absence of oxygen (Zhou et al. 2012). Molecular oxygen increased the ET activity of bc_1 by up to 82%, depending on the structural integrity and activity of the complex (Figure 19.6a, Table 19.1). It should be stressed that this enhanced activity is clearly independent of the production of superoxide, especially at low oxygen concentration, as both high- and low-potential chains are simultaneously stimulated (Figure 19.6b and c). Similarly, it has recently been reported that molecular oxygen enhances the ET activity of bc_1 from *R. capsulatus* by 42%. Even under antimycin A-inhibited conditions, oxygen-dependent enhancement of pre-steady-state bc_1 activity was observed (Muller et al. 2002; Zhou et al. 2012). The oxygen-dependent and oxygen-independent ET activities in bc_1 share the same mechanistic origin (rate-limiting step) because they have the same activation energy, pH profile, and ionic strength dependency (Zhou et al. 2012).

How molecular oxygen modulates bc_1 activity remains largely unknown. Nevertheless, it is important to pinpoint the exact step that is regulated by oxygen in the reaction coordinates of bc_1. Toward this end, the pre-steady-state rates of reduction of hemes b_L and b_H in bc_1 by QH_2 were determined under aerobic and anaerobic conditions, using a stopped-flow apparatus, and the results showed that the oxygen enhances the rate of low-potential reduction compared to that in the absence of oxygen. This enhancement of ET activity is sensitive to the Q_P site inhibitor stigmatellin but insensitive to the Q_N site inhibitor antimycin (Zhou et al. 2012). Thus, the site of dioxygen action is likely the Q_P site, which is further supported by the determination of the pre-steady-state rates of the reduction of hemes b_L and b_H in the heme b_H (H111N) or b_L (H198N) knockout mutant complexes (Yang et al. 2008; Zhou et al. 2012).

19.4.3 INTERACTION OF BC_1 COMPLEXES WITH CYTOSOLIC PROTEINS

Although structural and functional studies of mitochondrial bc_1 have been intensive, investigations into the interactions of bc_1 with other components of the ET chain and with soluble enzymes of the mitochondrial matrix have been limited. Such investigations are particularly lacking for bc_1 complexes of photosynthetic purple bacteria. In mitochondria, components of the ET chain form supercomplexes in the inner membrane (Schagger and Pfeiffer 2000). Complex II (succinate–ubiquinone oxidoreductase) is the only enzyme that participates in both the tricarboxylic acid (TCA) cycle and the electron transport chain (Oyedotun and Lemire 2004). Other enzymes of the TCA cycle are known to couple to the ET chain through their product NADH at complex I (Averet et al. 2002; Fukushima et al. 1989; Ovadi et al. 1994). However, Wang et al. (2010) showed that mitochondrial bc_1 interacts with malate dehydrogenase (MDH), aconitase (ACON), and aspartate transaminase by a novel coprecipitation method coupled with matrix-assisted laser desorption ionization (MALDI).

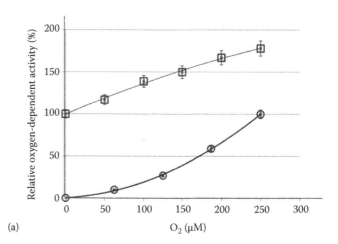

(a)

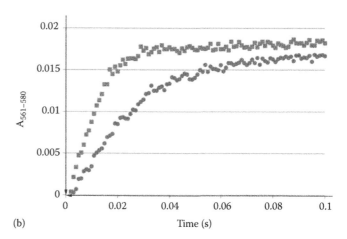

(b)

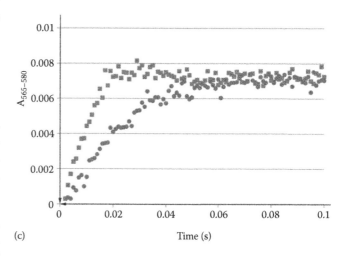

(c)

FIGURE 19.6 Oxygen-dependent ET and superoxide generation activities of cyt bc_1. (a) Oxygen-dependent ET and superoxide generation activities of *Rsbc₁*. The curve with open squares represents the net increase of cyt bc_1 activity, and the curve with open circles is the superoxide-generating activity from the oxidation of ubiquinol by *Rsbc₁* under various oxygen concentrations. (b) Time trace of heme b_L reduction by *Rsbc₁* under aerobic (filled squares) and anaerobic (filled circles) conditions. Reductions of heme b_H were determined from the increase at $A_{561-580}$. (c) Time trace of heme b_L reduction, determined from the increase at $A_{565-580}$.

TABLE 19.1

Electron Transfer Activities of Various bc_1 Preparations in the Presence or Absence of Oxygen

	Specific Activity[b] (μmol c reduced/min/nmol b)	
Preparation[a]	$+O_2$	$-O_2$
$Rsbc_1$	3.0	2.0
$Rsbc_1$ + IV	4.2	2.3
$Rs\Delta$IV	0.6	0.5
c_1-14Gly-IV-6His	3.8	2.1
$Mtbc_1$	22.0	16.0

[a] c_1-14Gly-IV-6His, $Rsbc_1$ with subunit IV fused to cyt c_1 C-terminus through 14 glycine residues; IV, recombinant, purified subunit IV; $Mtbc_1$, mitochondrial bc_1; $Rsbc_1$, wild-type, four-subunit complex; $Rsbc_1$ + IV, wild-type complex with added IV; $Rs\Delta$IV, $Rsbc_1$ lacking subunit IV.

[b] Specific activity refers to the antimycin-sensitive electron transfer activity.

In a recent effort to identify Interactions of bacterial bc_1 with cytosolic components and using a procedure specifically designed to identify interacting partners of bacterial bc_1, we found evidence of interaction of $Rsbc_1$ with UspA, a member of the large family of universal stress proteins of bacteria, archaea, fungi, and plants highly abundant in the cytosol of growth-arrested cells (Kvint et al. 2003). Purified, recombinant UspA enhances the ET activity of $Rsbc_1$ by ~40% but decreases its superoxide-generating activity by >60% (Su, PhD thesis, 2013, Oklahoma Statue University). Interestingly, the effect of UspA on the ET activity of $Rsbc_1$ is oxygen dependent; no enhancement in activity was observed in the absence of oxygen. These results, though preliminary, led us to suggest a regulatory function of UspA in cellular energy conservation. The capability of UspA to increase the ET activity and decrease superoxide generation by bc_1 could explain why this protein is expressed under stress conditions.

19.5 ORIGIN OF SUPEROXIDE GENERATION

It has long been recognized that during mitochondrial respiration, there is a continuous release of electrons that have escaped from the ET chain. They react with molecular oxygen to form $O_2^{\cdot-}$, which are progenitors of reactive oxygen species (ROS) (Boveris and Chance 1973; Loschen et al. 1974). ROS are well known as one of the toxicities mediating oxidative stress (Andreyev et al. 2005; Barja 2004; Van Remmen and Richardson 2001), causing damage to DNA/RNA, proteins, carbohydrates, and lipids and leading to aging, cancer, and inflammation (Moro et al. 2005; Sadek et al. 2003; Shigenaga et al. 1994). Isolated mitochondria under steady-state respiration (state 4) generate 0.6–1.0 nmol of H_2O_2/min/mg protein, accounting for about 2% of O_2 uptake under physiological conditions (Boveris et al. 1972). Production of $O_2^{\cdot-}$ during mitochondrial respiration is closely related to mitochondrial coupling efficiency; more $O_2^{\cdot-}$ is produced when the membrane potential of mitochondria is high (Korshunov et al. 1997;

Rottenberg et al. 2009). Two segments of the respiratory chain have been demonstrated to be responsible for the generation of $O_2^{\cdot-}$ in mitochondria. One is the NADH–Q oxidoreductase (complex I), and the other is cyt bc_1. The generated $O_2^{\cdot-}$ is subsequently dismutated to H_2O_2 either spontaneously or by the action of superoxide dismutases (McCord and Fridovich 1969). In the past, information concerning sites of mitochondrial generated $O_2^{\cdot-}$ was mostly acquired from studies using intact heart mitochondria with selected ET inhibitors and by measuring the H_2O_2 concentration in the suspending medium (Nohl and Jordan 1986; Turrens et al. 1985).

While the physiological role of this low amount of superoxide remains a focus of extensive investigations, its generation by mitochondrial cyt bc_1 has been carefully characterized. ROS generated by the cyt bc_1 has more impact on the cytosolic concentration of ROS (Muller et al. 2004). The production of $O_2^{\cdot-}$ by bc_1 is greatly enhanced when the complex is inhibited by the Q_N site inhibitor antimycin (Sun and Trumpower 2003), when the membrane is fully potentiated (Rottenberg et al. 2009; Zhang et al. 1998a), or when structural integrity of the complex is compromised (Yin et al. 2010). Mechanistically, identifying the source of electron leakage has been a centerpiece of ROS research on the bc_1 complex, and several competing hypotheses have been proposed. One hypothesis proposed that superoxide is the result of electron leakage from SQ_P to oxygen during the Q_P site reaction (Turrens et al. 1985). A second proposed site of electron donors to generate $O_2^{\cdot-}$ is the reduced heme b_L because of its low redox potential (Nohl and Jordan 1986; Zhang et al. 1998a), which implies a reversed electron flow from b_L to molecular oxygen. From a structural point of view, the $O_2^{\cdot-}$ generation by bc_1 depends on where oxygen is located in the structure and how many molecules of oxygen there are within the structure, which could be dependent on the structural integrity of the bc_1 complex (Yin et al. 2010).

Most of the early observations on production of ROS by mitochondrial bc_1 can be reproduced in bacterial bc_1 (Lee et al. 2011b; Sarewicz et al. 2010; Yang et al. 2008; Yin et al. 2010), making them ideal model systems for investigating the mechanism of ROS generation by bc_1. An added advantage of using bacterial bc_1 for mechanistic studies of ROS production is the fact that the amount of ROS generated by bacterial bc_1 is higher, making detection easier.

19.5.1 MUTATIONS THAT SUPPRESS OR ENHANCE SUPEROXIDE GENERATION IN BACTERIAL BC_1

It has been reported that superoxide generation is reciprocally related to the ET activity. Bovine cyt bc_1 has 10 times the ET activity but only one-tenth of the activity for superoxide generation compared with $Rsbc_1$ (Yin et al. 2009). Mutations resulting in lower ET activity are always accompanied by higher superoxide generation (Table 19.2). The Δ-subunit IV mutant of $Rsbc_1$ or three-subunit $Rsbc_1$, in which the only supernumerary subunit IV was deleted, has 3.5 times higher superoxide generation activity than that of the wild type. When the

TABLE 19.2
Summary of ET Activity and Superoxide Production by Various bc_1 Complex Preparations

Cyt bc_1 Preparations	Relative ET Activities	Relative Superoxide Production −Ant A	+Ant A	Ref
$Btbc_1$[a]	826 (20.0)[b]	11 (0.008)[c]		Zhang et al. 1998a
wt$Rsbc_1$	100 (2.42)	100 (0.07)	329 (0.23)[d]	Yin et al. 2010
3sub-$Rsbc_1$ (3 subunits)	26 (0.63)	357 (0.25)	600 (0.42)	Yin et al. 2010
3sub-$Rsbc_1$ + wtIV	96 (2.32)	100 (0.07)	329 (0.23)	Yin et al. 2010
3sub-$Rsbc_1$ + IV(Y81A)	84 (2.03)	129 (0.09)	386 (0.27)	Yin et al. 2010
3sub-$Rsbc_1$ + IV(R84E)	58 (1.41)	229 (0.16)	486 (0.34)	Yin et al. 2010
3sub-$Rsbc_1$ + IV(81-84)A	38 (0.92)	300 (0.21)	543 (0.38)	Yin et al. 2010
IV-c_1 fused $Rsbc_1$	165	70	–	Zhou et al. 2012
wt$Rcbc_1$[e]	100	100	530	Lee et al. 2011b
Y302A$Rcbc_1$[e]	15.4	730	520	Lee et al. 2011b
Y302F$Rcbc_1$[e]	77.4	570	450	Lee et al. 2011b
$Scbc_1$[f]	100	100	–	Wenz et al. 2007
F129K$Scbc_1$[f]	6.8	200	–	Wenz et al. 2007

[a] Numbers for bovine cyt bc_1 are in reference to wt$Rsbc_1$.
[b] Numbers in parentheses are specific activity in µmol cyt c reduction/min/nmol of cyt b.
[c] Numbers in parentheses are XO unit defined as chemiluminescence generated by 1 U of xanthine oxidase.
[d] Numbers are in reference to wild-type $Rsbc_1$ in the absence of antimycin A.
[e] Numbers are in reference to wild-type $Rcbc_1$.
[f] Numbers are in reference to wild-type yeast bc_1.

subunit IV was reintegrated into the complex, both the ET and superoxide generation activities returned to the same level as that of the wild-type complex (Tso et al. 2006; Yin et al. 2009). The subunit IV–cyt c_1 fusion reduces superoxide production to 70% of the wild-type level, while ET activity is 60% higher (Su et al. 2012). Mutations introduced to conserved residues often lead to changes in superoxide activity. For example, mutations introduced to Y302 of *R. capsulatus* cyt b, the equivalent of Y278 in human and Y279 in bovine cyt b, all have decreased ET activity and an increased rate of superoxide generation (Lee et al. 2011b). In particular, the Y302C mutant

covalently modifies the ISC of ISP, resulting in impaired catalytic function and enhanced superoxide generation. Similarly, the M183L mutant in cyt c_1 of $Rcbc_1$ was unable to grow photosynthetically due to cyt c_1 inactivation (Gray et al. 1992), but higher superoxide generation also was reported (Borek et al. 2008). Many other examples showing an inverse relationship between ET activity and the amount of superoxide generated by mutant bc_1 can be found in the literature (Sarewicz et al. 2010; Wenz et al. 2007). Interestingly, structurally compromised bacterial bc_1 by proteinase K treatment displays a similar relationship (Figure 19.7) (Yin et al. 2010).

19.5.2 Ubisemiquinol as a Source for Superoxide Formation

The Q-cycle mechanism of cyt bc_1 function mandates a bifurcated ET at the Q_P site, which suggests the presence of a possible transient $SQ_P^{\bullet}$ at the Q_P site during the ET reaction. This transient $SQ_P^{\bullet}$ has been hypothesized to be the source of electrons for $O_2^{\bullet-}$ (Figure 19.8a), implying that bc_1 would have higher ET activity under anaerobic than aerobic conditions, as no electron leakage would occur in the absence of oxygen. While this was true when activity of succinate–cyt c oxidoreductase was determined using succinate as a substrate (Zhang et al. 1998a), the effect of oxygen on the activity of ubiquinol–cyt c oxidoreductase is more complicated; different results have been reported (Muller et al. 2003). One of the problems with this hypothesis is that under physiological conditions, the time span for this $SQ_P^{\bullet}$ species to exist is extremely short, if it exists at all, as evidenced by the failure

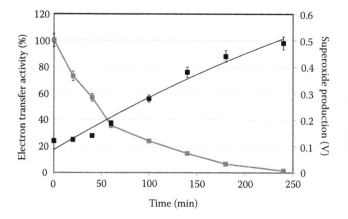

FIGURE 19.7 Inverse relationship between ET activity and superoxide generation activity of $Rsbc_1$. Samples of $Rsbc_1$ incubated with proteinase K were taken at given time points for measurements of ET activity (gray) and superoxide generation activity (black).

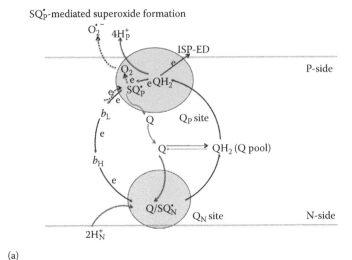

(a)

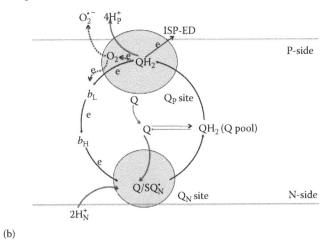

(b)

FIGURE 19.8 Mechanistic models for the generation of superoxide by cyt bc_1. (a) Semiquinol radical-mediated superoxide generation by cyt bc_1. (b) Oxygen-mediated superoxide generation by cyt bc_1.

in detecting this species under physiologically relevant conditions. Indeed, the postulated ubisemiquinone species were only detected under antimycin-inhibited and anaerobic conditions and never accumulated to a significant level (Cape et al. 2007; Zhang et al. 2007).

19.5.3 REDUCED HEME B_L AS A SOURCE FOR SUPEROXIDE FORMATION

It has also been proposed that reduced b_L heme has a sufficiently low midpoint potential to be the electron donor for O_2^- formation (Figure 19.8a). This mechanism gained indirect experimental support by the observation that superoxide production is maximal when the quinol pool is partially oxidized (Drose and Brandt 2008), which was interpreted as backward transfer of an electron from reduced heme b_L to an oxidized quinone at the Q_p site to form an $SQ_p^•$. This $SQ_p^•$ then serves as the electron source for superoxide formation. Since this model

also requires the presence of an $SQ_p^•$, it suffers from the paradoxical absence of observed $SQ_p^•$.

19.5.4 ROLE OF OXYGEN IN MEDIATING SUPEROXIDE PRODUCTION

A more recent proposed mechanism was based on the observations that the ET rate increases in bc_1 in proportion to the oxygen tension (Yin et al. 2010; Zhou et al. 2012) and that bc_1 is still operational in pre-steady-state kinetics even in the absence of heme b_L (Yang et al. 2008), suggesting that oxygen trapped in bc_1 could serve as the receptor for stray electrons and a source of superoxide. Since the mutant $Rsbc_1$ that lacks heme b_L produces a similar amount of O_2^- to that of the antimycin A-inhibited wild-type complex, the heme b_L is certainly unable to serve as the source of electrons for superoxide production in this case. This mechanism, as it was proposed, involves a direct transfer of an electron from ubiquinol to oxygen for superoxide production by bc_1 (Figure 19.8b). Clearly, this model deviates from the classic semiquinone intermediate hypothesis in that it removes the requirement for a stabilized $SQ_P^{\bullet}$ as an electron donor. According to this mechanism, O_2 is able to bind to a hydrophobic niche in the vicinity of the Q_P site and acts as an electron acceptor. Together with the high-potential acceptor, ISC, this bound O_2 accepts an electron and a proton resulting from the bifurcated oxidation of QH_2 to produce protonated superoxide $\left(O_2^{\bullet-}\right)$, which escapes to a hydrophilic environment (intermembrane space or matrix) and becomes deprotonated to produce $O_2^{\bullet-}$, as the pKa for $O_2^{\bullet}H$ is approximately 4.8 in a hydrophilic environment. Based on this mechanism, only a small amount of $O_2^{\bullet}H$ would leak to become superoxide anions under normal circumstances, because O_2, in this case, is also involved in the bifurcated oxidation of QH_2 by mediating ET from QH_2 to heme b_L. The observation that oxygen increased the ET activity of bc_1 supports this hypothesis.

19.6 CONCLUDING REMARKS

The cyt bc_1 complex is one of the best-characterized respiratory and photosynthetic complexes, both structurally and mechanistically, thanks in large part to approximately 15 years of intensive post-3D-structure studies. Despite the controversial issues discussed in this article, it is reasonable to say that most structure/function questions concerning cyt bc_1 have been resolved. On the other hand, how the activity of the bc_1 complex is regulated in relation to other cellular activities is very much uncharted territory, deserving future exploration. Considering the vital role of cyt bc_1 in cellular respiration and photosynthesis, it is conceivable that many cellular factors may influence the structure and function of the bc_1 complexes. Mutations in cyt bc_1 complexes and defects in cellular machineries that assemble this complex have been linked to various pathogenic conditions, the mechanisms of which have yet to be determined. The cyt bc_1 complex has been an important target of chemical intervention for disease control, both in humans and in crops. The rise of inevitable drug resistance demands further investigation into the underlying mechanism(s) and development of new therapeutics.

ACKNOWLEDGMENTS

The authors thank George Leiman for editorial assistance during the preparation of this manuscript. This research was supported by the Intramural Research Program of the Center for Cancer Research, National Cancer Institute, National Institutes of Health.

ABBREVIATIONS

bc_1: complex III, ubiquinol–cytochrome c oxidoreductase, or cytochrome bc_1
b_L: low-potential heme or b_{566}
b_H: high-potential heme or b_{562}
Btbc$_1$: *Bos taurus* bc_1
cyt: cytochrome

ET: electron transfer
H_2O_2: hydrogen peroxide
ISC: iron–sulfur cluster
ISP: iron–sulfur protein
ISP-ED: extrinsic domain of ISP
Mtbc$_1$: mitochondrial cyt bc_1
$O_2^{\bullet-}$: superoxide anion
P_f: inhibitor: inhibitor that fixes ISP-ED conformation
P_m: inhibitor: inhibitor that makes ISP-ED mobile
Q_N site: ubiquinone reduction site
Q_P site: ubiquinol oxidation site
Rsbc$_1$: bc_1 from *Rhodobacter sphaeroides*
QH_2: ubiquinol
Q: ubiquinone
Rcbc$_1$: bc_1 from *Rhodobacter capsulatus*
Rsbc$_1$: bc_1 from *Rhodobacter sphaeroides*
ROS: reactive oxygen species
$SQ_P^{\bullet}$: ubisemiquinone anion at Q_P site
TCA: tricarboxylic acid

REFERENCES

Andrews KM, Crofts AR, and Gennis RB (1990) Large-scale purification and characterization of a highly active four-subunit cytochrome bc_1 complex from *Rhodobacter sphaeroides*. *Biochemistry* 29: 2645–2651.

Andreyev AY, Kushnareva YE, and Starkov AA (2005) Mitochondrial metabolism of reactive oxygen species. *Biochemistry Biokhimiia* 70: 200–214.

Averet N, Aguilaniu H, Bunoust O, Gustafsson L, and Rigoulet M (2002) NADH is specifically channeled through the mitochondrial porin channel in Saccharomyces cerevisiae. *Journal of Bioenergetics and Biomembranes* 34: 499–506.

Barja G (2004) Free radicals and aging. *Trends in Neurosciences* 27: 595–600.

Bechmann G, and Weiss H (1991) Regulation of the proton/electron stoichiometry of mitochondrial ubiquinol:cytochrome c reductase by the membrane potential. *European Journal of Biochemistry* 195: 431–438.

Berry EA, Zhang Z, Bellamy HD, and Huang L (2000) Crystallographic location of two Zn2+-binding sites in the avian cytochrome bc_1 complex. *Biochimica et Biophysica Acta* 1459: 440–448.

Berry EA, Huang L, Saechao LK, Pon NG, Valkova-Valchanova M, and Daldal F (2004) X-ray structure of *Rhodobacter capsulatus* cytochrome bc_1: Comparison with its mitochondrial and chloroplast counterparts. *Photosynthesis Research* 81: 251–275.

Borek A, Sarewicz M, and Osyczka A (2008) Movement of the iron–sulfur head domain of cytochrome bc(1) transiently opens the catalytic Qo site for reaction with oxygen. *Biochemistry* 47: 12365–12370.

Boveris A, and Chance B (1973) The mitochondrial generation of hydrogen peroxide. General properties and effect of hyperbaric oxygen. *Biochemical Journal* 134: 707–716.

Boveris A, Oshino N, and Chance B (1972) The cellular production of hydrogen peroxide. *Biochemical Journal* 128: 617–630.

Brandt U (1996) Bifurcated ubihydroquinone oxidation in the cytochrome bc_1 complex by proton-gated charge transfer. *FEBS Letters* 387: 1–6.

Brandt U, and von Jagow G (1991) Analysis of inhibitor binding to the mitochondrial cytochrome c reductase by fluorescence quench titration. *European Journal of Biochemistry* 195: 163–170.

Brandt U, and Trumpower B (1994) The protonmotive Q cycle in mitochondria and bacteria. *Critical Reviews in Biochemistry and Molecular Biology* 29: 165–197.

Bruel C, di Rago J, Slonimski PP, and Lemesle-Meunier D (1995) Role of the evolutionarily conserved cytochrome b tryptophan 142 in the ubiquinol oxidation catalyzed by the bc_1 complex in the yeast Saccharomyces cerevisiae. *The Journal of Biological Chemistry* 270: 22321–22328.

Cape JL, Bowman MK, and Kramer DM (2007) A semiquinone intermediate generated at the Qo site of the cytochrome bc_1 complex: Importance for the Q-cycle and superoxide production. *Proceedings of the National Academy of Sciences of the United States of America* 104: 7887–7892.

Chen YR, Usui S, Yu CA, and Yu L (1994) Role of subunit IV in the cytochrome b-c1 complex from *Rhodobacter sphaeroides*. *Biochemistry* 33: 10207–10214.

Crofts AR (2004) The cytochrome bc_1 complex: Function in the context of structure. *Annual Review of Physiology* 66: 689–733.

Crofts AR, and Wang Z (1989) How rapid are the internal reactions of the ubiquinol:cytochrome c2 oxidoreductase? *Photosynthesis Research* 22: 69–87.

Crofts AR, Meinhardt SW, Jones KR, and Snozzi M (1983) The role of the quinone pool in the cyclic electron-transfer chain of *Rhodopseudomonas sphaeroides*: A modified q-cycle mechanism. *Biochimica et Biophysica Acta* 723: 202–218.

Crofts AR, Barquera B, Bechmann G, Guergova M, Salcedo-Hernandez R, Hacker B, Hong S, Gennis RB, and Mathis P (1995) *Photosynthesis: From light to biosphere* (The Netherlands: Kluwer Academic Publications).

Crofts AR, Shinkarev VP, Kolling DR, and Hong S (2003) The modified Q-cycle explains the apparent mismatch between the kinetics of reduction of cytochromes c1 and bH in the bc_1 complex. *The Journal of Biological Chemistry* 278: 36191–36201.

Crofts AR, Lhee S, Crofts SB, Cheng J, and Rose S (2006) Proton pumping in the bc_1 complex: A new gating mechanism that prevents short circuits. *Biochimica et Biophysica Acta* 1757: 1019–1034.

Crofts AR, Holland JT, Victoria D, Kolling DR, Dikanov SA, Gilbreth R, Lhee S, Kuras R, and Kuras MG (2008) The Q-cycle reviewed: How well does a monomeric mechanism of the bc(1) complex account for the function of a dimeric complex? *Biochimica et Biophysica Acta* 1777: 1001–1019.

Darrouzet E, Valkova-Valchanova M, and Daldal F (2000a) Probing the role of the Fe-S subunit hinge region during Q(o) site catalysis in *Rhodobacter capsulatus* bc(1) complex. *Biochemistry* 39: 15475–15483.

Darrouzet E, Valkova-Valchanova M, Moser CC, Dutton PL, and Daldal F (2000b) Uncovering the [2Fe2S] domain movement in cytochrome bc(1) and its implications for energy conversion. *Proceedings of the National Academy of Sciences of the United States of America* 97: 4567–4572.

de Vries S, Albracht SP, and Leeuwerik FJ (1979) The multiplicity and stoichiometry of the prosthetic groups in QH2: Cytochrome c oxidoreductase as studied by EPR. *Biochimica et Biophysica Acta* 546: 316–333.

di Rago JP, and Colson AM (1988) Molecular basis for resistance to antimycin and diuron, Q-cycle inhibitors acting at the Qi site in the mitochondrial ubiquinol–cytochrome c reductase in Saccharomyces cerevisiae. *The Journal of Biological Chemistry* 263: 12564–12570.

Ding H, Robertson DE, Daldal F, and Dutton PL (1992) Cytochrome bc_1 complex [2Fe-2S] cluster and its interaction with ubiquinone and ubihydroquinone at the Qo site: A double-occupancy Qo site model. *Biochemistry* 31: 3144–3158.

Ding H, Moser CC, Robertson DE, Tokito MK, Daldal F, and Dutton PL (1995) Ubiquinone pair in the Qo site central to the primary energy conversion reactions of cytochrome bc_1 complex. *Biochemistry* 34: 15979–15996.

Drose S, and Brandt U (2008) The mechanism of mitochondrial superoxide production by the cytochrome bc_1 complex. *The Journal of Biological Chemistry* 283: 21649–21654.

Esser L, Quinn B, Li Y, Zhang M, Elberry M, Yu L, Yu CA, and Xia D (2004) Crystallographic studies of quinol oxidation site inhibitors: A modified classification of inhibitors for the cytochrome bc_1 complex. *Journal of Molecular Biology* 341: 281–302.

Esser L, Gong X, Yang S, Yu L, Yu CA, and Xia D (2006) Surface-modulated motion switch: Capture and release of iron–sulfur protein in the cytochrome bc_1 complex. *Proceedings of the National Academy of Sciences of the United States of America* 103: 13045–13050.

Esser L, Elberry M, Zhou F, Yu CA, Yu L, and Xia D (2008) Inhibitor-complexed structures of the cytochrome bc_1 from the photosynthetic bacterium *Rhodobacter sphaeroides*. *The Journal of Biological Chemistry* 283: 2846–2857.

Fukushima T, Decker RV, Anderson WM, and Spivey HO (1989) Substrate channeling of NADH and binding of dehydrogenases to complex I. *The Journal of Biological Chemistry* 264: 16483–16488.

Garland PB, Clegg RA, Doxer D, Downie JA, and Haddock BA (1975) Proton-translocating nitrate reductase of Escherichia coli. In *Electron transfer chains and oxidative phosphorylation*, E. Quagliatrello, S. Papa, F. Falmieri, E.C. Slater, and N. Siliprandi, eds. (Amsterdam, The Netherlands: North-Holland Publishing Co.), pp. 351–358.

Gennis RB, Barquera B, Hacker B, Van Doren SR, Arnaud S, Crofts AR, Davidson E, Gray KA, and Daldal F (1993) The bc_1 complexes of *Rhodobacter sphaeroides* and *Rhodobacter capsulatus*. *Journal of Bioenergetics and Biomembrane* 25: 195–209.

Giachini L, Francia F, Veronesi G, Lee DW, Daldal F, Huang LS, Berry EA, Cocco T, Papa S, Boscherini F, and Venturoli G (2007) X-ray absorption studies of Zn2+ binding sites in bacterial, avian, and bovine cytochrome bc_1 complexes. *Biophysical Journal* 93: 2934–2951.

Gray KA, Davidson E, and Daldal F (1992) Mutagenesis of methionine-183 drastically affects the physicochemical properties of cytochrome c_1 of the bc_1 complex of *Rhodobacter capsulatus*. *Biochemistry* 31: 11864–11873.

Gurung B, Yu L, Xia D, and Yu CA (2005) The iron–sulfur cluster of the Rieske iron–sulfur protein functions as a proton-exiting gate in teh cytochrome bc(1) complex. *The Journal of Biological Chemistry* 280: 24895–24902.

Hong S, Ugulava N, Guergova-Kuras M, and Crofts AR (1999) The energy landscape for ubihydroquinone oxidation at the Q(o) site of the bc(1) complex in *Rhodobacter sphaeroides*. *The Journal of Biological Chemistry* 274: 33931–33944.

Huang LS, Cobessi D, Tung EY, and Berry EA (2005) Binding of the respiratory chain inhibitor antimycin to the mitochondrial bc_1 complex: A new crystal structure reveals an altered intramolecular hydrogen-bonding pattern. *Journal of Molecular Biology* 351: 573–597.

Hunte C, Koepke J, Lange C, Rossmanith T, and Michel H (2000) Structure at 2.3 A resolution of the cytochrome bc(1) complex from the yeast *Saccharomyces cerevisiae* co-crystallized with an antibody Fv fragment. *Structure* 15: 669–684.

Hunte C, Palsdottir H, and Trumpower BL (2003) Protonmotive pathways and mechanisms in the cytochrome bc_1 complex. *FEBS Letters* 545: 39–46.

Hunter SH (1950) Anaerobic and aerobic growth of purple bacteria (athiorhodaceae) in chemically defined media. *Journal of General Bicrobiology* 4: 286–293.

Iwata S, Lee JW, Okada K, Lee JK, Iwata M, Rasmussen B, Link TA, Ramaswamy S, and Jap BK (1998) Complete structure of the 11-subunit bovine mitochondrial cytochrome bc_1 complex. *Science* 281: 64–71.

Junemann S, Heathcote P, and Rich PR (1998) On the mechanism of quinol oxidation in the bc_1 complex. *The Journal of Biological Chemistry* 273: 21603–21607.

Kim H, Xia D, Yu CA, Xia JZ, Kachurin AM, Zhang L, Yu L, and Deisenhofer J (1998) Inhibitor binding changes domain mobility in the iron–sulfur protein of the mitochondrial bc_1 complex from bovine heart. *Proceedings of the National Academy of Sciences of the United States of America* 95: 8026–8033.

Kleiner D (1974) The effect of Zn2+ ions on mitochondrial electron transport. *Archives of Biochemistry and Biophysics* 165: 121–125.

Kleiner D, and von Jagow G (1972) On the inhibition of mitochondrial electron transport by Zn(2+) ions. *FEBS Letters* 20: 229–232.

Kleinschroth T, Castellani M, Trinh CH, Morgner N, Brutschy B, Ludwig B, and Hunte C (2011) X-ray structure of the dimeric cytochrome bc(1) complex from the soil bacterium Paracoccus denitrificans at 2.7-A resolution. *Biochimica et Biophysica Acta* 1807: 1606–1615.

Korshunov SS, Skulachev VP, and Starkov AA (1997) High protonic potential actuates a mechanism of production of reactive oxygen species in mitochondria. *FEBS Letters* 416: 15–18.

Kurisu G, Zhang H, Smith JL, and Cramer WA (2003) Structure of the cytochrome b6f complex of oxygenic photosynthesis: Tuning the cavity. *Science* 302: 1009–1014.

Kvint K, Nachin L, Diez A, and Nystrom T (2003) The bacterial universal stress protein: Function and regulation. *Current Opinion in Microbiology* 6: 140–145.

Lee DW, El Khoury Y, Francia F, Zambelli B, Ciurli S, Venturoli G, Hellwig P, and Daldal F (2011a) Zinc inhibition of bacterial cytochrome bc(1) reveals the role of cytochrome b E295 in proton release at the Q(o) site. *Biochemistry* 50: 4263–4272.

Lee DW, Selamoglu N, Lanciano P, Cooley JW, Forquer I, Kramer DM, and Daldal F (2011b) Loss of a conserved tyrosine residue of cytochrome b induces reactive oxygen species production by cytochrome bc_1. *The Journal of Biological Chemistry* 286: 18139–18148.

Link TA (1997) The role of the 'Rieske' iron sulfur protein in the hydroquinone oxidation (Q(P)) site of the cytochrome bc_1 complex. The 'proton-gated affinity change' mechanism. *FEBS Letters* 412: 257–264.

Link TA, Haase U, Brandt U, and von Jagow G (1993) What information do inhibitors provide about the structure of the hydroquinone oxidation site of ubihydroquinone: Cytochrome c oxidoreductase? *Journal of Bioenergetics and Biomembrane* 25: 221–232.

Liu X, Yu CA, and Yu L (2004) The role of extra fragment at the C-terminal of cytochrome b (Residues 421–445) in the cytochrome bc_1 complex from *Rhodobacter sphaeroides*. *The Journal of Biological Chemistry* 279: 47363–47371.

Ljungdahl PO, Pennoyer JD, Robertson DE, and Trumpower BL (1987) Purification of highly active cytochrome bc_1 complexes from phylogenetically diverse species by a single chromatophaphic procedure. *Biochimica et Biophysica Acta* 891: 227–241.

Loschen G, Azzi A, Richter C, and Flohe L (1974) Superoxide radicals as precursors of mitochondrial hydrogen peroxide. *FEBS Letters* 42: 68–72.

Ma H-W, Yang S, Yu L, and Yu C-A (2008) Formation of engineered intersubunit disulfide bond in cytochrome bc_1 complex disrupts electron transfer activity in the complex. *Biochimica et Biophysica Acta (BBA)—Bioenergetics* 1777: 317–326.

Mather MW, Yu L, and Yu CA (1995) The involvement of threonine 160 of cytochrome b of *Rhodobacter sphaeroides* cytochrome bc_1 complex in quinone binding and interaction with subunit IV. *The Journal of Biological Chemistry* 270: 28668–28675.

McCord JM, and Fridovich I (1969) Superoxide dismutase. An enzymic function for erythrocuprein (hemocuprein). *The Journal of Biological Chemistry* 244: 6049–6055.

Millett F, and Durham B (2009) Chapter 5 Use of ruthenium photooxidation techniques to study electron transfer in the cytochrome bc_1 complex. *Methods in Enzymology* 456: 95–109.

Moro MA, Almeida A, Bolanos JP, and Lizasoain I (2005) Mitochondrial respiratory chain and free radical generation in stroke. *Free Radical Biology & Medicine* 39: 1291–1304.

Mouncey NJ, Gak E, Choudhary M, Oh J, and Kaplan S (2000) Respiratory pathways of *Rhodobacter sphaeroides* 2.4.1(T): Identification and characterization of genes encoding quinol oxidases. *FEMS Microbiology Letters* 192: 205–210.

Muller F, Crofts AR, and Kramer DM (2002) Multiple Q-cycle bypass reactions at the Qo site of the cytochrome bc_1 complex. *Biochemistry* 41: 7866–7874.

Muller FL, Roberts AG, Bowman MK, and Kramer DM (2003) Architecture of the Qo site of the cytochrome bc_1 complex probed by superoxide production. *Biochemistry* 42: 6493–6499.

Muller FL, Liu Y, and Van Remmen H (2004) Complex III releases superoxide to both sides of the inner mitochondrial membrane. *The Journal of Biological Chemistry* 279: 49064–49073.

Musatov A, and Robinson NC (1994) Detergent-solubilized monomeric and dimeric cytochrome bc_1 isolated from bovine heart. *Biochemistry* 33: 13005–13012.

Nohl H, and Jordan W (1986) The mitochondrial site of superoxide formation. *Biochemistry Biophysics Research Communication* 138: 533–539.

Obungu VH, Amyot S, Wang Y, and Beattie DS (1998) Amino acids involved in the putative movement of the iron–sulfur protein of the bc_1 complex during catalysis. *FASEB Journal* 12: A1394.

Oh JI, and Kaplan S (2001) Generalized approach to the regulation and integration of gene expression. *Molecular Microbiology* 39: 1116–1123.

Ohnishi T, and Trumpower BL (1980) Differential effects of antimycin on ubisemiquinone bound in different environments in isolated succinate. cytochrome c reductase complex. *The Journal of Biological Chemistry* 255: 3278–3284.

Orme-Johnson NR, Hansen RE, and Beinert H (1974) Electron paramagnetic resonance–detectable electron acceptors in beef heart mitochondria. Ubihydroquinone–cytochrome c reductase segment of the electron transfer system and complex mitochondrial fragments. *The Journal of Biological Chemistry* 249: 1928–1939.

Osyczka A, Moser CC, Daldal F, and Dutton PL (2004) Reversible redox energy coupling in electron transfer chains. *Nature* 427: 607–612.

Osyczka A, Moser CC, and Dutton PL (2005) Fixing the Q cycle. *Trends in Biochemical Sciences* 30: 176–182.

Oudshoorn P, Van Steeg H, Swinkels BW, Schoppink P, and Grivell LA (1987) Subunit II of yeast QH2:cytochrome-c oxidoreductase. Nucleotide sequence of the gene and features of the protein. *European Journal of Biochemistry* 163: 97–103.

Ovadi J, Huang Y, and Spivey HO (1994) Binding of malate dehydrogenase and NADH channelling to complex I. *Journal of Molecular Recognition: JMR* 7: 265–272.

Oyedotun KS, and Lemire BD (2004) The quaternary structure of the Saccharomyces cerevisiae succinate dehydrogenase. Homology modeling, cofactor docking, and molecular dynamics simulation studies. *The Journal of Biological Chemistry* 279: 9424–9431.

Pemberton JM, Horne IM, and McEwan AG (1998) Regulation of photosynthetic gene expression in purple bacteria. *Microbiology* 144 (Pt 2): 267–278.

Purvis DJ, Theiler R, and Niederman RA (1990) Chromatographic and protein chemical analysis of the ubiquinol–cytochrome c2 oxidoreductase isolated from *Rhodobacter sphaeroides*. *The Journal of Biological Chemistry* 265: 1208–1215.

Rajagukguk S, Yang S, Yu CA, Yu L, Durham B, and Millett F (2007) Effect of mutations in the cytochrome b ef loop on the electron-transfer reactions of the Rieske iron–sulfur protein in the cytochrome bc(1) complex. *Biochemistry* 46: 1791–1798.

Robertson DE, Ding H, Chelminski PR, Slaughter C, Hsu J, Moomaw C, Tokito M, Daldal F, and Dutton PL (1993) Hydroubiquinone–cytochyrome c2 oxidoreductase from *Rhodobacter capsulatus*: Definition of a minimal, functional isolated preparation. *Biochemistry* 32: 1310–1317.

Rottenberg H, Covian R, and Trumpower BL (2009) Membrane potential greatly enhances superoxide generation by the cytochrome bc1 complex reconstituted into phospholipid vesicles. *The Journal of Biological Chemistry* 284: 19203–19210.

Sadek HA, Nulton-Persson AC, Szweda PA, and Szweda LI (2003) Cardiac ischemia/reperfusion, aging, and redox-dependent alterations in mitochondrial function. *Archives of Biochemistry and Biophysics* 420: 201–208.

Salerno JC (1984) Cytochrome electron spin resonance line shapes, ligand fields, and components stoichiometry in ubiquinol–cytochrome c oxidoreductase. *The Journal of Biological Chemistry* 259: 2331–2336.

Sarewicz M, Borek A, Cieluch E, Swierczek M, and Osyczka A (2010) Discrimination between two possible reaction sequences that create potential risk of generation of deleterious radicals by cytochrome bc(1). Implications for the mechanism of superoxide production. *Biochimica et Biophysica Acta* 1797: 1820–1827.

Schagger H, and Pfeiffer K (2000) Supercomplexes in the respiratory chains of yeast and mammalian mitochondria. *EMBO Journal* 19: 1777–1783.

Schagger H, Link TA, Engel WD, and von Jagow G (1986) Isolation of the eleven protein subunits of the bc1 complex from beef heart. *Methods in Enzymology* 126: 224–237.

Schagger H, Hagen T, Roth B, Brandt U, Link TA, and von Jagow G (1990) Phospholipid specificity of bovine heart bc1 complex. *European Journal of Biochemistry* 190: 123–130.

Schagger H, Brandt U, Gencic S, and von Jagow G (1995) Ubiquinol–cytochrome-c reductase from human and bovine mitochondria. *Methods in Enzymology* 260: 82–96.

Shigenaga MK, Hagen TM, and Ames BN (1994) Oxidative damage and mitochondrial decay in aging. *Proceedings of the National Academy of Sciences of the United States of America* 91: 10771–10778.

Skulachev VP, Chistyakov VV, Jasaitis AA, and Smirnova EG (1967) Inhibition of the respiratory chain by zinc ions. *Biochemistry Biophysics Research Communication* 26: 1–6.

Snyder CH, Gutierrez-Cirlos EB, and Trumpower B (2000) Evidence for a concerted mechanism of ubiquinol oxidation by the cytochrome bc1 complex. *The Journal of Biological Chemistry* 275: 13535–13541.

Stroebel D, Choquet Y, Popot JL, and Picot D (2003) An atypical haem in the cytochrome b(6)f complex. *Nature* 426: 413–418.

Su T, Esser L, Xia D, Yu CA, and Yu L (2012) Generation, characterization and crystallization of a cytochrome c(1)-subunit IV fused cytochrome bc(1) complex from *Rhodobacter sphaeroides*. *Biochimica et Biophysica Acta* 1817: 298–305.

Sun J, and Trumpower BL (2003) Superoxide anion generation by the cytochrome bc1 complex. *Archives of Biochemistry and Biophysics* 419: 198–206.

Suwanto A, and Kaplan S (1989a) Physical and genetic mapping of the *Rhodobacter sphaeroides* 2.4.1 genome: Genome size, fragment identification, and gene localization. *Journal of Bacteriology* 171: 5840–5849.

Suwanto A, and Kaplan S (1989b) Physical and genetic mapping of the *Rhodobacter sphaeroides* 2.4.1 genome: Presence of two unique circular chromosomes. *Journal of Bacteriology* 171: 5850–5859.

Thony-Meyer L (1997) Biogenesis of respiratory cytochromes in bacteria. *Microbiology and Molecular Biology Reviews: MMBR* 61: 337–376.

Tian H, Yu L, Mather MW, and Yu CA (1997) The involvement of serine 175 and alanine 185 of cytochrome b of *Rhodobacter sphaeroides* cytochrome bc1 complex in interaction with iron–sulfur protein. *The Journal of Biological Chemistry* 272: 23722–23728.

Tian H, Yu L, Mather MW, and Yu CA (1998) Flexibility of the neck region of the rieske iron–sulfur protein is functionally important in the cytochrome bc1 complex. *The Journal of Biological Chemistry* 273: 27953–27959.

Tian H, White S, Yu L, and Yu CA (1999) Evidence for the head domain movement of the Rieske iron–sulfur protein in electron transfer reaction of the cytochrome bc1 complex. *The Journal of Biological Chemistry* 274: 7146–7152.

Trumpower BL (1976) Evidence for a protonmotive Q cycle mechanism of electron transfer through the cytochrome b-c1 complex. *Biochemistry Biophysics Research Communication* 70: 73–80.

Trumpower BL (2002) A concerted, alternating sites mechanism of ubiquinol oxidation by the dimeric cytochrome bc(1) complex. *Biochimica et Biophysica Acta* 1555: 166–173.

Trumpower BL, and Gennis RB (1994) Energy transduction by cytochrome complexes in mitochondrial and bacterial respiration: The enzymology of coupling electron transfer reactions to transmembrane proton translocation. *Annual Review of Biochemistry* 63: 675–716.

Tso SC, Shenoy SK, Quinn BN, and Yu L (2000) Subunit IV of cytochrome bc1 complex from *Rhodobacter sphaeroides*. Localization of regions essential for interaction with the three-subunit core complex. *The Journal of Biological Chemistry* 275: 15287–15294.

Tso SC, Yin Y, Yu CA, and Yu L (2006) Identification of amino acid residues essential for reconstitutive activity of subunit IV of the cytochrome bc(1) complex from *Rhodobacter sphaeroides*. *Biochimica et Biophysica Acta* 1757: 1561–1567.

Turrens JF, Alexandre A, and Lehninger AL (1985) Ubisemiquinone is the electron donor for superoxide formation by complex III of heart mitochondria. *Archives of Biochemistry and Biophysics* 237: 408–414.

Van Remmen H, and Richardson A (2001) Oxidative damage to mitochondria and aging. *Experimental Gerontology* 36: 957–968.

Vermeglio A, and Joliot P (1999) The photosynthetic apparatus of *Rhodobacter sphaeroides*. *Trends in Microbiology* 7: 435–440.

von Jagow G, and Link TA (1986) Use of specific inhibitors on the mitochondrial bc1 complex. *Methods in Enzymology* 126: 253–271.

Wang Q, Yu L, and Yu CA (2010) Cross-talk between mitochondrial malate dehydrogenase and the cytochrome bc_1 complex. *The Journal of Biological Chemistry* 285: 10408–10414.

Wenz T, Covian R, Hellwig P, Macmillan F, Meunier B, Trumpower BL, and Hunte C (2007) Mutational analysis of cytochrome b at the ubiquinol oxidation site of yeast complex III. *The Journal of Biological Chemistry* 282: 3977–3988.

Wikstrom M, Krab K, and Saraste M (1981) Proton-translocating cytochrome complexes. *Annual Review of Biochemistry* 50: 623–655.

Xia D, Yu CA, Kim H, Xia JZ, Kachurin AM, Zhang L, Yu L, and Deisenhofer J (1997) Crystal structure of the cytochrome bc_1 complex from bovine heart mitochondria. *Science* 277: 60–66.

Xia D, Esser L, Yu L, and Yu CA (2007) Structural basis for the mechanism of electron bifurcation at the quinol oxidation site of the cytochrome bc_1 complex. *Photosynthesis Research* 92: 17–34.

Xia D, Esser L, Elberry M, Zhou F, Yu L, and Yu CA (2008) The road to the crystal structure of the cytochrome bc_1 complex from the anoxigenic, photosynthetic bacterium *Rhodobacter sphaeroides*. *Journal of Bioenergetics and Biomembrane* 40: 485–492.

Xia D, Esser L, Tang WK, Zhou F, Zhou Y, Yu L, and Yu CA (2013) Structural analysis of cytochrome bc_1 complexes: Implications to the mechanism of function. *Biochimica et Biophysica Acta* 1827: 1278–1294.

Xiao K, Yu L, and Yu CA (2000) Confirmation of the involvement of protein domain movement during the catalytic cycle of the cytochrome bc_1 complex by the formation of an inter-subunit disulfide bond between cytochrome b and the iron–sulfur protein. *The Journal of Biological Chemistry* 275: 38597–38604.

Xiao KH, Liu XY, Yu CA, and Yu L (2004) The extra fragment of the iron–sulfur protein (residues 96–107) of *Rhodobacter sphaeroides* cytochrome bc(1) complex is required for protein stability. *Biochemistry* 43: 1488–1495.

Xiao YM, Esser L, Zhou F, Li C, Zhou YH, Yu CA, Qin ZH, and Xia D (2014) Studies on inhibition of respiratory cytochrome bc_1 complex by the fungicide pyrimorph suggest a novel inhibitory mechanism. *PLoS One* 9: e93765.

Yang S, Ma HW, Yu L, and Yu CA (2008) On the mechanism of quinol oxidation at the QP site in the cytochrome bc_1 complex: Studied using mutants lacking cytochrome bL or bH. *The Journal of Biological Chemistry* 283: 28767–28776.

Yin Y, Tso SC, Yu CA, and Yu L (2009) Effect of subunit IV on superoxide generation by *Rhodobacter sphaeroides* cytochrome bc(1) complex. *Biochimica et Biophysica Acta* 1787: 913–919.

Yin Y, Yang S, Yu L, and Yu CA (2010) Reaction mechanism of superoxide generation during ubiquinol oxidation by the cytochrome bc_1 complex. *The Journal of Biological Chemistry* 285: 17038–17045.

Yu L, and Yu CA (1982) Isolation and properties of the cytochrome B-C1 complex from *Rhodopseudomonas sphaeroides*. *Biochemistry Biophysics Research Communication* 108: 1285–1292.

Yu L, and Yu CA (1987) Identification of cytochrome b and a molecular weight 12K protein as the ubiquinone-binding proteins in the cytochrome b-c1 complex of a photosynthetic bacterium *Rhodobacter sphaeroides* R-26. *Biochemistry* 26: 3658–3664.

Yu L, and Yu CA (1991) Essentiality of the molecular weight 15,000 protein (subunit IV) in the cytochrome b-c1 complex of *Rhodobacter sphaeroides*. *Biochemistry* 30: 4934–4939.

Yu CA, Nagoaka S, Yu L, and King TE (1980) Evidence of ubisemiquinone radicals in electron transfer at the cytochromes b and c1 region of the cardiac respiratory chain. *Archives of Biochemistry and Biophysics* 204: 59–70.

Yu L, Mei QC, and Yu CA (1984) Characterization of purified cytochrome b-c1 complex from *Rhodopseudomonas sphaeroides* R-26. *The Journal of Biological Chemistry* 259: 5752–5760.

Yu L, Tso SC, Shenoy SK, Quinn BN, and Xia D (1999) The role of the supernumerary subunit of *Rhodobacter sphaeroides* cytochrome bc_1 complex. *Journal of Bioenergetics and Biomembranes* 31: 251–257.

Yun C, Beci R, Crofts AR, Kaplan S, and Gennis RB (1990) Cloning and DNA sequencing of the fbc operon encoding the cytochrome bc_1 complex from *Rhodobacter sphaeroides*. *European Journal of Biochemistry* 194: 399–411.

Zhang L, Yu L, and Yu CA (1998a) Generation of superoxide anion by succinate–cytochrome c reductase from bovine heart mitochondria. *The Journal of Biological Chemistry* 273: 33972–33976.

Zhang Z, Huang L, Shulmeister VM, Chi YI, Kim KK, Hung LW, Crofts AR, Berry EA, and Kim SH (1998b) Electron transfer by domain movement in cytochrome bc_1. *Nature* 392: 677–684.

Zhang H, Osyczka A, Dutton PL, and Moser CC (2007) Exposing the complex III Qo semiquinone radical. *Biochimica et Biophysica Acta* 1767: 883–887.

Zhou F, Yin Y, Su T, Yu L, and Yu CA (2012) Oxygen dependent electron transfer in the cytochrome bc(1) complex. *Biochimica et Biophysica Acta* 1817: 2103–2109.

Zhu J, Egawa T, Yeh S, Yu L, and Yu CA (2007) Simultaneous reduction of iron–sulfur protein and cytochrome bL during ubiquinol oxidation in cytochrome bc_1 complex. *Proceedings of the National Academy of Sciences of the United States of America* 104: 4864–4869.

20 Photosynthetic Apparatus in Cyanobacteria and Microalgae

Isabella Moro, Nicoletta La Rocca, and Nicoletta Rascio

CONTENTS

20.1 Introduction 349
20.2 Photosynthetic Apparatus in Cyanobacteria 350
    20.2.1 Cyanobacterial PSII 350
        20.2.1.1 PSII Core 350
        20.2.1.2 Oxygen-Evolving Complex 352
        20.2.1.3 Phycobilisomes 352
    20.2.2 Cytochrome b_6f 353
    20.2.3 Cyanobacterial PSI 354
        20.2.3.1 The IsiA Complex 355
    20.2.4 Linear and Cycling Electron Flow 356
20.3 Green Oxyphotobacteria 358
    20.3.1 Prochlorophyte Chl *a/b*-Binding Proteins 358
20.4 *Acaryochloris marina* and *Halomicronema hongdechloris* 359
20.5 Microalgal Photosystems 359
    20.5.1 Photosystems of Red Microalgae 360
        20.5.1.1 Photosystem II 360
        20.5.1.2 Phycobilisomes 360
        20.5.1.3 Photosystem I 361
    20.5.2 Photosystems of Green Microalgae 361
        20.5.2.1 LHCII 362
        20.5.2.2 LHCI 362
    20.5.3 Photosystems of Chlorophyll *c*-Containing Microalgae 362
        20.5.3.1 FCP Complex in Diatoms 363
        20.5.3.2 acpPC and PCP Complexes in Dinoflagellates 363
20.6 Conclusions 364
Acknowledgments 364
References 364

20.1 INTRODUCTION

About 2.4 billion years (Gyr) ago, primitive prokaryotic photoautotrophs (cyanobacteria) acquired the ability to photo-oxidize water [1], giving rise to the oxygen-evolving photosynthesis, an event of paramount importance regarded as the *Big Bang* of the evolution of life on earth [2]. Between 2.4 and 2.0 Gyr, the oxygenic photosynthesis carried out by these microorganisms led to the gradual increase in atmospheric oxygen (to 1–2%) responsible for the *Great Oxygenation Event* (also called *Oxygen Catastrophe*), which caused the extinction of anaerobic organisms unable to adapt to the new oxidizing environment but also triggered the evolution of the aerobic organisms, thus profoundly affecting the diversity of life. The emergence of photosynthetic eukaryotes, 17–18 Gyr ago [3], and the increased photosynthetic productivity of algae and, subsequently, land plants accounted for the gradual

rise of the oxygen level up to the value (~20%) observed in today's atmosphere [4]. Essentially all the oxygen present in the aerobic environments on earth derives from the oxygenic photosynthesis that currently accounts for ~1.6×10^{14} kg O_2 every year [5].

In both prokaryotic and eukaryotic photoautotrophs, the oxygenic photosynthesis relies on two multiprotein complexes, namely, photosystem II (PSII) and photosystem I (PSI), linked by the protein complex of cytochrome b_6f (Cyt b_6f) and by small mobile electron carriers (plastoquinone and plastocyanin or cytochrome c_6) [6]. These complexes are embedded in membranes (thylakoid membranes) lining flattened sacs with an aqueous lumen (thylakoids), located inside the cytoplasm in cyanobacteria and inside the chloroplasts in eukaryotic photoautotrophs. The two photosystems operate in series to move electrons from water to NADP+ by using the light energy harvested through specific *antenna* systems. The light energy-dependent electron transfer results in generation of a proton motive force across the thylakoid membrane, which drives the synthesis of ATP via the ATP synthase, a fourth membrane-bound protein complex [7]. In this way, reducing potential (NADPH) and chemical energy (ATP) are provided for the production of organic compounds (*photosynthates*, mostly carbohydrates), which, together with the O_2 released from the water photo-oxidation, are indispensable for sustaining the life on earth.

20.2 PHOTOSYNTHETIC APPARATUS IN CYANOBACTERIA

The cyanobacterial cells (Figure 20.1a–c) exhibit an internal thylakoid system organized as a series of roughly parallel double-membrane layers distributed within the cytoplasm according to a species-specific topology [8,9]. The uniqueness of these prokaryotic organisms, which can perform oxygenic photosynthesis and respiration simultaneously, is that their thylakoid membranes, in addition to the components of the photosynthetic electron transport chain, house also most of the respiratory electron transport complexes (Figure 20.2). The photosynthetic PSI and PSII as well as the respiratory type I NAD(P)H dehydrogenase (NDH-1) and the succinate dehydrogenase (SDH) are harbored by these membranes together with the Cyt b_6f complex that functions in both photosynthetic and respiratory electron flows. The mobile electron carriers plastoquinone and plastocyanin or cytochrome c_{553} (usually referred to as cytochrome c_6) are also common to photosynthetic and respiratory pathways [10]. Moreover, two respiratory terminal oxidases are localized in the thylakoid membrane: a quinol oxidase (Cyd) that transfers electrons to

O_2 directly from reduced plastoquinones and a cytochrome *c* oxidase (COX) that receives electrons from Cyt b_6f via a soluble redox carrier (plastocyanin or cytochrome c_6) [11]. It is noteworthy that, unlike eukaryotic microalgae that possess an equal amount of PSI and PSII, cyanobacteria exhibit PSI/PSII ratios ranging from 3 to over 5 [12]. It has been suggested that this high quantity of PSI might serve in light to compete effectively with COX for electrons, thus maximizing the number of electrons that can be used for photosynthesis [13].

Although the coexistence of photosynthesis and respiration in the same membrane offers possibility for versatile electron transport routes, it also requires control systems to match the electron transport modes with the physiological needs of the cell [14]. It has been shown, for instance, that in normal conditions, respiratory complexes, such as NDH-1 and SDH, are clustered into functional *islands* (100–300 nm) in the thylakoid membranes to provide a way to preferentially channel the mobile electron carriers to the final respiratory acceptor located in close proximity in the same functional island [10,14]. This suggests that, in the thylakoid membranes, a spatial separation between photosynthetic and respiratory activities is maintained, at least under some conditions [10].

20.2.1 CYANOBACTERIAL PSII

The PSII is the membrane-spanning pigment–protein complex that functions as a light-driven water:plastoquinone oxidoreductase. PSII is the only known biological system able to catalyze the water oxidation, which is one of the most energetically demanding reaction in nature ($\Delta E_m = 1.23$ V) [15]. PSII is arranged as a membrane intrinsic core complex, made up of a reaction center (RC) and an inner light-harvesting antenna, linked to an extrinsic oxygen-evolving complex (OEC) and to a large outer antenna system. In spite of the enormous phylogenetic and ecological diversity achieved by the oxygenic photosynthetic organisms during their evolutive time course from the early ancestors, the PSII core complex is remained remarkably conserved structurally and functionally [16].

20.2.1.1 PSII Core

The crystal structure of the PSII core complex, solved at a resolution from 3.8 to 1.9 Å in some cyanobacteria, such as *Thermosynechococcus elongatus* and *T. vulcanus* [17–20], showed that the photosystem is present in nature in a dimeric form. Each monomer is composed of 17 intrinsic protein subunits, named Psb proteins, that comprehend the four major proteins of RC and inner antennae (Figure 20.3a) together with a series of other minor polypeptides (< 15 kDa) involved in assembly, stabilization, and dimerization of the core complex. The RC of a monomer is made up of a heterodimer of D1(PsbA)/D2 (PsbD) homologous proteins (32 and 34 kDa, respectively) that provide the scaffold for six chlorophyll *a* (Chl *a*), two pheophytin (Pheo: Chl *a* devoid of Mg), two β-carotene (β-Car), and two major plastoquinone (PQ, referred to as Q_A and Q_B) molecules, and for one non-heme iron and one bicarbonate ion, many of which involved in the

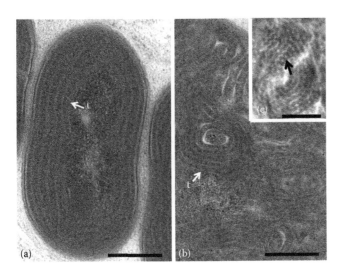

FIGURE 20.1   (a) Transmission electron micrographs of cyanobacterial cells. (a) Cell of the filamentous cyanobacterium *Spirulina* sp. with thylakoids (t) running parallelly to the cell surface. (b) Particular of an *Anabaena* sp. cell with thylakoids (t) irregularly distributed in the cytoplasm. (c) Thylakoids of *Chroococcus* sp. showing phycobilisomes (arrow) on the membrane surfaces. (Bars = 500 nm).

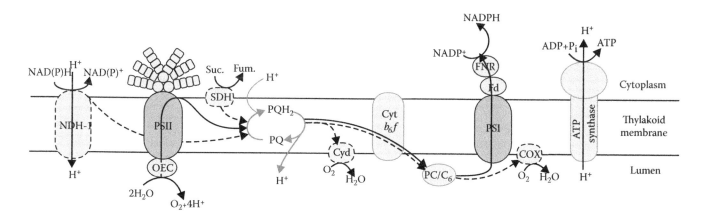

FIGURE 20.2 Schematic view of photosynthetic and respiratory complexes in the thylakoid membrane of cyanobacteria. Continuous outlines indicate the photosynthetic complexes, dashed outlines the respiratory complexes, and gray outlines the shared complexes and the soluble electron transporters. Photosynthetic and respiratory electron transport ways are drawn by continuous and dashed lines, respectively. C_6, cytochrome c_6; COX, cytochrome c oxidase; Cyd, quinol oxidase; Cyt b_6f, cytochrome b_6f; Fd, ferredoxin; FNR, ferredoxin-NADP$^+$ reductase; Fum, fumarate; NDH-1, type I NAD(P)H dehydrogenase; OEC, oxygen-evolving complex; PC, plastocyanin; PQ, plastoquinone; PSI, photosystem I; PSII, photosystem II; SDH, succinate dehydrogenase; Suc, succinate. See text for detailed explanation.

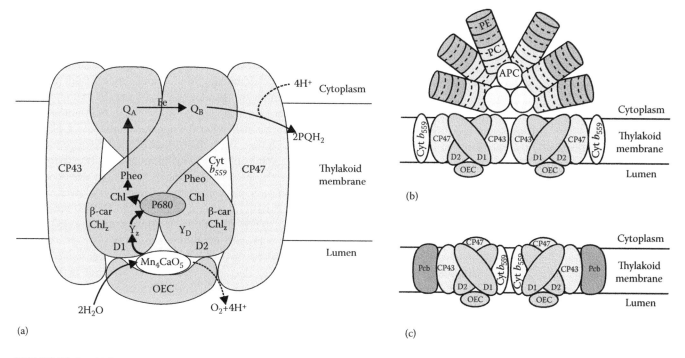

FIGURE 20.3 (a) Scheme showing the major intrinsic proteins of a PSII core monomer, and the OEC with the metal cluster carrying out the water oxidation. The electron flow from water to plastoquinone and the cofactors involved in the photochemical charge separation and in the electron transport chain are represented. β-car, β-carotene; Chl, accessory chlorophyll; Chl$_Z$, peripheral chlorophyll; Cyt b_{559}, cytochrome b_{559}; P680, primary electron donor; Pheo, pheophytin; Q$_A$ and Q$_B$, plastoquinones; PQH$_2$, plastoquinol; Y$_Z$ and Y$_D$, tyrosines 161. (b) Scheme of a cyanobacterial hemidiscoidal phycobilisome bound to the dimeric PSII core. APC, allophycocyanin; PC, phycocyanin; PE, phycoerythrin. (c) Dimeric PSII core of a green oxyphotobacterium with intrinsic Pcb (prochlorophyte Chl a/b binding) proteins instead of a phycobilisome as peripheral antenna. See text for detailed explanation.

photochemical reaction of charge separation and in electron transport chain. A peculiarity of the cyanobacterial RC is the possibility to have distinct D1 protein isoforms, with few sequence changes, codified by a small family of *PsbA* genes that differently expresses depending on the environmental cues [21]. In particular, the D1:2 form, known as *stress form*, is synthesized under excess light and controls the efficiency

of the PSII in this stressful condition [22]. In contrast to cyanobacteria, only one *PsbA* gene encoding a single form of D1 exists in the eukaryotic photoautotrophs [23].

The inner light-harvesting antenna of the PSII core consists of the CP43 (PsbC) and CP47 (PsbB) proteins, which belong to a family of six-helix transmembrane chlorophyll/carotenoid binding proteins [24] and harbor 13 Chls a/4 β-Cars and 16 Chls *a*/5 β-Cars, respectively. The two pigment–protein complexes, enclosing on opposite sides of the D1/D2 heterodimer, are able to absorb light and to funnel the excitation energy, also collected from the large outer antenna, to the pigments linked to D1/D2 proteins in the RC [20]. Two peripheral Chls *a* (Chl$_{ZD1}$ and Chl$_{ZD2}$) mediate the energy transfer from the inner antenna to the four excitonically coupled central Chls *a*, which comprehend two accessory Chls (Chl$_{D1}$ and Chl$_{D2}$) and a special pair of Chls (ChlP$_{D1}$ and ChlP$_{D2}$), known as P680 (due to the peak of the lowest-energy absorption at 680 nm). P680 is regarded as the primary electron donor that, in the excited electronic state (P680*), carries out the photochemical reaction of charge separation by transferring an electron to a nearby pheophytin (Pheo$_{D1}$), with the involvement of the accessory Chl$_{D1}$, to form the radical pair [P680$^+$, Pheo$^-$]. The charge separation is then stabilized by migration of the electron from the primary acceptor Pheo$^-$ to a first plastoquinone (Q$_A$), fixed to the Q$_A$ site on D2. Although both D1 and D2 harbor an accessory Chl *a* and a Pheo molecule, only Chl$_{D1}$ and Pheo$_{D1}$ are active in the electron transfer from P680* to Q$_A$. The acquired electron proceeds then from Q$_A$ to another plastoquinone (Q$_B$), bound at the Q$_B$ site on D1, which becomes a reduced semiquinone (PQ$^-$) [25]. This transfer is favored by the non-heme iron sat in between the two plastoquinone molecules [26]. After a second photochemical turnover, the Q$_B$, fully reduced to plastoquinol (PQ$^-$), captures two protons from the cytoplasm and leaves the RC as PQH$_2$, moving to the plastoquinone pool in the lipid bilayer of the thylakoid membrane and allowing an oxidized plastoquinone to bind to the empty Q$_B$ site [25]. This last event is positively affected by a bicarbonate ion bound to the non-heme iron [27] and seems also to involve in some way the Qc, a recently discovered third plastoquinone [19], whose precise role is still under investigation [28]. The PSII RC of all oxygenic photoautotrophs also contains a cytochrome (cyt b_{559}) composed of two heme-bridged polypeptides (PsbE and PsbF) forming a (α,β) heterodimer. The unique peculiarity of cyt b_{559} is that it covers a broad range of redox potentials, going from a high (HP) (from +310 to +400 mV) to an intermediate (IP) (from +125 to +240 mV) up to a low (LP) (from −40 to +80 mV) potential form [29]. The cyt b_{559} does not participate in the primary electron transport from P680 to Q$_B$, but it plays effective roles in PSII protection. It is involved in a photoprotective cyclic electron flow (CEF) within the RC [30], and it also displays some enzymatic functions to counteract the accumulation of the most harmful reactive oxygen species (ROS) connected to the PSII activity [29].

20.2.1.2 Oxygen-Evolving Complex

On the lumenal side of thylakoid membrane, the intrinsic PSII core links the OEC (Figure 20.3a) made up of some extrinsic proteins and of the metal cluster, composed of 4Mn and 1Ca bridged by five oxygen atoms, that carries out the light-driven water oxidation [20]. The catalytic Mn$_4$CaO$_5$, which has ligands on D1 and CP43, is also associated with four molecules of H$_2$O, some of them likely serving as a substrate for the oxidative reaction [20].

The primary cation radical (P680$^+$), created by the photochemical charge separation, shows the highest redox potential

observed in a biological system (+1250 mV) [31], which makes it able to take out an electron from water (E_m = +820 mV). This occurs through an indirect way that involves an intermediate redox-active tyrosine (D1-Tyr 161 known as Y$_z$) and the Mn$_4$CaO$_5$ cluster that act as an oxidizing equivalent accumulator [32]. According to the *classic* Kok model [33], referred to as *Kok cycle* or *S-state cycle*, the 4Mn of the catalytic cluster can proceed through five oxidation states (S$_0$>S$_1$>S$_2$>S$_3$>S$_4$). In a very simplified scheme, each P680$^+$ reduction after a charge separation removes, via the Y$_z$, one electron from the Mn cluster that advances to the next higher oxidized state, so that four sequential photochemical reactions drive the Mn cluster from the S$_0$ to the S$_4$ state, with accumulation of four oxidizing equivalents. In this last state, the catalytic cluster acquires the capability to oxidize two H$_2$O molecules recovering the four electrons (returning to the S$_0$ state) and releasing four H$^+$ and one O$_2$ into the thylakoid lumen. The fast turnover of water oxidation also involves a chloride ion (Cl$^-$) that seems to act in facilitating the proton removal from the Mn$_4$CaO$_5$ cluster [20].

The catalytic heart of the OEC is surrounded by extrinsic proteins interacting with a variety of PSII core subunits [34]. The essential role of these proteins, which do not ligate directly the Mn$_4$CaO$_5$ cluster, is to support the water oxidation and the O$_2$ evolution by stabilizing the catalytic cluster, shielding it from the aqueous lumenal environment and providing channels for the entry of water and the exit of protons and oxygen [5]. Although the mechanism of water splitting has remained unchanged during the timescale of photoautotroph evolution, the composition of the OEC proteins has undergone some changes in different eukaryotic groups with respect to the prokaryotic ancestors. Three major extrinsic proteins are present in the cyanobacterial OEC: PsbO (33 kDa), PsbU (12 kDa), and PsbV (15 kDa), which is also referred to as cytochrome c_{550}. Among these, PsbO is the predominant protein maintained in every type of oxygenic photoautotrophs. It is also known as manganese stabilizing protein (MSP) due to its recognized function in Mn cluster stabilization [32]. Recently, two other proteins, named PsbP-like and PsbQ-like, have been found as components of the cyanobacterial OEC. These proteins, also referred to as CyanoP and CyanoQ, exhibit homology to the PsbP and PsbQ proteins, which have replaced the PsbU and PsbV subunits in the OEC of green algae and higher plants, of which they are considered prokaryotic precursors [34].

20.2.1.3 Phycobilisomes

The main light-harvesting antenna of the cyanobacterial PSII is the phycobilisome (PBS), a highly structured extramembrane supercomplex (Figure 20.1c) made up of phycobiliproteins (PBP), which absorb light in the range of 490–650 nm, complementing the absorption of Chl *a* (with peaks at ~440 and ~670 nm) [35]. The PBPs, namely, phycocyanin (PC), phycoerythrocyanin (PEC), phycoerythrin (PE), and allophycocyanin (APC), are polypeptides bearing one to three units of linear tetrapyrrole chromophores (bilins), which are phycocyanobilin (PCB) and phycoerythrobilin (PEB), covalently bound to Cys residues [36]. The PBPs are composed of homologous α and β subunits bound to each other to form the supercomplex building blocks. The α/β units, called *monomers*, are assembled into stable ring-shaped trimers $(\alpha,\beta)_3$ with two of them attached, in turn, to each other in a hexameric disk $(\alpha,\beta)_6$.

The typical *hemidiscoidal* model of PBS (Figure 20.3b), found in many cyanobacteria, exhibits hexamers of PE (or PEC) and PC stacked to form six or eight peripheral rods that radiate out, like a fan, from a central core consisting of two to five cylinders of APC trimers [35]. The PBP examers are spatially arranged to produce a directional energy flow toward the final sink, which is usually the PSII core. PE or PEC disks that absorb blue-yellow (490–550 nm) and orange (550–620 nm) light, respectively, are located at the distal part of the rods, while those of PC that absorb red light (620–630 nm) are located at their proximal part. The rods are connected to cylinders of APC that, absorbing red light of longer wavelengths (650–670 nm), can collect the excitation energy from the peripheral rods and funnel it very efficiently to the Chl *a* of PSII [35]. In some cases, the PBS peripheral rods consist exclusively of PC, with the distal exameric disks having an energy level higher than the proximal ones [35]. Linker polypeptides, usually colorless, comprehending rod, rod core, core, and core–membrane linkers, are required to connect adjacent PBPs and to joint PBS and thylakoid membrane, in order to govern the supercomplex assembly and to optimize the absorbance characteristics of PSPs and the excitation energy flow to the intrinsic PSII core [37]. The polypeptide mainly involved in mediating the interaction of PBS with the thylakoid membrane is thought to be the core–membrane linker ApcE, also known as L_{cm} or *anchor* polypeptide. ApcE, which is the only bilin (PC)-attached linker protein, together with a form of APC, namely APC-B, have been postulated as terminal long-wavelength emitters that transfer excitation energy from PBS to the Chl *a* of PSII

[38,39]. Two copies of both ApcE and APC-B are present per PBS core, each of them plausibly binding a monomer of the dimeric PSII [35,38]. The PBS–PSII interaction, however, is rather weak, and PBSs can diffuse on the thylakoid membrane surface decoupling from a PSII and attaching to another PSII or also to a PSI. This occurs, for instance, during the events of state transitions that maintain the balance of the absorbed light energy between the two photosystems under changing light conditions [40]. Interestingly, a unique type of PBS, devoid of the central core and preferentially linked to the PSI, has been recently found in cyanobacteria. This *rod-type* PBS is mostly composed of hexameric PC units and is connected to the thylakoid membrane, or directly to PSI complexes, through the rod linker protein named CpcL, which is involved in the efficient energy transfer from PBS to photosystem core [35,41,42].

20.2.2 CYTOCHROME b_6f

The cytochrome b_6f (Cyt b_6f) is the membrane-spanning multiprotein complex, which provides the electronic connection between the photosystem RCs operating as a plastoquinone:plastocyanin (or cytochrome c_6) oxidoreductase. The complex, whose structure and function have been maintained almost identical in cyanobacteria and eukaryotic photoautotrophs [43,44], couples the electron flow to a proton translocation across the thylakoid membrane, from cytoplasm, usually referred to as stroma, into the lumen. The consequent H^+ gradient contributes, together with the resulting ΔpH, to establish the electrochemical potential gradient (the *proton motive force*) required for the ATP synthesis [25]. The crystal structure of the native Cyt b_6f from the cyanobacteria *Mastigocladus laminosus* and *Nostoc* sp. PCC7120 [45–47] revealed a symmetric homodimeric complex composed of two monomers each consisting of four major (Figure 20.4)

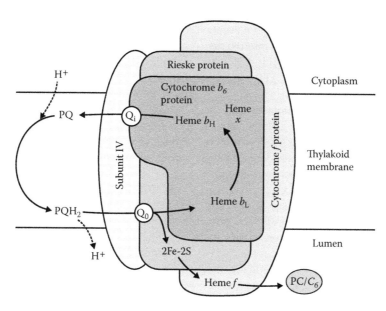

FIGURE 20.4 Scheme of the cytochrome b_6f monomer showing the four major proteins. The electron flow from reduced plastoquinone (PQH_2) to plastocyanin (PC) or cytochrome c_6 (C_6) and the redox-active prosthetic groups involved in electron transfer are represented. See text for detailed explanation.

and four minor subunits (known as Pet proteins). The three major subunits that bind the redox-active prosthetic groups are the cytochrome b_6 (PetB, 24 kDa), which bears two b-type hemes, one with higher (b_H) the other with lower (b_L) redox potential; the Rieske iron–sulfur protein (PetC, 19 kDa), with a [2Fe–2S] cluster; and the c-type cytochrome (PetA, 32 kDa), called f from $frons$ (Latin for leaf). The fourth major protein, named subunit IV (PetD, 17 kDa), lacks a redox-active prosthetic group, but is required for the catalytic activity of the complex. Together with the cytochrome b_6 subunit, it participates in the binding sites for oxidized and reduced plastoquinones, which are located in a lipophilic *quinone exchange cavity* between the two monomers [43,44,48]. The four small subunits (3–4 kDa), arranged as a *picket fence* at the lateral boundary of the monomer, may have essentially a structural role [44].

The Cyt b_6f complex drives the electron transfer between the reduced plastoquinol (PQH_2) coming from PSII and the electron carrier (plastocyanin or cytochrome c_6) that will move to PSI. The two-electron oxidation of plastoquinol occurs at a lumenal PQH_2 binding site (Q_o) on the p-side (electropositive side) of the quinone exchange cavity and results in the release of two protons into the thylakoid aqueous lumen and in the detachment of oxidized PQ from Q_o. One of the two electrons is transferred, via the high potential chain of Rieske iron–sulfur protein and cyt f, to the lumenal mobile carrier (plastocyanin or cytochrome c_6). The second electron is translocated across the membrane, through the two heme groups (b_L and b_H, also defined b_p and b_n) of cytochrome b_6, to a stromal Q_i site on the n-side (electronegative side) of the quinone exchange cavity, where a bound oxidized plastoquinone (PQ) becomes a reduced semiquinone (PQ^-). Following a second PQH_2 oxidation at the Q_o site and a second electron supply to the semiquinone at the Q_i site, two protons are taken up from the stroma and the fully reduced plastoquinone (PQ^-) is released as PQH_2 into the membrane lipid bilayer to join the plastoquinone pool. Through these sequential events, referred to as Q-cycle, $2H^+$ pass from the stroma into the thylakoid lumen per each electron moving from PSII to PSI, with generation of a proton motive force useful for ATP synthesis [25,43].

Although the Cyt b_6f structure bears similarities to the analogous respiratory Cyt bc_1 complex, the complement of prosthetic groups in the photosynthetic complex is strikingly different since each monomer comprehends one Chl a bound to the subunit IV, one β-Car bound to small subunits, and also a fourth heme group (named heme x or c_i or c_n) covalently bound to the b_6 polypeptide nearby the Qi site as an additional ubiquitous prosthetic group [44,45,48–50]. The presence of Chl a and β-Car in the Cyt b_6f complex that does not require the light capture poses questions about the role of these unexpected pigments. It has been suggested that Chl a, as an intrinsic component of the complex, may exert a structural role and may also be involved in sensing the redox state of plastoquinone that triggers the process of photosystem state transition [51]. However, it is known that the light absorption can induce triplet excited state of Chl ($^3Chl^*$), which transfers its excitation energy to O_2 generating the very harmful singlet oxygen $\left(^1O_2^*\right)$, unless it is quenched [44]. Although in the Cyt b_6f complex the β-Car is too far from the Chl a for a direct quenching of $^3Chl^*$, it might be involved in scavenging the O_2^* produced [52]. Finally, the unique extra heme x, which lies near the heme b_n on the n-side of the Cyt b_6f complex, is assumed to participate essentially in a cycling electron flow around the PSI [44,45].

As already written, the electron transfer from the Cyt b_6f to the PSI can involve either the copper protein plastocyanin or the iron protein cytochrome c_6. Most cyanobacteria and green microalgae are able to produce both of these soluble carriers, with the protein synthesized depending on the environmental availability on the constituent metal [53].

20.2.3 CYANOBACTERIAL PSI

The PSI is the large pigment–protein complex that spans the thylakoid membrane catalyzing the light-induced electron transfer from reduced plastocyanin or cytochrome c_6 at the lumenal side to the oxidized ferredoxin or flavodoxin on the opposite stromal side of the membrane. The crystal structure of PSI from *T. elongatus*, solved at a resolution of 2.5 Å, revealed that the photosystem is organized in nature as a clover-leaf shaped homotrimeric core complex [54,55], which comprehends the RC and an inner antenna system. PSI trimers have been found in almost every cyanobacteria analyzed, although also tetrameric and dimeric PSI have been recently noticed in some cyanobacterial strains, such as *Nostoc* sp. PCC7120 [56] and *Chroococcidiopsis* sp. TS-821 [57]. Each PSI monomer is made up of 12 proteins (known as Psa), with two major intrinsic subunits (PsaA and PsaB) (Figure 20.5a) surrounded by 10 minor polypeptides, for some of which the function is clearly established. PsaA and PsaB are large (~83 kDa) homologous polypeptides forming a heterodimer harboring most cofactors of the photochemical charge separation and electron transport chain (six Chls a, two phylloquinones, and one of three [4Fe–4S] clusters), which represent the hearth (the RC) of the complex. In addition, the PsaA/PsaB dimer contains ~100 Chl a and 22 β-Car molecules fulfilling the function of an inner light-harvesting antenna that transfers very efficiently the excitation energy to the primary electron donor. The latter is a special heterodimeric pair of Chl a and Chl a' (the C13^2 epimer of Chl a) designed P700 due to the peak of the lowest-energy absorption at 700 nm [58].

The primary electron donor (P700) in the excited electronic state (P700*) carries out the photochemical charge separation transferring an electron to the primary acceptor A_0 (a monomeric Chl a) via another accessory Chl a (A). In this reaction, the PSI generates the most negative redox potential (−1000 mV) in nature [59]. The electron flow proceeds then from A_0 to A_1, a phylloquinone, also known as vitamin K_1, and from that to the proximal [4Fe–4S] cluster (Fx) located near the stromal surface of the PsaA/PsaB heterodimer. Chlorophylls and phylloquinones of the RC are arranged in two quasi-symmetrical branches, both of them able to transfer the electron from P700 to Fx, albeit at different rates [54,60]. Two further [4Fe–4S]

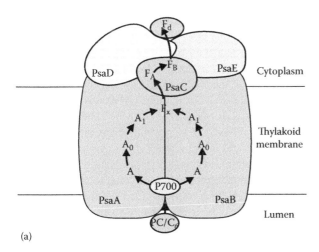

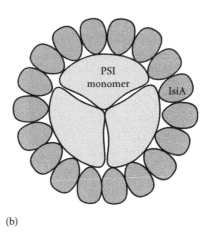

(a)   (b)

FIGURE 20.5   (a) Scheme of a PSI core monomer showing the two major intrinsic proteins and the extrinsic polypeptides involved in electron transport and ferredoxin anchoring. The electron flow from plastocyanin (PC) or cytochrome c_6 (C_6) to ferredoxin (Fd) and the cofactors involved in the photochemical charge separation and in the electron transport chain are represented. A and A_0, chlorophylls; A_1, phylloquinone; Fx, Fa, and Fb, 4Fe–4S clusters; P700, primary electron donor. (b) Scheme of the PSI–IsiA supercomplex, with 18 IsiA units around the trimeric PSI core. See text for detailed explanation.

clusters (F_A and F_B), coordinated by the extrinsic stromal polypeptide PsaC, then transfer the electron from Fx to the soluble ferredoxin, an iron–sulfur [2Fe–2S] protein, or, in iron-deficiency conditions, to the FMN-containing flavodoxin. The successive electron transfer from these carriers to the last acceptor $NADP^+$, through the soluble flavoprotein ferredoxin (or flavodoxin)–$NADP^+$ reductase (FNR), will lead finally to the production of NADPH. The PsaC and the surrounding PsaD and PsaE are the stromal subunits involved together in the transient anchoring of both ferredoxin and flavodoxin to the PSI core [61], while the docking site of the electron donors (plastocyanin or cytochrome c_6) for the reduction of P700$^+$ is provided by the lumenal surface of the PsaA/PsaB heterodimer [58].

The capability of PSI to drive the light-dependent electron flow is astonishing, as shown by a quantum yield (i.e., number of photons absorbed/number of charge separations) near 1.0, which makes this photosystem the most efficient nanophotochemical machine in nature [62]. It is noteworthy that the basic structure of the PSI core, like that of PSII, has been conserved in the eukaryotic photoautotrophs, despite more than 1 billion years of evolution. Apart from change of some minor polypeptides, the most distinctive feature of the eukaryotic PSI core is its existence in the thylakoid membrane as a monomeric instead of trimeric complex [62,63].

A striking characteristic of almost all PSI complexes is the presence in the light-harvesting antenna of a number of *red* Chls (excitonically coupled dimers or trimers of Chl *a*) absorbing light at longer and less energetic wavelengths than P700 [54,64]. The occurrence of long-wavelength Chls (LWCs) with absorption peaks from 708–719 up to 735 nm is unique to the cyanobacterial PSI core [65–67]. In eukaryotic photoautotrophs, the LWCs absorbing light at wavelengths over 700 nm are located mostly in the peripheral antenna of PSI [64]. No red chlorophylls are present in the PSI core of the green microalga *Chlamydomonas reinhardtii* [68]. The LWCs

of cyanobacterial PSI are assumed to fulfil an efficient uphill energy transfer to bulk Chls and then to P700, thus extending the spectral range of light-harvesting and increasing the absorbance cross section of the photosystem to facilitate the light energy capture in low light conditions [67,69].

20.2.3.1   The IsiA Complex

The cyanobacterial PSI does not possess a constitutive peripheral antenna system, while a large light-harvesting complex (LHC) is induced under iron-deficiency conditions [70,71]. Although iron is the most abundant element on earth, its concentration in aquatic environments, particularly in open oceans, is limited due to the low solubility of Fe^{3+} in slightly alkaline water [72]. The low iron bioavailability can greatly limit the photosynthetic activity of photoautotroph organisms affecting primarily the PSI that needs iron–sulfur clusters for the electron transport. In cyanobacteria, the PSI/PSII ratio, which is usually 3, reaching the value of 5.5 in *Arthrospira platensis* [12], undergoes a sharp decrease under iron-limitation stress [72], but the reduction of the PSI level is paralleled by the induction of the IsiA (iron starvation inducible) protein. The IsiA (36 kDa) is a six-helix transmembrane Chl *a*-binding protein belonging to the so-called *core complex* antenna family, the most well-known members of which are the CP43 and CP47 antenna proteins of the PSII core [24,73]. The protein induced by iron deficiency, also known as CP43', associates with the PSI to form a PSI–IsiA supercomplex (Figure 20.5b), where an antenna ring of 18 IsiA units, each of them binding 16–17 Chls *a*, encircles the trimeric PSI core [74,75]. The IsiA, which can become the most abundant Chl-binding protein in the cell [76], increases the absorption cross section of PSI, doubling the light-harvesting capacity of the photosystem [75] and compensating for the lowered PSI level in response of iron deficiency. The efficiency of energy transfer from the IsiA ring toward the PSI core is very high and probably involves some Chl *a* molecules bound to minor PSI polypeptides. In particular, three Chls *a* coordinated by the intrinsic subunit PsaJ are supposed to provide the excitonic link for this efficient energy transfer [58]. The IsiA proteins bind also four carotenoids (two ß-Car, one zeaxanthin, and one echinenone), which play a role as light-harvesting pigments [77], but also function in photosystem protection against the excess absorbed energy as quenchers of the harmful Chl excited states [78].

20.2.4 LINEAR AND CYCLING ELECTRON FLOW

The pathway of electron transfer from water as an initial electron donor to NADP$^+$ as a terminal electron acceptor, which involves both PSII and PSI working in series, is named linear electron flow (LEF) or Z-scheme. The LEF couples the final production of NADPH with the synthesis of ATP through a proton motive force generated across the thylakoid membrane. More precisely, the splitting of 2H$_2$O releases four H$^+$ into the thylakoid lumen and provides four electrons, whose transport through the LEF leads to the synthesis of 2NADPH and the translocation across the membrane of 8 H$^+$ protons by the action of the Q-cycle. In this way, a total of 12 H$^+$ are accumulated into the thylakoid lumen [79]. This electrochemical proton gradient powers the phosphorylation (photophosphorylation) of ADP to ATP through the ATP synthase, which is a multiprotein complex, very similar to the one operating in respiration. The thylakoidal ATP synthase is composed of two subcomplexes connected to each other by a stalk, namely, an integral subcomplex (CF$_o$) containing a proton-conducing ring formed by 14 c-subunits, and an extrinsic subcomplex (CF$_1$) containing the ATP catalytic site [4]. On the basis of the rotational catalysis model [80], the synthesis of 3 ATP requires the passage through the CF$_o$ ring of a number of protons corresponding to the c-subunits (i.e., 14). Thus, the linear flow of four electrons, coupled to 12 H$^+$ released into the thylakoid lumen, yields 2NADPH and 2.57 ATP [79].

An optimal photosynthetic performance requires the fine-tuning between the light energy conversion through the photosystems and its use by metabolic reactions. This makes the balance between phosphorylating (ATP) and reducing (NADPH) power

a very important parameter [79]. The calculated value of the ATP/NADPH ratio generated by the LEF is about 1.28, which is not sufficient to support the CO$_2$ fixation through the Calvin–Benson–Bassham cycle (that requires a ratio of 1.5) as well as the other metabolic processes depending on photosynthesis. Thus, the oxygenic photoautotrophs have developed alternative electron flow pathways to provide extra ATP in order to match the ATP/NADPH supply with the metabolic demand [79].

The CEF around the PSI is regarded as the main pathway involved in additional ATP production [81]. During CEF, electrons are returned from the acceptor side of PSI to the intersystem electron transport chain at the level of plastoquinone or Cyt *b$_6$f*. The CEF pathway gives rise to a trans-thylakoidal proton gradient (proton motive force) useful for photophosphorylation, but it does not generate reducing power, thus re-equilibrating the ATP/NADPH balance [79,82].

Cyanobacteria seem to have at least two major routes of CEF around the PSI [10,83] (Figure 20.6). One of them, the NADPH:PQ oxidoreductase route, involves electron transfer from ferredoxin (via FNR and NADPH) to plastoquinone through a particular type I NAD(P)H oxidoreductase (NDH-1) complex (defined as NDH-CEF), which possesses specific subunits essential for coupling it to the photosynthetic electron flow [84,85]. The electron transfer from ferredoxin to plastoquinone via a ferredoxin:plastoquinone oxidoreductase (FQR) is the second route of CEF, also indicated as *antimycin A-sensitive route* due to its sensitivity to this electron transfer inhibitor [86]. It has been proposed that the still elusive FQR might involve the peculiar heme *x* of the Cyt *b$_6$f*. The presence of this extra heme on the *n*-side of the complex, near the Q$_i$ site, in a strong electronic coupling with the heme *b$_H$* and at a position accessible from the cytoplasm, strongly suggests that the ferredoxin might be able to transfer electrons via the heme *x* to an oxidized quinone bound to the Q$_i$ site of the Cyt *b$_6$f* [25,45,50]. The FQR-mediated CEF requires the participation of the PGR5 (proton gradient regulation 5) protein, an enzyme component that might have a regulatory function on the redox poising of the electron transfer chain [79,86]. In fact, whereas the NDH-1-dependent CEF is coupled to the metabolism demand, operating primarily to meet the request of additional ATP synthesis, the major role of FQR-mediated

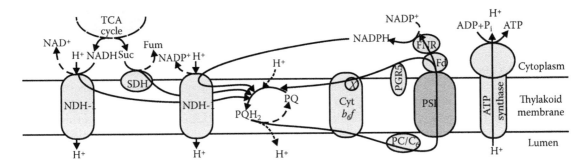

FIGURE 20.6 Schematic view of the cyanobacterial electron transport chains operating in the CEF around the PSI. The supply of electrons via respiratory complexes is also shown. C_6, cytochrome c_6; Cyt *b$_6$f*, cytochrome *b$_6$f*; Fd, ferredoxin; FNR, ferredoxin-NADP$^+$ reductase; Fum, fumarate; NDH-1, type I NAD(P)H dehydrogenase; PGR5, proton gradient regulation 5 protein; PSI, photosystem I; PC, plastocyanin; PQ, plastoquinone; SDH, succinate dehydrogenase; Suc, succinate; TC cycle, tricarboxylic acid cycle. See text for detailed explanation.

CEF is in recycling excess reducing power under conditions (such as high light or limiting electron sinks) that provoke high NADPH/ATP ratios [86].

A further route of CEF that can occur in the thylakoid membranes of cyanobacteria relies on the interaction of the coexisting respiratory and photosynthetic electron transport chains through common mobile electron carriers. In this pathway, electrons coming from sugar breakdown via tricarboxylic acid (TCA) cycle are supplied to plastoquinone by respiratory complexes (NDH-1 and SDH) and then transferred to PSI via Cyt b_6f and plastocyanin (or cytochrome c_6), with generation of a transmembrane proton gradient useful for ATP synthesis but with no net change of reducing power [14,87]. This peculiar route of cell redox-balance regulation is triggered by changes in the redox state of plastoquinone pool and involves the rearrangement of the respiratory complexes within the thylakoid membranes [14]. Under conditions of NADPH/ATP ratios requiring the CEF activation, the respiratory complexes (NDH-1 and SDH), previously clustered in discrete functional *islands*, become much more evenly distributed in the membrane, thus greatly enhancing the probability of a closer association with the PSI and of the electron transfer from them to the photosystem [10,83]. The pathway of signal transduction linking plastoquinone redox state to the redistribution of the electron transport complexes in the thylakoid membrane is still unknown [10].

Two major routes of CEF have been proposed also for eukaryotic microalgae (Figure 20.7). The first one involves the electron transfer from NADPH to plastoquinone through the activity of a NADPH dehydrogenase. However, many eukaryotic algae, including most of green microalgae, lack a chloroplast type I NDH complex, although their ability to reduce plastoquinone nonphotochemically is well documented [88]. In *C. reinhardtii*, it has been found that the plastoquinone reduction is mediated by a type II NAD(P)H dehydrogenase (NDH-2), a monomeric enzyme with a noncovalently bound flavin as the prosthetic group, localized in the thylakoid membranes [88]. Alternative to the NDH-2 route, there is the antimycin A-sensitive one that drives the reduction of plastoquinone pool via a putative FQR complex [89]. Also for the eukaryotic microalgae, the hypothesis exists that the electron flow from ferredoxin to plastoquinone might be mediated directly by the Cyt b_6f complex via the peculiar extra heme x [50]. Recently, two gene products common to all photosynthetic eukaryotes have been proposed as essential components of the FQR complex: the PGR5 protein, inherited from cyanobacteria, and the eukaryotic PGR5-like1 (PGRL1) protein [81]. The PGR5/PGRL1/Fd-dependent CEF seems to be required mostly under conditions leading to the ferredoxin overreduction [90].

A certain function in balancing the chloroplast ATP/NADPH budget may also be ascribed in eukaryotic microalgae to the so-called *malate valve*, which is based on metabolic

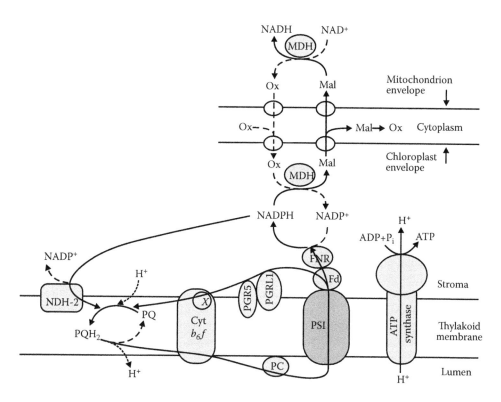

FIGURE 20.7 Schematic view of the electron transport chains operating in the CEF around the PSI in *C. reinhardtii*. The metabolic chloroplast/cytoplasm/mitochondrion interactions of the *malate valve* pathway are also included in the scheme. Cyt b_6f, cytochrome b_6f; Fd, ferredoxin; FNR, ferredoxin-NADP$^+$ reductase; Mal, malate; MDH, malate dehydrogenase; NDH-2, type II NAD(P)H dehydrogenase; Ox, oxaloacetate; PGR5, proton gradient regulation 5 protein; PGRL1, PGR5-like1 protein; PQ, plastoquinone; PSI, photosystem I; PC, plastocyanin. See text for detailed explanation.

interactions among the chloroplast, cytoplasm, and mitochondrion and occurs through malate/oxaloacetate shuttles operating in both the organelle envelopes [91]. In the malate valve, oxaloacetate, which entered the chloroplast from the cytosol via the malate/oxaloacetate shuttle, is reduced to malate by a stromal light-activated malate dehydrogenase, which uses the electrons of photosynthetic NADPH. The malate, exported from the chloroplast through the same shuttle, can be oxidized back to oxaloacetate in the cytosol or move to the mitochondrion via the malate/oxaloacetate shuttle of the organelle, serving as a substrate for light-enhanced dark respiration [89]. This indirect export of excess reducing equivalents, which does not interfere with the ATP production, can contribute to balance the ATP/NADPH ratio in the chloroplast [92].

20.3 GREEN OXYPHOTOBACTERIA

Cyanobacteria with PBSs as peripheral LHCs, known as blue-green cyanobacteria, represent the vast majority of prokaryotic species performing oxygenic photosynthesis (referred to as oxyphotobacteria). However, *green* forms of cyanobacteria from different natural environments, which use intrinsic membrane chlorophyll *a/b* binding proteins instead of extrinsic PBPs to collect light for photosynthesis, also exist. The extant known *green cyanobacteria* or *green oxyphotobacteria*, formerly named prochlorophytes, belong to the genera *Prochlorococcus, Prochloron,* and *Prochlorotrix*, which are polyphyletic within the cyanobacterial radiation [93].

Prochlorococcus, whose representative species is *P. marinus*, is the most abundant genus of the smallest (0.5–0.7 μm) known photosynthetic organisms in the open oceans, where it is ubiquitously distributed from 40°N to 40°S [94]. The free-living *Prochlorococcus* and the closest relative blue-green cyanobacterium *Synechococcus* can contribute, together, over 60% of the primary production of the oligotrophic oceans [95,96]. Genetically and physiologically distinct ecotypes of *Prochlorococcus*, co-occurring in the tropical and subtropical gyres, can be distinguished in two major divisions, occupying different light niches in the water column: the high-light (HL)-adapted ecotypes, thriving the upper well-illuminated layer, and the low-light (LL)-adapted ecotypes, growing in the deeper (up to −200 m) dimly illuminated euphotic zone [94,97,98]. The coccoid *Prochloron*, whose type species is *P. didemni*, is found in tropical and subtropical waters as obligate endosymbiont of marine tunicates [99,100], although the *Prochloron* capability to survive, at least for some time, in free-living form has been hypothesized [101]. The oxyphotobacteria of the genus *Prochlorotrix*, which includes the species *P. hollandica* [102] and *P. scandica* [103], are unique among the green cyanobacteria, being filamentous free-living organisms whose presence is restricted to the eutrophic freshwaters or brackish habitats of North Europe [104].

20.3.1 PROCHLOROPHYTE CHL *A/B*-BINDING PROTEINS

In all the green oxyphotobacteria, the major peripheral antenna system is provided by intrinsic membrane proteins defined as

Pcb (prochlorophyte Chl *a/b*-binding) proteins (30–38 kDa; Figure 20.3c), each of them binding 15 Chls with Chl *a*/Chl *b* ratios depending on the organism and strain [24]. The Pcbs share no relatedness with the three-helix light-harvesting chlorophyll-binding proteins of the eukaryotic LHC superfamily. They belong, instead, to the six-helix transmembrane proteins of the *core complex* antenna family, which includes the CP43 and CP47 complexes of the PSII core and the iron starvation-inducible (IsiA) protein [105].

LL-adapted ecotypes of *Prochlorococcus* possess eight different Pcb proteins harboring Chls with high *b/a* ratios [92], which form large constitutive antennae for both photosystems [70,106]. Eight Pcb subunits, with four distributed on each side of the dimeric core, form a Pcb–PSII supercomplex [106], while a ring of 18 Pcb subunits around the trimeric core gives rise to a giant Pcb–PSI supercomplex similar to the IsiA–PSI supercomplex induced in blue-green cyanobacteria by iron deficiency [70,71]. Peculiarly, *Prochlorococcus* is the unique photosynthetic organism that uses divinil Chl *a* (DVChl *a*) and divinil Chl *b* (DVChl *b*) instead of the monovinil Chls (MVChls) as major antenna pigments [94]. Since the DVChls harvest blue light more efficiently than MVChls, *Prochlorococcus* is allowed to sustain photosynthesis under the deep-sea water column, where blue is the predominant light [107]. In addition, LL-adapted strains also possess a blue light-absorbing phycoerythrin (PE) that links phycourobilin (PUB) and phycoerythrobilin (PEB) as chromophores and is connected to the thylakoid membranes [108,109]. Although this constitutive PE may function as an antenna pigment, its contribution to the total light-harvesting capacity of *Prochlorococcus* is very low due to its tiny amount in the cell [110].

The HL-adapted ecotypes of *Prochlorococcus* show only the constitutive PSII antenna, formed by subunits of a single Pcb harboring Chls with a very low *b/a* ratio [93]. The PSI, instead, is *naked* plausibly for the uselessness of an additional antenna for this photosystem in the upper well-illuminated layer of ocean waters [106]. These ecotypes possess also an unusual form of phycoerythrin with the sole β polypeptide linking one PEB as a chromophore [111]. This peculiar β PE, however, is a green light-absorbing pigment without any light-harvesting function but with a possible role as a photoreceptor [111].

A large Pcb–PSII supercomplex formed by 10 units of a single Pcb protein, with five flanking each side of the dimeric PSII core, is present in *Prochloron didemni*, which does not show any antenna complex around the PSI trimeric core [112].

Three different types of Pcb proteins (PcbA-C) form supramolecular complexes with both PSI and PSII in *Prochlorotrix hollandica* [113]. The major supercomplex of this freshwater filamentous oxyphotobacterium is the Pcb–PSI, which consists of 18 PcbC subunits encircling the trimeric PSI core [113–115]. PcbA and PcbB, instead, associate with the PSII to form a Pcb–PSII supercomplex including 14 Pcb subunits with seven likely arranged in a double arc on each flank of the dimeric core [113].

In addition to Chl *a* and Chl *b*, the Pcb proteins of green oxyphotobacteria harbor carotenoids, which are mainly zeaxanthin together with α-carotene in *Prochlorococcus*, β-carotene in *Prochloron*, and α- and β-carotene in *Prochlorotrix* [94,115]. Differently from chlorophylls, whose primary function is to greatly enhance the total light-harvesting capacity of photosystems, the Pcb-linked carotenoids seem to play an essential role as photoprotective pigments against the harmful excess light by quenching efficiently the chlorophyll excited states [116].

20.4 *ACARYOCHLORIS MARINA* AND *HALOMICRONEMA HONGDECHLORIS*

Acaryochloris marina is the only known oxyphotobacterium that uses Chl *d* as the predominant (over 95%) photosynthetic pigment [117,118]. This is an unusual form of chlorophyll, which absorbs light above 700 nm *in vivo*, allowing *A. marina* to thrive in niche environments enriched in far-red light and to use longer light wavelengths (700–740 nm) not exploited by organisms with other Chls (namely, Chl *a*, *b*, or *c*) [119]. *A. marina* strains are widely distributed in most saline and freshwater environments on earth, where they reside as epibionts beneath the lower layer of coral-reef ascidians, as free-living epiphytes on undersides of red, brown, and green macroalgae or in microbial mat communities [120,121].

Two different systems are used by *A. marina* to capture light: an integral Chl *d*-binding light-harvesting antenna and an external PBP complex. This makes *A. marina* able to exploit visible (500–600 nm), in addition to infrared (700–740 nm), light for photosynthesis [122]. In *A. marina*, large quantities of Chl *d* and small amounts of Chl *a*, together with zeaxanthin and α-carotene [118], are bound to two Pcb homologous proteins (PcbA and PcbC), which functionally associate with a photosystem core [123]. A giant PcbA–PSII supercomplex in *A. marina* cells consists of 16 copies of PcbA subunits flanking two dimeric PSII cores arranged end to end, with four Pcb proteins for each PSII monomer [124]. No Pcb–PSI association, instead, occurs in *A. marina* in no-limiting environmental conditions [124]. By contrast, cells exposed to iron deficiency are able to assemble a giant PcbC–PSI supercomplex similar to the IsiA–PSI supercomplex of blue-green cyanobacteria [71] with 18 PcbC subunits surrounding the trimeric PSI core [125]. In these cells, the poor iron supply, which causes a drop in the PSI level, induces the synthesis of the PcbC leading to the assemblage of the antenna that enhances the light-harvesting capacity of PSI compensating for the decrease in the PSI/PSII ratio [125]. *A. marina* also possesses a peculiar *rod-type* PBS consisting of three homoexamers of $(\alpha,\beta)_6$ PC and one heteroexamer formed by $(\alpha,\beta)_3$ PC/$(\alpha,\beta)_3$ APC [126]. The PBS rods are closely associated to the PSII complexes to which excitation energy is efficiently transferred [127]. In this way, a single PSII is simultaneously connected with both the integral membrane-bound (Pcb) and the external water-soluble (PBP) light-harvesting antennae [127].

It is noteworthy that in *A. marina* (as unique oxygenic phototroph), the Chl *d* is present not only in the peripheral antennae but also in PSII and PSI RCs, where it replaces Chl *a*. In

particular, a homodimer of Chl *d* absorbing maximally at 713 nm is the primary electron donor of PSII (designated P713), while the primary electron donor of PSI (designated P740) is a heterodimer of Chl *d* and Chl *d'* (the 13^2 epimer of Chl *d*) absorbing maximally at 740 nm [128,129]. In *A. marina*, the redox potentials of these primary electron donors, despite the lower photon energy of their excitation wavelengths, are very similar to those of blue-green cyanobacteria, and the energy-storage efficiency of the light reactions *in vivo* is comparable to, or even higher than, that in Chl *a* utilizing species [122,130,131]. This reveals that the solar spectrum used by oxyphotobacteria expands over the visible portion, and that the photosynthesis can be viable in even redden light environments [122].

The extension of oxygenic photosynthesis into the infrared region has recently been confirmed by the isolation from a stromatolite of the filamentous cyanobacterium *Halomicronema hongdechloris* that, in addition to Chl *a* and PBSs, contains Chl *f* as an accessory photopigment [132]. Chl *f* is the most red-shifted chlorophyll found in an oxygenic photoautotroph. This pigment allows the photosynthetic absorbance region to be extended up to 750 nm [122,133]. The photophysical and photochemical functions of Chl *f*, which can constitute ~10% of the total Chls in *H. hongdechloris*, are still under investigation, although its involvement in the light-harvesting systems has been clearly established [134,135].

20.5 MICROALGAL PHOTOSYSTEMS

The uptake by a eukaryotic nonphotosynthetic host of an ancient cyanobacterium, afterward genetically integrated and reduced to a cell organelle, represented the primary endosymbiosis event that led eukaryotic photoautotrophs with *primary* chloroplasts (glaucophytes, red algae, and chlorophytes [green algae, mosses, ferns, and higher plants]) to evolve since ~1.8 Gyr ago [3]. The ensuing secondary endosymbioses between eukaryotic hosts and engulfed green or red algae gave rise to a spread of photoautotrophs with the so-called *secondary* or *second-hand* plastids, whose acquisition had a major impact on photosynthetic organism diversity, evolution, adaptive strategies, and global ecology [136]. Chlorophytes and organisms with secondary plastids of green algal origin, such as chlorarachniophytes and euglenids, represent the *green lineage* of photosynthetic eukaryotes, while the *red lineage* comprehends red algae and organisms, such as cryptophytes, diatoms, most photosynthetic dinoflagellates, and related algal groups, collectively termed Chromalveolates, whose secondary plastids originated from a red alga [137,138]. The algae of the red lineage represent over 50% of the presently described protist species [139].

A further event, which occurred in the evolutionary history of plastids, was the tertiary endosymbiosis, which involved some lineages of photosynthetic dinoflagellates. In this case, a dinoflagellate ancestor that, as chromalveolate, already had a secondary plastid of red algal origin engulfed the cell of another chromalveolate, replacing the ancestral plastid that was no longer photosynthetically active with a tertiary organelle acquired from the algal host [137]. Furthermore, in some other dino-flagellates, the ancestral secondary plastid was replaced with another secondary plastid derived from a green alga through a so-called *serial secondary endosymbiosis* [137,138].

During the time course of evolution from prokaryotic cyanobacteria to the wide range of eukaryotic photoautotrophs, the major components of the intrinsic core complexes of PSI and PSII, with the inner antennae and the RC proteins harboring the cofactors of charge separation and electron transport chain, remained largely conserved [59,140], whereas some changes occurred in the minor polypeptides of the two photosystems and in composition of the extrinsic proteins of the OEC [34]. Drastic modifications, instead, were undergone by the peripheral components of the photosynthetic machinery fulfilling the function of light capture and excitation energy transfer to the photosystem RCs to drive the oxygenic photosynthesis. In particular, the antenna system based on bulky extrinsic phycobilisomes was replaced mostly by a mosaic of membrane-intrinsic LHCs that led the different groups of photosynthetic organisms to optimize the light-harvesting capacity and to adapt to different photic environments [141].

The LHCs belong to an extended chlorophyll-binding protein superfamily (named LHC superfamily), which, in addition to LHC components devoted to light capture, also comprehends LHC-like members mostly involved in photosystem protection against the dangerous excess light [142]. Among these are the eukaryotic early light-induced proteins (ELIPs) of green microalgae, the LHC-stress related (LHCSR and LHCX) proteins of green microalgae and diatoms [143,144], and the ancestral high light-inducible proteins (HLIPs) already present in cyanobacteria [73,145–147]. It is thought that the original role of the extended LHC superfamily may have actually been photoprotection rather than light harvesting [148]. All the LHC members present in photosynthetic eukaryotes, with the primary function to absorb light, span the thylakoid membranes, possess three evolutionarily conserved α-helical transmembrane regions connected by both stroma and lumen-exposed loops [141,142], and share chlorophyll-binding domains named *LHC motifs* [149]. These proteins also bind carotenoids as photosynthetic pigments, thus expanding the range of absorbed light, which is important especially in aquatic environments [150].

Different microalgal groups are characterized by the specific pigments associated with the LHCs. In red algae, the LHCs (referred to as LHCR, with *R* for Rhodophyta) bind Chl *a* and zeaxanthin as major carotenoids; in green algae, the pigments bound to LHCs (named CAB, for Chl *a/b*) are Chl *a*, Chl *b*, and lutein as most abundant carotenoids; while in chromalveolates, the LHCs (known as CAC, for Chl *a/c*) bind Chl *a*, Chl *c*, and different xanthophylls, such as fucoxanthin (in diatoms) or peridinin (in dinoflagellates), as dominant carotenoids [141,151]. Recently, a new type of light-harvesting antenna proteins, belonging to the LHC superfamily, has been discovered [142]. These new three-helix transmembrane proteins, termed *red lineage CAB-like proteins* (RedCAPs), are found in red algae and in algae (such as diatoms) with secondary chloroplasts of red algal origin [142,152].

20.5.1 PHOTOSYSTEMS OF RED MICROALGAE

20.5.1.1 Photosystem II

The red algae are evolutionarily one of the most primitive algal groups originated from a primary endosymbiotic event. Their plastids, enveloped by a double membrane, exhibit a photosynthetic apparatus that represents a transitional state between cyanobacteria and photosynthetic eukaryotes. The PSII, in fact, is similar to the cyanobacterial photosystem since it contains PBSs as primary light-harvesting antenna, while the PSI possesses an intrinsic chlorophyll-binding LHC, like the other eukaryotic algae [153]. The PSI and PSII complexes are distributed in membranes of thylakoids singly arranged in the stroma, as it occurs in cyanobacteria [154].

The composition of the dimeric PSII core complex in red algae is similar to that of the cyanobacterial PSII, with D1, D2, Cyt b_{559}, CP43, and CP47 as major polypeptides [155]. Also the OEC bound on the lumenal side of the PSII is rather similar to that of cyanobacteria, as it contains the extrinsic proteins PsbO (33kDa), PsbU (9 kDa), and PsbV (cytochrome c_{550}, 15 kDa) common to the prokaryotic complex. However, the OEC of red algae also possesses a unique 20 kDa protein (named PsbQ') not found in other photosynthetic organisms [156]. The red algal PsbQ' shows some similarities to the PsbQ-like (Cyano Q) protein from cyanobacteria as well as to the PsbQ (17 kDa) protein of the OEC from green algae and higher plants of which it is considered an ancestral form [156].

20.5.1.2 Phycobilisomes

Like in cyanobacteria, the stromal surface of thylakoid membranes in red algal plastids is covered by PBSs (Figure 20.8a), each of them likely associated to one PSII dimer [157]. The PBSs are made up of a tricylindrical core of trimeric $(\alpha\beta)_3$

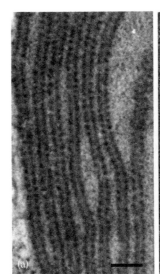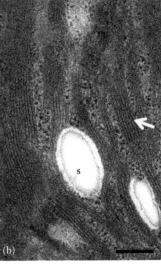

FIGURE 20.8 Transmission electron micrographs of microalgal chloroplast details. (a) Thylakoids from a red chloroplast with phycobilisomes on the membrane surfaces. Bar = 500 nm. (b) Chloroplast of the green microalga *C. reinhardtii* showing single and stacked (arrow) thylakoids. s, starch; bar = 200 nm.

APC, peripheral rods containing exameric $(\alpha\beta)_6$ PC or PC/PE, and colorless linker polypeptides, which mediate the highly ordered structure of the supramolecular antenna complex that also possesses ApcE and APC-B as terminal emitters [154,158]. In red algal PBSs, the PC can be present as C-PC containing only phycocyanobilin (PCB) and R-PC that links both PCB and phycoerythrobilin (PEB). The PE, which is the predominant antenna pigment in most red microalgae, shows two types of covalently attached chromophores, PEB and phycourobilin (PUB), and can be classified in different classes based on absorption properties [154,158]. Two different spectral types of PE are present at the periphery of PBS rods: the outer B-PE carrying an additional chromophorylated γ-subunit linked with PEB and PUB, and the inner b-PE lacking it [154,159,160]. This γ subunit, which acts as a linker responsible for the association between B-PE and b-PE in the rod elements, contributes to the light absorption of PE, but also participates in the regulation of energy transfer along the PBS [160].

Two morphological types of PBSs exist in different red algal species: a smaller hemidiscoidal and a larger hemispherical (also described as hemiellipsoidal) supercomplex [154]. The two PBSs exhibit the same central core but differ in the number of peripheral rods that in the hemispherical type are more than 10 and radiate into diverse directions in space giving rise to a very large dome-shaped PBS. In *Porphyridium cruentum*, for instance, the measured PBS showed the overall size of 60 × 41 × 34 nm (length × width × height) [157]. Although PBSs are the major LHCs of PSII, they may transfer excitation energy also to PSI due to their crowding distribution on the thylakoid membrane and to the small spacing between the neighboring photosystems in the same membrane [161].

20.5.1.3 Photosystem I

The red algal PSI differs from the cyanobacterial photosystem as it exists in a monomeric instead of a trimeric form and possesses a Chl *a*-binding LHC (referred to as LHCI) as LHC associated to the central core [153]. The LHCI is made up of a species-dependent number (from 3 to 6) of proteins, termed Lhcr [162], each of them binding an average of 8 Chl *a* and 0.5 β-Car molecules. The Lhcr polypeptides bind also 3–4 molecules of zeaxanthin that does not function as a photosynthetic accessory pigment, being rather required for structural stability [163]. The PSI and its LHC antenna give rise to a supercomplex (PSI-LHCI) (Figure 20.9a) with the Lhcr proteins forming a crescent along the monomeric PSI core at the side of the minor polypeptides PsaF/PsaJ [68,153]. Also, proteins of the RedCAP family have been found in the red algal chloroplasts, with a potential role in light harvesting [142].

Recently, Busch et al. [162] showed that the PSI of the primitive red alga *Cyanidioschyzon merolae* in addition to the LHCI possesses a firmly linked PBS subcomplex containing PC but no APC as the light-harvesting antenna. This finding led the authors to emphasize that the PSI system employed by red algae for sunlight capturing may represent an evolutionary intermediate between cyanobacteria and other eukaryotic photoautotrophs [162].

20.5.2 Photosystems of Green Microalgae

Like the red algae, also the green algae, as well as higher plants, possess chloroplasts derived from a primary endosymbiotic event. The organelles of this green lineage have a two-membrane envelope and contain an inner membrane system formed by bands of stacked thylakoids (grana) and thylakoids running singly in the stroma (Figure 20.8b).

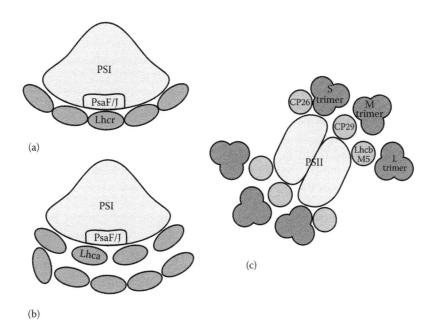

FIGURE 20.9 Scheme of the PSI–LCHI supercomplex in red (a) and in green (b) microalgae, and of the PSII–HLCII supercomplex in green microalgae (c). See text for detailed explanation.

In chloroplasts of the model organism *C. reinhardtii*, the appressed thylakoids comprise more than 70% of the entire thylakoid system [164]. However, unlike the chloroplasts of higher plants, the thylakoids of this green microalga seem to be weakly stacked, and the two photosystems do not exhibit a defined lateral segregation but are rather distributed through the thylakoid membranes, similarly to what occurs in other microalgal classes [164].

In green algae, like in higher plants, both the monomeric PSI and the dimeric PSII have intrinsic peripheral antenna systems formed by CAB members of the LHC superfamily [141]. Moreover, the composition of the PSII-associated OEC is rather different from that of cyanobacterial and red algal complexes. The *green algal-type* OEC lacks the PsbU and PsbV polypeptides and, besides the ubiquitous PsbO, contains the PsbP (23 kDa) and PsbQ (17 kDa) proteins derived from homologous precursors in ancient cyanobacteria (CyanoP and CyanoQ) and red algae (PsbQ') [34,156] through considerable genetic modifications [165]. A fourth still enigmatic protein, referred to as PsbR (10 kDa), not found in cyanobacteria and red algae, is also present in the green algal OEC [34].

20.5.2.1 LHCII

The peripheral PSII LHCs (LHCIIs) of green microalgae (Figure 20.9c), mostly studied in *C. reinhardtii*, consist of two different types of highly homologous CAB proteins (designated Lhcb), which comprehend nine major LHCII proteins (majLHCII, around 28 kDa) and two minor (CP26 and CP29) polypeptides. Most of the majLHCII proteins (namely, LhcbM1-4 and LhcbM6-9) are organized *in vivo* as heterotrimers that give rise to the major LHCII antenna complex, while the CP26 (Lhcb4) and CP29 (Lhcb5) are monomeric proteins, which form minor LHCII antennae [166,167]. Also a majLHCII protein (LhcbM5) is present in a monomeric form [168]. In green microalgae, as well as in higher plants, each protein of the majLHCII binds 14 Chls (8 Chl *a* and 6 Chl *b*) and 4 xanthophylls (1 neoxanthin, 1 violaxanthin, and 2 luteins that can be partly substituted by loroxanthin) [169,170]. The proteins of the minor antennae CP26 and CP29 bind 9 and 13 Chls, respectively, with higher Chl *a/b* ratios, and 2 or 3 xanthophylls (lutein, neoxanthin, and violaxanthin) [171,172]. The monomeric LHCII proteins are located next to CP43 and CP47, at the interface between the PSII core complex and the major LHCII antenna and function in the efficient energy transfer from the majLHCII trimers toward the photosystem RC [141]. Three types of majLHCII trimers, named strong (S), moderate (M), and loose (L), can be distinguished on the basis of the association degree with the PSII core [173]. Experimental evidence shows that the green algal photosystem can form a PSII–LHCII supercomplex with up to six majLHCII trimers per dimeric core (the $C_2S_2M_2L_2$) [174] and that a number of additional (*extra*) majLHCII trimers, present in the thylakoid membrane, might be located in between the PSII supercomplexes enlarging the photosystem light-harvesting capacity [170].

20.5.2.2 LHCI

The CAB proteins of the PSI peripheral light-harvesting antenna (LHCI) of green algae, as well as of higher plants, are known as Lhca proteins (20–24 kDa). These highly homologous proteins, nine of them (designated Lhca1 through 9) identified in *C. reinhardtii* [175], bind 10 Chls, with an average Chl *a*/Chl *b* ratio of 2.2 [176–178], together with 2 or 3 carotenoids (lutein, violaxanthin, and β-carotene) [179,180]. Some Lhca complexes, named *red Lhca*, peculiarly harbor *red-shifted spectral forms* of chlorophylls, which absorb light at wavelengths above 700 nm and have emission maxima between 707 and 715 nm [180]. The biological function of these long wavelength Chls (LWC) is still under debate. They might increase the absorption cross section of PSI in low light conditions [181], but might also have a photoprotective role under excess light, due to their low energy level that allows to focus the excitation energy in Lhca domains containing carotenoids with high efficiency in chlorophyll triplet quenching [182,183].

A recently published model of *C. reinhardtii* PSI–LHCI supercomplex shows 9 Lhcas arranged in a double half-moon-shaped row along the monomeric PSI core (Figure 20.9b). Like in red microalgae, the LHCI crescent is located at the side of the minor polypeptides PsaF/PsaJ [68,178]. However, different Lhca/core stoichiometries, giving from 6 to 14 Lhca subunits, have been proposed in literature for the PSI–LHCI [177,184–187]. This might be related to the finding that the green algal PSI–LHCI supercomplex possesses a high flexibility and that the size and composition of the LHCI can be modulated in response to changing environmental conditions [141].

20.5.3 Photosystems of Chlorophyll *c*-Containing Microalgae

The Chl c-containing microalgae, also known as chromalveolates, have plastids derived from a secondary endosymbiotic incorporation of a red microalga in a eukaryotic host [137,138]. The plastids of this *red algal lineage* are characteristically enveloped by 3 or 4 membranes and show thylakoids not differentiated into distinct granal and stromal regions but usually arranged in bands of three (Figure 20.10a,b), with photosystems rather homogeneously distributed [188].

Like the other eukaryotic photoautotrophs, these microalgae show monomeric PSI and dimeric PSII with conserved intrinsic core components. They have a *red algal-type* OEC exhibiting the four extrinsic proteins PsbO, PsbU, PsbV, and PsbQ' [189]. The photosystems possess membrane-bound LHCs highly homologous to the LHCs of green microalgae, but with distinct pigment composition and a stronger reliance on carotenoids for the light gathering [190]. In these peripheral antennae, the accessory chlorophyll is a form of Chl *c* (namely, Chl c_1 or Chl c_2 instead of Chl *b*, and lutein is replaced by unique forms of carbonyl carotenoids, such as fucoxanthin in diatoms and peridinin in most dinoflagellates. The copresence in the LHCs of Chl *c* with a strong light absorption at 460 nm

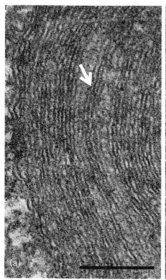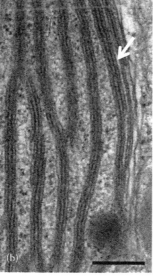

FIGURE 20.10 Transmission electron micrographs of two chromoalveolate chloroplasts. In both the diatom *P. tricornutum* (a) and the eustigmatophycean *Nannochloropsis gaditana* (b), the parallel thylakoids are mostly arranged in bands of three (arrows). Bars = 200 nm.

and xanthophylls with light absorption up to 540 nm clearly extends the light-harvesting capacity of these microalgae in the blue-green part of the solar spectrum, which dominates in marine environments [191,192]. In the antenna complexes, both Chl c and carotenoids transfer excitation energy to Chl a. The efficiency of the energy transfer from fucoxanthin and peridinin to Chl a is exceptionally high, approaching 100% [150]. Owing to the prominence of carotenoids as main light-harvesting pigments, these integral LHCs are usually called the FCP (fucoxanthin–chlorophyll protein) complex [190] and acpPC (Chl a–Chl c_2–peridinin protein complex) [193].

20.5.3.1 FCP Complex in Diatoms

The FCPs of diatoms are proteins with a molecular weight ranging from 17 to 23 kDa [190], which generally bind about 8 Chls a, 2 Chls c, and 8 fucoxanthin molecules [194]. Minor amounts of diadinoxanthin, a carotenoid involved in the photoprotective xanthophyll cycle, are also bound to the FCPs [195], giving rise to a Chl/Car ratio near 1 [192]. Nowadays, three major groups of FCPs (termed I, II, and III) can be distinguished for sequence and function in diatoms [196]. The FCPs of group I (called LHCF proteins) represent *classical* proteins that fulfil the function of light harvesting [197]. The FCPs of group II (called LHCR proteins), which also function as an antenna, are polypeptides closely related to red algal LHCI proteins [198]. The FCPs of group III (called LHCX proteins), instead, are proteins with high similarity to LHCSR proteins of *C. reinhardtii*, essentially involved in photosystem protection against surplus light [144]. The concentrations of LHCF and LHCR proteins are comparable in the diatom chloroplasts, while the concentration of the LHCX proteins is much lower [199].

In the centric diatom *Cyclotella meneghiniana*, two major FCP complexes (FCPa and FCPb) differing in oligomeric state and polypeptide composition have been characterized [190]. FCPa is a trimeric complex built of LHCF and LHCX proteins that plays a role in both light harvesting and photoprotection. The higher oligomeric (hexameric or nonameric) FCPb complex, instead, is composed of LHCF proteins and has the sole light-harvesting function [190,200]. Although a specific association of FCPs to one of the two photosystems is still controversial, some evidence suggests that FCPa might be associated with PSII, while FCPb might be weakly bound to PSI [198]. By contrast, only a single antenna fraction composed of LHCF polypeptides, basically organized as trimers, has been signaled in the pennate diatom *Phaeodactylum tricornutum* [196,201]. Different trimeric FCP pools, however, seem to be associated with the two photosystems [202]. Besides the major peripheral FCPs, diatoms also possess a minor antenna complex tightly connected to the PSI [195,198,203–205]. This specific FCP complex contains more diadinoxanthin and less fucoxanthin and is almost exclusively composed of LHCR proteins closely related to those of red algal LHCI [195]. Furthermore, RedCAP proteins with a possible role of light-harvesting antenna have recently been identified in centric and pennate diatoms [142,152]. It has been suggested that the RedCAPs might be associated to the PSI together with some FCPs [205] or bound at the periphery of FCP complexes shared between PSI and PSII [202].

20.5.3.2 acpPC and PCP Complexes in Dinoflagellates

Some photosynthetic dinoflagellates, whose chloroplasts evolved from a chromalveolate engulfed through a tertiary endosymbiotic event (such as *Gymnodinium* and *Dinophysis*, which acquired the organelle from a haptophyte and a cryptomonad, respectively) [206,207], have LHCs with a fucoxanthin derivative as the main carotenoid [206]. In contrast, dinoflagellates with secondary chloroplasts of red algal origin possess antenna systems based on the carotenoid peridinin as the key light-harvesting pigment [206]. Two different types of primary antennae can be distinguished in these microalgae: the already mentioned acpPC and a peridinin–Chl a protein (PCP) complex.

The acpPC (also known as APC or iAPC) is a membrane-bound intrinsic complex whose proteins (19–20 kDa in size) belong to the LHC superfamily and have sequence similarity to diatom FCPs [193]. Each protein binds 7 Chl a, 4 Chl c_2, and 11 peridinin molecules, together with 2 molecules of the photoprotective carotenoid diadinoxanthin, with a Chl/Car ratio of ~1 [208]. The energy transfer from peridinin and Chl c_2 to Chl a is very efficient, approaching 100% [209]. Both PSI and PSII possess an integral antenna formed by acpPCs. However, how these light-harvesting systems are organized to give rise to PSI and PSII supercomplexes is still an open question that requires further research [193].

The PCP (also indicated as sPCP) complex is a unique extrinsic water-soluble antenna encased within the aqueous thylakoid lumen, with no similarity to any known LHC [193]. The sPCP complex can consist of homodimers of a shorter

(14–16 kDa) polypeptide or monomers of a longer (32–35 kDa) polypeptide, depending on the species [210]. The 2 Å resolved structure of the sPCP complex from *Amphidinium carterae* revealed a trimer of long monomeric subunits each of them holding 2 Chl *a* and 8 peridinin molecules [211]. A similar organization has been suggested for the sPCP complex of the symbiotic dinoflagellate *Symbiodinium* sp. [212]. The peridinin of this extrinsic water-soluble antenna conveys its excitation energy, with an efficiency near 100%, to the Chl *a*, which, in turn, passes the excitation energy to the membrane-bound LHCs [212,213]. The peridinin also plays an excellent photoprotective role, being able to quench the triplet excitation of Chl *a* [214,215].

20.6   CONCLUSIONS

About 2.5 Gyr ago, the acquired ability of primitive cyanobacteria to carry out the oxygenic photosynthesis, by using water as the electron donor, led to the appearance and subsequent increase of oxygen in the atmosphere. This event profoundly affected the diversity of life, triggering the evolution of the aerobic organisms. A first endosymbiosis between a eukaryotic nonphotosynthetic host and an ancient cyanobacterium gave rise to microalgae with primary chloroplasts. Secondary, serial secondary, and tertiary endosymbioses between a range of eukaryotic hosts and different eukaryotic microalgae accounted then for the diversity, evolution, and ecology of the photosynthetic organisms.

In both prokaryotic and eukaryotic photoautotrophs, the oxygenic photosynthesis relies on two photosystems: PSII able to take out electrons from water and PSI able to give electrons to NADP$^+$. During the time course of evolution, the PSI and PSII intrinsic cores, made up of the RCs, harboring the cofactors of charge separation (photochemical reaction) and electron transport chain, and of the inner LHCs, have remained largely conserved. On the contrary, drastic modifications were undergone by the components of the peripheral antennae devoted to the environmental light absorption. This already occurred in some cyanobacteria, which replaced the phycobiliproteins with forms of chlorophylls (namely, Chl *a*, Chl *b*, Chl *d*, and Chl *f*) as light-harvesting pigments. Also in eukaryotic photoautotrophs, with the exception of red algae, the phycobiliproteins were lost and the major light-harvesting function was fulfilled by specific forms of chlorophylls (Chl *a*, Chl *b*, or Chl *c*) and carotenoids (such as fucoxanthin and peridinin).

The evolution of a variety of pigment–protein antenna complexes, with different light absorption features, allowed cyanobacteria and microalgae to photosynthesize by exploiting a wide range of light wavelengths in the visible up to the infrared region of the solar spectrum.

ACKNOWLEDGMENTS

The authors thank Matteo Simonetti of the Department of Biology, University of Padova, for his precious computer support in the production of the figures.

REFERENCES

1. A. Bekker, H. D. Holland, P.-L. Wang, D. Rumble, H. J. Stein, J. L. Hannah, L. L. Coetzee, and N. J. Beukes, *Nature, 427*: 117 (2004).
2. J. Barber, *Photosynth. Res., 80*: 137 (2004).
3. B. Rasmussen, I. R. Fletcher, J. J. Brocks, and M. R. Kilburn, *Nature, 455*: 1101 (2008).
4. M. F. Hohman-Marriot, and R. E. Brankenship, *Annu. Rev. Plant Biol., 62*: 515 (2011).
5. A. Williamson, B. Conlan, W. Hillier, and T. Wydrzynski, *Photosynth. Res., 107*: 71 (2011).
6. J. Whitmarsh, and Govindjee, in *Concepts in Photobiology: Photosynthesis and Photomorphogenesis* (G. S. Singhal, G. Renger, S. K. Sopory, K.-D. Irrgang, and Govindjee eds.) Kluwer Academic Publishers, Dordrecht, 1999, p. 11.
7. R. E. McCarty, Y. Evron, and E. A. Johnson, *Annu. Rev. Plant Physiol. Plant Mol. Biol., 51*: 83 (2000).
8. I. Moro, N. Rascio, N. la Rocca, M. Di Bella, and C. Andeoli, *Algol. Stud., 123*: 1 (2007).
9. K. Sciuto, C. Andreoli, N. Rascio, N. la Rocca, and I. Moro, *Cladistics, 28*: 357 (2012).
10. C. W. Mullineaux, *Biochim. Biophys. Acta, 1837*: 403 (2014).
11. D. J. Lea-Smith, N. Ross, M. Zori, D. S. Bendall, J. S. Dennis, S. A. Scott, A. G. Smith, and C. J. Howe, *Plant Physiol., 162*: 484 (2013).
12. M. G. Rakhimberdieva, M. G. Boikenko, V. A. Karapetyan, and I. N. Standichuk, *Biochemistry, 40*: 15780 (2001).
13. W. F. J. Wermaas, *Encycl. Life Sci.*, Nature Publshing Group, London, 2001, p. 245.
14. L.-N. Liu, S. J. Bryan, F. Huang, J. Yu, P. J. Nixon, P. R. Rich, and C. W. Mullineaux, *Proc. Natl. Acad. Sci. U.S.A., 109*: 2678 (2012).
15. K. Linke, and F. M. Ho, *Biochim. Biophys. Acta, 1837*: 14 (2014).
16. D. J. Vinyard, G. M. Ananyev, and G. C. Dismukes, *Annu. Rev. Biochem., 82*: 577 (2013).
17. A. Zouni, H. T. Witt, J. Kern, P. Fromme, N. Krauss, W. Saenger, and P. Orth, *Nature, 409*: 739 (2001).
18. N. Kamiya, and J.-R. Shen, *Proc. Natl. Acad. Sci. U.S.A., 100*: 98 (2003).
19. A. Guskov, J. Kern, A. Gabdulkhakov, M. Broser, A. Zouni, and W. Saenger, *Nat. Struct. Mol. Biol., 16*: 334 (2009).
20. Y. Umena, K. Kawakami, J.-R. Shen, and N. Kamiya, *Nature, 473*: 55 (2011).
21. P. B. Kós, Z. Deák, O. Cheregi, and I. Vass, *Biochim. Biophys. Acta, 1777*: 74 (2008).
22. J. Sander, M. Nowaczyk, J. Buchta, H. Dau, I. Vass, Z. Deák, M. Dorogi, M. Iwai, and M. Rögner, *J. Biol. Chem., 285*: 29851 (2010).
23. P. Mulo, C. Sicora, and E.-M. Aro, *Cell. Mol. Life Sci., 66*: 3697 (2009).
24. J. W. Murray, J. Duncan, and J. Barber, *Trends Plant Sci., 11*: 152 (2006).
25. N. Nelson, and A. Ben-Shem, *Nat. Rev., 5*: 1 (2004).
26. N. Msilini, M. Zaghdoudi, S. Govindachary, M. Lachaâl, Z. Ouerghi, and R. Carpentier, *Photosynth. Res., 197*: 247 (2011).
27. D. Shevela, J. J. Eaton-Rye, J.-R. Shen, and Govindjee, *Biochim. Biophys. Acta, 1817*: 1134 (2012).
28. M. D. Lambreva, D. Russo, F. Polticelli, V. Scognamiglio, A. Antonacci, V. Zobnina, G. Campi, and G. Rea, *Curr. Prot. Pept. Sci., 15*: 285 (2014).
29. P. Pospíšil, *J. Photochem. Photobiol. B: Biol., 104*: 341 (2011).
30. K. E. Shinopoulos, and G. W. Brudvig, *Biochim. Biophys. Acta, 1817*: 66 (2012).

31. F. Rappaport, A. Boussac, D. A. Force, J. Peloquin, M. Brynda, M. Sugiura, S. Un, R. D. Britt, and B. A. Diner, *J. Am. Chem. Soc.*, *131*: 4425 (2009).
32. I. L. McConnell, *Photosynth. Res.*, *98*: 261 (2008).
33. B. Kok, B. Forbush, and M. McGloin, *Photochem. Photobiol.*, *11*: 457 (1970).
34. T. M. Bricker, J. L. Roose, R. D. Fagerlund, L. K. Frankel, and J. J. Eaton-Rye, *Biochim. Biophys. Acta*, *1817*: 121 (2012).
35. M. Watanabe, and M. Ikeuchi, *Photosynth. Res.*, *116*: 265 (2013).
36. R. MacColl, *J. Struct. Biol.*, *124*: 311 (1998).
37. L. N. Liu, X. L. Chen, Y. Z. Zhang, and B. C. Zhou, *Biochim. Biophys. Acta*, *1708*: 133 (2005).
38. A. A. Arteni, G. Ajlani, and E. J. Boekema, *Biochim. Biophys. Acta*, *1787*: 272 (2009).
39. X. Gao, T.-D. Wei, N. Zhang, B.-B. Xie, H.-N. Su, X.-Y. Zhang, X.-L. Chen, B.-C. Zhou, Z.-X. Wang, J.-W. Wu, and Y.-Z. Zhang, *Mol. Microbiol.*, *85*: 907 (2012).
40. C. W. Mullineaux, *Photosynth. Res.*, *95*: 175 (2008).
41. K. Kondo, Y. Ochiai, M. Katayama, and M. Ikeuchi, *Plant Physiol.*, *144*: 1200 (2007).
42. M. Watanabe, D. A. Semchonok, M. T. Webber-Birungi, S. Ehira, K. Kondo, R. Narikawa, M. Ohmori, E. J. Boekema, and M. Ikeuchi, *Proc. Natl. Acad. Sci. U.S.A.*, *111*: 2512 (2014).
43. J. F. Allen, *Trends Plant Sci.*, *9*: 130 (2004).
44. W. A. Cramer, H. Zhang, J. Yan, G. Kurisu, and J. L. Smith, *Biochemistry*, *43*: 5921 (2004).
45. G. Kurisu, H. Zhang, J. L. Smith, and W. A. Cramer, *Science*, *302*: 1009 (2003).
46. D. Baniulis, E. Yamashita; J. P. Whittelegge, A. I. Zatsman, M. P. Hendrich, A. S. Hasan, C. M. Ryan, and W. A. Cramer, *J. Biol. Chem.*, *284*: 9861 (2009).
47. D. Baniulis, H. Zhang, T. Zakharova, S. S. Hasas, and W. A. Kramer, *Methods Mol. Biol.*, *684*: 65 (2011).
48. W. A. Cramer, and H. Zhang, *Biochim. Biophys. Acta*, *1757*: 339 (2006).
49. D. Stroebel, Y. Chroquet, J.-L. Popot, and D. Picot, *Nature*, *426*: 413 (2003).
50. D. Baniulis, E. Yamashita, H. Zhang, S. S. Hasan, and W. A. Cramer, *Photochem. Photobiol.*, *84*: 1349 (2008).
51. A. de Lacroix de Lavallette, G. Finazzi, and F. Zito, *Biochemistry*, *47*: 5259 (2008).
52. M. Sang, F. Ma, J. Xie, X.-B. Chen, K.-B. Wang, X.-C. Qin, W.-D. Wang, J.-Q. Zhao, L.-B. Li, J.-P. Zhang, and T.-Y. Kuang, *Biophys. Chem.*, *146*: 7 (2010).
53. M. A. De La Rosa, J. A. Navarro, A. Díaz-Quintana, B. De la Cerda, F. P. Molina-Heredia, A. Balme, P. Murdoch, I. Díaz-Moreno, R. V. Durán, and M. Hervás, *Bioelectrochemistry*, *55*: 41 (2002).
54. P. Jordan, P. Fromme, H. T. Witt, O. Klukas, W. Saenger, and N. Krauß, *Nature*, *411*: 909 (2001).
55. R. K. Le, B. J. Harris, I. J. Iwuchukwu, B. D. Bruce, X. Cheng, S. Quian, W. T. Heller, H. O'Neill, and P. D. Frymier, *Arch. Biochem. Biophys.*, *550–551*: 50 (2014).
56. M. Watanabe, H. Kubota, H. Wada, R. Narikawa, and M. Ikeuchi, *Plant Cell Physiol.*, *52*: 162 (1011).
57. Y. Li, Y. Lin, P. C. Loughlin, and M. Chen, *Front. Plant Sci.*, *5*: 63 (2014).
58. I. Grotjohann, and P. Fromme, *Photosynth. Res.*, *85*: 51 (2005).
59. A. Amunts, O. Drory, and N. Nelson, *Nature*, *447*: 58 (2007).
60. M. Guergova-Kuras, B. Bourdeaux, A. Joliot, P. Joliot, and K. Redding, *Proc. Natl. Acad. Sci. U.S.A.*, *98*: 4437 (2001).
61. M. Hervás, J. A. Navarro, and M. A. De La Rosa, *Acc. Chem. Res.*, *36*: 798 (2003).
62. A. Amunts, and N. Nelson, *Plant Physiol. Biochem.*, *46*: 228 (2008).
63. N. Nelson, and F. Yokum, *Annu. Rev. Plant Physiol.*, *57*: 521 (2006).
64. K. Gibasiewicz, A. Szrajner, J. A. Ihalainen, M. Germano, J. P. Dekker, and R. van Grondelle, *J. Phys. Chem. B, 109*: 21180 (2005).
65. L.-O. Pålsson, C. Flemming, B. Gobets, R. van Grondelle, J. P. Dekker, and E. Schlodder, *Biophys. J.*, *74*: 2611 (1998).
66. M. Brecht, H. Studier, A. F. Eli, F. Jelezko, and R. Bittl, *Biochemistry*, *46*: 799 (2007).
67. E. Schlodder, M. Hussels, M. Çetin, N. V. Karapetyan, and M. Brecht, *Biochim. Biophys. Acta*, *1807*: 1423 (2011).
68. A. Busch, and M. Hippler, *Biochim. Biophys. Acta*, *1807*: 864 (2011).
69. N. V. Karapetyan, Y. V. Bolychevtseva, N. P. Yurina, I. V. Terekhova, V. V. Shubin, and M. Brecht, *Biochemistry (Mosc.)*, *79*: 213 (2014).
70. T. S. Bibby, J. Nield, F. Partensky, and J. Barber, *Nature*, *413*: 590 (2001).
71. E. J. Boekema, A. Hifney, A. E. Yakushevska, M. Piotrowski, W. Keegstra, S. Berry, K.-P. Michel, E. K. Pistorius, and J. Kruip, *Nature*, *412*: 745 (2001).
72. N. Yeremenko, R. Kouril, J. A. Ihalainen, S. D'Haene, N. van Oosterwijk, E. G. Andrizhiyevskaya, W. Keegstra, H. L. Dekker, M. Hagemann, H. J. Boekema, H. C. P. Matthijs, and J. P. Dekker, *Biochemistry*, *43*: 10308 (2004).
73. B. R. Green, in *Light-Harvesting Antennas in Photosynthesis* (B. R. Green, and W. W. Parson, eds.) Kluwer Academic Publishers, Dordrecht, 2003, p. 129.
74. T. S. Bibby, J. Nield, and J. Barber, *Nature*, *412*: 743 (2001).
75. E. G. Andrizhiyevskaya, T. M. E. Schwabe, M. Germano, S. D'Haene, J. Kruip, R. van Grondelle, and J. P. Dekker, *Biochim. Biophys. Acta*, *1556*: 265 (2002).
76. T. J. Ryan-Keogh, A. I. Macey, A. M. Cockshutt, C. M. Moore, and T. S. Bibby, *J. Phycol.*, *48*: 145 (2012).
77. R. Berera, I. H. M. van Stokkum, J. T. M. Kennis, R. van Grondelle, and J. P. Dekker, *Chem. Phys.*, *373*: 65 (2010).
78. R. Berera, I. H. M. van Stokkum, S. D'Haene, J. T. M. Kennis, R. van Grondelle, and J. P. Dekker, *Biophys. J.*, *96*: 2261 (2009).
79. D. M. Kramer, and J. R. Evans, *Plant Physiol.*, *155*: 70 (2011).
80. P. D. Boyer, *Annu. Rev. Biochem.*, *66*: 717 (1997).
81. G. Perlier, D. Tolleter, E. Billon, and L. Courmac, *Photosynth. Res.*, *106*: 19 (2010).
82. J. Nogales, S. Gudmundsson, E. M. Knight, B. O. Palsson, and I. Thiele, *Proc. Natl. Acad. Sci. U.S.A.*, *109*: 11431 (2012).
83. C. W. Mullineaux, *Front. Plant Sci.*, *5*: 7 (2014).
84. N. Battchilova, L. Wei, L. Du, L. Bersanini, E.-M. Aro, and W. Ma, *J. Biol. Chem.*, *286*: 36992 (2011).
85. D. Schwarz, H. Schubert, J. Georg, W. R. Hess, and M. Hagemann, *Plant Physiol.*, *163*: 1191 (2013).
86. N. Yeremenko, R. Jeanjean, P. Prommeenate, V. Krasikov, P. J. Nixon, W. F. J. Vermaas, M. Havaux, and H. C. P. Mattijs, *Plant Cell Physiol.*, *46*: 1433 (2005).
87. J. D. Rochaix, *Biochim. Biophys. Acta*, *1807*: 878 (2011).
88. F. Jans, E. Mignolet, P. A. Houyoux, P. Cardol, B. Ghysels, S. Cuiné, L., Cournac, G. Peltier, C. Remacle, and F. Franck, *Proc. Natl. Acad. Sci. U.S.A.*, *105*: 20546 (2008).
89. P. Cardol, G. Forti, and G. Finazzi, *Biochim. Biophys. Acta*, *1807*: 912 (2011).
90. X. Johnson, J. Steinbeck, R. M. Dent, H. Takahashi, P. Richaud, S. I. Ozawa, L. Houille-Vernes, D. Petroutsos, F. Rappaport, A. R. Grossman, K. K. Niyogi, M. Hippler, and J. Alric, *Plant Physiol.*, *165*: 438 (2014).

91. K. Noguchi, and K. Yoshida, *Mitochondrion*, 8: 87 (2008).

92. R. Scheibe, J. E. Backhausen, V. Emmerlich, and S. Holtgrefe, *J. Exp. Bot.*, *416*: 1481 (2005).

93. F. Partensky, and L. Garczarek, in *Photosynthesis in Algae* (A. W. D. Larkum, S. E. Douglas, and J. A. Raven eds.) Kluwer Academic Publishers, Dordrecht, 2003, p. 29.

94. F. Partensky, W. R. Hess, and D. Vaulot, *Microbiol. Mol. Biol. Rev.*, *63*: 106 (1999).

95. H. Liu, H. A. Nolla, and L. Campbell, *Aquat. Microb. Ecol.*, *12*: 39 (1997).

96. C. B. Field, M. J. Behrenfeld, J. T. Randerson, and P. Falkowski, *Science*, *281*: 237 (1998).

97. W. R. Hess, G. Rocap, C. S. Ting, F. Larimer, S. Stilwagen, J. Lamerdin, and S. W. Chisholm, *Photosynth. Res.*, *70*: 71 (2001).

98. C. C. Thompson, G. G. Z. Silva, N. M. Vieira, R. Edwards, A. C. P. Vicente, and F. B. Thompson, *Microb. Ecol.*, *66*: 752 (2013).

99. R. A. Lewin, *Phycologia*, *16*: 217 (1977).

100. E. Hirose, X. Turon, S. López-Legentil, P. K. Erwin, and M. Hirose, *Sistem. Biodiv. 10*: 435 (2012).

101. J. Münchhoff, E. Hirose, T. Maruyama, M. Sunairi, B. P. Burns, and B. A. Neilan, *Environ. Microbiol.*, *9*: 890 (2007).

102. T. Burger-Wiersma, M. Veenhuis, H. J. Korthals, C. C. M. van de Wiel, and L. R. Mur, *Nature*, *320*: 262 (1986).

103. A. V. Pinevich, O. M. Skulberg, H. C. P. Mattijs, H. Schubert, E. Willen, O. V. Gavrilova, and N. Velichko, *Microbios*, *100*: 159 (1999).

104. A. Pinevich, N. Velichko, and N. Ivanikova, *Front. Microbiol.*, *3*: 173 (2012).

105. M. Chen, and T. S. Bibby, *Photosynth. Res.*, *86*: 165 (2005).

106. T. S. Bibby, I. Mary, J. Nield, F. Partensky, and J. Barber, *Nature*, *424*: 1051 (2003).

107. H. Itoh, and A. Tanaka, *Proc. Natl. Acad. Sci. U.S.A.*, *108*: 18014 (2011).

108. S. Penno, L. Campbell, and W. R. Hess, *J. Phycol.*, *36*: 723 (2000).

109. J. Wiethaus, A. W. U. Busch, T. Dammeyer, and N. Frankenberg-Dinkel, *Eur. J. Cell Biol.*, *89*: 1005 (2010).

110. C. Steglich, C. W. Mullineaux, K. Teuchner, W. R. Hess, and H. Lokstein, *FEBS Lett.*, *553*: 79 (2003).

111. C. Steglich, N. Frankenberg-Dinkel, S. Penno, and W. R. Hess, *Environ. Microbiol.*, *7*: 1611 (2005).

112. T. S. Bibby, J. Nield, M. Chen, A. W. D. Larkum, and J. Barber, *Proc. Natl. Acad. Sci. U.S.A.*, *100*: 9050 (2003).

113. V. A. Boichenko, A. V. Pinevich, and I. N. Stadnichuk, *Biochim. Biophys. Acta*, *1767*: 801 (2007).

114. L. Bumba, F. Prasil, and F. Vácha, *Biochim. Biophys. Acta*, *1708*: 1 (2005).

115. M. Herbstová, R. Litvín, Z. Gardian, J. Kimenda, and F. Vácha, *Biochim. Biophys. Acta*, *1797*: 89 (2010).

116. M. Dunchan, M. Herbstová, M. Fuciman, Z. Gardian, F. Vácha, and T. Polívka, *J. Phys. Chem.*, *114*: 9275 (2010).

117. H. Miyashita, H. Ikemoto, N. Kurano, K. Adachi, M. Chihara, and S. Miyachi, *Nature*, *383*: 402 (1996).

118. H. Miyashita, K. Adachi, N. Kurano, H. Ikemoto, M. Chihara, and S. Miyachi, *Plant Cell Physiol.*, *38*: 274 (1997).

119. M. Kühl, M. Chen, P. J. Ralph, U. Schreiber, and A. W. D. Larkum, *Nature*, *433*: 820 (2005).

120. Y.-W. Chan, A. Nenninger, S. J. H. Clokie, N. H. Mann, D. J. Scanlan, A. L. Whitworthm and M. R. J. Clokie, *FEMS Microbiol. Evol.*, *61*: 65 (2007).

121. R. Mohr, B. Voß, M. Schliep, T. Kurz, I. Maldener, D. G. Adams, A. D. W. Larkum, M. Chen, and W. R. Hess, *IMSE J.*, *4*: 1456 (2010).

122. M. Chen, and R. E. Blankenship, *Trends Plant Sci.*, *16*: 427 (2011).

123. M. Chen, R. G. Hiller, C. J. Howe, and A. W. D. Larkum, *Mol. Biol. Evol.*, *22*: 21 (2005).

124. M. Chen, T. S. Bibby, J. Nield, A. W. D. Larkum, and J. Barber, *FEBS Lett.*, *579*: 1306 (2005).

125. M. Chen, T. S. Bibby, J. Nield, A. Larkum, and J. Barber, *Biochim. Biophys. Acta*, *1708*: 367 (2005).

126. C. Theiss, F. J. Schmitt, J. Pieper, C. Nganou, M. Grehn, M. Vitali, R. Olliges, H. J. Eichler, and H. J. Eckert, *J. Plant Physiol.*, *168*: 1473 (2011).

127. M. Chen, M. Floetenmeyer, and T. S. Bibby, *FEBS Lett.*, *583*: 2535 (2009).

128. T. Tomo, T. Okubo, S. Akimoto, M. Yokono, H. Miyashita, T. Tsuchiya, T. Noguchi, and M. Mimuro, *Proc. Natl. Acad. Sci. U.S.A.*, *104*: 7283 (2007).

129. T. Tomo, Y. Kato, T. Suzuki, S. Akimoto, T. Okubo, T. Noguchi, K. Hasegawa, T. Tsuchiya, K. Tanaka, M. Fukuya, N. Dohmae, T. Watanabe, and M. Mimuro, *J. Biol. Chem.*, *283*: 18198 (2008).

130. T. Tomo, S. I. Allakhverdiev, and M. Mimuro, *J. Photochem. Photobiol., B: Biol.*, *104*: 333 (2011).

131. S. P. Mielke, N. Y. Kiang, R. E. Blankenship, M. R. Gunner, and D. Mauzerall, *Biochim. Biophys. Acta*, *1807*: 1231 (2011).

132. M. Chen, Y. Li, D. Birch, and R. D. Willows, *FEBS Lett.*, *586*: 3249 (2012).

133. M. Chen, M. Schliep, R. D. Willows, Z. L. Cai, B. A. Neilan, and H. Scheer, *Science*, *329*: 1318 (2010).

134. M. Li, D. A. Semchonok, E. J. Boekema, and B. D. Bruce, *Plant Cell*, *26*: 1230 (2014).

135. T. Tomo, T. Shinoda, M. Chen, S. I. Allakhverdiev, and S. Akimoto, *Biochim. Biophys. Acta*, *1873*: 1484 (2014).

136. P. G. Falkowski, M. E. Katz, A. H. Knoll, A. Quigg, J. A. Raven, O. Schofield, and F. J. Taylor, *Science*, *305*: 354 (2004).

137. P. J. Keeling, *Phil. Trans. R. Soc. B*, *365*: 729 (2010).

138. B. R. Green, *Photosynth. Res.*, *107*: 103 (2011).

139. T. Chavalier-Smith, in *Organelles, Genomes and Eukaryotic Evolution* (R. P. Hirt, and D. Horner eds.) Taylor and Francis, London, 2004, p. 71.

140. J. Nield, O. Kruse, J. Ruprecht, P. da Fonseca, C. Buchel, and J. Barber, *J. Biol. Chem.*, *275*: 27940 (2000).

141. J. A. D. Neilson, and D. G. Durnford, *Photosynth. Res.*, *106*: 57 (2010).

142. J. Engelken, H. Brinkmann, and I. Adamska, *BMC Evol. Biol.*, *10*: 233 (2010).

143. H. Teramoto, A. Ishii, Y. Kimura, K. Hasegawa, S. Nakazawa, T. Makamura, S. Higashi, M. Watanabe, and T. Ono, *Plant Cell Physiol.*, *47*: 419 (2006).

144. G. Peers, T. B. Truong, E. Ostendorf, A. Busch, D. Elrad, A. R. Grossman, M. Hippler, and K. K. Niyogi, *Nature*, *462*: 518 (2009).

145. C. Funk, and W. Vermaas, *Biochemistry*, *38*: 9397 (1999).

146. M. Heddad, and I. Adamska, *Comp. Funct. Genom.*, *3*: 504 (2002).

147. O. Kilian, A. S. Steunou, A. R. Grossman, and D. Bhaya, *Mol. Plant*, *1*: 155 (2008).

148. M. H. Montane, and K. Kloppstech, *Gene*, *258*: 1 (2000).

149. S. Jansson, *Trends Plant Sci.*, *4*: 1360 (1999).

150. M. Mimuro, and S. Akimoto, in *Photosynthesis in Algae* (A. W. D. Larkum, S. E. Douglas, and J. A. Raven eds.) Kluwer Academic Publishers, Dordrecht, 2003, p. 335.

151. G. E. Hoffman, M. V. Sanchez Puerta, and C. F. Delwiche, *BMC Evol. Biol.*, *11*: 101 (2011).

152. S. Sturm, J. Engelken, A. Gruber, S. Vugrinec, P. Kroth, I. Adamska, and J. Lavaud, *BMC Evol. Biol.*, *13*: 159 (2013).

153. Z. Gardian, L. Bumba, A. Schrofel, M. Herbstova, J. Nebesarova, and F. Vácha, *Biochim. Biophys. Acta*, *1767*: 725 (2007).

154. H.-N. Su, B.-B. Xie, X.-Y. Zhang, B.-C. Zhou, and Y.-Z. Zhang, *Photosynth. Res.*, *106*: 73 (2010).
155. L. Bumba, H. Havelková-Doušová, M. Hušák, and F. Vácha, *Eur. J. Biochem.*, *271*: 2967 (2004).
156. H. Ohta, T. Suzuki, M. Ueno, A. Okumura, S. Yoshihara, J.-R. Shen, and I. Enami, *Eur. J. Biochem.*, *270*: 4156 (2003).
157. A. A. Arteni, L.-N. Liu, T. J. Aartsma, Y.-Z. Zhang, B.-C. Zhou, and E. J. Boekema, *Photosynth. Res.*, *95*: 169 (2008).
158. C. M. Toole, and F. C. T. Allnut, in *Photosynthesis in Algae* (A. V. D. Larkum, S. E. Douglas, and J. A. Raven, eds.) Kluwer Academic Publishers, Dordrecht, 2003, p. 305.
159. L. Talarico, N. Rascio, F. Dalla Vecchia, and G. Maranzana, *Plant Biosyst.*, *132*: 87 (1998).
160. L.-N. Liu, A. T. Elmalk, T. J. Aartsma, J.-C. Thomas, G. E. M. Lamers, B.-C. Zhou, and Y.-Z. Zhang, *PLoS One*, *3*: e3134 (2008).
161. L.-N. Liu, T. J. Aartsma, J.-C. Thomas, G. E. M. Lamers, B.-C. Zhou, and Y.-Z. Zhang, *J. Biol. Chem.*, *283*: 34946 (2008).
162. A. Busch, J. Nield, and M. Hippler, *Plant J.*, *62*: 886 (2010).
163. B. Grabowski, S. Tan, F. X. Cunningham Jr., and E. Gantt, *Photosynth. Res.*, *63*: 85 (2000).
164. N. R. Bertos, and S. P. Gibbs, *J. Phycol.*, *34*: 1009 (1998).
165. K. Ifuku, K. Ido, and F. Sato, *J. Photochem. Photobiol., B: Biol.*, *104*: 158 (2011).
166. J. Minagawa, and Y. Takahashi, *Photosynth. Res.*, *82*: 241 (2004).
167. A. Koziol, T. Borza, K.-I. Ishida, P. Keeling, R. W. Lee, and D. G. Durnford, *Plant Physiol.*, *143*: 1802 (2007).
168. H. Takahashi, M. Iwai, Y. Takahashi, and J. Minagawa, *Proc. Natl. Acad. Sci. U.S.A.*, *103*: 477 (2006).
169. Z. Liu, H. Yan, K. Wang, T. Kuang, J. Zhang, L. Gui, X. An, and W. Chang, *Nature*, *428*: 287 (2004).
170. B. Drop, M. Webber-Birungi, A. K. N. Yadav, A. Filipowicz-Szymanka, F. Fusetti, E. J. Boekema, and R. Croce, *Biochim. Biophys. Acta*, *1837*: 63 (2014).
171. R. Croce, G. Canino, F. Ros, and R. Bassi, *Biochemistry*, *41*: 7334 (2002).
172. X. Pan, M. Li, T. Wan, L. Wang, C. Jia, Z. Hou, X. Zhao, J. Zhang, and W. Chang, *Nat. Struct. Mol. Biol.*, *18*: 309 (2011).
173. E. J. Boekema, H. van Roon, F. Calkoen, R. Bassi, and J. P. Dekker, *Biochemistry*, *38*: 2233 (1999).
174. R. Tokutsu, N. Kato, K. H. Bui, T. Ishikawa, and J. Minagawa, *J. Biol. Chem.*, *287*: 31574 (2012).
175. D. Elrad, and A. R. Grossman, *Curr. Genet.*, *45*: 61 (2004).
176. R. Croce, T. Morosinotto, S. Castelletti, J. Breton, and R. Bassi, *Biochim. Biophys. Acta*, *1556*: 29 (2002).
177. J. Kargul, J. Nield, and J. Barber, *J. Biol. Chem.*, *278*: 16135 (2003).
178. B. Drop, M. Webber-Birungi, F. Fusetti, R. Kouril, K. E. Redding, E. J. Boekema, and R. Croce, *J. Biol. Chem.*, *286*: 44874 (2011).
179. V. H. R. Schmid, S. Potthast, M. Wiener, V. Bergauer, H. Paulsen, and S. Storf, *J. Biol. Chem.*, *277*: 37397 (2002).
180. M. Mozzo, M. Mantelli, F. Passarini, S. Caffarri, R. Croce, and R. Bassi, *Biochim. Biophys. Acta*, *1797*: 212 (2010).
181. A. Rivadossi, G. Zucchelli, F. M. Garlaschi, and R. C. Jennings, *Photosynth. Res.*, *60*: 209 (1999).
182. D. Carbonera, T. Agostini, T. Morosinotto, and R. Bassi, *Biochemistry*, *44*: 8337 (2005).
183. R. Croce, M. Mozzo, T. Morosinotto, A. Romeo, R. Hienerwadel, and R. Bassi, *Biochemistry*, *46*: 3846 (2007).
184. M. Germano, A. Yakusshevska, W. Keegstra, H. van Gorkom, J. P. Dekker, and E. J. Boekema, *FEBS Lett.*, *525*: 121 (2002).
185. J. Kargul, M. V. Turkina, J. Nield, S. Benson, A. V. Vener, and J. Barber, *FEBS J.*, *272*: 4797 (2005).

186. Y. Takahashi, T. A. Yasui, E. J. Stauber, and M. Hippler, *Biochemistry*, *43*: 7816 (2004).
187. E. J. Stauber, A. Busch, B. Naumann, A. Stavos, and M. Hippler, *Proteomics*, *9*: 398 (2009).
188. A. M. Pyszniak, and S. P. Gibbs, *Protoplasma*, *166*: 208 (1992).
189. I. Enami, T. Suzuki, O. Tada, Y. Nakada, K. Nakamura, A. Tohri, H. Ohta, I. Inoue, and J.-R. Shen, *FEBS J.*, *272*: 5020 (2005).
190. C. Büchel, *Biochemistry*, *42*: 13027 (2003).
191. A. W. D. Larkum, in *Photosynthesis in Algae* (A. W. D. Larkum, S. E. Douglas, and J. A. Raven eds.) Kluwer Academic Publishers, Dordrecht, 2003, p. 277.
192. E. Papagiannakis, I. H. M. van Stokkum, H. Fey, C. Büchel, and R. van Grondelle, *Photosynth. Res.*, *86*: 241 (2005).
193. L. Boldt, D. Yellowlees, and W. Leggat, *PLoS One*, *7*: e47456 (2012).
194. L. Premvardhan, B. Robert, A. Beer, and C. Büchel, *Biochim. Biophys. Acta*, *1797*: 1647 (2010).
195. B. Lepetit, D. Volke, M. Gilbert, C. Wilhelm, and R. Goss, *Plant Physiol.*, *154*: 1905 (2010).
196. J. Joshi-Deo, M. Schmidt, A. Gruber, W. Weisheit, M. Mittag, P. G. Kroth, and C. Büchel, *J. Exp. Bot.*, *61*: 3079 (2010).
197. B. R. Green, in *Light Harvesting Antennas in Photosynthesis* (B. R. Green, and W. W. Parson eds.) Kluwer Academic Publishers, Dordrecht, 2003, p. 1293.
198. T. Veith, J. Brauns, W. Weisheit, M. Mittag, and C. Büchel, *Biochim. Biophys. Acta*, *1787*: 905 (2009).
199. B. Lepetit, R. Goss, T. Jakob, and C. Wilhelm, *Photosynth. Res.*, *111*: 245 (2012).
200. A. Beer, K. Gundermann, J. Beckmann, and C. Büchel, *Biochemistry*, *45*: 13046 (2006).
201. B. Lepetit, D. Volke, M. Szábo, R. Hoffmann, G. Garab, C. Wilhelm, and R. Goss, *Biochemistry*, *46*: 9813 (2007).
202. K. Gundermann, M. Schmidt, W. Weisheit, M. Mittag, and C. Büchel, *Biochim. Biophys. Acta*, *1827*: 303 (2013).
203. T. Veith, and C. Büchel, *Biochim. Biophys. Acta*, *1767*: 1428 (2007).
204. I. Grouneva, A. Rokka, and E.-M. Aro, *J. Proteome Res.*, *109*: 5338 (2011).
205. Y. Ikeda, A. Yamagishi, M. Komura, T. Suzuki, N. Dohmae, Y. Shibata, S. Itoh, H. Koike, and K. Satoh, *Biochim. Biophys. Acta*, *1827*: 529 (2013).
206. T. Tengs, O. J. Dahlberg, K. Shalchian-Tabrizi, D. Klaveness, K. Rudi, C. F. Delwiche, and K. S. Jakobsen, *Mol. Biol. Evol*, *17*: 718 (2000).
207. P. J. Keeling, *Am. J. Bot.*, *91*: 1481 (2004).
208. R. G. Hiller, P. M. Wrench, A. P. Gooley, G. Shoebridge, and J. Breton, *Photochem. Photobiol.*, *57*: 125 (1993).
209. T. Polívka, I. H. M. van Stokkum, D. Zigmantas, R. van Grondelle, V. Sundström, and R. G. Hiller, *Biochemistry*, *45*: 8516 (2006).
210. D. G. Durnford, in *Photosynthesis in Algae* (A. W. D. Larkum, S. E. Douglas, and J. A. Raven eds.) Kluwer Academic Publishers, Dordrecht, 2003, p. 63.
211. E. Hofmann, P. Wrench, F. Sharples, R. Hiller, W. Welte, and K. Diederichs, *Science*, *272*: 1788 (1996).
212. J. Jiang, H. Zhang, Y. Kang, D. Bina, C. S. Lo, and R. E. Blankenship, *Biochim. Biophys. Acta*, *1817*: 83 (2012).
213. A. Damjanovic, T. Ritz, and K. Schulten, *Biophys. J.*, *79*: 1695 (2000).
214. D. M. Niedzwiedzki, J. Jiang, C. S. Lo, and R. E. Blankenship, *J. Phys. Chem.*, *117*: 11091 (2013).
215. M. Di Valentin, C. E. Tait, E. Salvadori, L. Orian, A. Polimeno, and D. Carbonera, *Biochim. Biophys. Acta*, *1837*: 85 (2014).

21 Excess Light and Limited Carbon
Two Problems with Which Cyanobacteria and Microalgae Cope

Nicoletta La Rocca, Isabella Moro, and Nicoletta Rascio

CONTENTS

21.1 Light Underwater 369
21.2 Dangerous Effects of Excess Light 370
21.3 Nonphotochemical Excess Energy Dissipation in Antenna Systems 370
    21.3.1 NPQ in Cyanobacteria 371
        21.3.1.1 Orange Carotenoid Protein 371
        21.3.1.2 High Light-Inducible Proteins and Iron Starvation-Inducible Proteins 372
    21.3.2 NPQ in Microalgae 372
        21.3.2.1 Xanthophyll Cycles 373
21.4 Photoprotection Mechanisms in PSII RC 375
21.5 Alternative Electron Transport Pathways 375
    21.5.1 Alternative Electron Transport Pathways in Cyanobacteria 375
    21.5.2 Alternative Electron Transport Pathways in Microalgae 377
    21.5.3 Water–Water Cycle 378
21.6 State Transitions 378
21.7 Recovery of the PSII 380
21.8 Limited Carbon Availability 381
    21.8.1 Dual Function of Rubisco 381
21.9 Cyanobacterial CCM 382
    21.9.1 Inorganic Carbon Uptake 383
    21.9.2 Carboxysomes 385
21.10 CCM in Microalgae 385
    21.10.1 CCM in Green Microalgae 385
        21.10.1.1 Ci Transmembrane Transport 386
        21.10.1.2 Pyrenoid 387
    21.10.2 CCM in Diatoms 387
    21.10.3 CCM in Dinoflagellates 389
    21.10.4 CCM in Other Microalgae 389
21.11 Conclusions 390
Acknowledgments 390
References 390

21.1 LIGHT UNDERWATER

The absorptive, polarizing, and scattering properties of water, dissolved matters, and suspended particles (including cells of marine microorganisms), the intermittent changes in incoming solar radiation, like those caused, for instance, by a broken cloud cover, and the strong fluctuations in downwelling irradiance, due to sunlight rays reflection by surface waves, make the underwater light field an ever-changing environment. Moreover, the farthest penetration of blue light into seawater, which absorbs first the longer and less energetic wavelength in infrared, red, orange, and yellow regions of the solar spectrum, leads to a progressive dominance of the blue-green (400–500 nm) wavebands with depth [1].

The phytoplanktonic cyanobacteria and microalgae, whose photosynthetic activity accounts for about 50% of the global productivity of the whole biosphere [2,3], are incapable of doing directional movements. They are passively transported within the water column by tides, streams, and turbulent mixing and are subjected to rapid changes in the underwater light climate as they move also vertically in the water column [3].

The environmental factors that can mostly affect growth and productivity of these photosynthetic microorganisms are nutrient availability, temperature, and light. However, the first two factors are more likely to vary on daily or seasonal scales, while fluctuations in light intensity can occur with very high frequency (seconds or minutes) and huge amplitude (from darkness to full sunlight) [3]. Sudden changes in the underwater light regime and, in particular, a sharp rise of irradiance or other environmental conditions that lead to an excessive light are potentially unsafe for the phytoplanktonic organisms because they can cause a decrease in photosynthetic efficiency with a significant impact on the cell fitness [4].

21.2 DANGEROUS EFFECTS OF EXCESS LIGHT

It is well known that, although being essential for photosynthesis, light can become a harmful and even lethal factor for photosynthetic organisms [5]. This occurs in natural environments when the organisms are exposed to a too strong illumination or even under low light intensities when unfavorable conditions, such as reduced CO_2 availability or other nutritional deficiencies, limit the capability of assimilative processes to efficiently utilize the chemical energy (ATP) and reducing power (NADPH) produced by the photosystem activity. Under these conditions, the light absorbed by pigment molecules, such as chlorophylls and phycobiliproteins, creates an excess excitation, which induces the formation of dangerous reactive oxygen species (ROS) that can greatly photodamage the photosystems, leading to a decline of photosynthetic activity broadly termed *photoinhibition*.

In oxygenic photoautotrophs, the major site of photoinhibition is the reaction center (RC) of the photosystem II (PSII), which is the most light sensitive component of the photosynthetic apparatus [6]. An excess light energy absorbed by PSII gives rise to the over-reduction of the plastoquinone pool in the hydrophobic phase of the thylakoid membrane and to the lack of reducible plastoquinone (PQ) molecules that makes the PQ binding site (Q_B) at the PSII RC unoccupied. This causes the stabilization as semiquinone $\left(Q_A^-\right)$ of the PQ bound to the Q_A site of the RC, with the block of the forward electron transport [6]. Consequently, after the photochemical reaction of charge separation, a charge recombination from the reduced primary electron acceptor pheophytin (Pheo$^-$) to the oxidized primary electron donor P680 (P680$^+$) can occur. This leads to formation of the triplet excited state of P680 (^{3}P680*) that interacts with the molecular oxygen generating the singlet oxygen ($^1O_2$), which is the most harmful ROS for a photosynthetic cell [7]. In addition, other ROS (such as O_2^-, H_2O_2, and OH^-) can also derive from the O_2 reduction through reduced electron carriers at the acceptor side of PSII [8].

In a *classic* scheme of photoinhibition, the ROS, mostly $^1O_2$, directly affect PSII components, in particular, the D1 protein of the RC, triggering its oxidative damage that causes the photosystem inactivation [9]. However, according to another proposed scheme, the primary event of PSII photodamage occurs at the photosystem donor side through the disruption by UV-B radiation of the water splitting Mn_4CaO_5 cluster of

the oxygen evolving complex (OEC) [10]. The ensuing lack of electron supply to the RC leads to accumulation of long-lived highly oxidizing radicals, such as P680$^+$ and Yz$^+$ (the radical of the redox-active D1-Tyr 161), that prompt the degradation of both the proteins (D1 and D2) forming the heterodimeric RC [11,12]. Subsequently, the excess light absorbed by light-harvesting chlorophylls leads to generation of ROS (in particular, $^1O_2$) that engender the harmful effects by inhibiting the *de novo* synthesis of damaged proteins, essential for the PSII recovery [13,14]. In both schemes, the ROS accumulated within the PSII under excess light energy are directly or indirectly responsible for the PSII photodamage that, if not repaired, causes a *chronic* irreversible photoinhibition [14].

The PSI is more tolerant than PSII against high light stress, but it becomes damaged when an excessive electron flow from PSII exceeds the capability of the PSI electron acceptors to cope with the electrons, resulting in formation of very harmful ROS, which can destroy the iron–sulfur centers and the RC proteins [15,16]. This oxidative photodamage is particularly dangerous being practically permanent, due to the extremely slow recovery of the PSI [15].

Cyanobacteria and microalgae, like the other organisms carrying out the oxygenic photosynthesis, have evolved an array of protective mechanisms to cope with the dangerous effects of light on PSII, which comprehend: nonphotochemical dissipation of excess absorbed energy in the light-harvesting systems, elimination of the harmful radical states in the RC, activation of alternative routes able to remove electrons from the photosynthetic electron transport chain, balance of light energy absorption by the two photosystems, and fast repair of photodamaged proteins of RC to recover the PSII activity.

The mechanisms that control the PSII activity and regulate the electron flow throughout the electron transfer chain are also able to provide a PSI photoprotection avoiding the over-reduction of the electron acceptors and the formation of ROS on its reducing side [15,16].

The manifold strategies evolved to safeguard the integrity and functionality of the photosynthetic machinery make cyanobacteria and microalgae able to live and grow under a wide range of light conditions by responding promptly to quantity/quality variations of the incident radiation in the inhabited environments [17,18].

21.3 NONPHOTOCHEMICAL EXCESS ENERGY DISSIPATION IN ANTENNA SYSTEMS

A ubiquitous mechanism for balancing the input and utilization of light energy in photosynthesis to avoid light-induced damage consists in dissipating the excess absorbed energy as heat at the PSII antenna level, with a consequent lower amount of energy funneled to the RC. This harmless thermal dissipation of absorbed light energy, which occurs on the timescale of seconds to minutes, results in a decrease in chlorophyll fluorescence monitored as non-photochemical quenching (NPQ). Under the general NPQ term, numerous processes with distinct underlying mechanisms are included, which operate in cyanobacteria and microalgae to regulate the light harvesting in response to light conditions in different habitats [19].

21.3.1 NPQ IN CYANOBACTERIA

21.3.1.1 Orange Carotenoid Protein

Unlike eukaryotic microalgae, which exhibit intrinsic antenna complexes, most cyanobacteria have a major extramembrane light-harvesting system. This consists of phycobilisomes (PBSs), which are highly structured supercomplexes of phycobiliproteins (Figure 21.1), where peripheral rods of phycoerythrin (PE) and phycocyanin (PC) transfer the excitation energy to a central core of allophycocyanin (ACP) preferentially linked to the PSII in the thylakoid membrane [20]. Thus, in cyanobacteria, the NPQ cannot be triggered by a pH drop in the thylakoid lumen induced by high light conditions, as occurs in microalgae (see page 430). Instead, it relies on the activation of a soluble orange carotenoid protein (OCP), which contains a single noncovalently bound carotenoid (3′-hydroxyechinenone) [21] and is localized in the cytoplasmic interthylakoid region corresponding to the PBS site of the photosynthetic membranes [22].

The OCP protein (35 kDa) operates as a photoreceptor responding to blue-green light (450–550 nm) and is the only known photoactive protein with a carotenoid as photoresponsive chromophore [23,24]. Under high light conditions (Figure 21.1), the blue-green light absorption by 3′-hydroxyechinenone induces structural changes in both carotenoid and protein causing the conversion of the inactive orange form of OCP (OCP^o) to the active red form (OCP^r) that can bind to the ACP core of PBS [24]. Due to the very low quantum yield (~ 0.03), the OCP^o to OCP^r conversion occurs only under high light conditions (with higher OCP^r levels at stronger illuminations), making the OCP a photoactive protein able to sense the light intensity and the photoinhibitory state [25].

The OCP^r–PBS interaction permits the thermal dissipation of excess absorbed energy through the charge [26] or energy [27] transfer from an excited bilin chromophore of ACP (called APC_{660} due to its fluorescence maximum) to the OCP^r carotenoid, which acts as an effective quencher [26,28]. In this way, the OCP operates as a light sensor, signal propagator, and energy quencher in an NPQ mechanism that can prevent more than 80% of the PBS excitation from reaching the photosystem RC [24,26]. In cells no longer under high light regime (Figure 21.1), the OCP^r detaches from the PBS core and spontaneously reverts back to the inactive OCP^o, with the recovery of the antenna fluorescence. However, the OCP-dependent NPQ turnoff depends on the presence of a protein termed *fluorescence recovery protein* (FRP) [23]. FRP is a 13 kDa protein attached to the thylakoid membrane in a dimeric active or a tetrameric inactive form [29]. The dimeric FRP is constitutively active and can interact with the OCP exhibiting a higher affinity to OCP^r than to OCP^o. The OCP^r reconversion into the inactive orange form requires its dissociation from the strongly attached PBS. The FRP–OCP interaction seems to promote the OCP^r detachment from the PBS and to accelerate its reversion back to the OCP^o form, leading to a fast regain of a fully functional antenna [23].

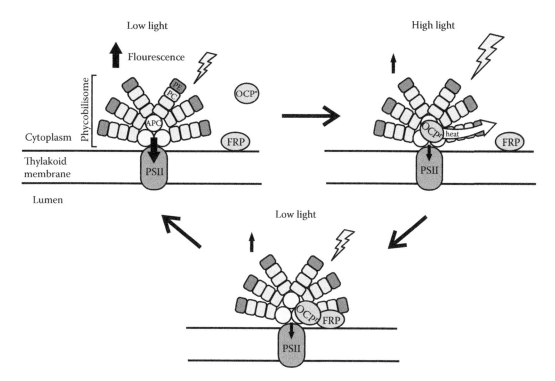

FIGURE 21.1 Schematic drawing of the OCP-dependent NPQ in cyanobacteria. (APC, allophycocyanin; FRP, fluorescence recovery protein; PC, phycocyanin; PE, phycoerythrin; PSII, photosystem II; OCP^o, inactive orange form; OCP^r, active red form). See text for detailed explanation.

The OCP-dependent NPQ operates exclusively in cyanobacteria with PBSs as main light-harvesting antenna. Neither green oxyphotobacteria (formerly Prochlorophyta) lacking PBS nor eukaryotic microalgae, including Rhodophyta having PBSs, possess OCP proteins [19].

21.3.1.2 High Light-Inducible Proteins and Iron Starvation-Inducible Proteins

Several other photoprotective mechanisms operate in cyanobacteria to decrease the amount of energy arriving at the RCs by thermal dissipation. Two of these mechanisms rely on the high light-inducible proteins (HLIP) and iron starvation-inducible (IsiA) proteins.

The HLIPs, also called small CAB-like proteins (SCPs), are integral membrane single-helix proteins belonging to the LHC-like family of chlorophyll and carotenoid binding proteins [30]. These proteins, mostly associated with PSII [31,32], are present in low amounts under low light and accumulate transiently upon cell exposure to high light or to other stressful conditions that result in absorption of excess excitation energy by the photosynthetic apparatus [31,33,34]. Genes encoding HLIPs have been identified in all sequenced cyanobacteria, with the highest gene number found in genomes of marine strains adapted to high light intensities [35]. The cyanobacterial HLIPs, which are considered primitive LHC-like polypeptides [36], seem to play different functions in the cell. They are mainly involved, through direct or indirect means, in nonphotochemical dissipation of excess absorbed light energy within antenna complexes [37]. Moreover, they may function as pigment carriers, binding chlorophylls released during the PSII turnover so as to protect the cyanobacterium from the potentially phototoxic effects of these free pigments [38].

The IsiA protein, also known as CP43', is an integral membrane six-helix protein belonging to the so-called *core-complex* antenna family of chlorophyll-binding proteins, which also includes the CP43 and CP47 inner antennae of the PSII core [39,40]. This protein accumulates in cyanobacteria under iron stress but also under high light conditions [41], playing a double role in enhancing the light-harvesting ability of PSI and in providing photoprotection to PSII [41,42].

The iron concentration in aquatic environments, particularly in open oceans, can become limited, due to the low solubility of Fe^{3+} in slightly alkaline water [41]. The low iron bioavailability affects primarily the PSI, which has three iron–sulfur clusters in the electron transport chain that operates in transferring electrons from the primary electron donor P700 of the RC to the stromal electron acceptor ferredoxin [41]. This causes a decrease in the PSI/PSII ratio, which in cyanobacteria is usually around 3 [41], and exposes the photosynthetic apparatus to photooxidative damage due to the inability of PSI to cope with the electrons arriving from PSII. However, the reduction of the PSI level is paralleled by the induction of the IsiA protein that associates with the PSI to form a PSI–IsiA supercomplex where an antenna ring of 18 IsiA units (or more), each of them binding 16–17 Chls *a*, encircles the trimeric PSI core [43,44]. The IsiA increases the absorption cross section of PSI, doubling the light-harvesting capacity of the core complex [44] and compensating for the lowered PSI level in response to iron deficiency. In this way, a higher activity of PSI is maintained, which reduces the excitation pressure on PSII and hence the oxidative stress [45]. The IsiA, which also accumulates in high light conditions, can form large IsiA aggregates detached from PSI, whose number greatly increases at the rise of light intensity also under iron-sufficient conditions [42]. These *empty* IsiA aggregates are involved in protection of cyanobacterial cells against the deleterious effects of light giving rise to very effective fluorescence quenching complexes [46]. It has been suggested that they can associate with PSII [47], and there is evidence that IsiA can quench excess PSII excitation [48]. The IsiA units contain chlorophyll *a*, zeaxanthin, and the echinenone, which is likely the carotenoid that quenches the chlorophyll excited state energy through an energy transfer mechanism [49]. Interestingly, this is the same type of carotenoid that operates in the OCP-dependent NPQ. The empty IsiA complexes can also thermally dissipate the energy coming from PBSs through the interaction with free PBSs that diffuse on the thylakoid membrane surface [50]. The energy transfer from PBSs to IsiA protects the PSII as it reduces the photosystem antenna size, lowering the energy that arrives at the RC [44]. Moreover, the quenching activity of the IsiA complexes can counteract the risk of cell photodamage due to free PBSs [50].

The cyanobacteria lacking PBS do not accumulate IsiA under iron stress and/or high light conditions. These green oxyphotobacteria have light-harvesting systems that rely on membrane intrinsic chlorophyll-binding proteins, known as Pcb (prochlorophyte Chl *a/b* binding) proteins, which belong to the same *core-complex* antenna family of the IsiA protein [51]. The Pcb proteins harbor chlorophyll *a*, chlorophyll *b*, and zeaxanthin as dominant carotenoid. Different species of Pcb proteins form either a ring around the PSI trimer, giving rise to a Pcb–PSI supercomplex (similarly to the IsiA–PSI supercomplex), or aggregate in complexes flanking the PSII dimers [52–55]. Like the IsiA, the Pcb proteins, in addition to the light-harvesting function, can play a role in photoprotection by quenching chlorophyll excited states under excess light conditions [56]. However, while in IsiA the candidate quencher is echinenone, in Pcb, the dissipation of excess excitation energy seems to be achieved by zeaxanthin through a charge transfer mechanism [56].

21.3.2 NPQ in Microalgae

In most eukaryotic microalgae, with integral membrane chlorophyll-binding proteins belonging to the LHC superfamily as major light-harvesting antennae, the NPQ is triggered by the strong acidification of the thylakoid lumen that occurs when the chemical energy generated by the photosynthetic light reactions exceeds the capacity of the assimilation reactions. In these organisms, the NPQ of excess energy relies on specific carotenoids involved in so-called *xanthophyll cycles* and on a particular type of LHC protein named LHCSR (LHC stress related) protein [19].

The LCHSR protein (formerly known as LI818 protein) is an ancient member of the LHC superfamily whose orthologs are widely distributed throughout photosynthetic algal taxa, except red algae that have PBS as major light-harvesting antenna [57]. The LHCSR protein is present in green microalgae, such as *Chlamydomonas reinhardtii* and the primitive prasinophyte *Ostreococcus tauri* [57]. It is worth noting that LHCSR relatives, referred to as LHCX proteins [58], are found also in chromalveolate algae, such as diatoms, whose secondary plastid is derived from a red alga [59]. This suggests that in these organisms, the *LHCX* gene has been acquired laterally from a prasinophyte-like alga through a cryptic endosymbiotic event preceding the symbiosis with the red alga [57,60]. In contrast to what happens for the *LHC* genes of light-harvesting complexes, the *LHCSR* gene expression is low under limiting light conditions, while it greatly increases in high light along with the enhanced capability of NPQ [61–64]. LHCSR is a pigment-binding protein that forms stable and specific complexes with chlorophylls and carotenoids of the xanthophyll cycle involved in quenching the chlorophyll excited states [61,62].

The LHCSR of green microalgae functions as a sensor of thylakoid lumen acidification through the protonation of acidic residues on the lumen facing side that enhances its constitutive energy dissipating activity [62]. Thus, LHCSR fulfils the double function of sensor for lumenal pH and of effective energy quencher [62]. It has been suggested that in *C. reinhardtii*, the protonate LHCSR can interact with (and induce quenching in) other LHC proteins of the PSII light-harvesting system [62,65]. Moreover, the LHCSR can also quench the excitation energy of detached LHCII components moving from PSII to PSI during the state transition (see page 437) [64].

Unlike the LHCSR of green microalgae, the LHCX of diatoms does not have lumenal acidic amino acids and plausibly misses the role of pH sensor. However, it shares with LHCSR the capacity to amplify the quenching activity [66]. So far, it is not clear how this protein functions in the NPQ process. It has been suggested that LHCX could act by binding the quenching carotenoid diatoxanthin [61] or by inducing conformational changes in the fucoxanthin-chlorophyll proteins (FCPs) of the PSII antenna or, again, by influencing the connectivity between the FCP and the photosystem core [67].

21.3.2.1 Xanthophyll Cycles

The NPQ of the antenna chlorophyll excited states involves specific carotenoids belonging to the so-called *xanthophyll cycles* found in microalgae, namely, the violaxanthin cycle in chlorophytes and the diadinoxanthin cycle in diatoms and other chromalveolates [68].

The violaxanthin cycle (Figure 21.2) consists of a two-step de-epoxidation that converts the di-epoxy violaxanthin into the epoxy-free zeaxanthin via the mono-epoxy antheraxanthin. The de-epoxidative reaction occurs in high light at the lumenal side of the thylakoid membrane and is catalyzed by

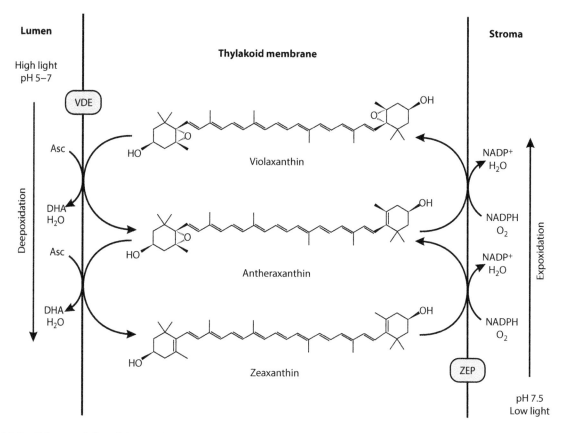

FIGURE 21.2 Scheme of the violaxanthin cycle. (Asc, ascorbate; DHA, dehydroascorbate; VDE, violaxanthin de-epoxidase; ZED, zeaxanthin epoxidase.) See text for detailed explanation.

the violaxanthin de-epoxidase (VDE) enzyme, which exhibits a pH optimum of 5.2 [69]. The decrease in lumenal pH (below ~6) due to strong light conditions induces the binding to the thylakoid membrane of the previously soluble VDE, which becomes activated and carries out the violaxanthin de-epoxidation by using ascorbate as the electron donor [70]. Low light intensities or darkness reverts the two-step reaction leading to the re-epoxidation of zeaxanthin into violaxanthin through the constitutive activity of the zeaxanthin epoxidase (ZEP) enzyme, which has a pH optimum of 7.5 [70]. ZEP is located at the stromal side of thylakoid membrane and utilizes O_2 and NADPH as cosubstrates to carry out the epoxidative reaction [68,71]. The zeaxanthin formed under high light conditions is the active quencher of the chlorophyll excitation energy, although the mechanistic details of the zeaxanthin-dependent energy dissipation are still under debate [68]. According to the current view, the zeaxanthin could interact with a chlorophyll *a*, giving rise to a quenching complex leading to the thermal dissipation of the chlorophyll excitation energy through a Chl–Car charge transfer mechanism [72]. A peculiar violaxanthin cycle occurs in members of *Prasinophyceae* (a primitive class of green microalgae), with only the first step of de-epoxidation and the antheraxanthin that replaces zeaxanthin as a quencher in the thermal dissipation of excitation energy [73].

Although the violaxanthin cycle has been found in all tested green microalgae, its contribution to the overall NPQ differs among different species and cannot be so significant due to the activity of other mechanisms, such as the state transitions, involved along with the xanthophyll cycle in the protection of photosynthetic apparatus against photoinhibition [74]. Conversely, in diatoms and in other chromalveolates, which do not rely on the state transition quenching, the xanthophyll cycle, namely, diadinoxanthin cycle, represents the most important mechanism acting in nonphotochemical dissipation of excessively absorbed excitation energy [68]. These organisms can display an NPQ capacity up to five times higher than that of green microalgae [75], and this superior NPQ capacity has been attributed, at least in part, to their specific xanthophyll cycle [76].

The diadinoxanthin cycle (Figure 21.3) comprises the one-step de-epoxidation of the mono-epoxy diadinoxanthin to form the quenching xanthophyll diatoxanthin. The reaction is triggered by the transthylakoidal ΔpH produced under high light conditions, and is reversed in low light or darkness [68]. Microalgae performing the diadinoxanthin cycle can also contain xanthophylls of the violaxanthin cycle. However, these latter pigments are intermediate products of the pathway that leads to the synthesis of the diadinoxanthin cycle carotenoids, whose concentrations drastically increase under high light conditions [77]. The enzymes involved in the diadinoxanthin cycle are the lumenal diadinoxanthin de-epoxidase (DDE), which has a pH optimum near 5.5 and, like the VDE, requires ascorbate as electron donor, and the stromal diatoxanthin epoxidase (DEP), showing the same cosubstrate requirement of ZEP and a pH optimum of 7.5 [71,78]. DEP is completely inhibited during high light illumination by the light-driven ΔpH, while in low light, it carries out the back reaction of epoxidation that rapidly removes the diatoxanthin-dependent NPQ [79]. Despite the basic importance of the xanthophyll cycle in diatoms and in other chromalveolates, the mechanistic aspects of the diatoxanthin-dependent energy dissipation have not been elucidated yet [67,68].

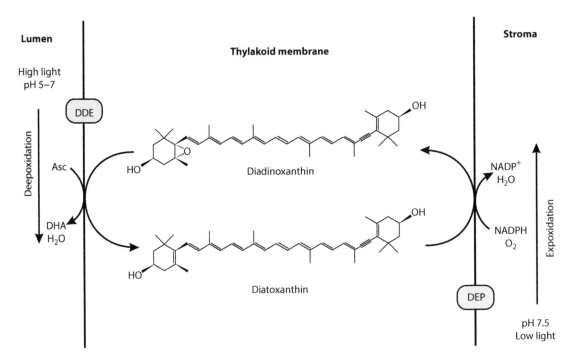

FIGURE 21.3 Scheme of the diadinoxanthin cycle. (Asc, ascorbate; DDE, diadinoxanthin de-epoxidase; DEP, diatoxanthin epoxidase; DHA, dehydroascorbate.) See text for detailed explanation.

The activity of a xanthophyll cycle in red algae is still not clear, although the presence of zeaxanthin, antheraxanthin, and violaxanthin has been noticed in some species [80]. As for red algae, having PBSs as major light-harvesting system, a peculiar mechanism to prevent photodamage of PSII RC due to an excess excitation energy has been shown in the model microalga *Porphyridium cruentum*. This mechanism is triggered by a strong green light and involves the energetic decoupling of phycoerythrin (PE) within the PBS [81,82]. Two different spectral types of PE are present at the periphery of PBS rods: the outer B-PE that carries an additional chromophoric γ-subunit, and the inner b-PE that lacks it. The γ-subunit is preferentially sensitive to the intense light and plays a central role in photoprotection. Under low light, this polypeptide essentially acts as a linker responsible for the association between B-PE and b-PE in the rod elements. Conversely, the γ-subunit sensitized by a strong light decouples the outer B-PE from the inner b-PE. In this way, the photon flow from PBS rods to the PBS core and then to the PSII RC is lowered, and a proportion of excitation energy is dissipated in the form of B-PE fluorescence [81,82].

21.4 PHOTOPROTECTION MECHANISMS IN PSII RC

A ubiquitous strategy of oxygenic photoautotrophs to protect the PSII RC against oxidative photodamage consists in preventing the formation of the harmful radical pair ^{3}[P680$^+$ Pheo$^-$], whose charge recombination results in triplet chlorophyll (^{3}P680*) and subsequent highly dangerous singlet oxygen ($^1O_2$), by favoring the competitive pathway of direct nonradiative [P680$^+$ Q$_A^-$] radical pair recombination. The prevailing of the safe route is obtained through the increase in the plastoquinone (Q$_A$) redox potential derived from subtle conformational changes of the RC D$_2$ protein (which links Q$_A$), triggered by the Q$_A$ over-reduction [83–85].

A major role in PSII RC photoprotection is also played in cyanobacteria by an alternative D1 protein of the RC [86]. It is known, in fact, that cyanobacteria possess different copies of D1, encoded by a family of genes differentially expressed depending on the environmental conditions [87]. In cells growing under low light, the predominant copy of D1 is the D1:1, while upon shift of cells to high light, the D1:1 is replaced by the D1:2 copy that can reach 70% of the total D1 proteins [86,88]. D1:2, the so-called *stress form*, differs from D1:1 mainly for the exchange of Gln-130 with Glu [89]. The Gnl-130 is hydrogen-bonded to the pheophytin, and the amino acid substitution causes an increase in pheophytin redox potential, which results in a higher rate of the indirect nonradiative route of [P680$^+$ Pheo$^-$] radical pair recombination, which avoids the repopulation of the excited ^{3}P680* and the formation of harmful singlet oxygen [86,90]. However, it has been found that the PSII RC is also protected against the damaging $^1O_2$ eventually produced by the scavenging activities of tocopherol and plastoquinol [91,92].

Moreover, a body of evidence demonstrates that a cyclic electron flow within the RC is another mechanism engaged by

oxygenic photoautotrophs to protect the PSII under conditions (such as high light or low CO_2 availability) making the electron transport carriers over-reduced [93,94]. This cyclic electron flow requires the plastoquinone (Q$_A$) in reduced form and a high ΔpH across the thylakoid membrane. The low lumenal pH hampers the water splitting reaction in the OEC, limiting the electron transfer to P680$^+$. However, an electron can arrive to P680$^+$ from the plastoquinone Q$_B$ (PQH$_2$), reduced by Q$_A$, via the cytochrome b_{559} (cyt b_{559}) present in the PSII RC. The unique peculiarity of cyt b_{559} is that it covers a broad range of redox potentials, going from a high (HP) (from +310 to +400 mV) to a low (LP) (from –40 to +80 mV) potential form [95]. In the photoprotective electron flow, it operates in the high redox potential form [96,97]. The cyt b_{559}, which in this form is maintained in the reduced state, can give an electron to the P680$^+$ through a series of redox intermediates, principally the peripheral Chlz and the β-carotene bound to the D2 protein [98], and it can be then re-reduced by the PQH$_2$. A cyclic electron flow via P680 > Q$_A$ > Q$_B$ > cyt b_{559} > D2 Chlz/β-carotene > P680$^+$ occurs in this way, allowing the charge separation capacity of the PSII RC to be maintained and the excess energy accumulated in P680 to be safely dissipated as heat by the sequential oxidoreductions [96,97].

The mechanisms of the excess light energy quenching in PSII RC, which complement those operating in the antenna complexes, belong to the oldest photoprotective mechanisms of the photoautotrophs, as the evolution of the photosystem RCs preceded that of the light-harvesting systems [85].

21.5 ALTERNATIVE ELECTRON TRANSPORT PATHWAYS

The engagement of multiple alternative electron transport pathways to remove electrons from the photosynthetic electron transport chain is a further system to prevent the dangerous effects of the electron carrier over-reduction. The exploitation of these switches, also known as *electron valves*, to alleviate the electron pressure on the PSII in conditions of excess excitation, allows cyanobacteria and microalgae to cope with fast-changing light intensities.

21.5.1 ALTERNATIVE ELECTRON TRANSPORT PATHWAYS IN CYANOBACTERIA

A role of electron valve for photoprotecting photosynthetic machinery has been recently assigned in cyanobacteria to flavodiiron proteins (FDPs) [99–103]. The FDPs (also referred to as A-type flavoproteins, Flvs) comprise a family of proteins widespread among strict and facultative anaerobic bacteria and archea, where they act against oxygen and/or nitric oxide by transferring electrons to these toxic compounds [104]. Two redox centers are common to all FDPs: the β-lactamate-like domain, containing a nonheme catalytic diiron center, and the flavodoxin-like domain, which harbors a flavin mononucleotide (FMN) moiety [105]. The FDPs found in cyanobacteria, as well as in green algae and in mosses and lycophytes, possess an extra flavin-reductase domain [99], which enables the oxidation of NAD(P)H and the reduction of a substrate within the same enzyme [103]. This makes the FDPs a unique family of cyanobacterial proteins involved in electron transfer processes [102]. Studies carried out on the model organism *Synechocystis* sp. PCC 6803 revealed the presence of four FDPs (Flv1–4) whose gene expression was greatly enhanced by inorganic carbon limitation and high light conditions [99]. All these FPDs are involved in safeguarding the photosystems against oxidative damage.

Flv1 and Flv3 are heterodimeric soluble proteins that provide protection to PSI by acting as powerful electron sinks (Figure 21.4). Due to their NAD(P)H: oxygen oxidoreductase activity, the Flv1/Flv3 proteins acquire electrons from the acceptor side of PSI delivering them to O_2, which is reduced to harmless H_2O in a process referred to as Mehler-like reaction [101,102]. Similarly to the real Mehler reaction (see page 436), the Flv1/Flv3-mediated reaction gives the excess of electrons to molecular oxygen, but it differs from the Mehler one for no concomitant production of dangerous ROS [101,102,106]. Under severe conditions of inorganic carbon starvation and high light intensities, more than 50% of the electrons released from water splitting by PSII, and arrived to the acceptor site of PSI, can be directed to O_2 via the Flv1/Flv3 activity [101,102].

Flv2 and Flv4 are FDP proteins specific of cyanobacteria, where they play a crucial role in protection of PSII against photoinhibition caused by high light. They form soluble heterodimers with high affinity to membranes in presence of cations [99]. The Flv2/Flv4 acts as a very effective electron sink at the acceptor side of PSII (Figure 21.4), preventing the over-reduction of plastoquinone pool and the formation of harmful singlet oxygen [100,103]. The function of Flv2/Flv4 is coordinated with that of a thylakoid membrane intrinsic protein (the so-called Sll0218), whose level also increases in high light [100]. According to the model proposed by Zhang et al. [100], the Sll0218 stabilizes the PSII dimer and induces slight changes in the photosystem complex. These changes facilitate the opening of the alternative electron transfer

route from PSII to a still unknown acceptor via the Flv2/Flv4 heterodimer, which in light is associated with the thylakoid membrane (due to the increased concentration of Mg^{2+} on the cytoplasmic membrane surface). Intact phycobilisomes with a correct energy transfer to PSII are required for the Flv2/Flv4-mediated electron transfer mechanism, which can absorb up to 30% of the electrons originating from PSII [103]. Differently from the Flv1/Flv3 proteins that are present in all cyanobacteria, the Flv2/Flv4 proteins are proper to β-cyanobacteria from freshwater and coastal marine environments (see page 441) [103].

Some other electron valves have been found to operate in cyanobacteria as alternative electron sinks to prevent dangerous over-reduction of the electron transport chain (Figure 21.4). They comprise the bidirectional hydrogenase (Hox) [107,108] and the terminal oxidases COX (an aa_3-type cytochrome c oxidase complex) and Cyd (a cytochrome bd–quinol oxidase complex), which are located in the thylakoid membranes and extract electrons from the intersystem electron transport chain upstream of PSI [109]. The Hox is a [Ni–Fe]-hydrogenase that can be reduced by ferredoxin and flavodoxin [108] and that uses electrons taken from the acceptor side of PSI to reduce, in turn, protons to H_2. This hydrogenase, which is inhibited by O_2, is active under anaerobic or microaerobic conditions and might be important as alternative sink for electrons from Flv1/3 when O_2 is scarcely available [110]. The Cyd oxidase seems to operate mainly in preventing the over-reduction of plastoquinone pool under excess activity of PSII by transferring to oxygen the electrons directly arrived from the reduced plastoquinone (plastoquinol). The COX, which can receive electrons from the Cyt b_6f complex via soluble redox carriers (plastocyanin or cytochrome c_6), possibly functions as terminal electron sink under conditions of low PSI activity [109].

A further plastoquinol terminal oxidase (PTOX) has been found in some α-cyanobacteria from oligotrophic marine and open-ocean waters, such as *Synechococcus*, *Prochlorococcus*

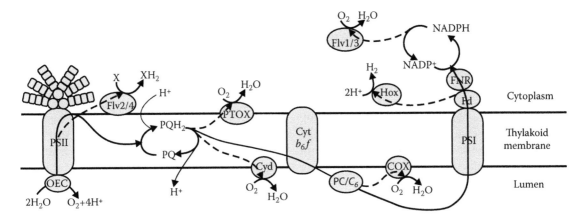

FIGURE 21.4 Schematic view of the linear electron flow (------) from water to NADP$^+$ and of the alternative electron transport pathways (—) acting as *electron valves* in cyanobacteria. (C_6, cytochrome c_6; COX, aa_3-type cytochrome c oxidase; Cyd, cytochrome bd-quinol oxidase; Cyt b_6f, cytochrome b_6f complex; Hox, [Ni-Fe]-hydrogenase; Fd, ferredoxin; Flv1/3 and Flv2/4, flavodiiron proteins; FNR, ferredoxin-NADP reductase; OEC, oxygen evolving complex; PC, plastocyanin; PQ, plastoquinone; PSI, photosystem I; PSII, photosystem II; PTOX, plastoquinol terminal oxidase; X, unknown electron acceptor.) See text for detailed explanation.

marinus, and *Acaryochloris marina* strains [111,112]. Interestingly, the *Synechococcus* strains that possess PTOX do not synthesize either IsiA or Pbc proteins as additional light-harvesting antennae around the PSI in iron-depleted conditions that lower the relative abundance of this photosystem. Thus, the PTOX-mediated electron transfer from plastoquinol to oxygen (Figure 21.4) provides an electron valve to keep PSII oxidized and to avoid the damaging consequences of a compromise PSI level in the iron-poor oceanic environment in which excess light excitation can commonly occur [111]. Moreover, the engagement of PTOX, which uses two iron atoms as opposed to the Cyt b_6f complex and PSI requiring a total of 18 iron atoms, may help the strains that possess this terminal oxidase to conserve iron under limiting conditions [111]. The electron flows from PSII to oxygen via PTOX, as well as via Cyd and COX-dependent alternative electron routes, create a sort of water-to-water pseudocycle around PSII that contributes to conserving energy in the form of trans-thylakoidal ΔpH useful for ATP synthesis [111].

21.5.2 Alternative Electron Transport Pathways in Microalgae

Also the eukaryotic organisms exhibiting oxygenic photosynthesis possess the photoprotective PTOX (in this case referred to as plastid terminal oxidase), acquired through the endosymbiotic events that led to the chloroplast formation [112]. PTOX is present in microalgae of both *green lineage* (green microalgae and algae with secondary chloroplasts derived from a symbiotic event with a green microalga) and *red lineage* (red microalgae and algae with secondary chloroplasts originated from a symbiotic event with a red microalga). In an oceanic strain of the green microalga *Ostreococcus* with a reduced PSI content likely due to iron deficiency, for instance, the diversion of electrons from plastoquinol to oxygen via PTOX was able to counterbalance the increased PSII photosensitivity, leading to a water-to-water pseudocycle like the one noticed in

cyanobacteria [113]. Moreover, it has been found that the iron limitation can induce the PTOX transcription in diatoms [112].

PTOX is a key component of the thylakoid electron transport chain called *chlororespiration*, which consists in the nonphotochemical reduction of plastoquinone pool and in its subsequent reoxidation at the expense of O_2 [114]. In such a pathway (Figure 21.5), a plastidial NAD(P)H dehydrogenase complex [115] carries out the plastoquinone reduction by using electrons from stromal NAD(P)H, while the PTOX is the plastoquinol oxidase that pulls out electrons from the plastoquinone pool to reduce O_2 to H_2O [112]. A major role of chlororespiration is in protecting the photosynthetic machinery against photooxidative stress. In environmental conditions, such as restricted availability of CO_2 or excess light, leading to a fully reduced PSI acceptor side, the remotion of stromal reductants via chlororespiration can act as an electron safety valve that helps to decrease the electron pressure on PSI acceptors, preventing the ROS production by the Mehler reaction and therefore protecting PSI from photoinhibition [116]. Moreover, the chlororespiration gives rise to a proton gradient across the thylakoid membrane that can contribute to ATP synthesis [117]. In some diatoms, the chlororespiration can participate to the xanthophyll cycle induction [118]. This is because the proton gradient created across the thylakoid membrane can be enough to activate the DDE enzyme that, although having the pH optimum at 5.5, already operates at almost neutral pH [78]. The same proton gradient also inhibits the DEP enzyme with the maintenance of the active quencher diatoxanthin [118]. However, the involvement of chlororespiration in xanthophyll cycle was noticed in *Phaeodactylum tricornutum* but not confirmed in other diatoms, suggesting that species-specific differences can exist in the patterns of NPQ regulation [118].

In several species of unicellular green algae, including the model organism *C. reinhardtii*, a further safety mechanism employed in photoprotection as electron-pressure valve relies on the [Fe–Fe]-hydrogenase, an oxygen-sensitive enzyme,

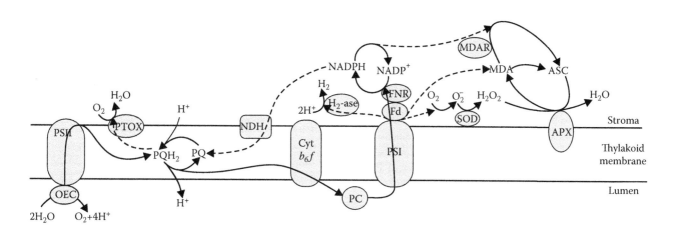

FIGURE 21.5 Schematic view of the linear electron flow (– – – –) from water to NADP$^+$ and of the alternative electron transport pathways (—) acting as *electron valves* in eukaryotic microalgae. (APX, ascorbate peroxidase; Asc, ascorbate; H$_2$-ase, [Fe-Fe]-hydrogenase; Cyt b_6f, cytochrome b_6f complex; Fd, ferredoxin; FNR, ferredoxin-NADP reductase; MDA, monodehydroascorbate; MDAR, monodehydroascorbate reductase; NDH, NAD(P)H dehydrogenase; OEC, oxygen evolving complex; PC, plastocyanin; PQ, plastoquinone; PSI, photosystem I; PSII, photosystem II; PTOX, plastid terminal oxidase; SOD, superoxide dismutase.) See text for detailed explanation.

which, like the cyanobacterial [Ni–Fe] enzyme, diverts electrons from the photosynthetic electron transport chain at the level of ferredoxin, catalyzing the reduction of protons to H_2 [119]. In *C. reinhardtii*, this photo-dependent reaction occurs in metabolic conditions, such as sulfur deficiency, that lead to *anaerobic oxygenic photosynthesis*, in which the anaerobiosis is maintained by consumption of photosynthetically generated oxygen through cell respiration [120].

21.5.3 WATER–WATER CYCLE

An alternative electron flow, regarded as an important system adopted by all the oxygenic photoautotrophs to prevent PSII photodamage under over-reduction of PSI electron acceptors, consists in diverging part of the electrons arrived at the PSI acceptor side toward the molecular oxygen [121] (Figure 21.5). This direct photoreduction of O_2 by reduced electron transport components (mainly reduced ferredoxin) of PSI, referred to as *Mehler reaction*, leads to formation of superoxide anion $\left(O_2^-\right)$, which is quickly disproportionated to H_2O_2 and O_2 by a thylakoid membrane attached form of the ubiquitous metalloenzyme superoxide dismutase (SOD) [122–124].

Although the electron flow to oxygen is useful for counteracting the PSII photodamage, it produces ROS that can attack and degrade some target components of the photosynthetic machinery, among which the iron–sulfur centers of PSI and the cyt$f$ of Cyt b_6f complex [125]. Thus, the prompt scavenging of H_2O_2, after that of O_2^-, becomes necessary to safeguard an efficient photosynthesis. In eukaryotic microalgae, this essential event is catalyzed by ascorbate peroxidase (APX) enzymes (Figure 21.5) that reduce H_2O_2 to H_2O by using ascorbate as electron donor [122]. The reaction gives rise to monodehydroascorbate, which can be reduced back to ascorbate through electrons from photoreduced ferredoxin or by the stromatic monodehydroascorbate reductase that takes electrons from NADPH [125]. Cyanobacteria do not possess ascorbate peroxidases, relying their antioxidant system on catalase–peroxidase enzymes and peroxiredoxins [126,127]. However, in cyanobacteria, the photoreduction of O_2 does not seem to produce ROS, because it essentially occurs through the so-called *Mehler-like* reaction based on the NAD(P)H: oxygen oxidoreductase activity of flavodiiron proteins [101,102,106]. This kind of proteins is also present in several green microalgae, where their possible function in photosynthesis-associated electron transfer reactions remains to be established [128].

The alternative electron transport route giving rise to water as final product is known as *water–water cycle* because the oxygen derived from the photooxidation of water at PSII is reduced back to water at PSI [121]. It is also called *pseudo-cyclic electron flow* since neither NADPH production nor O_2 evolution occurs, but only ATP synthesis, despite the involvement of both photosystems [129]. The water–water cycle can be active also in normal growth conditions playing a secondary role in extra supply of ATP to photosynthesis. However, it operates mainly under excess light or low CO_2 availability as a safety valve that allows the PSI excessive reducing power to be dissipated [128,130,131].

21.6 STATE TRANSITIONS

PSII and PSI have distinct light absorption properties, and their excitation degree can become unbalanced by quantitative/qualitative changes of environmental light conditions, exposing the photosynthetic machinery to a risk. Under light preferentially absorbed by PSII, for instance, the excessive excitation pressure on the photosystem can induce photoinhibition and damage to the RC [6]. This makes the capability to balance quickly the absorbed light energy between the two photosystems of basal importance for photosynthetic organisms in order to maintain the photosynthetic efficiency under changing light environments. The equalization of PSII and PSI excitation is achieved in cyanobacteria, red, and green microalgae on a timescale of minutes through a mechanism called *state transitions*, which consist in shuttling the energy of the main light-harvesting complexes (PBS or LHCII) between the two photosystems [132,133]. Namely, light conditions favoring the overexcitation of PSII induce a transition to state 2, in which more energy of the light-harvesting complexes is diverted from PSII to PSI, whereas lights that cause excess excitation of PSI induce a transition to state 1 in which more energy absorbed by the light-harvesting complexes is transferred to PSII [132]. Although this kind of energy redistribution is a shared strategy, the molecular mechanism of state transitions is quite different between cyanobacteria and red algae (with PBSs) and green algae (with LHCII), due to the different composition of their major light-harvesting complexes [134].

The cyanobacterial state transitions are regulated mostly by *mobile PBSs*, which are antenna complexes diffusing rapidly on the thylakoid membrane surface [135]. In state 1, the PBSs are transiently associated to PSII, while in state 2, a large number of PBSs move away from PSII toward PSI [136]. The cyanobacterial cells can sense any imbalance of the photosystem excitation, activating immediately the state transition response, which is triggered by the redox state of the plastoquinone pool, like in the other oxygenic photoautotrophs [134,137]. Plastoquinone oxidation induces state 1, whereas reduction induces state 2. It has been proposed that the sensing of the plastoquinone redox state involves the Chl a, which is present in the Cyt b_6f complex and whose phytyl tail protrudes into the Q_o site that binds the reduced plastoquinone (PQH_2) [138]. However, the pathway of signal transduction that links a change in plastoquinone redox state to a change in PBS coupling to photosystems has not been entirely clarified [139], although a crucial role in state transition has been ascribed to some components of the PBS core [140]. A strongly conserved membrane protein designated RpaC (regulator of phycobilisome association C) is specifically required for state transitions, likely as a binding factor that regulates the stability of the PBS–PSII interaction [141]. The synthesis of RpaC only in low light conditions led to suggest that in cyanobacteria, the major role of state transitions was in maximizing the light-harvesting efficiency to face light deficiency [134]. However, the state transitions have been found to have also a protective function in high light conditions [142]. In cells exposed to excess light, in fact, a change in light energy distribution between the two photosystems occurs, with the increase in energy supply to PSI in order to avoid photodamage. This kind of state transition depends on an alternative PSI subunit named PsaK2, which is present only in high light conditions and is involved in the energy transfer from PBS to PSI [142].

Although the PBS mobility has been proved to be the predominant mechanism to achieve the light state transitions in cyanobacteria, an *energy spillover* in which PSII transfers its excess excitation to PSI has also been proposed as a strategy to balance the energy distribution between the two photosystems [143]. The energy spillover mechanism, which occurs in state 2, presumes a strong coupling between PBS and PSII and a transient coupling between PSII Chl *a* and PSI Chl *a* with the PBS > PSII > PSI energy flow. This pathway is correlated with a reversible monomerization of the trimeric PSI that increases the PSI mobility in the thylakoid membrane enhancing the probability for the two photosystems to meet each other to realize the energy transfer from PSII to PSI, which works as a deeper energy trap [144,145]. Recently, the formation of a PBS–PSII–PSI megacomplex has been noticed in cyanobacteria [146].

The energy spillover from PBS to PSI via PSII is the prevalent mechanism utilized by red microalgae, including the unicellular *P. cruentum*, to share the light excitation energy between the two photosystems in the rapid adaptation following changes in environmental light conditions [147,148].

In green microalgae, the state transitions, mainly studied in *C. reinhardtii*, involve the relocation between PSII and PSI of the LHCII, the large peripheral antenna of PSII made up of major trimeric (majLHCII) and minor monomeric (CP26 and

CP29) complexes [149] of chlorophyll-binding proteins belonging to the LHC (light-harvesting complex) superfamily [150].

In these organisms, the state transitions are regulated by the plastoquinone pool redox state through the reversible phosphorylation of antenna components by the so-called Stt7 kinase, a thylakoidal enzyme firmly associated with both LHCII and Cyt b_6f complexes [151]. Under preferential excitation of PSII, a favored docking of the reduced plastoquinone to the Q_o site of the Cyt b_6f complex, perceived through the linked Chl *a* [138], causes the activation of the Stt7 kinase [152,153]. The enzyme phosphorylates proteins of the LHCII, a fraction of which moves from PSII to PSI to balance the light excitation energy between the two photosystems [154]. Conversely, the preferential excitation of PSI, which leads to the oxidized plastoquinone pool, inactivates the Stt7 and reverts the process, with the LHCII that turns back to PSII thanks to the dephosphorylation carried out by a thylakoidal phosphatase of the PP2C-type family [155]. So far, it is not known whether the phosphatase is constitutively active or whether it is regulated by the redox state of plastoquinone pool [155]. The phosphorylation carried out by Stt7, which occurs on polypeptides of a number of the trimeric majLHCIIs, the monomeric CP26 and CP29, and a monomeric majLHCII (named LhcbM5), triggers the undocking of the peripheral antenna from PSII and the lateral migration in the thylakoid membrane [156]. According to the model proposed by Iwai et al. [157] (Figure 21.6), the undocked phospho-LHCII polypeptides (both trimeric and monomeric) aggregate to form large complexes, which are in energy-dissipative state

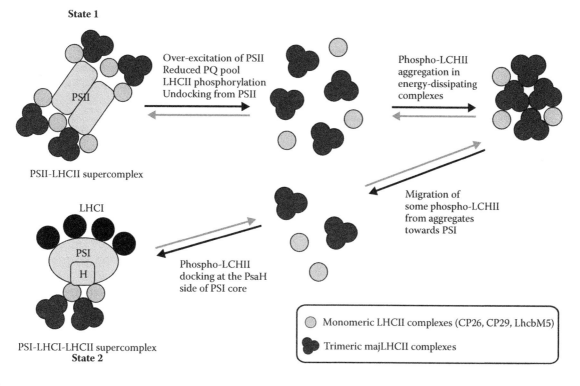

FIGURE 21.6 Schematic drawing of the state transition model in *C. reinhardtii* as proposed by Iwai et al. [157]. The black arrows show the steps of state 1 to state 2 transition after phosphorylation of LHCII complexes. The gray arrows indicate the state 2 transition back to state 1 after the LHCII complex dephosphorylation. See text for detailed explanation.

to suppress the deleterious effects of the excess energy. A certain number of these phospho-LHCIIs migrate then from the aggregated toward the PSI, giving rise to the supercomplex PSI–LHCI–LHCII of state 2. The monomeric LHCII complexes, and in particular CP29, seem to play a pivotal role in this shuttle, acting as linkers between the trimeric majLHCIIs and the PSI on the PsaH subunit of the photosystem core at the opposite side of the light-harvesting antenna LHCI [156,158]. During the transition back to state 1, the dephosphorylated LHCII polypeptides dissociate from PSI and associate with the aggregates from which they move back to PSII [157].

It has been suggested that about 80% of the trimeric majLHCII and a quantity of CP26, CP29, and LhcbM5 dissociate from PSII during the transition from state 1 to state 2, causing a disassembly of the PSII–LHCII supercomplex [159,160]. However, it has been shown that, although the state transition greatly reduces the antenna size of the PSII, its effect on the PSI antenna is rather small [160]. This because a predominant part of the dissociated phospho-LHCII remains as energy-dissipative aggregates detached from PSI [157,161]. This gains sense considering that the transition to state 2 is a short-term photoprotective strategy that in high light requires the immediate reduction of the PSII antenna size to avoid photodamage, but does not need the concomitant increase in the PSI light-harvesting capacity, which would even lead to PSI damage [162].

In *C. reinhardtii* chloroplasts, the state transition, besides the function in energy distribution between photosystems, seems to play an important role in regulating the ATP and NADPH production in response to the metabolic demand. Under conditions causing unbalanced ATP/NADPH ratios with ATP depletion, a nonphotochemical plastoquinone reduction occurs by electron transfer from stromal reductants via the thylakoidal type II NAD(P)H dehydrogenase. This event, added to the photochemical reduction of the plastoquinone pool, mimics an overexcitation of PSII and triggers, also in low light conditions, the transition to state 2, which favors the light-harvesting of PSI [163]. The increased energy absorption by PSI leads to an enhanced cyclic electron flow around the photosystem, which permits an additional ATP synthesis that re-equilibrates the ATP/NADPH stoichiometry required for the assimilative metabolism [163]. Recent data indicate that the enhancement of this cyclic electron flow is accompanied by the appearance of a supercomplex containing PSI, LHCI, and LHCII, together with Cyt b_6f and FNR (ferredoxin-NADP reductase) [164].

21.7 RECOVERY OF THE PSII

Both visible light and UV-B radiation can interfere with the PSII activity by damaging the proteins of the RC. The photoinhibitory effects of visible light depend on the over-reduction of the photosystem acceptor side that promotes the formation of ROS, mainly singlet oxygen, which induce the oxidative damage of D1 protein of the RC. The UV-B radiation, instead, acts by disrupting the Mn_4CaO_5 cluster involved in water oxidation, with the consequent accumulation of highly oxidizing cations, which cause the photochemical cleavage of both D1 and D2 proteins of the heterodimeric RC [11,12]. Irrespective of the underlying mechanism of photoinhibition, the recovery of the PSII needs necessarily the replacement of damaged proteins with newly synthesized functional copies. This occurs in thylakoid membranes through a tightly regulated repair process [165,166]. Although the photodamage by visible light requires mostly the substitution of D1, while the one caused by UV-B radiation needs the change of both D1 and D2, the same repair system is likely used under the two conditions, with few specific differences [167].

The current data suggest that the scheme of PSII repair in cyanobacteria as well as in eukaryotic microalgae includes (Figure 21.7): (1) monomerization of the dimeric PSII complex and its partial disassembly to permit access to the damaged proteins; (2) degradation and removal of the damaged subunits; (3) incorporation of the newly synthesized subunits into the subcomplex; and (4) reassembly of the fully functional dimeric PSII [165]. The partial disassembly of the PSII starts from the detachment of the lumenal extrinsic proteins of the OEC that triggers the monomerization of the dimeric PSII core. Subsequently, the monomeric PSII (composed of the RC and the inner antennae CP43 and CP47) undergoes the disjunction of CP43 from RC with formation of the RC–CP47 complex [165] in which degradation and replacement of the damaged protein take place [168]. Two families of thylakoid-bound proteases are mainly implicated in dismantling and remotion of D1 protein damaged by visible light: the ATP-dependent FtsH family and the ATP-independent Deg family [169–171]. After UV-B-induced damage, instead, only the FtsH enzyme is required for the proteolytic removal of both D1 and D2 proteins [167]. The remotion of damaged proteins is synchronized with synthesis and incorporation of the newly synthesized copies into the RC–CP47 complex [11,172]. The subsequent rebinding of RC–CP47 to CP43 gives rise to a monomeric PSII that can dimerize by the reattachment to the assembled lumenal extrinsic components of the OEC. A number of auxiliary factors have been reported to assist the PSII repair [165,166,173], which in cyanobacteria may be carried out in defined thylakoid regions called *thylakoid centers* [165], while in *C. reinhardtii*, it takes place in sites dispersed through the thylakoid system [174].

Interestingly, chlorophylls released during the PSII damage are temporarily stored and reutilized upon the repair and reassembly of the photosystem complex. This occurs through the pigment binding with proteins, which act as *emergency* scavengers that prevent the chlorophyll degradation and avoid the formation of harmful singlet oxygen from the free pigment molecules [175]. In cyanobacteria, the role of photoprotective temporary chlorophyll reservoirs is played by members of the HLIP family [176], while in chloroplasts of eukaryotic microalgae, related proteins belonging to the LHC-like family, such as early light-induced proteins (ELIPs), high-intensity light-inducible LHC-like 4 protein (LHC4), stress-enhanced proteins (SEPs), and one helix proteins (OHPs), may be involved in a similar role [150,177].

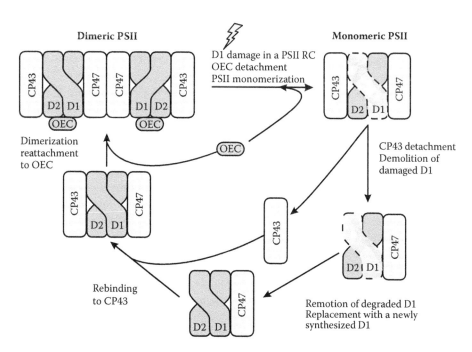

FIGURE 21.7 Scheme of the damage/repair cycle of the PSII RC in cyanobacteria and microalgae. (D1/D2, heterodimeric RC; CP43/CP47, inner antenna complexes; OEC, oxygen evolving complex.) See text for detailed explanation.

It must be considered that the damage/repair cycle of PSII RC does not occur only under photoinhibitory conditions but is also observed in normal growth conditions. The PSII activity, in fact, leads always to formation of ROS that are involved in damage of the D1 protein, which is characterized by a very short half-life [178]. Thus, degradation/resynthesis of D1, defined as a *suicide protein* that sacrifices itself to protect the rest of PSII against oxidative damage, is an unavoidable event that takes place under light at any intensity [179,180]. Since ROS accumulation and D1 damage linearly depend on the photon fluence rate, the irreversible photoinhibition of PSII occurs when high light conditions lead the rate of D1 degradation to exceed that of protein resynthesis and replacement in the damaged RC [181].

21.8 LIMITED CARBON AVAILABILITY

The ability to supply the Calvin–Benson–Bassham (CBB) cycle with CO_2 concentrations sufficient to sustain high rate of photosynthesis is crucial for photoautotrophic organisms. However, aquatic environments can vary dramatically with respect to the availability of CO_2, which often becomes severely limiting [182]. This is essentially due to (1) CO_2 diffusion 10,000 times slower in water than in air; (2) high HCO_3^-/CO_2 ratios, which make HCO_3^- the predominant form of inorganic carbon (Ci) especially in the pH range of 7–8.3 frequent in water bodies; and (3) high variability of total $Ci(CO_2 + HCO_3^-)$ concentrations, which are sensible to the pH values, lowering at neutral or acidic pH. In addition, fluctuations of environmental levels of O_2 can furthermore hinder the exploitation of available CO_2 for photosynthesis, due to the double carboxylase/oxygenase activity of the key enzyme Rubisco that makes CO_2 and O_2 competitive substrates [183].

The capability to overcome low environmental availability of Ci has been developed by cyanobacteria and microalgae through *CO_2 concentrating mechanisms* (CCMs), which elevate the Ci content inside the cells and lead to accumulation of CO_2 around the active site of Rubisco, thus enabling the enzyme carboxylase activity and warranting the efficient CO_2 fixation [184].

21.8.1 DUAL FUNCTION OF RUBISCO

The enzyme Rubisco (D-ribulose 1,5-bisphosphate [RuBP] carboxylase/oxygenase) is the most abundant protein in the biosphere. By its activity, the oxygenic photoautotrophs fix annually, via the CBB cycle, 10^{11} tons of CO_2 into organic biomass, with marine microorganisms accounting for nearly half of the CO_2 fixed [185]. However, Rubisco is not a perfect catalyst, suffering for low turnover rates and for its inevitable tendency to confuse the substrate of photosynthesis (CO_2) with the product (O_2). Thus, the enzyme, in addition to the RuBP carboxylation to produce two molecules of 3-phosphoglicerate (3PGA) entering the CBB cycle, also catalyzes the RuBP oxygenation. This last reaction gives rise to a molecule of 2-phosphoglicolate (2PG), which is the starting point of the wasteful photorespiratory (C2) cycle, that leads to a net carbon loss, hindering the carbon biomass productivity of photosynthetic organisms [186].

Due to the dual function of Rubisco, the efficiency with which CO_2 is able to compete with O_2 in the catalytic site depends on the inherent ability of the enzyme to discriminate between the two substrates. This Rubisco property is represented by the CO_2/O_2 specificity factor (Ω):

$$\Omega = K_o V_c / K_c V_o$$

where V_c and V_o are the maximal velocities of carboxylation and oxygenation, and K_c and K_o are the values of Michaelis constants for CO_2 (K_{CO2}) and O_2 (K_{O2}), respectively. The ratio of carboxylation to oxygenation rates is defined by the product of the specificity factor (Ω) and the ratio of CO_2 and O_2 concentrations at the active site of the enzyme [185,187].

Four distinct forms (I, II, III, and IV) of Rubisco exist in nature, based on difference in the primary sequence of the catalytic component [188]. Cyanobacteria and microalgae, with exception of dinoflagellates, possess different subtypes (named IA-D) of the form I Rubisco (referred to as L_8S_8 Rubisco) composed of eight large (L, ~50 kDa) and eight small (S, ~15 kDa) subunits organized as a core of four catalytic L dimers (L_2) capped on top and bottom by four regulatory S subunits. The Rubisco of dinoflagellates, instead, is a prokaryotic-type form II (referred to as L_2 Rubisco) composed of a simple L dimer [188,189].

Although the catalytic process is uniformly conserved, the substrate specific factor can differ substantially among the Rubisco forms and also among the different photosynthetic organisms. Cyanobacteria and green microalgae have type IB Rubiscos with low and intermediate values of CO_2 specificity factors (Ω = ~35–40 and ~60, respectively) [187]. The highest CO_2 specificity factors are shown by type ID Rubiscos from microalgae of red lineage, such as diatoms (Ω = ~110) and red algae (Ω = ~130–160) [185,190], while the lowest values are exhibited by the enzymes of dinoflagellates (Ω = ~35) [191]. Interestingly, an inverse correlation between specificity factor and turnover rate (V_c for carboxylation) has been noticed, with low specificities coupled to high turnover numbers and vice versa [185].

The competition of CO_2 and O_2 for the Rubisco catalytic site makes clear that higher values of Ω favor the RuBP carboxylation but cannot eliminate its oxygenation and the subsequent wasteful photorespiratory process, which remains common to all the oxygen-producing photoautotrophs [192]. However, in cyanobacteria as well as in most microalgae, the oxygenase activity of Rubisco is greatly lowered or almost totally repressed through the huge increase in CO_2 concentrations near the active site obtained by the effective action of the evolved CCMs [184,193].

21.9 CYANOBACTERIAL CCM

Cyanobacteria are the photosynthetic organisms inhabiting the widest range of ecological environments among which are the aquatic ones, such as open oceans, marine coasts, estuaries, freshwaters, and ponds [183].

Cyanobacteria first appeared around 2.4 billion years ago and carried out the oxygenic photosynthesis in an atmosphere that, until about 1.8 billion years ago, maintained CO_2 levels over 100-fold higher and O_2 levels much lower than today [194,195]. Thus, the ancient cyanobacteria did not require any CCM to achieve efficient photosynthesis under these conditions. However, the earth experienced dramatic change in evolutionary climate and atmospheric CO_2/O_2 ratios throughout the past 1.8 billion years. In particular, a marked drop in CO_2 concentration and an almost doubling in O_2 level, which

occurred around 400 million years ago [194], may have led the CO_2 to be a limiting carbon resource and the wasteful oxygenase activity of Rubisco to become a significant problem for photosynthetic organisms [183]. These environmental conditions may have represented the driving force for the evolution of the cyanobacterial CCM needed to face high photorespiration and low-efficiency carbon gain [196].

The cyanobacterial CCM rely on active uptake systems for both CO_2 and HCO_3^- to accumulate and retain Ci in the cytosol; intracellular conversion of CO_2 to HCO_3^-; and, finally, efficient conversion of the high HCO_3^- concentrations to CO_2 in carboxysomes, which are protein-bound microcompartments (Figure 21.8) that enclose the majority of the cellular Rubisco and where CO_2 fixation takes place [197]. This mechanism, which allows CO_2 to concentrate up to 1000-fold around the active site of Rubisco, is likely the most effective CCM of any photosynthetic organism [184,198]. The CCM enables the cells to achieve a saturated rate of CO_2 fixation at less than 10–15 μM exogenous CO_2, despite the low affinity and selectivity for CO_2 of the cyanobacterial form IB Rubisco having a CO_2 specificity factor (Ω) around 35–40 and a K_m for CO_2 exceeding 300 μM [190]. CCMs have been found in all the cyanobacteria so far examined, suggesting that this is an obligatory mechanism for survival in most natural environments [196].

The development of CCM possibly allowed ancient cyanobacteria to retain a Rubisco with low affinity for CO_2, but with a high rate of catalytic turnover per unit protein [199]. Due to CCM, this form of Rubisco can operate near V_{max} with a smaller nitrogen investment in enzyme [196]. The higher efficiency in nitrogen use and the photorespiration knocking down gained by CCM confer to cyanobacteria metabolic advantages that outweigh the energetic cost required for active accumulation of environmental Ci [200].

The extant cyanobacteria can be divided into two major CCM groups based on the form I Rubisco employed in CO_2 fixation [188]. Species containing the form IA Rubisco are referred to as α-cyanobacteria, while β-cyanobacteria are those having the form IB Rubisco [197,201]. The two

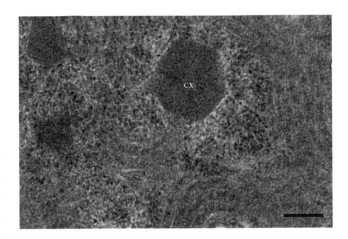

FIGURE 21.8 Transmission electron microscope detail of a cyanobacterial cell showing a polyhedric carboxysome (cx). (Bar = 0.2 μm).

cyanobacterial groups also possess carboxysomes ultrastructurally similar but proteomically different [202]. The so-called α-carboxysomes are peculiar to form IA Rubisco-containing bacteria, such as α-cyanobacteria and some chemosynthetic proteobacteria [203,204]. The microcompartments referred to as β-carboxysomes, instead, are found in β-cyanobacteria exhibiting form IB Rubisco [201].

The α-cyanobacteria are mostly oceanic strains that occupy aquatic habitats relatively constant in pH value (~8.3) and Ci levels (around 2 mM), with HCO_3^- as a dominant form of Ci, being the values of HCO_3^-/CO_2 ratios over 100 [183]. They often live in low light–low nutrient deep environments that severely limit the growth rate (as low as one cell division per day) of cyanobacterial populations [182,196]. Nevertheless, slow-growing α-cyanobacteria, such as *Prochlorococcus* spp. and *Synechococcus* spp., that predominate the oceanic waters covering near 75% of our *blue planet* surface, are highly abundant and productive on a global scale [205,206]. They contribute a very significant fraction (over 60%) of the ocean productivity that, in turn, accounts for nearly 50% of the net primary productivity of the whole biosphere [2,3].

Most β-cyanobacteria, conversely, are forms from freshwater ecosystems, with some coastal marine and estuarine strains [183]. These species inhabit a range of ecological niches that are much more variable in pH values and Ci levels than open ocean waters, so that they can experience substantial diurnal and/or seasonal change in Ci availability [183].

21.9.1 Inorganic Carbon Uptake

The first phase of cyanobacterial CCM consists in active uptake of both CO_2 and HCO_3^- by specific transport systems. So far, a total of three HCO_3^- transporters and two CO_2 uptake systems have been discovered in the examined cyanobacteria. These different CCM components vary in net affinity for Ci and in flux rate, with some of them being constitutively expressed and others genetically induced under conditions of Ci limitation [207–211]. The cyanobacterial Ci transporters are inactive in darkness but are activated in a few seconds upon illumination, probably through specific events of phosphorylation [182,212]. Generally, freshwater and estuarine strains of β-cyanobacteria, which have to face the greatest variability in environmental Ci levels, exhibit the most Ci transporters, whereas oceanic strains of α-cyanobacteria, which can rely on more constant levels of available Ci, tend to have the fewest transporters [182,183] (Figure 21.9).

The three cyanobacterial HCO_3^- transporters are referred to as BCT1, SbtA, and BicA.

BCT1 is an inducible transporter belonging to the ATP binding cassette (ABC) family of transporters, which are energized by ATP [213]. This is the only HCO_3^- primary uniporter found in cyanobacteria [214]. It shows a high photosynthetic affinity for HCO_3^- (around 15 μM) and a medium flux rate [215], and is strongly induced under severe Ci limitation [208,216,217]. BCT1 is generally present in β-cyanobacteria [183], while marine and oceanic α-cyanobacteria, such as *Prochlorococcus* strains, do not possess it [218].

SbtA is an inducible Na^+-dependent HCO_3^- transporter [219] that displays a very high photosynthetic affinity for HCO_3^- (around 2 μM) and a relatively low flux rate [220]. This kind of transporter, the abundance of which greatly increases under Ci limitation, has been found in β-cyanobacteria, while there is not a clear evidence of its presence in α-cyanobacteria [196]. Weak homologues of SbtA have been found in α-cyanobacteria, but their potential activity as HCO_3^- transporters is still devoid of confirmation [196].

BicA is another Na^+-dependent HCO_3^- transporter without sequence similarity to SbtA. Differently from SbtA, it exhibits a relatively low photosynthetic affinity for HCO_3^- and a high flux rate [220], and is widespread in both β- and α-cyanobacteria [182]. Homologues of BicA, with transport affinities for HCO_3^- ranging from 74 to 353 μM, were noticed in different cyanobacterial strains [220]. The BicA, whose expression is largely constitutive in nature, seems to be the only recognizable candidate for HCO_3^- uptake in oceanic α-cyanobacteria [182].

Both SbtA and BicA carry out a Na^+/HCO_3^- symport driven by the standing electrochemical Na^+ gradient (inward directed) maintained via Na^+/H^+ antiporters, which are known to operate in cyanobacterial cells [221–223]. The fact that marine and oceanic α-cyanobacteria exhibit only a Na^+-dependent HCO_3^- transport, lacking the ATP-dependent BCT1, suggests the strategy of energizing the HCO_3^- uptake by exploiting the driving force of the electrochemical Na^+ gradient rather than spending the ATP required by the ABC transporter [218].

The cyanobacterial active CO_2 uptake is based on two specialized type I NADPH dehydrogenase (NDH-1) complexes, referred to as $(NDH-1)_3$ and $(NDH-1)_4$, which possess alternative subunits with CO_2 hydration activity [224–228]. These complexes cannot be called true transporters, but rather active CO_2 uptake facilitators. They, in fact, operate as *vectorial carbonic anhydrases*, catalyzing the unidirectional conversion to HCO_3^- of CO_2 that passively entered the cell mostly through plasma membrane water channels (aquaporins) [229]. Both the CO_2 uptake systems reside largely, if not exclusively, in the thylakoid membrane [228,230,231], where a $\Delta\mu H^+$ generated by the photosynthetic electron transport can directly energize the conversion of CO_2 to HCO_3^- via the formation of alkaline microdomains on the membrane cytoplasmic face [207,226,229]. In this way, the intracellular HCO_3^- concentration is enhanced and the inward diffusion gradient for CO_2 is maintained. PxcA, a plasma membrane H^+ extrusion pump, is an ancillary component of the CO_2 uptake systems. Its activity is essential during the initial phase of CO_2 uptake (up to 30 s after illumination) when the ejection from the cytoplasm of H^+ produced by the CO_2 conversion to HCO_3^- is essential for maintaining the internal pH value [196].

As regards the two specialized NADPH dehydrogenases, $(NDH-1)_4$ is a constitutively expressed complex involved in a lower affinity CO_2 uptake (~10 μM), while $(NDH-1)_3$ is a higher affinity CO_2 uptake (1–2 μM) complex inducible under limiting Ci conditions [224,232]. Both the CO_2 uptake systems can usually be found in β-cyanobacteria, whereas the oceanic α-*Synechococcus* strains possess only the constitutive

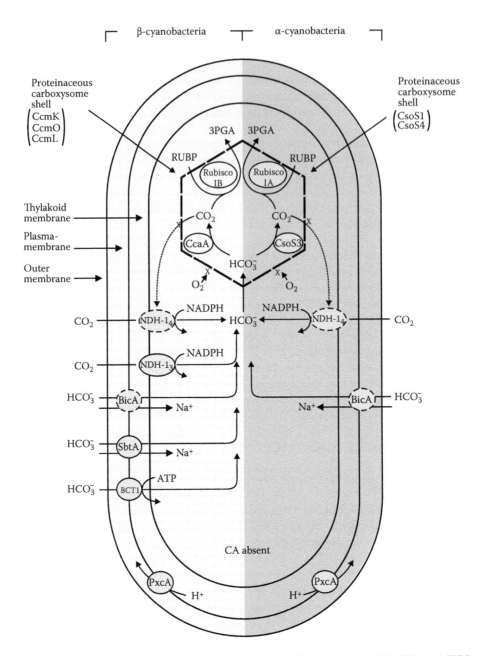

FIGURE 21.9 Schematic diagram illustrating the CCM components in α- and β-cyanobacteria. The CO_2 and HCO_3^- transporters with a dished outline are constitutive, while those with a continuous outline are induced under low Ci conditions. (CcaA and CsoS3, β-carbonic anhydrases; 3PGA, 3 phosphoglycerate; PxcA, H$^+$ extrusion pump; RUBP, ribulose-1,5-bisphosphate.) See text for detailed explanation.

$(NDH-1)_4$ complex and the α-*Prochlorococcus* strains lack any system of active CO_2 uptake [196,218]. This confirms that α-cyanobacteria living in habitats with rather constant Ci availability, such as the oceanic waters, rely on CCM based on a few constitutive low-affinity Ci uptake systems, although some evidence exists that in low Ci conditions, their system affinity can be in some way improved [233,234]. Conversely, β-cyanobacteria, which occupy habitats subjected to fluctuations of available Ci, have the highest capability to cope with environmental Ci deficiencies by up-regulating the expression of high affinity Ci (both CO_2 and HCO_3^-) uptake systems. The transcription of low Ci inducible genes occurs in as little as 15 min [208–210,235], while the completion of

the induced high-affinity uptake systems is achieved in a few hours [209,236,237].

Two subfamilies of LysR-type transcription factors, known as CmpR and CcmR, regulate the expression of Ci responsive genes [216,235,238–241]. Generally, CmpR functions as an activator of the BCT1 genes under Ci limitation, while CcmR acts as a repressor of the $NDH-1_3$ and SbtA genes in high Ci conditions [182,216,235,238,242]. Interestingly, the activity of these LysR-type transcription factors depends on some metabolic intermediates, among which are 2-phosphoglicolate (2PG) and NADP$^+$. More precisely, the 2PG, whose concentration arises under limiting Ci from the Rubisco oxygenase reaction, behaves as a primary effector that increases the CmpR inductive activity on BCT1 gene expression [239]. Conversely, the $NADP^+$ is an effector that greatly enhances the CcmR repressive action. Thus, the decrease in $NADP^+$ levels in low Ci conditions, due to the diminished NADPH consumption by CBB cycle, leads to de-repression of the CcmR regulated genes of NDH-1$_3$ and SbtA [240]. In this way, changes in the concentration of metabolites from pathways depending on Ci supply can act as useful signals to modulate the expression of genes involved in cell adaptation to environmental Ci availability [241].

21.9.2 Carboxysomes

Through the Ci uptake systems, cyanobacteria actively accumulate in cytoplasm high HCO_3^- amounts, which are maintained far for equilibrium partly for the absence of carbonic anhydrase activity [241]. The HCO_3^- moves then into the carboxysomes (Figure 21.9) where a carbonic anhydrase catalyzes its conversion to CO_2 that is provided to Rubisco at up to 50 mM [243]. In this way, carboxylation activity and high photosynthetic rates are warranted in spite of the low CO_2 affinity of the enzyme.

Carboxysomes, which carry out the final stages of CCM, are icosahedral bodies (90–250 nm in diameter) bounded by a thin proteinaceous shell (3–6 nm) that encapsulates the enzymatic components: Rubisco and carbonic anhydrase [244,245]. The α- and β-carboxysomes, typical of α- and β-cyanobacteria, show the same ultrastructural organization but have a different protein complement. The α-carboxysome proteins are named CsoS (CsoS1-4), while the β-carboxysome ones are the Ccm (CcmKLMNO) and CcaA [246–248]. Two of these proteins, namely, CsoS3 (also known as CsoSCA) and CcaA, are the β-carbonic anhydrases that supply Rubisco, tightly packed in the carboxysome core, with CO_2 [248,249].

A basic role played by the proteinaceous shell is to minimize the leakage of this CO_2 from carboxysome, providing a gas diffusional barrier that also impedes the entry of O_2. The major shell constituents are the CsoS1 and the CcmK and CcmO proteins. All of them belong to the BMC (Bacterial Micro Compartment) protein family and exist as hexagonally shaped hexamers whose edges interlock to build the thin sheets that form the flat facets of the icosahedral carboxysome shell [250–353]. Other two proteins, CsoS4 and CcmL, instead, are pentamers that form the vertices of the shell [254]. Each polymeric unit of these shell proteins exhibits a narrow central pore (4–7 nm in diameter) with a positive charge, which selectively favors the diffusion across the shell of negatively charged substrates and products (such as HCO_3^-, RuBP, 3PGA and 2PG) compared with uncharged CO_2 and O_2 [246,250,251,253]. Moreover, it has to be pointed out that the function of carboxysome shell against the CO_2 leakage is supplemented by the effective CO_2 uptake systems (NDH-1)$_{3/4}$ that actively recycle the CO_2 escaped from the carboxysome back into HCO_3^- before it leaves the cell [182]. A possible action in preventing the outward leakage of CO_2 from the cell has also been suggested for periplasmic carbonic anhydrases (named EcaA and EcaB) [196,198].

In addition to the main shell proteins, a structural role has also been assigned to other carboxysomal components, in particular, to the CcmM protein of β-carboxysomes that functions as an essential scaffold. It contains two regions homologous to γ-carbonic anhydrase and Rubisco small subunit that allow CcmM to interact with both CcaA and Rubisco. CcmM is the focal point for a bicarbonate dehydration complex (made of CcaA–CcmM–CcmN), which is positioned on the inner surface of the shell, warranting the CO_2 supply to the near Rubisco [245,255]. Moreover, in some β-carboxysomes that lack CcaA, the CcmM may also act as a catalytically active carbonic anhydrase [245,248,255–257]. Carboxysomes are present in all cyanobacteria, and their allocation is actively controlled in dividing cells to ensure that the daughter cells inherit a similar number of these basic CCM components [258].

21.10 CCM IN MICROALGAE

Most microalgae living in aquatic ecosystems with frequent CO_2 limitation have evolved CCMs that enable them to accumulate Ci inside the cell in excess of environmental concentrations to minimize the wasteful photorespiratory pathway and to increase the photosynthetic performance [197,259,260]. However, unlike cyanobacteria, which have a basic constitutive form of CCM also expressed in cells growing at high and even hypernormal Ci concentrations [182], the eukaryotic microalgae activate CCMs only under low Ci conditions, relying solely on CO_2 diffusion from the external medium at nonlimiting Ci concentrations [197,261].

Diverse CCMs have been discovered in different groups of microalgae, which can vary in complexity but, like the cyanobacterial CCM, are all based on energized Ci uptake systems and carbonic anhydrase activity to maintain high CO_2 levels around the Rubisco that in most of these microorganisms is concentrated in a specific plastidial microcompartment called pyrenoid (Figure 21.10) [262]. The microalgal CCMs have evolved after the primary endosymbiosis, plausibly in periods of low Ci availability due to a decrease in environmental CO_2 concentrations. Interestingly, no homology between the active HCO_3^- uptake systems of different microalgal groups has been found, suggesting that these photoautotrophs may have recruited a large variety of HCO_3^- transporters through diverse ancestral symbiotic hosts, evolving specific systems for the Ci acquisition in case of need [263]. Unlike the well-characterized cyanobacterial CCM, less is known about many aspects of the CCMs that most microalgal organisms can use to face the low environmental Ci [197,264,265]. However, new interesting information has recently been gained on CCM of green microalgae and some other dominant algal groups, such as diatoms and dinoflagellates [261,264].

21.10.1 CCM in Green Microalgae

The green microalgae can take up both CO_2 and HCO_3^- from the aquatic environment. In these eukaryotic cells, the Ci species must cross three barriers, namely, the plasma

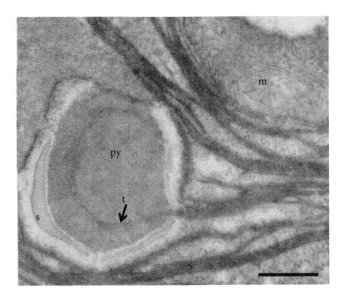

FIGURE 21.10 Transmission electron microscope detail of a green microalgal cell showing the chloroplast with a pyrenoid crossed by a thylakoid. (m, mitochondrion; s, starch; py, pyrenoid; t, thylakoid; bar = 0.5 μm.)

membrane, the inner membrane of the two-membrane envelope of the primary chloroplast, and the thylakoidal membrane, before reaching the lumen of a transpyrenoid thylakoid, which is the site devoted to CO_2 supply to the nearby packaged Rubisco [266]. Thus, the whole CCM requires Ci transport systems localized to each membrane together with associated CA activities in different subcellular compartments [266–269]. This CCM leads to a CO_2 concentration around Rubisco up to 40-fold of the ambient one

[270], which is much lower than the concentration obtained by the inductive cyanobacterial CCM.

The understanding of CCM functional components in green microalgae has been gained largely from research on the model organism *C. reinhardtii* [261]. To date, some cross-membrane Ci channels or transporters [265], some specific carbonic anhydrases [267], and, in addition, a plastidic complex of soluble proteins [271] have been recognized as CCM candidates in *C. reinhardtii* (Figure 21.11). All these candidates are under control of the transcriptional regulator CIA5 (also known as CCM1 or CCM master switch), whose expression is independent of Ci concentration [272–274]. The master regulator CIA5 is involved in the signal transduction mechanism sensing the changes in CO_2 availability and operates as a transcription inducer of Ci-responsive genes essential to the function of CCM in Ci-limiting conditions [273].

21.10.1.1 Ci Transmembrane Transport

Two low Ci-inducible transporters, referred to as HLA3 and LCI1, are predicted to be located in *C. reinhardtii* plasma membrane [275,276]. HLA3 is a HCO_3^- transporter of ABC type, whose gene transcription undergoes a 40-fold increase in cells shifted to a Ci-limiting condition [277,278]. HLA3 homologues have been found in several other green microalgae [279]. LCI1 represents a peculiar low Ci-inducible plasma membrane transporter that, according to Ohnishi et al. [276], may be involved in both CO_2 and HCO_3^- uptake. A Rhesus-like (RHP1) protein, with similarity to Rhesus proteins of human red blood membrane [280], has also been found in *C. reinhardtii* plasma membrane [281]. Although a role of RHP1 as a channel facilitating the CO_2 diffusion in low Ci conditions has not been ascertained so far, such a role cannot be ruled out [265]. An extracellular α-carbonic anhydrase (CAH1),

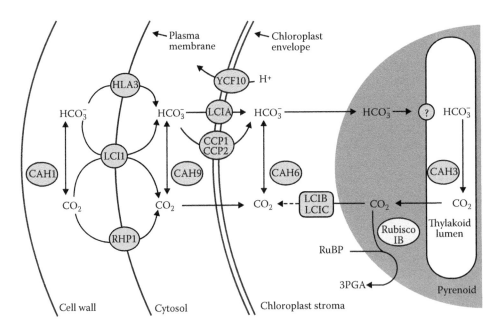

FIGURE 21.11 Schematic diagram illustrating the CCM components in the green microalga *C. reinhardtii*. (CCP1/2, HLA3, LCI1, LCIA, RHP1, Ci transporters; CAH1, CAH3, CAH6, CAH9, carbonic anhydrases; LCIB/C, soluble protein complex; 3PGA, 3 phosphoglycerate; RUBP, ribulose-1,5-bisphosphate; YFC10, H+ extrusion pump.) See text for detailed explanation.

whose expression greatly increases when the CCM is induced [268,278,282], has been suggested to operate in facilitating the CO_2 influx into the cell [261].

LCIA, CCP1, and CCP2 are three chloroplast envelope proteins proposed as putative transporters implicated in CCM. LCIA (also named NAR1.2) is a protein belonging to the formate/nitrite transporter family, which has been reported to transport HCO_3^- into the chloroplast, although with a rather low affinity [274,283]. Among the CCM genes, the one encoding LCIA is the most Ci-responsive, with a 4000-fold increased transcription under Ci limitation [278]. CCP1 and CCP2 are two closely related proteins exhibiting sequence similarity to the large family of mitochondrial carrier proteins involved in the transport of various metabolites across the inner membrane. Experimental evidence suggests that, rather than in Ci transport, CCP1/2 might play a role in translocation of metabolic intermediates important in Ci-limiting conditions [284]. Also genes encoding these proteins are highly responsive to Ci. Transcripts of *CCP1*, for instance, undergo a 2000-fold increase in low Ci conditions [278]. The Ci accumulated inside the alkaline stroma of chloroplast then moves across the thylakoid membrane as HCO_3^- through a yet unidentified transporter [285].

An integral protein of the chloroplast envelope inner membrane, named YFC10, acts as an ancillary component of CCM with a role similar to that of the cyanobacterial PxcA. YFC10, in fact, is a proton pump that seems to be required to maintain the alkaline stromal pH through the extrusion of H^+ generated inside the chloroplast by the hydration to HCO_3^- of Ci entered as CO_2 [265].

21.10.1.2 Pyrenoid

Pyrenoids (Figure 21.10) are electron-dense protein inclusions found in the chloroplast of most green microalgae, as well as in other eukaryotic algae and in hornworts [286]. They are specialized microcompartments, usually surrounded by a starch sheath [287,288], that play a fundamental role in functioning of the CCM.

The amorphous pyrenoid matrix is composed predominantly (more than 90%) of tightly packaged form IB Rubisco [261], which shows intermediate values of CO_2 specificity factor ($\Omega = \sim 50$–60) and a K_m for CO_2 around 30 μM [190]. The functional pyrenoid requires the correct Rubisco package [289] and the presence of other components, among which a recently identified protein named CIA6 [290] and the Rubisco activase that maintains the Rubisco activity for maximal CO_2 fixation [291]. The pyrenoid matrix is penetrated by a network of thylakoid tubules [287,292] containing the lumenal α-carbonic anhydrase CAH3 [269,275,293]. This enzyme performs a critical function in the acidic thylakoid lumen catalyzing the rapid dehydration of the accumulated HCO_3^- to CO_2 that moves through the thylakoid membrane into the pyrenoid for assimilation by Rubisco. CAH3 is a constitutive lumenal carbonic anhydrase whose level does not rise in cells grown under limiting Ci. In this case, however, CAH3 undergoes a phosphorylation that causes a change in its distribution. The enzyme, which in normal conditions is located also in

nonpyrenoid thylakoids, concentrates in the pyrenoid crossing thylakoids and exhibits a five- to sixfold increased activity [294].

Unlike cyanobacterial carboxysomes, pyrenoids are not delimited by a protein shell with the function of hampering the CO_2 leakage. However, a barrier for the necessary hindrance to CO_2 escape is furnished by the LCIB/LCIC complex [271]. LCIB and the homologous LCIC are soluble proteins forming high-molecular-weight complexes (350 kDa) that under nonlimiting Ci are dispersed through the entire chloroplast stroma [271]. In low Ci conditions, the LCIB/LCIC transcripts greatly increase and the complexes relocalize quickly (within 1 h) to the region surrounding the pyrenoid [271], giving rise to a structural diffusion barrier that minimizes the leakage of CO_2 from the pyrenoid matrix maintaining high CO_2 concentrations for efficient fixation by Rubisco. The escaped CO_2, however, can be recaptured and reconverted to HCO_3^- by the stromal β-carbonic anhydrase CAH6, which also localizes itself around the pyrenoid [271,295]. CAH6 is a constitutive enzyme, whose expression is slightly upregulated under low CO_2 conditions [295].

The pyrenoid is a basic component of the CCM, and all pyrenoid-containing green microalgae are able to activate mechanisms to concentrate Ci inside the cells [197]. However, there are also microalgae that can express CCM despite the pyrenoid absence. A form of CCM, in fact, has been recognized in some pyrenoid-less strains of the green algal genus *Chloromonas* [296]. These algal strains exhibit a low CO_2-inducible high photosynthetic affinity for CO_2 correlated with increased carbonic anhydrase activity. They can accumulate intracellular Ci pools, although smaller than those of pyrenoid-containing microalgae [296].

21.10.2 CCM in Diatoms

Diatoms are predominant microalgae that play a major role in the global carbon cycle accounting for 40% of marine and 20% of total primary production [297]. These microorganisms, derived from a secondary endosymbiosis between a heterotrophic eukaryote and a red alga [298], possess a chloroplast surrounded by four membranes, with the innermost two corresponding to the organelle envelope and the outermost two formed by a specialized fraction of endoplasmic reticulum (named CER: chloroplast endoplasmic reticulum) [299]. Between CER and chloroplast envelope, there is a periplastidal compartment, which represents the remains of the red algal endosymbiont cytosol [300]. An elongated pyrenoid is present in the chloroplast (Figure 21.12).

Like the red algae and other algae of red lineage, diatoms have a form 1D Rubisco, which exhibits a high CO_2 specificity factor ($\Omega = \sim 110$) [191] and an intermediate affinity for CO_2, with a K_m (~ 28–40 μM) [301] that is higher than the usual CO_2 concentration in marine waters (10–15 μM) [302]. Diatoms can perform significant rates of CO_2 fixation in Ci-depleted environments because they adapt themselves to the low Ci availability through an efficient CCM, which elevates the CO_2 concentrations around the active site of Rubisco [264]. The

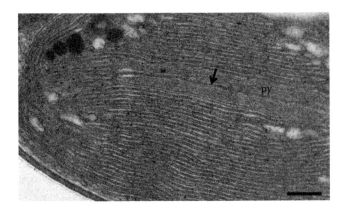

FIGURE 21.12 Transmission electron microscope detail of a *P. tricornutum* cell showing the chloroplast with an elongated pyrenoid (py) crossed by few thylakoids (arrow). (Bar = 0.2 μm).

CCM is based on the uptake of both CO_2 and HCO_3^- across the plasma membrane, the accumulation of Ci within the cell (two- to sixfold with respect to the extracellular concentration), and the intracellular carbon flux across the four membranes enclosing the secondary plastid as far as the pyrenoid where the confined Rubisco is supplied with CO_2. All this takes place with the critical cooperation of specific carbonic anhydrases located at different cell compartments [303,304].

The activity of CCM is induced/suppressed in a few hours under low/high CO_2 concentrations, respectively, indicating that the cells can sense the changes in environmental CO_2 availability as a signal for a fast response to the new CO_2 conditions [305]. Interestingly, recent findings suggest that diatoms can perceive the change in ambient CO_2 levels through a mammalian-type signalling pathway that involves cAMP as a second messenger and leads to the regulation of the expression of CCM genes [306,307].

Some molecular and functional aspects of diatom CCM are still poorly understood, and the available information has derived from studies on few organisms such as *P. tricornutum* and some species of *Thalassiosira*. As regards *P. tricornutum*, a CCM model termed *chloroplast pump* has recently been proposed [308,309]. According to the model (Figure 21.13), most of the environmental Ci enters the cell as CO_2 by diffusion across the high permeable plasma membrane, while HCO_3^- is taken up actively by membrane-embedded transporters transcribed specifically under low Ci conditions. All of these transporters (namely PtSLC4-2, and likely PtSLC4-1 and PtSLC4-4) show homology to the HCO_3^- transporters of the mammalian SLC4 (Solute Carrier 4) protein family [263]. The CO_2 that entered the cytoplasm is converted to HCO_3^- by one of the constitutive α-carbonic anhydrases localized within the four layered chloroplastic membrane system [303], and, ultimately, the HCO_3^- is actively transported into the chloroplast stroma plausibly through another putative PtSCL4-type carrier. The accumulated HCO_3^- diffuses toward the pyrenoid where it is rapidly converted to CO_2 by pyrenoidal β-carbonic anhydrases (PtCA1 and PtCA2), whose transcription is induced under CO_2 limited conditions [310]. Part of the CO_2 is fixed by Rubisco; a part escapes from pyrenoid and leaves chloroplast, but it is converted back to HCO_3^- by α-carbonic anhydrases of the periplastidal compartment [309]. Neither extracellular nor cytoplasmic carbonic anhydrases are present in *P. tricornutum* [304]. In the chloroplast pump model, the active transport of HCO_3^- into the organelle represents the major driving force for the CCM because it maintains an elevated CO_2 concentration (~60 μM) around the Rubisco [308] and lowers the CO_2 level in the cytoplasm, thus greatly enhancing the diffusive CO_2 influx into the cell [309].

The CCM model of *P. tricornutum* might be applicable to other diatoms having similar physiological characteristics [309]. However, the finding in *Thalassiosira* species of

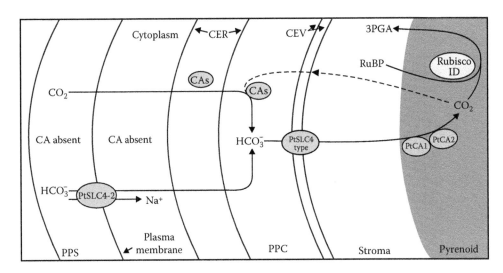

FIGURE 21.13 Schematic diagram illustrating the components of a proposed CCM model for the diatom *P. tricornutum*. (CER, chloroplast endoplasmic reticulum; CEV, chloroplast envelope; 3PGA, 3 phosphoglycerate; PPC, periplastidal compartment; PPS, periplasmic space; PtCA1, PtCA2, CAs, carbonic anhydrases; PtSLC4-2, PtSLC4-type, HCO_3^- transporters; RUBP, ribulose-1,5-bisphosphate.) See text for detailed explanation.

diverse carbonic anhydrases with different localizations in extra and intracellular compartments [304] strongly suggests the existence of biophysical CCMs with distinct architectures within this microalgal group.

A C_4 metabolism functioning in a single cell, like that found in some higher plants [311] and green macroalgae [312,313], has also been proposed for diatoms as a biochemical CCM able to increase the CO_2 import into the chloroplast [314]. According to a simplified route of diatom single-celled C_4, the phosphoenolpyruvate (C_3) carboxylation, occurring in the cytoplasm, would give rise to C_4 acids (oxalacetate and malate), which would be decarboxylated in the chloroplast stroma with production of CO_2 then fixed by Rubisco in the CBB cycle [197]. The hypothesis of a C_4-like activity in diatoms was supported by the predominance of C_4 compounds found in short-time ^{14}C labelling experiments in diatoms such as *Thalassiosira weissflogii* [315] and *Haslea ostrearia* [316] and by some effects of C_4 enzyme inhibition on photosynthesis [317,318]. However, despite the presence of a complete set of genes encoding the C_4 route enzymes in sequenced genomes of *P. tricornutum* and *T. pseudonana* [319,320], the biochemical analyses [315], the effects of key enzyme inhibition [321], and the cellular localization of carboxylating and decarboxylating enzymes [321,322] led to rather contradictory conclusions on the activity of a C_4 metabolism in these diatoms. At present, the existence of a C_4 route in some diatoms has been ascertained, although its direct role as biochemical CCM is still under debate [321,323].

21.10.3 CCM in Dinoflagellates

Dinoflagellates are important members of the phytoplankton communities from aquatic environments, in particular from marine waters where they can be present as free-living organisms occasionally forming conspicuous blooms (red tides), or as symbionts in a number of invertebrates, including giant clams (such as *Tridacne gigas*) and reef-building corals [324].

Two features make dinoflagellates unique among the oxygenic photoautotrophs. The first concerns the plastid, which, according to different lineages, can be derived from a secondary endosymbiosis with a red alga, from a serial secondary endosymbiosis with a green alga, or from a tertiary endosymbiosis with a chromalveolate (such as a diatom or a haptophyte) [298]. The second feature is the use as a carbon-fixing enzyme of form II Rubisco, the homodimeric enzyme formed by only large subunits (L_2) that they share with the purple anaerobic bacteria [325].

The dinoflagellate form II Rubisco exhibits CO_2 affinity and CO_2 specificity factor ($\Omega = \sim35$) significantly lower than the more common eukaryotic form I Rubisco [191]. This can greatly limit the net carbon fixation in the marine environments where, at 8–8.2 pH values, which approximate the pH found in open sea, most Ci is present in the form of HCO_3^- and only 10–20 µM (~1%) is available as CO_2 [326].

Although rather limited and often contradictory information is available on the Ci acquisition by dinoflagellates [197,264], their ability to increase the CO_2 supply to Rubisco by overcoming the low Ci availability through a CCM is shown by the Ci accumulation inside the cells from 5- to 70-fold relative to the environmental concentrations, depending on species and growth conditions [327–329].

Experimental data acquired on free living dinoflagellates support two possible CCM models [264]. The first relies on the CO_2 entry at the plasma membrane with [330] or without [331,332] the aid of an external carbonic anhydrase catalyzing the HCO_3^- dehydration. The second, found in red tide-forming species, is based on HCO_3^- uptake and intracellular carbonic anhydrase activity [333]. In most dinoflagellates, the pyrenoid is a major CCM component [334–336], although species lacking this intraplastid compartment, such as *Protoceratium reticulatum*, also exhibit an efficient CCM [332].

Dinoflagellates of the genus *Symbiodinium* (referred to as zooxanthellae) are symbionts of marine invertebrates, which rely on the microalgal photosynthate supply to satisfy over 90% of their energy requirement [337,338]. Interestingly, in these symbiotic systems, the zooxanthellae are assisted in acquiring the inorganic carbon from the surrounding seawater by host carbonic anhydrases, the level of which positively correlates with the zooxanthella number [339,340].

21.10.4 CCM in Other Microalgae

The ability to maintain high photosynthetic rates in spite of environmental Ci deficiency through concentration of CO_2 around Rubisco has been demonstrated in several microalgae belonging to other taxonomic groups. It is thought that nearly all marine phytoplankton can operate a CCM [197,264], although the specific mechanisms leading to the intracellular CO_2 accumulation are not always well understood. Some available information has been gained for a few worldwide-distributed bloom-forming marine species of Haptophyta such as the flagellate *Phaeocystis globosa* and the coccolithophorid *Emiliania huxleyi*.

As regards *P. globosa*, experimental data show that it can rely on both CO_2 and HCO_3^- as Ci sources for photosynthesis. In addition to direct HCO_3^- uptake, an external carbonic anhydrase, whose activity increases with decreasing CO_2 concentration, favors the dehydration of HCO_3^- to CO_2, which then diffuses across the plasma membrane [341]. The high carbon acquisition efficiency of *P. globosa*, which is comparable with that of diatoms, accounts for intracellular CO_2 concentrations up to 20 times higher than the external CO_2 and for the photosynthetic carbon fixation rate, which is maintained near CO_2 saturation at present-day CO_2 levels [264,342].

E. huxleyi is the most abundant coccolithophore, able to form blooms up to 8×10^6 km^2 [343]. This species operates a CCM responsive to CO_2 and HCO_3^- [344], but, interestingly, a number of differences have been found between strains from the northern and the southern hemisphere [345], which are genetically distinct [346]. Studies focused on northern hemisphere strains showed that *E. huxleyi* uses CO_2 as the primary Ci source for photosynthesis, although some discrepancies exist over the importance of HCO_3^- [347]. Furthermore, it does not possess external CO_2-regulated carbonic anhydrase activity and exhibits a CO_2 uptake system with low affinity [342]. These *E. huxleyi* strains show a low efficient CCM and photosynthetic carbon fixation rates well below CO_2 saturation at the present CO_2 levels [264,342]. This is also due to a substantial leakage of CO_2 from the cell across the highly permeable membranes [347]. Conversely, investigations on *E. huxleyi* strains from southern hemisphere [345] revealed a strong CCM activity, external carbonic anhydrases, and a high affinity for CO_2 in photosynthesis. Moreover, in *E. huxleyi* from the southern hemisphere, which includes both calcifying and noncalcifying isolates, it has also been found that HCO_3^- is the preferred Ci source for photosynthesis in noncalcifying strains while the calcifying ones exhibit a higher preference for CO_2 [345].

As regards the calcifying strains of *E. huxleyi*, it had previously been proposed that the process of calcification could act as an energy-efficient CCM by enhancing the intracellular concentration of CO_2 and then by using it for photosynthetic carbon fixation [348,349]. However, new evidences suggest that calcification and photosynthesis are not tightly linked in the microalga and that photosynthesis is not mechanistically dependent on calcification. Thus, calcification, which usually increases at low CO_2 conditions, is not involved in inorganic carbon delivering to Rubisco [344,350].

21.11 CONCLUSIONS

Sudden changes in light regime causing sharp rises of irradiance can be very harmful for the photosynthetic phytoplanktonic populations inhabiting aquatic environments. The absorption of excess light energy, in fact, leads to formation of dangerous ROS, which greatly damage the photosystems, in particular PSII, with the subsequent photoinhibition of photosynthetic activity.

Cyanobacteria and microalgae have evolved a plethora of strategies able to protect the PSII (and the entire photosynthetic apparatus) against the detrimental effects of excess light. These strategies essentially rely on (1) nonphotochemical dissipation of part of absorbed energy in light-harvesting complexes to reduce the energy funneled to RC, (2) activation of alternative electron transfer routes to divert electron from the photosynthetic electron transport chain, (3) balance of the light energy absorption by the two photosystems to maintain the photosynthetic efficiency under changing light environments, and, finally, (4) fast repair of photodamaged PSII to recover the photosystem activity.

The safeguard of a functional photosynthetic apparatus allows the production of ATP and NADPH required for the CO_2 fixation in the CBB cycle via the carboxylase activity of the enzyme Rubisco. However, an environmental problem that the phytoplanktonic organisms have to face is the availability of CO_2, which can become severely limiting due to its low diffusion rate in water and to the predominance of HCO_3^- as Ci form. Moreover, the dual (carboxylase/oxygenase) function of Rubisco forces the CO_2 to compete with O_2 for the enzyme.

Thus, the phytoplanktonic microorganisms have evolved constitutive and/or inductive CCMs to increase the Ci uptake into the cell and to allow the CO_2 to be accumulated at the catalytic site of Rubisco. The CCMs of cyanobacteria and green microalgae are well characterized, while less is known about CCMs of other microalgal groups, although it is thought that nearly all the phytoplanktonic species can operate a CCM.

The manifold strategies evolved by these photosynthetic microorganisms to cope with environmental hindrances, such as excess light and limited carbon availability, allow them to account for about 50% of the primary productivity of the whole biosphere.

ACKNOWLEDGMENTS

The authors thank Matteo Simonetti of the Department of Biology, University of Padova, for his precious computer support in the production of the figures.

REFERENCES

1. F. A. Depauw, A. Rogato, M. Rivera d'Alcalá, and A. Falciatore, *J. Exp. Bot.*, *63*: 1575 (2012).
2. C. B. Field, M. J. Behrenfeld, J. T. Randerson, and P. Falkowski, *Science*, *281*: 237 (1998).
3. H. L. MacIntire, T. M. Kana, and R. J. Geider, *Trends Plant Sci.*, *5*: 12 (2000).
4. J. Lavaud, R. F. Strzepek, and P. G. Kroth, *Limnol. Oceanogr.*, *52*: 1188 (2007).
5. S. B. Powles, *Annu. Rev. Plant Physiol.*, *35*: 15 (1984).
6. I. Vass, *Biochim. Biophys. Acta*, *1817*: 209 (2012).
7. C. Triantaphylides, M. Krischke, F. A. Hoeberichts, B. Kras, G. Gresser, M. Havaux, F. Van Breusegem, and M. J. Mueller, *Plant Physiol.*, *148*: 960 (2008).
8. P. Pospíšil, *Biochim. Biophys. Acta*, *1787*: 1151 (2009).
9. A. Krieger-Liszkay, C. Fufezan, and A. Trebst, *Photosynth. Res.*, *98*: 551 (2008).
10. N. Ohnishi, S. I. Allakhverdiev, S. Takahashi, S. Higashi, M. Watanabe, Y. Nishiyama, and N. Murata, *Biochemistry*, *44*: 8494 (2005).
11. L. Sass, C. Spetea, Z. Máté, F. Nagy, and I. Vass, *Photosynth. Res.*, *54*: 55 (1997).
12. S. Takahashi, and M. R. Badger, *Trends Plant Sci.*, *16*: 53 (2011).
13. Y. Nishiyama, S. I. Allakhverdiev, and N. Murata, *Physiol. Plant.*, *142*: 35 (2011).
14. N. Murata, S. I. Allakhverdiev, and Y. Nishiyama, *Biochim. Biophys. Acta*, *1817*: 1127 (2012).
15. K. Sonoike, *Physiol. Plant.*, *142*: 56 (2011).
16. M. Tikkanen, N. R. Mekala, and E.-M. Aro, *Biochim. Biophys. Acta*, *1837*: 210 (2014).
17. D. Kirilovsky, and C. A. Kerfeld, *Biochim. Biophys. Acta*, *1817*: 158 (2012).
18. N. La Rocca, K. Sciuto, A. Meneghesso, I. Moro, N. Rascio, and T. Morosinotto, *Physiol. Plant.*, *153*: 654 (2015).
19. K. K. Niyogi, and T. B. Truong, *Curr. Opin. Plant Biol.*, *16*: 307 (2013).
20. M. Watanabe, and M. Ikeuchi, *Photosynth. Res.*, *116*: 265 (2013).
21. C. Boulay, L. Abasova, C. Six, I. Vass, and D. Kirilovsky, *Biochim. Biophys. Acta*, *1777*: 1344 (2008).

22. A. Wilson, G. Ajlani, J.-M. Verbavatz, I. Vass, C. A. Kerfeld, and D. Kirilovski, *Plant Cell*, *18*: 992 (2006).

23. C. Boulay, A. Wilson, S. D'Haene, and D. Kirilovski, *Proc. Natl. Acad. Sci. U.S.A.*, *107*: 11620 (2010).

24. D. Kirilovski, and C. A. Kerfeld, *Photochem. Photobiol. Sci*, *12*: 1135 (2013).

25. A. Wilson, C. Punginelli, A. Gall, C. Bonetti, M. Alexandre, J.-M. Routaboul, C. H. Kerfeld, R. van Grondelle, B. Robert, J. T. M. Kennis, and D. Kirilovski, *Proc. Natl. Acad. Sci. U.S.A.*, *105*: 12075 (2008).

26. L. Tian, I. K. M. van Stokkum, R. B. M. Koehorst, A. Jongerius, D. Kirilovski, and H. Van Amerongen, *J. Am. Chem. Soc.*, *133*: 18304 (2011).

27. R. Berera, I. H. M. van Stokkum, M. Gwizdala, A. Wilson, D. Kirilovski, and R. Drondelle, *J. Phys. Chem. B*, *116*: 2568 (2012).

28. L. Tian, I. K. M. van Stokkum, R. B. M. Koehorst, and H. Van Amerongen, *J. Phys. Chem. B*, *117*: 11000 (2013).

29. M. Sutter, A. Wilson, R. L. Leverenz, R. Lopez-Igual, A. Thurotte, A. Salmeen, D. Kirilovski, and C. A. Kerfeld, *Proc. Natl. Acad. Sci. U.S.A.*, *110*: 10022 (2013).

30. C. Funk, and W. Vermaas, *Biochemistry*, *38*: 9397 (1999).

31. K. Promnares, J. Komenda, L. Bumba, J. Nebesarova, and F. Vacha, *J. Biol. Chem.*, *281*: 32705 (2006).

32. G. Kufryk, M. A. Hernandez-Prieto, T. Kieselbach, H. Miranda, W. Vermaas, and C. Funk, *Photosynth. Res.*, *95*: 135 (2008).

33. Q. He, N. Dolganov, O. Björkman, and A. R. Grossman, *J. Biol. Chem.*, *276*: 306 (2001).

34. A. K. Singh, L. M. McIntyre, and L. A. Sherman, *Plant Physiol.*, *132*: 1825 (2003).

35. D. Bhaya, A. Dufresne, D. Vaulot, and A. R. Grossman, *FEMS Microbiol. Lett.*, *215*: 209 (2002).

36. O. Kilian, A. S. Steunou, A. R. Grossman, and D. Bhaya, *Mol. Plant*, *1*: 155 (2008).

37. M. Havaux, G. Guedeney, Q. He, and A. R. Grossman, *Biochim. Biophys. Acta*, *1557*: 21 (2003).

38. H. Xu, D. Vavilin, C. Funk, and W. Vermaas, *J. Biol. Chem.*, *279*: 27971 (2004).

39. B. R. Green, in *Light-Harvesting Antennas in Photosynthesis* (B. R. Green, and W. W. Parson, eds.), Kluwer Academic Publishers, Dordrecht, 2003, p. 129.

40. J. W. Murray, J. Duncan, and J. Barber, *Trends Plant Sci.*, *11*: 152 (2006).

41. N. Yeremenko, R. Kouřil, J. A. Ihalainen, S. D'Haene, N. van Oosterwijk, E. G. Andrizhiyevskaya, W. Keegstra, H. L. Dekker, M. Hagemann, H. J. Boekema, H. C. P. Matthijs, and J. P. Dekker, *Biochemistry*, *43*: 10308 (2004).

42. M. Havaux, G. Guedeney, M. Hagemann, N. Yeremenko, H. C. P. Matthijs, and R. Jeanjean, *FEBS Lett.*, *579*: 2289 (2005).

43. T. S. Bibby, J. Nield, and J. Barber, *Nature*, *412*: 743 (2001).

44. E. G. Andrizhiyevskaya, T. M. E. Schwabe, M. Germano, S. D'Haene, J. Kruip, R. van Grondelle, and J. P. Dekker, *Biochim. Biophys. Acta*, *1556*: 265 (2002).

45. A. Wilson, C. Boulay, A. Wilde, C. A. Kerfeld, and D. Kirilovski, *Plant Cell*, *19*: 656 (2007).

46. J. A. Ihalainen, S. D'Haene, N. Yeremenko, H. van Roon, A. A. Arteni, E. J. Boekema, R. van Grondelle, H. C. P. Matthijs, and J. P. Dekker, *Biochemistry*, *44*: 10846 (2005).

47. M. Sarcina, and C. W. Mullineaux, *J. Biol. Chem.*, *279*: 36514 (2004).

48. S. Sandström, Y. I. Park, G. Öquist, and P. Gustafsson, *Photochem. Photobiol.*, *74*: 431 (2001).

49. R. Berera, I. H. M. Van Stokkum, S. D'Haene, J. T. M. Kennis, R. van Grondelle, and J. P. Dekker, *Biophys. J.*, *96*: 2261 (2009).

50. S. Joshua, S. Bailey, N. H. Mann, and C. W. Mullineaux, *Plant Physiol.*, *138*: 1577 (2005).

51. M. Chen, and T. S. Bibby, *Photosynth. Res.*, *86*: 165 (2005).

52. T. S. Bibby, I. Mary, J. Nield, F. Partensky, and J. Barber, *Nature*, *424*: 1051 (2003).

53. T. S. Bibby, J. Nield, M. Chen, A. W. D. Larkum, and J. Barber, *Proc. Natl. Acad. Sci. U.S.A.*, *100*: 9050 (2003).

54. F. Partensky, and L. Garczarek, in *Photosynthesis in Algae* (A. W. D. Larkum, S. E. Douglas, and J. A. Raven, eds.), Kluwer Academic Publishers, Dordrecht, 2003, p. 29.

55. V. A. Boichenco, A. V. Pinevich, and I. N. Stadnichuk, *Biochim. Biophys. Acta*, *1767*: 801 (2007).

56. M. Durchan, M. Herbstová, M. Fuciman, Z. Gardian, F. Vácha, and T. Polívka, *J. Phys. Chem.*, *114*: 9275 (2010).

57. G. Peers, T. B. Truong, E. Ostendorf, A. Busch, D. Elrad, A. R. Grossman, M. Hippler, and K. K. Niyogi, *Nature*, *462*: 518 (2009).

58. B. Lepetit, S. Sturm, A. Rogato, A. Gruber, M. Sachse, A. Falciatore, P. G. Knoth, and J. Lavaud, *Plant Physiol.*, *161*: 853 (2013).

59. P. J. Keeling, *J. Eukaryot. Microbiol.*, *56*: 1 (2009).

60. A. Moustafa, B. Beszteri, U. G. Mayer, C. Bowler, K. Valentin, and D. Bhattacharia, *Science*, *324*: 1724 (2009).

61. S.-H. Zhu, and B. R. Green, *Biochim. Biophys. Acta*, *1797*: 1449 (2010).

62. G. Bonente, M. Ballottari, T. B. Truong, T. Morosinotto, T. K. Ahn, G. R. Fleming, K. K. Niyogi, and R. Bassi, *PLoS Biol.*, *9*: e1000577 (2011).

63. S. Mou, X. Zhang, N. Ye, M. Dong, C. Liang, J. Miao, D. Xu, and Z. Zheng, *Extremophiles*, *16*: 193 (2012).

64. G. Allorent, R. Tokutsu, T. Roach, G. Peers, P. Gardol, J. Girard-Bascou, D. Seigneurin-Berny, D. Petroutsous, M. Kuntz, C. Breyton, F. Franck, F.-A. Wollman, K. K. Niyogi, A. Krieger-Liszkay, J. Minagawa, and G. Finazzi, *Plant Cell*, *25*: 545 (2013).

65. P. Ferrante, M. Ballottari, G. Bonente, G. Giuliano, R. Bassi, *J. Biol. Chem.*, *287*: 16276 (2012).

66. B. Bailleul, A. Rogato, A. de Martino, S. Coesel, P. Cardol, C. Bowler, A. Falciatore, and G. Finazzi, *Proc. Natl. Acad. Sci. U.S.A.*, *107*: 18214 (2010).

67. B. Lepetit, R. Goss, T. Jakob, and C. Wilhelm, *Photosynth. Res.*, *111*: 245 (2012).

68. R. Goss, and T. Jakob, *Photosynth. Res.*, *106*: 103 (2010).

69. E. E. Pfündel, M. Renganathan, A. M. Gilmore, H. Y. Yamamoto, and R. A. Dilley, *Plant Physiol.*, *106*: 1647 (1994).

70. C. E. Bratt, P. O. Arvidsson, M. Carlsson, and H. E. Akerlund, *Photosynth. Res.*, *45*: 169 (1995).

71. K. Büch, H. Stransky, and A. Hager, *FEBS Lett.*, *376*: 45 (1995).

72. N. E. Holt, D. Zigmantas, L. Valkunas, X. P. Li, K. K. Niyogi, and G. R. Fleming, *Science*, *307*: 433 (2005).

73. R. Goss, K. Böhme, and C. Wilhelm, *Planta*, *205*: 613 (1998).

74. J. Masojídek, J. Kopeský, M. Koblízek, and G. Torzillo, *Plant Biol.*, *6*: 342 (2004).

75. A. V. Ruban, J. Lavaud, B. Rousseau, G. Guglielmi, P. Horton, and A.-L. Etienne, *Photosynth. Res.*, *82*: 165 (2004).

76. C. Brunet, and J. Lavaud, *J. Plank. Res.*, *32*: 1609 (2010).

77. A. Shumann, R. Goss, T. Jakob, and C. Wilhelm, *Phycologia*, *46*: 113 (2007).

78. T. Jakob, R. Goss, and C. Wilhelm, *J. Plant Physiol.*, *158*: 383 (2001).

79. R. Goss, E. A. Pinto, C. Wilhelm, and M. Richter, *J. Plant Physiol.*, *163*: 1008 (2006).

80. N. Schubert, E. García-Mendoza, and I. Pacheco-Ruiz, *J. Phycol.*, *42*: 1208 (2006).

81. L.-N. Liu, A. T. Elmalk, T. J. Aartsma, J.-C. Thomas, G. E. M. Lamers, B.-C. Zhou, and Y.-Z. Zhang, *PLoS One*, *3*: e3134 (2008).

82. L.-N. Liu, T. J. Aartsma, J.-C. Thomas, B.-C. Zhou, and Y.-Z. Zhang, *PLoS One*, *4*: e5295 (2009).

83. H. Ishikita, and E.-W. Knapp, *J. Am. Chem. Soc.*, *127*: 14714 (2005).

84. C. Fufezan, M. C. Gross, M. Sjödin, A. W. Rutherford, A. Krieger-Liszkay, and D. Kirilowsky, *J. Biol. Chem.*, *282*: 12492 (2007).

85. A. G. Ivanov, P. V. Sane, V. Hurry, G. Öquist, and N. P. A. Huner, *Photosynth. Res.*, *98*: 565 (2008).

86. J. Sander, M. Nowaczyk, J. Buchta, H. Dau, I. Vass, Z. Deák, M. Dorogi, M. Iwai, and M. Rögner, *J. Biol. Chem.*, *285*: 29851 (2010).

87. P. B. Kós, Z. Deák, O. Cheregi, and I. Vass, *Biochim. Biophys. Acta*, *1777*: 74 (2008).

88. P. Mulo, C. Sicora, and E.-M. Aro, *Cell. Mol. Life Sci.*, *66*: 3687 (2009).

89. K. Cser, and I. Vass, *Biochim. Biophys. Acta*, *1767*: 233 (2007).

90. M. Sugiura, C. Azami, K. Koyama, A. W. Rutherford, F. Rappaport, and A. Boussac, *Biochim. Biophys. Acta*, *1837*: 139 (2014).

91. A. Krieger-Liszkay, and A. Trebst, *J. Exp. Bot.*, *57*: 1677 (2006).

92. J. Kruk, and A. Trebst, *Biochim. Biophys. Acta*, *1777*: 154 (2008).

93. O. Prasil, Z. Kolber, J. A. Berry, and P. Falkowski, *Photosynth. Res.*, *48*: 395 (1996).

94. C. Miyake, and A. Yokota, *Plant Cell Physiol.*, *42*: 508 (2001).

95. P. Pospíšil, *J. Photochem. Photobiol. B: Biol.*, *104*: 341 (2011).

96. C. Miyake, K. Kuniaki, Y. Kobayashi, and A. Yokota, *Plant Cell Physiol.*, *43*: 951 (2002).

97. K. E. Shinopoulos, and G. W. Brudvig, *Biochim. Biophys. Acta*, *1817*: 66 (2012).

98. C. A. Tracewell, and G. W. Brudvig, *Biochemistry*, *47*: 11559 (2008).

99. P. Zhang, Y. Allahverdyieva, M. Eisenhut, and E.-M. Aro, *PLoS One*, *4*: e5331 (2009).

100. P. Zhang, M. Eisenhut, A. M. Brandt, D. Carmel, H. M. Silén, I. Vass, Y. Allahverdyieva, T. A. Salminen, and E.-M. Aro, *Plant Cell*, *24*: 1952 (2012).

101. Y. Allahverdyieva, M. Ermakova, M. Eisenhut, P. Zhang, P. Richaud, M. Hagemann, L. Courmac, and E.-M. Aro, *J. Biol. Chem.*, *286*: 24007 (2011).

102. Y. Allahverdyieva, H. Mustila, M. Ermakova, L. Bersanini, P. Richaud, G. Ajlani, N. Battchikova, L. Cournac, and E. M. Aro, *Proc. Natl. Acad. Sci. U.S.A.*, *110*: 4111 (2013).

103. L. Bersanini, N. Battchikova, M. Jokel, A. Rehman, I. Vass, Y. Allahverdyieva, and E.-M. Aro, *Plant Physiol.*, *164*: 805 (2014).

104. J. B. Vicente, M. C. Justino, V. L. Conçalves, M. Saraiva, and M. Teixeira, *Methods Enzymol.*, *437*: 21 (2008).

105. C. Frazão, G. Silva, C. M. Gomes, P. Matias, R. Coehlo, L. Sieker, S. Macedo, M. Y. Liu, S. Oliveira, M. Teixeira, A. V. Xavier, C. Rodriguez-Pousada, M. A. Carrondo, and J. Le Gall, *Nat. Struct. Biol.*, *7*: 1041 (2000).

106. Y. Helman, D. Tchernov, L. Reinhold, M. Shibata, T. Ogawa, R. Schwarz, I. Ohad, and A. Kaplan, *Curr. Biol.*, *13*: 230 (2003).

107. J. Appel, S. Phunpruch, K. Steinmüller, and R. Schulz, *Arch. Microbiol.*, *173*: 333 (2000).

108. K. Gutekunst, X. Chen, K. Schreiber, U. Kaspar, S. Makam, and J. Appel, *J. Biol. Chem.*, *289*: 1930 (2014).

109. D. J. Lea-Smith, N. Ross, M. Zori, D. S. Bendall, J. S. Dennis, S. A. Scott, A. G. Smith, and C. J. Howe, *Plant Physiol.*, *162*: 484 (2013).

110. C. W. Mullineaux, *Front. Plant Sci.*, *5*: 7 (2014).

111. S. Bailey, A. Melis, K. R. M. Mackey, P. Cardol, G. Finazzi, G. van Dijken, G. M. Berg, K. Arrigo, J. Shrager, and A. Grossman, *Biochim. Biophys. Acta*, *1777*: 269 (2008).

112. A. E. McDonald, A. G. Ivanov, R. Bode, D. P. Maxwell, S. R. Rodermel, and N. P. Hüner, *Biochim. Biophys. Acta*, *1807*: 954 (2011).

113. P. Cardol, B. Bailleul, F. Rappaport, E. Derelle, D. Béal, C. Breyton, S. Bailey, F. A. Wollman, A. Grossman, H. Moreau, and G. Finazzi, *Proc. Natl. Acad. Sci. U.S.A.*, *105*: 7881 (2008).

114. G. Peltier, and L. Cournac, *Annu. Rev. Plant Biol.*, *53*: 523 (2002).

115. C. Desplats, F. Mus, S. Cuiné, E. Billon, L. Cournac, and G. Peltier, *J. Biol. Chem.*, *284*: 4148 (2009).

116. D. Rumeau, G. Peltier, and L. Cournac, *Plant Cell Environ.*, *30*: 1041 (2007).

117. S. Cruz, R. Goss, C. Wilhelm, R. Leegood, P. Horton, and T. Jakob, *J. Exp. Bot.*, *62*: 509 (2010).

118. I. Grouneva, T. Jakob, C. Wilhelm, and R. Goss, *Biochim. Biophys. Acta*, *1787*: 929 (2009).

119. S. T. Stripp, and T. Happe, *Dalton Trans.*, *45*: 9960 (2009).

120. A. Hemschemeier, A. Melis, and T. Happe, *Photosynth. Res.*, *102*: 523 (2009).

121. K. Asada, *Annu. Rev. Plant Physiol. Plant Mol. Biol*, *50*: 601 (1999).

122. D. R. Ort, and N. R. Baker, *Curr. Opin. Plant Biol.*, *5*: 193 (2002).

123. F. Wolfe-Simon, D. Grzebyk, O. Schofield, and P. G. Falkowski, *J. Phycol.*, *41*: 453 (2005).

124. B. Pryia, J. Premanandh, R. T. Dhanalakshmi, T. Seethalakshmi, I. Uma, D. Prabaharan, and G. Subramaniam, *BMC Genomics*, *8*: 435 (2007).

125. C. Miyake, and K. Asada, in *Photosynthesis in Algae* (A. W. D. Larkum, S. E. Douglas, and J. A. Raven, eds.), Kluwer Academic Publishers, Dordrecht, 2003, p. 183.

126. T. Stork, K.-P. Michel, E. K. Pistorius, and K.-J. Dietz, *J. Exp. Bot.*, *56*: 3193 (2005).

127. B. N. Tripathi, I. Bhatt, and K.-J. Dietz, *Protoplasma*, *235*: 3–15 (2009).

128. G. Peltier, D. Tolleter, E. Billon, and L. Cournac, *Photosynth. Res.*, *106*: 19 (2010).

129. J. F. Allen, *Trends Plant Sci.*, *8*: 15 (2003).

130. C. Hackenberg, A. Engelhardt, H. C. P. Matthijs, F. Wittink, H. Bauwe, A. Kaplan, and M. Hagemann, *Planta*, *230*: 625 (2009).

131. C. Miyake, *Plant Cell Physiol.*, *51*: 1951 (2010).

132. J. F. Allen, *Science*, *299*: 1530 (2003).

133. S. Lemeille, and J. D. Rochaix, *Photosynth. Res.*, *106*: 33 (2010).

134. C. W. Mullineaux, and E.-J. Emlyn-Jones, *J. Exp. Bot.*, *56*: 389 (2005).

135. S. Yang, R. Zhang, C. Hu, J. Xie, and J. Zhao, *Photosynth. Res.*, *99*: 99 (2009).

136. X. Xu, S. Yang, H. Xie, and J. Zhao, *Biochim. Biophys. Res. Comm.*, *422*: 233 (2012).

137. J. F. Allen, S. Santabarbara, C. A. Allen, and S. Puthiyaveetil, *PLoS One*, *6*: e26372 (2011).

138. A. de Lacroix de Lavallette, G. Finazzi, and F. Zito, *Biochemistry, 47*: 5259 (2008).

139. C. W. Mullineaux, *Photosynth. Res., 95*: 175 (2008).

140. K. Kondo, C. W. Mullineaux, and M. Ikeuchi, *Photosynth. Res., 99*: 217 (2009).

141. S. Joshua, and C. W. Mullineaux, *Biochim. Biophys. Acta, 1709*: 58 (2005).

142. T. Fujimori, Y. Hihara, and K. Sonoike, *J. Biol. Chem., 280*: 22191 (2005).

143. S. Federman, S. Malkin, and A. Scherz, *Photosynth. Res., 64*: 199 (2000).

144. H. Li, D. Li, S. Yang, J. Xie, and J. Zhao, *Biochim. Biophys. Acta, 1757*: 1512 (2006).

145. R. Zhang, J. Xie, and J. Zhao, *Photosynth. Res., 99*: 107 (2009).

146. H. Liu, H. Zhang, D. M. Niedzwiedzki, M. Prado, G. He, M. L. Gross, and R. E. Blankenship, *Science, 342*: 1104 (2013).

147. J. Biggins, and D. Bruce, *Photosynth. Res., 20*: 1 (1989).

148. M. Yokono, A. Murakami, and S. Akimoto, *Biochim. Biophys. Acta, 1807*: 847 (2011).

149. A. Koziol, T. Borza, K.-I. Ishida, P. Keeling, R. W. Lee, and D. G. Durnford, *Plant Physiol., 143*: 1802 (2007).

150. J. Engelken, H. Brinkmann, and I. Adamska, *BMC Evol. Biol., 10*: 233 (2010).

151. S. Lemeille, A. Willig, N. Depège-Fargeix, C. Delessert, R. Bassi, and J. D. Rochaix, *PLoS Biol., 7*: e45 (2009).

152. F. Zito, G. Finazzi, R. Delosme, W. Nitschke, D. Picot, and F.-A. Wollman., *EMBO J., 18*: 2961 (1999).

153. N. Nelson, and A. Ben-Shem, *Nat. Rev., 5*: 1 (2004).

154. S. Lemeille, M. V. Turkina, A. V. Vener, and J.-D. Rochaix, *Mol. Cell. Proteom., 9(6)*: 1281 (2010).

155. J.-D. Rochaix, S. Lemeille, A. Shapiguzov, I. Samol., G. Fucile, A. Willig, and M. Goldschidt-Clermont, *Phil. Trans. R. Soc. B., 367*: 3466 (2012).

156. H. Takahashi, M. Iwai, Y. Takahashi, and J. Minagawa, *Proc. Natl. Acad. Sci. U.S.A., 103*: 477 (2006).

157. M. Iwai, M. Yokono, N. Inada, and J. Minagawa, *Proc. Natl. Acad. Sci. U.S.A., 107*: 2337 (2010).

158. R. Tokutsu, M. Iwai, and J. Minagawa, *J. Biol. Chem., 284*: 7777 (2009).

159. M. Iwai, Y. Takahashi, and J. Minagawa, *Plant Cell, 20*: 2177 (2008).

160. C. Ünlü, B. Drop, R. Croce, and H. van Amerongen, *Proc. Natl. Acad. Sci. U.S.A., 111*: 3460 (2014).

161. G. Nagy, R. Ünnep, O. Zsiros, R. Tokutsu, K. Takizawa, L. Porcar, L. Moyet, D. Petroutsos, G. Garab, G. Finazzi, and J. Minagawa, *Proc. Natl. Acad. Sci. U.S.A., 111*: 5042 (2014).

162. M. Grieco, M. Tikkanen, V. Paakkarinen, S. Kangasjärvi, and E.-M. Aro, *Plant Physiol., 160*: 1896 (2012).

163. P. Cardol, J. Alric, J. Girard-Bascou, F. Franck, F.-A. Wollman, and G. Finazzi, *Proc. Natl. Acad. Sci. U.S.A., 106*: 15979 (2009).

164. M. Iwai, K. Takizawa, R. Tokutsu, A. Okamuro, Y. Takahashi, and J. Minagawa, *Nature, 464*: 1210 (2010).

165. P. J. Nixon, F. Michoux, J. Yu, M. Bohem, and J. Komenda, *Ann. Bot., 106*: 1 (2010).

166. J. Komenda, R. Sobotka, and P. J. Nixon, *Curr. Opin. Plant Biol., 15*: 245 (2012).

167. O. Cheregi, C. Sicona, P. B. Kós, M. Barker, P. J. Nixon, and I. Vass, *Biochim. Biophys. Acta, 1767*: 829 (2007).

168. M. Edelman, and A. K. Matoo, *Photosynth. Res., 98*: 609 (2008).

169. Z. Adam, A. Rudella, and K. J. Van Wijk, *Curr. Opin. Plant Biol., 9*: 234 (2006).

170. P. F. Huesgen, H. Schuhmann, and I. Adamska, *Physiol. Plant., 123*: 413 (2005).

171. D. A. Campbell, Z. Hossain, A. M. Cockshutt, O. Zhaxybayeva, H. Wu, and G. Li, *Photosynth. Res., 115*: 43 (2013).

172. K. Nath, A. Jajoo, R. S. Poudyal, R. Timilsina, Y. S. Park, E.-M. Aro, H. G. Nam, and C.-H. Lee, *FEBS Lett., 587*: 3372 (2013).

173. P. Mulo, S. Sirpiö, M. Suorsa, and E.-M. Aro, *Photosynth. Res., 98*: 489 (2008).

174. J. Uniacke, and W. Zerges, *Plant Cell, 19*: 3640 (2007).

175. R. K. Sihna, J. Komenda, J. Knoppova, M. Sedlarova, and P. Pospíšil, *Plant Cell Environ., 35*: 806 (2012).

176. D. Vavilin, D. Yao, and W. Vermaas, *J. Biol. Chem., 282*: 37660 (2007).

177. H. Teramoto, A. Ishii, Y. Kimura, K. Hasegawa, S. Nakazawa, T. Makamura, S. Higashi, M. Watanabe, and T. Ono, *Plant Cell Physiol., 47*: 419 (2006).

178. K. Apel, and H. Hirt, *Annu. Rev. Plant Biol., 55*: 373 (2004).

179. E.-M. Aro, I. Virgin, and B. Andersson, *Biochim. Biophys. Acta, 1443*: 113 (1993).

180. J. M. Andersen, and W. S. Chow, *Philos. Trans., R. Soc. Lond. B Biol. Sci., 357*: 1421 (2002).

181. S. Takahashi, and N. Murata, *Trends Plant Sci., 13*: 178 (2008).

182. G. D. Price, M. R. Badger, F. J. Woodger, and B. M. Long, *J. Exp. Bot., 59*: 1441 (2008).

183. M. R. Badger, G. D. Price, B. M. Long, and F. J. Woodger, *J. Exp. Bot., 57*: 249 (2006).

184. M. R. Badger, and G. D. Price, *J. Exp. Bot., 54*: 602 (2003).

185. M. van Lun, J. S. Hub, D. van der Spoel, and I. Andersson, *J. Am. Chem. Soc, 136*: 3165 (2014).

186. W. W. Cleland, T. J. Andrews, S. Gutteridge, F. C. Hartman, and G. H. Lorimer, *Chem. Rev., 98*: 549 (1998).

187. F. R. Tabita, *Photosynth. Res., 60*: 1 (1999).

188. F. R. Tabita, S. Satagopan, T. E. Hanson, N. E. Kreel, and S. S. Scott, *J. Exp. Bot., 59*: 1515 (2008).

189. R. J. Spreitzer, and M. E. Salvucci, *Annu. Rev. Plant Biol., 53*: 449 (2002).

190. G. G. Tcherkez, D. G. Farquhar, and T. J. Andrews, *Proc. Natl. Acad. Sci. U.S.A., 103*: 7246 (2006).

191. S. M. Whitney, and T. J. Andrews, *Aust. J. Plant Physiol., 25*: 131 (1998).

192. H. Bauwe, M. Hagemann, and A. R. Fernie, *Trends Plant Sci., 15*: 330 (2010).

193. M. Eisenhut, W. Ruth, M. Haimovich, H. Bauwe, A. Kaplan, and M. Hagemann, *Proc. Natl. Acad. Sci. U.S.A., 105*: 17199 (2008).

194. R. A. Berner, *Geochim. Cosmochim. Acta, 70*: 5653 (2006).

195. B. Rasmussen, I. R. Fletcher, J. J. Brocks, and M. R. Kilburn, *Nature, 455*: 1101 (2008).

196. G. D. Price, *Photosynth. Res., 109*: 47 (2011).

197. M. Giordano, J. Beardall, and J. A. Raven, *Annu. Rev. Plant Biol., 56*: 99 (2005).

198. E. V. Kupriyanova, M. A. Sinetova, S. M. Cho, Y.-I. Park, D. A. Los, and N. A. Pronina, *Photosynth. Res., 117*: 133 (2013).

199. M. R. Badger, T. J. Andrews, M. S. Whitney, M. Ludwig, D. C. Yellowlees, W. Leggat, and G. D. Price, *Can. J. Bot., 76*: 1052 (1998).

200. J. A. Raven, and W. J. Lucas, in *Inorganic Carbon Uptake by Aquatic Photosynthetic Organisms* (W. J. Lucas, and J. A. Berry, eds.), American Society of Plant Physiologist, Rockville, MD, 1985, p. 305.

201. M. R. Badger, D. Hanson, and G. D. Price, *Funct. Plant Biol., 29*: 161 (2002).

202. B. D. Rae, B. M. Long, M. R. Badger, and G. D. Price, *Microbiol. Mol. Biol. Rev., 77*: 357 (2013).

203. G. C. Cannon, and J. M. Shively, *Arch. Microbiol.*, *134*: 52 (1983).

204. G. C. Cannon, C. E. Bradburne, H. C. Aldrich, S. H. Baker, S. Heinhorst, and J. M. Shively, *Appl. Environ. Microbiol.*, *67*: 5351 (2001).

205. H. B. Liu, H. A. Nolla, and L. Campbell, *Aquat. Microbiol. Ecol.*, *12*: 39 (1997).

206. F. Partensky, W. R. Hess, and D. Vaulot, *Microbiol. Mol. Biol. Rev.*, *63*: 106 (1999).

207. A. Kaplan, and L. Reinhold, *Annu. Rev. Plant Physiol. Plant Mol. Biol.*, *50*: 539 (1999).

208. P. J. McGinn, G. D. Price, R. Maleszka, and M. R. Badger, *Plant Physiol.*, *132*: 218 (2003).

209. F. J. Woodger, M. R. Badger, and G. D. Price, *Plant Physiol.*, *133*: 2069 (2003).

210. F. J. Woodger, M. R. Badger, and G. D. Price, *Can. J. Bot.*, *83*: 698 (2005).

211. P. P. Zhang, N. Battchikova, T. Jansen, J. Appel, T. Ogawa, and E.-M. Aro, *Plant Cell*, *16*: 3326 (2004).

212. D. Sültemeyer, B., Klughammer, M. R. Badger, and G. D. Price, *Can. J. Bot.*, *76*: 954 (1998).

213. C. F. Higgins, *Res. Microbiol.*, *152*: 205 (2001).

214. T. Omata, G. D. Price, M. R. Badger, M. Okamura, S. Gohta, and T. Ogawa, *Proc. Natl. Acad. Sci. U.S.A.*, *96*: 13571 (1999).

215. T. Omata, Y. Takahashi, O. Yamaguchi, and T. Nishimura, *Funct. Plant Biol.*, *29*: 151 (2002).

216. H. L. Wang, B. L. Postier, and R. L. Burnap, *J. Biol. Chem.*, *279*: 5739 (2004).

217. F. J. Woodger, M. R. Badger, and G. D. Price, *Plant Physiol.*, *139*: 1959 (2005).

218. D. A. Bryant, *Proc. Natl. Acad. Sci. U.S.A.*, *100*: 9647 (2003).

219. M. Shibata, H. Katoh, M. Sonoda, H. Ohkawa, M., Shimoyama, H. Fukuzawa, A. Kaplan, and T. Ogawa, *J. Biol. Chem.*, *277*: 18658 (2002).

220. G. D. Price, F. J. Woodger, M. R. Badger, S. M. Howitt, and L. Tucker, *Proc. Natl. Acad. Sci. U.S.A.*, *101*: 18228 (2004).

221. M. Billini, K. Stamatakis, and V. Sophianopoulou, *J. Bacteriol.*, *190*: 6318 (2008).

222. G. S. Espie, and R. A. Kandasamy, *Plant Physiol.*, *104*: 1419 (1994).

223. R. Waditee, T. Hibino, T. Nakamura, A. Incharoensakdi, and T. Takabe, *Proc. Natl. Acad. Sci. U.S.A.*, *99*: 4109 (2002).

224. H. Ohkawa, H. B. Pakrasi, and T. Ogawa, *J. Biol. Chem.*, *275*: 31630 (2000).

225. M., Shibata, H. Ohkawa, T. Kaneko, H. Fukuzawa, S. Tabata, A. Kaplan, and T. Ogawa, *Proc. Natl. Acad. Sci. U.S.A.*, *98*: 11789 (2001).

226. S. Maeda, M. R. Badger, and G. D. Price, *Mol. Microbiol.*, *43*: 425 (2002).

227. T. Ogawa, and H. Mi, *Photosynth. Res.*, *93*: 69 (2007).

228. N. Battchikova, M. Eisenhut, and E. M. Aro, *Biochim. Biophys. Acta*, *1082*: 935 (2011).

229. D. Tchernov, Y. Helman, N. Keren, B. Luz, I. Ohad, L. Reinhold, T. Ogawa, and A. Kaplan, *J. Biol. Chem.*, *276*: 23450 (2001).

230. I. M. Folea, P. Zhang, M. M. Novaczyk, E.-M. Aro, and E. J. Boekema, *FEBS Lett.*, *582*: 249 (2008).

231. M. Xu, T. Ogawa, H. B. Pakrasi, and H. Mi, *Plant Cell Physiol.*, *49*: 994 (2008).

232. G. D. Price, S. Maeda, T. Omata, and M. R. Badger, *Funct. Plant Biol.*, *29*: 131 (2002).

233. M. Hassidim, N. Keren, I. Ohad, L. Reinhold, and A. Kaplan, *J. Phycol.*, *33*: 811 (1997).

234. K. A. Palinska, W. Laloui, S. Bédu, S. Loiseaux-de Goer, A. M. Castets, R. Rippka, and N. Tandeau de Marsac, *Microbiology*, *148*: 2405 (2002).

235. F. J. Woodger, D. A. Bryant, and G. D. Price, *J. Bacteriol.*, *189*: 3335 (2007).

236. T. Omata, and T. Ogawa, *Plant Physiol.*, *80*: 525 (1986).

237. J. W. Yu, G. B. Price, and M. R. Badger, *Austr. J. Plant Physiol.*, *21*: 185 (1994).

238. T. Omata, S. Gohta, Y. Takahashi, Y. Harano, and S. Maeda, *J. Bacteriol.*, *183*: 1891 (2001).

239. T. Nishimura, Y. Takahashi, O. Yamaguchi, H. Suzuki, S. Maeda, and T. Omata, *Mol. Microbiol.*, *68*: 98 (2008).

240. S. M. Daley, A. D. Cappell, M. J. Carrick, and R. L. Burnap, *PLoS One*, *7*: e41286 (2012).

241. R. L. Burnap, R. Nambudiri, and S. Holland, *Photosynth. Res.*, *118*: 115 (2013).

242. R. M. Figge, C. Cassier- Chauvat, F. Chauvat, and R. Cerff, *Mol. Microbiol.*, *39*: 455 (2001).

243. A. Kaplan, M. R. Badger, and J. A. Berry, *Planta 149*: 219 (1980).

244. C. V. Iancu, H. J. Ding, D. M. Morris, D. P. Dias, A. D. Gonzales, A. Martino, and G. J. Jensen, *J. Mol. Biol.*, *372*: 764 (2007).

245. B. M. Long, M. R. Badger, S. M. Whitney, and G. D. Price, *J. Biol. Chem.*, *282*: 29323 (2007).

246. T. O. Yeates, C. A. Kerfeld, S. Heinhorst, G. C. Cannon, and J. M. Shively, *Nat. Rev. Microbiol.*, *6*: 681 (2008).

247. S. Tanaka, M. R. Sawaya, M. Phillips, and T. O. Yeates, *Prot. Sci.*, *18*: 108 (2009).

248. G. C. Cannon, S. Heinhorst, and C. A. Kerfeld, *Biochim. Biophys. Acta*, *1804*: 382 (2010).

249. M. R. Sawaya, G. C. Cannon, S. Heinhorst, S. Tanaka, E. B. Williams, T. O. Yeates, and C. A. Kerfeld, *J. Biol. Chem.*, *281*: 7546 (2006).

250. C. A. Kerfeld, M. R. Sawaya, S. Tanaka, C. V. Nguyen, M. Phillis, M. Beeby, and T. O. Yeates, *Science*, *309*: 936 (2005).

251. Y. Tsai, M. R. Sawaya, G. C. Cannon, F. Cai, E. B. Williams, S. Heinhorst, C. A. Kerfeld, and T. O. Yeates, *PLoS Biol.*, *5*: 1345 (2007).

252. J. N. Kinney, S. D. Axen, and C. A. Kerfeld, *Photosynth. Res.*, *109*: 21 (2011).

253. G. S. Espie, and M. S. Kimber, *Photosynth. Res.*, *109*: 7 (2011).

254. S. Tanaka, C. A. Kerfeld, M. R. Sawaya, F. Cai, S. Heinhorst, G. C. Cannon, and T. O. Yeates, *Science*, *319*: 1083 (2008).

255. S. S.-W. Cot, A. K.-C. So, and G. S. Espie, *J. Bacteriol.*, *190*: 936 (2008).

256. B. M. Long, L. Tucker, M. R. Badger, and G. D. Price, *Plant Physiol.*, *153*: 285 (2010).

257. K. L. Peña, S. E. Castel, C. De Araujo, G. S. Espie, and M. S. Kimber, *Proc. Natl. Acad. Sci. U.S.A.*, *107*: 2455 (2010).

258. D. F. Savage, B. Afonso, A. H. Chen, and P. A. Silver, *Science*, *327*: 1258 (2010).

259. M. H. Spanding, *J. Exp. Bot.*, *59*: 1463 (2008).

260. J. A. Raven, C. S. Cockell, and C. L. De La Rocha, *Philos. Trans. R. Soc. B.*, *363*: 2641 (2008).

261. M. Meyer, and H. Griffiths, *J. Exp. Bot.*, *64*: 769 (2013).

262. J. A. Raven, M. Giordano, J. Beardall, and S. C. Maberly, *Philos. Trans. R. Soc. B.*, *367*: 493 (2012).

263. K. Nakajima, A. Tanaka, and Y. Matsuda, *Proc. Natl. Acad. Sci. U.S.A.*, *110*: 1767 (2013).

264. J. R. Reinfelder, *Annu. Rev. Mar. Sci.*, *3*: 291 (2011).

265. Y. Wang, D. Duanmu, and M. H. Spalding, *Photosynth. Res.*, *109*: 115 (2011).

266. A. Markelova, M. Sinetova, E. Kupriyanova, and N. A. Pronina, *Russ. J. Plant Physiol.*, 56: 761 (2009).

267. J. V. Moroney, and R. A. Ynalvez, *Eukaryot. Cell*, 6: 1251 (2007).

268. J. V. Moroney, Y. Ma, W. D. Frey, K. A. Fulisier, T. T. Pham, T. A. Simms, R. J. DiMario, J. Yang, and B. Mukherjee, *Photosynth. Res.*, 109: 133 (2011).

269. M. A. Sinetova, E. V. Kupriyanova, A. G. Markelova, S. I. Allakhverdier, and N. A. Pronina, *Biochim. Biophys. Acta*, 1817: 1248 (2012).

270. M. R. Badger, A. Kaplan, and J. A. Berry, *Plant Physiol.*, 66: 407 (1980).

271. T. Yamano, T. Tsujikawa, K. Hatano, S. I. Ozawa, Y. Takahashi, and H. Fukuzawa, *Plant Cell Physiol.*, 51: 1453 (2010).

272. H. Fukuzawa, K. Miura, K. Ishizaki, K. I. Kucho, T. Saito, T. Kohinata, and K. Ohyama, *Proc. Natl. Acad. Sci. U.S.A.*, 98: 5347 (2001).

273. Y. Xiang, J. Zhang, and D. P. Weeks, *Proc. Natl. Acad. Sci. U.S.A.*, 98: 5341 (2001).

274. K. Miura, T. Yamano, S. Yoshioka, T. Kohinata, Y. Inoue, F. Taniguchi, E. Asamizu, Y. Nakamura, S. Tabata, K. T. Yamato, K. Ohyama, and H. Fukuzawa, *Plant Physiol.*, 135: 1595 (2004).

275. D. Duanmu, Y. J. Wang, and M. H. Spanding, *Plant Physiol.*, 149: 929 (2009).

276. N. Ohnishi, B. Mukherjee, T. Tsujikawa, M., Yanase, H. Nakano, J. Moroney, and H. Fukuzawa, *Plant Cell*, 22: 3105 (2010).

277. D. Duanmu, A. R. Miller, K. M. Horken, D. P. Weeks, and M. H. Spanding, *Proc. Natl. Acad. Sci. U.S.A.*, 106: 5990 (2009).

278. A. J. Brueggeman, D. S. Gangadharaiah, M. F. Cserhati, D. Casero, D. P. Weeks, and I. Ladunga, *Plant Cell*, 24: 1860 (2012).

279. T. Yamano, and H. Fukuzawa, *J. Basic Microbiol.*, 49: 42 (2009).

280. S. Kustu, and W. Inwood, *Transfus. Clin. Biol.*, 13: 103 (2006).

281. C. Yoshihara, K. Inoue, D. Schichnes, S. Ruzin, W. Inwood, and S. Kutsu, *Mol. Plant*, 1: 1007 (2008).

282. S. Yoshioka, F. Tagiguchi, K. Miura, T. Inoue, T. Yamano, and H. Fukuzawa, *Plant Cell*, 16: 1466 (2004).

283. V. Mariscalc, P. Moulin, M. Orsel, A. J. Miller, E. Fernández, and A. Galván, *Protist*, 157: 421 (2006).

284. S. V. Pollock, D. L. Prout, A. C. Godfrey, S. D. Lemaire, and J. V. Moroney, *Plant Mol. Biol.*, 91: 505 (2004).

285. J. A. Raven, *Photosynth. Res.*, 106: 123 (2010).

286. J. C. Villareal, and S. S. Renner, *Proc. Natl. Acad. Sci. U.S.A.*, 109: 18873 (2012).

287. N. Rascio, G. Casadoro, and C. Andreoli, *Bot. Mar.*, 23: 25 (1980).

288. N. Rascio, G. Casadoro, and C. Andreoli, *Bot. Mar.*, 23: 467 (1980).

289. M. T. Meyer, T. Genkov, J. N. Skepper, J. Jouhet, and M. C. Mitchell, *Proc. Natl. Acad. Sci. U.S.A.*, 109: 19474 (2012).

290. Y. Ma, S. V. Pollock, Y. Xiao, K. Cunnusamy, and J. V. Moroney, *Plant Physiol.*, 156: 884 (2011).

291. S. V. Pollock, S. L. Colombo, D. L. Prout, A. C. Godfrey, and J. V. Moroney, *Plant Physiol.*, 133: 1854 (2003).

292. N. Rascio, G. Casadoro, and C. Andreoli, *Bot. Mar.*, 23: 31 (1980).

293. M. Mitra, C. B. Mason, Y. Xiao, R. A. Ynalvez, S. M. Lato, and J. V. Moroney, *Can. J. Bot.*, 83: 780 (2005).

294. A. Blanco-Rivero, T. Shutova, M. J. Román, A. Villarejo, and F. Martinez, *PLoS One*, 7: e49063 (2012).

295. M. Mitra, S. M. Lato, R. A. Ynalvez, Y. Xiao, and J. V. Moroney, *Plant Physiol.*, 135: 173 (2004).

296. E. Morita, T. Abe, M. Tsuzuki, S. Fujiwara, N. Sato, A. Hirata, K. Sonoike, and H. Nozaki, *Planta*, 204: 269 (1998).

297. P. G. Falkowski, and J. A. Raven, *Aquatic Photosynthesis*, University Press, Princeton, NJ, 2007, pp. 484.

298. P. J. Keeling, *Phil. Trans. R. Soc. B.*, 365: 729 (2010).

299. J. Prihoda, A. Tanaka, W. B. M. de Paula, J. F. Allen, L. Tirichine, and C. Bowler, *J. Exp. Bot.*, 63: 1543 (2012).

300. M. Peschke, D. Moog, A. Klingl, U. G. Maier, and F. Hempel, *Proc. Natl. Acad. Sci. U.S.A.*, 110: 10860 (2013).

301. S. M. Whitney, P. Balder, G. S. Hudson, and T. J. Andrews, *Plant J.*, 26: 535 (2001).

302. U. Riesebell, D. A. Wolf-Gladrow, and V. S. Smetacek, *Nature*, 361: 249 (1993).

303. M. Tachibana, A. E. Allen, S. Kikutani, Y. Endo, C. Bowler, and Y. Matsuda, *Photosynth. Res.*, 109: 205 (2011).

304. M. Samukawa, C. Shen, B. M. Hopkinson, and Y. Marsuda, *Photosynth. Res.*, 121: 235 (2014).

305. Y. Matsuda, T. Hara, and B. Colman, *Plant Cell Environ.*, 24: 611 (2001).

306. Y. Marsuda, K. Nakajima, and M. Tachibana, *Photosynth. Res.*, 109: 191 (2011).

307. N. Ohno, T. Inoue, R. Yamashiki, K. Nakajima, Y. Kitahara, M. Ishibashi, and Y. Matsuda, *Plant Physiol.*, 158: 499 (2012).

308. B. M. Hopkinson, C. L. Dupont, A. E. Allen, and F. M. M. Morel, *Proc. Natl. Acad. Sci. U.S.A.*, 108: 3830 (2011).

309. B. M. Hopkinson, *Photosynth. Res.*, 121: 223 (2014).

310. H. Harada, and Y. Matsuda, *Can. J. Bot.*, 83: 909 (2005).

311. G. E. Edwards, V. R. Franceschi, and E. V. Voznesenskaya, *Annu. Rev. Plant Biol.*, 55: 173 (2004).

312. J. B. Reiskind, and G. Bowes, *Proc. Natl. Acad. Sci. U.S.A.*, 88: 2883 (1991).

313. J. Xu, X. Fan, X. Zhang, D. Xu, S. Mou, S. Cao, Z. Zheng, J. Miao, and N. Ye, *PLoS One*, 7: e37438 (2012).

314. J. R. Reinfelder, A. M. L. Kraepiel, and F. M. M. Morel, *Nature*, 407: 996 (2000).

315. K. Roberts, E. Granum, R. C. Leegood, and J. A. Raven, *Plant Physiol.*, 145: 230 (2007).

316. M. Rech, A. Morant-Manceau, and G. Trembling, *Photosynthetica*, 46: 56 (2008).

317. J. R. Reinfelder, A. J. Milligan, and F. M. M. Morel, *Plant Physiol.*, 135: 2106 (2004).

318. P. J. McGinn, and F. M. M. Morel, *Plant Physiol.*, 146: 300 (2008).

319. E. V. Armbrust, J. Berges, C. Bowler, B. Green, and D. Martinez, *Science*, 306: 79 (2004).

320. C. Bowler, A. E. Allen, J. H. Badger, J. Grimwood, K. Jabbari, A. Kuo, U. Maheswari, C. Martens, F. Maumus, R. P. Otillar, E. Rayko, A. Salamov, K. Vandepoele, B. Beszteri, A. Gruber, M. Heijde, M. Katinka, T. Mock, K. Valentin, F. Verret, J. A. Berges, C. Brownlee, J.-P. Cadoret, A. Chiovitti, C. J. Choi, S. Coesel, A. De Martino, J. C. Detter, C. Durkin, A. Falciatore, J. Fournet, M. Haruta, M. J. J. Huysmann, B. D. Jenkins, K. Jiroutova, R. E. Jorgensen, Y. Joubert, A. Kaplan, N. Kröger, O. G. Kroth, J. La Roche, E. Lindquist, M. Lommer, V. Martin-Jézéquel, P. J. Lopez, S. Lucas, M. Mangogna, K. McGinnis, L. K. Medlin, A. Montsant, M.-P. Oudot-La Secq, C. Napoli, M. Obornik, M. Schnitzler Parker, J.-L. Petit, B. M. Porcel, N. Poulsen, M. Robison, L. Rychlewski, T. A. Rynearson, J. Schmutz, H. Shapiro, M. Siaut, M. Stanley, M. R. Sussman, A. R. Taylor, A. Vardi, P. von Dassow, W. Vyverman, A. Willis, L. S. Wyrwicz, D. S. Rokhsar, J. Weissenbach, E. V. Armbrust, B. R. Geen, Y. Van de Peer, and I. V. Grigoriev, *Nature*, 456: 239 (2008).

321. M. Haimovich-Dayan, N. Garfinkel, D. Ewe, Y. Marcus, A. Gruber, H. Wagner, P. G. Kroth, and A. Kaplan, *New Phytol.*, *197*: 177 (2013).

322. P. G. Kroth, A. Chiovitti, A. Gruber, V. Martin-Jézequél, T. Mock, M. S. Parker, M. S. Stanley, A. Kaplan, L. Caron, and T. Weber, *PLoS One*, *3*: e1426 (2008).

323. R. Tanaka, S. Kikutani, and A. Mahardika, *Photosynth. Res.*, *121*: 251 (2014).

324. R. K. Trench, in *Biology of the Dinoflagellates* (F. J. R. Taylor ed.), Blackwell Scientific Publishers, London, 1987, p. 530.

325. D. Morse, P. Salois, P. Markovic, and J. W. Hastings, *Science*, *286*: 1622 (1995).

326. P. J. Hansen, N. Lumdholm, and B. Rost, *Mar. Ecol. Prog. Ser.*, *334*: 63 (2007).

327. I. Berman-Frank, J. Erez, and A. Kaplan, *Can. J. Bot.*, *76*: 1043 (1998).

328. N. A. Nimer, C. Brownlee, and M. J. Merrett, *Plant Physiol.*, *120*: 105 (1999).

329. W. Leggat, M. R. Badger, and D. Yellowless, *Plant Physiol.*, *121*: 1247 (1999).

330. M. Lapointe, T. D. B. MacKenzie, and D. Morse, *Plant Physiol.*, *147*: 1427 (2008).

331. J. S. Dason, I. E. Huertas, and B. Colman, *J. Phycol.*, *40*: 285 (2004).

332. S. Ratti, M. Giordano, and D. Morse, *J. Phycol.*, *43*: 693 (2007).

333. B. Rost, K. U. Richter, U. Riebesell, and P. J. Hansen, *Plant Cell Environ.*, *29*: 810 (2006).

334. E. Schnepf, and M. Elbrächter, *Grana*, *38*: 81 (1999).

335. N. Nassoury, L. Fritz, and D. Morse, *Plant Cell*, *13*: 923 (2001).

336. N. Nassoury, Y. Wang, and D. Morse, *Traffic*, *6*: 548 (2005).

337. C. R. Fisher, W. K. Fitt, and R. K. Trench, *Biol. Bull.*, *169*: 230 (1985).

338. P. G. Falkowski, Z. Dubinsky, Z. Muscatine, and L. McCloskey, *Bioscience*, *43*: 606 (1993).

339. V. M. Weis, and S. S. Reynolds, *Physiol. Biochem. Zool.*, *72*: 307 (1999).

340. W. Leggat, E. M. Marendy, B. Baillie, S. M. Whitney, M. Ludwig, M. R. Badger, and D. Yellowless, *Funct. Plant Biol.*, *29*: 309 (2002).

341. J. T. M. Elzenga, H. B. A. Prins, and J. Stefels, *Limnol. Oceanogr.*, *45*: 372 (2000).

342. B. Rost, U. Riebesell, S. Burkhardt, and D. Sültemeyer, *Limnol. Oceanogr.*, *48*: 55 (2003).

343. T. S. Moore, M. D. Dowell, and B. A. Franz, *Rem. Sens. Environ.*, *117*: 249 (2012).

344. L. T. Bach, L. C. M. Mackinder, K. G. Schulz, G. Wheeler, D. C. Schroeder, C. Brownlee, and U. Riebesell, *New Phytol.*, *199*: 121 (2013).

345. S. Stojkovic, J. Beardall, and R. Matear, *J. Phycol.*, *49*: 670 (2013).

346. M. D. Iglesias-Rodriguez, P. R. Halloran, R. E. M. Rickaby, I. R. Hallm, E. Colmenero-Hidalgo, J. R. Gittins, D. R. H. Green, T. Tirrell, S. J. Gibbs, P. von Dassow, E. Rehm, E. V. Armbrust, and K. P. Boessenkool, *Science*, *320*: 336 (2006).

347. K. G. Shultz, B. Rost, S. Burkhardt, U. Riebesell, S. Thoms, and D. A. Wolf-Gladrow, *Geochim. Cosmochim. Acta*, *71*: 5301 (2007).

348. N. A. Nimer, and M. J. Merrett, *New Phytol.*, *121*: 173 (1992).

349. E. Paasche, *Phycologia*, *40*: 503 (2001).

350. N. Leonardos, B. Read, B. Thake, and J. R. Young, *J. Phycol.*, *45*: 1046 (2009).

22 Rethinking the Limitations of Photosynthesis in Cyanobacteria

Yehouda Marcus

CONTENTS

22.1 Introduction 397
    22.1.1 Photosynthetic Machinery 397
    22.1.2 CO_2 Concentrating Mechanism 398
    22.1.3 Limitations of Metabolic Flux 398
22.2 Limitations of Photosynthesis 399
    22.2.1 Rubisco 399
    22.2.2 Supply of Inorganic Carbon.. 400
    22.2.3 Reductive Phase 401
    22.2.4 RuBP Regeneration 401
22.3 Concluding Remarks 401
References... 402

22.1 INTRODUCTION

Cyanobacteria (blue-green algae), the progenitors of all eukaryotic photosynthetic organisms, are free-living or symbiotic prokaryotes that are found in a broad range of ecological habitats. Although cyanobacteria mainly populate aquatic (freshwater, marine, and hypersaline) environments in the water column or the benthic zone [1], they are also found in terrestrial environments in a variety of moderate and extreme conditions, including the most inhospitable and infertile habitats on Earth [2]. Cyanobacteria can inhabit cold and frozen zones [3] but are also found in hot springs [4] or on volcanic ash, dry rocks, [5] and even soil crusts in deserts, where they face exposure to high irradiance and extreme temperatures [6]. Cyanobacteria can live in nutrient-rich or nutrient-depleted waters and soils. This ability to grow in diverse habitats results primarily from metabolic and physiological flexibility. Although they are photoautotrophs that require only water, sunlight, CO_2, and certain minerals for growth, cyanobacteria have also preserved their heterotrophic metabolism, enabling them to live in darkness for extended periods of time and, in a number of strains, enabling a photomixotrophic mode of living. Many cyanobacterial species are facultative diazotrophs that fix gaseous nitrogen to a bioavailable form in the absence of fixed nitrogen in the environment. In fact, these species are the only nitrogen fixers found in aerobic environments [7].

Cyanobacteria were the first taxonomic group to use H_2O as a source of electrons for CO_2 assimilation, consequently evolving O_2 as a by-product. However, several strains have preserved the ability of their progenitors, the photosynthetic bacteria, to use alternative electron sources (e.g., the cyanobacterium *Oscilatoria* uses either H_2O or H_2S [8]). As a result of their metabolic flexibility, in many ecosystems, cyanobacteria are important or even unique primary producers that introduce inorganic carbon (Ci) and atmospheric nitrogen into the biosphere.

The evolutionary success of the cyanobacteria and their increase in biomass had a dramatic effect on the Earth's atmosphere and biosphere: cyanobacteria increased the O_2 concentration, which was negligible in the ancient atmosphere, and decreased the CO_2 concentration. On the one hand, this effect resulted in the extinction of many obligatory anaerobic organisms; on the other hand, it encouraged the evolution of aerobic life forms. Even currently, cyanobacteria comprise a substantial fraction of the Earth's primary producers and contribute up to 30% of the global oxygen production [9].

22.1.1 PHOTOSYNTHETIC MACHINERY

Although cyanobacteria are prokaryotes, their photosynthetic metabolism is similar to that of algae and higher plants. The photosynthetic light machinery in the thylakoids captures light, which is used to transfer electrons originating from H_2O oxidation through the electron transport chain to $NADP^+$, thereby generating the H^+ gradient that is later utilized for ATP synthesis. The reducing power (NADPH) and ATP produced by this machinery are used in the reductive pentose phosphate cycle (Calvin cycle) to fix CO_2 and convert the products into carbohydrates. In this pathway, 13 biochemical reactions catalyzed by 11 enzymes are integrated in an autocatalytic cycle that produces triose phosphates from CO_2 (Figure 22.1). The Calvin cycle is roughly divided into three phases. The primary reaction, catalyzed by ribulose-1,5-bisphosphate carboxylase/oxygenase (rubisco), is carboxylation of ribulose-1,5-bisphosphate (RuBP) and cleavage of this product, which forms two molecules of 3-phosphoglycerate (PGA). In addition, rubisco

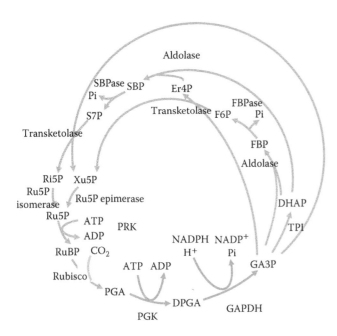

FIGURE 22.1 (**See color insert.**) Outline of the Calvin cycle. The Calvin cycle is divided into three phases: carboxylation, reduction, and regeneration (red, green, and blue arrows, respectively). DHAP, dihydoxy acetone phosphate; DPGA, 1,3-diphosphoglyceric acid; Er4P, erythrose-4-phosphate; F6P, fructose-6-phosphate; FBP, 1,6-fructose-6-phosphate; FBPase, fructose bisphosphatase; GA3P, glyceraldehyde-3-phosphate; GAPDH, GA3P dehydrogenase; PGA, 3-phosphoglyceric acid; PGK, phosphoglycerate kinase; PRK, phosphoribulose kinase; Ri5P, ribose-5-phosphate; Ru5P, ribulose-5-phosphate; rubisco, RuBP carboxylase/oxygenase; RuBP, 1,5-ribulose bisphosphate; S7P, sedoheptulose-7-phosphate; SBP, sedoheptulose-1,7-bisphosphate; SBPase, sedoheptulose bisphosphatase; TPI, triose phosphate isomerase; Xu5P, xylulose-5-phosphate.

catalyzes the oxygenation of RuBP, which is the primary reaction of photorespiration that produces one molecule of PGA and one of phosphoglycolate. In the next phase, PGA is reduced to glyceraldehyde-3-phosphate (GA3P) in two consecutive reactions catalyzed by phosphoglycerate kinase (PGK) and GA3P dehydrogenase (GAPDH) using ATP and NADPH. The GA3P is isomerized to dihydroxyacetone phosphate (DHAP) by triose phosphate isomerase (TPI). Finally, these triose phosphates are utilized in a series of reactions for regenerating the primary CO_2 acceptor, RuBP, or serve as substrates in carbohydrate biosynthesis, eventually becoming the precursors of all organic molecules (Figure 22.1) [10–12].

Because of the key role of the Calvin cycle in carbon metabolism, other metabolic pathways, growth, and development in cyanobacteria are affected by its constraints and regulation. Therefore, any attempt to improve photosynthesis, growth, or secondary metabolite and biofuel production in cyanobacteria requires analysis of the inherent limitations of carbon fixation.

22.1.2 CO_2 Concentrating Mechanism

The affinity of cyanobacterial rubisco to its substrate CO_2 is low (100–300 µM) [13,14]. Because rubisco is a bifunctional enzyme that catalyzes two competing reactions, i.e., carboxylation and oxygenation of RuBP, the apparent affinity of the enzyme to CO_2 is even lower under aerobic conditions. If the carboxylation reaction in cyanobacteria relied on diffusional CO_2 entry as it does in C_3 plants, the rate of carbon fixation in water equilibrated with air (approximately 10 µM) would be negligible (3–9% of the rate at saturating CO_2 concentrations).

In contrast to C_3 higher plants, cyanobacteria and algae possess a CO_2 concentrating mechanism (CCM) [15–18]. Ci transport into the cell is facilitated by CO_2 and HCO_3^- or CO_3^{-2} uptake systems [19–21]. Thus, for example, the cyanobacterium *Synechocystis* PCC 6803 has two CO_2 uptake systems (Cup) bound to the thylakoid membranes: one is constitutive, and the other is induced by low CO_2 concentrations. These two systems are considered to convert the CO_2 that diffuses into the cells to HCO_3^- in a mechanism that involves the thylakoidal NADP(H) dehydrogenase complex [21,22]. In addition, three HCO_3^- transporters are located on the cytoplasmic membrane: (1) a low Ci-induced, high-affinity HCO_3^- transporter, BCT1, which is encoded by the *cmp*ABCD operon and belongs to the bacterial ATP-binding cassette; (2) a low Ci-induced, medium-affinity, Na^+-dependent HCO_3^- transporter, SbtA; and (3) a low-affinity HCO_3^- transporter, BicA [16–18,20,23]. These transporters, through an influx higher than that of photosynthesis, elevate the internal Ci level up to 1000 times the concentration of external carbon [15]. Regardless of which Ci species are taken up, HCO_3^- and CO_2 are not present in equilibrium in the cyanobacterial cytoplasm, and the most abundant dissolved Ci species is presumably HCO_3^- [24,25]. The intracellular HCO_3^- in cyanobacteria is converted to CO_2 by a low quantity of carbonic anhydrase in the vicinity of rubisco, which resides within large proteinaceous bodies called carboxysomes [18,26]. Consequently, the CO_2 concentration at the carboxylation site is elevated. In addition, it has been shown that the carboxylase reaction in intact carboxysomes is less sensitive to O_2 inhibition than the reaction mediated by solubilized enzyme originating from carboxysomes [19]. These findings indicate that the carboxysome is a primitive organelle that compartmentalizes rubisco and carbonic anhydrase and that provides rubisco an environment enriched with CO_2 but presumably deficient in O_2. As a result, a high rate of photosynthesis and low rate of photorespiration are obtained at low external Ci concentrations [16,19,24].

22.1.3 Limitations of Metabolic Flux

The flux of metabolites through a sequence of biochemical reactions (metabolic pathway) is limited by the internal constraints (limitations) of the pathway, similarly to how liquid flow in a network of pipelines depends on the diameter of the pipelines. Flux limitations can result from the kinetic properties of enzymes and transporters; insufficient amounts of substrates, enzymes, and transporters; or the activity of regulatory agents. The constraints of the metabolic pathway are altered according to substrate availability and accumulation of intermediates and products. The flux limitations are regulated either by factors that coordinate the reaction rates within the specified pathway or coordinate with other metabolic pathways, or by factors that maintain homeostatic levels of key molecules and are subjected to regulation by developmental processes and external conditions.

Two approaches are generally taken to identify limitations in metabolic pathways. Traditionally, it was assumed that only nonequilibrium reactions, which are considered to be unidirectional processes, limit the metabolite flux [27]. Therefore, the free energy of the various reactions in a pathway was calculated by determining the concentrations of the substrates and products of each reaction, and the nonequilibrium processes were identified and considered to be rate-limiting steps [10,28]. In most metabolic pathways, the first unidirectional reaction is regulated. As there is no correlation between the free energy of a reaction and its rate, the extent of limitation imposed by a given reaction on the metabolite flux cannot be quantified using this approach. Alternatively, the flux limitation exerted by each variable (e.g., enzymes, substrates, and regulatory agents) could be quantified by metabolic control analysis [29,30]. In this analysis, indices that quantify the fractional change in rate (J) brought about by an infinitesimal and fractional change in the examined variable (Pi), in addition to other variables that are kept constant, are evaluated and defined as $\left(C_i^J\right)$ according to Equation 22.1:

$$C_i^J = \frac{\partial J/J}{\partial Pi/Pi} = \frac{\partial lnJ}{\partial lnPi} \qquad (22.1)$$

Control and response coefficients were determined by Kacser and Burns [29] for enzymes and external factors, respectively, that affect the metabolite flux through the pathway and elasticity coefficients identified for variables that have an effect on the reaction rate. The absolute values of C_i^J vary between 0, for variables that have no effect on the rate, and 1, for variables that proportionally affect the rate. In contrast to the thermodynamic approach, which assumes that the metabolite flux is bottlenecked by a single or a few nonequilibrium reactions, in metabolic control analysis, each reaction limits the metabolite flux to a certain degree depending on the value of the control and response coefficients [29]. To practically determine the control coefficient of an enzyme, its amount or activity should be varied by using specific inhibitors [31] or genetics techniques [32]. Not surprisingly, the development of molecular methods such as antisense technology [33], which enables the manipulation of enzyme concentrations, has advanced the control analysis of metabolic pathways in transformable organisms.

The discussion in Section 22.2 will focus on contemporary knowledge of the constraints of photosynthesis in cyanobacteria in relation to plants and algae.

22.2 LIMITATIONS OF PHOTOSYNTHESIS

22.2.1 Rubisco

Paradoxically, despite its central role in carbon fixation, rubisco is considered to be an inefficient enzyme [34] as a result of its inferior kinetic properties and antagonistic mode of action. Compared to the other enzymes of the Calvin cycle, rubisco has a low catalytic turnover (120–720 carboxylations per catalytic site per minute), and its affinity to the substrate CO_2 is also low (10–300 μM) relative to the atmospheric CO_2 concentration [13]. Moreover, due to the substrate ambiguity of rubisco, the carboxylation reaction is competitively inhibited by O_2 under ambient conditions, and fixed carbon is oxygenized and wasted in the photorespiratory pathway. Rubisco is also subject to inhibition by self-produced inhibitors or by RuBP binding prior to its activation [35]. Based on these inherent constraints of rubisco, RuBP carboxylation has been proposed to be the main rate-limiting step of carbon fixation [10,12,36]. Determination of the concentrations of Calvin cycle intermediates and calculation of the free energy changes (ΔG) of the 13 reactions involved in the pathway revealed four reactions, catalyzed by the enzymes rubisco, fructose and sedoheptulose bisphosphatase, and Ru-5-phosphate kinase, that are not in equilibrium and that may therefore limit the rate of photosynthesis [10]. Further analysis at various irradiance levels and CO_2 concentrations showed that rubisco is RuBP-saturated under light saturation and limiting CO_2 concentrations and that carboxylation capacity is the main limiting factor for photosynthesis under these conditions, whereas regeneration of RuBP constrains carbon fixation under conditions of light and CO_2 saturation [12,36,37]. This analysis demonstrated the central role of rubisco in controlling the carbon fixation rate.

Several approaches have been taken to examine the influence of rubisco abundance and activity on the rate of photosynthesis. Based on studies of the substantial changes in the rate of photosynthesis and rubisco activity observed throughout the day in the aerial leaves of the amphibious plant *Nuphar lutea* grown under solar irradiation, both parameters were correlated. Similar alterations were observed in rubisco activity and the rate of photosynthesis under light saturation (but not under light limitation), indicating that the limitation imposed by rubisco on the light-saturated rate of photosynthesis at natural conditions is extensive [38].

Using antisense RNA [33], the amount of rubisco was reduced in leaves of C_3 and C_4 plants, and the effect on photosynthesis rate was assessed. Whereas Hudson et al. [39] found that rubisco limited the rate of photosynthesis in *Nicotiana tabacum* (C_3 plant), Stitt and collaborators observed considerable variation in the photosynthetic limitations imposed by rubisco, depending on the irradiance levels during growth and measurement [32,40,41]. In *Flaveria bidentis* (C_4 plant), a high correlation was observed between rubisco content and the light-saturated rate of photosynthesis [42], although C_4 plants can presumably concentrate CO_2 at the carboxylation site of rubisco.

A series of rubisco mutants in the cyanobacterium *Synechocystis* PCC 6803 has been used to test the relationship between rubisco level, activity and catalytic turnover (K_{cat}) and photosynthetic rate [43]. These mutations reduced the K_{cat} of the enzyme by between 6-fold and 20-fold. As a result, the cells compensated for the low K_{cat} of the mutant enzyme by increasing its concentration. Nevertheless, a decrease of up to 75% was observed in the carboxylation capacity of the mutant cells. Despite this striking decrease in rubisco performance and capacity, the photosynthetic rate at saturating irradiance and limiting Ci concentrations, conditions under which rubisco is considered the main limiting factor, was hardly affected by the mutations, whereas the strongest effect of the mutations was seen in the photosynthetic rate at saturating irradiance and Ci concentrations, i.e., conditions under which photosynthesis is considered to be limited by RuBP regeneration [36]. The correlation between rubisco capacity and K_{cat} on the one hand and photosynthesis rate at saturating irradiance and Ci concentration on the other hand revealed that up to 90% reduction of the K_{cat} of the enzyme or a two-thirds reduction in the carboxylation capacity of the cells only slightly affected the photosynthesis rate. Further decline in rubisco performance and capacity, however, resulted in a decrease in the photosynthetic capacity of the cells. Interpolation of these data demonstrated that 20% of the wild-type carboxylation capacity is the minimal activity required to support photoautotrophic growth of this organism. These findings indicate that despite the low specificity factor* and affinity to CO_2 of the soluble enzyme, the activity of rubisco only minimally limits the rate of photosynthesis in *Synechocystis* [43].

Based on the common belief that rubisco is an inefficient enzyme whose performance is the main limitation in carbon fixation, rubisco has been targeted for improvement of kinetic properties to enhance the photosynthetic rate by genetic engineering [see Ref. 34 for a review]. Despite intensive efforts for several decades, an improved rubisco variant, i.e., a variant with better specificity, lower $K_m(CO_2)$ and high K_{cat}, has not been developed [34,44], presumably because of the mechanism of rubisco catalysis and its constraints. Because rubisco does not bind covalently to its gaseous substrates (CO_2 and O_2), elevation of its affinity to CO_2 or its specificity for CO_2 requires either tighter binding of the carboxylase transition state intermediate or an increase in the binding ratio of the carboxylase to oxygenase transition state intermediates to the enzyme. Tighter binding of the transition state intermediates may slow subsequent catalytic steps, which consequently reduces the K_{cat} [45,46]. Indeed, a comparison of the kinetic parameters of rubisco from diverse phylogenetic origins and of various rubisco mutants revealed an inverse relationship between the K_{cat} and specificity and a direct relationship between the K_{cat} and $K_m(CO_2)$ (i.e., as long as the $K_m(CO_2)$ decreases, the specificity increases, but the K_{cat} decreases) [13,46,47]. Hence, an improvement in $K_m(CO_2)$ or specificity results in a decline in the K_{cat} of the enzyme, and eventually, the effect on the rate of carboxylation is less pronounced. The underlying assumption in the attempt to improve rubisco is that the enzyme is not optimized for the ambient atmosphere; as such, its replacement with a more efficient catalyst would improve the rate of photosynthesis [48]. This assumption was eliminated by Savir et al. [47], who examined the effect of

the interplay between the kinetic parameters of rubisco and the net CO_2 fixation rate and found that the kinetic properties of the enzyme are nearly optimal with regard to the gaseous environment in which the enzyme resides. Accordingly, opportunities for further improvement of rubisco performance by genetic means are limited.

22.2.2 Supply of Inorganic Carbon

Because the Ci transport capability is induced in low CO_2-grown cyanobacteria, the limitation that the Ci transport exerts on the rate of photosynthesis was assessed in cells grown at various CO_2 concentrations. Kaplan et al. [15] found that the Ci influx limits the rate of photosynthesis in cells of the cyanobacterium *Anabaena variabilis* grown in high CO_2 (5% CO_2 in air) but not in low CO_2 (ambient air). Because Ci is transported against a large concentration gradient at a rate higher than the photosynthetic rate, it imposes a considerable energetic burden on the cells, and the consumption of ATP and NADPH required to motivate the Ci transport increases as a result [20,21,49]. The entry of charged Ci species, their accumulation at concentrations of up to 50 mM, and their interconversion affect the pH and the ion balance in the cytoplasm in two ways: (1) to maintain electroneutrality, the electric charge of the ionic Ci species must be electrically balanced by the presence of other ions with opposite charge; (2) conversion of CO_2 to HCO_3^- in the cytoplasm (and dehydration of the latter in the carboxysome) liberates or consumes protons in a stoichiometric ratio that should be neutralized or extruded to maintain the intracellular pH [50]. These pH-homeostatic mechanisms and ion compensatory movements at rates comparable to that of the Ci add to the energetic cost of the CCM.

Catalytic conversion of HCO_3^- to CO_2 in cyanobacteria does not occur in the cytoplasm and is apparently restricted to the carboxysome [24], whereas the reverse reaction is facilitated by the light-dependent thylakoidal NADP(H) dehydrogenase complex [21]. As a result, the cytoplasmic CO_2 concentration is maintained below the concentration that would be expected at equilibrium with HCO_3^-. This notion was demonstrated through expression of a foreign carbonic anhydrase in the cytoplasm of the cyanobacterium *Synechococcus* PCC 7942, which equilibrated the HCO_3^- and CO_2 concentrations, led to a CO_2 leak, and eventually led to a dissipation of the intracellular Ci pool. Consequently, the rate of photosynthesis of this mutant was low at ambient CO_2 concentration [25]. Two proteins in the cyanobacterial β-type carboxysome and one in the α-type carboxysome have been identified as carbonic anhydrase: (1) the product of the *ccaA* gene, which has been found in *Synechococcus* (*elongatus*) PCC 7942 and *Synechocystis* PCC 6803, and (2) the carboxysome shell protein *ccmM* and the product of the gene *csoSCA* (also known as *csoS3*) in the α-type carboxysome [26].

Although the activity of carbonic anhydrase is low in cyanobacteria [51], it is essential for photosynthesis and growth at ambient CO_2 concentration, as demonstrated using the permeable carbonic anhydrase inhibitor, ethoxyzolamide, or by mutations in the carboxysomal carbonic anhydrase, ccaA and

---

* A factor that determines the rate ratio of rubisco carboxylase to oxygenase reactions [67].

ccmM [51–53]. Evidently, the nonenzymatic conversion of HCO_3^- to CO_2 near the carboxylation site of rubisco cannot sustain photosynthesis; therefore, the rate of photosynthesis is limited by this reaction. This observation raises the question of whether overexpression of carboxysomal carbonic anhydrase enhances the rate of photosynthesis. Practically, such an experiment would be infeasible because of structural constraints of the carboxysomes. It has been predicted, however, that the amount of carbonic anhydrase in the carboxysome is optimized: low activity does not supply sufficient CO_2 for photosynthesis, whereas high activity leads to the leakage of CO_2 from the carboxysome [24,54].

Since other than rubisco, no other Calvin cycle enzymes are found in the carboxysome [18], RuBP penetrates into the carboxysome core, whereas PGA and phosphoglycolate diffuse out, presumably through pores in the carboxysome shell [26]. The movement of metabolites between the carboxysome and the cytoplasm has not yet been investigated; therefore, the limitation that such movement imposes on CO_2 fixation is unknown.

22.2.3 REDUCTIVE PHASE

In the reductive phase, 3-PGA formed by rubisco is converted to triose phosphate via two successive reactions using ATP and NADPH produced by the photosynthetic light reactions. Firstly, PGK catalyzes the phosphorylation of 3-PGA. The product, 1,3-diphosphoglycerate, is then reduced by GAPDH to form GA3P (Figure 22.1). Analysis of the limitation imposed by PGK on the rate of photosynthesis has not been reported thus far in plants, algae, or cyanobacteria, whereas the limitations imposed by GAPDH have been analyzed in tobacco plants using antisense RNA. Although GAPDH is one of the most regulated enzymes in the Calvin cycle [11,55], its limitation on the rate of photosynthesis is low (control coefficient < 0.2) [56].

Assuming that GA3P and PGA are in equilibrium, it could be shown that the concentration ratio of GA3P to PGA, termed the *assimilatory force*, equals the product of the mass action ratios of ATP and NADPH synthesis multiplied by the equilibrium constant of PGA reduction [57]. Heber et al. [28], who studied the relationship between the assimilatory force and PGA reduction in leaves and chloroplasts of higher plants, observed that PGA reduction was close to equilibrium and that the assimilatory force did not limit photosynthesis, even at low irradiance [28]. In contrast, in *Synechocystis* PCC 6803, it was found that the GA3P-to-PGA concentration ratio increased with increased irradiance, as long as light limited the rate of photosynthesis, indicating that in this organism, ATP or NADPH limits the rate of photosynthesis under light limitation [43].

ATP and NADPH limitation in cyanobacteria presumably results from the consumption of excess energy, in the form of ATP and NADPH, by the CCM [49], which would decrease the levels of ATP or NADPH, thereby limiting PGA reduction. ATP is also required for the formation of RuBP from ribulose-5-phosphate in the Calvin cycle (Figure 22.1). It is therefore likely that competition among PGA reduction, RuBP production, and Ci transport for limited resources of ATP and NADPH limits Ci uptake and carbon fixation [43].

22.2.4 RUBP REGENERATION

In this phase, triose phosphates are converted in a nonlinear sequence of reactions to the pentose phosphate RuBP, which serves as a CO_2 acceptor (Figure 22.1). Analysis of the metabolites in plants and algae revealed that three highly regulated reactions in this phase (hydrolysis of 1,6-fructose bisphosphate (FBP) and sedoheptulose 1,7-bisphosphate (SBP) and phosphorylation of ribulose-5-phosphate) are not in equilibrium [10]. Nevertheless, underexpression of two of the enzymes responsible for these reactions (fructose bisphosphatase and phosphoribulose kinase) using antisense technology had almost no effect on the rate of photosynthesis in the examined plants, as reflected by their low control coefficients [58–60]. By contrast, a decrease in the activity of sedoheptulose bisphosphatase, which catalyzes the nonequilibrium hydrolysis of SBP, reduced the level of RuBP and, consequently, the rate of photosynthesis, particularly at saturating irradiance and CO_2 concentrations [61,62]. Furthermore, overexpression of cyanobacterial sedoheptulose/fructose bisphosphatase* in tobacco enhanced the rate of photosynthesis [63]. Surprisingly, the reversible enzymes aldolase (which catalyzes the formation of FBP and SBP) and, in particular, transketolase (which catalyzes the formation of the pentose phosphates ribose-5-phosphate and xylulose-5-phosphate) (Figure 22.1), imposed a considerable limitation on the rate of photosynthesis at saturating irradiance and CO_2 concentrations, as revealed by the control coefficients [64,65]. In an alternate approach, Ma et al. [66] showed that overexpression of aldolase and TPI increased the rate of photosynthesis in the cyanobacterium *Anabaena* PCC 7120.

These findings raise the question of why, in contrast to the main assumption of the thermodynamic approach, equilibrium enzymes in the Calvin cycle were found to limit the rate of photosynthesis more than the highly regulated nonequilibrium enzymes in the control analyses discussed herein. This discrepancy could be explained in several ways: (1) the activity of a nonequilibrium enzyme is higher than that of an equilibrium enzyme because the former is available in excess; (2) a decrease in enzyme activity may convert a reaction at equilibrium to a nonequilibrium reaction; and (3) alterations in enzyme content are compensated for by other processes e.g., metabolic regulation, which maintains constant *in vivo* activity despite changes in enzyme availability. Thus, a decrease in rubisco content in antisense plants was accompanied by an increase in rubisco activation [40].

22.3 CONCLUDING REMARKS

Differences in the limiting factors for photosynthesis in cyanobacteria versus plants mainly result from the operation of the CCM in the former. On the one hand, the CCM in cyanobacteria elevates the CO_2 concentration at the carboxylation site of rubisco, thereby reducing the limitation of photosynthesis by rubisco; on the other hand, the excess energy in the

---

* In cyanobacteria, the same enzyme catalyzes both reactions [68].

form of ATP and NADPH consumed by the CCM limits PGA reduction and, consequently, the rate of photosynthesis.

REFERENCES

1. Gallon, J. R., Jones, D. A. and Page, T. S., 1996. *Trichodesmium*, the paradoxial diazotroph. *Arch. Hydrobiol. Suppl., Algol. Stud.*, 83: 215–243.
2. Whitton, B. A., 1992. Diversity, ecology and taxonomy of the cyanobacteria. In: *Photosynthetic prokaryotes*, eds. N. H. Mann and N. G. Carr. 1–51. New York, Plenum Press.
3. Tang, E. P. Y. and Vincent, W. F., 2002. Strategies of thermal adaptation by high-latitude cyanobacteria. *New Phytol.*, 142: 315–323.
4. Castenholz, R. W., 1973. Ecology of blue-green algae in hot springs. In: *The biology of blue-green algae*, eds. N. G. Carr and B. A. Whitton. 379–414. Oxford, Blackwell Scientific Publications.
5. Krumbein, W. E. and Jens, K., 1981. Biogenic rock varnishes of the Negev desert (Israel) an ecological study of iron and manganese transformation by cyanobacteria and fungi. *Oecologia*, 50: 25–38.
6. Dor, I. and Danin, A., 1996. Cyanobacterial desert crusts in the Dead Sea Valley, Israel. *Arch. Hydrobiol. Suppl. Algol. Stud.*, 83: 197–206.
7. Fay, P., 1992. Oxygen relations of nitrogen fixation in cyanobacteria. *Microbiol. Rev.*, 56: 340–373.
8. Padan, E., 1979. Facultative anoxygenic photosynthesis in cyanobacteria. *Annu. Rev. Plant Physiol.*, 30: 27–40.
9. Canfield, D. E., 2005. The early history of atmospheric oxygen: Homage to Robert M. Garrels. *Annu. Rev. Earth Planet. Sci.*, 33: 1–36.
10. Bassham, J. A. and Krause, G. H., 1969. Free energy changes and metabolic regulation in steady-state photosynthetic carbon reduction. *Biochim. Biophys. Acta*, 189: 207–221.
11. Leegood, R. C., 1990. Enzymes of the Calvin cycle. In: *Methods in plant biochemistry*, ed. P. J. Lea. 15–36. London, Academic Press.
12. Woodrow, I. E. and Berry, J. A., 1988. Enzymatic regulation of photosynthetic CO_2 fixation in C3 plants. *Annu. Rev. Plant Physiol. Plant Mol. Biol.*, 39: 533–594.
13. Badger, M. R. and Andrews, T. J., 1987. Co-evolution of Rubisco and CO_2 concentrating mechanisms. In: *Progress in photosynthesis research*, ed. J. Biggins. 601–609. Dordrecht, Nijhoff Publishers.
14. Marcus, Y., Altman-Gueta, H., Finkler, A. and Gurevitz, M., 2003. The dual role of cysteine 172 in redox regulation in Rubisco activity and degradation. *J. Bacteriol.*, 185: 1509–1517.
15. Kaplan, A., Badger, M. R. and Berry, J. A., 1980. Photosynthesis and the intracellular inorganic carbon pool in the blue-green alga *Anabaena variabilis*: Response to external CO_2 concentration. *Planta*, 149: 216–219.
16. Badger, M. R. and Price, G. D., 2003. CO_2 concentrating mechanisms in cyanobcteria: Molecular components, their diversity and evolution. *J. Exp. Bot.*, 54: 609–622.
17. Giordano, M., Beardall, J. and Raven, J. A., 2005. CO_2 concentrating mechanisms in algae: Mechanisms, environmental modulation, and evolution. *Annu. Rev. Plant Biol.*, 56: 99–131.
18. Kaplan, A., Hagemann, M., Kahlon, S. and Ogawa, T., 2008. Carbon acquisition by cyanobacteria: Mechanisms, comparative genomics and evolution. In: *The cyanobacteria: molecular biology, genomics and evolution*, eds. A. Herrero and E. Flores. 305–334. Norfolk, Caister Academic Press.
19. Marcus, Y., Berry, J. A. and Pierce, J., 1992. Photosynthesis and photorespiration in a mutant of the cyanobacterium *Synechocystis* PCC 6803 lacking carboxysomes. *Planta*, 187: 511–516.
20. Omata, T. et al., 1999. Identification of an ATP-binding cassette transporter involved in bicarbonate uptake in the cyanobacterium *Synechococcus* sp. strain PCC 7942. *Proc. Natl. Acad. Sci. U.S.A.*, 96: 13571–13576.
21. Ohkawa, H., Pakrasi, H. B. and Ogawa, T., 2000. Two types of functionally distinct NAD(P)H dehydrogenases in *Synechocystis* sp. strain PCC6803. *J. Biol. Chem.*, 275: 31630–31634.
22. Shibata, M. et al., 2001. Distinct constitutive and low-CO_2-induced CO_2 uptake systems in cyanobacteria: Genes involved and their phylogenetic relationship with homologous genes in other organisms. *Proc. Natl. Acad. Sci. U.S.A.*, 98: 11789–11794.
23. Price, G. D., Woodger, F. J., Badger, M. R., Howitt, S. M. and Tucker, L., 2004. Identification of a SulP-type bicarbonate transporter in marine cyanobacteria. *Proc. Natl. Acad. Sci. U.S.A.*, 101: 18228–18233.
24. Reinhold, L., Zviman, M. and Kaplan, A., 1989. A quantiative model for inorganic carbon fluxes and photosynthesis in cyanobacteria. *Plant Physiol. Biochem.*, 27: 945–954.
25. Price, G. D. and Badger, M. R., 1989. Expression of human carbonic anhydrase in the cyanobacterium *Synechococcus* PCC 7942 creates a high CO_2-requiring phenotype. Evidence for a central role for carboxysomes in the CO_2 concentrating mechanism. *Plant Physiol.*, 91: 505–513.
26. Yeates, T. O., Kerfeld, C. A., Heinhorst, A., Cannon, G. C. and Shively, J. M., 2008. Protein-based organelles in bacteria: Carboxysomes and related microcompartments. *Nat. Rev. Microbiol.*, 6: 681–691.
27. Krebs, H. A., 1969. The role of equilibria in the regulation of metabolism. *Curr. Top. Cell. Regul.*, 1: 45–58.
28. Heber, U., Neimanis, S., Dietz, K.-J. and Viil, J., 1986. Assimilatory power as a driving force in photosynthesis. *Biochim. Biophys. Acta*, 852: 144–155.
29. Kacser, H. and Porteus, J. W., 1987. Control of metabolism; what do we have to measure. *Trends Biochem. Sci.*, 12: 5–14.
30. Heinrich, R. and Rapoport, T. A., 1974. A linear steady state treatment of enzymatic chains. General properties, control and effector strength. *Eur. J. Biochem.*, 42: 89–95.
31. Groen, A. K., van Roermund, C. W. T., Vervoorn, R. C. and Tager, J. M., 1986. Control of gluconeogenesis in rat liver cells. Flux control coefficients of the enzymes in the gluconeogenic pathway in the absence and presence of glucagon. *Biochem. J.*, 237: 379–389.
32. Stitt, M. and Schulze, E. D., 1994. Does Rubisco control the rate of photosynthesis and plant growth? An exercise in molecular ecophysiology. *Plant Cell Environ.*, 17: 465–467.
33. Bourque, J. E., 1995. Antisense strategies for genetic manipulations in plants. *Plant Sci.*, 105: 125–149.
34. Spreitzer, R. E. and Salvucci, M. E., 2002. Rubisco: Structure, regulatory interactions, and possibilities for a better enzyme. *Annu. Rev. Plant Biol.*, 53: 449–475.
35. Cleland, W. W., Andrews, T. J., Gutteridge, S., Hartman, F. C. and Lorimer, G. H., 1998. Mechanism of Rubisco: The carbamate as general base. *Chem. Rev.*, 98: 549–561.
36. Farquhar, G. D., von Caemmerer, S. and Berry, J. A., 1980. A biochemical model of photosynthetic CO_2 fixation in leaves of C3 plants. *Planta*, 149: 78–90.
37. Laisk, A. and Oya, V. M., 1974. Photosynthesis of leaves subjected to brief impulses of CO_2. *Soviet Plant Physiol.*, 21: 928–935.

38. Snir, A., Gurevitz, M. and Marcus, Y., 2006. Alterations in Rubisco activity and in stomatal behavior induce a daily rhythm in photosynthesis of aerial leaves in the amphibious plant *Nuphar lutea*. *Photosynth. Res.*, 90: 233–242.

39. Hudson, G. S., Evans, J. R., von Caemmerer, S., Arvidsson, Y. B. C. and Andrews, T. J., 1992. Reduction of ribulose-1,5-bisphosphate carboxylase/oxygenase content by antisense RNA reduces photosynthesis in transgenic tobacco plants. *Plant Physiol.*, 98: 294–302.

40. Quick, W. P. et al., 1991. Decreased ribulose-1,5-bisphosphate carboxylase-oxygenase in transgenic tobacco transformed with "antisense" rbcS: I. Impact on photosynthesis in ambient growth conditions. *Planta*, 183: 542–554.

41. Stitt, M. et al., 1991. Decreased ribulose-1,5-bisphosphate carboxylase-oxygenase in transgenic tobacco transformed with "antisense" rbcS: II. Flux- control coefficients for photosynthesis in varying light, carbon dioxide, and air humidity. *Planta*, 183: 555–566.

42. Furbank, R. T., Chitty, J. A., von Caemmerer, S. and Jenkins, C. D., 1996. Antisense RNA inhibition of *RbcS* gene expression reduces rubisco level and photosynthesis in the C4 plant *Flaveria bidentis*. *Plant Physiol.*, 111: 725–734.

43. Marcus, Y., Altman-Gueta, H., Wolff, Y. and Gurevitz, M., 2011. Rubisco mutagenesis provides new insight into limitations on photosynthesis and growth in *Synechocystis* PCC6803. *J. Exp. Bot.*, 62: 4173–4182.

44. Whitney, S. M., Houtz, R. L. and Alonso, H., 2011. Advancing our understanding and capacity to engineer nature's CO_2-sequestering enzyme, Rubisco. *Plant Physiol.*, 155: 27–35.

45. Lorimer, G. H., Chen, Y.-R. and Hartman, F. C., 1993. A role for the ε-amino group of lysine-334 of ribulose-1,5-bisphosphate carboxylase in the addition of carbon dioxide to the 2,3-enediol(ate) of ribulose 1,5-bisphosphate. *Biochemistry*, 32: 9018–9024.

46. Tcherkez, G. G., Farquhar, G. D. and Andrews, T. J., 2006. Despite slow catalysis and confused substrate specificity, all ribulose bisphosphate carboxylases may be nearly perfectly optimized. *Proc. Natl. Acad. Sci. U.S.A.*, 103: 7246–7251.

47. Savir, Y., Noor, E., Milo, R. and Tlusty, T., 2010. Cross-species analysis traces adaptation of Rubisco toward optimality in a low-dimensional lands. *Proc. Natl. Acad. Sci. U.S.A.*, 107: 3475–3480.

48. Andrews, T. J. and Whitney, S. M., 2003. Manipulating ribulose bisphosphate carboxylase/oxygenase in the chloroplasts of higher plants. *Arch. Biochem. Biophys.*, 414: 159–169.

49. Raven, J. A., Beardall, J. and Giordano, M., 2014. Energy costs of carbon dioxide concentrating mechanisms in aquatic organisms. *Photosynth. Res.*, 121: 111–124.

50. Marcus, Y., 1997. Distribution of inorganic carbon among its component species in cyanobacterial: Do cyanobacteria in fact actively accumulate inorganic carbon? *J. Theor. Biol.*, 185: 31–45.

51. Marcus, Y., Schwarz, R., Friedberg, D. and Kaplan, A., 1986. High CO_2 requiring mutant of *Anacystis nidulans* R2. *Plant Physiol.*, 82: 610–612.

52. So, A. K. C., Van Spall, H. G. C., Coleman, J. R. and Espie, G. S., 1998. Catalytic exchange of ^{18}O from $^{13}C^{18}O$-labelled CO_2 by wild-type cells and *ecaA*, *ecaB*, and *ccaA* mutants of the cyanobacteria *Synechococcus* PCC7942 and *Synechocystis* PCC6803. *Can. J. Bot.*, 76: 1153–1160.

53. Price, G. D. and Badger, M. R., 1989. Ethoxyzolamide inhibition of CO_2-dependent photosynthesis in the cyanobacterium *Synechococcus* PCC79421. *Plant Physiol.*, 89: 44–50.

54. Reinhold, L., Kosloff, R. and Kaplan, A., 1991. A model for inorganic carbon fluxes and photosynthesis in cyanobacterial carboxysomes. *Can. J. Bot.*, 69: 984–988.

55. Tamoi, M., Miyazaki, T., Fukamizo, T. and Shigeoka, S., 2005. The Calvin cycle in cyanobacteria is regulated by CP12 via NAD(H)/NADP(H) ratio under light/dark conditions. *Plant J.*, 42: 504–513.

56. Price, G. D., Evans, J. R., von Caemmerer, S., Yu, J.-W. and Badger, M. R., 1995. Specific reduction of chloroplast glyceraldehehyde-3-phosphate dehydrogenase activity by antisense RNA reduces CO_2 assimilation via a reduction in ribulose bisphosphate regeneration in transgenic tobacco plants. *Planta*, 195: 369–378.

57. Arnon, D., Whatley, E. R. and Allen, M. B., 1958. Assimilatory power in photosynthesis: Photosynthetic phosphorylation by isolated chloroplasts is coupled with TPN reduction. *Science*, 127: 1026–1034.

58. Kossmann, J., Sonnewald, U. and Willmitzer, L., 1994. Reduction of the chloroplastic fructose-1,6-bisphosphatase in transgenic potato plants impairs photosynthesis and plant growth. *Plant J.*, 6: 637–650.

59. Paul, M. J. et al., 1995. Reduction in phosphoribulokinase activity by antisense RNA in transgenic tobacco: Effect on CO_2 assimilation and growth in low irradiance. *Plant J.*, 7: 535–542.

60. Muschak, M. et al., 1997. Gas exchange and ultrastructural analysis of transgenic potato plants expressing mRNA antisense construct targeted to the cp-fructose-1,6-bisphosphate phosphatase. *Photosynthetica*, 33: 455–465.

61. Harrison, E. P., Willingham, N. M., Lloyd, J. C. and Raines, C. A., 1998. Reduced sedoheptulose-1,7-bisphosphatase levels in transgenic tobacco lead to decreased photosynthetic capacity and altered carbohydrate accumulation. *Planta*, 204: 27–36.

62. Harrison, E. P., Olcer, H., Lloyd, J. C., Long, S. P. and Raines, C. A., 2001. Small decreases in SBPase cause a linear decline in the apparent RuBP regeneration rate, but do not affect Rubisco carboxylation capacity. *J. Exp. Bot.*, 52: 1779–1784.

63. Miyagawa, Y., Tamoi, M. and Shigeoka, S., 2001. Over-expression of a cyanobacterial fructose-1,6-/sedoheptulose-1,7-bisphosphatase in tobacco enhances photosynthesis and growth. *Nat. Biotechnol.*, 19: 965–969.

64. Haake, V., Zrenner, R., Sonnewlad, U. and Stitt, M., 1998. A moderate decrease of plastid aldolase activity inhibits photosynthesis, alters the levels of sugars and starch, and inhibits growth of potato plants. *Plant J.*, 14: 147–157.

65. Henkes, S., Sonnewald, U., Badur, R., Flachmann, R. and Stitt, M., 2001. A small decrease of plastid transketolase activity in antisense tobacco transformants has dramatic effects on photosynthesis and phenylpropanoid metabolism. *Plant Cell*, 13: 535–551.

66. Ma, W., Wei, L., Wang, Q., Shi, D. and Chen, H., 2007. Increased activity of the non-regulated enzymes fructose-1,6-bisphosphate aldolase and triosephosphate isomerase in *Anabaena* sp. strain PCC 7120 increases photosynthetic yield. *J. Appl. Phycol.*, 19: 207–213.

67. Laing, W. A., Ogren, W. L. and Hageman, R. H., 1974. Regulation of soybean net photosynthetic CO_2 fxation by the interaction of CO_2, O_2, and ribulose-1,5-bisphosphate carboxylase. *Plant Physiol.*, 54: 678–685.

68. Kaneko, T. et al., 1996. Sequence analysis of the genome of the unicellular cyanobacterium *Synechocystis* sp. strain PCC6803. II. Sequence determination of the entire genome and assignment of potential protein-coding regions. *DNA Res.*, 3: 109–136.

23 Induction and Relaxation of Bacteriochlorophyll Fluorescence in Photosynthetic Bacteria*

Péter Maróti

CONTENTS

23.1 Introduction 405
23.2 BChl Fluorescence—Properties and Detection...407
    23.2.1 Induction of Prompt Fluorescence..407
    23.2.2 Relaxation of Prompt Fluorescence...408
    23.2.3 Delayed Fluorescence 409
23.3 Fluorescence from Isolated RC 409
    23.3.1 Structure and Function of the Bacterial RC 409
        23.3.1.1 Donor-Side Reactions 410
        23.3.1.2 Acceptor-Side Reactions...410
    23.3.2 Free Energy and Protonation of Q_A 411
    23.3.3 Kautsky Effect of the RC 412
    23.3.4 Bionanotechnological Applications 413
23.4 RC-Controlled Fluorescence from Intact Bacteria 413
    23.4.1 Polyphasic Fluorescence Induction 414
        23.4.1.1 Photochemical Rise 414
        24.4.1.2 Triplet Quenching 415
        23.4.1.3 Slow Rising Phase...416
    23.4.2 Assembly of the Photosynthetic Apparatus Monitored by BChl Fluorescence 416
    23.4.3 Biosensing and Bioremediation of Heavy Metal Pollution in the Environment 418
23.5 Conclusions..418
Acknowledgments..419
List of Abbreviations...419
References...419

23.1 INTRODUCTION

The reaction center of bacterial photosynthesis drives a light-induced open-chain intraprotein electron transfer that is part of a larger-scale cyclic electron transfer connecting the cyt bc_1 complex via mobile electron carriers of cyt c and quinones (Q). The electron transfer can be tracked by kinetic changes of the bacteriochlorophyll (BChl) fluorescence originating either from the reaction center or from the antenna. Here, we will concentrate on induction and relaxation of fluorescence and on delayed fluorescence in the millisecond-to-minute time range. The fluorescence induction is decomposed into three (photochemical, triplet quenching, and slow) phases. The light intensity and redox reactions at the donor and acceptor sides as possible bottlenecks of the turnover of the reaction center are reviewed and analyzed critically. We demonstrate that (1) the

induction and relaxation of the yield of the BChl fluorescence are fast and convenient tools to monitor the rate-limiting steps in view of closing and reopening the bacterial reaction center upon (and after) continuous illumination, and (2) under physiological conditions in *Rhodobacter sphaeroides*, mainly, the donor-side reactions limit the turnover time to some tens of microseconds, (3) which progressively increases with longer illumination in accordance with finite pool sizes and structural confinements of reduced cyt c_2 and oxidized quinones in the periplasmic and membrane phases of the photosynthetic unit, respectively. Fluorescence techniques recently have gained a broad spectrum of applications in (1) easy and inexpensive determination of the photosynthetic capacity of bacteria under field conditions, (2) ontogenesis of the photosynthetic apparatus, and (3) production of novel systems and (nano)materials mimicking the natural processes of the bacterial photosynthesis.

---

* Dedicated to the memory of Colin A. Wraight (1945–2014).

The essential processes of photophysics, biochemistry, cell biology, and physiology in photosynthetic bacteria take place on wide ranges of spatial and temporal hierarchy from cells to atoms of key biomolecules and from femtoseconds to minutes, respectively (Figure 23.1). In purple bacterium *Rhodobacter* (*Rba.*) *sphaeroides*, the invagination of the cytoplasmic membrane at curved regions is followed by detachment of spherical vesicles housing the photosynthetic apparatus.[1] The light-harvesting (LH), reaction center (RC), cyt *bc*₁, and ATP synthase complexes of the purple photosynthetic bacteria and their organization in the intracytoplasmic membrane (ICM) have been characterized by high atomic resolution and an unprecedented level of biochemistry and physical chemistry.[2–5] Anoxygenic photobacteria all use some forms of bacteriochlorophyll (BChl) as a primary photosynthetic pigment, organized into LH and RC in the form of BChl–protein complexes. The different BChl aggregates of the peripherial (LH2) and core (LH1) antenna complexes absorb the light and funnel the electronic excitation energy to the RC. While the absorption takes place within a couple of femtoseconds (10^{-15} s), the migration and capture of the electronic excitation energy in the pigment bed and by the RC, respectively, occur in the (sub)picosecond (10^{-12} s) time

range. The excitation energy is converted to chemical energy via charge separation in the RC protein, which plays an essential role (1) in holding the redox states of the energy transducing machinery out of equilibrium by photochemistry, (2) in coupling the two-electron chemistry of the quinone reduction and the one-electron chemistry of the electron transfer chain, and (3) in initializing the proton translocation across the bioenergetic membrane. One of the key issues of present research is to reveal how the RC controls the light-activated biological electron (and proton) transfer.

A very small fraction of the excitation energy is dissipated as BChl fluorescence that can originate mainly from the antenna (intact cells) or directly from the RC (isolated RC in a detergent solution and, to a much smaller extent, from chromatophores or whole cells). The temporal changes of BChl fluorescence have been an invaluable source of information about photosynthetic processes. While the very fast (nanosecond-to-femtosecond) changes are related to excitation, energy transfer, and relaxation processes in the pigments, slower (microsecond-to-second) variations are connected to photochemical utilization of the absorbed energy. The initial slow changes of the BChl fluorescence upon rectangular-shaped excitation are called fluorescence induction, discovered more

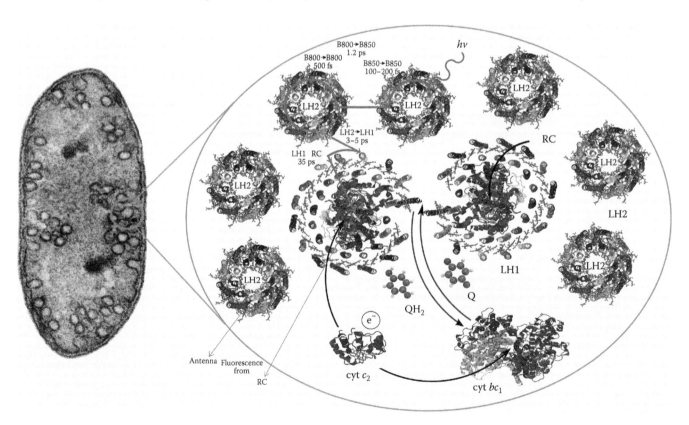

FIGURE 23.1 **(See color insert.)** Hierarchy of organization of the photosynthetic apparatus of *Rba. sphaeroides*: intact cell (left) and top-view model of the intracytoplasmic membrane with atomic resolution of the BChl–protein complexes of the light-harvesting (peripheral [LH2] and core [LH1]) antenna and the reaction center (RC) in supercomplex (dimeric) arrangement (right). The light energy (*h·ν*) absorbed in the pigment bed converts to electronic excitation energy and migrates to the BChl dimer of the RC. A tiny fraction of the electronic excitation energy will be reemitted as BChl fluorescence from the antenna or from RC and used to monitor the photosynthetic processes. The captured energy in the RC drives cyclic electron (e⁻) transfer between RC and cyt *bc*₁ complex with participation of mobile electron carriers of cyt *c*₂ in the aqueous periplasmic side and quinone (Q) in the hydrophobic layer of the membrane. (Adapted from Kis, M. et al., *Photosynth. Res.*, 122, 261–273, 2014.)

than 80 years ago.[6] The Kautsky effect launched an unprecedented development of our understanding about photosystem II electron transport[7–10] and became the basis of an exceptionally useful noninvasive diagnostic tool for determination of the photosynthetic capacity of organisms.[11–14]

Compared to the numbers and efforts of fluorescence induction studies on higher plants, relatively few investigations and a low level of instrumentation have been devoted to photosynthetic bacteria. Additionally, recent competent monographs and textbooks failed to include BChl fluorescence transients when chlorophyll (Chl) fluorescence induction of higher plants and algae was discussed.[15–17] Because of the significantly lesser complexity of the structure and function of bacteria, the dark–light transition of the BChl fluorescence is considerably simpler than that of Chl fluorescence in higher plants. This property can be utilized if the bacteria are taken as a model of the more advanced oxygenic systems. Similarly valuable information about the capture of light energy,[18,19] export or recombination of charges in RCs,[20] energetization of the membrane,[21] assembly of the photosynthetic apparatus[22] and marine ecosystem,[23–26] or heavy metal pollution in aqueous habitats[27,28] can be obtained in bacteria as in oxygenic organisms.

The purpose of this chapter is to survey the different BChl fluorescence methods used in research of photosynthetic bacteria. The current state of progress in this evolving field, with some selected unsolved problems, will be summarized, and insights will be given into the structure-based mechanisms. This comprehensive review has emerged from investigations on several different strains by a wide variety of methods, including fluorescence spectroscopy. As there are many distinctions between the different strains and experimental conditions, it is not enough to pool these data from different sources, because they will not fit into a coherent model. Besides making an attempt to find common points in the structure–function of the bacteria, we will call attention to the diversity of species and differences in operations, as well. The focus will be set on strain *Rba. sphaeroides*, but some references will be made to other species. Comparisons will be made between properties of the isolated RC and that embedded in an intact membrane and operating under natural conditions. We try to present a unified view of the RC mediated electron transfers and the assembly and organization of the photosynthetic apparatus.

23.2 BChl FLUORESCENCE— PROPERTIES AND DETECTION

23.2.1 INDUCTION OF PROMPT FLUORESCENCE

In contrast to the large variety of BChl–protein absorption bands in photosynthetic organisms, the fluorescence spectrum is characterized by a well-behaved bell shape as the fluorescence emission occurs from the lowest excited singlet state after vibrational relaxation. On one hand, the two antenna systems, LH1 (B875) and LH2 (B800–B850), are well separated in the room-temperature absorption spectrum of intact cells of *Rba. sphaeroides*; on the other hand, the fluorescence

band is not structured (Figure 23.2). The Stokes shift and the quantum yield, however, depend on the aggregation number and the environment of the BChl pigment. While the fluorescence spectrum of BChl *a* dissolved in toluene has a maximum at 790 nm and quantum yield of 0.2,[29] the fluorescence of the dimeric BChl form in the isolated RC is far-redshifted with a peak at 915 nm, and the quantum yield is as low as 4×10^{-4}.[30] The BChl fluorescence from the antenna is somewhat blueshifted with 890 nm peaks of emission.[31–33]

The quantum yield of BChl fluorescence can change during photosynthetic processes (shortly but not correctly named fluorescence induction) that can be formally treated as appearance and disappearance of different fluorescence quenchers. They are produced by various (light or dark) reactions and have different kinetics (lifetimes) and efficiencies. The major candidates for fluorescence quenchers in photosynthetic bacteria are the oxidized primary quinone, Q_A, and dimer, P^+, in the RC (photochemical quenching) and triplet states of BChl, ^{3}BChl, carotenoids, ^{3}Car, or oxygen ($^3O_2$), either in the RC or in the antenna (triplet quenching). Additionally, there also a couple of other factors that can modulate the quantum yield of BChl fluorescence. Compared to the major fluorescence quenchers, they are usually weak quenchers whose contribution in modulation of the BChl fluorescence yield is difficult to prove. The protonic and electric components of the light-induced membrane potential increase the nonradiative energy dissipation and therefore appear as fluorescence quenchers.[21,34–36] The oxidized quinone in the pool and oxidized BChl pigments, per se, can also quench the fluorescence of BChl in a nonphotochemical way.[37,38] Conformational changes may also influence the fluorescence yield by modifying the distances and interactions among the pigments.[8,9,39,40]

The principle of the most straightforward way to measure the BChl fluorescence induction goes back to the original method used by Kautsky and Hirsch in 1931.[6] They observed

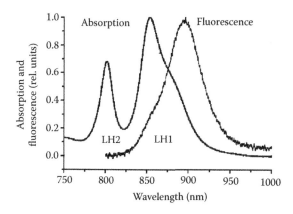

FIGURE 23.2 Steady-state far-red absorption and fluorescence spectra of light-harvesting BChl pigments in intact cells of *R sphaeroides* at room temperature. While the peripheral LH2 complex contains two types of BChls with well-separated absorption bands at 800 and 850 nm, the core complex LH1 has BChl of 875 nm band showing up as a shoulder in the spectrum. The fluorescence spectrum has a simple bell shape, and its peak is Stokes-shifted to 900 nm.

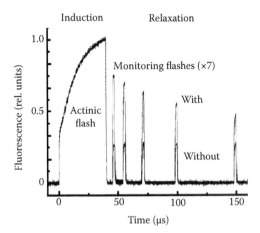

FIGURE 23.3 Principle of measurement of induction and relaxation of BChl fluorescence in intact cells of photosynthetic bacteria. The rectangular shape of actinic excitation from a high-power laser diode induces an increase of the BChl fluorescence yield (induction). After turning off the excitation, the fluorescence states of the cell are tested by a series of weak (attenuated by a factor of 7) and short flashes arranged in geometrical order in the dark. The ratio of the fluorescence signals evoked by monitoring flashes with and without preactivation will offer the kinetics of decay of the BChl fluorescence yield (relaxation).

the Chl red fluorescence of a leaf upon rectangular-shaped illumination (sunlight) through appropriate filters with their own eyes. The more precise detection of BChl fluorescence kinetics of photosynthetic bacteria requires excitation with a laser diode (or an array of light-emitting diodes [LEDs][12,41]) and solid-state photodetectors due to a longer wavelength of emission and higher time resolution (Figure 23.3).[42] The kinetics of the BChl fluorescence reflecting the $PQ_A \rightarrow P^+Q_A^-$ photochemical charge separation can be characterized by four essential parameters: the initial level (F_0), the lag phase, the (normalized) variable fluorescence ($F_v = [F_{max} - F_0]/F_0$, where F_{max} is the maximum fluorescence level), and the rate

constant of the fluorescence rise (k_p). F_0 is the fluorescence originating from unconnected centers ("dead" fluorescence), the lag phase is characteristic of the connectivity of the photosynthetic units (PSUs), F_v reflects the effectiveness of capture of the photons to photochemistry, and k_p is the product of the absorption cross section of the LH system at the wavelength (or band) of the excitation and the light intensity of the excitation.[27] Because of the increasing demand for kinetic Chl fluorometers in ecophysiology, the market has offered a rich selection of families of instruments, including the PEA (plant efficiency analyzer) by Hansatech Instruments Ltd (King's Lynn, Norfolk, United Kingdom, 2001), the PAM (pulse-amplitude modulation) by Walz (Effeltrich, Germany, 1986), and PSI from Photon System Instruments (Brno, Czech Republic). These commercial fluorometers use modulation methods with a series of short and nonsaturating flashes. The hardware and software are constructed primarily to study the photosynthetic capacity of higher plants but can be used to investigate anoxygenic photosynthetic bacteria with appropriate modifications, as well.[43]

23.2.2 RELAXATION OF PROMPT FLUORESCENCE

The yield of BChl fluorescence gradually decreases (relaxes) from a high level to a lower one after termination of the excitation tested by weak monitoring flashes in the dark (Figure 23.3). While the induction raises the yield from a dark level to a light-adapted level, the relaxation reverses this process but with significantly different kinetics (Figure 23.4). The qualitative interpretation of induction and relaxation of BChl fluorescence is similar and is based on the old and simple principle of photochemical trapping of excitons in the antenna: the BChl fluorescence level is low (F_0) if the RC is open and high (F_{max}) if it is closed.[44] The observed kinetics of relaxation reflects all light-induced redox reactions of the RC and donor/acceptor pools, which are responsible for closure and reopening of the RC for photochemistry.[45]

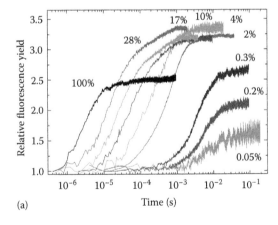
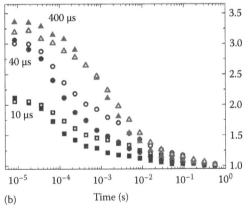

FIGURE 23.4 (a) BChl fluorescence induction and (b) relaxation in intact cells of *Rba. sphaeroides* at attenuated exciting light intensities (%) and after variable duration of excitation (10–400 μs), respectively. The (relative) fluorescence yield refers to that of the initial value, and note the logarithmic time scale. The induction kinetics is a complex function of the light intensity (a), and the relaxation becomes significantly slower at longer excitation but only slightly slower in the presence of terbutrine (closed symbols in b).

23.2.3 DELAYED FLUORESCENCE

Prompt fluorescence will be observed if the excited BChl is deactivated directly by emission of fluorescence. The lifetime of the prompt fluorescence can change from a couple of nanoseconds (intrinsic lifetime of BChl fluorescence) to (sub) picoseconds depending on the rates of various competing reactions (energy transfer, stabilization of separated charges, etc.). There is another way of emission when the energy of the excited state is transiently stored by intermediate (triplets, unrelaxed or relaxed charge-separated) states and the BChl* (P*) is repopulated thermally after definite delay (Figure 23.5). This type of emission is called delayed fluorescence (DF). The spectrum of the DF is indistinguishable from that of the prompt fluorescence, and the lifetime depends on the back-reaction kinetics of the intermediate steps. It can be as short as that of the prompt fluorescence (in this case, they can be hardly distinguished based on kinetic measurements) but will fall in the much larger (millisecond-to-second) time domain if the electron transfer reaches Q_A after charge separation. The slower the kinetics of the DF, the smaller will be its intensity as P* is repopulated from always deeper and deeper states. The exponential Boltzmann factor determines the DF intensity from the depth of the free energy gap between P* and the intermediate state. In the bacterial RC, this gap is much larger (about 860 meV) than that in green plants (about 450 meV), which makes the intensity of DF from bacterial systems extremely weak.

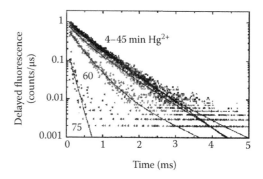

FIGURE 23.6 Decay of delayed fluorescence from whole cells of *Rba. sphaeroides* treated by 26 μM mercury ions for various durations (the numbers refer to minutes) after flash excitation (Nd:YAG laser at 532 nm, flash duration of 5 ns, and repetition rate of 1 Hz). The modified rate constants and amplitudes of the DF kinetics indicate the harmful effect of Hg^{2+} ions on photosynthetic bacteria in the early phase of destruction of the photosynthetic apparatus.

The measurement of DF in the bacterial RC requires special techniques mainly due to two factors: (1) the DF intensity is exceptionally low, and (2) the intensive prompt fluorescence cannot be spectrally separated from the DF as they have identical spectra. To overcome these difficulties, either a mechanical shutter should be used to separate prompt fluorescence and DF,[46–50] or the photodetector should be protected from oversaturation by electronic gating.[51] Photon counting mode is frequently applied to follow the drop of the DF intensity within two or three orders of magnitude with acceptable signal to noise (Figure 23.6).

To observe the Kautsky effect in photosynthetic organisms, two conditions are required at least: (1) the presence of fluorescence quenchers either of photochemical or nonphotochemical origin and (2) short- or long-range order of the components participating in BChl fluorescence. The ordered arrangement of the complexes within the photosynthetic membrane together with connectivity of the RC and some neighboring antenna complexes (PSU) assure the exchange of excitation energy and/or reducing equivalents, resulting in complex kinetics of induction/relaxation of BChl fluorescence. The fluorescence is indicative of the assembly and function of the RC and photosynthetic membrane.

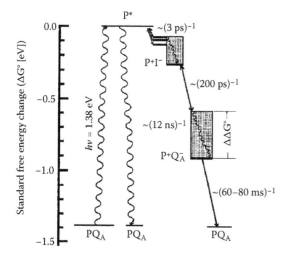

FIGURE 23.5 Forward and backward electron transfer between transient states in RC of *Rba. sphaeroides*. The absorbed photon ($h \cdot \nu$) excites the BChl dimer (P) to its first excited singlet state (P*). The de-excitation can occur either through a sequence of charge stabilization steps or through emission of prompt fluorescence (wavy arrow to the ground state). P* can be thermally repopulated from any of the transient charge-separated states, resulting in fluorescence emission with a delay in prompt fluorescence. The intensity and kinetics of the delayed fluorescence depends on the free energy difference between P* and the intermediate state and on the rate constants of the electron transfer, respectively.

23.3 FLUORESCENCE FROM ISOLATED RC

23.3.1 STRUCTURE AND FUNCTION OF THE BACTERIAL RC

Light-induced charge separation and subsequent electron transfer take place in the RC protein resolved with atomic precision (for *Rba. Sphaeroides*, see Ref. 52). Depending on the nature of the terminal electron acceptor species, type I and type II RCs can be defined. The type I RCs in green and brown sulfur bacteria (*Chlorobiaceae* and *Heliobacteriaceae*) use iron–sulfur complexes to transfer electrons one at a time from an intermediate quinone to a soluble ferredoxin. The type II RCs in all purple (sulfur-oxidizing *Chromatiaceae* and nonsulfur

Rhodobacteraceae) and green filamentous (*Chloroflexaceae*) bacteria export reducing equivalents in pairs, as quinol. All RCs house a homologous pattern of cofactors and protein arrangements with strong twofold symmetry (Figure 23.7). In purple photosynthetic bacteria, the RC is a heterodimer of similar L and M subunits that bind all the active cofactors: four BChls, two bacteriopheophytins (Bpheos), two quinones (Q), and a nonheme iron atom (Fe). The H subunit stabilizes the whole complex, is enriched with water molecules at the cytoplasmic side, and facilitates proton uptake coupled to the electron transfer. Although the two branches of cofactors show a high degree of rotational symmetry, the electron transfer from the primary donor P to the secondary quinone Q_B via $BChl_A$, BPh_{eoA}, and Q_A occurs exclusively through the A branch. The structural and compositional symmetry contrasts with additional functional asymmetry. Whereas the two quinones are chemically identical in many species, their properties are very different: Q_A is a tightly bound and one-electron redox couple, and Q_B is a reversibly bound and two-electron redox species. The light-induced charge transfer reactions in the RC are influenced by external factors, including the local electric fields of the phospholipids in the membrane.[53,54] In order to avoid the wasteful recombination of the separated charges, fast initial electron transfers take place on both the donor and acceptor sides. The products of the charge stabilization are the double-reduced QH_2 and the single-oxidized

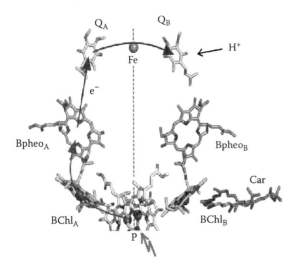

FIGURE 23.7 Arrangement of the pigments in the bacterial reaction center (*Rba. sphaeroides*). For clarity, the protein has been omitted from this picture. The dimer P, the monomeric BChl, Bpheo, and ubiquinone Q cofactors (sticks) are bound to the protein, form two membrane-spanning branches A (active) and B (inactive), and are arranged around a quasi-twofold rotational symmetry axis that is normal to the plane of the membrane and passes through the bacteriochlorophyll dimer (P) and the nonheme iron (Fe) midway between the primary (Q_A) and the secondary (Q_B) quinones. Mg atoms of BChl and nonheme Fe are shown as spheres. The light-induced charge separation evokes electron transfer from P to Q_B via monomeric $BChl_A$, $Bpheo_A$, and Q_A and uptake of the H^+ ions by Q_B^- via an extended H-bond network of water molecules and protonatable amino acids in the cytoplasmic phase.

cyt c^{3+} exported into the membrane and the aqueous periplasmic side of the RC, respectively.

23.3.1.1 Donor-Side Reactions

The immediate electron donor to the oxidized dimer is a *c*-type heme that can be (depending on strains) (1) a triheme (genus *Rhodovulum*[55]) or tetraheme (*Blastochloris* [*Bcl.*] *viridis*, *Rubrivivax* [*Rvx.*] *gelatinosus*, and *Thermochromatium tepidum*[56]) cyt *c* subunit of the RC or (2) a soluble (*Rba. sphaeroides*, *Rhodobacter* [*Rba.*] *capsulatus*, *Rhodopseudomonas* [*Rps.*] *palustris*, *Rhodospirillum* [*Rsp.*] *rubrum*, etc.) or membrane-bound monoheme cyt *c* (cyt c_y of *Rba. capsulatus*). In cyt subunits bound firmly to the RC of *Bcl. viridis*, the hemes extend linearly into the periplasmic space and are arranged alternatively according to the redox midpoints (high–low–high–low). The overall transit time of the electron in the subunit is about 10 μs. The electron transfer from the RC subunit and to the cyt bc_1 complex is mediated by soluble carriers (*c*-type cyt or high-potential iron–sulfur protein) with half-time between 100 and 400 μs, which is still much shorter than few milliseconds of rate-limiting reactions of the cyclic flow in the membrane.

The electron donation to P^+ in RC with no separate cyt subunit occurs by mobile cyt *c*. The molecular details of docking and binding and kinetics of electron transfer have been well established for *Rba. sphaeroides* (see review in Ref. 57). Due to the electrostatic steering effect, the binding rate of the reduced cyt *c* is higher ($k_{on} \approx 1 \cdot 10^9$ M^{-1} s^{-1}, close to the diffusion limit) than that of the oxidized form. As the dissociation rates ($k_{off} \approx 2 \cdot 10^3$ s^{-1}) are similar, cyt c^{3+} is more tightly bound (about fourfold) to the RC than cyt c^{2+}.[58,59] The product (oxidized cyt *c*) inhibits the fast exchange for reduced cyt *c* at the docking side of the RC.

23.3.1.2 Acceptor-Side Reactions

The quinone bound to the Q_B binding site of the RC is fully reduced to QH_2 by two successive photochemical turnovers and exchanged for new Q from the quinone pool. The quinone reduction cycle is depicted in Figure 23.8. The coupling of the electron transfer with protonation processes makes the acceptor-side reactions more complicated than those on the donor side. Mutational and structural studies showed that the protons were picked up from the aqueous cytoplasmic side at the "proton entry" point determined by His-H126, His-H128, and Asp-H124 and were delivered to Q_B through a relatively short chain of protonatable amino aids.[60] It was also observed that mutations at remote sites, which were, however, in H-bond connection with the acidic cluster at the Q_B site, could severely modify the pattern of proton uptake.[61] These investigations led to the extension of the concept of a limited proton delivery pathway and favored the more distributed locations of binding and delivery pathways of protons. According to this view, the significant part of the cytoplasmic site of the RC from Q_B to Q_A binding sites constitutes a network of H-bonds consisting of protonatable amino acids and water molecules that facilitates collective phenomena in protonation.[62,63]

A driving force assay on interquinone electron transfer demonstrated the sequential reduction of the quinone to

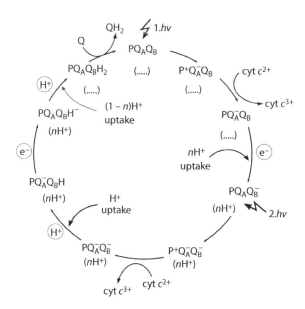

FIGURE 23.8 Quinone reduction turnover in the bacterial RC by proton-coupled light-induced electron transfer. The RC exports reduced (QH_2) and oxidized (cyt c^{3+}) species at the expense of photons and protons. Following the first saturating flash, charge separation occurs, the electron is transferred to Q_B, and the anionic semiquinone Q_B^- induces substoichiometric ($n < 1$) uptake of "Bohr protons" by the protonatable groups of the protein. The photooxidized dimer (P^+) is rereduced quickly by cyt c^{2+}, and the RC is prepared for a new charge separation by the second flash. The second electron transfer to Q_B is preceded by the uptake of a "chemical proton" by Q_B^-. After double reduction of Q_B, the second proton needed by QH_2 will be obtained partly from the bulk and partly from internal rearrangement of the Bohr protons. The two electrons and two H^+ ions are transferred to Q_B in a sequential mode.

quinol in the order of electron–proton–electron–proton.[64] The $Q_A^- Q_B^- \rightarrow Q_A Q_B^{2-}$ second interquinone electron transfer needs the uptake of the first (chemical) proton; therefore, the electron transfer is proton activated. The rate of the electron transfer is controlled by the $Q_A^- Q_B^-(\dots) \leftrightarrow Q_A^- Q_B^-(H^+)$ protonation equilibrium between the semireduced quinones.

It is worth mentioning that the redox interconversions of Q in the oxidizing (Q_o) and reducing (Q_i) sites of the cyt bc_1 complex follow both sequential and concerted patterns. At the Q_o binding site, the two electron transfer steps occur in a concerted way, involving two distinct acceptor chains. At the Q_i site, however, the reactions occur as a stepwise process, involving a singly reduced intermediate. Sequential proton release can be observed in the oxygen-evolving complex of photosystem II of green plants. The stoichiometry of the electrons and protons in the water cleavage process is close to even (1:1:1:1), although this has been disputed recently.[65]

23.3.2 FREE ENERGY AND PROTONATION OF Q_A

The unique advantage of the DF of the dimer is the possibility to determine the absolute value of the free energy of $P^+ Q_A^-$ relative to that of P^* by measuring the ratio of the intensities of the prompt and delayed emissions.[46,48,66] In the isolated RC

from *Rba. sphaeroides*, the amplitude of the DF is extremely small: $A_{DF}/A_{prompt} = 2.3 \times 10^{-16}$,[48] which indicates a considerable drop in the free energy of $P^+ Q_A^-$, amounting to $\Delta G^\circ_A = -910$ meV at pH 7.8. This low free energy level stabilizes the intermediate charge-separated state at the expense of a decrease in the yield of photochemical energy conversion.

The rate of electron transfer to and from Q_A in the RC depends on the driving force, which can be conveniently modified by changing the redox midpoint potential of the primary quinone. This can be achieved either by substitution of the native ubiquinone by quinones of different midpoints[48,67,68] or by changing the proteineous environment around Q_A by site-directed mutagenesis.[69] Low-potential quinones that replace the native ubiquinone decrease the free energy gap and therefore increase sensitively (in an exponential manner to $-\Delta\Delta G^0_A$) the intensity of DF. Simultaneously, the rate of the DF decay will also increase due to the introduction of an alternative (indirect) charge recombination pathway. The charge back reaction occurs directly (by a tunneling mechanism) and indirectly via an uphill reaction through the thermally excited $P^+ I^-$ state, depending on the free energy of $P^+ Q_A^-$. The direct path dominates if $G^\circ_A < -800$ meV.[70] The energetics of a great variety of low-potential quinones at the Q_A binding site was determined by DF measurements (Figure 23.9).

Single-site mutation of isoleucine to threonine at M265 increases DF by two orders of magnitude, revealing 120 mV lowering of the midpoint potential of Q_A.[66] The free energy gap between P^* and $P^+ Q_A^-$ depends not only on the midpoint of Q_A/Q_A^- but also on that of P/P^+. Several mutations were constructed close to the dimer whose midpoint potential increased systematically by introduction of H-bonds.[50,71] Mutations on the acceptor and donor sides control the energetics of the cofactors in the RC, which can be followed by measuring the DF of the dimer.

The free energy level of the quinone at the Q_A site, and thus the corresponding DF signal, should be sensitive to the protonation equilibrium either with the quinone itself or with the amino acids nearby. Fluorescence measurements can give information about the direct proton uptake by Q_A^- or about the pK_a values and positions of the neighboring amino acids participating in flash-induced H^+ ion binding. Earlier equilibrium redox potentiometry on native chromatophore membranes of *Rba. sphaeroides* showed strong pH dependence with a slope of -60 mV/pH unit up to $pK_a = 9.8$ for the reduced species, i.e., the reduction was coupled to the uptake of one proton per electron.[72] As the semiquinone is anionic, the protons are bound to the protein, rather than to the quinone, but coupled to Q_A reduction. In strong contrast with potentiometric redox titrations of Q_A, the DF showed very low (10–15 mV/pH unit) pH dependence in the isolated RC[47,48] and in chromatophores.[49] Several possibilities can be considered to solve the discrepancy. (1) Q_A might be reduced by disproportionation with QH_2 through the Q_B site, so the titration actually reflects the quinone pool, giving the -60 mV/pH unit dependence expected for the Q/QH_2 couple. However, the parameters necessary to achieve a strong pH dependence are not in good agreement with expected properties of Q_A and Q_B. (2) The time scale of

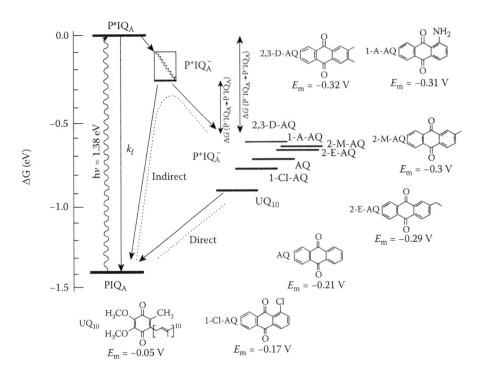

FIGURE 23.9 Direct and indirect pathways of charge recombination and origin of delayed fluorescence in bacterial RC where the native ubiquinone (UQ_{10}) at the Q_A binding site is replaced by a series of low-potential quinones (anthraquinone, AQ analogs). Upon decrease of the redox midpoint potential of the quinones, the $P^+Q_A^- \rightarrow PQ_A$ back reaction can proceed in an indirect way through P^+I^- (reduced $Bpheo_A$), and the free energy gap between P^* and $P^+Q_A^-$ decreases, resulting in increased delayed fluorescence.

the equilibrium potentiometric titrations allows the reduced state $\left(Q_A^-\right)$ to relax to a different conformation that is accompanied by stoichiometric H^+ binding. (3) As the Q_A is buried into the protein interior and not readily accessible for the redox mediators, the poor mediation could be a reasonable danger in the potentiometric measurements.

23.3.3 Kautsky Effect of the RC

The induction of the dimer fluorescence is unique in the isolated RC and displays different kinetics as observed routinely

in intact cells (Figure 23.10). Depending on the conditions, the intensity of the fluorescence decreases rather than increases during dark–light transition because P that is capable of emitting fluorescence will be converted to a nonfluorescence P^+ form progressively during the light excitation.[73,74] If, however, a fast external electron donor is added to the RC, an increase of the fluorescence intensity is experienced. Due to the fast electron donation, the amount of fluorescent P remains constant, but the primary quinone will be reduced progressively to Q_A^-, which is a weaker fluorescence quencher than its oxidized form Q_A. In case of known quenching properties of

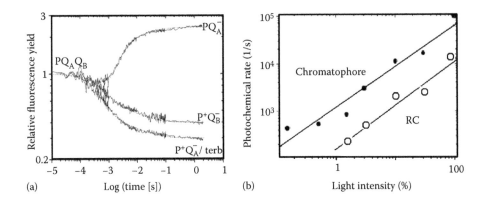

FIGURE 23.10 Induction of BChl dimer fluorescence from isolated bacterial RC of *Rba. sphaeroides*. (a) The kinetics are sensitive to the occupancy of the Q_B binding site by native quinone or by terbutrine, a potent interquinone electron transfer inhibitor, and to the fast external donor (ferrocene) to P^+. (b) Comparison of the photochemical rate constants versus (attenuated) light intensity measured in RCs isolated (no antenna) or embedded in chromatophore membrane (with antenna).

the species, the analysis of the fluorescence induction gives information about (light and dark) kinetics of redox conversion of the major cofactors in the RC. Additionally, the rate constant of the photochemical reaction reflects the absorption cross section of the apparatus. As the isolated RC does not have any harvesting pigments, the photochemical conversion is much (about two orders of magnitude) slower than that in chromatophores where the RC is embedded in the LH system. The amplification of the photochemical rates is characteristic of the antenna size and connection to the RC. These are essential features of the bacterial system, and BChl fluorescence is one of the few available assays for their determination.

Another interesting observation is the wavelength dependence of the induction and polarization of the fluorescence.[75] With narrow-band laser diodes, the accessory BChl (B, 800 nm), the upper excitonic state of the dimer (P_-, 810 nm), and the lower excitonic state of the special pair (P_+, 865 nm) were excited selectively. The fluorescence spectrum of the wild-type RC showed two bands centered at 850 nm (B) and 910 nm (P_-). While the monotonous decay of the fluorescence yield at 910 nm tracked the light-induced oxidation of the dimer, the kinetics of the fluorescence yield at 850 nm showed an initial rise before the drop. The anisotropy of the fluorescence excited at 865 nm (P_-) was very close to the limiting value (0.4) on the whole spectral range. The excitation of both B and P_- at 808 nm resulted in wavelength-dependent depolarization of the fluorescence from 0.35 to 0.24 in wild type and from 0.30 to 0.24 in the RC of a triple mutant (L131LH–M160LH–M197FH). The additivity law of the anisotropies of the fluorescence species accounts for the wavelength dependence of the anisotropy. The induction and anisotropies of the measured fluorescence can be explained by very fast energy transfer from $^1B^*$ to $^1P_-$ (either directly or indirectly by internal conversion from $^1P_+$) and to the oxidized dimer.

23.3.4 BIONANOTECHNOLOGICAL APPLICATIONS

The bacterial RC is a robust redox protein that can preserve the ability of light-induced charge separation under a wide range of conditions for a long time. It serves as a light-driven electronic device with a quantum conversion efficiency close to 100%[76] and can be an excellent target for research and utilization of the solar fuel.[77,78] In the presence of an appropriate pool of electron donors and acceptors, the RC can be considered as a continuous source of electrons upon illumination. This property is utilized in a diversity of redox-linked reactions mimicking the natural photosynthesis. Photoactivated biodevices are constructed by incorporation of the bacterial RC into artificial proteoliposomes.[79] The temporal and thermal stability of the membrane system is usually restricted but can be enhanced by the use of a complex membrane structure consisting of phospholipids, lipopolymers, and detergents[80] or by incorporation of LH complexes into the membrane.[81]

In addition to mimicking the primary processes of photosynthesis in aqueous membrane systems, another strategy also has been used by coupling the light-induced electron transfer from the RC to external devices.[82,83] By attachment of bacterial RCs to single-walled carbon nanotubes, a one-dimensional photosensitive electron conductor was constructed.[84] In a different approach, the RC was bound to a nickel layer on a gold surface through a polyhistidine tag at the end of the M subunit and incorporated into a photovoltaic cell. In this way, an enhanced photocurrent was generated through molecular relays with LH complexes.[85–87] By deposition of RC isolated from *Rba. sphaeroides* on the surface of indium tin oxide (ITO), a transparent conductive oxide, a nanocomposite was created where the RC preserved the photochemical activity.[88] In similar experiments, the bacterial RC was bound to porous silicon microcavities either through silane–glutaraldehyde chemistry or via a noncovalent peptide cross-linker and preserved the accessibility to external cofactors.[89] Although the quantum efficiency of the RC embedded into photo devices is generally low (~10%), these new types of biomaterials are useful models of optoelectronics. The nanoscale circuits integrated into solid-state electronics are attractive biotechnological applications, and the development of biodevices has become a challenging new field. Understanding what happens at the nanoscale could allow us to tailor-design materials to build better solar cells, batteries, nanoscale wires, and more.

23.4 RC-CONTROLLED FLUORESCENCE FROM INTACT BACTERIA

The photosynthetic apparatus of the bacteria in the ICM accommodates all the components necessary to convert light energy to electrochemical potential (Figure 23.1). The LH (antenna) complexes absorb photons and allow the transfer of electronic excitation to the RC protein. From the high degree of connectivity in the electronic excitation transfer, the dense packing of these complexes can be deduced.[90] In the RC, charge separation occurs between P and Q_B, which is further stabilized by subsequent electron transfers from the secondary donor (cyt c) in the periplasmic space to the pool of quinones (Q) in the hydrophobic domain of the membrane. Cytochrome c_2 and quinones act as electron transfer shuttles between two membrane-bound proteins, the RC and the cyt bc_1 complexes. The cyclic electron transfer through the RC is accompanied by net uptake of protons from the cytoplasm and transfer to the periplasm. The light-induced proton motive force completed by the cyt bc_1 complex drives ATP synthesis.

As there is an overwhelming excess (about two orders of magnitude) in the number of antenna BChls in relation to the BChl dimer in the RC, the observed fluorescence as a vast process during excitation energy migration will originate from the LH BChls. The RC, however, is a shallow photochemical trap for the excitation and can control the yield of antenna fluorescence by modification of the rates of de-excitations of the various pathways.[91] The different fluorescence assays of induction,[18,92,93] relaxation,[45] and DF[20] can be used to characterize the electron and proton transfer in intact cells.

23.4.1 Polyphasic Fluorescence Induction

The kinetics of BChl fluorescence in response to rectangular-shaped excitation has several phases in the time range from milliseconds to minutes in intact cells of *Rba. sphaeroides* (Figure 24.4). The dominating component is the (photochemical) rise attributed to charge separation in the RC, which is overlapped by additional phases due to electronic excitation and electron transfer and to other effects (e.g., conformational change) of smaller impact on the fluorescence yield.

23.4.1.1 Photochemical Rise

The charge separation is purely photochemical and follows the reciprocity law: the product of excitation light intensity and rate constant of the photochemical rise is constant.[94] At smaller and larger light intensities, the photochemical phase is overlapped by other phases due to electron transfer from RC to cyt bc_1 complex and (BChl or carotenoid [Car]) triplet quenching, respectively. The photochemical phase is less perturbed in wild-type bacteria when the rise time is about 1 ms, a kinetic borderline between a low level of exchange of electrons between RC and other membrane components and insignificant accumulation of triplet quenchers.

At a low level of light intensity exciting the RC, the slow photochemical accumulation of closed RC $\left(P^+Q_A^-\right)$ competes with reopening of the RC due to import and export of electrons on the donor and acceptor sides, respectively.[45] The bottleneck of the (cyclic) electron transport in the membrane is constituted by reactions of the cyt bc_1 complex. The turnover rate can be conveniently followed by monitoring the characteristic electrochromic changes of the carotenoids, which serve as sensitive endogenous probes (voltmeters) of the membrane electric potential. The method is useful for flash or continuous light excitations in intact cells.[92] The overall turnover time ranges from a few milliseconds to some tens of milliseconds. The actual value is determined by (1) the cyt bc_1/RC ratio and (2) the rates of electron transfer via mobile quinone and cyt c species. If the cyt bc_1/RC ratio is large (0.5 as in *Rba. sphaeroides* or 2.0 as in *Rba. capsulatus* grown at very high light intensity), and the quinone pool is reduced (as it is under anaerobic conditions), then the half-time of the flux will be as small as 1 ms. If, however, the bc_1/RC ratio is small (0.1 as in *Rsp. rubrum*), then much more time (some tens of milliseconds) will be required to rereduce the large amount of photooxidized RC. The discussion of the rates of diffusion-controlled redox reactions on the donor and acceptor sides needs deeper insight.

23.4.1.1.1 Donor-Side Limitation

Fluorescence induction and relaxation measurements indicate that the rate-limiting reaction is the rereduction of the oxidized dimer by a mobile electron carrier of reduced cyt c_2 in *Rba. sphaeroides*.[42] The simultaneous absorption change measurements of P/P^+, the effects of depletion of cyt c_2 (CYCAI mutant), and the negligible influence of the acceptor-side inhibitor (terbutryn) on the relaxation kinetics demonstrate that the relaxation is controlled dominantly by the donor-side reactions.[45,93] This can be attributed to the possible supramolecular organization of the photosynthetic apparatus,[95] which assures a relatively small pool of mobile electron donors to P^+ and can be exhausted shortly after the excitation. The longer the excitation, the slower the relaxation of the closed state of the RC, i.e., more time is needed for opening the RC. In photosynthetic bacteria with a cyt subunit bound to the RC (*Rvx. gelatinosus* or *Bcl. viridis*), the view is not so clear as the redox conditions and confinements of the quinone pool in the membrane (see Section 23.4.1.1.2) may play a determining role. If the quinone pool is highly reduced, then the Q_B binding site of the RC will be occupied mainly by QH_2, and the export of the reduced species will be severely limited. Additionally, it has been proposed that the quinone pool is a misnomer and each quinone molecule in the pool can only access a limited number of RCs.[96] These conditions can be made responsible for acceptor-side limitation in the relaxation of the BChl fluorescence of *Rba. sphaeroides* reported earlier.[22,97] For comparison, the relaxation of the Chl fluorescence in green plants is exclusively attributed to the quinone acceptor system because the rereduction of the oxidized Chl dimer $\left(P_{680}^+\right)$ by the secondary donor (tyrosine 161 of the D_1 subunit of photosystem II) is much faster (>1/μs) than any of the electron transfer reactions in the acceptor side.

23.4.1.1.2 Acceptor-Side Limitation

In all photosynthetic anoxygenic bacteria, ubiquinones are the mobile electron carrier between the RC and the cyt bc_1 complex. As they have a long isoprenoid tail, their mobility is limited to hydrophobic phases of the membrane and protein interfaces. The quinone/quinol exchange at the Q_B site can meet structural obstacles. It is well demonstrated that the ring of LH1 molecules surrounds the RC both in monomeric and in dimeric (S-shaped) association forms.[98] Both open and closed rings of LH1 were observed depending on the species and conditions. The encircled RC forms an apparent barrier against the exchange of the redox species.[99] It was discovered that the small PufX polypeptide of 80 amino acids was required for dimeric core complex formation and played a crucial role in other forms of long-range supraorganization of the PSU.[100,101] However, in the absence of high-resolution structures of the core complexes, its precise location and function are still under debate. The best resolution (4.8 Å) was obtained from the crystal structure of *Rps. palustris*, where the complete closure of the RC was prevented by a single transmembrane helix (W), whose structure is not known.[102] The PufX polypeptide can belong to the LH1 assembly at the dimer junction[98] or is situated between the LH1 ring and the Q_B site of the RC.[103] The export of the reduced quinone is brought about by fine tuning of the activities of all the key players in the membrane. The transport of Q/QH_2 through the ring of the core antenna should be controlled by the photochemistry of the RC; the flexibility of the LH1 complex; and the location, structure, and dynamics of the PufX polypeptide.[95,103,104]

The quinone reduction cycle in chromatophores takes about 1.6 ms, including 1.2 ms required for the slow binding-release steps.[96] The half-time required for the released quinol to arrive and bind at the oxidizing site of the cyt bc_1 complex (Q_0) takes 5–10 ms.[105] The overall cycling time of electrons in chromatophore was estimated to be 13 ms[106] and 22 ms.[107] Similarly to the mobile cyt c carriers on the donor side, the quinones in the supercomplex structure are also exposed to confinement,[96] which shortens the average distance of diffusion between the Q_B and Q_0 sites of the RC and cyt bc_1 complex, respectively.[108] However, it was found that the rate of diffusion was not the rate-limiting step of the electron transfer.[106] Recently, a theoretical treatment of the overall operation of the electron transfer was provided where the model considers the reverse electron flow from the ubiquinone pool.[109] The attempt is challenging, but it is hard to deduce kinetic data from this study.

23.4.1.1.3 Monitoring the Photosynthetic Capacity

There is a continuing interest in finding appropriate methods to monitor the photosynthetic capacity of different organisms *in vivo* or under field conditions.[14,21,97,107,110] The rate of the electron transfer characterizes properly the photosynthetic capacity of the photosynthetic organism. For higher plants, several techniques have been elaborated, among which the fluorescence induction of Chl fluorescence has proved to be the most convenient and widespread method (see, for recent review, Refs. 7, 112, and 113). Families of Chl fluorometers (PEA, PAM, and PSI; see Section 23.2.1) have found a large market in the fields of research, agriculture, and protection of the aquatic ecosystem.[24,113–115] In most types of PAM fluorometer, the rate of linear electron transfer (the number of electrons passing through photosystem II) in oxygenic organisms can be estimated routinely based on calibration between the fluorescence yield and the amount of fixed CO_2 in the absence of photorespiration (sometimes neglected in stress experiments).[116,117]

Unfortunately, a similar assay with comparable precision has not been worked out yet to determine the photosynthetic cyclic electron flow in bacteria, although there have been some encouraging attempts.[42,43,107] If the cyclic electron flow is driven by continuous low light, then the bottleneck can be shifted from the cyt bc_1 complex to the RC. The condition for the interchange is the commeasurable rates of light excitation (photochemistry) in the RC, k_I, and the electron transfer reactions in the cyt bc_1 complex, k_e. At a lower rate of light-induced charge generation than that of the efflux of charges in the RC, the function of the RC will determine the observed rate of the cyclic electron flow. The steady-state conditions can be utilized to determine the rate of the light-induced cyclic electron transfer from the measured fluorescence level, $F = F_{max} - (F_{max} - F_0) \cdot k_e/(k_I + k_e)$, where F_0 and F_{max} are the initial and maximum levels of fluorescence. If the stationary conditions are not fulfilled, the decomposition of the kinetics will be not straightforward, due to the combination of one-electron donor-side and two-electron acceptor-side reactions. Based on the quinone reduction cycle of the RC (Figure 23.8),

the temptation is large to construct and solve a set of kinetic equations (for higher plants, see Refs. 16, 119, and 120). In general, the solution has highly limited predictive power due to the large number of unknown or faintly known parameters (kinetic rate constants).

The main difficulty in determining the photosynthetic capacity originates from the lack of the proper assay to measure one of the final products (e.g., ATP) and the relationship between the yield of the BChl fluorescence and the amount of the product (calibration). Additionally, the interaction of the photosynthetic electron transport with the respiratory chain and the redox poise as well as the stoichiometry of the redox agents and protein complexes are also less elaborated. All vary significantly and depend on the species and growth conditions in a hardly predictable way. For example, under anaerobic conditions in the dark, the quinone pool is largely reduced, and neither the RC nor the cyt bc_1 complex can function optimally. Similarly, the oxidizing conditions are also not favorable. For optimum function, the ambient redox poise should be kept in the region of the redox midpoint potential of the ubiquinone pool (~100 mV). The stoichiometric ratios are also important parameters but depend on the strains, light intensity, oxygen tension, etc. Typically, the ubiquinones are in large excess ([UQ]/[RC], ~20–30), but fewer cyt bc_1 complex and mobile cyt c carriers are available ([cyt bc_1]/[RC], ~0.5, and [cyt c_2]/[RC], ~0.5) in *Rba. sphaeroides* grown under anaerobic photosynthetic conditions. Despite the difficulties, the calibration awaits elaboration, and efforts will be soon benefited by reliable and practical determination of the cyclic electron transfer in bacteria.

24.4.1.2 Triplet Quenching

Upon increase of the exciting light intensity, the photochemical rise of the BChl fluorescence is overlapped by a fast decrease due to appearance of (BChl and/or carotenoid) triplet (T) quenchers from singlet (S) states via a reversible light reaction in a closed RC: $Q_A^-[S] \underset{k_T}{\overset{k_I}{\rightleftharpoons}} Q_A^-[T]$ (Figure 23.11). The rate constant of production of T in the light is $k_I + k_T$, and that of disappearance in the dark is k_T. The lifetimes of the triplets are short, specific to the species, and atmosphere dependent if the triplet transfer to the triplet ground state of oxygen is energetically favored.[120] At room temperature, the lifetime of carotenoid triplet states is between 1 and 10 μs (3–5 μs[121] and 3.5 μs[122] in *Rsp. rubrum* and 8.4 μs[21] and 10 μs[123] in *Rba. sphaeroides*), and the decay time of the BChl triplet state is about 6 μs[121] in the RC and about 70 μs[121] or 50 μs[123] in the antenna. After a transient time (>1/[$k_I + k_T$]), stationary (BChl or Car) triplet level will be established and will not modify the kinetics of the fluorescence induction. The level of the quenched fluorescence will depend on the k_I/k_T ratio. In carotenoidless mutant strain R-26, the effect of BChl triplet quenching is more pronounced than in the wild-type strain. The fluorescence induction is an excellent assay to study the protective function of carotenoid triplets both in the antenna and in the RC against harmful oxidation (destruction) of BChl pigments upon intense illumination.[124,125]

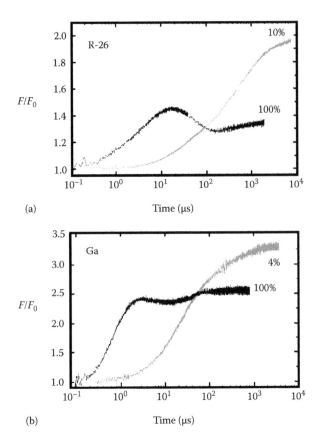

(a)

(b)

FIGURE 23.11 BChl fluorescence induction curves at high (100%) and attenuated (4% and 10%) exciting light intensities in intact cells of carotenoidless mutant R-26 (a) and wild-type (carotenoid-containing) Ga strains (b) of *Rba. sphaeroides*. The kinetics of fluorescence induction are considerably depressed by fluorescence quenching due to production of long-lived BChl (R-26) and short-lived carotenoid (Ga) triplets.

23.4.1.3 Slow Rising Phase

The maximum fluorescence level (F_{max}) cannot be reached after a single (monoexponential) rise predicted by simple photochemistry (Figure 23.12). Deviations from the mono-exponential kinetics can be observed (1) at the initial and (2) at the terminal ranges of the fluorescence rise. (1) The kinetics begins with a time delay (lag phase) attributed to the energetic coupling of the PSUs.[91,126] The extent of the initial lag phase can be used to estimate how densely the units are packed.[90] (2) Under appropriate conditions, the unusual behavior of the other end of the rising kinetics can be nicely visualized. The remaining ~20% of the rise to F_{max} slows down and behaves like a separate phase. Its origin is debated: in analogy with a similar effect of Chl induction in green plants,[9] multiple turnover of the RC, a thermal step after (or between) light reactions, heterogeneity of the antenna system, or conformational change might be made responsible for the observed slow phase of the fluorescence rise. Whatever the reason for the slow component, the bacterium cannot be brought to its maximum fluorescence state after short (laser-flash) excitation even if it is very bright and saturating from the point of view of closure of the RC.

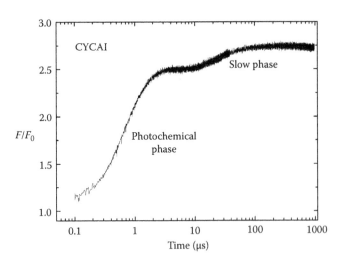

FIGURE 23.12 Separation of the photochemical and slow phases of the BChl fluorescence rise in intact cells of cytochrome c_2-deficient mutant (CYCAI) of photosynthetic bacterium *Rba. sphaeroides*. In the absence of an electron donor to P+, the short flash excitation assures single turnover of the RC and is unable to pull the fluorescence (yield) to the maximum level by a single (monoexponential) phase. The amplitude and rate constant of the rise of the slow component are about 5 and 50 times smaller than those of the photochemical phase, respectively.

Two selected problems related to (eco)physiology of photosynthetic bacteria will be discussed in the following sections to demonstrate the effectiveness of BChl fluorescence to answer practical questions.

23.4.2 Assembly of the Photosynthetic Apparatus Monitored by BChl Fluorescence

The bacterium *Rba. sphaeroides* is an ideal model system, and the light-induced cyclic electron flow provides an excellent tool for examining the biogenesis and assembly of photosynthetic membranes.[127] The bacterium shows remarkable versatility and metabolic elegance as it is capable of growth by aerobic and anaerobic respiration, fermentation, and anoxygenic photosynthesis.[128] Both photosynthesis and membrane development can be simultaneously studied, and significant problems in many areas of interest in cell biology, physiology, and bioenergetics can be addressed.

The formation of ICM can be induced by sudden modification of essential environmental factors. A convenient method is the control of the oxygen content of the culture. High oxygen tension under chemoheterotrophic growth conditions represses ICM formation but initiates a gratuitous induction of ICM assembly and BChl synthesis in the dark ("greening") by invagination of the cytoplasmic membrane (CM) together with the synthesis and assembly of the LH and RC complexes.[1,129] It was shown from fluorescence relaxation studies that the photosynthetic apparatus was assembled in the functional unit not in one step but in a sequential manner.[22,28,130] Initially, the RC-core–LH–PufX complex was built and inserted into CM in a form that was largely inactive in forward electron transfer. This was followed by the activation of the functional electron transfer together with the addition of the peripheral LH antenna (LH2) that started to surround the core complex. Finally, ICM was formed after invagination and vesicularization of the CM. A similar mechanism of sequential assembly has been observed for photosystems I and II in cyanobacteria. Additionally, the initial synthesis of the thylakoid membranes in plant proplastids may involve invagination of the inner envelope, into which thylakoid proteins were first inserted.[131]

A different method uses the development of the ICM during steady-state growing of the cells. After several generations, the division of the cells can be synchronized by different methods, including dilution of the cells in their late stationary growth phase,[132,133] starvation of the cells for phosphate or malate,[134] or exposure of the culture to a sequence of dark–light periods. In a steady-state synchronous culture, the number of cells increases stepwise as all of the cells are in the same stage of their development. Under the conditions of the experiment in Figure 23.13, the doubling time amounts about 3 h. The advantage of the synchronization is the possibility of direct tracking of the insertion of the photosynthetic apparatus

into the old and newly formed ICM. In sharp contrast to the stepwise increase of the population, the production, insertion into the ICM, and activation of the photosynthetic proteins (LH, RC, and cyt bc_1 complexes) are cell-cycle-independent processes, i.e., they follow not a stepwise but, rather, a continuous process.[135] However, the phospholipid synthesis[136] and the electrochromic signal due to the cyt bc_1 complex[20] show obvious cell-cycle dependence (Figure 23.13). The cell-cycle-dependent synthesis and accumulation of the phospholipids in the membrane are not unique to phototrophically grown cells. Reports were published about fluctuation in the rates of phospholipid synthesis and stepwise increase in total lipid content during the cell cycle of *Escherichia coli*.[137]

To understand the observed cell-cycle-dependent processes, we have to connect them to the major events of cell division. At the time of cell division, the area of the total membrane surface of the cell should increase significantly due to the resulting daughter cells with their own outer, cytoplasmic, and ICM membrane systems. This is a highly energy-consuming process that should influence the phospholipid synthesis of the whole cell. This is why a burst of phospholipid synthesis occurs prior to cell division, and the phospholipids will be inserted into the replicating ICM as it is being partitioned to daughter cells. At the time of cell division, however, the synthesis of the phospholipids is transiently interrupted, and no new phospholipids will be incorporated into the ICM. The uncoupling of protein and lipid incorporation into the ICM immediately preceding or at the time of cell division has widespread possible consequences,[138] among which we will concentrate now on the cell-cycle-specific changes of the fluidity and crowding (protein/phospholipid ratio) of the ICM bilayer. These changes will influence the supercomplex structure, the mobility of the species in the cyclic electron transfer, and the properties of the reporter molecules (BChl and carotenoids) in several ways. (1) The signaling species are usually bound to integral membrane proteins immersed in the ICM. For example, the yield of fluorescence is sensitive to the hydrophobicity of the surroundings: it is high when immersed in the membrane and becomes lower when partitioned in the aqueous phase. The process is controlled by the fluidity of the membrane. (2) The average distance between the PSUs will decrease, and therefore, the connectivity will increase upon crowding the membrane via an increased protein/phospholipid ratio. (3) The diffusion of electron carriers as part of the cyclic electron transport chain is faster at a high protein/phospholipid ratio, suggesting that the electron transport is more efficient when the membrane is crowded.[139] Indeed, the electrochromic change that is connected to the energetization of the membrane demonstrates a cell-cycle-dependent increase upon cultivation (Figure 23.13). The changes observed during the cell cycle can be attributed either to a shorter distance of diffusion (crowding) or to an increased diffusion coefficient (increased fluidity of the membrane), or to both effects.

The knowledge gained from synchronization of the bacterial culture and measurement of the electron transfer characteristics is only a small section of the great wealth of information that has been accumulated about the induction and assembly

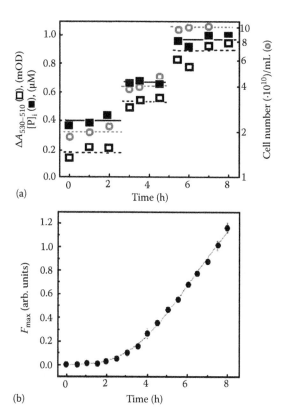

FIGURE 23.13 Cell-cycle-dependent increase of the cell number, the total phosphate content of the cells, and the amplitude of the electrochromic absorption change (a), and cell-cycle-independent increase of the maximum fluorescence of BChl induction (b) upon illumination of synchronized culture of *Rba. sphaeroides* wild type 2.4.1. The cell-cycle-dependent changes demonstrate the stepwise assembly of the membrane from phospholipids and its energetization upon division of the cell. (Adapted from Kis, M. et al., *Photosynth. Res.*, 122, 261–273, 2014.)

of the ICM and its connection to the photosynthetic apparatus. However, many other issues remain to be resolved, including better understanding of the formation, development, and function of ICM invaginations from the CM membrane during steady-state cell division: how do the essential constituents of the photosynthetic apparatus find their correct location, and how do the physical state and crowding of the membrane bilayer during cell division control the activity of the different components? The proteomics and dynamics of the ICM at the level of structural resolution in wide time scales (from femtoseconds to hours) constitute perspectives for the next few years.

23.4.3 BIOSENSING AND BIOREMEDIATION OF HEAVY METAL POLLUTION IN THE ENVIRONMENT

The electron transfer through the RC monitored by BChl fluorescence may contribute to useful applications in conservation of the environment including the protection of the biodiversity of aqueous habitats[113,140–143] and the monitoring and the remediation of heavy metal pollution.[144–146] As bacteria are involved in the transformation of metal compounds, they play an important role in the fate of the metal ion in the environment by modifying its mobility and bioavailability, and hence the intrinsic toxicity. Metal-related bacterial metabolism can be manifested by biosorption of metals into biomass or precipitation of ions. Purple photosynthetic bacteria are frequently used in these studies as they are very versatile microorganisms and can grow under different conditions: photoautotrophically or photoheterotrophically in the light, and chemotrophically in the dark under aerobic conditions. Some species are also able to grow by anaerobic respiration in the dark. Due to this versatility, they have been proved as highly promising candidates for bioremediation[147] and seem to act as sponges for the heavy metals,[148,149] including mercury accumulated mainly in waterways as a consequence of anthropogenic activities. Here, we will concentrate on recent developments in biosensing of environmental pollution by mercury (II) ions, which belong to the most toxic agents in nature. The rate of growth of *Rba. sphaeroides* culture is halved at 2 μM Hg^{2+} concentration,[27] which demonstrates orders-of-magnitude higher sensitivity than experienced for other heavy metal ions.[145]

The Hg^{2+} ions of the contaminated environment are adsorbed on the surface of the cell and transported into the cell by "open-gate" mechanism, i.e., by one of the physiological transport systems (e.g., magnesium uptake pathway). As the mercury ions have no physiological function, the bacterium tries to decrease the internal mercury concentration by the transient loss of the uptake system, the temporary increase of the efficiency of the efflux system, or the enzymatic detoxification encoded by the *mer* operon, which can be located on plasmids, chromosomes, or transposons.[150] Due to enzymatic detoxification, part of the Hg^{2+} ions is reduced by the mercuric reductase MerA, a glutathione reductase-related enzyme, converting the cation to metallic mercury.[151] Hg^0 is volatile and

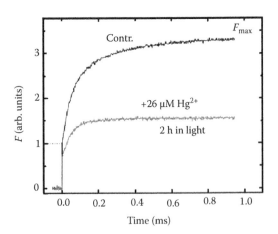

FIGURE 23.14 Induction of BChl fluorescence of photosynthetic bacterium *Rba. sphaeroides* wild type 2.4.1 in the exponential phase of development of the cells. The culture was untreated (control) or treated with 26 μM Hg^{2+} ions, and the sample was kept for 2 h in the illuminated culture before measurement. All fluorescence characteristics (the initial level F_0, the maximum level F_{max}, and the variable fluorescence F_v) demonstrate a significant drop upon mercury treatment.

leaves the cell by passive diffusion.[152] The mercury (II) ions being accumulated in the cell have strong affinity toward the thiol groups and express strong toxicity, resulting in the damage of the biosynthetic pathway of BChl and heme[153] and the photosynthetic apparatus.[27] The mercury (II) ion exerts the harmful effect on the photosynthetic machinery in a sequential manner. The RC is the primary target, and less sensitive molecular complexes are the cyt bc_1 and the LH system.[92] Although the RC protein can preserve its ability for light-induced charge separation up to high mercury concentration ($[Hg^{2+}]/[RC] < 500$), the internal electron transfer is impeded at a much lower mercury level. The most sensitive reaction is the Q_B-related electron transfer, which is blocked at a mercury concentration one order of magnitude smaller ($[Hg^{2+}]/[RC] < 50$) (Sipka et al., unpublished results). Any assay of the RC-controlled electron transfer is sensitive enough to detect the harmful effect of the mercury (II) ion on intact bacterial cells. The measurement of the fluorescence induction is a peculiarly adequate method to monitor the photosynthetic capacity of the bacteria even under field conditions[42] and has the advantage of fast (in the range of minutes) and sensitive detection of mercury in the early phase of pollution (Figure 23.14).[27]

23.5 CONCLUSIONS

Advances in BChl fluorescence kinetic techniques in combination with structural information during the past few years have vastly improved our knowledge on the mechanism of photosynthetic energy transduction and biological electron transfer. Here, we have only scratched the surface of the problems of cyclic electron transfer mediated by the bacterial RC. Further extensive characterization will certainly constitute the basis for new breakthroughs in organization and function of the photosynthetic complexes in the membrane. We can be sure that additional deeply buried secrets of the fundamental biological processes will be discovered in the near future. These findings will set the stage for engineering new biological functions and new materials based on light-induced redox chemistry mimicking the natural bacterial RC.

ACKNOWLEDGMENTS

Thanks to the programs of TAMOP 4.2.2.A-11/1KONV-2012-0060, 4.2.2.B, 4.2.2.D-15/1/KONV-2015-0024 and COST Action (CM1306) for financial support.

LIST OF ABBREVIATIONS

BChl: bacteriochlorophyll
DF: delayed fluorescence
P: bacteriochlorophyll dimer
Q_A: primary quinone
Q_B: secondary quinone
RC: reaction center

REFERENCES

1. Tucker, J.D., Siebert, C.A., Escalante, M., Adams, P.G., Olsen, J.D., Otto, C., Stokes, D.J. and Hunter, C.N., Membrane invagination in *Rhodobacter sphaeroides* is initiated at curved regions of the cytoplasmic membrane, then forms both budded and fully detached spherical vesicles, *Mol. Microbiol.*, 76, 833–847, 2010.
2. Bahatyrova, S., Frese, R.N., Siebert, C.A., Olsen, J.D., van der Werf, K.O., van Grondelle, R., Niederman, R.A., Bullough, P.A., Otto, C. and Hunter, C.N., The native architecture of a photosynthetic membrane, *Nature*, 430, 1058–1062, 2004.
3. Sener, M.K., Olsen, J.D., Hunter, C.N. and Schulten, K., Atomic-level structural and functional model of a bacterial photosynthetic membrane vesicle, *Proc. Natl. Acad. Sci. U.S.A.*, 104, 15723–15728, 2007.
4. Sener, M.K. and Schulten, K., From atomic-level structure to supramolecular organization in the photosynthetic unit of purple bacteria, in: Hunter, C.N., Daldal, F., Thurnauer, M. and Beatty, J.T., editors, *Advances in Photosynthesis and Respiration: The Purple Phototrophic Bacteria*, Springer, Dordrecht, pp. 275–294, 2009.
5. Cartron, M.L., Olsen, J.D., Sener, M., Jackson, P.J., Brindley, A.A., Qian, P., Dickman, M.J., Leggett, G.J., Schulten, K. and Hunter, C.N., Integration of energy and electron transfer processes in the photosynthetic membrane of *Rhodobacter sphaeroides*, *Biochim. Biophys. Acta*, 1837(10), 1769–1780, 2014.
6. Kautsky, H. and Hirsch, A., Neue Versuche zur Kohlensäureassimilation, *Naturwissenschaften*, 19, 964, 1931.
7. Stirbet, A. and Govindjee, Chlorophyll a fluorescence induction: A personal perspective of the thermal phase, the J–I–P rise, *Photosynth. Res.*, 113, 15–61, 2012.
8. Schansker, G., Tóth, S.C., Kovács, L., Holtzwarth, A.R. and Garab, G., Evidence for a fluorescence yield change driven by a light-induced conformational change within photosystem II during the fast chlorophyll a fluorescence rise, *Biochim. Biophys. Acta*, 1807, 1032–1043, 2011.
9. Schansker, G., Tóth, S.C., Holtzwarth, A.R. and Garab, G., Chlorophyll a fluorescence: Beyond the limits of the Q_A model, *Photosynth. Res.*, 120(1–2), 43–58, 2014.
10. Laisk, A. and Oja, V., Thermal phase and excitonic connectivity in fluorescence induction, *Photosynth. Res.*, 117, 431–448, 2013.
11. White, A.J. and Critchley, C., Rapid light curves: A new fluorescence method to assess the state of the photosynthetic apparatus, *Photosynth. Res.*, 59, 63–72, 1999.
12. Nedbal, L., Trtílek, M. and Kaftan, D., Flash fluorescence induction: A novel method to study regulation of Photosystem II, *J. Photochem. Photobiol., B*, 48, 154–157, 1999.
13. Rascher, U., Liebig, M. and Lüttge, U., Evalution of instant light-response curves of chlorophyll fluorescence parameters obtained with a portable chlorophyll fluorometer on site in the field, *Plant Cell Environ.*, 23, 1398–1405, 2000.
14. Schreiber, U., Pulse-amplitude-modulation (PAM) fluorometry and saturation pulse method: An overview, in: Papageorgiou, G. and Govindjee, editors, *Chlorophyll a Fluorescence: A Signature of Photosynthesis*, Advances in Photosynthesis and Respiration, Vol 19, Springer, Dordrecht, pp. 279–319, 2004.
15. Papageorgiou, G. and Govindjee, (eds). *Chlorophyll a Fluorescence: A Signature of Photosynthesis*, Advances in Photosynthesis and Respiration, Springer, Dordrecht, 2004.
16. Laisk, A., Nedbal, L. and Govindjee, (eds). *Photosynthesis in Silico: Understanding Complexity from Molecules to Ecosystems*, Springer, Dordrecht, 2009.
17. Hunter, C.N., Daldal, F., Thurnauer, M. and Beatty, J.T., (eds). *Advances in Photosynthesis and Respiration: The Purple Phototrophic Bacteria*, Springer, Dordrecht, 2009.
18. Trissl, H.-W., Antenna organization in purple bacteria investigated by means of fluorescence induction curves, *Photosynth. Res.*, 47, 175–185, 1996.
19. Trissl, H.-W., Law, C.J. and Cogdell, R.J., Uphill energy transfer in LH2-containing purple bacteria at room temperature, *Biochim. Biophys. Acta*, 1412, 149–172, 1999.
20. Asztalos, E. and Maróti, P., Export or recombination of charges in reaction centers in intact cells of photosynthetic bacteria, *Biochim. Biophys. Acta*, 1787, 1444–1450, 2009.
21. Bina, D., Litvin, R. and Vácha, F., Kinetics of in vivo bacteriochlorophyll fluorescence yield and the state of photosynthetic apparatus of purple bacteria, *Photosynth. Res.*, 99, 115–125, 2009.
22. Koblizek, M., Shih, J.D., Breitbart, S.I., Ratcliffe, E.C., Kolber, Z.S., Hunter, C.N. and Niederman, R.A., Sequential assembly of photosynthetic units in *Rhodobacter sphaeroides* as revealed by fast repetition rate analysis of variable bacteriochlorophyll a fluorescence, *Biochim. Biophys. Acta*, 1706, 220–231, 2005.
23. Yurkov, V. and Beatty, T., Isolation of aerobic anoxygenic photosynthetic bacteria from black smoker plume waters of the Juan de Fuca Ridge in the Pacific Ocean, *Appl. Environ. Microbiol.*, 64, 337–341, 1998.
24. Kolber, Z.S., van Dover, C.L., Niederman, R.A. and Falkowski, P.G., Bacterial photosynthesis in surface waters of the open ocean, *Nature*, 407, 177–179, 2000.
25. Kolber, Z.S., Plumley, F.G., Lang, A.S., Beatty, J.T., Blankenship, R.E., Van Dover, C.L., Vetriani, C., Koblizek, M., Rathgeber, C. and Falkowski, P.G., Contribution of aerobic photoheterotrophic bacteria to the carbon cycle in the ocean, *Science*, 292, 2492–2495, 2001.
26. Hohmann-Marriott, M.F. and Blankenship, R.E., Variable fluorescence in green sulfur bacteria, *Biochim. Biophys. Acta*, 1767, 106–113, 2007.

27. Asztalos, E., Italiano, F., Milano, F., Maróti, P. and Trotta, M., Early detection of mercury contamination by fluorescence induction of photosynthetic bacteria, *Photochem. Photobiol. Sci.*, 9, 1218–1223, 2010.

28. Kis, M., Asztalos, E., Sipka, G. and Maróti, P., Assembly of photosynthetic apparatus in *Rhodobacter sphaeroides* as revealed by functional assessments at different growth phases and in synchronized and greening cells, *Photosynth. Res.*, 122, 261–273, 2014.

29. Connolly, J.S., Samuel, E.B. and Janzen, A.F., Effects of solvent on the fluorescence properties of bacteriochlorophyll a, *Photochem. Photobiol.*, 36, 565–574, 1982.

30. Borisov, A.Y. and Godik, V.I., Energy transfer to the reaction centres in bacterial photosynthesis. II. Bacteriochlorophyll fluorescence lifetimes and quantum yields for some purple bacteria, *J. Bioenerg.*, 6, 515–523, 1972.

31. Zankel, K.L., Reed, D.W. and Clayton, R.K., Fluorescence and photochemical quenching in photosynthetic reaction centers, *Proc. Natl. Acad. Sci. U.S.A.*, 61, 1243–1249, 1968.

32. De Klerk, H., Govindjee, Kamen, M.D. and Lavorel, J., Age and fluorescence characteristics in some species of *Athiorhodaceae*, *Proc. Natl. Acad. Sci. U.S.A.*, 62, 972–978, 1969.

33. Zankel, K.L. and Clayton, R.K., Uphill energy transfer in a photosynthetic bacterium, *Photochem. Photobiol.*, 9, 7–15, 1969.

34. Sherman, L.A. and Cohen, W.S., Proton uptake and quenching of bacteriochlorophyll fluorescence in *Rhodopseudomonas spheroides*, *Biochim. Biophys. Acta*, 283, 54–66, 1972.

35. Gottfried, D.S., Stocker, J.W. and Boxer, S.G., Stark effect spectroscopy of bacteriochlorophyll in light-harvesting complexes from photosynthetic bacteria, *Biochim. Biophys. Acta*, 1059, 63–75, 1991.

36. Steiger, J. and Sauer, K., Electric field effects on the steady-state fluorescence emission of *Rb. sphaeroides* chromatophores, in: Mathis, M., editor, *Photosynthesis: From Light to Biosphere*, Kluwer Academic Publishers, Dordrecht, pp. 735–738, 1995.

37. Tóth, S.Z., Schansker, G. and Strasser, R.J., A non-invasive assay of the plastoquinone pool redox state based on the OJIP-transient, *Photosynth. Res.*, 93, 193–203, 2007.

38. Kingma, H., Duysens, L.N.M. and Van Grondelle, R., Magnetic field stimulated luminescence and a matrix model for energy transfer. A new method for determining the redox state of the first quinone acceptor in the reaction center of whole cells of *Rhodospirillum rubrum*, *Biochim. Biophys. Acta*, 725, 434–443, 1983.

39. Holmes, N.G. and Allen, J.F., Protein phosphorylation as a control for excitation energy transfer in *Rhodospirillum rubrum*, *FEBS Lett.*, 200, 144–148, 1986.

40. Ghosh, R., Tschopp, P., Ghosh-Eicher, S. and Bachofen, R., Protein phosphorylation in *Rhodospirillum rubrum*: Further characterization of the B873 kinase activity, *Biochim. Biophys. Acta*, 1184, 37–44, 1994.

41. Koblizek, M., Kaftan, D. and Nedbal, L., On the relationship between the non-photochemical quenching of the chlorophyll fluorescence and the Photosystem II light harvesting efficiency. A repetitive flash fluorescence induction study, *Photosynth. Res.*, 68, 141–152, 2001.

42. Kocsis, P., Asztalos, E., Gingl, Z. and Maróti, P., Kinetic bacteriochlorophyll fluorometer, *Photosynth. Res.*, 105, 73–82, 2010.

43. Ritchie, R.J. and Runcie, J.W., Photosynthetic electron transport in an anoxygenic photosynthetic bacterium *Afifella (Rhodopseudomonas) marina* measured using PAM fluorometry, *Photochem. Photobiol.*, 89, 370–383, 2013.

44. Vredenberg, W.J. and Duysens, L.N.M., Transfer of energy from bacteriochlorophyll to a reaction center during bacterial photosynthesis, *Nature*, 197, 355–357, 1963.

45. Asztalos, E., Sipka, G. and Maróti, P., Fluorescence relaxation in intact cells of photosynthetic bacteria: Donor and acceptor side limitations of reopening of the reaction center, *Photosynth. Res.*, 124(1), 31–44, 2015.

46. Arata, H. and Parson, W.W., Delayed fluorescence from *Rhodopseudomonas sphaeroides* reaction centers: Enthalpy and free energy changes accompanying electron transfer from P870 to quinones, *Biochim. Biophys. Acta*, 638, 201–209, 1981.

47. McPherson, P.H., Nagarajan, V., Parson, W.W., Okamura, M.Y. and Feher, G., pH-dependence of the free energy gap between DQ_A and $D^+Q_A^-$ determined from delayed fluorescence in reaction centers from *Rhodobacter sphaeroides* R-26, *Biochim. Biophys. Acta*, 1019, 91–94, 1990.

48. Turzó, K., Laczkó, G., Filus, Z. and Maróti, P., Quinone-dependent delayed fluorescence from reaction centers of photosynthetic bacteria, *Biophys. J.*, 79, 14–25, 2000.

49. Maróti, P. and Wraight, C.A., The redox midpoint potential of the primary quinone of reaction centers in chromatophores of *Rhodobacter sphaeroides* is pH independent, *Eur. Biophys. J.*, 37, 1207–1217, 2008.

50. Onidas, D., Sipka, G., Asztalos, E. and Maróti, P., Mutational control of bioenergetics of bacterial reaction center probed by delayed fluorescence, *Biochim. Biophys. Acta*, 1827, 1191–1199, 2013.

51. Filus, Z., Laczkó, G., Wraight, C.A. and Maróti, P., Delayed fluorescence from the photosynthetic reaction center measured by electronic gating of the photomultiplier, *Biopolymers*, 74(1–2), 92–95, 2004.

52. Koepke, J., Krammer, E.M., Klingen, A.R., Sebban, P., Ullmann, G.M. and Fritzsch, G., pH modulates the quinone position in the photosynthetic reaction center from *Rhodobacter sphaeroides* in the neutral and charge separated states, *J. Mol. Biol.*, 371, 396–409, 2007.

53. Pilotelle-Bunner, A., Beaunier, P., Tandori, J., Maróti, P., Clarke, R.J. and Sebban, P., The local electric field within phospholipid membranes modulates the charge transfer reactions in reaction centres, *Biochim. Biophys. Acta*, 1787, 1039–1049, 2009.

54. Gerencsér, L., Boros, B., Derrien, V., Hanson, D.K., Wraight, C.A., Sebban, P. and Maróti, P., Stigmatellin probes the electrostatic potential in the Q_B site of the photosynthetic reaction center, *Biophys. J.*, 108, 1–16, 2015.

55. Tsukatani, Y., Matsuura, K., Masuda, S., Shimada, K., Hiraishi, A. and Nagashima, K.V.P., Phylogenetic distribution of unusual triheme to tetraheme cyt subunit in the reaction center complex of purple photosynthetic bacteria, *Photosynth. Res.*, 79, 83–91, 2004.

56. Nogi, T., Hirano, Y. and Miki, K., Structural and functional studies on the tetraheme cyt subunit and its electron donor proteins: The possible docking mechanisms during the electron transfer reaction, *Photosynth. Res.*, 85, 87–99, 2005.

57. Axelrod, H., Miyashita, O. and Okamura, M.Y., Structure and function of the Cyt c_2: Reaction center complex from *Rhodobacter sphaeroides*, in: Hunter, C.N., Daldal, F., Thurnauer, M. and Beatty, J.T., editors, *Advances in Photosynthesis and Respiration: The Purple Phototrophic Bacteria*, Springer, Dordrecht, pp. 323–336, 2009.

58. Gerencsér, L., Laczkó, G. and Maróti, P., Unbinding of oxidized Cyt *c* from photosynthetic reaction center of *Rhodobacter sphaeroides* is the bottleneck of fast turnover, *Biochemistry*, 38(51), 16866–16875, 1999.

59. Larson, J.W. and Wraight, C.A., Preferential binding of equine ferricyt *c* to the bacterial photosynthetic reaction center from *Rhodobacter sphaeroides*, *Biochemistry*, 39, 14822–14830, 2000.

60. Okamura, M.Y., Paddock, M.L., Graige, M.S. and Feher, G., Proton and electron transfer in bacterial reaction centers, *Biochim. Biophys. Acta*, 1458, 148–163, 2000.

61. Tandori, J., Maróti, P., Alexov, E., Sebban, P. and Baciou, L., Key role of proline L209 in connecting the distant quinone pockets in the reaction center of *Rhodobacter sphaeroides*, *Proc. Natl. Acad. Sci. U.S.A.*, 99, 6702–6706, 2002.

62. Cheap, H., Tandori, J., Derrien, V., Benoit, M., de Oliveira, P., Köpke, J., Lavergne, J., Maróti, P. and Sebban, P., Evidence for delocalized anticooperative flash induced proton bindings as revealed by mutants at M266His iron ligand in bacterial reaction centers, *Biochemistry*, 46, 4510–4521, 2007.

63. Cheap, H., Bernad, S., Derrien, V., Gerencsér, L., Tandori, J., de Oliveira, P., Hanson, D.K., Maróti, P. and Sebban, P., M234Glu is a component of the proton sponge in the reaction center from photosynthetic bacteria, *Biochim. Biophys. Acta*, 1787, 1505–1515, 2009.

64. Graige, M.S., Paddock, M.L., Bruce, J.M., Feher, G. and Okamura, M.Y., Mechanism of proton-coupled electron transfer for quinone (Q_B) reduction in reaction centers of *Rb. sphaeroides*, *J. Am. Chem. Soc.*, 118, 9005–9016, 1996.

65. Cox, N., Retegan, M., Neese, F., Pantazis, D.A., Boussac, A. and Lubitz, W., Electronic structure of the oxygen evolving complex in photosystem II prior to O–O bond formation, *Science*, 345(6198), 804–808, 2014.

66. Rinyu, L., Martin, E.W., Takahashi, E., Maróti, P. and Wraight, C.A., Modulation of the free energy of the primary quinone acceptor (Q_A) in reaction centers from *Rhodobacter sphaeroides*: Contributions from the protein and protein–lipid (cardiolipin) interactions, *Biochim. Biophys. Acta*, 1655, 93–101, 2004.

67. Gunner, M.R. and Dutton, P.L., Temperature and $-\Delta G°$ dependence of the electron transfer from BPho⁻ to Q_A in reaction center protein from *Rhodobacter sphaeroides* with different quinones as Q_A, *J. Am. Chem. Soc.*, 111, 3400–3412, 1989.

68. Zhu, Z. and Gunner, M.R., Energetics of quinone-dependent electron and proton transfers in *Rhodobacter sphaeroides* photosynthetic reaction centers, *Biochemistry*, 1, 82–96, 2005.

69. Takahashi, E., Wells, T.A. and Wraight, C.A., Protein control of the redox potential of the primary acceptor quinone in reaction centers from *Rhodobacter sphaeroides*, *Biochemistry*, 40, 1020–1028, 2001.

70. Woodbury, N.W., Parson, W.W., Gunner, M.R., Prince, R.C. and Dutton, P.L., Radical-pair energetics and decay mechanisms in reaction centers containing anthraquinones, naphtoquinones or benzoquinones in place of ubiquinone, *Biochim. Biophys. Acta*, 851, 6–22, 1986.

71. Williams, J.C. and Allen, J.P., Directed modification of reaction centers from purple bacteria, in: Hunter, C.N., Daldal, F., Thurnauer, M. and Beatty, J.T., editors, *Advances in Photosynthesis and Respiration: The Purple Phototrophic Bacteria*, Springer, Dordrecht, pp. 337–353, 2009.

72. Prince, R.C. and Dutton, P.L., Protonation and the reducing potential of the primary electron acceptor, in: Clayton, R.K. and Sistrom, W.R. editors, *The Photosynthetic Bacteria*, Plenum Press, New York, pp. 439–453, 1978.

73. Osváth, S., Laczkó, G., Sebban, P. and Maróti, P., Electron transfer in reaction centers of *Rhodobacter sphaeroides* and *Rhodobacter capsulatus* monitored by fluorescence of the bacteriochlorophyll dimer, *Photosynth. Res.*, 47, 41–49, 1996.

74. Maróti, P., Kinetics and yields of bacteriochlorophyll fluorescence: Redox and conformation changes in reaction center of *Rhodobacter sphaeroides*, *Eur. Biophys. J.*, 37, 1175–1184, 2008.

75. Sipka, G. and Maróti, P., Induction and anisotropy of fluorescence of reaction center from photosynthetic bacterium *Rhodobacter sphaeroides*, *Photosynth. Res.*, 2015. ISSN: 0166-8595. DOI 10.1007/s11120-015-0096-y.

76. Wraight, C.A. and Clayton, R.K., The absolute quantum efficiency of bacteriochlorophyll photooxidation in reaction centers, *Biochim. Biophys. Acta*, 333, 246–260, 1973.

77. Gust, D., Moore, T.A. and Moore, A.L., Solar fuels via artificial photosynthesis, *Acc. Chem. Res.*, 42(12), 1890–1898, 2009.

78. Maróti, P. and Trotta, M., Artificial photosynthetic systems, in: Griesbeck, A., Oelgemöller, M. and Ghetti, F., editors, *CRC Handbook of Organic Photochemistry and Photobiology*, CRC Press, Boca Raton, FL, pp. 1289–1324, 2012.

79. Milano, F., Trotta, M., Dorogi, M., Fischer, B., Giotta, L., Agostiano, A., Maróti, P., Kálmán, L. and Nagy, L., Light induced transmembrane proton gradient in artificial lipid vesicles reconstituted with photosynthetic reaction centers, *J. Bioenerg. Biomembr.* 44, 373–384, 2012.

80. Laible, P.D., Kelly, R.F., Wasielewski, M.R. and Firestone, M.A., Electron transfer dynamics of photosynthetic reaction centers in thermoresponsive soft materials, *J. Phys. Chem. B*, 109, 23679–23686, 2005.

81. Kobayashi, M., Fujioka, Y., Mori, T., Terashima, H., Suzuki, H., Shimada, Y., Saito, T., Wang, Z.Y. and Nozawa, T., Reconstitution of photosynthetic reaction center complexes in liposomes and their thermal stability, *Biosci. Biotechnol. Biochem.*, 69, 1130–1136, 2005.

82. Lu, Y., Xu, J., Liu, B. and Kong, J., Photosynthetic reaction center functionalized nano-composite films: Effective strategies for probing and exploiting the photo-induced electron transfer of photosensitive membrane protein, *Biosens. Bioelectron.*, 22, 1173–1185, 2007.

83. Mackowski, S., Hybrid nanostructures for efficient light harvesting, *J. Phys. Condens. Matter*, 22, 193102, 2010.

84. Dorogi, M., Bálint, Z., Miko, C., Vileno, B., Milas, M., Hernádi, K., Forró, L., Váró, G. and Nagy, L., Stabilization effect of single-walled carbon nanotubes on the functioning of photosynthetic reaction centers, *J. Phys. Chem. B*, 110, 21473–21479, 2006.

85. Das, R., Kiley, P.J., Segal, M., Norville, J., Yu, A.A., Wang, L., Trammel, S.A., Reddick, L.E., Kumar, R., Stellacci, F., Lebedev, N., Schnur, J., Bruce, B.D., Zhang, S. and Baldo, M., Integration of photosynthetic protein molecular complexes in solid-state electronic devices, *Nano Lett.*, 4, 1079–1083, 2004.

86. Trammell, S.A., Wang, L., Zullo, J.M., Shashidhar, R. and Lebedev, N., Orientated binding of photosynthetic reaction centers on gold using Ni–NTA self-assembled monolayers, *Biosens. Bioelectron.*, 19, 1649–1655, 2004.

87. den Hollander, M.J., Magis, J.G., Fuchsenberger, P., Aartsma, T.J., Jones, M.R. and Frese, R.N., Enhanced photocurrent generation by photosynthetic bacterial reaction centers through molecular relays, light-harvesting complexes, and direct protein–gold interactions, *Langmuir*, 27, 10282–10294, 2011.

88. Szabó, T., Bencsik, G., Magyar, M., Visy, C., Gingl, Z., Nagy, K., Váró, G., Hajdu, K., Kozák, G. and Nagy, L., Photosynthetic reaction centers/ITO hybrid nanostructure, *Mater. Sci. Eng. C*, 33(2), 769–773, 2013.

89. Hajdu, K., Gergely, C., Martin, M., Cloitre, T., Zimányi, L., Tenger, K., Khoroshyy, P., Palestino, G., Agarwal, V., Hernádi, K., Németh, Z. and Nagy, L., Porous silicon/photosynthetic reaction center hybrid nanostructure, *Langmuir*, 28, 11866–11873, 2012.

90. de Rivoyre, M., Ginet, N., Bouyer, P. and Lavergne, J., Excitation transfer connectivity in different purple bacteria: A theoretical and experimental study, *Biochim. Biophys. Acta*, 1797(11), 1780–1794, 2010.

91. Lavergne, J. and Trissl, H.-W., Theory of fluorescence induction in photosystem II: Derivation of analytical expressions in a model including exciton-radical-pair equilibrium and restricted energy transfer between photosynthetic units, *Biophys. J.*, 68, 2474–2492, 1995.

92. Asztalos, E., Sipka, G., Kis, M., Trotta, M. and Maróti, P., The reaction center is the sensitive target of the mercury(II) ion in intact cells of photosynthetic bacteria, *Photosynth. Res.*, 112, 129–140, 2012.

93. Maróti, P., Asztalos, E. and Sipka, G., Fluorescence assay for photosynthetic capacity of bacteria, *Biophys. J.*, 104(2), 545a, 2013.

94. Maróti, P. and Asztalos, E., Calculation of connectivity of photosynthetic units in intact cells of *Rhodobacter sphaeroides*, in: Lu, C. editor, *Research for Food, Fuel and Future—15th International Conference on Photosynthesis*, pp. 27–31, 2012.

95. Lavergne, J., Vermeglio, A. and Joliot, P., Coupling between RC and Cyt bc_1 complex, in: Hunter, C.N., Daldal, F., Thurnauer, M. and Beatty, J.T., editors, *Advances in Photosynthesis and Respiration: The Purple Phototrophic Bacteria*, Springer, Dordrecht, pp. 509–536, 2009.

96. Comayras, F., Jungas, C. and Lavergne, J., Functional consequences of the organization of the photosynthetic apparatus in *Rhodobacter sphaeroides*: I. Quinone domains and excitation transfer in chromatophores and reaction center antenna complexes, *J. Biol. Chem.*, 280, 11203–11213, 2005.

97. Kolber, Z.S., Prasil, O. and Falkowski, P.G., Measurements of variable chlorophyll fluorescence using fast repetition rate techniques: Defining methodology and experimental protocols, *Biochim. Biophys. Acta*, 1367, 88–106, 1998.

98. Scheuring, S., The supramolecular assembly of the photosynthetic apparatus of purple bacteria investigated by high-resolution atomic force microscopy, in: Hunter, C.N., Daldal, F., Thurnauer, M. and Beatty, J.T., editors, *Advances in Photosynthesis and Respiration: The Purple Phototrophic Bacteria*, Springer, Dordrecht, pp. 941–952, 2009.

99. Qian, P., Bullough, P.A. and Hunter, C.N., Three-dimensional reconstruction of a membrane-bending complex—The RC-LH1–PufX core dimer of *Rhodobacter sphaeroides*, *J. Biol. Chem.*, 283(20), 14002–14011, 2008.

100. Frese, R.N., Olsen, J.D., Branvall, R., Westerhuis, W.H.J., Hunter, C.N. and van Grondelle, R., The long-range supraorganization of the bacterial photosynthetic unit: A key role for PufX, *Proc. Natl. Acad. Sci. U.S.A.*, 97(10), 5197–5202, 2000.

101. Siebert, C.A., Qian, P., Fotiadis, D., Engel, A., Hunter, C.N. and Bulloughet, P.A., Molecular architecture of photosynthetic membranes in *Rhodobacter sphaeroides*: The role of PufX, *EMBO J.*, 23(4), 690–700, 2004.

102. Roszak, A.W., Howard, T.D., Southall, J., Gardiner, A.T., Law, C.J., Isaacs, N.W. and Cogdell, R.J., Crystal Structure of the RC-LH1 core complex from *Rhodopseudomonas palustris*, *Science*, 302, 1969–1972, 2003.

103. Bullough, P.A., Quian, P. and Hunter, C.N., Reaction center–light-harvesting core complexes of purple bacteria, in: Hunter, C.N., Daldal, F., Thurnauer, M. and Beatty, J.T., editors, *Advances in Photosynthesis and Respiration: The Purple Phototrophic Bacteria*, Springer, Dordrecht, pp. 155–179, 2009.

104. Barz, W.P., Vermeglio, A., Francia, F., Venturoli, G., Melandri, B.A. and Oesterhelt, D., Role of the Puf-X protein in photosynthetic growth of *Rhodobacter sphaeroides* 2. Puf-X is required for efficient ubiquinone ubiquinol exchange between the reaction-center Q(B) site and the Cyt bc(1) complex, *Biochemistry*, 34(46), 15248–15258, 1995.

105. Crofts, A.R., Meinhardt, S.W., Jones, K.R. and Snozzi, M., The role of the quinone pool in the cyclic electron transfer chain of *Rhodopseudomonas sphaeroides*. A modified Q-cycle mechanism, *Biochim. Biophys. Acta*, 723, 202–218, 1983.

106. Geyer, T. and Helms, V., A spatial model of the chromatophore vesicles of *Rhodobacter sphaeroides* and the position of the cyt bc(1) complex, *Biophys. J.*, 913, 921–926, 2006.

107. Bina, D., Litvin, R. and Vácha, F., Absorption changes accompanying the fast fluorescence induction in the purple bacterium *Rhodobacter sphaeroides*, *Photosynth. Res.*, 105, 115–121, 2010.

108. Joliot, P., Joliot, A. and Vermeglio, A., Fast oxidation of the primary electron acceptor under anaerobic conditions requires the organization of the photosynthetic chain of *Rhodobacter sphaeroides* in supercomplexes, *Biochim. Biophys. Acta*, 1706, 204–214, 2005.

109. Klamt, S., Grammel, H., Straube, R., Ghosh, R. and Gilles, E.D., Modelling the electron transport chain of purple non-sulfur bacteria, *Mol. Syst. Biol.*, 4, 156, 2008.

110. Kramer, D.M., Robinson, H.R. and Crofts, A.R., A portable multi-flash kinetic fluorimeter for measurement of donor and acceptor reactions of Photosystem 2 in leaves of whole plants under field conditions, *Photosynth. Res.*, 26, 181–193, 1990.

111. Schreiber, U., Klughammer, C. and Kolbowski, J., Assessment of wavelength-dependent parameters of photosynthetic electron transport with a new type of multicolor PAM chlorophyll fluorometer, *Photosynth. Res.*, 113, 127–144, 2012.

112. Kalaji, H.M., Goltsev, V., Bosa, K., Allakhverdiev, S.I., Strasser, R.J. and Govindjee, Experimental in vivo measurements of light emission in plants: A perspective dedicated to David Walker, *Photosynth. Res.*, 114, 69–96, 2012.

113. Falkowski, P.G. and Raven, J.A., *Aquatic Photosynthesis*, 2nd edition, Princeton University Press, Princeton, New Jersey, 2007.

114. Hubas, C., Jesus, B., Pasarelli, C. and Jeanthon, C., Tools providing new insight into coastal anoxygenic purple bacterial mats, *Res. Microbiol.*, 162, 858–868, 2011.

115. Ritchie, R.J., Photosynthesis in the blue water lily (*Nymphaea caerulea* Saligny) using PAM fluorometry, *Int. J. Plant Sci.*, 173, 124–136, 2012.

116. Van Kooten, O. and Snel, J.F.H., The use of chlorophyll fluorescence nomenclature in plant stress physiology, *Photosynth. Res.*, 25, 147–150, 1990.

117. Beer, S. and Axelsson, L., Limitations in the use of PAM fluorometry for measuring photosynthetic rates of macroalgae at high irradiances, *Eur. J. Phycol.*, 39, 1–7, 2004.

118. Stirbet, A., Govindjee, Strasser, B. and Strasser, R.J., Chlorophyll a fluorescence induction in higher plants: Modelling and numerical simulation, *J. Theor. Biol.*, 193, 131–151, 1998.

119. Vredenberg, W.J., Kinetic analysis and mathematical modeling of primary photochemical and photoelectrochemical processes in plant photosystems, *BioSystems (Elsevier)*, 103, 138–151, 2011.

120. Rondonuwu, F.S., Taguchi, T., Fujii, R., Yokoyama, K., Koyama, Y. and Watanabe, Y., The energies and kinetics of triplet carotenoids in the LH2 antenna complexes as determined by phosphorescence spectroscopy, *Chem. Phys. Lett.*, 384, 364–371, 2004.

121. Monger, T.G., Cogdell, R.J. and Parson, W.W., Triplet states of bacteriochlorophyll and carotenoids in chromatophores of photosynthetic bacteria, *Biochim. Biophys. Acta*, 449, 136–153, 1976.

122. Rademaker, H., Hoff, A.J., van Grondelle, R. and Duysens, L.N., Carotenoid triplet yields in normal and deuterated *Rhodospirillum rubrum*, *Biochim. Biophys. Acta*, 592(2), 240–257, 1980.

123. Arellano, J.B., Melo, T.B., Fyfe, P.K., Cogdell, R.J. and Naqvi, K.R., Multichannel flash spectroscopy of the reaction centers of wild-type and mutant *Rhodobacter sphaeroides*: Bacteriochlorophyll B mediated interaction between the carotenoids triplet and the special pair, *Photochem. Photobiol.*, 79, 68–75, 2004.

124. Fraser, N.J., Hashimoto, H. and Cogdell, R.J., Carotenoids and bacterial photosynthesis: The story so far…, *Photosynth. Res.*, 70, 249–256, 2001.

125. Slouf, V., Chábera, P., Olsen, J.D., Martin, E.C., Qian, P., Hunter, C.N. and Polivka, T., Photoprotection in a purple phototrophic bacterium mediated by oxygen-dependent alteration of carotenoid excited-state properties, *Proc. Natl. Acad. Sci. U.S.A.*, 109, 8570–8575, 2012.

126. Joliot, P. and Joliot, A., Etudes cinétique de la réaction photochimique libérant l'oxygène au cours de la photosynthèse, *C. R. Acad. Sci. Paris*, 258, 4622–4625, 1964.

127. Sturgis, J.N. and Niederman, R.A., Organization and assembly of light-harvesting complexes in the purple bacterial membrane, in: Hunter, C.N., Daldal, F., Thurnauer, M. and Beatty, J.T., editors, *Advances in Photosynthesis and Respiration: The Purple Phototrophic Bacteria*, Springer, Dordrecht, pp. 253–273, 2009.

128. Asztalos, E., Kis, M. and Maróti, P., Aging photosynthetic bacteria monitored by absorption and fluorescence changes, *Acta Biol. Szeg.*, 54(2), 149–154, 2010.

129. Takemoto, J. and Lascelles, J., Coupling between bacteriochlorophyll and membrane protein synthesis in *Rhodopseudomonas sphaeroides*, *Proc. Natl. Acad. Sci. U.S.A.*, 70, 799–803, 1973.

130. Niederman, R.A., Structure, function and formation of bacterial intracytoplasmic membranes, in: Shively, J.M. editor, *Complex Intracellular Structures in Prokaryotes, Microbiology Monographs*, Vol. 2, Springer Verlag, Berlin, pp. 193–227, 2006.

131. Cline, K., Biogenesis of green plant thylakoid membranes, in: Green, B.R. and Parson, W.W., editors, *Light-Harvesting Antennas in Photosynthesis, Advances in Photosynthesis and Respiration*, Vol. 13, Kluwer Academic Publishers, Dordrecht, pp. 353–372, 2003.

132. Cutler, R.G. and Evans, J.E., Synchronization of bacteria by a stationary-phase method, *J. Bacteriol.*, 91(2), 469–476, 1966.

133. Lueking, D.R., Fraley, R.T. and Kaplan, S., Intracytoplasmic membrane synthesis in synchronous cell populations of *Rhodopseudomonas sphaeroides*, *J. Biol. Chem.*, 253, 451–457, 1978.

134. Myers, C.R. and Collins, M.L.P., Cell-cycle-specific fluctuation in cytoplasmic membrane composition in aerobically grown *Rhodospirillum rubrum*, *J. Bacteriol.*, 169(12), 5445–5451, 1987.

135. Wraight, C.A., Lueking, D.R., Fraley, R.T. and Kaplan, S., Synthesis of photopigments and electron transfer components in synchronous phototropic cultures of *Rhodopseudomonas sphaeroides*, *J. Biol. Chem.*, 253(2), 465–471, 1978.

136. Knacker, T., Harwood, J.L., Hunter, C.N. and Russell, N.J., Lipid biosynthesis in synchronized cultures of the photosynthetic bacterium *Rhodopseudomonas sphaeroides*, *Biochem. J.*, 229, 701–710, 1985.

137. Mozharov, A.D., Shchipakin, V.N., Fishov, I.L. and Evtodienko, Y.V., Changes in the composition of membrane phospholipids during the cell-cycle of *Escherichia coli*, *FEBS Lett.*, 186, 103–106, 1985.

138. Kiley, P.J. and Kaplan, S., Molecular genetics of photosynthetic membrane biosynthesis in *Rhodobacter sphaeroides*, *Microbiol. Rev.*, 52(1), 50–69, 1988.

139. Woronowicz, K., Olubanjo, O.B., Sung, H.C., Lamptey, J.L. and Niederman, R.A., The effects of protein crowding in bacterial photosynthetic membranes on the flow of quinone redox species between the photochemical reaction center and the ubiquinol–cyt c_2 oxidoreductase, *Photosynth. Res.*, 111(1–2), 125–138, 2012.

140. Koblízek, M., Masín, M., Ras, J., Poulton, A.J. and Prásil, O., Rapid growth rates of aerobic anoxygenic phototrophs in the ocean, *Environ. Microbiol.*, 9(10), 2401–2406, 2007.

141. Hojerová, E., Mašín, M., Brunet, C., Ferrera, I., Gasol, J.M. and Koblížek, M., Distribution and growth of aerobic anoxygenic phototrophs in the Mediterranean Sea, *Environ. Microbiol.*, 13(10), 2717–2725, 2011.

142. Medová, H., Boldareva, E.N., Hrouzek, P., Borzenko, S.V., Namsaraev, Z.B., Gorlenko, V.M., Namsaraev, B.B. and Koblížek, M., High abundances of aerobic anoxygenic phototrophs in saline steppe lakes, *FEMS Microbiol. Ecol.*, 76(2), 393–400, 2011.

143. Ferrera, I., Gasol, J.M., Sebastián, M., Hojerová, E. and Koblízek, M., Comparison of growth rates of aerobic anoxygenic phototrophic bacteria and other bacterioplankton groups in coastal Mediterranean waters, *Appl. Environ. Microbiol.*, 77(21), 7451–7458, 2011.

144. Gadd, G.M., Bioremedial potential of microbial mechanisms of metal mobilization and immobilization, *Curr. Opin. Biotech.*, 11, 271–279, 2000.

145. Giotta, L., Agostiano, A., Italiano, F., Milano, F. and Trotta, M., Heavy metal ion influence on the photosynthetic growth of *Rhodobacter sphaeroides*, *Chemosphere*, 62, 1490–1499, 2006.

146. Borsetti, F., Martelli, P.L., Casadio, R. and Zannoni, D., Metals and metalloids in photosynthetic bacteria: Interactions, resistance and putative homeostasis revealed by genome analysis, in: Hunter, C.N., Daldal, F., Thurnauer, M. and Beatty, J.T., editors, *Advances in Photosynthesis and Respiration: The Purple Phototrophic Bacteria*, Springer, Dordrecht, pp. 655–689, 2009.

147. Deng, X. and Jia, P., Construction and characterization of a photosynthetic bacterium genetically engineered for Hg^{2+} uptake, *Bioresour. Technol.*, 102, 3083–3088, 2011.

148. Italiano, F., Buccolieri, A., Giotta, L., Agostiano, A., Valli, L., Milano, F. and Trotta, M., Response of the carotenoidless mutant *Rhodobacter sphaeroides* growing cells to cobalt and nickel exposure, *Int. Biodeter. Biodegrad.*, 63, 948–957, 2009.

149. Italiano, F., D'Amici, G.M., Rinalducci, S., De Leo, F., Zolla, L., Gallerani, R., Trotta, M. and Ceci, L.R., The photosynthetic membrane proteome of *Rhodobacter sphaeroides* R-26.1 exposed to cobalt, *Res. Microbiol.*, 162(5), 520–527, 2011.

150. Reniero, D., Galli, E. and Barbieri, P., Cloning and comparison of mercury and organomercurial-resistance determinants from a *Pseudomonas stutzeri* plasmid, *Gene*, 166, 77–82, 1995.

151. Schiering, N., Kabsch, W., Moore, M.J., Distefano, M.D., Walsh, C.T. and Pai, E.F., Structure of the detoxification catalyst mercuric ion reductase from *Bacillus* sp. Strain RC607, *Nature*, 352, 168–172, 1991.

152. Silver, S. and Phung, L.T., Bacterial heavy metal resistance: New surprises, *Annu. Rev. Microbiol.*, 50, 753–789, 1996.

153. Pisani, F., Italiano, F., de Leo, F., Gallerani, R., Rinalducci, S., Zolla, L., Agostiano, A., Ceci, L.R. and Trotta, M., Soluble proteome investigation of cobalt effect on the carotenoidless mutant of *Rhodobacter sphaeroides*, *J. Appl. Microbiol.*, 106, 338–349, 2009.

24 Photosynthesis in Eukaryotic Algae with Secondary Plastids

Christian Wilhelm and Reimund Goss

CONTENTS

24.1 Introduction 425
24.2 Phylogeny and Morphology of Secondary Plastids 426
    24.2.1 Stramenopila 426
        24.2.1.1 Diatoms 426
        24.2.1.2 Eustigmatophytes 427
        24.2.1.3 Synchromophytes 427
    24.2.2 Alveolata 428
        24.2.2.1 Dinophyta 428
        24.2.2.2 Chromera 428
        24.2.2.3 Cryptophyta 428
        24.2.2.4 Haptophyta 429
24.3 Pigments, Pigment-Binding Complexes, and Their Macromolecular Organization 429
    24.3.1 FCPs of Diatoms 429
    24.3.2 Light-Harvesting Complexes of Other Groups 430
    24.3.3 Pigment Dynamics in Diatoms 431
    24.3.4 Pigment Dynamics in Other Groups 432
    24.3.5 Photosystems and Their Macromolecular Organization in the Membrane 433
24.4 Alternative Electron Flows 433
    24.4.1 Chlororespiration 433
    24.4.2 Cyclic Electron Flow 434
    24.4.3 Mehler Reaction 435
24.5 Carbon Assimilation 435
    24.5.1 Regulation of the Carbon Assimilation 436
    24.5.2 Organelle Interaction 436
24.6 Regulation of Photosynthesis 436
    24.6.1 Photoprotection in Diatoms 436
    24.6.2 Photoprotection in Other Groups 438
    24.6.3 Light Acclimation 439
Acknowledgments 439
References 440

24.1 INTRODUCTION

Eurkaryotic algae with secondary plastids can be found in the clade of photobiotic heterokonts, which consists of about 18 algal classes differing in cell structure and pigmentation. The major groups in this clade are the fucoxanthin-containing diatoms, the brown algae, chrysophytes (golden algae), the greenish Xanthophytes and Eustigmatophytes, and finally the phycobilin-containing Cryptophytes. Their plastids derived from a primitive red alga, which was engulfed by a nonphotosynthetic, animal-like host cell. During evolution, large parts of the genome of the autotrophic partner were transferred to the nucleus of the host by lateral gene transfer, which resulted in a sophisticated regulation of transcription and protein translocation apparatus in these cells. Since the host nucleus is phylogenetically far from that of green algae or higher plants, the regulation of the metabolic pathways is quite different from *green* cells. This chapter reviews the differences in metabolic regulation of the photosynthetic machinery and surprising modifications in the cellular compartmentalization of the major metabolic pathways in these cells. In this chapter, we concentrate on diatoms because, in this taxon, the present state of the art is much more advanced compared to the other groups. However, it has to be emphasized that, up to now, it is not clear if the metabolic situation found in diatom plastids can be extended to the other groups of the heterokontic clade.

24.2 PHYLOGENY AND MORPHOLOGY OF SECONDARY PLASTIDS

All chloroplasts are of prokaryotic origin by an endosymbiosis of a photosynthetic prokaryote and a eukaryotic host organism. The integration of the symbiotic genome by lateral gene transfer into the host nucleus and the subsequently established stable and controlled genetic interaction between the host nucleus and the symbiont led to the *semiautonomous* primary plastids, which can be found in green and red algae. The endosymbiotic engulfment of a eukaryotic organism already housing a plastid generated a new cellular organization containing two eukaryotic nuclei and one or more prokaryotic secondary plastids. The latter are surrounded by four membranes, two from the prokaryotic endosymbiont and two additional membranes derived from the cytoplasmic membranes of the first and second host organism. The nucleus of the first host organism degraded either totally or is still present as a minor genomic compartment called *nucleomorph* (Gillot et al. 1980). During evolution, secondary endosymbiotic events occurred at least three times. The uptake of a green alga led to the Chlorarachniophytes and to the Euglenoids, whereas the uptake of a red alga generated the secondary plastids in the Stramenopila (including 11 clades of autotrophic phototrophs: Raphidophytes, Phaeothamniophytes, Phaeophytes, Xanthophytes, Pinguiophytes, Chrysophytes, Synchromophytes, Eustigmatophytes, Dictyochophytes, Pelagophytes, and the Diatoms), the Alveolata (including two photosynthetic clades: the Chromerida and Dinophyta), the Haptophyta, and the Cryptophyta (each containing only one photosynthetic clade). This review will not discuss the phylogenetic traces in all these groups in detail. For details of the phylogeny of secondary plastids, the reader is referred to the recent review by Qiu et al. (2013). Here we concentrate on the diatoms, the Cryptophytes, the Eustigmatophytes, the Synchromophytes, the Dinoflagellates, and the Chromerophytes because from these groups, the information on genome organization, biochemistry, and physiology of photosynthetic components and reactions is studied best. From most of the other groups, the information is very scarce.

However, the consequences of the host–endosymbiont interaction cannot be overemphasized: secondary plastids are controlled by a host nucleus, which is phylogenetically far from a plant nucleus. For instance, in the genome of diatoms, two thirds of the genes are of animal and bacterial origin and only one third is plant-like (Falciatore and Bowler 2002). This deep phylogenetic difference with respect to the host genome becomes evident by the fact that, in secondary plastids, basic metabolic pathways are differently compartmentalized, regulated, or in some cases lacking or replaced by pathways that are not present in other photosynthetic cells (Wilhelm et al. 2006). In conclusion, secondary plastids are not simply differently colored plastids; instead they represent a unique photosynthetic organelle with special features and abilities unknown so far in green plastids.

24.2.1 STRAMENOPILA

24.2.1.1 Diatoms

Diatoms are the most important group within the Stramenopila. They fix up to 25% of the global carbon and can be found in all environments (marine, freshwater, ice, soil, or aerophytic). They are characterized by a carbon-poor silica shell (frustules) instead of a cellulosic cell wall. The frustules have two valves, a slightly larger epivalve and smaller hypovalve, and girdle bands, which are loops of silica inserted between the two valves. During division, each daughter cell receives one of the valves with the consequence that one daughter cell is always smaller. This pauperization in cell size leads to a critical volume when growth arrests and sexual reproduction are initiated. The diatoms can be divided into two major classes: the Pennales and the Centrales. The pennate diatoms show a strict bilaterally symmetrical pattern housing normally two large parietal brownish chloroplasts, whereas the centric are circular, mostly containing several up to many chloroplasts per cell. Since the cell wall is very brittle, it is extremely difficult to isolate intact chloroplasts.

The chloroplast is surrounded by four membranes, whereby the outer two form a fold of the endoplasmic reticulum. Therefore, the plastids are closely located to the nucleus, and the outer membranes form a continuum with the nuclear envelope. This close interaction of nucleus and plastids is discussed in the context of the chloroplast import of proteins, which are encoded in the nucleus and have to be transported across four membranes. In the light microscope, the plastids have a brownish color due to the high content of fucoxanthin, a green light-absorbing xanthophyll whose concentration exceeds that of chlorophyll that led to the brown color (pigments; see Section 24.3). The ultrastructure of the plastid (Figure 24.1) shows a large pyrenoid where the enzymes of the Calvin cycle form large supramolecular aggregates. Instead of starch, the storage product of diatoms is chrysolaminaran, which is a β-1,3 linked polyglucan. The

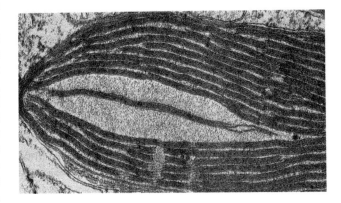

FIGURE 24.1 TEM of a section of the chloroplast from the diatom *Melosira* showing the triple membranes running in parallel with a central pyrenoid. (Reproduction of Figure 5.1. The fine structure of algal cells by J.D. Dodge. With permission.)

thylakoids are grouped in stacks of three; beneath the plastid envelope, one stack of three encloses all other thylakoids. In contrast to green plastids, where the membrane stacking of the thylakoids is found to depend on the light climate, the membrane topology in diatoms is fixed. Changes can be observed only in the amount of thylakoids per plastid but not in the numbers of stacks. In brown algae, the chloroplasts show a very similar architecture; the only difference is the pigmentation in the xanthophylls related to photoprotection (see Section 24.3).

24.2.1.2 Eustigmatophytes

The Eustigmatophytes are greenish because they do not contain any brownish pigment like fucoxanthin. Their major carotenoid is violaxanthin; in contrast to all other Stramenopila, they do not possess chlorophyll c. The name of this group is traced back to its very unique eyespot, which is, in contrast to other algal photoreceptors, not located in chloroplast but in the cytosol. Another unusual feature of the eustigmatophycean chloroplast is the lack of a girdle lamella, although the thylakoid arrangement as triple membranes is identical to that of diatoms or brown algae. Recently, the Eustigmatophytes have attracted high interest because one of the major genus *Nannochloropsis* can be genetically modified also by homologous recombination leading to stable and well-defined transformants (Kilian et al. 2011).

24.2.1.3 Synchromophytes

The Synchromophytes have been recently discovered by Horn et al. (2007) and confirmed as a new taxonomic class by Schmidt et al. (2012). The brownish amoeboid cells form a network of individual cells sitting inside a lorica and contain instead single chloroplasts called plastid complexes. Here, one complex consists of several plastids; each of them is surrounded by two membranes, whereas the whole complex is included in a fold of the endoplasmic reticulum. One complex can contain between 2 and 8 individual chloroplasts that share one common pyrenoid (Figure 24.2). Each chloroplast shows a typical heterokontic plastid organization with a girdle lamella and thylakoids running in three over the whole length of the membrane stack. Inside the cell also endocytobiotic bacteria could be found in a dividing stage, an indication that the bacteria can reproduce inside the cell. Their plasmids may support lateral gene transfer from the plastid to the nucleus. Cells with chloroplast complexes and endocytobiotic bacteria may represent a phylogenetic prototype for efficient lateral gene transfer. Normally the translocation of genes is lethal when the gene products may be functionally indispensable. However, if inside a plastid complex one unit remains functional, the chances for successful lateral gene transfer increase. This might be one of the mechanisms why the phylogenetic diversification inside the Stramenopiles runs faster than in all other organismic groups. The pigmentation of the plastids in the Synchromophyceae is the same as in diatoms.

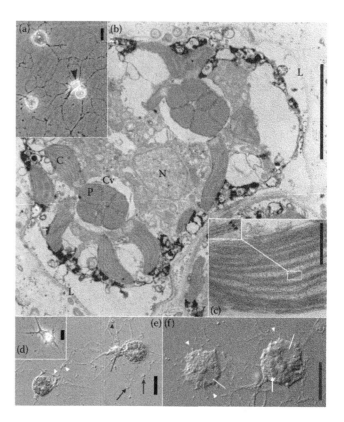

FIGURE 24.2 *Synchroma grande.* (a) Phase-contrast light microscopical image; scale bar: 20 mm. The circular main cell bodies of the three sessile amoebae surrounded by loricae. The reticulopodial strands of the cells are highly branched forming a meroplasmodium. After binary cell division of the sessile amoeba on the right, one daughter cell has hatched out of the lorica to become a migrating amoeba with a fusiform shape (arrowhead). (b) Transmission electron microscopical overview of a sessile amoeba; scale bar: 5 mm. Every chloroplast has a pigmented lobe (C), which is connected to the terminal pyrenoid (P). The pyrenoids of six neighboring chloroplasts are densely aggregated in the center of the radiating lobes and are surrounded by a capping vesicle (Cv). Two chloroplast aggregates representing multiplastidic secondary endosymbionts are visible in the cell. N: nucleus; L: lorica. (c) Detail of a pigmented chloroplast lobe containing longitudinally arranged lamellae of three adpressed thylakoids, which are marked by arrowheads in the enlarged insert; scale bar: 0.5 mm. (d) Migrating amoeba with spiny outline; phase contrast optics; scale bar: 20 mm. (e) Two sessile amoebae interconnected by reticulopodia with engulfed organisms or particles (white arrowheads). Aggregates and remnants of the digesting vacuoles containing the engulfed materials occur within (black arrowhead) or outside (black arrows) of the meroplasmodial network. A particularly large accumulation of digesting vacuoles (black line) occurs outside of the lorica in front of the ostiolum; differential interference contrast (DIC) optics; scale bar: 20 mm. (f) Two sessile amoebae interconnected by reticulopodia. White arrowheads point to the loricae. Main cell bodies have a somewhat irregularly stellar outline due to protuberances of the cytoplasm attached to the lorica. Within the cells, nuclei are marked with white lines. A chloroplast complex in which the pyrenoids are densely aggregated in the center of the radiating lobes is clearly visible (white arrow); DIC optics; scale bar: 20 mm. (Reproduced from Horn S et al., *Protist* 158, 277–293, 2007. With permission.)

24.2.2 ALVEOLATA

24.2.2.1 Dinophyta

The cell organization of Dinophyta is unique with respect to many features. The cell is covered by a species-specific *armor* built up by cellulose plates. In the interphase, the chromosomes remain highly condensed and the DNA is not bound to histones, and thus no nucleosomes are formed. The plastids derive from different endosymbionts, which led to a variety of plastid types in the dinophycean clade. The *true* dinophycean plastid (e.g., in *Peridinium cinctum*) is surrounded by three envelope membranes (Figure 24.3). Inside the chloroplast, the thylakoids run in stacks of 2–4, and the major light-harvesting pigment is peridinin, a brownish carotenoid. In a second type in dinophytes (e.g., in *Peridinium balticum*), chloroplasts can be found, which appear as reduced heterokont plastids. Here, we typically find a girdle lamella and thylakoid triple stacks with the major photosynthetic carotenoid fucoxanthin. A third type that can be found in Dinoflagellates originates from an endosymbiosis with a green alga. Here again, the plastid is surrounded by three membranes, but the thylakoid organization shows reduced grana and stroma differentiation. In that case, chlorophyll b and typical green algal carotenoids like lutein can be found. In some other cases, Dinoflagellates contain *kleptoplasts*, which are taken up but become digested after several cell cycles.

24.2.2.2 Chromera

The genus Chromera is the only photosynthetically active genus in the Chromerida, which contains many heterotrophic groups, e.g., the Plasmodium, the malaria pathogen. Although the host organisms in the Chromerida do not have close phylogenetic relationship to heterokontic algae, Obornik et al. (2011, 2012) showed that the chloroplast of

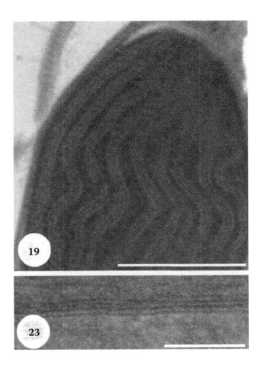

FIGURE 24.4 TEM of a section of the chloroplast of *Chromera velia* cells showing typical three lamellar structures of the thylakoids. (Reproduction from Obornik M et al., *Protist* 162, 1125–130, 2011. With permission.)

this group has many similarities with heterokontic algae (Figure 24.4). The chloroplast envelope consists of four membranes, and the thylakoids run in stacks of three. Although the similarity to heterokontic algae is quite evident, the biochemical composition and the functional organization of the Chromera plastid show some specific peculiarities, which give evidence of a specific diversification within the clade of the Apicomplexa.

24.2.2.3 Cryptophyta

The Cryptophyte algae are the only group housing red algae-derived secondary plastids, where remnants of the primary eukaryotic host nucleus are still present as nucleomorph. It is extremely reduced and contains only a few genes (Archibald et al. 2001). Interestingly, the gene transfer from mitochondrion-to-nucleus still occurs in all organisms with secondary plastids, whereas plastid-to-nucleus and nucleomorph-to-nucleus transfers do not. This explains why some essential genes remained in the nucleomorph. The chloroplast is again surrounded by four membranes with a special space between the second and third membrane. This space represents a reduced eukaryotic compartment because it contains not only the nucleomorph but also eukaryotic polysomes. Inside the plastid, the thylakoids always run as twins with a unique protein density in the lumen (Figure 24.5). This is due to the fact that the light-harvesting antenna is composed of phycobilins, which are not organized as phycobilisomes but instead of soluble complexes located in the lumen of the thylakoids. The chromophores are either phycoerythrin or phycocyanin.

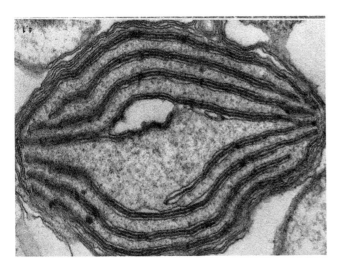

FIGURE 24.3 TEM of a section of the chloroplast from the dinoflagellate *Amphidinium britanicum* showing the thylakoid membranes running in stacks of two or three. (Reproduction of Figure 4.1. The fine structure of algal cells by J.D. Dodge. With permission.)

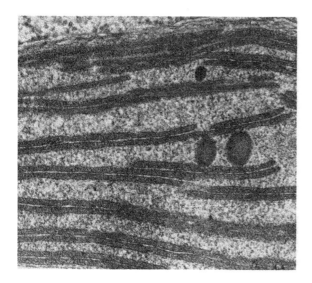

FIGURE 24.5 TEM of the thylakoid structure of the Cryptophyte *Chromonas mesostigmatica* showing the thylakoid membranes running as twins filled with electron dense material, which represents the phycobiliproteins inside the lumen. (Reproduction of Figure 4.4. The fine structure of algal cells by J.D. Dodge. With permission.)

Therefore, the cells appear either brownish-red or blue depending on the type of the chromophores. The width of the thylakoid lumen strongly depends on the amount of phycobilins, which is strictly controlled by the light intensity during growth. Under low light, the pigment content is high, and the volume of the thylakoid lumen is increased.

24.2.2.4 Haptophyta

The Haptophytes are an ecologically important phytoplankton group in the oceans; their contribution to global primary production is similar to that of diatoms. The cells have a brownish color due to their high content of fucoxanthin and some of its derivatives. They also contain several types of chlorophyll c; some of them are unusually quite hydrophobic due to a long chain alcohol as the side group. The ultrastructure of the plastid is quite similar to that of heterokontic algae; however, a girdle lamella is normally lacking, but the invariant triple thylakoid structure is also always present (Figure 24.6).

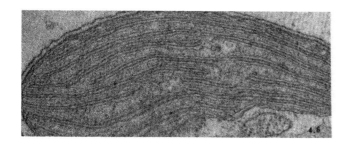

FIGURE 24.6 TEM of a section of the chloroplast from the haptophyte *Chrysochromulina* showing the thylakoid membranes running in stacks of two or three without a girdle lamella. (Reproduction of Figure 4.6. The fine structure of algal cells by J.D. Dodge. With permission.)

24.3 PIGMENTS, PIGMENT-BINDING COMPLEXES, AND THEIR MACROMOLECULAR ORGANIZATION

24.3.1 FCPs of Diatoms

A lot of work has been performed during recent years concerning the nature, structure, and function of the diatom antenna complexes, the fucoxanthin (Fx) chlorophyll proteins (FCP). Based on these works, it has become clear that the diatom antenna complexes can be divided into three main groups. The first group consists of the Lhcf proteins, which constitute the main peripheral light-harvesting complex, which serves as the antenna for both PSII and PSI (Lepetit et al. 2010; Grouneva et al. 2011; Gundermann et al. 2013; Nagao et al. 2013). The Lhcf proteins are encoded by a large number of genes, i.e., in *Phaeodactylum tricornutum* (Bowler et al. 2008) and in *Thalassiosira pseudonana* (Armbrust et al. 2004), 17 and 11 Lhcf genes, respectively, have been annotated so far. These FCP complexes typically bind 3–4 molecules of Chl a, 1 Chl c, and 3–4 Fx per apoprotein, while the DD/Dt cycle pigments are present in substoichiometric amounts (Beer et al. 2006). The high concentration of Fx molecules in the light-harvesting complexes in combination with their absorption of wavelengths up to 550 nm gives the diatoms their typical brown color (Figure 24.7). The second group of FCPs contains the Lhcr proteins. These proteins form light-harvesting complexes, which are specific for PSI (Veith et al. 2009; Lepetit et al. 2010; Grouneva et al. 2011; Ikeda et al. 2013). Like the Lhcf proteins, the Lhcr proteins are encoded by numerous genes. So far, both in *P. tricornutum* and *T. pseudonana*, 14 Lhcr genes have been found (Bowler et al. 2008; Armbrust et al. 2004). In contrast to the Lhcf proteins, the Lhcr proteins contain lower amounts of Fx but are enriched in the DD/Dt cycle pigments (Lepetit et al. 2008). However, precise pigment stoichiometries for the PSI-specific Lhcr proteins have

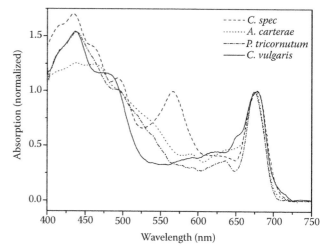

FIGURE 24.7 Absorption spectra of the cryptophyte *Cryptomonas spec.* (Chl a/phycoerythrin/phycocyanin), the dinoflagellate *Amphidinium carterae* (Chl a/peridinin), and the diatom *P. tricornutum* (Chl a/Chl c/fucoxanthin) in comparison to the absorption spectrum of the green alga *Chlorella vulgaris* (Chl a/Chl b).

not been reported so far. The enrichment of DD/Dt cycle pigments is accompanied by a higher de-epoxidation state, i.e., a higher concentration of Dt during high light illumination compared to the Lhcf-containing peripheral major antenna complexes (Lepetit et al. 2008). The third group of proteins consists of the Lhcx proteins. These proteins were identified as components of the peripheral antenna complexes as well as the PSI-specific antenna (Beer et al. 2006; Lepetit et al. 2010; Grouneva et al. 2011; Nagao et al. 2013). The number of genes that encode the Lhcx proteins is not as high as the number of genes for the Lhcf and Lhcr proteins. In *P. tricornutum* (Bowler et al. 2008) and in *T. pseudonana* (Armbrust et al. 2004), 4 and 6 Lhcx genes, respectively, have been annotated so far. The Lhcx proteins serve a special function in the light-harvesting system of diatoms. These proteins are essential for efficient photoprotection via the thermal dissipation of excessive excitation energy in the process of NPQ (Bailleul et al. 2010). Thus, they play a comparable role in NPQ as the PsbS protein in higher plants (Li et al. 2000) and the Lhcsr proteins in green algae (Peers et al. 2009; Bonente et al. 2011) and mosses (Alboresi et al. 2010; Pinnola et al. 2013). In contrast to the PsbS protein and in conjunction with the Lhcsr proteins, the Lhcx proteins are likely pigmented (Beer et al. 2006). In agreement with their important role in photoprotection, recent studies have shown that the Lhcx genes are activated during illumination with high light intensities, resulting in increasing concentrations of Lhcx proteins in high light exposed diatom cells (Nymark et al. 2009, 2013; Bailleul et al. 2010; Lepetit et al. 2013). Besides the three main groups of FCP proteins, recent studies have suggested that a fourth type of light-harvesting proteins is present in diatoms. These proteins are called RedCAPs and are related to the light-harvesting proteins of the red algae (Sturm et al. 2013).

With respect to the structure of the FCP complexes, there seem to be differences between the pennate and the centric diatoms. The centric diatoms, including *Cyclotella meneghiniana* and *T. pseudonana*, contain two different types of light-harvesting complexes (Büchel 2003; Beer et al. 2006; Lepetit et al. 2010; Nagao et al. 2013). The first FCP complex has been termed FCP-A and consists of trimers of Lhcf and Lhcx proteins. The second complex, the FCP-B, is oligomeric and is composed solely of Lhcf proteins. The basic unit of the peripheral FCP of pennate diatoms, including *P. tricornutum*, is the trimer, although, depending on the solubilization conditions, higher oligomeric states of the FCP can be isolated (Lepetit et al. 2007; Joshi-Deo et al. 2010; Nagao et al. 2013). Recent measurements have indicated that the trimers of *P. tricornutum* are composed of three major trimers, which themselves consist of only Lhcf proteins without a contribution of Lhcr and Lhcx proteins (Gundermann et al. 2013).

The FCP complexes are able to undergo a certain type of aggregation that is, however, not comparable to the macroaggregation of LHCII, which spans several thylakoid membranes (Schaller et al. 2014). For *C. meneghiniana*, it has been shown that the FCP-A complex is able to aggregate at low pH values and in the presence of the de-epoxidized DD/Dt cycle pigment Dt, as indicated by a pronounced quenching of the Chl a

fluorescence of the isolated FCP complexes (Gundermann and Büchel 2008; Miloslavina et al. 2009). High concentrations of the thylakoid membrane lipids prevented the FCP aggregation and thus the fluorescence quenching. In contrast to the FCP-A, the FCP-B was unable to aggregate at low pH values (Gundermann and Büchel 2008). Recent measurements have demonstrated that the isolated main FCP of *P. tricornutum* is also able to aggregate at low pH values, resulting in a reduction of the Chl a fluorescence of the complex (Schaller et al. 2014). In this study, the aggregation at low pH values was aided by high concentrations of the divalent cation Mg^{2+}. By comparing the aggregation behavior of the isolated FCP with the purified LHCII of higher plants, the study furthermore showed that the FCP aggregation is significantly less pronounced than that of LHCII.

24.3.2 LIGHT-HARVESTING COMPLEXES OF OTHER GROUPS

Brown algae (Phaeophyceae) contain FCP complexes that are comparable to the FCPs of diatoms (De Martino et al. 2000). The FCP proteins are encoded by six Lhcf genes in *Macrocystis pyrifera* (Apt et al. 1995) and *Laminaria saccharina* (Caron et al. 1996; De Martino et al. 2000). The Lhcf gene sequences show a high level of homology between the two species and also to expressed sequence tags of *Laminaria digitata* (De Martino et al. 2000), which indicates that the FCP complexes are very similar in all macrophytic brown algae. In contrast to diatoms, the antenna complexes of PSII and PSI seem to be identical. The FCP complexes bind Chl a and c and Fx as the major light-harvesting pigment (Pascal et al. 1998; De Martino et al. 2000). Since the brown algae use the VAZ cycle for photoprotection and not the DD/Dt cycle of diatoms, the FCPs possess binding sites for V (Katoh et al. 1989). Interestingly, in the FCP complexes, two pools of Fx molecules exist, which are characterized by different absorption wavelengths (Katoh et al. 1989). This has also been reported for the FCP complexes of diatoms where even three different Fx pools, termed blue, green, and red absorbing Fx, seem to exist (Szábo et al. 2008; Premvardhan et al. 2010). The ability of Fx molecules to absorb light in the blue and blue-green part of the spectrum in combination with their high amount in the FCP complexes leads to the brown color of the Phaeophyceae (Figure 24.7). In contrast to higher plants, the Lhcf proteins generally bind high concentrations of xanthophylls, i.e., the Chl-to-carotenoid ratio is around 1, whereas in green plastids, this ratio is typically higher than 3. Recent studies of LHC genes of brown algae have demonstrated that besides the Lhcf genes, also Lhcsr genes are present, which are likely to play an important role in the process of NPQ (Dittami et al. 2010; Konotchick et al. 2013; Ocampo-Alvarez et al. 2013).

The haptophytes contain membrane-intrinsic antenna complexes. The genes that are coding for the light-harvesting complexes suggest that Lhcf, Lhcr, and Lhcx proteins are present as it has been found for diatoms (Neilson and Durnford 2010; Hoffman et al. 2011). The antenna complexes bind Chl a, Chl c, most probably Fx, and DD since the haptophytes possess the DD/Dt cycle with comparable characteristics to the diatoms (Garrido and Zapata 1998). The pigment composition of the Haptophytes is complex, and eight different pigment types have been found, which seem to correspond with the phylogenetic trees for the group (Zapata et al. 2004). Differentiation of the eight pigment types was based on the distribution of nine Chl c pigments and five fucoxanthin derivates, including Chl c esterified with the membrane lipid MGDG and keto-derivatives of Fx.

The dinophytes are an interesting group with respect to their light-harvesting complexes. The pigmentation is different compared to the other groups, and peridinin, not Fx, represents the major light-harvesting carotenoid (Hoffman et al. 1996). The high concentration of peridinin bound to the light-harvesting complexes (see below) in conjunction with their light absorption in the blue-green and green part of the spectrum gives the haptophytes their brown color (Figure 24.7). In addition to the membrane light-harvesting complexes, the dinoflagellates contain a water-soluble Chl a peridinin complex (PCP), which is located in the thylakoid lumen (Haxo et al. 1976; Prezelin and Haxo 1976; Hoffman et al. 1996). The PCP complexes are very diverse with respect to their genes, their pigmentation, and spectroscopic characteristics. In general, two forms of PCP exist: the monomeric PCP (Iglesias-Prieto et al. 1991; Hoffman et al. 1996; Weis et al. 2002) and the homodimer (Iglesias-Prieto et al. 1991; Weis et al. 2002). In *Amphidinium carterae*, three copies of the monomeric PCP associate to form a trimer (Hoffman et al. 1996; Wormke et al. 2008). In this dinoflagellate, the PCP complexes are isolated in two different forms, namely, the main form PCP (MFPCP) with a peridinin-to-Chl a ratio of 8 to 2 (Hoffman et al. 1996) and the high salt PCP (HSPCP), which binds 6 peridinin and 2 Chl a (Schulte et al. 2009). Both the MFPCP and the HSPCP have been crystallized and studied by x-ray analysis (Hoffman et al. 1996; Schulte et al. 2009). The highest peridinin-to-Chl a ratio has been found in the PCP of *Alexandrium cohorticula*, which contains 12 peridinin per 2 Chl a molecules (Ogata et al. 1994). The membrane-intrinsic light-harvesting complexes of dinoflagellates are supposed to bind 7 Chl a, 4 Chl c, 10 to 12 peridinin, and 2 molecules of DD (Hiller et al. 1993). Until now, nothing is known about the contribution of Lhcf and Lhcr proteins to the native complexes.

The light-harvesting complexes of *Chromera velia*, which belongs to the chromerophyta, have been investigated in recent studies (Pan et al. 2012; Tichy et al. 2013; Mann et al. 2014). The results of these studies indicate that two types of light-harvesting complexes are present in this alga (Tichy et al. 2013). One LHC, termed CLH, is comparable to the FCP complexes of diatoms with the exception that the *C. velia* binds only Chl a and isofucoxanthin as main light-harvesting xanthophyll. Although the isofucoxanthin concentration of the CLH is not as high as the fucoxanthin concentration of the diatom or brown algal FCP, i.e., a Chl a-to-isofucoxanthin ratio of around 2.5 has been reported (Tichy et al. 2013), the *C. velia* cells are of a brownish color. The second LHC is of the Lhcr type. It is associated with PSI and is related to the red

algal PSI antenna. Recently, evidence was presented that the PSI-specific LHC is more tightly associated with PSI under conditions of a high light cultivation of *C. velia* cells (Mann et al. 2014).

The cryptophyta contain phycobilisomes, which are located in the thylakoid lumen and membrane-intrinsic light-harvesting complexes (Lichtlé et al. 1980, 1987; Ingram and Hiller 1983). The membrane-intrinsic antenna binds Chl a, Chl c2, and alloxanthin. The complexes are built up of only Lhcr proteins; Lhcf and Lhcsr proteins seem to be missing (Bathke et al. 1999). Several studies have presented evidence that both PSII and PSI possess a specific light-harvesting complex (Rhiel et al. 1987; Hiller et al. 1992; Bathke et al. 1999). In PSI, the PSI core complex is surrounded by six to eight monomers of the Lhc proteins (Kereïche et al. 2008). The color of the cryptophyta is caused by the cryptomonad-type phyco-erythrin and phycocyanin, which are bound to the phycobilisomes. Dominance of phycoerythrin leads to a red color of the cells, while in the presence of high amounts of phycocyanin, the cells are blue-green (Figure 24.7).

24.3.3 Pigment Dynamics in Diatoms

Diatoms contain the so-called diadinoxanthin (DD)/diatoxanthin (Dt) cycle as the main xanthophyll cycle (XC; Hager and Stransky 1970; Stransky and Hager 1970). The DD/Dt cycle comprises a forward reaction from DD to Dt, which takes place during high light illumination, and a backward reaction during low light periods or darkness, which reverts Dt back to DD (Figure 24.8). The forward reaction that consists of a removal of the one epoxy group of DD is catalyzed by the enzyme DD de-epoxidase (DDE). The backward reaction reintroduces the epoxy group into Dt and is catalyzed by the Dt epoxidase (DEP). In addition, the diatoms also contain the violaxanthin (V), antheraxanthin (A), and zeaxanthin (Z) cycle (Lohr and Wilhelm 1999) that is typical for higher plants and green and brown algae (Yamamoto et al. 1962; Hager 1967; Hager and Stransky 1970). In this cycle (Figure 24.8), the di-epoxy xanthophyll V is converted to the epoxy-free Z, via the intermediate A, in a two-step de-epoxidation sequence by the enzyme V de-epoxidase (VDE). The VAZ cycle is reverted by the enzyme Z epoxidase (ZEP), which reintroduces the two epoxy groups, again with A as the intermediate. The XC enzymes of diatoms show differences to the enzymes of higher plants and green algae (Figure 24.8). The pH optimum of the DDE is shifted toward higher pH values (Jakob et al. 2001), and the concentration of the cosubstrate ascorbate, which is needed for full activity, is lower (Grouneva et al. 2006). In addition and in contrast to higher plants, the DEP is fully inactivated under conditions of a light-driven proton gradient (Mewes and Richter 2002; Goss et al. 2006a). As a consequence, the DD/Dt cycle of diatoms is able to react faster and with a higher capacity to sudden increases in the light intensity as the VAZ cycle of higher plants. While the DD/Dt cycle is involved in the photoprotection of the diatom photosynthetic apparatus via nonphotochemical quenching of the chlorophyll (Chl) a fluorescence (Olaizola et al. 1994;

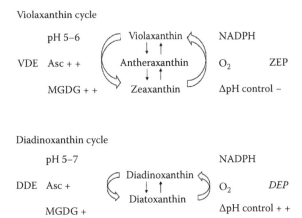

Violaxanthin cycle

FIGURE 24.8 Reaction sequence and enzymes of the violaxanthin (VAZ) and diadinoxanthin (DD/Dt) cycle. The VAZ cycle is present in vascular plants and green and brown algae; the DD/Dt cycle is found in the algal classes Bacillariophyceae, Xanthophyceae, Haptophyceae, and Dinophyceae. The figure also shows the cofactor requirements of the enzymes catalyzing the de-epoxidation reaction (VDE and DDE) and the epoxidation reaction (ZEP and DEP), respectively. Symbols behind the cofactors indicate whether high (+ +) or low (+) concentrations of the respective substrates are needed for high enzyme activity. The figure also addresses the important observation that the proton gradient inhibits Dt epoxidation (high ΔpH control, + +), whereas Zx epoxidation is unaffected by the presence of the transmembrane ΔpH (–). It also depicts the pH range of the thylakoid lumen, where VDE and DDE activity can be observed. VDE: violaxanthin de-epoxidase, DDE: diadinoxanthin de-epoxidase, ZEP: zeaxanthin epoxidase, DEP: diatoxanthin epoxidase, Asc: ascorbate, MGDG: monogalactosyl-diacylglycerole. (Reproduction of Figure 1 from Goss R, Jakob T, *Photosynth Res* 106, 103–122, 2010. With kind permission.)

Lavaud et al. 2002a; Goss et al. 2006a) and via the detoxification of ROS (Lepetit et al. 2010), the VAZ cycle in diatoms serves another important role. The VAZ cycle pigments only amount to significant concentrations when the diatom cells are illuminated with very high light intensities for several hours (Lohr and Wilhelm 1999). Under these conditions, the Ch biosynthesis is halted, but carotenoids are further synthesized. However, carotenoid biosynthesis is blocked on the level of the de-epoxidized, photoprotective pigments Dt and Z during high light exposure. Following the pigment conversions after a shift from high light to low light conditions, it became clear that the VAZ cycle pigments in diatoms serve as precursor pigments for the DD/Dt pigments and finally for the synthesis of the main light-harvesting xanthophyll Fx (Lohr and Wilhelm 1999, 2001). Based on a detailed analysis of the conversion kinetics, it was proposed that V is a common precursor of the carotenoids with an allenic or acetylenic group, and that during the sequence of events, V is converted to DD and finally to Fx (Lohr and Wilhelm 2001).

24.3.4 PIGMENT DYNAMICS IN OTHER GROUPS

In contrast to diatoms, brown algae do not possess the pigments DD and Dt and thus the DD/Dt cycle. Instead, they utilize

the VAZ cycle (Figure 24.8) that is typically found in higher plants and green algae (Hager and Stransky 1970). In general, the de-epoxidation and epoxidation reactions in brown algae show comparable features to those of higher plants, but some important differences have additionally been reported. Like in higher plants, the de-epoxidation reaction, i.e., the conversion of V to A and Z, is faster than the epoxidation reaction, i.e., the back reaction from Z to A and V. However, the kinetics of the epoxidation reaction seems to depend on the acclimation status of the blades of the macrophytic brown algae (García-Mendoza and Colombo-Pallotta 2007). In *M. pyrifera*, a fast epoxidation reaction was observed in blades near the water surface, which were exposed to high light intensities. In these high light acclimated blades, the ratio between the de-epoxidation and the epoxidation reaction exhibited a value of around 2. In low light acclimated blades that were situated in deeper water layers, the epoxidation was slow, thus resembling more closely the value of around 10 that is normally observed for the ratio of the de-epoxidation to the epoxidation reaction in higher plants and green algae (Siefermann and Yamamoto 1975; Goss et al. 2006b). Furthermore, and like in higher plants and green algae, the epoxidation seems not to be inhibited by the presence of the light-driven proton gradient. However, other studies have indicated that under certain conditions, i.e., darkness in combination with anoxia or high temperatures, the epoxidation in brown algae may be blocked (Fernandez-Marin et al. 2011). The insensitivity of the epoxidation to the light-driven proton gradient is in contrast to the situation in diatoms that are characterized by a very fast epoxidation reaction that is completely blocked during illumination by ΔpH (Mewes and Richter 2002; Goss et al. 2006a).

Comparable characteristics of a fast epoxidation reaction could be observed in haptophytes (Goss et al. 2006a), which like the diatoms use the DD/Dt cycle (Figure 24.8). In the haptophyte *Prymnesium parvum*, a fast conversion of DD to Dt was observed during high light illumination. In low light phases following the high light exposure, Dt was converted back to DD with fast kinetics. Although the kinetics of the epoxidation reaction were not as fast as in diatoms, the back reaction of the DD/Dt cycle was significantly enhanced compared to that of higher plants and green algae. Like in diatoms, it was observed that the epoxidase of the haptophyte could be inhibited by the light-driven proton gradient (Goss et al. 2006a). The presence of the VAZ cycle pigments (Figure 24.8) in haptophytes that are cultivated under prolonged high light illumination is another characteristic of the haptophytes that is comparable to the diatoms (Lohr and Wilhelm 1999). However, since it has been shown that the VAZ cycle pigments serve as precursor pigments for the biosynthesis of the DD/Dt cycle pigments (Lohr and Wilhelm 1999, 2001), their presence in the cells of haptophytes seems only logical.

Like the brown algae, the chrysophytes and the eustigmatophyceae contain a functional VAZ cycle (Figure 24.8; Hager and Stransky 1970). In addition, DD was detected in some ice algae belonging to the chrysophytes (Tanabe et al.

2011). In contrast to the VAZ cycle pigments, where light-induced conversion of V to A and Z was detected, DD was not convertible to Dt in these algae. This observation is puzzling since in diatoms the DDE is able to de-epoxidize both substrates (Lohr and Wilhelm 1999), and also for the isolated VDE of higher plants, the conversion of both DD and V has been reported (Yamamoto and Higashi 1978; Goss 2003). Dinophytes exhibit the DD/Dt cycle (Figure 24.8; Brown et al. 1999), and thus, it has to be expected that under stimulation with high light intensities, the VAZ cycle pigments are also present. Recently, an interesting VAZ cycle (Figure 24.8) has been reported in *C. velia,* which belongs to the chromerophyta (Kotabová et al. 2011). In contrast to higher plants and green and brown algae, the second step of both de-epoxidation and epoxidation reaction, i.e., the conversion of A to Z and A to V, respectively, seems to be so fast that the intermediate A cannot be observed in the intact cells. Finally, the cryptophytes seem to completely lack a XC.

24.3.5 PHOTOSYSTEMS AND THEIR MACROMOLECULAR ORGANIZATION IN THE MEMBRANE

The thylakoid membranes of higher plants show a clear separation into grana and stroma membranes and a heterogeneous distribution of the photosystems (Dekker and Boekema 2005). PSII is enriched in the appressed membranes of the grana, whereas PSI is localized in the stroma membranes and the margin regions of the grana membranes. During recent years, it has become clear that the LHCII plays an important role in the formation of the grana membranes. Through the interaction with the main membrane lipid, MGDG LHCII is able to form large macroaggregates, the so-called LHCII macrodomains (Simidjiev et al. 1998, 2000; Garab and Mustardy 1999). In addition, the electrostatic interactions between LHCII complexes in opposing thylakoid membranes, in conjunction with the high Mg^{2+} concentration in the chloroplast stroma, lead to a tight packaging of the thylakoids in the granum (Chow et al. 1991; Kirchhoff et al. 2003). The LHCII macroaggregates span several grana membranes and form supercomplexes that are characterized by excitonic coupling and thus excitation energy transfer between complexes located in different thylakoid membranes (Barzda et al. 1996). Thylakoid membranes of diatoms, on the other hand, do not show a comparable heterogeneity and are characterized by regular stacks of three membranes (Gibbs 1962, 1970). Recent results have indicated that the absence of grana formation, i.e., the ability to tightly pack thylakoid membranes, may be caused by various factors. The first important difference to thylakoid membranes of higher plants and green algae lies in the lipid composition of the diatom thylakoids. Whereas higher plant thylakoid membranes are dominated by the neutral galactolipids MGDG and DGDG, the diatom membranes are enriched in the negatively charged lipids SQDG and PG (Goss et al. 2009; Lepetit et al. 2012). This leads to large membrane areas with a negative surface charge, and it can be expected that these membrane areas show a strong electrostatic repulsion. In addition, the FCP of diatoms has been shown to be significantly less sensitive

to Mg^{2+} ions and low pH values compared to the LHCII of higher plants (Schaller et al. 2014). This leads to a strongly reduced capacity of the FCP to form large aggregates, i.e., macrodomains, as they are typical for the LHCII and needed for grana formation. Despite the lack of grana and stroma formation in diatom thylakoid membranes, heterogeneity of the membrane has been proposed (Lepetit et al. 2012). The model for the typical stacks of three membranes takes into account the high concentrations of negatively charged lipids and two other factors: the enrichment of isolated FCP complexes with the neutral galactolipid MGDG (Lepetit et al. 2010) and the inhibition of the DD to Dt conversion by the negatively charged lipid SQDG (Goss et al. 2009). With regard to the enrichment of the FCP with MGDG, it has been shown that MGDG forms a lipid shield around the FCP complexes, which incorporates a large part of the pool of DD/Dt cycle pigments. With respect to the inhibition of the DD de-epoxidation, it has become clear that SQDG, even in low concentrations, completely inhibits the enzyme DDE, which performs the conversion of DD to Dt. Based on these experimental results, the model (Lepetit et al. 2012) predicts that the inner membranes of the stacks of three are enriched in PSII and FCP complexes, which are surrounded by an MGDG shield that harbors a large part of the DD/Dt cycle pigments. The inner membranes are the places where the DD to Dt conversion is taking place. Together with the exclusion of SQDG from the inner membranes, the DD de-epoxidation can lead to lateral aggregation of FCP complexes, which is, however, different from a macroaggregation, thus inducing the nonphotochemical quenching of Chl a fluorescence (NPQ). According to the model, the outer membranes are enriched in PSI, the ATP synthases, and the negatively charged SQDG, which has been shown to interact with the ATP synthase (Pick et al. 1985). In conclusion, the recent results on the lipid and protein composition of diatom thylakoids indicate that a certain heterogeneity and patch-like areas might be present in the membranes. However, the heterogeneity is far from being as pronounced as in the grana and stroma membranes of higher plants.

24.4 ALTERNATIVE ELECTRON FLOWS

24.4.1 CHLORORESPIRATION

In addition to the linear electron flow, several additional electron transport pathways are present in the thylakoid membrane of diatoms. To differentiate these electron transport pathways from linear electron flow, the electron transports have been termed alternative electron transport pathways. One of these pathways, which are usually observed during periods of prolonged darkness where the light-driven electron flows are not present, is the chlororespiratory pathway. Chlororespiration was first observed in green algae (Bennoun 1982; Wilhelm and Duval 1990), but since then, evidence for chlororespiratory electron flow has also been found for higher plants and diatoms (Caron et al. 1987; Ting and Owens 1993). In contrast to linear electron flow where electrons from the OEC at PSII are transferred to PSI via the $Cytb_6/f$ complex, where they finally lead to the reduction of NADP⁺, the chlororespiratory electron flow starts from the reduced NADPH+H⁺. The electrons derived from NADPH+H⁺ are used to reduce the plastoquinone pool by the enzyme NADPH+H⁺ dehydrogenase (Ndh). In a second step, the PQ pool is oxidized via a terminal oxidase that reduces molecular oxygen. To differentiate this enzyme from the mitochondrial oxidase, it has been termed plastidal terminal oxidase (PTOX). The presence of a chlororespiratory electron pathway in diatoms leads to the reduction of the PQ pool and the establishment of a proton gradient over the thylakoid membrane during periods of darkness (Bennoun 1982; Wilhelm and Duval 1990). It thus enables a slow but persistent DD to Dt de-epoxidation during prolonged dark phases (Jakob et al. 1999), a process that is normally driven by high light illumination. In contrast to general belief (Peltier et al. 1987), recent measurements have shown that, in diatoms, chlororespiration is also active during actinic illumination (Eisenstadt et al. 2008; Grouneva et al. 2009). Furthermore, evidence has been gathered that in the centric and pennate diatoms, the chlororespiratory pathways are characterized by profound differences (Grouneva et al. 2009). Experiments with *P. tricornutum* and *C. meneghiniana* that were illuminated in the presence of the PSII inhibitor DCMU showed that the pennate *P. tricornutum* was able to generate a proton gradient under these conditions, which led to DD to Dt conversion and NPQ. Illumination of the centric *C. meneghiniana*, on the other hand, did not lead to DD de-epoxidation and the induction of NPQ, thus indicating the absence of ΔpH. It was proposed (Figure 24.9) that in the presence of DCMU, chlororespiratory electron flow was responsible for the generation of the proton gradient in *P. tricornutum* (Grouneva et al. 2009). This electron flow most likely involves a donation of electrons from stromal sources to the PQ pool, which is then oxidized by the PTOX. The generation of the proton gradient is realized by a proton-translocating NADPH+H⁺ dehydrogenase, i.e., a type I Ndh as it is found in higher plants (Sazanov et al. 1998). As evidenced by determinations of the PSI re-reduction kinetics, in *C. meneghiniana*, the PQ pool is not oxidized via the PTOX, but the electrons are shuttled to PSI with high priority and are used to reduce NADP⁺ (Grouneva et al. 2009). In addition, chlororespiration in *C. meneghiniana* likely utilizes a type II Ndh to reduce the PQ pool. In contrast to the type I Ndh, this enzyme is unable to translocate protons across the thylakoid membrane (Peltier and Cournac 2002; Jans et al. 2008), and consequently, the establishment of a proton gradient in the presence of DCMU is missing in the centric diatom.

24.4.2 Cyclic Electron Flow

During normal operation, PSII drives the electron transport from oxygen to plastoquinone at the PSII acceptor site and from the reduced quinones into the photosynthetic electron transport chain, where finally the PSI-triggered reduction of NADP⁺ is taking place. In addition to the well-known cyclic electron transport around PSI, a second cyclic electron transport has been revealed. This cyclic electron flow around PSII

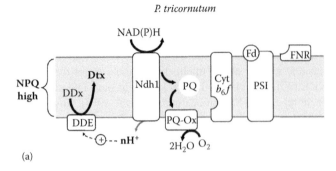

P. tricornutum

(a)

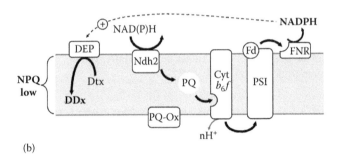

C. meneghiniana

(b)

FIGURE 24.9 Model depicting the influence of alternative electron pathways on the activity of the diadinoxanthin cycle in two diatoms during a period of low light (LL) illumination in the presence of DCMU after a preceding period of high light (HL) illumination in the absence of DCMU. (Modified from Grouneva I et al., *Biochim Biophys Acta* 1787, 929–938, 2009.) DCMU prevents the electron donation from PSII to the PQ pool. In *P. tricornutum* (a), there is evidence of electron transport from stromal electron donors via the PQ-pool to a PQ-oxidase. In conjunction with a proton translocating type 1 Ndh, this chlororespiratory electron transport generates a proton gradient across the thylakoid membrane, which activates the DDE and simultaneously suppresses the activity of the DEP. In *C. meneghiniana* (b), alternative electrons from stromal sources are preferentially donated to PSI, which finally contributes to an NADPH production and enables the rapid Dt epoxidation. No indications of a proton gradient were found in *C. meneghiniana*, which points to the involvement of a type 2 Ndh that is not coupled to a direct proton translocation across the thylakoid membrane. For further details and references, refer to the text. (Reproduction of Figure 3 from Goss R, Jakob T, *Photosynth Res* 106, 103–122, 2010. With kind permission.)

has been found in higher plants and the green alga *Chlorella pyrenoidosa* and presumably involves the PSII subunit Cytb₅₅₉ (Whimarsh and Pakrasi 1996). A comparable PSII cyclic electron flow has also been observed in the diatom *P. tricornutum* and has been interpreted as a photoprotection mechanism against supersaturating light intensities (Lavaud et al. 2002b, 2007) because, at the acceptor site of PSII, it helps to avoid the oversaturation of the electron carriers after PS II, i.e., the plastoquinones and the plastoquinone pool. At the donor site of PSII, it competes with the oxidation of water and the subsequent O₂ evolution (Geel et al. 1997), thus relieving the PSII when the OEC is temporarily inactive. Interestingly, it has been observed that, in diatoms, the amount of cyclic electron transport around PSII can be higher than in green algae (Lavaud et al. 2002b, 2007). Since the induction of the cyclic PSII electron transport is extremely fast, it has been proposed that diatoms, which are located in a habitat with sudden changes in the light intensity, particularly rely on this photoprotective mechanism (Feikema et al. 2006; Lavaud et al. 2007). With regard to the mechanism of the alternative cyclic electron flow, it has been demonstrated that the electrons are diverted from the linear electron transport via the plastoquinone Q_B at the PSII acceptor site (Lavaud et al. 2007). It is likely that the redox state of the PQ pool controls the activity of the PSII cycle electron transport (Feikema et al. 2006) and that the pathway of the electrons around PSII involves the $Cytb_{559}$ subunit (Feikema et al. 2006; Kaminskaya et al. 2007). It is believed but yet unknown that additional electron donors are located between $Cytb_{559}$ and P_{680}. The tyrosine residue Y_D, which was originally thought to be involved in the donation of electrons to the PSII reaction center (Lavaud et al. 2002b), is, according to recent measurements (Feikema et al. 2006), not a component of the cyclic PSII pathway.

24.4.3 MEHLER REACTION

Another alternative electron flow is the reduction of oxygen to the superoxide radical during the photosynthetic electron flow. This electron flow has been termed the Mehler reaction (for a review on the Mehler reaction in plants and algae, see Badger et al. 2000; for a recent review on plants, see Driever and Baker 2011). Molecular oxygen can interact with reduced FeS-X centers associated with the PSI core subunits PsaA and PsaB (Asada 1999) or with stromal components that accept electrons from PSI, like reduced ferredoxin (Furbank and Badger 1983). The superoxide anion radical is then dismutated to hydrogen peroxide and finally converted to water by ascorbate peroxidase. Since the electrons transported in the Mehler reaction result from the water-splitting at PSII and the deactivation of the superoxide anion radical generates water, this electron flow has been termed water–water cycle (Asada 1999). Although many studies have gained experimental evidence for an O_2 consumption during illumination in many algal species, the unequivocal assignment of this process to the Mehler reaction is difficult. However, through the use of $^{18}O_2$, Claquin et al. (2004) were able to demonstrate that in the diatom *Cylindrotheca fusiformis*, 50–60% of the light-stimulated oxygen uptake was caused by the Mehler reaction. Later, also using $^{18}O_2$ and mass spectrometry, it was shown that the two diatoms, *T. pseudonana* and *Nitzschia epithemioides*, exhibited significant rates of oxygen photoreduction due to the Mehler reaction at saturating light intensities (Waring et al. 2010). High light cultivation of the cells increased the capacity of the Mehler reaction, and interestingly, *T. pseudonana* showed five times higher rates of Mehler reaction in comparison to *N. epithemioides*. The higher rates of Mehler reaction in the high light-grown cultures were accompanied by significant increases in the enzymes used to detoxify the reactive oxygen species (ROS) produced by the Mehler reaction. The Mehler reaction is not restricted to diatoms, and recently, it has been reported that in the dinoflagellate *Symbiodinium* sp., a symbiotic dinoflagellate of cnidaria, the Mehler reaction is responsible for the photoreduction of oxygen (Roberty et al. 2014). In this alga, the Mehler reaction is the main alternative electron flow at the onset and during the steady state of illumination and acts as an important photoprotective mechanism.

24.5 CARBON ASSIMILATION

The only pathway to assimilate CO_2 in eukaryotic algae is the reductive pentose pathway (Calvin cycle). The initial reaction is catalyzed by the Rubisco and is the rate limiting step in the photosynthetic apparatus. Due to its slow turnover rate of about 0.3 s, the carboxylation capacity is regulated by the protein amount. This is the reason why Rubisco represents about 50% of total cell proteins. During evolution, at least four different Rubisco types have evolved. The type I B enzyme consists of eight large and eight small subunits and originates from cyanobacteria (Tabita et al. 2008). This enzyme is found in green algae and plants. In red algae and all plastids derived from a secondary endocytobiosis with these algae, Rubisco ID is present, which is phylogenetically related to purple bacteria (e.g., Ralstonia) having similar enzymatic key features (turnover rate or substrate specificity). However, secondary plastids in the Alveolates (typically Dinophyta and Chromera) fix CO_2 by using Rubisco type II (Morse et al. 1995). This enzyme consists of only a large subunit, and its turnover rate is three times faster. Both features together reduce the protein demand per active catalytic center to one fourth. However, the enzyme was not really successful in evolution due to its low affinity to CO_2. These cells can grow only under either high CO_2 or low oxygen and strongly depend on an efficient CO_2 concentrating mechanism (CCM).

Under natural conditions, the availability of inorganic carbon can vary dramatically over the period of a day due to strong pH oscillations in the water body by the uptake of Ci and nitrate. pH values up to 9 can be observed very often. Under this condition, free CO_2 is negligible, and the only source for Ci is the bicarbonate. Therefore, many phytoplankton algae contain transporters for free CO_2, biocarbonate, and the enzyme carboanhydrase (CA), which catalyzes the equilibrium between free CO_2 and carbonic acid. Beside this biophysical mechanism, a biochemical CCM can be used by fixing bicarbonate via phosphoenolpyruvate-carboxylase (PEPC) producing a C4 metabolite, which can be decarboxylated in close vicinity to the Rubisco to increase the local concentration of Ci. In higher plants, C4 photosynthesis is associated with a spatial separation of the prefixation via PEPC and the Ci assimilation via Rubisco to prevent a futile cycle. C4 photosynthesis has also been observed in single-cell diatoms (Reinfelder et al. 2000), at least in *Thalassiosira weissflogii*. In the model diatom *P. tricornutum*, this pathway is not very likely because the enzyme pyruvate-phosphate dikinase, which delivers PEP for the PEPC reaction, can be inactivated without any effect on photosynthesis (Haimovic-Dayan et al. 2013). In this case, an incomplete C4 pathway may have a function in photoprotection. Most of the algal plastids contain pyrenoids where the assimilated carbon is stored either as starch (e.g., in Dinoflagellates) or as chrysolaminaran (in diatoms). Pyrenoids are embedded in a shield of highly concentrated Rubisco protein (Raven et al. 2012) together with the fructose-1,6-bisphosphate aldolase (FBA), which is necessary for sequestering the assimilate between storage and the replenishment of the Calvin cycle with carbon skeletons for the primary acceptor RubP. Moroney and Ynalvez (2007) proposed that one function of the pyrenoids is to separate the Rubisco from the CA in the stroma of the chloroplast.

24.5.1 Regulation of the Carbon Assimilation

The regulation of the Calvin cycle in secondary plastids differs fundamentally from what is known for green chloroplasts. Here the Calvin cycle becomes activated in light by the redox control of the thioredoxin system. In light, ferredoxin becomes reduced due to the light-driven electron flow, and thioredoxin then reduces free S–S bonds in target enzymes, which leads to conformational changes and enzyme activation. However, in diatoms, the key enzymes of the Calvin cycle are not a target for the thioredoxin system with the exception of the FBPase and the phosphogylcerate kinase. Instead, in the pyrenoid, a CA is activated by reduced thioredoxin. This indicates that the CCM is under redox control in diatoms but not the Calvin cycle (Matsuda and Kroth 2014). This hypothesis is supported by the fact that, in diatoms, the genes for the Rubisco–activase could not be found; instead, the CbbX genes have been detected. The gene products of the CbbX genes are related to a protein having Rubisco–activase activity in the α-protobacterium *Rhodobacter sphaeroides* (Mueller-Cajar et al. 2011).

Since in green plants the plastid houses the reductive pentose phosphate cycle (PPC; Calvin cycle) as well as the oxidative PPC, a strict regulation is needed to prevent a futile cycle. Therefore, the oxidative PPC has to be active in the dark to deliver the carbon skeletons for the biosynthesis of nucleotides, which cannot be imported from the cytosol, because the plastid envelope is impermeable for nucleotides in green plastids. However, in diatoms, the oxidative PPC is lacking in the chloroplast, and the membrane is permeable for nucleotides (for review, see Kroth et al. 2008). This different compartmentalization of the major metabolic pathways asks for different regulatory elements. Recently, it has been shown by Mekhalfi et al. (2014) that GAPDH interacts in the light with Fd-NADP reductase and the small CP12 protein. Therefore, GAPDH is active in the light and inhibited in the dark, thus rerouting reductants to other dark-active reductive processes in the diatom plastid.

24.5.2 Organelle Interaction

Recently, Smith et al. (2012) published a comparative analysis of those genes whose products regulate carbon flow from photosynthesis to the major macromolecular pools in three different diatoms: *T. pseudonana*, *P. tricornutum*, and *Fragilariopsis cylindrus*. Unexpectedly they found that all diatoms have conserved the lower half of glycolysis in the mitochondria, whereas the key regulatory enzymes of the upper half of glycolysis are located in the cytosol. A second unexpected result was that between these three different species, basic differences with respect to the genomic translocation of key components of basic metabolic pathways can be found. Finally, the genome analysis revealed that the metabolite transport between the different intracellular compartments does not always follow the typical plant-like pattern and that the differences observed play a crucial role in understanding differences in carbon flux regulation in diatoms compared to other algae. Most of the available data indicate that, in diatoms, the mitochondria play a much stronger regulatory role in metabolic flux control as in green algal cells, and this might be the reason for their high ecological adaptability. This suggestion was recently supported by metabolome analysis in *P. tricornutum* cultures, which had been shifted from blue to red light (Jungandreas et al. 2014). The authors observed that under these shift conditions, growth is nearly arrested, although carbon is still assimilated by the photosynthetic machinery. However, under this condition, proteins are no longer formed, and all metabolites produced are funneled to carbohydrates. Instead of a redox control of carbon partitioning, Wilhelm et al. (2014) hypothesize that blue light receptors like aureochrome function as metabolic switches by influencing the enzyme activities, which regulate carbon partitioning between chloroplasts and mitochondria. However, it must be emphasized that this picture is restricted to diatoms and that the knowledge on the regulation of C-partitioning by plastid–mitochondrial interaction is not yet known in other secondary plastids.

24.6 REGULATION OF PHOTOSYNTHESIS

24.6.1 Photoprotection in Diatoms

The main photoprotection mechanism of diatoms is the so-called nonphotochemical quenching (NPQ; for reviews on NPQ in higher plants, see Horton et al. 2008; Jahns and Holzwarth 2012). This process describes the thermal dissipation of that part of the excitation, which cannot be used for photosynthesis under conditions of a high light illumination. The fast induction of NPQ after exposure to high light intensities makes this photoprotection mechanism valuable for diatom cells (Lavaud et al. 2002a; Goss et al. 2006a; Lavaud 2007; Lavaud and Goss 2014). This is especially true for diatoms in the open waters of the oceans where the light conditions may change rapidly on a time scale of seconds or minutes (Lavaud 2007). Diatom cells can be exposed to very high light intensities at the water surface at one moment and in the next moment drift to deeper water layers with near darkness. With regard to the kinetics of the NPQ onset, it has been observed that NPQ in diatoms consists of different components with different induction and relaxation kinetics. In the centric diatom *C. meneghiniana*, three different components of NPQ have been observed (Grouneva et al. 2008). The first component is induced very rapidly within less than a minute. This component shows the same kinetics as the establishment of the light-driven proton gradient and relies on the presence of preformed Dt. The fast NPQ component in diatoms is comparable to a similar component in higher plants, which has been termed transient NPQ (Finazzi et al. 2004; Kalituho et al. 2007). The second component is induced within a time span of several minutes and can be correlated with the main de-epoxidation of DD to Dt. The third component describes a stable NPQ in a dark period following high light illumination and is caused by the slow epoxidation of Dt back to DD due to limitations in the availability of the cosubstrate of the epoxidation reaction, NADPH+H+ (Goss et al. 2006a). It is reasonable to believe that the complexity of NPQ in diatoms with different induction and relaxation kinetics provides the diatom cell with a photoprotection mechanism that is able to cope with different illumination scenarios. The very fast induction of the transient NPQ is suited for cells that are situated in the open waters where they experience rapid changes of the light conditions. The slower induction of the main NPQ component and the slow relaxation kinetics of the third NPQ

component provide the ability to cope with longer periods of high light illumination as they may be experienced in the open ocean, but as they are typical for diatoms that are located in the coastal regions.

With regard to the molecular mechanism of NPQ in diatoms, several recent models have been put forward (Goss and Jakob 2010; Gundermann and Büchel 2012; Lavaud and Goss 2014; Goss and Lepetit 2015). These models (Figure 24.10) describe the factors and the sequence of events that are needed to induce the structural changes of the antenna system, which transform the FCP complexes from a light-harvesting system to a system that effectively dissipates the excessive excitation energy as heat. The models predict the existence of two different quenching sites within PSII (see also Miloslavina et al. 2009). The first quenching site, termed Q1, consists of FCP complexes, which have dissociated from the PSII core. After the dissociation, these complexes form aggregates (Gundermann and Büchel 2012; Schaller et al. 2014), which are characterized by a shift of the Chl a fluorescence emission from 680 to 700 nm (Miloslavina et al. 2009; Lavaud and Lepetit 2013). It has been proposed that nonphotochemical quenching at the Q1 site is independent of the light-driven

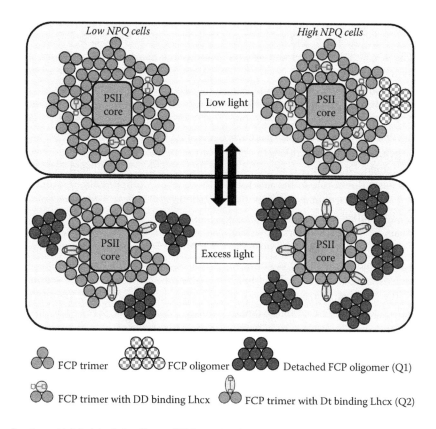

FIGURE 24.10 (See color insert.) Model of the diatom NPQ mechanism in PSII of low and high NPQ cells. After transfer from low light intensities to excessive light conditions, several structural rearrangements are taking place in the PSII antenna, which are reversed by transferring the cells back to low light. The conversion of DD to Dt at the Lhcx protein within the FCP trimer leads to the formation of the Q2 site. The concomitant conformational change of the trimer promotes an expulsion of adjacent trimers, which form oligomeric complexes, which then can build up Q1. The extent of the structural changes upon high light acclimation depends on the pre-acclimation conditions of the cells, which directly influence the initial amount of DD/Dt and Lhcx proteins as well as the presence of oligomeric antenna complexes due to an altered thylakoid membrane lipid composition. (Reproduction of Figure 3 of Goss R, Lepetit B, *J Plant Physiol* 172, 13–32, 2015. With kind permission.)

conversion of DD to Dt (Miloslavina et al. 2009). The second quenching site Q2 is composed of FCP complexes, which stay attached to the PSII core complex (Miloslavina et al. 2009). In contrast to Q1, quenching at site Q2 seems to depend on the presence of Dt. With respect to the sequence of events during the establishment of the two quenching sites, one of the recent models predicts that Q2 is formed first (Goss and Lepetit 2015). The conversion of DD to Dt in the FCP complexes at the Q2 site is thought to induce a conformational change of the Dt-binding FCP complexes. This conformational change may include the Lhcx proteins, which have been shown to be important for NPQ (Bailleul et al. 2010; Lepetit et al. 2013), at the Q2 site or the FCP trimers themselves. This conformational change may then induce the dissociation of a large part of the FCP trimers that under low light conditions form the FCP PSII supercomplex together with the PSII core complex. In addition, the dissociation of the FCP trimers might render the FCPs more sensitive to protonation by the light-driven proton gradient, thus facilitating the aggregation of the complexes (Gundermann and Büchel 2012; Schaller et al. 2014). Although a lot of data on the mechanism of NPQ have been gathered recently, some central points are still unclear. The protein composition of both the Q1 and the Q2 site is still an open question, i.e., it is unclear which of the Lhcf and Lhcx proteins form the two quenching sites. In addition, the nature of the actual quenching site is still unknown, and it has to be clarified if Chl a–Chl a or Chl a–xanthophyll interactions lead to the conversion of excitation energy into heat.

Besides NPQ, the diatoms contain other important photoprotection mechanisms. The cyclic electron flow around PSII (Lavaud et al. 2002b, 2007) seems to be another important mechanism, which helps the diatoms to prevent the overexcitation of PSII and thus the generation of harmful ROS. The de-epoxidized DD/Dt cycle pigment Dt not only is involved in the process of NPQ but also has a direct role in the inactivation of ROS (Lepetit et al. 2010). In particular, diatoms that are acclimated to high light intensities possess an increased pool of DD/Dt cycle pigments. A significant part of DD/Dt molecules in these algae is not bound to the FCP antenna complexes but is located in the free lipid phase of the membrane. Comparable to what has been observed before for Z in higher plants (Havaux et al. 2007), the lipid solubilized Dt serves as antioxidant and detoxifies ROS that are generated in the photosynthetic electron transport under supersaturating light conditions. Deactivation of ROS by Dt helps to prevent damage to the thylakoid membrane lipids, the embedded photosynthetic pigment protein complexes, and the photosynthetic pigments.

Photoprotection by state transitions, on the other hand, seems to be missing in diatoms. State transitions describe a process where, after a phosphorylation, a part of the LHCII dissociates from the PSII core complex and connects to PSI (for a recent review, see Minagawa 2011). This mechanism, which is controlled by the binding of reduced plastoquinone to the Cytb6/f complex, is used by the plant to compensate for imbalances in the excitation of the two photosystems. While in higher plants the main function of state transitions is the avoidance of overexcitation of PSII or PSI, in green algae,

state transitions also occur as a reaction to high light intensities (Garcia-Mendoza et al. 2002). The absence of state transitions in diatoms was first indicated by Owens (1986) who found that the diatom FCP complexes do not preferentially excite PSII or PSI. In contrast, the absorbed excitation energy is quite equally distributed between the two photosystems. This equal distribution is caused by the main peripheral FCP complex, which serves as an antenna for both photosystems. In addition, due to the lack of thylakoid membrane heterogeneity, i.e., grana and stroma formation as in higher plants, and the regular thylakoid arrangement in stacks of three, PSII and PSI are almost equally distributed in the thylakoid membrane of diatoms (Pyszniak and Gibbs 1992). Thus, the need for a specific mechanism to counterbalance differences in the photosystem excitation is absent in diatoms and other algal groups, which are characterized by a regular thylakoid membrane arrangement.

24.6.2 Photoprotection in Other Groups

Although the photoprotection of brown algae is not as extensively studied as that of the diatoms, recent studies have provided new insight into the process of NPQ, especially in the macrophytic brown algae (García-Mendoza and Colombo-Pallotta 2007; García-Mendoza et al. 2011; Li et al. 2014). The analysis of NPQ in the brown algae is especially interesting since these algae contain the VAZ cycle of higher plants and green algae on the one hand, and the FCP antenna complexes that are typical for the diatoms on the other. NPQ in the brown algae depends on the presence of the de-epoxidized VAZ cycle pigments A and Z (Benet et al. 1994; Uhrmacher et al. 1995; Harker et al. 1999). In comparison to diatoms and in contrast to higher plants and green algae, the single presence of the light-driven proton gradient does not suffice to induce NPQ, which is established only by the combined action of these two components (García-Mendoza and Colombo-Pallotta 2007; García-Mendoza et al. 2011; Li et al. 2014). The extent of NPQ in the macrophytic brown algae depends on the concentration of the VAZ cycle pigments (Ocampo-Alvarez et al. 2013), i.e., very high NPQ values can be observed in high light acclimated blades that contain a large pool of VAZ cycle pigments, whereas low light acclimated blades with low concentrations of V, A, and Z are characterized by a significantly lower capacity for NPQ. The importance of special light-harvesting proteins for the establishment of NPQ that is found in higher plants, green algae, and diatoms is also valid for brown algae. Recent studies have suggested that Lhcx/Lhcsr proteins are needed for enhanced NPQ, possibly by providing binding sites for the active Z (Ocampo-Alvarez et al. 2013). In line with this suggestion, it has been demonstrated that in *M. pyrifera*, some Lhcsr genes show a higher expression under high light conditions (Konotchick et al. 2013).

The NPQ of the DD/Dt cycle containing haptophytes shows characteristics that are comparable to those of the diatoms. In the haptophyte *P. parvum*, NPQ is correlated with the amount of Dt (Goss et al. 2006a; Dimier et al. 2009). The role of the proton gradient seems to be restricted to the induction of the DD to Dt de-epoxidation, and it is likely that ΔpH alone cannot induce NPQ (Goss et al. 2006a). However, for *Emiliania huxleyi*, it has been reported that the extent of NPQ correlates with the epoxidized DD and not Dt (Harris et al. 2005). In chrysophytes, NPQ depends on the de-epoxidation reactions of the VAZ cycle (Lichtlé et al. 1995; Dimier et al. 2009). In those chrysophytes where DD is additionally present, it cannot be converted to Dt (Tanabe et al. 2011), which implies that DD and Dt are not involved in NPQ but merely serve as precursor pigments in carotenoid biosynthesis (Lohr and Wilhelm 1999, 2001).

The NPQ of the eustigmatophyceae depends on the operation of the VAZ cycle, and the main function of the light-driven proton can be seen in the activation of the VDE (Gentile and Blanch 2001; Cao et al. 2013). The similarities to the NPQ mechanism in diatoms and brown algae are further strengthened by the finding that in the genome of *Nannochloropsis salina*, Lhcsr sequences were detected and genes coding for the PsbS protein of higher plants were missing (Vieler et al. 2012).

In dinophytes, the Dt-dependent NPQ seems to be located in the membrane-bound LHC-related antenna complex, the acpCP (Iglesias-Prieto and Trench 1997). Other mechanisms that contribute to NPQ but are not correlated with the DD/Dt cycle seem to be induced when the full capacity of the XC-dependent NPQ is reached. These NPQ mechanisms consist of the dissociation of the membrane intrinsic acpCP and the water-soluble peridinin chlorophyll protein complex (PCP), which is located at the lumenal side of the thylakoid membrane, from the PSII core complex, thereby relieving part of the excitation pressure on the PSII reaction center (Hill et al. 2012). Until today, gene sequences of Lhcsr proteins were only found in one dinophyte, namely, *Karlodinium micrum* (Patron et al. 2006).

In the chromerophyte *C. velia*, the single presence of the light-induced proton gradient is insufficient to induce NPQ, and the de-epoxidation of V to Z is needed (Kotabová et al. 2011). A recent study investigated in close detail the acclimation of *C. velia* to high light intensities (Mann et al. 2014). It was observed that high light-grown *C. velia* cultures showed a significantly increased pool of VAZ cycle pigments. Comparable to the situation in diatoms, the high light synthesized V, A, and Z were not active in NPQ. Isolation of the *C. velia* pigment protein complexes showed that the XC pigments were associated with the chromera light-harvesting complex (CLH). However, as revealed by spectroscopic measurements, the newly synthesized V, A, and Z were not bound to the protein but located in a lipid shield surrounding the antenna complexes as it has been observed for the diatoms (Lepetit et al. 2010). In addition, the study (Mann et al. 2014) showed the association of Lhcr proteins with PSI, which is also typical for the diatom PSI. The authors concluded that light acclimation of *C. velia* shows features that are comparable to those of diatoms.

The cryptophytes, although lacking a functional XC, are characterized by a fast NPQ, which depends on the establishment of the light-driven proton gradient (Kana et al. 2012).

Consequently, NPQ in *Rhodomonas salina* is of the qE type (high-energy-state quenching), which in higher plants and green algae is induced by ΔpH. The thermal dissipation of excitation energy is located in the membrane-intrinsic LHC antenna but not in the phycobiliproteins, which are additionally present in these algae.

24.6.3 Light Acclimation

Algal cells generally acclimate to changing light conditions to improve their photophysiological performance. Under high light, the cells decrease the chlorophyll content per cell, which goes in parallel with a reduction in antenna size and higher light saturated photosynthetic capacity. During light acclimation, the cells undergo a complete reorganization of the proteome, which is regulated by redox-controlled gene expression. In green plastids, the plastoquinone pool functions as a general switch to modulate gene expression and protein translation via redox signals inside the plastid and by retrograde signaling to nuclear encoded genes (Foyer et al. 2012; Szechynska-Hebda and Karpinski 2013). However, in the plastids derived from secondary symbiosis, the redox state of the plastid not only is under light control but also depends on plastid–mitochondrial interaction. This leads to the fact that, in the dark, the plastoquinone pool is not completely oxidized and, therefore, the redox change of the pool cannot be used for light intensity-mediated signaling. Nevertheless, all heterokontic plastids show comparable light acclimation reactions as green plastids, e.g., HL-mediated chlorophyll reduction, higher photosynthetic capacity, and improved photoprotection (Wilhelm et al. 2014). Recently, it was shown that, in diatoms, light acclimation depends on the presence of blue light (Schellenberger-Costa et al. 2013) and that the blue-light receptor aureochrome 1A is involved in the signal transduction from light sensing to HL phenotype (Schellenberger-Costa et al. 2013). This indicates that at least in diatoms, a light sensor is involved in light intensity acclimation. Nevertheless, recent work of Lepetit et al. (2014) showed that also in diatoms, the redox state of the plastoquinone pool is an important factor to alter the gene expression pattern. Recent data from a metabolome analysis (Jungandreas et al. 2014) showed that blue light directly alters the redox state of the metabolome, which may indirectly influence gene expression. The work has been reviewed by Wilhelm et al. (2014), and an overview on the mode of actions of photoreceptors in diatoms is given by Fortunato et al. (2015). Until now, the regulatory influence of the photoreceptors on the development and function of secondary plastids is not yet studied on the molecular or physiological level.

ACKNOWLEDGMENTS

The authors thank Dr. Torsten Jakob for the original absorption spectra used for Figure 24.7. The authors thank the Deutsche Forschungsgemeinschaft (DFG) for financial support (Go 818/7-1 and Wi 764/15-1).

REFERENCES

Alboresi A, Gerotto C, Giacometti GM, Bassi R, Morosinotto T. *Physcomitrella patens* mutants affected on heat dissipation clarify the evolution of photoprotection mechanisms upon land colonization. *Proc Natl Acad Sci U S A* 107, 11128–11133, 2010.

Apt KE, Clendennen SK, Powers DA, Grossman AR. The gene family encoding the fucoxanthin chlorophyll proteins from the brown alga *Macrocystis pyrifera*. *Mol Gen Genet* 246, 455–464, 1995.

Archibald JM, Cavalier-Smith T, Maier U, Douglas S. Molecular chaperones encoded by a reduced nucleus: The cryptomonad nucleomorph. *J Mol Evol* 52, 490–501, 2001.

Armbrust EV, Berges JA, Bowler C et al. The genome of the diatom *Thalassiosira* pseudonana: Ecology, evolution, and metabolism. *Science* 306, 79–86, 2004.

Asada K. The water-water cycle in chloroplasts: Scavenging of active oxygens and dissipation of excess photons. *Annu Rev Plant Physiol Plant Mol Biol* 50, 601–639, 1999.

Badger MR, von Caemmerer S, Ruuska S, Nakano H. Electron flow to oxygen in higher plants and algae: Rates and control of direct photoreduction (Mehler reaction) and Rubisco oxygenase. *Philos Trans R Soc Lond B Biol Sci* 355, 1433–1446, 2000.

Bailleul B, Rogato A, de Martino A et al. An atypical member of the light-harvesting complex stress-related protein family modulates diatom responses to light. *Proc Natl Acad Sci U S A* 107, 18214–18219, 2010.

Barzda V, Istokovics A, Simidjiev I, Garab G. Structural flexibility of chiral macroaggregates of light-harvesting chlorophyll a/b pigment-protein complexes. Light-induced reversible structural changes associated with energy dissipation. *Biochemistry* 35, 8981–8985, 1996.

Bathke L, Rhiel E, Krumbein WE, Marquardt J. Biochemical and Immunochemical Investigations on the light-harvesting system of the cryptophyte *Rhodomonas* sp.: Evidence for a photosystem I specific antenna. *Plant Biol* 1, 516–523, 1999.

Beer A, Gundermann K, Beckmann J, Büchel C. Subunit composition and pigmentation of fucoxanthin-chlorophyll proteins in diatoms: Evidence for a subunit involved in diadinoxanthin and diatoxanthin binding. *Biochemistry* 45, 13046–13053, 2006.

Benet H, Bruss U, Duval J-C, Kloareg B. Photosynthesis and photoinhibition in protoplasts of the marine brown alga *Laminaria saccharina*. *J Exp Bot* 45, 211–220, 1994.

Bennoun P. Evidence for a respiratory chain in the chloroplast. *Proc Natl Acad Sci U S A* 79, 4352–4356, 1982.

Bonente G, Ballotari M, Truong TB et al. Analysis of LhcSR3, a protein essential for feedback de-excitation in the green alga *Chlamydomonas reinhardtii*. *PLoS Biol* 9, e1000577, 2011.

Bowler C, Allen AE, Badger JH et al. The *Phaeodactylum* genome reveals the evolutionary history of diatom genomes. *Nature* 456, 239–244, 2008.

Brown BE, Ambarsari I, Warner ME, Fitt WK, Dunne RP, Gibb SW, Cummings DG. Diurnal changes in photochemical efficiency and xanthophyll concentrations in shallow water reef corals: Evidence for photoinhibition and photoprotection. *Coral Reefs* 18, 99–105, 1999.

Büchel C. Fucoxanthin-chlorophyll proteins in diatoms: 18 and 19 kDa subunits assemble into different oligomeric states. *Biochemistry* 42, 13027–13034, 2003.

Cao S, Zhang X, Xu D et al. A transthylakoid proton gradient and inhibitors induce a non-photochemical fluorescence quenching in unicellular algae *Nannochloropsis sp.* *FEBS Lett* 587, 1310–1315, 2013.

Caron L, Berkaloff C, Duval JC, Jupin H. Chlorophyll fluorescence transients from the diatom *Phaeodactylum tricornutum-* relative rates of cyclic phosphorylation and chlororespiration. *Photosynth Res* 11, 131–139, 1987.

Caron L, Douady D, Quinet-Szely M, de Göer S, Berkaloff C. Gene structure of a chlorophyll a/c-binding protein from a brown alga: Presence of an intron and phylogenetic implications. *J Mol Evol* 43, 270–280, 1996.

Chow WS, Miller C, Anderson JM. Surface charges, the heterogeneous lateral distribution of the two photosystems, and thylakoid stacking. *Biochim Biophys Acta* 1057, 69–77, 1991.

Claquin P, Kromkamp JC, Martin-Jezequel V. Relationship between photosynthetic metabolism and cell cycle in a synchronized culture of the marine alga *Cylindrotheca fusiformis* (Bacillariophyceae). *Eur J Phycol* 39, 33–41, 2004.

Dekker JP, Boekema EJ. Supramolecular organization of thylakoid membrane proteins in green plants. *Biochim Biophys Acta* 1706, 12–39, 2005.

De Martino A, Douady D, Quinet-Szely M, Rousseau B, Crépineau F, Apt K, Caron L. The light-harvesting antenna of brown algae. Highly homologous proteins encoded by a multigene family. *Eur J Biochem* 267, 5540–5549, 2000.

Dimier C, Giovanni S, Ferdinando T, Brunet C. Comparative ecophysiology of the xanthophyll cycle in six marine phytoplanktonic species. *Protist* 160, 397–411, 2009.

Dittami S, Michel G, Collén J, Boyen C, Tonon T. Chlorophyllbinding proteins revisited—A multigenic family of light-harvesting and stress proteins from a brown algal perspective. *Evol Biol* 10, 365, 2010.

Driever SM, Baker NR. The water–water cycle in leaves is not a major alternative electron sink for dissipation of excess excitation energy when CO_2 assimilation is restricted. *Plant Cell Environ* 34, 837–846, 2011.

Eisenstadt D, Ohad I, Keren N, Kaplan A. Changes in the photosynthetic reaction centre II in the diatom *Phaeodactylum tricornutum* result in non-photochemical fluorescence quenching. *Environ Microbiol* 10, 1997–2007, 2008.

Falciatore A, Bowler C. Revealing the molecular secrets of marine diatoms. *Annu Rev Plant Biol* 53, 109–130, 2002.

Feikema WO, Marosvölgyi MA, Lavaud J, van Gorkom HJ. Cyclic electron transfer in photosystem II in the marine diatom *Phaeodactylum tricornutum*. *Biochim Biophys Acta* 1757, 829–834, 2006.

Fernandez-Marin B, Miguez F, Becerril J, Garcia-Plazaola J. Activation of violaxanthin cycle in darkness is a common response to different abiotic stresses: A case study in *Pelvetia canaliculata*. *Plant Biol* 11, 181, 2011.

Finazzi G, Johnson GN, Dall'Osto L, Joliot P, Wollman F-A, Bassi R. A zeaxanthin-independent nonphotochemical quenching mechanism localized in the photosystem II core complex. *Proc Natl Acad Sci U S A* 101, 12375–12380, 2004.

Fortunato AE, Annunziata R, Juaber M, Bouly J-P, Falciatore A. Dealing with light: The widespread and multitasking cryptochrome/photolyase family in photosynthetic organisms. *J Plant Physiol* 172, 42–54, 2015.

Foyer C, Neukermans J, Queval G, Noctor G, Harbonson J. Photosynthetic control of electron transport and the regulation of gene expression. *J Exp Bot* 63, 1637–1661, 2012.

Furbank RT, Badger MR. Oxygen exchange associated with electron transport and photophosphorylation in spinach thylakoids. *Biochim Biophys Acta* 723, 400–409, 1983.

Garab G, Mustardy L. Role of LHCII-containing macrodomains in the structure, function and dynamics of grana. *Aust J Plant Physiol* 26, 649–658, 1999.

Garcia-Mendoza E, Matthijs HC, Schubert H, Mur LR. Non-photochemical of chlorophyll fluorescence in *Chlorella fusca* acclimated to constant and dynamic light conditions. *Photosynth Res* 74, 303–315, 2002.

García-Mendoza E, Colombo-Pallotta MF. The giant kelp *Macrocystis pyrifera* presents a different nonphotochemical quenching control than higher plants. *New Phytol* 173, 526–536, 2007.

García-Mendoza E, Ocampo-Alvarez H, Govindjee. Photoprotection in the brown alga *Macrocystis pyrifera*: Evolutionary implications. *J Photochem Photobiol B: Biology* 104, 377–385, 2011.

Garrido JL, Zapata M. Detection of new pigments from *Emiliania huxleyi* (Prymnesiophyceae) by high-performance liquid chromatography, liquid chromatography-mass spectrometry, visible spectroscopy, and fast atom bombardment mass spectrometry. *J Phycol* 34:70–78, 1998.

Geel C, Versluis W, Snel JFH. Estimation of oxygen evolution by marine phytoplankton from measurement of the efficiency of photosystem II electron flow. *Photosynth Res* 51, 61–70, 1997.

Gentile M-P, Blanch HW. Physiology and xanthophyll cycle activity of *Nannochloropsis gaditana*. *Biotechnol Bioeng* 75, 1–12, 2001.

Gibbs SP. The ultrastructure of the chloroplasts of algae. *J Ultrastruct Res* 7, 418–435, 1962.

Gibbs SP. The comparative ultrastructure of the algal chloroplast. *Ann N Y Acad Sci* 175, 454–473, 1970.

Gillot, MA, Gibbs, SP. The cryptomonad nucleomorph—Its ultrastructure and evolutionary significance. *J Phycol* 16, 558–568, 1980.

Goss R. Substrate specificity of the violaxanthin de-epoxidase of the primitive green alga *Mantoniella squamata* (Prasinophyceae). *Planta* 217, 801–812, 2003.

Goss R, Jakob T. Regulation and function of xanthophyll cycle-dependent photoprotection in algae. *Photosynth Res* 106, 103–122, 2010.

Goss R, Lepetit B. Biodiversity of NPQ. *J Plant Physiol* 172, 13–32, 2015.

Goss R, Pinto EA, Wilhelm C, Richter M. The importance of a highly active and ΔpH-regulated diatoxanthin epoxidase for the regulation of the PS II antenna function in diadinoxanthin cycle containing algae. *J Plant Physiol* 163, 1008–1021, 2006a.

Goss R, Lepetit B, Wilhelm C. Evidence for a rebinding of antheraxanthin to the light-harvesting complex during the epoxidation reaction of the violaxanthin cycle. *J Plant Physiol* 163, 585–590, 2006b.

Goss R, Nerlich J, Lepetit B, Schaller S, Vieler A, Wilhelm C. The lipid dependence of diadinoxanthin de-epoxidation presents new evidence for a macrodomain organization of the diatom thylakoid membrane. *J Plant Physiol* 166, 1839–1854, 2009.

Grouneva I, Jakob T, Wilhelm C, Goss R. Influence of ascorbate and pH on the activity of the diatom xanthophyll cycle-enzyme diadinoxanthin de-epoxidase. *Physiol Plant* 126, 205–211, 2006.

Grouneva I, Jakob T, Wilhelm C, Goss R. A new multicomponent NPQ mechanism in the diatom *Cyclotella meneghiniana*. *Plant Cell Physiol* 49, 1217–1225, 2008.

Grouneva I, Jakob T, Wilhelm C, Goss R. The regulation of xanthophyll cycle activity and of non-photochemical fluorescence quenching by two alternative electron flows in the diatoms *Phaeodactylum tricornutum* and *Cyclotella meneghiniana*. *Biochim Biophys Acta* 1787, 929–938, 2009.

Grouneva I, Rokka A, Aro EM. The thylakoid membrane proteome of two marine diatoms outlines both diatom-specific and species-specific features of the photosynthetic machinery. *J Prot Res* 10, 5338–5353, 2011.

Gundermann K, Büchel C. The fluorescence yield of the trimeric fucoxanthin-chlorophyll-protein FCPa in the diatom *Cyclotella meneghiniana* is dependent on the amount of bound diatoxanthin. *Photosynth Res* 95, 229–235, 2008.

Gundermann K, Büchel C. Factors determining the fluorescence yield of fucoxanthin-chlorophyll complexes (FCP) involved in non-photochemical quenching in diatoms. *Biochim Biophys Acta* 1817, 1044–1052, 2012.

Gundermann K, Schmidt M, Weisheit W, Mittag M, Büchel C. Identification of several sub-populations in the pool of light harvesting proteins in the pennate diatom *Phaeodactylum tricornutum*. *Biochim Biophys Acta* 1827, 303–310, 2013.

Hager A. Untersuchungen über die lichtinduzierten, reversiblen Xanthophyllumwandlungen an *Chlorella* und *Spinacia oleracea*. *Planta* 74, 148–172, 1967.

Hager A, Stransky H. Das Carotinoidmuster und die Verbreitung des lichtinduzierten Xanthophyllcyclus in verschiedenen Algenklassen. *Archiv Mikrobiol* 73, 77–89, 1970.

Haimovic-Dayan M, Garfinkel N, Ewe D et al. The role of C4 metabolism in the marine diatom *Phaeodactylum tricornutum*. *New Phytol* 191, 175–188, 2013.

Harker M, Berkaloff C, Lemoine Y et al. Effects of high light and desiccation on the operation of the xanthophyll cycle in two marine brown algae. *Eur J Phycol* 34, 35–42, 1999.

Harris GN, Scanlan DJ, Geider RJ. Acclimation of *Emiliania huxleyi* (Prymnesiophyceae) to photon flux density. *J Phycol* 41, 851–862, 2005.

Havaux M, Dall'Osto I, Bassi R. Zeaxanthin has enhanced antioxidant capacity with respect to all other xanthophylls in Arabidopsis leaves and functions independent of binding to PSII antennae. *Plant Physiol* 145, 1506–1520, 2007.

Haxo FT, Kycia JH, Somers GF, Bennett A, Siegelman HW. Peridinin–chlorophyll *a* proteins of the dinoflagellate *Amphidinium carterae* (Plymouth 450). *Plant Physiol* 57, 297–303, 1976.

Hill R, Larkum A, Prášil O, Kramer D, Szabó M, Kumar V, Ralph P. Light-induced dissociation of antenna complexes in the symbionts of scleractinian corals correlates with sensitivity to coral bleaching. *Coral Reefs* 31, 963–975, 2012.

Hiller RG, Scaramuzzi CD, Breton J. The organisation of photosynthetic pigments in a cryptophyte alga: A linear dichroism study. *Biochim Biophys Acta* 1102, 360–364, 1992.

Hiller RG, Wrench PM, Gooley AP, Shoebridge G, Breton J. The major intrinsic light-harvesting protein of Amphidinium: Characterization and relation to other light-harvesting protein. *Photochem Photobiol* 57, 125–131, 1993.

Hoffman GE, Sanchez Puerta MV, Delwiche CF. Evolution of light-harvesting complex proteins from Chl c-containing algae. *BMC Evol Biol* 11, 101, 2011.

Hofmann E, Wrench PM, Sharples FP, Hiller RG, Welte W, Diederichs K. Structural basis of light harvesting by carotenoids: Peridinin-chlorophyll-protein from *Amphidinium carterae*. *Science* 272, 1788–1791, 1996.

Horn S, Ehlers K, Fritzsch G, Gil-Rodriguez M, Wilhelm C, Schnetter R. *Synchroma grande* spec nov (Synchromophyceae class nov. Heterokontophyta): An amoemoid marine alga with unique plastid complexes. *Protist* 158, 277–293, 2007.

Horton P, Johnson MP, Perez-Bueno ML, Kiss AZ, Ruban AV. Photosynthetic acclimation: Does the dynamic structure and macro-organisation of photosystem II in higher plant grana membranes regulate light harvesting states? *FEBS J* 275, 1069–1079, 2008.

Iglesias-Prieto R, Trench RK. Acclimation and adaptation to irradiance in symbiotic dinoflagellates. II. Response of chlorophyll–protein complexes to different photon-flux densities. *Mar Biol* 130, 23–33, 1997.

Iglesias-Prieto R, Govind NS, Trench RK. Apoprotein composition and spectroscopic characterization of the water-soluble peridinin chlorophyll alpha-proteins from 3 symbiotic dinoflagellates. *Proc R Soc Lond B Biol Sci* 246, 275–283, 1991.

Ikeda Y, Yamagishi A, Komura M et al. Two types of fucoxanthin-chlorophyll-binding proteins I tightly bound to the photosystem I core complex in marine centric diatoms. *Biochim Biophys Acta* 1827, 529–539, 2013.

Ingram K, Hiller RG. Isolation and characterization of a major chlorophyll a/c2 light-harvesting protein from a Chroomonas species (Cryptophyceae). *Biochim Biophys Acta* 722, 310–319, 1983.

Jahns P, Holzwarth AR. The role of the xanthophyll cycle and of lutein in photoprotection of photosystem II. *Biochim Biophys Acta* 1817, 182–193, 2012.

Jakob T, Goss R, Wilhelm C. Activation of diadinoxanthin de-epoxidase due to a chlororespiratory proton gradient in the dark in the diatom *Phaeodactylum tricornutum*. *Plant Biol* 1, 76–82, 1999.

Jakob T, Goss R, Wilhelm C. Unusual pH-dependence of diadinoxanthin de-epoxidase activation causes chlororespiratory induced accumulation of diatoxanthin in the diatom *Phaeodactylum tricornutum*. *J Plant Physiol* 158, 383–390, 2001.

Jans F, Mignolet E, Houyoux P et al. A type II NAD(P)H dehydrogenase mediates light-independent plastoquinone reduction in the chloroplast of Chlamydomonas. *Proc Natl Acad Sci U S A* 105, 20546–20551, 2008.

Joshi-Deo J, Schmidt M, Gruber A, Weisheit W, Mittag M, Kroth PG, Buchel C. Characterization of a trimeric light-harvesting complex in the diatom *Phaeodactylum tricornutum* built of FcpA and FcpE proteins. *J Exp Bot* 61, 3079–3087, 2010.

Jungandreas A, Schellenberger-Costa B, Jakob T, von Bergen M, Baumann S, Wilhelm C. The acclimation of *Phaeodactylum tricornutum* to blue and red light does not influence the photosynthetic light reaction but strongly disturbs the carbon allocation pattern. *PLoS One* 9, e99727, 2014.

Kalituho L, Beran KC, Jahns P. The transiently generated nonphotochemical quenching of excitation energy in Arabidopsis leaves is modulated by zeaxanthin. *Plant Physiol* 143, 1861–1870, 2007.

Kaminskaya O, Shuvalov VA, Renger G. Two reaction pathways for transformation of high potential cytochrome b559 of PSII into the intermediate potential form. *Biochim Biophys Acta* 1767, 550–558, 2007.

Kana R, Kotabova E, Sobotka R, Prasil O. Nonphotochemical quenching in cryptophyte alga *Rhodomonas salina* is located in chlorophyll a/c antennae. *PLoS One* 7, e29700, 2012.

Katoh H, Mimuro M, Takaichi S. Light-harvesting particles isolated from a brown alga, *Dictyota dichotoma*. A supramolecular assembly of fucoxanthin–chlorophyll–protein complexes. *Biochim Biophys Acta* 976, 233–240, 1989.

Kereïche S, Kouril R, Oostergetel GT et al. Association of chlorophyll a/c2 complexes to photosystem I and photosystem II in the cryptophyte *Rhodomonas* CS24. *Biochim Biophys Acta* 1777, 1122–1128, 2008.

Kilian O, Benemann C, Niyoki K, Vick B. High-efficiency homologous recombination in the oil-producing alga Nannochloropsis sp. *Proc Natl Acad Sci U S A* 108, 21265–21269, 2011.

Kirchhoff H, Hinz HR, Rosgen J. Aggregation and fluorescence quenching of chlorophyll a of the light-harvesting complex II from spinach in vitro. *Biochim Biophys Acta* 1606, 105–116, 2003.

Konotchick T, Dupont CL, Valas RE, Badger JH, Allen AE. Transcriptomic analysis of metabolic function in the giant kelp, *Macrocystis pyrifera*, across depth and season. *New Phytol* 198, 398–407, 2013.

Kotabová E, Kaňa R, Jarešová J, Prášil O. Non-photochemical fluorescence quenching in *Chromera velia* is enabled by fast violaxanthin de-epoxidation. *FEBS Lett* 585, 1941–1945, 2011.

Kroth PG, Chiovitti A, Gruber A et al. A model for carbohydrate metabolism in the diatom Phaeodactylum tricornutum deduced from comparative whole genome analysis. *PLoS One* 3, e1426, 2008.

Lavaud J. Fast regulation of photosynthesis in diatoms: Mechanisms, evolution and ecophysiology. *Funct Plant Sci Biotechnol* 1, 267–287, 2007.

Lavaud J, Lepetit B. An explanation for the inter-species variability of the photoprotective non-photochemical chlorophyll fluorescence quenching in diatoms. *Biochim Biophys Acta* 1827, 294–302, 2013.

Lavaud J, Goss R. The peculiar features of non-photochemical fluorescence quenching in diatoms and brown algae. In: Demmig-Adams B, Adams III WW, Garab G, Govindjee (eds.) *Non-Photochemical Quenching and Thermal Energy Dissipation in Plants, Algae and Cyanobacteria, Series: Advances in Photosynthesis and Respiration*, Springer, Dordrecht, pp. 421–443, 2014.

Lavaud J, Rousseau B, Etienne AL. In diatoms, a transthylakoid proton gradient alone is not sufficient to induce a non-photochemical fluorescence quenching. *FEBS Lett* 523, 163–166, 2002a.

Lavaud J, Van Gorkom HJ, Etienne AL. Photosystem II electron transfer cycle and chlororespiration in planktonic diatoms. *Photosynth Res* 74, 51–59, 2002b.

Lavaud J, Strzepek RF, Kroth PG. Photoprotection capacity differs among diatoms: Possible consequences on the spatial distribution of diatoms related to fluctuations in the underwater light climate. *Limnol Oceanogr* 52, 1188–1194, 2007.

Lepetit B, Volke D, Szabo M, Hoffmann R, Garab GZ, Wilhelm C, Goss R. Spectroscopic and molecular characterization of the oligomeric antenna of the diatom *Phaeodactylum tricornutum*. *Biochemistry* 46, 9813–9822, 2007.

Lepetit B, Volke D, Szabo M, Hoffmann R, Garab G, Wilhelm C, Goss R. The oligomeric antenna of the diatom *P. tricornutum*-localization of diadinoxanthin cycle pigments. In: Allen JF, Gantt E, Golbeck JH, Osmond B (eds.) *Photosynthesis. Energy from the Sun.* Springer, Dordrecht, pp. 277–280, 2008.

Lepetit B, Volke D, Gilbert M, Wilhelm C, Goss R. Evidence for the existence of one antenna-associated, lipid-dissolved, and two protein-bound pools of diadinoxanthin cycle pigments in diatoms. *Plant Physiol* 154, 1905–1920, 2010.

Lepetit B, Goss R, Jakob T, Wilhelm C. Molecular dynamics of the diatom thylakoid membrane under different light conditions. *Photosynth Res* 111, 245–257, 2012.

Lepetit B, Sturm S, Rogato A et al. High light acclimation in the secondary plastids containing diatom *Phaeodactylum tricornutum* is triggered by the redox state of the plastoquinone pool. *Plant Physiol* 161, 53–865, 2013.

Li X, Björkman O, Shih C, Grossman AR, Rosenquist M, Jansson S, Niyogi K. A pigment-binding protein essential for the regulation of photosynthetic light-harvesting. *Nature* 403, 391–395, 2000.

Li XM, Zhang QS, Tang YZ, Yu YQ, Liu HL, Li LX. Highly efficient photoprotective responses to high light stress in *Sargassum thunbergii* germlings, a representative brown macroalga of intertidal zone. *J Sea Res* 85, 491–498, 2014.

Lichtlé C, Jupin H, Duval JC. Energy transfers from photosystem II to photosystem I in *Cryptomonas rufescens* (Cryptophyceae). *Biochim Biophys Acta* 591, 104–112, 1980.

Lichtlé C, Duval JC, Lemoine Y. Comparative biochemical, functional and ultra-structural studies of photosystem particles from a Cryptophycea: *Cryptomonas rufescens*; isolation of an active phycoerythrin particle. *Biochim Biophys Acta* 894, 76–90, 1987.

Lichtlé C, Arsalane W, Duval JC, Passaquet C. Characterization of the light-harvesting complex of *Giraudyopsis stellifer* (Chrysophyceae) and effects of light stress. *J Phycol* 31, 380–387, 1995.

Lohr M, Wilhelm C. Algae displaying the diadinoxanthin cycle also possess the violaxanthin cycle. *Proc Natl Acad Sci U S A* 96, 8784–8789, 1999.

Lohr M, Wilhelm C. Xanthophyll synthesis in diatoms: Quantification of putative intermediates and comparison of pigment conversion kinetics with rate constants derived from a model. *Planta* 212, 382–391, 2001.

Mann M, Hoppenz P, Jakob T, Weisheit W, Mittag M, Wilhelm C, Goss R. Unusual features of the high light acclimation of *Chromera velia*. *Photosynth Res* 122, 159–169, 2014.

Matsuda Y, Kroth P. Carbon fixation in diatoms. In: Hohmann-Marriott MF, editor. *The Structural Basis of Biological Energy Generation, Advances in Photosynthesis and Respiration*, Springer, Netherlands. vol. 39, pp. 335–362, 2014.

Mekhalfi M, Puppo C, Avilan L, Lebrun R, Mansuelle P, Maberly, SC, Gontero B. Glyceraldehyde-3-phosphate dehydrogenase is regulated by ferredoxin-NADP reductase in the diatom *Asterionella formosa*. *New Phytol* 203, 414–423, 2014.

Mewes H, Richter M. Supplementary ultraviolet-B radiation induces a rapid reversal of the diadinoxanthin cycle in the strong light-exposed diatom *Phaeodactylum tricornutum*. *Plant Physiol* 130, 1527–1535, 2002.

Miloslavina Y, Grouneva I, Lambrev PH, Lepetit B, Goss R, Wilhelm C, Holzwarth AR. Ultrafast fluorescence study on the location and mechanism of non-photochemical quenching in diatoms. *Biochim Biophys Acta* 1787, 1189–1197, 2009.

Minagawa J. State transitions-The molecular remodeling of photosynthetic supercomplexes that controls energy flow in the chloroplast. *Biochim Biophys Acta* 1807, 897–905, 2011.

Moroney JV, Ynalvez RA. Proposed carbon dioxide concentrating mechanism in Chlamydomonas reinhardtii. *Eukaryot Cell* 6, 1251–1259, 2007.

Morse D, Salois P, Markovic P, Hastings JW. A nuclear encoded form II Rubisco in Dinoflagelles. *Science* 268, 1622–1624, 1995.

Mueller-Cajar O, Stotz M, Wendler P, Hartl FU, Bracher A, Hayer-Hartl M. Structure and function of the AAA+ protein CbbX, a red-type Rubisco activase. *Nature* 479, 194–199, 2011.

Nagao R, Takahashi S, Suzuki T, Dohmae N, Nakazato K, Tomo T. Comparison of oligomeric states and polypeptide compositions of fucoxanthin chlorophyll a/c-binding protein complexes among various diatom species. *Photosynth Res* 117, 281–288, 2013.

Neilson JAD, Durnford DG. Structural and functional diversification of the light-harvesting complexes in photosynthetic eukaryotes. *Photosynth Res* 106, 57–71, 2010.

Nymark M, Valle KC, Brembu T, Hancke K, Winge P, Andresen K, Johnsen G, Bones AM. An integrated analysis of molecular acclimation to high light in the marine diatom *Phaeodactylum tricornutum*. *PLoS One* 4, e7743, 2009.

Nymark M, Valle KC, Hancke K et al. Molecular and photosynthetic responses to prolonged darkness and subsequent acclimation to re-illumination in the diatom *Phaeodactylum tricornutum*. *PLoS One* 8, e58722, 2013.

Oborník M, Modrý D, Lukeš M, Cernotíková-Stříbrná E, Cihlář J, Tesařová M, Kotabová E, Vancová M, Prášil O, Lukeš J. Morphology, ultrastructure and life cycle of Vitrella brassicaformis n. sp., n. gen., a novel chromerid from the Great Barrier Reef. *Protist* 163, 306–323, 2012.

Obornik M, Vancova M, Lai D, Janouskovec J, Keeling P, Lukes J. Morphology and ultrastructure of multiple life cycle stages of the photosynthetic relative of apicomplexa, *Chromera velia*. *Protist* 162, 1125–1130, 2011.

Ocampo-Alvarez H, García-Mendoza E, Govindjee. Antagonist effect between violaxanthin and de-epoxidated pigments in nonphotochemical quenching induction in the qE deficient brown alga *Macrocystis pyrifera*. *Biochim Biophys Acta* 1827, 427–437, 2013.

Ogata T, Kodama M, Nomura S, Kobayashi M, Nozawa T, Katoh T, Mimuro M. A novel peridinin–chlorophyll a protein (PCP) from the marine dinoflagellate *Alexandrium cohorticula*: A high pigment content and plural spectral forms of peridinin and chlorophyll *a*. *FEBS Lett* 356, 367–371, 1994.

Olaizola M, LaRoche J, Kolber Z, Falkowski PG. Non-photochemical fluorescence quenching and the diadinoxanthin cycle in a marine diatom. *Photosynth Res* 41, 357–370, 1994.

Owens TG. Light-harvesting function in the diatom *Phaeodactylum tricornutum*: II. Distribution of excitation energy between the photosystems. *Plant Physiol* 80, 739–746, 1986.

Pan H, Slapeta J, Carter D, Chen M. Phylogenetic analysis of the light-harvesting system in *Chromera velia*. *Photosynth Res* 111, 19–28, 2012.

Pascal AA, Caron L, Rousseau B, Lapouge K, Duval JC, Robert B. Resonance Raman spectroscopy of a light-harvesting protein from the brown alga *Laminaria saccharina*. *Biochemistry* 37, 2450–2457, 1998.

Patron NJ, Waller RF, Keeling PJ. A tertiary plastid uses genes from two endosymbionts. *J Mol Biol* 357, 1373–1382, 2006.

Peers G, Truong TB, Ostendorf E et al. An ancient light-harvesting protein is critical for the regulation of algal photosynthesis. *Nature* 462, 518–521, 2009.

Peltier G, Cournac L. Chlororespiration. *Annu Rev Plant Biol* 53, 523–550, 2002.

Peltier G, Ravenel J, Vermeglio A. Inhibition of a respiratory activity by short saturating flashes in Chlamydomonas: Evidence for a chlororespiration. *Biochim Biophys Acta* 893, 83–90, 1987.

Pick U, Gounaris K, Weiss M, Barber J. Tightly bound sulpholipids in chloroplast CF0-CF1. *Biochim Biophys Acta* 808, 415–420, 1985.

Pinnola A, Dall'Osto L, Gerotto C, Morosinotto T, Bassi R, Alboresi A. Zeaxanthin binds to light-harvesting complex stress-related protein to enhance nonphotochemical quenching in *Physcomitrella patens*. *Plant Cell* 25, 3519–3534, 2013.

Premvardhan L, Robert B, Beer A, Büchel C. Pigment organization in fucoxanthin chlorophyll a/c2 proteins (FCP) based on resonance Raman spectroscopy and sequence analysis. *Biochim Biophys Acta* 1797, 1647–1656, 2010.

Prezelin BB, Haxo FT. Purification and characterization of peridinin–chlorophyll *a* proteins from marine dinoflagellates *Glenodinium* sp. and *Gonyaulax polyedra*. *Planta* 128, 133–141, 1976.

Pyszniak AM, Gibbs SP. Immunocytochemical localization of photosystem I and the fucoxanthin–chlorophyll a/c light-harvesting complex in the diatom *Phaeodactylum tricornutum*. *Protoplasma* 166, 208–217, 1992.

Qiu H, Yoon HS, Bhattacharya D. Algal endosymbionts as vectors of horizontal gene transfer in photosynthetic eukaryotes. *Front Plant Sci*, 4, 366, 2013.

Raven JA, Giordano M, Beardall J, Maberly SC. Algal evolution in relation to atmospheric CO$_2$: Carboxylases, carbon-concentrating mechanisms and carbon oxidation cycles. *Philos Trans R Soc Lond B* 367, 493–507, 2012.

Reinfelder JR, Kraepiel AML, Morel FFM. Unicellular C4 photosynthesis in a marine diatom. *Nature* 407, 996–999, 2000.

Rhiel E, Mörschel E, Wehrmeyer W. Characterization and structural analysis of a chlorophyll a/c light-harvesting complex and of photosystem I particles isolated from thylakoid membranes of *Cryptomonas maculata* (Cryptophyceae). *Eur J Cell Biol* 43, 82–92, 1987.

Roberty S, Bailleul B, Berne N, Franck F, Cardol P. PSI Mehler reaction is the main alternative photosynthetic electron pathway in *Symbiodinium* sp., symbiotic dinoflagellate of cnidarians. *New Phytol* 204, 81–91, 2014.

Sazanov LA, Burrows PA, Nixon PJ. The plastid ndh genes code for an NADH-specific dehydrogenase: Isolation of a complex I analogue from pea thylakoid membranes. *Proc Natl Acad Sci U S A* 95, 1319–1324, 1998.

Schaller S, Richter K, Wilhelm C, Goss R. Influence of pH, Mg^{2+}, and lipid composition on the aggregation state of the diatom FCP in comparison to the LHCII of vascular plants. *Photosynth Res* 119, 305–317, 2014.

Schellenberger-Costa B, Sachse M, Jungandreas A et al. Aureochrome 1a is involved in the photoacclimation of the diatom *Phaeodactylum tricornutum*. *PLoS One* 8, e74451, 2013.

Schmidt M, Horn S, Flieger K, Ehlers K, Wilhelm C, Schnetter R. *Synchroma pusillum* sp. Nov. And other new algal isolates with chloroplast complexes confirm the Synchromophyceae (Ochrophyta) as a widely distributed group of amoeboid algae. *Protist* 163, 544–559, 2012.

Schulte T, Sharples FP, Hiller RG, Hofmann E. X-ray structure of the high-salt form of the peridinin–chlorophyll a-protein from the dinoflagellate *Amphidinium carterae*: Modulation of the spectral properties of pigments by the protein environment. *Biochemistry* 48, 4466–4475, 2009.

Siefermann D, Yamamoto H. NADPH and oxygen-dependent epoxidation of zeaxanthin in isolated chloroplasts. *Biochem Biophys Res Commun* 62, 456–461, 1975.

Simidjiev I, Barzda V, Mustardy L, Garab G. Role of thylakoid lipids in the structural flexibility of lamellar aggregates of the isolated light-harvesting chlorophyll a/b complex of photosystem II. *Biochemistry* 37, 4169–4173, 1998.

Simidjiev I, Stoylova S, Amenitsch H et al. Self-assembly of large, ordered lamellae from non-bilayer lipids and integral membrane proteins in vitro. *Proc Natl Acad Sci U S A* 97, 1473–1476, 2000.

Smith SR, Abbriano RM, Hildebrand M. Comparative analysis of diatom genomes reveals substantial differences in the organization of carbon partitioning pathways. *Algal Res* 1, 2–16, 2012.

Stransky H, Hager A. Das Carotinoidmuster und die Verbreitung des lichtinduzierten Xanthophyllzyklus in verschiedenen Algenklassen. II: Xanthophyceae. *Arch Microbiol* 71, 164–190, 1970.

Sturm S, Engelken J, Gruber A, Vugrinec S, Kroth PG, Adamska I, Lavaud J. A novel type of light-harvesting antenna protein of red algal origin in algae with secondary plastids. *Evol Biol* 13, 159–173, 2013.

Szábo M, Lepetit B, Goss R, Wilhelm C, Mustardy L, Garab G. Structurally flexible macro-organization of the pigment-protein complexes of the diatom *Phaeodactylum tricornutum*. *Photosynth Res* 95, 237–245, 2008.

Szechynska-Hebda M, Karpinski S. Light intensity-dependent retrograde signalling in higher plants. *J Plant Physiol* 170, 1501–1516, 2013.

Tabita FR, Satagopan S, Hanson TE, Kreel, NE, Scott SS. Distinct form I, II, III, IV Rubisco proteins from the three kingdoms of life provide clues about Rubisco evolution and structure/function relationships. *J Exp Bot* 59, 1515–1524, 2008.

Tanabe Y, Shitara T, Kashino Y, Hara Y, Kudoh S. Utilizing the effective xanthophyll cycle for blooming of *Ochromonas smithii* and *O. itoi* (Chrysophyceae) on the snow surface. *PLoS One* 6, e14690, 2011.

Tichy J, Gardian Z, Bina D et al. Light harvesting complexes of *Chromera velia*, photosynthetic relative of apicomplexan parasites. *Biochim Biophys Acta* 1827, 723–729, 2013.

Ting CS, Owens TG. Photochemical and nonphotochemical fluorescence quenching processes in the diatom *Phaeodactylum tricornutum*. *Plant Physiol* 101, 1323–1330, 1993.

Uhrmacher S, Hanelt D, Nultsch W. Zeaxanthin content and the degree of photoinhibition are linearly correlated in the brown alga *Dictyota dichotoma*. *Mar Biol* 123, 159–165, 1995.

Veith T, Brauns J, Weisheit W, Mittag M, Büchel C. Identification of a specific fucoxanthin-chlorophyll protein in the light harvesting complex of photosystem I in the diatom *Cyclotella meneghiniana*. *Biochim Biophys Acta* 1787, 905–912, 2009.

Vieler A, Wu G, Tsai C-H et al. Genome, functional gene annotation, and nuclear transformation of the heterokont oleaginous alga *Nannochloropsis oceanica* CCMP1779. *PLoS Genet* 8, e1003064, 2012.

Waring J, Klenell M, Underwood GJC, Baker NR. Light-induced responses of oxygen photoreduction, reactive oxygen species production and scavenging in two diatom species. *J Phycol* 46, 1206–1217, 2010.

Weis VM, Verde EA, Reynolds WS. Characterization of a short form peridinin–chlorophyll protein (PCP) cDNA and protein from the symbiotic dinoflagellate *Symbiodinium muscatineu* (dinophyceae) from the sea anemone *Anthopleura elegantissima* (cnidaria). *J Phycol* 38, 157–163, 2002.

Whimarsh J, Pakrasi HB. Form and function of cytochrome b-559. In: Ort DR, Yocum CF (eds.) *Oxygenic Photosynthesis: The Light Reactions*, Kluwer Academic Publishers, Dordrecht, pp. 249–264, 1996.

Wilhelm C, Duval J. Fluorescence induction kinetics as a tool to detect a chlororespiratory activity in the prasinophycean alga, *Mantoniella squamata*. *Biochim Biophys Acta* 1016, 197–202, 1990.

Wilhelm C, Büchel C, Fisahn J et al. The regulation of carbon and nutrient assimilation in diatoms is significantly different from green algae. *Protist* 157, 91–124, 2006.

Wilhelm C, Jungandreas A, Jakob T, Goss R. Light acclimation in diatoms: From phenomenology to mechanisms. *Mar Gen* 16, 5–15, 2014.

Wormke S, Mackowski S, Schaller A, Brotosudarmo TH, Johanning S, Scheer H, Brauchle C. Single molecule fluorescence of native and refolded peridinin–chlorophyll-protein complexes. *J Fluoresc* 18, 611–617, 2008.

Yamamoto HY, Higashi RM. Violaxanthin de-epoxidase. Lipid composition and substrate specificity. *Arch Biochem Biophys* 190, 514–522, 1978.

Yamamoto H, Nakayama T, Chichester C. Studies on the light and dark interconversions of leaf xanthophylls. *Arch Biochem Biophys* 97, 168–173, 1962.

Zapata M, Jeffrey SW, Wright SW, Rodríguez F, Garrido JL, Clementson L. Photosynthetic pigments in 37 species (65 strains) of Haptophyta: Implications for oceanography and chemotaxonomy. *Mar Ecol Prog Ser* 270, 83–102, 2004.

Section VII

Photosynthesis in Higher Plants

25 Current Methods in Photosynthesis Research

Jutta Papenbrock, Yelena Churakova, Bernhard Huchzermeyer, and Hans-Werner Koyro

CONTENTS

25.1 Introduction 447
25.2 Structure and Function of the Thylakoid Membrane 448
    25.2.1 Absorption of Light Energy 448
    25.2.2 Thylakoid Membrane 448
    25.2.3 Photosynthetic Efficiency..449
    25.2.4 Distribution of Activation Energy between PS_{II} and PS_I 449
    25.2.5 Xanthophyll Cycle 449
    25.2.6 ATP Synthesis 450
    25.2.7 Photorespiration 450
    25.2.8 ROS Production..450
    25.2.9 Electron Sinks 451
    25.2.10 Futile Cycles..451
    25.2.11 C_4/CAM..452
25.3 Noninvasive Methods Allowing Monitoring of Plant Performance....................................452
    25.3.1 Gas Exchange and Assimilation of CO_2 452
    25.3.2 Chlorophyll Fluorescence and Plant Performance in Photosynthesis......................452
    25.3.3 Method 453
    25.3.4 Chlorophyll Fluorescence Parameters and Their Meaning 453
    25.3.5 What to Expect from Chlorophyll Fluorescence Measurements 454
    25.3.6 Examples of PAM Fluorimetry Combined with Gas Exchange Measurements of Stressed Plants 454
    25.3.7 Limits of Chlorophyll Fluorescence Monitoring of Photosynthesis 455
    25.3.8 Determination of Chlorophyll Fluorescence Parameters as Potential Stress Markers 456
    25.3.9 Case Study 1: *Lablab purpureus*, Drought Stress..456
    25.3.10 Case Study 2: *Euphorbia tirucalli*, Drought Stress 456
    25.3.11 Case Study 3: *Balanites aegyptiaca*, Drought Stress..457
    25.3.12 Case Study 4: Drought Stress, Several Species Investigated by Several Groups............................457
    25.3.13 Case Study 5: *Tripolium pannonicum*, Nutrient Deficiency 457
    25.3.14 Case Study 6: Brassica napus, Nutrient Deficiency and Biotic Stress 458
    25.3.15 Alternative Methods for Analysis of Plant Stress Response..................................458
25.4 Summary 458
References..459

25.1 INTRODUCTION

Currently, functional traits of plants are widely analyzed to better understand how plants respond to environmental factors. For this purpose, the identification of protocols allowing for the effective evaluation of plant growth and performance is necessary. As photosynthesis plays a key role in plant metabolism, it is logical to conclude that there is a correlation between respective parameters and plant growth and development.

We will provide information on traits that can be measured *in vitro* and explain how these traits are mirrored by chlorophyll fluorescence parameters, thus allowing for *in vivo* monitoring. An important aspect of this is to estimate sensitivity of methods. Using case studies as examples, we will explain the limits of the presented methods. For instance, such limits can be found in breeding experiments, when plants are able to adapt to environmental stress while grown in the presence of only mild stress.

Global climate change and the need to feed a growing world population require improving a detailed understanding of plant adaptations to their environment. There is already a huge body of information available. Methods and the focus of individual investigations have changed during the last decades (see reviews for reference: Long et al. 2004; Ashraf and Harris 2005; Khan and Weber 2008; Hu and Xiong 2014). Several indicators of plant performance have been identified, and reliability of crop yield prediction based on such traits has been tested (Collins et al. 2008). It became obvious that plants differ in their strategies to adapt to their environment. In this context, it was questioned whether effects of individual traits are additive and independent or function in a sequential manner (Fritsche-Neto and Borém 2012).

While early research aimed to describe environmental effects on the whole-plant or plant-organ level, genomic, proteomic, and metabolomic approaches were used more recently (Montfort 1937; Waisel 1972; Flowers et al. 1977; Greenway and Munns 1980; Koyro et al. 2011, 2013; Debez et al. 2012; Kosová et al. 2013). Early experiments focused on field experiments with individual crop plant species or typical ecosystems (Montfort 1937), while currently, plants are preferentially analyzed in the laboratory, where the effect on plant performance of individual environmental factors can be analyzed (Baker 2008). Moreover, in many cases, plant growth has been monitored for short periods of time rather than during a whole life cycle. Recent approaches require additional data evaluation and interpretation because patterns of enzyme activities and metabolites, for instance, change during growth and development. Adverse environmental conditions, when applied at a moderate degree, have been shown to retard growth and development (Chinnusamy et al. 2007). Therefore, subsequent to any changes in growth conditions, a huge number of metabolites and proteins have to be identified to change in abundance (Zhang et al. 2012). Discriminating between changes caused by retardation of development and those that are due to plant stress response requires a lot of analytical work.

We know that laboratory conditions are imperfect, but they allow for the growth of plants under defined conditions, thus allowing scientists the ability to compare data created by different teams in different regions (and their otherwise different environments) (Gil et al. 2014). For the comparison of whole-plant and general physiology traits, standard methods have been suggested (Pérez-Harguindeguy et al. 2013). However, no such agreement currently exists for biochemical and -omics parameters. This is especially contested concerning stress analysis, where it is questioned which of the traits are sensitive and meaningful, and which of them relate to only a few of the stress adaptation strategies expressed in individual plant species (Claeys et al. 2014).

In this chapter, we are focusing on traits related to primary reactions in photosynthesis, e.g., absorption of light energy and immediate reactions of photosynthesis. Specifically, we are concentrating on noninvasive methods that allow for monitoring the performance of individual plants. This way, plant stress response can be observed during a certain period of time, while excluding variations of signals due to differences in the sensitivity of individual plants. However, depending on their respective strategies of how to adapt to an environment, plant species differ in their stress-induced expression of these traits (Chaerle et al. 2007; Koyro et al. 2014b).

25.2 STRUCTURE AND FUNCTION OF THE THYLAKOID MEMBRANE

25.2.1 Absorption of Light Energy

Any pigment is characterized by its ability to absorb the energy of light quanta to reach an activated state. Subsequently, the absorbed energy will be released as heat. Chlorophyll is special as it is more stable in its activated state. Thus, it allows several alternative energy-consuming reactions to occur in addition to heat release (Butler 1978). These reactions are as follows (Baker 2008): (1) Chlorophyll in its activated state can function as an electron donor due to its negative redox potential. Photosynthetic electron transport is among the best analyzed redox chains. (2) Activated chlorophyll can release part of the activation energy in the form of fluorescence light quanta. (3) Chlorophyll can interact with other pigments of the thylakoid membrane by transfer of activation energy.

As the portion of heat released from activated chlorophyll can be assumed to be constant under physiological conditions, changes in chlorophyll fluorescence can be used as an indicator of changes in other energy-consuming pathways, i.e., an increase in one of the activities will reduce the share of the other ones. Since an increase in any of the energy-consuming pathways reduces chlorophyll fluorescence intensity, the term *fluorescence energy quenching* has been introduced. Energy consumption by photosynthesis (photosynthetic quenching [qQ]) (see Table 25.1 for the nomeclature used) is differentiated from quenching by other energy-consuming pathways, such as nonphotochemical quenching (NPQ) (Baker 2008; Kalaji et al. 2014).

25.2.2 Thylakoid Membrane

Photosynthetic open-chain electron transport takes place at the chloroplast thylakoid membranes. The application of specific herbicides along with redox compounds helps dissect the electron transport chain and its biochemical and physiological functions. Inhibitors of electron transport (Tischer and Strotmann 1977), uncouplers destroying the transmembrane concentration gradient of protons and the membrane potential (Izawa and Good 1972), and inhibitors of the chloroplast coupling factor (CF_0CF_1) (Huchzermeyer 1982) all help in understanding how these compounds interact in photosynthesis (Trebst 2007). With the exception of mobile components (pools of plastoquinone [PQ], plastocyanin, ferredoxin), redox partners are bound to three protein complexes, namely, photosystem II (PS_{II}), the cytochrome complex (cyt b_6/f), and photosystem I (PS_I). Interaction among thylakoid proteins leads to clustering, as well as to structuring of the thylakoid membranes, including the formation of grana stacks and intergrana domains. Most of the PS_{II} complexes are found inside grana structures, while PS_I, along with the most abundant thylakoid protein, the chloroplast F-type ATPase (CF_0CF_1), is found in unstacked regions of the thylakoid membranes (Andersson and Anderson 1980;

TABLE 25.1

Parameters Used for the Analysis of Chlorophyll Fluorescence Analysis

Nomenclature

F_m	Maximal fluorescence
F'_m	Maximal fluorescence yield induced by a short saturating light flash during the exposure of the leaf to actinic light
F_o	Minimal fluorescence
F'_o	Fluorescence yield obtained in the leaf when exposed to actinic light following maximal oxidation of the reaction centers of photosystem II
F_v	Maximum variable fluorescence, defined as $F_m - F_o$
F'_v	Variable fluorescence from light-adapted leaf
F_v/F_o	A value that is proportional to the activity of the water splitting complex on the donor side of PS_{II}
F_v/F_m	Maximum quantum efficiency of PS_{II} photochemistry
F'_v/F'_m	PS_{II} maximum efficiency
ETR	Index of electron transport rate (ETR = $Y_{PSII} \times PPFD \times 0.5 \times a$)
PPFD	Photosynthetically active photon flux density
$\Delta F/F'_m$	Photochemical efficiency
Y_{PSII}	Effective quantum yield of PS_{II}
Y(NPQ)	Photoprotective nonphotochemical quenching, NPQ $\left(NPQ = F_m/F'_m - 1\right)$
Y(NO)	All other nonphotoprotective nonphotochemical quenching
qE	High-energy-state quenching/energy-dependent quenching
qI	Photoinhibition
qQ	Photochemical fluorescence quenching
qT	Apparent state transition

Pfeiffer and Krupinska 2005). In fact, PS_{II} is thought to initiate membrane stacking by protein-to-protein interaction (Chow et al. 2005). It is generally agreed upon that the patchiness of protein distribution in thylakoids is responsible for significantly controlling the light-use efficiency of photosynthetic electron transport. How PS_{II}- and light harvesting complex-mediated membrane stacking is controlled remains a matter of discussion (Kirchhoff et al. 2007; Fristedt et al. 2009; Goral et al. 2012). Proven interactions of stroma proteins, such as Rubisco and enzymes of the Calvin cycle, with the CF_0CF_1 complex have been discussed in terms of an improved metabolite flow, but regulation mechanisms controlling such aggregations, and the implications of this observation, remain obscure (Huchzermeyer et al. 1986; Süss et al. 1993; Wedel et al. 1997; Michelet et al. 2013). Such aggregates are quite unstable and tend to escape from analysis in cell extracts, when protein concentrations and concentrations of solutes are far below *in vivo* conditions. While individual steps, such as electron transfer among redox partners, and photophosphorylation of ADP plus phosphate to form ATP could be analyzed *in vitro*, regulation of activities, as well as structure–function relations, require analysis of an intact system. Thus, fine-tuning of energy use has to be investigated under *in vivo* conditions (Anderson et al. 1995; Eckardt 2011).

This context becomes clear when analyzing plant performance under suboptimal growth conditions. Plants can adapt

to their environment by modifying turnover rates of enzymes and, thus, establish a new metabolic equilibrium (Møller 2001). Such adaptive responses have been analyzed by measuring metabolic and growth activity of plants under high light intensity, drought stress, or salt stress (Chaves et al. 2009).

25.2.3 Photosynthetic Efficiency

Like any other pigment, light-activated chlorophyll is characterized by a half-life time. If none of the alternative reactions occur, absorbed light energy will be released as heat (Rosenqvist and van Kooten 2003). Therefore, it is obvious that the efficiency of open-chain electron transport depends on the availability of electron acceptors. Electron transport from activated chlorophyll to its electron acceptor, as well as to those following downstream in the reaction chain, has to take place faster as compared to any of the energy-wasting side reactions (Harbinson and Rosenqvist 2003). This also applies to the *end products* of photosynthetic electron transport, NADPH and ATP. Side reactions can be limited, if $NADP^+$ and ADP are efficiently recycled by reactions of the Calvin cycle and other (redox) energy-consuming reactions (Koyro et al. 2014a).

25.2.4 Distribution of Activation Energy between PS_{II} and PS_I

Exact tuning of the activities of PS_{II} and PS_I is a prerequisite of open-chain electron transport (Horton et al. 1996). Equal probability of light energy absorption events is tuned by adjacent light-harvesting complexes. Modification of this neighborhood is achieved by protein phosphorylation/dephosphorylation under the control of the PQ/PQH_2 ratio (Bennett 1983; Allen and Pfannschmidt 2000; Rochaix et al. 2012). It can be shown that protein kinase activity not only leads to reorientation of light-harvesting complexes but also affects general thylakoid membrane structure, including patterns of membrane stacking (Anderson and Aro 1994; Fristedt et al. 2009). These observations are summarized by the term *state transitions*. State transitions affect the efficiency of photosynthesis and, thus, chlorophyll fluorescence, as will be discussed in Section 25.3.4 (Anderson et al. 2008; Chuartzman et al. 2008).

25.2.5 Xanthophyll Cycle

Thylakoid membranes can be understood as devices converting the energy of absorbed light quanta to other forms of energy (Mitchell 1961; Dilley 1972; Horton et al. 1996). A proton gradient is built by the concerted action of the water splitting system and PQ pool-mediated electron transport from PS_{II} to cyt b_6/f (Takizawa et al. 2007). Acidification of the thylakoid lumen (and concomitant ion movements) induces conformational changes in membrane proteins and changes in protein interactions, and modifies catalytic activities of enzymes (Jagendorf 1972; 2002; Jajoo et al. 2012). One important aspect significantly affecting the energy efficiency of photosynthesis is the ΔpH-controlled variation of distances between chlorophylls and helper pigments, such as xanthophylls, and other carotenoids (Osto et al. 2012). When activation energy transfer between chlorophylls and xanthophylls is intensified and enzymes of the xanthophyll cycle are activated by an internal acid pH of the thylakoids, a significant portion of absorbed light energy will subsequently be released as heat (Havaux and Niyogi 1999; Matsubara et al. 2008; Johnson et al. 2009). This event can be monitored as a reduction in chlorophyll fluorescence, and it is one important component of NPQ of chlorophyll fluorescence (see Section 25.3.4) (Pérez-Bueno et al. 2008; Johnson et al. 2012).

25.2.6 ATP Synthesis

According to the alternating catalytic site model of ATP synthesis (Boyer 2000), the three catalytic sites of the chloroplast ATPase undergo a three-step sequence of conformational states. Nucleotide binding to the open state of a catalytic site induces a spontaneous closure of that site, and ATP can be spontaneously formed from bound phosphate and ADP. The product can be released by subsequent protein protonation (Fromme and Gräber 1990). The Boyer model proposes that the conformational change of only one site is allowed if simultaneous changes take place at the two other sites of the enzyme complex (Boyer 1993).

Depending on experimental conditions, any of the conformational changes can become rate limiting, and catalytic sites can idle between the states of ATP synthesis and hydrolysis (Groth and Junge 1993). Therefore, apparent changes in substrate affinity, different patterns of enzyme-bound substrate molecules, and changes in energy use efficiency (H+/ATP ratios) have been measured (Evron and McCarty 2000; Richter et al. 2000; Hisabori et al. 2002; Turina et al. 2003). *In vivo* catalytic activity of the enzyme is light controlled (Girault et al. 1998). It has been found that proton-induced ATP release from the thioredoxin-mediated reduced state readily occurs when the energy of the transmembrane ΔpH equals the hydrolysis energy of ATP. However, the concentration gradient of protons has to be more than tenfold higher to induce a similar effect on the oxidized enzyme (Turina et al. 2003; Richter et al. 2005). This may explain why experimental conditions can affect both light-use efficiency and steady-state level of chlorophyll fluorescence.

25.2.7 Photorespiration

NADPH and ATP are consumed by reactions of the Calvin cycle, e.g., assimilation of CO_2 and recycling of the primary acceptor molecule, ribulose 1,5-bisphosphate. In all plants, the initial reaction is catalyzed by the enzyme Rubisco. At a certain percantage, depending on the temperature and availability of substrates, Rubisco can use O_2 as a substrate instead of CO_2. This results in the formation of a phosphoglycolate plus a phosphoglycerate instead of two molecules of phosphoglycerate (Eckardt 2005). The C_2 compound has to be

removed to prevent the buildup of eventually toxic concentrations (Eisenhut et al. 2008). The reaction sequence consuming two molecules of phosphoglycolate to form phosphoglycerate plus CO_2 was named *photorespiration* because CO_2 release takes place inside mitochondria and because the enzyme Rubisco, which initiates the reaction sequence, is exclusively active in the light (Zhang and Portis 1999).

The catalytic pathway of photorespiration takes place in three compartments: chloroplasts, peroxisomes, and mitochondria. It includes transamination and deamination reactions. Released ammonium is transported to the chloroplast to be refixated by the glutamine oxoglutarate aminotransferase system. In summary, this pathway (1) significantly contributes to apparent photosynthetic quenching of chlorophyll fluorescence and (2) helps reduce the internal O_2 concentration in leaves (Farquhar et al. 1980; Wingler et al. 2000). This can be important under conditions when stomata are closed in the presence of high light intensity and the CO_2/O_2 ratio is decreasing (Figure 25.1).

25.2.8 ROS Production

Lacking sufficient amounts of electron acceptors, chlorophyll will stay in its light-activated state for a prolonged period of time, and the probability of electron transfer to molecular oxygen will increase. Reactive oxygen species (ROS) will be formed not only by electron transfer from chlorophyll of PS_{II} but also from other intermediates characterized by a sufficiently negative redox potential. In general, ROS originates from metabolic activities in different cell compartments, especially in peroxisomes, mitochondria, and chloroplasts, but might be also generated owing to environmental stresses such as high light intensity, drought, salinity, or heavy metals (Mittler 2002; Hossain et al. 2012; Anjum et al. 2014).

Low concentrations of ROS are essential, as they function as signaling molecules. Retrograde signaling from chloroplasts to the nucleus is discussed in the literature (Suzuki et al. 2011). On the other hand, all plants are prepared, to a specific

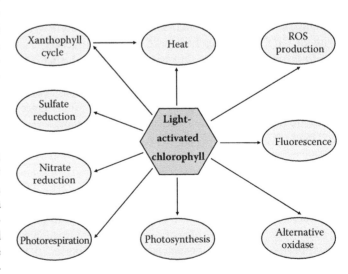

FIGURE 25.1 Flow of redox energy in chloroplasts.

extent, to control ROS concentration and detoxify radicals. A prominent ROS detoxification pathway includes ascorbic acid and the glutathione pool (Apel and Hirt 2004; Moradi and Ismail 2007) and results in a significant consumption of electrons from the photosynthetic electron transport chain. Thus, ROS detoxification results in significant apparent photosynthetic quenching of chlorophyll fluorescence (Moradi and Ismail 2007; Chen and Gallie 2008).

Nevertheless, in the presence of high light intensities, the probability of destructive events increases. It has been observed that proteins of the reaction centers become inactivated. For instance, half-lives of the D1 core protein of PS_{II} in the range of 15–30 min have been found (Aro et al. 1992; Murata et al. 2007). Parallel decrease of the chlorophyll concentration in leaves reduces the risk of ROS production by activated chlorophyll while affecting photosynthetic capacity. Respective light-initiated reaction sequences have been named *photoinhibition*. They contribute to NPQ of chlorophyll fluorescence and can be distinguished from other quenching effects by their slower kinetics (Holt et al. 2004; Zaks et al. 2012).

Plants are equipped, to varying degrees, with several systems to detoxify ROS in order to prevent the buildup of high, eventually toxic concentrations (Scandalios 1993; Vertuani et al. 2004). In this respect, molecules having antioxidative potential can be discriminated from enzyme-catalyzed reaction sequences (Gill and Tuteja 2010).

Photosynthetic adaptation is a common strategy in order to counteract an excess of electrons. Plant species adapted to different light regimes show differential photosynthetic characteristics. These differences can be based on changes at any level of photosynthetic metabolism, such as organization of thylakoid membranes or creation of electron and CO_2 sinks by overproduction of secondary metabolites. Plant species adapted to high light often have a high light-compensation point, light-saturation point, and maximal photosynthetic rate (Givnish 1988; Hölscher et al. 2006; Wong et al. 2012). In addition, the same plant species grown under different light environments were often able to acclimate to the habitat by morphological and physiological changes (Griffin et al. 2004; Aleric and Kirkman 2005; Huang et al. 2007; Zhang et al. 2007; Dai et al. 2009; Wong et al. 2012).

25.2.9 Electron Sinks

Ferredoxin accepts electrons from PS_I. In open-chain electron transport, the electron will be transferred to finally reduce $NADP^+$ and generate NADPH as a reducing coenzyme. But ferredoxin is not part of the thylakoids and, similar to NADPH, can function as a distributor of electrons inside the chloroplasts. While thioredoxins control the light-dependent state of catalytic activity of their target enzymes (see Section 25.2.6) (Buchanan and Balmer 2005), nitrate reduction, sulfate reduction, and cyclic electron transport around PS_I are major consumers of electrons from reduced ferredoxin in higher plants (Beevers and Hageman 1969; Gandin et al. 2014).

It has been discussed in the literature that cyclic electron transport leading to cyclic photophosphorylation is necessary

to provide the *extra ATP* needed to balance the requirements of the Calvin cycle. Thus, it is obvious that with respect to total PS_I activity, the share of cyclic electron transport has to be significant (Hosler and Yocum 1987; Foyer et al. 2012). In organic compounds, nitrogen is always found in its reduced state, and nitrate is the dominant form of nitrogen uptake. Calculating C/N ratios of organic matter leads to the assumption that about one-third of the electrons delivered by ferredoxin will be used, directly or indirectly, for nitrate reduction. In line with the calculations, it was observed that nitrate fertilization helps alleviate stress when plants are suffering from overreduction of photosynthetic redox reactions (Masclaux-Daubresse et al. 2010).

More recently, alternative pathways of the photosynthetic electron transport chain have become the focus of research. It was found, for instance, that in contrast to *Arabidopsis thaliana*, the halophyte *Thellungiella halophila* shows significant chloroplast alternative oxidase activity. When grown at hyperosmotic salinity, alternative oxidase in the halophyte will consume up to 30% of the electrons released from PS_{II} (Stepien and Johnson 2009).

In our hands, preliminary data from experiments with beans and beets indicate that ethylene activates alternative oxidase activity (Nahar, personal communication). With respect to monitoring photosynthesis by means of chlorophyll fluorescence measurement, it has to be pointed out that respective signals will be subsumed as "photosynthetic quenching," i.e., the method does not allow discrimination between photosynthetic electron transport leading to CO_2 fixation and electron transport ending up in one of these sinks (McDonald et al. 2011).

25.2.10 Futile Cycles

It was obvious that photosynthesis follows the same principle in all plants. In more detailed investigations, it was found that there are significant differences expressed among plant species, such as CO_2 concentrating reactions as well as side reactions, in addition to the main route of energy and substrate flow. Those reaction sequences that either dissipate absorbed light energy or lead to a release of CO_2 have been called futile cycles (Paul and Pellny 2003). Such reactions were understood to act as safety valves when reactions involved in light energy capture occur at a higher rate compared to energy-consuming reactions.

While the interaction of metabolic pathways of chloroplasts and mitochondria and concomitant exchange of metabolites is well understood in terms of photorespiration, other pathways still need clarification (Paul and Pellny 2003). This especially applies to pathways involving acetate as a substrate. Differences in acetate production and its transmembrane mobility have been observed when comparing developing and mature leaves (Heintze et al. 1990). Several pathways in all major cellular compartments lead to the production of compounds that help plants to survive under environmental stress (high light intensity, drought, etc.). Respective compounds may show osmotic activity, stabilize protein structures, or act as a ROS scavenger (Krasensky and Jonak 2012). In any case, production of these substances consumes energy and contributes to photosynthetic quenching of chlorophyll fluorescence. On the other hand, with the exception of osmotically active substances, production of these substances will not directly contribute to plant growth.

Significant changes in energy supply were observed when monitoring light energy on the surface of leaves during a sunny day, especially if periodical shading events occurred. It was discussed (Leong and Anderson 1984) that optimal photosynthesis rates can be reached, if chloroplasts are adapted to low actinic light while being equipped with systems allowing for the waste of surplus energy. When comparing leaves from plants adapted to shade and to high levels of sunlight, anatomical, physiological, and biochemical features corroborating this idea were identified (Leong and Anderson 1984).

Under conditions when dissipation of surplus energy was not sufficient, ROS production, along with typical membrane damage, was observed. In experiments, such conditions occured (1) when the export of assimilated carbon via the phosphate translocator was limited by phosphate starvation (Heineke et al. 1989; Dietz and Heilos 1990) or (2) in the presence of herbicides (Rutherford and Krieger-Lizkay 2001).

The observation that herbicide-mediated inhibition of electron transfer can lead to ROS production initiated a search for "herbicide protectors," to use these herbicides in cultures of ROS sensitive crops. However, early investigations were complicated by the fact that some of the protectors were stimulating herbicide degradation or herbicide export rather than interfering with ROS production or promoting ROS detoxification (Hatzios 1991; Beckie and Tardif 2012).

25.2.11 C₄/CAM

C_4 plants and CAM plants possess CO_2 concentrating mechanisms allowing the Calvin cycle to work efficiently, though leaf temperature and the internal CO_2/O_2 ratio would be suboptimal for photosynthesis in a C_3 plant (Sage 2002). In most C_4 plants, the water splitting system is inhibited in bundle sheath chloroplasts, and Rubisco is active in an environment depleted of O_2. Thus, photorespiration rate is reduced in C_4 plants (Woo et al. 1970; Sage et al. 2012). CO_2, along with reduction equivalents, is imported into the bundle sheath by means of the C_4 component produced inside mesophyll chloroplasts (Sage et al. 2012). Analysis of chlorophyll fluorescence data from C_4 plants is complicated by several aspects. In some plants, such as maize, the C_3 pathway can significantly contribute to overall CO_2 fixation (Preiss et al. 1994). Activities of PS_{II} in mesophyll chloroplasts and PS_I in the bundle sheath are not strictly coupled. Fluorescence signals from the two chloroplast species may vary and contribute, to varying degrees, to the overall fluorescence signal. Correction for the PS_I fluorescence signal in C_4 plants is more complicated than in C_3 plants. In general, a two-wavelength method for signal correction is suggested (Klughammer and Schreiber

1998), but the reliability of this method has not, as of yet, been sufficiently tested in stress experiments.

25.3 NONINVASIVE METHODS ALLOWING MONITORING OF PLANT PERFORMANCE

25.3.1 GAS EXCHANGE AND ASSIMILATION OF CO₂

In addition to monitoring plant growth parameters, gas exchange measurements were the first noninvasive method allowing scientists to describe plant performance in different environments (Meyer and Genty 1998). Transpiration rate, production of oxygen, and consumption of CO_2 could be measured in parallel. These data correlated well with plant growth and production of biomass. Stomatal conductance was the dominant parameter apparently controlling all of these factors (Lawlor and Fock 1975). In early experiments, the technique did not allow for the measurement of biochemical reactions inside leaves. These limitations were overcome by the introduction of infrared spectroscopy and measurement of $\delta^{13}C$ values of produced sugars, though the latter approach is an invasive method (Meyer and Genty 1998; Gilbert et al. 2011). Meanwhile, it became obvious that gas exchange measurement is an ideal method adding necessary information to the second noninvasive method: monitoring of chlorophyll fluorescence. Interpretation of gas exchange measurement on its own is hampered by the fact that the share of futile cycles (respiration, photorespiration, etc.) is not immediately obvious. Only an apparent, though physiologically relevant, rate of photosynthesis can be measured.

25.3.2 CHLOROPHYLL FLUORESCENCE AND PLANT PERFORMANCE IN PHOTOSYNTHESIS

Analysis of adaptation mechanisms, especially responses to changing light intensity, requires high time resolution as well as analysis of energy use efficiency. These requirements can be met by monitoring chlorophyll fluorescence. Though the release of fluorescence energy is a minor reaction compared to the shares of the other two reactions, monitoring chlorophyll fluorescence provides information about disturbances within the other pathways. Therefore, measuring chlorophyll fluorescence became an important tool for noninvasive analysis of photosynthesis (Buschmann et al. 2001; Oxborough 2004).

Fluorescence measurements can be performed on chloroplast suspensions as well as on leaves of intact plants and can be carried out in the laboratory as well as in the field (Susila et al. 2004; Gottardini et al. 2014). Quite often, such measurements are performed in parallel with gas exchange measurements. For this purpose, sensors can be fixed inside leaf clips, thus allowing synchronous measurement of gas exchange and fluorescence (Biehler et al. 1997). This way, information on CO_2 uptake, O_2 release, light absorption, and electron transport activity can be aligned to receive a complete picture of photosynthetic activity. By collecting these data, bottlenecks in leaf physiology and biochemistry may be more easily identified. However, in Section 25.3.8, we will provide recent data showing that plants are still keeping some secrets.

25.3.3 METHOD

In early experiments, it had been observed that chilling-sensitive plants exhibit a sudden undershoot in chlorophyll fluorescence immediately after saturating light pulses. This signal turned out to be a reliable criterion for the onset of chilling stress (Larcher and Neuner 1989).

Chlorophyll fluorescence can provide information about the changes in photosynthetic performance of an individual plant. But it is important to minimize, for instance, differences in absorbance occurring among leaves as well as leaf areas. Leaf anatomy, as well as leaf pigment content, can vary a lot among plant species, especially when comparing those growing in shade and high-sunlight environments (Boardman 1977). Relative measurements using the same leaf have proven to provide the most reliable results (Pfündel 1998). When comparability between measurements can be guaranteed, diurnal changes in energy distribution may be measurable. Such measurements can be performed with C_3, C_4, or CAM plants (Pfündel 1998; Hastilestari et al. 2013). Using this method, energy demand of photosynthetic pathways may be compared, and the contribution of otherwise futile reaction sequences to the protection from high light-mediated damage may be investigated. In young leaves, this patchiness may be due to differences in cell development. In mature leaves, there may be lateral variation of stomatal conductance (Omasa and Takayama 2003). Similar differences can be found within individual plants as well when comparing sun- and shade-exposed leaves of trees.

It is well documented that chlorophyll fluorescence imaging allows screening for spatial differences in photosynthetic activity. In such experimental approaches, demonstrable gradients in activity across the leaf area of sun and shade leaves are exhibited (Lichtenthaler et al. 2007). Applying chlorophyll fluorescence imaging may provide additional information, especially in experiments analyzing plant stress responses. Application of stress to only one-half of a leaf (by shading, for instance) is a widely used approach (Omasa et al. 1987; Hideg 2008).

In experiments, temporal changes in the distribution of chlorophyll fluorescence have been observed (Meyer and Genty 1998). Using chlorophyll fluorescence imaging in combination with measuring local CO_2 assimilation activity, a linear relationship was found between the mean photochemical yield of PS_{II}, calculated from images of the photochemical yield of PS_{II}, and the average quantum yield of linear electron transport. This linearity holds true for a wide range of plant species exposed to different treatments and conditions (Genty and Meyer 1995).

There are two major sources of experimental error for quantitative analysis of PS_{II} fluorescence: (1) chlorophyll fluorescence of PS_I can contribute up to 30% of total fluorescence in C_3 plants and 40% in C_4 plants, and (2) PQ can significantly contribute to chlorophyll quenching (Oxborough 2004). Recently, a dual pulse-amplitude-modulation (PAM)

instrument was developed that overcomes these limitations by measuring PS_{II} chlorophyll fluorescence (excitation at 460 or 620 nm, emission > 700 nm) and PS_I dual wavelength absorbance with a sample wavelength of 830 nm and a reference wavelength of 870 nm. Therefore, the quantum yields of photochemical energy conversion and nonphotochemical energy dissipation in PS_{II} and PS_I can be investigated separately (Schreiber and Klughammer 2008).

Using light curves, the dependence of PS_{II} electron transport on actinic light intensity can be estimated (Oxborough 2004). The performances of plants adapted to different light intensities can be compared using this method, despite being grown in different environments. Fluorescence measurement under field conditions allows for the estimation of their flexibility in adapting to their environment, even if the leaf anatomy of experimental plants makes it difficult to use conventional methods, such as gas exchange measurement (Demmig-Adams et al. 2006).

25.3.4 CHLOROPHYLL FLUORESCENCE PARAMETERS AND THEIR MEANING

The most frequently applied chlorophyll fluorescence ratio is F_v/F_m (the efficiency of excitation capture by "open" PS_{II} reaction centers; regular values, 0.74–0.85) since it is easy and fast to determine. A decline in F_v/F_m indicates a decline in the quantum yield of PS_{II} photochemistry and a disturbance in or damage to the photosynthetic apparatus (Pfündel 1998; Demmig-Adams et al. 2012). It cannot really be judged from an F_v/F_m ratio as is whether the decline in F_v/F_m is due to photoinhibition of PS_{II} units (qI) or other causes. On the other hand, F_v/F_m is a relatively inert ratio where stress-induced changes are detected rather late. A much more sensitive ratio, F_v/F_o, was first recommended by Lichtenthaler and Buschmann (1984). The quantum yield of noncyclic electron transport was found to be directly proportional to the product of the photochemical fluorescence quenching (qQ) and (F_v/F_m) (Genty et al. 1989).

As the rate of noncyclic electron transport increases with light intensity, an increase in high-energy-state quenching (qE), which results from the increased energization of the thylakoid membrane, will occur concomitant with a decrease in qQ (Genty et al. 1989).

In order to fully appreciate the physiological significance of non-photochemical fluorescence quenching (NPQ), it is necessary to understand the underlying processes with different relaxation times, ranging from a few minutes to several hours. The kinetics of relaxation can be used to distinguish them. To attain a full understanding of the contribution of fast- and slow-relaxing quenching, it is necessary to perform relaxation analysis (Walters and Horton 1991; Ruban and Horton 1995). In such an experiment, quenching is allowed to relax while F_m is recorded at regular intervals. As each estimation of F_m requires a saturating light flash, it is important that the intervals between flashes are long enough to allow relaxation, before the next flash is applied (Maxwell and Johnson 2000).

A graph of log(F_m) against time is then produced. The value that would have been attained, if only slowly relaxing quenching had been present, can be estimated by extrapolation of data points: A line is drawn from the end of the relaxation back to the time when actinic light was removed (Maxwell and Johnson 2000). ETR is the index of the electron transport rate. It is calculated from the equation ETR = Y_{PSII} × PPFD × 0.5 × a (Björkman and Deming 1987). Though this equation contains several assumptions, ETR in most experiments correlates linearly with rates of O_2 evolution, if the measured values are corrected for O_2 exchange in the dark. This value is measured immediately after turning off actinic light. The theoretical maximal value of O_2 evolution is O_2/ETR = 0.25 (Beer et al. 2000).

A complete quenching analysis is quite time-consuming. To reduce the time needed for this type of measurement, a faster procedure was developed, called rapid light curves. Rapid light curves are plots of ETR versus actinic irradiance applied for 10 s. Large changes in maximum ETR were observed when leaves were shifted from dark to moderate light or from dark to photoinhibitory light and vice versa. Maximum ETR was low when following long-term dark adaptation but increased to maximum levels within 8–15 min of illumination. It took more than 3 h, however, to return irradiated leaves back to the fully dark-adapted state. Quenching analysis of rapid light curves revealed large qE development in long-term dark-adapted leaves, accounting for a low ETR. Leaves photoinhibited for 3 h had a similarly reduced ETR. In photoinhibited leaves, however, photoinhibition (qI) was largely responsible for ETR reduction (White and Critchley 1999).

25.3.5 What to Expect from Chlorophyll Fluorescence Measurements

Plant species have been found to differ in stress response strategies. Accordingly, different types of stress-mediated modifications of chlorophyll fluorescence signals can be expected.

1. Activation of futile cycles, for instance, will increase the photosynthetic electron transport rate (stimulate ETR and qQ), though plant growth (production of biomass) will be inhibited. Thus, an analysis of chlorophyll fluorescence data will show an apparent increase in light energy use efficiency, while measurements based on biomass production will give the opposite result.
2. Improved dissipation of absorbed energy by energy transfer to the xanthophyll cycle will reduce energy use efficiency. Increased stress perception will be paralleled by an apparent reduction in the efficiency of electron transport. The reason for this inhibition will become obvious when a correlation with energy-dependent quenching (qE) is analyzed.
3. If plants preferentially adapt to stress by stimulation of their ROS-scavenging capacity, NPQ increases with stress perception as well, or an apparent increase

of photosynthetic quenching is observed. Typically, a spontaneous increase of photoinhibition (qI) will be observed if applied stress exceeds a threshold value.
4. Analysis of chlorophyll fluorescence data will be most complicated if stress response involves changes in the shares of PS_{II} and PS_I activities, as a correction for the PS_I background signal has to be modified prior to standard data analysis. These changes in PS_{II}/PS_I activity can be expected if (a) light quality (spectral distribution of absorbed light quanta) is changed or (b) stress-induced changes of metabolic activity lead to changes in the NADP⁺/NADPH ratio. This type of stress response is indicated by apparent state transition (qT). Recently, a dual-PAM instrument was developed that overcomes these limitations, and nonphotochemical energy dissipation in PS_{II} and PS_I can now be investigated separately (Schreiber and Klughammer 2008).

In most cases, it can be observed that stress response and adaptation to a changed environment results in a mixed response, and the type of response may vary with incubation time under stress. Nevertheless, the above-mentioned chlorophyll fluorescence data aid in indicating which of the other invasive and noninvasive methods should be used to better understand a plant's specific stress response.

25.3.6 Examples of PAM Fluorimetry Combined with Gas Exchange Measurements of Stressed Plants

The capability of photosynthetic adaptation and acclimation is specific to species and stress type. Photoprotective processes are induced by plants, subsequently reducing the formation of ROS during photosynthetic activity by achieving a low ratio of ETR to carbon assimilation rate A. The theoretical optimum is defined as a ratio of 4 µmol electrons m⁻² s⁻¹ per µmol CO_2 m⁻² s⁻¹. Thereby, plants under environmental conditions have to deal with an excess of electrons and might slightly suffer oxidative stress due to limitations of stomatal conductance, as proposed by Salazar-Parra et al. (2012). However, no oxidative stress was indicated by chlorophyll fluorescence measurements due to an increase in thermal dissipation (e.g., qI and qE) or an alteration of the fluorescence quenching (Kramer et al. 2004). Interestingly, limitation by reduced stomatal conductance seems to cause an elevated ETR/A ratio (da Silva et al. 2011). On the other hand, plants are also able to reduce the ETR/A ratio when extracellular CO_2 concentration increases (Kao et al. 2003; Schmidt 2014). The alteration of ETR might be a potential stress tolerance mechanism, described in experiments with *Chenopodia quinoa* Willd. (Schmidt 2014). Measurements of A/Ci curves (i.e., CO_2 assimilation rate (A) as a function of intercellular CO2 concentration (Ci)) reveal an elevation of apparent electron transport rate (ETR) with increasing extracellular CO_2 concentration and constant light intensity. The elevated ratio of the ETR to the maximum rate of net CO_2 assimilation under environmental conditions (i.e.,

1500 PAR, 400 ppm CO_2) decreased to the optimum level of 4 μmol electrons m^{-2} s^{-1} per μmol CO_2 m^{-2} s^{-1}, thus balancing the excess of electrons. However, a photosynthetic adaption related to thermal dissipation was not shown in quinoa plants (Schmidt 2014).

In species less tolerant than quinoa, such as *Jatropha curcas* (L.), it was also shown that the photosynthetic resistance of plants subjected to hyperosmotic salinity could be much less pronounced (da Silva et al. 2011). Under these conditions, salt stress caused reduction in leaf gas exchange parameters such as CO_2 fixation, stomatal conductance, and transpiration, but not in the photochemical efficiency of PS_{II}, as measured by chlorophyll fluorescence. After 14 days under salt stress, plants showed an ETR/A ratio fourfold higher than the control. This difference was maintained even after the recovery period. Thus, in this species and under these conditions, the salt-induced ionic toxicity effects were capable of inducing acute photosynthetic damage (photochemistry and gas exchange) due to stomatal and biochemical limitations.

In addition, further suboptimal growth conditions can overstrain the ability to compensate for photosynthetic stresses. *Alnus formosana* (Burkill) Makino and the fern *Pyrrosia lingua* Berecht & J. Presl., adapted or acclimated to high light, always had higher light-saturation points and maximal photosynthetic rates. However, photosynthetic capacity varied in the materials within a broad range (maximal photosynthetic rate ranging from 2 to 23 μmol CO_2 m^{-2} s^{-1}). The ratio of ETR to gross photosynthetic rate (PG) was similar for *A. formosana*

and *P. lingua* when measured under constant temperature, but the PPFD varied in both species grown under 100% sunlight and measured at different seasonal temperatures (15°C, 20°C, 25°C, and 30°C). The ETR/PG ratio increased with increasing temperature (Wong et al. 2012).

25.3.7 LIMITS OF CHLOROPHYLL FLUORESCENCE MONITORING OF PHOTOSYNTHESIS

In several studies, the responses of plants to abiotic stress were demonstrated by *in vitro* assays to assess stress tolerance. As was recently reported by Claeys et al. (2014), in most studies, very high stress levels are used, and criteria such as germination, plant survival, or the development of visual symptoms, such as bleaching, are measured. However, by tracking sensitive parameters such as shoot growth over a range of stress levels, rather than traditional, less sensitive parameters such as germination, visible stress symptoms, or root growth, a more accurate picture may be obtained of the stress sensitivity of, for instance, an ecotype or a transgenic line. In order to handle a high number of samples, it has to be tested whether the analysis of chlorophyll fluorescence parameters as a sensitive noninvasive method gives reliable results during the application of mild stress conditions.

The following six case studies demonstrate the difficulties of analyzing photosynthetic parameters under stress conditions. Examples of the influence of abiotic stresses (such as drought, salt, and nutrient limitation) and biotic stress (such as

TABLE 25.2
Summary of Stress Parameters and Chlorophyll Fluorescence Results Collected in the Selected Case Studies

Species	Condition	Strength of Stress	Reduction in Biomass (%)	F_v/F_m
Lablab purpureus	Drought	Moderate	26%	0.79 to 0.825
Euphorbia tirucalli	Drought	Mild	20%	0.73
		Moderate	40%	0.70
Balanites aegyptiaca	Drought	Moderate	65%	0.856 to 0.792
		Severe	8%	0.849 to 0.73
Balanites aegyptiaca	Salt	Moderate	81%	0.747 to 0.741
		Severe	7%	0.580 to 0.538
Tripolium pannonicum	Nutrients Nitrate	Mild	0% 100 ± 12.9	0.78 ± 0.04
		Moderate	64.3% 35.7 ± 13.6	0.83 ± 0.00
Brassica napus	Nutrients Sulfate	Mild	12% 0.47 ± 0.09	0.752 to 0.831
		Moderate	16% 0.45 ± 0.13	0.714
Brassica napus	Fungus	Mild	22% 0.41 ± 0.11	0.799 to 0.815
	Nutrients Sulfate + fungus	Mild	37% 0.34 ± 0.07	0.574 to 0.815

fungal infection) on different plant species will be discussed. In all cases, analyses of biometrical parameters, such as the determination of the gain in biomass in control and stressed plants, showed highly significant differences among conditions (summarized in Table 25.2). However, in several cases, photosynthetic parameters obtained by PAM imaging did not reveal significant differences or only revealed differences when under severe stress conditions.

25.3.8 Determination of Chlorophyll Fluorescence Parameters as Potential Stress Markers

It is theorized that chlorophyll fluorescence parameters are very sensitive markers in the identification and quantification of stress effects. Early stress symptoms measured shortly after exposure in the complex activity of the photosystems are especially thought to be easily determined by the noninvasive analysis of PAM fluorescence. In our laboratory, several species were treated with different kind of stresses and different levels of stress. Our main aim was to identify differences in the stress tolerance among genotypes. Therefore, we expected small but significant differences. In our experiments, we were interested in the application of mild or moderate stress to plants because our focus is on analyzing plants actively reacting and adapting to stress to regain homeostasis as soon as possible.

25.3.9 Case Study 1: *Lablab purpureus*, Drought Stress

One experimental plant was the herbaceous plant *Lablab purpureus* L., a potential protein-delivering species from the Fabaceae family grown in Africa and India. Due to climate change, farmers are interested in the cultivation of more drought-tolerant cultivars. Different genotypes from Africa and India were screened in the greenhouse for their drought tolerance under moderate stress conditions. Several methods were applied to analyze and quantify the influence of the stress, such as biometrical data, stomata opening, and PAM imaging. In this context, the main focus is on the suitability of chlorophyll fluorescence in drought tolerance screenings.

Chlorophyll fluorescence parameters were continuously monitored alongside the application of mild stress for a period of only several days up to 5 weeks, but no correlation to growth parameters was observed (Guretzki and Papenbrock 2013, 2014). The measurements of F_v/F_m, ETR, NPQ/4, Y(II), Y(NO), and Y(NPQ) showed almost no significant differences among the treatments. Some of the drought-stressed groups showed even better values compared to the control groups of the same genotypes. For example, F_v/F_m is considered to be a fast-measuring parameter for plant stress, and for nonstressed C_3 plants, a value of about 0.83 (Björkman and Demmig 1987) is expected. This approximate value was obtained in most cases, with one exception: *L. purpureus,* in both control and drought groups of the genotypes. Chlorophyll fluorescence does not appear to be sensitive enough to detect early symptoms of drought stress, at least in the herbaceous plant *L. purpureus.*

25.3.10 Case Study 2: *Euphorbia tirucalli,* Drought Stress

Perhaps the analysis of chlorophyll fluorescence parameters is more useful for more drought-tolerant plants with different morphology and adaptations to drought stress, like succulence or reduction of the leaf blade. Due to their morphology, for some plant species, the analysis of stomatal closure, as previously recommended, is technically not manageable. In this case, the analysis of chlorophyll fluorescence parameters was successfully performed with one member from the Euphorbiaceae family. In this investigation, the identification of drought-tolerant species and selection of the most tolerant genotypes of *Euphorbia tirucalli* L. was researched (Hastilestari et al. 2013). *E. tirucalli* is a a shrub or a small tree that can grow up to 7–12 m high that bears pencillike branches. The peculiar nature of the branches gave rise to the common name *pencil euphorb*. It is native to tropical areas in Africa; can be cultivated on marginal, arid land; and might be a good alternative source of bioenergy. Interestingly, this plant has two different photosynthetic pathways. In the leaves, it shows C_3 metabolism, and in the stem, CAM, a very important adaptation to arid conditions. Moreover, this plant is special in its capacity to produce high rates of secondary metabolites such as isoprenyl compounds. Thus, the plant creates an efficient carbon and electron sink.

In our experiments, two *E. tirucalli* genotypes, called Morocco and Senegal, according to their origin, were cultivated in a greenhouse at different volumetric water contents (VWCs, given in %: 25%, 15%, 10%, and 5%) for 8 weeks to induce mild and moderate drought stress. During this period of time, the performance of plants was continuously analyzed by several methods. Among them were biometrical parameters and PAM imaging to quantify the physiological status of stressed plants in comparison to the controls. Several parameters of photosynthesis were influenced by water shortage in the *E. tirucalli* genotypes Morocco and Senegal. Quantum efficiency of genotypes Morocco and Senegal in the photosystems of leaves and stems decreased linearly with water limitation over 8 weeks. Stems with CAM in both genotypes showed higher quantum efficiency than in C_3 leaves. Quantum efficiency of Morocco leaves for all VWCs (%)was in the range of 0.757–0.605. These values were higher than those for genotype Senegal (0.758–0.579) at similar VWCs. Genotype Morocco also had higher values at stem level (0.780–0.643) than genotype Senegal (0.780–0.616). In the leaves of both genotypes, there was no significant difference between different VWCs in the first 3 weeks of the experiment, but there was a significant difference beginning from week 4 onward. When considering stems, however, the genotypes performed differently. In genotype Morocco, significant differences between VWCs started to develop in week 5, while in genotype Senegal, changes started in week 4. Cessation of photosynthesis was supported by a decline in F_v/F_m along with prolonged drought in both genotypes. The decline of F_v/F_m was more pronounced at lower VWCs, whereby VWC 5% showed a lower decline. The decrease of F_v/F_m at high water limitation has been related to a decline in the function of primary photochemical reactions, primarily involving inhibition of PS_{II}, which is located in the thylakoid membrane system (Souza et al. 2003). The F_v/F_m values of leaves and stems were not significantly different during the first 3 weeks of the experiments, during which stress symptoms, such as leaf senescence, did not appear yet. After prolonged stress, F_v/F_m values in the stems of both genotypes were higher than in the leaves. In summary, the evaluation of the quantum efficiency could be successfully applied for the differentiation of genotypes with respect to drought tolerance (Hastilestari et al. 2013).

25.3.11 CASE STUDY 3: *BALANITES AEGYPTIACA*, DROUGHT STRESS

For further comparison, we would like to report experimental results using the drought-tolerant, multipurpose tree *Balanites aegyptiaca* L. It is a xerophytic tree found in tropical and nontropical areas in north and west Africa and west Asia (Khamis and Papenbrock 2014). It shows anatomical leaf adaptations to drought, such as thick waxy cuticles, numerous sunken stomata on both leaf surfaces, and thicker leaf laminae. A thick layer of trichomes on the leaves of *B. aegyptiaca* reflects excess light and forms a protective shield against high irradiance in the arid habitat. Leaf traits are among the major adaptive features for drought tolerance in this species (Elfeel et al. 2007). Six genotypes were investigated by the application of noninvasive methods, such as stomatal conductance and chlorophyll fluorescence. The impact of mild and moderate water limitation on chlorophyll fluorescence was examined through photochemical efficiency $\left(\Delta F/F_m\right)$ and quantum efficiency (F_v/F_m) during cultivation for 4 weeks at different VWCs (35%, 25%, 5%). The measurements showed that there were no significant differences among the genotypes per treatment, though on the fourth week, there was some variation, though not significant, in the mean average of photochemical efficiency for all treatments, with values between 0.439 and 0.603. The mean average for quantum efficiency ranged from 0.721 to 0.865. For mild stress conditions, these are impressive differences; however, due to the large standard deviation, the statistical analysis did not reveal significant differences. In the same experiments, results from stomatal conductance measurements clearly revealed significant differences among genotypes and treatments (Khamis et al. in preparation). Therefore, due to the quantification of small differences among genotypes after application of mild and moderate drought stress on *B. aegyptiaca*, we could not identify a suitable parameter of chlorophyll fluorescence photosynthesis.

25.3.12 CASE STUDY 4: DROUGHT STRESS, SEVERAL SPECIES INVESTIGATED BY SEVERAL GROUPS

Similar results were also obtained in other studies analyzing water limitation effects on plants. For example, in a study with the leafy rosette plant *A. thaliana* (L.) Heynh., drought stress

was initiated by completely withholding water. A change in the measured values of F_v/F_m, NPQ, and Y(II) occurred only after long-term (more than 10 days) drought stress. ETR and Y(NO) measurements behaved similarly. Initially, there was no impact of stress, and then both factors reflected strong signs of stress (Woo et al. 2008). In *Phaseolus vulgaris* L., only a slight decrease of F_v/F_m was found 7 days after stopping irrigation (Miyashita et al. 2005). Another study with *A. thaliana* compared the behavior of the photosynthetic system under mild, moderate, and severe drought stress. In this experiment, water was withheld until the soil water content reached 66–68% for mild drought stress, 50–52% for moderate drought stress, and 43–45% for severe drought stress, in comparison to the soil water content of the control group. Severe drought stress caused the strongest changes in comparison to the control group. But mild drought stress led moderate drought stress in regard to larger modifications in photosynthesis. It was concluded that the response of the plant matched with the Threshold for Tolerance Model (Sperdouli and Moustakas 2012). According to this model, tolerance mechanisms are started with lag time or induced by threshold concentrations (Barcelo and Poschenrieder 2002). Moderate stress caused less damage to the plant, because stress adaptation processes and repair mechanisms started in the plant, whereas during mild drought conditions, the stress threshold was not reached. Therefore, the plants were more affected under mild drought stress, which was reflected by greater altered chlorophyll fluorescence values in comparison to the moderate group (Lichtenthaler 1998; Sperdouli and Moustakas 2012). Classifying the correct strength of drought stress is important for the characterization of drought tolerance in different genotypes so that chlorophyll fluorescence measurements are used more efficiently. Often, the differences between the unstressed and stressed plants are too low to make definitive statements about the drought tolerance of the tested genotypes. In conclusion, chlorophyll fluorescence is only appropriate in examining the impact of severe drought stress conditions or in recovery experiments (Guretzki and Papenbrock 2013).

There is the potential for new measurement protocols to provide more reliable data about early symptoms of drought stress in the photosynthetic system. Burke et al. (2010) measured F_v/F_m at two time points by harvesting leaf punches. The chlorophyll fluorescence measurement then loses the advantage of being a nondestructive method. The closing of the stomata is the main reason for changes in photosynthesis under mild to moderate drought stress, as summarized by Medrano et al. (2002).

25.3.13 CASE STUDY 5: *TRIPOLIUM PANNONICUM*, NUTRIENT DEFICIENCY

In another set of experiments, the effect of another type of abiotic stress, nutrient deficiency, was analyzed by measuring chlorophyll fluorescence parameters. We are specifically interested in the cultivation of salt-tolerant plants as new crop plants and would like to optimize the cultivation conditions. Nitrogen is one of the most important nutrients in plant production. In seawater, as well as natural salt marshes, the nitrate concentration is low in comparison to farmland. For many species, a strong correlation exists between total leaf nitrogen and CO_2 assimilation at high irradiance (Evans 1989), and consequently, fluorescence seems to be a suitable method to analyze the effects of different nitrogen supplies. The halophyte *Tripolium pannonicum* was grown hydroponically with different concentrations of NO_3-N (between 1 and 100 mg NO_3-N per liter) and 220 mM NaCl. Only nitrogen content in leaves declined with decreasing nitrogen concentration in the nutrient solution. Plant uptake of nitrogen exhibited a similar tendency, but without strong differences. Only the plants treated with 1 mg NO_3-N per liter differed significantly from the others. In this grouping, less biomass was gained; less nitrogen and phosphorus were taken in; and chlorophyll, carotenoid, and nitrogen contents were much lower than in the other treatment groups. Contrary to the other results, maximum quantum efficiency of PS_{II} photochemistry was slightly higher at lower NO_3-N concentrations in the nutrient solution (Buhmann et al. 2015). Our results were in agreement with results reported before: nitrogen content has to reach very low levels before F_v/F_m is affected (Baker and Rosenqvist 2004). Therefore, the results of chlorophyll fluorescence analysis are not conclusive with respect to moderate nutrient deficiency.

However, it is likely that changes in the status of many other nutrients in leaves will have little effect on fluorescence characteristics. For example, a reduction of sulfur levels in sugar beet leaves had to reach starvation levels before any changes in F_q/F_m, F_q/F_v, and F_v/F_m were observed (Kastori et al. 2000). Consequently, more careful studies of the effects of deficiency of specific nutrients on leaf fluorescence characteristics are required before using fluorescence parameters to screen for nutrient deficiencies.

25.3.14 CASE STUDY 6: *BRASSICA NAPUS*, NUTRIENT DEFICIENCY AND BIOTIC STRESS

To understand the connection between sulfur metabolism and pathogen attack, plants of different *Brassica napus* varieties were grown with a full supply of sulfur, and after a certain period of time, a sulfur deficiency was applied followed by a *Verticillium longisporum* infection. The detection of early symptoms of pathogen attack is one major field of application of chlorophyll fluorescence measurements as stress indicators, as was described by Berger et al. (2007). The plants were analyzed by PAM imaging with the aim to detect very early nutrient stress and/or infection symptoms (Weese et al. 2015). Chlorophyll fluorescence measurements were done with noninfected and infected plants grown with different sulfur concentrations. Independently of the sulfur supply, the F_v/F_m values of the four varieties investigated ranged from 0.714 to 0.831, indicating photosynthetic activity in the same range when grown at higher or lower sulfur concentrations. Baker and Rosenqvist (2004) concluded that the F_v/F_m ratio was not sensitive to sulfur

stress until starvation levels were reached. Under full sulfur supply, two of the four varieties performed better (around 5%), with respect to quantum yield, than the others. A slight decrease of the photosynthetic yield was observed under sulfur deprivation, but in one variety that performed best under full sulfur supply, sulfur limitation led to strongly decreasing values of the quantum yield of PS_{II}. On the other hand, the quantum yield of one variety remained almost constant. Infection with the fungus *V. longisporum* led to a significant decrease of the photosynthetic yield, ranging from approximately 4.5% to 41%. The variety performing best with respect to quantum yield under full supply and worst under sulfur limitation was not significantly influenced by infection (Weese et al. 2015). These results implicated that for this one variety, the sulfur limitation led to worsening values of the quantum yield of PS_{II}, and the infection with the fungus remained unnoticed. This was also supported by other chlorophyll fluorescence parameters, such as $\left(\Delta F/F_m \right)$. In this case the variety seemed to suffer sulfur starvation because the quantum yield was only sensitive to sulfur stress until starvation levels were reached (Kastori et al. 2000; Baker and Rosenqvist 2004). A higher infection rate without killing the plants appeared impossible. Obviously, sulfur limitation influenced chlorophyll fluorescence more than fungal infection.

25.3.15 ALTERNATIVE METHODS FOR ANALYSIS OF PLANT STRESS RESPONSE

Thermal imaging appears to harness significant potential for the visualization of biotic and abiotic stress in plants. Thermography relies on imaging the long-wave (thermal) radiation emitted by the subject as an indicator of leaf temperature, which itself is indicative of changes in leaf transpiration. The technique to visualize plant stress via thermal radiation is a reliable tool for drought stress, as was shown by others (Jones 2007) and by our own experiments (Guretzki and Papenbrock 2013, 2014). It is possible to use thermography in an active mode by following temperature kinetics after changes in incident radiation (Chaerle et al. 2007). For other stress factors, such as biotic stress, it might be less reliable, because the correlation of stress response with leaf temperature is not linear, in addition to not being well understood. Recent publications describe the potential of utilizing thermal imaging in various experiment types, ranging from those attempting to minimize effects of environmental variability for a fast and detailed analysis of *A. thaliana* mutants to those that involve screening canopy structures outdoors (Chaerle et al. 2007; Costa et al. 2013).

25.4 SUMMARY

Advanced physiological, biochemical, and molecular biological methods allow description of plant performance in photosynthesis in a more detailed way. We can measure not only biomass production and shoot and root growth, but also the synthesis of individual metabolites, activity of individual enzymes, and expression of relevant genes (Seki et al. 2001; Chaerle et al. 2007; Shulaev et al. 2008; Kosová et al. 2013; Koyro et al. 2013; Pérez-Harguindeguy et al. 2013). But in most cases, the latter types of measurements require working on plant extracts.

When monitoring plant stress response or in approaches aiming at defining optimal crop production regimes, such destructive approaches require starting experiments with high numbers of plants because a set of plants will have to be harvested (destroyed) for each of the measuring intervals. Moreover, these methods result in a high experimental error, because data from individual plants are compared. Another problem is that enzyme activities measured *in vitro* do not necessarily resemble *in vivo* activities. Enzyme activities measured in plant extracts resemble the maximal turnover rate of the tested enzyme, rather than its catalytic activity under physiological conditions inside the cell. Heterogeneity of the analyzed plant material requires statistical analysis of experimental data. Noninvasive methods overcome this problem because monitoring can be performed by repetitive analysis of an individual plant (Pfündel 1998). It may be expected that effects of stress factors on growth and development are immediately visible. Among the noninvasive methods, chlorophyll fluorescence monitoring has attracted much attention, because the analyzed parameter is directly linked to photosynthesis, i.e., the technique can monitor the basis of energy use and biomass production (Oxborough 2004).

Over the last few decades, methods have been developed to distinguish the responses of fluorescence signals caused by different reactions taking place in chloroplasts and thylakoid membranes (Baker 2008). It has been proven by alignment of data from destructive methods and chlorophyll fluorescence that the respective data interpretation is correct. But, until now, not all of the biochemical stress responses that have been identified by destructive methods can be discriminated by means of chlorophyll fluorescence measurement. The parameter *photosynthetic quenching*, for instance, summarizes all factors that affect photosynthetic electron transport rate.

There is an ever-growing demand for crop plants with improved stress tolerance. These plants are expected to produce a more reliable crop yield despite "extreme events," such as short drought or intense illumination periods, during a growth season. Breeding for such crop lines requires testing plant responses in the presence of only minor environmental stress. Under stressful conditions, all plants are capable of adapting to the adverse situation. They will adjust physiological and biochemical functions and, as a result, adopt a new homeostasis at cellular and tissue levels. Nevertheless, there will be at least minor differences in energy use efficiency and, thus, productivity.

We therefore have compared the sensitivity of chlorophyll fluorescence parameters to those measured by other methods. In agreement with results published by other teams, we found high sensitivity in parameters such as shoot growth and dry weight production (Claeys et al. 2014; Pérez-Harguindeguy et al. 2013). On the other hand, it became obvious that chloroplast fluorescence measurement in experiments with some plant species was not sufficiently sensitive. As documented in case studies 1, 3, 4, 5, and 6, the measured fluorescence parameters were not sufficiently sensitive to measure stress responses at an early treatment stage.

In principle, there may be two reasons responsible for the failure of the method:

1. A complex leaf anatomy, heterogeneity of the leaf surface, and patchiness of metabolic activity in the leaves may cause problems when identifying the analyzed area in repetitive measurements. There will be a pronounced mean variation of data, and minor stress-induced effects will not be detected as significant stress responses (see Case Study 3). Some experiments encounter problems with the exact measurement of F_o. This results in data scattering, as well as the aforementioned data variations (Lichtenthaler et al. 2007). In other cases, meaningful results can be achieved by measuring other fluorescence parameters (Kalaji et al. 2014).

2. As stated, fluorescence parameters do not allow for discrimination between all of the biochemical pathways resulting in chlorophyll fluorescence quenching (Harbinson and Rosenqvist 2003; Kalaji et al. 2014). Therefore, no significant stress response may be visible if plants can use alternative electron or carbon sinks, characterized by kinetic parameters similar to the one of photosynthesis. *E. tirucalli*, for instance, is capable of producing high amounts of isoprenoids and further secondary metabolites. Respective pathways are competing for assimilated carbon with energy metabolism and plant growth. Therefore, minor changes in sugar consumption of the latter pathways can be compensated by increased synthesis of secondary compounds. For these reasons, moderate stress could not be detected in respective experiments (see Case Study 2).

In conclusion, measuring chlorophyll fluorescence parameters is an elegant, noninvasive method to monitor severe stress effects in plants. This method is recommended for laboratory experiments as well as field studies, especially if high numbers of samples have to be tested or if space in the greenhouse is incapable of supporting destructive experimental approaches. But, if possible, further methods should be used in parallel to check sensitivity and reliability.

REFERENCES

Aleric, K. M. and Kirkmann, L. K. 2005. Growth and photosynthetic responses of the federally endangered shrub, *Lindera melissifolia* (Lauraceae), to varied light environments. *Am. J. Bot.* 92:682–689.

Allen, J. F. and Pfannschmidt, T. 2000. Balancing the two photosystems: Photosynthetic electron transfer governs transcription of reaction centre genes in chloroplasts. *Phil. Trans. R. Soc. Lond.* 355:1351–1359.

Anderson, J. M. and Aro, E.-M. 1994. Grana stacking and protection of photosystem II in thylakoid membranes of higher plant leaves under sustained high irradiance: An hypothesis. *Photosynth. Res.* 41:315–326.

Anderson, J. M., Chow, W. S., and Park, Y.-I. 1995. The grand design of photosynthesis: Acclimation of the photosynthetic apparatus to environmental cues. *Photosynth. Res.* 46:129–139.

Anderson, J. M., Chow, W. S., and De Las Rivas, J. 2008. Dynamic flexibility in the structure and function of photosystem II in higher plant thylakoid membranes: the grana enigma. *Photosynth. Res.* 98:575–587.

Andersson, B. and Anderson, J. M. 1980. Lateral heterogeneity in the distribution of chlorophyll-protein complexes of the thylakoid membranes of spinach chloroplasts. *Biochim. Biophys. Acta* 593:427–440.

Anjum, N. A., Aref, I. M., Duarte, A. C., Pereira, E., Ahmad, I., and Iqbal, M. 2014. Glutathione and proline can coordinately make plants withstand the joint attack of metal(loid) and salinity stresses. *Front. Plant Sci.* 5:662.

Apel, K. and Hirt, H. 2004. Reactive oxygen species: Metabolism, oxidative stress, and signal transduction. *Annu. Rev. Plant Biol.* 55:373–399.

Aro, E.-M., Kettunen, R., and Tyystjärvi, E. 1992. ATP and light regulate D1 protein modification and degradation Role of D1* in photoinhibition. *FEBS Lett.* 297:29–33.

Ashraf, M. and Harris, P. J. C. 2005. *Abiotic Stresses. Plant Resistance through Breeding and Molecular Approaches.* New York: The Haworth Press.

Baker, N. R. 2008. Chlorophyll fluorescence: A probe of photosynthesis *in vivo. Ann. Rev. Plant Biol.* 59:89–113.

Baker, N. R. and Rosenqvist, E. 2004. Applications of chlorophyll fluorescence can improve crop production strategies: An examination of future possibilities. *J. Exp. Bot.* 55:1607–1621.

Barcelo, J. and Poschenrieder, C. 2002. Fast root growth responses, root exudates, and internal detoxification as clues to the mechanisms of aluminium toxicity and resistance: A review. *Environ. Exp. Bot.* 48:75–92.

Beckie, H. J. and Tardif, F. J. 2012. Herbicide cross resistance in weeds. *Crop Prot.* 35:15–28.

Beer, S., Larsson, C., Poryan, O., and Axelsson, L. 2000. Photosynthetic rates of *Ulva* (Chlorophyta) measured by pulse amplitude modulated (PAM) fluorometry. *Eur. J. Phycol.* 35:69–74.

Beevers, L. and Hageman, R. H. 1969. Nitrate reduction in higher plants. *Annu. Rev. Plant Physiol.* 20:495–422.

Bennett, J. 1983. Regulation of photosynthesis by reversible phosphorylation of the light harvesting chlorophyll a/b protein. *Biochem. J.* 212:1–13.

Berger, S., Benediktyová, Z., Matous, K., Bonfig, K., Mueller, M. J., Nedbal, L., and Roitsch, T. 2007. Visualization of dynamics of plant-pathogen interaction by novel combination of chlorophyll fluorescence imaging and statistical analysis: Differential effects of virulent and avirulent strains of *P. syringae* and of oxylipins on *A. thaliana. J. Exp. Bot.* 58: 797–806.

Biehler, K., Haupt, S., Beckmann, J., Fock, H., and Becker, T. W. 1997. Simultaneous CO_2- and $^{16}O_2/^{18}O_2$-gas exchange and fluorescence measurements indicate differences in light energy dissipation between the wild type and the phytochrome-deficient *aurea* mutant of tomato during water stress. *J. Exp. Bot.* 48:1439–1449.

Björkman, O. and Demmig, B. 1987. Photon yield of O_2 evolution and chlorophyll fluorescence characteristics at 77 K among vascular plants of diverse origins. *Planta* 170:489–504.

Boardman, N. K. 1977. Comparative photosynthesis of sun and shade plants. *Annu. Rev. Plant Physiol.* 28:355–377.

Boyer, P. D. 1993. The binding change mechanism for ATP synthase—Some probabilities and possibilities. *Biochim. Biophys. Acta* 1140:215–250.

Boyer, P. D. 2000. Catalytic site forms and controls in ATP synthase catalysis. *Biochim. Biophys. Acta* 1458:252–262.

Buchanan, B. B. and Balmer, Y. 2005. Redox regulation: A broadening horizon. *Annu. Rev. Plant Biol.* 56:187–220.

Buhmann, A., Waller, U., Wecker, B., and Papenbrock, J. 2014. Optimization of culturing conditions and selection of species for the use of halophytes as biofilter for nutrient-rich saline water. *Agr. Water Manage.* 149:102–114.

Burke, J. J., Franks, C. D., Birow, G., and Xin, Z. 2010. Selection system for the stay-green drought tolerance trait in sorghum germplasm. *Agron. J.* 102:1118–1122.

Buschmann, C., Langsdorf, G., and Lichtenthaler, H. K. 2001. Imaging of the blue, green, and red fluorescence emission of plants: An overview. *Photosynthetica* 38:483–491.

Butler, W. L. 1978. Energy distribution in the photochemical apparatus of photosynthesis. *Annu. Rev. Plant Physiol.* 29:345–378.

Chaerle, L., Leinonen, I., Jones, H. G., and van der Straeten, D. 2007. Monitoring and screening plant populations with combined thermal and chlorophyll fluorescence imaging. *J. Exp. Bot.* 58:773–784.

Chaves, M. M., Flexas, J., and Pinheiro, C. 2009. Photosynthesis under drought and salt stress: Regulation mechanisms from whole plant to cell. *Ann. Bot.* 103:551–560.

Chen, Z. and Gallie, D. R. 2008. Dehydroascorbate reductase affects non-photochemical quenching and photosynthetic performance. *J. Biol. Chem.* 283:21347–21361.

Chinnusamy, V., Zhu, J., and Zhu, J.-K. 2007. Cold stress regulation of gene expression in plants. *Trends Plant Sci.* 12:444–451.

Chow, W. S., Kim, E.-H., Horton, P., and Anderson, J. M. 2005. Granal stacking of thylakoid membranes in higher plant chloroplasts: The physicochemical forces at work and the functional consequences that ensue. *Photochem. Photobiol. Sci.* 4:1081–1090.

Chuartzman, S. G., Nevo, R., Shimoni, E. et al. 2008. Thylakoid membrane remodeling during state transitions in *Arabidopsis. Plant Cell* 20:1029–1039.

Claeys, H., Van Landeghem, S., Dubois, M., Maleux, K., and Inzé, D. 2014. What is stress? Dose-response effects in commonly used in vitro stress assays. *Plant Physiol.* 165:519–527.

Collins, N. C., Tardieu, F., and Tuberosa, R. 2008. Quantitative trait loci and crop performance under abiotic stress: Where do we stand? *Plant Physiol.* 147:469–486.

Costa, J. M., Grant, O. M., and Chaves, M. M. 2013. Thermography to explore plant-environment interactions. *J. Exp. Bot.* 64:3937–3949.

da Silva, E. N., Ribeiro, R. V., Ferreira-Silva1, S. L., Viégas, R. A., and Silveira, J. A. G. 2011. Salt stress induced damages on the photosynthesis of physic nut young plants. *Sci. Agric. (Piracicaba, Braz.)* 68:62–68.

Dai, Y., Shen, Z., Liu, Y., Wang, L., Hannaway, D., and Lu, H. 2009. Effects of shade treatments on the photosynthetic capacity, chlorophyll fluorescence, and chlorophyll content of *Tetrastigma hemsleyanum* Diels et Gilg. *Environ. Exp. Bot.* 65:177–182.

Debez, A., Braun, H.-P., Pich, A., Taamalli, W., Koyro, H. W., Abdelly, C., and Huchzermeyer, B. 2012. Proteomic and physiological responses of the halophyte Cakile maritima to moderate salinity at the germinative and vegetative stages. *J. Proteomics* 75:5667–5694.

Demmig-Adams, B., Ebbert, V., Zarter, C. R., and Adams, W. W. 2006. Characteristics and species-dependent employment of flexible versus sustained thermal dissipation and photoinhibition. In *Photoprotection, Photoinhibition, Gene Regulation, and Environment, Advances in Photosynthesis and Respiration*, eds. B. Demmig-Adams, W. W. Adams, and A. K. Mattoo, vol. 21, 39–48. Dordrecht Springer.

Demmig-Adams, B., Cohu, C. M., Muller, O., and Adams, W. W. 2012. Modulation of photosynthetic energy conversion efficiency in nature: From seconds to seasons. *Photosynth. Res.* 113:75–88.

Dietz, K.-J. and Heilos, L. 1990. Carbon metabolism in spinach leaves as affected by leaf age and phosphorus and sulfur nutrition. *Plant Physiol.* 93:1219–1225.

Dilley, R. A. 1972. Ion transport. H⁺, K⁺, Mg²⁺ exchange phenomena. *Methods Enzymol.* 24:68–74.

Eckardt, N. A. 2005. Photorespiration revisited. *Plant Cell* 17:2139–2141.

Eckardt, N. A. 2011. Fine-tuning photosynthesis: Structural basis of photoprotective energy dissipation. *Plant Cell* 23:1189.

Eisenhut, M., Ruth, W., Haimovich, M., Bauwe, H., Kaplan, A., and Hagemann, M. 2008. The photorespiratory glycolate metabolism is essential for cyanobacteria and might have been conveyed endosymbiontically to plants. *Proc. Natl. Acad. Sci. U.S.A.* 105:17199–17204.

Elfeel, A. A., Warrag, E. I., and Musnad, H. A. 2007. Response of *Balanites aegyptiaca* (L.) Del. seedlings from varied geographical source to imposed drought stress. *Discov. Innov.* 18:319–325.

Evans, J. R. 1989. Photosynthesis and nitrogen relationships in leaves of C_3 plants. *Oecologia* 78:9–19.

Evron, Y. and McCarty, R. E. 2000. Simultaneous measurement of ΔpH and electron transport in chloroplast thylakoids by 9-aminoacridine fluorescence. *Plant Physiol.* 124:407–414.

Farquhar, G. D., von Caemmerer, S., and Berry, J. A. 1980. A biochemical model of photosynthetic CO_2 assimilation in leaves of C_3 species. *Planta* 149:78–90.

Flowers, T. J., Troke, P. F., and Yeo, A. R. 1977. The mechanism of salt tolerance in halophytes. *Annu. Rev. Plant Physiol.* 28:89–121.

Foyer, C. H., Neukermans, J., Queval1, G., Noctor, G., and Harbinson, J. 2012. Photosynthetic control of electron transport and the regulation of gene expression. *J. Exp. Bot.* 63:1637–1661.

Fristedt, R., Willig, A., Granath, P., Crévecoeur, M., Rochaix, J.-D., and Vener, A. V. 2009. Phosphorylation of photosystem II controls functional macroscopic folding of photosynthetic membranes in Arabidopsis. *Plant Cell* 21:3950–3964.

Fritsche-Neto, R. and Borém, A. 2012. *Plant Breeding for Abiotic Stress Tolerance*. Berlin: Springer.

Fromme, P. and Gräber, P. 1990. Uni-site catalysis in thylakoids. The influence of membrane energization on ATP hydrolysis and ATP-Pi exchange. *FEBS Lett.* 269:247–251.

Gandin, A., Denysyuk, M., and Cousins, A. B. 2014. Disruption of the mitochondrial alternative oxidase (AOX) and uncoupling protein (UCP) alters rates of foliar nitrate and carbon assimilation in *Arabidopsis thaliana*. *J. Exp. Bot.* 65:3133–3142.

Genty, B. and Meyer, S. 1995. Quantitative mapping of leaf photosynthesis using imaging. *Aust. J. Plant Physiol.* 22:277–284.

Genty, B., Briantais, J. M., and Baker, N. R. 1989. The relationship between the quantum yield of photosynthetic electron transport and quenching of chlorophyll fluorescence. *Biochim. Biophys. Acta* 990:87–92.

Gil, R., Boscaiu, M., Lull, C., Bautista, I., Lidón, A., and Vicente, O. 2014. Are soluble carbohydrates ecologically relevant for salt tolerance in halophytes? *Funct. Plant Biol.* 40:805–818.

Gilbert, A., Silvestre, V., Robins, R. J., Tcherkez, G., and Remaud, G. S. 2011. A 13C NMR spectrometric method for the determination of intramolecular δ13C values in fructose from plant sucrose samples. *New Phytol.* 191:579–588.

Gill, S. S. and Tuteja, N. 2010. Reactive oxygen species and antioxidant machinery in abiotic stress tolerance in crop plants. *Plant Physiol. Biochem.* 48:909–930.

Girault, G., Berger, G., and Zimmermann, J.-L. 1998. Nucleotide-CF1 interactions and current views on the catalytic mechanism. *Photosynth. Res.* 57:253–266.

Givnish, T. J. 1988. Adaptation to sun vs. shade: A whole-plant perspective. *Aust. J. Plant Physiol.* 15:63–92.

Goral, T. K., Johnson, M. P., Duffy, C. P. D., Brain, A.-P. R., Ruban, A. V., and Mullineaux, C. W. 2012. Light-harvesting antenna composition controls the macrostructure and dynamics of thylakoid membranes in *Arabidopsis*. *Plant J.* 69:289–301.

Gottardini, E., Cristofori, A., Cristofolini, F., Nali, C., Pellegrini, E., Busotti, F., and Ferretti, M. 2014. Chlorophyll-related indicators are linked to visible ozone symptoms: Evidence from a field study on native *Viburnum lantana* L. plants in northern Italy. *Ecol. Indic.* 39:65–74.

Greenway, M. and Munns, R. 1980. Mechanisms of salt tolerance in non-halophytes. *Annu. Rev. Plant Physiol.* 31:149–190.

Griffin, J. J., Ranney, T. G., and Pharr, D. M. 2004. Photosynthesis, chlorophyll fluorescence, and carbohydrate content of Illicium taxa grown under varied irradiance. *J. Am. Soc. Hort. Sci.* 129:46–53.

Groth, G. and Junge, W. 1993. Proton slip of the chloroplast ATPase: Its nucleotide dependence, energetic threshold, and relation to an alternating site mechanism of catalysis. *Biochemistry* 32:8103–8111.

Guretzki, S. and Papenbrock, J. 2013. Comparative analysis of methods analyzing effects of drought on herbaceous plants. *J. Appl. Bot. Food Chem.* 86:47–54.

Guretzki, S. and Papenbrock, J. 2014. Characterization of *Lablab purpureus* (L.) Sweet regarding drought tolerance, trypsin inhibitor activity and cyanogenic potential for selection in breeding programs. *J. Agron. Crop Sci.* 200:24–35.

Harbinson, J. and Rosenqvist, E. 2003. An introduction to chlorophyll fluorescence. In *Practical Applications of Chlorophyll Fluorescence in Plant Biology*, eds. J. R. De Ell and P. M. A. Toivonen, 1–29. New York: Springer Science + Business Media.

Hastilestari, B. R., Mudersbach, M., Tomala, F. et al. 2013. *Euphorbia tirucalli* L.—Comprehensive characterization of a drought tolerant plant with a potential as biofuel source. *PLoS One* 8:e63501.

Hatzios, K. K. 1991. An overview of the mechanisms of action of herbicide safeners. *Z. Naturforsch.* 46c:819–827.

Havaux, M. and Niyogi, K. K. 1999. The violaxanthin cycle protects plants from photooxidative damage by more than one mechanism. *Proc. Natl. Acad. Sci. U.S.A.* 96:8762–8767.

Heineke, D., Stitt, M., and Heldt, H. W. 1989. Effects of inorganic phosphate on the light dependent thylakoid energization of intact spinach chloroplasts. *Plant Physiol.* 91:221–226.

Heintze, A., Gorlach, J., Leuschner, C., Hoppe, P., Hagelstein, P., Schulze-Siebert, D., and Schultz, G. 1990. Plastidic isoprenoid synthesis during chloroplast development. Change from metabolic autonomy to a division-of-labor stage. *Plant Physiol.* 93:1121–1127.

Hideg, É. A. 2008. Comparative study of fluorescent singlet oxygen probes in plant leaves. *Cent. Eur. J. Biol.* 3:273–284.

Hisabori, T., Konno, H., Ichimura, H., Strotmann, H., and Bald, D. 2002. Molecular devices of chloroplast F(1)-ATP synthase for the regulation. *Biochim. Biophys. Acta* 1555:140–146.

Hölscher, D., Leuschner, C., Bohman, K., Hagemeier, M., Juhrbandt, J., and Tjitrosemito, S. 2006. Leaf gas exchange of trees in old-growth and young secondary forest stands in Sulawesi, Indonesia. *Trees* 20:278–285.

Holt, N. E., Fleming, G. R., and Niyogi, K. K. 2004. Toward an understanding of the mechanism of nonphotochemical quenching in green plants. *Biochemistry* 43:8281–8289.

Horton, P., Ruban, A. V., and Walters, R. G. 1996. Regulation of light harvesting in green plants. *Annu. Rev. Plant Physiol. Plant Mol. Biol.* 47:665–684.

Hosler, J. P. and Yocum, C. F. 1987. Regulation of cyclic photophosphorylation during ferredoxin-mediated electron transport. *Plant Physiol.* 83:965–969.

Hossain, M. A., Piyatida, P., Teixeira da Silva, J. A., and Fujita, M. 2012. Molecular mechanism of heavy metal toxicity and tolerance in plants: central role of glutathione in detoxification of reactive oxygen species and methylglyoxal and in heavy metal chelation. *J. Bot.* 2012:872875.

Hu, H. and Xiong, L. 2014. Genetic engineering and breeding of drought-resistant crops. *Annu. Rev. Plant Biol.* 65:715–741.

Huang, J., Boerner, R. E. J., and Rebbeck, J. 2007. Ecophysiological responses of two herbaceous species to prescribed burning, alone or in combination with overstory thinning. *Am. J. Bot.* 94:755–763.

Huchzermeyer, B. 1982. Energy transfer inhibition induced by nitrofen. *Z. Naturforsch.* 37c:787–792.

Huchzermeyer, B., Löhr, A., and Willms, I. 1986. A direct interaction between photosystem I and chloroplast coupling factor. *Biochem. J.* 234:217–220.

Izawa, S. and Good, N. E. 1972. Inhibition of photosynthetic electron transport and photophosphorylation. *Meth. Enzymol.* 24:355–377.

Jagendorf, A. T. 1972. Two-stage phosphorylation techniques: Light-to-dark and acid-to-base procedures. *Methods Enzymol.* 24:103–113.

Jagendorf, A. T. 2002. Photophosphorylation and the chemiosmotic perspective. *Photosynth. Res.* 73:233–241.

Jajoo, A., Szabo, M., Zsiros, O., and Garab, G. 2012. Low pH induced structural reorganization in thylakoid membranes. *Biochim. Biophys. Acta* 1817:1388–1391.

Johnson, M. P., Pérez-Bueno, M. L., Zia, A., Horton, P., and Ruban, A. V. 2009. The zeaxanthin-independent and zeaxanthin-dependent qE components of nonphotochemical quenching involve common confoemational changes within the photosystem II antenna in *Arabidopsis*. *Plant Physiol.* 149:1061–1075.

Johnson, M. P., Zia, A., and Ruban, A. V. 2012. Elevated ΔpH restores rapidly reversible photoprotective energy dissipation in *Arabidopsis* chloroplasts deficient in lutein and xanthophyll cycle activity. *Planta* 235:193–204.

Jones, H. G. 2007. Monitoring plant and soil water status: Established and novel methods revisited and their relevance to studies of drought tolerance. *J. Exp. Bot.* 58:119–130.

Kalaji, H. M., Schansker, G., Ladle, R. J. et al. 2014. Frequently asked questions about in vivo chlorophyll fluorescence: Practical issues. *Photosynth. Res.* 122:121–158.

Kao, W. Y., Tsai, T. T., and Shih, C. N. 2003. Photosynthetic gas exchange and chlorophyll a fluorescence of three wild soybean species in response to NaCl treatments. *Photosynthetica* 41:415–419.

Kastori, R., Plesnicar, M., Arsenijevic-Maksimovic, I., Petrovic, N., Pankovic, D., and Sakac, Z. 2000. Photosynthesis, chlorophyll fluorescence and water relations in young sugar beet plants as affected by sulfur supply. *J. Plant Nutr.* 23:1037–1049.

Khamis, G. and Papenbrock, J. 2014. Newly established drought-tolerant plants as renewable primary products as source of bioenergy. *Emir. J. Food Agric.* 26:1057–1066.

Khan, M. A. and Weber, D. J. 2008. *Ecophysiology of High Salinity Tolerant Plants*. Dordrecht: Springer.

Kirchhoff, H., Haase, W., Haferkamp, S., Schott, T., Borinski, M., Kubitscheck, U., and Rögner, M. 2007. Structural and functional self-organization of photosystem II in grana thylakoids. *Biochim. Biophys. Acta* 1767:1180–1188.

Klughammer, C. and Schreiber, U. 1998. Measuring P700 absorbance changes in the near infrared spectral region with a dual wavelength pulse modulation system. In *Photosynthesis: Mechanism and Effects*, ed. G. Grab, vol. 5, 4357–4360. Dordrecht: Kluwer Academic Publishers.

Kosová, K., Vitámvás, P., Urban, M. O., and Prásil, I. T. 2013. Plant proteome responses to salinity stress—Comparison of glycophytes and halophytes. *Funct. Plant Biol.* 40:775–786.

Koyro, H.-W., Geissler, N., Seenivasan, R., and Huchzermeyer, B. 2011. Plant stress physiology: Physiological and biochemical strategies allowing plants/crops to thrive under ionic stress. In *Handbook of Plant and Crop Stress*, ed. M. Pessarakli, 1051–1094. Boca Raton, FL: CRC Press.

Koyro, H.-W., Zörb, C., Debez, A., and Huchzermeyer, B. 2013. The effect of hyperosmotic salinity on protein patterns and enzyme activities of halophytes. *Funct. Plant Biol.* 40:487–507.

Koyro, H.-W., Huchzermeyer, B., and Zörb, C. 2014a. Salt stress tolerance: Stress effect on metabolism and development of plants. In *Handbook of Plant & Crop Physiology*, ed. M. Pessarakli, 3rd ed., 487–507. Boca Raton, FL: CRC Press.

Koyro, H.-W., Lieth, H., Gul, B. et al. 2014b. Importance of the diversity in between halophytes to agriculture and land management in arid and semiarid countries. In *Sabkha Ecosystems: Volume IV: Cash Crop Halophyte and Biodiversity Conservation. Tasks for Vegetation Science*, eds. M. A. Khan et al., 175–198. Dordrecht: Springer.

Kramer, D. M., Johnson, G., Kiirats, O., and Edwards, G. E. 2004. New fluorescence parameters for the determination of QA redox state and excitation energy fluxes. *Photosynth. Res.* 79:209–218.

Krasensky, J. and Jonak, C. 2012. Drought, salt, and temperature stress-induced metabolic rearrangements and regulatory networks. *J. Exp. Bot.* 63:1593–1608.

Larcher, W. and Neuner, G. 1989. Cold-induced sudden reversible lowering of in vivo chlorophyll fluorescence after saturating light pulses: A sensitive marker for chilling susceptiblity. *Plant Physiol.* 89:740–742.

Lawlor, D. and Fock, H. 1975. Photosynthesis and photorespiratory CO_2 evolution of water-stressed sunflower leaves. *Planta* 126:247–258.

Leong, T.-Y. and Anderson, J. M. 1984. Adaptation of the thylakoid membranes of pea chloroplasts to light intensities. II. Regulation of electron transport capacities, electron carriers, coupling factor (CF1) activity and rates of photosynthesis. *Photosynth. Res.* 5:117–128.

Lichtenthaler, H. K. 1998. The stress concept in plants: An introduction. *Ann. N. Y. Acad. Sci.* 851:187–198.

Lichtenthaler, H. K. and Buschmann, C. 1984. Beziehungen zwischen Photosynthese und Baumsterben. *Allg. Forst. Z.* 39:12–16.

Lichtenthaler, H. K., Ac, A., Marek, M. V., Kalina, J., and Urban, O. 2007. Differences in pigment composition, photosynthetic rates and chlorophyll fluorescence images of sun and shade leaves of four tree species. *Plant Physiol. Biochem.* 45:577–588.

Long, S. P., Ainsworth, E. A., Rogers, A., and Ort, D. R. 2004. Rising atmospheric carbon dioxide: Plants FACE the future. *Annu. Rev. Plant Biol.* 55:591–628.

Masclaux-Daubresse, C., Daniel-Vedele, F., Dechorgnat, J., Chardon, F., Gaufichon, L., and Suzuki, A. 2010. Nitrogen uptake, assimilation and remobilization in plants: challenges for sustainable and productive agriculture. *Ann. Bot.* 105:1141–1157.

Matsubara, S., Krause, G. H., Seltmann, M., Virgo, A., Kursar, T. A., Jahns, P., and Winter, K. 2008. *Plant Cell Environ.* 31:548–561.

Maxwell, K. and Johnson, G. N. 2000. Chlorophyll fluorescence—A practical guide. *J. Exp. Bot.* 51:659–668.

McDonald, A. E., Ivanov, A. G., Bode, R., Maxwell, D. P., Rodermel, S. R., and Hüner, N. P. A. 2011. Flexibility in photosynthetic electron transport: the physiological role of plastoquinol terminal oxidase (PTOX). *Biochim. Biophys. Acta* 1807:954–967.

Medrano, H., Escalona, J. M., Bota, J., Gulías, J., and Flexas, J. 2002. Regulation of photosynthesis of C_3 plants in response to progressive drought: Stomatal conductance as a reference parameter. *Ann. Bot.* 89:895–905.

Meyer, S. and Genty, B. 1998. Mapping intercellular CO_2 mole fraction (Ci) in *Rosa rubiginosa* leaves fed with abscisic acid by using chlorophyll fluorescence imaging. *Plant Physiol.* 116:947–957.

Michelet, L., Zaffagnini, M., Morisse, S. et al. 2013. Redox regulation of the Calvin-Benson cycle: Something old, something new. *Front. Plant Sci.* 4:470.

Mitchell, P. 1961. Coupling of phosphorylation to electron and hydrogen transfer by a chemi-osmotic type of mechanism. *Nature* 191:144–148.

Mittler, R. 2002. Oxidative stress, antioxidants and stress tolerance. *Trends Plant Sci.* 7:405–410.

Miyashita, K., Tanakamaru, S., Maitani, T., and Kimura, K. 2005. Recovery responses of photosynthesis, transpiration, and stomatal conductance in kidney bean following drought stress. *Environ. Exp. Bot.* 53:205–214.

Møller, I. M. 2001. Plant Mitochondria and oxidative stress: Electron transport, NADPH turnover, and metabolism of reactive oxygen species. *Annu. Rev. Plant Physiol. Plant Mol. Biol.* 52:561–591.

Montfort, C. 1937. Die Trockenresistenz der Gezeitenpflanzen und die Frage der Übereinstimmung von Standort und Vegetation. *Ber. Deut. Bot. Ges.* 55:85–95.

Moradi, F. and Ismail, A. M. 2007. Responses of photosynthesis, chlorophyll fluorescence and ROS-scavenging systems to salt stress during seedling and reproductive stages in rice. *Ann. Bot.* 99:1161–1173.

Murata, N., Takahashi, S., Nishiyama, Y., and Allakhverdiev, S. I. 2007. Photoinhibition of photosystem II under environmental stress. *Biochim. Biophys. Acta* 1767:414–421.

Omasa, K. and Takayama, K. 2003. Simultaneous measurement of stomatal conductance, non-photochemical quenching, and photochemical yield of photosystem II in intact leaves by thermal and chlorophyll fluorescence imaging. *Plant Cell Physiol.* 44:1290–1300.

Omasa, K., Shimazaki, K.-I., Aiga, I., Larcher, W., and Onoe, M. 1987. Image analysis of chlorophyll fluorescence transients for diagnosing the photosynthetic system of attached leaves. *Plant Physiol.* 84:748–752.

Osto, L. D., Holt, N. E., Kaligotla, S. et al. 2012. Zeaxanthin protects plant photosynthesis by modulating chlorophyll triplet yield in specific light-harvesting antenna subunits. *J. Biol. Chem.* 287:41820–41834.

Oxborough, K. 2004. Imaging of chlorophyll a fluorescence: Theoretical and practical aspects of an emerging technique for the monitoring of photosynthetic performance. *J. Exp. Bot.* 55:1195–1205.

Paul, M. J. and Pellny, T. K. 2003. Carbon metabolite feedback regulation of leaf photosynthesis and development. *J. Exp. Bot.* 54:539–547.

Pérez-Bueno, M. L., Johnson, M. P., Zia, A., Ruban, A. V., and Horton, P. 2008. The Lhcb protein and xanthophyll composition of the light harvesting antenna controls the ΔpH-dependency of non-photochemical quenching in Arabidopsis thaliana. *FEBS Lett.* 582:1477–1482.

Pérez-Harguindeguy, N., Díaz, S., Garnier, E. et al. 2013. New handbook for standardised measurement of plant functional traits worldwide. *Aust. J. Bot.* 61:167–234.

Pfeiffer, S. and Krupinska, K. 2005. New insights in thylakoid membrane organization. *Plant Cell Physiol.* 46:1443–1451.

Pfündel, E. 1998. Estimating the contribution of photosystem I to total leaf chlorophyll fluorescence. *Photosynth. Res.* 56:185–195.

Preiss, M., Koopmann, E., Meyer, G., Koyro, H. W., and Schultz, G. 1994. Malate as additional substrate for fatty acid synthesis in a C_4-plant type developed by salt stress from a C_3-plant type maize. A screening for malate as substrate for fatty acid synthesis in chloroplasts. *J. Plant Physiol.* 143:544–549.

Richter, M. L., Hein, R. and Huchzermeyer, B. 2000. Important subunit interactions in the chloroplast ATP synthase. *Biochim. Biophys. Acta* 1458:326–342.

Richter, M. L., Samra, H. S., He, F., Giessel, A. J., and Kuczera, K. K. 2005. Coupling proton movement to ATP synthesis in the chloroplast ATP synthase. *J. Bioenerg. Biomembr.* 37:467–473.

Rochaix, J. D., Lemeille, S., Shapiguzov, A., Samol, I., Fucile, G., Willig, A., and Goldschmidt-Clermont, M. 2012. Protein kinases and phosphatases involved in the acclimation of the photosynthetic apparatus to a changing light environment. *Philos. Trans. R. Soc. Lond. B Biol. Sci.* 367:3466–3474.

Rosenqvist, E. and van Kooten, O. 2003. Chlorophyll fluorescence: A general description and nomenclature. In *Practical Applications of Chlorophyll Fluorescence in Plant Biology*, eds. J. R. DeEll and P. M. A. Toivonnen, 31–77. New York: Springer.

Ruban, A. V. and Horton, P. 1995. An investigation of the sustained component of nonphotochemical quenching of chlorophyll fluorescence in Isolated chloroplasts and leaves of spinach. *Plant Physiol.* 108:721–726.

Rutherford, A. W. and Krieger-Lizkay, A. 2001. Herbicide-induced oxidative stress in photosystem II. *Trends Biochem. Sci.* 26:648–653.

Sage, R. F. 2002. Variation in the k_{cat} of Rubisco in C_3 and C_4 plants and some implications for photosynthetic performance at high and low temperature. *J. Exp. Bot.* 53:609–620.

Sage, R. F., Sage, T. L., and Kocacinar, F. 2012. Photorespiration and the evolution of C_4 photosynthesis. *Annu. Rev. Plant Biol.* 63:19–47.

Salazar-Parra, C., Aguirreolea, J., Sanchez-Diaz, M., Irigoyen, J. J., and Morales, F. 2012. Climate change (elevated CO_2, elevated temperature and moderate drought) triggers the antioxidant enzymes' response of grapevine cv. Tempranillo, avoiding oxidative damage. *Physiol. Plant.* 144:99–110.

Scandalios, J. G. 1993. Oxygen stress and superoxide dismutases. *Plant Physiol.* 101:7–12.

Schmidt, J. 2014. Impact of the soil amendment biochar on the copper tolerance of *Chenopodium quinoa* Willd.—An ecotoxicological study. Unpublished master's thesis, Justus-Liebig-University, Giessen, Germany.

Schreiber, U. and Klughammer, C. 2008. Non-photochemical fluorescence quenching and quantum yields in PS I and PS II: Analysis of heat-induced limitations using Maxi-Imaging-PAM and Dual-PAM-100. *PAM Appl. Notes* 1:15–18.

Seki, M., Narusaka, M., Abe, H. et al. 2001. Monitoring the expression pattern of 1300 *Arabidopsis* genes under drought and cold stresses by using a full-length cDNA microarray. *Plant Cell* 13:61–72.

Shulaev, V., Diego Cortes, D., Miller, G., and Mittler, R. 2008. Metabolomics for plant stress response. *Physiol. Plant.* 132:199–208.

Souza, R. P., Machado, E. C., Silva, J. A. B., Lagôa, A. M. M. A., and Silveira, J. A. G. 2003. Photosynthetic gas exchange, chlorophyll fluorescence and some associated metabolic changes in cowpea (*Vigna unguiculata*) during water stress and recovery. *Environ. Exp. Bot.* 51:45–56.

Sperdouli, I. and Moustakas, M. 2012. Spatio-temporal heterogeneity in *Arabidopsis thaliana* leaves under drought stress. *Plant Biol.* 14:118–128.

Stepien, P. and Johnson, G. N. 2009. Contrasting responses of photosynthesis to salt stress in the glycophyte *Arabidopsis* and the halophyte *Thellungiella*: Role of the plastid terminal oxidase as an alternative electron sink. *Plant Physiol.* 149:1154–1165.

Susila, P., Lazar, D., Ilik, P., Tomek, P., and Naus, J. 2004. The gradient of exciting radiation within a sample affects relative heights of steps in the fast chlorophyll a fluorescence rise. *Photosynthetica* 42:161–172.

Süss, K.-H. Arkona, C., Manteuffel, R., and Adler, K. 1993. Calvin cycle multienzyme complexes are bound to chloroplast coupling factor CF_1. *Proc. Natl. Acad. Sci. U.S.A.* 90:5514–5518.

Suzuki, N., Koussevitzky, S., Mittler, R., and Miller, G. 2011. ROS and redox signalling in the response of plants to abiotic stress. *Plant Cell Environ.* 35:259–270.

Takizawa, K., Cruz, J. A., Kanazawa, A., and Kramer, D. M. 2007. The thylakoid proton motive force in vivo. Quantitative, non-invasive probes, energetics, and regulatory consequences of light-induced pmf. *Biochim. Biophys. Acta* 1767:1233–1244.

Tischer, W. and Strotmann, H. 1977. Relationship between inhibitor binding by chloroplasts and inhibiton of photosynthetic electron transport. *Biochim. Biophys. Acta* 460:113–125.

Trebst, A. 2007. Inhibitors in the functional dissection of the photosynthetic electron transport system. *Photosynth. Res.* 92:217–224.

Turina, P., Samoray, D., and Gräber, P. 2003. H+/ATP ratio of proton transport-coupled ATP synthesis and hydrolysis catalysed by CF0F1-liposomes. *EMBO J.* 22:418–426.

Vertuani, S., Angusti, A., and Manfredini, S. 2004. The antioxidants and pro-antioxidants network: An overview. *Curr. Pharm. Des.* 10:1677–1694.

Waisel, Y. 1972. *The Biology of Halophytes.* London: Academic Press.

Walters, R. G. and Horton, P. 1991. Resolution of components of non-photochemical fluorescence quenching in barley leaves. *Photosynth. Res.* 27:121–133.

Wedel, N., Soll, J., and Paap, B. 1997. CP12 provides a new mode of light regulation of Calvin cycle actiovitiy in higher plants. *Proc. Natl. Acad. Sci. U.S.A.* 94:10479–10484.

Weese, A., Papenbrock, J., and Riemenschneider, A. 2015. *Brassica napus* L. varieties show a broad variability in their morphology, physiology and metabolite levels in response to sulfur limitations and to pathogen attack. *Front. Plant Sci.* 6:9.

White, A. J. and Critchley, C. 1999. Rapid light curves: A new fluorescence method to assess the state of the photosynthetic apparatus. *Photosynth. Res.* 59:63–72.

Wingler, A., Lea, P. J., Quick, W. P., and Leegood, R. C. 2000. Photorespiration: Metabolic pathways and their role in stress protection. *Phil. Trans. R. Soc. Lond.* 355:1517–1529.

Wong, S.-L., Chen, C.-W. Huang, H.-W., and Weng, J.-H. 2012. Using combined measurements of gas exchange and chlorophyll fluorescence to investigate the photosynthetic light responses of plant species adapted to different light regimes. *Photosynthetica* 50:206–214.

Woo, K. C., Anderson, J. M., Boardman, N. K., Downton, W. J. S., Osmond, C. B., and Thorne, S. W. 1970. Deficient photosystem II in agranal bundle sheath chloroplasts of C_4 plants. *Proc. Natl. Acad. Sci. U.S.A.* 67:18–25.

Woo, N. S., Badger, M. R., and Pogson, B. J. 2008. A rapid, non-invasive procedure for quantitative assessment of drought survival using chlorophyll fluorescence. *Plant Methods* 4:27.

Zaks, J., Amarnath, K., Kramer, D. M., Niyogi, K. K., and Fleming, G. R. 2012. A kinetic model of rapidly reversible nonphotochemical quenching. *Proc. Natl. Acad. Sci. U.S.A.* 109:15757–15762.

Zhang, N. and Portis, A. R. Jr. 1999. Mechanism of light regulation of Rubisco: A specific role for the larger Rubisco activase isoform involving reductive activation by thioredoxin-f. *Proc. Natl. Acad. Sci. U.S.A.* 96:9438–9443.

Zhang, S. B., Hu, H., Xu, K., Li, Z. R., and Yang, Y. P. 2007. Flexible and reversible responses to different irradiance levels during photosynthetic acclimation of *Cypripedium guttatum*. *J. Plant Physiol.* 164:611–620.

Zhang, H., Han, B., Wang, T., Chen, S., Li, H., Zhang, Y., and Dai, S. 2012. Mechanisms of plant salt response: insights from proteomics. *J. Proteome Res.* 11:49–67.

26 Forage Crops and Their Photosynthesis

M. Anowarul Islam and Albert T. Adjesiwor

CONTENTS

26.1 Introduction 465
26.2 Photosynthetic Pathways and Physiology in Forage Crops 466
26.3 Photosynthetic Pigments and Their Roles in Forage Productivity and Quality 466
26.4 Factors Influencing the Photosynthetic Process 468
26.5 Management Considerations to Improve Photosynthesis Efficiency and Forage Productivity ... 469
    26.5.1 Grazing and Haying Management 469
    26.5.2 Irrigation and Moisture Management 470
    26.5.3 Nitrogen Management 471
    26.5.4 Forage Mixtures 471
26.6 Conclusions 471
References 471

26.1 INTRODUCTION

Forages and grassland agriculture play an important role in the economy of the nation and worldwide. Photosynthesis affects plant growth and productivity, forage quality, and stand persistence. In the presence of water and carbon dioxide (CO_2), plants are able to convert solar energy into usable chemical energy through a series of reactions, collectively called photosynthesis. Photosynthesis forms the basis of life on earth. Carbon fixation serves as the primary source of carbon for soil microorganisms and macroorganisms, and higher animals including humans. Not only does photosynthesis provide the needed carbon, but also, the oxygen produced as a result of plant photosynthesis is vital to our survival on this planet. Forage plants, just like other plants, fix the necessary carbon, which serves as food source for both wildlife and livestock. Humans are the ultimate beneficiaries of photosynthesis. Everything from the cloth we wear, the oxygen we breathe, the milk we drink, to the steak we enjoy is a product indirectly obtained from the photosynthesis of plants, including forages. Owing to the critical role played by photosynthesis to our existence on this planet, this subject has received much attention.

Forage crops, defined by the United States Department of Agriculture (USDA) as "plants grown for haying or grazing," play a significant role in the livestock industry and thus are a major contributor to the economy of many countries. In 2014, more than 126 million metric tons of hay was harvested in the United States alone, accounting for more than US$19 billion in terms of value (USDA-NASS 2015). Forage crops are often grouped into grasses, legumes, forbs (herbaceous nonlegume broadleaves), and grasslike plants. They are grown either in monocultures or in mixtures. Growing forages in mixtures often requires careful selection of species, mainly because shading of one species by the other often reduces light interception and, thus, photosynthesis and dry matter production are greatly compromised.

Photosynthesis in forages is becoming increasingly important because the options for increasing yield in agricultural crops are limited. Therefore, understanding and exploiting the physiology of photosynthesis in forages presents a major prospect for increasing yield (Morgan et al. 2002). However, improved crop yields through manipulation of photosynthetic process require understanding of the control mechanisms (Paul and Foyer 2001). There are reports that the changing global climate in the form of elevating CO_2 levels and increasing temperature will have profound effects on agricultural crops, including forage. For example, the projected increase in atmospheric CO_2 content is thought to impact photosynthesis positively (Allen and Prasad 2004), and rising temperatures favor warm-weather crops in cold regions (Lara and Andreo 2011). However, plants with different photosynthetic pathways respond differently to levels of CO_2. Photosynthesis in C_3 plants will be improved more than that of C_4 plants with rising levels of CO_2 because of CO_2 saturation in the photosynthetic pathway of C4 plants (Allen and Prasad 2004). Understanding the photosynthetic systems of forages and how they respond to environmental cues will be important in exploiting strategies to improve yield and quality for forages. Additionally, grasslands play a very important role in the global carbon cycle, and this necessitates steps to better understand how photosynthetic pathways will respond to some projected environmental cues (Ghannoum et al. 2000). This chapter is focused on photosynthetic processes of forages and factors that affect photosynthesis in forages. Additionally, management options that affect photosynthesis and options to exploit these factors for improved photosynthesis and biomass production in forages are discussed.

26.2 PHOTOSYNTHETIC PATHWAYS AND PHYSIOLOGY IN FORAGE CROPS

Scientists have long anticipated that photosynthesis in higher plants involves a series of processes. This series of processes was confirmed by various researchers in the 1960s (Ranson and Thomas 1960; Calvin and Bassham 1962; Kortschalk et al. 1965; Hatch and Slack 1966) and early 1970s (Downton 1970). Plants are able to assimilate carbon through either the C_3, C_4, or crassulacean acid metabolism (CAM) photosynthetic pathways. Evolution is thought of as the main factor producing the major photosynthetic variants in plants (Ehleringer et al. 1991; Connor et al. 2011). Plants have evolved from algae with the C_3 photosynthetic pathway (Connor et al. 2011), making C_3 plants the most dominant (about 85–95%) among all plant species (Ehleringer et al. 1991; Ashraf and Harris 2013). Forage plants predominantly utilize either the C_3 or C_4 pathways. However, there is evidence that some grasses utilize both C_3 and C_4 pathways. A survey of the main grasses in North America showed that within the *Panicum* genera of the Panicoidae, the subgenus *Dichanthelium* exhibits the C_3 pathway, while some plants in the *Eupanicum* subgenus have both C_3 and C_4 pathways (Waller and Lewis 1979). There is no known major forage plant exhibiting CAM photosynthesis. However, identifying or breeding for forage plants with the CAM pathway will be a major step toward extreme drought resistance because plants exhibiting CAM are well adapted to xeric conditions.

Plants exhibiting the C_4 pathway differ from C_3 types in a variety of ways (Table 26.1). However, the key difference is the type of acid formed at the initial process of CO_2 fixation. In C_3 forages, a five-carbon sugar, ribulose bisphosphate (RuBP), accepts CO_2 in the presence of the photosynthetic enzyme called rubisco to form an unstable product, which then splits, forming a three-carbon phosphoglyceric acid (PGA) (Connor et al. 2011). In C_4 forages, the enzyme phosphoenolpyruvate (PEP) mediates the CO_2 fixation process to form a four-carbon acid (oxalic acid) (Volenec and Nelson 2007; Connor et al. 2011). Cool-season forage grasses, some forbs, and all legumes exhibit the C_3 photosynthetic pathway (Volenec and Nelson 2007). Examples of C_3 forages include alfalfa (*Medicago sativa* L.), clovers (*Trifolium* spp.), bromegrasses (*Bromus* spp.), timothy (*Phleum pratense* L.), orchard grass (*Dactylis glomerata* L.), tall fescue (*Schedonorus arundinaceus* [Schreb.] Dumort., nom. cons.), crested wheatgrass (*Agropyron cristatum* [L.] Gaertn.), reed canarygrass (*Phalaries arundinacea* L.), and ryegrass (*Lolium perenne* L.). Warm-season grasses (C_4) include panic grass (*Panicum amarum* Ell.), switchgrass (*Panicum virgatum* L.), Rhodes grass (*Chloris gayana* Kunth), Bahia grass (*Paspalum notatum* Flüggé), Indian grass (*Sorgastrum nutans* [L.] Nash), Bermuda grass (*Cynodon dactylon* [L.] Pers.), big bluestem (*Andropogon gerardii* Vitman), maize (*Zea mays* L.), and sorghum (*Sorghum bicolor* L.).

26.3 PHOTOSYNTHETIC PIGMENTS AND THEIR ROLES IN FORAGE PRODUCTIVITY AND QUALITY

Photosynthetic pigments are substances with varying chemical structures occurring as porphyrin pigments, carotenoids, anthocyanins, and flavones (Costache et al. 2012). In higher plants, including forages, the main photosynthetic pigments can be grouped into *chlorophylls* and *carotenoids* (Boyer 1990). Both pigments occur in the *chloroplast* and are bound to proteins in the *thylakoids* (Boyer 1990). Several attempts have been made in the past to identify the major forms of chlorophylls. Stokes (1864) performed spectrophotometric measurements as far back as the 1860s in which he speculated that green plants may have more than one type of chlorophyll. In an attempt to separate the different types of chlorophylls alluded to in previous studies, Sorby (1873)

TABLE 26.1
Physiological and Agronomic Performance That Separates C_3 and C_4 Metabolism in Forages

Factor	C_3 Forages	C_4 Forages	Reference(s)
Leaf anatomy	Lack well-defined parenchymatic bundle sheaths, starch grains mainly within mesophyll	Well-developed parenchymatic bundle sheaths with high concentration of starch and chloroplast	Bisalputra et al. (1969)
CO_2 compensation	High (37–70 ppm)	Low (0–10 ppm)	Downton and Tregunna (1968); Black (1971)
Light saturation	CO_2 uptake becomes saturated at 0.838×10^4 to 1.676×10^4 J/m²	Continued CO_2 uptake to nearly full sunlight (6.285×10^4 to 7.542×10^4 J/m²)	Waller and Lewis (1979)
Optimum temperature for CO_2 uptake	10–25°C	30–40°C	Black (1971)
Photosynthetic rate	Relatively slow	Twofold to threefold greater	Waller and Lewis (1979)
Carbon loss through photorespiration	High (23–30%)	No net loss	Volenec and Nelson (2007)
Photosynthetic efficiency	Relatively low	Relatively high	Waller and Lewis (1979)
Translocation rate	Relatively slow	Relatively rapid	Volenec and Nelson (2007)
Water use efficiency	Low, due to high resistance to CO_2	High, due to low resistance to CO_2	Volenec and Nelson (2007)
Forage digestibility	Relatively rapid	Relatively slow	Moser and Jennings (2007)

used several solvents to partition the chlorophylls extracted from olive algae, *Fucus* or *Laminaria*. He found that the algae contain the *blue chlorophyll*, which is now known to be chlorophyll *a*, but lack the *yellow chlorophyll*, which is now known as chlorophyll *b*. In addition to the *a* and *b* chlorophylls, a third chlorophyll was present in the algae, which Sorby called *chlorofuscine*, now called chlorophyll c_1 and c_2. Recent studies have confirmed that chlorophyll exists as *a*, *b*, or *c* (Costache et al. 2012; Sumanata et al. 2014). In addition to leaves, stems, petioles, leaf sheaths, as well as some green pods contain photosynthetic pigments and thus are able to intercept light energy and photosynthesize. However, leaves remain the main organs for photosynthesis and forage production (Misra and Misra 1981).

Pigments are present in various types and amounts. Though the type and amount of pigments present in plants vary based on the type of species (Costache et al. 2012) and environment (Sumanata et al. 2014), chlorophyll *a* and *b* are the most abundant form of pigments in plants (Boyer 1990). The main difference between chlorophyll *a* and *b* is the presence of a methyl group on position 3 (ring 3) of chlorophyll *a*, whereas chlorophyll *b* has an aldehyde group. Chlorophyll *a* and *b* are present in a ratio (*a*/*b*) of approximately 3:1 (Boyer 1990). A decrease in the chlorophyll *a*/*b* ratio is a key change

that occurs during leaf senescence. Additionally, a change in *a*/*b* ratio occurs when forage plants are growing under shade. The ratio is therefore a good indication of the ability of forage plants to survive shading. For example, hoary tick clover (*Desmodium canescens* L.) and crown vetch (*Coronilla varia* L.) are shade tolerant, but switchgrass (*P. virgatum* L.) and prairie cordgrass (*Spartina pectinata* Bosc ex Link) are shade intolerant (Van Sambeek et al. 2007). Aside from the important role chlorophylls play in photosynthesis and forage productivity, they have important bearing on the quality of forages, which in turn determines animal preference. To this end, attempts have been made to evaluate the effect of *greenness* of forage on grazing preferences. Both wildlife and domestic livestock were found to prefer green *Panicum maximum* Jacq. (Treydte et al. 2013). Since greenness is an indication of the approximate concentration of chlorophyll, animal preference for green forage shows that chlorophyll concentration has a relationship with forage quality. A recent study by Christianson and Creel (2015) has shown that the concentration of photosynthetic pigments was highly correlated with forage nitrogen, digestibility, and neutral detergent fiber (Table 26.2). Since protein is predominantly made up of nitrogen, the results by Christianson and Creel (2015) are an indication that photosynthetic pigment concentration can be used

TABLE 26.2
Correlations (*r*) among Photosynthetic Pigments and Physicochemical Components of Graminoid Samples Commonly Consumed by Elk (*Cervus elaphus*) in Winter and Spring and All Forage Types across Both Seasons

Forage Component	Chlorophyll	Carotenoids	Nitrogen	NDF	GE
Winter[a]					
Carotenoids	**0.969** ($F_{[1,103]} = 1595$)				
Nitrogen	**0.722** ($F_{[1,103]} = 112.2$)	**0.726** ($F_{[1,103]} = 114.7$)			
NDF	**−0.694** ($F_{[1,103]} = 95.74$)	**−0.706** ($F_{[1,103]} = 95.74$)	**−0.642** ($F_{[1,106]} = 74.28$)		
GE	−0.04 ($F_{[1,102]} = 0.17$)	−0.04 ($F_{[1,102]} = 0.17$)	−0.070 ($F_{[1,105]} = 0.51$)	**0.214** ($F_{[1,105]} = 5.02$)	
IVDMD	**0.231** ($F_{[1,103]} = 5.79$)	**0.231** ($F_{[1,103]} = 5.79$)	**0.355** ($F_{[1,106]} = 15.25$)	−0.105 ($F_{[1,106]} = 1.19$)	0.150 ($F_{[1,105]} = 2.41$)
Spring[b]					
Carotenoids	**0.988** ($F_{[1,16]} = 657.5$)				
Nitrogen	**0.974** ($F_{[1,14]} = 257.9$)	**0.955** ($F_{[1,14]} = 143.9$)			
NDF	**−0.85** ($F_{[1,13]} = 33.76$)	**−0.841** ($F_{[1,13]} = 31.38$)	**−0.762** ($F_{[1,18]} = 24.95$)		
GE	**0.608** ($F_{[1,12]} = 7.05$)	**0.622** ($F_{[1,12]} = 7.57$)	0.282 ($F_{[1,17]} = 1.47$)	0.013 ($F_{[1,17]} = 0.00$)	
IVDMD	**0.876** ($F_{[1,14]} = 46.03$)	**0.863** ($F_{[1,14]} = 40.91$)	**0.768** ($F_{[1,19]} = 27.37$)	**0.768** ($F_{[1,18]} = 41.84$)	0.134 ($F_{[1,17]} = 0.31$)
All Forages[c]					
Carotenoids	**0.908** ($F_{[1,199]} = 934.4$)				
Nitrogen	**0.596** ($F_{[1,197]} = 108.6$)	**0.535** ($F_{[1,197]} = 78.96$)			
NDF	**−0.486** ($F_{[1,195]} = 60.14$)	**−0.625** ($F_{[1,195]} = 124.8$)	**−0.403** ($F_{[1,234]} = 45.45$)		
GE	**0.307** ($F_{[1,193]} = 20.11$)	**0.506** ($F_{[1,193]} = 66.34$)	0.126 ($F_{[1,228]} = 3.69$)	**−0.704** ($F_{[1,228]} = 223.6$)	
IVDMD	**0.311** ($F_{[1,197]} = 21.03$)	**0.312** ($F_{[1,197]} = 21.26$)	**0.456** ($F_{[1,237]} = 62.27$)	**−0.384** ($F_{[1,234]} = 40.51$)	**0.155** ($F_{[1,228]} = 5.61$)

Source: Adapted from Christianson, D., and S. Creel, *Can. J. Zool.*, 93, 51–59, 2015. With permission.

Note: Significant (*P* < 0.05) correlations are shown in boldface. GE, gross energy; IVDMD, *in vitro* dry matter digestibility; NDF, neutral detergent fiber.

[a] Winter graminoids.
[b] Spring grazing patches.
[c] Both winter and spring.

to predict the crude protein content of forages, an important quality parameter. This effect of greenness on forage quality might explain the seasonality in the conception and birth timings in wildlife. Birth timing in the African buffalo (*Syncerus caffer*) was found to be highly correlated with greenness of vegetation (Ryan et al. 2007). The authors elucidated that the improved protein levels, which normally occur 1 month after the first green flush following the onset of rains, might be a trigger for conception in the African buffalo.

26.4 FACTORS INFLUENCING THE PHOTOSYNTHETIC PROCESS

Photosynthesis, just like any other biological process, is influenced by a large number of factors. This is often the case because enzymes are involved in the process, and thus, factors affecting the enzymes will definitely affect the process. Photosynthesis is one of the main plant processes that are very sensitive to heat damage (Seemann et al. 1984). To this end, photosynthetic response and adaptation of plants to temperature have received enormous scientific investigation and attention (Berry and Björkman 1980; Badger et al. 1982; Seemann et al. 1984; Yamasaki et al. 2002; Hikosaka et al. 2006; Yamori et al. 2014). The effect of temperature on photosynthesis is sigmoidal. This sigmoidal relationship is due to the optimum temperature requirements of enzymes (e.g., rubisco and PEP) that mediate the CO_2 fixation process. The level of rubisco was reported to decline with increasing temperature (Yamasaki et al. 2002). Photosynthetic dependence on temperature is also influenced by factors such as the intercellular CO_2 pressure and light intensity (Berry and Björkman 1980) or activation energy of ribulose-1,5-bisphosphate (RuBP) (Hikosaka et al. 2006). At increased intercellular CO_2 pressure or light intensity, the effect of temperature on photosynthesis is more pronounced (Berry and Björkman 1980). Hikosaka et al. (2006) argued that shifts in optimum temperature requirements for photosynthesis at ambient CO_2 concentration are more influenced by the activation energy of the maximum rate of RuBP carboxylation. Ambient temperature has a profound effect on photosynthesis in forages and plants in general. However, forage plants differ in their tolerance to temperature. The C_4 forage species are more adapted to warm conditions compared to their C_3 counterparts.

The availability of photosynthetically active radiation (PAR) is of utmost importance in plant photosynthesis. Apparent photosynthesis in *L. perenne* was found to have increased with increasing light intensity (Wilson and Cooper 1969). The extent of the effect of light intensity on forages differs from one photosynthetic pathway to the other. The C_3 plants thrive relatively well under low light intensity compared to C_4 plants (Ashraf and Harris 2013). Shade is particularly important in forages. This is because most forages are grown in mixtures rather than in monocultures. There is overwhelming evidence that grass–legume forage mixtures often result in transgressive overyielding (Berdahl et al. 2001; Mwangi et al. 2004; Nyfeler et al. 2009; Giambalvo et al. 2011; Sanderson et al. 2013; Sturludóttir et al. 2014). However,

FIGURE 26.1 (See color insert.) Grass–legume mixture study at the University of Wyoming Sheridan Research and Extension Center, Sheridan, Wyoming, in 2014. Different mixtures of forage legumes and grass were used. Forage species includes alfalfa (*Medicago sativa* L.), bird's-foot trefoil (*Lotus corniculatus* L.), sainfoin (*Onobrychis viciifolia* Scop.), and meadow bromegrass (*Bromus biebersteinii* Roem. & Schult.). Forage production and forage quality varied among mixtures. (Picture taken by M. Anowarul Islam.)

this yield increase is only possible when the species grown are compatible (Figure 26.1). In other words, the species must be able to tolerate shade enough so as to produce appreciable dry matter. Most often than not, some forages (e.g., *P. virgatum* and *S. pectinata*) are described as shade intolerant. This is a principal allusion to the fact that their photosynthetic rates are greatly reduced when grown in mixtures. This phenomenon makes species selection an important consideration in the establishment of mixtures. Paul and Foyer (2001) opined that under shade conditions, leaves modify their chloroplast protein as well as the photosynthetic pigments in order to optimize light interception and light use efficiency. Within a single growing season, growth in the form of tillering within grass plant canopies turns sun leaves to shade leaves, and this places a morphological constraint on the response of grasses to light (Wang et al. 2012). It thus appears that the ability of forage plants to adapt these physiological and morphological strategies will determine their productivity when grown in a mixture.

The CO_2 concentration in the atmosphere has a profound effect on photosynthesis. Therefore, understanding the effects of projected increases in CO_2 concentration on photosynthetic rates in C_3 and C_4 forages is essential for appreciating future carbon cycles and forage productivity. Theoretically, as the atmospheric CO_2 concentration increases, the net photosynthesis is expected to increase because of reduced photorespiration (Lara and Andreo 2011). Though it is generally expected that the rising CO_2 concentration in the atmosphere will increase photosynthesis in C_3 plants more than that of C_4 (Lara and Andreo 2011), the extent of this stimulation varies with temperature and among species and cultures (Allen and Prasad 2004). In a meta-analysis, it was found that an increase in CO_2 combined with an 18% increase in the leaf area of soybean resulted in a 59% increase in canopy photosynthetic rate (Ainsworth et al. 2002). A long-term (10 years) study investigating the effects of elevated CO_2 on photosynthetic rate of *L. perenne*, a C_3 perennial forage grass, however, showed that the projected increases in yield in C_3 plants could not be realized. Makino and Mae (1999) elucidated that prolonged exposure to elevated CO_2 levels reduces initial photosynthesis stimulation in many plant species and, in many cases, results in decreased photosynthesis. These findings cast doubts on the prolonged effects the projected increase in elevated CO_2 will have on forages. In a short-term study, Leakey et al. (2006) found that in the absence of moisture stress, elevated CO_2 had no effect on the photosynthesis of *Z. mays*, a C_4 grass. From the foregoing, it is apparent that the photosynthetic response of species to elevated CO_2 varies, and these responses are often dependent on other environmental factors such as moisture and temperature. This necessitates research focused on specific important forage crops and taking into account as many environmental factors as possible for a clear understanding of future effects of atmospheric CO_2 elevation on forages.

Water, being a raw material for photosynthesis, affects the process considerably in all phases (Ashraf and Harris 2013). Under water-stressed conditions, dehydration of the mesophyll cells and disruption of photosynthetic pigments reduce plant photosynthesis (Anjum et al. 2011). Reduction in leaf area, an important factor in light interception, often occurs during moisture-stressed conditions and this reduces photosynthetic productivity (Nogués and Baker 2000). The photosynthetic advantage of C_4 grasses over C_3 is often lost under moisture-stressed conditions (Taylor et al. 2011). The chlorophyll content of leaves generally declines under moisture-stressed conditions (Singer et al. 2003). This often results in a reduction in the photosynthetic rate of forages.

Both quality and quantity of light are important factors regulating the photosynthetic process. Therefore, the potential of dry matter production in forage crops depends on their ability to respond to variations in light energy (Eagles and Östgård 1971). It is generally expected that all things being equal, the longer the exposure of plants to quality light, the higher the photosynthesis and carbon accumulation. However, this response is dependent on several factors, especially temperature (Heide et al. 1985a) and leaf expansion pattern and arrangement (Eagles and Östgård 1971). Longer photoperiods have been found to enhance photosynthesis and dry matter accumulation in forage crops. For example, Eagles and Östgård (1971) reported higher relative growth rates (RGRs) in *D. glomerata* under 24 h light compared to 8 h when temperatures were moderately low (5°C and 15°C). However, at higher temperatures, longer photoperiods had a negative effect on dry matter production (Eagles and Östgård 1971). In *Poa pratensis* L., Heide et al. (1985a) observed that the main increases in RGR occurred with a photoperiod of 14–18 h at temperatures of 9°C and 21°C. Similar results were obtained when *P. pratensis* and *Bromus inermis* Leyss were exposed to continuous light at low temperatures (≤15°C) (Heide et al. 1985b). Although this could be due to adaptation

of *P. pratensis* and *B. inermis* to high latitude and cool temperatures (because they are C_3 grasses), it appears that the response to the photoperiod is intricately linked to temperature. Heide et al. (1985b), however, explained that increases in RGR were caused mainly by higher leaf area ratios at longer photoperiods and not directly net photosynthesis.

Carbohydrate (CHO) reserves are nonstructural carbohydrates (NSC) present predominantly as sucrose and fructosan in temperate grasses and as sucrose and starch in tropical grasses (White 1973). They are particularly important in the production of new leaves after forage grazing or clipping. This is because after forage leaves have been harvested, active photosynthesis is very low, and plants have to rely heavily on the stored CHO for growth and survival. CHO reserves are important not only in the production of photosynthetic tissues but also in respiration. A direct relationship between CHO reserves and respiration rate has been reported (Azcón-Bieto et al. 1983). This direct relationship is an indication that the rate of CO_2 uptake might be affected by the level of NSC. Azcón-Bieto et al. (1983) concluded in a study that the rate of CO_2 uptake is highly dependent on the level of respiratory substrates in leaves. The theory of the role of accumulation of photosynthates on the rate of photosynthesis has been put forward over a century ago (Boussingault 1868). The theory seeks to explain that the rate of photosynthesis is a function of the balance between carbon fixation (photosynthesis) and the rate of use. It is therefore logical to assume that as the level of CHO reserves increases, negative feedback will eventually reduce the rate of photosynthesis. Downregulation of photosynthesis as a result of photosynthate accumulation has been reported in several other crops (Paul and Driscoll 1997; Paul and Pellny 2003; Nebauer et al. 2011). Paul and Foyer (2001) elucidated that in the presence of an imbalance in the CHO content of sources and sinks, accumulation of CHO in leaves can reduce the expression of photosynthetic genes and ultimately lead to accelerated leaf senescence. This phenomenon is of utmost importance in forages where fresh plant tissues are desired to obtain high-quality hay.

26.5 MANAGEMENT CONSIDERATIONS TO IMPROVE PHOTOSYNTHESIS EFFICIENCY AND FORAGE PRODUCTIVITY

The efficiency of photosynthesis and improvement of forage productivity can be enhanced by improved management practices. Better understanding and improved knowledge of photosynthesis processes can be useful in realizing how forage crops grow in order to manage forage crops for optimum productivity and greater stand persistence. This information is very useful for forage managers and livestock producers to manage pastures and hay for improved forage productivity and increased animal performance.

26.5.1 GRAZING AND HAYING MANAGEMENT

Grazing management affects the photosynthesis of forage crops and grasslands through the removal of photosynthetic tissues (mostly leaves and green stems). Grazing or clipping too close to the soil surface, in other words, overremoval of photosynthetic tissues, reduces photosynthesis. If this comes at a time when environmental conditions are not favorable, for example, close (<4–6 weeks) to winter months, stands may be damaged or lost. Grazing or haying must be done to allow plants enough time to accumulate photosynthates (NSC) in their organs (mainly roots) for stand survival and prolonged productivity. There are, however, contrasting reports on the effects of herbivory and/or defoliation on productivity of forage grasslands. Some stimulatory (Georgiadis et al. 1989) as well as negative (Ferraro and Oesterheld 2002; Chen et al. 2005) effects have been reported. According to Noy-Meir (1993), for grazing to have an immediate negative effect on subsequent growth, the following conditions must exist:

1. Reduction of photosynthetic leaf area
2. Removal of active apical meristems that act as sinks for photosynthates and produce new shoot tissues
3. Loss of nutrients for growth stored in the shoot

Noy-Meir (1993) also explained that for grazing to have a stimulatory effect on growth, the following mechanisms might be involved:

1. Increase in light intensity to remaining leaves and thus in photosynthetic rate per unit available leaf area and mass, and in the ratio of photosynthesis to respiration and senescence
2. Improved supply of water and nutrients to remaining leaves, resulting in increased growth rate or longer growing period
3. Increased photosynthetic efficiency and reduced senescence of young regrowth leaves compared with older removed leaves
4. Increased amount of photosynthates allocated to newer regrowth and actively photosynthesizing leaves at the expense of nonphotosynthetic roots, stems, and sometimes inflorescences
5. Activation of dormant meristems in response to removal of apical meristems and/or modified light conditions, increasing the number of sinks for vegetative and reproductive growth

It is apparent from the assessments of Noy-Meir (1993) that the stimulatory effect of grazing on grasslands is dependent on numerous factors that are often nonexistent. However, undamaged leaves remaining after defoliation of *Solidago altissima* L., a native perennial forb in the United States, had higher specific leaf area, and mass-based photosynthesis was stimulated (Meyer 1998). Because the photosynthetic capacity of leaves declines with age, it is rational to assume that removal of old leaves during grazing can increase the rate of photosynthesis as a result of the active participation of younger leaves. Following severe defoliation, in rotationally grazed swards, photosynthetic rate and rate of production of new leaves increased rapidly (Parsons and Penning 1988). This shows that grazed

swards are reliant on CHO reserves to produce photosynthetic tissues. Consequently, shortening the regrowth period reduces the amount of dry matter available for harvest (Parsons and Penning 1988). Parsons and Penning (1988) explain that severe defoliation in itself is not as harmful to swards that are rotationally grazed, but steps must be taken to allow for enough regrowth period. This is an allusion to the fact that the timing of the severe defoliation might also be an issue. Grazing must be done at a time of the season when plants have sufficient growing days to enable accumulation of NSC reserves. Height at grazing of swards also affects net productivity. Dry matter productivity per unit area declined in swards below 2.5 cm in height above the soil surface (Grant et al. 1983).

Grazing was found to have altered the effects of temperature on leaf photosynthesis (Shen et al. 2013). Although Shen et al. (2013) attributed this effect to modification of the physical environment of grassland plants, their findings suggest that grazing effects on photosynthesis should be looked at critically in the wake of the warming climate. Another problem with heavy defoliation in pastures that is often overlooked is the dominance of weeds. In the presence of weeds, overgrazing of pastures might give weeds a competitive advantage because weeds tend to either avoid or tolerate herbivory (Olson and Wallender 1999). The productivity of pastures can be greatly reduced, or they can even be rendered unusable, due to the dominance of weeds and shading of desirable species, which reduces their photosynthetic activity.

26.5.2 Irrigation and Moisture Management

Owing to the important role water plays in the photosynthetic process, irrigation management is important in enhancing photosynthesis. A deficit in moisture often leads to stomatal closure, an adaptation to reduce water loss. However, this adaptation affects CO_2 intake and, consequently, photosynthesis. High-moisture-stress conditions can cause loss of chlorophyll in plants (Nayyar and Gupta 2006; Hussein et al. 2013), but the extent of damage might be higher in C_3 plants compared to C_4 (Nayyar and Gupta 2006). Loss of chlorophyll has dire consequences on photosynthesis because it is the main pigment for photosynthesis. Irrigation modifies the microclimate of plants. For example, crop canopy temperature changes during irrigation (Cavero et al. 2009). Changes in leaf temperatures could have negative or stimulatory effects on photosynthesis. If leaf temperatures were too high prior to irrigation, lower canopy temperatures (than optimum) emanating from irrigation could stimulate photosynthesis. However, reduction in temperatures below optimum requirements could have negative effects on photosynthesis (Urrego-Pereira et al. 2013a).

The effect of irrigation on canopy temperature is an indication that cool- (C_3) and warm-season (C_4) forages will have differential responses. Sprinkler irrigation was found to have decreased net photosynthesis in *Z. mays* on 80% of the days measurements were taken, and this translated into 19% average reduction in net photosynthesis. However, *M. sativa* photosynthesis was almost unaffected by the sprinkler irrigation (Urrego-Pereira et al. 2013a). Although the researchers attribute the differential response to differences in leaf wettability of *M. sativa* and *Z. mays*, and the differences in temperature adaptations of the two plant species, it appears that the wettability of leaves might have influenced CO_2 diffusion rates. It must be noted that overhead irrigation might also influence CO_2 diffusion into leaves. This is because the rate of diffusion of CO_2 is 10,000 times slower in water than in air (Evans and Caemmerer 1996). Thus, for forages having pubescent and hydrophilic leaf surfaces (e.g., *Bromus biebersteinii* Roem. & Schult.), overhead irrigation can reduce the rate of diffusion of CO_2, especially during irrigation. However, transpiration can also be reduced during irrigation, which has a positive effect of photosynthesis. Up to a 36% decrease in transpiration was observed in *Z. mays* canopies subjected to sprinkler irrigation (Urrego-Pereira et al. 2013b).

26.5.3 NITROGEN MANAGEMENT

Nutrient management is one of the major recommended agronomic practices in crop production. Among the essential macronutrient and micronutrients in forage production, nitrogen is the most limiting and expensive. Probably one of the major known deficiency symptoms of nitrogen is yellowing of leaves. This perhaps is enough justification of the important role nitrogen plays in the formation of photosynthetic pigments (mainly chlorophyll) and the effects nitrogen deficiency might have on photosynthesis (Boussadia et al. 2009) and yield of forages, especially grasses. Evans (1989) opined that the photosynthetic capacity of leaves has a close relationship with nitrogen content. This relationship observed by Evans (1989) could be because the proteins in the Calvin cycle and *thylakoids*, which are essential to photosynthesis, are major components of leaf nitrogen.

Owing to the important effect of nitrogen on photosynthesis and biomass production in forages, grasses are often planted in mixtures with N_2-fixing legumes to supply the needed nitrogen. However, legumes must be inoculated with the right *Rhizobium* strain to ensure adequate N_2 fixation. The amount of nitrogen fixed by legumes differs based on the legume species, compatibility of the plants in the mixtures, and several other factors beyond the scope of this chapter. Therefore, N_2 fixation and transfer potential must be taken into account when using legumes as a nitrogen source to ensure that nitrogen deficiency does not occur. Synthetic nitrogen fertilizers are often used as the sole source of nitrogen for plants or as supplements to N_2-fixing legumes in forage production systems in order to supply the needed amount of nitrogen for photosynthesis and dry matter production. Nitrogen application to a stay-green *Z. mays* increased the net photosynthesis rate of leaves (Li et al. 2012). In *S. bicolor*, nitrogen deficiency reduced leaf area, chlorophyll, and net photosynthesis and consequently resulted in reduced biomass production (Zhao et al. 2005).

26.5.4 FORAGE MIXTURES

Forage mixtures, including grasslands, often contain two or more species. Mutual shading is thus common, and this often reduces the productivity of less shade-intolerant species. In pastures, mutual shading can reduce photosynthetic rates of species, especially toward the end of the regrowth period (Braga et al. 2008). Row spacing must be taken into consideration in establishing plant mixtures as this affects resource competition. It is common practice to seed mixtures above recommended rates to compensate for stand thinning, common in some forage species (e.g., *M. sativa*). However, this affects light interception, and thus, photosynthesis will very likely be reduced. *Bromus tectorum* L. plants grown in full sunlight produced higher amount of biomass, had a higher number of tillers and leaves, and partitioned more photosynthates to roots relative to plants growing in shade (Pierson et al. 1990). This response is not uncommon in desirable agricultural crops, including forages. The modification of light quality arising from neighboring plants often triggers *shade avoidance syndrome*, which often reduces yield (Page et al. 2010). The subject of shade avoidance is, however, beyond the scope of this discussion. Shading, which reduced photosynthetic photo flux to 30%, resulted in 57% reduction in yield of *Brachiaria decumbens* Stapf., a C_4 tropical forage grass (Gómez et al. 2012).

26.6 CONCLUSIONS

Photosynthesis is the primary source of carbon on earth. Forages fix carbon, which is used by animals and humans, who are the ultimate beneficiaries. Photosynthetic pigments are not only important in carbon fixation but also good indicators of forage quality. Similar to any biological process, photosynthesis in forages is affected by innumerable factors, spanning from limitations imposed by the environment to management issues. To enhance photosynthesis and forage productivity and stand persistence, management issues must be given priority since most environmental constraints are not easy to control. Grazing and haying management, adequate soil moisture, nutrient management (especially nitrogen), and careful selection of species to be grown in mixtures have the potential to improve carbon fixation and dry matter production in forage crops.

REFERENCES

Ainsworth, E.A., P.A. Davey, C.J. Bernacchi, O.C. Dermody, E.A. Heaton, D.J. Moore, P.B. Morgan, S.L. Naidu, H. Yoora, X. Zhu, P.S. Curtis, and S.P. Long. 2002. A meta-analysis of elevated [CO2] effects on soybean (*Glycine max*) physiology, growth and yield. *Global Change Biology* 8:695–709.

Allen, L.H. Jr. and P.V.V. Prasad. 2004. Crop responses to elevated carbon dioxide. In Goodman, R.M. *Encyclopedia of Plant and Crop Science*. Marcel Dekker, New York. pp. 346–348. CRC Press, Boca Raton, FL.

Anjum, S.A., X. Xie, L. Wang, M.F. Saleem, C. Man, and W. Lei. 2011. Morphological, physiological and biochemical responses of plants to drought stress. *African Journal of Agricultural Research* 9:2026–2032.

Ashraf, M. and P.J.C. Harris. 2013. Photosynthesis under stressful environments: An overview. *Photosynthetica* 51:163–190.

Azcón-Bieto, J., H. Lambers, and D.A. Day. 1983. Effect of photosynthesis and carbohydrate status on respiratory rates and the involvement of the alternative pathway in leaf respiration. *Plant Physiology* 72:598–603.

Badger, M.R., O. Björkman, and P.A. Armond. 1982. An analysis of photosynthetic response and adaptation to temperature in higher plants: Temperature acclimation in desert evergreen *Nerium oleander* L. *Plant, Cell and Environment* 5:85–99.

Berdahl, J.D., J.F. Karn, and J.R. Hendrickson. 2001. Dry matter yields of cool-season grass monocultures and grass–alfalfa binary mixtures. *Agronomy Journal* 93:463–467.

Berry, J. and O. Björkman. 1980. Photosynthetic response and adaptation to temperature in higher plants. *Annual Review of Plant Physiology* 31:491–543.

Bisalputra, T., W.J.S. Downton, and E.B. Tregunna. 1969. The distiction and ultrastructure of chloroplasts in leaves differing in photosynthetic carbon metabolism. I. Wheat, *Sorghum*, and *Aristida* (Gramineae). *Canadian Journal of Botany* 47:15–21.

Black, C.C. 1971. Ecological implications of dividing plants into groups with distinct photosynthetic production capacities. *Advances in Ecological Research* 7:87–114.

Boussadia, O., K. Steppe, H. Zgallai, S.B. El Hadj, M. Braham, R. Lemeur, and M.C. Van Labeke. 2009. Effects of nitrogen deficiency on leaf photosynthesis, carbohydrate status and biomass production in tow olive cultivars 'Meski' and 'Koroneiki'. *Scientia Horticulturae* 123:336–342.

Boussingault, J.B. 1868. *Agronomie, chimie agricole et physiologie.* Gauthier-Villars, Paris.

Boyer, R.F. 1990. Isolation and spectrophotometric characterization of photosynthetic pigments. *Biochemistry Education* 18:203–206.

Braga, G.J., C.G.S. Pedreira, V.R. Herling, P.H. de C. Luz, and C.G. de Lima. 2008. Herbage allowance effects on leaf photosynthesis and canopy light interception in palisade grass pastures under rotational stocking. *Tropical Grasslands* 42:214–223.

Calvin, M. and J.A. Bassham. 1962. *The photosynthesis of carbon compounds.* W.A. Benjamin, New York.

Cavero, J., E.T. Medina, M. Puig, and A. Martínez-Cob. 2009. Sprinkler irrigation changes maize canopy micro climate and crop water use status, transpiration, and temperature. *Agronomy Journal* 101:854–864.

Chen, S.P., Y.F. Bai, G.H. Lin, Y. Liang, and X.G. Han. 2005. Effects of grazing on photosynthetic characteristics of major steppe species in the Xilin Basin, Inner Mongolia, China. *Photosynthetica* 43:559–565.

Christianson, D. and S. Creel. 2015. Photosynthetic pigments estimate diet quality in forage and feces of elk (*Cervus elaphus*). *Canadian Journal of Zoology* 93:51–59.

Connor, D.J., R.S. Looms, and K.G. Cassman. 2011. *Crop ecology: Productivity and management in agricultural systems,* 2nd edition. Cambridge University Press, Cambridge.

Costache, M.A., G. Campeanu, and G. Neata. 2012. Studies concerning the extraction of chlorophyll and total carotenoids from vegetables. *Romanian Biotechnological Letters* 17:7702–7708.

Downton, W.J.S. 1970. Preferential C_4-dicarboxylic acid synthesis, the post illumination CO_2 burst, carboxyl transfer step, and grana configurations in plants with C_4-photosynthesis. *Canadian Journal of Botany* 48:1795–1800.

Eagles, C.F. and O. Østgård. 1971. Variation in growth and development in natural populations of *Dactylis glomerata* from Norway and Portugal. I. Growth analysis. *Journal of Applied Ecology* 8:367–381.

Ehleringer, J.R., R.F. Sage, L.B. Flanagan, and R.W. Pearcy. 1991. Climate change and the evolution of the C_4 photosynthesis. *Trends in Ecology and Evolution* 6:95–99.

Evans, J.R. 1989. Photosynthesis and nitrogen relationships in leaves of C_3 plants. *Oecologia* 78:9–19.

Evans, J.R. and S. von Caemmerer. 1996. Carbon dioxide diffusion inside leaves. *Plant Physiology* 110:339–346.

Ferraro, D.O. and M. Oesterheld. 2002. Effect of defoliation on grass growth: A quantitative review. *Oikos* 98:125–133.

Georgiadis, N.J., R.W. Ruess, S.J. McNaughton, and D. Western. 1989. Ecological conditions that determine when grazing stimulates grass production. *Oecologia* 81:316–322.

Ghannoum, O., S. Von Caemmerer, L.H. Ziska, and J.P. Conroy. 2000. The growth response of C_4 plants to rising atmospheric CO_2 partial pressure: A reassessment. *Plant, Cell and Environment* 23:931–942.

Giambalvo, D., P. Ruisi, G. Miceli, A.S. Frenda, and G. Amato. 2011. Forage production, N uptake, N2 fixation, and N recovery of berseem clover grown in pure stand and in mixture with annual ryegrass under different managements. *Plant and Soil* 342:379–391.

Gómez, S., O. Guenni, and L.B. de Guenni. 2012. Growth, leaf photosynthesis and canopy light use efficiency under differing irradiance and soil N supplies in the forage grass *Brachiaria decumbens* Stapf. *Grass and Forage Science* 68:395–407.

Grant, S.A., G.T. Barthram, L. Torvell, J. King, and H.K. Smith. 1983. Sward management, lamina turnover and tiller production density in continuously stocked *Loliuum perenne*-dominated swards. *Grass and Forage Science* 38:333–344.

Hatch, M.D. and C.R. Slack. 1966. Photosynthesis by sugar-cane leaves. A new carboxylation reaction and the pathway of sugar formation. *Biochemistry Journal* 101:103–111.

Heide, O.M., M.G. Bush, and L.T. Evans. 1985a. Interaction of photoperiod and gibberellin on growth and photosynthesis of high-latitude *Poa pratensis*. *Physiologia Plantarum* 65:135–145.

Heide, O.M., R.K.M. Hay, and H. Baugeröd. 1985b. Specific day-length effects on leaf growth and dry-matter production in high-latitude grasses. *Annals of Botany* 55:579–586.

Hikosaka, K., K. Ishikawa, A. Borjigidai, O. Muller, and Y. Onoda. 2006. Temperature acclimation of photosynthesis: Mechanisms involved in the changes in temperature dependence of photosynthetic rate. *Journal of Experimental Botany* 57:291–302.

Hussein, M.M., H.M. Mehanna and S.M. El-Lethy. 2013. Water deficit and foliar fertilization and their effect on growth and photosynthetic pigments of Jatropha plants. *World Applied Sciences Journal* 27:454–461.

Kortschalk, H.P., C.E. Hartt, and G.O. Burr. 1965. Carbon dioxide fixation in sugar-cane leaves. *Plant Physiology* 40:209–213.

Lara, M.V. and C.S. Andreo. 2011. C_4 plants adaptation to high levels of CO_2 and drought environments. In: Shanker, A. (Ed.) *Abiotic stress in plants—Mechanisms and adaptations.* InTech, Rijeka, Croatia, pp. 415–428.

Leakey, A.D.B., M. Uribelarrea, E.A. Ainsworth, S.L. Naidu, A. Rogers, D.R. Ort, and S.P. Long. 2006. Photosynthesis, productivity, and yield of maize are not affected by open-air elevation of CO_2 concentration in the absence of drought. *Plant Physiology* 140:779–790.

Li, G., Z.-S. Zhang, H.-Y. Gao, P. Lui, S.-T. Dong, J.-W. Zhang, and B. Zhao. 2012. Effects of nitrogen on photosynthetic characteristics of leaves from two different stay-green corn (*Zea mays* L.) varieties at the grain-filling stage. *Canadian Journal of Plant Science* 92:671–680.

Makino, A. and T. Mae. 1999. Photosynthesis and plant growth at elevated levels of CO_2. *Plant Cell Physiology* 40:999–1006.

Meyer, G.A. 1998. Pattern of defoliation and its effect on photosynthesis and growth of Goldenrod. *Functional Ecology* 12:270–279.

Misra, M.K. and B.N. Misra. 1981. Seasonal changes in leaf area index and chlorophyll in an Indian grassland. *Journal of Ecology* 69:797–805.

Morgan, P.W., S.A. Finlayson, K.L. Childs, J.E. Mullet, and W.L. Rooney. 2002. Opportunities to improve adaptability and yield in grasses: Lessons from sorghum. *Crop Science* 42:1791–1799.

Moser, L.E. and J.A. Jennings. 2007. Grass and legume structure and morphology. In: Barnes, R.F., C.J. Nelson, K.J. Moore, and M. Collins (Eds.) *Forages: The science of grassland agriculture*, Vol. II, 6th Edition. Blackwell Publishing, Ames, IA, pp. 15–35.

Mwangi, D.M., G. Cadisch, W. Thorpe, and K.E. Giller. 2004. Harvesting management options for legumes intercropped in Napier grass in the central highlands of Kenya. *Tropical Grassland* 38:234–244.

Nayyar, H. and D. Gupta. 2006. Differential sensitivity of C_3 and C_4 plants to water deficit stress: Association with oxidative stress and antioxidants. *Environmental and Experimental Botany* 58:106–113.

Nebauer, S.G., B. Renau-Morata, J.L. Guardiola, and R.V. Molina. 2011. Photosynthesis down-regulation precedes carbohydrate accumulation under sink limitation in citrus. *Tree Physiology* 31:169–177.

Nogués, S. and N.R. Baker. 2000. Effects of drought on photosynthesis in Mediterranean plants grown under enhanced UV-B radiation. *Journal of Experimental Botany* 51:1309–1317.

Noy-Meir, I. 1993. Compensating growth of grazed plants and its relevance to the use of rangelands. *Ecological Applications* 3:32–34.

Nyfeler, D., O. Huguenin-Elie, M. Suter, E. Frossard, J. Connolly, and A. Lüscher. 2009. Strong mixture effects among four species in fertilized agricultural grassland led to persistent and consistent transgressive over yielding. *Journal of Applied Ecology* 46:683–691.

Olson, B.E. and R.T. Wallander. 1999. Carbon allocation in *Euphorbia esula* and neighbors after defoliation. *Canadian Journal of Botany* 77:1641–1647.

Page, E.R., M. Tollernaar, E.A. Lee, L. Lukens, and C.J. Swanton. 2010. Shade avoidance: An integral component of crop–weed competition. *Weed Research* 50:281–288.

Parsons, A.J. and P.D. Penning. 1988. The effect of the duration of regrowth on photosynthesis, leaf death and the average rate of growth in a rotationally grazed sward. *Grass and Forage Science* 43:15–27.

Paul, M.J. and S.P. Driscoll. 1997. Sugar repression of photosynthesis: The role of carbohydrates in signaling nitrogen deficiency through source: Sink imbalance. *Plant, Cell and Environment* 20:110–116.

Paul, M.J. and C.H. Foyer. 2001. Sink regulation of photosynthesis. *Journal of Experimental Botany* 52:1383–1400.

Paul, M.J. and T.K. Pellny. 2003. Carbon metabolite feedback regulation of leaf photosynthesis and development. *Journal of Experimental Botany* 54:539–547.

Pierson, E.A., R.N. Mack, and R.A. Black. 1990. The effect of shading on photosynthesis, growth, and regrowth following defoliation for *Bromus tectorum*. *Oecologia* 84:534–543.

Ranson, S.C. and M. Thomas. 1960. Crassulacean acid metabolism. *Annual Review of Plant Physiology* 11:81–110.

Ryan, S.J., C.U. Knechtel, and W.M. Getz. 2007. Ecological cues, gestation length, and birth timing in African buffalo (*Syncerus caffer*). *Behavioral Ecology* 18:635–644.

Sanderson, M.A., G. Brink, R. Stout, and L. Ruth. 2013. Grass–legume proportions in forage seed mixtures and effects on herbage yield and weed abundance. *Agronomy Journal* 105:1289–1297.

Seemann, J.R., J.A. Berry, and W.J.S. Downton. 1984. Photosynthetic response and adaptation to high temperature in desert plants. *Plant Physiology* 75:364–368.

Shen, H., S. Wang, and Y. Tang. 2013. Grazing alters warming effects on leaf photosynthesis and respiration in *Gentiana straminea*, an alpine forb species. *Journal of Plant Ecology* 6:418–427.

Singer, S.M., Y.I. Helmy, A.N. Karas, and A.F. Abou-Hadid. 2003. Influences of different water stress treatments on growth, development and production of snap bean (*Phaseolus vulgaris* L.). *Acta Horticulturae (International Society for Horticultural Science)* 614:605–611.

Sorby, H.C. 1873. On comparative vegetable chromatology. In: Wellburn, A.R. 1994. The spectral determination of chlorophylls *a* and *b*, as well as total carotenoids, using various solvents with spectrophotometers of different resolution. *Journal of Plant Physiology* 144:307–313.

Stokes, G.G. 1864. On the supposed identity of biliverdin with chlorophyll, with remarks on the constitution of chlorophyll. In: Wellburn, A.R. 1994. The spectral determination of chlorophylls *a* and *b*, as well as total carotenoids, using various solvents with spectrophotometers of different resolution. *Journal of Plant Physiology* 144:307–313.

Sturludóttir, E., C. Brophy, G. Bélanger, A.-M. Gustavsson, M. Jørgensen, T. Lunnan, and Á. Helgadóttir. 2014. Benefits of mixing grasses and legumes for herbage yield and nutritive value in Northern Europe and Canada. *Grass and Forage Science* 69:229–240.

Sumanata, N., C.I. Haque, J. Nishika, and R. Suprakash. 2014. Spectrophotometric analysis of chlorophylls and carotenoids from common fern species by using various extracting solvents. *Research Journal of Chemical Sciences* 4:63–69.

Taylor, S.H., B.S. Ripley, F.I. Woodward, and C.P. Osborne. 2011. Drought limitation of photosynthesis differs between C_3 and C_4 grasses in a comparative experiment. *Plant, Cell and Environment* 34:65–75.

Treydte, A.C., S. Baumgartner, I.M.A. Heitkönig, C.C. Grant, and W.M. Getz. 2013. Herbaceous forage and selection patterns by Ungulates across varying herbivore assemblages in a South African savanna. *PLoS One* 8:e82831.

Urrego-Pereira, Y.F., A. Martínez-Cob, V. Fernández, and J. Cavero. 2013a. Daytime sprinkler irrigation effects on net photosynthesis of maize and alfalfa. *Agronomy Journal* 105:1515–1528.

Urrego-Pereira, Y., J. Cavero, E.T. Medina, and A. Martínez-Cob. 2013b. Microclimate and physiological changes under a center pivot system irrigating maize. *Agricultural Water Management* 119:19–31.

USDA-NASS. 2015. *Crop values 2014 summary*. United States Department of Agriculture–National Agriculture Statistics Service. Washington, DC. 49 pp.

Van Sambeek, J.W., N.E. Navarrete-Tindall, H.E. Garrett, C.-H. Lin, R.L. McGraw, and D.C. Wallace. 2007. Ranking the shade tolerance of forty-five candidate groundcovers for agroforestry plantings. *The Temperate Agroforester* 15:1–10.

Volenec, J.J. and C.J. Nelson. 2007. Physiology of forage plants. In: Barnes, R.F., C.J. Nelson, K.J. Moore, and M. Collins (Eds.) *Forages: The science of grassland agriculture*, Vol. II, 6th Edition. Blackwell Publishing, Ames, IA, pp. 37–52.

Waller, S.S. and J.K. Lewis. 1979. Occurrence of C_3 and C_4 photosynthetic pathways in North American grasses. *Journal of Range Management* 32:12–28.

Wang, D., M.W. Maughan, J. Sun, X. Feng, F. Miguez, D. Lee, and M.C. Dietze. 2012. Impact of nitrogen allocation on growth and photosynthesis of Miscanthus (*Miscanthus × giganteus*). *Global Change Biology Bioenergy* 4:688–697.

White, L.M. 1973. Carbohydrate reserves of grasses: A review. *Journal of Range Management* 26:13–18.

Wilson, D. and J.P. Cooper. 1969. Effect of light intensity and CO_2 on apparent photosynthesis and its relationship with leaf anatomy in genotypes of *Lolium perenne* L. *New Phytology* 68:627–644.

Yamasaki, T., T. Yamakawa, Y. Yamane, H. Koike, K. Satoh, and S. Katoh. 2002. Temperature acclimation of photosynthesis and related changes in photosystem II electron transport in winter wheat. *Plant Physiology* 128:1087–1097.

Yamori, W., K. Hikosaka, and D.A. Way. 2014. Temperature response of photosynthesis in C_3, C_4, and CAM plants: Temperature acclimation and temperature adaptation. *Photosynthesis Research* 119:101–117.

Zhao, D., K.R. Reddy, V.G. Kakani, and V.R. Reddy. 2005. Nitrogen deficiency effects on plant growth, leaf photosynthesis, and hyperspectral reflectance properties of sorghum. *European Journal of Agronomy* 22:391–403.

27 Photosynthetic Competition between Forest Trees

Piotr Robakowski

CONTENTS

27.1 Introduction 475
    27.1.1 Definition of Photosynthetic Competition 476
27.2 Competition between Trees 477
27.3 Competition for Light 477
27.4 Competition for Water and Nutrients... 479
27.5 Competitive Interactions between Native and Invasive Trees 480
27.6 Importance of Photosynthetic Competition for Evolution of Plants.. 483
References... 484

27.1 INTRODUCTION

Plants compete for light, water, and nutrients, all of which are used for photosynthesis. Long-lived trees have withstood the pressure of competitors, which resulted in the occurrence of many morphological and physiological adaptations. They allow trees to capture resources and improve photosynthesis more efficiently than their neighbors. Photosynthetic competition between trees was defined as the ability of a tree to maintain or increase its net CO_2 assimilation rate at the expense of a reduction of neighboring-tree photosynthesis. A way to win photosynthetic competition may consist in picking up or lowering availability of resources for neighbors to reduce photosynthetic capacity of competitors. Thus, the competition for light, water, and nutrients can be regarded as an integral part of photosynthetic competition.

At a young age, during intensive growth, the competition for light influences photosynthetic and photomorphogenic processes. Shade-intolerant tree species usually have a high growth rate, high leaf mass-to-area ratio, high nitrogen content in leaves, great net CO_2 assimilation rate, maximal carboxylation rate, and maximal electron transfer rate, which give them an advantage in photosynthetic competition under high light. In contrast, shade-tolerant trees show a low leaf mass-to-area ratio, large leaf area, thin epidermis, and high chlorophyll concentration in leaves, allowing them to use dissipated light energy filtered through a canopy for photosynthesis. Seedlings of shade-tolerant trees growing under the canopy or in gaps are adapted to compete for light provided by sun flecks.

Tree species characterized by higher water and nutrient uptake or greater photosynthetic water and nitrogen use efficiency are more predisposed to win the photosynthetic competition. The ability to regulate gas exchange with stomata and the capacity of water and nutrient transport with vessels or tracheids affect the photosynthetic performance of trees.

The photosynthetic competition can be regarded as a result of competition for resources, adaptation and acclimatization of the photosynthetic apparatus, and photosynthetic performance modified by site and biotic conditions.

Between an invasive and native species, the competition can be more intense than between native species. An invader is often morphologically and physiologically adapted to be a superior competitor over a native species. Usually, an invader has higher photosynthetic capacity, or it is able to reduce drastically the availability of resources used for photosynthesis by its native competitor. However, a higher photosynthetic capacity is not a guarantee to win the competition.

The functional groups of trees differ in their ability to maximize resource uptake and to protect their photosynthetic apparatus toward an inhibition of photosynthesis. The winner of the photosynthetic competition has to adapt the photosynthetic structures and processes to competitive interactions with neighbors. The photosynthetic competition exerts a selection pressure promoting the species that are able to maintain or increase their photosynthetic efficiency at the expense of their competitors, and under stress conditions, they effectively invest more resources into defensive mechanisms to protect the photosynthetic functions.

The role of some morphological and physiological adaptations that occur in plants is to help them to compete efficiently with neighbors. The overarching aim is to take advantage over competitors and win in fighting for space and site resources such as light, water, and nutrients. Plants are able to modify their competitive behavior in response to the changing environment and interactions with neighbors, but also, they can change growth conditions or influence other living organisms (bacteria, mycorrhizal fungi, insects, or even mammals) to protect themselves against herbivores and enhance their own performance at their neighbors' expense (Weir 2007; Pierik et al. 2013). Plant behavior triggered by the presence of other plants remains an intriguing question.

Plants can optimize competition in at least one of three ways: (1) reduction of competitive encounters by avoiding their neighbors, (2) enhancement of their competitive effects by aggressively confronting their neighbors, or (3) tolerance of the competitive effects of their neighbors (Novoplansky 2009). In a natural environment, all of these can be observed, but also, some intermediate strategy is possible (Grime 1977).

Plants perceive their competitors based on differences in electromagnetic radiation and metabolic concentrations and fluxes (Pierik et al. 2013). They capture light and are able to recognize its intensity (King et al. 1978; Cymerski and Kopcewicz 1994), which can be modified by other plants. In the forest, solar light is filtered by leaves of dominant trees in the canopy before reaching the lower stand layers and forest floor vegetation (Vézina and Boulter 1966). In the shade of a canopy of forest trees, the ratio of red to far-red radiation is lower than in the open (Messier and Bellefleur 1988). A degree of shading can be a cue to acclimate the photosynthetic apparatus to new irradiance conditions modified by swifter growing competitors. Moreover, plants can communicate and compete using volatiles emitted by leaves and root exudates, which affect directly their competitors or change the growth conditions in a way to inhibit seeds' germination, growth, biomass allocation, and photosynthesis of neighbors (Einhellig 1995; Mallik 2008; Wu et al. 2009; Pierik et al. 2013).

Interactions between plants may cause changes in photosynthetic capacity, giving an advantage to some competitors with a higher ability to acclimate or higher photosynthetic plasticity over the others. The photosynthetic competition is a biological *filter* that causes natural selection among individuals belonging to the same species and among different species (Lambers et al. 1998). Thus, the species having a greater capacity to adapt their photosynthetic apparatus to the changing environment can gain advantages over competitors, and it is more plausible that they will attain evolutionary success. In a short time scale, it is profitable for a plant to acclimate photosynthesis to a stressor, e.g., high irradiance, nutrient deficit, or low temperature. The species can differ in plasticity of their photosynthetic acclimation to different stressors, which may also influence the results of the competition between them (Valladares et al. 2000).

Competition is a process that is difficult to detect directly; thus, in most studies, only the outcomes of competition, for example, biomass allocation, growth, and photosynthetic rates, were measured. For practical and technical reasons, the measurements of growth or photosynthesis are undertaken only once at the arbitrary assigned end point of the experiments. However, during the vegetative season, the changing environmental conditions affect individual growth and photosynthesis, and also interactions among plants. Therefore, so as not to miss the dynamics of competition, measurements should be repeated through time (Trinder et al. 2013). Temporal dynamics particularly concern photosynthesis and photosynthetic competition, which may vary instantaneously but also within a season or within longer periods in the function of changes in light, temperature, humidity, and availability of nutrients. A species-specific ability to compete with neighbors can be modified by long-lasting plant–plant interactions under changing environmental conditions, which lead to the development of the response strategies, e.g., shade avoidance strategy (Pierik et al. 2013).

27.1.1 Definition of Photosynthetic Competition

Plants growing in intraspecific or interspecific communities differ in growth rate, biomass production, phenology, mortality, net CO_2 assimilation, and respiration rates as well as other phenotypical characteristics. However, all these traits of an individual plant may be acquired due to differences in the capacity to exploit environmental resources, and not the interaction between plants. To distinguish competition from adaptation or acclimatization to a site or biotic impact (e.g., predation or herbivores), it was defined as the tendency of neighboring plants to utilize the same quantum of light, ion of a mineral nutrient, molecule of water, or volume of space (Grime 1974, 1977). Consequently, photosynthetic competition can be understood as the ability of a plant to maintain or increase net CO_2 assimilation rate at the cost of a decrease in net CO_2 assimilation rate of a neighboring plant. A way for a plant to win the photosynthetic competition may consist in picking up more light, nutrients, or water, which are indispensable for photosynthesis, and in lowering availability of these resources for neighbors. To gain victory over a competitor, plants differently adapt and thus increase their ability to capture light energy and photosynthetic substrates, or they only decrease availability of photosynthetic substrates for their neighbors. Species, ecotypes, or individuals may differ in their capacity to compete for resources, which can also be modified by the effects of environmental factors on photosynthesis. It can be expected that alterations in key photosynthetic parameters such as quantum yield of photosystem II (PSII) efficiency (Φ_{PSII}), maximal net CO_2 assimilation rate (A_{max}), maximal carboxylation rate (V_{cmax}), and maximal electron transfer rate (J_{max}) reflect not only a photosynthetic response of plants to changing site conditions but also a response to the presence of competitors. Adaptation or acclimatization to environmental changes together with short-lasting responses to instantaneous climate changes or nutrient and water pulses may overlap and mask long-term trends in photosynthetic adaptation to the impact of competition on photosynthesis. In contrast, changes in morphological traits, such as formation of additional tissue and organs or growth rate reduction, usually need more time to occur and can be connected more easily with the adaptation to the given environmental conditions than any changes in photosynthesis.

A distinctive characteristic of competitive plants is their mechanisms of phenotypic response, which maximize the capture of resources. These mechanisms are advantageous only if a plant is growing in an environment where light, water, and nutrients are intensively absorbed and the availability of the photosynthetic substrates can be sustained (Grime 1977). Under stress, which inhibits photosynthesis, however, a successful competitor should be able to protect its photosynthetic apparatus through structural and physiological adaptations toward unfavorable conditions. An example of such an adaptation is the photoprotective mechanism to dissipate excessive energy as heat in the xanthophyll cycle developed in evergreen plants in response to the stress of high light and low temperature (Adams and Demmig-Adams 1994; Niyogi 1999; Adams et al. 2004).

Among all taxonomic groups of plants, long-lived trees are exposed to changing site conditions, interactions with different plant species, and thus, competition for long periods. They have developed advantageous species-specific life strategies encompassing photosynthetic structures and processes.

27.2 COMPETITION BETWEEN TREES

The intensity of competition between forest trees is contingent on distance (spacing) between individuals, species composition, proportion of stand mixture, and availability of site resources. In managed forests, seedlings are planted with the given spacing; therefore, at the beginning of growth, the competition for resources is not intense. In abundant natural regeneration of trees, there is less space between seedlings; thus, in contrast to a managed forest, young individuals compete for light and other resources from the beginning of growth. Differences in growth and photosynthetic capacity among individuals originate from ontogenetic traits and microsite conditions. When the space between trees is filled with crowns and the canopy is dense, competition for light and other resources needed for dynamic growth determines all life processes in trees. In thickets and small polewood stages, the growth rate of trees is highest, and competition between them, especially for light, is also the most intense during the whole life of a tree. In high polewood and mature stands, the growth rate and fierceness of competition diminish, and spacing among individuals increases because of trees' mortality.

In monospecific, even-aged stands, trees differ in height, and they compete mostly for light needed for photosynthesis. In these stands, individual growth dynamics decide success or defeat. Light-requiring species, e.g., Scotch pine (*Pinus sylvestris* L.), characterized by high values of the light compensation point of photosynthesis grow slowly, and usually live not far under the canopy. In case of monospecific stands of shade-tolerant trees, e.g., silver fir (*Abies alba* Mill.), the populations are more differentiated in height and, even in deep shade, can survive longer with low height increments. In mixed stands, trees compete within the same species as well as with other species differing in shade tolerance, where species-specific genetic traits play a greater role. At a young age, growth and photosynthetic competition for light eliminate the greatest number of trees when compared with the number of individuals dying of environmental stressors, except for mass wind damage or pest outbreaks (Barnes et al. 1998).

27.3 COMPETITION FOR LIGHT

Light is indispensable for photosynthetic and photomorphogenic processes in plants, which are able to perceive light intensity and quality. They identify the red/far-red ratio, which enables them not only to *measure* the length of day and night but also to recognize their neighbors (King et al. 1978; Cymerski and Kopcewicz 1994; Pierik et al. 2013).

Seedlings and advance trees' regeneration may occur in contrasting light environments depending on genetically founded species-specific light requirements. Pioneer, early successional trees are adapted to full sunlight and grow in open areas; e.g., after a clear-cut or in postfire areas, mid-successional species are tolerant of moderate shade and often occur in gaps, and late-successional, conservative, shade-tolerant species may grow in small gaps or under the canopy of mature trees. In full sunlight, a successful competitor is characterized by a high relative growth rate and great leaf area expansion above its neighbors, high Φ_{PSII}, high A_{max} and dark respiration rates (R_d), high nitrogen content in leave, and high biomass allocation to above-ground organs (Balandier et al. 2006). On the contrary, shade-tolerant species are able to survive in light-limiting conditions and respond positively to the canopy openings through increases in growth dynamics and photosynthesis rates. Leaves of these species are structurally and physiologically adapted to maintain photosynthesis even in extremely low light conditions. A low leaf mass to area ratio (LMA), thin epidermis, and high chlorophyll concentration in leaves allow them to use energy provided by sun flecks for photosynthesis (Grassi and Bagnaresi 2001; Wyka et al. 2007; Houter and Pons 2014). Stomatal opening in response to increased light intensity was consistently more rapid in four shade-tolerant species of North American trees, which used more efficiently brief periods of high illumination by sun flecks to increase net CO_2 assimilation rate (Woods and Turner 1971). However, under the same light conditions, shade-tolerant trees show lower photosynthetic capacity than light-demanding species. Sun flecks last from seconds up to minutes and may provide from about 30% to 80% of light energy, which is absorbed by leaves (Küppers et al. 1996). The characteristics of leaves adapted to deep shade, a medium-size gap in the stand, or full light are compared in Table 27.1.

In the forest, light is supplied from above the canopy formed by dominant species; thus, individuals that situate their crowns above those of neighbors benefit directly from increased photosynthetic rates and indirectly by reducing the growth of those neighbors via shade. In stands where seedlings recruit under the canopy and small gaps, light is a limiting factor, and the most competitive species are characterized by the adaptations to shade that confer greater survivorship and growth (Craine and Dybzinski 2013).

The natural regeneration of many species of trees finds the best conditions for growth and photosynthesis in gaps with higher irradiance than under the canopy (but not so high as in the open), low temperature amplitudes, and high humidity. Abundant young and advance regeneration compete for light, which is the main factor limiting photosynthesis and growth in gaps. Tree species differ not only in shade tolerance but also in the ability to acclimate to increased gap size and irradiances. Shade-tolerant conifers *Picea rubens* (Sarg.) and *Abies balsamea* (L.) Mill. regeneration occurred in gaps of different

TABLE 27.1

Structural, Physiological, and Biochemical Leaf Characteristics Allowing Trees to Optimize Intensity of Photosynthesis and Compete for Light in the Different Light Environments: Strong Shade, Forest Gap of Medium Size, and Full Light

Leaf Characteristics	Shade	Medium Gap	Full Light
Morphological			
Leaf position	Horizontal	Horizontal	Vertical
Dry weight	Low	Higher	Highest
Lamina thickness	Low	Higher (high plasticity)	Highest
Leaf area	Largest	Large	Small
Leaf mass-to-area ratio	Low	Higher (high plasticity)	Highest
Stomatal density	Low	–	Higher
Leaf punch strength	Low	Higher	Highest
Anatomical			
Epidermis	Thinnest	Thin	Thick
Palisade parenchyma	Thin, large spaces among cells	–	Thickest, cells well packed,
Spongy parenchyma	Similar	–	Similar
Sclerenchyma layer	Thin	–	Thick
Ultrastructural			
Chloroplast per area	Few	–	Many
Number of grana per chloroplast section	Low	High	–
Chloroplast section surface	Low	High	
Appressed-to-nonappressed grana membrane ratio	High	Low	Low
Biochemical			
Chlorophyll per leaf dry mass	High	Low	Low
Chlorophyll per leaf area	High	Low	Low
Chlorophyll a/b ratio	Low	Low	High
Carotenoids per dry mass	Low	Low	High
Carotenoids per area	Low	Higher	Highest
Nitrogen per dry mass	Similar	Similar	Similar
Nitrogen per area	Low	Low	High
Rubisco per area	Low	–	High
Photosynthetic Parameters			
Maximal net CO_2 assimilation rate per area	Low	Higher	Highest
Maximal carboxylation rate per area	Low	Higher	Highest
Maximal electron transfer rate per area	Low	Higher	Highest
Effective quantum yield of PSII photochemistry	Low	Higher	Highest
Light compensation point of photosynthesis	Low	Low	High
Saturation light level of photosynthesis	Low	Low	High
Dark respiration per leaf dry mass	Similar	–	Similar
Dark respiration per leaf area	Low	Low	High

Source: Lambers, H. et al., Plant physiological ecology, Springer-Verlag, New York, 1998; Grassi, G., and Bagnaresi, U., *Tree Physiology*, 21, 959–967, 2001; Kierzkowski, D. et al., *Polish Journal of Ecology*, 55(4), 821–825, 2007; Wyka, T. et al., *Tree Physiology*, 27, 1293–1306, 2007; Houter, N.C., and Pons, T.L., *Oecologica*, 175, 37–50, 2014.

sizes. Competition between seedlings of these species depended on a gap size: *P. rubens* successfully competed with *A. balsamea* in irregular gaps of intermediate size (100–300 m²) and in an unharvested control. However, in the shade of small gaps and in the high irradiance of large gaps *P. rubens* was unable to surpass *A. balsamea* even if leaf area-based A_{max} of spruce was higher. This was due to the higher leaf mass-based A_{max} and water use efficiency in photosynthesis (WUE) of *A. balsamea*. These results indicated that structural leaf traits expressed as lower LMA, higher growth dynamics, and higher plasticity of acclimation to the light environment gave an advantage to *A. balsamea* over *P. rubens* (Dumais and Prévost 2014).

Photosynthetic characteristics (V_{cmax}, J_{max}, leaf nitrogen, chlorophyll concentrations) and nitrogen partitioning to light harvesting, bioenergetics, and Rubisco were studied along a light gradient in crowns of four temperate deciduous species differing in shade tolerance (ranging from the most shade intolerant to strongly tolerant species): *Populus tremula* L., *Fraxinus excelsior* L., *Tilia cordata* Mill., and *Corylus avellana* L. There was evidence in leaves that structural adaptations to irradiance expressed in LMA were more important than between-species differences in biochemistry (Niinemets et al. 1998). It is generally believed that a greater size of plants gives them an advantage in the competition for light. However, the importance of size-asymmetric light competition is contingent on the species-specific light requirements and successional status. The pioneer woody species in early secondary succession differed in some morphological and photosynthetic traits, but their ability for light capture, light use efficiency, and photosynthetic rate per unit plant mass, unexpectedly, were not significantly different (van Kuijk et al. 2008). Shade-intolerant, pioneer tree species are usually characterized by high growth dynamics. In the young stand formed by these species (e.g., monospecific stands of *P. sylvestris* L.), the competition for light is intensive and size asymmetric, i.e., only the individuals who have their crowns in the canopy can survive. At the older stage of the development of a stand, the competition for light is low and symmetric. In contrast, even in a monospecific forest dominated by a shade-tolerant species (e.g., *A. alba* Mill.), individuals of different size may find their light niches. Under the canopy of shade-tolerant trees, in deep shade, young shade-tolerant saplings of *A. alba* and *Fagus sylvatica* (L.) were able to use dissipate light or direct light from sun flecks for photosynthesis (Robakowski and Antczak 2008). The interspecific and intraspecific light competition among the shade-tolerant trees is size asymmetric.

27.4 COMPETITION FOR WATER AND NUTRIENTS

When two species with pronounced differences in water use share a common space, the competition for water may be crucial for their photosynthetic machinery and survival, especially under drought conditions. *F. sylvatica* L. and *Quercus petraea* Matt. (Liebl.), economically important forest tree species, differ in drought tolerance. The former is more sensitive to water stress, and the latter is more tolerant to periodic water deficits. At the same values of stomatal conductance to water vapor (g_s), the 30-year-old *F. sylvatica* showed lower photosynthetic capacity than *Q. petraea* (Aranda et al. 2000). This was due to the more pronounced stomatal closure in *Fagus* leaves when compared with those of *Quercus*. These results indicated the interspecific differences in sensitivity to drought, which were reflected in g_s and A_{max}. In the southernmost limit of these species, ranges where they were threatened by drought, a higher tolerance to water deficit of *Q. petraea* gave it an advantage over *F. sylvatica*. Furthermore, it is projected that drought may become more intense as a result of climate change and will enhance the

species-specific differences in photosynthetic response to water deficit.

Photosynthesis was found to be less inhibited by drought than growth (Epron and Dreyer 1993; Dreyer 1997). A 5-year study on effects of drought on adult trees of *F. sylvatica* and *Q. petraea* in a mixed stand in Germany confirmed that a seasonal water deficit had only weak or no influence on A_{max} except for a severe drought in August, which reduced A_{max} in *Fagus* by 30%. This decrease in photosynthesis coincided with summer drought but also with high air temperatures. Therefore, it was impossible to identify the main reason for reduction of stomatal conductance. However, the two species differed in photosynthetic capacity: *Quercus* showed around twofold higher A_{max} values during midsummer compared to *Fagus* leaves (14–18 versus 7–10 μmol m^{-2} s^{-1}). This difference can be attributed to significantly lower stomatal conductance in *Fagus* leaves (Leuschner et al. 2001). Although in this study, a drought effect could not be observed in *Quercus* leaves, the late frosts remarkably decreased A_{max} in this species and might influence the interspecific photosynthetic competition. *Q. petraea* showed higher photosynthetic capacity and stomatal conductance than *F. sylvatica* even under severe drought. Leuschner et al. (2001) found that leaf conductance, stem hydraulic conductivity, A_{max}, stem diameter, and fine root growth were more susceptible to drought in *Quercus* than in *Fagus* leaves. Despite the lower values of its ecophysiological characteristics, *Fagus* usually outcompeted *Quercus* in the study site in Germany. This was probably due to the higher shade tolerance of *Fagus* and its high efficiency of light capture due to its leaf orientation and crown density. The importance of drought to photosynthetic competition between these species was higher in the southern limit of their geographical range in Spain, where more severe stress of drought occurred compared with the study site in Germany (Aranda et al. 2000).

Macroelements and microelements are indispensable for photosynthesis. Kazada et al. (2004) compared growth, A_{max}, and nutrients concentration in leaves of *F. sylvatica*, *Q. petraea*, and *Acer pseudoplatanus* saplings growing along the light gradient from the forest edge into the closed canopy. In the year of planting, the A_{max} in *A. pseudoplatanus* was the highest, but during the next 5 years, it was the lowest compared with *Fagus* and *Quercus*. This abrupt decrease in A_{max} was observed simultaneously with lower nitrogen concentration in *Acer* leaves than in the leaves of both other species. The differences in A_{max} among the study species were related to nitrogen nutrition. This relationship was significant only in some years for *Fagus* and *Quercus* but was observed during the whole experiment (6 years) for *Acer*. The plants had the highest nutrient concentration in shade and lower LMA compared with saplings growing at the edge of the forest. To increase light harvesting, saplings growing in shade produced thinner leaves, which enabled them to distribute nitrogen across a larger area (Niinemets et al. 1998) (Table 27.1). However, the results of Kazada et al. indicated that a large leaf area might not compensate for the effect of nutrient deficit on photosynthesis for nutrient-demanding species, which had to compete for nutrients with mature trees and other plants in vegetal cover under the canopy shade.

In an unmanaged forest dominated by *Pinus densiflora* Siebold and Zucc. in the canopy, intensive competition occurred between this species and understory vegetation for water and nutrients (Kume et al. 2003). The values of A_{max}, g_s, and ($\delta^{13}C$) isotope of carbon 13 in *Pinus* needles were smaller in the unmanaged forest compared with the managed forest, where the understory vegetation was removed. The highest WUE in the unmanaged forests was at the cost of low photosynthetic nitrogen use efficiency (PNUE). The lower PNUE and larger C/N ratio in the unmanaged stand indicated that nitrogen restriction in needles was higher in the unmanaged stands than in the managed stands. Kume et al. (2003) suggested that by the comparison of competition for light, the advantages of taller plants are not obvious in the acquisition of water and nutrients. However, the competition observed at the level of roots influenced remarkably photosynthetic capacity and efficiency of needles.

Nutrient availability is important for the success of invasive tree species, and it can affect competition between a native and exotic plants. Yuan et al. (2013) studied interspecific competition between an exotic tree (or shrub) *Rhus typhina* L. and a native tree *Quercus acutissima* Carr. under three soil nitrogen/phosphorous (N/P) ratios: 5, 15, and 45 (N-limited, basic N and P supply, and P-limited conditions, respectively). Potted plants were grown in a greenhouse in monocultures or species mixtures. A_{max} for *R. typhina* was higher in all N/P treatments compared with *Q. acutissima*, although leaf nitrogen concentration and leaf phosphorous concentration were higher in *Q. acutissima* than in *R. typhina* leaves. In basic and P-limited conditions, A_{max} of the native species was lower in the mixtures than in the monocultures. With the basic N and P supply, A_{max} for *R. typhina* was higher in the competition treatment compared with monocultures. In competition, the greater PNUE and photosynthetic phosphorous use efficiency (PPUE) gave an advantage to *R. typhina* over *Q. acutissima* independently of the N/P treatments. P limitation reduced the PNUE of the study species, and PPUE was lowest in the N-limited conditions. The interspecific differences between the species growing in competition resulted from their strategies of nutrient allocation to the photosynthetic machinery or nonphotosynthetic leaf tissues. *Q. acutissima* invested more N and P in leaf structural traits and defensive mechanisms against herbivores at the cost of PNUE and PPUE. In contrast, *R. typhina* showing low LMA and high A_{max}, PNUE, PPUE, and relative growth rate was able to outcompete *Q. acutissima*. Thus, when an exotic and native species are compared, a greater investment and maintaining of biomass and nutrients in the *R. typhina* photosynthetic apparatus allowed this exotic species to acclimate better to different N/P ratios and gave it an advantage over *Q. acutissima* (Yuan et al. 2013).

27.5 COMPETITIVE INTERACTIONS BETWEEN NATIVE AND INVASIVE TREES

Competition between an invasive and native plant species can be more intense than locally growing native species. An invader is often morphologically and physiologically adapted to be a superior competitor over a native species. The former often gets an advantage due to high growth rate, mass seed production, high ecological plasticity, allocation of resources to photosynthetic tissue, and also high photosynthetic capacity (Walck et al. 1999; Mangla et al. 2011; Yuan et al. 2013; Kuehne et al. 2014). It has been observed that moderately shade-tolerant *Quercus rubra* L. natural regeneration is inhibited throughout its native North American range, whereas in Central Europe, it is often more abundant and valuable for silviculture. Northern red oak was introduced to European forests from North America and adapted well to the new site conditions. It competes successfully with native moderately shade-tolerant *Quercus robur* L. and shade-tolerant *A. pseudoplatanus* L. and *Carpins betulus* L., due to its specific photosynthetic traits. Under closed canopies and in small canopy gaps, *Q. rubra* showed high photosynthetic capacity and low R_d, light compensation point, and quantum efficiency of photosynthesis when compared with the native species. Superior total leaf area and high A_{max} in red oak regardless of canopy type led to the highest total plant A_{max}. Interestingly, the photosynthetic advantage of this exotic species was not reflected in its height growth, which suggested that higher net CO_2 assimilation rates were related to an investment into increased leaf area at the cost of height growth. Thus, the regeneration success of red oak in Europe is related to an enhanced photosynthetic performance relative to associated shade-tolerant species (Kuehne et al. 2014). However, plasticity of photosynthetic parameters was not higher in *Q. rubra* compared with *A. pseudoplatanus* and *C. betulus*, suggesting that species-specific traits were more effective in successful competition than plasticity.

In European forests, *Prunus serotina* (Ehrh.) is also an exotic woody species originated from North America, which is much more invasive compared with *Q. rubra* (Marquis 1990). It was observed that the natural regeneration of *P. serotina* and *Q. petraea* Matt. (Liebl.) co-occurs under the canopies of Scotch pine stands (Photo 27.1a,b). In most of the stands, the natural regeneration of cherry was more abundant and dynamic; therefore, it had the advantage over oak, forming the undergrowth and second storey (Robakowski and Bielinis 2011). However, in some pine stands, it was oak that formed the undergrowth and second layer under the pine canopy, and cherry was present as an admixture. The question was addressed about causes of different outcomes of the competition between these species. Black cherry is well known for its high growth dynamics and allelopathic properties, which can be helpful in winning the competition with oak (Marquis 1990; Vetter 2000; Koutika et al. 2007). Additionally, when compared with oak, cherry produces more frequently and a greater amount of seeds, which can germinate in the second or third year after seeding (Marquis 1990). The life strategy of *Q. petraea* is opposite that of *P. serotina*: at a young age, it is a slow-growing tree investing more resources into the development of roots, and it produces a lower amount of seeds each 4–6 years (Mosandl and Kleinert 1998). In favorable light conditions, at a young age, cherry is able to attain

PHOTO 27.1 Natural regeneration of native sessile oak (*Quercus petraea* Matt. [Liebl.] and invasive black cherry (*Prunus serotina* Ehrh.) growing in a gap in Scotch pine (*Pinus sylvestris* L.) forest. (a) A seedling of sessile oak under the shade of a black cherry crown. The seedlings compete for light provided by sun flacks passing through the dense canopy of a small pole stand. (b) Competition for space and site resources between sessile oak (in the center) and more numerous black cherry seedlings (western Poland, managed experimental Scotch pine forest). (Photo from *P. Robakowski*.)

fivefold greater height than oak within one vegetative season (Robakowski and Bielinis 2011). The photosynthetic strategy of this species consists in an investment of resources into fast development of a large assimilative area, although it depends strongly on the light environment. In low light (from 6% to 10% of full irradiance), growth and photosynthesis of cherry were inhibited. The response of cherry to the light environment with regard to height growth, biomass production, and leaf area increase was much more dynamic compared with oak. It was hypothesized that photosynthesis (A_{max}) of *P. serotina* seedlings growing in competition with those of *Q. petraea* would increase at the cost of its competitor. This photosynthetic competition would encompass the competition for light, water, and nutrients, and additionally, it might be affected by allelochemical interactions between oak and cherry. However, in natural conditions, it is extremely difficult to separate the effects of different forest site factors from an effect of the competitors' neighborhood on photosynthesis.

To at least partially control the environmental conditions, a pot experiment was established, with *Q. petraea* and *P. serotina* seedlings growing in one of three light treatments (10%, 25%, and 100% of full irradiance) and in four combinations of competition and mulching with black cherry leaves (Q, three seedlings of oak; P, three seedlings of cherry; Q+L, three seedlings of oak mulched with black cherry leaves; Q+P+L, three oak seedlings and six cherry seedlings with mulching). There was evidence that after 4 months of acclimation to the experimental conditions, the light environment influenced most remarkably the growth and photosynthesis of both species. A_{max} and nitrogen content in leaves were also contingent on the effect of between-species competition and mulching with *P. serotina* leaves. For both species, net CO_2 assimilation rate was lowest in low light (10% of full irradiance) and did not differ between 25% and 100% (Figure 27.1). Interestingly, the effect of competition and mulching on A_{max} was significant in invasive *P. serotina* (Figure 27.1a) but not in the *target Q. petraea* seedlings (Figure 27.1b). For cherry, A_{max} was higher in the monoculture (P) than in the mixture with oak (Q+P+L) independently of the light treatment. This decrease in A_{max} of *Prunus* might result from the interspecific competition for resources with oak but also from the intraspecific competition among cherry seedlings and an effect of mulching with cherry leaves. The physiological mechanism of differences in A_{max} between P and Q+P+L cannot be elucidated with the changes in nitrogen concentration in *Prunus* leaves (Figures 27.2a and 27.3a), but rather, with the limitation of CO_2 assimilation by the lower stomatal conductance in leaves of cherry seedlings growing in competition with oak compared with those in monoculture ($P = 0.04$; P, probability). In both species, A_{max} and g_s were lowest in low light, but the absolute differences in the values of these parameters between the light treatments were higher in *P. serotina* compared with *Q. petraea*, indicating that the former had higher photosynthetic plasticity than the latter. High photosynthetic plasticity can help a species to win the competition for light and other resources needed for photosynthesis with a species characterized by a lower photosynthetic plasticity.

Nitrogen concentration expressed per leaf dry mass in *Prunus* was higher in mid light compared with high light, but it did not differ from low light (Figure 27.2a). There was no significant effect of competition and mulching on nitrogen concentration in cherry leaves. In contrast, nitrogen concentration in *Q. petraea* leaves depended not only on light but also on competition and mulching interacting with light (Figure 27.2b). The lowest nitrogen concentration was shown in leaves of oak seedlings growing in high light in Q+P+L, and the highest concentration was in mid light and Q+L. Nitrogen concentration in oak leaves decreased with greater light and competition with cherry. However, mulching with cherry leaves (Q+L) did not induce a reduction of nitrogen concentration; on the contrary, in mid light, it was even higher than in the control (Q) under low light and in all combinations acclimated to full light. When nitrogen content was expressed per leaf area, in *Prunus*, light acclimation was evident, but there was no effect of competition (Figure 27.3a). Oak showed

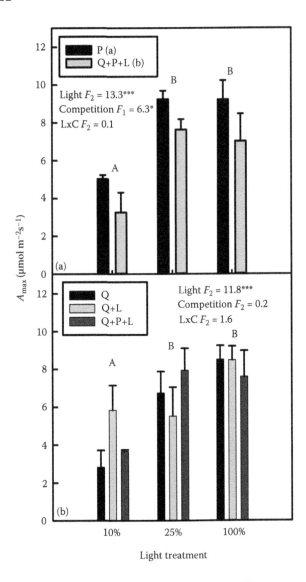

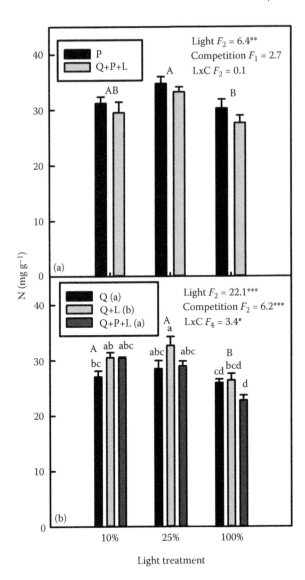

FIGURE 27.1 Maximal net CO_2 assimilation rate (A_{max}, means ± SE) of *Quercus petraea* and *Prunus serotina* seedlings growing in pots under one of three light treatments (10%, 25%, or 100% of full irradiance) in four combinations of competition and mulching with *Prunus* leaves: Q, three seedlings of *Q. petraea*; P, three seedlings of *P. serotina*; Q+L, three seedlings of *Q. petraea* with mulching; Q+P+L, three seedlings of *Q. petraea* and six seedlings of *P. serotina* with mulching. The measurements of gas exchange were conducted at the end of August 2011 after 4 months of acclimation to the experimental conditions. The results of two-factorial analysis of variance with the interaction between the light treatments and combinations of competition are given at $P < 0.05$. The same capital letters above columns indicate a lack of significant differences among the light treatments and the small letters in the legend inserted in the plot between the combinations of seedlings ($n = 6$, n, number of seedlings).

the highest nitrogen content in Q+L and the lowest in Q+P+L in high light. These results suggest that the expected variation of nitrogen concentration in oak leaves with the light environment can be modified by the competition with black cherry, and also, these results did not allow the exclusion of a nutritional or positive allelopathic effect that might be induced by mulching with cherry leaves.

FIGURE 27.2 Nitrogen concentration (means ± SE) expressed per dry mass in leaf of (a) *Prunus serotina* and (b) *Quercus petraea* acclimated to 10%, 25%, or 100% of full irradiance and one of four combinations of competition. The results of two-factorial analysis of variance and Tukey's test are shown ($n = 6$). The same small letters above columns indicate a lack of significant differences among the combinations within the light treatments. For further explanations, see Figure 27.1.

The interspecific competition between the young seedlings of *Q. petraea* and *P. serotina* was more remarkable in a non-light-limiting environment (100% of full irradiance) and in 25% compared with the low-light treatment (10%), supporting the hypothesis that the competition becomes more severe under full light access. However, the photosynthetic competition reflected by the differences in A_{max} between P and Q+P+L was observed in all light treatments. The study species showed different strategies of changes in nitrogen content in leaves. In *Prunus* leaves, it depended on the light environment, without a significant effect of competition. In *Quercus* leaves, nitrogen content was affected not only by the light regime but also, to a lesser extent, by competition and mulching. Although there was evidence that the species

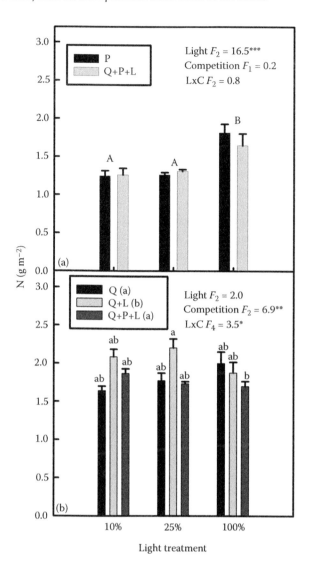

FIGURE 27.3 Nitrogen content (means ± SE) expressed per leaf area of (a) *Prunus serotina* and (b) *Quercus petraea* acclimated to 10%, 25%, or 100% of full light and to one of four combinations of competition. For further explanations, see Figure 27.1 ($n = 6$).

an individual tree and its neighbors on photosynthesis may stimulate or modify evolutionary processes such as adaptation and developing of optimal life strategies. The photosynthetic competition reflects the competition for the resources used as the substrates in photosynthetic processes. According to the optimization theory, the use of resources such as light, nitrogen, and water is most effective when it leads to a maximization of daily carbon gain. The optimization of photosynthesis in populations of trees can be analyzed at the level of an individual but also at that of a whole stand. Due to different interactions among trees, photosynthesis of a population of trees (a whole stand) cannot be a simple sum of net assimilation rates of individuals, as the performance of one individual depends on stature, leaf area, leaf distribution, and photosynthesis of its neighbors (Anten 2005). Maximization of photosynthesis of an individual may be in conflict with maximization of the whole stand's photosynthesis. Additionally, it can be challenging to distinguish adaptations induced by the habitat conditions from those caused by competition and other interactions among plants, e.g., allelopathic relations.

In the forest environment, light availability is the main factor limiting photosynthesis and growth of young trees until the attainment of maturity by a stand. The leaf area index (LAI) and leaf orientation in the canopy depend on light conditions: Under a low-light regime, LAI should be low and leaves horizontally oriented to reduce self-shading. Under a high-light regime, the value of LAI is high, and leaves are steeply inclined (Saeki 1960). The structural and physiological adaptations of shade-tolerant trees to photosynthesize in shade under the canopy of dominant tree species characterized by high relative growth rate, rapid leaf expansion, and large leaf area resulted from an ability to survive and develop in spite of a reduced light level and changes in spectrum after light passing through leaves of the canopy. By increasing their leaf area, plants increase the amount of light captured, but photosynthesis is influenced by many factors, in particular, by total concentration and an allocation of nitrogen to different photosynthetic structures and processes (Hikosaka and Terashima 1996). Under low light conditions (10% of full irradiance), shade-tolerant and late-successional *A. alba* (Mill.) was able to invest more nitrogen in light-harvesting complexes compared with the high-light treatment (Robakowski et al. 2003). This adaptation to the low-light environment may be an advantage in the competition for light, with fewer shade-tolerant tree species growing under a dense canopy or in gaps. It seems that an increased capacity for light capture can be particularly advantageous for the species, which use light from sun flecks in a great proportion for photosynthesis, as in the case of *A. alba* and *F. sylvatica* L. (Robakowski and Antczak 2008). At the canopy level, plants have LAI values greater than the predicted optima for maximal whole-stand PNUE and have leaf nitrogen contents lower than optimal. It was found that the evolutionary stable LAI values are always greater than the optimal LAI values for canopy photosynthesis (Anten 2005). Generally, trees growing in competition in a stand overinvest in height growth and produce larger leaf area, and leaves have more horizontal orientation than optimal.

differed in growth dynamics and photosynthetic traits and oak was often surpassed by cherry, they grew together under the canopy of Scotch pine forests, and both showed good condition. The study species of trees were evolutionarily close. The kin selection theory suggests that individuals will be more competitive with strangers compared to relatives (Dudley et al. 2013), which is concomitant with the observations of interactions between *Q. petraea* and *P. serotina* in the experimental and natural conditions.

27.6 IMPORTANCE OF PHOTOSYNTHETIC COMPETITION FOR EVOLUTION OF PLANTS

Usually, trees grow in associations of individuals of the same and/or other species. Effects of different interactions between

Brodribb et al. (2012) studied the history of competition between two evolutionarily different functional groups: conifer and angiosperm trees. Angiosperms developed during the Cretaceous at the expense of gymnosperm diversity and abundance, except for the conifers, which remain dominant in the canopy of many forest ecosystems. Photosynthetic structural and physiological adaptations of their photosynthetic apparatus can explain to some extent the conifer's advantages over angiosperms in some biomes. Although the conifers have generally lower A_{max} (~16 μmol m^{-2} s^{-1}) compared with woody angiosperms (~30 μmol m^{-2} s^{-1}) they are successful in many biomes, except for lowland equatorial land forest. Thus, the higher value of A_{max} does not guarantee an advantage over a competitor. The light environment in this forest promotes creation of large leaves with low mass per unit area, which are characterized by high intensity of photosynthesis and high productivity. On a global scale, however, the conifers show a higher LMA ratio than angiosperm trees (Reich et al. 1997). The mean value of LMA for the conifers is above twofold higher (227 g/cm^2) than for the angiosperms (106 g/cm^2) (Brodribb et al. 2012). Among the conifers, the family *Pinaceae* has attained particularly great evolutionary success, which is illustrated by the large geographical ranges of the *Pinus* species in the northern boreal forests. This is due to their tracheid-bearing vascular system, which is tolerant to freeze–thaw embolism. Additionally, in some conifers (*Pinus*, *Abies*), A_{max}, V_{cmax}, and J_{max} can be equivalent to or greater than those of associated angiosperm trees (Wullschleger 1993; Robakowski et al. 2003; Brodribb et al. 2012). In boreal forests of the Northern Hemisphere, under strong shade, *A. alba* (Mill.) and *Taxus baccata* (L.) are able to intercept dissipated light more efficiently than shade-intolerant trees due to flat needles. Similarly, in the Southern Hemisphere, *Podocarpaceae* formed flat leaves and shoots to increase light harvesting in the deep shade of tropical forests and compete successfully with angiosperms (Brodribb et al. 2012). There was evidence that leaves of evergreen angiosperms had higher photosynthetic capacity compared with their coniferous counterparts. At a common leaf nitrogen concentration, mean A_{mass} of angiosperms was 55% higher than that of conifers. Angiosperms are characterized by higher photosynthetic nitrogen use efficiency (PNUE) than conifers thanks to higher stomatal conductance and lower LMA, respectively (Lusk et al. 2003). It was hypothesized that low stomatal density and stomata structure of conifers have not been followed by a decreasing CO_2 concentration in air. Conifer xylem typically has lower specific conductivity than angiosperm xylem due to a greater hydraulic resistance in narrow-diameter tracheids than in vessels of angiosperms.

The role of photosynthetic competition in plant evolution may be, at least to some extent, elucidated by the *growth–differentiation balance theory*. This theory allows us to predict allocation of resources to growth or to defensive mechanisms against stressors under the changing environment. When any environmental factor inhibits growth more than it decreases photosynthesis, a plant may invest the resources into secondary metabolites and structural and functional differentiation processes (Matyssek et al. 2005a,b; Leudemann et al. 2009).

Under stress conditions, the importance of photosynthetic competition increases as it may exert a strong selection pressure promoting the species, which are able to maintain or increase their photosynthetic capacity and efficiency at the expense of their competitors and invest more resources into defensive mechanisms. The structural adaptations of the photosynthetic apparatus to increase light-harvesting efficiency, stomatal density adjustment to improve the regulation of gas exchange, molecular and biochemical adaptations to increase photosynthetic capacity, and development of root system to increase water and nutrients absorption can be regarded as the life strategy, which can give an advantage to one tree species over the others in photosynthetic competition and allow it to achieve evolutionary success.

REFERENCES

Adams W. W., Demmig-Adams B. 1994. Carotenoid composition and down regulation of photosystem II in three conifer species during the winter. *Physiologiae Plantarum* 92: 451–458.

Adams W. W., Zarter C. R., Ebbert V., Demmig-Adams B. 2004. Photoprotective strategies of overwintering evergreens. *Bio-Science* 54(1): 41–49.

Anten N. P. R. 2005. Optimal photosynthetic characteristics of individual plants in vegetation stands and implications for species coexistence. *Annals of Botany* 95: 495–506.

Aranda I., Gil L., Pardos J. A. 2000. Water relations and gas exchange in *Fagus sylvatica* L. and *Quercus petraea* (Mattuschka) Liebl. In a mixed stand at their southern limit of distribution in Europe. *Trees* 14: 344–352.

Balandier P., Collet C., Miller J. H., Reynolds P. E., Zedaker S. M. 2006. Designing forest vegetation management strategies based on the mechanisms and dynamics of crop tree competition by neighbouring vegetation. *Forestry* 79(1): 3–27.

Barnes B. V., Zak D. R., Denton S. R. Spurr S. H. 2012. *Forest Ecology*, 4th edition. John Wiley & Sons, New York, pp. 386–406.

Brodribb T. J., Pittermann J., Coomes D. A. 2012. Elegance versus speed: Examining the competition between conifer and angiosperm trees. *International Journal of Plant Sciences* 173(6): 673–694.

Craine J. M., Dybzinski R. 2013. Mechanisms of plant competition for nutrients, water and light. *Functional Ecology* 27: 833–840.

Cymerski M., Kopcewicz J. 1994. Labile phytochrome and photoperiodic flower induction in *Pharbitis nil* Chois. The irreversible phytochrome hypothesis. *Acta Societatis Botanicorum Poloniae* 63(3–4): 275–278.

Dreyer E. 1997. Photosynthesis and drought in forest trees. In: Rennenberg H., Eeschrich W., Ziegler H. (Eds.), *Trees—Contribution to Modern Tree Physiology*. Backhuys, Leiden, pp. 215–238.

Dudley S. A., Murphy G. P., File A. L. 2013. Mechanisms of plant competition. Kin recognition and competition in plants. *Functional Ecology* 27: 898–906.

Dumais D., Prévost M. 2014. Physiology and growth of advance *Picea rubens* and *Abies balsamea* regeneration following different canopy openings. *Tree Physiology* 34(2): 194–204.

Einhellig F. A. 1995. Mechanism of action of allelochemicals in allelopathy. In: Inderjit, K. M. M. Dakshini, F. A. Einhellig (Eds.), *Allelopathy, Organisms, Processes and Applications*. ASC Symp., Ser. 582. American Chemical Society, Washington, DC, pp. 96–116.

Epron D., Dreyer E. 1993. Long-term effects of drought on photosynthesis of adult oak trees (*Quercus petraea* (Matt.) Liebl. and *Quercus robur* L.) in a natural stand. *New Phytologist* 125: 381–389.

Grassi G., Bagnaresi U. 2001. Foliar morphological and physiological plasticity in *Picea abies* and *Abies alba* saplings along a natural light gradient. *Tree Physiology* 21: 959–967.

Grime J. P. 1974. Vegetation classification by reference to strategies. *Nature* 250: 26–31.

Grime J. P. 1977. Evidence for the existence of three primary strategies in plants and its relevance to ecological and evolutionary theory. *The American Naturalist* 111(982): 1169–1194.

Hikosaka K., Terashima I. 1996. Nitrogen partitioning among photosynthetic components and its consequences in sun and shade plants. *Functional Ecology* 10: 335–343.

Houter N. C., Pons T. L. 2014. Gap effects on leaf traits of tropical rainforest trees differing in juvenile light requirement. *Oecologica* 175: 37–50.

Kazada M., Salzer J., Schmid I., Von Wrangell P. 2004. Importance of mineral nutrition for photosynthesis and growth of *Quercus petraea*, *Fagus sylvatica* and *Acer pseudoplatanus* planted under Norway spruce canopy. *Plant and Soil* 264: 25–34.

Kierzkowski D., Samardakiewicz S., Robakowski P. 2007. Variation in ultrastructure of chloroplasts of needles of silver fir (*Abies alba* Mill.) saplings growing under the canopies of diverse tree species. *Polish Journal of Ecology* 55(4): 821–825.

King R. W., Vince-Prue D., Quail P. H. 1978. Light requirement, phytochrome and photoperiodic induction of flowering of *Pharbitis nil* Chois. *Planta* 141: 15–22.

Koutika L.-S., Vanderhoeven S., Chapuis-Lardy L., Dassonville N., Meerts P. 2007. Assessment of changes in soil organic matter after invasion by exotic plant species. *Biology and Fertility of Soils* 44: 331–341.

Kuehne Ch., Nosko P., Horwath T., Bauhus J. 2014. A comparative study of physiological and morphological seedling traits associated with shade tolerance in introduced red oak (*Quercus rubra*) and native hardwood tree species in southwestern Germany. *Tree Physiology* 34: 184–193.

Kume A., Satomura T., Tsuboi N., Chiwa M., Hanba Y. T., Nakane K., Horikoshi T., Sakugawa H. 2003. Effects of understory vegetation on the ecophysiological characteristics of an overstory pine, *Pinus densiflora*. *Forest Ecology and Management* 176: 195–203.

Küppers M., Timm H., Orth F., Stegemann J., Stöber R., Schneider H., Paliwal K., Karunaichamy K. S. T. K., Oritz R. 1996. Effects of light environment and successional status on light fleck use by understory trees of temperate and tropical forests. *Tree Physiology* 16: 69–80.

Lambers H., Chapin F. S., III, Pons T. L. 1998. Plant physiological ecology. Springer-Verlag, New York.

Leudemann G., Matyssek R., Winkler J. B., Grams T. E. E. 2009. Contrasting ozone x pathogen interaction as mediated through competition between juvenile European beech (*Fagus sylvatica*) and Norway spruce (*Picea abies*). *Plant Soil* 323: 47–60.

Leuschner Ch., Backes K., Hertel D., Schipka F., Schmitt U., Terborg O., Runge M. 2001. Drought responses at leaf, stem and fine root levels of competitive *Fagus sylvatica* L. and *Quercus petraea* (Matt.) Liebl. trees in dry and wet years. *Forest Ecology and Management* 149: 33–46.

Lusk Ch. H., Wright I., Reich P. B. 2003. Photosynthetic differences contribute to competitive advantage of evergreen angiosperm trees over evergreen conifers in productive habitats. *New Phytologist* 160: 329–336.

Mallik U. A. 2008. Allelopathy in forested ecosystems. In: Zeng R. S., Mallik U. A., Luo M. S. (Eds.), *Allelopathy in Sustainable Agriculture and Forestry*. Springer, New York, pp. 363–386.

Mangla S., Sheley R. L., James J. J., Radosevich S. R. 2011. Intra and interspecific competition among invasive and native species during early stages of plant growth. *Plant Ecology* 212: 531–542.

Marquis A. D. 1990. *Prunus serotina* Ehrh. Black Cherry. Silvics of North America: 2. In: Burns R. M., Honkala B. H. (Eds.), *Hardwoods—Agriculture Handbook 654*. U.S. Department of Agriculture, Forest Service, Washington, DC.

Matyssek R., Agerer R., Ernst D., Munch J. C., Oßwald W., Pretzsch H., Priesack E., Schnyder H., Treutter D. 2005a. The plant's capacity and regulating resource demand. *Plant Biology* 7: 560–580.

Matyssek R., Schnyder H., Ernst D., Munch J. C., Oßwald W., Pretzsch H., Priesack E., Schnyder H., Treutter D. 2005b. Resource allocation in plants. The balance between resource sequestration and retention. *Plant Biology* 7: 557–559.

Messier C., Bellefleur P. 1988. Light quantity and quality on the forest floor of pioneer and climax stage in a birch–beech–sugar maple stand. *Canadian Journal of Forest Research* 18: 615–622.

Mosandl R., Kleinert A. 1998. Development of oaks (*Quercus petraea* (Matt.) Liebl.) emerged from bird-dispersed seeds under old-growth pine (*Pinus sylvestris* L.) stands. *Forest Ecology and Management* 106: 35–44.

Niinemets Ü., Kull O., Tenhunen J. D. 1998. An analysis of light effects on foliar morphology, physiology, and light interception in temperate deciduous woody species of contrasting shade tolerance. *Tree Physiology* 18: 681–696.

Niyogi K. K. 1999. Photoprotection revisited: Genetic and molecular approaches. *Annual Review of Plant Physiology* 50: 333–359.

Novoplansky A. 2009. Picking battles wisely: Plant behavior under competition. *Plant, Cell and Environment* 32: 726–741.

Pierik R., Mommer L., Voesenek L. A. C. J. 2013. Molecular mechanisms of plant competition: Neighbour detection and response strategies. *Functional Ecology* 27: 841–853.

Reich P. B., Walters M. B., Ellsworth D. S. 1997. From tropics to tundra: Global convergence in plant functioning. *Proceedings of the National Academy of Sciences of the United States of America* 94: 13730–13734.

Robakowski P., Antczak P. 2008. Ability of silver fir and European beech saplings to acclimate photochemical processes to the light environment under different canopies of trees. *Polish Journal of Ecology* 56(1): 3–16.

Robakowski P., Bielinis E. 2011. Competition between sessile oak (*Quercus petraea*) and black cherry (*Prunus serotina*): Dynamics of seedlings growth. *Polish Journal of Ecology* 59(2): 297–306.

Robakowski P., Montpied P., Dreyer E. 2003. Plasticity of morphological and physiological traits in response to different levels of irradiance in seedlings of silver fir (*Abies alba* Mill). *Trees* 17: 431–441.

Saeki T. 1960. Interrelationships between leaf amount, light distribution and total photosynthesis in a plant community. *Botanical Magazine Tokyo* 73: 55–63.

Trinder C. J., Brooker R. W., Robinson D. 2013. Mechanisms of plant competition. Plant ecology's guilty little secret: Understanding the dynamics of plant competition. *Functional Ecology* 27: 918–829.

Valladares F., Martinez-Ferri E., Balaguer L., Perez-Corona E., Manrique E. 2000. Low leaf-level response to light and nutrients in Mediterranean evergreen oaks: A conservative resource-use strategy? *New Phytologist* 148: 79–91.

van Kuijk M., Anten N. P. R., Oomen R. J., van Bentum D. W., Werger M. J. A. 2008. The limited importance of size-asymmetric light competition and growth of pioneer species in early secondary forest succession in Vietnam. *Oecologia* 157: 1–12.

Vetter J. 2000. Plant cyanogenic glycosides. *Toxicon* 38: 11–36.

Vézina P. E., Boulter D. W. K. 1966. The spectral composition of near ultraviolet and visible radiation beneath forest canopies. *Canadian Journal of Botany* 44: 1267–1284.

Walck J. L., Baskin J. M., Baskin C. C. 1999. Relative competitive abilities and growth characteristics of a narrowly endemic and a geographically widespread *Solidago* species (*Asteraceae*). *American Journal of Botany* 86(6): 820–828.

Weir T. 2007. The role of allelopathy and mycorrhizal associations in biological invasions. *Allelopathy Journal* 20(1): 43–50.

Woods D. B., Turner N. C. 1971. Stomatal response to changing light by four tree species of varying shade tolerance. *New Phytologist* 70: 77–84.

Wu A.-P., Yu H., Gao S.-Q., Huang Z.-Y., He W.-M., Miao S.-L., Dong M. 2009. Differential belowground allelopathic effects of leaf and root of *Mikania micrantha*. *Trees* 23: 11–17.

Wullschleger S. D. 1993. Biochemical limitations to carbon assimilation in C3 plants—A retrospective analysis of the A/Ci curves from 109 species. *Journal of Experimental Botany* 44: 907–920.

Wyka T., Robakowski P., Żytkowiak R. 2007. Leaf acclimation to contrasting irradiance in juvenile evergreen and deciduous trees. *Tree Physiology* 27: 1293–1306.

Yuan Y., Guo W., Ding W., Du N., Luo Y., Liu J., Xu F., Wang R. 2013. Competitive interaction between the exotic plant *Rhus typhina* L. and the native tree *Quercus acutissima* Carr. in Northern China under different soil N: P ratios. *Plant Soil* 372: 389–400.

28 Functional Traits and Plasticity of Plants

Elena Masarovičová, Mária Májeková, and Ivana Vykouková

CONTENTS

28.1 Introduction 487
28.2 Plant Functional Traits 487
    28.2.1 History of Functional Traits and Functional Ecology 487
    28.2.2 Different Approaches to the Assessment of Functional Traits 488
    28.2.3 Intraspecific Trait Variability 489
    28.2.4 Plant Ecological Strategies 490
        28.2.4.1 Leaf Economics Spectrum 491
        28.2.4.2 Plant Water Stress and Drought Tolerance 492
        28.2.4.3 Plant Height 492
        28.2.4.4 Other Important Dimensions of Plant Ecological Strategies 492
28.3 Plasticity 493
    28.3.1 History of Phenotypic Plasticity and Adaptation of Plants to Environmental Conditions ... 493
    28.3.2 Adaptive Plasticity of Plants in Relationship to the Environment 494
    28.3.3 Phenotypic Plasticity and Evolution of Plants 495
    28.3.4 Morphological, Anatomical, and Physiological Plasticity of Woody Species 495
    28.3.5 Relationship between Plasticity of Whole Plant and Leaf Traits 496
    28.3.6 Plant Plasticity and Succession of Plant Populations 497
    28.3.7 Climatic Conditions and Plant–Plant Interaction 498
    28.3.8 Phenotypic Plasticity and Climate Changes 499
28.4 Concluding Remarks 500
Glossary 500
Acknowledgments 501
References 501

28.1 INTRODUCTION

The aim of comparative plant ecology is to know how functional traits vary among plant species in different habitats, under variable climatic conditions, and to what extent this variation has adaptive character. Plant species vary in morphological, anatomical, and physiological traits, despite their shared key functional purpose of gas exchange (photosynthesis, respiration, transpiration). This capacity of a given species to express different phenotype in different environments is known as phenotypic plasticity. Plasticity is an important aspect of evolution, development, and function of plants in their environments, and therefore, at present, it has been widely recognized as a significant mode of plant functional trait diversity. Variation in leaf traits can be found across species, among populations of the same species, and even between organs produced by a single plant. Similar modifications of leaf structure and form in response to the environment appear at each of these levels (across or within species, populations, or individuals).

Fundamental questions for evolutionary ecologists under global climatic change are how plant species will respond to these new conditions and what mechanisms will be involved in this process. Thus, the understanding of phenotypic plasticity will be crucial for predicting change in species distribution, community composition, and plant production under global climatic change (cf. Chaturvedi and Raghubanshi 2013; Gratani 2014). In this chapter, we therefore focus on characterizing functional traits and plasticity of the plants in relationship to climatic conditions, evolution of species, and formation of plant populations. Attention is also paid to these characteristics from the aspect of their importance for comparative plant ecology as a fundamental scientific discipline.

28.2 PLANT FUNCTIONAL TRAITS

28.2.1 HISTORY OF FUNCTIONAL TRAITS AND FUNCTIONAL ECOLOGY

Plant functional traits enable scientists to link processes observed on plant individuals to plant population dynamics, species' distribution patterns and their coexistence, community assembly, and finally, ecosystem functioning. Traits have been used by evolutionists since Darwin (1859) as surrogates for organismal performance. Over the last few decades, progress in community ecology (Grime 1974; McGill et al. 2006) and ecosystem ecology (Chapin 1993; Lavorel and Garnier 2002) has forced the concept of trait beyond these original boundaries. Functional ecology has slowly emerged as a novel and progressive way to respond to the arising issues of changing climate and increased diversity loss (Calow 1987; Keddy 1992). Ecologists have searched for ways to target these questions first with a more categorical approach of plant functional groups (Noble and Gitay 1996; Lavorel et al. 1997; McIntyre and Lavorel 2001). Functional groups represented groups of plant species that had the same function on the level of an individual, similar response to the environmental conditions, and/or similar effect on the ecosystems, and shared the similar set of functional traits (McIntyre and Lavorel 2001). Although functional groups represented a first necessary step toward functional ecology, there were several drawbacks to this approach, for example, no exact measurements and regional contingency of plant functional groups.

Therefore, ecologists recognized the need for a shift toward continuous traits, which are relatively easy to measure and which would be directly linked to organismal performance (Diaz and Cabido 2001). At this point, several questions were raised such as which functional traits were important for individual performance of plants, as well as for scaling up from plant individuals to the communities and further on to the ecosystem functioning. Several lists of key plant functional traits have been proposed. One of the first was an L–H–S scheme that focused on the three traits related to important dimensions of variation between species (Westoby 1998). Those were leaf traits (L), plant height (H), and seed mass (S). A more extensive common core list of traits for functional ecology was then presented by Weiher et al. (1999). In order to make the results on functional traits comparable across the globe, efforts have been made to standardize traits measurements (Garnier et al. 2001, 2007; Cornelissen et al. 2003; Pérez-Harguindeguy et al. 2013). With both the list of core traits and standardized measurement protocols, functional ecology moved quickly onward, and many local and regional studies were conducted. An even further step onward was research that aimed at

1. The search for global patterns in plant functional traits, especially along different climatic gradients and biogeographic regions (McIntyre et al. 1999; Niinemets 2001; Diaz et al. 2004; Reich and Oleksyn 2004; Wright et al. 2004; Moles et al. 2005)
2. The identification of the suites of quantitative traits to explain ecosystem functioning (e.g., Eviner and Chapin 2003) or plant strategies (e.g., Vile et al. 2006)
3. The assessment of functional diversity based on the clustering of plant traits (e.g., Petchey and Gaston 2002)

However, in order to compare the local results, to make the general patterns visible, and to draw conclusions on ecological mechanisms in plants, compilation of the data on functional traits from across the whole globe and their accessibility for all ecologists have been necessary. With this purpose in mind, several trait databases have been built (this process has been continuing up to now). The very first has been the trait database on the species common to the British Isles (Grime et al. 1988). BiolFlor database (Kuhn et al. 2004) has been built mainly as a tool for plant invasion ecology. LEDA database (Kleyer et al. 2008) has compiled a large amount of trait data on northwest European flora. CLO-PLA is a database for clonal traits (Klimešová and de Bello 2009). The largest database so far has been the TRY database (Kattge et al. 2011), the purpose of which has been to literally *try* and compile trait data from all around the world. With the emergence of trait databases, a boom of studies exploded bringing a whole new perspective to functional ecology. An increasing number of studies showed evidence that plant functional traits influenced the success or failure of species in a given environment (Cornwell and Ackerley 2009), the processes that affect their abundance and rarity in a region (Mouillot et al. 2007), the mechanisms that rule successional replacements (Kahmen and Poshold 2004), and the mechanisms that rule species' response to climate change (Soudzilovskaia et al. 2013). Many more studies have been conducted, and therefore the general aim nowadays is to identify the leading dimensions among plant functional traits (Laughlin 2014; Reich 2014) and incorporate the knowledge gained during the past 20 years into the more theoretical mechanisms of plant ecology (Adler et al. 2006; Mayfield and Levine 2010).

28.2.2 DIFFERENT APPROACHES TO THE ASSESSMENT OF FUNCTIONAL TRAITS

The definition of the terms *trait* and *functional trait* remained for a long time ambiguous, and their usage was sometimes confusing. For example, the term *trait* was used for measurements on an individual level (such as leaf traits or traits describing stature of plants and their seed mass) and on a plot level (such as standing biomass or vegetation cover; Weiher et al. 1999; Petchey et al. 2004). Another confusing example was the usage of *trait* for soil conditions or microbial phosphorus (Eviner 2004). In order to unify the definition of functional traits and to make the concept of functional traits stronger and unambiguous, Violle et al. (2007) presented a clear definition of functional traits: "... any morphological, physiological and phenological traits which impact fitness indirectly via their effects on growth, reproduction and survival, the three components of individual performance." These authors further suggested that traits should be used only used at the level of individuals without any additional information required from the environment or at any other level of organization (e.g., community). This definition of the term *functional trait*, sometimes referred to as *trait for short*, has been widely accepted.

A hierarchy in the functional concept was extended from Arnold's *morphology, performance, fitness* paradigm (Arnold 1983, proposed for animals) to plant ecology

(Ackerly et al. 2000). The hierarchy was proposed as follows (Violle et al. 2007):

1. *Performance* was substituted by *performance traits* that affect fitness directly and the three performance traits were growth, reproduction, and survival:
   1.1 Growth, assessed as vegetative biomass that represents the net cumulated outcome of all growth and loss processes
   1.2 Reproduction, assessed as a reproductive output measured by the seed biomass and/or the number of seeds produced
   1.3 Plant survival, assessed as a binary variable (an individual is dead or alive in the environmental condition considered) or a probability
2. *Morphology* was substituted by *functional traits* that affect fitness indirectly. They were defined as any morphological, physiological, and phenological traits measured at the level of individual.

Hodgson et al. (1999) divided functional traits into *hard* and *soft* traits. *Hard* traits were characterized as those that captured the function of interest, but which were either difficult or expensive to measure. In a sense, they could be taken as an analogy to the *performance traits* introduced by Violle et al. (2007). *Soft* traits were characterized as surrogates of the function of interest that were less difficult and/or expensive to obtain. Soft traits could be used to access hard traits. This dichotomy is nowadays rarely used in functional ecology, because the difficulty and expense of a measurement might depend on one's amount of help and funding available.

Lavorel and Garnier (2002) proposed to distinguish *functional effect traits* and *functional response traits*. Functional effect traits are species traits that impact ecosystem functioning. An example is leaf dry matter content and its effect on the leaves' decomposition rate and nutrients cycling (Fortunel et al. 2009). Functional response traits are traits that vary consistently in response to changes in environmental factors. The following are some examples: (1) life forms and leaf traits for response to climate (MacGillivray and Grime 1995; Niinemets 2001), (2) specific leaf area (SLA) and leaf chemical composition for response to soil nutrients availability (Poorter and de Jong 1999; Hodgson et al. 2011), and (3) life cycle, plant height, architecture, and seed traits for response to disturbance (Bond and Midgley 1995; McIntyre and Lavorel 2001; de Bello et al. 2005).

Environmental tolerances to shade, drought, heavy metals, or herbivory, as well as habitat and ecological preferences (e.g., Ellenberg's indices; Ellenberg 1988), are sometimes also called *traits* (e.g., frost tolerance; McGill et al. 2006). However, according to the definition given by Violle et al. (2007), this should not be so, since external, i.e., environmental, variables are required for their definition. Therefore, these authors proposed that the response of the whole-organism performance to an environmental variable should be called *ecological performance*, defined as "… the optimum and/or the breadth of distribution of performance traits along an environmental gradient." Ecological performance is dependent on the simultaneous and coordinated response of multiple functional traits to environmental factors. However, if environmental tolerances are measured as quantitative variables, they could fall into the characteristic of a response trait.

28.2.3 Intraspecific Trait Variability

Functional traits measured on the level of individuals are affected by both variability within species (intraspecific variability) and variability between species (interspecific variability). Although some traits are generally not variable within species, for example, the life form and ability to fix nitrogen, most quantitative traits are highly variable within species (Westoby et al. 2002; Cornelissen et al. 2003; Albert et al. 2010). Intraspecific trait variability has been neglected in trait-based community ecology for several years, as the basic assumptions were that interspecific trait variability is more important as intraspecific variability (McGill et al. 2006). There were also two main constraints to assess intraspecific trait variability: (1) more extensive datasets needed for its evaluation, and (2) the methodological problems to disentangle the intraspecific and interspecific trait variability. The first problem has become less pronounced with the building of the extensive datasets of functional trait measures and also thanks to better connection and cooperation between scientists worldwide. There have been also several attempts to solve the methodological issues on disentangling the intraspecific and interspecific trait variability (de Bello et al. 2011; Lepš et al. 2011; Violle et al. 2012).

Intraspecific trait variability represents the overall variability of trait values of a given species and is an expression of a species response to changes in environmental conditions through both adaptation and acclimation (Geber and Griffen 2003). The latter is also referred to as phenotypic plasticity and is that part of intraspecific trait variability that reflects the trait variability resulting from environmental heterogeneity in time, space, or during individual lifetime (Coleman et al. 1994). However, as intraspecific trait variability results from both mechanisms and their interaction (Scheiner and Lyman 1991), it should be taken as a whole and not restricted to just one of the mechanisms (Coleman et al. 1994).

Intraspecific trait variability also results from different organizational levels and thus can be decomposed into three components, which reflect the hierarchy of organization in ecological systems: (1) within-individual variability, (2) between-individual variability, and (3) population-level variability (Bolnic et al. 2003; Albert et al. 2011). (1) Within-individual variability can arise from spatial heterogeneity, such as shaded versus sun leaves (Richardson et al. 2001); from temporal variability, such as differences in ontogeny (Coleman et al. 1994), phenology, seasonality, and climate change; or from differences in individual history, such as former disease or disturbance that has affected a particular organ (Albert et al. 2010). (2) Between-individual variability, i.e., variability within a particular population, can arise from the coexistence of different genotypes or from differential

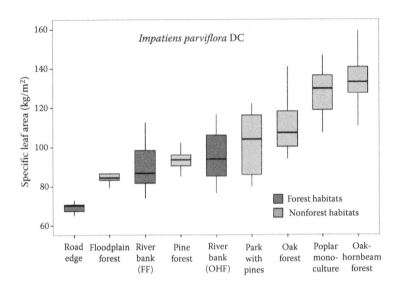

FIGURE 28.1 **(See color insert.)** The intraspecific variability in SLA in an invasive species *I. parviflora* DC measured in nine populations inhabiting different environmental conditions. These represent the habitats that the species invaded throughout the last 30 years and in which they have established a viable population, often monodominant. The habitats differ in light, nutrient, and water availability. FF: floodplain forest; OHF: oak-hornbeam forest.

response to resource availability (nutrients, water, light), stress (e.g., dry conditions), disturbance (e.g., grazing or mowing), or biotic interactions (competition or facilitation; Albert et al. 2011). (3) Population-level variability reflects the differences in trait values between populations of a given species (in space or time). Similarly to between-individual variability, population-level variability can be due to different genotypic composition of populations as well as due to differences in plastic response to the prevailing environmental conditions (e.g., climatic or disturbance; Sandquist and Ehleringer 1997). Some plant species are more variable, i.e., plastic, than others. The species that have a wide ecological valency are usually the more plastic ones with higher intraspecific trait variability. Examples are many invasive species for which their high plastic response to different environmental conditions enables them to colonize novel habitats. An example of an invasive species with a very high intraspecific trait variability (mainly on the population level) is *Impatiens parviflora* DC, a species native to Asia and invasive in Europe (Figure 28.1; Majekova et al. unpublished data).

28.2.4 PLANT ECOLOGICAL STRATEGIES

A common definition of a strategy is a plan of action designed to achieve a major aim. The major aim of (plant) species is to grow, reproduce, and survive. Therefore, according to Westoby et al. (2002), ecological strategy is "… the manner in which species secure carbon profit during vegetative growth and ensure gene transmission into the future." Plant species all compete for the same resources of light, water, CO_2, and mineral nutrients, and thus they need to differ in the ways of acquiring and managing these resources. The different ways in which plants make their living are expressed through different ecological strategies that can be captured

and quantified by several dimensions of variation between species (Westoby et al. 2002; Wright et al. 2004; Laughlin 2014; Reich 2014). These dimensions can be expressed through different costs and benefits, i.e., trade-offs, related to particular plant organs or whole-plant traits (Figure 28.2), and through the correlations between them (Westoby et al. 2002; Laughlin 2014). Westoby (1998) first proposed a pragmatic leaf–height–seed (LHS) plant strategy scheme, as these three simple traits (SLA, height, and seed mass) influence plant persistence, establishment, and dispersal. Later, Westoby and colleagues added leaf area (Westoby et al. 2002) and wood density and root traits (Westoby and Wright 2006) as potentially important plant strategy dimensions. Laughlin (2014) summarized the known dimensions of plant ecological strategies and tested their correlations in order to allocate those potentially most important. To focus on the independent, i.e., uncorrelated, dimensions in plant strategies is important in order to acquire unique information about plant function and in order to maximize the understanding of diversity patterns and ecosystem processes (Wright et al. 2007; Laughlin 2014).

Some of the plant ecological strategy dimensions are not very well defined, and their costs and benefits are not well known. An example of this is the leaf size–twig size spectrum (Westoby et al. 2002) or the trade-off behind clonal spread of species (Klimešová and Klimeš 2007). The lack of information results mainly from difficulties with the quantification of traits related to these strategies, and therefore sufficient empirical support is still missing. On the other hand, some dimensions are easily quantified, and the trade-offs behind them are more clear and straightforward. An example is the *leaf economics spectrum* (Wright et al. 2004) that have been very popular among plant scientists in the last ten years and have gained much empirical support from local to global scale. It represents a functional trade-off between a resource

Height
competition for light

Leaf
photosynthesis,
gas exchange

Flowers
sexual reproduction

Stem
transport and storage of
nutrients and water, support

Seeds
fecundity, dispersal,
germination, establishment

Root system
absorption, transport, and
storage of nutrients and water

FIGURE 28.2 **(See color insert.)** Different plant organs and plant height reflect different functions that are important for plant fitness. (Modified from Laughlin, D. C., *Journal of Ecology* 102:186–193, 2014.)

acquisition strategy and a resource conservation strategy in plants mediated through leaf traits. Recently, Reich (2014) suggested an extension of the economics spectrum from leaf to whole plant. He proposed a single *fast–slow* plant economics spectrum that combined leaves, stems, and roots as a key feature of the plant universe that could help to explain individual ecological strategies.

28.2.4.1 Leaf Economics Spectrum

Leaves are the most studied plant organs, and the leaf economics spectrum is the best-known dimension of plant ecological strategy. It describes a trade-off between maximum rate of carbon acquisition and leaf life span. In other words, it describes a trade-off between a resource acquisition strategy related to the ability to photosynthesize more and grow more rapidly, and a resource conservation strategy related to the ability to store resources and cope with environments with low resources (Westoby et al. 2002; Wright et al. 2004).

Individual leaf economic traits are quite simple and correlated among each other (Wright et al. 2004; Laughlin 2014). The latter implies that a single dimension accounts for the majority of variation between them (the main concept of the leaf economics spectrum; Wright et al. 2004). The simple traits involved in leaf economics spectrum are

1. LMA, leaf mass per area, which measures the leaf dry-mass investments per unit area of light-intercepting leaf. LMA is made up of lamina depth multiplied by tissue density (Witkowski and Lamont 1991).

2. SLA, specific leaf area, is 1/LMA and represents the ratio of leaf area to its dry weight (Garnier et al. 2001).

3. LDMC, leaf dry matter content, represents leaf dry mass divided by the fresh mass (Garnier et al. 2001).

4. Leaf thickness measurement can be problematic due to variation in leaf morphology, the differences in thickness within individual leaves, and the relatively small dimensions. It can be effectively estimated using SLA and LDMC (Vile et al. 2005).

5. Leaf nitrogen is important for the proteins needed in the photosynthetic process, especially Rubisco. Most nitrogen is fixed by plants from the atmosphere (Lambers et al. 1998).

6. Leaf phosphorus is found in nucleic acids, lipid membranes, and bioenergetic molecules such as ATP. Phosphorus derives from weathering of soil minerals at a site (Lambers et al. 1998).

7. Leaf life span describes the duration of a leaf's life (Wright et al. 2004).

Species with lower SLA have higher tissue density and/or thicker laminas (Witkowski and Lamont 1991; Wilson et al. 1999). Low-SLA species tend to achieve longer average leaf life span in a variety of habitats (Reich et al. 1997; Wright and Westoby 2002), suggesting that longer leaf life span requires more structural strength. Lower SLA protects against damage, and species with longer leaf life span should also invest more to protect themselves against herbivores. Already thicker and tougher leaves are the most common form of defense (Coley 1983), but long leaf life span may also be correlated with greater relative allocation to defensive compounds, such as tannins or phenols (Coley 1988). Species with high SLA tend to have higher photosynthetic capacity, as well as generally higher leaf N concentrations (Reich et al. 1997) and shorter diffusion paths from stomata to chloroplasts (Parkhurst 1994). High SLA with high photosynthetic capacity and generally faster turnover of leaves is associated with a more flexible response to the spatial patchiness of light and soil resources (Grime 1994), which gives this strategy short-term advantage over the low-SLA strategy. However, species with low SLA and longer life span have more long-term advantages such as (1) longer mean residence time of nutrients (Aerts and Chapin 2000), (2) slow decomposition of low-SLA litter that may restrict opportunities for potentially fast-growing competitors, and (3) accumulation of greater total leaf mass over time (Cornelissen et al. 1999). By comparing the costs and benefits of these two opposing strategies, it seems that longer leaf life span has clear advantages over short leaf life span. However, over a longer period of time, leaves suffer damage from herbivores and pathogens (Coley and Barone 1996), are colonized by epiphylls (Coley et al. 1993), and can be overshadowed by newly produced leaves. Taking the advantages and disadvantages of both strategies into consideration, they enable a coexistence of a wide variety of species (Westoby et al. 2002). There are other leaf properties that are not strongly correlated with leaf economics traits and less explored, for example, minimum water potential (Ackerly 2004), leaf surface area (Pierce et al. 2013), hydraulic conductance (Sack et al. 2003), and vein density (Sack and Scoffoni 2013).

28.2.4.2 Plant Water Stress and Drought Tolerance

Cell turgor loss has been considered as the classic indicator of plant water stress, having impacts on cellular structural integrity, metabolism, and whole-plant performance (Kramer and Boyer 1995; McDowell 2011). Consequently, the leaf water potential at turgor loss point (or wilting point; πtlp) has been recognized as a main physiological determinant of the response of plants to water stress. The πtlp is considered as a *hard* trait that quantifies leaf and plant drought tolerance most directly, because a more negative πtlp represents the case when leaf remains turgid and maintains its function (Sack et al. 2003; Lenz et al. 2006). Plants with low, i.e., more negative, πtlp tend to maintain stomatal conductance, hydraulic conductance, photosynthetic gas exchange, and growth at lower soil water potential, which is especially important when droughts occur during the growing season (Sack et al. 2003; Mitchell et al. 2008). The πtlp is thus a trait quantifying the ability to *tolerate* drought rather than to *avoid* drought.

Plants with high, i.e., less negative, πtlp tend to prefer conditions where water is not a limiting factor and species develop a less conservative strategy, with the ability to move and store water well and with the capacity to achieve higher growth rates. As a disadvantage, species with this strategy are then more vulnerable to drought and wilting in terms of their πtlp (Reich 2014). πtlp as a trait has potential for quantifying

drought tolerance in plants (Niinemets 2001; Brodribb and Holbrook 2003; Lenz et al. 2006). For example, Bartlett et al. (2012) found that πtlp was strongly correlated with water availability within and across biomes. These authors also found that leaf mass per area (LMA) was poorly correlated to drought tolerance, stressing the importance of the πtlp as an indicator of drought tolerance relative to other plant traits. On a more local scale, Savage and Cavender-Bares (2012) found that for a group of co-occurring willows, a set of traits that included πtlp varied in parallel with species abundances along a local soil moisture gradient, suggesting that species with traits well matched to specific locations along the moisture gradient dominated those locations.

28.2.4.3 Plant Height

Plant height is an easily measured whole-plant trait that has been used by many ecologists as a key trait in understanding plant strategies (e.g., Westoby 1998; Weiher et al. 1999; Reich 2014). It has been classically related to plant competitiveness for light (Keddy and Shipley 1989), such that being taller than the neighboring species has a competitive advantage of better access to light and higher photosynthesis. The costs of being tall include higher investments in stem and supporting structures, as well as disadvantages in transport of water to height (Westoby et al. 2002). In an ideal case, if a species lived without any neighbors, it would grow close to the ground and thus would maximize the benefits of photosynthesis over the costs into stem investments and maintenance. However, such species would be, in real situations, outcompeted by a taller species that spent some part of its resources on nonproductive stem tissues in order to grow a little higher. This species would be then outcompeted by another slightly taller species, and so forth (Westoby et al. 2002). Therefore, plant height is one of the most plastic traits among individuals of the same species, dependent on the specific set of abiotic and biotic conditions in which a given individual grows.

Plant height encompasses several strategies that are to be considered together when trying to understand its costs and benefits: the upper limit of height, the rate at which species grow upward, and the duration over which stems persist at the maximum height (Iwasa et al. 1985). The upper limit of height is the point in which the costs fully deplete current photosynthesis. The rate at which species grow upward needs to be considered mainly over cycles of disturbance and subsequent growth. Disturbances such as mowing in meadows, cutting in forests, or natural calamities open canopy, so that light becomes available also closer to the ground and enables the shorter species to coexist with the taller species (Westoby et al. 2002).

28.2.4.4 Other Important Dimensions of Plant Ecological Strategies

Stems provide structural support and transport of water and nutrients, and they can be very important for defense against herbivores and for storage of resources, especially throughout unfavorable conditions. Stem density represents a trade-off between the efficiency of hydraulic conductivity and resistance to drought- or freezing-induced cavitation (Hacke et al. 2001). It also reflects a trade-off between growth rate and survival (Wright et al. 2010). More specifically in woody plants, wood density is related to resistance to decay, storage capacity, and mechanical strength (Chave et al. 2009), while bark thickness is related to defense against fire, pests, or pathogens (Paine et al. 2010).

Roots are perhaps the least studied plant organs, because of the difficulty of their measurements in the field. There have been, however, several studies that have focused solely on root traits (Kutschera and Lichtenegger 1982, 1992), and our understanding of root function is nowadays improving (Eshel and Beeckman 2013). Root traits such as specific root length or tissue density may represent a trade-off between growth rate and life span. Root traits may influence the plant's ability to reach nutrient-rich patches with their fine absorptive roots. The empirical evidence is, however, ambiguous whether fine root traits are correlated in their functioning with leaf traits, i.e., whether roots reflect a *whole-plant economics spectrum* (Craine and Lee 2003; Tjoelker et al. 2005; Reich 2014).

Flowering phenology is a key component of plant function, as it affects plant interactions with pollinators (Hegland et al. 2009). Flowering phenology as a trait is one of the most plastic in time during a season, as well as particularly sensitive to global change (Fitter and Fitter 2002). Although onset of flowering was included in the first common core list of traits (Weiher et al. 1999), it has received little scientific attention in the subsequent years. One reason may be because the onset of flowering, its duration, and the timing of pollination are all contingent upon local environmental conditions and therefore difficult to compare across regions (Mouradov et al. 2002).

Seeds have different abilities of dispersal, germination, and seedling establishment (Grubb 1977). There is a well-defined trade-off between seed size and seed output that has been related to the competition–colonization trade-off (Westoby et al. 2002). Recent studies, however, indicate that seed mass reflects rather a trade-off between stress tolerance and fecundity (Muller-Landau 2010; Lonnberg and Eriksson 2013). Seed mass and seed shape also influence persistence in the seed bank (Thompson et al. 1993; Moles et al. 2000), and thus may reflect the ability of species to buffer against competition during unfavorable conditions (Angert et al. 2009).

Life-history traits are usually categorical whole-plant traits such as the life form (*sensu* Raunkiaer 1934), growth form (e.g., forbs, grasses, and graminoids), the occurrence of vegetative reproduction, or the capacity for clonal growth (Klimešová and Klimeš 2007). Life history can also include continuous traits, such as life span. Categorical traits are more stable and less plastic and are useful for the first division, e.g., to grasses and forbs, as these may have phylogenetically inherited different responses to environmental conditions. There are many plant traits that can be measured (Pérez-Harguindeguy et al. 2013), and therefore many new important dimensions of variation among species and correlations among traits to be unraveled. Plant functional ecology is a relatively young science that has still much more to offer for our understanding of the plant world and the relationships within it.

28.3 PLASTICITY

28.3.1 HISTORY OF PHENOTYPIC PLASTICITY AND ADAPTATION OF PLANTS TO ENVIRONMENTAL CONDITIONS

Natural environments inevitably vary both spatially and temporally. According to the classic evolution model, organisms accommodate that variation by means of natural selection, which through evolutionary time matches specific genotypes and environments. Considering a simple relationship of genotype to phenotype, this powerful model provides a genetic mechanism for adaptive phenotypic changes in populations. In one of the pioneer papers (Sultan 1995), the author focused on a second major mode of adaptation, one which is becoming particularly well understood in plants: the capacity of a single genotype to produce different, functionally appropriate phenotypes in different environments, or *adaptive phenotypic plasticity*. Based on phenotypic plasticity, adaptation occurs through individual development and physiology as well as through change in population gene frequencies. Thus, a single genotype may be able to maintain function and hence reproductive fitness under a variety of environmental conditions. For this reason, individual adaptive response has important implications for our understanding of natural selection and evolutionary diversification (Forsman 2015).

From an evolutional view, the phenotype is considered a result of environmental interference superimposed on the *inner reality* of the genes, which are seen as the basis of all evolutionarily meaningful variations (Sultan 1992). According to this model, adaptive diversity is best studied in a uniform environment, where genetic variation will be most clearly revealed. Alternatively, the genotype may be understood as a developmental system that will produce one of a number of possible phenotypes depending on its environmental circumstances. To the extent that the various phenotypes are functionally adaptive to the environments in which they are produced, the individual's phenotypic repertoire in itself comprises a significant mode of adaption to environments (e.g., Bradshaw 1965). This view of the phenotype as determined jointly by genotype and environment implied before 50 years ago quite a different approach, one designed to reveal the genotype's entire repertoire of responses.

According to Scheiner (1993), plasticity of plants reveals two distinct approaches to this problem: One is an approach that used the term *phenotypic plasticity* to denote all phenotypic changes across environments (e.g., De Jong and Stearns 1991). These authors used the terms *phenotypic plasticity* and *norm of reaction* interchangeably. In this study, plasticity is simply a neutral metric of phenotypic differences in various environments, of unknown and possibly little functional significance. This approach thus provided information as to the amount and pattern of phenotypic variability, but did not address the selective impact of that variability (Zhang and Lechowicz 1994). The other approach is one that conceptually distinguished aspects of phenotypic response to environment that are functionally adaptive from those that are developmentally or biochemically inevitable. This distinction is essential if we wish to know whether the phenotypic responses of genotypes may constitute adaptations. Thus, adaptive phenotypic plasticity could be defined as phenotypic response to an environment that enhances plant function and therefore fitness of the plant in that particular environment (Sultan 1987). Therefore, the argument that a given phenotypic response is adaptive (functionally appropriate) must rest on engineering principles and ecophysiological interpretation (Sultan and Bazzaz 1993a). For instance, the rate of photosynthesis per unit leaf area will inevitably be drastically reduced under low irradiance, such that plants grown in reduced light intensity produce only a fraction of the biomass of plants grown in high irradiance. In the case of light-reduced plants, then, the functionally adaptive plastic response would be to maximize photosynthetic surface area relative to biomass, thereby increasing light interception per gram of plant tissue. In other words, under reduced light conditions, the values of SLA, as one from the plant functional traits, increase.

It should be stressed that the plant physiological and ecophysiological papers provide a strong basis on which to evaluate the functional significance of changes in specific morphological, allocation, and metabolic traits. However, a given trait response can simultaneously reflect both inevitable and adaptive aspects. For instance, plants of *Calamagrostis canadensis* produce fewer and shorter rhizomes under low-temperature conditions of the soil. This results in preferential clonal expansion into more favorable sites and more extensive exploitation of those sites once entered (Macdonald and Lieffers 1993). According to Schlichting (1989a,b), adaptive interpretation of trait plasticity is also complicated by the fact that fitness of plants results from many aspects of the phenotype. Therefore, it is often necessary to consider interactions among sets of functionally related traits in order to evaluate the fitness effect of change in a particular trait. In some cases, how a trait is defined affects the interpretation of plastic response. Adaptive interpretation of plastic response is particularly complex when traits that interact in their effect on fitness, such as leaf size and number, show different amounts or directions of response to environmental changes. An ecophysiological perspective suggested that the functional importance of a given trait to plant fitness will indeed depend on environmental conditions. For example, relative biomass allocation to root tissues can strongly correlate with fitness when soil moisture or nutrients are in short supply, but not in more favorable soil environments (Sultan and Bazzaz 1993a). The environmental specificity of such correlation can suggest explanation in trait definition and in adaptive prediction. For instance, the correlation of photosynthetic water use efficiency (WUE; the ratio of the photosynthetic rate to the transpiration rate) with plant fitness varied not only in strength but also in sign depending on environmental conditions and in the number of taxa. One explanation for this paradoxical result is the fact that an increase in the above-mentioned ratio can be produced either by stomatal limitation of photosynthesis with water loss or by an increased photosynthetic rate (Donovan and Ehleringer 1994). Thus, depending on the strength of drought stress, these alternative physiological changes can have different adaptive consequences.

28.3.2 Adaptive Plasticity of Plants in Relationship to the Environment

One of the difficulties that arise immediately in research of plasticity is the interpretation of the phenotypic responses to environment. As a result of physical and biochemical effects on metabolic and developmental processes, plants in unfavorable or resource-poor environments inevitably show reduced growth. Hence, although phenotypes produced in suboptimal environments may include alterations that adaptively maximize function in those environments, they will necessarily reflect growth limits as well (Sultan and Bazzaz 1993a). For this reason, all of the phenotypic changes associated with different environmental conditions cannot be assumed to represent adaptive plastic adjustment. It should be stressed that major abiotic factors such as irradiance, water availability, nutrient content in the soil, and pH together with biotic factors, such as density and size of neighboring plants, represent key environmental pressures to which plants respond phenotypically. However, it is not always clear which environmental variable is most important, i.e., limiting, for plant growth and survival.

Numerous studies in a broad range of plant species and life forms have shown adaptive plasticity in many traits such as biomass allocation, morphology features, defense biochemical processes, metabolic pathways, physiological parameters, etc. A study of norms of reactions in some plants provided examples of adaptive response of individual genotypes to several environmental factors in growth traits important to resource acquisition. In the earlier experiments of Sultan and Bazzaz (1993a), 10 genotypes from an old-field population of *Polygonum persicaria* were cloned and grown in the glasshouse at three light levels, covering the range of light availabilities measured in the field (from 100% to 8% of full summer sun). Other environmental factors were held constant at favorable levels. It was found that total plant biomass was reduced by 98% in clones grown at very low irradiance (variant 8% of full summer sun) compared with those given full (100%) of sun irradiance. In addition, under moderately and severely reduced light conditions, genotypes doubled and tripled their proportional allocation of biomass to leaf tissue. Light-deprived plants also expressed great plasticity in leaf morphology by increasing SLA as a very sensitive leaf functional trait. This thin spreading of leaf tissue is known to enhance light-harvesting efficiency under conditions of low irradiance. The combined effect of these phenotypic changes under moderate and low irradiance was to dramatically increase leaf area relative to leaf biomass (SLA). Furthermore, all 10 genotypes survived and produced viable achenes across the full range of irradiance. Thus, these phenotypic responses were associated with a very broad range of tolerance for variability in light conditions within individual genotypes. The phenotypes of light-deprived plants of *P. persicaria* plants thus showed inevitable growth reductions as well as alterations that maximized photosynthetic surface area relative to plant biomass.

In order to interpret these changes as plastic adaptations to low irradiance, it is necessary to determine that they occurred specifically in response to low irradiance conditions rather than as a general *stress phenotype* produced in any suboptimal environment. To investigate this question of resource specificity, a second experiment was performed (Sultan and Bazzaz 1993b) in which genotypes from the same population were grown at four soil moisture treatments: severe drought (daily wilting), moderate drought (occasionally wilting), field capacity (soil both moist and aerated), and flooded (pots submerged to soil level). The other environmental factors such as irradiance, temperature, air humidity, and soil nutrients were maintained at constant, favorable levels. It is known that one specific response appropriate to drought stress would be to increase biomass allocation into the roots relative to vegetative tissues of shoots, which maximizes the supply of water available to those tissues. It was found that all of the *P. persicaria* genotypes showed this plastic adjustment to water limitation, increasing root-to-shoot biomass ratio by 70% in moderately dried soil, and by more than 100% in extremely dry soil relative to plants of the same genotypes given ample soil moisture. In contrast to clones of closely related genotypes deprived of light, plants subjected to drought did not alter SLA or change tissue allocation to leaves compared with well-watered plants. Phenotypic responses were thus specific to the limiting resource, in this case soil water supply (soil humidity).

28.3.3 Phenotypic Plasticity and Evolution of Plants

In the field of phenotypic plasticity and evolution of plants, scientists moved from variation for plasticity being considered as a nuisance in evolutionary studies to it being the primary target of investigations that use an array of methods, including quantitative and molecular genetics, as well as of several approaches that model the evolution of plant plastic responses. In modern biological literature, it is traced how plants are developed and interact with their environment, including gene-by-environment interactions. However, the problem was that many biologists retained some misconceptions about the nature of plasticity, especially its relationship with the genetics of an organism. At present, phenotypic plasticity is acknowledged as an important concept in modern evolutionary thinking. There is genetic variation in nature for plastic responses of the plants (genetic variation for plant plasticity; for details, see Pigliucci 2005). According to this author, the commonplace observation of genetic variation for plasticity within populations implies that it can evolve by responding to natural selection, which, in turn, suggests that adaptive phenotypic plasticity occurs in natural populations.

Bossdorf and Pigliucci (2009) examined genetic variation for thigmomorphogenesis (phenotypic changes by which plants react to mechanical stress, e.g., wind) within and among natural populations of the *Arabidopsis thaliana*. Wind significantly affected plant growth and phenology, and there was genetic variation for some aspects of plasticity to wind among *A. thaliana* populations. These authors found that phenotypic traits were organized into three distinct and, to a large degree, statistically independent covariance modules associated with plant size, phenology, and growth form, respectively. These phenotypic modules differed in their responsiveness to wind, in the degree of genetic variability for plasticity, and in the extent to which plasticity affected fitness of the plants. Therefore, it is likely that thigmomorphogenesis in this species evolves quasi-independently in different phenotypic modules. Nevertheless, there must also be limits to the evolution of adaptive plasticity.

Other than the possibility of genetic and/or developmental constraints, the idea has been advanced that there might be several types of measurable costs to maintaining plasticity, as well as limits to the ability of an organism of being adaptively plastic (De Witt 1998). Although important, the conceptual difference between costs and limits is often neglected: costs result in a decrease in fitness even when an optimal phenotype is expressed, whereas limits exist in the failure to express an optimal phenotype to begin with. Costs of plasticity were initially difficult to detect, although later research has found them in a variety of systems (e.g., Relyea 2002). Research into the costs of plasticity is still in its *early stages*, but is both theoretically important and empirically challenging, and should also become a major area of future investigation. Pigliucci (2005) emphasized that future research on the macroevolutionary consequences of plasticity must document instances of both genetic assimilation and phenotypic accommodation, map them in a phylogenetic context, and devise empirical approaches to study them.

28.3.4 Morphological, Anatomical, and Physiological Plasticity of Woody Species

The environment can induce changes in the individual's behavior at a morphological and/or physiological level (Price et al. 2003), and such changes may be crucial to survival in heterogeneous and variable conditions (Zunzunegui et al. 2011; Figure 28.3). For certain morphological traits, phenotypic plasticity has been shown to reflect genetic correlation relatively well, and traits belonging to the same suite of characters are more highly genetically and phenotypically correlated than traits from different suites (Waitt and Levin 1998). The selection for photosynthetic traits may often operate indirectly via correlation with other traits, emphasizing the importance of viewing the phenotype as an integrated function of growth, morphology, life history, and physiology (Arntz and Delph 2001). The timing of plant development can itself be plastic (Sultan 2000), and many phenotypic responses to environmental stress factors may be the consequence of growth reduction due to resource limitations (Dorn et al. 2000). Differences among species and population may reflect selective pressures on plasticity, different limitations acting upon the maximization of plasticity, or a combination of both (Valladares et al. 2007). The potential plastic response

FIGURE 28.3 High morphological variability in leaves of *Leaucanthemum vulgare* Lam.

of a given plant trait may be large, but the observed plasticity may be lowered by resource limitations or environmental stress factors.

Morphological, anatomical, and physiological plasticity may have a different role in plant adaptation to environmental changes. In particular, plasticity for physiological and life-history traits may allow plants to grow and reproduce in spatially or temporally variable environments (for details, see Gratani 2014). In our earlier papers, we studied the growth parameters (Masarovičová and Minarčic 1984), carbon dioxide exchange (photosynthesis, respiration; Masarovičová 1988), chlorophyll content, stomata apparatus (Masarovičová and Minarčic 1985), and anatomical and ultrastructural characteristics (Masarovičová and Minarčic 1980) in the leaves of European beech (Fagus sylvatica L.) seedlings or saplings grown under different light conditions. Moreover, we compared these morphological, anatomical, and physiological features in the sun and shade leaves of tall F. sylvatica trees grown in the natural forest stand (Masarovičová and Štefančík 1990). Special attention was paid to the different growing phases such are spring shoots (first growth phase) and summer shoots (second growth phase) grown under different microclimatic (light and temperature) conditions. This phenomenon is known as a polycyclic growth. This is especially conspicuous for woody plants where the periodicity of shoot (and root) growth is regularly encountered (e.g., Masarovičová 1992). In general, sun leaves, as well as the leaves of summer shoots, showed significantly higher rates of photosynthesis, photorespiration,

and dark respiration, as well as photosynthetic productivity. Specific leaf mass, mean leaf area, stomata density and size, and chlorophyll content per unit dry mass were also significantly different in both types of leaves. The importance of these findings for annual carbon gain of this Slovakian distinguished forest trees was also presented. A similar comparative study on another type of woody plants, oaks (*Quercus robur* and *Q. petraea*), which are characterized by great intraspecific variability, was published, and as loessial woody species, they hybridize interspecifically (for details, see Masarovičová 1991). It was found that mean and maximal daily net photosynthetic rates, shoot length, leaf area, stomatal density, and chlorophyll contents were significantly higher in shoots of the second growth phase. Intraspecific differences in leaf shape within each shoot growth phase were also determined.

Phenotypic plasticity in 19 functional traits (7 morphological and 12 physiological) in tree and shrub species across the five study sites in a tropical dry forest, showing variable soil moisture content, was recently studied by Chaturvedi and Raghubanshi (2013). The authors found that the plasticity in functional traits significantly varied across the study sites. The plasticity in functional traits also differed significantly across species. All traits under study affected the relative growth rate of the tree and shrub species directly or indirectly. However, the strength of effect is determined by an environmental parameter, and in the case of tropical dry forest, soil water availability is the important parameter. Plasticity in functional traits due to changes in environmental parameters explained the variation in relative growth rate. Stepwise multiple regression indicated that more than 80% variability in relative growth rate can be explained by canopy cover, leaf area index, SLA, and leaf intrinsic WUE alone. The first three variables represent quantity of photosynthetic surface, and the last represents water use economy of a species. All these parameters are also significantly modulated by soil moisture availability. An important point to note here is that photosynthetic rate is not an important parameter to determine relative growth rate in tropical dry forest where water economy and extended period of leaflessness are critical.

28.3.5 RELATIONSHIP BETWEEN PLASTICITY OF WHOLE PLANT AND LEAF TRAITS

As has already been mentioned, phenotypic plasticity is the ability of a genotype to produce distinct phenotypes when exposed to different environments throughout its ontogeny (e.g., Pigliucci 2005). Phenotypic plasticity has been shown to have significant evolutionary consequences (Murren et al. 2005). In the review by Navas and Garnier (2002), plasticity (or phenotypic plasticity) was defined as the ability of an organism to adjust its performance by altering its morphology and/or physiology in response to varying environmental conditions. However, the authors stressed that, in practice, this is difficult to evaluate, and any change in a plant trait has been generally called a plastic adjustment. Morphological plasticity plays an important role in resource acquisition of plants, and variations in size and placement of resource-acquiring organs such as leaves and roots are of major importance for plant adjustment to resource availability.

In plant ecology, the theory of *modules* and *metamers* was also adapted some years ago (e.g., White 1979). According to this theory, plants consist of a set of metamers composed of a section of stem, including its leaf and associated axillary meristem and roots. Metamers resulting from the development of meristems are assembled in modules that are aggregated to form a plant. Due to this integrated morphological organization, the plasticity of the whole plant depends on plastic responses of both the number and individual characters of metamers. A potential consequence of this organization is that the whole plant and metamer vegetative traits could differ in magnitude of plastic response to varying resources, with these differences resulting in a *hierarchy of plasticities* among traits at the plant level. Such a hierarchy exists for reproductive traits since the number of propagules produced by a plant varies more among environments than propagule dimension does. If such a hierarchy exists, some traits would be more important than others for a plant to adapt to a changing environment. In an effort to answer this question, Navas and Garnier (2002) published a paper in which they tested whether (1) changes in growth forms of *Rubia peregrine* recorded in nature could be reproduced by manipulating nutrient, light, and water availability; (2) these changes could be related to a hierarchy of plasticity among whole plant and aerial metamer traits; and (3) this hierarchy is consistent for different resources that may constrain growth. The authors found that assessing the plasticity of a species in response to a change in environment strongly depends on which trait is measured. If only one resource varies, then traits that relate to the capture and/or use of that particular resource (or a trait known to be closely related to such a trait) are good candidates to estimate the plasticity of a species, since those changes are likely to be of functional values. Similar statement has already been published by Schlichting in 1989 (1989b). Navas and Garnier (2002) found in their study with *Rubia peregrina* that mean internode length was relevant for evaluating plant response to light, but not for nutrient and water supply. It seems that other metamer traits, such as root hair length or specific root length, would have probably been more relevant. The last authors mentioned that characterizing the plasticity of a species to multiple changes in the environment is more difficult, since those changes are not linked to a change in a single function. Based on their findings, it was suggested that, in such cases, owing to their consistent responsiveness to both aerial and soil resources, allocation-related traits (such as the proportional allocation to roots) are of more general value than aerial metamer traits to assess the plasticity of a species.

In general, ecologists have categorized plant species according to plant functional types and have also identified several continuous plant functional traits that vary in predictable ways along environmental gradients. Functional types are widely used in global climate models to group species according to their function in the community or ecosystem, e.g., C_3 and C_4 grasses, herbs and shrubs, deciduous or evergreen woody species, and N-fixing legumes. Plant functional traits are those that help describe the ecology of species using easily quantified variables such as seed size, plant height, leaf life span, leaf mass per area (SLA), etc. Nicotra et al. (2010) advocated that plant functional traits should have priority for the investigation of (adaptive) phenotypic plasticity and identification of molecular and genetic mechanisms across species. Adaptive plasticity in functional traits likely assists to rapid adaptation to new environmental conditions. Indeed, many remarkable instances of plasticity entail plant response to biotic factors of the environment such as herbivores, bacteria, other different pathogens, mycorrhizal relations, etc. However, this aspect of plant plasticity would demand a separate chapter in this handbook.

28.3.6 PLANT PLASTICITY AND SUCCESSION OF PLANT POPULATIONS

There are many examples in outdoors of the acclimation of photosynthesis, respiration, and biomass allocation to environmental factors such as irradiance and water or nutrient supply. A high capacity of plants to this acclimation reflects a phenotypic plasticity for a specific trait. However, a relatively small plasticity for one trait may result from a large plasticity in other traits. For example, the low morphological plasticity (stem length) of an alpine species *Stellaria longipes* is a consequence of a high physiological plasticity (ethylene production). Both mentioned traits are directly related to the same environmental factor (wind stress), and the expressed phenotype has a direct effect on the plant's fitness (for details, see Lambers et al. 2008). In addition, a large morphological plasticity in biomass allocation between roots and leaves in response to nutrient supply or irradiance results in a low plasticity of the plant's growth rate, so that this varies relatively little between different environments. A high plasticity allows a genotype (plant species) to maintain dominance in spatially or temporally variable environments by enabling them to continuously explore new patches that have not been depleted, thus sustaining resource capture and maintaining fitness. By contrast, in habitats of predictably low resource supply, plant production would be restricted to a continuously low level, and a strategy of conservation of captured resources, associated with low growth, would be favored. Such a contention is hard to verify, in view of the fact that greater plasticity for one trait is made possible by smaller plasticity for another. There are certainly convincing examples of greater plasticity associated with competitive ability in a particular environment. In 1984, Küppers had already stressed that late-successional species tend to have a great potential for adjustment of their photosynthetic characteristics to shade than do early-successional species. Similarly, a further case is the response of stem elongation to shade light. Low irradiance also suppresses branching in dicotyledonous species and enhances tillering in grasses like perennial ryegrass *Lolium perenne* or Italian ryegrass *Lolium multiflorum*.

This plastic response is important in coping with plant neighbors. To confirm the importance of the phytochrome system for the perception of neighboring plants, Ballaré (1999) used transgenic plants of *Nicotiana tabacum* overexpressing a phytochrome gene. These transgenic plants showed a dramatically smaller response to the red/far red ratio of radiation and to neighboring plants. In a stand of such transgenic plants, the small plants of the population were rapidly suppressed by the neighbors. These results indicated that a high degree of plasticity in morphological parameters plays an important role in the competition with surrounding plants. In general, it appears that fast-growing species from high-resource environments are more plastic for some traits, such as photosynthetic characteristics and the rate of stem elongation in response to shade, surrounding plants, and wind. A large phenotypic plasticity for various plant traits (e.g., photosynthetic characteristics, nutrient acquisition, or stem elongation) may also contribute to competitive success. In addition, competitive advantage may be based on a profitable association with another organism, such as symbiotic N_2-fixing microorganism and mycorrhizal fungus.

A few recent studies (e.g., Al Hayek et al. 2014) examined consequences of ecotypic differentiation within alpine foundation species for community diversity and their feedback for the fitness of foundation species. Additionally, no study has quantified ecotypic differences in competitive effects in the field and in controlled conditions to disentangle genetic from plasticity effects in foundation/subordinate species interactions. The above-mentioned authors focused on a subalpine community of the French Pyrenees including two phenotypes of a cushion-forming species, *Festuca gautieri*. It was found that trait differences across habitats had both genetic and plasticity bases, with stronger contribution of the latter base. Field experiments showed higher competition within loose than tight phenotypes. In contrast, shadehouse results showed higher competitive ability for tight phenotypes. However, as changes in interactions across habitats were due to environmental effects without changes in cushion effects, the authors argued that heritable and plastic changes in competitive effects maintain high subordinate species diversity through decreasing competition. High reproduction cost for loose cushions when hosting subordinates highlighting the occurrence of community feedback was confirmed. The removal of subordinate species within loose cushions significantly increased the flower production of *F. gautieri*. Loose cushions in which subordinate species were removed produced twice the number of inflorescences as control loose cushions. These findings suggested that phenotypic differentiation within foundation species may cascade on subordinate species diversity through heritable and plastic changes in the competitive effects of foundation species, and that community feedback may affect the fitness of the foundation species.

Presented results brought additional support to the emerging perspective that diffuse within-trophic level biotic interactions might have important evolutionary consequences at the community level. However, in an evolutionary context, it is crucial to quantify not only the costs but also the benefits of

hosting other species (cf. Bronstein 2009). Moreover, the presence of ecotypic variation and ecotypic differentiation within species also presents a notable challenge for forecasting the effect of climate change on plant vegetation. Local adaptation plays a prominent role in ecological and evolutionary processes and is not rare. Ecotypic differentiation occurs even over short distances, such as along topographic or elevation gradients (cf. Liancourt and Tielbörger 2009).

28.3.7 CLIMATIC CONDITIONS AND PLANT–PLANT INTERACTION

Future climate modifications are likely to affect community composition and structure not only directly through abiotic changes but also indirectly by modifying the direction and intensity of species interactions, including competition. Understanding indirect effects of climate modification may be particularly important for plant communities where the direction and intensity of plant–plant interactions vary along environmental gradients. Based on the *stress gradient hypothesis*, competition is thought to rule at the benign end of stress gradients, while facilitation may occur when approaching the harsher end of a gradient, where it eventually wanes (Michalet et al. 2006). Although much attention has been paid to the effects of climate change on plant communities, only a few field-based experimental studies have explicitly induced a test for the role of plant–plant interactions in this context. Liancourt et al. (2013) in a three-year experiment set in the Mongolian steppe examined the response of the common grass *Festuca lenensis* to manipulated temperature and water while controlling for topographic variation, plant–plant interactions, and ecotypic differentiation. Plant survival and growth responses to a warmer, drier climate varied within the landscape. Response to simulated increased precipitation occurred only in the absence of neighbors, demonstrating that plant–plant interactions can supersede the effects of climate change. *F. lenensis* also showed evidence of local adaptation in populations that were only 300 m apart. Plant individuals from the steep and dry upper slope showed a higher stress/drought tolerance, whereas those from the more productive lower slope showed a higher biomass production and a greater ability to cope with competition. Moreover, the response of this species to increased precipitation was ecotype specific, with water addition benefiting only the least stress-tolerant ecotype from the lower slope origin.

This multifaceted approach illustrated the importance of placing climate change experiments within a realistic ecological and evolutionary framework. Existing sources of variation impacting plant performance may buffer or obscure climate change effects. Liancourt et al. (2013) emphasized that climate change will not produce consistent consequences across the landscape even for the same species and also that existing abiotic and biotic sources of variation in plant performance may buffer or even obscure climate change effects. The *stress gradient hypothesis* appears to offer a solid framework for evaluating the contingent effects of changes in a single resource such as water. Nevertheless, predicting plant responses to all concurrent climate change effects appears more challenging than expected.

28.3.8 Phenotypic Plasticity and Climate Changes

Phenotypic plasticity is usually defined as a property of individual genotypes to produce different phenotypes when exposed to different environmental conditions. It could also be characterized as a change in the phenotype expressed by a single genotype in different environments. Phenotypic plasticity could itself be under genetic control and therefore subjected to selective pressure. However, according to Pigliucci et al. (2006), some important points need to be considered for the purposes of the following discussion. First, not all phenotypic plasticity is adaptive (in the evolutionary sense of improving the organism's survival or reproduction). Some traits are plastic because of unavoidable constraints imposed by the biochemistry, physiology, or developmental biology of the organism. Second, plasticity may be expressed at the behavioral, biochemical, physiological, or developmental levels. While all these phenomena share the fundamental biological property of being part of the genotype-specific repertoire of environmentally induced phenotypes, there are significant differences in the degree of reversibility of different kinds of plasticity. Biochemical and physiological responses can be reversed over short timescales, and developmental plasticity tends to be irreversible or takes longer to be reversed. Third, the type and degree of plasticity are specific to individual traits and environmental conditions. The same trait may be plastic in response to, for example, changes in temperature, but not to nutrients, and a certain trait may be plastic in response to temperature while other traits are not plastic. Finally, there seems to be abundant genetic variation for a variety of plastic responses in natural populations, which makes possible the evolution of plasticity by natural selection and other mechanisms. Thus, it could be concluded that phenotypic plasticity is a common property of the reaction norm of a genotype (for a given trait, within a certain range of environmental conditions). Plasticity is what makes possible the appearance of an environmentally induced novel phenotype, and a process of selection on the expression of such phenotype in a new environment may end up *fixing* (genetically assimilating) it by altering the shape of the reaction norm.

Plants are exposed to heterogeneity in the environment where new stress factors (i.e., climate change, land use change, and invasiveness) are introduced and where interspecific and intraspecific difference may reflect resource limitation and/or environmental stress factors. Phenotypic plasticity is considered one of the major means by which plants can cope with environmental factor variability. Nevertheless, the extent to which phenotypic plasticity may facilitate survival under environmental condition changes still remains largely unknown because results are sometimes controversial (Gratani 2014). Thus, it is important to identify plant functional traits in which plasticity may play a determinant role in plant response to global change as well as on the ecological consequences at an ecosystem level for the competition between wild and invasive species, considering that species with greater adaptive plasticity may be more likely to survive in novel environmental conditions (Davidson et al. 2011). Therefore, in the near future, it will be important to increase long-term studies on natural populations in order to understand plant response to environmental factor fluctuations including climate change. According to the above-mentioned author, there is the necessity to analyze variations at phenotypic and genetic levels for the same species, and in particular, for endemic and rare species because these could have drastic effects at an ecosystem level. Moreover, climate changes have been shown to affect abundance and distribution of plant species, as well as plant community composition (e.g., Menzel et al. 2006). Under rapid climate change, phenotypic plasticity rather than genetic diversity is likely to play a crucial role in allowing plants to persist in their environment. Different responses to climate occur not only between populations throughout a species range but also between co-occurring individuals within a population (for details, see Gratani 2014).

One way by which plants can respond to climate changes is through environmentally induced shifts in phenotype (phenotypic plasticity). Understanding of these phenotypic responses is crucial for prediction of changing climate effects on both native species and crops. Based on ecological, evolutionary, physiological, and molecular biology perspectives, it will be possible to provide directives for perspective research as well as interdisciplinary dialogue on the relevance of phenotypic plasticity under changing climate. Some years ago, Jump and Penuelas (2005) argued that plastic responses to rapid climate change are less important than adaptation in the geographic range of distribution. These authors declared that the failure to expand beyond current limits demonstrated that species' adaptive potential has been largely exhausted and that phenotypic plasticity will be an unimportant factor because impulse for plastic responses in the first place might no longer be *reliable* in changed climates. Afterward, Lande (2009) asserted that plasticity can both provide a buffer against rapid climate change and assist in more rapid adaptation. Plasticity is a characteristic of a given trait in response to a given environmental stimulus, rather than a characteristic of an organism as a whole. Likewise, some responses are examples of adaptive plasticity, providing a fitness benefit, whereas others are inevitable responses to physical processes or resource limitations. Both adaptive and non-adaptive plasticity will play a role in the context of plant responses to climate change (van Kleunen and Fischer 2005).

At present (e.g., Nicotra et al. 2010), there is general acceptance that a high level of genetic variation within natural populations improves the potential to face, resist, and adapt to novel biotic and abiotic environmental changes including the tolerance of climatic change. A portion of this genetic variation determines the ability of plants to sense changes in the environment and produce a plastic response (Figure 28.4). In the context of rapid climate change, phenotypic plasticity can be a crucial determinant of plant responses, both short and long term. The above-mentioned authors discussed in detail how new developments in understanding of signaling cascades and

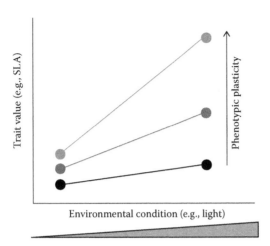

FIGURE 28.4 Schematic showing the different phenotypic plasticity of species as a response to particular environmental conditions. Upper line depicts the greatest phenotypic plasticity, middle line the least. (Modified from Nicotra, A. B. et al., *Trends in Plant Science* 15: 684–692, 2010.)

epigenetics in particular hold promise for interdisciplinary approaches to understanding the evolution of plant plasticity and for predicting how plasticity will influence the responses of native plants and crops to climate change. It could be concluded that there is increasing evidence of the importance of plasticity in the plant (adaptive, genotypic, phenotypic, morphological, and physiological plasticity, respectively) under climate change. It is important to discuss the potential roles of the mentioned plant plasticity in determining plant response to changing climate as well as effects of climate change in a way that is accessible and relevant to ecologists, physiologists, and molecular biologists.

According to Vitasse et al. (2010), phenotypic plasticity allows large shifts in the timing of phenology within one single generation and drives phenotypic variability under environmental changes; thus, it will enhance the inherent adaptive capacities of plants against future changes of climate. Mid- and high-elevation populations should experience a longer growing season with climate warming. These authors found for beech seedlings (*F. sylvatica*) that climate change could reduce population fitness at low elevation by shortening the period of growing season mainly due to earlier leaf senescence, whereas it could increase population fitness at high elevation by increasing the growing season length and consequently also by the growth. For oak seedlings (*Quercus petraea*), it was observed that all populations tend to inhabit climates colder than the one corresponding to the optimum in their growing season length, even though the population observed at the lowest elevation is currently close to its optimum. Overall, a high phenological plasticity was found for both species. Timing of leaf senescence exhibited hyperbolic trends for beech due to earlier senescence at the lowest elevation garden and no or slight trends for oak. There was no difference in the magnitude of phenological plasticity

among population from different elevations. For both species, the growing season length increased to reach maximum values at about 10–13°C of annual temperature according to the population. Since the magnitude of phonological plasticity is high for all the tested populations, they are likely to respond immediately to temperature variations in terms of leaf phenology. Consequently, the mid- to high-elevation populations are likely to experience a longer growing season with climate warming. The result found suggested that climate warming could lengthen the growing season of all populations over the altitudinal gradient, although the low-elevation populations, especially of beech, may experience accelerated senescence and shorter growing season due to drought and other climate changes associated with warming. Thus, this study is in line with some previous studies (e.g., Rweyongeza et al. 2007) showing that at high elevation, current changes in climate could drive conditions toward the optimal range for population growth while negatively affecting fitness in the southern part of the species range and at low elevation (e.g., Rehfeldt et al. 2002).

28.4 CONCLUDING REMARKS

Plant response to their immediate environment is particularly flexible because plants can vary the number as well as the size of parts, and can differently allocate and reallocate resources to various tissues. Ecological distribution of plants may be partly determined by the capacity for individual adaptation shown by its individual species. Population divergence and ultimately allopatric speciation (through geographic isolation) under local selection pressures may be alleviated by this property of individual plasticity. Plants undoubtedly provide excellent biological systems for study of phenotypic plasticity, because they may be replicated clonally and subjected to environmental manipulations. To really understand the process of evolutionary change, it is essential to study the relationships between plants and their environment on an evolutionary timescale and in terms of both individual developmental and physiological responses. Target of plant phenotypic response will thus also enlarge the view of developmental mechanisms, genetic diversity, and evolutionary diversification.

GLOSSARY

Acclimation: Set of physiological (not genetic) responses on environmental conditions. Change of certain environmental factors can improve survival of the plant.

Adaptability: Ability to accommodate variable environment. However, the measure of tolerance cannot be exceeded.

Adaptation: Process by which the plants during evolution accommodate (or resist) environmental condition through genetically conditioned morphological and physiological changes. The term *adaptation* has a wider meaning than the term *adjustment*.

Adaptive plasticity: Phenotypic plasticity that increases the global fitness of a genotype.

Adjustment: Realignment of features and performance of the plants with environmental conditions so that they can survive.

Fitness: The fitness of a plant individual is taken as the relative abundance and success of its genes (often measured as the number of surviving offspring) over multiple generations. In many cases, especially with large or long-lived species, direct estimates of fitness are not feasible, and total biomass, seed number, biomass, or growth rate of a single generation are used as proxies.

Functional traits: Any trait that is directly or indirectly affecting individual performance and fitness of species.

Functional types: Categorical assessments enabling plant species to be grouped according to functional position in a community (or ecosystem) and their use of resources. These plants have similar responses to the environmental factors and/or similar effects on ecosystems and biomes. The plants of one functional group have a set of similar key functional traits. For example, classification based on growth form (e.g., herb, grass, shrub), photosynthetic pathway, leaf longevity, or nitrogen-fixing status.

Genome (genotypic) plasticity: A change in genome structure or organization associated with environmental signals, leading to the evolution of new phenotypes, might result from mutational hotspots, genome expansion, transposable elements, or somatic recombination.

Genotype: To report to a genotype, we do so in a population genetic sense, not in reference to a molecular sequence of a single gene, but to the complete genome.

Intraspecific trait variability: The overall variability of trait values expressed by individuals within a species.

Morphological plasticity: A change in morphology (shape and size) of plant organs (mainly leaf and root) that is important for utilization of environmental resources (e.g., water, nutrients, light, and temperature).

Phenotype: The appearance or characteristics of an organism resulting from both genetic and environmental influences (genotype + environment = phenotype). All organisms have a phenotype, not just those expressing a mutation in a given gene of interest.

Phenotypic plasticity: The range of phenotypes a single genotype can express as a function of its environment. Originally this feature was apprehended in the context of ontogeny of the plant. At present, phenotypic plasticity includes changes occurring in the adult plants as their behavior under certain environmental conditions.

Physiological plasticity: A change in physiological processes associated with environmental changes, leading to the formation of new phenotypes.

Plasticity: Ability of a single genotype to exhibit a range of different phenotypes in response to variation in the environment. In response to variable environmental conditions, the plant can adjust its features

through morphological traits and/or physiological processes.

Resilience (elasticity): Ability of plant to return to the original state after action, perturbation, or disturbance. Plants (ecosystems) resist the damage, and after deviation, fast recovery (regeneration) occurred.

Trait: Any morphological, physiological, phenological, or behavioral characteristic that can be measured at the individual level.

ACKNOWLEDGMENTS

This contribution was financially supported by the Grant Agency VEGA, with grant no. 1/0218/14, and by the Slovak Research and Development Agency, project no. APVV-0866-12.

REFERENCES

Ackerly, D. 2004. Functional strategies of chaparral shrubs in relation to seasonal water deficit and disturbance. *Ecological Monographs* 74: 25–44.

Ackerly, D. D., Dudley, S. A., Sultan, S. E., Schmitt, J., Coleman, J. S., Linder, C. R. et al. 2000. The evolution of plant ecophysiological traits: Recent advances and future directions. *Bioscience* 50: 979–995.

Adler, P. B., HilleRisLambers, J., Kyriakidis, P. C., Guan, Q. F. and Levine, J. M. 2006. Climate variability has a stabilizing effect on the coexistence of prairie grasses. *Proceedings of the National Academy of Sciences of the United States of America* 103: 12793–12798.

Aerts, R. and Chapin, F. S. 2000. The mineral nutrition of wild plants revisited: A reevaluation of processes and patterns. *Advances in Ecological Research* 30: 1–67.

Albert, C. H., Thuiller, W., Yoccoz, N. G., Soudant, A., Boucher, F., Saccone, P. and Lavorel, S. 2010. Intraspecific functional variability: Extent, structure and sources of variation. *Journal of Ecology* 98: 604–613.

Albert, C. H., Grassein, F., Schurr, F. M., Vieilledent, G. and Violle, C. 2011. When and how should intraspecific variability be considered in trait-based plant ecology? *Perspectives in Plant Ecology, Evolution and Systematics* 3: 217–225.

Al Hayek, P., Touzard, B., Le Bagousse-Pinguet, Y. and Michalet, R. 2014. Phenotypic differentiation within a foundation grass species correlates with species richness in a subalpine community. *Oecologia* 176: 533–544.

Angert, A. L., Huxman, T. E., Chesson, P. and Venable, D. L. 2009. Functional tradeoffs determine species coexistence via the storage effect. *Proceedings of the National Academy of Sciences of the United States of America* 106: 11641–11645.

Arnold, S. J. 1983. Morphology, performance and fitness. *American Zoologist* 23: 347–361.

Arntz, M. A. and Delph, L. F. 2001. Pattern and process: Evidence for the evolution of photosynthetic traits in natural populations. *Oecologia* 127: 455–467.

Ballaré C. L. 1999. Keeping up with the neighbours: Phytochrome sensing and other signalling mechanisms. *Trends in Plant Sciences* 4: 97–102.

Bartlett, M. K., Scoffoni, C. and Sack, L. 2012. The determinants of leaf turgor loss point and prediction of drought tolerance of species and biomes: A global meta-analysis. *Ecology Letters* 15: 393–405.

Bolnick, D. I., Svanback, R., Fordyce, J. A., Yang, L. H., Davis, J. M., Hulsey, C. D. and Forister, M. L. 2003. The ecology of individuals: Incidence and implications of individual specialization. *American Naturalist* 161: 1–28.

Bond, W. J. and Midgley, J. J. 1995. Kill thy neighbor—An individualistic argument for the evolution of flammability. *Oikos* 73: 79–85.

Bossdorf, O. and Pigliucci, M. 2009. Plasticity to wind is modular and genetically variable in *Arabidopsis thaliana*. *Evolutionary Ecology* 23(5): 669–685.

Bradshaw, A. D. 1965. Evolutionary significance of phenotypic plasticity in plants. *Advances in Genetics* 13: 115–155.

Brodribb, T. J. and Holbrook, N. M. 2003. Stomatal closure during leaf dehydration, correlation with other leaf physiological traits. *Plant Physiology* 132: 2166–2173.

Bronstein, J. L. 2009. The evolution of facilitation and mutualism. *Journal of Ecology* 97: 1160–1170.

Calow, P. 1987. Towards a definition of functional ecology. *Functional Ecology* 1: 57–61.

Chapin, F. S. III. 1993. Functional role of growth forms in ecosystem and global processes. In: Ehleringer, J. R. and Field, C. B. (eds.), *Scaling Physiological Processes. Leaf to Globe.* Academic Press, San Diego, pp. 287–312.

Chaturvedi, R. K. and Raghubanshi, A. S. 2013. Phenotypic plasticity in functional traits of woody species in tropical forest. In: Valentino, J. B. and Harrelson, P. C. (eds.), *Phenotypic Plasticity: Molecular Mechanisms, Evolutionary Significance and Impact on Speciation.* Nova Science Publishers, Hauppauge, NY, pp. 35–66.

Chave, J., Coomes, D., Jansen, S., Lewis, S. L., Swenson, N. G. and Zanne, A. E. 2009. Towards a worldwide wood economics spectrum. *Ecology Letters* 12: 351–366.

Coleman, J. S., McConnaughay, K. D. M. and Ackerly, D. D. 1994. Interpreting phenotypic variation in plants. *Trends in Ecology and Evolution* 9: 187–191.

Coley, P. D. 1983. Herbivory and defensive characteristics of tree species in a lowland Tropical forest. *Ecological Monographs* 53: 209–229.

Coley, P. D. 1988. Effect of plant-growth rate and leaf lifetime on the amount and type of anti-herbivore defence. *Oecologia* 74: 531–536.

Coley, P. D. and Barone, J. A. 1996. Herbivory and plant defenses in tropical forests. *Annual Review of Ecology and Systematics* 27: 305–335.

Coley, P. D., Kursar, T. A. and Machado, J. L. 1993. Colonization of tropical rain-forest leaves by epiphylls—Effect of site and host leaf lifetime. *Ecology* 74: 619–623.

Cornelissen, J. H. C., Pérez-Harguindeguy, N., Diaz, S., Grime, J. P., Marzano, B., Cabido, M., Vendramini, F. and Cerabolini, B. 1999. Leaf structure and defence control litter decomposition rate across species and life forms in regional floras on two continents. *New Phytologist* 143: 191–200.

Cornelissen, J. H. C., Lavorel, S., Garnier, E., Diaz, S., Buchmann, N., Gurvich, D. E., Reich, P. B., ter Steege, H., Morgan, H. D., van der Heijden, M. G. A., Pausas, J. G. and Poorter, H. 2003. A handbook of protocols for standardised and easy measurement of plant Functional traits worldwide. *Australian Journal of Botany* 51: 335–380.

Cornwell, W. K. and Ackerly, D. D. 2009. Community assembly and shifts in plant trait distributions across an environmental gradient in coastal California. *Ecological Monographs* 79: 109–126.

Craine, J. M. and Lee, W. G. 2003. Covariation in leaf and root traits for native and non native grasses along an altitudinal gradient in New Zealand. *Oecologia* 134: 471–478.

Darwin, C. 1859. On the *Origin of Species.* John Murray, London.

Davidson, A. M., Jennions, M. and Nicotra, A. B. 2011. Do invasive species show higher phenotypic plasticity than native species and, if so, is it adaptive? A meta-analysis. *Ecology Letters* 14: 419–431.

de Bello, F., Lepš, J. and Sebastia, M. T. 2005. Predictive value of plant traits to grazing along a climatic gradient in the Mediterranean. *Journal of Applied Ecology* 42: 824–833.

de Bello, F., Lavorel, S., Albert, C. H., Thuiller, W., Grigulis, K., Doležal, J., Janeček, S. and Lepš, J. 2011. Quantifying the relevance of intraspecific trait variability for functional diversity. *Methods in Ecology and Evolution* 2: 163–174.

De Jong, G. and Stearns, S. C. 1991. Phenotypic plasticity and the expression of genetic variation. In: Dudley, E. C. (ed.), *The Unity of Evolutionary Biology, vol. II. International Congress of Systematic and Evolutionary Biology IV.* Dioscorides Press, Portland, OR, pp. 707–718.

De Witt, T. J. 1998. Costs and limits of phenotypic plasticity. *Trends in Ecology and Evolution* 13: 77–81.

Diaz, S. and Cabido, M. 2001. Vive la difference: Plant functional diversity matters to ecosystem processes. *Trends in Ecology and Evolution* 16: 646–655.

Diaz, S., Hodgson, J. G., Thompson, K., Cabido, M., Cornelissen, J. H. C., Jalili, A. et al. 2004. The plant traits that drive ecosystems: Evidence from three continents. *Journal of Vegetation Science* 15: 295–304.

Donovan, L. A. and Ehleringer, J. R. 1994. Potential for selection on plants of water-use efficiency as estimated by carbon isotope discrimination. *American Journal of Botany* 81: 927–935.

Dorn, L. A., Pyle, E. H. and Schmitt, J. 2000. Plasticity to light cues and resources in Arabidopsis thaliana: Testing for adaptive value and costs. *Evolution* 54: 1982–1994.

Ellenberg, H. 1988. *Vegetation Ecology of Central Europe*, 4th Ed. Cambridge University Press, Cambridge.

Eshel, A. and Beeckman, T. 2013. *Plant Roots: The Hidden Half.* CRC Press, New York.

Eviner, V. T. 2004. Plant traits that influence ecosystem processes vary independently among species. *Ecology* 85: 2215–2229.

Eviner, V. T. and Chapin, F. S. 2003. Functional matrix: A conceptual framework for predicting multiple plant effects on ecosystem processes. *Annual Review of Ecology, Evolution and Systematics* 34: 455–485.

Fitter, A. H. and Fitter, R. S. R. 2002. Rapid changes in flowering time in British plants. *Science* 296: 1689–1691.

Forsman, A. 2015. Rethinking phenotypic plasticity and its consequences for individuals, populations and species. *Heredity* 115: 276–284.

Fortunel, C., Garnier, E., Joffre, R., Kazakou, E., Quested, H., Grigulis, K. et al. 2009. Leaf traits capture the effects of land use changes and climate on litter decomposability of grasslands across Europe. *Ecology* 90: 598–611.

Garnier, E., Shipley, B., Roumet, C. and Laurent, G. 2001. A standardized protocol for the determination of specific leaf area and leaf dry matter content. *Functional Ecology* 15: 688–695.

Garnier, E., Lavorel, S., Ansquer, P., Castro, H., Cruz, P., Dolezal, J. et al. 2007. Assessing the effects of land-use change on plant traits, communities and ecosystem functioning in grasslands: A standardized methodology and lessons from an application to 11 European sites. *Annals of Botany* 99: 967–985.

Geber, M. A. and Griffen, L. R. 2003. Inheritance and natural selection on functional traits. *International Journal of Plant Sciences* 164: 21–42.

Gratani, L. 2014. Plant phenotypic plasticity in response to environmental factors. *Advances in Botany* 2014: 17.

Grime, J. P. 1974. Vegetation classification by reference to strategies. *Nature* 250: 26–31.

Grime, J. P. 1994. The role of plasticity in exploiting environmental heterogeneity. In: Caldwell, M. M. and Pearcy, R. W. (eds.), *Exploitation of Environmental Heterogeneity by Plants: Ecophysiological Processes Above- and Below-Ground.* Academic Press, New York, pp. 1–19.

Grime, J. P., Hodgson, J. G. and Hunt, R. 1988. *Comparative Plant Ecology: A Functional Approach to Common British Species.* Unwin Hyman, London, UK.

Grubb, P. J. 1977. The maintenance of species-richness in plant communities: The importance of the regeneration niche. *Biological Reviews* 52: 107–145.

Hacke, U. G., Sperry, J. S., Pockman, W. T., Davis, S. D. and McCulloch, K. A. 2001. Trends in wood density and structure are linked to prevention of xylem implosion by negative pressure. *Oecologia* 126: 457–461.

Hegland, S. J., Nielsen, A., Lazaro, A., Bjerknes, A.-L. and Totland, O. 2009. How does climate warming affect plant-pollinator interactions? *Ecology Letters* 12: 184–195.

Hodgson, J. G., Montserrat-Marti, G., Charles, M., Jones, G., Wilson, P., Shipley, B. et al. 2011. Is leaf dry matter content a better predictor of soil fertility than specific leaf area? *Annals of Botany* 108: 1337–1345.

Hodgson, J. G., Wilson, P. J., Hunt, R., Grime, J. P. and Thompson, K. 1999. Allocating C-S-R plant functional types: A soft approach to a hard problem. *Oikos* 85: 282–294.

Iwasa, Y., Cohen, D. and Leon, J. A. 1985. Tree height and crown shape, as results of competitive games. *Journal of Theoretical Biology* 112: 279–297.

Jump, A. S. and Penuelas, J. 2005. Running to stand still: Adaptation and the response of plants to rapid climate change. *Ecology Letters* 8: 1010–1020.

Kahmen, S. and Poschlod, P. 2004. Plant functional trait responses to grassland succession over 25 years. *Journal of Vegetation Science* 15: 21–32.

Kattge, J., Diaz, S., Lavorel, S., Prentice, C., Leadley, P., Bonisch, G. et al. 2011. TRY—A global database of plant traits. *Global Change Biology* 17: 2905–2935.

Keddy, P. A. 1992. A pragmatic approach to functional ecology. *Functional Ecology* 6: 621–626.

Keddy, P. A. and Shipley, B. 1989. Competitive hierarchies in herbaceous plant communities. *Oikos* 54: 234–241.

Kleyer, M., Bekker, R. M., Knevel, I. C., Bakker, J. P., Thompson, K., Sonnenschein, M. et al. 2008. The LEDA Traitbase: A database of life-history traits of the Northwest European flora. Journal of Ecology 96: 1266–1274.

Klimešová, J. and de Bello, F. 2009. CLO-PLA: The database of clonal and bud bank traits of Central European flora. *Journal of Vegetation Science* 20: 511–516.

Klimešová, J. and Klimeš, L. 2007. Bud banks and their role in vegetative regeneration—A literature review and proposal for simple classification and assessment. *Perspectives in Plant Ecology Evolution and Systematics* 8: 115–129.

Kramer, P. J. and Boyer, J. S. 1995. *Water Relations of Plants and Soils.* Academic Press, San Diego, CA.

Kuhn, I., Durka, W. and Klotz, S. 2004. BiolFlor—A new plant-trait database as a tool for plant invasion ecology. *Diversity and Distributions* 10: 363–365.

Küppers, M. 1984. Carbon relations and competition between woody species in a central European hedgerow. I. Photosynthetic characteristics. *Oecologia* 64: 332–343.

Kutschera, L. and Lichtenegger, E. 1982. *Wurzelatlas mitteleuropäischer Grünlandpflanzen, Bd. 1. Monocotyledoneae,* Fischer, Stuttgart.

Kutschera, L. and Lichtenegger, E. 1992. *Wurzelatlas mitteleuropäischer Grünlandpflanzen, Bd. 2. Pteridophyta und Dicotyledoneae (Magnoliopsida).* Teil 1 Morphologie, Anatomie, Ökologie, Verbreitung, Soziologie, Wirtschaft, Fischer, Stuttgart.

Lambers, H., Chapin, F. S. and Pons, T. L. 1998. *Plant Physiological Ecology.* Springer, New York.

Lambers, H., Chapin, F. S. and Pons, T. L. 2008. *Plant Physiological Ecology,* 2nd Ed. Springer Science, Business Media, LLC, New York pp. 604.

Lande, R. 2009. Adaptation to an extraordinary environment by evolution of phenotypic plasticity and genetic assimilation. *Journal of Evolutionary Biology* 22(7): 1435–1446.

Laughlin, D. C. 2014. The intrinsic dimensionality of plant traits and its relevance to community assembly. *Journal of Ecology* 102: 186–193.

Lavorel, S. and Garnier, E. 2002. Predicting changes in community composition and ecosystem functioning from plant traits: Revisiting the Holy Grail. *Functional Ecology* 16: 545–556.

Lavorel, S., McIntyre, S., Landsberg, J. and Forbes, T. D. A. 1997. Plant functional classifications: From general groups to specific groups based on response to disturbance. *Trends in Ecology and Evolution* 12: 474–478.

Lenz, T. I., Wright, I. J. and Westoby, M. 2006. Interrelations among pressure-volume curve traits across species and water availability gradients. *Physiologia Plantarum* 127: 423–433.

Lepš, J., de Bello, F., Šmilauer, P. and Doležal, J. 2011. Community trait response to environment: Disentangling species turnover vs intraspecific trait variability effects. *Ecography* 34: 856–863.

Liancourt, P. and Tielbörger, K. 2009. Competition and a short growing season lead to ecotypic differentiation at the two extremes of the ecological range. *Functional Ecology* 23: 397–404.

Liancourt, P., Spence, L. A., Song, D. S., Lkhagva, A., Sharkhuu, A., Boldgiv, B., Helliker, B. R., Petraitis, P. S. and Casper, B. B. 2013. Plant response to climate change varies with topography, interactions with neighbors, and ecotype. *Ecology* 94: 444–453.

Lonnberg, K. and Eriksson, O. 2013. Rules of the seed size game: Contests between large seeded and small-seeded species. *Oikos* 122: 1080–1084.

MacDonald, S. E. and Lieffers, V. J. 1993. Rhizome plasticity and clonal foraging of Calamagrostis canadensis in response to habitat heterogeneity. *Journal of Ecology* 81: 769–776.

MacGillivray, C. W. and Grime, J. P. 1995. Genome size predicts frost-resistance in British herbaceous plants—Implications for rates of vegetation response to global warming. *Functional Ecology* 9: 320–325.

Masarovičová, E. 1988. Comparative study of growth and carbon uptake in Fagus sylvatica L. trees growing under different light conditions. *Biologia Plantarum* 30: 285–293.

Masarovičová, E. 1991. Leaf shape, stomata density and photosynthetic rate of the common oak leaves. *Biologia Plantarum* 33: 495–500.

Masarovičová, E. 1992. Morphological, physiological, biochemical and production characteristics of three oak species. *Acta Physiologiae Plantarum* 14: 99–106.

Masarovičová, E. and Minarčic, P. 1980. Qualitative and quantitative analysis of the Fagus sylvatica L. leaves. II. Ultrastructural characteristics and photorespiratory activity. *Biológia (Biology, Bratislava)* 35: 235–242.

Masarovičová, E. and Minarčic, P. 1984. Photosynthetic response and adaptation of Fagus sylvatica L. trees to light conditions. I. Growth of leaves, shoots and trees. *Biológia (Biology, Bratislava)* 39: 867–876.

Masarovičová, E. and Minarčič, P. 1985. Photosynthetic response and adaptation of *Fagus sylvatica* L. trees to light conditions: Leaf chlorophyll contents, leaf dry matter, specific leaf area and mass, stomatal density. *Biológia (Biology, Bratislava)* 40: 473–481.

Masarovičová, E. and Štefančík, L. 1990. Some ecophysiological features in sun and shade leaves of tall beech trees. *Biologia Plantarum* 32: 374–387.

Mayfield, M. M. and Levine, J. M. 2010. Opposing effects of competitive exclusion on the phylogenetic structure of communities. *Ecology Letters* 13: 1085–1093.

McDowell, N. G. 2011. Mechanisms linking drought, hydraulics, carbon metabolism, and vegetation mortality. *Plant Physiology* 155: 1051–1059.

McGill, B. J., Enquist, B. J., Weiher, E. and Westoby, M. 2006. Rebuilding community ecology from functional traits. *Trends in Ecology and Evolution* 21: 178–185.

McIntyre, S. and Lavorel, S. 2001. Livestock grazing in subtropical pastures: Steps in the analysis of attribute response and plant functional types. *Journal of Ecology* 89: 209–226.

McIntyre, S., Lavorel, S., Landsberg, J. and Forbes, T. D. A. 1999. Disturbance response in vegetation towards a global perspective on functional traits. *Journal of Vegetation Science* 10: 621–630.

Menzel, A., Sparks, T. H., Estrella, N., Koch, E., Aasa, A., Ahas, R. et al. 2006. European phenological response to climate change matches the warming pattern. *Global Change Biology* 12: 1–8.

Michalet, R., Brooker, R. W., Cavieres, L. A., Kikvidze, Z., Lortie, C. J., Pugnaire, F. I., Valiente-Banuet, A. and Callaway, R. M. 2006. Do biotic interactions shape both sides of the humped-back model of species richness in plant communities? *Ecology Letters* 9: 767–773.

Mitchell, P. J., Veneklaas, E. J., Lambers, H. and Burgess, S. S. O. 2008. Leaf water relations during summer water deficit: Differential responses in turgor maintenance and variation in leaf structure among different plant communities in south-western Australia. *Plant Cell and Environment* 31: 1791–1802.

Moles, A. T., Hodson, D. W. and Webb, C. J. 2000. Seed size and shape and persistence in the soil in the New Zealand flora. *Oikos* 89: 541–545.

Moles, A. T., Ackerly, D. D., Webb, C. O., Tweddle, J. C., Dickie, J. B. and Westoby, M. 2005. A brief history of seed size. *Science* 307: 576–580.

Mouillot, D., Mason, N. W. H. and Wilson, J. B. 2007. Is the abundance of species determined by their functional traits? A new method with a test using plant communities. *Oecologia* 152: 729–737.

Mouradov, A., Cremer, F. and Coupland, G. 2002. Control of flowering time: Interacting pathways as a basis for diversity. *Plant and Cell* 14: S111–S130.

Muller-Landau, H. C. 2010. The tolerance-fecundity trade-off and the maintenance of diversity in seed size. *Proceedings of the National Academy of Sciences of the United States of America* 107: 4242–4247.

Murren, C. J., Denning, W. and Pigliucci, M. 2005. Relationships between vegetative and life history traits and fitness in a novel field environment: impacts of herbivores. *Evolutionary Ecology* 19: 583–601.

Navas, M. L. and Garnier, E. 2002. Plasticity of whole plant and leaf traits in Rubia peregrina in response to light, nutrient and water availability. *Acta Oecologica* 23: 375–383.

Nicotra, A. B., Atkin, O. K., Bonser, S. P., Davidson, A. M., Finnegan, E. J., Mathesius, U., Poot, P., Purugganan, M. D., Richards, C. L., Valladares, F. and van Kleunen, M. 2010. Plant phenotypic plasticity in a changing climate. *Trends in Plant Science* 15: 684–692.

Niinemets, U. 2001. Global-scale climatic controls of leaf dry mass per area, density, and thickness in trees and shrubs. *Ecology* 82: 453–469.

Noble, I. R. and Gitay, H. 1996. A functional classification for predicting the dynamics of landscapes. *Journal of Vegetation Science* 7: 329–336.

Paine, C. E. T., Stahl, C., Courtois, E. A., Patino, S., Sarmiento, C. and Baraloto, C. 2010. Functional explanations for variation in bark thickness in tropical rain forest trees. *Functional Ecology* 24: 1202–1210.

Parkhurst, D. F. 1994. Diffusion of CO_2 and other gases inside leaves. *New Phytologist* 126: 49–479.

Pérez-Harguindeguy, N., Diaz, S., Garnier, E., Lavorel, S., Poorter, H., Jaureguiberry, P. et al. 2013. New handbook for standardised measurement of plant functional traits worldwide. *Australian Journal of Botany* 61: 167–234.

Petchey, O. L. and Gaston, K. J. 2002. Functional diversity (FD), species richness and community composition. *Ecology Letters* 5: 402–411.

Petchey, O. L., Hector, A. and Gaston, K. J. 2004. How do different measures of functional diversity perform? *Ecology* 85: 847–857.

Pierce, S., Brusa, G., Vagge, I. and Cerabolini, B. E. L. 2013. Allocating CSR plant functional types: The use of leaf economics and size traits to classify woody and herbaceous vascular plants. *Functional Ecology* 27: 1002–1010.

Pigliucci, M. 2005. Evolution of phenotypic plasticity: Where are we going now? *Trends in Ecology and Evolution* 20: 481–486.

Pigliucci, M., Murren, C. J. and Schlichting, C. D. 2006. Phenotypic plasticity and evolution by genetic assimilation. *Journal of Experimental Biology* 209: 2362–2367.

Poorter, H. and de Jong, R. 1999. A comparison of specific leaf area, chemical composition and leaf construction costs of field plants from 15 habitats differing in productivity. *New Phytologist* 143: 163–176.

Price, T. D., Qvarnstrom, A. and Irwin, D. E. 2003. The role of phenotypic plasticity in driving genetic evolution. *Proceedings of the Royal Society of London. Series B: Biological Sciences* 270: 1433–1440.

Raunkiaer, C. 1934. *The life forms of plants and statistical plant geografy.* Oxford University Press, Oxford, 632 p.

Rehfeldt, G. E., Tchebakova, N. M., Parfenova, Y. I., Wykoff, W. R., Kuzmina, N. A. and Milyutin, L. I. 2002. Intraspecific responses to climate in *Pinus sylvestris. Global Change Biology* 8: 912–929.

Reich, P. B. 2014. The world-wide 'fast-slow' plant economics spectrum: a traits manifesto. *Journal of Ecology* 102: 275–301.

Reich, P. B. and Oleksyn, J. 2004. Global patterns of plant leaf N and P in relation to temperature and latitude. *Proceedings of the National Academy of Sciences of the United States of America* 101: 11001–11006.

Reich, P. B., Walters, M. B. and Ellsworth, D. S. 1997. From tropics to tundra: Global convergence in plant functioning. *Proceedings of the National Academy of Sciences of the United States of America* 94: 13730–13734.

Relyea, R. A. 2002. Costs of phenotypic plasticity. *American Naturalist* 159: 272–282.

Richardson, A. D., Ashton, P. M. S., Berlyn, G. P., McGroddy, M. E. and Cameron, I. R. 2001. Within-crown foliar plasticity of western hemlock. Tsuga heterophylla, in relation to stand age. Annals of Botany 88: 1007–1015.

Rweyongeza, D. M., Yang, R. C., Dhir, N. K., Barnhardt, N. K. and Hansen, C. 2007. Genetic variation and climatic impacts on survival and growth of white spruce in Alberta, Canada. Silvae Genetica 56: 117–127.

Sack, L. and Scoffoni, C. 2013. Leaf venation: Structure, function, development, evolution, ecology and applications in the past, present and future. New Phytologist 198: 983–1000.

Sack, L., Cowan, P. D., Jaikumar, N. and Holbrook, N. M. 2003. The 'hydrology' of leaves: Co-ordination of structure and function in temperate woody species. Plant Cell and Environment 26: 1343–1356.

Sandquist, D. R. and Ehleringer, J. R. 1997. Intraspecific variation of leaf pubescence and drought response in Encelia farinosa associated with contrasting desert environments. New Phytologist 135: 635–644.

Savage, J. A. and Cavender-Bares, J. 2012. Habitat specialization and the role of trait lability in structuring diverse willow (genus Salix) communities. Ecology 93: S138–S150.

Scheiner, S. M. 1993. Genetics and evolution of phenotypic plasticity. Annual Review of Ecology and Systematics 24: 35–68.

Scheiner, S. M. and Lyman, R. F. 1991. The genetics of phenotypic plasticity. 2. Response to selection. Journal of Evolutionary Biology 4: 23–50.

Schlichting, C. D. 1989a. Phenotypic integration and environmental change. Bioscience 39: 460–464.

Schlichting, C. D. 1989b. Phenotypic plasticity in Phlox: II. Plasticity of character correlations. Oecologia 78: 496–501.

Soudzilovskaia, N. A., Elumeeva, T. G., Onipchenko, V. G., Shidakov, I.I., Salpagarova, F. S., Khubiev, A. B., Tekeev, D. K. and Cornelissen, J. H. C. 2013. Functional traits predict relationship between plant abundance dynamic and long-term climate warming. Proceedings of the National Academy of Sciences of the United States of America 110: 18180–18184.

Sultan, S. E. 1987. Evolutionary implications of phenotypic plasticity in plants. Evolutionary Biology 21: 127–176.

Sultan, S. E. 1992. Phenotypic plasticity and the Neo-Darwinian legacy. Evolution Trends in Plants 6: 61–71.

Sultan, S. E. 1995. Phenotypic plasticity and plant adaptation. Acta Botanica Neerlandica 44: 363–383.

Sultan, S. E. 2000. Phenotypic plasticity for plant development, function and life history. Trends in Plant Science 5: 537–542.

Sultan, S. E. and Bazzaz, F. A. 1993a. Phenotypic plasticity in Polygonum persicaria. I. Diversity and uniformity in genotypic norms of reaction to light. Evolution 47: 1009–1031.

Sultan, S. E. and Bazzaz, F. A. 1993b. Phenotypic plasticity in Polygonum persicaria. II. Norms of reaction to soil moisture, ecological breadth, and the maintenance of genetic diversity. Evolution 47: 1032–1049.

Thompson, K., Band, S. R. and Hodgson, J. G. 1993. Seed size and seed shape predict persistence in soil. Functional Ecology 7: 236–241.

Tjoelker, M. G., Craine, J. M., Wedin, D., Reich, P. B. and Tilman, D. 2005. Linking leaf and root trait syndromes among 39 grassland and savannah species. New Phytologist 167: 493–508.

Valladares, F., Gianoli, E. and Gómez, J. M. 2007. Ecological limits to plant phenotypic plasticity. New Phytologist 176: 749–763.

van Kleunen, M. and Fischer, M. 2005. Constraints on the evolution of adaptive phenotypic plasticity in plants. New Phytologist 166: 49–60.

Vile, D., Garnier, E., Shipley, B., Laurent, G., Navas, M. L., Roumet, C. et al. 2005. Specific leaf area and dry matter content estimate thickness in laminar leaves. Annals of Botany 96: 1129–1136.

Vile, D., Shipley, B. and Garnier, E. 2006. A structural equation model to integrate changes in functional strategies during old-field succession. Ecology 87: 504–517.

Violle, C., Navas, M. L., Vile, D., Kazakou, E., Fortunel, C., Hummel, I. and Garnier, E. 2007. Let the concept of trait be functional! Oikos 116: 882–892.

Violle, C., Enquist, B. J., McGill, B. J., Jiang, L., Albert, C. H., Hulshof, C., Jung, V. and Messier, J. 2012. The return of the variance: Intraspecific variability in community ecology. Trends in Ecology and Evolution 27: 244–252.

Vitasse, Y., Bresson, C. C., Kremer, A., Michalet, R. and Delzon, S. 2010. Quantifying phenological plasticity to temperature in two temperate tree species. Functional Ecology 24: 1211–1218.

Waitt, D. E. and Levin, D. A. 1998. Genetic and phenotypic correlation in plants: A botanical test of Cheverud's conjecture. Heredity 80: 310–319.

Weiher, E., van der Werf, A., Thompson, K., Roderick, M., Garnier, E. and Eriksson, O. 1999. Challenging Theophrastus: A common core list of plant traits for functional ecology. Journal of Vegetation Science 10: 609–620.

Westoby, M. 1998. A leaf-height-seed (LHS) plant ecology strategy scheme. Plant and Soil 199: 213–227.

Westoby, M., Falster, D. S., Moles, A. T., Vesk, P. A. and Wright, I. J. 2002. Plant ecological strategies: Some leading dimensions of variation between species. Annual Review of Ecology and Systematics 33: 125–159.

Westoby, M. and Wright, I. J. 2006. Land-plant ecology on the basis of functional traits. Trends in Ecology and Evolution 21: 261–268.

White, J. 1979. The plant as a metapopulation. Annual Review of Ecology and Systematics 10: 109–145.

Wilson, P. J., Thompson, K. and Hodgson, J. G. 1999. Specific leaf area and leaf dry matter content as alternative predictors of plant strategies. New Phytologist 143: 155–162.

Witkowski, E. T. F. and Lamont, B. B. 1991. Leaf specific mass confounds leaf density and thickness. Oecologia 88: 486–493.

Wright, I. J. and Westoby, M. 2002. Leaves at low versus high rainfall: Coordination of structure, lifespan and physiology. New Phytologist 155: 403–416.

Wright, I. J., Reich, P. B., Westoby, M., Ackerly, D. D., Baruch, Z., Bongers, F. et al. 2004. The worldwide leaf economics spectrum. Nature 428: 821–827.

Wright, I. J., Ackerly, D. D., Bongers, F., Harms, K. E., Ibarra-Manriquez, G., Martinez Ramos, M. et al. 2007. Relationships among ecologically important dimensions of plant trait variation in seven Neotropical forests. Annals of Botany 99: 1003–1015.

Wright, J. S., Kitajima, K., Kraft, N. J. B., Reich, P. B., Wright, I. J., Bunker, D. E. et al. 2010. Functional traits and the growth-mortality trade-off in tropical trees. Ecology 91: 3664–3674.

Zhang, J. and Lechowicz, M. J. 1994. Correlation between time of flowering and phenotypic plasticity in Arabidopsis thaliana. American Journal of Botany 81: 1336–1342.

Zunzunegui, M., Barradas, M. C. D., Ain-Lhout, F., Alvarez-Cansino, L., Esquivias, M. P. and Novo, F. G. 2011. Seasonal physiological plasticity and recovery capacity after summer stress in Mediterranean scrub communities. Plant Ecology 212: 127–142.

Section VIII

Photosynthesis and Plant/Crop Productivity
and Photosynthetic Products

29 Carbon Photoassimilation and Photosynthate Partitioning in Plants

Carlos M. Figueroa, Claudia V. Piattoni, Karina E.J. Trípodi, Florencio E. Podestá, and Alberto A. Iglesias

CONTENTS

29.1 Intracellular and Intercellular Partitioning of Photosynthates 509
29.2 Metabolism in the Leaf Cell...510
    29.2.1 Starch Metabolism in the Leaf Cell..510
        29.2.1.1 Starch Synthesis and Degradation in the Chloroplast 510
        29.2.1.2 Regulation of the Starch Metabolism in the Chloroplast..513
    29.2.2 Exchange of Metabolites between Chloroplast and Cytosol 514
    29.2.3 Synthesis of Sucrose and Sugar-Alcohols 515
        29.2.3.1 The Pathway of Sucrose Synthesis 515
        29.2.3.2 Sugar-Alcohols as Major Photosynthetic Products...517
    29.2.4 Respiratory Metabolism in the Leaf Cell 519
29.3 Transport of Carbon from the Leaf to Sink Tissues..520
    29.3.1 Sucrose Transporters 521
        29.3.1.1 SWEET Transporters..521
        29.3.1.2 SUT Transporters..522
    29.3.2 Mannitol and Glucitol Transporters 523
29.4 Metabolism in Heterotrophic Cells 524
    29.4.1 Metabolism of Sucrose and Sugar-Alcohols...524
        29.4.1.1 Sucrose Utilization in Nonphotosynthetic Tissues 524
        29.4.1.2 Utilization of Mannitol and Glucitol in Heterotrophic Tissues 526
    29.4.2 Metabolism for Starch Storage 527
    29.4.3 Metabolism for Lipid Storage 528
29.5 Concluding Remarks and Perspectives..529
Acknowledgments...529
References...529

29.1 INTRACELLULAR AND INTERCELLULAR PARTITIONING OF PHOTOSYNTHATES

The photosynthetic process involves the use of sunlight energy to convert (reduce) the inorganic atmospheric form of carbon (CO_2) into organic molecules. The primary photosynthates are carbohydrates, which then supply substrates to different metabolic pathways leading to the synthesis of other cell components, such as lipids, amino acids, proteins, and nucleic acids. Also, these photogenerated biomolecules are key to sustaining life on earth, as they serve as the organic feedstock to many other heterotrophic organisms (Cortassa et al. 2012). Higher plants perform oxygenic photosynthesis in green tissues (mainly in leaves) and more specifically in cells having chloroplasts. The entire process can be divided into two parts: the light phase and the synthetic phase. The light phase comprises steps mainly taking place at the thylakoid membrane of chloroplast to convert the electromagnetic energy from light into reducing power (NADPH) and chemical energy (ATP)

with the associated photolysis of water, according to the following general equation:

$$2\ H_2O + 3\ ADP + 3\ P_i + 2\ NADP^+ + light \rightarrow O_2 + 3\ ATP + 2\ NADPH + 2\ H^+$$

The synthetic phase develops in the chloroplast stroma where, via the Benson–Calvin cycle (BCC), ATP and NADPH (produced during the light phase) are used for carbon fixation and carbohydrates production (Cortassa et al. 2012; Iglesias and Podestá 2005; Rochaix 2011), with the general equation for synthesis of one molecule of hexose-phosphate (hexose-P):

$$6\ CO_2 + 18\ ATP + 12\ NADPH + 12\ H^+ \rightarrow hexose\text{-}P + 18\ ADP + 17\ P_i + 12\ NADP^+$$

The BCC is an autotrophic pathway operatively acting in the chloroplast stroma and comprising three stages (detailed

509

TABLE 29.1

Reactions and Enzymes Involved in the Three Stages of the Benson–Calvin Cycle

Stage	EC	Name	Reaction
Carboxylation	4.1.1.39	Rul-1,5bisP carboxylase/ oxygenase	Rul-1,5bisP + CO_2 → 2 3PGA
Reduction	2.7.2.3	3PGA kinase	3PGA + ATP ⇆ 1,3bisPGA + ADP
	1.2.1.13	Ga3P dehydrogenase	1,3bisPGA + NADPH + H^+ ⇆ Ga3P + $NADP^+$ + P_i
	5.3.1.1	Triose-P isomerase	Ga3P ⇆ DHAP
Regeneration	4.1.2.13	Aldolase	Ga3P + DHAP ⇆ Fru-1,6bisP
	3.1.3.11	Fru-1,6bisP 1-phosphatase	Fru-1,6bisP + H_2O ⇆ Fru-6P + P_i
	2.2.1.1	Transketolase	Fru-1,6bisP + Ga3P ⇆ Xul-5P + erythrose-4P
	4.1.2.13	Aldolase	erythrose-4P + DHAP ⇆ sedoheptulose-1,7bisP
	3.1.3.37	Sedoheptulose-1,7bisP 1-phosphatase	sedoheptulose-1,7bisP + H_2O → sedoheptulose-7P + P_i
	2.2.1.1	Transketolase	sedoheptulose-7P + DHAP ⇆ Rib-5P + Xul-5P
	5.3.1.6	Rib-5P isomerase	Rib-5P ⇆ Rul-5P
	5.1.3.1	Xul-5P epimerase	Xul-5P ⇆ Rul-5P
	2.7.1.19	Rul-5P kinase	Rul-5P + ATP → Rul-1,5bisP + ADP

in Table 29.1; see also Figure 29.1): (1) carboxylation, comprising a single reaction, catalyzed by ribulose-1,5-bis-phosphate (Rul-1,5bisP) carboxylase/oxygenase, where CO2 is combined with Rul-1,5bisP to produce two molecules of 3P-glycerate (3PGA); (2) reduction, a stage including reactions using ATP to convert 3PGA in 1,3-bisP-glycerate (1,3bisPGA) (mediated by P-glycerate kinase) followed by reduction of this intermediate to triose-P (namely glyceraldehyde-3P [Ga3P] interconverted with dihydroxyacetone-P [DHAP]) at the expenses of NADPH catalyzed by Ga3P dehydrogenase (Ga3PDHase); and (3) regeneration, involving 10 reactions with the interconversion of sugar-P of between 3 and 7 carbon atoms plus the consumption of ATP to regenerate the initial acceptor of CO_2 of the cycle, Rul-1,5bisP (Piattoni et al. 2014).

Triose-P and hexose-P are key metabolites of the BCC, since from them the photoassimilated carbon is derived into different metabolic fates (Figure 29.1). Thus, Fru-6P is a starting point for the pathway leading to biosynthesis of starch, a major transitory product of photosynthesis accumulated in the chloroplast

during the day. Besides, triose-P can be mobilized to the cytosol by interchange with inorganic orthophosphate (Pi) via the specific Pi/triose-P translocator present in the chloroplast envelope. The latter provides carbon for cytosolic metabolism, and it also constitutes an indirect export of ATP and NADPH. Once in the cytosol, carbohydrates are utilized to synthesize soluble sugars that are accumulated and/or used for transporting photosynthates to nonphotosynthetic parts of the plant through the phloem. Sucrose (Suc) is the main sugar produced in the cytosol and delivered between photosynthetic and heterotrophic cells in most plants; but many species (for example, those of the Rosaceae family) also synthesize sugar-alcohols for similar purposes. Suc or sugar-alcohols enter the cytosol of nonphotosynthetic cells to fuel different metabolic routes, including those taking place in plastids for carbon storage in the form of polysaccharides (starch) or lipids (triacylglycerides [TAGs]) (Cortassa et al. 2012; Iglesias and Podestá 2005; Piattoni et al. 2014).

The above described events establish that, in plants, the primary products of photosynthesis enter metabolic routes determining two levels of photoassimilates partitioning: intracellular and intercellular. In photosynthetic cells, the carbohydrates are partitioned between the chloroplast (where starch is accumulated in a relatively stationary form) and the cytosol (where Suc and sugar-alcohols constitute a mobile pool). In heterotrophic cells, partition starts in the cytosol and includes delivery of specific metabolites to plastids for synthesis of starch (Figure 29.2) or TAGs (Figure 29.3), which constitute long-term carbon and energy storage forms. The global view thus indicates that intercellular partition allows distribution of photoassimilates between source and sink tissues in a plant. Therefore, understanding the function and regulation of these metabolic and physiological processes is relevant, because they critically determine plant productivity. In the following, we detail the biochemical characteristics of carbohydrates metabolism and transport within and between the different plant cells.

29.2 METABOLISM IN THE LEAF CELL

29.2.1 STARCH METABOLISM IN THE LEAF CELL

29.2.1.1 Starch Synthesis and Degradation in the Chloroplast

During the light period, a part (~30%) of the surplus of photoassimilates is derived from the BCC to produce starch within the chloroplast. The polysaccharide thus synthesized is transitorily stored, and it mainly serves to supply carbon metabolites during the night (Cortassa et al. 2012; Iglesias and Podestá 2005). The starch metabolism in the chloroplast involves the reactions detailed in Table 29.2. Synthesis initiates from the BCC intermediate Fru-6P (Figure 29.1) that is converted to Glc-6P and then to Glc-1P by the consecutive action of Glc-6P isomerase and phosphoglucomutase. Then follows the step catalyzed by ADP-glucose pyrophosphorylase (ADPGlc PPase) that produces ADPGlc and inorganic pyrophosphate (PPi) from Glc-1P and ATP in the presence of Mg^{2+}. ADPGlc is the substrate specifically used by starch synthase to elongate

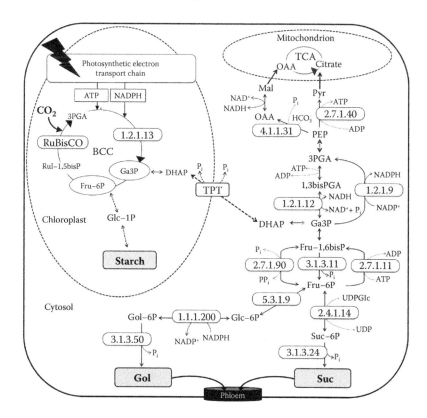

FIGURE 29.1 Photosynthetic metabolism and diagram of carbon flow in a plant autotrophic cell. Photoassimilated carbon is partitioned for transitory accumulation of starch within the chloroplast and for synthesis of Suc and/or sugar-alcohols (in this picture represented by Gol) in the cell cytosol.

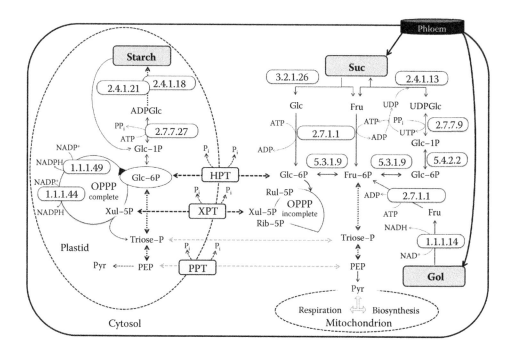

FIGURE 29.2 Carbon flow in a heterotrophic plant cell mainly accumulating starch for long-term carbon storage. In the diagram, it is considered a plant species using Suc and Gol as metabolites for carbon transport between photosynthetic and heterotrophic cells.

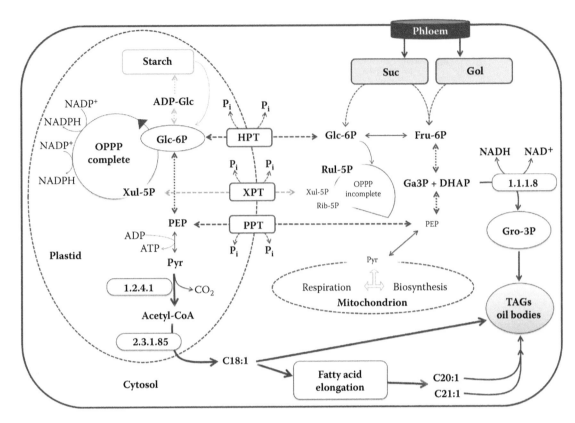

FIGURE 29.3 Diagram of the carbon flow in a heterotrophic plant cell mainly accumulating lipids (TAGs) for long-term storage. It is considered a plant species that transport photoassimilates between cells in the form of Suc and Gol.

TABLE 29.2

Reactions Involved in Starch Metabolism

EC	Name	Reaction
5.3.1.9	Glc-6P isomerase	Fru-6P ⇆ Glc-6P
5.4.2.2	Phosphoglucomutase	Glc-6P ⇆ Glc-1P
2.7.7.27	ADPGlc PPase	Glc-1P + ATP ⇆ ADPGlc + PPi
2.4.1.21	Starch synthase	ADPGlc + (α-1,4-glucan)$_n$ → (α-1,4-glucan)$_{n+1}$ + ADP
2.4.1.18	Branching enzyme	Linear α-1,4-polyglucan → α-1,4-polyglucan with α-1,6-linkage branch point
2.4.1.1	Starch phosphorylase	(α-1,4-glucan)$_n$ + Pi ⇆ (α-1,4-glucan)$_{n-1}$ + Glc-1P
3.2.1.1	α-Amylase	α-1,4-polyglucan + H$_2$O → maltose + linear and branched oligosaccharides + branched dextrins
3.2.1.2	β-Amylase	α-1,4-polyglucan + H$_2$O → maltose + dextrins
2.4.1.25	4-α-Glucanotransferase	(α-1,4-glucan)$_p$ + (α-1,4-glucan)$_q$ ⇆ (α-1,4-glucan)$_{p+q-1}$ + Glc
3.2.1.4	Debranching enzyme	α-1,4-polyglucan with α-1,6-linkage branch point → linear α-1,4-polyglucan

an α-1,4-polyglucan chain, and next the branching enzyme catalyzes the formation of the α-1,6-branch points of the amylopectin fraction of the polymer (Ballicora et al. 2003, 2004; Cortassa et al. 2012).

Concerning breakdown of chloroplastic starch, two routes are possible depending on the respective use of Pi or H$_2$O as substrates to cleave the α-1,4-glucan chain from its nonreducing end. A specific starch phosphorylase cuts α-1,4-glycosyl bonds (in polymers larger than maltotetraose) by phosphorolysis, whereas α- and β-amylase cleave by hydrolysis. The α-amylase is endoamylolytic and produces a mixture of linear and branched oligosaccharides, branched dextrins, maltotriose, maltose, and Glc. The exoamylolytic characteristic of β-amylase mainly renders maltose and dextrins. By either the phosphorolytic or the hydrolytic pathway, the complete degradation of starch requires other enzymes. Namely, the complementary action of a 4-α-glucanotransferase (or D-enzyme, D coming from disproportionating) and debranching enzyme respectively catalyzes condensation of two α-1,4-polyglucan chains and rearrangement of the polymer from branched to linear forms (Iglesias and Podestá 2005; Zeeman et al. 2010) (Table 29.2).

The described pathway for starch synthesis in chloroplasts involving the use of ADPGlc as the glycosyl donor is supported by numerous experimental data (Espada 1962; Heldt et al. 1977; Kossmann et al. 1994; Kruckeberg et al. 1989; Neuhaus and Stitt 1990). An early work demonstrated the presence of ADPGlc PPase in photosynthetic cells (Espada 1962). Later, it was demonstrated that isolated chloroplasts accumulate starch with the only requirement of CO_2 and light, thus showing no requirement of other cellular components (Heldt et al. 1977). Also, the relevance of Fru-6P as a starter intermediate became clear after studies with mutant plants having reduced activity of Glc-6P isomerase (Kruckeberg et al. 1989) or phosphoglucomutase (Neuhaus and Stitt 1990); as well as from experiments using antisense RNA strategy to produce transformed potato plants with lower amounts of Fru-1,6bisP 1Pase (Kossmann et al. 1994). All these works established a direct relationship where lower levels of Fru-6P in leaf cells result in lower accumulation of starch. Additionally, other enzymes of the malto-oligosaccharide metabolism (including D-enzyme) play critical roles determining the formation of the starch granule and the semicrystalline structure of the polymer (Zeeman et al. 2010).

29.2.1.2 Regulation of the Starch Metabolism in the Chloroplast

Starch synthesis in chloroplast is mainly modulated at the step of producing ADPGlc, as buttressed by conclusive results obtained from different experimental approaches. As extensively reviewed (see Ballicora et al. 2003, 2004; Preiss et al. 1991 and references therein), information derived from plant cell mutants and analysis on flux control strongly evidence that ADPGlc PPase is a rate-limiting enzyme and that levels of its activity regulatory properties are critical for the polysaccharide accumulation in green cells. Although the reaction catalyzed by ADPGlc PPase is reversible *in vitro* (Table 29.2), the presence of inorganic pyrophosphase and the use of the sugar nucleotide by starch synthase track the process to operate irreversibly *in vivo*.

Many studies report the structural, kinetic, and regulatory properties of ADPGlc PPase from cells performing oxygenic photosynthesis: cyanobacteria, green algae, and leaves of higher plants (Ballicora et al. 2004). The enzyme from eukaryotic organisms is composed by two subunits (α [~50 kDa] and β [~54 kDa]) that arrange an $\alpha_2\beta_2$ heterotetramer, whereas the cyanobacterial protein is homotetrameric (α_4). It has been proposed that the β subunit emerged after duplication of an ancestor gene, and it specialized during evolution to adapt allosteric regulation according to different demands related with plant species and tissue. This latter picture is strongly supported by studies performed in the characterization of the different hybrid proteins produced after expression of the six genes found in *Arabidopsis thaliana* that code for respective α (two genes) and β (four genes) subunits (Crevillen et al. 2003).

Different mechanisms concur to determine the regulation of ADPGlc PPase in green tissues. First, the enzyme is under transcriptional control (Koch 1996). Second, ADPGlc PPase from plants is mainly regulated by allosteric effects of 3PGA (activator) and Pi (inhibitor), which in a concerted manner (plus conditions of molecular crowding found in the stroma

environment) elicit ultrasensitive responses in the enzyme behavior (Aon et al. 2001; Ballicora et al. 2003, 2004). Activation by 3PGA involves an increase in V_{max} (between 5- and 100-fold) and also in the affinity toward substrates, which in combination with the cross-talk action established with the inhibitor Pi determine wide variations in the levels of the enzyme activity depending on the ratio 3PGA/Pi *in vivo*. Since 3PGA is the primary product of carbon fixation by the BCC (Table 29.1) and Pi is a critical metabolite in the photosynthetic process (Figure 29.1), the allosteric regulation of ADPGlc PPase is consistent with a physiological relevance. Under light conditions, the high 3PGA/Pi ratio in the chloroplast activates the enzyme, thus leading photoassimilates to produce starch. In the opposite scenario taking place in the dark, the inhibitory effect of Pi overrides to minimize the enzyme activity and block the polysaccharide synthesis (Ballicora et al. 2003, 2004; Iglesias and Podestá 2005; Piattoni et al. 2014; Preiss et al. 1991).

Another level of regulation of ADPGlc PPase associated to chloroplastic starch synthesis is given by post-translational redox modification (Ballicora et al. 2000; Fu et al. 1998; Smith and Stitt 2007). The enzyme from potato tuber undergoes activation (complementary to that produced by 3PGA) after reduction of an intermolecular disulfide bridge formed between α subunits in the $\alpha_2\beta_2$ heterotetramer. Chemical (dithiothreitol) and more efficiently biological (reduced thioredoxin *f* and *m*) redox compounds exert the reduction of the Cys12 linkage formed between the catalytic subunits (Ballicora et al. 2000; Fu et al. 1998). Oxidized thioredoxin reverses the process with recovery of the disulfide bridge in the enzyme. Although the molecular characterization of the reductive activating mechanism has been performed with the potato tuber enzyme, the cysteine residue involved is conserved in leaf ADPGlc PPases. Thus, the enzyme would be regulated on the basis of a light/dark cycle by redox modification mediated by the ferredoxin–thioredoxin system being operative in chloroplasts (Smith and Stitt 2007). During the day, reduced thioredoxin modifies the enzyme enhancing (in combination with the high 3PGA/Pi ratio) ADPGlc PPase activity to favor starch synthesis. At night, the process is reversed because thioredoxin is mostly oxidized (and the 3PGA/Pi ratio is low). Indeed, it has recently been demonstrated that these regulatory mechanisms are important for adjusting starch synthesis to day length in *A. thaliana* plants (Mugford et al. 2014).

In the last decade, it was suggested that in photosynthetic cells, the thioredoxin-dependent activation of starch synthesis was linked to the regulatory signal trehalose-6P (Tre-6P) (Kolbe et al. 2005; Lunn et al. 2006). However, Martins et al. (2013) recently argued against this hypothesis and proposed a new scenario where Tre-6P acts reducing the rate of starch degradation rather than promoting its synthesis. Although the underlying molecular mechanism is still unknown, it would operate in addition to the regulation of starch degradation by the circadian clock, which is responsible for adapting the rates of starch synthesis and degradation to the photoperiod (Smith and Stitt 2007; Stitt and Zeeman 2012). These studies importantly contribute to have a better comprehension of how starch metabolism is dynamically regulated. Further work is necessary to have a complete understanding of this central metabolic route in photosynthetic cells, especially in discovering the mechanisms that modulate breakdown of starch at the molecular level acting in coordination with its synthesis.

29.2.2   EXCHANGE OF METABOLITES BETWEEN CHLOROPLAST AND CYTOSOL

In plants, *de novo* synthesis of triose-P by the chloroplastic photosynthetic process generates carbon and energy sources for the whole plant during light periods. In order to deliver carbon for all the metabolic pathways, triose-P need to be partitioned between the different cell compartments. Chloroplasts and other plastids present in plants contain two concentric layers of membranes (called the envelope) that separate the stroma from the surrounding cytosol. The envelope has many functions and is of utmost importance in maintaining the global carbon flux, and the transporters for the exchange of metabolites between plastids and the cytoplasm are proteins implanted in its inner membrane. These proteins exchange metabolic precursors, intermediates, and end products, thus connecting the metabolism inside the organelle with that of the whole cell (Weber and Linka 2011). The type of transporters present in each plastid depends on the kind of cell and/or developmental stage. Each organ has different transporter proteins to interconnect not only the cells but also different cell compartments, which is critical to coordinating the entire metabolism. Intracellular metabolite partition is an example of how superior organisms have evolved to increment the control of metabolic pathways, allowing the simultaneous operation of metabolic fluxes competing for the same substrates and avoiding the occurrence of futile cycles (Lunn 2007).

Among the most studied and well-characterized membrane transporters are the plastidic phosphate translocators, which are homodimers of highly hydrophobic membrane proteins catalyzing the counter exchange of phosphorylated C3, C5, or C6 compounds with inorganic orthophosphate (Pi) (Weber et al. 2005). This antiport mode ensures that net transport of carbon is achieved without impairing phosphate balance between stroma and cytosol (Weber and Linka 2011). Inside the chloroplast, the main fate of triose-P is starch synthesis as a transitory reserve of carbon and energy. Alternatively, triose-P are transported to the cytosol via the triose-P transporter (TPT) found mainly in photosynthetic cells (Figure 29.1). TPT is specific for triose-P and 3PGA, and cannot transport phospho-enol-pyruvate (PEP), pentose-P, or hexose-P (Tegeder and Weber 2006; Toyota et al. 2006). In vascular plants, TPT function is to allocate a portion of the fixed carbon to be metabolized for producing hexose-P, and further Suc, by gluconeogenesis or used to obtain energy and the building blocks necessary for cell maintenance through glycolysis. As TPT also exchanges 3PGA, in photosynthetic cells, its combined action with the cytosolic nonphosphorylating Ga3PDHase (np-Ga3PDHase, EC 1.2.1.9) constitutes a shuttle system for

the export of NADPH photosynthetically generated from the chloroplast to the cytosol (Kelly and Gibbs 1973).

The energy-rich glycolytic intermediate PEP can also be delivered from the cytosol to plastids via the PEP/phosphate translocator (PPT). This transporter is specific for PEP (Weber and Linka 2011), and it represents the main route of PEP supply to plastids, since chloroplasts and most non-green plastids (except those from lipid-storing tissues) lack the ability to form this metabolite via a complete glycolytic pathway (Flugge et al. 2011) (Figures 29.2 and 29.3). Inside the plastids, PEP serves as a substrate for the pyruvate (Pyr) kinase generating Pyr and ATP. PEP and Pyr act as precursors for several pathways including fatty acid (FA) biosynthesis or the shikimate pathway, leading to a variety of important secondary compounds (Flugge 1999). Most plastid types can transport PEP across their envelope membranes. In C_3 plants, chloroplasts have a low PEP-transport activity, and non-green plastids also have the capability for PEP–Pi exchange. However, chloroplasts from C_4 plants show remarkably high rates of PEP transport. In these plants, the export of PEP is part of the C_4 photosynthetic carbon metabolism and has to keep pace with the rate of CO_2 assimilation (Fischer and Weber 2002). After induction of CAM, transcript amounts of genes coding for the PPT and the Glc-6P phosphate translocator (GPT) increase drastically, while TPT transcripts remain unchanged (Hausler et al. 2000).

The GPT belongs to the plastidic phosphate translocators family and it is preferentially expressed in non-green tissues. This translocator imports carbon skeletons in the form of Glc-6P into plastids for starch biosynthesis and for the oxidative pentose-phosphate pathway (OPPP), thus delivering reduction equivalents for FA biosynthesis or nitrite reduction. The GPT exhibits broad substrate specificity, accepting Glc-6P and triose-P (including 3PGA) as counter exchange substrates for Pi or even between them (Neuhaus and Emes 2000; Weber and Linka 2011; Weber et al. 2005). In the case of starch biosynthesis, the imported Glc-6P is converted in the stroma to ADPGlc, the precursor of starch synthesis (Neuhaus and Emes 2000). The pentose-P/Pi translocator (XPT) was the last of these carrier proteins discovered in vascular plants, and it counter-exchanges triose-P, xylulose-5-phosphate (Xul-5P), Rub-5P, and erythrose-4-phosphate for Pi (Eicks et al. 2002). The proposed function of XPT is the transport of pentose-P for incorporation into the plastidial OPPP and the BCC because of the lack of the non-oxidative branch of the incomplete cytosolic OPPP (Weber and Linka 2011) (Figures 29.2 and 29.3).

With some exceptions, starch is the principal energy reserve compound accumulated in photosynthetic and heterotrophic tissues. Starch storage in leaves is transitory, and concentrations are highly variable during the day. During the night, the carbon and energy are provided through the breakdown of transitory starch and the export of the resulting maltose and Glc. Export of maltose, being the main product exported, is mediated by the maltose transporter MEX1, in addition to the plastid Glc transporter (pGlcT) (Ludewig and Flugge 2013). These nonphosphorylated-sugar transport systems are

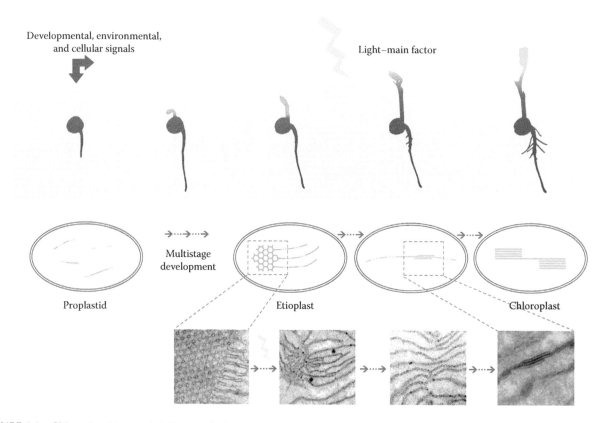

FIGURE 1.1 Chloroplast biogenesis in leaves of a hypogeal germinating plant coordinated with seedling development.

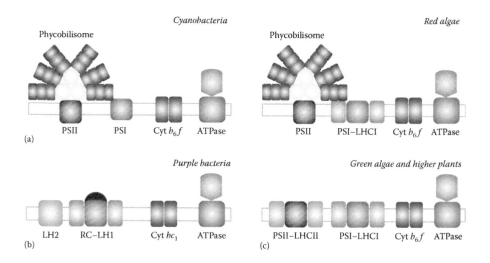

FIGURE 3.1 Variety of the photosynthetic apparatus in different photosynthetic organisms.

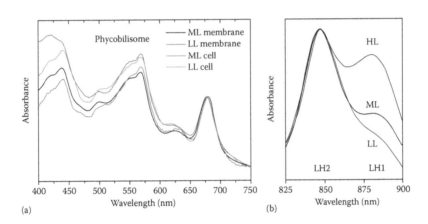

(a)

(b)

FIGURE 3.2 (a) Room-temperature absorption spectra of whole cells and isolated PBsome thylakoid membranes from the unicellular red alga *Porphyridium cruentum* grown under low light (LL, 6 W/m²) and moderate light (ML, 15 W/m²). (From Liu, L.N. et al., *J. Biol. Chem.*, 283, 34946–34953, 2008.) Spectra were normalized to the chlorophyll absorption band at 682 nm. Results showed that the amount of PBsomes (500–650 nm) is dependent upon the light intensity. (b) The absorption spectrum of purple photosynthetic membranes adapted to high light (HL), ML, and LL. The absorption ratios of LH2 (845 nm) to the core complex (LH1, 880 nm) are 1.16, 1.86, and 2.6, respectively. (From Liu, L.N. et al., *J. Mol. Biol.*, 393, 27–35, 2009; Scheuring, S., and J.N. Sturgis, *Science*, 309, 484–487, 2005.) Results indicated that the ratio of antenna to RC varies according to the light intensity.

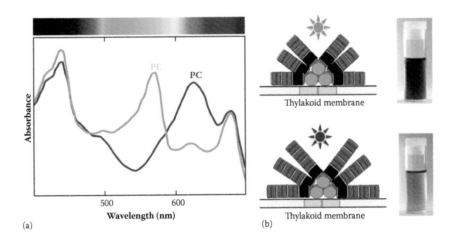

(a)

(b)

FIGURE 3.3 (a) Whole-cell absorption spectra of the cyanobacterium *Fremyella diplosiphon* cells grown in green and red light. The phycoerythrin (PE) and phycocyanin (PC) absorption peaks are indicated. (From Kehoe, D.M., and A. Gutu, *Annu. Rev. Plant Biol.*, 57, 127–150, 2006.) (b) Changes of *F. diplosiphon* PBsomes in composition and structure induced by green and red light.

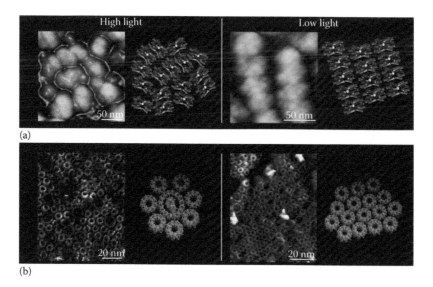

(a)

(b)

FIGURE 3.4 (a) AFM images presenting the adaptation of the organization of PBsomes on the red algal thylakoid membrane. (From Liu, L.N. et al., *J. Biol. Chem.*, 283, 34946–34953, 2008.) (b) AFM images showing the organization of photosynthetic membranes from the purple photosynthetic bacterium *Rhodospirillum photometricum* adapted to high light and low light. (From Liu, L.N., and S. Scheuring, *Trends Plant Sci.*, 18, 277–286, 2013; Scheuring, S., and J.N. Sturgis, *Science*, 309, 484–487, 2005.)

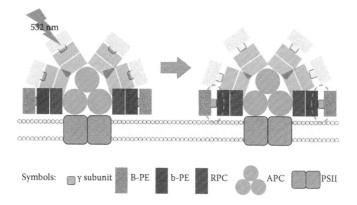

Symbols: γ subunit   B-PE   b-PE   RPC   APC   PSII

FIGURE 3.6 Schematic model of light-induced decoupling of PBsomes, indicating a possible photoprotective mechanism of the PBsome. Green open circles present the potential decoupling sites within the PBsome. B-PE, B-phycoerythrin; b-PE, b-phycoerythrin; RPC, R-phycocyanin.

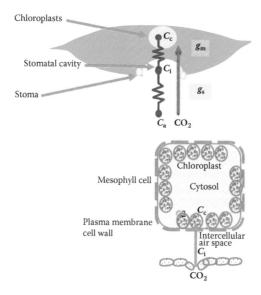

FIGURE 4.1 A model of components of a photosynthetic CO_2 diffusion pathway.

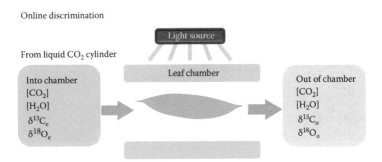

FIGURE 4.4 A model of measurement set up with leaf chamber.

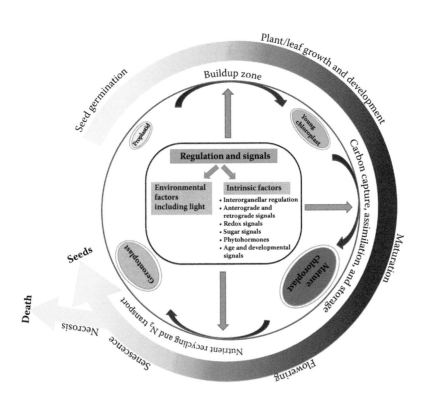

FIGURE 5.1 The major events associated with formation and demolition of chloroplast and associated regulatory mechanisms during leaf development and senescence. (Modified from Springer Science+Business Media: *Plastid Development in Leaves During Growth and Senescence*, ed. B. Biswal, K. Krupinska, and U.C. Biswal, 3–16, Advances in Photosynthesis and Respiration, vol. 36, The dynamic role of chloroplasts in integrating plant growth and development, 2013, Krupinska, K. et al.)

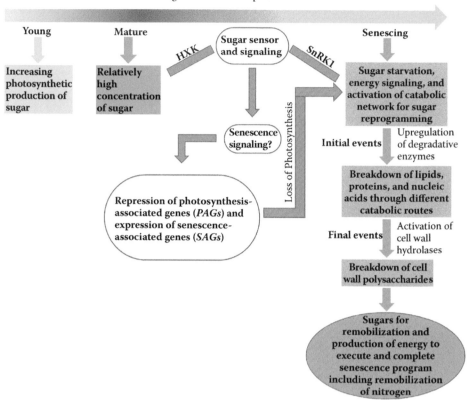

Progress of leaf development

Young   Mature   Senescing

HXK   Sugar sensor and signaling   SnRK1

Increasing photosynthetic production of sugar

Relatively high concentration of sugar

Sugar starvation, energy signaling, and activation of catabolic network for sugar reprogramming

Loss of Photosynthesis

Initial events   Upregulation of degradative enzymes

Senescence signaling?

Repression of photosynthesis-associated genes (*PAGs*) and expression of senescence-associated genes (*SAGs*)

Breakdown of lipids, proteins, and nucleic acids through different catabolic routes

Final events   Activation of cell wall hydrolases

Breakdown of cell wall polysaccharides

Sugars for remobilization and production of energy to execute and complete senescence program including remobilization of nitrogen

FIGURE 5.6 Changes in sugar reprogramming during leaf development and senescence. The figure shows variation in the concentration of sugar during development and involvement of regulatory mechanisms mediated by sugar sensors for induction of senescence and cellular sugar homeostasis. (Modified from Biswal, B. et al., Photosynthetic regulation of senescence in green leaves: Involvement of sugar signalling, in *Photosynthesis—Overviews on Recent Progress and Future Perspectives*, eds. S. Itoh, P. Mohanty, and K.N. Guruprasad, 245–260, IK International Publishing House Pvt, Ltd., New Delhi, India, 2012.)

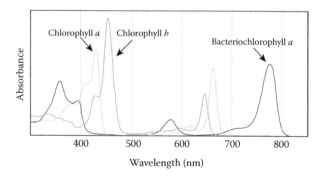

FIGURE 9.2 Absorption spectra of Chls. The absorption spectra of pigments dissolved in nonpolar solvents are shown for Chl *a* and *b* and Bchl *a*. Note that the spectra of these pigments show substantial shifts in absorbance *in vivo*, where they are associated with specific proteins.

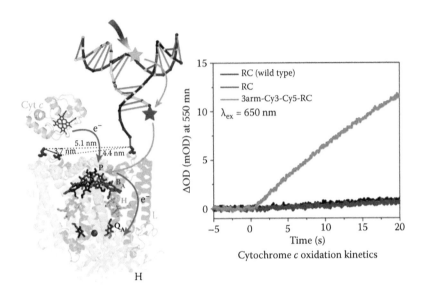

FIGURE 11.19 RC from the purple bacterium, *R. sphaeroides* 2.4.1, bioconjugated to a 3arm-DNA construct sequences. The two stars on strands 2 and 3 represent the positions of the two dye molecules. Because of the presence of three Cys residues, up to three 3arm-DNA junctions (and hence three pairs of dyes) can be conjugated to the RC. (Reprinted with permission from Dutta, P.K. et al., *Journal of the American Chemical Society*, 136(47): 16618–16625. Copyright 2014 American Chemical Society.)

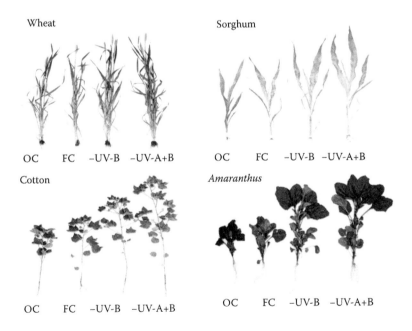

FIGURE 15.2 Photographs showing effect of exclusion of UV-B and UV-A+B on growth and biomass accumulation of wheat, cotton, *Amaranthus*, and sorghum plants.

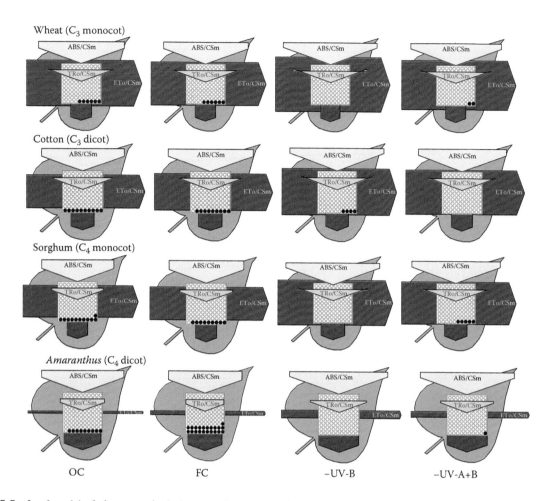

FIGURE 15.5 Leaf model of phenomenological energy fluxes per excited cross section (CS) of wheat, cotton, sorghum, and *Amaranthus* leaves showing the effect of exclusion of UV-B and UV-A+B.

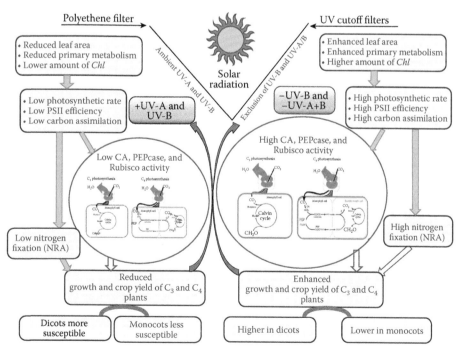

FIGURE 15.11 A model to summarize the growth, photosynthetic, and yield responses to solar UV-B and UV-A+B exclusion for C_3 and C_4 plants. NRA, nitrate reductase activity.

FIGURE 17.1 Limited K/N macronutrient study field views (irrigated June-bearing strawberry cv. *Mira*) initial flowering (a) and initial fruit ripening (b); strawberry plants at full blooming on June 3 (c), full bearing on June 26 (d), early maturity on July 2 (e), full maturity on July 13 (f); marketable fruit yields for the 50%-limited K/N treatments plot on July 13 (g); and higher marketable fruit yields for the 25%-limited K/N treatments plot determined on the same day (July 13) (h).

FIGURE 17.7 Strawberry nursery plant propagation and methods of determination of strawberry nursery productivity. Positioning of mother-stock crowns, runners, and daughter plants at first daughter-plant growth stage in the fields in mid-July (a) and overview positioning of strawberry nursery production systems in late September (b).

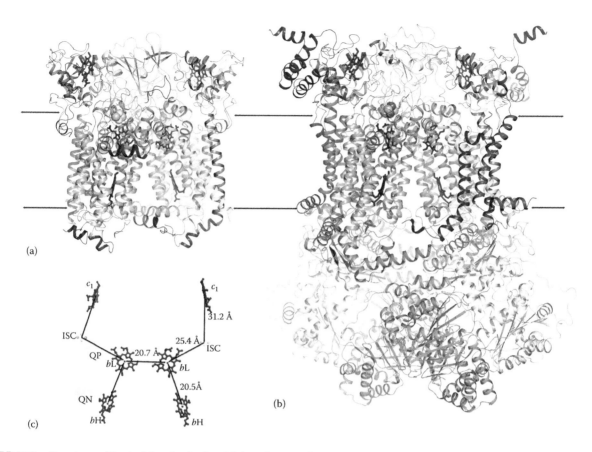

FIGURE 19.2 Structures of bacterial and mitochondrial cyt bc_1 complexes.

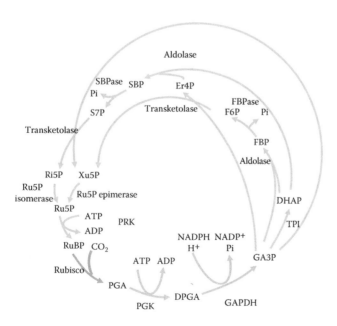

FIGURE 22.1 Outline of the Calvin cycle. The Calvin cycle is divided into three phases: carboxylation, reduction, and regeneration (red, green, and blue arrows, respectively).

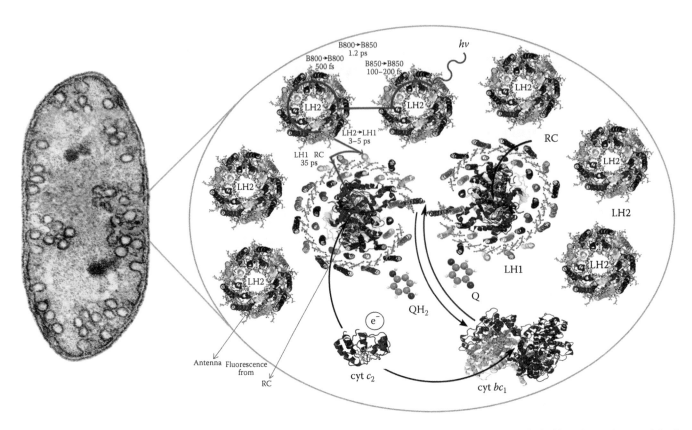

FIGURE 23.1 Hierarchy of organization of the photosynthetic apparatus of *Rba. sphaeroides*: intact cell (left) and top-view model of the intracytoplasmic membrane with atomic resolution of the BChl–protein complexes of the light-harvesting (peripheral [LH2] and core [LH1]) antenna and the reaction center (RC) in supercomplex (dimeric) arrangement (right).

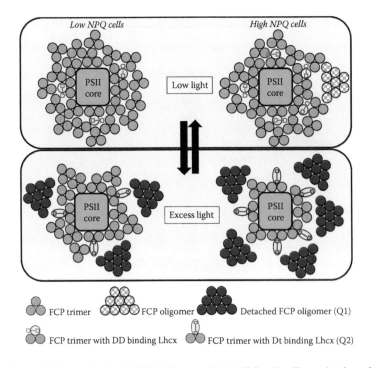

FIGURE 24.10 Model of the diatom NPQ mechanism in PSII of low and high NPQ cells. (Reproduction of Figure 3 of Goss R, Lepetit B, *J Plant Physiol* 172, 13–32, 2015. With kind permission.)

FIGURE 26.1 Grass–legume mixture study at the University of Wyoming Sheridan Research and Extension Center, Sheridan, Wyoming, in 2014. Different mixtures of forage legumes and grass were used. Forage species includes alfalfa (*Medicago sativa* L.), bird's-foot trefoil (*Lotus corniculatus* L.), sainfoin (*Onobrychis viciifolia* Scop.), and meadow bromegrass (*Bromus biebersteinii* Roem. & Schult.). Forage production and forage quality varied among mixtures. (Picture taken by M. Anowarul Islam.)

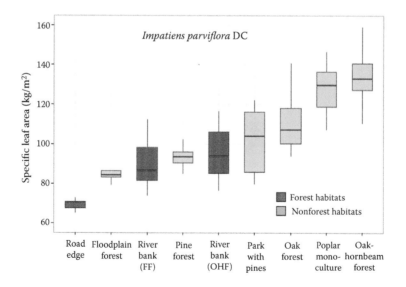

FIGURE 28.1 The intraspecific variability in SLA in an invasive species *I. parviflora* DC measured in nine populations inhabiting different environmental conditions. These represent the habitats that the species invaded throughout the last 30 years and in which they have established a viable population, often monodominant. The habitats differ in light, nutrient, and water availability. FF: floodplain forest; OHF: oak-hornbeam forest.

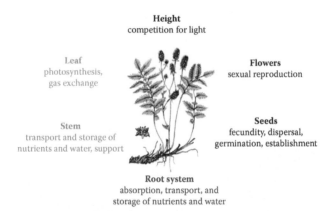

FIGURE 28.2 Different plant organs and plant height reflect different functions that are important for plant fitness. (Modified from Laughlin, D. C., *Journal of Ecology* 102:186–193, 2014.)

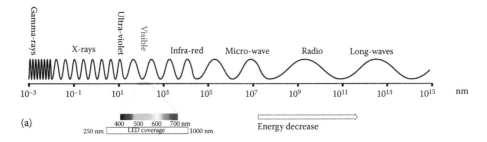

(a)

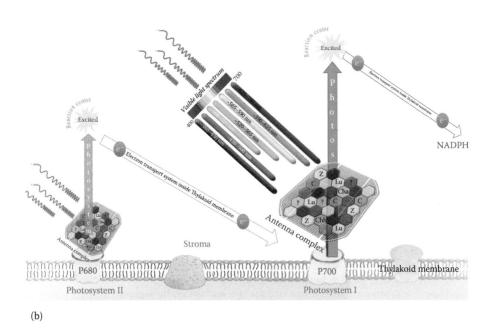

(b)

FIGURE 42.1 (a) The visible light is a small fraction of the full electromagnetic spectrum; (b) the antenna complexes located on thylakoid membrane consist of various light-harvesting molecules, which could only trap photon energy in the range of the visible spectrum light (~400 to ~700 nm) and send the photons through an energy transfer path to Chl *a* molecule as the main molecule in each RC. The antenna pigment molecules mainly include Chl *b*: chlorophyll *b*; Chl *a*: chlorophyll *a*; C: β-carotene; Z: zeaxanthin; and Lu: lutein.

mechanistically different from the others described above as they do not depend on the interchange of Pi or phosphorylated metabolites.

29.2.3 Synthesis of Sucrose and Sugar-Alcohols

Suc is the main molecule utilized by plants to transport carbon from source to sink tissues and constitutes an important carbon reservoir in certain species and tissues, like sugar cane (*Saccharum officinarum*) stems and sugar beet (*Beta vulgaris*) roots. Suc is not only important for carbon partition but also functions as a regulator for the expression of many genes (Koch 2004). In addition, it is thought that accumulation of the disaccharide protects the structure and function of macromolecules in plants exposed to cold, drought, and salt stress (Lunn and MacRae 2003). In addition to Suc, many plants produce, translocate, and store polyols, mainly mannitol (Mol) and glucitol (Gol, also known as sorbitol). In this section, we will describe the metabolic routes and the main regulatory mechanisms that allow the coordinated synthesis of these major metabolites in plants based on photosynthetic carbon availability.

29.2.3.1 The Pathway of Sucrose Synthesis

Suc synthesis occurs exclusively in the cytosol of plant cells by a metabolic route that utilizes triose-P imported from the plastid. This pathway includes the reactions detailed in Table 29.3 and Figure 29.1, starting with the production of Fru-1,6bisP from Ga3P and DHAP in a step catalyzed by aldolase, followed by hydrolysis of the product to Fru-6P by the cytosolic Fru-1,6bisP 1-phosphatase (cFru-1,6bisP 1-Pase). Fru-6P is transformed into Glc-6P and Glc-1P in two successive reactions catalyzed by Glc-6P isomerase and phosphoglucomutase, respectively. Glc-1P and UTP are then utilized by UDP-glucose pyrophosphorylase (UDPGlc PPase) to produce UDPGlc. Fru-6P and UDPGlc are the substrates of sucrose-6-phosphate synthase (Suc-6P Sase). Finally, Suc-6P is hydrolyzed by a specific phosphatase to release Suc, which could be either exported to other

tissues or transported to the vacuole (Lunn and Furbank 1999). Suc is continuously synthesized to provide carbon and energy to heterotrophic tissues. However, the building blocks necessary for its synthesis vary along the diurnal cycle. Leaf cells of many plant species (like *A. thaliana*, tobacco, and pea) store starch during the day and break it down during the following night. The resulting products (mainly maltose) are exported to the cytosol, where they are used for respiration or Suc synthesis (MacRae and Lunn 2012; Stitt et al. 2010).

29.2.3.1.1 Control of Suc Synthesis by cFru-1,6bisP 1-Pase

Several mechanisms act in an orchestrated way to regulate the synthesis of Suc. A key metabolic control point is the reaction catalyzed by cFru-1,6bisP 1-Pase (Figure 29.1), which is inhibited by the signal metabolite Fru-2,6bisP. The concentration of this molecule is strictly regulated by the bifunctional enzyme Fru-6P 2-kinase/Fru-2,6bisP 2-phosphatase (EC 2.7.1.105/EC 3.1.3.46), which is subject to allosteric regulation by metabolites (Markham and Kruger 2002) (Figure 29.4). Furumoto et al. (2001) showed that the enzyme from *A. thaliana* leaves is phosphorylated *in vivo*, with the level of phosphorylation being fluctuating along the diurnal cycle. These results support the view that such a modification could be important for modulating the synthesis/degradation of Fru-2,6bisP. Later, it was shown that the phosphorylated bifunctional enzyme binds to 14-3-3 regulatory proteins (Kulma et al. 2004), although this interaction does not change its kinetic parameters. Therefore, the physiological relevance of such post-translational modification remains unclear.

Fru-2,6bisP inhibits cFru-1,6bisP 1-Pase from spinach (*Spinacia oleracea*) leaves by decreasing the affinity for the substrate and increasing the sensitivity toward the inhibitors Pi and AMP (Herzog et al. 1984; Stitt et al. 1985). Therefore, high levels of Fru-2,6bisP lead to inhibition of Suc synthesis. The kinase activity of the enzyme responsible for the synthesis of this metabolite is inhibited by phosphorylated C3 molecules and PPi while it is activated by Pi, whereas the phosphatase activity degrading Fru-2,6bisP is decreased by the products Fru-6P and Pi (Markham and Kruger 2002) (see Figure 29.4). Therefore, high concentrations of triose-P in the cytosol (which are expected in the light) would have a dual role in promoting their use. First, triose-P are the precursors of Fru-1,6bisP; second, they indirectly activate cFru-1,6bisP 1-Pase by decreasing the level of the inhibitor Fru-2,6bisP. Additionally, high triose-P concentrations would increase the amount of hexose-P in the cytosol, which means more Fru-6P available for Suc synthesis as well as more Glc-6P (which activates Suc-6P Sase; see below). If the rate of Suc synthesis exceeds the capacity of the leaf to export the disaccharide, its accumulation will inhibit Suc-6P Sase. This would increase the levels of hexose-P and restrict the release of Pi in the cytosol, thus increasing the hexose-P/Pi ratio, which in turn promotes Fru-2,6bisP accumulation (see above) and inhibition of cFru-1,6bisP 1-Pase. Lower amounts of Pi in the cytosol would restrict interchange of triose-P with the chloroplast, which increases the 3PGA/Pi ratio in the stroma, leading to

---

TABLE 29.3

Reactions Involved in Sucrose Synthesis and Degradation

EC	Name	Reaction
4.1.2.13	Aldolase	Ga3P + DHAP ⇋ Fru-1,6bisP
3.1.3.11	cFru-1,6bisP 1-Pase	Fru-1,6bisP → Fru-6P
5.3.1.9	Glc-6P isomerase	Fru-6P ⇋ Glc-6P
5.4.2.2	Phosphoglucomutase	Glc-6P ⇋ Glc-1P
2.7.7.9	UDPGlc pyrophosphorylase	Glc-1P + UTP ⇋ UDPGlc + PPi
2.4.1.14	Suc-6P synthase	UDPGlc + Fru-6P ⇋ Suc-6P + UDP
3.1.3.24	Suc-6P phosphatase	Suc-6P → Suc + Pi
2.4.1.13	Suc Sase	UDPGlc + Fru ⇋ Suc + UDP
3.2.1.26	Invertase	Suc + H_2O → Glc + Fru

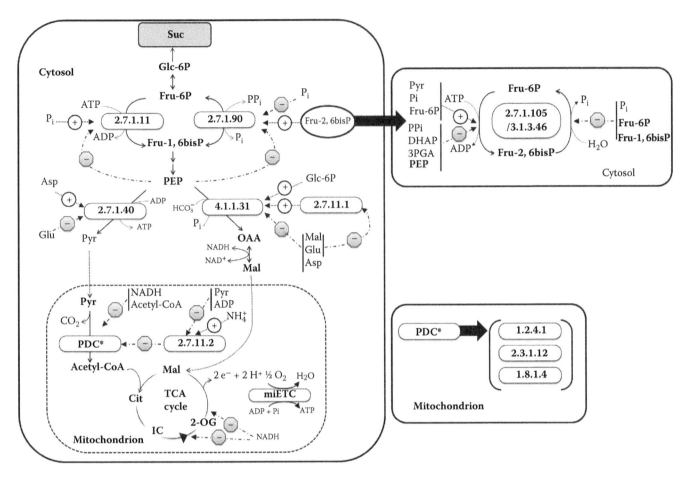

FIGURE 29.4 Metabolic regulation of the cytosolic events of plant respiration. The central role of PEP in the control of upstream reactions is key to understanding the unique mode in which the plant cell regulates cytosolic glycolysis. Plus signs denote enzyme activation; minus signs denote inhibition.

allosteric activation of ADPGlc PPase and starch synthesis (MacRae and Lunn 2012). This model has been tested by modification of the Fru-2,6bisP levels and enzymes regulated by this metabolite, thus confirming its significance for carbon partition in plants (Nielsen et al. 2004).

Interspecies variations in the properties of the cFru-1,6bisP 1-Pase allow the regulation of Suc synthesis to operate in systems where metabolite levels reach extreme values. In mesophyll cells of maize (*Zea mays*) leaves, where Suc is synthesized, DHAP levels are higher than in C_3 plants (Leegood 1985) due to the existence of a DHAP/3PGA shuttle that reduces 3PGA in the mesophyll and sends DHAP back to the bundle sheath cells, where carbon fixation by the BCC takes place. The shuttle is driven by metabolite concentration gradients (Leegood 1985; Stitt and Heldt 1985), and hence elevated levels of DHAP in the mesophyll. However, as noted above, an increase in DHAP would trigger a decrease in Fru2,6bisP, enhancement of cFru-1,6bisP 1-Pase activity, and a high rate of Suc synthesis. This situation could enable enough DHAP to be removed from the shuttle to prevent the C_4 cycle operation. However, since maize cFru-1,6bisP 1-Pase exhibits lower apparent affinity for Fru-1,6bisP than its C_3 counterparts, Suc synthesis in maize proceeds only when DHAP levels are high enough (Stitt and Heldt 1985).

29.2.3.1.2 Regulation of Suc-6P Sase

As mentioned above, Suc-6P Sase is modulated by Glc-6P (activator) and Pi (inhibitor). The activity of the enzyme changes along the diurnal cycle in response to light/dark transitions. During the light period, levels of hexose-P increase in the cytosol, while Pi decreases due to its transport into the chloroplast in exchange with triose-P. These changes result in an increase in the Glc-6P/Pi ratio, which in turn increases Suc-6P Sase activity (Gerhardt et al. 1987). In spinach leaves, the enzyme activity increases after illumination, declines gradually during the light period, continues decreasing in the dark, and recovers gradually during the night. Stitt et al. (1988) evidenced that these changes are not due to variations of the maximal activity (V_{max}), but they reflect a differential sensitivity to the inhibitor Pi in the selective assay (V_{sel}). It was later shown that regulation of Suc-6P Sase also involves protein phosphorylation (Huber and Huber 1992). The enzyme from spinach leaves is phosphorylated at three Ser residues: 158, 229, and 424, but only the first one seems to be important for the response to light/dark transitions.

Phosphorylation of the spinach leaf Suc-6P Sase at Ser158 (which is highly conserved among species) modifies the affinity for the substrate Fru-6P and the sensitivity for the effectors (Glc-6P and Pi) without altering the V_{max} (Toroser et al. 1999).

Therefore, changes in activity due to phosphorylation of this site are only observed when the enzyme is assayed at suboptimal concentration of substrates (V_{sel}). It was suggested that changes observed under different light/dark treatments are species specific (Huber et al. 1989). However, a detailed work on Suc-6P Sase from maize leaves showed that it behaves similarly to the enzyme from spinach leaves, i.e., light activation increases fivefold the apparent affinity for Fru-6P and UDPGlc without altering the V_{max} (Lunn and Hatch 1997). Furthermore, Suc-6P Sase from wheat (*Triticum aestivum*) leaves harvested under light conditions also showed no change in V_{max}, increased affinity toward Fru-6P and UDPGlc (four- and seven-fold, respectively), and decreased inhibition by Pi when compared to samples harvested from dark treated leaves (Trevanion et al. 2004). Conflicting results arose due to dissimilar interspecies enzyme kinetics and the use of fixed concentrations of substrates. Therefore, proper kinetic parameters are necessary before establishing correlations between light/dark transitions and the activation state of Suc-6P Sase.

The protein kinase responsible for phosphorylation of Suc-6P Sase at Ser158 is thought to be a SnRK1, which is further modulated by phosphorylation and metabolites (inhibition by Glc-6P), which adds another level of complexity to the regulation of Suc synthesis. The phospho-Ser158-Suc-6P Sase can be dephosphorylated by a type 2A protein phosphatase, which also seems to be regulated by metabolites (inhibition by Pi) and protein synthesis/degradation in response to changes in the light/dark regime (Winter and Huber 2000). It is worth noting that the Glc-6P/Pi ratio modulates Suc-6P Sase both directly and indirectly by regulation of the protein kinase and phosphatase. More recently, Wu et al. (2014) showed that Suc-6P Sase from carbon-starved *A. thaliana* seedlings is activated upon addition of external Suc through phosphorylation of a protein phosphatase, which in turn dephosphorylates Ser152 (analogous to Ser158 of the spinach leaves enzyme). Therefore, the underlying mechanisms that govern the cycle of Suc-6P Sase phosphorylation/dephosphorylation at Ser158 are still far from being completely understood.

Suc-6P Sase from spinach leaves is also phosphorylated at Ser229, after which it interacts (in the presence of Mg^{2+}) with 14-3-3 regulatory proteins (Toroser et al. 1998); however, the physiological significance of this process remains unclear. The third phosphorylation site on Suc-6P Sase from spinach leaves (Ser424) is well conserved in enzymes from other species, and it is thought to be involved in the response to osmotic stress (Toroser and Huber 1997). The spinach leaf enzyme is activated upon phosphorylation of Ser424 by a CDPK, which probably antagonizes with the inhibitory effect of phosphorylation at Ser158, thus allowing the production of Suc under conditions where it could be restricted (Winter and Huber 2000).

29.2.3.2 Sugar-Alcohols as Major Photosynthetic Products

Suc and starch are the primary photosynthetic products in most plants, although many species also produce sugar-alcohols as major photosynthates (Bieleski 1982). Polyols are

TABLE 29.4

Reactions Involved in the Synthesis and Degradation of Polyols

EC	Name	Reaction
5.3.1.8	Man-6P isomerase	Fru-6P $\leftrightarrows$ Man-6P
1.1.1.224	Man-6P reductase	Man-6P + NADPH + H$^+$ $\leftrightarrows$ Mol-1P + NADP$^+$
3.1.3.22	Mol-1P phosphatase	Mol-1P + H$_2$O $\rightarrow$ Mol + Pi
1.1.1.200	Ald-6P reductase	Glc-6P + NADPH + H$^+$ $\leftrightarrows$ Gol-6P + NADP$^+$
3.1.3.50	Gol-6P phosphatase	Gol-6P + H$_2$O $\rightarrow$ Gol + Pi
1.1.1.255	Mol dehydrogenase	Mol + NAD$^+$ $\leftrightarrows$ Man + NADH + H$^+$
2.7.1.1	Hexokinase	Man + ATP $\leftrightarrows$ Man-6P + ADP
1.1.1.14	Gol dehydrogenase	Gol + NAD$^+$ $\leftrightarrows$ Fru + NADH + H$^+$

important not only for carbon partition but also as protective molecules against certain types of abiotic stress, like salinity and drought (Loescher and Everard 2000). Even more, polyols are emerging as important compounds for the biofuel industry (Kunkes et al. 2008). Despite the importance of these metabolites, the occurrence and regulation of their metabolic pathways are far from being well understood. In this section, we will describe the enzymatic steps and regulatory mechanisms that lead to the production and use of Mol and Gol in different plant species. The reactions involved in the metabolism of sugar-alcohols are detailed in Table 29.4.

29.2.3.2.1 Biosynthesis of Mol

Mol synthesis, transport, and degradation have been studied in detail using celery (*Apium graveolens*) as a model plant. Studies performed with $^{14}CO_2$ demonstrated that, in celery, the newly fixed carbon in the soluble fraction is distributed in equal parts between Suc and Mol (Davis et al. 1988). This polyol is produced in photosynthetically active leaves and transported to sink tissues through the phloem along with Suc (Rumpho et al. 1983). Then, it is utilized in nonphotosynthetic tissues to support growth and maintenance (Stoop and Pharr 1992). The enzymes responsible for Mol synthesis are mannose-6P (Man-6P) isomerase, Man-6P reductase (Man-6P Rase), and Mol-1P phosphatase (Table 29.4). These enzymes were recovered together with phosphoenolpyruvate carboxylase (PEP Case) after differential centrifugation, which indicated their cytosolic localization. Man-6P isomerase and Man-6P Rase were also found in the cytosolic fraction by using Suc density gradient centrifugation (Rumpho et al. 1983) and immunological studies (Everard et al. 1993). The NADPH necessary for the reaction catalyzed by Man-6P Rase could be provided by the cytosolic np-Ga3PDHase (Gao and Loescher 2000; Gomez Casati et al. 2000; Rumpho et al. 1983), whose activity in celery leaves is higher than in plant species where Mol is not a photosynthetic product (Rumpho et al. 1983).

Man-6P Rase was first purified from celery leaves and showed to be very specific toward both substrates Man-6P and NADPH (Loescher et al. 1992), and the recombinant protein showed similar kinetic properties (Everard et al. 1997). The activity seems to be under tight transcriptional regulation during leaf development, with transcripts, protein, and activity levels peaking in mature leaves (Everard et al. 1997). In experiments where celery plants were irrigated with high NaCl concentrations, the level of Mol remained constant while Suc decreased significantly, which led to a higher Mol/Suc ratio. Under this situation, the amount of Man-6P Rase protein was unchanged, but its activity was higher than in nonstressed plants. Therefore, it was suggested that the enzyme could be under post-translational regulation (Everard et al. 1994); unfortunately, experimental evidences to support this hypothesis are still lacking.

By using short $^{14}CO_2$ pulses (between 20 and 120 s), it was possible to demonstrate that Man-6P and Mol-1P are early photosynthetic products (Loescher et al. 1992). Davis et al. (1988) demonstrated that after a 10 min $^{14}CO_2$ pulse and a 10 min chase period, the vast majority of labeled carbon in the neutral soluble fraction (72–95%) was found in Mol and Suc. Also, these authors showed that the Mol/Suc ratio increased as leaves matured, indicating that synthesis of Mol increases dramatically over that of Suc with leaf maturation. Interestingly, the activities of enzymes responsible for Mol biosynthesis were also higher in fully developed than in young leaves. This study clearly established that Mol synthesis and export capacities are interconnected (Davis et al. 1988). A more detailed study revealed that both Suc and Mol are exported in the light. However, the disaccharide is initially exported at a higher rate than Mol in the night; then, when the Suc pools are depleted, the sugar-alcohol becomes the predominant exported metabolite (Davis and Loescher 1990).

Synthesis of Mol in plants is not restricted to celery. This sugar-alcohol also constitutes a major photosynthate in olive (*Olea europaea*) trees, although this species mainly utilizes stachyose to export carbon from leaves. Flora and Madore (1993) showed that Suc, Mol, and galactinol were rapidly labeled in a short-term $^{14}CO_2$ pulse-chase experiment, whereas stachyose and raffinose were labeled more slowly. These authors also demonstrated that the amount of Mol and stachyose in leaves accounted for 30% and 5% of the total label, respectively, whereas the label in phloem sap was predominantly recovered as stachyose (50–60% of the total) and only marginal amounts of Mol were found (less than 5%). Another study confirmed that Mol is a predominant sugar in olive leaves and roots (41% and 29% of the total sugar content, respectively) (Cataldi et al. 2000), and it was postulated that this polyol could be an important osmoprotectant under high salinity conditions (Conde et al. 2007).

29.2.3.2.2 *Synthesis of Gol in Rosaceous Species*

In addition to plants producing Mol, there are important crop species from the Rosaceae family like apple (*Malus domestica*) and peach (*Prunus persica*) that use Gol to transport carbon from source to sink tissues (Loescher and Everard 2000). The set of synthesizing enzymes and the associated reactions are similar to those described above for Mol synthesis,

although the isomerization step is not necessary in this case. The enzymes involved in Gol synthesis are aldose-6P reductase (Ald-6P Rase) and Gol-6P phosphatase (Figure 29.1 and Table 29.4). Similarly to Mol, Gol is synthesized in fully developed leaves and transported to heterotrophic tissues. It is thought that these enzymes are located in the cytosol; this hypothesis is supported by the observation that sequences coding for Ald-6P Rase from several species have no predicted transit peptide (Loescher et al. 1982), but experimental evidences are still necessary to clarify this issue.

Early experiments performed with apple seedlings demonstrated that Gol and Suc accounted for nearly 70% of the label in the leaf water-soluble material after a 30 min $^{14}CO_2$ pulse in the light (Grant and Rees 1981). Then, Loescher et al. (1982) found 85–89% of the label in Gol when mature apple leaves were treated with $^{14}CO_2$ for 1–4 h. These works established a predominant role of Gol as a major photosynthate. Afterward, several authors investigated the expression and activity patterns of the Gol metabolizing enzymes in different rosaceous species. The initial studies on Ald-6P Rase were done with apple (Negm and Loescher 1981) and loquat (*Eriobotrya japonica*) (Hirai 1981) leaves. Both enzymes were very specific toward NADPH, and the apple enzyme showed some residual activity with other aldose-P. These results were confirmed with the enzyme purified from apple seedlings (Kanayama and Yamaki 1993) and in a more recent study performed with the highly purified recombinant protein from apple leaves (Figueroa and Iglesias 2010). The activity of Ald-6P Rase is inhibited nearly by 50% in the presence of 3 mM ATP (competitive inhibitor with respect to NADPH, K_i of 0.18 mM) (Kanayama and Yamaki 1993) and 4 mM Pi (mixed-type inhibitor with respect to Glc6P, K_i of 0.42 mM) (Zhou et al. 2003). The physiological relevance of ATP inhibition is difficult to explain. The inhibitory effect exerted by Pi is similar to that observed for Suc-6P Sase and ADPGlc PPase and fits well with the proposed role for Ald-6P Rase. As mentioned before, triose-P are exported from the chloroplast in exchange for Pi during light hours, which in turn would promote Gol synthesis by providing the substrate (Glc-6P) and releasing the inhibition exerted by Pi.

Gol-6P phosphatase has been poorly studied, and there is only one work dealing with the kinetic properties of the apple enzyme (Zhou et al. 2003). This phosphatase is highly specific for Gol-6P, and its activity is absolutely dependent on Mg^{2+}. Interestingly, both products of the reaction inhibited the enzyme, although with modest effects: 10 mM Pi reduced the activity by 82%, while 150 mM Gol produced a 40% inhibition (competitive, K_i of 109 mM). Interestingly, cFru-1,6bisP 1-Pase from apple leaves is also inhibited by Gol, although high concentrations (400 mM) are necessary to lower the activity by only 30% (Zhou and Cheng 2004). The modest effect of Gol turns doubtful the physiological relevance of this inhibition, but Nadwodnik and Lohaus (2008) demonstrated that such high concentrations are present in polyol synthesizing species like celery and peach. If the inhibitory control happens *in vivo*, Gol could be reducing its own synthesis, thus diverting the carbon flow to other products (like Suc and starch). Indeed, this hypothesis is consistent with increased starch levels in apple leaves with blocked carbon export by phloem girdling (Zhou and Quebedeaux 2003) and antisense inhibition of Ald-6P Rase (Cheng et al. 2005).

29.2.4 RESPIRATORY METABOLISM IN THE LEAF CELL

As in every living organism, respiration is a central process directed to meet the energy needs of the plant cell. Through the oxidation of photosynthetically generated metabolites, plants are able to generate ATP to drive energetically costly reactions and also provide the essential backbones of many biosynthetic processes. Respiration is also implicated in many other ways in different tissues and thus fulfills a plethora of functions not necessarily related with the classical main function of chemical power generation. Many of the pathways and reactions are identical to its counterpart in other organisms, but the presence of photosynthesis lends plant respiration many exclusive characteristics, both in its basic organization and in regulation as well. Most of the differences are based on the autotrophic nature of plants and are due to the effect that these organisms can synthesize their own respiratory substrates. As plants also possess heterotrophic tissues, respiration and its control are not uniform in different plant parts. As in most aerobic organisms, plant respiration consists in the generation, as a rule in the cytosol, of respiratory substrates followed by the complete oxidation of those compounds in the mitochondria. In a typical plant, both processes present well-differentiated characteristics, *oddities*, when compared to other strictly heterotrophic organisms. This section will cover the first part of the process, the production of cytosolic ATP and respiratory substrates and its control, highlighting those aspects that make plant respiration unique.

Plant respiration in the photosynthetic cell differs strongly from any other eukaryotic cell (Plaxton 1996; Plaxton and Podestá 2006). Instead of the classical Glc degradation ladder leading to Pyr, in the light, most of the carbon reaching PEP comes from triose-P exported from the chloroplast. At the same time, triose-P may be directed to Suc formation. PEP metabolism is also more complex, as there are at least two major pathways for its utilization: the universal substrate level phosphorylation of ADP by Pyr kinase (Pyr Kase, EC 2.7.1.40) that yields Pyr and ATP, and the carboxylation reaction leading primarily to oxaloacetate (OAA) and further on to malate and Asp (Figure 29.4). All four metabolites can be used as substrates for mitochondrial respiration. Other important differences upstream of PEP are the presence of the PPi-dependent phosphofructokinase (PPi-Fru-6P 1-Kase, EC 2.7.1.90) and the np-Ga3PDHase, whose role in the export of NADPH from the chloroplast to the cytosol has been described above (see Figure 29.1).

PEP can be used in at least four ways in the cytosol of plant cells: the canonical Pyr Kase-catalyzed substrate level phosphorylation (reaction 1), the carboxylation catalyzed by PEP carboxylase (PEP Case, EC 4.1.1.29) (reaction 2), and the reversible PEP carboxykinase (EC 4.1.1.32) and Pyr-Pi

dikinase (EC 2.7.9.1) reactions (reactions 3 and 4, respectively) (see also Figure 29.4):

1. $PEP + ADP \rightarrow Pyr + ATP$
2. $PEP + HCO_3^- \rightarrow OAA + Pi$
3. $PEP + ADP + HCO_3^- \leftrightarrows OAA + ATP$
4. $PEP + AMP + PPi \leftrightarrows Pyr + ATP + Pi$

The first two reactions are the main utilization ways and the best characterized so far. Plant cytosolic Pyr Kases (cPyr Kases) have been purified, characterized, and functionally analyzed from a number of species and in both photosynthetic and nonphotosynthetic tissues (Andre et al. 2007; Oliver et al. 2008; Plaxton and Podestá 2006; Zervoudakis et al. 1997). As in other organisms, its main contribution is the generation of respiratory Pyr, although this metabolite can also be used for amino acid synthesis. Also similar is the fact that in vascular plants, cPyr Kases are expressed in tissue-specific isoforms, with regulatory properties suited to the location (Hu and Plaxton 1996; Podestá and Plaxton 1991, 1994; Turner and Plaxton 2000). Plant cPyr Kases are inhibited by ATP; a common feature of that lends the enzyme the capacity to be modulated by the energy status of the cell (Podestá and Plaxton 1991, 1994). Nevertheless, in some cases, inhibition by ATP is too weak to be physiologically meaningful (Podestá and Plaxton 1991, 1994). cPyr Kases from photosynthetic, endosperm, cotyledon, or fruit tissue are inhibited by Glu and activated by Asp. Feedback inhibition of some plant cPyr Kases by Glu offers a means to control this enzyme and PEP Case by retroinhibition during active N-assimilation (Plaxton 1996; Plaxton and Podestá 2006). Asp, on the other hand, relieves Glu inhibition and thus behaves as an activator (Smith et al. 2000; Turner and Plaxton 2000). This allosteric control allows carbon flux to be directed either through PEP Case or cPyr Kase (Turner et al. 2005), permitting cPyr Kase to overcome Glu inhibition so ATP can be produced freely to meet the energy demands of protein synthesis.

The anapleurotic, nonphotosynthetic or C_3 form of PEP Case is also subject to fine control in plants. As outlined above, PEP Case control is highly coordinated with that of cPyr Kase, a consequence of the fact that both enzymes share a common substrate, and flux through each reaction determines the fate of carbon metabolism at that point. PEP Case's immediate product is OAA, which in turn can be converted to malate by reduction mediated by malate dehydrogenase (EC 1.1.1.37) or transaminated to Asp (reactions 5 and 6, respectively):

5. $OAA + NADH + H^+ \leftrightarrows malate + NAD^+$
6. $OAA + Glu \leftrightarrows Asp + 2\text{-}oxoglutarate$

Malate is a substrate for mitochondrial respiration, albeit of much lesser importance than Pyr. The role of PEP Case in the supply of carbon skeletons for FA production (combined with the activities of plastidic malate dehydrogenase and malic enzyme) is also not the most important (Plaxton and O'Leary 2012; Schwender et al. 2003). Therefore, PEP Case's main role under normal conditions is most probably an anapleurotic one, providing intermediates for the tricarboxylic acids (TCA) cycle and N fixation (Plaxton and Podestá 2006; Schwender et al. 2003). In certain stress situations, namely, Pi deficiency or anoxia, PEP Case may be key to circumvent cPyr Kase, the activity that can be curtailed by lack of adenylates (Plaxton and Podestá 2006). In fact, levels of PEP Case are increased in many situations that reflect this kind of stress (Plaxton and O'Leary 2012). In Pi-deprived plants, there is an increased export of malic and citric acids to the root environment with the purpose of scavenging Pi from the soil. Under this condition, a massive diversion of carbon flow is exerted to generate TCA intermediates through PEP Case.

In the last decade, an emerging body of evidence has offered a much more complex view of PEP Case biochemistry. In fact, PEP Cases can be composed by two different types of proteins, the plant-type PEP Cases (PTPC) and the bacterial-type PEP Cases (BTPC), owing to their sequence homologies with plant or bacterial PEP Cases (O'Leary et al. 2009; Park et al. 2012; Sánchez and Cejudo 2003). PTPCs, photosynthetic (C_4 or CAM) or nonphotosynthetic (C_3), display a high degree of sequence conservation, are encoded by small gene families, and express differentially in different tissues in response to local regulatory requirements (Sánchez and Cejudo 2003). BTPC genes, on the other hand, encode for a larger polypeptide that is distantly related to PTPCs (Plaxton and O'Leary 2012). Before the role of BTPC in plants was known, research on the biochemistry of anapleurotic PEP Cases (and photosynthetic isozymes as well) had shown in all cases inhibition by L-malate, activation by Glc-6P, and the existence of reversible phosphorylation that rendered the enzyme in an activated form less sensitive to feedback inhibition by malate (Chollet et al. 1996; Plaxton and O'Leary 2012; Plaxton and Podestá 2006; Podestá 2004; Podestá and Plaxton 1994; Zhang et al. 1995). Phosphorylation of PTPC occurred at a conserved Ser residue that was absent from bacterial PEP Case (Chollet et al. 1996). The native enzyme in general displays a homotetrameric structure of ~400 kDa in size (Chollet et al. 1996; Izui et al. 2004).

With respect to the conversion of Fru-6P into Fru-1,6bisP, the two enzymes that catalyze the glycolytic process of phosphorylation of Fru-6P, namely, the ATP-dependent phosphofructokinase (ATP-Fru-6P 1-Kase, EC 2.7.1.11) and the PPi-Fru-6P 1-Kase, are under tight metabolic regulation (Plaxton 1996; Plaxton and Podestá 2006; Turner and Plaxton 2003). Both enzymes are an example of the central role of PEP as a regulatory metabolite, as it has been repeatedly reported as a strong inhibitor of the phosphorylation of Fru-6P by either ATP or PPi. While Pi relieves the inhibition by PEP of plant cytosolic ATP-Fru-6P 1-Kases, it is a strong inhibitor of the PPi-Fru-6P 1-Kase (Plaxton and Podestá 2006; Turner and Plaxton 2003) (Figure 29.4). This feature may direct carbon preferentially through one of the two Fru-6P 1-Kases according to the Pi (and adenylate) status of the cell. PEP also exerts its regulatory influence by inhibiting the Fru-6P 2-kinase activity, thereby decreasing Fru-2,6bisP levels and

curtailing activation of the PPi-Fru-6P 1-Kase while releasing the inhibition of cFru-1,6bisP 1-Pase (Kulma et al. 2004). Fru-2,6bisP does not affect the ATP-Fru-6P 1-Kase in plants. The complex regulatory properties of Fru-6P 2-Kase/2-Pase have been suggested to reflect that its operation is involved in modulating carbon flow under conditions other than photosynthetic metabolism alone (Nielsen et al. 2004).

29.3 TRANSPORT OF CARBON FROM THE LEAF TO SINK TISSUES

The yield of a crop has a direct correspondence with the ability of the plant to allocate carbon from source leaves to the economically important sink tissue. In addition, growth and energy needs from other sink tissues must also be conveniently satisfied by the plant. It is generally accepted that solutes produced by photosynthesis generate sufficient hydrostatic pressure (or turgor) in mesophyll cells to drive transport from source to sink tissues through symplast. According to this mechanism, loading of Suc into the phloem at the source site increases the hydrostatic pressure and drives the subsequent flow (Wyse et al. 1985). Nevertheless, in the light of recent studies, this view has required further adjustments. The pathway through which Suc is finally loaded in the phloem has been subject to some controversy given that three types of phloem loading strategies were proposed: apoplastic, by polymer trapping, or by diffusion. The first two are also called active strategies, since they involve some form of energy, while the last kind of phloem loading is passive.

In the more common apoplastic loading, Suc is pumped from the mesophyll cell into the cell wall space by a recently described transporter type (SWEET—see Section 29.3.1.1) and subsequently retrieved on the companion cell-sieve element (CC-SE) complex through a membrane Suc transporter (SUT). This is a H^+ cotransporter and is coupled to the function of a H^+-ATPase, which energizes the membrane to drive Suc against a concentration gradient. In the second, less frequent active phloem loading, Suc diffuses via plasmodesmata to a specialized companion cell, the intermediary cell, and is subsequently converted into an oligosaccharide of the raffinose or stachyose family (with three or four hexose units, respectively). Energy is not required for transport but for oligosaccharide biosynthesis. In the passive transport, Suc is thought to move symplastically from one mesophyll cell to the next until it reaches the sieve tubes entirely through plasmodesmata. According to this model, assimilate concentration in the cytosol of mesophyll cells is high enough to impel long-distance transport by pressure flow (Figure 29.5).

In the past, there was a controversy since each proposed mechanism had their advocates, and research depicted arguments and evidence supporting exclusively one of them. Meanwhile, other groups thought of the problem as being species-dependent and proposed that both could be operative (Eschrich and Fromm 1994; Hawker et al. 1991). At the moment, a number of studies have shed light over this puzzling scene, and now it is rather clear that the occurrence of different mechanisms of phloem loading is related to several

Loading strategy	Apoplastic loading	Polymer trapping	Diffusion
Active or passive	Active	Active	Passive
Symplastic or apoplastic	Apoplastic	Symplastic	Symplastic
Type of companion cell	Transfer cells (TCs)	Intermediary cells (ICs)	Ordinary or simple companion cells (SC)
Number of plasmodesmata at MC-TC interface	Few	Abundant	Moderate
Schematic diagram of the loading strategy	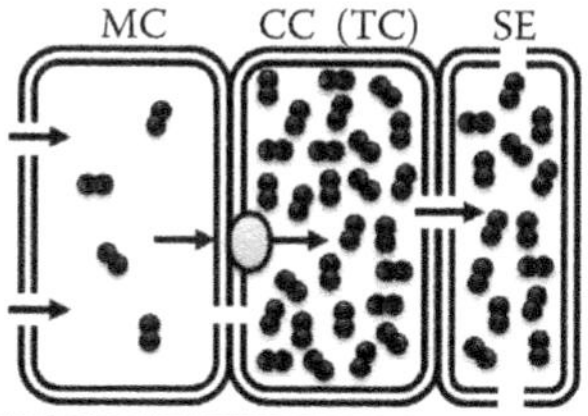	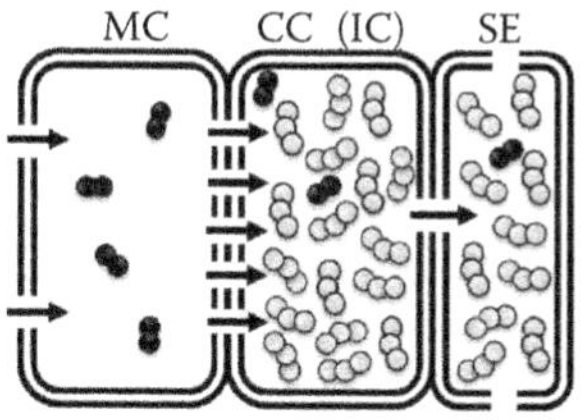	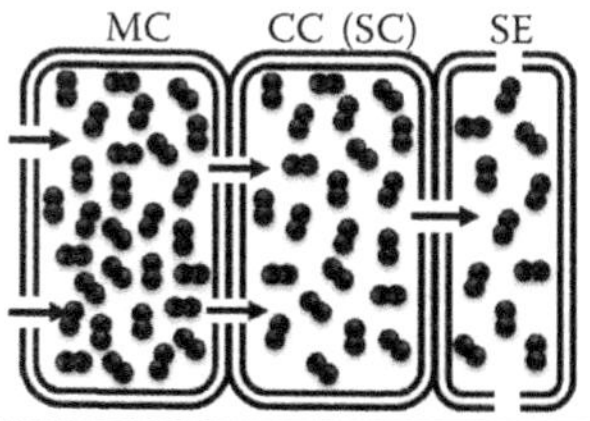
Transport sugar	Sucrose (●)	Sucrose (●), oligosaccharides (⊘)	Sucrose, often sugar alcohols

FIGURE 29.5 Features of loading strategies. (i) Apoplastic loading: sucrose is pumped from the apoplast into the transfer cell (TC, a companion cell) by a transporter (SUT, gray circle), reducing sucrose concentration in the mesophyll cell (MC). There are scarce plasmodesmata in the interface MC–TC. (ii) Polymer trapping: sucrose passively enters to intermediary cell (IC) through abundant plasmodesmata. At the IC, sucrose is converted to oligosaccharides (mostly raffinose or stachyose). (iii) Passive diffusion: sucrose passively diffuses from MC to a simple companion cell (SC) by means of a moderate plasmodesmata number. Assimilates concentration in minor veins is high in apoplastic or polymer trapping loaders, but slightly lower in passive loaders. (Adapted from De Schepper, V. et al., *J. Exp. Bot.* 64:4839–50, 2013.)

factors such as the growth form (herbs or trees), the frequency of plasmodesmata, and the ecophysiological context, among others. In fact, a survey of 45 species of different angiosperm families (Fu et al. 2011) has established a strong correlation between herbaceous plants with active phloem loading and trees with passive phloem loading. Interestingly, foliar Suc content was low in herbs but high in trees. It is also remarkable that the concentration of transport sugars in leaves is connected with the frequency of plasmodesmata. Species with high or intermediate number of plasmodesmata show high transport sugar content in leaves, while those species with small plasmodesmata frequency have low transport sugars (Rennie and Turgeon 2009).

Something difficult to conciliate with common sense is the fact that a tree can transport sugars over a few meters without the need for active transport, whereas herbs use this strategy more frequently (De Schepper et al. 2013). Ecophysiological considerations revealed evolutionary advantages for the rise of active phloem loading in most herbs: it maximizes the amount of soluble carbohydrates transported per unit of volume and avoids accumulation of nonstructural carbohydrates (NSC: Suc, starch, sugar-alcohols) in the leaf mesophyll. Several studies suggested that low levels of NSC in source tissues help to increase the growth capacity of the whole plant by preventing unnecessary carbohydrate accumulation and eluding the feedback inhibition of photosynthesis (Fondy and Geiger 1982; Stitt 2013; Turgeon 2010; Usadel et al. 2008). This trait is particularly relevant in herbaceous species, as they need to add new leaves continuously (high growth rate). In support of this, lower predawn NSC accumulation was measured in crop herbs than in trees (Turgeon 2010). In contrast, most trees produce leaves sporadically in springtime and use storage carbon for this purpose. A number of other interacting constraints are

probably involved; for instance, some researchers have suggested that a close symplastic coupling of the mesophyll and phloem makes transport vulnerable to water shortages during drought, which would affect herbaceous species more than woody plants (van Bel 2003).

It thus appears that the passive symplastic loading mode is the most primitive loading system, which evolved to active symplastic (polymer trapping) and apoplastic loading depending on the species requirements and growth conditions (De Schepper et al. 2013). The panorama seems to be straightforward in gymnosperms, although less studied than angiosperms. According to electronic micrographies, pre-phloem pathway and the sieve elements are continuous with symplasmic connections along the conducting elements. Anatomic differences between conducting elements of both groups of plants would suggest that Suc produced in the mesophyll cells reaches the phloem by passive symplastic loading, similar to what was proposed for many angiosperm tree species (Turgeon 2010; Turgeon and Medville 1998). It is worth emphasizing that phloem functions are different according to organ location. At least three components can be defined: collection phloem in source organs (minor veins), transport phloem (from sources to sinks), and release phloem in sink organs (Figure 29.6). Exchange with the xylem all along the path is also important (Dinant and Lemoine 2010).

29.3.1 Sucrose Transporters

29.3.1.1 SWEET Transporters

Recently, a new type of widespread sugar transporters was described. These proteins are not only important for carbon translocation but also play a key role under pathogen attack. The

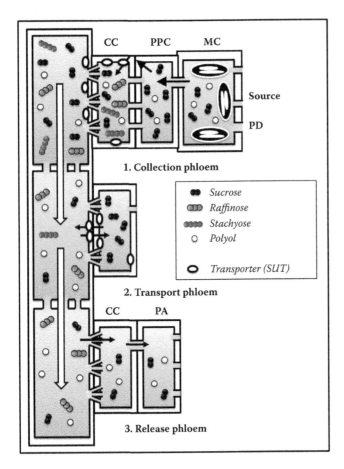

FIGURE 29.6 Schematic illustration of assimilates transport from source to sink. Sucrose and other sugars produced in mesophyll cells (MC) by carbon fixation are transferred through plasmodesmata to phloem parenchyma cells (PPC) and then translocated to the apoplast by SWEET proteins. Then, sucrose is taken up by SUT (gray circle) in the CC of the collection phloem (TC in apoplastic loaders) or transported by plasmodesmata to other companion cells (IC or SC), according to the loading strategy utilized. Subsequent movement of sucrose through transport phloem is likely by apoplastic loading, and sucrose release in sink organs is mostly a symplastic process. (Adapted from Dinant, S., and R. Lemoine, *C. R. Biol.* 333:307–19, 2010.)

SWEET (which stands for Sugars Will Eventually be Exported Transporters) family has 17 members in *A. thaliana* and a similar number in other relevant crops such as rice (*Oryza sativa*), tomato (*Solanum lycopersicum*), and soybean (*Glycine max*) (Chen et al. 2010; Lin et al. 2014). Some members of this large protein family (AtSWEET11 and AtSWEET12) are specifically involved in Suc efflux to the apoplast. The *A. thaliana* double-mutant *atsweet11,12* shows high levels of sugars and a starch excess phenotype at the end of the night (Chen et al. 2012). The finding of these transporters completed the model of apoplastic phloem loading, where Suc is transported from mesophyll to phloem parenchyma cells through plasmodesmata and the SWEET11 and 12 transporters, which are localized in these cells, efflux Suc to the apoplast. Then, the disaccharide is taken by SUT transporters localized in the plasma membrane of the CC–SE complex, thus concentrating

Suc in the sieve element (Baker et al. 2012). Expression of these transporters is upregulated upon infection by pathogenic bacteria, which benefit from the host's sugar secretion mechanism to support their growth. Indeed, rice plants where OsSWEET11 is not available are more resistant to infection by *Xanthomonas oryzae*, which produces a worldwide spread disease (Chen 2014). Therefore, genetic manipulation of these transporters could help in reducing the occurrence of several devastating diseases.

Other members of the SWEET family are involved in Glc and Fru transport in different tissues and compartments. For instance, natural variation in leaf Fru content was linked to AtSWEET17 (which exports Fru from the vacuole to the cytosol), and the *atsweet17* mutant showed stunted growth and lower seed yield. AtSWEET17 is highly conserved among plants; therefore, it was suggested that manipulation of this transporter might allow manipulation of carbon allocation in important crop species (Chardon et al. 2013). Both AtSWEET17 and its close paralog AtSWEET16 are mainly expressed in the vacuole of root cells, where they play an important role for maintaining cytosolic Fru homeostasis (Guo et al. 2014). More recently, Lin et al. (2014) reported that AtSWEET9 together with Suc-6P Sase are important factors for nectar secretion. These authors proposed a model where Suc is synthesized in the nectary parenchyma and then secreted to the extracellular space by AtSWEET9, where the disaccharide is converted into Glc and Fru by an apoplastic invertase. It is clear from these examples that discovery of the SWEET transporters in the last few years has been of exceptional importance to unravel the mechanisms of sugar translocation in different plant cells and compartments. Also, this knowledge opens a whole new scenario to develop innovative transgenic crops with improved carbon sinks.

29.3.1.2 SUT Transporters

In recent years, work with transgenic plants has yielded compelling evidence favoring the above description of loading mechanisms. Remarkable considerations deserve SUT transporters. Substantial findings have been done in solanaceous, the typical apoplastic loaders, in which the coupled activity of H+-ATPase and Suc transporters raises the disaccharide concentration greatly in the phloem with respect to surrounding cells. In celery, for instance, ratios from 4.5 to 40 have been observed (Nadwodnik and Lohaus 2008). The firsts SUT genes isolated were those present in spinach (SoSUT) and potato (StSUT) (Riesmeier et al. 1992, 1993, 1994). Antisense experiments rendered plant with bleaching leaves and reduction in root and tuber development, while leaves displayed considerable accumulation of Suc and starch. SUTs were localized in the phloem plasma membrane in these works, and subsequent studies characterized SUTs throughout the plant kingdom. Alternatively, in *Verbascum phoeniceum*, RNAi of galactinol synthase did not cause typical phenotype for apoplastic loaders. Since this enzyme is responsible for galactinol synthesis, which is a precursor for raffinose family oligomers, this was considered as a demonstration of symplastic polymer trapping operating in this herb (McCaskill and Turgeon

2007). A recent analysis of the anatomy of leaves uncovered a strong correlation between the polymer trapping strategy (with presence of intermediary cells) of phloem loading and species frequent in tropical and subtropical regions of the world (Davidson et al. 2011).

Outstandingly, an early anatomic observation of ultrastructure of vascular system in rice suggested that both apoplastic and symplastic phloem loading pathways were possible (Chonan et al. 1984). Confirmation of this came later with the identification of all SUTs present in rice and evaluation of mutant's phenotypes. SUTs are categorized into three major groups, all H$^+$-coupled symporters: type I (specific to plasma membrane of eudicots), type II (present in plasma membrane of all plants), and type III (present in vacuolar membrane of all plants). Members of type I are responsible for phloem loading or Suc import into sink organs. Type II SUTs for phloem loading are subdivided into two groups: type IIAB with a longer central loop, an ancestral form of the type II SUTs found in angiosperm, and the type IIB subgroup that is monocot specific. Type III SUTs are localized at the vacuolar membrane (tonoplast) and function in Suc transport from the vacuole to the cytosol (Reinders et al. 2012). In *A. thaliana*, type I SUTs have a wide range for substrate specificity, from Suc to several α- and ß-glucosides, with high affinities. On the other hand, type II SUTs cannot transport ß-glucosides, and type III SUTs have an intermediate substrate specificity between type I and II transporters. While rice type II SUT mutation did not produce changes, plants lacking OsSUT2 (type III SUT) displayed severe growth defects compared with wild-type and OsSUT2-complemented lines (Eom et al. 2012). Authors depicted a revised symplastic passive loading mechanism mediated by vacuolar Suc trapping, in which the tonoplast SUT is essential to regulate Suc flux to the phloem. In the *Populus* genus, down-regulation of tonoplastic SUT increased vacuolar Suc sequestration and lowered water uptake and physiological responsiveness to drought stress (Frost et al. 2012). Indeed, concentration of precursor for raffinose family oligomers was lower during water stress in tonoplast SUT-RNAi plants. Constitutive vacuolar Suc sequestration in transgenic plants affected source provisioning to expanding leaves. Previously described symptoms were not observed under normal water availability. Taken together, these results point out a species-dependent role for tonoplast SUT.

The activity of SUT is regulated at multiple levels. They are under hormonal control, and Suc itself can act as a hormone-like signal down-regulating enzyme activity (Kühn 2003). Transporter levels are enhanced by auxin-type hormones and light (Roblin et al. 1998). These authors have also shown that phosphorylation inhibits Suc transport, adding a further level to the complex regulation of the disaccharide movement. As proton symporters, they are also regulated by the proton motive force (i.e., membrane potential and pH gradient) (Carpaneto et al. 2005). Activity of StSUT1 is regulated by the redox potential and its dimerization status in the plasma membrane (Krugel et al. 2008). Different methods have shown that Suc transporter mRNA moves through phloem sap at least in five different species (*A. thaliana*, pumpkin, potato, tomato,

tobacco), but the relevance of this is still under investigation (Suc transporter regulation at the transcriptional, post-transcriptional, and post-translational level) (Liesche et al. 2011).

Increased Suc transport from source leaves to sink organs by manipulating the SUTs was proposed as an effective method to enhance crop productivity (Ainsworth and Bush 2011), especially in conditions where carbohydrate is in excess (Stitt 2013). If more Suc is transported for growth or storage, productivity would increase, since less of the disaccharide implies less product mediated inhibition of photosynthesis. As mentioned above, plants contain multiple Suc transporters with differing expression patterns, properties, and subcellular locations. The physiological function of specific SUTs has been inferred mostly from gene repression or attenuation, mutations, and expression patterns, while biochemical activity has been characterized predominantly by electrophysiology in *Xenopus laevis* oocytes and by Suc uptake assays into yeast. An exhaustive study was undertaken in *A. thaliana*, where researchers tested different members of SUT family in order to identify the high-level expression promoters as well as to assign the role of each SUT (Dasgupta et al. 2014). The most interesting finding was that overexpression of AtSUC1 in companion cells, under the regulation of a promoter that is not repressed by Suc concentration, did not result in improved growth, as expected. Quite the opposite, plants were smaller than wild type, even when they did show increased phloem loading and Suc transport. Surprisingly, these plants induced a phosphate deficiency response, although it was not clear if this was owing to insufficient Pi supply (more Suc requires more Pi to be metabolized) or to a perception of a carbon/phosphate imbalance.

Suc unloading in sink cells is also apoplastic in some tissues like the developing embryo or the endosperm, which lack of symplastic continuity with the parent plant (Hawker et al. 1991). However, in roots and expanding leaves, this process may be completely symplastic. Indeed, in grape berry, a switch from symplastic to apoplastic unloading was observed during fruit ripening (Zhang et al. 2006). The opposite, a change from apoplastic to symplastic unloading, was reported in potato tuber during tuberization (Zhang et al. 2006). The unloading of Suc at the sink and its subsequent transformation or storage in the vacuole provides the means of maintaining the steep concentration gradient needed to support the carbon flow. However, to what extent phloem loading of Suc is involved in the coupling between source and sink is still a matter of debate. Clearly, considerable progress has been achieved in the discernment of transport processes occurring at the whole plant. In spite of that, unsolved aspects of plant physiology that could hold important clues for the enhancement of plant breeding need further awareness.

29.3.2 Mannitol and Glucitol Transporters

The first eukaryotic polyol transporter to be studied was obtained from celery petiole phloem (Noiraud et al. 2001). These authors showed that AgMaT1 confers transgenic yeast cells the ability to grow in a Mol-supplemented medium. This H$^+$-Mol cotransporter could uptake other polyols like Gol and xylitol, although at a lower rate. Expression of AgMaT1 was high in mature leaves and lower in young leaves, indicating that it is responsible for Mol loading in celery phloem cells. A second Mol transporter (AgMaT2) was later described and proved to be expressed in the plasma membrane of the sieve elements, companion cells, and phloem parenchyma cells of both petiole and leaflet of mature leaves, but its expression was low in young leaves. These results agree with the proposed role of mature and young leaves in celery for Mol synthesis and utilization, respectively. Like AgMaT1, the specificity of AgMaT2 for Mol was low, and other polyols such as Gol and xylitol compete with Mol for transport (Juchaux-Cachau et al. 2007).

Gol transporters PmPLT1 and PmPLT2 from common plantain (*Plantago major*, a Gol-producing species) were cloned and localized in companion cells of the phloem in source leaves. These proteins cotransported H$^+$ and Gol, as it was reported for Mol transporters AgMaT1 and AgMaT2 from celery. The K_m values of PmPLT1 and PmPLT2 for Gol were higher than 10 mM, and both were able to transport Mol; therefore, they were classified as low-specificity and low-affinity Gol transporters (Ramsperger-Gleixner et al. 2004). Later, two high-affinity Gol transporters from apple (MdSOT3 and MdSOT5) were cloned and studied in detail. MdSOT3 was expressed only in mature leaves, and MdSOT5 was expressed in flowers and leaves of different developmental stages. Both transporters were localized in the phloem of minor veins from source leaves, thus suggesting that they play an important role for Gol phloem loading (Watari et al. 2004). Gol transporters (PcSOT1 and PcSOT2) were cloned from developing fruits and young leaves of sour cherry (*Prunus cerasus*), and both showed high sequence similarity to AgMaT1. As it was shown for apple Gol transporters, these H$^+$-Gol cotransporters are quite specific for Gol (uptake rates for Mol and Suc were 4% and 12% of that for Gol, respectively). PcSOT1 is expressed in young leaves and in the later stages of fruit development, when accumulation of soluble sugars (Glc, Fru, and Gol) is higher, whereas PcSOT2 is expressed in the early stages of fruit development, and it might be important when the fruit experiences the first increase in growth and/or during pit hardening (Gao et al. 2003). These data strongly suggest an apoplastic step for glucitol phloem unloading in apple sink tissues. Indeed, experimental evidence for such a mechanism was provided by Zhang et al. (2004). A monosaccharide transporter was localized in the plasma membrane of the sieve element and the parenchyma cells, with increasing amounts as fruit developed. Also, an H$^+$-ATPase was localized in the plasma membrane of the SE–CC complex. In addition, [^{14}C]Gol unloading suggested an energy-driven transport mechanism.

It seems that activity of polyol transporters is mainly regulated by controlling their spatiotemporal expression. However, a new regulatory mechanism for sugar uptake was recently described for apple fruits. Suc (MdSUT1) and Gol (MdSOT6) transporters (both located in the plasma membrane of phloem and parenchyma cells from different tissues) are activated by an endoplasmic reticulum-anchored cytochrome *b5* (MdCYB5). The

authors of this work proposed that interaction of these transporters with MdCYB5 increases the affinity toward their substrate. When sugar levels are restored, interaction with MdCYTB5 is released and the transporter affinity is lowered. This mechanism would ensure that a certain level of cytosolic sugar is maintained even under low carbon availability (Fan et al. 2009).

29.4 METABOLISM IN HETEROTROPHIC CELLS

29.4.1 METABOLISM OF SUCROSE AND SUGAR-ALCOHOLS

After phloem unloading, photosynthates are incorporated in the heterotrophic cell and used for maintenance and, depending on the tissue type, long-term carbon storage (mainly as starch or lipids). As previously described, Suc is the main molecule used for intercellular carbon translocation, although some species also utilize polyols to supply carbon to heterotrophic tissues. In these species, the pathways of both Suc and sugar-alcohols are interconnected at the hexose-P pool level. In this section, we will describe the main reactions leading to the utilization of carbon in nonphotosynthetic tissues and the regulatory mechanisms that control these pathways.

29.4.1.1 Sucrose Utilization in Nonphotosynthetic Tissues

Suc transported to heterotrophic tissues through the phloem is metabolized by two different enzymes, Suc synthase (Suc Sase) and invertases (β-D-fructofuranosidases) (Figure 29.2 and Table 29.3). Suc Sase is a cytosolic enzyme that produces UDPGlc and Fru from UDP and Suc, thus maintaining the energy of the glycosidic bond in the sugar–nucleotide molecule. UDPGlc is then utilized for cellulose and callose synthesis or by UDPGlc PPase (in the pyrophosphorolytic direction) to produce Glc-1P (Winter and Huber 2000), which enters the hexose-P pool and is further utilized for respiration and/or accumulation of storage products such as starch and TAGs, depending on the tissue type and developmental stage.

It was initially proposed that Suc Sase was not subject to any regulatory mechanism because its activity is not modulated by metabolites. However, it was later shown that this enzyme is under regulation by phosphorylation. Labeling studies performed with ^{32}P and detached maize leaves combined with TOF-MS and protein sequencing by Edman's degradation identified Ser15 as the phosphorylation site of Suc Sase 2 from maize leaves (Huber et al. 1996). Extracts obtained from the elongation zone incubated in the presence or absence of ATP showed that phosphorylation has no effect on the enzyme's V_{max} but increases the affinity for UDP and Suc. Activity measurements with subsaturating concentrations of UDPGlc and Fru showed no differences; therefore, it was concluded that the phosphorylation had no effect on the apparent affinities for these substrates. These authors also showed that maize Suc Sase 2 was *in vitro* phosphorylated by a partially purified CDPK from the same tissue (Huber et al. 1996). Similar results were obtained by Nakai et al. (1998) in a study performed with Suc Sase from mung bean (*Vigna radiata*). The authors cloned the sequence coding for this enzyme and introduced mutations at Ser11 (analogous to Ser15 from maize Suc Sase) to mimic phosphorylation (to Asp and Glu). The mutant enzymes showed an increased apparent affinity toward Suc; in addition, *in vitro* phosphorylation of the recombinant wild-type enzyme by a partially purified CDPK from soybean nodules showed an increase in the apparent affinity for Suc. None of these enzymes presented changes in the kinetics for UDPGlc or Fru (Nakai et al. 1998).

A third study on Suc Sase from tomato fruit showed that partially purified phosphorylated Suc Sase had a higher apparent affinity for UDP. Phosphorylation at the N-terminus Ser site was not sufficient to change the kinetics of this enzyme; however, an increase in the apparent affinity for UDP was observed in a multiphosphorylated isoform (Anguenot et al. 1999). Similar results were obtained with Suc Sase purified from Japanese pear (*Pyrus pyrifolia*) fruit (Tanase et al. 2002), where the phosphorylated enzyme showed a higher apparent affinity for UDP and no changes in the Suc kinetics, when compared with the enzyme treated with protein phosphatase 1. In this case, the phosphorylated enzyme also had decreased affinities for Fru and UDPGlc when compared with the phosphatase-treated enzyme. The phosphorylated isoform was more abundant in young fruits and decreased along with tissue development (Tanase et al. 2002), which agrees with the proposed role of Suc Sase in the degradation of Suc in heterotrophic tissues. Altogether, these results suggest that phosphorylation at Ser15 activates degradation of Suc and/or restricts its synthesis.

It was later demonstrated that phosphorylation of maize leaf Suc Sase also occurs at Ser170. This site is not involved in changes at the kinetic level, but it is responsible for targeting the enzyme to the 26S proteasome for degradation (Hardin et al. 2003), in a process mediated by ubiquitin conjugation (Hardin and Huber 2004). It was suggested that proteasome degradation of Suc Sase could be related with the significant decrease in protein abundance observed in the sink to source transition of maize leaves (Hardin et al. 2003). These authors also showed that both Ser15 and Ser170 are phosphorylated by a CDPK, whereas phosphorylation by SnRK1 is restricted to Ser15 (Hardin et al. 2003). In addition, they demonstrated that phosphorylation at Ser15 increased the activity of recombinant Suc Sase from maize leaves in the direction of Suc cleavage at acidic pH in the absence of the second phosphorylation site (Ser170) (Hardin et al. 2004), which agrees with previous reports (Huber et al. 1996; Nakai et al. 1998; Tanase et al. 2002) and reflects the increased affinity for Suc and/or UDP. More recently, Fedosejevs et al. (2014) described the phosphorylation at Ser11 (similar to Ser15 in the maize leaf enzyme) of Suc Sase from developing castor oil (*Ricinus communis*) seeds by a CDPK. Interestingly, elimination of Suc supply by excision of intact pods triggered dephosphorylation of Suc Sase *in planta*, which was correlated with diminished activity and immunoreactive polypeptides. Thus, phosphorylation of Ser11 in the castor oil seed enzyme seems to protect it from degradation rather than changing its kinetic properties.

Suc can also be degraded by invertases, which can be classified based on their optimal pH for activity (acidic, neutral,

and alkaline) and solubility (soluble and insoluble). They also differ in localization: soluble acid invertase is vacuolar, insoluble acid invertase is extracellular (apoplastic), and neutral/alkaline invertases are cytosolic (Winter and Huber 2000). Cytosolic invertases are minimally active in most systems, and cytosolic Suc is frequently transported into the vacuole for degradation (Koch 2004). Barratt et al. (2009) showed that the *A. thaliana* double-mutant *cinv1/cinv2* (which code for neutral/alkaline invertases) has a severe growth phenotype, whereas the quadruple mutant *sus1/sus2/sus3/sus4*, which only has Suc Sase activity in the phloem, shows no obvious phenotype. Therefore, it was concluded that the main pathway for Suc utilization in heterotrophic tissues involves cytosolic invertases. However, these results should be cautiously interpreted, because the growth phenotype could be due to abnormal root development, where these cytosolic invertases are preferentially expressed.

Cytosolic invertase is considered to play a maintenance role, degrading Suc when activities of Suc Sase and acid invertase are low (Winter and Huber 2000). The regulation of cytosolic invertase activity is not well understood, although there are some reports showing product inhibition. For instance, the purified enzyme from broad bean (*Vicia faba*) developing cotyledons was inhibited by Fru (30% inhibition at 10 mM Fru) (Ross et al. 1996) and that purified from chicory (*Cichorium intybus*) roots was strongly inhibited by Fru and Glc with similar $I_{0.5}$ values (~16 mM) (Van den Ende and Van Laere 1995). Neutral and alkaline invertases purified from carrot (*Daucus carota*) cell culture showed similar affinity for Suc. The alkaline invertase was highly specific for Suc and showed inhibition by Fru and Glc (Lee and Sturm 1996). However, because Fru and Glc accumulate in the vacuole, the physiological relevance of the inhibition by these hexoses is still not clear.

In addition to Suc, acid invertases accept other substrates such as stachyose and raffinose, although with a reduced cleavage efficiency. These enzymes share several enzymatic properties; both are glycosylated proteins and possess a high sequence identity. Thus, it is thought that acidic invertases have a different evolutionary origin than neutral/alkaline invertases (Roitsch and Gonzalez 2004). It has been proposed that apoplastic invertase plays a key role for assimilate uptake in sink tissues by generating a Suc gradient that mobilizes the disaccharide from source tissues (Eschrich 1980), which is particularly important in tissues where plasmodesmatal connections between cells are absent. This situation is observed in developing seeds and pollen, where the involvement of cell wall invertases on Suc unloading was observed. During early seed development in barley, apoplastic invertase is highly expressed together with a hexose transporter, which ensures a high Suc uptake due to their combined action (Weschke et al. 2003). Interestingly, expression of extracellular invertases is promoted by glucose (one of the products of its reaction), which leads to a feed-forward mechanism that enhances the flow of assimilates to heterotrophic tissues when carbon is available (Roitsch et al. 1995).

Transport of Suc to the vacuole for degradation by vacuolar invertase is the predominant path in expanding tissues.

This has the advantage of producing two osmotically active molecules from one, which promotes vacuolar expansion. Transgenic tomato fruits constitutively expressing an antisense soluble acid invertase gene showed increased Suc levels and a 30% reduction in size compared to wild-type fruits. Thus, Suc accumulating fruits appear to acquire less water during development with the concomitant reduction in expansion (Klann et al. 1996). The vacuolar invertase from darnel (false wheat, *Lolium temulentum*) leaves is inhibited by the product Fru (95% inhibition at 10 mM Suc and 20 mM Fru, noncompetitive inhibition, K_i of 2.5 mM), although the physiological relevance of this inhibition is not known (Walker et al. 1997). Conversely, it is well known that expression of vacuolar invertases is regulated by sugars, hormones, and several environmental factors (Koch 2004).

It has been long known that apoplastic and vacuolar invertases are regulated by a proteinaceous inhibitor, although their sequence only became available in the last decade. In recent years, several authors have characterized these inhibitors from different sources, including *A. thaliana* (Link et al. 2004) and tomato (Reca et al. 2008). It has been hypothesized that such inhibitors might help in controlling carbon partition in important crop species. Indeed, ectopic overexpression of a tobacco vacuolar invertase inhibitor in potato led to a 75% reduction in the production of cold-induced hexoses in tubers, with no significant changes in starch quality and quantity (Greiner et al. 1999). Similar results were recently obtained by McKenzie et al. (2013), who directed the expression of the potato vacuolar invertase inhibitor INH2α to potato tubers. Interestingly, they also showed that production of acrylamide during fry tests was also reduced in the transgenic lines, thus confirming the biotechnological application of these plants.

29.4.1.2 Utilization of Mannitol and Glucitol in Heterotrophic Tissues

As previously mentioned (see Section 29.2.3.2.1), celery transports Mol from mature leaves to young developing tissues, where it is unloaded and incorporated into the hexose-P pool after a series of reactions (Table 29.4). Mol is first converted into Man in a NAD-dependent reaction catalyzed by Mol dehydrogenase (Mol DHase), then phosphorylated by hexokinase, and finally transformed into Fru-6P by the activity of Man-6P isomerase. It is worth noting that the activity of the latter in celery petioles is higher than that observed in other plant species (Stoop and Pharr 1994b), suggesting that it plays an important role in Mol catabolism.

Mol DHase was first purified from celery roots, and it was shown that its product was Man and not Fru, as it was observed for the enzyme from many fungi and bacteria. Several metabolites were tested, but only NADH inhibited the enzyme; indeed, the activity was modulated by the NAD^+/NADH ratio (Stoop and Pharr 1992). The activity of Mol DHase in Mol-containing celery suspension cultures was higher in comparison to those supplemented with Suc or Man. Additionally, cultures transferred from Man to Mol showed increased Mol DHase activity and vice versa, i.e., its activity was decreased when cultures were transferred from Mol to Man, which confirmed the importance of this enzyme for the sugar-alcohol utilization (Stoop and Pharr 1993). Interestingly, celery plants exposed to salt stress showed increased activity of Man-6P Rase and decreased Mol DHase activity. Therefore, Mol accumulation in these plants seems to be regulated by elevated synthesis and lower utilization rates in sink tissues (Stoop and Pharr 1994a).

Expression of the gene encoding Mol DHase in cultured celery cells was repressed by sugars. Glc-grown cells had little Mol DHase activity but showed an important increase in transcripts, protein, and enzyme activity after starvation, whereas replenishment of Glc resulted in opposite changes (Prata et al. 1997). Sugar repression of Mol DHase expression could be beneficial under certain stress conditions by promoting sugar utilization and restricting Mol degradation, thus increasing its concentration to act as an osmoprotectant. Consistent with its proposed physiological role, Mol DHase has been localized in the cytosol of nonphotosynthetic cells but also in the nucleus, although its function in this compartment is not clear (Yamamoto et al. 1997; Zamski et al. 1996). This enzyme was also found in the apoplast in response to salicylic acid, which resembles a pathogen attack. The authors suggested that secretion of Mol DHase could help the plant to metabolize Mol secreted by certain pathogens in response to the plant's reactive oxygen species (Blackburn et al. 2010; Williamson et al. 1995).

Plants that produce Gol as a primary photosynthetic product (like apple and peach) have an extensive gene family coding for Gol dehydrogenase (Gol DHase) (Velasco et al. 2010; Verde et al. 2013), which converts Gol into Fru in an NAD-dependent reaction (Figure 29.2 and Table 29.4). Fru is then phosphorylated by hexokinase and directly enters the hexose-P pool. The expression pattern of the gene encoding Gol DHase was studied in apple (Nosarszewski et al. 2004; Park et al. 2002; Yamada et al. 1999), peach (Yamada et al. 2001), and loquat (Bantog et al. 2000) fruits. These works showed that activity of the enzyme is high in immature fruits, then declines transiently, and peaks again with fruit maturation; therefore, it was suggested that control of Gol DHase activity during fruit development occurs mainly at the transcriptional level. Accordingly with its proposed role in developing tissues, Gol DHase transcripts were also found in young apple leaves (Park et al. 2002). More recently, Nosarzewski and Archbold (2007) demonstrated that genes coding for Gol DHase in apple fruit are differentially expressed in seed and cortex tissues during early development. Furthermore, these authors showed that the seed enzyme accounts for most of the whole fruit Gol DHase activity. These data are in good agreement with the idea that Gol produced in photosynthetically active leaves is used to transport carbon to developing and sink tissues, where it is utilized to support growth and maintenance.

It was proposed that Gol transported to peach fruits has a critical role for development because the conversion of the sugar-alcohol into Fru is directly associated with the fruit growth rate. A clear advantage of Gol oxidation over Suc degradation is the net production of NADH (see Figure 29.2 and reactions in Tables 29.3 and 29.4), which could increase growth efficiency and promote a reduction in the respiration costs (Morandi et al. 2008). Similarly to Mol DHase, it was assumed that Gol DHase is a cytosolic enzyme. A recent study performed with leaves and fruits from apple confirmed the enzyme's localization, although it was also found in plastids. Gol is also present in plastids; thus, it was suggested that Gol DHase could play a key role for starch metabolism by providing hexoses for the synthesis of the polyglucan. Another hypothesis is that the enzyme helps in maintaining the osmotic balance of the chloroplast by regulating the concentration of Gol in the stroma (Wang et al. 2009).

Gol DHase has been purified from apple callus (Negm and Loescher 1979) and fruit (Yamaguchi et al. 1994) and Japanese pear fruit (Oura et al. 2000). These studies showed that the enzyme is specific for NAD[+], but it has certain degree of promiscuity toward the sugar-alcohol. Similar results were obtained with the recombinant enzymes from plum (*Prunus salicina*) (Guo et al. 2012), tomato (Ohta et al. 2005), and *A. thaliana* (Aguayo et al. 2013); however, the information obtained from species that normally do not accumulate Gol is difficult to extrapolate to those where the polyol is a major photosynthate. Recently, Hartman et al. (2014) demonstrated that Gol DHase from peach fruits is regulated by thioredoxin *h* from the same tissue. These authors proposed a metabolic scenario where Gol DHase is inactivated under oxidative stress conditions, thus leading to accumulation of Gol, which could play a key role as a free radical scavenger under such environment. Once the redox balance is reestablished, thioredoxin *h* would reduce the enzyme to allow the normal carbon flux into Fru. This is the first report showing that an enzyme from Gol metabolism is regulated at the post-translational level, which opens a completely new field of study on carbon partitioning in polyol synthesizing species.

As mentioned above, sugar-alcohols play a key role for both carbon partitioning and abiotic stress resistance; however, the studies involving enzymes from Mol and Gol metabolisms in plants are limited compared to those dealing with the synthesis and degradation of Suc and starch. Therefore, we need to improve our effort to study the enzymes involved in such pathways to unravel the underlying regulatory mechanisms. In this way, we would be able to better understand not only the physiology of sugar-alcohols metabolism but also a widely distributed stress tolerance mechanism in plants.

29.4.2 Metabolism for Starch Storage

Starch granules in heterotrophic cells of plants localize in non-green plastids. In certain reserve tissues and organs of many species (i.e., fruits, seeds, tubers, roots), the polysaccharide is actively synthesized in the developmental stage to accumulate to levels of up to ~50–90% of the dry weight (Smith et al. 1995). This constitutes a long-term storage of carbohydrates that is degraded during germination (seeds, tubers) or maturation (fruits) to supply demands of carbon and energy. Depending on relative amounts of amylase and amylopectin as well as of other structural characteristics (i.e., degree of phosphorylation), amyloplastic starch granules vary in shape, size, and physical properties in different

plant species (Ball and Morell 2003; Smith et al. 1995). This structural variety can be used to fingerprint the identity of the botanical origin of the polyglucan by microscopic examination of the granules.

The biochemical characteristics of heterotrophic tissues plants have been far less studied than those determined for the green counterparts, and thus some controversies have arisen concerning the flow of photoassimilates in nonphotosynthetic cells. This is particularly true for starch metabolism in amyloplasts, which has been less characterized than in chloroplasts. It is clear that the enzymatic reactions involved in the synthesis and degradation of the polyglucan in amyloplasts are same as those detailed in Table 29.2 for the pathway in chloroplasts (Iglesias and Podestá 2005). However, debate is currently open with respect to whether the metabolism of synthesis (from Glc-6P) occurs entirely within the plastid or if ADP-Glc is produced in the cytosol (Denyer et al. 1996; Pozueta-Romero et al. 1991). Figure 29.2 details the flow of carbon in heterotrophic cells in a model where synthesis of the polyglucan is confined into the plastid, according to what has been demonstrated in tissues of many plant species (as reviewed by Ballicora et al. 2003, 2004; Cortassa et al. 2012; Iglesias and Podestá 2005; Preiss et al. 1991; Tiessen et al. 2002). As studied by Keeling et al. (1988), Glc-6P (rather than triose-P) would be the metabolite actively translocated through the plastidic envelope to serve as the starting substrate for starch elongation. This scenario is compatible with the absence of Fru-1,6bisP 1-Pase in amyloplasts, which blocks the passage of triose-P to hexose-P. Concerning the route proposing that ADPGlc originates in the cytosol and then is translocated to the plastid for starch build-up, it would be an alternative operating in cereal endosperm (Kleczkowski 1996), particularly at certain developmental stages of the reserve tissue.

The distinctive structure and composition of starch granules are related with the characteristics of some of the enzymes involved in the metabolism of the polysaccharide. Worthy of mention is the occurrence of multiple isoforms of branching enzyme and starch synthesis (including soluble and granule-bound forms). As exhaustively analyzed (Ball and Morell 2003; Smith et al. 1997), the distinctive kinetic properties of these isozymes would critically determine the size of the polyglucan chains to be elongated, the degree of ramification, and relative contents of amylase and amylopectin in the entire polymer. In addition, the involvement of a protein named R1 in phosphorylating starch (Blennow et al. 2002) was demonstrated to be relevant. Phosphorylation is the only covalent modification present at different amounts (between 0.2% and 0.4% depending on the plant species) in the natural polymeric carbohydrate (mainly in amylopectin). The degree of phosphorylation is determinant for the fine structure and rheological properties of the starch, which establishes that this issue is of high relevance for the industrial uses of the polymer. R1 from plants has been characterized as a 120 kDa protein active as α-glucan water dikinase (EC 2.7.9.4):

$$\text{ATP} + \alpha\text{-glucan} + H_2O \leftrightarrows \text{AMP} + \text{phospho-}\alpha\text{-glucan} + Pi$$

and thus mono-phospho-esterifying Glc units in amylopectin by transferring the β-phosphate from ATP to positions C6 and C3 of the carbohydrate (Ritte et al. 2002). Studies using antisense strategy to generate potato mutants deficient in R1 indicated that the functionality of this protein is conclusive for both starch biosynthesis and degradation.

With respect to regulation of starch metabolism in amyloplasts, the picture is similar to that described above for the route in chloroplasts. ADPGlc PPase is the main target of regulation for the synthetic pathway, and the enzyme from heterotrophic cells has similar structural, kinetic, and regulatory properties than that from leaves (Ballicora et al. 2003, 2004; Cortassa et al. 2012; Iglesias and Podestá 2005), with some differences in the fine regulation of the activity established by differential expression of isoforms of the α and β subunits (Ballicora et al. 2004; Crevillen et al. 2003). In fact, the first characterization of the redox chemical modification related to modulation was performed with ADPGlc PPase from potato tubers (Ballicora et al. 2000; Fu et al. 1998). On the other hand, studies performed so far indicate that the association between ADPGlc PPase regulation and levels of starch accumulated in relation with the signal molecule Tre-6P would take place also in plant heterotrophic tissues (Stitt 2013).

29.4.3 Metabolism for Lipid Storage

In oily seeds, main carbon reserves are stored as oils instead of starch. Seeds like sunflower (*Helianthus annuus*) and soybean, and fruits like olives and avocado (*Persea americana*), are clear examples of plants accumulating oils with importance in the agronomic field (Taiz and Zeiger 2010). One of the advantages of lipid storage is that carbon is in a more reduced form than in carbohydrates, thus releasing more than twice the energy during catabolism; therefore, more reduction equivalents are needed during lipid synthesis (Baud and Lepiniec 2010; Theodoulou and Eastmond 2012). Carbon long-term storage as lipids takes place in nonphotosynthetic tissues in the form of TAGs, and reserves accumulation occurs in the cytosol (either in cotyledons or endosperm) as oil bodies or oleosomes. These structures are spherical organelles composed of a phospholipid monolayer with the TAGs in the interior and stabilized by the presence of oleosins, proteins embedded in the lipid membrane (Baud and Lepiniec 2010; Taiz and Zeiger 2010).

Lipid synthesis in plants requires the cooperation of different organelles (Figure 29.3). TAGs are assembled from glycerol-3-phosphate (Gro-3P) and acylated FAs in the endoplasmic reticulum through a series of reactions known as the *Kennedy pathway*. In plants, the *de novo* synthesis of FAs occurs in the plastid, unlike other organisms where the process takes place in the cytosol (Ohlrogge and Jaworski 1997). As FAs synthesis happens in heterotrophic tissues, carbon skeletons, energy, and reduction equivalents need to be imported from the cytosol. As described above, Suc represents the major form by which photosynthetically assimilated carbon is transported into sink tissues, thus providing the precursors for lipid synthesis (Figure 29.3). Two glycolytic intermediates are the building blocks for TAGs synthesis. Gro-3P

is produced from DHAP by the Gro-3P dehydrogenase in the cytosol, while Pyr provides the acetyl-CoA for FAs synthesis inside the plastid (Baud and Lepiniec 2010). Synthesis of FAs from acetyl-CoA requires ATP and NADPH, and thus these metabolites have to be imported or generated within plastids. Carbon skeletons also have to be imported into plastids as Glc-6P, Pyr, PEP, triose-P, or malate (Ludewig and Flugge 2013).

The intermediate used for FA synthesis depends on the plant species being considered. Due to the complexity and redundancy of plant metabolic pathways leading to the production of precursors, the proportion of carbon fluxes providing carbon and energy for FA synthesis through each pathway are hard to be addressed (Baud and Lepiniec 2010). Metabolic flux analysis has shown that in Brassicaceae (*Brassica napus* or *A. thaliana*), imported Suc is metabolized through cytosolic glycolysis until PEP, then imported into the plastid, and converted to Pyr inside the organelle (Schwender et al. 2003). In non-green heterotrophic embryos, like sunflower, Suc metabolism only goes through hexose-P in the cytosol, then hexose-P are imported into the plastid, and any further metabolism to Pyr seems to take place inside this organelle (Baud and Lepiniec 2010).

Independently of the intermediate delivered inside the plastid, all compounds are transported and converted by the steps needed of plastidic glycolysis until Pyr. The next step is the production of acetyl-CoA by decarboxylation of Pyr, a reaction catalyzed by the Pyr dehydrogenase complex that resides inside the plastid (Baud and Lepiniec 2010; Schwender et al. 2003). After acetyl-CoA is produced, the first committed step in the pathway of FAs synthesis is catalyzed by the acetyl-CoA carboxylase (AC Case) (Bates et al. 2013). Assembly of FAs occurs on acyl carrier protein (ACP) via a cycle of four reactions that elongates the acyl chain by two carbons each time. The cyclic condensation of acetyl-CoA proceed by a set of enzymes that are thought to be forming a complex collectively referred to as FA synthase (FA Sase), which includes the AC Case and is the strategic control point (Taiz and Zeiger 2010). As in yeast, animals, and bacteria, plant AC Case is highly regulated and is a key control point over the flux of carbon into FAs. In addition to control by phosphorylation, redox status, and protein interactions, feedback on AC Case by 18:1-ACP has recently been described (Bates et al. 2013).

FA synthesis has a high demand of energy and reducing power, and many pathways contribute to ATP and NADPH production. It has been reported that green seeds can also use light to supply NADPH and ATP, which allows a *bypass* of glycolysis via Rul-1,5bisP carboxylase activity and enzymes using pentose-P. This alternative pathway is more carbon efficient, resulting in 20% more acetyl-CoA available for oil synthesis, and also does not require reductant supply from the OPPP (Bates et al. 2013). However, in non-green seeds (where cytosolic and plastidic metabolisms are connected at the hexose-P level), the plastidic OPPP is the main pathway producing NADPH, and glycolysis is the one providing ATP (Baud and Lepiniec 2010; Schwender et al. 2003).

Once FAs are synthesized, they are transported and the assembly of TAGs takes place in the endoplasmic reticulum from Gro-3P and free FAs (Baud and Lepiniec 2010). Although there is evidence for a channeled pathway of free FAs through the plastid envelope, it is still unknown whether there are specific transporters for this process (Bates et al. 2013). During germination, oil seeds metabolize TAGs reserves by β-oxidation to obtain the energy and carbon required for growth and development until photosynthesis turns active. In the germinating process, the glyoxylate cycle plays a critical function to effectively mobilize carbon skeletons from oils to carbohydrates by gluconeogenesis (Taiz and Zeiger 2010).

TAGs coming from vegetable oils are primarily used for alimentary purposes, but they hold considerable potential for a wide range of uses depending upon the physicochemical properties conferred by their constituent FAs. As crude oil supplies decline, vegetable oils are gaining interest as substitutes for petroleum-derived fuels, lubricants, and specially chemicals (Cahoon et al. 2007). These industrial applications prompted researchers to engineer plants in order to increase carbon flux into TAG synthesis or produce new types of FAs. However, the efforts made until today have not been completely successful. This presumes that a greater understanding of FA synthesis and its regulation is needed to reach these goals.

29.5 CONCLUDING REMARKS AND PERSPECTIVES

This chapter describes the many faceted ways in which plants are able to gather their own components from the environment and distribute them through the whole organism. Expanding our knowledge in the field is of paramount importance in a world of fixed size with an ever-growing population that will impose an also ever-growing burden on food production. In short, plants sustain life on the planet by efficiently assimilating carbon and mobilizing it to the agronomically important sink organs. Assessing the organization, regulation, and coordination of all the involved processes is mandatory to recreate a green revolution that allows all humans access to at least the minimal food requirements based on the manipulation of selective parts of the biosynthetic machinery to improve plant efficiency and productivity. It is also true that in the few years since the first edition of this book was written, the notion that a climate change is well on its way is uncontestable. Under these circumstances, research must also consider what those changes will be, how they will affect crops, and how plant metabolism can be tailored to withstand such an abrupt change. Certain optimism is acceptable on this matter because of the manifest ability of plants to adapt to environmental changes, something that has been occurring since they first became terrestrial organisms.

ACKNOWLEDGMENTS

Work in our laboratories was granted by ANPCyT, CONICET, and UNL. All the authors are members of the Investigator Career from CONICET.

REFERENCES

Aguayo, M. F., D. Ampuero, P. Mandujano et al. 2013. Sorbitol dehydrogenase is a cytosolic protein required for sorbitol metabolism in *Arabidopsis thaliana*. *Plant Sci*. 205–6:63–75.

Ainsworth, E. A. and D. R. Bush. 2011. Carbohydrate export from the leaf: A highly regulated process and target to enhance photosynthesis and productivity. *Plant Physiol*. 155:64–9.

Andre, C., J. E. Froehlich, M. R. Moll, and C. Benning. 2007. A heteromeric plastidic pyruvate kinase complex involved in seed oil biosynthesis in arabidopsis. *Plant Cell* 19:2006–22.

Anguenot, R., S. Yelle, and B. Nguyen-Quoc. 1999. Purification of tomato sucrose synthase phosphorylated isoforms by Fe(III)-immobilized metal affinity chromatography. *Arch. Biochem. Biophys*. 365:163–9.

Aon, M. A., D. F. Gomez-Casati, A. A. Iglesias, and S. Cortassa. 2001. Ultrasensitivity in (supra)molecularly organized and crowded environments. *Cell Biol. Int*. 25:1091–9.

Baker, R. F., K. A. Leach, and D. M. Braun. 2012. SWEET as sugar: New sucrose effluxers in plants. *Mol. Plant* 5:766–8.

Ball, S. G. and M. K. Morell. 2003. From bacterial glycogen to starch: Understanding the biogenesis of the plant starch granule. *Annu. Rev. Plant Biol*. 54:207–33.

Ballicora, M. A., J. B. Frueauf, Y. Fu, P. Schurmann, and J. Preiss. 2000. Activation of the potato tuber ADP-glucose pyrophosphorylase by thioredoxin. *J. Biol. Chem*. 275:1315–20.

Ballicora, M. A., A. A. Iglesias, and J. Preiss. 2003. ADP-glucose pyrophosphorylase, A regulatory enzyme for bacterial glycogen synthesis. *Microbiol. Mol. Biol. Rev*. 67:213–25.

Ballicora, M. A., A. A. Iglesias, and J. Preiss. 2004. ADP-glucose pyrophosphorylase: A regulatory enzyme for plant starch synthesis. *Photosynth. Res*. 79:1–24.

Bantog, N. A., K. Yamada, N. Niwa, K. Shiratake, and S. Yamaki. 2000. Gene expression of NAD+-dependent sorbitol dehydrogenase and NADP+-dependent sorbitol-6-phosphate dehydrogenase during development of loquat (*Eriobotrya japonica* Lindl.) fruit. *J. Japan Soc. Hort. Sci*. 69:231–6.

Barratt, D. H., P. Derbyshire, K. Findlay et al. 2009. Normal growth of Arabidopsis requires cytosolic invertase but not sucrose synthase. *Proc. Natl. Acad. Sci. U.S.A*. 106:13124–9.

Bates, P. D., S. Stymne, and J. Ohlrogge. 2013. Biochemical pathways in seed oil synthesis. *Curr. Opin. Plant Biol*. 16:358–64.

Baud, S. and L. Lepiniec. 2010. Physiological and developmental regulation of seed oil production. *Prog. Lipid Res*. 49: 235–49.

Bieleski, R. L. 1982. Sugar alcohols. In *Plant Carbohydrates I: Intracellular Carbohydrates*, eds. F. A. Loewus, and W. Tanner, 158–92. Berlin: Springer-Verlag.

Blackburn, K., F. Y. Cheng, J. D. Williamson, and M. B. Goshe. 2010. Data-independent liquid chromatography/mass spectrometry (LC/MS^E) detection and quantification of the secreted *Apium graveolens* pathogen defense protein mannitol dehydrogenase. *Rapid Commun. Mass Spectrom*. 24:1009–16.

Blennow, A., T. H. Nielsen, L. Baunsgaard, R. Mikkelsen, and S. B. Engelsen. 2002. Starch phosphorylation: A new front line in starch research. *Trends Plant Sci*. 7:445–50.

Cahoon, E. B., J. M. Shockey, C. R. Dietrich et al. 2007. Engineering oilseeds for sustainable production of industrial and nutritional feedstocks: Solving bottlenecks in fatty acid flux. *Curr. Opin. Plant Biol*. 10:236–44.

Carpaneto, A., D. Geiger, E. Bamberg et al. 2005. Phloem-localized, proton-coupled sucrose carrier ZmSUT1 mediates sucrose efflux under the control of the sucrose gradient and the proton motive force. *J. Biol. Chem*. 280:21437–43.

Cataldi, T. R., G. Margiotta, L. Iasi et al. 2000. Determination of sugar compounds in olive plant extracts by anion-exchange chromatography with pulsed amperometric detection. *Anal. Chem.* 72:3902–7.

Chardon, F., M. Bedu, F. Calenge et al. 2013. Leaf fructose content is controlled by the vacuolar transporter SWEET17 in Arabidopsis. *Curr. Biol.* 23:697–702.

Chen, L. Q. 2014. SWEET sugar transporters for phloem transport and pathogen nutrition. *New Phytol.* 201:1150–5.

Chen, L. Q., B. H. Hou, S. Lalonde et al. 2010. Sugar transporters for intercellular exchange and nutrition of pathogens. *Nature* 468:527–32.

Chen, L. Q., X. Q. Qu, B. H. Hou et al. 2012. Sucrose efflux mediated by SWEET proteins as a key step for phloem transport. *Science* 335:207–11.

Cheng, L., R. Zhou, E. J. Reidel, T. D. Sharkey, and A. M. Dandekar. 2005. Antisense inhibition of sorbitol synthesis leads to up-regulation of starch synthesis without altering CO_2 assimilation in apple leaves. *Planta* 220:767–76.

Chollet, R., J. Vidal, and M. H. O'Leary. 1996. Phosphoenolpyruvate carboxylase: A ubiquitous, highly regulated enzyme in plants. *Annu. Rev. Plant Physiol. Plant Mol. Biol.* 47:273–98.

Chonan, N., H. Kawahara, and T. Matsuda. 1984. Ultrastructure of vascular bundles and fundamental parenchyma in relation to movement of photosynthate in leaf sheath of rice. *Jpn. J. Crop Sci.* 53:435–44.

Conde, C., P. Silva, A. Agasse et al. 2007. Utilization and transport of mannitol in *Olea europaea* and implications for salt stress tolerance. *Plant Cell Physiol.* 48:42–53.

Cortassa, S., M. A. Aon, A. A. Iglesias, J. C. Aon, and D. Lloyd. 2012. *An Introduction to Metabolic and Cellular Engineering*, 2nd ed. Singapore: World Scientific Publishing.

Crevillen, P., M. A. Ballicora, A. Merida, J. Preiss, and J. M. Romero. 2003. The different large subunit isoforms of *Arabidopsis thaliana* ADP-glucose pyrophosphorylase confer distinct kinetic and regulatory properties to the heterotetrameric enzyme. *J. Biol. Chem.* 278:28508–15.

Dasgupta, K., A. S. Khadilkar, R. Sulpice et al. 2014. Expression of sucrose transporter cDNAs specifically in companion cells enhances phloem loading and long-distance transport of sucrose but leads to an inhibition of growth and the perception of a phosphate limitation. *Plant Physiol.* 165:715–31.

Davidson, A., F. Keller, and R. Turgeon. 2011. Phloem loading, plant growth form, and climate. *Protoplasma* 248:153–63.

Davis, J. M. and W. H. Loescher. 1990. [^{14}C]-assimilate translocation in the light and dark in celery (*Apium graveolens*) leaves of different ages. *Physiol. Plant.* 79:656–62.

Davis, J. M., J. K. Fellman, and W. H. Loescher. 1988. Biosynthesis of sucrose and mannitol as a function of leaf age in celery (*Apium graveolens* L.). *Plant Physiol.* 86:129–33.

De Schepper, V., T. De Swaef, I. Bauweraerts, and K. Steppe. 2013. Phloem transport: A review of mechanisms and controls. *J. Exp. Bot.* 64:4839–50.

Denyer, K., F. Dunlap, T. Thorbjornsen, P. Keeling, and A. M. Smith. 1996. The major form of ADP-glucose pyrophosphorylase in maize endosperm is extra-plastidial. *Plant Physiol.* 112:779–85.

Dinant, S. and R. Lemoine. 2010. The phloem pathway: New issues and old debates. *C. R. Biol.* 333:307–19.

Eicks, M., V. Maurino, S. Knappe, U. I. Flugge, and K. Fischer. 2002. The plastidic pentose phosphate translocator represents a link between the cytosolic and the plastidic pentose phosphate pathways in plants. *Plant Physiol.* 128:512–22.

Eom, J. S., S. B. Choi, J. M. Ward, and J. S. Jeon. 2012. The mechanism of phloem loading in rice (*Oryza sativa*). *Mol. Cells* 33:431–8.

Eschrich, W. 1980. Free space invertase, its possible role in phloem unloading. *Ber. Dtsch. Bot. Ges.* 93:363–78.

Eschrich, W. and J. Fromm. 1994. Evidence for two pathways of phloem loading. *Physiol. Plant.* 90:699–707.

Espada, J. 1962. Enzymic synthesis of adenosine diphosphate glucose from glucose 1-phosphate and adenosine triphosphate. *J. Biol. Chem.* 237:3577–81.

Everard, J. D., V. R. Franceschi, and W. H. Loescher. 1993. Mannose-6-phosphate reductase, a key enzyme in photoassimilate partitioning, is abundant and located in the cytosol of photosynthetically active cells of celery (*Apium graveolens* L.) source leaves. *Plant Physiol.* 102:345–56.

Everard, J. D., R. Gucci, S. C. Kann, J. A. Flore, and W. H. Loescher. 1994. Gas exchange and carbon partitioning in the leaves of celery (*Apium graveolens* L.) at various levels of root zone salinity. *Plant Physiol.* 106:281–92.

Everard, J. D., C. Cantini, R. Grumet, J. Plummer, and W. H. Loescher. 1997. Molecular cloning of mannose-6-phosphate reductase and its developmental expression in celery. *Plant Physiol.* 113:1427–35.

Fan, R. C., C. C. Peng, Y. H. Xu et al. 2009. Apple sucrose transporter SUT1 and sorbitol transporter SOT6 interact with cytochrome b5 to regulate their affinity for substrate sugars. *Plant Physiol.* 150:1880–901.

Fedosejevs, E. T., S. Ying, J. Park et al. 2014. Biochemical and molecular characterization of RcSUS1, a cytosolic sucrose synthase isozyme phosphorylated *in vivo* at serine-11 in developing castor oil seeds. *J. Biol. Chem.* 289:33412–24.

Figueroa, C. M. and A. A. Iglesias. 2010. Aldose-6-phosphate reductase from apple leaves: Importance of the quaternary structure for enzyme activity. *Biochimie* 92:81–8.

Fischer, K. and A. Weber. 2002. Transport of carbon in non-green plastids. *Trends Plant Sci.* 7:345–51.

Flora, L. and M. Madore. 1993. Stachyose and mannitol transport in olive (*Olea europaea* L.). *Planta* 189:484–90.

Flugge, U. I. 1999. Phosphate translocators in plastids. *Annu. Rev. Plant Physiol. Plant Mol. Biol.* 50:27–45.

Flugge, U. I., R. E. Hausler, F. Ludewig, and M. Gierth. 2011. The role of transporters in supplying energy to plant plastids. *J. Exp. Bot.* 62:2381–92.

Fondy, B. R. and D. R. Geiger. 1982. Diurnal pattern of translocation and carbohydrate metabolism in source leaves of *Beta vulgaris* L. *Plant Physiol.* 70:671–6.

Frost, C. J., B. Nyamdari, C. J. Tsai, and S. A. Harding. 2012. The tonoplast-localized sucrose transporter in *Populus* (PtaSUT4) regulates whole-plant water relations, responses to water stress, and photosynthesis. *PLoS One* 7:e44467.

Fu, Y., M. A. Ballicora, J. F. Leykam, and J. Preiss. 1998. Mechanism of reductive activation of potato tuber ADP-glucose pyrophosphorylase. *J. Biol. Chem.* 273:25045–52.

Fu, Q., L. Cheng, Y. Guo, and R. Turgeon. 2011. Phloem loading strategies and water relations in trees and herbaceous plants. *Plant Physiol.* 157:1518–27.

Furumoto, T., M. Teramoto, N. Inada et al. 2001. Phosphorylation of a bifunctional enzyme, 6-phosphofructo-2-kinase/fructose-2,6-bisphosphate 2-phosphatase, is regulated physiologically and developmentally in rosette leaves of *Arabidopsis thaliana*. *Plant Cell Physiol.* 42:1044–8.

Gao, Z. and W. H. Loescher. 2000. NADPH supply and mannitol biosynthesis. Characterization, cloning, and regulation of the non-reversible glyceraldehyde-3-phosphate dehydrogenase in celery leaves. *Plant Physiol.* 124:321–30.

Gao, Z., L. Maurousset, R. Lemoine et al. 2003. Cloning, expression, and characterization of sorbitol transporters from developing sour cherry fruit and leaf sink tissues. *Plant Physiol.* 131:1566–75.

Gerhardt, R., M. Stitt, and H. W. Heldt. 1987. Subcellular metabolite levels in spinach leaves: Regulation of sucrose synthesis during diurnal alterations in photosynthetic partitioning. *Plant Physiol.* 83:399–407.

Gomez Casati, D. F., J. I. Sesma, and A. A. Iglesias. 2000. Structural and kinetic characterization of NADP-dependent, non-phosphorylating glyceraldehyde-3-phosphate dehydrogenase from celery leaves. *Plant Sci.* 154:107–15.

Grant, C. R. and T. a. Rees. 1981. Sorbitol metabolism by apple seedlings. *Phytochemistry* 20:1505–11.

Greiner, S., T. Rausch, U. Sonnewald, and K. Herbers. 1999. Ectopic expression of a tobacco invertase inhibitor homolog prevents cold-induced sweetening of potato tubers. *Nat. Biotechnol.* 17:708–11.

Guo, Z. X., T. F. Pan, K. T. Li et al. 2012. Cloning of NAD-SDH cDNA from plum fruit and its expression and characterization. *Plant Physiol. Biochem.* 57:175–80.

Guo, W. J., R. Nagy, H. Y. Chen et al. 2014. SWEET17, a facilitative transporter, mediates fructose transport across the tonoplast of Arabidopsis roots and leaves. *Plant Physiol.* 164:777–89.

Hardin, S. C. and S. C. Huber. 2004. Proteasome activity and the post-translational control of sucrose synthase stability in maize leaves. *Plant Physiol. Biochem.* 42:197–208.

Hardin, S. C., G. Q. Tang, A. Scholz et al. 2003. Phosphorylation of sucrose synthase at serine 170: Occurrence and possible role as a signal for proteolysis. *Plant J.* 35:588–603.

Hardin, S. C., H. Winter, and S. C. Huber. 2004. Phosphorylation of the amino terminus of maize sucrose synthase in relation to membrane association and enzyme activity. *Plant Physiol.* 134:1427–38.

Hartman, M. D., C. M. Figueroa, C. V. Piattoni, and A. A. Iglesias. 2014. Glucitol dehydrogenase from peach (*Prunus persica*) fruits is regulated by thioredoxin *h*. *Plant Cell Physiol.* 55:1157–68.

Hausler, R. E., N. H. Schlieben, P. Nicolay et al. 2000. Control of carbon partitioning and photosynthesis by the triose phosphate/phosphate translocator in transgenic tobacco plants (*Nicotiana tabacum* L.). I. Comparative physiological analysis of tobacco plants with antisense repression and overexpression of the triose phosphate/phosphate translocator. *Planta* 210:371–82.

Hawker, J., C. Jenner, and C. Niemietz. 1991. Sugar metabolism and compartmentation. *Funct. Plant Biol.* 18:227–37.

Heldt, H. W., C. J. Chon, and D. Maronde. 1977. Role of orthophosphate and other factors in the regulation of starch formation in leaves and isolated chloroplasts. *Plant Physiol.* 59:1146–55.

Herzog, B., M. Stitt, and H. W. Heldt. 1984. Control of photosynthetic sucrose synthesis by fructose 2,6-bisphosphate: III. Properties of the cytosolic fructose 1,6-bisphosphatase. *Plant Physiol.* 75:561–5.

Hirai, M. 1981. Purification and characteristics of sorbitol-6-phosphate dehydrogenase from loquat leaves. *Plant Physiol.* 67:221–4.

Hu, Z. and W. C. Plaxton. 1996. Purification and characterization of cytosolic pyruvate kinase from leaves of the castor oil plant. *Arch. Biochem. Biophys.* 333:298–307.

Huber, J. L. and S. C. Huber. 1992. Site-specific serine phosphorylation of spinach leaf sucrose-phosphate synthase. *Biochem. J.* 283(Pt 3):877–82.

Huber, S. C., T. H. Nielsen, J. L. A. Huber, and D. M. Pharr. 1989. Variation among species in light activation of sucrose-phosphate synthase. *Plant Cell Physiol.* 30:277–85.

Huber, S. C., J. L. Huber, P. C. Liao et al. 1996. Phosphorylation of serine-15 of maize leaf sucrose synthase. Occurrence *in vivo* and possible regulatory significance. *Plant Physiol.* 112:793–802.

Iglesias, A. A. and F. E. Podestá. 2005. Photosynthate formation and partitioning in crop plants. In *Handbook of Photosynthesis*, ed. M. Pessarakli, 525–45. Boca Raton, FL: CRC Press.

Izui, K., H. Matsumura, T. Furumoto, and Y. Kai. 2004. Phosphoenolpyruvate carboxylase: A new era of structural biology. *Annu. Rev. Plant Biol.* 55:69–84.

Juchaux-Cachau, M., L. Landouar-Arsivaud, J. P. Pichaut et al. 2007. Characterization of AgMaT2, a plasma membrane mannitol transporter from celery, expressed in phloem cells, including phloem parenchyma cells. *Plant Physiol.* 145:62–74.

Kanayama, Y. and S. Yamaki. 1993. Purification and properties of NADP-dependent sorbitol-6-phosphate dehydrogenase from apple seedlings. *Plant Cell Physiol.* 34:819–23.

Keeling, P. L., J. R. Wood, R. H. Tyson, and I. G. Bridges. 1988. Starch biosynthesis in developing wheat grain: Evidence against the direct involvement of triose phosphates in the metabolic pathway. *Plant Physiol.* 87:311–19.

Kelly, G. J. and M. Gibbs. 1973. A mechanism for the indirect transfer of photosynthetically reduced nicotinamide adenine dinucleotide phosphate from chloroplasts to the cytoplasm. *Plant Physiol.* 52:674–6.

Klann, E. M., B. Hall, and A. B. Bennett. 1996. Antisense acid invertase (TIV1) gene alters soluble sugar composition and size in transgenic tomato fruit. *Plant Physiol.* 112:1321–30.

Kleczkowski, L. A. 1996. Back to the drawing board: Redefining starch synthesis in cereals. *Trends Plant Sci.* 112:363–4.

Koch, K. E. 1996. Carbohydrate-modulated gene expression in plants. *Annu. Rev. Plant Physiol. Plant Mol. Biol.* 47:509–40.

Koch, K. 2004. Sucrose metabolism: Regulatory mechanisms and pivotal roles in sugar sensing and plant development. *Curr. Opin. Plant Biol.* 7:235–46.

Kolbe, A., A. Tiessen, H. Schluepmann et al. 2005. Trehalose 6-phosphate regulates starch synthesis via posttranslational redox activation of ADP-glucose pyrophosphorylase. *Proc. Natl. Acad. Sci. U.S.A.* 102:11118–23.

Kossmann, J., U. Sonnewald, and L. Willmitzer. 1994. Reduction of chloroplastic fructose-1,6-bisphosphate in transgenic potato plants impairs photosynthesis and plant growth. *Plant J.* 6:637–50.

Kruckeberg, A. L., H. E. Neuhaus, R. Feil, L. D. Gottlieb, and M. Stitt. 1989. Decreased-activity mutants of phosphoglucose isomerase in the cytosol and chloroplast of *Clarkia xantiana*. Impact on mass-action ratios and fluxes to sucrose and starch, and estimation of flux control coefficients and elasticity coefficients. *Biochem. J.* 261:457–67.

Krugel, U., L. M. Veenhoff, J. Langbein et al. 2008. Transport and sorting of the *Solanum tuberosum* sucrose transporter SUT1 is affected by posttranslational modification. *Plant Cell* 20:2497–513.

Kühn, C. 2003. A comparison of the sucrose transporter systems of different plant species. *Plant Biol.* 5:215–32.

Kulma, A., D. Villadsen, D. G. Campbell et al. 2004. Phosphorylation and 14-3-3 binding of Arabidopsis 6-phosphofructo-2-kinase/fructose-2,6-bisphosphatase. *Plant J.* 37:654–67.

Kunkes, E. L., D. A. Simonetti, R. M. West et al. 2008. Catalytic conversion of biomass to monofunctional hydrocarbons and targeted liquid-fuel classes. *Science* 322:417–21.

Lee, H. S. and A. Sturm. 1996. Purification and characterization of neutral and alkaline invertase from carrot. *Plant Physiol.* 112:1513–22.

Leegood, R. C. 1985. The intercellular compartmentation of metabolites in leaves of *Zea mays* L. *Planta* 164:163–71.

Liesche, J., U. Krugel, H. He et al. 2011. Sucrose transporter regulation at the transcriptional, post-transcriptional and post-translational level. *J. Plant Physiol.* 168:1426–33.

Lin, I. W., D. Sosso, L. Q. Chen et al. 2014. Nectar secretion requires sucrose phosphate synthases and the sugar transporter SWEET9. *Nature* 508:546–9.

Link, M., T. Rausch, and S. Greiner. 2004. In Arabidopsis thaliana, the invertase inhibitors AtC/VIF1 and 2 exhibit distinct target enzyme specificities and expression profiles. *FEBS Lett.* 573:105–9.

Loescher, W. H. and J. D. Everard. 2000. Regulation of sugar alcohol biosynthesis. In *Photosynthesis: Physiology and Metabolism*, eds. R. C. Leegood, T. D. Sharkey, and S. von Caemmerer, 275–99. Dordrecht: Springer.

Loescher, W. H., G. C. Marlow, and R. A. Kennedy. 1982. Sorbitol metabolism and sink-source interconversions in developing apple leaves. *Plant Physiol.* 70:335–9.

Loescher, W. H., R. H. Tyson, J. D. Everard, R. J. Redgwell, and R. L. Bieleski. 1992. Mannitol synthesis in higher plants: Evidence for the role and characterization of a NADPH-dependent mannose 6-phosphate reductase. *Plant Physiol.* 98:1396–402.

Ludewig, F. and U. I. Flugge. 2013. Role of metabolite transporters in source-sink carbon allocation. *Front. Plant Sci.* 4:231.

Lunn, J. E. 2007. Compartmentation in plant metabolism. *J. Exp. Bot.* 58:35–47.

Lunn, J. E. and R. T. Furbank. 1999. Tansley Review No. 105. Sucrose biosynthesis in C4 plants. *New Phytol.* 143:221–37.

Lunn, J. E. and M. D. Hatch. 1997. The role of sucrose-phosphate synthase in the control of photosynthate partitioning in *Zea mays* leaves. *Funct. Plant Biol.* 24:1–8.

Lunn, J. E. and E. MacRae. 2003. New complexities in the synthesis of sucrose. *Curr. Opin. Plant Biol.* 6:208–14.

Lunn, J. E., R. Feil, J. H. Hendriks et al. 2006. Sugar-induced increases in trehalose 6-phosphate are correlated with redox activation of ADPglucose pyrophosphorylase and higher rates of starch synthesis in *Arabidopsis thaliana*. *Biochem. J.* 397:139–48.

MacRae, E. and J. E. Lunn. 2012. Photosynthetic sucrose biosynthesis: An evolutionary perspective. In *Photosynthesis*, eds. J. J. Eaton-Rye, B. C. Tripathy, and T. D. Sharkey, 675–702. Dordrecht: Springer.

Markham, J. E. and N. J. Kruger. 2002. Kinetic properties of bifunctional 6-phosphofructo-2-kinase/fructose-2,6-bisphosphatase from spinach leaves. *Eur. J. Biochem.* 269:1267–77.

Martins, M. C., M. Hejazi, J. Fettke et al. 2013. Feedback inhibition of starch degradation in Arabidopsis leaves mediated by trehalose 6-phosphate. *Plant Physiol.* 163:1142–63.

McCaskill, A. and R. Turgeon. 2007. Phloem loading in *Verbascum phoeniceum* L. depends on the synthesis of raffinose-family oligosaccharides. *Proc. Natl. Acad. Sci. U.S.A.* 104:19619–24.

McKenzie, M. J., R. K. Chen, J. C. Harris, M. J. Ashworth, and D. A. Brummell. 2013. Post-translational regulation of acid invertase activity by vacuolar invertase inhibitor affects resistance to cold-induced sweetening of potato tubers. *Plant Cell Environ.* 36:176–85.

Morandi, B., L. Corelli Grappadelli, M. Rieger, and R. Lo Bianco. 2008. Carbohydrate availability affects growth and metabolism in peach fruit. *Physiol. Plant.* 133:229–41.

Mugford, S. T., O. Fernandez, J. Brinton et al. 2014. Regulatory properties of ADP glucose pyrophosphorylase are required for adjustment of leaf starch synthesis in different photoperiods. *Plant Physiol.* 166:1733–47.

Nadwodnik, J. and G. Lohaus. 2008. Subcellular concentrations of sugar alcohols and sugars in relation to phloem translocation in *Plantago major*, *Plantago maritima*, *Prunus persica*, and *Apium graveolens*. *Planta* 227:1079–89.

Nakai, T., T. Konishi, X. Q. Zhang et al. 1998. An increase in apparent affinity for sucrose of mung bean sucrose synthase is caused by in vitro phosphorylation or directed mutagenesis of Ser11. *Plant Cell Physiol.* 39:1337–41.

Negm, F. B. and W. H. Loescher. 1979. Detection and characterization of sorbitol dehydrogenase from apple callus tissue. *Plant Physiol.* 64:69–73.

Negm, F. B. and W. H. Loescher. 1981. Characterization and partial purification of aldose-6-phosphate reductase (alditol-6-phosphate:NADP 1-oxidoreductase) from apple leaves. *Plant Physiol.* 67:139–42.

Neuhaus, H. E. and M. J. Emes. 2000. Nonphotosynthetic metabolism in plastids. *Annu. Rev. Plant Physiol. Plant Mol. Biol.* 51:111–40.

Neuhaus, H. E. and M. Stitt. 1990. Control analysis of photosynthate partitioning: Impact of reduced activity of ADPglucose pyrophosphorylase or plastid phosphoglucomutase on the fluxes to starch and sucrose in Arabidopsis. *Planta* 182:445–54.

Nielsen, T. H., J. H. Rung, and D. Villadsen. 2004. Fructose-2,6-bisphosphate: A traffic signal in plant metabolism. *Trends Plant Sci.* 9:556–63.

Noiraud, N., L. Maurousset, and R. Lemoine. 2001. Identification of a mannitol transporter, AgMaT1, in celery phloem. *Plant Cell* 13:695–705.

Nosarzewski, M. and D. D. Archbold. 2007. Tissue-specific expression of sorbitol dehydrogenase in apple fruit during early development. *J. Exp. Bot.* 58:1863–72.

Nosarszewski, M., A. M. Clements, A. B. Downie, and D. D. Archbold. 2004. Sorbitol dehydrogenase expression and activity during apple fruit set and early development. *Physiol. Plant.* 121:391–8.

O'Leary, B., S. K. Rao, J. Kim, and W. C. Plaxton. 2009. Bacterial-type phosphoenolpyruvate carboxylase (PEPC) functions as a catalytic and regulatory subunit of the novel class-2 PEPC complex of vascular plants. *J. Biol. Chem.* 284:24797–805.

Ohlrogge, J. B. and J. G. Jaworski. 1997. Regulation of fatty acid synthesis. *Annu. Rev. Plant Physiol. Plant Mol. Biol.* 48:109–36.

Ohta, K., R. Moriguchi, K. Kanahama, S. Yamaki, and Y. Kanayama. 2005. Molecular evidence of sorbitol dehydrogenase in tomato, a non-Rosaceae plant. *Phytochemistry* 66:2822–8.

Oliver, S. N., J. E. Lunn, E. Urbanczyk-Wochniak et al. 2008. Decreased expression of cytosolic pyruvate kinase in potato tubers leads to a decline in pyruvate resulting in an *in vivo* repression of the alternative oxidase. *Plant Physiol.* 148:1640–54.

Oura, Y., K. Yamada, K. Shiratake, and S. Yamaki. 2000. Purification and characterization of a NAD+-dependent sorbitol dehydrogenase from Japanese pear fruit. *Phytochemistry* 54:567–72.

Park, S. W., K. J. Song, M. Y. Kim et al. 2002. Molecular cloning and characterization of four cDNAs encoding the isoforms of NAD-dependent sorbitol dehydrogenase from the Fuji apple. *Plant Sci.* 162:513–19.

Park, J., N. Khuu, A. S. M. Howard, R. T. Mullen, and W. C. Plaxton. 2012. Bacterial- and plant-type phosphoenolpyruvate carboxylase isozymes from developing castor oil seeds interact *in vivo* and associate with the surface of mitochondria. *Plant J.* 71:251–62.

Piattoni, C. V., C. M. Figueroa, V. E. Perotti, F. E. Podesta, and A. A. Iglesias. 2014. Biochemistry and physiology of carbon partitioning in crop plants. In *Handbook of Plant and Crop Physiology*, ed. M. Pessarakli, 193–215. Boca Raton, FL: CRC Press.

Plaxton, W. C. 1996. The organization and regulation of plant glycolysis. *Annu. Rev. Plant Physiol. Plant Mol. Biol.* 47:185–214.

Plaxton, W. C. and B. O'Leary. 2012. The central role of phosphoenolpyruvate metabolism in developing oilseeds. In *Seed Development: OMICS Technologies toward Improvement of Seed Quality and Crop Yield*, eds. G. K. Agrawal, and R. Rakwal, 279–304. Dordrecht: Springer.

Plaxton, W. C. and F. E. Podestá. 2006. The functional organization and control of plant respiration. *Crit. Rev. Plant Sci.* 25:159–98.

Podestá, F. E. 2004. Plant glycolysis. In *Encyclopedia of Plant and Crop Science*, ed. T. Goodman. New York: Marcel Dekker.

Podestá, F. E. and W. C. Plaxton. 1991. Kinetic and regulatory properties of cytosolic pyruvate kinase from germinating castor oil seeds. *Biochem. J.* 279:495–501.

Podestá, F. E. and W. C. Plaxton. 1994. Regulation of carbon metabolism in germinating *Ricinus communis* cotyledons. II. Properties of phosphoenolpyruvate carboxylase and cytosolic pyruvate kinase associated with the regulation of glycolysis and nitrogen assimilation. *Planta* 194:406–17.

Pozueta-Romero, J., M. Frehner, A. M. Viale, and T. Akazawa. 1991. Direct transport of ADPglucose by an adenylate translocator is linked to starch biosynthesis in amyloplasts. *Proc. Natl. Acad. Sci. U.S.A.* 88:5769–73.

Prata, R., J. D. Williamson, M. A. Conkling, and D. M. Pharr. 1997. Sugar repression of mannitol dehydrogenase activity in celery cells. *Plant Physiol.* 114:307–14.

Preiss, J., K. Ball, B. Smith-White et al. 1991. Starch biosynthesis and its regulation. *Biochem. Soc. Trans.* 19:539–47.

Ramsperger-Gleixner, M., D. Geiger, R. Hedrich, and N. Sauer. 2004. Differential expression of sucrose transporter and polyol transporter genes during maturation of common plantain companion cells. *Plant Physiol.* 134:147–60.

Reca, I. B., A. Brutus, R. D'Avino et al. 2008. Molecular cloning, expression and characterization of a novel apoplastic invertase inhibitor from tomato (*Solanum lycopersicum*) and its use to purify a vacuolar invertase. *Biochimie* 90:1611–23.

Reinders, A., A. B. Sivitz, and J. M. Ward. 2012. Evolution of plant sucrose uptake transporters. *Front. Plant Sci.* 3:22.

Rennie, E. A. and R. Turgeon. 2009. A comprehensive picture of phloem loading strategies. *Proc. Natl. Acad. Sci. U.S.A.* 106:14162–7.

Riesmeier, J. W., L. Willmitzer, and W. B. Frommer. 1992. Isolation and characterization of a sucrose carrier cDNA from spinach by functional expression in yeast. *EMBO J.* 11:4705–13.

Riesmeier, J. W., B. Hirner, and W. B. Frommer. 1993. Potato sucrose transporter expression in minor veins indicates a role in phloem loading. *Plant Cell* 5:1591–8.

Riesmeier, J. W., L. Willmitzer, and W. B. Frommer. 1994. Evidence for an essential role of the sucrose transporter in phloem loading and assimilate partitioning. *EMBO J.* 13:1–7.

Ritte, G., J. R. Lloyd, N. Eckermann et al. 2002. The starch-related R1 protein is an alpha-glucan, water dikinase. *Proc. Natl. Acad. Sci. U.S.A.* 99:7166–71.

Roblin, G., S. Sakr, J. Bonmort, and S. Delrot. 1998. Regulation of a plant plasma membrane sucrose transporter by phosphorylation. *FEBS Lett.* 424:165–8.

Rochaix, J. D. 2011. Reprint of: Regulation of photosynthetic electron transport. *Biochim. Biophys. Acta* 1807:878–86.

Roitsch, T. and M. C. Gonzalez. 2004. Function and regulation of plant invertases: Sweet sensations. *Trends Plant Sci.* 9:606–13.

Roitsch, T., M. Bittner, and D. E. Godt. 1995. Induction of apoplastic invertase of *Chenopodium rubrum* by D-glucose and a glucose analog and tissue-specific expression suggest a role in sink-source regulation. *Plant Physiol.* 108:285–94.

Ross, H. A., D. McRae, and H. V. Davies. 1996. Sucrolytic enzyme activities in cotyledons of the faba bean (developmental changes and purification of alkaline invertase). *Plant Physiol.* 111:329–38.

Rumpho, M. E., G. E. Edwards, and W. H. Loescher. 1983. A pathway for photosynthetic carbon flow to mannitol in celery leaves: Activity and localization of key enzymes. *Plant Physiol.* 73:869–73.

Sánchez, R. and F. J. Cejudo. 2003. Identification and expression analysis of a gene encoding a bacterial-type phosphoenolpyruvate carboxylase from arabidopsis and rice. *Plant Physiol.* 132:949–57.

Schwender, J., J. B. Ohlrogge, and Y. Shachar-Hill. 2003. A flux model of glycolysis and the oxidative pentosephosphate pathway in developing *Brassica napus* embryos. *J. Biol. Chem.* 278:29442–53.

Smith, A. M. and M. Stitt. 2007. Coordination of carbon supply and plant growth. *Plant Cell Environ.* 30:1126–49.

Smith, A. M., K. Denyer, and C. R. Martin. 1995. What controls the amount and structure of starch in storage organs? *Plant Physiol.* 107:673–7.

Smith, A. M., K. Denyer, and C. Martin. 1997. The synthesis of the starch granule. *Annu. Rev. Plant Physiol. Plant Mol. Biol.* 48:67–87.

Smith, C. R., V. L. Knowles, and W. C. Plaxton. 2000. Purification and characterization of cytosolic pyruvate kinase from *Brassica napus* (rapeseed) suspension cell cultures: Implications for the integration of glycolysis with nitrogen assimilation. *Eur. J. Biochem.* 267:4477–85.

Stitt, M. 2013. Systems-integration of plant metabolism: Means, motive and opportunity. *Curr. Opin. Plant Biol.* 16:381–8.

Stitt, M. and H. W. Heldt. 1985. Generation and maintenance of concentration gradients between the mesophyll and bundle sheath in maize leaves. *Biochim. Biophys. Acta, Bioenergetics* 808:400–14.

Stitt, M. and S. C. Zeeman. 2012. Starch turnover: Pathways, regulation and role in growth. *Curr. Opin. Plant Biol.* 15:282–92.

Stitt, M., B. Herzog, and H. W. Heldt. 1985. Control of photosynthetic sucrose synthesis by fructose 2,6-bisphosphate: V. Modulation of the spinach leaf cytosolic fructose 1,6-bisphosphatase activity *in vitro* by substrate, products, pH, magnesium, fructose 2,6-bisphosphate, adenosine monophosphate, and dihydroxyacetone phosphate. *Plant Physiol.* 79:590–8.

Stitt, M., I. Wilke, R. Feil, and H. W. Heldt. 1988. Coarse control of sucrose-phosphate synthase in leaves: Alterations of the kinetic properties in response to the rate of photosynthesis and the accumulation of sucrose. *Planta* 174:217–30.

Stitt, M., J. Lunn, and B. Usadel. 2010. Arabidopsis and primary photosynthetic metabolism—More than the icing on the cake. *Plant J.* 61:1067–91.

Stoop, J. M. and D. M. Pharr. 1992. Partial purification and characterization of mannitol: Mannose 1-oxidoreductase from celeriac (*Apium graveolens* var. *rapaceum*) roots. *Arch. Biochem. Biophys.* 298:612–19.

Stoop, J. and D. M. Pharr. 1993. Effect of different carbon sources on relative growth rate, internal carbohydrates, and mannitol 1-oxidoreductase activity in celery suspension cultures. *Plant Physiol.* 103:1001–8.

Stoop, J. and D. M. Pharr. 1994a. Mannitol metabolism in celery stressed by excess macronutrients. *Plant Physiol.* 106:503–11.

Stoop, J. M. H. and D. M. Pharr. 1994b. Growth substrate and nutrient salt environment alter mannitol-to-hexose partitioning in celery petioles. *J. Am. Soc. Hort. Sci.* 119:237–42.

Taiz, L. and E. Zeiger. 2010. *Plant Physiology*, 5th ed. Sunderland, MA: Sinauer Associates.

Tanase, K., K. Shiratake, H. Mori, and S. Yamaki. 2002. Changes in the phosphorylation state of sucrose synthase during development of Japanese pear fruit. *Physiol. Plant.* 114:21–6.

Tegeder, M. and P. M. Weber. 2006. Metabolite transporters in the control of plant primary metabolism. In *Control of Primary Metabolism in Plants*, eds. W. C. Plaxton, and M. T. McManus. Oxford: Blackwell Publishing.

Theodoulou, F. L. and P. J. Eastmond. 2012. Seed storage oil catabolism: A story of give and take. *Curr. Opin. Plant Biol.* 15:322–8.

Tiessen, A., J. H. Hendriks, M. Stitt et al. 2002. Starch synthesis in potato tubers is regulated by post-translational redox modification of ADP-glucose pyrophosphorylase: A novel regulatory mechanism linking starch synthesis to the sucrose supply. *Plant Cell* 14:2191–213.

Toroser, D. and S. C. Huber. 1997. Protein phosphorylation as a mechanism for osmotic-stress activation of sucrose-phosphate synthase in spinach leaves. *Plant Physiol.* 114:947–55.

Toroser, D., G. S. Athwal, and S. C. Huber. 1998. Site-specific regulatory interaction between spinach leaf sucrose-phosphate synthase and 14-3-3 proteins. *FEBS Lett.* 435:110–14.

Toroser, D., R. McMichael, Jr., K. P. Krause et al. 1999. Site-directed mutagenesis of serine 158 demonstrates its role in spinach leaf sucrose-phosphate synthase modulation. *Plant J.* 17:407–13.

Toyota, K., M. Tamura, T. Ohdan, and Y. Nakamura. 2006. Expression profiling of starch metabolism-related plastidic translocator genes in rice. *Planta* 223:248–57.

Trevanion, S. J., C. K. Castleden, C. H. Foyer et al. 2004. Regulation of sucrose-phosphate synthase in wheat (*Triticum aestivum*) leaves. *Funct. Plant Biol.* 31:685–95.

Turgeon, R. 2010. The role of phloem loading reconsidered. *Plant Physiol.* 152:1817–23.

Turgeon, R. and R. Medville. 1998. The absence of phloem loading in willow leaves. *Proc. Natl. Acad. Sci. U.S.A.* 95:12055–60.

Turner, W. L. and W. C. Plaxton. 2000. Purification and characterization of cytosolic pyruvate kinase from banana fruit. *Biochem. J.* 352(Pt 3):875–82.

Turner, W. L. and W. C. Plaxton. 2003. Purification and characterization of pyrophosphate- and ATP-dependent phosphofructokinases from banana fruit. *Planta* 217:113–21.

Turner, W. L., V. L. Knowles, and W. C. Plaxton. 2005. Cytosolic pyruvate kinase subunit composition, activity, and amount in developing castor and soybean seeds, and biochemical characterization of the purified castor seed enzyme. *Planta* 222:1051–62.

Usadel, B., O. E. Blasing, Y. Gibon et al. 2008. Global transcript levels respond to small changes of the carbon status during progressive exhaustion of carbohydrates in Arabidopsis rosettes. *Plant Physiol.* 146:1834–61.

van Bel, A. J. 2003. Transport phloem: Low profile, high impact. *Plant Physiol.* 131:1509–10.

Van den Ende, W. and A. Van Laere. 1995. Purification and properties of a neutral invertase from the roots of *Cichorium intybus*. *Physiol. Plant.* 93:241–8.

Velasco, R., A. Zharkikh, J. Affourtit et al. 2010. The genome of the domesticated apple (*Malus* x *domestica* Borkh.). *Nat. Genet.* 42:833–9.

Verde, I., A. G. Abbott, S. Scalabrin et al. 2013. The high-quality draft genome of peach (*Prunus persica*) identifies unique patterns of genetic diversity, domestication and genome evolution. *Nat. Genet.* 45:487–94.

Walker, R. P., A. L. Winters, and C. J. Pollock. 1997. Purification and characterization of invertases from leaves of *Lolium temulentum*. *New Phytol.* 135:259–66.

Wang, X. L., Y. H. Xu, C. C. Peng, R. C. Fan, and X. Q. Gao. 2009. Ubiquitous distribution and different subcellular localization of sorbitol dehydrogenase in fruit and leaf of apple. *J. Exp. Bot.* 60:1025–34.

Watari, J., Y. Kobae, S. Yamaki et al. 2004. Identification of sorbitol transporters expressed in the phloem of apple source leaves. *Plant Cell Physiol.* 45:1032–41.

Weber, A. P. and N. Linka. 2011. Connecting the plastid: Transporters of the plastid envelope and their role in linking plastidial with cytosolic metabolism. *Annu. Rev. Plant Biol.* 62:53–77.

Weber, A. P., R. Schwacke, and U. I. Flugge. 2005. Solute transporters of the plastid envelope membrane. *Annu. Rev. Plant Biol.* 56:133–64.

Weschke, W., R. Panitz, S. Gubatz et al. 2003. The role of invertases and hexose transporters in controlling sugar ratios in maternal and filial tissues of barley caryopses during early development. *Plant J.* 33:395–411.

Williamson, J. D., J. M. Stoop, M. O. Massel, M. A. Conkling, and D. M. Pharr. 1995. Sequence analysis of a mannitol dehydrogenase cDNA from plants reveals a function for the pathogenesis-related protein ELI3. *Proc. Natl. Acad. Sci. U.S.A.* 92:7148–52.

Winter, H. and S. C. Huber. 2000. Regulation of sucrose metabolism in higher plants: Localization and regulation of activity of key enzymes. *Crit. Rev. Biochem. Mol. Biol.* 35:253–89.

Wu, X., K. Sklodowski, B. Encke, and W. X. Schulze. 2014. A kinase-phosphatase signaling module with BSK8 and BSL2 involved in regulation of sucrose-phosphate synthase. *J. Proteome. Res.* 13:3397–409.

Wyse, R., D. Briskin, and B. Aloni. 1985. Regulation of carbon partitioning in photosynthetic tissue. In *Proceedings of the Eighth Annual Symposium on Plant Physiology*, eds. R. L. Heath, and J. Preiss, 231–53. Baltimore, MD: Waverly Press.

Yamada, K., H. Mori, and S. Yamaki. 1999. Gene expression of NAD-dependent sorbitol dehydrogenase during fruit development of apple (*Malus pumila* Mill. var. *domestica* Schneid.). *J. Japan Soc. Hort. Sci.* 68:1099–103.

Yamada, K., N. Niwa, K. Shiratake, and S. Yamaki. 2001. cDNA cloning of NAD-dependent sorbitol dehydrogenase from peach fruit and its expression during fruit development *J. Hort. Sci. Biotechnol.* 76:581–7.

Yamaguchi, H., Y. Kanayama, and S. Yamaki. 1994. Purification and properties of NAD-dependent sorbitol dehydrogenase from apple fruit. *Plant Cell Physiol.* 35:887–92.

Yamamoto, Y. T., E. Zamski, J. D. Williamson, M. A. Conkling, and D. M. Pharr. 1997. Subcellular localization of celery mannitol dehydrogenase. A cytosolic metabolic enzyme in nuclei. *Plant Physiol.* 115:1397–403.

Zamski, E., Y. T. Yamamoto, J. D. Williamson, M. A. Conkling, and D. M. Pharr. 1996. Immunolocalization of mannitol dehydrogenase in celery plants and cells. *Plant Physiol.* 112:931–8.

Zeeman, S. C., J. Kossmann, and A. M. Smith. 2010. Starch: Its metabolism, evolution, and biotechnological modification in plants. *Annu. Rev. Plant Biol.* 61:209–34.

Zervoudakis, G., C. D. Georgiou, G. Mavroidis, G. Kokolakis, and K. Angelopoulos. 1997. Characterization of purified leaf cytosolic pyruvate kinase from the C4 plant *Cynodon dactylon*. *Physiol. Plant.* 101:563–9.

Zhang, X. Q., B. Li, and R. Chollet. 1995. In vivo regulatory phosphorylation of soybean nodule phosphoenolpyruvate carboxylase. *Plant Physiol.* 108:1561–902.

Zhang, L. Y., Y. B. Peng, S. Pelleschi-Travier et al. 2004. Evidence for apoplasmic phloem unloading in developing apple fruit. *Plant Physiol.* 135:574–86.

Zhang, X. Y., X. L. Wang, X. F. Wang et al. 2006. A shift of Phloem unloading from symplasmic to apoplasmic pathway is involved in developmental onset of ripening in grape berry. *Plant Physiol.* 142:220–32.

Zhou, R. and L. Cheng. 2004. Biochemical characterization of cytosolic fructose-1,6-bisphosphatase from apple (*Malus domestica*) leaves. *Plant Cell Physiol.* 45:879–86.

Zhou, R. and B. Quebedeaux. 2003. Changes in photosynthesis and carbohydrate metabolism in mature apple leaves in response to whole plant source-sink manipulation. *J. Am. Soc. Hort. Sci.* 128:113–19.

Zhou, R., L. Cheng, and R. Wayne. 2003. Purification and characterization of sorbitol-6-phosphate phosphatase from apple leaves. *Plant Sci.* 165:227–32.

Section IX

Photosynthesis and Plant Genetics

30 Genetic Factors Affecting Photosynthesis

A.K. Joshi and Shree P. Pandey

CONTENTS

30.1 Introduction 539
30.2 Morphophysiological Factors: Genetic Basis 540
    30.2.1 Erect Leaves 540
    30.2.2 Crop Canopy and Leaf Area Index 541
    30.2.3 Chlorophyll Content 541
    30.2.4 Awns 542
30.3 C_3 versus C_4 Plants 542
    30.3.1 Heritability 542
    30.3.2 Heterosis 542
    30.3.3 Inheritance 543
    30.3.4 C_3–C_4 Hybridization 543
30.4 Nuclear Genes of Photosynthetic Apparatus 543
    30.4.1 Some Early Research 544
    30.4.2 Transcription and Translation 544
    30.4.3 Regulation of Ribulose Bisphosphate Carboxylase/Oxygenase 545
    30.4.4 Rubisco Activase 546
    30.4.5 Light-Regulated Nuclear Genes 546
30.5 Extranuclear Control Chloroplasts 547
    30.5.1 Inheritance 547
    30.5.2 Plastid Genome Organization 548
    30.5.3 Genes for Photosynthesis-Related Proteins 548
        30.5.3.1 Cytochrome b_6/f Complex 548
        30.5.3.2 Genes for PS I 548
        30.5.3.3 Genes for ATP Synthase Complex 549
        30.5.3.4 Other Genes in Chloroplast Genome 549
    30.5.4 Chloroplast Gene Expression 549
        30.5.4.1 Post-Transcriptional Regulation 550
        30.5.4.2 Translational Regulation 550
30.6 Photosynthesis and Plant Stress Responses 551
    30.6.1 Abiotic Stress 551
    30.6.2 Biotic Stress 551
    30.6.3 Growth-Defense Paradigm 552
30.7 Photosynthetic Strategy in Crops 552
    30.7.1 Manipulation of Photosynthetic Area 553
    30.7.2 Breeding for C_4 Mechanism 553
    30.7.3 Changes in Biochemical Activities 554
    30.7.4 Rubisco and Crop Improvement 554
    30.7.5 Future Prospects 554
References 555

30.1 INTRODUCTION

Photosynthesis, the process by which light energy is captured by the plant canopy and transformed into chemical energy, is the key to dry matter production and hence to yield of crop plants. All crop production that ultimately depends on photosynthesis [1] in turn depends on (1) the size and spatial orientation of light intercepting green area, (2) the duration of this area in active state, and (3) the specific rate of photosynthesis per unit green area [2]. The capture and use of solar energy by plants occur at different organizational levels and are influenced by various environmental factors whose effects can be ameliorated by genetically variable elements.

Historically there have not been strong interactions between photosynthesis research and other fields of plant science mainly due to two reasons [3]: (1) most techniques and tools developed to examine photosynthesis were unique to itself, and (2) there is a belief that CO_2 fixation and carbohydrate production are the only functions of photosynthesis, with carbohydrates the only link between photosynthesis and other biological phenomena [3]. However, recent research has established that photosynthesis is closely related to many other physiological processes [3] such as controlling the redox state of cells, regulating enzyme activity and other cellular processes [4,5], and generating reactive oxygen species, which are regulatory factors for many biological processes [6,7]. Precursor molecules of chlorophyll that act as a chloroplast-derived signal are involved in cell cycle regulation [8]. Photosynthesis research now employs the methods and tools of molecular biology and genetics, which are central methods for plant science, and therefore, there is better understanding about photosynthesis in crop plants [3].

The transition from an etiolated seedling to a fully green plant is a dramatic developmental event as any other in the life cycle of higher plants [9]. During this process, the expression of many genes is affected in many different ways, where light is also required as a signal for transcription of many photosynthetic genes in higher plants (for review, see Refs. [9–11]).

Like all other traits, photosynthesis is also influenced by genetic and environmental factors. Genes control various characteristics of plants that affect photosynthesis through light capture. The differences in C_3 and C_4 plants are also the consequence of genetic changes during evolution. Regulation of photosynthesis requires close cooperation of the genetic elements of plant cell that, in eukaryotes, are localized not only in the nucleus but in the cytoplasm (plastids) as well. The mechanisms controlling plastid development and gene expression in higher plants that used to be poorly understood [12,13] are now understood with enhanced precision [14].

Photosynthesis comprises an extremely complex series of biochemical, physiological, anatomical, and morphological features that have to be precisely matched and timed if biomass production of the crop is to increase and become consistent [2,3]. Evidence so far suggests the existence of substantial interspecific and intraspecific genetic variability in the rate of photosynthesis, but its utilization has not been satisfactory [15,16]. Breeding achievements were largely the outcome of changes in plant development and morphology and not through direct changes in photosynthetic rate or properties [15]. However, the scope for optimizing photosynthesis does exist, and thus, manipulation of the efficiency of photosynthesis along with respiration could be a useful strategy for enhancing yield of crops [3,17,18].

There is considerable variation in photosynthetic rates per unit leaf area between C_3 species [18]. This is an overlooked and untapped resource yet has huge potential to identify natural mechanisms that have evolved to allow plants to survive in different environments. Systematic analysis of plants from different geographic regions, using high-throughput *in vivo* photosynthesis tools, has the potential to identify novel targets for manipulation in crop plants. Development of photosynthetic capacity also differs in the same species when grown

under different environmental conditions; this can result in changes in morphology and anatomy and is termed *developmental acclimation*. In addition, some plant species are able to alter photosynthetic capacity in fully mature leaves, displaying a dynamic acclimatory response. A recent study revealed that developmental and dynamic acclimations are distinct processes and that dynamic acclimation may have a role in increasing the fitness of plants in natural environments [19]. However, most studies focus on plants grown in constant conditions or after a shift to different conditions, e.g., low to high light. There have been a few studies where plants have been exposed to fluctuating conditions. Elucidation of the mechanisms that are employed by plants during environmental fluctuations can provide novel targets for improving crop yield. This is particularly relevant given that many crop species lack any dynamic range to respond to this variation [20].

30.2 MORPHOPHYSIOLOGICAL FACTORS: GENETIC BASIS

Several morphological factors play an important role in influencing the rate of photosynthesis [18,21]. Manipulation of these morphological traits can be used to identify a crop ideotype [22]. An increase in the rate of photosynthesis can be achieved by identifying genotypes possessing improved photosynthetic apparatus, while assimilate partitioning can be brought about by genetic manipulation of growth functions of organs for endogenous regulation of physiological processes. Selection for improved photosynthetic activity and the identification of forms with high combining ability for photosynthetic characters have been reported [23].

Photosynthesis is a function of total leaf area, leaf duration, and total intercepted radiation by the canopy [24,25]. These clearly suggest that various morphological factors in plants influence photosynthesis mainly by influencing the canopy light interception. Thus, genetic variation for photosynthetic efficiency may be compensated for by morphological changes in total leaf area and light interception properties [16,26]. A number of studies have demonstrated the presence of genetic variation in photosynthetic traits of crops as well as interaction of photosynthetic phenotypes with their environment [27].

30.2.1 ERECT LEAVES

Crop varieties with erect leaves generally exhibit higher photosynthetic rates and in some cases greater economic yield due to better light interception [21,23]. However, no yield advantage for erect leaf was reported in barley [28]. In rice, the short, stiff upright leaf habit is governed by a single gene [29] having pleiotropic effects toward the reduced number of shorter internodes and the increased number in shorter panicles [30]. Erectness is reported to be associated with higher yields in densely planted fields [31] and also with reduced photoinhibition under high light intensities [32]. Later reports [33] also suggest that new rice cultivars with very erect leaves were associated with increased light capture for photosynthesis, nitrogen storage for grain filling, and also grain yields [34]. An efficient photomodel of maize inbred with erect leaf was also proposed [35].

30.2.2 CROP CANOPY AND LEAF AREA INDEX

Morphological traits influencing crop canopy are easily measurable and include leaf characters such as length, width, rigidity, and angle, and in cereals, tiller angle [21]. These traits that affect photosynthetic rate through light interception and correlated physiological, developmental, and morphological changes [1] usually possess high narrow-sense heritability [36,37]. Considerable additive genetic variation has been reported in rye grass for light transmission coefficient (K), which gives a quantifiable measure of the light trapping capabilities of a crop [37,38]. In wheat, some accessions of wild emmer types have been reported to possess high photosynthetic capacity with no significant reduction in leaf size [39].

Optimum leaf area index allows greater light penetration and hence is considered more desirable for photosynthesis. The components of leaf area index possess highly significant additive gene effects in wheat [40] and beans [41]. In pea, a gene (*st*) reduces stipule size, while another gene (*af*) substitutes tendrils for leaflets [42]. Cotton plants having okra-shaped leaves, which is due to a dominant gene mutation (L^0L^0), possess relatively greater photosynthetic rates owing to better light interception [43]. In *Cicer arietinum*, ovate leaf size is dominant over lanceolate shape and is controlled by a single gene (*Ovlt*) [44]. In a study by Ghatge [44], leaf and leaflet size were found to be under the control of three genes, two of which were supplementary while the other was inhibitory. Leaf shape in gherkin (*Cucumis* spp.) is controlled by a gene (*nt*) having incomplete dominance [45]. Leaf width in sunflower was reported to be under the control of four genes, while leaf length by at least three. For both traits, the estimates of broad sense heritability were high, but those for narrow-sense heritability was low [46]. Abzalov and Fatkullaeva [47] found that two genes (O^l and In^l) control the shape of leaf blade in cotton. The allele for sinuate leaf shape (*c*) in *Matthiola incana* was found to be recessive to the allele for normal entire leaf (C) [48].

Recent advances have revealed that various aspects of leaf growth and morphology such as leaf shape, size, and flatness, among many, are regulated by transcription factors (TFs) that are ultimately regulated by the small-RNAs (smRNAs). Regulatory smRNAs, in plants classified in two broad groups of microRNAs and small-interfering RNAs (siRNAs), are a family of noncoding RNAs. In eukaryotes, these range between 18 and 30 nucleotides (nt), whereas in bacteria, these may range up to several hundred nt [49]. Although several features govern their action, these smRNAs regulate their target in a sequence-specific manner [50] in a process termed as RNA-interference (RNAi) or post-transcriptional gene silencing (PTGS). These target complementary sequences in the mRNAs to block accumulation of proteins by either degrading the message (transcriptional cleavage; preferred mode in plants) or inhibiting translation [51]. The miRNAs

are synthesized as primary transcripts, folded into double-stranded (dsRNA) stem loops, in the nucleus via RNA-Pol II, whereas the siRNAs mature from dsRNAs that are generated in the cytoplasm with the help of a family of RNA-directed RNA polymerases (RdRs). These dsRNAs are processed by the Dicer-like proteins (DCLs), and finally, the mature smRNAs are loaded on the RNA-induced-silencing complex (RISC) with the help of the Argonaute (AGO) protein effectors [52,53]. Leaf morphology has been shown to be under the genetic control of several of these smRNA genes such as the AGO7 proteins through the action of the trans-acting siRNAs [54].

The recent information on the molecular players playing an important role in leaf morphophysiology suggests crucial regulation during development. Leaf initiation, leaf polarity, phase transition, growth, differentiation, and senescence are regulated by miRNAs via several TFs [55]. For instance, the establishment of polarity requires positional information along the radial axis. In Arabidopsis, two HD-ZIPIII TF genes, PHABULOSA (PHB) and PHAVOLUTA (PHV), govern this information [56]. The miR165/166 module targets the PHB TFs [57]. Similarly, the plant-specific TF family of GROWTH REGULATING FACTORS (GRF) regulates leaf growth. These are under the control of miR396 network that also includes the involvement of TCP4 TFs as well as the ASYMMETRIC LEAVES 1 (AS1) and AS2 TFs [58]. Genetic interactions among TFs, miRNAs, RdR6, SGS3, and AGO7 during leaf development are nearly established. Similarly, miR319-TCP TF network regulates aspects of leaf growth and senescence [59].

30.2.3 CHLOROPHYLL CONTENT

The rate of net photosynthesis in crop plants declines after certain age that varies both between and within species [60]. Several genes alter chlorophyll loss in crop plants. In soybean, genes at $d_1 + d_2$ loci prevent leaf yellowing [61–64]. In contrast, a recessive gene (y_3) causes early yellowing of leaves. Using isogenic soybean lines in the background of variety Clark, Guiamet et al. [63] observed that *Cyt G* and $d_1 d_2$ do not delay the decline in photosynthetic rate during monocarpic senescence relative to Clark, but $G d_1 d_2$ delays it. The inheritance pattern of the *B* gene conferring chlorophyll deficiency in *Cucurbita moschata* showed variable expression of the gene B_2, and a number of combinations of *B* genes were found to control chlorophyll expression [65].

It has been demonstrated that chlorophyll deficiency in a chlorophyll-deficient mutant of cowpea was due to a single recessive gene [66,67]. Sen and Bhowal [68] proposed that a monogenic multiple allelic series controlled the level of chlorophyll production. Kohle [69], in crosses involving *Vigna* subspecies, concluded that foliage color was controlled by two complementary genes and that light-green foliage was dominant to dark-green foliage.

Occurrence of stay-green trait has been reported in different crops [70] including wheat [71–73]. In durum wheat, a stay-green mutant has been associated with increased leaf area, rate, and duration of grain filling and photosynthetic competence [74]. It is also reported that green and viable leaves significantly contribute photosynthates to developing grain [75]. Since there is a strong association between the duration of photosynthetically active leaf area and grain yield [76], selection for stay-green is expected to have a significant implication in productivity of wheat particularly under harsh environments [56], including those that are heat stressed [77]. Genetic variation for *stay-green spike*' affecting photosynthesis was reported in durum wheat [78].

In *Sorghum bicolor*, this trait is influenced by a major gene that exhibits varying levels of dominant gene action and allows the plants to retain their leaves in the active photosynthetic state when subjected to water stress [79]. In Festuca, the stay-green nuclear gene (*sid*) inhibits chlorophyll loss and the degradation of some thylakoid compounds but does not interfere with other senescence processes [80].

On the other hand, when RdR2-silenced lines were subjected to high UV-B levels under controlled and natural environments in the wild tobacco plants, they suffered from significant loss of chlorophyll resulting in fitness consequences with reduced stalk length, dry mass, and seed production [81]. The reduced phenolics in the leaves of RdR2-silenced plants may cause effective loss of sunscreens for leaves, resulting in very rapid degradation of chlorophyll under high UV-B environments, reducing photosynthesis, biomass, and seed yield. These findings may be critical for designing crop-engineering strategies for high elevations subjected to climate change.

30.2.4 AWNS

The significant contribution of awns toward photosynthesis has been known for a very long time [82]. Awns provide a large surface area favorably placed for both light interception and CO_2 uptake, which may double the net photosynthetic rate of wheat ears [83]. Awns in wheat are the result of the interaction between the awns promoting genes present on group 2 chromosomes [84] and the awns suppressing genes Hd, B_1, B_2, and B_3 present on the short arm of chromosome 4B, the long arm of 5A, 6B, and 1D, respectively [84,85]. Among these, B_1 is the most potent followed by B_2, Hd, and B_3 [86]. Awns and glumes that are discrete organs of the spike display variable photosynthetic capacity based on their morphology, development, and metabolic capacity [87] and are likely to be under independent genetic control. This is important in view of the fact that spike photosynthesis can contribute substantially to grain filling [87].

30.3 C₃ VERSUS C₄ PLANTS

On the basis of certain inherent differences concerning photosynthesis, crop plants in general can be grouped into C_3 and C_4 plants. Most of the cultivated crops (e.g., wheat, rice, barley, oat, legumes, etc.) are C_3 plants where photosynthesis occurs through Calvin cycle and losses due to photorespiration are much higher. This is largely due to the enzyme Rubisco [18],

which not only is an inefficient enzyme with a low turnover number but also catalyzes two competing reactions: carboxylation and oxygenation [88]. The oxygenation reaction directs the flow of carbon through the photorespiratory pathway resulting in losses of between 25% and 30% of the carbon fixed [18]. Therefore, reducing the Rubisco oxygenase reaction has been considered to have the potential to increase carbon assimilation significantly [89].

The C_4 plants (e.g., sorghum, pearl millet, maize, sugarcane, etc.) possess approximately 50% higher photosynthetic efficiency through organic acid metabolism in the bundle sheath cells with a very low (about 10% of C_3 plants) rate of photorespiration [90]. The most important characteristic of C_4 plants is the presence of two types of chloroplasts in their leaves. The combined action of these chloroplasts, which differ in their structure as well as function (photochemical properties), leads to enhanced efficiency of carbon fixation. Consequently, the photosynthetic rates differ markedly between these two types of plant species. About 15 genera contain both C_3 and C_4 species, but none of these includes important crops [91]. Also no major C_3 crop has been shown to possess C_4 traits [21].

30.3.1 HERITABILITY

Heritability estimates for photosynthesis have been determined in a number of C_3 and C_4 crops, for example, maize [92–94], rice [95], cotton [96], soybean [97–99], peas [100], sunflower, and wheat [101]. In the C_3 plants, heritability estimates reported were moderate to high in rice [95]; low to high in tobacco [102]; and moderate [98] and low to moderate [99] in soybean. Similarly in maize, the realized heritabilities were found quite low for photosynthesis during the reproductive growth stage [92]. During the vegetative stage, the estimates of narrow-sense heritabilities and realized heritabilities were much higher [93].

30.3.2 HETEROSIS

Various reports on the extent of heterosis for photosynthesis indicate that a considerable range of heterosis exists among C_3 and C_4 plants. For example, in wheat, heterosis ranges from almost zero [103] to high [104–106]. In cotton, the intraspecific and interspecific (*G. hirsutum* × *G. barbadense*) hybrids exhibited considerable heterosis in photosynthetic rates [104]. Heterosis in cucumber ranged from 7.2% to 33% [107], while it was reported to be high in tomato [108]. Similarly, in maize, a significant amount of heterosis for photosynthesis has been observed [109–112]. Offerman and Peterhansel [113] argued that there is ample evidence that heterosis is associated with increased rates of photosynthesis, and the underlying biochemical principles are better understood. The importance of epigenetic chromatin modifications in heterosis is now established, and the first direct links between epigenetic changes and improved photosynthesis have also been demonstrated enabling possibility of epigenetic engineering [113].

30.3.3 INHERITANCE

Inheritance studies on photosynthesis have been largely based on studies related to photosynthetic rate as well as component traits that influence it. It has been suggested that the rate of photosynthesis decreases with an increase in the ploidy level [23]. For example, in wheat, a relatively higher rate of photosynthesis was observed in diploid wheat *T. aegilopoides var. boecticum* in comparison to hexaploid *T. aestivum var. erythrospermum* [114]. In another study, the photosynthetic rate was highest in diploid followed by tetraploid and hexaploid wheats [115].

Photosynthesis, being a complex process, is governed by a number of genes. There are reports that suggest that the rate of photosynthesis is inherited quantitatively, for example, Ojima [97] and Wiebold et al. [99] in soybean and Hobbs and Mahon [116] in pea. On the other hand, a single locus was found to influence the photosynthetic rate in rice where the lower rate was dominant over the higher rate [117]. Dominance for a low photosynthetic rate has also been noticed in dry bean [118]. In maize, net-photosynthesis rate, chlorophyll contents, and sucrose phosphate synthase activity were found to be controlled by two major genes [94].

The role of wheat D genome in the expression of net photosynthesis has been analyzed [119,120]. Haour-Lurton and Planchon [121] found that both arms of chromosome 3 D and short arm of chromosome 6 D control the major mechanism of photosynthesis under saturating light conditions. Bobo et al. [122] observed that the control of photosystem II (PS II) in wheat is under the control of 3 D chromosomes, and this might be related to bread wheat productivity. The role of cytoplasmic factors has been documented in detail in various crops (see Sections 32.4 and 32.5; while the inheritance aspects are covered in Sections 32.4 and 32.6).

30.3.4 C_3–C_4 HYBRIDIZATION

The first $C_3 \times C_4$ hybrids were produced by crossing *Atriplex rosea* (C_4) and *Atriplex prostrata* (C_3, formerly termed *A. patula* ssp. *hastata* and *A. triangularis*; [123], and *A. rosea* and *A. glabriuscula* (C_3) [124,125]). Although none of the hybrids showed complete C_4 characteristics, some exhibited certain anatomical and biochemical features quite close to C_4 plants [126]. Hybridization involving *A. rosea* (C_4) and *A. glabriuscula* or *A. hortensis* or *A. patula* (all C_3) led them to conclude that the genetic differences between them were quite narrow, and the inheritance of photosynthetic pathway is largely under the control of nuclear genes. The occurrence of wide segregation in F_2 and F_3 indicated the role of a smaller number of genes in the inheritance of Kranz anatomy [127].

In *Panicum*, hybrids involving C_3 and C_3–C_4 species were developed by Brown et al. [128]. The F_1's showed intermediate behavior for anatomical and biochemical characteristics. Hybridization of C_3 and C_4 types in *Flaveria* has shown greater promise for increasing photosynthetic efficiency [129]. Many of the *Flaveria* F_1 hybrids were sterile [130], and even if F_2 hybrids were generated, the problems associated with chromosome abnormalities and pairing persisted, and mapping populations could not be formed [125,131]. This led to abandonment of all hybrid studies [132].

For C_4 pathway, *Setaria viridis* is considered a potential candidate [131–133]. Recently, Oakley et al. [132] successfully crossed the C_3 species *A. prostrata* with the C_4 species *A. rosea* to produce F_1 and F_2 hybrids. They found that C_4 metabolic cycle was disrupted in the hybrids, but one hybrid exhibited a C_4 metabolic cycle. The anatomy of the hybrids resembled that of C_3–C_4 intermediate species using a glycine shuttle to concentrate CO_2 in the bundle sheath. They proposed that the progeny of these hybrids will be carried forward and are expected to further segregate C_3 and C_4 traits and will help in the discovery of C_4 genes using high-throughput methods.

At present, it is much easier to interpret patterns observed in $C_3 \times C_4$ hybrid due to substantial progress made in our understanding of photosynthetic methodology, theoretical models of C_3 and C_4 photosynthesis, and structural adaptations that enhance C_4 function [132,134,135]. New predictions from theoretical models of C_4 photosynthesis also facilitate interpretation of $C_3 \times C_4$ hybridization results [134,136]. The current progress enables us to address the degree to which $C_3 \times C_4$ hybrids express the physiology of C_3, C_4, or C_3–C_4 intermediate species [132,135].

30.4 NUCLEAR GENES OF PHOTOSYNTHETIC APPARATUS

The genetic factors influencing character(s) in crop plants are located in nuclear as well as cytoplasmic genome. The two genetic systems interact to regulate various processes associated with plant growth and development. Regulation of photosynthetic machinery also requires close cooperation of nuclear and cytoplasmic genetic systems in a plant cell [137]. The chloroplast genome, though it plays a key role in the regulation of photosynthesis, is not large enough to provide complete information for this complex process. Hence, a considerable part of the genetic information necessary for chloroplast development and activity is localized in the nucleus [138]. The diversity of plastid types being controlled by the developmental program of the plant also indicates that there is a significant flow of information between the nuclear and the chloroplast genome of a cell [139]. The role of nuclear genes in photosynthesis through chloroplast (and leaf) differentiation also emerges from the observation that in chloroplast mutants, leaf morphology is unaffected, but a small percentage of mutants show alterations in leaf development [140]. Now, it is known that photosynthesis together with other plastid functions requires the products of several hundred genes, of which only about 120 are present in the approximately 150 kb chloroplast genome; all other plastid proteins are expressed from nuclear genes [139]. In higher plants, all the enzymes that have been characterized in the chlorophyll biosynthetic pathway (e.g., δ-amino b levulinic acid to protochlorophyll [ide]) are encoded in the nuclear genome [141]. The 7Ag.7DL translocation [142,143] in wheat is associated with reduced floral abortion and improved utilization of photosynthetic capacity in high-yield environments [144].

30.4.1 SOME EARLY RESEARCH

Based on their studies on *C. reinhardtii*, Levine and Goodenough [145] suggested that 8 linkage groups out of 16 affecting chloroplast properties are localized in the nucleus. However, this was contradicted by McVittie and Davies [146] who related all the 16 linkage groups with nuclear genome. The influence of nuclear genome on the regulation of some specific chloroplast ribosomal proteins was demonstrated by Bogorad et al. [147,148]. Similarly, 50s subunits of chloroplast ribosomal proteins in *Nicotiana* were reported to be inherited by genes present in nuclear genome [149].

It has been well known for a long time that chlorophyll biosynthesis is controlled mainly by the nuclear genome [138]. Experimental mutagenesis and induction of the chlorophyll mutations has been widely used in investigating the genetic control systems of chloroplast biogenesis. Chlorophyll mutants in barley have been described a long time ago [150]. In *chlorina*-type pea mutants, 15 loci were found to control plastid pigment formation [151]. Analysis of mutants compromised in plastid-targeted proteins forms a major component of strategy for deciphering plastid function; its application in past has been limited due to lethality of mutants [152]. A substantial part of lamellar protein of chloroplasts is coded by the nuclear genome, as reviewed by Pogson and Albrecht [153].

Pigment aggregation and packing in photosynthetic membranes are determined by lamellar proteins, the synthesis of which is under the dual control of nucleus and plastid genes [154]. In *Arabidopsis thaliana*, 60 non-allelic genes control chlorophyll biosynthesis through their influence on chloroplast structure and function [152,155]. In a nuclear mutant of *Vicia faba* having a block in photosystem I (PS I), chloroplasts were incapable of NADP photoreduction [156]. Similarly, mutations in nuclear genes have been studied in *Scenedesmus obliquus* for organization of PS I [157]; PS II in pea [158]; and PS I and PS II in cotton [159].

Studies from many laboratories led to the conclusion that the earlier belief that nuclear genes were regulated exclusively at the transcriptional level and chloroplast genes at the post-transcriptional level is oversimplified view [137]. Gene expression is complex and involves several steps and genes, both nuclear and that of chloroplast, starting with the transcription of a gene or operon into a pre-mRNA that is processed into a mature mRNA molecule by splicing and editing [137].

30.4.2 TRANSCRIPTION AND TRANSLATION

There is a close cooperation between the transcription–translation system of chloroplasts, nucleus, and cytoplasm [160–162]. Induction of chloroplast DNA replication and plastid RNA polymerases is controlled by the nucleus and is formed in the cytoplasm [148,163]. Nasyrov [138], presenting a generalized scheme of genetic regulation of chloroplast biogenesis and photosynthesis, suggested that a large amount of the genetic information localized in the nucleus is translated on the 80s ribosomes, and these gene products regulate and are directly involved in chloroplast biosynthesis. Also, the chloroplast number is positively correlated with gene dosage [164]. On the other hand, nuclear genome replication and transcription depend, in turn, on chloroplast gene products [165,166].

Various polypeptides identified in PS II, photosystem PS I, the cytochrome b_6/f complex, and the coupling factor are encoded in the chloroplast as well as nuclear DNA [167–169]. The polypeptides encoded in the nucleus are commonly synthesized as a larger precursor on cytosolic ribosomes, and they are subsequently imported into the chloroplast [170]. The most studied nucleus-encoded chloroplast proteins are the Chl a/b-binding proteins (CAB protein) and SSU, and their photosynthetic pathways and functional regulation have been discussed in a number of reviews [171–177]. At least three *Arabidopsis* loci were identified as necessary for coupling the expression of some nuclear genes to the functional state of the chloroplast [178]. Recently, the environmental, systemic, and metabolic signals triggering the stoichiometry adjustments of ATP synthase and the cytochrome b_6f complex were reported by Schöttler et al. [179].

In higher plants, PS I consists of at least 13 subunits, including PS I-A through PS I-L and PS I-N. The genes encoding these polypeptides are designated *psa A* through *psa L* and *psa N*, respectively, and are located in both the nuclear and chloroplast genomes [180–185]. The PS I-H subunit is a peripheral membrane protein of about 10 kDa encoded by a nuclear gene, *psa H*. Several chloroplast DNAs and a genome clone for PS I-H have been isolated and analyzed in higher plants [186–189] and in *Chlamydomonas* [190]. However, little is known about the genomic organization of the genes encoding PS I-H. Nakamura and Obokata [191] identified three structurally different *psa H* genes in the nuclear genome of *Nicotiana sylvestris* designated *psa Ha*, *psa Hb*, and *psa Hc*, and all the three genes are expressed in young leaves. Each gene has two introns: one between sequences encoding a transit peptide and the N-terminal acidic domain, and the other between the N-terminal domain and a central hydrophobic domain. The occurrence of two isoforms of PS I-H in *N. sylvestris* [192] led these workers to conclude that the *psa H* products may be subjected to post-translational modifications.

Research done in the past decade indicates the presence of a plastid-derived signal that specifically influences the expression of nuclear genes for plastid proteins [193]. Although the nature of the plastid derived signal (s) is still not known, it has been suggested that the signal is not directly related to photosynthesis [194–196]. The results obtained in tobacco indicate that plastids control nuclear gene expression via different and gene-specific cis-regulatory elements [193].

Differences in gene regulatory mechanisms between monocot and dicot cells have been found in several studies [197–199]. The promoter of a monocot gene for the small subunit of ribulose 1,5-bisphosphate carboxylase is active in monocot cells but not in dicot cells. Conversely, a dicot promoter remains active in dicot cells but not in monocot cells

[198]. Yamamoto et al. [200] demonstrated that the promoter of pine cab 6 gene is also active in a monocot cell in a light-independent but cell-type-specific manner.

Waters et al. [201] reported that, in several land plants, GOLDEN2-LIKE (GLK) TFs are required for chloroplast development, that GLK proteins influence photosynthetic gene expression independently of the phyB signaling pathway, and that the two GLK genes (glk 1 and glk 2) are differentially responsive to plastid retrograde signals. They suggested that GLK genes help to coregulate and synchronize the expression of a suite of nuclear photosynthetic genes and thus act to optimize photosynthetic capacity in varying environmental and developmental conditions.

Many TFs, belonging to different families (e.g., MYB, bZIP, and DREB), have been related to abiotic stress responses; however, only a few are known to regulate the expression of photosynthesis-related genes in response to stress [202]. Several of them belonging to the MYB family play an important role in both stomatal and nonstomatal responses by regulation of stomatal numbers and sizes, and metabolic components, respectively [202]. Parlitz et al. [203] studied the ability of A. thaliana leaves to regreen after dark incubation with the aim to identify TFs that are involved in the regulation of early dark-induced senescence and regreening. Leaves recovered in two days shading, but longer periods of darkness result in irreversible senescence. Large-scale qRT-PCR analysis of 1872 TF genes revealed that 649 of them are regulated in leaves during normal development upon shading or re-illumination. Leaf shading triggered upregulation of 150 TF genes, some of which are involved in controlling senescence. Of those, 39 TF genes were upregulated after two days in the dark and regained preshading expression level after two days of re-illumination. A larger number of 422 TF genes were down-regulated upon shading.

Recently, Imam et al. [204] analyzed the gene targets for four TFs: FnrL, PrrA, CrpK, and MppG (RSP_2888), which are known or predicted to control photosynthesis in Rhodobacter sphaeroides. They identified 52 operons under direct control of FnrL, illustrating its regulatory role in photosynthesis, iron homeostasis, nitrogen metabolism, and regulation of sRNA synthesis. Using global gene expression analysis combined with ChIP-seq, they mapped the regulons of PrrA, CrpK, and MppG. They concluded that the Rrf2 family protein, MppG, plays an important role in photopigment biosynthesis as part of an incoherent feed-forward loop with PrrA.

While investigating photosynthesis and nitrogen availability in the accumulation of biomass, An et al. [205] cloned for the first time the GATA TF PdGNC from the fast-growing poplar clone NE-19. The GATA TF family plays important roles in chloroplast development and nitrogen metabolism. The overexpression results from Arabidopsis under high nitrate, sufficient nitrate, and low nitrate indicated that PdGNC improved photosynthetic capacity and plant growth under low nitrate levels; thus, it could potentially be used in transgenic breeding to improve nitrate utilization and plant growth rates under limited nitrogen conditions [205].

30.4.3 Regulation of Ribulose Bisphosphate Carboxylase/Oxygenase

A typical example of complementation of nuclear and cytoplasmic genes in chloroplast biogenesis of eukaryotic algae and higher plants is the regulation of the synthesis of ribulose 1,5-bisphosphate carboxylase/oxygenase (Rubisco), a key regulatory enzyme of photosynthetic carbon assimilation. Rubisco is the most abundant protein on earth, to which 50% of leaf nitrogen is allocated. It is the only enzyme that catalyzes the carbon assimilation for plant growth. Due to Rubisco's slow catalytic rate, low affinity for CO_2, and capacity to use O_2 for photorespiration, this enzyme has been regarded as notoriously inefficient [206], as this severely compromises photosynthetic carbon assimilation capacity of plants.

Using genetic analysis of reciprocal interspecific tobacco hybrids and mutants, it was shown that chloroplast RNA coded for the large subunit of the enzyme while nuclear genes controlled the formation of the small subunit [149,207]. Nasyrov [138] believed that the small subunit of Rubisco serves as an inducer of the transcription of the large subunit of mRNA. Multiple copies of the gene for the small subunit in the nucleus [170,208] appear to match the multiple chloroplast genomes. Regulation involves proteolysis in the organelle of newly imported small subunits when the large subunit is unavailable for assembly [209].

Transgenic tobacco (Nicotiana tabacum) plants that express RbeS antisense RNAs have been characterized as a tool to gain insight into regulatory interactions between the nucleus cytoplasm and chloroplast [210]. Jiang and Rodermel [211] reported that large subunit protein production is controlled post-transcriptionally in the mutants. They also reported that antisense plants do not adapt to genetic alterations in Rubisco content by significantly altering nuclear and chloroplast gene expression.

Higher plant Rubisco, being one of the largest enzymes with molecular weight of 560 kDa, has a hexadecameric L_8S_8 structure, comprising of eight large units and eight small units. Wealth of structures have now been reviewed [206]. A classic α/β-barrel is formed by the C-terminal domain of the large subunit, where the residues in the loop between α-helix and β-strand (at the mouth of the α/β-barrel) contribute to the active site. Based on the kinetic experiments from 100 species, it is clear that significant variations exist in the catalytic properties of Rubisco in land plants, indicating variations in photosynthetic capacities among plant species [212,213]. This natural variation may be harnessed, in principle, to improve Rubisco performance and thus crop productivity.

Rubisco activity is highly regulated in plant cells [214] by the carbamylation of critical active site residues such as a lysine, which forms a catalytically active complex with the help of Mg^{2+}. Another mode of regulating Rubisco activity is with tight binding of the naturally occurring, low molecular weight sugar-phosphate inhibitors. These inhibitors bind to the active site of the enzyme and prevent either carbamylation and/or binding of the substrate. A number of sugar-phosphate inhibitors have been characterized [215], of which

2-carboxy-D-arabinitol 1-phosphate (CA1P), found only in the chloroplast, formed during low irradiation and darkness, is the best characterized naturally occurring inhibitor that binds to carbamylated Rubisco active site. Rubisco-catalyzed reactions, during daytime, may misfire products such as D-glycero-2,3-diulose-1.5-bisphosphate (PDBP). These *misfired* products could also lock the carbamylated sites of Rubisco [215]. Removal of the tightly bound inhibitors is performed by the enzyme, Rubisco activase (RCA), and hydrolysis of ATP (for CA1P dephosphorylation is achieved with the help of CA1P phosphatase, CA1Pase). Tight binding of inhibitors to Rubisco may prevent its degradation during stress conditions because of an oxidized chloroplast environment.

30.4.4 RUBISCO ACTIVASE

RCA represents >5% of the total leaf protein, making it an abundant protein. It is a catalytic chaperone that belongs to the AAA+ (ATPases associated with a variety of cellular activities) superfamily of proteins. RCA remodels the configuration of Rubisco to release the latter's active sites from inhibitors, which makes it essential for activation and maintenance of Rubisco activity. By enforcing conformational changes in Rubisco and changing its activation state, RCA thus modulates photosynthesis by modulating the rate of electron transport to rate of CO_2 fixation. RCA has been detected in all the plant species in two isoforms, α and β [213,216]. Both forms can hydrolyze ATP and activate Rubisco. In many species, two isoforms are formed as a result of alternate splicing of the product from a single gene, whereas in some others (cotton, soybean), two separate genes may encode two isoforms. Nevertheless, proper redox regulation is needed for functioning of the holoenzyme (including regulation of Rubisco activation under irradiation).

Expression of RCA gene is long known to exhibit a strong circadian rhythm. Light induction of photosynthesis is correlated to modulation of Rubisco activation by RCA; RCA is sensitive to inhibition by ADP [216]. During the increase in irradiation, the induction of photosynthetic C assimilation is regulated by RCA [217]. Under fluctuating light conditions, the rate of photosynthetic induction and its translation into C-gain for plants may be under RCA's control. Taking these together, during the transition of low irradiation to high irradiation, photosynthesis induction rates are regulated by RCA via regulating the Rubisco activation states, which in turn is regulated by RCA's sensitivity to ADP. Under fluctuating light conditions, altering the regulatory properties of RCA may increase the rate of photosynthetic induction, thus enhancing photosynthetic performance [215,216].

30.4.5 LIGHT-REGULATED NUCLEAR GENES

A number of light-regulated nuclear genes are known that influence changes in transcription, mRNA, and protein abundance in higher plants [9–11]. Among these light-regulated genes, *Rbcs* and *Lhc* genes have been studied the most [11,218–222]. Multiple phytochrome genes have been identified in several species [223]. Sato [224] and Tomizawa et al. [223], who were working on pea, have documented in detail the multiple transcripts for a single phytochrome I gene. The *Arabidopsis* genome has been found to contain four to five phytochrome-related gene sequences [225], while genomic clones for three phytochrome loci have been isolated from maize [226]. In rice [227] and *Zucchini* [228], the existence of other genes encoding type II phytochrome has been suggested. In Petunia, the chloroplast proteins, the chlorophyll a/b binding (*cab*) proteins, and the small subunit of Rubisco are encoded by nuclear genes, which require light for expression. However, there is variation in the expressions among specific classes of *cab genes* [229].

Studies done so far suggest that light responsiveness in higher plants is the outcome of interaction between nuclear factors and cis-acting regulatory elements, which are generally located within the 5' upstream regions of light-regulated genes [11,221,230]. There are two types of DNA-binding proteins involved in light responsiveness [11]: one that interacts with the light-responsive regulatory elements (LREs) on the upstream region of photosynthetic genes, e.g., *Cyt1* factor binding to Box II and Box III of Rbc S-3A gene promoter of pea [231–233], and the other that modulates the level of expression conferred by LRE, e.g., ASF-2 and 3AF1 [233,234].

The cis-acting elements involved in light regulation of the nuclear gene (*Gap b*) encoding the B subunit of chloroplast glyceraldehyde 3-phosphate dehydrogenase have been characterized in *A. thaliana* [235]. Kwon et al. [235] showed that a 664-bp *Gap B* promoter fragment is sufficient to confer light induction and organ-specific expression of the *Escherichia coli* β glucuronidase reporter gene (*Gus*) in transgenic tobacco (*N. tabacum*) plants. Deletion analysis indicated that the −261 to −173 upstream region of the *Gap B* gene containing four direct repeats with the consensus sequence 5'-ATGAA (A/G) A-3' (Gap Boxes) was essential for light induction. Using gel mobility shift assays, Kwon et al. [235] identified a nuclear factor from tobacco plants that binds specifically to the repeated elements within regions that are essential for light regulation of Gap A and Gap B genes of *A. thaliana*. This binding factor was found distinct from the GT-1 factors, which binds to the Box II and Box III within the LRE of the *Rbc S-3A* gene of pea encoding the small subunit of Rubisco [231–233,236,237].

Pfannschmidt et al. [238] reported that photosynthetic control of chloroplast gene expression indicates an evolutionary explanation: the redox signal-transduction pathway can be short, the response rapid, and the control direct. They showed that the redox state of plastoquinone also controls the rate of transcription of genes encoding reaction-center apoproteins of PS I and PS II. As a result of this control, the stoichiometry between the two photosystems changes in a way that counteracts the inefficiency produced when either photosystem limits the rate of the other. In eukaryotes, these reaction-center proteins are encoded universally within the chloroplast [238]. Tyagi and Gaur [239] reviewed light regulation of nuclear photosynthetic genes in higher plants. They reported that signaling components can either directly influence the binding of light regulatory *trans*-acting factors to *cis*-acting elements present in a photosynthetic gene promoter or modulate their activity by various means to facilitate transcription in response to light. Some *cis*-acting elements, designated as light-responsive regulatory elements (LREs), show a high degree of conservation among photoresponsive nuclear genes in plants. Overall, gene regulation seems to involve interplay of several *cis*-acting elements and regulatory factors [239].

30.5 EXTRANUCLEAR CONTROL CHLOROPLASTS

Chloroplasts, the vital cell organelles for photosynthesis, are semiautonomous bodies having their own DNA. The mature chloroplasts are composed of membranous grana (thylakoids) and the stroma. The components of genetic machinery of chloroplasts are found in the stroma. Despite having their own DNA, the majority of chloroplast components are encoded by nuclear genes. For photosynthesis, genetic information flows from several hundred genes localized both in nucleus and chloroplasts. In order to integrate photosynthetic activity in cell metabolism, it is possible that in the course of evolution, chloroplasts acquired a double genetic subordination to the nuclear genome and to their own semiautonomous one [138]. The use of chloroplasts to study photosynthesis and the intricacy of photosynthetic complexes has yielded new information in the control of organic gene expression and the communication of different genomes in eukaryotic cells [139].

30.5.1 INHERITANCE

Chloroplast DNA is usually inherited maternally [240]. This fact was observed by Correns [241] and Baur [242] in the first decade of this century. Correns [241] observed it for pigmentation in the progenies of normal and variegated mutants of *Mirabilis jalapa* and postulated the principle of maternal or cytoplasmic inheritance (*Status albomaculatus*). Baur [242] demonstrated biparental inheritance (*Status para albomaculatus*) in *Pelargonium zonale* where both parents contributed to plastid inheritance. These early observations of non-Mendelian inherited mutant phenotypes affecting chloroplasts in higher plants suggested the existence of genetic material in chloroplasts [243].

Although the inheritance of chloroplast DNA genome in angiosperms is maternal in general, exceptions of paternal or biparental inheritance have been reported in several angiosperm taxa [244–251]. Paternal plastid DNA (pt DNA) inheritance in angiosperms is a rare event and has been demonstrated in intraspecific hybrids of *Medicago sativa* [247,252] and in interspecific hybrids between *Daucus muricatus* and *D. carota ssp. sativus* [253]. Recently, it has been demonstrated in the hybrids of *Zantedeschia odorata* × *Z. aethiopica* [250] and two crosses of *Stellaria porsildii* and *S. longifolia* [249]. In contrast to the angiosperms, the majority of gymnosperms have paternal pt DNA inheritance, e.g., the genera *Pinus* [254], *Larix* [255], *Pseudotsuga* [256], and *Picea* [257]. On the basis of cytological and physiological mechanisms, which

cause uniparental or biparental inheritance, Hagemann and Schroder [258] defined four plant types: *Lycopersicon* type, *Solanum* type, *Triticum* type, and *Pelargonium* type. Out of these, only the *Pelargonium* type exhibited biparental plastid inheritance, while the other three showed uniparental plastid transmission. The absence and presence of plastid DNA (pt DNA) in generative or sperm cells generally correlated with maternal and biparental pt DNA inheritance, respectively [259]. Based upon detection of pt DNA in generative or sperm cells, Corriveau and Coleman [259] reported that 18% of 235 species studied were potentially capable of biparental pt DNA inheritance. Similarly, out of 398 species studied, 27% were reported to possess potential for biparental pt DNA inheritance [260]. The percentage of progeny containing biparental pt DNA varies from species to species. If more than 5% of the progeny contain biparental pt DNA, it is defined as regular biparental pt DNA inheritance, whereas less than 5% of progeny with biparental pt DNA is defined as occasional biparental pt DNA inheritance [261]. In hybrids containing biparental plastids, the two types of plastids usually segregate, with different tissues containing a single plastid type. This is called *plastid sorting out*, which is usually apparent by the plant having green and white sectors in leaves. This type of variegation has been observed in *Pelargonium* [262] and *Stellaria* [249], and in the *hybrids* of *Zantedeschia odorata* × *Z. aethiopica* [250]. Further evidence for this comes from restriction fragment length polymorphism (RFLP) analysis using DNA from different shoots of one *Medicago* plant [247].

The inheritance of plastid DNA is usually determined through the use of a parent having a plastome-encoded chlorophyll deficiency in reciprocal crosses. Biparental pt DNA inheritance is detected when green and white sectored leaves appear in the progeny. By this method, biparental pt DNA inheritance has been reported in several genera including *Pelargonium* [263,264], Medicago [265,266], and *Oneothera* [267]. Recently, RFLP with closed pt DNA as a probe has been used to study pt DNA inheritance in a number of cases such as Douglas fir [256], Larix [255], *Medicago sativa* [247], *Pinus banksiana* [254], alfalfa [252], *Zantedeschia* [250], and *Stellaria* [249].

Different factors play significant roles in controlling pt DNA inheritance in different angiosperm species, for example, the paternal genome in *Petunia hybrida* [268] and *Pelargonium* species [264,269] or the maternal [252] or paternal genome [265] in *Medicago* species, the plastome in *Oneothera* species [267], and plastome–genome incompatibility in *Zantedeschia* [250].

McCauley et al. [270] studied inheritance of cpDNA in *Silene vulgaris* and suggested that chloroplast DNA (cpDNA) is maternally inherited in the majority but not all angiosperm species. The mode of inheritance of cpDNA is a critical determinant of its molecular evolution and of its population genetic structure. Likewise, analyzing the inheritance of the chloroplast genome in *Passiflora* by investigating both interspecific and intraspecific crosses, it was concluded that of 11 crosses, 2 intraspecific crosses demonstrated maternal inheritance, whereas the 9 interspecific crosses had paternal inheritance

[271]. All interspecific crosses had primarily paternal inheritance, whereas all intraspecific crosses had primarily maternal inheritance [271].

30.5.2 Plastid Genome Organization

Since the discovery of the existence of genetic material in chloroplasts, researchers have demonstrated the presence of a unique DNA species in the chloroplasts of higher plants and algae, which codes for several chloroplast proteins. Plastid genome in vascular plants and algae is a circular DNA molecule with an average contour length of about 50 μm and in general are 120–180 kbp in size [272]. A dominant feature of most chloroplast DNA is the presence of two large inverted repeats ranging from 10 to 80 kbp dividing the chloroplast chromosome into two unequal parts: a large single copy region (70–85 kbp) and a small single copy region (10–20 kbp). Due largely to the variation in the sizes of inverted repeats, the chloroplast DNA in different species varies from 120 to 180 kbp [273]. Genes coding for rRNAs and tRNAs were the first to be identified on the chloroplast genome through hybridization of unlabeled chloroplast DNA and 32p-labeled chloroplast rRNA [274,275] and I-labeled chloroplast tRNA [276].

The organization and composition of DNA within photosynthetic and nonphotosynthetic plastids are similar [274,277]. The DNA molecules are associated with the internal thylakoid membrane in the aggregate of 10–20 DNA molecules [278]. In wheat, the DNA copies per plastid have been found to increase during chloroplast development [279]. Chloroplast DNA is also reported to exist in rings or in a more dispersed arrangement [278–280]. About 140 open reading frames (ORFs) have been found to exist on both the strands of chloroplast DNA, e.g., in tobacco [273], which code either for a polypeptide of more than 30 amino acids or for ribosomal and transfer RNAs.

In the 1980s, the available information about the plastid genomes and genes indicated that the plastid genome codes for more than 100 genes [217], which could be grouped into three types: (1) genes involved in photosynthesis, (2) genes for chloroplast gene expression, and (3) putative sequence for an NADH oxidoreductase complex and other unidentified genes [281–283]. The understanding about plastid genomes and genes grew quite rapidly, and the complete sequences of the plastid genomes to be first known were in *N. tabacum* [283], *Marchantia polymorpha* [282], and *Oryza sativa* [284]. In the past few years, substantial information has been gathered about the physical and genetic structure and the regulatory mechanism of gene expression of chloroplast genome. Prior to 2004, only seven published crop chloroplast genomes were available. This number increased to 23 in the next two years [285]. In the same year, the first complete chloroplast genome sequence for a member of the Rutaceae and Sapindales was reported [286]. Recently, chloroplast genome (871 bp in length) of a medicinal plant *Datura stramonium* was reported [287]. So far, approximately 400 complete cp genome sequences have been sequenced in GenBank [287]. Most of these are from economically important plants, such

as *Solanum lycopersicum*, *O. sativa*, and *N. tabacum*. On the other hand, only a few cp genome sequences are reported for medicinal plants such as *Salvia miltiorrhiza* and *Panax ginseng* [287]. These studies provide valuable information for phylogenetic and cp genetic engineering for future use to mankind.

30.5.3 Genes for Photosynthesis-Related Proteins

Much information is now available about the chloroplast genes coding for protein that are directly involved in photosynthesis such as those coding for a large subunit of Rubisco (*rbcL*), polypeptides of PS I and II, cytochrome b_6/f complex, genes coding for ATP synthase complex, etc. Chloroplast genome codes for the large subunit of *rbcL* [207,288] that is composed of 8 large and 8 small subunits. The *rbcL* gene in higher plants and *Chlamydomonas* is smaller (1.8 kb) and contains no introns, while in *Euglena*, it is relatively larger (6.5 kb) and contains 9 introns each of 0.5 kb [289].

All the 12 polypeptides of the oxygen-evolving apparatus of PS II (P 680 core complex) are coded by the chloroplast genome in the large single copy region. In contrast, the light-harvesting part of PS II (chlorophyll a/b-binding proteins) is encoded in the nucleus. Several PS II genes coded by plastid DNA are known. These are *PsbA* [275,290–293], *PsbB* [294], *PsbC* [295], *PsbD* [295–297], *PsbE* and *F* [298], *PsbG* [299], *PsbH* [300,301], *PsbI* and *J* [302–305], *PsbK* [306], *PsbL* [307], and *PsbM* and *N* [308]. Among these, the association of PsbG [309] and *PsbI* with PS II has not been established [273].

30.5.3.1 Cytochrome b_6/f Complex

Of the various components of cytochrome b_6/f complex, the genes for cyt f (*Pet A*), Cyt b6 (*Pet B*), and subunit IV (*Pet D*) are coded by chloroplast genome and were first identified in spinach chloroplast genome by Alt et al. [169]. The cytochrome b_6f complex is the smallest redox-active complex of the photosynthetic electron transport chain. Its structure has been elucidated in the cyanobacteria *Mastigocladus laminosus* [310] and *Nostoc* sp. PCC 7120 [311], and in the eukaryotic alga *Chlamydomonas reinhardtii* [312]. The cytochrome b_6f complex functions as a dimer with a total molecular mass of ~220 kDa, and each monomer consists of eight different subunits [311,313]. Eight subunits of the cytochrome b_6f complex is reported. The small M subunit (PETM) and the Rieske 2Fe2S protein (PETC) are nuclear encoded, while the other six are encoded in the chloroplast genome (plastome) [179]. Except for the FNR and the small chloroplast-encoded L subunit (PetL), which is located at the periphery of the complex and only plays a role in complex stability [179], all other subunits are essential for the function and accumulation of the cytochrome b_6f complex [314].

30.5.3.2 Genes for PS I

Of the 13 polypeptides constituting the core complex of PS I, only 5 (*Psa A*, *Psa B*, *Psa C*, *Psa I*, and *Psa J*) are present in chloroplast genome [180]. The genes *Psa A* and *Psa B* first detected in the large single copy region of spinach chloroplast [315] and sequenced in maize [316] constitute an operon along with a ribosomal gene *rps* 14 [316,317]. The gene *Psa C* was first detected in tobacco [318], *Psa I* in barley [319], and *Psa J* in cyanobacterium [320].

Photosystem stoichiometry adjustments involve regulation of genes of both PS I and PS II reaction centers. However, in *Arabidopsis* and pea (*Pisum sativum*) chloroplasts, only PS I genes seem to be regulated at the level of transcription. Chloroplast sensor kinase (CSK) is reported to communicate the redox state of plastoquinone to the chloroplast transcriptional apparatus and thus initiates an appropriate change in photosystem stoichiometry [321]. Downstream events involving interactions between CSK and candidates for its cognate transcriptional regulator involve plastid transcription kinase (PTK), which is a protein kinase that catalyzes phosphorylation of proteins that serve as regulatory cofactors of the chloroplast RNA polymerase [322,323]. Another interaction partner of CSK is chloroplast sigma factor-1 (SIG-1) [324], which is phosphorylated under PQ oxidizing conditions, when incident light favors PS I (light 1) [324]. Phosphorylated SIG-1 represses transcription at psa (PS I reaction center) promoters while efficiently transcribing psb (PS II) promoters [324]. Puthiyaveetil and Allen [321] proposed that CSK is the SIG-1 kinase since CSK function and SIG-1 phosphorylation have the same gene expression phenotype, which is suppression of psa genes [321,324]. In addition, CSK and SIG-1 interact *in vivo* in yeast, and oxidized PQ, the signal for sigma factor phosphorylation, is also the signal that promotes the kinase activity of CSK.

30.5.3.3 Genes for ATP Synthase Complex

The genes coding for six polypeptides of ATP synthase, three each belonging to the outer surface of thylakoid membrane, CFI ($\alpha,\beta\epsilon$), and within the thylakoid membrane, CFO (I, III, and IV), are located on the chloroplast genome [325–328]. These chloroplast encoded genes are the component of the two operons (large and small), which are represented by several thousand base pairs [273]. The larger operon is composed of genes *atpI*, *atpH*, and *atpF*, which have been characterized in spinach and pea [328,329]. The smaller operon consists of *atpB* (498 codons) and *atpE* (137 codons) genes and has been sequenced in maize [330] and spinach [290].

A. thaliana carries two genes that appear to code ATP synthase γ subunits, ATPC1 and ATPC2 (At4G04640 and At1G15700), located on chromosomes 4 and 1, respectively [331,332], and showing 73% sequence similarity [331]. Both *Arabidopsis* γ homologues contain the domain responsible for the redox regulation of the enzyme [331]. Using a series of mutants, it was shown that both these subunits can support photosynthetic ATP synthesis *in vivo* with similar specific activities, but that in wild-type plants, only γ_1 is involved in ATP synthesis in photosynthesis [333]. The γ_1-containing ATP synthase shows classical light-induced redox regulation, whereas the mutant expressing only γ_2-ATP synthase (gamma exchange-revised ATP synthase, *gamera*) shows equally high ATP synthase activity in the light and dark [333].

The *pgr5* mutant of *A. thaliana* suffers from an impaired thylakoid lumen acidification and cannot properly induce photosynthetic control at the cytochrome b_6f complex. The molecular basis of the impaired lumen acidification in *pgr5* has been suggested to result either from a reduced cyclic electron flow [334,335] or from an impaired metabolic feedback inhibition and deregulation of ATP synthase activity [336]. In response to changing metabolic ATP and NADPH demands during environmental fluctuations and leaf development, strong adjustments of the cytochrome b_6f complex and ATP synthase are observed [179]. These changes in complex content and activity closely correlate with linear electron flux and leaf assimilation capacity, suggesting that the cytochrome b_6f complex and ATP synthase are the predominant points of photosynthetic flux control [179]. A precise coregulation of ATP synthase and the cytochrome b_6f complex is essential to ensure an optimized balance between light utilization for photosynthesis and growth and harmless dissipation of excess light by photoprotective mechanisms [179].

30.5.3.4 Other Genes in Chloroplast Genome

Genes for NADH dehydrogenase and initiation and elongation factors are also coded by chloroplast genome. However, many chloroplast DNA nucleotide sequences studied so far have unknown functions. It has been suggested that many of these might be coding for proteins related to chloroplast organization and function [273]. Nucleotide sequences of chloroplast DNA coding for amino acid sequences resembling the components of NADH dehydrogenase in the respiratory chain of human mitochondria have been reported in tobacco [283] and other plants [282,284,337,338].

30.5.4 Chloroplast Gene Expression

Chloroplasts are derived from small undifferentiated proplastids present in meristematic cells. The genes for chloroplast development and differentiation reside both in chloroplast as well as nucleus. Besides, the genes for photosynthesis are also present in these two cell components. Thus, chloroplast gene expression needs coordination with nuclear gene expression at various stages of growth and development [139,177,281,339,340]. Early efforts in this direction placed emphasis on the transcription of genes for photosynthetic proteins and tRNAs, which recently shifted to post-transcriptional and translational regulatory mechanisms [139]. Two component regulatory systems comprising redox sensors (electron carriers that initiate control of gene expression upon oxidation and reduction) and redox response regulators (DNA binding proteins that modify gene expression due to the action of redox sensors) that respond to changes in redox potential have recently been discovered in bacteria [341].

In the chloroplasts of higher plants, genes can be transcribed by three different RNA polymerases. Two RNA polymerases, called RPOTp and RPOTmp, are nuclear encoded, monomeric, and evolutionarily related to bacteriophage RNA polymerases. The third RNA polymerase, called the *plastid-encoded RNA polymerase* (PEP), consists of several chloroplast-encoded subunits and is related to eubacterial RNA polymerases (reviewed in Ref. [342]). In higher plants, its activity is regulated by at least six nuclear-encoded sigma factors (reviewed in Refs. [343–345]). Additionally, >10 other nuclear-encoded regulatory proteins associate with the PEP and, together, they constitute the *transcriptionally active chromosome complex*. They may have structural and regulatory functions, thus enabling nuclear control of chloroplast gene expression [346].

30.5.4.1 Post-Transcriptional Regulation

Transcription has been shown to be an important regulatory component for chloroplast gene expression. However, discrepancies between transcription rates and accumulation of messages during the development of chloroplasts suggest that post-transcriptional controls, including RNA processing and stability, play major roles in chloroplast gene regulation [347–352].

The similarity in the relative initiation rates of chloroplast genes coupled with significant differences in their RNA populations [347,348,353] suggests that relative promoter strength controls most chloroplast transcription units. In mustard, it was found that protein from etioplasts and chloroplasts binds differentially to the promoter region of *psbA*, indicating that a *trans* acting factor may play an important role in the regulation of gene expression in plastids [354].

In chloroplasts, the post-transcriptional regulation of differential mRNA accumulation may occur at the level of mRNA maturation and/or at the level of mRNA stability. In spinach, inverted repeats in 3′ untranslated region, which flank plastid protein coding genes in both mono and polycistronic transcription units, are processing signals for precise mRNA 3′ end formation [355,356] and are also required for stabilization of 5′ mRNA segments [355,357–359]. It has been shown that, in spinach, the half-lives of *psbA* and *rbcL* mRNAs are relatively longer than the known half-lives of bacterial mRNA, and for both, mRNA polysomic association is critical to facilitating specific mRNA degradation [359].

Differential RNA processing and splicing of pre-mRNAs also constitute integral post-transcriptional control steps [347–349,351,352]. A well-known example of complex processing of plastid mRNAs is the chloroplast operon *psbB–psbH–pet B–pet D* (encoding proteins for PS II and the cytochrome b₆f complex) and *rps2–atp1–atp H–atp F–atp A* (for the ribosomal protein S₂ and subunit of the ATP synthase) [329,360–362].

Chloroplast RNA-binding proteins (RBPs) were first purified from tobacco and are postulated to be involved in the post-transcriptional processing of chloroplast RNAs [363]. Recently, a spinach chloroplast RBP was shown *in vitro* to be required for processing of the 3′ ends of chloroplast messages [364]. Similarly, a cDNA clone from *Arabidopsis* encoding a chloroplast RBP that possesses same structural organization as all known chloroplast RBPs and is capable of binding single stranded DNA has been hypothesized to participate in post-transcriptional events of chloroplast gene expression [365].

Despite their structural homology, the genes encoding chloroplast RBPs from various species exhibit distinct patterns of expression. For example, in tobacco, three of the five genes express substantially higher mRNA levels in leaves than in roots [363]; in spinach, expression of RBP gene (28 rnp) is highest in leaves followed by cotyledons and stems [364]; the maize RBP message accumulates only in leaves [366], while the expression of RBP gene (*Atrbp*) in *Arabidopsis* is highest in leaves and lowest in roots [340]. The different patterns of expression may reflect different functions of each member, which is supported by the fact that each member contains a highly diverged, unique acidic region. The function of these acidic domains is unknown, but they have been suggested to be involved in protein–protein interaction [367,368]. It has been demonstrated that post-transcriptional processing and splicing of pre-mRNAs are complex events and require many ribonucleotide proteins such as the heterogeneous nuclear ribonucleoprotein splicing complex [369]. Therefore, it has been suggested that the functional specification of each chloroplast RBP may be achieved in part by interacting with other processing proteins through acidic domains [340].

Genes encoding proteins with only one RNA-binding domain have been isolated from maize [370], tobacco [371], and *Arabidopsis* [365]. The *Arabidopsis* chloroplast RBP gene (*Atrbp 33*) is nuclear-encoded and has been shown to encode as many as nine messages [340]. Multiple messages from a single RBP gene are common in other organisms and are produced through a combination of alternative transcription start sites and differential splicing (for review, see Refs. [369,372]). This is a mechanism by which a single gene can encode multiple proteins with specific functions in RNA processing, thereby indicating the need for a much lesser number of individual genes [340].

Most plastid-encoded genes are cotranscribed and form polycistronic mRNAs and may also contain introns [179]. In general, polycistronic transcripts undergo a complex post-transcriptional maturation, but in a few cases, individual nucleotides are modified via RNA editing (reviewed in Refs. [345,373,374]).

30.5.4.2 Translational Regulation

Translational and post-translational regulation of chloroplast genes are quite complex and also require the role of both nuclear and chloroplast genomes. The importance of translational and post-translational regulation in plastid gene expression is evident from the lack of correspondence between plastid mRNA and proteins levels in different growth conditions (light or dark), tissues (e.g., chromoplasts in fruits), and mutants that affect photosynthetic functions [139].

In vascular plants, there are a number of reports on translational control in plastids [350], which suggest that chloroplast ribosomes are quite similar to bacterial ribosomes [139]. Evidence is also available to indicate that translation of certain plastid mRNAs is controlled directly or indirectly by light. For example, this has been observed for the translation of the large subunit protein of Rubiscos [375] and D I reaction center protein of PS II [376]. It has been suggested that chlorophyll a, the pigment associated with PS I and II reaction center and light-harvesting proteins, may be needed to release an inhibition of chain elongation [377]. Thus, biosynthesis of chlorophyll a may also be a regulatory factor in translation of specific plastid mRNAs [139].

After the formation of a mature mRNA, it is translated by the 70S ribosomes of the chloroplast, which often requires additional transcript-specific nuclear-encoded translation initiation factors. For instance, in green algae *C. reinhardtii*, gene expression is controlled at the level of transcript stability and translation initiation, while transcriptional regulation and mRNA maturation appear to be less important [378–381]. However, in contrast to higher plants, green algae contain only the eubacterial-type RNA polymerase with only a single sigma factor, indicating that substantial differences in the complexity of transcriptional regulation may exist between algae and higher plants [382,383].

30.6 PHOTOSYNTHESIS AND PLANT STRESS RESPONSES

In their agroecological and ecological habitats, plants constantly face a plethora of abiotic (such as high light, water deficiency/flooding, cold, high temperature, salinity, etc.) as well as biotic (such as pathogens, herbivores) stresses. These stresses negatively affect the productivity of plants due to reduced growth and photosynthetic efficiency. An important question for discussion is whether the decrease in photosynthesis is an adaptive mechanism to tide over stress, thus contributing to tolerance mechanism of plants, providing enough plasticity, or whether it is a consequence of stress leading to severe fitness losses.

30.6.1 ABIOTIC STRESS

Abiotic stresses are highly limiting factors in crop productivity in agricultural systems, where decrease in photosynthesis is a well-established phenomenon. Such reduction in photosynthetic rates may be due to decreased CO_2 availability caused by diffusion limitations through the stomata and the mesophyll [384,385], stomatal inhibitions, suppression of Rubisco activity, reductions of Rubisco levels, RCA activity, and sensitivity of PS II [386]. Secondary effects, namely, oxidative stress, occur and are mostly present under multiple stress conditions [387] and can seriously affect photosynthetic machinery [388].

Photosynthesis is among the primary processes to be affected by drought and salinity [389]. Both salt and drought stresses lead to down-regulation of photosynthetic genes [389]. However, compared to drought, salt stress affected more genes and more intensely, possibly reflecting the combined effects of dehydration and osmotic stress in salt-stressed plants [389]. Salt stress may also lead to damage in PS II in wheat and barley [386,390]. In rice, SUV3 dual (DNA and RNA) helicase, upon overexpression, has been discovered to provide salinity stress tolerance by antioxidant machinery and improving photosynthesis [391]. In soybean, salinity stress

decreases the photosynthetic ratio; Rubisco activity decreases because of the decrease in Rubisco activation states; salinity negatively affects the carboxylation process. High salt stress also profoundly impacts accumulation of RCA protein [392]. Due to roles detailed in Section 32.4, Rubisco has been regarded as an ideal protein to study the role of stresses on photosynthesis metabolism.

Heat/cold-dependent increase in active oxygen species in light promotes oxidative modification of critical residues on Rubisco, tagging it for degradation [214]. Oxidative stress reduces transcript accumulation of the Rubisco small subunit. Rubisco turnover may be an important step in photosynthetic response of plants under abiotic stress. For example, under drought conditions, expression of a large subunit is negatively affected due to the increase in expression of glutamine synthase. An earlier hypothesis that Rubisco activity during heat shock may be maintained with the help of molecular chaperone RCA has now been critically questioned. Biochemical evidences show that, on one hand, Rubisco is an exceptionally heat-stable protein, whereas RCA is equally heat liable. It has been shown that elevated temperatures inhibit photosynthesis because of temperature sensitivity of RCA. Compelling evidences support the hypothesis that during elevated temperature stress, net photosynthesis is limited by Rubisco activity, which is governed by the activity of RCA. RCA expression during heat stress may be regulated post-transcriptionally [393,394].

30.6.2 BIOTIC STRESS

Like abiotic stresses, biotic stresses such as infection from pathogens and attack from insect herbivores also decrease photosynthesis [395,396]. Metagenomics approaches employing microarray studies on 22 biotic damages on eight plant species confirm that down-regulation of photosynthesis is a universal phenomenon when any biotic agent—insects, fungi, bacteria, or viruses—damages plants. During this process, transcripts of genes of light reaction, carbon reduction cycle, and pigment synthesis are reduced irrespective of the type of biotic attack [396].

When wheat plants are attacked by the Blast pathogen, net carbon assimilation rates, stomatal conductance, and transpiration rate are reduced and photosynthesis is severely impaired [397]. These reductions in photosynthetic efficiency could not be attributed to diffusive limitations; rather biochemical changes such as reduced Rubisco activity were regarded as major factors [397]. During infection of pathogens, major reduction in pathogenic proteins is recorded. Reducing photosynthesis has been regarded as one of the effective defense mechanisms during pathogen attack [398]. During pathogen attack (such as bacterial citrus cancer), the decrease in the expression of genes responsible for photosynthetic efficiency such as Rubisco, RCA, and ATP synthase and the increase in NADPH dehydrogenase accumulation are recorded [398]. Such reduction in photosynthetic proteins is also believed to act in defense by reducing the nutrient content of leaves for feeding of biotrophic pathogens. Moreover, the Rubisco small subunit has been implicated in tobamovirus movement and plant antiviral defenses [399].

Attack from insect species also reduces photosynthetic capacity of plants. Indeed, reductions in photosynthetic capacities during herbivore attack may be much more than what would be predicted due to removal of canopy area by herbivores [395]. This may be attributed to reduction of accumulations of proteins involved in photosynthesis. Degradation of Rubisco in response to attack from herbivore and pathogen as well as chilling is documented [400–403]. Further, RCA has been shown to be one of the primary proteins down-regulated during herbivore attack [404]. Studies of knocking-down expression of Rubisco and RCA independently in wild tobacco plants had different effects on herbivory during attack from generalists and specialists [405]. While RCA-silenced plants were susceptible to both generalists and specialists, Rubisco-silenced plants were susceptible to specialists. It has been concluded that plant herbivory is influenced by the mechanism of reduction of photosynthesis rather than just the magnitude [405].

30.6.3 GROWTH-DEFENSE PARADIGM

It is believed that biotic stresses divert allocation of resources from growth to energetically demanding processes of production of defenses, which may be targeted toward resistance to stress or tolerance mechanisms. Here, down-regulation of photosynthesis is a conserved feature. Reduced photosynthesis may reduce plant growth and may also limit resource availability for attacking agents. On the other hand, resource-intensive (carbon [C], nitrogen [N], ATP, etc.) production of defense metabolites would require robust photosynthesis demands. Such tradeoffs between stress–photosynthesis–growth and defense that requires transition from growth-oriented to defense-oriented metabolism presumably evolved to maximize plant fitness [406].

Growth–defense tradeoffs have been interpreted in terms of carbon-nutrient balance hypothesis as well as growth differentiation balance hypothesis. Broadly, these hypothesize that growth is governed by limitations in C and/or N, while photosynthesis remains unchanged: during reduced growth, the more abundant resource is invested in defense. They also presume that growth–defense tradeoffs should be observed under resource-abundant situations. Contrary to this, trade-offs are found between photosynthesis and defense [407].

Optimal defense (OD) theory is the best experimentally supported hypothesis explaining plant defense strategies. The OD theory states that because costly plant resources may be invested in growth and fitness-related traits or in defense-related traits, plant defenses, although adaptive, have allocation costs, thus favoring evolution of inducible defenses to minimize this tradeoff [408,409]. These tradeoffs are thought to be more stringent in resource-limited conditions, especially for N. One of the major sources of N is thought to be Rubisco, which is thought to also serve as N-storage protein [410]. Decrease in Rubisco activity has been associated with *dramatic changes of nitrate metabolism* in cultivated tobacco

when plants are grown under unstressed conditions [411]. On the other hand, ^{15}N-pulse labeling experiments in wild tobacco species subjected to herbivory stimulus show that N required for defense metabolites does not come from Rubisco but rather from recently assimilated N [412,413]. On the other hand, consistent with resource-based growth–defense tradeoff, silencing of RCA in wild tobacco impairs photosynthesis, growth, as well as jasmonic acid-isoleucine-based herbivore defenses: RCA may attenuate JA-Ile signaling during herbivory [414]. Indeed, JA-signaling, which imposes significant costs on plant, has been implicated in temporal dynamics and suppression of photosynthesis and growth [406]. Crosstalk of phytohormones is emerging as a major component in the balancing act of growth–defense tradeoffs [415]. RCA is proposed to play additional roles of mediating growth-defense tradeoffs and allowing plants to anticipate constraints in resources before they actually happen when plants are under attack [414].

30.7 PHOTOSYNTHETIC STRATEGY IN CROPS

An adequate food supply is the first necessity for a decent life [416]. Global food production will need to increase by more than 50% before 2050 to satisfy the food and fuel demands of an increasing population [417]. During the last century, crop yields, especially those of cereals, have increased dramatically. This increase over the past few decades has been attributed primarily to an efficient partitioning of resources to the economic parts and not to the increase in the biomass of crops [418]. Now, it is being argued by many that productivity of some crops has attained a plateau and that further improvement may be difficult to achieve by presently available technologies and breeding methods [418]. It has been suggested that breeding progress might be accelerated considerably if physiological, biochemical, and morphological characteristics were used as selection criteria [419]. Although more than 90% of crop biomass is derived from photosynthetic products, increasing photosynthetic capacity and/or efficiency has not yet been properly addressed by crop breeders [417]. The new green revolution will need identification of specific targets that will directly improve leaf photosynthesis [417].

Photosynthetic efficiency in crop plants is the primary component of dry matter productivity. The final biological or economical yield can theoretically be increased either by increasing the amount of photosynthesis, by reducing wasteful respiration, or by optimizing assimilate partitioning [17]. Photosynthesis, being influenced by many biochemical, physiological, anatomical, and morphological factors, needs perfect tuning of all these factors for an enhanced and consistent biomass production in crops. However, most attempts by plant breeders to select for an increased rate of photosynthesis have so far failed to result in enhanced biomass production [2,420].

Photosynthetic efficiency can be improved by increasing the efficiency of this process either at the level of light capture or at the level of CO_2 fixation [421]. Plant breeders started working for increased photosynthetic efficiency for increasing plant biomass and productivity after it was realized that considerable genetic variability exists for photosynthetic efficiency between and within species. Heat stability of RCA has been shown to be an important determinant of heat stability of photosynthesis [422]. Researchers have tried to manipulate photosynthetic efficiency by following hybridization both within and between C_3 and C_4 plants.

There is strong evidence for occurrence of genetic variation in both photosynthetic efficiency and capacity in cereals [423]. Basic research in photosynthesis has also confirmed that substantial improvements are theoretically possible [424,425]. In fact, to achieve a significant increase in yield in crops like wheat, anthesis biomass and photosynthetic provision of carbon to the filling grain must be improved [426].

Increasing photosynthetic capacity and efficiency is an important objective of wheat yield consortium (WYC) wherein the effort is being focused on phenotypic selection for photosynthetic capacity and efficiency, capturing the photosynthetic potential of spikes, optimizing canopy photosynthesis and photosynthetic duration, chloroplast CO_2 pumps, optimizing RuBP regeneration, improving the thermal stability of RCA, and replacement of large subunit Rubisco [423]. Yamori [417] suggested seven ways to enhance photosynthesis in plants:

1. Improving the Rubisco function by improving Rubisco catalytic activity and altering Rubisco amount per leaf area
2. Increasing the thermostability of RCA to sustain Rubisco activity at high temperature
3. Enhancing CO_2 concentration around Rubisco to maximize catalytic rate and minimize photorespiration by turning C_3 plants into C_4 plants, and installing algal or cyanobacterial carbon-concentrating mechanisms (CCM) into C_3 plants, redesigning photorespiratory metabolism, and improving CO_2 transfer pathways via stomata and/or mesophyll cells
4. Enhancing chloroplast electron transport rate
   i. Improving whole chain electron transport
   ii. Modifying light-harvesting systems
5. Enhancing enzyme activity of the Calvin cycle (e.g., SBPase)
6. Enhancing the capacity of metabolite transport processes and carbon utilization
7. Other researches such as QTL analyses for manipulation of mitochondrial respiration and improving photosynthesis under fluctuating light conditions

30.7.1 MANIPULATION OF PHOTOSYNTHETIC AREA

Most of the successful attempts for increasing photosynthetic efficiency have been toward changing the light-intercepting area of crops and duration of this area in the active stage. This has been successfully achieved in cereals (e.g., wheat and rice), pulses (e.g., pea), and other crops (e.g., sugar beet) as well. In comparison to old varieties, modern varieties of wheat possess more erect and a reduced number of leaves (in Australia) and wider leaf blades (in North America), while in barley, modern

varieties possess reduced leaf number (in the UK) (see review by Feil [419]). Modern Indian wheat varieties also possess relatively greater photosynthetic efficiency through increased flag leaf area and optimization of orientation and the number of leaves. Among the most popular varieties of wheat in India, HUW 234 possesses a relatively larger flag leaf area, while HUW 206 variety has slowly senescing leaves, which remain green until a very late stage. Similarly, modern rice varieties having erect and slowly senescing leaves are good examples of increased photosynthetic efficiency. Pea varieties having increased tendrillar growth (e.g., Indian HUP-2 variety) present another example of better light interception capacity of crop varieties. Selection of *Lolium perenne* genotypes for leaf and tiller insertion angle and relatively more rigid leaves resulted in a 30% increase in dry matter production [427]. The selected genotypes in perennial ryegrass were also found to show considerable reduction in respiratory losses under field conditions with a consequent 13–20% increase in dry matter production [428,429]. Recently, Yamori [417] suggested that there is a possibility to improve photosynthesis by altering the Rubisco amount per leaf area.

30.7.2 BREEDING FOR C_4 MECHANISM

Crops of C_4 species are about 40% [430] to 50% [431] more efficient than those of C_3. Therefore, incorporation of the C_4 characteristic was suggested for a long time as a possible mechanism to reduce photorespiratory losses and enhance photosynthetic efficiency of C_3 plant species. It was continued to be advocated in many crops with significant research on genetic engineering of C_4 features into rice (*O. sativa*) owing to expected potential to increase crop productivity [134,432,433]. However, from the beginning, hybridization of C_3 and C_4 plants instead of showing expected results has shown the genetic and physiological complexities of incorporating C_4 systems into C_3 crops [91,434,435]. Nevertheless, hybridization of the C_3 and C_4 types in some species such as *Flaveria* pointed toward possible scope of improving crop biomass through enhanced photosynthetic efficiency [129]. Around 20 years ago, while reviewing the progress in the use of interspecific and intergeneric plant hybrids for transferring the C_4 photosynthetic pathway into C_3 species, Brown and Bouton [130] suggested that C_4 photosynthesis appears to be a combination of independently inherited traits, and converting C_3 species into functional C_4 types would require more changes than possible. However, researchers continued to believe that increases in photosynthesis via CO_2 enrichment can improve crop yield [18,89,436]. Since CO_2 absorbed by plants is finally diffused to chloroplast stroma to be fixed by Rubisco and the primary resistances for CO_2 diffusion were considered to be at the stomata and at mesophyll cells, improving mesophyll CO_2 conductance via overexpressing aquaporin [437,438] as well as stomatal CO_2 conductance via manipulating stomatal characteristics [439,440] was proposed as a target for enhancing CO_2 levels around Rubisco. On the other hand, a variety of strategies for introducing CO_2-concentrating approaches into C_3 plants were also under way. These approaches aim to improve productivity by introducing C_4-like features into rice (C_4 rice) [441], by introducing cyanobacterial bicarbonate transporters (CO_2/HCO_3–transporter) into the C_3 chloroplast membranes in higher plants [442], or by engineering new pathways into the chloroplast that bypass photorespiration and release CO_2 directly into the chloroplast stroma [443]. Although a huge challenge, the possibility to engineer C_4 photosynthesis into C_3 plants is considered a reality in the future [431]. This hope is based on the fact that all key enzymes present in C_4 plants are also present in C_3, although the expression levels are much lower [444]. The C_4 plants have less dense topology, higher robustness, better modularity, and higher CO_2 and radiation use efficiency, which provide important basis for engineering C_4 photosynthesis into C_3 plants. In this transition, systems biology will play a critical role in many aspects, including identification of key regulatory elements controlling the development of C_4 features and viable routine toward C_4 using constraint-based modeling approach [445] partly demonstrated by Wang et al. [431] by comparing constraint-based metabolic networks of C_3 and C_4 plants.

30.7.3 Changes in Biochemical Activities

The manipulation of carboxylase/oxygenase ratio, which shows considerable variability between several photosynthetic organisms [446], has been suggested to be a promising way to enhance photosynthetic efficiency [447]. However, efforts done so far to improve the ratio within the species have been unsuccessful [447]. Identification of tobacco mutants for long-term survival under low CO_2 or high O_2 has not shown any changes in the carboxylase/oxygenase ratio or in the CO_2 compensation point [448–450].

Another alternative suggested in this direction has been the selection of plants with an increased level of Rubisco [16], which was later found to be not a very useful biochemical marker for yield [451,452]. Deckard et al. [453] compared old and new wheat varieties for the activity of several enzymes including Rubisco, which determine the efficiency of light utilization. The correlation coefficients between year of release of North Dakotan hard red spring wheat cultivars and the activity of glucose-6-dehydrogenase, oxidase, malate dehydrogenase, phosphofructokinase, and Rubisco were generally low. It was also observed that, in newer cultivars, the maximum protease activity in flag leaf blades during grain filling is 4–6 days later than older cultivars, which cause a delayed onset of leaf senescence.

30.7.4 Rubisco and Crop Improvement

Rubisco still continues to be the only enzyme in plants that can assimilate atmospheric carbon. Equally, Rubisco is the limiting step in assimilating atmospheric C (CO_2) under ambient concentrations and high light conditions. This makes Rubisco the most attractive target for improving photosynthetic efficiency for crops. It has been suggested that optimizing Rubisco function and its regulation by, e.g., RCA, would improve photosynthetic performance. A significant amount of natural variation exists in the catalytic property of Rubisco from diverse land plants. This variation present in nature may potentially be exploited in future for crop use. Nuclear transformation of Rubisco has been challenging. Modifying Rubisco's catalytic properties by engineering chloroplast encoded L-subunits followed by chloroplast transformation has initially shown success [213,215]. Apart from Rubisco, manipulating and optimizing the ancillary protein, RCA, for thermal tolerance may also potentially improve Rubisco performance. Thus, optimizing Rubisco function and regulation may help in engineering crops that are adaptable to a wide range of environmental conditions.

Rubisco is known to have a low catalytic rate, a primary factor explaining its high concentration in C_3 plant leaves. Therefore, improving the Rubisco performance via quality control and/or quantity control is an obvious target for both increasing the photosynthetic performance and nitrogen use efficiency. Recently, the introduction of a C_4-Rubisco small subunit (*RbcS*) gene from sorghum into rice successfully produced chimeric Rubisco with a greater catalytic turnover rate of Rubisco (*kcat*) in the transgenic rice [454]. Also, single residues controlling enzymatic properties of Rubisco have been identified and successfully engineered to produce greater Rubisco proteins in *Flaveria* species from C_3 to C_4 catalysis [212]. Moreover, it has been reported that antisense reductions of Rubisco improved the photosynthetic rate at high CO_2 concentrations in rice [455], as it may be possible to reallocate a large amount of nitrogen from Rubisco to other photosynthetic components (e.g., Calvin cycle enzymes, electron transport systems). Thus, many attempts have been made to improve Rubisco function by genetic manipulation.

30.7.5 Future Prospects

Although the work done so far suggests that manipulation of the efficiency of photosynthesis can be a useful strategy for increasing plant yield [17], the prospects do not seem to be very bright at the moment [2]. Nevertheless, the advances made in the direction of nuclear and chloroplast genes have raised high hopes for a breakthrough not very distant from now. The main target in this direction would be the genetic manipulation of CO_2 fixing enzymes, especially Rubisco and the introduction of C_4 photosynthesis in important C_3 crops like rice and wheat. Also the likely use of modern tools such as antisense technology to obtain desired results in the near future cannot be ruled out. Recent advances in chloroplast transformation have raised hopes for future molecular manipulations. With the possibility of achieving large increases in yield through introducing a C_4-like mechanism in crops, a major new project was initiated (funded by the Bill and Melinda Gates Foundation) to transfer C_4 characteristics into rice (http://beta.irri.org/projects15/c4rice). This project will utilize a wide array of approaches including a phenotypic screen for C_4-*ness* applied to rice and sorghum mutants combined with introducing other necessary genes into rice using genetic transformation [432,456].

REFERENCES

1. Evans LT: Physiological basis of crop yield. In: *Crop Physiology: Some Case Histories* (Evans LT ed.). Cambridge, UK: Cambridge University Press; 1975: 327–356.
2. Kuckuck H, Kobabe G, Wenzel G: *Fundamentals of Plant Breeding.* Berlin, Heidelberg: Springer-Verlag; 1991: 175–176.
3. Tanaka A, Makino A: Photosynthetic research in plant science. *Plant and Cell Physiology* 2009, **50**(4):681–683.
4. Buchanan BB, Balmer Y: Redox regulation: A broadening horizon. *Annual Review of Plant Biology* 2005, **56**:187–220.
5. Hisabori T, Motohashi K, Hosoya-Matsuda N, Ueoka-Nakanishi H, Romano PG: Towards a functional dissection of thioredoxin networks in plant cells. *Photochemistry and Photobiology* 2007, **83**(1):145–151.
6. Wagner D, Przybyla D, Op den Camp R, Kim C, Landgraf F, Lee KP, Wursch M, Laloi C, Nater M, Hideg E et al.: The genetic basis of singlet oxygen-induced stress responses of Arabidopsis thaliana. *Science* 2004, **306**(5699):1183–1185.
7. Beck CF: Signaling pathways from the chloroplast to the nucleus. *Planta* 2005, **222**(5):743–756.
8. Kobayashi Y, Kanesaki Y, Tanaka A, Kuroiwa H, Kuroiwa T, Tanaka K: Tetrapyrrole signal as a cell-cycle coordinator from organelle to nuclear DNA replication in plant cells. *Proceedings of the National Academy of Sciences of the United States of America* 2009, **106**(3):803–807.
9. Thompson WF, White MJ: Physiological and molecular studies of light-regulated nuclear genes in higher-plants. *Annual Review of Plant Physiology and Plant Molecular Biology* 1991, **42**:423–466.
10. Tobin EM, Silverthorne J: Light regulation of gene-expression in higher-plants. *Annual Review of Plant Physiology and Plant Molecular Biology* 1985, **36**:569–593.
11. Gilmartin PM, Sarokin L, Memelink J, Chua NH: Molecular light switches for plant genes. *Plant Cell* 1990, **2**(5):369–378.
12. Kusumi K, Inada H, Kawabata S, Iba K, Nishimura M: Chlorophyll deficiency caused by a specific blockage of the C-5-pathway in seedlings of virescent mutant rice. *Plant and Cell Physiology* 1994, **35**(3):445–449.
13. Pyke KA: Plastid division and development. *The Plant Cell Online* 1999, **11**(4):549–556.
14. Lepistö A, Rintamäki E: Coordination of plastid and light signaling pathways upon development of Arabidopsis leaves under various photoperiods. *Molecular Plant* 2012, **5**(4): 799–816.
15. Austin RB: New opportunities in breeding. *Hortscience* 1988, **23**(1):41–45.
16. Nelson CJ: Genetic associations between photosynthetic characteristics and yield—Review of the evidence. *Plant Physiology and Biochemistry* 1988, **26**(4):543–554.
17. Azcon-Bieto J, Caballero A: *Plant Breeding: Principles and Prospects.* London: Chapman and Hall; 1993.
18. Raines CA: Increasing photosynthetic carbon assimilation in C3 plants to improve crop yield: Current and future strategies. *Plant Physiology* 2011, **155**(1):36–42.
19. Athanasiou K, Dyson BC, Webster RE, Johnson GN: Dynamic acclimation of photosynthesis increases plant fitness in changing environments. *Plant Physiology* 2010, **152**(1):366–373.
20. Murchie E, Pinto M, Horton P: Agriculture and the new challenges for photosynthesis research. *New Phytologist* 2009, **181**(3):532–552.
21. Wilson D: Breeding for morphological and physiological traits. In: *Plant Breeding II* (Frey KJ ed.). New Delhi: Kalyani Publishers; 1981.
22. Donald CM: Breeding of crop ideotypes. *Euphytica* 1968, **17**(3):385–403.
23. Gupta US: *Crop Improvement—Physiological Attributes.* New Delhi: Oxford and IBH Publishing; 1992.
24. Monteith JL: Does light limit crop production? In: *Physiological Processes Limiting Plant Productivity* (Johnson CB ed.). London: Butterworths; 1981: 23–38.
25. Bugbee BG, Salisbury FB: Exploring the limits of crop productivity. 1. Photosynthetic efficiency of wheat in high irradiance environments. *Plant Physiology* 1988, **88**(3):869–878.
26. Elmore CD: The paradox of no correlation between leaf photosynthetic rates and crop yield. In: *Predicting Photosynthesis for Ecosystem Models* (Hesketh JD, Jones JW eds.). Boca Raton, FL: CRC Press; 1980: 155–167.
27. Flood PJ, Harbinson J, Aarts MG: Natural genetic variation in plant photosynthesis. *Trends in Plant Science* 2011, **16**(6):327–335.
28. Tungland L, Chapko LB, Wiersma JV, Rasmusson DC: Effect of erect leaf angle on grain-yield in barley. *Crop Science* 1987, **27**(1):37–40.
29. Aquino RC, Jennings PR: Inheritance and significance of dwarfism in an indica rice variety. *Crop Science* 1966, **6**(6):551–554.
30. Morishima H, Oka HI, Chang TT: Analysis of genetic variations in plant type of rice. 1. Estimation of indices showing genetic plant types and their correlations with yielding capacity in a segregating population. *Japan Journal of Breeding* 1967, **17**:73–84.
31. Sinclair TR, Sheehy JE: Erect leaves and photosynthesis in rice. *Science* 1999, **283**:1455.
32. Murchie EH, Chen Y-Z, Hubbart S, Peng S, Horton P: Interactions between senescence and leaf orientation determine in situ patterns of photosynthesis and photoinhibition in field-grown rice. *Plant Physiology* 1999, **119**(2):553–564.
33. Morinaka Y, Sakamoto T, Inukai Y, Agetsuma M, Kitano H, Ashikari M, Matsuoka M: Morphological alteration caused by brassinosteroid insensitivity increases the biomass and grain production of rice. *Plant Physiology* 2006, **141**(3):924–931.
34. Sakamoto T, Morinaka Y, Ohnishi T, Sunohara H, Fujioka S, Ueguchi-Tanaka M, Mizutani M, Sakata K, Takatsuto S, Yoshida S: Erect leaves caused by brassinosteroid deficiency increase biomass production and grain yield in rice. *Nature Biotechnology* 2006, **24**(1):105–109.
35. Radenović ČN, Babić M, Delić N, Hojka ZM, Stanković G, Trifunović B, Ristanović D, Selaković DM: Photosynthetic properties of erect leaf maize inbred lines as the efficient photo-model in breeding and seed production. *Genetika* 2003, **35**(2):85–97.
36. Wallace D, Ozbun J, Munger H: Physiological genetics of crop yield. *Advances in Agronomy* 1972, **24**:97–146.
37. Ackerly DD, Dudley SA, Sultan SE, Schmitt J, Coleman JS, Linder CR, Sandquist DR, Geber MA, Evans AS, Dawson TE et al.: The evolution of plant ecophysiological traits: Recent advances and future directions. *BioScience* 2000, **50**(11):979–995.
38. Rhodes I: Relationship between productivity and some components of canopy structure in ryegrass (Lolium Spp). 4. Canopy characters and their relationship with sward yields in some intra population selections. *Journal of Agricultural Science* 1975, **84**:345–351.
39. Carver BF, Nevo E: Genetic diversity of photosynthetic characters in native populations of Triticum dicoccoides. *Photosynthesis Research* 1990, **25**(2):119–128.

40. Hsu P, Walton PD: Inheritance of morphological and agronomic characters in spring wheat. *Euphytica* 1970, **19**(1):54–60.

41. Duarte R, Adams M: Component interaction in relation to expression of a complex trait in a field bean cross. *Crop Science* 1963, **3**(3):185–186.

42. Snoad B: Preliminary assessment of leafless peas. *Euphytica* 1974, **23**(2):257–265.

43. Karami E, Weaver JB: Dry-matter production, yield, photosynthesis, chlorophyll content and specific leaf weight of cotton in relation to leaf shape and color. *Journal of Agricultural Science* 1980, **94**:281–286.

44. Ghatge RD: Genic relationship of leaf/leaflet shape inheritance with leaf/leaflet size in chickpea Cicer arietinum L. *Annals of Agricultural Research* 1992, **13**:228–234.

45. Koch P, Costa C: Inheritance of plant and fruit characters in gherkin. *Horticultura Brasileira* 1991, **9**:73–77.

46. Marinkovic R, Joksimovic J, Dozet B: Genetics of leaf length and width in sunflower (*Helianthus annus* L.). *Helia* 1993, **16**:31–34.

47. Abzalov MF, Fatkullaeva GN: Studies of the anthocyanin pigmentation genetics and its relationship with the leaves shape in cotton Gossypium-hirsutum L. *Genetika* 1993, **29**(8):1356–1365.

48. Ecker R, Barzilay A, Osherenko E: Linkage relationships of genes for leaf morphology and double flowering in Matthiola-incana. *Euphytica* 1994, **74**(1–2):133–136.

49. Pandey SP, Winkler JA, Li H, Camacho DM, Collins JJ, Walker GC: Central role for RNase YbeY in Hfq-dependent and Hfq-independent small-RNA regulation in bacteria. *BMC Genomics* 2014, **15**(1):121.

50. Srivastava PK, Moturu TR, Pandey P, Baldwin IT, Pandey SP: A comparison of performance of plant miRNA target prediction tools and the characterization of features for genome-wide target prediction. *BMC Genomics* 2014, **15**(1):348.

51. Tang G, Reinhart BJ, Bartel DP, Zamore PD: A biochemical framework for RNA silencing in plants. *Genes & Development* 2003, **17**(1):49–63.

52. He L, Hannon GJ: MicroRNAs: Small RNAs with a big role in gene regulation. *Nature Reviews Genetics* 2004, **5**(7):522–531.

53. Pandey SP, Moturu TR, Pandey P: Roles of small RNAs in regulation of signaling and adaptive responses in plants. In: *Recent Trends in Gene Expression* (Mandal SS ed.). Hauppauge, NY: Nova Science Publishers; 2013: 107–132.

54. Adenot X, Elmayan T, Lauressergues D, Boutet S, Bouché N, Gasciolli V, Vaucheret H: DRB4-dependent *TAS3* trans-acting siRNAs control leaf morphology through AGO7. *Current Biology* 2006, **16**(9):927–932.

55. Pulido A, Laufs P: Coordination of developmental processes by small RNAs during leaf development. *Journal of Experimental Botany* 2010, **61**(5):1277–1291.

56. Reynolds M: Physiological approaches to wheat breeding. In: *FAO Plant Production and Protection Series.* (Curtis BC, Rajaram S, Gomezmacpherson H eds.) Rome: Food and Agriculture Organization; 2002.

57. Mallory AC, Reinhart BJ, Jones-Rhoades MW, Tang G, Zamore PD, Barton MK, Bartel DP: MicroRNA control of PHABULOSA in leaf development: Importance of pairing to the microRNA 5′ region. *The EMBO Journal* 2004, **23**(16):3356–3364.

58. Mecchia MA, Debernardi JM, Rodriguez RE, Schommer C, Palatnik JF: MicroRNA miR396 and *RDR6* synergistically regulate leaf development. *Mechanisms of Development* 2013, **130**(1):2–13.

59. Palatnik JF, Allen E, Wu X, Schommer C, Schwab R, Carrington JC, Weigel D: Control of leaf morphogenesis by microRNAs. *Nature* 2003, **425**(6955):257–263.

60. Aslam M, Hunt LA: Photosynthesis and transpiration of flag leaf in 4 spring-wheat cultivars. *Planta* 1978, **141**(1):23–28.

61. Terao H: Maternal inheritance in the soyabean. *American Naturalist* 1918, **52**(613): 51–56.

62. Reese PF, Boerma HR: Additional genes for green seed coat in soybean. *Journal of Heredity* 1989, **80**(1):86–88.

63. Guiamet JJ, Teeri JA, Nooden LD: Effects of nuclear and cytoplasmic genes altering chlorophyll loss on gas-exchange during monocarpic senescence in soybean. *Plant and Cell Physiology* 1990, **31**(8):1123–1130.

64. Bernard RL, Weiss MG: Qualitative genetics. In: *Soybeans: Improvement, Production, and Uses*, vol. 1. (Caldwell BE et al. eds.). Madison, WI: American Society of Agronomy; 1973: 117–154.

65. Shifris O: *Report—Cucurbit Genetics Cooperative*, vol. 16. North Carolina: CGC; 1993: 64–67.

66. Saunders A: Inheritance in the cowpea III: Mutations and linkages. *South African Journal of Agricultural Science* 1960, **3**:327–348.

67. Kirchhoff WR, Hall AE, Roose ML: Inheritance of a mutation influencing chlorophyll content and composition in cowpea. *Crop Science* 1989, **29**(1):105–108.

68. Sen NK, Bhowal J: Genetics of Vigna sinensis (L.) savi. *Genetica* 1962, **32**(1):247–266.

69. Kohle A: Genetic studies in Vigna. *Poona Agricultural. College Magazine* 1970, **59**:126–137.

70. Thomas H, Howarth CJ: Five ways to stay green. *Journal of Experimental Botany* 2000, **51**(Suppl 1):329–337.

71. Joshi A, Kumari M, Singh V, Reddy C, Kumar S, Rane J, Chand R: Stay green trait: Variation, inheritance and its association with spot blotch resistance in spring wheat (Triticum aestivum L.). *Euphytica* 2007, **153**(1–2):59–71.

72. Ahlawat S, Chhabra AK, Behl R, Bisht S: Genotypic divergence analysis for stay green characters in wheat (Triticum aestivum L. em. Thell). *The South Pacific Journal of Natural and Applied Sciences* 2008, **26**(1):73–81.

73. Rehman A, Habib I, Ahmad N, Hussain M, Khan MA, Farooq J, Ali MA: Screening wheat germplasm for heat tolerance at terminal growth stage. *Plant Omics* 2009, **2**(1):9–19.

74. Spano G, Di Fonzo N, Perrotta C, Platani C, Ronga G, Lawlor D, Napier J, Shewry P: Physiological characterization of 'stay green' mutants in durum wheat. *Journal of Experimental Botany* 2003, **54**(386):1415–1420.

75. Thorne G: Distribution between parts of the main shoot and the tillers of photosynthate produced before and after anthesis in the top three leaves of main shoots of Hobbit and Maris Huntsman winter wheat. *Annals of Applied Biology* 1982, **101**(3):553–559.

76. Rawson H, Hindmarsh J, Fischer R, Stockman Y: Changes in leaf photosynthesis with plant ontogeny and relationships with yield per ear in wheat cultivars and 120 progeny. *Functional Plant Biology* 1983, **10**(6):503–514.

77. Kumari M, Pudake R, Singh V, Joshi AK: Association of staygreen trait with canopy temperature depression and yield traits under terminal heat stress in wheat (Triticum aestivum L.). *Euphytica* 2013, **190**(1):87–97.

78. Abbad H, El Jaafari S, Bort J, Araus J: Comparative relationship of the flag leaf and the ear photosynthesis with the biomass and grain yield of durum wheat under a range of water conditions and different genotypes. *Agronomie* 2004, **24**:19–28.

79. Walulu RS, Rosenow DT, Wester DB, Nguyen HT: Inheritance of the stay green trait in sorghum. *Crop Science* 1994, **34**(4):970–972.

80. Thomas H: Sid—A Mendelian locus controlling thylakoid membrane disassembly in senescing leaves of Festuca pratensis. *Theoretical and Applied Genetics* 1987, **73**(4): 551–555.

81. Pandey SP, Baldwin IT: Silencing RNA-directed RNA polymerase 2 increases the susceptibility of Nicotiana attenuata to UV in the field and in the glasshouse. *The Plant Journal* 2008, **54**(5):845–862.

82. Harlan HV, Hulton HFE: Development of barley kernels in normal and clipped spikes and the limitations of awnless and hooded varieties. *Journal of the Institute Brewing*, 1920, **26**(12):639–641.

83. Evans LT, Rawson HM: Photosynthesis and respiration by flag leaf and components of ear during grain development in wheat. *Australian Journal of Biological Sciences* 1970, **23**(2):245–254.

84. Sears ER: Aneuploids of common wheat. *Missouri Agricultural Experiment Station Research Bulletin* 1954, **572**:P. 59.

85. Tsunewaki K: Comparative gene analysis of common wheat and its ancestral species. III. Glume hairiness. *Genetics* 1966, **53**(2):303.

86. Worland AJG, Gale MD, Law CN: Wheat genetics. In: *Wheat Breeding—Its Scientific Basis* (Lupton FGH ed.). London: Chapman & Hall; 1987: 129–172.

87. Tambussi EA, Bort J, Guiamet JJ, Nogués S, Araus JL: The photosynthetic role of ears in C3 cereals: Metabolism, water use efficiency and contribution to grain yield. *Critical Reviews in Plant Sciences* 2007, **26**(1):1–16.

88. Portis Jr. AR, Parry MA: Discoveries in Rubisco (Ribulose 1, 5-bisphosphate carboxylase/oxygenase): A historical perspective. *Photosynthesis Research* 2007, **94**(1):121–143.

89. Long SP, Zhu XG, Naidu SL, Ort DR: Can improvement in photosynthesis increase crop yields? *Plant, Cell & Environment* 2006, **29**(3):315–330.

90. Zelitch I, Day PR: The effect on net photosynthesis of pedigree selection for low and high rates of photorespiration in tobacco. *Plant Physiology* 1973, **52**(1):33–37.

91. Bojorkman O: CO_2 *Metabolism and Plant Productivity.* Baltimore, MD: University Park Press; 1976.

92. Crosbie TM, Mock JJ, Pearce RB: *Agronomie Abstr.* Madison, WI: Am Soc Agron; 1977: 52.

93. Crosbie TM, Mock JJ, Pearce RB: Variability and selection advance for photosynthesis in Iowa stiff stalk synthetic maize population. *Crop Science* 1977, **17**(4):511–514.

94. Wang Y, Yang S, Irfan M, Zhang C, Sun Q, Wu S, Lin F: Genetic analysis of carbon metabolism-related traits in maize using mixed major and polygene models. *Australian Journal of Crop Science* 2013, **7**(8):1205–1211.

95. McDonald D, Stansel J, Gilmore E: *Breeding Researches in Asia and Oceania.* New Delhi: Indian Society of Genetics & Plant Breeding; 1974: 1067–1073.

96. Rakhmankulov S, Gaziyants S: Intensity of the functioning of the photosynthetic apparatus of cotton interspecific hybrids in relation to economic productivity. *Sel'skokhozyaistvennaya Biologiya* 1980, **15**(3):374–378.

97. Ojima M: Improvement of photosynthetic capacity in soybean variety. *Japan Agricultural Research Quarterly* 1974, **8**:6–12.

98. Harrison S, Boerma H, Ashley D, Corbin F: Heritability of canopy photosynthetic capacity and its relationship to seed yield. In: *World Soybean Research Conference II, 1979: Abstracts: 1980.* Boulder, CO: Westview Press; 1980.

99. Wiebold JW, Shibles R, Green DE: Selection for apparent photosynthesis and related leaf traits in early generations of soybeans. *Crop Science* 1981, **21**:969–973.

100. Mahon JD, Hobbs SLA: Selection of peas for photosynthetic CO2 exchange-rate under field conditions. *Crop Science* 1981, **21**(4):616–621.

101. Khodadadi M, Dehghani H, Fotokian MH, Rain B: Genetic diversity and heritability of chlorophyll content and photosynthetic indexes among some Iranian wheat genotypes. *Journal of Biodiversity and Environmental Sciences* 2014, **4**:12–23.

102. Matsuda T: Studies on the breeding of high yield variety in air-cured tobacco. IV. Inheritance of apparent photosynthetic rate, rate of photorespiration and rate of respiration. *Utsunomiya tab Shikenjo Hokoku* 1978, **16**:9–18.

103. Muramoto H, Hesketh J, El-Sharkawy M: Relationships among rate of leaf area development, photosynthetic rate, and rate of dry matter production among American cultivated cottons and other species. *Crop Science* 1965, **5**:163–166.

104. Bhatt JG, Rao MRK: Heterosis in growth and photosynthetic rate in hybrids of cotton. *Euphytica* 1981, **30**(1):129–133.

105. Singh KB, Kandola HS: Heterosis in wheat. *Indian Journal of Genetics and Plant Breeding* 1969, **29**(1):53–61.

106. Barbosa-Neto JF, Sorrells ME, Cisar G: Prediction of heterosis in wheat using coefficient of parentage and RFLP-based estimates of genetic relationship. *Genome* 1996, **39**(6):1142–1149.

107. Brezhnev DD, Tagmazyan IA: A study of photosynthesis and gibberellin content in cucumbers in relation to the phenomenon of heterosis. *Byulleten Vsesoyuznogo Ordena Lenina Instituta Rastenievodstva Imeni NI Vavilova* 1973, **36**:54–57.

108. Stanev V, Angelov M, Tsonev T, Danilov Z: Photosynthetic rate and chlorophyll content in leaves of tomato hybrids with heterosis for yield. *Fiziologiya Rasteni (Sofia)* 1984, **10**:32–39.

109. Heichel GH, Musgrave RB: Varietal differences in net photosynthesis of Zea mays L. *Crop Science* 1969, **9**(4):483–486.

110. Albergoni F, Basso B, Pe E, Ottaviano E: A study of photosynthesis and gibberellin content in cucumbers in relation to the phenomenon of heterosis. Photosynthetic rate in maize—Inheritance and correlation with morphological traits. *Maydica* 1983, **28**(4):439–448.

111. Akita S, Mochizuki N, Yamada M, Tanaka I: Variations of heterosis in leaf photosynthetic activity of maize (Zea-mays-L) with growth-stages. *Japanese Journal of Crop Science* 1986, **55**(4):404–407.

112. Mehta H, Sarkar KR: Heterosis for leaf photosynthesis, grain yield and yield components in maize. *Euphytica* 1992, **61**(2):161–168.

113. Offermann S, Peterhansel C: Can we learn from heterosis and epigenetics to improve photosynthesis? *Current Opinion in Plant Biology* 2014, **19**:105–110.

114. Singh MK, Tsunoda S: Photosynthetic and transpirational response of a cultivated and a wild-species of triticum to soil-moisture and air humidity. *Photosynthetica* 1978, **12**(3): 280–283.

115. Austin RB, Morgan CL, Ford MA: Flag leaf photosynthesis of Triticum-aestivum and related diploid and tetraploid species. *Annals of Botany* 1982, **49**(2):177–189.

116. Hobbs SLA, Mahon JD: Inheritance of chlorophyll content, ribulose-1,5-bisphosphate carboxylase activity, and stomatal-resistance in peas. *Crop Science* 1985, **25**(6):1031–1034.

117. Hayashi K, Yamamoto T, Nakagahra M: Genetic control for leaf photosynthesis in rice, Oryza sativa L. *Japanese Journal of Breeding* 1977, **27**:49–56.

118. Izhar S, Wallace DH: Studies of physiological basis for yield differences. 3. Genetic variation in photosynthetic efficiency of Phaseolus vulgaris L. *Crop Science* 1967, **7**(5):457–460.

119. Planchon C, Fesquet J: Effect of the D-genome and of selection on photosynthesis in wheat. *Theoretical and Applied Genetics* 1982, **61**(4):359–365.

120. Watanabe N, Kawajiri T, Nishikawa K: Contribution of D-genome to the photosynthetic oxygen evolution in mesophyll protoplasts isolated from leaves of wheat seedlings. *Plant Physiology and Biochemistry* 1988, **26**(4):421–425.

121. Haour-Lurton B, Planchon C: Role of D-genome chromosomes in photosynthesis expression in wheats. *Theoretical and Applied Genetics* 1985, **69**(4):443–446.

122. Bobo MS, Planchon C, Morris R: Chromosome-3D influences photosystem-II quantum efficiency in winter-wheat. *Crop Science* 1992, **32**(4):958–961.

123. Kadereit G, Mavrodiev EV, Zacharias EH, Sukhorukov AP: Molecular phylogeny of Atripliceae (Chenopodioideae, Chenopodiaceae): Implications for systematics, biogeography, flower and fruit evolution, and the origin of C4 photosynthesis. *American Journal of Botany* 2010, **97**(10):1664–1687.

124. Björkman O, Gauhl E, Nobs M: Comparative studies of Atriplex species with and without β-carboxylation photosynthesis and their first-generation hybrid. *Carnegie Institution Washington Yearbook*. Washington, DC: 1969, **68**:620–633.

125. Osmond CB, Bjorkman O, Anderson DJ: *Physiological Processes in Plant Ecology. Toward a Synthesis with Atriplex.* Berlin: Springer Verlag; 1980.

126. Bjorkman O, Nobs M, Pearcy R, Boynton J, Berry J: *Photosynthesis and Photo-Respiration.* Sydney: Wiley-Interscience; 1971.

127. Bjorkman O, Troughton J, Nobs M: Basic mechanisms in plant morphogenesis. *Grookhaven Symposia in Biology* 1974, **25**:206–226.

128. Brown RH, Bouton JH, Evans PT, Malter HE, Rigsby LL: Photosynthesis, morphology, leaf anatomy, and cytogenetics of hybrids between C-3 and C-3/C-4 panicum species. *Plant Physiology* 1985, **77**(3):653–658.

129. Huber WE, Brown RH, Bouton JH, Sternberg LO: CO$_2$ exchange, cytogenetics, and leaf anatomy of hybrids between photosynthetically distinct Flaveria species. *Plant Physiology* 1989, **89**(3):839–844.

130. Brown RH, Bouton JH: Physiology and genetics of interspecific hybrids between photosynthetic types. *Annual Review of Plant Physiology and Plant Molecular Biology* 1993, **44**:435–456.

131. Covshoff S, Burgess SJ, Kneřová J, Kümpers BM: Getting the most out of natural variation in C4 photosynthesis. *Photosynthesis Research* 2014, **119**(1–2):157–167.

132. Oakley JC, Sultmanis S, Stinson CR, Sage TL, Sage RF: Comparative studies of C3 and C4 Atriplex hybrids in the genomics era: Physiological assessments. *Journal of Experimental Botany* 2014, **65**(13): 3637–3647.

133. Li P, Brutnell TP: Setaria viridis and Setaria italica, model genetic systems for the Panicoid grasses. *Journal of Experimental Botany* 2011, **62**(9):3031–3037.

134. von Caemmerer S, Evans JR: Enhancing C3 photosynthesis. *Plant Physiology* 2010, **154**(2):589–592.

135. Sage TL, Busch FA, Johnson DC, Friesen PC, Stinson CR, Stata M, Sultmanis S, Rahman BA, Rawsthorne S, Sage RF: Initial events during the evolution of C4 photosynthesis in C3 species of Flaveria. *Plant Physiology* 2013, **163**(3):1266–1276.

136. Ubierna N, Sun W, Kramer DM, Cousins AB: The efficiency of C4 photosynthesis under low light conditions in Zea mays, Miscanthus x giganteus and Flaveria bidentis. *Plant, Cell & Environment* 2013, **36**:365–381.

137. Pfannschmidt T: Chloroplast redox signals: How photosynthesis controls its own genes. *Trends in Plant Science* 2003, **8**(1):33–41.

138. Nasyrov YS: Genetic-control of photosynthesis and improving of crop productivity. *Annual Review of Plant Physiology and Plant Molecular Biology* 1978, **29**:215–237.

139. Gruissem W: Chloroplast gene-expression—How plants turn their plastids on. *Cell* 1989, **56**(2):161–170.

140. Reiter RS, Coomber SA, Bourett TM, Bartley GE, Scolnik PA: Control of leaf and chloroplast development by the Arabidopsis gene pale cress. *Plant Cell* 1994, **6**(9):1253–1264.

141. Reinbothe S, Reinbothe C: The regulation of enzymes involved in chlorophyll biosynthesis. *European Journal of Biochemistry* 1996, **237**(2):323–343.

142. Reynolds M, Calderini D, Condon A, Rajaram S: Physiological basis of yield gains in wheat associated with the Lr19 translocation from Agropyron elongatum. In: *Wheat in a Global Environment*. The Netherlands: Springer; 2001: 345–351.

143. Miralles D, Resnicoff E, Carretero R: Yield improvement associated with Lr19 translocation in wheat. In: *Scale and Complexity in Plant Systems Research: Gene-Plant-Crop Relationships* (Spiertz JHJ, Struik PC, van Laar HH eds.). The The Netherlands: Springer; 2007: 171–178.

144. Foulkes MJ, Slafer GA, Davies WJ, Berry PM, Sylvester-Bradley R, Martre P, Calderini DF, Griffiths S, Reynolds MP: Raising yield potential of wheat. III. Optimizing partitioning to grain while maintaining lodging resistance. *Journal of Experimental Botany* 2011, **62**(2):469–486.

145. Levine RP, Goodenough UW: The genetics of photosynthesis and of the chloroplast in Chlamydomonas reinhardtii. *Annual Review of Genetics* 1970, **4**:397–408.

146. McVittie A, Davies DR: Location of Mendelian linkage groups in Chlamydomonas reinhardtii. *Molecular and General Genetics* 1971, **112**(3):225–228.

147. Mets L, Bogorad L: Altered chloroplast ribosomal-proteins associated with erythromycin resistant mutants in 2 genetic systems of Chlamydomonas reinhardtii. *Proceedings of the National Academy of Sciences of the United States of America* 1972, **69**(12):3779–3783.

148. Bogorad L: Genetic and evolutionary relationships between plastids and the nuclear-cytoplasmic system. In: *Genetic Aspects of Photosynthesis* (Nasyrov YS, Kvitko KV, Šesták Z eds.). The Hague: Dr. Junk; 1975.

149. Wildman SG, Chaen K, Gray JC, Kung SD, Kwanjueen P, Sakano K: *Evolution of Ferredoxin and Fraction I Protein in the Genus Nicotiana.* Columbus, OH: The Ohio State University Press; 1975.

150. Holm G: Chlorophyll mutations in barley. *Acta Agriculturae Scandinavica* 1954, **4**(1):457–471.

151. Blixt S: Linkage studies in Pisum. 9. Linkage of gene Chi5 in chromosome. 2. *Agri Hortique Genetica* 1968, **26**(1–2):87.

152. Savage LJ, Imre KM, Hall DA, Last RL: Analysis of essential Arabidopsis nuclear genes encoding plastid-targeted proteins. *PLoS One* 2013, **8**(9):e73291.

153. Pogson BJ, Albrecht V: Genetic dissection of chloroplast biogenesis and development: An overview. *Plant Physiology* 2011, **155**(4):1545–1551.

154. Aliev KA, Muzafarova S, Radzhabov H, Kholmatova M, Ulugbekova G, Nasyrov YS: Actual problems of genetics of

photosynthesis. In: *Genetic Aspects of Photosynthesis* (Nasyrov YS, Kvitko KV, Šesták Z eds.). The Hague: Dr. Junk; 1975: 133–145.

155. Kasyanenko AG, Nasyrov YS: O deĭstvii geneticheskikh faktorov na fotosinteticheskiĭ apparat Arabidopsis thaliana (L.) HEYNH. [Action of genetic factors on photosynthetic apparatus of Arabidopsis thaliana (L.) HEYNH.]. *Fiziologiya Rasteni* 1968, **15**:422–429.

156. Heber U, Gottschalk W: Die Bestimmung des genetisc fixierten Stoffsechselblockes einer Photosynthese-XMutante von *Vicia faba. Zeitschrift für Naturforschung* 1963, **Teil. B. 18b**:36–44.

157. Römer S, Humbeck K, Senger H: Analysis of the molecular organization of photosystem I during light-dependent chloroplast differentiation in mutant C-6D of Scenedesmus obliquus. *Botanica Acta* 1990, **103**(2):155–161.

158. Stummann BM, Henningsen KW: Nuclear pea mutants deficient in chlorophyll b and the major polypeptide of the light-harvesting chlorophyll a/b protein of photosystem II. *Photosynthesis Research* 1984, **5**(4):275–292.

159. Yakubova MM, Rubin A, Khramova GA, Matorin D: Hill reaction and delayed fluorescence in mutants of Gossypium hirsutum. In: *Genetic Aspects of Photosynthesis.* (Nasyrov YS, Sestaked Z eds.) The Hague: Springer; 1975: 263–269.

160. Herrmann F: Struktur und Funktion der genetischen Information in den Plastiden. II. Untersuchung der photosynthesedefekten Plastommutante alba-1 von Antirrhinum majus L. *Photosynthetica* 1971:258–266.

161. Anderson JM, Levine RP: The relationship between chlorophyll-protein complexes and chloroplast membrane polypeptides. *Biochimica et Biophysica Acta* 1974, **357**(1):118–126.

162. Genge S, Pilger D, Hiller RG: The relationship between chlorophyll b and pigment-protein complex II. *Biochimica et Biophysica Acta* 1974, **347**(1):22–30.

163. Parthier B, Krauspe R, Munsche D, Wollgiehn R: The biogenesis of chloroplasts. In: *The Chemistry and Biochemistry of Plant Proteins.* (Harborne JB, van Sumere CF eds.) New York: Academic Press; 1975: 167–210.

164. Butterfass T: Control of plastid division by means of nuclear DNA amount. *Protoplasma* 1973, **76**(2):167–195.

165. Hoober JK, Stegeman WJ: *Genetics and Biogenesis of Mitochondria and Chloroplasts.* Columbus, OH: The Ohio State University Press; 1975.

166. Wolf G, Rimpau J: Evidence for cytoplasmic control of gene-expression in higher-plants. *Nature* 1977, **265**(5593): 470–472.

167. von Wettstein D: *Chloroplast and Nucleus: Concerted Interplay between Genomes of Different Cell Organelles.* The Emil Heitz lecture. Berlin: Springer; 1981.

168. von Wettstein D, Poulsen C, Oliver RP, Gough SP, Kannangara CG, Moller BL, Simpson D: Genetics: New frontiers. In: *Proceedings of the XV th Int Congr of Genetics* 1984:267–285.

169. Alt J, Westhoff P, Sears BB, Nelson N, Hurt E, Hauska G, Herrmann RG: Genes and transcripts for the polypeptides of the cytochrome b6/f complex from spinach thylakoid membranes. *EMBO Journal* 1983, **2**(6):979–986.

170. Coruzzi G, Broglie R, Lamppa G, Chua N-H: Expression of nuclear genes encoding the small subunit of ribulose-1, 5-bisphosphate carboxylase. In: *Structure and Function of Plant Genomes* (Ciferri O ed.). New York: Springer; 1983: 47–59.

171. Brecht E: The light-harvesting chlorophyll a/b-protein complex-II of higher-plants—Results from a 20-year research period. *Photobiochemistry and Photobiophysics* 1986, **12**(1–2):37–50.

172. Thorner JP, Peter GF, Chitnis PR, Nechushtai R, Vainstein A: *Light-Energy Transduction in Photosynthesis: Higher Plant and Bacterial Models,* vol. 54. Rockville, MD: American Society of Plant Physiologists; 1988.

173. Manzara T, Gruissem W: Organization and expression of the genes encoding ribulose-1,5-bisphosphate carboxylase in higher-plants. *Photosynthesis Research* 1988, **16**(1–2):117–139.

174. Dean C, Pichersky E, Dunsmuir P: Structure, evolution, and regulation of rbcs genes in higher-plants. *Annual Review of Plant Physiology and Plant Molecular Biology* 1989, **40**:415–439.

175. Keegstra K: Transport and routing of proteins into chloroplasts. *Cell* 1989, **56**(2):247–253.

176. Keegstra K, Olsen LJ, Theg SM: Chloroplastic Precursors and their transport across the envelope membranes. *Annual Review of Plant Physiology and Plant Molecular Biology* 1989, **40**:471–501.

177. Taylor WC: Regulatory interactions between nuclear and plastid genomes. *Annual Review of Plant Physiology and Plant Molecular Biology* 1989, **40**:211–233.

178. Susek RE, Ausubel FM, Chory J: Signal-transduction mutants of Arabidopsis uncouple nuclear CAB and RBCS gene-expression from chloroplast development. *Cell* 1993, **74**(5): 787–799.

179. Schöttler MA, Tóth SZ, Boulouis A, Kahlau S: Photosynthetic complex stoichiometry dynamics in higher plants: Biogenesis, function, and turnover of ATP synthase and the cytochrome b 6 f complex. *Journal of Experimental Botany* 2015, **66**(9):2373–2400.

180. Scheller HV, Moller BL: Photosystem-I polypeptides. *Physiologia Plantarum* 1990, **78**(3):484–494.

181. Herrmann RG, Oelmüller R, Bichler J, Schneiderbauer A, Steppuhn J, Wedel N, Tyagi AK, Westhoff P: The thylakoid membrane of higher plants: Genes, their expression and interaction. In: *Plant Molecular Biology,* vol. 2 (Harrmann RG, Larkins BA eds.). New York: Springer; 1991: 411–427.

182. Okkels JS, Scheller HV, Svendsen I, Moller BL: Isolation and characterization of a cDNA clone encoding an 18-kDa hydrophobic photosystem-I subunit (Psi-L) from barley (Hordeum vulgare L). *Journal of Biological Chemistry* 1991, **266**(11):6767–6773.

183. Bryant DA: Molecular biology of photosystem I. In: *The Photosystems: Structure, Function and Molecular Biology* (Barber J ed.). Amsterdam: Elsevier Science Publishers; 1992.

184. Golbeck JH: Structure and function of photosystem-I. *Annual Review of Plant Physiology and Plant Molecular Biology* 1992, **43**:293–324.

185. Knoetzel J, Simpson DJ: The primary structure of a cDNA for PsaN, encoding an extrinsic lumenal polypeptide of barley photosystem-I. *Plant Molecular Biology* 1993, **22**(2):337–345.

186. Okkels JS, Scheller HV, Jepsen LB, Moller BL: A cDNA clone encoding the precursor for a 10.2 kDa photosystem-I polypeptide of barley. *FEBS Letters* 1989, **250**(2):575–579.

187. Steppuhn J, Hermans J, Nechushtai R, Herrmann GS, Herrmann RG: Nucleotide-sequences of cDNA clones encoding the entire precursor polypeptide for subunit-VI and of the plastome-encoded gene for subunit-VII of the photosystem-I reaction center from spinach. *Current Genetics* 1989, **16**(2):99–108.

188. de Pater S, Hensgens LA, Schilperoort RA: Structure and expression of a light-inducible shoot-specific rice gene. *Plant Molecular Biology* 1990, **15**(3):399–406.

189. Hayashida N, Izuchi S, Sugiura M, Obokata J: Nucleotide-sequence of cDNA clones encoding the PS I-H subunit of photosystem-I in tobacco. *Plant and Cell Physiology* 1992, **33**(7):1031–1034.

190. Franzen LG, Frank G, Zuber H, Rochaix JD: Isolation and characterization of cDNA clones encoding photosystem-I subunits with molecular masses 11.0, 10.0 and 8.4 kDa from Chlamydomonas-reinhardtii. *Molecular & General Genetics* 1989, **219**(1–2):137–144.

191. Nakamura M, Obokata J: Organization of the psaH gene family of photosystem-I in Nicotiana-sylvestris. *Plant and Cell Physiology* 1994, **35**(2):297–302.

192. Obokata J, Mikami K, Hayashida N, Nakamura M, Sugiura M: Molecular heterogeneity of photosystem-I—psaD, psaE, psaF, psaH, and psaL are all present in isoforms in Nicotiana Spp. *Plant Physiology* 1993, **102**(4):1259–1267.

193. Bolle C, Sopory S, Lubberstedt T, Klosgen RB, Herrmann RG, Oelmuller R: The role of plastids in the expression of nuclear genes for thylakoid proteins studied with chimeric beta-glucuronidase gene fusions. *Plant Physiology* 1994, **105**(4):1355–1364.

194. Palomares R, Herrmann RG, Oelmuller R: Posttranscriptional and posttranslational regulatory steps are crucial in controlling the appearance and stability of thylakoid polypeptides during the transition of etiolated tobacco seedlings to white-light. *European Journal of Biochemistry* 1993, **217**(1):345–352.

195. Oelmuller R, Dietrich G, Link G, Mohr H: Regulatory factors involved in gene-expression (subunits of ribulose-1,5-bisphosphate carboxylase) in mustard (Sinapis alba L) cotyledons. *Planta* 1986, **169**(2):260–266.

196. Oelmuller R, Levitan I, Bergfeld R, Rajasekhar VK, Mohr H: Expression of nuclear genes as affected by treatments acting on the plastids. *Planta* 1986, **168**(4):482–492.

197. Keith B, Chua NH: Monocot and dicot pre-messenger-RNAs are processed with different efficiencies in transgenic tobacco. *EMBO Journal* 1986, **5**(10):2419–2425.

198. Schaffner AR, Sheen J: Maize rbcS promoter activity depends on sequence elements not found in dicot rbcS promoters. *Plant Cell* 1991, **3**(9):997–1012.

199. Luan S, Bogorad L: A rice cab gene promoter contains separate cis-acting elements that regulate expression in dicot and monocot plants. *The Plant Cell Online* 1992, **4**(8):971–981.

200. Yamamoto N, Tada Y, Fujimura T: The promoter of a pine photosynthetic gene allows expression of a beta-glucuronidase reporter gene in transgenic rice plants in a light-independent but tissue-specific manner. *Plant and Cell Physiology* 1994, **35**(5):773–778.

201. Waters MT, Wang P, Korkaric M, Capper RG, Saunders NJ, Langdale JA: GLK transcription factors coordinate expression of the photosynthetic apparatus in Arabidopsis. *The Plant Cell Online* 2009, **21**(4):1109–1128.

202. Saibo NJ, Lourenço T, Oliveira MM: Transcription factors and regulation of photosynthetic and related metabolism under environmental stresses. *Annals of Botany* 2009, **103**(4):609–623.

203. Parlitz S, Kunze R, Mueller-Roeber B, Balazadeh S: Regulation of photosynthesis and transcription factor expression by leaf shading and re-illumination in *Arabidopsis thaliana* leaves. *Journal of Plant Physiology* 2011, **168**(12):1311–1319.

204. Imam S, Noguera DR, Donohue TJ: Global analysis of photosynthesis transcriptional regulatory networks. *PLoS Genetics* 2014, **10**(12):e1004837.

205. An Y, Han X, Tang S, Xia X, Yin W: Poplar GATA transcription factor PdGNC is capable of regulating chloroplast ultrastructure, photosynthesis, and vegetative growth in Arabidopsis under varying nitrogen levels. *Plant Cell, Tissue and Organ Culture (PCTOC)* 2014, **119**(2):313–327.

206. Spreitzer RJ, Salvucci ME: Rubisco: Structure, regulatory interactions, and possibilities for a better enzyme. *Annual Review of Plant Biology* 2002, **53**(1):449–475.

207. Chan PH, Wildman SG: Chloroplast DNA codes for primary structure of large subunit of fraction I protein. *Biochimica et Biophysica Acta* 1972, **277**(3):677–680.

208. Dunsmuir P, Bedbrook J: Chlorophyll a/b binding proteins and the small subunit of ribulose bisphosphate carboxylase are encoded by multiple genes in petunia. In: *Structure and Function of Plant Genomes* (Ciferri O ed.). New York: Springer; 1983: 221–230.

209. Schmidt GW, Mishkind ML: Rapid degradation of unassembled ribulose 1, 5-bisphosphate carboxylase small subunits in chloroplasts. *Proceedings of the National Academy of Sciences of the United States of America* 1983, **80**(9):2632–2636.

210. Rodermel SR, Abbott MS, Bogorad L: Nuclear-organelle interactions: Nuclear antisense gene inhibits ribulose bisphosphate carboxylase enzyme levels in transformed tobacco plants. *Cell* 1988, **55**(4):673–681.

211. Jiang CZ, Rodermel SR: Regulation of photosynthesis during leaf development in rbcS antisense DNA mutants of tobacco. *Plant Physiology* 1995, **107**(1):215–224.

212. Whitney SM, Sharwood RE, Orr D, White SJ, Alonso H, Galmés J: Isoleucine 309 acts as a C4 catalytic switch that increases ribulose-1, 5-bisphosphate carboxylase/oxygenase (Rubisco) carboxylation rate in Flaveria. *Proceedings of the National Academy of Sciences of the United States of America* 2011, **108**(35):14688–14693.

213. Parry MA, Andralojc PJ, Scales JC, Salvucci ME, Carmo-Silva AE, Alonso H, Whitney SM: Rubisco activity and regulation as targets for crop improvement. *Journal of Experimental Botany* 2013, **64**(3):717–730.

214. Parry MA, Keys AJ, Madgwick PJ, Carmo-Silva AE, Andralojc PJ: Rubisco regulation: A role for inhibitors. *Journal of Experimental Botany* 2008, **59**(7):1569–1580.

215. Carmo-Silva E, Scales JC, Madgwick PJ, Parry MA: Optimizing Rubisco and its regulation for greater resource use efficiency. *Plant, Cell & Environment* 2015, **38**:1817–1832.

216. Carmo-Silva AE, Salvucci ME: The regulatory properties of Rubisco activase differ among species and affect photosynthetic induction during light transitions. *Plant Physiology* 2013, **161**(4):1645–1655.

217. Mott KA, Woodrow IE: Modelling the role of Rubisco activase in limiting non-steady-state photosynthesis. *Journal of Experimental Botany* 2000, **51**:399–406.

218. Dean C, Vandenelzen P, Tamaki S, Dunsmuir P, Bedbrook J: Differential expression of the 8 genes of the petunia ribulose bisphosphate carboxylase small subunit multi-gene family. *EMBO Journal* 1985, **4**(12):3055–3061.

219. Giuliano G, Pichersky E, Malik VS, Timko MP, Scolnik PA, Cashmore AR: An evolutionarily conserved protein-binding sequence upstream of a plant light-regulated gene. *Proceedings of the National Academy of Sciences of the United States of America* 1988, **85**(19):7089–7093.

220. Gidoni D, Brosio P, Bondnutter D, Bedbrook J, Dunsmuir P: Novel cis-acting elements in Petunia Cab gene promoters. *Molecular & General Genetics* 1989, **215**(2):337–344.

221. Manzara T, Carrasco P, Gruissem W: Developmental and organ-specific changes in promoter DNA-protein interactions in the tomato rbcS gene family. *Plant Cell* 1991, **3**(12):1305–1316.

222. Quail PH: Phytochrome—A light-activated molecular switch that regulates plant gene-expression. *Annual Review of Genetics* 1991, **25**:389–409.

223. Tomizawa K, Nagatani A, Furuya M: Phytochrome genes—Studies using the tools of molecular-biology and photomorphogenetic mutants. *Photochemistry and Photobiology* 1990, **52**(1):265–275.

224. Sato N: Nucleotide-sequence and expression of the phytochrome gene in Pisum-sativum—Differential regulation by light of multiple transcripts. *Plant Molecular Biology* 1988, **11**(5):697–710.

225. Sharrock RA, Quail PH: Novel phytochrome sequences in Arabidopsis-thaliana—Structure, evolution, and differential expression of a plant regulatory photoreceptor family. *Genes & Development* 1989, **3**(11):1745–1757.

226. Christensen AH, Quail PH: Structure and expression of a maize phytochrome-encoding gene. *Gene* 1989, **85**(2):381–390.

227. Kay SA, Keith B, Shinozaki K, Chye ML, Chua NH: The rice phytochrome gene—Structure, autoregulated expression, and binding of GT-1 to a conserved site in the 5′ upstream region. *Plant Cell* 1989, **1**(3):351–360.

228. Lissemore JL, Colbert JT, Quail PH: Cloning of cDNA for phytochrome from etiolated cucurbita and coordinate photoregulation of the abundance of 2 distinct phytochrome transcripts. *Plant Molecular Biology* 1987, **8**(6):485–496.

229. Stayton MM, Brosio P, Dunsmuir P: Photosynthetic genes of Petunia (Mitchell) are differentially expressed during the diurnal cycle. *Plant Physiology* 1989, **89**(3):776–782.

230. Schindler U, Cashmore AR: Photoregulated gene-expression may involve ubiquitous DNA-binding proteins. *EMBO Journal* 1990, **9**(11):3415–3427.

231. Green PJ, Kay SA, Chua NH: Sequence-specific interactions of a pea nuclear factor with light-responsive elements upstream of the rbcS-3a gene. *EMBO Journal* 1987, **6**(9):2543–2549.

232. Green PJ, Yong MH, Cuozzo M, Kanomurakami Y, Silverstein P, Chua NH: Binding-site requirements for pea nuclear-protein factor GT-1 correlate with sequences required for light-dependent transcriptional activation of the rbcS-3a gene. *EMBO Journal* 1988, **7**(13):4035–4044.

233. Lam E, Chua NH: GT-1 binding-site confers light responsive expression in transgenic tobacco. *Science* 1990, **248**(4954):471–474.

234. Lam E, Kanomurakami Y, Gilmartin P, Niner B, Chua NH: A metal-dependent DNA-binding protein interacts with a constitutive element of a light-responsive promoter. *Plant Cell* 1990, **2**(9):857–866.

235. Kwon HB, Park SC, Peng HP, Goodman HM, Dewdney J, Shih MC: Identification of a light-responsive region of the nuclear gene encoding the B-subunit of chloroplast glyceraldehyde-3-phosphate dehydrogenase from Arabidopsis-thaliana. *Plant Physiology* 1994, **105**(1):357–367.

236. Gilmartin PM, Memelink J, Hiratsuka K, Kay SA, Chua NH: Characterization of a gene encoding a DNA-binding protein with specificity for a light-responsive element. *Plant Cell* 1992, **4**(7):839–849.

237. Perisic O, Lam E: A tobacco DNA-binding protein that interacts with a light-responsive box-II element. *Plant Cell* 1992, **4**(7):831–838.

238. Pfannschmidt T, Nilsson A, Allen JF: Photosynthetic control of chloroplast gene expression. *Nature* 1999, **397**(6720):625–628.

239. Tyagi AK, Gaur T: Light regulation of nuclear photosynthetic genes in higher plants. *Critical Reviews in Plant Sciences* 2003, **22**(5):417–452.

240. Sears BB: Elimination of plastids during spermatogenesis and fertilization in the plant kingdom. *Plasmid* 1980, **4**(3):233–255.

241. Correns C: Vererbungsversuche mit blass (gelb)-grünen und buntblättrigen Sippen bei Mirabilis jalapa, Urtica pilulifera und Lunaria annua. *Z Indukt Abstamm Vererbungsl* 1909, **1**:291–329.

242. Baur E: "Varietates albomarginatae hort" von *Palargonium zonale*. *Z Indukt Abstamm-Vererbungsl* 1909, **1**:330–351.

243. Bedbrook JR, Kolodner R: Structure of chloroplast DNA. *Annual Review of Plant Physiology and Plant Molecular Biology* 1979, **30**:593–620.

244. Metzlaff M, Borner T, Hagemann R: Variations of chloroplast DNAs in the genus Pelargonium and their biparental inheritance. *Theoretical and Applied Genetics* 1981, **60**(1):37–41.

245. Medgyesy P, Pay A, Marton L: Transmission of paternal chloroplasts in Nicotiana. *Molecular & General Genetics* 1986, **204**(2):195–198.

246. Lee DJ, Blake TK, Smith SE: Biparental inheritance of chloroplast DNA and the existence of heteroplasmic cells in alfalfa. *Theoretical and Applied Genetics* 1988, **76**(4):545–549.

247. Schumann CM, Hancock JF: Paternal inheritance of plastids in Medicago-sativa. *Theoretical and Applied Genetics* 1989, **78**(6):863–866.

248. Cruzan MB, Arnold ML, Carney SE, Wollenberg KR: cpDNA inheritance in interspecific crosses and evolutionary inference in Louisiana irises. *American Journal of Botany* 1993, **80**(3):344–350.

249. Chong DKX, Chinnappa CC, Yeh FC, Chuong S: Chloroplast DNA inheritance in the Stellaria longipes complex (Caryophyllaceae). *Theoretical and Applied Genetics* 1994, **88**(5):614–617.

250. Yao JL, Cohen D, Rowland RE: Inheritance and plastome genome incompatibility in interspecific hybrids of Zantedeschia (Araceae). *Theoretical and Applied Genetics* 1994, **88**(2):255–260.

251. McKinnon GE, Vaillancourt RE, Tilyard PA, Potts BM: Maternal inheritance of the chloroplast genome in Eucalyptus globulus and interspecific hybrids. *Genome* 2001, **44**(5):831–835.

252. Masoud SA, Johnson LB, Sorensen EL: High transmission of paternal plastid DNA in alfalfa plants demonstrated by restriction fragment polymorphic analysis. *Theoretical and Applied Genetics* 1990, **79**(1):49–55.

253. Boblenz K, Nothnagel T, Metzlaff M: Paternal inheritance of plastids in the genus Daucus. *Molecular & General Genetics* 1990, **220**(3):489–491.

254. Wagner DB, Govindaraju DR, Yeatman CW, Pitel JA: Paternal chloroplast DNA inheritance in a diallel cross of jack pine (Pinus-banksiana Lamb). *Journal of Heredity* 1989, **80**(6):483–485.

255. Szmidt AE, Alden T, Hallgren JE: Paternal inheritance of chloroplast DNA in Larix. *Plant Molecular Biology* 1987, **9**(1):59–64.

256. Neale DB, Wheeler NC, Allard RW: Paternal inheritance of chloroplast DNA in Douglas-fir. *Canadian Journal of Forest Research-Revue Canadienne De Recherche Forestiere* 1986, **16**(5):1152–1154.

257. Stine M, Keathley D: Paternal inheritance of plastids in Engelmann spruce x blue spruce hybrids. *Journal of Heredity* 1990, **81**(6):443–446.

258. Hagemann R, Schroder MB: The cytological basis of the plastid inheritance in angiosperms. *Protoplasma* 1989, **152**(2–3):57–64.

259. Corriveau JL, Coleman AW: Rapid screening method to detect potential biparental inheritance of plastid DNA and results for over 200 angiosperm species. *American Journal of Botany* 1988, **75**(10):1443–1458.

260. Harris SA, Ingram R: Chloroplast DNA and biosystematics—The effects of intraspecific diversity and plastid transmission. *Taxon* 1991, **40**(3):393–412.

261. Smith S: Biparental inheritance of organelles and its implications in crop improvement. *Plant Breeding Reviews* 1989, **6**:361–393.

262. Metzlaff M, Pohlheim F, Borner T, Hagemann R: Hybrid variegation in the genus Pelargonium. *Current Genetics* 1982, **5**(3):245–249.

263. Tilney-Bassett RAE: The control of plastid inheritance in Pelargonium II. *Heredity* 1973, **30**:1–13.

264. Tilney-Bassett RAE, Almouslem A: Variation in plastid inheritance between pelargonium cultivars and their hybrids. *Heredity* 1989, **63**:145–153.

265. Smith SE: Influence of parental genotype on plastid inheritance in Medicago-sativa. *Journal of Heredity* 1989, **80**(3):214–217.

266. Smith SE, Bingham ET, Fulton RW: Transmission of chlorophyll deficiencies in Medicago-sativa—Evidence for biparental inheritance of plastids. *Journal of Heredity* 1986, **77**(1):35–38.

267. Chiu WL, Stubbe W, Sears BB: Plastid inheritance in Oenothera—Organelle genome modifies the extent of biparental plastid transmission. *Current Genetics* 1988, **13**(2):181–189.

268. Cornu A, Dulieu H: Pollen transmission of plastid-DNA under genotypic control in Petunia-hybrida Hort. *Journal of Heredity* 1988, **79**(1):40–44.

269. Tilney-Bassett R, Almouslem A, Amoatey H: Complementary genes control biparental plastid inheritance in Pelargonium. *Theoretical and Applied Genetics* 1992, **85**(2–3):317–324.

270. McCauley DE, Sundby AK, Bailey MF, Welch ME: Inheritance of chloroplast DNA is not strictly maternal in Silene vulgaris (Caryophyllaceae): Evidence from experimental crosses and natural populations. *American Journal of Botany* 2007, **94**(8):1333–1337.

271. Hansen AK, Escobar LK, Gilbert LE, Jansen RK: Paternal, maternal, and biparental inheritance of the chloroplast genome in Passiflora (Passifloraceae): Implications for phylogenetic studies. *American Journal of Botany* 2007, **94**(1):42–46.

272. Whitfeld PR, Bottomley W: Organization and structure of chloroplast genes. *Annual Review of Plant Physiology and Plant Molecular Biology* 1983, **34**:279–310.

273. Tyagi A, Kelkar N, Kapoor S, Maheshwari S: The chloroplast genome: Genetic potential and its expression. In: *Photosynthesis: Photoreactions to Plant Productivity* (Abrol YP, Mohanty P, Govindjee eds.). The Netherlands: Springer; 1993: 3–47.

274. Thomas J, Tewari K: Conservation of 70S ribosomal RNA genes in the chloroplast DNAs of higher plants. *Proceedings of the National Academy of Sciences of the United States of America* 1974, **71**(8):3147–3151.

275. Bedbrook JR, Kolodner R, Bogorad L: Zea-mays chloroplast ribosomal-rna genes are part of a 22,000 base pair inverted repeat. *Cell* 1977, **11**(4):739–749.

276. Steinmetz A, Mubumbila M, Keller M, Burkard G, Weil JH, Driesel AJ, Crouse EJ, Gordon K, Bohnert HJ, Herrmann RG: *Mapping of tRNA Genes on the Circular DNA Molecule of* Spinacia oleracea *Chloroplasts.* Amsterdam: Biomedical Press, Elsevier; 1978.

277. Gounaris K, Barber J, Harwood JL: The thylakoid membranes of higher-plant chloroplasts. *Biochemical Journal* 1986, **237**(2):313–326.

278. Herrmann R, Possingham J: Plastid DNA—The plastome. In: *Chloroplasts* (Reinert J ed.). Berlin Heidelberg: Springer; 1980: 45–96.

279. Miyamura S, Nagata T, Kuroiwa T: Quantitative fluorescence microscopy on dynamic changes of plastid nucleoids during wheat development. *Protoplasma* 1986, **133**(1):66–72.

280. Kowallik K, Herrmann R: Structural and functional aspects of the plastome. II. DNA regions during plastid development. *Portugaliae Acta Biologica Serie A* 1974, **14**:111–126.

281. Mullet JE: Chloroplast development and gene-expression. *Annual Review of Plant Physiology and Plant Molecular Biology* 1988, **39**:475–502.

282. Ohyama K, Fukuzawa H, Kohchi T, Shirai H, Sano T, Sano S, Umesono K, Shiki Y, Takeuchi M, Chang Z et al.: Chloroplast gene organization deduced from complete sequence of liverwort Marchantia-polymorpha chloroplast DNA. *Nature* 1986, **322**(6079):572–574.

283. Shinozaki K, Ohme M, Tanaka M, Wakasugi T, Hayashida N, Matsubayashi T, Zaita N, Chunwongse J, Obokata J, Yamaguchishinozaki K et al.: The complete nucleotide-sequence of the tobacco chloroplast genome—Its gene organization and expression. *EMBO Journal* 1986, **5**(9):2043–2049.

284. Hiratsuka J, Shimada H, Whittier R, Ishibashi T, Sakamoto M, Mori M, Kondo C, Honji Y, Sun CR, Meng BY et al.: The complete sequence of the rice (Oryza-sativa) chloroplast genome—Intermolecular recombination between distinct transfer-RNA genes accounts for a major plastid DNA inversion during the evolution of the cereals. *Molecular & General Genetics* 1989, **217**(2–3):185–194.

285. Daniell H, Lee S-B, Grevich J, Saski C, Quesada-Vargas T, Guda C, Tomkins J, Jansen RK: Complete chloroplast genome sequences of Solanum bulbocastanum, Solanum lycopersicum and comparative analyses with other Solanaceae genomes. *Theoretical and Applied Genetics* 2006, **112**(8):1503–1518.

286. Bausher MG, Singh ND, Lee S-B, Jansen RK, Daniell H: The complete chloroplast genome sequence of Citrus sinensis (L.) Osbeck var 'Ridge Pineapple': Organization and phylogenetic relationships to other angiosperms. *BMC Plant Biology* 2006, **6**(1):21.

287. Yang Y, Yuanye D, Qing L, Jinjian L, Xiwen L, Yitao W: Complete chloroplast genome sequence of poisonous and medicinal plant Datura stramonium: Organizations and implications for genetic engineering. *PLoS One* 2014, **9**(11):e110656.

288. Coen DM, Bedbrook JR, Bogorad L, Rich A: Maize chloroplast DNA fragment encoding large subunit of ribulose bisphosphate carboxylase. *Proceedings of the National Academy of Sciences of the United States of America* 1977, **74**(12):5487–5491.

289. Koller B, Gingrich JC, Stiegler GL, Farley MA, Delius H, Hallick RB: 9 introns with conserved boundary sequences in the euglena gracilis chloroplast ribulose-1,5-bisphosphate carboxylase gene. *Cell* 1984, **36**(2):545–553.

290. Zurawski G, Bohnert HJ, Whitfeld PR, Bottomley W: Nucleotide-sequence of the gene for the Mr 32,000 thylakoid membrane-protein from Spinacia-oleracea and Nicotiana debney predicts a totally conserved primary translation product of Mr 38,950. *Proceedings of the National Academy of Sciences of the United States of America-Biological Sciences* 1982, **79**(24):7699–7703.

291. Erickson JM, Rahire M, Rochaix JD: Chlamydomonas reinhardtii gene for the 32000 mol wt protein of photosystem-II contains 4 large introns and is located entirely within the chloroplast inverted repeat. *EMBO Journal* 1984, **3**(12): 2753–2762.

292. Karabin GD, Farley M, Hallick RB: Chloroplast gene for Mr 32000 polypeptide of photosystem-II in Euglena-gracilis is interrupted by 4 introns with conserved boundary sequences. *Nucleic Acids Research* 1984, **12**(14):5801–5812.

293. Keller M, Stutz E: Structure of the Euglena-gracilis chloroplast gene (psbA) coding for the 32-kDa protein of photosystem-II. *FEBS Letters* 1984, **175**(1):173–177.

294. Morris J, Herrmann RG: Nucleotide-sequence of the gene for the P680 chlorophyll-a apoprotein of the photosystem-ii reaction center from spinach. *Nucleic Acids Research* 1984, **12**(6):2837–2850.

295. Alt J, Morris J, Westhoff P, Herrmann RG: Nucleotide-sequence of the clustered genes for the 44-kd chlorophyll-a apoprotein and the 32kd-like protein of the photosystem-II reaction center in the spinach plastid chromosome. *Current Genetics* 1984, **8**(8):597–606.

296. Holschuh K, Bottomley W, Whitfeld PR: Structure of the spinach chloroplast genes for the D2 and 44-kd reaction-center proteins of photosystem-II and for tRNASer (UGA). *Nucleic Acids Research* 1984, **12**(23):8819–8834.

297. Rasmussen OF, Bookjans G, Stummann BM, Henningsen KW: Localization and nucleotide-sequence of the gene for the membrane polypeptide-D2 from pea chloroplast DNA. *Plant Molecular Biology* 1984, **3**(4):191–199.

298. Herrmann RG, Alt J, Schiller B, Widger WR, Cramer WA: Nucleotide-sequence of the gene for apocytochrome B-559 on the spinach plastid chromosome—Implications for the structure of the membrane-protein. *FEBS Letters* 1984, **176**(1):239–244.

299. Steinmetz AA, Castroviejo M, Sayre RT, Bogorad L: Protein PSII-G—An additional component of photosystem-II identified through its plastid gene in maize. *Journal of Biological Chemistry* 1986, **261**(6):2485–2488.

300. Westhoff P, Farchaus JW, Herrmann RG: The gene for the M-R 10,000 phosphoprotein associated with photosystem-II is part of the psbB operon of the spinach plastic chromosome. *Current Genetics* 1986, **11**(3):165–169.

301. Hird SM, Dyer TA, Gray JC: The gene for the 10 kDa phosphoprotein of photosystem-II is located in chloroplast DNA. *FEBS Letters* 1986, **209**(2):181–186.

302. Cushman JC, Christopher DA, Little MC, Hallick RB, Price CA: Organization of the psbE, psbF, orf38, and orf42 gene loci on the Euglena-gracilis chloroplast genome. *Current Genetics* 1988, **13**(2):173–180.

303. Cantrell A, Bryant DA: Molecular-cloning and nucleotide-sequence of the psaA and psaB genes of the cyanobacterium Synechococcus sp. PCC-7002. *Plant Molecular Biology* 1987, **9**(5):453–468.

304. Kawaguchi H, Fukuda I, Shiina T, Toyoshima Y: Dynamical behavior of psb gene transcripts in greening wheat seedlings. I. Time course of accumulation of the pshA through psbN gene transcripts during light-induced greening. *Plant Molecular Biology* 1992, **20**(4):695–704.

305. Yoo K-Y, Park D-S, Tae G-S: Molecular cloning and characterization of the psbL and psbJ genes for photosystem II from *Panax ginseng*. *Journal of Plant Biology* 2004, **47**(3):203–209.

306. Murata N, Miyao M, Hayashida N, Hidaka T, Sugiura M: Identification of a new gene in the chloroplast genome encoding a low-molecular-mass polypeptide of photosystem-II complex. *FEBS Letters* 1988, **235**(1–2):283–288.

307. Webber AN, Hird SM, Packman LC, Dyer TA, Gray JC: A photosystem-II polypeptide is encoded by an open reading frame cotranscribed with genes for cytochrome b-559 in wheat chloroplast DNA. *Plant Molecular Biology* 1989, **12**(2):141–151.

308. Ikeuchi M, Koike H, Inoue Y: N-terminal sequencing of low-molecular-mass components in cyanobacterial photosystem-II core complex—2 components correspond to unidentified open reading frames of plant chloroplast DNA. *FEBS Letters* 1989, **253**(1–2):178–182.

309. Gray JC, Webber AN, Hird SM, Willey DL, Dyer TA : *Current Research in Photosynthesis*, vol. III. Netherlands: Kluwer Academic Publishers; 1990: 461–468.

310. Kurisu G, Zhang H, Smith JL, Cramer WA: Structure of the cytochrome b6f complex of oxygenic photosynthesis: Tuning the cavity. *Science* 2003, **302**(5647):1009–1014.

311. Baniulis D, Yamashita E, Whitelegge JP, Zatsman AI, Hendrich MP, Hasan SS, Ryan CM, Cramer WA: Structure-function, stability, and chemical modification of the cyanobacterial cytochrome b6f complex from Nostoc sp. PCC 7120. *Journal of Biological Chemistry* 2009, **284**(15):9861–9869.

312. Stroebel D, Choquet Y, Popot JL, Picot D: An atypical haem in the cytochrome b(6)f complex. *Nature* 2003, **426**(6965): 413–418.

313. Cramer WA, Zhang H, Yan J, Kurisu G, Smith JL: Transmembrane traffic in the cytochrome b6f complex. *Annual Review of Biochemistry* 2006, **75**:769–790.

314. Hojka M, Thiele W, Toth SZ, Lein W, Bock R, Schottler MA: Inducible repression of nuclear-encoded subunits of the cytochrome b6f complex in tobacco reveals an extraordinarily long lifetime of the complex. *Plant Physiology* 2014, **165**(4):1632–1646.

315. Westhoff P, Alt J, Nelson N, Bottomley W, Bunemann H, Herrmann RG: Genes and transcripts for the P700-chlorophyll-alpha-apoprotein and subunit-2 of the photosystem-I reaction center complex from spinach thylakoid membranes. *Plant Molecular Biology* 1983, **2**(2):95–107.

316. Fish LE, Kuck U, Bogorad L: 2 partially homologous adjacent light-inducible maize chloroplast genes encoding polypeptides of the P700 chlorophyll-a protein complex of photosystem-I. *Journal of Biological Chemistry* 1985, **260**(3):1413–1421.

317. Sugiura M: The chloroplast chromosomes in land plants. *Annual Review of Cell Biology* 1989, **5**:51–70.

318. Hayashida N, Matsubayashi T, Shinozaki K, Sugiura M, Inoue K, Hiyama T: The gene for the 9 kd polypeptide, a possible apoprotein for the iron-sulfur center-A and center-B of the photosystem-I complex, in tobacco chloroplast DNA. *Current Genetics* 1987, **12**(4):247–250.

319. Scheller HV, Okkels JS, Hoj PB, Svendsen I, Roepstorff P, Moller BL: The primary structure of a 4.0-kDa photosystem-I polypeptide encoded by the chloroplast psaI gene. *Journal of Biological Chemistry* 1989, **264**(31):18402–18406.

320. Koike H, Ikeuchi M, Hiyama T, Inoue Y: Identification of photosystem-I components from the cyanobacterium, Synechococcus vulcanus by N-terminal sequencing. *FEBS Letters* 1989, **253**(1–2):257–263.

321. Puthiyaveetil S, Allen JF: Transients in chloroplast gene transcription. *Biochemical and Biophysical Research Communications* 2008, **368**(4):871–874.

322. Baginsky S, Tiller K, Pfannschmidt T, Link G: PTK, the chloroplast RNA polymerase-associated protein kinase from mustard (Sinapis alba), mediates redox control of plastid in vitro transcription. *Plant Molecular Biology* 1999, **39**(5):1013–1023.

323. Ogrzewalla K, Piotrowski M, Reinbothe S, Link G: The plastid transcription kinase from mustard (Sinapis alba L.). *European Journal of Biochemistry* 2002, **269**(13):3329–3337.

324. Shimizu M, Kato H, Ogawa T, Kurachi A, Nakagawa Y, Kobayashi H: Sigma factor phosphorylation in the photosynthetic control of photosystem stoichiometry. *Proceedings of the National Academy of Sciences of the United States of America* 2010, **107**(23):10760–10764.

325. Westhoff P, Alt J, Nelson N, Herrmann RG: Genes and transcripts for the ATP synthase CF0 subunit-1 and subunit-2 from spinach thylakoid membranes. *Molecular & General Genetics* 1985, **199**(2):290–299.

326. Alt J, Winter P, Sebald W, Moser JG, Schedel R, Westhoff P, Herrmann RG: Localization and nucleotide-sequence of the gene for the ATP synthase proteolipid subunit on the spinach plastid chromosome. *Current Genetics* 1983, **7**(2):129–138.

327. Bird CR, Koller B, Auffret AD, Huttly AK, Howe CJ, Dyer TA, Gray JC: The wheat chloroplast gene for CF0 subunit-I of ATP synthase contains a large intron. *EMBO Journal* 1985, **4**(6):1381–1388.

328. Hennig J, Herrmann RG: Chloroplast ATP synthase of spinach contains 9 nonidentical subunit species, 6 of which are encoded by plastid chromosomes in 2 operons in a phylogenetically conserved arrangement. *Molecular & General Genetics* 1986, **203**(1):117–128.

329. Hudson GS, Mason JG, Holton TA, Koller B, Cox GB, Whitfeld PR, Bottomley W: A gene-cluster in the spinach and pea chloroplast genomes encoding one CF1 and 3 CF0 subunits of the H+-ATP synthase complex and the ribosomal protein-S2. *Journal of Molecular Biology* 1987, **196**(2):283–298.

330. Krebbers ET, Larrinua IM, Mcintosh L, Bogorad L: The maize chloroplast genes for the beta-subunit and epsilon-subunit of the photosynthetic coupling factor CF1 are fused. *Nucleic Acids Research* 1982, **10**(16):4985–5002.

331. Inohara N, Iwamoto A, Moriyama Y, Shimomura S, Maeda M, Futai M: Two genes, atpC1 and atpC2, for the gamma subunit of Arabidopsis thaliana chloroplast ATP synthase. *Journal of Biological Chemistry* 1991, **266**(12):7333–7338.

332. Legen J, Miséra S, Herrrmann RG, Meurer J: Map positions of 69 Arabidopsis thaliana genes of all known nuclear encoded constituent polypeptides and various regulatory factors of the photosynthetic membrane: A case study. *DNA Research* 2001, **8**(2):53–60.

333. Kohzuma K, Dal Bosco C, Kanazawa A, Dhingra A, Nitschke W, Meurer J, Kramer DM: Thioredoxin-insensitive plastid ATP synthase that performs moonlighting functions. *Proceedings of the National Academy of Sciences of the United States of America* 2012, **109**(9):3293–3298.

334. Nandha B, Finazzi G, Joliot P, Hald S, Johnson GN: The role of PGR5 in the redox poising of photosynthetic electron transport. *Biochimica et Biophysica Acta* 2007, **1767**(10):1252–1259.

335. Joliot P, Johnson GN: Regulation of cyclic and linear electron flow in higher plants. *Proceedings of the National Academy of Sciences of the United States of America* 2007, **108**(32):13317–13322.

336. Avenson TJ, Cruz JA, Kanazawa A, Kramer DM: Regulating the proton budget of higher plant photosynthesis. *Proceedings of the National Academy of Sciences of the United States of America* 2005, **102**(27):9709–9713.

337. Neng BY, Matsubayashi T, Wakasugi T, Shinozaki K, Sugiura M, Hirai A, Mikami T, Kishima Y, Kinoshita T: Ubiquity of the genes for components of a NADH dehydrogenase in higher plant chloroplast genomes. *Plant Science* 1986, **47**:181–184.

338. Schantz R, Bogorad L: Maize chloroplast genes ndhD, ndhE, and psaC—Sequences, transcripts and transcript pools. *Plant Molecular Biology* 1988, **11**(3):239–247.

339. Rochaix JD: Posttranscriptional steps in the expression of chloroplast genes. *Annual Review of Cell Biology* 1992, **8**:1–28.

340. Cheng SH, Cline K, Delisle AJ: An Arabidopsis chloroplast RNA-binding protein gene encodes multiple messenger-RNAs with different 5′ ends. *Plant Physiology* 1994, **106**(1):303–311.

341. Allen JF: Redox control of gene expression and the function of chloroplast genomes—An hypothesis. *Photosynthesis Research* 1993, **36**(2):95–102.

342. Liere K, Weihe A, Borner T: The transcription machineries of plant mitochondria and chloroplasts: Composition, function, and regulation. *Journal of Plant Physiology* 2011, **168**(12):1345–1360.

343. Schweer J, Türkeri H, Kolpack A, Link G: Role and regulation of plastid sigma factors and their functional interactions during chloroplast transcription. Recent lessons from Arabidopsis thaliana. *European Journal of Cell Biology* 2010, **144**:1924–1935.

344. Lerbs-Mache S: Function of plastid sigma factors in higher plants: Regulation of gene expression or just preservation of constitutive transcription? *Plant Molecular Biology* 2011, **76**(3–5):235–249.

345. Lyska D, Meierhoff K, Westhoff P: How to build functional thylakoid membranes: From plastid transcription to protein complex assembly. *Planta* 2013, **237**(2):413–428.

346. Steiner S, Schroter Y, Pfalz J, Pfannschmidt T: Identification of essential subunits in the plastid-encoded RNA polymerase complex reveals building blocks for proper plastid development. *Plant Physiology* 2011, **157**(3):1043–1055.

347. Deng XW, Stern DB, Tonkyn JC, Gruissem W: Plastid run-on transcription—Application to determine the transcriptional regulation of spinach plastid genes. *Journal of Biological Chemistry* 1987, **262**(20):9641–9648.

348. Mullet JE, Klein RR: Transcription and RNA stability are important determinants of higher-plant chloroplast RNA Levels. *EMBO Journal* 1987, **6**(6):1571–1579.

349. Deng XW, Gruissem W: Constitutive transcription and regulation of gene-expression in non-photosynthetic plastids of higher-plants. *EMBO Journal* 1988, **7**(11):3301–3308.

350. Brawerman G: Messenger-RNA decay—Finding the right targets. *Cell* 1989, **57**(1):9–10.

351. Salvador ML, Klein U, Bogorad L: Light-regulated and endogenous fluctuations of chloroplast transcript levels in chlamydomonas—Regulation by transcription and RNA degradation. *Plant Journal* 1993, **3**(2):213–219.

352. Salvador ML, Klein U, Bogorad L: 5′ sequences are important positive and negative determinants of the longevity of chlamydomonas chloroplast gene transcripts. *Proceedings of the National Academy of Sciences of the United States of America* 1993, **90**(4):1556–1560.

353. Gruissem W, Callan K, Lynch J, Manzara T, Meighan M, Narita J, Piechulla B, Sugita M, Thelander M, Wanner L: Plastid and nuclear gene expression during tomato fruit formation. *Tomato Biotechnology (USA)* 1987, 239–249.

354. Eisermann A, Tiller K, Link G: In vitro transcription and DNA-binding characteristics of chloroplast and etioplast extracts from mustard (Sinapis-alba) indicate differential usage of the psbA promoter. *EMBO Journal* 1990, **9**(12):3981–3987.

355. Stern DB, Jones H, Gruissem W: Function of plastid messenger-RNA 3′ inverted repeats—RNA stabilization and gene-specific protein-binding. *Journal of Biological Chemistry* 1989, **264**(31): 18742–18750.

356. Stern DB, Gruissem W: Control of plastid gene-expression—3′ inverted repeats act as messenger-RNA processing and stabilizing elements, but do not terminate transcription. *Cell* 1987, **51**(6):1145–1157.

357. Stern DB, Radwanski ER, Kindle KL: A 3' stem/loop structure of the Chlamydomonas chloroplast atpB gene regulates messenger-RNA accumulation in vivo. *Plant Cell* 1991, **3**(3):285–297.

358. Adams CC, Stern DB: Control of messenger-RNA stability in chloroplasts by 3' inverted repeats—Effects of stem and loop mutations on degradation of psbA messenger-RNA in vitro. *Nucleic Acids Research* 1990, **18**(20):6003–6010.

359. Klaff P, Gruissem W: Changes in chloroplast messenger-RNA stability during leaf development. *Plant Cell* 1991, **3**(5):517–529.

360. Cozens AL, Walker JE, Phillips AL, Huttly AK, Gray JC: A 6th subunit of ATP synthase, an F0 component, is encoded in the pea chloroplast genome. *EMBO Journal* 1986, **5**(2):217–222.

361. Rock CD, Barkan A, Taylor WC: The maize plastid psbB-psbF-petB-petD gene-cluster—Spliced and unspliced petB and petD RNAs encode alternative products. *Current Genetics* 1987, **12**(1):69–77.

362. Westhoff P, Herrmann RG: Complex RNA maturation in chloroplasts—The psbB operon from spinach. *European Journal of Biochemistry* 1988, **171**(3):551–564.

363. Li YQ, Sugiura M: 3 distinct ribonucleoproteins from tobacco chloroplasts—Each contains a unique amino terminal acidic domain and 2 ribonucleoprotein consensus motifs. *EMBO Journal* 1990, **9**(10):3059–3066.

364. Schuster G, Gruissem W: Chloroplast messenger-RNA 3' end processing requires a nuclear-encoded RNA-binding protein. *EMBO Journal* 1991, **10**(6):1493–1502.

365. Delisle AJ: RNA-binding protein from Arabidopsis. *Plant Physiology* 1993, **102**(1):313–314.

366. Cook WB, Walker JC: Identification of a maize nucleic acid-binding protein (NBP) belonging to a family of nuclear-encoded chloroplast proteins. *Nucleic Acids Research* 1992, **20**(2):359–364.

367. Preugschat F, Wold B: Isolation and characterization of a Xenopus laevis C-protein cDNA—Structure and expression of a heterogeneous nuclear ribonucleoprotein core protein. *Proceedings of the National Academy of Sciences of the United States of America* 1988, **85**(24):9669–9673.

368. Barzvi D, Shagan T, Schindler U, Cashmore AR: RNP-T, a ribonucleoprotein from Arabidopsis-thaliana, contains 2 RNP-80 motifs and a novel acidic repeat arranged in an alpha-helix conformation. *Plant Molecular Biology* 1992, **20**(5): 833–838.

369. Bandziulis RJ, Swanson MS, Dreyfuss G: RNA-binding proteins as developmental regulators. *Genes & Development* 1989, **3**(4):431–437.

370. Ludevid MD, Freire MA, Gomez J, Burd CG, Albericio F, Giralt E, Dreyfuss G, Pages M: RNA-binding characteristics of a 16 kDa glycine-rich protein from maize. *Plant Journal* 1992, **2**(6):999–1003.

371. Hirose T, Sugita M, Sugiura M: CDNA structure, expression and nucleic acid-binding properties of 3 RNA-binding proteins in tobacco—Occurrence of tissue-specific alternative splicing. *Nucleic Acids Research* 1993, **21**(17):3981–3987.

372. Smith CWJ, Patton JG, Nadalginard B: Alternative splicing in the control of gene-expression. *Annual Review of Genetics* 1989, **23**:527–577.

373. Bollenbach TJ, Schuster G, Portnoy V, Stern DB: Processing, degradation, and polyadenylation of chloroplast transcripts. *Topics in Current Genetics* 2007, **19**:175–211.

374. Schmitz-Linneweber C, Barkan A: RNA splicing and RNA editing in chloroplasts. *Topics in Current Genetics* 2007, **19**:213–248.

375. Berry JO, Carr JP, Klessig DF: Messenger-RNAs encoding ribulose-1,5-bisphosphate carboxylase remain bound to polysomes but are not translated in amaranth seedlings transferred to darkness. *Proceedings of the National Academy of Sciences of the United States of America* 1988, **85**(12):4190–4194.

376. Fromm H, Devic M, Fluhr R, Edelman M: Control of psbA gene-expression—In mature Spirodela chloroplasts light regulation of 32-kd protein-synthesis is independent of transcript level. *EMBO Journal* 1985, **4**(2):291–295.

377. Klein RR, Mason HS, Mullet JE: Light-regulated translation of chloroplast proteins. 1. Transcripts of psaA-psaB, psbA, and rbcL are associated with polysomes in dark-grown and illuminated barley seedlings. *Journal of Cell Biology* 1988, **106**(2):289–301.

378. Choquet Y, Wostrikoff K, Rimbault B, Zito F, Girard-Bascou J, Drapier D, Wollman FA: Assembly-controlled regulation of chloroplast gene translation. *Biochemical Society Transactions* 2001, **29**(Pt 4):421–426.

379. Eberhard S, Drapier D, Wollman FA: Searching limiting steps in the expression of chloroplast-encoded proteins: Relations between gene copy number, transcription, transcript abundance and translation rate in the chloroplast of Chlamydomonas reinhardtii. *The Plant Journal* 2002, **31**(2):149–160.

380. Raynaud C, Loiselay C, Wostrikoff K, Kuras R, Girard-Bascou J, Wollman FA, Choquet Y: Evidence for regulatory function of nucleus-encoded factors on mRNA stabilization and translation in the chloroplast. *Proceedings of the National Academy of Sciences of the United States of America* 2007, **104**(21):9093–9098.

381. Boulouis A, Raynaud C, Bujaldon S, Aznar A, Wollman FA, Choquet Y: The nucleus-encoded trans-acting factor MCA1 plays a critical role in the regulation of cytochrome f synthesis in Chlamydomonas chloroplasts. *Plant Cell* 2011, **23**(1):333–349.

382. Carter ML, Smith AC, Kobayashi H, Purton S, Herrin DL: Structure, circadian regulation and bioinformatic analysis of the unique sigma factor gene in Chlamydomonas reinhardtii. *Photosynthesis Research* 2004, **82**(3):339–349.

383. Bohne AV, Irihimovitch V, Weihe A, Stern DB: Chlamydomonas reinhardtii encodes a single sigma70-like factor which likely functions in chloroplast transcription. *Current Genetics* 2006, **49**(5):333–340.

384. Flexas J, Bota J, Loreto F, Cornic G, Sharkey T: Diffusive and metabolic limitations to photosynthesis under drought and salinity in C3 plants. *Plant Biology* 2004, **6**(3):269–279.

385. Flexas J, Diaz-Espejo A, Galmes J, Kaldenhoff R, Medrano H, Ribas-Carbo M: Rapid variations of mesophyll conductance in response to changes in CO2 concentration around leaves. *Plant, Cell & Environment* 2007, **30**(10):1284–1298.

386. Mehta P, Jajoo A, Mathur S, Bharti S: Chlorophyll a fluorescence study revealing effects of high salt stress on Photosystem II in wheat leaves. *Plant Physiology and Biochemistry* 2010, **48**(1):16–20.

387. Chaves M, Oliveira M: Mechanisms underlying plant resilience to water deficits: Prospects for water-saving agriculture. *Journal of Experimental Botany* 2004, **55**(407):2365–2384.

388. Ort DR: When there is too much light. *Plant Physiology* 2001, **125**(1):29–32.

389. Chaves M, Flexas J, Pinheiro C: Photosynthesis under drought and salt stress: Regulation mechanisms from whole plant to cell. *Annals of Botany* 2009, **103**(4):551–560.

390. Kalaji HM, Bosa K, Kościelniak J, Żuk-Gołaszewska K: Effects of salt stress on photosystem II efficiency and CO_2 assimilation of two Syrian barley landraces. *Environmental and Experimental Botany* 2011, **73**:64–72.

391. Tuteja N, Sahoo RK, Garg B, Tuteja R: OsSUV3 dual helicase functions in salinity stress tolerance by maintaining photosynthesis and antioxidant machinery in rice (Oryza sativa L. cv. IR64). *The Plant Journal* 2013, **76**(1):115–127.

392. Chen P, Yan K, Shao H, Zhao S: Physiological mechanisms for high salt tolerance in wild soybean (Glycine soja) from Yellow River Delta, China: photosynthesis, osmotic regulation, ion flux and antioxidant capacity. *PLoS One* 2013, **8**(12):e83227.

393. Vargas-Suárez M, Ayala-Ochoa A, Lozano-Franco J, García-Torres I, Díaz-Quiñonez A, Ortíz-Navarrete VF, Sánchez-de-Jiménez E: Rubisco activase chaperone activity is regulated by a post-translational mechanism in maize leaves. *Journal of Experimental Botany* 2004, **55**(408):2533–2539.

394. DeRidder BP, Salvucci ME: Modulation of Rubisco activase gene expression during heat stress in cotton (Gossypium hirsutum L.) involves post-transcriptional mechanisms. *Plant Science* 2007, **172**:246–254.

395. Zangerl A, Hamilton J, Miller T, Crofts A, Oxborough K, Berenbaum M, De Lucia E: Impact of folivory on photosynthesis is greater than the sum of its holes. *Proceedings of the National Academy of Sciences of the United States of America* 2002, **99**(2):1088–1091.

396. Bilgin DD, Zavala JA, Zhu J, Clough SJ, Ort DR, DeLUCIA E: Biotic stress globally downregulates photosynthesis genes. *Plant, Cell & Environment* 2010, **33**(10):1597–1613.

397. Debona D, Rodrigues FÁ, Rios JA, Martins SCV, Pereira LF, DaMatta FM: Limitations to photosynthesis in leaves of wheat plants infected by Pyricularia oryzae. *Phytopathology* 2014, **104**(1):34–39.

398. Garavaglia BS, Thomas L, Gottig N, Zimaro T, Garofalo CG, Gehring C, Ottado J: Shedding light on the role of photosynthesis in pathogen colonization and host defense. *Communicative & Integrative Biology* 2010, **3**(4):382–384.

399. Zhao J, Liu Q, Zhang H, Jia Q, Hong Y, Liu Y: RuBisCO small subunit is involved in Tobamovirus movement and Tm2²-mediated extreme resistance. *Plant Physiology* 2013, **161**(1):374–383.

400. Hahlbrock K, Bednarek P, Ciolkowski I, Hamberger B, Heise A, Liedgens H, Logemann E, Nürnberger T, Schmelzer E, Somssich IE: Non-self recognition, transcriptional reprogramming, and secondary metabolite accumulation during plant/pathogen interactions. *Proceedings of the National Academy of Sciences of the United States of America* 2003, **100**(Suppl 2):14569–14576.

401. Hermsmeier D, Schittko U, Baldwin IT: Molecular interactions between the specialist herbivore Manduca sexta (Lepidoptera, Sphingidae) and its natural host Nicotiana attenuata. I. Large-scale changes in the accumulation of growth-and defense-related plant mRNAs. *Plant Physiology* 2001, **125**(2):683–700.

402. Hui D, Iqbal J, Lehmann K, Gase K, Saluz HP, Baldwin IT: Molecular interactions between the specialist herbivore Manduca sexta (lepidoptera, sphingidae) and its natural host Nicotiana attenuata: V. microarray analysis and further characterization of large-scale changes in herbivore-induced mRNAs. *Plant Physiology* 2003, **131**(4):1877–1893.

403. Yan SP, Zhang QY, Tang ZC, Su WA, Sun WN: Comparative proteomic analysis provides new insights into chilling stress responses in rice. *Molecular & Cellular Proteomics* 2006, **5**(3):484–496.

404. Giri AP, Wünsche H, Mitra S, Zavala JA, Muck A, Svatoš A, Baldwin IT: Molecular interactions between the specialist herbivore Manduca sexta (Lepidoptera, Sphingidae) and its natural host Nicotiana attenuata. VII. Changes in the plant's proteome. *Plant Physiology* 2006, **142**(4):1621–1641.

405. Mitra S, Baldwin IT: Independently silencing two photosynthetic proteins in Nicotiana attenuata has different effects on herbivore resistance. *Plant Physiology* 2008, **148**(2):1128–1138.

406. Attaran E, Major I, Cruz J, Rosa B, Koo A, Chen J, Kramer D, He SY, Howe G: Temporal dynamics of growth and photosynthesis suppression in response to jasmonate signaling. *Plant Physiology* 2014, **165**(3):1302–1314.

407. Massad TJ, Dyer LA, Vega G: Costs of defense and a test of the carbon-nutrient balance and growth-differentiation balance hypotheses for two co-occurring classes of plant defense. *PLoS One* 2012, **7**(10):e47554.

408. McKey D: The distribution of secondary compounds within plants. In: *Herbivores: Their Interaction with Secondary Plant Metabolites.* (Rosenthal GA, Berenbaum MR eds.) New York: Academic Press; 1979: 55–133.

409. Coley PD, Bryant JP, Chapin FS, 3rd: Resource availability and plant antiherbivore defense. *Science* 1985, **230**(4728):895–899.

410. Millard P: The accumulation and storage of nitrogen by herbaceous plants. *Plant Cell Environment* 1988, **11**:1–8.

411. Matt P, Krapp A, Haake V, Mock HP, Stitt M: Decreased Rubisco activity leads to dramatic changes of nitrate metabolism, amino acid metabolism and the levels of phenylpropanoids and nicotine in tobacco antisense RBCS transformants. *The Plant Journal* 2002, **30**(6):663–677.

412. Ullmann-Zeunert L, Stanton MA, Wielsch N, Bartram S, Hummert C, Svatoš A, Baldwin IT, Groten K: Quantification of growth–defense trade-offs in a common currency: Nitrogen required for phenolamide biosynthesis is not derived from ribulose-1, 5-bisphosphate carboxylase/oxygenase turnover. *The Plant Journal* 2013, **75**(3):417–429.

413. Stanton MA, Ullmann-Zeunert L, Wielsch N, Bartram S, Svatoš A, Baldwin IT, Groten K: Silencing ribulose-1, 5-bisphosphate carboxylase/oxygenase expression does not disrupt nitrogen allocation to defense after simulated herbivory in Nicotiana attenuata. *Plant Signaling & Behavior* 2013, **8**(12):e27570.

414. Mitra S, Baldwin IT: RuBPCase activase (RCA) mediates growth–defense trade-offs: Silencing RCA redirects jasmonic acid (JA) flux from JA-isoleucine to methyl jasmonate (MeJA) to attenuate induced defense responses in Nicotiana attenuata. *New Phytologist* 2014, **201**(4):1385–1395.

415. Huot B, Yao J, Montgomery BL, He SY: Growth-defense trade-offs in plants: A balancing act to optimize fitness. *Molecular Plant* 2014, **7**(8):1267–1287.

416. Borlaug NE: Increasing and stabilizing food production. In: *Plant Breeding 2* (Frey KJ ed.). New Delhi: Kalyani; 1981.

417. Yamori W: Improving photosynthesis to increase food and fuel production by biotechnological strategies in crops. *Journal of Plant Biochemistry & Physiology* 2013, **1**:113.

418. Medrano H, Vadell J: Photosynthesis improvement as a way to increase crop yield. In: *Photosynthesis: Photoreactions to Plant Productivity* (Abrol YP, Mohanty P, Govindjee eds.). The Netherlands: Springer; 1993: 571–582.

419. Feil B: Breeding progress in small grain cereals—A comparison of old and modern cultivars. *Plant Breeding* 1992, **108**(1):1–11.

420. Richards RA: Selectable traits to increase crop photosynthesis and yield of grain crops. *Journal of Experimental Botany* 2000, 51:447–458.

421. Neyra CA: *Biochemical Basis of Plant Breeding*, vol. I. Carbon Metabolism. Boca Raton, FL: CRC Press; 1985.

422. Salvucci ME, Crafts-Brandner SJ: Inhibition of photosynthesis by heat stress: The activation state of Rubisco as a limiting factor in photosynthesis. *Physiologia Plantarum* 2004, **120**(2):179–186.

423. Reynolds M, Bonnett D, Chapman SC, Furbank RT, Manès Y, Mather DE, Parry MA: Raising yield potential of wheat. I. Overview of a consortium approach and breeding strategies. *Journal of Experimental Botany* 2011, **62**(2):439–452.

424. Parry M, Madgwick P, Carvalho J, Andralojc P: Prospects for increasing photosynthesis by overcoming the limitations of Rubisco. *Journal of Agricultural Science-Cambridge* 2007, **145**(1):31–34.

425. Zhu X-G, Long SP, Ort DR: Improving photosynthetic efficiency for greater yield. *Annual Review of Plant Biology* 2010, **61**:235–261.

426. Parry MA, Reynolds M, Salvucci ME, Raines C, Andralojc PJ, Zhu X-G, Price GD, Condon AG, Furbank RT: Raising yield potential of wheat. II. Increasing photosynthetic capacity and efficiency. *Journal of Experimental Botany* 2011, **62**(2):453–467.

427. Rhodes I: The relationship between productivity and some components of canopy structure in ryegrass (Lolium spp.). *The Journal of Agricultural Science* 1973, **80**(1):171–176.

428. Wilson D: Response to selection for dark respiration rate of mature leaves in Lolium-perenne and its effects on growth of young plants and simulated swards. *Annals of Botany* 1982, **49**(3):303–312.

429. Wilson D, Jones JG: Effect of selection for dark respiration rate of mature leaves on crop yields of Lolium-Perenne cv. S23. *Annals of Botany* 1982, **49**(3):313–320.

430. Monteith JL: Reassessment of maximum growth-rates for C3 and C4 crops. *Experimental Agriculture* 1978, **14**(1):1–5.

431. Wang C, Guo L, Li Y, Wang Z: Systematic comparison of C3 and C4 plants based on metabolic network analysis. *BMC Systems Biology* 2012, **6**(Suppl 2):S9.

432. Hibberd JM, Sheehy JE, Langdale JA: Using C$_4$ photosynthesis to increase the yield of rice—Rationale and feasibility. *Current Opinion in Plant Biology* 2008, **11**(2):228–231.

433. Taniguchi Y, Ohkawa H, Masumoto C, Fukuda T, Tamai T, Lee K, Sudoh S, Tsuchida H, Sasaki H, Fukayama H: Overproduction of C4 photosynthetic enzymes in transgenic rice plants: An approach to introduce the C4-like photosynthetic pathway into rice. *Journal of Experimental Botany* 2008, **59**(7):1799–1809.

434. Sweetlove LJ, Last RL, Fernie AR: Predictive metabolic engineering: A goal for systems biology. *Plant Physiology* 2003, **132**(2):420–425.

435. Gutiérrez RA, Shasha DE, Coruzzi GM: Systems biology for the virtual plant. *Plant Physiology* 2005, **138**(2):550–554.

436. Zhu X-G, de Sturler E, Long SP: Optimizing the distribution of resources between enzymes of carbon metabolism can dramatically increase photosynthetic rate: A numerical simulation using an evolutionary algorithm. *Plant Physiology* 2007, **145**(2):513–526.

437. Hanba YT, Shibasaka M, Hayashi Y, Hayakawa T, Kasamo K, Terashima I, Katsuhara M: Overexpression of the barley aquaporin HvPIP2; 1 increases internal CO2 conductance and CO2 assimilation in the leaves of transgenic rice plants. *Plant and Cell Physiology* 2004, **45**(5):521–529.

438. Tsuchihira A, Hanba YT, Kato N, Doi T, Kawazu T, Maeshima M: Effect of over expression of radish plasma membrane aquaporins on water-use efficiency, photosynthesis and growth of Eucalyptus trees. *Tree Physiology* 2010, **30**(3):417–430.

439. Tanaka Y, Sugano SS, Shimada T, Hara-Nishimura I: Enhancement of leaf photosynthetic capacity through increased stomatal density in Arabidopsis. *New Phytologist* 2013, **198**(3):757–764.

440. Kusumi K, Hirotsuka S, Kumamaru T, Iba K: Increased leaf photosynthesis caused by elevated stomatal conductance in a rice mutant deficient in SLAC1, a guard cell anion channel protein. *Journal of Experimental Botany* 2012, **63**(15):5635–5644.

441. Covshoff S, Hibberd JM: Integrating C$_4$ photosynthesis into C$_3$ crops to increase yield potential. *Current Opinion in Biotechnology* 2012, **23**(2):209–214.

442. Price GD, Badger MR, von Caemmerer S: The prospect of using cyanobacterial bicarbonate transporters to improve leaf photosynthesis in C3 crop plants. *Plant Physiology* 2011, **155**(1):20–26.

443. Kebeish R, Niessen M, Thiruveedhi K, Bari R, Hirsch H-J, Rosenkranz R, Stäbler N, Schönfeld B, Kreuzaler F, Peterhänsel C: Chloroplastic photorespiratory bypass increases photosynthesis and biomass production in Arabidopsis thaliana. *Nature Biotechnology* 2007, **25**(5):593–599.

444. Kajala K, Covshoff S, Karki S, Woodfield H, Tolley BJ, Dionora MJA, Mogul RT, Mabilangan AE, Danila FR, Hibberd JM: Strategies for engineering a two-celled C4 photosynthetic pathway into rice. *Journal of Experimental Botany* 2011, **62**(9):3001–3010.

445. Zhu XG, Shan L, Wang Y, Quick WP: C4 Rice–An ideal arena for systems biology research. *Journal of Integrative Plant Biology* 2010, **52**(8):762–770.

446. Jordan DB, Ogren WL: Species variation in the specificity of ribulose-biphosphate carboxylase-oxygenase. *Nature* 1981, **291**(5815):513–515.

447. Gutteridge S: Limitations of the primary events of CO2 fixation in photosynthetic organisms—The structure and mechanism of Rubisco. *Biochimica et Biophysica Acta* 1990, **1015**(1):1–14.

448. Medrano H, Primomillo E: Selection of Nicotiana-tabacum haploids of high photosynthetic efficiency. *Plant Physiology* 1985, **79**(2):505–508.

449. Zelitch I: Selection and characterization of tobacco plants with novel O-2-resistant photosynthesis. *Plant Physiology* 1989, **90**(4):1457–1464.

450. Delgado E, Azconbieto J, Aranda X, Palazon J, Medrano H: Leaf photosynthesis and respiration of high CO$_2$-grown tobacco plants selected for survival under CO$_2$ compensation point conditions. *Plant Physiology* 1992, **98**(3):949–954.

451. Rocher JP, Prioul JL, Lecharny A, Reyss A, Joussaume M: Genetic-variability in carbon fixation, sucrose-P-synthase and ADP glucose pyrophosphorylase in maize plants of differing growth-rate. *Plant Physiology* 1989, **89**(2):416–420.

452. Loza-Tavera H, Martínez-Barajas E, Sánchez-de-Jiménez E: Regulation of ribulose-1,5-bisphosphate carboxylase expression in second leaves of maize seedlings from low and high yield populations. *Plant Physiology* 1990, **93**(2):541–548.

453. Deckard EL, Busch RH, Kofoid KD: In: *Exploitation of physiological and genetic variability to enhance crop productivity* (Harper JE, Schrader LE, Howell RW eds.). Rockville, MD: American Soc Plant Physiology; 1985: 46–54.

454. Ishikawa C, Hatanaka T, Misoo S, Miyake C, Fukayama H: Functional incorporation of sorghum small subunit increases the catalytic turnover rate of Rubisco in transgenic rice. *Plant Physiology* 2011, **156**(3):1603–1611.

455. Makino A, Shimada T, Takumi S, Kaneko K, Matsuoka M, Shimamoto K, Nakano H, Miyao-Tokutomi M, Mae T, Yamamoto N: Does decrease in ribulose-1,5-bisphosphate carboxylase by antisense RbcS lead to a higher N-use efficiency of photosynthesis under conditions of saturating CO2 and light in rice plants? *Plant Physiology* 1997, **114**(2):483–491.

456. Furbank RT, von Caemmerer S, Sheehy J, Edwards G: C4 rice: A challenge for plant phenomics. *Functional Plant Biology* 2009, **36**(11):845–856.

Section X

*Photosynthetic Activity Measurements
and Analysis of Photosynthetic Pigments*

31 Quantifying Growth Nondestructively Using Whole-Plant CO$_2$ Exchange Is a Powerful Tool for Phenotyping

Evangelos D. Leonardos and Bernard Grodzinski

CONTENTS

31.1 Introduction 571
31.2 Growth Analysis 572
    31.2.1 Destructive Analysis of Biomass Gain 572
    31.2.2 Nondestructive Analysis 573
        31.2.2.1 Chamber and System Design 573
        31.2.2.2 Net Carbon Exchange Rate (NCER) and Net Carbon Gain (NCG) 573
31.3 Agricultural and Ecological Case Studies 574
    31.3.1 Optimizing Greenhouse Control Systems 574
        31.3.1.1 Light 575
        31.3.1.2 CO$_2$ Concentration 576
        31.3.1.3 Temperature 577
    31.3.2 Ethylene and Plant Photosynthesis 578
    31.3.3 Leaf Morphology and Canopy Architecture 579
    31.3.4 An *Arabidopsis* Case Study 580
31.4 Conclusions 581
References 581

31.1 INTRODUCTION

Growth rate and productivity in terms of biomass gain are major plant traits during phenotyping. Plant growth and productivity are frequently quantified on the basis of dry matter accumulation that is dependent on photosynthesis. Carbon, oxygen, and hydrogen represent approximately 96% of the dry mass (Epstein, 1972) with net C reduction having been derived from photosynthetic fixation of CO$_2$ via the C$_3$ cycle (Woodrow and Berry, 1988; Geiger and Servaites, 1994; Raines, 2003). Many researchers continue to measure photosynthetic responses of single leaves alone, even though it is difficult to correlate quantitatively whole-plant growth and productivity to the photosynthetic rate of a single laminar structure (Elmore, 1980; Gifford and Jenkins, 1982; Bunce, 1986; Pereira, 1994; Leonardos and Grodzinski, 2001, 2011). Most crops achieve less than 50% of their photosynthetic potential due to mutual shading, limitations of CO$_2$, and nutrient and water supply during plant development (Bunce, 1986; Ort and Baker, 1988; Hay and Walker, 1989; Lawlor, 1995; Horton, 2000; Murchie et al., 2009; Zhu et al., 2010). In addition, respiratory losses occur from all tissues and constitute an important limitation to plant C gain during the life cycle (Zelitch, 1982; Lambers, 1985; Amthor, 2000). The balance of daytime C gain and nighttime C loss from the whole plant determines the rate of daily C accumulation, which subsequently controls plant growth and development (Penning de Vries et al., 1974; McCree, 1986; Dutton et al., 1988; Amthor, 1989). Plant biomass production can be correlated to whole-plant net CO$_2$ exchange rates if the duration of the photosynthetically active canopy is known and nighttime respiratory losses are assessed (Bate and Canvin, 1971; Charles-Edwards and Acock, 1977; Christy and Porter, 1982; Zelitch, 1982; Bugbee, 1992; Wheeler et al., 1994). In addition, as will be discussed later in this chapter, nonlaminar structures such as the petioles, stems, branches, and reproductive organs can be photosynthetically active and can significantly contribute to daily C gain (Aschan and Pfanz, 2003; Leonardos et al., 2014).

Two different experimental approaches have been used to estimate growth and productivity. The traditional approach involves measurement of phenotype traits such as dry weight following destructive harvests (Evans, 1972). However, the emphasis in this chapter is more on the value of nondestructive procedures based on continual whole-plant CO$_2$ exchange measurements for quantifying biomass gain and growth rates. There are advantages and disadvantages of both approaches to understanding the effects of environmental factors, plant pathogen interactions, and gene regulation on whole-plant productivity. The primary advantages of quantifying whole-plant gas exchange are that these analyses (1) are made nondestructively, (2) require fewer plants than do traditional destructive analytical procedures, (3) permit direct and accurate comparisons of photosynthetic C gain and nighttime respiratory C loss in real time that can be correlated with development, and (4) individual plants can be followed throughout their life cycle. For example, if the plant is an individual of value in a breeding program, its phenotype traits can be monitored without its destruction. A problem with net C gain estimated by CO_2 exchange alone is that partitioning of reduced C, N, and S compounds and ultimately the growth, form, and development of sinks still need to be determined. Quantitative analysis of assimilate allocation and development of sinks often relies on additional experimental approaches requiring invasive sampling procedures.

Recently, with the prominence of plant genetics and phenotyping, there has been a growing interest in quantifying plant growth noninvasively (Tackenberg, 2007; Golzarian et al., 2011; Bellasio et al., 2012; Dhondt et al., 2013; Fiorani and Schurr, 2013; Paulus et al., 2014). Advances in technologies such as 2D and 3D imaging are providing important plant growth metrics that help define important anatomical and architectural features of the plant canopy. Merging of these noninvasive techniques with whole-plant gas exchange analysis can be used to further understand how primary metabolic processes such as the photosynthesis and respiratory activities of source and sink tissues and their daily patterns of net C gain are controlling development patterns during the life cycle (Leonardos et al., 2014). Furthermore, most infrared gas analyzers offer the means to quantify not only CO_2 exchanges but also H_2O exchanges providing additional valuable phenotyping parameters such as the canopy transpiration and water use efficiency (Leonardos et al., 2014).

31.2 GROWTH ANALYSIS

31.2.1 Destructive Analysis of Biomass Gain

Traditional measurements of plant dry mass require destructive harvest of the plants (Evans, 1972; van der Werf, 1996). Although large numbers of samples are generally required for statistical accuracy, the great advantages of conventional plant growth analysis are that it provides accurate measurements of whole-plant biomass and the opportunity to determine dry matter allocation to developing sinks (Hunt, 1982; Poorter and Garnier, 1996). Samples collected during destructive plant growth analysis are available for more detailed analyses of other variables such as leaf area, organ size, chemical composition, metabolites, proteins, and genes. A historic advantage of destructive analyses is that, although sample handling can be very labor intensive, for assessing general growth patterns, the procedures do not require sophisticated equipment. Modern equipment such as leaf area meters, flat-bed scanners, and electronic balances greatly enhance the speed and accuracy with which measurements can be made and data analyzed. Destructive analytical procedures have been applied extensively in agricultural and ecological studies. Readers

are directed to several key references for more detailed discussions of these traditional methodologies (Evans, 1972; Causton and Venus, 1981; Hunt, 1982, 1990; Beadle, 1993; Poorter and Garnier, 1996; van der Werf, 1996).

Two important parameters that are frequently determined following conventional destructive harvests are dry mass and leaf area (Evans, 1972; Roberts et al., 1987; Hunt, 1990). Total dry weight is a measurement of photosynthetic accumulation of biomass corrected for respiratory loss over time (Evans, 1972). Leaf area measurements provide a means of expressing the photosynthetic potential. Normally the time intervals are several days or weeks. Another important calculation that is derived from dry weight data is the relative growth rate (RGR), which, simply defined, is the increase in dry weight per unit dry weight, per unit time.

Plant productivity is significantly correlated with RGR in many species (van Andel and Biere, 1990). In both agricultural and ecological studies, it is valuable to compare the RGR of different species (Grime and Hunt, 1975; Poorter, 1990; van Andel and Biere, 1990) during development, as influenced by exposure to different environmental conditions (Evans, 1972; Causton and Venus, 1981). Grime and Hunt (1975) conducted extensive examinations of the RGR of 132 species of flowering plants from contrasting habitats. Between 2 and 5 weeks after germination, RGR ranged from 0.22 to 2.20 g g^{-1} week^{-1}. Herbaceous species tended to have higher RGR than woody species. Sun plants had higher RGR than shade adapted species. In a study of 24 C_3 species with varying RGR, Poorter et al. (1990) reported that short-term rates of shoot net photosynthesis, dark respiration, and root respiration were all positively correlated with RGR on a dry weight basis. Fast-growing species fixed more CO_2 per unit total plant dry weight than slow-growing species. In addition, fast-growing plants allocated a lower percentage of their fixed C to shoot and root respiration and more C to leaves.

Daily growth rate is the balance between daytime dry weight gain (primarily from C fixation) and nighttime dry weight loss as a result of dark respiration. Diurnal patterns of plant dry weight change can only be detected by destructive growth analysis when sampling is frequent (e.g., hourly) and large number of plants are sacrificed (Hunt, 1980; Wickens and Cheeseman, 1988). Seedlings of the grass *Holcus lanatus* were grown in full or limited nutrient medium and harvested hourly for dry weight measurement over a 3 day period (Hunt, 1980). A diurnal pattern of dry weight change was distinguished using regression analysis, with the maximum growth occurring during the period of illumination. The RGR in the dark was not significantly different from zero, suggesting that the plants did not respire during the nighttime. However, random variability in the primary dry weight data was high. Wickens and Cheeseman (1988) also applied destructive analysis techniques in a short-term study with seedlings of *Spergular marina* L. and *Lactuca sativa* L. grown in nutrient solutions in controlled environments. It was found that the RGR was higher during the nighttime than during the daytime, which the authors reasoned was unrealistic. However, measurement of whole-plant CO_2 exchange has been adopted as a more sensitive method to measure small changes in dry matter accumulation occurring during short-term studies.

31.2.2 Nondestructive Analysis

Unlike destructive analyses that can be relatively inexpensive, nondestructive analyses based on net C exchange rate (NCER) generally require specialized equipment for the measurement of CO_2 fluxes and environmentally controlled plant holding chambers.

31.2.2.1 Chamber and System Design

There have been many gas exchange systems developed for measuring whole-plant biomass accumulation based on analysis of CO_2 exchange. It is beyond the scope of this chapter to describe these systems and materials used in their construction in any detail. The reader is directed to specific references (Bate and Canvin, 1971; Hand, 1973; Louwerse and Eikhoudt, 1975; Acock et al., 1977; Mortensen, 1982; McCree, 1986; Dutton et al., 1988; Bugbee, 1992; Corey and Wheeler, 1992; Hand et al., 1992; Lawlor et al., 1993; Long and Hällgren, 1993; Poorter and Welschen, 1993; Garcia et al., 1994; Nederhoff and Vegter, 1994; Stasiak et al., 1998; van Iersel and Bugbee, 2000; Wheeler et al., 2002; Baker et al., 2014). Although each system may vary in the degree of automation and complexity of its design, whole-plant CO_2 exchange systems have two basic components: (1) the plant holding chamber with its associated environmental control systems and (2) the gas mixing and analysis systems. In addition, specialized hardware and software are used to integrate environmental control with data collection and analysis.

The size of assimilation chambers varies greatly depending on the number of plants and the characteristics of the canopy that are being investigated. For example, many of the systems developed to measure whole-plant gas exchange in *Arabidopsis* are only big enough to include plants at the rosette stage (Donahue et al., 1997; Dodd et al., 2004; Tocquin and Perilleux, 2004; http://www.licor.com/env/products/photosynthesis/chambers/6400-17.html), whereas analyses of the older, bigger plants with well-developed inflorescences yield important data on the role of the reproductive shoot canopy on final productivity (Earley et al., 2009; Leonardos et al., 2014). The light source will also vary depending on the style and the objective of the experiment. In chambers designed to enclose a portion of the canopy in field (Louwerse and Eikhoudt, 1975; Garcia et al., 1994; Baker et al., 2014) or in the greenhouse (Lake, 1966; Mortensen, 1982; Dutton et al., 1988; Hand et al., 1992; Lawlor et al., 1993; Nederhoff and Vegter, 1994), natural radiation alone may be used. Artificial irradiation, usually supplied from overhead lights (e.g., high-intensity discharge lamps), is a common feature of many laboratory systems. In addition, novel systems have been designed with the capacity to provide inner canopy lighting (Stasiak et al., 1998). Today there is a move in many research facilities including ours to use new light-emitting diodes (LEDs) since the quality of this technology has improved greatly in the last decade, and these artificial lights do not generate the heat

of the older high-intensity discharge lamps (Mitchell et al., 2012). Materials commonly used to construct the chambers and the gas analysis system may affect plant growth, development, and gas exchange (Tibbitts et al., 1977; Bloom et al., 1980; Knight, 1992; Long and Hällgren, 1993; Wheeler et al., 2002). For example, plastics can release volatile hydrocarbons such as C_2H_4 gas, which can alter plant development, canopy architecture, and net CO_2 exchange (see Section 31.3.2).

The CO_2 level inside the chamber will either decrease due to net photosynthesis or increase in the dark as a result of respiration. The rate of change in CO_2 levels depends on chamber volume, canopy size, environmental conditions, and the design of the gas analysis system. The gas mixing and analysis system may be described *closed*, *semiclosed*, or *open* (Bugbee, 1992; Mitchell, 1992; Long and Hällgren, 1993). A *closed* system is one that is completely isolated from outside air and the chamber air is recycled after analysis. A *closed* system is not suitable for long-term whole-plant gas exchange or growth studies because CO_2 is rapidly depleted during photosynthesis. An example of a *semiclosed* system is one in which the plant chamber is closed to the outside atmosphere, but CO_2 is added to compensate for the depletion of CO_2, which occurs during plant photosynthesis (Dutton et al., 1988; Mitchell, 1992; Wheeler, 1992). In an *open* system, there is a continuous flow of air through the chamber and gases are not recycled to the plants (Bugbee, 1992; Mitchell, 1992; Miller et al., 1996; Donahue et al., 1997). The selection of the most suitable system for measuring whole-plant growth and productivity depends on the research objectives.

31.2.2.2 Net Carbon Exchange Rate (NCER) and Net Carbon Gain (NCG)

Daily C balance of the whole plant includes the daytime net photosynthesis and the nighttime respiration, which can be constantly monitored over a 24 h day/night period (Figure 31.1).

Daytime net C gain (C_d) is the integrated NCER during the day:

$$C_d = \sum_i^m (NCER_i . t_i)$$

where $NCER_i$ is the whole-plant net photosynthetic rate over a period of time (t_i).

The total nighttime respiratory C loss of whole plant is integrated as

$$C_n = \sum_j^n (NCER_{dj} . t_j)$$

where $NCER_{dj}$ is the whole-plant dark respiration rate during a period of time (t_j). The whole-plant daily net C gain (ΔC) is calculated as

$$\Delta C = C_d - C_n$$

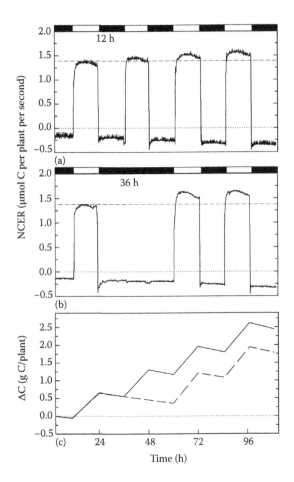

FIGURE 31.1 Diurnal patterns of NCER of greenhouse peppers (*Capsicum annum* 'Cubico') maintained at (a) 12 h/12 h day/night, (b) an extended dark period of 36 h, and (c) daily C gain (ΔC) calculated from a (solid line) and from b (broken line). In both sets of plants, CO_2 concentration during the experiment was maintained at 350 μl L^{-1}, and irradiance was 1150 μmol m^{-2} s^{-1} PAR during the 12 h daytime period. Temperature was 22°C during the daytime and 18°C at night. (Adapted from Watts B., The Effects of Temperature and CO_2 Enrichment on Growth and Photoassimilate Partitioning in Peppers (*Capsicum annuum L.*). Guelph, Canada: University of Guelph, 1995. With permission.)

NCER and therefore net daily C gain can be expressed on different metrics in order to normalize for differences among individual plants. For example, it can be expressed on a per plant basis, on a per surface area basis (leaf or plant), on a per weight basis (fresh or dry), or on a per volume basis.

The plant RGR (e.g., increase in dry mass per unit of dry mass per unit of time) can be estimated from measuring NCERs over a period of time (e.g., ΔC) and by destructive determination of the plant tissue C content and the total plant dry mass (Ho, 1976; Dutton et al., 1988).

The NCER of a common greenhouse sweet pepper shows a clear diurnal pattern (Figure 31.1a). The significance of nighttime C losses on daily C gain (ΔC) is clearly illustrated in Figure 31.1c. Panels a and b show the NCER of two similar populations (i.e., A and B) of pepper plants. The only difference between the two populations was the A population was maintained in a 12:12 h day/night regimes throughout the experiment, whereas the B population was subjected to a 36 h uninterrupted dark period, which corresponded to 24–60 h into the experiment. During the first dark period of the experiment, $NCER_d$ (negative values) of the two populations were similar. The rates of whole-plant net photosynthesis (positive values) were also similar (approximately 1.3 μmol C $plant^{-1}$ s^{-1}) during the first 12 h light period. Thus, at the end of the first 24 h period, the ΔC was virtually identical in the two populations (Figure 31.1c). Stated in conventional terms used in destructive growth analysis, these data show that RGRs of the two populations were the same.

During the next 3-day period (24–72 h), population A increased its daytime NCER and nighttime $NCER_d$, consistent with increases in net photosynthesis and dark respiration as plants increased in size and new sinks developed. In comparison, population B, which was maintained in total darkness for a 36 h period, lost biomass (Figure 31.1c). During the extended dark period, there was a reduction in specific leaf weight, which corresponded with a reduction in stored reserves of sucrose and starch (Watts, 1995). Figure 31.1c shows the effect of darkness on productivity. Interestingly, however, as a result of the extended dark period in population B, net photosynthesis on the third day increased more dramatically than that of the control population (Figure 31.1a). By the end of the fourth day, leaf starch reserves were replenished (Watts, 1995). The data in Figure 31.1 represent a study of a relatively simple environmental perturbation in which only the length of a single dark period was altered. As many as 50–100 times more plants would have been required to obtain a similar data set if a destructive growth analysis protocol had been used.

31.3 AGRICULTURAL AND ECOLOGICAL CASE STUDIES

In agricultural and ecological studies, light intensity, CO_2 level in the atmosphere, temperature, and nutrient availability are all important environmental variables affecting source and sink development that determine net CO_2 exchange. Plant growth regulators and pathogens affect sink–source relationships in part by altering canopy architecture, development, and allocation of assimilates within the plant. Integrated analyses of development of sources and sinks are required for a full appreciation of the value of whole-plant NCER measurements for studying growth and productivity.

31.3.1 OPTIMIZING GREENHOUSE CONTROL SYSTEMS

In contrast to agricultural field production systems, modern greenhouses provide an opportunity for control of both the aerial and root environments of high value crops (Porter and Grodzinski, 1985; Hanan, 1998). Our whole-plant gas analysis system was initially designed to develop environmental algorithms that could be used to predict growth of greenhouse crops and to design a larger commercial scale test in which productivity and crop yield could be assessed (Dutton et al., 1988; Leonardos et al., 1994).

Whole-plant NCER response to irradiance (I), CO$_2$ concentration, and temperature (T) can be expressed by the following polynomial function:

$$NCER = \beta_1 + \beta_2 I + \beta_3 CO_2 + \beta_4 T + \beta_5 I^2 + \beta_6 CO_2^2 + \beta_7 T^2 + \beta_8 I*T$$
$$+ \beta_9 I*CO_2 + \beta_{10} T*CO_2 + \beta_{11} I*T*CO_2$$

where NCER is in units of μmol CO$_2$ m^{-2} s^{-1}, I is in units of μmol photon m^{-2} s^{-1} photosynthetically active radiation (400–700 nm, PAR), CO$_2$ unit is μl L^{-1}, T is in units of °C, and β_1–β_{11} are coefficients.

There is a strong interaction among the influence of irradiance, CO$_2$ concentration, and temperature on whole-plant net photosynthesis, daily C gain, and growth. Irradiance is the most important determinant of whole-plant net photosynthesis followed by CO$_2$ and temperature. In a woody ornamental species, roses, irradiance, CO$_2$, and temperature accounted for 70%, 20%, and 5%, respectively, of the variance in whole-plant NCER (Jiao et al., 1991c). In an herbaceous C$_3$ crop, *Alstroemeria*, irradiance accounted for almost 60% of the variation in whole-plant NCER, whereas CO$_2$ and temperature accounted for 23% and 14%, respectively (Leonardos et al., 1994). This crop is more sensitive to fluctuations in temperature due to its large rhizomes. The polynomial equation helps to describe how environmental variables affect net C gain due to photosynthesis. By knowing the duration of the photoperiod, as well as the respiration rate of the plants at different night temperatures, one can predict how daily C gain and growth will be affected in controlled environments (Jiao et al., 1991c; Leonardos et al., 1994).

31.3.1.1 Light

Light intensity and quality inside a canopy fluctuates dramatically due to time of day, cloud cover, canopy density, and season (Hay and Walker, 1989; Björkman and Demming-Adams, 1994; Pearcy and Pfitsch, 1994; Hanan, 1998). As illustrated in Figure 31.1, the most dramatic changes in C gain occur diurnally. Due to mutual shading within the canopy, differences in leaf position, orientation, age, dark respiration of different organs, and sun and shade tolerance adaptations, whole-plant NCER is a better measurement of whole-plant growth response to light than that obtained from single leaf studies (Duncan, 1971; Acock et al., 1978; Russell et al., 1989; Reddy et al., 1991; Warren Wilson et al., 1992; Leonardos et al., 1994; Pons and Poorter, 2014). Figure 31.2a shows whole-plant photosynthesis of greenhouse roses. Leaf photosynthesis is saturated at lower light intensities than are required to saturate NCER of whole plants. The maximum rate of leaf photosynthesis is much higher than that of whole plants. Furthermore, the light compensation point (LCP; i.e., the light intensity at which C gain and C loss are balanced) is lower for the leaves than for whole plants. Increasing the irradiance

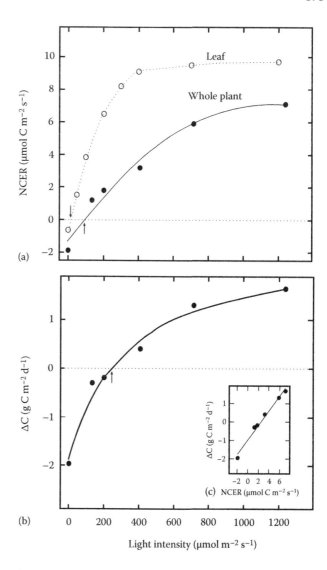

FIGURE 31.2 Effect of light intensity on (a) NCER of a leaf (broken line, open symbols) and a whole plant (solid line, solid symbols), (b) whole-plant net C gain (ΔC), and (c) the relationship between plant ΔC and plant NCER of a greenhouse rose, *Rosa hybrida* 'Samantha'. Values of ΔC were calculated from plant NCER measurements for a 10 h/14 h day/night period. The arrows in panels a and b indicate LCPs of leaf photosynthesis, of whole-plant photosynthesis, and of whole-plant C gain.

from 0 to 1200 μmol m^{-2} s^{-1} PAR resulted in a marked increase in whole-plant NCER (Figure 31.2a) and ΔC (Figure 31.2b) calculated for a 24 h period consisting of a 10 h day and a 14 h night. ΔC was linearly proportional to daytime whole-plant NCER (Figure 31.2b). We define the LCP for ΔC as that light intensity required during daytime hours to sustain photosynthetic C gain, which will balance nighttime (dark) respiratory C losses. The LCP for ΔC (Figure 31.2b) was greater than the LCP for the whole-plant NCER (Figure 31.2a). The difference between the LCP of the whole-plant NCER (Figure 31.2a) and that of ΔC (Figure 31.2b) was primarily due to the duration of the night period and the magnitude of nighttime respiration.

In a different experiment in which rose plants were either irradiated over a 12 h light period or continuously for 24 h with half the irradiance, but with the same total radiant energy input of 17.6 mol m^{-2}, net C gains during the periods of illumination were identical (Jiao et al., 1991a). In both cases, approximately 1.8 g C m^{-2} was assimilated in the light period. However, when the plants exposed to a 12 h daytime period were placed in the dark for 12 h, ΔC over the 24 h period was reduced to 1 g C m^{-2}. The length of light period not only affected the total canopy C assimilation during the day but also influenced C loss through respiration in the subsequent dark period. Commercially grown greenhouse roses are frequently provided with artificial lighting at night to offset nighttime respiratory loss, even though the irradiance levels achieved at the canopy level are well below those achieved when natural sunlight is available.

31.3.1.2 CO$_2$ Concentration

At present suboptimal atmospheric CO$_2$ (approximately 400 µl L^{-1}) and inhibitory O$_2$ (21%) levels, rates of CO$_2$ assimilation in C$_3$ plants are limited by CO$_2$ availability (Zelitch, 1992; Bowes, 1993; Sims and Pearcy, 1994; Drake et al., 1997). C$_3$ species grown under present atmospheric conditions lose as much as 40% of CO$_2$ assimilated as a result of photorespiration, which is regulated by CO$_2$ and O$_2$ concentrations and temperature (Zelitch, 1992; Keys and Leegood, 2002). There are two direct effects of increasing CO$_2$ concentration on increasing net photosynthesis (Kramer, 1981; Grodzinski, 1992). One is the direct increase in the primary substrate CO$_2$ for carboxylation. Stomata tend to close in response to high CO$_2$; however, the increase in gradient between atmospheric CO$_2$ and leaf internal CO$_2$ concentration under CO$_2$ enrichment offsets the inhibiting effect of stomatal closure, resulting in higher rates of CO$_2$ fixation. The increase in the CO$_2$ concentration at the site of fixation in the chloroplast has a second direct effect on carboxylation efficiency of the chloroplast. Oxygenase activity of Rubisco is reduced, and the flow of C to the glycolate pathway (i.e., photorespiration) is proportionally reduced. The benefits of CO$_2$ enrichment can come directly from the enhanced photosynthetic rate per unit leaf area or indirectly as a more long-term consequence of an increased total plant leaf area and altered pattern of carbon partitioning among developing sinks (Kramer, 1981; Grodzinski, 1992; Garcia et al., 1994). One of the consequences of CO$_2$ enrichment of young tomato seedlings, for example, is an increase in the allocation of assimilates to the sinks such as the roots (Woodrow et al., 1987). Healthy root establishment is a fundamental objective during transplant production. Many bedding plants are grown commercially in greenhouses under CO$_2$ enrichment (normally 1000–2000 µl L^{-1}) to establish vigorous root systems, which will improve the degree of hardiness of these transplants when they are exposed to field conditions (Porter and Grodzinski, 1985).

In greenhouse production systems, daytime CO$_2$ enrichment is commonly used to stimulate growth and enhance crop yield (Kimball and Idso, 1983; Porter and Grodzinski, 1985; Hanan, 1998). Typical leaf and whole-plant net photosynthetic responses of an herbaceous C$_3$ greenhouse cut flower crop, *Alstroemeria*, to varying levels of CO$_2$ are shown in

Figure 31.3. The major differences between leaf and whole-plant CO$_2$ exchange was the higher rate of leaf gas exchange when comparisons were made at the same CO$_2$ concentration (Leonardos et al., 1994). Leaf NCER was 18 µmol CO$_2$ m^{-2} s^{-1} at 1500 µl L^{-1} CO$_2$ under 1000 µmol m^{-2} s^{-1} PAR, whereas whole-plant NCER was 9 µmol CO$_2$ m^{-2} s^{-1} at 1500 µl L^{-1} CO$_2$ under 1200 µmol m^{-2} s^{-1} PAR. The lower rate of CO$_2$ fixation by whole-plant NCER was primarily due to mutual shading and respiratory activity of sinks. Nevertheless, CO$_2$ enrichment marginally reduces the LCP in some crops (Nederhoff and Vegter, 1994) and substantially increases the optimum irradiance for conversion efficiency as well as the maximum conversion efficiency (Warren Wilson et al., 1992; Caporn et al., 1994). Quantum yields of C$_3$ leaves are dependent on CO$_2$ concentration, leaf temperature, and O$_2$, whereas quantum yields of C$_4$ leaves are independent of these factors (Ehleringer and Bjorkman, 1977; Ehleringer et al., 1997).

Although the extent of photosynthetic and growth responses to CO$_2$ enrichment varies with plant species and depends on other environmental variables including light, water, nutrients, and temperature, increases in photosynthesis, growth, and productivity have been observed in nearly all C$_3$ species tested (Kimball and Idso, 1983; Porter and Grodzinski, 1985; Wittwer, 1985; Long, 1991; Bowes, 1993; Kirschbaum, 1994; Dippery et al., 1995; Leakey et al., 2009a). In C$_4$ species, a natural mechanism of CO$_2$ enrichment in the bundle sheath cells exists reducing photorespiration (Hatch, 1987; Grodzinski, 1992; Bowes, 1993). Net photosynthesis is usually much higher in C$_4$ than in C$_3$ plants at ambient CO$_2$ (Zelitch, 1992; Leonardos and Grodzinski, 2000). When there is a positive growth response to elevated CO$_2$ by C$_4$ plants, that response is usually less than that observed among C$_3$ plants (Kimball and Idso, 1983; Porter and Grodzinski, 1985; Bowes, 1993; Dippery et al., 1995; Ehleringer et al., 1997; Kellogg et al., 1999; Sage and Monson, 1999; Wand et al.,

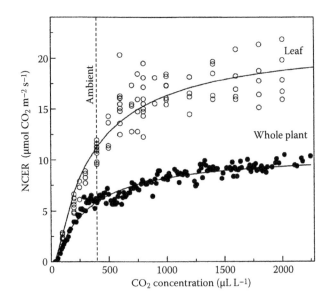

FIGURE 31.3 Effect of CO$_2$ concentration on leaf (open symbols) and plant (closed symbols) NCER of *Alstroemeria* sp. 'Jacqueline'.

2001). The CO_2 exchange response of CAM plants to CO_2 enrichment also depends on species, developmental stage, and environmental factors such as light, temperature, nutrients, and water availability (Osmond and Bjorkman, 1975; Huerta and Ting, 1988; Hogan et al., 1991; Cui and Nobel, 1994). Under conditions of water stress, stomata are a major limitation to C assimilation in CAM, C_4, and C_3 plants (Sage and Monson, 1999). In C_3 plants at ambient CO_2, stomatal limitation is about 30% of the total limitation of leaf photosynthesis (Jones, 1983). Elevated levels of atmospheric CO_2 increase the CO_2 gradient between the atmosphere and the fixation site of the chloroplasts, and high CO_2 generally reduces stomatal conductance and increases water use efficiency in C_3 and C_4 plants (Lawlor and Mitchell, 1991; Wolfe, 1995; Ghannoum et al., 2001; Leakey et al., 2009a; Leonardos et al., 2014).

During long-term exposures to elevated CO_2, photosynthesis on a leaf and on a whole-plant basis is altered in a species-specific manner (Grodzinski, 1992; Bowes, 1993; Gifford, 1995). In roses grown at 1000 µl L^{-1} CO_2 for several weeks, whole-plant net photosynthesis was identical to that of plants grown under ambient CO_2 conditions, indicating no inhibiting effect of long-term CO_2 enrichment. Similarly, lettuce plants grown under CO_2 enrichment showed no decrease in canopy photosynthesis under high CO_2 (Caporn et al., 1994). Nevertheless, in some herbaceous species even in canopies open to the light and with healthy developing root systems, prolonged exposure to high CO_2 results in reduced rates of mature leaf photosynthesis when comparisons were made at ambient CO_2 and O_2 (Porter and Grodzinski, 1985; Sage et al., 1989). These observations can, in part, be explained by the fact that under CO_2 enrichment, plant growth is enhanced and new sinks (e.g., leaves) (Kramer, 1981; Grodzinski, 1992) place a heavy demand on the key nutrients such as N (Porter and Grodzinski, 1984, 1985; Sage et al., 1989; Besford et al., 1990; Grodzinski, 1992; Bowes, 1993; Lawlor, 2002; Stitt et al., 2002; Paul and Pellny, 2003). In some species grown at high CO_2, a reduction in leaf photosynthesis can be attributed to a reduction in key enzymes associated with the fixation of CO_2 (e.g., Rubisco; Porter and Grodzinski, 1984; Sage et al., 1989; Grodzinski, 1992; Bowes, 1993). These enzymes are a major source of N, and their levels tend to be reduced during senescence as N is reallocated to growing sinks. As mentioned above, the magnitude of photorespiration relative to that of photosynthesis is reduced at high CO_2 (Grodzinski, 1992; Zelitch, 1992; Bowes, 1993). However, there is no reduction in glycolate oxidase activity, a key enzyme of the photorespiratory pathway at high CO_2 (Porter and Grodzinski, 1984; Zelitch, 1992). The evidence is that dark respiratory processes increase in plants grown at elevated levels of CO_2. For example, enhanced whole-plant dark respiration following growth under CO_2 enrichment has been observed in several species (Leakey et al., 2009b; Leonardos et al., 2014) and can, in part, be attributed to the increased level of carbohydrates present in leaf tissue of plants grown at high CO_2 levels (Amthor, 1989; Thomas and Griffin, 1994). In wheat, the number of mitochondria in the leaf mesophyll appears to increase during growth at high CO_2 (Robertson et al., 1995). These observations

together serve to illustrate that during CO_2 enrichment, there will be profound changes in both photosynthetic and respiratory CO_2 fluxes as developmental processes are generally accelerated compared to growth at ambient CO_2 (Grodzinski, 1992). However, the effects of short-term and long-term high CO_2 on respiration rates need to be carefully examined in view of limitations and errors that can arise using CO_2 gas exchange systems (Jahnke, 2001; Jahnke and Krewitt, 2002; Hunt, 2003; Long and Bernacchi, 2003).

Studies with several greenhouse crops show daytime starch accumulation at high CO_2 (Madsen, 1968; Madore and Grodzinski, 1985; Leonardos et al., 1996; Jiao and Grodzinski, 1998). Furthermore, the increase in photoassimilate storage supports an enhanced nighttime carbon export rate from the leaves, which, in part, explains the faster growth rate of plants exposed to elevated daytime CO_2 (Grange, 1985; Grodzinski, 1992; Geiger and Servaites, 1994; Watts, 1995; Jiao and Grodzinski, 1998; Grimmer and Komor, 1999). Storage of carbohydrates may compete with leaf and root growth and reduce the maximum growth rate (Chapin et al., 1990). The increase in stored carbohydrates (e.g., starch levels) during long-term CO_2 enrichment represents a problem in equating net CO_2 exchange rates obtained nondestructively to growth and development. Daytime and nighttime CO_2 exchanges are dependent on the partitioning and the allocation of the stored reserves such as starch. Starch in the source leaves definitely represents biomass accumulation (Schulze and Schulze, 1994; Thomas and Griffin, 1994). However, growth and development requires further partitioning and allocation of these reserves to developing sinks such as the roots and reproductive structures. In a survey of 42 C_3, C_3–C_4 intermediate, and C_4 photosynthetic types, the linear relationship between leaf photosynthesis and C export at ambient CO_2 breaks down at elevated CO_2 (Grodzinski et al., 1998; Leonardos and Grodzinski, 2000, 2002; Lalonde et al., 2003). Of the leaf parameters tested, C export correlated best with whole-plant RGR obtained noninvasively using whole-plant CO_2 exchange analysis (Leonardos and Grodzinski, 2001).

31.3.1.3 Temperature

One of the advantages of using whole-plant NCER as a tool to study growth is that the effects of temperature during the dark can be differentiated from those in the light. Light and CO_2 are environmental parameters that primarily affect photosynthesis and photorespiration of the leaves. However, temperature affects all aspects of metabolism, growth, and development of all organs.

Leaf net photosynthesis of most C_3 plants has an optimal temperature range of 20–35°C at ambient CO_2 level and saturating light (Berry and Bjorkman, 1980). Leaf photorespiration increases sharply at temperatures above 30°C due to decreases in CO_2 solubility (Ku and Edwards, 1979) and in CO_2/O_2 specificity of Rubisco (Jordan and Orgen, 1984; Brooks and Farquhar, 1985; Bernacchi et al., 2002). In addition, as we have outlined elsewhere, temperature can dramatically alter C export rates from leaves (Jiao and Grodzinski, 1996, 1998; Leonardos et al., 1996; Leonardos and Grodzinski, 2002;

Leonardos et al., 2003). Thus, temperature moderates source–sink relationships by affecting fluxes of metabolites as well as photosynthetic and respiratory metabolism more directly.

Because CO_2 enrichment reduces photorespiration in C_3 plants, the optimal temperature for whole-plant photosynthesis is usually shifted a few degrees higher than at ambient CO_2 (Woo and Wong, 1983; Jiao et al., 1991b; Long, 1991; Kirschbaum, 1994). Both the LCP (Jiao et al., 1991b) and the CO_2 compensation point (Nilwik, 1980) of whole plants increase with a rise in temperature because respiration from all tissues is enhanced. The effect of temperature on C is the balance of its effects on photosynthesis, photorespiration, and dark respiration. Both daytime and nighttime temperatures are important in relation to plant daily C gain. In white clover plants, ΔC at 30/10°C (day/night) was higher than plants maintained at 30/20°C (McCree and Amthor, 1982).

The importance of nighttime temperature on daily net C gain during greenhouse rose production is shown in Table 31.1. Roses were maintained at either 27/27°C (day/night) or at 27/17°C (day/night). The respiration rate of rose plants maintained at 27°C during the night was twice that of plants maintained at 17°C during the night. Daily net C gain maintained at 27/17°C was 50% higher than that of plants maintained at the same 27/27°C day/night temperature. Dark respiration rate is more sensitive to changes in temperature than photosynthesis. Respiration rate generally increases exponentially with increasing temperature (Johnson and Thornley, 1985) with a Q_{10} of about 2 (Lambers, 1985). However, the rate varies with developmental stage of specific tissues. For example, in a flowering rose shoot, the respiration rate of the flower bud on a dry weight basis is three to four times higher than that of leaves and accounts for half of the total respiratory C loss from the shoot (Jiao et al., 1991b).

In greenhouses, root zone temperatures can be controlled by bench heating and cooling systems. Lower root zone temperatures stimulate flowering in *Alstroemeria*, which alters the growth and development pattern of the whole plant. The NCER of *Alstroemeria* is very sensitive to changes in aerial and root zone temperature (Leonardos et al., 1994). The optimal temperature for leaf photosynthesis under ambient CO_2 level and saturating light is about 20°C, whereas that of whole-plant NCER is only 10–12°C, which, in part, reflects the metabolism of the rhizomes. Whole-plant gas exchange measurements have also been used to discriminate between

growth and maintenance respiration and how these processes relate to C use efficiency (van Iersel and Seymour, 2000; van Iersel, 2003).

31.3.2 Ethylene and Plant Photosynthesis

The use of plant growth regulators and herbicides for controlling vegetative growth of agricultural crops has increased dramatically in recent decades (Davis and Curry, 1991; Abeles et al., 1993; Hanan, 1998). Ethylene is a natural plant growth regulator that is produced in nonphotosynthetic organs (e.g., flowers and fruits) as well as in photosynthetic leaf tissues (Abeles et al., 1993). Because of our interest in CO_2 enrichment in closed environments, we began to investigate the relationship between C_2H_4 and CO_2 gas exchange in photosynthetic tissue (Grodzinski, 1984, 1992; Porter and Grodzinski, 1985). The stimulatory effect of high CO_2 levels on C_2H_4 release from photosynthetic tissue during short-term exposures (one to eight hours) has been demonstrated in intact plants, in detached leaves, and in excised leaf tissue (Dhawan et al., 1981; Grodzinski et al., 1982, 1983; Kao and Yang, 1982; Philosoph-Hadas et al., 1986). The CO_2 levels that affect C_2H_4 release from leaf tissue during short-term incubations (i.e., 50–5000 μl L^{-1}; Woodrow and Grodzinski, 1994) parallel those encountered by leaf tissue in closed greenhouse environments or in tissue cultures (Woodrow et al., 1990). Active photosynthesis under high irradiance can deplete the CO_2 to below ambient levels in protected environments (Porter and Grodzinski, 1985; Hanan, 1998). Therefore, CO_2 is added to supplement growth. Interestingly, predicted future global CO_2 concentrations also fall within this range (Grodzinski, 1992; Bowes, 1993). Our early studies showed that C_2H_4 release from C_3 and C_4 leaf tissue during short-term exposures to varying CO_2 are different (Grodzinski et al., 1982, 1983). Long-term exposure to elevated CO_2 concentrations modifies endogenous C_2H_4 metabolism and affects plant growth and development (Grodzinski, 1992; Woodrow and Grodzinski, 1994). Prolonged growth at high CO_2 results in a persistent increase in the rate of endogenous C_2H_4 release, which can, only in part, be attributed to an increase in the endogenous pools of C_2H_4 pathway intermediates (Woodrow and Grodzinski, 1994). During acclimation to high CO_2, leaves appear to have higher levels of ethylene-forming enzyme activity (Philosoph-Hadas et al., 1986; Woodrow

TABLE 31.1
Effect of Night Temperature on Whole-Plant Daily C Gain (ΔC) of Greenhouse Roses (*Rosa hybrida* 'Samantha') Maintained at 12:12 h Day/Night

Temperature (°C) Day/Night	NCER	NCER$_d$	C_d	C_n	ΔC	C_n/C_d
	(μmol C m^{-2} s^{-1})		(g C m^{-2})		(%)	
27/27	6.3 ± 0.35	2.6 ± 0.16	3.3 ± 0.15	1.5 ± 0.09	1.8 ± 0.16	45 ± 2.4
27/17	6.9 ± 0.35	1.3 ± 0.07	3.5 ± 0.16	0.8 ± 0.05	2.7 ± 0.17	23 ± 1.3

Source: Adapted from Jiao J et al., *Canadian Journal of Plant Science* 71:235–243, 1991.

and Grodzinski, 1994). Photosynthetically active young leaves contribute most of the C_2H_4 emanating from the canopy (Woodrow and Grodzinski, 1994; Wheeler et al., 1996; Stasiak, 2002).

All lower and higher plant tissues produce C_2H_4, which can elicit a wide range of biochemical and morphological responses (Abeles et al., 1993). For example, leaf ontogenesis and maturation are correlated with changing rates of both C_2H_4 emanation and sensitivity to exogenously supplied C_2H_4 (Osborne, 1991; Woodrow et al., 1990; Woodrow and Grodzinski, 1994). Ethylene can modify leaf and whole-plant photosynthesis (Kays and Pallas, 1980; Taylor and Gunderson, 1986; Woodrow and Grodzinski, 1989; Woodrow et al., 1989). Vegetative growth measured nondestructively as net C gain was reduced by 50% within 24 h of C_2H_4 exposure in *Lycopersicon esculentum* L. (Woodrow and Grodzinski, 1989) and 35% in *Xanthium strumarium* L. (Woodrow et al., 1989). Similar results were obtained with destructive analysis of tomato (Woodrow et al., 1988) and corn (Cliquet et al., 1991). The observed decrease was attributed to well-known morphological responses exhibited by these plants when treated with C_2H_4 (Crocker et al., 1932; Woodrow et al., 1988). The reductions in whole-plant NCER were attributed to (1) epinastic changes in leaf angle (i.e., light interception patterns) (Woodrow and Grodzinski, 1989; Woodrow et al., 1989) and (2) alteration of sink–source relations (Woodrow et al., 1988). For example, when the leaves that showed C_2H_4 induced epinasty were repositioned with respect to the overhead light source in the analysis chamber, a NCER comparable to that of the untreated plants was observed. The reduction in C gain associated with C_2H_4 is an indirect effect of C_2H_4 on canopy photosynthesis since it is a consequence of C_2H_4-induced epinastic responses, which alter the orientation of the leaves and light interception (Woodrow et al., 1988, 1989; Woodrow and Grodzinski, 1989).

The role of C_2H_4 in regulating the CO_2 exchange of varying plant density was further tested by treating model canopies of tomato seedlings with C_2H_4. Plants were exposed to a 12:12 h, day/night regime, during which canopy NCER was measured after treatment with C_2H_4 (Grodzinski et al., 1996). The critical leaf area index (LAI) at which 95% of the maximum rate of canopy photosynthesis was achieved corresponded to a value of about 5. When a well-developed (i.e., dense) canopy, with a LAI of about 6, was treated with C_2H_4, there was no change in the photosynthesis of the stand. However, when the LAI was only 4, treatment with C_2H_4 resulted in a 20% decrease in canopy photosynthesis. These studies with model canopies support our earlier conclusion that the effects of C_2H_4 on photosynthesis and C gain (Woodrow et al., 1988, 1989; Woodrow and Grodzinski, 1989) can be ascribed to classical hormonal responses such as epinastic development, which result in altered light interception within the canopy.

Light interception is a major determinant of canopy photosynthesis. Endogenously produced C_2H_4 can accumulate to physiologically active levels in plant canopies (Abeles et al., 1993). For example, C_2H_4 concentrations sufficient to stimulate premature cotton boll abscission have been documented

in field cotton canopies (Heiman et al., 1971). In closed greenhouse environments in which crops are growing, C_2H_4 levels of 10–15 ppb have been detected (Woodrow and Grodzinski, 1994), which can be attributed to production by the plant tissue. In ongoing closed environment studies with lettuce, wheat, and soybean canopies at the National Aeronautics and Space Administration (United States; Corey and Wheeler, 1992; Wheeler et al., 1996) and at the University of Guelph (Stasiak, 2002), C_2H_4 levels of up to 100 ppb in air samples have been detected. The productivity of wheat and rice is also affected by low concentration of ethylene that might accumulate in closed environments (Klassen and Bugbee, 2002). Collectively, these observations support the view that C_2H_4 from the plants accumulates in closed environments and can modify canopy architecture, light trapping, and whole-plant photosynthesis. Canopy density (i.e., LAI) and light interception patterns are important factors in determining the extent to which exposure to this growth regulator will alter daily C gain (Stasiak et al., 1998; Stasiak, 2002).

31.3.3 Leaf Morphology and Canopy Architecture

Leaf anatomy and canopy architecture markedly influence canopy light interception and photosynthesis especially during the plant's life cycle. For example, in near-isogenic cotton lines, variations in leaf morphologies (i.e., size, and shape) resulted in different LAI and leaf dry weight (Wells et al., 1986). The genotypic variation in LAI of different lines caused differences in light penetration through the canopy, integrated canopy photosynthesis, and limit yield. We used whole-plant gas exchange to better understand photosynthesis and productivity of pea mutants with very different leaf anatomy. "Afila" mutants of peas, in which a single gene modification results in replacement of the laminar-shaped leaflets by cylindrically shaped tendrils (Marx, 1987; Gould et al., 1989; Côté et al., 1992), have high plant NCER in the light because more light penetrates the leaf canopy than is the case with conventional leafy cultivars. In the semi-leafless "afila" phenotypes, the tendrils and laminar-shaped stipules accounted for approximately 60% and 40% respectively, of the total plant photosynthesis, even though on a chlorophyll or area basis, the tendrils were predicted to account for only 30% of whole-plant photosynthesis. These values were derived from two different sets of experiments. In one set of experiments, $^{14}CO_2$ was supplied to whole plants for 1 min after which the plants were rapidly killed to prevent translocation of ^{14}C-labelled assimilates. The $^{14}CO_2$ fixed in the different photosynthetic organs was measured following destructive analysis. In a parallel experiment, similar values for the contributions of leaflets, stipules, and tendrils to plant CO_2 exchange were determined by measuring whole-plant NCER before and after surgical removal of the tendrils or the stipules or the leaflets (Côté and Grodzinski, 1999). These experiments demonstrate the importance of tendril structures in peas and the heterogeneous nature of leaf canopies. They also serve to underscore the need for a greater degree of resolution in monitoring gas exchanges from different parts of the canopy if we are to fully describe how canopy architecture contributes to canopy photosynthesis, growth, and development through the season.

31.3.4 An *Arabidopsis* Case Study

Among the vascular plants frequently studied, *Arabidopsis thaliana* has a unique distinction because it was the plant whose genome was sequenced first (Arabidopsis Genome Initiative, 2000). As a model organism for studying genetic modifications, there have been numerous reports on the photosynthesis of the leaves and growth of the rosette leaf canopy in different environmental conditions including those where the plants were grown in an elevated CO_2 condition (Lake, 2004; Teng et al., 2009). However, few studies show how the diel patterns of whole-plant gas exchange that include both photosynthesis and dark respiration affect daily carbon gain. Notably, Dodd et al. (2005), using analyses of whole-plant C-exchanges, showed that in selected transgenics of *Arabidopsis* in which the circadian clock matches the photoperiod, both net C-fixation and growth of the rosette stage plants are enhanced at ambient CO_2. Earley et al. (2009) used measurements of whole-plant C-fixation that were taken at one time during the daytime but at several stages during the plants' life cycle to investigate the importance of the inflorescence tissues as well as the rosette leaves to lifetime C assimilation. They concluded from a study of ecotypes with different patterns of vegetative and reproductive architectures that the inflorescences during reproductive development were contributing from 36% to 93% to total C-gain when the plants were grown under similar environmental conditions including ambient CO_2 levels. As discussed in this chapter, it is well known that plant growth depends on the balance between the primary processes in both source and sink tissues (McCree, 1986; Dutton et al., 1988; Leonardos and Grodzinski, 2011). Recently, using whole-plant CO_2 analysis, we were able to confirm that the inflorescence tissues, largely those nonlaminar in form, were major contributors to daily C-gain, growth, and final plant productivity in terms of the seed number and oil content when the *A. thaliana* plants were cultured under both CO_2 conditions (Dahal et al., 2014; Leonardos et al., 2014).

In many species, nonlaminar structures in the canopy such as petioles and stems are photosynthetically active (Aschan and Pfanz, 2003; Wittmann et al., 2006; Cernusak and Hutley, 2011; Hua et al., 2012; Simbo et al., 2013). It is well known that among C_3 species, the nonlaminar structures such as the petioles and stem respire CO_2 and also can, via a PEP carboxylase, refix that CO_2 in a partial C_4 photosynthetic system (Hibberd and Quick, 2002). Under elevated CO_2, plants develop larger canopies and generally have higher respiration (Leonardos and Grodzinski, 2011). We have been interested for some time in the possibilities of controlling glycolytic and respiratory metabolism with the view to better regulating sink strength and the demand for photoassimilates from the photosynthetically active source tissues. When plants are exposed to high CO_2 levels, the problems in achieving greater growth often are associated with the imbalance between source strength and sink strength. A working hypothesis is that many

growth processes are dependent on rapid rates of respiration. Respiration is an anabolic process that provides energy as well as many essential metabolites needed for biosynthetic processes and growth in all tissues in the plant (Nunes-Nesi et al., 2011; Weraduwage et al., 2011). Using *Arabidopsis* as a model transgenic system, members of our team were the first to demonstrate that at ambient levels, transgenics with either constitutive or seed-specific partial repression of mitochondrial pyruvate dehydrogenase kinase (mt*PDHK*) had higher leaf respiration and greater seed weight and oil content (Marillia et al., 2003). Interestingly, at elevated CO_2, two constitutively repressed mt*PDHK* lines, 10'4 and 3'1, having altered branching, inflorescence architecture, and silique number compared to controls, show the greatest enhancement of plant growth and oil production (Weraduwage, 2013; Dahal et al., 2014). Because we had no knowledge of how a mutation might alter a central respiratory process via pyruvate metabolism, we choose to use whole-plant gas analysis as a phenotyping tool. With the mutation expressed constitutively, one might be altering net gas exchanges in petioles or stem tissues as well as leaves. For example, an enhancement of pyruvate flux through modification of mt*PDHK* could theoretically alter the capacity for a C_4-like refixation of CO_2 from respiratory processes in the nonlaminar canopy structures such as the stems and branches as suggested by Hibberd and Quick (2002). The primary objectives of the study by Leonardos et al. (2014) involved determining whole-plant respiration, photosynthesis, and daily C gain in *Arabidopsis* and assessing the contribution of the rosette leaves and of the inflorescence tissues to whole-plant gas exchanges throughout the life cycle at both ambient and high CO_2 conditions.

When sink strength was altered by modifying mitochondrial pyruvate metabolism, increases in productivity occurred in both conditions. Respiration and photosynthesis during canopy transition from its rosette leaf form to the highly branched inflorescence stages were quantified. Diel patterns of leaf and whole-plant gas exchange were measured at four critical developmental stages for controls (wild-type (WT) and plasmid (pBI121)) and two transgenics (3'1 and 10'4) *Arabidopsis* lines having partial constitutive repression of mt*PDHK* grown at ambient CO_2 and elevated CO_2. In Figure 31.4, and C), only data for the wild-type control plants are shown. It suffices here to note that, in all plants tested, when the reproductive inflorescence structures were well developed, they had higher photosynthesis and respiration than they did at the rosette leaf stage. In fact, the inflorescences were contributing over 90% of photosynthesis and daily C gain. Interestingly, when the gas exchange data were expressed on a plant basis, gas exchange and ΔC were 700–900% higher and water-use-efficiency was double when the inflorescence sustained daily growth (Figure 31.4a).

When gas exchanges were expressed on a dry-matter basis, photosynthesis and respiration and ΔC were 50–75% lower late in the life cycle reflecting the fact that the plants were bigger as the inflorescence was well developed. Interestingly, on a surface-area basis, the whole-plant gas exchanges and ΔC were remarkably constant during development, further reflecting

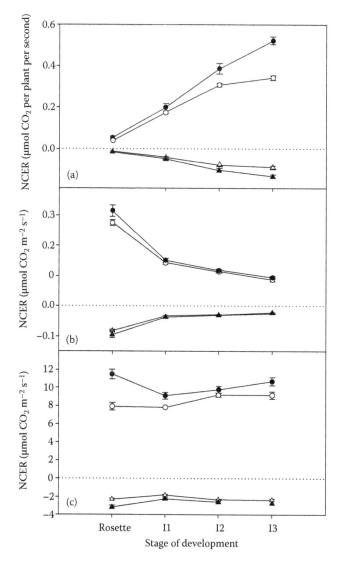

FIGURE 31.4 Whole-plant NCER of *Arabidopsis* plants at different stages of development. Measurements were taken at the rosette stage and the inflorescence first, second, and third stages (I1, I2, and I3). During the rosette stage (28–32 days after sowing [DAS]), the rosette had reached about 30% of its maximum leaf area and the inflorescence was just visible. During I1 (44–48 DAS), the rosette reached its maximum leaf area, and flowers and siliques were prominent on the main inflorescence. During I2 (54–58 DAS), secondary and tertiary inflorescences were well developed also. During I3 (64–68 DAS), the inflorescence had reached its maximum branching, with little flowering remaining and most siliques developing and seeds ripening. Positive NCER values represent photosynthesis (circles), whereas negative values represent dark respiration (triangles). Measurements were taken at ambient CO_2 (40 Pa; open symbols) or elevated CO_2 (80 Pa; closed symbols) on plants acclimated to those conditions. Data are on per plant (a), dry mass (b), and area (c) bases.

the high photosynthetic strength (i.e., source strength) and carbon-use efficiency of the inflorescence tissues (Figure 31.4c). Although not shown here, Leonardos et al. (2014) noted that all *Arabidopsis* lines had greater photosynthesis, respiration, ΔC, and water use efficiency under elevated CO_2. However, at the rosette stage, the wild-type controls responded

more to the elevated CO_2 levels, whereas, during the more mature inflorescence life cycle stage (I2 and I3), it was the transgenics that responded more to the higher amounts of available CO_2 (Leonardos et al. 2014). Taken together, this case study with *Arabidopsis* demonstrates the need for measuring many key physiological traits such as whole-plant photosynthesis, respiration, and ΔC to assess how source and sink strengths might -be changing during the life cycle of the plant. This experimental approach is especially instructive following genetic modification of a primary metabolic process that might be affecting glycolytic and respiratory metabolism, or the capacity to refix respired CO_2 through the operation of a PEP carboxylase.

31.4 CONCLUSIONS

It is naive to suggest that any single technique such as classical destructive growth analysis or a less invasive analysis based on CO_2 gas exchange can provide all the data necessary to describe the complex sequence of events that occur during plant development. Further advances in imaging and remote sensing procedures will undoubtedly facilitate a more complete analysis of plant biomass accumulation and canopy development. Many experimental approaches need to be applied simultaneously at each stage of the vegetative and the reproductive development of a crop to fully assess primary gas exchanges, source–sink interactions, and their impact on productivity.

Although there are limitations to all procedures used to investigate whole-plant growth and productivity, measurements of CO_2 exchange made in real time provide valuable information for many levels of inquiry. Analysis of diel patterns of CO_2 gas analysis is by no means limited to studies of photosynthetic and respiratory metabolism. As suggested decades ago, the applications of using gas signatures to quantify growth are endless (Penning de Vries et al., 1974; McCree, 1986; Dutton et al., 1988). Numerous case studies within our group alone demonstrate the value of gas exchange in obtaining algorithms for modelling and optimizing productivity of crops in controlled environments (Jiao et al., 1991c; Leonardos et al., 1994; Stasiak et al., 1998); assessing the role of canopy architecture and form on productivity (Woodrow and Grodzinski, 1989; Woodrow et al., 1989; Côté and Grodzinski, 1999); correlating whole-plant RGR of natural photosynthetic variants with specific leaf functions such as C-fixation or export (Leonardos and Grodzinski, 2001); assessing plant pathogen interactions (Johnstone et al., 2005a,b); investigating acclimation processes such as hardening of overwintering perennials (Savitch et al., 2002; Leonardos et al., 2003); and evaluating the impact of specific gene alterations on growth (Grodzinski et al., 1999; Leonardos et al., 2014).

REFERENCES

Abeles FB, Morgan PW, Saltvein Jr. ME. 1993. *Ethylene in Plant Biology*. New York: Academic Press.
Acock B, Charles-Edwards DA, Hearn AR. 1977. Growth response of a chrysanthemum (Morifolium) crop to the environment. I. Experimental techniques. *Annals of Botany* 41:41–48.

Acock B, Charles-Edwards DA, Fitter DJ, Hand DW, Ludwig LJWWJ, Withers AC. 1978. The contribution of leaves from different levels within a tomato crop to canopy net photosynthesis: An experimental examination of two canopy models. *Journal of Experimental Botany* 29:815–827.

Amthor JS. 1989. *Respiration and Crop Productivity*. New York: Springer-Verlag.

Amthor JS. 2000. The McCree–de Wit–Penning de Vries–Thornley respiration paradigms: 30 years later. *Annals of Botany* 86:1–20.

Arabidopsis Genome Initiative, 2000. Analysis of the genome sequence of the flowering plant *Arabidopsis thaliana*. *Nature* 408:796–815.

Aschan G, Pfanz H. 2003. Non-foliar photosynthesis—A strategy of additional carbon acquisition. *Flora* 198:81–97.

Baker JT, Gitz III DC, Lascano RJ. 2014. Field evaluation of open system chambers for measuring whole canopy gas exchanges. *Agronomy Journal* 106:537–544.

Bate GC, Canvin DT. 1971. A gas-exchange system for measuring the productivity of plant populations in controlled environments. *Canadian Journal of Botany* 49:601–608.

Beadle CL. 1993. In: Hall DO, Scurlock JM, Bolhar-Nordenkampf HR, Leegood RC, Long SP, eds. *Photosynthesis and Production in a Changing Environment: A Field and a Laboratory Manual*. London: Chapman and Hall, p. 36.

Bellasio C, Olejníčková J, Tesař R, Šebela D, Nedbal L. 2012. Computer reconstruction of plant growth and chlorophyll fluorescence emission in three spatial dimensions. *Sensors* 12:1052–1071.

Bernacchi CJ, Portis AR, Nakano H, von Caemmerer S, Long SP. 2002. Temperature response of mesophyll conductance. Implications for the determination of Rubisco enzyme kinetics and for limitations to photosynthesis in vivo. *Plant Physiology* 130:1992–1998.

Berry JA, Björkman O. 1980. Photosynthetic response and adaptation to temperature in higher plants. *Annual Review of Plant Physiology and Plant Molecular Biology* 31:491–543.

Besford RT, Ludwig LJ, Withers AC. 1990. The greenhouse effect: Acclimation of tomato plants growing in high CO_2, photosynthesis and ribulose-1, 5-bisphosphate carboxylase protein. *Journal of Experimental Botany* 41:925–931.

Björkman O, Demming-Adams B. 1994. Regulation of photosynthetic light energy capture, conversion, and dissipation in leaves of higher plants. In: Schulze E-D, Caldwell MM, eds. *Ecophysiology of Photosynthesis*. Berlin: Springer-Verlag, pp. 17–47.

Bloom A, Mooney, Björkman O, Berry J. 1980. Materials and methods for carbon dioxide and water exchange analysis. *Plant, Cell and Environment* 3:371–376.

Bowes G. 1993. Facing the inevitable: Plants and increasing atmospheric CO_2. *Annual Review of Plant Physiology and Plant Molecular Biology* 44:309–332.

Brooks A, Farquhar GD. 1985. Effects of temperature on the CO_2/O_2 specificity of ribulose-1,5-bisphosphate carboxylase/oxygenase and the rate of respiration in the light. *Planta* 165:397–406.

Bugbee B. 1992. Steady-state canopy gas exchange: System design and operation. *HortScience* 27:770–776.

Bunce JA. 1986. Measurements and modeling of photosynthesis in field crops. *CRC Critical Reviews in Plant Sciences* 4:47–77.

Caporn SJM, Hand DW, Mansfield TA, Wellburn AR. 1994. Canopy photosynthesis of CO_2-enriched lettuce (*Lactuca sativa L.*). Response to short-term changes in CO_2, temperature and oxides of nitrogen. *New Phytologist* 126:45–52.

Causton DR, Venus JC. 1981. *The Biometry of Plant Growth*. London: Edward Arnold.

Cernusak LA, Hutley LB. 2011. Stable isotopes reveal the contribution of corticular photosynthesis to growth in branches of *Eucalyptus miniata*. *Plant Physiology* 155:515–523.

Chapin FSI, Schulze ED, Mooney HA. 1990. The ecology and economics of storage in plants. *Annual Review of Ecology and Systematics* 21:423–447.

Charles-Edwards DA, Acock B. 1977. Growth response of a chrysanthemum [morifolium] crop to the environment. II. A mathematical analysis relating photosynthesis and growth. *Annals of Botany* 41:49–58.

Christy AL, Porter CA. 1982. Canopy photosynthesis and yield in soybean. In: Govindjee, ed. *Photosynthesis Development, Carbon Metabolism, and Plant Productivity*. New York: Academic Press, pp. 449–511.

Cliquet J-B, Boutin J-P, Deleens E, Morot-Gaudry J-F. 1991. Ethephon effects on translocation and partitioning of assimilates in *Zea mays*. *Plant Physiology and Biochemistry* 29:623–630.

Corey KA, Wheeler RM. 1992. Gas exchange capabilities in NASA's plant biomass production chamber. *BioScience* 42:503–509.

Côté R, Grodzinski B. 1999. Improving light interception by selecting morphological leaf phenotypes: A case study using "alifa" pea mutant. *AES Technical Papers* ES-288.

Côté R, Jerrath JM, Posluszny U, Grodzinski B. 1992. Comparative leaf development of conventional and semileafless peas (*Pisum sativum*). *Canadian Journal of Botany* 70:571–580.

Crocker W, Zimmerman PW, Hitchcock AE. 1932. Ethylene-induced epinasty of leaves and the relation of gravity to it. *Contributions from Boyce Thompson Institute* 4:177–218.

Cui M, Nobel PS. 1994. Gas exchange and growth responses to elevated CO_2 and light levels in the CAM species *Opuntia ficus-indica*. *Plant, Cell and Environment* 17:935–944.

Dahal K, Weraduwage SM, Kane K, Rauf SA, Leonardos ED, Gadapati W, Savitch L et al. 2014. Enhancing biomass production and yield by maintaining enhanced capacity for CO_2 uptake in response to elevated CO_2. *Canadian Journal of Plant Science* 94:1075–1083.

Davis TD, Curry EA. 1991. Chemical regulation of vegetative growth. *Critical Reviews in Plant Science* 10:151–188.

Dhawan KR, Bassi PK, Spencer MS. 1981. Effects of carbon dioxide on ethylene production and action in intact sunflower plants. *Plant Physiology* 68:831–834.

Dhondt S, Wuyts N, Inzé D. 2013. Cell to whole-plant phenotyping: The best is yet to come. *Trends in Plant Science* 18:428–439.

Dippery JK, Tissue DT, Thomas RB, Strain BR. 1995. Effects of low and elevated CO_2 on C3 and C4 annuals. 1. Growth and biomass allocation. *Oecologia* 101:13–20.

Dodd AN, Parkinson K, Webb AAR. 2004. Independent circadian regulation of assimilation and stomatal conductance in the ztl-1 mutant of *Arabidopsis*. *New Phytologist* 162:63–70.

Dodd AN, Salathia N, Hall A, Kevei E, Toth R, Nagy F, Hibberd JM, Millar AJ, Webb AAR. 2005. Plant circadian clocks increase photosynthesis, growth, survival and competitive advantage. *Science* 309:630–633.

Donahue RA, Poulson ME, Edwards GE. 1997. A method for measuring whole plant photosynthesis in *Arabidopsis thaliana*. *Photosynthesis Research* 52:263–269.

Drake BG, Gonzales-Meler M, Long SP. 1997. More efficient plants: A consequence of rising atmospheric CO_2. *Annual Review of Plant Physiology and Plant Molecular Biology* 48:609–639.

Duncan WG. 1971. Leaf angles, leaf area and canopy photosynthesis. *Crop Science* 11:482–485.

Dutton R, Jiao J, Tsujita MJ, Grodzinski B. 1988. Whole plant CO_2 exchange measurements for nondestructive estimation of growth. *Plant Physiology* 86:355–358.

Earley EJ, Ingland B, Winkler J, Tonsor SJ. 2009. Inflorescences contribute more than rosettes to lifetime carbon gain in *Arabidopsis thaliana (Brassicaceae)*. *American Journal of Botany* 96:786–792.

Ehleringer J, Björkman O. 1977. Quantum yields for CO_2 [carbon dioxide] uptake in C3 and C4 plants: Dependence on temperature, CO_2, and O_2 [oxygen] concentration. *Plant Physiology* 59:85–90.

Ehleringer JR, Cerling TE, Helliker BR. 1997. C4 photosynthesis, atmospheric CO_2, and climate. *Oecologia* 112:285–299.

Elmore CD. 1980. The paradox of no correlation between leaf photosynthetic rates and crop yields. In: Hesketh JD, Jones JW, eds. *Predicting Photosynthesis for Ecosystem Models. Vol. II.* Boca Raton, FL: CRC Press, pp. 155–167.

Epstein E. 1972. *Mineral Nutrition of Plants: Principles and Perspectives.* New York: John Wiley & Sons.

Evans GC. 1972. *The Quantitative Analysis of Plant Growth.* Oxford: Blackwell Scientific Publications.

Fiorani F, Schurr U. 2013. Future scenarios for plant phenotyping. *Annual Review of Plant Biology* 64:267–91.

Garcia RL, Idso SB, Wall GW, Kimball BA. 1994. Changes in net photosynthesis and growth of *Pinus eldarica* seedlings in response to atmospheric CO_2 enrichment. *Plant, Cell and Environment* 17:971–978.

Geiger DR, Servaites JC. 1994. Diurnal regulation of photosynthetic carbon metabolism in C3 plants. *Annual Review of Plant Physiology and Plant Molecular Biology* 45:235–256.

Ghannoum O, von Caemmerer S, Conroy JP. 2001. Plant water use efficiency of 17 Australian NAD-ME and NADP-ME C4 grasses at ambient and elevated CO_2 partial pressure. *Australian Journal of Plant Physiology* 28:1207–1217.

Gifford RM. 1995. Whole plant respiration and photosynthesis of wheat under increased CO_2 concentration and temperature: Long-term vs short-term distinctions for modelling. *Global Change Biology* 1:385–396.

Gifford RM, Jenkins CLD. 1982. Prospects of applying knowledge of photosynthesis towards improving crop production. In: Govindjee, ed. *Photosynthesis Development, Carbon Metabolism, and Plant Productivity.* New York: Academic Press, pp. 419–457.

Golzarian MR, Frick RA, Rajendran K, Berger B, Roy S, Tester M, Lun DS. 2011. Accurate inference of shoot biomass from high-throughput images of cereal plants. *Plant Methods* 7:2–12.

Gould KS, Cutter EG, Young JPW. 1989. Does growth rate determine leaf form in *Pisum sativum*? *Canadian Journal of Botany* 67:2590–2595.

Grange RI. 1985. Carbon partitioning in mature leaves of pepper: Effects of daylength. *Journal of Experimental Botany* 36:1749–1759.

Grime JP, Hunt R. 1975. Relative growth-rate: Its range and adaptive significance in a local flora. *Journal of Ecology* 63:393–422.

Grimmer C, Komor E. 1999. Assimilate export by leaves of *Ricinus communis* L. growing under normal and elevated carbon dioxide concentrations: The same rate during the day, a different rate at night. *Planta* 209:275–281.

Grodzinski B. 1984. Enhancement of ethylene release from leaf tissue during glycolate decarboxylation: A possible role for photorespiration. *Plant Physiology* 74:871–876.

Grodzinski B. 1992. Carbon dioxide enrichment: Plant nutrition and growth regulation. *BioScience* 42:517–525.

Grodzinski B, Boesel I, Horton RF. 1982. Ethylene release from leaves of *Xanthium strumarium* L. and *Zea mays* L. *Journal of Experimental Botany* 33:344–354.

Grodzinski B, Boesel I, Horton RF. 1983. Light stimulation of ethylene release from leaves of *Gomphrena globosa* L. *Plant Physiology* 71:588–593.

Grodzinski B, Woodrow L, Leonardos ED, Dixon M, Tsujita MJ. 1996. Plant responses to short- and long-term exposures to high carbon dioxide levels in closed environments. *Advances in Space Research* 18:203–211.

Grodzinski B, Jiao J, Leonardos ED. 1998. Estimating photosynthesis and concurrent export rates in C3 and C4 species at ambient and elevated CO_2. *Plant Physiology* 117:207–215.

Grodzinski B, Jiao J, Knowles VL, Plaxton WC. 1999. Photosynthesis and carbon partitioning in transgenic tobacco plants deficient in leaf cytosolic pyruvate kinase. *Plant Physiology* 120:887–895.

Hanan JJ. 1998. *Greenhouses. Advanced Technology for Protected Horticulture.* New York: CRC Press.

Hand DW. 1973. A null balance method for measuring crop photosynthesis in an airtight day-lit controlled-environment cabinet. *Agricultural Meteorology* 12:259–270.

Hand DW, Clark G, Hannah MA, Thornley JHM, Warren Wilson J. 1992. Measuring the canopy net photosynthesis of glasshouse crops. *Journal of Experimental Botany* 43:375–381.

Hatch MD. 1987. C4 photosynthesis: A unique blend of modified biochemistry, anatomy and ultrastructure. *Biochimica et Biophysica Acta* 895:81–106.

Hay PKM, Walker AJ. 1989. *An Introduction to the Physiology of Crop Yield.* Harlow: Longman.

Heiman MD, Meredith FI, Gonzalez CL. 1971. Ethylene production in the cotton plant (*Gossypium hirsutum* L.) canopy and its effect on fruit abscission. *Crop Science* 11:25–27.

Hibberd, JM, Quick, WP. 2002. Characteristics of C4 photosynthesis in stems and petioles of C3 flowering plants. *Nature* 415:451–454.

Ho LC. 1976. Variation in the carbon/dry matter ratio in plant material. *Annals of Botany* 40:163–165.

Hogan KP, Smith AP, Ziska LH. 1991. Potential effects of elevated CO_2 and changes in temperature on tropical plants. *Plant, Cell and Environment* 14:763–778.

Horton P. 2000. Prospects for crop improvement through the genetic manipulation of photosynthesis: Morphological and biochemical aspects of light capture. *Journal of Experimental Botany* 51:475–485.

Hua W, Li R-J, Zhan G-M, Liu J, Li J, Wang X-F, Liu G-H, Wang H-Z, 2012. Maternal control of seed oil content in *Brassica napus*: The role of silique wall photosynthesis. *Plant Journal* 69:432–444.

Huerta AJ, Ting IP. 1988. Effects of various levels of CO_2 on the induction of Crassulacean acid metabolism in *Portulacaria afra* (L.) Jacq. *Plant Physiology* 88:183–188.

Hunt R. 1980. Diurnal progressions in dry weight and short-term plant growth studies. *Plant, Cell and Environment* 3:475–478.

Hunt R. 1982. *Plant Growth Curves.* London: Edward Arnold.

Hunt R. 1990. *Basic Growth Analysis.* London: Unwin Hyman.

Hunt S. 2003. Measurements of photosynthesis and respiration in plants. *Physiologia Plantarum* 117:314–325.

Jahnke S. 2001. Atmospheric CO_2 concentration does not directly affect leaf respiration in bean or poplar. *Plant, Cell and Environment* 24:1139–1151.

Jahnke S, Krewitt M. 2002. Atmospheric CO_2 concentration may directly affect leaf respiration measurement in tobacco, but not respiration itself. *Plant, Cell and Environment* 25:641–651.

Jiao J, Grodzinski B. 1996. The effect of leaf temperature and photorespiratory conditions on export of sugars during steady-state photosynthesis in *Salvia splendens. Plant Physiology* 111:169–178.

Jiao J, Grodzinski B. 1998. Environmental influences on photosynthesis and carbon export in greenhouse roses during development of the flowering shoot. *Journal of the American Society for Horticultural Science* 123:1081–1088.

Jiao J, Tsujita MJ, Grodzinski B. 1991a. Influence of temperature on net CO_2 exchange in roses. *Canadian Journal of Plant Science* 71:235–243.

Jiao J, Tsujita MJ, Grodzinski B. 1991b. Influence of irradiation and CO_2 enrichment on whole plant net CO_2 exchange in roses. *Canadian Journal of Plant Science* 71:245–252.

Jiao J, Tsujita MJ, Grodzinski B. 1991c. Optimizing aerial environments for greenhouse rose production utilizing whole-plant net CO_2 exchange data. *Canadian Journal of Plant Science* 71:253–261.

Johnson IR, Thornley JHM. 1985. Temperature dependence of plant and crop processes. *Annals of Botany* 55:1–24.

Johnstone M, Chatterton S, Sutton JC, Grodzinski B. 2005a. Net carbon gain and growth of bell peppers, *Capsicum annuum* 'cubico', following root infection by *Pythium aphanidermatum*. *Phytopathology* 95:354–361.

Johnstone M, Yu H, Liu W, Leonardos ED, Sutton J, Grodzinski B. 2005b. Physiological changes associated with pythium root rot in hydroponic lettuce. *Acta Horticulturae* 635:67–71.

Jones HG. 1983. *Plants and Microclimate*. Cambridge: Cambridge University Press.

Jordan DB, Ogren WL. 1984. The CO_2/O_2 specificity of ribulose-1,5-bisphosphate carboxylase/oxygenase. Dependence on ribulose bisphosphate concentration, pH and temperature. *Planta* 161:308–313.

Kao CH, Yang SF. 1982. Light inhibition of the conversion of 1-aminocyclopropane-1-carboxylic acid (ACC) to ethylene in leaves is mediated through carbon dioxide. *Planta* 155:261–266.

Kays SJ, Pallas Jr. JE. 1980. Inhibition of photosynthesis by ethylene. *Nature* 285:51–52.

Kellogg EA, Farnsworth EJ, Russo ET, Bazzaz F. 1999. Growth responses of C4 grasses of contrasting origin to elevated CO_2. *Annals of Botany* 84:279–288.

Keys AJ, Leegood RC. 2002. Photorespiratory carbon and nitrogen cycling: Evidence from studies of mutant and transgenic plants. In: Foyer CH, Noctor G, eds. *Photosynthetic Nitrogen Assimilation and Associated Carbon and Respiratory Metabolism*. Dordrecht: Kluwer Academic, pp. 115–134.

Kimball BA, Idso SB. 1983. Increasing atmospheric CO_2 effects on crop yield, water use and climate. *Agricultural Water Management* 7:55–72.

Kirschbaum MUF. 1994. The sensitivity of C3 photosynthesis to increasing CO_2: A theoretical analysis of its dependence on temperature and background CO_2 concentration. *Plant, Cell and Environment* 17:747–754.

Klassen SP, Bugbee B. 2002. Sensitivity of wheat and rice to low levels of atmospheric ethylene. *Crop Science* 42:746–753.

Knight SL. 1992. Constructing specialized plant growth chambers for gas exchange research: Considerations and concerns. *HortScience* 27:767–769.

Kramer PJ. 1981. Carbon dioxide concentration, photosynthesis, and dry matter production. *BioScience* 20:1201–1208.

Ku SB, Edwards GE. 1979. Oxygen inhibition of photosynthesis. I. Temperature dependence and relation to O_2/CO_2 solubility ratio. *Plant Physiology* 59:986–990.

Lake JV. 1966. Measurement and control of the rate of carbon dioxide assimilation by glasshouse crops. *Nature* 209:97–98.

Lake JA. 2004. Gas exchange: New challenges with *Arabidopsis*. *New Phytologist* 162:1–8.

Lalonde S, Tegeder M, Throne-Holst M, Frommer WB, Patrick JW. 2003. Phloem loading and unloading of sugars and amino acids. *Plant, Cell and Environment* 26:37–56.

Lambers H. 1985. Respiration in intact plants and tissues. *Encyclopedia of Plant Physiology. Vol. 18*. Berlin: Springer-Verlag, pp. 418–473.

Lawlor DW. 1995. Photosynthesis, productivity and environment. *Journal of Experimental Botany* 46:1449–1461.

Lawlor DW. 2002. Carbon and nitrogen assimilation in relation to yield: Mechanisms are key to understanding production systems. *Journal of Experimental Botany* 53:773–788.

Lawlor DW, Mitchell RAC. 1991. The effects of increasing CO_2 on crop photosynthesis and productivity: A review of field studies. *Plant, Cell and Environment* 14:807–818.

Lawlor DW, Mitchell RAC, Franklin J, Mitchell VJ, Driscoll SP. 1993. Facility for studying the effects of elevated carbon dioxide concentration and increased temperature on crops. *Plant, Cell and Environment* 16:603–608.

Leakey ADB, Ainsworth EA, Bernacchi CJ, Rogers A, Long SP, Ort DR. 2009a. Elevated CO_2 effects on plant carbon, nitrogen, and water relations: Six important lessons from FACE. *Journal of Experimental Botany* 60:2859–2876.

Leakey ADB, Xu F, Gillespie KM, McGrath JM, Ainsworth EA, Ort DR. 2009b. Genomic basis for stimulated respiration by plants growing under elevated carbon dioxide. *Proceedings of the National Academy of Sciences* 106:3597–3602.

Leonardos ED, Grodzinski B. 2000. Photosynthesis, immediate export and carbon partitioning in source leaves of C3, C3–C4 intermediate and C4 *Panicum* and *Flaveria* species at ambient and elevated CO_2 levels. *Plant, Cell and Environment* 23:839–851.

Leonardos ED, Grodzinski B. 2001. Correlating source leaf photosynthesis and export characteristics of C3, C3–C4 intermediate and C4 *Panicum* and *Flaveria* species with whole-plant relative growth rate. Proceedings for the 12th International Congress on Photosynthesis, PS2001, Brisbane, Australia, CSIRO Publishing.

Leonardos ED, Grodzinski B. 2002. Quantifying immediate C export from source leaves. In: Pessarakli M, ed. *Handbook of Plant and Crop Physiology (2nd edn)*. New York: Marcel Dekker, pp. 407–420.

Leonardos ED, Grodzinski B. 2011. Plant systems/ Photosynthesis and productivity of vascular plants in controlled and field environments. In: Moo-Young M, ed. *Comprehensive Biotechnology (2nd edn)*. Oxford, UK, 4:177–189.

Leonardos ED, Tsujita MJ, Grodzinski B. 1994. Net carbon dioxide exchange rates and predicted growth patterns in Alstroemeria "Jacqueline" at varying irradiances, carbon dioxide concentrations and air temperatures. *Journal of the American Society for Horticultural Science* 119:1265–1275.

Leonardos ED, Tsujita MJ, Grodzinski B. 1996. The effect of source or sink temperature on photosynthesis and 14C-partitioning in, and export from a source leaf of *Alstroemeria sp.* cv. Jacqueline. *Physiologia Plantarum* 97:563–575.

Leonardos ED, Savitch LV, Huner NPA, Öquist G, Grodzinski B. 2003. Daily photosynthetic and C-export patterns in winter wheat leaves during cold stress and acclimation. *Physiologia Plantarum* 117:521–531.

Leonardos ED, Rauf SA, Weraduwage SM, Marillia E-F, Taylor DC, Micallef BJ, Grodzinski B. 2014. Photosynthetic capacity of the inflorescence is a major contributor to daily-C-gain and the responsiveness of growth to elevated CO_2 in *Arabidopsis thaliana* with repressed expression of mitochondrial-pyruvate-dehydrogenase-kinase. *Environmental and Experimental Botany* 107:84–97.

Long SP. 1991. Modification of the response of photosynthetic productivity to rising temperature by atmospheric CO_2 concentrations: Has its importance been underestimated? *Plant, Cell and Environment* 14:729–739.

Long SP, Bernacchi CJ. 2003. Gas exchange measurements, what can they tell us about the underlying limitations to photosynthesis? Procedures and sources of error. *Journal of Experimental Botany* 54:2393–2401.

Long SP, Hällgren J-E. 1993. Measurements of CO_2 assimilation by plants in the field and the laboratory. In: Hall DO, Scurlock JM, Bolhar-Nordenkampf HR, Leegood RC, Long SP, eds. *Photosynthesis and Production in a Changing Environment: A Field and a Laboratory Manual*. London: Chapman and Hall, pp. 129–167.

Louwerse W, Eikhoudt JW. 1975. A mobile laboratory for measuring photosynthesis, respiration and transpiration of field crops. *Photosynthetica* 9:31–34.

Madore M, Grodzinski B. 1985. Photosynthesis and transport of 14C-labelled in a dwarf cucumber cultivar under CO_2 enrichment. *Journal of Plant Physiology* 121:59–71.

Madsen E. 1968. Effect of CO_2 concentration on the accumulation of starch and sugar in tomato leaves. *Physiologia Plantarum* 21:168–175.

Marillia E-F, Micallef BJ, Micallef M, Weninger A, Pedersen KK, Zou J, Taylor DC. 2003. Biochemical and physiological studies of *Arabidopsis thaliana* transgenic lines with repressed expression of the mitochondrial pyruvate dehydrogenase kinase. *Journal of Experimental Botany* 54:259–270.

Marx GA. 1987. A suite of mutants that modify pattern formation in pea leaves. *Plant Molecular Biology Reporter* 5:311–335.

McCree KJ. 1986. Measuring the whole-plant daily carbon balance. *Photosynthetica* 20:82–93.

McCree KJ, Amthor ME. 1982. Effects of diurnal variation in temperature on the carbon balances of white clover plants. *Crop Science* 22:822–827.

Miller DP, Howell GS, Flore JA. 1996. A whole-plant, open, gas-exchange system for measuring net photosynthesis of potted woody plants. *HortScience* 31:944–946.

Mitchell CA. 1992. Measurement of photosynthetic gas exchange in controlled environments. *HortScience* 27:764–767.

Mitchell CA, Both A-J, Bourget CM, Burr JF, Kubota C, Lopez RG, Morrow RC, Runkle ES. 2012. LEDs: The future of greenhouse lighting! *Chronica Horticulturae, a Publication of the International Society for Horticultural Science*, 52(1):6–12.

Mortensen LM. 1982. Growth responses of some greenhouse plants to environment. I. Experimental techniques. *Scientia Horticulturae* 16:39–46.

Murchie EH, Pinto M, Horton P. 2009. Agriculture and the new challenges for photosynthesis research. *New Phytologist* 181: 532–552.

Nederhoff EM, Vegter JG. 1994. Photosynthesis of stands of tomato, cucumber and sweet pepper measured in greenhouse under various CO_2-concentrations. *Annals of Botany* 73:353–361.

Nilwik HJM. 1980. Photosynthesis of whole sweet pepper plants 2. Response to CO_2 (carbon dioxide) concentration, irradiance and temperature as influenced by cultivation conditions. *Photosynthetica* 14:382–391.

Nunes-Nesi A, Araújo WL, Fernie AR. 2011. Targeting mitochondrial metabolism and machinery as a means to enhance photosynthesis. *Plant Physiology* 155:101–107.

Ort DR, Baker NR. 1988. Consideration of photosynthetic efficiency at low light as a major determinant of crop photosynthetic performance. *Plant Physiology and Biochemistry* 26: 555–565.

Osborne DJ. 1991. Ethylene in leaf ontogeny and abscission. In: Mattoo AK, Suttle JC, eds. *Ethylene*. Boca Raton, FL: CRC Press, pp. 193–214.

Osmond CB, Björkman O. 1975. Pathways of CO_2 fixation in the CAM plant *Kalanchoë daigremontiana*. II. Effects of O_2 and CO_2 concentration on light and dark CO_2 fixation. *Australian Journal of Plant Physiology* 2:155–162.

Paul MJ, Pellny TK. 2003. Carbon metabolite feedback regulation of leaf photosynthesis and development. *Journal of Experimental Botany* 54:539–547.

Paulus S, Behmann J, Mahlein A-K, Plumer L, Kuhlmann H. 2014. Low-cost 3D systems: Suitable tools for plant phenotyping. *Sensors* 14:3001–3018.

Pearcy RW, Pfitsch WA. 1994. The consequences of sun flecks for photosynthesis and growth of forest understory plants. In: Schulze E-D, Caldwell MM, eds. *Ecophysiology of Photosynthesis*. Berlin: Springer-Verlag, pp. 343–359.

Penning de Vries FWT, Brunsting AHM, van Larr HH. 1974. Products, requirements and efficiency of biosynthetic process: A quantitative approach. *Journal of Theoretical Biology* 45:377–399.

Pereira JS. 1994. Gas exchange and growth. In: Schulze E-D, Caldwell MM, eds. *Ecophysiology of Photosynthesis*. Berlin: Springer-Verlag, pp. 147–181.

Philosoph-Hadas S, Aharoni N, Yang SF. 1986. Carbon dioxide enhances the development of the ethylene-forming enzyme in tobacco leaf discs. *Plant Physiology* 82:925–929.

Pons TL, Poorter H. 2014. The effect of irradiance on the carbon balance and tissue characteristics of five herbaceous species differing in shade-tolerance. *Frontiers in Plant Science.* 5:1–14.

Poorter H. 1990. Interspecific variation in relative growth rate: On ecological causes and physiological consequences. In: Lambers H, Cambridge ML, Konings H, Pons TL, eds. *Causes and Consequences of Variation in Growth Rate and Productivity of Higher Plants*. The Hague: SPB Academic Publishing, pp. 45–68.

Poorter H, Garnier E. 1996. Plant growth analysis: An evaluation of experimental design and computational. *Journal of Experimental Botany* 47:1343–1351.

Poorter H, Welschen RAM. 1993. Variation in RGR underlying carbon economy. In: Hendry GAF, Grime JP, eds. *Methods in Comparative Plant Ecology, A Laboratory Manual*. London: Chapman & Hall, p. 107.

Poorter H, Remkes C, Lambers H. 1990. Carbon and nitrogen economy of 24 wild species differing in relative growth rate. *Plant Physiology* 94:621–627.

Porter MA, Grodzinski B. 1984. Acclimation to high CO_2 in bean. *Plant Physiology* 74:413–416.

Porter MA, Grodzinski B. 1985. CO_2 enrichment of protected crops. *Horticultural Reviews* 7:345–398.

Raines C. 2003. The Calvin cycle revised. *Photosynthesis Research* 75:1–10.

Reddy VR, Baker DN, Hodges HF. 1991. Temperature effects on cotton canopy growth, photosynthesis, and respiration. *Agronomy Journal* 83:699–704.

Roberts MJ, Long SP, Tieszen LL, Beadle CL. 1987. Measurement of plant biomass and net primary production. In: Coombs J, Hall DO, Long SP, Scurlock JMO, eds. *Techniques in Bioproductivity and Photosynthesis*. Oxford: Pergamon Press, pp. 1–19.

Robertson EJ, Williams M, Hardwood JL, Linsday JG, Leaver CJ, Leech RM. 1995. Mitochondria increase three-fold and mitochondrial proteins and lipid change dramatically in post-meristematic cells in young wheat leaves grown in elevated CO_2. *Plant Physiology* 108:469–474.

Russell G, Jarvis PG, Monteith JL. 1989. Absorption of radiation by canopies and stand growth. In: Russell G, Marshall B, Jarvis PG, eds. *Plant Canopies: Their Form and Functions.* Cambridge: Cambridge University Press, pp. 21–39.

Sage RF, Monson RK. 1999. *C4 Plant Biology.* New York: Academic Press.

Sage RF, Sharkey TD, Seamann JR. 1989. Acclimation of photosynthesis to elevated CO_2 in five C3 species. *Plant Physiology* 89:590–596.

Savitch LV, Leonardos ED, Krol M, Jansson S, Grodzinski B, Huner NPA, Öquist G. 2002. Two different strategies for light utilisation in photosynthesis in relation to growth and cold acclimation. *Plant, Cell and Environment* 25:761–771.

Schulze W, Schulze E-D. 1994. The significance of assimilatory starch for growth in *Arabidopsis thaliana* wild-type and starchless mutants. In: Schulze E-D, Caldwell MM, eds. *Ecophysiology of Photosynthesis.* Berlin: Springer-Verlag, pp. 123–131.

Simbo DG, Van den Bilcke N, Samson R. 2013. Contribution of corticular photosynthesis to bud development in African baobab (*Adansonia digitata* L.) and Castor bean (*Ricinus communis* L.) seedlings. *Environmental and Experimental Botany* 95:1–5.

Sims DA, Pearcy RW. 1994. Scaling sun and shade photosynthetic acclimation of *Alocasia macrorrhiza* to whole-plant performance—I. Carbon balance and allocation at different daily photon flux densities. *Plant, Cell and Environment* 17:881–887.

Stasiak MA. 2002. Effect of inner canopy irradiation on plant productivity in a sealed environment. PhD Thesis, Department of Plant Agriculture. Guelph, Canada: University of Guelph.

Stasiak MA, Cote R, Dixon MAD, Grodzinski B. 1998. Increasing plant productivity in closed environments with inner canopy illumination. *Life Support Biosphere Sciences* 5:175–182.

Stitt M, Muller C, Matt P, Gibon Y, Carillo P, Morcuende R, Scheible W-R, Knapp A. 2002. Steps towards an integrated view of nitrogen metabolism. *Journal of Experimental Botany* 53:959–970.

Tackenberg O. 2007. A new method for non-destructive measurement of biomass, growth rates, vertical biomass distribution and dry matter content based on digital image analysis. *Annals of Botany* 99:777–783.

Taylor GE, Gunderson CA. 1986. The response of foliar gas exchange to exogenously applied ethylene. *Plant Physiology* 82:653–657.

Teng N, Jin B, Wang Q, Hao H, Ceulemans R, Kuang T, Lin J. 2009. No detectable maternal effects of elevated CO_2 on *Arabidopsis thaliana* over 15 generations. *PLoS ONE* 4(6): e6035.

Thomas RB, Griffin KL. 1994. Direct and indirect effects of atmospheric carbon dioxide enrichment on leaf respiration of *Glycine max* (L.) Merr. *Plant Physiology* 104:355–361.

Tibbitts TW, McFarlane JC, Krizek DT, Berry WL, Hammer PA, Hodgson RH, Langhans RW. 1977. Contaminants in plant growth chambers. *HortScience* 12:310–311.

Tocquin P, Périlleux C. 2004. Design of a versatile device for measuring whole plant gas exchanges in Arabidopsis thaliana. *New Phytologist* 162:223–229.

van Andel J, Biere A. 1990. Ecological significance of variability in growth rate and plant productivity. In: Lambers H, Cambridge ML, Konings H, Pons TL, eds. *Causes and Consequences of Variation in Growth Rate and Productivity of Higher Plants.* The Hague: SPB Academic Publishing, p. 257.

van der Werf A. 1996. Growth analysis and photoassimilate partitioning. In: Zamski E, Schaffer AA, eds. *Photoassimilate Distribution in Plants and Crops: Source–Sink Relationships.* New York: Marcel Dekker, pp. 1–20.

van Iersel MW. 2003. Carbon use efficiency depends on growth respiration, maintenance respiration, and relative growth rate. A case study with lettuce. *Plant, Cell and Environment* 26:1441–1449.

van Iersel MW, Bugbee B. 2000. A multiple chamber, semicontinuous, crop carbon dioxide exchange system: Design, calibration, and data interpretation. *Journal of the American Society for Horticultural Science* 125:86–92.

van Iersel MW, Seymour L. 2000. Growth respiration, maintenance respiration, and carbon fixation of vinca (*Catharanthus roseus* L. G. Don.): A time series analysis. *Journal of the American Society for Horticultural Science* 125:702–706.

Wand SJE, Midgley GF, Stock WD. 2001. Growth responses to elevated CO_2 in NADP-ME, NAD-ME and PCK C4 grasses and a C3 grass from South Africa. *Australian Journal of Plant Physiology* 28:13–25.

Warren Wilson J, Hand DW, Hannah MA. 1992. Light interception and photosynthetic efficiency in some glasshouse crops. *Journal of Experimental Botany* 43:363–373.

Watts B. 1995. The Effects of Temperature and CO_2 Enrichment on Growth and Photoassimilate Partitioning in Peppers (*Capsicum annuum* L.). Guelph, Canada: University of Guelph.

Wells R, Meredith Jr. WR, Williford JR. 1986. Canopy photosynthesis and its relationship to plant productivity in near-isogenic lines differing in leaf morphology. *Plant Physiology* 82:635–640.

Weraduwage SM. 2013. Harnessing the anabolic properties of dark respiration to enhance sink activity at high CO_2 using *Arabidopsis thaliana* l. with partially suppressed mitochondrial pyruvate dehydrogenase kinase (Ph.D. thesis). University of Guelph, Guelph, Canada.

Weraduwage SM, Micallef BJ, Grodzinski B, Taylor DC, Marillia E-F. 2011. Plant systems/roles of dark respiration in plant growth and productivity. In: Moo-Young, M. ed. *Comprehensive Biotechnology, vol. 4 (2nd edn).* Oxford, UK: Elsevier, pp. 191–207.

Wheeler RM. 1992. Gas-exchange measurements using a large, closed plant growth chamber. *HortScience* 27:777–780.

Wheeler RM, Mackowiak CL, Sager JC, Yorio NC, Knott WM, Berry WL. 1994. Growth and gas exchange by lettuce stands in a closed, controlled environment. *Journal of the American Society for Horticultural Science* 119:610–615.

Wheeler RM, Peterson BV, Sager JC, Knott WM. 1996. Ethylene production by plants in a closed environment. *Advances in Space Research* 18:193–196.

Wheeler RM, Stutte GW, Subbarao GV, Yorio NC. 2002. Plant growth and human life support for space travel. In: Pessarakli M, ed. *Handbook of Plant and Crop Physiology (2nd edn).* New York: Marcel Dekker, pp. 925–941.

Wickens LK, Cheeseman JM. 1988. Application of growth analysis to physiological studies involving environmental discontinuities. *Physiologia Plantarum* 73:271–277.

Wittmann C, Pfanz H, Loreto F, Centritto M, Pietrini F, Alessio G, 2006. Stem CO_2 release under illumination: Corticular photosynthesis, photorespiration or inhibition of mitochondrial respiration? *Plant Cell and Environment* 29:1149–1158.

Wittwer SH. 1985. Carbon dioxide levels in the biosphere: Effects on plant productivity. *CRC Critical Reviews in Plant Science* 2:171–198.

Wolfe DW. 1995. Physiological and growth response to atmospheric carbon dioxide concentration. In: Pessarakli M, ed. *Handbook of Plant and Crop Physiology.* New York: Marcel Dekker, p. 223.

Woo KC, Wong SC. 1983. Inhibition of CO_2 assimilation by supraoptimal CO_2: Effect of light and temperature. *Australian Journal of Plant Physiology* 10:75–85.

Woodrow IE, Berry JA. 1988. Enzymatic regulation of photosynthetic CO_2 fixation in C3 plants. *Annual Review of Plant Physiology and Plant Molecular Biology* 39:533–594.

Woodrow L, Grodzinski B. 1989. An evaluation of the effects of ethylene on carbon assimilation in *Lycopersicon esculentum Mill. Journal of Experimental Botany* 40:361–368.

Woodrow L, Grodzinski B. 1994. Ethylene exchange in *Lycopersicon esculentum Mill.* Leaves during short- and long-term exposures to CO_2. *Journal of Experimental Botany* 44:471–480.

Woodrow L, Liptay A, Grodzinski B. 1987. The effects of CO_2 enrichment and ethephon application on the production of tomato transplants. *Acta Horticulturae* 201:133–140.

Woodrow L, Thompson RG, Grodzinski B. 1988. Effects of ethylene on photosynthesis and partitioning in tomato *Lycopersicon esculentum Mill. Journal of Experimental Botany* 39:667–684.

Woodrow L, Jiao J, Tsujita MJ, Grodzinski B. 1989. Whole plant and leaf steady state gas exchange during ethylene exposure in *Xanthium strumarium L. Plant Physiology* 90:85–90.

Woodrow L, Jiao J, Tsujita MJ, Grodzinski B. 1990. Photoautotrophic systems: Ethylene and carbon dioxide interactions from the callus to the canopy. In: Flores HE, Arteca RN, Shannon JC, eds. *Polyamines and Ethylene: Biochemistry, Physiology, and Interactions.* Rockville, MD: American Society of Plant Physiologists, pp. 91–100.

Zelitch I. 1982. The close relationship between net photosynthesis and crop yield. *BioScience* 32:796–802.

Zelitch I. 1992. Control of plant productivity by regulation of photorespiration. *BioScience* 42:510–517.

Zhu XG, Long SP, Ort DR. 2010. Improving photosynthetic efficiency for greater yield. *Annual Review of Plant Biology* 61:235–261.

32 Nonubiquitous Carotenoids in Higher Plants

Presence, Role in Photosynthesis, and Guidelines for Identification

Raquel Esteban and José Ignacio García-Plazaola

CONTENTS

32.1 Pigment Composition of Higher Plant Chloroplasts: Ubiquitous and Unusual Carotenoids 589
32.2 Low-Light Carotenoids 591
    32.2.1 *Trans*-Neoxanthin 591
    32.2.2 α-Carotene 591
    32.2.3 Lutein Epoxide 593
    32.2.4 Lactucaxanthin 594
32.3 High-Light Carotenoids 594
    32.3.1 Xanthophyll Esters 594
    32.3.2 Red *Retro*-Carotenoids: Escholtzxanthin and Rhodoxanthin 595
32.4 HPLC Determination of Nonubiquitous Carotenoids 596
32.5 Concluding Remarks 597
Acknowledgments 597
Abbreviations 597
References 597

32.1 PIGMENT COMPOSITION OF HIGHER PLANT CHLOROPLASTS: UBIQUITOUS AND UNUSUAL CAROTENOIDS

More than 600 carotenoids and carotenoid-derived species have been found in the chloroplasts of plants and also of other organisms, including algae, fungi, and bacteria. In plants, this group of 40-C polyene molecules (tetraterpenes) are synthesized and accumulated in plastids, where they fulfill a multitude of roles: from the ecological function (e.g., attraction of pollinators) to the cellular level (e.g., retrograde signaling) (Havaux 2013). Despite this astonishing functional and structural diversity, the chloroplast pigment composition of the green lineage of the plant kingdom (which includes green algae, hornworts, liverworts, mosses, and vascular plants) is remarkably constant. Thus, with a few exceptions, functional chloroplasts from all species analyzed so far contain six *ubiquitous* carotenoids: neoxanthin (Neo), violaxanthin (Vio), antheraxanthin (Ant), zeaxanthin (Zea), lutein (L), and β-carotene (β-Car), together with the main two chlorophylls, Chl *a* and Chl *b* (Young et al. 1997; Esteban et al. 2015).

Depending on the presence or absence of oxygen in the hydrocarbon structure, carotenoids can be separated into two major groups: xanthophylls (Neo, Vio, Ant, Zea, and L) and carotenes (β-Car), respectively. In photosynthetic tissues, carotenoids are rarely free in thylakoids or lipid droplets. Indeed, they typically appear in the formation of chlorophyll–carotenoid–protein complexes. Such is the case for β-Car, which is bound to the reaction centers of both photosystems (Cazzaniga et al. 2012), while xanthophylls are present in light-harvesting systems, where they occupy specific binding sites (Formaggio et al. 2001). Biosynthetically, all these carotenoids are produced from the linear lycopene through two main biosynthetic routes: the β-route (β-branch), which involves the β-cyclation of both extremes of lycopene, and the α-route (α-branch), which implies one β-ring and one ε-rings (Figure 32.1). Carotenoids display a plethora of functions in photosynthetic plant tissues: light harvesting, thermal energy dissipation, triplet Chl quenching, singlet oxygen scavenging, and membrane stabilization (Havaux and Niyogi 1999).

However, this *eight-basic-pigment* composition is supplemented in some species and under some circumstances by the presence of other *nonubiquitous* carotenoids. Some of them partly replace *ubiquitous* carotenoids at their binding sites, while others show a distinctive location. Some of them are only induced by specific environmental conditions (mainly related to the growth irradiance), while others have a constitutive character, their presence being phylogenetically determined. In view of all that has been mentioned so far, one may predict that although uncommon or nonubiquitous, these

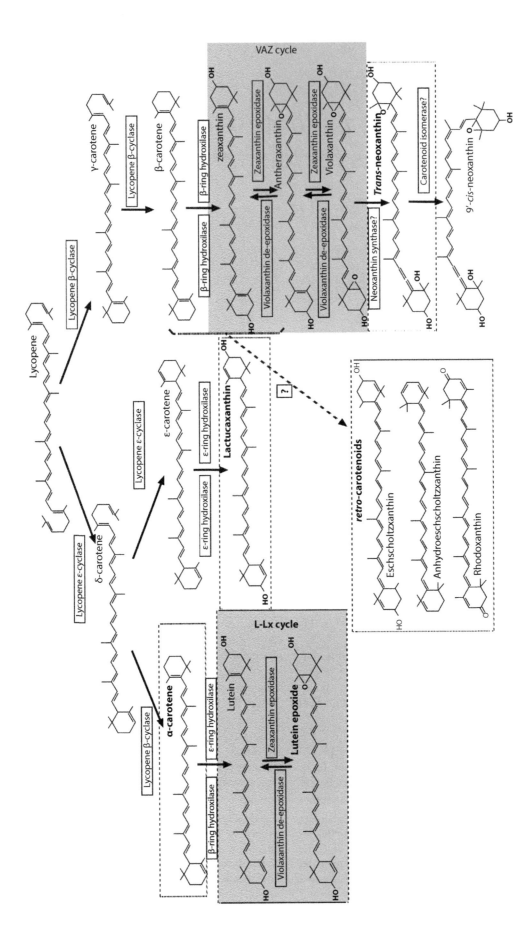

FIGURE 32.1 Carotenoid biosynthesis from lycopene showing hypothetical and confirmed pathways for the nonubiquitous carotenoids addressed in this chapter: *trans*-Neo, α-Car, Lx, Lac, and red *retro*-carotenoids.

TABLE 32.1

Occurrence of Nonubiquitous Carotenoids Described in the Present Chapter

Light Environment	Nonubiquitous Carotenoids	Frequency	Where?	When?	Role
Shade environment	*Trans*-neoxanthin	Rare	LHCs (N1)?	Nonfoliar tissues	Unknown
	α-Carotene	Common	Core complexes	Constitutive, low light	Light harvesting
	Lactucaxanthin	Rare	LHCs (L1, L2)	Constitutive	Similar to lutein?
	Lutein epoxide	Common	LHCs (L2, V1)	Woody plants	Second xanthophyll cycle
Sun environment	Xanthophyll esters	Common	Plastoglobules	Stress periods, senescence	Unknown
	Red *retro*-carotenoids	Rare	Plastoglobules	Stress periods	Filters, ROS quenchers

Note: The frequency of these carotenoids, together with the answers to the three major questions addressed in this chapter, is shown: (1) where they are located, (2) when they occur, and (3) their presumed role/the reason for their presence.

unusual carotenoids could be, indeed, important components for the functionality of the chloroplast or/and could have a key role in plant acclimation to variable stress situations, at the leaf and whole-plant levels. This chapter will describe, therefore, the presence of these nonubiquitous carotenoids (under which circumstances they occur). Specifically, it will concentrate on the occurrence of the six groups of nonubiquitous carotenoids more frequently found in photosynthetic tissues of vascular plants: lactucaxanthin (Lac), lutein epoxide (Lx), *trans*-neoxanthin (*trans*-Neo), α-carotene (α-Car), xanthophyll esters (XEs), and the group of the red *retro*-carotenoids (which includes escholtzxanthin, anhydroescholtzxanthin, and rhodoxanthin). With the exception of the last two groups (grouped as high-light carotenoids—see Section 32.3), low-light-acclimated leaves are typically enriched in the presence of such nonubiquitous carotenoids (thereby low-light carotenoids—see Section 32.2). Three major questions will be addressed on the natural occurrence of each compound in the forthcoming sections (with the limitations of the available information) (Table 32.1): (1) Where are they located (i.e., plastoglobules [PGs], thylakoid, protein–pigment complexes, etc.)? (2) When do they occur (i.e., constitutive, seasonal, stress-induced, etc.)? (3) What is their presumed role or the reason for their presence (i.e., energy collectors, photoprotection, etc.)? Some tips for their identification and quantification in routinely high-performance liquid chromatography (HPLC) measurements will also be provided (Section 32.4).

32.2 LOW-LIGHT CAROTENOIDS

32.2.1 *Trans*-Neoxanthin

Neo is a xanthophyll exclusively found in the green lineage of the plant kingdom. From an evolutionary point of view, its presence is associated with the appearance of Chl *b* (Takaichi and Mimuro 1998). Biosynthetically, *trans*-Neo is synthesized from *trans*-Vio through the action of a Neo synthase, being subsequently transformed to *cis*-Neo by the action of an isomerase (North et al. 2007), which is also able to catalyze the reverse reaction (Figure 32.1). Curiously, the usual configuration of

carotenoids in chloroplast is *trans*, Neo being the only carotenoid regularly present in *cis* configuration. As a consequence of this particular configuration, it is supposed to be specifically bound to the N1 binding site of light-harvesting proteins of photosystem II (PSII) that preferentially binds 9-*cis*-5,6-epoxy xanthophylls (Formaggio et al. 2001). Although this site is basically occupied by 9′*cis*-Neo, in photosynthetic fruits or stems, this xanthophyll can be replaced by another unusual carotenoid: 9′*cis*-Vio (Snyder et al. 2004; Esteban et al. 2010b). Recent findings have also demonstrated that *cis*-Neo has a specific role in photoprotection as a scavenger of the superoxide anion (Dall'Osto et al. 2007). Apart from its function in the light-harvesting antennae, *cis*-Neo, together with *cis*-Vio, is a substrate for abscisic acid (ABA) synthesis, an important hormone implicated in plant developmental processes.

The occurrence of *trans*-Neo is, then, unusual; however, its natural accumulation has been documented in chromoplasts of roots, petals, and fruits (Takaichi and Mimuro 1998) and likewise, its presence has been convincingly demonstrated in chloroplasts (Li and Walton 1990; Strand et al. 2000; Kruk et al. 2005). From these studies, some general patterns of its presence in chloroplasts can be depicted. Thus, while in green leaves, 9′*cis*-Neo typically represents more than 90% of the total Neo content (Strand et al. 2000), the accumulation of *trans*-Neo seems to be favored in plants growing under low light, as is the case in etiolated cabbage heads (Kruk et al. 2005), photosynthetic stems of avocado (Esteban et al. 2010b), and bean leaves, where it can represent up to 40% of total Neo (Li and Walton 1990). *Arabidopsis* plants, where the *ABA-4* gene (encodes a membrane protein involved in ABA biosynthesis) was constitutively expressed, also showed enhanced levels of *trans*-Neo (North et al. 2007). Apart from these few studies, the proportion of *cis*-Neo and *trans*-Neo isomers has been very rarely reported, complicating the task of identifying differential functional roles or specific protein locations for both isomers.

32.2.2 α-Carotene

α-Car is a carotenoid containing one β-ring and one ε-ring, and it is synthesized through the α-branch of the carotenoid biosynthetic pathway, via the cyclization of lycopene by the successive action of two enzymes, lycopene ε-cyclase and lycopene β-cyclase (Figure 32.1) (Cunningham et al. 1996). In chloroplasts, it is supposed to be bound to the core complexes in both photosystems, PSI and PSII, together with (or replacing) β-Car (Yamamoto and Bassi 1996). α-Car is found in significant amounts in deep orange-yellow fruits (e.g., apricot, cantaloupe, mango, nectarine, peach, papaya); roots; and tubers (e.g., pumpkin, sweet potato, carrot, butternut squash) (Pennington and Fisher 2010; Khoo et al. 2011). Hydroxylation of α-Car gives rise to L, the most abundant carotenoid in photosynthetic tissues, α-Car being an obligate intermediate on L biosynthesis. The single most striking fact is that in green tissues of most of the species, α-Car is not ubiquitous (contrary to L, which is always present), and when it is naturally present, it accumulates in low amounts (typical concentrations on chlorophyll basis, ~10 mmol/mol Chl $a + b$). Overall, its presence is not rare in leaves, because it can be detected in a considerable number of species and families (Thayer and Björkman 1990; Demmig-Adams and Adams 1992). Extremely high contents (>50 mmol/mol Chl $a+b$) have indeed been described in several species covering a wide range of taxonomical groups, such as *Tetraselmis chui* Butcher (microalgae) (Garrido et al. 2009), *Cladophora glomerata* Kütz (macroalgae) (Choo et al. 2005), *Picea abies* (L.) H. Karst, *Pinus contorta* Douglas ex Loudon, *Pinus sylvestris* L. (Pinaceae) (Kronfuss et al. 1998; Savitch et al. 2002; Tegischer et al. 2002; Tausz et al. 2005), and several angiosperms, such as *Paullinia bracteosa* Radlk, *Dieffenbachia longispatha* Engl. & Krause,

Piper cordulatum C.DC. (Koninger et al. 1995), *Virola sebifera* Aubl. (Matsubara et al. 2008, 2009), *Connarus williamsii* Britton, *Helicornia irrasa* Lane ex R.R. Sm., *Calathea lutea* Schult., *Costus villosissimus* Jacq., *Cryosophila warscewiczii* (H.Wendl.) Bartlett, *Gustavia superb* (Kunth) O. Berg, *Miconia argéntea* (Sw.) DC., *Anacardium excelsum* L. (Krause et al. 2004), *Fagus sylvatica* L. (García-Plazaola et al. 2001), and *Ligustrum lucidum* Aiton (Esteban et al. 2009b).

α-Car concentrations are far from being stable. Indeed, in α-Car species (species with α-Car presence), the equilibrium between the two carotenes of the biosynthetic pathway (α-Car and β-Car) and between the corresponding carotenoids in each of the branches (α-branch and β-branch carotenoids) may swing during acclimation to changes in light environments (Thayer and Björkman 1990; Matsubara et al. 2009). As well, α-Car contents have been reported to change between seasons, with lower levels of α-Car during winter in conifer species (Adams and Demmig-Adams 1994) and during summer in the bay laurel (unpublished data) (Figure 32.2c). Indeed, this balance is determined by the differential activity of the cyclases of the biosynthetic pathway (Cunningham et al. 1996), as occurs under low light, when the ε-cyclases are upregulated (Hirschberg 2001), leading to an enhanced content of α-branch carotenoids. Therefore, these enzymes (cyclases) play a strategic function in controlling adjustments of leaf carotenoid composition under acclimation processes. This is exemplified in tropical forests pioneer plants, in which β-Car concentration was enhanced with gap size, whereas

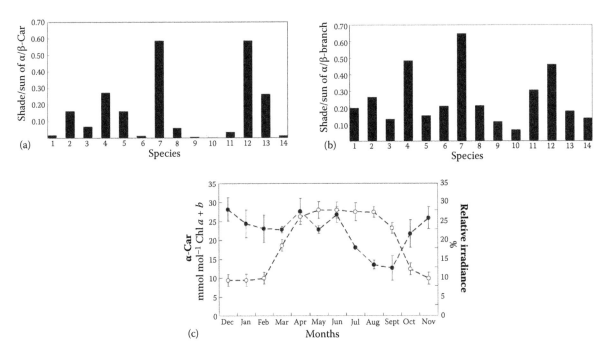

FIGURE 32.2   Relative increase on (a) α/β-Car and (b) α/β-branch when shade leaves are compared with sun leaves of the same temperate forest species: 1, *Quercus pirenaica* Willd.; 2, *Rubus ulmifolius* Schott; 3, *Acer pseudoplatanus* L.; 4, *Quercus petraea* (Matt.) Liebl.; 5, *Viburnun lantana* L.; 6, *Castanea sativa* Mill.; 7, *Corylus avellana* L.; 8, *Cornus sanguínea* L.; 9, *Betula celtiberica* Rothm. & Vasc.; 10, *Populus tremula* L.; 11, *Alnus glutinosa* (L.) Gaertn.; 12, *Fagus sylvatica* L.; 13, *Quercus robur* L.; and 14, *Quercus ilex* L. (c) Seasonal changes in α-Car pool (closed circles) on chlorophyll basis (mmol/mol Chl $a + b$) of a laurel tree (*Laurus nobilis*) growing on the Basque Country University campus. The tree was shaded by a building from October to March, and the canopy received quite low total daily light fluxes until spring. Open circles represent the relative irradiance. Each point is the mean of three to four replicates ± standard error (SE).

α-Car content decreased (Krause et al. 2001). In general, α-Car has been detected in leaves of shade-tolerant species, leading to a high α/β ratio under shade or low light, which is usually reversed in sun leaves (e.g., Thayer and Björkman 1990; Demmig-Adams and Adams 1992; Koninger et al. 1995; Demmig-Adams 1998; Krause et al. 2004; Matsubara et al. 2009) (Figure 32.2a, b). However, not all shade plants accumulate α-Car; such is the case for *Vinca minor* L., which can grow in extremely deep shade without accumulating α-Car (Demmig-Adams 1998). Undeniably, when α-Car is present, its concentration is invariably higher in shade leaves in contrast with that in sun leaves (Figure 32.2). Based on these data, an important role as light-harvesting pigment in shade acclimation has been suggested (Krause et al. 2001; Caliandro et al. 2013). Interestingly, both branches (α-branch and β-branch) have reactive oxygen species (ROS)-scavenging capacities in leaves (Havaux et al. 2007; Johnson et al. 2007); hence, the regulation of the α-branch and β-branch in the acclimation to variable environments may be crucial for the upregulation of carotenoid biosynthesis and for leaf photoprotection, and therefore, for the whole-plant level (Caliandro et al. 2013).

32.2.3 Lutein Epoxide

Lutein-5,6-epoxide is a xanthophyll generated from L by epoxidation of the β-ring, presumably by the action of a zeaxanthin epoxidase (ZEP) (García-Plazaola et al. 2007) (Figure 32.1). This carotenoid, although nonubiquitous in the plant kingdom, participates in an important light-dependent xanthophyll cycle (Lx-L cycle) (Bungard et al. 1999) with key functions for the whole plant and, together with the violaxanthin, antheraxanthin, zeaxanthin (VAZ) cycle, comprises the two xanthophyll cycles present in higher plants.

In this cycle, Lx undergoes a de-epoxidation to L driven by light (frequently day periods or changes in light conditions) that occurs simultaneously with de-epoxidation of Vio in the VAZ cycle. The newly formed L from Lx is denoted as ΔL, to differentiate from the constitutive L. The ΔL augments the already existing pool of L in plants (L + ΔL) (Nichol et al. 2012). Under low light or darkness (usually night periods or deep-shade situations), ΔL may eventually be epoxidized back to Lx by the enzyme ZEP, leading to a *completed* Lx-L cycle. In the majority of Lx species, however, once ΔL is formed, it remains for days or even weeks, with no overnight recovery of the initial Lx pool, leading to a *truncated* Lx-L cycle (Esteban and García-Plazaola 2014). This truncated cycle (de-epoxidation of Lx to ΔL) usually occurs in response to sudden changes in environmental conditions (severe environmental stress) or developmental stage (budbreak, floral development) (Matsubara et al. 2011).

The Lx-L cycle pigments compete for the same binding sites in chloroplast pigment–protein complexes with VAZ cycle pigments. In monomeric light-harvesting complex (LHC), Lx is bound to the internal L2 site, while both the peripheral V1 and the internal binding site L2 of trimers can be occupied by Lx (Matsubara et al. 2007). When illuminated, ΔL formed from Lx de-epoxidation replaces Lx in L2 in monomeric

antenna proteins and in V1 and L2 in trimers, displacing VAZ pigments. While de-epoxidation of Lx to ΔL has also been observed in PSI (Matsubara et al. 2007), the precise location of these pigments has not been established.

Lx is, in general, present in unstressed shaded tissues (e.g., inner-canopy long-lived leaves or light-protected tissues as buds) (Matsubara et al. 2003, 2005; García-Plazaola et al. 2004, 2007). However, it can also be found in sun leaves of several species, such as *Ocotea foetens* L., *Laurus azorica* (Seub.) Franco (Esteban et al. 2007), *Virola elongata* Warb. (Matsubara et al. 2009), and the parasitic plants *Cuscuta reflexa* Roxb. and *Cassytha* sp. (Snyder et al. 2005; Close et al. 2006). It has been detected in all plant organs, in petals (Tai and Chen 2000; Mélendez-Martínez et al. 2006), in floral receptacles (Miller et al. 2009), in fruits (Rabinowitch et al. 1975; Razungles et al. 1996; Gandul-Rojas et al. 1999; Esteban et al. 2010b), in stems (Bungard et al. 1999; Esteban et al. 2010b), in seeds (Edelenbos et al. 2001), in cotyledons (Esteban et al. 2009a), and in leaves. Lx is present in a broad range of rather unrelated plant taxa (Esteban et al. 2009b; Matsubara et al. 2009) but is absent in green algae (at least in the ones analyzed), with the exception of Antarctic alga *Koliella antarctica*, recently described (La Rocca et al. 2015). It is rare among bryophytes and frequent in several unrelated groups of gymnosperms (Pinaceae and Gingkoaceae) and angiosperms (Myristicaceae, Laruaraceae, Fagaceae, Fabaceae, Ericaceae, Cornaceae, Verbenaceae, and Caprifoliaceae) (García-Plazaola et al. 2004; Matsubara et al. 2008, 2009; Esteban et al. 2009b). Despite this wide distribution, Lx is more frequent in woody species, and it is in these key woody species where research has been centered in recent years. Examples of these species are *Inga* sp. (Fabaceae) (Matsubara et al. 2005, 2007, 2008), *Umbellularia californica* (Nees) Nutt., *Laurus nobilis* L. (Lauraceae) (Esteban et al. 2007), *Persea americana* Mill. (Lauraceae) (Förster et al. 2009, 2011; Matsubara et al. 2013), and *O. foetens* (Lauraceae) (Esteban et al. 2010a). Only one herbaceous species (nonparasite) has been investigated so far: *Cucumis sativus* L. (Esteban et al. 2009a; Matsubara et al. 2009), together with some holoparasites (Bungard et al. 1999; Matsubara et al. 2003; Snyder et al. 2005; Close et al. 2006). Curiously, differences in Lx content among taxa can be considerable: concentrations lower than 10 mmol Lx/mol Chl *a* + *b* are commonly seen among angiosperms, while concentrations exceeding the latter value are relatively rare (Esteban and García-Plazaola 2014).

The function of Lx will presumably vary depending on the plant organ (e.g., stem, leaf, flower, seed, and fruit) under consideration and is determined by the type of cycle in each of the species (truncated or completed). In chromoplasts (typically typified in petals and pericarps), it plays a role in biocommunication as a visual cue to pollinators and for seed dispersal (Mélendez-Martínez et al. 2006). Conversely, when Lx is present in photosynthetically active thylakoids, the completed Lx-L cycle provides fine-tuning for the adjustment of photoprotective energy dissipation complementary to the VAZ cycle. Equally, the truncated Lx-L cycle (most of the species) may act as a rapid switch for the photosynthetic apparatus from a photosynthetically highly efficient state (Lx accumulated under low light or darkness has been correlated with efficient light harvesting by PSII) (Matsubara et al. 2007) to a photoprotected one. Upon light, the newly formed ΔL is thought to have a role in the regulation of thermal energy dissipation (Garcia-Plazaola et al. 2003; Matsubara et al. 2008; Esteban et al. 2010a; Förster et al. 2011), reflected in the rapid development of non-photochemical quenching (NPQ) (Matsubara et al. 2008; Jia et al. 2013).

Overall, Lx-L cycle activity provides a complementary, more dynamic and precise mechanism of photoprotection than the VAZ cycle in fluctuating environmental conditions as in forests, where plants are exposed to sudden irradiance changes and light-limited environments. The Lx-L cycle enables plants to maximize carbon gain while preventing photosynthetic damage in one fell swoop.

32.2.4 Lactucaxanthin

Lac is a xanthophyll that contains two hydroxylated ε-end groups at both extremes of the molecule, whereas L has one ε-group and one β-group and Zea has two β-hydroxylated end-groups (Figure 32.1). Its presence has been reported in chloroplasts of a few species belonging to the three major angiosperm clades: the genus *Lactuca* (Siefermann-Harms et al. 1981) and *Sonchus* in the asterids, the genus *Euonymus* (Demmig-Adams and Adams 1996) in the rosids, and the genus *Saxifraga* (unpublished results) in the core eudicots. Its presence in these unrelated taxa suggests a polyphyletic origin of this carotenoid. Despite its scarce presence across the phylogenetic tree of higher plants, when this xanthophyll is present, its concentration is high, ranging from 40 to 70 mmol/mol Chl *a+b* (Adams et al. 1992; Demmig-Adams 1998).

Reconstitution experiments have shown that Lac can be bound to LHCs; however, this requires the presence of L bound to L1 to stabilize the complexes (Phillip et al. 2002). It has been shown that Lac has a high affinity for N1 and L2 binding sites. In these sites, Lac may substitute Neo and L, respectively (Phillip and Young 1995). However, the species analyzed so far do not show a reduction in Neo content, suggesting that *in vivo*, most (if not all) of the Lac pool is bound in the L2 site. In contrast, with Neo, there is a significant decrease in the L pool in plants possessing Lac (Demmig-Adams and Adams 1996; Pogson et al. 1996). In fact, this decrease is proportional to the accumulation of Lac, leading to a stable Lac+L pool in leaves (Demmig-Adams and Adams 1996; Pogson et al. 1996; Lizaraso et al. 2010; Baslam et al. 2013). Consistently, with Lac location in L2, ratios of L to Lac range between 1:1 and 4:1 (Adams et al. 1992; Phillip and Young 1995, 2006; Pogson et al. 1996; Demmig-Adams 1998). However, Lac content is never higher than that of L, except in plants treated with inhibitors of cyclases, in which it may reach a ratio of 1:4 (Phillip and Young 2006), substituting not only L molecules but also Neo at N1.

A specific role for Lac, different from that of L, has not been proposed. However, field studies (Adams et al. 1992; Demmig-Adams and Adams 1996; Demmig-Adams 1998;

Baslam et al. 2013) show that Lac content is higher in plants or leaves developed under low light. Light-induced differences in expression/activity of lycopene ε-cyclase and β-cyclase can be responsible for that trend (Caliandro et al. 2013). Nevertheless, a specific role of Lac in plant performance under low light has not been ruled out. Interestingly, despite its sparse presence across the evolutionary tree of angiosperms, when it is present, it always replaces L, suggesting that its function could be the same.

32.3 HIGH-LIGHT CAROTENOIDS

32.3.1 Xanthophyll Esters

Contrasting with functional chloroplasts, in chromoplasts, xanthophylls are frequently esterified by fatty acids, thanks to the presence of hydroxyl groups (–OH) (Biswal 1995). These XEs are frequently found in yellow- or orange-colored organs such as fruits (Hornero-Méndez and Mínguez-Mosquera 2000; Solovchenko et al. 2010), flowers (Ariizumi et al. 2014), or tubers (Fernández-Orozco et al. 2013), where they accumulate in the hydrophobic core of specialized structures called PGs (Bréhélin et al. 2007). These globular osmiophilic lipid–protein structures are present in the stroma of plastids and actively involved in several metabolic processes, including carotenoid metabolism (Bréhélin et al. 2007). XE formation is catalyzed by the action of a still-uncharacterized xanthophyll acyl transferase (Ariizumi et al. 2014).

However, the presence of XEs is not restricted to fruits, flowers, or tubers, as their presence has also been reported during autumn senescence in leaves of some deciduous trees such as *F. sylvatica*, *Cornus sanguinea* L., and *Populus tremula* L. (García-Plazaola and Becerril 2001; García-Plazaola et al. 2003). In senescing chloroplasts (also called gerontoplasts), lipids released from dismantling thylakoids contribute to the formation of PG (Biswal 1995), while breakdown of protein–pigment complexes leads to pigment release. Then, while chlorophylls and β-Car are degraded, most of the xanthophylls remain until the last stages of senescence. Once they are free from pigment–protein complexes (Biswal 1995), they can be esterified, forming XEs. When present in deciduous trees, XEs are responsible for the bright autumn leaf coloration (Figure 32.3), thereby potentially contributing to the optical masking of chlorophyll and its catabolites.

Apart from their presence in senescing leaves, XEs are not usually found in green, photosynthetically active leaves. However, in some species, particularly evergreens, they can accumulate transiently during unfavorable periods (Figure 32.3). Thus, their accumulation in PGs (Biswal 1995) has been described during periods of drought (Barry et al. 1992) or chilling stress (Hormaetxe et al. 2004, 2007). In fact, the presence and size of PGs increase reversibly during stress exposure in parallel to XEs. The proteomic and metabolic study of PGs strongly suggests that these organelles are not only a storage compartment for lipophilic components, and XEs could be involved in some of their functions (Ytterberg et al. 2006). However, the presence and functional role of XEs in functional chloroplasts rarely have been evaluated, and it

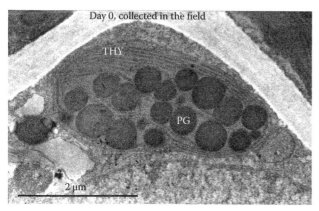
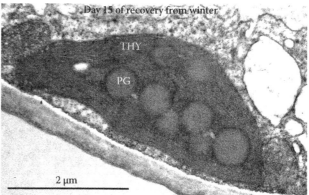

FIGURE 32.3 Morphological changes in chloroplasts of *Buxus sempervirens* during the process of stress recovery. When leaves accumulate red carotenoids, the number of plastoglobules increase, and most thylakoids disappear or lose their structure (left). After recovery, red pigments disappear in parallel with thylakoid reorganization (right). (Photographs were taken by Unai Artetxe, University of the Basque Country.)

is still uncertain whether the accumulation of these modified xanthophylls confers any advantage or they are just a waste product.

32.3.2 RED *RETRO*-CAROTENOIDS: ESCHOLTZXANTHIN AND RHODOXANTHIN

The prefix *retro-* identifies those carotenoids in which double bonds are shifted in the polyene chain by one position relative to most carotenoids. These pigments are relatively uncommon in nature, particularly in the case of photosynthetic tissues. To date, very few *retro*-carotenoids have been detected in functional leaves: escholtzxanthin, anhydroescholtzxanthin, and rhodoxanthin. Very little is known about their biosynthesis, although it has been proposed that, at least in the case of rhodoxanthin, they derive from the β-branch carotenoids: Ant, Zea, or β-Car (Han et al. 2003). All *retro*-carotenoids share in common a long-chain chromophore, which displaces the absorbance maximum to longer wavelengths, generating a characteristic red color. Apart from this color shift, the presence of long chromophores is particularly well suited to quench ROS (Hirayama et al. 1994; Bouvier et al. 1998).

Accumulation of red *retro*-carotenoids occurs in photosynthetic tissues, in parallel with the transformation of chloroplasts in PG-enriched plastids (Merzylak et al. 2005; Silva-Cancino et al. 2012). This observation, together with the fact that escholtzxanthin and rhodoxanthin fail to support *in vitro* antenna protein reassembly (Phillip et al. 2002), suggests that these pigments only accumulate in PGs, where they are presumably dissolved in the lipid matrix of these spherical bodies.

PGs usually form during stress periods, as a consequence of the hydrolysis of chloroplast membrane lipids (Davidi et al. 2014). As all the environmental stresses induce oxidative stress in plants (Fanciullino et al. 2014), and oxidative signals control *retro*-carotenoid synthesis, as occurs *in vivo* in *Capsicum annuum* L. chromoplasts (Bouvier et al. 1998), chloroplast reddening is linked to stress situations. These organelles cannot be considered true chromoplasts, since they

are not specialized for carotenoid storage as occurs in petals and flowers (Vishnevetsky et al. 1999). Besides, the process of differentiation is fully reversible once the environmental conditions return to a nonstressful situation (Silva-Cancino et al. 2012).

Leaf reddening during stress periods is a widespread phenomenon, usually related to the vacuolar accumulation of anthocyanins or betacyanins but also to the accumulation of carotenoids in plastids (Hughes et al. 2011). In the infrequent case of the accumulation of *retro*-carotenoids, reddening is mainly due to the presence of rhodoxanthin. This is the case in gymnosperms (*Cryptomeria japonica* [L.f.] D. Don, *Taxodium distichum* [L.] Rich., *Taxus cuspidata* Siebold & Zucc., *Thuja occidentalis* L., *Thuja plicata* Donn ex D.Don) (Weger et al. 1993; Ida et al. 1995; Han et al. 2003; Maslova et al. 2009) but also some angiosperms, as in the case of the drought-induced reddening in *Aloe vera* (L.) Burm.f. (Diaz et al. 1990) or *Aloe arborescens* Miller (Merzylak et al. 2005), and even cryptogams, such as the horsetails *Equisetum variegatum* Schleich. ex F.Weber & D.Mohr and *Equisetum scirpoides* Michx. (Petrov et al. 2010). As mentioned before for the case of Lac, the scattered taxonomic distribution of rhodoxanthin also suggests a polyphyletic origin. On the other hand, the shrub *Buxus sempervirens* L. and other species of the genus *Buxus* represent a unique example of accumulation of several *retro*-carotenoids: escholtzxanthin, anhydroescholtzxanthin, and monoanhydroescholtzxanthin (Hormaetxe et al. 2004) (Figure 32.4).

Two environmental factors have been identified as inducers of *retro*-carotenoid accumulation: chilling temperatures and water stress, both combined with high irradiance (Koiwa et al. 1986; Han et al. 2004; Hormaetxe et al. 2007). Specifically, chloroplasts containing red PGs occupy the upper mesophyll layer (Hormaetxe et al. 2005), suggesting that these pigments, attenuating visible light, might provide shielding to lower mesophyll layers. This light-filtering mechanism has been experimentally demonstrated for rhodoxanthin (Han et al. 2003) and escholtzxanthin (Hormaetxe et al. 2005). However, their photoprotective role as powerful antioxidants cannot be ruled out.

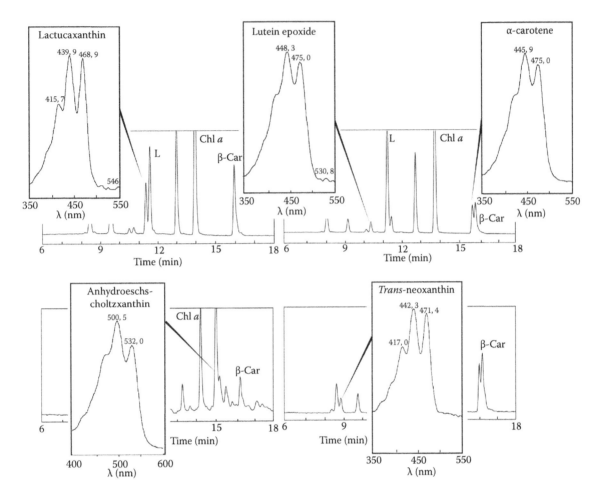

FIGURE 32.4 Representative examples of chromatograms showing the relative positions and absorption spectra of nonubiquitous carotenoids. Upper left: pigment analysis from *Saxifraga hirsuta* L. leaf, where lactucaxanthin elutes immediately before lutein. Upper right: pigment analysis from *Ocotea foetens* leaf, where lutein epoxide elutes after antheraxanthin and α-carotene before β-carotene. Bottom left: pigment analysis from *B. sempervirens* leaf showing the presence of anhydroeschscholtzxanthin and several xanthophyll esters. Bottom right: pigment analysis from *Tilia cordata* Mill. leaf where *trans*-neoxanthin elutes after *cis*-neoxanthin. For easier interpretation of chromatograms, note that the retention times and positions of lutein (L), chlorophyll *a* (Chl *a*), and β-carotene (β-Car) are indicated.

32.4 HPLC DETERMINATION OF NONUBIQUITOUS CAROTENOIDS

As the main topic of this review is nonubiquitous carotenoids, their determination, analytical identification, and quantification may cause difficulties for researchers unfamiliar with them. In this section, therefore, several tips will be provided for these cases, taking into account the absorption spectra for each of the carotenoids (Figure 32.4). Note that the correct identification and quantification requires the use of standards of known concentration. If not commercially available, standards can be purified from reference material (see details in this section). At least the correct identification of these carotenoids can be performed on the basis of the comparison of chromatographic and spectroscopic properties of samples from well-known examples of each compound (provided in the following). Description of chromatographic properties is done on the basis of a previously described HPLC protocol (García-Plazaola and Esteban 2012), but carotenoid separation is remarkably similar in other reverse-phase HPLC methods.

1. *Trans*-Neo: It is detected as Neo; however, the spectrum differs from the *cis*-Neo (the main Neo in plants), and the retention time is usually longer (Figure 32.4). Shade-tolerant trees, such as *Tilia cordata* Mill., are enriched on this isomer.

2. α-Car: It is detected as a peak that elutes immediately before β-Car (Figure 32.4). To confirm its presence, carrot can be used, but in general, the double peak formed by both carotene isomers (α and β) can be easily detected in many shade-tolerant and understory species.

3. Lx: The Lx peak is detected close to (usually before) the peak of Ant (Figure 32.4). Germination of commercially available avocado seeds produces shade leaves that can be used to confirm its presence. Disappearance or decrease of this peak after illumination confirms its presence (indicating the de-epoxidation of Lx to ΔLx).

4. Lac: The peak is detected close to (usually before) L (Figure 32.4). Lettuce leaves contain significant

amounts of this xanthophyll and can be used to confirm its presence. The absorbance spectrum is remarkably different from other xanthophylls.

5. XEs: Usually, they are detected on chromatograms by the presence of dozens of overlapping peaks. Each peak corresponds to an ester formed with a unique combination of a xanthophyll and a fatty acid. Esterification with fatty acids does not alter the visible spectrum, but it moves the retention time because of the decrease in the degree of polarity. Saponification with KOH breaks the ester, releasing xanthophylls that can be analyzed, but the only way to characterize the exact composition of XEs is coupled to a mass detector.

6. Red *retro*-carotenoids: They can be easily detected by the presence of unusual peaks in the chromatograms, which have their maxima at around 500 nm. Red fruits of yew (*Taxus baccata* L.) are a good source of rhodoxanthin.

32.5 CONCLUDING REMARKS

Nonubiquitous carotenoids, although not universal among kingdom Plantae, are not rare at all. Indeed, low-light carotenoids, such as α-Car and Lx, are more widespread than initially considered when their presence in photosynthetic tissues was discovered. Frequently associated with a specific acclimation syndrome or a taxonomic group, the physiological function of most of them (if any) remains elusive. Future research will probably lengthen the list of nonubiquitous carotenoids, opening new doors to exciting scientific findings.

ACKNOWLEDGMENTS

The authors acknowledge the support of research grant UPV/EHU-GV IT-624-13. R.E. received a JAE-Doc-2011-046 contract from the Spanish Research Council (CSIC) of the programme «Junta para la Ampliación de Estudios» co-financed by the European Social Fund. We are indebted to colleagues at University of the Basque Country (OB, UA, AH, BFM) for creating a pigment database, from which we obtained data on α-Car species for this chapter.

ABBREVIATIONS

ABA: abscisic acid
Ant: antheraxanthin
α-Car: α-carotene
β-Car: β-carotene
HPLC: high-performance liquid chromatography
ΔL: newly formed lutein from lutein epoxide
L: lutein
Lac: lactucaxanthin
LHC: light-harvesting complex
Lx: lutein epoxide
Neo: neoxanthin
PG: plastoglobules
PSII: photosystem II
ROS: reactive oxygen species
THY: thylakoids
***Trans*-Neo:** trans-neoxanthin
VAZ: violaxanthin, antheraxanthin, zeaxanthin cycle
Vio: violaxanthin
XE: xanthophyll esters
Zea: zeaxanthin
ZEP: zeaxanthin epoxidase

REFERENCES

Adams WW, Demmig-Adams B. Carotenoid composition and down regulation of photosystem II in three conifer species during the winter. *Physiol. Plant.* 1994; 92: 451–458.

Adams WW, Volk M, Hoehn A, Demmig-Adams B. Leaf orientation and the response of the xanthophyll cycle to incident light. *Oecologia* 1992; 90: 404–410.

Ariizumi T, Kishimoto S, Kakami R, Maoka T, Hirakawa H, Suzuki Y, Ozeki Y, Shirasawa K, Bernillon S, Okabe Y, Moing A, Asamizu E, Rotham C, Ohmiya A, Ezura H. Identification of the carotenoid modifying gene *PALE YELLOW PETAL 1* as an essential factor in xanthophyll esterification and yellow flower pigmentation in tomato (*Solanum lycopersicum*). *Plant J.* 2014; 79: 453–465.

Barry P, Evershed RP, Young A, Prescott MC, Britton G. Characterization of carotenoid acyl esters produced in drought-stressed barley seedling. *Phytochemistry* 1992; 31: 3163–3168.

Baslam M, Esteban R, García-Plazaola JI, Goicoechea N. Effectiveness of arbuscular mycorrhizal fungi (AFM) for inducing the accumulation of major carotenoids, chlorophylls and tocopherol in green and red leaf lettuces. *Appl. Microbiol. Biotechnol.* 2013; 97: 3119–3128.

Biswal B. Carotenoid catabolism during leaf senescence and its control by light. *J. Photochem. Photobiol.* 1995; 30: 3–13.

Bouvier F, Backhaus RA, Camara B. Induction and control of chromoplast-specific carotenoid genes by oxidative stress. *J. Biol. Chem.* 1998; 273: 30651–30659.

Bréhélin C, Kessler F, van Wijk KJ. Plastoglobules: Versatile lipoprotein particles in plastids. *Trends Plant Sci.* 2007; 12: 260–266.

Bungard RA, Ruban AV, Hibberd JM, Press MC, Horton P, Scholes JD. Unusual carotenoid composition and a new type of xanthophyll cycle in plants. *Proc. Natl. Acad. Sci. U.S.A.* 1999; 96: 1135–1139.

Caliandro R, Nagel KA, Kastenholz B, Bassi R, Li Z, Niyogi KK, Pogson BJ, Schurr U, Matsubara S. Effects of altered α- and β-branch carotenoid biosynthesis on photoprotection and whole-plant acclimation of *Arabidopsis* to photo-oxidative stress. *Plant Cell Environ.* 2013; 36: 438–453.

Cazzaniga S, Li Z, Niyogi KK, Bassi R, Dall'Osto L. The Arabidopsis szl1 mutant reveals a critical role of β-carotene in photosystem I photoprotection. *Plant Physiol.* 2012; 159: 1745–1758.

Choo KS, Nilsson J, Pedersen M, Snoeijs P. Photosynthesis, carbon uptake and antioxidant defence in two coexisting filamentous green algae under different stress conditions. *Mar. Ecol. Prog. Ser.* 2005; 292: 127–138.

Close DC, Davidson NJ, Davies NW. Seasonal fluctuations in pigment chemistry of co-occurring plant hemi-parasites of distinct form and function. *Environ. Exp. Bot.* 2006; 58: 41–46.

Cunningham FX, Pogson B, Sun Z, McDonald KA, DellaPenna D, Gantt E. Functional analysis of the beta and epsilon lycopene cyclase enzymes of *Arabidopsis* reveals a mechanism for control of cyclic carotenoid formation. *Plant Cell* 1996; 8: 1613–1626.

Dall'Osto L, Cazzaniga S, North H, Marion-Poll A, Bassi R. The *Arabidopsis* aba4-1 mutant reveals function for neoxanthin in protection against photooxidative stress. *Plant Cell* 2007; 19: 1048–1064.

Davidi L, Shimoni E, Khozin-Goldberg I, Zamir A, Pick U. Origin of β-carotene-rich plastoglobuli in *Dunaliella bardawil*. *Plant Physiol*. 2014; 164: 2139–2156.

Demmig-Adams B. Survey of thermal energy dissipation and pigment composition in sun and shade leaves. *Plant Cell Physiol*. 1998; 39: 474–482.

Demmig-Adams B, Adams WW III. Carotenoid composition in sun and shade leaves of plants with different life forms. *Plant Cell Environ*. 1992; 15: 411–419.

Demmig-Adams B, Adams WW III. Chlorophyll and carotenoid composition in leaves of *Euonymus kiautschovicus* acclimated to different degrees of light stress in the field. *Aust. J. Plant Physiol*. 1996; 23: 649–659.

Diaz M, Ball E, Lüttge U. Stress-induced accumulation of the xanthophyll rhodoxanthin in leaves of *Aloe vera*. *Plant Physiol. Biochem*. 1990; 28: 679–682.

Edelenbos MLP, Christensen LP, Grevsen K. HPLC determination of chlorophyll and carotenoid pigments in processed green pea cultivars (*Pisum sativum* L.). *J. Agric. Food Chem*. 2001; 49: 4768–4774.

Esteban R, García-Plazaola JI. Involvement of a second xanthophyll cycle in non-photochemical quenching of chlorophyll fluorescence: The lutein epoxide story. In: Demmig-Adams B, Adams WW III, Garab G, Govindjee, eds. *Non-Photochemical Fluorescence Quenching and Energy Dissipation in Plants, Algae, and Cyanobacteria. Advances in Photosynthesis and Respiration*. Dordrecht, The Netherlands, Springer, 2014: pp. 277–295.

Esteban R, Jiménez ET, Jiménez MS, Morales D, Hormaetxe K, Becerril JM, Garcia-Plazaola JI. Dynamics of violaxanthin and lutein epoxide xanthophyll cycles in Lauraceae tree species under field conditions. *Tree Physiol*. 2007; 27: 1407–1414.

Esteban R, Becerril JM, García-Plazaola JI. Lutein epoxide cycle, more than just a forest tale. *Plant Signal. Behav*. 2009a; 4: 342–344.

Esteban R, Olano JM, Castresana J, Fernandez-Marin B, Hernandez A, Becerril JM, Garcia-Plazaola JI. Distribution and evolutionary trends of photoprotective isoprenoids (xanthophylls and tocopherols) within the plant kingdom. *Physiol. Plant*. 2009b; 135: 379–389.

Esteban R, Matsubara S, Jiménez MS, Morales D, Brito P, Lorenzo R, Fernández-Marín B, Becerril JM, García-Plazaola JI. Operation and regulation of the lutein epoxide cycle in seedlings of *Ocotea foetens*. *Funct. Plant Biol*. 2010a; 37: 859–869.

Esteban R, Olascoaga B, Becerril JM, Garcia-Plazaola JI. Insights into carotenoid dynamics in non-foliar photosynthetic tissues of avocado. *Physiol. Plant*. 2010b; 40: 69–78.

Esteban R, Barrutia O, Artetxe U, Fernández-Marín B, Hernández A, García-Plazaola JI. Internal and external factors affecting photosynthetic pigment composition in plants: A meta-analytical approach. *New Phytol*. 2014; 206: 268–280.

Fanciullino AL, Bidel LPR, Urban L. Carotenoid responses to environmental stimuli: Integrating redox and carbon control into a fruit model. *Plant Cell Environ*. 2014; 37: 273–289.

Fernández-Orozco R, Gallardo-Guerrero L, Hornero-Méndez D. Carotenoid profiling in tubers of different potato (*Solanum* sp.) cultivars: Accumulation of carotenoids mediated by xanthophyll esterification. *Food Chem*. 2013; 141: 2864–2872.

Formaggio E, Cinque G, Bassi R. Functional architecture of the major light-harvesting complex from higher plants. *J. Mol. Biol*. 2001; 314: 1157–1166.

Förster B, Osmond CB, Pogson BJ. De novo synthesis and degradation of Lx and V cycle pigments during shade and sun acclimation in avocado leaves. *Plant Physiol*. 2009; 149: 1179–1195.

Förster B, Osmond CB, Pogson BJ. Lutein from deepoxidation of lutein epoxide replaces zeaxanthin to sustain an enhanced capacity for nonphotochemical chlorophyll fluorescence quenching in avocado shade leaves in the dark. *Plant Physiol*. 2011; 156: 393–403.

Gandul-Rojas B, Cepero MRL, Mínguez-Mosquera MI. Chlorophyll and carotenoid patterns in olive fruits, *Olea europaea* cv. Arbequina. *J. Agric. Food Chem*. 1999; 47: 2207–2212.

García-Plazaola JI, Becerril JM. Seasonal changes in photosynthetic pigments and antioxidants in beech (*Fagus sylvatica*) in a Mediterranean climate: Implications for tree decline diagnosis. *Aust. J. Plant Physiol*. 2001; 28: 225–232.

García-Plazaola JI, Esteban R, Prometheus wiki contributors. Determination of chlorophylls and carotenoids by HPLC, Version 10, 2012. Available at http://prometheuswiki.publish.csiro.au/tikiindex.php?page=Determination+of+chlorophylls+and+carotenoids+by+HPLC.

García-Plazaola JI, Hernández A, Becerril JM. Antioxidant and pigment composition during autumnal leaf senescence in woody deciduous species differing in their ecological traits. *Plant Biol*. 2003; 5: 557–566.

García-Plazaola JI, Hormaetxe K, Hernández A, Olano JM, Becerril JM. The lutein epoxide cycle in vegetative buds of woody plants. *Funct. Plant Biol*. 2004; 31: 815–823.

García-Plazaola JI, Matsubara S, Osmond CB. The lutein epoxide cycle in higher plants: Its relationships to other xanthophyll cycles and possible functions. *Funct. Plant Biol*. 2007; 34: 759–773.

Garrido JL, Rodriguez F, Zapata M. Occurrence of loroxanthin, loroxanthin decenoate, and loroxanthin dodecenoate in *Tetraselmis* species (Prasinophyceae, Chlorophyta). *J. Phycol*. 2009; 45: 366–374.

Han Q, Shinohara K, Kakubari Y, Mukai Y. Photoprotective role of rhodoxanthin during cold acclimation in *Cryptomeria japonica*. *Plant Cell Environ*. 2003; 26: 715–723.

Han Q, Katahata S, Kakubari Y, Mukai Y. Seasonal changes in the xanthophyll cycle and antioxidants in sun-exposed and shaded parts of crown of *Cryptomeria japonica* in relation to rhodoxanthin accumulation during cold acclimatization. *Tree Physiol*. 2004; 24: 609–616.

Havaux M. Carotenoid oxidation products as stress signals in plants. *Plant J*. 2013; 79: 597–606.

Havaux M., Niyogi K. The violaxanthin cycle protects plants from photo-oxidative damage by more than one mechanism. *Proc. Natl. Acad. Sci. U.S.A*. 1999; 96: 8762–8767.

Havaux M., Dall'Osto L, Bassi R. Zeaxanthin has enhanced antioxidant capacity with respect to all other xanthophylls in *Arabidopsis* leaves and functions independent of binding to PSII antennae. *Plant Physiol*. 2007; 145: 1506–1520.

Hirayama O, Nakamura K, Hamada S, Kobayasi Y. Singlet oxygen quenching ability of naturally occurring carotenoids. *Lipids* 1994; 29: 149–150.

Hirschberg J. Carotenoid biosynthesis in flowering plants. *Curr. Opin. Plant Biol*. 2001; 4: 210–218.

Hormaetxe K, Hernández A, Becerril JM, García-Plazaola JI. Role of red carotenoids in photoprotection during winter acclimatization in *Buxus sempervirens* leaves. *Plant Biol*. 2004; 6: 325–332.

Hormaetxe K, Becerril JM, Fleck I, Pinto M, García-Plazaola JI. Functional role of (retro)-carotenoids as passive light filters in the leaves of *Buxus sempervirens* L.: Increased protection of photosynthetic tissues? *J. Exp. Bot*. 2005; 56: 2629–2636.

Hormaetxe K, Becerril A, Hernández A, Esteban R, García-Plazaola JI. Plasticity of photoprotective mechanisms of *Buxus sempervirens* L. leaves in response to extreme temperatures. *Plant Biol.* 2007; 9: 59–68.

Hornero-Méndez D, Mínguez-Mosquera MI. Xanthophyll esterification accompanying carotenoid overaccumulation in chromoplast of *Capsicum annum* ripening fruits is a constitutive process and useful for ripeness index. *J. Agric. Food Chem.* 2000; 48: 1617–1622.

Hughes NM. Winter leaf reddening in "evergreen" species. *New Phytol.* 2011; 190: 573–581.

Ida K, Masamoto K, Maoka T, Fujiwara Y, Takeda S, Hasegawa E. The leaves of the common box, *Buxus sempervirens* (Buxaceae), become red as the level of a red carotenoid, anhydroeschscholtzxanthin, increases. *J. Plant Res.* 1995; 108: 369–376.

Jia H, Förster B, Chow WS, Pogson BJ, Osmond CB. Decreased photochemical efficiency of photosystem II following sunlight exposure of shade-grown leaves of avocado: Because of, or in spite of, two kinetically distinct xanthophyll cycles? *Plant Physiol.* 2013; 161: 836–852.

Johnson MP, Havaux M, Triantaphylidès C, Ksas B, Pascal AA, Robert B, Davison PA, Ruban AV, Horton P. Elevated zeaxanthin bound to oligomeric LHCII enhances the resistance of Arabidopsis to photooxidative stress by a lipid-protective, antioxidant mechanism. *J Biol. Chem.* 2007; 282: 22605–22618.

Khoo HE, Prasad KN, Kong KW, Jiang Y, Ismail A. Carotenoids and their isomers: Color pigments in fruits and vegetables. *Molecules* 2011; 16: 1710–1738.

Koiwa H, Ikeda T, Yoshida Y. Reversal of chromoplasts to chloroplasts in *Buxus* leaves. *Bot. Mag. (Tokyo)* 1986; 99: 233–240.

Koninger M, Harris GC, Virgo A, Winter K. Xanthophyll-cycle pigments and photosynthetic capacity in tropical forest species: A comparative field study on canopy, gap and understory plants. *Oecologia* 1995; 104: 280–290.

Krause GH, Koroleva OY, Dalling JW, Winter K. Acclimation of tropical tree seedlings to excessive light in simulated tree-fall gaps. *Plant Cell Environ.* 2001; 24: 1345–1352.

Krause GH, Grube E, Koroleva OY, Barth C, Winter K. Do mature shade leaves of tropical tree seedlings acclimate to high sunlight and UV radiation? *Funct. Plant Biol.* 2004; 31: 743–756.

Kronfuss G, Polle A, Tausz M, Havranek WM, Wieser G. Effects of ozone and mild drought stress on gas exchange, antioxidants and chloroplast pigments in current-year needles of young Norway spruce *Picea abies* (L.) Karst. *Trees-Struct. Funct.* 1998; 12: 482–489.

Kruk J. Occurrence of chlorophyll precursors in leaves of cabbage heads—The case of natural etiolation. *J. Photochem. Photobiol.* 2005; 80: 187–194.

La Rocca N, Sciuto K, Meneghesso A, Moro I, Rascio N, Morosinotto T. Photosynthesis in extreme environments: Responses to different light regimes in the Antarctic alga *Koliella Antarctica*. *Physiol. Plant.* 2014; 153: 654–667.

Li Y, Walton C. Violaxanthin is an abscisic acid precursor in water-stressed dark-grown bean leaves. *Plant Physiol.* 1990; 92: 551–559.

Lizaraso K, Fernández-Marín B, Becerril JM, García-Plazaola JI. Ageing and irradiance enhance Vitamin E content in green edible tissues from crop plants. *J. Sci. Food Agric.* 2010; 90: 1994–1999.

Maslova TG, Mamushina NS, Sherstneva OA, Bubolo LS, Zubkova EK. Seasonal structural and functional changes in the photosynthetic apparatus of evergreen conifers. *Russ. J. Plant Physiol.* 2009; 56: 607–615.

Matsubara S, Morosinotto T, Bassi R, Christian AL, Fischer-Schliebs E, Luttge U, Orthen B, Franco AC, Scarano FR, Forster B, Pogson BJ, Osmond CB. Occurrence of the lutein-epoxide cycle in mistletoes of the Loranthaceae and Viscaceae. *Planta* 2003; 217: 868–879.

Matsubara S, Naumann M, Martin R, Nichol C, Rascher U, Morosinotto T, Bassi R, Osmond B. Slowly reversible de-epoxidation of lutein-epoxide in deep shade leaves of a tropical tree legume may "lock in" lutein-based photoprotection during acclimation to strong light. *J. Exp. Bot.* 2005; 56: 461–468.

Matsubara S, Morosinotto T, Krause H, Seltmann M, Winter K, Osmond B, Jahns P, Bassi R. Achieving better light harvesting in the shade: Accumulation of lutein epoxide increases light-harvesting efficiency in shade leaves of *Inga* species. *Plant Phys.* 2007; 144: 926–941.

Matsubara S, Krause GH, Seltmann M, Virgo A, Kursar TA, Jahns P, Winter K. Lutein epoxide cycle, light harvesting and photoprotection in species of the tropical tree genus Inga. *Plant Cell Environ.* 2008; 31: 548–555.

Matsubara S, Krause GH, Aranda J, Virgo A, Beisel KG, Jahns P, Winter K. Sun–shade patterns of leaf carotenoid composition in 86 species of neotropical forest plants. *Funct. Plant Biol.* 2009; 36: 20–36.

Matsubara S, Chen YC, Caliandro R, Govindjee, Clegg RM. Photosystem II fluorescence lifetime imaging in avocado leaves: Contributions of the lutein-epoxide and violaxanthin cycles to fluorescence quenching. *J. Photochem. Photobiol. B* 2011; 104: 271–284.

Matsubara S, Förster B, Waterman M, Robinson SA, Pogson BJ, Gunning B, Osmond B. From ecophysiology to phenomics: Some implications of photoprotection and shade–sun acclimation *in situ* for dynamics of thylakoids *in vitro*. *Phil. Trans. R. Soc. Lond. B* 2013; 367: 3503–3514.

Mélendez-Martínez AJ, Britton G, Vicario IM, Heredia FJ. HPLC analysis of geometrical isomers of lutein epoxide isolated from dandelion (*Taraxacum officinale* F. Weber ex Wiggers). *Phytochemistry* 2006; 67: 771–777.

Merzylak MN, Solovchenko A, Pogosyan S. Optical properties of rhodoxanthin accumulated in *Aloe arborescens* Mill. Leaves under high-light stress with special reference to its photoprotective function. *Photochem. Photobiol. Sci.* 2005; 4: 333–340.

Miller R, Watling JR, Robinson SA. Functional transition in the floral receptacle of the sacred lotus (*Nelumbo nucifera*): From thermogenesis to photosynthesis. *Funct. Plant Biol.* 2009; 36: 471–480.

Nichol CJ, Pieruschka R, Takayama K, Förster B, Kolber Z, Rasher U, Grace J, Robinson SA, Pogson B, Osmond B. Canopy conundrums: Building on the biosphere 2 experience to scale measurements of inner and outer canopy photoprotection from the leaf to the landscape. *Funct. Plant Biol.* 2012; 39: 1–24.

North HM, De Almeida A, Boutin JP, Frey A, To A, Botran L, Sotta B, Marion-Poll A. The Arabidopsis ABA-deficient mutant *aba4* demonstrates that the major route for stress-induced ABA accumulation is via neoxanthin isomers. *Plant J.* 2007; 50: 810–824.

Pennington JAT, Fisher RA. Food component profiles for fruit and vegetable subgroups. *J Food Compos. Anal.* 2010; 23: 411–418.

Petrov KA, Sofronova VE, Chepalov VA, Perk AA, Maksimov KH. Seasonal changes in the content of photosynthetic pigments in perennial grasses of cryolithic zone. *Russ. J. Plant Physiol.* 2010; 57: 181–188.

Phillip D, Young AJ. Occurrence of the carotenoid lactucaxanthin in higher plant LHCII. *Photosynth. Res.* 1995; 43: 273–282.

Phillip D, Young AJ. Preferential inhibition of the lycopene ε-cyclase by the substituted triethylamine compound MPTA in higher plants. *J. Plant Physiol.* 2006; 163: 383–391.

Phillip D, Hobe S, Paulsen H, Molnar P, Hashimoto H, Young AJ. The binding of xanthophylls to the bulk light-harvesting complex of photosystem II of higher plants. *J. Biol. Chem.* 2002; 277: 25160–25169.

Pogson B, McDonald KA, Truong M, Britton G, DellaPenna D. *Arabidopsis* carotenoid mutants demonstrate that lutein is not essential for photosynthesis in higher plants. *Plant Cell* 1996; 8: 1627–1639.

Rabinowitch HDP, Budowski P, Kedar N. Carotenoids and epoxide cycles in mature-green tomatoes. *Planta* 1975; 122: 91–97.

Razungles AJ, Babic I, Sapis JC, Bayonove CL. Particular behaviour of epoxy xanthophylls during veraison and maturation of grape. *J. Agric. Food Chem.* 1996; 44: 3821–3825.

Savitch LV, Leonardos ED, Krol M, Jansson S, Grodzinski B, Huner NPA, Oquist G. Two different strategies for light utilization in photosynthesis in relation to growth and cold acclimation. *Plant Cell Environ.* 2002; 25: 761–771.

Siefermann-Harms D, Hertzberg S, Borch G, Liaaen-Jensen S. Lactucaxanthin, an ε,ε-carotene-3,3′-diol from *Lactuca sativa*. *Phytochemistry* 1981; 20: 85–88.

Silva-Cancino MC, Esteban R, Artetxe U, García-Plazaola JI. Patterns of spatio-temporal distribution of winter chronic photoinhibition in leaves of three evergreen Mediterranean species with contrasting acclimation responses. *Physiol. Plant.* 2012; 144: 289–301.

Snyder AM, Clark BM, Robert B, Ruban AV, Bungard RA. Carotenoid specificity of light-harvesting complex II binding sites: Occurrence of 9-cis-violaxanthin in the neoxanthin-binding site in the parasitic angiosperm *Cuscuta reflexa*. *J. Biol. Chem.* 2004; 279: 5162–5168.

Snyder AM, Clark BM, Bungard RA. Light-dependent conversion of carotenoids in the parasitic angiosperm *Cuscuta reflexa* L. *Plant Cell Environ.* 2005; 28: 1326–1333.

Solovchenko AE, Merzlyak MN, Pogosyan SI. Light-induced decrease of reflectance provides an insight in the photoprotective mechanisms of ripening apple fruit. *Plant Sci.* 2010; 178: 281–288.

Strand A, Kvernberg K, Karlse AM, Liaaen-Jenses S. Geometrical *E/Z* isomers of (*6R*)- and (*6S*)-neoxanthin and biological implications. *Biochem. Syst. Ecol.* 2000; 28: 443–455.

Tai CY, Chen BH. Analysis and stability of carotenoids in the daylily (*Hemerocallis disticha*) as affected by various treatments. *J. Agric. Food Chem.* 2000; 58: 5962–5964.

Takaichi S, Mimuro M. Distribution and geometric isomerism of neoxanthin in oxygenic phototrophs: 9′-*Cis*, a sole molecular form. *Plant Cell Physiol.* 1998; 39: 968–977.

Tausz M, Loffler S, Posch S, Monschein S, Grill D, Katzel R. Do photoprotective pigments and antioxidants in needles of *Pinus sylvestris* relate to high N or water availability at field plots in a dry year? *Phyton-Ann. Rei Bot. A* 2005; 45: 107–116.

Tegischer K, Tausz M, Wieser G, Grill D. Tree- and needle-age-dependent variations in antioxidants and photoprotective pigments in Norway spruce needles at the alpine timberline. *Tree Physiol.* 2002; 22: 591–596.

Thayer SS, Björkman O. Leaf xanthophyll content and composition in sun and shade determined by HPLC. *Photosynth. Res.* 1990; 23: 331–343.

Vishnevetsky M, Ovadis M, Vainstein A. Carotenoid sequestration in plants: The role of carotenoid-associated proteins. *Trends Plant Sci.* 1999; 4: 232–235.

Weger HG, Silim SN, Guy RD. Photosynthetic acclimation to low temperature by western cedar seedlings. *Plant Cell Environ.* 1993; 16: 711–718.

Yamamoto HY, Bassi R. Carotenoids: Localization and function. In: Ort DR, Yocum CF, eds. *Oxygpp. enic Photosynthesis: The Light Reactions.* Dordrecht, Kluwer, 1996: pp. 539–563.

Young AJ, Phillip D, Savill J. Carotenoids in higher plant photosynthesis. In: Pessaraki M, ed. *Handbook of Photosynthesis.* New York, Taylor and Francis, 1997: pp. 575–596.

Ytterberg AJ, Peltier JB, van Wijk KJ. Protein profiling of plastoglobules in chloroplasts and chromoplasts. A surprising site for differential accumulation of metabolic enzymes. *Plant Physiol.* 2006; 140: 984–997.

Section XI

Photosynthesis and Its Relationship with Plant Nutrient Elements

33 Role of Phosphorus in Photosynthetic Carbon Assimilation and Partitioning

Anna M. Rychter, Idupulapati M. Rao, and Juan Andrés Cardoso

CONTENTS

33.1 Introduction 603
33.2 Short-Term *In Vitro* Effects of Pi Deprivation...605
    33.2.1 Phosphate Translocators 605
    33.2.2 Regulation of Photosynthesis..606
    33.2.3 Starch Biosynthesis...607
    33.2.4 Sucrose Biosynthesis 607
33.3 Long-Term *In Vivo* Effects of Pi Deprivation..608
    33.3.1 Plant Growth Response and Phosphate Concentration...609
    33.3.2 Photosynthetic Machinery 610
    33.3.3 Carbon Metabolism 611
    33.3.4 Intracellular Pi Compartmentation...613
    33.3.5 Carbon Partitioning and Export 614
33.4 Recovery of Plants from Phosphate Deficiency..615
33.5 Interaction of Phosphate Deficiency with Environmental Factors Affecting Photosynthesis........616
33.6 Acclimation and Adaptation of Plants to Phosphate Deficiency 617
33.7 Conclusions and the Way Forward 617
Acknowledgments..618
References...618

33.1 INTRODUCTION

Phosphorus (P) is one of the essential mineral nutrients for plants and is required in many compounds in cells and organelles [1]. These compounds are associated with numerous components of metabolism (sugar phosphates, nucleic acids, nucleotides, coenzymes, phospholipids) and are closely associated with energy transfer (triphosphonucleotides) and genetic material (nucleic acids). The covalent ester bond between two P atoms is at a higher *energy level* than the covalent bonds between many other kinds of atoms. That is, it takes more energy for these compounds to be synthesized, and conversely, they release more energy when either they are hydrolyzed or they participate in alternative reactions such as P addition to other molecules. Plants must have P for plant growth and development. Limited inorganic phosphate (Pi) supply results in numerous perturbations in plant growth and development and strongly affects plant yields [2].

Photosynthesis is the primary physiological process whereby CO_2 diffuses down a concentration gradient from the atmosphere, through the epidermis and into the chloroplasts, where energy derived photochemically is used to assimilate CO_2 in the formation of organic compounds (Figure 33.1). In algae and higher plants, there is only one primary carboxylating mechanism, which results in the net synthesis of carbon compounds. The photosynthetic carbon reduction (PCR) cycle is common to all plants (C_3, C_4, and crassulacean acid metabolism [CAM] plants). Nonetheless, C_4 and CAM plants have auxiliary mechanisms of carbon fixation [1].

During photosynthesis, carbon is fixed through the PCR cycle in the chloroplast and is then exported to the cytosol as triose phosphate (triose-P). The triose-P is then converted to sucrose in the cytosol, releasing Pi, which is then available to further allow export of triose-P from the chloroplast. If there is any restriction of sucrose synthesis in the cytosol, it will lead to a decreased export of triose-P from the chloroplast, so more photosynthate is retained in the stroma for conversion to starch (Figure 33.1). Chloroplastic starch degradation may be closely related to internal factors in the cell such as the supply and demand of carbon substrates. Orthophosphate (Pi), together with CO_2 and H_2O, is a primary substrate of photosynthesis [3] according to the following overall equation:

$$3CO_2 + 6H_2O + Pi \xrightarrow{hv} triose-P + H_2O + 3O_2.$$

thin the chloroplast, Pi is involved in photophosphorylation, as a proton gradient is discharged through an ATPase into the chloroplast stroma. In the stroma, ATP is consumed by the PCR cycle. Nine molecules of Pi are consumed for every three molecules of CO_2 fixed and three molecules of O_2 evolved. Eight molecules of Pi are released in the PCR

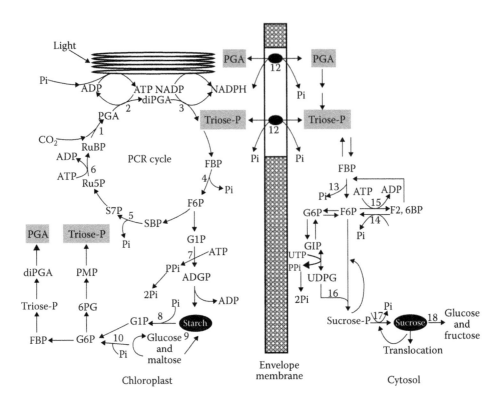

FIGURE 33.1 Simplified model depicting the reactions of photosynthetic carbon metabolism in which Pi has a regulating function or in which energy-rich phosphates and the corresponding phosphate esters are involved. Because of these functions, strict compartmentation and regulation of the Pi level in the metabolic pool are essential for photosynthesis in leaf cells. Fixed carbon inputs and reducing equivalents converge in the PCR cycle. Two major branch points of the PCR cycle lead to the production of starch in the chloroplast and the export of triose-P to the cytosol through the Pi translocator, located on the inner envelope of the chloroplast membrane. Synthesis of sucrose in the cytosol is linked to the release of Pi that is returned to the stroma through the Pi translocator in exchange for triose-P. The dashed arrows indicate possible feedback mechanisms. The reactions are catalyzed by enzymes numbered as follows: 1, Rubisco; 2, PGA kinase; 3, NADP-G3P dehydrogenase; 4, FBPase; 5, SBPase; 6, Ru5P kinase; 7, ADPG PPase; 8, phosphorylase; 9, b-amylase; 10, hexokinase; 11, NADP-G6P dehydrogenase; 12, Pi translocator; 13, FBPase; 14, F2,6BPase; 15, F6P-2-kinase; 16, SPS; 17, SPPase; and 18, invertase.

cycle, and the remaining molecule of Pi is incorporated into triose-P, which is transported to the cytosol in exchange for imported Pi. Sucrose synthesis in the cytosol releases Pi and thereby recycles Pi. Four molecules of Pi must enter the chloroplast for every molecule of sucrose synthesized in the cytosol. Adequate supply of Pi is essential for the assimilation of photosynthetic carbon in plants [3], and there has been a great deal of interest for the past three decades related to the idea that the level of Pi in plant tissues may regulate various aspects of photosynthesis and the flow of carbon between starch and sucrose biosynthesis [4–16]. In addition, it was proposed that Pi may be involved in the partitioning of photosynthates between plant parts [17–21].

The rate of photosynthesis is dependent on the ATP/reductant (NADPH, NADH, and ferredoxin) balance, which can be stabilized by extrachloroplastic compartments such as mitochondria [22]. At the whole plant level, photosynthesis is regulated by sink demand [23]. In P-deficient plants, low sink strength imposes the primary limitation on photosynthesis [15]. Therefore, the response of photosynthesis to phosphate limitation is a *whole plant* one and depends on the dynamic interactions between sink and source tissues [15,23]. The decrease in phosphate concentration due to limited Pi supply from the growth medium involves several

changes not only in the photosynthetic process but also in glycolysis, respiration, and nitrogen metabolism, which affect the rate of net photosynthesis. Metabolic aspects of the phosphate-starvation response were reviewed by Plaxton and Carswell [13] and more recently the molecular ones and networks with regard to Pi sensing and signaling in plants [24,25].

Plant productivity depends on photosynthesis. Inadequate supply of Pi limits photosynthesis because of its large demand for adenylate energy and the role of phosphorylated intermediates in the PCR cycle [14]. The inhibition of photosynthesis due to Pi deprivation results from both short- and long-term effects of Pi on photosynthetic carbon assimilation and carbon partitioning processes [12]. In this chapter, we review the research progress that contributed to our present understanding of the role of Pi in photosynthetic carbon assimilation and partitioning. To illustrate the effects of Pi deprivation on photosynthesis and partitioning of photosynthates, a simplified outline is presented of the short-term *in vitro* effects of Pi deprivation, followed by long-term *in vivo* effects of Pi deprivation, the recovery of plants from P deficiency, interaction of phosphate deficiency with environmental factors affecting photosynthesis, and the acclimation and adaptive responses of plants to P deficiency.

33.2 SHORT-TERM *IN VITRO* EFFECTS OF Pi DEPRIVATION

The evidence for a crucial role of Pi in the regulation of photosynthesis arose from the studies of photosynthetic induction. It was demonstrated that in isolated chloroplasts, the induction period is due to a need to build up the pool sizes of the intermediates of the PCR cycle [26,27]. The interrelationships between Pi and induction, together with the demonstration that isolated chloroplasts require Pi for the continuation of photosynthesis, led to the concept that C_3 chloroplast is not a fully self-sufficient photosynthetic organelle [28].

Experimental observations on the photosynthetic induction period have led to the view that the chloroplast produces triose-P (glyceraldehyde-3-phosphate [G-3-P] and dihydroxyacetone phosphate [DHAP]), which it exchanges for Pi from the cytoplasm of the cell [29,30]. Subsequent research work indicated that light activation of key enzymes [31–33] may be involved along with the autocatalytic build-up of metabolites [34] to overcome the lag period in photosynthetic CO_2 fixation [35]. However, experimental verification of these hypotheses with intact wheat leaves suggested that light activation of enzymes might not be a limiting factor during photosynthetic induction [36].

Studies of the short-term effects of Pi on photosynthesis, based on *in vitro* experiments, have shown the inhibition of triose-P export from the chloroplast to the cytosol through the Pi translocator leading to the build-up of starch and a decrease in the rate of photosynthesis [35,37–39]. It was demonstrated that in isolated chloroplasts, the increase in Pi concentration in incubation medium up to 1 mM stimulated net photosynthesis and lowered starch production, whereas low Pi concentration in external medium increased starch synthesis despite a low photosynthetic rate [40–42]. Low supply of Pi might restrict photophosphorylation, which should lead to increased energization of the thylakoid membrane, decreased electron flow, and associated inhibition of photosynthesis. At high Pi supply, triose-P export competes with ribulose 1,5-bisphosphate (RuBP) regeneration, and the rate of photosynthesis can be diminished.

Optimal photosynthesis of isolated chloroplasts requires a finely balanced concentration of Pi in the cytosol [43]. This optimal concentration may be maintained by transport to and from the vacuole and by metabolic processes causing changes in the rate of sucrose synthesis [19,43]. Over the short term, low Pi in the cytosol decreases the export of triose-P from the chloroplast, which leads to the inhibition of sucrose synthesis in the cytosol [8,35,44,45].

33.2.1 PHOSPHATE TRANSLOCATORS

In higher plants, photosynthesis is compartmentalized in the chloroplast, which is bounded by the envelope membranes that serve both as a barrier separating the chloroplast stroma from the cytoplasm and a bridge enabling rapid exchange of specific metabolites between the two (Figure 33.1) [46–48]. The outer envelope membrane is nonspecifically permeable to all molecules, both charged and uncharged. The impermeability of the inner envelope membrane to hydrophilic solutes such as Pi, phosphate esters, dicarboxylates, and glucose is overcome by translocators that catalyze specific transfer of metabolites across the envelope [47,48]. The energy-transducing thylakoid membranes, located within the chloroplasts, are distinct from the envelope membranes.

The mechanism by which external Pi influences photosynthesis has been attributed to the operation of the Pi translocator, an antiport located in the inner membrane of the chloroplast envelope that facilitates a rapid counter-exchange of Pi, triose-P, and 3-phosphoglyceric acid (PGA) [40,47,48]. The major flow of metabolites across the chloroplast envelope is mediated by the Pi translocator, which enables the specific transport of Pi and phosphorylated compounds such that photosynthetically fixed carbon in the form of triose-P can be exported from the stroma to the cytosol in a one-to-one stoichiometric and obligatory exchange for Pi [49]. The Pi released during biosynthetic processes is shuttled back through the Pi translocator into the chloroplasts for the formation of ATP catalyzed by the thylakoid ATPase [50].

If triose-P is regarded as the end product of the PCR cycle (Figure 33.1), then one molecule of Pi must be made available for incorporation into triose-P for every three molecules of CO_2 fixed. Some Pi will be released within the stroma as triose-P is utilized for starch synthesis, but starch synthesis is usually slower (by a factor of 3–4) than maximal CO_2 fixation. Virtually all the remaining Pi must enter the chloroplast in exchange for exported triose-P [47–49]. In the short term, a sudden decrease in the Pi concentration in the cytosol of photosynthetic mesophyll cells will have a direct effect on the triose-P and Pi exchange between the chloroplast and the cytosol, decreasing the availability of Pi in the chloroplast and thus decreasing the production of ATP needed in the turnover of the PCR cycle.

Triose phosphate/phosphate translocator (TPT) was the first phosphate transporter to be cloned from plants [51]. The activity of TPT is closely associated with photosynthetic carbon metabolism, and the expression of the TPT gene is observed only in photosynthetic tissues [42]. Its importance for *in vivo* communication between chloroplast and cytosol was demonstrated in transgenic potato plants with reduced expression of the TPT at both RNA and protein levels due to antisense inhibition [52]. Four different groups of Pi transporters have been described so far in plastids, and one among them is phosphoenolpyruvate (PEP)/phosphate transporter, which transports Pi out of the chloroplast into cytosol under most physiological conditions [53].

Versaw and Harrison [54] described a low-affinity Pi transporter PHT2;1, H^+/Pi symporter, located in the inner envelope of the chloroplast. The identification of the null mutant of *Arabidopsis thaliana*, pht2;1-1, revealed that the PHT2;1 transporter affects Pi allocation and modulates Pi-starvation responses including the expression of genes and the translocation of Pi within leaves [54]. The presence of several transporters indicates highly controlled transport of phosphate into and out of the chloroplast.

The synthesis of sucrose from triose-P is believed to make the major contribution to the recycling of Pi (Figure 33.1). Sucrose synthesis releases Pi due to the action of a phosphatase, and rapid export of sucrose from the cytoplasm will make Pi available as fast as the plant can synthesize triose-P; little or none will be available for storage within the stroma as starch. If the demand for sucrose by growing sinks is less, however, excess triose-P would be stored as starch and the rate of photosynthesis possibly diminished.

Another important function of the Pi translocator is to link intra- and extra-chloroplast pyridine nucleotide and adenylate systems through shuttles involving the exchange of DHAP and PGA. Photosynthetically produced ATP and NADPH are not directly available to the extra-chloroplastic compartments due to the low permeability of the inner envelope membrane to these compounds in mature tissue. The Pi translocator provides an indirect shuttle system for transferring ATP and NADPH to the cytoplasm involving exchange of triose-P and PGA. This shuttle can operate in either direction depending on the redox potential of the pyridine nucleotides in the cytoplasm and stroma [47].

Gerhardt et al. [55] observed asymmetric distribution of DHAP and 3-PGA across the chloroplast envelope in spinach leaves and suggested that the Pi translocator may be kinetically limiting *in vivo*. The reduction of TPT activity *in vivo* by antisense repression of chloroplast TPT resembles the situation of chloroplasts performing photosynthesis under Pi limitation [40]. To examine more specifically the role of the Pi translocator in assimilate partitioning in photosynthetic tissues, Barnes et al. [56] transformed tobacco plants with sense and antisense constructs of a cDNA encoding the tobacco Pi translocator. Although the transformed plants showed a 15-fold variation in Pi translocator activity, the growth and development and the rate of photosynthesis showed no consistent differences between antisense and sense transformants. In contrast, the distribution of assimilate between starch and sugar had been altered with no change in the amount of sucrose in leaves, suggesting a homeostatic mechanism for maintaining sucrose concentrations in the leaves at the expense of glucose and fructose. However, in potato plants, antisense repression of the triose-P translocator affected carbon partitioning as chloroplasts isolated from such plants showed reduced import of Pi, reduced rate of photosynthesis, and change in carbon partitioning into starch at the expense of sucrose and amino acids [52]. Published evidence indicates that TPT exerts a considerable control on the rate of both CO_2 assimilation and sucrose biosynthesis [42].

33.2.2 REGULATION OF PHOTOSYNTHESIS

Since Pi, triose-P, and PGA are exchanged through the Pi translocator, changes in the Pi concentration outside the chloroplast could affect the PCR cycle indirectly by altering the amount of intermediates within the chloroplast. Pi might also have direct effects on PCR cycle enzymes through the level of activation. Heldt et al. [57] indicated that Pi is required for light activation of ribulose 1,5-bisphosphate carboxylase/

oxygenase (Rubisco). Later, Bhagwat [58] showed that Pi is an activator of Rubisco. However, Machler and Nösberger [59] showed that although the activity of Rubisco decreased with decreased stromal Pi concentration, they believed this to be an indirect effect mediated through the changes in stromal pH.

The activation of fructose-1,6-bisphosphatase (FBPase) [60] and of sedoheptulose 1,7-bisphosphatase (SBPase) [59] is strongly inhibited by Pi concentrations in the range of 5–10 mM. Pi inhibited the PCR cycle turnover in thiol-activated stromal extracts. This inhibition was due primarily to effects on the SBPase [61]. Another PCR cycle enzyme, the light-activated form of ribulose-5-phosphate kinase (Ru5Pkinase), is inhibited by the monovalent ionic species of Pi [62]. The decrease in the concentration of stromal Pi, which occurs upon illumination, is therefore likely to enhance the activity of the PCR cycle.

The reduction in photosynthetic rate occurs when cytoplasmic Pi is decreased. For example, when Pi is sequestered in the cytoplasm by mannose [63] or glycerol [64], reduction of photosynthetic rate might be explained in terms of an endproduct inhibition [65]. This end-product inhibition could be due to high concentrations of triose-P. Because the properties of the Pi translocator dictate that the total Pi (inorganic plus organic) within the chloroplast is relatively constant [49], high triose-P is automatically coupled with low Pi, which in turn could limit photosynthesis [5,9,44].

The consumption of Pi as a substrate of photosynthesis [28] could decrease photosynthesis by a direct effect of low stromal Pi concentration on Rubisco [57]. Low stromal Pi concentration, together with the accumulation of triose-P, might influence the activation state of Rubisco by various mechanisms [5]. Rubisco could be inactivated by the buildup of various intermediates, for example, ribose-5-phosphate [66,67] and other chloroplast metabolites [68]; or it may be inactivated by the build-up of PGA [66]. Another possibility is that the pH of the stroma could be changed [69,70].

Alternatively, inhibition of photosynthesis might occur due to a drop in the ATP/ADP ratio [71]. A decrease in stromal Pi concentration could diminish the rate of photophosphorylation and thereby reduce the rate of carbon fixation because of the sensitivity of the PCR cycle to the ATP/ADP quotient. Such a reduction is readily demonstrated with isolated chloroplasts photosynthesizing in a medium containing suboptimal Pi concentrations. The reduced concentration of Pi leads to a reduction in ATP/ADP, which could restrict the activity of Rubisco activase and therefore Rubisco carbamylation [72].

Robinson and Giersch [73] determined the concentration of Pi in the stroma of isolated chloroplasts during photosynthesis under Pi-limited and Pi-saturated conditions. These authors used colorimetric and ^{32}P labeling techniques and found that when chloroplasts are illuminated in the absence of added Pi, photosynthesis declines rapidly due to Pi depletion in the stroma, which was estimated to be 1.4 mM by the colorimetric method and 0.2 mM by ^{32}P high-performance liquid chromatography. With optimal concentrations of Pi added to the medium, the stromal Pi concentration was estimated to be 2.6 and 1.6 mM with the colorometric and ^{32}P methods, respectively. This study demonstrated that any decrease in the supply of Pi from the medium leads to a rapid decrease in stromal Pi to the point where photophosphorylation may become Pi-limited, decreasing the rate of photosynthesis.

33.2.3 Starch Biosynthesis

The important role of Pi in starch synthesis stems from the work of Preiss [4,74]. The author showed that ADP-glucose pyrophosphorylase (ADPG PPase), the key regulatory enzyme for starch synthesis, is stimulated by high triose-P/Pi levels. In the chloroplast, the concentration of these effector molecules was postulated to vary due to the physiological conditions to which the plant was exposed [4]. It has been shown that starch synthesis is greatly increased in those plant species where mannose-phosphate accumulates as a result of mannose feeding, which serves to lower the cytoplasmic Pi concentration [75].

A specific effect of Pi ions is exerted through the control of the distribution of newly fixed carbon between starch synthesis in the chloroplasts and the transfer of triose-P to the cytoplasm followed by synthesis of sucrose [49]. In isolated chloroplasts, low Pi slows photosynthesis and shifts the flow of carbon toward starch [49]. In some leaves, mannose feeding produces the same effect by sequestering Pi, as an abnormal hexokinase reaction becomes linked to oxidative phosphorylation [63,75]. Low levels of phosphate and high levels of sugars in phosphate-limited plants will lead to increased levels of ADPG PPase transcript, which could contribute to increase in starch accumulation [76].

The starch deposited in the chloroplasts is usually degraded during the subsequent night period (Figure 33.1). An increased stromal Pi level favors starch breakdown [77]. Glucose-1-phosphate, the product of phosphorylytic starch degradation, is transformed through the oxidative pentose phosphate pathway [78,79] and also through phosphofructokinase [80] to triose-P or further to PGA [49,81,82]. The Pi translocator catalyzes the export of these phosphate esters into the cytosol.

The influence of Pi concentrations outside the chloroplast on the steady-state concentrations of various stromal metabolites and the corresponding rates of CO_2 fixation and starch production was determined using a kinetic model [83] based on control theory [84]. This kinetic analysis indicated that PGA and Pi play an important role in regulating starch synthesis and that ATP, glucose-1-P, and fructose-6-P make significant contributions. Since these metabolites are either substrates or effectors of the ADPG PPase, the analysis is consistent with the view that Pi is a negative effector and PGA is a positive effector of ADPG synthesis and that the PGA/Pi ratio therefore regulates starch synthesis [74].

33.2.4 Sucrose Biosynthesis

Sucrose is a major product of photosynthesis. In many plants, it is the main form in which carbon is translocated through the phloem of the vascular system from the leaf to other parts of the plant, but sucrose and other sugars may also be isolated and stored in vacuoles in the mesophyll cells. Sucrose is not merely a crucial sugar of vascular plants but is preeminently the sugar of vascular plants [85]. The rate of sucrose synthesis is a function of the carbon fixation rate, the chemical partitioning of carbon between starch and sucrose, and the rate of sucrose export from the leaf [86]. Several processes may be involved in regulating the movement of carbon from the chloroplast to the vascular tissue [87]. It is not possible in this review to present a complete analysis.

Sucrose formation occurs exclusively in the cytoplasm [86]. Substantial progress has been made in elucidating the biochemical mechanisms that control sucrose formation in leaves [8,9,85,88,89]. The cytosolic sucrose formation pathway starts with triose-P exported from the chloroplast, which are converted to hexose phosphate (hexose-P) and ultimately to sucrose (Figure 33.1). The key enzymes involved in the synthesis of sucrose from triose-P are cytoplasmic FBPase and sucrose-phosphate synthase (SPS) [45,88,90–94]. It is recognized that there are at least two key aspects of the regulation of the pathway of sucrose biosynthesis: (1) control of cytosolic FBPase by the regulatory metabolite fructose-2, 6-bisphosphate (F2,6BP) [45]; and (2) control of SPS activity by allosteric effectors and protein phosphorylation [86,88,94]. Although control of the sucrose biosynthesis pathway is shared between cytosolic FBPase and SPS, it appears that SPS probably exerts more of a limitation to the maximal rate of sucrose synthesis than does FBPase [93]. However, it was found that decreased expression of these two enzymes in antisense *Arabidopsis* lines has different consequences for photosynthetic carbon metabolism [95]. In transformants with decreased expression of SPS, there was a slight inhibition of sucrose synthesis, there was no accumulation of phosphorylated intermediates, and carbon partitioning was not redirected to starch. This indicates that decreased expression of SPS triggers compensatory responses that favor sucrose synthesis, which included an increase in the UDP-glucose/hexose-P ratio and a decrease in pyrophosphate concentration. Strand et al. [95] conclude that these responses are presumably triggered when sucrose synthesis is decreased both in light and dark conditions. Decreased expression of cytosolic FBPase represented a passive response to the lower rate of sucrose synthesis and led to accumulation of phosphorylated intermediates, Pi limitation of photosynthesis, and high rates of starch synthesis [95]. Work on *Arabidopsis* showed that loss of the two major SPS isoforms in leaves limited sucrose synthesis without changing carbon partitioning in favor of starch during the light period, yet limiting starch degradation during the dark period [96].

Regulation of FBPase received increased attention with the discovery of F2,6BP in plants [45]. The extensive studies of Stitt et al. [8,45,91–93,97–99] showed that F2,6BP plays a key regulatory role in sucrose biosynthesis. In plants, the level of F2,6BP responds to changes in light, specific metabolites, sugars, and CO_2. F2,6BP is a potent inhibitor of cytoplasmic FBPase and sensitizes FBPase to the effects of FBP and Pi. F2,6BP decreases when triose-P becomes available for sucrose synthesis, and it increases when hexose-P accumulates in the cytosol. The response of the cytosolic FBPase to a rising supply of triose-P has been described in a semi-empirical model [8,100]. This model predicts how cytosolic FBPase activity responds to a rising rate of photosynthesis and relates closely with the actual response of sucrose synthesis *in vivo*.

UDP-glucose pyrophosphorylase (UDPG PPase) is an important enzyme producing UDP-glucose for sucrose synthesis in leaves. The UDP-ase encoding gene of *A. thaliana* was suggested as a possible regulatory entity that is closely involved in the readjustment of plant response to environmental signaling [101]. In *Arabidopsis* mutants (pho 1–2) impaired in Pi status, *Ugp* was found to be upregulated by conditions of phosphate deficiency [99]. Ciereszko et al. [102] concluded that under Pi deficiency, UGP-ase represents a transcriptionally regulated step in sucrose synthesis/metabolism, and that it is involved in homeostatic mechanisms for adjusting to the nutritional status of the plant.

Huber and coworkers have documented the role of SPS in the regulation of photosynthetic sucrose synthesis and partitioning in leaves [87,89,93,103–108]. SPS is minimally regulated at three levels. The steady-state level of the SPS enzyme protein is regulated developmentally during leaf expansion [106]. There are two distinct mechanisms to control the enzyme activity of the SPS protein: (1) allosteric control by G6P (activator) and Pi (inhibitor) and (2) protein phosphorylation (covalent modification). These two mechanisms are often referred to as *fine* and *coarse* controls, respectively. There are apparent differences among species in the properties of SPS that may reflect different strategies for the control of carbon partitioning [107]. The importance of SPS in the regulation of carbon partitioning in leaves has been confirmed using recombinant DNA technology [109]. However, SPS is not the only determinant of the rate of sucrose synthesis. In some cases, the growth rate of the whole plant is correlated closely with SPS activity in leaves [110].

Sucrose synthesis is a Pi-liberating process (net reaction, 4 triose-P + $3H_2O$ = 1 sucrose + 4Pi). The liberation of Pi in the cytoplasm during sucrose synthesis favors continued triose-P export from the chloroplast by counter-exchange through the Pi translocator. Thus, under conditions that favor sucrose synthesis, triose-P molecules are partitioned away from the starch biosynthetic pathway that resides in the chloroplast. If sucrose synthesis in the cytoplasm is reduced, triose-P remains within the chloroplast for starch synthesis. The resulting increase in PGA within the chloroplast stroma (high PGA/Pi ratio) also favors starch synthesis by allosterically activating the starch-synthesizing enzyme ADPG PPase [72,111].

Pi may be involved in determining the proportion of the flux of photosynthetically fixed carbon between starch synthesis and export from the chloroplast [112]. As an inhibitor of SPS and cytosolic FBPase [113] and an activator of fructose-6-phosphate-2-kinase [114], Pi plays a critical role in regulating the rate of sucrose synthesis. When sucrose synthesis in the cytosol is restricted, there can indeed be substantial changes of the stromal Pi in leaves [42]. The rate of sucrose synthesis may also have an indirect control over

the synthesis and accumulation of starch in leaves. Cytosolic FBPase and SPS, when acting in coordination with the Pi translocator, may represent an important link between sink demand and rates of carbon partitioning into starch and sucrose [115,116]. A change of partitioning does not necessarily imply that the rate of photosynthesis has been inhibited [9]. However, Pieters et al. [15] found that low sink strength lowers sucrose synthesis and restricts the recycling of Pi back to the chloroplast, thus limiting the rate of net photosynthesis.

Four lines of evidence suggest that short-term availability of Pi in the cytosol may restrict sucrose synthesis and can limit the maximal rate of photosynthesis in saturating light and CO_2 [8]. The first approach is based on the manipulation of leaf material through Pi or mannose feeding [3,7]. A second approach is based on observations that the net rate of CO_2 assimilation does not always increase in C_3 plants when the O_2 concentration is decreased from 21% to 2% to suppress photorespiration, which is generally known as O_2 *insensitivity* [5,117,118]. A third approach involves using a brief interruption of photosynthesis to transiently increase the Pi level in the cytoplasm of the leaf [119]. A fourth line of evidence comes from the study of photosynthetic oscillations that can be triggered by increasing the CO_2 or lowering O_2 [4], or by a short period in the dark [119]. These oscillations are decreased when Pi is supplied to leaves and are increased when mannose is supplied to sequester Pi. As sucrose is the major end product of photosynthesis, it is likely that a restriction in sucrose synthesis can limit photosynthesis through short-time limitation of Pi in the cytosol.

33.3 LONG-TERM *IN VIVO* EFFECTS OF Pi DEPRIVATION

The view that Pi is an important regulator of the rate of photosynthesis and of the partitioning of triose phosphates between starch biosynthesis and sucrose biosynthesis is to a large extent based on research carried out with *in vitro* systems involving the use of isolated chloroplasts, enzyme systems, and protoplasts, and with detached leaves or leaf disks fed with mannose to induce Pi deprivation. All of these studies point to the fact that the concentration of Pi in the cytosol versus that in the chloroplast is what potentially controls the intracellular flow and distribution of triose-P, and possibly of the rate of photosynthesis itself.

Studies of long-term limitations of Pi on photosynthesis and carbon partitioning based on *in vivo* experiments using low-phosphate (low P) plants have shown that the inhibition of photosynthesis was to a large extent due to limitations imposed on the PCR cycle in terms of RuBP regeneration [6,18,120–131], while the changes in carbon partitioning could be influenced in part by the relative capacities of the enzymes involved in starch and sucrose metabolism [131]. Experimental work performed by Pieters et al. [15] showed that during Pi deficiency, low rates of sucrose synthesis due to low demand from sinks limit Pi recycling to chloroplast and restrict photosynthesis.

33.3.1 Plant Growth Response and Phosphate Concentration

Long-term P deficiency greatly affects the plant growth processes at subcellular, cellular, and whole organ levels of organization [1,25]. The growth of several plant species tested was greatly reduced by P deficiency. Leaf area, leaf number, and shoot dry matter per plant were found to be more sensitive to P deficiency than photosynthetic rate per unit leaf area [18,19,127,129,132,133]. Effects of P deficiency were similar in C_3 (sunflower and wheat) and C_4 (maize) species [127]. In Pi withdrawal experiments of the range of C_3, C_{3-4} intermediate, C_4 annual, and perennial monocotyledons and dicotyledons species, it was shown that C_3 and C_4 species had similar photosynthetic P use efficiency, but the growth of C_3 species was more affected by Pi supply than C_4 species. Moreover, leaf photosynthetic rates were not correlated with growth response [134]. These results indicated that the relative growth rate (RGR) decreased before any significant effect on photosynthesis [134]. Growth analysis of maize field crops under P deficiency supported the idea that P deficiency affects plant growth, especially leaf growth, earlier and to a greater extent than photosynthesis per unit leaf area [135]. Jacob and Lawlor [127] showed that the extreme P deficiency reduced plant height by 52%, leaf area per plant by 95%, and shoot dry weight per plant by 93% in sunflower (Table 33.1). The respective reductions were 57%, 89%, and 90% in maize and 53%, 91%, and 93% in wheat. P-deficient leaves contained more and smaller cells per unit leaf area. The mean cell volume and specific leaf weight were reduced to a smaller extent by P deficiency.

A typical response to phosphate deficiency is the increase in the root mass/shoot mass ratio resulting from the decrease in shoot growth and the increase in root growth [136–138].

The increase in root elongation and growth is probably a plant response to low Pi in the surrounding medium [139–143]. However, this response is not a rule as shown in *Arabidopsis* plants, where Pi deficiency resulted in a short primary root phenotype as a result of limited division of meristematic cells, early cellular differentiation, and reduction of cell elongation zone [144]. From the studies on bean plants, it was found that the RGR of phosphate-deficient roots was higher only at the beginning of phosphate starvation, and after 2 weeks (with severe P deficiency), RGR was significantly lower as a result of decreasing ATP concentration in the roots [145].

To assess the importance of increased carbon allocation to roots for the adaptation of plants to low P availability, Nielsen et al. [146] constructed carbon budgets for four common bean genotypes with contrasting adaptation to low P availability in the field (*P efficiency*). They found that P-efficient genotypes allocated a larger fraction of their biomass to root growth, especially under low-P conditions. They also found that efficient genotypes had lower rates of root respiration than inefficient genotypes, which enabled them to maintain greater root biomass allocation than inefficient genotypes without increasing overall root carbon costs. Hogh-Jensen et al. [147] tested the influence of P deficiency on growth and nitrogen fixation of white clover plants. Their results indicated that nitrogen fixation did not limit the growth of clover plants experiencing P deficiency. Low-P status in plant tissue induced changes in the relative growth of roots, nodules, and shoots rather than changes in nitrogen and carbon uptake rates per unit mass or area of these organs.

The extent to which plant growth might be affected by P supply may depend on the sink–source status of the examined plant and how this is regulated [148]. The reduction in shoot biomass production in low-P plants may be attributed to a lower rate of leaf expansion, which may be induced by

TABLE 33.1

Effect of Extreme P Deficiency on Plant Characteristics of Sunflower, Maize, and Wheat

Plant Growth Characteristics	Treatment	Sunflower	Maize	Wheat
Plant height (cm)	Control	46	103	68
	P-deficient	22	44	32
Leaf area per plant (cm²)	Control	895	708	232
	P-deficient	41	79	21
Shoot dry matter per plant (g)	Control	6.0	4.0	2.8
	P-deficient	0.4	0.4	0.2
Number of cells/m² leaf area (×10⁷)	Control	723	152	98
	P-deficient	967	172	105
Mean cell volume (pl)	Control	27	43	144
	P-deficient	19	38	131
A_{mes}/A_{leaf}[a]	Control	31	9	14
	P-deficient	34	10	13
Specific leaf weight (g fresh wt./m²)	Control	235	145	204
	P-deficient	222	121	201

Source: Jacob J, Lawlor DW, *J. Exp. Bot.* 42:1003–1011, 1991.

[a] A_{mes}, mesophyll surface area; A_{leaf}, leaf surface area.

lower hydraulic conductance of the root system and a lower leaf water potential [18,19,149,150]. Using experimental and simulation techniques, Rodriguez et al. [151] identified the existence of direct effects of P deficiency on individual leaf area expansion. Chiera and Rufty [152] found that expansion of soybean leaves under P stress was limited by the number of cell divisions, which would imply control of cell division by a common regulatory factor within the leaf canopy. The reduction in leaf expansion in low-P sugar beet plants was associated with a 30% increase in leaf dry weight per unit area. Only 9% (or less) of the increase in dry weight in low-P leaves was due to starch [126]. Most of the remainder of the increase in dry weight may be attributed to other structural carbohydrates (e.g., cellulose and hemicelluloses). But our knowledge is limited regarding the effects of P on cell wall properties, especially those affecting cell division and cell wall expansion.

Jacob and Lawlor [127] reported that the extreme P deficiency in nutrient solution not only diminishes plant growth but also drastically reduces the total and inorganic P contents of leaves of sunflower, maize, and wheat (Table 33.2). The concentration of Pi in the leaf water decreased as the Pi content per unit leaf area decreased. Soluble protein content was lower in P-deficient leaves of all the three species, while chlorophyll content was reduced in sunflower and maize. Studies on rice plants under short-term (<1 day) to long-term Pi deficiency (>7 days) revealed that both shoots and roots showed a gradual decrease in Pi content as the length of the treatment increased [138]. This was consistent with other studies under long-term Pi starvation [153–156].

Under Pi deficiency, the concentration of Pi in leaves depends mainly on the transport from the roots and mobilization of stored phosphate from older leaves [157]. Short-term phosphate starvation tends to maintain constant cytoplasmic Pi concentration at the expense of the vacuolar pool [158]. To regulate Pi homeostasis, plants develop signaling mechanisms [159]. It was recognized since many years that restriction of P, nitrate, or sulfur influences transpiration, stomatal conductance, and root hydraulic conductivity. The experiments with *Lotus japonicus* indicated that roots are capable of monitoring the nutrient content of the solution in the root apoplasm and of initiating responses that anticipate by hours or days any metabolic changes resulting from nutrient deficiency [160]. Recent studies have shown that the sensory mechanisms that monitor environmental Pi and transmit the nutritional signal to adjust root development involve complex antagonistic interactions between external Pi and Fe bioavailability [161].

Reduced hydraulic conductance resulting from phosphate deficiency may affect the distribution of phosphate and nitrate ions between shoot and root [162]. In phosphate-sufficient bean plants, the equal Pi distribution between shoot and root was noted, whereas in plants grown on a Pi-deficient solution, almost 70% was partitioned to the shoot [163]. It seems that during moderate phosphate deficiency, the leaf Pi pool remains relatively more stable mainly due to the possible effect of Pi recycling processes [164].

33.3.2 PHOTOSYNTHETIC MACHINERY

Phosphorus deprivation has been associated with the restriction of ATP production by altering the light-reaction component of photosynthesis [165]. Several investigations have indicated that Pi deprivation involve nonstomatal factors that have significant effects on photosynthesis [19,166–168].

Studies conducted with isolated chloroplasts, thylakoid membranes, and pigment systems have shown that the primary processes of light reactions of photosynthesis and photosynthetic electron transport were relatively little affected by long-term Pi deprivation [6,38,121]. However, it was shown that phosphate availability might change thylakoid membrane lipid composition by replacing some phospholipids for galactolipid digalactosyldiacylglycerol [169,170]. Significant changes in plasma membrane phospholipid composition were also observed in bean roots during prolonged phosphate deficiency [163].

TABLE 33.2
Effect of Extreme P Deficiency on the Leaf Composition in Sunflower, Maize, and Wheat Plants

Plant Growth Characteristics	Treatment	Sunflower	Maize	Wheat
Total P content (mmol/m^2)	Control	7.20	4.80	5.90
	P-deficient	1.81	0.51	0.97
Pi concentration (mmol/m^2)	Control	1.65	0.58	0.65
	P-deficient	0.21	0.11	0.21
Concentration of Pi in leaf tissue (mol/m^3)	Control	7.81	4.51	4.00
	P-deficient	1.20	1.01	1.21
Total chlorophyll (g/m^2)	Control	0.56	0.42	0.50
	P-deficient	0.48	0.26	0.54
Total soluble protein (g/m^2)	Control	12.2	5.8	6.28
	P-deficient	6.7	1.1	5.83

Source: Jacob J, Lawlor DW, *J. Exp. Bot.* 42:1003–1011, 1991.

Investigations on changes in photochemical apparatus organization and function in relation to leaf P status in sugar beet revealed the following: low-P leaves exhibited increased levels of chlorophyll/area, PSI/area, LHCP/area, Cyt-b_{563}/area, and Cyt-f/area, while PSII, Cyt-b_{559}, and Q per area were not much affected [121]. PSII electron transport was slightly decreased per area, while PSI electron transport was slightly increased so that the ratio of PSII/PSI is decreased. The expression of certain genes encoding proteins involved in PSI, PSII, Rubisco, and chlorophyll a/b-binding proteins is restrained following Pi deficiency [171–173]. The proteomic analysis performed by Zhang et al. [168] showed that Pi starvation included changes in the levels of proteins involved in the light harvesting complex, the Calvin cycle, and CO_2 fixation.

It is generally believed that the results from in vitro studies with external supplies of artificial electron donors and acceptors and possibly damaged or atypical membranes may not always represent the in vivo situation. Light scattering and modulated chlorophyll a fluorescence have been successfully employed by several research workers as experimental probes for analyzing the state of the photosynthetic apparatus in vivo [174]. Measurements of in vivo fluorescence at room temperature and fluorescence at 77 K suggested that the low-P leaves had less mobility of the antenna, which may be due to (1) the enhanced phosphatase activity leading to dephosphorylation of the antenna and (2) the fact that the large proton gradient may promote dephosphorylation [175]. However, low-P leaves, to overcome this difficulty, developed a larger permanent antenna [121].

Using modulated chlorophyll a fluorescence techniques, the effects of extreme P deficiency during growth in the in vivo photochemical activity of PSII were determined in leaves of sunflower and maize [176]. In both species, long-term P deficiency decreased the efficiency of excitation energy capture by open PSII reaction centers, the photochemical quenching coefficient of PSII fluorescence, and the in vivo quantum yield of PSII photochemistry, and increased the nonphotochemical dissipation of excitation energy. Observations from PSII fluorescence from intact leaves suggested that P deficiency causes photo-inhibition of PSII. Furthermore, their calculations showed that there was a relatively higher rate of electron transport across PSII per net CO_2 assimilated in extreme P-deficient leaves. Most of these photosynthetic electrons that are not used for CO_2 reduction are diverted to photorespiration leading to proportionately more photorespiration and less CO_2 fixation in P-deficient leaves [177]. Photorespiration can alleviate Pi deficiency and reduce photo-inhibition by consuming the surplus energy produced by photosynthesis and accelerating the recycling of phosphorus, which, in turn, reduces the impact of low phosphorus stress [178].

The important role of photorespiration for supporting photosynthesis when isolated chloroplasts were incubated at a low Pi level was shown by Usuda and Edwards [43]. Heber et al. [179] proposed that photorespiration substantially increases Pi availability for photosynthesis in the leaves of spinach. Unicellular green algae, Chlorella vulgaris, was used to study the effect of low-phosphate supply on glycolate metabolism [180]. P deficiency did not change chlorophyll concentration, but with subsequent medium alkalization, dissolved inorganic carbon increased the photosynthetic O_2 evolution and intra-chloroplast oxygen concentration resulting in enhancement of glycolate production [181]. The study of post-illumination burst (PIB) of CO_2, which is interpreted as short-lived continuation of photorespiration in dark, indicated that the photorespiratory potential activity of P-deficient bean leaves is enhanced [182]. The importance of photorespiratory metabolism in Pi balance in bean plants under moderate phosphate deficiency was also suggested by Kondracka and Rychter [164], but the elucidation of its role needs further studies.

Plesnicar et al. [130] evaluated the efficiency of PSII photochemistry and electron transport, and light utilization capacity of sunflower leaves grown under sub- to supraoptimal Pi supply conditions. The apparent quantum yield (based on the initial slope of the relationship between photon flux density and rate of O_2 evolution) and the maximum (light and CO_2-saturated) rates of photosynthesis were the highest with the plants that were grown in optimal (0.5 mol m^{-3} Pi and 1.0 mol m^{-3}) Pi concentrations in nutrient solution. The photosynthetic efficiency was decreased by sub- or supraoptimal supply of Pi in nutrient solution. They suggested that the processes associated with nonphotochemical energy dissipation could modify the efficiency with which the reaction centers can capture and utilize excitation energy during Pi limitation of photosynthesis. This down-regulation of the efficiency of PSII photochemistry by nonphotochemical energy was attributed to the adjustment of the rate of photochemistry to match that of photosynthetic carbon metabolism in order to avoid overexcitation of the PSII reaction centers.

33.3.3 Carbon Metabolism

Several studies have shown that P deficiency in leaves decreases the rate of net CO_2 assimilation by intact leaves of C_3 and C_4 plants. This decline in net photosynthesis with long-term inadequate supply of Pi may result from a decrease in the conductance of CO_2 from the atmosphere to the chloroplasts; from a detrimental effect on the photosynthetic mechanism (mesophyll activity) itself; or from a combination of the two. It is often associated with decreases in Rubisco activity, RuBP concentration, rate of RuBP regeneration, stomatal conductance, and an increase in mesophyll resistance [6,123,127,128,183,184].

Phosphorus deficiency reduced the rate of photosynthesis in leaves by reducing the carboxylation efficiency and apparent quantum yield [6,124,130] by its influence on leaf metabolism, and also by decreasing leaf conductance [19,123]. Jacob and Lawlor [127] analyzed the effects of P deficiency on stomatal and mesophyll limitations of photosynthesis in sunflower, maize, and wheat plants. They found that stomatal conductance did not restrict the CO_2 diffusion rate; rather the metabolism of the mesophyll was the limiting factor. This was shown by poor carboxylation efficiency and decreased apparent quantum yield for CO_2 assimilation, both of which contributed to the increase in relative mesophyll limitation of photosynthesis in P-deficient leaves.

Brooks [6] attempted to determine which aspects of photosynthetic metabolism are affected when spinach plants are grown with inadequate P supply. P deficiency caused reductions in Rubisco activity, RuBP regenerating capacity, and quantum yield. The reduction in quantum yield was accompanied by changes in chlorophyll fluorescence of PSI and PSII measured at 77 K. The levels of RuBP and PGA were significantly reduced than the control leaves, while the response of photosynthesis to low [O_2] was similar to control leaves, indicating that the photosynthesis is not limited by triose-P utilization. Dietz and Foyer [7] also observed decreased levels of phosphorylated metabolites in leaves as a result of P deficiency. The decrease in phosphorylated sugar levels was also observed in roots despite the increased sugar concentrations, which indicates that sugar phosphorylation may be limited by lower activity of fructokinase and hexokinase [185].

Rao and Terry [19] explored the changes in the activity of PCR cycle enzymes in relation to leaf Pi status. Low-P leaves exhibited increased levels in total activity of Rubisco, FBPase, and Ru5PKinase, while the activity of PGA kinase, G-3-P-dehydrogenase, transketolase, and FBP aldolase decreased. The percentage light activation of Rubisco, PGA kinase, G-3-P-dehydrogenase, FBPase, SBPase, and R5PKinase was lower in low-P leaves (Table 33.3). Jacob and Lawlor [128] have also shown that P deficiency decreased the RuBP content of the leaf more than it decreased Rubisco. They suggested that the decreased specific activity of Rubisco found in Pi-deficient sunflower leaves is a consequence of the decreased ratio of RuBP to RuBP binding sites observed in such leaves allowing inhibitors to bind to the active sites of the enzyme.

It has been shown that long-term inadequate supply of Pi decreases the rate of photosynthesis by limiting the capacity for regeneration of RuBP, although decreased activation of Rubisco may play a part [6,19,121,122]. Rao et al. [123] measured a number of metabolites in low-P leaves, including RuBP, PGA, triose-P, FBP, F6P, G6P, adenylates, nicotinamide nucleotides, and Pi (Table 33.3). They suggested that RuBP regeneration in moderately P-deficient leaves is limited by decreased supply of carbon due to increased diversion of assimilated carbon for starch synthesis rather than by the decreased supply of ATP.

What are the precise metabolic control points that diminish regeneration of RuBP in P-deficient leaves? Several factors, including the initial activity of PCR cycle enzymes, the supply of ATP and NADPH, and the availability of fixed carbon, all affect the RuBP regeneration capacity of leaves. At moderate P-deficient conditions, RuBP regeneration of sugar beet leaves may be limited by the supply of Ru5P and the initial activity of the Ru5P kinase [123,131]. The conditions necessary to alter the RuBP pool size by this mechanism need to be clearly understood. According to Jacob and Lawlor [175], it is more probable that a deficiency of ATP in severely Pi-deficient leaves slows down the PCR cycle activity and thus decreases the regeneration of ATP. They found marked reductions in the amounts of ATP, ADP, and oxidized pyridine nucleotides per unit leaf area in extremely Pi-deficient sunflower and maize leaves (Table 33.4).

As pointed out by Noctor and Foyer [22], a small change in the ratio of ATP and NADPH production during photosynthesis relative to the ratio of their consumption has an impact on cell adenylate and redox status. In bean leaves, during moderate phosphate deficiency, the net photosynthetic rate was lower and the concentration of NADPH increased; the ratio of NAD(P)H/NAD(P) also increased [186]. At the same time, leaf ATP concentration was reduced by 50% [187]. The reduction in leaf ATP concentration was comparable in light and dark periods. The determinations of ATP in leaf extracts during the light period reflect chloroplastic, mitochondrial, and cytosolic pools of ATP, whereas the leaf extracts from the dark period reflect mainly cytosolic and mitochondrial ATP pools. The ATP produced during photophosphorylation may be immediately utilized in the chloroplasts to support CO_2 fixation and chloroplast synthetic processes [188]. ATP synthesized in mitochondria can be transported to cytosol to support cytosolic reactions connected with sucrose synthesis [188]. Therefore, small differences between light and dark concentrations of ATP in phosphate-deficient leaves may reflect the determination of only the cytosolic pool being strongly dependent on the efficiency of mitochondrial ATP production [187]. It was found by Rychter's group that the efficiency of mitochondrial ATP production in bean plants during phosphate deficiency is lower due to increased participation of a cyanide-resistant, alternative pathway (AOX) [187–191],

TABLE 33.3

Effect of Low-P Treatment on the Percent Light Activation of Certain PCR Cycle Enzymes and Pool Sizes of Sugar Phosphates in Leaves of 5-Week-Old Sugar Beet Plants

PCR Cycle Enzymes and Metabolites	Control	Low P
Light Activation of PCR Cycle Enzymes (%)		
Rubisco	82	73
PGA kinase	78	65
NADP-G3PD	34	10
FBPase	33	39
SBPase	82	82
Ru5P kinase	34	23
Pool Size of Sugar Phosphates (μmol/m²)		
RuBP	66	32 (48)[a]
PGA	125	38 (30)
Triose-P	21	10 (48)
FBP	27	18 (67)
F6P	18	2 (11)
G6P	4	7 (16)

Source: Rao IM, Terry N, *Plant Physiol.* 90:814–819, 1989; Rao IM et al., *Plant Physiol.* 90:820–826, 1989.

[a] Figures in parentheses represent percentage of control values.

TABLE 33.4

Effect of P Deficiency on Adenylates and Nicotinamide Nucleotides of the Third Fully Expanded Leaves of Sunflower and Maize Grown at P-Sufficient (10 mM Pi) or P-Deficient (0 mM) Conditions (Values Indicate Pool Sizes in µmol/m²)

Leaf Metabolites	Sunflower		Maize	
	P-Sufficient	P-Deficient	P-Sufficient	P-Deficient
	Adenylates			
ATP	19.8	8.9	22.4	5.7
ADP	13.5	6.5	14.5	8.6
AMP	7.4	6.5 NS	9.1	8.8 NS
Total	40.7	21.9	46.0	23.1
ATP/ADP	1.5	1.4	1.5	0.7
	Nicotinamide Nucleotides			
NAD⁺	13.9	5.9	19.7	11.5
NADP⁺	12.6	7.1	16.8	9.8
NADH	3.2	4.4 NS	7.3	5.2
NADPH	4.5	4.1	8.9	9.4 NS
Total	34.2	21.4	52.7	35.9
NADPH/ NADP⁺	0.36	0.58	0.53	0.96

Source: Jacob J, Lawlor DW, *Plant Cell Environ.* 16:785–795, 1993.
Note: NS = not significant at $p < 0.05$.

which bypasses two respiratory chain phosphorylation sites. Determination of actual participation of AOX and ATP efficiency of respiratory chain phosphorylation in bean, tobacco, and *Gliricidia sepium* leaves revealed that during prolonged phosphate deficiency, AOX expression is species-dependent and is not observed in tobacco or *G. sepium* [192].

The rates of photosynthesis in C_3 plants have been modeled on Rubisco kinetics and the supply of CO_2, RuBP, and Pi [5,10,72,193–196]. It seems clear that at all levels, regulation is serving to maximize efficiency while striving to avoid damage to the photochemical apparatus [193]. In general, nonlimiting processes of photosynthesis are regulated to balance the capacity of limiting processes [194]. When photosynthesis is limited by the capacity of Rubisco, the activities of electron transport and Pi regeneration are down-regulated so that the rate of RuBP regeneration matches the rate of RuBP consumption by Rubisco. Similarly, when photosynthesis is limited by electron transport or Pi regeneration, the activity of Rubisco is down-regulated to balance the limitation in the rate of RuBP regeneration.

It is important to understand that several parameters interact and a change in any one will result in a change in the activity of the others [84,197]. When the activity or level of any one of the components is reduced (Ru5P kinase or RuBP), that component temporarily assumes an increased importance until equilibrium is restored. The enzymes of the PCR cycle, the pool sizes of sugar phosphates, along with the flux of ATP and NADPH, interacting as a system, share control over the

rate of photosynthesis. None of these system elements control the rate but all regulate jointly. It is the self-regulated lowering of the RuBP pool and not the inability to regenerate it faster that is a major factor in restoring and maintaining metabolic balance [196].

33.3.4 INTRACELLULAR PI COMPARTMENTATION

In order to prove that Pi regulation of photosynthesis occurs *in vivo*, it will be essential to demonstrate that cytosolic and chloroplastic Pi concentrations vary sufficiently to bring about changes in the flow and distribution of triose-P within the cell. There are practical problems to overcome in determining Pi compartmentation between chloroplast, cytoplasm, and vacuole. An additional problem is that there may be an internal Pi buffering mechanism. For example, if a mechanism for regulated transport of Pi across the tonoplast membrane was present, then the vacuole could act as a Pi reserve for the cytosol. More general evidence for a cytosolic Pi buffering mechanism arises from studies on P-deficient plants, which appear to maintain the cytosolic Pi level at the expense of vacuolar Pi [159,198].

Methods have been developed for the assay of subcellular metabolite levels using leaf protoplasts. The protoplasts were ruptured by passage through a nylon net or a capillary tube. This was followed by immediate filtration of the particles (formed after rupture of the protoplasts) through a layer of silicone oil [71,199,200] or a combination of membrane filters [201]. Unfortunately, it has proved experimentally difficult to accurately determine chloroplastic and cytosolic Pi concentrations. Part of the problem relates to the presence, in the leaf cell vacuole, of a comparatively large amount of Pi [202], which masks the much smaller amount present in the cytosol. Furthermore, protoplasts are of limited value since their carbohydrate metabolism is almost certainly affected by the lack of sucrose export to the phloem.

³¹P-nuclear magnetic resonance (³¹P-NMR) spectroscopy can provide information on the relative concentration of Pi in the different cellular compartments [202]. A characteristic feature of the ³¹P-NMR spectra of most plants tissues is the detection of two clearly resolved Pi signals, assigned to the cytoplasmic and vacuolar pools. *In vivo* ³¹P-NMR provides an important method for studying the interaction between the two pools under different physiological conditions. With ³¹P-NMR spectra, it is possible to determine the absolute concentrations of Pi in the cytosol and the vacuole and thus to assess the extent to which the Pi distribution across the tonoplast reaches electrochemical equilibrium under different nutritional conditions [203–205].

The ³¹P-NMR technique has been applied extensively in studies of chloroplasts, protoplasts, cell suspensions, leaves, and roots [17,18,124,157,206–217]. Foyer and Spencer [17] determined the intracellular distribution of Pi in barley leaves grown under different Pi regimes. They showed large differences in the vacuolar Pi content between the plants grown at different levels of P supply. In contrast, the cytosolic Pi level was similar in the leaves of plants grown at 1 and 25 mM

Pi. Based on these data, they suggested that in leaves as in isolated cells, the cytoplasmic Pi level is maintained constant as far as is possible, while the vacuolar Pi pool is allowed to fluctuate in order to buffer the Pi in the cytoplasm [206,207]. Several studies suggest the role of the vacuole in homeostasis of the cytoplasmic Pi concentration. Under different external phosphate levels, the cytoplasmic phosphate concentration remains relatively stable at the expense of the vacuolar pool, which decreases under Pi deficiency [158,159]. The mechanisms that control Pi transport from and to the vacuole are not clear, but changes in cytosolic and vacuolar Pi concentrations are considered as a signal for triggering different starvation response systems [159].

Using ^{31}P-NMR, Lauer et al. [124] determined P compartmentation in leaves of reproductive soybean as affected by P supply in nutrient solution (Table 33.5). As the concentration of P in nutrient solution increased from 0.05 to 0.45 mM, the vacuolar P pool size increased relative to cytoplasmic and hexose monophosphate P pools. Under low-P supply (0.05 mM), cytoplasmic P pool size was greatly reduced at full flower and full seed growth stages. This study indicated that the cytoplasmic P pool and leaf carbon metabolism dependent on it are buffered by the vacuolar P pool until the late stages of reproductive growth of soybeans.

Kerr et al. [104] found that the rates of net fixation of carbon, assimilate export, and net starch accumulation are not constant in continuous light. Since cytoplasmic concentrations of key regulatory metabolites such as F2,6BP and Pi could fluctuate as photosynthetic rates change [98], it may be possible that changes in intracellular Pi compartmentation could alter endogenous rhythms of photosynthesis and SPS activity.

33.3.5 CARBON PARTITIONING AND EXPORT

The partitioning of photosynthate between starch and sucrose appears to be strictly regulated at both genetic and biochemical levels [4]. There is a distinct interspecific variation in the ratio of starch: sucrose synthesized in leaves of different species [90,115]. This genetically determined predisposition allows classification of plants as high (e.g., soybean), intermediate (e.g., spinach), or low (e.g., barley) starch formers. P deficiency increased the starch synthesis relative to sucrose in soybean, spinach, maize, and barley leaves, although the accompanying limitation on photosynthetic capacity varied considerably between the species [17,172]. Usuda and Shimogawara [218] measured carbon fixation, carbon export, and carbon partitioning in maize seedlings in the early morning and at noon in P-adequate and P-deficient leaves (Table 33.6). P deficiency caused marked reductions in carbon fixation and carbon export and changed the partitioning of fixed carbon between starch and sucrose.

Long-term P deficiency causes increased starch concentrations in organs of several plant species [18,126,133]. These elevated starch concentrations in P-deficient plants may result from increased partitioning of photosynthetically fixed carbon into starch at the expense of sucrose synthesis in leaves [18,123,172] and decreased starch utilization in plant organs during the dark phase of the diurnal cycle [133,172]. These results suggested that the accumulation of starch while decreasing sucrose production enables leaves to maintain Pi levels under low phosphorus stress.

Accumulation of high starch concentration in leaves and stems and decreased starch utilization in the dark in P-deficient soybean plants indicated that growth was restricted to a greater degree than photosynthetic capacity [133]. However,

---

TABLE 33.5
^{31}P-NMR Determination of P Compartmentation in Leaves of Reproductive Soybeans as Affected by P Nutrition

Growth Stage	Phosphate Pools[a]	P Supply to Plants (mM)			
		0.05	0.10	0.20	0.45
		Pool Size (mM)			
Full flower	HMP	0.54	2.11	5.75	7.65
	P_c	0.23	0.87	3.56	8.32
	P_v	<0.05	<0.10	3.50	8.01
Full pod	HMP	0.42	0.81	1.24	8.98
	P_c	0.23	0.69	0.93	7.59
	P_v	<0.025	<0.005	0.51	13.56
Full seed	HMP	<0.01	0.39	0.78	4.10
	P_c	<0.01	0.21	0.72	5.63
	P_v	<0.01	<0.05	<0.10	7.65

Source: Lauer MJ et al., *Plant Physiol.* 89:1331–1336, 1989.

[a] HMP, hexose monophosphate; P_c, cytoplasmic inorganic phosphate; P_v, vacuolar inorganic phosphate.

TABLE 33.6
Carbon Fixation, Carbon Export, and Carbon Partitioning between Starch and Sucrose in the Middle Part of the Third Leaves of 18- or 19-Day-Old Maize Plants

Measurement Period	Measurement[a]	Treatment	
		Control	Low P
Early morning	Carbon fixed[b]	293	110
	Carbon exported[b]	221	81.9
	Carbon partitioning[c]	0.191	0.051
	Carbon fixed	214	112
	Carbon exported	158	103
	Carbon partitioning	0.253	0.103

Source: Usuda H, Shimogawara K, *Plant Cell Physiol.* 32:497–504, 1991.

[a] Measurement conditions were 1400 μmol/m^2/s PAR and 33 Pa ambient CO_2 concentration.

[b] Matom/m^2/2 h.

[c] Carbon partitioning was expressed as a ratio of carbon atom accumulated in starch to carbon atom accumulated in sucrose (including transported sucrose).

in barley plants, omission of Pi from the growth medium resulted in increase in fructan concentration while little or no effect on starch, sucrose, glucose, and fructose was observed, which indicates that, in some plants, the mechanism for carbon partitioning into fructans is more sensitive toward low-P conditions than the mechanism for carbon partitioning into starch [219].

The work of Qiu and Israel [20] addressed the issue of whether increased starch accumulation is the cause or the result of decreased growth in P-deficient soybean plants. During onset of P deficiency, significant decreases in RGR and in day and night leaf elongation rate occurred before or at the same time as significant increases in stem, leaf, and root starch concentrations. Based on these data, they concluded that disruption of metabolic functions associated with growth impairs utilization of available nonstructural carbohydrate in plants adjusting to P-deficiency stress.

Pieters et al. [15] studied the importance of sink demand on photosynthesis limitation during low-Pi conditions. The source/sink ratio was altered by darkening of all but two source leaves and compared to fully illuminated leaves of tobacco plants grown in Pi-sufficient and Pi-deficient conditions. They concluded that in tobacco plants grown in phosphate-deficient conditions, low demand for assimilate (low sink strength) is the primary reason for photosynthesis limitation. Pi deficiency drastically decreased RuBP content in the Pi-deficient leaves and hence the rate of photosynthesis. This decrease was the result of end-product limitation since decreased sucrose synthesis restricted Pi recycling to chloroplast, thereby limiting ATP synthesis and RuBP regeneration.

In P-deficient sugar beet leaves, large accumulations of not only starch but also sucrose and glucose were observed. This accumulation was associated with a marked reduction in carbon export from the leaves [126]. P deficiency also increased the levels of starch, sucrose, and glucose of petioles, storage root, and fibrous roots of sugar beet [131]. In contrast to sugar beet, P deficiency in soybean leaves caused a significant decrease in sucrose concentration together with a decrease in the activity of SPS [18,20]. The apparent carbon export rate from leaves was also restricted in soybean, but the assimilate transport to stems and roots exceeded assimilate utilization in these organs, which implies that carbohydrate availability was not the primary factor limiting the growth of nonphotosynthetic organs of P-deficient plants [20]. De Groot et al. [16] investigated growth and dry mass partitioning in tomato as affected by P nutrition and light. They found that at mild P limitation, transport and utilization of assimilates in growth, not the production of assimilates, result in an increase in starch accumulation, and at severe P limitation, the production of assimilates is limited.

In bean leaves, sucrose concentration increased, but light-promoted accumulation of sucrose was lower than in control leaves [164]. It is consistent with the observation of enhanced sucrose translocation from shoots to roots during phosphate deficiency [21,220–222]. The increase in soluble sugars in bean roots is believed to be not only the result of greater assimilate transport from leaves to roots but also higher

hydrolysis of sucrose [223] and decrease in hexose phosphorylation [185]. Typical responses to phosphate-deficiency stress in root meristematic tissue include increase in sugar concentration, increase in the size of the vacuolar compartment, and changes in factors that control the rate of respiration [224].

33.4 RECOVERY OF PLANTS FROM PHOSPHATE DEFICIENCY

Several researchers tested the reversibility of the long-term Pi-deprivation effects on plant processes such as Pi transport, photosynthesis, carbon partitioning, and growth. Leaf Pi levels of P-deficient plants raised markedly when the Pi supply was increased to spinach [225], potato [226], barley [227], maize, and soybean [228] due to an enhanced P uptake system. Obviously, a transport system with a large capacity for Pi uptake was induced in the root system when the plants were deprived of Pi. This system may catalyze a rapid accumulation of Pi in the leaves once the Pi availability is improved [214]. Based on the comparison of the results of the long-term experiment with those of the short-term uptake experiments, Jungk et al. [228] concluded that plants markedly adapt P uptake kinetics to their P status.

Increased Pi supply to low-P plants should increase leaf RuBP and should eliminate the inhibition of photosynthesis. It should also lower the pool sizes of storage carbohydrates (mainly starch) due to the recovery in leaf expansion and plant growth. Understanding the changes involved in the reversibility of low-P effects is important in predicting long-term plant growth and yield because of the varying sink strengths during plant development. The ability to reduce accumulations of starch and to relieve photosynthetic inhibition can significantly restore photosynthetic rates and increase the amount of photosynthate available for the actively growing sinks.

The changes in photosynthesis and carbon partitioning induced by low-P treatment could be due both to structural modifications induced by long-term phosphate stress and to metabolic changes accommodating the shortage of Pi as a reactant in biochemical pathways. These effects may be distinguished since the latter should be readily reversible when the supply of Pi is restored. The effects of P deficiency on photosynthesis were shown to be rapidly reversible with the resupply of P to the P-deficient plants or Pi feeding [6,7,20,120,131,225].

Brooks [6] reported that when low-P spinach plants were returned to nutrient solutions with adequate Pi, the percentage activation of Rubisco, amounts of RuBP and PGA, quantum yield, and maximal RuBP regeneration rate were increased within 24 h. The rapid increase in leaf RuBP and other sugar phosphates, which occurred as a consequence of increased Pi supply to low-P plants, substantiates the claim that the photosynthesis in low-P leaves was limited by RuBP regeneration [123].

Rao and Terry [131] monitored changes in photosynthesis, carbon partitioning, and plant growth in sugar beet by increasing the Pi supply to low-P plants. Within 72 h of increased Pi supply, low-P plants developed very high leaf blade Pi concentrations (up to sixfold of control levels). This dramatic increase in leaf blade Pi concentration was associated with a rapid increase in leaf sugar phosphates (especially RuBP), ATP, and total adenylates, which led to the rapid recovery (within 4 h) of the rate of photosynthesis. Increased Pi supply to low-P plants also decreased the amount of carbon accumulation in leaf blades in the form of starch, sucrose, and glucose, but this decrease was found to be slower than the recovery of photosynthesis. These results suggest that the effects of low P on photosynthetic machinery and the partitioning of fixed carbon are reversible. The rapid recovery of photosynthesis may be attributed to the lack of marked effects of low P on the structure and function of the photosynthetic membrane system [115]. Rao et al. [120] measured the changes in light scattering *in vivo* during photosynthetic induction with variation in the leaf Pi status. Light scattering was markedly increased during photosynthetic induction in low-P leaves. This effect was reversible, disappearing within 24 h after P resupply.

Compared to the recovery of photosynthesis, the recovery in leaf expansion and other plant growth parameters were found to be slower in sugar beet [125]. When P-deficient soybean plants were supplied with adequate P, starch concentrations in leaves and stems decreased to the levels of P-sufficient plants within 3 days [20]. Thus, starch stored in leaves and stems is ready to be utilized in the synthesis of structural biomass during the time required for activation and development of additional photosynthetic capacity.

In the context of whole plant growth, plants may have developed the ability to buffer photosynthetic metabolism against decreases in P supply using Pi stored in the vacuoles. The poor correlations between short-term measurements of photosynthetic rate and long-term plant growth [229] may be due to the buffering power of the vacuoles [151]. Therefore, the primary influence of P deficiency on plant growth may be through a reduction in leaf expansion rather than through a marked reduction in photosynthetic capacity.

33.5 INTERACTION OF PHOSPHATE DEFICIENCY WITH ENVIRONMENTAL FACTORS AFFECTING PHOTOSYNTHESIS

Although P is involved in several photosynthetic processes, little is known about the interactive effects between P deficiency and environmental factors controlling photosynthetic rate (e.g., ambient CO_2 concentration, water availability, temperature, salinity concentration, irradiance level, etc.). Some plants experience higher growth rates under elevated ambient CO_2. Therefore the raise of ambient CO_2 concentration will concomitantly increase the amount of P required to sustain increased growth rates [21,143,233–235]. Low availability of P in soil may indirectly affect photosynthetic responses under elevated CO_2 through reductions in plant growth and lower carbohydrate sink/source balance [235–237]. In cottonwood (*Populus deltoides*), low-P supply limited carbon assimilation rates under elevated CO_2 concentration [235]. This limitation in photosynthesis was primarily driven by the effects on

Rubisco activity through the effects on Rubisco content and stomatal conductance [235]. Similar results for the interaction of low Pi and elevated CO_2 were found for white lupin (*Lupinus albus*) [238]. In cotton plants, elevated CO_2 reduced diffusional limitations to photosynthesis at various Pi levels, yet it failed to reduce the nonstomatal limitations to photosynthesis in phosphorus-deficient plants [238]. However, work on potato showed an opposite effect with a lack of interactive effects between CO_2 enrichment and P deficiency on leaf gas exchange parameters [239].

In tomato plants, low Pi significantly increased the susceptibility to chill-induced inhibition of photosynthesis [240]. The reduction of photosynthetic capacity was attributed to the dysfunction of Calvin cycle and limitations in Pi supply to the chloroplasts, leading to a down-regulation of photochemical activity in PSII. Phosphate-fed grapevine leaves subjected to low temperature had an increase of 71–80% on photosynthetic rate [241]. Conversely, *Arabidopsis* mutants with decreased or increased shoot Pi concentrations showed that low Pi triggers cold acclimatization of photosynthetic carbon metabolism leading to an increase in Rubisco expression, changes in Calvin cycle enzymes, and increased expression of enzymes of sucrose biosynthesis [242]. These results suggest that low-Pi levels resulting from low rates of sucrose synthesis can induce long-term changes in photosynthesis at the level of gene expression.

In a study conducted to test the effect of light intensity and phosphorus nutrition on the phosphate-mining capacity of white lupin, it was shown that higher light intensity increased the net photosynthetic rate under both P-deficient and P-sufficient conditions [243]. Under exposure to high light and P deficiency, knockout mutants of *Arabidopsis* lacking the transcription factor PHR1 (PHOSPHATE STARVATION RESPONSE I) showed a reduction of the quantum efficiency of the PSII and degradation of the core units of PSII, which resulted in irreversible photodamage [244], thus indicating that PHR1 is needed to avoid permanent damage of photosystems during high light conditions.

Under combined stress factors of salinity (100 mM) and low P supply, there were no additive effects on leaf gas exchange parameters of barley plants [245]. Results indicated that changes in reduction in gas exchange rates were predominantly due to low P rather than due to salt exposure. Another study found that 100 mM NaCl addition to P-deficient *Hordeum marinum* plants significantly restored the plant growth and improved CO_2 assimilation rate [245].

Increasing P availability in soil could improve tolerance to drought conditions due to increased osmolyte production that aid to the maintenance of leaf water potential [246,247], and reduction in water loss via leaf transpiration and increase in instantaneous transpiration efficiency (CO_2 assimilation rate/transpiration) [248]. However, other studies showed that increase in P supply did not affect osmotic potential, stomatal conductance, or transpiration rate in plants [249,250]. A study using bean genotypes showed that foliar Pi supply did not affect net CO_2 assimilation rate, stomatal conductance, or chlorophyll fluorescence parameters during drought stress;

however, a positive effect of an extra foliar Pi supply before drought stress was shown on the photosynthesis after a period of rewatering [251].

Phosphate deficiency stress, as most if not all stresses, involves also oxidative stress. Oxidative stress results from imbalance between reactive oxygen species (ROS) production and antioxidant activity (ROS homeostasis). In nonphotosynthetic tissues, phosphate deficiency induced mild oxidative stress [149,168,230–232,252–254]. In the leaves, the main source of ROS production is photorespiration and photosynthetic electron transport chain [255]. Recent work of Zhang et al. [168] combining physiological determinations with proteomic analysis indicated oxidative stress in maize leaves under Pi deprivation. They showed that an increased nonphotochemical quenching (i.e., a thermal dissipation process) is accompanied by an increased concentration of O_2^- in leaf extracts. During phosphate deficiency, increased rate of photorespiration was observed in several studies (e.g., see Refs. [164,177,179,182]). In maize, the increase in photorespiration rate was accompanied by increased leaf ROS level [168]. Under low Pi, maize leaves showed also an increased activity of the ascorbate–glutathione cycle and increased synthesis of ascorbic acid and several other peroxiredoxins, but despite increased antioxidant defense system, higher malonyldialdehyde content, an indicator of lipid membrane oxidative damage, was observed [168]. ROSs are part of a general stress response and may act as a signal [255], yet the link between ROS and Pi deficiency is far from being elucidated.

33.6 ACCLIMATION AND ADAPTATION OF PLANTS TO PHOSPHATE DEFICIENCY

Deficiency of phosphate in the growth medium creates a stress condition for growing plants. Growth under Pi-limiting conditions triggers a cluster of acclimation responses at the morphological, physiological, biochemical, and molecular levels. These responses include reduced plant growth, altered root system architecture, and secretion of organic acids, phosphatases, and nucleases to increase Pi acquisition [137–139,156,256]. Plants can achieve tolerance to stress either by adaptation or by acclimation. Adaptation refers to heritable modifications, whereas acclimation refers to nonheritable modifications in metabolism and morphology of plant that is subjected to stressful conditions [257]. Both terms are often confusing in the literature. It is important to note that many researchers describe acclimation process and refer to it as adaptation to phosphate deficiency.

The effect of phosphate deficiency on photosynthesis depends on the capability of the plant metabolism to acclimate to low internal Pi supply. The acclimation of plants to P deficiency is complex and involves integrated cellular, tissue, and whole plant responses. Plants acclimate to P stress by changes in the pattern of growth, changes in the activity of Pi transport system, and changes in the physiological and metabolic activities. Changes in the pattern of growth and root architecture can be achieved by the increase in extension rates of roots, root hairs, and lateral root formation [157]. In some plants

(e.g., maize, common bean), aerenchyma development in root tissue is induced under low soil P and considered to enhance soil exploration and P acquisition [258]. Some plant species develop the cluster or proteoid roots, releasing organic acids and phosphatases to growth media or form the symbiotic associations of roots with mycorrhizae. All those responses are presumed to enhance Pi acquisition from the soil and involve altered gene expression. In acclimation of plants to low-Pi environment, over 100 genes may be involved, and the expression of some of those genes was described by Abel et al. [259] and Raghothama [159]. Sensing a low-Pi environment involves not only the changes in Pi uptake and transport system [260] but also remobilization of phosphate from roots and older leaves to growing leaves to maintain the rate of net photosynthesis [158,261]. Products involved in sulfolipid biosynthesis, phospholipid degradation, and starch biosynthesis increase to promote Pi use [165,168]. A number of proteins involved in hormone and organic acid synthesis are regulated to promote Pi absorption and mobilization [168,262,263]. This included an increased abundance of proteins related to energy metabolism and phosphorylation, accumulation of stress and defense-related proteins through scavenging of ROS, and proteins involved in ethylene biosynthesis [167,262,263].

The metabolic changes that occur in response to P stress may be part of an acclimation of plants to low-P environments. This physiological and metabolic adjustment increases the amount of Pi available for photosynthesis and other essential physiological functions [13]. In photosynthetic carbon metabolism, Pi is liberated during the synthesis of carbohydrates, organic acids, and amino acids, and during photorespiration. In different plant species, under Pi deficiency, some of the above-mentioned reactions may be enhanced and thereby temporarily serve as an additional Pi source [129,164,177,264]. Also, an enhanced activity of PEP carboxylase and changes in PEP metabolism were observed [252–254,265]. Kondracka and Rychter [164] indicated the crucial role of PEP carboxylase, PEP metabolism, and enhanced amino acid synthesis for Pi recirculation during photosynthesis under moderate phosphate deficiency in bean leaves. It seems that the extent of acclimation of plants to low-phosphate conditions depends on individual plant species and serves primarily to maintain the rate of net photosynthesis through internal Pi recycling processes.

33.7 CONCLUSIONS AND THE WAY FORWARD

The pioneering work of Walker and colleagues demonstrated that the isolated chloroplast requires a continuous supply of Pi in order to sustain photosynthesis. The Pi imported into the chloroplasts from the cytosol in exchange for triose-P and the Pi released from metabolic intermediates in the chloroplast stroma is available for photophosphorylation, which generates ATP for utilization in the PCR cycle. Thus, an adequate supply and internal cycling of Pi in the cell are essential for the regeneration of RuBP in the PCR cycle, which is a major limitation to maintain the rate of photosynthesis under Pi deprivation. The view that Pi supply is maintained *in vivo* by sucrose synthesis within the cytosol has been strengthened by substantial experimental evidence. The subcellular compartmentation of reactions and the resulting conservation of stromal and cytosolic Pi play an important role in the regulation of photosynthesis and carbon partitioning in leaves. Further, the rapid recovery of photosynthesis after P resupply to low-P leaves provides the direct evidence for the Pi regulation of photosynthesis *in vivo*.

Inadequate supply of Pi to plants limits the rate of photosynthesis due to both short- and long-term influences of Pi on the development of the photosynthetic machinery and metabolism. In the short term, low Pi might restrict photophosphorylation, which should lead to increased energization of the thylakoid membrane, decreased electron flow, and associated inhibition of photosynthesis. Inadequate supply of Pi over the long term decreases the rate of photosynthesis by limiting the capacity for regeneration of RuBP in the PCR cycle. However, the precise mechanisms that control RuBP regeneration under Pi deprivation are yet to be elucidated.

The research reviewed here suggests the following:

1. Pi deprivation does not directly affect photosynthetic electron transport.
2. Pi deprivation reduces photosynthesis through the limitation of RuBP regeneration and not through Rubisco.
3. RuBP regeneration may be limited by the supply of ATP and by increased partitioning of sugar phosphates to starch and sucrose synthesis.
4. Pi deprivation affects leaf area the most and photosynthesis to a lesser extent.
5. Pi deprivation diminishes carbon export more than the rate of photosynthesis; carbon accumulates in leaves of Pi-deprived plants.
6. Pi-deprivation effects on photosynthesis and carbon partitioning are reversible.
7. Sink strength imposes the most important regulatory role on photosynthesis *in vivo* during phosphate deficiency.

During the last 15 years, the use of *Arabidopsis* mutants with increased or decreased Pi level in the shoots and transgenic plants with altered gene expression served as powerful tools for studying the *in vivo* effect of Pi on photosynthetic carbon metabolism. Phosphate concentration in the leaves depends strongly on long- and short-distance transport processes and the efficiency of the uptake process [53]. The expression of genes encoding high-affinity root phosphate transporters is regulated by the phosphate status of the plant [262]. Overexpressing genes encoding high-affinity phosphate transporters may be one of the strategies for increasing Pi uptake and in consequence leaf Pi concentration. Chloroplast phosphate transporter (PHT2;1) may be a key component in coordinating Pi acquisition and also Pi allocation toward the demands of photosynthetic carbon metabolism [54].

Phosphite (Phi), the analog of phosphate, is known to interfere with many Pi-starvation responses and could serve

as an interesting tool to study plant responses to phosphate starvation. Varadarajan et al. [266] provided molecular evidence that Phi suppresses expression of several Pi starvation-induced genes. They suggest that suppression of multiple Pi-starvation responses by Phi may be due to inhibition of primary Pi-starvation response mechanisms and therefore could serve as a tool in dissecting the Pi starvation-induced molecular changes [266]. More recently, it has been shown that Phi interferes with Pi signaling at the local and systemic plant level, thus indicating that Pi acts as a signal to report its own availability [25].

The goal of this chapter was not to make an exhaustive review of all the work carried out so far on Pi regulation of photosynthesis, but rather to evaluate the role of Pi in the regulation of photosynthetic carbon assimilation and partitioning and to point out where our understanding is limited. It is clear that the Pi concentration in the cytosol is what potentially controls the rate of photosynthesis *in vivo* and partitioning of photoassimilates between starch and sucrose. Even though our knowledge from isolated chloroplasts provides substantial basis for the role for Pi in the control of photosynthesis, and undoubted importance of Pi to the life of *higher* plants, advanced theories concerning the mechanisms of Pi control of photosynthesis *in vivo* remain to be fully tested experimentally.

ACKNOWLEDGMENTS

The authors greatly appreciate the partial financial support from the CGIAR Research Programs on Grain Legumes, and Livestock & Fish.

REFERENCES

1. Bieleski R, Ferguson IB. Physiology and metabolism of phosphate and its compounds. In: Bieleski R, Ferguson IB, eds. *Encyclopedia of Plant Physiology.* New York: Springer-Verlag, 1983:422–449.
2. Moorby J, Besford RT. Mineral nutrition and growth. In: Laüchli A, Bieleski RL, eds. *Encyclopedia of Plant Physiology.* New York: Springer-Verlag, 1983:481–527.
3. Sivak MN, Walker DA. Photosynthesis in vivo can be limited by phosphate supply. *New Phytol.* 1986; 102:499–512.
4. Preiss J. Regulation of the biosynthesis and degradation of starch. *Ann. Rev. Plant Physiol.* 1982; 33:431–454.
5. Sharkey TD. Photosynthesis in intact leaves of C₃ plants: Physics, physiology and rate limitations. *Bot. Rev.* 1985; 51: 53–105.
6. Brooks A. Effects of phosphorus nutrition on ribulose-1,5-biophosphate carboxylase activation, photosynthetic quantum yield and amounts of some Calvin-cycle metabolites in spinach leaves. *Aust. J. Plant Physiol.* 1986; 13:221–237.
7. Dietz KJ, Foyer C. The relationship between phosphate status and photosynthesis in leaves—Reversibility of the effects of phosphate deficiency on photosynthesis. *Planta* 1986; 167:376–381.
8. Stitt M, Huber S, Kerr P. Control of photosynthetic sucrose synthesis. In: Hatch MD, Boardman NK, eds. *The Biochemistry of Plants. A Comprehensive Treatise.* New York: Academic Press, 1987:327–409.

9. Stitt M, Quick W. Photosynthetic carbon partitioning: Its regulation and possibilities for manipulation. *Physiol. Plant* 1989; 77:633–641.

10. Woodrow IE, Berry JA. Enzymatic regulation of photosynthetic CO_2, fixation in C_3 plants. *Ann. Rev. Plant Physiol. Plant Mol. Biol.* 1988; 39:533–594.

11. Terry N, Rao IM. Nutrients and photosynthesis: Iron and phosphorus as case studies. In: Porter JR, Lawor DW, eds. *Plant Growth: Interactions with Nutrition and Environment.* Cambridge: Cambridge University Press, 1991:55–79.

12. Rao IM. Role of phosphorus in photosynthesis. In: Pessarakli M, ed. *Handbook of Photosynthesis.* New York: Marcel Dekker, 1996:173–193.

13. Plaxton WC, Carswell MC. Metabolic aspects of the phosphate starvation response in plants. In: Lerner R, ed. *Plant Responses to Environmental Stresses. From Phytohormones to Genome Reorganization.* New York: Marcel Dekker, 1999:349–372.

14. Rao IM, Terry N. Photosynthetic adaptation to nutrient stress. In: Yunus M, Pathre U, Mohanty P, eds. *Probing Photosynthesis: Mechanism, Regulation and Adaptation.* London: Taylor & Francis, 2000:379–397.

15. Pieters AJ, Paul MJ, Lawlor DW. Low sink demand limits photosynthesis under Pi deficiency. *J. Exp. Bot.* 2001; 52:1083–1091.

16. De Groot CC, Marcelis LFM, Van den Boogaard R, Lambers H. Growth and dry-mass partitioning in tomato as affected by phosphorus nutrition and light. *Plant Cell Environ.* 2001; 24:1309–1317.

17. Foyer C, Spencer C. The relationship between phosphate status and photosynthesis in leaves. Effects on intracellular ortho-phosphate distribution, photosynthesis and assimilate partitioning. *Planta* 1986; 167:369–375.

18. Fredeen AL, Rao IM, Terry N. Influence of phosphorus nutrition on growth and carbon partitioning in glycine max. *Plant Physiol.* 1989; 89:225–230.

19. Rao IM, Terry N. Leaf phosphate status, photosynthesis, and carbon partitioning in sugar-beet I. Changes in growth, gas-exchange, and Calvin cycle enzymes. *Plant Physiol.* 1989; 90:814–819.

20. Qiu J, Israel DW. Carbohydrate accumulation and utilization in soybean plants in response to altered phosphorus nutrition. *Physiol. Plant.* 1994; 90:722–728.

21. Ciereszko I, Gniazdowska A, Mikulska M, Rychter AM. Assimilate translocation in bean plants (*Phaseolus vulgaris* L.) during phosphate deficiency. *J. Plant Physiol.* 1996; 149:343–348.

22. Noctor G, Foyer CH. Homeostasis of adenylate status during photosynthesis in a fluctuating environment. *J. Exp. Bot.* 2000; 51:347–356.

23. Paul MJ, Foyer CH. Sink regulation of photosynthesis. *J. Exp. Bot.* 2001; 52:1383–1400.

24. Li K, Xu C, Li Z, Zhang K, Yang A, Zhang J. Comparative proteome analyses of phosphorus responses in maize (*Zea mays* L.) roots of wild-type and a low-P-tolerant mutant reveal root characteristics associated with phosphorus efficiency. *Plant J.* 2008; 55:927–939.

25. Zhang Z, Liao H, Lucas WJ. Molecular mechanisms underlying phosphate sensing, signaling, and adaptation in plants. *J. Integr. Plant Biol.* 2014; 56:192–220.

26. Walker DA. Photosynthetic induction. In: Akoyonoglou G, ed. *Photosynthesis IV. Regulation of Carbon Metabolism.* Philadelphia: Balaban, 1981:189–202.

27. Lilley R, Chon C, Mosbach A, Heldt H. The distribution of metabolites between spinach chloroplasts and medium during photosynthesis in vitro. *Biochim. Biophys. Acta A* 1977; 460:259–272.

28. Walker DA, Herold A. Can chloroplasts support photosynthesis unaided? In: Fugita Y, Fatoh S, Shibita K, Miyachu S, eds. *Plant Cell Physiol. Special Issue. Photosynthetic Organelles: Structure and Function.* Tokyo: Japanese Society of Plant Physiologists and Centre for Academic Publication, 1977:295–310.

29. Giersch C, Heber U, Kaiser G, Walker DA, Robinson SP. Intracellular metabolite gradients and flow of carbon during photosynthesis of leaf protoplasts. *Arch. Biochem. Biophys.* 1980; 205:246–259.

30. Robinson SP, Walker DA. Photosynthetic carbon reduction cycle. In: Stumpf PK, Conn EE, eds. *The Biochemistry of Plants: A Comprehensive Treatise.* New York: Academic Press, 1981:193–236.

31. Stitt M, Wirtz W, Heldt H. Metabolite levels during induction in the chloroplast and extrachloroplast compartments of spinach protoplasts. *Biochim. Biophys. Acta* 1980; 593:85–102.

32. Marques IA, Anderson LE. Changing kinetic properties of fructose-1,6-bisphosphatase from pea chloroplasts during photosynthetic induction. *Plant Physiol.* 1985; 77:807–810.

33. Marques I, Ford D, Muschinek G, Anderson LE. Photosynthetic carbon metabolism in isolated pea chloroplasts: Metabolite levels and enzyme activities. *Arch. Biochem. Biophys.* 1987; 252:458–466.

34. Leegood RC, Walker DA. Regulation of fructose-1,6-biphosphatase activity in intact chloroplasts. Studies of the mechanism of inactivation. *Biochim. Biophys. Acta* 1980; 593:362–370.

35. Leegood RC, Walker DA. Regulation of the Benson–Calvin cycle. In: Barber J, Barber NR, eds. *Photosynthetic Mechanisms and the Environment.* New York: Elsevier, 1985: 188–258.

36. Kobza J, Edwards GE. The photosynthetic induction response in wheat leaves: Net CO_2 uptake, enzyme activation, and leaf metabolites. *Planta* 1987; 171:549–559.

37. Giersch C, Robinson SP. Regulation of photosynthetic carbon metabolism during phosphate limitation of photosynthesis in isolated spinach-chloroplasts. *Photosynth. Res.* 1987; 14:211–227.

38. Furbank RT, Foyer CH, Walker DA. Regulation of photosynthesis in isolated spinach-chloroplasts during ortho-phosphate limitation. *Biochim. Biophys. Acta* 1987; 894:552–561.

39. Heineke D, Stitt M, Heldt HW. Effects of inorganic phosphate on the light dependent thylakoid energization of intact spinach chloroplasts. *Plant Physiol.* 1989; 91:221–226.

40. Flugge UI. Phosphate translocation in the regulation of photosynthesis. *J. Exp. Bot.* 1995; 46:1317–1323.

41. Flugge UI. Metabolite transporters in plastids. *Curr. Opin. Plant Biol.* 1998; 1:201–205.

42. Flugge UI. Phosphate translocators in plastids. *Ann. Rev. Plant Physiol. Plant Mol. Biol.* 1999; 50:27–45.

43. Usuda H, Edwards GE. Influence of varying CO_2 and ortho-phosphate concentrations on rates of photosynthesis, and synthesis of glycolate and dihydroxyacetone phosphate by wheat chloroplasts. *Plant Physiol.* 1982; 69:469–473.

44. Sharkey TD, Vanderveer PJ. Stromal phosphate concentration is low during feedback limited photosynthesis. *Plant Physiol.* 1989; 91:679–684.

45. Stitt M. Fructose-2,6-bisphosphate as a regulatory molecule in plants. *Ann. Rev. Plant Physiol. Plant Mol. Biol.* 1990; 41:153–185.

46. Heber U, Walker D. The chloroplast envelope, barrier or bridge? *Trends Biochem. Sci.* 1979; 4:252–256.

47. Heber U, Heldt HW. The chloroplast envelope: Structure, function, and role in leaf metabolism. *Ann. Rev. Plant Physiol.* 1981; 32:139–168.

48. Flugge UI, Heldt HW. The phosphate-triose phosphate-phosphoglycerate translocator of the chloroplast. *Trends Biochem. Sci.* 1984; 9:530–533.

49. Heldt HW, Chon CH, Maronde D, Herold A, Stankovic AZ, Walker DA, Kraminer A, Kirk MR, Heber U. Role of ortho-phosphate and other factors in the regulation of starch formation in leaves and isolated chloroplasts. *Plant Physiol.* 1977; 59:1146–1155.

50. Flugge UI, Heldt HW. Metabolite translocators of the chloroplast envelope. *Ann. Rev. Plant Physiol. Plant Mol. Biol.* 1991; 42:129–144.

51. Flugge UI, Fischer K, Gross A. Molecular-cloning and in vitro expression of the chloroplast phosphate translocator protein. *Biol. Chem.* 1989; 370:643–644.

52. Riesmeier JW, Flugge UI, Schulz B, Heineke D, Heldt HW, Willmitzer L, Frommer WB. Antisense repression of the chloroplast triose phosphate translocator affects carbon partitioning in transgenic potato plants. *Proc. Natl. Acad. Sci. U.S.A.* 1993; 90:6160–6164.

53. Rausch C, Bucher M. Molecular mechanisms of phosphate transport in plants. *Planta* 2002; 216:23–37.

54. Versaw WK, Harrison MJ. A chloroplast phosphate transporter, PHT2;1, influences allocation of phosphate within the plant and phosphate-starvation responses. *Plant Cell* 2002; 14:1751–1766.

55. Gerhardt R, Stitt M, Heldt HW. Subcellular metabolite levels in spinach leaves. Regulation of sucrose synthesis during diurnal alternation in photosynthetic partitioning. *Plant Physiol.* 1987; 83:399–407.

56. Barnes SA, Knight JS, Gray JC. Alteration of the amount of the chloroplast phosphate translocator in transgenic tobacco affects the distribution of assimilate between starch and sugar. *Plant Physiol.* 1994; 106:1123–1129.

57. Heldt HW, Chon CJ, Lorimer H. Phosphate requirement for the light activation of ribulose-1,5-biphosphate carboxylase in intact spinach chloroplasts. *FEBS Lett.* 1978; 92:234–240.

58. Bhagwat AS. Activation of spinach ribulose 1,5-bisphosphate carboxylase by inorganic phosphate. *Plant Sci. Lett.* 1981; 23:197–206.

59. Machler F, Nösberger J. Influence of inorganic-phosphate on photosynthesis of wheat chloroplasts. 2. Ribulose bisphosphate carboxylase activity. *J. Exp. Bot.* 1984; 35:488–494.

60. Charles SA, Halliwell B. Properties of freshly purified and thiol-treated spinach chloroplast fructose bisphosphatase. *Biochem. J.* 1980; 185:689–693.

61. Furbank R, Lilley R. Effects of inorganic phosphate on the photosynthetic carbon reduction cycle in extracts from the stroma of pea chloroplasts. *Biochim. Biophys. Acta* 1980; 592:65–75.

62. Gardemann A, Stitt M, Heldt HW. Control of CO_2 fixation. Regulation of spinach ribulose-5-phosphate kinase by stromal metabolite levels. *Biochim. Biophys. Acta* 1983; 722:51–60.

63. Harris GC, Cheesbrough JK, Walker DA. Effects of mannose on photosynthetic gas exchange in spinach leaf discs. *Plant Physiol.* 1983; 71:108–111.

64. Leegood RC, Labate CA, Huber SC, Neuhaus HE, Stitt M. Phosphate sequestration by glycerol and its effects on photosynthetic carbon assimilation by leaves. *Planta* 1988; 176:117–126.

65. Azcon-Bieto J. Inhibition of photosynthesis by carbohydrates in wheat leaves. *Plant Physiol.* 1983; 73:681–686.

66. Hatch AL, Jensen RG. Regulation of ribulose-1,5-bisphosphate carboxylase from tobacco: Changes in pH response and affinity for CO_2 and Mg^{2+} induced by chloroplast intermediates. *Arch. Biochem. Biophys.* 1980; 205:587–594.

67. Jordan DBC, Ogren WL. Binding of phosphorylated effectors by active and inactive forms of ribulose-1,5-biphosphate carboxylase. *Biochemistry* 1983; 22:3410–3418.

68. Badger MR, Lorimer GH. Interaction of sugar phosphates with the catalytic site of ribulose-1,5-bisphosphate carboxylase. *Biochemistry* 1981; 20:2219–2225.

69. Enser U, Heber U. Metabolic regulation by pH gradients. *Biochim. Biophys. Acta* 1980; 592:577–591.

70. Flugge UI, Freisl M, Heldt HW. The mechanism of the control of carbon fixation by the pH in the chloroplast stroma. *Planta* 1980; 149:48–51.

71. Robinson SP, Walker DA. The control of 3-phospho-glycerate reduction in isolated chloroplasts by the concentrations of ATP, ADP and 3-phosphoglycerate. *Biochim. Biophys. Acta* 1979; 545:528–536.

72. Portis AR. Regulation of ribulose 1,5-bisphosphate carboxylase oxygenase activity. *Ann. Rev. Plant Physiol. Plant Mol. Biol.* 1992; 43:415–437.

73. Robinson SP, Giersch C. Inorganic-phosphate concentration in the stroma of isolated-chloroplasts and its influence on photosynthesis. *Aust. J. Plant Physiol.* 1987; 14:451–462.

74. Preiss J. Regulation of the C_3 reductive cycle and carbohydrate synthesis. In: Tolbert NE, Preiss J, eds. *Regulation of Atmospheric CO_2 and O_2 by Photosynthetic Carbon Metabolism.* New York: Oxford University Press, 1994:93–102.

75. Chen-She S-H, Lewis DH, Walker DA. Stimulation of photosynthetic starch formation by sequestration of cytoplasmic orthophosphate. *New Phytol.* 1975; 74:383–392.

76. Nielsen TH, Krapp A, Roper-Schwarz U, Stitt M. The sugar-mediated regulation of genes encoding the small subunit of Rubisco and the regulatory subunit of ADP glucose pyrophosphorylase is modified by phosphate and nitrogen. *Plant Cell Environ.* 1998; 21:443–454.

77. Steup M. Starch degradation. In: Davies DD, ed. *The Biochemistry of Plants.* New York: Academic Press, 1988:255–296.

78. Stitt M, ap Rees T. Estimation of the activity of the oxidative pentosephosphate pathway in pea chloroplasts. *Phytochemistry* 1980; 19:1583–1585.

79. Stitt M, Rees AA. Carbohydrate breakdown by chloroplasts of *Pisum sativum. Biochim. Biophys. Acta* 1980; 627:131–143.

80. Kelly GJ, Latzko E. Chloroplast phosphofructokinase. *Plant Physiol.* 1977; 60:290–294.

81. Stitt M, Heldt HW. Physiological rates of starch breakdown in isolated intact spinach chloroplasts. *Plant Physiol.* 1981; 68:755–761.

82. Dennis DT, Miernyk JA. Compartmentation of non-photosynthetic carbohydrate metabolism. *Ann. Rev. Plant Physiol.* 1982; 33:27–50.

83. Pettersson G, Ryde-Pettersson U. Metabolites controlling the rate of starch synthesis in chloroplast of C_3 plants. *Eur. J. Biochem.* 1989; 179:169–172.

84. Kacser H. Control of metabolism. In: Davies DD, ed. *The Biochemistry of Plants.* New York: Academic Press, 1987:39–67.

85. ap Rees T. Sucrose metabolism. In: Lewis DH, ed. *Storage Carbohydrates in Vascular Plants.* Cambridge: Cambridge University Press, 1984:53–73.

86. Winter H, Huber SC. Regulation of sucrose metabolism in higher plants: Localization and regulation of activity of key enzymes. *Crit. Rev. Biochem. Mol. Biol.* 2000; 35:253–289.

87. Wardlaw IF. The control of carbon partitioning in plants. *New Phytol.* 1990; 116:341–381.

88. Huber SC, Huber JL. Role of sucrose-phosphate synthase in sucrose metabolism in leaves. *Plant Physiol.* 1992; 99:1275–1278.

89. Scott P, Lange AJ, Pilkis SJ, Kruger NJ. Carbon metabolism in leaves of transgenic tobacco (*Nicotiana tabacum* L.) containing elevated fructose 2,6-bisphosphate levels. *Plant J.* 1995; 7:461–469.

90. Huber SC. Role of sucrose-phosphate in partitioning of carbon in leaves. *Plant Physiol.* 1983; 71:818–821.

91. Stitt M, Wirtz W, Heldt HW. Regulation of sucrose synthesis by cytoplasmic fructosebisphosphatase and sucrose phosphate synthase during photosynthesis in varying light and carbon-dioxide. *Plant Physiol.* 1983; 72:767–774.

92. Stitt M, Gerhardt R, Kurzel B, Heldt HW. A role for fructose 2,6-bisphosphate in the regulation of spinach leaves. *Plant Physiol.* 1983; 72:1139–1141.

93. Stitt M. Control analysis of photosynthetic sucrose synthesis—Assignment of elasticity coefficients and flux-control coefficients to the cytosolic fructose 1,6-bisphosphatase and sucrose phosphate synthase. *Philos. Trans. R. Soc. (Lond.) Ser. B—Biol. Sci.* 1989; 323:327–338.

94. Huber SC, Huber JL. Role and regulation of sucrose-phosphate synthase in higher plants. *Ann. Rev. Plant Physiol. Plant Mol. Biol.* 1996; 47:431–444.

95. Strand A, Zrenner R, Trevanion S, Stitt M, Gustafsson P, Gardestrom P. Decreased expression of two key enzymes in the sucrose biosynthesis pathway, cytosolic fructose-1,6-bisphosphatase and sucrose phosphate synthase, has remarkably different consequences for photosynthetic carbon metabolism in transgenic *Arabidopsis thaliana*. *Plant J.* 2000; 23:759–770.

96. Volkert K, Debast S, Voll LM, Voll H, Schießl I, Hofmann J, Schneider S, Börnke F. Loss of the two major leaf isoforms of sucrose-phosphate synthase in *Arabidopsis thaliana* limits sucrose synthesis and nocturnal starch degradation but does not alter carbon partitioning during photosynthesis. *J. Exp. Bot.* 2014; 65:5217–5229.

97. Stitt M, Cseke C, Buchanan BB. Regulation of fructose 2,6-bisphosphate concentration in spinach leaves. *Eur. J. Biochem.* 1984; 143:89–93.

98. Stitt M, Herzog B, Heldt HW. Control of photosynthetic sucrose synthesis by fructose-2,6-bisphosphate. Coordination of CO_2 fixation and sucrose synthesis. *Plant Physiol.* 1984; 75:548–553.

99. Stitt M, Herzog B, Heldt HW. Control of photosyntheti sucrose synthesis by fructose 2,6-bisphosphate. V. Modulation of the spinach leaf cytosolic fructose 1,6-bisphosphatase activity in vitro by substrate, products, pH, magnesium, fructose 2,6-bisphosphate, adenosine-monophosphate, and dihydroxyacetone phosphate. *Plant Physiol.* 1984; 79:590–598.

100. Stitt M, Gerhardt, R, Wilke I, Heldt HW. The contribution of fructose-2,6-bisphosphate to the regulation of sucrose synthesis during photosynthesis. *Physiol. Plant.* 1987; 69:377–386.

101. Cierszko I, Johansson H, Kleczkowski L. Sucrose and light regulation of a cold-inducible UDP-glucose pyrophosphorylase gene via a hexokinase-independent and abscisic acid-insensitive pathway in *Arabidopsis*. *Biochem. J.* 2001; 354:67–72.

102. Cierszko I, Johansson H, Hurry V, Kleczkowski LA. Phosphate status affects the gene expression, protein content and enzymatic activity of UDP-glucose pyrophosphorylase in wild-type and pho mutants of *Arabidopsis*. *Planta* 2001; 212:598–605.

103. Doehlhert DC, Huber SC. Phosphate inhibition of spinach leaf sucrose phosphate synthase as affected by glucose-6-phosphate and phosphoglucoisomerase. *Plant Physiol.* 1984; 76:250–253.

104. Kerr PS, Rufty TW, Huber SC. Endogenous rhythms in photosynthesis, sucrose phosphate synthase activity, and stomatal-resistance in leaves of soybean (*Glycine max* L. Merr.). *Plant Physiol.* 1985; 77:275–280.

105. Kerr PS, Huber SC. Coordinate control of sucrose formation in soybean leaves by sucrose-phosphate synthase and fructose-2,6-bisphosphate. *Planta* 1987; 170:197–204.

106. Walker JL, Huber SC. Regulation of sucrose-phosphate synthase activity in spinach leaves by protein level and covalent modifications. *Planta* 1989; 177:116–120.

107. Huber SC, Huber JL, Pharr DM. Variation among species in light activation of sucrose-phosphate synthase. *Plant Cell Physiol.* 1989; 30:277–285.

108. Huber SC. Biochemical mechanism for regulation of sucrose accumulation in leaves during photosynthesis. *Plant Physiol.* 1989; 91:656–662.

109. Worrell AC, Bruneau JM, Summerfelt K, Boersig M, Voelker TA. Expression of a maize sucrose phosphate synthase in tomato alters leaf carbohydrate partitioning. *Plant Cell* 1991; 3:1121–1130.

110. Rocher JP, Prioul JL, Lecharny A, Reyss A, Joussaume M. Genetic variability in carbon fixation, sucrose-P-synthase and ADP glucose pyrophosphorylase in maize plants of differing growth rate. *Plant Physiol.* 1989; 89:416–420.

111. Preiss J, Levi C. Starch biosynthesis and degradation. In: Preiss J, ed. *The Biochemistry of Plants: A Comprehensive Treatise. Carbohydrates: Structure and Function.* New York: Academic Press, 1980:371–423.

112. Woodrow IEM, Walker DA. Regulation of photosynthetic carbon metabolism. The effect of inorganic phosphate on stromal sedoheptulose-1,7-bisphosphatase. *Eur. J. Biochem.* 1983; 132:121–126.

113. Harbron S, Foyer CH, Walker DA. The purification and properties of sucrose phosphate synthetase from spinach leaves: The involvement of this enzyme and fructose bisphosphatase in the regulation of sucrose biosynthesis. *Arch. Biochem. Biophys.* 1981; 212:237–246.

114. Cseke C, Buchanan BB. An enzyme synthesizing fructose-2,6-bis-phosphate occurs in leaves and is regulated by metabolic effectors. *FEBS Lett.* 1983; 155:139–142.

115. Huber SC. Interspecific variation in activity and regulation of leaf sucrose-phosphate synthase. *Z. Pflanzenphysiol.* 1981; 102:443–450.

116. Rufty TW, Huber SC. Changes in starch formation and activities of sucrose phosphate synthase and cytoplasmic fructose-1,6-bisphosphatase in response to source-sink alternations. *Plant Physiol.* 1983; 72:474–480.

117. Sharkey TD, Stitt M, Heineke D, Gerhardt R, Raschke K, Heldt HW. Limitation of photosynthesis by carbon metabolism. II. O_2 insensitive CO_2 uptake results from limitation of triose phosphate utilization. *Plant Physiol.* 1986; 81:1123–1129.

118. Leegood RC, Furbank RT. Stimulation of photosynthesis by 2-percent oxygen at low-temperatures is restored by phosphate. *Planta* 1986; 168:84–93.

119. Stitt M. Limitation of photosynthesis by carbon metabolism. I. Evidence for excess electron transport capacity in leaves carrying out photosynthesis in saturating light and CO_2. *Plant Physiol.* 1986; 81:1115–1122.

120. Rao IM, Abadia J, Terry N. Leaf phosphate status and photosynthesis in vivo: Changes in light-scattering and chlorophyll fluorescence during photosynthetic induction in sugar-beet leaves. *Plant Sci.* 1986; 44:133–137.

121. Abadia J, Rao IM, Terry N. Changes in leaf phosphate status have only small effects on the photochemical apparatus of sugar beet leaves. *Plant Sci.* 1987; 50:49–55.

122. Brooks A, Woo KC, Wong SC. Effects of phosphorus nutrition on the response of photosynthesis to CO_2 and O_2, activation of ribulose bisphosphate carboxylase and amounts of ribulose bisphosphate and 3-phosphoglycerate in spinach leaves. *Photosynth. Res.* 1988; 15:133–141.

123. Rao IM, Arulanantham AR, Terry N. Leaf phosphate status, photosynthesis and carbon partitioning in sugar beet II. Diurnal changes in sugar phosphates, adenylates, and nicotinamide nucleotides. *Plant Physiol.* 1989; 90:820–826.

124. Lauer MJ, Blevins DG, Sierzputowska-Gracz H. ^{31}P-Nuclear magnetic resonance determination of phosphate compartmentation in leaves of reproductive soybeans (*Glycine max* L.) as affected by phosphate nutrition. *Plant Physiol.* 1989; 89:1331–1336.

125. Fredeen AL, Raab TK, Rao IM, Terry N. Effects of phosphorus nutrition on photosynthesis in *Glycine max*. *Planta* 1990; 181:399–405.

126. Rao IM, Fredeen AL, Terry N. Leaf phosphate status, photosynthesis, and carbon partitioning in sugar beet III. Diurnal changes in carbon partitioning and carbon export. *Plant Physiol.* 1990; 92:29–36.

127. Jacob J, Lawlor DW. Stomatal and mesophyll limitations of photosynthesis in phosphate deficient sunflower, maize and wheat plants. *J. Exp. Bot.* 1991; 42:1003–1011.

128. Jacob J, Lawlor DW. Dependence of photosynthesis of sunflower and maize leaves on phosphate supply, ribulose-1,5-bisphosphate carboxylase oxygenase activity, and ribulose-1,5-bisphosphate pool size. *Plant Physiol.* 1992; 98:801–807.

129. Rao IM, Fredeen AL, Terry N. Influence of phosphorus limitation on photosynthesis, carbon allocation and partitioning in sugar beet and soybean grown with a short photoperiod. *Plant Physiol. Biochem.* 1993; 31:223–231.

130. Plesnicar M, Kastori R, Petrovic N, Pankovic D. Photosynthesis and chlorophyll fluorescence in sunflower (*Helianthus annuus* L) leaves as affected by phosphorus-nutrition. *J. Exp. Bot.* 1994; 45:919–924.

131. Rao IM, Terry N. Leaf phosphate status, photosynthesis, and carbon partitioning in sugar-beet IV. Changes with time following increased supply of phosphate to low-phosphate plants. *Plant Physiol.* 1995; 107:1313–1321.

132. Lynch J, Lauchli A, Epstein E. Vegetative growth of the common bean in response to phosphorus nutrition. *Crop Sci.* 1991; 31:380–387.

133. Qiu J, Israel DW. Diurnal starch accumulation and utilization in phosphorus-deficient soybean plants. *Plant Physiol.* 1992; 98:316–323.

134. Halsted M, Lynch J. Phosphorus responses of C-3 and C-4 species. *J. Exp. Bot.* 1996; 47:497–505.

135. Plenet D, Etchebest S, Mollier A, Pellerin S. Growth analysis of maize field crops under phosphorus deficiency I. Leaf growth. *Plant Soil* 2000; 223:117–130.

136. Reymond M, Svistoonoff S, Loudet O, Nussaume L, Desnos T. Identification of QTL controlling root growth response to phosphate starvation in *Arabidopsis thaliana*. *Plant Cell Environ.* 2006; 29:115–125.

137. Jiang CF, Gao XH, Liao L, Harberd NP, Fu XD. Phosphate starvation, root architecture and anthocyanin accumulation responses are modulated by the gibberellin-DELLA signaling pathway in *Arabidopsis*. *Plant Physiol.* 2007; 145: 1460–1470.

138. Secco D, Jabnoune M, Walker H, Shou H, Wu P, Poirier Y, Whelan J. Spatio-temporal transcript profiling of rice roots and shoots to phosphate starvation and recovery. *Plant Cell.* 2013; 25:4285–4304.

139. Rao IM, Borrero V, Ricaurte J, Garcia R, Ayarza MA. Adaptive attributes of tropical foliage species to acid soils III. Differences in phosphorus acquisition and utilization as influenced by varying phosphorus supply and soil type. *J. Plant Nutr.* 1997; 20:155–180.

140. Rao IM, Friesen DK, Osaki M. Plant adaptation to phosphorus-limited tropical soils. In: Pessarakli M, ed. *Handbook of Plant and Crop Stress*. New York: Marcel Dekker, 1999:61–96.

141. Mollier A, Pellerin S. Maize root system growth and development as influenced by phosphorus deficiency. *J. Exp. Bot.* 1999; 50:487–497.

142. Ciereszko I, Janonis A, Kociakowska M. Growth and metabolism of cucumber in phosphate-deficient conditions. *J. Plant Nutr.* 2002; 25:1115–1127.

143. Vance CP, Uhde-Stone C, Allen DL. Phosphorus acquisition and use: Critical adaptations by plants for securing a nonrenewable resource. *New Phytol.* 2003; 157:423–447.

144. Chacón-López A, Ibarra-Laclette E, Sanchéz Calderón L, Gutiérez-Alanís D, Herrea-Estrella L. Global expression pattern comparison between low phosphorus insensitive 4 and WT Arabidopsis reveals an important role of reactive oxygen species and jasmonic acid in the root tip response to phosphate starvation. *Plant Signal. Behav.* 2011; 6:382–392.

145. Gniazdowska A, Mikulska M, Rychter AM. Growth, nitrate uptake and respiration rate in bean roots under phosphate deficiency. *Biol. Plant* 1998; 41:217–226.

146. Nielsen K, Eshel A, Lynch JP. The effect of phosphorus availability on the carbon economy of contrasting common bean (*Phaseolus vulgaris* L.) genotypes. *J. Exp. Bot.* 2001; 52:329–339.

147. Hogh-Jensen H, Schjoerring JK, Soussana J-F. The influence of phosphorus deficiency on growth and nitrogen fixation of white clover plants. *Ann. Bot.* 2002; 90:745–753.

148. Stitt M. Rising CO_2 levels and their potential significance for carbon flow in photosynthetic cells. *Plant Cell Environ.* 1991; 14:741–762.

149. Radin JW, Eidenbock MP. Carbon accumulation during photosynthesis in leaves of nitrogen- and phosphorus-stressed cotton. *Plant Physiol.* 1986; 82:869–871.

150. Hart AL, Greer DH. Photosynthesis and carbon export in white clover plants grown at various levels of phosphorus supply. *Physiol. Plant.* 1988; 73:46–51.

151. Rodriguez D, Zubillaga MM, Ploschuk EL, Keltjens WG, Goudriaan A, Lavado RS. Leaf area expansion and assimilate production in sunflower (*Helianthus annuus* L.) growing under low phosphorus conditions. *Plant Soil* 1998; 202:133–147.

152. Chiera J, Rufty T. Leaf initiation and development in soybean under phosphorus stress. *J. Exp. Bot.* 2002; 53:473–481.

153. Wang C, Ying S, Huang H, Li K, Wu P, Shou H. Involvement of OsSPX1 in phosphate homeostasis in rice. *Plant J.* 2009; 57:895–904.

154. Secco D, Baumann A, Poirier Y. Characterization of the rice PHO1 gene family reveals a key role for OsPHO1;2 in phosphate homeostasis and the evolution of a distinct clade in dicotyledons. *Plant Physiol.* 2010; 152:1693–1704.

155. Jia H, Ren H, Gu M, Zhao J, Sun S, Zhang X, Chen J, Wu P, Xu G. The phosphate transporter gene OsPht1;8 is involved in phosphate homeostasis in rice. *Plant Physiol.* 2011; 156:1164–1175.

156. Rouached H, Stefanovic A, Secco D, Bulak Arpat A, Gout E, Bligny R, Poirier Y. Uncoupling phosphate deficiency from its major effects on growth and transcriptome via PHO1 expression in *Arabidopsis*. *Plant J.* 2011; 65:557–570.

157. Schachtman DP, Reid RJ, Ayling SM. Phosphorus uptake by plants. From soil to cell. *Plant Physiol.* 1998; 116:447–453.

158. Mimura T, Sakano K, Shimmen T. Studies on distribution, retranslocation and homeostasis of inorganic phosphate in barley leaves. *Plant Cell Environ.* 1996; 19:311–320.

159. Raghothama KG. Phosphate acquisition. *Ann. Rev. Plant Physiol. Plant Mol. Biol.* 1999; 50:665–693.

160. Clarkson DT, Carvajal M, Henzler T, Waterhouse RN, Smyth AJ, Cooke, DT, Steudle E. Root hydraulic conductance; diurnal aquaporin expression and the effects of nutrient stress. *J. Exp. Bot.* 2000; 51:61–70.

161. Abel S. Phosphate sensing in root development. *Curr. Opin. Plant. Biol.* 2011; 14:303–309.

162. Radin JW, Mathews MA. Water transport properties of cortical cells in roots of nitrogen- and phosphorus-deficient cotton seedlings. *Plant Physiol.* 1989; 89:264–268.

163. Gniazdowska A, Szal B, Rychter AM. The effect of phosphate deficiency on membrane phospholipid composition of bean (*Phaseolus vulgaris* L.) roots. *Acta Physiol. Plant.* 1999; 21:263–269.

164. Kondracka A, Rychter AM. The role of Pi recycling processes during photosynthesis in phosphate-deficient bean plants. *J. Exp. Bot.* 1997; 48:1461–1468.

165. Zhang H, Huang Y, Ye X, Shi L, Xu F. Genotypic differences in phosphorus acquisition and the rhizosphere properties of *Brassica napus* in response to low phosphorus stress. *Plant Soil.* 2009; 320:91–102.

166. Reich PB, Oleksyn J, Wright IJ. Leaf phosphorus influences the photosynthesis–nitrogen relation: A cross-biome analysis of 314 species. *Oecologia* 2009; 160:207–212.

167. Suriyagoda LD, Lambers H, Ryan MH, Renton M. Effects of leaf development and phosphorus supply on the photosynthetic characteristics of perennial legume species with pasture potential: Modelling photosynthesis with leaf development. *Funct. Plant. Biol.* 2010; 37:713–725.

168. Zhang K, Liu H, Tao P, Chen H. Comparative proteomic analysis provide new insights into low phosphorus stress responses in maize leaves. *PLoS One* 2014; 9:e98215.

169. Essigmann B, Guler S, Narang RA, Linke D, Benning C. Phosphate availability affects the thylakoid lipid composition and the expression of SQD1, a gene required for sulfolipid biosynthesis in *Arabidopsis thaliana. Proc. Natl. Acad. Sci. U.S.A.* 1998; 95:1950–1955.

170. Andersson X, Stridh MH, Larsson KE, Liljenberg C, Sandelius AS. Phosphate-deficient oat replaces a major portion of the plasma membrane phospholipids with the galactolipid digalactosyldiacylglycerol. *FEBS Lett.* 2003; 537:128–132.

171. Morcuende R, Bari R, Gibon Y, Zheng W, Pant BD, Bläsing O, Usadel B, Czechowski T, Udvardi MK, Stitt M, Scheible W-R. Genome-wide reprogramming of metabolism and regulatory networks of *Arabidopsis* in response to phosphorus. *Plant Cell Environ.* 2007; 30:85–112.

172. Calderon-Vazquez C, Ibarra-Laclette E, Caballero-Perez J, Herrera-Estrella L. Transcript profiling of *Zea mays* roots reveals gene responses to phosphate deficiency at the plant- and species-specific levels. *J. Exp. Bot.* 2008; 59:2479–2497.

173. Hammond JP, Broadley MR, Bowen HC, Spracklen WP, Hayden RM, White PJ. Gene expression changes in phosphorus deficient potato (*Solanum tuberosum* L.) leaves and the potential for diagnostic gene expression markers. *PLoS One* 2011; 6:e24606.

174. Krause GH, Weis E. Chlorophyll fluorescence and photosynthesis: The basics. *Ann. Rev. Plant Physiol. Plant Mol. Biol.* 1991; 42:313–349.

175. Horton P. Interactions between electron transport and carbon assimilation: Regulation of light harvesting. In: Briggs WR, ed. *Photosynthesis.* New York: Alan R. Liss, 1989:393–406.

176. Jacob J, Lawlor DW. In vivo photosynthetic electron- transport does not limit photosynthetic capacity in phosphate-deficient sunflower and maize leaves. *Plant Cell Environ.* 1993; 16:785–795.

177. Jacob J, Lawlor DW. Extreme phosphate deficiency decreases the in vivo CO_2/O_2 specificity factor of ribulose 1,5-bisphosphate carboxylase-oxygenase in intact leaves of sunflower. *J. Exp. Bot.* 1993; 44:1635–1641.

178. Singh P, Kumar P, Abrol Y, Naik M. Photorespiratory nitrogen cycle—A critical evaluation. *Physiol. Plant.* 1986; 66:169–176.

179. Heber U, Viil J, Neimanis S, Mimura T, Dietz KJ. Photoinhibitory damage to chloroplasts under phosphate deficiency and alleviation of deficiency and damage by photorespiratory reactions. *Z. Naturforsc. J. Biosci.* 1989; 44:524–536.

180. Kozlowska B, Maleszewski S. Low-level of inorganic orthophosphate in growth-medium increases metabolism and excretion of glycolate by chlorella-vulgaris cells cultivated under air conditions. *Plant Physiol. Biochem.* 1994; 32:717–721.

181. Kozlowska-Szerenos B, Zielinski P, Maleszewski S. Involvement of glycolate metabolism in acclimation of *Chlorella vulgaris* cultures to low phosphate supply. *Plant Physiol. Biochem.* 2000; 38:727–734.

182. Hauschild T, Ciereszko I, Maleszewski S. Influence of phosphorus deficiency on post-irradiation burst of CO_2 from bean (*Phaseolus vulgaris* L.) leaves. *Photosynthetica* 1996; 32:1–9.

183. Sicher RC, Kremer DF. Effects of phosphate deficiency on assimilate partitioning in barley seedlings. *Plant Sci.* 1988; 57:9–17.

184. Usuda H, Shimogawara K. Phosphate deficiency in maize. 2. Enzyme activities. *Plant Cell Physiol.* 1991; 32:1313–1317.

185. Rychter AM, Randall DD. The effect of phosphate deficiency on carbohydrate metabolism in bean roots. *Physiol. Plant.* 1994; 91:383–388.

186. Juszczuk IM, Rychter AM. Changes in pyridine nucleotide levels in leaves and roots of bean plants (*Phaseolus vulgaris* L.) during phosphate deficiency. *J. Plant Physiol.* 1997; 151:399–404.

187. Mikulska M, Bomsel JL, Rychter AM. The influence of phosphate deficiency on photosynthesis, respiration and adenine nucleotide pool in bean leaves. *Photosynthetica* 1998; 35:79–88.

188. Krömer S. Respiration during photosynthesis. *Ann. Rev. Plant Physiol. Plant Mol. Biol.* 1995; 46:45–70.

189. Rychter AM, Mikulska M. The relationship between phosphate status and cyanide-resistant respiration in bean roots. *Physiol. Plant.* 1990; 79:663–667.

190. Rychter AM, Chauveau M, Bomsel JL, Lance C. The effect of phosphate deficiency on mitochondrial activity and adenylate levels in bean roots. *Physiol. Plant.* 1992; 84:80–86.

191. Juszczuk IM, Wagner AM, Rychter AM. Regulation of alternative oxidase activity during phosphate deficiency in bean roots (*Phaseolus vulgaris*). *Physiol. Plant.* 2001; 113:185–192.

192. Gonzalez-Meler MA, Giles L, Thomas RB, Siedow JN. Metabolic regulation of leaf respiration and alternative pathway activity in response to phosphate supply. *Plant Cell Environ.* 2001; 24:205–215.

193. Foyer CH, Furbank R, Harbinson J, Horton P. The mechanism contributing to photosynthetic control of electron transport by carbon assimilation in leaves. *Photosynth. Res.* 1990; 25:83–100.

194. Sage RF. A model describing the regulation of ribulose-1,5-bisphosphate carboxylase, electron-transport, and triose phosphate use in response to light-intensity and CO_2 in C-3 plants. *Plant Physiol.* 1990; 94:1728–1734.

195. Bowes G. Facing the inevitable: Plants and increasing atmospheric CO_2. *Ann. Rev. Plant Physiol. Plant Mol. Biol.* 1993; 44:309–332.

196. Geiger DR, Servaites JC. Diurnal regulation of photosynthetic carbon metabolism in C3 plants. *Ann. Rev. Plant Physiol. Plant Mol. Biol.* 1994; 45:235–256.

197. Servaites JC, Shieh WJ, Geiger DR. Regulation of photosynthetic carbon reduction cycle by ribulose bisphosphate and phosphoglyceric acid. *Plant Physiol.* 1991; 97:1115–1121.

198. Bieleski RL. Phosphate pools, phosphate transport, and phosphate availability. *Ann. Rev. Plant Physiol.* 1973; 24:225–252.

199. Hampp R. Rapid separation of the plastid, mitochondrial, and cytoplasmic fractions from intact leaf protoplasts of *Avena*. *Planta* 1980; 150:291–298.

200. Wirtz W, Stitt M, Heldt HW. Enzymic determination of metabolites in the subcellular compartments of spinach protoplasts. *Plant Physiol.* 1980; 66:187–193.

201. Lilley RMC, Stitt M, Mader G, Heldt HW. Rapid fractionation of wheat leaf protoplasts using membrane filtration. *Plant Physiol.* 1982; 70:965–970.

202. Hamp R, Goller M, Zeigler H. Adenylate levels, energy charge and phosphorylation potential during dark–light and light–dark transition in chloroplasts, mitochondria, and cytosol of mesophyll protoplasts from *Avena sativa* L. *Plant Physiol.* 1982; 69:448–455.

203. Ratcliffe RG. In vivo NMR studies of higher plants and algae. *Adv. Bot. Res.* 1994; 20:43–123.

204. Lee RB, Ratcliffe RG, Southon TE. ^{31}P-NMR measurement of the cytosolic and vacuolar Pi content of mature roots: Relationship with phosphorus status and phosphate fluxes. *J. Exp. Bot.* 1990; 41:1063–1078.

205. Lee RB, Ratcliffe RG. Subcellular distribution of inorganic phosphate and levels of nucleoside triphosphate in mature maize roots at low external phosphate concentrations: Measurements with ^{31}P-NMR. *J. Exp. Bot.* 1993; 44:587–598.

206. Foyer CH, Walker D, Spencer C, Mann B. Observations on the phosphate status and intracellular pH of intact cells, protoplasts and chloroplasts from photosynthetic tissue using phosphorus-31 nuclear magnetic resonance. *Biochem. J.* 1982; 202:429–434.

207. Rebeille F, Blingy R, Martin J-B, Douce R. Relationship between the cytoplasm and vacuole phosphate pool in *Acer pseudoplatanus* cells. *Arch. Biochem. Biophys.* 1983; 225: 143–148.

208. Waterton JC, Bridges IA, Irwing MP. Intracellular compartmentation detected by ^{31}P-NMR in intact photosynthetic wheat-leaf tissue. *Biochim. Biophys. Acta* 1983; 763:315–320.

209. Lee RB, Ratcliffe RG. Phosphorus nutrition and the intracellular if inorganic phosphate in pea root tips: A quantitative study using ^{31}P-NMR. *J. Exp. Bot.* 1983; 34:1222–1224.

210. Mitsumori F, Ito I. Phosphorus-31 nuclear magnetic resonance studies of photosynthesizing *Chlorella*. *FEBS Lett.* 1984; 174:248–252.

211. Bligny R, Foray M, Roby C, Douce R. Transport and phosphorylation of choline in higher plant cells. Phosphorus-31 nuclear magnetic resonance studies. *J. Biol. Chem.* 1989; 264:4888–4895.

212. Loughman BC, Ratcliffe RG, Southon TE. Observations on the cytoplasmic and vacuolar orthophosphate pools in leaf tissues using in vivo ^{31}P-NMR spectroscopy. *FEBS Lett.* 1989; 242:279–284.

213. Lundberg P, Weich RG, Jensen P, Vogel HG. Phosphorus-31 and nitrogen-14 NMR studies of the uptake of phosphorus and nitrogen compounds in the marine microalgae *Ulva lactuca*. *Plant Physiol.* 1989; 89:1380–1387.

214. Bligny R, Gardestrom P, Roby C, Douce R. ^{31}P NMR studies of spinach leaves and their chloroplasts. *J. Biol. Chem.* 1990; 256:1319–1326.

215. Mimura T, Dietz K-J, Kaiser W, Schramm MJ, Kaiser G, Heber U. Phosphate transport across biomembranes and cytosolic phosphate homeostasis in barley leaves. *Planta* 1990; 180:139–146.

216. Hentrich S, Hebeler M, Grimme LH, Leibfritz D, Mayer A. ^{31}P-NMR Saturation-transfer experiments in *Chlamydomonas reinhardtii*—Evidence for the NMR visibility of chloroplastidic-Pi. *Eur. Biophys. J., Biophys. Lett.* 1993; 22:31–39.

217. Lee RB, Ratcliffe RG. Nuclear magnetic resonance studies of the location and function of plant nutrients in vivo. *Plant Soil* 1993; 155/156:45–55.

218. Usuda H, Shimogawara K. Phosphate deficiency in maize. 1. Leaf phosphate status, growth, photosynthesis and carbon partitioning. *Plant Cell Physiol.* 1991; 32:497–504.

219. Wang C, Tillberg J-E. Effects of short-term phosphorus deficiency on carbohydrate storage in sink and source leaves of barley (*Hordeum vulgare*). *New Phytol.* 1997; 136:131–135.

220. Cakmak I, Hengeler C, Marschner H. Partitioning of shoot and root dry matter and carbohydrates in bean plants suffering from phosphorus, potassium and magnesium deficiency. *J. Exp. Bot.* 994; 45:1245–1250.

221. Ciereszko I, Milosek I, Rychter AM. Assimilate distribution in bean plants (*Phaseolus vulgaris* L.) during phosphate limitation. *Acta Soc. Bot. Poloniae* 1999; 68:269–273.

222. Ciereszko I, Farrar JF, Rychter AM. Compartmentation and fluxes of sugars in roots of *Phaseolus vulgaris* under phosphate deficiency. *Biol. Plant* 1999; 42:223–231.

223. Ciereszko I, Zambrzycka A, Rychter A. Sucrose hydrolysis in bean roots (*Phaseolus vulgaris* L.) under phosphate deficiency. *Plant Sci.* 1998; 133:139–144.

224. Wanke M, Ciereszko I, Podbielkowska M, Rychter AM. Response to phosphate deficiency in bean (*Phaseolus vulgaris* L.) roots. Respiratory metabolism, sugar localization and changes in ultrastructure of bean root cells. *Ann. Bot.* 1998; 82:809–819.

225. Dietz KJ. Recovery of spinach leaves from sulfate and phosphate deficiency. *J. Plant Physiol.* 1989; 134:551–557.

226. Cogliatti DH, Clarkson DT. Physiological-changes and phosphate-uptake by potato plants during development of, and recovery from phosphate efficiency. *Physiol. Plant.* 1983; 58:287–294.

227. Drew MC, Saker LR, Barber SA, Jenkins W. Changes in the kinetics of phosphate and potassium absorption in nutrient-deficient barley roots measured by solution-depletion technique. *Planta* 1984; 1984:490–499.

228. Jungk A, Asher CJ, Edwards DG, Mayer D. Influence of phosphate status on phosphate uptake kinetics of maize (*Zea mays*) and soybean (*Glycine max*). *Plant Soil* 1990; 124:175–182.

229. McGraw JB, Wulf RD. The study of plant growth: A link between the physiological ecology and population biology of plants. *J. Theor. Biol.* 1983; 103:21–28.

230. Shin R, Berg H, Schachtman DP. Reactive oxygen species and root hairs in *Arabidopsis* root response to nitrogen, phosphorus and potassium deficiency. *Plant Cell Physiol.* 2005; 46:1350–1357.

231. Tran HT, Plaxton WC. Proteomic analysis of alterations in the secretome of *Arabidopsis thaliana* suspension cells subjected to nutritional phosphate deficiency. *Proteomics* 2008; 8:4317–4326.

232. Torabi S, Wissuwa M, Heidari M, Naghavi MR, Gilany K, Hajirezaei MR, Omidi M, Yazdi-Samadi B, Ismail AM, Salekdeh GH. A comparative proteome approach to decipher the mechanism of rice adaptation to phosphorous deficiency. *Proteomics* 2009; 9:159–170.

233. Körner C. Plant CO_2 responses: An issue of definition, time and resource supply. *New Phytol.* 2006; 172:393–411.

234. Lewis JD, Ward JK, Tissue DT. Phosphorus supply drives nonlinear responses of cottonwood (*Populus deltoides*) to increases in CO_2 concentration from glacial to future concentrations. *New Phytol.* 2010; 187:438–448.

235. Tissue DT, Lewis JD, Niinemets Ü. Photosynthetic responses of cottonwood seedlings grown in glacial through future atmospheric [CO_2] vary with phosphorus supply. *Tree Physiol.* 2010; 30:1361–1372.

236. Field CB, Matson PA, Mooney HA. Responses of terrestrial ecosystems to a changing atmosphere. A resource-based approach. *Annu. Rev. Ecol. Syst.* 1992; 23:201–235.

237. Lewis JD, Griffin KL, Thomas RB, Strain BR. Phosphorus supply affects the acclimation of photosynthesis in loblolly pine to elevated carbon dioxide. *Tree Physiol.* 1994; 14:1229–1244.

238. Campbell CD, Sage RE. Interactions between the effects of atmospheric CO_2 content and P nutrition on photosynthesis in white lupin (*Lupinus albus* L.). *Plant Cell Environ.* 2006; 29:844–853.

239. Flesiher DH, Wang Q, Timlin DJ, Chun J-A, Reddy VR. Response of potato gas exchange and productivity to phosphorus deficiency and carbon dioxide enrichment. *Crop Sci.* 2011; 52:1803–1815.

240. Zhou YH, Wu JX, Zhu LJ, Shi K, Yu JQ. Effects of phosphorus and chilling under low irradiance on photosynthesis and growth of tomato plants. *Biol. Plant.* 2009; 53:378–382.

241. Hendrickson L, Chow WS, Furbank RT. Low temperature effects on grapevine photosynthesis: The role of inorganic phosphate. *Funct. Plant Biol.* 2004; 31:789–801.

242. Hurry V, Strand A, Furbank R, Stitt M. The role of inorganic phosphate in the development of freezing tolerance and the acclimatization of photosynthesis to low temperature is revealed by the pho mutants of *Arabidopsis thaliana*. *Plant J.* 2000; 24:383–396.

243. Cheng L, Tang X, Vance CP, White PJ, Zhang F, Shen J. Interactions between light intensity and phosphorus nutrition affect the phosphate mining capacity of white lupin (*Lupinus albus* L.). *J. Exp. Bot.* 2014; 65:2995–3003.

244. Nilsson L, Lundmark M, Jensen PE, Nielsen TH. The *Arabidopsis* transcription factor PHR1 is essential for adaptation to high light and retaining functional photosynthesis during phosphate starvation. *Physiol. Plant.* 2012; 144:35–47.

245. Zribi OT, Abdelly C, Debez A. Interactive effects of salinity and phosphorus availability on growth, water relations, nutritional status and photosynthetic activity of barley (*Hordeum vulgare* L.). *Plant Biol.* 2011; 13:872–880.

246. Shubhra, Dayal J, Goswami CL, Munjal R. Influence of phosphorus application on water relations, biochemical parameters and gum content in cluster bean under water deficit. *Biol. Plant.* 2004; 48:445–448.

247. Lambers H, Shane MW, Cramer MD, Pearse SJ, Veneklaas EJ. Root structure and functioning for efficient acquisition of phosphorus: Matching morphological and physiological traits. *Ann. Bot.-Lond.* 2006; 98:693–713.

248. Jin J, Lauricella D, Armstrong R, Sale P, Tang C. Phosphorus application and elevated CO_2 enhance drought tolerance in field pea grown in a phosphorus-deficient vertisol. *Ann. Bot.-Lond.* 2014; doi:10.1093/aob/mcu209.

249. Nelsen CE, Safir GR. Increased drought tolerance of mycorrhizal onion plants caused by improved phosphorus nutrition. *Planta* 1982; 154:407–413.

250. Graciano C, Guiamet JJ, Goya JF. Impact of nitrogen and phosphorus fertilization on drought responses in *Eucalyptus grandis* seedlings. *Forest Ecol. Manag.* 2005; 212:40–49.

251. dos Santos MG, Ribeiro RV, de Oliveira RF, Machado EC, Pimentel C. The role of inorganic phosphate on photosynthesis recovery of common bean after a mild water deficit. *Plant Sci.* 2006; 170:659–664.

252. Plaxton WC, Tran HT. Metabolic adaptations of phosphate-starved plants. *Plant Physiol.* 2011; 156:1006–1015.

253. Juszczuk I, Malusa E, Rychter AM. Oxidative stress during phosphate deficiency in roots of bean plants (*Phaseolus vulgaris* L.). *J. Plant Physiol.* 2001; 158:1299–1305.

254. Malusa E, Laurenti E, Juszczuk I, Ferrari RP, Rychter AM. Free radical production in roots of *Phaseolus vulgaris* subjected to phosphate deficiency stress. *Plant Physiol. Biochem.* 2002; 40:963–967.

255. Foyer CH, Noctor G. Redox sensing and signalling associated with reactive oxygen in chloroplasts, peroxisomes and mitochondria. *Physiol. Plant* 2003; 119:355–364.

256. Chiou TJ, Lin SI. Signaling network in sensing phosphate availability in plants. *Annu. Rev. Plant Biol.* 2011; 62:185–206.

257. Bray EA, Bailey-Serres J, Weretilnyk E. Responses to abiotic stresses. In: Buchanan BB, Gruissem W, Jones RL, eds. *Biochemistry and Molecular Biology of Plants*. Rockville, MD: American Society of Plant Physiology, 2000:1158–1203.

258. Lynch JP. Root phenes for enhanced soil exploration and phosphorus acquisition: Tools for future crops. *Plant Physiol.* 2011; 156:1041–1049.

259. Abel S, Ticconi C, Delatorre CA. Phosphate sensing in higher plants. *Physiol. Plant.* 2002; 115:1–8.

260. Smith FW. The phosphate uptake mechanism. *Plant Soil* 2002; 245:105–114.

261. Mimura T. Homeostasis and transport of inorganic phosphate in plants. *Plant Cell Physiol.* 1995; 36:1–7.

262. Yao Y, Sun H, Xu F, Zhang X, Liu S. Comparative proteome analysis of metabolic changes by low phosphorus stress in two *Brassica napus* genotypes. *Planta* 2011; 233:523–537.

263. Chen Z, Cui Q, Liang C, Sun L, Tian J, Liao H. Identification of differentially expressed proteins in soybean nodules under phosphorus deficiency through proteomic analysis. *Proteomics* 2011; 11:4648–4659.

264. Dietz K-J, Heilos L. Carbon metabolism in spinach leaves as affected by leaf age and phosphorus and sulfur metabolism. *Plant Physiol.* 1990; 93:1219–1225.

265. Duff SMG, Moorhead GBG, Lefebvre DD, Plaxton WC. Phosphate starvation inducible "bypasses" of adenylate and phosphate-dependent glycolytic enzymes in *Brassica nigra* suspension cells. *Plant Physiol.* 1989; 90:1275–1278.

266. Varadarajan DK, Karthikeyan AS, Matilda PD, Raghothama KG. Phosphite, an analog of phosphate, suppresses the coordinated expression of genes under phosphate starvation. *Plant Physiol.* 2002; 129:1232–1240.

Section XII

Photosynthesis under Environmental
Stress Conditions

34 Photosynthesis in Plants under Stressful Conditions

Rama Shanker Dubey

CONTENTS

34.1 Introduction 629
34.2 Effects of Stressful Conditions on Photosynthesis 631
    34.2.1 Salinity 631
        34.2.1.1 Stomatal Closure and CO_2 Diffusion 631
        34.2.1.2 Chloroplast Structure and Pigment Composition 632
        34.2.1.3 Photosystems, Photochemical Activities, and Chlorophyll Fluorescence 633
        34.2.1.4 Carboxylation 634
        34.2.1.5 Level of Photosynthates 634
    34.2.2 Drought 635
        34.2.2.1 Leaf Area and Stomatal Conductance 635
        34.2.2.2 Ultrastructural Changes in Chloroplasts 636
        34.2.2.3 Chlorophyll Fluorescence and Photochemical Reactions 636
        34.2.2.4 Carboxylation under Drought 637
        34.2.2.5 Levels of Carbohydrates and Related Enzymes 638
    34.2.3 Heat Stress 638
    34.2.4 Chilling 639
    34.2.5 Anaerobiosis 640
    34.2.6 Air Pollutants 640
    34.2.7 Metal Toxicity 642
34.3 Stress-Induced ROS Production and Damage to Photosynthetic Apparatus 643
34.4 Role of TFs and MAP Kinases in Photosynthetic Metabolism under Stresses 643
34.5 Concluding Remarks 644
References 645

34.1 INTRODUCTION

Abiotic stressful conditions of the environment such as soil salinity, drought, flood, heat, cold, anaerobiosis, gaseous pollutants, radiation, and high levels of metals in the soil are major constraints to global crop productivity. Extremes of these conditions cause alterations in a wide range of physiological, biochemical, and molecular processes in plants, affecting growth and metabolism of plants and ultimately leading to decrease in crop yield. To achieve sustainability in food grain production, in order to feed growing population, it is essential to ensure yield stability in stress-prone environments.

Photosynthesis is the key physiological process of all green plants and provides the energy source for plant metabolism and is directly related to productivity in terms of primary production of biomass and yield. Though growth of a plant is controlled by a range of physiological, biochemical, and molecular processes, photosynthesis is the key process that contributes substantially to the plant growth and productivity. The process of photosynthesis involves two key events: the light reactions and dark reactions. In the light

reactions, light energy is converted into ATP and NADPH and oxygen is evolved, whereas in the dark reactions, CO_2 is assimilated to carbohydrates by utilizing the products of light reactions, ATP, and NADPH (Dulai et al. 2011). Among the products of photosynthesis, (1) the high energy reduced form of organic carbon (carbohydrate) is related to yield, whereas (2) molecular oxygen supports the life of all organisms on this planet. The abiotic stresses severely affect photosynthesis in growing plants by influencing different components and phases associated with photosynthesis such as decreasing utilization of light energy (Plaut 1995), altering pigment composition and causing destruction in fine structure of chloroplasts (Behera et al. 2002), impairing photophosphorylation and ATP synthesis (Medrano et al. 2002), affecting activity of photosystem II (PSII) (Sharma and Hall 1992), and decreasing stomatal conductance leading to closure of stomata and decreased stomatal control of CO_2 diffusion (Flexas et al. 2002), causing alteration in the amount and activity of the key enzyme of carbon assimilation, ribulose bisphosphate carboxylase/oxygenase (Rubisco) (Allen et al. 2000), and leading to overproduction of reactive oxygen species (ROS) that in turn causes oxidative damage to biomolecules (Saibo et al. 2009). Many signal components such as transcription factors (TFs) and protein kinases have been identified and functionally characterized that mediate the responses of stress and regulate the expression of genes related to stomatal functioning and photosynthetic metabolism (Saibo et al. 2009).

A specific environmental stress may affect any particular event associated with photosynthesis, whereas the effects of many stresses may be commonly affecting the same parameter. For example, chilling of plants leads to disorganization of thylakoids, disruption of membrane integrity leading to solute leakage and osmotic imbalance, cellular dehydration, decline in Rubisco activity, disruption in circadian regulation of key photosynthetic enzymes, and inhibition in the translocation of carbohydrates (Ting et al. 1991; Thomashow 1999). Heat stress increases membrane fluidity, which leads to disorganization of chloroplast thylakoid membranes, dissociation of

PSII complex, destacking of grana lamellae, and inactivation of Rubisco (Crafts-Brandner and Salvucci 2002). Waterlogged conditions result in anaerobic environment, decreased nutrient absorption, reduced stomatal conductance, and decreased levels of ATP and chlorophylls (Taiz and Zeiger 1998). On the other hand, salinity and drought stresses cause increase in cellular osmotic concentrations, resulting in lower leaf water potential, disruption of homeostasis and ion distribution in cells, decreased stomatal conductance, stomatal closure, altered chlorophyll fluorescence, photoinhibition of PSII, impaired ATP synthesis and ribulose-1,5-bisphosphate (RuBP) regeneration, conformational changes in membrane-bound ATPase enzyme complex, as well as decrease in both activity and concentration of Rubisco enzyme (Plaut 1995; Flexas and Medrano 2002). It has been observed that under salinity and drought stresses, reduced mesophyll conductance is an equally important cause of lower CO_2 diffusion and that plasma membrane aquaporins are closely related to mesophyll

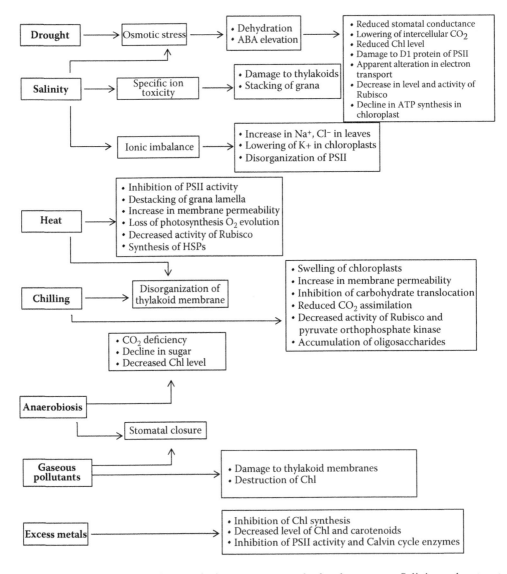

FIGURE 34.1 Effects of abiotic stresses on photosynthetic components and related processes. Salinity and water stresses have many effects in common as both cause dehydration and increase in the level of ABA. Heat and chilling cause disorganization of thylakoid membranes. The primary effect of anaerobiosis and gaseous pollutants is stomatal closure.

conductance (Saibo et al. 2009). In fact, the effects of salinity and drought on photosynthesis are primarily attributed to the stomatal limitations for diffusion of gases, which ultimately alter photosynthesis (Chaves et al. 2009). Salt stress in addition to osmotic effects exerts specific ion effects due to Na^+ and Cl^- penetrating in the chloroplasts, leading to ion toxicity and resulting in nutritional imbalance due to competition of salt ions and nutrients (Plaut 1995).

Polluting gases such as SO_2, NO_2, H_2S, and O_3 enter leaves and inhibit stomatal movements (Taiz and Zeiger 1998). Excess levels of many metals like Cd^{2+}, Ni^{2+}, Pb^{2+}, Cu^{2+}, and Hg^{2+} within the plant tissues directly affect PSII activity, alter photosynthetic partitioning, and inhibit Rubisco activity (Krupa et al. 1993; Quartacci et al. 2000; Verma and Dubey 2001; Carpentier 2002). The overall impact of various environmental stresses on different components of photosynthetic process may be described according to the scheme presented in Figure 34.1. This chapter presents our current status of knowledge related to the effects of different environmental stresses on the individual components of photosynthesis such as pigments, photosystems, components of electron transport system, alterations in the activities of enzymes involved in CO_2 assimilation, and changes in gas exchange characteristics in crop plants. In addition, stress-induced effects on production of ROS that cause damage to photosynthetic apparatus and enzymes, as well as various signaling components involved in stress-induced damage and adaptation in plants, are discussed.

34.2 EFFECTS OF STRESSFUL CONDITIONS ON PHOTOSYNTHESIS

The key abiotic stressful conditions of the environment to which plants are generally exposed include soil salinity, drought, heat, chilling, anaerobiosis, excessive levels of metals in the soil, gaseous pollutants, etc. Photosynthetic efficiency of plants is drastically reduced under these stresses. The extent of the effect depends on the plant species, the developmental stage of the plant, as well as the type, intensity, and duration of stress.

34.2.1 SALINITY

Salinity of the soil is one of the most important abiotic stresses that challenges global sustainable agriculture. Worldwide cultivable land is shrinking at an alarming rate due to salinization of the soil. Salinity limits plant growth and productivity and causes severe loss in yield of crops in vast areas of the world and more especially in arid and semiarid regions (Taiz and Zeiger 1998).

Accumulation of soluble salts from poor-quality irrigation water, irrigation of soil with saline water, and improper or restricted drainage system to flush out accumulated salts often lead to a high level of salt buildup in the soil. The problem of salinity is increasing in recent years due to global warming, with the consequent rise in sea level and increase in storm incidences, particularly in coastal areas (Moradi and Ismail 2007). The predominant salts in saline soils are chlorides and sulfates

of Na^+, Mg^{2+}, and Ca^{2+}. NaCl contributes substantially to salinity due to its exceptionally high solubility. Plants growing in saline environments suffer injury due to osmotic stress, specific ion toxicities, and ionic imbalance (Taiz and Zeiger 1998). Osmotic stress results due to lowering of soil water potential. Specific ion toxicities result due to accumulation of injurious concentrations of Na^+, Cl^-, or SO_4^{2-} in the cells. Ionic imbalance or nutritional imbalance results in salt-stressed plants due to competition of salt ions with the nutrients. Due to a large amount of soluble salts found in saline soils, plants cannot absorb the necessary nutrients required for their growth.

Salinity affects photosynthesis due to reduction in stomatal conductance as well as reduced CO_2 diffusion through the stomata, decreased intercellular partial pressure of CO_2 in leaves, reduction in chlorophyll content (Downton et al. 1985), changes in ultrastructure of chloroplasts (Salama et al. 1994; Suleyman et al. 2002), decreased photochemical and carboxylation reactions (Sharma and Hall 1992), increased level of soluble sugars in the tissues (Dubey and Singh 1999), and reduced content of organic acids (Sanchez et al. 2007).

34.2.1.1 Stomatal Closure and CO_2 Diffusion
Salinity stress decreases leaf turgor, which in turn leads to closure of stomata and reduced stomatal conductance (Chaves et al. 2009). Stomatal closure is often a rapid initial response to salt stress. Decreased stomatal conductance under salt stress results due to combined effects of osmotic stress as well as Na^+ toxicity (Plaut 1995). It is suggested that a small decline in stomatal conductance in leaves of salt-stressed plants may provide protection to the plants by allowing plant water saving and improving plant water-use efficiency (Chaves et al. 2009). The stomatal closure observed under salt stress appears to be a result of accumulation of abscisic acid (ABA) in leaves of salt-stressed plants (Taiz and Zeiger 1998). Rasmuson and Anderson (2002) observed that in cheat grass (*Bromus tectorum* L.), salinity led to stunted growth through reduced leaf initiation and expansion and reduced photosynthetic rates, primarily due to stomatal limitation. Besides reduced CO_2 diffusion through the stomata, a reduced mesophyll conductance to CO_2 (g_m) is also observed in salt-stressed plants. Alteration in mesophyll conductance under salt stress appears to be linked to alterations in physical structure of the intercellular spaces and changes in membrane permeability (Chaves et al. 2009). Many studies suggest that g_m is depressed under salt stress, and it can be induced by exogenous application of ABA (Chaves et al. 2009). Flexas et al. (2004) concluded that salt stress predominantly affected diffusion of CO_2 in the leaves through a decrease in stomatal and mesophyll conductances, and not due to the biochemical capacity to assimilate CO_2.

Salinity invariably leads to stomatal closure and decreases intercellular CO_2 concentration (C_i) in leaves (Plaut 1995). The immediate effect of salinity is mainly osmotic, and continuous exposure further leads to specific ion toxicities. However, varied observations have been reported regarding accumulation of salinity ions in leaf tissues and alteration in the rate of photosynthesis (Downton et al. 1985; Plaut et al. 1989). According to Downton et al. (1985), in spinach leaves, stomatal conductance and intercellular partial pressure of CO_2 decreased due to salinity, but this had little effect on photosynthetic rate although it improved water use efficiency (WUE).

Rice (*Oryza sativa* L.) cultivars differing in salt tolerance when stressed for 1 week with 60 and 120 mmol NaCl showed substantial reduction in carbon assimilation rate and stomatal conductance (Dionisio-Sese and Tobita 2000). Similarly, in pepper plants, CO_2 assimilation decreased under 100–150 mM NaCl treatment but not under 50 mM (Plaut 1995). In isolated mesophyll cells of cowpea leaves, the CO_2 fixation rate decreased by 30% in the medium containing 130 mM NaCl, whereas under 173 mM NaCl, photosynthetic rate was severely and irreversibly inhibited (Plaut et al. 1989). These observations suggest that under higher salinity level, the osmotic effect and the concentration of salinity ions as well as duration of exposure also become important in determining the gas exchange rate, the CO_2 concentration in leaves, and, in turn, the rate of photosynthesis. Certain workers have pointed out the possible involvement of carbonic anhydrase and aquaporins in the regulation of g_m; however, the role of these components in an apparent reduced g_m under salinity is not well understood (Chaves et al. 2009).

34.2.1.2 Chloroplast Structure and Pigment Composition

High concentrations of Na^+ and Cl^+, which accumulate in the chloroplasts under salinity stress, cause damage to thylakoid membranes (Omoto et al. 2010). Salinity leads to destruction of fine structure and swelling of chloroplasts, instability of pigment protein complex (Lapina and Popov 1999), degradation of chlorophylls (Downton and Grant 1985; Lapina and Popov 1999), and alteration in the content and composition of carotenoids (Lapina and Popov 1999). Chloroplasts isolated from leaves of salt-stressed spinach plants showed about 80% of the photosynthetic capacity compared to chloroplasts from control leaves (Downton and Grant 1985). Salinity-induced swelling of thylakoid membranes appears to be due to a change in the ionic composition of the stroma (Salama et al. 1994). Under salinity, plants accumulate higher levels of Na^+ and Cl^- ions within the chloroplasts, which leads to shrinking of thylakoid membranes (Rottenberg 1977) and stacking of adjacent membranes in grana (Barber 1982). In salinized barley, wheat, and pea plants, a marked loosening between the chlorophyll and the protein was observed in the chloroplasts (Lapina and Popov 1999).

It is observed that under salt stress, breakdown of chlorophyll (Chl) occurs due to the increased level of the toxic cation Na^+ within the chloroplasts (Yang et al. 2011). Various workers have observed decreased level of Chl pigments in salt-stressed spinach, sunflower, alfalfa, wheat, and castor bean plants (Downton et al. 1985; Salama et al. 1994; Lapina and Popov 1999; Ashraf and Harris 2013). Salinity treatment of 60–240 mM NaCl for 10 days caused significant reduction in the contents of Chl *a*, Chl *b*, and carotenoids in bean (*Vicia faba* L.) seedlings (Amira and Qados 2011). Downton et al. (1985) observed that leaves of spinach plants grown under 200 mM NaCl contained about 73% of the chlorophyll per unit area of control plants. Salt-induced decline in the level of leaf Chls could be due to its impaired biosynthesis or increased degradation (Ashraf and Harris 2013). Some workers attribute the reduced level of Chls in salt-stressed plants mainly due to the destruction of Chl *a*, which is supposed to be more sensitive to salinity than chlorophyll *b* (Reddy and Vora 1986). Experiments using sunflower callus and whole plants have shown that the important precursors of Chl synthesis, i.e., glutamate and 5-aminolaevulinic acid (ALA), decreased in salt-stressed calli and leaves, indicating that salt stress affects more markedly Chl biosynthesis than Chl breakdown (Akram and Ashraf 2011).

Mature trees of *Prunus salicina*, acclimated to salinity under field conditions, showed reduced leaf Chl content, which was apparently related to increased leaf chloride content and decreased CO_2 assimilation capacity (Ziska et al. 1990). In such trees, if leaf chloride level exceeded 0.25 mol kg^{-1} dry weight, a significant reduction in Chl content as well as visual leaf damage was apparent (Ziska et al. 1990). Cultivars of crop species differing in salt tolerance when grown under saline conditions show different degrees of reduction in chlorophyll level. Chlorophyll in salt-tolerant cultivars is more effectively protected against the deleterious effects of Na^+ because such plants show higher accumulation of vacuolar Na^+ and osmolytes like putrescine and quaternary ammonium compounds in the chloroplasts (Salama et al. 1994).

Seedlings of rice cultivars differing in salt tolerance, when raised under increasing levels of NaCl salinity in sand culture experiments, showed significant decrease in the level of Chls with greater decrease in salt-sensitive cultivars than the tolerants (Singh and Dubey 1995). An assessment of total Chl level (Chl *a* + *b*) in rice seedlings of differing salt tolerance, raised under increasing levels of NaCl salinity over 5- to 20-day growth period, indicated that in salt-tolerant *cvs*. CSR-1 and CSR-3 with moderate salinity level of 7 dS m^{-1} NaCl, almost no change occurred in total Chl level, whereas with a higher salinity level of 14 dS m^{-1} NaCl, a marked decline in Chl level was observed (Singh and Dubey 1995). On the other hand, in salt-sensitive rice *cvs*. Ratna and Jaya, with increase in salinity, a concomitant decline in total Chl level was seen. Salt-stressed seedlings of tolerant rice cultivars maintain higher level of total Chls compared to the sensitive ones under similar level of salinization (Singh and Dubey 1995). Similar to Chl *a*, decreased level of Chl *b* is noticed in plants grown in salinized medium (Tewari and Singh 1991; Salama et al. 1994; Singh and Dubey 1995). Salama et al. (1994) noted an increase in Chl *a*/Chl *b* ratio in salt-sensitive wheat plants due to salinity, although separately the levels of Chl *a* as well as that of Chl *b* decreased. Salinity induces genotype-specific change in the level of carotenoids (Singh and Dubey 1995). Under 14 dS m^{-1} NaCl level of salinization, seedlings of salt-sensitive rice cultivars showed more decline in the level of carotenoids compared to the tolerants (Singh and Dubey 1995). It is suggested that salt stress causes an increase in zeaxanthin content and degradation of β-carotene, which are apparently involved in protection against photoinhibition (Sharma and Hall 1992).

34.2.1.3 Photosystems, Photochemical Activities, and Chlorophyll Fluorescence

When plants are subjected to salinity and other stresses under high and/or persistent excess light, photoinhibition occurs. Salinity stress enhances the susceptibility of plants to photoinhibition, which leads to the formation of toxic singlet oxygen in chloroplasts and degradation of the quinone-binding protein, now known as D1 protein in the PSII complex and is caused by excess light (Kyle et al. 1984). Impairment of D1 results in disruption of the light-dependent separation of charge between P680 and pheophytin a, and this phenomenon is associated with interruption of the transport of electrons that is medicated by PSII, ultimately leading to decreased photosynthetic activity (Suleyman et al. 2002). Under natural conditions, in the field, salt stress very often occurs in combination with prolonged light excess, and it has been observed that the combination of light and salt stress is synergistic in inactivating PSII (Suleyman et al. 2002). It is suggested that under stress, the photoprotective mechanisms compete with photochemistry for the absorbed energy, leading to a decrease in quantum yield of PSII (Chaves et al. 2009).

According to Kyle et al. (1984), the 32 kDa D_1 protein, which is one of the two reaction center proteins of PSII, is the primary site of damage due to photoinhibition. The level of photoinhibition can be determined by the extent of damage and repair of D_1 protein (Ohad et al. 1984). Photodamaged PSII is repaired in a process involving the rapid turnover of D_1, with degradation of damaged D_1 and subsequent light-dependent synthesis of precursor to D_1 termed as pre-D_1. The damaged D_1 is replaced by newly synthesized pre-D_1 (Suleyman et al. 2002). In wheat plants, it was shown by Mishra et al. (1991) that the inhibition of protein synthesis including the D_1 protein and closure of stomata by salt stress are responsible for the exacerbation of photoinhibition under salt stress. Mehta et al. (2010) observed that in wheat (*Triticum aestivum*) plants, the donor side of the PSII was damaged more than the acceptor side due to salt stress (0.1–0.5 M NaCl) and that the salt-induced damage to PSII was reversible.

Chloroplasts isolated from salt-stressed seedlings of rice cultivars of differing salt tolerance, however, showed different levels of electron transport activities compared to control grown plants (Singh and Dubey 1995). Results of an experiment conducted to examine electron transport reactions in chloroplasts isolated from 20-day grown seedlings of a salt-tolerant rice *cv*. CSR-1 and a sensitive *cv*. Ratna are shown in Table 34.1. As is evident from the table, about 53% decrease in whole-chain electron transport activity and 71% decrease in PSII activity can be seen in chloroplasts isolated from 14 dS m^{-1} NaCl grown seedlings of salt-sensitive *cv*. Ratna compared to electron transport reactions in chloroplasts isolated from nonsalinized seedlings of this cultivar. The extent of inhibition in electron transport activities due to salinity was less in the tolerant cultivar than in the sensitive one.

Wang et al. (2009), while conducting experiments using a salt-sensitive rice *cv*. Peta and a salt-tolerant *cv*. Pokkali, showed that 200 mM NaCl stress inhibited electron transport activity of PSII more than that of PSI, and the degree of inhibition of electron transport activity was lower in the salt-tolerant cultivar than the sensitive one. Furthermore, these workers observed that PSII particles of *cv*. Pokkali contained more 33 and 43 kDa polypeptides than *cv*. Peta and that PSII particles after NaCl treatment showed deficiency of 23 kDa polypeptides.

Alteration in the photochemical activity of salinity-exposed plants appears to be possibly due to more absorption of potentially toxic ions Na$^+$ and Cl$^-$ in these species, which could penetrate the chloroplasts and exert its adverse effects (Salama et al. 1994). Another possible explanation appears to be salt stress-induced photodamage to PSII; however, the mechanism by which salt stress enhances the photodamage to PSII remains unclear (Suleyman et al. 2002).

Chlorophyll fluorescence measurements have revealed that under salt stress, nonphotochemical quenching increases, whereas the electron transport rate (ETR) decreases. In sorghum plants, it was observed by Netondo et al. (2004) that the maximum quantum yield of PSII, the ratio of variable

TABLE 34.1
Effect of Increasing Levels of NaCl Salinity on Electron Transport Reactions in Chloroplasts Isolated from Rice Seedlings Grown for 20 days under 7 and 14 dS m^{-1} NaCl Salinity Levels

Rice Cultivar	1. Assay	Salinity		
		Control (without NaCl)	7 ds m^{-1} NaCl	14 ds m^{-1} NaCl
CSR-1 (T)	PS I+II	430.90 ± 30.60	370.40 ± 28.00	295.60 ± 22.60
	Whole chain (H$_2$O MV)	286.19 ± 27.80	213.09 ± 23.70	145.18 ± 20.20
	PSII			
	(H$_2$O Pd$_{ox}$)			
Ratna (S)	PS I+II	384.84 ± 42.50	280.50 ± 32.80	162.50 ± 28.50
	Whole chain (H$_2$O MV)	250.84 ± 36.80	129.30 ± 24.60	72.58 ± 18.40
	PSII			
	(H$_2$O Pd$_{ox}$)			

Note: Reaction rates are expressed as mmol H$_2$O consumed or evolved mg^{-1} chlorophyll per h^{-1}. Values are mean ± SD based on three independent observations. T and S in parentheses indicate salt-tolerant and -sensitive rice cultivar, respectively.

fluorescence to maximal fluorescence (F_v/F_m), the photochemical quenching coefficient (q_P), and the ETR significantly decreased, but the nonphotochemical quenching coefficient (q_N) increased substantially under saline conditions. The F_v/F_m ratio is an important parameter, which determines the maximum quantum efficiency of PSII. Larcher et al. (1990), while examining the combined effects of salt and temperature stresses on chlorophyll fluorescence characteristics of cowpea (*Vigna unguiculata* L.) plants, observed that appreciable differences between controls and the various salt levels could be seen in only a few of the fluorescence characteristics. These workers observed that the fluorescence indicators such as F_v/F_m, q_P, and q_N remained practically unaffected, whereas the peak of the induction transient, F_p (expressed as a fraction of F_m), was 20% higher for salt-stressed plants than for controls. Similarly, in rice cultivars differing in salinity tolerance, F_v/F_m was almost not affected by salt stress, whereas q_N increased in sensitive cultivars with increasing level of salt stress (Dionisio-Sese and Tobita 2000).

Mishra et al. (1991), while examining the effects of salt and light stress on wheat plants, observed that with NaCl treatments, intrinsic chlorophyll fluorescence level (F_0) did not change, whereas gradual reduction in variable chlorophyll fluorescence (F_v) occurred, which was a result of a decrease in maximal fluorescence (F_m) upon salt treatment. However, no significant difference in the F_v/F_m ratio could be observed between salt-treated and control plants (Mishra et al. 1991). A decrease in room temperature fluorescence of Chls associated with PSII was observed in salt-stressed sorghum plants, which appeared to be due to photoinhibition of PSII activity (Sharma and Hall 1992). When three rice cultivars contrasting in their salinity tolerance during the vegetative and reproductive stages were subjected to salinity stress, the actual quantum yield of PSII did not change substantially, whereas the ETR decreased and q_N increased significantly, and it was consistent in both stages despite the longer exposure to salinity during the reproductive stage (Moradi and Ismail 2007). These observations suggest that salinity may reduce photosynthesis through its indirect effect on the photosynthetic apparatus.

34.2.1.4 Carboxylation

The key enzyme in the photosynthetic carbon reduction in all plants is Rubisco (EC 4.1.1.39). Its level as well as carboxylating capacity decreases in plants subjected to salt stress (Seemann and Sharkey 1986). One of the most prominent effects of salt stress is the stomata closure, which leads to a lower concentration of intercellular CO_2, which in turn causes deactivation of Rubisco (Moradi and Ismail 2007; Chaves et al. 2009). Supply of CO_2 to Rubisco is impaired under salinity stress. Besides Rubisco activity, salt-induced osmotic stress adversely affects the activities of the stroma enzymes involved in CO_2 reduction and also the enzymes involved in regeneration of the Rubisco substrate, RuBP (Ashraf and Harris 2013). In bean (*Phaseolus vulgaris*) plants, 100 mM NaCl reduced photosynthetic efficiency, which was a consequence of decreased Rubisco activity and decrease in pool size of RuBP (Seemann and Sharkey 1986). In winged

bean (*Phosphocarpus tetragonolobus*) plants, NaCl salinity decreased the activities of Rubisco as well as phosphoenolpyruvate carboxylase (PEPCase, EC 4.1.1.31), and also the rate of photosynthetic CO_2 fixation (Rajmane and Karadge 1986). In leaves of 7-day-old barley seedlings, grown in the presence of 100 mM NaCl, Rubisco level was only about 20% of the control plants (Miteva et al. 1992). Miteva et al. (1992) suggested that in barley plants, NaCl salinity inhibited the synthesis of total soluble protein with a more pronounced inhibition of Rubisco synthesis. It is suggested that the reduction in the amount of Rubisco protein under salt stress could be due to the effect of salt at the level of transcription, translation, or gene regulation (Ziska et al. 1990).

Rubisco isolated from many plant species appears to be sensitive to NaCl (Flowers et al. 1977). It is believed that under salt stress conditions, compatible solutes like proline and its analogs accumulate in the cytoplasm/chloroplasts and provide possible protection to this enzyme against osmotic and toxic effects of salinity (Solomon et al. 1994). Under *in vitro* conditions, the proline-related analogs N-methyl-L-proline and N-methyl-trans-4-hydroxy-L-proline have been shown to ameliorate the inhibition of the activity of Rubisco by NaCl (Solomon et al. 1994). Though it is observed that salinity decreases the activity of Rubisco in many plant species (Miteva et al. 1992), NaCl-adapted plants of *Tamarix jordanis* showed production of higher level of Rubisco as well as of compatible solutes (Solomon et al. 1994). Higher content of Rubisco protein in salt-adapted plants might contribute toward better adaptation of plants to salinity (Takabe et al. 1988). The activity of PEPCase has been shown to rise considerably in salt-stressed plants compared to control grown plants (Miteva and Vaklinova 1989). NaCl-stressed barley plants showed four times higher PEPCase activity than unstressed plants (Miteva and Vaklinova 1989). PEPCase isolated from many halophytes like *Suaeda monoica*, *Chloris gayana*, and *Cakile maritima* was shown to be not only a salt-tolerant enzyme but also a salt-requiring enzyme (Shomer-Ilan et al. 1985).

34.2.1.5 Level of Photosynthates

The principal end products of leaf photosynthesis are starch and sucrose. It has been observed that plants under salinity stress show higher starch content and accumulate soluble sugars especially sucrose (Flowers et al. 1977). The accumulation of photosynthates starch and sugars under saline conditions is mainly attributed to the impaired carbohydrate utilization as respiration rate decreases at high salinity levels (Das et al. 1990). The accumulation of soluble sugars under stressful condition of salinity might contribute to a favorable osmotic potential and render a protective role to biomolecules (Flowers et al. 1977).

The responses of NaCl salinity on the level of starch and sugars depends on the plant organs as well as the genotypes of plants studied (Dubey 1982; Lawlor 2002). Rice genotypes of differing salt tolerance accumulate varying levels of sugars in plant parts when subjected to saline stress. An examination of the levels of total, reducing, and nonreducing sugars in shoots of the seedlings of salt-sensitive *cvs*. Ratna and Jaya

TABLE 34.2

Levels of Nonreducing, Reducing, and Total Sugars (mg^{-1} g dry wt) in Shoots of 20-day Grown Nonsalinized (Control) and Salt-Stressed (14 dS m^{-1} NaCl) Seedlings of Salt-Sensitive Rice *cvs.* Ratna and Jaya, and Salt-Tolerant *cvs.* CSR-1 and CSR-3

| Rice Cultivar | Nonreducing Sugars | | Reducing Sugars | | Total Sugars | |
	Control (without NaCl)	14 dS m^{-1} NaCl	Control (without NaCl)	14 dS m^{-1} NaCl	Control (without NaCl)	14 dS m^{-1} NaCl
Ratna (S)	6.0	26.2	8.1	17.8	14.1	44.0
Jaya (S)	8.2	19.5	11.5	30.2	19.7	49.7
CSR-1 (T)	5.8	12.2	10.2	14.1	16.0	26.3
CSR-1 (T)	7.1	15.2	9.8	18.6	16.9	33.8

Note: Values are mean based on three independent determinations. T and S in parentheses indicate salt-tolerant and -sensitive cultivar, respectively.

and tolerant *cvs.* CSR-1 and CSR-3 grown in the presence of 14 dS m^{-1} NaCl indicated that, in both sets of cultivars, salinity caused increase in the level of sugars with more increase in the sensitive rice cultivars than in the tolerants (Table 34.2). It was observed that at 14 dS m^{-1} NaCl salinity level, shoots of sensitive rice cultivars maintained about 2.52–3.14 times total sugar level compared to nonsalinized seedlings.

34.2.2 DROUGHT

Water deficit for prolonged period leads to the state of drought stress in growing plants. Plants are often subjected to periods of soil and atmospheric water deficits during their life cycle. Photosynthesis is a key plant process that is adversely affected by drought. Drought reduces plant growth and affects photosynthesis by reducing leaf area, enhancing stomatal closure, decreasing water status in the leaf tissues, decreasing CO_2 availability caused by diffusion limitations through the stomata and the mesophyll, causing ultrastructural changes in chloroplasts, affecting electron transport reactions impairing ATP synthesis, reducing the rate of CO_2 assimilation and RuBP generation, and altering the level of photosynthates in the tissues. Drought causes an imbalance in the hormone level in plants. Due to alteration in hormonal balance, concentrations of many key enzymes of photosynthesis decline in water-stressed plants.

34.2.2.1 Leaf Area and Stomatal Conductance

As a result of decrease in the water content of the leaves, cells shrink, cell volume decreases, and the solutes within the cell become more concentrated. The plasma membrane becomes thicker and compressed resulting in inhibition of cell expansion. Leaf area as well as size of individual leaves and the number of total leaves are reduced under water stress. Decreasing relative water content (RWC) and water potential of leaves progressively decrease stomatal conductance, leading to decline in CO_2 molar fraction in chloroplasts, decreased CO_2 assimilation, and reduced rate of photosynthesis (Lawlor 2002).

Stomatal closure is considered as the primary cause of reduced photosynthesis under mild to moderate drought stress

in C_3 and C_4 plants (Chaves et al. 2011). Stomatal closure leads to decrease in CO_2 intake by mesophyll cells, leading to decreased intercellular CO_2 partial pressure (C_i), decreased chloroplastic CO_2 concentration (C_c), and thereby decreased CO_2 assimilation and net photosynthesis (Flexas et al. 2002). In C_4 plants, though photosynthesis saturates at much lower CO_2 concentrations than C_3 plants, C_4 photosynthesis can be even more sensitive to drought than C_3 photosynthesis, in spite of the larger photosynthetic capacity and WUE, because C_4 plants have limited capacity for photorespiration to provide alternative sinks of electrons under drought (Chaves et al. 2011). C_4 and crassulacean acid metabolism (CAM) plants are best adapted to arid environments because they have higher WUE than that of C_3 plants.

Under drought, guard cells, when exposed to the atmosphere, in air of low humidity, lose water too rapidly by evaporation causing the stomata to close by a mechanism called hydropassive closure (Taiz and Zeiger 1998). In a different mechanism of stomatal closure, called hydroactive closure, the whole leaf gets dehydrated under water stress and increased synthesis of ABA takes place in mesophyll cells and it accumulates in the chloroplast (Taiz and Zeiger 1998). Stored ABA is then released to the chloroplast (cell wall space) from where it reaches to the guard cells through the transpiration stream. Redistribution of stored ABA from the mesophyll chloroplasts to the apoplasts initiates the closure of stomata (Cornish and Zeevaart 1985). Under water-deficit conditions in the soil, messengers from the root system like root drying or increased delivery of ABA from root to leaves via transpiration stream also induce stomatal closure (Taiz and Zeiger 1998). Under drought, stomatal closure is caused mainly due to the action of ABA. High ABA level causes an increase in cytosolic Ca^{2+} and activation of plasma membrane-localized anion channels (Ashraf and Harris 2013). Both stomatal conductance and mesophyll conductance (g_m) get reduced under drought. The contributions of stomatal and nonstomatal limitations to photosynthesis under drought depend on the intensity and duration of stress being imposed. Under prolonged and intensified water deficit, limitations of nonstomatal processes become more important, especially due to lowered g_m (Chaves et al. 2011).

34.2.2.2 Ultrastructural Changes in Chloroplasts

Drought stress causes substantial damage to photosynthetic pigments and deterioration of thylakoid membranes. A reduction in Chl content is generally observed under drought stress with greater reduction of Chl *b* than that of Chl *a* (Ashraf and Harris 2013). It is suggested that decline in Chl content under drought may be attributed to accelerated breakdown of Chl rather than its slow synthesis (Ashraf and Harris 2013).

Under drought stress, chloroplast volume decreases, permanent adhesions occur within the grana, partitions become thinner, lipid droplets increase in number and size, many thylakoid proteins are oxidatively damaged, and structural changes occur in light-harvesting chlorophyll protein complexes (Maroti et al. 1984; Tambussi et al. 2000; Gussakovsky et al. 2002). Maroti et al. (1984), while investigating ultrastructural changes in chloroplasts of different plant species due to drought, observed that contraction of stroma, swelling, and blistering of thylakoids were characteristic features of the chloroplasts of CAM succulent plant *Sedum sexangulare* and mesophyll chloroplasts of C$_4$ sclerophyllous plant *Testuca vaginata*. These workers noted that under naturally induced drought, the chloroplasts elongated and contracted along the cell wall, and stroma aggregated and was found along the inside surface of the envelope. The effect of drought on *S. sexangulare* chloroplast was marked by shriveling of the cells and chloroplasts with a decrease in the size of electron dense granules and the electron density of the whole cytoplasm (Maroti et al. 1984).

It was suggested by Poljakoff-Mayber (1981) that swelling, distortion of stroma and grana lamellae regions, and the appearance of lipid droplets were common features of chloroplasts in conditions of drought. Decrease in the volume of cells and chloroplasts has been noted by other workers in plant tissues undergoing dehydration due to long drought (Kaiser and Heber 1981; Sharkey and Murray 1982). It is regarded that stromal aggregation under drought is a reversible process as the normal structure is restored after drought recovery, whereas the accumulation of lipid droplets in the intrathylakoidal space may play an adaptive role during drought conditions (Maroti et al. 1984). An examination of the structural changes of bundle sheath and mesophyll chloroplasts of a C$_4$ plant *T. vaginata* that underwent water stress suggests that chloroplasts of bundle sheath cells are more resistant to drought than those of mesophyll cells (Maroti et al. 1984).

It has been shown that, in chloroplasts, chiral macroaggregate formation of the light-harvesting chlorophyll *a/b* pigment protein complexes (LHCIIs) occurs, which is involved in the lateral separation of the two photosystems, protects the photosynthetic apparatus against photoinhibitory damage, and plays an important role in the structure and function of the chloroplast (Gussakovsky et al. 2002). The imposition of moderate water deficit (water potential of −1.3 MPa) to pea leaves led to a 75% inhibition of photosynthesis and to increases in zeaxanthin, malondialdehyde, oxidized proteins, and mitochondrial, cytosolic, and chloroplastic superoxide dismutase activities (Ormaetxe et al. 1998), whereas severe water deficit

(−1.9 MPa) almost completely inhibited photosynthesis and decreased the levels of chlorophylls, β-carotene, neoxanthin, etc. Imposition of −2.0 MPa water stress on 4-week-old wheat plants led to oxidative damage of 68, 54, 41, and 24 kDa thylakoid polypeptides and accumulation of many cross-linked high molecular weight proteins with the substantial decrease in photosynthetic electron transport activity (Tambussi et al. 2000).

34.2.2.3 Chlorophyll Fluorescence and Photochemical Reactions

Drought stress causes characteristic changes in Chl *a* fluorescence kinetics and hence damages PSII reaction center (Zhang et al. 2011). Many studies have shown that drought damages oxygen evolving center (OEC) coupled with PSII and degrades D1 polypeptide leading to the inactivation of the PSII reaction center (Ashraf and Harris 2013). The damage caused to PSII under drought is associated with light-induced oxidative stress. ROS are produced within the thylakoids when photon absorption exceeds the rate of photon utilization (Foyer and Noctor 2000).

In vivo chlorophyll fluorescence of dark-adapted leaf, which can be elicited by very dim light beam modulated at high frequency, represents the minimum chlorophyll fluorescence (F_0) and is not significantly modified by water stress (Havaux et al. 1986). After illumination with nonmodulated white light of higher intensity, the fluorescence increases rapidly to a peak point. The fluorescence between F_0 and peak point is termed as variable fluorescence (F_v). Maximum fluorescence (F_m) can be induced by a short pulse of saturating white light. The value of F_v reflects the reduction of the primary electron acceptor Q_A of PSII. In the oxidized state, Q_A quenches fluorescence. Quenching of F_v reflects the working of entire photosynthetic process, more especially primary photochemical events, and it depends on reduction–oxidation of Q_A, light-induced proton gradient across the thylakoid membrane, or the light energy distribution between the two photosystems. Depending on the degree of oxidation or reduction of the electron transport chain, F_v is quenched or enhanced (Zrust et al. 1988). On the acceptor side of PSII, the quinone and plastoquinone pools are possibly responsible for fluorescence quenching (Zrust et al. 1988). Knowing the values of F_0, F_v, and F_m, the value of the photochemical component of fluorescence quenching (q_Q) can be calculated as suggested by Schreiber et al. (1986) as $q_Q = (F_m − F_v)/(F_m − F_0)$.

Drought causes drastic changes in the different modulated fluorescence levels, resulting in severe reduction in the q_Q value (Havaux et al. 1988). Dehydration of leaves or extreme water losses caused alterations in chlorophyll fluorescence in many plant species (Havaux 1992). At extreme water deficit, fluorescence changes are more pronounced. In oak leaves, change in chlorophyll fluorescence was detected only when water-deficit values exceeded 30% (Plaut 1995). At more severe dehydration, increase in nonphotochemical quenching was observed, whereas photochemical quenching remained unaffected (Plaut 1995). Seven potato genotypes grown under drought showed decline in variable fluorescence (F_v) of leaves with a concomitant decrease in net photosynthetic rate (P_N) (Zrust et al. 1988). In potatoes, total dry matter production in water-stressed plants could be correlated with F_v, which is a measure of the capacity of primary photochemical event (Zrust et al. 1988). It was suggested by Zrust et al. (1988) that decreasing values of F_v under water deficit indicate diminishing photosynthetic activity in potato leaves and that F_v values provide a method for the study of changes in the photosynthetic capacity of the potatoes in response to drought. In a bryophyte *Physcomitrella patens*, when osmotic stress was imposed with 0.4 M sorbitol, chloroplast development, Chl content, and the thylakoid protein composition were affected, and the photosynthetic apparatus of the moss responded to oxidative stress by increasing nonphotochemical quenching (Azzabi et al. 2012).

It has been shown that photochemical activities decline under drought, with PSII activity being more drought sensitive than PSI (Nogues and Alegre 2002). Both the degree of dehydration as well as the duration of dehydration imposed influence PSII activities (Liu et al. 2014). In two native Mediterranean plants, rosemary (*Rosmarinus officinalis* L.) and lavender (*Lavandula stoechas* L.) imposition of drought led to the decrease in the relative quantum efficiency of PSII photochemistry and decreased the efficiency of energy capture by open PSII reaction centers. These events were associated with down-regulation of electron transport (Nogues and Alegre 2002). Similarly, in drought-stressed wheat leaves, kinetics of the Hill reaction activity declined significantly (Behera et al. 2002). In a study on metabolic consumption of photosynthetic electron transport in tomato plants, Haupt-Herting and Fock (2002) observed down-regulation of PSII under drought. Liu et al. (2014) observed that papaya leaves, when dehydrated in air, showed decline in the PSII activity with a concomitant decline in the leaf RWC. These workers suggested that the degree of dehydration and its duration both play important roles in damage to photosynthetic apparatus.

The observed inhibition of PSII activity under drought might not be due to the direct effect of stress on photochemical activity but to photoinhibition (Havaux 1992). When leaves of *Lycopersicon esculentum*, *Solanum tuberosum*, and *Solanum ingrum* plants were illuminated with intense white light at 25°C, photoinhibition damage of PSII was more pronounced in drought-stressed leaves compared to undesiccated controls (Havaux 1992). In tomato and potato, drought created by treatment of intact leaves did not significantly alter the PSII functioning in dark- and light-adapted leaf samples (Havaux 1992). In these plants, PSII was shown to be drought resistant.

Impaired photophosphorylation, decreased ATP synthesis by the enzyme ATP synthase, and loss of ATP content have been observed in plants subjected to drought (Lawlor 2002). Among the different events during photophosphorylation, electron transport activity and uncoupling or thylakoid energization are not affected under drought, whereas the possible effect of drought appears to be decreased ATP synthesis by the chloroplastic enzyme ATP synthetase (Lawlor 2002).

Drought condition retards chloroplastic ATP synthesis. At low RWC of leaves, inhibition in ATP synthesis occurs due to progressive inactivation or loss of ATP synthase activity resulting from increased Mg^{2+} concentration in chloroplasts (Lawlor 2002). Under drought, Mg^{2+} concentration increases in the chloroplasts. Sunflower plants grown at high Mg^{2+} level in the nutrient medium showed lower photosynthetic rate compared to control grown plants (Rao et al. 1987). Decreased ATP content affects cell metabolism substantially, limits RuBP biosynthesis, and decreases photosynthetic potential of plants under drought (Lawlor 2002). Though certain reports suggest the importance of stomatal closure in restricting the supply of CO_2 under mild drought, many studies suggest that photosynthesis may be more directly limited by nonstomatal factors such as the effect of drought on the ATP synthase activity, leading to a restricted ATP supply (Lawlor 2002). In fact, the activity of the enzyme ATP synthase is retarded and sometimes inhibited depending on the severity of drought stress (Farooq et al. 2009).

34.2.2.4 Carboxylation under Drought

Drought stress causes decrease in CO_2 assimilation due to stomatal closure and thereby restricting CO_2 entry into leaves (Parry et al. 2002). Some studies also suggest that decreased photosynthetic capacity under drought results from impaired regeneration of ribulose-1,5-bisphosphate as well as decreased availability and activity of CO_2 assimilating enzyme Rubisco (Parry et al. 2002). The amount of Rubisco in leaves is controlled by the rate of its synthesis and degradation. In many plant species such as tomato, Arabidopsis, and rice, a rapid decrease in the abundance of steady-state level of Rubisco small subunit (*rbc* S) transcripts has been observed under drought, which indicates decreased synthesis of this enzyme under drought (Parry et al. 2002; Srivastava 2002). Dehydration has a direct effect on Rubisco activity (Parry et al. 2002). In fact, Rubisco activity is modulated *in vivo* either by reaction with CO_2 and Mg^{2+}, leading to carbamylation of a lysine residue at the catalytic site that is essential for activity, or by the binding of inhibitors within the catalytic site, leading to inhibition of enzyme activity (Parry et al. 2002). In tobacco plants, it has been shown that drought leads to decrease in Rubisco activity, and this decrease is due to the presence of greater amounts of tight-binding Rubisco inhibitors in droughted leaves (Parry et al. 2002). Loss of Rubisco activity has been suggested as a rapid and very early response of drought stress in soybean (Majumdar et al. 1991). Both increased severity and duration of drought cause decrease in Rubisco activity and its protein content in sunflower and wheat plants (Parry et al. 2002). It is suggested that severe drought conditions limit photosynthesis due to a decrease in the activities of Rubisco, phosphoenol pyruvate carboxylase (PEPCase), NADP-malic enzyme (NADP-ME), fructose-1,6-bisphosphatase (FBPase), and pyruvate orthophosphate dikinase (PPDK) (Farooq et al. 2009). Under severe drought, carboxylation efficiency of Rubisco declines, and it acts more as oxygenase than carboxylase (Farooq et al. 2009).

According to Berkowitz and Gibbs (1983), drought causes acidification of chloroplast stroma, which also might contribute to inhibited Rubisco activity. Depending on the extent of accumulation of osmolytes as well as transport of ions and low molecular weight osmolytes, changes in chloroplast/protoplast volume occur, leading to alteration in the behavior of Rubisco (Kaiser 1987). It has been specifically shown that drought leads to limited RuBP regeneration and that RuBP concentration decreases in droughted plants (Flexas et al. 2002). This decreased level of RuBP might be due to progressive down-regulation of metabolic processes in mesophyll cells under drought and might be one of the factors for decreased Rubisco activity and thereby decreased photosynthetic efficiency under drought (Flexas et al. 2002).

34.2.2.5 Levels of Carbohydrates and Related Enzymes

Drought decreases the photosynthetic rate, disrupts carbohydrate metabolism, and alters the level of two end products of photosynthesis: starch and sucrose in the cells. Due to low carbon supply under drought conditions, chloroplastic starch may be remobilized to provide carbon in favor of more sucrose synthesis (Harn and Daie 1992). This is presumably due to increased activity of acid invertase under drought (Kim et al. 2000). Sucrose accumulation in the leaves may hamper the rate of sucrose export to the sink organs. The rate of sucrose synthesis is regulated by the two enzymes, cytosolic FBPase and sucrose phosphate synthase (SPS), which are subject to various types of metabolic regulations. Activities of these two enzymes decline in water-stressed leaves. In drought-stressed leaves of sugar beet (*Beta vulgaris* L.), when water potential decreased from −0.8 to −4.3 MPa, activities of FBPase and SPS declined and also starch content declined by 10%, whereas about threefold increase in sucrose level was observed (Harn and Daie 1992). In bean leaves, drought caused a decline in the partitioning ratio of starch/sucrose with no change in FBPase activity, whereas SPS activity was reduced by 60% (Vassey et al. 1991). Vassey et al. (1991), while estimating SPS activity in leaves of bean plants subjected to osmotic stress, observed that a mild water stress of −0.9 MPa reduced SPS activity by 50%, and this effect was a consequence of the inhibition of photosynthesis caused by stomatal closure. Drought causes decrease in photosynthesis and the consumption of assimilates in expanding leaves, as a result of which the amount of photosynthate exported from leaves decreases (Taiz and Zeiger 1998). Drought limits the size of the source and sink tissues and affects phloem loading. The assimilate translocation and dry matter portioning are also impaired; however, the effects vary with the plant species, stage, duration, and severity of drought (Farooq et al. 2009).

34.2.3 Heat Stress

Photosynthesis is one of the most temperature-sensitive processes in plants. High temperature causes physiological, biochemical, and molecular changes in plants and reduces photosynthesis. Most plants show extensive physiological and

biochemical adaptation to the large environmental range of temperature. However, when temperatures exceed the normal growing range of plants, heat injury takes place. Most of the crop plants generally grow in the 15–45°C temperature range. An increase in temperature of 10–15°C above normal growth temperature leads to disorganization of chloroplast thylakoid membranes, dissociation of PSII light-harvesting complex, destacking of grana lamellae, separation of nonbilayer lipids of thylakoid membranes, loss of photosynthetic O_2 evolution, as well as denaturation and inactivation of many enzymes. Even a moderate degree of heat stress slows the growth of the whole plant and causes decline in photosynthetic rate much faster than respiratory rate (Taiz and Zeiger 1998).

PSII is more sensitive to elevated temperatures than PSI (Berry and Bjorkmann 1980). In intact pea leaves with high-temperature treatment, PSII activity was inhibited or down-regulated, whereas PSI activity was stimulated (Havaux et al. 1991). In dark-adapted pea leaves, heat treatment caused inhibition of photosynthetic oxygen evolution and decrease in photochemical energy storage, which were correlated with a marked loss of variable PSII chlorophyll fluorescence emission, whereas the capacity of cyclic electron flow around PSI increased (Havaux et al. 1991). In cold-adapted C_4 *Atriplex sabulosa* plants, electron transport activity of PSII was more sensitive to high temperature than in heat-adapted C_4 *Tidestromia obiongifolia* plants (Bjorkman et al. 1980). In both of these species, decline in CO_2 assimilation under heat stress paralleled with decline in PSII activity. Experiments have shown that thermal inactivation of PSII is due to extraction of divalent ions Ca^{2+} and Mn^{2+} from the oxygen-evolving complex of PSII as well as due to dissociation of the 32 kDa extrinsic polypeptide that is involved in the stabilization of the Mn cluster (Carpentier 1999).

High temperature raises membrane fluidity, causes peroxidation and lateral diffusion of membrane lipids, and increases membrane permeability leading to decreased proton gradient formation across the thylakoid membrane (Berry and Bjorkmann 1980). Proton leakiness of the thylakoid membranes occurs at high temperatures resulting in the impairment of ATP synthesis (Yamori et al. 2014). Disorganization of chloroplast thylakoid membrane takes place at higher temperatures (Carpentier 1999). With the rise in temperature, increased fluidity of the membrane lipids takes place, and the strength of hydrogen bonds and electrostatic interactions between polar groups of proteins within the aqueous phase of the membrane decrease (Taiz and Zeiger 1998). As a result, integral membrane proteins tend to associate more strongly with lipid phase, and nonbilayer lipids of the thylakoid membrane form aggregates of cylindrical inverted micelles (Gounaris et al. 1983).

Synthesis of heat shock proteins (HSPs) occurs in plants exposed to high temperatures, and it has been observed that in the soluble portion of the chloroplasts, almost 19 small molecular weight HSPs (sHSPs) are synthesized; these sHSPs play an important role in photosynthesis and whole plant thermotolerance (Heckathorn et al. 2002). Chloroplast sHSPs are regarded as the most abundant and heat responsive of the plastid HSPs (Heckathorn et al. 2002). *In vivo* and *in vitro* experiments from *Agrostis stolonifera* genotypes indicate that chloroplast sHSPs associate with thylakoids and protect PSII during heat stress, possibly by stabilizing the O_2-evolving complex (Heckathorn et al. 2002). The expression of HSPs at high temperature is important for protein folding and assembly, stabilization of proteins and membranes, and cellular homeostasis at high temperature (Yamori et al. 2014).

Moderate heat stress causes a reduction in the activity of carboxylating enzyme Rubisco. The level and activation state of Rubisco decrease with rise in temperature. In many plant species, the Rubisco activation state decreases at high temperature (Yamori et al. 2014). Maintenance of a high-activation state of Rubisco is important for high-temperature acclimation of plants (Yamori et al. 2014). In maize plants, temperatures exceeding 32.5°C caused decline in the activation state of Rubisco, and the enzyme was nearly completely inactivated at 45°C (Crafts-Brandner and Salvucci 2002). Under heat stress, reduction in photosynthetic capacity is more associated with disorganization of chloroplast membrane and uncoupling of energy transfer mechanism in chloroplasts than the inactivation of enzymes.

34.2.4 CHILLING

Plants growing in tropical and subtropical regions often exhibit sensitivity to chilling temperatures. In a variety of plant species, exposure to temperatures between 0°C and 15°C causes chilling injury, leading to increased membrane permeability, inhibition of chlorophyll synthesis, swelling of chloroplasts, inhibition in the activities of photosystems, and increased susceptibility to photoinhibition, with eventually decreased photosynthetic ability (Yu et al. 2002; Wang and Guo 2005). Chilling inhibits light-dependent reactions in thylakoid membranes and reduces CO_2 assimilation in chloroplast stroma (Wang and Guo 2005). The activity of light-dependent reaction appears to be more sensitive to chilling than the dark reaction. The extent of injury due to chilling depends on the duration of chilling, irradiance level, relative humidity, and the plant species (Ting et al. 1991). Crop species like maize, bean, rice, tomato, cucumber, sweet potato, and cotton are chilling-sensitive. Even in the same crop species, certain cultivars are chilling-sensitive and others are tolerant (Ting et al. 1991). Chilling injury occurs in sensitive plant species at temperatures that are too low for normal growth, but not so low that ice formation could take place (Taiz and Zeiger 1998). In chilling-sensitive rice plants, the reaction centers and antenna systems are damaged more severely due to chilling leading to the lower photosynthetic rate compared to similarly stressed tolerant plants (Wang and Guo 2005). Nonphotochemical quenching has been suggested to be a self-protective mechanism that plays an important role in protecting photosynthetic apparatus against photoinhibition under chilling (Wang and Guo 2005).

Photosynthesis is significantly lower at temperatures around 10°C compared to that at 20–25°C (Laing et al. 2002). In chilling-sensitive plants, the lipids in the bilayer have a high percentage of saturated fatty acid chains, and such membranes tend to solidify into a semicrystalline state at a temperature well above 0°C (Taiz and Zeiger 1998). In high-yielding Indica rice varieties, chilling leads to decline in photosynthetic rate, swelling of chloroplasts, and accumulation of starch grains within the chloroplasts (Park and Tsunoda 1979). When black alder (*Alnus glutinosa*) seedlings fertilized with different doses of nitrate were exposed to 2.5 h of night time chilling temperatures of −1–4°C, net photosynthesis declined by 17% for plants receiving low nitrate fertilizer (0.36 mM) and by 19% for plants receiving high nitrate fertilizer (7.14 mM). It was suggested that in black alder, chilling stimulated stomatal closure only at a high nitrate level and that the major impact of chilling on photosynthesis involved interference with biochemical functions (Vogel and Dawson 1991).

Light intensity influences the effect of low temperature on photosynthesis. At high light intensities and chilling, PSII is uniquely damaged and photoinhibition of photosynthesis takes place (Yu et al. 2002). Photoinhibition occurs at low temperatures even at low light intensities, leading to PSII injury (Lapina and Popov 1999). Photoinhibition as well as decline in CO_2 assimilation have been reported in many plants following chilling treatment. In *Zea mays*, when either lamina of the second leaf or the whole plant was subjected to chilling treatment, significant photoinhibition of PSII occurred (Nie et al. 1992). Plants exposed to low temperatures for a longer period show sustained down-regulation of PSII complexes with low intrinsic efficiency of PSII electron transport (F_v/F_m) (Matsubara et al. 2002). In intact leaves of an Australian mistletoe *Amyema miquelli*, efficiency of excitation energy transfer from light-harvesting pigments to Chl *a* molecules in PSII core complexes was markedly reduced in winter (Matsubara et al. 2002). Chilling leads to degradation of photosynthetic pigments, which is more pronounced in chilling-sensitive species like *Cucumis sativa* and maize compared to chilling-tolerant species like *Pisum sativum* (Lapina and Popov 1999). Sensitization of photosynthesis to photoinhibition at low temperature appears to be due to decreased activity of oxygen-scavenging enzymes, slowdown of physiological processes, and the inhibition of PSII repair cycle (Ivashuta et al. 2002). It has been shown that presence of a large proportion of *cis*-unsaturated fatty acids in phosphatidyl glycerol (PG) of chloroplast membranes is correlated to chilling resistance in plants (Ariizumi et al. 2002). Transgenic rice seedlings showing 29.4% and 32% *cis*-unsaturated fatty acids compared to wild-type seedlings with 19.3% fatty acids had improved chilling tolerance (Ariizumi et al. 2002).

In chilling-sensitive plant species, low-temperature exposure causes significant loss of the activity of certain carbon reduction cycle enzymes like Rubisco, sedoheptulose-1,7-bisphosphatase (SBPase), and chloroplastic FBPase. In *Z. mays* genotypes, growth at 14°C resulted in a 75% decrease in Rubisco activity and a 50% decrease in the activity of the C_4 enzyme, NADP-malate dehydrogenase, compared to plants grown at 24°C, whereas no change was observed in the activity of PEPCase (Stamp 1987). An exposure of 5°C and 7°C of the subtropical fruit tree mango (*Mangifera indica* L.) led to substantial decline in CO_2 assimilation, an increase in stomatal limitation, and lower Rubisco activity (Allen et al. 2000). Similarly, in herbaceous chilling-sensitive crop tomato, overnight chilling caused severe disruption in the circadian regulation of key photosynthetic enzymes, leading to dysfunction of photosynthesis (Allen et al. 2000).

Accumulation of starch, sucrose, and hexoses is observed in plants exposed to low temperature (Guy et al. 1992; Yuanyuan et al. 2009). The most abundant and most commonly accumulated sugar is sucrose, which may accumulate up to tenfold in certain plants (Guy et al. 1992). In spinach leaves, when plants were transferred from 25°C to 5°C conditions, a sudden increase in the sucrose level was observed with a concomitant increase in the activity of its biosynthetic enzyme SPS; however, the activities of the enzymes sucrose synthase and invertase remained unaffected (Guy et al. 1992). As a cryoprotectant, sucrose accumulation has adaptive significance for cold-exposed plants (Guy et al. 1992). Accumulation of starch under low-temperature treatment is mainly due to production of this photosynthate in excess of its needs (Guy et al. 1992). Besides, the levels of a few oligosaccharides increase during cold acclimation in cereals and woody plants (Yuanyuan et al. 2009). Galactosylsucrose oligosaccharides such as raffinose and stachyose also accumulate during seasonal cold acclimation in woody plants (Yuanyuan et al. 2009). Soluble sugars serve as osmoprotectants and nutrients, and also contribute to the increase in the cryostability of cellular membranes by interacting with the lipid bilayer and protecting the cells from damage caused by cold stress (Yuanyuan et al. 2009). Higher sugar concentrations might also enhance leaf senescence and down-regulate the process of photosynthesis (Yuanyuan et al. 2009).

Liu et al. (2012) observed that tomato (*S. lycopersicum*) plants when exposed to a low temperature of 4°C for 3 days showed the expression of many TFs, post-translational proteins, metabolic enzymes, and photosynthesis-related genes, which were specifically modulated. These specific modulations appear to play a vital role in conferring cold tolerance in tomato.

34.2.5 ANAEROBIOSIS

Plants growing under natural conditions are often exposed to soil waterlogging. Due to excessive rain, flood, and poor drainage, soil gets water-logged, physicochemical properties of the soil is altered, and oxygen gets depleted from the bulk water, leading to anaerobiosis. Due to O_2 restriction, plants show depressed growth, reduced photosynthesis, and severe losses in yield (Taiz and Zeiger 1998). The level of carbohydrates and their translocation get affected (Parent et al. 2008). Plants show reduction in stomatal conductance and decreased root hydraulic activity (Parent et al. 2008). In some plant species like pea and tomato, flooding leads to stomatal closure without significant change in leaf water potential (Taiz and Zeiger 1998). It is believed that oxygen shortage in roots stimulates ABA production, and movement of ABA to leaves can account for the stomatal closure (Zhang and Zhang 1994).

Plants growing under anaerobic conditions show reduced stomatal aperture, a decrease in leaf chlorophyll content, early senescence, and reduced leaf area, which ultimately contribute to reduced photosynthesis (Parent et al. 2008). Other factors like CO_2 deficiency in water, low irradiances in muddy water, settling of silt on the leaves, as well as slow diffusion of gases in water also contribute to decreased photosynthesis (Setter et al. 1989). Slow diffusion results in restriction of CO_2 influx during photosynthesis. Complete submergence is a common feature associated with low-land rice crop in Southeast Asian flood plains, where deep water and floating rice cultivars are grown (Lambers and Seshu 1982). Setter et al. (1989), while examining the effect of submergence on photosynthetic capacity of rice cultivars, observed that due to stagnation of water, supply of CO_2 to the chloroplasts was restricted, and this was the prime reason for decreased photosynthesis of plants. CO_2 enrichment of water increased the rate of photosynthesis. Long period of submergence makes the leaves chlorotic, and chloroplasts lose the capacity to fix CO_2 (Lambers and Seshu 1982). Concentration of soluble sugars decreases in plants after submergence; a greater decrease is observed in submergence-sensitive cultivars than the tolerant ones (Setter et al. 1989). O_2 deficiency tends to accelerate breakdown of carbohydrates, and therefore, a high rate of photosynthesis is required in submerged plant parts in order to compensate for the carbon loss (Setter et al. 1989). Anaerobic conditions reduce metabolic activity and cause reduced translocation of photoassimilates from source leaves to sink roots with a marked decline in sugar level. Maintenance of adequate photosynthetic activity and requisite level of sugars in roots are important adaptations of flooding-tolerant plants (Parent et al. 2008).

34.2.6 AIR POLLUTANTS

Many gases such as CO_2, CO, SO_2, NO, NO_2, H_2S, HF, ozone, ammonia, and ethylene as well as many hydrocarbons, which are emitted in the atmosphere, serve as air pollutants and in excess concentrations inhibit plant growth and show deleterious effect on photosynthesis. Among these, SO_2, fluorides, ammonia, and particulate matters are local pollutants. Ozone, a major oxidant, which is produced in the atmosphere during a complex reaction involving nitrogen oxides (NO, NO_2), hydrocarbons, and sunlight in a chain of atmospheric events, is considered as one of the most potent phytotoxic air pollutants. Due to higher concentration of CO_2 and other *greenhouse gases* in the atmosphere, increased absorption of infrared radiation takes place (Taiz and Zeiger 1998), which is posing a serious threat of global warming. This may ultimately have serious impact on plant health.

SO_2 enters leaves through stomata and causes stomatal closure. It gets dissolved in the cells and produces bisulfite and sulfite ions; the latter is toxic for the cell (Taiz and Zeiger 1998). Fluoride also enters the leaves mainly through the stomatal openings. Particulate matters such as cement dust, magnesium-lime dust, and carbon soot get deposited on the surface of leaves and inhibit normal photosynthesis.

Elevated CO_2 levels cause stomatal closure and reduce uptake of other pollutants (Taiz and Zeiger 1998). NO or NO_2 also reaches the cells through stomata and, when present in air in concentration greater than 0.1 $\mu l\ L^{-1}$, inhibits photosynthesis. The concentration of the polluting gases varies depending on location, direction of wind, rainfall, sunlight, humidity, temperature, etc. (Taiz and Zeiger 1998). Table 34.3 shows common air pollutants, visible morphological changes that occur in plants due to these pollutants, and the associated metabolic alterations.

Ozone is the main pollutant in the oxidant smog complex. It is present in high concentrations in urban and nearby areas. Ozone reduces the rate of photosynthesis by damaging photosynthetic apparatus and causes great losses in crop yield. High-ground ozone level has been reported in many areas of the world due to industrial emissions. Ozone binds to plasma membrane, and the regulation of stomatal aperture by guard cells is disturbed. Both SO_2 and ozone inhibit the translocation of photosynthetic products via a disturbed phloem loading due to inactivation of the plasmalemma-bound ATPase, which ultimately leads to increased starch accumulation and finally bleaching of the photosynthetic pigments (Dominy and Heath 1985). Due to its highly reactive nature, ozone damages chloroplast envelope and thylakoid membranes, disrupts chemiosmotic balance, and causes decrease in the level and inhibition in the activity of the carboxylating enzyme Rubisco (Quartacci et al. 2000; Guidi et al. 2002). The impairment of carboxylation is regarded as the primary effect of ozone on photosynthesis (Guidi et al. 2002). Presence of ozone promotes photoinhibition even when the light intensity is moderate (Guidi et al. 2002). Decomposition of ozone spontaneously in aqueous medium within the cell or its reaction with a number of compounds such as phenolics and other organic molecules produces ROS such as superoxide anion $\left(O_2^-\right)$, singlet oxygen $(^1O_2{}^*)$, hydroxyl radical (OH), and peroxides that denature proteins, damage nucleic acids, and cause peroxidation of membrane lipids (Taiz and Zeiger 1998). Singh et al. (2009), while investigating the effects of elevated ozone on electron transport, carbon fixation, stomatal conductance, and pigment concentrations in soybean (*Glycine max* L.), observed that dark reactions of photosynthesis were more sensitive than light reactions to ozone stress. These workers observed that the reduction in photosynthesis due to ozone was a result of damage to the photosynthetic apparatus, leading to accumulation of internal CO_2 and stomatal closure (Singh et al. 2009).

Global climate change has led to the depletion of the ozone layer in both poles resulting in the penetration of increasing amounts of ultraviolet-B (UV-B) radiation to the earth's surface. It is a matter of great concern to both terrestrial and aquatic plants, which are the primary producers supporting life on earth. UV-B is injurious to photosynthetic apparatus and inhibits photosynthesis in both C_3 and C_4 plants (Ormond and Hale 1995). Due to UV-B radiation, damage to PSII occurs, marked by an increase in variable chlorophyll fluorescence (Stapleton et al. 1997). In rice and pea leaves, the quantum yield of photosynthetic oxygen evolution decreased with a concomitant decrease in the ratios of variable to maximum chlorophyll fluorescence yield due to UV-B radiation (He et al. 1994; Ormond and Hale 1995). Destruction of chlorophyll and carotenoids occurs due to UV-B radiation in sensitive plant species (He et al. 1994). It has been observed that UV-B radiation suppresses the expression and synthesis of

TABLE 34.3
Common Air Pollutants and Their Effects on Photosynthesis

Pollutants	Morphological Changes	Metabolic Alterations
1. SO_2 and derivatives	1. Chlorophyll bleaching 2. Leaf discoloration 3. Stomatal closure 4. Growth retardation	1. Alteration in $FAD/FADH_2$ and NAD^+ functions 2. Decrease in ATP pool 3. Peroxidation of thylakoid membranes 4. Inhibition in translocation of photosynthetic products
2. NO and NO_2	1. Change in leaf color 2. Growth retardation	1. Reaction with olefins 2. Peroxidation of membrane lipids
3. Elevated CO_2	1. Stomatal closure 2. Growth retardation 3. Abscission	1. Reduced uptake of nutrients 2. Decreased root permeability
4. Ozone	1. Decreased stomatal conductance 2. Bleaching of photosynthetic pigments 3. Damage to chloroplast envelope and photosynthetic apparatus 4. Abscission	1. Splitting of olefinic bonds and reaction with thiols 2. Oxidation of glutathione and proteinic –SH gps 3. Inhibition of lipid synthesis in mitochondria and microsomes 4. Inactivation of several key enzymes 5. Inactivation of α-1-proteinase inhibitor 6. Inactivation of plasmalemma-bound ATPase 7. Uncoupling of photophosphorylation
5. Peroxides and PAN (peroxyacetyl nitrate)	1. Epinasty 2. Necrosis of leaves 3. Browning 4. Early ripening 5. Abscission	1. Reaction with NADPH 2. Lipid peroxidation 3. Acetylation of amines 4. Reaction with thiols of enzymes

photosynthetic proteins Rubisco large (*rbc* L) and small (*rbc* S) subunits and chlorophyll *a/b*-binding proteins (Mackerness and Jordan 1999). The extent of down-regulation is dependent on the severity of UV-B exposure. UV-B exposure damages photosynthetic apparatus, and maximum damage is seen during the senescence phase of development. Macromolecules of the cell such as DNA, RNA, and proteins due to their strong absorption at 280–315 nm are more prone to damage due to UV radiations.

34.2.7 METAL TOXICITY

In areas with high anthropogenic pressures, contamination of soil with metals has become a major global environmental problem, leading to losses in agricultural yield. Nonessential metals for plants, which contaminate soil environment, include Cd, Pb, Hg, Cr, and As, whereas many essential metals, which are micronutrients for plants but accumulate in high concentrations in the soil due to anthropogenic activities, include Fe, Cu, Ni, Zn, etc. (Dubey 2010). High levels of these metals in the soil environment may arise due to wastes from metal work industries, mining, presence of naturally occurring ore bodies, exhausts from automobiles, power stations, exploitation of natural resources, agricultural runoff including waste water irrigation, waste disposal, airborne pollution, etc. (Dubey 2010). Most of these metals show serious adverse effects on growth and metabolic processes in plants and cause reduction in chlorophyll content, degeneration of chloroplasts, disorganization of chloroplast thylakoids, and reduction in photosynthesis.

Cadmium, a long-range transported heavy metal pollutant, inhibits the synthesis of chlorophylls and carotenoids and affects the ultrastructure of developing chloroplasts in many plant species (Muthuchelian et al. 1988; Bhardwaj and Mascarehas 1989). Exposure of 7-day-old etiolated *Vigna sinensis* L (savi) leaf segments to heavy metals Cu^{2+} and Cd^{2+} for 24 h caused inhibition in the synthesis of chlorophylls, with more inhibition of Chl *a* than of Chl *b* (Muthuchelian et al. 1988). Similarly, Pb^{2+} alters photosynthetic pigment composition and disturbs the geranial structure of chloroplasts (Carpentier 2002). Pb^{2+} reduces the concentrations of total in rice (*O. sativa* L.) plants. When rice seedlings were raised in sand cultures containing nutrient solutions supplemented with 500 or 1000 μM $Pb(NO_3)_2$, the level of chlorophylls (Chl *a+b*) declined in the seedlings compared to controls (Verma and Dubey 2002). Ni^{2+} reduces pigment content in various photosynthetic organisms and affects both photosystems (Carpentier 2002). Both Cd and Pb cause reduced content of photosynthetic pigments and distort the shape of stomata (Srivastava et al. 2014). Srivastava et al. (2014) observed that when rice seedlings were grown in sand cultures and exposed to 150 μM $Cd(NO_3)_2$ or 600 μM $Pb(CH_3COO)_2$ individually or in combination for 8–16 days, a significant reduction in the level of photosynthetic pigments Chl *a*, Chl *b*, and carotenoids was observed. Scanning electron microscope (SEM) imaging of leaf stomata from Cd- and Pb-treated plants revealed that Pb caused more distortion in the shape of guard cells than Cd (Srivastava et al. 2014).

Most of the heavy metals preferentially inhibit PSII activity. In chloroplasts isolated from Cd^{2+}-treated *T. aestivum* seedlings, 70% decline in oxygen evolution and inhibition in PSII-mediated electron transport were observed (Bhardwaj and Mascarenhas 1989). It is suggested that Cd^{2+} affects electron transport on the oxidizing site of PSII (Bhardwaj and Mascarenhas 1989). Cadmium also reduces the turnover rate of the D1 protein of the reaction center of PSII (Geiken et al. 1998). Cu^{2+}, which is an integral part of plastocyanin, inhibits the electron transport at a site connecting both PSII and PSI. Excess copper induces changes in the lipid composition and fluidity of PSII-enriched membranes in wheat (Quartacci et al. 2000). Under *in vitro* conditions, high Cu (II) levels significantly modify the oxygen-evolving complex of PSII by dissociating the Mn cluster and associated cofactors in PSII-enriched oxygenic and non-oxygenic thylakoid membranes (Yruela et al. 2000). Hg^{2+} inhibits both photosystems; the inhibition in PSI is reported at the donor side beyond the cytochrome *b/f* complex, whereas PSII is affected on both donor and acceptor sides (Carpentier 200). Tripathy et al. (1981) demonstrated that Ni^{2+} affected both photosystems with greater effect on PSII than on PSI. Hg^{2+} binds to thylakoid membrane proteins, reacts directly with plastocyanin, replaces copper, and alters the enzyme ferredoxin:NADP-reductase by reacting with the –SH group (Carpentier 2002). Mn^{2+} toxicity reduces photosynthesis in rice bean seedlings due to peroxidative impairment of thylakoid membrane function (Subramanyam and Rathore 2000). Aluminum (Al) toxicity to plants is a prevalent problem in acid soils. Al together with kinetin delays the loss of pigment and protein contents and the activities of PSII and PSI in detached wheat primary leaves (Subhan and Murthy 2000).

Cadmium and Ni cause decline in CO_2 fixation rates and have pronounced effects on the Calvin cycle enzymes (Krupa et al. 1993; Carpentier 2002). A reduced CO_2 assimilation rate in *Helianthus annus* plants subjected to Cd(II) treatment, in addition to reduced Rubisco activity, photochemical quenching, and quantum efficiency of PSII, was observed (Cagno et al. 2001). In pigeon pea (*Cajanus cajan* L.) plants, due to Cd and Ni treatments, *in vivo* CER decreased and marked inhibition in the activities of the Calvin cycle enzymes occurred (Sheoran et al. 1990). Rice plants grown over a 30-day period in nutrient solution containing increasing copper levels ranging from 0.002 to 6.25 mg L^{-1} showed a progressive decline in Rubisco activity (Lidon and Henriques 1991). It is concluded that, in rice plants, Cu-led inhibition in photosynthetic activity is primarily due to decreased Rubisco activity (Lidon and Henriques 1991).

Arsenic (As) is a ubiquitous element present in the environment, and it is a significant contaminant of soils and groundwater in many regions of the world. The contents of both Chl *a* and *b* decrease in leaves of plants grown in soil with high arsenic concentrations (Sharma et al. 2014). In rice plants, chlorophyll contents decreased significantly with a concomitant reduction in growth and yield of the plants, suggesting that arsenic toxicity affects the photosynthesis, which ultimately results in the reduction in growth and yield of rice (Rahman et al. 2007).

Heavy metals affect photosynthetic partitioning within the different organs of plants. Cd toxicity in rice limits the availability of the photoassimilate sucrose in the cells by favoring its enhanced degradation due to invertase and sucrose synthase activities (Verma and Dubey 2001). Calvin cycle reactions are slowed down in metal-exposed plants, and limitation of ATP and NADPH consumption occurs, which leads to inhibition of photosynthetic electron transport (Krupa et al. 1993). These observations suggest that inhibition in photosynthesis in metal-exposed plants is due to inhibition of synthesis of chloroplast pigments, reduced activity of the electron transport system, as well as the Calvin cycle enzymes.

34.3 STRESS-INDUCED ROS PRODUCTION AND DAMAGE TO PHOTOSYNTHETIC APPARATUS

All environmental stresses such as salinity, drought, chilling, heat, metal toxicity, gaseous pollutants, UV-B radiation, etc. cause enhanced production of ROS. These ROS include free radicals such as superoxide anion $\left(O_2^-\right)$ and hydroxyl radical ($^\bullet OH$), as well as nonradical molecules like hydrogen peroxide (H_2O_2), singlet oxygen ($^1O_2$), etc. (Sharma et al. 2012). In plants, electron transport activities of chloroplasts substantially contribute to overall production of ROS in the cell. ROS are continuously produced in the chloroplasts by partial reduction of O_2 molecules. Though within the chloroplasts, ROS are generated from several locations, but ETCs in PSI and PSII are the key sources of ROS production. Among the two photosystems, the major site of O_2^- production is the thylakoid membrane-bound primary electron acceptor of PSI, where only due to one electron transfer onto O_2, the first and moderately reactive ROS O_2^- is produced (Gill and Tuteja 2010). ROS are always formed by the inevitable leakage of electrons onto O_2 due to disruption of cellular homeostasis. When the level of ROS exceeds the defense mechanisms, a cell is said to be in a state of *oxidative stress*. The overproduced ROS during stresses can pose a threat to cells and more especially thylakoid membranes and associated proteins causing peroxidation of lipids, oxidation of proteins, damage to nucleic acids, and enzyme inhibition, and ultimately leading to death of the cells (Sharma et al. 2012). Chloroplasts are especially at risk of oxidative damage due to the abundance of the photosensitizers and polyunsaturated fatty acids (PUFAs) in the chloroplast envelope. The highly reactive $^1O_2$ can be produced via triplet chlorophyll (Chl) formation in the antenna system and in the reaction center of PSII, whereas the Chl triplet state can react with $^3O_2$ to give very highly destructive ROS $^1O_2^-$ (Krieger-Liszkay 2005). Due to the presence of ROS-producing centers such as triplet Chl, ETC in PSI and PSII, chloroplasts serve as a major site of ROS production.

The overproduced ROS during stresses cause oxidation of proteins leading to their covalent modifications. Protein oxidation events are mainly irreversible, and formation of protein carbonyls is widely used as a marker of protein oxidation. Oxidation of many amino acids such as Arg, His, Lys, Pro, Thr, and Trp gives free carbonyl groups, and the function of such oxidized proteins gets altered. Various stresses lead to the carbonylation of proteins within the chloroplasts altering their function. Oxidized proteins are more efficiently degraded, and therefore, proteolytic activity invariably increases under stresses. When *Arabidopsis thaliana* plants were treated with 300 mM NaCl for 72 h under continuous illumination, and oxidatively modified proteins were analyzed using a carbonyl-targeted proteomics approach, it was observed that 17 types of proteins were modified by two more in chloroplasts and other organelles of stressed plants compared to the nonstressed plants (Mano et al. 2014). In mung bean (*P. vulgaris*) seedlings, Cd toxicity elevates the level of lipid peroxides and causes decrease in the chlorophyll level, which appears to be due to peroxide-mediated degradation (Somashekaraiah et al. 1992). Carotenoids scavenge ROS and protect chloroplast membranes and chlorophyll molecules from light-dependent oxidative damage during stressful conditions (Ashraf and Harris 2013).

Using intact thylakoids from spinach leaves, it has been shown that the damage to PSII under heat stress is due to the direct action of ROS produced during heat treatment (Yamashita et al. 2008). Heat treatment of 40°C for 30 min to spinach thylakoid membranes caused cleavage of D1 protein of PSII, aggregation of the D1 protein with the polypeptides D2 and CP43, and release of three extrinsic proteins: PsbO, –P, and –Q (Yamashita et al. 2008). These events were suppressed due to the addition of ROS scavenger sodium ascorbate showing that heat stress induced lipid peroxidation in thylakoid membranes under aerobic conditions (Yamashita et al. 2008).

34.4 ROLE OF TFs AND MAP KINASES IN PHOTOSYNTHETIC METABOLISM UNDER STRESSES

Transcriptional and post-transcriptional regulations play an important role in environmental stress-induced metabolic alterations in plants. TFs are essential for the regulation of the gene expression. Abiotic stress-induced alterations in photosynthetic responses require differential expression of genes, regulated by specific TFs. A number of TFs have been identified that regulate directly or indirectly the expression of various genes related to the process of photosynthesis. TFs are proteins with a DNA domain that binds to the *cis*-acting elements present in the promoter of a target gene and either induces (activators) or represses (repressors) the activity of the RNA polymerase and regulates gene expression (Saibo et al. 2009). Depending on their DNA-binding domain, TFs have been grouped into different families, and a group of genes controlled by a particular TF is called a regulon (Saibo et al. 2009). Regulons associated with plant responses to abiotic stresses are either ABA dependent or ABA independent. A large number of genes involved in both stomatal and nonstomatal processes associated with photosynthesis are differentially expressed under stresses, and the TFs modulate the expression of these genes.

Two myeloblastosis oncogene (MYB)-type TFs have been identified, which are regulators of stomatal movements. AtMYB60, a R2R3-MYB gene of Arabidopsis, was specifically expressed in guard cells, leading to increase in stomatal aperture, and the expression was regulated by light, ABA, and drought (Cominelli et al. 2005). Similarly AtMYB61, another member of the *A. thaliana* family of R2R3-MYB TFs, was also shown to be specifically expressed in the guard cells, where its transcription was repressed by light, leading to decrease in stomatal aperture (Liang et al. 2005). While AtMYB60 causes stomatal opening by the ABA-dependent signaling pathway, AtMYB61 mediates closure of stomatal aperture by an ABA-independent signaling pathway. This suggests that AtMYB60 and AtMYB61 have distinct expression patterns and functions (Saibo et al. 2009). It is regarded that stomatal development is related to environmental conditions, and five TFs have been identified, which are essential for cell transitions leading to stomatal formation. However, the mechanisms of how environmental conditions regulate stomatal development are not fully understood (Saibo et al. 2009).

The genes associated with nonstomatal processes of photosynthesis such as those encoding chlorophyll *a/b*-binding (CAB) proteins, PSI and PSII subunits, oxygen-evolving complex, and Rubisco subunits are under the control of a transcriptional regulatory network that responds to external stimuli (Saibo et al. 2009). Two MYB-like TFs have been isolated from barley, HvMCB1 and HvMCB2, which bind specifically to CAB promoters from barley and wheat and are required for maximal CAB gene expression. The transcription of both genes *HvMCB1* and *HvMCB2* is regulated under salt, osmotic, and oxidative stress (Churin et al. 2003). A TF LONG HYPOCOTYL 5 (HY5), a bZIP-type protein, controls the expression of Chl *a/b* binding protein 2 (CAB2) and regulates the expression of the gene for the Rubisco small subunit (RbcS1A) (Lee et al. 2007). Another TF, OsMYB4, when expressed, causes accumulation of glycine betaine, which stabilizes Rubisco under saline conditions (Khafagy et al. 2009). Similarly, in maize, two TFs DOF1 and DOF2 have been identified, DOF1 being an activator of C4-PEPC gene transcription whereas DOF2 is its repressor (Yanagisawa and Sheen 1998).

It has been shown in CAM plants that transcriptional regulation mediates the expression of CAM-specific genes leading to enhanced expression under drought and saline conditions (Saiba et al. 2009). Both *cis*-acting DNA sequences and *trans*-acting factors play a vital role in transcriptional activation of CAM-specific genes under salinity (Saiba et al. 2009). In CAM plants, increased synthesis of the specific isozymes of PEPCase and NAD-dependent glyceraldehyde-3-phosphate dehydrogenase occurs under salinity due to enhanced transcription of the genes *Ppcl* and *Gapl*, which have binding sites for the MYB class of TFs (Schaeffer et al. 1995). This suggests that, in CAM plants, increased synthesis of CAM-specific enzymes is mediated via MYB-type TFs. TFs regulating several stress responsive genes have been used as important targets for engineering abiotic stress tolerance with improved photosynthetic efficacy in crop plants. The results are more encouraging compared to a single gene encoding a single protein. Many crop plants have been transformed with TFs, and these plants are tolerant to stresses with improved photosynthetic parameters (Saibo et al. 2009).

Mitogen-activated protein kinases (MAPKs) regulate a wide range of cellular metabolic processes by effectively serving as components in the signal transduction pathway. MAPK cascades include a MAP kinase kinase kinase (MAPKKK), a MAP kinase kinase (MAPKK), and a MAP kinase (MAPK), which work in concert. Activated MAPKs move from cytoplasm to the nucleus, and phosphorylate TFs leading to alterations in expression of genes (Zhang et al. 2011). Increased expression of MAP kinase enzyme cascade has been shown in plants exposed to cold, salinity, and drought stresses (Ashraf and Harris 2013). Transgenic maize plants constitutively expressing tobacco MAPKKK (NPK1) resulted in enhanced tolerance to drought with a concomitant high rate of photosynthesis, suggesting the role of NPK1 in protection of photosynthetic machinery under stresses (Shou et al. 2004; Ashraf and Harris 2013).

34.5 CONCLUDING REMARKS

Photosynthesis is essentially the only mechanism of energy input into the living world and represents a dominant physiological process in plants that is highly sensitive to environment. Various abiotic stresses like salinity, drought, heat, chilling, anaerobiosis, polluting gases, and excess metals in the soil limit photosynthetic efficacy of crops resulting in decreased productivity. Despite the extensive studies conducted on the effects of various stressful conditions on different photosynthetic parameters in growing plants, our knowledge is still incomplete to specify the precise mechanisms related to stress recognition and sequence of events triggered, thereby leading to alteration in various components of photosynthetic process under the particular stress. Various steps involved in the overall process of photosynthesis associated with conversion of transient energy of a photon into stable chemical energy like sucrose and other photosynthates within the photosynthetic apparatus are so tightly linked that any impairment at a particular step would influence the complete series of events ultimately limiting photosynthesis. Our scientific knowhow and devices are to be advanced to exactly identify and monitor the slightest change occurring due to a stressful condition on the different photosynthetic parameters.

Salinity and drought are major constraints to crop productivity in arid and semiarid tropics because they affect the otherwise most productive agricultural lands. Extremes of high and low temperatures cause heat and chilling injury in plants. Plants growing in flooded soils often suffer with the problem of anaerobiosis. Industrialization has led to increased content of various metals in the soil. Therefore, to produce crop plants with better photosynthetic efficacy and showing tolerance to different stressful conditions, it is essential to identify various regulatory processes and limitations to photosynthesis under the stresses and the stepwise details of various stomatal and nonstomatal photosynthetic events influenced by these stresses, and to examine the adaptive mechanisms by which plants would show tolerance to these stresses. Abiotic stress-tolerant crop plants should maintain optimal photosynthetic efficiency, avoiding the energy imbalance caused due to stresses. In nature, plants are often exposed to multiple stresses, and the combination of these stresses coactivates various TFs and signaling pathways and causes complex responses related to photosynthesis. Therefore, interaction of many stresses occurring together on photosynthetic parameters needs to be examined to produce plants tolerant to multiple stresses.

Several classes of genes have been identified to engineer crop plants tolerant to a particular stress or multiple stresses with better yield and photosynthetic efficiency. Overexpression of genes-encoding enzymes that synthesize osmoprotectants and regulate metabolic pathways leading to salinity and drought adaptation has helped in producing salt- and drought-tolerant transgenic plants with better photosynthetic efficacy. Transgenic plants overexpressing mitochondrial superoxide dismutase (Mn-SOD) show improved tolerance to drought, freezing, and many herbicides. Tolerance to oxidative stress is being realized as an important factor in providing tolerance to a wide range of environmental stresses.

Many TFs regulate the expression of photosynthesis-related genes. Some TFs have been identified, which regulate stomatal aperture under stresses, and some regulate expression of genes coding for components of photosynthetic carbon metabolism under stresses. Certain transgenic plants have been successfully produced that overexpress TFs and show tolerance to abiotic stresses with enhanced photosynthetic efficiency.

Stress tolerance is a multigenic phenomenon, and therefore identification, characterization, and assessment of many more complex mechanisms involving interplay of TFs, signaling pathways, and gene products, which govern complex traits like WUE, stomatal conductance, ability to exclude salt, and maintenance of optimal photochemical and carboxylation reactions under stresses, are essential. With the advent of biotechnological tools and rapid advancements being made to explore the details of photosynthetic phenomenon under stresses, it is likely that, in near future, many more specific genes associated with stress tolerance would be identified to produce many transgenic crop plants with improved tolerance to stressful environments with improved photosynthetic efficiency.

REFERENCES

Akram, N.A. and M. Ashraf. 2011. Improvement in growth, chlorophyll pigments and photosynthetic performance in salt-stressed plants of sunflower (*Helianthus annuus* L.) by foliar application of 5-aminolevulinic acid. *Agrochimica* 55:94–104.

Allen, D.J., K. Ratner, Y.E. Giller, E.E. Gussakovsky, Y. Shahak, and D.R. Ort. 2000. An overnight chill induces a delayed inhibition of photosynthesis at mid day in mango (*Mangifera indica* L.). *J Exp Bot* 51:1893–1902.

Amira, M.S. and A. Qados. 2011. Effect of salt stress on plant growth and metabolism of bean plant *Vicia faba* (L.). *J Saudi Soc Agric Sci* 10:7–15.

Ariizumi, S., S. Kishitani, R. Inatsugi, I. Nishida, N. Murata, and K. Toriyama. 2002. An increase in unsaturation of fatty acids in phosphatidyl glycerol from leaves improves the rates of photosynthesis and growth at low temperatures in transgenic rice seedlings. *Plant Cell Physiol* 43:751–758.

Ashraf, M. and P.J.C. Harris. 2013. Photosynthesis under stressful environments: An overview. *Photosynthetica* 51:163–190.

Azzabi, G., A. Pinnola, N. Betterle, R. Bassi, and A. Alboresi. 2012. Enhancement of non-photochemical quenching in the Bryophyte *Physcomitrella patens* during acclimation to salt and osmotic stress. *Plant Cell Physiol* 53:1815–1825.

Barber, J. 1982. Influence of surface changes on thylakoid structure and function. *Annu Rev Plant Physiol* 33:261–295.

Behera, R.K., P.C. Mishra, and N.K. Choudhury. 2002. High irradiance and water stress induce alterations in pigment composition and chloroplast activities of primary wheat leaves. *J Plant Physiol* 159:967–973.

Berkowitz, G.A. and M. Gibbs. 1983. Reduced osmotic potential inhibition of photosynthesis: Site specific effects of osmotically induced stromal acidification. *Plant Physiol* 72:1100–1109.

Berry, J. and O. Bjorkmann. 1980. Photosynthetic response and adaptation to temperature in higher plants. *Annu Rev Plant Physiol* 31:491–543.

Bhardwaj, R. and C. Mascarenhas. 1989. Cadmium induced inhibition of photosynthesis in vivo during development of chloroplast in *Triticum aestivum* L. *Plant Physiol Biochem (India)* 16:40–48.

Bjorkman, O., M.R. Badger, and P.A. Armond. 1980. Response and adaptation of photosynthesis to high temperature. In: *Adaptation of Plants to Water and High Temperature Stress*, eds. N.C. Turner and P.J. Kramer, 233–249. New York: Wiley.

Cagno, R.D., L. Guidi, L.D. Gara, and G.F. Soldatini. 2001. Combined cadmium and ozone treatments affect photosynthesis and ascorbate dependent defenses in sunflower. *New Phytol* 151:627–636.

Carpentier, R. 1999. Effect of high temperature stress on the photosynthetic apparatus. In: *Handbook of Plant and Crop Stress*, ed. M. Pessarakli, 337–348. New York: Marcel Dekker.

Carpentier, R. 2002. The negative action of toxic divalent cations on the photosynthetic apparatus. In: *Handbook of Plant and Crop Physiology*, ed. M. Pessarakli, 763–772. New York: Marcel Dekker.

Chaves, M.M., J. Flexas, and C. Pinheiro. 2009. Photosynthesis under drought and salt stress: Regulation mechanisms from whole plant to cell. *Ann Bot* 103:551–560.

Churin, Y., E. Adam, L. Kozma-Bognar, F. Nagy, and T. Borner. 2003. Characterization of two Myb-like transcription factors binding to CAB promoters in wheat and barley. *Plant Mol Biol* 52:447–462.

Cominelli, E., M. Galbiati, A. Vavasseur, L. Conti, T. Sala, M. Vuylsteke et al. 2005. A guard-cell-specific MYB transcription factor regulates stomatal movements and plant drought tolerance. *Curr Biol* 15:1196–1200.

Cornish, K. and J.A.D. Zeevaart. 1985. Movement of abscisic acid into the apoplast in response to water stress in *Xanthium strumarium* L. *Plant Physiol* 78:623–626.

Crafts-Brandner, S.J. and M.E. Salvucci. 2002. Sensitivity of photosynthesis in a C4 plant, maize, to heat stress. *Plant Physiol* 129:1773–1780.

Das, N., M. Misra, and A.N. Mishra. 1990. Sodium chloride salt stress induced metabolic changes in callus cultures of pearl millet (*Pennisetum americanum* L. Leeke): Free solute accumulation. *J Plant Physiol* 137:244–246.

Dionisio-Sese, M.L. and S. Tobita. 2000. Effects of salinity on sodium content and photosynthetic responses of rice seedlings differing in salt tolerance. *J Plant Physiol* 157:54–58.

Dominy, P.J. and R.L. Heath. 1985. Inhibition of the K[+] stimulated ATPase of the plasmalemma of pinto bean leaves by ozone. *Plant Physiol* 77:43–45.

Downton, W.J.S., W.J.R. Grant, and S.P. Robinson. 1985. Photosynthetic and stomatal responses of spinach leaves to salt stress. *Plant Physiol* 78:85–88.

Dubey, R.S. 1982. Biochemical changes in germinating rice seeds under saline stress. *Biochem Physiol Pflanzen* 177:523–535.

Dubey, R.S. 2010. Metal toxicity, oxidative stress and antioxidative defense system in plants. In: *Reactive Oxygen Species and Antioxidants in Higher Plants*, ed. S. Dutta Gupta, 177–203. Boca Raton, FL: CRC Press.

Dubey, R.S. and A.K. Singh. 1999. Salinity induces accumulation of soluble sugars and alters activity of sugar metabolizing enzymes in rice plants. *Biol Plant* 42:233–239.

Dulai, S., I. Molnár, and M. Molnár-Láng. 2011. Changes of photosynthetic parameters in wheat/barley introgression lines during salt stress. *Acta Biol Szeged* 55:73–75.

Farooq, M., A. Wahid, N. Kobayashi, D. Fujita, and S.M.A. Basra. 2009. Plant drought stress: Effects, mechanisms and management. *Agron Sustain Dev* 29:185–212.

Flexas, J. and M. Medrano. 2002. Drought inhibition of photosynthesis in C3 plants: Stomatal and non-stomatal limitations revisited. *Ann Bot* 89:183–189.

Flexas, J., J. Bota, J.M. Escalona, B. Sampol, and H. Medrano. 2002. Effects of drought on photosynthesis in grapevines under field conditions: An evaluation of stomatal and mesophyll limitations. *Funct Plant Biol* 29:461–471.

Flexas, J., J. Bota, F. Loreto, G. Cornic, and T.D. Sharkey. 2004. Diffusive and metabolic limitations to photosynthesis under drought and salinity in C3 plants. *Plant Biol* 6:269–279.

Flowers, T.J., P.F. Troke, and A.R. Yeo. 1977. The mechanism of salt tolerance in halophytes. *Annu Rev Plant Physiol* 28:89–121.

Foyer, C.H. and G. Noctor. 2000. Oxygen processing in photosynthesis: Regulation and signalling. *New Phytol* 146:359–388.

Geiken, B., J. Masojidek, M. Rizzuto, M.L. Pompili, and M.T. Giardi. 1998. Incorporation of [35S] methionine in higher plants reveals that stimulation of the D1 reaction centre II protein turn over accompanies tolerance to heavy metal stress. *Plant Cell Environ* 21:1265–1273.

Gill, S.S. and N. Tuteja. 2010. Reactive oxygen species and antioxidant machinery in abiotic stress tolerance in crop plants. *Plant Physiol Biochem* 48:909–930.

Gounaris, K., A.P.R. Brain, P.J. Quinn, and W.P. Williams. 1983. Structural and functional changes associated with heat induced phase separations of non bilayer lipid in chloroplast thylakoid membranes. *FEBS Lett* 153:47–52.

Guidi, L., E. Degl'Innocenti, and G.F. Soldatini. 2002. Assimilation of CO_2, enzyme activation and photosynthetic electron transport in bean leaves, as affected by high light and ozone. *New Phytol* 156:377–388.

Gussakovsky, E.E., B.A. Salakhutdinov, and Y. Shahak. 2002. Chiral macroaggregates of LHCII detected by circularly polarized luminescence in intact pea leaves are sensitive to drought stress. *Funct Plant Biol* 29:955–963.

Guy, C.L., J.L.A. Huber, and S.C. Huber. 1992. Sucrose phosphate synthase and sucrose accumulation at low temperature. *Plant Physiol* 100:502–508.

Harn, C. and J. Daie. 1992. Regulation of the cytosolic fructose-1,6-bisphosphatase by post-translational modification and protein level in drought-stressed leaves of sugar beet. *Plant Cell Physiol* 33:763–770.

Haupt-Herting, S. and H.P. Fock. 2002. Oxygen exchange in relation to carbon assimilation in water stressed leaves during photosynthesis. *Ann Bot* 89:851–859.

Havaux, M. 1992. Stress tolerance of photosystem II *in vivo*. Antagonistic effects of water, heat and photoinhibition stresses. *Plant Physiol* 100:424–432.

Havaux, M., O. Canaani, and S. Malkin. 1986. Photosynthetic responses of leaves to water stress expressed by photoacoustic and related methods. *Plant Physiol* 82:827–839.

Havaux, M., M. Ernez, and R. Lannoye. 1988. Correlation between heat tolerance and drought tolerance in cereals demonstrated by rapid chlorophyll fluorescence tests. *J Plant Physiol* 133:555–560.

Havaux, M., H. Greppin, and R.J. Strasser. 1991. Functioning of photosystem I and II in pea leaves exposed to heat stress in the presence or absence of light. *Planta* 186:88–98.

He, J., L.K. Huang, and M.I. Witecross. 1994. Chloroplast ultrastructure changes in *Pisum sativum* associated with supplementary ultraviolet (UV-B) radiation. *Plant Cell Environ* 17:771–775.

Heckathorn, S.A., S.L. Ryan, J.A. Baylis, D. Wang, E.W. Hamilton, L. Cundiff, and D.S. Luthe. 2002. In vivo evidence from an *Agrostis stolonifera* selection genotype that chloroplast small heat shock proteins can protect photosystem II during heat stress. *Funct Plant Biol* 29:933–944.

Ivashuta, S., R. Imai, K. Uchiyama, M. Gau, and Y. Shimamoto. 2002. Changes in chloroplast FtsH-like gene during cold acclimation in alfalfa (*Medicago sativa*). *J Plant Physiol* 159:85–90.

Kaiser, W.M. 1987. Effects of water deficit on photosynthetic capacity. *Physiol Plant* 71:142–149.

Kaiser, W.M. and U. Heber. 1981. Photosynthesis under osmotic stress: Effect of high solute concentrations on the permeability properties of the chloroplast envelope and on activity of stroma enzymes. *Planta* 153:423–429.

Khafagy, M.A., A.A. Arafa, and M.F. El-Banna. 2009. Glycinebetaine and ascorbic acid can alleviate the harmful effects of NaCl salinity in sweet pepper. *Aust J Crop Sci* 3:257–267.

Kim, J.Y., A. Mahé, J. Brangeon, and J.L. Prioul. 2000. A maize vacuolar invertase, *IVR2*, is induced by water stress: Organ/tissue specificity and diurnal modulation of expression. *Plant Physiol* 124:71–84.

Krieger-Liszkay, A. 2005. Singlet oxygen production in photosynthesis. *J Exp Bot* 56:337–346.

Krupa, Z., A. Siedlecka, W. Maksymiec, and T. Baszynski. 1993. In vivo response of photosynthetic apparatus of *Phaseolus vulgaris* to nickel toxicity. *J Plant Physiol* 142:664–668.

Kyle, D.J., I. Ohad, and C.J. Arntzen. 1984. Membrane protein damage and repair: Selective loss of a quinone-protein function in chloroplast membranes. *Proc Natl Acad Sci U S A* 81:4070–4074.

Laing, W.A., D.H. Greer, and B.D. Campbell. 2002. Story responses of growth and photosynthesis of five C3 pasture species to elevated CO_2 at low temperatures. *Funct Plant Biol* 29:1089–1096.

Lambers, D.H.R. and D.V. Seshu. 1982. Some ideas on breeding procedures and requirements for deep water rice improvement. *Proceedings of the 1981 International Deepwater Rice Workshop*, IRRI, Los Banos, 29–44.

Lapina, I.P. and B.A. Popov. 1999. Photosynthetic pigment metabolism in plants during stress. In: *Handbook of Plant and Crop Stress*, ed. M. Pessarakli, 527–543. New York: Marcel Dekker.

Larcher, W., J. Wagner, and A. Thammathaworn. 1990. Effect of superimposed temperature stress on in vivo chlorophyll fluorescence of *Vigna unguiculata* under saline stress. *J Plant Physiol* 136:92–102.

Lawlor, D.W. 2002. Limitation to photosynthesis in water-stressed leaves: Stomata vs metabolism and the role of ATP. *Ann Bot* 89:275–294.

Lee, J., K. He, V. Stolc, L. Horim, F. Pablo, G. Ying et al. 2007. Analysis of transcription factor HY5 genomic binding sites revealed its hierarchical role in light regulation of development. *Plant Cell* 19:731–749.

Liang, Y.K., C. Dubos, I.C. Dodd, G.H. Holroyd, A.M. Hetherington, and M.M. Campbell. 2005. AtMYB61, an R2R3-MYB transcription factor controlling stomatal aperture in *Arabidopsis thaliana*. *Curr Biol* 15:1201–1206.

Lidon, F.C. and F.S. Henriques. 1991. Limiting step on photosynthesis of rice plants treated with varying copper levels. *J Plant Physiol* 138:115–118.

Liu, H., B. Ouyang, J. Zhang, T. Wang, H. Li, Y. Zhang et al. 2012. Differential modulation of photosynthesis, signaling, and transcriptional regulation between tolerant and sensitive tomato genotypes under cold stress. *PLoS One* 7(11):e50785.

Liu, J.H., T. Peng, and W.S. Dai. 2014. Critical cis-acting elements and interacting transcription factors: Key players associated with abiotic stress responses in plants. *Plant Mol Biol Reporter* 32:303–317.

Mackerness, S.A.H. and B.R. Jordan. 1999. Changes in gene expression in response to ultraviolet B-induced stress. In: *Handbook of Plant and Crop Stress*, 2nd ed., ed. M. Pessarakli, 749–768. New York: Marcel Dekker.

Majumdar, S., S. Ghosh, B.R. Glick, and E.B. Dumbroff. 1991. Activities of chlorophyllase, phosphoenolpyruvate carboxylase and ribulose-1,5-bisphosphate carboxylase in the primary leaves of soybean during senescence and drought. *Physiol Plant* 81:473–480.

Mano, J., M. Nagata, S. Okamura, T. Shiraya, and T. Mitsui. 2014. Identification of oxidatively modified proteins in salt-stressed *Arabidopsis*: A carbonyl-targeted proteomics approach. *Plant Cell Physiol* 55:1233–1244.

Maroti, I., Z. Tuba, and M. Csik. 1984. Changes of chloroplast ultrastructure and carbohydrate level in Festuca, Achillea and Sedum during drought and after recovery. *J Plant Physiol* 116:1–10.

Matsubara, S., A. Gilmore, M.C. Ball, J.M. Anderson, and C.B. Osmond. 2002. Sustained down regulation of photosystem II in mistletoes during winter depression of photosynthesis. *Funct Plant Biol* 29:1157–1169.

Medrano, H., J.M. Escalona, J. Bota, J. Gulias, and J. Flexas. 2002. Regulation of photosynthesis of C_3 plants in response to progressive drought: Stomatal conductance as a reference parameter. *Ann Bot* 89:895–905.

Mehta, P., A. Jajoo, S. Mathur, and S. Bharti. 2010. Chlorophyll a fluorescence study revealing effects of high salt stress on photosystem II in wheat leaves. *Plant Physiol Biochem* 48:16–20.

Mishra, S.K., D. Subahmanyam, and G.S. Singhal. 1991. Interrelationship between salt and light stress on the primary process of photosynthesis. *J Plant Physiol* 138:92–96.

Miteva, T.S. and S.G. Vaklinova. 1989. Salt stress and activity of some photosynthetic enzymes. *C R Acad Bulg Sci* 42:87–89.

Miteva, T.S., N.Z. Zhelev, and L.P. Popova. 1992. Effect of salinity on the synthesis of ribulose-1,5-bisphosphate carboxylase/ oxygenase in barley leaves. *J Plant Physiol* 140:46–51.

Moradi, F. and A.M. Ismail. 2007. Responses of photosynthesis, chlorophyll fluorescence and ROS-scavenging systems to salt stress during seedling and reproductive stages in rice. *Ann Bot* 99:1161–1173.

Muthuchelian, K., S. Maria, V. Rani, and K. Paliwal. 1988. Differential action of Cu^{2+} and Cd^{2+} on chlorophyll biosynthesis and nitrate reductase activity in *Vigna sinensis* L. *Indian J Plant Physiol* 31:169–173.

Netondo, G.W., J.C. Onyango, and E. Beck. 2004. Sorghum and salinity: II. Gas exchange and chlorophyll fluorescence of sorghum under salt stress. *Crop Sci* 44:806–811.

Nie, G.Y., S.P. Long, and N.R. Baker. 1992. The effects of development and suboptimal growth temperatures on photosynthetic capacity and susceptibility to chilling dependent photoinhibition in *Zea mays*. *Physiol Plant* 85:554–560.

Nogues, S. and L. Alegre. 2002. An increase in water deficit has no impact on the photosynthetic capacity of field grown Mediterranean plants. *Funct Plant Biol* 29:621–630.

Ohad, I., D.J. Kyle, and C.J. Arntzen. 1984. Membrane protein damage and repair: Removal and replacement of inactivate 32-kilodalton polypeptide in chloroplast membranes. *J Cell Biol* 99:481–485.

Omoto, E., M. Taniguchi, and H. Miyake. 2010. Effects of salinity stress on the structure of bundle sheath and mesophyll chloroplasts in NAD-malic enzyme and PCK type C4 plants. *Plant Prod Sci* 13:169–176.

Ormaetxe, I.I., P.R. Escuredo, C.A. Igor, and M. Becana. 1998. Oxidative damage in pea plants exposed to water deficit or paraquat. *Plant Physiol* 116:173–181.

Ormond, D.P. and B.A. Hale. 1995. Physiological responses of plants and crops to ultraviolet-B radiation stress. In: *Handbook of Plant and Crop Physiology*, ed. M. Pessarakli, 761–770. New York: Marcel Dekker.

Parent, C., N. Capelli, A. Berger, M. Crevecoeur, and J.F. Dat. 2008. An overview of plant responses to soil water logging. *Plant Stress* 2:20–27.

Park, I.K. and S. Tsunoda. 1979. Effect of low temperature on chloroplast structure in cultivars of rice. *Plant Cell Physiol* 20:1449–1453.

Parry, M.A.J., P.J. Andralojc, S. Khan, P.J. Lea, and A.J. Keys. 2002. Rubisco activity: Effects of drought stress. *Ann Bot* 89:833–839.

Plaut, Z. 1995. Photosynthesis in plants/crops under water and salt stress. In: *Handbook of Plant and Crop Physiology*, ed. M. Pessarakli, 587–603. New York: Marcel Dekker.

Plaut, Z., C.M. Grieve, and E. Federman. 1989. Salinity effects on photosynthesis in isolated mesophyll cells of cowpea leaves. *Plant Physiol* 91:493–499.

Poljakoff-Mayber, A. 1981. Ultrastructural consequences of drought. In: *Physiology and Biochemistry of Drought Resistance in Plants*, eds. L.G. Paleg and D. Aspinall, 389–403. Sydney: Academic Press.

Quartacci, M.F., C. Pinzino, C.L.M. Sgherri, F.D. Vecchia, and F. Navari-Izzo. 2000. Growth in excess copper induces changes in the lipid composition and fluidity to PSII enriched membranes in wheat. *Physiol Plant* 108:87–93.

Rahman, M.A., H. Hasegawa, M.M. Rahman, M.N. Islam, M.A.M. Miah, and A. Tasmin. 2007. Effect of arsenic on photosynthesis, growth and yield of five widely cultivated rice (*Oryza sativa* L.) varieties in Bangladesh. *Chemosphere* 67:1072–1079.

Rajmane, N.A. and B.A. Karadge. 1986. Photosynthesis and photorespiration in winged bean (*Psophocarpus tetragonolobus* L.) grown under saline conditions. *Photosynthetica* 20:139–145.

Rao, I.M., R.E. Sharp, and J.S. Boyer. 1987. Leaf magnesium alters photosynthetic response to low water potentials in sunflower. *Plant Physiol* 84:1214–1219.

Rasmuson, K.E. and J.E. Anderson. 2002. Salinity affects development, growth and photosynthesis in cheat grass. *J Range Manag.* 55:80–87.

Reddy, M.P. and A.B. Vora. 1986. Changes in pigment composition, hill reaction activity and saccharide metabolism in Bajra (*Pennisetum typhoids* H.) leaves under NaCl salinity. *Photosynthetica* 20:50–55.

Rottenberg, H. 1977. Proton and ion transport across the thylakoid membranes. In: *Photosynthesis I. Encyclopedia of Plant Physiology*, N.S. Vol. 5, eds. A. Trebst and M. Avron, 338–349. Berlin: Springer-Verlag.

Saibo, N.J.M., T. Lourenço, and M.M. Oliveira. 2009. Transcription factors and regulation of photosynthetic and related metabolism under environmental stresses. *Ann Bot* 103:609–623.

Salama, S., S. Trivedi, M. Busheva, A.A. Arafa, G. Garab, and L. Erdei. 1994. Effects of NaCl salinity on growth, cation accumulation, chloroplast structure and function in wheat cultivars differing in salt tolerance. *J Plant Physiol* 144:241–247.

Sanchez, D.H., M.R. Siahpoosh, U. Roessner, M. Udvardi, and J. Kopka. 2007. Plant metabolomics reveals conserved and divergent metabolic responses to salinity. *Physiol Plant* 132:209–219.

Schaeffer, H.J., N.R. Forsthoefel, and J.C. Cushman. 1995. Identification of enhancer and silencer regions involved in salt-responsive expression of crassulacean acid metabolism (CAM) genes in the facultative halophyte *Mesembryanthemum crystallinum*. *Plant Mol Biol* 28:205–218.

Schreiber, U., U. Schliwa, and W. Bilger. 1986. Continuous recording of photochemical and non-photochemical chlorophyll fluorescence quenching with a new type of modulated fluorometer. *Photosyn Res* 10:51–62.

Seemann, J.R. and T.D. Sharkey. 1986. Salinity and nitrogen effects on photosynthesis, ribulose-1,5-bisphosphate carboxylase and metabolic pool size in *Phaseolus vulgaris* L. *Plant Physiol* 82:555–560.

Setter, T.L., I. Waters, I. Wallace, P. Bhekasut, and H. Greenway. 1989. Submergence of rice. I. Growth and photosynthetic response to CO_2 enrichment of flood water. *Aust J Plant Physiol* 16:251–263.

Sharkey, T.D. and R.B. Murray. 1982. Effect of water stress on photosynthetic electron transport, photophosphorylation and metabolite levels of *Xanthium stumarium* mesophyll cells. *Planta* 156:199–206.

Sharma, P.K. and D.O. Hall. 1992. Changes in carotenoid composition and photosynthesis in sorghum under high light and salt stresses. *J Plant Physiol* 140:661–666.

Sharma, P., A.B. Jha, and R.S. Dubey. 2014. Arsenic toxicity and tolerance mechanisms in crop plants. In: *Handbook of Plant and Crop Physiology*, ed. M. Pessarakli, 3rd ed., 733–783. New York: CRC Press.

Sheoran, I.S., H.R. Singal, and R. Singh. 1990. Effect of cadmium and nickel on photosynthesis and the enzymes of photosynthetic carbon reduction cycle in pigeon pea (*Cajanus cajan* L). *Photosynth Res* 23:345–351.

Shomer-Ilan, A., D. Moualem-Beno, and Y. Waisel. 1985. Effect of NaCl on the properties of phosphoenol pyruvate carboxylase from *Suaeda monoica* and *Chloris gayana*. *Physiol Plant* 65:72–78.

Shou, H., P. Bordallo, and K. Wang. 2004. Expression of the *Nicotiana* protein kinase (NPK1) enhanced drought tolerance in transgenic maize. *J Exp Bot* 55:1013–1019.

Singh, A.K. and R.S. Dubey. 1995. Changes in chlorophyll a and b contents and activity of photosystem I and II in rice seedlings induced by NaCl. *Photosynthetica* 31:489–499.

Singh, E., S. Tiwari and M. Agrawal. 2009. Effects of elevated ozone on photosynthesis and stomatal conductance of two soybean varieties: A case study to assess impacts of one component of predicted global climate change. *Plant Biol* 11:101–108.

Solomon, A., S. Beer, Y. Waisel, G.P. Jones, and L.G. Paleg. 1994. Effect of NaCl on the carboxylating activity of Rubisco from *Tamerix Jordonis* in the presence and absence of proline-related compatible solutes. *Physiol Plant* 90:198–204.

Somashekaraiah, B.V., K. Padmaja, and A.R.K. Prasad. 1992. Phytotoxicity of cadmium ions on germinating seedlings of mung bean (*Phaseolus vulgaris*): Involvement of lipid peroxides in chlorophyll degradation. *Physiol Plant* 85:85–89.

Srivastava, L.M. 2002. Abscisic acid and stress tolerance in plants In: *Plant Growth and Development: Hormones and Environment*, ed. L.M. Srivastava, 381–412. New York: Academic Press.

Srivastava, R.K., P. Pandey, R. Rajpoot, A. Rani, and R.S. Dubey. 2014. Cadmium and lead interactive effects on oxidative stress and antioxidative responses in rice seedlings. *Protoplasma* 251:1047–1065.

Stamp, P. 1987. Photosynthetic traits of maize genotypes at constant and at fluctuating temperatures. *Plant Physiol Biochem* 25:729–733.

Stapleton, A.E., C.S. Thornber, and V. Walbot. 1997. UV-B component of sunlight causes measurable damage in field grown maize: Developmental and cellular heterogeneity of damage and repair. *Plant Cell Environ* 20:279–290.

Subhan, D. and S.D.S. Murthy. 2000. Synergistic effect of $AlCl_3$ and kinetin on chlorophyll and protein contents and photochemical activities in detached wheat primary leaves during dark incubation. *Photosynthetica* 38:211–214.

Subramanyam, D. and V.S. Rathore. 2000. Influence of manganese toxicity on photosynthesis in rice bean (*Vigna umbellata*) seedlings. *Photosynthetica* 38:449–453.

Suleyman, I.A., Y. Nishiyama, S. Miyairi, H. Yamamoto, N. Inagaki, Y. Kaneasaki and M. Murata. 2002. Salt stress inhibits the repair of photodamaged photosystem II by suppressing the transcription and translation of *psb A* genes in *Synechocystis*. *Plant Physiol* 130:1443–1453.

Taiz, L. and E. Zeiger. 1998. *Plant Physiology*, 2nd ed. Redwood City, CA: Benjamin Cummings.

Takabe, T., A. Incharoensakdi, K. Arakawa, and S. Yokota. 1988. CO_2 fixation rate and Rubisco content increase in the halotolerant cyanobacterium *Aphanotheca halophytica* grown in high salinities. *Plant Physiol* 88:1120–1124.

Tambussi, E.A., C.G. Bartoli, J. Beltrano, J.J. Guiamet, and J.C. Araus. 2000. Oxidative damage to thylakoid proteins in water-stressed leaves of wheat (*Triticum aestivum*). *Physiol Plant* 108:398–404.

Tewari, T.N. and B.B. Singh. 1991. Stress studies in lentil (*Lens esculenta* Moench) II. Sodicity induced changes in chlorophyll, nitrate and nitrite reductase, nucleic acids, proline, yield and yield component in lentil. *Plant Soil* 136:225–230.

Thomashow, M.F. 1999. Plant cold acclimation: Freezing tolerance genes and regulatory mechanisms. *Annu Rev Plant Physiol Plant Mol Biol* 50:571–599.

Ting, C.S., T.G. Owens, and D.W. Wolfe. 1991. Seedling growth and chilling stress effects on photosynthesis in chilling sensitive and chilling tolerant cultivars of *Zea mays*. *J Plant Physiol* 137:559–564.

Tripathy, B.C., B. Bhatia, and P. Mohanty. 1981. Inactivation of chloroplast photosynthetic electron-transport activity by Ni^{2+}. *Biochim Biophys Acta* 638:217–224.

Vassey, T.L., W.P. Quick, T.D. Sharkey, and M. Stitt. 1991. Water stress, carbon dioxide and light effects on sucrose phosphate synthase activity in *Phaseolus vulgaris*. *Physiol Plant* 81:37–44.

Verma, S. and R.S. Dubey. 2001. Effect of cadmium on soluble sugars and enzymes of their metabolism in rice. *Biol Plant* 44:117–123.

Verma, S. and R.S. Dubey. 2002. Influence of lead toxicity on photosynthetic pigments, lipid peroxidation and activities of antioxidant enzymes in rice plants. *Indian J Agril Biochem* 15:17–22.

Vogel, C.S. and J.O. Dawson. 1991. Nitrate reductase activity, nitrogenase activity and photosynthesis of black alder exposed to chilling temperatures. *Physiol Plant* 82:551–558.

Wang, G. and Z. Guo. 2005. Effects of chilling stress on photosynthetic rate and chlorophyll fluorescence parameters in seedlings of two rice cultivars differing in cold tolerance. *Rice Sci* 12:187–191.

Wang, R.L., C. Hua, F. Zhou, and Q.-C. Zhou. 2009. Effects of NaCl stress on photochemical activity and thylakoid membrane polypeptide composition of a salt-tolerant and a salt-sensitive rice cultivar. *Photosynthetica* 47:125–127.

Yamashita, A., N. Nijo, P. Pospíšil, N. Morita, D. Takenaka, R. Aminaka et al. 2008. Reactive oxygen species are responsible for the damage to photosystem II under moderate heat stress. *J Biol Chem* 283:28380–28391.

Yamori, W., K. Hikosaka, and D.A. Way. 2014. Temperature response of photosynthesis in C3, C4, and CAM plants: Temperature acclimation and temperature adaptation. *Photosynth Res* 119:101–117.

Yanagisawa, S. and J. Sheen. 1998. Involvement of maize Dof zinc finger proteins in tissue-specific and light-regulated gene expression. *Plant Cell* 10:75–89.

Yang, J.Y., W. Zheng, Y. Tian, Y. Wu, and D.W. Zhou. 2011. Effects of various mixed salt-alkaline stresses on growth, photosynthesis, and photosynthetic pigment concentrations of *Medicago ruthenica* seedlings. *Photosynthetica* 49:275–284.

Yruela, I., M. Alfonso, M. Baron, and R. Picorel. 2000. Copper effect on the protein composition of photosystem II. *Physiol Plant* 110:551–557.

Yu, J., Y. Zhaou, L. Houang, and D. Allen. 2002. Chill induced inhibition of photosynthesis: Genotypic variation within *Cucumis sativus*. *Plant Cell Physiol* 43:1182–1188.

Yuanyuan, M., Z. Yali, L. Jiang, and S. Hongbo. 2009. Roles of plant soluble sugars and their responses to plant cold stress. *African J Biotech* 8:2004–2010.

Zhang, J. and X. Zhang. 1994. Can early wilting of old leaves account for much of the ABA accumulation in flooded pea plants. *J Exp Bot* 45:1335–1342.

Zhang, Y.-L., Y.-Y. Hu, H.-H. Luo, W.-S. Chow, and W.-F. Zhang. 2011. Two distinct strategies of cotton and soybean differing in leaf movement to perform photosynthesis under drought in the field. *Functional Plant Biol* 38:567–575.

Ziska, L.H., J.R. Seemann, and T.M. DeJong. 1990. Salinity induced limitations on photosynthesis in *Prunus salicina*, a deciduous tree species. *Plant Physiol* 93:864–870.

Zrust, J., K. Vacek, J. Hala, I. Janackova, F. Adamec, M. Ambroz et al. 1988. Influence of water stress on photosynthesis and variable chlorophyll fluorescence of potato leaves. *Biol Plant* 36:209–214.

35 Effects of Salinity and Drought Stress on Photosynthesis, Growth, and Development of Ornamental Plants

Hyun-Sug Choi, Xiaoya Cai, and Mengmeng Gu

CONTENTS

35.1 Introduction 651
35.2 Salinity and Drought Stresses 651
35.3 Effects of Salinity Stress on Ornamental Plants 652
    35.3.1 Effects of Salinity Stress on Plant Growth and Development 652
        35.3.1.1 Effects of Salinity on Photosynthesis 653
    35.3.2 Salinity-Tolerant Mechanisms 653
35.4 Effects of Drought Stress on Ornamental Plants 654
    35.4.1 Effects of Drought Stress on Plant Growth and Development 654
        35.4.1.1 Effects of Drought Stress on Photosynthesis 654
    35.4.2 Drought-Tolerant Mechanisms 655
35.5 Salinity and Drought Tolerance in Ornamental Plants 655
35.6 Coping with Salinity and Drought Stress in Ornamental Plants 656
References 658

35.1 INTRODUCTION

Ornamental plants could potentially face abiotic stresses (e.g., salinity and drought stress) at both production and postproduction stages (i.e., in the landscape). In both production and landscape situations, abiotic stresses, such as salinity and drought, adversely affect growth and development by causing morphological, physiological, biochemical, and molecular changes. The effects of these stresses range from the molecular level to the whole-plant level, from possible damage to tolerance and hardening mechanisms (Beck et al. 2004). Responses to salinity and drought stresses vary significantly among ornamental plants. Plants may change leaf orientation, transpiration rate, lipid membrane composition, and osmotic adjustment for short-term acclimation, or morphological changes may occur for long-term adaptation. With the paramount importance of the role of water quantity and quality in the life of plants, the following discussion aims to evaluate the effects of salinity and drought stresses on photosynthesis, growth, and development of ornamental plants.

35.2 SALINITY AND DROUGHT STRESSES

Generally, soil salinity could develop in both primary (natural) and secondary (induced) ways. Primary salinity is naturally formed from long-term weathering of the parent materials, which contain mineral nutrients. These parent materials are considered as beginning sources from which to dissolve and release considerable amounts of solutes into soil. Primary salinity also comes from the deposition of ocean salts, which contain inorganic ions carried by winds or rains. Secondary salinity is a major cause of increased soil solute concentrations, and it is facilitated by heavy application of irrigation and fertilizers. With increased population and industrial development, alternative water sources, such as reclaimed water, are becoming commonly used to irrigate urban landscapes and for ornamental plant production (Niu and Rodriguez 2008). However, alternative water sources often contain large amounts of inorganic nutrients, which could cause salt damage in sensitive plants. Waterlogged conditions caused by heavy irrigation application could prevent salts from leaching out of the soil profile. Excessive application of soluble fertilizers or compost could increase secondary soil salinity too. It is a common problem for greenhouse production with closed systems in Japan and the Netherlands, where soil gradually becomes dry with poor infiltration (Pessarakli 1991). In arid and semiarid regions, high soil salinity is also a result of low rainfall and high evapotranspiration.

The production of many ornamental plants is in containers and not in soil. Artificial substrate, instead of soil, is often used in container production. Except in the case of some coir substrates, most other substrate components themselves do not pose salinity issues. Salinity in container production is mainly introduced through irrigation water with high electrical conductivity (EC). To simplify the issue, the term "soil" is used only to refer to soil in the ground or container substrate.

The most common salinity problems are caused by soluble salts, such as NaCl, $CaCl_2$, $MgCl_2$, and KCl, which could negatively affect plant growth and development when a high level of salts accumulates in the root zone. Dissolved salts in irrigation water contain cations and anions. Sodium (Na^+) and chloride (Cl^-) are usually the most prevalent ions in saline water, followed by Ca^{2+}, Mg^{2+}, HCO_3^-, PO_4^{3-}, and NO_3^- (Charley and McGarity 1964). The pH of saline soils ranged between 7.0 and 8.5, and the exchangeable sodium percentage (ESP) is less than 15%. Saline soil with EC higher than 4 dS/m is considered to problematic for ornamental plant production, and EC ranges of different types of water are shown in Table 35.1 (Pitman and Läuchli 2002).

Climate changes, including global warming, are becoming a critical factor to limit water resources for plants grown in many regions of the world. Water supplies have been rapidly shrinking during drought in many states in the United States, including Georgia, Texas, and California. Due to decreasing water resources and increasing population and urbanization, water conservation and development of more efficient irrigation systems are critical in greenhouse and landscape water management (Nicolas et al. 2008; Niu et al. 2006). Landscape water use restrictions are also becoming more common in many municipalities to conserve water; these exacerbate the effects of drought stress on plant establishment and survival (Niu et al. 2008).

In nature, plants can be subjected to either short-term drought stress or long-term drought stress. Water deficit can affect plants in different ways, such as stomatal closure, growth inhibition, and reduced transpiration rates and photosynthesis (Yordanov et al. 2003). Drought stress causes substantial loss of plant growth and productivity, and it can be categorized into three groups of stress severity, mild, moderate, and desiccated. Mild drought stress causes slight loss of water through transpiration, which results in partial stomatal closure and lower net photoassimilates but minimal effects on plant performance. Moderate stress is considered as more extensive water deficit for plant growth, and plants have reduced turgor pressure and malfunction of cell enzyme activities. Desiccated drought stress causes water deficit in the

protoplasm and plasmolysis through loss of turgor pressure and metabolism disturbance (Shao et al. 2008).

35.3 EFFECTS OF SALINITY STRESS ON ORNAMENTAL PLANTS

35.3.1 EFFECTS OF SALINITY STRESS ON PLANT GROWTH AND DEVELOPMENT

Inhibition of plant growth due to soil salinity is a result of osmotic and ionic effects. There are three major constraints for plants grown on saline soil: (1) high soil osmotic pressure decreases soil water potential, which limits plants' ability to absorb water from soil; (2) excessive uptake of Na^+ and Cl^- causes ion toxicity in plants; and (3) alternations in nutrient availability results in nutrient imbalance, which depresses the uptake of other mineral nutrients (Gorham 2007). High soil salinity could also disturb membrane integrity and function, causing nutrition deficiency symptoms in plant growth (Grattan and Grieve 1999).

The concept of the *two-phase growth response to salinity* has been developed by Munns (2002, 2005). After exposure to salinity, the first phase of growth reduction happens quickly (within minutes) due to the osmotic effects in cellular water relations. After the initial decrease in leaf growth, there is a gradual recovery of the growth rate until a new steady state is reached (Munn 2002). The second phase takes days, weeks, or months due to salt accumulation in leaves, causing salinity toxicity in plants (salinity-specific effect). Salinity toxicity primarily occurs in the older leaves with Na^+ and Cl^- buildup in the transpiring leaves, causing nutrient deficiency and declined protein synthesis. Based on this two-phase concept, the initial growth reduction for both salinity-sensitive and salinity-tolerant plants is caused by an osmotic effect, while in the second phase, salinity-sensitive and salinity-tolerant plants are determined by their ability to prevent salt from accumulating in transpiring leaves to toxic levels (Munns et al. 2006).

Many studies have reported the reduction of visual quality, growth, and development under salinity stress. Under mild salinity stress conditions, growth of leaves and stems may be inhibited, while the roots may continue to grow and elongate (Hsiao and Xu 2000). After long exposure to moderate and high salinity levels, many plant species showed severe salinity damage symptoms of scorching and necrosis around the leaf margins and, ultimately, defoliation, causing plant death eventually (Marschner 1995). Leaf injury and plant death are probably caused by high salt load in leaves that exceeds the capacity of salt compartmentalization in the vacuoles, which results in salt buildup in the cytoplasm to toxic levels (Munns et al. 2006). Roses (*Rosa hybrida* L.) are an economically important ornamental plant, and they had a 25–50% decrease in shoot growth at EC between 2 and 3 dS/m (Bernstein et al. 1972). In a study by Cabrera et al. (2009), salinity stress negatively affected biomass, cut flower production, and foliage quality of "Red France," "Manetti," "Natal Briar," and "Dr. Huey" rose cultivars. Shoot growth reduction has been

TABLE 35.1
Classification of Water Quality Based on Soil Salinity

Water Designation	Total Dissolved Salts (mg/L)	Salinity (EC, dS/m at 25°C)
Freshwater	<500	<0.6
Slightly brackish	500–1000	0.6–1.5
Brackish	1000–2000	1.5–3.0
Moderately saline	2000–5000	3.0–8.0
Saline	5000–10,000	8.0–15.0
Highly saline	10,000–35,000	15.0–45.0

Source: Pitman, M.G., and Läuchli, A., Global impact of salinity and agricultural ecosystems, pp. 3–20, In: Läuchli, A. and Lüttge, U. (eds.), *Salinity: Environment—Plants—Molecules*, Kluwer Academic Publishers, Dordrecht, 2002.

observed in angelonia (*Angelonia angustifolia*), "Calico" ornamental pepper (*Capsicum annuum*), helenium (*Helenium amarum*), licorice plant (*Helichrysum petiolatum*), and plumbago (*Plumbago auriculata*) under high salinity levels (Niu et al. 2010b). Niu et al. (2012a) found that shoot growth and plant height of zinnia plants (*Zinnia marylandica*) was reduced at elevated salinity. In native wildflowers, mealy cup sage (*Salvia farinacea*), Hooker's evening primrose (*Oenothera elata*), and plains zinnia (*Zinnia grandiflora*) had reduced shoot dry weight as EC increased from 1.5 to 7.3 dS/m (Niu et al. 2012b). Increasing soil salinity can also delay flowering. High salinity resulted in a decrease in energetic reserves, preventing the development of flower structure and causing a delay in anthesis of *Iris hexagona* (Van Zandt and Mopper 2002). In two bulbous species, *Hippeastrum hybridum* and *Ornithogalum arabicum*, salinity stress caused a significant reduction of bulb, leaf, and root weight (Shilo et al. 2002). In a study by Cai et al. (2014a), the number of flowers of rose cultivars "Caldwell Pink," "Carefree Delight," "Marie Pavie," and "The Fairy" decreased as salinity level increased to 4.0 or 8.0 dS/m.

35.3.1.1 Effects of Salinity on Photosynthesis

Photosynthesis is the primary process affected by salinity stress (Munns et al. 2006). Salinity stress increases the permeability of thylakoids to Na^+ and Cl^-, resulting in inhibition of photosystems I and II (PSI and PSII) (Gilmour et al. 1982). Kirst (1989) reported that the electron transport stage between PSI and PSII was significantly affected by ionic stress in macroalgae. In *Porphyra*, *Ulva*, and *Enteromorpha* species, salinity stress mostly affected the site between plastoquinone and P700 (Wiltens et al. 1978). Satoh et al. (1983) found that high salinity stress affects three sites in the photosynthetic apparatus, causing impairment in photoactivation of electron flow on the reducing side of PSI, electron flow on the water side of PSII, and the transfer of light between the pigment complexes. In *Spirulina platensis*, high salinity stress caused a decrease in PSII due to the loss of thylakoid membrane protein (Sudhir et al. 2005). The impairment of electron transport in chloroplast and mitochondria caused by high NaCl concentration could result in cellular accumulation of reactive oxygen species (ROS), such as $^1O_2$, H_2O_2, O_2^-, and $HO\cdot$, which caused cellular oxidative stress and cellular damage (Foyer and Noctor 2003). Salinity stress decreases the internal CO_2 concentration (C_i) and slows down CO_2 reduction by the Calvin cycle, resulting in depletion of the oxidized $NADP^+$. With the depletion of the final electron acceptor ($NADP^+$) in PSI, the leakage of electrons to O_2, forming O_2^-, increases (Hsu and Kao 2003). With decreased reactions of Calvin cycles under salinity stress, more H_2O_2 is generated in the peroxisome through photorespiration in C_3 plants (Leegood et al. 1995). Accumulation of ROS also contributed to the activation of cell membrane-bound NADPH oxidase and apoplastic diamine oxidase under salinity stress (Cross et al. 1999). High ROS accumulation caused lipid degradation or lipid peroxidation in the membrane, resulting in destruction of pigments, such as chlorophyll, carotenoid, and xanthophyll in the mesophyll

cells (Dong et al. 2001; Tsai et al. 2004). The decreased chlorophyll content could be identified by low SPAD readings, which is the leaf greenness index, under salinity stress. Saleh (2012) found that chlorophyll SPAD and leaf chlorophyll a and b contents were significantly reduced at elevated salinity levels due to their adverse effects on membrane stability. In *Grevilea*, salinity stress significantly reduced protochlorophyll, chlorophylls, and carotenoids (Kennedy and De Fillippis 1999). Parida et al. (2002) also reported the reduction of leaf chlorophyll a, chlorophyll b, and carotenoid in *Bruguiera parviflora*. The reduced SPAD readings under severe salinity stress have also been reported in garden roses, azalea, crape myrtles, cherry rootstocks, and a group of 10 herbaceous ornamental plants (Cabrera 2003, 2009; Cai et al. 2014b; Niu et al. 2007a, 2012c; Sotiropoulos et al. 2006). With reduced diffusivity of mesophyll conductance to CO_2, the production of carboxylation is limited under salinity stress; thus, the photosynthesis is inhibited (Farooq et al. 2009). The reduction of net photosynthetic rate under salinity stress has been reported in *Argyranthemum coronopifolium*, Texas mountain laurel (*Sophora secundiflora*), green ash (*Fraxinus pennsylvanica*), *Phillyrea latifolia*, and garden roses (Cai et al. 2014b; Herralde et al. 1998; Niu et al. 2010a; Pezeshki and Chambers 1986; Tattini et al. 2002).

35.3.2 Salinity-Tolerant Mechanisms

Plants have developed various mechanisms to tolerate salinity stress. Under salinity stress conditions, plants can reduce the adverse effects by intercellular compartmentalization, ion transport and uptake, maintaining a relatively high cytosolic potassium/sodium (K^+/Na^+) ratio, activation of antioxidant enzymes and synthesis of antioxidant compounds, and biosynthesis of osmoprotectants and compatible solutes (Gupta and Huang 2014). Ion compartmentalization is an important mechanism for plant growth under salinity stress. Excess salts are restricted in the vacuole or compartmentalized in different tissues to prevent salt accumulation in the cytoplasm (Zhu 2003). Plants can tolerate salinity stress by maximizing Na^+ efflux from the root and its recirculation out of the shoot. Potassium plays an important role in maintaining turgor within cells, and it is essential for protein synthesis, enzyme activation, and stomatal movement. With a high cytosolic potassium/sodium (K^+/Na^+) ratio, the uptake of Na^+ is restricted (Mengel and Kirkby 2001). The K^+/Na^+ ratio has been considered a useful parameter to evaluate plant salt tolerance (Cakmak 2005). Due to the formation of ROS under salinity stress, plants have developed antioxidant defense mechanisms to protect them against oxidative stress, including enzymatic (superoxide dismutase, catalase, ascorbate peroxidase, glutathione reductase, monodehydroascorbate reductase, dehydroascorbate reductase, glutathione peroxidase, and guaiacol peroxidase) and nonenzymatic (ascorbic acid, glutathione, phenolic compounds, alkaloids, and non-protein amino acids) antioxidant defense systems to scavenge ROS (Gill and Tuteja 2010). Compatible solutes mainly include proline, glycine betaine, sugar, and polyols. The compatible organic solute accumulations significantly contribute to osmotic adjustment (Hasegawa et al. 2000). Osmotic adjustment is an important mechanism of salinity tolerance in plants because it contributes to the maintenance of water uptake and cell turgor, allowing stomatal opening, photosynthesis, and cell expansion (Serraj and Sinclair 2002). In addition, salinity stress increases abscisic acid (ABA) and cytokinins (Thomas et al. 1992). ABA has been shown to alleviate the inhibitory effect of NaCl on phytosynthesis and growth (Popova et al. 1995). Gupta et al. (1998) also found that ABA-inducible genes are important in salt-tolerant mechanisms in rice. With decreased photosynthesis under salinity stress, shifting the C_3 carbon fixation pathway to the more water-efficient crassulacean acid metabolism (CAM) pathway could help a plant increase water use efficiency, which allows plants to reduce transpiratory water loss (Cushman et al. 1989). Zhu and Meinzer (1999) found that salt-tolerant plant species *Atriplex lentiformis* can shift from the C_3 pathway to the more water-efficient C_4 carbon fixation pathway.

Through understanding a plant's salinity-tolerant mechanisms, plant responses to salinity stress are categorized into the tolerant halophytes and susceptible glycophytes. Halophytes are well adapted to a saline environment, and they have the ability to minimize the entry of salts into plants and minimize the cytoplasm salinity concentrations. In halophytes, large amounts of ions are contained within vacuoles. Most glycophytes grow well in a nonsaline environment, and they have a relatively poor ability to tolerate salinity stress (Munns 2002).

35.4 EFFECTS OF DROUGHT STRESS ON ORNAMENTAL PLANTS

35.4.1 EFFECTS OF DROUGHT STRESS ON PLANT GROWTH AND DEVELOPMENT

Plant growth is accomplished through cell division, cell enlargement, and differentiation (Farooq et al. 2008). Under water deficiency, cell elongation of higher plants is inhibited by reduced turgor pressure (Nonami 1998). Plant growth, height, and leaf area were reduced by impaired mitosis, cell elongation, and expansion under drought (Hussain et al. 2008). Consequently, irreversible damages in physiological function occurred with impaired mitosis and limited cell elongation and expansion, which caused reduction of plant growth and development (Mahajan and Tuteja 2005; Zhu 2002).

Drought stress increases production of ABA hormone in plant roots and shoots, and induces complete stomatal closure in leaves, resulting in reduction of net photoassimilation. ABA could mediate many stress responses to help plant survival under stress conditions, and it is defined as a stress hormone (Zhang et al. 2006). Many studies have reported the involvement of ABA in mediating drought stress. ABA induces genes that encode enzymes and protein involved in cellular dehydration tolerance, which could help plants regulate water status under drought stress (Zhu 2002). Zhang et al. (1987) found that ABA produced in dehydrated roots is transported to the

xylem to regulate stomatal opening and leaf growth in the shoots. Sharp et al. (2000) reported that enhanced ABA in roots during drought stress inhibits ethylene production and helps maintain plant shoot and root growth under low water potential. Under prolonged soil drying, accelerated ABA concentration in the xylem causes severe stomatal inhibition and leaf wilting (Zhang and Davies 1989). ABA also has been found to increase generation of O_2^- and H_2O_2, therefore inducing activities of antioxidant enzymes such as superoxide dismutase, catalase, ascorbate peroxidase, and glutathione reductase in plant tissues (Jiang and Zhang 2001).

Drought-induced growth reduction has been reported in many ornamental plant species. Reduced shoot dry weight and total leaf area were found under moisture stress conditions in bonfire salvia (*Salvia splendens* F. Sellow), which was due to the loss of turgor during dry-down cycles (Eakes 1991). In bigtooth maples (*Acer grandidentatum* Nutt.), drought-stressed plants were less efficient in accumulating dry matter than well-irrigated plants (Bsoul et al. 2006). In Big Bend bluebonnet (*Lupinus bavardii*), cut raceme yield, shoot and root dry weight decreased as substrate moisture content decreased to 12% (Niu et al. 2007b). In potted miniature roses, severe drought stress significantly reduced both leaf number and total leaf area by 40% compared to well-irrigated plants (Williams et al. 1999). The reduction of total production of flower stems was observed in "Eurored" rose under water deficit conditions (Bolla et al. 2010). Cai et al. (2012) found that shoot growth and flower number of rose cultivars "Radrazz," "Belinda's Dream," "Old Blush," and "Marie Pavie" were reduced by drought stress. In snapdragon (*Antirhinum majus* L.), the imposed drought stress significantly reduced all growth parameters and flower yield compared to non-stressed plants (Asrar et al. 2012). In a study of ornamental herbaceous perennials, mealy sage (*Salvia farinacea* Benth. "Henry Duelberg") had reductions in plant height and leaf area, while fan flower (*Scaevola aemula* R. Br. "New Wonder") had a reduction in shoot dry weight under drought stress (Starman and Lombardini 2006). In "Raspberry Ice" bougainvillea (*Bougainvillea spectabilis*), drought stress reduced plant height and width but increased flowering (Ma and Gu 2012).

35.4.1.1 Effects of Drought Stress on Photosynthesis

Photosynthesis is one of the most sensitive physiological processes to environmental stresses (Huang 2004), which cause impaired photosynthetic machinery and changes in photosynthetic pigments and components (Anjum et al. 2003; Wahid and Rasul 2005). Drought stress lowers C_i and inhibits photosynthetic enzymes such as ATP synthase, phosphoenolpyruvate carboxylase, NADP-malic enzyme, fructose-1,6-bisphosphatase, and pyruvate orthophosphate dikinase, thereby limiting acceptor sites of ribulose-1,5-bisphospate carboxylase/oxygenase (Rubisco) and causing reduction in net photosynthetic rate (Cornic et al. 1992; Farooq et al. 2009; Haupt-Herting and Fock 2000; Tezara et al. 1999). Drought stress may also inhibit photosynthetic electron transport through PSII (Chen and Hsu 1995). *In vivo* studies showed that drought stress caused damage to the oxygen-evolving complex of PSII (Lu and Zhang 1999) and to the PSII reaction centers associated with D1 protein (Cornic 1994). Photosynthetic pigments play an important role in harvesting light and production of reducing powers. Reduced photosynthetic pigment contents under drought stress significantly decreased photosynthesis in sunflower (Reddy et al. 2004). Under drought stress, the chlorophyll a/b ratio and the chlorophyll/carotenoid ratio change (Farooq et al. 2009), resulting in reduction of chlorophyll a and b (Ashraf et al. 1994). Reduction of chlorophyll content under drought stress has been reported in cotton, *Catharanthus roseus*, sunflower, and *Vaccinium myrtillus* (Jaleel et al. 2008; Kiani et al. 2008; Massacci et al. 2008; Tahkokorpi et al. 2007). The reduction of chlorophyll b is generally greater than chlorophyll a (Jaleel et al. 2009). Carotenes are critical in plant antioxidant defense systems, and they are susceptible to oxidative destruction (Jaleel et al. 2009). With excessive ROS accumulation under drought stress, the antioxidant enzymes, such as β-carotenes, superoxide dismutase, catalase peroxidase, ascorbate peroxidase, and glutathione reductase, can alleviate oxidative damages (Jaleel et al. 2009). Havaux (1998) found that β-carotene is an effective antioxidant to protect photochemical process, and it prevents the generation of singlet oxygen by direct quenching of triplet chlorophyll (Farooq et al. 2009). The loss of balance between the ROS production and the antioxidant defense under drought stress causes oxidative stress in some important photosynthesis components, such as photosynthetic enzymes, stomatal oscillations, and adenosine triphosphate synthesis (Farooq et al. 2009).

Both stomatal and nonstomal limitations contribute to reduced photosynthesis under drought stress (Shangguan et al. 1999). Stomatal limitation of photosynthesis under drought stress causes changes in photosynthetic reactions (Zlatev and Yordanov 2004). By controlling stomatal movement under water deficit conditions, drought-tolerant plants have improved ability to maintain the functionality of the photosynthetic machinery (Lawlor 1995). Nonstomal limitations of photosynthesis includes reduced carboxylation efficiency (Jia and Gray 2004), reduced ribulose-1,5-bisphosphate (RuBP) generation (Tezara and Lawlor 1995), reduced amount of functional Rubisco (Kanechi et al. 1995), and inhibited functional activity of PSII, which could result in damages in the primary photochemical and biochemical processes (Lawlor 2002).

35.4.2 DROUGHT-TOLERANT MECHANISMS

Plants have developed various tolerance mechanisms under drought stress, and these mechanisms help plants adapt to water deficit conditions through changes at morphological, physiological, and molecular levels (Farooq et al. 2009). A shortened plant life cycle or growing season allows plants to reproduce before drought stress occurs, which is referred as drought escape. Drought-tolerant plants limit transpiration rate through reduced leaf size, and they develop an extensive and deep root system to maintain high water potential at drought stress (Pallardy 2008). Ludlow and Muchow (1990)

reported that glaucousness on leaves or waxy bloom is considered a desirable trait for drought tolerance, which helps plants maintain high tissue water potential under drought stress. Osmotic adjustment could improve tissue water status by actively accumulating inorganic solutes (calcium, potassium, chloride ion, etc.) and compatible organic solutes (soluble sugars, sugar alcohols, proline, etc.) in the cytoplasm, thereby attracting water into the cell and maintaining turgor pressure in water-limiting environments (Farooq et al. 2009). The antioxidant defense system is important for scavenging excessive ROS under drought stress, including both enzymatic and nonenzymatic components (Farooq et al. 2009). In clones of *Coffea canephora*, drought tolerance might be associated with enhanced activity of antioxidant enzymes, which protects plants against oxidative stress (Lima et al. 2002). With possible membrane disruption at drought stress, maintenance of membrane integrity and stability under water stress is becoming an important component for the evaluation of drought tolerance (Bajji et al. 2002). Aquaporins are important in facilitating and regulating passive exchange of water across the membrane, and they can increase water permeability by regulating hydraulic conductivity of the membrane (Maurel and Chrispeels 2001). In transgenic tobacco, plant vigor was significantly improved by overexpression of the plasma membrane aquaporin (Aharon et al. 2003). Synthesis of transcription factors and stress proteins is also an important drought-tolerant mechanism (Farooq et al. 2009). Improvement of plant drought tolerance by introduction of a novel dehydration-responsive element-binding gene transcriptional factor has been found in groundnut and rice (Mathur et al. 2004; Yamaguchi-Shinozaki and Shinozaki 2004). Changing the expression levels of late embryogenesis abundant/dehydrin-type genes and molecular chaperones under drought stress could protect the cellular proteins from denaturation (Mahajan and Tuteja 2005). Heat shock proteins are found to improve plant drought tolerance by serving as molecular chaperones to prevent protein denaturation during stress (Gorantla et al. 2006). In addition, membrane-stabilizing proteins and late embryogenic abundant proteins are responsible for conferring drought tolerance, and they increase water binding capacity and protect the partner protein from degradation under drought stress (Farooq et al. 2009).

35.5 SALINITY AND DROUGHT TOLERANCE IN ORNAMENTAL PLANTS

Both salinity and drought stresses induce osmotic stress. When the soluble salt levels in the soil solution are high enough to limit water uptake, physiological drought occurs. Plant adaptation to salinity and drought stresses is related to maintenance of plant water potential through partial closure of stomata and reduced leaf development. The demand for salinity and drought-tolerant ornamental plants is increasing in arid and semiarid regions as a lower quantity and quality of water is used for landscape irrigation, and it is the primary solution for the design of sustainable landscaping.

The relative salinity tolerance based on the growth and physiological responses to elevated salinity levels has been evaluated in many studies among multiple ornamental plant species. In hybrid Japanese *Limonium*, there was little or no effect of salinity stress on stem yield and length of flowering stems, and they can tolerate salinity up to 11.5 dS/m (Shilo et al. 2002). In a study by Niu et al. (2007c), Texas bluebonnet (*Lupinus texensis* Hook.) was more salinity tolerant than Big Bend bluebonnet (*Lupinus havardii* Wats.), with less shoot growth reduction and visual injury at salinity up to 7.6 dS/m. At 3.7 and 5.7 dS/m, *L. texensis* had lower shoot Na^+ and Cl^- concentrations than *L. havardii*, which may cause the lower leaf water potential in *L. havardii* than in *L. texensis* (Niu et al. 2007c). This is in accordance with the findings of Wu et al. (2001) that salinity-tolerant plants had less salt accumulation in leaf tissue than less salinity-tolerant plants among 10 landscape plant species. As salinity increased from 1.4 to 6.4 dS/m, salinity-tolerant rose cultivars "Little Buckaroo," "Sea Foam," and "Rise N Shine" had little visual damage and shoot dry weight reduction (Niu et al. 2013). In a study by Cai et al. (2014b), drought-tolerant rose cultivars had higher shoot growth, flower number, and leaf SPAD readings at a high salinity level (EC at 10.0 dS/m). In 10 herbaceous perennials and ground covers, SPAD readings were not affected by salinity treatment in salinity-tolerant species *Gaillardia aristata* Pursh, *Lantana × hybrida*, *Lonicera japonica* Thunb., and *Verbena macdougalii* Heller, while less salinity-tolerant species *Lantana montevidensis* (Spreng.) Brig. and *Glandularia × hybrida* (Gronland & Rumpler) G.L. Nesom & Pruski had reduced SPAD readings at elevated salinity stress (Niu et al. 2007a). In two native landscape woody ornamentals, Texas mountain laurel (*Sophora secundiflora*) was more salinity tolerant than Mexican redbud (*Cercis canadensis* var. *mexicana*), with less reduction in shoot growth and elongation at EC of 3.0 and 6.0 dS/m (Niu et al. 2010a). With higher photosynthesis, stomatal conductance, osmotic adjustment, and chlorophyll retention as well as restricted translocation of Na^+ from root to shoot at elevated salinity stress, *Ziziphus rotundifolia* is more salinity tolerant than *Ziziphus nummularia* (Meena et al. 2003). Ornamental plants have a wide range of salinity tolerance, which varies among species, environmental conditions, and substrate. The aesthetic value is the primary assessment of salinity tolerance in ornamental plants in landscapes, while plant growth is also an important parameter.

Reduction in plant photosynthetic rate under drought stress is one of the primary tolerance mechanisms protecting plants from desiccation (Chaves 1991). In a study on begonia (*Begonia semperflorens* L.), photosynthesis and stomatal conductance decreased as substrate moisture content decreased to 13.6% (by volume) (Miralles-Crespo and van Iersel 2011). This is accordance with a study on four bedding plants, "Bonfire Red" salvia (*Salvia splendens* Sell ex Roem. & Schult), "Cooler Peppermint" vinca (*Catharanthus roseus* L.G. Don.), "Lavender White" petunia (*Petunia × hybrida* Hort ex. Vilm.), and "Cherry" impatiens (*Impatiens wallerana* Hook.), in which plants had the lowest photosynthetic rates at 9% substrate moisture content (Nemali and van Iersel 2008).

In a study by Gu et al. (2007), four birch genotypes (*Betula* L.) had reduced net photosynthesis, stomatal conductance, and transpiration rate at water deficit conditions. In potted miniature roses, reduced photosynthesis and stomatal conductance at water deficit conditions can be restored to rates comparable with control plants after rewatering (Williams et al. 1999). Cai et al. (2012) also reported reduced photosynthesis and stomatal conductance at drought stress in four garden rose cultivars. Drought-tolerant species of *Eucalyptus* and *Nothofagus* had greater photosynthesis and stomatal conductance at decreasing substrate moisture content than less drought-tolerant species (Ngugi et al. 2004). Similar results were found with rose and oleander (*Nerium oleander* L.), wherein drought tolerant cultivars/clones under water deficit conditions had greater gas exchange rates than those less tolerant to drought (Niu et al. 2008; Niu and Rodriguez 2009). Hagidimitriou and Pontikis (2005) also reported that drought-tolerant cultivars of olive (*Olea europaea* L.) trees can maintain higher photosynthesis than less tolerant cultivars at drought stress.

Drought-tolerant plants could maintain acceptable visual quality, growth, and productivity under stress conditions (Wang et al. 2003). Under drought treatment (18% substrate moisture content), there was no reduction of shoot dry weight in agastache (*Agastache urticifolia* [Beth.]), ornamental pepper (*Capsicum annuum* L. "Black Pearl"), or vinca (*Catharanthus roseus* L.G. Don "Titan") compared with control plants, and they were considered drought tolerant (Niu et al. 2006). In three herbaceous Australian ornamental species (*Orthosiphon aristatus*, *Dianella revolute* "Breeze," *Ptilotus nobilis*), *D. revolute* and *P. nobilis* were considered drought tolerant by maintaining acceptable aesthetic performance under water deficit conditions (Kjelgren et al. 2009). In four rose rootstocks, drought-tolerant cultivar *Rosa × fortuniana* had little reduction in shoot growth under drought stress, while *R. odorata* was the most sensitive to drought stress with the greatest growth reduction (Niu and Rodriguez 2009). In a study on *Phaseolus vulgaris*, drought-tolerant cultivars "Carioca" and "Ouro Negro" had higher tissue water retention capacity and growth under water deficit conditions (Franca et al. 2000). In 17 species of bedding plants and Kentucky bluegrass (*Poa pratensis* L.) investigated, petunia (*Petunia × hybrida* hort. ex E. Vilm. "Merlin White") and mock verbena (*Glandularia* J.F. Gmel. "Imagination") were considered to be drought tolerant, and they were able to maintain acceptable growth and visual quality with little or no supplemental irrigation (Henson et al. 2006).

35.6 COPING WITH SALINITY AND DROUGHT STRESS IN ORNAMENTAL PLANTS

As water shortage and poor water quality are critical problems for agriculture in many regions of the world, the selection and use of salinity- and drought-tolerant plants becomes increasingly important for the development of sustainable landscapes. Some management practices have been utilized to improve plant performance under salinity and drought stresses in ornamental plant production.

Improving plant responses to salinity and drought stress can be achieved in many ways. Stress-tolerant plants are selected in areas where water sources are limited, and they have a strong ability to survive under conditions with limited water. By understanding plants' responses to salinity and drought stress and their tolerant mechanisms, it is possible to produce salinity- and drought-tolerant cultivars/genotypes by production of genetically modified or transgenic plants (Farooq et al. 2009). Many newly tolerant plant species have been developed in classical and molecular ways to reduce the adverse effects of abiotic stresses on plant aesthetic performance. Arbuscular mycorrhizal (AM) organisms can increase salinity and drought tolerance in host plants (Augé 2001; Evelin et al. 2009). Under severe salinity and drought stress, the spore number of AM fungi around the root zone increased, resulting in an increase in water uptake and tissue hydration (Augé 2001; Evelin et al. 2009). AM organisms can also increase photosynthesis by increasing cation nutrients and osmolyte accumulation under water stress conditions (Augé 2001; Evelin et al. 2009). Preconditioning plants with saline water has been reported to increase plant salinity tolerance in *Calceolaria hybrida*, *Calendula officinalis*, and *Petunia × hybrida* (Fornes et al. 2007). Replacing less salinity-tolerant plants with halophytes *Suaeda maritime* and *Sesuvium portulacastrum* can mitigate the effects of salinity stress in some areas, because halophytes can cause great reduction of soil salinity by accumulating high salts in their leaf tissue (Ravindran et al. 2007). Gramineae species are recommended to eliminate solute-rich soil solutions during noncultivation periods, and they can develop extensive root systems to decrease salinity concentrations. These green cover crops can also help to improve soil physical properties by improving water infiltration and soil aeration.

A proper amount of fertilizer should be applied to synchronize suitable growth of ornamental plants without significantly increasing soil salinity. Avoiding overfertilization can reduce soil salinity and maintain water potential in the root zone. Excessive accumulation of ions such as Na^+ and Cl^- can lead to potassium (K^+) and calcium (Ca^{2+}) deficiency. Potassium is a major osmoticum, and it plays an important role in osmotic adjustment, stomatal movement, protein synthesis, enzyme activation, and restriction of Na^+ uptake. Application of K fertilizer reduces adverse effects of salinity by regulating a desirable K^+/Na^+ ratio in cytoplasm, which helps plants maintain photosynthetic capacity under high salinity conditions. Increasing K application has been reported to increase leaf expansion and crop production in *Citrus tangerina* under elevated salinity (Anac et al. 1997). Calcium application can also reduce salinity toxicity in *Solanum lycopersicum* plants by replacing Na^+ in the plasma membrane (Song and Fujiyama 1996), which can facilitate higher K^+/Na^+ selectivity.

An efficient irrigation system is important to ameliorate the effects of salinity stress on plants. An irrigation system for a landscape needs to apply enough water with a lower amount of salts to leach out and dilute the accumulated soluble salts on the soil surface, especially on clay soil profiles with low hydraulic conductivity. Implementation of a drainage program can leach salts from the landscape root zone, resulting in a decrease in soil salinity. Sprinkler irrigation is the most common method of irrigating landscape ornamental plants. However, this type of irrigation system increases the risk of salt accumulation on leaves, causing foliar salinity injury even at relatively low salinity levels. Adjusting the sprinkler header to direct the water flow below the leaf canopy can prevent foliar injury. Under high salinity levels, a drip or flood irrigation system was recommended to reduce the adverse effect of salt spray on landscape plants (Jordan et al. 2001). The drip irrigation system delivers water around the drip lines, which cannot completely eliminate salt accumulation in areas without receiving irrigation water. Proper irrigation scheduling is needed to increase water use efficiency and decrease salt accumulation through monitoring the evapotranspiration, soil moisture, and soil salinity.

Foliar application of plant growth regulators, both natural and synthetic, could improve plant growth and development under various abiotic stresses (Farooq et al. 2009). Under severe drought stress conditions, exogenous ABA applications delay wilting in many bedding plants (Waterland et al. 2010). Application of cytokinin and gibberellins mediated plant responses to drought by closing stomata in many flowering plants, and it caused cell division and enlargement, thereby increasing plant growth and development (Waterland et al. 2010). Exogenous application of a low concentration of salicylic acid was found to increase plant drought tolerance by increasing antioxidant enzymes, nitrate metabolism enzymes, and ABA contents in leaves (Hayat et al. 2005). In addition, exogenous application of jasmonic acid, brassinolide, uniconazole, and methyl jasmonate can also improve plant drought tolerance by increasing activities of superoxide dismutase, catalase and ascorbate peroxidase, and ABA, as well as total carotenoid contents (Li et al. 1998).

In addition to plant growth regulators, other exogenous applications, such as osmoprotectants (proline, trehalose, fructan, mannitol, glycine betaine, and others), silicon, and K, are important for plants' adaptation to drought environments (Benlloch-Gonzalez et al. 2010; Egilla et al. 2005; Hattori et al. 2005; Zhu 2002). The osmoprotectants help plants improve drought tolerance by mediating osmotic adjustment and protecting subcellular structures in stressed plants (Farooq et al. 2009). Application of glycine betaine or proline was effective in ameliorating the adverse effects of abiotic stresses including salinity and drought stress on many plant species (Ashraf and Foolad 2007), including bedding plant petunia (Yamada et al. 2005). Exogenous application of silicon has been mostly used for field crop production under drought stress, which increased the capacity for water transport and reduced oxidative stress in sorghum (*Sorghum bicolor*) (Hattori et al. 2005). Exogenous K fertilizer application has been studied to increase stomatal conductance and net photosynthesis under drought stress in *Hibiscus rosa-sinensis* (Egilla et al. 2005) and *Helianthus annuus* plants (Benlloch-Gonzalez et al. 2010) with K deficiency.

With water becoming more and more limited and water of low quality being used more and more in agriculture sectors other than food production, salinity and drought issues will probably become prevalent or accompany ornamental plant production and landscapes, thus affecting plant performance during production and in landscapes. Understanding the effects of salinity and drought stresses on photosynthesis, probably the most important physiological process, of ornamental plants could help combat issues in ornamental plant production and landscape management.

REFERENCES

Aharon, R., Y. Shahak, S. Wininger, R. Bendov, Y. Kapulnik, and G. Galili. 2003. Overexpression of a plasma membrane aquaporins in transgenic tobacco improves plant vigor under favorable growth conditions but not under drought or salt stress. *Plant Cell* 15:439–447.

Anac, D., B. Okur, C. Cilic, U. Aksoy, Z. Hepaksoy, S. Anac, M.A. Ul, and F. Dorsan. 1997. Potassium fertilization to control salinization effect. pp. 370–377. In: Johnson, A.E. (ed.). *Food Security in the WANA Region, the Essential Need for Balanced Fertilization.* International Potash Institute, Basel.

Anjum, F., M. Yaseen, E. Rasul, A. Wahid, and S. Anjum, 2003. Water stress in barley (*Hordeum vulgare* L.). I. Effect on chemical composition and chlorophyll contents. *Pakistan J. Agric. Sci.* 40:45–49.

Ashraf, M. and M.R. Foolad. 2007. Roles of glycine betaine and proline in improving plant abiotic stress resistance. *Environ. Exp. Bot.* 59:206–216.

Ashraf, M.Y., A.R. Azmi, A.H. Khan, and S.A. Ala. 1994. Effect of water stress on total phenols, peroxidase activity and chlorophyll content in wheat (*Triticum aestivum* L.) genotypes under soil water deficits. *Acta Physiol. Plant.* 16:185–191.

Asrar, A.A., G.M. Abdel-Fattah, and K.M. Elhindi. 2012. Improving growth, flower yield, and water relations of snapdragon (*Antirhinum majus* L.) plants grown under well-watered and waterstress conditions using arbuscular mycorrhizal fungi. *Photosynthetica* 50:305–316.

Augé, R.M. 2001. Water relations, drought and vesicular–arbuscular mycorrhizal symbiosis. *Mycorrhiza* 11:3–42.

Bajji, M., J. Kinet, and S. Lutts. 2002. The use of the electrolyte leakage method for assessing cell membrane stability as a water stress tolerance test in durum wheat. *Plant Growth Regul.* 36:61–70.

Beck, E.H., R. Heim, and J. Hansen. 2004. Plant resistance to cold stress: Mechanisms and environmental signals triggering frost hardening and dehardening. *J. Biosci.* 29:449–459.

Benlloch-Gonzalez, M., J. Romera, S. Cristescu, F. Harren, J.M. Fournier, and M. Benlloch. 2010. K+ starvation inhibits waterstress induced stomatal closure via ethylene synthesis in sunflower plants. *J. Exp. Bot.* 61:1139–1145.

Bernstein, L., L.E. Francois, and R.A. Clark. 1972. Salt tolerance of ornamental shrubs and ground covers. *J. Am. Soc. Hort. Sci.* 97:550–556.

Bolla, A., D. Voyiatzis, M. Koukourikou-Petridou, and D. Chimonidou. 2010. Photosynthetic parameters and cut-flower yield of rose 'Eurored' (H.T.) are adversely affected by mild water stress irrespective of substrate composition. *Sci. Hort.* 126:390–394.

Bsoul, E., R.S. Hilaire, and D.M. VanLeeuwen. 2006. Bigtooth maples exposed to asynchronous cyclic irrigation show provenance differences in drought adaptation mechanisms. *J. Am. Soc. Hort. Sci.* 131:459–468.

Cabrera, R.I. 2003. Growth, quality and nutrient responses of azalea hybrids to salinity. *Acta Hort.* 609:241–245.

Cabrera, R.I. 2009. Revisiting the salt tolerance of crape myrtles (*Lagerstroemia spp.*). *Arboric. Urban Fores.* 35:129–134.

Cabrera, R.I., A.R. Solís-Pérez, and J.J. Sloan. 2009. Greenhouse rose yield and ion accumulation responses to salt stress as modulated by rootstock selection. *HortScience* 44:2000–2008.

Cai, X., T. Starman, G. Niu, C. Hall, and L. Lombardini. 2012. Response of selected garden roses to drought stress. *HortScience* 47:1400–1403.

Cai, X., G. Niu, T. Starman, and C. Hall. 2014a. Response of six garden roses to salt stress. *Sci. Hort.* 168:27–32.

Cai, X., Y. Sun, T. Starman, C. Hall, and G. Niu. 2014b. The response of 18 Earth-Kind® roses to salt stress. *HortScience* 49:544–549.

Cakmak, I. 2005. The role of potassium in alleviating detrimental effects of abiotic stresses in plants. *J. Plant Nutr. Soil Sci.* 168:521–530.

Charley, J.L. and J.W. McGarity. 1964. High soil nitrate levels in patterned saltbush communities. *Nature* 201:1351–1352.

Chaves, M.M. 1991. Effects of water deficits on carbon assimilation. *J. Expt. Bot.* 42:1–16.

Chen, Y.H. and B.D. Hsu. 1995. Effect of dehydration on the electron transport of Chlorella. An in vivo fluorescence study. *Photosynth. Res.* 46:295–299.

Cornic, G. 1994. Drought stress and high light effects on leaf photosynthesis. pp. 297–313. In: Baker, N.R. and J.R. Boyer (eds.). *Photoinhibition of Photosynthesis: From Molecular Mechanisms to the Field.* Bios Scientific, Oxford.

Cornic, G., J. Ghashghaie, B. Genty, and J.M. Briantais. 1992. Leaf photosynthesis is resistant to a mild drought stress. *Photosynthetica* 27:295–309.

Cross, A.R., R. Erichson, B.A. Ellis, and J.T. Curnutte. 1999. Spontaneous activation of NADPH oxidase in cell-free system: Unexpected multiple effects of magnesium ion concentrations. *Biochem. J.* 8:229–233.

Cushman, J.C., G. Meyer, C.B. Michalowski, J.M. Schmitt, and H.J. Bohnert. 1989. Salt stress leads to differential expression of two isogenes of PEP Case during CAM induction in the common Ice plant. *Plant Cell* 1:715–725.

Dong, H.L., S.K. Young, and B.L. Chin. 2001. The inductive responses of the antioxidant enzymes by salt stress in the rice (*Oryza sativa* L.). *J. Plant Physiol.* 158:737–745.

Eakes, D.J., R.D. Wright, and J.R. Seiler. 1991. Moisture stress conditioning effects on *Salvia splendens* "Bonfire". *J. Am. Soc. Hort. Sci.* 116:716–719.

Egilla, J.N., F.T. Davies, and T.W. Boutton. 2005. Drought stress influences leaf water content, photosynthesis, and water use efficiency of Hibiscus rosa-sinensis at three potassium concentrations. *Photosynthetica* 43:135–140.

Evelin, H., R. Kapoor, and B. Giri. 2009. Arbuscular mycorrhizal fungi in alleviation of salt stress: A review. *Ann. Bot.* 104:1263–1280.

Farooq, M., T. Aziz, S.M.A. Basra, M.A. Cheema, and H. Rehamn. 2008. Chilling tolerance in hybrid maize induced by seed priming with salicylic acid. *J. Agron. Crop Sci.* 194:161–168.

Farooq, M., A. Wahid, N. Kobayashi, D. Fujita, and S.M.A. Basra. 2009. Plant drought stress: Effects, mechanisms and management. *Agron. Sustain. Dev.* 29:185–212.

Fornes, F., R.M. Belda, C. Carrión, V. Noguera, P. García-Agustín, and M. Abad. 2007. Pre-conditioning ornamental plants to drought by mean of saline water irrigation as related to salinity tolerance. *Sci. Hort.* 113:52–59.

Foyer, C.H. and G. Noctor. 2003. Redox sensing and signaling associated with reactive oxygen in chloroplasts, peroxisomes and mitochondria. *Physiol. Plant.* 119:355–364.

Franca, M.G.C., A.T.P. Thi, C. Pimentel, R.O.P. Rossiello, F.Y. Zuily, and D. Laffray. 2000. Differences in growth and water relations among *Phaseolus vulgaris* cultivars in response to induced drought stress. *Environ. Exp. Bot.* 43:227–237.

Gill, S.S. and N. Tuteja. 2010. Reactive oxygen species and antioxidant machinery in abiotic stress tolerance in crop plants. *Plant Physiol. Biochem.* 48:909–930.

Gilmour, D.J., M.F. Hipkins, and A.D. Boney. 1982. The effect of salt stress on the primary processes of photosynthesis in *Dunaliella tertiolecta. Plant Sci. Lett.* 26:325–330.

Gorantla, M., P.R. Babu, V.B.R. Lachagari, A.M.M. Reddy, R. Wusirika, J.L. Bennetzen, and A.R. Reddy. 2006. Identification of stress-responsive genes in an indica rice (*Oryza sativa* L.) using ESTs generated from drought-stressed seedlings. *J. Exp. Bot.* 58:253–265.

Gorham, J. 2007. Sodium. pp. 569–583. In: Barker, A.V. and D.J. Pilbeam (eds.). *Handbook of Plant Nutrition*. CRC Press, Boca Raton, FL.

Grattan, S.R. and C.M. Grieve. 1999. Salinity–mineral nutrient relations in horticultural crops. *Sci. Hort.* 78:127–157.

Gu, M., C.R. Rom, J.A. Robbins, and D.M. Oosterhuis. 2007. Effect of water deficit on gas exchange, osmotic solutes, leaf abscission, and growth of four birch genotypes (*Betula L.*) under a controlled environment. *HortScience* 42:1383–1391.

Gupta, B. and B. Huang, B. 2014. Mechanism of salinity tolerance in plants: Physiological, biochemical, and molecular characterization. *Int. J. Genomics* 2014:701596.

Gupta, S., M.K. Chattopadhyay, P. Chatterjee, B. Ghosh, and D.N. SenGupta. 1998. Expression of abscisic acid–responsive element binding protein in salt tolerant indica rice (*Oryza sativa* L. cv. Pokkali). *Plant Mol. Biol.* 137:629–637.

Hagidimitriou, M. and C.A. Pontikis. 2005. Seasonal changes in CO_2 assimilation in leaves of five major Greek olive cultivars. *Sci. Hort.* 104:11–24.

Hasegawa, P.M., R.A. Bressan, J.K. Zhu, and H.J. Bohnert. 2000. Plant cellular and molecular responses to high salinity. *Plant Physiol. Plant Mol. Biol.* 51:463–499.

Hattori, T., S. Inanaga, A. Hideki, A. Ping, M. Shigenori, L. Miroslava, and A. Lux. 2005. Application of silicon enhanced drought tolerance in *Sorghum bicolor. Physiol. Plant.* 123:459–466.

Haupt-Herting, S. and H.P. Fock. 2000. Exchange of oxygen and its role in energy dissipation during drought stress in tomato plants. *Physiol. Plant.* 110:489–495.

Havaux, M. 1998. Carotenoids as membrane stabilizers in chloroplasts. *Trends Plant Sci.* 3:147–151.

Hayat, S., Q. Fariduddin, B. Ali, and A. Ahmad. 2005. Effect of salicylic acid on growth and enzyme activities of wheat seedlings. *Acta Agron. Hung.* 53:433–437.

Henson, D.Y., S.E. Newman, and D.E. Hartley. 2006. Performance of selected herbaceous annual ornamentals grown at decreasing levels of irrigation. *HortScience* 41:1481–1486.

Herralde, F.D., C. Biel, R. Savé, M.A. Morales, A. Torrecillas, J.J. Alarcón, and M.J. Sánchez-Blanco. 1998. Effect of water and salt stress on the growth and gas exchange and water relations in *Argyranthemum coronopifolium* plants. *Plant Sci.* 139:9–17.

Hsiao, T.C. and L.K. Xu. 2000. Sensitivity of growth of roots versus leaves to water stress: Biophysical analysis and relation to water transport. *J. Exp. Bot.* 25:1595–1616.

Hsu, S.Y. and C.H. Kao. 2003. Differential effect of sorbitol and polyethylene glycol on antioxidant enzymes in rice leaves. *Plant Growth Regul.* 39:83–90.

Huang, B. 2004. Recent advances in drought and heat stress physiology of turfgrass: A review. *Acta Hort.* 661:185–192.

Hussain, M., M.A. Malik, M. Farooq, M.Y. Ashraf, and M.A. Cheema. 2008. Improving drought tolerance by exogenous application of glycine betaine and salicylic acid in sunflower. *J. Agron. Crop Sci.* 194:193–199.

Jaleel, C.A., R. Gopi, B. Sankar, M. Gomathinayagam, and R. Panneerselvam. 2008. Differential responses in water use efficiency in two varieties of *Catharanthus roseus* under drought stress. *Comp. Rend. Biol.* 331:42–47.

Jaleel, C.A., P. Manivannan, A. Wahid, M. Farooq, H.J. Al-Juburi, R. Somasundaram, and R. Panneerselvam. 2009. Drought stress in plants: A review on morphological characteristics and pigments composition. *Int. J. Agric. Biol.* 11:100–105.

Jia, Y. and V.M. Gray. 2004. Interrelationships between nitrogen supply and photosynthetic parameters in *Vicia faba* L. *Photosynthetica* 41:605–610.

Jiang, M. and J. Zhang. 2001. Effect of abscisic acid on active oxygen species, antioxidative defense system and oxidative damage in leaves of maize seedlings. *Plant Cell Physiol.* 42:1265–1273.

Jordan, L.A., D.A. Devitt, R.L. Morris, and D.S. Neuman. 2001. Foliar damage to ornamental trees sprinkler-irrigated with reuse water. *Irrig. Sci.* 21:17–25.

Kanechi, M., E. Kunitomo, N. Inagaki and S. Maekawa. 1995. Water stress effects on ribulose-1,5-bisphosphate carboxylase and its relationship to photosynthesis in sunflower leaves. pp. 597–600. In: Mathis, M. (ed.). *Photosynthesis: From Light to Biosphere*. Kluwer Academic Publisher, Dordrecht.

Kennedy, B.F. and L.F. De Fillippis. 1999. Physiological and oxidative response to NaCl of the salt tolerant *Grevillea ilicifolia* and the salt sensitive *Grevillea arenaria. J. Plant Physiol.* 155:746–754.

Kiani, S.P., P. Maury, A. Sarrafi, and P. Grieu. 2008. QTL analysis of chlorophyll fluorescence parameters in sunflower (*Helianthus annuus* L.) under well-watered and water-stressed conditions. *Plant Sci.* 175:565–573.

Kirst, G.O. 1989. Salinity tolerance of eukaryotic marine algae. *Annu. Rev. Plant Physiol. Plant Mol. Biol.* 40:21–53.

Kjelgren, R., L. Wang, and D. Joyce. 2009. Water deficit stress responses of three native Australian ornamental herbaceous wildflower species for water-wise landscape. *HortScience* 44:1358–1365.

Lawlor, D.W. 1995. Effects of water deficit on photosynthesis. pp. 129–160. In: Smirnoff, N. (ed.). *Environment and Plant Metabolism*. Bios Scientific, Oxford.

Lawlor, D.W. 2002. Limitation of photosynthesis in water-stressed leaves. Stomatal metabolism and the role of ATP. *Ann. Bot.* 89:871–885.

Leegood, R.C., P.J. Lea, M.D. Adcock, and R.E. Hausler. 1995. The regulation and control of photorespiration. *J. Exp. Bot.* 46:1397–1414.

Li, L., J. Van Staden, and A.K. Jager. 1998. Effects of plant growth regulators on the antioxidant system in seedlings of two maize cultivars subjected to water stress. *Plant Growth Regul.* 25:81–87.

Lima, A.L.S., F.M. DaMatta, H.A. Pinheiro, M.R. Totola, and M.E. Loureiro. 2002. Photochemical responses and oxidative stress in two clones of *Coffea canephora* under water deficit conditions. *Environ. Exp. Bot.* 47:239–247.

Lu, C. and J. Zhang. 1999. Effects of water stress on photosystem II photochemistry and its thermostability in wheat plants. *J. Exp. Bot.* 50:1199–1206.

Ludlow, M.M. and R.C. Muchow. 1990. A critical evaluation of traits for improving crop yields in water-limited environments. *Adv. Agron.* 43:107–153.

Ma, S. and M. Gu. 2012. Effects of water stress and selected plant growth retardants on growth and flowering of "Raspberry Ice" Bougainvillea (*Bougainvillea spectabilis*). *Acta Hort.* 937:237–242.

Mahajan, S. and N. Tuteja. 2005. Cold, salinity and drought stresses: An overview. *Arch. Biochem. Biophys.* 444:139–158.

Marschner, H. 1995. *Mineral Nutrition of Higher Plants*, 2nd ed. Academic Press, San Diego, CA.

Massacci, A., S.M. Nabiev, L. Pietrosanti, S.K. Nematov, T.N. Chernikova, K. Thor, and J. Leipner. 2008. Response of the photosynthetic apparatus of cotton (*Gossypium hirsutum*) to the onset of drought stress under field conditions studied by gas-exchange analysis and chlorophyll fluorescence imaging. *Plant Physiol. Biochem.* 46:189–195.

Mathur, P.B., M.J. Devi, R. Serraj, K. Yamaguchi-Shinozaki, V. Vadez, and K.K. Sharma. Evaluation of transgenic groundnut lines under water limited conditions. *Int. Archis. Newslett.* 24:33–34.

Maurel, C. and M.J. Chrispeels. 2001. Aquaporins: A molecular entry into plant water relations. *Plant Physiol.* 125:135–138.

Meena, S.K., N.K. Gupta, S. Gupta, S.K. Khandelwal, and E.V.D. Sastry. 2003. Effect of sodium chloride on the growth and gas exchange of young *Ziziphus* seedling rootstocks. *J. Hortic. Sci. Biotechnol.* 78:454–457.

Mengel, K. and E.A. Kirkby. 2001. *Principles of Plant Nutrition*, 5th ed. Kluwer Academic Publishers, Dordrecht.

Miralles-Crespo, J. and M.W. van Iersel. 2011. A calibrated time domain transmissometry soil moisture sensor can be used for precise automated irrigation of container-grown plants. *HortScience* 46:889–894.

Munns, R. 2002. Comparative physiology of salt and water stress. *Plant Cell Environ.* 25:239–250.

Munns, R. 2005. Genes and salt tolerance: Bringing them together. *New Phytol.* 167:645–663.

Munns, R., R.A. James, and A. Läuchli. 2006. Approaches to increasing the salt tolerance of wheat and other cereals. *J. Exp. Bot.* 57:1025–1043.

Nemali, K.S. and M.W. van Iersel. 2008. Physiological responses to different substrate water contents: Screening for high water-use efficiency in bedding plants. *J. Am. Soc. Hort.* 133:333–340.

Ngugi, M.R., D. Doley, and M.A. Hunt. 2004. Physiological responses to water stress in *Eucalyptus cloeziana* and *E. argophloia* seedlings. *Trees (Berl.)* 18:381–389.

Nicolas, E., T. Ferrandez, J.S. Rubio, J.J. Alarcon, and M.J. Sanchez-Blanco. 2008. Annual water status, development, and flowering patterns for *Rosmarinus officinalis* plants under different irrigation conditions. *HortScience* 43:1580–1585.

Niu, G. and D.S. Rodriguez. 2008. Responses of growth and ion uptake of four rose rootstocks to chloride- or sulfate-dominated salinity. *J. Am. Soc. Hort. Sci.* 133:663–669.

Niu, G. and D.S. Rodriguez. 2009. Growth and physiological responses of four rose rootstocks to drought stress. *J. Am. Soc. Hort. Sci.* 134:202–209.

Niu, G., D.S. Rodriguez, and Y.T. Wang. 2006. Impact of drought and temperature on growth and leaf gas exchange of six bedding plant species under greenhouse conditions. *HortScience* 41:1408–1411.

Niu, G., D.S. Rodriguez, and L. Aguiniga. 2007a. Growth and landscape performance of ten herbaceous species in response to saline water irrigation. *J. Environ. Hort.* 25:204–210.

Niu, G., D.S. Rodriguez, L. Rodriguez, and W. Mackay. 2007b. Effect of water stress on growth and flower yield of Big Bend bluebonnet. *HortTechnology* 17:557–560.

Niu, G., D.S. Rodriguez, L. Aguiniga, and W. Mackay. 2007c. Salinity tolerance of *Lupinus havardii* and *Lupinus texensis*. *HortScience* 42:526–528.

Niu, G., D.S. Rodriguez, and W. Mackay. 2008. Growth and physiological responses to drought stress in four oleander clones. *J. Am. Soc. Hort. Sci.* 133:188–196.

Niu, G., D.S. Rodriguez, and M. Gu. 2010a. Salinity tolerance of *Sophora secundiflora* and *Cercis canadensis* var. *Mexicana*. *HortScience* 45:424–427.

Niu, G., D.S. Rodriguez, and T. Starman. 2010b. Response of bedding plants to saline water irrigation. *HortScience* 45:628–636.

Niu, G., M. Wang, D.S. Rodriguez, and D. Zhang. 2012a. Response of zinnia plants to saline water irrigation. *HortScience* 47:793–797.

Niu, G., D.S. Rodriguez, and C. McKenney. 2012b. Response of selected wildflower species to saline water irrigation. *HortScience* 47:1351–1355.

Niu, G., W. Xu, D. Rodriguez, and Y. Sun. 2012c. Growth and physiological responses of Maize and Sorghum genotypes to salt stress. *ISRN Agron.* 2012:145072.

Niu, G., T. Starman, and D. Byrne. 2013. Responses of growth and mineral nutrition of garden roses to saline water irrigation. *HortScience* 48:756–761.

Nonami, H. 1998. Plant water relations and control of cell elongation at low water potentials. *J. Plant Res.* 111:373–382.

Pallardy, S.G. 2008. *Physiology of Woody Plants*, 3rd ed. Academic Press, San Diego, CA.

Parida, A., A.B. Das, and P. Das. 2002. NaCl stress causes changes in photosynthetic pigments, proteins and other metabolic components in the leaves of a true mangrove, *Bruguiera parviflora*, in hydroponic cultures. *J. Plant Biol.* 45:28–36.

Pessarakli, M. 1991. Dry matter yield, nitrogen-15 absorption, and water uptake by green bean under sodium chloride stress. *Crop Sci.* 31:1633–1640.

Pezeshki, S.R. and J.L. Chambers. 1986. Effect of soil salinity on stomatal conductance and photosynthesis of green ash (*Fraxinus pennsylvanica*). *Can. J. For. Res.* 16:569–573.

Pitman, M.G. and A. Läuchli. 2002. Global impact of salinity and agricultural ecosystems. pp. 3–20. In: Läuchli, A. and U. Lüttge (eds.). *Salinity: Environment—Plants—Molecules.* Kluwer Academic Publishers, Dordrecht.

Popova, L.P., Z.G. Stoinova, and L.T. Maslenkova. 1995. Involvement of abscisic acid in photosynthetic process in *Hordeum vulgare* L. during salinity stress. *J. Plant Growth Regul.* 14:211–218.

Ravindran, K.C., K. Venkatesan, V. Balakrishnan, K.P. Chellappan, and T. Balasubramanian. 2007. Restoration of saline land by halophytes for Indian soils. *Soil Biol. Biochem.* 39:2661–2664.

Reddy, A.R., K.V. Chaitanya, and M. Vivekanandan. 2004. Drought-induced responses of photosynthesis and antioxidant metabolism in higher plants. *J. Plant Physiol.* 161:1189–1202.

Saleh, B. 2012. Effect of salt stress on growth and chlorophyll content of some cultivated cotton varieties grown in Syria. *Commun. Soil Sci. Plant Anal.* 43:1976–1983.

Satoh, K., C.M. Smith, and D.C. Fork. 1983. Effects of salinity on primary processes of photosynthesis in the red alga *Porphyra perforata*. *Plant Physiol.* 73:643–647.

Serraj, R. and T.R. Sinclair. 2002. Osmolyte accumulation: Can it really help increase crop yield under drought conditions? *Plant Cell Environ.* 25:333–341.

Shangguan, Z., M. Shao, and J. Dyckmans. 1999. Interaction of osmotic adjustment and photosynthesis in winter wheat under soil drought. *J. Plant Physiol.* 154:753–758.

Shao, H., L. Chu, C.A. Jaleel, and C. Zhao. 2008. Water-deficit stress-induced anatomical changes in higher plant. *C. R. Biol.* 331:215–225.

Sharp, R.E., M.E. LeNoble, M.A. Else, E.T. Thorne, and F. Gherardi. 2000. Endogenous ABA maintains shoot growth in tomato independently of effects on plant water balance: Evidence for an interaction with ethylene. *J. Exp. Bot.* 51:1575–1584.

Shilo, R., M. Ding, D. Pasternak, and M. Zaccai. 2002. Cultivation of cut flower and bulb species with saline water. *Sci. Hort.* 92:41–54.

Song, J.Q. and H. Fujiyama. 1996. Ameliorative effect of potassium on rice and tomato subjected to sodium salinization. *Soil Sci. Plant Nutr.* 42:493–501.

Sotiropoulos, T.E.I., N. Therios, D. Almaliotis, I. Papadakis, and K.N. Dimassi. 2006. Response of cherry rootstocks to boron and salinity. *J. Plant Nutr.* 29:1691–1698.

Starman, T. and L. Lombardini. 2006. Growth, gas exchange, and chlorophyll fluorescence of four ornamental herbaceous perennials during water deficit conditions. *J. Am. Soc. Hort. Sci.* 131:469–475.

Sudhir, P.R., D. Pogoryelov, L. Kovacs, G. Garab, and S.D.S. Murthy. 2005. The effects of salt stress on photosynthetic electron transport and thylakoid membrane proteins in the cyanobacterium *Spirulina platensis*. *J. Biochem. Mol. Biol.* 38:481–485.

Tahkokorpi, M.K., K. Taulavuori, K. Laine, and E. Taulavuori. 2007. After effects of drought-related winter stress in previous and current year stems of Vaccinium myrtillus L. *Environ. Exp. Bot.* 61:85–93.

Tattini, M., G. Montagni, and M.L. Traversi. 2002. Gas exchange, water relations and osmotic adjustment in *Phillyrea latifolia* grown at various salinity concentrations. *Tree Physiol.* 22:403–412.

Tezara, W. and D.W. Lawlor. 1995. Effects of water stress on the biochemistry and physiology of photosynthesis in sunflower. pp. 625–628. In: Mathis, M. (ed.). *Photosynthesis: From Light to Biosphere*. Kluwer Academic Publisher, Dordrecht.

Tezara, W., V.J. Mitchell, S.D. Driscoll, and D.W. Lawlor. 1999. Water stress inhibits plant photosynthesis by decreasing coupling factor and ATP. *Nature* 401:914–917.

Thomas, J.C., E.F. McElwain, and H.J. Bohnert. 1992. Convergent induction of osmotic stress responses. *Plant Physiol.* 100:416–423.

Tsai, Y.C., C.Y. Hong, L.F. Liu, and C.H. Kao. 2004. Relative importance of Na+ and Cl– in NaCl-induced antioxidant systems in roots of rice seedlings. *Physiol. Plant.* 122:86–94.

Van Zandt, P.A. and S. Mopper. 2002. Delayed and carry-over effects of salinity on flowering in *Iris hexagona* (Iridaceae). *Am. J. Bot.* 89:364–383.

Wahid, A. and E. Rasul. 2005. Photosynthesis in leaf, stem, flower and fruit. pp. 479–497. In: Pessarakli, M. (ed.). *Handbook of Photosynthesis*, 2nd ed. CRC Press.

Wang, W.X., B. Vinocur, and A. Altman. 2003. Plant responses to drought, salinity and extreme temperatures: Towards genetic engineering for stress tolerance. *Planta* 218:1–14.

Waterland, N.L., C.A. Campbell, J.J. Finer, and M.L. Jones. 2010. Abscisic acid application enhances drought stress tolerance in bedding plants. *HortScience* 45:409–413.

Williams, M.H., E., Rosenqvist, and M. Buchhave. 1999. Response of potted miniature roses (*Rosa × hybrida*) to reduced water availability during production. *J. Hort. Sci. Biotechnol.* 74:301–308.

Wiltens, J., U. Schreiber, and W. Vidaver. 1978. Chlorophyll fluorescence induction: An indicator of photosynthetic activity in marine algae undergoing desiccation. *Can. J. Bot.* 56:2787–2794.

Wu, L., X. Guo, and A. Harivandi. 2001. Salt tolerance and salt accumulation of landscape plants irrigated by sprinkler and drip irrigation systems. *J. Plant Nutr.* 24:1473–1490.

Yamada, M., H. Morishita, K. Urano, N. Shiozaki, K. Yamaguchi-Shinozaki, K. Shinozaki, and Y. Yoshiba. 2005. Effects of free proline accumulation in petunias under drought stress. *J. Exp. Bot.* 56:1975–1981.

Yamaguchi-Shinozaki, K. and K. Shinozaki. 2004. Improving drought and cold stress tolerance in transgenic rice. *Proceedings of World Rice Research Conference*, Tsukuba, Japan, 5–7 November.

Yordanov, I., V. Velikova, and T. Tsonev. 2003. Plant responses to drought and stress tolerance. *Bulg. J. Plant Physiol. Special Issue.* 187–206.

Zhang, J. and W.J. Davies. 1989. Sequential responses of whole plant water relations towards prolonged soil drying and the mediation by xylem sap ABA concentrations in the regulation of stomatal behavior of sunflower plants. *New Phytol.* 113:167–174.

Zhang, J., U. Schurr, and W.J. Davies. 1987. Control of stomatal behavior by abscisic acid which apparently originates in roots. *J. Exp. Bot.* 38:1174–1181.

Zhang, J., W. Jia, J. Yang, and A.M. Ismail. 2006. Role of ABA in integrating plant responses to drought and salt stresses. *Field Crop Res.* 97:111–119.

Zhu, J.K. 2002. Salt and drought stress signal transduction in plants. *Annu. Rev. Plant Biol.* 53:247–273.

Zhu, J.K. 2003. Regulation of ion homeostais under salt stress. *Curr. Opin. Plant Biol.* 6:441–445.

Zhu, J.K. and F.C. Meinzer. 1999. Efficiency of C-4 photosynthesis in *Atriplex lentiformis* under salinity stress. *Aust. J. Plant Physiol.* 26:79–86.

Zlatev, Z. and I. Yordanov. 2004. Effects of soil drought on photosynthesis and chlorophyll fluorescence in common bean plants. *Bulg. J. Plant Physiol.* 30:3–18.

36 Photosynthetic Strategies of Desiccation-Tolerant Organisms

Beatriz Fernández-Marín, Andreas Holzinger, and José Ignacio García-Plazaola

CONTENTS

36.1 Definitions of Desiccation Tolerance and Occurrence among Photosynthetic Organisms 663
    36.1.1 Photosynthetic Organisms and Water Restriction 663
    36.1.2 Definitions of Desiccation Tolerance: What Is It and What Is It Not? 664
36.2 Living without Water 665
    36.2.1 General Mechanisms of Desiccation Tolerance 665
    36.2.2 Glassy State 665
    36.2.3 Contrasting Strategies: Poikilochlorophyllous versus Homoiochlorophyllous Species 667
36.3 Chloroplast Changes upon Desiccation/Rehydration Cycles 667
    36.3.1 Chloroplast Ultrastructural Changes 667
    36.3.2 Changes in Photosynthetic Membranes, Pigments, and Proteins 669
36.4 Carbon Balance 669
36.5 Photoprotection in Desiccation-Tolerant Organisms 672
    36.5.1 Light Avoidance 672
    36.5.2 Excess Energy Dissipation as Heat 672
    36.5.3 Antioxidative Response 673
36.6 The Peculiar Case of Poikilochlorophyllous Plants 673
    36.6.1 Chloroplast Ultrastructural Changes in Poikilochlorophyllous Plants 674
    36.6.2 Changes in Photosynthetic Pigments in Poikilochlorophyllous Plants 674
    36.6.3 Gas Exchange in Poikilochlorophyllous Plants 674
36.7 Ecological and Evolutionary Significance of Desiccation Tolerance in Photosynthetic Organisms 675
    36.7.1 Desiccation Tolerance in Green Algae 675
    36.7.2 Desiccation Tolerance in Terrestrial Plants 675
Acknowledgments 676
References 676

36.1 DEFINITIONS OF DESICCATION TOLERANCE AND OCCURRENCE AMONG PHOTOSYNTHETIC ORGANISMS

36.1.1 PHOTOSYNTHETIC ORGANISMS AND WATER RESTRICTION

Water restriction is one of the major world limitations for growth and survival of photosynthetic organisms, carbon assimilation and growth being the primary processes affected by water scarcity. Photosynthesis is very sensitively and negatively affected by the decrease in the availability of water within the cell for three major reasons: (1) water is essential to maintain cellular functions and integrity, (2) water molecules are the electron donors for the photosynthetic electron transport chain, and (3) in vascular plants, an additional side effect of water limitation is usually a restriction in CO_2 reaching the chloroplasts. To cope with these effects, photosynthetic organisms employ three different strategies: (1) escape, (2) avoidance, and (3) tolerance (Franks 2011). The first consists in the

early completion of the life cycle, together with the formation of resistance propagules such as spores or seeds, before dehydration of tissues leads to irreversible damage. The second, widespread in most vascular plants (tracheophytes), consists in the delay of desiccation by imposing structural barriers (stomata, cuticle) to water loss. A side effect of this strategy is that it also imposes limitations to CO_2 availability in the chloroplasts, decreasing photosynthetic rates. These limitations are caused by stomatal closure, by changes in mesophyll conductance to CO_2 (Chaves et al. 2009), and at later stress stages, by metabolic restrictions such as those induced by decreases in Rubisco content and activity (Galmés et al. 2013). In parallel to these processes, a series of mechanisms acting on the maintenance of cell turgor and of suitable conditions for metabolic activity are activated.

The third strategy (tolerance) is shown by those organisms that are able to equilibrate their water content with that of the air, interrupt their photosynthetic activity (and the rest of metabolism), and then resume normal metabolic activity after favorable water conditions have returned: these are the so-called

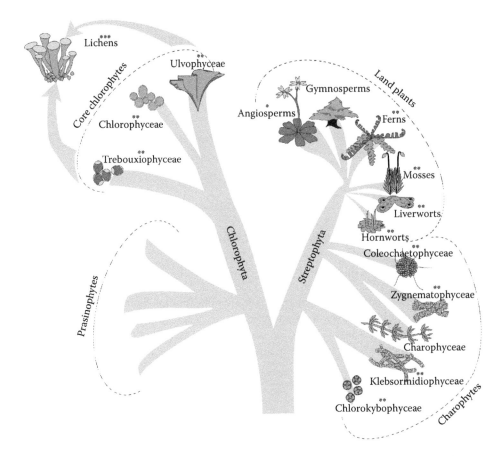

FIGURE 36.1 Evolutionary tree of green-photosynthetic organisms showing the frequency of desiccation-tolerance strategy in photosynthetic tissues. Asterisks indicate the frequency of desiccation tolerance within each phylogenetic group: *, infrequent; **, frequent; ***, very frequent. No asterisk indicates no desiccation tolerance described in photosynthetic tissues for the group.

desiccation-tolerant (DT) organisms. This group includes virtually all lichens, many terrestrial and upper intertidal algae, numerous bryophytes, some pteridophytes, and very few angiosperms (around 250 species only, described up to now) and is still unknown in gymnosperms (Alpert 2006) (Figure 36.1).

36.1.2 Definitions of Desiccation Tolerance: What Is It and What Is It Not?

The intuitive concept of a DT organism is obvious: one that is able to recover its function after being extremely dry. However, it is difficult to condense this straightforward idea into a unique quantitative physiological definition throughout the entire taxonomic (from algae to angiosperms) and functional (from seeds or pollen to leaves) ranges. Several approaches have been employed to define the limits of desiccation tolerance. As a general statement, it could be affirmed that DT organisms are able to equilibrate their water content with that of the air (at relative humidity [RH] ≤50%) and then resume normal metabolic activity after rewatering (Gaff 1971). This dehydration process leads to a loss of water of more than 90% of their relative water content (RWC), which is equivalent to reach a water potential of −100 MPa or even lower (e.g., Proctor 2001; Alpert 2005, 2006). The absolute water content (gH₂O/g dry mass [DM]) can also be used as an estimator of the dryness threshold. Those organisms able to recover after

drying to 0.1 gH₂O/g DM or lower would be considered DT (Vertuci and Farrant 1995; Gaff 1997; Alpert 2005). An alternative to measurements of the RWC is to use the lowest value of protoplasmic water potential at which a tissue can survive. With this criterion, DT tissues are those that can reach below −25 MPa (down to −160 MPa) and still recover, while desiccation-sensitive (DS) tissues are unable to recover after reaching −25 MPa (or even less negative) water potentials (Gaff and Oliver 2013). Nevertheless, it should be pointed out that most of these thresholds have been established for plants, while the same consensus has still not been reached for algae.

It is noteworthy to highlight here that DT greatly differs from drought tolerance strategy. Thus, while drought-tolerant plants cope with water loss in order to maintain proper cell turgor and functioning during stress and are not able to stand protoplasmic water potentials beyond −22 MPa (Gaff and Oliver 2013), DT plants maintain extremely low intracellular water content and experience the water deficit as a transient stress condition from which they escape by switching on additional mechanisms for toleration of further tissue dehydration. Cellular functions are then suspended, and the tissue goes into a *quiescent state*. The tissue can stay in this quiescent state for short to extremely long periods (hours to even years) and then restart normal functions when remoistened. From this point of view, desiccation tolerance can be viewed as a strategy of *drought avoidance* (Tuba et al. 1998; Proctor 2000).

Also, desiccation tolerance must not be confused with the term *poikilohydry*, sometimes misunderstood in the literature. Poikilohydry refers to the ability of an organism to passively change its water content in response to water availability, eventually equilibrating its water content with that of the ambient (Kappen and Valladares 2007), which does not necessarily mean it is able to survive a severe dehydration event. Thus, DT organisms could be considered poikilohydric, but not all poikilohydric organisms are DT (Proctor and Tuba 2002; Kranner et al. 2008).

36.2 LIVING WITHOUT WATER

36.2.1 GENERAL MECHANISMS OF DESICCATION TOLERANCE

Some of the general mechanisms that DT organisms display during dehydration are common to drought tolerance, i.e., stomatal closure (in tracheophytes), synthesis of osmolytes and drought-induced proteins (such as late embryogenesis abundant [LEA] proteins, dehydrins, and early light-induced proteins [ELIPs]), etc. (Figure 36.2). Nevertheless, the ability of DT organisms to recover from severe loss of cell turgor entails more specific mechanisms constitutively expressed, or triggered below a certain cell water potential (Dinakar and Bartels 2013; Yobi et al. 2013). Desiccation implies two main types of biochemical/physical damage that tolerant organisms have to prevent and/or repair: those associated with the disruption of ultrastructures (including loss of membrane integrity and changes in configuration of macromolecules such as enzymes) (Hoekstra et al. 2001) and those related to oxidative stress due to the accumulation of agents such as free radicals (Dinakar et al. 2012). Consequently, DT organisms have developed several mechanisms to assure cellular integrity upon low water content (for a summary, see Figure 36.2). Those mechanisms can be constitutively present in the organism or triggered during dehydration. Generally, DT organisms subjected to rapid and frequent desiccation/rehydration cycles in their natural habitats seem to base their DT strategy on constitutive mechanisms and on the mechanism of repair activated during rehydration (i.e., most algae, bryophytes, and lichens), while most vascular plants posses a greater *third battery* of mechanisms triggered during desiccation (Oliver et al. 1997; Rascio and La Rocca 2005).

Three main strategies are found in DT organisms to cope with the mechanical stress (Figure 36.2): (1) the accumulation of low-molecular-weight osmoprotectants (sugars and/or polyols and proteins) that allow the preservation of turgor, and/or membrane and macromolecule structure (Farrant 2000), (2) the division of the central vacuole into small vacuoles that provide higher mechanical stability (Rascio and LaRocca 2005; Proctor et al. 2007), and (3) the flexibilization of cell walls that then shrink and blend during the loss of turgor without irreparable damages in the plasma membrane to cell wall interaction and cell-to-cell connection (plasmodesma) (Moore et al. 2009). These strategies can appear separately (i.e., some Chlorophyta algae base their protective mechanism on osmolyte accumulation, while some Streptophyta members do on

highly flexible cell walls) (Holzinger et al. 2011; Holzinger and Karsten 2013) but can also be simultaneously found in the same cell (i.e., the DT moss *Polytrichum formossum* shows a blend of flexible cell walls and the central vacuole divided into numerous smaller vacuoles, which provide higher mechanical stability, during drying) (Proctor et al. 2007).

On the other hand, the antioxidant mechanisms of DT organisms play such an important role in assuring survival during the dry state and also during the initial steps of rehydration when an oxidative burst accompanies the resumption of metabolic activity (Kranner et al. 2002; Kranner and Birtik 2005) that the collapse of the antioxidant system determines the longevity of tissues in the desiccated state (Kranner et al. 2002). Thus, an efficient antioxidant system together with the inhibition of senescence programs induced by severe dehydration guarantees survival after drying (Kranner et al. 2002; Griffiths et al. 2014). Among the main antioxidant molecules, glutathione and the carotenoid zeaxanthin (Z) seem to play crucial roles in the tolerance of photosynthetic tissues to desiccation since the content of these two molecules generally increases during dehydration even when considering species of phylogenetically diverse groups such as bryophytes, lichens, and angiosperms (Augusti et al. 2001; Kranner et al. 2002, 2008; Fernandez-Marin et al. 2010, 2011a). Specific antioxidant responses closely related to chloroplast metabolism are discussed extensively in Sections 36.3.2 and 36.5.3.

In sum, DT mechanisms are lead to minimize membrane and protein damage, mechanical stress (due to the loss of turgor), and oxidative stress (linked to water restriction) to levels that are reparable during rehydration. Strong similarities in these mechanisms between brophytes and algae recently have been revealed by transcriptomics analyses suggesting a common origin in the development of the DT strategy (Holzinger et al. 2014).

36.2.2 GLASSY STATE

When cytoplasm reduces its water content to below $0.8gH_2O/g$ dry weight (DW), the viscosity increases exponentially and becomes extremely high, and it enters a glassy state (Leprince and Hoekstra 1998). A glass is defined as an amorphous metastable state that resembles a solid, brittle material but retains the disorder and physical properties of the liquid state (Franks et al. 1991). In other words, a glass is a highly viscous liquid with the physical properties of a solid but the thermodynamic properties of a liquid (Leprince and Buitink 2007). In that state, the relative position of one molecule to another is random, comparable to that in the liquid state (Angell 1998). Due to the high viscosity of the glass, macromolecules are immobilized to a certain extent in a rigid matrix, which limits molecular mobility (Leprince and Buitink 2007). Thus, formation of intracellular glasses slows down deteriorative reactions (although they cannot be totally stopped) (Leprince and Buitink 2007) and prevents the denaturation of proteins. Although there is only little direct evidence of glass formation in dry photosynthetic tissues (Fernandez-Marin et al. 2013), the glassy state is generally considered a crucial stage in the long-term preservation of dry tissues. The formation of the glass depends on water

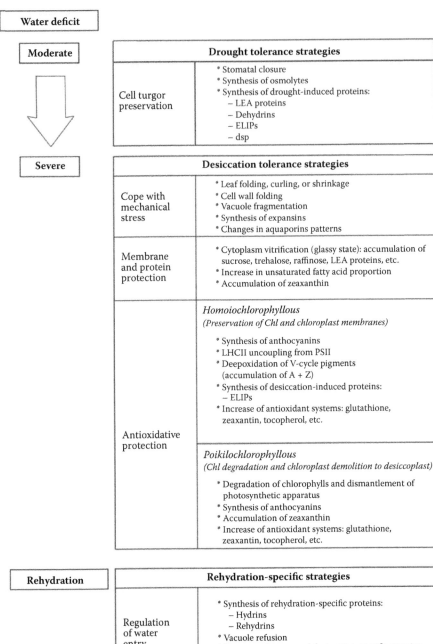

FIGURE 36.2 Summary of general mechanisms of desiccation tolerance in photosynthetic tissues in accordance with the severity of the water deficit. Mechanisms with special relevance during the rehydration are also shown.

content, temperature, and chemical composition of the matrix (i.e., cellular compartments but also intercellular spaces). Although sugars were initially supposed to be the main molecules responsible for intracellular glass formation, it is currently thought that some proteins (particularly LEA proteins) may also play a crucial role in glass formation (Leprince and Buitink 2007). The interaction of these special sugars and proteins with other cytoplasmic components such as salts, organic acids, and amino acids can also intervene in the properties of intracellular glasses (Buitink and Leprince 2004).

36.2.3 CONTRASTING STRATEGIES: POIKILOCHLOROPHYLLOUS VERSUS HOMOIOCHLOROPHYLLOUS SPECIES

Under water limitation the risk of oxidative damage increases in photosynthetic organisms. Light absorption by chlorophylls does not cease during the loss of water, and the energy that cannot be properly transferred to the electron transport chain in the chloroplast will lead to the overproduction of reactive oxygen species (ROS) (Farrant 2007). Even then, most DT organisms retain their chlorophylls and thylakoids during the desiccation, which allow them to rapidly resume photosynthesis after rewatering (Figure 36.3). These organisms are called homoiochlorophyllous and include all DT algae and bryophytes and most vascular plants. To reduce the risk for photooxidative stress during the dry periods, a limited group of DT plants (the so-called poikilochlorophyllous plants) have developed a contrasting strategy by which the chloroplast ultrastructure is dismantled and chlorophylls are degraded during the loss of water (Figure 36.3). Poikilochlorophylly is a scarce strategy among DT organisms that implies a longer period after rewatering until photosynthetic activity is completely restored (several days, usually). It has hitherto been described in a few genera among monocots only, i.e., *Eragrostis* or *Pleurostima* (Van der Willigen et al. 2001; Aidar et al.

2014), and has been studied in depth in the genus *Xerophyta* from the family Velloziaceae (Hallam and Luff 1980; Tuba et al. 1993a,b; Sherwin and Farrant 1998; Ingle et al. 2008; Georgieva et al. 2011; Pérez et al. 2011; Beckett et al. 2012; Solymosi et al. 2013; etc.). Considering that the vast majority of DT photosynthetic organisms show the homoiochlorophyllous strategy (which implies the retention of chlorophylls and thylakoidal organization during the dry stage) and due to the particularities of their photosynthetic mechanisms, this chapter is focused in the processes of homoiochlorophyllous organisms. Nevertheless, the main changes related to chloroplast and photosynthesis in poikilochlorophyllous plants are summarized in Section 36.6.

36.3 CHLOROPLAST CHANGES UPON DESICCATION/REHYDRATION CYCLES

36.3.1 CHLOROPLAST ULTRASTRUCTURAL CHANGES

Methodological difficulties in studying ultrastructure in desiccated tissues limit the amount of available information regarding desiccation-related changes in chloroplasts. However, some general patterns can be extracted from the literature. As is also the case with the general responses to desiccation, the rate of dehydration can greatly influence the extent

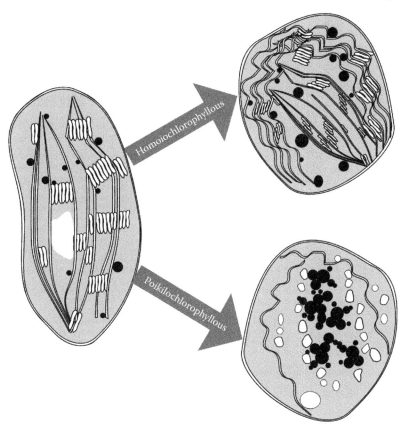

FIGURE 36.3 Ultrastructural changes occurring during desiccation in chloroplast of homoiochlorophyllous (upper right) and poikilochlorophyllous (lower right) DT photosynthetic organisms. Both strategies can lead to spherical chloroplast rich in plastoglobuli (black balls). The increase in the number of plastoglobuli is particularly noticeable in poikilochlorophyllous plants (lower right), which is thought to be related to the dismantling of photosynthetic membranes. Although a kind of reorganization of thylakoidal membranes can be observed in some homoiochlorophyllous species, their photosynthetic membranes remain in a fully functional conformation during the dry state.

of ultrastructural modifications to the chloroplasts. During desiccation, chloroplasts usually change from an oval to a spherical shape independently of the poikilochlorophyllous or homoiochlorophyllous strategy of the species (Figure 36.3). This trend has been observed in photosynthetic tissues of DT angiosperms (Sherwin and Farrant 1996; Ingle et al. 2008), pteridophytes (Bergtrom et al. 1982), and mosses (Proctor et al. 2007) and is frequently accompanied by the movement of chloroplasts from a peripheral to a central location within the cell (Nagy-Deri et al. 2011). Inner membranes are maintained with their thylakoidal organization (Figures 36.4 and

36.5), although some variations in the pattern of grana stacking can be observed (Fernandez-Marin et al. 2013). Also, an increasing number of plastoglobules (lipophilic bodies) has been described during the desiccation of terrestrial microalgae (Holzinger et al. 2011; Karsten and Holzinger 2012) and bryophytes (Fernandez-Marin et al. 2013). These electrodense bodies not only are a storage structure for dismantling of thylakoids but also are furthermore involved in several metabolic functions (Ytterberg et al. 2006). Noticeably, the accumulation of a dense substance in the thylakoidal lumen has been observed in angiosperms (Georgieva et al. 2010) and

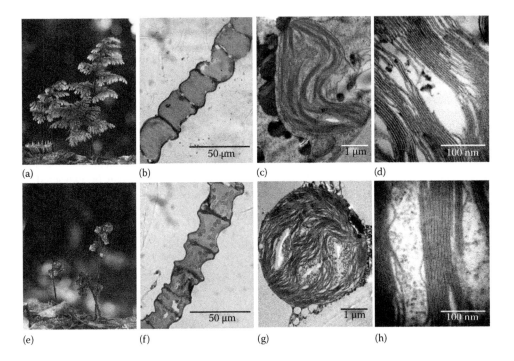

(a)   (b)   (c)   (d)

(e)   (f)   (g)   (h)

FIGURE 36.4 Characterization of morphological changes at macroscopic and microscopic levels experienced during desiccation by the DT ferns of the Hymenophyllaceae family (filmy ferns). (a–d) Hydrated plant. (e–h) Desiccated plant. (a and e) At the whole-plant level, note the curling of the fronds. (b and f) Details at cellular level: note the folding of the flexible cell walls in an optic microscope image of the monostromatic frond. (c–g) Spherification of the chloroplast and reorganization of thylakoids. (d) Detail of thylakoid stacking that accumulates a dark substance in the lumen due to desiccation (h). (Photographs kindly provided by U. Artetxe and J.I. García-Plazaola.)

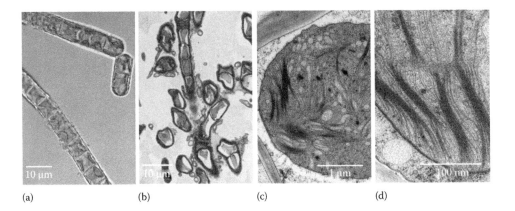

(a)   (b)   (c)   (d)

FIGURE 36.5 Characterization of morphological changes at the microscopic level experienced during desiccation by the streptophycean green alga *Klebsormidium crenulatum*. (a) Hydrated cells in liquid medium; (b–d) desiccated cells. In c and d, note the remanent thylakoidal organization in the chloroplast in the dry state. (a, b: Reprinted from Holzinger, A. et al., *J. Phycol.*, 47, 591–602, 2011. With permission. c, d: Micrographs of A. Holzinger.)

pteridophytes (Figure 36.4h). Although this substance is still unidentified (it is probably related to the accumulation of some phenolic compounds), its accumulation during desiccation has been proposed as a mechanism to protect thylakoid membranes against oxidative damage during desiccation and initial steps of rehydration (Georgieva et al. 2010). Accumulation of phenolic compounds (i.e., anthocyanins) in the chloroplast has already been reported in DT angiosperms (Koonjul et al. 2000). Also, an increase in the quantity of starch deposits can occur during the desiccation (Gasulla et al. 2013b).

36.3.2 CHANGES IN PHOTOSYNTHETIC MEMBRANES, PIGMENTS, AND PROTEINS

Even when homoiochlorophyllous plants retain their photosynthetic apparatus (proteins and pigments), desiccation triggers some changes in their content and composition. Thus, membranes are the first targets of damage during desiccation, and their protection is essential to maintain metabolic functions. Lipid composition of thylakoidal membranes determines their functionality and integrity under stress conditions (Murata et al. 1992; Sahsah et al. 1998). Very recently, Gasulla et al. (2013a) have revealed that, at least in angiosperms, thylakoidal lipid composition undergoes significant changes during the desiccation, although total lipid content remains constant. Mainly, monogalactosyldiacylglycerol is removed from thylakoids upon drying, which instead accumulate oligogalactolipids and phosphatidylinositol. This change in lipid composition seems to enhance the proportion of bilayer-forming lipids that may help in membrane and protein stabilization (Gasulla et al. 2013a). Fatty acid composition of thylakoids could be differential in DT and DS organisms, although more work is needed in that direction before a general conclusion can be drawn, i.e., oleic and linoleic acids were higher and linolenic acid lower in the isolated chloroplasts of two DT angiosperm than in the DS pea plant (Koonjul et al. 2000). Despite noticeable changes in lipid composition of chloroplast membranes, thylakoidal pigment–protein complexes are remarkably stable during desiccation and rehydration (Deng et al. 2003).

Transcriptomic and proteomic studies have shown that the expression of many protective genes is induced during the process of desiccation (Moore et al. 2009). Remarkable among plastidial proteins is the upregulation of the ELIP genes in response to desiccation, a process that has been observed in angiosperms (Alamillo and Bartels 2001), lycophytes (Iturriaga et al. 2006), and mosses (Zeng et al. 2002). ELIPs are a family of low-molecular-weight proteins able to bind chlorophyll and carotenoids (Adamska 1997). Desiccation-induced ELIPs likely protect photosynthetic tissues against photooxidative stress (Hutin et al. 2003) by allowing the accumulation of protective carotenoids such as Z and lutein (Alamillo and Bartels 2001; Dinakar et al. 2012; Gaff and Oliver 2013). In the streptophytic green alga *Klebsormidium*, a strong upregulation of plastidic transcripts was described upon severe desiccation (Holzinger et al. 2014). Four of the six upregulated transcripts with a putative function are similar to plastidic proteins: two upregulated transcripts are possibly involved in nonphotochemical quenching (ELIP1, 35.9-fold upregulated; NPQ4 (PsbS), 35.1-fold upregulated); another interesting contig shows similarity to a plastidic glucose transporter (GLT1, 30.1-fold upregulated) (Holzinger et al. 2014). High-light protection in *Klebsormidium* has recently also been demonstrated by a genomic approach (Hori et al. 2014). Hori et al. (2014) describe a cyclic electron flow activity in *Klebsormidium flaccidum*, which is triggered by 18 NADH oxidoreductase subunits that constitute an NADH dehydrogenase-like complex (NDH).

Thus, chlorophyll content and chlorophyll *a/b* ratio can slightly decrease during desiccation (Degl'Innocenti et al. 2008), while some carotenoids increase. Particularly remarkable is the case of Z, whose increase in drying tissues has been reported for an elevated number of species and taxonomic groups (Kranner et al. 2002; Stepigova et al. 2008; Fernandez-Marin et al. 2009, 2010, 2011a). The use of metabolic inhibitors suggests that (2) the enzyme responsible for Z synthesis under this conditions is violaxanthin deepoxidase (VDE) and not β-carotene hydroxylase and (2) the enzyme Z epoxidase, which converts Z back to violaxanthin, could be inhibited during desiccation (Fernandez-Marin et al. 2009, 2011b). Thus, VDE catalyzes the formation of Z from violaxanthin, in a process directly activated by desiccation and independent of light (which was previously thought to be the only factor able to activate the enzyme) (Fernandez-Marin et al. 2009). Interestingly, VDE is still active at very low intracellular water content (in the rubbery state), being inactivated only in the glassy state. Long-term storage of dry tissues from the DT moss *Syntrichia ruralis* with different contents of Z has shown that Z accumulation correlates with the recovery of photochemical efficiency during subsequent rehydration (Fernandez-Marin et al. 2013). Considering the important functions of Z as a powerful antioxidant (Havaux et al. 2007) in the preservation of thylakoid structure (Havaux 1998) and in the reorganization of light harvesting complex II (LHCII) under stressful conditions (Johnson et al. 2011), the results obtained with *Syntrichia* strongly suggest that Z accumulated during desiccation must play a prominent role in the preservation of the integrity of thylakoidal membranes and photosynthetic apparatus (Fernandez-Marin et al. 2013).

36.4 CARBON BALANCE

DT organisms have adapted to alternation of photosynthetically active and inactive episodes that match periods of water availability in their habitats. Alterations in photosynthesis during desiccation/rehydration cycles follow different patterns among different phylogenetic groups of DT organisms, but generally, net CO_2 assimilation (A_N) is more sensitive than photochemical activity to water loss (Deltoro et al. 1998; Georgieva et al. 2007), and that respiration that can endure longer at low cell water content (Schwab et al. 1989; Tuba et al. 1996a). In tracheophytes (vascular plants), the decrease in A_N starts relatively soon during the dehydration, below 90% RWC (Peeva and Cornic 2009), and even earlier in poikilochlorophyllous species, at RWC >90%, with a pronounced decrease below 80% (Van der Willigen et al. 2001; Beckett et al. 2012) (Figure 36.6) (see also Section 36.6 for details on poikilochlorophyllous plants). Stomatal closure and the consequent limitation in CO_2 availability for Rubisco seem to be the main determinants of that decrease observed in tracheophytes (Tuba et al. 1996b; Georgieva et al. 2007; Degl'Innocenti et al. 2008). In contrast to that, different behavior is found in bryophytes and lichens: they are able to keep a positive net assimilation long during dehydration until RWC close to 0 (see, for instance, Deltoro et al. 1998) (Figure 36.6). High RWC in most bryophytes and lichens induces a decrease in the A_N due to increasing resistance to CO_2 diffusion of the external water, adhered to the tissue (Tuba et al. 1996a; Deltoro et al. 1998; Lange et al. 2001; Royles et al. 2013). Although few works have assessed respiration in green algae during desiccation due to technical difficulties (Holzinger 2009; Holzinger and Karsten 2013), photosynthesis shows a similar behavior to that found in lichens and bryophytes at high water contents. Maximum A_N occurs around 98% RH, and A_N decreases both at saturation (100% RH) and during dehydration, having the lower limit for carbon assimilation at 68% RH (reviewed in Holzinger and Karsten 2013). However, in some aeroterrestrial algae, 100% RH is needed for optimum photosynthesis and growth (Häubner et al. 2006).

At lower water contents, it has been suggested that proteins involved in the Calvin cycle are more affected by desiccation than proteins of the photosynthetic electron transport chain (Dinakar et al. 2012). In accordance with that, proteomic works in DT species of lichenic algae, bryophytes, and angiosperms have recently revealed that several proteins related to carbon fixation (mainly Rubisco and associated proteins) decrease their abundance during desiccation (Oliver et al. 2011) because of their degradation (Gasulla et al. 2013b; Cruz de Carvalho et al. 2014) and the downregulation of the expression of photosynthesis-related genes (Rascio and

LaRocca 2005). A crucial role for the enzyme Rubisco activase in the regulation of Rubisco activity during dehydration/rehydration cycles also has been suggested but still has not been studied in DT organisms (Dinakar et al. 2012). On the other hand, the transcriptomic analysis of the aerotrerrestrial alga *Klebsormidium* shows that the expression of the majority of antenna and carbon assimilation transcripts is upregulated during desiccation, which seems to be involved in the preparation for next rehydration (Holzinger et al. 2014).

Most of the DT species known among tracheophytes show a C_3 metabolism. Nevertheless, at least three C_4 species (*Sporobolus stapfianus*, *Tripogon loliiformis*, and *Tripogon minimus*) and a crassulacean acid metabolism (CAM) (*Blossfeldia liliputiana*) have hitherto been reported (Proctor and Pence 2002; Martinelli et al. 2007; see Niinemets et al. 2014 for C_4 species list). Interestingly, among cryptogams, where the desiccation-tolerance strategy is much more widespread; also, carbon concentration mechanisms (CCMs) are more common: many lichens (Badger et al. 1993; Smith and Griffiths 1996; Lange et al. 1999) and intertidal and terrestrial algae contain CCM. Additionally, multiple gains and losses of the pyrenoid have occurred during the evolution of both Chlorophyta and Streptophyta lineages (Meyer and Griffiths 2013; Kroth 2015). The presence of CCM could represent an advantageous strategy complementary to DT considering that the considerable thickness of cell walls in many DT species together with the migration of chloroplasts to the center of the cell during loss of cell turgor could increase the resistance of the tissue to CO_2 assimilation (reducing mesophyll conductance). Future research in that direction may shed light on the trade-offs between photosynthesis and tolerance to desiccation throughout the evolution of photosynthetic organism.

In sum, optimal conditions for a positive A_N occur in a narrow range of water availability, light, and temperature conditions. In addition, maximal A_N is not particularly high in most DT organisms (Table 36.1). Furthermore, respiration decays

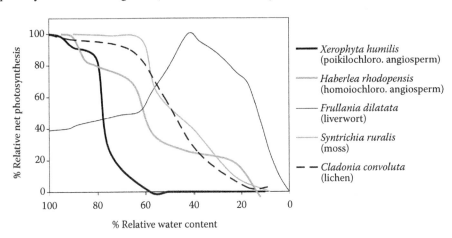

FIGURE 36.6 Carbon net assimilation (net photosynthesis) as a function of tissue relative water content (RWC) during the dehydration process in DT organisms representing different taxonomic groups (poikilochlorophyllous and homoiochlorophyllous angiosperms, mosses, liverworts and lichens). Note the difference in the water content level at which each group ceases the carbon assimilation: poikilochlorophyllous plants (i.e., *Xerophyta*) reach an A_N of 0 at the highest RWC (>50%). (Idealized model based on data from Tuba, Z. et al., *New Phytol.*, 133, 353–361, 1996; Deltoro, V.I. et al., *Planta* 207, 224–228, 1998; Peeva, V., and Cornic, G., *Env. Exp. Bot.*, 65, 310–318, 2009; Beckett, M. et al., *Plant Cell Environ.*, 35, 2061–2074, 2012.)

TABLE 36.1

Examples of Maximal Net CO_2 Assimilation (A_N max) in Desiccation-Tolerant Organisms of Different Taxonomic Groups

Species	A_N max	RWC (%)	PPFD (μmol m^{-2} s^{-1})	Reference
		Green Algae		
Chlorophyta				
Apatococcus lobatus	3.7 μmol CO_2 m^{-2} s^{-1}		20,000 lux	Bertsch 1966
Streptophyta				
Klebosormidium crenulatum	0.5 μmol CO_2 m^{-2} s^{-1}	100	600	Holzinger et al. 2014
		Lichens		
Cyanolichens				
Collema cristatum	2.8 μmol CO_2 m^{-2} s^{-1}	600 WC (% DM)	1000–2000	Lange et al. 2001
Chlorolichens				
Cladonia convoluta	15 μmol CO_2 kg^{-1} s^{-1}	100	300	Tuba et al. 1996a
Cladonia furcata	13 μmol CO_2 kg^{-1} s^{-1}	100	300	Tuba et al. 1996a
Lecanora muralis	6 μmol CO_2 m^{-2} s^{-1}		1200	Lange et al. 2001
Verrucaria baldensis	0.45 μmol CO_2 m^{-2} s^{-1}			Tetriach and Geletti 1997
		Bryophytes		
Mosses				
Syntrichia ruralis	15 μmol CO_2 kg^{-1} s^{-1}	90	300	Tuba et al. 1996a
Fissidens serrulatus	13 μmol CO_2 g^{-1} h^{-1}		200–800	Gabriel and Bates 2003
Liverworts				
Frullania dilatata	80 μg CO_2 g^{-1} DM min^{-1}	40	140	Deltoro et al. 1998
Pellia endiviifolia	55 μg CO_2 g^{-1} DM min^{-1}	65	140	Deltoro et al. 1998
		Pteridophytes		
Selaginella bryopteris	1 μmol CO_2 m^{-2} s^{-1}	100	300	Pandey et al. 2010
		Angiosperms		
Homoiochlorophyllous species				
Craterostigma plantagineum	14 μmol CO_2 m^{-2} s^{-1}	100	2000	Dinakar et al. 2012
	(10 μmol CO_2 m^{-2} s^{-1})		(500)	Dinakar et al. 2012
Haberlea rhodopensis	6 μmol CO_2 m^{-2} s^{-1}		500	Peeva and Cornic 2009
Ramonda nathaliae	6 μmol CO_2 m^{-2} s^{-1}		1000	Rakic et al. 2015
Ramonda serbica	0.5 μmol CO_2 m^{-2} s^{-1}	97	600	Degl'Innocenti et al. 2008
	6 μmol CO_2 m^{-2} s^{-1}		1000	Rakic et al. 2015
Poikilochlorophyllous Species				
Eragrostis nindensis	5.5 μmol CO_2 m^{-2} s^{-1}	90	2000	Van der Willigen et al. 2001
Reaumuria soongorica	30 μmol CO_2 m^{-2} s^{-1}			Xu et al. 2008
Xerophyta humilis	12 μmol CO_2 m^{-2} s^{-1}	100	700	Beckett et al. 2012
Xerophyta scarbida	20 μmol CO_2 m^{-2} s^{-1}	100	1500	Pérez et al. 2011

Note: Different units are shown when transformation to μmol CO_2 m^{-2} s^{-1} was not possible from the original data. Relative water content (RWC) (or water content [WC] when RWC was not provided) and light conditions (photosynthetic photon flux density [PPFD]) during the measurements at which the A_N max values were obtained, and the original sources (references), are shown.

later than assimilation during dehydration, and high respiration rates are established during the first stage of rehydration, i.e., initial peak of respiration after rewatering can reach three-times-higher values than at steady state in lichens (Sundberg et al. 1999). In that context of short and intermittent periods of positive A_N, together with the relatively low maximal A_N, desiccation tolerance is usually related to slow growth rates. For lichens and bryophytes, this short temporal window of positive A_N can be limited to a few hours during the sunrise through daily wetting and drying cycles (Lange et al. 2006).

The success of a DT photosynthetic organism relies on its ability to compensate for the loss of carbon (by respiration) during dry or dark periods with sufficient A_N during hydrated episodes under illumination, which must assure growth and reproduction. A low compensation point (<50 μmol photons m^{-2} s^{-1}; and even much lower in green algae, ~5 μmol photons m^{-2} s^{-1}) (Karsten et al. 2010) and low dark respiration (<2 μmol CO_2 m^{-2} s^{-1}) during hydrated periods could also conform an efficient strategy that has been observed in some DT tracheophytes from the *Ramonda* genus (Rakic et al. 2015) and bryophytes (Van der Poorten and Goffinet 2009). As an example of their metabolic efficiency, it has been shown for the moss *Tortula princeps* that A_N during 6 days of rain (around 200 mmol CO_2/m^2) was enough to compensate for carbon loss due to respiration during 69 days of dry conditions (Graham et al. 2006). Additionally, some recent works on lichens have shown that a wet period in the dark following a photosynthetically active period can improve thallus growth, probably due to enhancement in the conversion of photoassimilates into thallus extension (Bidussi et al. 2013).

36.5 PHOTOPROTECTION IN DESICCATION-TOLERANT ORGANISMS

36.5.1 LIGHT AVOIDANCE

Dehydration directly affects the electron transport chain of the thylakoidal membranes since water molecules are the primary electron donors. The loss of water, in the presence of light, progressively increases the risk of chlorophyll overexcitation, which can generate oxygen radicals and reactive species (ROS) (Heber and Shuvalov 2005). In addition, oxygen diffusion is higher in desiccated tissues than in hydrated ones. DT photosynthetic organisms must deal with these hazardous conditions during the desiccation process but also during long dry periods, in which despite the unavailability of water in the tissues, chlorophylls can still absorb light energy. Rehydration, far from being the end of the stress, is related to an oxidative burst induced by the resumption of metabolic activity (Minibayeva and Beckett 2001). To counteract these effects, DT organisms upregulate photoprotection mechanisms in response to desiccation (Kranner et al. 2002; Farrant et al. 2003). Most of these mechanisms are common with DS plants and algae, but others are triggered only by desiccation.

Tissue folding is frequently an unavoidable mechanical consequence of dehydration, but it also represents the first barrier of photoprotection during dehydration as it implies a substantial reduction in the exposed photosynthetic area (Schwab et al. 1989; Barták et al. 2006; Georgieva et al. 2009; Rafsanjani et al. 2015). Curling of photosynthetic tissues can be found in representatives of all groups of DT organisms (i.e., *Lobaria pulmonaria* among lichens, *Polytrichum* or *Syntrichia* among mosses, *Selaginella*, *Polypodium* or *Asplenium* among pteridophytes, and *Myrothamnus* or *Haberlea* among angiosperms) (see, for instance, curling of fronds from filmy ferns in Figure 36.4e). In some species, such as the fern *Ceterach officinarum*, frond curling combined with the presence of a layer of brown scales or trichomes in the abaxial epidermis reduces almost completely chlorophyll excitation in the desiccated state.

The accumulation of light-screening molecules in external layers of tissues, such as anthocyanins in the epidermis of angiosperms (i.e., *Craterostigma wilmsii*) (Sherwin and Farrant 1998), astaxanthin in terrestrial green algae (Bidigare et al. 1993), and purple-colored iron-complexed phenolics in the green alga *Zygogonium ericetorum* (Aigner et al. 2013) and secondary fungal metabolites (parietin, melanin, rhizocarpic acid, etc.) in the cortex of lichens (Solhaug et al. 2010), reduce the incidence of light to chloroplasts. Particularly, in lichens, desiccation induces important changes in optical properties caused by the generation of air-filled spaces within the cortex, which may increase the efficiency of light screening up to 60–80% for blue light (Solhaug et al. 2010).

36.5.2 EXCESS ENERGY DISSIPATION AS HEAT

In most species, the absorption of light by chlorophylls cannot be totally avoided with the strategies described above, and the dissipation as heat of the excess of energy absorbed by chlorophylls (thermal dissipation) becomes the most important photoprotection mechanism. This process is present in all photosynthetic organisms studied up to now, and it can be easily estimated by the decrease in the maximal fluorescence emission of chlorophyll a (F_m), the so-called nonphotochemical quenching of chlorophyll fluorescence (NPQ). This parameter can be used as a proxy to study the thermal energy dissipation, and frequently, the term NPQ is used as synonym for thermal dissipation. Nevertheless, the reader should be aware that different processes other than thermal dissipation such as chloroplast movements or changes in optical properties (Bilger 2014) can also reduce the fluorescence yield, affecting the NPQ.

NPQ, more than a single mechanism, represents the sum of a large spectrum of processes (García-Plazaola et al. 2012). The most deeply characterized form of NPQ is the ΔpH-dependent one, which is activated upon illumination thanks to the generation of a transthylakoidal proton gradient coupled to electron transport. It also requires the presence of a specific protein: PsbS in vascular plants (Li et al. 2000) and streptophycean green algae (Gerotto and Morosinotto 2013), LHSCR in green algae (Peers et al. 2009), and both proteins in bryophytes (Alboresi et al. 2010) and the Ulvophyceae (Mou et al. 2013). Formation of NPQ is also amplified by the presence of Z (Morosinotto and Bassi 2014). This form of NPQ is also active in DT plants and algae and contributes to photoprotection during desiccation, mainly before the activation of sustained forms of NPQ. Recently, a Z-specific NPQ mechanism has been described in terrestrial Chlorophyceae, Klebsormidiophyceae, and Trebouxiophyceae species but not found in the terrestrial species *Tetracystis aeria* (Chlorophyceae) (Quaas et al. 2015) and *Cylindrocystis* sp. (Zygnematophyceae) (Lunch et al. 2013). In general, the xanthophyll cycle pool size might also be related to other factors in Zygnematophyceae, like the geographical distribution, as shown for the genus *Cosmarium* (Stamenković et al. 2014).

In addition to the ΔpH-dependent NPQ, in most DT, the dehydration itself is thought to be responsible for the development of the so-called desiccation-induced NPQ (NPQ_{DT}) (Heber et al. 2006, 2010; Nabe et al. 2007; Wieners et al. 2012; Bilger 2014). The NPQ_{DT} is not present in vascular DT plants (ferns and angiosperms) but is widespread in all the other groups of DT organisms where it is easily recognizable by a characteristic decrease in basal fluorescence (F_0) and a loss of variable fluorescence (F_v) in the desiccated state (Heber et al. 2000; Bilger 2014). Contrasting with the light-induced NPQ, the NPQ_{DT} does not require either the generation of a proton gradient or Z, desiccation being the sole triggering factor involved in its activation (Heber 2012). The inhibition of NPQ_{DT} by glutaraldehyde, which produces insoluble cross-linking aggregates, suggests that a conformational change in photosystem II (PSII) structure is involved in the activation of such a mechanism, which implies a shift from a light-harvesting configuration to a dissipative one (Heber 2008). There, dissipation may occur in the antenna or in the reaction center of PSII (Heber 2012). Contrasting with PSII, photosystem I (PSI) seems to be unable to activate a comparable NPQ_{DT}. As a consequence PSI, is able to operate even under severe desiccation when activity of PSII (fluorescence levels) is already negligible (Veerman et al. 2007; Gao et al. 2014, 2015). Under illumination, PSI seems to be inactivated in DT organisms (but not in DS ones) as part of their particular strategy (Nabe et al. 2007). Phosphorylation of PSII and LHCII during desiccation relates to the rearrangement of thylakoidal proteins and the formation of a supercomplex (together with PSI subunits), which recently has been suggested as a photoprotective mechanism in DT green macroalgae (Gao et al. 2014, 2015). Independently of the phylogenetic origin, the development of NPQ seems to follow a similar pattern among DT organisms (García-Plazaola et al. 2012 and references therein). Briefly, NPQ increases during desiccation and remains at a high value during the whole dry period. During the initial phase of rehydration, NPQ can transiently undergo a further increase, due to the increasing demand for photoprotection occurring during the reestablishment of metabolic activity and the reorganization of the photosynthetic apparatus. After rehydration has been completed, initial lower values of NPQ progressively recover. Nevertheless, factors such as the photosynthetic photon flux density (PPFD) and the speed of dehydration influence the NPQ, leading to certain variability even among individuals of the same species (Hamerlynck et al. 2002; Gasulla et al. 2009; Heber et al. 2010). Also, growing conditions affect the speed and the amplitude of NPQ development, i.e., samples of the lichen *Lobaria pulmonari* collected from an oak forest showed higher and faster NPQ than a parallel set of samples collected from a beech (more shady and humid) forest (Fernandez-Marin et al. 2010). Some metabolites can also play an important role in the development of the NPQ. As an example, it recently has been proposed that the availability of D-arabitol but not of other sugars in lichens enhances the NPQ of the photobiont (Kosugi et al. 2013).

36.5.3 Antioxidative Response

Antioxidative mechanisms play such a crucial role in DT organisms that their survival through a desiccated period can be directly related to their antioxidant status (Kranner et al. 2002). As stated in Sections 36.3 and 36.2.1, the antioxidant role of Z is of paramount importance in desiccated chloroplasts. Z is accumulated during desiccation and kept at high contents even during long periods of desiccation, until it decreases again during the next rehydration (Kranner et al. 2002; Fernandez-Marin et al. 2013). Together with it, another lipophilic antioxidant, α-tocopherol, whose function partly overlaps with that of Z (Havaux et al. 2000, 2005), is also upregulated in some species (including mosses, ferns, and angiosperms) in response to desiccation events (Tausz et al. 2001; Fernandez-Marin et al. 2011a). Similarly, accumulation of γ-tocopherol has been reported during desiccation of the DT moss *Tortula ruraliformis* (Seel et al. 1992). The tocopherol content, however, can also progressively decrease during prolonged desiccation (Kranner et al. 2002).

Some hydrophilic antioxidants also play fundamental protective roles under desiccation in photosynthetic tissues. This is particularly the case for glutathione, whose reduced form's (GSH) proportion is a determinant of the survival rate in the desiccated state (Kranner et al. 2002). A general response in DT plants is to protect chloroplasts by increasing the total amount of GSH during dehydration (Rascio and LaRocca 2005). After this initial period, GSH decreases with prolonged dry episodes, and then it recovers during rehydration (Kranner et al. 2002). Apart from its well-known antioxidant role, it has been suggested that GSH is involved in a thiol–disulfide cycle that protects proteins from oxidative damage and the irreversible formation of disulfide bonds (Kranner et al. 2008). Recent studies highlight also the potential role of other low-molecular-weight thiols, such as glutamyl amino acids, in the preservation of dry photosynthetic tissues (Yobi et al. 2013).

36.6 THE PECULIAR CASE OF POIKILOCHLOROPHYLLOUS PLANTS

There is a particular group of DT plants, the so-called poikilochlorophyllous (only described in monocots; see Section 36.2.3), in which chlorophyll is degraded during desiccation in order to avoid damages to the photosynthetic apparatus (Figure 36.2). This mechanism involves also the disorganization of chloroplast thylakoids, which recover their normal organization and chlorophyll content after rewatering (Farrant et al. 1999) (Figure 36.5). Poikilochlorophylly represents an important advantage as it prevents oxidative damages that may occur during the dehydration process and during the dry state, when photodynamic chlorophyll molecules become a very powerful source of ROS (see Section 36.1.1). However, the metabolic cost of this strategy is that chloroplast must be rebuilt during rehydration, by the *de novo* synthesis of all photosynthetic pigments and the entire thylakoid system (Sherwin and Farrant 1996). Consequently, the resumption of normal photosynthetic activity can only take place several days after rehydration, while only a few hours (or even minutes in the case of nonvascular plants) are needed for most homoiochlorophyllous organisms.

36.6.1 Chloroplast Ultrastructural Changes in Poikilochlorophyllous Plants

Chloroplasts of poikilochlorophyllous plants progressively lose their thylakoidal structure during desiccation (Figure 36.3). Grana stacking is lost, and thylakoidal membranes are disorganized and broken. In parallel, an increasing number of plastoglobuli are accumulated. As a result, in the dry state, only a few poorly developed inner membranes are found in the plastids (Sherwin and Farrant 1996; Solymosi et al. 2013) (Figure 36.3). Inner membrane material can be stored in hexagonal, paracrystalline prolamellar bodies and lamellar prothylakoids (Solymosi et al. 2013) or in vesicles (Sherwin and Farrant 1996). Plastids with these properties found in the dry leaves of poikilochlorophyllous species are called desiccoplasts (Tuba et al. 1993a,b), or xeroplasts (Ingle et al. 2008). Desiccation of poikilochlorophyllous plants is also accompanied by a downregulation of the expression of genes that encode important photosynthetic proteins such as the D1 subunit of the PSII core complex and the oxygen-evolving complex (Ingle et al. 2007).

During rehydration, the plastids slowly recover their normal ultrastructure in a process that usually takes several days (≥48 h) (Sherwin and Farrant 1996). Thylakoidal membranes can start their reassembly in the dark, while the formation of grana depends on light and correlates with the light-dependent synthesis of Lhcb2 protein (Ingle et al. 2008). This observation is in agreement with the fact that grana stacking, generally in plants, is mediated by the adhesion between Lhcb proteins (reviewed in Pribil et al. 2014). Reconstitution of photosynthetic membranes during rehydration is also accompanied by an increase in total fatty acid content and by a higher proportion of unsaturated fatty acids (Georgieva et al. 2011).

36.6.2 Changes in Photosynthetic Pigments in Poikilochlorophyllous Plants

Regulated chlorophyll and carotenoid degradation starts when RWC decreases to approximately below 80% (Van der Willigen et al. 2001). Chlorophylls can be completely (Tuba et al. 1994; Aidar et al. 2014) or almost completely degraded (<10% of the content of fully hydrated leaves) (Pérez et al. 2011), whereas around 20% of the carotenoid content found in hydrated leaves is retained in desiccoplasts (Tuba et al. 1994). Thus, carotenoids can appear as the only photosynthetic pigments remaining in the dry leaves (Tuba et al. 1993a). In particular, Z content increases during desiccation (Beckett et al. 2012), while violaxanthin, neoxanthin, and β-carotene decrease (Beckett et al. 2012). As described for homoiochlorophyllous species in Sections 36.2.1, 36.3.2, and 36.5.3, the role of Z in poikilochlorophyllous plants could be

strongly related to the protection of the membranous structures in which thylakoids are converted during the desiccation. Z could protect those membranes against oxidation and destabilization, considering the highly antioxidant activity of this xanthophyll even when free in the membranes (Havaux et al. 2007; Dall'Osto et al. 2010).

During the rehydration, the resynthesis of chlorophyll starts in the first 12 h, but final content is not reached until 72–96 h after rewatering (Pérez et al. 2011). Interestingly, the chlorophyll a/b ratio increases progressively during the rehydration, which has been interpreted as a faster rise in the PSII reaction centers with respect to the antenna complexes (Pérez et al. 2011). In some homoiochlorophyllous species, anthocyanin content of leaves follows an inverse pattern to chlorophyll content, being accumulated during desiccation, retained in the dry state, and reduced during rehydration (Sherwin and Farrant 1998).

36.6.3 Gas Exchange in Poikilochlorophyllous Plants

During desiccation, photosynthesis rapidly decreases when RWC reaches values below 80%, most probably due to the photochemical limitation that would affect the energy-dependent regeneration of Rubisco enzyme (Beckett et al. 2012) and the breakdown of photosynthetic proteins, such as the components of PSII (Ingle et al. 2007).

CO_2 assimilation starts several hours after rehydration, at RWC ≥70%, when the chloroplast ultrastructure has recovered and the chlorophylls have been resynthesized (Van der Willigen et al. 2001). Net CO_2 assimilation can start within the first 24 h of rehydration, but normal rates are not reached until 1 or 2 days after (around 72 h after rewatering) (Tuba et al. 1994). Respiration, by contrast, is restored much earlier, almost immediately after rewatering (Van der Willigen et al. 2001). This fast reactivation of mitochondria (which retain their intact ultrastructure in the dry state) is thought to assure the energy supply needed for the rebuilding of chloroplast ultrastructure and synthesis of photosynthetic components during rehydration (Van der Willigen et al. 2001).

The activity of Rubisco, more than its content, seems to be altered during desiccation/rehydration cycles (Pérez et al. 2011). The enzyme is thought to adopt a special protective (aggregated) conformation during the dry state that is reverted into an active conformation during rehydration (Pérez et al. 2011). The process could be regulated by redox status modifications (Pérez et al. 2011).

To the best of our knowledge, only one paper, by Beckett et al. (2012), has so far reported about emissions of volatile organic compounds (VOCs) by poikilochlorophyllous species. In this work, it is shown that *Xerophyta humilis* emits isoprene during the first steps of desiccation (reaching a maximum at around 85% RWC) and in the last steps of rehydration (with a maximum at RWC >90%). With a different emission pattern, hexanal was also detected. Thus, hexanal emission increased with desiccation, peaking at 35% RWC, but was not emitted during rehydration. Beckett et al. (2012) surmise that isoprene could be involved in membrane protection during moderate levels of desiccation, while hexanal emission could be related to processes of thylakoid membrane disassembly into membranous vesicles.

36.7 ECOLOGICAL AND EVOLUTIONARY SIGNIFICANCE OF DESICCATION TOLERANCE IN PHOTOSYNTHETIC ORGANISMS

Toward the evolution of photosynthetic organisms, the conquering of land represented an explosion in terms of biodiversity and functional specifications. In the earliest terrestrial photosynthetic organisms, the investment in mechanisms to deal with light excess, temperature variability, and, particularly, desiccation represented the key to success. Interestingly, the acquisition of desiccation tolerance, although likely initially related to the transition from water to terrestrial environments, has been lost and regained multiple times, leading to the scattered but widespread presence across the evolutionary tree of eukaryotic phototrophs. In this final section of the chapter, we review the phylogenetically diverse groups in which desiccation tolerance has been described in photosynthetic tissues, together with the most plausible ecological restrictions that constrain the presence of DT species to certain habitats.

36.7.1 DESICCATION TOLERANCE IN GREEN ALGAE

Within green algae, it became obvious over the past decade that phylogenetic relationships allowed us to clearly separate between the chlorophytic lineage (containing the core chlorophytes with Ulvophyceae, Chlorophyceae, and Trebouxiophyceae) and the streptophytic lineage with all charophytes (e.g., Leliaert et al. 2012; Friedl and Rybalka 2012) (Figure 36.1). Within the latter lineage of organisms, the sisters to land plants are expected (e.g., Wodniok et al. 2011; Becker 2012). Interestingly, desiccation tolerance of the vegetative forms (and we are here excluding tolerant permanent stages, i.e., zygospores) is realized in both the chlorophytic and streptophytic lineage. In the chlorophytic lineage, almost all "core chlorophytes" have DT members (Figure 36.1). Particular attention regarding DT has been given to the chlorophycean genera *Desmococcus*, *Apatococcus*, *Coccomyxa* (Lüttge and Büdel 2010), *Bracteacoccus*, *Scenedesmus*, and *Chlorosarcinopsis* (Gray et al. 2007) and to the Trebouxiophycean genera *Trebouxia* (Lüttge and Büdel 2010), *Botryococcus* (Demura et al. 2014), *Chlorella*, and *Myrmecia* (Gray et al. 2007). Moreover, these core chlorophytes contain genera that tend to lichenization, leading to most of the known lichens with eukaryotic photobionts, e.g., members of the Trebouxiophyceae, and even Ulvophyceae, particularly Trentepohliales (Ong et al. 1992; Lüttge and Büdel 2010), which were recently phylogenetically characterized (Hametner et al. 2014).

Within the streptophytic lineage (Streptophyta), at least in four of the six green algal classes (Chlorokybophyceae, Zygnematophyceae, Klebsormidiophyceae, and Coleochaetaceae), DT vegetative forms have been found, and some of their genera, extensively studied, i.e., *Zygnema* (McLean and Pessoney 1972; Holzinger et al. 2009; Pichrtová et al. 2014a), *Zygogonium* (Holzinger et al. 2010; Stancheva et al. 2014), *Klebsormidium* (Karsten et al. 2010; Karsten and Holzinger 2012), *Interfilum* (Karsten et al. 2014), and *Colechaete* (Graham et al. 2012) (Figure 36.1). There are no DT Charophyceae known. This desiccation sensitivity likely prevented further steps in terrestrialization of the group, despite their being morphologically more advanced when compared to the Zygnematophyceae, which are believed to be sisters to land plants (e.g., Wodniok et al. 2011). Within the Zygnematophyceae, the situation concerning desiccation tolerance is complex; mostly, the Zygnematales (including genera like *Zygnema*, *Cylindrocystis*, and *Mesotaenium*) contain DT species, and mostly, these organisms are additionally tolerant to cold temperatures. In contrast, members of the Desmidiales depend on strictly oligotrophic waters as found in peat bogs but do not show DT mechanisms.

It appeared to be beneficial for the initial events of terrestrialization within this group that DT mechanisms were realized. Several of these organisms might not be termed *DT* in the strict sense but are only DT after some hardening processes (i.e., akinete formation) (e.g., McLean and Pessoney 1972; Pichrtová et al. 2014b; Herburger et al. 2015). The acclimation process might go along with an accumulation of storage compounds and an increase in osmotic strength (Kaplan et al. 2013). Within the morphologically advanced members of *Coleochaete*, DT is reported (e.g., Graham et al. 2012). Klebsormidiophyceae have been found to tolerate a certain level of desiccation in the vegetative state (e.g., Karsten et al. 2010; Karsten and Holzinger 2012) and some of the physiological requirements (Kaplan et al. 2012) and transcriptomic changes after severe desiccation have been recently described (Holzinger et al. 2014).

Most of the green algae with a DT strategy live in terrestrial habitats, which is the case for the chlorophytic as well as the streptophytic lineage. While many examples from, for example, the Trebouxiophyta are described as "soil crust algae" (e.g., Ettl and Gärtner 1994; Karsten and Holzinger 2014), there are more uncommon examples, where terrestrial life forms have only recently been discovered. For example, within the Ulvophyceae, in most cases strictly marine organisms, there are exceptions like the terrestrial Trentepohliales (e.g., López-Bautista 2002) and Cladophorales (Rindi et al. 2006).

36.7.2 DESICCATION TOLERANCE IN TERRESTRIAL PLANTS

One of the terrestrialized branches of streptophyta evolved into the first plants 600 mya (http://www.timetree.org). As a consequence, desiccation tolerance in plants is considered to be an ancestral trait, which contributed to the establishment of plants on terrestrial habitats but constrained independence from water. Thus, desiccation tolerance is widespread in basal groups, and it has been experimentally demonstrated in the three groups of bryophytes: mosses, liverworts, and hornworts (Wood 2007) (Figure 36.1). Bryophytes are poikilohydric, and water uptake and loss occur rapidly and without specialized organs. Contrasting with this group, desiccation tolerance is much more restricted in vascular plants (ferns and angiosperms) and even absent in gymnosperms (Alpert 2005). Despite the existence of only a few vascular plants that can be considered DT, this trait is conserved in spores and seeds. Among ferns, desiccation tolerance is present in both the gametophyte and the sporophyte generations (Watkins et al. 2007), being more frequent than in angiosperms (Proctor and Pence 2002). Particularly relevant is the case of filmy ferns (Hymenophyllaceae) (Figure 36.4), which have developed desiccation tolerance in constantly moist forests in which desiccation is rarely experienced (Proctor 2012) and water saving mechanisms are unnecessary. Among basal angiosperms, desiccation tolerance is not present (Oliver et al. 2005), but after the diversification of angiosperms, desiccation tolerance of vegetative tissues has re-evolved at least 10 times (Oliver et al. 2000) in habitats with intermittent water availability, such as rocky outcrops. Considering the monophyletic origin of plants, it has been proposed that all these forms of desiccation tolerance derive from an ancestral form of desiccation tolerance proper to bryophytes (Rascio and LaRocca 2005).

Overall, terrestrialization of green algae and cyanobacteria is an evolutionary event that has repeatedly occurred in fluctuating environments. This is the case with biological crusts, temporary pools, and intertidal habitats. It has been proposed that the generation of temporary pools of freshwater during glaciation events (snowball stages) in the Proterozoic period might have been fundamental for the diversification and colonization of terrestrial habitats by streptophytic algae (Becker 2012). Additionally, the early process of lichenization (600 mya) (Retallack 2012) has been one of the most successful processes in the independence from water, which has had remarkable and conspicuous success, generating a particularly resistant symbiosis, which even dominates the landscape in certain biocenosis in tropical and polar deserts, and tundra and alpine regions, where higher plants are not able to survive. Desiccation tolerance has also evolved from the sea in some Ulvophyceaen macroalgal species occupying the upper intertidal belt.

Irrespective of the conditions that have lead to the evolution of DT, when desiccated, phototrophs are particularly resistant to environmental stress and, for example, can restore photosynthetic activity even after decades of dry storage (Alpert 2000). Particularly remarkable is the case of desiccated lichens, whose photobionts are more tolerant to photoinhibition when living in symbiosis (Kosugi et al. 2009) and do not show symptoms of damage even after prolonged exposition to the extreme conditions of outer space that are lethal for bacteria and other microorganisms (Sancho et al. 2007).

Independently of the monophyletic or polyphyletic nature of the presence of desiccation tolerance across the evolutionary tree of green photosynthetic eukaryotes, when this character is not useful (because of the development of stenohydric or homeohydric habits), it is lost. For example, even among lichens (in general considered extremely DT), there are species such as *Pseudocyphellaria dissimilis*, a shade-adapted lichen from the rainforests of New Zealand, that are highly sensitive to desiccation (Green et al. 1991). The costs/benefits of desiccation tolerance justify its labile character. In essence, the basic economics equation of a DT organism states that carbon assimilation during hydrated periods must be higher than carbon costs of being DT together with carbon losses during desiccated periods (see also Section 36.4). This strategy can be considered opportunistic in the sense that these organisms take advantage of water scarcity in some of the most extreme habitats of the biosphere.

ACKNOWLEDGMENTS

The authors acknowledge financial support by the European Union to BFM (Marie Curie Action FP7-PEOPLE-2012-IEF 328370 "MELISSA") and support to JIGP for a visiting professorship by Salvador de Madariaga Programme and by the University of Innsbruck. The present work has also been supported by Austrian Science Fund (FWF) grants P 24242-B16 and I 1951-B16 to AH.

REFERENCES

Adamska, I. 1997. ELIPs—Light-induced stress proteins. *Physiol Plant* 100:794–805.

Aidar, S.T., Meirelles, S.T., Oliveira, R.F., Chaves, A.R.M., Fernandes-Junior, P.I. 2014. Photosynthetic response of poikilochlorophyllous desiccation-tolerant *Pleurostima purpurea* (Velloziaceae) to dehydration and rehydration. *Photosynthetica* 52:124–133.

Aigner, S., Remias, D., Karsten, U., Holzinger, A. 2013. Unusual phenolic compounds contribute to the ecophysiological performance in the purple-colored green alga *Zygogonium ericetorum* (Zygnematophyceae, Streptophyta) from a high-alpine habitat. *J Phycol* 49:648–660.

Alamillo, J., Bartels, D. 2001. Effects of desiccation on photosynthesis pigments and the ELIP-like dsp 22 protein complexes in the resurrection plant *Craterostigma plantagineum*. *Plant Sci* 160:1161–1170.

Alboresi, A., Gerotto, C., Giacometti, G.M., Bassi, R., Morosinotto, T. 2010. *Physcomitrella patens* mutants affected on heat dissipation clarify the evolution of photoprotection mechanisms upon land colonization. *Proc Natl Acad Sci U S A* 107:11128–11133.

Alpert, P. 2000. The discovery, scope, and puzzle of desiccation tolerance in plants. *Plant Ecol* 151:5–17.

Alpert, P. 2005. The limits and frontiers of desiccation-tolerant life. *Int Comp Biol* 45:685–695.

Alpert, P. 2006. Constraints of tolerance: Why are desiccation tolerant organisms so small or rare? *J Exp Bot* 209:1575–1584.

Angell, A. 1998. Liquid landscape. *Nature* 393:521–523.

Augusti, A., Scartazza, A., Navari-Izzo, F., Sgherri, C.L.M., Stevanovic, B., Brugnoli, E. 2001. Photosystem II photochemical efficiency, zeaxanthin and antioxidant contents in the poikilohydric *Ramonda serbica* during dehydration and rehydration. *Photosynth Res* 67:79–88.

Badger, M.R., Pfanz, H., Budel, B., Heber, U., Lange, O.L. 1993. Evidence for the functioning of photosynthetic CO_2 concentrating mechanisms in lichens containing green algal and cyanobacterial photobionts. *Planta* 191:57–70.

Barták, M., Solhaug, A., Vrablikova, H., Gauslaa, Y. 2006. Curling during desiccation protects the foliose lichen *Lobaria pulmonaria* against photoinhibition. *Oecology* 149:553–560.

Becker, B. 2012. Snow ball earth and the split of streptophyta and chlorophyta. *Trends Plant Sci* 18:180–183.

Beckett, M., Loreto, F., Velikova, V. et al. 2012. Photosynthetic limitations and volatile and non-volatile isoprenoids in the poikilochlorophyllous resurrection plant *Xerophyta humilis* during dehydration and rehydration. *Plant Cell Environ* 35:2061–2074.

Bergtrom, G., Schaller, M., Eickmeter, W.G. 1982. Ultrastructural and biochemical bases of resurrection in the drought-tolerant vascular plant, *Selaginella lepidophylla*. *J Ultrastruct Res* 78:269–282.

Bertsch, A. 1966. CO_2 Gaswechsel der Grünalge *Apatococcus lobatus*. *Planta* 70:46–72.

Bidigare, R.R., Ondrusek, M.E., Kennicutt, M.C., Iturriaga, R., Harvey, H.R., Hoham, R.W., Macko, S.A. Evidence for a photoprotective function for secondary carotenoids of snow algae. *J Phycol* 29:427–434.

Bidussi, M., Gauslaa, Y., Solhaug, A. 2013. Prolonging the hydration and active metabolism from light periods into nights substantially enhances lichen growth. *Planta* 237:1359–1366.

Bilger, W. 2014. Desiccation-induced quenching of chlorophyll fluorescence in cryptogams. In *Non-photochemical quenching and energy dissipation in plants, advances in photosynthesis and respiration*, vol. 40, eds. B. Demmig-Adams, G. Garab, W. Adams III, Govindjee, 409–420. Dordrecht: Springer.

Buitink, J., Leprince, O. 2004. Glass formation in plant anhydrobiotes: Survival in the dry state. *Cryobiology* 48:215–228.

Chaves, M.M., Flexas, J., Pinheiro, C. 2009. Photosynthesis under drought and salt stress: Regulation mechanisms from the whole plant to cell. *Ann Bot* 103:551–560.

Cruz de Carvalho, R., Bernardes da Silva, A., Soares, R., Almeida, A., Coelho, A.V., Marques da Silva, J., Branquinho, C. 2014. Differential proteomics of dehydration and rehydration in bryophytes: Evidence towards a common desiccation tolerance mechanism. *Plant Cell Environ* 37:1499–1515.

Dall'Osto, L., Cazzaniga, S., Havaux, M., Bassi, R. 2010. Enhanced photoprotection by protein-bound vs free xanthophyll pools: A comparative analysis of chlorophyll *b* and xanthophyll biosynthesis mutants. *Mol Plant* 3:576–593.

Degl'Innocenti, E., Guidi, L., Stevanovic, B., Navari, F. 2008. CO_2 fixation and chlorophyll *a* fluorescence in leaves of *Ramonda serbica* during a dehydration–rehydration cycle. *J Plant Physiol* 165:723–733.

Deltoro, V.I., Calatayud, A., Gimeno, C., Abadía, A., Barreno, E. 1998. Changes in chlorophyll a fluorescence, photosynthetic CO_2 assimilation and xanthophyll cycle interconversions during dehydration in desiccation-tolerant and intolerant liverworts. *Planta* 207:224–228.

Demura, M., Ioki, M., Kwachi, M., Nakajima, N., Watanabe, M.M. 2014. Desiccation tolerance of *Botryococcus braunii* (Trebouxiophyceae, Chlorophyta) and extreme temperature tolerance of dehydrated cells. *J Appl Phycol* 26:49–53.

Deng, X., Hu, Z.A., Wang, H.X., Wen, X.G., Kuang, T.Y. 2003. A comparison of photosynthetic apparatus of the detached leaves of the resurrection plant *Boea hygrometrica* with its non-tolerant relative *Chirita heterotrichia* in response to dehydration and rehydration. *Plant Sci* 165:851–861.

Dinakar, C., Bartels, D. 2013. Desiccation tolerance in resurrection plants: New insights from transcriptome, proteome, and metabolome analysis. *Front Plant Sci* 4:482.

Dinakar, C., Djilianov, D., Bartels, D. 2012. Photosynthesis in desiccation tolerant plants: Energy metabolism and antioxidative stress defence. *Plant Sci* 182:29–41.

Ettl, H., Gärtner, G. 1994. *Syllabus der Boden Luft und Flechtenalgen*. Stuttgart: Gustav Fischer.

Farrant, J.M. 2000. A comparison of mechanisms of desiccation-tolerance among three angiosperm resurrection plant species. *Plant Ecol* 151:29–39.

Farrant, J. 2007. Mechanisms of desiccation tolerance in angiosperm resurrection plants. In *Plant desiccation tolerance*, eds. M.A. Jenks, A.J. Wood, 51–90. Ames, IA: Blackwell Publishing.

Farrant, J.M., Cooper, K., Kruger, L.A., Sherwin, H.W. 1999. The effect of drying rate on the survival of three desiccation-tolerant angiosperm species. *Ann Bot* 84:371–379.

Farrant, J.M., Van der Willigen, C., Loffell, D.A., Bartsch, S., Whittaker, A. 2003. An investigation into the role of light during desiccation of three angiosperm resurrection plants. *Plant Cell Environ* 26:2175–1286.

Fernandez-Marin, B., Balaguer, L., Esteban, R., Becerril, J.M., García-Plazaola, J.I. 2009. Dark induction of the photoprotective xanthophyll cycle in response to dehydration. *J Plant Physiol* 166:1734–1744.

Fernandez-Marin, B., Becerril, J.M., García-Plazaola, J.I. 2010. Unravelling the roles of desiccation-induced xanthophyll cycle activity in darkness: A case study in *Lobaria pulmonaria*. *Planta* 231:1335–1342.

Fernandez-Marin, B., Míguez, F., Becerril, J.M., García-Plazaola, J.I. 2011a. Dehydration-mediated activation of the xanthophyll cycle in darkness: Is it related to desiccation tolerance? *Planta* 243:579–588.

Fernandez-Marin, B., Míguez, F., Becerril, J.M., García-Plazaola, J.I. 2011b. Activation of violaxanthin cycle in darkness is a common response to different abiotic stresses: A case study in *Pelvetia canaliculata*. *BMC Plant Biol* 11:181.

Fernandez-Marin, B., Kranner, I., San Sebastián, M. et al. 2013. Evidence for the absence of enzymatic reactions in the glassy state. A case study of xanthophyll cycle pigments in the desiccation-tolerant moss *Syntrichia ruralis*. *J Exp Bot* 64:3013–3043.

Franks, S.J. 2011. Plasticity and evolution in drought avoidance and escape in the annual plant *Brassica rapa*. *New Phytol* 190:249–257.

Franks, F., Hatley, R.H.M., Mathias, S.F. 1991. Materials science and the production of shelf-stable biologicals. *Bio Pharm* 4:38–42.

Friedl, T., Rybalka, N. 2012. Systematics of the green algae: A brief introduction to the current status. *Progr Bot* 73:259–280.

Gabriel, R., Bates, J.W. 2003. Responses of photosynthesis to irradiance in bryophytes of the Azores laurel forest. *J Bryol* 25:101–105.

Gaff, D.F. 1971. Desiccation-tolerant flowering plants in Southern Africa. *Science* 3:1033–1034.

Gaff, D.F. 1997. Response of desiccation tolerant "resurrection" plants to water stress. In *Mechanisms of environmental stress resistance in plants*, ed. R.K. Basra, 43–58. London: Harwood Academic Publishers.

Gaff, D.F., Oliver, M.J. 2013. The evolution of desiccation tolerance in angiosperm plants: A rare yet common phenomenon. *Funct Plant Biol* 40:315–328.

Galmés, J., Aranjuelo, I., Medrano, H., Flexas, J. 2013. Variations in Rubisco content and activity under variable climatic factors. *Photosynth Res* 117:73–90.

Gao, S., Zheng, Z., Gu, W., Xie, X., Huan, L., Pang, G., Wang, G. 2014. Photosystem I shows a higher tolerance to sorbitol-induced osmotic stress than photosystem II in the intertidal macro-algae *Ulva prolifera* (Chlorophyta). *Physiol Plant* 152:380–388.

Gao, S., Gu, W., Xiong, Q. et al. 2015. Desiccation enhances phosphorylation of PSII and affects the distribution of protein complexes in the thylakoid membrane. *Physiol Plant* 153:492–502.

García-Plazaola, J.I., Esteban, R., Fernandez-Marin, B., Kranner, I., Porcar-Castell, A. 2012. Thermal energy dissipation and xanthophyll cycles beyond the Arabidopsis model. *Photosynth Res* 113:89–103.

Gasulla, F., Gómez de Nova, P., Esteban-Carrasco, A., Zapata, J.M., Barreno, E., Guéra, A. 2009. Dehydration rate and time of desiccation affect recovery of the lichenic algae *Trebouxia erici*: Alternative and classical protective mechanisms. *Planta* 231:195–208.

Gasulla, F., Dorp, K., Dombrink, I., Zahringer, U., Gisch, N., Dormann, P., Bartels, D. 2013a. The role of lipid metabolism in the acquisition of desiccation tolerance in *Craterostigma plantagineum*: A comparative approach. *Plant J* 75:726–741.

Gasulla, F., Jain, R., Barreno, E., Guera, A., Balbuena, T., Thelen, J.J., Oliver, M.J. 2013b. The response of *Asterochloris erici* (Ahmadjian) Skaloud et Peksa to desiccation: A proteomic approach. *Plant Cell Environ* 36:1363–1378.

Georgieva, K., Szigeti, Z., Sarvari, E. et al. 2007. Photosynthetic activity of homoiochlorophyllous desiccation tolerant plant *Haberlea rhodopensis* during dehydration and rehydration. *Planta* 225:955–964.

Georgieva, K., Röding, A., Büchel, C. 2009. Changes in some thylakoid membrane proteins and pigments upon desiccation of the resurrection plant *Haberlea rhodopensis*. *J Plant Physiol* 166:1520–1528.

Georgieva, K., Sarvari, E., Keresztes, A. 2010. Protection of thylakoids against combined light and drought by a lumenal substance in the resurrection plant *Haberlea rhodopensis*. *Ann Bot* 105:117–126.

Georgieva, K., Ivanova, A., Doncheva, S., Petkova, S., Stefanov, D., Péli, E., Tuba, Z. 2011. Fatty acid content during reconstitution of the photosynthetic apparatus in the air-dried leaves of *Xerophyta scabrida* after rehydration. *Biol Plant* 55:581–585.

Gerotto, C., Morosinotto, T. 2013. Evolution of photoprotection mechanisms upon land colonization: Evidences of PSBS-dependent NPQ in late Streptophyte algae. *Physiol Plant* 149:583–598.

Graham, E.A., Hamilton, M.P., Mishler, B.D., Rundel, P.W., Hansen, M.H. 2006. Use of a networked digital camera to estimate net CO₂ uptake of a desiccation-tolerant moss. *Int J Plant Sci* 167:751–758.

Graham, L.A., Arancibia-Avila, P., Taylor, W.A., Strother, P.K., Cook, M.E. 2012. Aeroterrestrial *Coleochaete* (Streptophyta, Coleochaetales) models early plant adaptation to land. *Am J Bot* 99:130–144.

Gray, D.W., Lewis, L.A., Cardon, Z.G. 2007. Photosynthetic recovery following desiccation of desert green algae (Chlorophyta) and their aquatic relatives. *Plant Cell Environ* 30:1240–1255.

Green, T.G.A., Kilian, E., Lange, O.L. 1991. Pseudocyphellaria dissimilis: A desiccation-sensitive highly shade-adapted lichen from New Zealand. *Oecologia* 85:498–503.

Griffiths, C.A., Gaff, D.F., Neale, A.D. 2014. Drying without senescence in resurrectino plants. *Front Plant Sci* 5:36.

Hallam, N.D., Luff, S.E. 1980. Fine structural changes in the mesophyll tissue of the leaves of *Xerophyta villosa* during desiccation. *Bot Gaz* 141:173–179.

Hamerlynck, E.P., Csintalan, Z., Nagy, Z., Tuba, Z., Goodin, D., Henebry, G.M. 2002. Ecophysiological consequences of contrasting microenvironments on the desiccation tolerant moss *Tortula ruralis*. *Oecologia* 131:498–505.

Hametner, C., Stocker-Wörgötter, E., Grube, M. 2014. New insights into diversity and selectivity of trentepohlialean lichen photobionts from the extratropics. *Symbiosis* 63:31–40.

Häubner, N., Schumann, R., Karsten, U. 2006. Aeroterrestrial microalgae growing in biofilms on facades—Response to temperature and water stress. *Microb Ecol* 51:285–293.

Havaux, M. 1998. Carotenoids as membrane stabilizers in chloroplasts. *Trends Plant Sci* 3:147–151.

Havaux, M., Bonfi, J.-P., Lütz, C., Niyogi, K.K. 2000. Photodamage of the photosynthetic apparatus and its dependence on the leaf developmental stage in the *npq1 Arabidopsis* mutant deficient in the xanthophyll cycle enzyme violaxanthin de-epoxidase. *Plant Physiol* 124:273–284.

Havaux, M., Eymery, F., Porfirova, S., Rey, P., Dörmann, P. 2005. Vitamin E protects against photoinhibition and photooxidative stress in *Arabidopsis thaliana*. *Plant Cell* 17:3451–3469.

Havaux, M., Dall'Osto, L., Bassi, R. 2007. Zeaxanthin has enhanced antioxidant capacity with respect to all other xanthophylls in Arabidopsis leaves and functions independent of binding to PS II antennae. *Plant Physiol* 145:1506–1520.

Heber, U. 2008. Photoprotection of green plants: A mechanism of ultra-fast thermal energy dissipation in desiccated lichens. *Planta* 228:641–650.

Heber, U. 2012. Conservation and dissipation of light energy in desiccation-tolerant photoautotrophs, two sides of the same coin. *Photosynth Res* 113:5–13.

Heber, U., Shuvalov, V.A. 2005. Photochemical reactions of chlorophyll in dehydrated photosystem II: Two chlorophyll forms (680 and 700 nm). *Photosynth Res* 84:85–91.

Heber, U., Bilger, W., Bligny, R., Lange, O.L. 2000. Phototolerance of lichens, mosses and higher plants in an alpine environment: Analysis of photoreactions. *Planta* 211:770–780.

Heber, U., Lange, O.L., Shuvalov, V.A. 2006. Conservation and dissipation of light energy as complementary processes: Homoiohydric and poikilohydric autotrophs. *J Exp Bot* 57: 1211–1223.

Heber, U., Bilger, W., Turk, R., Lange, O.L. 2010. Photoprotection of reaction centres in photosynthetic organisms: Mechanisms of thermal energy dissipation in desiccated thalli of the lichen *Lobaria pulmonaria*. *New Phytol* 185:459–470.

Herburger, K., Lewis, L.A., Holzinger, A. 2015. Photosynthetic efficiency, desiccation tolerance and ultrastructure in two phylogenetically distinct strains of alpine *Zygnema* sp. (Zygnematophyceae, Streptophyta): Role of pre-akinete formation. *Protoplasma* 252:571–589.

Hoekstra, F.A., Golovina, E.A., Buitink, J. 2001. Mechanisms of plant desiccation tolerance. *Trends Plant Sci* 6:1360–1385.

Holzinger, A. 2009. Desiccation tolerance in green algae: Implications of physiological adaptation and structural requirements. In *Algae: Nutrition, pollution control and energy sources*, ed. K.N. Hagen, pp. 41–56. New York: Nova Science Publishers.

Holzinger, A., Karsten, U. 2013. Desiccation stress and tolerance in green algae: Consequences for ultrastructure, physiological and molecular mechanisms. *Front Plant Sci* 4:327.

Holzinger, A., Roleda, M.Y., Lütz, C. 2009. The vegetative arctic green alga *Zygnema* is insensitive to experimental UV exposure. *Micron* 40:831–838.

Holzinger, A., Tschaikner, A., Remias, D. 2010. Cytoarchitecture of the desiccation-tolerant green alga *Zygogonium ericetorum*. *Protoplasma* 243:15–24.

Holzinger, A., Lütz, C., Karsten, U. 2011. Desiccation stress causes structural and ultrastructural alterations in the aeroterrestrial green alga *Klebsormidium crenulatum* (Klebsormidiophyceae, Streptophyta) isolated from an alpine soil crust. *J Phycol* 47: 591–602.

Holzinger, A., Kaplan, F., Blaas, K., Zechmann, B., Komsic-Buchmann, K., Becker, B. 2014. Transcriptomics of desiccation tolerance in the streptophyte green alga *Klebsormidium* reveal a land plant-like defense. *PLoS One* 9:10.

Hori, K., Maruyama, F., Fujisawa, T. et al. 2014. *Klebsormidium flaccidum* genome reveals primary factors for plant terrestrial adaptation. *Nat Commun* 5:3978.

Hutin, C., Nussaume, L., Moise, N., Moya, I., Kloppstech, K., Havaux, M. 2003. Early light-induced proteins protect Arabidopsis from photooxidative stress. *Proc Natl Acad Sci U S A* 100:4921–4926.

Ingle, R.A., Schmidt, U.G., Farrant, J.M., Thomson, J.A., Mundree, S.G. 2007. Proteomic analysis of leaf proteins during dehydration of the resurrection plant *Xerophyta viscosa*. *Plant Cell Environ* 30:435–446.

Ingle, R.A., Collett, H., Cooper, K., Takahashi, Y., Farrant, J.M., Illing, N. 2008. Chloroplast biogenesis during rehydration of the resurrection plant *Xerophyta humilis*: Parallels to the etioplast-chloroplast transition. *Plant Cell Environ* 31:1813–1824.

Iturriaga, G., Cushman, M.A.F., Cushman, J.C. 2006. An EST catalogue from the resurrection plant *Selaginella lepidophylla* reveals abiotic stress-adaptive genes. *Plant Sci* 170:1173–1184.

Johnson, M.P., Goral, T.K., Duffy, C.D.P., Brain, A.P.R., Mullineaux, C.W., Ruban, A.V. 2011. Photoprotective energy dissipation involves the reorganization of photosystem II light-harvesting complexes in the grana membranes of spinach chloroplasts. *Plant Cell* 23:1468–1479.

Kaplan, F., Lewis, L.A., Wastian, J., Holzinger, A. 2012. Plasmolysis effects and osmotic potential of two phylogenetically distinct alpine strains of *Klebsormidium* (Streptophyta). *Protoplasma* 249:789–804.

Kaplan, F., Lewis, L.A., Herburger, K., Holzinger, A. 2013. Osmotic stress in the arctic and antarctic green alga *Zygnema* sp. (Zygnemtales, Streptophyta): Effects on photosynthesis and ultrastructure. *Micron* 44:317–330.

Kappen, L., Valladares, F. 2007. Opportunistic growth and desiccation tolerance: The ecological success of poikilohydrous autotrophs. In *Functional plant ecology*, ed. F. Pugnaire, 7–66. Boca Ratón, FL: CRC Press.

Karsten, U., Holzinger, A. 2012. Light, temperature and desiccation effects on photosynthetic activity and drought-induced ultrastructural changes in the green alga *Klebsormidium dissectum* (Streptophyta) from a high alpine soil crust. *Microb Ecol* 63:51–63.

Karsten, U., Holzinger, A. 2014. Green algae in alpine biological soil crust communities: Acclimation strategies against ultraviolet radiation and dehydration. *Biodivers Conserv* 23:1845–1858.

Karsten, U., Lütz, C., Holzinger, A. 2010. Ecophysiological performance of the aeroterrestrial green alga *Klebsormidium crenulatum* (Charophyceae, Streptophyta) isolated from an alpine soil crust with an emphasis on desiccation stress. *J Phycol* 46:1187–1197.

Karsten, U., Herburger, K., Holzinger, A. 2014. Dehydration, temperature, and light tolerance in members of the aeroterrestrial green algal genus *Interfilum* (Streptophyta) from biogeographically different temperate soils. *J Phycol* 50:804–816.

Koonjul, P.K., Brandt, W.F., Lindsey, G.G., Farrant, J.M. 2000. Isolation and characterisation of chloroplasts from *Myrothamnus flabellifolius*. *J Plant Physiol* 156:584–594.

Kosugi, M., Arita, M., Shizuma, R., Moriyama, Y., Kashino, Y., Koike, H., Satoh, K. 2009. Responses to desiccation stress in lichens are different from those in their photobionts. *Plant Cell Physiol* 50:879–888.

Kosugi, M., Miyake, H., Ymamkawa, H. et al. 2013. Arabitol provided by lichenous fungi enhances ability to dissipate excess light energy in a symbiotic green alga under desiccation. *Plant Cell Physiol* 54:1316–1325.

Kranner, I., Birtic, S. 2005. A modulating role for antioxidant in desiccation-tolerance. *Integr Comp Biol* 45:734–740.

Kranner, I., Beckett, R.P., Wornik, S., Zorn, M., Pfeifhofer, H.W. 2002. Revival of a resurrection plant correlates with its antioxidant status. *Plant J* 31:13–24.

Kranner, I., Beckett, R.P., Hochman, A., Nash, T.H. 2008. Desiccation-tolerance in lichens: A review. *Bryologist* 111: 576–593.

Kroth, P.G. 2015. The biodiversity of carbon assimilation. *J Plant Physiol* 172:76–81.

Lange, O.L., Green, T.G.A., Reichenberger, H. 1999. The response of lichen photosynthesis to external CO_2 concentration and its interaction with thallus water-status. *J Plant Physiol* 154:157–166.

Lange, O.L., Green, T.G.A., Heber, U. 2001. Hydration-dependent photosynthetic production of lichens: What do laboratory studies tell us about field performance? *J Exp Bot* 52:2033–2042.

Lange, O.L., Green, T.G.A., Melzer, B., Meyer, A., Zellner, H. 2006. Water relations and CO_2 exchange of the terrestrial lichen *Teloschistes capensis* in the Namib fog desert: Measurements during two seasons in the field and under controlled conditions. *Flora* 201:268–280.

Leliaert, F., Smith, D.R., Moreau, H., Herron, M.D., Verbruggen, H., Delwiche, C.F., De Clerck, O. 2012. Phylogeny and molecular evolution of green algae. *Crit Rev Plant Sci* 31:1–46.

Leprince, O., Hoekstra, F.A. 1998. The responses of cytochrome redox state and energy metabolism to dehydration support a role for cytoplasmic viscosity in desiccation tolerance. *Plant Phys* 118:1253–1264.

Leprince, O., Buitink, J. 2007. The glassy state in dry seeds and pollen. In *Plant desiccation tolerance*, eds. M.A. Jenks, A.J. Wood, 193–214. Oxford: Blackwell Publishing.

Li, X.P., Björkman, O., Shih, C., Grossman, A.R., Rosenquist, M., Jansson, S., Niyogi, K.K. 2000. A pigment-binding protein essential for regulation of photosynthetic light harvesting. *Nature* 403:391–395.

López-Bautista, J.M., Waters, D.A., Chapman, R.L. 2002. The Trentepohliales revisited. *Constancea* 83. California: University and Jepson Herbaria, P.C. Silva Festschrift.

Lunch, C.K., LaFountain, A.M., Thomas, S., Frank, H.A., Lewis, L.A., Cardon, Z.G. 2013. The xanthophyll cycle and NPQ in diverse desert and aquatic green algae. *Photosynth Res* 115:139–151.

Lüttge, U., Büdel, B. 2010. Resurrection kinetics of photosynthesis in desiccation-tolerant terrestrial green algae (Chlorophyta) on tree bark. *Plant Biol* 123:437–444.

Martinelli, T., Whittaker, A., Masclaux-Daubresse, C., Farrant, J.M., Loreto, F., Vazzana, C. 2007. Evidence for the presence of photorespiration in desiccation-sensitive leaves of the C4 'resurrection' plant *Sporobolus stapfianus* during dehydration stress. *J Exp Bot* 58:3929–3939.

McLean, R.J., Pessoney, G.F. 1972. Formation and resistance of akinetes of *Zygnema*. In *Contributions in phycology*, eds. B.C. Parker, R.M. Brown, 145–152. Lawrence, KS: Allen Press.

Minibayeva, F., Beckett, P. 2001. High rates of extracellular super-oxide production in bryophytes and lichens, and an oxidative burst in response to rehydration following desiccation. *New Phytol* 152:333–341.

Meyer, M., Griffiths, H. 2013. Origins and diversity of eukaryotic CO_2-concentrating mechanisms: Lessons for the future. *J Exp Bot* 64:769–786.

Moore, J.P., Le, N.T., Brandt, W.F., Driouich, A., Farrant, J.M. 2009. Towards a systems-based understanding of plant desiccation tolerance. *Trends Plant Sci* 14:110–117.

Morosinotto, T., Bassi, R. 2014. Molecular mechanisms for activation of non-photochemical fluorescence quenching: From unicellular algae to mosses and higher plants. In *Non-photochemical quenching and energy dissipation in plants, advances in photosynthesis and respiration*, vol. 40, eds. B. Demmig-Adams, G. Garab, W. Adams III, Govindjee, 315–331. Dordrecht: Springer.

Mou, S., Zhang, X., Dong, M. et al. 2013. Photoprotection in the green tidal alga *Ulva prolifera*: Role of LHCSR and PsbS proteins in response to high light stress. *Plant Biol* 15:1033–1039.

Murata, N., Ishizaki-Nishizawa, O., Higashi, S., Hayashi, H., Tasaka, Y., Nishida, L. 1992. Genetically engineered alteration in the chilling sensitivity of plants. *Nature* 356: 710–713.

Nabe, H., Funabiki, R., Kashino, Y., Koike, H., Satoh, K. 2007. Responses to desiccation stress in bryophytes and important role of dithiothreitol-insensitive non-photochemical quenching against photoinhibition in dehydrated states. *Plant Cell Physiol* 48:1548–1557.

Nagy-Deri, H., Peli, E.R., Georgieva, K., Tuba, Z. 2011. Changes in chloroplast morphology of different parenchyma cells in leaves of *Haberlea rhodopensis* Friv. during desiccation and following rehydration. *Photosynthetica* 49:119–126.

Niinemets, Ü., Keenan, T.F., Hallik, L. 2014. A worldwide analysis of within-canopy variations in leaf structural, chemical and physiological traits across plant functional types. *New Phytol* 205:973–993.

Oliver, M.J., Wood, A.J., O'Mahony, P. 1997. How some plants recover from vegetative desiccation: A repair based strategy. *Acta Physiol Plant* 19:419–425.

Oliver, M.J., Tuba, Z., Mishler, B.D. 2000. The evolution of vegetative desiccation tolerance in land plants. *Plant Ecol* 151:85–100.

Oliver, M.J., Velten, J., Mishler, B.D. 2005. Desiccation tolerance in bryophytes: A reflection of the primitive strategy for plant survival in dehydrating habitats? *Integr Comp Biol* 45:788–799.

Oliver, M.J., Jain, J., Balbuena, T.S., Agrawal, G., Gasulla, F., Thelen, J.J. 2011. Proteome analysis of leaves of the desiccation-tolerant grass, *Sporobolus stapfianus*, in response to dehydration. *Phytochemistry* 72:1273–1284.

Ong, B.L., Lim, M., Wee, Y.C. 1992. Effects of desiccation and illumination on photosynthesis and pigmentation of the edaphic population of *Trentepohlia odorata* (Chlorophyta). *J Phycol* 28:768–772.

Pandey, V., Ranjan, S., Deeba, F., Pandey, A.K., Singh, R., Shirke, P.A., Pathre, U.V. 2010. Desiccation-induced physiological and biochemical changes in resurrection plant, *Selaginella bryopteris*. *J Plant Physiol* 167:1351–1359.

Peers, G., Truong, T.B., Ostendorf, E. et al. 2009. An ancient light-harvesting protein is critical for the regulation of algal photosynthesis. *Nature* 462:518–521.

Peeva, V., Cornic, G. 2009. Leaf photosynthesis of *Haberlea rhodopensis* before and during drought. *Env Exp Bot* 65: 310–318.

Pérez, P., Rabnecz, G., Laufer, Z., Gutierrez, D., Tuba, Z., Martinez-Carrasco, R. 2011. Restoration of photosystem II photochemistry and carbon assimilation and related changes in chlorophyll and protein contents during the rehydration of desiccated *Xerophyta scabrida* leaves. *J Exp Bot* 62:895–905.

Pichrtová, M., Hájek, T., Elster, J. 2014a. Osmotic stress and recovery in field populations of *Zygnema* sp. (Zygnematophyceae, Streptophyta) on Svalbard (High Arctic) subjected to natural desiccation. *FEMS Microbiol Ecol* 89:270–280.

Pichrtová, M., Kulichová, J., Holzinger, A. 2014b. Nitrogen limitation and slow drying induce desiccation tolerance in conjugating green algae (Zygnematophyceae) from polar habitats. *PLoS One* 9:11.

Pribil, M., Labs, M., Leister, D. 2014. Structure and dynamics of thylakoids in land plants. *J Exp Bot* 65:1955–1972.

Proctor, M.C.F. 2000. The bryophyte paradox: Tolerance of desiccation, evasion of drought. *Plant Ecol* 151:41–49.

Proctor, M. 2001. Patterns of desiccation tolerance and recovery in bryophytes. *Plant Growth Regul* 35:147–156.

Proctor, M. 2012. Light and desiccation responses of some Hymeophyllaceae (filmy ferns) from Trinidad, Venezuela and New Zealand: Poikilohydry in a light-limited but low evaporation ecological niche. *Ann Bot* 109:1019–1026.

Proctor, M.C.F., Pence, V.C. 2002. Vegetative tissues: Bryophytes, vascular resurrection plants and vegetative propagules. In: *Desiccation and survival in plants. Drying without dying*, eds. M. Black, H.W. Blank, 207–238. Oxon: CABI Publishing.

Proctor, M.C.F., Tuba, Z. 2002. Poikilohydry and homoihydry: Antithesis or spectrum of possibilities? *New Phytol* 156:327–349.

Proctor, M.C.F., Ligrone, R., Duckett, J.G. 2007. Desiccation tolerance in the moss *Polytrichum formosum*: Physiological and fine-structural changes during desiccation and recovery. *Ann Bot* 99:75–93.

Quaas, R., Berteotti, S., Ballottari, M., Flieger, K., Bassi, R., Wilhelm, C., Goss, R. 2015. Non-photochemical quenching and xanthophyll cycle activities in six green algal species suggest mechanistic differences in the process of excess energy dissipation. *J Plant Physiol* 172:92–103.

Rafsanjani, A., Brule, V., Western, T.L., Pasini, D. 2015. Hydro-responsive curling of the resurrection plant *Selaginella lepidophylla*. *Sci Rep* 5:8064.

Rakic, T., Gajic, G., Lazarevic, M., Stevanovic, B. 2015. Effects of different light intensities, CO_2 concentrations, temperatures and drought stress on photosynthetic activity in two paleo-endemic resurrection plant species *Ramonda serbica* and *R. nathaliae*. *Environ Exp Bot* 109:63–72.

Rascio, N., LaRocca, N. 2005. Resurrection plants: The puzzle of surviving extreme vegetative desiccation. *Crit Rev Plant Sci* 24:209–225.

Retallack, G.J. 2012. Ediacaran life on land. *Nature* 493:89–92.

Rindi, F., López-Bautista, J.M., Sherwood, A.R., Guiry, M.D. 2006. Morphology and phylogenetic position of *Spongiochrysis hawaiiensis* gen. et sp. nov., the first known terrestrial member of the order Cladophorales (Ulvophyceae, Chlorophyta). *Int J Syst Evol Microbiol* 56:913–922.

Royles, J., Ogee, J., Wingate, L., Hodgson, D.A., Convey, P., Griffiths, H. 2013. Temporal separation between CO_2 assimilation and growth? Experimental and theoretical evidence from the desiccation-tolerant moss *Syntrichia ruralis*. *New Phytol* 197:1152–1160.

Sahsah, Y., Campos, P., Gareil, M., Zuily-Fodil, Y., Pham-Thi, A.T. 1998. Enzymatic degradation of polar lipids in *Vigna unguiculata* leaves and influence of drought stress. *Physiol Plant* 104:577–586.

Sancho, L.G., de la Torre, R., Horneck, G. et al. 2007. Lichen survive in space: Results from the 2005 lichens experiment. *Astrobiology* 7:443–454.

Schwab, K.B., Schreiber, U., Heber, U. 1989. Response of photosynthesis and respiration of resurrection plants to desiccation and rehydration. *Planta* 177:217–227.

Seel, W., Hendry, G., Lee, J. 1992. Effects of desiccation on some activated oxygen processing enzymes and antioxidants in mosses. *J Exp Bot* 43:1031–1037.

Sherwin, H., Farrant, J. 1996. Differences in rehydration of three desiccation-tolerant angiosperm species. *Ann Bot* 78:703–710.

Sherwin, H., Farrant, J. 1998. Protection mechanisms against excess light in the resurrection plants *Craterostigma wilmsii* and *Xerophyta viscosa*. *Plant Growth Regul* 24:203–210.

Smith, E.C., Griffiths, H. 1996. The occurrence of the chloroplast pyrenoid is correlated with the activity of a CO_2-concentrating mechanism and carbon isotope discrimination in lichens and bryophytes. *Planta* 198:6–16.

Solhaug, K.A., Larsson, P., Gauslaa, Y. 2010. Light screening in lichen cortices can be quantified by chlorophyll fluorescence techniques for both reflecting and absorbing pigments. *Planta* 231:1003–1011.

Solymosi, K., Tuba, Z., Boddi, B. 2013. Desiccoplast–etioplast–chloroplast transformation under rehydration of desiccated poikilochlorophyllous *Xerophyta humilis* leaves in the dark and upon subsequent illumination. *J Plant Physiol* 170:583–590.

Stamenković, M., Bischof, K., Hanelt, D. 2014. Xanthophyll cycle pool size and composition in several *Cosmarium* strains (Zygnematophyceae, Streptophyta) are related to their geographic distribution patterns. *Protist* 165:14–30.

Stancheva, R., Hall, J.D., Herburger, K., Lewis, L.A., McCourt, R.M., Sheath, R.G., Holzinger, A. 2014. Phylogenetic position of *Zygogonium ericetorum* (Zygnematophyceae, Charophyta) from a high alpine habitat and ultrastructural characterization of unusual aplanospores. *J Phycol* 50:790–803.

Stepigova, J., Gauslaa, Y., Cempirkova-Vrablikova, H., Solhaug, K.A. 2008. Irradiance prior to and during desiccation improves the tolerance to excess irradiance in the desiccated state of the old forest lichen *Lobaria pulmonaria*. *Photosynthetica* 46:286–290.

Sundberg, B., Ekblad, A., Nasholm, T., Palmqviest, K. 1999. Lichen respiration in relation to active time, temperature, nitrogen and ergosterol concentrations. *Funct Ecol* 13:119–125.

Tausz, M., Hietz, P., Briones, O. 2001. The significance of carotenoids and tocopherols in photoprotection of seven epiphytic fern species of a Mexican cloud forest. *Aust J Plant Physiol* 28:775–783.

Tetriach, M., Geletti, A. 1997. CO_2 exchange of the endolithic lichen *Verrucaria baldensis* from karst habitats in northern Italy. *Oecologia* 111:515–522.

Tuba, Z., Lichtenthaler, H.K., Csintalan, Z., Pocs, T. 1993a. Regreening of desiccated leaves of the poikilochlorophyllous *Xerophyta scabrida* upon rehydration. *J Plant Physiol* 142:103–108.

Tuba, Z., Lichtenthaler, H.K., Maroti, I., Csintalan, Z. 1993b. Resynthesis of thylakoids and functional chloroplasts in the desiccated leaves of the poikilochlorophyllous plant *Xerophyta scabrida* upon rehydration. *J Plant Physiol* 142:742–748.

Tuba, Z., Lichtenthaler, H.K., Csintalan, Z., Nagy, Z., Szente, K. 1994. Reconstitution of chlorophylls and photosynthetic CO_2 assimilation upon rehydration of the desiccated poikilochlorophyllous plan *Xerophyta scabrida* (Pax) Th. Dur. Et Schinz. *Planta* 192:414–420.

Tuba, Z., Csintalan, Z., Proctor, M.C.F. 1996a. Photosynthetic responses of a moss, *Tortula ruralis*, ssp. *ruralis*, and the lichens *Cladonia convoluta* and *C. furcata* to water deficit and short periods of desiccation, and their ecophysiological significance: A baseline study at present-day CO_2 concentration. *New Phytol* 133:353–361.

Tuba, Z., Lichtenthaler, H.K., Csintalan, Z., Nagy, Z., Szente, K. 1996b. Loss of chlorophylls, cessation of photosynthetic CO_2 assimilation and respiration in the poikilochlorophyllous plant *Xerophyta scabrida*. *Physiol Plant* 96:383–388.

Tuba, Z., Proctor, M.C.F., Csintalan, Z. 1998. Ecophysiological responses of homoiochlorophyllous and poikilochlorophyllous desiccation-tolerant plants: A comparison and an ecological perspective. *Plant Growth Reg* 24:211–217.

Van der Poorten, A., Goffinet, B. 2009. *Introduction to bryophytes*. Cambridge: Cambridge University Press.

Van der Willigen, C., Pammenter, N.W., Mundree, S., Farrant, J. 2001. Some physiological comparisons between the resurrection grass, *Eragrostis nindensis*, and the related desiccation-sensitive species, *E. curvula*. *Plant Growth Reg* 35:121–129.

Veerman, J., Vasil'ev, S., Paton, G.D., Ramanauskas, J., Bruce, D. 2007. Photoprotection in the lichen *Parmelia sulcata*: The origins of desiccation-induced fluorescence quenching. *Plant Physiol* 145:997–1005.

Vertuci, C.W., Farrant, J.M. 1995. Acquisition and loss of desiccation tolerance. In *Seed development and germination*, J. Kigel, G. Galili, 237–272. New York: Marcel Dekker.

Watkins, J.E., Mack, M.C., Sinclair, T.R., Mulkey, S.S. 2007. Ecological and evolutionary consequences of desiccation tolerance in tropical fern gametophytes. *New Phytol* 176:708–717.

Wieners, P.C., Mudimu, O., Bilger, W. 2012. Desiccation-induced non-radiative dissipation in isolated green lichen algae. *Photosynth Res* 113:239–247.

Wodniok, S., Brinkmann, H., Glöckner, G., Heidel, A.J., Philippe, H., Melkonian, M., Becker, B. 2011. Origin of land plants: Do conjugating green algae hold the key? *BMC Evol Biol* 11:104.

Wood, A.J. 2007. The nature and distribution of vegetative desiccation-tolerance in hornworts, liverworts and mosses. *Bryologist* 110:163–167.

Xu, D.H., Li, J.H., Fang, X.W., Wang, G., Su, P.X. 2008. Photosynthetic activity of poikilochlorophyllous desiccation tolerant plant *Reaumuria soongorica* during dehydration and re-hydration. *Photosynthetica* 46:547–551.

Yobi, A., Wone, B.W.M., Xu, W. et al. 2013. Metabolomic profiling in *Selaginella lepidophylla* at various hydration states provides new insights into the mechanistic basis of desiccation tolerance. *Mol Plant* 6:369–385.

Ytterberg, A.J., Peltier, J.B., van Wijk, K.J. 2006. Protein profiling of plastoglobules in chloroplasts and chromoplasts. A surprising site for differential accumulation of metabolic enzymes. *Plant Physiol* 140:984–997.

Zeng, Q., Chen, X., Woods, A.J. 2002. Two early light-inducible protein (ELIP) cDNAs from the resurrection plant *Tortula ruralis* are differentially expressed in response to desiccation, rehydration, salinity and high light. *J Exp Bot* 53:1197–1205.

37 Drought Tolerance of Photosynthesis

Katya Georgieva and Gergana Mihailova

CONTENTS

37.1 Introduction 683
37.2 Homoichlorophylly and Poikilochlorophylly 684
    37.2.1 Changes in the Photosynthetic Pigments..684
37.3 Ultrastructural Changes in Chloroplasts 685
    37.3.1 PDT Plants...685
    37.3.2 HDT Plants...685
37.4 Effect of Drought on Photosynthesis...685
    37.4.1 Photosynthetic Activity during Dehydration and Rehydration of PDT Plants...........685
    37.4.2 Photosynthetic Activity during Dehydration and Rehydration of HDT Plants 686
        37.4.2.1 Effect of Drought on Net Photosynthesis 686
        37.4.2.2 Effect of Drought on Photochemical Activity 687
        37.4.2.3 Changes in Photosynthesis-Related Proteins...688
        37.4.2.4 Changes in Photosynthesis-Related Genes 689
37.5 Protection of the Photosynthetic Apparatus upon Desiccation 689
37.6 Effect of High Light and High Temperature during Desiccation...691
37.7 Summary 691
References...692

37.1 INTRODUCTION

Drought is one of the major environmental factors that inhibits many metabolic processes and constrains plant growth and crop productivity. The ongoing global warming and current climate changes are enlarging the land areas where plants experience water deficit. Understanding the responses of plants to their external environment is of importance with respect to basic research, but it is also an attractive target for improving stress tolerance [1]. Thus, our understanding of the drought adaptation mechanisms is of importance to meet the goal of increased plant productivity under the projected critical global scenarios that are related to water availability.

The most severe form of water deficit is desiccation, when most of the protoplasmic water is lost and only a very small amount of tightly bound water remains in the cell. Plants are very sensitive to desiccation during the vegetative phase of their life cycle, and very few plants acquire desiccation tolerance in the vegetative tissues. These include a small group of angiosperms, termed *resurrection plants* [2], which are capable of surviving water loss to an air-dry state. Resurrection plants are mostly poikilohydrous, which means that their water content adjusts with the relative humidity in the environment. They are able to stay in the dehydrated state until water becomes available and allows them to rehydrate and to resume full physiological activities [3–5]. In this small group of plants, the mature leaves, roots, and shoots can lose up to 95% of their water. Vegetative desiccation tolerance in angiosperms is comparatively rare, with approximately 300–400 species being reported as desiccation tolerant [6,7]. It has

been suggested that desiccation tolerance is connected with size limitation, since all examples of desiccation tolerant flowering plants do not exceed a certain height [8]; perhaps the largest known resurrection plant is the small woody shrub *Myrothamnus flabellifolia* [9]. Most resurrection plants are herbaceous plants. Resurrection plants are found in ecological niches with limited seasonal water availability, preferentially on rocky outcrops in semiarid and arid countries, and a rich diversity is found in Southern Africa [6,7,10]. Resurrection species are found also in the Balkans (*Haberlea rhodopensis, Ramonda* spp.), China (*Boea hygrometrica*), Australia (*Sporobolus* and *Eragrostis* spp.), North and Central America (*Tortula ruralis*), and South America (*Pleurostima purpurea*) [11,12]. Resurrection plants have been identified within the angiosperms both among monocotyledonous and dicotyledonous plants, but no desiccation tolerant gymnosperms or trees have been reported yet [5,13,14].

Desiccation tolerance can be achieved either by mechanisms that are based on the protection of cellular integrity or mechanisms that are based on the repair of desiccation- or rehydration-induced cellular damage [12,15,16]. Bryophytes are considered as *fully desiccation-tolerant* plants [17] and can withstand very fast drying. The tolerance to desiccation of these plants is based on cellular protection upon dehydration coupled with the repair of cell damage during rehydration [18,19]. The vascular species, designated as *modified desiccation-tolerant* plants, can survive desiccation only if the drying rate is slow. In these plants, the tolerance relies mainly on cellular protection during dehydration [12,20].

The physical properties of the photosynthetic apparatus are of crucial importance in desiccation-tolerant plants. The photosynthetic apparatus is very sensitive and liable to injury and needs to be maintained or quickly repaired upon rehydration [21].

37.2 HOMOICHLOROPHYLLY AND POIKILOCHLOROPHYLLY

Down-regulation of photosynthesis during dehydration of resurrection plants is achieved by one of two mechanisms termed *poikilochlorophylly* and *homoiochlorophylly*. The homoiochlorophyllous desiccation-tolerant (HDT) plants retain their photosynthetic apparatus and chlorophyll during drying, whereas the poikilochlorophyllous desiccation-tolerant (PDT) plants dismantle their photosynthetic apparatus and lose chlorophyll on drying [22–24]. Many PDT plants are monocots, and the best studied are three African shrubs in the Velloziaceae, *Xerophyta scabrida*, *Xerophyta humilis*, and *Xerophyta viscosa*, and a member of the Liliceae, *Borya nitida* Labill., from Western Australia. The main advantage of poikilochlorophylly is to limit photo-oxidative damage and not having to maintain the photosynthetic apparatus intact through long inactive periods of desiccation [23–25]. The poikilochlorophyllous desiccation tolerance strategy has evolved in habitats where the plants remain in the desiccated state for 8–10 months. However, as the photosynthetic apparatus has to be reassembled on rehydration, recovery time is generally longer in these species [26]. The HDT plants, mostly dicotyledonous species and ferns, are generally adapted to more rapid alternations of wet and dry periods than the PDT species.

37.2.1 CHANGES IN THE PHOTOSYNTHETIC PIGMENTS

The PDT plant *X. viscosa* lose almost all chlorophyll on drying and the carotenoid content declines by 60% [22]. Upon rehydration, very little of these pigments are reconstituted within the first 24 h, but thereafter, pigment content increases, reaching control levels after 120 h. The study on *X. scabrida*, stored 5 years in an air-dry state, also shows that the chlorophyll content is completely restored after 120 h of rehydration [27]. Regardless of the similar relative water content (RWC) of leaves following 72 and 120 h of rehydration, the chlorophyll content of the former is lower by 50%. The chlorophyll breakdown in PDT plants depends on the rate of water loss and light intensity during desiccation. If *B. nitida* is desiccated rapidly, it does not have the time to break down chlorophyll and loses viability [28], and the same is true for *X. humilis* [29]. *X. humilis* leaves lose virtually all chlorophyll when dried in the light, but only half when dried in the dark [26]. Moreover, the carotenoid content is reduced by 50% upon desiccation in the light but only 10% in the dark. These pigments are recovered upon rehydration in both treatments. So, the chlorophyll loss seems to be a result of photo-oxidation under natural circumstances [30]. This suggests that PDT plants in general probably do not decompose their chlorophyll enzymatically, but rather do not invest in preserving it through the dry state [31].

The extent of chlorophyll loss in HDT plants varies from species to species and may be influenced by environmental factors. *M. flabellifolia*, a short woody shrub from southern Africa, loses half of its chlorophyll and approximately 25% of carotenoids on drying [32]. In *M. flabellifolia*, the recovery of carotenoid levels was slower (72 h) than the recovery of chlorophyll (24 h). The chlorophyll and carotenoid concentration of leaves of *Craterostigma wilmsii* decline by approximately 30% and 20%, respectively, during drying, regardless of the drying rate [22,32]. The chlorophyll content of *C. wilmsii* recovers almost to control levels after 45 h.

No significant change in the chlorophyll content of detached leaves from HDT Gesneriaceae plants *B. hygrometrica* [33] and *H. rhodopensis* [34] is detected during dehydration. When *H. rhodopensis* plants, growing under low irradiance in their natural habitat, are desiccated to air-dry state at a similar light intensity (about 30 μmol m^{-2} s^{-1}) and 23/20°C day/night temperature, the chlorophyll content decrease by 10–15% [35,36]. However, not only dehydration of plants at high temperature (38/30°C) increases the rate of water loss threefold, but also the chlorophyll content decline by 30% in air-dry state [36]. Desiccation of these shade plants at light intensity of 350 μmol m^{-2} s^{-1} results in 30% reduction of chlorophyll content [37]. Similar changes in chlorophyll content are observed in dried leaves of plants, growing on sun-exposed limestone rocks and, thus, receiving full sunlight of about 1500–1700 μmol m^{-2} s^{-1} [38]. Following rehydration, chlorophyll content recovers reaching the control values.

The photosynthetic machinery reacts to dehydration with significant changes in the xanthophyll cycle pigments. Desiccation-induced zeaxanthin accumulation is observed in many resurrection plants [39–42]. Casper et al. [43] hypothesized that under dehydrating conditions, even low light levels (50 μmol photons m^{-2} s^{-1}) become excessive and zeaxanthin-related photoprotection is engaged in the resurrection plant *Selaginella lepidophylla*. It has been suggested that the xanthophyll cycle, in addition to its participation in the nonphotochemical energy quenching, might have a direct antioxidant action by enhancing the tolerance of thylakoid membranes to lipid peroxidation [44]. According to Havaux et al. [45], the antioxidant activity of zeaxanthin, distinct from NPQ, can occur in the absence of PSII light-harvesting complexes. In addition, Alamillo and Bartels [39] suggested that dsp 22, localized in the PSII, could be one of the proteins synthesized under the desiccation stress to maintain an adequate hydrophobic environment that enables the stabilization of the zeaxanthin produced under these conditions. Furthermore, Fernández-Marín et al. [46] showed that violaxanthin can be de-epoxidized into zeaxanthin, and zeaxanthin epoxidized back to violaxanthin in the total absence of light, parallel to dehydration–rehydration cycles in the desiccation-tolerant fern *Ceterach officinarum*. The inhibition of de-epoxidation by dithiothreitol suggests that violaxanthin de-epoxidase is the enzyme responsible for the conversion of violaxanthin into zeaxanthin during dark desiccation. It has been proposed that dark formation of zeaxanthin during desiccation could be understood as one of those mechanisms that started in the dehydration phase, and dryness itself may be an environmental stimulus sufficient to trigger the activation of the xanthophyll cycle, independent of light.

37.3 ULTRASTRUCTURAL CHANGES IN CHLOROPLASTS

37.3.1 PDT PLANTS

The loss of protoplasmic water results in considerable anatomical and ultrastructural reorganization of the leaf tissue [22,47]. Ultrastructural observations of mesophyll cells from dried plants of *X. humilis* show that the plasma membrane remains appressed to the cell wall and the outer membranes of organelles are intact [32]. The integrity of most of the cell structure is maintained during drying of *B. nitida* with the exception of plastids [48]. The dismantling of the photosynthetic apparatus is proposed as a strictly organized protective mechanism, rather than damage to be repaired after rehydration [24]. The thylakoid system within the chloroplasts is completely replaced by small groups of plastoglobuli and osmophilic, stretched lipid material [22,49]. The elongated osmophilic structures, termed *desiccoplasts* [22], appear to occupy the positions previously occupied by the thylakoids. During drying, the central vacuole divided into a number of smaller ones filled with relatively electron dense material [50–52]. The replacement of water in vacuoles with nonaqueous substances would serve to maintain volume within cytoplasm and reduce the extent of plasmalemma withdrawal from the cell wall during drying and thus minimize mechanical stress [31].

Ingle et al. [53] defined six distinct stages in chloroplast biogenesis during rehydration of *X. humilis*. The first change is observed within 3 h of rehydration when many of the smaller membrane-bound vesicles start to elongate to form the precursors to thylakoid membranes (stage 2). Stage 3 is characterized by the elongation of chloroplasts and by the first appearance of starch bodies. The small, elongated membrane-bound vesicles are assembled into single thylakoid membranes in stage 4. Grana are first visible in stage 5 and are assembled from three to four thylakoid membranes. Further stacking of thylakoids occur, and by stage 6, grana were more frequent and far thicker, with more than seven thylakoids stacked together. While thylakoid reassembly in *X. humilis* is independent of light, the formation of grana is found to be light dependent [53]. A few chloroplasts with clear grana (stage 5) are first detectable at 12 h following rehydration. The proportions of chloroplasts in stages 5 and 6 steadily increase with time following rehydration in the light.

37.3.2 HDT PLANTS

The subcellular organization of mesophyll cells from hydrated leaves of *C. wilmsii* and *M. flabellifolia* are typical of most other angiosperms, having a single central vacuole with cytoplasm occurring peripherally and adjacent to the cell wall [31]. Thylakoid membranes are clearly defined and some starch is present. The thylakoid membranes of *M. flabellifolia* have a unique form of stacking, referred to as a *staircase* arrangement

[22]. This feature is not specific to resurrection species in general, and it has been suggested that such an arrangement allows a more effective absorption of incoming radiation [54].

Cells from dry leaves of *C. wilmsii* have folded walls, and considerable plasma membrane withdrawal from the cell wall has occurred [26,32,55]. Vicré et al. [55] have shown that wall folding is accompanied by changes in wall architecture and chemistry, and it is a controlled process. The central vacuole disappears, and there are numerous small vacuoles in the cytoplasm and no starch is evident. Chloroplasts become rounded, but all membranes appear intact and thylakoid stacks are still clearly visible. In the dry state, the thylakoids of *M. flabellifolia* retain their staircase arrangement. The chloroplasts appear ovate, and there is no starch present. There is no plasmalemma withdrawal from the cell wall and vacuoles containing some relatively electron dense material. The electron dense appearance of the extracytomic space might suggest the presence of materials, possibly similar to those in the vacuoles. This would contribute considerably to the stabilization of the subcellular milieu in these species [31].

The inner membranes are maintained in dried organelles of *Boea hygroscopica*, where no thylakoid swelling or breakage has taken place, but an apparent change in the membrane system assemblage could be noticed [56]. Most thylakoids are appressed to form large stacks. The increased thylakoid stacking in dried chloroplasts could be a consequence of membrane and/or environmental changes leading to a weakening of the repulsive force between the membrane surfaces [57].

During desiccation of *H. rhodopensis* and *Ramonda serbica*, rounding and dislocation of the chloroplast to the middle part of the cell is observed [58–61]. In addition, the size of the chloroplasts is smaller compared with control plants both in palisade and spongy parenchyma [61]. At the air-dry state, the outer chloroplast envelope is disturbed [58,62], and the starch grains disappear from the chloroplast [59]. Upon rehydration, the chloroplasts progressively assume the shape and thylakoid distribution typical of control tissue.

37.4 EFFECT OF DROUGHT ON PHOTOSYNTHESIS

Photosynthesis is very sensitive to water deficit. The loss of photosynthetic capacity after dehydration could be triggered by several factors, such as the closure of stomata, decrease in intracellular CO_2 concentration, degradation of pigments, loss of function of photosystems, or destruction of photosynthetic structure [33].

37.4.1 PHOTOSYNTHETIC ACTIVITY DURING DEHYDRATION AND REHYDRATION OF PDT PLANTS

In poikilochlorophyllous DT plants, the photosynthetic activity declines at the initial stage of dehydration at relatively high water content [31,63]. For example, net photosynthesis of *X. humilis* decreases rapidly below 80% RWC and ceases by 57% RWC, when chlorophyll content decreases about 50%

[64]. It has been shown that the main photosynthetic limitation is photochemical, and it develops simultaneously with the onset of chlorophyll degradation in dehydrating leaves. Beckett et al. [64] suggested that photochemical impairment in dehydrating leaves is due to the loss of functional PSII reaction centers. Furthermore, the photochemical efficiency is not homogeneous across the *X. humilis* leaves. The basal part of the leaves continues to have a high photochemical efficiency until the tissues are dehydrated, but the apical part is largely impaired during early stages of dehydration.

The inhibition of photosynthetic activity in *Xerophyta* species during desiccation correlates with a reduction of PSII proteins [65,66]. The abundance of several chloroplast proteins involved in photosynthesis is significantly decreased as a result of desiccation of *X. viscosa*: PsbO and PsbP (two components of the oxygen evolving complex of PSII), the PSII stability factor HCF136, Lhcb2 (a component of the light-harvesting antennae), PsbS (a component of PSII that plays a role in nonphotochemical quenching), the α-subunit of the F-ATPase, and the Calvin cycle enzyme transketalose [66]. Of these, only HCF136 is significantly lower already at 65% RWC, which correlates with the reported decline of the maximum quantum efficiency of PSII, estimated by the ratio of F_v/F_m [31]. It was proposed that reduced levels of HCF136 protein in *X. viscosa* during drying may be one component of the shutdown of photosynthesis [66]. At RWC of 30%, PsbP and Lhcb2 are no longer detectable suggesting that *de novo* synthesis of these two proteins would be required during rehydration for resumption of PSII activity. Thus, poikilochlorophylly in *X. viscosa* involves the breakdown of photosynthetic proteins during dismantling of the thylakoid membranes. The observation that psbP and psbO mRNA levels also decline in the related species *X. humilis* at RWCs below 50% suggests that at least some of the observed decrease in protein levels is caused by down-regulation of gene expression [65].

The rate of dehydration is very important for recovery of the photosynthetic function after rehydration. Differences between rapid and slow drying could be related to differences in the extent of breakdown of the photosynthetic apparatus. Slowly dried leaves of *X. humilis* lost more than 50% of their chlorophyll at 50% RWC resulting in an initial decline in quantum efficiency of PSII [32]. Further inhibition in PSII activity is due to further chlorophyll loss and thylakoid dismantling. In contrast, rapidly dried leaves keep PSII activity at lower water contents since much of their chlorophyll is retained and thylakoid membranes are not dismantled during drying. However, rapidly dried leaves are unable to recover PSII activity upon rehydration, suggesting that irreversible damage may have occurred to the electron transport system. Farrant et al. [32] proposed that during rapid drying, there is insufficient time for complete induction of protection mechanisms, resulting in damage during dehydration, which is exacerbated on rehydration.

Chlorophyll–fluorescence measurements show that initial recovery of the photosystems is extraordinarily rapid. *X. viscosa* recovers photochemical activity before chlorophyll levels reach their pre-desiccation levels, and before full water content is attained [22]. After 24 h of rehydration (60% RWC), about 10% of the chlorophyll is recovered, but there is a 60% recovery of F_v/F_m. Forty-eight hours of rehydration (80% RWC) leads to 50% chlorophyll recovery and 90% recovery of F_v/F_m. In *Xerophyta* species, the amounts of chlorophyll and rates of CO_2 assimilation reach normal values in 72 h [22,23]. Similarly, Ingle et al. [53] detected net CO_2 assimilation after 12 h of rehydration of *X. humilis* and prior to any increase in chlorophyll content. They suggested that the residual chlorophyll present in desiccated tissue (<1 mg g DW^{-1}) is sufficient for the resumption of photosynthetic activity during rehydration.

Rehydration of dry leaves from *X. humilis* in the presence of transcriptional and translational inhibitors reveals that the initial partial recovery of the electron transport system of PSII is possible without *de novo* gene transcription [50]. Thus, most of the mRNAs for chlorophyll biosynthesis and for components required for PSII activity appear to be stored in the dry leaves. These mRNAs together with the protein synthetic apparatus must be stabilized and maintained in the dry state and reactivated upon rehydration. However, new transcription is required after approximately 18 h of rehydration of this species for complete recovery of the photosystem function. Furthermore, recovery on rehydration does require the synthesis of new proteins [50].

Investigation on the role of light for the transcription of eight *psb* genes encoding subunits of PSII (*psb A, D, O, P, R, S, Tn, Y*) showed that with the exception of *psbA* and *psbD*, water is the primary signal for mRNA accumulation, although the rate of synthesis is delayed in the absence of light [53]. While both *psbA* and *psbD* mRNA are stably stored in desiccated tissue, synthesis of the D1 protein is light-dependent and is first detectable 12 h after rehydration, correlating with the resumption of photosynthetic activity. Light is required for the synthesis of two other chloroplast proteins, one of the LHCII proteins (Lhcb2) and digalactosyldiacylglycerol synthase 1 (DGD1), playing important roles in the stability and activity of PSII and in the stacking of thylakoid membranes. The synthesis of Lhcb2 correlated with the formation of granal stacks in rehydrating plants.

37.4.2 PHOTOSYNTHETIC ACTIVITY DURING DEHYDRATION AND REHYDRATION OF HDT PLANTS

As has been already mentioned, one advantage that homoiochlorophyllous plants possess is the ability to resume photosynthesis promptly upon rehydration. Thus, in these resurrection plants, the photosynthetic apparatus must be maintained in a recoverable condition throughout the dehydration process. It has been proposed that the switch-off of photosynthesis in homoiochlorophyllous plants is likely to be a programmed process that involves some kind of protective mechanisms [33].

37.4.2.1 Effect of Drought on Net Photosynthesis

The rate of photosynthesis turned out to be very sensitive to desiccation, and it decreases during the initial phase of water loss. Experimental evidence suggests that the net CO_2 assimilation starts to decline between RWC of 80% and 75% and ceases by 55% in *Myrothamnus flabellifolius* and 40% in *C. wilmsii* [31]. It has been shown that dehydration of *H. rhodopensis* up to 50% leaf RWC reduces the CO_2 uptake by 50%, and it is strongly inhibited when the RWC drops to 20% [35,62]. After desiccation to 5% RWC, there is no net CO_2 assimilation rate, but the illuminated desiccated leaves have shown respiration activity. Actually, it was found that fully hydrated *H. rhodopensis* plants show a low rate of leaf net CO_2 uptake (4–6 μmol m^{-2} s^{-1}) under saturating photosynthetic photon flux densities in normal air, which could be partly explained by a very low mesophyll CO_2 conductance [67]. Stomatal limitation is generally accepted to be the main reason of reduced photosynthesis under drought stress [68]. The extent of stomatal control of photosynthesis is demonstrated in *Ramonda mykoni* leaves, which can easily be stripped of the lower epidermis. After stripping, CO_2 uptake decreases more gradually than when photosynthesis is under stomatal control [69]. Schwab et al. [69] showed that up to a water loss of 50% from the leaves, photosynthesis is limited by stomatal closure and not by inhibition of reactions of the photosynthetic apparatus. In addition, the decline in CO_2 conductance (both stomatal and mesophyll conductances) from ambient air to the carboxylating site inside the chloroplasts explains the reduction in CO_2 assimilation in *H. rhodopensis* when RWC decreases to about 40% [67]. The further lowering and inhibition of net CO_2 assimilation is caused by the decrease in both stomatal conductance and photochemical activity [35,67]. Moreover, the reduction in photosynthetic rate during desiccation correlates with the leaf folding of *C. wilmsii* and *M. flabellifolius* [31] and the incurving of the fronds of *Selaginella bryopteris* [70]. The assimilation ceases when folding is nearing completion.

In response to water stress, the diffusional limitations through stomata and mesophyll result in decreased CO_2 availability for Rubisco and consequently in decreased activity of the key enzyme in C3 metabolism. Experimental data showed that while the activity of Rubisco remains similar to those in the controls during the first stages of dehydration of *R. serbica*, it is undetectable in dry leaves (7% RWC) [71]. In contrast to the strong reduction in CO_2 assimilation rate during dehydration of *R. serbica* and *H. rhodopensis*, a more gradual decrease in carboxylating activity of Rubisco is observed [71–73]. The quantity of Rubisco in leaves is controlled by the rate of synthesis and degradation of the enzyme, even in stressful environments. The decrease in the activity of Rubisco is more related to the activity of Rubisco activase and ATP/ADP ratio than to the changes at the protein level [74]. Transcripts encoding the small subunit of Rubisco in *C. plantagineum* are reported to be down-regulated in response to water stress [75].

An induction of crassulacean acid metabolism (CAM) under extreme desiccation of *H. rhodopensis* is reported [76]. CAM cycling exhibited a different degree under well-watered conditions and extreme desiccation, and it proved an efficient mechanism of saving water [72,77]. Furthermore, phosphoenolpyruvate carboxylase (PEPC) transcripts related to C_4/CAM-type photosynthetic metabolism are substantially induced in the water-deficient samples [78].

The decrease in net CO_2 assimilation during cellular dehydration correlates with the decrease in cellular ATP levels [74]. Although the photosynthesis is down-regulated, cellular ATP is not fully depleted because respiration is less sensitive to the loss of water during wilting in the resurrection plants. Until leaf water content is reduced by about 70%, dark respiration in *Craterostigma plantagineum* leaves is independent of water stress [69]. Even at 15% RWC, about 30% of maximal respiration rates persist. There is full recovery of respiration in *C. plantagineum* within 2 h of rewatering when leaf water content reaches 20–25%.

It was observed that the sensitivity of photosynthesis to dehydration is similar in resurrection plants and nonresurrection C_3 plants [67,69]. The most marked difference is apparent in the recovery from desiccation. The ability of resurrection plants to fully recover photosynthesis after rewatering is likely linked to protective mechanisms that maintain the integrity of the photosynthetic machinery during desiccation [10,79]. Upon rehydration, plants regain water content rapidly, and the stomatal conductance increased faster compared to CO_2 assimilation rate [60,71,80]. CO_2 fixation is not restored until the RWC of *R. serbica* reaches values of 50%, and it still remains lower than in the controls at 84% RWC, whereas the stomatal conductance and Rubisco activity strongly increase at 50% RWC [71]. Similarly, CO_2 assimilation slowly increases following rehydration of *H. rhodopensis* [60,72,80]. After 24 h of rehydration, it is 15% of the control value and almost full recovery is observed after 7 days. Recovery of the CO_2 assimilation suggests that the photosynthetic apparatus had not been damaged by desiccation per se.

37.4.2.2 Effect of Drought on Photochemical Activity

Associated with the decrease in photosynthetic rate observed during dehydration of resurrection plants, there is also a decrease in the photochemical activity. However, the extent and rate of inhibition of photochemical activity differ between different resurrection plants. The resurrection fern *S. lepidophylla* shows a very steep decline in the maximum quantum efficiency of PSII, estimated by the ratio F_v/F_m, below RWC of about 75% followed by a rapid decline below RWC of 30% [81]. In contrast, the F_v/F_m ratio is unaffected in the beginning of dehydration of *R. serbica* and strongly decreases at the threshold RWC of about 30–40% [79]. Interestingly, it was found that the reduction of the RWC to about 40% declines the photochemical efficiency of PSII by 40% and 60% in *Ramonda nathaliae* and *R. serbica*, respectively [82]. In addition, it was shown that at the beginning of the desiccation of *C. wilmsii* [32] and *H. rhodopensis* [34–36,67,83], F_v/F_m is also relatively insensitive to desiccation, and this ratio declines when RWC drops below 20%. The decrease in variable fluorescence is mostly attributable to a strong decrease in the maximum chlorophyll fluorescence, F_m, accompanied by a moderate decrease in the ground chlorophyll fluorescence, F_o [35]. The maximum photochemical activity of PSII is almost fully inhibited at 10% RWC, indicating inhibition of the activity of the PSII reaction centers.

Moreover, it was found that while the PSII activity in *C. wilmsii* is not affected by the rate of drying, in leaves of *M. flabellifolius*, F_v/F_m begins to decline at a higher water content in slowly dried leaves (50% RWC) than in rapidly dried leaves (20% RWC). Actually, it was observed that rapidly dried leaves are unable to recover PSII activity, indicating damage to that system. According to Farrant et al. [32], at slow drying to 50% RWC, leaves of *M. flabellifolius* has reorientated through 90°, and adaxial surfaces, which retains chlorophyll, are shaded from light. Abaxial surfaces appeared brown and thylakoid membranes have become distended. Such changes could limit PSII activity. Rapidly dried leaves do not complete leaf folding and chlorophyll-retaining adaxial surfaces remain exposed to light, enabling PSII activity to continue at low (20% RWC) water contents.

Chlorophyll fluorescence analysis indicated that the maximum photochemical efficiency of PSII, given by F_v/F_m, is less affected by increasing water deficit than the quantum yield of PSII linear electron transport during illumination, ΦPSII [35,60,79]. This parameter is equal to the product of photochemical quenching (q_P) and the efficiency of excitation capture by open PSII reaction centers (F_v/F_m) [84]. The inhibition of ΦPSII during dehydration of *H. rhodopensis* leaves is mainly due to a decrease in the proportion of open PSII reaction centers [34,71,85]. A reduction in the total number of operating PSII centers in severely dehydrated *H. rhodopensis* plants is also confirmed by thermoluminescence measurements [34,35,86]. Furthermore, it was proposed that the drying-induced disconnection of the two photosystems in *Haberlea* explains the lower electron transport rate in the light-adapted state [87].

Among the thylakoid reactions, electron donation to PSII from water splitting complex appears to be affected by water stress [69]. The oxygen-evolving complex of thylakoid membranes is the most sensitive part of the photosynthetic apparatus to different stress treatments including dehydration. Dehydration of *H. rhodopensis* up to 8% RWC results in a considerable decline of oxygen yield and a well-expressed loss of oscillations, indicating that the number of oxygen-evolving centers from grana regions is reduced and that the centers still evolving oxygen does not work properly [62,88]. Furthermore, the dehydration leads to reduction of the number of oxygen-evolving centers in S_0 state, which is probably related to the reduced electron transport activity and hindered deactivation of oxidized states of oxygen-evolving centers.

PSI activity, measured by far-red induced P700 oxidation, increases in moderately dehydrated *H. rhodopensis* leaves (RWC up to 50%) but is strongly reduced when RWC drops to 20% [34,36]. The increased PSI signal could be a result of the desiccation-induced changes in leaf optics. Perhaps light scattering increases with desiccation. On the other hand, the enhanced P700 oxidation correlates with the limitation in electron flow from PSII, i.e., decreased PSII electron transport activity.

The desiccation-tolerant species all show complete recovery of photochemical activity, indicating that chloroplasts become fully functional upon rehydration. Similarly to dehydration, the rate of recovery depends on plant species. The effect of rehydration on the recovery of photochemical activity is measurable already after 3 h of rehydration of *H. rhodopensis* and 24 h after the rehydration the photochemical activity of the previously desiccated leaves reaches the control values [34,83]. When leaves of *R. serbica* are allowed to take up water, they show a fast recovery of photosynthetic activity [79]. Most of the recovery of F_v/F_m and ΦPS II is accomplished in about 20–25 h. *C. wilmsii* recovers water content very rapidly and F_v/F_m reaches control levels by 32 h [22]. Photochemical activity is restored before the plant is fully hydrated and before chlorophyll content reaches control levels. The rehydration of leaves of *M. flabellifolia* is slower than in *C. wilmsii*, possibly because of the time to recover hydraulic conductance in the xylem vessels in the former [22]. There is a delay in the resumption of photochemical activity during initial rehydration despite recovery of chlorophyll content. The photochemical activity is recovered after 55 h of rehydration, and like *C. wilmsii*, this occurs before the plant is fully rehydrated.

37.4.2.3 Changes in Photosynthesis-Related Proteins

A proteome study of the resurrection plants revealed regulation of proteins related to photosynthesis, energy and sugar metabolism, detoxification and scavenging of reactive oxygen species, and stress response proteins [89,90]. From a total of 223 proteins that are detected and analyzed in *B. hygrometrica* leaves, 35% show increased abundance in dehydrated leaves, 5% are induced in rehydrated leaves, and 60% show decreased or unchanged abundance in dehydrated and rehydrated leaves [89]. The programmed regulation of protein expression triggered by changes of water status was suggested.

Dehydration and subsequent rehydration induce alterations in a number of proteins related to photosynthesis, including the Rubisco large subunit, chl a/b binding protein, and oxygen-evolving complex [78,89,91]. However, while these proteins are down-regulated in desiccated *Sporobolus stapfianus* [91] and *Selaginella tamariscina* [90], some accumulation of Rubisco large subunit and the 23 kDa protein of oxygen-evolving complex are accumulating in the dehydrated leaves of *B. hygrometrica* [89]. Since the 23 kDa polypeptide has the function of stabilizing the Mn_4O_x cluster, it was proposed that its abundance in dehydrated and rehydrated leaves of *B. hygrometrica* may help to maintain the structure of oxygen-evolving complex during dehydration and the regain of function during rehydration [89].

In contrast to desiccation-sensitive plants, the thylakoid pigment–protein complexes have been shown to be highly stable during desiccation and rehydration of *B. hygrometrica* leaves [33] and *H. rhodopensis* [35]. Preserving the integrity of the photosynthetic apparatus of resurrection plants upon desiccation is important for the fast recovery after rehydration. Desiccation of *H. rhodopensis* to air-dried state at very low light irradiance lead to a little decrease in the level of D1, D2, Psb S, and Psa A/B proteins in thylakoids [41]. Alamillo and Bartels [39] have shown that the D1 protein levels are low in *C. plantagineum* leaves desiccated in the light but remain high in samples desiccated in the dark. The D1 protein levels are inversely related to the accumulation of desiccation-related protein (dsp) 22.

Proteins of thylakoid membranes decrease about 30% as a result of 7 days dehydration of *B. hygroscopica* (80% RWC) and remain unchanged up to 9% RWC but increase again to control values after 48 h after rehydration [56]. Dehydrated plants show lesser amounts of lipids and the lipid-to-protein ratio decreases during dehydration. It was suggested that the decrease in thylakoid proteins observed during dehydration may be associated with degradation of lipoprotein thylakoid structure [92]. EPR measurements of spin-labeled proteins with 3-maleimido proxyl show the presence of three different groups of proteins with different mobility in thylakoid membranes. The rotational correlation time of groups 1 and 2 increases with dehydration and decreases upon rehydration, whereas group 3 shows little changes [56].

37.4.2.4 Changes in Photosynthesis-Related Genes

Experimental evidence suggests that most changes in gene expression occur during the dehydration phase in desiccation-tolerant higher plants [93], while in desiccation-tolerant mosses such as *T. ruralis*, the protein synthesis is most apparent during rehydration [94].

Down-regulation of photosynthesis-related genes observed in *H. rhodopensis* [78] as well as in *C. plantagineum* [95] is a common response to desiccation. It was shown that the gene encoding the small subunit of Rubisco (*rbcS*) is strongly expressed in unstressed *C. plantagineum* leaves, but the transcript rapidly disappears during desiccation [75,95]. Rubisco activase transcript is abundant in control plants, suggesting a constitutive priming to respond to dehydration. Around 15 h after rehydration of the *C. plantagineum*, the level of the rbcS mRNA corresponds to that found in unstressed leaves, indicating that a 15-h rehydrated plant has the transcriptional potential similar to that of unstressed plants [75]. Furthermore, transcripts of light-harvesting complex lhcb1 and the other photosynthesis-related genes are highly represented under normal conditions and rehydration, but repressed during desiccation of *H. rhodopensis* [78]. In contrast to the general decrease in Rubisco, PEPC transcripts are substantially induced in dehydrated *H. rhodopensis*, possibly indicating a drought-induced switch from C_3- to C_4-type photosynthesis and possible CAM idling [78]. Although no carbon is being gained during CAM idling, this could be an important eco-physiological adaptation to maintain photosystem stability and survive prolonged periods without water.

A novel nuclear gene family expressed in response to dehydration has been identified in leaves of *C. plantagineum* [93] that encodes plastid-targeted proteins (CpPTP). These proteins have the ability to interact with the plastid DNA, suggesting a role in the down-regulation of chloroplast-encoded genes that occurs upon drying.

The early light-induced proteins (ELIPs) are nuclear-encoded proteins, which are associated with thylakoid membranes in the chloroplast. A number of desiccation-related genes encoding ELIPs have been isolated, including the *dsp22*

gene from *C. plantagineum* [96]. The dsp22 transcript is not detected in unstressed plants but is one of the most abundant desiccation-induced transcripts [97]. Similarly, this transcript is detectable only in desiccation-tolerant tissue of *S. stapfianus*, which has experienced severe drought stress, but the transcript accumulation appears to be significantly greater in *C. plantagineum* than *S. stapfianus* [98].

Interestingly, transgenic tobacco that ectopically expresses two dehydration responsive genes encoding late embryogenesis abundant proteins (designated as BhLEA1 and BhLEA2) from the resurrection plant *B. hygrometrica* is more tolerant to drought compared to the wild type [99]. Immunoblot analysis showed that the stability of proteins related to photosynthesis like ribulose-bisphosphate carboxylase large subunit (RbcL), light-harvesting complex of PSII (LHCII), and the 33 kDa protein of the oxygen-evolving complex (PsbO) are less degraded in transgenic plants under drought stress.

37.5 PROTECTION OF THE PHOTOSYNTHETIC APPARATUS UPON DESICCATION

During dehydration, an imbalance between CO_2 assimilation rate and PSII activity is observed, i.e., CO_2 assimilation is strongly inhibited in moderately dehydrated leaves when PSII activity is not significantly influenced [35,81]. This suggests that photosynthetic electrons are used for other photochemical processes under the stomatal limitation of photosynthesis. Both increased photorespiration and increased allocation of electrons to oxygen under water stress have been suggested as possible mechanisms by which excess photochemical energy can be dissipated and high quantum yield of PSII electron transport can be maintained, thus protecting PSII from damage [100]. Moreover, the large amount of chlorophyll molecules, which remain during desiccation of homoiochlorophyllous resurrection plants, could be a source for potentially harmful singlet oxygen production. Resurrection plants have many mechanisms to reduce reactive oxygen species (ROS) formation and quench their activity and thus to minimize photo-oxidative damage [101]. These include leaf folding to reduce the surface area exposed to light and heat, and the accumulation of anthocyanins or other pigments that mask chlorophylls, reflect light, and can also act as antioxidants [23,26,31,102].

The decrease in leaf surface during drought is due to a reduction in leaf cell volume as already seen in *C. wilmsii* and *M. flabellifolius* [31], where the volume occupied by the cytoplasm decreases by a similar amount to that of the cell wall boundary [31]. This behavior minimizes mechanical stress, as the plasmodesmata and the plasmalemma membrane remain intact [31]. This is in contrast to nonresurrection plants where a small decrease in leaf surface of only 20% leads to a mechanical stress that ruptures membranes [67].

Farrant et al. [26] have shown that *C. wilmsii* did not survive drying under irradiation if the leaves were prevented from folding, despite protection from increased anthocyanin and sucrose. Leaf surface area decreases and an extensive shrinkage and some folding of *H. rhodopensis* leaves also occur due to loss of turgor. Leaf folding is observed when the

RWC is lower than 50%, while shrinkage correlates with the time point when the leaves lose most of their water content. It has been estimated that leaves of *H. rhodopensis* drop to about 60% of their original area [37]. Moreover, during dehydration of *H. rhodopensis*, the blue and green fluorescence emissions increase. The blue and green fluorescence has been generally attributed to cell-wall-bound ferulic acid, and thus this increase may be connected to the shrinkage of leaves. Accumulation of dense luminal substance (possibly phenolics) in the thylakoid lumen is demonstrated and is proposed as a mechanism protecting the thylakoid membranes of *H. rhodopensis* during desiccation and during recovery [60]. Some phenolics are considered to be antioxidants [103] and ROS scavengers [104]. In *H. rhodopensis*, during drying and rehydration under illumination, the extent of loss and reappearance of photosynthetic activity correlates strongly with the presence or absence of dense luminal substance.

Shutdown of photosynthesis during the early phase of drying and the decline in PSII activity in severely desiccated leaves could represent a protective mechanism from toxic oxygen production in order to maintain the membrane integrity and to ensure protoplast survival [31,81,105]. Water deficit results in a reduction of net photosynthetic rate and starch content in leaf tissues. There is a rapid conversion of starch into sugars, the most common of which are sucrose and trehalose, and they have been proposed to contribute toward the maintenance of turgor during stress and the prevention of protein denaturation and membrane fusions in the cell [70,106].

An additional mechanism to reduce the amount of ROS is by nonradiative dissipation of excessive excitation energy. Simultaneously with the reduction of electron-transport rate through PSII, the fraction of excitation energy dissipated as heat in the PSII antenna increased markedly. The energy-dependent component of nonphotochemical quenching (NPQ) shows an increase in dehydrating *R. serbica* leaves when RWC drops to near 30%, whereas further decreases in RWC below these values cause a decrease in NPQ [79]. Hence, it can be concluded that NPQ has a major role in preventing photoinhibition in the early stage of dehydration, while at extremely low RWCs, other mechanisms may become important to avoid photodamage.

The integration of classical analysis of chlorophyll fluorescence with a detailed analysis of energy partitioning [107] showed a readjustment in the function and extent of the different components of energy management in PSII depending on the degree of dehydration of *H. rhodopensis* [108]. Under well-hydrated conditions, the highest fraction of absorbed energy is used for photochemistry (Φ_{PSII}), while the contribution of thermal dissipation *via* light-independent thermal dissipation ($\Phi_{f,D}$) is approximately 30% of the absorbed light. The energy fraction channeled *via* photoprotective nonphotochemical quenching (Φ_{NPQ}) is low (about 6% of the absorbed light), and the contribution of photoinactivated PSIIs (Φ_{NF}) is negligible (less than 0.1% of the absorbed light). At an early stage of dehydration, the contribution of the light-dependent thermal dissipation, in terms of both Φ_{NF} and Φ_{NPQ}, rises to 2% and 10%, while the energy fraction used for photochemistry

slightly decreases in comparison to the predehydration fraction. Severe plant dehydration leads to a pronounced reduction of energy utilization for photochemistry and of the energy fraction dissipated by Φ_{NPQ}, and resulted in a strong increase in Φ_{NF}. The energy fraction dissipated *via* the fluorescence, and constitutive thermal dissipation ($\Phi_{f,D}$) is almost constant over the development of dehydration; its contribution under severe dehydration is likely to change in a compensatory manner. Thus, under these severe dehydration conditions, energy is mainly allocated to light-independent thermal dissipation ($\Phi_{f,D}$) and light-dependent thermal dissipation associated with photoinactivated PSII (Φ_{NF}). Those pathways together dissipate more than 94% of total light absorbed.

Xanthophyll cycle pigments play an important role in the protection of photosynthesis and could be crucial for the dehydration tolerance in the resurrection angiosperms [33]. Indeed, zeaxanthin accumulation is observed in the leaves of many resurrection plants during desiccation [33,39,40], even at low light intensity [41,43] and in the dark [42,46]. It has been suggested that dark formation of zeaxanthin during desiccation could be understood as one of those mechanisms that starts in the dehydration phase, and dryness itself may be an environmental stimulus that is sufficient to trigger the activation of the xanthophyll cycle, independent of light [42,46]. However, violaxanthin de-epoxidase activity does not take place continuously during drying, but only during the last period of desiccation when almost all water content has been lost [109]. Moreover, it was suggested that the xanthophyll cycle might have a direct antioxidant action by enhancing the tolerance of thylakoid membranes to lipid peroxidation [44].

A determinant role in defense mechanisms from free radicals is played by the activity of antioxidant systems, whose upregulation during dehydration is a common event in resurrection plants [26,31,40,102,103,110–112]. However, what appears to be a distinguishing feature of the functioning of these antioxidants in resurrection plants is the ability to maintain their antioxidant potential in the dry state such that the same antioxidants can be utilized during the early stages of rehydration, thus protecting against the ROS stress associated with reconstitution of full metabolism (reviewed in Refs. [5,14,113]). Farrant et al. [114] have shown that the antioxidant enzymes retain the ability to detoxify ROS even at RWC of <10%, suggesting that there is some protection of these proteins that prevents their denaturation and maintains the native state in dry conditions. In addition, the accumulation of antioxidants like ascorbate, glutathione, and α-tocopherol is observed upon dehydration [31].

Different classes of desiccation-induced proteins seem to be involved in maintenance of chloroplast stability of homoiochlorophyllous resurrection plants in the dried state [19,96,98]. One of these is a chloroplast-localized, desiccation-related protein found in leaves of *C. plantagineum* [39,97] and *S. stapfianus* [98] that shows high similarities with ELIPs.

Overall, desiccation-tolerant plants must be able to limit the damage caused by desiccation to a repairable level, maintain the physiological integrity, and mobilize mechanisms upon rehydration that repair damage caused during desiccation and subsequent rehydration.

37.6 EFFECT OF HIGH LIGHT AND HIGH TEMPERATURE DURING DESICCATION

Upon drying, angiosperm tissues must be protected against a number of stresses brought about by, or in association with, extreme water loss. Drought is often associated with high temperature and high radiance, which has a strong impact on the vitality of plants.

The presence of high light intensities during dehydration can be extremely damaging to photosynthetically active tissues [102]. Studies on the resurrection fern *Polypodium polypodioides* [115] and on the resurrection moss *T. ruralis* [116] have shown that there is more damage, and recovery times are longer, when plants are dried under high light compared to low light conditions. However, irradiation with the very bright light of the sun during the summer months is found to be lethal to desiccated leaves of *P. polypodioides* even when they are curled [115]. In general, shade plants are more vulnerable to photoinhibition than sun plants. *H. rhodopensis* taken from a shaded environment is very sensitive to photoinhibition [117]. Increasing the light intensity from 50 to 120 μmol m^{-2} s^{-1} sharply decreases the quantum yield of PSII photochemistry, measured in *H. rhodopensis* leaves, and it is almost 90% reduced after treatment at 350 μmol m^{-2} s^{-1}. Exposure of well-watered *H. rhodopensis* plants to 350 μmol m^{-2} s^{-1} for 19 days leads to a strong reduction of rates of net photosynthesis and transpiration, contents of chlorophyll and carotenoids, as well as PSII activity. Furthermore, desiccation of plants at irradiance of 350 μmol m^{-2} s^{-1} induces irreversible changes in the photosynthetic apparatus, and leaves (except the youngest ones) do not recover after rehydration [37]. In order to elucidate limiting parameters for recovery, and the reason for the irreversible damage of the photosynthetic apparatus due to dehydration at high irradiance, changes in the structure and function of the photosynthetic apparatus of *H. rhodopensis* are investigated during desiccation and subsequent rehydration under a medium light irradiance (100 μmol m^{-2} s^{-1}) versus a low one (30 μmol m^{-2} s^{-1}). Regardless of the similar rate of water loss, dehydration at 100 μmol m^{-2} s^{-1} decreases the quantum efficiency of PSII photochemistry, and particularly the CO$_2$ assimilation rate, more rapidly than at 30 μmol m^{-2} s^{-1}. Dehydration induces accumulation of stress proteins in leaves under both light intensities. The appearance of dense luminal substances in the thylakoid lumen during desiccation and recovery under low light is suggested to protect the leaves from oxidative damage. The disappearance of the dense luminal substance during desiccation under high light correlates with the oxidative damage and ceases the recovery of photosynthesis upon rewatering [60].

In contrast to high sensitivity of *H. rhodopensis* to photoinhibition, comparatively well-expressed thermostability was found [117]. Dehydration of plants at high temperature increases the rate of water loss and has a more detrimental effect than either drought or high temperature alone [36].

Exposure of *H. rhodopensis* to 38/30°C day/night temperature for a week decreases the quantum efficiency of PSII, and the rate of photosynthetic oxygen evolution is reduced by about 35%. High-temperature treatment reduces PSI activity only slightly (by 8%), confirming the previous results that PSI is more heat resistant than PSII [118,119]. In addition, it was shown that the activity of superoxide dismutase and catalase slightly decreases as a result of heat treatment, while the activity of ascorbate peroxidase and guaiacol peroxidase are not affected [112]. The photochemical activity of PSII and PSI and photosynthetic rate decrease rapidly when plants are desiccated at high temperature compared to optimal temperature. Furthermore, some reduction in the amount of the main PSI and PSII proteins is observed especially in severely desiccated leaves. It was clearly showed that desiccation of the *H. rhodopensis* at high temperature has more damaging effects than desiccation at optimal temperature, and recovery is slower. But the damage is limited to a level where repair is still possible, and thus plants fully recover after rehydration. Increased thermal energy dissipation together with higher proline and carotenoid content in the course of desiccation at 38°C compared to desiccation at 23°C probably help in overcoming the stress. The study on isolated thylakoids demonstrates increased distribution of excitation energy to PSI as a result of high-temperature treatment of *H. rhodopensis*, which is enhanced upon desiccation [62]. In addition, the surface charge density of thylakoid membranes isolated from plants desiccated at 38°C is higher in comparison with these at 23°C, which is in agreement with the decreased membrane stacking, confirmed by electron microscopy data. Dehydration leads to a decrease in amplitudes of oxygen yields and to loss of the oscillation pattern. Rehydration results in partial recovery of the amplitudes of flash oxygen yields as well as of population of S$_0$ state in plants desiccated at 23°C. However, it is not observed in plants dehydrated at 38°C.

The most common molecular response of plants submitted to heat stress is the expression of heat shock proteins. Using antibodies rose against two sunflower small heat shock proteins (HSP17.6 and HSP17.9), Alamillo et al. [120] detected immunologically related proteins in unstressed vegetative tissues from the resurrection plant *C. plantagineum*. In whole plants, further accumulation of these polypeptides was induced by heat shock (2 h at either 37°C or 42°C) or water stress.

37.7 SUMMARY

The photosynthetic apparatus is very sensitive to drought stress, and its acclimation to unfavorable conditions is of primary importance for survival of plants. Resurrection plants use two strategies for down-regulation of photosynthesis during dehydration. PDT species dismantle their photosynthetic apparatus and lose all of their chlorophyll during drying and thus minimizing photo-oxidative damage. HDT strategy is based on the preservation of the integrity of the photosynthetic apparatus by protective mechanisms. The rate of photosynthesis in HDT plants decreases during the initial phase of water loss due to stomata closure, and excess energy from excited chlorophyll molecules is transferred to oxygen causing a rapid production of ROS. The first system that limits the formation of ROS consists of reducing the light–chlorophyll interaction by leaf folding and of synthesizing large amounts of anthocyanins that mask chlorophylls, reflect light, and can also act as antioxidants. The efficiency of antioxidant defense system and stabilization of the cellular environment by accumulation of carbohydrate, osmolytes, and LEA proteins play a crucial role in acquisition of desiccation tolerance. Indeed, the activity of antioxidant enzymes and the pools of reduced ascorbate and glutathione increase during drying and remain high in fully dehydrated leaves. An additional mechanism to reduce the amount of ROS is by nonradiative dissipation of excessive excitation energy. Desiccation-induced zeaxanthin accumulation is observed in many resurrection plants. Zeaxanthin has important functions in energy dissipation as heat as well as in membrane stabilization and antioxidant protection. A huge variety of genes and their products are upregulated in resurrection plants during drought stress. Changes in gene expression result in morphological and physiological adaptations, which enable survival in a desiccated state. Many dehydration-induced genes from several resurrection plants have been characterized, although few have been introduced into desiccation-sensitive plants and tested for improving drought tolerance, but the obtained results are promising. More work is necessary to define gene functions and dissect the complex regulation of gene expression in a wider range of species.

REFERENCES

1. Ramanjulu S, Bartels D. Drought- and desiccation-induced modulation of gene expression in plants. *Plant Cell Environ.* 2002; 25:141–151.
2. Gaff DF. Desiccation-tolerant flowering plants in Southern Africa. *Science.* 1971; 174(4013):1033–1034.
3. Vertucci CW, Farrant JM. Acquisition and loss of desiccation tolerance. In: Kigel J, Galili G, eds. *Seed Development and Germination.* New York: Marcel Dekker, 1995:237–271.
4. Bartels D. Desiccation tolerance studied in the resurrection plant *Craterostigma plantagineum. Integr. Comp. Biol.* 2005; 45(5):696–701.
5. Farrant JM. Mechanisms of desiccation tolerance in angiosperm resurrection plants. In: Jenks MA, Wood AJ, eds. *Plant Desiccation Tolerance.* Wallingford: CABI, 2007: 51–90.
6. Porembski S, Barthlott W. Granitic and gneissic outcrops (inselbergs) as centers of diversity for desiccation-tolerant vascular plants. *Plant Ecol.* 2000; 151:19–28.
7. Moore JP, Farrant JM. A system based molecular biology analysis of resurrection plants for crop and forage improvement in arid environments. In: Tuteja N, Gill SS, Tiburcio AF, Tuteja R, eds. *Improving Crop Resistance to Abiotic Stress,* vols. 1–2. Weinheim: Wiley-VCH, 2012:399–418.
8. Bewley JD, Krochko JE. Desiccation-tolerance. In: Lange OL, Nobel PS, Osmond CB, Ziegler H, eds. *Physiological Plant Ecology II. Encyclopedia of Plant Physiology,* vol. 12/B. Berlin: Springer, 1982:325–378.
9. Sherwin HW, Pammenter NW, February E, Willigen CV, Farrant JM. Xylem hydraulic characteristics, water relations and wood anatomy of the resurrection plant *Myrothamnus flabellifolius* Welw. *Ann. Bot.* 1998; 81:567–575.
10. Bartels D, Salamini F. Desiccation tolerance in the resurrection plant *Craterostigma plantagineum.* A contribution to the study of drought tolerance at the molecular level. *Plant Physiol.* 2001; 127(4):1346–1353.
11. Alpert P. Constraints of tolerance: Why are desiccation-tolerant organisms so small or rare? *J. Exp. Biol.* 2006; 209(9):1575–1584.
12. Oliver MJ, Tuba Z, Mishler BD. The evolution of vegetative desiccation tolerance in land plants. *Plant Ecol.* 2000; 151:85–100.
13. Alpert P, Oliver MJ. Drying without dying. In: Black M, Pritchard HW, eds. *Desiccation and Survival in Plants: Drying Without Dying,* vol. 3. Wallingford: CABI, 2002:3–43.
14. Moore JP, Le NT, Brandt WF, Driouich A, Farrant JM. Towards a systems-based understanding of plant desiccation tolerance. *Trends Plant Sci.* 2009; 14(2):110–117.
15. Bewley JD. Physiological aspects of desiccation tolerance. *Annu. Rev. Plant Physiol.* 1979; 30(1):195–238.
16. Bewley JD, Oliver MJ. Desiccation tolerance in vegetative plant tissues and seeds: Protein synthesis in relation to desiccation and a potential role for protection and repair mechanisms. In: Somero GN, Osmond CB, Bolis CA, eds. *Water and Life.* Berlin: Springer, 1992:141–160.
17. Oliver MJ, Bewley J. Desiccation-tolerance in plant tissues. A mechanistic overview. *Hortic. Rev.* 1997; 18:171–214.
18. Oliver MJ, Velten J, Wood AJ. Bryophytes as experimental models for the study of environmental stress-tolerance: *Tortula ruralis* and desiccation-tolerance in mosses. *Plant Ecol.* 2000; 151:73–84.
19. Rascio N, La Rocca N. Resurrection plants: The puzzle of surviving extreme vegetative desiccation. *Crit. Rev. Plant Sci.* 2005; 24:209–225.
20. Cooper K, Farrant JM. Recovery of the resurrection plant *Craterostigma wilmsii* from desiccation: Protection versus repair. *J. Exp. Bot.* 2002; 53(375):1805–1813.
21. Godde D. Adaptations of the photosynthetic apparatus to stress conditions. In: Lerner HR, ed. *Plant Responses to Environmental Stresses. From Phytohormones to Genome Reorganization.* New York: Marcel Dekker, 1999:449–474.
22. Sherwin HW, Farrant JM. Differences in rehydration of three desiccation tolerant angiosperm species. *Ann. Bot.* 1996; 78:703–710.
23. Tuba Z, Proctor MCF, Csintalan Z. Ecophysiological responses of homoiochlorophyllous and poikilochlorophyllous desiccation-tolerant plants: A comparison and an ecological perspective. *Plant Grow. Regul.* 1998; 24:211–217.
24. Proctor MCF, Tuba Z. Poikilohydry and homoihydry: Antithesis or spectrum of possibilities? *New Phytol.* 2002; 156:327–349.
25. Toldi O, Tuba Z, Scott P. Vegetative desiccation tolerance: Is it a goldmine for bioengineering crops? *Plant Sci.* 2009; 176(2):187–199.
26. Farrant JM, Bartsch S, Loffell D, van der Willigen C, Whittaker A. An investigation into the effects of light on the desiccation of three resurrection plants species. *Plant Cell Environ.* 2003; 26:1275–1286.
27. Georgieva K, Ivanova A, Doncheva S, Petkova S, Stefanov D, Péli E, Tuba Z. Changes in fatty acid content during reconstitution of photosynthetic apparatus of poikilochlorophyllous and air dried *Xerophyta scabrida* leaves following rehydration. *Biol. Plant.* 2011; 55(3):581–585.

28. Gaff DF, Churchill DM. *Borya nitida* Labill.—An Australian species in the Liliaceae with desiccation-tolerant leaves. *Aust. J. Bot.* 1976; 24:209–224.

29. Cooper K. The effect of drying rate on the resurrection species *Craterostigma wilmsii* (homoiochlorophyllous) and *Xerophyta humilis* (poikilochlorophyllous). MSc thesis, University of Cape Town, South Africa, 2001.

30. Tuba Z, Smirnoff N, Csintalan Z, Szente K, Nagy Z. Respiration during slow desiccation of the poikilochlorophyllous desiccation tolerant plant *Xerophyta scabrida* at present day CO_2 concentration. *Plant Physiol. Biochem.* 1997; 35:381–386.

31. Farrant JM. A comparison of mechanisms of desiccation-tolerance among three angiosperm resurrection plant species. *Plant Ecol.* 2000; 151:29–39.

32. Farrant JM, Cooper K, Kruger LA, Sherwin HW. The effect of drying rate on the survival of three desiccation-tolerant angiosperm species. *Ann. Bot.* 1999; 84:371–379.

33. Deng X, Hu ZA, Wang HX, Wen XG, Kuang TY. A comparison of photosynthetic apparatus of the detached leaves of the resurrection plant *Boea hygrometrica* with its non-tolerant relative *Chirita heterotrichia* in response to dehydration and rehydration. *Plant Sci.* 2003; 165:851–861.

34. Georgieva K, Maslenkova L, Peeva P, Markovska Y, Stefanov D, Tuba Z. Comparative study on the changes in photosynthetic activity of the homoiochlorophyllous desiccation-tolerant *Haberlea rhodopensis* and desiccation sensitive spinach leaves during desiccation and rehydration. *Photosynth. Res.* 2005; 85:191–203.

35. Georgieva K, Szigeti Z, Savarti E, Gaspar L, Maslenkova L, Peeva V, Peli E, Tuba Z. Photosynthetic activity of homoiochlorophyllous desiccation tolerant plant *Haberlea rhodopensis* during dehydration and rehydration. *Planta.* 2007; 225:955–964.

36. Mihailova G, Petkova S, Büchel C, Georgieva K. Desiccation of the resurrection plant *Haberlea rhodopensis* at high temperature. *Photosynth. Res.* 2011; 108:5–13.

37. Georgieva K, Lenk S, Buschmann C. Responses of the resurrection plant *Haberlea rhodopensis* to high irradiance. *Photosynthetica.* 2008; 46:208–215.

38. Sárvári É, Mihailova G, Solti Á, Keresztes Á, Velitchkova M, Georgieva K. Comparison of thylakoid structure and organization in sun and shade *Haberlea rhodopensis* populations under desiccation and rehydration. *J. Plant Physiol.* 2014; 171(17):1591–1600.

39. Alamillo J, Bartels D. Effects of desiccation on photosynthesis pigments and the ELIP-like dsp 22 protein complexes in the resurrection plant *Craterostigma plantagineum. Plant Sci.* 2001; 160:1161–1170.

40. Kranner I, Beckett RP, Wornik S, Zorn M, Pfeifhofer HW. Revival of a resurrection plant correlates with its antioxidant status. *Plant J.* 2002; 31(1):13–24.

41. Georgieva K, Röding A, Büchel C. Changes in some thylakoid membrane proteins and pigments upon desiccation of the resurrection plant *Haberlea rhodopensis. J. Plant Physiol.* 2009; 166:1520–1528.

42. Fernández-Marín B, Míguez F, Becerril JM, García-Plazaola JI. Dehydration-mediated activation of the xanthophyll cycle in darkness: Is it related to desiccation tolerance? *Planta.* 2011; 234(3):579–588.

43. Casper C, Eickmeier WG, Osmond CB. Changes of fluorescence and xanthophyll pigments during dehydration in the resurrection plant *Selaginella lepidophylla* in low and medium light intensities. *Oecologia.* 1993; 94(4):528–533.

44. Niyogi KK. Photoprotection revisited: Genetic and molecular approaches. *Annu. Rev. Plant Physiol. Plant Mol. Biol.* 1999; 50:333–359.

45. Havaux M, Dall'Osto L, Bassi R. Zeaxanthin has enhanced antioxidant capacity with respect to all other xanthophylls in Arabidopsis leaves and functions independent of binding to PSII antennae. *Plant Physiol.* 2007; 145(4):1506–1520.

46. Fernández-Marín B, Balaguer L, Esteban R, Becerril JM, García-Plazaola JI. Dark induction of the photoprotective xanthophyll cycle in response to dehydration. *J. Plant Physiol.* 2009; 166(16):1734–1744.

47. Moore JP, Nguema-Ona E, Chevalier L, Lindsey GG, Brandt WF, Lerouge P, Farrant JM, Driouich A. Response of the leaf cell wall to desiccation in the resurrection plant *Myrothamnus flabellifolius. Plant Physiol.* 2006; 141(2):651–662.

48. Gaff DF, Zee SY, O'Brien TP. The fine structure of dehydrated and reviving leaves of *Borya nitida* Labill.—A desiccation-tolerant plant. *Aust. J. Bot.* 1976; 24(2):225–236.

49. Tuba Z, Lichtenthaler HK, Maroti I, Csintalan Z. Resynthesis of thylakoids and functional chloroplasts in the desiccated leaves of the poikilochlorophyllous plant *Xerophyta scabrida* upon rehydration. *J. Plant Physiol.* 1993; 142(6):742–748.

50. Dace H, Sherwin HW, Illing N, Farrant JM. Use of metabolic inhibitors to elucidate mechanisms of recovery from desiccation stress in the resurrection plant *Xerophyta humilis. Plant Grow. Regul.* 1998; 24(3):171–177.

51. Tuba Z, Lichtenthalter HK, Csintalan Z, Nagy Z, Szente K. Reconstitution of chlorophylls and photosynthetic CO_2 assimilation upon rehydration of the desiccated poikilochlorophyllous plant *Xerophyta scabrida* (Pax) Th. Dur. et Schinz. *Planta.* 1994; 192:414–420.

52. Hedderson N, Balsamo RA, Cooper K, Farrant JM. Leaf tensile properties of resurrection plants differ among species in their response to drying. *South Afr. J. Bot.* 2009; 75:8–16.

53. Ingle RA, Collett H, Cooper K, Takahashi Y, Farrant JM, Illing N. Chloroplast biogenesis during rehydration of the resurrection plant *Xerophyta humilis*: Parallels to the etioplast–chloroplast transition. *Plant Cell Environ.* 2008; 31(12):1813–1824.

54. Wellburn FAM, Wellburn AR. Novel chloroplasts and unusual cellular ultrastructure in the "resurrection" plant *Myrothamnus flabellifolia* Welw. (Myrothamnaceae). *Bot. J. Linn. Soc.* 1976; 72:51–54.

55. Vicré M, Lerouxel O, Farrant J, Lerouge P, Driouich A. Composition and desiccation-induced alterations of the cell wall in the resurrection plant *Craterostigma wilmsii. Physiol. Plant.* 2004; 120:229–239.

56. Navari-Izzo F, Quartacci MF, Pinzino C, Rascio N, Vazzana C, Sgherri C. Protein dynamics in thylakoids of the desiccation-tolerant plant *Boea hygroscopica* during dehydration and rehydration. *Plant Physiol.* 2000; 124:1427–1436.

57. Barber J. Influence of surface charges on thylakoid structure and function. *Annu. Rev. Plant Physiol.* 1982; 33(1):261–295.

58. Markovska Y, Tsonev T, Kimenov G, Tutekova A. Physiological changes in higher poikilohydric plants—*Haberlea rhodopensis* Friv. and *Ramonda serbica* Panc. during drought and rewatering at different light regimes. *J. Plant Physiol.* 1994; 144:100–108.

59. Markovska Y, Tutekova A, Kimenov G. Ultrastructure of chloroplasts of poikilohydric plants *Haberlea rhodopensis* Friv. and *Ramonda serbica* Panc. during recovery from desiccation. *Photosynthetica.* 1995; 31:613–620.

60. Georgieva K, Sárvári E, Keresztes A. Protection of thylakoids against combined light and drought by a lumenal substance in the resurrection plant *Haberlea rhodopensis. Ann. Bot.* 2010; 105(1):117–126.

38 Photosynthesis under Heat Stress

Muhammad Farooq, Abdul Rehman, Abdul Wahid, and Kadambot H.M. Siddique

CONTENTS

38.1 Introduction 697
38.2 Leaf Senescence and Photosynthetic Pigments 697
38.3 Photosynthetic Apparatus 697
38.4 Cyclic and Noncyclic Electron Flow 698
38.5 Stomatal Oscillations 698
38.6 Photosynthetic Enzymes and Photorespiration 699
    38.6.1 Photosynthetic Enzymes 699
    38.6.2 Rubisco Activation 699
    38.6.3 Photorespiration 699
38.7 Conclusions 700
References 700

38.1 INTRODUCTION

Plant growth and development is dependent on a plant's ability to convert light energy into usable chemical energy during a process termed *photosynthesis*. However, temperatures above that optimum for growth (heat stress) may injure and/or irreversibly damage photosynthetic machinery, pigments, and metabolites including photosynthetic enzymes, as well as the stomata (Xu et al. 1995; Sairam et al. 2000; Farooq et al. 2011). Heat stress disrupts the integrity of thylakoid membranes, thus inhibiting the activities of membrane-associated electron carriers and enzymes (Rexroth et al. 2011), which reduces the rate of photosynthesis.

In this chapter, we describe the influence of heat stress on different components of photosynthesis, including photosynthetic pigments, photosystems, and the electron transport system, as well as the activities of different enzymes involved in carbon assimilation.

38.2 LEAF SENESCENCE AND PHOTOSYNTHETIC PIGMENTS

Progressive loss of green leaves is termed leaf senescence (Noodén 1988). High temperature accelerates senescence-related metabolic changes in plants (Paulsen 1994; Nawaz et al. 2013). Heat stress substantially reduces chlorophyll content by accelerating its degradation and/or reducing its rate of biosynthesis (Table 38.1, Figure 38.1) (Dutta et al. 2009; Efeoglu and Terzioglu 2009; Reda and Mandoura 2011). Reduced chlorophyll biosynthesis under heat stress is caused by reduced activities of the enzymes involved in its biosynthesis (Reda and Mandoura 2011), such as 5-aminolevulinate dehydratase (ALAD), the first enzyme in the pyrrole biosynthetic pathway (Tewari and Tripathy 1999; Mohanty et al. 2006); protochlorophyllide (Pchlide), oxidoreductase and porphobilinogen deaminase (Tewari and Tripathy 1998).

38.3 PHOTOSYNTHETIC APPARATUS

The photosynthetic machinery is sensitive to heat stress (Berry and Björkman 1980). In many plants, even brief exposure to moderate heat stress causes oxidative damage to the photosynthetic apparatus (Havaux 1993; Crafts-Brandner and Salvucci 2000). Heat stress strongly affects the integrity of the thylakoid membrane (Pfeiffer and Krupinska 2005) and disturbs photosystems I and II (PSI and PSII) and the oxygen-evolving complex (OEC), thus affecting phosphorylation (Dias and Lidon 2009; Rexroth et al. 2011). The adverse effects of short-term mild heat stress are reversible; however, long-term exposure may permanently impair the photosynthetic apparatus (Sharkey and Zhang 2010).

Heat stress causes ultrastructural changes such as swelling, expansion of the matrix zone, and loosening of the lamella in the chloroplast. Moreover, heat stress may unstack grana and induce the formation of inverted cylindrical micelles (Zhang et al. 2014). High temperatures induce changes in protein structure, which split the non-bilayer-forming lipids, thus unstacking packed membranes of granum (Fristedt et al. 2009). High temperatures can also inhibit the acceptor side of PSII electron transport, a bound quinone (QA) to reduced form of plastoquinone (PQH_2) between PSII and PSI being the most sensitive part of the electron transport chain (Yan et al. 2013). Furthermore, heat stress inactivates the OEC and reaction center, and dissociates the light-harvesting complex, which inhibits electron transport (Tang et al. 2007; Xue et al. 2011).

Thylakoid membrane stability depends on stabilization of unsaturated fatty acids in membranes (Raison et al. 1982). However, heat stress induces the production of reactive oxygen species (ROS) across the thylakoid membrane (Takahashi and Murata 2006), causing peroxidation of membrane lipids, which accelerates the denaturation of thylakoid proteins, thus intensifying electron leakage from membranes (Xu et al. 2006). Thus, ROS production hinders protein synthesis and PSII repair (Takahashi and Murata 2006).

TABLE 38.1

Influence of Heat Stress on Photosynthetic Pigments and Related Enzymes

Plant Species	Temperature	Effect on Photosynthetic Pigments/Enzymes	Reduction Relative to Control (%)	References
Wheat	45°C	Reduced total chlorophyll contents	11.36	Efeoglu and Terzioglu (2009)
Wheat	42°C	Reduced porphobilinogen deaminase activity	41	Tewari and Tripathy (1998)
Wheat	42°C	Reduced Pchlide synthesis activity	65	Tewari and Tripathy (1998)
Wheat	42°C	Reduced Pchlide oxidoreductase activity	90	Tewari and Tripathy (1998)
Cucumber	42°C	Inhibited chlorophyll biosynthesis	60	Tewari and Tripathy (1998)
Festuca arundinacea	35°C/30°C	Reduced chlorophyll contents	28.2–47.5	Cui et al. (2006)
Creeping bent grass	35°C/30°C	Reduced chlorophyll contents	51	Veerasamy et al. (2007)
Tomato	38°C	Reduced chlorophyll florescence	12–23	Murkowski (2001)

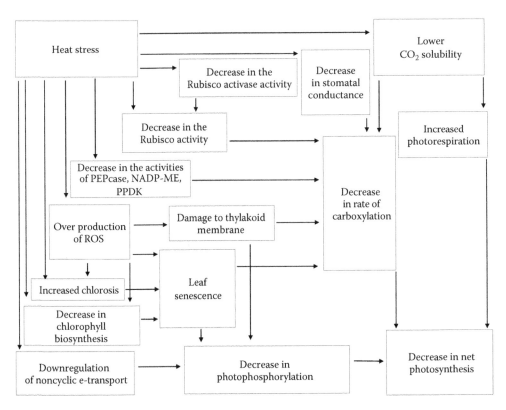

FIGURE 38.1 Influence of heat stress on photosynthesis. Heat stress accelerates leaf senescence and damages photosynthetic machinery and pigments. Heat stress also induces production of reactive oxygen species (ROS) across the thylakoid membrane, which causes oxidative damage. Severe heat stress limits photosynthesis due to a decline in activities of ribulose-1,5-bisphosphate carboxylase/oxygenase (Rubisco), phosphoenolpyruvate carboxylase (PEPCase), nicotinamide adenine dinucleotide phosphate malic enzyme (NADP-ME) and pyruvate phosphate dikinase (PPDK). Rubisco activity is also affected through heat-induced reduction in the activity of Rubisco activase. With increases in ambient temperature, stomatal conductance and CO_2 solubility in water decrease, which not only reduces carboxylation directly but also directs more electrons to form ROS and promotes photorespiration. Moreover, noncyclic electron transport is downregulated to match the reduced requirements of NADPH production, thus reducing ATP synthesis. (Modified from Farooq, M. et al., *Crit. Rev. Plant Sci.*, 30, 1–17, 2011.)

38.4 CYCLIC AND NONCYCLIC ELECTRON FLOW

Heat stress triggers the cyclic flow of electrons around PSI (Schrader et al. 2004). However, noncyclic electron transport is reduced, which substantially reduces NADPH (Sharkey and Schrader 2006) as electrons are diverted from the NADPH pool to the plastoquinone pool (Bukhov et al. 2005). This diversion of electron movement may help to protect PSII from

heat-induced damage by increasing zeaxanthin synthesis and reducing the size of light-harvesting antennae present on PSII (Sharkey and Schrader 2006).

38.5 STOMATAL OSCILLATIONS

The effect of temperature on stomatal regulation is important due to its effect on CO_2 influx from the atmosphere to the leaves (Xu et al. 2004). Heat stress accelerates transpiration, causing stomatal closure, which indirectly affects CO_2 fixation (Xu et al. 2004).

38.6 PHOTOSYNTHETIC ENZYMES AND PHOTORESPIRATION

38.6.1 PHOTOSYNTHETIC ENZYMES

Photosynthetic enzymes are sensitive to heat stress (Table 38.2); ribulose-1,5-bisphosphate carboxylase/oxygenase (Rubisco) is the principal enzyme involved in CO_2 fixation, regulating carboxylation during photosynthesis (Ogren 1984). Variation in daytime temperatures alters the photosynthetic rate by modifying the rate of ribulose-1,5-bisphosphate (RuBP) generation and changing the carboxylation site of Rubisco (Kattge and Knorr 2007). For instance, Rubisco activity in *Alhagi sparsifolia* increased with temperature to a maximal point at 34°C, after which it declined (Xue et al. 2011).

High temperature increases the CO_2 concentration in plants; elevated levels of CO_2 disturb the active state of Rubisco (Salvucci et al. 1986). Higher CO_2 levels, even in normal plants, limit photosynthesis by increasing RuBP consumption by Rubisco (Crafts-Brandner and Salvucci 2000). Reduced CO_2 levels limit the activity of photosynthetic enzymes. For instance, Xu et al. (2004) observed increased activity of phosphoenolpyruvate carboxylase (PEPcase) and reduced activities of Rubisco in photosynthetically active nonleaf parts of wheat (awns, glumes, peduncles, etc). A rapid decline in Rubisco activity was observed in cotton exposed to thermal stress for 12 days at 34°C/17°C (Crafts-Brandner and Law 2000) as heat stress decreased the light-saturated CO_2 exchange rate (Law and Crafts-Brandner 1999). Increases in temperature accelerate the catalytic activity of Rubisco. However, the ability of Rubisco to act as oxygenase and its lower affinity for CO_2 inhibit increases in the photosynthetic rate with rising temperature (Salvucci and Crafts-Brandner 2004a).

38.6.2 RUBISCO ACTIVATION

Rubisco activase (RCA) regulates the activity of Rubisco. It is a cytosol-synthesized chloroplast protein that aids removal of inhibitory sugar phosphates from active sites of Rubisco and thus regulates its activity (Salvucci and Ogren 1996). Therefore, the RCA-driven step in photosynthesis is critical for carboxylation of Rubisco (Salvucci and Ogren 1996; Spreitzer and Salvucci 2002).

RCA is thermolabile (Salvucci et al. 2001), such that reduced photosynthesis at moderate heat stress levels has been correlated with deactivation of Rubisco (Law and Crafts-Brandner 1999). For instance, Kobza and Edwards (1987) observed deactivation of Rubisco under mild heat stress in wheat. This indicates that the principal limitation to photosynthesis at moderate elevated temperatures is heat-induced Rubisco deactivation (Crafts-Brandner and Salvucci 2000). Heat stress also causes loss of RCA capacity to regulate the activity and efficiency of Rubisco (Salvucci and Crafts-Brandner 2004a,b). For instance, exposing wheat to 40°C altered the abundance of smaller and larger subunits of Rubisco and RCA under both dark and light conditions (Demirevska-Kepova et al. 2005). Higher-temperature stress from 450°C to 60°C inhibits the activity of Rubisco (Li et al. 2002) and is associated with the denaturation of RCA at high temperature (Salvucci et al. 2001). Moreover, the carboxylation capacity of Rubisco at elevated temperatures is limited to further generation of by-products other than RuBP, and Rubisco deactivation occurs at high temperature due to limited RCA activity (Salvucci and Crafts-Brandner 2004a).

In higher plants, Rubisco is relatively stable with heat stress (Feller et al. 1998; Crafts-Brandner and Salvucci 2000), while RCA loses stability under mild heat stress. Deactivation of Rubisco during mild heat stress has been observed in some studies and is possibly due to the deleterious effect of high temperature on chloroplast reactions (Velikova et al. 2012). It has been reported that inhibition of the carboxylation site of Rubisco is the most sensitive part of the photosynthetic machinery (Law and Crafts-Brandner 1999; Crafts-Brandner and Salvucci 2000).

38.6.3 PHOTORESPIRATION

Inhibition of photosynthesis due to heat stress is often attributed to increased rates of photorespiration (Lea and Leegood 1999). At high temperature, the solubility of O_2 increases relative to CO_2 in leaves, which reduces the efficiency

TABLE 38.2
Influence of Heat Stress on Carboxylation Enzymes

Plant Species	Temperature	Effect on Carboxylation Enzymes Activity	Reduction Relative to Control (%)	References
Wheat	40°C	Reduced Rubisco activation	78	Feller et al. (1998)
Maize	45°C	Malate-inhibited PEPCase activity	66	Crafts-Brandner and Salvucci (2002)
Maize	45°C	Complete inactivation of Rubisco	–	Crafts-Brandner and Salvucci (2002)
Maize	45°C	Pyruvate phosphate dikinase insensitive to heat stress	–	Crafts-Brandner and Salvucci (2002)
Cotton	50°C	Reduction in active sites of Rubisco	75	Crafts-Brandner and Salvucci (2000)
Cotton	40°C	Reduced Rubisco activation	34	Feller et al. (1998)
Tobacco	50°C	Reduction in active sites of Rubisco	75	Crafts-Brandner and Salvucci (2000)
Pea	>50°C	Malate-inhibited PEPCase activity	30	Chinthapalli et al. (2003)

Note: PEPCase, phosphoenolpyruvate carboxylase; Rubisco, ribulose-1,5-bisphosphate carboxylase/oxygenase.

of CO_2-concentrating mechanisms (Salvucci and Crafts-Brandner 2004a; Li and Liu 2007).

Rubisco is the key enzyme that catalyzes photorespiration and photosynthesis in plants, the rates of which are highly dependent on oxygenase and carboxylase activities of Rubisco, respectively (Laing et al. 1974). However, at high temperature, CO_2 concentration is the major limitation to the carboxylase activity of Rubisco. Similarly, heat stress decreases photosynthetic efficiency due to loss of CO_2 as a result of photorespiration (Jordan and Ogren 1984). Under heat stress, oxygenation of Rubisco overtakes carboxylation by forming phosphoglycolate, which needs to be reconverted to phosphoglycerate by the loss of CO_2 fixed during the light reaction using energy produced during photosynthesis. Increased oxygenase activity of Rubisco limits photosynthesis in C_3 plants by photorespiration (Sharkey 1988). Moreover, moderate heat stress (30–42°C) reduces photosynthesis due to less production of RuBP due to reduced activities of enzymes in the OEC and damage to electron transport in PSII (Salvucci and Crafts-Brandner 2004b). Similarly, temperature sensitivity of photosynthesis increases at higher CO_2 levels due to reduced CO_2 diffusion, thus increasing photorespiration (Sage and Coleman 2001; Sun et al. 2013).

38.7 CONCLUSIONS

With the changing climate, the incidence of heat stress is increasing, which limits growth and productivity in plants by disturbing the process of photosynthesis. Heat stress affects the photosynthetic capacity of plants by disturbing metabolic activities and by damaging photosynthetic machinery and enzymes. Heat stress damages the thylakoid membrane and PSII, thus disturbing electron transport and phosphorylation, which limits CO_2 fixation. Moreover, heat stress affects the enzymes involved in CO_2 fixation; Rubisco is the key enzyme and is regulated by another enzyme, RCA. However, thermal stress can inactivate both these enzymes, thus increasing photorespiration.

REFERENCES

Berry, J.A. and O. Björkman. 1980. Photosynthetic response and adaptation to temperature in higher plants. *Annual Review of Plant Physiology* 31:491–543.
Bukhov, N.G., T.G. Dzhibladze, and E.A. Egorova. 2005. Elevated temperatures inhibit ferredoxin-dependent cyclic electron flow around photosystem I. *Russian Journal of Plant Physiology* 52:578–583.
Chinthapalli, B., J. Murmu, and A.S. Raghavendra. 2003. Dramatic difference in the responses of phosphoenolpyruvate carboxylase to temperature in leaves of C_3 and C_4 plants. *Journal of Experimental Botany* 54:707–714.
Crafts-Brandner, S. and R. Law. 2000. Effect of heat stress on the inhibition and recovery of the ribulose-1,5-bisphosphate carboxylase/oxygenase activation state. *Planta* 212:67–74.
Crafts-Brandner, S.J. and M.E. Salvucci. 2000. Rubisco activase constrains the photosynthetic potential of leaves at high temperature and CO_2. *Proceedings of the National Academy of Sciences of the United States of America* 97:13430–13435.

Crafts-Brandner, S.J. and M.E. Salvucci. 2002. Sensitivity of photosynthesis in a C_4 plant, maize, to heat stress. *Plant Physiology* 129:1773–1780.
Cui, L., J. Li, Y. Fan, S. Xu, and Z. Zhang. 2006. High temperature effects on photosynthesis, PSII functionality and antioxidant activity of two *Festuca arundinacea* cultivars with different heat susceptibility. *Botanical Studies* 47:61–69.
Demirevska-Kepova, K., R. Holzer, L. Simova-Stoilova, and U. Feller. 2005. Heat stress effects on ribulose-1,5-bisphosphate carboxylase/oxygenase, Rubisco binding protein and Rubisco activase in wheat leaves. *Biologia Plantarum* 49:521–525.
Dias, A.S. and F.C. Lidon. 2009. Evaluation of grain filling rate and duration in bread and durum wheat, under heat stress after anthesis. *Journal of Agronomy and Crop Science* 195:137–147.
Dutta, S., S. Mohanty and B.C. Tripathy. 2009. Role of temperature stress on chloroplast biogenesis and protein import in pea. *Plant Physiology* 150:1050–1061.
Efeoglu, B. and S. Terzioglu. 2009. Photosynthetic responses of two wheat varieties to high temperature. *EurAsian Journal of BioSciences* 3:97–106.
Farooq, M., H. Bramley, J.A. Palta, and K.H.M. Siddique. 2011. Heat stress in wheat during reproductive and grain-filling phases. *Critical Reviews in Plant Sciences* 30:1–17.
Feller, U., S.J. Crafts-Brandner, and M.E. Salvucci. 1998. Moderately high temperatures inhibit ribulose-1,5-bisphosphate carboxylase/oxygenase (Rubisco) activase-mediated activation of Rubisco. *Plant Physiology* 116:539–546.
Fristedt, R., A. Willig, P. Granath, M. Crèvecoeur, J.-D. Rochaix, and A.V. Vener. 2009. Phosphorylation of photosystem II controls functional macroscopic folding of photosynthetic membranes in Arabidopsis. *Plant Cell* 21:3950–3964.
Havaux, M. 1993. Rapid photosynthetic adaptation to heat stress triggered in potato leaves by moderately elevated temperatures. *Plant, Cell and Environment* 16:461–467.
Jordan, D.B. and W.L. Ogren. 1984. The CO_2/O_2 specificity of ribulose 1,5-bisphosphate carboxylase/oxygenase dependence on ribulose bisphosphate concentration, pH and temperature. *Planta* 161:308–313.
Kattge, J. and W. Knorr. 2007. Temperature acclimation in a biochemical model of photosynthesis: A reanalysis of data from 36 species. *Plant, Cell and Environment* 30:1176–1190.
Kobza, J. and G.E. Edwards. 1987. Influences of leaf temperature on photosynthetic carbon metabolism in wheat. *Plant Physiology* 83:69–74.
Laing, W.A., W.L. Ogren, and R.H. Hageman. 1974. Regulation of soybean net photosynthetic CO_2 fixation by the interaction of CO_2, O_2 and ribulose-1,5-bisphosphate carboxylase. *Plant Physiology* 54:678–685.
Law, R.D. and S.J. Crafts-Brandner. 1999. Inhibition and acclimation of photosynthesis to heat stress is closely correlated with activation of ribulose-1,5-bisphosphate carboxylase/oxygenase. *Plant Physiology* 120:173–182.
Lea, P.J. and R.C. Leegood. 1999. *Plant Biochemistry and Molecular Biology*. Chichester: John Wiley & Sons.
Li, M. and H. Liu. 2007. Carbon isotope composition of plants along an altitudinal gradient and its relationship to environmental factors on the Qinghal-Tibet Plateau. *Polish Journal of Ecology* 55:67–78.
Li, G., H. Mao, X. Ruan, Q. Xu, Y. Gong, X. Zhang, and N. Zhao. 2002. Association of heat induced conformational change with activity loss of Rubisco. *Biochemical and Biophysical Research Communications* 290:1128–1132.

Mohanty, S., B. Grimm, and B.C. Tripathy. 2006. Light and dark modulation of chlorophyll biosynthetic genes in response to temperature. *Planta* 224:692–699.

Murkowski, A. 2001. Heat Stress and spermidine: Effect on chlorophyll fluorescence in tomato plants. *Biologia Plantarum* 44:53–57.

Nawaz, A., M. Farooq, S.A. Cheema, and A. Wahid. 2013. Differential response of wheat cultivars to terminal heat stress. *International Journal of Agriculture and Biology* 15:1354–1358.

Noodén, L.D. 1988. The phenomenon of senescence and aging. In: *Senescence and Aging in Plants*, L.D. Noodén and A.C. Leopold, eds., pp. 2–50. San Diego, CA: Academic Press.

Ogren, W.L. 1984. Photorespiration pathways, regulation, and modification. *Annual Review of Plant Physiology* 35:415–442.

Paulsen, G.M. 1994. High temperature responses of crop plants. In: *Physiology and Determination of Crop Yield*, K.J. Boote, I.M. Bennet, T.R. Sinclair and G.M. Paulsen, eds., pp. 365–389. Madison, WI: American Society of Agronomy.

Pfeiffer, S. and K. Krupinska. 2005. Chloroplast ultrastructure in leaves of *Urtica dioica* L. analyzed after high-pressure freezing and freeze-substitution and compared with conventional fixation followed by room temperature dehydration. *Microscopy Research and Technique* 68:368–376.

Raison, J.K., J.K. Roberts, and J.A. Berry. 1982. Correlations between the thermal stability of chloroplast (thylakoid) membranes and the composition and fluidity of their polar lipids upon acclimation of the higher plant, *Nerium oleander*, to growth temperature. *Biochimica et Biophysica Acta (BBA)— Biomembranes* 688:218–228.

Reda, F. and H.M.H. Mandoura. 2011. Response of enzymes activities, photosynthetic pigments, proline to low or high temperature stressed wheat plant exogenous proline or cysteine. *International Journal of Academic Research* 3:108–116.

Rexroth, S., C.W. Mullineaux, E.D. Sendtko, M. Rögner, and F. Koenig. 2011. The plasma membrane of the cyanobacterium *Gloeobacter violaceus* contains segregated bioenergetic domains. *Plant Cell* 23:2379–2390.

Sage, R.F. and J.R. Coleman. 2001. Effects of low atmospheric CO_2 on plants: More than a thing of the past. *Trends in Plant Science* 6:18–24.

Sairam, R.K., G.C. Srivastava and D.C. Saxena. 2000. Increased antioxidant activity under elevated temperatures: A mechanism of heat stress tolerance in wheat genotypes. *Biologia Plantarum* 43:245–251.

Salvucci, M.E. and S.J. Crafts-Brandner. 2004a. Relationship between the heat tolerance of photosynthesis and the thermal stability of rubisco activase in plants from contrasting thermal environments. *Plant Physiology* 134:1460–1470.

Salvucci, M.E. and S.J. Crafts-Brandner. 2004b. Inhibition of photosynthesis by heat stress: The activation state of Rubisco as a limiting factor in photosynthesis. *Physiologia Plantarum* 120:179–186.

Salvucci, M.E. and W.L. Ogren. 1996. The mechanism of Rubisco activase: Insights from studies of the properties and structure of the enzyme. *Photosynthesis Research* 47:1–11.

Salvucci, M.E., A.R. Portis, and W.L. Ogren. 1986. Light and CO_2 response of ribulose-1,5-bisphosphate carboxylase/oxygenase activation in Arabidopsis leaves. *Plant Physiology* 80:655–659.

Salvucci, M.E., K.W. Osteryoung, S.J. Crafts-Brandner, and E. Vierling. 2001. Exceptional sensitivity of Rubisco activase to thermal denaturation *in vitro* and *in vivo*. *Plant Physiology* 127:1053–1064.

Schrader, S.M., R.R. Wise, W.F. Wacholtz, D.R. Ort, and T.D. Sharkey. 2004. Thylakoid membrane responses to moderately high leaf temperature in pima cotton. *Plant, Cell and Environment* 27:725–735.

Sharkey, T. 1988. Estimating the rate of photorespiration in leaves. *Physiologia Plantarum* 73:147–152.

Sharkey, T.D. and S.M. Schrader. 2006. High temperature stress. In: *Physiology and Molecular Biology of Stress Tolerance in Plants*, K.V.M. Rao, A.S. Raghavendra and K.J. Reddy, eds., pp. 101–129. Dordrecht: Springer.

Sharkey, T.D. and R. Zhang. 2010. High temperature effects on electron and proton circuits of photosynthesis. *Journal of Integrative Plant Biology* 52:712–722.

Spreitzer, R.J. and M.E. Salvucci. 2002. Rubisco: Structure, regulatory interactions, and possibilities for a better enzyme. *Annual Review of Plant Biology* 53:449–475.

Sun, Z., K. Hüve, V. Vislap, and U. Niinemets. 2013. Elevated (CO_2) magnifies isoprene emissions under heat and improves thermal resistance in hybrid aspen. *Journal of Experimental Botany* 64:5509–5523.

Takahashi, S. and N. Murata. 2006. Glycerate-3-phosphate, produced by CO_2 fixation in the Calvin cycle, is critical for the synthesis of the D_1 protein of photosystem II. *Biochimica et Biophysica Acta* 1757:198–205.

Tang, Y., X. Wen, Q. Lu, Z. Yang, Z. Cheng, and C. Lu. 2007. Heat stress induces an aggregation of the light-harvesting complex of photosystem II in spinach plants. *Plant Physiology* 143:629–638.

Tewari, A.K. and B.C. Tripathy. 1998. Temperature-stress-induced impairment of chlorophyll biosynthetic reactions in cucumber and wheat. *Plant Physiology* 117:851–858.

Tewari, A.K. and B.C. Tripathy. 1999. Acclimation of chlorophyll biosynthetic reactions to temperature stress in cucumber (*Cucumis sativus* L.). *Planta* 208:431–437.

Veerasamy, M., Y. He, and B. Huang. 2007. Leaf Senescence and protein metabolism in creeping bentgrass exposed to heat stress and treated with cytokinins. *Journal of the American Society for Horticultural Science* 4:467–472.

Velikova, V., T.D. Sharkey, and F. Loreto. 2012. Stabilization of thylakoid membranes in isoprene-emitting plants reduces formation of reactive oxygen species. *Plant Signaling and Behavior* 7:139–141.

Xu, Q., A.Q. Paulsen, J.A. Guikema, and G.M. Paulsen. 1995. Functional and ultrastructural injury to photosynthesis in wheat by high temperature during maturation. *Environmental and Experimental Botany* 35:43–54.

Xu, X.L., Y.H. Zhang, and Z.M. Wang. 2004. Effect of heat stress during grain filling on phosphoenolpyruvate carboxylase and ribulose-1,5-bisphosphate carboxylase/oxygenase activities of various green organs in winter wheat. *Photosynthetica* 42:317–320.

Xu, S., J.L. Li, and X.Q. Zhang. 2006. Effects of heat acclimation pretreatment on changes of membrane lipid peroxidation, antioxidant metabolites, and ultrastructure of chloroplasts in two cool-season turfgrass species under heat stress. *Environmental and Experimental Botany* 56:274–285.

Xue, W., X.Y. Li, L.S. Lin, Y.J. Wang, and L. Li. 2011. Effects of elevated temperature on photosynthesis in desert plant *Alhagi sparsifolia* S. *Photosynthetica* 49:435–447.

Yan, K., P. Chen, H. Shao, and C. Shao. 2013. Dissection of photosynthetic electron transport process in sweet sorghum under heat stress. *PLoS One* 8(5):62100.

Zhang, J., X.D. Jiang, T.L. Li, and X.J. Cao. 2014. Photosynthesis and ultrastructure of photosynthetic apparatus in tomato leaves under elevated temperature. *Photosynthetica* 52:430–436.

Section XIII

Photosynthesis in the Past, Present, and Future Prospective

39 Evolutionary Ecology of s-Triazine-Resistant Plants

Pleiotropic Photosynthetic Reorganization in the Chloroplast Chronomutant

Jack Dekker

CONTENTS

39.1 Introduction 705
39.2 Nature of s-Triazine-Resistant Plants 706
39.3 Structure–Function Change in s-Triazine-Resistant Plants 706
    39.3.1 Structural Change in s-Triazine-Resistant Plants 706
    39.3.2 Functional Change in s-Triazine-Resistant Plants 707
        39.3.2.1 Resistance to s-Triazine Herbicides 707
        39.3.2.2 Differences in Carbon Assimilation Efficiency 707
        39.3.2.3 Complex *psbA* Mutant Phenotype 708
        39.3.2.4 *psbA* Mutant Pleiotropic Reorganization 708
39.4 Chlorophyll Fluorescence in R and S 709
39.5 Carbon Assimilation in R and S 709
39.6 Temperature Effects on Photosynthetic Function in R and S 710
    39.6.1 Diurnal Temperature Effects 710
    39.6.2 Controlled Leaf Temperature Effects 714
39.7 Rubisco Activity in R and S 715
39.8 Photosynthetic Regulation in R and S 717
39.9 Evolutionary Ecology of s-Triazine-Resistant Plants 719
    39.9.1 Pleiotropy in R 719
    39.9.2 R Adaptation to the Environment and Regulation of Carbon Assimilation 720
References 720

39.1 INTRODUCTION

The nature of s-triazine-resistant (R) plants is a complex adaptive photosynthetic system arising as a consequence of a single base-pair lesion in the chloroplast *psbA* gene. The D-1 protein product of *psbA* is a key element of photosystem II electron transport. Altered electron transport in R causes a pleiotropic cascade of self-reorganization of interacting, interdependent functional traits. Order in this highly conserved photosynthetic system is emergent: a new homeostatic equilibrium among the plastids, cells, and organs of the whole-plant R phenotype emerges.

The nature of R plants is how local photosynthetic opportunity space-time is seized and exploited relative to that of the susceptible (S) wild type. The altered D-1 protein product of the *psbA* gene has been regarded as less photosynthetically efficient in the R biotypes of several species. Past studies have shown lower carbon assimilation (*A*) and whole-plant yields in R relative to S; in other comparisons, R was greater than S; in

still others, R and S were comparable. The equivocal nature of our understanding of how photosynthesis differs between R and S, and how it is regulated, led us to focus experimental efforts on the chronobiology of both phenotypes.

A differential pattern of leaf disk chlorophyll fluorescence (LCF) between S and R *Brassica napus* was observed: R is a chronomutant. A phase shift in LCF maxima occurred in the daily light–dark cycle, supporting the hypothesis that alterations in chloroplast structure conferring s-triazine resistance imply altered temporal behavior of photosynthetic activity.

Studies were conducted observing diurnal patterns of *A* during development. Younger R plants had greater photosynthetic rates early and late in the diurnal, while those of S were greater during midday as well as the photoperiod as a whole. As *B. napus* plants aged, differences in *A* between biotypes increased: only during the late diurnal period was R greater. As *B. napus* began reproductive development, a reversal of photosynthetic differences occurred. R carbon assimilation was greater than S early, midday, and for the whole day period; S was superior to R only late in the day. These results indicate a more complex pattern of photosynthetic carbon assimilation than previously reported. The photosynthetic superiority of a biotype is a function of the time of day and the age of the plant.

R plants also differ in stomatal function from S plants. Total conductance to water vapor and intercellular CO_2 partial pressure in R was either equal to or greater than S over the lifetime of those plants, with the possible exception of some atypical episodes late in ontogeny and late in the light period. As a consequence of these phenomena, leaves of R plants were either the same temperature or cooler than leaves of S plants for the entire lives of both biotypes. At lower leaf and air temperatures, R and S leaves function in a similar way. As air temperature increases, their responses diverge, allowing R plants to compensate for their sensitivity to high temperature. R leaves generally are cooler, and total conductance to water is greater than in S, probably due to greater stomatal aperture size. As a result, at higher air temperatures, R leaves photosynthesize at cooler leaf temperatures closer to the optimal for both biotypes. In addition to other pleiotropic effects, R plants appear to be stomatal mutants. They constitute a model system to study regulation of stomatal function and the relationship between environmental cues and stomatal behavior.

Studies were conducted to test if there is any regulatory role for ribulose-1,5-bisphosphate carboxylase/oxygenase (Rubisco) in photosynthetic carbon assimilation in R and S *B. napus*. Rubisco percent activation and initial activity may account for R and S carbon assimilation differences during midday. Differences in *A* early and late in the day were not accounted for by differences in Rubisco (initial, total, % activation).

Studies were conducted to determine the response of R and S to different temperatures and gas atmospheres with infrared gas analysis and pulse amplitude-modulated chlorophyll fluorescence techniques. Photosynthetic regulation can be separated into three categories based on these studies. The first category is Rubisco-limited photosynthesis. When carbon assimilation was Rubisco limited, there was little difference between R and S biotypes. A second category, feedback-limited photosynthesis, was most evident at 15°C, less so at 25°C. The third category, photosynthetic electron transport-limited photosynthesis, was evident at 25°C and 35°C. R exhibited much more electron transport-limited carbon assimilation at 25°C and 35°C than did S. At 15°C, neither R nor S exhibited electron transport-limited carbon assimilation. The primary limitation to photosynthesis changes with changes in leaf temperature: electron transport limitations in R may be significant only at higher temperatures.

The reorganized R plant interacts with the environment in a different manner than does S. Under environmental conditions highly favorable to plant growth, S often has an advantage over R. Under certain less favorable conditions to plant growth and stressful conditions, R can be at an advantage over S. It can be envisioned that there were environmental conditions prior to the introduction of s-triazine herbicides in which R had an adaptive advantage over the more numerous S individuals in a population of a species. Under certain conditions, R might have exploited a photosynthetic niche underutilized by S. These conditions may have occurred in less favorable environments and may have been cool (or hot), low-light conditions interacting with other biochemical and diurnal plant factors early and late in the photoperiod, as well as more complex physiological conditions late in the plant's development. R survival and continuity could have been favored under these conditions.

39.2 NATURE OF s-TRIAZINE-RESISTANT PLANTS

The nature of s-triazine-resistant plants (R) is a complex adaptive photosynthetic system. Complex photosynthetic adaptation of the whole-plant phenotype arises in the dynamic network of interacting components behaving in parallel: the psbA gene, the D-1 protein, chloroplast ultrastructure, the R cell, and leaf tissues and organs. Order in this system is emergent: coherent structures and functional traits appear during the process of self-organization of the entire photosynthetic process. The nature of R plant self-organization is a pleiotropic consequence of a single base-pair lesion in the chloroplast psbA gene. This single mutation affects multiple phenotypic traits in a cascade from gene to global R metapopulation behavior. The R product of the psbA gene, the D-1 protein, is a key functional element of photosystem II (PSII) electron transport. Altered electron transport in R leads to a profound reorganization of interacting, interdependent functional units in the chloroplast. Change in the highly conserved photosynthetic apparatus leads to a new homeostatic equilibrium among the plastids, cells, tissues, organs, and the whole-plant R phenotype. The nature of R plants is the pleiotropic consequence of how and when the new phenotype seizes and exploits local photosynthetic opportunity space-time relative to that of the susceptible (S) wild type.

39.3 STRUCTURE–FUNCTION CHANGE IN s-TRIAZINE-RESISTANT PLANTS

39.3.1 STRUCTURAL CHANGE IN s-TRIAZINE-RESISTANT PLANTS

s-Triazine resistance in higher plants is due to a single base-pair mutation to the psbA chloroplast gene (Hirschberg and McIntosh 1983). The R variant has a single nucleotide base-pair mutation of guanine for adenine. This mutation results in substitution of a serine codon (AGT) for a glycine codon (GGT) at position 264 in the polypeptide product. The codon 264 change in the maternally inherited, cytoplasmic, psbA gene causes a change in its product, the transmembrane D-1 protein, a key functional element in PSII electron transport. This single base-pair mutation results in a profound decrease in the ability of the protein to bind s-triazines that induce toxic responses.

Several changes in thylakoid lipid chemical composition in R have been observed (Pillai and St. John 1981).

R phospholipids had higher linolenic acid concentrations and lower levels of oleic and linoleic fatty acids (Dekker 1993; Lemoine et al. 1986). R plants overall were richer in unsaturated fatty acids, had higher proportions (and quantitatively greater amounts) of monogalactosyl diglyceride and phosphatidyl choline, and had lower proportions of digalactosyl diglyceride and phosphatidyl choline, compared to S (Burke et al. 1982). As a consequence of a higher proportion of appressed thylakoids, R plants had a greater proportion of Δ3-*trans*-hexadecenoic acid in phosphatidylglycerol (Boardman 1977).

Although the leaf anatomy of R and S is similar, many plastid ultrastructural characters are different (Dekker 1993). R has decreased plastid starch content and increased grana stacking (and the associated characters of lower chlorophyll (chl) *a/b* ratio, increased chl *a/b* light-harvesting complex, and lower P700 chl *a* and chloroplast coupling factor amounts) (Mattoo et al. 1984). The resistant mutant had a larger volume of the chloroplast as grana lamellae and more thylakoids per granum. Vaughn reported that several triazine-resistant species, including *B napus*, have a shade-adapted chloroplast ultrastructure (Vaughn 1986; Vaughn and Duke 1984). These plastid ultrastructural changes in R are similar to those found in shade-adapted leaves (Boardman 1977; Dekker 1993; Dekker and Burmester 1992).

A single base-pair substitution at codon 264 in the *psbA* chloroplast gene in the highly conserved photosynthetic apparatus (PS) leads to a cascade of changes in the plants' morphology, physiology, and ecological reaction to its immediate environment, a nonintuitive result. With the genetic lesion, a dynamic reorganization of interacting, interdependent plant functional units occurs, a new homeostatic equilibrium among parts. This pleiotropic cascade of phenotypic changes that are stimulated by *psbA* gene codon 264 mutation is not predictable from genomic information.

39.3.2 Functional Change in s-Triazine-Resistant Plants

39.3.2.1 Resistance to s-Triazine Herbicides

The herbicide atrazine was introduced commercially around 1960 for broadleaf weed control in maize (*Zea mays*) production fields. s-Triazine resistance (R) in higher plants was first discovered in common groundsel (*Senecio vulgaris*) in 1968 in response to protracted exposure to simazine and atrazine in nursery stock (Ryan 1970). Since that time, s-triazine resistance has been confirmed in many species throughout the world. The rapid increase in R weed species and locations has been associated with sole reliance on s-triazine herbicides for weed control, simple selection by a single herbicide. In Iowa, for example, four species rapidly appeared: kochia (*Kochia scoparia*) (Dekker et al. 1987), giant foxtail (*Setaria faberi*) (Thornhill and Dekker 1993), common lamb's-quarters (*Chenopodium album*) (Dekker and Burmester 1989) and Pennsylvania smartweed (*Polygonum pensylvanicum*) (Dekker et al. 1991).

Herbicide resistance is conferred by a marked decrease in atrazine binding to the D-1 protein encoded by the chloroplast

(maternally inherited) *psbA* gene mutant (Hirschberg and McIntosh 1983). Studies revealed that this type of herbicide resistance was a consequence of reduced herbicide binding to the 32 kD chloroplast protein, D-1, in the resistant (R) biotype (Pfister et al. 1981; Steinback et al. 1981). s-triazine herbicides act by binding to the D-1 protein in susceptible (S) species in the chloroplast thylakoid membranes and thereby inhibiting PSII electron transport. The s-triazine resistance mechanism is not physiological; R is not a function of differential uptake, translocation, metabolism, accumulation, or membrane permeability (Radosevich 1970).

39.3.2.2 Differences in Carbon Assimilation Efficiency

Since the first discovery of s-triazine resistance (R) in higher plants, the altered D-1 protein product of the psbA gene has been regarded as less photosynthetically efficient in those R biotypes of the species (Dekker 1993). s-Triazine-resistant plants have been shown to have a decreased quantum efficiency of CO_2 assimilation compared to s-triazine-susceptible plants (S) (Holt et al. 1981) and have been generally regarded as less fit than S plants (Gressel and Segel 1978). This decreased quantum efficiency of CO_2 assimilation in R compared to the S wild type is credited to an altered redox state of PSII quinone acceptors and a shift in the equilibrium constant between Q_A^- and Q_B in favor of Q_A^- (Arntzen et al. 1979).

Several studies have shown lower photosynthetic carbon assimilation rates (*A*) in R *Amaranthus hybridus* (Ahrens and Stoller 1983) and *Senecio vulgaris* (Holt et al. 1981; Ort et al. 1983) relative to S. Beversdorf et al. (1988) found lower R whole-plant yields in field evaluations of *Brassica napus*. Additionally, van Oorshot and van Leeuwen (1984) found that *A* in S *Amaranthus retroflexus* was greater than that in R, while R and S *A* were comparable in *Chenopodium album*, *Polygonum lapathifolium*, *Poa annua*, *Solanum nigrum*, and *Stellaria media*. R biotypes of *Phalaris paradoxa* have been found to be photosynthetically superior to their S counterparts (Schonfeld et al. 1987). Jansen et al. (1986) observed that R *Chenopodium album* chloroplasts had lower electron transport rates between water and plastoquinone compared to S, yet no differences were found in the rate or quantum yield of whole-chain electron transport, or in *A*, between R and S. These inconsistent responses by R and S biotypes have led several to conclude that the change conferring R is not necessarily directly linked to inferior photosynthetic function (Holt and Goffner 1985; Jansen et al. 1986; Schonfeld et al. 1987).

An assessment of these and other studies reveals several possible reasons why different responses may have been observed. They include pleiotropic reorganization of the R chloroplast and the dynamic interrelationship between components of photosynthesis; the role of environment in altering responses; genetic factors, such as differences between biotypes and genome interactions within a biotype; and the possibility of an unnamed factor controlling photosynthesis (Arntzen and Duesing 1983; Duesing and Yue 1983). Comparison of different photosynthetic responses by R is further complicated by the many different environmental and biological conditions under which they were conducted. These include changes in plant species; plant age; plant uniformity; plant tissue; temperature; photosynthetic photon flux density (PPFD); the degree of experimental environmental control (field, glasshouse, or controlled environment chambers); and the diurnal light period length and variation of conditions. Few systematic studies have been conducted to separate these factors.

A mitigating factor in comparing R and S photosynthetic responses has been the use of model systems in which other genes besides the mutation to *psbA* have differed (e.g., Holt et al. 1981). Inferences from these studies have been confounded because they relied on a nonisogenic model system. McClosky and Holt (1990) have suggested that nuclear genome differences may compensate for differences in productivity between nonisogenic R and S selections and that detrimental effects may be attenuated by interactions of plastid and nuclear genomes (Stowe and Holt 1988). Many of these limitations may have been overcome in studies with nearly isonuclear biotypes of *B. napus* (Beversdorf and Hume 1984; Dekker and Burmester 1990a,b; Jursinic and Pearcy 1988; Vaughn 1986).

39.3.2.3 Complex *psbA* Mutant Phenotype

Decreases in electron transport function in the chloroplast have been believed to be the cause of decreased carbon assimilation rates and plant productivity in many reports (Dekker 1993). What is less clear in the literature is whether this change in D-1 structure and electron transport function directly modifies whole-leaf photosynthesis and plant productivity or only indirectly influences these functions (Holt and Goffner 1985). The dynamic nature of these responses has led several to conclude that the primary effect of R is complex,

involves more than one aspect of photosynthesis, and can be mitigated by other processes in the system (Dekker and Sharkey 1992; Ireland et al. 1988; McClosky and Holt 1990). For example, it has been pointed out that decreased electron transport of Q_A^- to Q_B in R is more rapid than the normally rate-limiting oxidation of plastoquinol (Barber 1983; Ort et al. 1983), while other studies indicate this step may be rate limiting (Jursinic and Pearcy1988).

39.3.2.4 *psbA* Mutant Pleiotropic Reorganization

The genetic change in R plants leads to a profound, pleiotropic reorganization of structural and functional units in the chloroplast (Dekker and Burmester 1993) (Table 39.1). This adaptive reorganization of photosynthetic components in the chloroplast may be a compensatory mechanism to maintain a functional interaction of the PSII complex lipids and proteins (Pillai and St. John 1981). This pleiotropic cascade includes both structural (Burke et al. 1982; Mattoo et al. 1984) and functional changes (Arntzen et al. 1979; Pillai and St. John 1981; Tranel and Dekker 2002; Vaughn 1986). The dynamic nature of the chloroplast to reach a new, markedly different, structural and functional equilibrium in response to the mutation of a key plastidic gene has been observed previously (Hugly et al. 1989; Kunst et al. 1989). This profound pleiotropic cascade of functional and structural changes conferred by changes in the D-1 protein could imply that the amino acid substitution is close to a primary functional and structural source of photosynthetic regulation. Mattoo et al. (1984) have suggested that the rapid anabolism–catabolism rate of the D-1 protein could serve as a signal resulting in the reorganization of membranes around the PSII complex. This dynamic reorganization has consequences for evaluating and understanding regulatory effects of electron transport in carbon assimilation (Dekker 1993).

TABLE 39.1
Cascade of Pleiotropic Effects at the Plastid, Cell, Tissue, Organ, Plant, Population, and Community Levels of Plant Organization Consequential to the Mutation of the *psbA* Chloroplast Gene

Gene	Plastid/Cell	Tissue/Organ	Plant	Population	Community
psbA chloroplast gene codon 264 mutation	Altered chloroplast D-1 protein: structural changes in PSII electron transport	Differential R/S diurnal pattern of photosynthetic carbon assimilation and chlorophyll fluorescence	s-Triazine resistance Higher R carbon assimilation early and late in the day	Enrichment of R biotypes Enhanced intraspecific photosynthetic phenotype diversity	Enrichment of R species Enhanced interspecific photosynthetic niche diversity
		Increased stomatal aperture opening	Higher carbon assimilation in older R plants Cooler leaf temperature		
	Increased thylakoid membrane stacking	Shade-adapted leaf morphology	Enhanced resistance to low-temperature stress Enhanced photosynthetic efficiency in lower light conditions		
	Increased thylakoid grana lamellae fatty acid unsaturation	Greater low-temperature lipid fluidity	Increased seed dormancy		

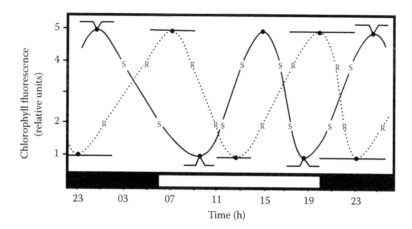

FIGURE 39.1 Diurnal oscillations in s-triazine-resistant and s-triazine-susceptible *Brassica napus* leaf disk chlorophyll fluorescence; relative daily maxima and minima combined over 3 days of a controlled environment experiment; variations in LCF amplitude removed. R, resistant means; S, susceptible means.

The equivocal nature of our knowledge of how photosynthesis differs between R and S, and how it is regulated, led us to focus experimental efforts on chronobiological understandings of the R phenotype under more dynamic, but closely controlled, growth conditions (Dekker and Burmester 1993). In particular, we have observed other pleiotropic effects in R: differential patterns of both carbon assimilation (*A*) and chlorophyll fluorescence over the course of the light–dark diurnal cycle. Subsequently, we observed that R plants have markedly different leaf and stomatal responses to temperature.

39.4 CHLOROPHYLL FLUORESCENCE IN R AND S

Variations in leaf disk chlorophyll fluorescence (LCF; chl *a*; terminal, F_t) intensity with time of day were studied to test the hypothesis that alterations in chloroplast structure that confer s-triazine resistance (R) may also imply altered temporal organization of chlorophyll fluorescence activities (Dekker and Westfall 1987a,b). Two periods of reduced photosynthetic efficiency occurred in the daily light–dark cycle in both biotypes (Figure 39.1). The times these occurred during the diurnal differed between biotypes. A phase shift in LCF maxima occurred between R and S biotypes. The R biotype was less photosynthetically efficient than S early and late in the light period. The S biotype was less efficient in the middle of the light and dark periods of the diurnal. This differential pattern of LCF supports the hypothesis that s-triazine resistance chloroplast alterations also imply an alteration in the temporal organization of chloroplast physiological function: the s-triazine-resistant phenotype is a chronomutant.

39.5 CARBON ASSIMILATION IN R AND S

Insights about the chronobiology of R and S chlorophyll fluorescence stimulated observations of how carbon assimilation changes during the light period and how it changes over the life of the two biotypes. Studies were conducted testing the

hypothesis that the mutation to the *psbA* plastid gene that confers s-triazine resistance also results in an altered diurnal pattern of photosynthetic carbon assimilation relative to that of the susceptible biotype.

Studies were conducted observing changes in the diurnal patterns of photosynthetic carbon assimilation (*A*) in R relative to that of the susceptible S wild type over the ontogenetic development of the plant (Dekker and Burmester 1992). *A* approximately tracked the increasing and decreasing diurnal light levels (Figure 39.2).

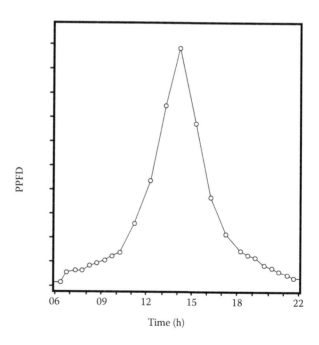

FIGURE 39.2 Changes in photosynthetic photon flux density (PPFD; µmol quanta m^{-1} period^{-1}) with time (hour of the day) in separate experiments on 3- to 9 1/2-leaf *Brassica napus* plants; periods of the diurnal: early (0060–1000h), midday (1000–1800h), and late (1800–2150h). (From Dekker, J.H., and Burmester, R.G., *Plant Physiol.*, 100, 2052–2058, 1992.)

Younger (3- to 4-leaf) R plants had greater photosynthetic rates early and late in the diurnal light period, while those of S were greater during midday as well as the photoperiod as a whole. A differential pattern of *A* was observed in 3- to 3 1/2-leaf plants: early and late in the photoperiod, R carbon assimilation exceeded that of S, while S carbon assimilation exceeded that of R during the midday period (Figure 39.3, Table 39.2). The same differential pattern of *A* between R and S was observed in 4-leaf plants.

As *B. napus* plants aged, differences in *A* between S and R became greater. At the 5 1/2- to 6-leaf growth stage, S leaf *A* exceeded R during midday and was similar to R early, while only during the late period was R greater (Figure 39.4, Table 39.2). For both biotypes, maximum leaf *A* was reached during this stage of development.

S continued to assimilate more carbon than R in 6 1/2- to 7 1/2-leaf plants, plants near the end of the vegetative phase of development, for most times of the day (Figure 39.5, Table 39.2). *A* in R equaled that in S late in the day.

As *B. napus* began the reproductive phase of development, 8 1/2- to 9 1/2-leaf plants, a nearly complete reversal of photosynthetic differences observed in previous growth stages occurred. R carbon assimilation was greater than S early, midday, and for the whole day period, while S (for the first time in its development) was superior to R later in the day (Figure 39.6, Table 39.2). An erratic pattern of *A*, especially in S plants, was observed late in the photoperiod. *A* increased for a time, while PPFD was incrementally decreasing. It is

unclear why this occurred. Overall, *A* at this stage of plant development was considerably less (S ca. 65% less; R ca. 45% less) than the vegetative-phase plants of 1 week previous and was the lowest *A* in their ontogeny. Also, this erratic pattern could be a function of plants overcoming limitations imposed during the preceding high-light, midday period, or erratic responses to the onset of senescence.

These relative photosynthetic relationships change in several ways with ontogeny and are summarized in Table 39.2. These results support a generalized model of carbon assimilation during the diurnal light period with changes in development (Figure 39.7, Table 39.2).

As the plants age during the vegetative phase of development, S gradually assimilates more than R in the early (Figure 39.7a) and then in the late (Figure 39.7c) part of the light period. At the end of the vegetative phase of development, R is photosynthetically inferior to S at most times of the day and is never better. This relationship between the two biotypes dramatically changes with the onset of the reproductive phase (8 1/2- to 9 1/2-leaf) of plant development: R is now superior to S during all periods (Figure 39.7b) of the diurnal light period with the exception of the late part of the day.

As both R and S plants developed from 3 to 9 1/2 leaves, they assimilated more carbon on an all-day basis until the 5 1/2- to 6-leaf stage, when *A* was the greatest, and then *A* declined through the reproductive stage when the oldest (8 1/2- to 9 1/2-leaf) plants had the lowest *A* (of Figure 39.7d). Differences observed in *A* at that later developmental stage could be a function of R losing photosynthetic competence at a slower rate than S. This quality of R could have important implications in agriculture and the critical seed development period: possibly, greater *A* late in development could result in greater accumulated carbon for seed development, a factor that might partially overcome reduced photosynthetic carbon assimilation earlier in the vegetative phase of development.

These studies indicate a more complex pattern of photosynthetic carbon assimilation than previously observed. The photosynthetic superiority of one biotype relative to the other was a function of the time of day and the age of the plant. The lower rate of R photosynthetic carbon assimilation can be overcome in some instances by certain environmental and developmental conditions. These results support the hypothesis tested: *psbA* plastid gene mutation conferring R also confers a different diurnal pattern of photosynthetic function than that in S. This work is also consistent with the pattern of differential chlorophyll fluorescence (F_t) during the diurnal previously reported (Dekker and Westfall 1987a,b).

39.6 TEMPERATURE EFFECTS ON PHOTOSYNTHETIC FUNCTION IN R AND S

39.6.1 DIURNAL TEMPERATURE EFFECTS

The diurnal pattern of carbon assimilation differs between R and S at different air temperatures and stages of development: R is greater than S at higher air temperature and late in ontogeny (Table 39.2). When carbon *A* was greater in R, S leaf

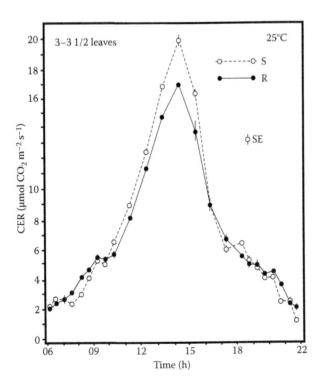

FIGURE 39.3 Changes in carbon assimilation (*A*; μmol CO_2 m^{-2} s^{-1}) in 3- to 3 1/2-leaf s-triazine-resistant (R; ●) and s-triazine-susceptible (S; ○) *Brassica napus* plants with time (hour of the day); leaf temperature, 25°C, *n* = 9. SE, ± standard error of the mean.

TABLE 39.2

Summary of Ontogenic (3- to 9 1/2-Leaf Plants) Effects on Photosynthetic Carbon Assimilation, Leaf Temperature, Total Conductance to Water Vapor, and Leaf Intercellular CO_2 Partial Pressure in s-Triazine-Resistant (R) and s-Triazine-Susceptible (S) Biotypes at 25°C Air Temperature with Changes within a Diurnal Photoperiod

Leaf Stage	Early 0600–1000 h	Midday 1000–1800 h	Late 1800–2150 h	All Day 0600–2150 h
Photosynthetic Carbon Assimilation (A; µmol CO_2 m^{-2} s^{-1})				
3–3 1/2	**R > S**	S > R	**R > S**	S > R
4	**R > S**	S > R	**R > S**	S > R
5 1/2–6	S = R	S > R	**R > S**	S > R
6 1/2–7 1/2	S > R	S > R	S = R	S > R
8 1/2–9 1/2	**R > S**	**R > S**	S > R	**R > S**
Leaf Temperature (°C)				
3–3 1/2	S = R	S = R	S = R	S = R
4	S = R	S > R	S = R	S > R
5 1/2–6	S > R	S > R	S = R	S > R
6 1/2–7 1/2	S > R	S > R	S = R	S > R
8 1/2–9 1/2	S > R	S > R	S = R	S > R
Total Conductance to Water Vapor (g; mmol m^{-2} s^{-1})				
3–3 1/2	S = R	S = R	**R > S**	**R > S**
4	**R > S**	**R > S**	S = R	**R > S**
5 1/2–6	**R > S**	S = R	**R > S**	**R > S**
6 1/2–7 1/2	S = R	S = R	S > R	S = R
8 1/2–9 1/2	**R > S**	**R > S**	S > R	**R > S**
Leaf Intercellular CO_2 Partial Pressure (C_i; µL)				
3–3 1/2	S = R	S = R	S = R	S = R
4	**R > S**	**R > S**	S = R	**R > S**
5 1/2–6	**R > S**	**R > S**	S = R	**R > S**
6 1/2–7 1/2	S = R	S = R	S > R	S = R
8 1/2–9 1/2	S = R	**R > S**	S > R	S = R

Source: Dekker, J.H., and Burmester, R.G., *Plant Physiol.*, 100, 2052–2058, 1992.

Note: Organized by leaf stages within photosynthetic parameters. **R > S** or **R < S** in bold type is significantly different with F-test comparison.

temperatures were similar to R (except 8 1/2–9 1/2 leaves), and S stomatal conductance (g) and C_i were less than or equal to R (Table 39.3) (Dekker and Burmester 1992). When carbon assimilation was greater in S, R leaf temperatures and stomatal conductances were less than or similar to S (except 8 1/2–9 1/2 leaves, late), and S C_i was less than or equal to R (except 8 1/2–9 1/2 leaves, late). Regardless of differences or similarities in A between the two biotypes, S leaf temperatures were greater than or equal to R (R was never greater than S), and R stomatal conductances and C_i were greater than or equal to S (except 6 1/2–7 1/2 and 8 1/2–9 1/2 leaves, late).

S plants assimilated more carbon over the entire day than R plants at 10°C, 15°C, and 25°C air temperatures but assimilated less than R at 35°C air temperature (Dekker and Burmester 1993) (Tables 39.4 and 39.5, Figures 39.8 and 39.9). During the early part of the light period, A in R plants was either similar (15°C air temperature) or greater (25° and 35°C) compared to that in S. During the midday part of the light

period, A in S was greater than that at 15° and 25°C air temperatures, while A in R was greater than that in S at 35°C. A in R was greater than in S at all three air temperatures at some times of the diurnal.

In most instances, leaf temperatures and g were similar in R and S at 15°C air temperature. At 25°C, S leaf temperatures were greater than or equal to R. At 35°C, R leaf temperatures were less than those of S during all parts of the light period (Table 39.4, Figure 39.10). At 25°C and 35°C, g in R plants was almost always greater than that of S (Table 39.4, Figure 39.11).

Carbon assimilation in R was greater than that in S at higher temperature and late in ontogeny. This contrasts with other studies in which S carbon assimilation was greater than R when leaf temperatures where held constant. When leaf temperature is not controlled, greater stomatal conductance and leaf cooling in R can overcome these disadvantages.

In addition to the differences in carbon assimilation, R plants also differ in stomatal function from S plants (Dekker

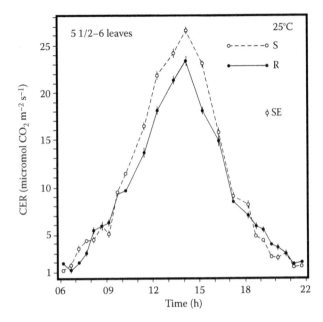

FIGURE 39.4 Changes in carbon assimilation (A; μmol CO_2 m^{-2} s^{-1}) in 5 1/2- to 6-leaf s-triazine-resistant (R; ●) and s-triazine-susceptible (S; ○) *Brassica napus* plants with time (hour of the day); leaf temperature, 25°C; $n = 9$. SE, ± standard error of the mean.

and Burmester 1992; Dekker et al. 1990). Total conductance to water vapor and intercellular CO_2 partial pressure in R was either equal to or greater than S over the lifetime of those plants, with the possible exception of some atypical episodes late in ontogeny and late in the light period. As a consequence of these phenomena, leaves of R plants were either the same temperature or cooler compared to leaves of S plants for the entire lives of both biotypes. The physiological or biochemical

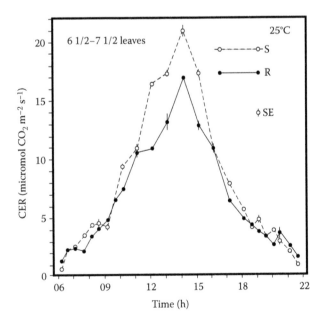

FIGURE 39.5 Changes in carbon assimilation (A; μmol CO_2 m^{-2} s^{-1}) in 6 1/2- to 7 1/2-leaf s-triazine-resistant (R; ●) and s-triazine-susceptible (S; ○) *Brassica napus* plants with time (hour of the day); leaf temperature, 25°C; $n = 9$. SE, ± standard error of the mean.

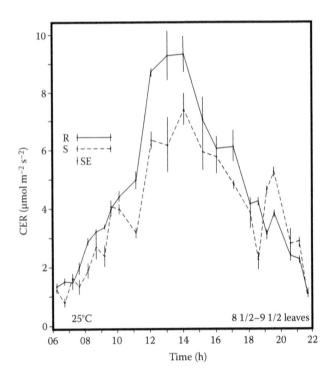

FIGURE 39.6 Changes in carbon assimilation (A; μmol CO_2 m^{-2} s^{-1}) in 8 1/2- to 9 1/2-leaf s-triazine-resistant (R; ●) and s-triazine-susceptible (S; ○) *Brassica napus* plants with time (hour of the day); leaf temperature, 25°C; $n = 9$. SE, ± standard error of the mean.

linkage connecting the primary R defect and stomatal function is not apparent but could provide an interesting experimental model system. These results on altered, cooler, leaf temperatures and stomatal function also have been observed in previous studies under quite different conditions (Dekker and Sharkey 1992). In all instances in these experiments, both

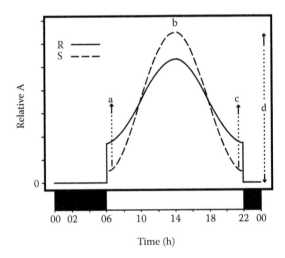

FIGURE 39.7 Model of the relationship between relative carbon assimilation rate (A) and the time of the day in s-triazine-resistant (R) and s-triazine-susceptible (S) *Brassica napus*. (a) Changes in relative A during the early portion of the diurnal. (b) Changes in relative A during the midday portion of the diurnal. (c) Changes in relative A during the late portion of the diurnal. (d) Changes in relative A during ontogeny.

TABLE 39.3

Summary of Ontogenic (3–9 1/2 Leaf Plants) Effects on Photosynthetic Carbon Assimilation (A; µmol CO_2 m^{-2} s^{-1}), Leaf Temperature (°C), Total Conductance to Water Vapor (g; mmol m^{-2} s^{-1}), and Leaf Intercellular CO_2 Partial Pressure (C_i; µL) in s-Triazine-Resistant (R) and s-Triazine-Susceptible (S) Biotypes at 25°C Air Temperature with Changes within a Diurnal Photoperiod

Leaf Stage	Parameter	Early 0600–1000 h	Midday 1000–1800 h	Late 1800–2150 h	All Day 0600–2150 h
3–3 1/2	A	**R > S**	S > R	**R > S**	S > R
	Leaf temp	S = R	S = R	S = R	S = R
	g	S = R	S = R	**R > S**	**R > S**
	C_i	S = R	S = R	S = R	S = R
4	A	**R > S**	S > R	**R > S**	S > R
	Leaf temp	S = R	S > R	S = R	S > R
	g	**R > S**	**R > S**	S = R	**R > S**
	C_i	**R > S**	**R > S**	S = R	**R > S**
5 1/2–6	A	S = R	S > R	**R > S**	S > R
	Leaf temp	S > R	S > R	S = R	S > R
	g	**R > S**	S = R	**R > S**	**R > S**
	C_i	**R > S**	**R > S**	S = R	**R > S**
6 1/2–7 1/2	A	S > R	S > R	S = R	S > R
	Leaf temp	S > R	S > R	S = R	S > R
	g	S = R	S = R	S > R	S = R
	C_i	S = R	S = R	S > R	S = R
8 1/2–9 1/2	A	**R > S**	**R > S**	S > R	**R > S**
	Leaf temp	S > R	S > R	S = R	S > R
	g	**R > S**	**R > S**	S > R	**R > S**
	C_i	S = R	R > S	S > R	S = R

Source: Dekker, J.H., and Burmester, R.G., *Plant Physiol.*, 100, 2052–2058, 1992.

Note: Organized by photosynthetic parameters within leaf stage. **R > S** or **R < S** in bold type is significantly different with F-test comparison.

TABLE 39.4

Changes in Carbon Assimilation Rate and Conductance to Water Vapor in Four Leaf s-Triazine-Resistant (R) and s-Triazine-Susceptible (S) *Brassica napus* Plants Grown at 15°C, 25°C, and 35°C Air Temperatures within a Diurnal Period

Air Temp	Early 0600–1000 h	Midday 1000–1800 h	Late 1800–2150 h	All Day 0600–2150 h
Photosynthetic Carbon Assimilation (A; µmol CO_2 m^{-2} s^{-1})				
15°C	R = S	S > R	**R > S**	S > R
25°C	**R > S**	S > R	**R > S**	S > R
35°C	**R > S**	**R > S**	**R > S**	**R > S**
Leaf Temperature (°C)				
15°C	R = S	**R > S**	R = S	R = S
25°C	R = S	S > R	R = S	S > R
35°C	S > R	S > R	S > R	S > R
Total Conductance to Water Vapor (g; mol m^{-2} s^{-1})				
15°C	R = S	R = S	**R > S**	R = S
25°C	**R > S**	**R > S**	R = S	**R > S**
35°C	**R > S**	**R > S**	**R > S**	**R > S**

Note: **R > S** or **R < S** in bold type is significantly different with F-test comparison.

TABLE 39.5

Comparison of Carbon Assimilation (Carbon Exchange Rate [CER]; μmol CO$_2$ m^{-2} s^{-1}) in s-Triazine-Resistant (R) and s-Triazine-Susceptible (S) *Brassica napus* Plants at 3–4 (Figure 39.2) and 8 1/2–9 1/2 Leaves (Figure 39.5) at 25°C, and 3–4 Leaves at 35°C (Figure 39.10) Air Temperatures

Temp	Leaf Stage	Early 0600–1000 h	Midday 1000–1800 h	Late 1800–2150 h	All Day 0600–2150 h
25°C	3–4	**R > S**	S > R	**R > S**	S > R
	8 1/2–9 1/2	**R > S**	**R > S**	S > R	**R > S**
35°C	3–4	**R > S**	**R > S**	**R > S**	**R > S**

Note: **R > S** or **R < S** in bold type is significantly different with F-test comparison.

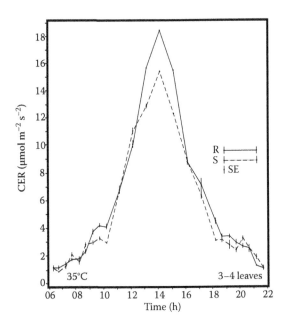

FIGURE 39.8 Changes in carbon assimilation (CER; μmol CO$_2$ m^{-2} s^{-1}) in 3- to 4-leaf s-triazine-resistant (R) and s-triazine-susceptible (S) *Brassica napus* plants with time (hour of the day); leaf temperature, 35°C; *n* = 9. SE, ± standard error of the mean.

R and S plants were functioning at or near (26–29°C) their photosynthetic temperature optima of ca. 28°C (Dekker and Sharkey 1992).

39.6.2 CONTROLLED LEAF TEMPERATURE EFFECTS

At 15°C, *A* in R and S was similar, but as leaf temperature increased to 35°C, S assimilated increasingly more carbon than R (Figure 39.12) (Dekker and Burmester 1993). For both biotypes, the optimal temperature for *A* was ca. 26–27°C.

Associated with increases in leaf temperature was a greater total conductance to water vapor in R relative to S under similar temperature conditions (Figure 39.13) (Dekker and Burmester 1993).

At lower leaf and air temperatures (15°C), R and S leaves function in a similar way (Dekker and Burmester 1993). But, as the temperature of their environment increases, their responses diverge considerably. If leaf temperature is controlled externally, R leaves assimilate considerably less carbon than S leaves, especially at hyperoptimal temperatures (e.g., 35°C).

If leaf temperature is not directly controlled, and both R and S are immersed in an identical air temperature environment,

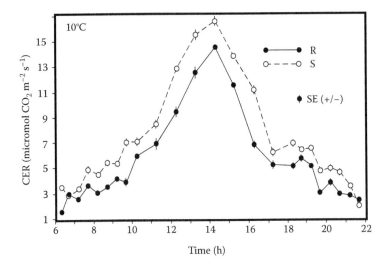

FIGURE 39.9 Changes in carbon assimilation (CER; μmol CO$_2$ m^{-2} s^{-1}) in s-triazine-resistant (R) and s-triazine-susceptible (S) *Brassica napus* plants with time (hour of the day); leaf temperature, 10°C. SE, ± standard error of the mean.

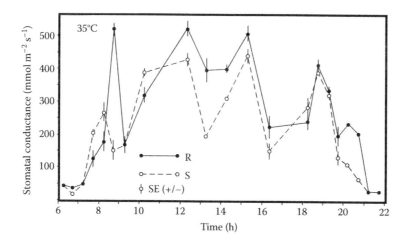

FIGURE 39.10 Change in leaf temperature at constant 35°C air temperature in s-triazine-resistant (R) and s-triazine-susceptible (S) *Brassica napus* plants with time (hour of the day). SE, ± standard error of the mean.

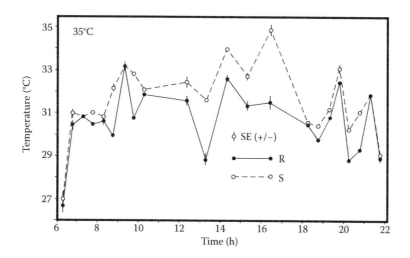

FIGURE 39.11 Stomatal conductance of water vapor versus diurnal time of R and S at 35°C air temperature.

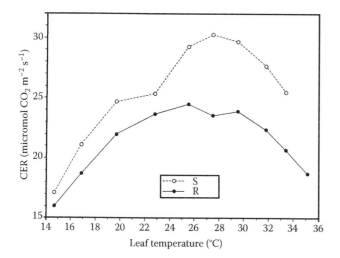

FIGURE 39.12 Carbon exchange rate (CER; μmol CO_2 m⁻² s⁻¹) versus leaf temperature (°C) in s-triazine-resistant (R) and s-triazine-susceptible (S) *Brassica napus*.

their responses to the same air temperature are different (Dekker and Burmester 1993). R leaves generally are cooler, and total conductance to water is greater than in S, probably due to greater stomatal aperture size. The consequence of this is that at higher air temperatures (e.g., 35°C), R leaves photosynthesize at cooler leaf temperatures, leaf temperatures closer to the optimal for both biotypes. In this way, R plants compensate for their high temperature sensitivity.

In addition to other pleiotropic effects of the mutation to the psbA gene in R, R plants appear to be stomatal mutants. R and S *B. napus* biotypes may constitute a good model system to study regulation of stomatal function and the relationship between environmental cues and stomatal behavior.

39.7 RUBISCO ACTIVITY IN R AND S

Studies were conducted to test if there is any regulatory role for ribulose-1,5-bisphosphate carboxylase/oxygenase (Rubisco) in the differential photosynthetic carbon assimilation previously

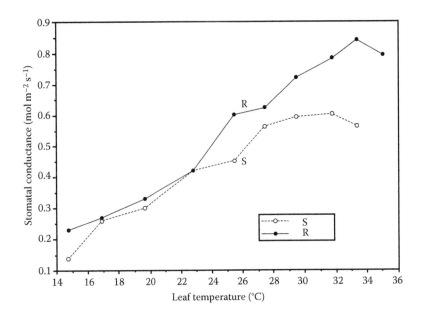

FIGURE 39.13 Stomatal conductance to water vapor (mol m^{-2} s^{-1}) versus leaf temperature (°C) in s-triazine-resistant (R) and s-triazine-susceptible (S) *Brassica napus*.

reported between the R and S *B. napus* (Dekker et al. 1990; Mazen and Dekker 1990). Diurnal oscillations in initial, total, and percent Rubisco activation (initial/total) were studied in two biotypes. Plants were evaluated under dynamic, changing light, conditions representative of a typical daily light cycle in the field (Figure 39.1).

Initial activity started to increase early in the day and reached its peak at midday, reaching a similar rate to that of the total (Figures 39.14 and 39.15, Table 39.6). By late afternoon, initial activity declined to values similar to those of the morning and remained constant overnight. Total Rubisco activity for both R and S remained constant, or decreased

slightly, during the light period (Figure 39.15, Table 39.6). At the very early and late portion of the day, and overnight, total activity was about twice that of the initial activity at those times.

Rubisco percent activity (% activation; initial/total) (Figure 39.16) and initial activity (Figure 39.14) may account for the carbon assimilation differences observed between R and S during the midday (high-light) period: their diurnal rise and fall corresponded to carbon exchange rate (CER) at those times (Table 39.6). CER differences early and late in the day (low light) were not accounted for by differences in any of the Rubisco activity indices (initial, total, % activation).

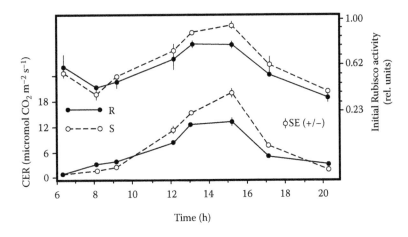

FIGURE 39.14 Carbon exchange rate (CER; μmol CO$_2$ m^{-2} s^{-1}) and initial Rubisco activity (relative units) versus diurnal time (hour of the day).

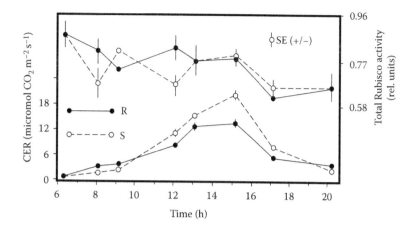

FIGURE 39.15 Carbon exchange rate (CER; μmol CO_2 m^{-2} s^{-1}) and total Rubisco activity (relative units) versus diurnal time (hour of the day).

TABLE 39.6

Rubisco (Initial, Total, % Activation, Initial/Total) and Carbon Assimilation (μmol CO_2 m^{-2} s^{-1}) of R and S *Brassica napus* during the Daily Diurnal Period (Figures 39.13 through 39.15)

	Early	Midday	Late
Rubisco: initial	R = S	S > R (1300–1400h)	R = S
Rubisco: total		No consistent relationship between R and S total activity	
Rubisco: % activation	R = S	S > R (1100–1600h)	R = S
Carbon assimilation	R = S	S > R (1200–1700h)	R = S

Source: Mazen, A.M.A., and Dekker, J., *Weed Sci. Soc. Am. Abstr.*, 30, 163, 1990.

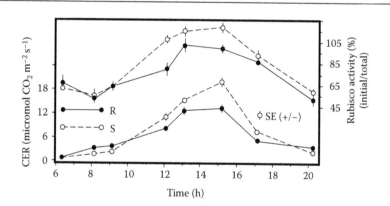

FIGURE 39.16 Carbon exchange rate (CER; μmol CO_2 m^{-2} s^{-1}) and % Rubisco activation (initial/total; relative units) versus diurnal time (hour of the day).

39.8 PHOTOSYNTHETIC REGULATION IN R AND S

Studies were conducted to determine the response of two biotypes of *B. napus* (triazine resistant [R] and susceptible [S]) to different temperatures and gas atmospheres (Dekker and Sharkey 1990, 1992). The response of photosynthetic carbon assimilation and chlorophyll fluorescence quenching to changes in intercellular CO_2 partial pressure (C_i), O_2 partial pressure, and leaf temperature (15–35°C) in R and S were examined to determine the effects of the changes in the resistant biotype on the overall process of photosynthesis in intact leaves. Observations included photosynthetic carbon assimilation (A) using infrared gas analysis techniques and pulse amplitude–modulated chlorophyll fluorescence. Plants were evaluated with strict leaf temperature control and leaf gases (N_2, CO_2, O_2, H_2O).

Several observations were made. Both variants responded to increasing availability of CO_2 (Figure 39.17). S was insensitive to very low atmospheres of O_2 at all temperatures evaluated (15–35°C) (Figure 39.17). S [O_2] insensitivity could indicate that A in that biotype is limited by feedback from sucrose synthesis. R was insensitive to very low O_2 atmospheres at 15°C but was very sensitive to low O_2 at 25°C. R is limited by electron transport limitations at 25°C and therefore responds to low [O_2] (removal of limitations imposed by photorespiration). Regulation in R changes markedly at 15°C, where the limitation to photosynthesis becomes similar to that of S, feedback limitation from sucrose synthesis.

From these observations, photosynthesis can be separated into three categories based on chlorophyll fluorescence quenching and responses to intercellular partial O_2 pressure (pO_2) and C_i (Dekker and Sharkey 1992) (Table 39.7). The first category of photosynthetic response occurs when photosynthetic CO_2 assimilation increases with C_i and q_P (photochemical chlorophyll fluorescence quenching) increases or q_N (nonphotochemical chlorophyll fluorescence quenching) decreases, indicating increasing electron transport with C_i. This category is called Rubisco-limited photosynthesis. When carbon assimilation was Rubisco, limited there was little difference between the resistant and susceptible biotypes. Rubisco activity parameters were similar between the two biotypes.

A second category, called feedback-limited photosynthesis, was evident at 15°C and 25°C above 300 mbar C_i. Characteristics of this category include constant or declining carbon assimilation rates as C_i increased, no pO_2 response or

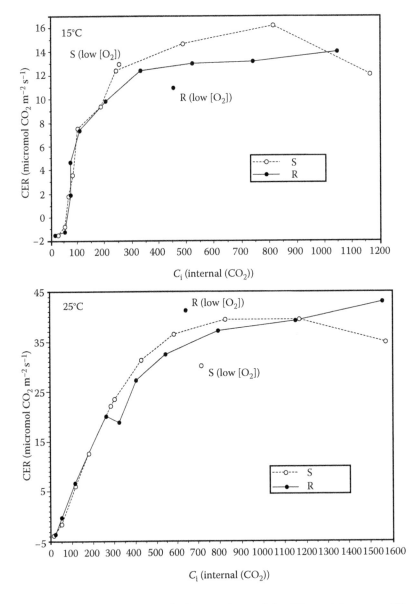

FIGURE 39.17 Carbon exchange rate (CER; mmol CO_2 m^{-2} s^{-1}) versus intercellular CO_2 partial pressure (C_i) at 15°C (top) and 25°C (bottom) in s-triazine-resistant (R) and s-triazine-susceptible (S) *Brassica napus* leaves; low [O_2], 700 μbar intercellular O_2 partial pressure.

TABLE 39.7

R and S Carbon Assimilation Rate (A) Limitations and Regulation Change with Leaf Temperature Changes

Leaf Temp	Carbon Assimilation	O₂ Sensitivity	Limitations/Regulation
25–35°C	S > R		Electron transport rate is limited
25°C		S [O₂] insensitive	Rubisco feedback inhibition from sucrose synthesis
		R responsive to [O₂]	Electron transport is limited
15°C		S [O₂] insensitive	Rubisco feedback inhibition from sucrose synthesis
		R [O₂] insensitive	
	R = S		Electron transport rate is not limited

O_2 stimulation of the carbon assimilation rate, and decline of q_P with increasing C_i. At 25°C, q_N also increased with C_i. This condition has been called feedback-limited photosynthesis (Sharkey 1990) and was most evident at 15°C, less so at 25°C. The importance of feedback at low temperature has been reported before (Sage and Sharkey 1987).

The third category, photosynthetic electron transport-limited photosynthesis, was evident at 25°C and 35°C at moderate to high CO_2. Characteristics of this category include the carbon assimilation rate increasing with C_i consistent with C_i suppression of photorespiration, O_2 inhibition of carbon assimilation, and q_P and q_N independent of C_i. This is believed to represent photosynthetic electron transport-limited photosynthesis (Sharkey 1990). The resistant biotype exhibited much more electron transport-limited carbon assimilation at 25°C and 35°C than did the susceptible biotype. At 15°C, neither the resistant nor the susceptible biotype exhibited electron transport-limited carbon assimilation.

These observations reveal the increasing importance of photosynthetic electron transport in controlling the overall rate of photosynthesis in the resistant biotype as temperature increases. These observations indicate that the primary limitation to photosynthesis changes with changes in leaf temperature and that electron transport limitations in R may be significant only at higher temperatures (Dekker and Sharkey 1990, 1992). At low temperatures (e.g., 15°C), R carbon assimilation rates responded to changing temperatures much quicker than S, a form of cold tolerance in R conferred by increased lipid fluidity and polarity. At low temperature, when the response curves of carbon assimilation to C_i indicated little or no electron transport limitation, carbon assimilation was similar in the resistant and susceptible biotype. With increasing temperature, more and more electron transport-limited carbon assimilation was observed, and a greater difference between resistant and susceptible biotypes was observed. These observations are consistent with previous reports that resistant biotypes are more sensitive to high temperature (Ducruet and Ort 1988; Havaux 1989). Our interpretation of our results is that this is caused by the increasing importance of photosynthetic electron transport in controlling the overall rate of photosynthesis as temperature increases. In summary, the temperature sensitivity of the effect of triazine resistance in *B. napus* is accounted for by the temperature dependence

of electron transport limitations to carbon assimilation, with no need to invoke an additional temperature sensitivity in the resistant biotype.

39.9 EVOLUTIONARY ECOLOGY OF s-TRIAZINE-RESISTANT PLANTS

39.9.1 PLEIOTROPY IN R

A single base-pair substitution at codon 264 in the *psbA* chloroplast gene in the highly conserved photosynthetic apparatus leads to a cascade of changes in the plant's morphology, physiology, and ecological reaction to its immediate environment, a nonintuitive result. With the genetic lesion, a dynamic reorganization of interacting, interdependent plant functional units occurs, a new homeostatic equilibrium among parts. The equivocal nature of our understanding of how photosynthesis differs between R and S, and how it is regulated, has led us to focus experimental efforts on chronobiological, environmental, and regulatory understandings of R under static and dynamic, closely controlled, growth conditions.

Many plant species exhibit an endogenous rhythm of carbon assimilation and stomatal function once entrained in a photoperiod. This rhythm is regulated to some extent independently of the plant's direct response to PPFD (Browse et al. 1981). Work in our laboratory indicated a consistent, differential pattern of chl a fluorescence (F_t), carbon assimilation, leaf temperature, total conductance to water vapor (g), and leaf intercellular CO_2 partial pressure (C_i) (Dekker 1993; Dekker and Burmester 1992; Dekker and Westfall 1987a,b) between S and R *B. napus* over the course of a diurnal light period: R is a chronomutant. Pleiotropic changes in R result in chloroplast and whole-leaf morphology similar to that of low-light, dark-adapted plants. The acquisition of shade-adapted morphology in R is not a plastic response of the phenotype to environment. A consistent, differential pattern of many photosynthetic functions was observed between R and S *B. napus* over the course of a diurnal light period. Photosynthetic superiority of one biotype over another was a function of the time of day, the age of the plant, and the temperature of the environment.

R plants varied in their relative advantage (or disadvantage) over S in terms of carbon assimilation as they aged, and in response to temperature. Carbon assimilation in R was much lower than that in S at high leaf temperatures (e.g., 35°C) when the leaf temperature was closely controlled. These results are consistent with those of others (Ducuret and Ort 1988; Gounaris and Barber 1983; Havaux 1989). When leaf temperature was not directly controlled but air temperature was, R carbon assimilation exceeded that of S at relatively high temperatures (e.g., 35°C air temperature) (Dekker and Burmester 1990a,b, 1992). In both experimental conditions, R leaf stomatal conductances were usually greater than in S. R possesses greater heat tolerance than S due to leaf cooling from greater stomatal conductance and leaf intercellular CO_2 partial pressure. As a consequence, at all important physiological temperatures (10–35°C), R leaves are cooler than S leaves. Stomatal function differentially regulates carbon assimilation in these two biotypes.

R also possesses tolerance to cool temperatures and to changes in temperature, which is conferred by pleiotropic membrane lipid changes. It is hypothesized that these ontogenetic and diurnal patterns of differential photosynthesis may be a consequence of correlative diurnal fluctuations in fatty acid biosynthesis and the dynamic changes in membrane lipids over the course of the light–dark daily cycle, changes in leaf membrane lipids with age (Lemoine et al. 1986), or microenvironmental temperature influences (Dekker and Burmester 1990; Ducruet and Lemoine 1985; Ireland et al. 1988). At relatively cool temperatures it has been hypothesized that the change in lipid saturation of chloroplast membranes could confer cold tolerance to R plants, resulting in greater carbon assimilation rates in R under those conditions (Pillai and St. John 1981).

R biotypes may have a selective, adaptive advantage over S in certain unfavorable ecological niches independent of the presence of s-triazine herbicides: in cool, low-light environments early and late in the day; at high temperatures; and late in the plant's development.

39.9.2   R Adaptation to the Environment and Regulation of Carbon Assimilation

Regulation of photosynthesis in R and S is controlled by many different factors. Limitations in electron transport in R are not the only critical factors in yield losses at the whole-plant level. The pleiotropic effects observed in R result in a new equilibrium between functional and structural components. It is this new dynamic pleiotropic reorganization that regulates carbon assimilation in R. Electron transport limitations are only one possible regulatory point in the photosynthetic pathway leading from light harvesting and the photolysis of water; through ribulose bisphosphate carboxylase/oxygenase; to starch/sucrose biosynthesis, translocation, and utilization. Carbon flux through the leaf is regulated at many points. Electron transport, even in R, is not the only critical regulatory step. In fact, Dekker and Sharkey (1992) have shown that the primary limitation to photosynthesis changes with changes in leaf temperature and that electron transport limitations in R may be significant only at higher temperatures.

The reorganized R plants interact with the environment in a different way than does S. It is this that causes the functional result observed. Under environmental conditions highly favorable to plant growth, S often has an advantage over R. Under certain less favorable conditions to plant growth and stressful conditions, R can be at an advantage over S.

It can be envisioned that there were environmental conditions prior to the introduction of s-triazine herbicides in which R had an adaptive advantage over the more numerous S individuals in a population of a species. Under certain conditions, R might have exploited a photosynthetic niche underutilized by S. These conditions may have occurred in less favorable environments and may have been cool (or hot), low-light conditions interacting with other biochemical and diurnal plant factors early and late in the photoperiod, as well as more complex physiological conditions late in the plant's development. Under these conditions, R survival and continuity could have been ensured at a higher frequency of occurrence than that due to the mutation rate of the *psbA* plastid gene alone, independent of the existence of a postulated plastome mutator (Duesing and Yue 1983).

REFERENCES

Ahrens, W.H. and E.W. Stoller. 1983. Competition, growth rate and CO_2 fixation in triazine-susceptible and resistant smooth pigweed (*Amaranthus hybridus*). *Weed Science* 31:438–444.

Arntzen, C.J. and J.H. Duesing. 1983. Chloroplast-encoded herbicide resistance. In: *Advances in Gene Technology: Molecular Genetics of Plants and Animals*, pp. 273–294, K. Downey, R.W. Voellmy, F. Ahmad and J. Schultz (Eds.). Academic Press, New York.

Arntzen, C.J., D.L. Ditto and P.E. Brewer. 1979. Chloroplast membrane alterations in triazine-resistant *Amaranthus retroflexus* biotypes. *Proceedings of the National Academy of Sciences of the United States of America* 76:278–282.

Barber, J. 1983. Photosynthetic electron transport in relation to thylakoid membrane composition and organization. *Plant, Cell and Environment* 6:311–322.

Beversdorf, W.D. and D.J. Hume. 1984. OAC Triton spring rapeseed. *Canadian Journal of Plant Science* 64:1007–1009.

Beversdorf, W.D., D.J. Hume and M.J. Donnelly-Vanderloo. 1988. Agronomic performance of triazine-resistant and susceptible reciprocal spring canola hybrids. *Crop Science* 28:932–934.

Boardman, N.K. 1977. Comparative photosynthesis of sun and shade plants. *Annual Review of Plant Physiology* 28:355–377.

Browse, J., P.G. Roughan and C.R. Slack. 1981. Light control of fatty acid synthesis and diurnal fluctuations of fatty acid composition in leaves. *Biochemical Journal* 196:347–354.

Burke, J.J., R.F. Wilson and J.R. Swafford. 1982. Characterization of chloroplasts isolated from triazine-susceptible and triazine-resistant biotypes of *Brassica campestris* L. *Plant Physiology* 70:24–29.

Dekker, J. 1993. Pleiotropy in triazine resistant *Brassica napus*: Leaf and environmental influences on photosynthetic regulation. *Zeitschrift Naturforschung* 48c:283–287.

Dekker, J. and R. Burmester. 1989. Mutant weeds of Iowa: s-triazine resistant plastids in *Chenopodium album* L. *Journal of the Iowa Academy of Science* 96:61–64.

Dekker, J. and R.G. Burmester. 1990a. Differential diurnal photosynthetic function in a psbA plastid gene chronomutant of *Brassica napus* L. In: *Chronobiology: Its Role in Clinical Medicine, General Biology and Agriculture*; Pt. B., Cellular and Molecular Mechanisms, pp. 243–251, D. Hayes, J.E. Pauly and R.J. Reiter (Eds.). Wiley-Liss, New York.

Dekker, J. and R.G. Burmester. 1990b. Differential pleiotropy in a psbA gene mutant of *Brassica napus* implies altered temporal photosynthesis and thermal tolerance. *Zeitschrift fur Naturforschung* 45c:474–477.

Dekker, J.H. and R.G. Burmester. 1992. Pleiotropy in triazine resistant *Brassica napus*: Ontogenetic and diurnal influences on photosynthesis. *Plant Physiology* 100:2052–2058.

Dekker, J.H. and R.G. Burmester. 1993. Pleiotropy in triazine resistant *Brassica napus*: Differential stomatal and leaf responses to the environment. In: *Research in Photosynthesis*, Vol. IV, pp. 631–634, N. Murata (Ed.). Kluwer Academic Publishers, Dordrecht.

Dekker, J. and T. Sharkey. 1990. Pleiotropy in triazine resistant *Brassica*: II. Differential responses to [CO₂] and [O₂]. *Plant Physiology Supplement* 93:207.

Dekker, J. and T.D. Sharkey. 1992. Regulation of photosynthesis in triazine resistant and susceptible *Brassica napus*. *Plant Physiology* 98:1069–1073.

Dekker, J. and B. Westfall. 1987a. Circadian oscillations in chlorophyll fluorescence in a triazine-resistant chronomutant of *Brassica napus*. In: *Progress in Clinical and Biological Research*. Advances in Chronobiology Pt. A., pp. 81–93, J.E. Pauly and L.E. Schering (Eds.). Alan R. Liss, New York.

Dekker, J.H. and B. Westfall. 1987b. A temporal phase mutation of chlorophyll fluorescence in triazine resistant *Brassica napus*. *Zeitschrift fur Naturforschung* 42c:135–138.

Dekker, J., R. Burmester, K.C. Chi and L. Jensen. 1987. Mutant weeds of Iowa: s-triazine resistant plasids in *Kochia scoparia*. *Iowa Journal of Research* 62:183–188.

Dekker, J., R. Burmester and T. Sharkey. 1990. Pleiotropy in triazine resistant *Brassica*: IV. Differential responses to temperature. *Plant Physiology Supplement* 93:209.

Dekker, J., R. Burmester and J. Wendel. 1991. Mutant weeds of Iowa: S-triazine resistant *Polygonum pensylvanicum* L. *Weed Technology* 5:211–213.

Ducruet, J.M. and Y. Lemoine. 1985. Increased heat sensitivity of the photosynthetic apparatus in triazine-resistant biotypes from different plant species. *Plant Cell Physiology* 26:419–429.

Ducruet, J.M. and D.R. Ort. 1988. Enhanced susceptibility of photosynthesis to high leaf temperature in triazine-resistant *Solanum nigrum* L. evidence for photosystem II D1 protein site of action. *Plant Science* 56:39–48.

Duesing, J.H. and S. Yue. 1983. Evidence for a plastome mutator (cpm) in triazine-resistant *Solanum nigrum*. *Weed Science Society of America Abstracts* 23:191.

Gounaris, K. and J. Barber. 1983. Monogalactosyldiacylglycerol: The most abundant polar lipid in nature. *Trends in Biochemical Sciences* 8:378–381.

Gressel, J. and L.A. Segel. 1978. The paucity of genetic adaptive resistance of plants to herbicides: Possible biological reasons and implications. *Journal of Theoretical Biology* 75:349–371.

Havaux, M. 1989. Comparison of atrazine-resistant and -susceptible biotypes of *Senecio vulgaris* L.: Effects of high and low temperatures on the *in vivo* photosynthetic electron transfer in intact leaves. *Journal of Experimental Botany* 40:849–854.

Hirschberg, J. and L. McIntosh. 1983. Molecular basis of herbicide resistance in *Amaranthus hybridus* L. *Science* 222:1346–1349.

Holt, J.S. and D.P. Goffner. 1985. Altered leaf structure and function in triazine-resistant common groundsel (*Senecio vulgaris*). *Plant Physiology* 79:699–705.

Holt, J.S., A.J. Stemler and S.R. Radosevich. 1981. Differential light responses of photosynthesis by triazine-resistant and triazine-susceptible *Senecio vulgaris* biotypes. *Plant Physiology* 67:744–748.

Hugly, L. Kunst, J. Browse and C. Somerville. 1989. Enhanced thermal tolerance of photosynthesis and altered chloroplast ultrastructure in a mutant of *Arabidopsis* deficient in lipid desaturation. *Plant Physiology* 90:1134–1142.

Ireland, J.S., P.S. Telfer, P.S. Covello, N.R. Baker and J. Barber. 1988. Studies on the limitations to photosynthesis in leaves of the atrazine-resistant mutant of *Senecio vulgaris* L. *Planta* 173:459–467.

Jansen, M.A.K., J.H. Hobe, J.C. Wesselius and J.J.S. van Rensen. 1986. Comparison of photosynthetic activity and growth performance in triazine-resistant and susceptible biotypes of *Chenopodium album*. *Physiologie Vegetale* 24:475–484.

Jursinic, P.A. and R.W. Pearcy. 1988. Determination of the rate limiting step for photosynthesis in a nearly isonuclear rapeseed (*Brassica napus* L.) biotype resistant to atrazine. *Plant Physiology* 88:1195–1200.

Kunst, L., J. Browse and C. Somerville. 1989. Altered chloroplast structure and function in a mutant of *Arabidopsis* deficient in plastid glycerol-3-phosphate acyltransferase activity. *Plant Physiology* 90:846–853.

Lemoine, Y., J.P. Dubacq, G. Zabulon and J.M. Ducruet. 1986. Organization of the photosynthetic apparatus from triazine-resistant and -susceptible biotypes of several plant species. *Canadian Journal of Botany* 64:2999–30007.

Mattoo, A.K., J.P. St. John and W.P. Wergin. 1984. Adaptive reorganization of protein and lipid components in chloroplast membranes as associated with herbicide binding. *Journal of Cell Biochemistry* 24:163–175.

Mazen, A.M.A. and J. Dekker. 1990. Diurnal oscillations in ribulose bisphosphate activity in triazine resistant and susceptible *Brassica napus*. *Weed Science Society of America Abstracts* 30:163.

McClosky, W.B. and J.S. Holt. 1990. Triazine resistance in *Senecio vulgaris* parental and nearly isonuclear backcrossed biotypes is correlated with reduced productivity. *Plant Physiology* 92:954–962.

Ort, D.R., W.H. Ahrens, B. Martin and E. Stoller. 1983. Comparison of photosynthetic performance in triazine-resistant and susceptible biotypes of *Amaranthus hybridus*. *Plant Physiology* 72:925–930.

Pfister, K., K. Steinback, G. Gardner, and C. Arntzen. 1981. Photoaffinity labeling of a herbicide receptor protein in chloroplast membranes. *Proceedings of the National Academy of Sciences of the United States of America* 78:981–985.

Pillai, P. and J.B. St. John. 1981. Lipid composition of chloroplast membranes from weed biotypes differentially sensitive to triazine herbicides. *Plant Physiology* 68:585–587.

Radosevich, S.R. 1970. Mechanism of atrazine resistance in lambsquarters and pigweed. *Weed Science* 25:316–318.

Ryan, G.F. 1970. Resistance of common groundsel to simazine and atrazine. *Weed Science* 18:614–616.

Sage, R.F. and T.D. Sharkey. 1987. The effect of temperature on the occurrence of O₂ and CO₂ insensitive photosynthesis in field grown plants. *Plant Physiology* 84:658–664.

Schonfeld, M., T. Yaacoby, O. Michael and B. Rubin. 1987. Triazine resistance without reduced vigor in *Phalaris paradoxa*. *Plant Physiology* 83:329–333.

Sharkey, T.D. 1990. Feedback limitation of photosynthesis and the physiological role of ribulose bisphosphate carboxylase carbamylation. *Botanical Magazine–Tokyo Special Issue* 2:87–105.

Steinback, K., K.L. McIntosh, L. Bogorad, and C. Arntzen. 1981. Identification of the triazine receptor protein as a chloroplast gene product. *Proceedings of the National Academy of Sciences of the United States of America* 78:7463–7467.

Stowe, A.E. and J.S. Holt. 1988. Comparison of triazine-resistant and -susceptible biotypes of *Senecio vulgaris* and their F1 hybrids. *Plant Physiology* 87:183–189.

Thornhill, R. and J. Dekker. 1993. Mutant weeds of Iowa: V. S-triazine resistant giant foxtail (*Setaria faberii* Hermm.). *Journal of the Iowa Academy of Science* 100:13–14.

Tranel, D. and J. Dekker. 2002. Differential seed germinability in triazine-resistant and -susceptible giant foxtail (*Setaria faberii*). *Asian Journal of Plant Sciences* 1(4):334–336.

van Oorshot, J.L.P. and P.H. van Leeuwen. 1984. Comparison of the photosynthetic capacity between intact leaves of triazine-resistant and -susceptible biotypes of six weed species. *Zeitschrift fur Naturforschung* 39c:440–442.

Vaughn, K.C. 1986. Characterization of triazine-resistant and susceptible isolines of canola (*Brassica napus* L.). *Plant Physiology* 82:859–863.

Vaughn, K.C. and S.O. Duke. 1984. Ultrastructural alterations to chloroplasts in triazine-resistant weed biotypes. *Physiologia Plantarum* 62:510–520.

40 Next Evolution of Agriculture
A Review of Innovations in Plant Factories

Brandon F. Merrill, Na Lu, Toshitaka Yamaguchi, Michiko Takagaki,
Toru Maruo, Toyoki Kozai, and Wataru Yamori

CONTENTS

40.1 Introduction 723
40.2 Plant Factories 724
40.3 Merits/Demerits of Plant Factories...726
    40.3.1 Plant Factories with Sunlight...726
    40.3.2 Plant Factories with Sunlight and Supplemental Light 727
    40.3.3 Plant Factories with Artificial Lighting...727
40.4 Control Systems + Innovations...729
    40.4.1 Temperature 729
    40.4.2 Humidity 731
    40.4.3 CO$_2$...732
    40.4.4 Light..732
        40.4.4.1 Light and Photosynthesis 733
        40.4.4.2 Supplemental Lighting in Plant Factories...734
        40.4.4.3 Light Quality and Light Source 735
40.5 Conclusion 737
Acknowledgments...737
References..737

40.1 INTRODUCTION

Agriculture is the foundation of modern society. The ability of a small portion of society to produce food for the whole allows humanity the leniency to study the arts and sciences, which have led to our current era. It has also supported humanity's growth far beyond populations naturally attainable; by mid-2013, the world population reached approximately 7.2 billion people. Even assuming that fertility levels will continue to decline, the world population is still expected to reach 9.6 billion in 2050 and 10.9 billion in 2100, according to the Food and Agriculture Organization (FAO) medium-variant projection (United Nations 2013). In addition to worldwide population growth, per capita food consumption will also increase. The daily caloric intake of the population overall will rise, though countries currently with high caloric intake will level off or slightly decrease consumption. In 2007, worldwide food consumption per person was about 2770 kcal per person per day; by 2050, this will increase to a worldwide average of just above 3000 kcal per person per day (Alexandratos and Bruinsma 2012).

In order for agriculture to sustain this growth, production increases must improve year on year. This can be done by either extending farmlands to produce more crops on more land or improving yields on current farmlands. However, in some countries, expanding farmland is not viable. Pragmatically speaking, island countries, small countries, and countries with suboptimal soils would be competing with the residential sector for land. Other factors that influence the choice between expansion and yield increases should also be considered: resource consumption—particularly the water supply—and seasonal weather limitations on the growing season. From an environmental perspective, a focus on agriculture expansion would cause greater greenhouse gas emissions than a focus on agriculture intensification (Godfray et al. 2014). All this considered, yield improvements should be prioritized. Plant factories provide the opportunity to improve overall productivity while addressing the considerations mentioned.

Plant factories are a new facility for growing plants under a controlled environment (e.g., temperature, light, CO$_2$ concentration, and nutrient source) and are able to obtain high-yield and high-quality production year-round. Plant factories have become popular in Japan, a country with limited available land where farmland expansion is not viable. East Asia is one of two areas in the world where farmland expansion will not occur, and all yield increases will need to come from crop intensification (Table 40.1). In fact, in 2012, Japan had already exceeded its suitable farmland—using about 12.5% of its land area for agriculture even though only 11.6% of its land is classified as *suitable as farmland* (World Bank 2014a,b). Even if the farmland area increases, it will likely

TABLE 40.1

Increases in Agricultural Yield from Farmland Expansion and Intensification

	Arable Land Expansion		Increases in Cropping Intensity		Yield Increases	
	1961–2007	2005/2007–2050	1961–2007	2005/2007–2050	1961–2007	2005/2007–2050
All Developing Countries	23	21	8	6	70	73
Sub-Saharan Africa	31	20	31	6	38	74
Near East/North Africa	17	0	22	20	62	80
Latin America and the Caribbean	40	40	7	7	53	53
South Asia	6	6	12	2	82	92
East Asia	28	0	–6	15	77	85
World	14	10	9	10	77	80

Source: Data were modified from Alexandratos, N., and Bruinsma, J., World agriculture towards 2030/2050, the 2012 revision, ESA working paper, FAO, Rome, 2012.

Note: Data were derived from 34 crops grown within 105 different countries, grown in rainwater-fed and irrigated cultivation conditions. Growth in production is due to increases in crop yield and land expansion, which resulted in a greater harvested area. Around 80% of the projected growth in developing countries would come from intensification of yield (73% increases) and higher cropping intensities (6% increase). Of note, North Africa and East Asia are not expected to have any farmland expansion to assist in yield improvements.

not significantly increase the overall production—leaving Japan and similar countries in need of yield improvements through crop intensification. Over the past 10 years, Japan has actually decreased the acreage of several important crops; in 2004, tomatoes claimed 13,100 ha, and lettuce 21,800 ha, but by 2013, tomatoes retained only 12,100 ha, while lettuce retained only 20,900 ha (Ministry of Agriculture, Forestry and Fisheries of Japan [MAFF] 2013). However, as long as yields are good, a reduction in space is acceptable. Indeed, since 2004, Japan has increased its tomato yield by 8% and its lettuce yield by 14% (MAFF 2013). The typical Japanese farm, as of 2013, produces 10.2 tons of tomatoes per 10 ares in the spring (1/10 of a hectare), though a sparse 4.25 tons in the autumn. In regard to lettuce, yields are 2.6 tons per 10 ares in the spring and rise to 3.1 tons in the fall and winter (MAFF 2013). However, these improvements are modest when compared to the potential yields derived from plant factories.

Plant factories can produce yields significantly higher than currently obtainable yields in traditional agriculture. The technology in the plant factories that utilize sunlight was developed in the Netherlands and other parts of Europe. Dutch-type plant factories have achieved an average of 70 tons of tomatoes per 10 ares. Recently, plant factories in Chiba University in Japan have achieved around 50 tons of tomatoes per 10 ares (Japan Plant Factory Association [JPFA] 2013), while the normal range of tomato yields in Japanese plant factories is within 25–50 tons per 10 ares. That is roughly two to five times higher production over traditional farming. Various plant management styles and technologies are being researched throughout Japan to further improve all aspects of plant factory production. Lettuce growth within plant factories using artificial light has achieved much greater yields than tomatoes. Plant factories have developed cultivation

systems to grow many levels of lettuce vertically, rather than expanding horizontally as a farm would. The plant factory in Chiba University in Japan can produce 197 tons of lettuce per 10 ares—a yield increase 75 times greater than field production (JPFA 2013). The technology for plant factories using artificial light is in progress in Asia, especially in Japan. Plant factories achieve these increases through resource optimizations and higher degrees of control over the plant environments. One impressive result of these optimizations is the water usage; the tomato yields mentioned were achieved using only 10% of the irrigation water that a field would use, while lettuce saves even more, using only a few percent of the water usage in a field (Kozai 2013a). In this chapter, various types of plant factories, as well as various new plant management styles and technologies, are summarized.

40.2 PLANT FACTORIES

Plant factory is an umbrella term for several types of facilities that grow plants. These facilities are categorized based on their main light energy source for plants, with three current types: (1) greenhouses using sunlight (typical Dutch-type greenhouse), (2) greenhouses that supplement sunlight with artificial lighting, and (3) closed-growth rooms with fully artificial lighting. In this chapter, these are referred to as (1) plant factories with sunlight, (2) plant factories with sunlight and supplemental light, and (3) plant factories with artificial lighting, respectively (Figure 40.1). Plant factories with sunlight make use of the sunlight, as per traditional agriculture, to grow plants. However, there are inconsistencies with sunlight if one is growing year-round, most notably with the day length. Therefore, some plant factories install artificial lighting in the greenhouse to create a more consistent light environment for the plants. The management of these light

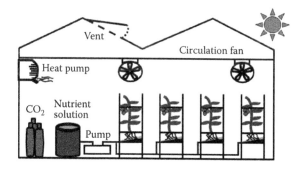

(a)

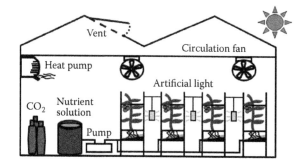

(b)

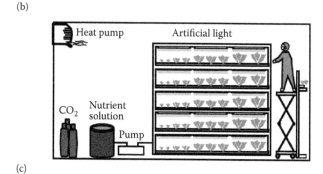

(c)

FIGURE 40.1 Three current types of plant factories: (a) plant factories with sunlight, the typical Dutch-type greenhouse; (b) plant factories with sunlight and supplemental light; and (c) plant factories with artificial lighting.

systems requires balancing the energy cost of the lights with the benefit toward plant growth. Along the same lines, plant factories with artificial lighting completely do away with sunlight, using solely artificial lights, in order to bring a much more consistent light environment to plants. However, in each of these systems, there are two foundational components in plant factories: environment-isolating structures, like greenhouses, as well as soilless systems.

All plant factories require a structure to separate them from the external environment. In the case of plant factories with sunlight and those with sunlight and supplemental light, the structure is a greenhouse for sunlight to enter and reach the plants. In the case of plant factories with artificial lighting, sunlight is not utilized, and the structure can be a solid, well-insulated building. Either of these structures presents numerous benefits for growing plants. The structures separate the plants from the exterior environment so that crops are protected from adverse weather conditions such as heavy rains or strong winds. In addition, they act as the primary barrier to pests, significantly reducing the number of pests within the plant factory. As the structure surrounds the plants, it can provide support for them and enable better use of the vertical space above them if cultivation methods allow it. Most importantly, these structures create a separated interior environment; within these, the environment can be manipulated and controlled to remain suitable for plants. Rather than only extending the harvesting season as had been done in the past, plant factories use equipment to control the environment to an extent that allows continued cultivation throughout the year. In order to do this, plant factories manage the temperature and airflow; regulate carbon dioxide levels and humidity; and exclude, minimize, or otherwise manage pests and diseases. Lighting is also controlled, though to differing degrees of management depending upon the type of plant factory in use. In addition, plant nutrition can be optimized through the use of soilless systems.

Soilless systems are the result of innovations in plant nutrition and nutrient delivery systems, which remove the need for soil to grow plants. Soilless systems can be separated into two broad categories: aggregate culture and liquid culture (Olympios 1999). The first category uses materials such as rock wool, cocopeat, oasis, or other solid materials to anchor the plant roots and absorb the nutrient solution, like a sponge, for it to be readily available to the roots. The latter category does not use such materials, and the nutrient solution directly flows around the roots. Both have strengths and weaknesses, but both use soluble fertilizers to dissolve the minerals into the water, creating a nutrient solution, which is then delivered to the plant roots. The complexities between minerals and soil are removed from these systems, allowing for more precise formulations of minerals in the nutrient solutions, and therefore, nutrient compositions can be customized for different crop needs (Gericke 1929; Hanan 1998; Hoagland and Arnon 1950). In plant factories, these soilless systems are all closed loop, meaning that any unused nutrient solution is recirculated back into the system. In traditional intensive agriculture farms, both fertilizer and water will leach into the soil, causing groundwater pollution and the waste of said resources (Thompson et al. 2007). New techniques such as subirrigation and drip irrigation can lower this leachate but do not eliminate it. However, soilless systems can recapture the unused nutrient solution, preventing groundwater pollution as well as allowing the unused solution to be recirculated into the system (Gorbe and Calatayud 2010; Grewal et al. 2011). By monitoring the nutrient solution, growers can analyze the recirculating solution to gauge the amount of water and nutrients used by plants and alter any water and nutrient additions into the system. In some circumstances, it may be preferable to the grower to discard the solution in the system and start afresh. In such situations, plant factories can run the nutrient solution through the system to remove all plant-usable nutrients from the solution and then flush the water out of the system (Massa et al. 2010). The management may assist the grower in controlling the solution as well as preventing the groundwater pollution caused by flushing nutrients into the soil.

Technologies and various equipment within plant factories further improve the environmental control in order to optimize plant growth. These plant-growth-optimized environments create higher yields while simultaneously pushing resource-use efficiencies and thus reducing waste (Kim et al. 2000). The complex interactions required to create such optimized environments also require computer control systems to assist growers in fine-tuning and automating the components in the greenhouse. The purpose of this chapter is to introduce the concept of plant factories, to reflect on recent advancements made in this field, and to consider the potential research focuses that could further improve plant factories in the future.

40.3   MERITS/DEMERITS OF PLANT FACTORIES

As with any technology, plant factories have disadvantages alongside their many merits. Many of these are common to all plant factories; however, due to plant factories having three styles, there are benefits and weaknesses belonging to the individual styles as well. This section will summarize these common strengths and weaknesses first before covering the type-dependent ones. In order to make the explanations easier, this section will focus on the two most popular crops in plant factories: tomatoes for sunlight and supplemental-light factories, and lettuce for artificial-light factories.

As mentioned in the introduction to plant factories, the structure surrounding the growing area provides a primary defense against pests and diseases. However, the environment within the plant factory is as suitable for pests as it is for plants, and it is inevitable for some pests to attempt entry into plant factories. In plant factories that are more open to the exterior environment, as sunlight and supplemental-light plant factories are, integrated pest management (IPM) techniques are used to minimize pest levels and pest damage while also minimizing pesticide use. IPM will not be discussed in detail in this chapter, but in essence, it is a combination of management techniques and living organisms to reduce pest populations to levels that ensure produce will be safe and sellable. In closed systems like artificial-light plant factories, extraordinary effort is put into preventing pest entry into the growing room, creating a clean room where the plants are grown. Thus, in these systems the crops are spared from pests and do not need any pesticides at all (Kozai 2013b). In either situation, plant factories have the potential to minimize the use of dangerous chemicals, a benefit for making high-production agriculture more environmentally friendly. Plant factories also enable CO_2 supplementation, which is not viable in the field, due to gas diffusion. In plant factories, the CO_2 can be injected into the growing area, resulting in increased plant growth. The management of CO_2 changes depending on which system is used, but each can increase and utilize CO_2 above the ambient levels in the atmosphere.

An additional benefit to the closed environment created by plant factories is the resource-use improvements. Generally speaking, most plants retain only 1–5% of the water they absorb, the rest being transpired. This means that roughly 95% of the water uptake by plants ends up as water vapor in the air. In open fields, this water is diffused into the atmosphere, but plant factories isolate this water vapor due to the structure around the plants. Depending on the technology used by the plant factories, a portion of the air moisture can be captured and recycled into the irrigation system. With such a high percentage of the water used becoming water vapor, this capture-and-recycle method is a significant boon to water-use efficiency. Additionally, there is potential to manage irrigation water in correlation to the level of transpiration, optimizing the amount of water used to only what is necessary for plants at any one time.

Common to all plant factories is the higher capital investment for equipment. To enable the highly controlled environments, various technologies must be purchased and used. All of this equipment also means higher operational and production costs, especially in the typical off-season as the environment must be highly managed to maintain optimal conditions. Structural components block rainwater from entering the growing area (a necessary sacrifice for maintaining a closed environment), which prevents rain from being a free source of irrigation. The rainwater can be captured and stored for later use, but this extra effort requires further infrastructure costs and is an additional cost to achieve something nature normally provides for free. The interactions of the equipment and control also necessitate computer control as the complexity without control systems would restrict management to only the highest echelon of skilled growers. Even so, control systems are not yet perfected and are expensive, and grower expertise is still necessary. The expertise needed to manage a plant factory is a different skill set from normal agriculture and requires more training, and currently, there is a limited plant factory–trained workforce. Expertise with plant physiology is also necessary, as soilless systems give much greater freedom of nutrient control than typical agriculture—nutrient solutions must be calculated and customized for each crop (Gorbe and Calatayud 2010). The high efficiency of water and nutrient use requires the nutrient solution to be recirculated, which adds cost and complexity to the system as this water must first be disinfected (and preferably analyzed) before it can be added back into the system.

40.3.1   Plant Factories with Sunlight

Plant factories with sunlight use sunlight as the light energy source for plants. As such, these plant factories do not need to pay for the costs of this light energy as it is freely available. This style uses various technologies for environmental control: heaters, CO_2 generators or injectors, nutrient management systems, air circulation fans, ventilation, heat and shade curtains, and coolers. Many of the technologies in plant factories with sunlight have been used in greenhouse environments prior to the rise of plant factories, providing a base of reference for management strategies. Plant factories with sunlight are also partially open to the outside environment by way of ventilation. This openness provides an inexpensive way to control temperatures by allowing heat to escape the greenhouse through the vents; opening the roof vents allows the rising heat to passively ascend out of the greenhouse (Hanan 1998). The ventilation also improves airflow by allowing wind to enter primarily through the side vents. Temperature can be solely passively managed as discussed, or it can be used in combination with active controls like heaters/coolers, shade curtains, and circulation fans.

However, the openness of plant factories with sunlight is also a weakness. These openings are entry points for pests and insects, and screens must be purchased and attached along any possible point of entry. In order to minimize pesticide use, IPM techniques are used to minimize pest populations. Various preventative techniques are used; however, pests inevitably find an entry point. Thus, plant factories with sunlight require more direct and involved IPM methods. Another weakness of plant factories with sunlight is similar to conventional agriculture: control depends upon the location of the plant factory. Control cannot be completely standardized as some locations need more winter heating or more summer cooling (Sethi and Sharma 2008). In comparison to the other plant factory types, a pitfall for sunlight plant factories is the inability to control light (aside from blocking it), and they must depend entirely on the sun. In situations where light levels are not optimal, sunlight plant factories can only adjust management to align with the change. This weakness is the reason why supplemental-light plant factories were created.

40.3.2 PLANT FACTORIES WITH SUNLIGHT AND SUPPLEMENTAL LIGHT

Plant factories with sunlight and supplemental light were born from the desire to have more control over lighting. These plant factories resemble plant factories with sunlight in every respect, except that they employ secondary light sources. These lights are installed to provide supplemental lighting when sunlight is suboptimal. When there is a reduction in available light plant yields also decrease (Heuvelink 2005). In plant factories with sunlight, these reductions are unavoidable. However, by installing supplemental lighting, these lower-light circumstances can be averted to ensure stable yields. The most obvious light changes occur with the seasons where the day length shortens in the winter. By using these secondary lights to restore the day length, plants are better able to regulate flowering (Heuvelink 2005). In many plants, flowering decreases during the winter due to less available lighting. By adding extra light, flowers can better develop, leading to more fruit produced. Another circumstance where supplemental lighting could be beneficial is during poor or cloudy weather, where light intensity fluctuates or decreases, since these suboptimal conditions could affect photosynthesis and result in reductions in plant biomass. In these situations, these secondary lights supplement the existing sunlight to maintain consistent light requirements for better yields (Dorais and Gosselin 2002). However, there are reasons for supplemental lighting beyond fluctuations of sunlight. As plants grow, the lower leaves of plants are always in the shadows of the upper leaves, decreasing their ability to perform

photosynthesis. Supplemental lighting can be positioned to provide lighting to these leaves and prevent performance loss due to lower lighting (Heuvelink 2005). All of these methods increase potential photosynthesis in the crop and contribute to better yields. From a business perspective, while the increased yields are beneficial, the stability in yields is equally important. Traditionally, to maintain yields of a crop in the winter, more space was necessary to cultivate the crop because of the decrease in production. However, supplemental lighting can create stable, year-round production without expanding the growing area by maintaining the necessary light intensity (Hanan 1998). Without the need to overestimate crop size to account for winter production decreases, growers can save land space for other uses.

However, while these plant factories are able to provide more consistency, the supplemental lights require energy for power. In exchange for the mentioned benefits, energy costs rise. Research on supplemental light is still needed to find the best management strategies; as it is not a natural occurrence, there are no models we can learn from in nature. Timing of lighting, duration, position in crop, distance from canopy, light intensity, and light wavelength are all in need of continued research and investigation. Until more expertise in this technique is gained, supplemental-light strategies will remain unoptimized and raise production costs and energy usage more than necessary. It is important to balance these energy costs with the yield increases, or supplemental lighting becomes a burden rather than a beneficial technology.

40.3.3 PLANT FACTORIES WITH ARTIFICIAL LIGHTING

Plant factories with artificial lighting completely replace sunlight with artificial sources of light. As sunlight is no longer needed, these factories can be completely closed off from the outside environment. For these plant factories, the building needs to be much more highly insulated than typical facilities. Buildings generally require an air exchange rate of 30% per hour; plant factories with artificial lighting are insulated to the point that these air exchanges are closer to 1% per hour (Kozai 2013b). This isolation from the outside means that, more so than plant factories with sunlight, outdoor weather has no effect on the interior growing area. A snowstorm or a heat wave will influence temperatures in plant factories with sunlight but have no effect, either in temperature or humidity, in these plant factories with artificial lighting. This isolation means that in these plant factories, there is much more control over the environment. Humidity control, a function of plant transpiration, can be more easily managed due to the closed nature of these plant factories. While field crops cannot viably capture water vapor and plant factories with sunlight can only capture a portion due to the loss through ventilation, plant factories with artificial lighting are so insulated that water vapor cannot escape. This makes it much easier to capture all this water vapor for reuse as irrigation. By capturing all the water vapor, plant factories with artificial lighting achieve around 50 times greater water-use efficiency compared to sunlight plant factories

TABLE 40.2
Resource-Use Efficiencies in Artificial Light Plant Factories versus Sunlight Plant Factories

Use Efficiency	CPPS	Greenhouse with Ventilators Closed and Enriched CO_2	Greenhouse with Ventilators Open	Theoretical Maximum Value for CPPS
Water-use efficiency	0.95–0.98	N/A	0.02–0.03	1.00
CO_2-use efficiency	0.87–0.89	0.4–0.6	N/A	1.00
Light-use efficiency of the lighting modules	0.027	–	–	About 0.10
Light-use efficiency by the plants	0.032–0.043	N/A	0.017	About 0.10
	0.05		0.003–0.032	
Electricity-use efficiency	0.007	–	–	About 0.04

Source: Data from Kozai, T., *Proc. Jpn. Acad. Ser. B Phys. Biol. Sci.*, 89, 447–461, 2013.

Note: Water-use efficiency (WUE) is over 30 times better in a closed plant production system (CPPS) over greenhouse production. CO_2-use efficiency (CUE) can double that of greenhouses as well. Because of the artificial lighting in CPPS, light-use efficiency by plants (LUE_P) increased over that in greenhouses due to the ability to control lighting. However, due to the artificial lighting, new variables are introduced: light-use efficiency of the lighting modules (LUE_L) and electricity-use efficiency (EUE). The final column shows that improvements can still be made in CPPS, particularly in the efficiency of electrical use.

(Table 40.2; Kozai 2013b). When compared to field cultivation, the water-use efficiency is even greater. In the same way that humidity cannot significantly escape the factory, CO_2 is equally well contained. Without the worry of diffusion to the exterior atmosphere, the CO_2 concentration can be raised to higher levels than other plant factories. CO_2 within these plant factories can easily quadruple ambient concentrations as the buildings' insulation prevents gas losses. The degree of isolation from the outside also serves as a greater barrier for the entry of pests and disease. By extension, pesticide use is significantly reduced as well. These plant factories put an increased emphasis on keeping the growing room void of pests, attempting to create a clean room.

At the time of this writing, plant factories with artificial lighting are primarily growing leafy vegetables such as lettuce. The compact size of these vegetables means that very little vertical space is necessary for their production and creates the opportunity to use the extra vertical space more effectively. These plant factories create shelves for the plants, with each shelf spaced sufficient for the height of the mature plant. Each shelf receives proper light intensity and airflow, and nutrients are delivered via a nutrient solution. This technique means that for the same space of land, plant factories with artificial lighting have multiple *fields* of crops, significantly increasing the potential yields in the same acreage as a farm. Since plant factories with artificial lighting are isolated from the exterior environment, environmental control can be standardized. Regardless of where the plant factory is built, the growing area of any artificial-light plant factory can be maintained with the same parameters and guidelines (Kozai 2013a). While one weakness of the plant factories with sunlight is its inability to provide complete standardization of environmental control, plant factories with artificial lighting can achieve such standardization because it is not restricted by external factors.

Plant factories with artificial lighting have numerous advantages due to being so closed; however, these plant factories tend to have greater energy inputs. As the free energy from sunlight has been completely excluded from these systems, new energy must be input into the system, causing a much higher cost of lighting in these factories. Typically, fluorescent lighting is used as the light source, due to its wide range in the light spectrum. However, these lights are not the most efficient in power usage, and the electricity cost from these represents nearly 25% of the production costs (Kozai 2013b). A wide variety of leafy vegetables can be grown together under fluorescent lighting as it has a broad light spectrum. However, there is another light source increasingly becoming popular: light-emitting diodes (LEDs). These lights are about 60% more power efficient, but require expertise to optimize the wavelength spectrum for a particular crop and are initially more expensive (Kozai 2013b). LEDs have a very narrow spectrum and, as such, do not support a variety of plants grown together. Each plant will have its own preferred spectrum, requiring a customized spectrum for each to achieve quality production. However, depending on the crop growing in the plant factory, light intensity may be very low. Mushrooms, which typically grow in the dark, would consequently have better efficiencies within artificial-light plant factories as the running cost of lighting is minimal. High-density cultivation within these plant factories offsets the costs of lighting, but a reduction in energy costs is necessary for their future. Another difficulty with these plant factories comes from capital investment: the highly insulated building requires higher construction costs to insulate effectively and efficiently. Finally, while this type of plant factory has very low levels of pests or disease, it is critical to prevent any entry of either. The environment is essentially void of competition, and should a pest gain entry, it would quickly explode in population and require significant time and effort to purge. The same applies for diseases, especially those that can transmit in the nutrient solution. The connectedness of the nutrient solution with all the plants in a shelf, combined with the high density of plants in these systems, means that a disease could spread rapidly if not quickly controlled. As in all plant factories, the need to disinfect nutrient solutions is not optional.

While there are numerous weaknesses in plant factories, it is fortunate that many of these weaknesses can be solved with continued research and improvement in management expertise. Several of the current issues already have research underway. In many instances, there are ongoing research projects that present several possible solutions for the issue. Current technologies and innovations for plant factories will be summarized in the following section.

40.4 CONTROL SYSTEMS + INNOVATIONS

Food production is determined by plant growth, with photosynthesis being the driving mechanism. It has been shown that photosynthetic reaction is affected by various environmental conditions such as light intensity (Yamori et al. 2010a, 2015), temperature (Yamori et al. 2005, 2006, 2010b), CO_2 concentration (Yamori et al. 2005; Yamori and von Caemmerer, 2009), and so on. By controlling the environment (e.g., light, temperature, humidity, airflow, CO_2 concentration, water, and nutrient solution), plant factories can achieve higher crop yield and higher-quality production. To do this, plant factories make use of various technologies to manage each environmental variable. Current environmental control is not perfect, but new technologies and new techniques are constantly undergoing research to improve the situation. Each environmental factor will be separated into its own subsection for ease of understanding, but it is vital to understand that these factors generally interact with each other and control of each is not isolated, as the subsections might suggest.

40.4.1 Temperature

Early protected agriculture often revolved around offsetting the changing seasons, particularly regarding changing temperatures. As a result, over several centuries, techniques to warm or cool the plant canopy have been polished.

Sunlight is the primary source of energy for plants. However, sunlight also creates large amounts of heat. In greenhouses, which are transparent and allow the sunlight to enter, heat also enters the greenhouse and continues to build up because it is trapped inside by the structure. One of the simplest methods to remove heat is to simply add an escape route via ventilation (Hanan 1998). Heat will naturally rise and escape through the vent openings, replaced by cooler air. Shade curtains can also be used to slow the heat buildup in the greenhouse by simply blocking some light from entering. Used together, shade curtains and vents are enough to manage the temperature in milder climates that do not rise above 33°C (Sethi and Sharma 2007). Both methods work well together and are cost-effective, leading to their wide adoption.

However, once passing that 33°C limit, additional technology is necessary.

Water is an effective heat sink. Within a cubic meter of space, just 1 g of water can evaporate and absorb enough heat to reduce the air temperature by 2.5°C (Sethi and Sharma 2007). This efficiency, in addition to the relative ease of obtaining water, has made it popular for use in cooling. Two common technologies already make use of this concept: fog-cooling systems (Figure 40.2) and pad-and-fan systems (Figure 40.3).

Pad-and-fan systems work by using large fans on one side of the growing area to pull air through the room (Figure 40.3). On the far side is a porous, water-saturated pad where outside air can enter. The fan pulls air through the pad, cooling the outside air as it passes through the water and pulls the now-cold air through the growing area. It then exhausts this air outside once it reaches the fans (Hanan 1998). This form of cooling achieves 5–10°C temperature reductions (Chandra et al. 1989; Kittas et al. 2003), though computer analysis indicates an opportunity for further temperature improvements (Landsberg et al. 1979). The downside to this technology is that there is a temperature gradient in the growing room as the air warms up over time—in a 60 m long greenhouse, there was an 8°C temperature gradient from the pad side to the fan side (Kittas et al. 2003). Finally, due to the limitations of directly cooling air via water, these systems cannot lower

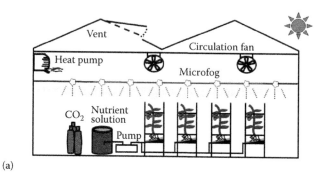

(a)

(b) (c) (d)

FIGURE 40.2 Typical fog-cooling system. These systems spray a very fine fog above the plant canopy. The fog uses the sun's energy to evaporate, removing heat from the greenhouse. The particle size of the fog is important to prevent plants from getting wet. In the picture shown, the Japanese company DAISEN uses 20–30 μm sized fog, sprayed at the high pressure of 4–6 MPa (the fogging nozzles are designed by Japanese company IKEUCHI). Control is based on the VPD within the growing area. (a) Microfog system, (b) nozzle, (c) without microfog, and (d) with microfog.

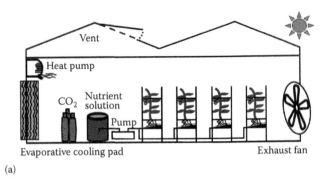

(a)

(b)   (c)

FIGURE 40.3 Typical pad-and-fan system. Along one wall are water-saturated pads through which air can pass. On the opposing side, large fans pull air through the pads and through the greenhouse. The water in the pads cools the air passing through it and pulls the cool air through the plant canopy. In the picture, the Japanese company ZEN-NOH keeps single-truss tomato plants cool by pulling the colder air from the pads into the growing area. (a) Pad-and-fan system, (b) evaporative cooling pad, and (c) exhaust fan.

temperatures past wet bulb temperatures. In environments that need to reduce temperatures further, combining the system with other cooling methods is advisable (Davies 2005).

As mentioned, pad-and-fan systems are limited by the wet bulb temperature. Thus, lowering the wet bulb temperature would enable improved cooling efficiencies. An experiment designed for this purpose proposed using a desiccant in the system to dry the input air. A new pad with a desiccant in it is installed in front of the cooling pad so that air must first pass through this pad to dry via passing through the desiccant before being cooled. The drier air lowers the wet bulb temperature, which allows for improved cooling in the evaporative pads. The result of this system is a successful reduction in temperature, lowering the temperature 5°C more than the typical pad-and-fan setup (Davies 2005). Another improvement to the system was to the build material. An experiment undertaken in Saudi Arabia sourced local materials to replace the commercial pads (made of Aspen wood excelsior fiber). Of the tested materials, luffa gourds showed improved efficiencies, improving the cooling efficiency over the commercial pad from 49.1% to 55.1% efficiency. It also demonstrated improved resistance to cooling degradation caused by salt or mold buildup. The luffa gourds were also price competitive with the Aspen fiber (Al-Sulaiman 2002). Both of these experiments prove that pad-and-fan systems still have room for optimizations and improvements.

The other method that makes use of water for cooling is fog cooling, which directly removes the heat buildup in the growing area. It does so via fine water particles that are sprayed above the canopy and absorb the heat in the air as they evaporate. This method of cooling can reduce temperatures by 3–5°C while maintaining temperature uniformity within 5°C (Arbel et al. 1999; Montero et al. 1989). The water particle size is critical, as larger water droplets increase the risk of wet foliage and consequent fungus growth. To prevent this, water droplets for fog cooling are between 2 and 60 nm in diameter (Katsoulas et al. 2001). While the overall cooling of fog systems is lower than pad-and-fan systems, the temperature uniformity is desirable. Research has been focusing on improving the efficiency of this technology to make the cooling effects more in line with or better than pad-and-fan systems. Several ideas have been investigated to improve pad-and-fan systems.

Fog cooling is not a complex idea, but optimal use requires complex environmental interactions. As such, most of the improvements in fog cooling recently have been in control and automation (Figure 40.2). Because of the complexities in fog cooling, coming from the interaction of various environmental factors, it was often used inefficiently. However, a model has been developed to reliably predict greenhouse parameters and to improve the control of fogging times and intervals (Abdel-Ghany and Kozai 2006). This model allows a grower to manage fog cooling based on environmental data to optimize fog-cooling usage. To allow for a more automated approach, logic controllers have been designed to account for the interactions between temperature and relative humidity (RH) and use this to control fogging durations to prevent wasteful use when conditions for use are poor (Hayashi et al. 2007). While temperature and RH are important, other methodologies have focused on vapor pressure deficit (VPD) instead of RH (Baille et al. 2006; Garcia and Medrano 2011; Meca et al. 2006). These experiments all focused on time and duration of fog application based on a set pressure from the system. However, it has been shown that the fogging pressure could affect efficiency. High-pressure fogging (6.89 MPa) was 28% more efficient in cooling than low-pressure cooling (405 kPa), and it had 64% better evaporation efficiency (Katsoulas et al. 2001). The improvement in evaporation efficiency results in lower water usage. In prior systems, only 50% of the fog contributed to temperature control, meaning half was wasted. To improve the water-use efficiency, another experiment demonstrated that the use of fans could improve water utilization. Upward-blowing fans were added to the fogging system to increase evaporation (and thus cooling efficiency). The result of this addition showed an improvement in the water evaporation ratio by 1.5 times over the norm (Toida et al. 2006). In addition to water savings, evaporation improvements also contribute to a more uniform temperature. An evolution of the variable-fogging research, variable ventilation was used with variable fog. Rather than opening vents to set positions, this system opened vents by various degrees in response to ventilation needs. The synergy between these to variable controls demonstrated 36% water savings and 30% reduction in electricity usage (Linker et al. 2011). As these types of control techniques improve, the efficiency of fog cooling will likely continue to become more efficient as well.

While cooling is important in warmer seasons, heating is necessary for cold seasons. A cost-effective method for maintaining a warm growing room is to use a heat curtain. The curtain essentially traps heat within the greenhouse, slowing the outward radiation of heat. In Colorado State in the United States, heat curtains showed a 30% reduction in energy consumption (Huang and Hanan 1976), which could be increased to 60% with better material choice for the curtain (Rebuck et al. 1977). However, in cold environments, where a heat curtain is not sufficient, active heating is necessary. Two technologies under active research that have been replacing more traditional forms of heating, such as oil heaters, are hot-water heating and heat pumps.

Hot-water heating fixes one major flaw of traditionally used oil or gas heating: heating distribution. The traditional heaters generally provide heat from above, which is inefficient. Hot water is provided near ground level, allowing the heat to rise into the plant canopy. Simple implementations of this are water-filled barrels that are placed throughout the greenhouse to absorb heat during the day and radiate the heat into the canopy at night. This passive heating method can raise nighttime temperatures by 8°C (Sethi and Sharma 2008). For a low-cost method, this is effective, but there are more effective methods, such as hot-water piping. Water is stored in a reservoir and distributed throughout the room via pipes along the floor. During the day, sunlight warms the water pipes, and this stored heat is radiated from the pipes overnight. This method has shown an 11°C warmer environment (Levav and Zamir 1987). These water heating systems can be actively heated, but the solar heating reduces energy costs while still providing effective heating.

The second heating source that has been rising in popularity is heat pumps (sometimes called heat exchangers). Early forms of heat pumps had poor coefficients of performance (COPs), or the efficiency to move heat. Around 10 years ago, COPs were about 4; today, these values are closer to 8 (Tong et al. 2010). Heat pumps move heat via ground-to-air, solar assisted ground-to-air, and air-to-air systems (Aye et al. 2010; Benli 2011; Chai et al. 2010; Ozgener 2010; Tong et al. 2010). Air-to-air heat pumps, specifically, were found to be quite efficient. They use 25–65% less energy and have a 1.3–2.6 times higher energy-use efficiency when compared to traditional kerosene burners. Additionally, CO_2 emissions were reduced by 59–79% (Tong et al. 2012). These improvements have enabled heat pumps to become a viable alternative to gas heaters.

Cooling and heating management changes when sunlight is excluded from the system. Fully artificial plant factories use various light sources to replace the sun. These lights are also the primary heat source. The highly insulated growing room retains this heat, allowing temperatures to rise as the light continues to produce heat. As long as the photoperiods in the system are managed effectively, cooling technology takes precedence in these facilities as the heat cannot be

expelled passively. Heat pumps are commonly used here as they can collect water vapor in the air while performing their main task of temperature control (Kozai 2013b). Temperature control and humidity control are closely linked, so many of the technologies for controlling temperature are also used for manipulating humidity.

40.4.2 Humidity

Humidity is measured in two values: RH and VPD. Control can use either of these measurements, so it is necessary to understand what each is. The former is a measure of the water vapor in the air as a percentage of the maximum, and the latter, a measure of the difference between saturation and the current water vapor pressure. Humidity has a direct relationship with temperature: higher temperatures reduce humidity by allowing more moisture in the air, and similarly, lowering temperatures increases humidity due to lower moisture capacity. As such, balancing control between the two parameters becomes necessary. Humidity control uses several technologies: heating, ventilation, heat pumps, and air circulation fans for decreasing humidity; fog-cooling and/or pad-and-fan cooling systems are used for raising humidity.

If humidity is too high, the simplest method is the same as for temperature. Opening vents allows water vapor to exit the greenhouse and thereby reduces humidity. Alternatively, introducing heat into the greenhouse will reduce humidity by increasing the water vapor capacity of the air. The heating could be from an active heating source or by opening the heat curtain during the night. At night, the humidity within the growing area continues to rise due to plant activity, while the air above the curtain cools down and therefore is less humid. By opening the curtain, the hot air rises, and the colder air falls into the growing area. The colder, dryer air warms up and absorbs moisture, decreasing humidity (Heuvelink 2005). Using heat pumps also dehumidifies the air because the cooling plate within the heat pump will have water vapor condense on it while the heat pump is on (Li et al. 2012). Increasing the humidity is either passive or active: waiting for transpiration to naturally raise humidity (passive) or using one of the two main cooling systems for plant factories (active).

Plants naturally transpire water, at a rate dependent on light intensity, airflow, and particularly VPD. If humidity is low and circumstances permit, it would be simplest to allow transpiration to slowly increase humidity. However, if humidity must be raised sooner, both pad-and-fan and fogging systems can raise the humidity. When air enters through the pad in a pad-and-fan system, the cooled air also has a higher moisture content, and this humidity will be pulled through the greenhouse. But a humidity gradient will exist, just like a temperature gradient does. Fogging provides a more uniform environment, with variance of 20% throughout the greenhouse (Arbel et al. 1999). With the aforementioned innovations in fog systems, it is likely that the uniformity can be further refined.

In regard to artificial-light plant factories, plants continue to transpire, and humidity will reach 100% if not controlled. Heat pumps are used to capture the water vapor while also cooling the room. However, if the heat pump stops, humidity will continue to rise until 100% (Kozai 2013b). As such, the rotating light schedule mentioned earlier allows the heat pumps to constantly run to offset the heat from the light source and to continue to capture humidity.

40.4.3 CO_2

CO_2 is essential for plant growth, being the source of carbon for plants. Increasing CO_2 concentrations in the air generally has beneficial effects on crop yield (Long and Zhu 2006; Zhu et al. 2007). The benefit of CO_2 enrichment within the greenhouse was found to be a 22% or more increase in yield (Challa and Schapendonk 1986), depending on the climate. Climates are involved because of the use of ventilation. In locations where temperatures are mild, the greenhouse vents can be closed, and CO_2 enrichment can simply be injected; in locations where temperatures require ventilation usage, CO_2 enrichment becomes more complicated. With ventilation, CO_2 interacts with the outside environment, allowing diffusion of CO_2 outward. For this reason, common control strategies revolve around maintaining an equilibrium with the outside, ambient CO_2 concentration. However, even in greenhouses that are well ventilated, CO_2 concentrations within a greenhouse have been shown to drop to well below ambient levels due to plant uptake (Pascale and Maggio 2008). For yield increases, even ambient-level CO_2 control would require some CO_2 injection. For this purpose, *null-balance CO_2 enrichment* within a greenhouse has been proposed (Figure 40.4): only when sensors show lower levels than ambient will CO_2 be injected into the greenhouse, thereby improving photosynthesis but minimizing CO_2 loss. For more aggressive CO_2 enrichment, ventilation rates are an obstacle that must

be resolved. When the vents open, the CO_2 will diffuse into the outside environment. Therefore, research on capturing the CO_2 prior to its diffusion is being worked on. As the research progresses, production trials will be run, and in the future, aggressive control of CO_2 within greenhouse may be a possibility. Much higher CO_2 enrichment in warmer locations where ventilation traditionally took priority over CO_2 enrichment could then accomplish both.

In contrast to sunlight plant factories, artificial-light plant factories do not need to worry about diffusion of CO_2 to the outside environment. Without fear of the gas being lost to the outside environment, CO_2 enrichment is typically above 1000 ppm (Kozai 2013b). The increase in CO_2 input as well as the lower passively lost CO_2 increases the usage efficiency. When compared to a sunlight greenhouse with its ventilation closed and using CO_2 enrichment, the CO_2-use efficiency in the artificial-light plant factories is about 1.8 times greater (Kozai 2013b).

40.4.4 Light

Light is a major factor for plant photosynthesis and plant morphological development. Sunlight provides the energy for plant processes to begin. Light influences plant growth and development over the plant lifetime via light characteristics, including light intensity, light sum over time, lighting timing, photoperiod, light quality, light direction, and light distribution over the plant. Improving light usage by plants would therefore be a likely candidate for improving yields. For light in plant factories with sunlight, this generally means finding a way to improve light incidence within the plant canopy or supplementing light via another source. In plant factories with artificial lighting, sunlight is completely replaced

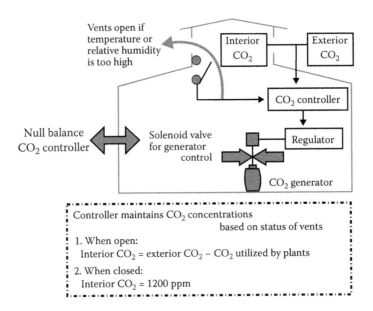

FIGURE 40.4 The null-balance CO_2 enrichment is a method to minimize CO_2 losses when vents are open. When vents are open, the CO_2 is maintained at ambient; when closed, CO_2 is injected to about 1000 ppm. (Modified from Japan Plant Factory Association [JPFA], Convention report: Operation test of integrated environmental control for plant factories, Specified Nonprofit Corporation: JPFA, Chiba, Japan, 2013, in Japanese.)

by artificial light sources. Thus, research is being done for both the improvement of light usage in plants as well as the improvement of the alternative light sources.

40.4.4.1 Light and Photosynthesis

Maximizing the photosynthetic potential of plants is one method for improving yields. Understanding the intricacies of photosynthesis brings opportunities to optimize the environment for plant growth. The efficiency of photosynthesis, or the process to take sunlight, water, and CO_2 and create energy, has been well studied and found to have a maximum theoretical efficiency of around 4.5% (Archer and Barber 2004; Barber and Tran 2013; Blankenship et al. 2011; Bolton and Hall 1991). However, plants do not naturally reach that efficiency, usually creating only about 1% dry matter (well below the 4.5% limitation), due to the complexities of photosynthesis and its interactions with the environment. Suboptimal environmental conditions, light interception by the plants, and nutrient and water uptake all reduce the potential for photosynthesis (Barber and Tran 2013). The nature of plant factories is to control these factors, thus creating the opportunity to improve photosynthetic efficiency.

In tandem with the research on the theoretical level, the commercial production of plants has also shown ways to improve photosynthesis potential. Many studies have investigated the general relationship between light and plant growth in production environments (Challa and Schapendonk 1983; Downs 1985). Commercial production data have also been reliable and consistent enough that a general rule has been established for the light usage of the major vegetable crops. If light levels drop by 1%, production will also decrease by about 1% (Heuvelink 2005). Figure 40.5 shows this linear relationship between the yield and the light interception of the plant canopy. In sunlight plant factories, the sunlight cannot be controlled, so cultural techniques are necessary to increase light-use efficiency. Measuring the leaf area index (LAI; the leaf area divided by the ground area occupied by these leaves) is valuable for understanding light usage. For tomato plants, an LAI of 3.0 (three times as much leaf area compared to

ground area) will intercept about 90% of the light entering the greenhouse (Heuvelink 2005). This type of knowledge allows growers to prepare environments to maximize light incidence and improve photosynthesis in the greenhouse. On the other side of the spectrum, there are times in which too much light is undesirable and must be managed. In these situations where light intensity is too harsh and will damage the crop (either through direct lighting harming the plant or due to the increased temperature), shade cloth can be utilized (Hanan 1998). Shade cloth simply blocks a portion of the light from reaching the plant, which may decrease potential yields, but protects the crop from damage. These methods do not control lighting, however, but rather simply, managing it.

Common methods for improving photosynthesis revolve around increasing light interception in the plant canopy. As is known, the upper canopy in plants is typically responsible for the majority of photosynthesis and carbon assimilation. Specifically, in tomato plants, the uppermost part of the canopy fixes 66% of the carbon assimilation, though it only accounts for about a quarter of the total leaf area (Acock and Charles-Edwards 1978). This benefit is important because the lower leaves in the plant canopy, which are shaded by the upper leaves, naturally have a lower capacity for photosynthesis than the upper leaves. It was originally thought that the lower leaves are aging and the loss in photosynthetic potential was due to the aging process. However, research has demonstrated that leaf aging is not a major detractor in photosynthesis but, rather, a result of the constantly lower light intensity due to leaf shading (Pettersen et al. 2010; Trouwborst and Hogewoning 2011). To examine this, plants were grown horizontally in order for all leaves, lower or upper, to receive the same light intensity from the sun. During these experiments, the older leaves maintained competitive levels of photosynthesis with the younger leaves. However, if the leaves were shaded for several days, this potential would permanently drop to the traditionally seen levels. Therefore, white plastic is usually used to cover the floor of the sunlight plant factory. Owing to the reflections of the white plastic, 50–80% of photosynthetically active radiation (PAR) light can be rebounded

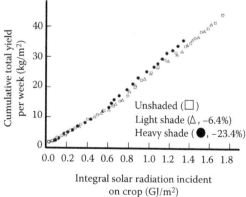

FIGURE 40.5 The relationship between cumulative total yield (kg/m²) and cumulative solar radiation incident on the crop (GJ/m²) in successive weeks in unshaded (□), light-shade (Δ; −6.4%), or heavy-shade (●; −23.4%) treatments. (Modified from Cockshull et al. 1992. Reprinted from Heuvelink, 2005. With permission.)

back to the plant canopy (Heuvelink 2005) resulting in the increment of light incidence in the canopy and improving the overall photosynthesis. In addition, by using diffusion glass or diffusion film, the light distribution can be improved both horizontally and vertically, improving overall photosynthesis. Research on different degrees of light diffusion on tomato plants shows that 71% of light being diffused increased the calculated crop photosynthesis by 7.2%. This effect was mainly attributed to a more uniform horizontal and vertical photosynthetic photon flux density (PPFD) distribution in the crop. The research also found that tomato plants acclimated to the high level of diffusion by maintaining a higher photosynthetic capacity in the leaves within the middle of the crop as well as a higher LAI. Moreover, diffused light resulted in lower leaf temperatures and less photoinhibition at the top of the canopy when global irradiance was high (Li et al. 2014). Thus, providing more light to the lower plant canopy and creating uniform light distribution in the canopy present opportunities to improve photosynthesis.

Research has been done to improve light usage via the improvement of plant physiological processes by biotechnology. Some of the plant processes with promising research are changes in the Calvin cycle, the chloroplast electron transport system, and Rubisco and Rubisco activase (Yamori 2013). The Calvin cycle is a procedure that uses CO_2 to create sugars. Research on the Calvin cycle has found bottlenecks in the process that offer possibilities for optimization (Raines 2011). Chloroplast electron transport is where energy molecules are created (ATP and NADPH) and used during the various processes of photosynthesis. The overexpression of protein involved in electron transfer (i.e., plastocyanin) improved the electron transport system and increased plant biomass (Chida et al. 2007; Pesaresi et al. 2009). Overexpression of Rubisco activase, which is the regulatory protein for the activation of Rubisco, was found to benefit photosynthesis under the natural fluctuating light conditions (Yamori et al. 2012). Further efforts to improve the plant functions related to photosynthesis could culminate in new plant breeds with higher yield capacities based solely on the plants' abilities, before the benefits of optimized environments are factored in.

40.4.4.2 Supplemental Lighting in Plant Factories

For more control over lighting, local light sources are necessary. Using supplementary lighting (or completely replacing sunlight with artificial lights) would allow for more control over the light received by plants and boost photosynthesis. This gives both plant factories with sunlight and supplemental light and plant factories with artificial lighting promise for greater yields if lighting can be done properly.

Using supplementary lighting allows for more stable light environments. In plant factories, the supplemental lighting is usually applied on high-wire crops like tomato, cucumber, and paprika; leafy vegetables like lettuce; as well as floriculture plants like roses and chrysanthemum. Theoretically, optimal lighting is expected to be used for certain crops to gain maximum production, but in practice, the cost and the limitation of other environmental factors have to be considered when

supplemental-light technology is applied. That means the supplemental lighting must be used as efficiently as possible. For example, in tomato production, the light input must be connected to weight gain of the salable product rather than to increasing too much vegetative growth. Thus, identifying the light timing, position, and photoperiod is important for the efficient utilization of the energy.

Supplemental light is mainly used during the time of year when ambient light intensity may limit plant growth. The timing in application of supplemental lighting varies with locations, seasons, and market demands. In high-latitude countries, the daytime is short and the light level is very low in the wintertime. Normally, during this season, the price of vegetables is also high because of the difficulty in producing vegetables outside. This is a good period to apply supplemental lighting on high-value crops. In low-latitude countries, during low-light-level seasons or cloudy days and rainy days, supplemental lighting is also needed for stable production and better quality of production. Adding supplementary light in those conditions provides a more consistent lighting environment year-round and translates into more stable yields. In experiments to supplement light in the winter, when light intensity is low, tomato yields increased by 70–106% (Dorais and Gosselin 2002). Even in summer, applying supplemental light within the plant canopy can still benefit plant growth and plant production. The average amount of solar radiation for the location should be investigated. This will give an idea of the range of solar radiation conditions at the site. By simulating the desired light of certain plants and calculating the actual sunlight they receive (which changes depending on the season/weather/day), estimates can be made for how much supplemental lighting should be applied for the desired production.

Lighting position is important because the supplemental lighting should be applied on an area where the leaves can convert as much of the energy as possible to achieve the desired production. There is research showing that tomato leaves at different positions transport assimilates to different parts of the plant (Shishido et al. 1991). This means the light should be used at the position where the leaves can absorb light and then contribute the most to production. For example, the young leaves of tomato plants are more active in photosynthesis than old leaves. Keeping the younger leaves warm and receiving more light is beneficial for the plant's development. As such, the most common way to use supplemental light is to mount lamps above the plants, called top lighting. For high-wire crops and high-plant-density cultivation crops (e.g., single-truss tomato in high plant density), increasing light incidence inside the middle or lower level of the canopy, where there is a lack of light, can also increase plant production (Grimstad 1987; Hovi et al. 2004; Gunnlaugsson and Adalsteinsson 2006; Ménard et al. 2006; Hovi-Pekkanen and Tahvonen 2008; Trouwborst et al. 2010; Lu et al. 2012a,b). Combining the two mounting methods, top lighting and interlighting, would be more efficient than solely applying supplemental light from above (Figure 40.6).

Photoperiod is also an important factor that affects plant growth. For some crops, and especially for the vegetative

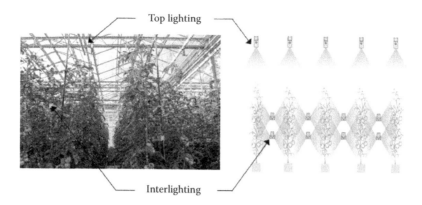

FIGURE 40.6 Two supplemental-light mounting methods: top lighting and interlighting (double line). The left side is a practical example in high-wire tomato plants. The right side is the sketch of this type of supplemental-lighting application. (Reprinted from Philips Horticulture LED Solutions. With permission.)

growth phase, a linear relationship exists between total amount of light received and plant growth (Lu et al. 2012a,b). There are two ways to increase the total light amount on the plant: to increase light intensity during certain periods and to increase lighting hours with certain light intensity. The strategy used should be determined based on the conditions of each location and cultivation system. To reduce the initial investment, growers prefer to use supplemental lighting as long as possible per day rather than using more lamps to maximize production. Thus, the effective hours of lighting were studied in several cases. For tomato plants under long photoperiods (over 17 h) and with continuous light treatment via supplemental lighting, leaf chlorosis was observed after several days of treatment. However, under almost 24 h of natural lighting in Finland, tomato plants do not show these negative symptoms (Heuvelink and Dorais 2005). An extended photoperiod of 18 h and 24 h had no significant effect on the leaf area of greenhouse tomato but did increase leaves' dry weight (Dorais et al. 1996). However, under a high light density, tomato leaf area was smaller in plants grown under a 24 h photoperiod than in the 14 h lighting treatment, but specific leaf weight was higher (Demers et al. 1998). In a semiclosed plant factory, the photoperiod is mainly influenced by sunlight, but in a fully artificial-light plant factory, the photoperiod is independent of sunlight. In order to make the operation of a supplemental lighting system as economical as possible, this system is expected to be operated during periods of the day with off-peak electricity rates (e.g., 10:00 p.m.–6:00 a.m.). It is not necessarily important for light to be continuously used for the plant to reach its desired light integral; sometimes, interval light can be used to meet the same cumulative integral every day.

40.4.4.3 Light Quality and Light Source

Light quality refers to the wavelength reaching the plant surface. Solar radiation from 400 to 700 nm is designated as PAR. Photosynthetic organisms are able to use this spectral range in the process of photosynthesis. The morphology and physiology of plants are strongly influenced by red (600–700 nm) and blue light (400–500 nm). In the photosynthetic photosystem within photosynthetic organisms, the existing pigments are chlorophylls and carotenoids. The two most important absorption peaks of chlorophyll are located in the red and blue regions from 625 to 675 nm and from 425 to 475 nm, respectively. Carotenoids, such as xanthophylls and carotenes, are located in the chloroplast plastid organelles on plant cells and absorb light mainly in the blue region. Red light is most efficient in light production because red light costs less energy per photon and plants absorb this energy effectively for photosynthesis; however, to achieve a better plant morphology, red light alone is not enough. Much research has found that blue light is indispensable to achieve a balanced morphology of most crop plants through the mediation of the cryptochrome family of photoreceptors. Blue light is more inefficient in terms of energy but also drives photosynthesis processes, especially on plant morphological formation and secondary metabolism products. There are many studies about the effects of red and blue light as well as different red-and-blue combinations on different crop types in the past years (Massa et al. 2008). Besides red and blue light, green light, ultraviolet (UV) light, and far-red light also have been used for research. Green light (500–600 nm) is commonly thought to be less effective in plants as most plants reflect green light. However, several studies reported that the addition of green light in combination with red and blue LEDs enhanced plant growth, since leaves in the lower layer of the canopy would be able to use the transmitted green light in photosynthesis (Klein 1992; Smith 1993; Kim 2004). Terashima et al. (2009) reported that green light can be absorbed by the lower layer of chloroplasts and drive leaf photosynthesis more efficiently than red light when sunflower was irradiated by strong white light. In addition, a high intensity of green light showed promoted shoot growth (Johkan et al. 2012). UV light also has been used to increase anthocyanin and carotenoid levels in lettuce (Caldwell and Britz 2006; Tsormpatsidis and Henbest 2008; Li and Kubota 2009). In summary, white light is needed as work light, but it is not always essential for the plant. Therefore, it is important to know that the exact wavelength and each ratio are inside the light spectrum.

There are some important points that need to be noted when selecting a light source for horticulture use: first, to select the required/correct light spectrum for a certain crop; second, high transformation efficiency, which is the efficiency of a light source in transforming electrical energy into optical energy; third, limiting any losses due to reflectors, ballasts, or external cooling; and last, good light maintenance over the service life, such as constant light intensity and long lifetime. The following four types of lamp are commonly used in plant factories for plant production.

Metal-halide lamps are preferred in semiclosed plant factories. The metal-halide lamp belongs to the category of high-intensity discharge (HID) lamps. The emission of visible radiation is based on the luminescent effect. The inclusion of metal halides during manufacturing allows, to a certain extent, the optimization of the spectral quality of the radiation emitted. Metal-halide lamps can be used in plant cultivation to totally replace daylight or for partially supplementing it during periods of lower availability. The high PAR output per lamp, the relatively high percentage of blue radiation (around 20%), and the electrical efficiency at approximately 25% make metal-halide lamps a viable option for year-round crop cultivation (Brown et al. 1995; Schuerger et al. 1997).

The high-pressure sodium (HPS) lamp is also a type of HID lamp. Compared to metal-halide lamps, HPS lamps produce 40–50% more mean lumens per watt. The factors of high radiant emission, high PAR emission, and high electrical efficiency have allowed the use of HPS lamps as supplemental lighting sources for supporting vegetative growth in a cost-effective way during wintertime in northern latitudes. Electrical efficiencies of HPS lamps are typically within 30–40%. Approximately 40% of the input energy is converted into photons inside the PAR region and almost 25–30% into far red and infrared (Simpson 2003). Metal-halide lamps' lifetimes are between 10,000 and 15,000 h, and the lifetimes of HPS lamps are between 12,000 and 24,000 h. Both metal-halide lamps and HPS lamps release a lot of heat, so a certain distance from lamp to plants is required during installation.

Fluorescent lamps are commonly utilized in plant-growth applications in closed-type plant factories. The electro-optical energy conversion of fluorescent lamps is more efficient in comparison to incandescent lamps. White fluorescent lamp lights have a mixture of red, green, and blue colors (also a small amount of UV light and far-red light) that attempts to imitate sunlight as closely as possible. Therefore, fluorescent lamps are frequently used for total substitution of natural daylight radiation in closed-growth environments. High-frequency, electric ballast fluorescent lamps can achieve electrical efficiency values of around 20–30% and have typical lifetimes of around 12,000 h (Simpson 2003).

LEDs can play a variety of roles in horticultural lighting. LED-based lighting systems have a number of advantages over lamps currently used in horticulture. For example, LED-based lighting systems save more energy and have a longer lifetime as well as controllable spectral output compared to conventional lamp types. The spectral output of an LED lighting system can be matched to plant photoreceptors and optimized to provide maximum production without wasting energy on nonproductive wavelengths and heat levels. The ability to dynamically combine the spectral output can also be used to influence plant morphology (Heo et al. 2002). LED lighting systems can be configured to produce very high light levels, but even at high light outputs, they can be operated in close proximity to plant tissue due to the low radiant heat output when cooled properly. There are some comparisons between high-efficiency HPS lamps and latest developed LED modules showing that the LED top-lighting systems use significantly less energy, saving up to 20–46% at comparable light intensity levels (PHILIPS Horticulture LED Solutions 2014b). LED bulbs and diodes have an outstanding operational lifetime expectation of up to 100,000 h. However, for horticulture use, another important consideration is the photon flux maintenance of a LED lighting system. The high-photon-flux maintenance LED systems are preferred for growing plants. Because most plants are also sensitive to intercepted light intensity and rapid light attenuation from LED lamps, these may reduce plant production and quality.

LED illumination also showed beneficial effects against pests and disease. Blue light has been shown to significantly suppress *Botrytis cinerea*, commonly called gray mold (Kook et al. 2013), and has lowered the leaf blight index, as well as fruit damage, from *Helicoverpa armigera* worms in tomatoes (Xu et al. 2012). However, using LEDs as a direct control requires much more research, as it is not a natural occurrence, and thus examples in nature are minimal. Complex reactions may or may not occur and need careful investigation. Some of the possible interactions with LED lighting include reactions to the visual stimuli of the LEDs, damaging effects from the light wavelength, and effects on internal physiological processes (Johansen et al. 2011). These intricacies prevent quick results and require time to discover viable techniques for use. However, pest monitoring has shown clear and positive results from the use of light spectrums. Research has shown that nocturnal insects are generally attracted to high-UV light fixtures (Shimoda and Honda 2013). As a result, black-light bulbs (BLB) have been used for monitoring in various industries. They have also been shown to have synergy with other monitoring methodologies. A good example is shown in trials of red flour beetles (*Tribolium castaneum*). Using pheromones to capture the beetles in traps, 1% of the test population was captured. When UV lighting was used, more beetles were attracted to the area, which increased the chance for the pheromones to lure the beetles into the traps—resulting in a 20% capture rate (Duehl et al. 2011). However, UV lighting is not the only effective spectrum. LED spectrums were tested for their effectiveness in attracting leaf miners (*Liriomyza trifolii*), and it was found that green LEDs were most effective. Green LEDs attracted the insects 1.4 times more successfully than UV lighting, while yellow LEDs resulted in a close improvement of 1.35 times as well (Kim and Lee 2014). It is clear that light plays a part in insect responses, and LEDs present the chance to produce much more precise light spectrums than prior technology. The separation of the spectrum allows researchers to better test insect responses to specific wavelengths. In time, research should uncover insect responses and create novel ways to integrate LEDs into IPM strategies.

40.5 CONCLUSION

Plant factories are currently under heavy research, and new innovations are steadily solving their weaknesses. Plant factories with sunlight and/or supplemental lighting are becoming more efficient with fog-cooling systems, allowing for a more uniform environment and more predictable plant growth. Alternative uses for LEDs are allowing for better-quality produce from plant factories, and the advances in LED efficiencies will make plant factories with artificial lighting even more desirable. The production advantage of plant factories overshadows traditional agriculture methods, and the fast-paced innovation will only widen the gap.

ACKNOWLEDGMENTS

This work was supported by the Japan Science and Technology Agency, PRESTO (to W.Y.), and in part by a Grant-in-Aid for Scientific Research from the Japan Society for the Promotion of Science (Scientific Research no. 25891005 to W.Y.).

REFERENCES

Abdel-Ghany, A. M. and T. Kozai. 2006. Dynamic modeling of the environment in a naturally ventilated, fog-cooled greenhouse. *Renewable Energy* 31: 1521–1539.

Acock, B. and D. A. Charles-Edwards. 1978. The contribution of leaves from different levels within a tomato crop to canopy net photosynthesis: An experimental examination of two canopy models. *Journal of Experimental Botany* 29: 815–827.

Alexandratos, N. and J. Bruinsma. 2012. World agriculture towards 2030/2050, The 2012 Revision. Rome: FAO, ESA Working Paper.

Al-Sulaiman, F. 2002. Evaluation of the performance of local fibers in evaporative cooling. *Energy Conversion and Management* 43: 2267–2273.

Arbel, A., O. Yekutieli, and M. Barak. 1999. Performance of a fog system for cooling greenhouses. *Journal of Agricultural Engineering Research* 2: 129–136.

Archer, M. D. and J. Barber. 2004. Photosynthesis and photoconversion. In *Molecular to Global Photosynthesis*, eds. E. Heuvelink, M. D. Archer, and J. D. Barber. London: Imperial College Press.

Aye, L., R. J. Fuller, and A. Canal. 2010. Evaluation of a heat pump system for greenhouse heating. *International Journal of Thermal Sciences* 1: 202–208.

Baille, A., M. M. González-Real, J. C. Gazquez et al. 2006. Effects of different cooling strategies on the transpiration rate and conductance of greenhouse sweet pepper crops. *Acta Horticulturae: International Symposium on Greenhouse Cooling* 719: 463–470.

Barber, J. and P. D. Tran. 2013. From natural to artificial photosynthesis. *Journal of the Royal Society* 10: 20120984.

Benli, H. 2011. Energetic performance analysis of a ground-source heat pump system with latent heat storage for a greenhouse heating. *Energy Conversion and Management* 52: 581–589.

Blankenship, R. E., D. M. Tiede, J. Barber et al. 2011. Comparing photosynthetic and photovoltaic efficiencies and recognizing the potential for improvement. *Science* 332: 805–809.

Bolton, J. R. and D. O. Hall. 1991. The maximum efficiency of photosynthesis. *Photochemistry and Photobiology* 53: 545–548.

Brown, C. S., A. C. Schuerger, and J. C. Sager. 1995. Growth and photomorphogenesis of pepper plants under red light-emitting diodes with supplemental blue or far-red lighting. *Journal of the American Society for Horticultural Science* 120: 808–813.

Caldwell, C. R. and S. J. Britz. 2006. Effect of supplemental ultraviolet radiation on the carotenoid and chlorophyll composition of green house-grown leaf lettuce (*Lactuca sativa* L.) cultivars. *Journal of Food Composition and Analysis* 19: 637–644.

Chai, L., C. Ma, Y. Zhang et al. 2010. Energy consumption and economic analysis of ground source heat pump used in greenhouse in Beijing. *Transactions of the Chinese Society of Agricultural Engineering* 26: 249–254.

Challa, H. and A. Schapendonk. 1983. Quantification of effects of light reduction in greenhouses on yield. *Acta Horticulturae, III International Symposium on Energy in Protected Cultivation* 148: 501–510.

Challa, H. and A. H. C. M. Schapendonk. 1986. Dynamic optimization of CO_2 concentration in relation to climate control in greenhouses. In *Carbon Dioxide Enrichment of Greenhouse Crops*, Vol. I, eds. H. Z. Enoch and B. A. Kimball. Boca Raton, FL: CRC Press.

Chandra, P., J. Singh, and G. Majumdar. 1989. Some results of evaporative cooling of a plastic greenhouse. *Journal of Agricultural Engineering ISAE* 26: 274–280.

Chida, H., A. Nakazawa, H. Akazaki et al. 2007. Expression of the algal cytochrome c6 gene in Arabidopsis enhances photosynthesis and growth. *Plant & Cell Physiology* 48: 948–957.

Cockshull, K. E., C. J. Graves, and C. R. J. Cave. 1992. The influence of shading on yield of glasshouse tomatoes. *Journal of Horticultural Science* 67: 11–24.

Davies, P. 2005. A solar cooling system for greenhouse food production in hot climates. *Solar Energy* 79: 661–668.

Demers, D. A., M. Dorais, C. H. Wien et al. 1998. Effects of supplemental light duration on greenhouse tomato (*Lycopersicon esculentum* Mill.) plants and fruit yields. *Scientia Horticulturae* 74: 295–306.

Dorais, M. and A. Gosselin. 2002. Physiological response of greenhouse vegetable crops to supplemental lighting. *Acta Horticulture* 580: 59–67.

Dorais, M., S. Yelle, and A. Gosselin. 1996. Influence of daylength on photosynthate partitioning and export in tomato and pepper plants. *New Zealand Journal of Crop and Horticultural Science* 24: 29–37.

Downs, S. J. 1985. Irradiance and plant growth in greenhouses during winter. *HortScience* 20: 1125.

Duehl, A. J., L. W. Cohnstaedt, R. T. Arbogast et al. 2011. Evaluating light attraction to increase trap efficiency for Tribolium castaneum (Coleoptera: Tenebrionidae). *Journal of Economic Entomology* 104: 1430–1435.

Garcia, M. and E. Medrano. 2011. Climatic effects of two cooling systems in greenhouses in the Mediterranean area: External mobile shading and fog system. *Biosystems Engineering* 108: 133–143.

Gericke, W. 1929. Aquaculture, a means of crop production. *American Journal of Botany* 16: 862.

Godfray, H., J. Charles and T. Garnett. 2014. Food security and sustainable intensification. *Philosophical Transactions of the Royal Society B: Biological Sciences* 369(1639): 20120273.

Gorbe, E. and A. Calatayud. 2010. Optimization of nutrition in soilless systems: A review. *Advances in Botanical Research* 53: 193–245.

Grewal, H., B. Maheshwari, and S. Parks. 2011. Water and nutrient use efficiency of a low-cost hydroponic greenhouse for a cucumber crop: An Australian case study. *Agricultural Water Management* 98: 841–846.

Grimstad, S. O. 1987. Supplementary lighting of early tomatoes after planting out in glass and acrylic greenhouses. *Scientia Horticulturae* 33: 189–196.

Gunnlaugsson, B. and S. Adalsteinsson. 2006. Interlight and plant density in year-round production of tomato at northern latitudes. *Acta Horticulturae* 711: 71–75.

Hanan, J. 1998. *Greenhouses: Advanced Technology for Protected Horticulture*. Boca Raton, FL: CRC Press.

Hayashi, M., E. Goto, and T. Kozai. 2007. Experimental verification of control logic for operation of a fog cooling system for a naturally ventilated greenhouse. *Environmental Control in Biology* 45: 47–58.

Heo, J., C. Lee, D. Chakrabarty et al. 2002. Growth responses of marigold and salvia bedding plants as affected by monochromic or mixture radiation provided by a light-emitting diode (LED). *Plant Growth Regulation* 38: 225–230.

Heuvelink, E. 2005. Developmental processes. In *Tomatoes*, ed. E. Heuvelink, 53–84. Wallingford, UK: CABI Publishing.

Heuvelink, E. and M. Dorais. 2005. Crop growth and yield. In *Tomatoes*, ed. E. Heuvelink, 85–144. Wallingford: CABI Publishing.

Hoagland, D. R. and D. I. Arnon. 1950. The water culture method for growing plants without soil. *California Agricultural. Experimental Station. Circular* 347: 32–35.

Hovi, T., J. Näkkilä, and R. Tahvonen. 2004. Interlighting improves production of year-round cucumber. *Scientia Horticulturae* 102: 283–294.

Hovi-Pekkanen, T. and R. Tahvonen. 2008. Effects of interlighting on yield and external fruit quality in year-round cultivated cucumber. *Scientia Horticulturae* 116: 152–161.

Huang, K. and J. Hanan. 1976. Theoretical analysis of internal and external covers for greenhouse heat conservation. *Hortscience* 11: 582–583.

Japan Plant Factory Association (JPFA). 2013. *Convention Report: Operation Test of Integrated Environmental Control for Plant Factories*. Chiba, Japan: Specified Nonprofit Corporation, JPFA (in Japanese).

Johansen, N. S., I. Vänninen, D. M. Pinto et al. 2011. In the light of new greenhouse technologies: 2. Direct effects of artificial lighting on arthropods and integrated pest management in greenhouse crops. *Annals of Applied Biology* 159: 1–27.

Johkan, M., K. Shoji, and F. Goto. 2012. Effect of green light wavelength and intensity on photomorphogenesis and photosynthesis in Lactuca sativa. *Environmental and Experimental Botany* 75: 128–133.

Katsoulas, N., A. Baille, and C. Kittas. 2001. Effect of misting on transpiration and conductances of a greenhouse rose canopy. *Agricultural and Forest Meteorology* 106: 233–247.

Kim, H.-H. 2004. Green-light supplementation for enhanced lettuce growth under red- and blue-light-emitting diodes. *HortScience* 39: 1617–1622.

Kim, M. and H. Lee. 2014. Attractive effects of American serpentine leafminer, *Liriomyza trifolii* (Burgess), to light-emitting diodes. *Journal of Insect Behavior* 27: 127–132.

Kim, H., C. Chun, T. Kozai et al. 2000. The potential use of photoperiod during transplant production under artificial lighting conditions on floral development and bolting, using spinach as a model. *HortScience* 35: 43–45.

Kittas, C., T. Bartzanas, and A. Jaffrin. 2003. Temperature gradients in a partially shaded large greenhouse equipped with evaporative cooling pads. *Biosystems Engineering* 85: 87–94.

Klein, R. M. 1992. Effects of green light on biological systems. *Biological Reviews* 67: 199–284.

Kook, H., S. Park, Y. Jang et al. 2013. Blue LED (light-emitting diodes)-mediated growth promotion and control of Botrytis disease in lettuce. *Acta Agriculturae Scandinavica, Section B—Soil & Plant Science* 63: 271–277.

Kozai, T. 2013a. Plant factory in Japan- current situation and perspectives. *Chronica Horticulturae* 53: 8–11.

Kozai, T. 2013b. Resource use efficiency of closed plant production system with artificial light: Concept, estimation and application to plant factory. *Proceedings of the Japan Academy. Series B, Physical and Biological Sciences* 89: 447–461.

Landsberg, J., B. White, and M. Thorpe. 1979. Computer analysis of the efficacy of evaporative cooling for glasshouses in high energy environments. *Journal of Agricultural Engineering Research* 24: 29–39.

Levav, N. and N. Zamir. 1987. Utilization of solar energy in greenhouse. In *Greenhouse Heating with Solar Energy, REU Technical Series 1*, ed. C. von Zabeltitz, 178–185. Rome: FAO, ENEA.

Li, Q. and C. Kubota. 2009. Effects of supplemental light quality on growth and phytochemicals of baby leaf lettuce. *Environmental and Experimental Botany* 67: 59–64.

Li, M., T. Kozai, G. Niu et al. 2012. Estimating the air exchange rate using water vapour as a tracer gas in a semi-closed growth chamber. *Biosystems Engineering* 113: 94–101.

Li, T., E. Heuvelink, T. A. Dueck et al. 2014. Enhancement of crop photosynthesis by diffuse light: Quantifying the contributing factors. *Annals of Botany* 114: 145–156.

Linker, R., M. Kacira, and A. Arbel. 2011. Robust climate control of a greenhouse equipped with variable-speed fans and a variable-pressure fogging system. *Biosystems Engineering* 110: 153–167.

Long, S. and X. Zhu. 2006. Can improvement in photosynthesis increase crop yields? *Plant, Cell & Environment* 29: 315–330.

Lu, N., T. Maruo, M. Johkan et al. 2012a. Effects of supplemental lighting within the canopy at different developing stages on tomato yield and quality of single-truss tomato plants grown at high density. *Environmental Control in Biology* 50: 1–11.

Lu, N., T. Maruo, M. Johkan et al. 2012b. Effects of supplemental lighting with light-emitting diodes (LEDs) on tomato yield and quality of single-truss tomato plants grown at high planting density. *Environmental Control in Biology* 50: 63–74.

Massa, G. D., H.-H. Kim, R. M. Wheeler et al. 2008. Plant productivity in response to LED lighting. *HortScience* 43: 1951–1956.

Massa, D., L. Incrocci, and R. Maggini. 2010. Strategies to decrease water drainage and nitrate emission from soilless cultures of greenhouse tomato. *Agricultural Water Management* 97: 971–980.

Meca, D., J. Lopez, J. Gazquez et al. 2006. Evaluation of two cooling systems in parral type greenhouses with pepper crops: Low pressure fog system verses whitening. *Acta Horticulturae, International Symposium on Greenhouse Cooling* 719: 515–519.

Ménard, C., M. Dorais, T. Hovi et al. 2006. Developmental and physiological responses of tomato and cucumber to additional blue light. *Acta Horticulturae* 711: 291–296.

Ministry of Agriculture, Forestry and Fisheries (MAFF). 2013. Statistics of Agriculture, Forestry and Fishery. Available at http://www.maff.go.jp/j/tokei/sokuhou/yasai_syutou13/index .html (in Japanese).

Montero, J., A. Anton, C. Biel et al. 1989. Cooling of greenhouses with compressed air fogging nozzles. *Acta Horticulturae, II Workshop on Greenhouse Construction and Design* 281: 199–210.

Muneer, S., E. Kim, J. Park et al. 2014. Influence of green, red and blue light emitting diodes on multiprotein complex proteins and photosynthetic activity under different light intensities in lettuce leaves (*Lactuca sativa* L.). *International Journal of Molecular Sciences* 15: 4657–4670.

Olympios, C. M. 1999. Overview of soilless culture: Advantages, constraints, and perspectives for its use in Mediterranean Countries. *Cahiers Options Méditerranéennes* 31: 307–324.

Ozgener, O. 2010. Use of solar assisted geothermal heat pump and small wind turbine systems for heating agricultural and residential buildings. *Energy* 35: 262–268.

Pascale, S. and A. Maggio. 2008. Plant stress management in semiarid greenhouse. *Acta Horticulturae, International Workshop on Greenhouse Environmental Control and Crop Production in Semi-Arid Regions* 797: 205–215.

Pesaresi, P., M. Scharfenberg, M. Weigel et al. 2009. Mutants, overexpressors, and interactors of Arabidopsis plastocyanin isoforms: Revised roles of plastocyanin in photosynthetic electron flow and thylakoid redox state. *Molecular Plant* 2: 236–248.

Pettersen, R. I., S. Torre, and H. R. Gislerød. 2010. Effects of leaf aging and light duration on photosynthetic characteristics in a cucumber canopy. *Scientia Horticulturae* 125: 82–87.

Philips Horticulture LED Solutions. 2014a. Available at http://www.lighting.philips.com/pwc_li/main/shared/assets/downloads/pdf/horticulture/leaflets/cl-g-led_interlighting-en.pdf.

Philips Horticulture LED Solutions. 2014b. Available at http://www.lighting.philips.com/pwc_li/main/shared/assets/downloads/pdf/horticulture/leaflets/cl-g-toplighting-en.pdf.

Raines, C. A. 2011. Increasing photosynthetic carbon assimilation in C3 plants to improve crop yield: Current and future strategies. *Plant Physiology* 155: 36–42.

Rebuck, S., R. Aldrich, and J. White. 1977. Internal curtains for energy conservation in greenhouses. *Transactions of the ASAE* 20: 732–734.

Schuerger, A. C., C. S. Brown, and E. C. Stryjewski. 1997. Anatomical features of pepper plants (*Capsicum annuum* L.) grown under red light-emitting diodes supplemented with blue or farred light. *Annals of Botany* 79: 273–282.

Sethi, V. P. and S. K. Sharma. 2007. Survey of cooling technologies for worldwide agricultural greenhouse applications. *Solar Energy* 81: 1447–1459.

Sethi, V. P. and S. K. Sharma. 2008. Survey and evaluation of heating technologies for worldwide agricultural greenhouse applications. *Solar Energy* 82: 832–859.

Shimoda, M. and K. Honda. 2013. Insect reactions to light and its applications to pest management. *Applied Entomology and Zoology* 48: 413–421.

Shishido, Y., C. J. Yun, T. Yuhashi et al. 1991. Changes in photosynthesis, translocation and distribution of 14C-assimilates during leaf development and the rate of contribution of each leaf to fruit growth in tomato. *Journal of the Japanese Society for Horticultural Science* 59: 771–779 (in Japanese with English summary).

Simpson, R. S. 2003. *Lighting Control: Technology and Applications.* Oxford: Taylor & Francis.

Smith, H. 1993. Sensing the light environment: The functions of the phytochrome family. In *Photomorphogenesis in Plants*, eds. R. E. Kendrick and G. H. M. Kronenberg, 2nd Ed., 377–416. Dordrecht: Kluwer Academic.

Terashima, I., T. Fujita, T. Inoue et al. 2009. Green light drives leaf photosynthesis more efficiently than red light in strong white light: Revisiting the enigmatic question of why leaves are green. *Plant Cell Physiology* 50: 684–697.

Thompson, R. B., C. Martínez-Gaitan, M. Gallardo et al. 2007. Identification of irrigation and N management practices that contribute to nitrate leaching loss from an intensive vegetable production system by use of a comprehensive survey. *Agricultural Water Management* 89: 261–274.

Toida, H., T. Kozai, K. Ohyama. 2006. Enhancing fog evaporation rate using an upward air stream to improve greenhouse cooling performance. *Biosystems Engineering* 93: 205–211.

Tong, Y., T. Kozai, N. Nishioka et al. 2010. Greenhouse heating using heat pumps with a high coefficient of performance (COP). *Biosystems Engineering* 106: 405–411.

Tong, Y., T. Kozai, N. Nishioka et al. 2012. Reductions in energy consumption and CO_2 emissions for greenhouses heated with heat pumps. *Applied Engineering in Agriculture* 28: 401–406.

Trouwborst, G. and S. Hogewoning. 2011. The influence of light intensity and leaf age on the photosynthetic capacity of leaves within a tomato canopy. *Journal of Horticultural Science and Biotechnology* 86: 403–407.

Trouwborst, G., J. Oosterkamp, S. W. Hogewoning et al. 2010. The responses of light interception, photosynthesis and fruit yield of cucumber to LED-lighting within the canopy. *Physiologia Plantarum* 138: 289–300.

Tsormpatsidis, E. and R. Henbest. 2008. UV Irradiance as a major influence on growth, development and secondary products of commercial importance in Lollo Rosso lettuce "Revolution" grown under polyethylene films. *Environmental and Experimental Botany* 63: 232–239.

United Nations. 2013. World population prospects: The 2012 revision, Volume I: Comprehensive tables. Available at http://esa.un.org/wpp/Documentation/pdf/WPP2012_Volume-I_Comprehensive-Tables.pdf.

World Bank. 2014a. Agricultural land (% of land area) from World Development Indicators (WDI) database. Available at http://data.worldbank.org/indicator/AG.LND.AGRI.ZS?display=map.

World Bank. 2014b. Arable land (% of land area) from World Development Indicators (WDI) database. Available at http://data.worldbank.org/indicator/AG.LND.ARBL.ZS?display=map.

Xu, H., Q. Xu, F. Li et al. 2012. Applications of xerophytophysiology in plant production—LED blue light as a stimulus improved the tomato crop. *Scientia Horticulturae* 148: 190–196.

Yamori, W. 2013. Improving photosynthesis to increase food and fuel production by biotechnological strategies in crops. *Journal of Plant Biochemistry & Physiology* 1: 1–3

Yamori, W. and S. von Caemmerer. 2009. Effect of Rubisco activase deficiency on the temperature response of CO_2 assimilation rate and Rubisco activation state: Insights from transgenic tobacco with reduced amounts of Rubisco activase. *Plant Physiology* 151: 2073–2082.

Yamori, W., K. Noguchi, and I. Terashima. 2005. Temperature acclimation of photosynthesis in spinach leaves: Analyses of photosynthetic components and temperature dependencies of photosynthetic partial reactions. *Plant, Cell & Environment* 28: 536–547.

Yamori, W., K. Noguchi, Y. T. Hanba et al. 2006. Effects of internal conductance on the temperature dependence of the photosynthetic rate in spinach leaves from contrasting growth temperatures. *Plant and Cell Physiology* 47: 1069–1080.

Yamori, W., J. R. Evans, and S. von Caemmerer. 2010a. Effects of growth and measurement light intensities on temperature dependence of CO_2 assimilation rate in tobacco leaves. *Plant, Cell & Environment* 33: 332–343.

Yamori, W., K. Noguchi, K. Hikosaka et al. 2010b. Phenotypic plasticity in photosynthetic temperature acclimation among crop species with different cold tolerances. *Plant Physiology* 152: 388–399.

Yamori, W., C. Masumoto, H. Fukayama et al. 2012. Rubisco activase is a key regulator of non-steady-state photosynthesis at any leaf temperature and, to a lesser extent, of steady-state photosynthesis at high temperature. *The Plant Journal: For Cell and Molecular Biology* 71: 871–880.

Yamori, W., T. Shikanai, and A. Makino. 2015. Photosystem I cyclic electron flow via chloroplast NADH dehydrogenase-like complex performs a physiological role for photosynthesis at low light. *Scientific Reports* 5: 13908.

Zhu, X.-G., E. de Sturler, and S. P. Long. 2007. Optimizing the distribution of resources between enzymes of carbon metabolism can dramatically increase photosynthetic rate: A numerical simulation using an evolutionary algorithm. *Plant Physiology* 145: 513–526.

41 Strategies for Optimizing Photosynthesis with Biotechnology to Improve Crop Yield

Wataru Yamori, Louis J. Irving, Shunsuke Adachi, and Florian A. Busch

CONTENTS

41.1 Introduction 741
41.2 Improving Photosynthetic Performance via Improvements in Rubisco Kinetics and N Allocation........................... 741
    41.2.1 Rubisco Specificity and Carboxylation Rate 741
    41.2.2 Rubisco Concentration and N Allocation.. 743
    41.2.3 Opportunities for Improving Rubisco Performance... 744
41.3 Increasing the Thermotolerance of Rubisco Activase to Maintain Higher Rubisco Activity under High Temperatures... 745
41.4 Increasing the CO_2 Concentration around Rubisco.. 745
41.5 Enhancing the Rate of Chloroplast Electron Transport in the Thylakoid Membranes.............................. 747
41.6 Enhancing the Activity of Calvin–Benson Cycle Enzymes... 748
41.7 Enhancing the Capacity of Phloem Carbohydrate Transport and Carbon Utilization............................. 750
41.8 Increasing Crop Yield by Optimizing Photosynthesis on the Canopy Level.. 750
41.9 Improving Photosynthetic Performance by QTL Analysis with Utilizing Natural Genetic Resources 752
41.10 Conclusion.. 754
Acknowledgments... 754
References.. 754

41.1 INTRODUCTION

The world's population is projected to increase to 9 billion by 2050 (Roberts 2011), while the availability of arable land will decrease as environmental degradation and destruction grow worse. Meeting the food demand of this growing population will be a major challenge, and a food shortage is projected. Therefore, it is imperative that we produce crops with improved quality and quantity using conventional selection and breeding, or genetic engineering, in order to meet the growing food demand through sustainable agricultural systems (Ashraf and Akram 2009).

In the past, large improvements in grain yield were achieved during the green revolution (Khush 2001); however, present rates of increase in crop yields are insufficient to keep pace with the rapid rates of global population growth. Thus, there is an urgent need to increase crop productivity beyond existing yield potentials to address the challenge of food security (Yamori 2013). Advances in biotechnology have led to increased opportunities to rapidly improve the crop production, for example, by facilitating the transfer of desired characteristics from other species, which is impossible through conventional plant breeding. Biotechnology for crop improvement has become a promising strategy to solve the food crisis.

So far, various crops have been engineered for enhanced resistance to various stresses such as herbicides, viruses, and abiotic stresses. The proportion of biomass partitioned into grain has already been brought close to its theoretical maximum thanks to plant breeding; thus, a fundamental limitation on plant productivity that has to be optimized for increasing crop yield is photosynthesis (Long et al. 2006). Photosynthesis is the process of converting light energy to chemical energy using CO_2 and water, and of storing it as carbohydrates. Enhancing photosynthetic capacity of plants is a promising approach to increase crop productivity. This chapter summarizes the various approaches by genetic engineering applied to enhance photosynthetic capacity and plant production. The targets considered for possible candidates include Rubisco, Rubisco activase, enzymes of Calvin–Benson cycle, CO_2 and carbohydrate transport, as well as photosynthetic electron transport (Figure 41.1). In addition, other areas that promise scope for photosynthetic improvement, such as altering canopy architecture or mining natural genetic variations for identifying new targets, are described.

41.2 IMPROVING PHOTOSYNTHETIC PERFORMANCE VIA IMPROVEMENTS IN RUBISCO KINETICS AND N ALLOCATION

41.2.1 RUBISCO SPECIFICITY AND CARBOXYLATION RATE

Rubisco (ribulose-1,5-bisphosphate carboxylase/oxygenase; EC 4.1.1.39) is the enzyme responsible for the fixation of CO_2 in

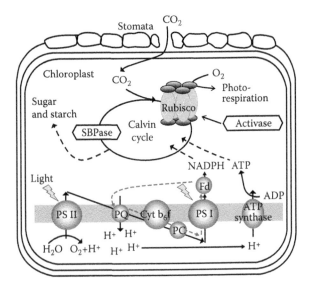

FIGURE 41.1 Scheme of C_3 photosynthetic reactions. Light energy captured by chlorophyll drives electron transport, which is used to reduce NADP+ to NADPH. Electron transport drives the translocation of H+ from the stroma into the lumen, generating a H+ electrochemical gradient and driving the production of ATP. ATP and NADPH produced by these reactions are now used to fix CO_2 (photosynthesis) into carbohydrates and to react with O_2 (photorespiration). Rubisco catalyzes the assimilation of CO_2 in the carboxylation reaction of the Calvin–Benson cycle in the chloroplast. Reactions of the Calvin–Benson cycle utilize NADPH and ATP to produce triose phosphates, which are required for the synthesis of carbohydrates and then finally of sugar and/or starch. The NADPH and ATP are also used in a range of other metabolic activities (e.g., nitrogen and sulfur metabolism, lipid and pigment synthesis) in the chloroplast. Rubisco: ribulose-1,5-bisphosphate carboxylase/oxygenase, SBPase: sedoheptulose-1,7-bisphosphatase, PS II: photosystem II, PQ: plastoquinone, Cyt b_6f: cytochrome b_6/f complex, PC: plastocyanin, PS I: photosystem I, Fd: ferredoxin.

photosynthesis. It has a low catalytic efficiency, able to fix only around 2–4 CO_2 molecules per second in higher plants; thus, 20–30% of nitrogen in terrestrial C_3 plant leaves is invested in Rubisco to maximally exploit the light environment (Galmés et al. 2014a). Given Rubisco's low catalytic efficiency, and the relatively high proportion of leaf nitrogen for which it accounts, numerous investigators have noted a strong positive correlation between leaf nitrogen content and photosynthetic rates (Evans 1989; Makino et al. 1997; Wright et al. 2004).

Rubisco can fix both carbon dioxide and oxygen in photosynthesis and photorespiration, respectively. Photosynthetic carbon fixation produces two molecules of phosphoglycerate (PGA) for every carbon fixed, while photorespiration produces one PGA and one phosphoglycolate (PGO). PGO must be recycled to PGA, with the loss of CO_2 and ammonia. Although the released CO_2 may be refixed by the chloroplasts, and the ammonia reassimilated in the leaves (Morris et al. 1988; Busch et al. 2013), photorespiration is generally considered to be a wasteful reaction.

To understand the reason Rubisco's oxygenase activity exists, we must consider the enzyme's evolutionary history.

Rubisco evolved around 3.5 Ga ago in an environment devoid of oxygen, and thus devoid of selection pressure against its oxygenase activity. The depletion of reduced iron and other chemicals in the oceans by photosynthetically evolved oxygen took around 1.3 Ga and led to an increase in molecular oxygen both in the oceans and release of free O_2 into the atmosphere (Schopf 2014), although by this time, despite the new selection pressure, Rubisco's oxygenase activity apparently could not be eliminated. Following the great oxygenation of 2.2 Ga, atmospheric O_2 concentrations remained below 10% of current atmospheric levels until approximately 800 Ma (Payne et al. 2009), while CO_2 levels remained 10–20 times present levels until the evolution of large, woody plants during the Devonian.

The present atmospheric partial pressure of O_2 is approximately 500 times higher than that of CO_2. Despite the lower solubility of oxygen than CO_2, oxygen levels may be 25- to 30-fold higher than CO_2 levels at Rubisco's catalytic site. Thus, over evolutionary time, as CO_2 levels have fallen and oxygen levels increased, Rubisco has evolved to discriminate between CO_2 and O_2, and modern C_3 Rubisco has a discrimination ratio (specificity) of around 100:1. However, increases in Rubisco's CO_2 affinity correlate with declines in the carboxylation rate, requiring an increased proportion of leaf nitrogen to be allocated to Rubisco, in order to maintain photosynthetic rates (Galmés et al. 2014a). This may have negative implications for whole plant nitrogen use efficiency, since approximately 50% of nitrogen in new leaves derives from the breakdown of Rubisco and other soluble proteins from senescent leaves (Mae and Ohira 1981; Mae 1986).

Drought causes stomatal closure. As CO_2 in intercellular gas spaces is rapidly depleted under high light conditions, drought induced stomatal closure leads to an increase in photorespiration as the O_2/CO_2 ratio increases rapidly. Galmés et al. (2014b) found that *Limonium* species in which drought had little effect on CO_2 availability at Rubisco's active site had lower CO_2 specificities than species in which drought significantly reduced CO_2 availability. As in other studies, a negative relationship existed between maximum carboxylation rate and specificity. As specificity increased, biomass accumulation increased, despite decreases in photosynthetic rates. Presumably this increase in biomass production was due to decreases in photorespiratory carbon losses.

Although generally seen as wasteful, photorespiration is thought to act as a mechanism by which excess light energy can be used, reducing photodamage to the chloroplast (André 2011). Even under moderate drought conditions, stomatal closure quickly depletes leaf CO_2 levels, while sucrose phosphate synthase activity is repressed, which may lead to feedback repression of photosynthesis (Vassey et al. 1991). In the absence of photorespiration, a lack of capacity for NADPH consumption would cause increases in both Mehler reaction-generated oxygen radicals and nonphotochemical quenching (Makino et al. 2002; Medrano et al. 2002). Given that most plants are probably under moderate stress most of the time, genetically engineered plants lacking oxygenase activity might exhibit higher levels of photodamage, reduced leaf lifespans, and reduced net carbon balance. It would seem that engineering an oxygenase-free Rubisco may not be viable within a C_3 architecture.

41.2.2 RUBISCO CONCENTRATION AND N ALLOCATION

Photosynthetic rates at ambient CO_2 levels correlate with leaf Rubisco concentration (Makino 2003). In grasses, Rubisco content increases rapidly during leaf emergence, reaching a maximum around full leaf expansion followed by a gradual decline until leaf death (Figure 41.2; Mae et al. 1983; Irving and Robinson 2006). In rice, even under high exogenous N supply conditions, approximately half of the nitrogen for new leaf growth comes from the remobilization of proteins from older leaves (Mae and Ohira 1981); thus, degradation of Rubisco in older leaves is necessary for the growth of new leaves. On the other hand, as leaf photosynthetic rates correlate with leaf Rubisco content (Makino et al. 1985), we might sensibly conclude that in order to maximize the photosynthetic rates of individual leaves, Rubisco degradation should be retarded, or synthesis maintained throughout the leaf lifespan. However, this would come at the expense of reduced rates of new leaf production.

Previous investigators have noted a correlation between leaf N content and the light environment of the canopy, allowing efficient distribution of nitrogen between leaves in a manner that increases canopy photosynthesis, compared to a uniform N distribution (Hirose and Werger 1987; Anten et al. 1995). This has been taken to suggest that plants have specific mechanisms for regulating leaf N content in a light-dependent manner (Grindlay 1997). Hikosaka (1996) grew the vine *Ipomoea tricolor* horizontally to prevent leaf shading, separating the effects of leaf age and light availability on nitrogen allocation, since older leaves tend to be lower in most canopies. At low N levels, older leaves contained significantly lower N levels than younger leaves, suggesting that leaf age may be a more important factor than light availability in determining leaf N contents. However, at high N levels, these differences disappeared, suggesting some plasticity under ideal conditions.

Hikosaka and Terashima (1995) demonstrated that the optimal partitioning of nitrogen into different photosynthetic components (e.g., Calvin–Benson cycle enzymes, chlorophyll and electron transport chain components) varies depending upon the light and nitrogen supply to the leaf. As leaves are part of a dynamic canopy, first exposed to full sunlight and with high levels of N influx, but later shaded with reduced N influx, we should expect to see changes in the balance of different protein levels through time, and perhaps inefficiencies, due to the leaves' limited ability for recycling and repartitioning nitrogen. At medium and high light levels (i.e., younger leaves at the top of the canopy), the optimal strategy

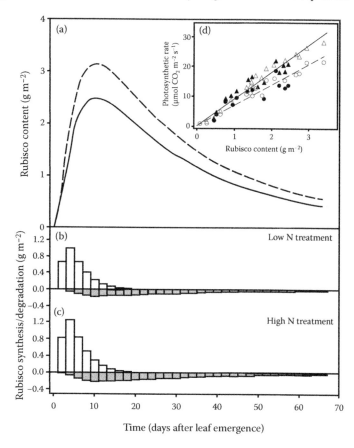

FIGURE 41.2 Rubisco content (a) and synthesis and degradation (b and c) through time in the twelfth leaf of rice plants fed with 1 or 3 mM NH_4NO_3 (solid and dashed lines, respectively) modelled using the model of Irving and Robinson (2006) with an a value of 3, and a b value of 0.004, and (d) the photosynthetic rate at 1 mM (closed symbols) or 3 mM (open symbols), at 350 ppm CO_2 and 500 mmol photons m^{-2} s^{-1} (circles) or 2000 μmol photons m^{-2} s^{-1} (triangles). (Data remodeled from Makino A et al., *Planta* 166:414–420, 1985.)

is for leaves to invest a large fraction of their nitrogen into Rubisco, particularly in nitrogen-replete plants. However, at lower light intensities (i.e., older leaves, deeper in the canopy), and at lower leaf N levels, a lower proportion of nitrogen should be partitioned to Rubisco. In this context, the vesicular export of Rubisco from chloroplasts during senescence can be understood (Izumi et al. 2010). As the production of Rubisco-containing bodies is suppressed by leaf carbohydrates, leaves lower in the canopy, which receive less light and produce less carbohydrate, may use leaf carbohydrate status as a signal to remobilize nitrogen for new leaf production.

At low CO_2 levels, such as those that have dominated for the last 400,000 years (Petit et al. 1999), Rubisco's carboxylation capacity is not saturated (Makino et al. 1983), and Rubisco content is understood to limit photosynthesis (Makino et al. 1988). In accordance with this, a recent paper demonstrated 32% and 15% biomass increases in Rubisco overexpressing rice grown at 280 and 400 ppm (ambient) CO_2, respectively, while at 1200 ppm CO_2 levels, no promotion of biomass production was noted. Conversely, in Rubisco antisense plants, biomass accumulation was significantly retarded compared to controls irrespective of CO_2 levels (Sudo et al. 2014). Plants grown at elevated CO_2 (e.g., in free-air CO_2 enrichment [FACE] experiments) generally exhibit reduced leaf Rubisco concentrations, relative to controls (Moore et al. 1999). In Arabidopsis grown at 1000 ppm CO_2, Rubisco protein content decreased by 34%, and transcript abundances by 38% and 60% from *rbcS* and *rbcL*, respectively (Cheng et al. 1998), although these results are not typical for all species. For example, Rubisco levels in pea, spinach, parsley, and maize are relatively unaffected by CO_2 elevation (Moore et al. 1998). Reduced N investment in Rubisco may allow greater biomass production and more rapid plant development.

Anten et al. (1995) demonstrated that in order to maximize photosynthetic rates, nitrogen should be remobilized from older leaves to newer, younger leaves, higher in the canopy. Hikosaka and Terashima (1995) further showed that Rubisco should be a prime target of nitrogen remobilization from older to younger leaves. If anything, Anten et al. (1995) show us that nitrogen should be remobilized more rapidly from older leaves than it actually is, and the gradient in canopy N levels should be steeper. We might posit various explanations for this, for example, that the optimal N distribution is further shaped by the energy costs associated with leaf growth, protein synthesis, and degradation, or that the maintenance of N in older leaves represents some form of insurance against grazing damage. Either way, increasing plant growth rates seems to require increasing Rubisco levels, which seems difficult without substantially increasing N supply, or increasing CO_2 levels at the site of carbon fixation.

41.2.3 Opportunities for Improving Rubisco Performance

Improving plant photosynthesis has been a long-held goal of plant molecular biologists. As noted, along with its carboxylase activity, Rubisco also functions as an oxygenase

in photorespiration, costing the plant carbon, nitrogen, and energy, but providing a degree of protection against photodamage. Photorespiratory carbon losses are a particular problem for C_3 plants, where primary assimilation of carbon dioxide is distributed among leaf mesophyll cells. While C_3 plants dominate in temperate environments, under drought conditions, C_4 plants have an advantage. In C_4 plants, CO_2 is assimilated by PEP carboxylase and shuttled to bundle sheath cells surrounding the leaf veins, where the CO_2 is liberated and subsequently refixed by Rubisco. PEP carboxylase has a high affinity for CO_2, allowing the maintenance of high CO_2 levels at the active site of Rubisco, thus suppressing Rubisco's oxygenase activity. Suppression of photorespiration, for example, by growing C_3 plants at reduced oxygen partial pressures, has long been known to lead to significant enhancement of both CO_2 fixation rates and growth (Björkman 1966; Björkman et al. 1968).

C_4 Rubiscos have maximum carboxylation rates up to double those in C_3 plants allowing them to invest less nitrogen in Rubisco (Ghannoum et al. 2005). The highest maximum carboxylation rates, and thus lowest specificities, are found in cyanobacteria, which can take advantage of both the lower solubility of oxygen in water compared with CO_2 (an effect that is exacerbated with depth) and carbon dioxide concentrating mechanisms (Raven 2000). A recent perspective (von Caemmerer et al. 2012) is optimistic about the possibility of the first C_4 prototype rice plants within a few years. Although the technical challenges are great, we should consider that C_4 has evolved more than 62 times in various flowering plant groups, including 24 times in the grasses alone (Studer et al. 2014).

In *Flaveria*, a single mutation (M309I) was key in reducing specificity and increasing the maximum carboxylation rate (Whitney et al. 2011), suggesting that the switch from C_3-type to C_4-type Rubisco may be relatively simpler than previously considered. Conversely, a recent paper using a modelling approach to explore the effect on mutations on the stability and activity of Rubisco found that 98.9% of amino acid substitutions had a moderate or highly destabilizing effect, while 1.1% of mutations were found to have a stabilizing effect (Studer et al. 2014). Most of these stabilizing mutations were close to the active site and generally would have a negative effect on enzyme function. Of ancestral mutations, 91.5% were noted to have had minor effects on enzyme stability, suggesting that stability is a strong constraint on enzyme evolution. The evolution of C_4 Rubisco required the presence of a strongly destabilizing mutation, which could then be followed up by a series of stabilizing mutations. The authors describe Rubisco as having evolved on a small island of stability, which raises fundamental questions about how far we can improve C_3 Rubisco in a stepwise fashion.

Recently, Lin et al. (2014) successfully produced transgenic tobacco plants, containing functional *Synechococcus elongatus* Rubisco. *S. elongatus* Rubisco has a high catalytic rate—approximately 12 CO_2 molecules can be fixed per second, compared to 3 molecules per second for host tobacco Rubisco. The transgenic plants contained only 12–18% of the Rubisco levels found in wild-type tobacco plants, and *S. elongatus* Rubisco has a low affinity for CO_2 (Km value of 200 µM compared to 6–12 µM for C_3 angiosperms); thus, they had to be grown at 9000 ppm CO_2 (Lin et al. 2014). It is clear that *Synechococcus* Rubisco could not be used in C_3 plants under current environmental conditions; however, if expressed in a C_4 architecture, or a C_3 plant with a CCM, it might significantly increase photosynthetic rates, allowing the reallocation of significant amounts of N for new leaf growth.

Rubisco evolved in a high CO_2, low O_2 environment around 3.5 billion years ago. For 85% of its history, it has evolved in an environment with oxygen levels less than 10% of present levels, where its oxygenase activity would have presented little issue. However, in the last 500 million years, the evolution of large, woody plants, and the resulting vast increase in atmospheric oxygen, prompted the evolution of increased specificity for CO_2. This increased specificity comes at the cost of the maximum carboxylation rate, requiring plants to partition up to one-third of leaf nitrogen to Rubisco. The evolution of C_4 plants with higher maximum carboxylation rates allowed a reduced nitrogen allocation to Rubisco, which could be invested in new biomass production. Clearly, a path to increase the yield of many major C_3 crops would be the development of C_4 varieties, using cyanobacterial Rubisco, which might break the deadlock between nitrogen investment in photosynthetic proteins in living leaves, and the production of new leaves and increased biomass.

41.3 INCREASING THE THERMOTOLERANCE OF RUBISCO ACTIVASE TO MAINTAIN HIGHER RUBISCO ACTIVITY UNDER HIGH TEMPERATURES

The Rubisco catalytic sites must first be activated for CO_2 fixation to take place (Figure 41.1). This requires the carbamylation of a lysine residue at the Rubisco catalytic site, allowing the binding of Mg^{2+} and ribulose-1,5-bisphosphate (RuBP). Rubisco activase facilitates carbamylation and the maintenance of Rubisco activity by removing inhibitors such as tight-binding sugar phosphates from Rubisco catalytic sites in an ATP-dependent manner (Spreitzer and Salvucci 2002; Portis 2003; Parry et al. 2008).

In many plant species, Rubisco activation state decreases at high temperatures (Crafts-Brandner and Salvucci 2000; Salvucci and Crafts-Brandner 2004a; Yamori et al. 2006, 2014; Yamori and von Caemmerer 2009). It has been reported that Rubisco deactivation at high temperature could be due to the insufficient activity of Rubisco activase to keep pace with the faster rates of Rubisco inactivation at high temperature because of its thermolability (Salvucci and Crafts-Brandner 2004b). Reduced levels of Rubisco activase resulted in decreased photosynthetic rate at high temperature using mutants/transgenic plants in Arabidopsis (Salvucci et al. 2006), rice (Yamori et al. 2012), and tobacco (Yamori and von Caemmerer 2009). It has been reported that overexpression of maize Rubisco activase in rice stimulated Rubisco activation

state and photosynthetic rate at high temperature (Yamori et al. 2012). In addition, transgenic Arabidopsis expressing thermotolerant Rubisco activase isoforms generated by either gene shuffling technology (Kurek et al. 2007) or chimeric Rubisco activase constructs (Kumar et al. 2009) exhibited greater photosynthetic rates, greater biomass production, and increased seed yields, compared with the wild-type plants. This suggests that Rubisco activase activity is a major limiting factor for photosynthesis under high temperature, and that engineering of Rubisco activase would be an efficient strategy to improve crop yield under high temperatures.

An additional vital role of Rubisco activase in the regulation of photosynthesis under fluctuating light has been reported recently (Yamori et al. 2012). At any leaf temperature, the speed by which Rubisco activation state and photosynthetic rate change following an increase in light intensity changed markedly faster in plants overexpressing maize Rubisco activase compared to wild-type rice. In addition, the parameter changed more slowly in antisense plants with reduced levels of Rubisco activase. Thus, a selective enhancement of Rubisco activase capacity would be also advantageous in natural environments where irradiance often fluctuates. Recently, the structure of Rubisco activase has been determined, providing insight into its interactions with Rubisco (Stotz et al. 2011) and its counterpart CbbX in red algae (Mueller-Cajar et al. 2011). The structural information coupled with the knowledge of regulation in Rubisco activase would contribute to improve its thermostability and catalytic properties.

41.4 INCREASING THE CO_2 CONCENTRATION AROUND RUBISCO

Rubisco is a dual-function enzyme that can use either CO_2 or O_2 to carboxylate or oxygenate its primary substrate RuBP, two processes that have been termed photosynthesis and photorespiration (Bauwe et al. 2010). While photosynthesis results in a net fixation of CO_2, the photorespiratory pathway requires ATP and releases previously fixed CO_2. The rate of photorespiration is affected by the concentration of CO_2 in the chloroplast (C_c) relative to the O_2 concentration and increases with increasing temperature. At current atmospheric CO_2 concentrations and a temperature of 30°C, the rate of photorespiratory CO_2 release from the mitochondria is approximately 25% of the rate of net CO_2 assimilation (Sàge et al. 2012). Increasing C_c, and thereby minimizing photorespiration, is therefore a promising target to increase the rate of photosynthesis in crops. In C_3 plants, C_c is determined not only by the atmospheric CO_2 concentration but also by the resistance for CO_2 to diffuse from the outside of the leaf to the chloroplast (Figures 41.1 and 41.3). CO_2 diffusion to the chloroplast can be influenced by modifying the conductance through the stomata (g_s) to the intercellular air space (IAS), either by increasing the stomatal density (Tanaka et al. 2013) or by preventing stomatal closure (Kusumi et al. 2012). Both approaches resulted in increased rates of net CO_2 assimilation, albeit at the cost of higher transpiration rates and lower water-use efficiency.

An alternative approach addresses the other major diffusion resistance for CO_2 from the intercellular air space into the mesophyll cell chloroplasts. In contrast to modifying g_s, increasing mesophyll conductance (g_m) does not negatively affect water-use efficiency. A large proportion of g_m has been shown to be dependent on cell wall thickness, meaning that CO_2 diffusion can potentially be improved by modifying plants to have smaller mesophyll cells with thinner cell walls (Terashima et al. 2011). Plants with smaller cells not only can afford to have thinner cell walls but also result in an increased mesophyll surface area. A larger surface area exposed to the IAS can be occupied by more chloroplasts pressing against the cell walls, which again aids CO_2 diffusion into the chloroplast (Evans et al. 1994), while at the same time impairs diffusion of photorespiratory CO_2 out into the IAS (Figure 41.3; Busch et al. 2013).

The second important component of g_m involves the diffusion of CO_2 through the plasma and chloroplast membranes (Evans et al. 2009), and a number of approaches are under way to increase the chloroplastic CO_2 concentration in C_3 plants by increasing the membrane permeability for CO_2. Aquaporins that are permeable to CO_2 are proteins that assist the CO_2 diffusion through the membranes by providing pores through which CO_2 can be channeled (Kaldenhoff 2012). It has been shown that the disruption to the aquaporin *AtPIP1;2* gene limits CO_2 transport across the membrane (Heckwolf et al. 2011), while the overexpression of different aquaporin genes resulted in increased g_m (Hanba et al. 2004;

Flexas et al. 2006). Modifying the membrane permeability for bicarbonate $\left(HCO_3^-\right)$ can also facilitate the transport of CO_2 across the membrane. Adding cyanobacterial HCO_3^- transporters to the chloroplast envelope of C_3 plants would provide a parallel route for inorganic carbon to enter the chloroplast, in addition to the diffusion of dissolved CO_2 (Price et al. 2011, 2013). Lastly, the enzyme carbonic anhydrase (CA) plays a role in facilitating the diffusion of CO_2 throughout the aqueous phase of the chloroplast stroma by interconverting between CO_2 and HCO_3^- and providing a more even distribution of CO_2 throughout the chloroplast (Evans et al. 2009). It has been suggested that the amount of CA found in plants is somewhat limiting the conductance in the stroma of C_3 crops, opening up the possibility for improving this aspect by molecular engineering (Makino et al. 1992; Tholen and Zhu 2011).

In addition to decreasing the resistance for CO_2 to diffuse into the chloroplast from the IAS, advances have been made to engineer plants that can make better use of the CO_2 released from photorespiration. The introduction of a bacterial pathway for glycolate metabolism into the chloroplasts of *Arabidopsis thaliana* creates a photorespiratory bypass that releases CO_2 directly into the chloroplast and thereby increases C_c (Kebeish et al. 2007; Maier et al. 2012). However, this approach might be limited to minor benefits in high yielding C_3 crop plants that already trap a high proportion of photorespiratory CO_2 by having a high chloroplast cover of the cell wall space facing the IAS (Busch et al. 2013).

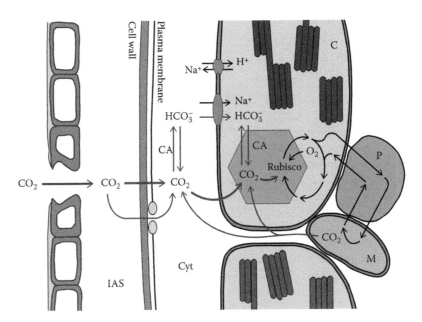

FIGURE 41.3 Schematic representation of mechanisms for concentrating CO_2 around Rubisco. The diagram shows CO_2 moving from the outside of the leaf through the stomatal pore to the intercellular air space (IAS), from where it diffuses through the cell wall and plasma membrane into the cytosol. CO_2 permeable aquaporins may aid the transport of CO_2 into the cytosol (Cyt) of the mesophyll cell. The transport of CO_2 into the chloroplast (C) could be improved by introducing a cyanobacterial HCO_3^- transporter into the chloroplast envelope, creating an additional pathway for CO_2 entering the chloroplast. The addition of a Rubisco and CA-containing carboxysome (represented by the hexagonal structure) could further increase the CO_2 concentration around Rubisco, minimizing the rate of photorespiration. Photorespiration involves peroxisomes (P) and mitochondria (M) and results in the release of CO_2. Capturing this CO_2, either by minimizing the diffusion out into the IAS or by engineering a pathway that results in the release of photorespiratory CO_2 directly into the chloroplast, also could increase the CO_2 concentration around Rubisco. (Adapted from Price GD et al., *Plant Physiology* 155: 20–26, 2011.)

A substantial increase in the CO_2 concentration around Rubisco may be achieved by installing a carbon concentrating mechanism (CCM) in C_3 crop plants. Cyanobacteria have evolved a single-cell CCM, in which Rubisco is encapsulated in a cellular compartment known as carboxysome (Price et al. 2011). In the carboxysomes, the CO_2 concentration is enriched up to 1000-fold, significantly reducing the rate of photorespiration. This is achieved by the conversion of CO_2 to HCO_3^-, which can be accumulated in the cell, by CA at the cyanobacterial thylakoid membrane, combined with the CA-mediated CO_2 release in the carboxysome. A fully functioning cyanobacterial-like CCM in the C_3 chloroplast would require the introduction of HCO_3^- transporters, adjustments in the expression of chloroplastic CA in order to allow HCO_3^- accumulation, and the establishment of a Rubisco- and CA-containing compartment, such as the carboxysome (Figure 41.3) (Price et al. 2011, 2013).

C_4 plants evolved a two-cell CCM, where CO_2 is initially fixed in the mesophyll cells by the enzyme phosphoenolpyruvate carboxylase (PEPC) to produce a C_4 acid. The organic acid diffuses to the specialized gas-tight bundle-sheath cells, where it is decarboxylated, resulting in significantly increased CO_2 concentrations around Rubisco. In addition to higher yields, the benefits of this photosynthetic pathway include an improved nitrogen- and water-use efficiency (Sage 2004). Currently, considerable efforts are undertaken to engineer features of the complex C_4 pathway into C_3 crops like rice (Covshoff and Hibberd 2012; von Caemmerer et al. 2012). Challenges associated with this approach include morphological adjustments, such as the establishment of a Kranz(-like) anatomy, as well as introducing the C_4 biochemistry into C_3 leaves (Covshoff and Hibberd 2012).

Another CCM that is based on a prefixation step of CO_2 into C_4 acids is the crassulacean acid metabolism (CAM). Here, the initial CO_2 fixation step by PEPC and the final fixation by Rubisco are not spatially separated like in C_4 plants, but both steps are located in the same cell type and are separated temporally. CAM plants take up CO_2 during the night and store it as C_4 acid in the vacuole, with the decarboxylation step and subsequent CO_2 fixation by Rubisco occurring during the day (Dodd et al. 2002). By shifting the CO_2 uptake to the nighttime and keeping stomata largely closed during the day, the plant can achieve high rates of photosynthesis with a water-use efficiency superior to both C_3 and C_4 plants. Early stages of attempting to move the CAM metabolism into C_3 plants have been reported, with the aim of improving plant productivity under water-limited conditions (Borland et al. 2014; DePaoli et al. 2014).

41.5 ENHANCING THE RATE OF CHLOROPLAST ELECTRON TRANSPORT IN THE THYLAKOID MEMBRANES

ATP and NADPH generated during photosynthetic electron transport in the thylakoid membranes are used to power photosynthetic carbon reduction. In a future high CO_2 world, the rate of CO_2 assimilation would be limited by the rate of RuBP regeneration in the Calvin–Benson cycle (Farquhar et al. 1980), which in turn is limited by the chloroplast electron transport capacity (Yamori et al. 2011). The cytochrome b_6/f complex has a unique role in chloroplast electron transport (Figure 41.1), as it can act in both linear electron transport (production of ATP and NADPH) and cyclic electron transport (ATP generation only) (Yamori et al. 2015). There is a strong linear relationship between chloroplast electron transport rate and cytochrome b_6/f complex content at any leaf temperature (Yamori et al. 2011). Thus, it could be a suitable target for genetic manipulation to improve photosynthesis and thus plant yield.

Figure 41.4 shows the Z-scheme of photosynthetic electron transfer from water to $NADP^+$ and approximate estimated times for various steps in the scheme. It has been proposed that photosynthetic electron transport could be mostly limited by the reduction and reoxidation of the diffusible components (i.e., plastoquinone, plastocyanin, and ferredoxin) that interconnect the membrane-bound complexes (Yamori et al. 2008). Previous experiments with antisense lines have shown that even a moderate reduction of the amounts of chloroplastic ferredoxin NADP(H) oxidoreductase (FNR), which catalyzes the terminal reaction of the photosynthetic electron transport chain by transferring electrons from reduced ferredoxin to $NADP^+$, has a negative impact on photosynthetic rate under both low and high light conditions (Hajirezaei et al. 2002). However, overexpression of FNR (Rodriguez et al. 2007) or ferredoxin (Yamamoto et al. 2006) did not increase photosynthesis or plant growth in tobacco, irrespective of growth light conditions. Electron transfer between the cytochrome b_6/f complex and photosystem I is exclusively mediated by plastocyanin in higher plants, whereas in many algae, it is mediated by cytochrome $c6$. Variations in plastocyanin levels have been reported to coincide with variations in photosynthetic electron transport activity (Burkey 1994; Burkey et al. 1996; Schöttler et al. 2004), leading to the conclusion that the plastocyanin pool size is limiting for photosynthetic electron transport. Two homologous plastocyanin isoforms are encoded by the genes *PETE1* and *PETE2* in the nuclear genome of *A. thaliana*. An analysis of knockout plants for *PETE1* and *PETE2* in *A. thaliana* showed that the plastocyanin content can be significantly decreased without apparent changes in the photosynthetic rate, suggesting that the concentration of plastocyanin is not limiting for photosynthetic electron transport rate (Pesaresi et al. 2009). However, overexpression of either *PETE1* or *PETE2* results in an increase in biomass production (Pesaresi et al. 2009). Thus, there is a discrepancy between experimental data of knockout and overexpression lines. The introduction of a parallel electron carrier between cytochrome b_6/f complex and photosystem I through the expression of an algal cytochrome $c6$ gene in *A. thaliana* resulted in an enhancement of electron transport rate and thus NADPH and ATP production, leading to enhanced plant growth (Chida et al. 2007). This important evidence indicates that plastocyanin could be a suitable target for genetic manipulation to improve electron transport and thus plant yield.

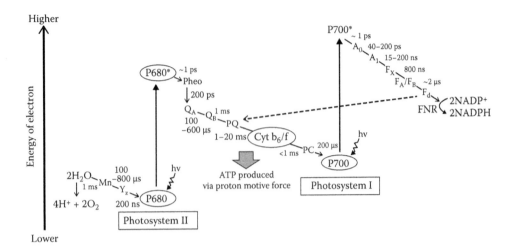

FIGURE 41.4 Scheme of photosynthetic electron transport from H_2O to $NADP^+$, leading to the release of O_2, the reduction of NAD^+ to NADPH, and production of ATP via proton motive force. Estimated times for various steps in the scheme are also noted on the figure. Mn: Mn cluster; Yz: tyrosine-161 on D1 protein; P680: the reaction center of photosystems II (PS II); P680*: the excited P680; Pheo: the primary electron acceptor of PS II; Q_A: the primary plastoquinone electron acceptor of PS II; Q_B: the secondary plastoquinone electron acceptor of PS II; PQ: plastoquinone pool; Cyt b_6/f: cytochrome b_6/f complex; PC: plastocyanin; P700: the reaction center of photosystem I (PS I); P700*: the excited P700; A_0: primary electron acceptor of PS I; A_1: secondary electron acceptor of PS I; F_X, F_A, and F_B: 3 different iron sulfur centers; F_d: ferredoxin; FNR: ferredoxin-NADP reductase. The straight lines show linear electron transport, while the dashed line shows cyclic electron transport around PS I. (Modified from Govindjee, Chlorophyll a fluorescence: A bit of basics and history, In: G.C. Papageorgiou, Govindjee (Eds.), *Chlorophyll a Fluorescence: A Signature of Photosynthesis*, Kluwer Academic Publishers, Dordrecht, pp. 1–42, 2004.)

There have been other reports that documented an enhancement of plant biomass by genetic manipulation of photosynthetic electron transport. In the plant cell, NADP is mainly located in the chloroplast, where $NADP^+$ functions as the final electron acceptor of the photosynthetic electron transport chain (Wigge et al. 1993). NAD kinase regulates NAD(H)/NADP(H) balance through its catalysis of NAD phosphorylation in the presence of ATP (Kawai and Murata 2008). In *A. thaliana*, one of the NADK isoforms localized in the chloroplast (NADK2; Chai et al. 2005) catalyzes a key step in the regulation of the NAD/NADP ratio (Kawai and Murata 2008). It succeeded in enhancing electron transport and CO_2 assimilation rates by overexpression of Arabidopsis chloroplastic *NADK* (*AtNADK2*) in rice (Takahara et al. 2010). Moreover, the controlled up-regulation of *Arabidopsis* chlorophyllide *a* oxygenase (CAO), involved in chlorophyll *b* biosynthesis, in tobacco has been shown to enhance electron transport rates, CO_2 assimilation, and plant biomass by co-modulating the expression of different thylakoid membrane proteins (Biswal et al. 2012). In addition, plants with a mutation in TAP38, an enzyme involved in dephosphorylation of light-harvesting complex of photosystem II, exhibited increased photosynthetic electron flow, leading to enhancement of plant growth under low light conditions (Pribil et al. 2010).

In situations where the electron transport rate is limited by the amount of available light that can be absorbed by the plant, increased light harvesting might boost photosynthetic productivity. Land plants utilize chlorophyll *a* and *b*, which absorb light in the visible part of the solar spectrum of about 400–700 nm. Chlorophyll *d* used by *Acaryochloris* (Miyashita et al. 1996) and chlorophyll *f* discovered recently

in the cyanobacterial communities of stromatolites (Chen et al. 2010) have red-shifted absorption spectra that enable their host organisms to perform oxygenic photosynthesis at the much longer wavelengths of 700–750 nm, which are inaccessible to other organisms. Introducing these chlorophylls into higher plants to supplement or replace the existing chlorophylls could potentially increase the amount of usable photon flux by up to 19% (Chen and Blankenship 2011). Since the photosynthetic electron transport chain provides energy and reducing equivalents for the reduction of fixed CO_2 to carbohydrates in the Calvin–Benson cycle as well as for nitrogen assimilation and other processes, genetic manipulation of photosynthetic electron transport could be a candidate to improve the entire photosynthetic system, and thus plant yield.

41.6 ENHANCING THE ACTIVITY OF CALVIN–BENSON CYCLE ENZYMES

The Calvin–Benson cycle utilizes ATP and NADPH from photosynthetic electron transport to fix CO_2 into carbon skeletons that are used for sucrose and starch production (Figure 41.1). The Calvin–Benson cycle also supplies intermediates to a lot of other pathways in the chloroplast, including the shikimate pathway for the biosynthesis of amino acids and lignin, isoprenoid biosynthesis, and precursors for nucleotide metabolism and cell wall synthesis (Lichtentahler 1999). This cycle comprises 11 different enzymes, catalyzing 13 reactions, and is initiated by Rubisco (Figure 41.5; Raines 2003). Four enzymes are regulated by thioredoxins: GAPDH, FBPase, SBPase, and PRK. Two of the 11 enzymes catalyze reversible reactions: aldolase and transketolase.

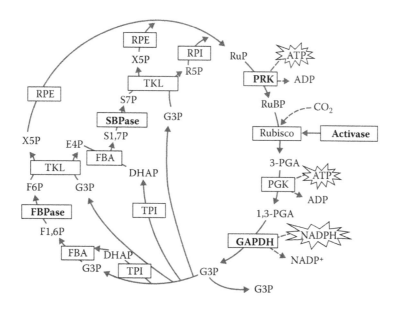

FIGURE 41.5 Scheme of the Calvin–Benson cycle. The Calvin–Benson cycle is the primary pathway for carbon fixation, and this cycle has 13 reaction steps catalyzed by 11 enzymes in the chloroplasts in higher plants. <Enzymes> Rubisco: ribulose-1,5-bisphosphate carboxylase/oxygenase; PGK: phosphoglycerate kinase; GAPDH: glyceraldehyde-3-phosphate dehydrogenase; TPI: triose phosphate isomerase; FBA: fructose-1,6-bisphosphate aldolase; FBPase: fructose-1,6-bisphosphatase; TKL: transketolase; SBPase: sedoheptulose-1,7-bisphosphatase; RPE: ribulose-5-phosphate 3-epimerase; RPI: ribose-5-phosphate isomerase; PRK: phosphoribulokinase. <Metabolites> RuBP: ribulose-1,5-bisphosphate; 3-PGA: 3-phosphoglycerate; 1,3-PGA: 1,3-bisphosphoglycerate; G3P: glyceraldehyde-3-phosphate; DHAP: dihydroxy-acetone phosphate; F1,6P: fructose-1,6-bisphosphate; F6P: fructose-6-phosphate; X5P: xylulose-5-phosphate; E4P: erythrose-4-phosphate; S1,7P: sedoheptulose-1,7-bisphosphate; S7P: sedoheptulose-7-phosphate; R5P: ribose-5-phosphate; RuP: ribulose-5-phosphate. Four enzymes are regulated by thioredoxins: GAPDH, FBPase, SBPase, and PRK. Some proteins that control the activity of Calvin–Benson cycle enzymes are also regulated by thioredoxins (i.e., Rubisco activase and CP12, which forms a complex with PRK and GAPDH). (Modified from Michelet, L. et al., *Frontiers in Plant Science* 4: 470, 2013.)

To understand how metabolic fluxes in photosynthesis reactions are determined, metabolic flux control analysis has been examined (Woodrow 2009). Enzymes are assigned a control coefficient (between 0 and 1), which is a measure of the degree to which the flux is sensitive to changes in enzyme concentration. Flux control analyses using antisense plants for various Calvin–Benson cycle enzymes revealed a high control coefficients for Rubisco activity on photosynthetic rate at ambient CO_2 concentrations and also sedoheptulose-1,7-bisphosphatase (SBPase) at high CO_2 concentrations, but low control coefficients for the other Calvin–Benson cycle enzymes (Raines 2011). It has been reported that the photosynthetic rate at high light is sensitive to SBPase activity (Harrison et al. 1998; Ölcer et al. 2001). Moreover, photosynthesis and biomass production were improved by overexpression of cyanobacterial fructose-1,6-bisphosphatase and sedoheptulose-1,7-bisphosphatase (FBPase/SBPase) in tobacco (Miyagawa et al. 2001). The subsequent research separated the effects of FBPase and SBPase by enhancing the activity of each one separately (Tamoi et al. 2006), showing a 25% rise in photosynthetic rate in response to increases in SBPase of between 60% and 330%. A smaller but significant 15% rise in A_{sat} was observed in response to a 130% rise in FBPase. Therefore, they concluded that the photosynthetic rate was apparently sensitive to both FBPase and SBPase activity at all irradiances. It has been reported that photosynthesis and biomass production are also improved by overexpression of SBPase

in rice (Feng et al. 2007) and in tobacco (Lefebvre et al. 2005). In addition, transgenic tobacco plants overexpressing SBPase from *A. thaliana* exhibited enhanced photosynthetic rate and biomass production when grown at 585 ppm CO_2 in FACE experiments (Rosenthal et al. 2011). These reports provide evidence that SBPase is a major limiting factor for the Calvin–Benson cycle and thus photosynthesis, and that engineering of SBPase would be an efficient strategy to improve photosynthetic rate and crop yield especially in a future high CO_2 world. The impact of FBPase on photosynthesis has not been brought to light yet, since the study of transgenic potato plants with reduced amounts of FBPase revealed that FBPase has a small control coefficient (Kossman et al. 1994). Further studies are needed to clarify the impact of FBPase on photosynthesis and plant growth.

There is a view among researchers in flux control analysis that enzymes catalyzing the reversible reactions of photosynthesis should not be flux limiting. However, nonregulated enzymes catalyzing reversible reactions, aldolase and transketolase, exerted significantly higher flux control for photosynthesis. In the first study, the photosynthetic rate of antisense potato plants with varying amounts of plastid aldolase was examined (Haake et al. 1998, 1999). Moderate reductions in aldolase activity were shown to cause significant reductions in photosynthetic rate and plant growth. Moreover, transgenic *Arabidopsis* and tobacco plants overexpressing plastidial aldolase exhibited enhanced photosynthetic rate, seed yield, and biomass production (Hatano-Iwasaki and Ogawa 2012; Uematsu et al. 2012). Also, experimental evidence has shown that small decreases in plastid transketolase activity in antisense tobacco transformants have dramatic effects on photosynthesis (Henkes et al. 2001). These studies confirm the rate-limiting nature of plastid aldolase and transketolase, and reveal its role in determining carbon flux in the Calvin–Benson cycle.

Taken together, the manipulation of SBPase and aldolase (and possibly transketolase) levels in chloroplasts could increase photosynthetic capacity and could be used as a novel biotechnological tool to produce crops with high yield. Since increased accumulation of foliar carbohydrates has been shown to limit photosynthetic rate (Leakey et al. 2009), it should be noted that carbohydrate export from the source (leaves) to its utilization in sink tissues (seed, stem, roots, etc.) could also be an important target for enhancing photosynthesis that has remained unexplored.

41.7 ENHANCING THE CAPACITY OF PHLOEM CARBOHYDRATE TRANSPORT AND CARBON UTILIZATION

Accumulation of nonstructural carbohydrates in mature *Phaseolus vulgaris* leaves is known to cause photosynthetic repression (Araya et al. 2006). Although in several species soluble sugars suppress the expression of photosynthetic genes (Pego et al. 2000), Araya et al. (2006) noted a decrease in the Rubisco large subunit concentration only at the final time point measured, and no difference in LHCII levels between control and plants fed with 20 mM sucrose solution, suggesting that any repression was not due to differences in leaf photochemistry. Conversely, Cheng et al. (1998) noted a 35–40% decrease in *rbcL* transcript—the gene encoding the Rubisco large subunit—in Arabidopsis plants grown at elevated CO_2.

Conventionally, it is considered that the main drivers of stomatal aperture are water availability, light intensity, vapor pressure deficit, and intercellular CO_2 concentration. However, recent evidence suggests that leaf carbon balance, which is at least partially mediated by phloem loading of carbohydrates produced in photosynthesis, can have an effect at levels comparable to hydraulic limitation (Nikinmaa et al. 2013). Assimilate transport in the phloem requires a hydrostatic gradient between photoassimilate sources (i.e., leaves) and sinks (e.g., roots and growing tissues). Phloem draws water from the xylem at sugar loading sites, thus competing directly with leaf tissues for water. High levels of sugar loading can lead to a decrease in flux rates due to increases in phloem sap viscosity (Hölttä et al. 2009), thus leading to a buildup in assimilate levels in the leaves, photosynthetic suppression, stomatal closure, and photorespiration. Conversely, high transpiration rates may also decrease water availability for phloem transport, leading to leaf carbohydrate accumulation (Hölttä et al. 2006).

Under drought conditions, sugars accumulate in the root phloem sap to counter the osmotic potential of the xylem.

Decreased sugar gradient between the leaves and the roots, and increased phloem sap viscosity, further decreases carbohydrate efflux from the leaves (Nikinmaa et al. 2013; Sevanto 2014). Carbohydrate build-up in leaves has been shown to reduce the production of Rubisco-containing bodies, which are important in the degradation of Rubisco in senescent leaves (Izumi et al. 2010), presumably since carbohydrate build-up is a signal that the leaf has sufficient light available for photosynthesis. Thus, carbohydrate build-up in leaves may delay leaf senescence, although it also tends to repress photosynthesis. Leaf-derived carbohydrates are also required for the production of new tissues (Lattanzi et al. 2005; Carvalho et al. 2006), but it is unclear whether carbon is limiting for leaf production for most plants under natural conditions.

Clearly, the loading and efficient transport of carbohydrates by the phloem have implications, both positive in terms of the potential for increased leaf lifespan and negative in terms of stomatal limitation, and potentially increased photorespiration. However, the balances of these factors are far from clear. Furthermore, it is presently unclear how much genetic variability exists for phloem transport characteristics, making a clear evaluation for its potential in improving crop yields impossible.

41.8 INCREASING CROP YIELD BY OPTIMIZING PHOTOSYNTHESIS ON THE CANOPY LEVEL

Improving photosynthesis to increase food production ultimately means maximizing the photosynthetic efficiency of a crop canopy rather than the individual plant. Canopy photosynthetic CO_2 uptake is integrated across the plant from the sunlit leaves that are above light saturation to the shaded leaves that operate below their full capacity. The ideal crop plant architecture should optimize the light-use efficiency at the leaf level while maximizing the amount of absorbed solar radiation by the canopy per unit ground area (Zhu et al. 2010). Several ways have been proposed for how to alter plant architecture and biochemistry in order to achieve the highest conversion efficiency of solar radiation to biomass. This includes distributing the solar radiation more evenly throughout the canopy as well as matching the photosynthetic capacity at different layers within the canopy to the available light (Zhu et al. 2010). Two main approaches are employed to distribute sunlight more evenly throughout the canopy: adjusting the leaf angle of the plant and modifying the leaf chlorophyll content, both with the aim to reduce the light absorption of the leaves at the top of the canopy and to optimize the light penetration to the leaves lower in the canopy (Figure 41.6).

Light absorption by the canopy varies with, among other things, the angle of the leaves and the leaf area per unit ground, or leaf area index (LAI). The leaf angle, at which the interception of daily incident solar radiation is maximized, depends on LAI. Early in the growing season when LAI is low, maximum light interception is achieved in canopies with horizontal leaves covering a large proportion of the ground

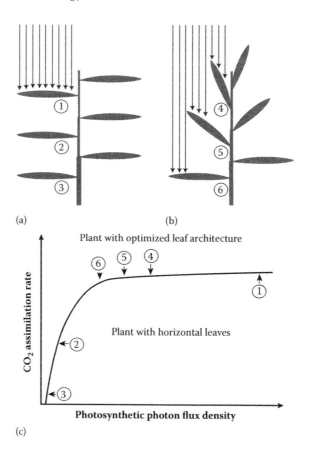

FIGURE 41.6   The effect of plant architecture on the CO_2 assimilation rate of leaves in a crop canopy. (a) Plants with mostly horizontal leaves absorb most of the incoming solar radiation in the upper layer of the canopy. While these leaves have to dissipate the majority of the absorbed photosynthetic photon flux density as heat, the shaded layers of the lower canopy operate below light saturation. (b) A plant with optimized leaf architecture and more erect leaves in the upper canopy. A larger fraction of incoming solar radiation can penetrate to the lower canopy, bringing all leaves of the canopy closer to the minimum PPFD, at which CO_2 assimilation is saturated (c). Similarly, a reduced amount of chlorophyll in the upper canopy facilitates a more even distribution of solar radiation throughout the canopy. (Modified from Long SP et al., *Plant, Cell & Environment* 29: 315–330, 2006.)

area. However, for canopies with an LAI above 2, horizontal leaves at the upper layers of the canopy would be exposed to light intensities above saturation and at the same time shade the leaves in the lower layers (Long et al. 2006). An arrangement where the plant has leaves at a more vertical angle in the upper canopy, transitioning to horizontal leaves in the lower canopy, allows for a more even distribution of the incident solar radiation later in the growing season. Erect leaves that increase light capture and improve light distribution in dense canopies have been shown to increase yield in several species (Lambert and Johnson 1978; Sakamoto et al. 2006). Although our knowledge of the mechanisms that control plant architecture are still rudimentary, some progress has been made on determining what mechanisms influence leaf angle. The control of leaf and branch angle is a complex quantitative trait, and genes involved have been identified in several species, including rice (Li et al. 2007), maize (Zhang et al. 2014), and *Arabidopsis* (Yoshihara et al. 2013).

Similar to the effects of plant architecture, a canopy that reduces light absorption in the top layers by minimizing the chlorophyll antennae will let more radiation penetrate to the lower layers and thereby distribute the light more evenly

throughout the canopy (Melis 2009; Ort et al. 2011). This has been shown to work effectively in algae mass cultures, where cells with truncated light-harvesting chlorophyll antenna were able to significantly increase biomass production compared to normally pigmented cells by having improved light-use efficiency (Melis et al. 1999). In crop species, promising results come from chlorophyll *b*-deficient soybean cultivars that only contain about half as much chlorophyll as their wild-type counterparts (Pettigrew et al. 1989). Even though the canopy extinction coefficients were similar between the mutant and wild-type plants, the chlorophyll-depleted cultivars outperformed the wild type by increasing the light penetration into the canopy.

Other changes to the plant architecture could also prove to be beneficial for maximizing the crop yield. For example, the wheat *tiller inhibition* (*tin*) mutant has a reduced number of tillers and fewer sterile tillers (Kebrom and Richards 2013). These plants use less water early in the growing season as compared to near-isogenic wild-type plants, which leaves them with more water available during the grain-filling period. When grown under water deficit, *tin* lines had larger and more grains per spikelet, offsetting the lower number of spikelets and producing a higher grain yield overall (Mitchell et al. 2013). It is conceivable that yield benefits might also be achieved under favorable conditions when the *tin* lines are grown at higher planting densities than current practice (Kebrom and Richards 2013; Mitchell et al. 2013).

41.9 IMPROVING PHOTOSYNTHETIC PERFORMANCE BY QTL ANALYSIS WITH UTILIZING NATURAL GENETIC RESOURCES

It is well recognized that there is wide intra- and interspecific variation in the rate of leaf photosynthesis. The utilization of these variations may be one of the important strategies for improving crop photosynthesis. Recent advances in molecular genetics have provided a lot of DNA markers, which facilitate identification of quantitative trait loci (QTL) for important agronomic traits (Yamamoto and Yano 2008). QTL analysis should be a useful approach to understand the genetic basis of the natural variation of leaf photosynthesis and will contribute to designing future breeding programs (Flood et al. 2011).

Most agronomic traits, such as plant height, seed number, and, of course, photosynthetic traits, are expressed in continuous values rather than a few discrete classes. These traits are called quantitative traits and are often controlled by multiple genetic factors. QTL analysis provides information about the number and location of QTLs on the genomes and the phenotypic effect of each QTL (Tanksley 1993; Figure 41.7). It is also possible to identify genes that underlie QTLs by high-resolution mapping, namely, map-based cloning (Yano 2001). It is expected that QTL analysis can find genes that are difficult to identify by mutant analysis (Ashikari and Matsuoka 2006). QTL analysis needs a segregating population consisting of different individuals. F_2 and BC_1F_1 populations are most frequently used in genetic analysis because of the relatively short time needed for their development, while populations such as recombinant inbred lines (RILs), backcrossed inbred lines (BILs), and chromosome segment substitution lines (CSSLs) are used in studies that aim to identify QTLs for complex traits (Yamamoto et al. 2014). Currently, genome-wide association studies (GWAS), which do not rely on a controlled bi-parental segregating population, are conducted to identify QTLs with the availability of large numbers of genetic markers such as single-nucleotide polymorphisms (SNPs) (Myles et al. 2009).

There are two options to identify QTLs associated with leaf photosynthesis. The first is direct measurements of leaf CO_2 assimilation rate of a segregating population. An initial report of QTLs for CO_2 assimilation rate is shown by Herve et al. (2001), who used sunflower RILs for genetic analysis. Subsequently, Teng et al. (2004) found two QTLs for CO_2

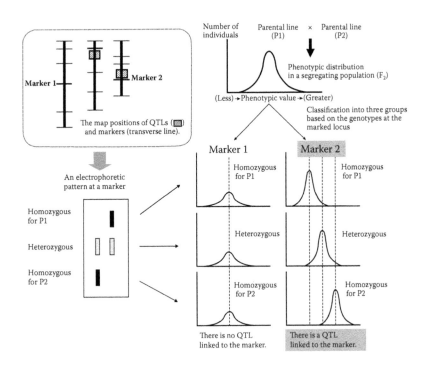

FIGURE 41.7 Mapping of quantitative trait loci. The phenotypic values for a particular trait of interest are evaluated using a segregating population such as F_2. Subsequently, genotypes of DNA markers that are evenly spaced in the genome are analyzed. When no QTL locates near the marker, the average of the phenotypic value will be similar among the three groups that are classified based on the genotypes at the marker locus (Marker 1). When a QTL locates near the marker, the average of the phenotypic value will be significantly different among the groups (Marker 2). QTL analysis applies this analysis to all the DNA markers that are used and evaluates the number and location of QTLs and the phenotypic effect of each QTL. (Modified from Yamamoto T, Yano M, *Kagaku* 77, 607–613, 2007.)

assimilation rate using double haploid (DH) lines derived from a cross between *japonica* and *indica* rice cultivars. Hu et al. (2009) and Gu et al. (2012) found several QTLs for rice photosynthesis using RILs and BILs lines, respectively, under both water-limited and well-watered conditions. Another option to identify QTLs is to evaluate phenotypic values of traits that are related to CO_2 assimilation rate. Among those, the SPAD value is the simplest nondestructive index for chlorophyll content (Takai et al. 2013a). Chlorophyll fluorescence is also used to identify QTLs for several photosynthesis-related traits (Yang et al. 2007; Guo et al. 2008; Liang et al. 2010; Czyczyło-Mysza et al. 2011; Kasajima et al. 2011). Furthermore, Rubisco and nitrogen content as well as carbon isotope discrimination ($\Delta^{13}C$), which reflects the ratio of leaf internal to external CO_2 concentration (C_i/C_a), have also been used for QTL analysis (Ishimaru et al. 2001; Laza et al. 2006; Takai et al. 2009; Khaembah et al. 2013).

The identification of the causal genes underlying QTLs for leaf photosynthesis is a major goal in these studies. Takai et al. (2013b) conducted high-resolution mapping for CO_2 assimilation and identified the *Green for Photosynthesis* (*GPS*) gene that corresponds to the difference between an *indica* and a *japonica* rice cultivar. The analysis showed that the *indica*-type *GPS* gene increases CO_2 assimilation rate by increasing Rubisco content, which might be attributed to enhancement of mesophyll numbers and leaf thickness. The sequencing analysis revealed that *GPS* encodes a plant-specific protein that may be involved in polar auxin transport (Qi et al. 2008) and is totally different from genes encoding components of plant photosynthetic apparatus.

DNA marker-assisted selection enables us to combine multiple QTLs, namely, QTL pyramiding. Adachi et al. (2014) combined two genetic regions from an *indica* rice in the genetic background of a *japonica* rice. The research showed that the pyramided line had significantly increased CO_2 assimilation rates compared with *japonica* rice (Figure 41.8). Furthermore, it has been suggested that pyramiding of several QTLs has the potential to increase CO_2 assimilation rate to even higher levels than those of parental lines (Adachi et al. 2013). Therefore, QTL pyramiding could be a promising approach for improving crop photosynthesis.

QTL analysis for leaf photosynthesis has become popular, although the number of studies for photosynthesis remains much lower than those for other important agronomic traits such as seed number, flowering time, and disease resistance (Yamamoto et al. 2009). Two things should be considered to facilitate QTL identification for leaf photosynthesis. The first is to enhance the measurement efficiency of gas exchange and related traits as these measurements are still labor-intensive and time-consuming (Adachi et al. 2011). The second is to decrease measurement variability because photosynthesis parameters are often influenced by environmental conditions and plant-to-plant variation that prevent the accurate evaluation of QTLs (Tanksley 1993).

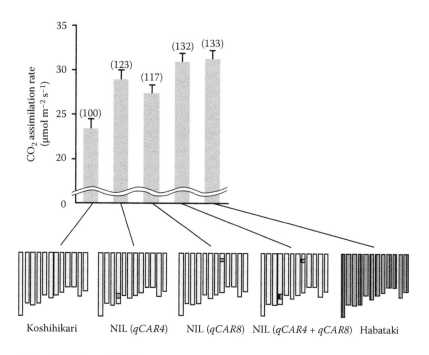

FIGURE 41.8 QTL pyramiding for CO_2 assimilation rate. Koshihikari is a commercial *japonica* cultivar. Habataki is a high-yielding *indica* cultivar. The NILs have single or double homozygous regions of Habataki in the genetic background of Koshihikari. CO_2 assimilation rates of flag leaves were measured at a photosynthetic photon flux density of 2000 µmol m^{-1} s^{-1}, an ambient CO_2 concentration of 370 µmol mol^{-1}, and an air temperature of 30°C. Values in parentheses are percentages relative to Koshihikari. (Modified from Adachi S et al., *Journal of Experimental Botany* 65: 2049–2056, 2014.)

The development of high-throughput measurement systems with high accuracy is a key factor for comprehensive understanding of the genetic basis of natural variation in photosynthesis.

41.10 CONCLUSION

The present rate of increase in crop yields is insufficient to keep pace with the rapid growth rates of the global population. Development of crops with increased yield and stress tolerance is essential to meet the future food and energy demands. Therefore, novel paths need to be explored to generate more efficient plants with higher production for the future. Enhancement of leaf photosynthetic capacity would provide one attractive avenue to drive increases in crop yields, since plant growth depends largely on its photosynthesis (Yamori 2013).

One could argue that plant photosynthesis cannot be improved, since its evolution has already perfected it. However, natural selection, which maximizes total fitness of the individual rather than agronomic yield, has shaped plant photosynthesis for life in environments that differ considerably from the resource-rich settings with little competition from other plants provided by modern agriculture (Leister 2012). The mutational constraints and selection pressures faced by higher plants, along with elements of historical contingency, have left us with a photosynthetic pathway that is far from perfect, but instead has been optimized to deal with the prevailing high O_2 and low CO_2 environment. The controlled conditions of the modern food production system force us to think about maximizing yield from the perspective of the entire crop canopy rather than of a single plant. The extent of our understanding of photosynthetic processes clearly indicates that enough scope has been left for improvements of this ancient and critical biological pathway to achieve our goals of sustainable food production.

ACKNOWLEDGMENTS

W.Y. was supported by the Japan Science and Technology Agency, PRESTO, by a Grant for Basic Science Research Projects from The Sumitomo Foundation, and in part by a Grant-in-Aid for Scientific Research from the Japan Society for the Promotion of Science (Scientific Research No. 25891005). F.B. was supported by the ARC Centre of Excellence for Translational Photosynthesis. S.A. was supported by the Japan Science and Technology Agency, PRESTO, and in part by a Grant-in-Aid for Scientific Research from the Japan Society for the Promotion of Science (Postdoctoral Fellowships).

REFERENCES

Adachi S, Nito N, Kondo M, Yamamoto T, Arai-Sanoh Y, Ando T, Ookawa T, Yano M, Hirasawa T (2011) Identification and characterization of genomic regions on chromosomes 4 and 8 that control the rate of photosynthesis in rice leaves. *Journal of Experimental Botany* 62: 1927–1938.

Adachi S, Nakae T, Uchida M, Soda K, Takai T, Oi T, Yamamoto T et al. (2013) The mesophyll anatomy enhancing CO_2 diffusion is a key trait for improving rice photosynthesis. *Journal of Experimental Botany* 64: 1061–1072.

Adachi S, Baptista LZ, Sueyoshi T, Murata K, Yamamoto T, Ebitani T, Ookawa T, Hirasawa T (2014) Introgression of two chromosome regions for leaf photosynthesis from an *indica* rice into the genetic background of a *japonica* rice. *Journal of Experimental Botany* 65: 2049–2056.

André M (2011) Modelling $^{18}O_2$ and $^{16}O_2$ unidirectional fluxes in plants: II. Analysis of Rubisco evolution. *Biosystems* 103: 252–264.

Anten NPR, Schieving F, Werger MJA (1995) Patterns of light and nitrogen distribution in relation to whole canopy carbon gain in C_3 and C_4 mono- and dicotyledonous species. *Oecologia* 101: 504–513.

Araya T, Noguchi K, Terashima I (2006) Effects of carbohydrate accumulation on photosynthesis differ between sink and source leaves of *Phaseolus vulgaris* L. *Plant and Cell Physiology* 47: 644–652.

Ashikari M, Matsuoka M (2006) Identification, isolation and pyramiding of quantitative trait loci for rice breeding. *Trends in Plant Science* 11: 344–350.

Ashraf M, Akram NA (2009) Improving salinity tolerance of plants through conventional breeding and genetic engineering: An analytical comparison. *Biotechnology Advances* 27: 744–752.

Bauwe H, Hagemann M, Fernie AR (2010) Photorespiration: Players, partners and origin. *Trends in Plant Science* 15: 330–336.

Biswal AK, Pattanayak GK, Pandey SS, Leelavathi S, Reddy VS, Govindjee, Tripathy BC (2012) Light intensity-dependent modulation of chlorophyll b biosynthesis and photosynthesis by overexpression of chlorophyllide *a* oxygenase in tobacco. *Plant Physiology* 159: 433–449.

Björkman O (1966) The effect of oxygen concentration on photosynthesis in higher plants. *Physiologia Plantarum* 19: 618–633.

Björkman O, Heisey WM, Nobs M, Nicholson F, Hart RW (1968) Effect of oxygen concentration on dry matter production in higher plants. *Carnegie Institute Year Book*. Carnegie Institute of Washington, Washington, DC, pp. 228–245.

Borland AM, Hartwell J, Weston DJ, Schlauch KA, Tschaplinski TJ, Tuskan GA, Yang X, Cushman JC (2014) Engineering crassulacean acid metabolism to improve water-use efficiency. *Trends in Plant Science* 19: 327–338.

Burkey KO (1994) Genetic variation of photosynthetic electron transport in barley: Identification of plastocyanin as a potential limiting factor. *Plant Science* 98: 177–187.

Burkey KO, Gizlice Z, Carter TE (1996) Genetic variation in soybean photosynthetic electron transport capacity is related to plastocyanin concentration in the chloroplast. *Photosynthesis Research* 49: 141–149.

Busch FA, Sage TL, Cousins AB, Sage RF (2013) C_3 plants enhance rates of photosynthesis by reassimilating photorespired and respired CO_2. *Plant, Cell & Environment* 36: 200–212.

Carvalho DD, Irving LJ, Carnevalli RA, Hodgson J, Matthew C (2006) Distribution of current photosynthate in two guinea grass (*Panicum maximum* Jacq.) cultivars. *Journal of Experimental Botany* 57: 2015–2024.

Chai MF, Chen QJ, An R, Chen YM, Chen J, Wang XC (2005) NADK2, an Arabidopsis chloroplastic NAD kinase, plays a vital role in both chlorophyll synthesis and chloroplast protection. *Plant Molecular Biology* 59: 553–564.

Chen M, Blankenship RE (2011) Expanding the solar spectrum used by photosynthesis. *Trends in Plant Science* 16: 427–431.

Chen M, Schliep M, Willows RD, Cai Z-L, Neilan BA, Scheer H (2010) A red-shifted chlorophyll. *Science* 329: 1318–1319.

Cheng SH, Moore BD, Seemann JR (1998) Effects of short and long-term elevated CO_2 on the expression of ribulose-1,5-bisphosphate carboxylase/oxygenase genes and carbohydrate accumulation in leaves and *Arabidopsis thaliana* (L.) Heynh. *Plant Physiology* 116: 715–723.

Chida H, Nakazawa A, Akazaki H, Hirano T, Suruga K, Ogawa M, Satoh T et al. (2007) Expression of the algal cytochrome c6 gene in Arabidopsis enhances photosynthesis and growth. *Plant & Cell Physiology* 48: 948–957.

Covshoff S, Hibberd JM (2012) Integrating C_4 photosynthesis into C_3 crops to increase yield potential. *Current Opinion in Biotechnology* 23: 209–214.

Crafts-Brandner SJ, Salvucci ME (2000) Rubisco activase constrains the photosynthetic potential of leaves at high temperature and CO_2. *Proceedings of the National Academy of Sciences of the United States of America* 97: 13430–13435.

Czyczyło-Mysza I, Marcińska I, Skrzypek E, Chrupek M, Grzesiak S, Hura T, Stojałowski S, Myśków B, Milczarski P, Quarrie S (2011) Mapping QTLs for yield components and chlorophyll a fluorescence parameters in wheat under three levels of water availability. *Plant Genetic Resources* 9: 291–295.

DePaoli HC, Borland AM, Tuskan GA, Cushman JC, Yang X (2014) Synthetic biology as it relates to CAM photosynthesis: Challenges and opportunities. *Journal of Experimental Botany* 65: 3381–3393.

Dodd AN, Borland AM, Haslam RP, Griffiths H, Maxwell K (2002) Crassulacean acid metabolism: Plastic, fantastic. *Journal of Experimental Botany* 53: 569–580.

Evans JR (1989) Photosynthesis and nitrogen relationships in leaves of C_3 plants. *Oecologia* 78(1): 9–19.

Evans JR, von Caemmerer S, Setchell BA, Hudson GS (1994) The relationship between CO_2 transfer conductance and leaf anatomy in transgenic tobacco with a reduced content of Rubisco. *Australian Journal of Plant Physiology* 21: 475–495.

Evans JR, Kaldenhoff R, Genty B, Terashima I (2009) Resistances along the CO_2 diffusion pathway inside leaves. *Journal of Experimental Botany* 60: 2235–2248.

Farquhar GD, von Caemmerer S, Berry JA (1980) A biochemical model of photosynthetic CO_2 assimilation in leaves of C_3 species. *Planta* 149: 78–90.

Feng L, Wang K, Li Y, Tan Y, Kong J, Li H, Zhu Y (2007) Overexpression of SBPase enhances photosynthesis against high temperature stress in transgenic rice plants. *Plant Cell Report* 26: 1635–1646.

Flexas J, Ribas-Carbó M, Hanson DT, Bota J, Otto B, Cifre J, McDowell N, Medrano H, Kaldenhoff R (2006) Tobacco aquaporin NtAQP1 is involved in mesophyll conductance to CO_2 in vivo. *The Plant Journal* 48: 427–439.

Flood PJ, Harbinson J, Aarts MGM (2011) Natural genetic variation in plant photosynthesis. *Trends in Plant Science* 16: 327–335.

Galmés J, Kapralov MV, Andralojc PJ, Conesa MÀ, Keys AJ, Parry MAJ, Flexas J (2014a) Expanding knowledge of the Rubisco kinetics variability in plant species: Environmental and evolutionary trends. *Plant, Cell and Environment* 37: 1989–2001.

Galmés J, Andralojc PJ, Kapralov MV, Flexas J, Keys AJ, Molins A, Parry MAJ, Conesa MÀ (2014b) Environmentally driven evolution of Rubisco and improved photosynthesis and growth within the C_3 genus *Limonium* (Plumbaginaceae). *New Phytologist* 203: 989–999.

Ghannoum O, Evans JR, Chow WS, Andrews TJ, Conroy JP, von Caemmerer S (2005) Faster Rubisco is the key to superior nitrogen use efficiency in NADP-Malic enzyme relative to NAD-Malic enzyme C_4 grasses. *Plant Physiology* 137: 638–650.

Govindjee (2004) Chlorophyll a fluorescence: A bit of basics and history. In *Chlorophyll a Fluorescence: A Signature of Photosynthesis*, eds. Papageorgiou GC, Govindjee. Kluwer Academic Publishers, Dordrecht, pp. 1–42.

Grindlay DJC (1997) Towards and explanation of crop nitrogen demand based on the optimization of leaf nitrogen per unit area. *Journal of Agricultural Science* 128: 377–396.

Gu J, Yin X, Struik PC, Stomph TJ, Wang H (2012) Using chromosome introgression lines to map quantitative trait loci for photosynthesis parameters in rice (*Oryza sativa* L.) leaves under drought and well-watered field conditions. *Journal of Experimental Botany* 63: 455–469.

Guo PG, Baum M, Varshney RK, Graner A, Grando S, Ceccarelli S (2008) QTLs for chlorophyll and chlorophyll fluorescence parameters in barley under post-flowering drought. *Euphytica* 163: 203–214.

Haake V, Zrenner R, Sonnewald U, Stitt M (1998) A moderate decrease of plastid aldolase activity inhibits photosynthesis, alters the levels of sugars and starch, and inhibits growth of potato plants. *The Plant Journal* 14: 147–157.

Haake V, Geiger M, Walch-Liu P, Engels C, Zrenner R, Stitt M (1999) Changes in aldolase activity in wild-type potato plants are important for acclimation to growth irradiance and carbon dioxide concentration, because plastid aldolase exerts control over the ambient rate of photosynthesis across a range of growth conditions. *The Plant Journal* 17: 479–489.

Hajirezaei M-R, Peisker M, Tschiersch H, Palatnik JF, Valle EM, Carrillo N, Sonnewald U (2002) Small changes in the activity of chloroplastic NADP⁺-dependent ferredoxin oxidoreductase lead to impaired plant growth and restrict photosynthetic activity of transgenic tobacco plants. *The Plant Journal* 29: 281–293.

Hanba YT, Shibasaka M, Hayashi Y, Hayakawa T, Kasamo K, Terashima I, Katsuhara M (2004) Overexpression of the barley aquaporin HvPIP2;1 increases internal CO_2 conductance and CO_2 assimilation in the leaves of transgenic rice plants. *Plant & Cell Physiology* 45: 521–529.

Harrison EP, Willingham NM, Lloyd JC, Raines CA (1998) Reduced sedohepulose-l,7-bisphosphatase levels in transgenic tobacco lead to decreased photosynthetic capacity and altered carbohydrate accumulation. *Planta* 204: 27–36.

Hatano-Iwasaki A, Ogawa K (2012) Biomass production is promoted by increasing an aldolase undergoing glutathionylation in *Arabidopsis thaliana*. *The International Journal of Developmental Biology* 6: 1–8.

Heckwolf M, Pater D, Hanson DT, Kaldenhoff R (2011) The *Arabidopsis thaliana* aquaporin AtPIP1;2 is a physiologically relevant CO_2 transport facilitator. *The Plant Journal* 67: 795–804.

Henkes S, Sonnewald U, Badur R, Flachmann R, Stitt M (2001) A small decrease of plastid transketolase activity in antisense tobacco transformants has dramatic effects on photosynthesis and phenylpropanoid metabolism. *The Plant Cell* 13: 535–551.

Herve D, Fabre F, Berrios EF, Leroux N, Al Chaarani G, Planchon C, Sarrafi A, Gentzbittel L (2001) QTL analysis of photosynthesis and water status traits in sunflower (*Helianthus annuus* L.) under greenhouse conditions. *Journal of Experimental Botany* 52: 1857–1864.

Hikosaka K (1996) Effects of leaf age, nitrogen nutrition and photon flux density on the organization of the photosynthetic apparatus in leaves of a vine (*Ipomoea tricolor* Cav) grown horizontally to avoid mutual shading of leaves. *Planta* 198: 144–150.

Hikosaka K, Terashima I (1995) A model of the acclimation of photosynthesis in the leaves of C_3 plants to sun and shade with respect to nitrogen use. *Plant, Cell and Environment* 18: 605–618.

Hirose T, Werger MJA (1987) Maximizing daily canopy photosynthesis with respect to the leaf nitrogen allocation pattern in the canopy. *Oecologia* 72: 520–526.

Hölttä T, Vesala T, Sevanto S, Perämäki M, Nikinmaa E (2006) Modeling xylem and phloem water flows in trees according to cohesion theory and Münch hypothesis. *Trees—Structure and Function* 20: 67–78.

Hölttä T, Mencuccini M, Nikinmaa E (2009) Linking phloem function to structure: Analysis with a coupled xylem—Phloem transport model. *Journal of Theoretical Biology* 259: 325–337.

Hu SP, Zhou Y, Zhang L, Zhu XD, Li L, Luo LJ, Liu GL, Zhou QM (2009) Correlation and quantitative trait loci analyses of total chlorophyll content and photosynthetic rate of rice (*Oryza sativa*) under water stress and well-watered conditions. *Journal of Integrative Plant Biology* 51: 879–888.

Irving LJ, Robinson D (2006) A dynamic model of Rubisco turnover in cereal leaves. *New Phytologist* 169: 493–504.

Ishimaru K, Kobayashi N, Ono K, Yano M, Ohsugi R (2001) Are contents of Rubisco, soluble protein and nitrogen in flag leaves of rice controlled by the same genetics? *Journal of Experimental Botany* 52: 1827–1833.

Izumi M, Wada S, Makino A, Ishida H (2010) The autophagic degradation of chloroplasts via Rubisco-containing bodies is specifically linked to leaf carbon status, but not nitrogen status in Arabidopsis. *Plant Physiology* 154: 1196–1209.

Kaldenhoff R (2012) Mechanisms underlying CO_2 diffusion in leaves. *Current Opinion in Plant Biology* 15: 276–281.

Kasajima I, Ebana K, Yamamoto T, Takahara K, Yano M, Kawai-Yamada M, Uchimiya H (2011) Molecular distinction in genetic regulation of nonphotochemical quenching in rice. *Proceedings of the National Academy of Sciences of the United States of America* 138: 13835–13840.

Kawai S, Murata K (2008) Structure and function of NAD kinase and NADP phosphatase: Key enzymes that regulate the intracellular balance of NAD(H) and NADP(H). *Bioscience, Biotechnology, and Biochemistry* 72: 919–930.

Kebeish R, Niessen M, Thiruveedhi K, Bari R, Hirsch HJ, Rosenkranz R, Stäbler N, Schönfeld B, Kreuzaler F, Peterhänsel C (2007) Chloroplastic photorespiratory bypass increases photosynthesis and biomass production in *Arabidopsis thaliana*. *Nature Biotechnology* 25: 593–599.

Kebrom TH, Richards RA (2013) Physiological perspectives of reduced tillering and stunting in the tiller inhibition (tin) mutant of wheat. *Functional Plant Biology* 40: 977–985.

Khaembah EN, Irving LJ, Thom, ER, Faville MJ, Easton HS, Matthew C (2013) Leaf Rubisco turnover in a perennial ryegrass (*Lolium perenne* L.) mapping population: Genetic variation, identification of associated QTL, and correlation with plant morphology and yield. *Journal of Experimental Botany* 64: 1305–1316.

Khush GS (2001) Green revolution: The way forward. *Nature Reviews Genetics* 2: 815–822.

Kossmann J, Sonnewald U, Willmitzer L (1994) Reduction of the chloroplastic fructose-1,6-bisphosphatase in transgenic potato plants impairs photosynthesis and plant growth. *The Plant Journal* 6: 637–650.

Kumar A, Li C, Portis Jr AR (2009) *Arabidopsis thaliana* expressing a thermostable chimeric Rubisco activase exhibits enhanced growth and higher rates of photosynthesis at moderately high temperatures. *Photosynthesis Research* 100: 143–153.

Kurek I, Chang TK, Bertain SM, Madrigal A, Liu L, Lassner MW, Zhu G (2007) Enhanced thermostability of Arabidopsis Rubisco activase improves photosynthesis and growth rates under moderate heat stress. *Plant Cell* 19: 3230–3241.

Kusumi K, Hirotsuka S, Kumamaru T, Iba K (2012) Increased leaf photosynthesis caused by elevated stomatal conductance in a rice mutant deficient in SLAC1, a guard cell anion channel protein. *Journal of Experimental Botany* 63: 5635–5644.

Lambert RJ, Johnson RR (1978) Leaf angle, tassel morphology, and the performance of maize hybrids. *Crop Science* 18: 499–502.

Lattanzi FA, Schnyder H, Thornton B (2005) The sources of carbon and nitrogen supplying leaf growth. Assessment of the role of stores with compartmental models. *Plant Physiology* 137: 383–395.

Laza MR, Kondo M, Ideta O, Barlaan E, Imbe T (2006) Identification of quantitative trait loci for $\delta^{13}C$ and productivity in irrigated lowland rice. *Crop Science* 46: 763–773.

Leakey AD, Ainsworth EA, Bernacchi CJ, Rogers A, Long SP, Ort DR (2009) Elevated CO_2 effects on plant carbon, nitrogen, and water relations: Six important lessons from FACE. *Journal of Experimental Botany* 60: 2859–2876.

Lefebvre S, Lawson T, Zakhleniuk OV, Lloyd JC, Raines CA, Fryer M (2005) Increased sedoheptulose-1,7-bisphosphatase activity in transgenic tobacco plants stimulates photosynthesis and growth from an early stage in development. *Plant Physiology* 138: 451–460.

Leister D (2012) How can the light reactions of photosynthesis be improved in plants? *Frontiers in Plant Science* 3: 199.

Li P, Wang Y, Qian Q, Fu Z, Wang M, Zeng D, Li B, Wang X, Li J (2007) *LAZY1* controls rice shoot gravitropism through regulating polar auxin transport. *Cell Research* 17: 402–410.

Liang Y, Zhang KP, Zhao L, Liu B, Meng QW, Tian JC, Zhao SJ (2010) Identification of chromosome regions conferring dry matter accumulation and photosynthesis in wheat (*Triticum aestivum* L.). *Euphytica* 171: 145–156.

Lichtentahler HK (1999) The 1-deoxy-d-xylulose-5-phosphate pathway of isoprenoid biosynthesis in plants. *Annual Review of Plant Physiology and Plant Molecular Biology* 50: 47–65.

Lin MT, Occhialini A, Andralojc PJ, Parry MAJ, Hanson MR (2014) A faster Rubisco with potential to increase photosynthesis in crops. *Nature* 513: 547–550.

Long SP, Zhu X-G, Naidu SL, Ort DR (2006) Can improvement in photosynthesis increase crop yields? *Plant, Cell & Environment* 29: 315–330.

Mae T (1986) Partitioning and utilization of nitrogen in rice plants. *Japan Agricultural Research Quarterly* 20: 115–120.

Mae T, Ohira K (1981) The relationship of nitrogen related to leaf growth and senescence in rice plants (*Oryza sativa* L.). *Plant and Cell Physiology* 22: 1067–1074.

Mae T, Makino A, Ohira A (1983) Changes in the amounts of ribulose bisphosphate carboxylase synthesized and degraded during the life-span of rice life (*Oryza sativa* L.). *Plant Physiology* 24: 1079–1086.

Maier A, Fahnenstich H, von Caemmerer S, Engqvist MKM, Weber APM, Fluegge U-I, Maurino VG (2012) Transgenic introduction of a glycolate oxidative cycle into *A. thaliana* chloroplasts leads to growth improvement. *Frontiers in Plant Science* 3: 38.

Makino A (2003) Rubisco and nitrogen relationships in rice: Leaf photosynthesis and plant growth. *Soil Science and Plant Nutrition* 49: 319–327.

Makino A, Mae T, Ohira K (1983) Photosynthesis and ribulose 1,5-bisphosphate carboxylase in rice leaves. *Plant Physiology* 73: 1002–1007.

Makino A, Mae T, Ohira K (1985) Photosynthesis and ribulose 1,5-bisphosphate carboxylase/oxygenase in rice leaves from emergence through senescence. Quantitative analysis by carboxylation/oxygenation and regeneration of ribulose 1,5-bisphosphate. *Planta* 166: 414–420.

Makino A, Mae T, Ohira K (1988) Differences between wheat and rice in the enzymic properties of ribulose-1,5-bisphosphate carboxylase/oxygenase and the relationship to photosynthetic gas exchange. *Planta* 174: 30–38.

Makino A, Sakashita H, Hidema J, Mae T, Ojima K, Osmond B (1992) Distinctive responses of ribulose-1,5-bisphosphate carboxylase and carbonic anhydrase in wheat leaves to nitrogen nutrition and their possible relationships to CO_2-transfer resistance. *Plant Physiology* 100: 1737–1743.

Makino A, Sato T, Nakano H, Mae T (1997) Leaf photosynthesis, plant growth and nitrogen allocation in rice under different irradiances. *Planta* 203: 390–398.

Makino A, Miyake C, Yokota A (2002) Physiological functions of the water-water cycle (Mehler reaction) and the cyclic electron flow around PSI in rice leaves. *Plant and Cell Physiology* 43: 1017–1026.

Medrano H, Escalona JM, Bota J, Gulías J, Flexas J (2002) Regulation of photosynthesis of C_3 plants in response to progressive drought: Stomatal conductance as a reference parameter. *Annals of Botany* 89: 895–905.

Melis A (2009) Solar energy conversion efficiencies in photosynthesis: Minimizing the chlorophyll antennae to maximize efficiency. *Plant Science* 177: 272–280.

Melis A, Neidhardt J, Benemann JR (1999) *Dunaliella salina* (Chlorophyta) with small chlorophyll antenna sizes exhibit higher photosynthetic productivities and photon use efficiencies than normally pigmented cells. *Journal of Applied Phycology* 10: 515–525.

Michelet L, Zaffagnini M, Morisse S, Sparla F, Perez-Perez ME, Francia F, Danon A et al. (2013) Redox regulation of the Calvin-Benson cycle: Something old, something new. *Frontiers in Plant Science* 4: 470.

Mitchell JH, Rebetzke GJ, Chapman SC, Fukai S (2013) Evaluation of reduced-tillering (tin) wheat lines in managed, terminal water deficit environments. *Journal of Experimental Botany* 64: 3439–3451.

Miyagawa Y, Tamoi M, Shigeoka S (2001) Overexpression of a cyanobacterial fructose-1,6-/sedoheptulose-1,7-bisphosphatase in tobacco enhances photosynthesis and growth. *Nature Biotechnology* 19: 965–969.

Miyashita H, Ikemoto H, Kurano N, Adachi K, Chihara M, Miyachi S (1996) Chlorophyll *d* as a major pigment. *Nature* 383: 402.

Moore BD, Cheng SH, Rice J, Seemann JR (1998) Sucrose cycling, Rubisco expression, and prediction of photosynthetic acclimation to elevated atmospheric CO_2. *Plant, Cell and Environment* 21: 905–915.

Moore BD, Cheng SH, Sims D, Seemann JR (1999) The biochemical and molecular basis for photosynthetic acclimation to elevated atmospheric CO_2. *Plant, Cell and Environment* 22: 567–582.

Morris PF, Layzell DB, Canvin DT (1988) Ammonia production and assimilation in glutamate synthase mutants of *Arabidopsis thaliana*. *Plant Physiology* 87: 148–154.

Mueller-Cajar O, Stotz M, Wendler P, Hartl FU, Bracher A, Hayer-Hartl M (2011) Structure and function of the AAA+ protein CbbX, a red-type RuBisCO activase. *Nature* 479: 194–199.

Myles S, Peiffer J, Brown PJ, Ersoz ES, Zhang Z, Costich DE, Buckler ES (2009) Association mapping: Critical considerations shift from genotyping to experimental design. *Plant Cell* 21: 2194–2202.

Nikinmaa E, Hölttä T, Hari P, Kolari P, Mäkelä A, Sevanto S, Vesala T (2013) Assimilate transport in phloem sets conditions for leaf gas exchange. *Plant, Cell and Environment* 36: 655–669.

Ölcer H, Lloyd JC and Raines CA (2001) Photosynthetic capacity is differentially affected by reductions in sedoheptulose-1,7-bisphosphatase activity during leaf development in transgenic tobacco plants. *Plant Physiology* 125: 982–989.

Ort DR, Zhu X, Melis A (2011) Optimizing antenna size to maximize photosynthetic efficiency. *Plant Physiology* 155: 79–85.

Parry MAJ, Keys AJ, Madgwick PJ, Carmo-Silva AE, Andralojc PJ (2008) Rubisco regulation: A role for inhibitors. *Journal of Experimental Botany* 59: 1569–1580.

Payne JL, Boyer AG, Brown JH, Finnegan S, Kowalewski M, Krause RA, Lyons SK, McClain CR, McShea DW, Novack-Gottshall PW, Smith FA, Stempien JA, Wang SC (2009) Two-phase increase in the maximum size of life over 3.5 billion years reflects biological innovation and environmental opportunity. *Proceedings of the National Academy of Sciences of the United States of America* 106: 24–27.

Pego JV, Kortstee AJ, Huijser C, Smeekens SCM (2000) Photosynthesis, sugars and the regulation of gene expression. *Journal of Experimental Botany* 51: 407–416.

Pesaresi P, Scharfenberg M, Weigel M, Granlund I, Schroder WP, Finazzi G, Rappaport F et al. (2009) Mutants, overexpressors, and interactors of *Arabidopsis* plastocyanin isoforms: Revised roles of plastocyanin in photosynthetic electron flow and thylakoid redox state. *Molecular Plant* 2: 236–248.

Petit JR, Jouzel J, Raynaud D, Barkov NI, Barnola JM, Basile I, Bender M et al. Saltzman E, Stievenard M (1999) Climate and atmospheric history of the past 420,000 years from the Vostok ice core, Antarctica. *Nature* 399: 429–436.

Pettigrew WT, Hesketh JD, Peters DB, Woolley JT (1989) Characterization of canopy photosynthesis of chlorophyll-deficient soybean isolines. *Crop Science* 29: 1025–1029.

Portis Jr AR (2003) Rubisco activase: Rubisco's catalytic chaperone. *Photosynthesis Research* 75: 11–27.

Pribil M, Pesaresi P, Hertle A, Barbato R, Leister D (2010) Role of plastid protein phosphatase TAP38 in LHCII dephosphorylation and thylakoid electron flow. *PLoS Biology* 8: e1000288.

Price GD, Badger MR, von Caemmerer S (2011) The prospect of using cyanobacterial bicarbonate transporters to improve leaf photosynthesis in C_3 crop plants. *Plant Physiology* 155: 20–26.

Price GD, Pengelly JJL, Forster B, Du J, Whitney SM, von Caemmerer S, Badger MR, Howitt SM, Evans JR (2013) The cyanobacterial CCM as a source of genes for improving photosynthetic CO_2 fixation in crop species. *Journal of Experimental Botany* 64: 753–768.

Qi J, Qian Q, Bu Q, Li S, Chen Q, Sun J, Liang W et al. Chen J, Chen M, Li C (2008) Mutation of the rice *Narrow leaf1* gene, which encodes a novel protein, affects vein patterning and polar auxin transport. *Plant Physiology* 147: 1947–1959.

Raines CA (2003) The Calvin cycle revisited. *Photosynthesis Research* 75: 1–10.

Raines CA (2011) Increasing photosynthetic carbon assimilation in C_3 plants to improve crop yield: Current and future strategies. *Plant Physiology* 155: 36–42.

Raven JA (2000) Land plant biochemistry. *Philosophical Transactions of the Royal Society of London B* 355: 833–846.

Roberts L (2011) 9 billion? *Science* 333: 540–543.

Rodriguez RE, Lodeyro A, Poli HO, Zurbriggen M, Palatnik JF, Tognetti VB, Tschiersch H et al. Transgenic tobacco plants overexpressing chloroplastic ferredoxin-NADP(H) reductase display normal rates of photosynthesis and increased tolerance to oxidative stress. *Plant Physiology* 143: 639–649.

Rosenthal D, Locke A, Khozaei M, Raines C, Long S, Ort D (2011) Over-expressing the C_3 photosynthesis cycle enzyme sedoheptulose-1-7 bisphosphatase improves photosynthetic carbon gain and yield under fully open air CO_2 fumigation (FACE). *BMC Plant Biology* 11: 123.

Sage RF (2004) The evolution of C_4 photosynthesis. *New Phytologist* 161: 341–370.

Sage RF, Sage TL, Kocacinar F (2012) Photorespiration and the evolution of C_4 photosynthesis. *Annual Review of Plant Biology* 63: 19–47.

Sakamoto T, Morinaka Y, Ohnishi T, Sunohara H, Fujioka S, Ueguchi-Tanaka M, Mizutani M, Sakata K, Takatsuto S, Yoshida S, Tanaka H, Kitano H, Matsuoka M (2006) Erect leaves caused by brassinosteroid deficiency increase biomass production and grain yield in rice. *Nature Biotechnology* 24: 105–109.

Salvucci ME, Crafts-Brandner SJ (2004a) Relationship between the heat tolerance of photosynthesis and the thermal stability of Rubisco activase in plants from contrasting thermal environments. *Plant Physiology* 134: 1460–1470.

Salvucci ME, Crafts-Brandner SJ (2004b) Inhibition of photosynthesis by heat stress: The activation state of Rubisco as a limiting factor in photosynthesis. *Physiologia Plantarum* 120: 179–186.

Salvucci ME, DeRidder BP, Portis AR (2006) Effect of activase level and isoform on the thermotolerance of photosynthesis in Arabidopsis. *Journal of Experimental Botany* 57: 3793–3799.

Schopf JW (2014) Geological evidence of oxygenic photosynthesis and the biotic response to the 2400–2200 Ma "great oxidation event". *Biochemistry* 79: 165–177.

Schöttler MA, Kirchhoff H, Weis E (2004) The role of plastocyanin in the adjustment of the photosynthetic electron transport to the carbon metabolism in tobacco. *Plant Physiology* 136: 4265–4274.

Sevanto S (2014) Phloem transport and drought. *Journal of Experimental Botany* 65: 1751–1759.

Spreitzer RJ, Salvucci ME (2002) Rubisco: Interactions, associations and the possibilities of a better enzyme. *Annual Review of Plant Biology* 53: 449–475.

Stotz M, Mueller-Cajar O, Ciniawsky S, Wendler P, Hartl FU, Bracher A, Hayer-Hartl M (2011) Structure of green-type RuBisCO activase from tobacco. *Nature Structural & Molecular Biology* 18: 1366–1370.

Studer RA, Christin PA, Williams MA, Orengo CA (2014) Stability-activity tradeoffs constrain the adaptive evolution of RubisCO. *Proceedings of the National Academy of Sciences of the United States of America* 111: 2223–2228.

Sudo E, Suzuki Y, Makino A (2014) Whole plant growth and N utilization in transgenic rice plants with increased or decreased Rubisco content under different CO2 partial pressures. *Plant and Cell Physiology* 55: 1905–1911.

Takahara K, Kasajima I, Takahashi H, Hashida SN, Itami T, Onodera H, Toki S, Yanagisawa S, Kawai-Yamada M, Uchimiya H (2010) Metabolome and photochemical analysis of rice plants overexpressing Arabidopsis NAD kinase gene. *Plant Physiology* 152: 1863–1873.

Takai T, Ohsumi A, San-Oh Y, Laza MRC, Kondo M, Yamamoto T, Yano M (2009) Detection of a quantitative trait locus controlling carbon isotope discrimination and its contribution to stomatal conductance in *japonica* rice. *Theoretical and Applied Genetics* 118: 1401–1410.

Takai T, Ohsumi A, Arai Y, Iwasawa N, Yano M, Yamamoto T, Yoshinaga S, Kondo M (2013a) QTL analysis of leaf photosynthesis in rice. *Japan Agricultural Research Quarterly: JARQ* 47: 227–235.

Takai T, Adachi S, Taguchi-Shiobara F, Sanoh-Arai Y, Iwasawa N, Yoshinaga S, Hirose S et al. (2013b) A natural variant of *NAL1*, selected in high-yield rice breeding programs, pleiotropically increases photosynthesis rate. *Scientific Report* 3: 2149.

Tamoi M, Nagaoka M, Miyagawa Y, Shigeoka S (2006) Contribution of fructose-l,6-bisphosphatase and sedoheptulose-l,7-bisphosphatase to the photosynthetic rate and carbon flow in the Calvin cycle in transgenic plants. *Plant & Cell Physiology* 47: 380–390.

Tanaka Y, Sugano SS, Shimada T, Hara-Nishimura I (2013) Enhancement of leaf photosynthetic capacity through increased stomatal density in Arabidopsis. *New Phytologist* 198: 757–764.

Tanksley SD (1993) Mapping polygenes. *Annual Review of Genetics* 27: 205–233.

Teng S, Qian Q, Zeng D, Kunihiro Y, Fujimoto K, Huang D, Zhu L (2004) QTL analysis of leaf photosynthetic rate and related physiological traits in rice (*Oryza sativa* L.). *Euphytica* 135: 1–7.

Terashima I, Hanba YT, Tholen D, Niinemets Ü (2011) Leaf functional anatomy in relation to photosynthesis. *Plant Physiology* 155: 108–116.

Tholen D, Zhu XG (2011) The mechanistic basis of internal conductance: A theoretical analysis of mesophyll cell photosynthesis and CO_2 diffusion. *Plant Physiology* 156: 90–105.

Uematsu K, Suzuki N, Iwamae T, Inui M, Yukawa H (2012) Increased fructose 1,6-bisphosphate aldolase in plastids enhances growth and photosynthesis of tobacco plants. *Journal of Experimental Botany* 63: 3001–3009.

Vassey TL, Quick WP, Sharkey TD, Stitt M (1991) Water stress, carbon dioxide, and light effects on sucrose-phosphate synthase activity in *Phaseolus vulgaris*. *Physiologia Plantarum* 81: 37–44.

von Caemmerer S, Quick WP, Furbank RT (2012) The development of C_4 rice: Current progress and future challenges. *Science* 336: 1671–1672.

Whitney SM, Sharwood RE, Orr D, White SJ, Alonso H, Galmés J (2011) Isoleucine 309 acts as a C_4 catalytic switch that increases ribulose-1,5-bisphosphate carboxylase/oxygenase (Rubisco) carboxylation rate in *Flaveria*. *Proceedings of the National Academy of Sciences of the United States of America* 108: 14688–14693.

Wigge B, Krömer S, Gardeström P (1993) The redox levels and subcellular distribution of pyridine nucleotides in illuminated barley leaf protoplasts studied by rapid fractionation. *Physiologia Plantarum* 88: 10–18.

Woodrow IE (2009) Flux control analysis of the rate of photosynthetic CO_2 assimilation. In *Photosynthesis in Silico*, eds. Laiska A, Nedbal L, Govindjee. Springer, Dordrecht, pp. 349–360.

Wright IJ, Reich PB, Westoby M, Ackerly DD, Baruch Z, Bongers F, Cavender-Bares J (2004) The worldwide leaf economics spectrum. *Nature* 428: 821–827.

Yamamoto T, Yano M (2007) Application of the rice genome information to the molecular breeding. *Kagaku* 77: 607–613.

Yamamoto T, Yano M (2008) Detection and molecular cloning of genes underlying quantitative phenotypic variations in rice. In *Rice Biology in the Genomics Era*, eds. Hirano HY, Sano Y, Hirai A, Sasaki T. Springer, Berlin, Heidelberg, pp. 295–308.

Yamamoto H, Kato H, Shinzaki Y, Horiguchi S, Shikanai T, Hase T, Endo T et al. (2006) Ferredoxin limits cyclic electron flow around PSI (CEF-PSI) in higher plants: Stimulation of CEF-PSI enhances non-photochemical quenching of Chl fluorescence in transplastomic tobacco. *Plant & Cell Physiology* 47: 1355–1371.

Yamamoto T, Yonemaru J, Yano M (2009) Towards the understanding of complex traits in rice: Substantially or superficially? *DNA Research* 16: 141–154.

Yamamoto T, Uga Y, Yano M (2014) Genomics-assisted allele mining and its integration into rice breeding. In *Genomics of Plant Genetic Resources*, Volume 2. Crop Productivity, Food Security and Nutritional Quality, eds. Tuberosa R, Graner A, Frison E. Springer, Dordrecht, pp. 251–265.

Yamori W (2013) Improving photosynthesis to increase food and fuel production by biotechnological strategies in crops. *J Plant Biochem & Physiol* 1: 1–13.

Yamori W, Shikanai T, Makino A (2015) Photosystem I cyclic electron flow via chloroplast NADH dehydrogenase-like complex performs a physiological role for photosynthesis at low light. *Scientific Reports* 5: 13908.

Yamori W, von Caemmerer S (2009) Effect of Rubisco activase deficiency on the temperature response of CO_2 assimilation rate and Rubisco activation state: Insights from transgenic tobacco with reduced amounts of Rubisco activase. *Plant Physiology* 151: 2073–2082.

Yamori W, Suzuki K, Noguchi K, Nakai M, Terashima I (2006) Effects of Rubisco kinetics and Rubisco activation state on the temperature dependence of the photosynthetic rate in spinach leaves from contrasting growth temperatures. *Plant, Cell & Environment* 29: 1659–1670.

Yamori W, Noguchi K, Kashino Y, Terashima I (2008) The role of electron transport in determining the temperature dependence of the photosynthetic rate in spinach leaves grown at contrasting temperatures. *Plant & Cell Physiology* 49: 583–591.

Yamori W, Takahashi S, Makino A, Price GD, Badger MR, von Caemmerer S (2011) The roles of ATP synthase and the cytochrome b_6/f complexes in limiting chloroplast electron transport and determining photosynthetic capacity. *Plant Physiology* 155: 956–962.

Yamori W, Masumoto C, Fukayama H, Makino A (2012) Rubisco activase is a key regulator of non steady-state photosynthesis at any leaf temperature and, to a lesser extent, of steady-state photosynthesis at high temperature. *The Plant Journal* 71: 871–880.

Yamori W, Hikosaka K, Way DA (2014) Temperature response of photosynthesis in C_3, C_4 and CAM plants: Temperature acclimation and temperature adaptation. *Photosynthesis Research* 119: 101–117.

Yang DL, Jing RL, Chang XP, Li W (2007) Quantitative trait loci mapping for chlorophyll fluorescence and associated traits in wheat (*Triticum aestivum*). *Journal of Integrative Plant Biology* 49: 646–654.

Yano M (2001) Genetic and molecular dissection of naturally occurring variation. *Current Opinion in Plant Biology* 4: 130–135.

Yoshihara T, Spalding EP, Iino M (2013) *AtLAZY1* is a signaling component required for gravitropism of the *Arabidopsis thaliana* inflorescence. *The Plant Journal* 74: 267–279.

Zhang J, Ku LX, Han ZP, Guo SL, Liu HJ, Zhang ZZ, Cao LR, Cui XJ, Chen YH (2014) The *ZmCLA4* gene in the *qLA4-1* QTL controls leaf angle in maize (*Zea mays* L.). *Journal of Experimental Botany* 65: 5063–5076.

Zhu X-G, Long SP, Ort DR (2010) Improving photosynthetic efficiency for greater yield. *Annual Review of Plant Biology* 61: 235–261.

42 Efficiency of Light-Emitting Diodes for Future Photosynthesis

Soleyman Dayani, Parisa Heydarizadeh, and Mohammad Reza Sabzalian

CONTENTS

42.1 Introduction 761
42.2 Electroluminescence: Efficiency in Converting Electricity to Optical Power for Photosynthesis 761
42.3 LED: A Tool for Innovative Studies on Photosynthesis 763
    42.3.1 Response of Higher Plants to LED Lighting 772
    42.3.2 Productivity of Algae under LED Lighting 773
42.4 Will LEDs Dominate the Future of Artificial Lighting for Photosynthesis? 774
42.5 Conclusions and Perspectives 777
References...777

42.1 INTRODUCTION

Oxygenic photosynthesis is a biophysicochemical process that converts carbon dioxide into organic compounds using light as a source of energy. It occurs in plants, algae, and cyanobacteria, but not in archaea. Photosynthesis uses water as a source of electrons, releasing oxygen as a waste product. Photosynthesis confers autotrophy to organisms and is the only natural process, allowing for the creation of food from simple and abundant compounds. Therefore, this process is vital for all aerobic life forms on earth because, in addition to maintaining normal levels of oxygen in the atmosphere, photosynthetic products directly or indirectly constitute the ultimate source of energy in food. Regardless of the type of photosynthetic organism, this process takes place according to the same scheme (Merchant and Sawaya 2005). The photosynthetic apparatus is composed of four macrocomplexes, namely, water-oxidizing photosystem II (PSII), cytochrome b6f, photosystem I (PSI), and H+-translocating ATP synthase (CF0F1) (Nelson and Ben-Shem 2004). They supply ATP and NADPH for the synthesis of many essential compounds, such as carbohydrates, for autotrophic growth.

Photosystems are composed of light-harvesting complexes (LHCs) and reaction centers (RCs). The harvested energy is transferred to the RC and is used to drive an electron flow within the membrane hosting the photosynthetic apparatus, the so-called thylakoid membrane. In prokaryotic photosynthetic organisms, such as cyanobacteria, thylakoid membranes are in direct contact with the cytosol, whereas in eukaryotic photosynthetic organisms, thylakoids are separated from the cytosol by envelopes, together forming a unique cell compartment, the chloroplast. A photosystem is a pigment–protein complex composed of a RC and a LHC. Two families of pigments are found in LHCs, namely, tetrapyrroles and carotenoids (Table 42.1). The presence of open-chain tetrapyrroles (phycobilins) is restricted to the phycobilisomes in cyanobacteria, glaucophytes, and red algae.

As assumed by Barber and Andersson (1992) in their paper "Too Much of a Good Thing: Light Can Be Bad for Photosynthesis," the amount of photons reaching the photosynthetic apparatus should be strictly regulated; otherwise, the photosynthetic process is completely affected. Besides, the wavelength conveys an important signal for living organisms. This information is sensed by different types of photoreceptors (Cheng et al. 2004). The pigment moiety of photoreceptors allows the receptor to extract the specific information related to the intensity of environmental light constraints from the incoming natural white light. This information is used to develop the adequate response.

In the following sections, we have reviewed the literature concerning the impact of light color provided by LED lights on plant and algal photosynthesis and productivity.

42.2 ELECTROLUMINESCENCE: EFFICIENCY IN CONVERTING ELECTRICITY TO OPTICAL POWER FOR PHOTOSYNTHESIS

The conventional artificial lighting around us is mainly based on three different technologies: fluorescent, incandescent, and high-intensity discharge (HID). However, there is another rapidly developing lighting source, which, in less than two decades of its practical presence, has shown to be a promising lighting technology for future. This technology is termed *solid-state lighting technology*, also known as semiconductor light-emitting diodes (LEDs) (Girón González 2012). The term *solid state* indicates that there are no argon and mercury (as in fluorescent) nor heating thin tungsten filament (as in incandescent) nor metal salts and gas (as in low/high sodium pressure or HIDs) for producing light. LEDs emit light when an electrical current passes through some specific material, usually an electronic diode, which is more like a computer chip

TABLE 42.1

Main Chlorophyll and Carotenoid Types in the Various Taxons of Photosynthetic Organisms

Pigment Type	Cyanobacteria	Glaucophytes	Red Algae	Brown Algae	Diatoms	Green Algae	Land Plants
Phycobilisomes	+	+	+	−	−	−	−
Chl *a*	+ Except in *Acaryochloris marina* and related taxa	+	+	+	+	+	+
Chl *b*	+ Except in Prochlorophytes	−	−	−	−	+	+
Chl *c*	−	−	−	+	+	−	−
Chl *d*	*Acaryochloris marina* and related taxa	−	−	−	−	−	−
Chl *f*	Only in filamentous cyanobacteria from Stromatolites	−	−	−	−	−	−
β-carotene	+	+	Unicellular	+	+	+	+
Fucoxanthin	−	−	−	+	+	−	−
Diadinoxanthin	−	−	−	Traces	+	−	−
Diatoxanthin	−	−	−	Traces	+	−	−
Violaxanthin	−	−	+	+	+	+	+
Lutein	−	−	Macrophytes	−	−	+	+
Zeaxanthin	Depends on species	+	+	+	Traces	+	+
Echinenone	Depends on species	−	−	−	−	−	−
Myxoxanthophyll	Depends on species	−	−	−	−	−	−
Canthaxanthin	Anabaena	−	−	−	−	−	−
Xanthophyll cycle	−	−	−	Traces	+	−	−

Note: +, Presence and −, absence.

(Lin et al. 2015). LEDs can have peak emission wavelengths from UV-C (~250 nm) to infrared (~1000 nm) (Bourget 2008), which is well beyond the light visible spectrum limits (~390 to ~700 nm) (Starr et al. 2010).

In electroluminescent devices like LEDs, the semiconductor material such as gallium arsenide (GaAs), indium phosphide (InP), and gallium nitride (GaN) as impurities are deposited in layers to form a p–n (positive–negative) junction. When positive external voltage is applied across contacts to the p- and n-type regions, electron and hole recombination leads to the released energy in the form of light. The electroluminescence phenomenon was discovered in the early 1920s following the development of solid-state semiconductor diodes (Lossev 1924). Later, the first practical LED, which could emit visible red light, was developed in General Electric Company by Holonyak, the father of the light-emitting diode, in 1961 (Lin et al. 2015). The first LEDs were of very low power and only used as an indicator. It required another 50 years to develop a real LED lighting system. In 1970s, yellow, green, and orange LEDs were introduced, and the technology led to improving the brightness of LEDs by tenfold (Bourget 2008). After almost another decade, in 1994, Nichia Corporation introduced the first high-brightness blue LED (Nakamura et al. 1991, 1996). Eventually, having all the three primary light spectra (colors) developed, white LED was designed in 1996 (Krames et al. 2007). The primary LEDs used low power electricity (lower than 1 W), and development

of high-power LED lights (1 W or greater) in 1999 was the final major advance in LED technology, which led to a boom in demand for LEDs in both general and professional communities. Since the 1960s, the efficiency and light output of LEDs have been doubled approximately every 36 months (Steele 2007; Lin et al. 2015).

Numerous advantages are associated with LEDs over other conventional lights. The general LED benefits include higher energy efficiency (LEDs use less electric power [watt] for generating one unit of light [lumen]); low LED chip energy costs; no sensitivity to low temperatures; no sensitivity to high humidity; and extremely low heat emission (Singh et al. 2014). There are no high touch temperatures, no fragile glass bulbs to break, and no hazardous materials, e.g., mercury (RoHS compliant), used in LEDs. Their average life span is roughly about 50,000 h, which is currently about 20 times more than that of an incandescent bulb (Kelly 2004). The small size of LEDs enables flexible lighting designs as needed in photobioreactors used to grow algae (Kim et al. 2015). LEDs could be integrated into digital control systems, which facilitate the application of high-performance and user-defined lighting programs such as varying intensity or spectral composition (Yeh and Chung 2009). The environmental impacts of LEDs should also be considered. LEDs are very durable, safe, and generally considered environmentally-friendly (Darko et al. 2014). These advantages could be linked with low electricity consumption, which is directly correlated with lower fossil fuel burning, decreased CO_2 and sulfur oxide emissions from power plants, and less nuclear wastes deposited into the environment. The reduction in mercury-containing lamp production and disposal is another promising environmental effect of LEDs (Schubert and Kim 2005).

There are, however, some drawbacks for widespread LED applications at present. LEDs are resistant to very low temperatures, but they are adversely affected by high temperatures, as is the case for almost all electronic chips. Therefore, LED lamps are commercially offered in packages, which include heat dissipation elements such as heat sinks, cooling fins, and fans. This imposes a considerably higher capital cost for commercial LED applications. They also consume DC electrical power, and it requires instruments to convert regular AC electrical power to DC. The energy cost and maintenance issues related to these appendixes lessen the economic soundness of LEDs (Olle and Viršile 2013). Another issue is that because LEDs emit light in a narrow and directional manner, covering a wide area by LED light requires installing a large number of LED lamps. This limitation is addressed by placing glasses or lenses to scatter the light beam, but it could reduce the optical effectiveness of LEDs. Finally, LEDs can be constructed to emit UVB and UVC if designed for that purpose. Therefore, care should be taken under circumstances of using such wavelengths as they might damage skin and eyes.

In spite of the few technical limitations, it is expected that mass production of LEDs, which is encouraged by the increasing general and industrial demands for energy-efficient lightings, and technological developments in optoelectronics, will significantly reduce the capital and operating costs and further enhance the overall light efficiency of LEDs in the near future (Haitz and Tsao 2011; Pinho et al. 2012). In recently developed organic LEDs (OLEDs) or polymer LEDs (PLEDs), the emitting layer is a film of organic compound or a polymer instead of inorganic compounds, which emits light in response to an electric current (Kamtekar et al. 2010). The new LEDs have the potential to generate light with high efficiency in electricity conversion up to 100%, providing a basis for what is called the ultimate light source.

42.3 LED: A TOOL FOR INNOVATIVE STUDIES ON PHOTOSYNTHESIS

The phenomenon of converting carbon dioxide and water under the influence of light energy (photons) into carbohydrate and oxygen gas in plants has been known since the mid-nineteenth century (Huzisige and Ke 1993; Singhal 1999). It was eventually termed *photosynthesis* by Charles Reid Barnes in 1893 (Gest 2002). Today, our understanding of this fascinating process is still expanding. Sunlight comprises a plethora of wavelengths, of which only a small fraction is the spectrum of visible light (Figure 42.1a). As we know, the light-absorbing complexes or antenna complexes in thylakoid membrane of plant cells act as solar cells, which can only perceive energy from particular wavelengths in visible spectrum and convert it into electrons. The reason for the specificity of plant photosynthesis to this small splice of the full electromagnetic spectrum

lies with the type of light receptors located on the antenna. Briefly, there is one chlorophyll *a* molecule in the center of each antenna complex as the core pigment. Chlorophyll *a* is a green pigment and plays a central role in photosynthesis. The fact that an object is green means when it is illuminated by a white light source, it absorbs blue and red light spectra and reflects the green wavelength. The other known pigments or light receptors on the antenna are mainly lutein, zeaxanthin, β-carotene, and chlorophyll *b* (Figure 42.1b) depending on the species (Table 42.1), each of which has a specific frequency absorption peak. The occurrence of different types of receptors in the antenna is not arbitrary. This pigment variation has roles such as providing extra light-harvesting nodes for trapping a wider range of spectra that leads to higher efficiency, and acts as a protecting shield against occasional damaging effects of sunlight on the photosynthetic apparatus. So, when it comes to studying the light effect on photosynthesis in plants and other photosynthetic organisms, we are literally confined to a small splice of light wavelengths ranging from ~400 to ~700 nm.

By the 1990s, the fact that red and blue colors are the primary light wavelengths involved in photosynthesis had been established; however, little was known about the independent effect of various light wavelengths on plants. One of the main obstacles that hindered research progress in this field was the lack of lighting instruments that can reliably emit light at a defined and narrow wavelength. It seems that this unique characteristic of LEDs was even neglected by the first scientists who studied the effect of red-light LEDs on plants (Bula et al. 1991). In fact, they were more impressed by the energy efficiency of the emerging lighting technology rather than its monochromatic emitting capability. However, the many advantages of LED lights over other conventional synthetic light sources in addition to its true spectral control were enough to catch the attention of plant biologists very soon. It is not surprising that one of the main applications of LEDs in plant research still continues to be the study of light component effect on photosynthesis (Table 42.2).

The first LEDs had small chips that used less than 100 mW, while today, LEDs used in car lighting system are capable of dissipating up to 10 W of power. The range of light power offered by LED manufacturing companies is now expanding from below 1 W to more than 100 W and light intensity ranging from below 10 to 100,000 $\mu mol\ m^{-2}\ s^{-1}$. This range of light power and intensity of LEDs broadly exceeds the plant photosynthetic pigment limits for more efficient photosynthesis without suffering from excessive sun irradiation.

The unique specification of LED technology that expedited the integration of LED lights into plant growth and photosynthesis research, as mentioned earlier, was its capability of true spectral control (Morrow 2008). In other words, the LEDs are capable of emitting wavelengths that almost exactly match plant photoreceptors. This characteristic facilitates the implementation of LEDs with desired spectral ranges that are involved in plant photosynthetic responses. It also provides the chances of independent control and fine-tuning of light intensity and spectral quality (Folta et al. 2005). Furthermore,

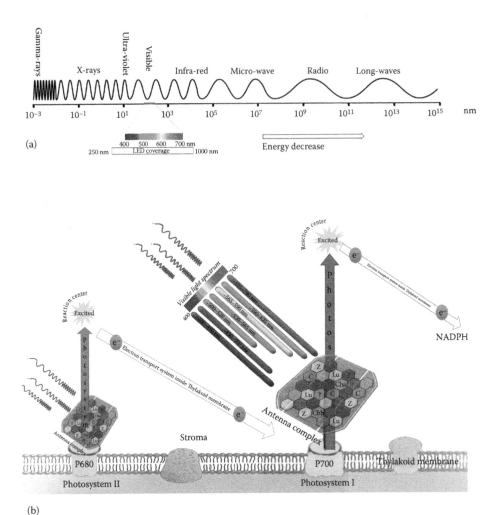

FIGURE 42.1 **(See color insert.)** (a) The visible light is a small fraction of the full electromagnetic spectrum; (b) the antenna complexes located on thylakoid membrane consist of various light-harvesting molecules, which could only trap photon energy in the range of the visible spectrum light (~400 to ~700 nm) and send the photons through an energy transfer path to Chl *a* molecule as the main molecule in each RC. The antenna pigment molecules mainly include Chl *b*: chlorophyll *b*; Chl *a*: chlorophyll *a*; C: β-carotene; Z: zeaxanthin; and Lu: lutein.

LED lamps have relatively cool photon-emitting surfaces and can be used for intracanopy photosynthesis to eliminate the negative effects of shading. This will enhance the productivity of closed-canopy crops in greenhouses (Trouwborst et al. 2010).

The idea of developing space-based plant growth systems as part of the space program in the 1990s encouraged the experiments with LEDs on plants (Bula et al. 1991). At that time, practically only red (660 nm) LEDs were available. Since then, experiments about LED lighting effects on photosynthesis, which directly influences the growth and developmental mechanisms in plants, have been going on for more than two decades now. Although red wavelength was close to Pr phytochrome absorption peak and seemed to be the only beneficial spectrum for photosynthesis-related responses, further experiments proved the indispensable necessity of the blue spectrum (400–500 nm) for enhanced photosynthetic processes and normal plant growth (Hoenecke et al. 1992). Some authors believe that the era of LED research in the plant

kingdom started when the first practical high-intensity blue LEDs were introduced (Kubota et al. 2011).

The fact that red and blue lights were the most physiologically important lights has led some scientists to believe that by removing the other seemingly unprofitable wavelengths like green and yellow, they could save energy while improving plant growth. However, it was then proved that green light may be used for photosynthesis, but this assumption soon lost its credibility as some plants grown under optimal solely red and blue lights looked purplish gray, and it was difficult to evaluate the plant health. The cure for such undesirable effects of wavelength deficiencies was to provide the missing light wavelengths together with red and blue, like projecting a maximum of 20% green light on affected plants (Massa et al. 2008; Mitchell 2012). It is now well established that other light spectral components also have significant influence on normal plant growth and development (Kim et al. 2004a). It, however, should not escape attention that elucidation of such intricate effects of light wavelengths on plants owes its credit

TABLE 42.2
LED Light Effects on Higher Plants

Plant	Tissue/Genus	LED Wavelength	Scope	References
Potato	Stem cuttings	Red and blue	The effect of light and hormones on morphogenesis (root/shoot).	Aksenova et al. 1994
	In vitro plantlets	Red (660–690 nm)	Increase in shoot length and chlorophyll content of the plantlets. Red light affected morphology rather than growth rate.	Miyashita et al. 1994
Broccoli	Sprouting broccoli (*Brassica oleacea* var. *italica*)	Red (627 nm)/blue (470 nm)	Significant positive effect of short-duration blue light on phytochemical compound accumulation.	Kopsell et al. 2014
	Sprouting broccoli (*Brassica oleacea* var. *italica*)	Red (627 nm)/blue (470 nm)	Increase in pigments, glucosinolates, and mineral elements.	Kopsell and Sams 2013
	Postharvest (*Brassica oleracea* L. var. *italica*)	Red and blue	Under red LED light, senescence and yellowing process after harvest delayed and ethylene production and reduction of ascorbate (AsA) were suppressed. Blue LED light did not significantly affect the senescence process after harvest.	Ma et al. 2014
Chrysanthemum	Plantlets (*Dendranthema grandiflorum*)	Blue LED, red LEDs, red + blue LEDs (RB), red + far-red LEDs (RFr), blue + far-red LEDs (BFr)	Net photosynthetic rate, growth, and leaf stomata.	Kim et al. 2004b
Parsley	Green tissues and sprouts (*Petroselinum crispum*, "Miller")	Red 640 nm, supplemented with 450, 660, and 735 nm	Content of carbohydrate, nitrate, vitamin C, phenolic compounds, antioxidant activity.	Urbonavičiūtė et al. 2008
	Preharvest (*Petroselinum* "plain leaved")	Red 638 nm (supplemental)	Accumulation of vitamin C, carbohydrates, enhancement of free radical binding activity, activity of nitrate-reducing enzymes.	Bliznikas et al. 2012
Kale	*Brassica oleracea* L. var. Acephala D.C.	730, 640, 525, 440, 400 nm	Selective increase in accumulation of chlorophylls, carotenoids, glucosinolates.	Lefsrud et al. 2008
Spinach	Whole plant	Red (660 nm, 620–700 nm)	Dry matter production and plant leaf area under LED were less and smaller than those under fluorescent control.	Yanagi and Okamoto 1994
	Preharvest (*Spinacia oleracea* "Géant d'hiver")	Red 638 nm (supplemental)	Metabolic unbalance, accumulation of low amounts of antioxidant compounds, diminishing the nutritional value.	Bliznikas et al. 2012
	Spinach (*Spinacea oleracea* L. cv. Nordic IV)	Red (660 nm), supplemented: FL blue	Improving the plant growth.	Yorio et al. 2001
Ginseng	Roots (*Panax ginseng* C. A. Mayer)	Red (630 nm); blue (465 nm)	Effect of light on metabolic processes of adventitious roots.	Park et al. 2013
	Roots (*Panax ginseng* Meyer)	380, 450, 470, 660 nm	Increased contents of ginsenosides in raw ginseng under 450 and 470 nm (blue) LED.	Park et al. 2012
Wheat	Wheat (*Triticum aestivum* L., cv. "USU Super Dwarf")	Blue (400–500 nm), red (600–700 nm), far red (700–800 nm)	Photomorphogenesis, photosynthesis, and seed yield.	Goins et al. 1997
	Root shoot	Red, blue	Root and shoot interaction in the greening of wheat seedlings.	Tripathy and Brown 1995
	Sprouted seeds; wheat (*Triticum aestivum* L.)	638 nm, 455 nm, 638 nm, 669 nm, 731 nm (basal components); basal + 385 nm, basal + 510 nm, basal + 595 nm	Antioxidant properties; basal components with supplemental green (510 nm) light could improve the antioxidant properties of sprouted seeds.	Samuolienė et al. 2011d
	Green tissue and sprouts; wheat grass (*Triticum aestivum* L. cv. "Pirvinta")	Red 640 nm supplemented with 450, 660, and 735 nm	Content of carbohydrate, nitrate, vitamin C, phenolic compounds, antioxidant activity.	Urbonavičiūtė et al. 2008

(Continued)

TABLE 42.2 (CONTINUED)
LED Light Effects on Higher Plants

Plant	Tissue/Genus	LED Wavelength	Scope	References
Strawberry	Strawberry "Akihime" shoot explants	70% red + 30% blue	Higher LED shoot and root fresh weight of plantlets; the number of leaves and the total of shoot and root fresh weight of plantlets under LEDs and control were equal. LED light for *in vitro* culture of plantlets contributed to improving the plant growth in acclimatization.	Duong et al. 2000
	Harvested fruits; *Fragaria × ananassa* "Strawberry Festival"	Blue (455 nm) Red (668 nm) FR (755 nm)	Light wavelengths modulation of volatile organic compounds.	Colquhoun et al. 2013
Lilium	*In vitro* regeneration of bulblets; *Lilium oriental* hybrid "Pesaro"	Red, blue, red + blue	Red + blue developed a higher number of bulblets per explant. Bulblets were bigger in size, and their fresh and dry weight and dry matter percentages were high.	Lian et al. 2002
Cyclamen	Flowering; *Cyclamen persicum* Mill. cv. "Dixie White"	Red, blue, red + blue	Light quality and photoperiod on flowering; flowering and growth of cyclamen can be controlled by lights.	Heo et al. 2003
Radish	Radish (*Saxa*)	455, 640, 660, 735 nm	Photosynthesis and control of plant growth by circadian manipulation of a relatively weak far red.	Tamulaitis et al. 2005
	Green vegetables and sprouts; leafy radish (*Raphanus sativus* L. cv. "Tamina")	Red 640 nm, supplemented with 450, 660, and 735 nm	Content of carbohydrate, nitrate, vitamin C, phenolic compounds, and the antioxidant activity.	Urbonavičiūtė et al. 2008
	Sprouted seed; radish (*Raphanus sativus* L.)	638, 455, 638, 669, and 731 nm (basal components); basal + 385 nm; basal + 510 nm, basal + 595 nm	Antioxidant properties.	Samuolienė et al. 2011d
	Radish (*Raphanus sativus* L. cv. Cherriette)	Red (660 nm), supplemented: FL blue	Improving the plant growth.	Yorio et al. 2001
Oat	Seedlings (*Avena sativa* cv Seger)	Red	Oat seedlings could detect IR LED radiation and showed differences in growth and gravitropic response.	Johnson et al. 1996
Pea	Seedlings, stem, leaves (*Pisum sativum* L.)	Red (625–630 nm); blue (465–470 nm)	Red induced b-carotene expression, antioxidant activity, and blue improved seedling weight and chlorophyll induction.	Wu et al., 2007
	(*Pisum sativum* L.)	Red (660 nm)	Photosynthetic apparatus.	Topchiy et al. 2005
Grape	Shoot and root growth; hybrid franc (Vitisriparia 9 *V. vinifera* cv. Cabernet Sauvignon), Ryuukyuuganebu (*V. ficifolia* var. *ganebu*), Kadainou R-1 (Ryuukyuuganebu 9 *V. vinifera* cv. Muscat of Alexandria)	Blue (480 nm); red (660 nm)	Growth and morphogenesis Higher rooting % and higher root numbers	Poudel et al. 2008
Lettuce	Red leaf lettuce; *Lactuca sativa* cv. Outredgeous	Red (640 nm), blue (440 nm), green (530 nm), far red (730 nm)	Morphology increased both accumulation of bioprotective compounds and total yield.	Stutte et al. 2009
	Lettuce (*Lactuca sativa* cv. Grand rapids)	Red (638 nm); supplementary	Substantial reduction of nitrate concentration, increased concentration of nutritionally valuable carbohydrates, variation in content of vitamin C.	Samuolienė et al. 2009
	Lettuce cultivars (*Lactuca sativa* L.)	Red (638 nm); supplementary	Enhanced antioxidant properties, increased content of alpha-carotene and phenolic compounds, enhanced free radical scavenging activity.	Zukauskas et al. 2009

(Continued)

TABLE 42.2 (CONTINUED)
LED Light Effects on Higher Plants

Plant	Tissue/Genus	LED Wavelength	Scope	References
Lettuce	Preharvest stage, baby lettuce; miniature variety of Romaine Little Gem "Thumper," curly baby leaf "Multibaby"	Red (638 nm); supplementary	High correlation between ascorbic acid and glucose content under supplemental red LED; increase in total phenolics, tocopherols, sugars, antioxidant capacity.	Samuolienė et al. 2012b
	Preharvest; red leaf "Multired 4," green leaf "Multigreen 3," light green leaf "Multiblond 2"	Red (638 nm); supplementary	Reduced content of nitrate and increased contents of total phenolic compounds and free-radical scavenging activity in two varieties.	Samuolienė et al. 2011c
	Lettuce (*Lactuca sativa* L. cv Banchu Red Fire)	Green (510, 524, 532 nm)	Promoting active plant growth; photosynthetic rate under green LED light at higher PPF was dramatically higher.	Johkan et al. 2012
	Red leaf "Multired 4," green leaf "Multigreen 3," light green leaf "Multiblond 2"	Blue (455, 470 nm); green (505, 530 nm); supplementary	Reduction in nitrates, increased concentration of nutritionally valuable carbohydrates.	Samuolienė et al. 2011b
	Red leaf "Multired 4," green leaf "Multigreen 3," light green leaf "Multiblond 2"	Blue 455/470 nm, green 505/530 nm, supplementary	Antioxidant properties in lettuce depended on multicomponent exposure of variety, light quality, or seasonality.	Samuolienė et al. 2012a
	Lettuce (*Lactuca sativa* L.)	Blue, red, blue + red	Blue light promoted the growth of plants after transplanting. High shoot and root biomasses, high content of photosynthetic pigments, high antioxidant activities in seedlings before transplanting. Seedling morphology affected under blue LED light.	Johkan et al. 2010
	Baby leaf lettuce (*Lactuca sativa* L. cv. Red Cross)	UV-A, blue, green, red, far-red: (18, 130, 130, 130, 160 μmol m^{-2} s^{-1}); supplementary	Enhanced nutritional value and growth.	Li and Kubota 2009
	Lettuce (*Lactuca sativa* L. var. *acephala* Dill. cv. LolloRossa)	Blue (450–500 nm); green (500–600 nm); red (650–700 nm); 70% red, 20% green, 10% blue	Efficiency of photosynthesis, photoinhibition process by changes in malonedialdehyde (MDA) concentration, peroxidase (POX) activity, and hydrogen peroxide (H_2O_2) content. More effective functioning of PSII at 70% red, 20% green, and 10% blue light composition.	Ilieva et al. 2010
	Lettuce; cv. Boston head	–	Comparison of high-pressure sodium (HPS) lamps and LED lights. Growth, dry mass, overall yield.	Martineau et al. 2012
	Lettuce varieties (*Lactuca sativa* L.)	Blue, red, blue/red	Effects of light quality and photosynthetic photon flux (PPF) on growth and morphogenesis.	Yanagi et al. 1996
	Lettuce seedlings (*Lactuca sativa* L. cv. Okayama-saradana)	Red/blue ratios	Growth and morphogenesis.	Okamoto et al. 1996
	Lettuce (*Grand Rapids*)	455, 640, 660, 735 nm	Photosynthesis and control of plant growth by circadian manipulation of a relatively weak far red.	Tamulaitis et al. 2005
	Green vegetables and sprouts (*Lactuca sativa* cv. "Grand rapids")	Red 640 nm, supplemented with 450, 660,735 nm	Content of carbohydrate, nitrate, vitamin C, phenolic compounds, and the antioxidant activity.	Urbonavičiūtė et al. 2008
	Lettuce seedlings (*Lactuca sativa* "Grand Rapids")	Red (660 nm)	Extended hypocotyls and elongated cotyledons under red-only LED. Blue fluorescence combination decreased growth.	Hoenecke et al. 1992
	Lettuce (*Lactuca sativa* L. var. *capitata*)	Red/blue Red/blue/white	Growth, development, and edible quality. Nitrate content under RBW LED was significantly lower. Leaves sugar content under RBW lights was significantly higher. Leaves chlorophyll, carotenoid, and protein showed no significant differences.	Lin et al. 2013

(Continued)

TABLE 42.2 (CONTINUED)
LED Light Effects on Higher Plants

Plant	Tissue/Genus	LED Wavelength	Scope	References
Lettuce	Lettuce; red leaf (*Lactuca sativa* L. "Sunmang"), green leaf lettuce (*Lactuca sativa* L. "Grand Rapid TBR")	Various combinations of blue (456 nm), red (655 nm)	Leaf shape, plant growth, accumulation of antioxidant phenolic compounds. Ratio of blue to red light is important for morphology, growth, phenolic compounds, and antioxidant properties.	Son and Oh 2013
	Lactuca sativa	Blue (460 nm)/+ Red (635, 660 nm)	Photocontrol of basic physiological processes.	Tarakanov et al. 2012
	(*Lactuca sativa* L. cv. Waldmann's Green)	Red (660 nm), supplemented: FL blue	Improving the plant growth.	Yorio et al. 2001
	Lactuca sativa	Green (500–600 nm), red and blue LEDs	Addition of green LED to (R + B) enhanced plant growth.	Kim et al. 2004a
	Lactuca sativa	Green light (500–600 nm), red and blue	Effects of green light under controlled environment.	Kim et al. 2005
	Seedling; *Lactuca sativa*	Red, blue, red + blue	The effects of light quality on the growth and chloroplast ultrastructure.	Zhang et al. 2010
	Seeds, lettuce, "Green Oak Leaf" lettuce	Red, blue, red + blue	Morphology, biomass and pigment, vitamin C contents.	Chen et al. 2014
	(*Lactuca sativa* var. *crispa* L., "Frillice")	Red-orange (630 nm), blue (460 nm), yellow (594 nm)	Evaluation of growth performance.	Pinho et al. 2007
	(*Lactuca sativa* L. "Rollo Rosa")	Red, blue, red + blue	Early growth and inorganic elements in leaf lettuce.	Shin et al. 2012
Petunia	Cut flowers (*Petunia* × *hybrida* cv. "Mitchell Diploid" [MD])	Blue (455 nm), red (668 nm), FR (755 nm)	Light wavelength modulation of volatile organic compounds.	Colquhoun et al. 2013
Blueberry	Harvested fruits; *Vaccinium corymbosum* "Scintilla"	Blue (455 nm), red (668 nm), FR (755 nm)	Light wavelength modulation of volatile organic compounds.	Colquhoun et al. 2013
Primula	Seedling; *Primula vulgaris*	100% red, 100% blue,	Higher flower buds and less days to flowering.	Sabzalian et al. 2014
Treasure flower	Seedling; *Gazania splendens*	70% red + 30% blue, or		
Stock	Seedling; *Matthiola incana*	100% white LED		
Mint	*Mentha spicata* (spear mint), *Mentha piperita* (pepper mint), *Mentha longifolia* (horse mint)	100% red, 100% blue, 70% red + 30% blue, or 100% white LED	Increased essential oil content, plant photosynthesis, fresh weight under 70/30% red-blue LED.	Sabzalian et al. 2014
Eucalyptus	*Eucalyptus citriodora*	80% red + 20% blue	Improved growth of plants in acclimatization.	Nhut et al. 2002
Marigold	Seedling; marigold (*Tagetes erecta* L. cv. Orange Boy)	Blue (440 nm), red (650 nm), far red (720 nm); supplementary	Growth and morphogenesis.	Heo et al. 2002
	Seedling; *Calendula officinalis*	100% red, 100% blue, 70% red + 30% blue, or 100% white LED	Higher flower buds and less days to flowering.	Sabzalian et al. 2014
	French marigold; *Tagetes patula*	Blue (460 nm)/+ Red (635, 660 nm)	Photocontrol of basic physiological processes.	Tarakanov et al. 2012
Salvia	Seedling; Salvia (*Salvia splendens* F. Sello ex Ruem&Schult. cv. Red Vista)	Blue (440 nm), red (650 nm), far red (720 nm); supplementary	Growth and morphogenesis.	Heo et al. 2002
Pepper	Leaves, stem (*Capsicum annuum* L. cv., Hungarian Wax)	Red (660 nm), far red (700/735–800 nm), blue (350–550 nm); supplementary	Effects of light quality on plant anatomy of leaves and stems.	Schuerger et al. 1997
	(*Capsicum annum* L. cv. "Hungarian Wax" pepper)	Red (660 nm), far red (735 nm)	Growth and photomorphogenesis.	Brown et al. 1995
	Sweet pepper variety "Reda"	455, 470, 505, 530 nm; supplementary (HPS)	Increased leaf area, fresh and dry weight, and photosynthetic pigment content.	Samuolienė et al. 2011a

(Continued)

TABLE 42.2 (CONTINUED)
LED Light Effects on Higher Plants

Plant	Tissue/Genus	LED Wavelength	Scope	References
Pepper	Fruit, chili peppers (*Capsicum annuum* L. cv. Cheonyang)	Red/+blue	Pungency, fruit color, primary and secondary metabolites (capsaicinoids).	Gangadhar et al. 2012
	Sweet pepper; *Capsicum annum* L. cv. Ferrari; *Capsicum annum* L. cv. Viper	80% 630 nm; 20% 460 nm	Growth, yield, and quality.	Stadler 2011
Calla lilies	Plantlet, *Zantedeschia jucunda* "Black Magic"	Red/blue	Improved growth and tuber formation.	Jao et al. 2005
Banana	Shoot explants	100% red, 90% red + 10% blue, 80% red + 20% blue, 70% red + 30% blue, 100% blue	Improved acclimatization, growth of plantlets under LED (80% red: 20 blue).	Nhut et al. 2000
Kudzu	Leaves (*Pueraria lobata* (Willd) Ohwi.)	Red light (656±5 nm)	Photosynthesis, stomatal conductance, isoprene.	Tennessen et al. 1994
Basil	*Clove Basil; Ocimum gratissimum*	Blue (460 nm)/+ Red (635, 660 nm)	Photocontrol of basic physiological processes.	Tarakanov et al. 2012
	Basil seed/seedling (*Ocimum basilicum* L.)	100% red, 100% blue, 70% red + 30% blue, or 100% white LED	Improved growth rate.	Sabzalian et al. 2014
	Green tissue and sprouts (*Ocimum basilicum* L.)	Red 640 nm, supplemented with 450, 660, and 735 nm	Content of carbohydrate, nitrate, vitamin C, phenolic compounds, and the antioxidant activity.	Urbonavičiūtė et al. 2008
Dills	Green tissue and sprouts (*Anethum graveolens* L. cv. "Szmaragd")	Red 640 nm, supplemented with 450, 660, and 735 nm	Content of carbohydrate, nitrate, vitamin C, phenolic compounds, and antioxidant activity.	Urbonavičiūtė et al. 2008
	Preharvest (*Anethum graveolens* "Mammouth")	Red (638 nm), supplemental	Accumulation of vitamin C, carbohydrates, enhancement of free radical binding nitrate-reducing enzymes activity.	Bliznikas et al. 2012
Rocket	Preharvest; rocket (*Eruca sativa* "Rucola")	Red (638 nm), supplemental	Metabolic unbalance, accumulation of low amounts of antioxidant compounds, diminishing the nutritional value.	Bliznikas et al. 2012
Coleus flower	*Coleus blumei* (also *Plectranthus scutellarioides*)	Blue (460 nm)/+ Red (635, 660 nm)	Photocontrol of basic physiological processes.	Tarakanov et al. 2012
Cymbidium	Plantlets	Super bright, blue	–	Tanaka et al. 1998
Marjoram	Green tissue and sprouts (*Majorana hortensis*, Moench.)	Red (640 nm), supplemented with 450, 660, and 735 nm	Content of carbohydrate, nitrate, vitamin C, phenolic compounds, antioxidant activity.	Urbonavičiūtė et al. 2008
	Marjoram (*Majorana hortensis*, Moench.)	Red (638 nm)	Substantial reduction of nitrate concentration, increased concentration of nutritionally valuable carbohydrates, variation in vitamin C.	Samuolienė et al. 2009
Onions	Green onions (*Allium cepa*, L. cv. Lietuvosdidieji)	Red (638 nm)	Substantial reduction of nitrate concentration, increased concentration of nutritionally valuable carbohydrates, variation in vitamin C.	Samuolienė et al. 2009
	Preharvest; green onions (*Allium cepa* "White Lisbon")	Red (638 nm), supplemental	Enhanced antioxidant and nutritional properties.	Bliznikas et al. 2012
	Onion leaves (*Allium cepa* L.)	Red (640 nm), supplemented with 450, 660, and 735 nm	Content of carbohydrate, nitrate, vitamin C, phenolic compounds, and antioxidant activity.	Urbonavičiūtė et al. 2008
Mustard	*Mustard greens; Brassica juncea*	Blue (460 nm)/+ Red (635, 660 nm)	Photocontrol of basic physiological processes.	Tarakanov et al. 2012
	(*Brassica juncea* L., "Red Lion")	455, 638, 665, 731 nm	Growth, nutritional quality, and antioxidant properties.	Samuolienė et al. 2013
	Preharvest; white mustard (*Sinapis alba* "Yellow mustard")	Red (638 nm), supplemental	Enhanced antioxidant and nutritional properties.	Bliznikas et al. 2012

(Continued)

TABLE 42.2 (CONTINUED)
LED Light Effects on Higher Plants

Plant	Tissue/Genus	LED Wavelength	Scope	References
Cabbages	Seedlings, cabbages (*Brassica oleracea* var. capitata L.) "Kinshun" (green leaf), "Red Rookie" (red leaf)	470, 500, 525, 660 nm	Growth of seedlings and the biosynthesis of plant pigments in leaves.	Mizuno et al. 2009
	Nonheading Chinese cabbage (*Brassica campestris* L.)	Blue (460 nm) Blue + red (11.1% + 88.9%) Red (660 nm)	Growth and vitamin C, soluble protein, sucrose, soluble sugar, starch, and pigment concentrations.	Li et al. 2012
	Chinese cabbage (*Brassica chinensis* L.)	Red (650 nm), blue (470 nm), supplemented: HPS	Growth, content of sugar, protein, photosynthetic pigments, and chlorophyll fluorescence parameters.	Avercheva et al. 2009
Tomato	Tomato (*S. lycopersicum* L., cv. MomotaroNatsumi)	White, red, blue	Tomato yield and quality of tomato under single-truss tomato production system.	Lu et al. 2012
	Tomato seedling ("Komeett")	100% red: 0% blue, 96% red: 4% blue, 84% red: 16% blue	Growth and morphological response.	Hernández and Kubota 2012
	Tomato seedlings	Blue (450 nm), red (660 nm)	Morphogenesis, photosynthesis, stem elongation, and flowering.	Nanya et al. 2012
	Transplants; tomato hybrid "Magnus" F_1	455, 470, 505, 530 nm; supplementary (HPS)	Increased leaf area, fresh and dry weight, and photosynthetic pigments content.	Samuolienė et al. 2011a
	Seedlings, cherry tomato (*Solanum lycopersicum* Mill.)	Red, blue, orange, green, red + blue (RB), red + blue + green (RBG)	Regulation of chloroplast ultrastructure, cross-section anatomy of leaves, and morphology of stomata.	Liu et al. 2011
	Tomato (*S. lycopersicum*), two cultivars	–	Comparison of intracanopy LED towers and overhead high-pressure sodium lamps for supplemental lighting of greenhouse-grown tomatoes.	Gómez et al. 2013
	Harvested fruits; tomatoes (*S. lycopersicum*, M82)	Blue (455 nm), red (668 nm), FR (755 nm)	Light wavelength modulation of volatile organic compounds.	Colquhoun et al. 2013
	Tomato (*Lycopersicon esculentum* "Trust")	Blue (455 nm), supplemented: HPS	Comparison of developmental and physiological changes in plants grown under HPS with different levels of blue light, and to evaluate the use of blue LEDs for extended and inner canopy lighting.	Ménard et al. 2005
	Seedling; grafted tomato (*Lycopersicon esculentum* Mill. cv. Momotaro; rootstock: cv. Helper M)	Red and blue LEDs, different red/blue PPFD ratios (100/0: R100, 97/3: R97B3, and 50/50: R50B50)	Growth, ribulose-1,5-bisphosphate carboxylase/oxygenase content, chlorophyll, and carbohydrate content.	Kaneko-Ohashi et al. 2004
	Tomato	Blue (450 nm), red (660 nm), white	Stimulating tomato crop canopy to induce xerophytophysiological regulations.	Xu et al. 2012
	Young tomato plants	Red/blue, 300, 450, and 550 µmol m^{-2} s^{-1}	Growth and leaf development were studied under different light intensities.	Fan et al. 2013
	Seedlings; six tomato (*Solanum lycopersicum*) cultivars ("Maxifort," "Komeett," "Success," "Felicity," "ShevaSheva," and "Liberty")	80% red + 20% blue; 95% red + 5% blue; 100% red; supplementary	Increased hypocotyl diameter, epicotyl length, shoot dry weight, leaf number, and leaf expansion.	Gómez and Mitchell 2015
	Seedlings	Red, blue, red + blue	The effects of light quality on the growth and chloroplast ultrastructure.	Zhang et al. 2010
Cucumber	Cucumber (*Cucumis sativus*, Hybrid "Mandy F1")	447, 638, 669, 731 nm	Optimization of light spectra for cucumber growth and evaluating after-effect on yield under LED lights.	Brazaitytė et al. 2009

(Continued)

TABLE 42.2 (CONTINUED)
LED Light Effects on Higher Plants

Plant	Tissue/Genus	LED Wavelength	Scope	References
Cucumber	Transplant; cucumber hybrid "Mandy"	455, 470, 505, 530 nm; supplemented: HPS	Increased leaf area, fresh and dry weight, decreased hypocotyls elongation, and enhanced development. Changes in synthesis of photosynthetic pigments and chlorophyll *a* and *b* ratio in leaves.	Novičkovas et al. 2010
	Cucumber (*Cucumis sativus*)	Red/blue	Leaf photosynthesis, morphology, and chemical composition.	Hogewoning et al. 2010
	Transplant; cucumber hybrid "Mirabelle" F_1	455, 470, 505, 530 nm; supplementary (HPS)	Increased leaf area, fresh and dry weight, and photosynthetic pigment content.	Samuolienė et al. 2011a
	Cucumber (*Cucumis sativus* "Bodega")	Blue (455 nm), supplemented: HPS	Comparison of developmental and physiological changes in plants grown under HPS with different levels of blue light, and to evaluate the use of blue LEDs for extended and inner canopy lighting.	Ménard et al. 2005
Lentil	Sprouted seed; lentil (*Lens esculenta* Moenh.)	638, 455, 638, 669, 731 nm (basal components); basal + 385 nm; basal + 510 nm, basal + 595 nm	Antioxidant properties; basal components with supplemental green (510 nm) light could improve the antioxidant properties of sprouted seeds.	Samuolienė et al. 2011d
	Lentil seed/seedling (*Lens culinaris* Medic)	100% red, 100% blue, 70% red + 30% blue, or 100% white LED.	Improved growth rate.	Sabzalian et al. 2014
Barley	Leaves; barley cv. Geungang	Red (660–670 nm), far red (730–740 nm), blue (470–475 nm), blue red (470–670 nm), green (505 nm), yellow (590–595 nm), white	Antioxidant activity of extracts of barley leaves.	Lee et al. 2010
	Barley grass (*Hordeum vulgare* L. cv. "Aura")	Red (640 nm), supplemented with 450, 660, and 735 nm	Content of carbohydrate, nitrate, vitamin C, phenolic compounds, and the antioxidant activity.	Urbonavičiūtė et al. 2008
Citrus	Postharvest fruit; *Satsuma mandarin* (Citrus unshiu Marc.)	Blue (470 nm), red (660 nm)	Accumulation of carotenoids (β-Cryptoxanthin) and expression of genes related to carotenoid biosynthesis; fruits.	Ma et al. 2012
	Postharvest fruit; *Satsuma mandarin* (Citrus unshiu Marc.)	Red (660 nm), supplemented: ethylene gas	Carotenoid accumulation and carotenogenic gene expression in the flavedo of citrus fruit.	Ma et al. 2015
Kohlrabi	(*Brassica oleracea* var. *gongylodes*, "Delicacy Purple")	455, 638, 665, 731 nm	Growth, nutritional quality, and antioxidant properties.	Samuolienė et al. 2013
Red Pak Choi	*B. rapa* var. *chinensis*, "Rubi F1")			
Tatsoi	(*Brassica rapa* var. *rosularis*)			
King protea	Plantlets; *Protea cynaroides* L.	Red (630 nm), blue (460 nm), red + blue, supplementary	Growth and morphogenesis, and concentrations of endogenous phenolic compounds.	Wu and Lin 2012
Tartary buckwheat	Sprouts, Tartary buckwheat (*Fagopyrum tartaricum* Gaertn.)	White (380 nm), blue (470 nm), red (660 nm)	Carotenoid biosynthetic gene expression levels and carotenoid accumulation.	Tuan et al. 2013
Rice	Rice leaves (*Oryza sativa* cv. Ilmi)	Blue, green, red, white	Metabolite profiling, antioxidant activities.	Jung et al. 2013
Rapeseed	Plantlets, rapeseed (*Brassica napus* L., cv. Westar)	Blue, red, blue + red (3:1, 1:1, 1:3)	Effects of different light qualities on the growth and morphogenesis.	Li et al. 2013
Rose	Rose plants "Akito"	Red (645 nm), red (680 nm)	Photosynthesis and the related optical properties of young reddish leaves and middle-age green leaves, quantifying the spectral dependence of photosynthesis at the canopy level.	Paradiso et al. 2011

(Continued)

TABLE 42.2 (CONTINUED)
LED Light Effects on Higher Plants

Plant	Tissue/Genus	LED Wavelength	Scope	References
Radicchio	*Cichorium intybus* L. subsp. *intybus* (Foliosum group) cv. Bianca di Milano	Blue (450–500 nm), green (500–600 nm), red (650–700 nm); 70% red, 20% green, 10% blue	Efficiency of photosynthesis, photoinhibition process by changes in malonedialdehyde (MDA) concentration, peroxidase (POX) activity, and hydrogen peroxide (H$_2$O$_2$) content. More effective functioning of PSII at 70% red, 20% green, and 10% blue light composition.	Ilieva et al. 2010
Cotton	Upland cotton, plantlets (*Gossypium hirsutum* L.)	Blue, blue + red (B:R = 3:1, 1:1, 1:3), red	Effects of different LED light sources on explants growth for tissue culture.	Li et al. 2010
Arabidopsis thaliana	*Arabidopsis thaliana*	Red (630 nm), blue (470 nm)	Studying photosynthesis using controlled pulses from a pulse width modulation (PWM) light dimming system equipped with LEDs.	Shimada and Taniguchi 2011

to lighting technologies with monochromatic emitting capability like LEDs.

LEDs have revolutionized the way we look at the light components and their effects on different plant morphological, physiological, and developmental aspects. Photosynthesis, however, seems to be benefited more significantly in terms of elucidating the hidden or less attended effects of auxiliary wavelengths and also developing strategies for improving the light-harvesting component and consequently the core photosynthesis apparatus efficiency. Thanks to LED technology, new horizons have been opened up to researchers to more delicately investigate the light effects on plants, particularly the photosynthesis phenomena in model plants (Shimada and Taniguchi 2011). Today, the rapid development of higher intensity and more economic LED lights has led to designing laboratory-scale controlled environment plant growth facilities equipped with LED lights (Sabzalian et al. 2014). They provide a wider scientific population with the opportunity to conduct innovative experiments on the influence of light spectra on plants.

42.3.1 Response of Higher Plants to LED Lighting

The body of LED research that has been conducted on plants since the beginning of the twenty-first century was broadly focused on two main subjects: plant growth aspects resulting from plant photosynthetic apparatus performance, and investigating the influence of monochromicatic light on plant secondary metabolite indices. Various aspects of plant physiology and morphology were extensively studied. Table 42.2 offers a comprehensive list of studies on the effects of LED light spectra on many cultivated crop plants. The annual leafy plants (i.e., vegetables and flowers) have been the primary study subjects under LED lighting, with lettuces holding the top rank in terms of history (Bula et al. 1991) and popularity. However, there are a few exceptions, including postharvest treatment of fruits (Ma et al. 2012, 2015) and suggesting LEDs as a light source for plant tissue culture (Nhut et al. 2000; Li et al. 2010). There are no reports at present, to our knowledge, regarding the investigation of LED light effects

on woody plants or perennials. Noticeably, almost all of the experiments were conducted inside controlled growth environment facilities, where many of the biotic and abiotic influencers on plant growth (e.g., gas concentrations, temperature, humidity, optimal light flux, nutrition, water condition, pest and diseases, etc.) were kept at optimal levels. In practice, the result of such research could not be generalized and applied to natural outdoor or even commercial indoor greenhouse cultivation practices (Pinho et al. 2007). However, it should be noted that although the high capital cost of LED lighting could be considered as the main obstacle for open-space or outdoor agricultural applications, as long as our knowledge in this field is still growing, such studies under controlled conditions are inevitable to evaluate the intricate and sole effects of LED light wavelengths on plants in the absence of any other known external or internal influences.

Because LEDs provide nearly monochromatic irradiation, they constitute the tool of choice to manipulate lighting conditions and study their impact in terms of plant growth and development. In fact, changes in the blue/red LED ratio may modulate the biochemical composition of the photosynthetic apparatus (Matsuda et al. 2004) as well as steady-state (Kinoshita et al. 2001) and dynamic (Kosvancova-Zitova et al. 2009) photosynthetic performances. For instance, the reduction of light spectra in blue and near UV regions is a key signal for leaf senescence and consequently for the cessation of photosynthetic activity (Causin et al. 2006). Also, the manipulation of the B/R ratio using LED setup improved the root number, fresh weight, and chlorophyll content in *Tripterospermum japonicum* (Moon et al. 2006) and *Mentha* spp. (Sabzalian et al. 2014).

More efficient absorption of the blue photons than red ones by leaves leads to stimulation of the electron transport rate. This is resulted from a large enhancement of ΔpH formation through thylakoid membranes (Matsuda et al. 2004; Kosvancova-Zitova et al. 2009), which in turn triggers the activation of stress-response mechanisms such as the xanthophyll cycle (Roháček et al. 2014). More tightly linked to photosynthesis, the B/R ratio may control the activity of several enzymes (Thomas 1981), and the migration of chloroplast in response to changes in irradiance level and stomatal opening (Briggs and Christie 2002), which are the mechanisms allowing the modulation of CO_2 uptake by the leaf. However, despite significant advances in understanding of the molecular mechanisms involving photoreceptors of blue and/or red lights and signaling, there is still little information about the mechanisms of monochromatic light action(s) on plant physiology (Darko et al. 2014; Lokhandwala et al. 2015).

The growing knowledge of light wavelength specificity effects on plant photosynthesis and other photosensing processes has encouraged designing delicate comparative studies on interspecies/intraspecies response variations under different LED lights (Samuoliené et al. 2011a, 2013; Sabzalian et al. 2014) and gene expression patterns (Ma et al. 2012, 2015; Wu and Lin 2012). The plant intracanopy or intercanopy lighting systems were also proposed for compensating the self-shading effects in some crop plants. The LED lights in this system are placed throughout leaf canopies, which resulted in biomass increase and elimination of premature senescence of lower leaves inside crop canopies (Ménard et al. 2005; Massa et al. 2007; Gómez et al. 2013).

LEDs have the capacity to emit high-intensity monochromatic light over a wide spectrum. However, the variation of light spectra used in plant research is still limited, including red, blue, and to a lesser extent, yellow (Pinho et al. 2007; Lee et al. 2010), green, and far-red (Tamulaitis et al. 2005; Stutte et al. 2009; Ilieva et al. 2010; Johkan et al. 2012), independently or in mixed ratios. Lee et al. (2010) investigated the effect of blue, green, yellow, red, far-red, and a few combinations of them on the antioxidant content of barley leaves. However, it should be noted that the visible spectrum of light limits (~390 to ~700 nm) is known to be responsible for most of the physiological phenomena, including photosynthesis in plants (Starr et al. 2010), and based on this fact, it might not seem rational to investigate the wavelengths outside this limit.

42.3.2 Productivity of Algae under LED Lighting

There is no generally accepted definition of algae, which constitutes a very diverse group of eukaryotic organisms. Some authors consider that algae have chlorophyll as their primary photosynthetic pigment and lack a sterile covering of cells around their reproductive cells (Lee 2008), while other authors exclude all prokaryotes (Nabors 2004) and therefore do not consider cyanobacteria as algae (Allaby 1992). Algae constitute a polyphyletic group (Nabors 2004) since they do not share a common ancestor, and although their plastids seem to originate from cyanobacteria (Keeling 2004), they were acquired in different ways. Green algae are examples of algae that have primary chloroplasts derived from endosymbiotic cyanobacteria. Diatoms are examples of algae with secondary chloroplasts derived from an endosymbiotic red alga (Palmer et al. 2004; Solymosi 2012).

Algae have photosynthetic machinery ultimately derived from cyanobacteria that produce oxygen as a by-product of photosynthesis. The role of algae in the biosphere is tremendous as they are responsible for more than 40% of the organic matter produced in the oceans, which is more than all terrestrial rainforests combined (Field et al. 1998; Granum et al. 2005; Bowler et al. 2010). In addition, microalgae play a crucial role in the formation of crude oil deposits under ocean floors, which are a rich natural source of fossil fuel (Shukla and Mohan 2012).

The growth of the human population and rapid expansion of cities into the countryside contribute to an accelerated depletion of natural resources (Alexandratos and Bruinsma 2012), increasing the prices of raw materials based on these resources in markets and potentially changing the earth's climate. To overcome difficulties in supplying the growing demand and reducing the cost of these resources, an intensive search for alternative pharmaceutical, food, high-value molecules (HVM), and energy sources, including *via* genetic engineering and nanotechnology, has started (Chisti 2008; Gordon and Seckbach 2012; Mimouni et al. 2012). Among the alternative natural sources, algae are considered reliable candidates for their efficient capacity for photosynthesis. In fact, algae can become biotechnological tools because they are capable of producing many high-value biocompounds such as polysaccharides, lipids, polyunsaturated fatty acids, pigments, and biofuels (Mimouni et al. 2012; Heydarizadeh et al. 2013; Hudek et al. 2014). Despite the considerable progress made in microalgal bioreactor technology (Cogne et al. 2011; Vasseur et al. 2012), the use of algae as cell factories still remains in its infancy. We believe that it is mainly due to the wide gaps between the knowledge of microalga like diatom biochemistry and physiology, especially where responses to stress conditions are concerned. A deep knowledge of algal stress physiology is of particular importance for any biotechnological manipulation because the production of high-value compounds often results from metabolic shifts that are induced by stressful conditions including high light intensities. For example, when facing abiotic stresses, the microalga starts to accumulate pigments (astaxanthine: Kopecky et al. 2000; β-carotene: Seyfadabi et al. 2010) or proteins (Seyfadabi et al. 2010; for reviews, see Lemoine and Schoefs 2010; Gacheva and Gigova 2014).

While a small number of algae can perform heterotrophic growth, the majority of algae relies on photosynthesis for their development, a phenomenon that requires light to proceed. Despite the fact that algae evolved in the presence of this particular light source and have a dedicated set of pigments to harvest photons (Cheng et al. 2004), sunlight full spectrum is not optimized for physiological purposes due to additional damaging rays such as ultraviolet (UV) and infrared (IR) photons that can damage the cellular structure (Lee and Palsson 1996; Holzinger and Lutz 2005). These photons are especially harmful when the light intensity is strong (PAR > 750 μmol m^{-2} s^{-1}), a condition that can easily be met outdoors on a sunny day, during which the PAR can reach a value as high as 2000 μmol m^{-2} s^{-1}. Therefore, it is tempting to speculate that monochromatic illumination by favorable wavelength peaks could be beneficial for algal growth and productivity.

Microalgae are unicellular organisms, so one can postulate that the impact of light quality on photosynthesis will directly influence their growth. It is generally thought that blue (420–470 nm) or red (650–680 nm) light is beneficial to algal cells (Schulze et al. 2014) (Table 42.3). In fact, the wavelength that is the most beneficial is probably taxon dependent (Table 42.3). For instance, using light bulbs emitting blue wavelength at 475 nm improved the growth of the green microalgae *Chlorella fusca* and *Chlorella vulgaris* biomasses as compared to red (650 nm) or white artificial lights (Wilhelm et al. 1985; Blair et al. 2014). Conversely, Kubin et al. (1983) observed that *C. vulgaris* had better productivity under green than blue or red irradiation. Wang et al. (2007) found that red LED (620–645 nm) was more effective than blue LED (460–475 nm) for *Spirulina* growth. In the case of the red seaweed, *Gracilaria tikvahiae*, white fluorescence lighting was more effective to produce tissue mass than monochromatic light from LED with the lowest growth rate obtained under blue LED (Kim et al. 2015) (Table 42.3). This was not completely unexpected as Dring and Lüning (1985) earlier reported the low photosynthetic quantum yield in this wavelength region, compared with green or red light. A similar result was reported with a mutant lacking phycoerythrin. It was shown that the combination of monochromatic LEDs in a dichromatic lighting (R + B, R + G, G + B; ratio = 1) or trichromatic lighting (R + G + B: 40/40/20%, R + G + B: 40/20/40%, R + G + B: 20/40/40%) allowed a growth rate of *G. tikvahiae* similar to that obtained with fluorescent tubes (Kim et al. 2015) (Table 42.4). Nevertheless, when the economic cost (based on electricity consumption) is taken into account, more biomass is obtained with LED illumination. The impact of light color can also be modulated by light intensity. For instance, the growth rate of the diatom *Haslea ostearia* is similar under blue, white, red, and far-red lights under high light intensity (100 μmol m^{-2} s^{-1}), whereas this microalga grows faster under blue light with low light intensity (20 μmol m^{-2} s^{-1}).

Most of the papers dealing with algae have been dedicated to the photosynthetic responses to monochromatic light delivered from conventional lighting systems (Grotjohann and Kowallik 1989; Senge and Senger 1990; Bader et al. 1992). For instance, it was shown that blue light from conventional lighting stimulates photosynthesis in most brown seaweeds when already saturated with red light (Dring, 1988; Schmid and Dring 1992, 1993), but not in red seaweeds, which are less efficient in light harvesting (Dring and Lüning 1985). If the change in the photosynthetic activity will probably be reflected in cell growth, one cannot still ignore the effects of monochromatic light on the cell cycle by itself. For instance, it was demonstrated that under red light, *Tetraselmis* occurs preferentially as swarmer, whereas under blue or green light, encyst cells largely prevail (*Tetraselmis gracilis*: Aidar et al. 1994; *T. suecica*; Abiusi et al. 2014). This topic is, however, beyond the scope of this review, and interested readers are requested to read the related literature (e.g., Huysman et al. 2010).

One of the major biotechnological potentials of microalgae is its high lipid production capacity for either health or biofuel purposes (Mimouni et al. 2012; Heydarizadeh et al. 2013; Ulmann et al. 2014). It is now well established that abiotic stresses such as lowering the temperature could modify the lipid content (Pasquet et al. 2014). Shu et al. (2012) reported a higher lipid content in *Chlorella* sp. illuminated with blue LED compared to red LED (Table 42.3). Testing the effect of light color provided by LED, Das et al. (2011) found little variation in the total fatty acid in *Nannochloropsis* sp. according to the light wavelength, whereas they observed a modification of the fatty acid composition. A similar observation has been made with *T. suecica* (Abiusi et al. 2014). They also reported that the total amount of lipid was increased under blue light, while red light lowered the fatty acid content (Table 42.3).

Light wavelength also influences the orientation of carbon metabolism within the cell. In a recent paper, Jungandreas et al. (2014) demonstrated that shifting the LED light quality from red to blue triggers fast and drastic changes in the metabolic profile. For instance, blue light (emitted by conventional lighting) irradiated microalgae contained higher chlorophyll content (Senger 1987). This stimulation was not observed in *G. tikvahiae* for which the Chl content was strongly reduced under blue LED (Kim et al. 2015) or in *T. suecica* for which no change in the biochemical composition was observed under different lightings (Abiusi et al. 2014). In *T. suecica*, the total Chl content was independent from the light wavelength used for growth, but the Chl *a*/Chl *b* ratio was higher under the red LED irradiation (Abiusi et al. 2014).

Pigments are among the prominent high-value molecules produced by microalgae (Mimouni et al. 2012; Heydarizadeh et al. 2013). Interestingly, blue LED triggered the accumulation of open tetrapyrroles phycoerythrin and phycocyanin, which are the typical red pigments in red seaweed *G. tikvahiae* (Kim et al. 2015). The application of blue light (delivered by conventional lighting) increased the production of marrenine in the diatom *Haslea ostrearia* (Mouget et al. 2005) and astaxanthin in the green alga *Haematococcus pluvialis* (Park and Lee 2001). In *T. suecica*, the level of carotenoid content remained unaffected by the irradiation wavelength (Abiusi et al. 2014).

The wavelength shift from red to blue light also favored protein synthesis, whereas the shift from blue to red light favored carbohydrate synthesis (Jungandreas et al. 2014). However, Abiusi et al. (2014) showed that green-light grown *T. suecica* contained more carbohydrates than cells grown under other light wavelengths. Also, the amount of proteins was not significantly lower in *T. suecica* (Abiusi et al. 2014) and *Cyclotella caspia* (Aidar et al. 1994) grown under red LED light than other lights, whereas a slight increase was observed in *T. gracilis* grown under the same type of illumination (Aidar et al. 1994).

42.4 WILL LEDs DOMINATE THE FUTURE OF ARTIFICIAL LIGHTING FOR PHOTOSYNTHESIS?

The world's population is increasing at a rapid pace. This requires more energy and materials to support food production, among many other needs. While the world population is increasing, natural resources are vanishing due to mostly

TABLE 42.3

Monochromatic Lighting Effects on Different Taxa of Algae

Family	Genus	Species	Illumination	Biomass Growth	Lipid Content	Chl	Phycoerythrin	Marrenine	References
						Pigments			
Prasinophyceae	*Tetraselmis*	*suecica* F&M-M33	R (635), G (552), B (448), W (halide lamp)	R and W	B increased the composition, but R decreased the fatty acid content	No effect, but R affected the Chla/b ratio		Nr	Abiusi et al. 2014
Chlorophyceae	*Chlamydomonas*	*reinhardtii*	Blue	Delayed cell division					Oldenhof et al. 2006
Coscinodiscophyceae	*Skeletonema*	*costatum*		Enhanced with G					Oh et al. 2008
Chlorophyceae	*Haematococcus*	*pluvialis*	Blue						Park and Lee 2001
Cyanobacteria	*Spirulina*	*platensis*	R (620–645), B (460–470), W (380–760) (LED)	R > B or W	–	–	Nr		Wang et al. 2007
Chlorophyceae	*Chlorella*	*vulgaris*	B (475), R (650), W (bulb)	B > R or W	–	–	Nr		Blair et al. 2014
Chlorophyceae	*Chlorella*	sp.	Blue, red	–	Blue > red	–	Nr		Shu et al. 2012
Eustigmatophyceae	*Nannochloropsis*	sp.	B (470), R (680), W	–	B > R or W + modification in the fatty acid composition	–	Nr		Das et al. 2011
	Gracilaria	*tikvahiae* (strain G-R1-ST1)	B (465–470), G (515–520), R (625–630), W (fluorescent tubes)	W > R = G (79%) > B (34%)	–	R = G = W > B	Higher under blue LED		Kim et al. 2015
	Gracilaria	*tikvahiae* – green mutant (strain G-R1-ST3)	B (465–470), G (515–520), R (625–630), W (fluorescent tubes)	W > R = G > B	–	–			Kim et al. 2015
Diatoms	*Haslea*	*ostrearia*	Blue fluorescent tubes					Enhanced production under blue (20 µE)	Mouget et al. 2005

Note: – = not measured; Nr = not relevant.

TABLE 42.4

Combination of Monochromatic Lighting Effect on Algae

| | | | Factor | | |
| | | | Biomass Growth | Lipid Content | References |
Genus	Species	Illumination			
Gracilaria	*tikvahiae* (strain G-RI-ST1)	B (465–470), G (515–520), R (625–630), W (fluorescent tubes)	R + B, R + G, G + B = W	–	Kim et al. 2015
Gracilaria	*tikvahiae* (strain G-RI-ST1)	B (465–470), G (515–520), R (625–630), W (fluorescent tubes)	R + G + B = W	–	Kim et al. 2015

low efficient or inefficient agricultural production systems. Electricity is becoming the major energy source for industry, houses, and modern agriculture. A considerable portion (about 19%) of the electricity produced in the world is used for lighting purposes, with about 75% of the global lighting still using old and inefficient solutions. This amount of energy requires about 1800 million barrels of oil or 642 power plants at a cost of $170 billion, which eventually ends in roughly 670 million tons of CO_2 gas deposited into the atmosphere (PFZK report, PHILIPS, 2014). So, together with changing the human energy consumption culture and norms, and replacing the planet's natural resources with renewable safer energy sources, it is inevitable to devise and implement more efficient lighting technologies. The LED light has shown to be a promising solution.

The agricultural energy consumption is no exception to the global energy problem and is already becoming an issue due to the need for more efficiency out of our farms. The capacity of global agricultural lands is determined; however, this valuable commodity of the human race is adversely affected by the negative consequences of global climate change, drought, salinity, erosion, and low-performance agricultural practices. Although the current scale of indoor agricultural production is not yet comparable to that of outdoors, it is expected that indoor food production solutions will boom in the near future to compensate for the deceasing productivity of common outdoor agricultural practices. Fortunately, the scientific community and the food production private sector around the world have already addressed the issue by focusing their efforts on developing solutions including optimizing climatic conditions, creating production automations, and challenging the lighting systems for improving the performance and efficiency of the indoor agricultural production. The study of photosynthesis is at the heart of these programs, since it is the most important physiological phenomenon inside the plants. Light is the major influencer of this phenomenon, and considerable efforts are being taken to better understand and improve the light effects on plants. Since its practical inception in the 1990s, LED technology has been considered a potential solution for future lighting. Today, this still developing technology is at

the center of a torrent of research in agriculture. The overall results of the literature strongly indicate that LED light can dramatically increase plant performance leading to higher yields (Table 42.1). A Japanese company is developing a LED vertical greenhouse, which is 100 times more productive than traditional methods with 40% less power consumption, 80% less food waste, and 99% less water usage than an outdoor field (MIRAI, CO., Japan). Similar solutions are being experienced and developed in other parts of the world.

Our world is becoming smarter, more digitized and connected, which together can bring more efficiency and productivity. The agriculture industry can benefit from such advancements toward improving indoor production solutions through automation and more delicate environmental condition control. LEDs are completely compatible with all the three criteria. As mentioned earlier, an LED is more like a computer chip than a light bulb. They can emit light at literally any narrow spectrum in any desired direction and at any power. Therefore, having understood the most favorable light requirements for a plant through its different stages of growth, one could simply design a computer program for the plants' lighting needs connected to LED lights to be sure that the plants will receive the most optimum, sufficient, and economic amount of light.

Today, this belief that LEDs will be a very important light source, perhaps a dominant one, in the future of lighting is widely accepted among the scientific community and the private sector. However, the capital cost of LED implementation is still a major drawback. Some companies like Philips and Konica-Minolta have put tremendous efforts to develop solutions for lowering the LED production costs and improving its characteristics. Among the different types of LEDs, the OLED or organic LED is highlighted. OLED is the same kind of technology used in some ultrathin TVs and smart phones and is expected to be the future of commercial LEDs in the market. Nevertheless, the blue OLEDs are still relatively inefficient. The Nobel Prize in physics in 2014 was awarded jointly to three Japanese scientists who discovered the blue LED (Amano et al. 1986; Nakamura et al. 1995), which clearly indicates both the importance and complications of developing such technology.

The technical and physiological advantages of the LED technology for future agriculture at this time are solely highlighted for indoor practices. There are no reports published regarding LED outdoor experiments on plants or feasibility studies. The LED contribution to sustainable agriculture, whether for indoor or outdoor practices, through minimizing carbon dioxide emissions and protecting the environment while guaranteeing a reliable and efficient solution for year-around food production surely will be one of the key elements for its future success. Some experts believe that by improving the productivity of future LED greenhouses per square meter, more families can grow their own vegetables in their home backyard greenhouse; hence, a seemingly small LED-lighted space could produce much higher yields than a conventional one. This is of utmost interest for alleviating hunger and famine among impoverished communities because the food can be grown locally inside small-scale highly efficient LED greenhouses fitted to people's needs.

The unique productivity, technological, physiological, and environmental advantages of LEDs over other conventional lightings in sustainable agriculture cannot be ignored. However, LEDs may change the future of indoor or outdoor agriculture, provided that the main drawback of its current high capital cost will be satisfactorily addressed. The technological advancements and economies of scale undoubtedly will get to a certain point that mass production will bring the costs down and a greater range of affordable options for LED farming will be available.

42.5 CONCLUSIONS AND PERSPECTIVES

The plant yield is a product of photosynthesis, and sunlight is the energy source for photosynthesis in plants. When humans inhabited new lands on the globe with less favorable climatic conditions, like colder or warmer temperatures, dimmer or shorter periods of sunlight exposure, or water and soils being in capable of nurturing the plant life, humans decided to devise ways to compensate for the shortages of nature with artificial supplementations. The light was given special attention, and humans soon learned that they could supplement and even replace the sunlight ration of the day with a synthetic illumination source, which used electricity to emit visible spectrum light, and this illumination source is known as a *lamp*. Since then, numerous innovations and discoveries have been introduced in the lighting industry aimed at lowering the costs, improving the light intensity and lamp durability, and eventually achieving more efficiency. LED technology is the latest in this course. LED performance has now increased up to ~300 lm/W and more, which outperforms most of the other current lighting technologies. LED design and production capability have reached a point at which a variety of different light colors and power outputs are being developed, which could efficiently fit into any lighting application. LEDs perhaps will dominate the future of artificial lighting. However, for crop production under LED lighting, although there is evidence that this technology has many advantages, there are some uncertainties that could affect LED application and their adoption at present.

The LED is a maturing technology and hence its capital cost compared to common lighting solutions is very high. At present, the less risky idea for their utilization in agriculture could be the application of LEDs as a complement for the available light systems to promote plant growth and to improve the value of production by stimulating early flowering or profuse flowering in plants. The LED as an advanced technology for its digital nature can be fine-tuned to provide the plant with seasonal and daily light requirement and be optimized for interplant and intraplant canopy light distribution as well. It is also believed that there is still a need for developing new designs and fabrications of LED lighting systems and to customize the spectral composition and device configuration of LEDs. The future of LED lighting in agriculture would be presumably more oriented toward specific enhancement of crop responses to lights for specific goals like higher ornamental quality, faster and more productive plant propagation, postharvest solutions, and the production of valuable primary and secondary metabolites.

REFERENCES

Abiusi, F., G. Sampietro, G. Marturano, N. Biondi, L. Rodolfi, M. D'Ottavio, and M. R. Tredici. 2014. Growth, photosynthetic efficiency, and biochemical composition of *Tertraselmis suecica* F&M-M33 grown with LEDs of different colors. *Biotechnol Bioeng* 111: 956–964.

Aidar, E., S. M. F. Gianesellagalvao, T. C. S. Sigaud, C. S. Asano, T. H. Liang, K. R. V. Rezende, M. K. Oishi, F. J. Aranha, G. M. Milani, and M. A. L. Sandes. 1994. Effects of light quality on growth, biochemical composition and photosynthetic production in *Cyclotella caspia* Grunow and *Tetraselmis gracilis* (Kylin) Butcher. *J Exp Mar Biol Ecol* 180: 175–187.

Aksenova, N. P., T. N. Konstantinova, L. I. Sergeeva, I. Macháčková, and S. A. Golyanovskaya. 1994. Morphogenesis of potato plants in vitro. I. Effect of light quality and hormones. *J Plant Grow Regul* 13: 143–146.

Alexandratos, N. and J. Bruinsma. 2012. *World Agriculture towards 2030/2050: The 2012 Revision*. Rome: ESA Working Paper, FAO.

Allaby, M. 1992. Algae. *The Concise Dictionary of Botany*. Oxford: Oxford University Press.

Amano, H., N. Sawaki, I. Akasaki, and Y. Toyoda. 1986. Metalorganic vapor phase epitaxial growth of a high quality GaN film using an AlN buffer layer. *Appl Phys Lett* 48: 353–355.

Avercheva, O., Y. A. Berkovich, A. Erokhin, T. Zhigalova, S. Pogosyan, and S. Smolyanina. 2009. Growth and photosynthesis of Chinese cabbage plants grown under light-emitting diode-based light source. *Russ J Plant Physiol* 56: 14–21.

Bader, K., G. Schmid, G. Ruyters, and W. Kowallik. 1992. Blue-light enhanced respiratory activity under photosynthesis conditions in Chlorella; a mass spectrometric analysis. *Z Naturforsch C: J Biosci* 47: 881–888.

Barber, J. and B. Andersson. 1992. Too much of a good thing: Light can be bad for photosynthesis. *Trends Biochem Sci* 17: 61–66.

Blair, M., B. Kokabian, and G. Gude. 2014. Light and growth medium on *Chlorella vulgaris* biomass production. *J Environ Chem Eng* 2: 665–674.

Bliznikas, Z., A. Žukauskas, G. Samuoliene, A. Viršile, A. Brazaityte, J. Jankauskiene, P. Duchovskis, and A. Novičkovas. 2012. Effect of supplementary pre-harvest LED lighting on the antioxidant and nutritional properties of green vegetables. *Acta Hort* 939: 85–91.

Bourget, C. M. 2008. An introduction to light-emitting diodes. *HortScience* 43: 1944–1946.

Bowler, C., A. Vardi, and A. E. Allen. 2010. Oceanographic and biogeochemical insights from diatom genomes. *Ann Rev Mar Sci* 2: 333–363.

Brazaitytė, A., P. Duchovskis, A. Urbonavičiūtė, G. Samuolienė, J. Jankauskienė, A. Kasiulevičiūtė-Bonakėrė, Z. Bliznikas, A. Novičkovas, K. Breivė, and A. Žukauskas. 2009. The effect of light-emitting diodes lighting on cucumber transplants and after-effect on yield. *Zemdirbyste–Agric* 96: 102–118.

Briggs, W. R. and J. M. Christie. 2002. Phototropins 1 and 2: Versatile plant blue-light receptors. *Trends Plant Sci* 7: 204–210.

Brown, C. S., A. C. Schuerger, and J. C. Sager. 1995. Growth and photomorphogenesis of pepper plants under red light-emitting diodes with supplemental blue or far-red lighting. *J Am Soc Hortic Sci* 120: 808–813.

Bula, R., R. Morrow, T. Tibbitts, D. Barta, R. Ignatius, and T. Martin. 1991. Light-emitting diodes as a radiation source for plants. *HortScience* 26: 203–205.

Causin, H. F., R. N. Jauregui, and A. J. Barneix. 2006. The effect of light spectral quality in leaf senescence and oxidative stress in wheat. *Plant Sci* 171: 24–33.

Chen, X.-L., W.-Z. Guo, X.-Z. Xue, L.-C. Wang, and X.-J. Qiao. 2014. Growth and quality responses of "Green Oak Leaf" lettuce as affected by monochromic or mixed radiation provided by fluorescent lamp (FL) and light-emitting diode (LED). *Sci Hort* 172: 168–175.

Cheng, M., J. Chory, and C. Fankhauser. 2004. Light signal transduction in higher plants. *Annu Rev Genet* 38: 87–117.

Chisti, Y. 2008. Biodiesel from microalgae beats bioethanol. *Trends Biotechnol* 26: 126–131.

Cogne, G., M. Rugen, A. Bockmayr, M. Titica, C. G. Dussap, J. F. Cornet, and J. Legrand 2011. A model-based method for investigating bioenergetic processes in autotrophically growing eukaryotic microalgae: Application to the green algae *Chlamydomonas reinhardtii*. *Biotechnol Progr* 27: 631–640.

Colquhoun, T. A., M. L. Schwieterman, J. L. Gilbert, E. A. Jaworski, K. M. Langer, C. R. Jones, G. V. Rushing, T. M. Hunter, J. Olmstead, and D. G. Clark. 2013. Light modulation of volatile organic compounds from petunia flowers and select fruits. *Postharvest Biol Technol* 86: 37–44.

Darko, E., P. Heydarizadeh, B. Schoefs, and M. R. Sabzalian. 2014. Photosynthesis under artificial light: The shift in primary and secondary metabolism. *Philos Trans R Soc Lond B: Biol* 369: 20130243.

Das, P., W. Lei, S. S. Aziz, and J. P. Obbard. 2011. Enhanced algae growth in both phototrophic and mixotrophic culture under blue light. *Biores Technol* 102: 3883–3887.

Dring, M. J. 1988. Photocontrol of development in algae. *Annu Rev Plant Physiol Plant Mol Biol* 39: 157–174.

Dring, M. J. and K. Lüning. 1985. Action spectra and spectral quantum yield of photosynthesis in marine macroalgae with thin and thick thalli. *Mar Biol* 87: 119–129.

Duong, T. N., T. Takamura, H. Watanabe, and M. Tanaka. 2000. Light emitting diodes (LEDs) as a radiation source for micropropagation of strawberry. In *Transplant Production in the 21st Century*. Netherlands: Springer, pp. 114–118.

Fan, X.-X., Z.-G. Xu, X.-Y. Liu, C.-M. Tang, L.-W. Wang, and X.-L. Han. 2013. Effects of light intensity on the growth and leaf development of young tomato plants grown under a combination of red and blue light. *Sci Hort* 153: 50–55.

Field, C. B., M. J. Behrenfeld, J. T. Randerson, and P. G. Falkowski. 1998. Primary production of the biosphere: Integrating terrestrial and oceanic components. *Science* 281: 237–240.

Folta, K. M., L. L. Koss, R. McMorrow, H.-H. Kim, J. D. Kenitz, R. Wheeler, and J. C. Sager. 2005. Design and fabrication of adjustable red-green-blue LED light arrays for plant research. *BMC Plant Biol* 5: 17.

Gacheva, G. V. and L. G. Gigova. 2014. Biological activity of microalgae can be enhanced by manipulating the cultivation temperature and irradiance. *Cent Eur J Biol* 9: 1168–1181.

Gangadhar, B. H., R. K. Mishra, G. Pandian, and S. W. Park. 2012. Comparative study of color, pungency, and biochemical composition in chili pepper (*Capsicum annuum*) under different light-emitting diode treatments. *HortScience* 47: 1729–1735.

Gest, H. 2002. History of the word photosynthesis and evolution of its definition. *Photosynth Res* 73: 7–10.

Girón González, E. 2012. *LEDs for General and Horticultural Lighting*. Aalto University, School of Electrical Engineering, Espoo, Finland.

Goins, G., N. Yorio, M. Sanwo, and C. Brown. 1997. Photomorphogenesis, photosynthesis, and seed yield of wheat plants grown under red light-emitting diodes (LEDs) with and without supplemental blue lighting. *J Exp Bot* 48: 1407–1413.

Gómez, C. and C. A. Mitchell. 2015. Growth responses of tomato seedlings to different spectra of supplemental lighting. *HortScience* 50: 112–118.

Gómez, C., R. C. Morrow, C. M. Bourget, G. D. Massa, and C. A. Mitchell. 2013. Comparison of intracanopy light-emitting diode towers and overhead high-pressure sodium lamps for supplemental lighting of greenhouse-grown tomatoes. *HortTechnology* 23: 93–98.

Gordon, R. and J. Seckbach. 2012. *The Science of Algal Fuels: Phycology, Geology, Biophotonics, Genomics and Nanotechnology*. Dordrecht: Springer.

Granum, E., J. A. Raven, and R. C. Leegood. 2005. How do marine diatoms fix 10 billion tons of inorganic carbon per year? *Can J Bot* 83: 898–908.

Grotjohann, N. and W. Kowallik. 1989. Influence of blue light on the activity of phosphofructokinase in *Chlorella kessleri*. *Physiol Plant* 75: 43–46.

Haitz, R. and J. Y. Tsao. 2011. Solid-state lighting: 'The case' 10 years after and future prospects. *Phys Status Solidi (a)* 208: 17–29.

Heo, J., C. Lee, D. Chakrabarty, and K. Paek. 2002. Growth responses of marigold and salvia bedding plants as affected by monochromic or mixture radiation provided by a light-emitting diode (LED). *Plant Grow Regul* 38: 225–230.

Heo, J. W., C. W. Lee, H. Murthy, and K. Y. Paek. 2003. Influence of light quality and photoperiod on flowering of Cyclamen persicum Mill. cv. 'Dixie White'. *Plant Grow Regul* 40: 7–10.

Hernández, R. and C. Kubota. 2012. Tomato seedling growth and morphological responses to supplemental LED lighting red: Blue ratios under varied daily solar light integrals. In *VII International Symposium on Light in Horticultural Systems* 956, pp. 187–194.

Heydarizadeh, P., I. Poirier, D. Loizeau, L. Ulmann, V. Mimouni, B. Schoefs, and M. Bertrand. 2013. Plastids of marine phytoplankton produce bioactive pigments and lipids. *Marine Drugs* 11: 3425–3471.

Hoenecke, M., R. Bula, and T. Tibbitts. 1992. Importance of 'blue' photon levels for lettuce seedlings grown under red-light-emitting diodes. *HortScience* 27: 427–430.

Hogewoning, S. W., G. Trouwborst, H. Maljaars, H. Poorter, W. van Ieperen, and J. Harbinson. 2010. Blue light dose–responses of leaf photosynthesis, morphology, and chemical composition of *Cucumis sativus* grown under different combinations of red and blue light. *J Exp Bot* 61: 3107–3117.

Holzinger, A. and C. Lutz. 2005. Algae and UV irradiation: Effects on ultrastructure and related metabolic functions. *Micron* 37: 190–197.

Hudek, K., L. C. Davis, J. Ibbini, and L. Erickson. 2014. Commercial products from algae. In *Algal Biorefineries*, vol. 1, Bajpai, R., Prokop, A., and Zappi, M. Eds. Netherlands: Springer, pp. 275–295.

Huysman, M. J. J., C. Martens, K. Vandepoele, J. Gillard, E. Rayko, M. Heijde, C. Bowler, D. Inzé, Y. Van de Peer, L. De Veylder, and W. Vyverman. 2010. Genome-wide analysis of the diatom cell cycle unveils a novel type of cyclins involved in environmental signaling. *Genome Biol* 11:R17.

Huzisige, H. and B. Ke. 1993. Dynamics of the history of photosynthesis research. *Photosynth Res* 38: 185–209.

Ilieva, I., T. Ivanova, Y. Naydenov, I. Dandolov, and D. Stefanov. 2010. Plant experiments with light-emitting diode module in Svet space greenhouse. *Adv Space Res* 46: 840–845.

Jao, R.-C., C.-C. Lai, W. Fang, and S.-F. Chang. 2005. Effects of red light on the growth of Zantedeschia plantlets in vitro and tuber formation using light-emitting diodes. *HortScience* 40: 436–438.

Johkan, M., K. Shoji, F. Goto, S.-N. Hashida, and T. Yoshihara. 2010. Blue light-emitting diode light irradiation of seedlings improves seedling quality and growth after transplanting in red leaf lettuce. *HortScience* 45: 1809–1814.

Johkan, M., K., Shoji, F. Goto, S. Hahida, and T. Yoshihara. 2012. Effect of green light wavelength and intensity on photomorphogenesis and photosynthesis in Lactuca sativa. *Env Exp Bot* 75: 128–133.

Johnson, C. F., C. S. Brown, R. M. Wheeler, J. C. Sager, D. K. Chapman, and G. F. Deitzer. 1996. Infrared light-emitting diode radiation causes gravitropic and morphological effects in dark-grown oat seedlings. *Photochem Photobiol* 63: 238–242.

Jung, E. S., S. Lee, S.-H. Lim, S.-H. Ha, K.-H. Liu, and C. H. Lee. 2013. Metabolite profiling of the short-term responses of rice leaves (*Oryza sativa* cv. Ilmi) cultivated under different LED lights and its correlations with antioxidant activities. *Plant Sci* 210: 61–69.

Jungandreas, A., B. Schellenberger Costa, T. Jakob, M. von Bergen, S. Baumann, and C. Wilhelm. 2014. The acclimation of *Phaeodactylum tricornutum* to blue and red light does not influence the photosynthetic light reaction by strongly disturbs the carbon allocation pattern. *PLoS One* 9: e99727.

Kamtekar, K. T., A. P. Monkman, and M. R. Bryce. 2010. Recent advances in white organic light-emitting materials and devices (WOLEDs). *Adv Mater* 22: 572–582.

Kaneko-Ohashi, K., K. Fujiwara, Y. Kimura, R. Matsuda, and K. Kurata. 2004. Effects of red and blue LEDs low light irradiation during low temperature storage on growth, ribulose-1,5-bisphosphate carboxylase/oxygenase content, chlorophyll content and carbohydrate content of grafted tomato plug seedlings. *Env Control Biol (Japan)* 42: 65–73.

Keeling, P. J. 2004. Diversity and evolutionary history of plastids and their hosts. *Am J Bot* 91: 1481–1493.

Kelly, T. L. 2004. *Solid State Lighting: Strategies for a Brighter Future*. Cambridge, MA: Massachusetts Institute of Technology.

Kim, H.-H., G. D. Goins, R. M. Wheeler, and J. C. Sager. 2004a. Green-light supplementation for enhanced lettuce growth under red-and blue-light-emitting diodes. *HortScience* 39: 1617–1622.

Kim, S.-J., E.-J. Hahn, J.-W. Heo, and K.-Y. Paek. 2004b. Effects of LEDs on net photosynthetic rate, growth and leaf stomata of chrysanthemum plantlets in vitro. *Sci Hort* 101: 143–151.

Kim, H.-H., R. M. Wheeler, J. C. Sager, G. Gains, and J. Naikane. 2005. Evaluation of lettuce growth using supplemental green light with red and blue light-emitting diodes in a controlled environment-a review of research at Kennedy Space Center. In *V International Symposium on Artificial Lighting in Horticulture* 711, pp. 111–120.

Kim, J. B., Y. Mao, G. Kraemer, and C. Yarish. 2015. Growth and pigment content of *Gracilaria tikvahiae* McLachlan under fluorescent and LED lighting. *Aquaculture* 436: 52–57.

Kinoshita, T., M. Doi, N. Suetsugu, T. Kagawa, M. Wada, and K. Shimazaki. 2001. Phot1 and phot2 mediate blue light regulation of stomatal opening. *Nature* 414: 656–660.

Kopecky, J., B. Schoefs, D. Stys, K. Loest, and O. Pulz. 2000. Microalgae as a source for secondary carotenoid production. A screening study. *Arch Hydrobiol Suppl Algol Stud* 98: 153–167.

Kopsell, D. A. and C. E. Sams. 2013. Increases in shoot tissue pigments, glucosinolates, and mineral elements in sprouting broccoli after exposure to short-duration blue light from light emitting diodes. *J Am Soc Hortic Sci* 138: 31–37.

Kopsell, D. A., C. E. Sams, T. C. Barickman, and R. C. Morrow. 2014. Sprouting broccoli accumulate higher concentrations of nutritionally important metabolites under narrow-band light-emitting diode lighting. *J Am Soc Hortic Sci* 139: 469–477.

Kosvancova-Zitova, M., O. Urban, M. Navratil, V. Spunda, T. M. Robson, and M. V. Marek. 2009. Blue radiation stimulates photosynthetic induction in *Fagus sylvatica* L. *Photosynthetica* 47: 388–398.

Krames, M. R., O. B. Shchekin, R. Mueller-Mach, G. O. Mueller, L. Zhou, G. Harbers, and M. G. Craford. 2007. Status and future of high-power light-emitting diodes for solid-state lighting. *J Disp Technol* 3: 160–175.

Kubin, S., E. Borns, J. Doucha, and U. Seiss. 1983. Light absorption and production rate of *Chlorella vulgaris* in light of different spectral composition. *Biochem Physiol Pflanz* 178: 193–205.

Kubota, C., P. Chia, Z. Yang, and Q. Li. 2011. Applications of far-red light emitting diodes in plant production under controlled environments. In *International Symposium on Advanced Technologies and Management Towards Sustainable Greenhouse Ecosystems: Greensys* 952, pp. 59–66.

Lee, R. E. 2008. *Phycology*. Cambridge: Cambridge University Press.

Lee, C. G. and B. O. Palsson. 1996. Photoacclimation of *Chlorella vulgaris* to red light from light-emitting diode leads to autospore release following each cellular division. *Biotechnol Progr* 12: 249–256.

Lee, N. Y., M.-J. Lee, Y.-K. Kim, J.-C. Park, H.-K. Park, J.-S. Choi, J.-N. Hyun, K.-J. Kim, K.-H. Park, and J.-K. Ko. 2010. Effect of light emitting diode radiation on antioxidant activity of barley leaf. *J Korean Soc Appl Biol Chem* 53: 685–690.

Lefsrud, M. G., D. A. Kopsell, and C. E. Sams. 2008. Irradiance from distinct wavelength light-emitting diodes affect secondary metabolites in kale. *HortScience* 43: 2243–2244.

Lemoine, Y. and B. Schoefs. 2010. Secondary ketocarotenoid astaxanthin biosynthesis in algae: A multifunctional response to stress. *Photosynth Res* 106: 155–177.

Li, Q. and C. Kubota. 2009. Effects of supplemental light quality on growth and phytochemicals of baby leaf lettuce. *Env Exp Bot* 67: 59–64.

Li, H., Z. Xu, and C. Tang. 2010. Effect of light-emitting diodes on growth and morphogenesis of upland cotton (*Gossypium hirsutum* L.) plantlets in vitro. *Plant Cell Tiss Org Cult* 103: 155–163.

Li, H., C. Tang, Z. Xu, X. Liu, and X. Han. 2012. Effects of different light sources on the growth of non-heading Chinese cabbage (Brassica campestris L.). *J Agric Sci* 4: 262.

Li, H., C. Tang, and Z. Xu. 2013. The effects of different light qualities on rapeseed (*Brassica napus* L.) plantlet growth and morphogenesis in vitro. *Sci Hort* 150: 117–124.

Lian, M.-L., H. Murthy, and K.-Y. Paek. 2002. Effects of light emitting diodes (LEDs) on the in vitro induction and growth of bulblets of *Lilium* oriental hybrid 'Pesaro'. *Sci Hort* 94: 365–370.

Lin, K.-H., M.-Y. Huang, W.-D. Huang, M.-H. Hsu, Z.-W. Yang, and C.-M. Yang. 2013. The effects of red, blue, and white light-emitting diodes on the growth, development, and edible quality of hydroponically grown lettuce (*Lactuca sativa* L. var. capitata). *Sci Hort* 150: 86–91.

Lin, C.-C., K.-J. Chen, D.-W. Lin, H.-V. Han, W.-C. Lai, J.-J. Huang, T.-C. Lu, S.-J. Chang, and H.-C. Kuo. 2015. Light emitting diodes. In *The Current Trends of Optics and Photonics*, Lee, C.-C., Ed. Netherlands: Springer, pp. 179–234.

Liu, X. Y., S. R. Guo, Z. G. Xu, X. L. Jiao, and T. Tezuka. 2011. Regulation of chloroplast ultrastructure, cross-section anatomy of leaves, and morphology of stomata of cherry tomato by different light irradiations of light-emitting diodes. *HortScience* 4 6: 217–221.

Lokhandwala, J., H. C. Hopkins, A. Rodriguez-Iglesias, C. Dattenböck, M. Schmoll, and B. D. Zoltowski. 2015. Structural biochemistry of a fungal LOV domain photoreceptor reveals an evolutionary conserved pathway integrating light and oxidative stress. *Structure* 23: 116–125.

Lossev, O. 1924. Oscillating crystals. *Wireless World Radio Rev* 271: 93–96.

Lu, N., T. Maruo, M. Johkan, M. Hohjo, S. Tsukagoshi, Y. Ito, T. Ichimura, and Y. Sshinohara. 2012. Effects of supplemental lighting with light-emitting diodes (LEDs) on tomato yield and quality of single-truss tomato plants grown at high planting density. *Env Control Biol* 50: 63–74.

Ma, G., L. Zhang, M. Kato, K. Yamawaki, Y. Kiriiwa, M. Yahata, Y. Ikoma, and H. Matsumoto. 2012. Effect of blue and red LED light irradiation on β-Cryptoxanthin accumulation in the flavedo of *Citrus* fruits. *J Agric Food Chem* 60: 197–201.

Ma, G., L. Zhang, C. K. Setiawan, K. Yamawaki, T. Asai, F. Nishikawa, S. Maezawa, H. Sato, N. Kanemitsu, and M. Kato. 2014. Effect of red and blue LED light irradiation on ascorbate content and expression of genes related to ascorbate metabolism in postharvest broccoli. *Postharvest Biol Technol* 94: 97–103.

Ma, G., L. Zhang, M. Kato, K. Yamawaki, Y. Kiriiwa, M. Yahata, Y. Ikoma, and H. Matsumoto. 2015. Effect of the combination of ethylene and red LED light irradiation on carotenoid accumulation and carotenogenic gene expression in the flavedo of citrus fruit. *Postharvest Biol Technol* 99: 99–104.

Martineau, V., M. Lefsrud, M. T. Naznin, and D. A. Kopsell. 2012. Comparison of light-emitting diode and high-pressure sodium light treatments for hydroponics growth of Boston lettuce. *HortScience* 47: 477–482.

Massa, G. D., J. C. Emmerich, R. C. Morrow, C. M. Bourget, and C. A. Mitchell. 2007. Plant-growth lighting for space life support: A review. *Gravit Space Res* 19: 19–30.

Massa, G. D., H.-H. Kim, R. M. Wheeler, and C. A. Mitchell. 2008. Plant productivity in response to LED lighting. *HortScience* 43: 1951–1956.

Matsuda, R., K. Ohashi-Kaneko, K. Fujiwara, E. Goto, and K. Kurata. 2004. Photosynthetic characteristics of rice leaves grown under red light with or without supplemental blue light. *Plant Cell Physiol* 45: 1870–1874.

Ménard, C., M. Dorais, T. Hovi, and A. Gosselin. 2005. Developmental and physiological responses of tomato and cucumber to additional blue light. In *V International Symposium on Artificial Lighting in Horticulture* 711, pp. 291–296.

Merchant, S. and M. R. Sawaya. 2005. The light reactions: A guide to recent acquisitions for the picture gallery. *Plant Cell* 17: 648–663.

Mimouni, V., L. Ulmann, V. Pasquet, M. Mathieu, L. Picot, G. Bougaran, J.-P. Cadoret, A. Morant-Manceau, and B. Schoefs. 2012. The potential of microalgae for the production of bioactive molecules of pharmaceutical interest. *Curr Pharm Biotechnol* 13: 2733–2750.

Mitchell, C. 2012. Plant lighting in controlled environments for space and earth applications. In *VII International Symposium on Light in Horticultural Systems* 956, pp. 23–36.

Miyashita, Y., Y. Kitaya, T. Kozai, and T. Kimura. 1994. Effects of red and far-red light on the growth and morphology of potato plantlets in vitro: Using light emitting diode as a light source for micropropagation. *Env Effect Control Plant Tiss Cult* 393: 189–194.

Mizuno, T., W. Amaki, and H. Watanabe. 2009. Effects of monochromatic light irradiation by LED on the growth and anthocyanin contents in leaves of cabbage seedlings. In *VI International Symposium on Light in Horticulture* 907, pp. 179–184.

Moon, H. K., S. Y. Park, Y. W. Kim, and C. S. Kim. 2006. Growth of Tsuru-rindo (*Tripterospermum japonicum*) cultured in vitro under various sources of light-emitting diode (LED) irradiation. *J Plant Biol* 49: 174–179.

Morrow, R. C. 2008. LED lighting in horticulture. *HortScience* 43: 1947–1950.

Mouget, J., P. Rosa, C. Vachoux, and G. Tremblin. 2005. Enhancement of marennine production by blue light in the diatom Hasleaostrearia. *J Appl Phycol* 17: 437–445.

Nabors, M. W. 2004. *Introduction to Botany*. San Francisco, CA: Pearson Education.

Nakamura, S., T. Mukai, and M. Senoh. 1991. High-power GaN pn junction blue-light-emitting diodes. *Jap J Appl Phys* 30: L1998.

Nakamura, S., M. Senoh, N. Iwasa, and S.-I. Nagahama. 1995. High-power InGaN single-quantum-well-structure blue and violet light-emitting diodes. *Appl Phys Lett* 67: 1868–1870.

Nakamura, S., T. Mukai, and N. Iwasa. 1996. Multilayer elements with indium gallium nitride on clad layers, dopes for pn junctions. US Patent 5578839 A.

Nanya, K., Y. Ishigami, S. Hikosaka, and E. Goto. 2012. Effects of blue and red light on stem elongation and flowering of tomato seedlings. In *VII International Symposium on Light in Horticultural Systems* 956, pp. 261–266.

Nelson, N. and A. Ben-Shem. 2004. The complex architecture of oxygenic photosynthesis. *Nat Rev Mol Cell Biol* 5: 971–982.

Nhut, D. T., L. Hong, H. Watanabe, M. Goi, and M. Tanaka. 2000. Growth of banana plantlets cultured in vitro under red and blue light-emitting diode (LED) irradiation source. In *International Symposium on Tropical and Subtropical Fruits* 575, pp. 117–124.

Nhut, D., T. Takamura, H. Watanabe, A. Murakami, K. Murakami, and M. Tanaka. 2002. Sugar-free micropropagation of *Eucalyptus citriodora* using light-emitting diodes (LEDs) and film-rockwool culture system. *Env Control Biol* 40: 147–155.

Novičkovas, A., A. Brazaitytė, P. Duchovskis, J. Jankauskienė, G. Samuolienė, A. Virsilė, R. Sirtautas, Z. Bliznikas, and A. Zukauskas. 2010. Solid-state lamps (LEDs) for the short-wavelength supplementary lighting in greenhouses: Experimental results with cucumber. In *XXVIII International Horticultural Congress on Science and Horticulture for People (IHC2010)* 927, pp. 723–730.

Oh, S. J., D. I. Kim, T. Sajima, Y. Shimasaki, Y. Matsuyama, Y. Oshima, T. Honjo, and H. S. Yang. 2008. Effects of irradiance of various wavelengths from light-emitting diodes on the growth of the harmful dinoflagellate *Heterocapsa circularisquama* and the diatom *Skeletonema costatum*. *Fish Sci* 74: 137–145.

Okamoto, K., T. Yanagi, and S. Kondo. 1996. Growth and morphogenesis of lettuce seedlings raised under different combinations of red and blue light. In *II Workshop on Environmental Regulation of Plant Morphogenesis* 435, pp. 149–158.

Oldenhof, H., V. Zachleder, and H. Van Den Ende. 2006. Blue and red-light regulation of the cell cycle in *Chlamydomonas reinhardtii* (Chlorophyta). *Eur J Phycol* 41: 313–320.

Olle, M. and A. Viršile. 2013. The effects of light-emitting diode lighting on greenhouse plant growth and quality. *Agric Food Sci* 22: 223–234.

Palmer, J. D., D. E. Soltis, and M. W. Chase. 2004. The plant tree of life: An overview and some points of view. *Am J Bot* 91: 1437–1445.

Paradiso, R., E. Meinen, J. F. H. Snel, P. De Visser, W. Van Ieperen, S. W. Hogewoning, and L. F. M. Marcelis. 2011. Spectral dependence of photosynthesis and light absorptance in single leaves and canopy in rose. *Sci Hort* 127: 548–554.

Park, E. K. and C. G. Lee 2001. Astaxanthin production by *Haematococcus pluvialis* under various light intensities and wavelengths. *J Microbiol Biotechnol* 11: 1024–1030.

Park, S. U., D.-J. Ahn, H.-J. Jeon, T. R. Kwon, H.-S. Lim, B.-S. Choi, K.-H. Baek, and H. Bae. 2012. Increase in the contents of ginsenosides in raw ginseng roots in response to exposure to 450 and 470 nm light from light-emitting diodes. *J Ginseng Res* 36: 198.

Park, S.-Y., J. G. Lee, H. S. Cho, E. S. Seong, H. Y. Kim, C. Y. Yu, and J. K. Kim. 2013. Metabolite profiling approach for assessing the effects of colored light-emitting diode lighting on the adventitious roots of ginseng ("Panax ginseng" CA Mayer). *Plant Omics* 6: 224.

Pasquet, V., L. Ulmann, V. Mimouni, F. Guihéneuf, B. Jacquette, A. Morant-Manceau, and G. Tremblin. 2014. Fatty acids profile and temperature in the cultured marine diatom *Odontella aurita*. *J Appl Phycol* 26: 2265–2271.

Pinho, P., L. Särkkä, E. Tetri, R. Tahvonen, and L. Halonen. 2007. Evaluation of lettuce growth under multi-spectral-component supplemental solid state lighting in greenhouse environment. *Int Rev Electr Eng* 2: 854–860.

Pinho, P., K. Jokinen, and L. Halonen. 2012. Horticultural lighting—Present and future challenges. *Lighting Res Technol* 44: 427–437.

Poudel, P. R., I. Kataoka, and R. Mochioka. 2008. Effect of red-and blue-light-emitting diodes on growth and morphogenesis of grapes. *Plant Cell Tiss Org Cult* 92: 147–153.

Roháček, K., M. Bertrand, B. Moreau, B. Jaquette, C. Caplat, B. Moreau, A. Morant-Manceau, and B. Schoefs. 2014. The relaxation of the nonphotochemical chlorophyll fluorescence quenching in diatoms: Kinetics, components and mechanisms. *Philos Trans R Soc Lond B: Biology* 369: 20130241.

Sabzalian, M. R., P. Heydarizadeh, M. Zahedi, A. Boroomand, M. Agharokh, M. R. Sahba, and B. Schoefs. 2014. High performance of vegetables, flowers, and medicinal plants in a red-blue LED incubator for indoor plant production. *Agron Sustain Dev* 34: 879–886.

Samuoliené, G., A. Urbonavičiūtė, P. Duchovskis, Z. Bliznikas, P. Vitta, and A. Žukauskas. 2009. Decrease in nitrate concentration in leafy vegetables under a solid-state illuminator. *HortScience* 44: 1857–1860.

Samuoliené, G., A. Brazaitytė, R. Sirtautas, A. Novičkovas, and P. Duchovskis. 2011a. Supplementary red-LED lighting affects phytochemicals and nitrate of baby leaf lettuce. *J Food Agric Env* 9: 271–274.

Samuoliené, G., A. Brazaitytė, R. Sirtautas, A. Novičkovas, and P. Duchovskis. 2011b. The effect of supplementary LED lighting on the antioxidant and nutritional properties of lettuce. In *International Symposium on Advanced Technologies and Management Towards Sustainable Greenhouse Ecosystems: Greensys 2011* 952, pp. 835–841.

Samuoliené, G., A., Brazaitytė, P. Duchovskis, A. Viršilé, J. Jankauskienė, R. Sirtautas, A. Novičkovas, S. Sakalauskienė, and J. Sakalauskaitė. 2011c. Cultivation of vegetable transplants using solid-state lamps for the short-wavelength supplementary lighting in greenhouses. In *International Symposium on Advanced Technologies and Management Towards Sustainable Greenhouse Ecosystems: Greensys 2011* 952, pp. 885–892.

Samuoliené, G., A. Urbonavičiūtė, A. Brazaitytė, G. Šabajevienė, J. Sakalauskaitė, and P. Duchovskis. 2011d. The impact of LED illumination on antioxidant properties of sprouted seeds. *Cent Eur J Biol* 6: 68–74.

Samuoliené, G., R. Sirtautas, A. Brazaitytė, and P. Duchovskis. 2012a. LED lighting and seasonality effects antioxidant properties of baby leaf lettuce. *Food Chem* 134: 1494–1499.

Samuoliené, G., R. Sirtautas, A. Brazaitytė, A. Viršilé, and P. Duchovskis. 2012b. Supplementary red-LED lighting and the changes in phytochemical content of two baby leaf lettuce varieties during three seasons. *J Food Agric Env* 10: 701–706.

Samuoliené, G., A. Brazaitytė, J. Jankauskienė, A. Viršilé, R. Sirtautas, A. Novičkovas, S. Sakalauskienė, J. Sakalauskaitė, and P. Duchovskis. 2013. LED irradiance level affects growth and nutritional quality of Brassica microgreens. *Cent Eur J Biol* 8: 1241–1249.

Schmid, R. and M. J. Dring. 1992. Circadian rhythm and fast responses to blue light of photosynthesis in Ectocarpus (Pheaophyta, Ectocarpales). I. Characterization of the rhythm and the blue-light response. *Planta* 187: 60–66.

Schmid, R. and M. J. Dring. 1993. Rapid, blue-light-induced acidifications at the surface of Ectocarpus and other marine macroalgae. *Plant Physiol* 101: 907–913.

Schubert, E. F. and J. K. Kim. 2005. Solid-state light sources getting smart. *Science* 308: 1274–1278.

Schuerger, A. C., C. S. Brown, and E. C. Stryjewski. 1997. Anatomical features of pepper plants (*Capsicum annuum* L.) grown under red light-emitting diodes supplemented with blue or far-red light. *Ann Bot* 79: 273–282.

Schulze, P. S. C., L. A. Barreira, H. G. C. Pereira, J. A. Perales, and J. C. S. Varela. 2014. Light emitting diode (LEDs) applied to microalgal production. *Trends Biotechnol* 32: 422–430.

Senge, M. and H. Senger. 1990. Functional changes in the photosynthetic apparatus during light adaptation of the green alga *Chlorella fusca*. *Photochem Photobiol B* 8: 63–71.

Senger, H. 1987. Chlorophyll biosynthesis in algae. In *Blue Light Responses: Phenomena and Occurrence in Plants and Microorganisms*, vol. 1, Senger, H. Ed. Boca Raton, FL: CRC Press, pp. 76–85.

Seyfadabi, J., Z. Ramezanpour, and A. Khoeyi. 2010. Protein, fatty acid and pigment content of *Chlorella vulgaris* under different light regimes. *J Appl Phycol* 23: 721–726.

Shimada, A. and Y. Taniguchi. 2011. Red and blue pulse timing control for pulse width modulation light dimming of light emitting diodes for plant cultivation. *J Photochem Photobiol B: Biology* 104: 399–404.

Shin, Y. S., M. J. Lee, E. S. Lee, J. H. Ahn, J. H. Lim, H. J. Kim, H. W. Park, Y. G. Um, S. D. Park, and J. H. Chai. 2012. Effect of LEDs (light emitting diodes) irradiation on growth and mineral absorption of lettuce (*Lactuca sativa* L. "Lollo Rosa"). *J Bio-Environ Control* 21: 180–185.

Shu, C. H., C. H. Tsai, W. H. Liao, K. Y. Chen, and H. C. Huang. 2012. Effects of light quality on the accumulation of oil in a mixed culture of *Chlorella* sp. and *Saccharomyces cerevisiae*. *J Chemtechnol Biotechnol* 87: 601–607.

Shukla, S. K. and R. Mohan. 2012. The contribution of diatoms to worldwide crude oil deposits. In *The Science of Algal Fuels: Phycology, Geology, Biophotonics, Genomics and Nanotechnology*; Gordon, R., Seckbach, J., Eds. Dordrecht: Springer, pp. 355–382.

Singh, D., C. Basu, M. Meinhardt-Wollweber, and B. Roth. 2014. LEDs for energy efficient greenhouse lighting. arXiv:1406.3016.

Singhal, G. 1999. *Concepts in Photobiology: Photosynthesis and Photomorphogenesis*. New York: Springer Science & Business Media.

Solymosi, K. 2012. Plastid structure, diversification and interconversions I. *Algae Curr Chem Biol* 6: 167–186.

Son, K.-H. and M.-M. Oh. 2013. Leaf shape, growth, and antioxidant phenolic compounds of two lettuce cultivars grown under various combinations of blue and red light-emitting diodes. *HortScience* 48: 988–995.

Stadler, C. 2011. Effects of lighting time and lighting source on growth, yield and quality of greenhouse sweet pepper. Available at http://www.lbhi.is/sites/default/files/gogn/vidhengi/thjonusta/utgefid_efni/RitLbhi/34_effects_of_lightning_-_rit_lbhi_nr._34.pdf.

Starr, C., C. Evers, and L. Starr. 2010. *Biology: Concepts and Applications Without Physiology*. Stamford, CT: Cengage Learning.

Steele, R. V. 2007. The story of a new light source. *Nat Photon* 1: 25–26.

Stutte, G. W., S. Edney, and T. Skerritt. 2009. Photoregulation of bioprotectant content of red leaf lettuce with light-emitting diodes. *HortScience* 44: 79–82.

Tamulaitis, G., P. Duchovskis, Z. Bliznikas, K. Breive, R. Ulinskaite, A. Brazaityte, A. Novičkovas, and A. Žukauskas. 2005. High-power light-emitting diode based facility for plant cultivation. *J Phys D: Appl Phys* 38: 3182.

Tanaka, M., T. Takamura, H. Watanabe, M. Endo, T. Yanagi, and K. Okamoto. 1998. In vitro growth of Cymbidium plantlets cultured under superbright red and blue light-emitting diodes (LEDs). *J Hort Sci Biotechnol* 73: 39–44.

Tarakanov, I., O. Yakovleva, I. Konovalova, G. Paliutina, and A. Anisimov. 2012. Light-emitting diodes: On the way to combinatorial lighting technologies for basic research and crop production. In *VII International Symposium on Light in Horticultural Systems* 956, pp. 171–178.

Tennessen, D. J., E. L. Singsaas, and T. D. Sharkey. 1994. Light-emitting diodes as a light source for photosynthesis research. *Photosynth Res* 39: 85–92.

Thomas, B. 1981. Specific effects of blue light on plant growth and development. In *Plants and the Daylight Spectrum*, Smith, H., Ed. London: Academic Press, pp. 453–459.

Topchiy, N., S. Sytnik, O. Syvash, and O. Zolotareva. 2005. The effect of additional red irradiation on the photosynthetic apparatus of *Pisum sativum*. *Photosynthetica* 43: 451–456.

Tripathy, B. C. and C. S. Brown. 1995. Root-shoot interaction in the greening of wheat seedlings grown under red light. *Plant Physiol* 107: 407–411.

Trouwborst, G., J. Oosterkamp, S. W. Hogewoning, J. Harbinson, and W. van Ieperen. 2010. The responses of light interception, photosynthesis and fruit yield of cucumber of LED-lighting within the canopy. *Physiol Plant* 138: 289–300.

Tuan, P. A., A. A. Thwe, Y. B. Kim, J. K. Kim, S.-J. Kim, S. Lee, S.-O. Chung, and S. U. Park. 2013. Effects of white, blue, and red light-emitting diodes on carotenoid biosynthetic gene expression levels and carotenoid accumulation in sprouts of tartary buckwheat (*Fagopyrum tataricum* Gaertn.). *J Agric Food Chem* 61: 12356–12361.

Ulmann, L., V. Mimouni, V. Blanckaert, V. Pasquet, B. Schoefs, and B. Chénais. 2014. The polyunsaturated fatty acids from microalgae as potential sources for health and disease. In *Polyunsaturated Fatty Acids: Sources, Antioxidant Properties, and Health Benefits*, Angel Catalá A., Ed. New York: Nova Publishers, pp. 23–44.

Urbonavičiūtė, A., G. Samuolienè, A. Brazaitytė, R. Ulinskaitė, J. Jankauskienė, P. Duchovskis, and A. Zukauskas. 2008. The possibility to control the metabolism of green vegetables and sprouts using light emitting diode illumination. *Sodininkystè ir Daržininkystè* 27: 83–92.

Vasseur, C., G. Bougaran, M. Garnier, J. Hamelin, C. Leboulanger, M. Le Chevanton, B. Mostajir, B. Sialve, J. P. Steyer, and E. Fouilland. 2012. Carbon conversion efficiency and population dynamics of a marine algaebacteria consortium growing on simplified synthetic digestate first step in a bioprocess coupling algal production and anaerobic digestion. *Bioresour Technol* 119: 79–87.

Wang, C., C. Fu, and Y. Liu. 2007. Effects of using light-emitting diodes on the cultivation of *Spirulina platensis*. *Biochem Eng* 37: 21–25.

Wilhelm, C., P. Krämer, and A. Wild. 1985. Effect of different light qualities on the ultrastructure, thylakoid membrane composition and assimilation metabolism of *Chlorella fusca*. *Plant Physiol* 64: 359–364.

Wu, H.-C. and C.-C. Lin. 2012. Red light-emitting diode light irradiation improves root and leaf formation in difficult-to-propagate *Protea cynaroides* L. plantlets in vitro. *HortScience* 47: 1490–1494.

Wu, M.-C., C.-Y. Hou, C.-M. Jiang, Y.-T. Wang, C.-Y. Wang, H.-H. Chen, and H.-M. Chang. 2007. A novel approach of LED light radiation improves the antioxidant activity of pea seedlings. *Food Chem* 101: 1753–1758.

Xu, H.-L., Q. Xu, F. Li, Y. Feng, F. Qin, and W. Fang. 2012. Applications of xerophytophysiology in plant production—LED blue light as a stimulus improved the tomato crop. *Sci Hort* 148: 190–196.

Yanagi, T. and K. Okamoto. 1994. Utilization of super-bright light emitting diodes as an artificial light source for plant growth. In *III International Symposium on Artificial Lighting in Horticulture* 418, pp. 223–228.

Yanagi, T., K. Okamoto, and S. Takita. 1996. Effects of blue, red, and blue/red lights of two different PPF levels on growth and morphogenesis of lettuce plants. In *International Symposium on Plant Production in Closed Ecosystems* 440, pp. 117–122.

Yeh, N. and J.-P. Chung. 2009. High-brightness LEDs—Energy efficient lighting sources and their potential in indoor plant cultivation. *Renew Sustain Energy Rev* 13: 2175–2180.

Yorio, N. C., G. D. Goins, H. R. Kagie, R. M. Wheeler, and J. C. Sager. 2001. Improving spinach, radish, and lettuce growth under red light-emitting diodes (LEDs) with blue light supplementation. *HortScience* 36: 380–383.

Zhang, H., Z.-G. Xu, J. Cui, A.-S. Gu, and Y.-S. Guo. 2010. Effects of light quality on the growth and chloroplast ultrastructure of tomato and lettuce seedlings. *Chin J Appl Ecol* 4: 23.

Zukauskas, A., Z. Bliznikas, K. Breivė, A. Novičkovas, G. Samuolienė, A. Urbonavičiūtė, A. Brazaitytė, J. Jankauskienė, and P. Duchovskis. 2009. Effect of supplementary pre-harvest LED lighting on the antioxidant properties of lettuce cultivars. In *VI International Symposium on Light in Horticulture* 907, pp. 87–90.

Index

Page numbers followed by f and t indicate figures and tables, respectively.

A

Abiotic stress; *see also* Stressful conditions
  application of glycine betaine or proline and, 657
  changes in the membrane organization under, 247
  effect on photosynthetic components, 630f
  environmental, 289–290
  genetic factors, 551
  under high temperature, 245
  to influence crop performance, 287
  in ornamental plants, 651
  resistance, 527
  responses of plants to, 455
  signaling during, 166
Abiotic stressful conditions; *see also* Stressful conditions
  effects on photosynthesis, 629–631, 630f
A/b ratio, chlorophylls, 669, 674
Abscisic acid (ABA)
  applications, 657
  in drought stress, 654
  salinity stress and, 654
Absolute water content, 664
Absorption
  of light energy, thylakoid membrane and, 448
Absorption spectra, *see* Chlorophylls/ bacteriochlorophylls, physicochemical properties of
  of chlorophyll (Chl), 174f
Acaryochloris marina, 96, 359, 762t; *see also* Cyanobacteria/microalgae, photosynthetic apparatus in
Acaryochloris spp., 748
A/C_c curve-fitting method, 66–67; *see also* Mesophyll conductance estimation
Acclimation, 489; *see also* Phenotypic plasticity
  developmental, 540
  dynamic, 540
  phosphate deficiency and, 617
Acer grandidentatum Nutt., 654
Acetazolamide, 195, 197
Acetonitrile
  cyclic voltammograms of Chls *a, b,* and *d* in, 138f
  square wave voltammograms of Chls *a, b, d, f,* and DV-Chl *a* in, 139f
Aconitase (ACON), 339
acpPC, *see* Allophycocyanins (APC)
Adaptation, 489
  phosphate deficiency and, 617
  plants, to environmental conditions, plasticity and, 493–494
Adaptive diversity, 493
Adaptive plasticity, 497; *see also* Plasticity
  environment condition and, 494–495
  phenotypic, 493, 494
Adenosine triphosphate (ATP), 202
ADP-glucose pyrophosphorylase (ADPGlc PPase), 510, 512, 513

Affinity Tags, 211
Agastache urticifolia, 656
7Ag.7DL translocation, 543–544
Aggregate culture, 725
AGO7 proteins, 541
Agriculture
  CO_2, 732, 732f
  control systems, 729–737
  cooling management, 729–731
  evolution of, 723–737
  fog-cooling systems, 729, 729f, 730
  heating management, 731
  humidity, 731–732
  innovations, 729–737
  light, 732–737, 733f, 735f
  overview, 723–724, 724t
  pad-and-fan systems, 729–731, 730f
  plant factories, *see* Plant factories
  temperature, 729–731, 729f, 730f
Air pollutants
  effects on photosynthesis, 640–642, 642t
ALAD, *see* 5-aminolevulinate dehydratase (ALAD)
ALB3 protein (ALBINO 3), 8
Aldose-6P reductase (Ald-6P Rase), 518
Algae, under LED lighting, 773–774, 775t, 776t
Alhagi sparsifolia, 699
Alkylthiols–gold surface, 211
Allium cepa, 769t
Allocation, nitrogen, 743–744, 743f
Allophycocyanins (APC)
  about, 351f
  in dinoflagellates, 363–364
  and orange carotenoid protein, 371
  phycobilisome (PBS) and, 352
Alternative electron transport pathways, *see* Light (excess)/carbon (limited)
Amaranthus hybridus, 707
Amaranthus retroflexus, 707
American Soil Classification System, 296
Amino acid sequences of precytochromes, 233f
5-aminolevulinate dehydratase (ALAD), 697
5-aminolevulinic acid (ALA) biosynthesis, 174
5-aminolevulinic acid (ALA) dehydratase (ALAD), 174
Ampere's law, 25
α-amylase, 512
Anaerobiosis
  effects on photosynthesis, 640
Analysis of Variance (ANOVA), 298–300, 298t
Anatomical plasticity, woody species, 495–496
Anethum graveolens, 769t
Angelonia, 653
Angelonia angustifolia, 653
Angiosperm seedlings, 182
Angular momentum, 27, 28f
Antenna proteins, 12
Antheraxanthin (Ant), 589
Anthocyanins, 672
Antimycin, 334
Antimycin A-sensitive route, 356

Antioxidant mechanisms
  dehydration and, 690
  of DT organisms, 665
  zeaxanthin and, 684
Antioxidants, 88, 88f
Antioxidative response, 673
Antirhinum majus L., 654
Apatococcus, 675
ApcE, core–membrane linker, 353
Apium graveolens (celery), 517
Apple *(Malus domestica)*, 518
Aquaporins, 746
Arabidopsis, 7, 8
  mutants/transgenic plants in, 745
  overexpression of, 748
Arabidopsis thaliana, 152, 161, 495, 513, 746, 747, 748, 749, 751, 772t
  growth analysis (case study), 580–581, 581f
Arbuscular mycorrhizal (AM) organisms, 657
Argonaute (AGO) protein, 541
Argyranthemum coronopifolium, 653
Artificial lighting
  future, LEDs and, 774, 776–777
  plant factories, 725, 725f, 727–729
    advantages, 728
    LEDs, 728
    shelves, 728
    vs. sunlight plant factories, 728t
Artificial photosynthesis
  basic architecture, 207–208, 208f, 209f
  photoconversion efficiency *vs.* structural complexity, 208, 209f
Asada–Halliwell–Foyer pathway, 166
Ascorbate, 163–164
Ascorbate oxidase (AO), 163
Ascorbate peroxidase (APX), 160f, 163, 166
Ascorbic acid (AA), 160f
Asplenium, 672
Atmospheric pressure chemical ionization (APCI)-mass spectra, 116, 120f
Atomic clocks of global positioning system, 37
Atomic force microscopy (AFM), 5, 57, 57f
ATP synthase (ATPase), 53, 54f, 81
ATP synthase complex, genes for, 549
ATP synthesis
  thylakoid membrane, 450
Atrazine, 707
Atriplex lentiformis, 654
Atriploid, 311
Avena sativa, 766t
Average wavelength of light, 38, 39
Avocado *(Persea americana)*, 528
Avoidance, water availability and, 663
Awns, genetic factors, 542

B

Back-crossed inbred lines (BILs), 752, 753
Bacterial-type PEP Cases (BTPC), 520
Bacteriochlorophyll (BChl) 663
  NMR spectra of, 126, 133–134, 134f
  81-OH -Chl *a* and, 100

Bacteriochlorophyll (BChl)
  a' and *g'*, 97–98
  g and *g'*, CD spectra of, 114
  molecular structures and carbon numbering of, 98f
  physicochemical properties of, *see* Chlorophylls/bacteriochlorophylls, physicochemical properties of
Bacteriochlorophyll (BChl) *a, b, g*
  about, 96, 98f, 99f
  and derivatives, 101, 102f
  and epimers, 109
Bacteriopheophytin (BPhe), 214f
  a and *b*, 98, 109, 110f
Balanites aegyptiaca
  drought stress (case study), 457
β-amylase, 512
Banana, 769t
Barley, 771t
Barnes, Charles Reid, 763
Basil, 769t
β-carotene (β-Car), 589
Begonia, 656
Begonia semperflorens L., 656
Benson-Calvin cycle (BCC), 161, 162, 509–510, 511f
  reactions and enzymes, 510t
Berry formation and marketable yields (case study), 293–294
Between-individual variability, 489–490
Bicarbonate removal, 197
Bienertia cycloptera, 311–313, 312f, 314f
Bienertia sinuspersici, 311–313
Big Bend bluebonnet, 654, 656
Bigness of binary photon, 29
Bigtooth maples, 654
Binary photon
  about, 23–34, 32t
  anatomy of, 30f
  blue-shifted, 39, 39f
  classes of, 32t
  properties of, 42–43
  test of, relativity of space and time, 36–42
  theory, 37
Biochemical activities, changes in, 554
Biochemistry/bioenergetics of photosynthesis, UVB on, *see* Ultraviolet-B radiation in photosynthesis determination
Biogenesis, chloroplast, 685
BiolFlor database, 488
Biological spectral weighting function (BSWF), 277
Biomass accumulation, leaf area and, 259–260, 260f
Biomass gain, destructive analysis of, 572–573
Biomimetic systems, 202
Biosynthesis of Chls and hemes, 175f
Biosynthetic enzymes
  complex formation of, 178–179
  subcellular localization of, 177–178
Biotechnology, photosynthesis and, 741–754
  Calvin–Benson cycle enzymes, activity of, 748–750, 749f
  chloroplast electron transport in thylakoid membranes, 747–748, 748f
  CO₂ concentration around Rubisco, 745–747, 746f
  optimizing, on canopy level, 750–752, 751f
  overview, 741
  phloem carbohydrate transport and carbon utilization, 750

photosynthetic performance, improving, 741–745
  by QTL analysis, natural genetic resources, 752–754, 752f, 753f
  Rubisco concentration and N allocation, 743–744, 743f
  Rubisco performance, opportunities for improving, 744–745
  Rubisco specificity and carboxylation rate, 741–743, 742f
  thermotolerance of Rubisco activase, 745
Biotic stress; *see also* Abiotic stress
  Brassica napus (case study), 458
  genetic factors, 551–552
Black-light bulbs (BLB), 736
Blossfeldia liliputiana, 670
Blueberry, 768t
Blue-green algae, *see* Cyanobacteria
Blue light, 735; *see also* Light
Blue-shifted binary photons, 39, 39f
Boea hygrometrica, 683, 684, 685
Boltzmann's constant, 19, 39
Born, Max, 22
Borya nitida, 684, 685
Botryococcus, 675
Botrytis cinerea, 736
Bougainvillea spectabilis, 654
Bracteacoccus, 675
Brassica campestris L., 770t
Brassica juncea, 769t
Brassica napus, 705, 707, 710, 717, 719, 771t
  nutrient deficiency and biotic stress (case study), 458
Brassica oleracea, 765t, 770t, 771t
Brassica rapa, 771t
Brassinolide, 657
Broadband UV sources, 276
Broccoli, 765t
Bruguiera parviflora, 653

C

CA, *see* Carbonic anhydrase
Cabbages, 770t
Cab genes, 546
Calamagrostis canadensis, 494
Calceolaria hybrida, 657
Calcium application, ornamental plants, 657
Calendula officinalis, 657, 768t
"Calico" ornamental pepper, 653
Calla lilies, 769t
Calvin–Benson cycle
  enzymes, activity of, 748–750, 749f
  photosynthetic electron transport, 748
  RuBP regeneration in, 747
Calvin cycle, 81, 88, 202
  as primary carboxylation mechanism, 308
  water content and, 670
CAM, *see* Crassulacean acid metabolism (CAM)
C₃ and C₄ carbon metabolism, 265–268, 268f
C₃ and C₄ plants; *see also* Solar ultraviolet (280–400 nm) exclusion on photosynthesis
  comparison between, 259b
  effects of solar UV exclusion, 259
  solar UV sensitivity of, 269–270, 269f
Canopy level, photosynthesis on, 750–752, 751f
Capsicum annuum, 653, 656, 768t–769t
Carbamylation, of lysine residue, 745
Carbohydrates
  drought conditions and levels of, 638

Carbohydrate transport, phloem carbon utilization and, 750
Carbon assimilation
  efficiency, differences in, 707–708
  phosphorus role in, 603–618
    long-term *in vivo* effects of Pi deprivation, 608–615
    overview, 603–604
    phosphate deficiency, *see* Phosphate deficiency
    short-term *in vitro* effects of Pi deprivation, 605–608
  in R and S biotypes, 709–710, 710f, 711t, 712f, 720
  Rubisco in R/S plants, 706
Carbon balance, 669–672, 670f, 671t
Carbon capturing and assimilation zone, 78
Carbon concentration mechanisms (CCMs), 670, 747
Carbon dioxide (CO₂)
  assimilation, 69, 70f, 452
    dehydration, 689
  availability, 663, 670
    R and S variants, 718
  plant growth and, 732, 732f
  small fruit crop responses to, 288
Carbon dioxide (CO₂) concentrating mechanisms (CCM)
  cyanobacteria
    about, 382–383
    carboxysomes, 385
    inorganic carbon uptake, 383–385, 384f
  in diatoms, 387–389
  in dinoflagellates, 389
  in green microalgae
    about, 385–386
    Ci transmembrane transport, 386–387
    pyrenoids, 387
Carbon dioxide (CO₂) concentration
  greenhouse control systems, 576–577, 576f
  Rubisco and, 745–747, 746f
Carbon dioxide (CO₂) diffusion
  salinity effects on, 631–632
Carbon dioxide (CO₂) exchange
  growth quantification using, *see* Growth analysis
Carbon electrodes, 211
Carbon fixation
  role of, 465
Carbon flow
  in heterotrophic plant cell, 510, 511f–512f
Carbonic anhydrase (CA)
  activity measured as ratio of CO₂ hydration, 194t
  and C₃ and C₄ carbon metabolism, 265–268
  near PSI, 196, 196f
  near PSII, 194–196, 194t, 195f
  of thylakoid of higher plants, *see* Thylakoid carbonic anhydrases
Carbon metabolism
  Pi deprivation and, 611–613, 612t, 613t
Carbon numbering
  of bacteriochlorophylls, 98f
  of chlorophylls, 97f
Carbon partitioning, 614–615, 614t
Carbon substrates, interface of RC with, 211
Carbon transport, from leaf to sink tissues, 520–524, 521f
  mannitol and glucitol transporters, 523–524
  sucrose transporters, 521–523, 522f
  SUT transporters, 522–523, 522f
  SWEET transporters, 521–522, 522f

Carbon utilization, phloem carbohydrate transport and, 750
Carboxylation
  under drought conditions, 637–638
  Rubisco, 700
  salinity effects on, 634
Carboxylation efficiency (CE), 291
Carboxylation rate, Rubisco specificity and, 741–743, 742f
Carboxysomes, 385
  defined, 747
α-carboxysomes, 383
α -carotene (α-Car), 591–593, 592f
  HPLC determination of, 596, 596f
β-carotene hydroxylase, 669
Carotenoids, 7, 206, 261
  biosynthesis from lycopene, 589, 590f
  content, desiccation and, 684
  degradation, 84; see also Photosynthetic pigments, degradation of
  high-light carotenoids, 594–595
    escholtzxanthin, 595
    red retro-carotenoids, 595
    rhodoxanthin, 595
    xanthophyll esters, 594–595
  low-light carotenoids, 591–594
    α-carotene, 591–593, 592f
    lactucaxanthin, 594
    lutein epoxide, 593–594
    trans-neoxanthin, 591
  nonubiquitous, see Nonubiquitous carotenoids
  pigment composition of, 589–591, 590f, 591t
  poikilochlorophyllous plants, 674
  types, 761, 762t
  ubiquitous, 589–591, 590f
Catharanthus roseus, 655, 656
C₄/CAM plants
  thylakoid membrane, 452
C₃–C₄ hybridization, 543
CCM, see Carbon dioxide (CO₂) concentrating mechanisms
Celery (Apium graveolens), 517
Cercis canadensis var. mexicana, 656
Ceterach officinarum, 672, 684
CFru-1,6bisP 1-Pase
  sucrose synthesis control by, 515–516, 516f
Chamber and system design, growth analysis and, 573
Chaperone, 82f
Charge (C), parity (P), sign of mass (M) theory, 33
Chemical robustness, 208
Chenopodiaceae family, C₄ photosynthesis in; see also Nontypical C₄ species, photosynthesis in
  Bienertia cycloptera/Bienertia sinuspersici, case of, 311–313, 312f
  Suaeda aralocaspica, case of, 313–314, 313f, 314f
Chenopodium album, 707
Chilling
  effects on photosynthesis, 639–640
Chl a/b binding protein, 688
Chlamydomonas reinhardtii, 178, 775t
Chl b reductase (CBR), 177
Chlide a oxygenase (CAO), 177
Chlorella, 675
Chlorella fusca, 774, 775t
Chlorella vulgaris, 774, 775t
Chlorenchyma cells, 311, 313

Chlorina-type pea mutants, 544
Chlorofluorocarbon (CFC), 257
Chlorokybophycaceae, 675
Chlorophyceae, 672, 775t
Chlorophyll (Chl)
  a/b ratio, 669, 674
  c₁ and c₂, 96, 97f
  chlorophyll a, 763
  content, genetic factors, 541–542
  deficiency, 541
  degradation, tracing path of, 82–84
  in diethylether, absorption properties of, 103t–107t
  fluorescence (R and S biotypes), 709, 709f
  heat stress and, 697
  light avoidance, 672
  loss
    in HDT plants, 684
    in PDT plants, 684
  molecular structures and carbon numbering of, 97f
  poikilochlorophyllous plants, 674
  types, 762t
Chlorophyll (Chl) a', b', d', f', DV-Chl a' and pheophytins, CD spectra of, 114, 117f, 118f
Chlorophyll (Chl) a, b, d, f and derivatives, 100–101, 101f
Chlorophyll (Chl) a, b, d, f and DV-Chl a
  about, 96, 97f
  NMR spectra of
    C-NMR spectra of, 122–124, 128t–129t
    H-NMR spectra of, 120–122, 124f, 125t–126t
    two-dimensional, 124, 126
    redox potentials of, 138–139, 138f, 139f, 140t–141t
Chlorophyll (Chl) a and a', 126
Chlorophyll (Chl) a and b, structures of, 174f
Chlorophyll (Chl) a' d' and DV-Chl a', 96–97, 100f
Chlorophyll (Chl) aΔ2,6-Phytadienol, 126
Chlorophyll (Chl) biosynthesis with chloroplast functionality, transcriptional coordination of, 183
Chlorophyll (Chl) biosynthetic center, 178
Chlorophyll–carotenoid–protein complexes, 589
Chlorophyll c-containing microalgae, photosystems of, see Microalgal photosystems
Chlorophyll (Chl) fluorescence, 151f, 262–264, 263f, 264t, 452–453
  drought effects on, 636–637
  measurements, 454
  monitoring, limits of, 455–456, 455t
  salinity effects on, 633–634
Chlorophyll (Chl) fluorescence method; see also Mesophyll conductance estimation
  constant J method, 67
  variable J method, 67
Chlorophyll (Chl) fluorescence parameters, 447
  meaning of, 453–454
  as potential stress markers, 456
Chlorophyllide (Chlide), 5
Chlorophyll metabolism regulation in plants
  about, 173–174, 174f
  Chl biosynthesis in plants
    ALA biosynthesis, 174
    Chl cycle, 177
    common steps, 174, 176
    Fe branch, 177

Mg branch, 176–177
  overview, 175f, 176t
  localization of enzymes, in Chl biosynthesis
    complex formation of biosynthetic enzymes, 178–179
    subcellular localization of biosynthetic enzymes, 177–178
  regulation of Chl biosynthesis
    posttranslational regulation, 184–186
    transcriptional regulation, 179–183
Chlorophyll-protein (CP) complexes
  appearance of, 10
Chlorophylls/bacteriochlorophylls, physicochemical properties of
  absorption spectra
    about, 102, 103f, 103t–107t, 107f
    BChl a /BChl b /BChl g and epimers, 109
    BPhe a/BPhe b, 109, 110f
    Chls a, b, d, f and DV-Chl a, 103, 107–108, 108f
    Phes a, b, d, f and divinyl pheophytin a, 108, 109f
    [Zn]-BChl a, 110, 110f, 111f
  BChl a, b, and g, 96, 98f, 99f
  BChl a' and g', 97–98
  Chl a, b, d, f and DV-Chl a, 96, 97f
  Chl a' d' and DV-Chl a', 96–97, 100f
  Chl c₁ and c₂, 96, 97f
  circular dichroism (CD) spectra
    BChl g and g', 114
    Chl a', b', d', f', DV-Chl a' and pheophytins, 114, 117f, 118f
    [Mg]- and [Zn]-BChl a, 116, 119f
  fluorescence spectra
    fluorescence lifetime of (B)Chls/(B)Phes, 111–114, 112t–113t, 115f
    fluorescence quantum yield, 111, 116f
    fluorescence wavelength maxima, 110, 111f
  high-performance liquid chromatography (HPCL)
    BChl a, b, g and derivatives, 101, 102f
    Chl a, b, d, f and derivatives, 100–101, 101f
    DV-Chl a, a', b and divinyl Phe a, 101–102
    normal phase/reversed-phase, 102f, 103f
  mass spectrometry (MS)
    APCI-mass spectra, 116, 120f
    FAB-mass spectra, 117–119, 121f
    MALDI-mass spectra, 119, 122f, 123f
  nuclear magnetic resonance (NMR) spectra
    BChl 663, 126, 133–134, 134f
    Chl a and a', 126
    Chls a, b, d, f and divinyl Chl a, 120–126, 125t–126t, 128t–129t, 130f–132f
    [Mg]-BChl a /[Zn]-BChl a/[2H]-BChl a (BPhe a), 134–138, 135f, 136–137t, 137f
  81-OH -Chl a and BChl a, 663, 100
  Phe a/BPhe a and b, 98
  redox potentials
    Chls a, b, d, f and DV-Chl a, 138–139, 138f, 139f, 140t–141t
    [M]-Chl a/[Zn]-BChl a /[2H]-BChl a, 139, 142, 142f
  [Zn]-BChl a, 100
Chlorophytes, 359
Chloroplast(s), 3, 77
  and cytosol, metabolites exchange between, 514–515
  envelope proteins, 387

extranuclear control, *see* Extranuclear
control chloroplasts
gene expression, 549–550
post-transcriptional regulation, 550
translational regulation, 550–551
higher plant, pigment composition of,
589–591, 590f, 591t
regulation of starch metabolism in, 513–514
starch synthesis and degradation in, 510–513
structure and pigment composition, salinity
effects on, 632
ultrastructural changes in, drought and, 636
Chloroplast biogenesis
coordinated with plant development, 3–4, 4f, 5f
gene expression changes during, 7–8
proteins/lipids/photosynthetic pigments,
changes in, 6–7, 6f
at structural level, 4–6
structure/function, correlation of
CP complexes, appearance of, 10
immunodetection of proteins, 12, 12t
membrane composition and structure,
mutual relations, 8
microscopic analysis, 8–10, 9f
photosynthetic apparatus, functional
changes of, 10, 12
Chloroplast development
changes during leaf senescence
photosynthetic pigments, degradation of,
82–84, 83f
primary photochemical reactions, loss
in, 84
Rubisco activity, decline in/enzyme
protein, loss of, 84–85
thylakoid complexes, disassembly of, 84
ultrastructural changes, 82, 83f
chloroplast biogenesis/senescence, signals
regulating (during leaf development)
role of ROS in signaling network of,
87–89, 88f
sugar signaling, 85–87, 86f
gene expression during, 79
overview, 77–79, 78f
thylakoids, biogenesis/stromal enzyme
complexes
ATP synthase, 81
cytochrome b_6f complex, 81
photosystem I (PSI), assembly of, 81
photosystem II (PSII), organization/
assembly of, 80–81
Rubisco, 81–82, 82f
thylakoid complexes, assembly of, 81
Chloroplast DNA (cpDNA)
inheritance of, 547–548
Chloroplast electron transport, in thylakoid
membranes, 747–748, 748f
Chloroplast electron transport rate to CO_2,
response to, 71
Chloroplast-linked retrograde signaling, 88
Chloroplasts changes, 667–669
biogenesis, stages, 685
photosynthetic membranes, desiccation/
rehydration cycles, 669
pigments, desiccation/rehydration cycles, 669
poikilochlorophyllous plants, 674
proteins, desiccation/rehydration cycles, 669
ultrastructural changes
desiccation/rehydration cycles, 667–669,
667f–668f
drought tolerance, 685
heat stress, 697

Chloroplast sensor kinase (CSK), 549
Chloroplast(s) stroma, 279
Chlorosarcinopsis, 675
Chromosome I/II (CI/CII), 328
Chromosome segment substitution lines
(CSSLs), 752
Chronomutant, 705
Chrysanthemum, 765t
Cicer arietinum, 541
Cichorium intybus, 772t
Circadian regulation, 57
Circular dichroism (CD) spectra, *see*
Chlorophylls/bacteriochlorophylls,
physicochemical properties of
Citrus, 771t
Citrus tangerina, 657
Citrus unshiu Marc., 771t
Climate changes
phenotypic plasticity and, 499–500, 500f
plant–plant interaction and, 498–499
Clock ticks of binary photons, 37
CLO-PLA database, 488
Closed loop, soilless systems, 725
Clove Basil, 769t
C_3 metabolism, 670
Coccomyxa, 675
Coffea canephora, 655
Colechaete, 675
Coleochaetaceae, 675
Coleochaete, 675
Coleus blumei, 769t
Coleus flower, 769t
Colimitation of reduced K/N inputs on
strawberry fruit productivity/mineral
nutrition quality, 295, 296f
Community ecology, 488
Competition, photosynthetic, *see* Photosynthetic
competition
Complementary chromatic adaptation (CCA), 56
Compton, Arthur, 22
Compton effect, 20, 34
Concentration
CO_2, Rubisco and, 745–747, 746f
Rubisco, 743–744, 743f
Conospermoid, 311
Constant J method, 67
CONSTITUTIVE PHOTOMORPHOGENIC1
(COP1), 179
Control by epistasy of synthesis (CES), 7
Controlled environment (CE), 276
Controlled leaf temperature effects, 714–715, 715f
Controlled movement of extrinsic domain of ISP,
334–336
Coproporphyrinogen III oxidase (CPOX), 176
Core complex antenna family, 355, 372
Corn–ryegrass–ryegrass–strawberry (CRRS)
system, 296
Coscinodiscophyceae, 775t
Cosmarium, 672
Cotton, 259–270, 279, 541, 542, 579, 616, 639,
699, 772t
Cottonwood (*Populus deltoides*), 616
Coulomb's law, 30n
Covalent binding, 210–211
Cover crops, 289, 657
CP43, *see* IsiA
CP complexes, *see* Chlorophyll-protein complexes
C_3 plants
vs. C_4 plants, genetic factors, 542–543
C_3–C_4 hybridization, 543
heritability, 542

heterosis, 542
inheritance, 543
C_4 plants, 670
breeding for, 553–554
vs. C_3 plants, genetic factors, 542–543
C_3–C_4 hybridization, 543
heritability, 542
heterosis, 542
inheritance, 543
Crassulacean acid-like metabolism (CAM) in
Portulaca genus, transition from C_4
photosynthesis to
about, 314–315
Portulaca grandiflora, case of, 318–319, 318f
Portulaca oleracea, case of, 315–318, 315f,
317f
Crassulacean acid metabolism (CAM), 307, 670,
687, 747
Craterostigma plantagineum, 687
Craterostigma wilmsii, 672, 684, 685, 687, 688
Crop canopy
architecture, plant growth and, 579–580
genetic factors, 541
Crop macronutrient stress, 289
Crop photosynthesis/transpiration, 288
Crop productivity
forage crops, photosynthesis efficiency and,
469–471; *see also* Forage crops
Crop rotation, 289
Crop yield
effect of solar UV exclusion on, 268–269,
269f
improving, *see* Biotechnology
Crystallization of Cyt bc_1, 330
Crystallographic structures of thylakoid
complexes, 89
Crystal structures of bacterial bc_1 from
R. sphaeroides
Cyt b subunit, structure of, 330–331, 331f,
332f
Cyt c_1 subunit, structure of, 331, 333
iron–sulfur protein subunit, structure of, 333
Cucumber, 770t–771t
Cucumis sativus, 770t–771t
Cucurbita moschata, 541
Cyamopsis plants, 269
Cyanidioschyzon merolae, 361
Cyanobacteria, 775t; *see also* Cytochrome
c_6-like proteins
about, 229
alternative electron transport pathways in,
375–377, 376f
photosynthesis/respiration in, 230–232
Cyanobacteria, NPQ in; *see also*
Nonphotochemical quenching (NPQ)
high light-inducible proteins (HLIP)/iron
starvation inducible (IsiA) proteins,
372
orange carotenoid protein, 371–372
Cyanobacteria CCM; *see also* Light (excess)/
carbon (limited)
about, 382–383
carboxysomes, 385
inorganic carbon uptake, 383–385, 384f
Cyanobacterial CYT c_6-like proteins, 232–236,
233f–236f, 236t
Cyanobacterial PSI, 354–356
Cyanobacterial PSII
oxygen-evolving complex (OEC), 352
phycobilisome (PBS), 352–353
PSII core complex, 350–352

Cyanobacteria/microalgae, photosynthetic
    apparatus in
  Acaryochloris marina, 359
  green oxyphotobacteria
    prochlorophyte Chl *a*/*b*-binding proteins,
        358–359
  Halomicronema hongdechloris, 359
  microalgal photosystems
    about, 359–360
    chlorophyll *c*-containing microalgae,
        photosystems of, 362–364
    green microalgae, photosystems of,
        361–362
    red microalgae, photosystems of, 360–361
  overview, 349–350
  photosynthetic apparatus in cyanobacteria
    about, 350, 351f
    cyanobacterial PSI, 354–356
    cyanobacterial PSII, 350–353
    cytochrome b_6f (Cyt b_6f), 353–354, 353f
    linear and cycling electron flow, 356–358,
        357f
Cyanobacterium *Nostoc* sp. PCC 7119, 233
Cyclamen, 766t
Cyclic electron flow (CEF), 352, 357
  heat stress and, 698
Cyclic voltammograms of Chls *a*, *b*, and *d*, 138f
Cyclotella caspia, 774
Cylindrocystis sp., 672, 675
CYT c_6-like proteins, *see* Cytochrome c_6-like
    proteins
Cytochrome, 203f, 223
  and ATP synthase, 55
Cytochrome *b*559, 246
Cytochrome bc_1 complex from photosynthetic
    purple bacteria
  genetic manipulation of, 328–329, 329f
  mechanism of bc_1 function in light of
        structural information
    controlled movement of extrinsic domain
        of ISP, 334–336
    experimental verification of SAMICS, 336
    Q-cycle mechanism/bifurcated ET at
        quinol oxidation site, 334, 335f
    ubisemiquinone free radical as reaction
        intermediate at quinol oxidation site,
        336–338, 337f
  overview, 327–328
  in photosynthetic bacteria growth, 328
  reactions catalyzed by, 328
  regulation of bc_1 activity in photosynthetic
        purple bacteria
    inhibition of, by zinc ions, 338
    interaction with cytosolic proteins,
        339–340
    oxygen on electron transfer activity of
        bc_1, 338–339, 340t
  structural analysis of
    about, 329–330
    crystal structures of bacterial bc_1 from
        R. sphaeroides, 330–333
    purification/crystallization/subunit
        composition of Cyt bc_1 from
        R. sphaeroides, 330
    subunit IV of *Rsbc*₁, role of, 333–334
  superoxide generation
    mutations suppressing/enhancing,
        340–341, 341t
    oxygen in mediating, 342–343
    reduced b_L heme as source of, 342
    ubisemiquinol as source for, 341–342, 342f

Cytochrome b_6f (Cyt b_6f), 353–354, 353f
Cytochrome b_6f complex, 81
Cytochrome b₆/f complex, 548, 747
Cytochrome *b* subunit, structure of, 330–331,
        331f, 332f
Cytochrome c_6-like proteins
  cyanobacterial, 232–236, 233f–236f, 236t
  green algae/higher plant CYT c_6-like
        proteins, 232
  overview, 229
  photosynthesis/respiration in cyanobacteria
    about, 230, 230s
    cytochrome c_M, 232
    and plastocyanin, 230–232
Cytochrome *c* oxidase (COX), 230, 236, 237s,
        350, 351f
Cytochrome c_1 subunit, structure of, 331, 333
Cytokinins, ornamental plants, 657
Cytosol
  and chloroplast, metabolites exchange
        between, 514–515
Cytosolic proteins, interaction of bc_1 complexes
        with, 339–340
Cytosolic Pyr Kases (cPyr Kases), 519

D

Daily growth rate, 572
Dark desiccation, 684
Dark-operative (light-independent)
        protochlorophyllide (Pchlide)
        reductase (DPOR), 174
Darwin, Charles, 487
Databases
  functional traits, 488
Daucus muricatus, 547
Daughter-plant phosphorus (DPP), 296–300,
        298t
De-epoxidation, 684
De-etiolation, 4
Deflection of starlight, 36f
Dehydroascorbate (DHA), 160f, 163, 374f
Dehydroascorbic acid reductase (DHAR), 166
Dendranthema grandiflorum, 765t
De novo gene transcription, 686
De novo synthesis, 673–674
Density functional theory (DFT), 114
Deoxyribonucleic nucleic acid (DNA), 288
Desiccation-induced NPQ (NPQ$_{DT}$), 673
Desiccation-induced proteins, 690
Desiccation/rehydration cycles, 665
  chloroplasts changes, 667–669
Desiccation-sensitive (DS) tissues, 664
Desiccation tolerance; *see also* Drought
        tolerance
  carbon balance, 669–672, 670f, 671t
  chloroplast changes, 667–669, 668f
  definition, 664–665
  ecological and evolutionary significance,
        675–676
  glassy state, 665–666
  in green algae, 675
  mechanisms, 665, 666f
  organisms, *see* Desiccation-tolerant (DT)
        organisms
  photoprotection, 672–673
  poikilochlorophyllous plants, 673–675
  poikilochlorophyllous vs.
        homoiochlorophyllous species, 667
  in terrestrial plants, 675–676
  vs. drought tolerance, 664

Desiccation-tolerant (DT) organisms
  defined, 663–664
  frequency, 664f
  photoprotection in, 672–673
Desiccoplasts, 674, 685
Desmococcus, 675
Destacking of thylakoids, 245
Destructive analysis, of biomass gain, 572–573
Development, of ornamental plants; *see also*
        Ornamental plants
  drought stress on, 654–655
  salinity stress on, 652–653
Developmental acclimation, 540
DGD1 (digalactosyldiacylglycerol synthase 1),
        686
D-glycero-2,3-diulose-1,5-bisphosphate (PDBP),
        546
Diadinoxanthin cycle, 374, 374f
Diadinoxanthin de-epoxidase (DDE), 374
Dianella revolute, 656
Diatoms, 773, 775t
  CCM in, 387–389
  FCP complex in, 363
Diatoxanthin epoxidase (DEP), 374
Dicer-like proteins (DCLs), 541
Diethylether
  absorption properties of chlorophylls in,
        103t–107t
  fluorescence properties of natural
        chlorophylls in, 112t–113t
Digalactosyldiacylglycerol (DGDG), 7, 224,
        245
Digalactosyldiacylglycerol synthase 1 (DGD1),
        686
Dihydroxyacetone-P (DHAP), 510
2,5-dihydroxybenzoic acid (DHB), 119, 123f
Dills, 769t
Dimer, 205
Dinoflagellates
  acpPC/PCP complexes in, 363–364
  CCM in, 389
Direct immobilization by laser printing, 210
Disulfide protein (PSSP), 160f
Dithiothreitol (DTT), 161
Diurnal temperature effects, 709f, 710–714,
        712f, 713t, 714f, 714t
Diversity of photosynthetic apparatus
        organization, 55
Divinyl chlorophyll (DV-Chl) *a*
  chlorophyll (Chl) *a*, *b*, *d*, *f* and, *see*
        Chlorophyll (Chl) *a*, *b*, *d*, *f* and
        DV-Chl *a*
Divinyl chlorophyll (DV-Chl) *a'*
  chlorophyll (Chl) *a'*, *b'*, *d'*, *f'* and/
        pheophytins, 114, 117f, 118f
  chlorophyll (Chl) *a' d'* and, 96–97, 100f
Divinyl chlorophyll (DV-Chl) *a*, *a'*, *b* and divinyl
        pheophytin *a*, 101–102
3,8-divinyl-chlorophyllide (DV-Chlide), 177
3,8-divinyl-protochlorophyllide (DV-Pchlide),
        177
Divinyl (DV)-reductase (DVR), 177
N-dodecyl-β-maltoside (DM), 194, 195f
N-dodecyl ß-d-maltoside (ß-DDM), 330
Doppler effect, 20, 37, 38, 42
Double quantum filtered correlation
        spectroscopy (DQF-COSY), 129,
        135f
Double-stranded (dsRNA), 541
D-1 protein, 705, 706
Drip irrigation, 657, 725

Drought
  avoidance, 664
  effects on photosynthesis, 635–638
    carbohydrates and related enzymes
        levels, 638
    carboxylation under, 637–638
    chlorophyll fluorescence and
        photochemical reactions, 636–637
    leaf area and stomatal conductance, 635
    ultrastructural changes in chloroplasts,
        636
  escape, defined, 655
  on leaf structure, 316
  stomatal closure and, 742
Drought stress
  Balanites aegyptiaca (case study), 457
  Euphorbia tirucalli (case study), 456–457
  Lablab purpureus (case study), 456
  several species (case study), 457
Drought stress, on ornamental plants, 654–655
  coping with, 656–658
  overview, 652
  photosynthesis, 654–655
  plant growth and development, 654–655
  tolerance
    mechanisms, 655
    overview, 655–656
Drought tolerance; *see also* Desiccation
        tolerance
  chloroplasts, ultrastructural changes, 685
  as dimension of plant ecological strategy, 492
  on HDT plants, photosynthetic activity,
        686–689
  high light/temperature, 691
  homoichlorophylly mechanism, 684–685
  leaf surface and, 689
  on PDT plants, photosynthetic activity,
        685–686
  on photosynthesis, 685–689
  poikilochlorophylly mechanism, 684–685
  vs. desiccation tolerance, 664
Dry weight (DW), 665
DS, *see* Desiccation-sensitive (DS) tissues
DT, *see* Desiccation-tolerant (DT) organisms
Dual function of Rubisco, 381–382
Dual-PAM fluorometer, 10
Dutch-type greenhouse, 724, 725f
DV-Chl, *see* Divinyl chlorophyll (DV-Chl)
DW, *see* Dry weight (DW)
Dynamic acclimations, 540

E

Early fruit formation, 290
Early light-induced proteins (ELIP), 360, 380,
        665, 689
Ecological performance
  defined, 489
Ecological significance, desiccation tolerance,
        675–676
Ecological strategy(ies), 490–493, 491f
  defined, 490
  drought tolerance, 492
  leaf economics spectrum, 491–492
  other dimensions, 492–493
  plant height, 492
  plant water stress, 492
Ecological systems
  organizational hierarchy in, 489–490
Economy
  forages and grassland agriculture and, 465

Ecosystem ecology, 488
Egeria densa, case of, 308–310, 309f
Electricity to optical power, converting, 761,
        762–763
Electroluminescence phenomenon, 761, 762–763
Electromagnetic wave theory, 18
Electromagnetism, laws of, 26f
Electron acceptor, 150, 209f
Electron donor, 209f
Electron flow, linear and cycling, 356–358, 357f
Electron paramagnetic resonance (EPR), 247,
        338
  spin trapping method, 224
Electron sinks
  thylakoid membrane, 451
Electron transfer (ET)
  absence of oxygen, 340t
  cascade, 204
  flux in photosynthesis/respiration, 237f
  hybrid interfaces for, *see* Nano-biohybrid (or
        biohybrid) photoconverter
  inverse relationship with superoxide
        generation activity of $Rsbc_1$, 341f
  in mitochondria, 328
  oxygen-dependent, 339f
  and superoxide production, 341t
Electron transport, 265f
Electron transport, chloroplast
  in thylakoid membranes, 747–748, 748f
Electron transport pathways, regulation of, 60
Electron valves, 375
Electrophoresis, bidimensional, 151f
Electrophysiological studies in *E. densa*, 308
Electrostatic absorption, 210
ELIP genes, 669
ELIPs, *see* Early light-induced proteins (ELIP)
Endogenous circadian rhythm, transcriptional
        regulation by, 182–183
Endosymbiotic theory, 224
Energizer for plant health, 288
Energy, 21
  dissipation, 265f, 672–673
Energy transfer, 206–207
  hybrid interfaces for, *see* Nano-biohybrid (or
        biohybrid) photoconverter
Enteromorpha spp., 653
Environmental abiotic stress factor, limited
        primary macronutrient supply as,
        289–290
Environmental conditions/factors
  adaptive plasticity of plants and, 494–495
  adverse conditions, 448
  phosphate deficiency and, 616–617
  plants adaptation to, plasticity and, 493–494
Environmental factor, agriculture
  CO_2, 732, 732f
  humidity, 731–732
  light, 732–737, 733f, 735f
  temperature, 729–731, 729f, 730f
Environmental temperature, 248
Enzyme activities, regulation of, 184
Enzymes; *see also specific entries*
  Calvin–Benson cycle, activity of, 748–750,
        749f
  drought conditions and levels of, 638
Epimerization, 98
Equipments, plant factories, 726
Eragrostis, 667
Eragrostis spp., 683
Erect leaves, 540–541
Eruca sativa, 769t

Escape, water availability and, 663
Escholtzxanthin, 595
Ethoxzolamide, 197
Ethylene
  plant photosynthesis and, 578–579
Etioplast model, 8, 9
Eucalyptus, 768t
Eucalyptus citriodora, 768t
Eucalyptus spp., 656
Euphorbia tirucalli
  drought stress (case study), 456–457
Eustigmatophyceae, 775t
Evolution
  agriculture, 723–737
  desiccation tolerance, 675–676
  s-triazine-resistant plants, 719–720
Evolution, of plants
  importance of photosynthetic competition
        for, 483–484
  phenotypic plasticity and, 495
Excess-energy hypothesis, 249
Excess light, dangerous effects of, 370
Excitation energy, 208f
Excitation energy trapping (TRo), 265
Expressed sequence tag (EST) sequencing,
        232
Extranuclear control chloroplasts, 547–551
  ATP synthase complex, genes for, 549
  chloroplast gene expression, 549–551
  cytochrome b_6/f Complex, 548
  genes for photosynthesis-related proteins,
        548–549
  inheritance, 547–548
  other genes, 549
  plastid genome organization, 548
  post-transcriptional regulation, 550
  PS I, genes for, 548–549
  translational regulation, 550–551
Extrinsic CA, 194

F

Facultative photosynthetic purple bacteria, 338
Fagopyrum tartaricum, 771t
Fan flower, 654
Fast atom bombardment (FAB)-mass spectra,
        117–119, 121f
Fast–slow plant economics spectrum, 491
Fatty acid (FA) biosynthesis, 514
Fecundity, 493
Feedback-limited photosynthesis, 706, 718–719
Ferredoxin, 160f
Ferredoxin–NADP oxidoreductase, 160f
Ferredoxin-NADP reductase (FNR), 351f
Ferredoxin-thioredoxin oxide-reductase (FTR),
        160f, 185
Ferrochelatase (FeCh), 177
Festuca gautieri, 498
Festuca lenensis, 498
Feynman diagram, 33f
Fitness, functional traits and, 488
Flaveria, 744
Flavin mononucleotide (FMN), 375
Flavodiiron (Flv), 60
Flavodiiron proteins (FDP), 375
Flavodoxin (Fld), 230
Flowering phenology, 493
Fluorescence decay, 215f
Fluorescence lifetime of (B)Chls/(B)Phes,
        111–114, 112t–113t, 115f
Fluorescence quantum yield, 111, 116f

Fluorescence recovery after photobleaching (FRAP), 58
Fluorescence recovery protein (FRP), 371f
Fluorescence spectra, *see* Chlorophylls/ bacteriochlorophylls, physicochemical properties of
Fluorescence wavelength maxima, 110, 111f
Fluorescent chlorophyll catabolites (FCC), 84
Fluorescent lamps, 736
Fluorescent lighting, plant factories, 728
Fluorescent (FLU) protein, 178
Fog-cooling systems, 729, 729f, 730
Forage crops, 465–471
    defined, 465
    factors, 468–469
    forage mixtures, 471
    grazing management, 469–470
    haying management, 469–470
    irrigation management, 470–471
    management, for photosynthesis efficiency and crop productivity improvement, 469–471
    moisture management, 470–471
    nitrogen management, 471
    nutrient management, 471
    overview, 465
    photosynthetic pathways and physiology in, 466, 466t
    photosynthetic pigments role in productivity and quality of, 466–468, 467t
    role in economy, 465
Forage mixtures, 471
Forster (or fluorescence) resonance energy transfer (FRET), 207, 213, 216
Fragaria × ananassa, 766t
Fraxinus pennsylvanica, 653
Free radicals, 88–89
Fucoxanthin–chlorophyll protein (FCP) complex in diatoms, 363
Fully desiccation-tolerant plants, 683
Functional change, s-triazine-resistant plants, 707–709
Functional changes in photosynthetic apparatus; *see also* Light-harvesting chlorophyll protein complexes of PSII (LHCII) for apparatus sensitivity
    under high light intensity, 249–250, 251f
    under high temperature
        LHCII organization for temperature sensitivity of photosynthetic apparatus, 247–248
        LHCII–PSII supercomplex, changes in, 246–247
        lipid composition on temperature sensitivity, 247
        thylakoid membranes, structural changes in, 245–246, 251f
    under low temperature
        functional changes, 248–249
        structural changes, 248
Functional ecology, 493
    history of, 487–488
    research aims, 488
Functional effect traits, 489
Functional response traits, 489
Functional traits, 487–493; *see also* Plasticity
    assessment approaches, 488–489
    BiolFlor database, 488
    CLO-PLA database, 488
    databases, 488
    defined, 488

fitness, 488
    hard traits, 489
    hierarchy, 488–489
    history of, 487–488
    intraspecific trait variability, 489–490, 490f
    LEDA database, 488
    morphology, 488, 489
    overview, 487
    performance, 488, 489
    plant ecological strategies, 490–493, 491f
        drought tolerance, 492
        leaf economics spectrum, 491–492
        other dimensions, 492–493
        plant height, 492
        plant water stress, 492
    research aims, 488
    soft traits, 489
    TRY database, 488
Futile cycles
    thylakoid membrane, 451–452

G

Gaillardia aristata, 656
Garden roses, 652–653, 771t
Gas exchange
    growth quantification using, *see* Growth analysis
    measurements, 67, 452
        with PAM fluorimetry, of stressed plants, 454–455
    parameters, 264–265, 265f, 266f
    poikilochlorophyllous plants, 674–675
GATA transcription factors, 181
Gazania splendens, 768t
Gene-by-environment interactions, 495
Gene expression
    changes during chloroplast biogenesis, 7–8
    during chloroplast development, 79
General Electric Company, 762
Genes
    inner reality of, 493
Genetic factors, 539–554
    C$_3$ vs. C$_4$ plants, 542–543
        C$_3$–C$_4$ hybridization, 543
        heritability, 542
        heterosis, 542
        inheritance, 543
    extranuclear control chloroplasts, 547–551
        ATP synthase complex, genes for, 549
        chloroplast gene expression, 549–551
        cytochrome b$_6$/f complex, 548
        genes for photosynthesis-related proteins, 548–549
        inheritance, 547–548
        other genes, 549
        plastid genome organization, 548
        post-transcriptional regulation, 550
        PS I, genes for, 548–549
        translational regulation, 550–551
    morphophysiological factors
        awns, 542
        chlorophyll content, 541–542
        crop canopy and leaf area index, 541
        erect leaves, 540–541
    nuclear genes of photosynthetic apparatus, 543–547
        early research on, 544
        light-regulated nuclear genes, 546–547
        ribulose bisphosphate carboxylase/ oxygenase, regulation of, 545–546

    rubisco activase, 546
        transcription, 544–545
        translation, 544–545
    overview, 539–540
    photosynthetic strategy in crops, 552–554
        breeding for C$_4$ mechanism, 553–554
        changes in biochemical activities, 554
        future prospects, 554
        manipulation of photosynthetic area, 553
        rubisco and crop improvement, 554
    plant stress responses
        abiotic stress, 551
        biotic stress, 551–552
        growth-defense paradigm, 552
Geranylgeranyl pyrophosphate (GGPP), 177
Geranylgeranyl reductase (GGR), 177
Gerontoplast, 78
Gesneriaceae plants, 684
Gibberellins, ornamental plants, 657
Ginseng, 765t
Glandularia × hybrida, 656
Glass, defined, 665
Glassy state, 665–666
Glc-6P phosphate translocator (GPT), 514
4-α-glucanotransferase, 512
Glucitol
    in heterotrophic tissues, 526–527
    transporters, 523–524
Glucose-1-phosphate, 607
Glucose 6-phosphate (Glc6P), 160f
Glu 1-semialdehyde (GSA), 174
Glutaredoxin (GRX), 160f, 163, 165
Glutathione (GSH), 163, 665, 673
Glutathione-dependent system, 164–165
Glutathione peroxidase (GPX), 160f
Glutathione reductase (GR), 160f, 163, 164
Glutathione S-transferase (GST), 160f, 166
Glutathionyl protein (PSSG), 160f
GluTR-binding protein (GluBP), 179
Glyceraldehyde-3P (Ga3P), 510
Glyceraldehyde 3-phosphate (Ga3P), 160f
Glyceraldehyde 3-phosphate dehydrogenase (GAPN), 160f, 162
Glycine betaine, 657
Glycoglycerolipids, 7
Golden2-like (GLK) transcription factors, 180–181
Gol synthesis, 518–519
Gossypium hirsutum, 772t
Gracilaria tikvahiae, 774, 775t, 776t
Grana formation, regulation of, 225; *see also* Quality control of photosystem II
Grana thylakoids, 83f
Grape, 290, 766t
Graphene-based electrodes, 211
Grasslands
    role in economy, 465
Gravitational force, 24, 25
Grazing management
    forage crops, 469–470
Green algae
    desiccation tolerance in, 675
    and higher plant CYT c$_6$-like proteins, 232
Green ash, 653
Greenhouse control systems, 574–578
    CO$_2$ concentration, 576–577, 576f
    light and, 575–576, 575f
    temperature and, 577–578, 578t
Greenhouses
    cooling, 729–731
    Dutch-type greenhouse, 724, 725f

heating, 731
   shade curtains, 729
   ventilation, 729
Greening process, 4
Green LEDs, 736; *see also* Light-emitting diodes
   (LEDs)
Green microalgae
   photosystems of, *see* Microalgal
      photosystems
Green microalgae, CCM in; *see also* Carbon
      dioxide (CO_2) concentrating
      mechanisms (CCM)
   about, 385–386
   Ci transmembrane transport, 386–387
   pyrenoids, 387
Green oxyphotobacteria, 358–359; *see*
      also Cyanobacteria/microalgae,
      photosynthetic apparatus in
Grevilea spp., 653
Growth
   assessment of, 489
Growth, of ornamental plants; *see also*
      Ornamental plants
   drought stress on, 654–655
   salinity stress on, 652–653
Growth analysis, 571–581
   agricultural and ecological case studies,
      574–581
      Arabidopsis (case study), 580–581, 581f
      ethylene and plant photosynthesis,
         578–579
      greenhouse control systems, optimizing,
         574–578
      leaf morphology and canopy architecture,
         579–580
   daily growth rate, 572
   destructive analysis of biomass gain,
      572–573
   nondestructive analysis, 573–574
      chamber and system design, 573
      net carbon exchange rate (NCER),
         573–574, 574f
      net carbon gain (NCG), 573–574, 574f
   overview, 571–572
   plant productivity and, 572
   relative growth rate (RGR), 572
Growth conditions, 448
Growth-defense paradigm
   genetic factors, 552
Growth rate and productivity, 571; *see also*
      Growth analysis
Growth regulators, 578

H

Haberlea, 672
Haberlea rhodopensis, 683, 684, 685, 687, 690,
   691
Haemtococcus pluvialis, 774, 775t
Halomicronema hongdechloris, 359; *see*
      also Cyanobacteria/microalgae,
      photosynthetic apparatus in
Haptophyta, 389
Hard traits, 489, 492
Haslea ostearia, 774, 775t
Haying management
   forage crops, 469–470
[2H]-bacteriochlorophyll (BChl) *a*, redox
      potentials of, 139, 142
[2H]-bacteriochlorophyll (BChl) *a* (BPhe *a*),
      NMR spectra of, 134

HDT, *see* Homoiochlorophyllous desiccation-
      tolerant (HDT) plants
Heat curtain, 731
Heat pumps/exchangers, 731
Heat shock proteins (HSP17.6 and HSP17.9),
   691
Heat stress
   damage effect of, 247
   effects on photosynthesis, 638–639
Heat stress, photosynthesis under
   cyclic/noncyclic electron flow, 698
   leaf senescence, 697, 698f
   photorespiration, 699–700
   photosynthetic apparatus, 697
   photosynthetic enzymes, 699, 699t
   photosynthetic pigments, 697, 698f, 698t
   Rubisco activation, 699
   stomatal oscillations, 698–699
   thylakoid membrane and, 697
Helenium, 653
Helenium amarum, 653
Helianthus annuus, 657
Helichrysum petiolatum, 653
Helicoverpa armigera, 736
Heliobacteria, 96
Heme biosynthetic pathway in angiosperms, 176t
Heme-copper-type terminal respiratory
      oxidases, 236
Hemeprotein, 231
Hemidiscoidal model of PBS, 353
Hemoprotein, 166
Herbicides, 578
   resistance to, 707
Heritability
   C_3 *vs.* C_4 plants, 542
Heterodimer, 206
Heteronuclear multiple-bond correlation
      (HMBC), 124, 126, 132f
Heteronuclear single-quantum coherence
      (HSQC), 124, 131f
Heterosis
   C_3 *vs.* C_4 plants, 542
Heterotrophic cells, metabolism in
   carbon flow in, 510, 511f–512f
   lipid storage metabolism, 528–529
   starch storage metabolism, 527–528
   sucrose and sugar-alcohols metabolism,
      524–527
Hexokinase (HXK), 86f, 87
Hexose-P, 514
   in BCC, 510
Hibiscus rosa-sinensis, 657
HID, *see* High-intensity discharge (HID) lamps
Higher leaf C_i, 291–293, 292f, 292t
   inhibiting leaf/fruit NO_3^- retention, 294–295
Higher plant CYT c_6-like proteins, green algae
      and, 232; *see also* Cytochrome c_6-like
      proteins
Higher plants, nonubiquitous carotenoids in
   high-light carotenoids, 594–595
      escholtzxanthin, 595
      red *retro*-carotenoids, 595
      rhodoxanthin, 595
      xanthophyll esters, 594–595
   HPLC determination of, 596–597, 596f
   low-light carotenoids, 591–595
      α-carotene, 591–593, 592f
      lactucaxanthin, 594
      lutein epoxide, 593–594
      trans-neoxanthin, 591
   pigment composition of, 589–591, 590f, 591t

Highest occupied molecular orbital (HOMO),
   139, 142
High-intensity discharge (HID) lamps, 736
High-light carotenoids, 594–595
   escholtzxanthin, 595
   red *retro*-carotenoids, 595
   rhodoxanthin, 595
   xanthophyll esters, 594–595
High light-inducible proteins (HLIP), 360, 372
High-light protection, 669
Highly ordered pyrolytic graphite (HOPG), 211
High-performance liquid chromatography
      (HPLC), *see* Chlorophylls/
      bacteriochlorophylls,
      physicochemical properties of
   nonubiquitous carotenoids, 596–597, 596f
High-pressure sodium (HPS) lamp, 736
High-SLA species, 492
Hippeastrum hybridum, 653
Histone acetyltransferase (HAT), 180
HMBC, *see* Heteronuclear multiple-bond
      correlation (HMBC)
Homoiochlorophyllous desiccation-tolerant
      (HDT) plants
   chloroplast changes, drought tolerance, 685
   defined, 684
   photosynthetic activity, drought on, 686–689
      net photosynthesis, 686–687
      photochemical activity, 687–688
      photosynthesis-related proteins, 688–689
Homoiochlorophyllous species
   defined, 667
   vs. poikilochlorophyllous, 667
Hooker's evening primrose, 653
Hordeum vulgare, 771t
Hot-water heating, 731
HPLC, *see* High-performance liquid
      chromatography (HPLC)
HPS, *see* High-pressure sodium (HPS) lamp
HSQC, *see* Heteronuclear single-quantum
      coherence (HSQC)
Humidity, 731–732
   relative humidity, 664, 730, 731
   vapor pressure deficit, 730, 731
Hybrid interfaces for electron/energy transfer
   artificial photosynthesis
      basic architecture, 207–208, 208f, 209f
      photoconversion efficiency *vs.* structural
         complexity, 208, 209f
   nano-biohybrid (or biohybrid) photoconverter
      about, 208–210, 209f
      hybrid interfaces for electron transfer,
         210–212
      hybrid interfaces for energy transfer,
         212–216
   overview, 201–202
   photosynthesis
      light-harvesting complexes, 205–207,
         206f, 207f
      photochemical energy conversion, 202, 202f
      Q-type RC, 202–205, 204f
      schematic representation of, 202f
4-hydoxy-α-cyanocinnamic acid (CHCA), 119,
   123f
Hydrilla verticillata, case of, 310–311
Hydrogen peroxide, 160f, 279
Hydroxyl radicals, 224
Hydroxymethylbilane (HMB), 174
7-hydroxymethyl Chl *a* reductase (HCAR), 177
3-hydroxy-2-picolinic acid (HPA), 119, 123f
Hypodermal cells, 313

I

IB Rubisco, 382, 384f, 387
Immunodetection of proteins, 12, 12t; *see also* Chloroplast biogenesis
Immunogold labeling, 319
  of FtsH proteases, 227f
Impatiens parviflora DC, 490
  intraspecific variability in SLA in, 490f
Impatiens wallerana Hook., 656
Impulse moment, 27
Incident photon flux density, 67
Indica rice, 753
Inheritance
  chloroplast DNA, 547–548
  C_3 *vs.* C_4 plants, 543
Inhibition of bacterial bc_1, by zinc ions, 338
Inner reality, of genes, 493
Innovations, agriculture, 729–737
Inorganic carbon (Ci) transmembrane transport, 386–387; *see also* Carbon dioxide (CO_2) concentrating mechanisms (CCM)
Inorganic carbon uptake, 383–385, 384f
Inorganic phosphate (Pi), 603
  long-term *in vivo* effects of deprivation, 608–615
    carbon metabolism, 611–613, 612t, 613t
    carbon partitioning and export, 614–615, 614t
    intracellular Pi compartmentation, 613–614, 614t
    photosynthetic machinery, 610–611, 610t
    plant growth response and phosphate concentration, 609–610, 609t
  photosynthetic carbon metabolism, 603, 604f
  short-term *in vitro* effects of deprivation, 605–608
    phosphate translocators, 605–606
    photosynthesis regulation, 606–607
    starch biosynthesis, 607
    sucrose biosynthesis, 607–608
Inorganic pyrophosphate (PPi), 510
In situ immunolocalization studies, 316
Integrated pest management (IPM), 726
Intercellular air space (IAS), 745, 746
Intercellular photosynthate partitioning, 509–510, 510t
Interfilum, 675
Intracellular photosynthate partitioning, 509–510, 510t
Intracellular Pi compartmentation, 613–614, 614t
Intramolecular proton transfer, 96
Intraspecific trait variability, 489–490, 490f
  between-individual variability, 489–490
  constraints, 489
  population-level variability, 489, 490
  in SLA in *Impatiens parviflora* DC, 490f
  within-individual variability, 489
Invasive trees
  and native trees, competitive interactions between, 480–483, 481f, 482f
Investment, plant factories, 726, 728
Ion compartmentalization, 653
IPM, *see* Integrated pest management (IPM)
Ipomoea tricolor, 743
Iris hexagona, 653
Iron starvation inducible (IsiA) complex, 355–356, 372
Iron-sulfur cluster (FeS-type RC), 203f
Iron-sulfur protein (ISP)
  controlled movement of extrinsic domain of, 334–336
  subunit, structure of, 333
Irrigation management
  forage crops, 470–471
Irrigation system, for ornamental plants, 657
IsiA (iron starvation inducible) complex, 355–356
Isomerization, 96
Isopentenyl pyrophosphate (IPP), 177
Isotope method, 68–69
Isotope ratio mass spectrometric analysis (IRMS), 69
ISP, *see* Iron-sulfur protein (ISP)
ISP-ED
  conformation switch, control mechanism for, 336
  mobility necessary for ET function of bc_1, 336

J

Japonica rice, 753
Jasmonic acid, 657

K

Kale, 765t
Kinetic energy, Einstein's equation for, 19
King protea, 771t
Klebsormidiophyceae, 675
Klebsormidium, 675
Klebsormidium flaccidum, 669
Klebsormidum, 669, 670
Kochia scoparia, 707
Kochioid, 311
Kohlrabi, 771t
Kranz anatomy, 307, 308, 311
Kranz-Suaedoid, 311
Kudzu, 769t

L

Lablab purpureus
  drought stress (case study), 456
Lactuca sativa, 766t–768t
Lactucaxanthin (Lac), 591, 594
  HPLC determination of, 596–597, 596f
LAI, *see* Leaf area index (LAI)
Lantana × hybrida, 656
Lantana montevidensis, 656
Late embryogenesis abundant (LEA) proteins, 665
LCF, *see* Leaf disk chlorophyll fluorescence (LCF)
LCHSR protein (formerly LI818 protein), 373
LDMC (leaf dry matter content), 491
LEA, *see* Late embryogenesis abundant (LEA) proteins
Leaching, 725
Leaf area
  and biomass accumulation, 259–260, 260f
  drought effects on, 635
Leaf area index (LAI), 579, 733, 734, 750–751
  components, 541
  genetic factors, 541
Leaf cell
  metabolites exchange between chloroplast and cytosol, 514–515
  respiratory metabolism in, 519–520
  starch metabolism in, 510–514, 511f–512f, 512t
  sucrose and sugar-alcohols synthesis, 515–519, 515t, 516f, 517t
Leaf Chamber Fluorometer (LI-COR), 67, 71
Leaf disk chlorophyll fluorescence (LCF), 705, 709, 709f
Leaf dry matter content (LDMC), 491
Leaf economics spectrum, as dimension of plant ecological strategy, 490, 491–492
  traits in, 491
Leaf folding, drought and, 689–690
Leaf–height–seed (LHS) plant strategy scheme, 490
Leaf intercellular CO_2 concentration (C_i)
  induced by limited K/N supply, 290
Leaf life span, 491
Leaf mass per area (LMA), 491, 492
Leaf model of phenomenological energy, 265f
Leaf morphology
  plant growth and, 579–580
Leaf nitrogen, 491
Leaf phosphorus, 491
Leaf polarity, 308
Leaf senescence, heat stress and, 697, 698f
Leaf size–twig size spectrum, 490
Leaf temperature (LT), 293, 293f
Leaf thickness measurement, 491
Leaf traits
  plasticity of whole plant and, 496–497
  variation in, 487
Leaf weight ratio (LWR), 260
LEDA database, 488
LEDs, *see* Light-emitting diodes (LEDs)
Lens culinaris, 771t
Lens esculenta, 771t
Lentil, 771t
Lettuce, 766t–768t
Lhcb2 protein, 674
LHCII, *see* Light harvesting complex II (LHCII)
L–H–S scheme, 488
Lichenization, 675
Licorice plant, 653
Life-history traits, 493
Light, 179
  desiccation and, 691
  photosynthesis and, 733–734, 733f
  photosynthetic competition for, 477–479, 478t
  plant factories and, *see* Lighting, plant factories
  quality and source, 735–737
  supplemental lighting, 734–735, 735f
Light, nature of
  binary photon
    about, 23–34, 32t
    properties of, 42–43
    test of, relativity of space and time, 36–42
  quantum mechanical photon/wave–particle duality, 17–23
  uncertainty principle, 34–36
Light absorption, by canopy, 750–751
Light avoidance, 672
Light (excess)/carbon (limited)
  alternative electron transport pathways
    in cyanobacteria, 375–377, 376f
    in microalgae, 377–378
    water-water cycle, 378
  CCM in microalgae
    in diatoms, 387–389
    in dinoflagellates, 389
    in green microalgae, 385–387
  cyanobacteria CCM
    about, 382–383
    carboxysomes, 385
    inorganic carbon uptake, 383–385, 384f

dangerous effects of excess light, 370
light underwater, 369–370
limited carbon availability
    dual function of Rubisco, 381–382
nonphotochemical excess energy dissipation
    in antenna systems
        about, 370–371
        NPQ in cyanobacteria, 371–372, 371f
        NPQ in microalgae, 372–375, 373f, 374f
photoprotection mechanisms in PSII RC, 375
recovery of PSII, 380–381
state transitions, 378–380, 379f
Light effects; see also Photosynthetic apparatus,
        adaptation/regulation of
    on photosynthetic machinery organization
        light-state transitions, 58–60, 59f
        photoadaptation of photosynthetic
            apparatus, 57–58, 57f
    on photosynthetic stoichiometry
        light-dark cycle, 57
        light intensity, 55–56, 55f
        light quality, 56–57, 56f
Light-emitting diodes (LEDs), 728, 736
    green LEDs, 736
    pest control and, 736
    yellow LEDs, 736
Light-emitting diodes (LEDs), for future
        photosynthesis, 761–777
    advantages of, 763
    artificial lighting, future of, 774, 776–777
    characteristic of, 763
    chlorophyll and carotenoid types, 762t
    electricity to optical power, converting, 761,
        762–763
    electroluminescence, 761, 762–763
    experiments about, 764
    higher plants to LED lighting
        effects, 765t–772t
        response of, 772–773
    overview, 761
    productivity of algae, 773–774, 775t, 776t
    technical and physiological advantages, 777
    tool for innovative studies, 763–774, 764f
    unique specification of, 763
Light energy
    absorption, thylakoid membrane and, 448
Light-harvesting chlorophyll protein complexes
        of PSI (LHCI), 223, 362
Light-harvesting chlorophyll protein complexes
        of PSII (LHCII), 223, 362
Light-harvesting chlorophyll protein complexes
        of PSII (LHCII) for apparatus
        sensitivity
    high light intensity, photosynthetic response
        to
        depending on temperature, 250–251
        structural/functional changes in
            photosynthetic apparatus under high
            light intensity, 249–250, 251f
    organization/functions of LHCII-PSII
        supercomplex
        lipids/xanthophylls, role of, 244–245
        overview, 243–244
        PSII core complex and OEC, 244
    temperature stress, photosynthetic response to
        structural/functional changes in
            photosynthetic apparatus under high
            temperature, 245–248, 251f
        structural/functional changes in
            photosynthetic apparatus under low
            temperature, 248–249, 251f

Light-harvesting complexes (LHCs), 7, 10, 328,
        329f, 761, see Photosynthesis
    composition of photosynthetic apparatus, 54
    and IsiA complex, 355–356
Light harvesting complex II (LHCII), 669, 673,
        686
Lighting, plant factories
    with artificial lighting, 725, 725f, 727–729
    position, 734
    with sunlight, 724, 725f, 726–727
    with sunlight and supplemental light, 724–
        725, 725f, 726–727, 734–735, 735f
Light intensity
    greenhouse control systems, 575–576, 575f
    plant factories, 728
Light intensity (high), structural changes in
        photosynthetic apparatus under,
        249–250, 251f
Light-intercepting area, manipulation of, 553
Light phase, 509
Light-regulated nuclear genes, 546–547
Light-responsive regulatory elements (LREs),
        546, 547
Light-screening molecules, 672
Light-state transitions, 58–60, 59f
Light underwater, 369–370
Liliceae, 684
Lilium, 766t
Limited carbon availability, 381–382
Limited K/N macronutrient supply (case study),
        see Small fruit crop responses
Limited primary macronutrient supply as
        environmental abiotic stress factor,
        289–290
Limiting step in photosynthesis at leaf
        temperatures; see also Mesophyll
        conductance estimation
    response of chloroplast electron transport
        rate to CO_2, 71
    response of CO_2 assimilation rate to CO_2,
        69, 70f
Limonium species, 742
Limonium spp., 656
Linear and cycling electron flow, 356–358, 357f
Linear electron flow (LEF), 356
Linear momentum, 21
Linolenic acid, desiccation and, 669
Lipid composition on temperature sensitivity, 247
Lipid peroxidation, 225
Lipids
    changes in, 6–7
    of higher-plant thylakoid membranes, 224
    and membranes, 279
Lipid storage
    metabolism for, 528–529
Liquid culture, 725
Liriomyza trifolii, 736
LMA (leaf mass per area), 491, 492
L-methionine, 160f
L-methionine sulfoxide, 160f
Lobaria pulmonari, 673
Lobaria pulmonaria, 672
Localization of enzymes, in Chl biosynthesis
    complex formation of biosynthetic enzymes,
        178–179
    subcellular localization of biosynthetic
        enzymes, 177–178
Lolium multiflorum, 497
Lolium perenne, 497
LONG HYPOCOTYL5 (HY5) transcription
        factors, 179–180

Long-term in vivo effects, of Pi deprivation,
        608–615
    carbon metabolism, 611–613, 612t, 613t
    carbon partitioning and export, 614–615,
        614t
    intracellular Pi compartmentation, 613–614,
        614t
    photosynthetic machinery, 610–611, 610t
    plant growth response and phosphate
        concentration, 609–610, 609t
Long wavelength Chls (LWC), 362
Lonicera japonica Thunb., 656
Lorentz transformation, 38
Low Chl accumulation A (LCAA), 177
Lowest unoccupied molecular orbital (LUMO),
        139, 142
Low-light carotenoids, 591–594
    α-carotene, 591–593, 592f
    lactucaxanthin, 594
    lutein epoxide, 593–594
    trans-neoxanthin, 591
Low-molecular-weight disulfide, 160f
Low-molecular-weight dithiols, 165
Low-molecular-weight thiol, 160f
Low-SLA species, 491–492
    advantages, 492
Lupinus bavardii, 654
Lupinus havardii, 656
Lupinus texensis, 656
Lutein (L), 589
Lutein epoxide (Lx), 591, 593–594
    HPLC determination of, 596
Lx-L cycle, 593
Lycopersicon esculentum, 770t
Lycopersicon type plants, 547

M

Magnesium (Mg) branch, 176–177
Magnesium Proto ME cyclase (MPCY), 177
Magnesium Proto methyltransferase (MPMT),
        177, 178
Magnetic moment, 32, 32t
Majorana hortensis, 769t
Malate, 519
Malate dehydrogenase (MDH), 339, 357f
Malic acid, 307
Malus domestica (apple), 518
Manganese stabilizing protein (MSP), 352
Mannitol
    in heterotrophic tissues, 526–527
    transporters, 523–524
Mannose-6P (Man-6P) isomerase, 517–518
Man-6P reductase (Man-6P Rase), 517–518
MAP kinases (MAPK)
    photosynthesis under stressful conditions
        and, 643–644
Marigold, 768t
Mass spectrometry (MS), see Chlorophylls/
        bacteriochlorophylls,
        physicochemical properties of
Matrix-assisted laser desorption ionization
        (MALDI)
    coprecipitation method with, 339
Matrix-assisted laser desorption/ionization
        (MALDI)-mass spectra, 119, 122f,
        123f
Matrix-assisted laser desorption/ionization-
        time of flight (MALDI-ToF)
        measurements, 214
Matter, defined, 23

Matthiola incana, 541, 768t
Maxwell's wave equation, 38
[M]-chlorophyll (Chl) *a*, redox potentials, 139, 142
Mealy cup sage, 653, 654
Mechanical stress, DT organisms, 665
Mechanisms, desiccation tolerance, 665, 666f
Medicago sativa, 547
Mehler reaction, 153, 378
Membrane-bound carbonic anhydrases; *see also* Thylakoid carbonic anhydrases
    about, 193–194
    carbonic anhydrase (CA)
        near PSI, 196, 196f
        near PSII, 194–196, 194t, 195f
Membrane composition and structure, mutual relations, 8; *see also* Chloroplast biogenesis
Membrane fluidity, 224
Membrane integrity, desiccation tolerance and, 665
Membrane stability, 248
Mentha longifolia, 768t
Mentha piperita, 768t
Mentha spicata, 768t
Mentha spp., 772
Mesophyll cells, 316, 318
Mesophyll conductance estimation
    CO_2 diffusional pathways, 65–66, 66f
    against environmental response
        limiting step in photosynthesis at leaf temperatures, 69–71, 70f
        short-term variations under variable environmental conditions, 69
    theory/method
        A/C_c curve-fitting method, 66–67
        chlorophyll fluorescence method, 67
        isotope method, 68–69
Mesotaenium, 675
Messenger RNA (mRNA), 232
Metal-chelating substance (MCS), 83
Metal-halide lamps, 736
Metalloproteins, 230
Metal toxicity
    effects on photosynthesis, 642–643
Metamers, 497
Methionine sulfoxide (MetSO), 167
Methionine sulfoxide reductase A-type (MSRA), 160f, 167
Methionine sulfoxide reductase B-type (MSRB), 160f, 167
Methyl jasmonate, 657
MEX1, maltose transporter, 514
Mexican redbud, 656
[Mg]-and [Zn]-bacteriochlorophyll (BChl) *a*, 116, 119f
[Mg]-bacteriochlorophyll (BChl) *a*, NMR spectra of, 134
Mg branch, 176–177
Microalgae, 773–774
    alternative electron transport pathways in, 377–378
Microalgae, NPQ in; *see also* Nonphotochemical quenching (NPQ)
    about, 372–373
    xanthophyll cycles, 373–375
Microalgal photosystems; *see also* Cyanobacteria/microalgae, photosynthetic apparatus in
    about, 359–360
    chlorophyll *c*-containing microalgae, photosystems of

about, 362–363
acpPC/PCP complexes in dinoflagellates, 363–364
FCP complex in diatoms, 363
green microalgae, photosystems of
    about, 361–362
    LHCI, 362
    LHCII, 362
red microalgae, photosystems of
    photosystem I, 361
    photosystem II, 360
    phycobilisomes, 360–361
MicroRNAs, 541
Microscopic analysis of chloroplast biogenesis stages, 8–10, 9f
Mint, 768t
Mirabilis jalapa, 547
MiR319-TCP TF network, 541
Mitochondria, 88
Mitochondrial complex I, 150
Mitochondrial heme biosynthesis, 178
M-nitrobenzyl alcohol, 121f
Mobile electron carriers, 350
Mobile PBsomes, 59
Modified desiccation-tolerant plants, 683
Modules, 497
Moisture management
    forage crops, 470–471
Molecular-level responses to UVB radiation, 280–281
Molecular oxygen, 160f, 339
Molecular structures
    of bacteriochlorophylls, 98f
    of chlorophylls, 97f
Mol-1P phosphatase, 517
Mol synthesis, 517–518
Moment of momentum, 27
Monodehydroascorbate (MDHA), 160f, 163
Monogalactosyldiacylglycerol (MGDG), 7, 224, 245
    desiccation and, 669
Monoisotopic mass, 116
Monomers, 353
Monomethylester (ME), 177
Monovinil Chls (MVChls), 358
Morphological plasticity, 496–497
    woody species, 495–496
Morphology, functional traits and, 488, 489
Morphophysiological factors (genetic basis)
    awns, 542
    chlorophyll content, 541–542
    crop canopy and leaf area index, 541
    erect leaves, 540–541
Mother-stock phosphorus (MSP), 296–300, 298t
Multiwalled carbon nanotubes (MWCNT), 211
Mustard, 769t
Mutation
    psbA gene, 708
Mutations suppressing/enhancing superoxide generation, 340–341, 341t
Myrmecia, 675
Myrothamnus, 672
Myrothamnus flabellifolia, 683, 684, 685, 687, 688

N

Nannochloropsis spp., 774, 775t
Nano-biohybrid (or biohybrid) photoconverter; *see also* Hybrid interfaces for electron/energy transfer

hybrid interfaces for electron transfer
    covalent binding, 210–211
    electrostatic absorption, 210
    interface of RC with carbon substrates, 211
    physisorption techniques, 210
    RC–NP interface, 212
    selective anchoring, 211
hybrid interfaces for energy transfer
    RC–NP interface, 212–213
    RC-organic interface, 213–216
Nanocondenser, 204
Nanocrystals (NC), 212
Native trees
    and invasive trees, competitive interactions between, 480–483, 481f, 482f
Natural chlorophylls in diethylether, fluorescence properties of, 112t–113t
Natural genetic resources, photosynthetic performance by QTL analysis and, 752–754, 752f, 753f
Naturally occurring chlorophylls, 96, 97f
NCER (net carbon exchange rate), 573–574, 574f
NCG (net carbon gain), 573–574, 574f
Ndh complex, *see* Thylakoid Ndh complex
Necrosis, 78
Negative mass, 23, 24f, 25f
Neoxanthin (Neo), 589
Nerium oleander L., 656
Net carbon exchange rate (NCER), 573–574, 574f
Net carbon gain (NCG), 573–574, 574f
Net CO_2 assimilation (A_N), 669–672, 670f, 671t
N-ethylmaleimide (NEM), 185
Net photosynthesis, drought on, 686–687
Nichia Corporation, 762
Nickel-nitrilotriacetic (Ni-NTA) affinity chromatography, 330
Nicotiana tabacum (transgenic tobacco), 498, 545
Nicotinamide adenine dinucleotide phosphate (NADPH), 202
    metabolism in cytosol, 161–163
    production in plastids, 159–161, 160f
Nicotinamide adenine dinucleotide phosphate (NADPH)-dependent Trx reductase (NTR), 185
Nicotinamide adenine dinucleotide phosphate (NADP)-malic enzyme (NADP-ME), 162
Nitrate reductase activity, 259
Nitrogen (N)
    allocation, Rubisco concentration and, 743–744, 743f
    cross-talk of sugar with, 87
    for plant growth, 288, 289
    in terrestrial plant leaves, 742
Nitrogen management
    forage crops, 471
Nitroxyl, 160f
NOESY, *see* Nuclear Overhauser and exchange spectroscopy
Non-adaptive plasticity, 499
Noncyclic electron flow, heat stress and, 698
Nondestructive analysis, 573–574
    chamber and system design, 573
    net carbon exchange rate (NCER), 573–574, 574f
    net carbon gain (NCG), 573–574, 574f
Nonfluorescent chlorophyll catabolites (NCC), 84

Noninvasive methods, plant performance
    monitoring and, 452–458
    alternative methods for plant stress response
        analysis, 458
    Balanites aegyptiaca, drought stress (case
        study), 457
    Brassica napus, nutrient deficiency and
        biotic stress (case study), 458
    chlorophyll fluorescence, 452–453
    chlorophyll fluorescence measurements, 454
    chlorophyll fluorescence monitoring, limits
        of, 455–456, 455t
    chlorophyll fluorescence parameters, and
        their meaning of, 453–454
    chlorophyll fluorescence parameters as
        potential stress markers, 456
    drought stress, several species (case study),
        457
    Euphorbia tirucalli, drought stress (case
        study), 456–457
    gas exchange and CO$_2$ assimilation, 452
    Lablab purpureus, drought stress (case
        study), 456
    method, 453
    PAM fluorimetry with gas exchange
        measurements of stressed plants,
        454–455
    Tripolium pannonicum, nutrient deficiency
        (case study), 457–458
Nonphosphorylating Ga3PDHase
    (np-Ga3PDHase), 514
Nonphotochemical excess energy dissipation in
    antenna systems, *see* Light (excess)/
    carbon (limited)
Nonphotochemical quenching (NPQ), 61
    of chlorophyll fluorescence, 672–673, 690
    in cyanobacteria
        high light-inducible proteins (HLIP)/iron
            starvation inducible (IsiA) proteins,
            372
        orange carotenoid protein, 371–372
    of fluorescence, 197
    in microalgae
        about, 372–373
        xanthophyll cycles, 373–375
Nontypical C$_4$ species, photosynthesis in
    in Chenopodiaceae family
        *Bienertia cycloptera/Bienertia
            sinuspersici*, case of, 311–313, 312f
        Suaeda aralocaspica, case of, 313–314,
            313f, 314f
        overview, 307–308
    in submersed aquatic plants of
        hydrocharitaceae family
        Egeria densa, case of, 308–310, 309f
        Hydrilla verticillata, case of, 310–311
    transition from, to crassulacean acid-like
        metabolism in *Portulaca* genus
        about, 314–315
        Portulaca grandiflora, case of, 318–319,
            318f
        Portulaca oleracea, case of, 315–318,
            315f, 317f
Nonubiquitous carotenoids, in higher plants
    high-light carotenoids, 594–595
        escholtzxanthin, 595
        red *retro*-carotenoids, 595
        rhodoxanthin, 595
        xanthophyll esters, 594–595
    HPLC determination of, 596–597, 596f
    low-light carotenoids, 591–594

α-carotene, 591–593, 592f
    lactucaxanthin, 594
    lutein epoxide, 593–594
    trans-neoxanthin, 591
    pigment composition of, 589–591, 590f, 591t
Norm of reaction, 493; *see also* Phenotypic
    plasticity
Nostoc sp. PCC 7119, 229, 233–236, 236t, 237s
Nothofagus spp., 656
NPQ, *see* Nonphotochemical quenching (NPQ)
NPQ$_{DT}$, *see* Desiccation-induced NPQ (NPQ$_{DT}$)
Nuclear-encoded genes for photosynthetic
    proteins, 280
Nuclear-encoded RNA polymerase (NEP), 79
Nuclear genes, of photosynthetic apparatus,
    543–547
    early research on, 544
    light-regulated nuclear genes, 546–547
    ribulose bisphosphate carboxylase/
        oxygenase, regulation of, 545–546
    rubisco activase, 546
    transcription, 544–545
    translation, 544–545
Nuclear magnetic resonance (NMR) spectra, *see*
    Chlorophylls/bacteriochlorophylls,
    physicochemical properties of
Nuclear *Ndh* genes, 151–153, 152f; *see also*
    Thylakoid Ndh complex
Nuclear Overhauser and exchange spectroscopy
    (NOESY), 124, 130f
Null-balance CO$_2$ enrichment, 732, 732f
Nursery productivity, 301
Nutrient deficiency
    Brassica napus (case study), 458
    Tripolium pannonicum (case study),
        457–458
Nutrient management
    forage crops, 471
Nutrients
    photosynthetic competition for, 479–480

O

Oat, 766t
Ocimum basilicum L., 769t
Ocimum gratissimum, 769t
OEC, *see* Oxygen-evolving complex (OEC)
Oenothera elata, 653
Olea europaea L., 656
Oleic acid, desiccation and, 669
Oligogalactolipids, desiccation and, 669
Olive, 656
One helix proteins (OHP), 380
Onions, 769t
Optical power, electricity to, 761, 762–763
Optimal defense (OD) theory, 552
Optomechanical counterforce hypothesis, 41
Orange carotenoid protein (OCP, 61, 371–372;
    see also Nonphotochemical
    quenching (NPQ)
Organelle, 77
Organic feedstock, 509
Ornamental plants, 651–658
    drought stress, effects, 654–655
        coping with, 656–658
        overview, 652
        photosynthesis, 654–655
        plant growth and development, 654–655
        tolerance, 655
    overview, 651

salinity and drought tolerance, 655–656
    salinity stress, effects, 651–654
        coping with, 656–658
        overview, 651–652
        photosynthesis, 653
        plant growth and development, 652–653
        tolerance, 653–654
Ornithogalum arabicum, 653
Orthosiphon aristatus, 656
Osmoprotectants, 657, 665
Osmotic adjustment, 654, 655
Oxidative damage in plants, systems for
    repairing, 166–167; *see also* Redox
    metabolism in photosynthetic
    organisms
Oxidative pentose-phosphate pathway (OPPP),
    159, 160f, 161, 514
Oxidative stress, desiccation tolerance, 665
Oxidized glutathione (GSSG), 160f
Oxidoreductase, 697
Oxygenation, Rubisco, 700
Oxygen-evolving complex (OEC), 194, 203f,
    243, 244, 688, 697
    about, 351f
    cyanobacterial PSII and, 352
Oxygenic photoautotrophs, 370
Oxygenic phototrophs, 177
Oxygen in mediating superoxide generation,
    342–343, 342f
Oxygen on electron transfer activity of bc_1,
    338–339, 340t
Oxygen uptake rate of redox proteins, 23t

P

Pad-and-fan systems, 729–731, 730f
Pair production of photon, 33
PAM fluorimetry
    with gas exchange measurements of stressed
        plants, 454–455
P-aminomethylbenzensulfonamide (mafenide),
    196
Panax ginseng, 765t
Paolillo model, 5
PAR, *see* Photosynthetically active radiation
    (PAR)
Parsley, 765t
Particulate theory of light, 18
Partitioning
    phosphorus role in, 603–618
        long-term *in vivo* effects of Pi
            deprivation, 608–615
        overview, 603–604
        phosphate deficiency, *see* Phosphate
            deficiency
        short-term *in vitro* effects of Pi
            deprivation, 605–608
    photosynthate, *see* Photosynthate
        partitioning
Paternal plastid DNA (pt DNA) inheritance, 547
Pauling electronegativity, 110, 139
PBG deaminase (PBGD), 174
PBsomes, *see* Phycobilisomes
Pchlide oxidoreductase (POR), 177
PDT, *see* Poikilochlorophyllous desiccation-
    tolerant (PDT) plants
Pea, 766t
Pelargonium type plants, 547
Pelargonium zonale, 547
Pentose-P, 514

Pentose-P/Pi translocator (XPT), 514
Pepper, 768t–769t
PEP/phosphate translocator (PPT), 514
Performance
    functional traits and, 488, 489
    photosynthetic, see Photosynthetic
        performance
    traits, 489
Performance index (PI), 263
Peridinin–Chl a protein (PCP) complex in
        dinoflagellates, 363–364
Peroxidase (PX), 153
Peroxiredoxin (PRX), 160f, 164, 165
Peroxiredoxin Q (PRXQ), 165
Persea americana (avocado), 528
Petroselinum crispum, 765t
Petunia, 768t
Petunia × hybrida, 656, 657, 768t
Phaeodactylum tricornutum, 377
Phalaris paradoxa, 707
Phaseolus vulgaris, 656, 750
ΔpH-dependent NPQ, 673
Phenotype, psbA gene mutant, 708
Phenotypic plasticity, 493; see also Plasticity
    adaptive, 493, 494
    climate changes and, 499–500, 500f
    evolution of plants and, 495
    leaf traits and, 496–497
    overview, 487
    plants adaptation to environmental
        conditions and, 493–494
Pheophorbide a oxygenase (PAO), 82–83, 83f
Pheophytin (Phe), chlorophyll (Chl) a', b', d', f',
        DV-Chl a' and, 114, 117f, 118f
Pheophytin (Phe) a, 98
Phillyrea latifolia, 653
Phloem carbohydrate transport, carbon
        utilization and, 750
Phosphate deficiency
    acclimation and adaptation of plants to, 617
    interaction with environmental factors,
        616–617
    plants recovery from, 615–616
Phosphate translocators, 605–606
Phosphatidylethanolamine (PE), 333
Phosphatidylglycerol (PG), 224, 245, 248
Phosphatidylinositol, desiccation and, 669
Phospho-enol-pyruvate (PEP), 514
Phosphoenolpyruvate, distal generation of, 314
Phosphoenolpyruvate carboxykinase (PEPCK), 317f
Phosphoenolpyruvate carboxylase (PEP Case),
        265, 266, 517, 519–520
Phosphoenolpyruvate carboxylase (PEPC), 309,
        313, 315, 317f, 687, 699
Phosphoglycerate (PGA), 700, 742
3-phosphoglycerate (3PGA), 160f
3-phosphoglyceric acid (PGA), 65, 605
Phosphoglycolate (PGO), 700, 742
Phospholipid hydroperoxide glutathione
        peroxidase (PHGPX), 163
Phosphorus (P); see also Inorganic phosphate (Pi)
    overview, 603
    for plant growth, 288
    role in carbon assimilation and partitioning,
        603–618
        long-term in vivo effects of Pi
            deprivation, 608–615
        overview, 603–604
        phosphate deficiency, see Phosphate
            deficiency

short-term in vitro effects of Pi
    deprivation, 605–608
Phosphorus macronutrient limitation/small fruit
        crop development (case study), see
        Small fruit crop responses
Phosphorylation, activity of Ndh complex by,
        154–155
Photoadaptation of photosynthetic apparatus,
        57–58, 57f
Photochemical activities
    drought on, 687–688
    salinity effects on, 633–634
Photochemical energy conversion, 202, 202f
Photochemical reactions
    drought effects on, 636–637
Photoconversion efficiency vs. structural
        complexity, 208, 209f
Photocycle, 204–205, 205f
Photoenzymes, 208
Photoinhibition, 249, 370
Photolysis, of water, 509
Photomorphogenesis, 4, 6f
    transcriptional regulation during, 182
Photons
    defined, 17
    energy of, 20
Photoperiod, 734–735
Photoprotection, 277; see also Photosynthetic
        apparatus, adaptation/regulation of
    nonphotochemical quenching/orange
        carotenoid protein, 61
    of PBsomes, 61, 61f
    in PSII RC, 375
Photoprotection, DT organisms, 672–673
    antioxidative response, 673
    energy dissipation as heat, 672–673
    light avoidance, 672
Photorespiration
    heat stress and, 699–700
    in high-compensation-point plants, 310
    thylakoid membrane, 450, 450f
Photosynthate partitioning, 509–529
    carbon transport from leaf to sink tissues,
        520–524, 521f
        mannitol and glucitol transporters,
            523–524
        sucrose transporters, 521–523, 522f
        SUT transporters, 522–523, 522f
        SWEET transporters, 521–522, 522f
    heterotrophic cells, metabolism in
        lipid storage metabolism, 528–529
        starch storage metabolism, 527–528
        sucrose and sugar-alcohols metabolism,
            524–527
    intercellular, 509–510, 510t
    intracellular, 509–510, 510t
    leaf cell
        metabolites exchange between
            chloroplast and cytosol, 514–515
        respiratory metabolism in, 519–520
        starch metabolism in, 510–514,
            511f–512f, 512t
        sucrose and sugar-alcohols synthesis,
            515–519, 515t, 516f, 517t
Photosynthates
    salinity effects on level of, 634–635, 635t
Photosynthesis; see also Cytochrome c₆-like
        proteins; Hybrid interfaces for
        electron/energy transfer
    artificial
        basic architecture, 207–208, 208f, 209f

photoconversion efficiency vs. structural
        complexity, 208, 209f
Benson–Calvin cycle (BCC), 509–510, 510t,
        511f
    with biotechnology, see Biotechnology,
        photosynthesis and
    on canopy level, optimizing, 750–752, 751f
    defined, 697
    efficiency of LEDs for, see Light-emitting
        diodes (LEDs)
    ethylene and, 578–579
    implications of, 201
    light-harvesting complexes
        energy transfer, 206–107
        spatial organization, 205–206, 206f
        spectroscopic properties, 206
    light phase, 509
    Ndh complex in, 153–154, 153f, 154f
    of ornamental plants; see also Ornamental
        plants
        drought stress on, 654–655
        salinity stress on, 653
    photochemical energy conversion, 202,
        202f
    Pi deprivation and regulation of, 606–607
    Q-type photosynthetic reaction centers
        electron transfer cascade, 204
        photocycle, 204–205, 205f
        spatial organization, 202, 204, 204f
        spectroscopic properties, 204
    research and other fields, 539–540
    and respiration in cyanobacteria
        about, 230, 230s
        cytochrome cₘ, 232
        and plastocyanin, 230–232
    role of, 465, 536
    synthetic phase, 509
Photosynthesis-associated genes (PAG), 85, 86f
Photosynthesis efficiency
    forage crops management for improved
        productivity, 469–471
Photosynthesis-related genes, drought on, 689
Photosynthesis-related proteins
    drought on, 688–689
    genes for, 548
Photosynthetically active radiation (PAR), 151f,
        280, 733–734, 735
Photosynthetic apparatus
    damage, stress-induced ROS production
        and, 643
    heat stress and, 697
    nuclear genes of, 543–547
        early research in, 544
        light-regulated nuclear genes, 546–547
        ribulose bisphosphate carboxylase/
            oxygenase, regulation of, 545–546
        rubisco activase, 546
        transcription, 544–545
        translation, 544–545
    protection upon desiccation, 689–691
Photosynthetic apparatus, adaptation/regulation
        of
    composition/organization of photosynthetic
        apparatus
        about, 53, 54f
        cytochrome and ATP synthase, 55
        diversity of photosynthetic apparatus
            organization, 55
        light-harvesting complexes (LHC), 54
        reaction centers (RC), 54–55
    electron transport pathways, regulation of, 60

light effects on photosynthetic machinery
    organization
    light-state transitions, 58–60, 59f
    photoadaptation of photosynthetic
        apparatus, 57–58, 57f
light effects on photosynthetic stoichiometry
    light–dark cycle, 57
    light intensity, 55–56, 55f
    light quality, 56–57, 56f
photoprotection
    nonphotochemical quenching/orange
        carotenoid protein, 61
    of PBsomes, 61, 61f
Photosynthetic apparatus, functional changes
    of, 10, 12; see also Chloroplast
    biogenesis
Photosynthetic area, manipulation of, 553
Photosynthetic bacteria growth, cytochrome bc_1
    complex in, 328
Photosynthetic carbon reduction cycle (PCR
    cycle), 317f, 603
Photosynthetic Chl–protein complexes, 183
Photosynthetic competition, 475–484
    defined, 476–477
    for evolution of plants, 483–484
    overview, 475–476
    between trees, 477
        interactions between native and invasive
            trees, 480–483, 481f, 482f
        for light, 477–479, 478t
        for nutrients, 479–480
        for water, 479–480
Photosynthetic efficiency
    genetic factors, 552–554
        breeding for C_4 mechanism, 553–554
        changes in biochemical activities, 554
        future prospects, 554
        manipulation of photosynthetic area,
            553
        rubisco and crop improvement, 554
    thylakoid membrane, 449
Photosynthetic electron transfer, 60, 159
Photosynthetic electron transport-limited
    photosynthesis, 706, 719
Photosynthetic enzymes
    heat stress and, 699, 699t
Photosynthetic machinery
    Pi deprivation and, 610–611, 610t
Photosynthetic membranes, desiccation/
    rehydration cycles, 669
Photosynthetic pathways
    forage crops, 466, 466t
Photosynthetic performance, improving
    by QTL analysis, natural genetic resources,
        752–754, 752f, 753f
    Rubisco kinetics, 741–745
        concentration and N allocation, 743–744,
            743f
        performance, opportunities for
            improving, 744–745
        specificity and carboxylation rate,
            741–743, 742f
Photosynthetic photon flux density (PPFD), 673,
    708, 709f, 734
Photosynthetic pigments, 260–262, 262t
    changes in, 6–7, 6f
    drought tolerance, 684–685
    heat stress and, 697, 698f, 698t
    poikilochlorophyllous plants, 674
    role in productivity and quality of forage
        crops, 466–468, 467t

Photosynthetic pigments, degradation of, 82–84,
    83f; see also Chloroplast development
    carotenoid degradation, 84
    chlorophyll degradation, tracing path of,
        82–84
    coordinated breakdown of pigment, 84
Photosynthetic regulation, R/S subtypes,
    717–719, 718f, 719t
Photosynthetic repression, cause, 750
Photosynthetic strategy in crops (genetic basis),
    552–554
    breeding for C_4 mechanism, 553–554
    changes in biochemical activities, 554
    future prospects, 554
    manipulation of photosynthetic area, 553
    rubisco and crop improvement, 554
Photosynthetic system, 41
Photosynthetic units (PSU)-based electron/
    energy transfer, see Hybrid interfaces
    for electron/energy transfer
Photosystem(s), 7; see also Microalgal
    photosystems
    of chlorophyll c-containing microalgae,
        362–364
    of green microalgae, 361–362
    of red microalgae, 360–361
    salinity effects on, 633–634
Photosystem I (PSI)
    assembly of, 81
    cyanobacterial, 354–356
    cyclic electron transport, 152, 153f
    genes for, 548–549
    heat stress and, 697
    photoprotection in DT organisms, 673
    PS_{II} and activation energy distribution
        between, 449
    and red microalgae, 361
    redox systems in plant cell, 160f
Photosystem II (PSII)
    core complex, 244, 350–352
    cyanobacterial
        oxygen-evolving complex (OEC), 352
        phycobilisome (PBS), 352–353
        PSII core complex, 350–352
    dehydration and, 690
    electron supply from, 153, 153f
    heat stress and, 697
    organization/assembly of, 80–81
    photoprotection in DT organisms, 673
    and PS_I activation energy distribution
        between, 449
    quality control of, see Quality control of
        photosystem II
    recovery of, 380–381
    and red microalgae, 360
    redox systems, 160f
Photosystem II (PSII) core phosphatase (PBCP),
    225
Phototransformation of violaxanthin, 244
Phototransistors, 208
Phycobilins, 761
Phycobilisomes (PBsomes)
    about, 54, 55f
    and cyanobacterial PSII, 352–353
    mobile, 59f
    photoprotection and, 61, 61f
    and photosystems of red microalgae,
        360–361
    supercomplexes of phycobiliproteins, 371
Phycocyanobilin (PCB), 352, 361
Phycoerythrins (PE), 54, 371f, 375

Phycoerythrobilin (PEB), 352, 361
Phycoerythrocyanin (PEC), 352
Phycourobilin (PUB), 361
Phylloquinone, 203f
Phylogenetic tree of cytochromes, 235f
Physiological plasticity, woody species, 495–496
Physisorption techniques, 210
Phytochromes (PHY), 177
Phytochromobilin synthase, 177
Phytohormones, 78
    signalings, 88–89
Phytol, 177
Pi, see Inorganic phosphate (Pi)
Picket fence, 354
Pigment composition, of higher plant
    chloroplast, 589–591, 590f, 591t
Pigments, photosynthetic
    desiccation/rehydration cycles, 669
    drought tolerance, 684–685
    heat stress and, 697, 698f, 698t
    poikilochlorophyllous plants, 674
Pisum sativum L., 766t
Planck's blackbody radiation law, 40, 41f
Planck's constant, 20, 27
Plant carboxylation, 290
Plant ecological strategies, 490–493, 491f
    defined, 490
    drought tolerance, 492
    leaf economics spectrum, 491–492
    other dimensions, 492–493
    plant height, 492
    plant water stress, 492
Plant factories
    aggregate culture, 725
    with artificial lighting, 725, 725f, 727–729
    capital investment, 726, 728
    CO_2 supplementation, 726
    defined, 724
    integrated pest management, 726
    liquid culture, 725
    merits/demerits of, 726–729, 728t
    soilless systems, 725
    structure, 725
    with sunlight, 724, 725f, 726–727, 729
    with sunlight and supplemental light,
        724–725, 725f, 727
    technologies and equipment, 726
Plant growth response
    Pi deprivation and, 609–610, 609t
Plant height
    as dimension of plant ecological strategy,
        492
Plant nutrient, 288
Plant performance
    indicators, 448
Plant performance monitoring, noninvasive
    methods, 452–458
    alternative methods for plant stress response
        analysis, 458
    Balanites aegyptiaca, drought stress (case
        study), 457
    Brassica napus, nutrient deficiency and
        biotic stress (case study), 458
    chlorophyll fluorescence, 452–453
    chlorophyll fluorescence measurements, 454
    chlorophyll fluorescence monitoring, limits
        of, 455–456, 455t
    chlorophyll fluorescence parameters, and
        their meaning of, 453–454
    chlorophyll fluorescence parameters as
        potential stress markers, 456

drought stress, several species (case study), 457
Euphorbia tirucalli, drought stress (case study), 456–457
gas exchange and CO_2 assimilation, 452
Lablab purpureus, drought stress (case study), 456
method, 453
PAM fluorimetry with gas exchange measurements of stressed plants, 454–455
Tripolium pannonicum, nutrient deficiency (case study), 457–458
Plant physiological responses, UVB and, 276–277
Plant–plant interaction
climatic conditions and, 498–499
Plant populations
succession of, plasticity and, 497–498
Plant productivity
RGR and, 572
Plants, evolution of
importance of photosynthetic competition for, 483–484
phenotypic plasticity and, 495
Plants adaptation
to environmental conditions, plasticity and, 493–494
Plants recovery
from phosphate deficiency, 615–616
Plant stress, defined, 289
Plant stress response analysis
alternative methods for, 458
Plant stress responses, genetic factors and
abiotic stress, 551
biotic stress, 551–552
growth-defense paradigm, 552
Plant survival
assessment of, 489
Plant-type PEP Cases (PTPC), 520
Plant water stress
as dimension of plant ecological strategy, 492
Plasmodesma, 665
Plasticity, 493–500; *see also* Functional traits
adaptive, environment condition and, 494–495
climatic conditions and plant–plant interaction, 498–499
defined, 496
hierarchy of plasticities, 497
history of, 493–494
non-adaptive, 499
overview, 487
phenotypic
climate changes and, 499–500, 500f
evolution of plants and, 495
plants adaptation to environmental conditions and, 493–494
succession of plant populations and, 497–498
of whole plant, leaf traits and, 496–497
of woody species, 495–496, 496f
Plastid-encoded RNA polymerase (PEP), 79
Plastid Glc transporter (pGlcT), 514
Plastidial proteins, desiccation and, 669
Plastidic glucose transporter, desiccation and, 669
Plastid nucleoids, 79
Plastids genome, 548
Plastid sorting out, 547
Plastid-targeted proteins (CpPTP), 689
Plastid transcription kinase (PTK), 549
Plastocyanin (Pc), 230–232, 355f
for photosynthetic electron transport, 747

Plastoglobuli, 82
Plastoquinol, 354
Plastoquinol terminal oxidase (PTOX), 376, 377
Plastoquinone (PQ), 58, 150, 246, 354, 697
Plectranthus scutellarioides, 769t
Pleiotropy
in R, 719–720
reorganization, *psbA* gene, 708–709, 708t
Pleurostima, 667
Pleurostima purpurea, 683
Plumbago auriculata, 653
^{31}P-nuclear magnetic resonance (^{31}P-NMR) spectroscopy, 613–614, 614t
Poa annua, 707
Poikilochlorophyllous desiccation-tolerant (PDT) plants
chloroplast changes, drought tolerance, 685
defined, 684
photosynthetic activity, drought on, 685–686
Poikilochlorophyllous species
chloroplast changes, desiccation tolerance, 674
defined, 667, 673
gas exchange in, 674–675
photosynthetic pigments, 674
vs. homoiochlorophyllous, 667
Z content, 674
Poikilohydry, 665
Polar plants, 308
Polyacrylamide gel, 195f
Polydimethyldiallylammonium chloride (PDDA), 210
Polyethene filter, 258f
Polygonum lapathifolium, 707
Polygonum pensylvanicum, 707
Polygonum persicaria, 494
Polypodium, 672
Polypodium polypodioides, 691
Poly(3-thiophene acetic acid), 212f
Polytrichum, 672
Polytrichum formossum, 665
Population-level variability, 489, 490
Populations, of plant
succession of, plasticity and, 497–498
Porphobilinogen deaminase, 697
Porphyra spp., 653
Porphyridium cruentum, 57, 361, 375
Portulaca grandiflora, 318–319, 318f
Portulaca oleracea, 315–318, 315f, 317f
Position, lighting, 734
Positive mass, 24f, 25f
Positrons, 33
Postillumination fluorescence, 151
Posttranscriptional control of *Ndh* gene expression, 154; *see also* Thylakoid Ndh complex
Post-transcriptional gene silencing (PTGS), 541
Post-transcriptional regulation
chloroplast gene expression, 550
Posttranslational regulation of Chl biosynthesis; *see also* Chlorophyll metabolism regulation in plants
enzyme activities, regulation of, 184
protein stability, regulation by, 185–186
redox regulation, 184–185
Potassium fertilizer, ornamental plants, 657
Potassium (K) for plants, 288
Potato, 765t

Power generation reduction/partitioning in plants, *see* Redox metabolism in photosynthetic organisms
Poynting vector, 31, 31f
PPFD, *see* Photosynthetic photon flux density (PPFD)
Prasinophyceae, 775t
Prasinophyceae, 374
Precorrin 2-dependent pathway, 174
Primary chloroplasts, 359
Primary macronutrients, 288–289
Primary photochemical reactions, loss in, 84
Primary photochemistry of PSII, 246
Primula vulgaris, 768t
Prochlorococcus, 358
Prochloron, 358
Prochlorophyte Chl *a/b*-binding proteins, 358–359; *see also* Cyanobacteria/microalgae, photosynthetic apparatus in
Prochlorotrix, 359
Productivity, of algae under LED lighting, 773–774, 775t, 776t
Prolamellar body (PLB)
feature of etioplasts, 4
Proline, 657
Proplastid model, 8
Proplastids, 5, 78
Protea cynaroides L., 771t
Protease-mediated degradation of plastid proteins, 89
Protein–glutathione mixed disulfides (PSSG), 164
Proteins
changes in, 6–7
desiccation/rehydration cycles, 669
glass formation, 666
in photosynthetic light and dark reactions, 278–279
Protein stability, regulation by, 185–186
Proteomic studies, desiccation, 669
Prothylakoids (PT), 4
Protochlorophyllide (Pchlide), 4, 5
Protogen oxidase (PPOX), 176, 178
Protoheme, 177
Proton gradient regulation 5 protein (PGR5), 357, 357f
Proton motif force (pmf), 336
Protons, 334
Protoplasmic water potential, 664
Protoplast, 313
Protoporphyrinogen IX (Protogen), 176
PsbA gene, 705, 706, 707
mutant phenotype, 708
pleiotropic reorganization, 708–709, 708t
PsbA mRNA, 686
Psb genes, 686
Pseudocyphellaria dissimilis, 676
Ptilotus nobilis, 656
πtlp (wilting point), 492
Pueraria lobata, 769t
Purification of Cyt bc_1, 330
Purple photosynthetic bacteria, 55
Pyrenoids, 387
Pyruvate (Pyr) kinase, 514
Pyruvate orthophosphate dikinase (PPDK), 310, 312, 313, 317f

Q

Q-cycle mechanism/bifurcated ET at quinol oxidation site, 334, 335f

Q-type photosynthetic reaction centers, *see* Photosynthesis
Quality and source, light, 735–737
Quality control of photosystem II
 grana formation, regulation of, 225
 role of thylakoid membranes in, 225, 226f
 thylakoid membranes
  structure and composition, 223–224
   thylakoid membrane lipids, 224
  thylakoid membranes, structure changes
   shrinkage and swelling of thylakoids, 224–225
   unstacking of thylakoid, 224
Quantitative trait loci (QTL) analysis
 photosynthetic performance by, natural genetic resources, 752–754, 752f, 753f
Quantum dots (QD), 213
Quantum electrodynamics (QED), 26n
Quantum electron transport, 263
Quantum mechanical photon, 17–23, 22; *see also* Light, nature of
Quantum of light, concept by Einstein, 21
Quiescent state, 664
Quinol oxidase (Cyd), 350, 351f
Quinol oxidation, 337f
Quinol oxidation site
 Q-cycle mechanism/bifurcated ET at, 334, 335f
 ubisemiquinone free radical as reaction intermediate at, 336–338, 337f
Quinone, 214f
 exchange cavity, 354

R

Radicchio, 772t
Radish, 766t
Rainwater, storage, 726
Ramonda genus, 672
Ramonda mykoni, 687
Ramonda nathaliae, 687
Ramonda serbica, 685, 687
Ramonda spp., 683
Rapeseed, 771t
Raphanus sativus L., 766t
"Raspberry Ice" bougainvillea, 654
Rbcs gene, 546
RbeS antisense RNAs, 545
RCA, *see* Rubisco activase (RCA)
Reaction center (RC), 97, 117, 761
 composition of photosynthetic apparatus, 54–55
 interface of, with carbon substrates, 211
 membrane intrinsic core complex, 350
Reaction center (RC)-NP interface, 212–213
Reaction center (RC)-organic interface, 213–216
Reactive nitrogen species (RNS), 166–167
Reactive oxygen free radicals, 88
Reactive oxygen species (ROS), 5, 629–630, 667, 672, 689, 697; *see also* ROS production
 and radicals, 279–280
 superoxide generation, 340
 thylakoid membrane and, 224
Reactive oxygen species (ROS) in signaling network; *see also* Sugar signaling
 cross-talk of, 87
 and phytohormone signalings, 88–89
 and retrograde signaling, 88
 ROS production, 87–88, 88f
Recombinant inbred lines (RILs), 752–753

Recovery, of plants
 from phosphate deficiency, 615–616
Recovery of PSII, 380–381
Red chlorophyll catabolite (RCC) production, 84
Red chlorophyll catabolite reductase (RCCR), 84
Red light, 735; *see also* Light
Red microalgae, photosystems of, *see* Microalgal photosystems
Redox metabolism in photosynthetic organisms
 oxidative damage in plants, systems for repairing, 166–167
 power generation reduction/partitioning in plants
  NADPH metabolism in the cytosol, 161–163
  NADPH production in plastids, 159–161, 160f
 redox molecules/enzymatic systems
  ascorbate, 163–164
  ascorbate peroxidase (APX), 166
  glutathione, 163
  glutathione-dependent system, 164–165
  glutathione *S*-transferases (GST), 166
  peroxiredoxin (PRX), 165
  superoxide dismutase (SOD), 165–166
  thioredoxin (TRX) system, 164
Redox potentials, *see* Chlorophylls/ bacteriochlorophylls, physicochemical properties of
Redox regulation, of Chl biosynthesis, 184–185
Red Pak Choi, 771t
Red *retro*-carotenoids, 591, 595
 HPLC determination of, 597
Red-shifted binary photons, 39, 39f
Reduced b_L heme as source of superoxide generation, 342
Reduced glutathione (GSH), 160f, 163
Rehydration/desiccation cycles, 665
 chloroplasts changes, 667–669
Relative growth rate (RGR), 572, 609; *see also* Growth analysis
Relative humidity (RH), 664, 730, 731; *see also* Humidity
Relative water content (RWC), 664, 670, 684, 685, 688
Relativity of space and time, 36–42
Reproduction
 assessment of, 489
Research
 functional ecology, 488
 nuclear genes, 544
 and other fields, 539–540
Research, current methods, 447–458
 noninvasive methods, plant performance monitoring and, 452–458
  alternative methods for plant stress response analysis, 458
  Balanites aegyptiaca, drought stress (case study), 457
  Brassica napus, nutrient deficiency and biotic stress (case study), 458
  chlorophyll fluorescence, 452–453
  chlorophyll fluorescence measurements, 454
  chlorophyll fluorescence monitoring, limits of, 455–456, 455t
  chlorophyll fluorescence parameters, and their meaning of, 453–454
  chlorophyll fluorescence parameters as potential stress markers, 456

drought stress, several species (case study), 457
  Euphorbia tirucalli, drought stress (case study), 456–457
  gas exchange and CO_2 assimilation, 452
  Lablab purpureus, drought stress (case study), 456
  method, 453
  PAM fluorimetry with gas exchange measurements of stressed plants, 454–455
  Tripolium pannonicum, nutrient deficiency (case study), 457–458
 overview, 447–448
 thylakoid membrane
  absorption of light energy, 448
  activation energy distribution between PS_{II} and PS_I, 449
  ATP synthesis, 450
  C_4/CAM, 452
  described, 448–449
  electron sinks, 451
  futile cycles, 451–452
  photorespiration, 450, 450f
  photosynthetic efficiency, 449
  ROS production, 450–451
  structure and function, 448–452
  xanthophyll cycle, 449–450
Respiration in cyanobacteria; *see also* Cytochrome c_6-like proteins
 about, 230, 230s
 cytochrome c_M, 232
 and plastocyanin, 230–232
Respiratory electron transport, 230, 351f
Respiratory metabolism
 in leaf cell, 519–520
Restriction fragment length polymorphism (RFLP) analysis, 547
Resurrection plants, 683
Retrograde mechanism, 8
Retrograde signaling, ROS and, 88
RGR, *see* Relative growth rate (RGR)
RH, *see* Relative humidity (RH)
Rhodobacter sphaeroides, 101, 545
Rhodospirillum rubrum, 101
Rhodoxanthin, 595
Ribose nucleic acid (RNA), 288
Ribose 5-phosphate (Rib5P), 160f
Ribulose-1,5-bisphosphate (RuBP), 699
Ribulose bisphosphate carboxylase/oxygenase regulation of, 545–546
Ribulose-1,5-bisphosphate carboxylase/ oxygenase, *see* Rubisco
Ribulose bisphosphate carboxylase-oxygenase (RuBisCO), 81, 82f, 307
Ribulose-1,5-bisphosphate (RuBP)), 65, 270, 278, 315, 468, 630, 637, 699
 regeneration, 66, 66f
Ribulose 5-phosphate (Rub5P), 160f
Rice, 771t
Rieske-type monooxygenase, 177
RNA-directed RNA polymerases (RdRs), 541
RNA-induced-silencing complex (RISC), 541
RNA-interference (RNAi), 541
RNA-Pol II, 541
Robust model of UV-photosynthetic response, 276
Rocket, 769t
Root traits, 493
ROS, *see* Reactive oxygen species (ROS)
Rosa × *fortuniana*, 656

Rosa hybrida L., 652
ROS production
    thylakoid membrane, 450–451
ROS production, stress-induced
    photosynthetic apparatus damage and, 643
Rubia peregrine, 497
Rubisco, 81–82, 82f, 266, 268f, 491, 545–546, 606
    activation, heat stress and, 699
    carbon assimilation in R/S plants, 706
    content, 663, 674
    crop improvement and, 554
    dual function of, 381–382
    oxygenation and carboxylation, 700
    R/S subtypes, activity in, 715–716, 716f, 717f, 717t
Rubisco, dual-function enzyme
    activase, thermotolerance of, 745
    CO_2 concentration around, 745–747, 746f
    kinetics
        concentration and N allocation, 743–744, 743f
        performance, opportunities for improving, 744–745
        specificity and carboxylation rate, 741–743, 742f
Rubisco activase (RCA), 546, 552, 670, 687, 699
Rubisco activity, decline in/enzyme protein, loss of, 84–85
Rubisco-containing bodies (RCB)
Rubisco degradation mechanism, 85
Rubisco-limited photosynthesis, 706, 718
Rub-5P, 514
RuBP (ribulose-1,5-bisphosphate), 65, 270, 278, 315, 468, 630, 637, 699
    regeneration, 66, 66f
RWC, *see* Relative water content (RWC)

S

S-adenosyl-L- methionine (SAM), 177
Salicylic acid, 657
Salinity, effects on photosynthesis, 631–635
    carboxylation, 634
    chloroplast structure and pigment composition, 632
    photosynthates level, 634–635, 635t
    photosystems, photochemical activities, and chlorophyll fluorescence, 633–634
    stomatal closure and CO_2 diffusion, 631–632
Salinity stress, on ornamental plants, 651–654
    coping with, 656–658
    overview, 651–652
    photosynthesis, 653
    plant growth and development, 652–653
    tolerance
        mechanisms, 653–654
        overview, 655–656
Salsoloid, 311
Salvia farinacea, 653, 654
Salvia splendens, 654, 656, 768t
SAMICS, *see* Surface-affinity modulated ISP conformation switch
Satsuma mandarin, 771t
Scaevola aemula, 654
Scarcity, water, *see* Water restriction
Scarecrow-like (SCL) transcription factors, 181
Scenedesmus, 675
Scenedesmus obliquus, 544
Scotomorphogenesis, 3, 6f
Secondary plastids, 359

Sedoheptulose-1,7-bisphosphatase (SBPase), 278, 749, 749f
Seed traits, 493
Selaginella, 672
Selaginella bryopteris, 687
Selaginella lepidophylla, 684, 687
Selaginella tamariscina, 688
Selective anchoring, 211
Self-assembled monolayer (SAM), 211
Semiconductor LEDs, defined, 761
Semiphotons, 23, 26f, 43
Semiquinol radical-mediated superoxide generation, 342f
Semiquinone radical, 334
Senecio vulgaris, 707
Senescence-associated genes (SAG), 82, 86f
Senescence-associated vacuole (SAV), 89
Senescence during leaf development, signals regulating, 85–89
Senescence induction
    depletion of sugars and, 85–86, 86f
    by excess sugars, 86–87
Sensitivity indices, 269
Sensors, sugar, 87
Sequence alignment of cytochromes, 234f
Serial secondary endosymbiosis, 360
Sesuvium portulacastrum, 657
Setaria faberi, 707
Shade curtains, 729
Shelves, plant factories, 728
Short-term *in vitro* effects, of Pi deprivation, 605–608
    phosphate translocators, 605–606
    photosynthesis regulation, 606–607
    starch biosynthesis, 607
    sucrose biosynthesis, 607–608
Shrinkage/swelling of thylakoids, 224–225
Sigma factor-1 (SIG-1), 549
Signals regulating chloroplast biogenesis/ senescence during leaf development
    role of ROS in signaling network of, 87–89, 88f
    sugar signaling, 85–87, 86f
Signal transduction mechanism, 87
Singlet oxygen, 225
Single-walled carbon nanotubes (SWCNT), 211, 212f
Sink tissues, carbon transport from leaf to, 520–524, 521f
    mannitol and glucitol transporters, 523–524
    sucrose transporters, 521–523, 522f
    SUT transporters, 522–523, 522f
    SWEET transporters, 521–522, 522f
Skeletonema costatum, 775t
Small fruit crop responses
    about, 287
    to carbon dioxide (CO_2)/primary macronutrients/crop rotation, 288–289
    limited K/N macronutrient supplies (case study)
        on berry formation and marketable yields, 293–294
        colimitation of reduced K/N inputs on strawberry fruit productivity/mineral nutrition quality, 295, 296f
        higher leaf C_i inhibiting leaf/fruit NO_3- retention, 294–295
        overview, 290–291, 291f
        resulting in higher leaf C_i, 291–293, 292f, 292t

limited primary macronutrient supply as environmental abiotic stress factor, 289–290
    phosphorus macronutrient limitation/small fruit crop development (case study)
        excessive P inputs prohibiting plant P/N uptake, 301
        overview, 296–297
        strawberry plant P optimization, 299–301
        strawberry plants and limited macronutrient P supplies in nursery production, 297–299, 298t
Small-interfering RNAs (siRNAs), 541
Small-RNAs (smRNAs)
    TFs regulated by, 541
Small subunit (SSU), 278
Smirnoff–Wheeler–Running pathway, 163
Snapdragon, 654
S-nitrosothiols (RSNO), 160f
ß-octyl glucopyranoside (ß-OG), 330
Sodium azide, 195
Sodium dodecyl sulfate polyacrylamide gel electrophoresis (SDS-PAGE), 150, 214–215, 333
Soft traits, 489
"Soil crust algae," 675
Soilless systems, 725
    aggregate culture, 725
    closed loop, 725
    liquid culture, 725
Soil nutrient limitation, 289
Soil salinity, *see* Salinity stress
Solanum lycopersicum, 657, 770t
Solanum nigrum, 707
Solanum type plants, 547
Solar energy, 159
Solar ultraviolet (280-400 nm) exclusion on photosynthesis
    on C_3 and C_4 plants, 259, 259b
    effect on crop yield, 268–269, 269f
    on photosynthetic performance
        C_3 and C_4 carbon metabolism, 265–268, 268f
        chlorophyll fluorescence, 262–264, 263f, 264t
        gas exchange parameters, 264–265, 265f, 266f
        leaf area and biomass accumulation, 259–260, 260f
        photosynthetic pigments, 260–262, 262t
    solar UV exclusion, 258–259
    solar UV sensitivity of C_3 and C_4 plants, 269–270, 269f
    stratospheric ozone depletion, 257–258
Solid-state lighting technology, defined, 761
Soluble electron carrier, 230
Sophora secundiflora, 653, 656
Soret bands, 107–108
Sorghum bicolor, 542, 657
Source, light, 735–737; *see also* Light
    selecting, 735–736
Specificity, Rubisco, 741–743, 742f
Specific leaf area (SLA), 489, 491, 494
    intraspecific variability in, 490f
Specific leaf weight (SLW), 260
Spinach, 765t
Spinacia oleracea, 765t
Spirulina platensis, 653, 774, 775t
Spodosols, 296
Sporobolus, 683
Sporobolus stapfianus, 670, 688

Spring constant of binary photon, 26, 26n
Sprinkler irrigation, for ornamental plants, 657
Square wave voltammogram (SWV), 138
  of Chls a, b, d, f, and DV-Chl a, 139f
Staircase arrangement, 685
Starch biosynthesis
  Pi deprivation and, 607
Starch metabolism
  in leaf cell, 510–514, 511f–512f, 512t
    reactions involved in, 512t
    regulation, in chloroplast, 513–514
    synthesis and degradation in chloroplast,
      510–513
Starch storage
  metabolism for, 527–528
State transitions, 378–380, 379f; see also Light
  (excess)/carbon (limited)
Stellaria longipes, 497
Stellaria media, 707
Stem density
  as dimension of plant ecological strategy,
    492–493
Stock, 768t
Stokes shift, 214
Stomatal closure
  salinity effects on, 631–632
Stomatal conductance
  drought effects on, 635
Stomatal oscillations, heat stress and, 698–699
Strategy(ies); see also Plant ecological
    strategies
  defined, 490
Stratospheric ozone depletion, 257–258; see
    also Solar ultraviolet (280-400 nm)
    exclusion on photosynthesis
Strawberry, 766t
Strawberry fruit productivity/mineral nutrition
    quality, colimitation of reduced K/N
    inputs on, 295, 296f
Strawberry marketable fruit yield, 296f
Strawberry nursery plant propagation, 297f
Strawberry physiological development, K
    and N macronutrient treatments
    on, 292t
Strawberry plant P optimization, 299–301
Strawberry plants and limited macronutrient
    P supplies in nursery production,
    297–299, 298t
Strawberry plants treatment (case study), 290,
    291f
Stress-enhanced proteins (SEP), 380
Stresses, salinity and drought, see Drought
    stress; Salinity stress
Stress form, 375
Stressful conditions, photosynthesis and,
    629–644
  air pollutants, 640–642, 642t
  anaerobiosis, 640
  chilling, 639–640
  drought, 635–638
    carbohydrates and related enzymes
      levels, 638
    carboxylation under, 637–638
    chlorophyll fluorescence and
      photochemical reactions, 636–637
    leaf area and stomatal conductance, 635
    ultrastructural changes in chloroplasts,
      636
  heat stress, 638–639
  metal toxicity, 642–643
  overview, 629–631

salinity, 631–635
  carboxylation, 634
  chloroplast structure and pigment
    composition, 632
  photosynthates level, 634–635, 635t
  photosystems, photochemical activities,
    and chlorophyll fluorescence, 633–634
  stomatal closure and CO₂ diffusion,
    631–632
  stress-induced ROS production and
    photosynthetic apparatus damage, 643
  TFs and MAP kinases role in, 643–644
Stress gradient hypothesis, 498–499
Stress-induced ROS production
  photosynthetic apparatus damage and, 643
Stress markers
  chlorophyll fluorescence parameters as, 456
Stress phenotype, 495
Stress tolerance, 493
S-Triazine-resistant (R) plants
  carbon assimilation, 709–710, 710f, 711t,
    712f, 720
  chlorophyll fluorescence, 709
  controlled leaf temperature effects, 714–715,
    715f
  diurnal temperature effects, 710–714
  evolutionary ecology, 719–720
  functional change in, 707–709
  nature, 706
  overview, 705–706
  photosynthetic regulation, 717–719, 718f, 719t
  pleiotropy in R, 719–720
  R adaptation, 720
  Rubisco activity, 715–716, 716f, 717f, 717t
  structural change in, 706–707
  temperature effects on, 710–715, 711t, 712f,
    713t, 714f
  vs. S plants, 705–706
String theory, 29
Stromal enzyme complexes, see Thylakoids,
    biogenesis
Strontium ion, 333
Structural change, s-triazine-resistant plants,
    706–707
Structural changes in photosynthetic apparatus;
    see also Light-harvesting chlorophyll
    protein complexes of PSII (LHCII)
    for apparatus sensitivity
  under high light intensity, 249–250, 251f
  under high temperature
    LHCII organization for temperature
      sensitivity, 247–248
    LHCII–PSII supercomplex, changes in,
      246–247
    lipid composition on temperature
      sensitivity, 247
    thylakoid membranes, structural changes
      in, 245–246, 251f
  under low temperature
    functional changes, 248–249
    structural changes, 248
Structure/specificity of Ndh complex, 151–153;
    see also Thylakoid Ndh complex
Suaeda aralocaspica, 313–314, 313f, 314f
Suaeda maritime, 657
Subcellular localization of biosynthetic
    enzymes, 177–178
Subirrigation, 725
Submersed aquatic macrophytes
  about, 308

Submersed aquatic plants of hydrocharitaceae
    family, C₄-like mechanism in;
    see also Nontypical C₄ species,
    photosynthesis in
  Egeria densa, case of, 308–310, 309f
  Hydrilla verticillata, case of, 310–311
Subunit composition of Cyt bc₁ from
    R. sphaeroides, 330
Subunit IV of Rsbc₁, role of, 333–334
Succinate dehydrogenase (SDH), 350, 351f
Suc-6P, 515
Suc-6P Sase
  regulation of, 516–517
Sucrose, 510
  metabolism in heterotrophic cells, 524–527
  in nonphotosynthetic tissues, 524–526
  synthesis in leaf cell, 515–519, 515t, 516f,
    517t
    control by cFru-1,6bisP 1-Pase, 515–516,
      516f
    pathway of, 515–517
    reactions involved in, 515t
    Suc-6P Sase regulation, 516–517
Sucrose biosynthesis
  Pi deprivation and, 607–608
Sucrose nonfermenting-1-related protein kinase
    1 (SnRK1), 86f, 87
Sucrose-phosphate synthase (SPS), 607, 608
Sucrose transporters, 521–523, 522f
Sugar
  glass formation, 666
  and hormones, interactions, 87
  sensors, 87
Sugar-alcohols
  metabolism in heterotrophic cells, 524–527
  as photosynthetic products, 517–519
    Gol synthesis, 518–519
    Mol synthesis, 517–518
  synthesis in leaf cell, 515–519, 515t, 516f,
    517t
Sugar signaling; see also Chloroplast
    development
  cross-talk of sugar, 87
  depletion of sugars/induction of senescence,
    85–86, 86f
  ROS in signaling network, 87
  senescence induction by excess sugars,
    86–87
  sugar sensors/signal transduction
    mechanism, 87
Suicide protein, 381
Sulfoquinovosyl diacylglycerol (SQDG), 224,
    250
Sunflower (Helianthus annuus), 528, 655
Sunlight into chemical energy, conversion, 328
Sunlight use, plant factories, 724, 725f, 726–727,
    729; see also Lighting, plant factories
  vs. artificial light, 728t
Sunyaev–Zel'dovich effect, 20
Superoxide dismutase (SOD), 153, 165–166
Superoxide generation, see Cytochrome bc₁
    complex from photosynthetic purple
    bacteria
Supplemental lighting, plant factories, 724–725,
    725f, 727, 734–735, 735f; see also
    Lighting, plant factories
Surface-affinity modulated ISP conformation
    switch (SAMICS)
  ISP-ED mobility necessary for ET function
    of bc₁, 336

mutagenesis in *bc*1 and presence of control mechanism for ISP-ED conformation switch, 336
    Q-cycle mechanism and, 335f
Susceptible (S) wild plants, 705
    vs. R plants, 705–706
SUT transporters, 522–523, 522f
SWEET transporters, 521–522, 522f
Symbiodinium, 389
Synechococcus elongatus, 744–745
Synechocystis, 8, 152
Synechocystis sp. PCC 6803, 229, 232
Synthetic phase, 509
Syntrichia, 672
Syntrichia ruralis, 669

T

Tagetes erecta, 768t
Tagetes patula, 768t
TAP38, mutation in, 748
Tartary buckwheat, 771t
Tatsoi, 771t
Technologies, plant factories, 726
Temperature; *see also* Heat stress
    agriculture and, 729–731, 729f, 730f
    controlled leaf temperature effects, 714–715
    deactivation of PSII, 246
    desiccation and, 691
    diurnal temperature effects, 709f, 710–714, 712f, 713t, 714f, 714t
    greenhouse control systems, 577–578, 578t
    Rubisco activity and, 745
Temperature sensitivity, LHCII organization for, 247–248
Temperature stress, photosynthetic response to structural/functional changes in photosynthetic apparatus
    under high temperature, 245–248, 251f
    under low temperature, 248–249, 251f
Ternary effect, 68
Terrestrialization, 675, 676
Terrestrial plants, desiccation tolerance in, 675–676
Tetracystis aeria, 672
Tetrahydrofuran, 142f
Tetrapyrrole biosynthetic pathway branches, 176
Tetrapyrroles, open-chain, 761
Tetraselmis gracilis, 774, 775t
Tetraselmis suecica, 774, 775t
Texas bluebonnet, 656
Texas mountain laurel, 653, 656
Theory of carbon isotope discrimination during photosynthesis, 68–69
Theory of relativity, 37
Thermotolerance, of Rubisco activase, 745
Thiol–disulfide proteins in plants, 164
Thiol protein (PSH), 160f
Thioredoxin (TRX), 160f, 164
Thioredoxin reductase (TRXR), 160f, 164
Thioredoxin reductase C (TRXRC), 160f
Third battery of mechanisms, 665
Thylakoid carbonic anhydrases
    membrane-bound carbonic anhydrases
        about, 193–194
        CA near PSI, 196, 196f
        CA near PSII, 194–196, 194t, 195f
    possible functions of, 197–198
    soluble lumenal carbonic anhydrase, 196–197
Thylakoid complexes

assembly of, 81
    disassembly of, 84
Thylakoid membranes, 547
    absorption of light energy, 448
    activation energy distribution between PS$_{II}$ and PS$_I$, 449
    ATP synthesis, 450
    biogenesis of, *see* Chloroplast biogenesis
    C$_4$/CAM, 452
    chloroplast electron transport in, 747–748, 748f
    described, 448–449
    electron sinks, 451
    futile cycles, 451–452
    photorespiration, 450, 450f
    photosynthetic efficiency, 449
    role of, in quality control of PSII, 225
    ROS production, 450–451
    structural changes in, 245–246, 251f
    structure and composition, 223–224
    structure and function, 448–452
    structure changes
        shrinkage and swelling of thylakoids, 224–225
        unstacking of thylakoid, 224
    thylakoid membrane lipids, 224
    xanthophyll cycle, 449–450
Thylakoid Ndh complex
    Ndh complex in photosynthesis, role of, 153–154, 153f, 154f
    nuclear *Ndh* genes/structure and specificity of Ndh complex, 151–153, 152f
    overview, 149
    reaction catalyzed by Ndh complex, 149–151, 150f, 151f
    regulation of
        activity of Ndh complex by phosphorylation, 154–155
        transcriptional/posttranscriptional control of *Ndh* gene expression, 154
Thylakoids
    DT organisms, 667
    ultrastructure, UVB in plants in, 279
Thylakoids, biogenesis; *see also* Chloroplast development
    ATP synthase, 81
    cytochrome b_6f complex, 81
    photosystem I (PSI), assembly of, 81
    photosystem II (PSII), organization/assembly of, 80–81
    Rubisco, 81–82, 82f
    thylakoid complexes, assembly of, 81
Tiller inhibition (tin) mutant, 751–752
Tissue folding, 672
Tobacco plants
    chlorophyll *b* biosynthesis, 748
    FBPase/SBPase in, 749–750
    transgenic, 744–745
Tolerance
    drought
        mechanisms of, 655
        overview, 655–656
    salinity
        mechanisms of, 653–654
        overview, 655–656
    water availability and, 663–664
Tomato, 770t
Tortula princeps, 672
Tortula ruraliformis, 673
Tortula ruralis, 689
Total dissolved solids (TDS), 294
Total dry weight, 572

Tracheophytes, 663
Trait for short, *see* Functional traits
Traits, 487–488, 489; *see also* Functional traits
    defined, 488
    hard, 489, 492
    intraspecific trait variability, 489–490, 490f
    soft, 489
Transcription, 544–545
Transcriptional control of *Ndh* gene expression, 154
Transcriptional regulation of Chl biosynthesis; *see also* Chlorophyll metabolism regulation in plants
    developmental and environmental cues
        endogenous circadian rhythm, transcriptional regulation by, 182–183
        photomorphogenesis, transcriptional regulation during, 182
        transcriptional coordination of Chl biosynthesis with chloroplast functionality, 183
    factors involved in Chl biosynthesis
        FAR-RED ELONGATED HYPOCOTYL3 (FHY3), 181
        GATA transcription factors, 181
        Golden2-like (GLK) transcription factors, 180–181
        LONG HYPOCOTYL5 (HY5), 179–180
        PHYTOCHROME-INTERACTING FACTOR (PIF), 180
        Scarecrow-like (SCL) transcription factors, 181
        TIMING OF CAB EXPRESSION1 (TOC1), 181–182
Transcription factors (TFs), 541
    photosynthesis under stressful conditions and, 643–644
Transcriptomic studies, desiccation, 669
Transgenic Ndh-defective tobacco, 154f
Transgenic tobacco (*Nicotiana tabacum*), 545
Translation, 544–545
Translational energy, 36f, 37
Translational regulation
    chloroplast gene expression, 550–551
Translocators, phosphate, 605–606
Transmembrane helix (TMH), 332
Transmission electron micrograph (TEM)
    of chromoalveolate chloroplasts, 363f
    cyanobacterial cell, 382f
    of cyanobacterial cells, 350f
    of microalgal chloroplast, 360f
    of *P. tricornutum* cell, 388f
Transmission electron microscopy (TEM)
    chloroplast biogenesis stages, 8, 9f, 11f
    PLB and, 5
    of thylakoid membranes, 225, 226f
Transmission spectra of UV cutoff filters/polyethene filter, 258f
Trans-neoxanthin (trans-Neo), 591
    HPLC determination of, 596
Transporters
    glucitol transporters, 523–524
    mannitol transporters, 523–524
    sucrose transporters, 521–523, 522f
    SUT transporters, 522–523, 522f
    SWEET transporters, 521–522, 522f
Treasure flower, 768t
Trebouxia, 675
Trebouxiophyceae, 675
Trebouxiophycean genera, 675

Trees, photosynthetic competition between, 477
    interactions between native and invasive trees, 480–483, 481f, 482f
        for light, 477–479, 478t
        for nutrients, 479–480
        for water, 479–480
Trentepohliales, 675
Tre-6P, 513
Tribolium castaneum, 736
Tricarboxylic acid (TCA), 277, 339
    cycle, 268
Tri-leaves, 301
Trimers, 244
Triose-P, 515, 603, 605
    in BCC, 510
    de novo synthesis of, 514
Triose phosphate/phosphate translocator (TPT), 514, 605
Tripeptide, 163
Tripogon loliiformis, 670
Tripogon minimus, 670
Tripolium pannonicum
    nutrient deficiency (case study), 457–458
Tripterospermum japonicum, 772
Triticum aestivum, 765t
Triticum type plants, 547
TRY database, 488
Turgor loss point, 492
Two-cysteine peroxiredoxin (2CysPRX), 160f
Two-dimensional NMR spectra; *see also* Chlorophylls/bacteriochlorophylls, physicochemical properties of
    HMBC, 124, 126
    HSQC, 124
    NOESY, 124
Two-step hypothesis, 249

U

Ubiquinol oxidation, 337f
Ubiquinones, 204f, 334
Ubiquitous carotenoids, 589–591, 590f, 591t
Ubisemiquinol as source for superoxide generation, 341–342, 342f
Ubisemiquinone free radical as reaction intermediate, 336–338, 337f
UDP-glucose pyrophosphorylase (UDPG PPase), 515, 608
Ultrastructural analysis, 10
Ultrastructural changes of chloroplasts, 82, 83f
Ultrastructure, chloroplast
    desiccation/rehydration cycles, 667–669, 667f–668f
    drought tolerance, 685
    heat stress and, 697
    poikilochlorophyllous plants, 674
Ultraviolet-B radiation
    as ecophysiological factor, 257
Ultraviolet-B radiation in photosynthesis determination
    on biochemistry/bioenergetics of photosynthesis
        general aspects, 277–278
        lipids and membranes, 279

proteins in photosynthetic light and dark reactions, 278–279
        reactive oxygen species (ROS) and radicals, 279–280
    issues on, 281t
    molecular-level responses to, 280–281
    outcome of UVB responses, factors affecting, 276
    overview, 275
    and plant physiological responses, 276–277
Ultraviolet cutoff filters, 258f
Ultraviolet sensitivity index (UV SI), 269, 270
Ulva spp., 653
Ulvophyceae, 675
Uncertainty principle, 34–36; *see also* Light, nature of
Undecyl hydroxy dioxobenzothiazole (UHDBT), 334
Uniconazole, 657
United States Department of Agriculture (USDA)
    forage crops defined by, 465
Unstacking of thylakoid, 224
Urogen III decarboxylase (UROD), 174
Uroporphyrinogen III (Urogen III) synthase (UROS), 174

V

Vaccinium corymbosum, 768t
Vaccinium myrtillus, 655
Vapor pressure deficit (VPD), 730, 731; *see also* Humidity
Variability
    between-individual variability, 489–490
    intraspecific trait variability, 489–490, 490f
    population-level variability, 489, 490
    within-individual variability, 489
Variable *J* method, 67
VDE, *see* Violaxanthin de-epoxidase (VDE)
Vector of acceleration, 24
Velloziaceae, 667, 684
Ventilation, greenhouses, 729
Verbena macdougalii Heller, 656
Violaxanthin (Vio), 244, 589, 684
Violaxanthin, antheraxanthin, zeaxanthin (VAZ) cycle, 593
Violaxanthin de-epoxidase (VDE), 197, 374, 669, 684
Virtual particles, 33
Volatile organic compounds (VOCs), 674

W

Water
    as heat sink, 729, 730
    photolysis of, 509
    photosynthetic competition for, 479–480
Water restriction, 663–664; *see also* Desiccation tolerance
Water-splitting complex, 278
Water stress
    as dimension of plant ecological strategy, 492
Water use efficiency (WUE), 290, 291, 293, 494
Water-water cycle, alternative electron transport pathways in, 378

Wave equation, 38
Wave packet, 18, 26n
Wave-particle duality, 17–23; *see also* Light, nature of
Wavicles, 27
Western blot analysis, 319
Wet bulb temperature, 730
Wheat, 765t
Wheat–ryegrass–ryegrass–strawberry (WRRS) system, 296, 298t
Wheat yield consortium (WYC), 553
Whole-plant economics spectrum, 493
Wilting point (πtlp), 492
Within-individual variability, 489
Wood density
    plant ecological strategy and, 493
Woody species
    morphological, anatomical, and physiological plasticity of, 495–496, 496f

X

Xanthophyll(s), 244–245, 589, 674, 684, 690
Xanthophyll cycle, 373–375; *see also* Nonphotochemical quenching (NPQ)
    thylakoid membrane, 449–450
Xanthophyll esters (XEs), 591, 594–595, 595f
    HPLC determination of, 597
Xerophyta, 667, 686
Xerophyta humilis, 674, 684, 685, 686
Xerophyta scabrida, 684
Xerophyta viscosa, 684, 686
Xeroplasts, 674
X-ray absorption fine-structure spectroscopy (XAFS), 338
X-ray crystallography, 206f
X-ray diffraction grating spectrometer, 20
Xylulose-5-phosphate (Xul-5P), 160f, 514

Y

Yellow LEDs, 736; *see also* Light-emitting diodes (LEDs)

Z

Zantedeschia jucunda, 769t
Zea mays, 707
Zeaxanthin (Zea), 589, 665, 684, 698
Zeaxanthin epoxidase (ZEP), 374, 593, 669
Zinnia grandiflora, 653
Zinnia marylandica, 653
Ziziphus nummularia, 656
Ziziphus rotundifolia, 656
[Zn]-bacteriochlorophyll (BChl) *a*
    about, 100
    CD spectra of, 116, 119f
    NMR spectra of, 134
    redox potentials of, 139, 142
Zooxanthellae, 389
Zygnema, 675
Zygnematophyceae, 672, 675
Zygogonium, 675
Zygogonium ericetorum, 672
Zymograms, 150f

Printed and bound by CPI Group (UK) Ltd, Croydon, CR0 4YY

17/10/2024

01775698-0016